ELEMENTS

Element	Symbol	Atomic Number	Element	Symbol	Atomic Number
Actinium	Ac	89	Mercury	Hg	80
Aluminum	Al	13	Molybdenum	Mo	42
Americium	Am	95	Neodymium	Nd	60
Antimony	Sb	51	Neon	Ne	10
Argon	Ar	18	Neptunium	Np	93
Arsenic	As	33	Nickel	Ni	28
Astatine	At	85	Niobium	Nb	41
Barium	Ba	56	(Columbrium)		
Berkelium	Bk	97	Nitrogen	N	7
Beryllium	Be	4	Nobelium	No	102
Bismuth	Bi	83	Osmium	Os	76
Boron	B	5	Oxygen	O	8
Bromine	Br	35	Palladium	Pd	46
Cadmium	Cd	48	Phosphorus	P	15
Calcium	Ca	20	Platinum	Pt	78
Californium	Cf	98	Plutonium	Pu	94
Carbon	C	6	Polonium	Po	84
Cerium	Ce	58	Potassium	K	19
Cesium	Cs	55	Praseodymium	Pr	59
Chlorine	Cl	17	Promethium	Pm	61
Chromium	Cr	24	Protactinium	Pa	91
Cobalt	Co	27	Radium	Ra	88
Copper	Cu	29	Radon	Rn	86
Curium	Cm	96	Rhenium	Re	75
Dysprosium	Dy	66	Rhodium	Rh	45
Einsteinium	Es	99	Rubidium	Rb	37
Erbium	Er	68	Ruthenium	Ru	44
Europium	Eu	63	Samarium	Sm	62
Fermium	Fm	100	Scandium	Sc	21
Fluorine	F	9	Selenium	Se	34
Francium	Fr	87	Silicon	Si	14
Gadolinium	Gd	64	Silver	Ag	47
Gallium	Ga	31	Sodium	Na	11
Germanium	Ge	32	Strontium	Sr	38
Gold	Au	79	Sulfur	S	16
Halfnium	Hf	72	Tantalum	Ta	73
Helium	He	2	Technetium	Tc	43
Holmium	Ho	67	Tellurium	Te	52
Hydrogen	H	1	Terbium	Tb	65
Indium	In	49	Thallium	Tl	81
Iodine	I	53	Thorium	Th	90
Iridium	Ir	77	Thulium	Tm	69
Iron	Fe	26	Tin	Sn	50
Krypton	Kr	36	Titanium	Ti	22
Lanthanum	La	57	Tungsten (Wolfram)	W	74
Lawrencium	Lr	103	Uranium	U	92
Lead	Pb	82	Vanadium	V	23
Lithium	Li	3	Xenon	Xe	54
Lutetium	Lu	71	Ytterbium	Yb	70
Magnesium	Mg	12	Yttrium	Y	39
Manganese	Mn	25	Zinc	Zn	30
Mendelevium	Md	101	Zirconium	Zr	40

Hawley's
Condensed Chemical
Dictionary

Hawley's Condensed Chemical Dictionary

ELEVENTH EDITION

Revised by

N. Irving Sax
and
Richard J. Lewis, Sr.

VNR VAN NOSTRAND REINHOLD
New York

Library of Congress Catalog Card Number 86-23333
ISBN 0-442-28097-1

Printed in the United States of America

Van Nostrand Reinhold
115 Fifth Avenue
New York, New York 10003

Van Nostrand Reinhold International Company Limited
11 New Fetter Lane
London EC4P 4EE, England

Van Nostrand Reinhold
480 La Trobe Street
Melbourne, Victoria 3000, Australia

Nelson Canada
1120 Birchmount Road
Scarborough, Ontario M1K 5G4, Canada

15 14 13 12 11 10 9 8 7 6 5 4

Library of Congress Cataloging-in-Publication Data

Condensed chemical dictionary.
  Hawley's condensed chemical dictionary.

  Rev. ed. of: The Condensed chemical dictionary.
10th ed./rev. by Gessner G. Hawley, 1981.
  1. Chemistry—Dictionaries. I. Hawley, Gessner
Goodrich, 1905–   II. Sax, N. Irving (Newton Irving)
III. Lewis, Richard J., Sr. IV. Title.
QD5.C5 1987   540'.3'21   86-23333
ISBN 0-442-28097-1

In fond memory
of
our good friend
Gessner G. Hawley

To
Pauline and Grace
for their unending support and assistance

Our gratitude to Alberta W. Gordon and Carol P. Wickell for their professional assistance with the production of this book

Introduction

The first edition of the *Condensed Chemical Dictionary* appeared in 1919, when the chemical industry in the United States was entering a huge expansion program as a result of World War I. The urgent need for such a reference book became apparent to Francis M. Turner, President of the Chemical Catalog Company, predecessor of the Reinhold Publishing Corporation. Under his supervision a succession of Editors developed and expanded the *Condensed Chemical Dictionary* to meet the growing needs of the chemical industries. Since his death this development has continued, with the result that the work has achieved worldwide recognition in its field.

The Condensed Chemical Dictionary is a unique publication. It is not a dictionary in the usual sense of a compilation of brief definitions, but rather a compendium of technical data and descriptive information covering many thousand chemicals and chemical phenomena, organized in such a way as to meet the needs of those who have only minutes to devote to any given substance or topic.

Three distinct types of information are presented: (1) descriptions of chemicals, raw materials, processes and equipment; (2) expanded definitions of chemical entities, phenomena, and terminology; and (3) description or identification of a wide range of trademarked products used in the chemical industries. Supplementing these are listings of accepted chemical abbreviations used in the literature, short biographies of chemists of historic importance, and descriptions or notations of the nature and location of many American technical societies and trade associations. In special cases editorial notes have been supplied where it was felt necessary to clarify or amplify a definition or description. A few entries written by specialists are acknowledged by use of the author's name.

In a work of this nature, selection of topics for inclusion can hardly fail to be influenced by current interests and developing concerns within the topic area. The growing importance to chemists, and to the general public as well, of environmental and health hazards, which came to the forefront so quickly in the 1960s was reflected in the Eighth Edition, which greatly increased its coverage of this aspect of chemistry. After that, the magnitude of the energy problem became uppermost in the thinking of a broad spectrum of engineers, chemists, and physicists, since it has certainly become one of the most important technical problems confronting this country. Both the Ninth and Tenth Editions, while retaining emphasis on environmental considerations, were expanded in the area of energy and its sources, as far as permitted by available information. The effort in the eleventh edition has been to provide condensed, authoritative, factually oriented statements and descriptions, and to resist prognostications as to the future potential of any particular energy source. At the same time, continuing attention has been devoted to common hazards, such as flammable and explosive materials, poisons, pesticides, carcinogens, corrosive agents, radioactive wastes, etc., in line with the practice followed in earlier editions, and with the increasing public concern over these matters.

In connection with certain classifications of substances, particularly pesticides and carcinogens, which have occasioned the most controversy, the statement "Use may be restricted" indicates that a state or local regulation may exist even though the product has not been officially banned, or that a definitive ruling on its use is pending. When a product has been banned outright, the statement "Use has been prohibited" is used. A number of disputed cases have arisen in recent years; though some have been definitely settled, others are still being evaluated or are in the process of litigation, for example, saccharin.

In such a work as this, in view of the many materials in various stages of evaluative testing, court procedures, appeals, hearings, etc., it is impossible to keep abreast of every development. The user should check the current status of any questionable products before making decisions that involve them (see also paragraph on Hazards, below).

Arrangement of Entries

The entries are listed in strict alphabetical order; that is, those comprised of two or more words are alphabetized as if they were a single word, e.g., "acidimetry" precedes "acid value," and "water-proofing agent" precedes "water softener." Many of the prefixes used in organic chemistry are disregarded in alphabetizing, since they are not considered an integral part of the name; these include ortho- (o-), meta- (m-), para- (p-), alpha (α), beta- (β), gamma- (γ), sec-, tert-, sym-, as-, uns-, cis-, trans-, endo-, exo-, d-, l-, dl-, n-, N-, O-, as well as all numerals denoting structure. However, there are certain prefixes that are an integral part of the name (iso-, di-, tri-, tetra-, cyclo-, bis-, neo-, pseudo-), and in these cases the name appears in its normal alphabetical position, e.g., dimethylamine under "D" and isobutane under "I." The same is true of mono- (used as little as possible) and of ortho-, and meta-, in inorganic compounds such as sodium orthophosphate.

Chemicals and Raw Materials

The information in the categories listed below is given for each substance in the sequence indicated; where entries are incomplete, it may be presumed that no reliable data were provided by the reference systems utilized.

Name: The commonly accepted name is the key entry. Terminological variations are indicated where necessary. In virtually all cases, the name is given in the singular number. A name having initial caps and enclosed in quotes is a trademark (TM). The superior numbers refer to the name of the manufacturer given in Appendix III.

Synonym: Alternate names (IUPAC and others), as well as trivial names, are indicated. Obsolete and slang names have been eliminated as far as possible. Most synonyms are entered independently and cross-referenced, but space limitation has not permitted complete consistency in this regard.

Chemical Abstract Service Registry Numbers (CAS): In this Eleventh Edition of the *Condensed Chemical Dictionary* in order to make it a part of the vast accumulation of chemical, physical, and toxicological data which presently exists in the scientific literature in general, we have for the first time included the American Chemical Society's Chemical Abstract Service Registry Numbers (CAS) for many of the chemical entries. This will make the Eleventh Edition harmonize with the worldwide scientific literature. It will permit use and comparison of data on a given material no matter under what synonym it might be published. It will, in fact, permit absolute identification of a compound with all of its synonyms.

Formula: The molecular (or atomic) formula is regularly given. Structural formulas are used in special cases of unusual importance or interest.

Properties: The properties typically given are: physical state, atomic number, atomic weight, valence, isotopes, odor, taste, density, boiling point (at 760 mm Hg unless otherwise stated), melting point (freezing point), refractive index, and solubility or miscibility. Various other properties are given where pertinent: flash point, autoignition point, electrical properties, tensile strength, hardness, expansion coefficient, etc.

Source or Occurrence: Geographical origin of metals, ores, essential oils, vegetable oils, and other natural products.

Derivation: The chemical reactions or other means of obtaining the product by current industrial methods are emphasized. Obsolete and "curiosity" methods have been largely eliminated.

Grade: Recognized grades as reported in the industrial literature, including technical, CP, USP, refined, reactor, semiconductor, etc.

Hazard: This category includes flammability, toxicity characterization, explosion risks, etc., based on authoritative data. Also given are the Threshold Limit Values (TLV) for workroom exposures established by the American Conference of Government Industrial Hygienists; and various rulings of the Food and Drug Administration.

It was not considered practicable to include occupational exposure recommendations made by the National Institute for Occupational Safety and Health or exposure standards established by the Occupational Safety and Health Administration.

The toxicity ratings are intended to be used only as indications of the industrial hazard presented by a given material, as most of them are based on tests made on laboratory animals. Qualified toxicologists or physicians should be consulted for specific evaluations, dosages, exposure times, and concentrations. For further information regarding these hazards, the reader is referred to the following entries: combustible material; flammable material; dust, industrial; corrosive material; oxidizing material; poison (1); toxicity; toxic materials; carcinogen.

Use: These are primarily large-scale applications. Because of the rapidity of change in the chemical industries and the difficulty of obtaining reliable current data, no attempt has been made to list uses in the order of their tonnage consumption.

The patent literature is not specifically represented.

Shipping Regulations: For some years, revised regulations for transportation labeling of hazardous materials have been issued annually by the International Air Transport Association of Geneva, Switzerland (IATA) and by the U.S. Department of Transportation (DOT). For this reason, it has been decided to discontinue listing of specific labeling requirements in this book, as has been customary throughout its publishing history. Though these have always been taken directly from the most recent available schedules, so many additions and changes in the official regulations are made each year that it is impossible to keep them up to date in a publication that extends over a four- or five-year period. Thus, any such endeavor has to be less than 100% accurate and may result in costly and even dangerous misinterpretations. Therefore, manufacturers and shippers of chemicals and related products should procure and utilize the "restricted articles," schedules established by the agencies referred to.

The presence of the notation "Shipping regulations" in the description of a substance indicates a strong probability that a labeling requirement exists; however, absence of this notation does not necessarily mean that the substance is unregulated. The only valid sources of this information are the current revision of the IATA Restricted Articles Regulations (for air transport) and the Hazardous Materials Regulations of the U.S. Department of Transportation. The former may be ordered from the International Air Transport Association, 26 Chemin de Joinville, Box 160, 1216 Cointrin, Geneva, Switzerland; the latter can be obtained from R. M. Graziano, Association of American Railroads, 1920 L St., N.W., Washington, D.C. 20036.

General Entries

It is likely that no two editors would completely agree about what general subjects should be included in a dictionary of this kind. The major subdivisions of matter directly involved with chemical reactions, the various states of matter, and the more important groups of compounds would almost certainly be regarded as essential; but beyond these, the area of selectivity widens rapidly. The topics either added or expanded by the present Editors were chosen chiefly because of their interest and importance, both industrial and biochemical, and secondarily because of the terminological confusion evidenced in the literature and in industrial practice. Regarding the latter, the reader is referred to the entries on gum, resin, pigment, dye, filler,

extender, reinforcing agent, homogeneous, and combustible materials. In some cases a position has been taken which may not be accepted by all, but which is defensible and certainly not arbitrary. Even Editors must acknowledge that the meanings and uses of terms often change illogically, and that such changes are usually irreversible.

Among the general entries are: important subdivisions of chemistry; short biographies of outstanding chemists of the past including all winners of the Nobel Prize for Chemistry; numerous group definitions (barbiturate, peroxide); major chemical and physicochemical phenomena (polymerization, catalysis); functional names (antifreeze, heat-exchange agent); terms describing special material forms (aerosol, foam, fiber); energy sources (solar cell, fuel cell, fusion); the more important chemical processes; and various types of machinery and equipment used in the process industries. No general entry is intended to be encyclopedic or definitive, but rather a condensation of essential information, to be supplemented by reference to specialized sources. To present all this in useful and acceptably complete form has been a challenging, though often frustrating task, which the Editors leave with the uneasy feeling that, like the breadcrumbs in the Hatter's butter, some mistakes may have got in as well.

Trademarks

Continuing the policy of previous editions, an essential component of the Dictionary comprises descriptions of several thousand proprietary industrial products. The information was either provided by the manufacturers of these materials or taken from announcements or advertisements appearing in the technical press. Each proprietary name is enclosed in quotation marks, is stated to be a trademark (or brand name), and is followed by a superscript number referring to the Numerical List of Manufacturers in Appendix III. From this, the address of the manufacturer can easily be found in the alphabetical List of Manufacturers contained in Appendix IV. We wish to thank the owners of these trademarks for making the information available. The space devoted to these is necessarily limited, as the constant proliferation of trademarked products makes it impossible to list more than a small fraction of them in a volume such as this.

The absence of a specific trademark designation does not mean that proprietary rights may not exist in a particular name. No listing, description, or designation in this book is to be construed as affecting the scope, validity, or ownership of any trademark rights that may exist therein. Neither the Editors nor the Publisher assumes any responsibility for the accuracy of any such description, or for the validity or ownership of any trademark.

Acknowledgments

A Request

Many corrections and suggestions have been made from readers during the long history of the earlier editions. The Editors have always tried to acknowledge these to the best of their ability. They have welcomed this correspondence, for it has been an important source of information about the ac-ceptance of the Dictionary by its readers. The present Editors and Publisher wish to encourage this reaction from the field, not only to permit corrections to be made in reprints, but also as a basis for preparing future editions. All letters addressed to the Publisher will be forwarded to the Editors.

N. Irving Sax
Richard J. Lewis, Sr.

Abbreviations

ACS	American Chemical Society	mg	milligram
atm	atmosphere	mg/m³, mg/m3	milligrams per cubic meter
ASTM	American Society for Testing and Materials	μCi/mL	microcuries per milliliter
		μg/m³	microgram per cubic meter
autoign temp	autoignition temperature	mi	Curie
aw	atomic weight	min	minimum, minute(s)
bp	boiling point	m-	meta
Btu	British thermal unit	mm	millimeter
C	degrees centigrade (Celsius)	mp	melting point
CAS	Chemical Abstracts Service registry number	Mw, mw	molecular weight
		N.D.	"New Drugs"
cc	cubic centimeter	NF	National Formulary grade of chemical
CC	closed cup		
C.I.	"Colour Index" (a standard British publication giving official numerical designations to colorants).	N.F.	"National Formulary"
		NIOSH	National Institute for Occupational Safety and Health
CL	ceiling level	nm	nanometers
CP	chemically pure: a grade designation signifying a minimum of impurities, but not 100% purity.	o-	ortho
		OC	Open Cup
		OSHA	Occupational Safety and Health Administration
cp	centipoise	p-	para
COC	Cleveland Open Cup	ppb	parts per billion
cu	cubic	ppm	parts per million
cu/ft/lb	cubic feet per pound	psi(a)	pounds per square inch (absolute)
d, D	density	%	percent
DOT	Department of Transportation	refr	refractive
DOTC	Department of Transportation Classification	sec	second
		sp vol	specific volume
e.g.	for example	TCC	Tagliabue Closed Cup
F	degrees Fahrenheit	TLV	threshold limit value
FCC	"Food Chemicals Codex"	TM	trademark
FDA	Food and Drug Administration	TOC	Tagliabue Open Cup
flash p	flash point	USAN	United States Adopted Name
fp	freezing point	USDA	U.S. Department of Agriculture
ft	feet	U.S.P.	United States Pharmacopeia
FTC	Federal Trade Commission (a consumer protection agency).	UV	ultraviolet
		vap press	vapor pressure
g	gram	wt/gal	weight per gallon
gal	gallon		
gL	grams per liter		
g/mL	grams per milliliter		**Greek Letters**
HOH, H₂O	water		
Hg	mercury	α	alpha
hr	hour	β	beta
i.e.	that is	ω	omega
L, l	liter	γ	gamma
lb(s)	pound (pounds)	μ	mu
lb/gal	pounds per gallon	Δ	delta

Contents

A

a.   See alpha.

A.   (1) Abbreviation for absolute temperature; (2) abbreviation for Angstrom.

AAAS.   Abbreviation of American Association for Advancement of Science.

abaca.   (Manila hemp).   The strongest vegetable fiber, obtained from the leaves of a tree of the banana family. The fibers are 4–8 feet long, light in weight, soft, lustrous, nearly white, and do not swell or lose strength when wet. Denier ranges from 300 to 500. Combustible, but self-extinguishing.
Sources: Philippines, Central America, Sumatra.
Use: Heavy cordage and twine, especially for marine use; manila paper.
See also hemp.

"abate."[57]   TM for insecticide. CAS: 3383-96-8.
[(CH$_3$O)$_2$ PSOC$_6$H$_4$]$_2$S.
Properties: Colorless crystals, mp 30C.
Hazard: Toxic by ingestion and inhalation. Cholinesterase inhibitor. TLV: 10 mg/m^3 of air.
Use: Pesticide.

Abel-Pensky.   A standard test for flash point.

abherent.   Any substance that prevents adhesion of a material to itself or to another material. It may be in the form of a dry powder (a silicate such as talc, mica, or diatomaceous earth); a suspension (bentonite-water); a solution (soap-water); or soft solid (stearic acid, tallow waxes). Abherents are used as dusting agents and mold washes in the adhesives, rubber and plastics industries. Fats and oils are used as abherents in the baking industry. Fluorocarbon resin coatings on metals are widely used on cooking utensils.
See also antiblock agent; dusting agent.

Abies Siberica oil.   See fir needle oil.

abietic acid.   (abietinic acid; sylvic acid).
C$_{19}$H$_{29}$COOH   (having a phenathrene ring system). A major active ingredient of rosin, where it occurs with other resin acids. The term is often applied to these mixtures, separation of which is not achieved in technical grade material.
Properties: Yellowish, resinous powder; mp 172–175C; optical rotation −106; soluble in alcohol, ether, chloroform, and benzene; insoluble in water. Combustible.
Derivation: Rosin, pine resin; tall oil.
Method of purification: Crystallization.
Grade: Technical.
Use: Abietates (resinates) of heavy metals as varnish driers; esters in lacquers and varnishes; fermentation industries; soaps.

"ABITOL."[266]   TM for a colorless, tacky, very viscous liquid; mixture of tetra-, di-, and dehydroabietyl alcohols made from rosin.
Use: Plasticizers, tackifiers, adhesive modifiers.

ablative material.   Any material that possesses a capability for rapidly dissipating heat from a substrate. Specialized ceramic tiles developed since 1980 for protection of the space shuttle have proved successful. The materials used are of two major types: (1) Fibers made from white silica, fused in an oven, cut into blocks, and coated with borosilicate glass; these are extremely efficient at temperatures up to 2300F. (2) An all-carbon composite (called reinforced carbon-carbon) made by laminating and curing layers of graphite fiber previously coated with a resin, which is pyrolized to carbon. The resulting tile is then treated with a mixture of alumina, silicon, and silicon carbide. Such composites are used for maximum-temperature (nose-cone) exposure up to 3000F. Both types are undamaged by the heat and are reusable. The tiles are adhered to the body of the spacecraft with a silicone adhesive. Ablative materials used on early spaceship trials were fluorocarbon polymers and glass-reinforced plastics, but these were wholly or partially decomposed during reentry.

abrasion.   Gradual erosion of the surface of a material both by physical forces (simultaneous cutting, shearing, and tearing) and by chemical degradation, chiefly oxidation. Temperature is a significant factor: friction may raise the temperature of the surface layers to the point where they become subject to chemical attack. Abrasion causes deterioration of many materials, especially of rubber (tire treads), where it can be offset by a high percentage of carbon black; other materials subjected to abrasion in their service life are textiles (laundering), leather and plastics

1

(shoe soles, belting), house paints and automobile lacquers (airborne dust, grit, etc.).
See also abrasive.

abrasive. A finely divided, hard, refractory material, ranging from 6 to 10 on the Mohs scale, used to reduce, smooth, clean, or polish the surfaces of other, less hard substances, such as glass, plastic, stone, wood, etc. Natural abrasive materials include diamond dust, garnet, sand (silica), corundum (aluminum oxide, emery), pumice, rouge (iron oxide), and feldspar; the more important synthetic types are silicon carbide, boron carbide, cerium oxide, and fused alumina. Abrasive in powder form may be (1) applied directly to the surface to be treated by mechanical pressure or compressed-air blast, as in cleaning building stone; (2) affixed to paper or textile backing after the particles have been coated with an adhesive; or (3) mixed with a bonding agent such as sodium silicate or clay, the particles being compressed into a wheel rotated by a power-driven shaft. A relatively recent development (1983) is an aluminum grinding wheel to which industrial diamonds are bonded with fluorocarbon polymer ("Teflon"). The process involves reaction of fluorine with the surfaces of the diamonds, chemical bonding of the fluorinated diamonds to the fluorocarbon, and further chemical bonding of the resulting material to the aluminum with application of heat and pressure.

abrasive, coated. See abrasive (2).

ABS. Abbreviation for (1) alkyl benzene sulfonate (detergent); (2) acrylonitrile-butadiene-styrene copolymer.
See ABS resin.

"absafil."[539] TM for an acrylonitrile-butadiene-styrene copolymer reinforced with glass fiber.
See also reinforced plastic.

abscisic acid. $C_{15}H_{20}O_4$. A plant growth regulator that promotes detachment of leaves and fruit.
Properties: Colorless crystals; mp 160C; sublimes at 120C; soluble in acetone, ether, chloroform; slightly soluble in water. Optically active.
Occurrence: In plants, fruits, and vegetables from which it can be extracted. Also made synthetically.
Use: In orchard sprays to facilitate fruit harvesting; defoliant; growth inhibitor.

absinthium. (wormwood). $(C_{30}H_{40}O_7)$.
An essential oil with intensely bitter taste due to presence of absinthin.

Hazard: Toxic by ingestion.
Use: A flavoring in liqueurs, vermouth.

absolute. (1) Free from admixture of other substances; pure. Example: absolute alcohol is dehydrated ethanol, 99% pure. (2) The pure essential oil obtained by double solvent extraction of flowers in the manufacture of perfumes.
See concrete (2). (3) Absolute temperature.

absolute temperature. The fundamental temperature scale used in theoretical physics and chemistry, and in certain engineering calculations such as the change in volume of a gas with temperature. Absolute temperatures are expressed either in degrees Kelvin or in degrees Rankine, corresponding respectively to the centigrade and Fahrenheit scales. Degrees Kelvin are obtained by adding 273 to the centigrade temperature (if above 0C) or subtracting the centigrade temperature from 273 (if below 0C). Degrees Rankine are obtained by subtracting 460 from the Fahrenheit temperature. The closest approach to absolute zero is −272C.

absorbent. (1) Any substance exhibiting the property of absorption, e.g., absorbent cotton, so made by removal of the natural waxes present.
See absorption (1).
(2) A material that does not transmit certain wavelengths of incident radiation.
See absorption (2).

absorption. (1) In chemical terminology, the penetration of one substance into the inner structure of another, as distinguished from adsorption, in which one substance is attracted to and held on the surface of another. Physicochemical absorption occurs between a liquid and a gas or vapor, as in the operation known as scrubbing in which the liquid is called an absorption oil; sulfuric acid, glycerol, and some other liquids absorb water vapor from the air under certain conditions. Physiological absorption takes place via porous tissues, such as the skin and intestinal walls, which permit passage of liquids and gases into the bloodstream.
See also adsorption; hygroscopic.
(2) In physical terminology, retention by a substance of certain wavelengths of radiation incident upon it, followed by either an increase in temperature of the substance or by a compensatory change in the energy state of its molecules. The UV component of sunlight is absorbed as the light passes through glass and some organic compounds, the radiant energy being transformed into thermal energy. The radiation-absorptive capacity of matter is utilized in analyti-

cal chemistry in various types of absorption spectroscopy.

(3) In physical chemistry, the ability of some elements to pick up or "capture" thermal neutrons produced in nuclear reactors as a result of fission. This is due to the large capture cross section of their atoms, which is measured in units called, "barns;" elements of particularly high neutron absorption capability are cadmium and boron.

absorption band. The range of wavelengths absorbed by a molecule; for example, absorption in the infrared band of 2.3 to 3.2 mʋ indicates the presence of OH and NH groups, while the 3.3 to 3.5 band indicates aliphatic structure. Atoms absorb only a single wavelength, producing lines, such as the sodium D line.
See also spectroscopy; resonance (2); UV absorber; excited state.

absorption oil.   See absorption (1).

absorption spectroscopy.   An important technique of instrumental analysis involving measurement of the absorption of radiant energy by a substance as a function of the energy incident upon it. Absorption processes occur throughout the electromagnetic spectrum, ranging from the gamma region (nuclear resonance absorption or the Mossbauer effect) to the radio region (nuclear magnetic resonance). In practice, they are limited to those processes that are followed by the emission of radiant energy of greater intensity than that which was absorbed. All absorption processes involve absorption of a photon by the substance being analyzed. If it emits the excess energy by emitting a photon of less energy than that absorbed, fluorescence or phosphorescence occurs, depending on the lifetime of the excited state.

The emitted energy is normally studied. If the source of radiant energy and the absorbing species are in identical energy states, i.e., in resonance, the excess energy is often given up by the nondirectional emission of a photon whose energy is identical with that absorbed.

Either absorption or emission may be studied, depending upon the chemical and instrumental circumstances. If the emitted energy is studied, the term "resonance fluorescence" is often used. However, if the absorbing species releases the excess energy in small steps by the process of intermolecular collision or some other mode, it is commonly understood that this phenomenon falls within the realm of absorption spectroscopy. The terms absorption spectroscopy, spectrophotometry, and absorptimetry are often used synonymously.

Most absorption spectroscopy is done in the ultraviolet, visible, and infrared regions of the electromagnetic spectrum.
See also emission spectroscopy; infrared spectroscopy.

ABS resin.   Any of a group of tough, rigid thermoplastics deriving their name from the three letters of the monomers which produce them.

acrylonitrile-butadiene-styrene resin.   Most contemporary ABS resins are true graft polymers consisting of an elastomeric polybutadiene or rubber phase, grafted with styrene and acrylonitrile monomers for compatibility, dispersed in a rigid styrene-acrylonitrile (SAN) matrix. Mechanical polyblends of elastomeric and rigid copolymers, e.g., butadiene-acrylonitrile rubber and SAN, historically the first ABS resins, are also marketed.

Varying the composition of the polymer by changing the ratios of the three monomers and use of other comonomers and additives results in ABS resins with a wide range of properties.
Properties: Dimensional stability over temperature range from −40 to +71C. Attacked by nitric and sulfuric acids, and by aldehydes, ketones, esters, and chlorinated hydrocarbons. Insoluble in alcohols, aliphatic hydrocarbons, mineral and vegetable oils. Processed by conventional molding and extrusion methods. D 1.04; tensile strength about 6500 psi; flexural strength 10,000 psi; good electrical resistance; water absorption 0.3–0.4%. Combustible, but slow-burning; flame retardants may be added. Can be vacuum-metallized or electroplated.
Grade: High-, medium-, and low-impact; molding and extrusion.
Use: Engineering plastics used for automobile body parts and fittings; telephones; bottles; heels; luggage; packaging; refrigerator door liners; plastic pipe and building panels (subject to local building codes); shower stalls; boats; radiator grills; machinery housings; business machines.
Note: Several trademarked types are "Cycolac," "Abson," Kralastic, Lustran. For further information, refer to Society of Plastics Industry, 355 Lexington Ave., New York, NY 10017.

abundance.   The relative amount (% by weight) of a substance in the earth's crust, including the atmosphere and the oceans.

(a) The abundance of the elements in the earth's crust is:

Rank	Element	% by wt.
1	Oxygen	49.2
2	Silicon	25.7
3	Aluminum	7.5
4	Iron	4.7
5	Calcium	3.4

Rank	Element	% by wt.
6	Sodium	2.6
7	Potassium	2.4
8	Magnesium	1.9
9	Hydrogen	0.9
10	Titanium	0.6
11	Chlorine	0.2
12	Phosphorus	0.1
13	Manganese	0.1
14	Carbon	0.09
15	Sulfur	0.05
16	Barium	0.05
	all others	0.51

(b) The percentages of inorganic compounds in the earth's crust, exclusive of water, are:

(1) SiO_2 55   (2) Al_2O_3 15   (3) $CaCO_3$ 8.8
(4) MgO 1.6   (5) Na_2O 1.6   (6) K_2O 1.9

(c) The most abundant organic materials are cellulose and its derivatives, and proteins. *Note*: On the universal scale, the most abundant element is hydrogen.

Ac.   Symbol for actinium; abbreviation for acetate.

AC.   Abbreviation for allyl chloride.

acacia gum.   See arabic gum.

acaricide.   A type of pesticide effective on mites and ticks (acarides).

"Accel."[123]   TM for a lactic acid starter culture for use in food processing.

accelerator.   (1) A compound, usually organic, that greatly reduces the time required for vulcanization of natural and synthetic rubbers, at the same time improving the aging and other physical properties. Organic accelerators invariably contain nitrogen, and many also contain sulfur. The latter type are called ultra-accelerators because of their greater activity. The major types include amines, guanidines, thiazoles, thiuram sulfides, and dithiocarbamates. The amines and guanidines are basic, the others acidic. The normal effective concentration of organic accelerators in a rubber mixture is 1% or less based on the rubber hydrocarbon present. Zinc oxide is required for activation, and in the case of acidic accelerators, stearic acid is required. The introduction of organic accelerators in the early twenties was largely responsible for the successful development of automobile tires and mechanical products for engineering uses. A few inorganic accelerators are still used in low-grade products, e.g., lime, magnesium oxide, and lead oxide. See also vulcanization; rubber.

(2) A compound added to a photographic developer to increase its activity, such as certain quaternary ammonium compounds and alkaline substances.

(3) A particle accelerator.

acceptability (foods).
See organoleptic.

acceptable risk.   A concept that has developed in recent years, especially in connection with toxic substances (insecticides, mercurials, carcinogens), food additives, air and water pollution, and related environmental concerns. It may be defined as a level of risk at which a seriously adverse result is highly unlikely to occur, "but at which one *cannot* prove whether *or not* there is 100% safety. It means living with reasonable assurance of safety and acceptable uncertainty." (Schmutz, J. F., C&EN, Jan. 16, 1978). Examples of acceptable risk that might be cited are diagnostic x-rays, fluoridation of water, and ingestion of saccharin in normal amounts.
The acceptability of the risks involved in nuclear power generation is controversial. The weight of the evidence has tended to shift toward the negative side since 1975 when an official safety study estimated the risk of a serious accident to be 1 in 20,000 years of reactor operation. An investigation made by the Oak Ridge National Laboratory based on data collected from 1969 to 1979 concluded that the risk of a major accident is 1 in 1000 years of reactor operation.

acceptor.   See donor.

acenaphthene.   (1,8-dihydroacenaphthalene; ethylenenaphthalene). $C_{10}H_6(CH_2)_2$ (a tricyclic compound).
Properties: White needles; d 1.024 (99/4C); fp 93.6C; bp 277.5C; refr index (100C) 1.6048. Soluble in hot alcohol; insoluble in water. Combustible.
Derivation: From coal tar.
Grade: Technical; 98%.
Use: Dye intermediate; pharmaceuticals; insecticide; fungicide; plastics.

acenaphthenequinone.   (1,2-acenapthenedione). $C_{10}H_6(CO)_2$ (a tricyclic compound).
Properties: Yellow needles; mp 261–263C; insoluble in water; soluble in alcohol.
Derivation: By oxidizing acenaphthene, using glacial acetic acid and sodium or potassium dichromate.
Grade: Technical.
Use: Dye synthesis.

acenocoumarin.   (3-(α-acetonyl-4-nitrobenzyl)-4-hydroxycoumarin). $C_{19}H_{15}NO_6$.
Properties: White, crystalline powder; tasteless

and odorless; mp 197C. Slightly soluble in water and organic solvents.
Use: Medicine (anticoagulant).

acephate. (acetylphosphoramidothioic acid ester). CAS: 30560-19-1. $C_4H_{10}NO_3PS$.
Properties: White crystals; mp 65C; soluble in water; slightly soluble in acetone and alcohol.
Hazard: Moderately toxic by ingestion.
Use: Insecticide.

acetadol. See aldol.

acetal. (diethylacetal; 1,1-diethoxyethane; ethylidenediethyl ether). CAS: 105-57-7. $CH_3CH(OC_2H_5)_2$.
Properties: Colorless, volatile liquid; agreeable odor; nutty aftertaste. Stable to alkalies, but readily decomposed by dilute acids. Forms a constant-boiling mixture with ethanol. Soluble in alcohol, ether, and water. D 0.831; bp 103–104C; vap press 20.0 mm (20C); flash p −5F (CC) (−20.5C); specific heat 0.520; refr index 1.38193 (20C); wt (lb/gal) 6.89; autoign temperature 446F (230C).
Derivation: Partial oxidation of ethanol the acetaldehyde first formed condensing with the alcohol.
Grades: Technical.
Hazard: Highly flammable. Dangerous fire risk. Explosive limits in air 1.65 to 10.4%. Moderately toxic and narcotic in high concentrations.
Use: Solvent; cosmetics; organic synthesis; perfumes; flavors.   See also acetal resin.

acetaldehyde. (acetic aldehyde; aldehyde; ethanal; ethyl aldehyde). CAS: 75-07-0. CH_3CHO.
Properties: Colorless liquid; pungent, fruity odor. D 0.783 (18/4C); bp 20.2C; mp −123.5C; vap press 740.0 mm (20C); flash p −40F (−40C) (OC); specific heat 0.650; refr index 1.3316 (20C); wt 650 lb/gal (20C); miscible with water, alcohol, ether, benzene, gasoline, solvent naphtha, toluene, xylene, turpentine, and acetone.
Derivation: (a) Oxidation of ethylene; (b) vapor phase oxidation of ethanol; (c) vapor-phase oxidation of propane and butane; (d) catalytic reaction of acetylene and water (chiefly in Germany).
Grade: Technical 99%.
Hazard: Highly flammable; toxic (narcotic). Dangerous fire, explosion risk. Explosive limits in air 4 to 57%. TLV: 100 ppm in air.
Use: Manufacture of acetic acid and acetic anhydride, n-butanol, 2-ethylhexanol, peracetic acid, aldol, pentaerythritol, pyridines, chloral, 1,3-butylene glycol, and trimethylolpropane; synthetic flavors.

acetaldehyde ammonia. See aldehyde ammonia.

acetaldehyde cyanohydrin. See lactonitrile.

acetal resin. (polyacetal). A polyoxymethylene thermoplastic polymer obtained by ionically initiated polymerization of formaldehyde + CH_2 to obtain a linear molecule of the type —O—CH_2—O—CH_2=CH_2—. Single molecules may have over 1500 —CH_2— units. As the molecule has no side chains, dense crystals are formed. Acetal resins are hard, rigid, strong, tough, and resilient; dielectric constant 3.7; dielectric strength 1200 volts/mil), 600 volts/mil (80-mil); dimensionally stable under exposure to moisture and heat, resistant to chemicals, solvents, flexing and creep, and have a high gloss and low friction surface. Can be chromium-plated, injection-molded, extruded, and blow-molded. Not recommended for use in strong acids or alkalies. They may be homopolymers or copolymers.
Properties: D 1.425; thermal conductivity 0.13 Btu/hr/sq ft/degree F/ft; coefficient of thermal expansion 4.5×10^{-5}/degree F; specific heat 0.35 Btu/lb/degree F; water absorption 0.41%/24 hour; tensile strength 10,000 psi; elongation 15%; hardness (Rockwell) R120; impact strength (notched) 1.4 ft-lb/inch; flexural strength 14,100 psi; shear strength 9500 psi. Combustible, but slow burning.
Use: An engineering plastic, often used as substitute for metals, as oil and gas pipes; automotive and appliance parts; industrial parts; hardware; communication equipment; aerosol containers for cosmetics.
See also "Delrin"; "Celcon."

acetamide. (acetic acid amine; ethanamide). CAS: 60-35-5. CH_3CONH_2.
Properties: Colorless, deliquescent crystals with a mousy odor. Soluble in water and alcohol; slightly soluble in ether; d 1.159; mp 80C; bp 223C; refr index 1.4274 (78.3C). Combustible.
Derivation: Interaction of ethyl acetate and ammonium hydroxide.
Grade: Technical; CP (odorless); intermediate; reagent.
Hazard: An experimental carcinogen.
Use: Organic synthesis (reactant, solvent, peroxide stabilizer); general solvent; lacquers; explosives; soldering flux; hygroscopic agent; wetting agent; penetrating agent.

acetamidine hydrochloride. $C_2H_6N_2HCl$.
Properties: Crystalline solid; slightly deliquescent; mp 166 C; soluble in water and alcohol; insoluble in acetone. Keep stoppered.
Derivation: Alcohol solution of acetonitrile + HCl + ammonia.
Hazard: Skin irritant. Moderately toxic by ingestion.

Use: Synthesis of pyrimidines and related groups of biochemically active compounds.

acetamido-. Prefix indicating the group CH_3CONH-. Also called acetamino- or acetylamino-.

3-acetamido-5-aminobenzoic acid.
Use: Intermediate in the manufacture of X-ray contrast media.

5-acetamido-8-amino-2-naphthalenesulfonic acid.
(acetyl-1,4-naphthalenediamine-7-sulfonic acid; acetylamino-1,6-Cleve's acid).
$C_{10}H_5NHCOCH_3)(NH_2)(SO_3H)$.
A reddish-brown paste.
Hazard: Toxic.
Use: Chemical intermediate; dyes.

8-acetamido-5-amino-2-naphthalenesulfonic acid.
(acetyl-1,4-naphthalenediamine-6-sulfonic acid; acetylamino-1,7-Cleve's acid).
$C_{10}H_5(NHCOCH)(NH_2)(SO_3H)$.   A paste.
Hazard: Toxic.
Use: Chemical intermediate; dyes.

p-acetamidobenzenesulfonyl chloride.
See N-acetylsulfanilyl chloride.

acetamidocyanoacetic ester. See ethyl acetamidocyanoacetate.

8-acetamido-2-naphthalenesulfonic acid magnesium salt. (acetyl-1,7-Cleve's acid).
$[C_{10}H_6(CH_3CONH)(SO_3)]_2Mg$.
Properties: Brownish-gray paste containing approximately 80% solids.
Use: Intermediate for dyes.

p-acetamidophenol. See p-acetylaminophenol.

"Acetamine."[28] TM for a group of azo dyes and developers made for application to acetate yarn, and especially suited to nylon.

acetamino-. See acetamido-.

acetaminophen. See p-acetylaminophenol.

acetanilide. (N-phenylacetamide).
CAS: 103-84-4. $C_6H_5NH(COCH_3)$.
Properties: White, shining crystalline leaflets or white, crystalline powder; odorless; stable in air; slightly burning taste; d 1.2105; mp 114–116C; bp 303.8C; soluble in hot water, alcohol, ether, chloroform, acetone, glycerol, and benzene. Flash p 345F (174C); combustible; autoign temperature 1015F (545C).
Derivation: Acetylation of aniline with glacial acetic acid
Grade: Technical; CP.

Hazard: Toxic by ingestion.
Use: Rubber accelerator; inhibitor in hydrogen peroxide; stabilizer for cellulose ester coatings; manufacture of intermediates (p-nitroaniline, p-nitroacetanilide; p-phenylenediamine); synthetic camphor; pharmaceutical chemicals; dyestuffs; precursor in penicillin manufacture; medicine (antiseptic); acetanisole.
See p-methoxyacetophenone.

acetate. (1) A salt of acetic acid in which the terminal hydrogen atom is replaced by a metal, as in copper acetate, $Cu(CH_3COO)_2$. (2) An ester of acetic acid where the substitution is by a radical as in ethyl acetate, $CH_3COOC_2H_5$. In cellulose acetate the hydroxyl radicals of the cellulose are involved in the esterification.
See also cellulose acetate; vinyl acetate.

acetate dye. One group comprises water-insoluble azo or anthraquinone dyes that have been highly dispersed to make them capable of penetrating and dyeing acetate fibers. A second class consists of water-insoluble amino azo dyes that are made water-soluble by treatment with formaldehyde and bisulfite. After absorption by the fiber the resulting sulfonic acids hydrolyze and regenerate the insoluble dyes.

acetate fiber. A manufactured fiber in which the fiber-forming substance is cellulose acetate. Where not less than 92% of the hydroxyl groups are acetylated, the term triacetate may be used as a generic description of the fiber (Federal Trade Commission). This fiber was formerly called "acetate rayon" or "acetate silk." The term "rayon" is not permissible for this type.
Properties: Thermoplastic; becomes tacky at 350F (176C). Soluble in acetone and glacial acetic acid; decomposed by concentrated solutions of strong acids and alkalies. Moisture absorption 6%. Tenacity (dry) 1.4 g/denier; (wet) about 1 g/denier. Elongation 50% dry, 40% wet. Combustible.
Use: Wearing apparel; industrial fabrics.
See also cellulose acetate; cellulose triacetate.

acetate fiber, saponified. Regenerated cellulose fibers obtained by complete saponification of highly oriented cellulose acetate fibers. Available in continuous filament form having a high degree of crystallinity and great strength.
Properties: Tensile strength (psi) 136,000–155,000; elongation 6%; d 1.5–1.6; moisture regain 9.6–10.7%; decomposes about 149C. Similar to cotton in chemical resistance, dyeing, and resistance to insects and mildew. Combustible.
Use: Cargo parachutes; typewriter ribbons; belts; webbing; tapes; carpet backing.

acetate film. A durable, highly transparent film with non-deforming characteristics, produced

from cellulose acetate resin. It is grease-, oil-, dust-, and air-proof and hygienic. Combustible. Available forms: Rolls and cut-to-size sheets.
Use: Laminates; support for photographic film; document preservation; pressure-sensitive tape; magnetic sound recording tape; window cartons and envelopes packaging.

acetate of lime. Commercial term for calcium acetate made from pyroligneous acid and milk of lime. There are brown and gray acetates of lime.
See also calcium acetate.

acetate process. See cellulose acetate.

acethydrazidepyridinium chloride. See Girard's "P" reagent.

acetic acid. (ethanoic acid; vinegar acid; methane-carboxylic acid). CAS: 64-19-7. (CH_3COOH). Glacial acetic acid is the pure compound (99.8% min), as distinguished from the usual water solutions known as acetic acid. 33rd highest-volume chemical produced in US (1985).
Properties: Clear, colorless liquid; pungent odor. Mp 16.63C; bp 118C (765 mm), 80C (202 mm); d 1.0492 (20/4C); wt/gal (20C) 8.64 lb; viscosity (20C) 1.22 cps; flash p (OC) 110F (43C); refr index 1.3715 (20C). Miscible with water, alcohol, glycerol, and ether; insoluble in carbon disulfide; autoign temperature 800F (426C). Combustible.
Derivation: (a) Liquid- and vapor-phase oxidation of petroleum gases (with catalyst); (b) oxidation of acetaldehyde; (c) reaction of methanol and carbon monoxide (with catalyst). This is the most cost-efficient method and has been in general use for some years; (d) fermentative oxidation of ethanol.
Grade: USP (glacial, 99.4 wt %, and dilute, 36–37 wt %); CP; technical (80; 99.5%); commercial (6, 28, 30, 36, 56, 60, 70, 80 and 99.5%); NF (diluted; 6.0 g/100 ml).
Hazard: Moderate fire risk. Pure acetic acid is moderately toxic by ingestion and inhalation, but dilute material is approved by FDA for food use. Strong irritant to skin and tissue. TLV: 10 ppm in air.
Use: Manufacturing of acetic anhydride, cellulose acetate, and vinyl acetate monomer; acetic esters; chloroacetic acid; production of plastics, pharmaceuticals, dyes, insecticides, photographic chemicals, etc., food additive (acidulant); latex coagulant; oil-well acidizer; textile printing.
See also vinegar.

acetic acid amine. See acetamide.

acetic acid, glacial. See acetic acid.

acetic aldehyde. See acetaldehyde.

acetic anhydride. (acetyl oxide; acetic oxide). CAS: 108-24-7. ($CH_3CO)_2O$.
Properties: Colorless, mobile, strongly refractive liquid; strong odor; d 1.0830 (20/20C); bp 139.9C; fp −73.1C; flash p 121F (49.4C) (CC). Autoign temperature 732F (385C); wt/gal (20C) 9.01 lbs. Miscible with alcohol, ether, and acetic acid; soluble in cold water; decomposes in hot water to form acetic acid. Combustible.
Derivation: (a) Oxidation of acetaldehyde with air or oxygen with catalyst; (b) by catalyzed thermal decomposition of acetic acid to ketone; (c) reaction of ethyl acetate and carbon monoxide; (d) from carbon monoxide and methanol.
Grade: C.P., technical (75, 85, 90–95%).
Hazard: Strong irritant and corrosive; may cause burns and eye damage. TLV: Ceiling 5 ppm in air. Moderate fire risk.
Use: Cellulose acetate fibers and plastics; vinyl acetate; dehydrating and acetylating agent in production of pharmaceuticals, dyes, perfumes, explosives; etc.; aspirin. Esterifying agent for food starch (5% max.)

acetic ester. See ethyl acetate.

acetic ether. See ethyl acetate.

acetic oxide. See acetic anhydride.

acetin. (monoacetin; glyceryl monoacetate). CAS: 102-76-1. $C_3H_5(OH)_2OOCCH_3$.
Acetin may also refer to glyceryl di- or triacetate, also known as diacetin and triacetin.
Properties: Colorless, thick liquid; hygroscopic; d 1.206 (20/4C); bp 158C (165 mm); 130C (3 mm); soluble in water, alcohol; slightly soluble in ether; insoluble in benzene. Combustible.
Derivation: By heating glycerol and strong acetic acid, distilling off the weak acetic acid formed and again heating with strong acetic acid and distilling.
Method of purification: Rectification.
Hazard: Moderately toxic; irritant.
Use: Tanning; solvent for dyes; food additive, gelatinizing agent in explosives.

acetoacetanilide. (acetylacetanilide). $CH_3COCH_2CONHC_6H_5$.
Properties: White, crystalline solid; mp 85C. Resembles ethyl acetoacetate in chemical reactivity. Slightly soluble in water; soluble in dilute sodium hydroxide, alcohol, ether, acids, chloroform, and hot benzene; d 1.26, flash p 325F (162.7C). Combustible.
Derivation: By reacting ethyl acetoacetate with

aniline, eliminating ethanol. Acetoacetanilide may also be prepared from aniline and diketene.
Grade: Technical.
Use: Organic synthesis; dyestuffs (intermediate in the manufacture of the dry colors generally referred to as Hansa and benzidine yellows).

acetoacet-o-anisidide.
$CH_3COCH_2CONHC_6H_4OCH_3$.
Properties: White, crystalline powder; mp 86.6C; d 1.1320 (86.6/20C); flash p 325F (162.7C) (OC). Combustible.
Use: Intermediate for azo pigments.

acetoacet-o-chloranilide.
$CH_3COCH_2CONHC_6H_4Cl$.
Properties: White, crystalline powder; mp 107C; bp decomposes; d 1.1920 (107/20C). Almost insoluble in water. Combustible; flash p 350F (176.6C) (OC).
Hazard: Toxic by ingestion.
Use: Intermediate for azo pigments.

acetoacet-p-chloranilide.
$CH_3COCH_2CONHC_6H_4Cl$.
Properties: White, crystalline powder; mp 133C; bp decomposes; flash p 320F (160C) (OC). Combustible. Very slightly soluble in water.
Hazard: Toxic by ingestion.
Use: Intermediate for azo pigments.

acetoacetic acid. (acetylacetic acid; diacetic acid; acetone carboxylic acid).
CH_3COCH_2COOH.
Properties: Colorless, oily liquid; soluble in water, alcohol, and ether; decomposes below 100C into acetone and carbon dioxide.
Hazard: Irritant to eyes and skin.
Use: Organic synthesis.

acetoacetic ester. See acetoacetate.

acetoacet-p-phenetidide.
$CH_3COCH_2CONHC_6H_4OCH_2CH_3$.
Properties: Crystalline powder; mp 108.5C; bp decomposes; d 1.0378 (108.5/20C); flash p (OC) 325F (162.7C). Combustible.
Hazard: Moderately toxic by ingestion.
Use: Intermediate for azo pigments.

acetoacet-o-toluidide.
$CH_3COCH_2CONHC_6H_4CH_3$.
Properties: Fine, white, granular powder; mp 106C; bp decomposes; d 1.062 (106C); slightly soluble in water. Flash p 320F (160C); Combustible.
Grade: Technical.
Hazard: Moderately toxic.
Use: Intermediate in the manufacture of Hansa and benzidine yellows.

acetoacet-p-toluidide.
$CH_3COCH_2CONHC_6H_4CH_3$.
Properties: White, crystalline powder; mp 93.0–96.0C; purity, 99% min.
Hazard: Moderately toxic.
Use: Light-fast yellow pigment intermediate; diazo coupler.

acetoacet-m-xylidide. (AAMX).
$(CH_3)_2C_6H_3NHCOCH_2COCH_3$.
Properties: White to light yellow crystals; mp 89–90C; d (20C) 1.238; water solubility (25C) 0.5%; flash p 340F (171C); Combustible.
Use: Intermediate for yellow pigments.

acetoaminofluorene. A pesticide. May not be used in food products or beverages (FDA).
Hazard: Toxic by ingestion.

p-acetoanisole. See p-methoxyacetophenone.

acetoglyceride. Usually an acetylated monoglyceride although commercial acetoglycerides will contain di- and triglycerides.
See acetostearin.

acetoin. See acetylmethylcarbinol.

acetol. See hydroxy-2-propanone.

acetomeroctol.
$CH_3COOHgC_6H_3(OH)C(CH_3)_2CH_2C(CH_3)_3$.
Properties: White solid; mp 155–157C; freely soluble in alcohol; soluble in ether or chloroform; sparingly soluble in benzene; insoluble in water.
Hazard: Toxic by ingestion.
Use: Medicine (antiseptic, solution 1:1000).

acetone. (dimethylketone; 2-propanone).
CAS: 67-64-1. CH_3COCH_3. 42nd highest-volume chemical produced in US (1985).

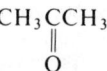

Properties: Colorless, volatile liquid; sweetish odor. Mp −94.3C; bp 56.2C; refr index (20C) 1.3591; d 0.792 (20/20C); wt/gal 6.64 lb (15C); flash p 15F (−9.4C) (OC). Autoign temperature 1000F (537C). Miscible with water, alcohol, ether, chloroform, and most oils.
Derivation: (a) Oxidation of cumene; (b) dehydrogenation or oxidation of isopropyl alcohol with metallic catalyst; (c) vapor phase oxidation of butane; (d) by-product of synthetic glyerol production.
Grade: Technical; CP; NF; electronic; spectrophotometric.

Hazard: Flammable; dangerous fire risk. Explosive limits in air 2.6-12.8%. TLV: 750 ppm in air. Narcotic in high concentrations. Moderately toxic by ingestion and inhalation.

Use: Chemicals (methyl isobutyl ketone; methyl isobutyl carbinol; methyl methacrylate; bisphenol-A); paint, varnish and lacquer solvent; cellulose acetate, especially as spinning solvent; to clean and dry parts of precision equipment; solvent for potassium iodide and permanganate; delusterant for cellulose acetate fibers; specification testing of vulcanized rubber products.

acetone bromoform.   See tribromo-tert-butyl alcohol.

acetonecarboxylic acid.   See acetoacetic acid.

acetone chloroform.   See chlorobutanol.

acetone cyanohydrin.   (α-hydroxyisobutyronitrile; 2-methyllactonitrile).   CAS: 75-86-5. $(CH_3)_2COHCN$.
Properties: Colorless liquid; bp 82C (23 mm); mp −20C; d 0.932 (19C); refr index 1.3996 (20C); flash p 165F (73.9C); soluble in alcohol and ether. Combustible; autoign temperature 1270F (685C).
Derivation: Condensing acetone with hydrocyanic acid.
Grade: Technical (97–98% pure).
Hazard: Toxic. Readily decomposes to hydrocyanic acid and acetone.
Use: Insecticides; intermediate for organic synthesis, especially methyl methacrylate.

acetonedicarboxyllic acid.   See β-ketoglutaric acid.

acetone oxime.   See acetoxime.

acetone semicarbazone.
$(CH_3)_2CNNHCONH_2$.   A chemical intermediate.
Properties: White powder; mp 188C.

acetone sodium bisulfite.   See sodium acetone bisulfite.

acetonitrile.   (methyl cyanide).   CAS: 75-05-8. CH_3CN.
Properties: Colorless, limpid liquid; aromatic odor; d 0.783; mp −41C; bp 82C; flash p 42F (5.56C). Soluble in water and alcohol; high dielectric constant; high polarity; strongly reactive.
Derivation: By-product of propylene-ammonia process for acrylonitrile.
Grade: Technical; nanograde; spectrophotometric.

Hazard: Flammable, dangerous fire risk. TLV: 40 ppm in air. Toxic action by skin absorption and inhalation.
Use: Solvent in hydrocarbon extraction processes, especially for butadiene; specialty solvent; intermediate; catalyst; separation of fatty acids from vegetable oils; manufacturing of synthetic pharmaceuticals.

acetonylacetone.   (1,2-diacetylethane; hexanedione-2,5; 2,5-diketohexane).   CAS: 110-13-4. $CH_3COCH_2CH_2COCH_3$.
Properties: Colorless liquid, soluble in water, d 0.9734 (20/20C), bp 192.2C, vap press 0.43 mm at 20C, fp −5.4C, flash p 185F (85C), bulk d 8.2 lb/gal (20C), autoign temperature 920F (493C). Combustible.
Derivation: By-product in the production of acetaldehyde from acetylene.
Grade: Technical.
Hazard: Irritant to eyes and skin.
Use: Solvent for cellulose acetate, roll-coating inks, lacquers, stains; intermediate for pharmaceuticals and photographic chemicals; electroplating.

acetonyl alcohol.   See hydroxy-2-propanone.

3-(α-acetonylbenzyl)-4-hydroxycoumarin.
See warfarin.

3-(α-acetonylfurfuryl)-4-hydroxycoumarin.
(sodium salt also used). A rodenticide.
Hazard: Highly toxic by ingestion and inhalation.

3-(α-acetonyl-4-nitrobenzyl)-4-hydroxycoumarin.
See acenocoumarin.

acetophenetidin.   (p-acetylphenetidin; acetophenetidide; phenacetin; p-ethoxyacetanilide).   CAS: 62-44-2. $CH_3CONHC_6H_4OC_2H_5$.
Properties: White crystals or powder; odorless and stable in air; soluble in alcohol, chloroform, and ether; slightly soluble in water; has slightly bitter taste; mp 135C. Derivation: By the interaction of p-phenetidin and glacial acetic acid, or of ethyl bromide and p-acetaminophenol.
Method of purification: Crystallization.
Grade: Technical; USP, as phenacetin.
Hazard: Toxic by ingestion.
Use: Medicine (analgesic); veterinary medicine.

acetophenone.   (phenylmethylketone; hypnone; acetylbenzene).   CAS: 98-86-2. $C_6H_5COCH_3$.

Properties: Colorless liquid with sweet, pungent odor and taste; bp 201.7; fp 19.7C; d 1.030 (20/20C); bulk d 8.56 lb/gal (20C); refr index 1.5363 (20C); flash p 180F (82.2C) (COC); slightly soluble in water; soluble in organic solvents and sulfuric acid. Combustible.

Derivation: (a) Friedel-Crafts process with benzene and acetic anhydride or acetyl chloride; (b) by-product from the oxidation of cumene; (c) oxidation of ethylbenzene.

Method of purification: Distillation and crystallization.

Grade: Technical, refined, perfumery.

Hazard: Narcotic in high concentrations, hypnotic.

Use: Perfumery, solvent, intermediate for pharmaceuticals, resins, etc.; flavoring, polymerization catalyst, organic synthesis.

acetophenone oxime.
Properties: Crystals.
Use: Antiozonant properties, antioxidant, antiskinning agent, piezoelectric properties, emulsifier-water/oleoresinous systems, end blocker, polymerization short stopper.

"Acetoquat."[400]   TM for a series of quaternary ammonium salts.

acetostearin.
$CH_3(CH_3)_{16}COOCH_2CHOHCH_2OOCCH_3$.
Acetylated glyceryl monostearate. Solid with peculiar combination of flexibility and nongreasiness. Derived from glyceryl monostearate or mixed glycerides by acetylation with acetic anhydride.
Use: Protective coating for food and as a plasticizer.

acetotoluidide.   See acetyl-o- or p-toluidine.

acetoxime.   (acetone oxime; 2-propanone oxime). $(CH_3)_2CNOH$.

$$CH_3 \diagdown \diagup CH_3$$
$$C$$
$$\parallel$$
$$N{-}OH$$

Properties: Colorless crystals; both basic and acidic in properties. Volatilizes in air; chloral-like odor; d 0.97 (20/20C); bp 136.3C; mp 61C; fairly easily hydrolyzed by dilute acids; soluble in alcohols, ethers, water. Combustible.

Derivation: Reaction of hydroxylamine in water solution with acetone, followed by ether extraction.

Grade: Technical.

Use: Organic synthesis (intermediate); solvent for cellulose ethers; primer for diesel fuels.

o-acetoxybenzoic acid.   See aspirin.

acetoxylation.   A method of synthesizing ethylene glycol in which ethylene is reacted with acetic acid in the presence of a catalyst, such as tellurium bromide, resulting in formation of mixed mono- and diacetates; this is followed by hydrolysis to ethylene glycol and acetic acid, with up to 95% yield of the glycol. It is thus considerably more efficient than the ethylene oxide method.

4-(p-acetoxyphenyl)-2-butanone.   See cue-lure.

acetozone.   See acetylbenzoyl peroxide.

"Acetulan."[493]   TM for a special fraction of acetylated lanolin alcohols.
Properties: Anhydrous; pale straw color; nearly odorless, low-viscosity liquid. Miscible with ethanol, mineral oil, and other common formulating materials; insoluble in water. Combustible.
Use: Hydrophobic penetrant, emollient, plasticizer, cosolvent, pigment dispersant, cosmetics.

acetylacetanilide.   See acetoacetanilide.

acetylacetic acid.   See acetoacetic acid.

acetylacetonatobis(ethylene)rhodium(1).
Properties: Orange crystal, air stable. Soluble in chloroform and ether.
Use: Reactive ethylene complex.

acetylacetonatodicarbonylrhodium(1).
Properties: Green crystal grades (dichroic red when crushed), air stable. Soluble in acetone, benzene, and chlorinated solvents.
Use: Homogeneous catalyst for hydroformylation reactions.

acetylacetone.   (diacetylmethane; pentanedione-2,4).   CAS: 123-54-6.
$CH_3COCH_2OCCH_3$.
Properties: Mobile, colorless or yellowish liquid. When cooled, solidifies to lustrous, pearly spangles. The liquid is affected by light, turning brown and forming resins; bp 140.5C, d 0.9753 (20/20C), bulk d 8.1 lb/gal, fp −23.5C, flash p 105F (40.5C) (TOC); soluble in water (acidified by hydrochloric acid); fairly soluble in neutral water; soluble in alcohol, chloroform, ether, benzene, acetone and glacial acetic acid. Combustible.
Derivation: By condensing ethyl acetate with acetone.
Hazard: Moderately toxic; moderate fire risk.
Use: Solvent for cellulose acetate, intermediate, chelating agent for metals, paint driers, lubricant additives, pesticides.

N-acetyl-l(+)-alanine. See alanine.

acetylamino-. See acetamido-.

p-acetylaminobenzenesulfonyl chloride. See N-acetylsulfanilyl chloride.

o-acetylaminobenzoic acid. See acetylanthranilic acid.

p-acetylaminophenol. (APAP; N-acetyl-p-aminophenol; acetaminophen; p-acetamidophenol; p-hydroxyacetanilide). CAS: 103-90-2. $CH_3CONHC_6H_4OH$.
Properties: Crystals, odorless, slightly bitter taste, d 1.293 (21/4C), mp 168C, slightly soluble in water and ether; soluble in alcohol; pH saturated aqueous solution 5.5–6.5.
Derivation: Interaction of p-aminophenol and an aqueous solution of acetic anhydride.
Use: Intermediate for pharmaceuticals and azo dyes, stabilizer for hydrogen peroxide, photographic chemicals, medicine (analgesic).

p-acetylaminophenyl salicylate. (phenetsal). $C_6H_4(NHCOCH_3)OOCC_6H_4OH$.
Properties: Fine, white, crystal scales; odorless; tasteless; soluble in alcohol, ether, and hot water; insoluble in light hydrocarbon solvents; decomposed by strong alkalies; mp 187–188C.
Derivation: By reducing p-nitrophenol salicylate to p-aminophenol salicylate and acetylating the latter.
Hazard: See aspirin.
Use: Medicine (analgesic).

p-acetylanisole. See p-methoxyacetophenone.

N-acetylanthranilic acid. (o-acetylaminobenzoic acid). $CH_3CONHC_6H_4COOH$.
Properties: Needles, plates, rhombic crystals (from glacial acetic acid); mp 185C, slightly soluble in water, soluble in hot alcohol, ether, and benzene. Combustible.
Derivation: Oxidation of o-acetyltoluidine with potassium permanganate in the presence of magnesium sulfate or potassium chloride.
Grade: Technical.
Use: Chemical (organic synthesis, anthranilic acid).

acetylation. Introduction of an acetyl radical ($CH_3CO—$) into the molecule of an organic compound having OH or NH_2 groups. The usual reagents for this purpose are acetic anhydride or acetyl chloride. Thus, ethanol (C_2H_5OH) may be converted to ethyl acetate ($C_2H_5OCOCH_3$). Cellulose is similarly converted to cellulose acetate by treatment with a mix containing acetic anhydride. Acetylation is commonly used to determine the number of hydroxyl groups in fats and oils.
See acetyl value.

acetylbenzene. See acetophenone.

acetyl benzoyl aconine. See aconitine.

acetyl benzoyl peroxide. (acetozone; benzozone). CAS: 644-31-5. $C_6H_5CO \cdot O_2 \cdot OCCH_3$.
Properties: White crystals; decomposed by water, alkaloids, organic matter, and some organic solvents; decomposes slowly and evaporates when gently heated and instantaneously (possibly explosively) if quickly heated, ground, or compressed; mp 36.6C; bp 130C (19 mm); moderately soluble in ether, chloroform, carbon teteachloride, and water; slightly soluble in mineral oils and alcohol. The commercial product is mixed with a neutral drying powder and contains 50% acetyl benzoyl peroxide.
Hazard: Toxic by ingestion. Strong irritant to skin and mucus membranes. Strong oxidizing agent; dangerous in contact with organic materials. Moderate explosion risk when shocked or heated.
Use: Medicine (active germicide); disinfectant; bleaching flour.

acetyl bromide. CH_3COBr. CAS: 506-96-7.
Properties: Colorless, fuming liquid; turns yellow in air; soluble in ether, chloroform and benzene; bp 81C; mp −96C; d 1.663 (16/4C).
Derivation: Interaction of acetic acid and phosphorus pentabromide.
Grade: Technical.
Hazard: Toxic by ingestion and inhalation; strong irritant to eyes and skin. Reacts violently with water or alcohol.
Use: Organic synthesis; dye manufacture.

α-acetylbutyrolactone. (α-acetobutyrolactone).
$$\overline{CH_2CH_2CH(OCCH_3)C(O)O}.$$
Properties: Liquid with ester-like odor, d 1.18–1.19 (20C), bp 142–144 (30 mm), partially soluble in water. Combustible.
Derivation: Sodium acetoacetate and ethylene oxide in absolute alcohol.
Hazard: Strong irritant to skin and mucus membranes.
Use: Organic synthesis.

acetyl carbinol. See hydroxy-2-propanone.

acetyl carbromal.   (N-acetyl-N-bromodiethyl-acetylurea).
$(C_2H_5)_2CBrCONHCONHCOCH_3$.
Properties: Crystals, slightly bitter taste, mp 109C. Slightly soluble in water; freely soluble in alcohol and ethyl acetate.
Hazard: Overdoses may be fatal.
Use: Medicine (sedative).

acetyl chloride.   (ethanoyl chloride).
CAS: 75-36-5. CH_3COCl.
Properties: Colorless, highly refractive, fuming liquid; strong odor; d 1.1051; mp −112C; bp 51–52C; flash p 40F (4.4C) (CC); soluble in ether, acetone, acetic acid.
Derivation: By mixing glacial acetic acid and phosphorus trichloride in the cold and heating a short time to drive off hydrochloric acid. The acetyl chloride is then distilled.
Hazard: Corrosive to skin and mucous membranes; toxic; strong eye irritant; flammable, dangerous fire risk. Reacts violently with water and alcohol.
Use: Organic preparations (acetylating agent); dyestuffs; pharmaceuticals.

acetyl choline.   (acetylethanoltrimethylammonium hydroxide).   CAS: 51-84-3.

$$CH_3-\underset{\underset{CH_3}{|}}{\overset{\overset{CH_3}{|}}{N}}-CH_2-CH_2-O-\overset{\overset{O}{\|}}{C}-CH_3$$

A derivative of choline; important because it acts as the chemical transmitter of nerve impulses in the autonomic system. It has been isolated and identified in brain tissue. The enzyme cholinesterase hydrolyzes acetylcholine into choline and acetic acid, and is necessary in the body to prevent acetylcholine poisoning.
See nerve gas.
Use: Medicine (as bromide and chloride); biochemical research.
See also cholinesterase inhibitor.

acetylcholinesterase.   See cholinesterase.

N-acetyl-L-cysteine (USAN).
$HSCH_2CH(NHCOCH_3)COOH$.   The N-acetyl derivative of the naturally occurring amino acid, l-cysteine.
Use: Medicine, biochemical research.

α-acetyldigitoxin (anhydrous). $C_{43}H_{66}O_{14}$.
Properties: Crystals; sparingly soluble in water, ether, petroleum ether; soluble in most organic solvents.

Derivation: Obtained by enzymatic hydrolysis of a digitalis extract.
Hazard: Toxic.
Use: Medicine (heart disease).

acetylene.   (ethyne).   CAS: 74-86-2.
$HC{\equiv}CH$.
Properties: Colorless gas, ethereal odor, d 0.91 (air=1), mp −81.8C (890 mm), bp −84C, soluble in alcohol and acetone; slightly soluble in water, flash p 0F (−17.7C) (CC), autoign temperature 635F (335C). An asphyxiant gas.
Derivation: (a) By the action of water on calcium carbide; (b) by cracking petroleum hydrocarbons with steam (Wulff process), or natural gas by partial oxidation (BASF process); (c) from fuel oil by modified arc process.
Grade: Technical, containing 98% acetylene and not more than 0.05% by volume of phosphine or hydrogen sulfide.
Hazard: Very flammable; dangerous fire risk; burns with intensely hot flame; explosive limits in air 2.5–80%. Forms explosive compounds with silver, mercury, and copper, which should be excluded from contact with acetylene in transmission systems. Copper alloys may be used with caution. Piping used should be electrically bonded and grounded.
Use: Manufacture of vinyl chloride; vinylidene chloride; vinyl acetate; acrylates; acrylonitrile; acetaldehyde; per- and trichloroethylene; cyclooctatetraene; 1,4-butanediol; carbon black; welding and cutting metals.

acetylene black.   The carbon black resulting from incomplete combustion of or thermal decomposition of acetylene.
See also carbon black.
Properties: High liquid adsorption, retention of high bulk volume, high electrical conductivity.
Use: Dry cell batteries, component of explosives, reinforcing agent in rubber and in thermal and sound insulation, gloss suppressor in paints, carburizing agent in hardening of steel, pigment in printing inks, filler in electroconductive polymers.

acetylene dichloride.   See sym-dichloroethylene.

acetylene hydrocarbon.   (alkyne).   One of a class of unsaturated hydrocarbons of the homologous series having the generic formula C_nH_{2n-2} and a structural formula containing a triple bond.

acetylene polymer.   See polyacetylene; cyclooctatetraene.

acetylene tetrabromide. (sym-tetrabromethane). CAS: 79-27-6. $CHBr_2CHBr_2$.
Properties: Yellowish liquid, soluble in alcohol and ether, insoluble in water, d 2.98–3.00, bp 239–242C with decomposition, 151C (54 mm), mp 0.1C, refr index 1.638. Combustible.
Derivation: Interaction of acetylene and bromine and subsequent distillation.
Method of purification: Rectification.
Grade: Technical.
Hazard: Irritant by inhalation and ingestion. TLV: 1 ppm in air.
Use: Separating minerals by specific gravity; solvent for fats, oils, and waxes; fluid in liquid gauges; solvent in microscopy.

acetylene tetrachloride. See sym-tetrachloroethane.

acetylenogen. Calcium carbide.

N-acetylethanolamine. (hydroxyethylacetamide). $CH_3CONHC_2H_4OH$.
Properties: Brown, viscous liquid; soluble in alcohol, ether, and water; d 1.122 (20/20C); boiling range 150–152C (5 mm); decomposes (10 mm); bulk d 9.34 lb/gal; refr index 1.4730 (20C); flash p 350F (176.6C); fp 15.8C. Combustible.
Grade: Technical (75% solution in water).
Use: Plasticizer for polyvinyl alcohol and for cellulosic and proteinaceous materials; humectant for paper products, glues, cork, and inks; textile conditioner.

acetylethanoltrimethylammonium hydroxide. See acetylcholine.

acetylferrocene. (ferrocenyl methyl ketone). $C_5H_5FeC_5H_4COCH_3$.
Properties: Orange crystals, mp 85–86C.
Use: Intermediate. See also ferrocene.

acetylformic acid. See pyruvic acid.

acetylide. A salt-like carbide formed by the reaction of acetylene and an alkali or alkaline-earth metal in liquid ammonia, or with silver, copper and mercury salts in aqueous solution. The latter are explosive when shocked or heated.
See also carbide; calcium carbide.

acetyl iodide. CH_3COI. CAS: 507-02-8.
Properties: Colorless, transparent, fuming liquid; suffocating odor; turns brown on exposure to air or moisture; soluble in ether and benzene; decomposed by water and alcohol; d 1.98; bp 105–108C; refr index 1.55.
Derivation: Reaction of acetic acid, iodine, and phosphorus; reaction of acetyl chloride and hydrogen iodide.
Method of purification: Distillation.
Grade: Technical.
Hazard: Strong irritant to eyes and skin.
Use: Organic synthesis.

acetylisoeugenol. (isoeugenol acetate). $C_6H_3(CHCHCH_3)(OCH_3)(OCOCH_3)$.
Properties: White crystals; spicy, clove-like odor; congealing point 77C; soluble in 27 parts of 95% alcohol. Combustible.
Method of purification: Crystallization.
Grade: Technical.
Use: Perfumery, particularly for carnation-type odors; flavoring.

acetyl ketene. See diketene.

acetylmethylcarbinol. (acetoin; 3-hydroxy-2-butanone; dimethylketol). CAS: 513-86-0. $CH_3COCHOHCH_3$.
Properties: Slightly yellow liquid or crystals (dimer), oxidizes gradually to diacetyl on exposure to air, d 1.016, bp 140–148C, mp 15C, soluble in alcohol, miscible with water, slightly soluble in ether. Combustible.
Derivation: Reduction of diacetyl.
Grade: Technical, FCC (as acetoin).
Use: Aroma carrier; preparation of flavors and essences.

acetyl nitrate. CH_3COONO_2.
Properties: Colorless, fuming liquid; hygroscopic; d 1.24.
Derivation: Reaction of acetic anhydride and nitrogen pentoxide.
Hazard: Corrosive to skin and mucous membranes. Explodes above 55C (130F) or in presence of mercuric oxide.
Use: Nitrating agent for organic compounds.

acetyl oxide. See acetic anhydride.

acetyl peroxide. (diacetyl peroxide). CAS: 110-22-5 $(CH_3CO)_2O_2$.
Properties: Colorless crystals, mp 30C, slightly soluble in cold water, soluble in alcohol and ether. Marketed as a 25% solution in dimethyl phthalate; flash p 113F(45C) (OC), fp approximately −8C, d 1.18 (20C). Flammable.
Hazard: (25% solution) Moderate fire risk; strong irritant to skin and eyes. The pure material is a strong oxidizer and severe explosion hazard; should not be stored after preparation, nor heated above 30C.
Use: Initiator and catalyst for resins.

acetylphenetidin. See acetophenetidin.

acetylphenol. See phenyl acetate.

N-acetyl-p-phenylenediamine. See p-aminoacet-anilide.

acetylpropionic acid. See levulinic acid.

acetyl propionyl. (2,3-pentanedione; methyl ethyl glyoxal; methyl ethyl diketone).
CAS: 600-14-6. $CH_3COCOCH_2CH_3$.
Properties: Yellow liquid, mp −52C, bp 106–110C, d 0.955–0.959 (15/4C), partly soluble in water. Combustible.
Grade: 99%.
Use: Flavors of butterscotch and chocolate type.

4-acetylresorcinol. (2,4-dihydroxyacetophe-none). $C_6H_3(OH)_2COCH_3$.
Properties: Light tan crystals, mp 146–148C. High absorptivity in UV region. Slightly soluble in water; soluble in most organic solvents except benzene and chloroform.
Use: UV absorber in plastics; dye intermediate; fungicide; plant growth promoter.

acetylsalicylic acid. See aspirin.

N-acetylsulfanilyl chloride. (p-acetamidobenzenesulfonyl chloride; p-acetylaminobenzenesulfonyl chloride). $(CH_3CONH)C_6H_4(SO_2Cl)$.
Properties: Light tan to brownish powder or fine crystals, mp 149C, soluble in chloroform and ethylene dichloride.
Hazard: Irritant to skin and mucous membranes.
Use: Intermediate in the manufacture of sulfa drugs.

2-acetylthiophene. (methyl-2-thienyl ketone).
$CH_3COC_4H_3S$.
Properties: Yellowish, oily liquid; mp 10–11C; bp 213.5C; 88–90C (10 mm); very slightly soluble in ether.
Use: Organic intermediate.

acetyl-o-toluidine. (o-acetotoluidide).
$CH_3CONHC_6H_4CH_3$.
Properties: Colorless crystals; mp 110C; bp 296C; d 1.168 (15C); soluble in alcohol, ether, benzene, chloroform, glacial acetic acid; slightly soluble in cold water; insoluble in hot water.
Derivation: By boiling glacial acetic acid with o-toluidine and distilling the product.
Grade: Technical.
Use: Organic synthesis.

acetyl-p-toluidine. (p-acetotoluidide).
$CH_3CONHC_6H_4CH_3$.
Properties: Colorless needles, mp 153C, bp 307C, d 1.212 (15/4C), slightly soluble in water; soluble

in alcohol, ether, ethyl acetate, glacial acetic acid.
Grade: Technical.
Use: Dyes.

acetyl triallyl citrate.
$CH_3COOC_3H_4(COOCH_2CH{:}CH_2)_3$.
Properties: Liquid, boiling range 142–143C (0.2 mm), d 1.140 (20C), refr index 1.4665 (25C), flash p 365F (185C). Combustible.
Use: Crosslinking agent for polyesters; monomer for polymerization.

acetyl tributyl citrate.
$CH_3COOC_3H_4(COOC_4H_9)_3$.
Properties: Colorless, odorless liquid; distillation range 172–174C (1 mm); pour p −60C; d 1.046 (25C); bulk d 8.74 lb/gal (25C); refr index 1.4408 (25C); viscosity 42.7 cps (25C); insoluble in water; flash p 400F (204C). Combustible.
Derivation: Esterification and acetylation of citric acid.
Grade: Technical.
Use: Plasticizer for vinyl resins.

acetyl triethyl citrate.
$CH_3COOC_3H_4(COOC_2H_5)_3$.
Properties: Colorless, odorless liquid; distillation range 131–132C (1 mm); pour p −47C; d 1.135 (25C); bulk d 9.47 lb/gal (25C); refr index 1.4386 (25C); viscosity 53.7 cps (25C); slightly soluble in water; flash p 370F (187C). Combustible.
Derivation: Esterification and acetylation of citric acid.
Grade: Technical.
Use: Plasticizer for cellulosics, particularly ethyl cellulose.

acetyl tri-2-ethylhexyl citrate.
$CH_3COOC_3H_4(COOC_8H_{17})_3$.
Properties: Liquid, bp 225C (1 mm), insoluble in water, flash p 430F (222C). Combustible.
Grade: Technical.
Use: Low-volatility plasticizer for vinyl resins.

N-acetyltryptophan. Available commercially as N-acetyl-L-tryptophan, mp 185–186C; N-acetyl-dl-triptophan, mp 205C.
Use: Nutrition and biochemical research; medicine.

acetyl valeryl. (heptadione-2,3).
$CH_3COCOC_4H_9$.
Yellow liquid, 92% pure. Combustible.
Use: Cheese, butter, and miscellaneous flavors.

acetyl value. The number of milligrams of potassium hydroxide required for neutralization of acetic acid obtained by the saponification of 1 g of acetylated fat or oil sample. Acetylation is carried out by boiling the sample with an equal amount of acetic anhydride, washing, and drying. Saponification values on the acetylated and on untreated fat are determined. From the results the acetyl value is calculated. It is a measure of the number of free hydroxyl groups in the fat or oil.

ACGIH. Abbreviation for American Conference of Governmental Industrial Hygienists. See Threshold Limit Values.

"Achromycin"[315]. TM for tetracycline hydrochloride.

acicular. Used in describing needle-shaped crystals or the particles in powders.

acid. One of a large class of chemical substances whose water solutions have one or more of the following properties; sour taste, ability to make litmus dye turn red and to cause other indicator dyes to change to characteristic colors, ability to react with and dissolve certain metals to form salts and ability to react with bases or alkalies to form salts. All acids contain hydrogen. In water, ionization or splitting of the molecule occurs so that some or most of this hydrogen forms H_3O^+ ions (hydronium ions), usually written more simply as hydrogen ions, H^+.

Acids are referred to as strong or weak according to the concentration of H^+ ion that results from ionization. Hydrochloric, nitric, and sulfuric are strong or highly ionized acids; acetic acid (CH_3COOH) and carbonic acid (H_2CO_3) are weak acids. Tenth normal hydrochloric acid is 100 times as acid (pH = 1) as tenth normal acetic acid (pH = 3). The pH range of acids is from 6.9 to 1.
See also pH.
When dealing with chemical reactions in solvents other than water, it is sometimes convenient to define an acid as a substance that ionizes to give the positive ion of the solvent. The common definitions of acid have been extended as more detailed studies of chemical reactions were made. The Lowry-Bronsted definition of an acid as a substance that can give up a proton is more useful in connection with an understanding of bases (see base). Perhaps the most significant contribution to the theory of acids was the electron-pair concept introduced by G. N. Lewis around 1915.
See Lewis electron theory.
The terms "hard" and "soft" acids and bases refer to the ease with which the electron orbitals can be disturbed or distorted. Hard acids have a high positive oxidation state and their valence electrons are not readily excited; soft acids and bases have little or no positive charge and easily excited

valence electrons. Hard acids combine preferentially with hard bases, and soft acids with soft bases. Soft acids tend to accept electrons and form covalent bonds more readily than hard acids. For example, the halogen acids arranged in a series by increasing atomic weight (and decreasing chemical activity) show a progression from hard (HF) to soft (HI).

A brief outline of the major groups of acids is as follows:
Inorganic
Mineral acids; sulfuric, nitric, hydrochloric, phosphoric. Hazard: All mineral acids are highly irritant and corrosive to human tissue.
Organic
Carboxylic (contain —COOH group)
aliphatic; acetic, formic
aromatic: benzoic, salicylic
Dicarboxylic (contain 2 —COOH groups)
oxalic, phthalic, sebacic, adipic
Fatty acids (contain —COOH group)
aliphatic: oleic, palmitic, stearic
aromatic: phenylstearic
Amino acids N-containing protein components
For further details, see Lewis acid, carboxylic acid, fatty acid, amino acid, and specific compounds.

1,2,4-acid. (1-amino-2-naphthol-4-sulfonic acid). $C_{10}H_5NH_2OHSO_3H$.
Properties: Pinkish-white to gray needles. Soluble in hot water, but almost insoluble in cold water.
Derivation: β-naphthol is nitrated to nitroso-β-naphthol by reaction with nitrous acid and the product treated with sodium bisulfite. Upon acidification the free sulfurous acid effects simultaneous reduction and sulfonation.
Use: Aniline dye intermediate.

1,8,2,4-acid. See Chicago acid.

acid amide. See amide.

acid ammonium sulfate. See ammonium bisulfate.

acid ammonium tartrate. See ammonium bitartrate.

acid anhydride. An oxide of a nonmetallic element or of an organic radical which is capable of forming an acid when united with water, or which can be formed by the abstraction of water from the acid molecule, or which can unite with basic oxides to form salts.

Acid Black 2.
Use: Hair color, reagent, biological stain.

acid butyl phosphate. See n-butyl acid phosphate.

acid calcium phosphate. See calcium phosphate, monobasic.

acid dye. An azo, triarylmethane, or anthraquinone dye with acid substituents such as nitro-, carboxy-, or sulfonic acid. They are most frequently applied in acid soluble to wool and silk, and no doubt combine with the basic groups of the proteins of those animal fibers. Orange II (C.I. 15510), black 10B, and acid alizarine blue B are examples.

acid ethylsulfate. See ethylsulfuric acid.

acid fungal protease. (fungal protease enzyme). Highly off-white powder.
Use: As a replacement for pepsin; chill-proofing agent for beer; in cereal treatment; feed supplement for baby pigs; rennet extender.

acid glaucine blue. See peacock blue.

acid, hard. See Lewis electron theory; acid.

acidic oxide. An oxide of a nonmetal, e.g., SO_2, CO_2, P_2O_5, SO_3, which form acids when combined with water.
See also acid anhydride.

acidimetry. The determination of the concentration of acid solutions or of the quantity of acid in a sample or mixture. This is usually done by titration with a solution of base of known strength (standard solution); an indicator is used to establish the end point.
See also pH.

acid lining. Silica brick lining used in steel-making furnaces.

acid magnesium citrate. See magnesium citrate, dibasic.

acid magnesium phosphate. See magnesium phosphate, monobasic.

acid methyl sulfate. See methylsulfuric acid.

acid mine drainage. (AMD). Water from both active and inactive coal mines which has become contaminated with sulfuric acid as a result of hydrolysis of ferric sulfate, the oxidation product of pyrite. This is a factor in water and stream pollution, which can be corrected by use of appropriate ion-exchange resins.

acid number. See acid value.

acid phosphatase. An enzyme found in blood serum which catalyzes the liberation of inorganic phosphate from phosphate esters. Optimum pH 5; is less active than alkaline phosphatase.
Use: Biochemical research.

acid phosphate. An acid salt of phosphoric acid such as NaH_2PO_4, $CaHPO_4$, etc. Also used to refer specifically to calcium phosphate monobasic, $Ca(H_2PO_4)_2$, or superphosphate of lime.

acid potassium oxalate. See potassium binoxalate.

acid potassium sulfate. See potassium bisulfate.

acid precipitation. (acid rain). Any form of precipitation (wet deposition) having a pH of 5.6 or less, the most important deleterious components being the sulfur dioxide and oxides of nitrogen either emitted as stack gases in highly industrialized areas or resulting from volcanic activity. The most sensitive sections of the US are the east and the extreme northwest; southeastern Canada is also affected. The average pH of precipitation in the New York area (1975–77) was 4.28; the extreme range of pH recorded in Glacier Park, Montana from 1972–76 was from 2.6 to 7.1 for an average of 5.78, due to natural causes. Acid precipitation is not only destructive to fish and other fresh water life, but also kills certain species of trees (especially spruce) and corrodes metal and cement structures. Increased industrial use of coal in recent years has been largely responsible for the incidence of acid rain, especially in the northeast.
See also dry deposition.

acid rain. See acid precipitation.

acid, soft. See Lewis electron theory; acid.

acidulant. Any of a number of acids (chiefly organic) either occurring naturally in fruits and vegetables or used as additives in food processing. They function in the following ways: (1) as bacteriostats in processed foods; (2) as aids to the sterilization of canned foods by lowering the pH; (3) as chelating agents for metal ions such as iron and copper which catalyze rancidity reactions in fats; (4) as flavor enhancers, by offsetting excessive sweetness by their tart taste. Commonly used acidulants are citric, acetic, fumaric, ascorbic, propionic, lactic, adipic, malic, sorbic, and tartaric acids.

"Acidulin."[100] TM for glutamic acid hydrochloride acid value. The number of milligrams of potassium hydroxide required to neutralize the free acids present in 1 g of oil, fat, or wax. The determination is made by titrating the sample in hot 95% ethanol using phenolphthalein as indicator.

Acid Yellow 9.   See 4-aminoazobenzene-3,4'-disulfonic acid.

acifluorfen, sodium salt.   (Blazer).
Use: Post emergent herbicide for soybeans, peanuts, and rice.

"Acintol."[252]   TM for a series of tall oils, crude and distilled, and tall oil derivatives such as fatty acids, rosins, tall oil heads, pitch. The derivatives are usually obtained by fractional distillation. Combustible
Use: Adhesives; cement; intermediates; degreasing compounds; emulsifiers; flotation agents; ink vehicles; leather chemicals; lubricants; metallic soaps; oil-well drilling muds; paints and varnishes; rubber chemicals; soaps and cleaners.

"ACL."[55]   TM for a series of solid, organic chlorine liberating compounds used in bleaches, cleansers, sanitizers, etc.

"Aclar."[50]   TM for a series of fluorohalocarbon films.
Properties: Useful properties from -200 to $+198C$.
Use: In packaging applications where a transparent, vapor, and/or gas barrier is required, as in packaging of foods for astronauts. Used in electronic and electrical applications because of insulating and heat-resistant properties. Extreme chemical resistance and ability to seal make it useful as a tank lining, etc.

"A-C-M."[299]   TM for a balanced mixture of ascorbic acid (vitamin C) and citric acid.
Use: As an antioxidant that protects flavor and prevents browning of fruits exposed to air. Used in home freezing and canning of fresh fruits.

"Acofor."[36]   TM for pale distilled tall oil fatty acids.
Properties: D 0.907 (25/25C); refr index 1.471 (20C); flash p 380F (193C) (OC), acid number 192; saponification number 194; unsaponifiable matter 2.5%; rosin acids 4.5%. Combustible.
Use: Paint and varnish; inks; soaps; disinfectants; textile oils; core oils, etc.

"Aconew."[36]   TM for pale distilled tall oil fatty acids with low rosin acid contents.

aconite.   (monkshood; wolfsbane; friar's cowl).
Hazard: An antipyretic drug; an alkaloid poison.

aconitic acid.   (propene-1,2,3-tricarboxylic acid). $H(COOH)C:C(COOH)CH_2(COOH)$.
Properties: White to yellowish, crystalline solid; mp approximately 195C with decomposition; soluble in water and alcohol. Combustible.
Derivation: (a) By dehydration of citric acid with sulfuric acid; (b) extraction from sugar cane bagasse, Aconitum napellus and other natural sources.
Use: Preparation of plasticizers and wetting agents; antioxidant; organic syntheses; itaconic acid manufacture; synthetic flavors.

aconitine.   (acetyl benzoyl aconine).
CAS: 302-27-2. $C_{34}H_{49}NO_{11}$.
Hazard: A highly toxic alkaloid, an antipyretic drug; readily absorbed by skin.

acraldehyde.   See acrolein.

"Acrawax."[73]   TM for a series of synthetic waxes supplied in solid and powdered forms.
Properties: Melting range from 83 to 143C.
Use: As antitack agents; flatting agents in paint; lubricant and mold release agents for butyl and neoprene elastomers; adhesives; rubber; plastics.

acridine (tricyclic).   CAS: 260-94-6. $C_{13}H_9N$.
Properties: Small, colorless needles. Soluble in alcohol, ether, or carbon disulfide; sparingly soluble in hot water. Sublimes at 100C, mp 111C, bp above 360C.
Derivation: (a) By extraction with dilute sulfuric acid from the anthracene fraction from coal tar and adding potassium dichromate. The acridine chromate precipitated is recrystallized, treated with ammonia and recrystallized. (b) Synthetically.
Hazard: Strong skin irritant.
Use: Manufacture of dyes; derivatives, especially acriflavine, proflavine; analytical reagent.

acridine orange.   (N,N,N',N'-tetramethyl-3,6-acridinediamine monohydrochloride).
CAS: 65-61-2. $C_{17}H_{19}N_3 \cdot HCl$.
Hazard: An in vitro mutagen.
Use: Selective biological stain for tumor cells, *in travitam*, and causes retardation of tumor growth.

acriflavine. $C_{14}H_{14}N_3Cl$.   A mixture of 3,6-diamino-10-methylacridinium chloride and 3,6-diaminoacridine.
Properties: Brownish or orange, odorless, granular powder. Soluble in 3 parts of water; incompletely soluble in alcohol; nearly insoluble in ether and chloroform. The aqueous solutions fluoresce

green on dilution. Also available as the hydro-chloride.
Use: Antiseptic and bacteriostat.

"Acrilan."[58] TM for an acrylic fiber.

"Acival."[496] TM for an aqueous acrylic emulsion.
Use: As a fabric finish, hand modifier, anticrack agent for pigment prints, non-woven binder, and upholstery backing.

"Acriviolet."[243] TM for dye mixture used as oral antiseptic.

acroleic acid. See acrylic acid.

acrolein. (2-propenal; acrylaldehyde; allyl alde-hyde; acraldehyde). CAS: 107-02-8. CH_2CHCHO.

$$
\begin{array}{c}
CH{=}O \\
| \\
CH \\
\| \\
CH_2
\end{array}
$$

Properties: Colorless or yellowish liquid; disagree-able, choking odor. Soluble in water, alcohol, and ether. Polymerizes readily unless inhibitor (hydroquinone) is added. Very reactive. Bp 52.7C, mp −87.0C, d 0.8427 (20/20C); bulk d 7.03 lb/gal (20C); flash p below 0F (−17C) (COC). Autoign temperature 532F (277C).
Derivation: (a) Oxidation of allyl alcohol or pro-pylene; (b) by heating glycerol with magnesium sulfate; (c) from propylene with bismuth-phos-phorus-molybdenum catalyst.
Method of purification: Rectification.
Grade: Technical.
Hazard: Very irritant to eyes and skin; toxic by inhalation and ingestion. TLV: 0.1 ppm in air. Dangerous fire risk. Explosive limits in air 2.8 to 31%.
Use: Intermediate for synthetic glycerol, polyure-thane and polyester resins, methionine, pharma-ceuticals; herbicide; warning agent in gases.

acrolein dimer. (2-formyl-3,4-dihydro-2H-pyran). $OCH{:}CHCH_2CH_2CHCHO$.
Properties: Liquid, d 1.0775 (20C), bp 151.3C, fp −100C, flash p 118F (47.7C) (OC), bulk d 8.96lb/gal (20/20C), soluble in water. Flamma-ble.
Hazard: Moderate fire risk.
Use: Intermediate for resins, pharmaceuticals, dye-stuffs.

"Acronal."[440] TM for dispersions, solutions and solids of acrylate homo- and copolymers.

ACR process. Abbreviation for advanced crack-ing reactor.
See ethylene (note).

"Acrylafil."[539] TM for styrene-acrylonitrile polymer with glass fiber reinforcement. Available with 35 and 40% glass fiber content.

acrylaldehyde. See acrolein.

acrylamide. CAS: 79-06-1. $CH_2CHCONH_2$.
Properties: Colorless, odorless crystals; mp 84.5C; bp 125C (25 mm); d 1.122 (30C); soluble in wa-ter, alcohol, acetone; insoluble in benzene, hep-tane. The solid is stable at room temperature but may polymerize violently on melting.
Derivation: Reaction of acrylonitrile with sulfuric acid (84.5%) and neutralization.
Grade: Technical (approximately 97% pure).
Hazard: Toxic by skin absorption. Irritant to skin and mucous membranes. TLV: 0.03 mg/m³ of air, suspected of carcinogenic potential for hu-mans, toxic by skin absorption.
Use: Synthesis of dyes, etc.; cross-linking agent; adhesives; paper and textile sizes, soil condition-ing agents; flocculants; sewage and waste treat-ment; ore processing; permanent press fabrics.

acrylate. (1) Any of several monomers used for the manufacture of thermosetting acrylic sur-face coating resins, e.g., 2-hydroxyethyl acrylate (HEA) and hydroxypropyl acrylate (HPA).
    (2) Polymer of acrylic acid or its esters, used in surface coatings, emulsion paints, paper and leather finishes, etc.
See also acrylic acid; acrylic resin.

acrylic acid. (acroleic acid; propenoic acid). CAS: 79-10-7. $H_2C{:}CHCOOH$.

$$
\begin{array}{c}
HO \\
|\| \\
H_2C{=}C{-}C{-}OH
\end{array}
$$

Properties: Colorless liquid; acrid odor. Polymer-izes readily. Miscible with water, alcohol, and ether; bp 140.9C; mp 12.1C; d 1.052 (20/20C), vap press 3.1 mm (20C); bulk d 8.6 lb/gal (20C); refr index 1.4224 (20C); flash p 130F (54.5C) (OC). Combustible.
Derivation: (a) Condensation of ethylene oxide with hydrocyanic acid followed by reaction with sulfuric acid at 320F; (b) acetylene, carbon monoxide, and water, with nickel catalyst; (c) propylene is vapor-oxidized to acrolein, which is oxidized to acrylic acid at 300C with molybde-num-vanadium catalyst; (d) hydrolysis of acryl-onitrile.

Grade: Technical (esterification and polymerization grades); glacial (97%).
Hazard: Irritant to skin. Toxic by inhalation. May polymerize explosively. TLV: 10 ppm in air.
Use: Monomer for polyacrylic and polymethacrylic acids and other acrylic polymers. See acrylic resin.

acrylic fiber. A manufactured fiber in which the fiber-forming substance is any long-chain synthetic polymer composed of at least 85% by weight of acrylonitrile units —$CH_2CH(CN)$— (Federal Trade Commission).
Properties: Tensile strength, 2 to 3 g/denier; water absorption 1.5 to 2.5%; d approximately 1.17.
Hazard: Fumes are toxic. Combustible.
Use: Modacrylic fibers; blankets; carpets.
See also modacrylic fiber; acrylic resin.

acrylic polymers. See acrylic resin.

acrylic resin. (acrylic fiber; nitrile rubber).
Thermoplastic polymers or copolymers of acrylic acid, methacrylic acid, esters of these acids, or acrylonitrile. The monomers are colorless liquids that polymerize readily in the presence of light, heat, or catalysts such as benzoyl peroxide; they must be stored or shipped with inhibitors present to avoid spontaneous and explosive polymerization.
See also acrylic acid, acrylonitrile, methyl methacrylate.
Properties: Acrylic resins vary from hard, brittle solids to fibrous, elastomeric structures to viscous liquids, depending on the monomer used and the method of polymerization. A distinctive property of cast sheet and extruded rods of acrylic resin is ability to transmit light.
Use: Bulk-polymerized: Hard, shatterproof, transparent or colored material (glass substitute, decorative illuminated signs, contact lenses, dentures, medical instruments, specimen preservation, furniture components). Suspension-polymerized: beads and molding powders (headlight lenses, adsorbents in chromatography, ion-exchange resins). Solution polymers: coatings for paper, textiles, wood, etc. Aqueous emulsions: adhesives, laminated structures, fabric coatings, nonwoven fabrics. Compounded prepolymers: exterior auto paints, applied by spray and baked. Acrylonitrile-derived acrylics are extruded into synthetic fibers and are also the basis of the nitrile family of synthetic elastomers.
See also acrylic acid, acrylonitrile, methyl methacrylate.

"Acryloid" Coating Resins.[23] TM for acrylic ester polymers in organic solvent solutions or in

100% solids form, water-white, and transparent. Films range from very hard to very soft.
Use: Exceptionally resistant surface coatings, such as heat-resistant and fumeproof enamels, vinyl and plastic printing; fluorescent coatings; clear and pigmented coatings on metals.

"Acryloid" Modifiers.[23] TM for thermoplastic acrylic polymers in powder form. Various grades facilitate processing or improve physical properties of rigid or semi-rigid polyvinyl chloride formulations.

"Acryloid" Oil Additives.[23] TM for acrylic polymers supplied in special oil solution or in diester lubricant.
Use: Viscosity-index improvement; pour-point depression of lubricating oils and hydraulic fluids; sludge dispersancy in lubricating and fuel oils.

"Acrylon."[65] TM for a group of acrylic rubbers outstanding in resistance to oil, grease, ozone, and oxidation.
Use: Gaskets and rubber parts for contact with oils and diester lubricants.

acrylonitrile. (propenenitrile; vinyl cyanide).
CAS: 107-13-1. $H_2C:CHCN$. 38th highest-volume chemical produced in US (1985).
Properties: Colorless, mobile liquid; mild odor; fp −83C; bp 77.3–77.4C; d 0.8004 (25C); flash p 32F (0C) (TOC). Soluble in all common organic solvents; partially miscible with water.
Derivation: (a) From propylene oxygen and ammonia with either bismuth phosphomolybdate or a uranium-based compounds as catalysts; (b) addition of hydrogen cyanide to acetylene with cuprous chloride catalyst; (c) dehydration of ethylene cyanohydrin.
Hazard: Toxic by inhalation and skin absorption. A carcinogen. Flammable, dangerous fire risk. Explosive limits in air 3 to 17%. TLV: 2 ppm, suspect of carcinogenic potential for humans.
Use: Monomer for acrylic and modacrylic fibers and high-strength whiskers; ABS and acrylonitrile styrene copolymers; nitrile rubber; cyanoethylation of cotton; synthetic soil blocks (acrylonitrile polymerized in wood pulp); organic synthesis; adiponitrile; grain fumigant; monomer for a semiconductive polymer that can be used like inorganic oxide catalysts in dehydrogenation of tert-butanol to isobutylene and water.

acrylonitrile-butadiene rubber. See nitrile rubber.

acrylonitrile-butadiene-styrene resin. See ABS resin.

acrylonitrile dimer. See methylene glutaronitrile.

acrylonitrile-styrene copolymer. A thermoplastic blend of acrylonitrile and styrene monomers having good dimensional stability and suitable for use in contact with foods. Among its numerous applications is that of bottles for soft drinks (TM "Cyclesafe"). FDA regulations limit the amount of acrylonitrile monomer that will be allocated to migrate from the container to the contents as 0.3 ppm.

acryloyl chloride. (acrylyl chloride).
$H_2CCHCOCl$.
Properties: Liquid; bp 75C.
Use: Monomer; intermediate.

"Acrysol."[23] (1) TM for aqueous solutions of sodium polyacrylate or other polymeric acrylic salts.
Use: Thickeners in paints, fabric coatings and backings; adhesives.
(2) TM for polyacrylic acid and copolymer products in aqueous solutions or dispersions. Some grades are solutions of sodium polyacrylate.
Use: Warp size for synthetic fibers, cotton and rayon; modifier of starch sizes.

ACS. Abbreviation for American Chemical Society.

"Actafoam."[511] TM for an activator-stabilizer for vinyl foams containing azodicarbonamide. Lead-free.

"Actane."[142] TM for fluoride-containing additive supplied in powder form for acid pickling solutions to dissolve siliceous films on metals as well as to assist in etching aluminum and other metals.

ACTH. (adrenocorticotropic hormone; corticotropin). CAS: 9002-60-2. One of the hormones secreted by the anterior lobe of the pituitary gland. It stimulates an increase in the secretion of the adrenal cortical steroid hormones. It is a polypeptide consisting of a 39-unit chain of amino acids, the sequence varying in certain positions with the biological species. ACTH was synthesized in 1960.
Properties: White powder with molecular weight approximately 3500. Freely soluble in water; soluble in 60–70% alcohol or acetone. Solutions are stable to heat.
Source: Extracted from whole pituitary glands of swine, sheep, and oxen. Normally isolated from swine.
Grade: Pure; USP as corticotropin injections. Units: Based on comparison with USP Corticotropin Reference Standard.
Hazard: May have damaging side effects.
Use: Medicine; biochemical research.

"Acti-dione."[519] TM for antibiotic cycloheximide, an agricultural fungicide.

"Actidip."[204] TM for a mildly alkaline powder having the property of refining crystal size in phosphate baths when used just prior to the phosphate.

actinide series. (actinoid series). The group of radioactive elements starting with actinium and ending with Element 105. All are classed as metals. Those with atomic number higher than 92 are called transuranic elements. The series includes the following elements; actinium, 89; thorium, 90; protoactinium, 91; uranium, 92; neptunium, 93; plutonium, 94; americium, 95; curium, 96; berkelium, 97; californium, 98; einsteinium, 99; fermium, 100; mendelevium, 101; nobelium, 102; and lawrencium 103. The isotopes of several of these elements are under study for possible use in such fields as radiography, neutron activation analysis, hydrology, and geophysics.

actinium. Ac. A radioactive metallic element; first member of the actinide series.
Properties: Atomic number 89; aw 227 (most stable isotope); silvery white metal, mp 1050C, bp (est) 3200C, oxidation state +3. Eleven radioactive isotopes; 227 has half-life of 21.8 years.
Derivation: Uranium ores; neutron bombardment of radium. Several compounds have been prepared. Available commercially at 98% minimum purity.
Hazard: Radioactive bone-seeking poison.
Use: Radioactive tracer (225 isotope).
See also lanthanum, to which actinium is closely similar.

actinomycin. A family of antibiotics produced by Streptomyces; reported to be active against *E. coli*, other bacteria, fungi, and to have cytostatic and radiomimetic activity. There are many forms of actinomycin; two of commercial importance are cactinomycin and dactinomycin.

activated alumina. See alumina, activated.

activated carbon. See carbon, activated.

activated sludge. See sewage sludge (2).

activation. The process of treating a substance or a molecule or atom by heat or radiation or the presence of another substance so that the first mentioned substance, atom or molecule will undergo chemical or physical change more rapidly or completely. Common types of activation are:
(1) Processing of carbon black, alumina, and other materials to impart improved adsorbent qualities.

Subjecting the material to steam or carbon dioxide at high temperatures is the method usually used.

See alumina, activated; carbon, activated.

(2) Heating or otherwise supplying energy to a substance (for example, ultraviolet or infrared radiation) to attain the necessary level of energy for the occurrence of a chemical reaction, or for emission of desired light waves, as in fluorescence or chemical lasers. The term excitation is also used.

(3) An important variation of (2) is the process of making a material radioactive by bombardment with neutrons, protons, or other nuclear particles.

See also activation analysis.

(4) Catalysis basically a process by which energy of activation for occurrence of a reaction is lowered by the presence of a non-reacting substance.

activation analysis. An extremely sensitive technique for identifying and measuring very small amounts of various elements. A sample is exposed to neutron bombardment in a nuclear reactor, for the purpose of producing radioisotopes from the stable elements. The characteristics of the induced radiations are sufficiently distinct that different elements in the sample can be accurately identified. The technique is particularly useful when concentration of the elements are too small to be measured by ordinary means. Trace elements have thus been determined in drugs, fertilizers, foods, fuels, glass, minerals, dusts, water, toxicants, etc.

activator. (1) A metallic oxide that promotes cross-linking of sulfur in rubber vulcanization. By far the most widely used is zinc oxide; in rubber mixes where no organic accelerator is used, oxides of magnesium, calcium or lead are effective.

(2) A fatty acid that increases the effectiveness of acidic organic accelerators; stearic acid is generally used, especially with thiazoles.

(3) A substance necessary in trace quantities to induce luminescence in certain crystals. Silver and copper are activators for zinc sulfide and cadmium sulfide.

See also initiator.

active amyl alcohol. See 2-methyl-1-butanol.

active carbon. See carbon, activated.

activity. (1) Chemical activity (thermodynamic activity): A quantity replacing actual molar concentration in mathematical expressions for the equilibrium constant so as to eliminate the effect of concentration on equilibrium constant. (2) Activity coefficient: A fractional number which when multiplied by the molar concentration of a substance in solution yields the chemical activity. This term gives an idea of how much interaction exists between molecules at higher concentration. (3) Activity of metals or elements: An active element will react with a compound of a less active element, to produce the latter as the free element, and the active element ends up in a new compound. Thus, magnesium, an active metal, will displace copper from copper sulfate to form magnesium sulfate and free metallic copper; chlorine will liberate iodine from sodium iodide and sodium chloride is formed. See electromotive series. (4) Activity product: The number resulting from the multiplication of the activities of slightly soluble substances. This is frequently called the solubility product. (5) Catalytic activity: See catalysis. (6) Optical activity: The existence of optical rotation in a substance. (7) Radioactivity activity coefficient.

See activity (2).

activity series. (displacement series; electromotive series). An arrangement of the metals in the order of their tendency to react with water and acids, so that each metal displaces from solution those below it in the series and is displaced by those above it. The arrangement of the more common metals is: K Na Mg Al Zn Fe Sn Pb H Cu Hg Ag Pt Au

"Acto."[51] TM for refined petroleum sodium sulfonate. Used as an oil-soluble emulsifier and surface-active agent.

"Actol."[243] TM for a series of polyoxypropylene diols, triols, and polyols. These vary in molecular weight from approximately 1,000 to 3,600; the diols and triols are almost insoluble in water, but the polyols are completely miscible with it.

Use: Urethane foams, elastomers, and coatings.

"Actox."[268] TM for a series of lead-free zinc oxides manufactured by the American Process (produced from zinc ore), and the French Process (produced from metal).

Available forms: Pellets and free flowing.

Use: Reinforcing agent and accelerator activator for rubber.

acyl. An organic acid group in which the OH of the carboxyl group is replaced by some other substituent (RCO—). Examples: acetyl, CH_3CO—; benzoyl, C_6H_5CO—.

ADA. Abbreviation for acetonedicarboxylic acid. See β-ketoglutaric acid.

1-adamantanamine hydrochloride. See amantadine hydrochloride.

adamantane. (sym-tricyclodecane). $C_{10}H_{16}$.
Has unique molecular structure consisting of
four fused cyclohexane rings.
Properties: White crystals, mp 270C (sublimes),
approximately 99% pure. Derivatives recently
developed (alkyl adamantanes) have potential
uses in imparting heat-, solvent-, and chemical
resistance to many basic types of plastics. Synthetic lubricants and pharmaceuticals are also
based on adamantane derivatives. Adamantane
diamine is used to cure epoxy resins.
See also "Symmetrel."

Adams, Roger. (1889–1971). An American chemist, born in Boston; graduated from Harvard
where he taught chemistry for some years. After
studying in Germany, he moved to the University
of Illinois in 1916 where he later became Chairman of the Department of Chemistry (1926–
1954). During his prolific career, he made this
department one of the best in the country, and
strongly influenced the development of industrial
chemical research in the US. His executive and
creative ability made him an outstanding figure
as a teacher, innovator, and administrator.
Among his research contributions were development of platinum-hydrogenation catalysts, and
structural determinations of chaulmoogric acid,
gossypol, alkaloids, and marijuana. He held
many important offices, including president of
the ACS and AAAS, and was a recipient of the
Priestley medal.

adamsite. See diphenylamine chloroarsine.

addition polymer. A polymer formed by the direct
addition or combination of the monomer molecules with one another. An example is the formation of polystyrene by stepwise combination of
styrene monomer units (approximately 1000 per
macromolecule).
See also polymerization.

additive. A nonspecific term applied to any substance added to a base material in low concentrations for a definite purpose. Additives can be
divided into two groups: (1) those which have
an auxiliary or secondary function (antioxidants
inhibitors, thickeners, plasticizers, flavoring
agents, colorants, etc.) and (2) those that are
essential to the existence of the end product (leavening agents in bread, curatives in rubber, blowing agents in cellular plastics, emulsifiers in mayonnaise, polymerization initiators in plastics, and
tanning agents in leather). It seems logical that
the latter group should be regarded less as additives than as base materials, since the end products could not exist without them. In any case,

a specific functional name is preferable to the
neutral term "additive."
See food additive.

adduct. See inclusion complex.

adenine. (6-aminopurine).   CAS: 73-24-5.
$C_5H_5N_5$.

Properties: White, odorless, microcrystalline powder with sharp salty taste; mp 360–365C (decomposes). Very slightly soluble in cold water; soluble in boiling water, acids, and alkalies; slightly
soluble in alcohol; insoluble in ether and chloroform. Aqueous solutions are neutral.
Occurrence: Ribonucleic acids and deoxyribonucleic acids, nucleosides, nucleotides, and many
important coenzymes.
Derivation: By extraction from tea; by synthesis
from uric acid; prepared from yeast ribonucleic
acid.
Use: Medicine and biochemical research.

adenosine. (adenine riboside; 9-β-D-ribofurano-
syl-adenine).   CAS: 58-61-7.
$C_5H_4N_5 \cdot C_5H_8O_4$.   The nucleoside composed
of adenine and ribose.
Properties: White, crystalline, odorless powder
with mild, saline, or bitter taste; mp 229C. Quite
soluble in hot water; practically insoluble in alcohol.
Derivation: Isolation following hydrolysis of yeast
nucleic acid.
Use: Biochemical research.

adenosine diphosphate. (5'-adenylphosphoric
acid; ADP; adenosine-5'-pyrophosphate;
adenosine diphosphoric acid).   CAS: 58-64-0.
$C_{10}H_{15}N_5O_{10}P_2$.   A nucleotide found in all
living cells and important in the storage of energy
for chemical reactions.
Derivation: (a) From adenosine triphosphate by
hydrolysis with the enzyme adenosinetriphos-
phatase from lobster or rabbit muscle; (b) by
yeast phosphorylation of adenosine.
Use: Biochemical research.

Commercially available as the sodium or barium salt of adenosine diphosphoric acid.
See adenosine diphosphate.

adenosine monophosphate. See adenylic acid.

adenosine phosphate (USAN). (5′-adenyldiphosphoric acid; ATP; adenosine triphosphate). CAS: 56-65-5. $C_{10}H_{16}N_5O_{13}P_3$. A nucleotide which serves as a source of energy for biochemical transformations in plants (photosynthesis) and also for many chemical reactions in the body, especially those associated with muscular activity, and with replication of cell components.
Properties: White, amorphous powder; odorless; faint sour taste. Soluble in water; insoluble in alcohol, ether, and organic solvents; stable in acidic solutions; decomposes in alkaline solution.
Derivation: Isolation from muscle tissue; yeast phosphorylation of adenosine.
Use: Biochemical research.
Commercially available as the disodium, dipotassium, and dibarium salts.

3′-adenylic acid. (yeast adenylic acid). $C_{10}H_{14}N_5O_7P$.
Properties: Crystals, mp 197C (decomposes). Almost insoluble in cold water; slightly soluble in boiling water. Gives quantitative yield of furfural when distilled with 20% hydrochloric acid.
Derivation: Extracted from nucleic acids of yeast; also made synthetically.
Use: Biochemical research.

5′-adenylic acid. (adenosine monophosphate; AA; adenosine phosphate; AMP; adenosine phosphoric acid).   CAS: 61-19-8.
$C_{10}H_{14}N_5O_7P$.   The monophosphoric ester of adenosine, i.e., the nucleotide containing adenine, d-ribose, and phosphoric acid. Adenylic acid is a constituent of many important coenzymes. Cyclic adenosine-3′,5′-monophosphate is designated by biochemists as cAMP

Properties: (muscle adenylic acid): Crystalline solid; mp 196–200C. Readily soluble in boiling water. Gives only traces of furfural when boiled with 20% hydrochloric acid.
Derivation: Extracted from muscle tissue; phosphorylation of adenosine.
Use: Biochemical research.

adhesion. The state in which two surfaces are held together by interfacial forces, which may consist of valence forces or interlocking action, or both. (ASTM)

adhesive. Any substance, inorganic or organic, natural or synthetic, that is capable of bonding other substances together by surface attachment. A brief classification by type is as follows:
  I. Inorganic
    1. Soluble silicates (water glass)
    2. Phosphate cements
    3. Portland cement (calcium oxide-silica)
    4. Other hydraulic cements (mortar, gypsum)
    5. Ceramic (silica-boric acid)
    6. Thermosetting powdered glasses ("Pyroceram")
  II. Organic
    1. Natural
      (a) Animal
          Hide and bone glue; fish glue
          Blood and casein glues
      (b) Vegetable
          Soybean starch cellulosics, rubber latex, and rubber-solvent (pressure-sensitive).
          Gums, terpene resins (rosin), mucilages
      (c) Mineral
          Asphalt, pitches, hydrocarbon resins
    2. Synthetic
      (a) Elastomer-solvent cements
      (b) Polysulfide sealants
      (c) Thermoplastic resins (for hot-melts)
          Polyethylene, isobutylene, polyamides, polyvinyl acetate
      (d) Thermosetting resins
          Epoxy, phenoformaldehyde, polyvinyl butyral, cyanoacrylates.
      (e) Silicone polymers and cements
For further information, refer to Case Western Reserve University in Cleveland, Ohio, which maintains a fundamental research center for adhesives and coatings.
See also following entries.

adhesive, high-temperature. (1) Organic polymers, e.g., polybenzimidazoles, that retain bond-

ing strength up to 260C for a relatively long time (500–1000 hours); strength drops rapidly above 260C, 80% being lost after 10 minutes at 535C. (2) Inorganic (ceramic), e.g., silica-boric acid mixtures or cermets produce bonds having high strength above 2000F adhesive lap-bond strengths can be over 2000 psi at 1000F. These adhesives are used largely for aerospace service, and metal/metal and glass/metal seals. A silicone cement is reported to have been used to adhere tiles to spacecraft.
See "RTV."

adhesive, hot-melt. A solid, thermoplastic material which quickly melts upon heating, and then sets to a firm bond on cooling. Most other types of adhesives set by evaporation of solvent. Hot-melt types offer the possibility of almost instantaneous bonding, making them well-suited to automated operation. In general, they are low-cost, low-strength products, but are entirely adequate for bonding cellulosic materials. Ingredients of hot-melts are polyethylene, polyvinyl acetate polyamides, hydrocarbon resins, as well as natural asphalts, bitumens, resinous materials, and waxes.
Use: Rapid and efficient bonding of low-strength materials, e.g., bookbinding, food cartons, side-seaming of cans, miscellaneous packaging applications.
See also sealant.

adhesive, rubber-based. (cement, rubber).
(1) A solution of natural or synthetic rubber in a suitable organic solvent, without sulfur or other curing agent; (2) a mixture of rubber (often reclaimed), filler, and tackifier (pine tar, liquid asphalt) applied to fabric backing (pressure-sensitive friction tape); (3) a room-temperature curing rubber-solvent-curative mixture, often made up in two parts which are blended just before use; (4) rubber latex especially for on-the-job repairing such as conveyor belts; (5) silicone rubber cement (see "RTV" and silicone).
Hazard: Those containing organic solvents, (1) and (3) above, are flammable.

adiabatic. Descriptive of a system or process in which no gain or loss of heat is allowed to occur.

adipic acid. (hexanedioic acid; 1,4-butanedicarboxylic acid).

$$CH_2CH_2CO_OH$$
$$CH_2CH_2COOH$$

$COOH(CH_2)_3COOH$.   48th highest-volume chemical produced in US (1985).

Properties: White, crystalline solid; mp 152C; bp 265C (100 mm); d 1.360 (20/4C); flash p 385F (196C) (CC). Slightly soluble in water, soluble in alcohol and acetone. Relatively stable. Combustible.
Derivation: Oxidation of cyclohexane, cyclohexanol, or cyclohexanone with air or nitric acid.
Grade: Technical; FCC.
Use: Manufacture of nylon and of polyurethane foams; preparation of esters for use as plasticizers and lubricants; food additive (acidulant), baking powders, adhesives.

"Adipol."[55]   TM for a series of adipate plasticizers.

adiponitrile. (1,4-dicyanobutane).
$NC(CH_2)_4CN$.
Properties: Water-white, odorless liquid; mp 1–3C; refr index 1.4369 (20C); bp 295C; flash p 200F (93.3C) (OC). Combustible. Slightly soluble in water, soluble in alcohol and chloroform.
Derivation: Chlorination of butadiene to dichlorobutylene, which is reacted with 35% sodium cyanide soluble to yield 1,4-dicyanobutylene which is hydrogenated to adiponitrile. Also by electroorganic synthesis from acrylonitrile.
Hazard: Toxic by ingestion and inhalation.
Use: Intermediate in the manufacture of nylon; organic synthesis.

"Adiprene."[28]   TM for a polyurethane rubber, the reaction product of diisocyanate and polyalkylene ether glycol. In its raw polymer form, it is a liquid of honey-like color and consistency which is compounded chemically (to polymerize it further) and converted into products by casting and other techniques.
See also polyurethane rubber.

adjuvant. A subsidiary ingredient or additive in a mixture (medicine, flavoring, perfume, etc.) which contributes to the effectiveness of the primary ingredient.

Adkins Catalyst. A catalyst containing copper chromite and copper oxide. It is used for the reduction of organic compounds usually at high temperatures and pressures. It is likewise used as a catalyst for dehydrogenation and for decarboxylation reactions.

Adkins-Peterson Reaction. Formation of formaldehyde by air-oxidation of methanol in the vapor phase over metal oxide catalysts. A 40% aqueous formaldehyde solution is obtained.

"ADMA."[313]   TM for a group of alkyldimethyl amines composed of even-numbered carbon chains from C_8 to C_{18}.

"Admex."[589] TM for a series of plasticizers consisting variously of epoxidized soybean oil, tallate esters, monomeric esters, and polyesters. Used in vinyl plastics.

admiralty metal. A nonferrous alloy containing 70–73% copper, 0.75–1.20% tin, remainder zinc. It offers good resistance to dilute acids and alkalies, seawater, and moist sulfurous atmospheres; d 8.53 (20C); liquidus temperature 935C; solidus temperature 900C.
Use: Condenser, evaporator, and heat exchanger tubes, plates, and ferrules.

adocain. A mixture of cocaine hydrochloride and adonidin (a glucoside from adonis vernalis).
Use: Heart stimulant and diuretic.

"Adofoam."[544] TM for oil-field additive used specifically in air-drilling and hydraulic fracturing to produce stable foam in fresh water, salt water, acid/water solutions, sulfur/water, and oil/water mixtures.

"Adogen."[589] TM for a series of fatty nitrogen chemicals including amines, amides, amine acetates and quaternary ammonium compounds. Available in various grades for specific applications in fabric softeners, ore separation, detergents, petroleum additives, corrosion inhibitors, bactericides, printing inks, antiblock and slip agents, waterproofing formulations and chemical intermediates.

"Adol."[589] TM for a series of industrial fatty alcohols, specifically cetyl, stearyl, and oleyl, available in a variety of grades for specific applications.

"Adomall."[544] TM for multi-value, water-fracturing additive for oil-field operations; kills bacteria, inhibits corrosion, lowers surface tension, reduces permeability damage, removes drilling mud from bore and fracture areas, and produces stable foam to return fracturing water from wells.

"Adomite."[544] TM for oil-field additives to reduce fluid loss during hydraulic fracturing. "Adomite Mark II" functions in oil-based fracturing fluids, "Adomite Aqua" in water-based fracture fluids.

"Adoquat."[544] TM for a quaternary ammonium salt used in water-flooding operations for secondary recovery of petroleum. It increases efficiency by inhibiting bacterial growth and reducing microbial plugging.

ADP. Abbreviation for (1) adenosine diphosphate, (2) ammonium dihydrogen phosphate.
See ammonium phosphate, monobasic.

adrenaline. (epinephrine). A hormone having a benzenoid structure.
$C_9H_{13}O_3N$.
It is obtained by extraction from the adrenal glands of cattle, and is also made synthetically. Its effect on body metabolism is pronounced, causing an increase in blood pressure and rate of heart beat. Under normal conditions, its rate of release into the system is constant; but emotional stresses such as fear or anger rapidly increase the output and result in temporarily heightened metabolic activity.
Hazard: Toxic by ingestion and injection.

adrenocorticotropic hormone. See ACTH; corticoid hormone.

"Adriamycin."[588] (doxorubicin hydrochloride). An antibiotic reported to be effective against such types of cancer as leukemia and cancers of the breast and bladder. It is made by fermentation of a soil fungus. Approved by FDA for clinical research, but is said to have deleterious side effects. Synthetic routes to adriamycin and its analogs have been developed.

adsorbent. A substance which has the ability to condense or hold molecules of other substances on its surface. Activated carbon, activated alumina, and silica gel are examples.

"Adsorbosil."[425] TM for adsorbents for use in chromatography. "Adsorbosil"-1 is a mixture of specially purified silica gel and 10% calcium sulfate designed as a thin-layer chromatography powder; "Adsorbosil" CAB is designed for column chromatography.

adsorption. Adherence of the atoms, ions, or molecules of a gas or liquid to the surface of another substance, called the adsorbent. The best-known examples are gas/solid and liquid/solid systems. Finely divided or microporous materials presenting a large area of active surface are strong adsorbents, and are used for removing colors, odors, and water vapor (activated carbon, activated alumina, silica gel). The attractive force of adsorption is relatively small, of the order of van der Waal's forces. When molecules of two or more substances are present, those of one substance may be adsorbed more readily than those of the others. This is called preferential adsorption.
See also absorption, chemisorption.

adsorption indicator. A substance used in analytical chemistry to detect the presence of a slight excess of another substance or ion in solution as the result of a color produced by adsorption of the indicator on a precipitate present in the solution. Thus, a precipitate of silver chloride will turn red in a solution containing even a minute excess of silver ion (silver nitrate solution), if fluorescein is present. In this example, fluorescein is the adsorption indicator.

"Advabrite."[230] TM for a series of optical brighteners giving a bright, blue-white fluorescence in very dilute solutions either in daylight or UV light.

"Advacar."[230] TM for a series of water-dispersible paint driers. Supplied in high metal concentration (calcium 4%, cobalt 6%, lead 24%, manganese 6%).

"Advacide."[230] TM for a series of fungicides and mildewcides for paints, etc., wood and fabric preservatives. "Advacide" TPLA is a free-flowing powder of triphenyl lead acetate containing 10% of a liquid aromatic hydrocarbon mixture, and used in antifouling paints.

"Advalite."[230] TM for a series of organic and organometallic compounds, specifically designed for the stabilization of vinyl coatings systems against heat and light degradation.

"Advawax."[230] TM for a series of wax modifiers. "P" = High molecular weight polybutene and paraffin wax. "M" = High molecular weight polybutene and amber-colored, microcrystalline wax. "280" = A hard, high-melting point synthetic wax for use in lacquers, varnishes, as a plastic lubricant and anti-blocking agent, as an extender or substitute for carnauba wax.

"Advawet."[230] TM for a series of wetting agents for emulsion paints, latex paints, and vinyl plastisols.

AEPD. Abbreviation for 2-amino-2-ethyl-1,3-propanediol.

aerate. To impregnate or saturate a material (usually a liquid) with air, or some similar gas. This is usually achieved by bubbling the air, through the liquid, or by spraying the liquid into air.

aerobic. Requiring air or oxygen.
See bacteria.

"Aero."[57] TM used as a combining form in naming a group of chemical products, e.g., "Aerofloat." They include the following:

case-hardening   metal heat-treating salts
mixture

catalysts   metallurgical additives
fertilizer additives   anticaking agents
flotation agents   sizing emulsions
flocculants   wetting agents
frothing/foaming   reinforcing agents
  agents

aerogel. Dispersion of a gas in a solid or a liquid. The reverse of an aerosol; flexible and rigid plastic foams are examples.
See foam, aerosol.

aerosol. A suspension of liquid or solid particles in a gas, the particles often being in the colloidal size range. Fog and smoke are common examples of natural aerosols; fine sprays (perfumes, insecticides, inhalants, anti-perspirants, paints, etc.) are man-made.

aerosols. Suspensions of various kinds may be formed by placing the components together with a compressed gas in a container (bomb). The pressure of the gas causes the mixture to be released as a fine spray (aerosol) or foam (aerogel) when a valve is opened. This technique is used on an industrial scale to spray paints and pesticides. It is also used in consumer items such as perfumes, deodorants, shaving cream, whipped cream, and the like. The propellant gas may be a hydrocarbon (propane, isobutane) or dimethyl ether. Admixture of 15% of methyl chloride with the hydrocarbons reduces their flammable risk, while water can be used with dimethyl ether. Carbon dioxide generated *in situ* is a more recently developed propellant, which does away with the flammability problem.

"Aerothene."[233] TM for a group of chlorinated solvents used as vapor pressure depressants, or with compressed gases to replace fluorocarbon propellant systems.

aerozine. A 1:1 mix of hydrazine and uns-dimethylhydrazine (UDMH).
Hazard: Flammable and explosive.
Use: Rocket fuels.
See also hydrazine.

AES. Abbreviation for Auger electron spectroscopy.
See spectroscopy.

AET. See aminoethylisothiourea dihydrobromide.

affinin. (N-isobutyl-2,6,8-decatrienamide). $C_{14}H_{23}NO$.
Properties: Yellowish, oily liquid; bp 163C (0.5 mm); mp 23C; refr index 1.52. Soluble in alcohol; insoluble in alkalies and acids.

Derivation: From *Heliopsis longipes*, or made synthetically.
Use: Insecticide activator.

affinity. The tendency of an atom or compound to react or combine with atoms or compounds of different chemical constitution. For example, paraffin hydrocarbons were so named because they are quite unreactive, the word "paraffin" meaning "very little affinity." The hemoglobin molecule has a much greater affinity for carbon monoxide than for oxygen. The free energy decrease is a quantitative measure of chemical affinity.

aflatoxins. A group of polynuclear molds (mycotoxins) produced chiefly by the fungus Aspergillus flavus; they are natural contaminants of a wide range of fruits, vegetables, and cereal grains.
Hazard: Highly toxic to many species of animals, including fish and birds. The B_2 and G_1 strains are known carcinogens. Aflatoxins fluoresce strongly under UV, are soluble in methanol, acetone, and chloroform, but only slightly soluble in water and hydrocarbon solvents. Prevention of mold growth is the most effective protection; removal or inactivation is possible by physical or chemical means (hand-sorting, solvent refining, etc.). Complete elimination of aflatoxins from foods is not feasible; FDA sets an upper limit of
20 ppb in foods and feeds, and 0.5 ppb in milk.
See also mycotoxin.

after-chromed dye. A dye that is improved in color or fastness by treatment with sodium dichromate, copper sulfate, or similar materials, after the fabrics are dyed.

Ag. Symbol for silver.

Agar. (agar-agar). A phycocolloid derived from red algae such as Gelidium and Gracilaria; it is a polysaccharide mixture of agarose and agaropectin.
Properties: Thin, translucent membranous pieces or pale buff powder. Strongly hydrophilic, it absorbs 20 times its weight of cold water with swelling; forms strong gels at approximately 40C.
Grade: Technical, USP, FCC.
Use: Microbiology and bacteriology (culture medium); antistaling agent in bakery products, confectionery, meats and poultry, gelation agent, desserts and beverages, protective colloid in ice cream, pet foods, health foods, laxatives, pharmaceuticals, dental impressions, laboratory reagents, photographic emulsions.
See also algae, alginic acid.

ageing. (1) *Deleterious*: Gradual deterioration of a material due to long exposure to the environment. Ageing characteristics of various materials:
(a) Vulcanized rubber and thermoplastic polymers lose strength and crack due to oxidation, sunlight, heat; this is retarded by antioxidants, e.g., phenyl-β-naphthylamine. For accelerated ageing tests see bomb.
(b) Foods: Spoilage and rancidity due to bacterial contamination, retarded by butylated hydroxyanisole and various propionates.
(c) Paints: Cracking, fading, chalking due to exposure to weather and photochemical degradation. Retarded by proper selection of vehicles and pigments. Accelerated weathering tests are used ("Weather-O-Meter").
(d) Metals: Rusting, pitting, and scaling due to corrosion, especially in moist acid and alkaline environments. Avoided by use of alloys in which noncorrosive metals are incorporated (stainless steel) or by plating or cladding the base metal with cromium or nickel. See corrosion, exposure testing. (2) *Beneficial*: Improvement of flavor by long storage. Cheeses develop a "sharp" flavor on ripening for 9–12 months; wines develop a "bouquet" after two or more years of storage, whiskeys stored in oaken casks for several years modify their flavor by extracting components of the wood. Tobacco is aged from 3 to 5 years after curing to remove unpleasant odors and improve smoking qualities.

Agent Orange. A toxic herbicide and defoliant containing 2,4,5-trichlorophenoxy-acetic acid (2,4,5-T) and 2,4-dichlorophenoxyacetic acid (2,4-D), with trace amounts of dioxin. Its use has been restricted.

age-resister. See antioxidant.

"Agerite."[69] TM for a series of antioxidants for rubber. Alba = p-benzyloxyphenol. DPPD = N,N'-diphenyl-p-phenylenediamine. Gel = A mixture of alkylated diphenylamines with selected petroleum wax. Geltrol = Phosphited polyalkyl polyphenol. Hipar = Mixture of phenyl-β-naphthylamine, isopropoxydiphenylamine, and diphenyl-p-phenylenediamine. HP = A blend of phenyl-β-naphthylamine and N,N'-diphenyl-p-phenylenediamine. Iso = p-isoproproxydiphenylamine. Powder = phenyl-β-naphthylamine. Resin = Aldol-α-naphthylamine. Resin D = A polymerized 1,2-dihydro-2,2,4-trimethylquinoline. Spar = Mixed mono-, di-, and tristyrenated phenols. Stalite and Stalite S = Mixture of alkylated diphenylamines. Superflex = Diphenylamine-acetone reaction product. Superlite = A mixture of polybutylated bisphenol A. White = syn-di-β-naphthyl-p-phenylenediamine.

agglomeration. (1) Combination or aggregation of colloidal particles suspended in a liquid into clusters or "flocs" of approximately spherical shape. It is usually achieved by neutralization of the electric charges which maintain the stability of the colloidal suspension. The terms flocculation and coagulation have a closely similar meaning.

(2) The food industry uses "agglomeration" in the sense of increasing the particle size of powdered food products. Because such powders tend to be hydrophobic because of the high surface tension of water, agglomeration causes them to be more readily dispersible in water--a process known as "instantizing." The agglomerates have varying degrees of open spaces (voids), and are loosely bound, foam-like structures. They are formed by mechanical means in chamber spray dryers, tubes, or fluidized beds, usually in the presence of moisture.
See also aggregation; agglutination; flocculation; coagulation.

agglutination. The combination or aggregation of particles of matter under the influence of a specific protein. The term is usually restricted to antigen-antibody reactions characterized by a clumping together of visible cells, such as bacteria or erythrocytes. The antigen is called an agglutinogen, and the antibody an agglutinin because of an apparent gluing or sticking action. See also aggregation.

aggregate. A collective term denoting any mixture of such particulates as sand, gravel, crushed stone, or cinders used in Portland cement formulations, road-building, paving compositions, animal husbandry, trickle filters, horticulture, etc.

aggregation. A general term describing the tendency of large molecules or colloidal particles to combine in clusters or clumps, especially in solution. When this occurs, usually as a result of removal of the electric charges by addition of an appropriate electrolyte, by the action of heat, or by mechanical agitation, the aggregates precipitate or separate from the dissolved state. Included in this term are the more specific terms agglutination, coagulation, flocculation, agglomeration, and coalescence.
See also these entries.

agitator. Any rotating device that induces motion in fluid mixtures over a wide range of viscosities, thus effecting uniform dispersion of their components. An important class of agitators comprises impellers which produce turbulent flow in liquids of low viscosity; their diameter is much less than that of the container. They may be either top-entering (vertical) or side-entering (at an angle of approximately 45 degrees);

medium-viscosity liquids are agitated by paddles attached to a central rotating member. Pastes and high-viscosity mixtures in which no turbulent flow is possible require agitators that closely fit the mixing chamber so as to provide the necessary shearing and squeezing action throughout the mass. These are kneading devices utilizing curved or helical rotors or sigma blades either single or double. Screw-type agitators permit continuous mixing by means of multiple shearing and blending action. So-called ribbon agitators are effective for dry powders, slurries, etc. A number of ingenious modifications and combinations of these types are widely used in the process industries. Most are available in laboratory sizes. See also impeller, mixing, kneader, screw.

aglucone. The nonsugar-like portion of a glucoside molecule.
See glycoside.

aglycone. A nonsugar hydrolytic product of a glycoside.
See glycoside.

agricultural chemical. (agrichemical). A chemical compound or mixture used to increase the productivity and quality of farm crops. Included are fertilizers, soil conditioners, fungicides, insecticides, herbicides, nematocides, and plant hormones. For further information, refer to National Agricultural Chemicals Association, Suite 900, 1155 15th St.,NW, Washington, D.C.

agricultural waste. See biomass, waste control, gasohol.

AIChE. Abbreviation for American Institute of Chemical Engineers.

air. A mixture (or solution) of gases, the composition of which varies with altitude and other conditions at the collection point. The composition of dry air at sea level is:

Substance	% by Wt.	% by Vol.
Nitrogen	75.53	78.00
Oxygen	23.16	20.95
Argon		0.93
Carbon dioxide		0.033*
Neon		0.0018
Helium		0.0005
Methane		0.0002
Krypton		0.0001
Nitrous oxide		0.000,05
Hydrogen		0.000,05
Xenon		0.000,008
Ozone		0.000,001

*The CO_2 content of air has increased 12 to 15% since 1900 due to combustion of fossil fuels.

See greenhouse effect.

The density of dry air is 1.29 g/L at 0C and 760 mm Hg. It is noncombustible, but will support combustion. Liquid air is air which has been subjected to a series of compression, expansion, and cooling operations until it liquefies.

Use: Source of oxygen, nitrogen, and rare gases; coolant; power source (compressed); cryogenic agent (liquid); particle classification; blowing agent (asphalt, soap, ice-cream mixes, whipped cream, etc.); flotation.

air classification. The separation of solid particles according to weight and/or size, by suspension in and settling from an air stream of appropriate velocity, as in air-floated clays and other particulate products.

air floatation. See air classification.

air gas. See producer gas.

air knife. See doctor knife.

"Airocel PK Foam Liquid."[270] TM for a protein-based, liquid concentration used to produce lightweight concrete, low density refractories, foamed ceramics, and other rigid inorganic foams. Also used as an air entraining agent to facilitate the pumping of concrete through pipe and flexible hose.

air pollution. (atmospheric pollution). Introduction of substances into the atmosphere that are not normally present therein and that have a harmful effect on man, animals, or plant life. Photosynthesis is significantly inhibited by air pollutants, especially in urban areas. The worst offenders are sulfur dioxide (which forms sulfurous acid on contact with water vapor), automotive emission products, metal dusts from smelters, coal smoke, and other particulates, formaldehyde and acrolein, and radioactive emanations. Control of these is exercised by the Environmental Protection Agency. As conventionally used, the term does not apply to interior air spaces such as industrial workrooms. TLVs (Threshold Limit Values) for the latter are established by the American Conference of Governmental Industrial Hygienists (ACGIH) and by OSHA.

See also environmental chemistry.

Akabori amino acid reactions. (1) Formation of aldehydes by oxidative decomposition of alpha-amino acids when heated with sugars. (2) Reduction of alpha-amino acids and esters by sodium amalgam and ethanolic hydrochloric acid to the corresponding alpha-amino aldehydes. (3) For-

mation of alkamines by heating mixtures of aromatic aldehydes and amino acids. No reaction was observed with tertiary amino groups.

"Akroflex" C.[28] TM for a rubber antioxidant containing 35% diphenyl-p-phenylenediamine $C_6H_4(NHC_6H_5)_2$ and 65% phenyl-α-naphthylamine ("Neozone" A).

Use: To improve the aging and service life of rubbers; anticrosslinking agent for SBR (styrene-butadiene-rubber).

"Aktoflo-S."[236] TM for a mixed oxyethylated phenol derivative with an optimum proportion of defoamant added. Added to emulsified-surfactant drilling muds, to reduce clay dispersion, and viscosity.

"Aktone."[285] TM for a rubber accelerator.

Used as a secondary accelerator, deodorizer, and secondary blowing agent in sponge compounds.

"Akweons."[152] TM for metal-processing oils.

Akweons 674 and 700 are wetting agents and corrosion inhibitors for use in acids; also exhibit fume-depressant properties. Akweons 250 is an acid fume depressant and pickling activator.

"Akwilox."[152] TM for high-density, brominated vegetable oils, manufactured for use in the soft drink industry.

Al. Symbol for aluminum.

alabaster. A fine-grained compact gypsum.

"Alacsan."[542] TM for a series of quaternary ammonium compounds.

Use: Cosmetics, germicides, algaecides, disinfectants, deodorants, and sanitizers.

"Alacstat" C-2.[542] TM for N,N-bis(2-hydroxyethyl alkylamine).

Use: Antistatic agent for polyolefins.

"Alamac."[590] TM for acetate salts of primary amines. Water-soluble or dispersible.

Use: Non-metallic ore flotation, corrosion inhibition.

"Alamide."[590] TM for a series of high molecular weight aliphatic amides (such as palmitamide and stearamide) produced by reacting ammonia with fatty acids.

Use: Intermediates for durable water repellents and finishes for textiles and paper; mold release and antiblocking agents.

"Alamine."[590] TM for a series of primary, secondary, and tertiary aliphatic amines, organic substituted ammonia derivatives. Soluble in a variety of organic solvents. Not appreciably soluble in water. Chain length from C_{12} to C_{18} with varying degrees of unsaturation.
Use: Corrosion inhibitors, ore flotation agents, textile finishing agents, asphalt antistripping agents, rubber compounding, color particle dispersion, petroleum product additives, chemical intermediates.

"Alanap."[248] TM for N-1-naphthylphthalamic acid.
Use: A herbicide.

alanine. (α-alanine; α-aminopropionic acid; 2-aminopropanoic acid).
$CH_3CH(NH_2)COOH$. A nonessential amino acid.
Properties: Colorless crystals, soluble in water, slightly soluble in alcohol, insoluble in ether, optically active. DL-alanine, mp 295C (decomposes), sublimes at 200C. L(+)-alanine, mp 297C (decomposes). D(-)-alanine, mp 295C (decomposes). L(+)-alanine hydrochloride; prisms, mp 204C (decomposes). L(+)-alanine, N-acetyl-, crystals, mp 116C. L(+)-alanine, N-benzoyl-, crystals, mp 152–154C.
Derivation: Hydrolysis of protein (silk, gelatin, zein), organic synthesis.
Grade: Reagent, technical.
Use: Microbiological research, biochemical research, dietary supplement.

β-alanine. (3-aminopropanoic acid; β-aminopropionic acid). $NH_2CH_2CH_2COOH$. A naturally occurring amino acid not found in protein.
Properties: White prisms, mp 198C (decomposes). Soluble in water, pH (50% sol) 6.0–7.3, slightly soluble in alcohol, insoluble in ether. Hydrochloride plates and leaflets, mp 122.5C. Platinichloride, yellow leaflets, mp 210C (decomposes).
Derivation: Addition of ammonia to β-propiolactone, other processes based on the reaction of ammonia with acrylonitrile, etc.
Use: Biochemical research; organic synthesis; production of calcium pantothenate; buffer in electroplating.

alanosine. $C_3H_7N_3O_4$.
Properties: Finely divided crystals, decomposes at 190C, optically active. Can be prepared in l-, d-, and dl forms. Insoluble in most organic solvents, slightly soluble in water.
Derivation: Fermentation of *Streptomyces alanosinicus*; also synthetically.
Use: Inhibitor of insect reproduction, antineoplastic, antibiotic.

β-alanylhistidine. See carnosine.

"Alar."[248] TM for a plant growth regulator (succinic acid- 2,2-dimethylhydrazide) which improves the color and texture of apples, grapes, and tomatoes; prevents premature dropping; growth retardant; multiple-flower stimulator.

alarmona. One of a class of cell-growth regulating metabolites (nucleotides) which enable bacteria to respond to metabolic and environmental changes. They are thought to act by controlling or affecting several biochemical reactions simultaneously, but the exact mechanism of their behavior has not been elucidated. One such nucleotide, discovered in 1982, is known as ZTP (5-amino-4-imidazole carboximide riboside- 5'-triphosphate).
See also nucleotide.

"Alathon."[28] TM for a polyethylene resin.
"Alathon" G-0530, designated as a reinforced polyethylene, contains 30% by weight of glass-fiber treated with a proprietary coupling agent that optimizes its reinforcing properties.

"Albigen."[440] TM for a water-soluble polymer used in the textile industry for stripping vat and other dyestuffs. Has no affinity for the fiber, promotes the stripping effect of alkaline hydrosulfite solutions.

"Albone."[28] TM for a series of hydrogen peroxide solutions which vary in hydrogen peroxide content from 35% to 90% by weight.
See hydrogen peroxide.

albumen. Commercial term for dried egg white used in the food industry.
See albumin, egg.

albumin. Any of a group of water-soluble proteins of wide occurrence in such natural products as milk (lactalbumin), blood serum, and eggs (ovalbumin). They are readily coagulated by heat and hydrolyze to α-amino acids or their derivatives.
See following entries.

albumin, egg. (ovalbumin). Chief protein occurring in egg white as a viscous, colorless fluid; it becomes an amorphous solid when dried which can be reconstituted with water. It is a heat-sensitive colloidal material which coagulates irreversibly at approximately 60C (140F). The dried product is available in commercial quantities.
Use: Protective colloid and emulsifying agent in bakery products (especially angel cake), clarifica-

tion of wines, adhesives, paper coatings, pharmaceuticals, enzyme activation, lithography, analytical reagent, antidote for mercury poisoning, mordant in dyes. *Note*: A recombinant DNA technique has made possible the formation of ovalbumin by the bacterium *E. coli*.

albumin, milk. (lactalbumin).   A component of skim milk protein (2 to 5%). Can be crystallized. Exact function is not known, but probably aids in stabilization of the fat particles.
See also milk.

ALCA.   Abbreviation for American Leather Chemists' Association.

alchemy.   The predecessor of chemistry, practiced from as early as 500 BC through the 16th century. Its two principal goals were transmutation of the baser metals into gold and discovery of a universal remedy. Modern chemistry grew out of alchemy by gradual stages.

alcian blue.
Properties: Greenish-black crystals with metallic sheen. Soluble in ethanol, cellosolve, ethylene glycol.
Use: Gelling agent for lubricating fluids, bacterial stain for histiocytes and fibroblasts.

Alcoa process.   A more efficient method of producing aluminum from bauxite which requires one-third less electric power than the Hall process. Alumina is reacted with chlorine, the resulting aluminum chloride yielding the metal and chlorine on electrolysis. No fluorine is required in the process. Prototype plants are under development.

alcohol.   A broad class of hydroxyl-containing organic compounds occurring naturally in plants and made synthetically from petroleum derivatives such as ethylene. Many are manufactured in tonnage quantities. The many types may be summarized as follows:
I.   Monohydric (1 OH group)
   1. Aliphatic
      (a) paraffinic (ethanol)
      (b) olefinic (allyl alcohol)
   2. Alicyclic (cyclohexanol)
   3. Aromatic (phenol, benzyl alcohol)
   4. Heterocyclic (furfuryl alcohol)
   5. Polycyclic (sterols)
II. Dihydric (2 OH groups): glycols and derivatives (diols)
III. Trihydric (3 OH groups): glycerol and derivatives
IV. Polyhydric (polyols) (3 or more OH groups)
Use: Organic synthesis, as solvents, detergents, beverages, pharmaceuticals, plasticizers, and fuels.

For further information, see monohydric, dihydric, trihydric, polyol, and specific alcohol.

alcohol, absolute.   See ethanol.

alcohol dehydrogenase.   An enzyme found in animal and plant tissue which acts upon ethanol and other alcohols producing acetaldehyde and other aldehydes.
Use: Biochemical research.

alcohol, denatured.   Ethanol to which another liquid has been added to make it unfit to use as a beverage (chiefly for tax reasons). In the US, it may be either Completely Denatured (CDA) (usually by the use of the toxic methanol) or Specially Denatured (SDA). The number of formulations authorized officially for making SDA have been reduced to four (Nos. 40, 40A, 40B, and 40C). They include the following denaturants: SDA 40B must contain brucine, brucine sulfate, or quassin plus tert-butanol; SDA 40A must contain sucrose octaacetate plus tert-butyl alcohol; SDA 40B must contain Bitrex and tert-butyl alcohol; SDA40C must contain only tert-butyl alcohol. For exact formulas, consult the Alcohol and Tobacco Tax Division of IRS, Washington, D.C.
Properties: See ethanol.
Hazard: Flammable, dangerous fire risk; TLV: 1000 ppm in air.
Use: Manufacture of acetaldehyde and other chemicals, solvents. antifreeze and brake fluids, fuels.

alcohol, grain.   Synonym for ethanol made from grain.

alcohol, industrial.   A mixture of 95% ethanol and 5% water, plus additives for denaturing or special solvent purposes.
See also alcohol, denatured.

alcoholysis.   A chemical reaction between an alcohol and another organic compound analogous to hydrolysis. The alcohol molecule decomposes to form a new compound with the reacting substance; the other reaction product being water. Both hydrolysis and alcoholysis may be considered as forms of solvolysis.
See also solvolysis.

alcohol, wood.   See methanol.

"Aldactone."[70]   TM for spironolactone .

"Aldaromes."[188]   TM for compositions of aromatic chemicals and essential oils used in embalming fluids and sprays to mask or cover the unpleasant odor of formaldehyde. Will not decompose when in contact with formaldehyde; gives clear solutions in 40% formaldehyde. Not water-soluble.

aldehyde. A broad class of organic compounds having the generic formula RCHO, and characterized by an unsaturated carbonyl group (C=O). They are formed from alcohols by either dehydrogenation or oxidation, and thus occupy an intermediate position between primary alcohols and the acids obtained from them by further oxidation. Their chemical derivation is indicated by the name: *al*(cohol) + *dehyd*(rogenation). Aldehydes are reactive compounds participating in oxidation, reduction, addition, and polymerization reactions. For specific properties, see individual compounds.

aldehyde ammonia. (acetaldehyde ammonia; 1-amino-ethanol). $CH_3CHOHNH_2$.

$$CH_3-CH\begin{array}{c}OH\\\\NH_2\end{array}$$

Properties: White, crystalline solid; stable in closed containers; resinifies on long exposure to air. Very soluble in water and alcohol. Mp 97C (partly decomposes).
Derivation: Action of acetaldehyde on ammonia.
Hazard: Irritant to eyes and skin; moderate fire risk.
Use: Accelerator for vulcanization of thread rubber, organic synthesis, source of acetaldehyde and ammonia.

aldehyde collidine. See 2-methyl-5-ethyl pyridine.

aldehydine. See 2-methyl-5-ethylpyridine.

Alder, Kurt. (1902-1958) A German chemist who won the Nobel prize for chemistry along with Otto Diels in 1950 for a project involving a practical method for making ring compounds from chain compounds by forcing them to combine with maleic anhydride. This is known as the Diels-Alder reaction which provided a method for synthesis of complex organic compounds. He had degrees from the Universities of Berlin and Kiel.

Alder-Rickert Rule. Adducts of 1,3-cyclohexadiene derivatives with acetylenedicarboxylic esters give on heating phthalate ester and ethylene. Similar adducts of cyclopentadiene revert on heating to starting materials (Retro-Diels-Alder).

Alder-Stein Rules. Set of rules governing the stereochemistry of the Diels-Alder reaction. The most important are: (1) The stereochemical relationship of groups attached to the diene and the dienophile is maintained in the product (cis-addition). (2) The product resulting from max accumulation of unsaturated centers in the transition state is favored (endo-rule).

aldicarb. $CH_3SC(CH_3)_2HC:NOCONHCH_3$.
Properties: Colorless crystals, mp 100C (212F); almost insoluble in water, slightly soluble in benzene and xylene; partly soluble in acetone and methylene chloride.
Hazard: Toxic by ingestion.
Use: Nematocide; insecticide.

"Aldo."[73] TM for a series of waxes, glyceryl monofatty acid esters, in technical, water-dispersible, and edible grades.
Use: Foods, plastics and rubber, cosmetics, lubricants, plasticizers, emulsion stabilizers.

aldol. (acetaldol; β-hydroxybutyraldehyde). $CH_3CHOHCH_2CHO$.
Properties: Water-white to pale yellow, syrupy liquid. Decomposes into crotonaldehyde and water on distillation under atmospheric pressure. Miscible with water, alcohol, ether, organic solvents; d 1.1098 (15.6/4C); bp 83C (20 mm); vapor press less than 0.1 mm (20C); specific heat 0.737; bulk d 9.17 lb/gal (20C). Flash p 150F (65.5C) (OC); fp below 0C. Autoign temperature 530F (276.6C). Combustible.
Derivation: By condensation of acetaldehyde in sodium hydroxide solution.
Grade: Technical (98%).
Hazard: Moderate fire risk.
Use: Synthesis of rubber accelerators and age resisters, perfumery, engraving, ore flotation, solvent, solvent mixtures for cellulose acetate, fungicides, organic synthesis, printer's rollers, cadmium plating, dyes, drugs, dyeing assistant, synthetic polymers.

aldolase. (zymohexase). An enzyme present in muscle involved in glycogenolysis and anaerobic glycolysis. It catalyzes production of dihydroxyacetone phosphate and phosphoglyceric aldehyde from fructose-1,6-diphosphate.
Use: Biochemical research.

aldol condensation. A reaction between two aldehyde or two ketone molecules in which the position of one of the hydrogen atoms is changed in such a way as to form a single molecule having one hydroxyl and one carbonyl group. Since such a molecule is partly an alcohol (OH group) and partly an aldehyde (CHO group) and represents a union of two smaller molecules, the reaction is called an ald-ol condensation. It can be repeated to form molecules of increasing molecular weight. The condensation of formaldehyde to sugars in plants, which on repetition builds up the more complex carbohydrate structures such as starch and cellulose, is thought to be a reaction of this type. It occurs most effectively in an alkaline medium.

aldo-α-naphthylamine condensate.
  Properties: Orange to dark red solid with characteristic odor; softens at 64C min; d 1.16, insoluble in water, gasoline; slightly soluble in alcohol and petroleum hydrocarbons; soluble in acetone, benzene, chloroform, and carbon disulfide.
  Use: Antioxidant in tire carcasses, tubes, insulating tape, black soles.

aldose. Any of a group of sugars whose molecule contains an aldehyde group and one or more alcohol groups. An example is glyceraldehyde ($HOCH_2 \cdot CHOH \cdot CHO$), specifically called an aldotriose because it contains three carbon atoms.

"Aldosperse."[73] TM for a series of polyethylene glycol monofatty acid esters.
  Properties: Yellow liquids to soft white solids (S-9); d approximately 1.0; mp less than 2C (0–9) to 24–29C (S-9). Soluble in most organic solvents. Combustible.
  Use: Antistatic agents, emulsifiers and detergents, wetting agents, dye assistants, thickening agents, viscosity stabilizers.

aldosterone. (electrocortin). $C_{21}H_{28}O_5$.

An adrenal cortical steroid hormone which is the most powerful mineralocorticoid. Probably the chief regulator of sodium, potassium, and chlorine metabolism; approximately 30 times as active as deoxycorticosterone.
  Properties: Crystals, mp 108–112C.
  Derivation: Isolated from adrenals; has been synthesized.
  Use: Medicine.

"Aldox."[204] TM for an acidic powdered compound used to deoxidize aluminum prior to spot welding or to desmut aluminum subsequent to etching.

aldrin. (HHDN).   CAS: 309-00-2.
  $C_{12}H_8Cl_6$.   The assigned common name for an insecticidal product containing 95% or more of 1,2,3,4,10,10-hexachloro-1,4,4a,5,8,8a-hexa-hydro-1,4,5,8,-endo, exo-dimethanonaphthalene. See also dieldrin and endrin.
  Properties: Brown to white, crystalline solid; mp 104–105.5C; insoluble in water; soluble in most organic solvents. Not affected by alkalies or dilute acids, compatible with most fertilizers, herbicides, fungicides and insecticides.
  Grade: Technical.
  Hazard: TLV 0.25 mg/m³, toxic effects from skin absorption.
  Use: Insecticide.

"Aldyl."[28] TM for high-strength polyethylene pipe composed principally of "Alathon."

ale. See brewing.

aleuritic acid. (9,10,16-trihydroxyhexadecanoic acid).   CAS: 533-87-9.
  $HOCH_2(CH_2)CH(OHCH(OH)(CH_2)COOH$.
  Properties: Yellowish solid extracted from shellac. Mw 304.48, mp 101C.
  Use: Perfumes.

alfin. A catalyst obtained from alkali alcoholates derived from a secondary alcohol which is a methyl carbinol and olefins possessing the grouping —CH=CH—CH₂, which may be part of a ring, as in toluene. The interaction of the alkali alcoholate (sodium isopropoxide) with the olefin halide (allyl chloride) gives a slurry of sodium chloride on which sodium isopropoxide and allyl sodium are adsorbed. This slurry is a typical alfin catalyst used to convert olefins into polyolefins. The elastomers produced are called alfin rubbers.

"Alfol."[544] TM for synthetic primary straight-chain alcohols made by Ziegler-type reaction of aluminum alkyls, ethane, and hydrogen. Lower alcohols (C_6 to C_{12}) are chiefly used in phthalate and other esters for use as vinyl plasticizers. Higher alcohols (C_{12} to C_{19}) are intermediates producing biodegradable surfactants, alcohol sulfates, alcohol ethoxylate nonionics, alcohol ether sulfate anionics, and amine-derived cationics.

"Alfrax."[280] TM for a series of refractory products composed principally of electrically fused aluminum oxide grain. Available as bonded refractories and refractory cements. Used as furnace linings.

algae. Chlorophyll-bearing organisms occurring in both salt- and freshwater, they have no flowers or seeds but reproduce by unicellular spores. They range in size from single cells to giant kelp over 100 feet long, and include most kinds of seaweed. There are four kinds of algae: brown, red, green, and blue-green. Blue-green algae are said to be the earliest form of life to appear on earth. The photosynthetic activity of algae accounts for the fact that over two-thirds of the world total of photosynthesis takes place in the oceans. Algae are harvested and used as food supplements (see carrageenan, agar), soil conditioners, animal feeds, and as a source of iodine; they also contain numerous minerals, vitamins, proteins, lipids, and essential amino acids. Alginic acid is another important derivative. Blue-green algae are water contaminants and are toxic to fish and other aquatic life. Phosphorus compounds in detergent wastes stimulate the growth of algae to such an extent that overpopulation at the water surface prevents light from reaching many of the plants; these decompose, removing oxygen and releasing carbon dioxide, thus making the water unsuitable for fish. Algae are being used in treatment of sewage and plant effluent in a proprietary flocculation process.
See also eutrophication, agar, biomass.

Algar-Flynn-Oyamada reaction. Alkaline hydrogen peroxide oxidation of o-hydroxyphenyl styryl ketones (chalcones) to flavonols via the intermediate dihyroflavonols.

"Algepon."[300] TM for series of dyeing, stripping, and discharge printing assistants for various applications in textile processing. Several types are each formulated for a specific function.

algicide. Chemical agent added to water to destroy algae. Copper sulfate is commonly employed for large water systems.

algin. A hydrophilic polysaccharide (phycocolloid or hydrocolloid) found exclusively in the brown algae. It is analogous to agar. The seaweed (giant kelp) is sea-harvested, water-extracted, and refined. US (California) and Great Britain are the chief producers.
See also alginic acid, alginate.

alginate. Any of several derivatives of alginic acid (e.g., calcium, sodium or potassium salts or propylene glycol alginate). They are hydrophilic colloids (hydrocolloids) obtained from seaweed. Sodium alginate is water-soluble but reacts with calcium salts to form insoluble calcium alginate.
Use: Food additive (thickener, stabilizer), yarns and fibers, medicine (first-aid dressings), meat substitute, high-protein food analogs.

alginic acid. $(C_6H_8O_6)_n$. A polysaccharide composed of β-d-mannuronic acid residues linked so that the carboxyl group of each unit is free, while the aldehyde group is shielded by a glycosidic linkage. It is a linear polymer of the mannuronic acid in the pyranose ring form.
Properties: White to yellow powder possessing marked hydrophilic colloidal properties for suspending, thickening, emulsifying, and stabilizing. Insoluble in organic solvents, slowly soluble in alkaline solutions. Absorbs up to 300 times its weight of water.
Grade: Refined (food), technical (commercial), NF (sodium alginate), FCC.
Use: Food industry as thickener and emulsifier; protective colloid in ice cream, tooth paste, cosmetics, pharmaceuticals, textile sizing, paper coatings, waterproofing agent for concrete, boiler water treatment, oil-well drilling muds, storage of gasoline as a solid.

alicyclic. A group of organic compounds characterized by arrangement of the carbon atoms in closed ring structures sometimes resembling boats, chairs, or even bird cages. These compounds have properties resembling those of aliphatics and should not be confused with aromatic compounds having the hexagonal benzene ring. Alicyclics are comprised of three subgroups: (1) cycloparaffins (saturated), (2) cycloolefins (unsaturated with two or more double bonds), and (3) cycloacetylenes (cyclynes) with a triple bond. The best-known cycloparaffins (sometimes called naphthenes) are cyclopropane, cyclohexane, and cyclopentane; typical of the cycloolefins are cyclopentadiene and cyclooctatetraene. Most alicyclics are derived from petroleum or coal tar. Many can be synthesized by various methods.
See also subgroups referred to above.

"Alidase."[70] TM for hyaluronidase.
See hyaluronic acid.

"Alipal."[307] TM for a series of anionic surfactants. CO-433. Sodium salt of sulfated nonylphenoxypoly(ethyleneoxy)ethanol; 28% active.
Use: Detergents, shampoos, scrub soaps, lime soap dispersant, emulsifier, antistatic agent.

aliphatic. One of the major groups of organic compounds characterized by straight- or branched-chain arrangement of the constituent

carbon atoms. Aliphatic hydrocarbons are comprised of three subgroups: (1) paraffins (alkanes), all of which are saturated and comparatively unreactive, the branched-chain types being much more suitable for gasoline than the straight-chain; (2) olefins (alkenes or alkadienes) which are unsaturated and quite reactive; (3) acetylenes (alkynes) which contain a triple bond and are highly reactive. In complex structures, the chains may be branched or cross-linked.
See also alicyclic aromatic chain.

"Aliquat."[590] TM for a group of fatty quaternary ammonium chlorides, stable in both acidic and alkaline media. Insoluble in water.
Use: Cationic textile softeners, corrosion inhibitors and flow control agents, phase transfer catalyst.

aliquot. A part which is a definite fraction of a whole; as aliquot samples for testing or analysis.

"Alite."[326] TM for a series of sintered metallic oxides.

"Alitrile."[259] TM for a series of high molecular weight aliphatic nitriles produced from fatty acids from tallow, hydrogenated tallow, coconut and soybean oils, and palmitic and stearic acids.
Use: Lubricating oil additives; plasticizers.

alizarin. (1,2-dihydroxyanthraquinone). $C_6H_4(CO)_2C_6H_2(OH)_2$.   Parent form of many dyes and pigments including mordants.

Properties: Orange-red crystals; brownish-yellow powder; soluble in aromatic solvents, hot methanol, and ether; sparingly soluble in water; moderately soluble in ethanol. C.I. 58000. Mp 289C, bp 430C (sublimable). Combustible.
Derivation: Anthracene is oxidized to anthraquinone; the sulfonic acid of which is then fused with caustic soda and potassium chlorate; the melt is run into hot water and the alizarin precipitated with hydrochloric acid. Occurs naturally in madder root.
Grade: Technical.
Use: Manufacture of dyes, production of lakes, indicators, biological stain.

alizarin blue. (C.I. 67410). $C_{17}H_9NO_4$.
Properties: Violet crystals, mp 268C, insoluble in water, soluble in glacial acetic acid and hot benzene.
Use: Indicator.

alizarin yellow R. (p-nitrophenylazosalicylate sodium; C.I. 14030).
$O_2NC_6H_4NNC_6H_3OHCOONa$.
Properties: Yellow-brown powder, soluble in water.
Use: Acid-base indicator, biological stain.

alkadiene. See diolefin.

alkali. Any substance which in water solution is bitter, more or less irritating or caustic to the skin and mucous membranes; turns litmus blue, and has a pH value greater than 7.0.
See also base, pH, alkali metal.
The alkali industry produces sodium hydroxide, sodium carbonate (soda ash), sodium chloride, salt cake, sodium bicarbonate and corresponding potassium compounds.

alkali blue. Class name for a group of pigment dry powders prepared by the phenylation of p-rosaniline or fuchsine, followed by drowning in hydrochloride acid, washing and sulfonating. Alkali blue on a weight basis has the highest tinting strength of all blue pigments. The press-cake may be "flushed" with vehicle to replace the water in the pulp.
Use: Printing inks, interior paints.

alkali cellulose. The product formed by steeping wood pulp with sodium hydroxide, the first step in the manufacture of viscose rayon and other cellulose derivatives.
See also carboxymethylcellulose.

alkali metal. A metal in Group 1A of the Periodic Table, i.e., lithium, sodium, potassium, rubidium, cesium, and francium. Except for francium, the alkali metals are all soft, silvery metals, which may be readily fused and volatilized; the melting and boiling points becoming lower with increasing atomic weight. The density increases with (but less rapidly than) the atomic weight, the atomic volume therefore becoming greater as the series is ascended. They are the most strongly electropositive of the metals. They react vigorously, at times violently, with water; within the group itself, the basicity increases with atomic weight, that of cesium being the greatest.

alkalimetry. The measurement of the concentration of bases or of the amount of free base present in a solution by titration or some other means of analysis.

alkaline earth. An oxide of an alkaline earth metal (lime).

alkaline-earth metals. Calcium, barium, strontium and radium (Group II A of Periodic Table). In general they are white and differ by shades of color or casts, malleable, extrudable, and machinable, and may be made into rods, wire or plate; less reactive than sodium and potassium and have higher melting and boiling points. See also specific entry.

alkaloid. A basic nitrogenous organic compound of vegetable origin. Usually derived from the nitrogen ring compounds pyridine, quinoline, isoquinoline, pyrrole, designated by the ending -ine. Though some are liquids, they are usually colorless, crystalline solids, having a bitter taste which combine with acids without elimination of water. They are soluble in alcohol, insoluble or sparingly soluble in water. Examples are atropine, morphine, nicotine, quinine, codeine, caffeine, cocaine, and strychnine.

alkane. See paraffin (1).

alkanesulfonic acid, mixed. RSO_3H (R is methyl, ethyl, propyl, mixed). Trade designation for a mixture of methane-, ethane-, and propane sulfonic acids. A strong nonoxidizing, nonsulfonating liquid acid which is thermally stable at moderately high temperatures.
Properties: Light amber liquid with sour odor, very corrosive, miscible with water and saturated fatty acids, mp below $-40C$, bp 120–140C (1 mm), d 1.38 (20C); pH (1% solution) 1.15.
Use: Catalyst; intermediate, reaction medium.

"Alkanol."[28] TM for a series of fatty alcohol-ethylene oxide condensation products used as nonionic surface-active agents in detergents, dispersing and emulsifying agents in paper, leather and textiles. These include grades OA, OE, OJ, OP, and HC. 189-S is a saturated hydrocarbon sodium sulfonate. B and BG are sodium alkyl-naphthalene sulfonates. Sulfur is tetrahydronaphthalene sodium sulfonate

alkanolamine. (alkylolamine). A compound such as ethanolamine $HOCH_2CH_2NH_2$, or triethanolamine, $(HOCH_2CH_2)_3N$, in which nitrogen is attached directly to the carbon of an alkyl alcohol. See specific compound.

alkene. See olefin.

"Alkor."[41] TM for a synthetic, furan-type resin cement which is acid- and alkali-proof and used as a mortar cement where temperatures do not exceed 380F.

alkoxyaluminum hydrides. (H_nAlOR_{3n}). A group of reducing agents especially useful in converting epoxides to alcohols. Derived by reaction of aluminum hydride with the corresponding alcohol in tetrahydrofuran.

"Alkyd Molding Compounds."[175] TM for a thermosetting plastic comprised of an unsaturated polyester (usually formulated with a diallyl phthalate cross-linking monomer), inorganic mineral fillers (clay, glass fiber, asbestos, etc.), and other modifiers.
Properties: High dimensional stability, good electrical resistivity, ease of molding at low pressures. Glass-reinforced type has high mechanical strength and impact resistance.
Forms available: Granular, putty (soft), glass-reinforced.
Use: Components of electrical systems, encapsulating compounds, electrical insulation.
See also alkyd resin.

alkyd resin. A thermosetting coating polymer, chemically similar to polyester resins, conventionally made by condensation, polymerization of a dihydric or polyhydric alcohol (ethylene glycol or glycerol) and a polybasic acid (phthalic anhydride), usually with a drying oil modifier. The process requires heating at 230–250C for up to 12 hours. A new and quite different method utilizes epoxy addition polymerization in which a mixture of glycidyl esters and organic acid anhydrides are heated with a metallic catalyst at 100C or less for only two to four hours. Cost and energy savings and improved application performance are realized by this process.
Use: Alkyd resins are used as vehicles in exterior house paints, marine paints and baking enamels. Molded alkyd resins are used for electrical components, distributor caps, encapsulation and a variety of similar applications.

alkyl. A paraffinic hydrocarbon group which may be derived from an alkane by dropping one hydrogen from the formula. Examples are methyl CH_3-, ethyl C_2H_5-, propyl $CH_3CH_2CH_2$-, isopropyl $(CH_3)_2CH_3$-. Such groups are often represented in formulas by the letter R and have the generic formula C_nH_{2n+1}.
See also aryl.

alkylaryl polyethyleneglycol ether. See isooctylphenoxypolyoxyethylene ethanol for a typical example of this class of compound.
Use: In surface-active agents.

alkylaryl sulfonate. An organic sulfonate of combined aliphatic and aromatic structure, e.g., alkylbenzene sulfonate.

alkylate. (1) A product of alkylation. (2) A term used in the petroleum industry to designate a branched-chain paraffin derived from an iso-paraffin and an olefin, e.g., isobutane reacts with ethylene (with catalyst) to form 2,2-dimethylbutane (neohexane). The product is used as a high-octane blending component of aviation and civilian gasolines. (3) In the detergent industry, the term is applied to the reaction product of benzene or its homologs with a long-chain olefin to form an intermediate, e.g., dodecylbenzene, used in the manufacture of detergents. It also designates a product made from a long-chain normal paraffin which is chlorinated to permit combination with benzene to yield a biodegradable alkylate. The adjectives "hard" and "soft" applied to detergents refer to their ease of decomposition by microorganisms.
See biodegradability, detergent.

alkylation. (1) The introduction of an alkyl radical into an organic molecule. This was one of the early chemical processes used in Germany to furnish intermediates for improved dyes, e.g., dimethylaniline. Other alkylation products are cumene, dodecylbenzene, ethylbenzene, and nonylphenol. (2) A process whereby a high-octane blending component for gasolines is derived from catalytic combination of an isoparaffin and an olefin.
See also alkylate (2) neohexane.

alkylbenzene sulfonate. (ABS). A branched-chain sulfonate type of synthetic detergent, usually a dodecylbenzene or tridecylbenzene sulfonate. Such compounds are known as "hard" detergents because of their resistance to breakdown by microorganisms. They are being replaced by linear sulfonates.
See alkyl sulfonate, linear; detergent; also sodium dodecylbenzene sulfonate.

alkyldimethylbenzylammonium chloride. General name for a quaternary detergent.
See for example benzalkonium chloride.

alkyl fluorophosphate. See diisopropyl fluorophosphate.

alkylolamine. See alkanolamine.

alkyl sulfonate. (linear alkylate sulfonate; LAS). A straight-chain alkylbenzene sulfonate, a detergent specially tailored for biodegradability. The linear alkylates may be normal or iso (branched at the end only), but are C_{10} or longer.
See Sodium dodecylbenzene sulfonate.

"Alkyltainer."[313] TM for custom-made sample container for aluminum alkyl compounds and similar pyrophoric chemicals, to facilitate safe handling.

alkyne. See acetylene hydrocarbon.

Allan-Robinson reaction. Preparation of flavones or isoflavones by condensing o-hydroxy-aryl ketones with anhydrides of aromatic acids and their sodium salts.

allantoin. (glyoxyldiureide; 5-ureidohydantoin). $C_4H_6N_4O_3$. The end product of purine metabolism in mammals other than man and other primates; it results from the oxidation of uric acid.
Properties: White to colorless, odorless, tasteless powder or crystals; mp 230C (decomposes); 1 g is soluble in 190 cc water or 500 cc alcohol; readily soluble in alkalies. Optically active forms are known.
Preparation: Produced by oxidation of uric acid. Also present in tobacco seeds, sugar beets, wheat sprouts.
Use: Biochemical research, medicine.

allele. One of two or more types of genes that may occur at a given position on a strand of DNA.

allelopathic chemical. Any of a wide range of natural herbicides of varying toxicity produced by many species of plants, as well as by soil microorganisms (bacteria, fungi). These compounds adversely affect other plants in the vicinity, either inhibiting germination and growth or killing them outright. They are extracted from the growing plant by leaching of its leaves, root exudates and decomposition of dead tissue. Examples of plants found to be sources of these toxic compounds are sunflowers, oats, and soybeans. Among the products that have been identified are amygdalin, caffeine, gallic acid, and arbutin. Many types of chemical structure are represented. Research is directed toward breeding and cultivation of allelopathic plants to utilize their weed-killing ability.

allene. (propadiene; dimethylenemethane). $H_2C:C:CH_2$.
Properties: Colorless gas, unstable, fp $-136.5C$, bp 34.5C. Can be readily liquefied.
Derivation: (a) Action of zinc dust on 2,3-dichloropropene, (b) pyrolysis of isobutylene, (c) electrolysis of potassium itaconate.
Use: Organic intermediate.

allergen. Any substance that acts in the manner of an antigen on coming into contact with body tissues by inhalation, ingestion, or skin adsorption. The allergen causes a specific reagin to be formed in the bloodstream. The ability to produce reagins in response to a given allergen is an inherited characteristic that differentiates an allergic from a nonallergic person. A reagin is actually an antibody. The specificity of the allergen-reagin reaction and its dependence on molecular configuration is similar to the antigen-antibody reaction.

The allergen molecule (often a protein such as pollen or wool) may be regarded as a key which precisely fits the corresponding structural shape of the reagin molecule. Allergies in the form of contact dermatitis can result from exposure to a wide range of plant products, some metals, and a few organic chemicals. Though they are alike in some ways antigen-antibody reactions protect the individual, whereas allergen-reagin reactions are harmful.
See also antigen-antibody.

allethrin. $C_{19}H_{26}O_3$. Generic name for 2-allyl-4-hydroxy-3-methyl-2-cyclopenten-1-one ester of chrysanthemummonocarboxylic acid. A synthetic insecticide structurally similar to pyrethrin and used in the same manner. For other synthetic analogs, see barthrin, cyclethrin, ethythrin, furethrin. Pyrethrin I differs in having a 2,4-pentadienyl group in place of the allyl of allethrin.
Properties: Clear, amber-colored, viscous liquid. D 1.005–1.015 (20/20C), refr index (20C) 1.5040. Insoluble in water; incompatible with alkalies; soluble in alcohol, carbon tetrachloride, kerosene and nitromethane. Combustible.
Derivation: Synthetically (glycerol, acetylene, and ethyl acetoacetate are the major raw materials).
Grade: 90%, technical (approximately 90% pure with 10% of isomers or related compounds), 20% technical, 2.5% technical.
Use: Insecticides; synergist.

"Allexcel."[342] TM for allethrin-containing insecticidal concentrations.

allicin. ($C_6H_{10}OS_2$). An antibacterial substance extracted from garlic (allium).
Properties: Yellow, oily liquid; sharp garlic odor; unstable; decomposes rapidly when heated; slightly soluble in water; very soluble in alcohol, benzene, and ether.
Use: Medicine.

alligatoring. Formation of cracks on the surface of thick paint layers, the underlying material remaining soft.

allo-. A prefix designating the more stable of two geometrical isomers (Chemical Abstracts).

allomaleic acid. See fumaric acid.

allomerism. A constancy of crystalline form or structure with a variation in chemical composition.
See also polyallomer.

alloocimene. (2,6-dimethyl-2,4,6-octatriene). $(CH_3)_2C(CH)_3CCH_3CHCH_3$.
Properties: Clear, almost colorless liquid. Boiling range (5–95%) 89–91C (20 mm), d 0.824 (15/15C), refr index 1.5278 (20C). Polymerizes and oxidizes readily. Combustible.
Derivation: Pyrolysis of α-pinene.
Use: Component of varnishes and a variety of polymers; fragrance.

allophanamide. See biuret.

allophanate. An unsaturated nitrogenous product made by reaction of an alcohol with two moles of isocyanic acid (a gas). Usually crystals, high-melting products that are easily isolated. Acid-sensitive and tertiary alcohols can be converted into allophanates.

D-allose. (β-D-allopyranose). $C_6H_{12}O_6$.
Properties: Crystals, mp 128C, mw 180.16. Soluble in water, insoluble in alcohol.
Derivation: Obtained from the leaves of *Protea Rubropilosa*.

allothreonine. See threonine.

allotrope. One of several possible forms of a substance.
See allotropy.

allotropy. (polymorphism). The existence of a substance in two or more forms which are significantly different in physical or chemical properties. The difference between the forms involves either (1) crystalline structure, (2) the number of atoms in the molecule of a gas, or (3) the molecular structure of a liquid. Carbon is a common example of (1), occurring in several crystal forms (diamond, carbon black, graphite) as well as several amorphous forms. Diatomic oxygen and triatomic ozone are instances of (2), and liquid sulfur and helium of (3). Uranium has three crystalline forms, manganese 4, and plutonium no less than 6. A number of other metals also have several allotropic forms which are often designated by Greek letters, e.g., alpha-, gamma-, and delta-iron.

alloxan. (mesoxalylurea). $C_4H_2O_4N_2 \cdot HOH$ and 4HOH.
Properties: White crystals, become pink on exposure to air; colorless, aqueous solution imparts

pink color to skin; mp 170C (decomposes) (various melting points in literature); soluble in water and alcohol.
Derivation: Oxidation of uric acid in acid solution.
Use: Biochemical research, cosmetics, organic synthesis.

alloy. A solid or liquid mixture of two or more metals, or of one or more metals with certain nonmetallic elements, as in carbon steels. The properties of alloys are often greatly different from those of the components. The purpose of an alloy is to improve the specific usefulness of the primary component--not to adulterate or degrade it. Gold is too soft to use without a small percentage of copper, the corrosion and oxidation resistance of steel is markedly increased by incorporation of from 15 to 18% of chromium and often a few percent of nickel (stainless steel). The presence of up to 1.5% carbon profoundly affects the properties of steels. Similarly, a low percentage of molybdenum improves the toughness and wear resistance of steel. The hundreds of special alloys available are instances of the tailor-made nature of alloys to meet specific operating conditions. Amorphous alloys for use in transformer coils are made by quick-quenching the melt.
See alloy, fusible, amalgam, superalloy.

alloy, fusible. (low-melting alloy; fusible metal). An alloy melting in the range of approximately 51–260C, usually containing bismuth, lead, tin, cadmium, or indium. Eutectic alloys are the particular compositions that have definite and minimum melting points compared with other mixtures of the same metals. The compositions of a few fusible alloys are given below:

System (C)	Composition	Eutectic Temperature
Cd-Bi	60 Bi-40 Cd	144
In-Bi	33.7 Bi-66.3 In	72
	67.0 Bi-33 In	109
Pb-Bi	56.5 Bi-43.5 Pb	125
Sn-Bi	58 Bi-42 Sn	139
Pb-Sn-Bi	52 Bi-16 Sn-32 Pb	96
Pb-Cd-Bi	52 Bi-8 Cd-40 Pb	92
Sn-Cd-Bi	54 Bi-20 Cd-26 Sn	102
In-Sn-Bi	58 Bi-17 Sn-25 In	79
Pb-Sn-Cd-Bi	*50 Bi-10 Cd-13.3 Sn-26.7 Pb	70
In-Pb-Sn-Bi	49.4 Bi-11.6 Sn-18 Pb-21 In	57
In-Cd-Pb-Sn-Bi	44.7 Bi-5.3 Cd-8.3 Sn-22.6 Pb-19.1 In	47

*Wood's metal.

alloy steel. A steel containing up to 10% of elements such as chromium, molybdenum, nickel, etc., usually with a low percentage of carbon. These added elements improve hardenability, wear resistance, toughness, and other properties. This term includes low-alloy steels in which the alloy content does not exceed 5%, but does not include stainless steel.
See also steel.

allylacetone. (5-hexene-2-one). $CH_2CHCH_2CH_2COCH_3$.
Properties: Colorless liquid, d 0.846 (20/4C), bulk d 6.99 lb/gal (20C), 5% to 95% distills between 127–129C, soluble in water and organic solvents.
Use: Intermediate in pharmaceutical synthesis, perfumes, fungicides, insecticides, fine chemicals.

allyl acrylate. $CH_2CHCOOCH_2CHCH_2$.
Liquid, bp 122–134C. Used as monomer for resins.

allyl alcohol. (2-propen-1-ol; propenyl alcohol). CAS: 107-18-6. $H_2C=CHCH_2OH$.
Properties: Colorless liquid with pungent mustard-like odor. Bp 96.9C, mp −129C, d 0.8520 (20/4C), bulk d 7.11 lb/gal (20C), refr index 1.4131, flash p 70F (21C) (TOC), autoign temp 713F (375C). Miscible with water, alcohol, chloroform, ether.
Derivation: (a) Hydrolysis of allyl chloride (from propylene) with dilute caustic, (b) isomerization of propylene oxide over lithium phosphate catalyst at 230–270C, (c) dehydration of propylene glycol.
Hazard: TLV 5 mg/m³, toxic effects from skin absorption.
Use: Esters for use in resins and plasticizers, intermediate for pharmaceuticals and other organic chemicals, manufacture of glycerol, and acrolein, military poison, herbicide.

allylamine. (2-propenylamine). $C_3H_5NH_2$.
Properties: Colorless to light yellow liquid, strong ammoniacal odor, attacks rubber and cork, bp 55–58C, d 0.759–0.761 (20/20C), refr index 1.4194 (22C). Soluble in water, alcohol, ether, and chloroform. Combustible.
Grade: CP, technical.
Use: Pharmaceutical intermediate, organic synthesis.

allyl bromide. (3-bromopropene; bromoallylene). $H_2C=CHCH_2Br$.
Properties: Colorless to light yellow liquid; irritating, unpleasant odor. D 1.398 (20/4C); mp −199C; bp 71.3C; refr index 1.4654; soluble in alcohol, ether, chloroform, carbon tetrachloride,

carbon disulfide; insoluble in water; flash p 30F (-1.1C), autoign temperature 563F (295C).
Grade: Technically pure (95% min purity via bromium titration).
Hazard: Strong irritant to skin and eyes; flammable; dangerous fire risk.
Use: Organic syntheses; preparation of resins, perfume intermediates.

allyl carbamate.
Use: Intermediate for fire proofing compound, emulsion and solvent polymers.

allyl chloride.   CAS: 107-5-1.
$H_2C=CHCH_2Cl$.
Properties: Colorless liquid with a disagreeable pungent odor. D 0.938 at 20C, mw 76.53, vapor press 295 at 20C, flash p (CC) −25F. Slightly soluble in water; miscible with alcohol, chloroform, ether, and petroleum ether.
Hazards: TLV: 1 ppm. Skin and eye irritant.
Use: Synthesis of allyl compounds.

allyl diglycol carbonate.   See diethylene glycol bis-(allyl carbonate).

4-allyl-1,2-diemthoxybenzene.   See methyl eugenol.

allylene.   See methylacetylene.

2-allyl-2-ethyl-1,3-propanediol.
$C_3H_5C(CH_2OH)_2CH_2CH_3$.   Off-white solid, freezing point 31C, d 0.976 (35C). Suggested as polymer additive and chemical intermediate.

allyl glycidyl ether.   (AGE).   CAS: 106-92-3.
$CH_2CHCH_2OCH_2=CH_2$.
Properties: Colorless liquid of pleasant odor. D 0.9698 at 20C, mw 114, freezing point -100C, bp 153.9C, vapor press 4.7 mm at 25C, flash p 135F (OC). Slightly soluble in water.
Hazards: TLV: 5 ppm. Skin and eye irritant.
Use: Resin intermediate, stabilizer of chlorinated compounds, vinyl resins, and rubber.

allyl hexanoate.   See allyl caproate.

1-allyl-4-hydroxybenzene.   See chavicol.

allylic rearrangements.   Migration of a C=C double bond in a three-carbon (allylic) system on treatment with nucleophiles under S_{N1} conditions (or under S_{N2} conditions when the nucleophilic attack takes place at the gamma- carbon).

allyl-α-ionone. $C_{16}H_{24}O$.   Yellow liquid with fruity odor, d 0.928–0.935 (25/25C). Stable, soluble in 70% alcohol. Made synthetically.

Properties: Combustible.
Use: A perfume and flavoring.

5-allyl-5-isobutylbarbituric acid. $C_{11}H_{16}N_2O_3$.
Properties: White, crystalline powder; odorless; slightly bitter taste; soluble in alcohol, ether, and chloroform; almost insoluble in water; mp 138–139C.
Use: Medicine (sedative).   See also barbiturate.

allyl isocyanate. C_3H_5NCO.
Properties: Colorless liquid, turns yellow on standing, d 0.935–0.945 (20C), flash p 110F (43.3C), bp 85.5–86C, freezing point less than −80C. Flammable.
Grade: Purity, 98% min.
Hazard: Fire risk. Toxic.
Use: Organic intermediate, cross-linking agent, polymer modifier.

allyl isothiocyanate.   (allyl isosulfocyanate; mustard oil; 2-propenyl isothiocyanate).
$H_2C=CHCH_2NCS$.
Properties: Colorless to pale yellow, oily liquid; pungent, irritating odor; sharp, biting taste. D 1.013–1.016 (25C), bp 152C, flash p 115F (46.1C), refr index 1.527; optically inactive. Soluble in alcohol, ether, carbon disulfide; slightly soluble in water. Flammable.
Derivation: Distillation of sodium thiocyanate and allyl chloride, or of seed of black mustard.
Grade: Technical, FCC.
Hazard: Toxic via ingestion, inhalation, skin contact; fire risk.
Use: Fumigant, ointments, and mustard plasters.

allyl mercaptan.   See allyl thiol.

allyl methacrylate. $CH_2C(CH_3)COOC_3H_5$.
Used as monomer and intermediate.

4-allyl-2-methoxyphenol.   See eugenol.

4-allyl-1,2-methylenedioxybenzene.   See safrole.

allyl pelargonate. $C_3H_5OOC(CH_2)_7CH_3$.
Properties: Liquid, fruity odor, bp 87–91C (3 mm), refr index 1.4332 (20.5C).
Combustible.
Use: Flavors and perfumes, polymers.

allyl propyl disulfide.   CAS: 2179-59-1.
$CH_2=CHCH_2S_2C_3H_7$.
Properties: Liquid with a pungent, irritating odor. Mw 148.16, d 0.9289 at 15C. Soluble in ether, carbon disulfide, and chloroform.
Occurrence: The chief volatile constituent of onion oil.
Hazards: TLV: 2 ppm. Eye irritant.

p-**allylphenol.**   See chavicol.

allyl resin.   A special class of polyester resins derived from esters of allyl alcohol and dibasic acids. Common monomers are allyl diglycol carbonate, also known as diethylene glycol bis(allyl carbonate), diallyl chlorendate, diallyl phthalate, diallyl isophthalate, and diallyl maleate. Polymerization occurs through the unsaturated allyl double bond to form thermosetting resins which are highly resistant to chemicals, moisture, abrasion, and heat; have low shrinkage and good electrical resistivity.
Use: As laminating adhesives and coatings, especially by impregnation of layered materials with prepolymers (called prepregs); allylic glass cloth, varnishes, applications requiring microwave transparency, encapsulation of electronic parts, vacuum impregnation of metal, casting of ceramics, molding compositions, heat-resistant furniture finishes.

allyl sulfide.   (diallyl sulfide; thioallyl ether). $(CH_2CHCH_2)_2S$.
Properties: Colorless liquid with garlic odor. Combustible. Bp 139C, d 0.888 (27/4C), refr index 1.4877 (27C). Insoluble in water; miscible with alcohol, ether, chloroform, and carbon tetrachloride.
Use: Component of artificial oil of garlic.

allyl thiol.   (allyl mercaptan; 2-propene-1-thiol). CH_2CHCH_2SH.
Properties: Water-white liquid (darkens on standing), strong garlic odor, d 0.925 (23/4C), bp 90C, insoluble in water, soluble in ether and alcohol. Combustible.
Use: Pharmaceutical intermediate, rubber accelerator intermediate.

allylthiourea.   (allylsulfocarbamide; thiosinamine; allyl sulfourea). $C_3N_5NHCSNH_2$.
Properties: White, crystalline solid; slight garlic odor; bitter taste; d 1.22; mp 78C. Soluble in water, ether, and solutions of borax, benzoates, urethane; insoluble in benzene; slightly soluble in 70% alcohol.
Derivation: Warming a mixture of equal parts of allyl isothiocyanate and absolute alcohol with an equal amount of 30% ammonia.
Use: Medicine, corrosion inhibitor, organic synthesis.

allyltrichlorosilane. $CH_2CHCH_2SiCl_3$.
Properties: Colorless liquid; pungent, irritating odor; bp 117.5C; d 1.217 (27C); refr index 1.487 (20C); flash p 95F (35C) (COC). Readily hydroyzed by moisture with the liberation by hydrochloric acid, polymerizes easily. Fire hazard.

Derivation: Reaction of allyl chloride with silicon (copper catalyst).
Use: Intermediate for silicones, glass fiber finishes.

4-allyl veratrole.   See methyl eugenol.

almond oil.   The volatile essential oil distilled from ground kernels of bitter almonds.
Use: Cosmetic creams, perfumes, liqueurs, food flavors (hydrocyanic acid-free).
See also amygdalin.
Note: "Bitter almonds contain amygdalin together with an enzyme that catalyzes its hydrolysis. When the kernels are ground and moistened, a volatile oil produced by the hydrolysis can be distilled from them consisting mainly of benzaldehyde and hydrocyanic acid. This is the oil of bitter almond used in pharmacy as a food flavor after removal of the hydrocyanic acid)." (Eckey, "Vegetable Fats and Oils.")

alnico.   Coined word for an alloy containing chiefly iron, aluminum, nickel, and cobalt which has outstanding properties as a permanent magnet.

aloin.   (barbaloin).   A mixture of active principles obtained from aloe. Varies in properties according to variety used.
Properties: Yellow crystals with bitter taste, darkens on exposure to air, none or slight odor of aloe, approximately 60% soluble in pyridine, slightly soluble in water and organic solvents.
Grade: Technical.
Use: Medicine, proprietary laxatives, electroplating baths, fermentation.

"Alox."[117]   TM for a series of oxygenated hydrocarbons derived from the controlled, liquid phase, partial oxidation of petroleum fractions. Each consists of mixtures of organic acids and hydroxy acids, lactones, esters, and unsaponifiable matter. Esters, amines, amides and metal soap derivatives are also available.
Use: Corrosion inhibitors, film-forming rust preventives, lubricity agents, emulsifiers.

"Aloxite."[280]   TM for aluminum oxide made by fusing materials high in alumina, such as bauxite, and for articles made therefrom.
Use: Abrasive grains and powders, grinding wheels, stones, razor hones, refractory cements, filter plates and tubes, diffuser plates and tubes, porous undergrain plates and coated abrasive products.

alpaca.   A natural fiber obtained from a South American animal similar to the llama. Properties resemble those of wool. Used for specialty clothing and also blended with polyester. Combustible.

"Alpco."[271] TM for a series of high-melting mineral waxes and resins.
Use: Carbon paper, inks, polishes, paper, plasticizers, surfactants, dispersants, casting waxes and surface coatings.

"Alperox" C.[154] TM for lauroyl peroxide (96.5% min.).

alpha (α). (1) A prefix denoting the position of a substituting atom or group in an organic compound. The Greek letters alpha, beta, gamma, etc., are usually not identical with the IUPAC numbering system 1, 2, 3, etc., since they do not start from the same carbon atom. However, alpha and beta are used with naphthalene ring compounds to show the 1 and 2 positions, respectively. Alpha, beta, etc., are also used to designate attachment to the side chain of a ring compound.
(2) Both a symbol and a term used for relative volatility in distillation.
(3) Symbol for optical rotation.
(4) A form of radiation consisting of helium nuclei.
See alpha particle.
(5) The major allotropic form of a substance, especially of metals, e.g., alpha-iron.

alpha-cellulose. The major component of wood and paper pulp. It is that portion of holocellulose that is insoluble in strong sodium hydroxide solution.
See also cellulose.

alpha particle. A helium nucleus emitted spontaneously from radioactive elements both natural and man-made. Its energy is in the range of 4 to 8 MeV, and is dissipated in a very short path, i.e., in a few centimeters of air or less than 0.005 mm of aluminum. It has the same mass (4) and positive charge (2) as the helium nucleus. Accelerated in a cyclotron alpha particles can be used to bombard the nuclei of other elements.
See also helium decay, radioactive.

"Alphazurine."[243] TM for triphenylmethane acid blues.

"Al Polymer."[216] TM for an aromatic polymer that can be cured at moderate temperatures to produce a poly(amide-imide). It is a single polymer system based on trimellitic anhydride. Supplied as a completely solid resin.
Use: As a complete 220 degree insulation system in wire enamels, dipping varnishes, impregnated cloth, glass and asbestos laminates, film and sleeving.

"Alsilox."[468] TM for a fusion product of 1% alumina, 65% litharge, and 34% silica, used in ceramics. Available in various particle sizes.

"Altax."[69] TM for 2,2'-dithiobisbenzothiazole (benzothiazyl disulfide); a rubber accelerator.

altheine. See asparagine.

alum. See aluminum ammonium sulfate, aluminum potassium sulfate, aluminum sulfate.

"Alumalith."[250] TM for amblygonite.

alum, burnt. (alum, dried). $AlK(SO_4)_2$.
Aluminum ammonium sulfate or aluminum potassium sulfate heated just sufficiently to drive off the water of crystallization.
Properties: White, odorless powder; sweetish taste. Absorbs moisture on exposure to air. Soluble in hot water and slowly soluble in cold water, insoluble in alcohol.
Use: Medicine (astringent).

alum, chrome. See chromium potassium sulfate.

alum, chrome ammonium. See chromium ammonium sulfate.

"Alumel."[166] TM for alloy consisting of approximately 94% nickel with small carefully controlled amounts of silicon, aluminum, and manganese.
Use: Chiefly used in thermocouples and lead wire.

"Alumex."[285] TM for a china clay; particle size: 55–60% minus 2 microns, 20–25% plus 5 microns, oil absorption 30 cc/100 g clay, pH 4.5–5.5.

alumina. See aluminum oxide and following entries.

alumina, activated. A highly porous, granular form of aluminum oxide having preferential adsorptive capacity for moisture and odor contained in gases and some liquids. When saturated, it can be regenerated by heat (176–315C). The cycle of adsorption and reactivation can be repeated many times. Granules range in size from powder (7 microns for chromatographic work) to pieces approximately 1–1/2 inches in diameter. Average density approximately 50 lb/cu ft.
Use: An effective desiccant for gases and vapors in the petroleum industry. It is also used as a catalyst or catalyst carrier in chromatography, and in water purification.
See also aluminum oxide.

alumina, fused. See aluminum oxide "Alundum."

alumina gel. See aluminum hydroxide gel.

alumina-silica fiber.
Properties: Amorphous structure, excellent resistance to all chemicals except hydrochloric acid and phosphoric acids and concentrated alkalies. Available in both short and long staple. Low heat conductivity, high thermal shock resistance. Tensile strength 400,000 psi, elastic modulus 16 million psi, upper temperature limit in oxidizing atmosphere 800C. Noncombustible.
Derivation: The short fiber type is made by blasting a stream of molten alumina and silica with a steam jet. Long staple is spun from a molten mixture of alumina and silica modified with zirconium.
Forms: Fibers, sheets, blankets.
Use: Nonwoven fabrics (short staple), woven fabric structures, cordage, thermal insulation, repair of furnace linings, piping molten metals, welding insulation (re-usable), insulation for rocket and space applications. *Note*: In finely divided form, alumina-silica is also used as a catalyst.

alumina trihydrate. (aluminum hydroxide; aluminum hydrate; hydrated alumina; hydrated aluminum oxide). $Al_2O_3 \cdot 3HOH$ or $Al(OH)_3$.
Properties: White crystalline powder, balls, or granules; d 2.42; insoluble in water; soluble in mineral acids and caustic soda. Releases water on heating.
Derivation: From bauxite. The ore is dissolved in strong caustic and aluminum hydroxide precipitated from the sodium aluminate solution by neutralization (as with carbon dioxide) or by autoprecipitation (Bayer process).
Grade: Technical, CP.
Use: Glass, ceramics, iron-free aluminum and aluminum salts, manufacture of activated alumina, base for organic lakes, flame retardants, mattress batting. Finely divided form (0.1–0.6 microns) used for rubber reinforcing agent, paper coating, filler, cosmetics.

aluminium. British spelling of aluminum.

aluminon. (aurine tricarboxylic acid, ammonium salt).
Properties: Mw 473.44, mp 220-5C (decomposes).
Use: As a reagent for aluminum in solution.

aluminosilicate. A compound of aluminum silicate with metal oxides or other radicals.
Use: Catalyst in refining petroleum, to soften water, and in detergents. The sodium compound, $Na_{12}[(AlO_2)_{12}(SiO_2)_{12}]xH_2O$, is typical.
See also zeolite, molecular sieve.

aluminum Al. (aluminium). Metallic element of atomic number 13. Group IIIA of the Periodic Table. Atomic weight 26.98154. Valence 3; no stable isotopes, monovalent in high-temperature compounds, i.e., AlCl and AlF. Most abundant metal in earths crust; third most abundant of all elements. Does not occur free in nature.
Properties: Silvery white, crystalline solid. Tensile strength (annealed) 6800 psi, cold-rolled 16,000 psi. D 2.708, mp 660C, bp 2450C. Forms protective coating of aluminum oxide approximately 50 angstroms thick, which makes it highly resistant to ordinary corrosion. Attacked by concentrated and dilute solutions of hydrochloric acid, hot concentrated sulfuric acid and perchloric acid. Also violently attacked by strong alkalies. Rapidly oxidized by water at 180C. Not attacked by dilute or cold concentrated sulfuric acid or concentrated nitric acid. Can ignite violently in powder form. Electrical conductivity approximately two thirds that of copper. Aluminum qualifies as both a light metal and a heavy metal, according to their respective definitions.
Derivation: From bauxite by Bayer process and subsequent electrolytic reduction by Hall process. There are several processes for obtaining ultrapure aluminum: (a) electrolytic (three-layer), (b) zone refining, and (c) chemical refining. Impurities as low as 0.2 ppm are possible.
  More efficient processes are the Alcoa and Toth processes which require much less electric power than the Hall process. Another method, using no electricity, involves heating a mixture of aluminum ores with a coal-derived fuel in a closed furnace. Still another process called calsintering, using fly ash as a source of alumina, has been described (1978). See calsintering.
Forms available: Structural shapes of all types, plates, rods, wire foil flakes, powder (technical and USP). Aluminum can be electrolytically coated and dyed by the anodizing process (see anodic coating); it can be foamed by incorporating zirconium hydride in molten aluminum and it is often alloyed with other metals or mechanically combined (fused or bonded) with boron and sapphire fibers or whiskers. Strengths up to 55,000 psi at 500C have been obtained in such composites. A vapor-deposition technique is used to form a tightly adherent coating from 0.2 to 1 mil thick on titanium and steel.
Hazard: Fine powder forms flammable and explosive mixtures in air. TLV: 10 mg/m³ of air, (soluble salts) 2 mg/m³ of air; (alkyls) 2 mg/m³ of air; (welding fumes) 5 mg/m³ of air.
Use: Building and construction, corrosion-resistant chemical equipment (desalination plants), die-cast auto parts, electrical industry (power transmission lines), photoengraving plates, per-

manent magnets, cryogenic technology, machinery and accessory equipment, miscible food-processing equipment, tubes for ointments, tooth paste, shaving cream, etc. Also as a powder in paints and protective coatings, as rocket fuel, ingredient of incendiary mixtures (thermite) and pyrotechnic devices, as a catalyst, foamed concrete vacuum metallizing and coating. Other uses are as foil in packaging, cooking, decorative stamping, and as flakes for insulation of liquid fuels.
See also aluminum alloy. For further information, refer to Aluminum Association, New York, N.Y.

aluminum acetate. A salt obtained by reaction of aluminum hydroxide and acetic acid with subsequent recrystallization. Its neutral form $Al(C_2H_3O_2)_3$ is a white, water-soluble powder used in solution as an antiseptic, astringent, and antiperspirant. Its basic form is $Al(C_2H_3O_2)_2OH$, also known as aluminum diacetate and aluminum subacetate. It is a crystalline solid, insoluble in water, used as a mordant in textile dyeing, as a flame retardant and waterproofing agent, and in manufacture of lakes and pigments. See also mordant rouge.

aluminum acetylacetonate. $Al(C_5H_7O_2)_3$.
Properties: Solid, mp 189C, bp 315C. Soluble in benzene and alcohol.
Use: Deposition of aluminum, catalyst.

aluminum alkyl. (Al trialkyl).　　Catalyst used in the Ziegler process.
Hazard: Pyrophoric liquid.
See triethylaluminum and triisobutylaluminum.

aluminum alloy. Aluminum containing variable amounts of manganese, silicon, copper, magnesium, lead, bismuth, nickel, chromium, zinc, or tin. A wide range of uses and properties are possible. Alloys may be obtained for casting or working, heat-treatable or non-heat-treatable, with a wide range of strength, corrosion resistance, machinability and weldability.
See also duralumin.

aluminum ammonium chloride. (ammonium aluminum chloride.　　$AlCl_3 \cdot NH_4Cl$.
Properties: White crystals, soluble in water. Mp 304C.
Use: Fur treatment.

aluminum ammonium sulfate. (ammonium alum; alum NF).　　$Al_2(SO_4)_3(NH_4)_2SO_4 \cdot 24H_2O$ or $AlNH_4(SO_4)_2 \cdot 12H_2O$.
Properties: Colorless crystals, odorless, strong astringent taste. Soluble in water, glycerol; insoluble in alcohol. D 1.645, mp 94.5C, bp loses 20 HOH at 120C.
Derivation: By crystallization from a mixture of ammonium and aluminum sulfates.
Method of purification: Recrystallization.
Grade: Technical, lump, ground, powdered, CP, NF, FCC.
Use: Mordant in dyeing, water and sewage purification, sizing paper, retanning leather, clarifying agent, food additive, manufacture of lakes and pigments, and fur treatment.

aluminum, anodized. See anodic coating.

aluminum antimonide. AlSb.
Properties: Crystalline solid, mp 1050C.
Derivation: Fusion of the elements followed by zone refining to purify.
Use: Semiconductor technology.

aluminum arsenide. AlAs.　　A semiconductor used in rectifiers, transistors, thermistors.
Hazard: Poisonous by ingestion.

aluminum borate. $Al_2O_3 \cdot B_2O_3$.
Properties: White, granular powder; decomposed by water.
Derivation: Interaction of aluminum hydroxide and boric oxide.
Grade: Technical, CP.
Use: Glass and ceramic industries, polymerization catalyst.

aluminum boride.
Properties: Powder; apparent bulk density (fully settled, light) 0.6–0.8 g/cc; dense, 1.2–1.4 g/cc; high neutron absorption.
Use: Nuclear shielding.

aluminum borohydride. $Al(BH_4)_3$.
Properties: Volatile, pyrophoric liquid; bp 44.5C; fp - 64.5C.
Derivation: (1) By reaction of sodium borohydride and aluminum chloride in the presence of a small amount of tributyl phosphate, (2) by reaction of trimethylaluminum and diborane.
Hazard: Ignites spontaneously in air, reacts violently with water.
Use: Intermediate in organic synthesis, jet fuel additive, reducing agent.

aluminum brass. An alloy containing 76% copper, 21.5 to 22.25% zinc, and 1.75 to 2.50% aluminum
Use: Condenser, evaporator and heat exchanger tubes and ferrules.

aluminum bromide. (a) $AlBr_3$; (b) $AlBr_3 \cdot 6HOH$.

Properties: White to yellowish, deliquescent crystals; exists as double molecules Al_2Br_6 in the vapor; soluble in alcohol, carbon disulfide, or ether. (a) D 3.01, mp 97.5C, bp 265C; (b) d 2.54, mp 93C (decomposes).
Derivation: (a) By passing bromine over heated aluminum; (b) reaction of hydrogen bromide with aluminum hydroxide.
Hazard: The anhydrous form reacts violently with water; corrosive to skin.
Uses (anhydrous): Bromination, alkylation, and isomerization catalyst in organic synthesis.

aluminum bronze. An alloy containing 88–96.1% copper, 2.3–10.5% aluminum, and small amounts of iron and tin. Characterized by high strength, ductility, hardness, and resistance to shock, fatigue, most chemicals and sea water. Powder: Also called gold bronze powder. An alloy of 90% copper and 10% aluminum reduced from leaf form to powder, polished mechanically, and coated with stearic acid. Available in the following grades: litho, molding, printing-ink, and radiator.
Use: A pigment in paints and inks.

aluminum-n-butoxide. $Al(OC_3H_9)_3$.
Properties: Yellow to white, crystalline solid; mp 101.5C (pure) and 88–96C (commercial); d 1.0251 (20C), bp 290–310C (30 mm). Soluble in aromatic, aliphatic, and chlorinated hydrocarbons.
Use: Ester exchange catalyst, defoamer ingredient, hydrophobic agent, intermediate.

aluminum calcium hydride. (calcium aluminum hydride). Al_2CaH_8.
Properties: Grayish solid. Soluble in tetrahydrofuran, insoluble in benzene.
Derivation: Reaction of $AlCl_3$ and CaH_2 in tetrahydrofuran.
Hazard: Flammable in contact with water and alcohols. Spontaneous ignition in moist air, store and handle in nitrogen.
Use: Versatile reducing agent.

aluminum carbide. Al_4C_3.
Properties: Yellow crystals or powder, decomposes in water with liberation of methane. D 2.36. Stable to 1400C.
Derivation: By heating aluminum oxide and coke in an electric furnace.
Grade: Technical.
Hazard: Dangerous fire risk in contact with moisture.
Use: Generating methane, catalyst, metallurgy, drying agent, reducing agent.

aluminum carbonate. A basic carbonate of variable composition; formula sometimes given as $Al_2O_3 \cdot CO_2$. White lumps or powder, insoluble in water, dissolves in hot hydrochloric acid or sulfuric acid. Formerly used as mild astringent, styptic. Normal aluminum carbonate $Al_2(CO_3)_3$ is not known as an individual compound.

"Aluminum Chelate."[134] TM for a group of compounds based on aluminum. BEA-1: Chemically modified aluminum secondary butoxide.
Properties: Pale yellow liquid, d 1.030 (21C), aluminum content 8.9–9.1%. PEA-1: Chemically modified aluminum isopropylate. Properties: Pale yellow liquid, d 1.035 (25C), aluminum content 9.3–10.0%. PEA-2: Chemically modified aluminum isopropylate. Properties: Pale yellow; soluble in aromatic, aliphatic, and chlorinated hydrocarbons; aluminum content 7.8–7.9%.
Use: Curing of epoxy, phenolic, castor oil alkyds, and high molecular weight polymers which are hydroxyl or carboxyl bearing; textile hydrophobing in solvent based systems; adhesion promotion.

aluminum chlorate. $Al(ClO_3)_3$.
Properties: Colorless crystals, deliquescent, soluble in water and alcohol.
Hazard: Powerful oxidizing material, keep out of contact with combustibles.
Use: Disinfectant, color control of acrylic resins.

aluminum chloride, anhydrous. $AlCl_3$
Properties: White or yellowish crystals, d 2.44 (25C), mp 190C (2.5 atm), sublimes readily at 178C, the vapor consists of double molecules Al_2Cl_6. Soluble in water.
Derivation: (a) By reaction of purified gaseous chlorine with molten aluminum, (b) by reaction of bauxite with coke and chlorine at approximately 875C. (This product is used to make the hydrate).
Impurities: (a) Ferric chloride-free aluminum, insolubles. Grade: Technical, reagent.
Hazard: Powerful irritant to tissue; moderately toxic by ingestion. Reacts violently with water evolving hydrogen chloride gas.
Use: Ethylbenzene catalyst, dyestuff intermediate, detergent alkylate, ethyl chloride, pharmaceuticals and organics (Friedel-Crafts catalyst), butyl rubber, petroleum refining, hydrocarbon resins, nucleating agent for titanium dioxide pigments.

aluminum chloride hydrate. $AlCl_3 \cdot 6HOH$.
Properties: White or yellowish, deliquescent, crystalline powder; nearly odorless; sweet astringent taste; d 2.4; mp decomposes. Soluble in water and alcohol. The water solution is acid.
Derivation: By crystallizing the anhydrous form from hydrochloric acid solution.
Grade: Technical, CP, NF.
Use: Pharmaceuticals and cosmetics, pigments,

roofing granules, special papers, photography, textiles (wool). Aluminum chloride solution 32 degrees Bé. Special grade of a solution containing only 0.005% iron as impurity, and having an acid reaction but containing no free acid. Use: Antiperspirants, roofing granules.
See also aluminum chloride (solution).

aluminum chlorohydrate. $[Al_2(OH)_5Cl]_x$.
An ingredient of commercial antiperspirant and deodorant preparations. Also used for water purification and treatment of sewage and plant effluent.

aluminum diacetate. See aluminum acetate.

aluminum diethyl monochloride. See diethylaluminum chloride.

aluminum diformate. (aluminum formate, basic). $Al(OH)(CHO_2)_2 \cdot HOH$.
Properties: White or gray powder. Soluble in water.
Derivation: Aluminum hydroxide is dissolved in formic acid and spray-dried. Solutions are also prepared by treating aluminum sulfate with formic acid, followed by lime.
Grade: Technical, solutions (12–20 degrees Bé).
Use: Waterproofing, mordanting, antiperspirants, tanning leather, improving wet strength of paper.

aluminum distearate. $Al(OH)(C_{18}H_{35}O_2)_2$.
Properties: White powder, mp 145C, d 1.009. Insoluble in water, alcohol, ether. Forms gel with aliphatic and aromatic hydrocarbons.
Use: Thickener in paints, inks, and greases. Water repellent, lubricant in plastics and cordage, in cement production.

aluminum ethylate. (aluminum ethoxide). $Al(OC_2H_5)_3$.
Properties: Colorless liquid which gradually solidifies, bp 200C (6 mm), mp 140C. Partly soluble in high-boiling organic solvents.
Derivation: Reaction of aluminum with ethanol, catalyzed by iodine and mercuric chloride.
Hazard: Strong irritant to eyes and skin.
Use: Reducing agent for aldehydes and ketones, polymerization catalyst.

aluminum ethylhexoate. (aluminum octoate).
A metallic salt of 2-ethylhexoic acid.
Use: A gelling agent for liquid hydrocarbons, and a paint additive.

aluminum fluoride, anhydrous. AlF_3.
Properties: White crystals, sublimes at approximately 1250C, d 2.882. Slightly soluble in water, insoluble in most organic solvents.

Derivation: (1) Action of hydrogen fluoride gas on alumina trihydrate; (2) reaction of hydrogen fluoride on a suspension of aluminum trihydrate, followed by calcining the hydrate formed; (3) reaction of fluosilicic acid on aluminum hydrate.
Grade: Technical.
Hazard: Strong irritant to tissue. TLV (as F): 2.5 mg/m³ of air.
Use: Production of aluminum to lower the melting point and increase the conductivity of the electrolyte, flux in ceramic glazes and enamels, manufacture of aluminum silicate, catalyst.

aluminum fluoride, hydrate. $AlF_3 \cdot 3.5HOH$.
Properties: White, crystalline powder. Slightly soluble in water.
Derivation: Action of hydrofluoric acid on alumina trihydrate and subsequent recovery by crystallization.
Grade: Technical, CP.
Hazard: See fluorine.
Use: Ceramics (production of white enamel).

aluminum fluosilicate. (aluminum silicofluoride). $Al_2(SiF_6)_3$.
Properties: White powder. Slightly soluble in cold water, readily soluble in hot water.
Grade: Technical.
Hazard: See fluorine.
Use: Artificial gems, enamels, glass.

aluminum formate. See aluminum triformate, and aluminum diformate.

aluminum formate, basic. See aluminum diformate.

aluminum formate, normal. See aluminum triformate.

aluminum formoacetate. $Al(OH)(OOCH)(OOCCH_3)$.
Properties: White powder, soluble in water and alcohol, used in textile water repellents.

aluminum hydrate. See alumina trihydrate.

aluminum hydride. AlH_3.
Properties: White to gray powder, decomposes at 160C (100C if catalyzed). Evolves hydrogen on contact with water.
Hazard: Dangerous fire and explosion risk.
Use: Electroless coatings on plastics, textiles, fibers, other metals, polymerization catalyst, reducing agent.

aluminum hydroxide. See alumina trihydrate.

aluminum hydroxide gel. (hydrous aluminum oxide; alumina gel). $Al_2O_3 \cdot xHOH$.
Properties: White gelatinous precipitate. Constants variable with the composition, d approxi-

mately 2.4. Insoluble in water and alcohol, soluble in acid and alkali. Noncombustible.
Derivation: By treating a solution of aluminum sulfate or chloride with caustic soda, sodium carbonate or ammonia, by precipitate from sodium aluminate solution, by seeding or acidifying (carbon dioxide is commonly used).
Grade: Technical, CP, USP (containing 4% Al_2O_3), NF (dried containing 50% Al_2O_3).
Use: Dyeing mordant, water purification, waterproofing fabrics, manufacture of lakes, filtering medium, chemicals (aluminum salts), lubricating compositions, manufacture of glass, sizing paper, ceramic glaze, antacid.

aluminum hydroxystearate.
$Al(OH)[OOC(CH_2)_{10}CHOH(CH_2)_5CH_3]_2$.
Properties: White powder, mp 155C, d 1.045. Less soluble in nonpolar compounds than other aluminum stearates and more soluble in polar compounds.
Use: Waterproofing of leather and cements, lubricant for plastics and ropes, paints and inks.

aluminum hypophosphite. $AlH_6O_6P_3$.
Properties: Crystalline solid, decomposes at 218C. Insoluble in water, soluble in hydrochloric acid, weak sulfuric acid, and sodium hydroxide solution.
Derivation: By heating an aluminum salt solution with sodium hypophosphite.
Hazard: Decomposes to toxic phosphine.
Use: Finishing agent for polyacrylonitrile fiber.

aluminum iodide. AlI_3 (anhydrous).
Properties: Brown-black, crystalline pieces (white when pure); mp 191C; bp 385C; d 3.9825. Soluble in alcohol, ether, carbon disulfide.
Derivation: Heating aluminum and iodine in a sealed tube.
Method of purification: Crystallization.
Grade: Technical.
Hazard: Reacts violently with water.
Use: Catalyst in organic synthesis.

aluminum isopropoxide. See aluminum isopropylate.

aluminum isopropylate. (aluminum isopropoxide). $Al(OC_3H_7)_3$.
Properties: White solid, d 1.035 (20C), mp 128–132C, bp 138–148C (10 mm). Soluble in alcohol, benzene; decomposes in water.
Derivation: From isopropanol and aluminum.
Grade: Distilled (purity approximately 100%).
Use: Dehydrating agent, catalyst, waterproofing textiles, organic synthesis, paints.

aluminum lactate. $C_9H_{15}AlO_9$.
Properties: Colorless, water-soluble powder.
Use: Fire foam.

aluminum metaphosphate. $Al(PO_3)_3$.
Properties: White powder, insoluble in water, mp approximately 1527C.
Use: As a constituent of glazes, enamels and glasses, and as a high temperature insulating cement.

aluminum monobasic stearate. See aluminum monostearate.

aluminum monopalmitate. See aluminum palmitate.

aluminum monostearate. (aluminum monobasic stearate). $Al(OH)_2[OOC(CH_2)_{16}CH_3]$.
Properties: Fine, white to yellowish-white powder; faint characteristic odor; mp 155C; d 1.020. Insoluble in water, alcohol, and ether. Forms a gel with aliphatic and aromatic hydrocarbons.
Derivation: Mixing solutions of a soluble aluminum salt and sodium stearate.
Grade: USP which describes it as a mixture of the monostearate and monopalmitate containing 14.5–16.0% Al_2O_3.
Use: Paints, inks, greases, waxes, thickening lubricating oils, waterproofing, gloss producer, stabilizer for plastics.

aluminum naphthenate.
Properties: Yellow substance of rubbery consistency with high thickening power. Combustible.
Derivation: Reaction of an aluminum salt with an alkali naphthenate in aqueous solution.
Use: Paint and varnish drier and bodying agent, detergent in lube oils, the solution in organic solvents has been proposed for insecticides and siccatives. See also soap (2).

aluminum nitrate. $Al(NO_3)_3 \cdot 9HOH$.
Properties: White crystals. Soluble in cold water, decomposes in hot water. Soluble in alcohol and acetone. Mp 73C, decomposes at 150C.
Derivation: Formed by the action of nitric acid on aluminum and crystallization.
Grade: Technical, CP (99.75%)
Hazard: Powerful oxidizing agent. Do not store near combustible materials.
Use: Textiles (mordant), leather tanning, manufacture of incandescent filaments, catalyst in petroleum refining, nucleonics, anticorrosion agent, antiperspirant.

aluminum nitride. AlN.
Properties: Crystalline solid, d 3.10, mp 2150C (4.3 atm), Mohs hardness 9+, decomposes when wet into aluminum hydroxide and ammonia.
Derivation: From coal and bauxite when heated in a stream of nitrogen.

Use: As semiconductor in electronics, nitriding of steel.

aluminum oleate. $Al(C_{18}H_{33}O_2)_3$.
Properties: Yellowish-white, viscous mass. Insoluble in water; soluble in alcohol, benzene, ether, oil, and turpentine. Combustible.
Derivation: By heating aluminum hydroxide, water, and oleic acid. The resultant mixture is filtered and dried.
Use: Waterproofing, drier for paints, etc.; thickener for lubricating oils, medicine, as lacquer for metals, lubricant for plastics, food additive.

aluminum orthophosphate. See aluminum phosphate.

aluminum oxalate. $Al_2C_6O_{12}$.
Properties: Finely divided solid (as hydrate), soluble in nitric and sulfuric acids, almost insoluble in water and alcohol.
Use: Mordant and dyeing assistant.

aluminum oxide. (alumina). Al_2O_3.
The mineral corundum is natural aluminum oxide, and emery, ruby, and sapphire are impure crystalline varieties. The mixed mineral bauxite is a hydrated aluminum oxide.
Properties: Vary according to the method of preparation. White powder, balls, or lumps of various mesh. D 3.4–4.0, mp 2030C, insoluble in water, difficultly soluble in mineral acids and strong alkali. Noncombustible.
See also alumina trihydrate; aluminum hydroxide gel.
Derivation: (a) Leaching of bauxite with caustic soda followed by precipitation of a hydrated aluminum oxide by hydrolysis and seeding of the solution. The alumina hydrate is then washed, filtered and calcined to remove water and obtain the anhydrous oxide.
See derivation under alumina trihydrate.
(b) Coal mine waste waters are used to get aluminum sulfate, which is then reduced to alumina.
Grade: Technical, CP, fibers, high purity, fused, calcined.
Hazard: Toxic by inhalation of dust.
Use: Production of aluminum, manufacture of abrasives, refractories, ceramics, electrical insulators, catalyst and catalyst supports, paper, spark plugs, crucibles and laboratory wares, adsorbent for gases and water vapors (see alumina activated), chromatographic analysis, fluxes, light bulbs, artificial gems, heat-resistant fibers, food additive (dispersing agent).
See also alumina, activated.

aluminum oxide, hydrated. See alumina trihydrate.

aluminum oxide, hydrous. See aluminum hydroxide gel.

aluminum palmitate. (aluminum monopalmitate). $Al(OH)_2(C_{16}H_{31}O_2)$.
Properties: White powder, mp 200C, d 1.072. Insoluble in alcohol and water, forms gel with hydrocarbons. Combustible.
Derivation: By heating aluminum hydroxide and palmitic acid and water. The resultant mixture is filtered and dried.
Use: Waterproofing leather, paper, textiles, thickening for lubricating oils, thickening or suspending agent in paints and inks, production of high gloss on leather and paper, ingredient of varnishes, lubricant for plastics, food additive.

aluminum paste. Aluminum powder ground in oil.
Use: Aluminum paints.

aluminum phosphate. (aluminum orthophosphate). $AlPO_4$.
Properties: White crystals. Insoluble in water and alcohol, slightly soluble in hydrochloric acid and nitric acid. D 2.566, mp 1500C.
Derivation: Interaction of solutions of aluminum sulfate and sodium phosphate.
Hazard: Solutions are corrosive to tissue.
Use: Ceramics, dental cements, cosmetics, paints and varnishes, pharmaceuticals, pulp and paper.

aluminum phosphide. AlP.
Properties: Dark gray or dark yellow crystals, d 2.85.
Hazard: Dangerous fire risk. It evolves phosphine.
Use: Insecticide, fumigant, semiconductor technology

aluminum picrate. $Al[(NO_2)_3C_6H_2O]_3$.
Hazard: Toxic by ingestion and inhalation; dangerous fire risk in contact with combustibles, severe explosion risk when shocked or heated. Strong oxidizer
Use: Explosive compositions.

aluminum potassium sulfate. (potash alum; alum NF; potassium alum). CAS: 7784-24-9. $Al_2(SO_4)_3 \cdot K_2SO_4 \cdot 24HOH$, sometimes written $AlK(SO_4)_2 \cdot 12HOH$.
Properties: White, odorless crystals having an astringent taste; d 1.75; mp 92C; bp loses 18 H_2O at 64.5C, anhydrous at 200C; soluble in water; insoluble in alcohol. Solutions in water are acid. Noncombustible.
Derivation: (a) From alunite, leucite or similar mineral. (b) Also derived by crystallization from a solution made by dissolving aluminum sulfate and potassium sulfate and mixing.

Grade: Technical, lump, ground, powdered, NF, FCC.

Use: Dyeing (mordant), paper, matches, paints, tanning agents, waterproofing agents, purification of water, aluminum salts, food additive, baking powder, astringent, cement hardener.

aluminum resinate. CAS: 61789-65-9.
Properties: Brown solid. Insoluble in water, soluble in oils.
Derivation: Heating together soluble aluminum salts and rosin.
Grade: Technical (fused, prepicitated).
Use: Drier for varnishes.
Hazard: Flammable, dangerous fire risk.

aluminum rubidium sulfate. (rubidium alum). $AlRb(SO_4)_2 \cdot 12HOH$.
Properties: Colorless crystals, soluble in hot water, insoluble in alcohol. D 1.867, mp 99C.

aluminum salicylate. $Al(C_6H_4OHCOO)_3$.
Properties: Reddish-white powder. Odorless. Soluble in dilute alkalies, insoluble in water, alcohol. Decomposed by acid.
Use: Medicine.

aluminum silicate. Any of the numerous types of clay which contain varying proportions of Al_2O_3 and SiO_2. Made synthetically by heating aluminum fluoride at 1000–2000C with silica and water vapor; the crystals or whiskers obtained are up to 1-cm long, have high strength, and are used in reinforced plastics.
Use: Same as for clay.
See also mullite, kaolin.

aluminum silicofluoride. See aluminum fluosilicate.

aluminum silicon. A light-weight alloy available as ingots or in powder form used for automotive parts, construction, etc. For new manufacturing method, see aluminum derivation.
Hazard: Powder is flammable, dangerous fire risk.

aluminum soaps. See aluminum oleate, aluminum palmitate, aluminum resinate, aluminum stearate. See also soap (2).

aluminum sodium chloride. See sodium tetrachloroaluminate.

aluminum sodium sulfate. (SAS; sodium aluminum sulfate; soda alum; alum).
$Al_2(SO_4)_3 \cdot Na_2SO_4 \cdot 24HOH$ or $AlNa(SO_4)_2 \cdot 12HOH$.
Properties: Colorless crystals, saline, astringent taste, effloresces in air. Soluble in water, insoluble in alcohol, d 1.675, mp 61C. Noncombustible.
Derivation: By heating a solution of aluminum sulfate and adding sodium chloride. The solution is allowed to cool with constant stirring. The alum meal deposited is washed with water and centrifuged.
Method of purification: Recrystallization.
Grade: Pure crystals, technical, CP, FCC.
Use: Textiles (mordant, waterproofing), dry colors, ceramics, tanning, paper size precipitant, matches, inks, engraving, sugar refining, water purification, medicine, confectionary, baking powders, food additive.

aluminum stearate. (aluminum tristearate). $Al(C_{18}H_{35}O_2)_3$.
Properties: White powder, d 1.070, mp 115C. Insoluble in water, alcohol, ether; soluble in alkali, petroleum, turpentine oil. Forms gel with aliphatic and aromatic hydrocarbons.
Derivation: Reaction of aluminum salts with stearic acid.
Grade: Technical.
Use: Paint and varnish drier, greases, waterproofing agent, cement additive, lubricants, cutting compounds, flatting agent, cosmetics and pharmaceuticals, defoaming agent in beet sugar and yeast processing.

aluminum subacetate. See aluminum acetate.

aluminum sulfate. (trade term for alum; pearl alum; pickle alum; cake alum; filter alum; papermakers' alum; patent alum).
(a) $Al_2(SO_4)_3$; (b) $(Al_2(SO_4)_3 \cdot 18HOH$.
Noncombustible 40th highest-volume chemical produced in the US. (1985)
Properties: White crystals, soluble in water (sweet taste), insoluble in alcohol, stable in air, d (a) 2.71, (b) 1.62, mp decomposes at 770C, (b) decomposes at 86.5C.
Derivation: (1) By treating pure kaolin or aluminum hydroxide or bauxite with sulfuric acid. The insoluble silicic acid is removed by filtration and the sulfate is obtained by crystallization. (2) Similarly, from waste coal-mining shale and sulfuric acid.
Grades; Iron-free, technical, CP, USP, FCC. A liquid form (49.7 H_2O) is also available.
Use: Sizing paper, lakes, alums, dyeing mordant foaming agent in fire foams, cloth fireproofing, white leather tannage, catalyst in manufacturing ethane, pH control in paper industry, waterproofing agent for concrete, clarifier for fats and oils, lubricating compositions, deodorizer and decolorizer in petroleum refining, sewage precipitating agent and for water purification, food additive.

aluminum sulfide. Al_2S_3.
Properties: Yellowish-gray lumps. Odor of hydrogen sulfide. Decomposes in moist air to hydrogen sulfide and a gray powder, d 2.02, mp 1100C.
Hazard: Irritant to skin and mucous membranes.
Use: To prepare hydrogen sulfide.

aluminum sulfocarbolate. See aluminum phenolsulfonate.

aluminum tartrate. $C_{12}H_{12}Al_2O_{18}$.
Properties: Crystalline powder, soluble in hot water and ammonia.
Use: Textile dyeing auxiliary.

aluminum thiocyanate. (aluminum sulfocyanate). $Al(SCN)_3$.
Properties: Yellowish powder. Soluble in water, insoluble in alcohol and ether.
Use: Mordant for textile dyeing, manufacturing pottery.

aluminum trialkyl. See triethylaluminum and triisobutylaluminum.

aluminum tributyl. See tributyl aluminum.

aluminum triethyl. See triethyl aluminum.

aluminum triformate. (aluminum formate, normal). $Al(HCOO)_3 \cdot 3HOH$.
Properties: White, crystalline powder. Soluble in hot water, slightly soluble in cold water.
Use: Textile (delustering rayon, mordanting, waterproofing, after-treatment of dyeings), paper (sizing), fur dyeing (mordant), and medicine.

aluminum trimethyl. See trimethyl aluminum.

aluminum triricinoleate. $Al(C_{17}H_{32}OHCOO)_3$.
Properties: Yellowish to brown plastic mass. Limited solubility in most organic solvents. Mp 95C. Combustible.
Derivation: Castor oil.
Use: Gelling agent, waterproofing, solvent-resistant lubricants.

aluminum tristearate. See aluminum stearate.

alum, N.F. May be either aluminum ammonium sulfate or aluminum potassium sulfate.

alum, papermakers. See aluminum sulfate.

alum, pearl. Specially prepared aluminum sulfate for the papermaking industry.

alum, pickle. Aluminum sulfate prepared to meet specifications of packers and preservers.

alum, porous. See aluminum sodium sulfate.

alum, potash. See aluminum potassium sulfate.

alum, rubidium. See rubidium sulfate.

alum, soda. See aluminum sodium sulfate.

"Alundum."[249] Al_2O_3. TM for a series of fused-alumina refractory and abrasive products. Fusion point 2000–2050C.
Grade: Grains, cements, and refractory shapes.
Use: (grains) Electrical insulation in radio and television tubes, (cements) embedding electrical resistors, metal melting applications, refractory brick setting (refractory shapes), high-temperature work as in furnaces, bricks, plates, muffles, tunnel kilns, tubes, laboratory ware.

alunite. (alum stone). CAS: 1344-28-1.
$KAl_3(OH)_6(SO_4)_3$. A naturally occurring basic potassium aluminum sulfate usually found with volcanic and other igneous rocks.
Occurrence: Utah, Arizona, California, Colorado, Nevada, Washington, Italy, Australia.
Use: Production of aluminum potassium compounds, millstones, substitute for bauxite in aluminum manufacture, decolorizing and deodorizing agent, fertilizer.

"Alusite" D.[446] TM for a 70% alumina brick with relatively low porosity, good resistance to mechanical abrasion and penetration by molten slags.
Use: In rotary lime, lime sludge and dolomite kilns, soaking pit curb walls, gas regenerators, nonferrous metallurgical refining furnaces, in reverberatory and brass melting furnaces and various lead furnaces.

Am. Symbol for americium.

Amadori rearrangement. Conversion of N-glycosides of aldoses to N-glycosides of the corresponding ketoses by acid or base catalysis.

amalgam. A mixture or alloy of mercury with any of a number of metals or alloys including cesium, sodium, tin, zinc, lithium, potassium, gold and silver as well as with some nonmetals. Dental amalgams are mixtures of mercury with a silver tin alloy. A sodium amalgam is formed in the preparation of pure sodium hydroxide by electrolysis of brine.
Use: Dental fillings, silvering mirrors, catalysis, analytical separation of metals, to facilitate application of active metals such as sodium, aluminum, and zinc in the preparation of titanium, etc., or in reduction of organic compounds.
See also sodium amalgam.

amanitin. (alpha) $C_{39}H_{54}N_{10}O_{13}S$.
(beta) $C_{39}H_{53}N_9O_{14}S$. Toxic principle from a species of mushroom (*Amanita phalloides*). The beta form has been obtained as acicular crystals which are soluble in water and methanol and ethanol.
Hazard: A poison! Ingestion may be fatal.

amantadine hydrochloride. (1-adamantanamine hydrochloride). $C_{10}H_{17} \cdot HCl$. A derivative of adamantane. An anti-viral drug. Also used in treatment of Parkinson's disease.
See "Symmetrel."

amaranth. (FDC Red No.2; Red Dye No.2).
CAS: 915-67-3.
$NaSO_3C_{10}H_5N=NC_{10}H_4(SO_3Na)_2OH$. An azo dye derived from naphthionic and R acids.
Properties: Dark red to purple powder, density approximately 1.50, soluble in water, glycerol, propylene glycol; insoluble in most organic solvents.
Hazard: An experimental carcinogen. May not be used in foods, drugs or cosmetics.
Use: Formerly a certified food and drug colorant. (Replaced for some applications by FDC Red No. 40.) Textile dye, color photography.

amatol. An explosive mixture of ammonium nitrate and TNT. The 50–50 mixture can be melted and poured for filling small shells, the 80% ammonium nitrate mixture is granular.
Hazard: Highly explosive. A powerful irritant to mucous membranes and by skin absorption.

amber. A polymerized fossil resin derived from an extinct variety of pine. Readily accumulates static electrical charge by friction; good electrical insulator.

ambergris. A waxy, opaque mass containing 80% cholesterol formed in intestinal tract of the sperm whale and found on beaches or afloat in the ocean.
Use: A fixative in perfumes, now largely replaced by synthetic products.

"Amberlac."[23] TM for modified alkyd-type resins for quick-drying lacquers.
Use: Metal primers, bottle cap coatings, food can coatings, appliance coatings.

"Amberlac."[165] TM for synthetic water soluble polymer. Colorless additive for film formers, improves scuff resistance, prevents sticking of thermoplastic coatings.
Use: High gloss paper coatings, coating of book covers, decorative papers, paperboard.

"Amberlite."[23] TM for several types of ion-exchange resins. Insoluble cross-linked polymers of various types in minute bead form. Strong acid, weak acid, strong base, and weak base forms, each having various grades differing in exchange capacity and porosity, for removing simple and complex cations and anions from aqueous and nonaqueous solutions. Reversible in action, can be regenerated.
Grade: Laboratory, liquid, nuclear, mixed bed, pharmaceutical.
Use: Water conditioning (softening and complete deionization), recovery and concentration of metals, antibiotics, vitamins, organic bases, catalysis, decolorization of sugar, manufacture of chemicals, neutralization of acid mine water drainage, analytical chemistry, water treatment in nuclear reactors, pharmaceuticals.

"Amberol."[23] TM for maleic-resin and resin-modified and unmodified phenolformaldehyde-type polymers in solid form. They react with various oils to produce fast drying, high gloss protective coatings and vehicles for printing inks.
Use: Varnishes, enamels, can liners, nitrocellulose sanding sealers, printing inks, tackifying and vulcanization of butyl rubber.

ambient temperature. The temperature of the environment in which an experiment is conducted or in which any physical or chemical event occurs.
See also room temperature.

"Ambitrol."[233] TM for coolants used in stationary industrial engines.

amblygonite. $Li(AlF)PO_4$ or $AlPO_4 \cdot LiF$.
A natural fluorophosphate of aluminum and lithium.
Properties: White to grayish-white. Contains up to 10.1% lithia, sometimes with partial replacement by sodium, d 3.01–3.09, Mohs hardness 6.
Occurrence: California, Maine, Connecticut, South Dakota, Germany, Norway, France.
Use: A source of lithium used in glazes and coatings.

ambomycin (USAN). An antibiotic produced by Streptomyces ambofaciens.

"Ambraloy-687."[324] TM for an aluminum-brass alloy containing arsenic as an inhibitor to increase its resistance to dezincification. Its nominal composition is copper 77%, zinc 20.96%, aluminum 2%, arsenic 0.04%.
Use: Principally as a condenser tube alloy where cooling water is salt or brackish, and where it

is resistant to the impinging action of turbulently flowing seawater containing air bubbles.

ambrettolide. (ω-6-hexadecenlactone; 6-hexadecenolide; 16-hydroxy-6-hexadecenoic acid; omega lactone). $C_{16}H_{28}O_2$.
Colorless liquid having powerful musk-like odor. Found in ambrette-seed oil.
Use: Flavoring, perfume fixative.

"Amerchol."[493] TM for a series of surface-active lanolin derivatives; most are soft solids.
Use: Emulsifiers and stabilizers for water and oil systems, emollients in pharmaceuticals and cosmetics.

American Association for the Advancement of Science (AAAS). Founded in 1848, the interests of this Association extend into all areas of natural and social science. Its leading publication is *Science*, established by Thomas Edison in 1880. It also publishes many symposium volumes based on papers presented at its annual meetings. One of its important activities in chemistry is sponsorship of the Gordon Research Conferences which originated in 1931 under the leadership of Dr. Neil E. Gordon; these have since been expanded to more than 30 technical conferences attended by chemists from many foreign countries. Society headquarters are located at 1333 H Street, NW, Washington, D.C. 20005.

American Carbon Society. The present name of the American Carbon Committee, a group incorporated in 1964 to operate the Biennial American Carbon Conferences. The committee also has sponsored the international journal *Carbon*.

American Chemical Society (ACS). The nationally chartered professional society for chemists in the US. One of the largest scientific organizations in the world, it was founded in 1874 and now has over 125,000 members. Its offices are at 1155 16th St., N.W., Washington, D.C. 20036. For further information, see Appendix IIB.
See also Chemical Abstracts.

American Conference of Governmental Industrial Hygienists (ACGIH). A group of scientists organized in 1938 for the purpose of determining standards of exposure to toxic and otherwise harmful materials in workroom air. The standards are revised annually and are available from the Secretary, ACGIH, 6500 Glenway Ave., Bldg. D-7, Cincinnati, Ohio, 45211.
See also Threshold Limit Value (TLV's).

American Institute of Chemical Engineers (AIChE). Founded in 1890, the AIChE is the largest society in the world devoted exclusively to the advancement and development of chemical engineering. Its official publication is *Chemical Engineering Progress*. It has over 50 local sections and many committees working in a wide range of activities. Its offices are located at 345 East 47th Street, New York, NY 10017.

American Institute of Chemists. Founded in 1923, the AIC is primarily concerned with chemists and chemical engineers as professional people rather than with chemistry as a science. Special emphasis is placed on the scientific integrity of the individual and on a code of ethics adhered to by all its members. It publishes a monthly journal, *The Chemist*.

American National Standards Institute (ANSI). A federation of trade associations, technical societies, professional groups, and consumer organizations which constitutes the US clearinghouse and coordinating body for voluntary standards activity on the national level. It eliminates duplication of standards activities and combines conflicting standards into single, nationally accepted standards. It is the US member of the International Organization for Standardization and the International Electrotechnical Commission. Over 1000 companies are members of the ANSI. One of its primary concerns is safety in such fields as hazardous chemicals, protective clothing, welding, fire control, electricity and construction operations, blasting, etc. Its address is 1430 Broadway, New York, NY, 10018.

American Petroleum Institute (API). Incorporated in 1919 under the laws of the District of Columbia, membership is approximately 15,000. The objects of the Institute as stated in its charter are: (1) to afford a means of cooperation with the government in all matters of national concern, (2) to foster foreign and domestic trade in American petroleum products, (3) to promote, in general, the interests of the petroleum industry in all its branches, (4) to promote the mutual improvement of its members and the study of the arts and sciences connected with the petroleum industry.

The activities of the Institute include the fields of standardization, design, care, and correct practice in the use of equipment, engineering and technology, fundamental research, safety and fire protection, industrial health, product labeling, waste disposal, testing methods and specifications, measuring, sampling and testing, nomenclature, metallurgy, corrosion prevention, pipeline, highway, waterway and railroad transportation, radio facilities, fuels and lubricants, agriculture, highways, aviation, vocational and personnel training, education and public relations, finance and accounting, statistics, petroleum reserves, taxation, legislation, and regulation. Its offices are located at 1220 L St., NW, Washington, D.C.

American Society for Metals (ASM).   Formally organized in 1935, this society actually had been active under other names since 1913 when the need for standards of metal quality and performance in the automobile became generally recognized. ASM publishes *Metals Review* and the famous "Metals Handbook," as well as research monographs on metals. It is active in all phases of metallurgical activity, metal research, education, and information retrieval. Its headquarters is at Metals Park, Ohio, 44073.

American Society for Testing and Materials (ASTM). This society, organized in 1898 and chartered in 1902, is a scientific and technical organization formed for "the development of standards on characteristics and performance of materials, products, systems and services, and the promotion of related knowledge." It is the world's largest source of voluntary consensus standards. The society operates via more than 125 main technical committees which function in prescribed fields under regulations that ensure balanced representation among producers, users, and general interest participants. Headquarters of the society is at 1916 Race St., Philadelphia, PA, 19103.

americium.   Am. A synthetic radioactive element of atomic number 95, a member of the actinide series. Atomic weight 241; 14 isotopes of widely varying half-life. Valence 3, but divalent, tetravalent, and higher valencies exist. Alpha and gamma emitter, forms compounds with oxygen, halides, lithium, etc. Metallic americium is silverwhite crystalline, d 13.6, mp approximately 100C. Half-life of Am 241 is 458 years.
Derivation: Multiple neutron capture in plutonium in nuclear reactors, plutonium isotopes yield Am 241 and Am 243 on beta decay. The 241 isotope is available in 100-gram quantities, the 243 isotope only in milligram quantities. The metal is obtained by reduction of Am trifluoride with barium in a vacuum at 1200C.
Hazard: A radioactive poison.
Use: Gamma radiography, radiochemical research, diagnostic aid, electronic devices.

"Amerlate" phosphorus.[493]   TM for the isopropyl ester of hydroxy, normal and branched chain acids of lanolin. A light yellow, soft solid that liquefies on contact with the skin. A hydrophilic emollient, moisturizer, conditioning agent, lubricant, pigment dispersant, nonionic auxiliary without emulsifier.

amethopterin.   See methotrexate.

amiben.   Generic name for 3-amino-2,5-dichlorobenzoic acid. $C_6H_2NH_2Cl_2COOH$.
Use: Herbicide or plant growth regulator.

amide.   A nitrogenous compound related to or derived from ammonia. Reaction of an alkali metal with ammonia yields inorganic amides, e.g., sodium amide ($NaNH_2$).
Organic amides are characterized by an acyl group ($-CONH_2$), usually attached to an organic group ($R=CONH_2$) and formamide ($HCONH_2$) are common examples. Carbamide (urea) is $CO(NH_2)_2$.
See also polyamide.

amidinomycin. $C_9H_{18}N_4O$.   An antibiotic which partially inhibits spore-forming bacteria.

4-amidino-1-(nitrosamino amidino)-1-tetrazene.
See tetrazene.

amidol.   See 2,4-diaminophenol hydrochloride.

amidopropylamine oxide.
Use: Foamer, foam booster, foam stabilizer, scour for household, cosmetic, and industrial applications.

aminacrine hydrochloride.   USAN name for 9-amino acridine hydrochloride.

amination.   The process of making an amine (RNH_2). The methods commonly used are (a) reduction of a nitro compound and, (b) action of ammonia on a chloro-, hydroxy-, or sulfonic acid compound.

amine.   A class of organic compounds of nitrogen that may be considered as derived from ammonia (NH_3) by replacing one or more of the hydrogen atoms with alkyl groups. The amine is primary, secondary, or tertiary depending on whether one, two, or three of the hydrogen atoms are replaced. All amines are basic in nature, and usually combine readily with hydrochloric or other strong acids to form salts.
See also fatty amine.

amine 220.   2-(8-Heptadecenyl)-2-imidazoline-1-ethanol). $C_{17}H_{33}C:NC_2H_4NC_2H_4OH$.
Properties: D 0.9330 (20/20C), bulk d 7.76 (20C) lb/gal, bp 235C (1 mm), flash p 465F (240C). Combustible.

Use: Demulsifier used particularly in the recovery of tar from water-gas process emulsions. A powerful cationic wetting agent. Useful in flotation processes involving siliceous minerals and the formation of emulsions and dispersions under acidic conditions.

amine 248. Dark-colored liquid or paste consisting of a non-volatile amine mixture with bis-(hexamethylene)triamine and its homologs as principal components. Disperses readily in water.
Use: Coagulant and flocculating agent, ingredient of antistripping agents for asphalt, corrosion inhibitor, anchoring agent, for making cationic amine salts.

amine absorption process. See Girbotol absorption process.

aminimide. Any of a group of nitrogen compounds derived by reaction of 1,1-dimethylhydrazine with an epoxide in the presence of an ester of a carboxylic acid. A number of types of epoxides and esters may be used providing a wide variety of products including short- and long-chain aliphatics and aromatics. Their major uses are considered to be in tire-cord dips to increase adhesion of cord to rubber, in soil-removing detergents (nonionic), in coating formulations, and in cosmetic creams and shampoos. They are stated to be biodegradable and without toxic hazard. They also have elastomer and cross-linking applications from adhesives, caulks, and sealants to foams and mechanical goods. As isocyanate precursors, aminimides can be prepared in a large number of structural variations. Bisaminimides are especially valuable for producing stable single-package prepolymer compositions for *in situ* generation of isocyanates in polyurethane applications.

p-aminoacetanilide. (4′-aminoacetanilide; N-acetyl-p-phenylenediamine).
$NH_2C_6H_4NHCOCH_3$.
Properties: Colorless or reddish crystals, soluble in alcohol and ether, slightly soluble in water. Mp: 162C, bp 267C. Combustible.
Derivation: Acetylation of p-phenylenediamine.
Hazard: Moderate toxicity by ingestion.
Use: Intermediate for azo dyes and pharmaceuticals.

aminoacetic acid. See glycine.

aminoacetophenetidide hydrochloride.
See phenocoll hydrochloride.

amino acid. An organic acid containing both a basic amino group (NH_2) and an acidic carboxyl group (COOH), thus they are amphoteric and exist in aqueous solution as dipolar ions. The 25 amino acids that have been established as protein constituents are alpha-amino acids (i.e., the —NH_2 group is attached to the carbon atom next to the —COOH group). Many other amino acids occur in the free state in plant or animal tissue. Twenty-two amino acids with structures identical with those that exist today have been identified in pre-Cambrian sedimentary rock, indicating an age of at least 3 million years. Amino acids have been created in the laboratory by passing an electric discharge through a mixture of ammonia, methane, and water vapor; it is believed that a similar reaction may have accounted for the original synthesis of amino acids on earth. See life, origin.
Amino acids can be obtained by hydrolysis of a protein; or they can be synthesized in various ways, especially by fermentation of glucose. An essential amino acid is one which cannot be synthesized by the body and is necessary for survival, namely, isoleucine phenylalanine, leucine, lysine, methionine, threonine, tryptophan, and valine. Nonessential amino acids (alanine, glycine, and about a dozen others) can be synthesized by the body in adequate quantities. Arginine and histidine are essential during periods of intensive growth.

All the essential and most of the nonessential amino acids have one or more asymmetric carbon atoms and are optically active.

Research in molecular biology has established the fact that heredity characteristics are, to a large extent, determined by the sequence of amino acids in the genes, which is programmed by deoxyribonucleic acid. This substance is composed of four amino acids and is itself a protein. Several amino acids have been synthesized by gene-splicing (recombinant DNA) methods. *Note*: Use of amino acids as fortification additives to foods; restricted by FDA to foods containing proteins.
See genetic code, deoxyribonucleic acid, chromatin, protein.

amino acid oxidase. See amino oxidase.

aminoamylene glycol. See 2-amino-2-ethyl-1,3-propanediol.

5-amino-2-aniliobenzenesulfonic acid.
See 4-amino-diphenylamine-2-sulfonic acid.

o-aminoanisole. See o-anisidine.

p-aminoanisole. See p-anisidine.

aminoanthraquinone. $C_6H_4(CO)_2C_6H_3NH_2$.
(tricyclic) (a) 1-amino, (b) 2-amino.
Properties: (a) Red, iridescent needles. (b) Red or orange-brown needles. Soluble in alcohol, chloroform, benzene and acetone; insoluble in water. Mp (a) 252C, (b) 302C, bp sublimes (both a and b).
Derivation: By reduction of nitroanthraquinones, or by the substitution of the amino radical direct for the sulfonic acid.
Use: Dye and pharmaceutical intermediate.

4-aminoantipyrine. (4-amino-1,5-dimethyl-2-phenyl-3-pyrazolone; 1,5-dimethyl-2-phenyl-4-aminopyrazolone). $C_{11}H_{13}N_3O$.
Properties: Mp 107–109C.
Use: Analytical reagent.

p-aminoazobenzene. (aniline yellow; phenyl-azoaniline). CAS: 60-09-3. $C_6H_5NNC_6H_4NH_2$.
Properties: Yellow to tan crystals. Soluble in alcohol and ether; slightly soluble in water. mp 126–128C, bp above 360C.
Derivation: (a) Heating diazoaminobenzene with aniline hydrochloride as catalyst. (b) Diazotization of a solution of aniline and aniline hydrochloride with hydrochloric acid and sodium nitrite.
Hazard: Suspected carcinogen.
Use: Dyes (chrysoidine, induline, solid yellow and acid yellow), insecticide.

4-aminoazobenzene-3,4′-disulfonic acid. (Acid Yellow 9). $C_6H_4(SO_3H)NNC_6H_3NH_2(SO_3H)$.
Properties: Bright, violet needles.
Use: For synthesizing dyes for dyeing wool.

p-aminoazobenzene hydrochloride. (aminoazobenzene salt). $C_6H_5NNC_6H_4NH_2 \cdot HCl$.
Properties: Steel-blue crystals. Soluble in alcohol; slightly soluble in water.
Derivation: By passing dry hydrogen chloride gas into a solution of aminoazobenzene.
Hazard: Suspected carcinogen.
Use: Dyes, coloring lacquers, intermediate.

aminoazobenzenemonosulfonic acid.
$NH_2C_6H_4NNC_6H_4SO_3H$.
Properties: Yellowish-white, microscopic needles. Barely soluble in water, almost insoluble in alcohol, ether, and chloroform.
Derivation: By sulfonating aminoazobenzene.
Hazard: Suspected carcinogen.
Use: Dyestuff manufacture.

o-aminoazotoluene. (Solvent Yellow; 3,2-amino-5-azotoluene; toluazotoluidine).

CAS: 97-56-3. $CH_3C_6H_4N_2C_6H_3NH_2CH_3$.
Properties: Reddish-brown to yellow crystals; soluble in alcohol, ether, oils, and fats; slightly soluble in water; mp 100–117C.
Derivation: From o-toluidine by treatment with nitrite and hydrochloric acid.
Hazard: Suspected carcinogen.
Use: Dyes, medicine.

6-p-(p-aminobenzamido)benzamido-1-naphthol-3-sulfonic acid.
$H_2NC_6H_4CONHC_6H_4CONHC_{10}H_5(OH)(SO_3H)$.
Gray paste containing approximately 35% solids.
Use: Intermediate.

aminobenzene. See aniline.

p-aminobenzenearsonic acid. See arsanilic acid.

2-amino-p-benzenedisulfonic acid. (aniline-2,5-disulfonic acid). $C_6H_3NH_2(SO_3H)_2 \cdot 4HOH$.
Properties: Crystals, very soluble in water and alcohol.
Derivation: Boiling sodium salt of 4-chloro-3-nitrobenzene sulfonate with sodium sulfite, resulting in formation of sodium-2-nitrobenzene disulfonate which is reduced with iron and acetic acid to aniline-2,5-disulfonic acid.
Use: Intermediate.

4-amino-m-benzenedisulfonic acid. (aniline-2,4-disulfonic acid). $C_6H_3NH_2(SO_3H)_2 \cdot 2HOH$.
Properties: Needles decompose when heated above 120C, very soluble in water and alcohol.
Derivation: By heating sulfanilic acid with fuming sulfuric acid at 170–180C.
Use: Dye intermediate.

m-aminobenzenesulfonic acid. See metanilic acid.

p-aminobenzenesulfonic acid. See sulfanilic acid.

m-aminobenzoic acid. $C_6H_4NH_2CO_2H$.
CAS: 99-05-8.
Yellowish or reddish crystals, sublimes easily, sweet taste. Slightly soluble in water, alcohol, and ether; mp 173C–174C.
Use: Dye intermediate.

o-aminobenzoic acid. See anthranilic acid.

p-aminobenzoic acid. (PABA).
CAS: 150-13-0. $NH_2C_6H_4CO_2H$.
Required by many organisms as a vitamin for growth, active in neutralizing the antibacteriostatic effect of some sulfonamide drugs.

Properties: Light buff, odorless crystals; white when pure; discolor on exposure to light and air; mp 186–187C. Sparingly soluble in cold water, soluble in hot water, glacial acetic acid, ethyl acetate. Unstable to ferric salts and oxidizing agents.
Derivation: Reduction of p-nitrobenzoic acid. Commercially available as the calcium, potassium, and sodium salts.
Food source: Widely distributed, especially in yeast.
Grade: Technical, NF.
Use: Dye intermediate, pharmaceuticals, nutrition, UV absorber in suntan lotions.

2-aminobenzothiazole. $C_6H_4NC(NH_2)S$ (bicyclic).
Use: As an azo dye intermediate and in photographic chemicals.

m-**aminobenzotrifluoride.** $CF_3C_6H_4NH_2$.
CAS: 98-16-8.
Properties: Colorless to oily yellow liquid.
Grade: Technical (88% min), purified (98% min).
Use: Pharmaceutical intermediate.

o-**aminobenzoylformic anhydride.** See isatin.

N-(p-**aminobenzoyl)glycine.** See p-aminohippuric acid.

o-**aminobiphenyl.** (o-phenylaniline; o-biphenylamine). CAS: 90-41-5. $C_6H_5C_6H_4NH_2$.
Properties: Colorless or purplish crystals, mp 49.3C, bp 299C, slightly soluble in water.
Derivation: Reduction of o-nitrobiphenyl.
Hazard: Toxic by ingestion, inhalation, and skin absorption. A carcinogen.
Use: Research, analytical chemistry.

1-aminobutane. See n-butylamine.

2-aminobutane. See sec-butylamine.

aminobutanoic acid. See aminobutyric acid.

2-amino-1-butanol. $CH_3CH_2CHNH_2CH_2OH$.
Properties: Colorless liquid, d 0.944 at 20/20C, mp −2C, bp 178C; at 10 mm = 79–80C, flash p 164F (73.3C), bulk wt 7.85 lb/gal (20C); pH

(0.1M aqueous solution) 11.11, refr index 1.453 (20C). Completely miscible in water at 20C, soluble in alcohols, corrosive to copper, brass, aluminum. Combustible.
Use: Emulsifying agent (in soap form) for oils, fats, and waxes, absorbent for acidic gases, organic synthesis.

aminobutyric acid. (aminobutanoic acid; GABA). CAS: 56-12-2. An unusual amino acid having the following isomers: alpha: $CH_3CH_2CH(NH_2)COOH$. Isolated from a bacterium (*Corynebacterium diphtheriae*). The dl form is a crystalline solid, mp 305C, soluble in water, slightly soluble in alcohol, insoluble in ether. The l(+) form is solid, mp 270C, sweetish taste, soluble in water. beta: $CH_2CH(NH_2)CH_2COOH$. The dl form is a tasteless solid, mp 190C, water soluble, insoluble in ether and alcohol. The d(-) form decomposes at 220C; gamma: (GABA) $H_2N(CH_2)_3COOH$. Obtained from bacteria, yeast, and plant life. Crystalline solid, mp 202C, soluble in water, insoluble in organic solvents. Decomposes to pyrrolidone and water on quick heating. This substance is reported to be a neurotransmitter which activates or retards nervous reactions in the cells of the brain, including the sense of pain. All three isomers have been synthesized by various reaction sequences, the first reported in 1880.

α-**aminocaproic acid.** See norleucine.

aminocaproic lactam. See caprolactam.

aminochlorobenzene. See chloroaniline.

4-amino-4′-chlorodiphenyl. $C_{12}H_{10}ClN$.
Properties: Crystalline solid, mp 128C, insoluble in water, soluble in acetone, benzene, glacial acetic acid, and alcohol.
Derivation: Chlorination of 4-nitrodiphenyl and reduction of the product with iron and hydrochloric acid in alcohol.
Use: Sulfur determination in various materials, e.g., rubber, coal, etc.

2-amino-4-chlorophenol. (p-chloro-o-aminophenol). $C_6H_3OHNH_2Cl$.
Properties: Light brown crystals, mp 138C (decomposes).
Derivation: Reduction of p-chloro-o-nitrophenol.
Use: Intermediate.

2-amino-4-chlorotoluene. 5-chloro-2-methylaniline (NH_2 = 1); 4-chloro-o-toluidine (CH_3 = 1). $ClNH_2C_6H_3CH_3$.
Properties: Off-white solid or light brown oil which tends to darken on storage, mp 20–22C.

Hazard: Toxic by ingestion and inhalation.
Use: Intermediate.

2-amino-5-chlorotoluene. $ClNH_2C_6H_3CH_3$.
Properties: Crystalline solid, mp 26–27C, sparingly soluble in water, soluble in dilute acids.
Use: Intermediate.

2-amino-6-chlorotoluene. 6-chloro-o-toluidine ($CH_3 = 1$); 3-chloro-2-methylaniline ($NH_2 = 1$). $ClNH_2C_6H_3CH_3$.
Properties: Liquid, mp 0–2C.
Hazard: Toxic by ingestion or inhalation.
Use: Intermediate.

4-amino-2-chlorotoluene. 2-chloro-p-toluidine ($CH_3 = 1$); 3-chloro-4-methylaniline ($NH_2 = 1$). $ClNH_2C_6H_3$.
Properties: Liquid, mp 21–24C.
Use: Intermediate.

4-amino-2-chlorotoluene-5-sulfonic acid. (Brilliant Toning Red Amine; permanent red 2B amine). $ClNH_2CH_3C_6H_2SO_3H$.
Properties: White to buff powder, essentially insoluble as free acid, soluble as sodium or ammonium salt.
Grade: 98.5% min purity.
Use: Intermediate for azo pigments.

5-amino-2-chlorotoluene-4-sulfonic acid. (Lake Red carbon Amine). $ClNH_2CH_3C_6H_2SO_3H$.
Properties: White to pink powder, essentially insoluble as free acid, soluble as sodium or ammonium salt.
Grade: 98.5% min purity.
Use: Intermediate for azo pigments.

m-amino-p-cresol methyl ether.   See 5-methyl-o-anisidine.

aminocyclohexane.   See cyclohexylamine.

l-amino-dehydrogenase.   See amino oxidase.

3-amino-2,5-dichlorobenzoic acid.   See amiben.

2-amino-4,6-dichlorophenol.   See 2,4-dichloro-6-aminophenol.

p-aminodiethylaniline. (N,N-diethyl-p-phenylenediamine; diethylaminoaniline). $(C_2H_5)_2NC_6H_4NH_2$.
Properties: Liquid, bp 260–262C, insoluble in water, soluble in alcohol and ether.
Derivation: Treatment of diethylaniline with nitrous acid and subsequent reduction.
Hazard: See aniline.

Use: Dye intermediate, source of diazonium compounds in diazo copying process.

p-aminodiethylaniline hydrochloride. $C_{10}H_{16}N_2 \cdot HCl$.
Properties: Colorless needles, soluble in water, alcohol; insoluble in ether.
Hazard: See aniline.
Use: Color photography.

2-amino-4,6-dihydroxypteridine.   See xanthopterin.

p-aminodimethylaniline. (dimethylaminoaniline; dimethyl-p-phenylenediamine). $(CH_3)_2NC_6H_4NH_2$.
Properties: Colorless, asbestos-like needles, stable in air when pure. If impure, the crystals liquefy. Soluble in water, alcohol, and benzene. Mp 41C, bp 257C.
Derivation: By reduction of p-nitrosodimethylaniline with zinc dust and hydrochloric acid.
Method of purification: Recrystallization from mixture of benzene and ligroin.
Hazard: Toxic by ingestion or inhalation of vapor.
Use: Base for production of methylene blue, photodeveloper, reagent for detection of hydrogen sulfide, reagent for cellulose, organic synthesis, reagent for certain bacteria.

aminodimethylbenzene.   See xylidine.

2-amino-4,6-dimethylpyridine. $(CH_3)_2C_5H_2NNH_2$.
Properties: Solid, mp 65.2–68.5C, bp 235C. Water-soluble.
Derivation: Prepared from 2-aminopyridine.
Use: Organic intermediate.

2-amino-4,6-dinitrophenol.   See picramic acid.

p-aminodiphenyl. (4-aminodiphenyl; p-xenylamine). CAS: 92-67-1. $C_6H_5C_6H_4NH_2$.
Properties: Colorless crystals. Mw 169.23, mp 53C, bp 302C.
Hazard: Human carcinogen. Toxic by ingestion, inhalation, skin absorption.
Use: Organic research.

p-aminodiphenylamine. (N-phenyl-p-phenylenediamine). $NH_2C_6H_4NHC_6H_5$.
Properties: Purple powder, mp 75C, insoluble in water, soluble in alcohol and acetone.
Derivation: Reduction of the coupling product of diazotized sulfanilic acid and diphenylamine.
Use: Dye intermediate, pharmaceuticals, photographic chemicals.

4-aminodiphenylamine-2-sulfonic acid. (5-amino-2-anilinobenzenesulfonic acid). $C_6H_5NHC_6H_3NH_2SO_3H$.
Properties: Needle-like crystals, barely soluble in water.
Derivation: From p-nitrodiphenylamine-o-sulfonic acid by reduction with iron and hydrochloric acid.
Use: Synthesis of dyestuffs.

aminodithioformic acid. See dithiocarbamic acid.

aminoethane. See ethylamine.

2-aminoethanesulfonic acid. See taurine.

2-aminoethanethiol. (cysteamine; mercaptamine; thioethanolamine). CAS: 60-23-5. $HSCH_2CH_2NH_2$.
Properties: Crystals with unpleasant odor, oxidizes on contact with air, mp 97C, soluble in water. Combustible.
Use: Medicine (believed to offer protection against radiation).

1-aminoethanol. See aldehyde ammonia.

2-aminoethanol. See ethanolamine.

2-(2-aminoethoxy)ethanol. ("Diglycolamine"; DGA). $NH_2CH_2CH_2OCH_2CH_2OH$.
Properties: Colorless, slightly viscous liquid with a mild amine odor. Miscible with water and alcohols, bp 221C, d 1.0572 (20/20C), flash p 260F (126.6C), fp −12.5C. Combustible.
Hazard: Strong irritant to tissue.
Use: Removal of acid components from gases, especially carbon dioxide and hydrogen sulfide from natural gas; intermediate.

aminoethoxyvinylglycine.
Use: Inhibitor of ethylene biosynthesis in plants.

1-[(aminoethyl)amino]-2-propanol. $C_5H_{14}N_2O$.
Properties: Thick, colorless liquid; bp 112C (11 mm Hg); d 0.984; refr index 1.743. Slight odor of ammonia.
Use: Curing epoxy resins.

aminoethylethanolamine. See hydroxyethyl-ethylenediamine.

4-aminoethylglyoxaline. See histamine.

4-(2-aminoethyl)imidazole. See histamine.

β-aminoethylisothiourea dihydrobromide. (2-(2-aminoethyl)-2-thiopseudourea dihydrobromide; AET). $C_3H_9N_3S\cdot2HBr$.
Properties: Crystals, hygroscopic, mp 194–195C.

Derivation: Thiourea is refluxed with 2-bromoethylamine hydrobromide in isopropanol.
Use: Enzyme activator, free radical detoxifier (believed to offer protection against radiation).

N-aminoethylpiperazine. CAS: 140-31-8.
$H_2NC_2H_4NCH_2CH_2NHCH_2CH_2$. An amine combining a primary, secondary, and tertiary amine in one molecule.
Properties: Liquid, d 0.9837, bp 222.0C, flash p 200F (93.3C), fp 17.6C. Soluble in water. Combustible.
Hazard: Strong irritant to tissue.
Use: Epoxy curing agent, intermediate for pharmaceuticals, anthelmintics, surface-active agents, synthetic fibers.

2-amino-2-ethyl-1,3-propanediol. (AEPD; aminoamylene glycol). $CH_2OHC(C_2H_5)NH_2CH_2OH$.
Properties: Solid or viscous liquid, mp 38C, soluble in water and alcohol.
Use: Emulsifying agent (in soap form) for oils, fats, and waxes, absorbent for acidic gases CO_2 and H_2S, organic synthesis.

2-aminoethylsulfuric acid. $NH_2CH_2CH_2OSO_3H$.
Properties: White, crystalline powder; noncorrosive; mp 274–280C; sinters at 274C and darkens without complete melting at 280C; d 1.782; bulk d 1.007. Soluble in water, insoluble in most organic solvents, pH (1% aqueous solution) 4.0 (20C), (5% aqueous solution) 3.3 (39C).
Use: Organic synthesis of ethyleneimine and various other compounds, amination of cotton.

3-amino-α-ethyl-2,4,6-triiodohydrocinnamic acid. See iopanoic acid.

4-aminofolic acid. See aminopterin.

aminoform. See hexamethylenetetramine.

amino-G acid. (2-naphthylamine-6,8-disulfonic acid; 7-amino-1,3-naphthalene-disulfonic acid). $C_{10}H_5(NH_2)_2(SO_3H)_2$.
Properties: White, crystalline solid; soluble in water.
Derivation: (a) Form G acid by heating sodium salt with ammonia and sodium bisulfite solution in an autoclave under pressure. (b) Sulfonation by beta-naphthylamine.
Use: Azo dye intermediate.

α-aminoglutaric acid. See glutamic acid.

amino-4-guanidovaleric acid. See arginine.

aminohexamethyleneimine.

$\overline{NH_2NCH_2CH_2CH_2CH_2CH_2CH_2}$.

Properties: Liquid, bp 170C, soluble in water and most organic solvents. Combustible.

Use: Intermediate for dyes, pharmaceuticals, and photographic chemicals.

2-aminohexanoic acid. See norleucine.

6-aminohexanoic acid. See aminocaproic acid.

p-aminohippuric acid. [N-(p-aminobenzoyl)-glycine; PAHA].
$NH_2C_6H_4CONHCH_2COOH$.

Properties: White, crystalline powder. Discolors on exposure to light. Soluble in alcohol and most organic solvents. Very soluble in dilute hydrochloric acid and alkalies. Forms a water-soluble sodium salt. Mp 197–199C.

Grade: USP.

Use: Medicine (diagnostic agent), intermediate.

aminohydroxybenzoic acids. See aminosalicyclic acids.

α-amino-β-hydroxybutyric acid. See threonine.

2-amino-2-hydroxymethyl-1,3-propanediol.
See tris(hydroxymethyl)aminomethane.

α-amino-β-hydroxypropionic acid. See serine.

α-amino-β-imidazolepropionic acid. See histidine.

α-aminoisocaproic acid. See leucine.

α-aminoisovaleric acid. See valine.

amino-J acid. (2-naphthylamine-5,7-disulfonic acid; 6-amino-1,3-naphthalene-disulfonic acid).
$C_{10}H_5(NH_2)_2(SO_3H)_2$.

Properties: Crystallizes in white lustrous leaflets from water and in long needles from hydrochloric acid solution.

Derivation: By sulfonation of either 2-naphthylamine-5-sulfonic acid or 2-naphthylamine-7-sulfonic acid.

Use: Azo dye intermediate.

2-amino-6-mercaptopurine.
Use: Pharmaceutical end product for treatment of leukemia.

aminomercuric chloride. See mercury, ammoniated.

aminomethane. See methylamine.

3-amino-4-methoxybenzanilide.
$CH_3OC_6H_3(NH_2)CONHC_6H_5$.

Properties: Gray powder.

Use: Dyes, pharmaceuticals and other organic chemicals.

1-amino-2-methoxy-5-methylbenzene.
See 5-methyl-o-anisidine.

m-(4-amino-3-methoxyphenylazo)benzenesulfonic acid. $H_2NC_6H_3(OCH_3)NNC_6H_4SO_3H$.

Properties: Maroon paste containing approximately 38% solids.

Use: Intermediate.

4-amino-4′-methyldiphenylamine-2-sulfonic acid.
(aminotoluidinobenzenesulfonic acid).
$CH_3C_6H_4NHC_6H_3NH_2SO_3H$.

Properties: Light to dark gray paste with characteristic odor.

Use: Intermediate.

4-amino-10-methylfolic acid. See methotrexate.

3-amino-5-methylisoxazole.
Use: Analogue of the nucleic acid constituent guanine.

2-amino-3-methylpentanoic acid. See isoleucine.

2-amino-2-methyl-1,3-propanediol. (AMPD; aminobutylene glycol; butanediolamine).
$CH_2OCH(CH_3)NH_2CH_2OH$. Corrosive to copper, brass, and aluminum.

Properties: Colorless crystals, soluble in water and alcohol, mp 110C.

Use: Emulsifying agent (in soap form) for oils, fats, and waxes; absorbent for acidic gases; organic synthesis; cosmetics.

2-amino-2-methyl-1-propanol. (isobutanolamine; AMP). $CH_3(CH_3)NH_2CH_2OH$.

Properties: Solid or viscous liquid, mp 30C, bp 165C, d 0.93, refr index 1.45.

Hazard: Toxic by ingestion.

Use: Emulsifying agent (in soap form) for oils, fats, and waxes, absorbent for acidic gases, organic synthesis, cosmetics.

2-amino-3-methylpyridine. (2-amino-3-picoline).

$\overline{N:C(NH_2)C(CH_3):CHCH:CH}$.

Properties: Liquid. Bp 221C, mp 29.5–33.3C, soluble in water.

Derivation: From 2-aminopyridine.

Hazard: Toxic by ingestion.

Use: Intermediate.

2-amino-4-methylpyridine.　(2-amino-4-picoline).

N:C(NH$_2$)CH:C(CH$_3$)CH:CH.
Properties: Crystals, bp 230.9C (115–117C at 11
mm), mp 96–99.0C. Sublimes on slow heating,
soluble in water and lower alcohols.
Derivation: Prepared from 2-aminopyridine.
Use: Intermediate, medicine.

2-amino-5-methylpyridine.　(2-amino-5-picoline).

N:C(NH$_2$)CH:CHC(CH$_3$):CH.
Properties: Crystals, bp 227.1C, mp 76.6C, soluble
in water.
Derivation: Prepared from 2-aminopyridine.
Use: Intermediate.

2-amino-6-methylpyridine.　(2-amino-6-picoline).

N:C(NH$_2$)CH:CHCH:C(CH$_3$).
Properties: Crystals, bp 214.4C, mp 43.7C, soluble
in water.
Derivation: Prepared from 2-aminopyridine.
Use: Intermediate.

2-amino-4-(methylthio)butyric acid.　See methionine.

α-amino-β-methylvaleric acid.　See isoleucine.

α-amino-γ-methylvaleric acid.　See leucine.

aminonaphthalensulfonic acid.　See naphthylaminesulfonic acid.

4-amino-1-naphthol.　(4-hydroxy-α-naphthyl-
amine).　C$_{10}$H$_9$NO.
Properties: Acicular crystals, soluble in water.
Derivation: Rearrangement of α-naphthylhydrox-
ylamine in acetone.
Use: Inhibitor of polymerization, chemical intermediate. Must be kept dry during storage to
avoid oxidation and discoloration.

aminonaphtholsulfonic acid.　Any of several sulfonated aromatic acids derived from naphthol or
naphthylamine and used as azo dye intermediates.

6-aminonicotinic acid.　(6-aminopyridine-3-car-
boxylic acid).　C$_6$H$_6$N$_2$O$_2$.
Properties: Crystals. Mp decomposes above 300C.

amino oxidase.　(l-amino acid oxidase; d-amino
acid oxidase; l-amino dehydrogenase).　An
enzyme which catalyzes the deamination of alpha-amino acids by dehydrogenation to keto acids and ammonia. Two types are recognized, acting on the d- and l-amino acids. Recent emphasis

has been on characterization of the d-amino oxidase, which is known to contain the flavin isoalloxazine as coenzyme. Both types are found in
animal tissue, especially in liver and kidney, as
well as in snake venom and certain bacteria.

2-amino-6-oxypurine.　See guanine.

aminoparathion.　O,O-diethyl-o,p-aminophenyl
phosphorothioate).　CAS: 3785-01-1.
Metabolite.

1-aminopentane.　See n-amylamine.

2-aminophenetole.　See o-phenetidine.

4-aminophenetole.　See p-phenetidine.

m-aminophenol　(m-hydroxyaniline).
　CAS: 591-27-5.　C$_6$H$_4$NH$_2$OH.

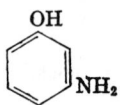

Properties: White crystals, mp 122C, soluble in
water, alcohol and ether.
Derivation: Fusion of m-sulfanilic acid with caustic soda and subsequent extraction of the melt
with ether.
Hazard: Toxic by ingestion.
Use: Dye intermediate, intermediate for p-amino-
salicylic acid.

o-aminophenol.　(o-hydroxyaniline).
　CAS: 95-55-6.　C$_6$H$_4$NH$_2$OH.
Properties: White crystals, turn brown with age,
mp 172–173C, sublimes on further heating. Soluble in cold water, alcohol, benzene, freely soluble
in ether.
Derivation: By reduction of o-nitrophenol mixed
with aqueous ammonia by means of a stream
of hydrogen sulfide. Also available as the hydrochloride.
Grade: Technical, 99% min.
Use: Dyeing furs and hair; dye intermediate for
azo and sulfur dyes, pharmaceuticals.

p-aminophenol.　(p-hydroxyaniline).
　CAS: 123-30-8.　C$_6$H$_4$NH$_2$OH.
Properties: White or reddish yellow crystals turn
violet on exposure to light, mp 184C (decomposes), soluble in water and alcohol.
Derivation: (a) By reduction of p-nitrophenol with
iron filings and hydrochloric acid; (b) by electrolytic reduction of nitrobenzene in concentrated
sulfuric acid and treatment with an alkali to free
the base. Also available as the hydrochloride.

Grade: Technical, photographic.

Use: Dyeing textiles, hair, furs, feathers, photographic developer, pharmaceuticals, antioxidants, oil additives.

4-amino-1-phenol-2,6-disulfonic acid.
$C_6H_2OHNH_2(SO_3H)_2$.
Properties: Fine needles. Soluble in water, slightly soluble in alcohol, insoluble in ether.
Derivation: Action of sulfur dioxide on p-nitrophenol.
Use: Dyes.

2-amino-1-phenol-4-sulfonic acid. (o-amino-phenol-p-sulfonic acid). $C_6H_3OHNH_2SO_3H$.
Properties: Brown crystals, fairly soluble in hot water, very soluble in alkaline solution, no mp, decomposes on heating.
Derivation: (a) Sulfonation and nitration of chlorobenzene followed by hydrolysis to phenol with caustic soda with subsequent reduction by sodium sulfide. (b) Sulfonation of o-aminophenol. (c) Sulfonation of phenol followed by nitration and reduction.
Use: Intermediate for dyes.

p-aminophenylarsonic acid.   See arsanilic acid.

m-aminophenylboronic acid hemisulfate.
(3-aminobenzeneboronic acid).
$H_2NC_6H_4B(OH)_2 \cdot 1/2H_2SO_4$.
Properties: Mw 186, mp above 300C.
Use: Adsorbant additive for the chromatographic separation of 3′-terminal polynucleotides from RNA.

1-amino-2-phenylethane.   See β-phenylethylamine.

o-aminophenylglyoxalic lactim.   See isatin.

p-aminophenylmercaptoacetic acid.
$NH_2C_6H_4SCH_2CO_2H$.
Properties: Mp 186–187C, insoluble in water, alcohol, benzene, chloroform; soluble in aqueous acid or alkali solutions.
Use: Synthetic intermediate for dyes and pharmaceuticals.

2-(p-aminophenyl)-6-methylbenzothiazole.   See dehydrothio-p-toluidine.

m-aminophenyl methyl carbinol.
$NH_2C_6H_4CH(OH)CH_3$.
Properties: Solid, d 1.12, bp 217.3C (100 min), mp 66.4C, soluble in water, flash p 315F (157C). Combustible.
Use: Carrier for dyeing synthetic fibers, intermediate for perfume, chemicals and pharmaceuticals.

1-(m-aminophenyl)-3-methyl-5-pyrazolone.

$\overline{NH_2C_6H_4NNC(CH_3)CH_2CO}$.
Properties: Light tan paste containing approximately 45% solids.
Use: Intermediate.

α-amino-β-phenylpropionic acid.   See phenylalanine.

aminophylline.   (3,7-dihydro-1,3-dimethyl-1H-purine-2,6-dione compounded with 1,2-ethanediamine (2:1)). $C_{16}H_{24}N_{10}O_4$.
Properties: White or slightly yellowish granular powder, slight ammonia odor, bitter taste. Mw 420.44.
Derivation: Prepared from theophylline and aqueous ethylenediamine.
Hazard: Cardiovascular and respiratory collapse.
Use: Small animal muscle relaxant, for heaves in horses, and diuretic in dogs with congestive heart failure.

aminopicoline.   See aminomethylpyridine.

aminoplast resin.   (amino resin).   A class of thermosetting resins made by the reaction of an amine with an aldehyde. The only aldehyde in commercial use is formaldehyde, and the most important amines are urea and melamine.
Use: Molding, adhesives, laminating, textile finishes, permanent-press fabrics, wash-and-wear apparel fabrics, protective coatings, paper manufacture, leather treatment, binders for fabrics, foundry sands, graphite resistors, plaster of paris fortification, foam structures, and ion-exchange resins.
See dimethylol urea, methylol urea, melamine resins, urea-formaldehyde resins.

2-aminopropane.   See isopropylamine.

2-aminopropanoic acid.   See alanine.

3-aminopropanoic acid.   See β-alanine.

1-amino-2-propanol.   See isopropanolamine.

2-amino-1-propanol.   (2-aminopropyl alcohol; β-propanolamine). $CH_3CH(NH_2)CH_2OH$.
Properties: Colorless to pale yellow liquid. Both l and dl forms are available. Dl-form: fish odor, bp 173–176C, freely soluble in water, alcohol, ether. l-form: refr index 1.4480–1.4495 (26C), distillation range approximately 114C at 100 mm Hg. Combustible.
Use: Organic synthesis and chemical intermediate.

3-amino-1-propanol.   (propanolamine).
$H_2NCH_2CH_2CH_2OH$.
Properties: Colorless liquid, mp 12.4C, bp 184–186C (168C), flash p 175F (79.4C), d 0.9786

(30C). Miscible with alcohol, water, acetone and chloroform. Combustible.
Grade: 99% pure.
Hazard: Irritant to tissue.
Use: Organic intermediate.

3-aminopropionitrile. $H_2NCH_2CH_2CN$.
Properties: Colorless liquid, bp 185C, refr index 1.44. May polymerize if stored in presence of air.
Derivation: Reaction of acrylonitrile with ammonia.
Use: Production of β-alanine and pantothenic acid.

2-aminopropyl alcohol. See 2-amino-1-propanol.

N-aminopropylmorpholine. (4-(3-aminopropyl) omorpholine).

$\overline{CH_2CH_2OCH_2CH_2NC_3H_6}NH_2$.
Properties: Colorless liquid, d 0.9872 (20/20C), bp 224.5C, flash p 220F (104.4C) (OC), fp −15C, soluble in water and alcohol. Combustible.
Hazard: Strongly irritant to tissue.
Use: Fiber synthesis; chemical intermediate.

γ-aminopropyltriethoxysilane.
$NH_2(CH_2)_3Si(OC_2H_5)_3$.
Properties: Liquid, bp 217C, d 0.94 (25C).
Use: Sizing of glass fibers for making laminates.

γ-aminopropyltrimethoxysilane.
$NH_2(CH_2)_3Si(OCH_3)_3$.
Properties: 100% active, water-white liquid, d 1.01; refr index 1.42.
Use: Glass fabric sizing, binder, adhesion promoter.

aminopterin. (4-aminofolic acid; aminopteroylglutamic acid). $C_{19}H_{20}N_8O_5 \cdot 2H_2O$. Differs slightly in structure from folic acid and antagonizes the utilization of folic acid by the body, an antimetabolite.
Properties: Occurs as clusters of yellow needles which are soluble in aqueous sodium hydroxide solutions.
Use: Medicine, rodenticide.

aminopteroylglutamic acid. See aminopterin.

6-aminopurine. See adenine.

2-aminopyridine. (α-pyridylamine).
CAS: 92-67-1. $C_5N_4NNH_2$.
Properties: White leaflets or large colorless crystals; fp 58.1C; bp 210.6C; soluble in water, alcohol, benzene, ether.

Hazard: Toxic. TLV: 0.5 ppm in air.
Use: Intermediate for antihistamines and other pharmaceuticals.

3-aminopyridine. (β-pyridylamine).
$C_6H_4NNH_2$.
Properties: White crystals; mp 64C; bp 250–252C; soluble in water, alcohol, benzene, ether.
Use: Intermediate in preparation of drugs and dyestuffs.

4-aminopyridine. $C_5H_4NNH_2$. CAS: 504-24-5.
Properties: Crystals, mp 158.9C, bp 273.5C, soluble in water.
Derivation: From 2-aminopyridine.
Grade: 95% (minimum).
Use: Intermediate.

4-aminosalicylic acid. (PASA; PAS; p-amino-salicylic acid; 4-amino-2-hydroxybenzoic acid).
CAS: 65-49-6 $NH_2C_6H_3(OH)COOH$.
Properties: White or nearly white bulky powder. Odorless or has slight acetous odor. Mp 150–151C (decomposes). Affected by light and air. Darkened solutions should not be used. Soluble in dilute sodium hydroxide and dilute nitric acid, slightly soluble in ether, practically insoluble in benzene, pH of (0.1% water soluble) 3.5. Water solutions decompose with evolution of carbon dioxide.
Derivation: From m-aminophenol and potassium bicarbonate, soluble under pressure.
Grade: USP.
Use: Medicine, industrial preservative.

5-aminosalicylic acid. (m-aminosalicylic acid; 5-amino-2-hydroxybenzoic acid).
CAS: 89-57-6. $NH_2C_6H_3(OH)COOH$.
Properties: White crystals, sometimes pinkish, mp 260–280C (decomposes), soluble in hot water or alcohol.
Derivation: From the corresponding nitrosalicylic acid by reduction.
Use: Dyes, intermediate.

α-aminosuccinamic acid. See asparagine.

aminosuccinic acid. See aspartic acid.

2-aminothiazole. (2-thiazylamine).
$SCHCHNCNH_2$.
Properties: Light yellow crystals, mp 90C, distills at 3 mm without decomposition. Slightly soluble in cold water, alcohol, and ether; soluble in hot water and dilute mineral acids.
Derivation: Chlorination of vinyl acetate and condensation with thiourea.
Use: Intermediate in synthesis of sulfathiazole, medicine (thyroid inhibitor).

α-amino-β-thiolpropionic acid.   See cysteine.

aminothiourea.   See thiosemicarbazide.

aminotoluene.   See m-, o-, or p-toluidine. See also benzylamine.

6-amino-s-triazine-2,4-diol.   See ammelide.

3-amino-1,2,4-triazole.   (amitrole).

CAS: 61-82-5.   NHNC(NH₂)NCH.
Properties: White, crystalline solid; mp 156–159C; soluble in water and alcohol.
Hazard: Use on food crops not permitted; TLV 0.2 mg/m³, suspected human carcinogen.
Use: Herbicide, defoliant.

4-amino-3,5,6-trichloropicolinic acid.   See picloram.

3-(3-amino-2,4,6-triiodophenyl)-2-ethylpropanoic acid.   See iopanoic acid.

aminourea hydrochloride.   See semicarbazide hydrochloride.

"Aminox."²⁴⁸   TM for a low-temperature reaction product of diphenylamine and acetone.
Properties: Light-tan powder; d 1.13; mp 85–95C; soluble in acetone, benzene, and ethylene dichloride; insoluble in water and gasoline.
Use: Antioxidant for nylon and light-colored rubber products.

aminoxylene.   See xylidine.

amitrole.   Generic name for 3-amino-1,2,4-triazole.

"Amizyme."¹¹⁴   TM for a series of enzyme preparations, high in dextrinizing or starch-liquefying properties. Available in tablet, powdered, or liquid form.
Derivation: Produced by growing pure microbial cultures on select media.
Use: Conversion of starch, used in paper coatings, adhesives, and textile sizes.

"Ammate."²⁸   TM for ammonium sulfamate in various grades.
Hazard: See ammonium sulfamate.
Use: Non-selective herbicide.

ammelide.   (6-amino-s-triazine-2,4-diol;

cyanuramide).   NC(OH)NC(OH)NC(NH₂).
Properties: Crystalline solid, mp decomposes, insoluble in alcohol, slightly soluble in hot water.

Similar to melamine and suggested for melamine-type (amino) resins.

ammeline.   (4,6-diamino-s-triazine-2-ol; cyanurdiamide).

NC(OH)NC(NH₂)NC(NH₂.
Properties: Crystalline solid, mp decomposes, insoluble in water and alcohol. A compound similar to melamine and used in melamine-type resins and in special high temperature lubricants.

ammine.   A coordination compound formed by the union of ammonia with a metallic substance in such a way that the nitrogen atoms are linked directly to the metal. Note the distinction from amines in which the nitrogen is attached directly to the carbon atom.
See cobaltammine and coordination compound.

ammonia, anhydrous.   NH₃.   CAS: 7664-41-7.
Third highest-volume chemical produced in US (1985).
Properties: Colorless gas (or liquid); sharp, intensely irritating odor; lighter than air; easily liquefied by pressure; bp −33.5C; fp −77C; vap press of liquid 8.5 atm (20C); sp vol 22.7 cu ft/lb (70C); d (liquid) 0.77 at 0C and 0.6819 at bp; very soluble in water, alcohol, ether; volume 22.7 cu ft/lb (70C); d of liquid 0.77 at 0C; 0.6819 at bp; very soluble in water, alcohol, and ether. Autoign temperature 1204F (650C). Combustible. Note: Ammonia is the first complex molecule to be identified in interstellar space, it has been observed in galactic dust clouds in the Milky Way and is believed to constitute the rings of the planet Saturn.
Derivation: From synthesis gas, a mixture of carbon monoxide, hydrogen, carbon dioxide, and nitrogen (from air) obtained by steam reforming or by partial combustion of natural gas (US), or from the action of steam on hot coke (Haber-Bosch process). The latter method is used in South Africa. After removal of the carbon oxides, the gas composition is adjusted to a ratio of 3 parts H₂ to 1 part N₂ and passed to the synthesis unit over a catalyst at pressures of about 300 atm and temperature of approximately 500C. The catalyst most widely used is produced by fusion of iron oxide (Fe₃O₄) containing aluminum oxide and potassium oxide as promoters, followed by reduction of the oxide. Chemisorption of the nitrogen on the catalyst surface is the rate-controlling step.
    Other methods include use of refinery off-gases, coke-oven gas, electrolytic hydrogen, and calcium cyanimide. Ammonia has been made experimentally using solar energy to activate the reaction

$$N_2 + 3H_2O \xrightarrow[\text{TiO}_2]{hv} 2NH_3 + 1.5O_2.$$

Ammonia is formed as an end product of animal metabolism by decomposition of uric acid.

Grade: Commercial 99.5%, refrigerant 99.97%.

Hazard: Inhalation of concentrated fumes may be fatal. TLV: 25 ppm in air. Moderate fire risk, explosive limits in air 16 to 25%. Forms explosive compounds in contact with silver or mercury.

Use: Fertilizers, either as such or in form of compounds, e.g., ammonium nitrate, manufacture of nitric acid, hydrazine hydrate, hydrogen cyanide, urethane, acrylonitrile, and sodium carbonate (by Solvay process); refrigerant, nitriding of steel, condensation catalyst, synthetic fibers, dyeing, neutralizing agent in petroleum industry, latex preservative, explosives, nitrocellulose, urea-formaldehyde, nitroparaffins, melamine, ethylenediamine, sulfite cooking liquors, fuel cells, rocket fuel, yeast nutrient, developing diazo films.

See also ammonium hydroxide; Haber, Fritz; synthesis gas.

ammonia, aromatic spirits. A mixture of 10% of ammonia in alcohol. Strong, suffocating odor.

Hazard: Irritant to mucous membranes. Flammable, keep tightly sealed.

Use: Medicine (respiratory stimulant).

ammonia-soda process. See Solvay process.

ammoniated mercury chloride. See mercury, ammoniated.

ammoniated ruthenium oxychloride. See ruthenium red.

ammoniated superphosphate. Fertilizer produced by mixing ammonia with superphosphate in the ratio of 5 parts to 100.

ammonio-cupric sulfate. See copper sulfate, ammoniated.

ammonio-ferric oxalate. See ferric ammonium oxalate.

ammonio-ferric sulfate. See ferric ammonium sulfate.

ammonium acetate. CAS: 631-61-8. $NH_4(C_2H_3O_2)$.

Properties: White, deliquescent, crystalline mass. Soluble in water, alcohol. Mp 114C, d 1.073. Combustible.

Derivation: By the interaction of glacial acetic acid and ammonia.

Use: Reagent in analytical chemistry, drugs, textile dyeing, preserving meats, foam rubbers, vinyl plastics, explosives.

ammonium acid carbonate. See ammonium bicarbonate.

ammonium acid fluoride. See ammonium bifluoride.

ammonium acid methanearsonate. $CH_3AsO(OH)ONH_4$. A post-emergent herbicide, available as "Ansar" 157, a clear solution containing 9.54% As.

ammonium acid phosphate. See ammonium phosphate, monobasic.

ammonium alginate. (ammonium polymannuronate). $C_6H_7O_6 \cdot NH_4$. A hydrophilic, colloid.

Properties: Filamentous, grainy, granular or powder, colorless or slightly yellow, and may have a slight characteristic smell and taste. Slowly soluble in water forming a viscous solution. Insoluble in alcohol.

Grade: Technical, FCC.

Use: Thickening agent and stabilizer in food products.

See also algin.

ammonium alum. See aluminum ammonium sulfate.

ammonium aluminum chloride. See aluminum ammonium chloride.

ammonium arsenate. CAS: 7784-44-3. $(NH_4)_2HAsO_4$.

Properties: White crystals or powder efflorescing in air with loss of ammonia, d 1.99. Soluble in water; decomposes in hot water.

ammonium aurin tricarboxylate. $C(C_6H_3OHCOONH_4)_2(C_6H_3(O)COONH_4)$ Forms colored lakes with aluminum, chromium, iron and beryllium.

ammonium benzenesulfonate. (ammonium sulfonate). $C_6H_5SO_3NH_4$.

Properties: Mp 271C (decomposes), d 1.34.

Grade: 35% solution in kerosene.

Hazard: Flammable.

ammonium benzoate. $C_6H_5COONH_4$.

Properties: White crystals or powder. Soluble in water, alcohol, and glycerol. Decomposes at 198C, d 1.260, sublimes at 160C.

Use: Medicine, latex preservative.

ammonium biborate. See ammonium borate.

ammonium bicarbonate. (ammonium acid carbonate; ammonium hydrogen carbonate).
CAS: 1066-33-7. NH_4HCO_3.
Properties: White crystals. Soluble in water, insoluble in alcohol. D 1.586, mp decomposes at 36 to 60C. Noncombustible.
Derivation: By heating ammonium hydroxide with an excess of carbon dioxide and evaporating.
Impurities: Ammonium carbonate.
Grade: Technical, CP, FCC.
Hazard: Evolves irritating fumes on heating to 35C.
Use: Production of ammonium salts, dyes, leavening agent for cookies, crackers, cream-puff doughs, fire-extinguishing compounds, pharmaceuticals, degreasing textiles, blowing agent for foam rubber, boiler scale removal, compost treatment.

ammonium bichromate. See ammonium dichromate.

ammonium bifluoride. (ammonium acid fluoride; ammonium hydrogen fluoride).
CAS: 1341-49-7. NH_4HF_2.
Properties: White crystals, deliquescent, d 1.211, soluble in water and alcohol.
Derivation: Action of ammonium hydroxide on hydrofluoric acid with subsequent crystallization.
Hazard: Corrosive to skin. TLV (as F): 2.5 mg/m^3 of air.
Use: Ceramics, chemical reagent, etching glass (white acid), sterilizer for brewery, dairy and other equipment; electroplating processing beryllium, laundry sour.

ammonium binoxalate. $(NH_4)HC_2O_4 \cdot H_2O$.
Properties: Colorless crystals. Soluble in water. D 1.556, decomposes on heating.
Derivation: Action of ammonium hydroxide on oxalic acid with subsequent crystallization.
Use: Analytical reagent, ink removal from fabrics.

ammonium biphosphate. See ammonium phosphate, monobasic.

ammonium bisulfate. (acid ammonium sulfate; ammonium hydrogen sulfate). NH_4HSO_4.
Properties: Colorless, deliquescent powder; mp 145C; d 1.79; soluble in water; insoluble in acetone and alcohol.
Use: Catalyst in organic synthesis, hair wave formulation.

ammonium bisulfide. See ammonium sulfide.

ammonium bitartrate. (acid ammonium tartrate). $(NH_4)HC_4H_4O_6$.
Properties: White crystals; soluble in water, acids, and alkalies; insoluble in alcohol; d 1.636.
Derivation: By the action of ammonium hydroxide on tartaric acid.
Use: Baking powder.

ammonium borate. (ammonium biborate). $NH_4HB_4O_7 \cdot 3H_2O$.
Properties: Colorless crystals, efflorescent with loss of ammonia. Soluble in water. D 2.38–2.95. Noncombustible.
Derivation: Action of ammonium hydroxide on boric acid with subsequent crystallization.
Hazard: Evolves irritating fumes especially when heated.
Use: Fireproofing compounds, electrical condensers, herbicide.

ammonium bromide. NH_4Br.
Properties: Colorless crystals or yellowish white powder, soluble in water and alcohol, somewhat hygroscopic. D 2.43, mp sublimes. Noncombustible.
Derivation: Action of hydrobromic acid on ammonium hydroxide with subsequent crystallization.
Grade: Technical, pure, CP, NF.
Use: Precipitating silver salts for photographic plates, medicine (for its bromide ion), analytical chemistry, process engraving, textile finishing, fire retardant, anticorrosive agents.

ammonium cadmium bromide. See cadmium ammonium bromide.

ammonium caprylate. (octanoic acid ammonium salt). $C_8H_{19}NO_2$.
Properties: Hygroscopic crystals, decompose at room temperature, mp approximately 75C. Hydrolyzes readily. Soluble in alcohol and glacial acetic acid, partly soluble in acetone, insoluble in benzene.
Use: Pesticide, photographic emulsions, chemical intermediate.

ammonium carbamate. $NH_4CO_2NH_2$.
Properties: White, rhombic, crystalline powder; very volatile; forms urea upon heating; soluble in water and alcohol. Sublimes at 60C, decomposes in air to evolve ammonia.
Derivation: Interaction of dry ammonia gas and carbon dioxide from ammonia liquor with ammonia and ammonium carbonate.
Grade: Technical.
Hazard: Evolves irritating fumes when heated.
Use: Fertilizer.

ammonium carbazotate. See ammonium picrate.

ammonium carbonate. (crystal ammonia; ammonium sesquicarbonate; hartshorn).
CAS: 506-87-6.
$(NH_4)HCO_3 \cdot (NH_4)CO_2NH_2$.
A mixture of ammonium acid carbonate and ammonium carbamate.
Properties: Colorless, crystalline plates or white powder; unstable in air, being converted into the bicarbonate. Strong odor of ammonia, sharp ammoniacal taste. Soluble in water, decomposes in hot water, yielding ammonia and carbon dioxide. Noncombustible.
Derivation: Ammonium salts are heated with calcium carbonate.
Method of purification: Sublimation.
Grade: Technical, lumps, cubes, powder, CP, NF, FCC.
Hazard: Evolves irritating fumes when heated.
Use: Ammonium salts, medicine (expectorant), baking powders, smelling-salts, fire-extinguishing compounds, pharmaceuticals, textiles (mordant), fermentation accelerator in wine manufacture, organic chemicals, ceramics, washing wool.

ammonium ceric nitrate. See ceric ammonium nitrate.

ammonium chlorate. CAS: 10192-29-7.
NH_4ClO_3.
Properties: Colorless or white crystals, water soluble.
Derivation: Reaction of ammonium chloride with sodium chlorate in solution.
Hazard: Spontaneous chemical reaction with reducing agents. Powerful oxidizer. When contaminated with combustible materials, it can ignite. Shock-sensitive, can detonate when exposed to heat or vibration, especially when contaminated.
Use: Explosives.

ammonium chloride. (sal ammoniac).
CAS: 12125-02-9. NH_4Cl.
Properties: White crystals; cool, saline taste; somewhat hygroscopic. Soluble in water, glycerol, slightly soluble in alcohol. Sublimes at 350C, d 1.54.
Derivation: (a) As a byproduct of the ammonia-soda process, (b) reaction of ammonium sulfate and sodium chloride solutions.
Grade: Technical (lumps or granulated), CP, USP, FCC.
Hazard: TLV (fume): 10 mg/m³ of air.
Use: Dry batteries, mordant (dyeing and printing), soldering flux, manufacturing of various ammonia compounds, fertilizer, pickling agent in zinc coating and tinning, electroplating, washing powders, melt-retarding snow treatment, resins and adhesive of urea-formaldehyde, bakery products.

ammonium chloroosmate. See ammonium hexachloroosmate.

ammonium chloroplatinate. See ammonium hexachloroplatinate.

ammonium chloroplatinite. (platinous ammonium chloride; platinum ammonium chloride).
CAS: 16919-58-7. $(NH_4)_2PtCl_4$.
Properties: Dark ruby red crystals, decomposes 140–150C, soluble in water, insoluble in alcohol.
Use: In photography.

ammonium chromate. CAS: 7788-98-9.
$(NH_4)_2CrO_4$.
Properties: Yellow crystals, soluble in cold water, insoluble in alcohol. D 1.866, mp decomposes.
Derivation: Addition of ammonium hydroxide to a soluble of ammonium bichromate, recovery by crystallization.
Impurities: Dichromates.
Grade: Technical, CP.
Hazard: Toxic by inhalation, strong irritant.
Use: Mordant in dyeing; photography (sensitizer for gelatin coatings), analytical reagent, catalyst, corrosion inhibitor.

ammonium chrome alum. See chromium ammonium sulfate.

ammonium chromium sulfate. See chromium ammonium sulfate.

ammonium citrate, dibasic. $(NH_4)_2HC_6H_5O_7$.
Properties: White granules, soluble in water, very slightly soluble in alcohol.
Use: Pharmaceuticals, rustproofing, cotton printing, plasticizer, analytically in determination of phosphate in fertilizer.

ammonium cobaltous phosphate. See cobaltous ammonium phosphate.

ammonium decaborate. See ammonium pentaborate.

ammonium dichromate. (ammonium bichromate). $(NH_4)_2Cr_2O_7$.
Properties: Orange needles, soluble in water and alcohol, d 2.152 (25C), decomposes with slight heating.
Derivation: Action of chromic acid on ammonium hydroxide with subsequent crystallization.
Hazard: Dusts and solutions are toxic, irritating to eyes and skin, dangerous fire risk. Strong oxidizing agent, may explode in contact with organic materials. Experimental carcinogen.
Use: Mordant for dyeing; pigments, manufactur-

ing of alizarin, chrome alum, oil purification, pickling, manufacture of catalysts, leather tanning, synthetic perfumes, photography, process engraving and lithography (sensitizer for photochemical insolubilization of albumin, etc.), chromic oxide, pyrotechnics.

ammonium dihydrogen phosphate. See ammonium phosphate monobasic.

ammonium dimethyldithiocarbamate. $(CH_3)_2NCS_2NH_4$.
Properties: Yellow crystals, soluble in water, decomposes in air.
Grade: 42% soluble in water.
Use: Fungicide.

ammonium dinitro-o-cresolate.
Hazard: Dangerous in contact with combustible materials. Strong oxidizing agent. Flammable.
Use: Herbicide.

ammonium dithiocarbamate. $CH_6N_2S_2$.
Properties: Yellow crystals, soluble in water, mp 99C (decomposes), d 1.45.
Derivation: Reaction of ammonia with carbon disulfide.
Use: Organic synthesis, especially heterocyclic compounds, analytical reagent.

ammonium fluoride. CAS: 12125-01-8. NH_4F.
Properties: White crystals, d 1.31, decomposed by heat, soluble in cold water.
Derivation: Interaction of ammonium hydroxide and hydrofluoric acid with subsequent crystallization.
Method of purification: Recrystallization.
Grade: Technical, CP.
Hazard: Corrosive to tissue. TLV (as F): 2.5 mg/m³ of air.
Use: Fluorides, analytical chemistry, antiseptic in brewing, etching glass, textile mordant, wood preservation, mothproofing.

ammonium fluosilicate. (ammonium silicofluoride). CAS: 1309-32-6. $(NH_4)_2SiF_6$.
Properties: White, crystalline powder; d 2.01; soluble in alcohol and water.
Hazard: Strong irritant to eyes and skin. TLV (as F): 2.5 mg/m³ of air.
Use: Laundry sours, mothproofing disinfectant in brewing industry, glass etching, light metal casting, electroplating.

ammonium formate. $HCOONH_4$.
Properties: Deliquescent, crystalline powder; d 1.26; mp 115C; soluble in water and alcohol.

Derivation: Reaction of formic acid with ammonia.
Use: Analytical chemistry (metal precipitant).

ammonium gluconate. $NH_4C_6H_{11}O_7$.
Properties: White powder, soluble in water, insoluble in alcohol. Optical rotation +11.6 (in water).
Preparation: From gluconic acid by neutralization with ammonia.
Use: Emulsifying agent for cheese and salad dressings.

ammonium glutamate. (monoammonium glutamate). See sodium glutamate.

ammonium hexachloroosmate. (ammonium chlorosmate; osmium ammonium chloride). $(NH_4)_2OsCl_6$.
Properties: Red powder. Contains 43.5% osmium. Soluble in alcohol, water.
Hazard: See osmium.

ammonium hexachloroplatinate. (ammonium chloroplatinate; platinic-ammonium chloride; platinic sal ammoniac; platinum ammonium chloride). $(NH_4)_2PtCl_6$.
Properties: Orange-red crystals or yellow powder. Slightly soluble in water, insoluble in alcohol, d 3.06, mp decomposes.
Grade: Technical, CP.
Use: Plating, platinum sponge.

ammonium hexafluorogermanate. $(NH_4)_2GeF_6$.
Properties: White crystalline solid, mp 380C (sublimes), soluble in cold water, insoluble in alcohol, d 2.564.

ammonium hexanitratocerate. See ceric ammonium nitrate.

ammonium hydrate. See ammonium hydroxide.

ammonium hydrogen carbonate. See ammonium bicarbonate.

ammonium hydrogen fluoride. See ammonium bifluoride.

ammonium hydrogen sulfate. See ammonium bisulfate.

ammonium hydrosulfide. See ammonium sulfide.

ammonium hydroxide. (ammonia solution; aqua ammonia; ammonium hydrate). NH_4OH.
Properties: Colorless liquid, strong odor. Concentration of solutions range up to approximately 30% ammonia.

Grade: Technical, CP, 16%, 20%, 26%, NF (strong), FCC.
Hazard: Liquid and vapor extremely irritating, especially to eyes.
Use: Textiles, manufacture of rayon, rubber, fertilizers, refrigeration, condensation polymerization, photography (development of latent images), pharmaceuticals, ammonia soaps, lubricants, fireproofing wood, ink manufacture, explosives, ceramics, ammonium compounds, saponifying fats and oils, organic synthesis, detergents, food additives, household cleansers.

ammonium hypophosphite. $NH_4N_2PO_2$.
Properties: Deliquescent crystals or white powder, decomposes when heated. Soluble in water and alcohol.
Hazard: Evolves flammable and toxic fumes on heating.
Use: Catalyst in nylon manufacture.
See phosphine.

ammonium ichthosulfonate. See ichthammol.

ammonium iodate. NH_4IO_3.
White, odorless, granular powder.
Hazard: Fire risk in contact with organic materials.
Use: Oxidizing agent.

ammonium iodide. NH_4I.
Properties: White, hygroscopic crystals or powder; soluble in water or alcohol; d 2.56; mp sublimes with decomposition. Affected by light.
Use: Iodides, medicine (expectorant), photography.

ammonium iron tartrate. See ferric ammonium tartrate.

ammonium laurate, anhydrous.
$C_{11}H_{23}COONH_4$.
Properties: Tan, wax-like material; free from ammonia odor. Soluble in ethanol, methanol, cottonseed oil and mineral oil. Soluble (hot) in naphtha, toluene, and vegetable oil, d 0.88 (25C), pH (5% dispersion) 7.6–7.8, mp 42–56C, neut value 120–125. Combustible.
Use: Production of oil-in-water emulsions with high oil content, cosmetics.

ammonium lignin sulfonate. See lignin sulfonate.

ammonium linoleate. $C_{17}H_{31}COONH_4$.
Properties: Yellow paste with ammoniacal odor. Soluble in water, ethanol, methanol. Emulsifies in naphtha, toluene, mineral oil and vegetable oil. D 1.1, pH (5% dispersion) 9.5–9.8, total solids 82%. Combustible.

Grade: Technical, 80%.
Use: Emulsifying agent for oils, waxes, and hydrocarbon solvents; surface tension reducer; detergent; water-repellent finishes.

ammonium metavanadate. (ammonium vanadate). NH_4VO_3.
Properties: White crystals, insoluble in saturated ammonium chloride solution slightly soluble in cold water, d 2.326, mp decomposes at 210C. Nonflammable.
Derivation: Alkali solutions of vanadium pentoxide and precipitation with ammonium chloride.
Use: For catalyst as vanadium pentoxide, dyes, varnishes, indelible inks, drier for paints and inks, photography, analytical reagent.

ammonium molybdate. (molybdic acid 85%). $(NH_4)_6Mo_7O_{24} \cdot 4H_2O$.
Properties: White, crystalline powder; soluble in water; insoluble in alcohol; d 2.27; mp decomposes. Nonflammable.
Derivation: Dissolving molybdenum trioxide in aqueous ammonia.
Grade: Technical, CP, reagent (contains 85% MoO_3).
Hazard: Irritant.
Use: Analytical reagent, pigments, catalyst for dehydrogenation and desulfurization in petroleum and coal technology, production of molybdenum metal, source of molybdate ions.

ammonium-12-molybdophosphate. (ammonium phosphomolybdate).
$(NH_4)_3PO_4 \cdot 12MoO_3 \cdot 3H_2O$, or $(NH_4)_3PMo_{12}O_{40} \cdot xH_2O$.
Properties: Yellow, crystalline powder; soluble in alkali; insoluble in alcohol and acids; very slightly soluble in water. Nonflammable.
Derivation: Interaction of ammonium molybdate and phosphoric and nitric acids.
Grade: 91% MoO_3.
Use: Reagent, ion exchange columns, photographic additives, imparting water resistance.

ammonium-12-molybdosilicate. (ammonium silicomolybdate). $(NH_4)_4SiMo_{12}O_{40} \cdot xH_2O$.
Properties: Crystalline, yellow granules; thermally stable. Only slightly soluble in water, ethanol, and ethyl acetate. Nonflammable.
Grade: Technical, reagent.
Use: Catalysts, reagents, in atomic energy as precipitants and inorganic ion-exchangers, in photographic processes as fixing agents and oxidizing agents, in plating processes as additive, and in plastics, adhesives, and cement industries for imparting water resistance.

ammonium nickel chloride. See nickel ammonium chloride.

ammonium nickel sulfate. See nickel ammonium sulfate.

ammonium nitrate. (Norway saltpeter). NH_4NO_3. 13th highest-volume chemical produced in US (1985).
Properties: Colorless crystals; soluble in water, alcohol, and alkalies. D 1.725, mp 169.6C, bp decomposes at 210C with evolution of nitrous oxide.
Derivation: Action of ammonia vapor on nitric acid.
Grade: Usually expressed in percent of nitrogen as 20.5% N, 33.5% N. FGAN is a fertilizer grade, prilled and usually coated with kieselguhr. Also available as an 83% solution. A temperature-stabilized grade is also available which inhibits breakdown of prills due to crystalline changes.
Hazard: May explode under confinement and high temperatures, but not readily detonated. Ventilate well. To fight fire, use large amounts of water. The material must be kept as cool as possible and removed from confinement and flooded with water in event of fire. Explodes more readily if contaminated with combustibles. Strong oxidizing agent. May be made resistant to flame and detonation by proprietary process involving addition of 5 to 10% ammonium phosphate.
Use: Fertilizer, explosives, especially as prills/oil mixture, pyrotechnics, herbicides and insecticides, manufacture of nitrous oxide, absorbent for nitrogen oxides, ingredient of freezing mixtures, oxidizer in solid rocket propellants, nutrient for antibiotics and yeast, catalyst.
See also explosives, high.

ammonium nitrate-carbonate mixtures. See calcium ammonium nitrate.

ammonium nitroso-β-phenylhydroxylamine. See cupferron.

ammonium oleate. $C_{17}H_{33}COONH_4$. An ammonium soap.
Properties: Yellow to brownish ointment-like mass, ammonia odor, decomposes on heating. Soluble in water and hot alcohol. Combustible.
Use: Emulsifying agent, cosmetics.

ammonium oxalate. $(NH_4)_2C_2O_4 \cdot H_2O$.
Properties: Colorless crystals, soluble in water, d 1.502, decomposed by heat.
Use: Analytical chemistry, safety explosives, manufacture of oxalates, rust and scale removal.

ammonium palmitate. $C_{15}H_{31}COONH_4$.
An ammonium soap.
Properties: Yellowish granules, mp 20C, soluble in water, hot alcohols and benzene. Combustible.

Derivation: Reaction of palmitic acid and ammonium carbonate.
Use: Thickening agent for petroleum-derived solvents, lubricants, etc.; waterproofing.

ammonium paratungstate. See ammonium tungstate.

ammonium pentaborate. (ammonium decaborate). $(NH_4)_2B_{10}O_{16} \cdot 8H_2O$.
Properties: Crystals or powder, soluble in water.
Use: Intermediate for boron chemicals, as a power-level control in atomic submarines.

ammonium perchlorate. (AP; APC). NH_4ClO_4.
Properties: White crystals. Soluble in water. D 1.95, mp decomposes on heating.
Derivation: Interaction of ammonium hydroxide, hydrochloric acid and sodium chlorate. Recovery by crystallization.
Hazard: Strong oxidizing agent, ignites violently with combustibles. Shock-sensitive, may explode when exposed to heat or by spontaneous chemical reaction. Sensitive, high explosive when contaminated with reducing materials. Skin irritant.
Use: Explosives, pyrotechnics, analytical chemistry, etching and engraving agent, smokeless rocket and jet propellant.

ammonium permanganate. NH_4MnO_4.
Properties: Crystal or powder form, having a metallic sheen in rich violet-brown or dark purple shades, soluble in water.
Hazard: May explode on shock or on exposure to heat. Toxic by ingestion, inhalation of dust or fume. Powerful oxidizer.

ammonium perrhenate. NH_4ReO_4
Properties: Colorless, weakly oxidizing solid. Stable to heat; decomposes at 365C; moderately soluble in hot water, slightly in soluble cold water.
Derivation: Liquid ion-exchange.
Hazard: Moderate fire risk in contact with reducers.

ammonium persulfate. $(NH_4)_2S_2O_8$.
Properties: White crystals; strong oxidizer; water soluble. D 1.98; mp decomposes.
Derivation: Electrolysis of concentrated solution of ammonium sulfate. Recovered by crystallization.
Hazard: Fire risk in contact with reducers.
Use: Oxidizer, bleaching agent; photography; etchant for printed circuit boards, copper; electroplating; manufacturing of other persulfates; deodorizing and bleaching oils; aniline dyes; preserving foods; depolarizer in batteries; washing infected yeast.

ammonium phosphate, dibasic. (ammonium phosphate, secondary; diammonium hydrogen phosphate; diammonium phosphate; DAP). $(NH_4)_2HPO_4$.
Properties: White crystals or powder, d 1.619, mildly alkaline in reaction, soluble in water, insoluble in alcohol. Noncombustible.
Derivation: Interaction of ammonium hydroxide and phosphoric acid in proper proportions.
Grade: Technical, CP, fertilizer, feed, dentifrice, highly purified for phosphors, FCC.
Use: Flameproofing of wood, paper, and textiles; coating vegetation to retard forest fires; to prevent afterglow in matches and smoking of candlewicks; fertilizer (high analysis phosphate type); plant nutrient solutions; manufacture of yeast, vinegar, and bread improvers; feed additive; flux for soldering tin, copper, brass, zinc; purifying sugar; in ammoniated dentrifices; halophosphate phosphors.

ammonium phosphate, hemibasic.
$NH_4H_2PO_4 \cdot H_3PO_4$.
Properties: White, crystalline material; somewhat hygroscopic. Strongly acid in reaction. Soluble in water. Noncombustible.
Use: Nutrient for truck gardens, yeast food, buffer for adjustment of pH values, metal cleaning.

ammonium phosphate, monobasic. (ammonium acid phosphate; ammonium biphosphate; ammonium dihydrogen phosphate; ammonium phosphate, primary). $NH_4H_2PO_4$.
Properties: Brilliant, white crystals or powder. Mildly acid in reaction. Moderately soluble in water, d 1.803. Noncombustible.
Derivation: Interaction of phosphoric acid and ammonia in proper proportions.
Grade: Technical, CP, FCC, single crystals.
Use: Fertilizers; flameproofing agent; to prevent afterglow in matches; plant nutrient solutions; manufacturing of yeast, vinegar, yeast foods, and bread improvers; food additive; analytical chemistry.

ammonium phosphite. (neutral ammonium phosphite). $(NH_4)_2HPO_3 \cdot H_2O$.
Properties: Colorless, crystalline mass. Hygroscopic. Soluble in water.
Grade: Technical.
Use: Chemical (reducing agent), lubricating grease (corrosion inhibitor).

ammonium phosphomolybdate. See ammonium-12-molybdophosphate.

ammonium phosphotungstate. (ammonium phosphowolframate).
$2(NH_4)_3PO_4 \cdot 24WO_3 \cdot xH_2O$.
Properties: White powder. Soluble in alkali, insoluble in acid, sparingly soluble in water.

Derivation: Interaction of ammonium tungstate, ammonium phosphate, and nitric acid.
Use: Chemical reagent, ion-exchange.

ammonium phosphowolframate. See ammonium phosphotungstate.

ammonium picrate. (ammonium carbazoate; ammonium picronitrate). CAS: 131-74-8.
$C_6H_2(NO_2)_3ONH_4$.
Properties: Yellow crystals, d 1.72, mp decomposes, slightly soluble in water and alcohol.
Hazard: A high explosive when dry, flammable when wet.
Use: Pyrotechnics, explosive compositions.

ammonium polymannuronate. See ammonium alginate.

ammonium polyphosphate. See urea-ammonium polyphosphate.

ammonium polysulfide. $(NH_4)_2S_x$.
Properties: Known only in solution, yellow, unstable, H_2S odor, decomposed by acids with evolution of hydrogen sulfide.
Derivation: Passing hydrogen sulfide into 28% ammonium hydroxide and dissolving an excess of sulfur in the resulting solution.
Hazard: Evolves toxic and flammable gas on contact with acids.
Use: Analytical reagent, insecticide spray.

ammonium reineckate. See Reinecke salt.

ammonium ricinoleate. $C_{17}H_{32}OHCOONH_4$.
Properties: White paste. Combustible.
Grade: Technical.
Use: Detergent, emulsifying agent.

ammonium salts. Salts formed by neutralization of ammonium hydroxide with acids. Usually white and water soluble; usually decomposed by heat into ammonia and the corresponding acid, which may also decompose. All ammonium salts liberate ammonia (NH_3) when heated with a strong base, e.g., sodium hydroxide or calcium hydroxide.

ammonium selenate. $(NH_4)_2SeO_4$.
Properties: Colorless crystals, d 2.194, soluble in water, insoluble in alcohol.
Use: Mothproofing agent.

ammonium selenite. $(NH_4)_2SeO_3 \cdot HOH$.
Properties: Colorless or slightly reddish crystals. Keep away from dust or light. Soluble in water.
Grade: Technical.
Use: Analysis (test for alkaloids), glass colorant.

ammonium sesquicarbonate. See ammonium carbonate.

ammonium silicofluoride. See ammonium fluosilicate.

ammonium silicomolybdate. See ammonium-12-molybdosilicate.

ammonium soap. A soap resulting from the reaction of a fatty acid with ammonium hydroxide. Has an appreciable vapor pressure of ammonia and decomposes on continued exposure leaving the fatty acid residue. Usually not sold as detergents, but used in toilet preparations and emulsions.

ammonium stearate. $C_{17}H_{35}COONH_4$.
Properties: Tan, wax-like solid, free from ammonia odor. Soluble in boiling water. Soluble in hot toluene, partly soluble in hot butyl acetate and ethanol, d 0.89 (22C), pH (3% dispersion) 7.6, mp 73–75C, neutralization value 70–80. Combustible.
Grade: Available as anhydrous solid or as paste.
Use: Vanishing creams, brushless shaving creams, other cosmetic products, waterproofing of cements, concrete, stucco, paper and textiles, etc.

ammonium sulfamate. CAS: 7773-06-0.
$NH_4OSO_2NH_2$.
Properties: White, hygroscopic solid; mp 130C; decomposes at 160C. Soluble in water and ammonia solution. Nonflammable.
Derivation: Hydrolysis of the product obtained when urea is treated with fuming sulfuric acid.
Hazard: Hot acid solutions when enclosed may explode. TLV: 10 mg/m³ of air.
Use: Flameproofing agent for textiles and certain grades of paper, weed and brush killer, electroplating, generation of nitrous oxide.

ammonium sulfate. CAS: 7783-20-2.
$(NH_4)_2SO_4$. 30th highest-volume chemical produced in US (1985).
Properties: Brownish-gray to white crystals according to degree of purity. Soluble in water, insoluble in alcohol and acetone, d 1.77, mp 513C with decomposition. Nonflammable.
Derivation: (a) Ammoniacal vapors from destructive distillation of coal react with sulfuric acid, followed by crystallization and drying. (b) Synthetic ammonia is neutralized with sulfuric acid. (c) Byproduct of manufacture of caprolactam. (d) From gypsum by reaction with ammonia and carbon dioxide.
Method of purifying: Recrystallization or sublimation.

Grade: Commercial, technical, CP, enzyme (no heavy metals), FCC.
Use: Fertilizers, water treatment, fermentation, fireproofing compositions, viscose rayon, tanning, food additive.

ammonium sulfate nitrate.
Properties: A double salt of approximately 60% ammonium sulfate and 40% ammonium nitrate, 26% nitrogen content. White to light gray granules, soluble in water.
Hazard: Oxidizer, dangerous in contact with organic materials.

ammonium sulfide. CAS: 12124-99-1.
$(NH_4)_2S$. The true sulfide is stable only in the absence of moisture and below 0C. The ammonium sulfide of commerce is largely ammonium bisulfide or hydrosulfide, NH_4HS.
Properties: Yellow crystals. Soluble in water, alcohol, and alkalies. Mp decomposes. Evolves hydrogen sulfide on contact with acids.
Grade: Technical, CP, liquid, 40–44%.
Hazard: Strong irritant to skin and mucous membranes.
Use: Textile industry, photography (developers), coloring brasses, bronzes, iron control in soda ash production, synthetic flavors.

ammonium sulfite. CAS: 10196-04-0.
$(NH_4)_2SO_3 \cdot H_2O$.
Properties: Colorless crystals; acrid, sulfurous taste. Hygroscopic, sublimes at 150C with decomposition, soluble in water, d 1.41.
Use: Chemical (intermediates, reducing agent), medicine, permanent wave solutions, photography, metal lubricants.

ammonium sulfocyanate. See ammonium thiocyanate.

ammonium sulfocyanide. See ammonium thiocyanate.

ammonium sulfonate. See ammonium benzenesulfonate.

ammonium sulforicinoleate.
Properties: Yellow liquid, soluble in alcohol. Combustible.
Grade: Technical.
Use: Medicine, furniture polish.

ammonium tartrate. CAS: 3164-29-2.
$(NH_4)_2C_4H_4O_6$.
Properties: White crystals, soluble in water and alcohol, d 1.601, decomposes on heating.
Use: Textile industry, medicine.

ammonium tetrathiocyanodiammonochromate. See Reinecke salt.

ammonium tetrathiotungstate. $(NH_4)_2WS_4$.
Properties: Orange-colored, crystalline powder; sensitive to heat; hydrogen sulfide odor; mp decomposes. Soluble in water, ammoniacal, and amine solutions.
Use: Source of high purity tungsten disulfide for catalysts, lubricants, semiconductors.

ammonium thiocyanate. (ammonium sulfocyanide; ammonium sulfocyanate).
CAS: 1762-95-4. NH_4SCN.
Properties: Colorless, deliquescent crystals; soluble in water, alcohol, acetone, and ammonia; d 1.3057; mp 149.6C; decomposes at 170C.
Derivation: By boiling an aqueous solution of ammonium cyanide with sulfur or polysulfides, or by the reaction of ammonia and carbon disulfide.
Grade: Technical, CP, 50–60% solution.
Use: Analytical chemistry; chemicals (thiourea); fertilizers; photography; ingredients of freezing solutions, especially liquid rocket propellants; fabric dyeing; zinc coating; weed killer and defoliant; adhesives; curing resins; pickling iron and steel; electroplating; temporary soil sterilizer; polymerization catalyst; separator of zirconium and hafnium, and of gold and iron.

ammonium thioglycolate. $HSCH_2COONH_4$.
Properties: Colorless liquid, repulsive odor. Evolves hydrogen sulfide. Combustible.
Use: Solutions of various strengths are used for hair waving and for hair removal.

ammonium thiosulfate. CAS: 1183-18-8.
$(NH_4)_2S_2O_3$.
Properties: White crystals decomposed by heat, very soluble in water; pH of 60% solution 6.5–7.0.
Grade: Pure crystals (97%), 60% photographic solution.
Use: Photographic fixing agent, especially for rapid development; analytical reagent; fungicide; reducing agent; brightener in silver plating baths; cleaning compounds for zinc-base die-cast metals; hair waving preparations; fog screens.

ammonium titanium oxalate. (titanium ammonium oxalate). $(NH_3)_2TiO(C_2O_4)_2$.
A water-soluble powder used as a mordant in dyeing cellulosic fibers, leather, etc.

ammonium tungstate. (ammonium wolframate; ammonium paratungstate).
$(NH_4)_6W_7O_{24} \cdot 6HOH$.   See also ammonium metatungstate.
Properties: White crystals; soluble in water; insoluble in alcohol.
Derivation: Interaction of ammonium hydroxide and tungstic acid with subsequent crystallization.

Use: Preparation of ammonium phosphotungstate and tungsten alloys.

ammonium valerate. (pentanoic acid, ammonium salt; valeric acid, ammonium salt).
$C_5H_{13}NO_2$.
Properties: Very hygroscopic crystals. Mp 108C, mw 119.16. Very soluble in water, alcohol, and ether.
Grades: Food and flavor codex.
Use: Flavoring material.

ammonium vanadate. See ammonium metavanadate.

ammonium wolframate. See ammonium tungstate.

ammonium zirconifluoride. See zirconium ammonium fluoride.

ammonium zirconyl carbonate.
$(NH_4)_3ZrOH(CO_3)_3 \cdot 2HOH$.   Available in aqueous solution, d 1.238 (24C). Stable up to approximately 68C; decomposes in dilute acids, alkalies.
Use: Ingredient in water repellents for paper and textiles, catalyst, stabilizer in latex emulsion paints, ingredient in floor wax to aid in resistance to detergents, lubricant in fabrication of glass fibers.

ammonobasic mercuric chloride. See mercury, ammoniated.

"Ammonyx."[328] TM for a series of quaternary ammonium chloride derivatives in which the substituents are methyl, ethyl, benzyl, stearyl, lauryl, phenyl, and cetyl groups. Some items are similar, i.e., isoquinolinium and pyridinium salts. These are all cationics. The "O" series are alkylamine oxides and are nonionics.
Use: Softeners, wetting agents, emulsifiers, some grades are germicides, fungicides, disinfectants, and slimicides.

"Ammo-Phos."[84] TM for high analysis ammonium phosphate-containing fertilizers.

amniote egg. The type of egg laid by reptiles and birds, having a nutritious yolk and a hard outer shell to protect the embryo from the dry environment. The amniote egg is named for the amnion, a sac that contains the embryo.

amobarbital. (5-ethyl-5-isoamylbarbituric acid).
$C_{11}N_{18}N_2O_3$.
Properties: White, crystalline powder; odorless with bitter taste. Mp 156–161C, solutions are

acid to litmus. Very slightly soluble in water, soluble in alcohol.

Grade: USP.

Hazard: May be habit forming drug of abuse.

Use: Medicine also as sodium salt (hypnotic).

amodiaquine hydrochloride.
$C_{20}H_{22}ON_3Cl \cdot 2HCl \cdot 2HOH$.
Properties: Yellow, odorless, bitter, crystalline solid. Mp 150–160C (decomposes), soluble in water, sparingly soluble in alcohol, very slightly soluble in benzene, chloroform, and ether; pH (1% solution) 4.0–4.8.

Grade: NF.

Use: Medicine (antimalarial).

amorphous. Noncrystalline, having no molecular lattice structure which is characteristic of the solid state. All liquids are amorphous, some materials that are apparently solid, such as glasses, or semisolid, such as some high polymers, rubber, and sulfur allotropes, also lack a definite crystal structure and a well-defined melting point. They are considered high-viscosity liquids. The cellulose molecule contains amorphous as well as crystalline areas. Carbon derived by thermal decomposition or partial combustion of coal, petroleum, and wood is amorphous (coke, carbon black, charcoal), though other forms (diamond, graphite) are crystalline. Amorphous metallic alloys for use as transformer coils are made by extremely rapid cooling of the molten mixture. They are composed of iron, nickel, phosphorus, and boron.

See also liquid, liquid crystal, glass, metallic.

amosite. A type of asbestos.
See asbestos.

AMP. (1) Abbreviation for 2-amino-2-methyl-1-propanol. (2) Abbreviation for adenosine monophosphate.
See adenylic acid.

A5MP. Abbreviation for adenosine-5-monophosphoric acid.
See adenylic acid (muscle adenylic acid).

"Ampco."[407] TM for a series of aluminum-iron-copper alloys containing 6–15% aluminum, 1.5–5.25% iron, balance copper. Resistant to fatigue, corrosionve, erosion, wear, and cavitation-pitting.
Use: For bushings, bearings, gears, slides, etc.

"Ampcoloy."[407] TM for a series of industrial copper alloys including low iron-aluminum bronzes, nickel-aluminum bronzes, tin bronzes, manganese bronzes, lead bronzes, beryllium-copper and high-conductivity alloys.

"Ampco-Trode."[407] TM for a series of aluminum-bronze arc-welding electrodes and filler rod containing 9.0–15.0% aluminum, 1.0–5.0% iron balance copper, for joining like or dissimilar metals and overlaying surfaces resistant to wear, corrosion, erosion, and cavitation-pitting.

AMPD. Abbreviation for 2-amino-2-methyl-1,3-propanediol.

amphetamine. (1-phenyl-2-aminopropane; methylphenethylamine; "Benzedrine").
$C_6H_5CH_2CH(NH_2)CH_3$.
Properties: Colorless, volatile liquid; characteristic strong odor and slightly burning taste; bp 200–203C (decomposes); flash p 80F (26.6C); soluble in alcohol and ether; slightly soluble in water.
Grade: Dextro-, dextrolevo-. Also available as phosphate and sulfate.
Hazard: Flammable, moderate fire risk. Basis of a group of hallucinogenic (habit-forming) drugs which affect the central nervous system. Sale and use restricted to physicians. Production limited by law.
Use: Medicine.

amphibole. A type of asbestos.
See asbestos.

amphiphilic. A molecule having a water-soluble polar head (hydrophilic) and a water-insoluble organic "tail" (hydrophobic), e.g., octyl alcohol, sodium stearate. Such molecules are necessary for emulsion formation and for controlling the structure of liquid crystals.
See also emulsion, liquid crystal.

ampholyte. A substance that can ionize to form either anions or cations and thus may act as either an acid or a base. An ampholytic detergent is cationic in acid media and anionic in base media. Water is an ampholyte.
See also amphoteric.

amphora catalyst. See catalyst, amphora.

amphoteric. Having the capacity of behaving either as an acid or a base. Thus aluminum hydroxide neutralizes acids with the formation of aluminum salts, $Al(OH)_3 + 3HCl \rightarrow AlCl_3 + 3HOH$, and also dissolves in strongly basic solutions to form aluminates $Al(OH)_3 + 3NaOH \rightarrow Na_3AlO_3 + 3HOH$. Amino acids and proteins are amphoteric, i.e., their molecules contain both an acid group (COOH) and a basic group (NH₂). Thus, wool can absorb both acidic and basic dyes.

amphotericin B. A polyene antifungal antibiotic.
$C_{47}H_{73}NO_{17}$.

Properties: Pale yellow, semicrystalline powder; mp >170C (gradual decomposition); insoluble in water; slightly soluble in methanol. Somewhat more soluble in dimethylsulfoxide.
Derivation: Fermentation with Sterptomyces nodosus. Commercially available as a deoxycholate complex.
Grade: USP.
Hazard: May have undesirable side effects.
Use: Medicine (meningitis treatment).

ampicillin (USAN). (6-D-α-aminophenyl-acetamido)penicillanic acid). $C_{16}H_{19}N_3O_4S$.
A semisynthetic antibiotic, active against some gram-negative infections.

amprolium. 1-[(4-Amino-2-propyl-5-pyrimidinyl)-methyl]-2-picolinium chloride).
$C_{14}H_{19}ClN_4$.   A cocidiostat used in veterinary medicine.

amprotropine phosphate. (phosphate of the *dl*-tropic acid   ester of 3-diethylamino-2,2-dimethyl-l-propanol). $C_{18}H_{29}NO_3 \cdot H_3PO_4$.
Properties: Bitter crystals. Mp 142–144C. Soluble in water, slightly soluble in alcohol.
Use: Medicine (antispasmodic).

"Ampvar."[41]   TM for synthetic-resin metal conditioner of the vinyl-phosphoric acid-zinc chromate type used to prepare metal surfaces for the application of corrosion-proof coatings.

amsonic acid. (4,4'-diamino-2,2'-stilbenedisulfonic acid). $C_{14}H_{14}N_2O_6S_2$.
Properties: Acicular crystals, slightly soluble in water.
Use: Manufacture of bleaching agents and organic dyes.

"Amthio."[50]   TM for an ammonium thiosulfate solution.
Properties: A reddish liquid fertilizer and soil conditioner used before planting. Contains 12% nitrogen and 26% sulfur, d 1.33, bulk wt approximately 11.1 lb/gal. Can be mixed and applied with many liquid fertilizer solutions or alone in irrigation water.
Hazard: Irritant. Avoid contact with skin or eyes.

amygdalic acid.   See mandelic acid.

amygdalin. (mandelonitrile-β-gentiobioside; amygdaloside). $C_6H_5CHCNOC_{12}H_{21}O_{10}$.
A glycoside found in bitter almonds, peaches, and apricots.
Properties: White crystals, bitter taste, mp 214–216C (anhydrous), soluble in water and alcohol, insoluble in ether.
See also almond oil (note).

amyl. The 5-carbon aliphatic group C_5H_{11}, also known as pentyl. Eight isomeric arrangements (exclusive of optical isomers) are possible. The amyl compounds occur (as in fusel oil) or are formed (as from the petroleum pentanes) as mixtures of several isomers and since their bp's are close and their other properties similar, it is neither easy nor usually necessary to purify them.
See following entries, especially amyl alcohol.

amyl acetate. (amylacetic ester; banana oil; pear oil).   CAS: 628-63-7. $CH_3COOC_5H_{11}$.
Commercial amyl acetate is a mixture of isomers the composition and properties depending upon the grade and derivation. The main isomers are isoamyl, normal, and secondary amyl acetates.
Properties: Colorless liquid, flash p 65 to 95F (18.3 to 35C) (CC) depending on grade, persistent banana-like odor. Autoign temperature approximately 714F (380C).
Derivation: Esterification of amyl alcohol (often fusel oil) with acetic acid and a small amount of sulfuric acid as catalyst.
Method of purification: Rectification.
Grade: Commercial (85–88%), high test (85–88%), technical (90–95%), pure (95–99%), special antibiotic grade. Amyl acetate is also sold by original source as from fusel oil, pentane, or Oxo process.
Hazard: Flammable; dangerous fire risk. Explosive limits in air 1.1 to 7.5%.
Use: Solvent for lacquers and paints, extraction of penicillin, photographic film, leather polishes, nail polish, warning odor, flavoring agent, printing and finishing fabrics, solvent for phosphors in fluorescent lamps.

n-amyl acetate.   CAS: 628-63-7.
$CH_3COOCH_2CH_2CH_2CH_2CH_3$.
Properties: Colorless liquid, bp 148.4C, mp −70.8C, d 0.879 (20/20C), wt/gal (20C) 7.22 lbs, flash p 77F (25C) (CC). Very slightly soluble in water miscible with alcohol and ether. Vapor heavier than air. Autoign temperature 714F (380C).
Derivation: Esterification of n-amyl alcohol with acetic acid.
Hazard: Flammable; dangerous fire risk. TLV: 100 ppm in air.
Use: See amyl acetate.

sec-amyl acetate.   CAS: 626-38-0.
$CH_3CO_2C_5H_{11}$.
Properties: Colorless liquid. May be mixture of secondary isomers. Distillation range 123–145C, mild odor, nonresidual, purity of ester content as amyl acetate 85–88%, d 0.862–0.866 (20/

20C), flash p 89F (31.6C) (CC), wt/gal (20C) approximately 7.19 lbs.

Derivation: Esterifiction of sec-amyl alcohol and acetic acid.

Grade: Technical.

Hazard: Flammable; high fire risk. Toxic. TLV: 125 ppm in air.

Use: Solvent for nitrocellulose and ethyl cellulose, cements, coated paper, lacquers, leather finishes, nail enamels, plastic wood, textile sizing, and printing compounds.

amylacetic ester.   See amyl acetate.

amyl acid phosphate. $(C_5H_{11})_2HPO_4$ and $C_5H_{11}H_2PO_4$.   A mixture of primary and amyl isomers. Water-white liquid, d 1.070–1.090, flash p 245F (118.3C) (COC). Insoluble in water, soluble in alcohol. Combustible.

Hazard: Strong irritant to tissue.

Use: Curing catalyst and accelerator in resins and coatings, stabilizer, dispersion agent, lubricating and antistatic agent in synthetic fibers.

amyl alchol.   (amyl hydrate).   Eight isomers of amyl alcohol, $C_5H_{11}OH$, are possible, exclusive of several optical isomers, and six are offered commercially. In addition, definite mixtures of the isomers are sold under a variety of names (unfortunately some of them identical with the names of the pure isomers) as well as fusel oil, a natural fermentation product. For descriptive data on the pure isomers, see: (1) n-amyl alcohol, primary (2) 2-methyl-1-butanol (active amyl alcohol from fusel oil) (3) isoamyl alcohol, primary (4) 2-pentanol (5) 3-pentanol (6) tert-amyl alcohol The other 2 isomers not described in detail are (7) sec-isoamyl alcohol (8) 2,2-dimethyl-1-propanol (1), (2), (3), and (8) are primary alcohols; (4), (5), and (7) are secondary alcohols; and (6) is a tertiary alcohol. (1), (4), and (5) are normal, and (2), (3), (6), (7), and (8) are branched-chain compounds. (2), (4), and (7) are asymmetric and have optically active forms.

amyl alcohol, primary.   A mixture of primary amyl alcohols made from normal butenes by the Oxo process is sold under this name. It consists of 60% primary n-amyl alcohol, 35% 2-methyl-1-butanol, and 5% 3-methyl-1-butanol.

Hazard: Flammable; moderate fire risk.

Use: A solvent.

n-amyl alcohol, primary.   (1-pentanol; n-butyl carbinol).   CAS: 71-41-0.
$CH_3(CH_2)_4OH$.
Properties: Colorless liquid, mild odor, bp 137.8C, fp −78.9C, d 0.812–0.819 (20/20C), wt/gal

(20C) 6.9 lbs, flash p 123F (50.5C) (OC). Autoign temperature 572F (300C). Slightly soluble in water; miscible with alcohol, benzene, and ether. Flammable.

Derivation: Fractional distillation of the mixed alcohols resulting from the chlorination and alkaline hydrolysis of pentane.

Grade: Technical, CP, 98%.

Hazard: LEL in air 1.2% by volume. Moderate fire risk.

Use: Raw material for pharmaceutical preparations; organic synthesis solvent.

amyl alcohol, primary active.   See 2-methyl-1-butanol.

n-sec-amyl alcohol.   See 2- and 3-pentanol.

sec-amyl alcohol, active.   See 2-pentanol.

tert-amyl alcohol.   (dimethylethylcarbinol; 2-methyl-2-butanol; amylene hydrate; tertpentanol).   CAS: 75-85-4.
$(CH_3)_2C(OH)CH_2CH_3$.
Properties: Colorless liquid, camphor odor and burning taste, d 0.81 (20/20C), fp −11.9C, bp 101.8C, refr index 1.4052 (20C), wt/gal 6.76 lbs, flash p 70F (21.2C) (OC). Slightly soluble in water, miscible with alcohol and ether, solutions neutral to litmus. Autoign temperature 819F (437C).

Derivation: Fractional distillation of the mixed alcohols resulting from the chlorination and alkaline hydroysis of pentanes.

Grade: Technical, CP, NF.

Hazard: Flammable; dangerous fire risk.

Use: Solvent, flotation agent, organic synthesis, medicine (sedative).

amyl alcohol, fermentation.   See fusel oil.

amyl aldehyde.   See n-valeraldehyde.

n-amylamine.   (pentylamine; 1-aminopentane). $C_5H_{11}NH_2$.
Properties: Colorless liquid, d 0.75 (20/20C), fp −55.0C, bp 104.4C, flash p 45F (7.2C) (OC). Soluble in water, alcohol, and ether.

Derivation: Reaction of ammonia and amyl chloride which gives a mixture of mono-, di-, and triamyl amines.

Grade: Technical.

Hazard: Flammable; dangerous fire risk. Strong irritant.

Use: Chemical intermediate, dyestuffs, rubber chemicals, insecticides, synthetic detergents, flotation agents, corrosion inhibitors, solvent, gasoline additive, pharmaceuticals.

sec-**amylamine.** (2-aminopentane).
$CH_3(CH_2)_2CH(CH_3)NH_2$.
Properties: Colorless liquid, bp 198F, d 0.7; flash p 20F (65.5C).
Hazard: Flammable; dangerous fire risk.
Use: See n-amylamine.

amylase. A class of enzymes which convert starch into sugars. Fungal and bacterial amylases from specific fungi and bacteria have been suggested for commercial fermentation processes.
Use: Textile desizing, conversion of starch to glucose sugar in syrups (especially corn syrups), baking (to improve crumb softness and shelf life), dry cleaning (to attack food spots and similar stains).
See also amylopsin, diastase, and ptyalin.

n-**amylbenzene.** (1-phenylpentane).
$C_6H_5CH_2(CH_2)_3CH_3$. Water-white liquid, mild odor, fp −75C, bp 205C, d 0.8585 (20/4C), flash p 150F (65.5C) (OC), Insoluble in water, soluble in hydrocarbons and coaltar solvents. Combustible.
Hazard: Irritant to skin and eyes, narcotic in high concentrations. Moderate fire risk.

sec-**amylbenzene.** (2-phenylpentane).
$CH_3CH(C_6H_5)CH_2CH_2CH_3$.
Properties: Clear liquid, fp −75C, bp 190.3C, d 0.861 (20/4C).
Hazard: Moderate fire hazard. Combustible. Irritant to skin and eyes, narcotic in high concentrations.
Grade: Pure, 99.0 mole %, technical, 95.0 mole %.
Use: Weed control, chemical intermediate.

amyl benzoate. See isoamyl benzoate.

amyl butyrate. See isoamyl butyrate.

amyl carbinol. See 1-hexanol.

n-**amyl chloride.** (1-chloropentane).
CAS: 543-59-9. $CH_3(CH_2)_3CH_2Cl$.
Properties: Colorless liquid. Bp 107.8C, fp −99C, d 0.883 (20/4C), refr index 1.4128 (20C), flash p 54F (12.2C) (OC). Miscible with alcohol and ether, insoluble in water. Autoign temperature 500F (260C).
Derivation: (a) Distillation of amyl alcohol with salt and sulfuric acid, (b) addition of hydrochloric acid to α-amylene.
Grade: Technical.
Hazard: Flammable; dangerous fire risk. lel 1.4%, uel 8.6%. May be narcotic in high concentrations.
Use: Chemical intermediate.

amyl chlorides, mixed.
Properties: Straw to purple-colored liquid, d 0.88 (20C), 95% distills between 85 and 109C, wt/gal 7.33 lb, refr index (20C) 1.406, insoluble in water, water azeotrope at 77–82C approximately 90% $C_5H_{11}Cl$, miscible with alcohol and ether. Flash p 38F (3.3C) (OC). Components: 1-chloropentane, bp 107.8C. 2-chloropentane, bp 96.7C. 3-chloropentane, bp 97.3C. 1-chloro-2-methylbutane, bp 99.9C. 1-chloro-3-methylbutane, bp 98.8C. 3-chloro-2-methylbutane, bp 93.0C. 2-chloro-2-methylbutane, bp 86.0C.
Derivation: Vapor phase chlorination of mixed normal pentane and isopentane.
Hazard: Flammable; dangerous fire risk. Explosive limits in air 1.4 to 8.6%. May be narcotic in high concentrations.
Use: Synthesis of other amyl compounds, solvent, rotogravure ink vehicles, soil fumigation.

α-**amylcinnamic alcohol.** (2-benzylidene-1-heptanol). $C_6H_5CH:C(CH_2OH)C_5H_{11}$.
Properties: Yellow liquid, floral odor, d 0.954–0.962 (25/25C), soluble in 3 parts 70% alcohol. Combustible.
Derivation: Synthetic.
Use: Perfumery, flavoring.

α-**amylcinnamic aldehyde.** (jasmine aldehyde; α-pentylcinnamaldehyde).
$C_6H_5CH:C(CHO)C_5H_{11}$.
Properties: Clear, yellow, oily liquid. Jasmine-like odor. Aldehyde content 97%., soluble in 6 volumes of 80% alcohol, d 0.962 to 0.968, refr index 1.554 to 1.559. Combustible.
Derivation: Synthetic.
Grade: Technical, FCC.
Use: Perfumery, flavoring.

6-n-**amyl-m-cresol.** $C_{12}H_{18}O$.
Properties: Liquid at room temperature, mp 24C, insoluble in water, soluble in acetone and ethanol.
Use: Prevention of molds, bactericide.

amyl-p-dimethylaminobenzoate. See "Escalol 506."

α-n-**amylene.** Legal label name for 1-pentene.

β-n-**amylene.** Legal label name for 2-pentene.

amylene dichloride. See dichloropentane.

amylene hydrate. See tert-amyl alcohol.

n-**amyl ether.** (diamyl ether). $C_{10}H_{22}O$.
Properties: Colorless liquid, bp 186C, fp −70C, d 0.783, flash p 135F (57C), autoign temperature

340F (171C), refr index 1.41. Insoluble in water, soluble in alcohol and ether.
Hazard: Narcotic in high concentration.
Use: General solvent for fats, oils, waxes, resins, etc.

amyl formate. $HCOOC_5H_{11}$.
Properties: Colorless liquid composed of a mixture of isomeric amyl formates with isoamyl formate in predominance. Plum-like odor. Less odoriferous and more active solvent than amyl acetate. It also has both a lower boiling point and a higher rate of evaporation. Miscible with oils, hydrocarbons, alcohols, ketones. Slightly soluble in water, d 0.880 to 0.885, bp 123.5C, flash p 80F (26.6C).
Grade: Technical, FCC.
Hazard: Flammable; dangerous fire risk. Toxic by ingestion and inhalation.
Use: Solvent for cellulose esters, resins, solvent mixtures, films and coatings, perfume for leather, flavoring.

n-amylfuroate. (amyl pyromucate). $C_4H_3OCO_2C_5H_{11}$.
Properties: Colorless oil, decomposes on standing, d 1.0335 (20/4C), bp 233C. Insoluble in water, soluble in alcohol. Combustible.
Derivation: Esterification of furoic acid.
Use: Perfumes, lacquers.

amyl hydrate.   See amyl alcohol.

amyl hydride.   See pentane.

"Amyliq."[173]   TM for starch-liquefying enzyme for sizes and adhesives.

amyl mercaptan.   Legal label name for pentane-thiol.

tert-amyl mercaptan.   See 2-methyl-2-butane-thiol.

amyl nitrate (mixed isomers). $C_5H_{11}NO_3$.
Properties: Colorless liquid, bp 145C, flash p 118F (47.8C), d 0.99 (20C), ethereal odor. Flammable.
Hazard: Oxidizing agent, moderate fire risk.
Use: Additive to increase cetane number of diesel fuels.

amyl nitrite. (isoamyl nitrite).
CAS: 1002-16-0. $(CH_3)_2CHCH_2CH_2NO_2$.
Properties: Yellowish liquid of peculiar ethereal, fruity odor, and pungent aromatic taste. Soluble in alcohol, almost insoluble in water. Decomposes on exposure to air, light, or water; d 0.865–0.875 (25C); bp 96–99C; autoign temperature 405F (207C).

Derivation: Interaction of amyl alcohol with nitrous acid.
Grade: NF, (75% min), technical.
Hazard: Flammable; dangerous fire risk, a strong oxidizer. Vapor may explode if ignited.
Use: Perfumes, diazonium compounds.

amyloglucosidase.   An enzyme used commercially to convert starches to dextrose.

"Amylon."[53]   TM for a high-amylose starch from high-amylose corn.

amylopectin.   The outer, almost insoluble portion of starch granules. It is a hexosan, a polymer of glucose, and is a branched molecule of many glucose units. It stains violet with iodine and forms a paste with water.

amylopsin. (animal diastase).   The starch-digesting enzyme of pancreatic juice, the most powerful enzyme of the digestive tract. It is an amylase which converts starches through the soluble-starch stage to various dextrins and maltose. Its acts in neutral, slightly acid, and slightly alkaline environments with an optimum pH of 6.3–7.2. It requires the presence of certain negative ions for activation.
Use: Biochemical research.

amylose.   The inner, relatively soluble portion of starch granules. Amylose is a hexosan, a polymer of glucose, and consists of long straight chains of glucose units joined by a 1,4-glycosidic linkage. It stains blue with iodine. Microcrystalline amylose is available chiefly as a food ingredient and dietary energy source.

o-sec-amylphenol. $C_5H_{11}C_6H_4OH$.
Properties: Clear, straw-colored liquid, d 0.955–0.971 (30/30C), initial bp over 235.0C, final bp below 250.0C, wt/gal 8.0 lbs, very slightly soluble in water, soluble in oil and organic solvents. Flash p 219F (OC). Combustible.
Use: Dispersing and mixing agent for paint pastes, antiskinning agent for paint, varnish and oleoresinous enamels, organic synthesis.

p-tert-amylphenol. $(CH_3)_2(C_2H_5)CC_6H_4OH$.
Properties: White crystals, mp 93C, bp 265–267C

(138C at 15 mm), slightly soluble in water, soluble in alcohol and ether. Flash p 232F (111C) (OC). Combustible.
Use: Manufacture of oil-soluble resins, chemical intermediate.

p-tert-amylphenyl acetate.
$C_5H_{11}C_6H_4OOCCH_3$.
Properties: Colorless liquid, d 0.996 (20C), boiling range 253–272C, fruity odor, flash p 240F (115.5C). Combustible.
Use: Perfumes, flavorings.

amyl propionate. $CH_3CH_2COOC_5H_{11}$.
An isoamyl isomer.
Properties: Colorless, high-boiling liquid; apple-like odor; d 0.869–0.873 (20/20C); wt/gal (20C) approximately 7.25 lbs; distillation range 135–175C; flash p 106F (41.1C) (OC); autoign temperature 712F (377C); miscible with most organic solvents. Flammable.
Derivation: By reacting amyl alcohol with propionic acid in the presence of sulfuric acid as a catalyst, followed by neutralization, drying, and distillation.
Hazard: Fire hazard.
Use: Perfumes, lacquers, flavors.

amyl pyromucate.   See n-amyl furoate.

amyl salicylate.   See isoamyl salicylate.

amyl sulfide.   See diamyl sulfide.

amyltrichlorosilane. $C_5H_{11}SiCl_3$.   A mixture of isomers.
Properties: Colorless to yellow liquid, bp 168C, d 1.137 (25/25C), refr index 1.4152 (20C), flash p 145F (62.8C) (COC), readily hydrolyzed by moisture with the liberation of hydrogen chloride. Combustible.
Derivation: By Grignard reaction of silicon tetrachloride and amyl magnesium chloride.
Hazard: Toxic and corrosive.
Use: Intermediate for silicones.

amyl valerate.   See isoamyl valerate.

amyl valerianate.   See isoamyl valerate.

"Amytal."[100]   TM for amobarbital (USP).

anabasine.   (neonicotine; 2-(3-pyridyl)piperidine).
$C_{10}H_{14}N_2$.   A naturally occurring alkaloid.
Properties: Colorless liquid, darkens on exposure to air, bp 270C, fp 9C, d 1.046 (20/20C), refr

index 1.5430 (20C). Miscible with water, soluble in alcohol and ether.
Derivation: (a) Extraction from Anabasis aphylla and Nicotiana glauca; (b) synthetic.
Use: Insecticide.

anacardic acid. $C_{22}H_{32}O_3$.   The chief component of cashew nutshell oil.

anacardium gum.   See cashew gum.

anaerobic.   Descriptive of a chemical reaction or a micro-organism that does not require the presence of air or oxygen. Examples are the fermentation of sugars by yeast and the decomposition of sewage sludge by anaerobic bacteria. It also applies to certain polymers that solidify when kept out of contact with air.
See also "Loctite", bacteria, botulism.

analcite.   (analcime).
$Na_2O \cdot Al_2O_3 \cdot 4SiO_2 \cdot 2HOH$.   A mineral, one of the zeolites.
Properties: Colorless, white, sometimes greenish-grayish, yellowish, or reddish white; hardness 5–5.5; d 2.22–2.29.
Occurrence: Europe, US, Nova Scotia.

analytical chemistry.   The subdivision of chemistry concerned with identification of materials (qualitative analysis) and with determination of the percentage composition of mixtures or the constituents of a pure compound (quantitative analysis). The gravimetric and volumetric (or "wet") methods (precipitation, titration, and solvent extraction) are still used for routine work, indeed new titration methods have been introduced, e.g., cryoscopic, pressure-metric (for reactions that produce a gaseous product), redox methods, and use of a F-sensitive electrode. However, faster and more accurate techniques (collectively called instrumental) have been developed in the last few decades. Among these are infrared, ultraviolet, and x-ray spectroscopy where the presence and amount of a metallic element is indicated by lines in its emission or absorption spectrum, colorimetry by which the percentage of a substance in soluble is determined by the intensity of its color, chromatography of various types by which the components of a liquid or gaseous mixture are determined by passing it through a column of porous material or on thin layers of finely divided solids, separation of mixtures in ion-exchange columns and radioactive tracer analysis. Optical and electron microscopy, mass spectrometry, microanalysis, nuclear magnetic resonance (NMR) and nuclear quadrupole resonance (NQR) spectroscopy all fall within the area of analytical chemistry. New and highly sophisticated techniques have been introduced in recent years, in many cases replacing traditional methods.

See also spectroscopy, nuclear magnetic resonance, NQR spectroscopy, chromatography instrumentation, fiber optical, supercritical fluid.

anaphylaxis. Abnormal reaction to a second injection of a foreign protein, e.g., penicillin. It is an extreme form of allergy which often has serious consequences (swelling of tissues) and has been known to be fatal.

anatase. (octahedrite). A natural crystallized form of titanium dioxide, d 3.8, refr index 2.5, mp 1560C.

"Anattene."[342] TM for annatto derivatives for use in coloring food-stuffs, i.e., cheeses, oranges.

andalusite. Al_2OSiO_4. A natural silicate of aluminum.
Properties: Gray, greenish, reddish, or bluish in color; d 3.1–3.2; hardness 7–7.5.
Occurrence: Massachusetts, Connecticut, California, Nevada, Europe, South Africa, Australia.
Use: Constituent of sillimanite refractories, spark plug insulators, laboratory ware, superrefractories.

androgen. A male sex hormone. The androgenic hormones are steroids and are synthesized in the body by the testis, the cortex of the adrenal gland, and, to slight extent, by the ovary.

androsterone. $C_{19}H_{30}O_2$. An androgenic steriod, metabolic product of testosterone. The international unit (IU) of adrogenic activity is defined as 0.1 mg androsterone

Properties: Crystalline solid, mp 185–185.5C, sublimes in high vacuum, dextrorotatory in solution, not precipitated by digitonin, practically insoluble in water, soluble in most organic solvents.
Derivation: Isolation from male urine, synthesis from cholesterol.
Use: Medicine, biochemical research.

Andrussov oxidation. Ammonia and methane are oxidized with air in the presence of platinum catalyst to form hydrogen cyanide. Side reactions are hydration of methane to carbon dioxide and hydrogen and oxidation of methane and ammonia to carbon monoxide and nitrogen. The reaction is strongly exothermic. The process has been elaborated wherever natural gas is abundant.

anesthetic. A chemical compound that induces loss of sensation in a specific part or all of the body. A brief classification of the more important agents is as follows:
A. General
  1. Hydrocarbons
     (a) Cyclopropane (USP). Effective in presence of substantial proportions of oxygen; flammable.
     (b) Ethylene (USP). Rapid anesthesia and rapid recovery; flammable.
  2. Halogenated hydrocarbons
     (a) Chloroform. Nonflammable. Its use is being abandoned because of its high toxicity.
     (b) Ethyl chloride. A gas at room temperature, liquefies at relatively low pressure. Applied as a stream from container directly on tissue. Sometimes used in gaseous form as inhalation type general anesthetic. Flammable.
     (c) Trichloroethylene. Toxic and flammable. Used as general anesthetic since 1934.
  3. Ethers
     (a) Ethyl ether (USP). First anesthetic used in surgery (1846), now largely replaced with less dangerous types. Highly flammable, explodes in presence of spark or open flame.
     (b) Vinyl ether. A liquid having many of the physiological properties of ethylene and ethyl ether. Highly flammable.
  4. Miscellaneous
     (a) Tribromoethanol. Basal anesthetic, supplemented by an inhalation type when general anesthesia is needed. Ingredient of "Avertin."
     (b) Nitrous oxide. Originally prepared by Priestley in 1772 (laughing gas); first used as anesthetic by Humphry Davy in 1800. Used (with oxygen) largely for dental surgery. Nonflammable.
     (c) Barbiturates.
B. Local
  1. alkaloids (cocaine)
  2. synthetic products (procaine group, e.g., "Novocain"); alkyl esters of aromatic acids (topical).
  3. quinine hydrochloride.

anethole. (anise camphor; p-methoxypropenyl-benzene; p-propenylanisole). $CH_3CH:CHC_6H_4OCH_3$.
Properties: White crystals with sweet taste, odor of oil of anise. Affected by light. Soluble in 8 volumes of 80% alcohol, 1 volume of 90% alcohol. Almost immiscible with water, d 0.983–0.987, refr index 1.557–1.561, optical rotation 0.08, mp 22–23C, distillation range 234–237C.
Derivation: By crystallization from anise or fennel oils, synthetically from p-cresol.
Grade: USP, technical, FCC.
Use: Perfumes, particularly for dentifrices, flavors, synthesis of anisic aldehyde, licorice candies, color photography (sensitizer in color-bleaching process), microscopy.

Anfinsen, Christian B. (1916-) An American biochemist who won the Nobel prize for chemistry in 1972. His work involved the molecular basis of evolution and the chemistry of enzymes. He worked with Moore and Stein. His PhD was granted from Harvard.

ANFO. A high explosive based on ammonium nitrate.
See also explosives, high.

angelic acid. (cis-2-methyl-2-butenoic acid; α-methyl-crotonic acid). $CH_3CH:C(CH_3)COOH$. The cis isomer of tiglic acid.
Properties: Colorless needles or prismatic crystals, spicy odor. Soluble in alcohol, ether, and hot water. D 0.9539 (76/4C), mp 45C, bp 185C, refr index 1.4434 (47C).
Derivation: From the root of Angelica archangelica or from the oil of Anthemis nobilis by distillation.
Use: Flavoring extracts.

angelica oil.
Properties: Essential oil, strong aromatic odor, spicy taste. Soluble in alcohol, d 0.853–0.918, optical rotation +16 to +41. Chief known constituents: phellandrene, valeric acid. Combustible.
Derivation: Distilled from the roots and seeds of Angelica archangelica found principally in Europe.
Grade: Technical, FCC.
Use: Preparation of liqueurs, perfumery.

"Angio-Conray."[329] TM for an 80% solution of sodium iothalamate used in diagnostic medicine.

angiotensin. (angiotonin; hypertensin). A peptide found in the blood, important in its effect on blood pressure. Both a decapeptide and an octapeptide are known. Their amino acid sequences and hence the complete structures have been established.

angiotonin. See angiotensin.

angstrom. A unit of length almost one one-hundred-millionth (10^{-8}) centimeter. The angstrom (abbreviation Å.) is now defined in terms of the wavelength of the red line of cadmium (6438.4696 Å). Used in stating distances between atoms, dimensions of molecules, wavelengths of short-wave radiation, etc.
See also nanometer.

anhydride. A chemical compound derived from an acid by elimination of a molecule of water. Thus sulfur trioxide (SO_3) is the anhydride of sulfuric acid (H_2SO_4), carbon dioxide (CO_2) is the anhydride of carbonic acid (H_2CO_3), phthalic acid $[C_6H_4(CO_2H)_2]$ minus water gives phthalic anhydride $[C_6H_4(CO_2)O]$. Not to be confused with anhydrous.

anhydrite. $CaSO_4$. A natural calcium sulfate usually occurring as compact granular masses and resembling marble in appearance. Differs from gypsum in hardness and lack of hydration.

anhydroenneaheptitol. (AEH; 4-hydroxy-2H-pyran-3,3,5,5(4H,6H)tetramethanol).
$OCH_2(CH_2OH)_2CHOHC(CH_2OH)_2CH_2$.
Use: Alkyd resins, rosin esters, urethane coatings and foams, surfactants, lubricating oil additives.

"Anhydrol."[214] TM for a gasoline-free solvent composed of 100 gal SD ethanol denatured with 10 gal isopropanol (90%) and 1 gal methyl isobutyl ketone.
Properties: (anhydrous grade) Bp 76.2–79.5C, d 0.7895–0.7935 (20/20C), lb/gal 6.6 (20C), flash p 54F (12.2C).
Grade: Anhydrous to 190 proof.
Hazard: Flammable; dangerous fire risk.
Use: Solvent in manufacture of printing inks, textile dyestuff solutions, window cleaners, synthetic detergents, aircraft de-icing fluids, and photographic chemicals.

anhydrous. Descriptive of an inorganic compound that does not contain water either adsorbed on its surface or combined as water of crystallization. Do not confuse with anhydride.

"Anhydrox."[236] TM for a compound to prevent or overcome anhydrite or gypsum contamination in drilling mud, by pretreatment of the mud to remove calcium and sulfate ions.

anidex. A synthetic fiber designated by the FTC as a cross-linked polyacrylate elastomer.

anileridine. (ethyl-1-(p-aminophenethyl)-4-phenylisonipecotate). $C_{22}H_{28}N_2O_2$.
Properties: White, odorless, crystalline powder. Oxidizes and darkens in air and on exposure to light. Exhibits polymorphism and of two crystalline forms observed, one melts at approximately 80C and the other at approximately 89C. Soluble in alcohol and chloroform. Very slightly soluble in water.
Grade: NF.
Hazard: Is addictive.
Use: Medicine (narcotic).

aniline. (aniline oil; phenylamine; aminobenzene). CAS: 62-53-3. $C_6H_5NH_2$.   One of the most important of the organic bases, the parent substance for many dyes and drugs.

Properties: Colorless, oily liquid; characteristic odor and taste; rapidly becomes brown on exposure to air and light. Vapors will contaminate foodstuffs and damage textiles. Soluble in alcohol, ether, and benzene; soluble in water; d 1.0235, solidifies at −6.2C, bp 184.4C, wt/gal 8.52 lbs (20C), refr index 1.5863 (20C), flash p 158F (70C) (CC), autoign temperature 1140F (615C). Combustible.
Derivation: (a) By catalytic vapor-phase reduction of nitrobenzene with hydrogen; (b) reduction of nitrobenzene with iron filings using hydrochloric acid as catalyst; (c) catalytic reaction of chlorobenzene and aqueous ammonia; (d) ammonolysis of phenol (Japan).
Grade: Commercial; CP.
Hazard: An allergen. Toxic if absorbed through the skin. TLV: 2 ppm in air.
Use: Rubber accelerators and antioxidants, dyes and intermediates, photographic chemicals (hydroquinone), isocyanates for urethane foams, pharmaceuticals, explosives, petroleum refining, diphenylamine, phenolics, herbicides, fungicides.

aniline acetate. $C_6H_5NH_2 \cdot CH_3COOH$.
Properties: Colorless liquid, becomes dark with age; on standing or heating is converted gradually to acetanilide; d 1.070–1.072; miscible with water and alcohol. Combustible.
Use: Organic synthesis.

aniline black. A black dye developed on cotton and other textiles from a bath containing aniline hydrochloride, an oxidizing agent (usually chromic acid) and a catalyzer (usually a vanadium or copper salt).
Hazard: See aniline.

aniline chloride. See aniline hydrochloride.

aniline-2,4-disulfonic acid. See 4-amino-m-benzenedisulfonic acid.

aniline-2,5-disulfonic acid. See 2-amino-p-benzenedisulfonic acid.

aniline dye. Any of a large class of synthetic dyes made from intermediates based upon or made from aniline. Most are somewhat toxic and irritating to eyes, skin, and mucous membranes. They are generally much less toxic than the intermediates from which they are derived.

aniline hydrochloride. (aniline salt; aniline chloride). $C_5H_5NH_2 \cdot HCl$.
Properties: White crystals, commercial product frequently greenish in appearance, darkens in light and air. Soluble in water, alcohol, and ether; d 1.2215, mp 198C, bp 245C.
Derivation: (a) By passing a current of dry hydrogen chloride gas into an ethereal solution of aniline; (b) neutralizing aniline at 100C with concentrated hydrochloric acid and subsequent crystallization.
Hazard: See aniline.
Use: Intermediates, dyeing and printing, aniline black.

aniline ink. A fast-drying printing ink used on kraft paper, cotton fabric, cellophane, polyethylene, etc. The name is due to the fact that original inks for this purpose were solutions of coal-tar dyes in organic solvents. Modern inks usually employ pigments rather than dyes and are of two types: spirit inks containing organic solvent as the vehicle, and emulsion inks in which water is the main vehicle.

aniline point. The lowest temperature at which equal volumes of aniline and the test liquid are miscible. Cloudiness occurs on phase separation. Used as a test for components of hydrocarbon fuel mixtures.

aniline salt. See aniline hydrochloride.

p-anilinesulfonic acid. See sulfanilic acid.

aniline yellow. See p-aminoazobenzene.

1-anilino-4-hydroxyanthraquinone.
$C_6H_4(CO)_2C_6H_2(OH)NHC_6H_5$ (tricyclic).
A chemical intermediate.

anilinophenol. See p-hydroxydiphenylamine.

animal black. See bone black, ivory black.

animal diastase. See amylopsin.

animal oil. See bone oil.

animal starch. See glycogen.

anion. An ion having a negative charge, anions in a liquid subjected to electric potential collect at the positive pole or anode. Examples are hydroxyl, OH^-, carbonate, CO_3^{--}, phosphate, PO_4^{---}.

o-anisaldehyde. (o-methoxybenzaldehyde; o-anisic aldehyde). CAS: 135-02-4. $C_6H_4(OCH_3)CHO$.
Properties: White to light tan solid; burned, slightly phenolic odor; bp 238C; mp (two crystalline forms) 38–39C and 3C; d (liquid) 1.1274 (25/25C); (solid) 1.258 (25/25C); refr index 1.5608 (20C), flash p 244F (117C); insoluble in water, soluble in alcohol. Combustible.
Grade: 95% (min).
Use: Intermediate.

p-anisaldehyde. (aubepine; p-anisic aldehyde; p-methoxybenzaldehyde). CAS: 123-11-5. $C_6H_4(OCH_3)CHO$.

CH₃O—⟨◯⟩—CHO

Properties: Colorless to pale yellow liquid having odor of hawthorn. Insoluble in water. Soluble in 5 volumes of 50% alcohol, d 1.119–1.122, refr index 1.570–1.572, mp OC, bp 248C. Combustible.
Derivation: Obtained from anethole or anisole by oxidation.
Grade: Liquid and crystals, latter being the disulfite compound.
Use: Perfumery, intermediate for antihistamines, electroplating, flavoring.

anise alcohol. See anisic alcohol.

anise camphor. See anethole.

anise oil. (anise seed oil; aniseed oil). See anethole.

anisic acid. (p-methoxybenzoic acid). $CH_3OC_6H_4COOH$.
Properties: White crystals or powder, d 1.385 (4C), mp 184C, bp 275–280C, soluble in alcohol and ether, almost insoluble in water.

Derivation: Oxidation of anethole.
Use: Repellent and ovicide.

anisic alcohol. (anisyl alcohol; anise alcohol; p-methoxybenzyl alcohol). $CH_3OC_6H_4CH_2OH$.
Properties: Solidifies at room temperature, floral odor. Soluble in 1 volume of 50% alcohol, insoluble in water, d 1.111–1.114, refr index 1.541–1.545, boiling range 255–265C. Combustible.
Derivation: Obtained from anisic aldehyde by reduction.
Grade: Technical, FCC.
Use: In perfumery for light floral odors, pharmaceutical intermediate, flavoring.

anisic aldehyde. See anisaldehyde.

o-anisidine. (o-methoxyaniline; o-amino-anisole). CAS: 90-04-0. $CH_3OC_6H_4NH_2$.
Properties: Reddish or yellowish oil, becomes brownish on exposure to air, volatile with steam, d 1.097 (20C), bp 225C, fp 5C; soluble in dilute mineral acid, alcohol, and ether; insoluble in water.
Derivation: (a) Reduction of o-nitroanisole with tin or iron and hydrochloric acid; (b) heating o-aminophenol with potassium methyl sulfate.
Method of purification: Steam distillation.
Grade: 99% (1% max moisture).
Hazard: Strong irritant. TLV: 0.1 ppm. Toxic when absorbed through the skin.
Use: Intermediate for azo dyes and for guaiacol.

p-anisidine (p-methoxyaniline; p-aminoanisole). CAS: 104-94-9 $CH_3OC_6H_4NH_2$.
Properties: Fused, crystalline mass; crystallizing point 57.2C (min); d 1.089 (55/55C); bp 242C; soluble in alcohol and ether; slightly soluble in water.
Derivation: (a) Reduction of p-nitroanisole with iron filings and hydrochloric acid; (b) methylation of p-aminophenol.
Grade: Technical.
Hazard: Strong irritant. TLV: 0.1 ppm.
Use: Azo dyestuffs, intermediate.

anisindione. (2-p-anisyl-1,3-indandione). $C_{16}H_{12}O_3$.
Properties: Pale yellow crystals, mp 156–157C.
Grade: ND.
Use: Anticoagulant (blood).

anisole. (methylphenyl ether; methoxybenzene). CAS: 100-66-3. $C_6H_5OCH_3$.
Properties: Colorless liquid, aromatic odor, soluble in alcohol and ether, insoluble in water, d

0.999 (15/15C), fp −37.8C, bp 155C, refr index 1.5170 (20C), flash p 125F (51.6C) (OC). Combustible.
Derivation: From sodium phenate and methyl chloride, heating phenol with methanol.
Use: Solvent, perfumery, vermicide, intermediate, flavoring.

anisomycin. $C_{14}H_{19}NO_4$.
Properties: Needles, mp 140C, slightly soluble in water, soluble in low-molecular-weight alcohols, ketones, and chloroform, slightly soluble in organic solvents. Water solutions are stable at room temperature over broad range of pH.
Use: Fungus inhibitor, mildew preventive in vegetables.

anisotropic. Descriptive of crystals whose index of refraction varies with the direction of the incident light. This is true of most crystals, e.g., calcite (Iceland spar); it is not true of isometric (cubic) crystals, which are isotropic.

anisoyl chloride. CAS: 100-07-2.
$CH_3OC_6H_4COCl$.
Properties: Clear crystals or amber liquid. Mp 22C, bp 262–263C, soluble in acetone and benzene, decomposed by water or alcohol, fumes in moist air.
Use: Intermediate for dyes and medicines.
Hazard: Solutions corrosive to tissue. Explosion risk when in closed containers due to pressure caused by decomposition at room temperature.

anisyl acetate. (p-methoxybenzyl acetate).
$CH_3OC_6H_4CH_2COCH_3$.
Properties: Colorless liquid, lilac-odor, soluble in 4 volumes of 60% alcohol. d 1.104–1.107, refr index 1.514–1.516. Combustible.
Derivation: Reaction of anisic alcohol with acetic anhydride.
Grade: Technical, FCC.
Use: Perfumery, flavoring.

anisylacetone. (Generic name for 4-(p-methoxyphenyl)-2-butanone).
$CH_3OC_6H_4C_2H_4COCH_3$.
Properties: A colorless to pale yellow liquid, mp 8C. Combustible.
Used in insect attractants, organic synthesis, flavoring.

anisyl alcohol. See anisic alcohol.

p-anisylchlorodiphenylmethane. (p-methoxytriphenylmethyl chloride). CAS: 14470-28-1.
$CH_3OC_6H_4C(C_{62}H_5)_2Cl$.
Properties: Mp 122−124C, mw 308.81.
Grades: Research.

Hazards: Skin and eye irritant.
Use: NH_2— protecting reagent for amino acids in oligonucleotide synthesis.

anisyl formate. (p-methoxybenzyl formate).
$CH_3OC_6H_4CH_2OCOH$.
Properties: Colorless liquid, lilac odor. Soluble in 5.5 volumes of 70% alcohol, d 1.139–1.141. Combustible.
Use: Perfumery, flavoring.

2-p-anisyl-1,3-indandione. See anisindione.

annatto. Vegetable dye containing ethyl bixin. $C_{27}H_{34}O_4$.
Derivation: From the seeds of Bixa orellana.
Occurrence: South America, West Indies, India.
Uses (as extract): Coloring margarine, sausage casings, etc.; food-product marking inks. For details, see regulations of Meat Inspection Division, USDA and FDA regulations.
See also bixin.

annealing. Maintenance of glass or metal at a specified temperature for a specific length of time (at least three days for plate glass) and then gradually cooling it at a predetermined rate. This treatment removes the internal strains resulting from previous operations and eliminates distortions and imperfections. A clearer, stronger, and more uniform material results.
See also tempering.

annellation. A chemical reaction in which one cyclic or ring structure is added to another to form a polycyclic compound.

"Ano."[203] TM for a series of dyestuffs used for coloring anodized aluminum.
See anodic coating.

anode. The positive electrode of an electrolytic cell, to which negatively charged ions travel when an electric current is passed through the cell. Such anodes are usually made of graphite or other form of carbon, although titanium has been successfully introduced in the chlor-alkali industry. In a primary cell (battery or fuel cell), the anode is the negative electrode.
See also cathode, electrode.

anode mud. Residue obtained from the bottom of a copper or other plating bath. In the electrolytic refining of copper, the anode mud contains the relatively inert metals platinum, silver, and gold and is usually collected and treated for the recovery of these metals and other rare elements.

anode process. See electrophoresis.

anodic (anodized) coating. The electrolytic treatment of aluminum, magnesium, and a few other metals as a result of which heavy, stable films of oxides are formed on their surfaces. A thin oxide film will form on an aluminum surface without special treatment on exposure of the metal to air. This provides excellent resistance to corrosion. This fact led to the development of electrochemical processes to produce much thicker and more effective protective and decorative coatings. The chief electrolytes used are sulfuric, oxalic, and chromic acids. The metal acts as the anode. Such anodic coatings are hard and have good electrical insulating properties. Their ability to absorb dyes and pigments makes it possible to obtain finishes in a complete range of colors, including black. The luster of the underlying metal gives them a metallic sheen; colorants can be used to reproduce the color of any metal with which the aluminum might be used. Anodized coatings can be used as preparatory treatments to electroplating; copper, nickel, cadmium, silver and iron have been successfully deposited over oxide coatings.
See also aluminum.

anomer. A specific kind of diastereoisomer (or epimer) occurring in some sugars and other substances having asymmetric carbon atoms.

"Anoplex."[85] TM for the sodium bisulfite complex of anisaldehyde.
$[C_6H_4(OCH_3)HC(OH)SO_3Na]$.
Properties: Dry, white powder; slightly soluble in water.
Use: Compounding of brighteners requiring a uniform dry solid bisulfite complex for electroplating.

ANPO. Abbreviation for α-naphthylphenyloxazole.

"Anstac-2M."[238] TM for an antistatic and cleaning agent for plastics, such as methyl methacrylates, vinyls and polystyrenes.
Use: In aircraft, sign, novelty, electrical, photographic, optical, and other industries.

antabuse.
Use: Drug for treatment of alcoholism. See tetraethylthiuram disulfide.

antacid. Any mildly alkaline substance such as sodium bicarbonate taken internally or in water solution to neutralize excess stomach acidity.

antagonist, structural. (antimetabolite). An organic compound that is structurally related to a biologically active substance (enzyme, nucleic acid, amino acid etc) and which acts as an inhibitor of its growth and development. Such biological antagonism exists between sulfa drugs and p-aminobenzoic acids, and also between histamine and a group of compounds collectively called antihistamines. One of the most important of these from an agricultural standpoint is imidazole which, together with similar compounds, is used to "antagonize" the metabolism of insects, especially those attacking fabrics; it is also being used in irrigation waters to protect plants from deleterious pests. Structural antagonists have important medical applications in the complex field of allergic disease.
See also antihistamine, anticoagulant, antigen-antibody.

"Antaron FC-34."[203] TM for a high-foaming, water-soluble, amphoteric surfactant with soap-like qualities, a complex fatty amido compound 40% active.
Use: Fulling agent and detergent for woolen and worsted fabrics; effective under neutral, acid, and alkaline conditions; recommended for use in bubble baths, detergents, in soaps for dedusting purposes.

"Antarox."[307] TM for a series of surfactants.
Use: Viscose spin bath additive, preventing accumulation of sludge in pipe lines, reels, etc. Prevents clogging of spinnerets; in manufacture of cellophane, prevents deposits from forming on extrusion slits and rollers, in the steel industry, to obtain cleaner sheets in the final wash of the reduced sheet during cold reduction.

antazoline. $C_{17}H_{19}N_3$.
Properties: White, odorless crystalline powder with bitter taste. Mp 120C. Sparingly soluble in alcohol and water, practically insoluble in benzene and ether.
Use: Medicine (antihistamine). Available as hydrochloride and phosphate.

anteiso-. Prefix denoting an isomer (usually a fatty acid or derivative) which has a single, simple branching attached at the third carbon from the end of a straight chain in distinction to an iso- compound where the attachment is to the second carbon from the end. For example isododecanoic acid would be $CH_3CH(CH_3)(CH_2)_8COOH$, while anteisododecanoic acid would be $CH_3CH_2CH(CH_3)(CH_2)_7COOH$.

anthelmintic. An agent used in veterinary medicine as a vermifuge.

anthelmycin. USAN name for an antibiotic substance produced by Streptomyces longissimus.

antheridiol. $C_{29}H_{42}O_5$.
Properties: Colorless, fine crystals; mp 250C; slightly soluble in water, soluble in warm methanol.
Use: A plant hormone having a specific sex function, it is secreted by certain water molds. It has been used to modify plant fertility. Said to be the first plant sex hormone to be discovered (1942).

anthocyanin. A flavonoid plant pigment which accounts for most of the red, pink, and blue colors in plants, fruits, and flowers. Water-soluble.
See also flavonoid.

"Anthomine."[300] TM for a dyeing assistant primarily for use in wool dyeing.

anthopyllite. $(Mg,Fe)_7Si_8O_{22}(OH)_2$. A natural magnesium-iron silicate.
See asbestos.

anthracene. CAS: 120-12-7.
$C_6H_4(CH)_2C_6H_4$.
Properties: Yellow crystals with blue fluorescence. Soluble in alcohol and ether, insoluble in water, d 1.25 (27/4C), mp 217C, bp 340C, fp 250F (121C) (CC). Combustible. It has semiconducting properties.

Derivation: (a) By salting out from crude anthracene oil and draining. The crude salts are purified by pressing and finally by the use of various solvents. Phenanthrene and carbazole are removed; (b) by distilling crude anthracene oil with alkali carbonate in iron retorts, the distillate containing only anthracene and phenanthrene. The latter is removed by carbon disulfide.
Method of purification: By sublimation with superheated steam or by crystallization from benzene followed by sublimation; for very pure crystals, zone melting of solid anthracene. Impurities: Phenanthrene, carbazole, and chrysene.
Grade: Commercial (90–95%), pure crystals.
Hazard: A carcinogen.
Use: Dyes, alizarin, phenanthrene, carbazole, anthraquinone, calico printing, a component of smoke screens, scintillation counting crystals, organic semi-conductor research.

anthracene oil. A coal-tar fraction boiling in the range 270–360C, a source of anthracene and similar aromatics. Also used as a wood preservative and pesticide, except on food crops.
Hazard: A carcinogen.

1,8,9-anthracenetriol. See anthralin.

anthracite. See coal.

anthragallic acid. See anthragallol.

anthragallol. (1,2,3-trihydroxyanthraquinone; anthragallic acid). $C_6H_4(CO)_2C_6H(OH)_3$.
Tricyclic.
Properties: Brown powder. Soluble in alcohol, ether, glacial acetic acid, slightly soluble in water and chloroform. Sublimes at 290C.
Derivation: Product of the reaction of benzoic, gallic, and sulfuric acids.
Use: Dyeing.

"Anthragen."[203] TM for a series of lake colors. Used for printing inks, wallpaper, coated paper, paint, rubber, and organic plastics.

"Anthralan."[203] TM for a series of acid dye-stuffs. Used on wool.

anthralin. (1,8,9-anthracenetriol; 1,8-dihydroxyanthranol). $C_{14}H_{10}O_3$.
Properties: Odorless, tasteless, yellow powder. Mp 176–181C. Filtrate from water suspension is neutral to litmus. Soluble in chloroform, acetone, benzene, and in solutions of alkali hydroxide; slightly soluble in alcohol, ether, and glacial acetic acid; insoluble in water. Combustible.
Derivation: By catalytic reduction of 1.8-dihydroxyanthraquinone with hydrogen at high pressure.
Grade: NF, (95%).
Hazard: Very irritating. Do not use on scalp or near eyes.
Use: Medicine (treatment of psoriasis).

anthranilic acid. (o-aminobenzoic acid).
CAS: 118-92-3. $C_6H_4(NH_2)(CO_2H)$.
Properties: Yellowish crystals; sweetish taste; soluble in hot water, alcohol, and ether. Mp 144–146C, sublimes. Combustible.
Derivation: Phthalimide plus an alkaline hypobromite solution.
Grade: Technical (95–98%), 99% or better.
Use: Dyes, drugs, perfumes, and pharmaceuticals.

anthranol. (9-hydroxyanthracene).
$C_{14}H_9OH$.
Properties: Crystals, mp 120C, soluble in organic solvents with a blue fluorescence. Changes in solution to anthrone. Combustible.
Use: Dyes.

anthranone. See anthrone.

"Anthrapole."[300] TM for a group of dye carriers or assistants for use in dyeing polyester fibers and blends. Active ingredients are aromatic esters, chlorinated hydrocarbons, and phenol derivatives. Emulsifying agents are incorporated.

anthrapurpurin. (1,2,7-trihydroxyanthraquinone; isopurpurin; purpurin red).
$C_6H_3OH(CO)_2C_6H_2(OH)_2$. Tricyclic.
Properties: Orange-yellow needles. Soluble in alcohol and alkalies, slightly soluble in ether and hot water. Mp 369C, bp 462C.
Derivation: By fusion of anthraquinonedisulfonic acid with caustic soda and potassium chlorate; the melt is run into hot water and the anthrapurpurin precipitated by hydrochloric acid.
Use: Dyeing, organic synthesis.

anthraquinone. CAS: 84-65-1.
$C_6H_4(CO)_2C_6H_4$.

Properties: Yellow needles. Soluble in alcohol, ether, and acetone; insoluble in water; d 1.419–1.438, mp 286C, bp 379–381C, flash p 365F (185C) (CC). Combustible.
Derivation: (a) By heating phthalic anhydride and benzene in the presence of aluminum chloride and dehydrating the product; (b) by condensation of 1,4-naphthoquinone with butadiene.
Method of purification: Sublimation.
Grade: Sublimed, 30% paste (sold on 100% basis), electrical 99.5%.
Use: Intermediate for dyes and organics, organic inhibitor, bird repellent for seeds. See also anthraquinone dye.

anthraquinone-1,5- and **1,8-disulfonic acids.**
(rho acid, chi acid) respectively.
$C_{14}H_8O_8S_2$.
Properties: (In their pure state) Slightly yellow to white crystals. The technical grade is grayish-white. Soluble in water and strong sulfuric acid. The 1,8-isomer is the more soluble. The 1,5-disulfonic acid melts with decomposition at 310–311C. The 1,8-isomer melts and decomposes at 293–294C.
Derivation: Anthraquinone is sulfonated with fuming sulfuric acid in the presence of mercury or mercuric oxide to a mixture of the 1,5- and 1,8-disulfonic acids which are separated by fractional crystallization.
Method of purification: Fractional crystallization from strong sulfuric acid or in form of their alkali salts from either acid or alkaline solutions.
Grade: Technical.
Use: Dyes.

anthraquinone dye. A dye whose molecular structure is based on anthraquinone.
$C_6H_4(CO)_2C_6H_4$.
The chromophore groups are $=C=O-$ and $=C=C=$. The benzene ring structure is important in the development of color. CI numbers from 58000 to 72999. These are acid or mordant dyes when OH or HSO_3 groups respectively are present. The anthraquinone dyes that can be reduced to an alkaline solution leuco (vat) derivative that has affinity for fibers and which can be reoxidized to the dye are known as anthraquinone vat dyes. They are largely used on cotton, rayon, and silk, and have excellent properties of color and fastness.

anthrarufin. See 1,5-dihydroxyanthraquinone.

anthrone. (anthranone; 9,10-dihydro-9-oxoanthracene). $C_{14}H_{10}O$. The keto is the more stable form of anthranol.
Properties: Colorless needles, mp 156C, insoluble in water, soluble in alcohol, benzene, and hot sodium hydroxide solutions.
Derivation: Reduction of anthraquinone with tin and hydrochloric acid.
Use: Rapid determination of sugar in body fluids, and of animal starch in liver tissue; general reagent for carbohydrates; organic synthesis.

anti-. (1) A prefix used in designating geometrical isomers in which there is a double bond between the carbon and nitrogen atoms. This prevents free rotation so that two different spatial arrangements of substituent atoms or groups are possible. When a given pair of these are on opposite sides of the double bond, the arrangement is called anti-; when they are on the same side, it is called syn-, as indicated below:

$$C_6H_5-\underset{HO-N}{\overset{\|}{C}}-H \qquad C_6H_5-\underset{N-OH}{\overset{\|}{C}}-H$$

anti   syn

These prefixes are disregarded in alphabetizing chemical names. (2) A prefix meaning "against" or "opposed to," as an antibody, antimalarial, etc.

antianxiety agent. See psychotropic drug.

antibiotic. A chemical substance produced by microorganisms that has the capacity in dilute solutions to inhibit the growth of other microorganisms or destroy them. Only approximately 20 out of several hundred known have proved generally useful in therapy. Those that are used must conform to FDA requirements. The most important groups of antibiotic-producing organisms are the bacteria, lower fungi or molds, and actinomycetes. These antibiotics belong to very diverse classes of chemical compounds. Most of the antibiotics produced by bacteria are polypeptides (such as tyrothricin, bacitracin, polymyxin). The penicillins are the only important antibiotics produced by fungi. Actinomycetes produce a wide variety of compounds (actinomycin, streptomycin, chloramphenicol, tetracycline). The antimicrobial activity (antibiotic spectrum) of antibiotics varies greatly; some are active only upon bacteria, others upon fungi, still others upon bacteria and fungi, some are active on viruses, some on protozoa, and some are also active on neoplasms. An organism sensitive to an antibiotic may upon continued contact with it develop resistance and yet remain sensitive to other antibiotics. Certain antibiotics are used as direct food additives to inhibit growth of bacteria and fungi; among these are nisin, pimaricin, nystatin, and tylosin. *Note*: A number of antibiotics have been restricted by FDA, both for direct use by humans and as additives to animal feeds. Among those that are in question are streptomycin, chloramphenicol, tetracycline, and penicillin.
See also penicillin, cephalosporin, plasmid, Waksman, Fleming.

antibody. See antigen-antibody.

antiblock agent. A substance, e.g., a finely divided solid of mineral nature which is added to a plastic mix to prevent adhesion of the surfaces of films made from the plastic to each other or to other surfaces. They are of particular value in polyolefin and vinyl films. The hard, infusible particles tend to roughen the surface and so maintain a small air space between adjacent layers of the film thus preventing adhesion. Silicate minerals are widely used for this purpose. Another type of antiblock function is performed by high-melting waxes, which bloom to the surface and form a layer that is harder than the plastic.

anticaking agent. An additive used primarily in certain finely divided food products that tend to be hygroscopic to prevent or inhibit agglomeration and thus maintain a free-flowing condition. Such substances as starch, calcium metasilicate, magnesium carbonate, silica, and magnesium stearate are used for this purpose in table salt, flours, sugar, coffee, whiteners, and similar products.

anticancer drug. See antineoplastic.

anticholinergic. A drug or pharmaceutical that inhibits the action of acetylcholine by reactivating the cholinesterase. Examples are atropine sulfate and pralidoxime iodide.

antichlor. Any product which serves to neutralize and remove hypochlorite or free chlorine after a bleaching operation. For many years the term was considered synonymous with sodium thiosulfate, but it may equally be applied to sodium disulfite or any other product used for the purpose.

anticoagulant. A complex organic compound often a carbohydrate that has the property of retarding the clotting or coagulation of blood. The most effective of these is heparin, which acts by interfering with the conversion of prothrombin to thrombin and by inhibiting the formation of thromboplastin. In addition to their specific clinical uses, anticoagulants have been applied with limited success to rodenticides (see warfarin). They are regarded as cumulative poisons, requiring multiple ingestions to be lethal. One type (diaphenadione) has been found to reduce blood cholesterol in experimental animals.

antidepressant. See psychotropic drug.

antidote. Any substance that inhibits or counteracts the effects of a poison which has entered the body by any route. Mild acids or alkalies (except sodium bicarbonate) exert neutralizing action if corrosive materials have been swallowed; for noncorrosive poisons, warm salt water or milk may be given to cause vomiting. Activated charcoal in water is effective in protecting the throat and stomach linings, except for corrosive poisons. Nothing should be administered by mouth if the subject is unconscious. Artificial respiration may be necessary. In no case should alcohol be used as an antidote. Atropine sulfate and pralidoxime iodide have been successfully used as antidotes for poisoning by cholinesterase inhibitors.

antiemetic agent. A compound usually classed as a pharmaceutical that inhibits or prevents nausea. A well-known type is dimenhydrinate ("Dramamine") which is useful to counteract motion sickness, nausea due to pregnancy, etc. A more recent antiemetic is a benzoquinolizine derivative used for nausea resulting from anesthesia ("Emete-con"[299]).

antienzyme. A substance present in the substrate which restricts or negates the catalytic activity of the enzyme on that substrate.

antifertility agent. A synthetic steroid sex hormone of the type normally produced by the body during pregnancy. Said to act by simulating the conditions of pregnancy and thus suppressing ovulation, which automatically prevents conception. The hormones used are basically of 2 kinds: (1) progestin (synthetic progesterone) and (2) synthetic estrogen. There are also several other derivatives all of which are much stronger than natural progesterone and estrogen when taken orally. The chemical modifications are necessary for oral potency.

Injectable contraceptives that remain effective over much longer periods than the oral type are also available. The active ingredient is medoxyprogesterone acetate. Experimental work is also being done with N,N'-octaethylenediamine bis-(dichloroacetamide) ("Fertilysin"). An estrogen-free type is available in England (chloromadinone acetate); it is said to be less effective, but with less tendency to blood clotting than estrogenic types.
Hazard: FDA requires that oral contraceptives carry a label warning of the tendency of these agents to form blood clots. There is also a possibility that they have other adverse side effects. *Note*: Proteins and peptides which act as enzyme inhibitors have been identified in the semen of some mammals.

antifoam agent. See defoaming agent.

antifouling paint. An organic coating formulated especially for use on the hulls and bottoms of ships, boats, buoys, pilings, and the like to protect them from attack by barnacles, teredos, and other marine organisms. The chief specific ingredient is a metallic naphthenate, e.g., copper naphthenate; mercury compounds are also used.

antifreeze. (1) Water additive. Any compound that lowers the freezing point of water. Both sodium chloride and magnesium chloride were once used, but their extreme corrosive properties made them a liability in automotive cooling systems. Methanol requires only 27% by volume for protection to −17.7C. Due to its tendency to evaporate rapidly at operating temperatures, its flammability, and low boiling point, 63.9C, it has been replaced by glycol derivatives which are relatively noncorrosive, nonflammable, have very low evaporation rate, and are effective heat-exchange agents. A concentration of 35% protects against freezing to −17.7C. Ethylene and propylene glycol antifreezes can be carried in an automotive cooling system for several years without damage, and are satisfactory coolants at summer operating temperatures. Methoxy propanol has been introduced as an antifreeze-coolant for diesel engines.
See also coolant.
(2) Gasoline additive. A proprietary preparation TM "Drygas".[580] consisting of methanol, isopropanol, or mixtures of these which lower the freezing point of water enough to inhibit ice formation in feed lines and carburetors. It is added directly to the gasoline.
Hazard: Poisonous. Flammable.

antigen-antibody. An antibody is a blood serum protein of the globulin fraction which is formed in response to introduction of an antigen. It has a molecular weight of approximately 160,000. An antigen is an infective organism (protein) with a molecular weight of at least 10,000; it is able to induce formation of an antibody in an organism into which it is introduced (by injection). Thus an animal is able to resist infections to which it has previously been exposed. The entire science of immunology is due to antigen-antibody reactions, the most outstanding feature of which is their specificity.

The antibodies produced in the bloodstream can react only with the homologous antigen or with those of a similar molecular structure. As a result, the animal can destroy a particular virus or bacterium and become immune. The specificity of antigens is due not so much to the molecule as a whole as to its configuration. Certain radicals (polar and quaternary ammonium groups) seem to "mate" with corresponding complementary structures in the antibody molecule. A precipitate or agglutinate is formed by the reaction which is analogous to colloidal (catalytic) reactions in some respects, e.g., surface configuration.
See also immunochemistry, Pasteur, allergen, antagonist structural.

antiglobulin. An agent used to coagulate globulin.

antihistamine. A synthetic substance structurally analogous to histamine whose presence in minute amounts prevents or counteracts the action of excess histamine formed in body tissues. These compounds are usually complex amines of various types and also have other physiological effects and medical uses. Examples are chlorpheniramine maleate, dimenhydrinate, diphenhydramine hydrochloride, imidazole, pheniramine maleate, pyrilamine maleate, thonzylamine hydrochloride, tripelennamine hydrochloride.
See also antagonist (structural).

antihypertensive agent. An organic compound having the property of lowering blood pressure in animals and man. Among the better-known types are the alkaloid reserpine and its derivative, syrosingopine, guanethidine sulfate, α-methyl dopa (α-methyl-3,4-dihydroxyphenylalanine) and hydralazine. They function by various nerve-blocking mechanisms involving structural antagonism. They should be taken only by prescription.

anti-inflammatory agent. Any of a number of drugs that prevent or inhibit inflammation of tissue. Most common of these is aspirin.

antiknock agent. Any of a number of organic compounds that increase the octane number of a gasoline when added in low percentages by reducing knock, especially in high-compression engines. Knock is caused by spontaneous oxidation reactions in the cylinder head resulting in loss of power and characteristic ignition noise. Branched-chain hydrocarbon gasolines ameliorate this problem and antiknock additives virtually eliminate it. Tetraethyllead, the most effective of these, has been used for many years but its contribution to air pollution has almost eliminated its use in automotive fuels. Lead-free gasolines (which contain only 0.05 gram per gallon) are now used in conjunction with catalytic converters. The antiknock agents used in them are nonmetallic compounds such as methyl-tert-butyl ether (MTBE) or a mixture of methanol and tert-butyl alcohol.
See also octane number, gasoline.

antilymphocytic serum. (ALS). An immunological suppressant for use in organ transplants. It acts to control the build-up of rejection factors in the blood which result from introduction of foreign organs into the body.
See also immunochemistry.

"Antilac."[165] TM for liquid antimony lactate containing 15% available antimony oxide. Recommended as a replacement for technical tartar emetic.

antimalarial agent. A natural or synthetic drug of the alkaloid type that is specific to combating malaria, a disease of the tropics. Most of the synthetic types are derivatives of 8-aminoquinoline (N-containing heterocyclic compounds) developed by research teams before and during World War II. Quinine has been the standard natural antimalarial drug for centuries. The first synthetic was pamaquin (1926), but it proved too toxic for more than limited use. Mepacrin ("Atabrin") was developed in 1932; it was used in the war and was found more effective than quinine, though it discolored the skin. Chloro-quin and pentaquine followed, the former being preferable to mepacrin with no skin discoloration. A more recently discovered nonheterocyclic called proguanil (1945) has the advantages of lower cost and toxicity than other antimalarials. See also chemotherapy.

antimatter. See antiparticle.

antimetabolite. See antagonist, structural, antihistamine, metabolite.

antimonial lead alloy. (hard lead). Lead containing from approximately 6 to 28% antimony. Common grades are as follows: (a) 15% antimony, resistant to sulfuric acid used in type metal; (b) national stock pile specification, 10.7–11.3% antimony; (c) battery grids 5–11% antimony; (d) battery terminals 4% antimony; (e) cable sheaths 1% antimony.

antimonic. A variation of the name antimony used for compounds in which the antimony has a valence of 5, e.g., antimony pentachloride, pentasulfide, etc.

antimonic acid. See antimony pentoxide.

antimonic anhydride. See antimony pentoxide.

antimonite. See stibnite.

antimonous. (antimonious). A variation of the name antimony used for compounds in which the antimony has a valence of 3, as in antimony tribromide, antimony trichloride, antimony trioxide, antimony trisulfide.

antimonous sulfide. See antimony trisulfide.

antimony. CAS: 7440-36-0. Sb (from Latin stibium). Metallic element of atomic number 51; Group VA of the Periodic Table. Two stable isotopes.
Properties: Aw 121.75, valences 3, 4, and 5; silver-white solid; mp 630.5C, bp 1635C, low thermal conductivity, Mohs hardness 3 to 3.5. Oxidized by nitric acid, not attacked by hydrochloric acid in absence of air, reacts with sulfuric acid and aqua regia. Combustible. A semiconductor.
Forms: Besides the stable metal, there are two allotropes—yellow crystals and amorphous black modifications.
Ores: Stibnite, kermasite, tetrahedrite, livingstonite, jamisonite.
Occurrence: Algeria, Bolivia, China, Mexico, South Africa, Peru, Yugoslavia.
Derivation: Reduction of stibnite with iron scrap; direct reduction of natural oxide ores. About half

the antimony used in the US is recovered from lead base battery scrap metal.

Grade: Up to 99.999% pure, technical, powder, commercial grade in 55-lb cakes $10 \times 10 \times 2.5$ inches.

Hazard: Use with adequate ventilation. Soluble salts are toxic. TLV (as Sb): 0.5 mg/m³ of air.

Use: Hardening alloy for lead, especially storage batteries and cable sheaths, bearing metal, type metal, solder, collapsible tubes and foil, sheet and pipe, semiconductor technology (99.999% grade), pyrotechnics.

antimony-124. Radioactive antimony isotope.

Properties: Half-life 60D, radiation= beta and gamma. The chemical form used is often antimony trichloride and oxychloride in hydrochloric acid solution.

Use: As a tracer, especially in solid state studies and marker of interfaces between products in pipe lines. The gamma ray has sufficient energy to eject neutrons from beryllium. Convenient portable neutron sources which may be reactivated in a nuclear reactor are made by irradiation of an antimony pellet encased in a beryllium shell.

Hazard: Radioactive poison.

antimony-125. Radioactive antimony isotope. Half-life 2.4Y, emits beta and gamma rays.

Hazard: Radioactive poison.

antimony black. Metallic antimony in the form of a fine powder produced by electrolysis or chemical action on an antimony salt solution. See also antimony trisulfide.

antimony bromide. See antimony tribromide.

antimony, caustic. See antimony trichloride.

antimony chloride. See antimony trichloride.

antimony chloride, basic. See antimony oxychloride.

antimonydichlorotrifluoride. $SbCl_2F_3$. A thick liquid stored in iron drums. Used as catalyst for fluorocarbon manufacture.

antimony fluoride. See antimony trifluoride.

antimony hydride. (stibine). CAS: 7803-52-3. SbH_3.

Properties: Colorless gas, mp −88C, bp −17C.

Derivation: Action of hydrogen chloride on antimony/metal compounds such as Zn_3Sb_2; also released by reduction of antimony compounds in hydrochloric acid solutions with zinc or other reducing metal.

Hazard: Toxic. TLV: 0.1 ppm in air.

antimony iodide. See antimony triiodide.

antimonyl. The radical or group SbO which occurs commonly in formulas of antimony compounds. Thus SbOCl is often named antimonyl chloride, and numerous other antimony compounds are sometimes named in a similar manner.

antimony lactate. CAS: 58164-88-8. $Sb(C_3H_5O_3)_3$.

Properties: Tan-colored mass, soluble in water.

Grade: Technical.

Hazard: Toxic. TLV (as Sb): 0.5 mg/m³ of air.

Use: Mordant in fabric dyeing.

antimony oxide. See antimony trioxide.

antimony oxychloride. (antimony chloride, basic; antimonyl chloride). SbOCl.

Properties: White powder, mp 170C (decomposes), soluble in hydrochloric acid and alkali tartrate solutions; insoluble in alcohol, ether, and water.

Derivation: Interaction of water and antimony chloride.

Hazard: Toxic. TLV (as Sb): 0.5 mg/m³ of air.

Use: Antimony salts, flameproofing textiles.

antimony pentachloride. (antimony perchloride). CAS: 7647-18-9. $SbCl_5$.

Properties: Reddish-yellow, oily liquid, offensive odor, hygroscopic. Solidifies by absorption of moisture. Decomposed by excess water into hydrochloric acid and antimony pentoxide. Soluble in an aqueous solution of tartaric acid in hydrochloric acid and chloroform. Mp 2.8C, d 2.34, bp 92C (30 mm).

Derivation: Action of chlorine on antimony powder.

Hazard: Corrosive, fumes in moist air, reacts strongly with organics. TLV (as Sb): 0.5 mg/m³ of air.

Use: Analysis (testing for alkaloids and cesium), dyeing intermediates, as chlorine carrier in organic chlorinations.

antimony pentafluoride. CAS: 7783-70-2. SbF_5.

Properties: Viscous, hygroscopic liquid, d 2.99 (23C), mp 7C, bp 149.5C, hydrolyzed by water, soluble in potassium fluoride (KF), liquid sulfur dioxide.

Derivation: Antimony pentachloride and anhydrous hydrogen fluoride.

Hazard: Corrosive to skin and tissue. TLV (as Sb): 0.5 mg/m³ of air.

Use: Catalyst and/or source of fluorine in fluorination reactions.

antimony pentasulfide. (antimony red; antimony persulfide; antimony sulfide golden).
CAS: 1315-04-4. Sb_2S_5.
Properties: Orange-yellow powder, odorless, insoluble in water, soluble in concentrated hydrochloric acid with evolution of hydrogen sulfide, soluble in alkali. Decomposes on heating.
Hazard: Flammable, dangerous fire risk near oxidizing materials. TLV (as Sb) 0.5 mg/m³ of air.
Use: Red pigment, rubber accelerator.

antimony pentoxide. (antimonic anhydride; antimonic acid; stibic anhydride).
CAS: 1314-60-9. Sb_2O_5.
Properties: White or yellowish powder, d 3.78, mp 450C, loses oxygen above 300C, slightly soluble in water, soluble in strong bases forming antimonates, insoluble in acids except concentrated hydrochloric acid.
Derivation: Action of concentrated nitric acid on the metal or the trioxide.
Use: Preparation of antimonates and other antimony compounds, flame retardant for textiles.

antimony persulfide. See antimony pentasulfide.

antimony potassium tartrate. (tartar emetic; potassium antimonyl tartrate; tartrated antimony).
$K(SbO)C_4O_6 \cdot 1/2HOH$.
Properties: Transparent, odorless crystals efflorescing on exposure to air, or white powder, sweetish metallic taste, d 2.6, at 100C loses all its water. Soluble in water, glycerol, insoluble in alcohol. Aqueous solution is slightly acid.
Derivation: By heating antimony trioxide with a solution of potassium bitartrate and subsequent crystallization.
Grade: Technical, crystals, powdered, CP, USP.
Hazard: Toxic. TLV (as Sb): 0.5 mg/m³ of air.
Use: Textile and leather mordant, medicine, insecticide.

antimony salt. (deHaens salt).   Mixture of antimony trifluoride and either sodium fluoride or ammonium sulfate.
Properties: White crystals, soluble in water.
Hazard: Toxic. TLV (as Sb): 0.5 mg/m³ of air.
Use: Dyeing and printing textiles.

antimony sodiate. See sodium antimonate.

antimony sulfate. (antimony trisulfate).
CAS: 7446-32-4. $Sb_2(SO_4)_3$.
Properties: White powder or lumps. Deliquescent, decomposes in water, d 3.62 (4C), flammable.
Hazard: A poison. TLV (as Sb): 0.5 mg/m³ of air.
Use: Matches, pyrotechnics.

antimony tribromide. (antimony bromide).
$SbBr_3$.
Properties: Yellow, deliquescent, crystalline mass. Soluble in carbon disulfide, hydrobromic acid, hydrochloric acid, ammonia. Decomposed by water, d 4.148, mp 96.6C, bp 280C.
Hazard: Toxic. TLV (as Sb): 0.5 mg/m³ of air.
Use: Analytical chemistry, mordant, manufacturing antimony salts.

antimony trichloride. (antimonous chloride; antimony chloride; caustic antimony).
CAS: 10025-91-9. $SbCl_3$.
Properties: Colorless, transparent, very hygroscopic, crystalline mass. Fumes slightly in air, soluble in alcohol, acetone, acids, with water forms antimony oxychloride, d 3.14, bp 223.6C, mp 73.2C.
Derivation: Interaction of chlorine and antimony or by dissolving antimony sulfide in hydrochloric acid.
Grade: Technical, CP.
Hazard: Corrosive liquid or solid. Very irritating to eyes, skin. TLV (as Sb): 0.5 mg/m³ of air.
Use: Antimony salts, bronzing iron, mordant, manufacturing lakes, chlorinating agent in organic synthesis, pharmaceuticals, fireproofing textiles, analytical reagent.

antimony trifluoride. (antimony fluoride).
CAS: 7783-56-4. SbF_3.
Properties: White to gray hygroscopic crystals. Mp 292C, d 4.58, soluble in water.
Grade: 99–100%.
Hazard: Strong irritant to eyes and skin.
Use: Porcelain, pottery, dyeing, fluorinating agent.

antimony triiodide. (antimony iodide).
SbI_3.
Properties: Red crystals. Volatile at high temperatures. Soluble in carbon disulfide, hydrochloric acid and solution of potassium iodide, insoluble in alcohol and chloroform, decomposes in water with precipitation of oxyiodide, d 4.768, mp 167C, bp 420C.
Derivation: Action of iodine on antimony.

antimony trioxide. (antimony white; antimony oxide).   CAS: 1309-64-4. Sb_2O_3.
Occurs in nature as valentinite.
Properties: White, odorless, crystalline powder; d 5.67; mp 655C. Insoluble in water, soluble in concentrated hydrochloric and sulfuric acids, strong alkalies. Amphoteric.
Derivation: Burning antimony in air, adding ammonium hydroxide to antimony chloride, directly from low-grade ores.
Grade: Technical, pigment.

Hazard: A suspected human carcinogen.

Use: Flameproofing of textiles, paper, and plastics (chiefly polyvinyl chloride); paint pigments; ceramic opacifier; catalyst; intermediate; staining iron and copper; phosphors; mordant; glass decolorizer.

antimony trisulfate. See antimony sulfate.

antimony trisulfide. (antimony orange; black antimony; antimony needles; antimonous sulfide; antimony sulfide). CAS: 1345-04-6. Sb_2S_3.

Properties: (a) Black crystals; (b) orange-red crystals. Insoluble in water, soluble in concentrated hydrochloric acid and sulfide solutions, d 4.562, mp 546C.

Derivation: (a) Occurs in nature as black crystalline stibnite. (b) As precipitated from solutions of salt of antimony, the trisulfide is an orange-red precipitate which is filtered dried and ground.

Grade: Technical.

Hazard: Explosion risk in contact with oxidizing materials.

Use: Vermilion or yellow pigment, antimony salts, pyrotechnics, matches, percussion caps, camouflage paints (reflects infrared radiation in same way as green vegetation), ruby glass.

antimycin A₁. ($C_{28}H_{40}O_9N_2$). An antibiotic substance said to have strong fungicidal properties.

Properties: Crystals; mp 139–140C; soluble in alcohol, ether, acetone, and chloroform; slightly soluble in benzene, carbon tetrachloride, and petroleum ether; insoluble in water.

Derivation: From Streptomyces.

Use: Active against a large group of fungi, but in general not against bacteria; possible insecticide and miticide.

antineutron. See antiparticle.

antineoplastic. A drug that inhibits the formation of tumors (neoplasms). Many of these are antibiotics used in treatment of cancer.

See, for example, adriamycin, bleomycin.

antioxidant. An organic compound added to rubber, natural fats and oils, food products, gasoline, and lubricating oils to retard oxidation, deterioration, rancidity, and gum formation, respectively. Rubber antioxidants are commonly of an aromatic amine type, such as di-β-naphthyl-p-phenylenediamine and phenyl-β-naphthylamine, 1% or less based on the rubber content of a mixture affords adequate protection. Many antioxidants are substituted phenolic compounds (butylated hydroxyanisole, di-tert-butyl-p-cresol, and propyl gallate). Food antioxidants are effective in very low concentration (not more than 0.01% in animal fats) and not only retard rancidity but protect the nutritional value by minimizing the breakdown of vitamins and essential fatty acids. Sequestering agents, such as citric and phosphoric acids, are frequently employed in antioxidant mixtures to nullify the harmful effect of traces of metallic impurities. Note: Max concentration of food antioxidants approved by FDA is 0.02%.

antiozonant. (antiozidant). A substance used to reverse or prevent the severe oxidizing action of ozone on elastomers both natural and synthetic. Among antiozonant materials used are petroleum waxes, both amorphous and microcrystalline, secondary aromatic amines (such as N, N-diphenyl-p-phenylenediamine), quinoline, and furan derivatives.

See also ozone.

antiparticle. (antimatter). Any of several species of subatomic particles that are identical in mass with electrons, protons, or neutrons, but opposite in electrical charge or (in the case of the neutron) in magnetic moment. Thus, a positron is an electron with a positive charge, an antiproton is a proton with a negative charge, an antineutron has no charge but has a magnetic moment opposite to that of a neutron. A photon has no antiparticle, since it has no particulate properties. When an antiparticle collides with its opposite particle, (e.g., a collision of an electron and a positron) both particles are annihilated and their masses are converted to photons of equivalent energy. The same is true of other subatomic particles (neutrinos, mesons, etc.) which are fantastically short-lived (of the order of billionths of a second). The quantum-mechanical concepts that led to the discovery of antiparticles include that of "strangeness" which is expressed numerically.

antiperspirant. Any substance having a mild astringent action which tends to reduce the size of skin pores and thus restrain the passage of moisture on local body areas. The most commonly used antiperspirant agent is aluminum chlorohydrate. Use of zirconium compounds in antiperspirant sprays has been virtually discontinued because of their suspected carcinogenicity, though they are permissible in creams. Antiperspirants exert a neutralizing action which gives them deodorant properties. The FDA classified them as drugs rather than as cosmetics.

antiproton. See antiparticle.

antipsychotic agent. See psychotropic drug.

antipyretic.
Any of a group of drugs that reduce fever or inflammation, e.g., aconite.

antipyrine. (phenazone; 2,3-dimethyl-1-phenyl-3-pyrazolin-5-one).   CAS: 60-80-0.
$(CH_3)_2(C_6H_5)C_3HN_2O$.
Properties: Colorless crystals or powder, odorless, slightly bitter taste, d 1.19, mp 110–113C, bp 319C. Soluble in water, alcohol, and chloroform; slightly soluble in ether.
Derivation: Condensation of methylphenylhydrazine and ethyl acetoacetate.
Method of purification: Crystallization.
Grade: Technical, NF.
Use: Medicine (analgesic), analytical reagent for nitrous acid, nitric acid, and iodine number.

antipyrine chloral hydrate. See chloral hydrate antipyrine.

antirheumatic. Any of various drugs used in the treatment of rheumatoid arthritis. Among these are certain gold salts, e.g., disodium aurothiomalate and gold sodium thiosulfate. Penicillamine is also reported to be effective. Cortisone is no longer widely used because of deleterious side effects.

antiscorbutic. Tending to prevent scurvy.
See ascorbic acid.

antiseptic. A substance applied to humans or animals that retards or stops the growth of microorganisms without necessarily destroying them, e.g., alcohol, boric acid and borates, certain dyes, as acriflavine, menthol, hydrogen peroxide, hypochlorites, iodine, mercuric chloride, and phenol. Many of these are corrosive and poisonous, and should be used with great caution. Among the newer antiseptics are hexachlorophene, which is toxic, and some quaternary ammonium compounds.
See also disinfectant, sanitizer, fumigant.

antiskinning agent. A liquid antioxidant used in paints and varnishes to inhibit formation of an oxidized film on the exposed surface in cans, pails, or other open containers.

antistatic agent. The marked tendency of thermoplastic polymers to accumulate static charges which result in adherent particles of dust and other foreign matter has required study of possible means of eliminating or reducing this property. The following have been tried. (1) Development of more electrically conductive polymers, e.g., tetracyanoquinodimethane. (2) Incorporation of additives which migrate to the surface of the plastic or fiber and modify its electrical properties. (Examples of these are fatty quaternary ammonium compounds, fatty amines, and phosphate esters.) Other types of antistatic additives are hygroscopic compounds, such as polyethylene glycols and hydrophobic slip additives, that markedly reduce the coefficient of friction of the plastic. (3) Copolymerization of an antistatic resin with the base polymer.

"Antistine" Phosphate.[305]   TM for antazoine phosphate.

antitussive. A medicinal preparation for suppressing coughs, often containing codeine. Chloroform is no longer permitted as an ingredient.

"Antox."[28]   TM for rubber antioxidant, a condensation product of butyraldehyde-aniline. Amber liquid.

"Antozite."[69,119]   TM for a series of antiozonants for use in natural and synthetic rubber.

"Antron."[28]   TM for nylon textile fibers in the form of continuous filament yarns and staple.

ANTU. (α-naphthylthiourea).   CAS: 86-88-4.
Hazard: TLV 0.3 mg/m³.

AOD process. Injection of a mixture of argon and oxygen into molten steel to reduce carbon impurities.

AP. Abbreviation for ammonium perchlorate.

ap-. A prefix denoting formation from or relationship to another compound, e.g., apomorhine.

APAP. Abbreviation for acetyl-p-amino-phenol.
See p-acetylaminophenol.

apatite. A natural calcium phosphate (usually containing fluorine) occurring in the earth's crust as phosphate rock. It is also the chief component of the bony structure of teeth.
Properties: Color variable, d 3.1–3.2, hardness 5, frequently in hexagonal crystals.
Occurrence: Eastern US, California, USSR, Canada, Europe.

Use: Source of phosphorus and phosphoric acid, manufacture of fertilizers, laser crystals.
See also fluoridation.

APC. Abbreviation for ammonium perchlorate.

apholate. Generic name for 2,2,4,4,6,6-hexakis(1-aziridinyl)-2,2,4,4,6,6-hexahydro-1,3,5,2,4,6-triazatriphosphorine .
$C_{12}H_{24}N_9P_3$.
Prevents reproduction of certain insects by inhibiting formation of DNA in eggs.
Use: Insect sterilant.

aphrodine. See yohimbine.

API. Abbreviation for American Petroleum Institute.

API gravity. A scale of measurement adopted by the API. It runs from 0.0 (equivalent to d 1.076) to 100.0 (equivalent to d 0.6112). The API values as used in the petroleum industry decrease as density increases.

"Apiezon."[431] TM for a series of hydrocarbon oils, greases, and waxes that are produced by molecular distillation and characterized by very low vapor pressure and good thermal stability. Used as lubricants and seals in high vacuum equipment and operations, as a stationary phase in gas chromatography. Combustible.

APO. See triethylenephosphoramide.

apocarotenal. Food color supplied in dark purplish-black beadlets. Vitamin A activity 120,000 units/gram. Dispersible in warm water. Approved for food use by FDA.

apparent density. See under density.

Appert, Nicolas. (1752–1841). A French pioneer in the science of food preservation. Though not a chemist, his work on application of heat to food products was in effect a form of home preserving which eventually developed into the canning industry. The idea of destroying bacteria by heat treatment was later applied more exhaustively by Pasteur.

apple acid. See malic acid.

apple oil. See isoamyl valerate.

applied research. The experimental investigation of a specific practical problem for the immediate purpose of creating a new product, improving an older one, or evaluating a proposed ingredient. The experimental program is set up to answer the question "What happens?" rather than "Why does it happen?" Examples are the determination of the value of a new rubber antioxidant, the substitution of one drying oil for another in a paint formulation, or the development of a new synthetic product. While applied research has produced the multitude of new materials that have revolutionized industry, agriculture, and medicine in the last 50 years or so, its achievements have usually been practical outgrowths of prior fundamental research.

aprotic solvent. A type of solvent which neither donates nor accepts protons. Examples: dimethylformamide, benzene, dimethyl sulfoxide.

APS. Abbreviation for appearance potential spectroscopy.
See spectroscopy.

aqua ammonia. See ammonium hydroxide.

"Aquadag."[46] TM for a dispersion of colloidal graphite in water. Used as a lubricant for dies, tools, and molds for metalworking, glassmaking, etc.; conductive coating.

"Aquagel."[236] TM for a gel-forming colloidal bentonite clay used in drilling muds.
See also bentonite.

"Aquapel."[266] TM for a series of alkylketene dimers prepared from long chain fatty acids. Used for paper sizing.

"Aquaprint."[293] TM for a resin-bonded pigment color for printing on textiles. The vehicle, an oil-in-water emulsion contains a water-insoluble binder which adheres to the fibers and anchors the color permanently to the cloth.

aqua regia. (nitrohydrochloric acid; chloronitrous acid; chlorazotic acid).
Properties: Fuming yellow, volatile, suffocating liquid.
Derivation: A mixture of nitric and hydrochloric acids, usually 1 part of nitric acid to 3 or 4 parts of hydrochloric acid.
Grade: Technical.
Hazard: A powerful oxidizer, toxic, corrosive liquid.
Use: Metallurgy, testing metals, dissolving metals (platinum, gold, etc).

"Aquarex."[28] TM for a series of wetting agents for elastomers. They act as stabilizers and mold lubricants.

"Aquarol."[300]   TM for water repellents of the wax-multivalent metal salt type for textiles.

"Aquasorb AR."[329]   TM for a phosphorus pentoxide (P_2O_5)-based desiccant.
Hazard: Powerful oxidizer and caustic.

Ar.   Symbol for argon approved by IUPAC.

ara-A.   (9-β-D-arabinofuranosyladenine; vidarabine).   CAS: 24356-66-9.   A biologically active pharmaceutical product having both antitumor and antiviral properties. It was originally prepared (1959) by chemical synthesis at Stanford Research Institute, and later isolated from a fermentation beer of *Streptomyces antibioticus*. It is commercially available under the TM "Vira-A". [330]

arabic gum.   (acacia gum).   CAS: 9000-01-5.
The dried water-soluble exudate from the stems of Acacia senegal or related species.
Properties: Thin flakes, powder, granules, or angular fragments; color white to yellowish white; almost odorless; mucilaginous consistency. Completely soluble in hot and cold water yielding a viscous solution of mucilage; insoluble in alcohol. The aqueous solution is acid to litmus. Combustible.
Occurrence: Sudan, West Africa, Nigeria.
Composition: A carbohydrate polymer, complex and highly branched. The central core or nucleus is D-galactose and D-glucuronic acid (actually the calcium, magnesium, and potassium salts) to which are attached sugars such as l-arabinose and l-rhamnose.
Grade: USP, FCC (both grades of acacia).
Use: Pharmaceuticals, adhesives, inks, textile printing, cosmetics, thickening agent and colloidal stabilizer in confectionery and food products, binding agent in tablets, emulsifier.

arabinogalactan.   A water-soluble polysaccharide extracted from timber of the western larch trees. It is a complex highly branched polymer of arabinose and galactose.
Properties: Dry, light tan powder, readily soluble in hot and cold water, both powder and solutions relatively stable. Combustible.
Use: Dispersing and emulsifying agent, lithography.

arabinose.   (pectin sugar; gum sugar).
$C_5H_{10}O_5$.   Both the d- and l-enantiomers occur naturally. l-Arabinose is common in vegetable gums, especially arabic.
Properties: White crystals, soluble in water and

$$
\begin{array}{c}
CHO \\
| \\
HCOH \\
| \\
HOCH \\
| \\
HOCH \\
| \\
CH_2OH
\end{array}
$$

glycerol, insoluble in alcohol and ether. Mp 158.5C, d 1.585 (20/4C). Combustible.
Use: Culture medium.

Ara-C.   Abbreviation for cytosine arabinoside.

arachidic acid.   (eicosanoic acid).
$CH_3(CH_2)_{18}COOH$.   A widely distributed but minor component of the fats of peanut oils and related plant species.
Properties: Shining, white, crystalline leaflets; soluble in ether; slightly soluble in alcohol; insoluble in water. Mp 75.4C, d 0.2840 (100/4C), bp 328C (decomposes), refr index 1.4250. Combustible.
Derivation: From peanut oil.
Grade: Technical, 99%.
Use: Organic synthesis, lubricating greases, waxes and plastics, source of arachidyl alcohol, biochemical research.

arachidonic acid.   (5,8,11,14-eicosatetraenoic acid).   CAS: 506-32-1.
$CH_3(CH_2)_4(CH:CHCH_2)_4(CH_2)_2COOH$.   A C_{20} unsaturated fatty acid. Combustible. An essential fatty acid.
Sources: Liver, brain, glandular organs, also made synthetically.
Use: Biochemical research, source of prostaglandins and other pharmacologically active compounds.
See also eicosanoid.

arachidyl alcohol.   (1-eicosanol).
$CH_3(CH_2)_{18}CH_2OH$.   A long-chain saturated fatty alcohol much like stearyl alcohol.
Properties: White, wax-like solid; mp 66.5C; bp 369C (220C at 3 mm); refr index 1.455; soluble in hot benzene. Combustible.
Derivation: Ziegler synthesis (trialkylaluminum process).
Grade: Technical, 99%.
Use: Lubricants, rubber, plastics, textiles, research.

arachin.   (arachine).   A protein from peanuts, a globulin containing arginine, histidine, lysine, cystine; yellow-green syrup, soluble in water and alcohol, insoluble in ether. Combustible.

aragonite. A form of calcium carbonate appearing in pearls.
See nacre.

aragonite needles. Slender crystals of the mineral aragonite that constitute most carbonate muds in the modern ocean. Some of the needles form by direct precipitation from sea water, and some by the collapse of the skeletons of organisms.

aralkonium chloride. $C_{21}H_{36}Cl_3N$.
Properties: Water-soluble solid, bitter taste.
Use: As sanitizer and deodorant.

aralkyl. See arylalkyl.

aramid. Generic name for a distinctive class of highly aromatic polyamide fibers which are characterized by their flame-retardant properties. Some types are also suitable for protective clothing, dust-filter bags, tire cord, and bullet-resistant structures. They are derived from p-phenylenediamine and terephthaloyl chloride.
See also "Nomex," "Kevlar," polyamide.

"Aramite."[248] TM for 2-(p-tert-butylphenoxy)-isopropyl-2-chloroethyl sulfite.
$(CH_3)_3CC_6H_4OCH_2CH(CH_3)OSOOC_2H_4Cl$.
Properties: Clear, light-colored oil; d 1.148–1.152 (20C); bp 175C (0.1 mm); very soluble in common organic solvents; insoluble in water; noncorrosive.
Grade: Technical (90% min), wettable powder, emulsifiable concentrate restricted to postharvest application on fruit trees.
Hazard: A carcinogen. Irritant to eyes and skin, toxic by ingestion.
Use: Antimicrobial agent, miticide.

"Aranox."[248] TM for p-(p-tolysulfonylamido)-diphenylamine.
$CH_3C_6H_4SO_2NHC_6H_4NHC_6H_5$.
Properties: Gray powder; d 1.32; mp 135C (min); soluble in acetone, benzene, and ethylene dichloride; insoluble in gasoline and cold water; slightly soluble in hot water or hot alkaline solutions.
Use: Antioxidant for light-colored rubber products.

"Arasan."[28] TM for seed disinfectants based on thiram.

"Arazate."[248] TM for zinc dibenzyl dithiocarbamate.

arbutin. (ursin). $C_{12}H_{16}O_7$. Available commercially in both natural and synthetic forms. Pure synthetic arbutin is hydroquinone-β-D-glucopyranoside.

OH

O—glucose

Properties (pure synthetic): White powder, mp 199–200C, soluble in water and alcohol, stable in storage.
Derivation: A glucoside found in the leaves of the cranberry, blueberry, and manzanita shrubs, and in the roots, trunks, and leaves of most pear species. Pure arbutin can be prepared synthetically from acetobromoglucose and hydroquinone in presence of alkali.
Use: Oxidation inhibitor, polymerization inhibitor, color stabilizer in photography, intermediate.

archaometry. Application of chemical and physical analytical methods to archaeology. Among those used are microanalytical methods, spectroscopic analysis, x-rays, and other types of nondestructive tests. For age determination, C-14 measurement (chemical dating) is one of the most valuable techniques.

arene. See aromatic.

Arens-van Dorp synthesis. The preparation of alkoxyethynyl alcohols from ketones and ethoxyacetylene. In the Isler modification, β-chlorovinyl ether is reacted with lithium amide to give lithium ethoxyacetylene which is then condensed with the ketone. This avoids the tedious preparation of ethoxyacetylene.

argentite. (silver glance). Ag_2S. Lead-gray to black or blackish-gray mineral. A natural silver sulfide. Contains 87.1% silver. Differs from other soft black minerals in cutting like wax. Soluble in nitric acid, d 7.2–7.36, Mohs hardness 2–2.5.
Occurrence: Nevada, Colorado, Montana, Mexico, Chile, Canada.
Use: An important ore of silver.

argentum. The Latin name for silver, hence the symbol Ag in chemical nomenclature.

arginase. An enzyme producing ornithine and urea by splitting arginine. It is found in liver.
Use: Biochemical research.

arginine. (guanidine aminovaleric acid; amino-4-guanidovaleric acid).

$$HN \diagdown \\ C-NCH_2CH_2CH_2CHCO_2H \\ H_2N \diagup \quad\quad\quad \underset{NH_2}{|}$$

An essential amino acid for rats, occurring naturally in the l(+) form. Available as the glutamate and hydrochloride.

Properties: Prisms from water containing two molecules of HOH, anhydrous plates from alcohol solution, dehydrates at 105C, decomposes at 244C, sparingly soluble in alcohol, insoluble in ether.

Derivation: Widely found in animal and plant proteins. It is precipitated as the flavianate from gelatin hydrolyzate in industry.

Use: Biochemical research, medicine, pharmaceuticals, dietary supplement.

argon. (Ar). A nonmetallic element of atomic number 18, in noble gas group of the Periodic System. Aw 39.948, in air to 0.94% by volume.

Properties: Colorless, odorless, tasteless, monatomic gas, it is not known to combine chemically with any element, but forms a stable clathrate with β-hydroquinone, fp $-189.3C$, bp $-185.8C$, d 1.38 (air = 1), sp vol 9.7 cu ft/lb (21.1C) (1 atm). Slightly soluble in water. Noncombustible; an asphyxiant gas.

Derivation: (a) By fractional distillation of liquid air. (b) By the treatment of atmospheric nitrogen with metals such as magnesium and calcium to form nitrides. (c) Recovery from natural gas oxidation bottoms-steam in ammonia plant. (d) Originally formed by radioactive decay of K-40.

Methods of purification: (a) Highly purified argon is obtained by passing the gas through a bed of titanium at 850C. (b) Synthetic zeolite molecular sieves separate oxygen from argon to give high purity gas.

Grade: Technical, highest purity (99.995%).

Use: Inert gas shield in arc welding, furnace brazing, plasma jet torches (with hydrogen), electric and specialized light bulbs (neon, fluorescent, sodium vapor, etc.), titanium and zirconium refining, flushing molten metals (steel) to remove dissolved gases, in Geiger-counting tubes, lasers, inert gas or atmosphere in miscible applications, decarburization of stainless steel (AOD process).

"Argyrol." TM for an organic compound of silver and a protein, used in medicine for its specific antiseptic and bacteriostatic action.

"Aridye."[293] TM for a product and process for printing colors on textiles using permanent and insoluble pigments suspended in an organic vehicle into which water is emulsified to give printing consistency. The vehicle contains a water-insoluble binder which adheres to the clot and anchors the color permanently to the fibers.

"Ariperm."[300] TM for a gas-fading inhibitor for application to dyed acetate fabrics. Protects the dyes from fading due to acid gases in the atmosphere.

"Arizole."[252] TM for a group of terpene products including anethole and sulfate pine oil. Used in perfumes, soaps, etc.

"Arkolene."[300] TM for textile wetting agents of the alkylarylsulfonate type, sodium or ammonium salts.

"Arkolube."[300] TM for textile lubricating and softening agents based on silicones, polyethylenes, and waxes.

arkose. A rock consisting primarily of sand-sized particles of feldspar. Most arkose accumulates close to the source area of the feldspar because feldspar weathers quickly to clay and seldom travels far.

"Arkotan."[300] TM for a synthetic tanning agent for leather. Ammonium salt of a naphthalenesulfonic acid complex.

"Arlacel."[89] TM for each of a series of nonionic emulsifiers for use in cosmetics and pharmaceuticals. They are fatty acid partial esters of polyols or polyol anhydrides.

"Arlex."[89] TM for noncrystallizing industrial humectant soluble containing 83% solids consisting of sorbitol and related polyhydric materials.

Use: For flexibilizing and moisture-conditioning in industrial applications including tobacco, glue compositions, cellulose products, etc.

"Armalon."[28] TM for TFE-fluorocarbon fiber felt and also for TFE-fluorocarbon resin-coated glass fabrics, tapes, and laminates.

"Armofos."[1] TM for sodium tripolyphosphate, anhydrous. $Na_5P_3O_{10}$.

Use: Sequestering agent for iron, calcium, and manesium ions; soap builder; detergent mixtures; deflocculator in drilling muds, paper, ceramics and textiles.

Armstrong's acid. (naphthalene-1,5-disulfonic acid). $C_{10}H_6(SO_3H)_2$.

Properties: White, crystalline solid; soluble in water.

Derivation: Sulfonation of naphthalene with fuming sulfuric acid at low temperature followed by separation from the 1,6-isomer.
Use: Dye intermediate.

Arndt-Eistert synthesis. Procedure for converting an acid to its next higher homolog.

"Arnel."[352] TM for an acetate fiber made from cellulose triacetate. It has a higher melting point, and is less soluble than cellulose acetate.
See acetate fiber, cellulose triacetate.

"Arnox"[245]. TM for a family of 1-component liquid and solid epoxy resins designed for compression and transfer molding, injection molding, filament winding and pultrusion.

aromatic. (arene). A major group of unsaturated cyclic hydrocarbons containing one or more rings, these are typified by benzene which has a 6-carbon ring containing three double bonds. The vast number of compounds of this important group derived chiefly from petroleum and coal tar are rather highly reactive and chemically versatile. The name is due to the strong and not unpleasant odor characteristic of most substances of this nature. Certain 5-membered cyclic compounds such as the furan group (heterocyclic) are analogous to aromatic compounds. *Note*: The term "aromatic" is often used in the perfume and fragrance industries to describe essential oils which are not aromatic in the chemical sense.

aromaticity. A stable electron shell configuration in organic molecules, especially those related to benzene.
See resonance, orbital theory.

aromatization. See hydroforming.

"Aromin."[51] TM for a highly aromatic solvent widely used as a carrier for chemical pesticides.

Arrhenius, Svante. (1859–1927) A native of Sweden, he won the Nobel prize in chemistry in 1903. He is best known for his fundamental investigations on electrolytic dissociation of compounds in water and other solvents, and for his basic equation stating the increase in the rate of a chemical reaction with rise in temperature:

$$\frac{d \ln k}{dT} = \frac{A}{RT^2}$$

in which, k is the specific reaction velocity, T is the absolute temperature, A is a constant usually referred to as the energy of activation of the reaction, and R is the gas law constant.

arsacetin. (sodium acetylarsanilate; sodium p-acetyl aminophenylarsonate). $CH_3CONHC_6H_4AsO(OH)ONa$.
Properties: White, crystalline powder; odorless; tasteless; free of arsenous or arsenic acid; solutions will admit of thorough sterilization. Soluble in cold water, but more so in warm water.
Use: Medicine (antisyphilitic).

arsanilic acid. (atoxylic acid; p-aminobenzene-arsonic acid; p-aminophenylarsonic acid). $C_6H_4 \cdot C_6H_8AsNO_3$.
Properties: White, crystalline powder; practically odorless; soluble in hot water; slightly soluble in cold water, alcohol, and acetic acid; insoluble in acetone, benzene, chloroform, and ether. Mp 232C.
Derivation: By condensing aniline with arsenic acid removing the excess of aniline by steam distillation in alkaline solution and setting the acid free by hydrochloric acid.
Hazard: Yields flammable vapors on heating above melting point. A poison.
Use: Arsanilates, manufacture of arsenical medicinal compounds such as arsphenamine, etc., veterinary medicine, grasshopper bait.

arsenic. As. CAS: 7440-38-2. A nonmetallic element of atomic number 33, group VA of Periodic Table, aw 74.9216, valence=2,3,5; no stable isotopes.
Properties: Silver-gray, brittle, crystalline solid that darkens in moist air. Allotropic forms: black, amorphous solid (β-arsenic), yellow, crystalline solid, d 5.72 (commercial product ranges from 5.6 to 5.9), mp 814C (36 atm), sublimes at 613C (1 atm), Mohs hardness 3.5, insoluble in water, caustic and nonoxidizing acids. Attacked by hydrochloric acid in presence of oxidant. Reacts with nitric acid. Low thermal conductivity; a semiconductor.
Derivation: Flue dust of copper and lead smelters from which it is obtained as white arsenic (arsenic trioxide) in varying degrees of purity. This is reduced with charcoal. The commercial grade is not made in US.
Grade: Technical, crude (90–95%), refined (99%), semiconductor grade 99.999%, single crystals.
Hazard: Carcinogen and mutagen. TLV OSHA standard for employee exposure is 10 μg/m3 of air. Respirators required for worker exposure to atmospheres of over 500 μg/m3. ACGIH TLV is 200 μg/m3 (arsenic and soluble compounds).
Uses (metallic form): Alloying additive for metals, especially lead and copper as shot, battery grids, cable sheaths, boiler tubes. High-purity (semiconductor) grade: used to make gallium arsenide

for dipoles and other electronic devices, doping agent in germanium and silicon solid state products, special solders, medicine.
See also arsenic trioxide.

arsenic acid. (orthoarsenic acid).
CAS: 7778-39-4. $H_3AsO_4 \cdot 1/2H_2O$.
Arsenic pentoxide is also sometimes called arsenic acid.
Properties: White, translucent crystals; soluble in water, alcohol, alkali, glycerol. D: 2–2.5, mp 35.5C, bp loses water at 160C.
Derivation: By digestion of arsenic with nitric acid.
Grade: Pure, technical, CP.
Use: Manufacture of arsenates, glass making, wood treating process, defoliant (regulated), desiccant for cotton, soil sterilant.

arsenical Babbitt.   See Babbitt metal.

arsenical nickel.   See niccolite.

arsenic anhydride.   See arsenic pentoxide.

arsenic, black.   (β-arsenic).   See arsenic.

arsenic bromide.   Legal label name for arsenic tribromide.

arsenic chloride.   See arsenic trichloride.

arsenic disulfide. (arsenic monosulfide; ruby arsenic; red arsenic glass; red arsenic sulfide; red arsenic).   CAS: 56320-22-0. As_2S_2 or AsS.   Occurs as mineral, realgar.
Properties: Orange-red powder, soluble in acids and alkalies, insoluble in water, d 3.4 to 2.6, mp 307C.
Deviation: By roasting arsenopyrite and iron pyrites and sublimation.
Grade: Technical.
Use: Leather industry, depilatory agent, paint pigment, shot manufacture, pyrotechnics, rodenticide, taxidermy.

arsenic hydride.   See arsine.

arsenic pentafluoride. AsF_5.   A gas, bp −52.8C, fp −79C. Readily hydrolyzed, soluble in alcohol and benzene.
Use: Doping agent in electroconductive polymers.

arsenic pentasulfide. As_2S_5.
Properties: Yellow or orange powder. Soluble in nitric acid and alkalies, insoluble in water, decomposes to sulfur and the trisulfide when heated.
Derivation: By precipitation from arsenic acid in a hydrochloric acid solution with hydrogen sulfide. It is filtered, then dried.

Grade: Technical.
Use: Paint pigments, light filters, other arsenic compounds.

arsenic pentoxide. (arsenic oxide; arsenic anhydride; arsenic acid).   CAS: 1303-28-2.
As_2O_5.
Properties: White, amorphous solid; deliquescent; forms arsenic acid in water. Soluble in water, alcohol; d 4.086, mp 315C.
Derivation: By action of oxidizing agent, such as nitric acid, on arsenious oxide.
Use: Arsenates, insecticides, dyeing and printing, weed killer, colored glass, metal adhesives.

arsenic sesquioxide.   See arsenic trioxide.

arsenic sulfide.   Legal label name for arsenic disulfide.

arsenic thioarsenate. $As(AsS_4)$.
Properties: Dry, free-flowing yellow powder, stable, high-melting. Insoluble in water and organic solvents, but soluble in aqueous caustics.
Use: Scavenger for certain oxidation catalysts and thermal protectant for metal-bonded adhesives and coating resins.

arsenic tribromide. (arsenic bromide; arsenious bromide; arsenous bromide).
CAS: 7784-33-0. $AsBr_3$.
Properties: Yellowish-white, hygroscopic crystals; d 3.54 (25C); mp 33C, bp 221C. Decomposed by water.
Derivation: Direct union of arsenic and bromine.
Use: Analytical chemistry, medicine.

arsenic trichloride. (arsenic chloride; arsenious chloride; arsenous chloride; caustic arsenic chloride; fuming liquid arsenic).
CAS: 7784-34-1. $AsCl_3$.
Properties: Colorless or pale yellow oil. Soluble in concentrated hydrochloric acid and most organic solvents, decomposed by water. Fumes in moist air. Bp 130.5C, fp −18C, d 2.163 (14/4C). Noncombustible.
Derivation: (a) By action of chlorine on arsenic; (b) by distillation of arsenic trioxide with concentrated hydrochloric acid.
Grade: Technical.
Hazard. Strong irritant to eyes and skin.
Use: Intermediate for organic arsenicals (pharmaceuticals insecticides), ceramics.   See arsenic.

arsenic trifluoride (arsenious fluoride).
CAS: 7784-35-2. AsF_3.
Properties: Mobile liquid which fumes in air.
Hazards: Extremely toxic.

Use: Fluorinating reagent, catalyst, ion implantation source and dopant.

arsenic trioxide. (crude arsenic; white arsenic; arsenious acid; arsenious oxide; arsenous anhydride). CAS: 1327-53-3. As_2O_3.
Properties: White, odorless, tasteless powder, slightly soluble in water, soluble in acids and alkalies, soluble in glycerol, d 3.865. Sublimes on heating.
Derivation: Smelting of copper and lead concentrates. Flue dust to which pyrite or galena concentrations are added, yields As_2O_3 vapor. Condensation gives product of varying purity called crude arsenic (90–95% pure). A higher-purity oxide called white arsenic is obtained by resubliming the crude As_2O_3 (99+% pure).
Hazard: A human carcinogen, a poison.
Use: Pigments, ceramic enamels, aniline colors, decolorizing agent in glass, insecticide, rodenticide, herbicide, sheep and cattle dip, hide preservative, wood preservative, preparation of other arsenic compounds.

arsenic trisulfide. (arsenious sulfide; arsenic sulfide[yellow]; arsenous sulfide; arsenic tersulfide). CAS: 1303-33-9. As_2S_3.
Properties: Yellow crystals or powder, changes to a red form at 170C, d 3.43, mp 300C, insoluble in water and hydrochloric acid, dissolves in alkaline sulfide solutions and nitric acid.
Derivation: Occurs in nature as the mineral orpiment. May be precipitated from arsenious acid solution by the action of hydrogen sulfide.
Grade: Technical, pigment, single crystals.
Use: Pigment, reducing agent, pyrotechnics, glass used for infrared lenses, semiconductors, hair removal from hides.

arsine. (arsenic hydride). CAS: 7784-42-1. AsH_3.
Properties: Colorless gas, fp −113.5C, bp −62C, decomposes 230C, soluble in water, slightly soluble in alcohol, alkalies.
Derivation: Reaction of aluminum arsenide with water or hydrochloric acid, electrochemical reduction of arsenic compounds in acid solutions.
Grade: Technical, 99% pure or in mixture with other gases.
Hazard: Poison by inhalation. TLV 0.05 ppm.
Use: Organic synthesis, military poison, doping agent for solid state electronic components.

arsphenamine. A specific for syphilis originally developed by Ehrlich, but no longer in use. It was a derivative of arsenic and benzene.
See Ehrlich.

"Artic."[28] TM for refrigeration grade of methyl chloride.

artificial cinnabar. See mercuric sulfide, red.

artificial snow. A copolymer of butyl and isobutyl methacrylate, often dispersed from an aerosol bomb or other atomizing device, used in decorative window displays, etc. Man-made snow is crystallized water vapor made by mechanical means.

"Arubren CP."[470] TM for a highly chlorinated aliphatic hydrocarbon compound used in rubber compounds to decrease flammability of vulcanizates.

arylalkyl. A compound containing both aliphatic and aromatic structures, e.g., alkyl benzenesulfonate. Also called aralkyl.

aryl. A compound whose molecules have the ring structure characteristic of benzene, naphthalene, phenanthrene, anthracene, etc, i.e., either the six-carbon ring of benzene or the condensed six-carbon rings of the other aromatic derivatives. For example, an aryl group may be phenyl C_6H_5 or naphthyl $C_{10}H_9$. Such groups are often represented in formulas by "R."
See also alkyl.

As. Symbol for arsenic.

as-. Abbreviation for asymmetrical, same as uns-.

ASA. Abbreviation for acrylic ester-modified styrene acrylonitrile terpolymer.
See also "Luran sulfur."

asarone. See 2,4,5-trimethoxy-1-propenylbenzene.

asbestine. A soft fibrous magnesium silicate. Used as a filler in paper, rubber, and plastics.

asbestos. CAS: 1332-21-4. A group of impure magnesium silicate minerals which occur in fibrous form. Colors: white, gray, green, brown. D 2.5. Noncombustible.
(1) Serpentine asbestos is the mineral chrysotile, a magnesium silicate. The fibers are strong and flexible so that spinning is possible with the longer fibers. A microcrystalline form, TM "Avibest," has been developed.
(2) Amphibole asbestos includes various silicates of magnesium, iron, calcium and sodium. The fibers are generally brittle and cannot be spun, but are more resistant to chemicals and to heat than serpentine asbestos.
(3) Amosite.
(4) Crocidolite.

Occurrence: Vermont, Arizona, California, North Carolina, Africa, Italy, Yukon, Quebec, Mexico,
Hazard: A carcinogen. Highly toxic by inhalation of dust particles. TLV (amosite) 0.5 fibers/cc more than 5 microns long; (chrysotile) 2 fibers/cc more than 5 microns long; (crocidolite) 0.2 fibers/cc more than 5 microns long; (other forms) 2 fibers/cc more than 5 microns long.
Use: Fireproof fabrics, brake lining, gaskets, roofing compositions, electrical and heat insulations, paint filler, chemical filters, reinforcing agent in rubber and plastics, component of paper dryer felts, diaphragm cells, cement reinforcement.
Note: A promising substitute for asbestos for cement reinforcement is glass fiber made from slate and limestone.

ascaridole.   1,4-peroxido-p-menthene-2).
$C_{10}H_{16}O_2$.
Properties: A liquid, naturally occurring peroxide, bp 84C (5 mm), d 1.011 (13/15C), refr index 1.4743 (20C).
Derivation: By vacuum distillation of chenopodium oil.
Hazard: Strong oxidizing agent, explodes on heating to 130C or in contact with organic acids.
Use: Initiator in polymerization, medicine.

ascorbic acid.   (l-ascorbic acid; vitamin C).
CAS: 5081-7. $C_6H_8O_6$.

$$\underset{O}{C}-\underset{OH}{C}=\underset{OH}{C}-\underset{H}{\overset{O}{\underset{|}{C}}}-\underset{OH}{\overset{H}{C}}-CH_2OH$$

A dietary factor which must be present in the diet of man to prevent scurvy. It cures scurvy and increases resistance to infection. Ascorbic acid presumably acts as an oxidation-reduction catalyst in the cell. It is readily oxidized; citrus juices should not be exposed to air for more than a few minutes before use.
Properties: White crystals (plates or needles), mp 192C, soluble in water, slightly soluble in alcohol, insoluble in ether, chloroform, benzene, petroleum ether, oils and fats, stable to air when dry.
Sources: Food source: acerola (West Indian cherry); citrus fruits; tomatoes; potatoes; green, leafy vegetables. Commercial sources: Synthetic product made by fermentation of sorbitol.
Units: One international unit is equivalent to 0.05 milligram of l-ascorbic acid. Grade: USP, FCC.
Use: Nutrition, color fixing, flavoring and preservative in meats and other foods, oxidant in bread doughs, abscission of citrus fruit in harvesting, reducing agent in analytical chemistry. The iron, calcium and sodium salts are available for biochemical research.

ascorbic acid oxidase.   An enzyme found in plant tissue which acts upon ascorbic acid in the presence of oxygen to produce dehydroascorbic acid.
Use: Biochemical research.

ascorbyl palmitate. $C_{22}H_{38}O_7$.   A white or yellowish-white powder having a citrus-like odor. Mp 116–117C; soluble in alcohol, animal and vegetable oils; slightly soluble in water.
Derivation: Palmitic and l-ascorbic acids.
Grade: FCC.
Use: Antioxidant for fats and oils, source of vitamin C, stabilizer, emulsifier.

-ase.   A suffix characterizing the names of many enzymes, e.g., diastase, cellulase, cholinesterase, etc. However, the names of some enzymes end in -in (i.e., pepsin, rennin, papain).

ash.   (1) In analytical chemistry, the residue remaining after complete combustion of a material. It consists of mineral matter (silica, alumina, iron oxide, etc.) the amount often being a specification requirement. (2) The end product of large-scale coal combustion as in power plants; now said to be the sixth most plentiful mineral in the US. It consists principally of fly ash, bottom ash, and boiler ash. Some of its values are recoverable and there are a number of industrial uses of fly ash, e.g., in cement products and road fill.
See also fly ash.

askarel.   A generic descriptive name for synthetic electrical insulating (dielectric) material which, when decomposed by the electric arc, evolves only nonexplosive gases or gaseous mixtures, i.e., chlorinated aromatic derivatives, particularly pentachlorodiphenyl and trichlorobenzene, but also including pentachlorodiphenyl oxide, pentachlorophenylbenzoate, hexachlorodiphenylmethane, pentachlorodiphenyl ketone, and pentachloroethylbenzene.   Nonflammable.
Use: Insulating medium in transformers, dielectric fluid.
See also dielectric, transformer oil.

ASM.   Abbreviation for American Society for Metals.

asparagic acid.   See aspartic acid.

l-asparaginase.   (l-asnase; colaspase; elspar).
CAS: 9015-68-3.   An enzyme used in the treatment of certain types of leukemia. Produced by biochemical activity of certain bacteria, yeasts, and fungi. Yields are in excess of 3500 units/g of source.

asparagin. (α-aminosuccinamic acid; β-aspara-gine; althein; aspartamic acid; aspartamide). $NH_2COCH_2CH(NH_2)COOH$. The β amide of aspartic acid, a nonessential amino acid, existing in the d(+)- and l(-)-isomeric forms as well as the dl-racemic mixture. l(-)-asparagine is the most common form.
Properties: l(-)-Asparagine monohydrate: White crystals, mp 234–235C, acid to litmus, nearly insoluble in ethanol, methanol, ether and benzene, soluble in acids and alkalies.
Derivation: Widely distributed in plants and animals both free and combined with proteins.
Use: Biochemical research, preparation of culture media, medicine.

asparaginic acid. See aspartic acid.

"Aspartame"[70]. $C_{14}H_{18}N_2O_5$. TM for a synthetic nonnutritive sweetener approved by FDA for tabletop use and as a packaged food additive. The US, Canada, and South Africa permit its use in carbonated beverages. A combination of aspartic acid and l-phenylalanine, it is said to be 200 times sweeter than sugar.
See also sweetener, non-nutritive.

aspartamic acid. See asparagine.

aspartamide. See asparagine.

aspartic acid. (asparaginic acid; asparagic acid; aminosuccinic acid). $COOHCH_2CH(NH_2)COOH$. A naturally occurring nonessential amino acid. The common form is l(+)-aspartic acid.

$$HO_2CCH_2CHCO_2H$$
$$|$$
$$NH_2$$

Properties: Colorless crystals, soluble in water, insoluble in alcohol and ether, optically active. dl-aspartic acid: mp 278–280C (decomposes), d 1.663 (12/12C). l(+)-aspartic acid: mp 251C. d(-)-aspartic acid: mp 269–271C (decomposes), d 1.6613.
Source: Young sugar cane, sugar beet molasses.
Derivation: Hydrolysis of asparagine, reaction of ammonia with diethyl fumarate.
Use: Biological and clinical studies, preparation of culture media, organic intermediate, ingredient of aspartame, detergents, fungicides, germicides, metal complexation. Available commercially as d(-)-, l(+)-, and dl-aspartic acid.

aspartocin. USAN for antibiotic produced by Streptomyces griseus.

aspergillic acid. (2-hydroxy-3-isobutyl-6-(1-methylpropyl)pyrazine-1-oxide). $C_{12}H_{20}N_2O_2$.

An antibiotic from strains of Aspergillus flavus.
Properties: Yellow crystals, mp 97C, insoluble in cold water, soluble in common organic solvents and dilute acids. Hydrochloride melts at 178C and is soluble in water.
Use: Antibiotic.

asphalt. (petroleum asphalt; Trinidad pitch; mineral pitch). CAS: 8052-42-4. A dark-brown to black cementitious material, solid or semisolid in consistency, in which the predominating constituents are bitumens which occur in nature as such or are obtained as residua in petroleum refining (ASTM). It is a mixture of paraffinic and aromatic hydrocarbons and heterocyclic compounds containing sulfur, nitrogen, and oxygen.
Properties: Black solid or viscous liquid, d approximately 1.0, soluble in carbon disulfide. Flash p 450F (132C), autoign temperature 900F (482C), solid, softens to viscous liquid at approximately 93C, penetration value (paving) 40–300 (roofing) 10–40. Good electrical resistivity. Combustible.
Occurrence: California, Trinidad, Venezuela, Cuba, Canada (Athabasca tar sands).
Hazard: Toxic by inhalation of fume. TLV (fume) 5 mg/m³.
Use: Paving and road-coating, roofing, sealing and joint filling, special paints, adhesive in electrical laminates and hot-melt compositions, diluent in low-grade rubber products, fluid loss control in hydraulic fracturing of oil wells, medium for radioactive waste disposal, pipeline and underground cable coating, rust-preventive hot-dip coatings, base for synthetic turf, water-retaining barrier for sandy soils, supporter of rapid bacterial growth in converting petroleum components to protein.
See also bacteria, protein, oil sands.

asphalt (blown). (mineral rubber; oxidized asphalt; hard hydrocarbon). Black, friable solid obtained by blowing air at high temperature through petroleum-derived asphalt with subsequent cooling. Penetration value 10–40, softening point 85 to 121C. Combustible.
Use: Primarily roofing, as diluent in low-grade rubber products and as thickener in oil-based drilling fluids.

asphalt (cut-back). A liquid petroleum product produced by fluxing an asphaltic base with suitable distillates. (ASTM).
Properties: Flash p 50F (10C) (OC).
Grade: Solution of residue from distillation in carbon tetrachloride, 99.5%.
Hazard: Flammable, dangerous fire hazard.
Use: Road surfaces.

asphaltene. A component of the bitumen in petroleums, petroleum products, malthas, asphalt cements and solid native bitumens, soluble in carbon disulfide but insoluble in paraffin naphthas. (ASTM). It is comprised of polynuclear hydrocarbons of molecular weight up to 20,000 joined by alkyl chains.

asphalt, liquid. See residual oil; asphalt (cutback).

asphalt (oxidized). See asphalt (blown).

asphalt paint. Asphaltic base in a volatile solvent with or without drying oils, resins, fillers, and pigments. Ground asbestos was frequently used as a component of heavy asphaltic paints for roofing and waterproofing purposes.
Hazard: Flammable, dangerous fire risk.

asphyxiant gas. A gas which has little or no positive toxic effect but which can bring about unconsciousness and death by replacing air and thus depriving an organism of oxygen. Among the so-called asphyxiant gases are carbon dioxide, nitrogen, helium, methane, and other hydrocarbon gases.

aspidospermine. $C_{22}H_{30}O_2N_2$.
Properties: White to brownish-yellow crystalline alkaloid. Mp 208C, bp 220C (1 to 2 mm), sublimes at 180C under reduced pressure. Soluble in fats and fixed oils, soluble in absolute alcohol, ether. Its sulfate and hydrochloride are soluble in water.
Use: Medicine (respiratory stimulant).

aspirin. (acetylsalicylic acid; o-acetoxybenzoic acid). CAS: 50-78-2. $CH_3COOC_6H_4COOH$.

Properties: White crystals or white, crystalline powder. Odorless, slightly bitter taste. Stable in dry air, slowly hydrolyzes in moist air to salicylic and acetic acids. Soluble in water, alcohol, chloroform, and ether; less soluble in absolute ether. Dissolves with decomposition in solutions of alkali hydroxides and carbonates. Mp 132–136C, bp 140C (decomposes).
Derivation: Action of acetic anhydride on salicylic acid.
Method of purification: Crystallization.
Grade: Technical, USP.

Hazard: An allergen, may cause local bleeding especially of the gums, 10-g dose may be fatal. May cause excessive biosynthesis of prostaglandins. Dust dispersed in air is serious explosion risk.
Use: Medicine (analgesic, anti-inflammatory, antipyretic).

"Aspon."[1] TM for a concentration of tetra-n-propyl dithionopyrophosphate a liquid insecticide assay. Determination of the content of a specific component of a mixture with no evaluation of other components. Such determinations are made on ores of various metals (especially precious metals), on pharmaceutical products to validate the amount of drug present in a given unit, and on organisms (bacteria) to determine their reactions to an antibiotic or insecticide. The latter procedure is called bioassay. Ores are assayed by heat fractionation; organic materials by solvent extraction and chemical separation.

assistant. A term loosely used in the textile industry for any chemical compound that aids in a processing step, e.g., scouring, dyeing, bleaching, finishing, etc.
See also auxiliary, dyeing assistant.

association. A reversible chemical combination due to any of the weaker classes of chemical bonding forces. Thus, the combination of two or more molecules due to hydrogen bonding as in the union of water molecules with one another or of acetic acid molecules with water molecules is called association; also, combination of water or solvent molecules with molecules of solute or with ions, i.e., hydrate formation or solvation. Formation of complex ions or chelates as copper ion with ammonia or copper ion with 8-hydroxyquinoline are other examples. Aqueous solutions of soaps or synthetic detergents are often called association colloids.

A-stage resin. (resole; one-step resin). An alkaline catalyzed thermosetting phenol-formaldehyde type resin consisting primarily of partially condensed phenol alcohols. At this stage, the product is fully soluble in one or more common solvents (alcohols, ketones) and is fusible at less than 150C. On further heating and without use of a catalyst or additive, the resin is eventually converted to the insoluble, infusible, cross-linked form (C-stage). The A-stage resin is a constituent of most commercial laminating varnishes, and is also used in special molding powders.
See B-stage resin, C-stage resin, novolak, phenolformaldehyde resin.

astatine. At.   Nonmetallic element of atomic number 85. Group VIIA of Periodic Table, aw 211. Heaviest member of the halogen family, has 20 isotopes, all radioactive; derived by α bombardment of bismuth. The two most stable isotopes have half-lives of approximately eight hours. Astatine occurs in nature to the extent of approximately one ounce in the entire earth's crust. Like iodine, it concentrates in the thyroid gland. Its use in medicine is still experimental.

ASTM. Abbreviation for American Society for Testing and Materials.

"Aston 108."[328] TM for a thermosetting polyamine, 20% active.
Use: Durable antistatic agents, softeners.

Aston, Francis William. (1877–1945) This noted English chemist and physicist carried out much of his work with J.J. Thomson at Cambridge. He was the pioneer investigator of isotopes, and his method of separating the lighter from the heavier atomic nuclei provided the technique that later developed into the mass spectroscope which utilizes a magnetic field for this purpose. Aston received the Nobel prize for this discovery in 1922, just three years after Rutherford performed the first transmutation of elements. Aston also correctly estimated the energy content of a hydrogen atom, and predicted the controlled release of this energy.

"Astracel."[203] TM for a group of fast dyes for union shades on polyester-carbon blended fabrics.

astrochemistry. Application of radioastronomy (microwave spectroscopy) to determination of the existence of chemical entities in the gas clouds of interstellar space and of elements and compounds in celestial bodies including their atmospheres. Such data were obtained from spectrographic study of the light from the sun and stars, from analysis of meteorites, and from actual samples from the moon. Hydrogen is by far the most abundant element in interstellar space, with helium a distant second. Over 25% of the elements, including carbon, have been identified as well as molecules of water, carbon monoxide, carbon dioxide, ammonia, ethane, methane, acetylene, formaldehyde, formic acid, methyl alcohol, hydrogen cyanide, and acetonitrile. When applied to the planets only the science is called chemical planetology.
See also nucleogenesis.

"Astrol."[203] TM for a group of fast alizarin direct blues.

asymmetry. A molecular structure in which an atom having four tetrahedral valences is attached to four different atoms or groups. The commonest cases involve the carbon atom, though they may exist also with other elements such as nitrogen and sulfur, i.e., lactic acid which contains one asymmetric carbon (indicated by *). In such cases, two optical isomers (l and d enantiomers) result which are nonsuperposable mirror images of each other.

$$HO-\overset{*}{\underset{CH_3}{C}}-H \qquad H-\underset{CH_3}{C}-OH$$
$$\text{L} \qquad\qquad \text{D}$$

Amino acids are also characterized by asymmetric carbons. Many compounds have more than one asymmetric carbon, e.g., tartaric acid, sugars, terpenes, etc. This results in the possibility of many optical isomers the number being determined by the formula 2_n, where n is the number of asymmetric carbons.
See also optical isomer, enantiomer, glyceraldehyde.

At. Symbol for astatine.

"Atabrine" Hydrochloride.[162] TM for quinacrine hydrochloride.

atactic. A type of polymer molecule in which substitutent groups or atoms are arranged randomly above and below the backbone chain of atoms when the latter are all in the same plane, as shown below.
See polymer, stereospecific.

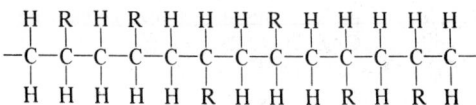

-ate. A suffix having two different meanings. (1) In inorganic compounds, it indicates a salt whose metal or radical is in the highest oxidation state as calcium sulfate, ammonium nitrate, etc. (2) In engineering terminology, it means "result of, " as in precipitate, condensate, alkylate, distillate, etc.

ATE. Abbreviation for aluminum triethyl.
See triethyl aluminum.

atenolol. (1-p-carbamoylmethylphenoxy-3-isopropylamino-2-propanol). $C_{14}H_{22}N_2O_3$.
Properties: Colorless crystals, mp 147C.
Use: An adrenergic blocker used in treatment of hypertension. FDA approved.

$$CH_3\!\!-\!\!CHNHCH_2CH(OH)CH_2O\!\!-\!\!\langle\bigcirc\rangle\!\!-\!\!CH_2\underset{\parallel O}{C}NH_2$$

"Atlac."[89]   TM for a series of polyester resins for use in reinforced plastics.

"Atlastavon."[41]   TM for synthetic-resin sheet lining of the plasticized polyvinyl chloride type used to protect steel tanks at temperatures up to 160F without brick sheathing. Good resistance to oxidizing acids.

ATM.   Abbreviation for aluminum trimethyl. See trimethyl aluminum.

atmosphere.   (1) The gaseous envelope that surrounds the earth. It is comprised of four major divisions: the troposphere (from sea level to approximately 10 km), the stratosphere (ozone region) which extends from approximately 10 to 50 km, the mesosphere (from approximately 50 to 100 km), and the thermosphere which ranges from approximately 100 to 1000 km or more. There are no sharp boundaries between the layers. The pressure drops rapidly as altitude increases (from 1 atm at sea level to 10^{-13} atm at 1000 km). The chemical entities and reactions that occur in these spheres are the subject of extensive research.

(2) The pressure exerted by the air at sea level (14.696 psi) which will support a column of mercury 760 mm high (approximately 30 in). This is standard barometric pressure, though it varies slightly with local meteorological conditions. It is often used to indicate working pressures of steam. The accepted abbreviation is atm.

(3) Any environmental gas or mix of gases, e.g., an atmosphere of nitrogen or an inert atmosphere.

atmosphere (controlled).   As used in the technology of food preservation and storage, a gaseous environment in which the concentration of oxygen, carbon dioxide, and nitrogen are held constant at a specific level the temperature also being controlled. Controlled atmosphere storage techniques are used on a commercial scale in the US.

atmospheric pollution.   See air pollution.

atom.   The smallest possible unit of an element, comprised of a nucleus containing one or more protons and (except hydrogen) two or more neutrons, and one or more electrons which revolve around it. The protons are positively charged, the neutrons have no charge, the electrons are negatively charged. As each atom contains the same number of protons as electrons, the atom is electrically neutral. Atoms in general are characterized by stability. One might wonder why the negatively charged electrons are not attracted into the positively charged nucleus in response to the law of opposite charges causing the atom to collapse on itself. That this does not occur is due to the nature of the electron which is not only a particle but also a standing electromagnetic wave. As explained by Dr. W. V. Houston, "The normal state of an atom is balance between the attraction of the nucleus for the electron wave and what might be called the elastic resistance to compression of the wave itself."

Atoms of the various elements differ in mass (weight), that is in the number of neutrons and protons and also in the number of electrons. Atoms of a given element are identical, but an element may have atoms of slightly different masses called isotopes. Individual atoms of uranium and thorium have been resolved at 5 Å units in the scanning electron microscope. Motion pictures of uranium atoms at magnification of 7.5 million times have been made at the Enrico Fermi Institute at University of Chicago.

Atoms of the same or different elements combine to form molecules. When the atoms are of two or more different elements these molecules are called compounds. Atoms remain essentially unchanged in chemical reactions except that some of the outermost electrons may be removed, shared, or transferred as occurs in oxidation, ionization, and chemical bonding. A few atomic species disintegrate as a result of nuclear changes and thus become radioactive. Heavy unstable atoms such as uranium-235 and plutonium can be split by bombardment with high-energy particles yielding tremendous energy. See also electron, proton, bonding, orbital theory, ionization, radioactivity, fission.

atomic absorption spectroscopy.   An analytical technique in which the substance to be analyzed is converted into an atomic vapor by spraying a solution into an acetylene-air flame. Some types of compounds require a reducing flame, such as acetylene-nitrous oxide. The absorbance at a selected wavelength is measured and compared with that of a reference substance. The absorbance measured is proportional to the concentration.

atomic energy.   See nuclear energy.

atomic hydrogen welding.   A method of welding in which hydrogen gas is passed through an arc between two tungsten electrodes. The arc breaks down the molecules to form atomic hydrogen.

The recombination of the atoms to form molecules and the combustion of the molecular hydrogen in atmospheric oxygen produce a flame temperature of 4000–5000C.

atomic number. The number of protons (positively charged mass units) in the nucleus of an atom, upon which its structure and properties depend. This number represents the location of an element in the Periodic Table. It is always the same as the number of negatively charged electrons in the shells. Thus, an atom is electrically neutral except in an ionized state, when one or more electrons have been gained or lost. Atomic numbers range from 1, for hydrogen, to 106, for the most recently discovered element.
See also Periodic Table, atomic weight (aw), mass number.

atomic weight. (aw). The average weight or mass of all the isotopes of an element as determined from the proportions in which they are present in a given element compared with the mass of the 12 isotope of carbon (taken as precisely 12.000), which is the official international standard. The true atomic weight of carbon when the masses of its isotopes are averaged is 12.01115; that of oxygen is 15.9994. The total mass of any atom is the sum of the masses of all its constituents (protons, neutrons, and electrons). Official atomic weight determinations are released periodically by the IUPAC.
See also atomic number, mass number.

atomic theory. See Dalton, John.

atomic volume. The atomic weight of an element divided by its density.

ATP. Abbreviation for adenosine triphosphate.

"Atpet."[89] TM for a series of emulsifiers used in conjunction with oil as corrosive inhibitors, solubilizers for production of soluble cutting oils, and water-block removal agents in oil well drilling.

atrazine. (2-chloro-4-ethylamino-6-isopropyl-amino-s-triazine). CAS: 1912-24-9.
Hazard: TLV 5 mg/m³.
Used as a herbicide, plant growth regulator, and weed-control agent for corn, etc., and for non-crop and industrial sites. Reported to inhibit photosynthesis of algae in streams.

atropine. (daturine). CAS: 51-55-8.
$C_{17}H_{23}NO_3$. An alkaloid obtained from species of Atropa, Datura, or Hyoscyamus.
Properties: White crystals or powder, optically inactive (but usually contains levorotatory hyoscyamine). Soluble in alcohol, ether, chloroform, and glycerol. Slightly soluble in water. Mp 114–116C.
Derivation: By extraction from Datura stramonium or by synthesis.
Grade: Technical, NF.
Use: Medicine (antidote for cholinesterase-inhibiting compounds, organophosphorus insecticides, nerve gases); artificial respiration may also be necessary.

attapulgite. $(MgAl)_5Si_8O_{22}(OH)_4 \cdot 4HOH$.
A hydrated aluminum-magnesium silicate, the chief ingredient of Fuller's earth.
See also clay.
Use: Drilling fluids, decolorizing oils, filter medium.

attar. (otto). An essential oil (fragrance) made by steam distillation of flowers especially roses.
See essential oil, perfume.

atto-. Prefix meaning 10^{-18} unit (symbol a), e.g., 1 ag = 1 attogram $= 10^{-18}$ g.

attrition mill. (burr mill). A grinding machine comprised essentially of two metal plates or discs with small projections (burrs). One plate may be stationary while another rotates or both may rotate in opposite directions. Feed enters through a hopper above the plates, and ground product emerges at the bottom. There are numerous variations in design.

Au. Symbol for gold, from Latin aurum.

auger. See screw.

"Aura."[108] TM for a powdered, chlorinated, alkaline polyphosphate detergent for mechanical dishwashing.

auramine. [4,4'-(imidocarbonyl)bis(N,N-dimethylaniline)]. CAS: 492-80-8.
$(CH_3)_2NC_6H_4(C:NH)C_6H_4N(CH_3)_2 \cdot HCl$.
Properties: Yellow flakes or powder; soluble in water, alcohol, and ether.
Use: Yellow dye for paper, textiles, leather; also an antiseptic; fungicide.

"Aurantiol"[227]. (methyl-N-3,7-dimethyl-7-hydroxyocylidene-anthranilate). $C_{18}H_{27}O_3N$.
TM for hydroxy-citronellal-methyl anthranilate Schiff base.

aureolin. See indian yellow.

"Aureomycin."[315] TM for chlortetracycline hydrochloride. An antibiotic. Must conform to FDA requirements.

"Auric."[28]   TM for a ferric oxide brown pigment.

auric and aurous compound.   See corresponding gold compound.

aurin (p-rosolic acid). $(C_6H_4OH)_2CC_6H_4O$.
A triphenylmethane derivative.
Properties: Reddish-brown pieces with greenish metallic luster; easily powdered; insoluble in water, benzene, and ether; soluble in alcohol.
Use: Indicator, dye intermediate.

austenite.   A component of steel, a nonmagnetic solid solution of carbon or ferric carbide in gamma iron. Very unstable below its critical temperature, but may be obtained in high carbon steels by rapid quenching from high temperatures. Addition of manganese and nickel lowers critical transition temperature and stable austenite may be obtained at room temperature. Characterized by a face-centered cubic lattice.

austenitic alloys.   (austenitic steels).   Alloys of iron, chromium, and nickel noted for their resistance to corrosion.

Australian bark.   See wattle bark.

autocatalysis.   A catalytic reaction induced by a product of the same reaction. This occurs in some types of thermal decomposition, in autoxidation, and in many biochemical systems, as when an enzyme activates its own precursor. See also autoxidation.

autoclave.   A chamber, usually of cylindrical shape, provided with a door or gate at one end which can be securely closed during operation. It is built heavily enough to accommodate steam pressures of considerable magnitude. It is used to effect chemical reactions requiring high temperature and pressure such as open-steam vulcanization of rubber. Sizes vary from laboratory units to production size, which may be over 50 ft long and three or more feet in diameter. The latter are provided with baffles to ensure equal distribution of the entering steam. Autoclaves are also used in certain sterilization processes.

autoignition point.   (autoign temperature).
The minimum temperature required to initiate or cause self-sustained combustion in any substance in the absence of a spark or flame. This varies with the test method. Some approximate autoign temperatures follow:

acetone	537.7C (1000F)
amyl acetate	398C (750F)
aniline	537.7C (1000F)
butane	430C (806F)
carbon disulfide	100C (212F)
ethyl ether	180C (356F)
phenol	715C (1319F)
toluene	537.7C (1000F)
pine shavings	265C (507F)
cotton batting	230C (446F)
magnesium powder	472C (883F)
nitrocellulose film	137C (279F)

See also flash point.

automatic control.   Maintenance of desired process conditions (temperature, pressure, etc.) by means of sensing devices which function either electromechanically (thermostat) or electronically (feedback). Is applicable to many operations and processes in the chemical industries, such as petroleum refining, evaporation, distillation, heat transfer, electroplating, calendering, extrusion, and many others. Automatic control is not identical with automation.
See also instrumentation.

automation.   Substitution of specially designed machines for manual labor in such mechanical operations as wrapping, packaging of small units, filling and capping bottles, sealing cans and containers, and materials feeding and proportioning. Automated procedures are much more efficient than manual and effect notable cost savings provided that the machinery is reliable. Do not confuse automation with automatic control.

automotive exhaust emission.   See air pollution.

autoxidation.   A spontaneous, self-catalyzed oxidation occurring in the presence of air, it usually involves a free-radical mechanism. It is initiated by heat, light, metallic catalysts, or free-radical generators. Industrial processes, such as manufacture of phenol and acetone from cumene, are based on autoxidation. Other instances are the drying of vegetable oils, the spoilage of fats, gum formation in lubricating oils, and the degradation of high polymers exposed to sunlight for long periods.
See also autocatalysis.

Auwers-Skita rule.   In its original form, the rule suggested that in cis-trans isomeric hydroaromatic compounds the cis had the higher density and refractive index and lower molecular refractivity. A more modern statement of the von Auwers- Skita rule (which has undergone several modifications since it was first enunciated) is that among alicyclic epimers not differing in dipole moment the isomer of highest heat content has the higher density, index of refraction, and boiling point.

Auwers synthesis. Expansion of coumarones to flavonols by treatment of 2-bromo-2-(α-bromobenzyl)coumarones with alcoholic alkali.

auxiliary. Any of a number of chemical compounds used in some phase of textile processing. They may be classified as follows: (1) fats, oils, and waxes; (2) starches, gums, and glues (sizing); (3) soaps and detergents; (4) inorganic chemicals (bleaching, mercerizing); (5) organic solvents; (6) special-purpose products (flameproofing, mildew-proofing, repellent and decorative coatings, and permanent press resins). They are sometimes also called assistants.

auxin. A natural or synthetic plant growth hormone that regulates longitudinal cell structure so as to permit bending of the stalk or stem in phototropic response. The natural materials are formed in small amounts in the green tips of growing plants, in root tips, and on the shaded side of growing shoots. 3-Indoleacetic acid is the most important natural auxin.
See also plant growth regulator.

auxochrome. A radical or group of atoms whose presence is essential in enabling a colored organic substance to be retained on fibers. The best examples are the groups —COOH, —SO$_3$H, —OH, and —NH$_2$.

"Av20."[507] TM for a vitreous 85% pure alumina used in ceramics. Impervious to gas, compressive strength 250,000 psi, maximum service temperature 2550F, hardness (Mohs) 8.5. Fabricated by dry press or extrusion methods.

"Avadex."[58] TM for a series of liquid or granular herbicides containing 2,3-dichloroallyl diisopropylthiocarbamate. Widely used to control growth of wild oats in agricultural crops.
Hazard: By ingestion and inhalation.

"Availaphos."[62] TM for a mineral supplement supplying phosphorus and calcium in readily available form for animal and poultry feeds.

aviation gasoline. See gasoline

"Avibest."[55] TM for a microcrystalline form of asbestos.

"Avicel."[261] TM for microcrystalline cellulose, a highly purified particulate form of cellulose. Par-

ticle size ranges from less than 1 to 150 microns (average varies with grade), density 1.55 (bulk density 0.3–0.5). Insoluble in dilute acids, organic solvents, oils, swells in dilute alkali. Dispersible in water to form stable gels or pourable suspensions. Adsorbs oily and syrupy materials.
Use: Aid to stabilization and emulsifiction, ingredient in foods, suspending agent, binder and hardening agent in tableting, separatory medium in column and thin-layer chromatography, pure cellulose raw material.

avidin. A protein occurring in egg white where it comprises approximately 0.2% of the total protein. It has the property of combining firmly with biotin and rendering it unavailable to organisms, since proteolytic enzymes do not destroy the avidin-biotin complex. Avidin loses its ability to combine with biotin when subjected to heat, hence cooked egg white does not lead to biotin deficiency.

"Avitene."[55] TM for a microcrystalline form of collagen.

"Avitex."[28] TM for a group of textile softeners lubricants and antistatic agents. Both anionic and cationic types.

"Avitone."[28] TM for a group of chemical compounds based on hydrocarbon sodium sulfonates that are used principally as softening lubricating and finishing agents for textiles, leather, and paper.

avocado oil. An edible oil high in unsaturated fatty acids.
Properties: D 0.91, acid value 1–7, saponification value 177–198, iodine value 71–95, fp 7–9C, refr index 1.461–1.465 (40C). Faint odor, bland taste, greenish color.
Use: Cosmetic creams, hair conditioners, suntan preparations, salad oils.

Avogadro's law. A principle stated in 1811 by the Italian chemist Amadeo Avogadro (1776–1856) that equal volumes of gases at the same temperature and pressure contain the same number of molecules regardless of their chemical nature and physical properties. This number (Avogadro's number) is 6.023×10^{23}. It is the number of molecules of any gas present in a volume of 22.41 L and is the same for the lightest gas (hydrogen) as for a heavy gas such as carbon dioxide or bromine. Avogadro's number is one of the fundamental constants of chemistry. It permits calculation of the amount of pure substance (mole), the basis of stoichiometric relationships. It also makes possible determination of how

much heavier a simple molecule of one gas is than that of another, as a result the relative molecular weights of gases can be ascertained by comparing the weights of equal volumes. Avogadro's number (conventionally represented by N' in chemical calculations) is now considered to be the number of atoms present in 12 grams of the carbon-12 isotope (one mole of carbon-12) and can be applied to any type of chemical entity.
See also mole.

"Avolin."[188] $C_6H_4(COOCH_3)_2$.   TM for a special perfume grade of dimethyl phthalate.

aza-.   Prefix indicating the presence of nitrogen in a heterocyclic ring.

3-azabicyclo(3,2,2)nonane. $C_8H_{15}N$.
Properties: White-tan solid, mp 180C (sublimes), partly soluble in water, solubility decreases with an increase in temperature. Readily soluble in alcohol, bulk d 4.67 lb/gal (20C).
Use: Intermediate for the preparation of pharmaceuticals and rubber chemicals.

8-azaguanine. (5-amino-1,4-dihydro-7H-1,2,3-triazolo[4,5-d]pyrimidin-7-1).
CAS: 134-58-7. $C_4H_4N_6O$.
Properties: Crystals from dilute aqueous sodium hydroxide. Insoluble in water, alcohol, and ether.
Grades: Refined.
Use: Inhibitor of purine synthesis.

"Azak."[266]   TM for an herbicide.
2,6-di-tert-butyl-p-tolyl methylcarbamate.

azathioprine.   (Imuran).   An immunosuppressive drug administered for the purpose of inhibiting the neutral tendency of the body to reject foreign tissues by one or more types of immunizing reactions, i.e., formation of leucocytes or antibodies. It has been used with some success in cases of kidney and liver transplants.

6-azauridine.   (6-azauracil riboside; as-triazine-3,-5(2H,4H)dione riboside).   CAS: 54-25-1.
$C_8H_{11}N_3O_6$.
Derivation: Microbiological fermentation.
Use: Research on cell formation and cancer.

azelaic acid.   (nonanedioic acid; 1,7-heptanedicarboxylic acid). $HOOC(CH_2)_7COOH$.
Properties: Yellowish to white, crystalline powder; mp 106C; bp 365C (decomposes); soluble in hot water, alcohol, and organic solvents.
Derivation: Oxidation of oleic acid by ozone.
Grade: Technical.
Use: Organic synthesis, lacquers, product of hydrotropic salts, alkyd resins, polyamides, polyes-

ter adhesives, low-temperature plasticizers, urethane elastomers.

azelaoyl chloride. $ClOC(CH_2)_7COCl$.
Properties: Bp 125–130C (3 mm). Slowly decomposes in cold water, soluble in hydrocarbons and ethers.
Use: Organic synthesis.

azeotrope.   See azeotropic mixture.

azeotropic distillation.   A type of distillation in which a substance is added to the mixture to be separated in order to form an azeotropic mixture with one or more of the components of the original mixture. The azeotrope or azeotropes thus formed will have boiling points different from the boiling points of the original mixture and will permit greater ease of separation.

azeotropic mixture.   (azeotrope).   A liquid mixture of two or more substances which behaves like a single substance in that the vapor produced by partial evaporation of liquid has the same composition as the liquid. The constant boiling mixture exhibits either a maximum or minimum boiling point as compared with other mixtures of the same substances.

azide.   Any of a group of compounds having the characteristic formula $R(N_3)_x$. R may be almost any metal atom, a hydrogen atom, a halogen atom, the ammonium radical, a complex ([CO(NH_3)_6], [Hg(CN)_2M] with M = Cu, Zn, Co, Ni) an organic radical like methyl, phenyl, nitrophenol, dinitrophenol, p-nitrobenzyl, ethyl nitrate, etc.), and a variety of other groups or radicals. The azide group has a chain structure rather than a ring structure. All the heavy metal azides as hydrogen azide and most if not all of the light metal azides (under appropriate conditions) are explosive. They should be handled with utmost care and protected from light, shock and heat. Many of the organic azides are also explosive.
See also lead azide, hydrogen azide.

aziminobenzene.   See 1,2,3-benaotriazole.

azine dye.   A class of dyes derived from phenazine.
$(C_6H_4)N_2(C_6H_4)$ (tricyclic).
The chromophore group may be $=C=N—$, but the color is probably due to the characteristic unsaturation of the benzene rings. The members of the group are quite varied in application. The nigrosines (Color Index 50415–50440) and safranines (Color Index 50200–50375) are examples of this group.
See also dye, synthetic.

azinphos methyl. (o,o-dimethyl-S-4-oxo-1,2,3-benzotriazin-3(4H)-yl methyl phosphorodithioate; "Guthion"). CAS: 86-50-0. $C_{10}H_{12}N_3O_3PS_2$.
Properties: Brown, waxy solid; mp 73C. Slightly soluble in water, soluble in most organic solvents.
Hazard: A poison, cholinesterase inhibitor. Absorbed by skin. TLV: 0.2 mg/m³ in air.
Use: Insecticide for fruit. Use may be restricted.

1-aziridineethanol. (N-(2-hydroxyethylethyleneimine). CAS: 1072-52-2. $C_2H_4NCH_2CH_2OH$.
Properties: Colorless liquid, bp 167.9C, flash p 185F (OC). Combustible. Handling: Inhibited with 1–3% dissolved sodium hydroxide.
Hazard: Irritant to skin and eyes.
Use: Chemical intermediate.

aziridine. A compound based on the ring structure.

CH₂NHCH₂.

See ethyleneimine which is aziridine itself, ethylethyleneimine, propyleneimine, polypropyleneimine, 1-aziridneethanol.

azlon. Generic name for a manufactured fiber in which the fiber-forming substance is composed of any regenerated naturally occurring protein (Federal Trade Commission). Proteins from corn, peanuts, and milk have been used. Azlon fiber has a soft hand, blends well with other fibers, and is used like wood. Combustible.

azobenzene. (diphenyldiimide; benzeneazobenzene). CAS: 103-33-3. $C_6H_5N_2C_6H_5$.
Properties: Yellow or orange crystals, mp 68C, bp 297C, d 1.203 (20/4C), soluble in alcohol and ether, insoluble in water. Combustible.
Derivation: Reduction of nitrobenzene with sodium stannite.
Hazard: Toxic; may cause liver injury.
Use: Manufacture of dyes and rubber accelerators, fumigant, acaricide.

azobenzene-p-sulfonic acid. $C_{12}H_{10}O_3N_2S$.
Properties: Orange crystals, mp 129C.
Use: Intermediate and reagent chemicals.

4,4'-azobis(4-cyanovaleric acid). CAS: 2638-94-0. [=NC(CH₃)(CN)CH₂CH₂CO₂H]₂.
Properties: Light sensitive. Mp 90C (decomposes), mw 280.28.
Grades: 80% solution in water.
Hazards: Toxic.

azobisdimethylvaleronitrile.
Properties: White, crystalline solid.

Use: Initiator for suspension polymerization of vinyl chloride and solution polmyerization of various monomers such as acrylonitrile, MMA, vinyl acetate.

1,1'-azobisformamide. (azodicarbonamide). $H_2NCONNCONH_2$.
Properties: Yellow powder, d 1.65 (20/20C), mp above 180C (decomposes), insoluble in common solvents, soluble in dimethyl sulfoxide. Hydrolyzes at high temperatures to nitrogen, carbon dioxide, and ammonia.
Derivation: From hydrazine.
Grade: Technical, FCC.
Use: Blowing agent for plastics and rubbers, maturing agent for flours.

azobisisobutyronitrile. $(CH_3)_2C(CN)NNC(CN)(CH_3)_2$.
Properties: White powder, mp 105C (decomposes), insoluble in water, soluble in many organic solvents and in vinyl monomers.
Hazard: Toxic by ingestion.
Use: Catalyst for vinyl polymerizations and for curing unsaturated polyester resins, blowing agent for plastics.

"Azocel."[552] TM for 1,1'-azobisformamide.

azodicarbonamide. See 1,1'-azobiformamide.

azodine. (benzeneazonaphthylethylenediamine). CAS: 136-40-3. $C_{18}H_{18}N_4$.
Properties: Red crystals, mp 107–108C.
Use: Reagent for rapid determination of penicillin in blood, urine and other media.

"Azodrin."[125] TM for dimethyl phosphate of 3-hydroxy-N-methyl-cis-crotonamide. $(CH_3O)_2P(O)OC(CH_3):CHC(O)NHCH_3$.
Also called monocrotophos.
Properties: Reddish brown solid with a mild ester odor, bp 125C, soluble in water and alcohol, almost insoluble in kerosene and diesel fuel. Commercially available as a water-miscible solution.
Hazard: Flammable, dangerous fire risk. Use may be restricted. Toxic via ingestion, inhalation, and skin absorption.
Use: Controls certain insects which attack cotton plants.

azo dye. Any of a broad series of synthetic dyes that have —N=N— as a chromophore group. They are produced from amino compounds by diazotization and coupling. Over half of the commercial dyestuffs are in this general category. By varying the chemical composition it is possible to produce acid, basic, direct, or mordant

dyes. This general group is subdivided as monoazo, disazo, trisazo, and tetrazo according to the number of —N=N— groups in the molecule. Examples are Chrysoidine Y, Bismarck Brown 2R, and Direct Green B.
See also dye, synthetic, azoic dye.

azo dye intermediate. Any of various sulfonated aromatic acids derived from α- and β-naphthol, naphthalene, and α- and β-napththylamine. Besides their systematic names, some are named after their discoverers while others have letter designations. All have a fused ring structure with amino, hydroxyl, or sulfonic groups at various locations. For details, see the following entries:

1,2,4-acid	hydrogen acid
amino-G acid	J acid
amino-J acid	Potassium acid
Armstrong's acid	Koch's acid
B acid	L acid
Broenner's acid	Laurent's acid
carbon acid	M acid
Casella's acid	Neville-Winter acid
Chicago acid (SS acid)	peri acid
chromotropic acid	R acid
Cleve's acid	R R acid (2R acid)
crocein (Bayer's acid)	Sulfur acid
gamma acid (F acid)	Schaeffer's acid
epsilon acid	Schoelkopf's acid
G acid	sultam acid
gamma acid	Tobias acid

azoic dye. There is no fundamental chromophoric difference between azo and azoic dyes. The differentiation is made to characterize a group of azo pigments which are precipitated within the cellulosic fiber by carrying out the dye coupling on the fiber. With the advent of the equally brilliant but more easily applied fiber-reactive dyes the azoics have lost some of their importance.

azophenylene. See phenazine.

"Azosol."[203] TM for a series of dyestuffs; soluble in organic solvents.
Use: Coloring spirit lacquers and spirit inks.

azosulfamide. (disodium 2-(4'-sulfamylphenyl-azo)-7-acetamido-1-hydroxynaphthalene-3,6-disulfonate). $C_{18}H_{14}N_4Na_2O_{10}S_3$.
Properties: Dark red, odorless, tasteless powder. Soluble in water with an intense red color, practically insoluble in organic solvents.
Use: Medicine (antibacterial).

azote. The French word for nitrogen (a = not, plus *zoo* = alive as in *zoon* = animal). Nitrogen-bearing compounds can be recognized by this root word as in such terms as azo, azide, azobenzene, carbazole, thiazole, etc. The derivation is due to the chemical inertness of nitrogen.

azotic acid. See nitric acid.

azoxybenzene. (diphenyldiazene oxide).
CAS: 495-48-7. $C_6H_5NO=NC_6H_5$.
Properties: Yellow crystals, d 1.16, mp 36C, soluble in alcohol, insoluble in water.
Use: Intermediate in organic synthesis.

azoxytoluidine. See diaminoazoxytoluene.

azulene. CAS: 275-51-4. $C_{10}H_7$.
Aromatic hydrocarbon with a 7-carbon ring fused to a 5-carbon ring.
Properties: Blue to greenish-black leaflets. Mw 128.19, mp 99-100C, bp 242C, decomposes at 270C, bp 115-135C at 10 mm, soluble in alcohol, ether, acetone.

azure blue. See cobalt blue.

azuresin.
Properties: Moist, irregular, dark blue or purple granules. Slightly pungent odor.
Derivation: Carbacrylic cation-exchange resin, in reversible combination with 3-amino-7-dimethyl-aminophenazathionium chloride. (azure A dye).
Grade: NF.
Use: Medicine (diagnostic test).

B

β. See beta.

B. Symbol for boron.

Ba. Symbol for barium.

Ba-137. See barium-137.

babassu oil. A nondrying, edible oil expressed from the kernels of the Babassu palm which grows in profusion in Brazil. Composition: 44% lauric acid, 15% myristic acid, 16% oleic acid, balance mixed acids. Usable in foods and soap-making, but supply is limited by cost of exploitation of large quantities potentially available. Combustible.

Babbitt metal. One of a group of soft alloys used widely for bearings. They have good bonding characteristics with the substrate metal, maintain oil films on their surfaces, and are nonseizing and antifriction. Used as cast, machined, or pre-formed bimetallic bearings in the form of a thin coating on a steel base; the main types are lead base, lead-silver base, tin base, cadmium base, and arsenical. The latter contains up to 3% arsenic.
Hazard: Dust is toxic by inhalation.

Babcock test. A rapid test for butterfat in milk introduced by Stephen M. Babcock in 1890 and now in world-wide use in the dairy industry.

B acid. (1-amino-8-naphthol-3,5-disulfonic acid). $C_{10}H_4NH_2OH(SO_3H)_2$.
Derivation: Sulfonation of 1-amino-8-naphthol-3-sulfonic acid.
Use: Azo dye intermediate.

bacillus. A type of bacteria characterized by a rod-like shape.

bacitracin. An antibiotic of polypeptide structure produced by the metabolic processes of *Bacillus subtilis*. It is effective against many Gram-positive bacteria, but ineffective against most aerobic Gram-negative bacteria.
Properties: White to pale buff, hygroscopic powder, odorless or with slight odor, bitter taste. Powder is stable to heat, but solutions deteriorate at room temperature. Freely soluble in water; soluble in alcohol, methanol, and glacial acetic acid; insoluble in acetone, chloroform, and ether. Solutions are neutral or slightly acid to litmus. Grade: USP, having a potency of approximately 40 to 50 units/mg.
Use: Medicine (antibacterial), feed supplement.

bacitracin methylene disalicylate.
Properties: White to gray-brown powder. Slight unpleasant odor, less bitter than bacitracin. Soluble in water, pyridine, ethanol; less soluble in acetone, ether, chloroform, benzene; pH of saturated aqueous soluble 3.5–5.0. Available also as the sodium salt.
Use: Antibiotic, feed additive.

bacteria. Microorganisms often composed of a single cell in the form of straight or curved rods (bacilli), spheres (cocci), or spiral structures. Their chemical composition is primarily protein and nucleic acid. Chlorophyll molecules are also present enabling bacteria to carry out photosynthesis. Some types called anaerobic are able to live and reproduce in the absence of air or oxygen, aerobic types require oxygen. Bacteria that can live either with or without oxygen are called facultative. Filamentous bacteria are related to blue-green algae. Molds that yield antibiotics (Actinomycetes) are of this type. Pathogenic bacteria are infectious organisms that cause such diseases as pneumonia, tuberculosis, syphilis, and typhus. The staining of bacteria for microscopic identification was originated by Koch, a German physician and bacteriologist (1843–1910). Bacteria are often classified as Gram-positive or Gram-negative. Food spoilage is often induced by bacterial contamination, but there are many beneficial types of bacteria in the body, e.g., intestinal flora that aid in metabolism. Bacteria rich in proteins can be produced by fermentation of animal wastes for feed supplements.
  The outstanding development in this field is the laboratory creation of bacteria by gene-splicing techniques. This noteworthy achievement has an enormous future potential in the chemical, agricultural, food, and pharmaceutical industries. The Supreme Court has ruled it to be a patentable invention.
See also recombinant DNA biotechnology.
Use: (1) Fermentation processes used in baking and the manufacture of alcohol, wine, vinegar, beer (yeast), and antibiotics (molds). See also (3).
  (2) Fixation of atmospheric nitrogen in the soil.

(3) Reaction with hydrocarbons (methane and other paraffins) to yield proteins (yeasts). (4) Purification of sewage sludge activated by bacteria (see sewage sludge). (5) Reaction with cellulose to form biopolymers and high-protein foodstuffs. (6) Reaction with waste materials (coal and cement dusts, gasworks effluent) to release plant nutrients for inexpensive fertilizers (USSR). (7) Precipitation and concentration of uranium and some other metals by compounds obtained from bacteria grown on carbonaceous materials such as lignin and cellulose. (8) Formation of azo compounds in soil treated with the herbicide propanil. (9) Synthesis of hormones by recombinant DNA methods (*E. coli*). See also insulin. (10) Miscellaneous reactions, e.g., oxidation of pentaerythritol to tris (hydroxymethyl)acetic acid; conversion of the sulfur in gypsum to elemental sulfur via hydrogen sulfide; clean-up of oil spills.
See also fermentation, virus, enzyme, biotechnology.

bactericide. (germicide).   Any agent that will kill bacteria, especially those causing disease. Bactericides vary greatly in their potency and specificity. They may be other organisms (bacteriophages), chemical compounds, or short-wave radiation.
See also virus, antibiotic, biocide.

bacteriophage.   A type of virus which attacks and destroys bacteria by surrounding and absorbing them.

bacteriostat.   A substance which prevents or retards the growth of bacteria. Examples are quaternary ammonium salts and hexachlorophene.
See also antiseptic.

baddeleyite. (zirconia). ZrO_2. A natural zirconium oxide.
Properties: Black, brown, yellow to colorless, streak white, luster submetallic to vitreous to greasy. Mp 2500–2950C. Highly resistant to chemicals, d 5.5–6.0.
Grade: Crude (53%, 73–75%), purified (98%).
Occurrence: Brazil, Ceylon.
Use: Corrosion- and heat-resistant applications; source of zirconium.

Badische acid. (2-naphthylamine-8-sulfonic acid). $C_{10}H_9NO_3S$.
Properties: Colorless needles, partially soluble in water and alkalies, slightly soluble in alcohol.
Derivation: Sulfonation of 2-naphthylamine.
Use: Azo dye intermediate.

Baekeland, L. H. (1863–1944). Born in Ghent, Belgium, did early research in photographic chemistry and invented Velox paper (1893). After working for several years in electrolytic research, he undertook fundamental study of the reaction products of phenol and formaldehyde which culminated in his discovery in 1907 of phenol-formaldehyde polymers originally called "Bakelite." The reaction itself had been investigated by Bayer in 1872, but Bakeland was the first to learn how to control it to yield dependable results on a commercial scale. The Bakelite Co. (now a division of Union Carbide) was founded in 1910.
See phenolformaldehyde resins.

Baekeland (Bakelite) process.   Condensation of phenol and formaldehyde to o-hydroxymethyl-phenol (Lederer-Manasse), which undergoes further arylation yielding a polymeric structure.

Baeyer-Drewson indigo synthesis.   Formation of indigos by an aldol addition of o-nitrobenzaldehydes to acetone, pyruvic acid, or acetaldehyde. Of interest mainly as a method of protecting o-nitrobenzaldehydes.

Baeyer-Villiger reactions.   The oxidation of aromatic, open chain, and cyclic ketones to esters and lactones by peracids.

baffle.   A flow-regulating device consisting of a perforated metal plate placed horizontally in liquid-mixing tanks, distillation columns, and the like to restrict or divert the passage of liquid, thus providing a uniformly dispersed flow. Baffles are also used in open-steam autoclaves to ensure even distribution of the entering steam.

bagasse.   A form of cellulose (biomass) derived as a byproduct of the crushing of sugar cane or guayule plants. Contains a high proportion of hemicellulose. After pulping with either soda or kraft cooking liquor, it can be made into a low grade of paper. It is also used in compressed form as an insulating board in construction; as a medium for growth of nutritive bacteria; in animal feeds; in manufacturing of furfural; and as on-site fuel for cane-sugar mills. In Hawaii, it is being used as a fuel for electric power generation.
See biomass.
Hazard: Dust is flammable in air.

baghouse. (bag filter).   A large-scale, dust-collecting device composed of a series of large cotton or nylon bags assembled in a heavy metal frame or housing. The bags may be as much as 10 ft high, each "bag" being made up of three units sewn together as one element approximately 18

inches in diameter. Discharge hoppers are located beneath the bags. A suction or blower system forces dust-laden air through an inlet port on one side of the frame just above the hopper space. It enters the bags, where it deposits its suspended solids, while the cleaned air is drawn through and leaves by an outlet port. A motor-driven shaker mechanism agitates the bags periodically, dislodging the accumulated layer of dust which falls into the hoppers. Installations of this type are often of impressive size, some containing over 300 bags.
See also "Nomex."

bait. An insecticide or rodenticide placed in such a way as to attract the pest. Arsenic compounds and Bordeaux mixture are typical insect baits. All types are highly toxic.
See pesticide.

"Bakelite."[214] TM for polyethylene, polypropylene, epoxy, phenolic, polystyrene, phenoxy, perylene, polysulfone, ethylene copolymers, ABS, acrylics, and vinyl resins and compounds.

Baker-Nathan effect. Effect originally observed in the reaction of p-substituted benzyl bromides with pyridine and other processes in which the observed rates are opposite to those predicted by the electron releasing inductive effect of alkyl groups, i.e., $CH_3 > CH_3CH_2 > (CH_3)_2CH > C(CH_3)_3$. To explain it, a type of electron delocalization involving sigma electrons was proposed, termed hyperconjugation, which manifests itself in systems in which a saturated carbon atom attached to an unsaturated carbon or one with an empty orbital bears at least one hydrogen atom.

Baker-Venkataraman rearrangement. Base-catalyzed rearrangement of o-acyloxyketones to β-diketones, important intermediates in the synthesis of chromones and flavones.

baking finish. A paint or varnish that requires baking at temperatures greater than 66C for the development of desired properties (ASTM). Such finishes are based on oil-modified alkyd, melamine, epoxy, nitrocellulose, or urea resins, or combinations of these. Baking is often done by infrared radiation producing high molecular weight coatings that are dense and tough.

baking powder. A synthetic leavening agent widely used in the baking industry. There are several types all of which are composed of a carbonate, a weak acid or acidic compound, and a filler. A typical composition is sodium bicarbonate, tartaric acid or monobasic calcium phosphate, and cornstarch. Ingredients sometimes used are ammonium carbonate and potassium bitartrate. Upon contact with moisture and heat the active ingredients react to evolve carbon dioxide which "raises" the dough in the early minutes of heat exposure, thus producing a stable solid foam. Wheat flour gluten is sufficiently elastic to retain the bubbles of carbon dioxide.

baking soda. See sodium bicarbonate.

BAL. Abbreviation for British Anti-Lewisite. See 2,3-dimercaptopropanol.

"Balan."[530] TM for benefin, a selective herbicide.

balance. (1) Exact equality of the number of atoms of various elements entering into a chemical reaction and the number of atoms of those elements in the reaction products. For example, in the reaction $NaOH + HCl \rightarrow NaCl + HOH$, the atoms in the input side are H[2], Na[1], O[1], and Cl[1]. Each of these is also present in the products, though in different combination. The atoms of catalysts (when present) do not enter into reactions and therefore are not involved. The balance of chemical reactions follows the law of conservation of mass.

The term "material balance" is used by chemical engineers in designing processing equipment. It denotes a precise list of all the substances to be introduced into a reaction and all those that will leave it in a given time, the two sums being equal.

(2) A precision instrument designed for weighing extremely small amounts of material with high accuracy. An analytical balance or microbalance for weights from about 1 g to 0.1 mg is standard equipment in chemical laboratories. Its essential feature is a one-piece metal beam (lever) pivoted on a knife-edge or flexure at its exact center (fulcrum) so that it is free to oscillate. From it are suspended two scale pans approximately two inches in diameter, each of which is also positioned on a knife-edge on the lower arms of the beam. Exact balance is indicated by a pointer attached to the beam. Either an aluminum rider or a chain and vernier is provided for maximum accuracy. Highly sophisticated balances operating electronically with built-in microprocessors have become available in recent years.

balata. See rubber, natural (note).

"Balco."[155] TM for an alloy of 70% nickel and 30% iron.
Properties: D 8.46, resistivity 20 ohm-mυ cm, tensile strength 70,000 psi.

Use: Square-loop laminated cores for magnetic amplifiers.

Baldwin rules for ring closure. Set of empirical rules, stereochemical in nature, predicting the relative facility of ring closure reactions.

ball clay. A clay that has good plasticity, strong bonding power, high refractoriness, and fires to a white or cream colored product. Used as bonding and plasticizing agent or chief ingredient of whiteware, porcelains, stoneware, terra cotta, glass refractories, floor and wall tile.
See also clay.

ball mill. A jacketed steel cylinder rotating on a horizontal axis and containing steel balls of varying diameter; the interior walls are usually equipped with baffle bars to impart a rolling and cascading action to the balls. The total weight of the balls may be 2000 lb or more. The grinding efficiency depends on the number of contacts between any two balls, thus the greater the number of balls the more effective the grinding action. The material is introduced through an opening in the center of the cylinder which is then hermetically closed. Discharge is by the same opening after replacing the cover plate with a grill to retain the balls. Ball mills can be adapted to continuous operation in which the feed enters at one end and is discharged at the other. Products ground are dry chemicals, paint pigments, etc.
See also pebble mill, jar mill.

Bally-Scholl synthesis. Formation of mesobenzanthrones by the action of glycerol, or a derivative, and sulfuric acid on anthraquinones or anthranols.

balsam. A resinous mixture of varying composition obtained from several species of evergreen trees or shrubs. Contains oleoresins, terpenes, and usually cinnamic and benzoic acids. All types are soluble in organic liquids and insoluble in water. Some have a penetrating, pleasant odor. They are combustible and in general non-toxic. The best-known types are:

(1) Peru balsam, from Central America, is a thick, viscous liquid (d 1.15) containing vanillin. Used in flavoring, chocolate manufacturing, as an ingredient in expectorants and cough syrups, and as a fragrance in shampoos and hair conditioners. A mild allergen. Shipped in drums.

(2) Tolu balsam, a plastic solid, is derived from a related tree in Colombia. Its uses are similar to Peru balsam. Source of tolu oil. Odorless.

(3) Copaiba balsam from Brazil and Venezuela (d 0.94–0.99) is a viscous liquid used in varnishes and lacquers as an odor fixative and in manufac-

ture of photographic paper. It is the source of copaiba oil.

(4) Balm of Gilead is from a Middle Eastern shrub is used in perfumery and medicine.

(5) Canada balsam from the North American balsam fir is a liquid, d 0.98, used in microscopy, in fine lacquers, as a flavoring, and as a fragrance.

(6) Benzoin resin (Benjamin gum).
See benzoin resin.

bamboo. A grass or plant native to southeast Asia having a rather high cellulose content which makes possible its use for specialty papers. Its fibers are longer than those of most other plants of this type, and are comparable to those of coniferous woods. Its composition is: total cellulose 58%, alpha cellulose 35%, pentosans 28%, lignin 23%. Also used for making light furniture, fishing rods, etc. Combustible.

Bamford-Stevens reaction. Formation of olefins by base-catalyzed decomposition of p-toluenesulfonylhydrazones of aldehydes and ketones.

banana oil. (1) (banana liquid) A solution of nitrocellulose in amyl acetate or similar solvent so termed because of its penetrating banana-like odor. (2) Synonym for amyl acetate.

Banbury mixer. A batch-type internal mixing machine, named after its inventor, which has been widely used in the rubber industry since 1920 for high-volume production. It will also accept plastic molding powders. Its chief feature is an enclosed barrel-shaped chamber in which two rotors with oppositely curved contours rotate rapidly on a horizontal axis, first masticating the rubber and then efficiently incorporating the dry ingredients. Both steam and water jacketing are provided. Batches may be up to 1000 lb. A plunger at the entrance port rides on top of the batch to furnish enough pressure for proper mixing. A hydraulically operated discharge gate is located below the mixing chamber.

band, absorption. See absorption band.

"Bandane."[316] TM for polychlorodicyclopentadiene isomers.
Used as an herbicide.

banded iron formation. An iron formation that consists of alternating iron-rich and iron-poor layers. Most rocks of this type are older than about two billion years.

"Ban-Kal."[204] TM for a nonfoaming liquid acid cleaner for dairies and food-processing plants.

"Banox."[108] TM for a series of dry, powdered, phosphate-type corrosion inhibitors. No. 1 is artificially colored. No. 1-P and WT are colorless.
Use: Refrigerator cars, refrigeration brine, cooling towers, and small water systems.

Banting, Sir Frederick. (1891–1941). A native of Ontario, Canada, Banting did his most important work in endocrinology. His brilliant research culminated in the preparation of the antidiabetic hormone which he called insulin, derived from the "isles of Langerhans" in the pancreas. He received the Nobel prize in medicine for this work together with MacLeod of the University of Toronto. In 1930, the Banting Institute was founded in Toronto. He was killed in an airplane crash.

"Banvel" D.[316] TM for an herbicide containing 2-methoxy-3,6-dichlorobenzoic acid (dimethylamine salt).

BAP. Abbreviation for benzyl-p-aminophenol.

"Barafene."[293] TM for çoatings used to prevent permeation of volatile ingredients, oils, and oxygen through polyolefin containers.

"Barafos."[236] TM for a polyphosphate compound for the treatment of drilling mud to reduce viscosity and gel strength.

"Baragel."[236] TM for a compound of purified bentonite and an organic base.
Use: Gelling agent for lubricating oils to prepare nonmelting greases.

"Barak."[28] TM for dibutylammonium oleate $(C_4H_9)_2NH_2COOC_{17}H_{33}$.
Properties: Translucent, light brown liquid. Combustible.
Use: To activate accelerators and improve processing of rubber and synthetic rubbers.

barban. Generic name for 4-chloro-2-butynyl-m-chlorocarbanilate.
$C_6H_4(Cl)NHCOOCH_2C:CCH_2Cl$.
Herbicide and plant growth regulator.

barberite. A nonferrous alloy containing 88.5% copper, 5% nickel, 5% tin, 1.5% silicon.
Properties: D 8.80, mp 1070C. It offers good resistance to sulfuric acid in all dilutions up to 60%, seawater, moist sulfurous atmospheres, and mine waters.

Barbier-Wieland degradation. Stepwise carboxylic acid degradation of aliphatic acids (particularly in sterol side chains) to the next lower homolog. The ester is converted to a tertiary alcohol that is dehydrated with acetic anhydride, and the olefin oxidized with chromic acid to a lower homologous carboxylic acid.

barbital. (diethylmalonylurea; diethylbarbituric acid; "Veronal"). CAS: 57-44-3. $C_8H_{12}N_2O_3$.

Properties: White crystals or powder, bitter taste, odorless, stable in air, mp 187–192C, soluble in hot water, alcohol, ether, acetone, and ethyl acetate.
Derivation: By the interaction of diethyl ester or diethylmalonic acid and urea.
Grade: Technical, CP.
Hazard: See barbiturates.
Use: Medicine (sedative), stabilizer for hydrogen peroxide. See also barbiturates.

barbiturate. A derivative of barbituric acid which produces depression of the central nervous system and consequent sedation. Used by prescription only for sedative and anesthetic purposes.
Hazard: Habit-forming. Several types, including amo-, seco-, and pentabarbital are under government restriction.

barbituric acid. (malonylurea; pyrimidinetrione; 2,4,6-tri-oxohexahydro pyrimidine).

$OCNHCOCH_2CONH \cdot 2HOH$.

Properties: White crystals, efflorescent, odorless, mp 245C with some decomposition, slightly soluble in water and alcohol, soluble in ether. Forms salts with metals.
Derivation: By condensing malonic acid ester with urea.
Grade: Technical.
Use: Preparation of barbiturates, polymerization catalyst, dyes.

"Barden."[285] TM for a group of hydrous aluminum silicates (sedimentary kaolins) from South Carolina.
Properties: D: 2.60, bulk d (aerated) 18–20 lb/cu ft, packed 35–40 lb/cu ft, creamy white, pH 4.5–5, air-floated, particle size 90% less than 2 microns.
Use: In pesticides, boxboard, flooring and tile ad-

hesives, fertilizers, roofing granules, putties, caulking compounds, etc.

Bardhan-Sengupta phenanthrene synthesis.
Formation of octahydrophenanthrene derivatives by cyclodehydration of derivatives of 2-(β-phenethyl)-1-cyclohexanol and consequent dehydration to phenanthrenes with selenium.

bar disintegrator. See cage mill.

Bardol[175]. TM for a coal-tar oil with aromatic content.
Properties: Dark-colored liquid, d 1.07–1.12 (25/25C), distillation at 300C 60% max, low viscous at 4.4C. Combustible.
Use: Swelling agent for natural and synthetic elastomers, dispersing agent for blacks and mineral fillers, tackifier, plasticizer.

Barfoed's reagent. Aqueous solution of copper acetate.
Use: To distinguish monosaccarides from disaccharides (red cuprous oxide forms in presence of glucose).

barite. ($BaSO_4$). Natural barium sulfate, barytes, heavy spar.

barium. CAS: 7440-39-3. Ba. Alkaline-earth element of atomic number 56, Group IIA of Periodic Table; aw 137.34; valence 2; 7 stable isotopes.
Properties: Silver-white, somewhat malleable metal, d 3.6, values for melting and boiling points are reported ranging from 704C to 850C for mp and from 1140C to 1637C for bp. The most acceptable values based on reliable original work appear to be mp 710C and bp 1500C. Extremely reactive, reacts readily with water, ammonia, halogens, oxygen, and most acids. Gives green color in flame. Extrudable and machinable.
Occurrence: Ores of barite and witherite are found in Georgia, Missouri, Arkansas, Kentucky, California, Nevada, Canada, Mexico.
Derivation: Reduction of barium oxide with aluminum or silicon in a vacuum at high temperature.
Forms: Rods, wire, plate, powder.
Grade: Technical, pure.
Hazard: Flammable (pyrophoric) at room temperature in powder form; store under inert gas, petroleum, or other oxygen-free liquid. When heated to approximately 200C in hydrogen, barium reacts violently forming BaH_2. TLV: For all soluble barium compounds, 0.5 mg/m^3 in air (as barium).
Use: Getter alloys in vacuum tubes, deoxidizer for copper, Frary's metal, lubricant for anode rotors in x-ray tubes, spark-plug alloys.

barium-137. Radioactive isotope of barium. See Cs-137.

barium acetate. CAS: 543-80-6.
$Ba(C_2H_3O_2)_2 \cdot H_2O$.
Properties: White crystals, soluble in water, insoluble in alcohol, d 2.02, mp decomposes.
Derivation: Acetic acid is added to a solution of barium sulfide. The product is recovered by evaporation and subsequent crystallization.
Grade: Technical, CP. See barium.
Use: Chemical reagent, acetates, textile mordant, catalyst manufacturing, paint and varnish driers.

barium aluminate. $3BaO \cdot Al_2O_3$.
Properties: Gray pulverized mass, soluble in water, acids. See barium.

barium azide. CAS: 18810-58-7. $Ba(N_3)_2$.
Crystalline solid, d 2.936, loses nitrogen at 120C, soluble in water, slightly soluble in alcohol.
Hazard: Explodes when shocked or heated. See also barium.
Use: High explosives.

barium binoxide. See barium peroxide.

barium borotungstate. (barium borowolframate).
$2BaO \cdot B_2O_3 \cdot 9WO_3 \cdot 18H_2O$.
Properties: Large, white crystals. Effloresces in air. Keep well stoppered! Soluble in water.
Hazard: A poison. TLV: See barium.
Use: Making borotungstates.

barium borowolframate. See barium borotungstate.

barium bromate. CAS: 13967-90-3.
$Ba(BrO_3)_2 \cdot H_2O$.
Properties: White crystals or crystalline powder. Slightly soluble in water, insoluble in alcohol, d 3.820, decomposes at 260C.
Derivation: By passing bromine into a soluble of barium hydroxide, barium bromide and barium bromate being formed which are separated by crystallization.
Grade: Pure, reagent.
Hazard: A poison. Moderate fire risk in contact with organic materials. TLV: See barium.
Analytical reagent, oxidizing agent, corrosion inhibitor.

barium bromide. $BaBr_2 \cdot 2H_2O$.
Properties: Colorless crystals. Soluble in water and in alcohol, d 3.852, mp (anhydrous) 847C.
Derivation: Interaction of barium sulfide and hydrobromic acid with subsequent crystallization.
Grade: Technical, CP.

Hazard: A poison. TLV: See barium.
Use: Manufacturing bromides, photographic compounds, phosphors.

barium carbonate. CAS: 513-77-9. $BaCO_3$.
Properties: White powder, found in nature as the mineral witherite. Insoluble in water, soluble in acids (except sulfuric), d 4.275, mp 174C at 90 atmospheres, 811C at one atmosphere.
Derivation: Precipitated barium carbonate is made by reaction of sodium carbonate or carbon dioxide with barium sulfide.
Grade: Technical, CP, reagent 99.5%.
Hazard: A poison. TLV: See barium.
Use: Treatment of brines in chlorine-alkali cells to remove sulfates, rodenticide, barium salts, ceramic flux, optical glass, case-hardening baths, ferrites, in radiation-resistant glass for color television tubes.

barium chlorate. CAS: 13477-00-4.
$Ba(ClO_3)_2 \cdot HOH$.
Properties: Colorless prisms or white powder. Soluble in water, d 3.179, mp 414C. Combustible.
Derivation: Electrolysis of barium chloride.
Grade: Technical, CP, reagent.
Hazard: A poison. TLV: See barium. Strong oxidizer, fire risk in contact with organic materials.
Use: Pyrotechnics, explosives, textile mordant, manufacture of other chlorates.

barium chloride. CAS: 10361-37-2.
$BaCl_2 \cdot 2H_2O$.
Properties: Colorless, flat crystals. Soluble in water, insoluble in alcohol, d 3.097, mp 960C (anhydrous). Combustible.
Derivation: (a) By the action of hydrochloric acid on barium carbonate or barium sulfide; (b) by heating a mixture of barium sulfate, carbon, and calcium chloride.
Grade: Technical (crystals or powdered), 99%, crystals, powdered, CP.
Hazard: Ingestion of 0.8 g may be fatal. TLV: see barium.
Use: Chemicals (artificial barium sulfate, other barium salts), reagents, lubrication oil additives, boiler compounds, textile dyeing, pigments, manufacture of white leather.

barium chromate. (lemon chrome; ultramarine yellow; baryta yellow; Steinbuhl yellow).
CAS: 10294-40-3. $BaCrO_4$.
Properties: Heavy, yellow, crystalline powder; soluble in acids; insoluble in water; d 4.498. Combustible.
Derivation: Interaction of barium chloride and sodium chromate. The precipitate is washed, filtered and dried.

Grade: Technical, CP.
Use: Safety matches, corrosion inhibitor in metal-joining compounds, pigment in paints, ceramics, fuses, pyrotechnics, metal primers, ignition control devices.   See also chrome yellow.

barium citrate. $Ba_3(C_6H_5O_7)_2 \cdot HOH$.
Properties: Grayish white crystalline powder. Soluble in water, hydrochloric and nitric acids.
Hazard: See barium.
Use: Manufacture of barium compounds, stabilizer for latex paints.

barium cyanide. CAS: 542-62-1. $Ba(CN)_2$.
Properties: White, crystalline powder. Soluble in water and alcohol.
Derivation: By the action of hydrocyanic acid on barium hydroxide with subsequent crystallization.
Hazard: See barium.
Use: Metallurgy, electroplating.

barium cyanoplatinite.   (platinum barium cyanide; barium platinum cyanide).
$BaPt(CN)_4 \cdot 4HOH$.
Properties: Yellow or green crystals, mp 100C (loses 2HOH), d 2.08, soluble in water, insoluble in alcohol.
Grade: CP.
Hazard: See barium and cyanides.
Use: X-ray screens.

barium dichromate.   (barium bichromate).
$BaCr_2O_7 \cdot 2H_2O$.
Properties: Brownish-red needles or crystalline masses. Soluble in acids, decomposed by water.
Hazard: See barium.

barium dioxide.   See barium peroxide.

barium di-o-phosphate.   See barium phosphate, secondary.

barium diphenylamine sulfonate.
$(C_6H_5NHC_6H_4SO_3)_2Ba$.
Properties: White crystals soluble in water.
Hazard: See barium.
Use: Indicator in oxidation-reduction titrations.

barium dithionate.   (barium hyposulfate).
$BaS_2O_6 \cdot 2HOH$.
Properties: Colorless crystals, soluble in hot water, slightly soluble in alcohol, d 4.536.
Derivation: Action of manganese dithionate on barium hydroxide.
Hazard: See barium.

barium diuranate.   See uranium-barium oxide.

barium ethylsulfate. $Ba(C_2H_5SO_4)_2 \cdot 2HOH$.
Properties: Colorless crystals, soluble in water and alcohol. Combustible.
Derivation: Interaction of barium hydroxide and ethylsulfuric acid.
Hazard: See barium.
Use: Organic preparations.

barium ferrite.
Grades: Powder.
Use: Permanent magnet material.

barium fluoride. CAS: 7787-32-8. BaF_2.
Properties: White powder. Sparingly soluble in water, d 4.828; mp 1354C.
Derivation: Interaction of barium sulfide and hydrofluoric acid followed by crystallization.
Grade: Technical, CP, single pure crystals, 99.98%.
Hazard: See barium.
Use: Ceramic flux, carbon brushes for electrical equipment, glass making, manufacture of other fluorides, crystals for spectroscopy, electronics, dry-film lubricants.

barium fluosilicate. (barium silicofluoride).
$BaSiF_6H$.
Properties: White, crystalline powder; insoluble in water.
Grade: Technical.
Hazard: See barium.
Use: Ceramics, insecticidal compositions.

barium fructose diphosphate. See fructose diphosphate, calcium, barium salts.

barium hexafluorogermanate. $BaGeF_6$.
White crystalline solid, mp approximately 665C, dissociates to barium fluoride and germanium fluoride, d 4.56.

barium hydrate. See barium hydroxide.

barium hydrosulfide. $Ba(SH)_2$.
Properties: Yellow crystals, hygroscopic, soluble in water.
Hazard: See barium.

barium hydroxide (anhydrous). $Ba(OH)_2$.
Available commercially.
See barium hydroxide hydrates, following.

barium hydroxide, monohydrate. (barium monohydrate). $Ba(OH)_2 \cdot HOH$.
Properties: White powder, soluble in dilute acids, slightly soluble in water.
Hazard: See barium.
Use: Manufacturing of oil and grease additive,

barium soaps and chemicals. Refining of beet sugar, alkalizing agent in water softening, sulfate removal agent in treatment of water and brine, boiler scale removal, dehairing agent, catalyst in manufacture of phenol-formaldehyde resins, insecticide and fungicide, sulfate controlling agent in ceramics, purifying agent for caustic soda, steel carbonizing agent, glass, refining edible oils, elastomer vulcanization.

barium hydroxide, octahydrate. (barium hydrate; barium octahydrate; caustic baryta).
$Ba(OH)_2 \cdot 8HOH$.
Properties: White powder or crystals. Absorbs carbon dioxide from air. Keep well stoppered! Soluble in water, alcohol, and ether, d 2.18, mp 78C (losing its water of crystallization), mp (anhydrous $Ba(OH)_2$) 408C.
Derivation: (a) By dissolving barium oxide in water with subsequent crystallization. (b) By precipitation from an aqueous solution of the sulfide by caustic soda. (c) By heating barium sulfide in earthenware retorts into which a current of moist carbonic acid is passed after which superheated steam is passed over the resulting heated carbonate.
Impurities: Iron and calcium in commercial grades.
Grade: Technical (crystals or anhydrous powder), CP, ACS reagent.
Hazard: See barium.
Use: Organic preparations, barium salts, analytical chemistry.   See also the monohydrate.

barium hydroxide pentahydrate. (barium pentahydrate). $Ba(OH)_2 \cdot 5HOH$.
Properties: Translucent, free-flowing, white flakes; approximately 65 lb/cu ft.
Hazard: See barium.
Use: Same as the octahydrate.

barium hypophosphite. $BaH_4(PO_2)_2$.
Properties: White, crystalline powder; odorless; soluble in water, insoluble in alcohol.
Hazard: See barium.
Use: Nickel plating.

barium hyposulfate. See barium dithionate.

barium hyposulfite. See barium thiosulfate.

barium iodate. $Ba(IO_3)_2$.
Properties: White, crystalline powder; slightly soluble in water, hydrochloric and nitric acids, insoluble in alcohol, d 5.23, mp decomposes at 476C.
Hazard: See barium.

barium iodide. $BaI_2 \cdot 2HOH$.
Properties: Colorless crystals, decomposes and reddens on exposure to air, soluble in water,

slightly soluble in alcohol, d 5.150, mp loses 2HOH and melts at 740C.
Derivation: Action of hydriodic acid on barium hydroxide or of barium carbonate on ferrous iodide solution.
Hazard: See barium.
Use: Preparation of other iodides.

barium manganate. (manganese green; Cassel green). $BaMnO_4$.
Properties: Emerald-green powder. Insoluble in water, decomposed by acids, d 4.85.
Hazard: See barium.
Use: Paint pigment.

barium mercury bromide. See mercuric barium bromide.

barium mercury iodide. See mercuric barium iodide.

barium metaphosphate. $Ba(PO_3)_2$.
Properties: White powder, slowly soluble in acids, insoluble in water.
Hazard: See barium.
Use: Glasses, porcelains, and enamels.

barium metasilicate. See barium silicate.

barium molydate. $BaMoO_4$.
Properties: White powder, slightly soluble in acids and water, absolute d 4.7 g/cc, approximate mp 1600C.
Grade: Crystal, 99.84% pure.
Hazard: See barium.
Use: Electronic and optical equipment, pigment in paints and other protective coatings.

barium monohydrate. See barium hydroxide, monohydrate.

barium monosulfide. See barium sulfide.

barium monoxide. See barium oxide.

barium nitrate. CAS: 10022-31-8. $Ba(NO_3)_2$.
Properties: Lustrous, white crystals; soluble in water; insoluble in alcohol; d 3.244; mp 575C.
Derivation: By the action of nitric acid on barium carbonate or sulfide.
Grade: Technical, crystals, fused mass or powder, CP.
Hazard: Strong oxidizing agent. See barium.
Use: Pyrotechnics (gives green light), incendiaries, chemicals (barium peroxide), ceramic glazes, rodenticide, electronics.

barium nitrite. $Ba(NO_2)_2 \cdot HOH$.
Properties: White to yellowish, crystalline powder. Soluble in alcohol, water; d 3.173, decomposes at 217C.

Hazard: See barium.
Use: Diazotization, corrosive inhibitor, explosives.

barium octahydrate. See barium hydroxide, octahydrate.

barium oxalate. $BaC_2O_4 \cdot HOH$.
Properties: White, crystalline powder; d 2.66; slightly soluble in water; soluble in dilute nitric or hydrochloric acid.
Hazard: See barium.
Use: Analytical reagent, pyrotechnics.

barium oxide. (barium monoxide; barium protoxide; calcined baryta). CAS: 1304-28-5. BaO.
Properties: White to yellowish-white powder, absorbs carbon dioxide readily from air. Soluble in acids and water, reacts violently with water to form the hydroxide. D 5.72, mp 1923C.
Derivation: Decomposition of carbonate at high temperature in presence of carbon, oxidation of barium nitrate.
Grade: Technical (regular grind) 208 lb/cu ft, technical fine grind (175 lb/cu ft), porous, carbide-free, 97%.
Hazard: Toxic by ingestion. See barium.
Use: Dehydrating agent for solvents, detergent for lubricating oils.

barium pentahydrate. See barium hydroxide pentahydrate.

barium perchlorate. $Ba(ClO_4)_2 \cdot 4H_2O$.
Properties: Colorless crystals, soluble in methanol and water, d 2.74, mp 505C.
Hazard: Oxidizer, fire and explosion risk in contact with organic materials. Toxic by ingestion. See barium.
Use: Manufacture of explosives, experimentally in rocket fuels.

barium permanganate. $Ba(MnO_4)_2$.
Properties: Brownish-violet crystals. Soluble in water.
Hazard: Oxidizing material. Fire and explosion risk in contact with organic materials. Toxic by ingestion. See barium.
Use: Strong disinfectant, manufacture of permanganates, depolarizing dry cells.

barium peroxide. (barium binoxide; barium dioxide; barium superoxide). CAS: 1304-29-6 BaO_2 and $BaO_2 \cdot 8HOH$.
Properties: Grayish-white powder, slightly soluble in water, d 4.96, mp 450C, decomposes 800C.
Derivation: By heating barium oxide in oxygen or air at approximately 1000F.
Grade: Technical, reagent.

Hazard: Oxidizing material. Fire and explosion risk in contact with organic materials. Keep cool and dry. Toxic by ingestion, skin irritant
Use: Bleaching, decolorizing glass, thermal welding of aluminum, manufacture of hydrogen peroxide, oxidizing agent, dyeing textiles.

barium phosphate, secondary. (barium di-o-phosphate). $BaHPO_4$.
Properties: White powder. Soluble in dilute nitric acid or dilute hydrochloric acid, slightly soluble in water, d 4.16.
Hazard: See barium.
Use: Flame retardant, phosphors.

barium phosphosilicate.
Use: Anticorrosive pigment for solvent-based epoxies and as auxiliary pigment for 1-package zinc-rich coatings. Also used in water-based coatings.

barium platinum cyanide. See barium cyanoplatinite.

barium potassium chromate. (Pigment E). $BaK(CrO_4)_2$.
Properties: Pale yellow pigment as compared with other chromate pigments, it has a low chloride and sulfate content and forms stronger, more elastic paint films, d 3.65.
Derivation: By a kiln reaction at 500C between potassium dichromate and barium carbonate.
Hazard: See barium.
Use: Component of anticorrosive paints for use on iron, steel and light metal alloys.

barium protoxide. See barium oxide.

barium pyrophosphate. $Ba_2P_2O_7$.
Properties: White powder, soluble in acids and ammonium salts, very slightly soluble in water.
Hazard: See barium.

barium selenide. BaSe.
Properties: Crystalline powder, d 5.0, decomposes in water.
Hazard: See barium.
Use: Semiconductors, photocells.

barium silicate. (barium metasilicate). $BaSiO_3$.
Properties: Colorless powder, d 4.4, bp 1604C, insoluble in water, soluble in acids.
Use: In ceramics.
Hazard: See barium.

barium silicide. $BaSi_2$.
Properties: Light-gray solid, evolves hydrogen on exposure to moisture.

Hazard: See barium.
Use: Metallurgy to deoxidize steel, etc.

barium silicofluoride. See barium fluosilicate.

barium-sodium niobate. A synthetic electro-optical crystal used to produce coherent green light in lasers, also to make such devices as electro-optical modulators and optical parametric oscillators. The crystal undergoes no optical damage from laser irradiation at high power levels.

barium stannate. $BaSnO_3 \cdot 3HOH$.
Properties: White, crystalline powder; sparingly soluble in water; readily in hydrochloric acid.
Hazard: See barium.
Use: Production of special ceramic insulations requiring dielectric properties.

barium stearate. $Ba(C_{18}H_{35}O_2)_2$.
Properties: White crystalline solid, insoluble in water or alcohol, mp 160C, d 1.145. Combustible.
Use: Waterproofing agent; lubricant in metalworking, plastics, and rubber; wax compounding; preparation of greases; heat and light stabilizer in plastics.

barium sulfate. (barytes (natural); blanc fixe (artificial, precipitated); basofor). $BaSO_4$.
Properties: White or yellowish, odorless, tasteless powder, soluble in concentrated sulfuric acid, d 4.25–4.5, particle size 2–25 microns, mp 1580C. Noncombustible.
Derivation: (a) By treating a solution of a barium salt with sodium sulfate (salt cake). (b) By-product in manufacture of hydrogen peroxide. (c) Occurs in nature as the mineral barite (Arkansas, Missouri, Georgia, Nevada, Canada, Mexico).
Grade: Technical, dry, pulp, bleached, ground, floated, natural, CP, USP, x-ray.
Use: Weighting mud in oil-drilling, paper coatings, paints, filler and delustrant for textiles, rubber, plastics and lithograph inks, base for lake colors, x-ray photography, opaque medium for gastrointestinal radiography, in battery plate expanders.

barium sulfide. (barium monosulfide; black ash). BaS.
Properties: Yellowish-green or gray powder or lumps. Soluble in water, decomposes to the hydrosulfide, d 4.25.
Derivation: Barium sulfate (crude barite) and coal are roasted in a furnace. The melt is lixiviated with hot water, filtered, and evaporated.
Impurities: Iron, arsenic.
Hazard: See barium.
Use: Dehairing hides, flame retardant, luminous

paints, barium salts, generating pure hydrogen sulfide.

barium sulfite. $BaSO_3$.
Properties: White powder, decomposed by heat. Soluble in dilute hydrochloric acid, insoluble in water.
Grade: Technical, CP.
Hazard: See barium.
Use: Analysis, paper manufacturing.

barium sulfocyanide.   See barium thiocyanate.

barium superoxide.   See barium peroxide.

barium tartrate. $BaC_4H_4O_6$.
Properties: White crystals, d 2.98, soluble in water, insoluble in alcohol.
Hazard: See barium.
Use: Pyrotechnics.

barium thiocyanate.   (barium sulfocyanide). $Ba(SCN)_2 \cdot 2HOH$.
Properties: White crystals. Soluble in water and in alcohol, deliquescent.
Derivation: By heating barium hydroxide with ammonium thiocyanate and subsequent crystallization.
Hazard: See barium.
Use: Making aluminum or potassium thiocyanates, dyeing, photography.

barium thiosulfate.   (barium hyposulfite). $BaS_2O_3 \cdot HOH$.
Properties: White crystalline powder, slightly soluble in water, insoluble in alcohol, d 3.5, decomposed by heat.
Hazard: See barium.
Use: Explosives, luminous paints, matches, varnishes, photography.

barium titanate. $BaTiO_3$.
Properties: Light gray-buff powder, mp 3010F, d 5.95, insoluble in water and alkalies, slightly soluble in dilute acids, soluble in concentrated sulfuric and hydrofluoric acids.
Use: Ferroelectric ceramics, single crystals either pure or doped with iron, are used in storage devices, dielectric amplifiers, and digital calculators.

barium tungstate.   (barium wolframate; barium white; tungsten white; wolfram white). $BaWO_4$.
Properties: White powder, insoluble in water, d 5.04.
Use: Pigment, in x-ray photography for manufacturing of intensifying and phosphorescent screens.

"Barium XA."[554]   TM for a product used by manufacturers of high quality tool steels. Elim-

inates chain type occlusions and degasifies the steel.

barium zirconate. $BaZrO_3$.
Properties: Light gray-buff powder, d 5.52, bulk d 140 lb/cu ft, mp 4550F, insoluble in water and alkalies, slightly soluble in acid.
Use: Manufacture of a white, easily colored silicone rubber compound having good heat stability at temperatures up to 500F; electronics.

barium zirconium silicate.   A complex of BaO, ZrO_2, SiO_2.
Properties: White powder, bulk d 118 lb/cu ft, mp 2800F, insoluble in water, alkalies, slightly soluble in acids, soluble in hydrofluoric acid.
Use: Production of electrical resistor ceramics, glaze opacifiers, stabilizer for colored ground coat enamels.

bark.   The cellulosic outer layer or cortex of trees and other woody plants. The bark of certain species such as oak, hemlock, etc., is a source of tannic acid; medicinal products, e.g., quercitrin and quillaja, are also derived from barks, especially cinchona from which quinine is obtained. Phenolic-rich bark extracts mixed with epichlorohydrin are reported useful as adhesive compounds. An unusual form of bark is cork from the oak species *Quercus suber*. In the pulp industry, bark is removed from logs with high-pressure jets of water.
See also hydraulic barking, cork, quinine.

barking, hydraulic.   See hydraulic barking.

barn.   A unit of measurement equal to 10^{-24} square centimeters for the cross-section target area of the nucleus of an atom.

"Barnesite."[88]   TM for a special rare earth for instrument lens polishing.

Barnett acetylation method.   Acetylation of hydroxy compounds such as cellulose with acid anhydrides in the presence of chlorine and sulfur dioxide. With cellulose, the process yields the diacetate below 65C and the triacetate above this temperature.

"Baroco."[236]   TM for a high yield clay compound for the preparation of drilling muds for use in formations containing moderate quantities of salt or other oil field electrolytes that may flocculate ordinary drilling muds.

"Baroid."[236]   TM for a weighting material made from selected barytes (barium sulfate ore). Added to drilling muds to increase unit weight thus increasing the hydrostatic head on the formations being drilled in deep wells to prevent the walls from caving.

barometric pressure. The pressure of the air at a particular point on or above the surface of the earth. At sea level, this pressure is sufficient to support a column of mercury approximately 29.9 inches in height (760 mm), equivalent to 14.7 pounds/square inch absolute (psia) or 1 atmosphere.

"Bar-O-Sil."[304] TM for a complex barium silicate vinyl stabilizer.
Properties: Fine white powder, d 2.5, refr index 1.5.
Use: Supplementary stabilizer for barium-cadmium and/or zinc types. Useful in film, sheeting, extrusions, and dispersion resin systems. Controls plating, hazing, crocking and fogging.

"Barosperse."[329] TM for a special barium sulfate formulation used in radiographic examinations of the gastrointestinal tract.

barrel finishing. Cleaning, smoothing, and polishing of metal or plastic items by mechanical friction obtained by placing them in drums or barrels which rotate on their horizontal axis. An abrasive medium and water are usually added. The barrels are six- or eight-sided, and often contain vertical dividers to make two or more compartments which can be individually loaded and unloaded. Such treatment is widely used for large-scale cleaning and burnishing of metal parts, which it finishes to exact dimensional tolerances much more economically than is possible by manual methods.

barrier, moisture. Any substance that is impervious to water or water vapor. Most effective are high polymer materials like vulcanized rubber, phenolformaldehyde resins, polyvinyl chloride, and polyethylene which are widely used as packaging films. The chief factors involved are polarity, crystallinity, and degree of cross-linking. Water-soluble surfactants and protective colloids increase the susceptibility of a film to water penetration. Any pigments and fillers must be completely wetted by the polymer. Properly formulated paints are effective moisture barriers.

barthrin. Generic name for a synthetic analog of allethrin described as the 6-chloropiperonyl ester of chrysanthemummonocarboxylic acid.
Use: As insecticide with applications similar to allethrin and other analogs as furethrin, ethythrin and cyclethrin. Relatively nontoxic to humans.

Barton, Derek H. R. (1918-) An English organic chemist who won the Nobel prize for chemistry in 1969 with Hassel. The field of conformational analysis in organic chemistry was initiated through his research in the terpene and steroid fields. He did extensive research in the area of carbanion autoxidations. He was instrumental in research concerning relationship of molecular rotation to structure in complex organic molecules. His education took place in London, France, and Ireland.

Barton reaction. Conversion of a nitrite ester to a γ-oximino alcohol by photolysis involving the homolytic cleavage of an N--O bond followed by hydrogen abstraction.

Bart reaction. (Scheller modification; Starkey modification). Formation of aromatic arsonic acids by treating aromatic diazonium compounds with alkali arsenites in the presence of cupric salts or powdered silver or copper; in the Scheller modification, primary aromatic amines are diazotized in the presence of arsenious chloride and a trace of cuprous chloride.

baryta, calcined. See barium oxide.

baryta, caustic. See barium hydroxide.

baryta water. A solution of barium hydroxide.

baryta yellow. See barium chromate.

barytes. See barium sulfate.

"Basacryl."[440] TM for a series of cationic dyestuffs for the dyeing and printing of polyacrylonitrile fiber.

basal metabolism. See metabolism.

"Basazol."[440] TM for dyes used in printing and dyeing paper composed of cellulosic fibers.

base. Any of a large class of compounds with one or more of the following properties: bitter taste, slippery feeling in solution, ability to turn litmus blue and to cause other indicators to take on characteristic colors, ability to react with (neutralize) acids to form salts. Included are both hydroxides and oxides of metals.
Water-soluble hydroxides such as sodium, potassium, and ammonium hydroxide, undergo ionization to produce hydroxyl ion (OH$^-$) in considerable concentration, and it is this ion that causes the previously mentioned properties common to bases. Such a base is strong or weak according to the fraction of the molecules which

break down (ionize) into positive ion and hydroxyl ion in the solution. Base strength in solution is expressed by pH. Common strong bases (alkalies) are sodium and potassium hydroxides, ammonium hydroxide, etc. These are caustic and corrosive to skin, eyes and mucous membranes. The pH range of basic solutions is from 7.1 to 14.

Modern chemical terminology defines bases in a broader manner. A Lowry-Bronsted base is any molecular or ionic substance that can combine with a proton (hydrogen ion) to form a new compound. A Lewis base is any substance that provides a pair of electrons for a covalent bond with a Lewis acid. Examples of such bases are hydroxyl ion and most anions, metal oxides, compounds of oxygen, nitrogen, sulfur with non-bonded electron pairs (such as water, ammonia, hydrogen sulfide).
For hard and soft bases, see Lewis electron theory.

BASF process. A process for producing acetylene by burning a mixture of low molecular weight hydrocarbons (as natural gas) with oxygen to produce a temperature of 1485C. The combustion products and cracked gases are quickly chilled by scrubbing with water and the acetylene is separated by distillation and solvent extraction from ethylene, carbon monoxide, hydrogen and other reaction products. The Sachsse process is similar.

basic. Descriptive of a compound that is more alkaline than other compounds of the same name, e.g., lead carbonate, basic, basic salt.

basic dichromate. See bismuth chromate.

basic fuchsin. (CI 42500). CAS: 569-61-9. A mixture of three parts pararosaniline acetate and one part pararosaniline hydrochloride.
Grades: Certifiable.
Use: For staining *tubercle bacillus* and in distinguishing between the *coli* and *aerogenes* types of bacteria in the Endo medium. Also used in the periodic acid-Schiff (PAS) method, the Feulgen stain, and in Gomoris aldehyde-function method for staining elastic tissue.

basic lining. A furnace lining containing basic compounds that decompose under furnace conditions to give basic oxides. The usual basic linings contain calcium and magnesium oxides or carbonates.

"Basicol."[188] TM for a series of essential oils intended as replacements for oils of lavender, geranium, lemon, pine, ylang ylang, neroli, and orris root.

basic oxide. An oxide which is a base or which forms a hydroxide when combined with water and/or which will neutralize acidic substances. Basic oxides are all metallic oxides, but there is a great variation in the degree of basicity. Some basic oxides such as those of sodium, calcium and magnesium combine with water with vigor or with relative ease and also neutralize all acidic substances rapidly and completely. The oxides of the heavy metals are only weakly basic, do not dissolve or react with water to any extent, and neutralize only the more strongly acidic substances. There is a gradual transition from basic to acidic oxides and certain oxides, such as aluminum oxide, show both acidic and basic properties.
See also base.

basic research. See fundamental research.

basic salt. A compound belonging in the category of both salt and base because it contains OH (hydroxide) or O (oxide) as well as the usual positive and negative radicals of normal salts. Among the best examples are bismuth subnitrate, often written $BiONO_3$, and basic copper carbonate, $Cu_2(OH)_2CO_3$. Most basic salts are insoluble in water and many are of variable composition.

basic slag. A slag produced in the manufacturing of steel. It contains a variable amount of tricalcium phosphate, calcium silicate, lime and oxides of iron, magnesium, and manganese. Used as a fertilizer for its phosphorus and lime.
See also slag.

basis metal. In electroplating the metal that is being coated constitutes the cathode and may be any of a large number of metals.

"Basogal phosphorus."[440] TM for leveling agent for vat dyeing.

bastnasite. An ore from which all nine of the lanthanide minerals (rare earths) are obtained. The only large deposit in the US is in southwest California.
See also monazite.

batch distillation. Distillation in which the entire sample of the material to be distilled, the charge, is placed in the still before the process is begun and product is withdrawn only from the condenser of the apparatus.

bating. In leather processing, the treatment of delimed skins with pancreatin or other tryptic enzyme to give a softer and smoother-grained product. The extent of bating varies from none for sole leather to 10 hours or more for soft kid skins. The chemical mechanism is not clearly defined.

batrachotoxinin A. $C_{24}H_{35}NO_5$. An isomeric component of batrachotoxin; strongest neurotoxin among venoms. It is a steroidal alkaloid. The A form is only 1/500th as strong as the complete venom, but is still as toxic as strychnine. It is found in the so-called poison dart frog of Colombia. Its structure has been elucidated; when synthesized it may prove useful in medicine.
See also snake venom.

battery. An electrochemical device that generates electric current by converting chemical energy to electrical energy. Its essential components are positive and negative electrodes made of more or less electrically conductive materials, a separating medium, and an electrolyte. There are four major types: (1) primary batteries (dry cells), which are not reversible and in which the anode (zinc) is the negative plate and the cathode (graphite) is the positive plate with ammonium chloride as electrolyte; (2) secondary or storage batteries, which are reversible and can be recharged and in which lead sponge is the negative plate (anode) and lead oxide the positive plate (cathode), with sulfuric acid as electrolyte; (3) nuclear and solar cells or energy converters; and (4) fuel cells. So-called superbatteries of high charge density have been developed using solid electrolytes of lithium-titanium dioxide and trilithium nitride in which lithium atoms are intercalated in the crystal structure. For further information, see dry cell, storage battery, voltaic cell, fuel cell, solar cell, intercalation.

battery acid. (electrolyte acid). Sulfuric acid of strengths suitable for use in storage batteries. Water-white, odorless, and practically free from iron.
Derivation: By diluting high-grade commercial sulfuric acid with distilled water to standard strengths.
Hazard: Corrosive to skin and tissue.
See sulfuric acid.
Use: Storage batteries.

battery limits. That portion of a chemical plant in which the actual processes are carried out, as distinguished from storage buildings, offices, and other subordinate structures called offsites.

batu. A variety of East India copal resin.
See East India.

Baudisch reaction. Synthesis of o-nitrosophenols from benzene or substituted benzenes, hydrox-

ylamine and hydrogen peroxide in the presence of copper salts.

Baumé. (Bé). An arbitrary scale of specific gravities devised by the French chemist Antoine Baumé and used in the graduation of hydrometers. The relations of specific gravity (at 60/60F) are as follows: Bé = 145−145/d for liquids heavier than water, Bé = 140/d - 130 for liquids lighter than water.

bauxite. A natural aggregate of aluminum-bearing minerals more or less impure in which the aluminum occurs largely as hydrated oxides. It is usually formed by prolonged weathering of aluminous rocks.
Contains 30–75% Al_2O_3, 9–31% HOH, 3–25% Fe_2O_3, 2–9%, SiO_2, 1–3% TiO_2.
Properties: White cream, yellow, brown, gray, or red, d 2–2.55, Mohs hardness 1–3, insoluble in water, decomposed by hydrochloric acid. Noncombustible.
Occurrence: Australia, Jamaica, France, Guiana, Guinea, US (Arkansas), Brazil.
Use: Most important ore of aluminum, aluminum chemicals, abrasives, aluminous cement, refractories, decolorizing and deodorizing agent, catalysts, filler in rubber, plastics, paints, cosmetics, hydraulic fracturing.
See also Bayer process, Hall process.
Note: Due to increasing cost of bauxite the use of other aluminum-containing minerals is under active investigation.

Bayer process. Process for making alumina from bauxite. The main use of alumina is in the production of metallic aluminum. Bauxite is mixed with hot concentrated sodium hydroxide, which dissolves the alumina and silica. The silica is precipitated and the dissolved alumina is separated from the solids, diluted, cooled, and then crystallized as aluminum hydroxide. The aluminum hydroxide is calcined to anhydrous alumina which is then shipped to reduction plants.
See also Hall process.

Bayer's acid. See crocein acid.

"Baygon."[181] TM for o-isopropoxyphenyl methylcarbamate.

bay oil. See myrcia oil.

"Bayol."[51] TM for technical-grade white mineral oils that are widely used where USP or NF quality is not required.

"Baytex."[181] TM for o,o-dimethyl (o-(4-(methylthio)-m-tolyl)phosphorothioate.
See fenthion.

BBO. See 2,5-dibiphenylyloxazole.

BBP. Abbreviation for butyl benzyl phthalate.

BCWL. Abbreviation for basic carbonate white lead.
See lead carbonate, basic.

"BDA."[233] TM for inhibited hydrochloric acid solution containing surfactants.
Use: In limestone and dolomite formations, in oil-well fracturing and acidizing.

Be. Symbol for beryllium.

Bé. Abbreviation for Baumé.

bead. (1) In a rubber-fabric composite (tires, transmission belts), the point at which the cut edges of the fabric meet after being folded over. A length of pure gum rubber, called a bead strip, is used to seal the joint. The bead must be removed from tires before reclaiming--an operation called debeading. (2) See microspheres.

beater. An open, oval tank into which digested paper pulp is fed together with water and other processing ingredients such as clay, rosin, pigments, etc. The resulting slurry (approximately 5% solids content) is strongly agitated by a rotating drum equipped with closely spaced, horizontal, fin-like projections which effectively disintegrate and macerate the wood fibers while the slurry is continuously circulated. The time of beating varies with the type and quality of the paper being made. Conical refiners (called Jordans) are often used to complete the beating operation for high-grade papers. The beating operation is critical in converting wood pulp to paper.

Bechamp reaction. Formation of p-amino or p-hydroxyphenylarsonic acids by heating aromatic amines or phenols with arsenic acid. The arsonylation requires an active hydrogen atom, and is practically limited to the benzene series. Apart from para substitution, a small amount of ortho arsonylation is observed, particularly when the para position is blocked.

"Beckacite."[36] TM for fumaric, maleic, and modified phenolic resins.

"Beckamine."[36] TM for urea-formaldehyde resins.

Beckmann rearrangement. The conversion of a ketone oxime to a substituted amide by an intermolecular rearrangement brought about by a catalyst. For example, the oxime of cyclohexa-none is converted into caprolactam with sulfuric acid as catalyst.

$$\underset{\substack{\|\\ \text{NOH}}}{R'C-R} \longrightarrow \underset{\substack{\|\\ R'N}}{RC-OH} \longrightarrow \underset{\substack{|\\ R\ N-H}}{R-C=O}$$

"Beckosol."[36] TM for alkyd resins used at coating vehicles.

Bacquerel, Henri. (1851–1908). A French physicist who shared the Nobel Prize in physics with the Curies for the discovery of the radioactivity of uranium salts. He also discovered the deflection of electrons by a magnetic field, as well as the existence and properties of gamma radiation.

BDTA.[522] See 3,3'4,4'-benzophenone tetracarboxylic dianhydride.

beer. See brewing.

beerstone. A deposit occurring during brewing operations on containers and consisting of calcium oxalate and organic material.

beeswax. Wax from the honeycomb of the bee. Beeswax consists largely of myricyl palmitate, cerotic acid and esters, and some high-carbon paraffins.
Properties: Brown or white (bleached) solid with faint odor, d 0.95, melting range 62–65C, insoluble in water, slightly soluble in alcohol, soluble in chloroform, ether, and oils. Combustible.
Grade: Technical, crude, refined, NF, FCC, USP (white).
Use: Furniture and floor waxes, shoe polishes, leather dressings, anatomical specimens, artificial fruit, textile sizes and finishes, church candles, cosmetic creams, lipsticks, adhesive compositions.

beet sugar. See sucrose

behenic acid. (docosanoic acid).
$CH_3(CH_2)_{20}COOH$. A saturated fatty acid, a minor component of the oils of the type of peanut and rapeseed.
Properties: Solid, mp 80.0C, bp 306C (60 mm), 265C (15 mm), d 0.8221 (100/4C), refr index 1.4270 (100C). Combustible.
Derivation: Occurs in bean oil, hydrogenated mustard oil, and rapeseed oil.
Grade: Technical, 99%.
Use: Cosmetics, waxes, plasticizers, chemicals, stabilizers.

behenone. $C_{22}H_{44}O$. An aliphatic ketone. Insoluble in water, inert, compatible with high-

melting waxes, fatty acids. Incompatible with resins, polymers, and organic solvents at room temperature but compatible with them at high temperature.
Use: As an antiblocking agent.

behenyl alcohol. (1-docosanol).
$CH_3(CH_2)_{20}CH_2OH$.   A long-chain, saturated fatty alcohol.
Properties: Colorless waxy solid; mp 71C, bp 180C (0.22 mm). Insoluble in water; soluble in ethanol, and chloroform. Combustible.
Derivation: Reduction of behenic acid with lithium aluminum hydride as catalyst.
Grade: Technical; 99%.
Use: Synthetic fibers; lubricants; evaporation retardant on water surfaces.

Beilstein, F. P. (1838–1906). A German chemist noted for his compilation "Handbuch der Organischen Chemie," the first edition of which appeared in 1880. A multi-volume compendium of the properties and reactions of organic compounds, it has been revised several times and remains a unique and fundamental contribution to chemical literature.

BEK. See butyl ethyl ketene.

belladonna. (deadly nightshade; banewort).
An herbaceous perennial bush (Atropa belladonna) of which the leaves and roots are used for their content of hyoscyamine and atropine.
Occurrence: Southern and central Europe, Asia Minor, Algeria; cultivated in North America, England, France.
Grade: Belladonna leaf, USP; belladonna root.
Hazard: Very toxic when high in atropine.
Use: Medicine (gastrointestinal relaxant).

bemberg. A cuprammonium rayon fiber. Flammable, not self-extinguishing.

"Bemul."[345]   TM for a practically odorless emulsifying agent; a pure white, edible glycerol monostearate in bead form; mp 58–59C, completely dispersible in hot water; completely soluble in alcohols and hot hydrocarbons.
Use: Pharmaceuticals, cosmetics, and foodstuffs; protective coating for edible hygroscopic powders, tablets, and crystals; pour-point depressant for lubricating oils; textile sizes; etc.

Benadryl. Proprietary name of diphenhydramine hydrochloride.

"Ben-A-Gel."[304]   TM for a highly beneficiated hydrous magnesium silicate for aqueous systems. Used for thickening or gelling water systems,

or as a suspension agent, and an emulsion stabilizer in oil-in-water emulsions.

Benary reaction. Action of Grignard reagents on enamino ketones or aldehydes yields β-substituted α,β-unsaturated ketones or aldehydes.

bench gas. See coal gas.

bendiocarb. $C_{11}H_{13}NO_4$.
Properties: White powder; mp 130C; slightly soluble in water.
Hazard: Poison by ingestion, skin absorption.
Use: Contact insecticide.

Benedict solution. A water solution of sodium carbonate, copper sulfate, and sodium citrate. The blue color changes to a red, orange, or yellow precipitate or suspension in the presence of a reducing sugar such as glucose, and is therefore used in testing for such materials, especially for urinalysis in the treatment of diabetes.
See Fehling's solution.

beneficiation. A process used in extractive metallurgy whereby an ore, either metallic or nonmetallic, is concentrated in preparation for further processing (smelting). Calcination is often an important step in beneficiation; others are physical separation of high-grade ore from impurities (gangue) by screening, washing, milling, or magnetic means. A process for removing sulfur from coal by chemical comminution has been developed.

benefin. (N-butyl-N-ethyl-α,α,α-tri-fluoro-2,6-dinitro-p-toluidine).
$C_6H_2(NO_2)_2CF_3NC_6H_{14}$.
Properties: Yellow-orange solid; mp 65–66.5C; bp 121–122C (0.5 mm). Slightly soluble in water, readily soluble in acetone and xylene.
Hazard: Highly toxic.
Use: Herbicide.

Benjamin gum. See benzoin resin.

benomyl. (methyl-1-(butylcarbamoyl)-2-benzimidazole-carbamate).   CAS: 17804-35-2.
Generic name for a post-harvest fungicide for peaches, apples, etc. Also used as oxidizer in sewage treatment.
Hazard: High toxicity by ingestion. TLV: 10 mg/m^3 of air.

bensulide. (N-(2-mercaptoethyl)benzenesulfonamide).   CAS: 741-58-2.
Use: Herbicide.

benthos. The botton-dwelling life of an ocean or freshwater environment.

"Bentone."[304] TM for organic derivatives of hydrous magnesium aluminum silicate minerals. Use: Gelling and pigment-suspending agents.

bentonite. A colloidal clay (aluminum silicate) composed chiefly of montmorillonite. There are two varieties: (1) sodium bentonite (Wyoming or western), which has high swelling capacity in water; and (2) calcium bentonite (southern), with negligible swelling capacity.
Properties: (Wyoming) Light to cream-colored impalpable powder; forms colloidal suspension in water, with strongly thixotropic properties.
Occurrence: Wyoming, Mississippi, Texas, Canada, Italy, USSR.
Use: Oil-well drilling fluids; cement slurries for oil-well casings; bonding agent in foundry sands and pelletizing of iron ores; sealant for canal walls; thickener in lubricating greases and fireproofing compositions; cosmetics; decolorizing agent; filler in ceramics, refractories, paper coatings; asphalt modifier; polishes and abrasives; food additive; catalyst support.
See also clay.

benzalacetone. See benzylidene acetone.

benzalazine. (benzylidene azine).
$C_6H_5CH:NN:CHC_6H_5$.
Properties: Yellow crystals; mp 91–93C; insoluble in cold water; soluble in benzene, hot alcohol.
Use: Stabilizer; polymerization catalyst; UV absorbent; reagent and intermediate.

benzal chloride. See benzyl dichloride.

benzaldehyde. (benzoic aldehyde; synthetic oil of bitter almond).　　CAS: 100-52-7.
C_6H_5CHO.

Properties: Colorless or yellowish, strongly refractive, volatile oil with odor resembling oil of bitter almond, and burning aromatic taste; oxidizes readily; miscible with alcohol, ether, fixed and volatile oils; slightly soluble in water. D 1.0415 (25/4C), refr index 1.5440–1.5464 at 20C, fp −56C; bp 178C. Flash p 145F (62.7C) (CC). Oxidizes in air to benzoic acid. Combustible. Autoign temperature 377F (191.6C).
Derivation: (a) Air oxidation of toluene with uranium or molybdenum oxides as catalysts; (b) reaction of benzyl dichloride with lime; (c) extraction from oil of bitter almond.

Impurities: Usually chlorides. Method of purification: Rectification.
Grade: Technical; NF. Note: The specifications, especially regarding impurities, vary considerably for the grades used for dye manufacture from those used in perfumery.
Hazard: Highly toxic.
Use: Chemical intermediate for dyes, flavoring materials, perfumes, and aromatic alcohols; solvent for oils, resins, some cellulose ethers, cellulose acetate and nitrate; flavoring compounds; synthetic perfumes; manufacturing of cinnamic acid, benzoic acid; pharmaceuticals; photographic chemicals.

benzaldehyde cyanohydrin. See mandelonitrile.

benzaldehyde green. See Malachite Green.

benzalkonium chloride. A mixture of alkyl dimethyl-benzylammonium chlorides of general formula $C_6H_5CH_2N(CH_3)_2RCl$ in which R is a mixture of the alkyls from C_8H_{17} to $C_{18}H_{37}$. It is a typical quaternary ammonium salt.
Properties: White or yellowish-white, amorphous powder or gelatinous pieces. Aromatic odor and very bitter taste; soluble in water, alcohol, or acetone; almost insoluble in ether; slightly soluble in benzene. Water solutions foam strongly when shaken and are alkaline to litmus.
Grade: USP.
Hazard: Highly toxic.
Use: Cationic detergent; surface antiseptic; fungicide.

benzamide. (benzoylamide).　　$C_6H_5CONH_2$.

Properties: Colorless crystals; mp 130C, bp 288C, d 1.341. Soluble in hot water, hot benzene, alcohol, and ether. Combustible.
Derivation: From benzoyl chloride and ammonia or ammonium carbonate.
Grades: Technical.
Use: Organic synthesis.

benzaminoacetic acid. See hippuric acid.

benzanilide. (benzoylaniline; phenylbenzamide). $C_6H_5NH(COC_6H_5)$.
Properties: White to reddish crystals and powder, related to acetanilide, containing benzoyl in place of acetyl radical. D 1.306; mp 160–162C. Soluble in alcohol; insoluble in water; slightly soluble in ether.

Derivation: From benzoic anhydride and aniline with sodium hydroxide.

Use: Intermediate in the synthesis of dyes, drugs and perfumes.

benzanthrone. $C_{17}H_{10}O$. A four-ring system.
Properties: Pale yellow needles; soluble in alcohol and other organic solvents. Mp 170C.
Derivation: (a) From anthranol and glycerol via condensation via sulfuric acid (anthranol is made from anthraquinone); (b) from anthracene in sulfuric acid solution by addition of glycerol and heating to 100–110C until the anthracene disappears. The reaction mass is then diluted with water, salted out and purified.
Method of purification: Crystallization from toluene.
Use: Dyes.

benzathine penicillin G. (N,N′-dibenzylethylenediamine dipenicillin G).
$2C_{16}H_{18}N_2O_4S \cdot C_{16}H_{20}N_2 \cdot 4H_2O$.
Properties: White, odorless, crystalline powder; slightly soluble in alcohol; almost insoluble in water; pH of a saturated solution is 4.5–7.5.
Grade: USP.
Use: Medicine (antibiotic).

benzazimide. See 4-ketobenzotriazine.

"Benzedrine."[71] TM for amphetamine sulfate.

benzene. CAS: 71-43-2. C_6H_6. 16th highest-volume chemical produced in US (1985).

I   II

III   IV

Structure: I. Complete ring showing all elements. II. Standard ring showing double bonds only. III. Simple ring without double bonds, with numerals indicating position of carbon atoms to which substituent atoms or groups may be attached (2 = ortho, 3 = meta, 4 = para). IV. Generalized structure with enclosed circle suggesting the resonance of this compound. This structure is now in general use. These structures are also referred to as the benzene nucleus.
Properties: Colorless to light-yellow, mobile, nonpolar liquid of highly refractive nature; aromatic odor; vapors burn with smoky flame; bp 80.1C; fp 5.5C, d 0.8790 (20/4C); wt/gal 7.32 lb; refr index 1.50110 at 20C, flash p 12F (−11C) (CC), surface tension 29 dynes/cm. Autoign temperature 1044F (562C). Miscible with alcohol, ether, acetone, carbon tetrachloride, carbon disulfide, acetic acid; slightly soluble in water.
Derivation: (a) Hydrodealkylation of toluene or pyrolysis of gasoline; (b) transalkylation of toluene by disproportionation reaction; (c) catalytic reforming of petroleum; (d) fractional distillation of coal tar.
Grade: Crude, straw color; motor; industrial pure (2C); nitration (1C); thiophene-free; 99 mole %; 99.94 mole %; nanograde.
Hazard: A carcinogen. Highly toxic. Flammable; dangerous fire risk. Explosive limits in air 1.5 to 8% by volume. TLV: 10 ppm in air.
Use: Manufacturing of ethylbenzene (for styrene monomer); dodecylbenzene (for detergents); cyclohexane (for nylon); phenol; nitrobenzene (for aniline); maleic anhydride; chlorobenzene; diphenyl; benzene hexachloride; benzene-sulfonic acid; as a solvent.
See also aromatic.

benzene azimide. See 1,2,3-benzotriazole.

benzeneazoanilide. See diazoaminobenzene.

benzeneazobenzene. See azobenzene.

benzeneazo-p-benzeneazo-β-naphthol. ("Sudan" III; tetraazobenzene-β-naphthol).
$C_{22}H_{16}ON_4$. A red dye; CI 26100.444.
Properties: Brown powder; mp 195C, insoluble in water; soluble in alcohol, oils, chloroform, glacial acetic acid.
Use: Coloring oils red; biological stain.

benzeneazonaphthylethylenediamine. See azodine.

benzenecarboxylic acid. See benzoic acid.

benzenediazonium chloride. $C_6H_5N(N)Cl$.
Properties: Ionic salt. Very soluble in water; insoluble in most organic solvents.
Hazard: Highly toxic. Can explode on heating.
Use: Dye intermediate.

benzene dibromide. See dibromobenzene.

benzene-o-dicarboxylic acid. See phthalic acid.

benzene-p-dicarboxylic acid. See terephthalic acid.

benzene hexachloride. (BHC). A commercial mixture of isomers of 1,2,3,4,5,6-hexachlorocyclohexane. An insecticide.
Hazard: The gamma isomer is highly toxic. Use may be restricted.
See also lindane.

benzenemonosulfonic acid.   See benzenesulfonic acid.

benzenephosphinic acid.   (phenylphosphinic acid). $C_6H_5H_2PO_2$.
Properties: Colorless crystals; mp 82–84C; d 1.376 (29C). Decomposes at 200C. Stable in air. Soluble in water, alcohol, acetone. Slightly soluble in ether; insoluble in benzene, hexane, CCl_4. Combustible.
Use: Antioxidant; intermediate for metallic salt formation; accelerator for organic peroxide catalysts.

benzenephosphonic acid.   (phenylphosphonic acid). $C_6H_5H_2PO_3$.
Properties: Colorless crystals. Mp 158C; d 1.475 (4C); decomposes at 275C; soluble in water, alcohol, CCl_4. Combustible.
Hazard: Highly toxic.
Use: Intermediate in antifouling paint agents; catalyst in organic reactions.

benzenephosphorus dichloride. $C_6H_5PCl_2$.
Properties: Highly reactive, colorless liquid. Mp −51C; bp 224.6C; d 1.315 (25C); refr index 1.5958 (25C). Soluble in common inert organic solvents; fumes in air; hydrolyzes in water.
Hazard: Highly corrosive to skin, tissue. Flammable.
Use: Organic synthesis, for derivation of plasticizers, polymers, antioxidants; oil additives.

benzenephosphorus oxydichloride. $C_6H_5POCl_2$.
Properties: Reactive colorless liquid. Mp 3.0C; bp 258C; d 1.197 (25C); refr index 1.5585 at 25C. Soluble in common inert organic solvents; hydrolyzes in water. Combustible.
Hazard: Strong irritant to skin, mucous membranes.
Use: Organic synthesis, for derivation of plasticizers, polymers, antioxidants, oil additives.

benzenesulfonic acid.   (benzenemonosulfonic acid; phenylsulfonic acid).   CAS: 42615-29-2. $C_6H_5SO_3H$.

Properties: Fine, deliquescent needles or large plates; mp 65–66C when anhydrous; with 1.5 molecules water, mp 43–44C; soluble in water, alcohol; slightly soluble in benzene; insoluble in ether and carbon disulfide.
Derivation: By reacting benzene with fuming sulfuric acid.
Hazard: Irritant to skin, eyes, mucous membranes.
Use: Manufacturing of phenol, resorcinol and other organic syntheses, and as a catalyst.

benzene-1,3,5-tricarboxylic acid chloride.
See trimesoyl trichloride.

benzenoid.   Any organic compound containing or derived from the benzene ring structure, e.g., phenol, nitrobenzene, anthracene, styrene. This large array of unsaturated compounds, derived chiefly from petroleum and coal tar, provides a broad base for the synthesis of polymers, dyes, intermediates.
See also aromatic.

benzenyl trichloride.   See benzotrichloride.

benzethonium chloride. $C_{27}H_{42}ClNO_2$.
A synthetic quaternary ammonium compound,
Properties: Colorless, odorless plates. Very bitter taste; mp 164–166C. Soluble in water, alcohol, acetone. Aqueous solution yields flocculent white precipitate with soap solutions.
Grade: NF.
Hazard: An oral poison.
Use: Antiseptic; cationic detergent.

benzhydrol.   (benzohydrol; diphenylmethanol; diphenylcarbinol).   CAS: 91-01-0. $(C_6H_5)_2CHOH$.

Properties: Needlelike, colorless crystals; mp 69C; bp 298C, 176C (13 mm). Slightly soluble in water, easily soluble in alcohol, ether, chloroform, and carbon disulfide. Combustible.
Derivation: Reduction of benzophenone with magnesium or zinc dust.
Use: Preparation of certain antihistamines; insecticide.

benzhydryl bromide.   See diphenylmethyl bromide.

benzhydryl chloride. $(C_6H_5)_2CHCl$.
Properties: Water-white to light straw-colored liquid; refr index 1.596; mp 16C; bp 140C (3 mm). Combustible.
Use: Organic synthesis.

benzidine.   (benzidine base; p-diaminodiphenyl).   CAS: 92-87-5. $NH_2(C_6H_4)_2NH_2$.

$$H_2N-\langle\rangle-\langle\rangle-NH_2$$

Properties: Grayish-yellow, white, or reddish gray crystalline powder; mp 127C; bp 400C; soluble in hot water, alcohol, ether; slightly soluble in cold water. Combustible. Also available as the hydrochloride.

Derivation: (a) By reducing nitrobenzene with zinc dust in alkaline solution followed by distillation; (b) by electrolysis of nitrobenzene, followed by distillation; (c) nitration of diphenyl followed by reduction of the product with zinc dust in alkaline solution, with subsequent distillation.

Grade: Technical (paste; powder 80–85%).

Hazard: Human carcinogen. Highly toxic by ingestion, inhalation, skin absorption.

Use: Organic synthesis; manufacture of dyes, especially of Congo red; detection of blood stains; stain in microscopy; reagent; stiffening agent in rubber compounding.

benzidinedicarboxylic acid. See diaminodiphenic acid.

benzidine dye. Any of a group of azo dyes derived from 3,3'-dichloro- benzidine; they include yellow and orange colors claimed to be light-fast and alkali-resistant. Congo Red is derived from benzidine and naphthionic acid.

Hazard: These compounds are highly toxic and carcinogens. Physical contact with them should be avoided.

benzidine rearrangment; semidine rearrangement. The acid-catalyzed rearrangement of hydrazobenzenes to 4,4'- diaminobiphenyls. If the hydrazobenzene contains a para substituent, the product is a p-aminodiphenylamine (semidine rearrangement).

benzidine sulfate. CAS: 531-86-2.
$C_{12}H_{12}N_2 \cdot H_2SO_4$.

Properties: White, crystalline powder. Soluble in ether; sparingly soluble in water, alcohol, dilute acids.

Derivation: Action of sulfuric acid or sodium sulfate on benzidine with subsequent recovery by precipitation.

Hazard: Poison by ingestion, skin absorption. A carcinogen.

Use: Organic synthesis.

benzil. (dibenzoyl). $C_6H_5CO \cdot COC_6H_5$.

Properties: Yellow needles. Soluble in alcohol, ether; insoluble in water. Mp 95C; bp 346–348C; d 1.521.

Derivation: From benzoin by oxidation with HNO_3.

Use: Organic synthesis; insecticide.

benzilic acid. (diphenylglycolic acid). $(C_6H_5)_2C(OH)COOH$.

Properties: White to tan powder with a characteristic odor; mp 148–151C; soluble in hot water, alcohol. Combustible.

Use: Chemical intermediate.

benzilic acid rearrangement. Rearrangement of benzyl to benzilic acid on treatment with base.

benzimidazole. (1,3-benzodiazole; azindole; benzoglyoxaline).   CAS: 51-17-2. $C_7H_6N_2$.

Properties: Tabular crystals. Mp 172–174C, mw 118.13, bp greater than 360C. Weak base sparingly soluble in cold water and ether. Freely soluble in alcohol, practically insoluble in benzene and petroleum ether.

benzine. The name "benzine" is archaic and misleading and should not be used. (ASTM Petroleum Definitions D-288.) Do not confuse with benzene.
See also ligroin.

benzocaine. See ethyl-p-aminobenzoate.

benzodihydropyrone. (dihydrocoumarin). $C_6H_8O_2$ (bicyclic).

Properties: White to light yellow, oily liquid with a sweet odor; congeals at 23C. Insoluble in water; soluble in alcohol, chloroform, ether. Combustible.

Use: Perfumery.

"Benzoflex."[316] TM for a series of plasticizers which are dibenzoate esters of dipropylene glycol or any of several polyethylene glycols.

Use: Primary plasticizer for vinyl resins; adhesive formulations; some grades in food packaging adhesives.

benzofuran. See coumarone.

benzoglycolic acid. See mandelic acid.

benzoguanamine. (2,4-diamino-6-phenyl-s-triazine). $C_6H_5C_3N_3(NH_2)_2$.

Properties: Crystals; d 1.40; mp 227–228C. Soluble in methyl"Cellosolve ", alcohol, dilute hydrochloric acid; partially soluble in dimethylformamide, acetone; practically insoluble in chloroform, ethyl acetate; insoluble in water, benzene, ether. Combustible.

Derivation: Benzonitrile and dicyandiamide in the presence of sodium and liquid ammonia.

Use: Thermosetting resins, resin modifiers; chemical intermediate for pesticides, pharmaceuticals, and dyestuffs.

benzohydrol. See benzhydrol.

benzoic acid. (carboxybenzene; benzenecarboxylic acid; phenylformic acid). CAS: 65-85-0. C_6H_5COOH. It occurs naturally in benzoin resin.

Properties: White scales or needle crystals with odor of benzoin or benzaldehyde; d 1.2659; mp 121.25C; bp 249.2C; partially sublimes at 100C; freely volatile in steam; flash p 250F (121.1C) (CC); soluble in alcohol, ether, chloroform, benzene, carbon disulfide, carbon tetrachloride, turpentine; slightly soluble in water. Combustible.

Derivation: (a) Decarboxylation of phthalic anhydride in the presence of catalysts; (b) chlorination of toluene to yield benzotrichloride, which is hydrolyzed to benzoic acid; (c) oxidation of toluene; (d) from benzoin resin.

Method of purification: Sublimation.

Grade: Technical; CP; USP; FCC.

Hazard: Moderately toxic by ingestion. Use restricted to 0.1% in foods.

Use: Sodium and butyl benzoates; plasticizers; benzoyl chloride; alkyd resins; food preservative; seasoning tobacco; flavors; perfumes; dentifrices; standard in analytical chemistry; antifungal agent.

benzoic aldehyde. See benzaldehyde.

benzoic anhydride. $(C_6H_5CO)_2O$.

Properties: Colorless prisms; d 1.198; mp 42C; bp 360C; refr index 1.576. Soluble in most organic solvents; insoluble in water.

Use: Dyes, intermediates, pharmaceuticals (benzoylating agent); organic synthesis.

benzoic trichloride. See benzotrichloride.

benzoin. (bitter almond-oil camphor; benzoylphenyl carbinol; 2-hydroxy-2-phenylacetophenone; phenylbenzoyl carbinol). CAS: 119-53-9 $C_6H_5CHOHCOC_6H_5$.

Properties: White or yellowish crystals; slight camphor odor; mp 137C; slightly soluble in water, ether; soluble in acetone, hot alcohol. Optically active. Combustible.

Derivation: Condensation of benzaldehyde in an alkaline cyanide solution.

Hazard: Highly toxic.

Use: Organic synthesis; intermediate; photopo-

lymerization catalyst. Note: Do not confuse with benzoin resin.

benzoin condensation. Cyanide-ion catalyzed condensation of aromatic aldehydes to give benzoins (acyloins).

α-benzoin oxime. (benzoin antioxime). $C_6H_5CHOHC:NOHC_6H_5$.

Properties: Solid; mp 150–152C.

Use: Organic intermediates and photographic chemicals; analytical reagent for determination of metals.

benzoin resin. (gum benzoin; Benjamin gum).

Properties: Reddish-brown globules; balsamic, vanilla-like odor; brittle at room temperature, but softened by heat. Soluble in warm alcohol and carbon disulfide; insoluble in water.

Source: Obtained from the Styrax tree in Southeast Asia and Sumatra. The Sumatran grade is higher-melting and only 75% soluble in alcohol.

Grade: Technical; tincture USP.

Constituents: Benzoic acid, cinnamic acid, vanillin.

Use: Source of benzoic acid; perfumery; cosmetics; medicine (antiseptic and expectorant). Note: Do not confuse with benzoin.

benzol. Obsolete name for benzene, no longer in approved use.

benzonitrile. (phenyl cyanide). CAS: 100-47-0. C_6H_5CN.

Properties: Colorless oil; almond-like odor; sharp taste; viscosity (100F) 1.054 centistokes; refr index 1.5289; soluble in boiling water, alcohol, ether; slightly soluble in cold water. D 1.0051; bp 190.7C; fp −13.1C.

Derivation: From benzoic acid by heating with lead thiocyanate.

Hazard: High toxicity; absorbed by skin.

Use: Manufacture of benzoguanamine; intermediate for rubber chemicals; solvent for nitrile rubber, specialty lacquers, and many resins and polymers, and for many anhydrous metallic salts.

benzophenol. See phenol.

benzophenone. (diphenylketone). CAS: 119-61-9. $(C_6H_5)_2CO$.

Properties: White prisms with sweet, rose-like odor. Partially soluble in alcohol, ether; soluble in chloroform; insoluble in water; mp 47.5C; bp 305C. Combustible.

Purification: Crystallization from alcohol.

Grade: Free from chlorine (FFC); also FCC.

Use: Organic synthesis; odor fixative; derivatives are used as ultraviolet absorbers; flavoring; soap fragrance; pharmaceuticals; polymerization inhibitor for styrene.

benzophenone oxide. See xanthone.

3,3′,4,4′-benzophenone tetracarboxylic dianhydride. (BTDA). $C_{17}H_6O_7$.
Properties: Free-flowing powder; mp 228C.
Use: Epoxy curing agent; heat-resistant polymers, specialty alkyd resins: polyesters and plasticizers.

benzopyrene. (3,4-benzpyrene).
CAS: [a] 50-32-8. $C_{20}H_{12}$. A polynuclear (five-ring) aromatic hydrocarbon. Found in coal tar, cigarette smoke, and in the atmosphere as a product of incomplete combustion.
Occurs as benzo[a]pyrene and benzo[e]pyrene.
Properties: (benzo[a]pyrene) Yellowish crystals; mp 179C; bp 310–312C (10 mm). Insoluble in water; slightly soluble in alcohol; soluble in benzene, toluene, xylene.
Hazard: Highly toxic; carcinogen by inhalation.

benzopyrone. See coumarin.

benzoquinone. See quinone.

benzosulfimide. See saccharin.

benzothiazole. C_6H_4SCHN (bicyclic).
Properties: Yellow liquid with unpleasant odor; d 1.246; refr index 1.637; bp 227C; slightly soluble in water; soluble in alcohol. Combustible.
Hazard: Highly toxic by ingestion.
Use: Derivatives used as rubber accelerators.

benzothiazolyl disulfide. See 2,2′-dithiobisbenzothiazole.

benzothiazyl-2-cyclohexylsulfenamide.
See N-cyclohexyl-2-benzothiazole- sulfenamide.

2-benzothiazyl-N,N-diethylthiocarbamyl sulfide.
(diethyldithiocarbamic acid-2-benzothiazoyl ester). $(C_6H_4SCN)SSCN(C_2H_5)_2$.
Properties: Free-flowing, light yellow to tan powder; d 1.27; mp 69C (min).
Use: Rubber accelerator.

benzothiazyl disulfide. See 2,2′-dithiobisbenzothiazole.

1,2,3-benzotriazole. (aziminobenzene; benzene azimide). $C_6H_4NHN_2$.
Properties: White to light tan, odorless, crystalline powder; b range 201–204C (15 mm); very stable toward acids and alkalies, and toward oxidation and reduction. Its basic characteristics are very weak but it forms stable metallic salts. Can exist in two tautomeric forms. Soluble in alcohol, benzene; slightly soluble in water.
Hazard: Highly toxic by ingestion; may explode under vacuum distillation.
Use: Photographic restrainer; chemical intermediate; derivatives are ultraviolet absorbers.

benzotrichloride. (toluene trichloride; benzenyl trichloride; benzoic trichloride; phenylchloroform). $C_6H_5CCl_3$.
Properties: Colorless to yellowish liquid; fumes in air; hydrolyzes in water; penetrating odor. Soluble in alcohol, ether; insoluble in water. D 1.38; bp 220C; fp −5C, refr index 1.5584.
Derivation: Chlorination of boiling toluene.
Method of purification: Rectification.
Hazard: Highly toxic by inhalation; fumes highly irritant.
Use: Synthetic dyes; organic synthesis.

benzotrifluoride. (toluene trifluoride; trifluoromethylbenzene). $C_6H_5F_3$.
Properties: Water-white liquid with aromatic odor. Bp 102.1C; fp −29.1C; d 1.1812 (25/4C); refr index 1.4146. Flash p 54F (12.2C) (CC). Miscible with alcohol, acetone, benzene, carbon tetrachloride, ether, n-heptane; insoluble in water.
Hazard: Highly toxic by inhalation. Flammable, dangerous fire risk. TLV (as F): 2.5 mg/m³ of air.
Use: Intermediate for dyes, and pharmaceuticals; solvent and dielectric fluid; vulcanizing agent; insecticides.

trans-β-benzoylacrylic acid.
$C_6H_5COCH:CHCOOH$.
Properties: Straw yellow needles or plates; mp 99C; soluble in most solvents but only slightly soluble in cold water and ligroin. Combustible.
Use: Reagent for characterizing phenols; intermediate in the manufacturing of bactericides, insecticides, surface-active agents, and the upgrading of drying oils.

N-benzoyl-l(+)-alanine. See alanine.

benzoylamide. See benzamide.

benzoylaminoacetic acid. See hippuric acid.

benzoylaniline. See benzanilide.

benzoyl chloride. CAS: 98-88-4. C_6H_5COCl.

Properties: Transparent, colorless liquid; pungent odor; vapor causes tears. D 1.2188; fp −0.5C; bp 197.2C; refr index 1.5536 (20C), flash p 162F (72.2C). Soluble in ether, carbon disulfide; decomposes in water. Combustible.

Derivation: (a) Interaction of benzoic acid and sulfuryl chloride; (b) benzotrichloride and water in the presence of zinc chloride; (c) phosphorus tri- or pentachloride and benzoic acid.

Grade: Technical, CP.

Hazard: Highly toxic, strong irritant to skin, eyes, mucous membranes, and via ingestion, inhalation.

Use: Introduction of benzoyl group; dye intermediates; benzoyl peroxide manufacturing; analytical reagent.

benzoyl-2,5-diethoxyaniline.
$C_6H_5CONHC_6H_3(OC_2H_5)_2$.
Properties: Gray pellets; mp 83–84C.
Hazard: Possibly toxic.
Use: Intermediate for pharmaceuticals, dyestuffs, and other organic chemicals.

benzoylferrocene.   (phenyl ferrocenyl ketone).
$C_5H_5FeC_5H_4COC_6H_5$.
Properties: Dark red crystalline solid; mp 107–108C.
Hazard: Possibly toxic.
Use: Intermediate.

benzoyl fluoride. C_6H_5COF.
Hazard: High toxicity; TLV (as F): 2.5 mg/m³ of air.
Use: Manufacturing of acyl and other fluorides.

benzoylglycin.   See hippuric acid.

benzoylglycocoll.   See hippuric acid.

benzoyl peroxide.   (dibenzoyl peroxide).
CAS: 94-36-0. $(C_6H_5CO)_2O_2$.

Properties: White, granular, crystalline solid; tasteless; faint odor of benzaldehyde. Active oxygen, approximately 6.5%. Soluble in nearly all organic solvents; slightly soluble in alcohols, vegetable oils; slightly soluble in water. Mp 103–105C; decomposes explosively above 105C. Autoign temperature 176F (80C); d 1.3340 (25C).
Grade: Technical, wet or dry; FCC.
Hazard: Highly toxic via inhalation. May explode spontaneously when dry <1% of water). Never mix unless at least 33% water is present. TLV: 5 mg/m³ of air. See also peroxides, organic.

Use: Bleaching agent for flour, fats, oils, and waxes; polymerization catalyst; drying agent for unsaturated oils; pharmaceutical and cosmetic purposes; rubber vulcanization without sulfur; burn out agent for acetate yarns; production of cheese; embossing vinyl flooring (proprietary).

benzoylphenyl carbinol.   See benzoin.

2-benzoylpyridine. $C_6H_5COC_5H_4N$.
Properties: Colorless liquid. Fp 42.7C; insoluble in water.
Grade: 98% (minimum).
Use: Organic synthesis.

4-benzoylpyridine. $C_6H_5COC_5H_4N$.
Properties: Colorless liquid. Fp 71.4C; insoluble in water.
Grade: 98% (minimum).
Use: Organic synthesis.

benzoylsulfonic imide.   See saccharin.

benzozone.   See acetyl benzoyl peroxide.

1,2-benzphenanthrene.   See chrysene.

benzpyrene.   See benzopyrene.

benzyl abietate. $C_{19}H_{29}COOCH_2C_6H_5$.
Properties: Nonvolatile, viscous liquid which resembles Canada balsam. Soluble in most anhydrous solvents.   See also balsam.

benzyl acetate.   (phenylmethyl acetate).
$C_6H_5CH_2OOCCH_3$.
Properties: Water-white liquid; floral odor. Soluble in alcohol, ether. Slightly soluble in water, soluble in alcohol. Flash p 216F (102.2C) (CC). D 1.059–1.062 (15C); bp 212C; refr index 1.5015–1.5035. Combustible. Autoign temperature 862F (460C).
Derivation: (a) By treating benzyl chloride with sodium acetate in various solvents; (b) by esterification of benzyl alcohol with acetic anhydride or acetic acid.
Method of purification: Distillation.
Grade: Free-from-chlorine grade which should have an ester content of 97% but for which lower grade material is sometimes substituted; technical grade which is not free from chlorine and for which ester content varies considerably; FCC.
Hazard: Highly toxic by inhalation; skin irritant.
Use: Artificial jasmine and other perfumes; soap perfume; flavoring; solvent and high boiler for cellulose acetate and nitrate, natural and synthetic resins; oils; lacquers; polishes; printing inks; varnish removers.

benzyl alcohol.   (α-hydroxytoluene; phenylmethanol; phenylcarbinol).   CAS: 100-51-6.
$C_6H_5CH_2OH$.

Properties: Water-white liquid; slight odor; sharp, burning taste. Bp 206C; flash p 220F (105C) (OC); d 1.040–1.050 (25/25C); refr index 1.5385–1.5405 (20C). somewhat soluble in water; miscible with alcohol, ether, chloroform. Combustible. Autoign temperature 817F (436C).

Derivation: (a) By hydrolysis of benzyl chloride; (b) from benzaldehyde by catalytic reduction or Cannizzaro reaction.

Method of purification: Distillation and chemical treatment.

Grade: Free from chlorine (FFC); technical; NF; textile; photographic; reagent; FCC.

Hazard: Highly toxic.

Use: Perfumes and flavors; photographic developer for color movie films; dyeing nylon filament, textiles and sheet plastics; solvent for dyestuffs, cellulose esters, casein, waxes, etc.; heat-sealing polyethylene films; intermediate for benzyl esters and ethers; bacteriostat; cosmetics, ointments, emulsions; ball point pen inks; stencil inks.

benzylamine.   (aminotoluene).
$C_6H_5CH_2NH_2$.

Properties: Light amber liquid; strongly alkaline reaction. Soluble in alcohol, ether, water. D 0.9813; bp 184.5C; refr index 1.540 at 20C; Combustible.

Derivation: From benzyl chloride and ammonia.

Hazard: Highly toxic, strong irritant to skin and mucous membranes.

Use: Chemical intermediate for dyes, pharmaceuticals, and polymers.

N-benzyl-p-aminophenol.   (BAP).
$C_6H_5CH_2NHC_6H_4OH$.

Properties: Light brown powder, melts between 84 and 90C; 96–99% pure; solubility 50% in anhydrous methanol, 50% in 95% ethanol, 0.06% in water; 0.1–0.5% in gasoline, varying with chemical nature of gasoline.

Use: In cracked gasoline, in concentration of 0.001–0.004% by weight to prevent gum formation.

2-benzylamino-1-propanol.
$C_6H_5NHCH(CH_2OH)CH_3$.

Properties: White to yellow solid. Both l and dl-forms are available. Mp (dl-form) 70–73C. Spe-

cific rotation (l-form) +38° to +44° (1.0% solution in alcohol) at 25C. Combustible.

benzylaniline. $C_6H_5NHCH_2C_6H_5$.
Properties: Colorless prisms. Soluble in alcohol, ether; insoluble in water. Mp 33C; bp 310C.
Use: Organic synthesis.

benzylbenzene.   See diphenylmethane.

benzyl benzoate.   CAS: 120-51-4.
$C_6H_5CH_2OOCC_6H_5$.
Properties: Water-white liquid; readily freezes. Sharp, burning taste and faint aromatic odor. Supercools easily. Insoluble in water, glycerol; soluble in alcohol, chloroform, ether. D 1.116–1.120 (25/25C); bp 325C; mp 18.8C; refr index 1.568–1.569 (20C). Flash p 298F (147.7C). Combustible.
Derivation: (a) By a Cannizzaro reaction from benzaldehyde; (b) by esterifying benzyl alcohol with benzoic acid; (c) by treating sodium benzoate with benzyl chloride. Method of purification: Distillation and crystallization.
Grade: Technical; USP; FCC.
Hazard: Irritant to eyes, skin.
Use: Fixative and solvent for musk in perfumes and flavors; medicine (external); plasticizer for nitrocellulose and cellulose acetate; miticide.

benzyl bromide.   (α-bromotoluene).
$C_6H_5CH_2Br$.
Properties: Clear, refractive liquid. Pleasant odor. Not easily hydrolyzed. Soluble in alcohol, benzene, ether; insoluble in water. A lachrymator. D 1.438 at 16C; bp 198–199C; fp −3.9C; vap d 5.8.
Derivation: (a) Bromination of toluene; (b) interaction of benzyl alcohol and hydrobromic acid.
Hazard: Highly toxic. Corrosive to skin and tissue.
Use: Making foaming and frothing agents; organic synthesis.

benzyl butyrate. $C_3H_7COOCH_2C_6H_5$.
Properties: Liquid; fruity odor; bp 240C; d 1.016 (17.5C); soluble in alcohol. Combustible.
Grade: Technical; FCC.
Use: Plasticizer; odorants; flavoring.

benzyl carbinol.   See phenethyl alcohol.

benzyl "Cellosolve."   See ethylene glycol monobenzyl ether.

benzyl chloride.   (α-chlorotoluene).
CAS: 100-44-7. $C_6H_5CH_2Cl$.
Properties: Colorless liquid; pungent odor. A lachrymator. D 1.090–1.111 (25/25C); fp −43C; bp 179C; refr index 1.5365 (25C), flash p 153F (67.2C) (OC). Combustible. Autoign tempera-

ture 1085F (525C). Soluble in alcohol, ether; insoluble in water.

Derivation: By passing chlorine over boiling toluene until it has increased 38% in weight. The product is washed with water and separated by fractional distillation.

Grade: Technical; CP; 95%; redistilled.

Forms: Anhydrous; stabilized (with aqueous sodium carbonate solution).

Hazard: Highly toxic, intense eye and skin irritant. TLV: 1 ppm in air.

Use: Dyes; intermediates; benzyl compounds; synthetic tannins; perfumery; pharmaceuticals; manufacture of photographic developer; gasoline gum inhibitors; penicillin precursors; quaternary ammonium compounds.

benzyl chlorocarbonate. (benzyl chloroformate). C_6H_5OCOCl.

Properties: Oily liquid with lachrymatory properties; odor of phosgene; decomposes above 100C; reacts with water to form hydrochloric acid.

Hazard: Highly toxic, emits very toxic phosgene fumes at 100C. Irritant to eyes.

Use: Peptide synthesis.

benzyl chloroformate. See benzyl chlorocarbonate.

o-**benzyl**-p-**chlorophenol.** (chlorophene, USAN; "Santophen"; septiphene; 4-chloro-α-phenyl-o-cresol). $C_6H_5CH_2C_6H_3OHCl$.

Properties: White to light tan or pink flakes; crystallizing point 45C min; d 1.202–1.206 (55/55C); odor slight phenolic. Insoluble in water; highly soluble in alcohol, other organic solvents; dispersible in aqueous media with the aid of soaps or synthetic dispersing agents; noncorrosive to most metals. Combustible.

Hazard: Highly toxic, an irritant.

Use: Active principle or enhancing agent for disinfectants.

benzyl cinnamate. (cinnamein). $C_9H_7O_2 \cdot C_7H_7$.

Properties: White crystals; aromatic odor; mp 39C; congeal point (min) 34C; bp 244C (25 mm); insoluble in water; soluble in alcohol.

Hazard: Moderately toxic.

Grade: Technical; FCC.

Use: Perfumery and flavors.

benzyl cyanide. (phenylacetonitrile; α-tolunitrile). $C_6H_5CH_2CN$.

Properties: Colorless oily liquid; aromatic odor; soluble in alcohol, ether; insoluble in water. D 1.0157; fp −24C; bp 230C; refr index 1.5211 (25C).

Derivation: Interaction of benzyl chloride and potassium cyanide.

Grade: Technical.

Hazard: Highly toxic, absorbed by skin. TLV (as CN): 5 mg/m³ of air.

Use: Organic synthesis; especially penicillin precursors.

benzyl dichloride. (benzylidene chloride; benzal chloride; chlorobenzal; α, α-dichlorotoluene). $C_6H_5CHCl_2$.

Properties: Colorless, oily liquid; faint aromatic odor; d 1.295 (16C); fp −16.1C; bp 207C; refr index 1.5502 (20C); soluble in alcohol, ether, dilute alkali; insoluble in water. Combustible.

Derivation: Chlorination of toluene, until two formula weights of chlorine are absorbed, in absence of catalysts but presence of light.

Hazard: Strong irritant and lachrymator.

Use: Dyes; manufacture of benzaldehyde and cinnamic acid.

benzyl ether. (dibenzyl ether).
CAS: 103-50-4. $(C_6H_5CH_2)_2O$.

Properties: Colorless, unstable liquid; bp 295C; d 1.001; flash p 275F (135C); refr index 1.54; insoluble in water; soluble in alcohol, ether, acetone.

Derivation: Benzaldehyde is reduced with cobalt complex, $[Co(CO)_4]_2$.

Hazard: Moderate by ingestion; skin irritant.

Use: Solvent; plasticizer for nitrocellulose.

N-benzylidiethanolamine.
$(C_6H_5CH_2N(C_2H_4OH)_2$.

Properties: Colorless to light yellow liquid; miscible with water. D 1.073; refr index 1.5345–1.5375; distilling range 240–255C. Combustible.

Use: Corrosive inhibitor; intermediate.

N-benzylidimethylamine. $C_6H_5CH_2N(CH_3)_2$.

Properties: Colorless to light yellow liquid. D 0.894 (27C); refr index 1.4985–1.5005 (25C); bp 180–182C; distilling range 65–68C (18 mm). Combustible.

Use: Intermediate, especially for quaternary ammonium compounds; dehydrohalogenating catalyst; corrosion inhibitor; acid neutralizer; potting compounds; adhesives; cellulose modifier.

benzyl disulfide. See dibenzyl disulfide.

N-benzylethanolamine.
$C_6H_5CH_2NH(C_2H_4OH)$.

Properties: Colorless to light yellow liquid. D 1.044 (27C); refr index 1.5400–1.5430; distillation range 240–255C. Combustible.

Use: Corrosion inhibitor; intermediate.

benzyl ethyl ether. $C_6H_5CH_2OC_2H_5$.

Properties: Colorless, oily liquid; aromatic odor; volatile in steam; insoluble in water; miscible

with alcohol, ether. Bp 185C; d 0.949; refr index 1.4955 (20C). Combustible.
Derivation: By boiling benzyl chloride with either sodium or potassium ethylate.
Hazard: Narcotic in high concentration; may be skin irritant.
Use: Organic synthesis; flavoring.

benzyl ethylsalicylate.
$C_6H_5CH_2OOCC_6H_4OC_2H_5$.
Used as fixative and solvent in perfumes.

benzyl fluoride. $C_6H_5CH_2F$.
Properties: Colorless liquid. Forms acicular crystals on prolonged cooling. D 1.022 at 25C; bp 139.8C (753 mm); fp −35C.
Derivation: By decomposing benzyltrimethylammonium fluoride.
Hazard: Very irritant. TLV (as F): 2.5 mg/m³ of air.
Use: Organic synthesis.

benzyl formate. $C_6H_5CH_2OOCH$.
Properties: Colorless liquid; fruity-spicy odor. Resembles benzyl acetate in many respects but differs in its greater volatility. D 1.083–1.087; refr index 1.511–1.513; bp 203C; miscible with alcohols, ketones, oils, aromatic, aliphatic and halogenated hydrocarbons; insoluble in water.
Hazard: May be narcotic in high concentration.
Use: Perfumes, flavoring; solvent for cellulose esters.

benzyl fumarate.
$C_6H_5CH_2OOCCH{=}CHCOOCH_2C_6H_5$.
Properties: White powder; mp 59C; bp 210C (5 mm); insoluble in water; soluble in alcohol, ether.
Derivation: Reaction of fumaric acid and benzyl alcohol.
Use: Spray deodorant.

benzylhexadecyldimethylammonium chloride.
See cetalkonium chloride.

benzylhydroquinone. See p-benzyloxyphenol.

benzylidene acetone. (benzalacetone; methyl styryl ketone; 4-phenyl-3-buten-2-one).
$C_6H_5CH{=}CHCOCH_3$.
Properties: Colorless crystals; odor of coumarin. Soluble in alcohol, ether, benzene, chloroform; insoluble in water. Mp 72C; congeal point 39C (min); bp 260–262C; refr index 1.5836 (46C), d 1.0377 (15/15C). Combustible.
Derivation: Condensation of benzaldehyde and acetone.
Use: Organic synthesis; perfumery; fixative; flavoring.

benzylidene azine. See benzalazine.

benzylidene chloride. See benzyl dichloride.

2-benzylidene-1-heptanol. See α-amylcinnamic alcohol.

benzyl iodide. $C_6H_5CH_2I$.
Properties: Colorless crystals or liquid. Soluble in alcohol, carbon disulfide, ether; insoluble in water. D 1.7335; mp 34.1C; bp decomposes.
Derivation: Interaction of benzyl chloride and hydriodic acid.
Hazard: Powerful irritant.

benzyl isoamyl ether. See isoamyl benzyl ether.

benzyl isobutyl ketone. See 4-methyl-1-phenyl-2-pentanone.

benzyl isoeugenol. (1-α-phenyl-4-propenylveratrole).
$CH_3CHCHC_6H_3(OCH_3)OCH_2C_6H_5$.
Properties: White, crystalline solid; floral odor of the carnation type. Soluble in alcohol, ether. Combustible.
Use: Perfumery; fixative.

N-benzylisopropylamine.
$C_6H_5CH_2NH(CH_3CHCH_3)$.
Properties: Colorless to yellow liquid; d 0.895 (25C); refr index 1.4995–1.5015 (25C). Combustible.
Use: Rust inhibitor; intermediate.

benzyl isothiocyanate. $C_6H_5CH_2NCS$.
Properties: Colorless to slightly yellow liquid; a lachrymator.
Hazard: Very irritant to tissues.
Use: Chemical intermediate.

benzyl mercaptan. See benzyl thiol.

benzylmethylamine. $C_6H_5CH_2NHCH_3$.
Properties: Colorless to light yellow liquid. D 0.936 (25C); refr index 1.5185–1.5220 (25C); distillation range 183–188C. Combustible.
Use: Organic synthesis.

N-benzyl-N,N-methylethanolamine.
$C_6H_5CH_2NCH_3(C_2H_4OH)$.
Properties: Colorless to light yellow liquid; d 1.006 (27C); refr index 1.5250–1.5270 (25C); distillation range 95–105C (2 mm). Combustible.
Use: Corrosive inhibitor; intermediate.

3-benzyl-4-methyl umbelliferone.
$C_6H_5CH_2CH_3C_9H_4O_3$.
Properties: Tan, crystalline powder; mp 255C min; slightly soluble in ethanol; insoluble in water.
Use: Optical whitening agent; intermediate.

p-**benzyloxyphenol.** (benzylhydroquinone; "Agerite Alba"). $C_6H_5CH_2OC_6H_4OH$.
Properties: Light tan powder; d 1.26; faint odor; mp 121–122C. Slightly soluble in water; practically insoluble in petroleum hydrocarbons; very soluble in benzene and alkalies. Combustible.
Use: Rubber antioxidant; stabilizer; polymerization inhibitor; chemical intermediate.

benzyl pelargonate. $C_6H_5CH_2OOCC_8H_{17}$.
Properties: Liquid; d 0.962 (15.5/15.5C); bp 315C; mild odor.
Use: In flavors and perfumes; bactericides and fungicides; organic synthesis.

p-**benzylphenol.** (4-hydroxydiphenylmethane). $C_6H_5CH_2C_6H_4OH$.
Properties: White crystals from ethanol; mp 84C; bp 320–322C. Soluble in ethanol, ether, chloroform, benzene, acetic acid, caustic alkalies; moderate solubility in hot water. Combustible.
Hazard: Toxic by ingestion.
Use: Antiseptic and germicide; organic synthesis.

benzyl phenylacetate.
$C_6H_5CH_2COOCH_2C_6H_5$.
Properties: Colorless liquid; honeylike odor. Soluble in alcohol; d 1.097–1.099; refr index 1.554–1.556. Combustible.
Use: Perfumery and flavors.

benzyl phenyl ketone. See deoxybenzoin.

benzyl propionate. $C_2H_5COOCH_2C_6H_5$.
Properties: Similar to benzyl acetate but has sweeter odor. Liquid. Bp 220C; d 1.036 (17.5C); insoluble in water. Combustible.
Grade: Technical; FCC.
Use: Perfumes; flavoring.

benzylpyridine. $C_6H_5CH_2C_5H_4N$.
Properties: Liquid. Bp 276.8C; mp 13.6C; d 1.061 (20C); refr index 1.5797. Insoluble in water. Combustible.
Hazard: Toxic by ingestion.

benzyl salicylate.
$C_6H_4(H_4OH)COOCH_2C_6H_5$.
Properties: Colorless liquid; faint sweet odor. Soluble in 9 vols of 90% alcohol. D 1.176–1.179; refr index 1.580–1.581; mp 24C min; bp 208C(26 mm). Combustible.
Derivation: Reaction of sodium salicylate and benzyl chloride.
Grade: Technical; FCC.
Use: Perfume fixative; solvent for synthetic musk; sun-screening lotions; soap odorant.

benzyl succinate. (dibenzyl succinate).
$C_6H_5CH_2OOCCH_2CH_2COOCH_2C_6H_5$.
Properties: White crystalline powder, almost tasteless. Soluble in alcohol, ether, chloroform, also

in fixed and volatile oils; insoluble in water. Mp 45C. Combustible.
Use: Medicine (anti-spasmodic).

benzyl sulfide. $(CH_2C_6H_5)_2S$.
Properties: Colorless plates. Soluble in alcohol, ether; insoluble in water. D 1.0712; mp 49C.
Derivation: Action of potassium sulfide on benzyl chloride and subsequent distillation.
Use: Organic synthesis.

benzyl thiocyanate. $C_6H_5CH_2CNS$.
Properties: Colorless crystals; mp 41C; bp 230C; insoluble in water; soluble in alcohol, ether.
Hazard: Strong irritant to skin, tissue. Moderate fire hazard;
Use: Insecticide.

benzyl thiol. (benzyl mercaptan; α-toluenethiol). $C_6H_5CH_2SH$.
Properties: Colorless liquid; bp 195C; insoluble in water; soluble in alcohol, carbon disulfide; flash p 158F (70C) (CC); d 1.05; strong odor; combustible.
Hazard: Toxic by inhalation and ingestion; irritant to tissue.
Use: Odorant; flavors.

2-**benzyl-6-thiouracil.** $C_6H_5CH_2C_3N_2OS$.
A drug intermediate.

benzyltrimethylammonium chloride.
$C_6H_5CH_2N(CH_3)_3Cl$.
Properties: A quaternary ammonium salt. Colorless crystals; stable up to 135C, above which benzyl chloride and trimethylamine are formed. Readily soluble in water, ethanol, and butanol; slightly soluble in butyl phthalate and tributyl phosphate. Properties of 60% solution: D 1.07 (20/20C); wt/gal 8.90 lb; fp below -50C.
Grade: 60–62% aqueous solution.
Use: Solvent for cellulose; gelling inhibitor in polyester resins; intermediate.

benzyltrimethylammonium hexafluorophosphate.
$C_6H_5CH_2N(CH_3)_3PF_6$.
Properties: Crystals; mp 160C.
Hazard: Toxic by ingestion, irritant to skin.

4-**benzyltrimethylammonium methoxide.**
$C_6H_5CH_2(CH_3)_3NOCH_3$.
Properties: A quaternary ammonium salt. Yellow liquid; decomposes on distillation.
Hazard: Toxic by ingestion, irritant to skin.
Use: Catalyst; organic soluble strong base.

benzyne. C_6H_4.
Properties: An unsaturated, cyclic hydrocarbon with a structure similar to benzene, in which

one of the double bonds is replaced by a triple bond. It may be prepared from benzenediazonium-2-carboxylate or from isatoic anhydride.

"Beraloy."[155] TM for beryllium-copper alloys supplied in two grades: "Beraloy" A (1.80–2.05% beryllium) and "Beraloy" D (1.60–1.80% beryllium). "Beraloy" A meets ASTM Specifications B-194-51T and B-197-51T.
Properties: High electrical conductivity; high resistance to fatigue, very low hysteresis or drift; easily formed when annealed; high strength and rigidity when heat treated; corrosion resistant.
Forms: Strip, round wire, flat wire.
Use: Diaphragms; springs; fabrication of lightweight, intricate parts.

berberine. $C_{20}H_{19}NO_5$.
Properties: White to yellow crystals; mp 145C (anhydrous); insoluble in water; soluble in ether, alcohol. Salts of berberine are: berberine bisulfate, berberine sulfate, and berberine hydrochloride. All three are yellow crystals, slightly soluble in water.
Derivation: From the root of Berberis vulgaris or Hydrastis canadensis.
Hazard: Toxic via ingestion, inhalation, skin absorption.
Use: Medicine (antipyretic).

Berg, Paul. (1926-) An American molecular biologist who won the Nobel prize for chemistry in 1980 with Sanger and Gilbert. Berg's research concerned the biochemistry of nucleic acid particularly regarding the recombinant DNA, that is a molecule containing DNA's from different species. His PhD was attained at Western Reserve, and he has been at Stanford University since 1970.

bergamot oil.
Properties: An essential oil. Brownish-yellow to green liquid; agreeable odor; bitter taste.
Use: In perfumery.

Bergius, Frederick. (1884-1949) A German chemist who won a Nobel prize in 1931 with Bosch for chemical high pressure methods. He invented a method of coal dust conversion into oil via pressurized hydrogen. He also invented a method for production of cattle feed and sugar from wood by hydrolysis. He was educated in Poland and Germany.

Bergius process. Formation of petroleum-like hydrocarbons by hydrogenation of coal at high temperatures and pressures (e.g., 450C and 300 atm) with or without catalysts; production of toluene by subjecting aromatic naphthas to cracking temperatures at 100 atm with a low partial pressure of hydrogen in the presence of a catalyst.

Bergius-Willstatter saccharification process.
Process for industrial production of fermentable sugar from wood by hydrolysis of tannin and xylan-free cellulose with 40-45% hydrochloric acid. The use of concentrated acid requires acid-resistant equipment and recovery of acid. The sugar produced must be rehydrolyzed prior to fermentation.

Bergmann azlactone peptide synthesis. Conversion of an acetylated amino acid and an aldehyde into an azlactone with an alkylene side chain, reaction with a second amino acid with ring opening and formation of an acylated unsaturated dipeptide, followed by catalytic hydrogenation and hydrolysis to the dipeptide.

Bergmann degradation. Stepwise degradation of polypeptides involving benzoylation, conversion to azides and treatment of the azides with benzyl alcohol; this treatment yields, via rearrangement to isocyanates, carbobenzoxy compounds which undergo catalytic hydrogenation and hydrolysis to the amide of the degraded peptide.

Bergmann-Zervas carbobenzoxy method. Formation of the N-carbobenzoxy derivative of an amino acid for use in peptide synthesis and liberation of the amino group at an appropriate stage of synthesis by hydrogenolysis of the activated CO bond.

beri-beri. A disease caused by a deficiency of vitamin B_1 in the diet.

berkelium. Bk. Synthetic radioactive element with atomic number 97 first produced (1949) as the 243 isotope by bombarding americium with helium ions in a cyclotron. The chemical properties of berkelium have been studied by tracer techniques and are similar to those of the other transuranium elements. Its oxidation behavior is similar to that of the rare earth cerium. It has a mp of 986C. There are 8 isotopes ranging from 243 to 250; the 249 isotope has been made by neutron bombardment of curium 244. Atomic weight uncertain; 249 is generally accepted. The following compounds have been identified by x-ray diffraction: berkelium dioxide (BkO_2), berkelium sesquioxide (Bk_2O_3), berkelium trifluoride (BkF_3), berkelium trichloride ($BkCl_3$), and berkelium oxychloride (BkOCl).

Berlin blue. Any of a number of the varieties of iron blue pigments.
See iron blue.

Berlin red. A red pigment consisting, essentially, of red iron oxide.

Berthelot, M. PE. (1827–1917). A French chemist who did fundamental work on the organic synthesis of hydrocarbons, fats, and carbohydrates. Opposed the then current idea that a "vital force" is responsible for synthesis. Did important work on explosives for French government. He was one of the first to prove that all chemical phenomena depend on physical forces that can be measured.

bertholite. A name given to chlorine when used as a poison gas.

Berthollet, Claude Louis. (1748–1822). A French chemist. Followed Lavoisier, but did not accept the latter's contention that oxygen is the characteristic constituent of acids. He was the first to propose chlorine as a bleaching agent. His essay on chemical physics (1803) was the first attempt to explain this subject. His speculations on stoichiometry, especially as regards relative masses of reacting atoms, profoundly affected later theories of chemical affinity.

beryl. $Be_3Al_2(SiO_3)_6$. Sometimes with replacement of beryllium by sodium, lithium, cesium. A natural silicate of beryllium and aluminum.
Hazard: A known carcinogen (OSHA).
See beryllium.

beryllia. See beryllium oxide.

beryllides. Intermetallic compounds made by chemically combining beryllium with such metals as zirconium and tantalum.

beryllium. CAS: 7440-41-7. Be. Metallic element of atomic number 4, group IIA of the periodic system. Atomic weight 9.0121. Valence 2; no stable isotopes.
Properties: A hard, brittle, gray-white metal; d 1.85; mp 1280C; soluble in acids (except nitric) and alkalies. Resistant to oxidation at ordinary temperatures. High heat capacity and thermal conductivity. It is the lightest structural metal known; can be fabricated by rolling, forging, and machining. Joining is chiefly by shrink-fitting; brazing and welding are difficult. Highly permeable to x-rays.
Occurrence: Beryl, the ore of beryllium, is found chiefly in South Africa, Zimbabwe, Brazil, Argentina, and India. Principal sources in US are Colorado, Maine, New Hampshire, and South Dakota. There are undeveloped deposits in Canada.
Derivation: The ore is converted to the oxide or hydroxide, then to the chloride or fluoride. The halide may be (a) reduced in a furnace by magnesium metal, or (b) reduced by electrolysis. (c) Liquid-liquid extraction with an organophosphate chelating agent can be used as a method of purification, or as an alternative process on the ore itself.
Grade: Technical, over 99.5% pure.
Forms: Hot-pressed or cold-pressed and sintered blocks; sheet (0.04 inch); tube; rods; wire; powder.
Hazard: A carcinogen (OSHA). Very high toxicity, especially by inhalation of dust. TLV: 0.002 mg/m³ of air.
Use: Structural material in space technology; moderator and reflector of neutrons in nuclear reactors; source of neutrons when bombarded with alpha particles; special windows for x-ray tubes; in gyroscopes, computer parts, inertial guidance systems; additive in solid propellant rocket fuels; beryllium-copper alloys.

beryllium acetate. CAS: 543-81-7.
$Be_4O(C_2H_3O_2)_6$.
Properties: White crystals; mp 285–286C, bp 330–331C; insoluble in water; hydrolyzed by hot water, dilute acids. Soluble in chloroform and other organic solvents. Can be crystallized from hot acetic acid in very pure form.
Hazard: Toxic by inhalation and ingestion. See also beryllium.
Use: Source of pure beryllium salts.

beryllium acetylacetonate. $Be(C_5H_7O_2)_2$.
Properties: Crystalline powder; slightly soluble in water; resistant to hydrolysis. A chelating nonionizing compound. Mp 108C; bp 270C; freely soluble in alcohol, ether; slightly soluble in water.
Hazard: Toxic. See also beryllium.

beryllium carbide. Be_2C.
Properties: Fine, hexagonal, hard, refractory crystals; attacked vigorously by strong, hot alkali solutions forming methane gas and alkali beryllate. D 1.91; decomposes above 2100C.
Derivation: By direct interaction of elemental beryllium and carbon; by reduction of beryllium oxide with carbon above 1500C.
Use: Nuclear reactor cores.

beryllium chloride. CAS: 7787-47-5. $BeCl_2$.
Properties: White or slightly yellow deliquescent crystals; sweetish taste. Mp 440C; bp 520C; d 1.90. Very soluble in water; soluble in alcohol, benzene, ether, carbon disulfide. Readily hydrolyzed.
Derivation: By passing chlorine over a mixture of beryllium oxide and carbon.
Hazard: Very toxic. See also beryllium.
Use: In dry form, as catalyst for organic reactions.

beryllium-copper. A precipitation-hardenable alloy; often also contains nickel or cobalt, and has relatively high electrical conductivity, high strength, and high hardness.

Properties: D 8.22. Tensile strength of heat-treated sheet 175,000 psi, elongation 5% in 2 inches; Brinell hardness 350; good electrical conductivity. Typical analysis: copper 97.4; beryllium 2.25; nickel 0.35.

Hazard: Avoid inhalation. See also beryllium.

Use: In electrical switch parts; watch springs; optical alloys; electronic equipment; valves and parts; spot-welding electrodes; nonsparking tools; springs and diaphragms; shims; cams; and bushings.

Note: A comparatively recent development is an 85 copper, 9 nickel, 6 tin alloy reported to be 15% stronger than Be-Cu.

beryllium fluoride. BeF_2.

Properties: Hygroscopic solid; mp 800C; d 1.986. Readily soluble in water; sparingly soluble in alcohol.

Derivation: By the thermal decomposition (900–950C) of ammonium beryllium fluoride.

Hazard: A known carcinogen. Toxic by inhalation and ingestion. TLV (as F): 2.5 mg/m³ of air.

Use: Production of beryllium metal by reduction with magnesium metal; nuclear reactors; glass manufacturing.

beryllium hydrate. See beryllium hydroxide.

beryllium hydride. BeH_2.

Properties: White solid. Reacts with water, dilute acids, methanol, liberating hydrogen. When heated to 220C it liberates hydrogen rapidly.

Hazards: For toxicity see beryllium. Fire risk when exposed to water, organic materials, and heat.

Use: Experimentally in rocket fuels.

beryllium hydroxide. (beryllium hydrate). CAS: 13327-32-7. $Be(OH)_2$.

Properties: White powder; decomposes to the oxide at 138C; soluble in acids, alkalies, insoluble in water.

Derivation: By precipitation with alkali from pure beryllium acetate, basic.

Grade: Technical.

Hazard: Very toxic. See also beryllium.

beryllium metaphosphate. $Be(PO_3)_2$.

Properties: White porous powder or granular material; has a high melting point; insoluble in water.

Hazard: Very toxic. See also beryllium.

Use: Raw material for special ceramic compositions; catalyst carrier.

beryllium nitrate. CAS: 13597-99-4. $Be(NO_3)_2$ $3H_2O$.

Properties: White to faintly yellowish, deliquescent mass; mp 60C; decomposes 100–200C; soluble in water, alcohol.

Derivation: Action of nitric acid on beryllium oxide, with subsequent evaporation and crystallization; reaction of beryllium sulfate with barium nitrate.

Grade: Technical; CP.

Use: Chemical reagent; gas mantle hardener.

Hazard: Very toxic. See also beryllium. Oxidizing material; dangerous fire risk.

beryllium nitride. Be_3N_2.

Properties: Hard, refractory, white crystals; mp 2200C; reacts with mineral acids to form the corresponding salts of beryllium and ammonia. Readily attached by strong alkali solutions, liberating ammonia.

Derivation: By heating beryllium metal powder in a dry, oxygen-free nitrogen atmosphere at 700–1400C. See also beryllium.

Use: Atomic energy; production of the radioactive carbon isotope C_{14} for tracer uses.

beryllium oxide. (beryllia). CAS: 1304-56-9. BeO.

Properties: White powder. A unique ceramic material. D 3.016; mp 2570C. Hardness (Mohs) 9. Soluble in acids and alkalies; insoluble in water. High electrical resistivity and thermal conductivity; transparent to microwave radiation; undamaged by nuclear radiation. High heat-stress resistance. Can be fabricated into finished shapes.

Derivation: By heating beryllium nitrate or hydroxide.

Grade: Technical; CP; pure; single crystals.

Hazard: Highly toxic by inhalation. See also beryllium. Keep container tightly closed and flush out after use.

Use: Electron tubes; resistor cores; windows in klystron tubes; transistor mountings; high-temperature reactor systems; additive to glass, ceramics and plastics; preparation of beryllium compounds; catalyst for organic reactions.

beryllium potassium fluoride. (potassium beryllium fluoride). $BeF_2 \cdot 2KF$.

Properties: White, crystalline masses. Soluble in water, insoluble in alcohol.

Hazard: Toxic by inhalation and ingestion. TLV (as F): 2.5 mg/m³ of air.

beryllium potassium sulfate. $BeSO_4K_2SO_4$.

Properties: Shiny crystals, insoluble in alcohol, soluble in water and concentrated potassium sulfate solution.

Use: Metal plating especially chromium and silver.
Hazard: See beryllium.

beryllium sodium fluoride. (sodium beryllium fluoride). $BeF_2 \cdot 2NaF$.
Properties: White, crystalline mass; soluble in water; mp approximately 350C.
Hazard: Toxic by inhalation and ingestion. TLV (as F): 2.5 mg/m³ of air.
Use: Making pure beryllium metal.

beryllium sulfate. CAS: 13510-49-1
$BeSO_4 \cdot 4H_2O$.
Properties: Colorless crystals, soluble in water, insoluble in alcohol, d 1.713, decomposes at 540C.
Hazard: A carcinogen (OSHA). See also beryllium.

Berzelius, J. J. (1779–1848). A native of Sweden, Berzelius was one of the foremost chemists of the 19th Century. He made many contributions to both fundamental and applied chemistry; he coined the words *isomer* and *catalyst*; classified minerals by chemical compound; recognized organic radicals which maintain their identity in a series of reactions; discovered selenium and thorium and isolated silicon, titanium, and zirconium; did pioneer work with solutions of proteinaceous materials which he recognized as being different from "true" solutions.

"Besk."[204] TM for a high-foaming heavy duty manual cleaner for dairies and food-processing plants.

Best, Charles H. (1899–1978). Born in Maine, Best was educated at the University of Toronto where he distinguished himself as a student of biochemistry. He collaborated with the late Dr. Frederick Banting in the isolation of the hormone insulin. He later became head of the insulin division of the Connaught Laboratories of the University as well as of the Banting and Best Research Institute. He also developed histaminase, an antiallergic enzyme, and the anticoagulant heparin.
See also Banting.

beta. (β). A prefix having meanings analogous to those of alpha.
   (1) It indicates (a) the position of a substituent atom or radical in a compound, (b) the second position in a naphthalene ring, or (c) the attachment of a chemical unit to the side-chain of an aromatic compound.
   (2) It refers to a secondary allotropic modification of a metal or compound.
   (3) It designates a type of radioactive decay.
See beta particle.

beta battery. Alternative name for the sodium/sulfur battery now under development.
See also storage battery, electrolyte.

"Betacote"[532a]. TM for a series of urethane prepolymers specifically designed for flame-proofing coatings on wood. Coatings 6 mils thick on Douglas fir have a flame-spread rating of 55.

betaine hydrochloride. (lycine hydrochloride).
CAS: 107-43-7. $C_5H_{11}O_2N \cdot HCl$.
Properties: Colorless crystals, mp 227–228C (decomposes), soluble in water and alcohol; insoluble in chloroform and ether. Aqueous solutions are strongly acid. Liberates hydrogen chloride at the mp.
Grade: Technical.
Use: Source of hydrogen chloride in solders and fluxes, organic synthesis.

betaine phosphate. $C_5H_{11}O_2N \cdot H_3PO_4$.
Properties: White, odorless granules; acid taste; mp 198–200C; very soluble in water.
Grade: Technical.
Use: Source of phosphoric acid.

"Betaprene".[36] TM for olefinic hydrocarbon resins used as coating vehicles.

"Betanox" Special.[248] TM for low-temperature reaction product of phenyl-β-naphthylamine and acetone.
Properties: Tan powder, d 1.16, mp above 120C, soluble in acetone, benzene, and ethylene dichloride, insoluble in water and gasoline.
Use: Antioxidant for wire insulation, tire treads, carcass, inner tubes, dark-colored footwear, proofing, and mechanical goods.

beta particle. A charged particle emitted from a radioactive atomic nucleus either natural or man made. The energies of beta particles range from 0 to 4 MeV. They carry a single charge; if this is negative the particle is identical with an electron, if positive it is a positron. Beta rays (streams of these particles) may cause skin burns and are harmful within the body. Protection to the skin can be afforded by a thin sheet of metal.
See also electron, decay, radioactive.

"Betasan"[1]. (n-β-o,o-diisopropyl dithiophosphoryl ethyl benzene sulfonamide). TM for a concentrated herbicide both liquid and granular.

betatron. An electromagnetic device for accelerating electrons (beta particles). Its action is similar in principle to that of an electric transformer in which the secondary windings are replaced with focusing magnets. The electrons travel around the core in a vacuum tube placed between the magnets. At each revolution around the core the electrons pick up the same energy as the voltage that would have been induced in one turn of wire at that point. The betatron can generate electron beams up to 320 MeV. Invented by D. W. Kerst in 1940, it is used chiefly for basic physical research.

Bettendorf's reagent. A reagent used for the detection of arsenic in presence of bismuth and antimony compounds. It consists of a concentrated solution of stannous chloride in fuming hydrochloric acid.
Hazard: Powerful tissue irritant.

Betterton-Kroll process. A process for obtaining bismuth and purifying desilverized lead that contains bismuth. Metallic calcium or magnesium is added to the molten lead to cause formation of high melting intermetallic compounds with bismuth. These separate as a surface scum and are skimmed off. The excess calcium and magnesium are removed from the lead by use of chlorine gas as mixed molten chlorides of lead or zinc. Bismuth of 99.995% purity is produced in this way.

Betti reaction. The reaction of aromatic aldehydes, primary aromatic or heterocylic amines and phenols leading to α-aminobenzylphenols.

Betts process. An electrolytic process for removing impurities from lead in which pure lead is deposited on a thin cathode of pure lead from an anode containing as much as 10% of silver, gold, bismuth, copper, antimony, arsenic, selenium, and other impurities. The electrolyte is lead fluosilicate and fluosilicic acid. The scrap anodes and the residues of impurities associated with them are either recast into anodes or treated to recover antimony, lead, silver, gold, bismuth, etc.

betula oil. See methyl salicylate.

"Beutene."[248] TM for a butyraldehyde-aniline reaction product.
Properties: Reddish-brown, free-flowing liquid; d 0.95; soluble in acetone, benzene, and ethylene dichloride; slightly soluble in gasoline; insoluble in water.
Use: Rubber accelerator.

BFE. Abbreviation for bromotrifluoroethylene.

BF₃-ether complex. See boron trifluoride etherate.

BFI powder. Proprietary preparation of bismuth formic iodide.
Use: Skin antiseptic.

BF₃-MEA. See boron trifluoride monoethylamine.

BF₃MeOH. See boron trifluoride-methanol.

BFPO. Abbreviation for bis(dimethylamino)-fluorophosphine oxide.
See dimefox.

BHA. Abbreviation for butylated hydroxyanisole.

BHC. Abbreviation for benzene hexachloride.

BHMT Amine. $NH_2(CH_2)_6NH(CH_2)_6NH_2$.
A liquid polyalkylene polyamine.
Used in asphalt additives and corrosive inhibitors.

BHT. Abbreviation for butylated hydroxytoluene.
See 2,6-di-tert-butyl-p-cresol.

Bi. Symbol for bismuth.

bi-. Prefix meaning two; di- is preferred in chemical nomenclature. Exceptions are bicarbonate, bisulfate, bitartrate, in which it indicates the presence of hydrogen in the molecule, e.g., $NaHCO_3$ (sodium bicarbonate).
See also bis-.

bibenzyl. See diphenylethane, symmetrical.

bicalcium phosphate. See calcium phosphate, dibasic.

bicarburetted hydrogen. See ethylene.

bicyclic. An organic compound in which two (only) ring structures occur. They may or may not be the same type of ring.
See naphthalene.

bicyclohexyl. (dicyclohexyl). $C_{12}H_{22}$.
Properties: Colorless, mobile liquid with pleasant odor. Bp 238.5C, fp 1 to 3C, d 0.883 (25/16C), wt/gal 7.37lb, refr index 1.480 (20C), flash p 165F (73.9C). Combustible. Autoign temperature 471F (244C).
Derivation: Hydrogenation of diphenyl.
Use: High-boiling solvent and penetrant.

"Bidrin."[125] (TM for dimethyl phosphate of 3-hydroxy-N,N-dimethyl-cis-crotonamide; dicrotophos).   CAS: 141-66-2.
$(CH_3O)_2P(O)OC(CH_3):CHC(O)N(CH_3)_2$.
Properties: Brown liquid with a mild ester odor,

bp 400C, miscible in water and xylene, slightly soluble in kerosene and diesel fuel. Commercially available water-miscible solution.
Hazard: Cholinesterase inhibitor. TLV: 0.25 mg/m^3 of air. Toxic by skin absorption.
Use: Insecticide.

biformin. $C_9H_6O_2$. An antibiotic produced by the fungus Polyporus biformis, reported to be active against various bacteria and fungi.

Biginelli reaction. Synthesis of tetrahydropyrimidinones by the acid catalyzed condensation of an aldehyde, a beta-keto ester, and urea.

"B-I-K."[248] TM for a surface-coated urea.
Properties: White powder, d 1.32, melting range 129–134C, soluble in water. Surface coating not soluble in water but is soluble in rubber. Slightly soluble in acetone, insoluble in benzene, gasoline and ethylene dichloride.
Use: Promoter for azodicarbonamide, a nitrogen blowing agent; activator for thiazoles, sulfenamides, and thiurams; odor reducer when used with nitrosoamine-type blowing agents.

"Bikalith."[88] TM for a series of lithium silicate ores including lepidolite, petalite, spodumene, and eucryptite.
Used in glass-making and ceramics.

bile acid. An acid found in bile (secretion of the liver). Bile acids are steroids having a hydroxyl group and a five carbon atom side chain terminating in a carboxyl group. Cholic acid is the most abundant bile acid in human bile. Others are deoxycholic and lithocholic acids. The bile acids do not occur free in bile but are linked to the amino acids, glycine and taurine. These conjugated acids are water-soluble. Their salts are powerful detergents and as such aid in the absorption of fats from the intestine.

bilirubin. (bilifulvin). $C_{33}H_{36}O_6N_4$.
Red coloring matter of bile. Also occurs in blood serum as decomposition product of hemoglobin.
Properties: Orange-red powder; mp 192C; soluble in acids, alkalies, chloroform, and benzene; insoluble in water; very slightly soluble in alcohol and ether.
Derivation: From bile pigment.
Use: Analytical chemistry, biochemical research.

bimetal. A type of thermometer in which the sensing element consists of two thin strips of metals having different expansion coefficients bonded together in a helical or spiral structure. The extent of deflection or bending induced by temperature change is indicated by a pointer on a dial. Reasonably accurate readings are obtained in this way, the range being from −185 to 425C. Bimetals are used in both laboratory and industry. See also thermometer.

binapacryl. Generic name for 2-sec-butyl-4,6-dinitrophenyl-3-methyl-2-butenoate.
$C_{15}H_{18}O_6N_2$.
Hazard: Toxic by ingestion and inhalation.
Use: Acaricide and fungicide.

binary. Descriptive of a system containing two and only two components. Such a system may be a chemical compound composed of two elements an element and a group (hydroxyl, methyl, etc.) or two groups, e.g., oxalic acid, it may also be a two-component solution or alloy.

bind. To exert a strong physiochemical attraction as often occurs between various proteins and water in hydrophilic gels, between organic dyes and fabrics, or between acids or bases and various chemical complexes.

binder. (1) The film-forming ingredient in paint, usually either a drying oil or a polymeric substance.
See also paint.
(2) In the food industry a material used in sausage manufacture that absorbs moisture at high temperatures, e.g., various flours, dried milk and soy protein.
(3) Any cementitious material that is soft at high temperatures and hard at room temperature used to hold dry powders or aggregate together, e.g., asphalt and sulfur in paving compositions and resins used in sand-casting.

binding energy. The energy that holds the protons together in an atomic nucleus. Since protons are positively charged, they exert strong mutually repellent forces and tremendous energy is required to keep them from flying apart. This energy is so great that it results in a slightly lower value for the mass of a nucleus taken as a whole than for the sum of its constituents taken individually. This phenomenon is of vast significance, for it means that a small fraction of mass has been converted into energy within the nucleus, as shown by Einstein's equivalence equation $E = mc^2$. Thus, when a U-235 nucleus (92 protons) is split as in the fission process a portion of its binding energy (equivalent to the mass difference) is released. It amounts to approximately 200 million electron volts per nucleus.
Binding energy may also be defined as the minimum energy required to dissociate a nucleus

into its component neutrons and protons. Neutron or proton binding energy is that required to remove a neutron or a proton from a nucleus; electron binding energy is that required to remove an electron from an atom or molecule.
See also mass defect, fission.

bioassay. See assay.

"Biobate."[173] TM for an enzymatic preparation for use in bating in the leather industry.

"Biocheck."[108] TM for a family of biocides, fungicides, and slimicides.
Use: Controlling and eliminating microbiological growth in pulp and paper mill water systems as well as for antibacterial papers.

biochemical oxygen demand. (BOD). A standardized means of estimating the degree of contamination of water supplies, especially those which receive contamination from sewage and industrial wastes. It is expressed as the quantity dissolved oxygen (in mg/L) required during stabilization of the decomposable organic matter by aerobic biochemical action. Determination of this quantity is accomplished by diluting suitable portions of the sample with water saturated with oxygen and measuring the dissolved oxygen in the mixture both immediately and after a period of incubation usually five days.
See also sewage sludge, biodegradability, dissolved oxygen (DO), and oxygen consumed (COD) as related terms.

biochemistry. Originally a subdivision of chemistry but now an independent science, biochemistry includes all aspects of chemistry that apply to living organisms. Thus, photochemistry is directly involved with photosynthesis and physical chemistry with osmosis--two phenomena that underlie all plant and animal life. Other important chemical mechanisms that apply directly to living organisms are catalysis, which takes place in biochemical systems by the agency of enzymes; nucleic acid and protein constitution and behavior, which is known to control the mechanism of genetics; colloid chemistry, which deals in part with the nature of cell walls, muscles, collagen, etc.; acid-base relations, involved in the pH of body fluids; and such nutritional components as amino acids, fats, carbohydrates, minerals, lipids and vitamins, all of which are essential to life. The chemical organization and reproductive behavior of microorganisms (bacteria and viruses) and a large part of agricultural chemistry are also included in biochemistry. Particularly active areas of biochemistry are nucleic acids, cell surfaces (membranes), enzymology, peptide hormones, molecular biology, and recombinant DNA.
See also biotechnology.

biocide. General name for any substance that kills or inhibits the growth of microorganisms such as bacteria, molds, slimes, fungi, etc. Many of them are also toxic to humans. Biocidal chemicals include chlorinated hydrocarbons, organometallics, halogen-releasing compounds, metallic salts, organic sulfur compounds, quaternary ammonium compounds, and phenolics.
See also antiseptic, disinfectant, fungicide, bactericide.

biocolloid. An aqueous colloidal suspension or dispersion produced by or within a living organism. Blood, milk, and egg yolk are examples.

biocomputer. A computer in which the silicon in the microchips has been replaced by a synthetic protein or polypeptide coated with a silver compound, the combination behaving as a metallic semiconductor. Such chips have been made experimentally, they have the potential of improving the storage capacity and operating efficiency of silicon chips substantially. The materials used in the experimental chips were polylysine on a glass substrate coated with an acrylate polymer and treated with silver nitrate.

bioconversion. Utilization of animal manures, garbage, and similar organic wastes for production of fuel gases by digestion, gasification, or liquefaction.
See also biogas, biomass.

biocytin. (epsilon-N-biotinyl-L-lysine).
$C_{16}H_{28}N_4O_4S$.
Properties: A naturally occurring complex of biotin isolated from yeast. Water-soluble crystals, mp 228.5C. It is believed to be an intermediate in the utilization of biotin by animal organisms.

biodegradability. The susceptibility of a substance to decompose by microorganisms, specifically the rate at which detergents and pesticides and other compounds may be chemically broken down by bacteria and/or natural environmental factors. Branched chain alkylbenzene sulfonates (ABS) are much more resistant to such decomposition than are linear alkylbenzene sulfonates (LAS) in which the long straight alkyl chain is readily attacked by bacteria. If the branching is at the end of a long alkyl chain (isoalkyls), the molecules are about as biodegradable as the normal alkyls. The alcohol sulfate anionic detergents and most of the nonionic detergents are biodegradable. Among pesticides the organophosphorus types while highly toxic are more biodegradable than DDT and its derivatives. Tests on a number of compounds gave results as follows: Easily biodegraded: n-propanol, ethanol, benzoic acid, benzaldehyde, ethyl acetate. Less easily biodegraded: ethylene glycol, isopropanol, o-cresol, diethylene glycol, pyridine, triethanolamine. Resistant to biodegration: aniline, methanol, monoethanolamine, methyl ethyl ketone, acetone. Additives that accelerate biodegradation of polyethylene, polystyrene and other plastics are available.

bioengineering. Application of the principles and methods of chemical engineering to biotechnology.

bioelectrochemistry. Application of the principles and techniques of electrochemistry to biological and medical problems. It includes such surface and interfacial phenomena as the electrical properties of membrane systems and processes, ion adsorption, enzymatic clotting, transmembrane pH and electrical gradients, protein phosphorylation, cells, and tissues.

bioethics. An interdisciplinary science for which research facilities were established in 1971 encompassing the ethical and social issues resulting from advances in medicine and the biosciences. Its scope includes a number of areas of importance to chemistry, e.g., reproductive and genetic phenomena, organ transplants, gerontology and antiaging techniques, biological warfare, contraception, etc. The Kennedy Institute at Georgetown University, Washington, D.C., is the chief center for information about this developing aspect of biomedical science.

bioflavonoid. A group of naturally occurring substances thought to maintain normal conditions in the walls of the small blood vessels. The bioflavonoids are widely distributed among plants, especially citrus fruits, black currants, and rose hips (hesperidin, rutin, quercitin). They have little or no medicinal value.

biogas. Methane ganerated from animal manure by bacterial anaerobic digestion. Small-scale units have been in use for some years, and the possibilities of utilizing the tremendous quantities of manure available in the US as an energy source have stimulated investigation of large-scale production. One installation utilizing a thermophilic fermentation technique at 55–60C has been operating in Florida since 1979, and another in Colorado since 1981. This energy source is also being exploited in China and India.
See also biomass.

biogeochemistry. A branch of geochemistry dealing with the interactions between living organisms and their mineral environment. It includes among other studies that of the effect of plants on weathering of rocks, of the chemical transformations that produced petroleum and coal, of the concentration of specific elements in vegetation at some time in the geochemical cycle (iodine in sea plants, uranium in some forms of decaying organic matter), and of the organic constituents of fossils.

biogenesis. See life, origin.

biogenic sediment. Sediment consisting of mineral grains that were once parts of organisms.

bioinorganic chemistry. Study of the mechanisms involved in the behavior of metal-containing molecules in living organisms, e.g., biological transport of iron, the effect of copper on nucleic acid and nucleoproteins, molybdenum and manganese complexes, etc.

bioluminescence. See chemiluminescence.

biomass. Any organic source of energy or chemicals that is renewable. Its major components are: (1) trees (wood) and all other vegetation; (2) agricultural products and wastes (corn, fruit, garbage ensilage, etc.); (3) algae and other marine plants; (4) metabolic wastes (manure, sewage); and (5) cellulosic urban waste. Conversion of these is performed in several ways: (1) by combustion (heat); (2) by fermentation (alcohol); (3) by gasification (synthesis gas); and (4) by anaerobic digestion (methane).

In terms of energy, wood is by far the most important component of biomass. It has become a significant source of industrial heat, e.g., in paper mills and power plants, and intensive cultivation of trees for this purpose is under way. Wood is also a potential source of alcohols; ethyl alcohol is produced from wood on large scale in Brazil as a gasoline substitute. Agricultural wastes are fermented or gasified to synthesis gas, manures and municipal waste yield methane (biogas) on digestion. In 1981, biomass supplied 3.5% of US energy requirements and this is expected to increase substantially.

biomaterial. Any material suitable for use as a surgical implant within the body to replace or support joints or tissues. They include such metals as aluminum, stainless steels, titanium, various forms of carbon, and especially plastics (polycarbonate, polyurethane, nylon, silicones). They have been used successfully in many areas of the body from hip and knee replacements to mastectomies. They must be compatible with the interior environment, noncorrosive and nondegradable, and duplicate as closely as possible the properties of the tissues they replace. A notable breakthrough was made in this field in 1982 when a complete artificial heart was implanted successfully in a living human. Its chief component was polyurethane, the base was aluminum, and the valves pyrolytic graphite together with polycarbonate and titanium. Other materials under consideration are styrene-butadiene copolymers for the diaphragms. The polyurethane used in the first implanted heart is TM "Biomer" produced by a division of Johnson & Johnson.

"Biomer". TM for a unique type of polyurethane used in heart implants.

"bioMeT 12."[288] TM for a rodent-repellent coating for cables in which the active ingredient is a mixture of tributyl tin salts. It has proved 95% effective in preventing destruction of telephone cables by rats and other rodents. Flexible, transparent, and effective for six months or more, it is applied mechanically over the plastic cable sheathing. Can also be used to protect other types of wiring, shipping containers, and similar products.

biomimetic chemistry. An interdisciplinary approach to biochemistry including both organic and inorganic aspects of this field. The term means imitation or mimicry of natural organic processes in living systems, and encompasses such subjects as enzyme systems, vitamin B_{12} and flavins, oxygen binding and activation, bioorganic mechanisms and nitrogen and small molecule fixation. The technique was utilized in the synthesis of the bleomycin molecule. A notable example of biomimetic chemistry is the development of model synthetic catalysts which imitate the action of natural enzymes. The behavior of chymotrypsin has been duplicated by a man-made catalyst that can accelerate certain reaction rates by the incredible factor of 100 billion.

"Biopal" VRO-20.[307] TM for an iodophor consisting of a solution of 20% available iodine in alkylphenoxypoly(ethyleneoxy)ethanol.
Properties: Brownish-black liquid, readily soluble in water. Slowly decreases in activity in the presence of light and should be packed in dark amber bottles or in suitably lined metal containers. Has both detergent and germicidal properties.
Use: Bactericide, sporocide, fungicide, and protozoacide in hard or soft water and at high or low temperatures.

biopolymer. A water-soluble polymer resulting from the action of bacteria (genus Xanthomonas)

on carbohydrates. The viscosity is almost as low as that of water. Chromium ion can be added to increase viscosity if desired. Such polymers are being used as viscosity builders in oil-well drilling muds and as thickeners and gel strength additives in aqueous media.
See also fermentation.

biophyl. A highly refined form of verxite in excess of 99% pure hydrobiotite (Mg-Fe-Al silicate).
Use: In foods and pharmaceuticals.

bioresmethrin. $C_{22}H_{26}O_3$ A synthetic insecticide of the pyrethrin type. It is biodegradable and has low toxicity, it is nonpersistent, and can act as a synergist.

biosynthesis. (1) Natural synthesis of organic compounds by plants and animals. For plants, this includes not only photosynthetic formation of carbohydrates and protein synthesis by means of nitrogen-fixing bacteria, but also a wide range of specialized organic compounds that are specific to individual species; many of these are poisonous. Animals and humans synthesize certain amino acids, hormones, cholesterol, etc. Some types (snakes, fish, toads, etc.) synthesize unique and powerful toxic principles.
(2) This term is also applied to such research achievements as synthesis of edible single-cell proteins by fermentation and gene-splicing techniques.
See protein, single-cell, recombinant DNA.

biota. A collective term for all the animals and plants of an ecosystem.

biotechnology. A definition prepared by a committee of British scientists in a report issued by the Organization for Economic Cooperation and Development (Paris) may be considered official and definitive. It states that biotechnology is "application of scientific and engineering principles to the processing of any organic or inorganic substance by biological agents to provide goods and services; the biological agents include a wide range of biological catalysts, particularly microorganisms, enzymes, and animal and plant cells." This involves commercial production of chemical compounds from either (1) renewable resources (biomass) or (2) nucleic acids (DNA). Examples of (1) are production of antibiotics, alcohols, and single-cell proteins by fermentation, and of (2) production of insulin, interferon, and synthetic bacteria by gene-splicing. Biotechnology appears to constitute a major technological revolution on a world-wide basis and may prove to be the most important development in the chemical industries since the plastics explosion of the 1930s.

biotin. (vitamin H; 2'-keto-3,4-imidazolido-2-tetrahydrothiophene-n-valeric acid). $C_{10}H_{16}N_2O_3S$. Biotin, frequently referred to as a member of the vitamin B complex, is necessary for the maintenance of health in animals and for growth of many microorganisms.

It influences fat metabolism, decarboxylation and carbon dioxide fixation, and deamination of some amino acids. It is closely related metabolically to pantothenic acid and folic acid. A biotin deficiency may be induced by ingestion of avidin, a raw-egg protein, because of the formation of a nonabsorbable biotin-avidin complex. Biotin is synthesized in the intestinal tract of humans; therefore, normally it is not essential in the diet.
Properties: White crystals, mp 230–232C, soluble in water and alcohol, insoluble in naphtha and chloroform, stable to heat, stable in neutral or acid solution, destroyed by strong alkali or oxidizing agents.
Sources: Egg yolk, kidney, liver, yeast, milk, molasses.
Units: Amounts are expressed in milligrams or micrograms of biotin. Grade: Practical, FCC.
Use: Medicine, nutrition.

N-biotinyl-l-lysine. See biocytin.

biotite. A component of igneous rocks and of soil similar to mica. It is a silicate of magnesium, iron, potassium, and aluminum.

biphenyl. See diphenyl.

o-**biphenylamine.** See o-aminobiphenyl.

o-**biphenyl biguanide.**
$NH_2(CNHNH)_2C_6H_4C_6H_5 \cdot H_2O$.
Properties: White to faintly pink powder, mp greater than 150C on dried material, ash less than 0.5%. Soluble in alcohol, "Carbitol," and "Cellosolve", very slightly soluble in water.
Use: Soap, antioxidant.

2,4'-biphenyldiamine. See 2,4'-diphenyldiamine.

Birch reduction. Reduction of aromatic rings by means of alkali metals in liquid ammonia to give mainly unconjugated dihydro derivatives.

birefringent. Descriptive of a type of crystal which separates an impinging light ray into two components that are at right angles to each other as a result two images appear each of which is caused by a light ray vibrating in only one direction (plane-polarized light). Such anisotropic crystals (Iceland spar) are used in nicol prisms.
See also nicol, anisotropic.

bis-. Prefix meaning "twice" or "again." Used in chemical nomenclature to indicate that a chemical grouping or radical occurs twice in a molecule, e.g., bisphenol A, where two phenolic groups appear $(CH_3)_2C(C_6H_5OH)_2$.
See also following entries.

N,N-bisacetoxethylaniline.
Grades: Brown liquid.
Hazard: Combustible.
Use: Coupling agent for disperse dyes for synthetic fibers.

2,2-bis(acetoxymethyl)propyl acetate.
$CH_3C(CH_2OOCCH_3)_2CH_2OOCCH_3$.
Properties: Colorless liquid, refr index 1.4359 (20C).
Use: As a plasticizer.

bisamides. General formula
$RCONHR'NHCOR$.
Properties: When R and R' have high molecular weight they are hard, light-colored waxes.

bis(2-aminoethyl)sulfide. $(H_2NCH_2CH_2)_2S$.
An ethyleneimine derivative.
Properties: Colorless liquid, fp 2.6C, bp 238C, d 8.7 lb/gal, refr index 1.5277, flash p 246F (118.9C), very soluble in water, benzene, and ethanol. Combustible.
Use: See ethyleneimine.

N,N-bis(3-aminopropyl)methylamine.
$CH_3N(C_3H_6NH_2)_2$.
Properties: Liquid, d 0.9307 (20/20C), bp 240C, fp −29.6C, flash p 220F (104.4C). Miscible in water. Combustible.
Hazard: Irritant.
Use: Chemical intermediate.

1,3-bis(2-benzothiazolylmercaptomethyl)urea.
$(C_6H_4NCS \cdot SCH_2NH)_2CO$.
Properties: Buff to light tan powder, mp 220C, d 1.38 (25C).
Use: Rubber accelerator.

p-bis[2-(5-p-biphenylyloxazoyl)]benzene.
(BOPOB). $C_{36}H_{25}O_2N_2$.
Properties: Shiny, yellow flakes; mp 327–328C; fluorescence peak 4400 Å; sparingly soluble in toluene.
Grade: Purified.
Use: Scintillation counter, wavelength shifter in liquid scintillators.

2,2-bis(p-bromophenyl)-1,1,1-trichloroethane.
$C_{14}H_9Br_2Cl_3$.   The bromine analog of DDT.
Hazard: Toxic by ingestion.
Use: Insecticide.

bis(tert-butylperoxy)-2,5-dimethylhexane.
$C_{16}H_{34}O_4$.
A cross-linking agent for polymers.
See "Varox."

Bischler-Napieralski reaction.   Cyclodehydration of β-phenethylamides to 3,4-dihydroisoquinoline derivatives by means of condensing agents such as phosphorous pentoxide or zinc chloride.

Bischler-Mohlau indole synthesis.   Formation of 2-substituted indoles by heating o-halogeno- or o-hydroxy-ketones with excess aniline via cyclization of the intermediate 2-arylaminoketone.

bis(2-chloroethoxy)methane.   See dichloroethyls formal.

4-[p-[bis(2-chloroethyl)amino]phenyl]butyric acid.
See chlorambucil.

5-[bis(2-chloroethyl)amino]uracil.   (uracil mustard). $\overline{\text{CONHCONHCHCN}}(C_2H_4Cl)_2$.
Properties: A cream-white, odorless, crystalline compound; moderately soluble in methanol and acetone.
Used in medicine.

bis(2-chloroethyl) ether.   See dichloroethyl ether.

bis(chloromethyl)ether.   (dichloromethyl ether).
CAS: 542-88-1. $(CH_2Cl)O(CH_2Cl)$.
Properties: Reported to form spontaneously from formaldehyde and chloride ions in moist air.
Hazard: Toxic by ingestion. A carcinogen. TLV: 0.001 ppm in air.
Use: Intermediate for ion-exchange resins; laboratory reagent.

1,1′-bischloromercuriferrocene.   [1,1-di(chloromercuri)ferrocene]. $(ClHgC_5H_4)_2Fe$.
Use: Inorganic polymers.

3,3-bis(chloromethyl)oxetane.   See "Penton."

bis(p-chlorophenoxy)methane.
$(ClC_6H_4O)_2CH_2$.
Properties: Solid, mp 65C, insoluble in water and oils, soluble in ether and acetone.
Use: Acaricide.

2,2-bis(p-chlorophenyl)-1,1-dichloroethane.
See TDE.

1,1-bis(p-chlorophenyl)ethanol.   See di(p-chlorophenyl)ethanol.

(1,1′-bis(p-chlorophenyl)-2,2,2-trichlorethanol.
(4,4′-dichloro-α-trichloromethylbenzhydrol; Kelthane). $CCl_3C(C_6H_4Cl)_2OH$.   An alcohol analog of DDT.
Hazard: Toxic by inhalation and ingestion.
Use: Miticide.

bis-cyclopentadienyliron.   See dicyclopentadienyliron.

bis(2,6-diethylphenyl)carbodiimide.
$(CH_2H_5)_2C_6H_3NCNC_6H_3(C_2H_5)_2$.
Properties: Light yellow to red-brown liquid with faintly acrid odor, d 1.007 (20/20C), bp 192–194C (0.4 mm), refr index 1.591 (23C). Soluble in organic solvents. Combustible.
Hazard: Toxic by inhalation. Damaging to eyes.
Use: Stabilizers in polyester and urethane systems, intermediate for textile chemicals and pharmaceuticals.

2,6-bis(dimethylaminomethyl)cyclohexanone.
$[(CH_3)_2NCH_2]_2C_6H_8O$.
Properties: D 0.95 (20C).
Use: Preservative for aqueous paint systems, casein, pigment dispersions, and adhesives.

2,6-bis(dimethylaminomethyl)cyclohexanone dihydrochloride. $C_{12}H_{24}N_2O \cdot 2HCl$.
Properties: Free-flowing, white to off-white crystalline salt.
Use: Preservative for aqueous systems, latex paints, adhesives coatings, wax emulsions, casein, and starch solution.

bis(1,3-dimethylbutyl)amine.
$[(CH_3)_2CH_2CH_2CH(CH_3)]_2NH$.
Properties: Liquid, d 0.772–0.778 (20/20C), distillation range 179.0–205C, bulk d 6.5 lb/gal, flash p 160F (71.1C). Combustible.

N,N′-bis(1,4-dimethylpentyl)-p-phenylenediamine.
(diheptyl-p-phenylenediamine).
$C_6H_4(CNH_7H_{15})_2$.
Properties: Amber to red liquid, d 0.90, fp 7.2C. Combustible.
Use: Gasoline antioxidant and sweetener.

bis(2-ethylhexyl)phthalate. $C_{24}H_{38}O_4$. Liquid used in vacuum pumps.

N,N-bis(1-ethyl-3-methylpentyl)-p-phenylenediamine. $C_6H_4(NHC_8H_{17})_2$.
Properties: Dark, reddish-brown liquid; d approximately 0.90; bulk d 7.5 lb/gal. Combustible.
Use: Antioxidant for polyunsaturated elastomers.

2,2-bis(p-ethylphenyl)-1,1-dichloroethane. See 1,1-dichloro-2,2-bis(p-ethylphenyl)ethane.

bisethylxanthogen. $(C_2H_5OCSS)_2$.
Properties: Yellow needles, onion-like odor, mp 28–32C, insoluble in water, freely soluble in benzene, ether, petroleum fractions.
Grade: 58% soluble in oil.
Use: Weed control, rubber accelerator, fungicide.

1,3-bis(3-glycidoxypropyl)tetramethyldisiloxane. $[OCH_2CHCH_2O(CH_2)_3Si(CH_2)_2]_2O$.
Properties: Liquid, d 0.99 (25C), refr index 1.4500 (25C), bp approximately 185C (2 mm). Soluble in acetone and benzene, insoluble in water. Flash p 300F (149C). Combustible.
Use: Chemical intermediate.

bishydroxycoumarin. [3,3'-methylenebis(4-hydroxycoumarin)dicoumarol]. $C_{19}H_{12}O_6$.
Properties: White or creamy-white crystalline powder, faint pleasant odor and slightly bitter taste, mp 287–293C. Readily soluble in solutions of fixed alkali hydroxides, slightly soluble in chloroform, almost insoluble in water, alcohol, and ether.
Derivation: (a) Originally extracted from spoiled sweet clover; (b) synthetically from methyl acetylsalicylate sodium, and formaldehyde.
Grade: USP.
Hazard: May cause hemorrhage.
Use: Anticoagulant for blood.

bis(1-hydroxycyclohexyl)peroxide. $C_6H_{10}(OH)_2O_2$.
Properties: Fine, white powder; mp 66–68C. Active oxygen 6.6% min.
Hazard: Dangerous fire risk in contact with organic materials. Strong oxidizing agent.
Use: Catalyst for polymerization of polyester resins.

N,N-bis(2-hydroxyethyl)alkylamine. Clear liquid used as an antistatic for blow molding applications for polyolefins. Approved for use in food packaging films.

bis(hydroxyethyl)butynediol ether.
$HO(CH_2)_2OCH_2C \cdot CCH_2C \cdot CCH_2O(CH_2)_2OH$.
Properties: Dark brown liquid, d 1.136 (25/15C),

solidifies below −15C, distillation range 116–235C (10 mm).
Hazard: May explode under alkaline conditions at high temperature.
Use: Intermediate for polyesters, plasticizers and plastics, nickel brightener in electroplating, corrosive inhibitor, pickling inhibitor prior to copper plating.

bis(hydroxyethyl) cocoamine oxide. A derivative coconut oil claimed to be useful as a gasoline additive to inhibit rust formation and icing of carburetors. Also used as foaming agent in shampoos, detergents, tooth pastes, and the like.

N,N-bis(hydroxyethyl)oleamide.
$CH_3(CH_2)_7HC:CH(CH_2)_7CON(CH_2CH_2OH)_2$.
A technical grade containing 25% excess amine. Light amber liquid with faint odor.
Used as a surface-active agent.

β-bishydroxyethyl sulfide. See thiodiglycol.

1,3-bishydroxymethylurea. See dimethylolurea.

4,4-bis(4-hydroxyphenyl)pentanoic acid. (diphenolic acid; DPA).
$CH_3C(C_6H_4OH)_2CH_2CH_2COOH$.
Properties: Light tan granules, mp 170–173C, d 1.30–1.32. Soluble in acetic acid, acetone, and ethanol; insoluble in benzene, carbon tetrachloride, and xylene. Slightly soluble in water.
Use: Paint formulations, coatings, and finishes.

bishydroxyphenyl sulfone. See dihydroxydiphenyl sulfone.

1,4-bis(2-hydroxypropyl)-2-methylpiperazine. $C_{11}H_{24}O_2N_2$.
Properties: Liquid, bp 145C (3 mm), d 1.0013 (25/25C), refr index 1.4803 (20C), flash p 300F (149C), miscible in water, odor-free. Combustible.
Use: Catalyst, chemical intermediate.

bis-intercalator. A unique type of natural antibiotics that have antitumor and antimicrobial properties. They function by interposing two symmetrical groups between the nucleotide bases of the DNA molecule. Some are quite toxic and may be mutagenic.
See also carzinophillin A.

2,4-bis(isopropylamino)-6-methoxy-s-triazine. (Gesafram; ontrach; promitol; prometon).
CAS: 1610-18-0. $C_{10}H_{19}N_5O$.
Properties: White, extruded pellets.
Hazard: Toxic by ingestion.
Use: Industrial herbicide.

bis-keto-triazine. ("Permafresh" 110). Water-white liquid, 40% active. Can be cured with magnesium chloride to provide chlorine resistance; compatible with optical brighteners. Used as a wash/wear finish for cellulosics and blends of cellulosics.

"Bismanol". (MnBi). An alloy or compound of bismuth and manganese which has an exceptionally high coercive force (3400 oersteds) and a high energy product. Produced by US Naval Ordnance Laboratory by powder metallurgy methods and used as a permanent magnet.

Bismarck Brown R. (toluene-2,4-diazo-bis-m-toluylenediamine hydrochloride; CI 21010). $CH_3C_6H_3[NNC_6H_2(CH_3)NH_2)_2] \cdot 2HCl$.
Properties: Dark brown powder, soluble in water and alcohol.
Derivation: Action of nitrous acid on toluylene diamine.
Use: Dye for wool and leather, biological stain.

Bismarck Brown Y. (4,4'-[m-phenylenebis-(azo)]bis(m-phenylenediamine)dihydrochloride; CI 21000). C_6H_4-1,3-$[NNC_6H_3$-2,4-$(NH_2)_2 \cdot HCl]_2$.
Properties: Black to brown powder. Soluble in water, insoluble in benzene and carbon tetrachloride.
Use: Dyeing textiles, biological stain.

"Bismate."[69] TM for bismuth dimethylidithiocarbamate.

bis(2-methoxyethoxy)ethyl ether. See dimethoxytetraglycol.

bis(1-methylamyl)sodium sulfosuccinate. (dihexyl sodium sulfosuccinate). $C_{16}H_{29}NaO_7S$.
Properties: White, waxy particles; readily soluble in hot water; slowly in cold water; soluble in benzene, carbon tetrachloride, acetone, and glycerol. Hydrolyzes in alkaline media.
Use: Surfactant, wetting agent.

N,N'-bis(1-methylheptyl)-p-phenylenediamine. $C_6H_4[NHCH(CH_3)C_6H_{13}]_2$.
Properties: Dark, reddish-brown liquid; d approximately 0.90 @ 26.6C; wt/gal 7.5 lb. Combustible.
Use: Antiozonant in polyunsaturated elastomers.

1,4-bis[2-(4-methyl-5-phenyloxazolyl)]benzene. (dimethyl-POPOP). $C_6H_4[C_3NO(CH_3)C_6H_5]_2$. Yellow needles with a bluish fluorescence, mp 231–234C. Used as a scintillation phosphor.

bismite. See bismuth trioxide.

bismuth. Bi. Metallic element of atomic number 83, Group VA of the periodic system. Atomic weight 208.9804. Valences 3,5; no stable isotopes; four naturally radioactive isotopes.
Properties: Brittle metal with reddish tinge. Soluble in nitric and hydrochloric acids. Highly diamagnetic (mass susceptibility -1.35×10^6). Expands 3.3% on solidification. Electrical resistivity higher in solid than in liquid state. Extrudable at 437F, not fabricable at room temperature, d 9.8 (20C), mp 271C, bp 1560C, Brinell hardness 7. Thermal conductivity (0.018 cal/sec/cc (250C)) is lowest of all metals except mercury. On heating it burns to form the oxide.
Source: (1) Metallurgical by-products (often lead bullion) obtained chiefly from smelting ores of lead, silver, copper, and gold; (2) ores used chiefly for their bismuth and one or two other metals as tin and tungsten.
Derivation: Debismuthizing of lead bullion by (a) fractional crystallization, (b) electrolytic (Betts) refining, or (c) addition of calcium or magnesium (Betterton-Kroll process) which removes bismuth.
Purification: By addition of molten caustic, zinc and finally chlorine (to make removable chlorides of the impurities). Impurities: Lead, iron, copper, arsenic, antimony, selenium.
Forms: Rods, wire, lump, powder.
Grade: 99.5+% pure, high purity (less than 10 ppm impurities), single crystals.
Hazard: Flammable in powder form.
Use: Pharmaceuticals and medicine, cosmetics (eye polish, lipstick), component of low-melting (fusible) alloys, catalyst in making acrylonitrile, additive to improve machinability of steels and other metals, coating selenium, thermoelectric materials, permanent magnets, semiconductors. See also bismuthinite and cosalite.

bismuth ammonium citrate.
Properties: Pearly, shining, transparent scales or white powder, slightly acid, metallic taste, composition varies. Soluble in water, slightly soluble in alcohol.
Derivation: Interaction of bismuth subnitrate, citric acid, and ammonium hydroxide.
Use: Medicine.

bismuth antimonide. BiSb. Single crystals used as semiconductors.

bismuth bromide. (bismuth tribromide). $BiBr_3$.
Properties: Yellow, crystalline powder. Hygroscopic. Decomposed by water with formation of bismuth oxybromide. Soluble in either hydrochloric acid (dilute) or solutions of potassium iodide, bromide, and chloride. Insoluble in alcohol, d 5.7, bp 453C, mp 218C.

bismuth bromide oxide.   See bismuth oxybromide.

bismuth carbonate, basic.   See bismuth subcarbonate.

bismuth chloride.   (bismuth trichloride).
$BiCl_3$.
Properties: White, very deliquescent crystals; volatilized by heat. Soluble in acids, insoluble in alcohol, decomposes in water to the oxychloride, d 4.56, mp 227C, bp decomposes at 300C.
Derivation: Action of hydrochloric acid on bismuth.
Use: Bismuth salts, catalyst.

bismuth chloride, basic.   See bismuth oxychloride.

bismuth chromate.   (basic dichromate).
$Bi_2O_3 \cdot 2CrO_3$.
Properties: Orange-red, amorphous powder; soluble in alkalies and acids; insoluble in water.
Derivation: Interaction of bismuth nitrate and potassium chromate.

bismuth citrate. $BiC_6H_5O_7$.
Properties: White powder, d 3.458, soluble in ammonia or alkali citrates, insoluble in water, slightly soluble in alcohol, mp decomposes.
Derivation: Boiling bismuth subnitrate with citric acid.
Use: Medicine.

bismuth dimethyldithiocarbamate.
CAS: 21260-46-8. $Bi[(CH_3)_2NC(S)S]_3$.
Properties: Lemon yellow powder, d 2.04, mp greater than 230C (decomposes), soluble in chloroform, slightly soluble in benzene, carbon disulfide, insoluble in water.
Use: Accelerator for natural rubber and SBR especially in cable covers and mechanical items.

bismuth ditannate.   See bismuth tannate.

bismuth ethyl chloride. $BiHC_2H_5Cl$.   White powder.
Hazard: Ignites spontaneously in air, dangerous fire risk. See also bismuth.

bismuth gallate, basic.   See bismuth subgallate.

bismuth glance.   See bismuthinite.

bismuth hydrate.   See bismuth hydroxide.

bismuth hydroxide.   (bismuth hydrate; bismuth oxyhydrate; bismuth trihydroxide; bismuth trihydrate; hydrated bismuth oxide).
$Bi(OH)_3$.

Properties: White, amorphous powder. Soluble in acids, insoluble in water, d 4.36.
Derivation: Action of sodium hydroxide on a solution of bismuth nitrate.
Use: Plutonium separation, hydrolysis of ribonucleic acid, absorbent.

bismuthinite.   (bismuth glance). Bi_2S_3.
May contain copper or iron.
Properties: Lead-gray mineral, often with yellow tarnish; metallic luster. Contains 81.2% bismuth, 18.8% sulfur, soluble in nitric acid, d 6.4–6.5, Mohs hardness 2.
Occurrence: Utah, Bolivia, Mexico.
Use: Ore of bismuth.

bismuth iodide.   (bismuth triiodide). BiI_3.
Properties: Grayish-black, metallic, glistening crystals. Soluble in alcohol, hydriodic acid, and potassium iodide solutions, insoluble in water; decomposes in hot water; d 5.778 (15C); sublimes at 438C.
Derivation: By the interaction of bismuth and iodine.
Hazard: Toxic by ingestion. See also bismuth.
Use: Analytical chemistry, manufacturing bismuth oxyiodide.

bismuth-β-naphthol. $Bi_2O_2(OH) \cdot C_{10}H_7O$.
Properties: Brown to gray powder, almost insoluble in water or other solvents.
Derivation: By treating a solution of sodium-β-naphtholate with acetic acid solution of bismuth nitrate and adding caustic soda solution to neutralize excess acid.
Use: Medicine.

bismuth nitrate.   (bismuth ternitrate; bismuth trinitrate). $Bi(NO_3)_3 \cdot 5HOH$.
Properties: Lustrous, clear, colorless, hygroscopic crystals, acid taste. Soluble in dilute nitric acid, alcohol, and acetone; slowly decomposed by water to the subnitrate; d 2.83; bp 75–80C (decomposes).
Derivation: Action of nitric acid on bismuth with subsequent recovery by evaporation and crystallization.
Hazard: Oxidizing material; fire risk near organic materials.
Use: Preparation of other bismuth salts, bismuth luster on tin, luminous paints and enamels, precipitation of alkaloids.

bismuth nitrate, basic.   See bismuth subnitrate.

bismuth oleate.
Properties: Yellowish-brown soft granular mass. Soluble in ether, insoluble in water.
Derivation: A mixture of bismuth trioxide and oleic acid.
Use: Catalyst in oxo process.

bismuth oxide. See bismuth trioxide.

bismuth oxide, hydrated. See bismuth hydroxide.

bismuth oxybromide. (bismuth bromide, basic; bismuth bromide oxide). BiBrO.
Properties: Dry powder, insoluble in water and alcohol, soluble in hydrochloric and hydrobromic acids.
Use: Cathodes for dry cells.

bismuth oxycarbonate. See bismuth subcarbonate.

bismuth oxychloride. (bismuth chloride, basic; bismuth subchloride). BiOCl.
Properties: White lustrous crystalline powder, d 7.717, soluble in acid, insoluble in water.
Derivation: By action of water on bismuth chloride, interaction of dilute nitric acid solution of bismuth nitrate with sodium chloride.
Use: Cosmetics, pigment, dry cell cathodes.

bismuth oxyhydrate. See bismuth hydroxide.

bismuth oxynitrate. See bismuth subnitrate.

bismuth pentafluoride. BiF_5.
Properties: Crystals, sublimes at 550C.
Hazard: Reacts violently with water and petrolatum above 50C. Strong irritant to eyes and skin.
Use: Fluorinating agent.

bismuth pentoxide. Bi_2O_5.   An acid anhydride, its salts have not been prepared in pure state. Made by oxidation of bismuth trioxide, giving a scarlet precipitate.

bismuth phosphate. $BiPO_4$.
Properties: Odorless crystals which do not melt when heated, d 6.32 (15C), soluble in nitric and hydrochloric acids, insoluble in acetic acid and alcohol, slightly soluble in water and weak acids.
Use: Plutonium recovery, optical glass.

bismuth potassium iodide. $BiI_3 \cdot 4KI$.
Properties: Red crystals, decomposed by water. Soluble in potassium iodide solution.
Use: Precipitation of vitamins and antibiotics from solution.

bismuth potassium tartrate. See potassium bismuth tartrate.

bismuth salicylate, basic. See bismuth subsalicylate.

bismuth selenide. (bismuth triselenide). Bi_2Se_4.

Properties: Black crystals, mp 710C, insoluble in water, decomposes on heating.
Use: Semiconductor technology.

bismuth stannate. $Bi_2(SnO_3)_3 \cdot 5HOH$.
Properties: Light colored crystals, Insoluble in water. Temperature of dehydration is approximately 140C.
Hazard: Toxic material; TLV (as tin): 2 mg/m^3 of air. See also bismuth.
Use: Component of ceramic capacitors, useful with barium titanate.

bismuth subcarbonate. (bismuth oxycarbonate; bismuth carbonate, basic). $(BiO)_2CO_3$ or $Bi_2O_3 \cdot CO_2 \cdot 1/2H_2O$.
Properties: White, odorless powder; tasteless; insoluble in water and alcohol; soluble in nitric or hydrochloric acid with effervescence; d 6.86. Stable in air but slowly affected by light.
Derivation: By adding ammonium carbonate to a solution of a bismuth salt.
Grade: Technical, CP, USP (90% Bi_2O_3 min).
Use: Bismuth compounds, cosmetics, opacifier in x-ray diagnosis, enamel fluxes, and ceramic glazes.

bismuth subchloride. See bismuth oxychloride.

bismuth subgallate. (basic bismuth gallate). $C_6H_2(OH)_3COOBi(OH)_2$.
Properties: Saffron-yellow powder, odorless and tasteless. Soluble in dilute alkalies, insoluble in water, alcohol, and ether. Stable in air but affected by light.
Derivation: Interaction of bismuth nitrate, glacial acetic acid, and gallic acid in aqueous solution.
Use: Medicine (treatment of alimentary canal).

bismuth subnitrate. (basic bismuth nitrate; bismuth oxynitrate). $4BiNO_3(OH)_2 \cdot BiO(OH)$.
Properties: White, heavy, slightly hygroscopic powder which shows acid to moistened litmus paper. Soluble in acids, insoluble in water and alcohol, d 4.928, mp 260C (decomposes).
Derivation: Hydrolysis of bismuth nitrate, filtering and drying.
Grade: Technical, CP, NF.
Use: Cosmetics, ceramic glazes, enamel fluxes.

bismuth subsalicylate. (basic bismuth salicylate). $Bi(C_7H_5O_3)_3Bi_2O_3$.
Properties: White, bulky, crystalline powder; tasteless; odorless; soluble in acids and alkalies; insoluble in water, alcohol, and ether. Stable in air but affected by light.
Derivation: By treating freshly prepared bismuth hydroxide with salicylic acid.
Use: Surface-coating plastics and copying paper.

bismuth sulfate. $Bi_2(SO_4)_3$.
Properties: White needles or powder. Contains approximately 68.5% bismuth, d 5.08 decomposes at 405C. Soluble in dilute hydrochloric or nitric acid, insoluble in alcohol, water.
Use: Analysis of metallic sulfates.

bismuth sulfide. (bismuth trisulfide). Bi_2S_3.
Properties: Blackish-brown powder. Soluble in nitric acid, insoluble in water, d 7.6–7.8, mp decomposes.
Derivation: (a) By melting bismuth and sulfur together. (b) By passing hydrogen sulfide into a soluble of a bismuth salt. (c) Occurs as the mineral bismuthinite.
Use: Manufacturing bismuth compounds.

bismuth tannate. (bismuth ditannate).
Properties: Light brownish-yellow powder containing approximately 36% bismuth, insoluble in water and alcohol, soluble in mineral acids.
Derivation: From freshly prepared bismuth hydroxide and tannin.
Use: Medicine (astringent).

bismuth telluride. (bismuth tritelluride).
CAS: 1304-82-1. Bi_2Te_3.
Properties: Gray, hexagonal platelets. Mp 573C, d 7.642.
Derivation: Stoichiometric combination of the elements.
Grade: Ingot, single crystals.
Hazard: Toxic. TLV: 10 mg/m³ of air (if selenium-doped 5 mg/m³ of air).
Use: Semiconductors for thermoelectric cooling and power generation applications.

bismuth tetraoxide. Bi_2O_4.
Properties: Heavy, yellowish-brown powder. Soluble in acids, insoluble in water, d 5.6, mp 305C.
Derivation: By further oxidation of bismuth trioxide.
Use: Lubricant for metal extrusion dies.

bismuth tribromide. See bismuth bromide.

bismuth trichloride. See bismuth chloride.

bismuth trihydrate. See bismuth hydroxide.

bismuth trihydroxide. See bismuth hydroxide.

bismuth triiodide. See bismuth iodide.

bismuth trinitrate. See bismuth nitrate.

bismuth trioxide. (bismuth oxide; bismuth yellow; bismite). Bi_2O_2.
Properties: Heavy, yellow powder. Soluble in acids, insoluble in water, d 8.8, mp 820C.

Derivation: Heating bismuth nitrate in air, ignition of bismuth hydroxide.
Use: Enameling cast iron ceramic and porcelain colors.

bismuth trisulfide. See bismuth sulfide.

bismuth tritelluride. See bismuth telluride.

bismuth yellow. See bismuth trioxide.

p-bis[2,5(5-α-naphthyloxazolyl)]benzene. (NOPON). $C_{32}H_{20}O_2N_2$.
Properties: Crystals, mp 215–217C.
Grade: Purified.
Use: Scintillation counting.

bis(3-nitrophenyl) disulfide. See nitrophenide.

bisphenol A. (4,4'-isopropylidenediphenol; 2,2-bis(4-hydroxyphenol)propane). $(CH_3)_2C(C_6H_4OH)_2$.

Properties: White flakes with a mild phenolic odor. Bp 220C (4 mm), mp 153C, d 1.195 (25/25C), flash p 175F (79.4C), insoluble in water, soluble in alcohol and dilute alkalies, slightly soluble in carbon tetrachloride. Combustible.
Derivation: Condensation reaction of phenol and acetone catalyzed by hydrochloric acid at 65C.
Use: Intermediate in manufacture of epoxy, polycarbonate, phenoxy, polysulfone and certain polyester resins, flame retardants, rubber chemicals, fungicide.

1,4-bis-2(5-phenyloxazoyl)benzene. (POPOP). $(C_6H_5HNO)_2C_6H_4$.
Properties: Light yellow, cottony needles; mp 245–246C; fluorescence max 4200 Å; solubilities (g/100 g at 25C): water 0.00, 95% ethanol 0.00, toluene 0.12, hexane 0.02. Combustible.
Grade: Purified.
Use: Band-shifter in scintillation counting.

bis(tetrachloroethyl) disulfide. $C_4H_2Cl_8S_2$.
Properties: Liquid, d 1.785 (23.3C), bp 185C (3 mm), soluble in benzene, hexane, ethanol. Combustible.
Hazard: Toxic.
Use: Agricultural chemicals, additives.

bis(tribromophenoxy)ethane.
Properties: White, crystalline powder.
Use: Flame retardant for many thermoplastic and thermoset systems.

bis(tri-n-butyltin) oxide.   CAS: 56-35-9.
$(C_4H_9)_3SnOSn(C_4H_9)_3$.
Properties: Slightly yellow liquid, bp 180C (2 mm), solidifies below -45C, d 1.17 (25C), flash p greater than 212F (100C) (TCC), viscosity 4.8 centistokes at 25C, almost insoluble in water, miscible with organic solvents. Forms compounds with cellulosic and lignin-containing materials not easily decomposed or dissolved in water. Combustible.
Derivation: Hydrolysis of tributyl tin chloride.
Hazard: Toxic via ingestion and inhalation. TLV (as tin): 0.1 mg/m³ of air.
Use: Fungicide and bactericide in underwater and antifouling paints, pesticide.

bis(trichlorosilyl)ethane.   (1,1,1,4,4,4-hexachloro-1,4-disilabutane). $Cl_3SiCH_2CH_2SiCl_3$.
Properties: Colorless liquid. Bp 202.9C, flash p 190F (COC). Readily hydrolyzed with liberation of hydrogen chloride. Combustible.
Derivation: Reaction of acetylene and trichlorosilane in presence of a peroxide catalyst.
Grade: Technical.
Hazard: Corrosive when exposed to moisture.
Use: Intermediate for silicones.

bistridecyl phthalate. $C_6H_4(COOC_{13}H_{27})_2$.
Properties: Liquid, d 0.9497 (25C), boiling range 280–290C (4 mm), fp −35C, flash p 485F (251C), refr index 1.483 (25C). Combustible.
Use: Primary plasticizer for most PVC resins.

bis(trimethylsilyl)trifluoroacetamide.
(BSTFA).   CAS: 21149-38-2.
$CF_3C[=NSi(CH_3)_3]OSi(CH_3)_3$.
Properties: bp 45–50C (14 mm), refr index 1.3839 (20C), mw 257.4.
Use: Preparation of volatile derivatives of a wide range of biologically active compounds for gas-liquid chromatographical analysis.

bithionol.   (bis[2-hydroxy-3,5-dichlorophenyl] sulfide; 2,2'-thiobis[4,6-dichlorophenol]).
CAS: 97-18-7. $HOCl_2C_6H_2SC_6H_2Cl_2OH$.
Properties: White or grayish-white, crystalline powder; mp 187C. Odorless or with slight aromatic or phenolic odor. Insoluble in water; freely soluble in acetone, alcohol, and ether; soluble in chloroform and dilute solution of fixed alkali hydroxides.
Grade: NF.
Hazard: Skin irritant, may not be used in cosmetics (FDA).
Use: Deodorant, germicide, fungistat, pharmaceuticals.

4,4'bi-o-tolylene diisocyanate.   See 3,3'-dimethyl-4,4'-biphenylene diisocyanate.

bitter almond oil.   See almond oil.

bittern.   The solution of bromides, magnesium and calcium salts that remains after sodium chloride has been crystallized by concentration of sea water or brines.

bitrex.   See denatonium benzoate.

"Bitumastic."[11]   TM for refined coal tar-based water-resistant protective coatings including hot-applied impermeable enamels and industrial coatings.

bitumen.   A mixture of hydrocarbons occurring both in the native state and as residue from California petroleum distillation. Soluble in carbon disulfide. Solid to viscous, semisolid liquid. Used in hot-melt adhesives, coatings, paints, sealants, roofing, and road-coating. Bitumens are found in asphalt, mineral waxes, and in lower grades of coal. Combustible.
See also asphalt, gilsonite, glance pitch, shale oil, oil sands.

bituminous coal.   See coal.

biuret.   (allophanamide; carbamylurea).
$NH_2CONHCONH_2 \cdot HOH$.
Properties: White needles, odorless, mp 190C (decomposes), soluble in water and alcohol, very slightly soluble in ether. Loses water of crystallization at approximately 110C.
Derivation: From urea by heat.
Method of purification: Crystallization.
Use: Analytical reagent, especially for proteins.

bixin. $C_{25}H_{30}O_4$.   A carotenoid obtained from seeds of *Bixa orellana*.
Properties: Orange crystals, decomposes at 217C.
Derivatives: Methyl ester (methyl bixin) $C_{26}H_{32}O_4$: blue crystals, mp 203C. Ethyl ester (ethyl bixin), $C_{27}H_{34}O_4$: Red crystals, mp 138C.
Use: The ethyl ester is used as a food coloring.
See also annatto.

Bk.   Symbol for berkelium.

black.   Any of several forms of finely divided carbon either pure or admixed with oils, fats, or waxes.
See acetylene black, animal black, bone black, carbon black, etc.

black, aniline.   See aniline black.

black antimony.   See antimony trisulfide.

black ash.   (1) (Papermaking) The product obtained by heating black liquor in furnaces. The

organic material is reduced to carbon. The alkaline components are leached out and used again in papermaking. The carbon may be treated to obtain activated carbon. (2)
See barium sulfide.

blackbody. In radiation physics, an ideal blackbody is a theoretical object that absorbs all the radiant energy falling upon it and emits it in the form of thermal radiation. The power radiated by a unit area of a blackbody is given by Planck's radiation law and the total power radiated is expressed by the Stefan-Boltzmann law.

black, bone. See bone black.

black cyanide. A mixture containing 45% calcium cyanide made from calcium cyanamid by heating it with sodium chloride and carbon.

black lead. See graphite.

Black Leaf 40. A pesticide consisting of a 40% solution of nicotine sulfate.

black liquor. (1) The liquor resulting from cooking pulpwood in an alkaline solution in the soda or sulfate (kraft) papermaking process. It is a source of lignin and tall oil and is said to be effective in removal of mercury from industrial effluents (USDA).
(2) Iron acetate liquor (black mordant).

black oil. See residual oil.

black phosphorus. See phosphorus.

black plate. Thin sheet steel obtained by rolling and usually used for containers. It is not coated with any metal, but a special lacquer or baked enamel finish is usually applied by the manufacturer.

black, platinum. See platinum black.

black powder. (blasting powder). A low explosive composed of potassium nitrate, charcoal, and sulfur. In some cases sodium nitrate is substituted for potassium nitrate. Typical proportions are 75%, 15%, and 10%. Gunpowder is probably the oldest variety.
Hazard: Sensitive to heat, deflagrates rapidly. Does not detonate but is a dangerous fire and explosion hazard.
Use: Time fuses for blasting and shell, in igniter and primer assemblies for propellants, pyrotechnics, mining, and blasting.

black rouge. See iron oxide, black.

black sand. A deposit of dark minerals with a high density found in stream beds and on beaches. Magnetite and ilmenite are usually present as also monazite and other minerals.

blackstrap. See molasses.

Bladen. (hexaethyl tetraphosphate).
Use: Insecticide.

Blaise reaction. Formation of β-oxoesters by treatment of α-bromocarboxylic esters with zinc in the presence of nitriles. The intermediate organozinc compound reacts with the nitrile and the complex is hydrolyzed with 30% potassium hydroxide.

Blanc (chloromethylation) reaction. Introduction of the CH_2Cl group into aromatic rings on treatment with formaldehyde and hydrochloric acid in the presence of zinc chloride.

Blanc reaction - Blanc rule. Cyclization of dicarboxylic acids on heating with acetic anhydride to give either cyclic anhydrides or ketones depending on the respective positions of the carboxyl groups: 1,4- and 1,5-diacids give anhydrides, while diacids in which the carboxy groups are in 1,6- or further removed positions give ketones.

blanc fixe. Precipitated barium sulfate.

blanch. To immerse vegetables or fruits in either hot water (80–100C) or steam in preparation for cooking or canning. The times and temperatures vary among processors; the higher the temperature the shorter the time required. The operation loosens the skin or peel when present, tends to remove the color, and decreases the volume. It also causes some loss of nutrient value, especially vitamins.

"Blancol."[307] TM for an anionic dispersing agent composed of the sodium salt of a sulfonated naphthalene condensate 90% active.
Use: Dispersing agent for pigments, earths, and solids in water; peptizing agent in insecticide formulations; in the paper industry for slime control, preventing coagulation of pitch, reducing two-sidedness to improve sizing, etc.; in the leather industry as a bleaching, dispersing, leveling, and neutralizing agent.

"Blancophor."[203] TM for optical whitening agents. FFG. A comarin derivative used as whitening agent for wool, nylon, acetate rayon, and mixed fibers. HS Brands. Stilbene derivatives used on cellulosic fibers, cotton and rayon fabrics, paper, and in household and industrial detergents.

"Blandol."[45] TM for white mineral oil (NF).
Use: Pharmaceutical and cosmetic formulations, plasticizers, paper penetrants, foam depressants.

blank. (1) A piece of material of any desired shape cut by a stamping die prepared for further processing. (2) See control (1).

blast furnace. A vertical coke-fired furnace used for smelting metallic ores, e.g., iron ore.

blast-furnace gas. Byproduct gas from smelting iron ore obtained by the passage of hot air over the coke in the blast furnaces. A typical gas will analyze 12.9% carbon dioxide, 26.3% carbon monoxide, 3.7% hydrogen, 57.1% nitrogen.
Hazard: Toxic by inhalation. See carbon monoxide.
Use: Heating blast-furnace stoves, boiler, or gas-engine fuel.

blasting agent. See black powder; ammonium nitrate; explosive, high, permissible, and low.

blasting gelatin. (SNG). A type of gelatinized dynamite containing approximately 7% of nitrocellulose.
Hazard: High explosive.

blasting powder. See black powder.

"B-L-E."[248] TM for high-temperature reaction product of diphenylamine and acetone.
Properties: Dark-brown, viscous liquid; d 1.087; soluble in acetone, benzene, and ethylene dichloride; insoluble in gasoline and water. Combustible.
Use: General-purpose rubber antioxidant.

bleach. To whiten a textile or paper by chemical action. Also the agent itself. Bleaching agents include hydrogen peroxide (the most common), sodium hypochlorite, sodium peroxide, sodium chlorite, calcium hypochlorite, hypochlorous acid, and many organic chlorine derivatives. Chlorinated lime is a bleaching powder used on an industrial scale. Household bleaching powders are sodium perborate and dichlorodimethylhydantoin.
Hazard: See calcium hypochlorite; lime, chlorinated. Some bleaching agents are toxic and strong oxidizing agents.

bleaching assistant. A material added to bleaching baths to secure more rapid and complete penetration of the bleach or improved regulation of the bleaching action, e.g., compounds of sulfonated oils and solvents, soluble pine oils, fatty alcohol salts, sodium silicate, sodium phosphate, magnesium sulfate, and borax.

bleach liquor. A solution of calcium hypochlorite and water.

bleed. (1) When a dye runs. (2) To release pressure gradually as via a valve.

blend. A uniform combination of two or more materials either of which could be used alone for the same purpose as the blend. For example, a fabric may be a blend of wool and nylon either of which is itself usable as fabric. Instances of materials that are often blended are:

plastics (polyblends)	grains
whiskeys	coffees
fabrics	paints
colors	tobaccos
metal powders	solvents
fertilizers	

See also mixture, mixing, kneading.

"Blendex."[525] TM for synthetic resinous products prepared from a variety of copolymer combinations. They are used to modify other polymers to attain a wide range of properties.

bleomycin. A glycopeptide antibiotic produced by *Streptomyces verticillus,* it functions as an antineoplastic and diagnostic agent. The molecule is exceedingly complex, but synthesis was achieved in 1982. It is a colorless to yellowish powder, soluble in water and methanol but insoluble in acetone and ether. It induces rupture of DNA strands.

blinding. (blister copper).
Properties: Copper (96–99% purity) produced by the reduction and smelting of copper ores. It has a blistered appearance probably caused by gas pockets. It is usually further refined electrolytically.

blister gas. See dibromodiethylsulfide.

blister packaging. A type of packaging used widely in the food and pharmaceutical industries consisting of a hollow cavity of various shapes and capacities in which the material is enclosed. Polyester and polyethylene resins are often used.

block. (1) Undesirable cohesion of films or layers of plastic.
See antiblock agent.
(2) A type of polymer.
See block polymer.

block polymer. A high polymer whose molecule is made up of alternating sections of one chemical composition separated by sections of a different

chemical nature or by a coupling group of low molecular weight. An example might be blocks of polyvinyl chloride interspersed with blocks of polyvinyl acetate. Such polymer combinations are made synthetically. They depend on the presence of an active site on the polymer chain which initiates the necessary reactions.
See also graft polymer, stereoblock polymer.

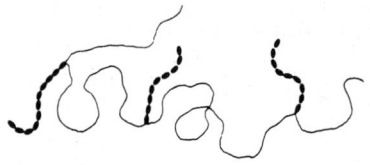

blood. A complex, liquid tissue of d 1.056 and pH 7.35–7.45. It is comprised of erythrocytes (red cells), leucocytes (white cells), platelets, plasma, proteins, and serum. The plasma fraction (55–70%) is whole blood from which the red and white cells and the platelets have been removed by centrifuging. Hemoglobin is a protein found in the erythrocytes. It contains the essential iron atom and functions as the transport agent for oxygen to the heart (artery) and of carbon dioxide from the heart (vein). Experimental work has been reported on the effectiveness of fluorocarbon compounds in carrying out the essential transport functions of blood, especially of the red cells.
Use: Plasma is used to restore liquid volume and thus osmotic pressure in the body where blood loss has been extensive. Animal blood is used as a component of adhesive mixtures. In dried or powder form it is a component of fertilizers poultry feeds and deer repellents.
See also hemoglobin, plasma, platelet, rhesus factor.

bloom. (1) A thin coating of an ingredient of a rubber or plastic mixture that migrates to the surface usually within a few hours after curing or setting. Sulfur bloom in vulcanized rubber products is most common; it is harmless but impairs the eye appeal of the product. Paraffin wax is often included purposely; when it migrates to the surface it provides an efficient barrier to sun-checking and oxidation.
  (2) A piece of steel made from an ingot.
  (3) An arbitrary scale for rating the strength of gelatin gels. When so used the word is capitalized.
  (4) Efflorescence of phytoplankton in sea water causing discoloration of the surface water.
See red tide.

blowing agent. A substance incorporated in a mixture for the purpose of producing a foam. One type decomposes when heated to processing temperature to evolve a gas, usually carbon dioxide, which is suspended in small globules in the mixture. Typical blowing agents of this kind are baking powder (bread and cake), sodium bicarbonate or ammonium carbonate (cellular or sponge rubber), halocarbons and methylene chloride in urethane, pentane in expanded polystyrene, hydrazine and related compounds in various types of foamed plastics. Another type is air used at room temperature as a blowing agent for rubber latex; it is introduced mechanically by whipping, after which the latex is coagulated with acid. Air is also used for this purpose in ice cream, whipped cream, and other food products, as well as in blown asphalt and blown vegetable oils.
See also foam.

blow molding. A technique for production of hollow thermoplastic products. It involves placing an extruded tube (parison) of the thermoplastic in a mold and applying sufficient air pressure to the inside of the tube to cause it to take on the conformation of the mold. Polyethylene is usually used but a number of other materials are adaptable to this method, e.g., cellulosics, nylons, polypropylene, and polycarbonates. It is an economically efficient process and is especially suitable for production of toys, bottles, and other containers as well as air-conditioning ducts and various industrial items. The method is not limited to hollow products; e.g., housings can be made by blowing a unit and sawing it along the parting line to make two housings.

blown asphalt. See asphalt, blown.

blown oil. (oxidized oil; base oil; thickened oil; polymerized oil). Vegetable and animal oils which have been heated and agitated by a current of air or oxygen. They are partially oxidized, deodorized, and polymerized by the treatment and are increased in density, viscosity and drying power. Common blown oils are castor, linseed, rape, whale and fish oils.
Use: Paints, varnishes, lubricants, and plasticizers.

blue copperas. See copper sulfate.

blue cross gas. See diphenylchloroarsine.

blue gas. See water gas.

blue lead. See lead sulfate, blue basic.

blueprint. See Turnbull's Blue.

blue verdigris. See copper acetate, basic.

blue vitriol. See copper sulfate.

blush. Precipitation of water vapor in the form of colloidal droplets on the surface of a varnish or lacquer film caused by lowering of the temperature immediately above the coated surface due to solvent evaporation. This results in unsightly graying of the dried film and can be avoided by use of a less volatile solvent.

board. See paperboard.

BOD. See biochemical oxygen demand.

body. (1) A non-specific term approximately synonymous with consistency or viscosity, usually descriptive of liquids, e.g., a heat-bodied oil (linseed oil which has been polymerized by heating). See also consistency.
(2) In biochemistry, an agglutinous substance present in the blood or tissues, e.g., antibody. See agglutination.
(3) An object having a unique physical property, e.g., blackbody.

Bodroux-Chichibabin aldehyde synthesis. Formation of aldehydes by treatment of orthoformates with Grignard reagents.

Bodroux-reaction. Formation of substituted amides by reaction of a simple aliphatic or aromatic ester with an amino magnesium halide obtained by treatment of a primary or secondary amine with a Grignard reagent at room temperature.

Boeseken's method. Determination of the relative configurations of the hydroxyl groups on the reducing and vicinal carbon atoms in a cyclic sugar. Boric acid forms complexes with cis hydroxyl groups on vicinal carbon atoms and the electrical conductivity of the solution is enhanced. Since there is no complex formation with trans-hydroxyl groups, no increase in conductivity is observed.

Bogert-Cook synthesis. Condensation of β-phenylethylmagnesium bromide with cyclohexanones followed by cyclodehydration of the tertiary alcohol with concentrated sulfuric acid with formation of octahydrophenanthrene derivatives and a small amount of spiran.

boghead coal. A variety of bituminous or subbituminous coal resembling cannel coal in appearance and behavior during combustion, characterized by a high percentage of algal remains and volatile matter. Upon distillation it gives exceptionally high yields of tar and oil (ASTM).

Bohn-Schmidt reaction. Hydroxylation of anthraquinones containing at least one hydroxyl group by fuming sulfuric acid or sulfuric acid and boric acid in the presence of a catlyst such as mercury.

boiled oil. See linseed oil, boiled.

boiler scale. A rocklike deposit occurring on boiler walls and tubes in which hard water has been heated or evaporated. Consists largely of calcium carbonate, calcium sulfate, or similar materials, depending on the mineral content of the water. Boiler scale decreases the rate of heat transfer through the boiler and tube walls resulting in increased heating costs and shortening of boiler life. Most boiler feed water is softened (treated to remove calcium and magnesium ions) before being used. Scale may be removed by treatment with ammonium bicarbonate solution. See also water, hard; zeolite.

boiling point. The temperature of a liquid at which its vapor pressure is equal to or very slightly greater than the atmospheric pressure of the environment. For water at sea level it is 212F (100C).

Boltzmann, Ludwig. (1844–1906). Born in Vienna, Boltzmann was interested primarily in physical chemistry and thermodynamics. His work has importance for chemistry because of his development of the kinetic theory of gases and the rules governing their viscosity and diffusion. The mathematical expression of his most important generalizations is known as Boltzmann's Law, still regarded as one of the cornerstones of physical science.

bomb. A small metal container which can contain gases or liquids under varying degrees of pressure. An aerosol bomb usually contains insecticides which are emitted as an atomized spray on release of pressure, the gases used being carbon dioxide, nitrous oxide, butane, etc. at relatively low pressure. An oxygen bomb is used for accelerated aging tests for rubber and plastic products; oxygen under high pressure is used. This device must be handled by a trained technician.

bombardment. Impingement upon an atomic nucleus of accelerated particles such as neutrons or deuterons for the purpose of inducing fission or of creating unstable nuclei which will thus become radioactive. This operation was first accomplished with positively charged particles in the cyclotron in the early 1940s and subsequently in nuclear reactors. Neutrons are commonly used in reactors because their lack of electrical charge permits easier penetration of the target nucleus. See also radioisotope, fission, fast atom bombardment.

"Bomyl."[50]   TM for dimethyl-3-hydroxyglutaconate dimethyl phosphate. ($C_9H_{15}O_8P$).
Used as an insecticide to control flies.
Properties: Liquid. Insoluble in water and kerosene. Corrosive to iron, steel, and brass. Stable when stored in glass. Nonflammable. Soluble in acetone, xylene, methanol, ethanol.
Hazard: Toxic by ingestion and inhalation.

BON.   Abbreviation for β-oxynaphthoic acid. See 3-hydroxy-2-naphthoic acid, BON Red.

"Bonaril."[233]   TM for a hydrolyzed polyacrylamide for use in foundry sands.

bond, chemical.   An attractive force between atoms strong enough to permit the combined aggregate to function as a unit. A more exact definition is not possible because attractive forces ranging upward from 0 to those involving more than 250 kcal/mole of bonds are known. A practical lower limit may be taken as 2–3 kcal/mole of bonds, the work necessary to break approximately 1.5×10^{24} bonds by separating their component atoms to infinite distance.

All bonds appear to originate with the electrostatic charges on electrons and atomic nuclei. Bonds result when the net coulombic interactions are sufficiently attractive. Different principal types of bonds recognized include metallic, covalent, ionic, and bridge.

Metallic bonding is the attraction of all the atomic nuclei in a crystal for the outer shell electrons which are shared in a delocalized manner among all available orbitals. Metal atoms characteristically provide more orbital vacancies than electrons for sharing with other atoms.

Covalent bonding results most commonly when electrons are shared by two atomic nuclei. Here the bonding electrons are relatively localized in the region of the two nuclei although frequently a degree of delocalization occurs when the shared electrons have a choice of orbitals. The conventional *single* covalent bond involves the sharing of two electrons. There may also be *double* bonds with four shared electrons, *triple* bonds with six shared electrons, and bonds of intermediate multiplicity.

Covalent bonds may range from *nonpolar*, involving electrons evenly shared by the two atoms, to extremely *polar*, where the bonding electrons are very unevenly shared. The limit of uneven sharing occurs when the bonding electron(s)

spends full time with one of the atoms. This makes this atom into a negative ion and leaves the other atom in the form of a positive ion. Ionic bonding is the electrostatic attraction among oppositely charged ions.

Bridge bonds involve compounds of hydrogen in which the hydrogen bears either a + or − charge. When hydrogen is attached by a polar covalent bond to one molecule it may attract another molecule bridging the two molecules together. If the hydrogen is + it may attract an electron pair of the other molecule. This is called a *protonic bridge*. If the hydrogen is − it may attract through a vacant orbital the nucleus of an atom of a second molecule. This called a *hydridic bridge*. Such bridges are at the lower range of bond strength but may be very significant in their effect on the physical properties of condensed states of those substances in which they are possible. R. T. Sanderson.
See also hydrogen bond.

"Bonderite."[62]   TM for chemical compositions for producing a corrosion-inhibiting finish on metals, preparing metal surfaces for the subsequent application of finish coats, conditioning metal surfaces to facilitate metal deformation operations.

"Bonderlube."[62]   TM for soap-like chemical composition for treating metal surfaces which have been pretreated with phosphatizing coating chemicals to form a lubricant layer adapted to cold forming and to retard the formation of rust.

"Bondogen."[69,496]   TM for oil soluble sulfonic acid of high molecular weight with a high-boiling hydrophilic alcohol and a paraffin oil.
Properties: Dark mahogany liquid, d 0.93, acid number 40–42. Combustible.
Use: Plasticizer and processing aid for elastomers.

bone ash.   An ash composed principally of tribasic calcium phosphate but containing minor amounts of magnesium phosphate, calcium carbonate, and calcium fluoride. Noncombustible.
Derivation: By calcining bones. A synthetic product is also available.
Use: Cleaning and polishing, ceramics, animal feeds. The better grades are used in coating molds for copper wire, bars, slabs, and other metals.

bone black.   (bone char; bone charcoal). Black pigment made by carbonizing bones. Carbon content is usually approximately 10%. Nonflammable in bulk.
Hazard: Flammable as suspended dust.
Use: Manufacturing activated carbon, decolorizing agent and filtering medium, cementation

reagent, adsorptive medium in gas masks, paint and varnish pigment, clarifying shellac, water purification.

bone china. Ceramic tableware of high quality in which a small percentage of bone ash is incorporated. Made chiefly in England.

bone meal. A product made by grinding animal bones. Raw meal is made from bones that have not been previously steamed. If pressure-steaming has been used, the meal is called "steamed." The fertilizer grade contains from 43–55% tricalcium phosphate, 20–25% phosphoric acid, and 4–5% ammonia. The feed grade according to Bureau of Animal Industry specifications must contain from 65–75% tricalcium phosphate and only approximately 2% ammonia. Much of the latter grade is imported.
Use: Fertilizer (raw); animal feeds (steamed).

bone oil. (animal oil; Dippel's oil; hartshorn oil). CAS: 8001-85-2.
Properties: Dark brown, fixed oil; repulsive odor. Soluble in water, d 0.900–0.980. Combustible. Chief constituents: Hydrocarbons, pyridine bases, and amines. Derivation: Destructive distillation of bones or other animal substance. Note distinction from bone tallow.
Grade: Technical.
Hazard: Evolves very toxic ammonium cyanide when heated to 180C.
Use: Organic preparations, source of pyrrole, denaturant for alcohol.

bone phosphate. (BPL). Phosphoric acid occurring in bones in the form of tribasic calcium phosphate.

bone seeker. An element or radioisotope that tends to lodge in the bones when absorbed into the body. Examples are fluorine, calcium, and strontium.

bone tallow. (bone fat). Fat obtained from animal bones by boiling in water, by treating with steam under pressure, or by solvent extraction. It is a glyceryl ester.

BON maroon. A calcium or manganese precipitated compound of 3-hydroxy-2-naphthoic acid and 2-naphthylamine-1-sulfonic acid.
See BON red.

BON red. Class name for a group of organic azo pigments made by coupling 3-hydroxy-2-naphthoic acid to various amines and forming the barium, calcium, strontium, or manganese salts. They have bright shades ranging from

yellow red to deep maroon, good light and heat resistance, non-bleeding in vehicles and solvents, and good opacity. They are widely used in printing inks, paints, enamels, lacquers, rubber, plastics, wall paper, textiles, floor coverings, and crayons.

Boord olefin synthesis. Regiospecific synthesis of olefins from aldehydes and Grignard reagents.

boort. (bort). See diamond, industrial.

BOPOB. See p-bis[2-(5-p-biphenylyloxazolyl)]benzene.

boracic acid. See boric acid.

boral. A composite material consisting of boron carbide crystals in aluminum with a cladding of commercially pure aluminum. Concentration of up to 50% boron carbide can be obtained.
Use: Reactor shields, neutron curtains, shutters for thermal curtains, safety rods, containers for fissionable material.
See also composite.

"Boran."[169] TM for diaminochrysazin used in the colorimetric determination of boron.

borane. One of a series of boron hydrides (compounds of boron and hydrogen). The simplest of these, BH_3, is unstable at atmospheric pressure and becomes diborane (B_2H_6) as gas at normal pressures. This is converted to higher boranes, i.e., penta-, deca-, etc., by condensation. This series progresses through a number of well-characterized crystalline compounds. Hydrides up to $B_{20}H_{26}$ exist. Most are not very stable and readily react with water to yield hydrogen. Many react violently with air. As a rule, they are highly toxic. Their properties have suggested investigation for rocket propulsion but they have not proved satisfactory for this purpose. There are also a number of organoboranes used as reducing agents in electroless nickel-plating of metals and plastics. Some of the compounds used are di- and triethylamine borane and pyridine borane.
See also carborane, diborane, pentaborane, organoborane.

borax. (sodium borate; tincal; borax decahydrate). $Na_2B_4O_7 \cdot 10H_2O$. A natural, hydrated, sodium borate found in salt lakes and alkali soils also the commercial name for sodium borate.
Hazard: Toxic. TLV: 5 mg/m³.

borax, anhydrous. (borax, dehydrated; sodium borate, anhydrous). CAS: 1303-96-4. $Na_2B_4O_7$.

Properties: White, free-flowing crystals; hygroscopic; forms partial hydrate in damp air; mp 741C; d 2.367; slightly soluble in cold water. Noncombustible.
Grade: Technical, 99% $Na_2B_4O_7$, standard, fine granular form, glass or fused.
Hazard: Toxic. TLV: 1 mg/m^3 in air.
Use: Manufacturing of glass, enamels and other ceramic products, herbicide.

borax pentahydrate. $Na_2B_4O_7 \cdot 5HOH$.
Properties: Free-flowing powder; begins to lose water of hydration at 122C; d 1.815. Noncombustible.
Grade: Crude, technical (99.5% $Na_2B_4O_7 \cdot 5H_2O$).
Hazard: Toxic. TLV: 1 mg/m^3 in air.
Use: Weed killer and soil sterilant, fungus control on citrus fruits (FDA tolerance 8 ppm of boron residue).

borazole. (borazine). $B_3N_3H_6$. Inorganic analog of benzene.
Properties: Colorless liquid, fp −58C, bp 53C, d 0.824 (0C), hydrolyzes to evolve boron hydrides.
Hazard: Dangerous fire risk. Toxic via ingestion and inhalation, strong irritant to tissue.

"Borazon."[245] TM for a boron nitride formed at very high pressure (85,000 atm) and temperatures (1800C) from mixtures of boron and nitrogen or from ordinary hexagonal boron nitride in the presence of catalysts such as lithium, calcium, magnesium, or their nitrides. Mohs hardness 10, d 3.48. Reacts extremely slowly with water; dissolves in fused sodium carbonate.
Use: See boron nitride.

Borcher's metal. A group of alloys of chromium with nickel and cobalt or of chromium and iron with a small proportion of molybdenum and/or silver or gold. Heat- and corrosion-resistant.
Use: Chemical apparatus, crucibles, pyrometer tubes, heat treating or annealing pots.

Bordeaux mixture. A fungicide and insecticide mixture made by adding slaked lime to a copper sulfate solution. It is either made by the user or bought as a powder ready for dissolving. Stabilizing agents are sometimes added to delay settling. Used especially for potato bugs and similar garden pests.
Hazard: Irritant and corrosive by ingestion.

"Bor-guard."[1] TM for a borax-propylene glycol-butynediol condensate. A corrosive inhibitor particularly suited for brake fluids of high boiling point also for antifreeze and hydraulic fluids.

boric acid. (boracic acid; orthoboric acid).
CAS: 10043-35-3. H_3BO_3.
Properties: Colorless, odorless scales or white powder; stable in air. D: 1.4347 (15C), mp indeterminate since it loses water in stages through metaboric acid, HBO_2, pyroboric acid, $H_2B_4O_7$, and to the oxide B_2O_3. Soluble in boiling water, alcohol, and glycerol. Noncombustible.
Derivation: (a) By adding hydrochloric or sulfuric acid to a solution of borax and crystallizing. (b) From weak borax brines by extraction with a kerosene solution of a chelating agent such as 2-ethyl-1,3-hexanediol or other polyols. Borates are stripped from the chelate by sulfuric acid.
Method of purification: Recrystallization.
Hazard: Toxic via ingestion. Use only weak solutions. Irritant to skin in dry form.
Grade: Technical, 99.9%, CP, USP.
Use: Heat-resistant (borosilicate) glass, glass fibers, porcelain enamels, boron chemicals, metallurgy (welding flux, brazing copper), flame retardant in cellulosic insulation, mattress batting and cotton textile products, fungus control on citrus fruits (FDA tolerance 8 ppm boron residue), ointment and eye wash (water solution only), nickel electroplating baths.

boric acid esters. (borate ester; trimethyl borate; tri-n-butyl borate; tricyclohexyl borate; tridodecylborate; tri-p-cresyl borates). Trihexylene glycol biborate compounds which are readily hydrolyzed to boric acid and the respective alcohols.
Properties: Colorless to yellow liquids, bp 230–350C. Combustible.
Use: Dehydrating agents, catalysts, sources of boric oxide, special solvents, stabilizers, plasticizers or adhesion additives to latex paints, ingredients of soldering and brazing fluxes.

orthoboric acid. See boric acid.

boric anhydride. See boric oxide.

boric oxide. (boric anhydride; boron oxide).
CAS: 1303-86-2. B_2O_3.
Properties: Colorless powder or vitreous crystals, slightly bitter taste. Soluble in alcohol and hot water; slightly soluble in cold water, d 2.46, mp approximately 450C, bp 1500C. Noncombustible.
Derivation: By heating boric acid.
Grade: Glass or fused form, powder, technical or high purity 99.99+%.
Hazard: TLV: 10 mg/m^3 in air.
Use: Production of boron, heat-resistant glassware, fire-resistant additive for paints, electronics, liquid encapsulation techniques, herbicide.

boride. An interstitial compound of boron and another metal (transition, alkaline-earth, or rare-earth). Such compounds are not stoichiometric, the boron atoms being linked together in zigzag chains, two-dimensional nets, or three-dimensional structures throughout the crystal.
Properties: Highly refractory, with mp from 2000 to 3000C; Mohs hardness from 8 to 10; thermally and electrically conductive; high chemical stability; do not react with hydrochloric or hydrofluoric acids but are attacked by hot alkali hydroxides; color varies from gray (transition-metal) to black (alkaline-earth) to blue (rare-earth).
Derivation: (1) Sintering mixtures of metal powder and boron at 2000C; (2) reduction of mixture of the metal oxide and boric oxide with aluminum, silicon, or carbon; (3) fused salt electrolysis; (4) vapor-phase deposition.
Use: High-temperature service as rocket nozzles, turbines, etc.

borneol. (bornyl alcohol; 2-camphanol; 2-hydroxycamphane). CAS: 507-70-0. $C_{10}H_{17}OH$.
Properties: White, translucent lumps; sharp, camphor-like odor; burning taste; soluble in alcohol and ether; insoluble in water. Optically active in natural form, racemic form made synthetically, d 1.011, mp 208C, bp 212C. Flammable.
Derivation: Natural form from a species of tree in Borneo, Sumatra, Synthesized from camphor by hydrogen reduction or from α-pinene.
Grade: Technical.
Hazard: Fire risk in presence of open flame.
Use: Perfumery, esters.

bornyl acetate. $C_{10}H_{17}OOCCH_3$.
Properties: Colorless liquid, solidifying to crystals at approximately 50F, piney-camphoraceous odor. Soluble in 3 volumes of 70% alcohol, miscible with 95% alcohol and with ether, d 0.980–0.984, refr index 1.463–1.465, mp 29C. Combustible.
Derivation: Interaction of borneol and acetic anhydride in the presence of formic acid.
Grade: Technical, FCC.
Use: Perfumery, flavoring, nitrocellulose solvent.

bornyl alcohol. See borneol.

bornyl formate. $C_{10}H_{17}OOCH$.
Properties: Colorless liquid having a piney odor, d 1.007–1.009. Combustible.
Grade: Technical.
Use: Perfuming of soaps, disinfectants and sanitary products, flavoring.

bornyl isovalerate. $C_{10}H_{17}OOC_5H_9$.
A constituent of valerian oil.

Properties: Limpid fluid, aromatic, valerian-like odor. Soluble in alcohol and ether, insoluble in water, d 0.951 (20C), bp 255–260C. Combustible.
Use: Medicine, essential oil intermediate, flavoring.

boroethane. See diborane.

boron. (B) Nonmetallic element of atomic number 5; Group IIIA of the Periodic Table. Atomic weight 10.81. Valence 3; two stable isotopes, 11 (approximately 81%) and 10 (approximately 19%).
Properties: Black, hard solid; brown, amorphous powder; crystals. Highly reactive. Soluble in concentrated nitric and sulfuric acids; insoluble in water, alcohol, and ether. High neutron absorption capacity. Amphoteric. A plant micronutrient, d 2.45, mp 2300C, Mohs hardness 9.3.
Sources: Borax, kernite, colemanite, ulexite.
Derivation: (a) By heating boric oxide with powdered magnesium or aluminum; (b) by vapor-phase reduction of boron trichloride with hydrogen over hot filaments (80–2000C); (c) electrolysis of fused salts.
Forms: Filament, powder, whiskers, single crystals.
Grade: Technical (90–92%), 99% pure, high-purity crystals.
Hazard: Dust ignites spontaneously in air, severe fire and explosion hazard. Reacts exothermally with metals above 900C; explodes with hydrogen iodide.
Use: Special-purpose alloys, cementation of iron, neutron absorber in reactor controls, oxygen scavenger for copper and other metals, fibers and filaments in composites with metals or ceramics, semiconductors, boron-coated tungsten wires, rocket propellant mixtures, high-temperature brazing alloys.
See also boron alloy, boron fiber, boron 10.

boron-10. Nonradioactive isotope of boron of mass number 10.
Properties: Has marked capacity for absorbing slow neutrons, emitting a high-energy alpha particle in the process.
Derivation: Constitutes approximately 19% of natural boron.
Forms available: Crystalline powder, dry amorphous powder, colloidal suspension of dry amorphous powder in oil; in boron trifluoride-calcium fluoride, in potassium borofluoride, in boron trifluoride ethyl etherate, in boric acid.
Use: Neutron counter, radiation shielding (in the form of boral), medicine.

boron alloy. A uniformly dispersed mixture of boron with another metal or metals. Ferroboron usually contains 15–25% boron, manganese boron usually 60–65% manganese.

Use: Degasifying and deoxidizing agents, to harden steel (in trace quantities) to increase conductivity of copper, turbojet engines.

boron bromide. See boron tribromide.

boron carbide. B_4C.
Properties: Hard, black crystals; d 2.6; Mohs hardness 9.3; mp 2350C; bp 3500C. Soluble in fused alkali, insoluble in water and acids. High capture cross-section for thermal neutrons.
Derivation: Heating boron oxide with carbon in an electric furnace.
Forms: Powder, crystals, rods, fibers, whiskers.
Hazard: Avoid inhalation of dust or particles.
Use: Abrasive powder, abrasion resister and refractory, control rods in nuclear reactors, reinforcing agent in composites for military aircraft, and other special applications.
See also boral.

boron chloride. See boron trichloride.

boron fiber. A vapor-deposited filament made by deposition of boron on a heated tungsten wire. Filament is 0.004 inch in diameter while the wire is only 0.0005 inch. Tensile strength 350,000 to 450,000 psi, elastic modules 55 million psi, upper temperature limit in oxidizing atmosphere 250C. Used in composites with epoxy resins for aircraft and space applications. They can be woven into fabrics.

boron fluoride. See boron trifluoride.

boron fuel. See rocket fuel.

boron hydride. See borane, diborane, decaborane, pentaborane.

boron nitride. BN.
Properties: White powder; particle diameter approximately 1 micron. Mp 3000C (sublimes). Graphite-like, hexagonal plate structure. High electrical resistance. Compressed at 10^6 psi, it becomes hard as diamond. Excellent heat-shock resistance, low mechanical strength, hygroscopic. Noncombustible.
Derivation: Heating a mixture of boric acid and tricalcium phosphate in ammonia atmosphere in an electric furnace.
Forms: Powder, compressed solid, fibers, whiskers.
Use: Refractory, furnace insulation, crucibles, rectifying tubes, dielectric, chemical equipment, self-lubricating bushings, molten metal pump parts, transistor and rectifier mounting wafers,

heatshield for plasma, nose-cone windows, heat-resistant fibers stable to 870C in oxidizing atmosphere for military composites, metalworking abrasive, high-temperature insulator, high-strength fibers. See also borazon.

boron nitride, pyrolytic. See "Boralloy."

boron oxide. See boric oxide.

boron phosphate. (Sometimes called borophosphoric acid). BPO_4.
Properties: White, non-hygroscopic crystals; d 1.873. Soluble in water, pH (1% solution) 2.0.
Use: Special glasses, ceramics, acid cleaner, dehydration catalyst.

boron phosphide. BP. A refractory, maroon powder; noncorrosive; Mohs hardness 9.5.
Derivation: Direct union of boron and phosphorus at approximately 1000C in a reducing atmosphere.
Hazard: Evolves toxic fumes in contact with water and acids. Ignites spontaneously at 390F (199C).

boron steel. See ferroboron.

boron tribromide. (boron bromide).
CAS: 10294-33-4. BBr_3.
Properties: Colorless, fuming liquid. Decomposed by alcohol and by water. D 1.69 (15C), bp 90C, fp −46C.
Derivation: (High purity.) Direct bromination of boron followed by rectification in quartz columns.
Grade: Technical, high purity.
Hazard: Corrosive to tissue. TLV: ceiling of 1 ppm in air. May explode when heated.
Use: Catalyst in organic synthesis manufacture of diborane.

boron trichloride. (boron chloride).
CAS: 10294-34-5. BCl_3.
Properties: Colorless fuming liquid. Decomposed by alcohol and by water. D 1.35 (25C), bp 12.5C, fp −107C. Reacts with hydrogen at 1200C.
Derivation: (a) Heating boric oxide and carbon with chlorine; (b) combining boric oxide with phosphorus pentachloride.
Grade: Technical (99%), CP (99.5%).
Hazard: Strong irritant to tissue. Fumes are corrosive and toxic.
Use: Catalyst in organic syntheses; source of many boron compounds; refining of alloys; soldering flux; making electrical resistors; extinguishing magnesium fires in heat-treating furnaces; manufacturing of diborane.

boron trifluoride. (boron fluoride).
CAS: 7637-07-2. BF_3.

Properties: Colorless gas, d 3.076 g/L, fp −126.8C, bp −101C, does not support combustion, soluble in cold water, hydrolyzes in hot water, soluble in concentrated sulfuric acid and most organic solvents. Easily forms double compounds such as that with ether known as boron trifluoride etherate or BF_3-ether complex.
Derivation: From borax and hydrofluoric acid or from boric acid and ammonium bifluoride. The complex formed is then treated with cold fuming sulfuric acid.
Grade: Pure (99% min).
Hazard: Toxic by inhalation, corrosive to skin and tissue. TLV: ceiling of 1 ppm in air.
Use: Catalyst in organic synthesis, production of diborane, instruments for measuring neutron intensity, soldering fluxes, gas brazing.

boron trifluoride etherate. (BF_3-ether complex). $CH_3CH_2O(BF_3)CH_2CH_3$.   A relatively stable coordination complex formed by the combination of diethyl ether with boron trifluoride, in which the boron atom is bonded to the oxygen of the ether.
Properties: Fuming liquid. Bp 259C (126C), flash p 147F (63.8C) (OC).
Combustible.
Hazard: Toxic by inhalation, corrosive to skin and tissue.
Use: Catalyst in organic synthesis.

boron trifluoride-methanol. (BF_3-MeOH).
A soluble of boron trifluoride in methanol.
Properties: Concentration 14 g/100 cc, d 0.90 (20C).
Use: As an esterification reagent for fats and oils.
Hazard: Moderate fire risk. Combustible.

boron trifluoride monoethylamine. (boron fluoride monoethylamine).
BF_3-MEA); BF_3-$C_2H_5NH_2$.
Properties: White to pale tan flakes, d 1.38, mp 88–90C. Soluble in furfuryl alcohol, polyglycol, acetone. Releases boron trifluoride at above 110C.
Hazard: Moderate fire risk. Combustible.
Use: Elevated temperature cure of epoxy resins.

borophosphoric acid. See boron phosphate.

borosilicate glass. See glass, heat-resistant.

"Boro-Silicon."[504] TM for fire and heat resistant, field castable elastomer with high hydrogen content. A solid material with resiliency to minimize impact due to secondary missile formation. Density 1.59 g/cc. Self. extinguishing.

"Boro-Spray."[88] TM for a crystalline product consisting chiefly of sodium pentaborate. Used

for spray applications to tree fruit and truck crops where boron deficiency is indicated.

borotungstic acid. (borowolframic acid). Various formulas and properties given.
Grade: Technical.
Use: Mineralogic assay.

Borsche-Drechsel cyclization. Acid-catalyzed rearrangement of cyclohexanone phenylhydrazone to tetrahydrocarbazole. Subsequent oxidation yields carbazole.

Bosch, Karl. (1874-1940) A German chemist who was the 1931 recipient of the Nobel prize with Friederick Bergius. In World War I, his catalyst study led to the production of synthetic gasoline. He also worked in the area of chemical high pressure methods. His research in ammonia synthesis aided in the manufacture of fertilizers and explosives. His PhD was awarded in Liepzig, Germany.

Bosch-Meiser urea process. Industrial process for formation of urea by reaction of carbon dioxide with ammonia at elevated temperatues and pressures.

"Botran."[519] TM for an agricultural fungicide 2,6-dichloro-4-nitroaniline.

botulism. An intense and often fatal poisoning caused by ingestion of the anaerobic bacterium *Clostridium botulinum*, a gram-positive bacillus which proliferates in many kinds of inadequately sterilized canned food products, i.e., those preserved at home. Death from respiratory paralysis occurs in from 30 to 65% of the cases. Heating to 80C or higher for 25 to 30 minutes before serving is effective protection against this powerful toxin. Extensive tests have indicated that meats such as bacon can be made resistant to botulism by treatment with a combination of potassium sorbate and sodium nitrite.

bound water. Water molecules that are tightly held by various chemical groups in a larger molecule. Carboxyl, hydroxyl, and amino groups are usually involved, hydrogen-bonding is often a factor. Proteins tend to bind water in this way and in meats it will remain unfrozen as low as −40C.

Bouveault aldehyde synthesis. Action of Grignard reagents on N,N-disubstituted formamides yields aldehydes.

Bouveault-Blanc reduction. Formation of alcohols by reduction of esters with sodium and an alcohol.

Boyle, Robert. (1627–1691). A native of Ireland, Boyle devoted his life to experiments in what was then called "natural philosophy," i.e., physical science. He was early influenced by Galileo. His interest aroused by a pump which had just been invented. Boyle studied the properties of air, on which he wrote a treatise (1660). Soon thereafter, he stated the famous law that bears his name (see next entry). Boyle's group of scientific enthusiasts was known as the "invisible college" and in 1663 it became the Royal Society of London. Boyle was one of the first to apply the principle which Francis Bacon had described as "the new method"—namely, inductive experimentation as opposed to the deductive method of Aristotle—and this became and has remained the cornerstone of scientific research. Boyle also investigated hydrostatics, desalination of sea water, crystals, electricity, etc. He approached but never quite stated the atomic theory of matter; however, he did distinguish between compounds and mixtures and conceived the idea of "particles" becoming associated to form molecules.

Boyle's Law. The volume of a sample of gas varies inversely with the pressure if the temperature remains constant. The relation is strictly true only for an imaginary perfect or ideal gas, but the law is satisfactory for practical calculations except when pressures are high or temperatures are approaching the liquefaction point. Van der Waal's equation is a refinement to take care of the inherent inaccuracy of Boyle's Law.
See also ideal gas.

"BPIC"[177]. TM for technical grade of terbutyl peroxy isopropyl carbonate a polymerization initiator for acrylic, ethylene, styrene, and other monomers and a cross-linking agent for silicone and ethylene propylene elastomers.

BPL. (1) Abbreviation for bone phosphate of lime.
See bone phosphate.
(2) Abbreviation for β-propiolactone.

"BPR"[55]. TM for insecticidal mixture containing varying proportions of pyrethrin, piperonyl butoxide, and rotenone in liquid or dust base.

Br. Symbol for bromine.

BRA. Abbreviation for β-resorcylic acid.

brackish water. Water that is lower in salinity than normal seawater and higher in. salinity than freshwater, ranging from 30 to 0.5 parts salt per 1,000. parts water.

Bradsher reaction. Acid catalyzed cyclodehydration of o-acyldiarylalkanes to polycyclic hydrocarbons and their heterocyclic analogs.

brake fluid. See hydraulic fluid.

branched chain. See chain.

bran oil. See furfural.

brasilin. (brazilin; Brazilwood extract). $C_{16}H_{14}O_5$. The crystalline, colorizing principle of Brazilwood.
Properties: White or pale yellow, rhombic needles from alcohol; turns orange in air or light. Soluble in water, alcohol, ether, and in alkali hydroxide soluble with a carmine-red color. Decomposes above 130C.
Use: Dyeing red and purple shades of wood, ink, textiles, etc. Acid-base indicator, turning yellow in acid and carmine red in alkali. Biological stain.

brass. Copper-zinc alloys of varying composition. Low-zinc brasses below 20% are resistant to stress-corrosion cracking and are easily formed. Red brass (15% zinc) is highly corrosion-resistant. Yellow brasses contain from 34 to 37% zinc and have good ductility and high strength and can withstand severe cold-working. Cartridge brass contains 30–33% zinc. Muntz metal (40% zinc) is primarily a hot-working alloy used where cold-forming operations are unnecessary. Some brasses also contain low percentages of other elements, e.g., manganese, aluminum, silicon, lead, and tin (admiralty metal, naval brass).
Hazard: Flammable in powder or finely divided form.
Use: Condenser tube plates, piping, hose nozzles and couplings, oil gauges, flow indicators, air cocks, drain cocks, marine equipment, jewelry, fine arts, stamping dies.
See also admiralty metal, aluminum brass, naval brass, red brass, yellow brass, Muntz metal, cartridge brass.

brassidic acid. (trans-13-docosenoic acid). $CH_3(CH_2)_7HC{=}CH(CH_2)_{11}COOH$.
An isomer of erucic acid.
Properties: White crystals. Mp 61–62C, bp 282C (30 mm), d 0.859, refr index 1.448 (57C). Insoluble in water, slightly soluble in alcohol, soluble in ether. Combustible.
Derivation: By treating erucic acid with nitrous acid (catalyst).

von Braun reaction. Reaction of tertiary amines with cyanogen bromide to form disubstituted cyanamides and an alkyl halide.

"Braze Bonding Agent."[69] TM for halogenated rubber derivatives and selected modifiers in solvent.
Properties: Deep red liquid, d 1.01, total solids 20–22%, flash p 93F (34C).
Hazard: Flammable, moderate fire risk.
Use: Cement to bond natural rubber, SBR or neoprene to steel.

Brazil wax. See carnauba wax.

brazing. A welding method in which a nonferrous filler alloy is inserted between the end or edges of the metals to be joined.
See also welding.

Bredt's rule. A restriction applying to bridged systems, which states that in some. bridged systems the branching points (the bridgeheads) cannot. be involved. in a double bond. As a corollary, reactions which should lead to such compounds will be hindered or will take a different course.

breeder. A particularly efficient type of nuclear reactor which is able to utilize the tremendous energy latent in U-238. This cannot be exploited in conventional (thermal) nuclear reactors which are fueled with enriched uranium or plutonium, for these eventually become depleted and must be replaced. The fuel used for the breeder reactor is a mixture of nonfissionable U-238 and plutonium-239 sealed in long thin hexagonal metal tubes which are in turn contained in cans called subassemblies. These constitute the reactor core. Around it are placed several layers of U-238, also in subassemblies. When criticality is reached, the unmoderated neutron flux from the core permeates the entire system and thus "breeds" fissionable plutonium-239 in the surrounding U-238. The amount of fissionable material thus made available is about 100 times as great as that obtainable with a conventional reactor since all the energy potential of the U-238 can be released. Twenty lbs of uranium has the potential of delivering approximately 52×10^6 kwh of electricity, only a small fraction of this would be extractable without breeding.
The breeder utilizes fast neutrons which are much more efficient than the slow (thermal) neutrons used in conventional reactors. Liquid sodium is the coolant in breeder reactors as it has no retarding effect on the neutrons, 2.9 neutrons per fission are produced in the breeder compared with only 2.4 in water-moderated reactors. It is this excess of neutrons that makes it possible for the fast breeder to produce more fuel than it consumes. Breeders have been operating on a commercial scale in several European countries for some years. The NRC has authorized construction of the Clinch River breeder, but funding was cancelled as a result of opposition by environmentalists and antinuclear groups. The only operating breeder reactor in the US is Argonne's EBR-11 in Idaho.

breeze. Coke particles less than one-half inch in diameter. This occurs to the extent of approximately 100 lbs/ton of coal processed.

"Brellin."[342] TM for gibberellic acids as plant growth-stimulating compounds.

Brestan. TM for triphenyltin acetate by Hoechst. AG, Postfach 80 03 20, D 6230 Frankfurt am Main 80, Federal Repulic of Germany.

"Bretol."[430] TM for cetyldimethylethyl ammonium bromide, a quaternary ammonium compound used in dental preparations and soldering fluxes.

bretonite. (iodoacetone). CH_3COCH_2I.
Properties: bp 102C.

"Brevon."[448] TM for vinyl resin films used for cocoon-type storage coatings.

brewing. The production of beer, ale, and malt liquors by a process involving a complex series of enzymatic reactions. The most important of these is the conversion of starch to a malt extract (wort), which in turn is fermented with yeast. Mashing is the preparation of wort from malt and cereals by enzymatic hydrolysis, after which the product is boiled with hops, which impart the characteristic taste and aroma of beer. The malt extract must contain the nutrients required for yeast growth. Mashing involves a complex interplay of chemical and enzymatic reactions that are not fully understood. Few changes have been made in the basic brewing processes for over a century, increased automatic operation and quality control techniques ensure a consistently good product.
See also fermentation, yeast, wort.

Brewster process. A method for the extraction of acetic acid from the acid distillate of the destructive distillation of wood. Isopropyl ether is used as the solvent for the acetic acid.
See Cochran process.

brick, refractory. A highly heat-resistant and nonconductive material used for furnace linings as in the glass and steel industries and other applications where temperatures above 1600C are involved. Some types are made of quartzite or high-silica clay (see firebrick), others of metallic ores such as chromite, magnesite, and zirconia.
See also refractory.

bridge. See bond, chemical; hydrogen bond.

brightener. A compound which when added to a nickel plating formulation of the Watts type (nickel sulfate and nickel chloride in a six to one ratio plus boric acid) will yield a bright reflective finish. There are two types: (1) naphthalene disulfonic acids, diphenyl sulfonates, aryl sulfonamides, etc., which give bright deposits on polished surfaces; and (2) metal ions having high hydrogen overvoltage in acid soluble (zinc, cadmium, selenium, etc., and unsaturated organic compounds such as thiourea, acetylene derivatives, azo dyes, etc.) which give mirror brightness as a result of their "leveling" action. Usually both types are used for maximum effectiveness. See also leveling (2), optical brightener.

bright stock. Lubricating oil of high viscosity obtained from residues of petroleum distillation by dewaxing and treatment with Fuller's earth or similar material. Sometimes also applied to viscous petroleum distillates.
Use: For blending with neutral oils in preparing automotive engine lubricating oils.

"Brij."[206] TM for a series of emulsifiers and wetting agents developed for use in emulsions of high alkalinity or acidity. They are polyoxyethylene ethers of higher aliphatic alcohols. Soluble in water and lower alcohols. Insoluble in coal-tar hydrocarbons.

brilliant crocein. (Crocein Scarlet MOO).
$C_6H_5N_2C_6H_4N_2C_{10}H_4OH(S)_3Na)_2$.
Properties: Light brown powder, cherry red solution in water. CI 17190.
Use: To dye wool and silk red from acid solution and cotton and paper with aid of a mordant. Also used for red lakes, biological stain.

brilliant green. (Malachite Green G).
$C_{27}H_{34}N_2O_4S$.
Properties: Yellow crystals, CI 42040. Soluble in water and alcohol.
Derivation: Condensation of benzaldehyde with diethylaniline in presence of sulfuric acid, followed by oxidation of the triphenylmethane product formed and conversion to sulfate.
Use: Dyeing textiles, inks, etc.; indicator; staining bacteria; antiseptic.
See also Malachite Green.

"Brilliant Toning Red."[141] TM for Permanent Red 2B azo pigments derived from β-hydroxynaphtholic acid.
Use: Printing inks, paints, enamels, lacquers, rubber, plastics, wallpaper, textiles, floor coverings, crayons, paper coatings.

brilliant toning red amine. See 4-amino-2-chlorotoluene-5-sulfonic acid.

brimstone. Lumps or blocks of sulfur obtained in refining of sulfur. It collects on floor of condensing chamber where it is cast into sticks. See sulfur.

brine. Any solution of sodium chloride and water, usually containing other salts also. The most industrially important brines are (1) in subterranean wells as in Michigan; (2) in desert lakes such as Great Salt Lake, Searles Lake, Salton Sea, and Dead Sea; and (3) the ocean. These are the sources of many inorganic chemicals such as soda ash, sodium sulfate, potassium chloride, bromine, chlorine, borax, etc. Brines are also used for the preservation and pickling of certain vegetables, meat curing and freezing mixtures. Concentration range from 3% (ocean) to 20% or more.
Large areas of sand and shale containing brines under high pressure exist along the Gulf Coast. These are reported to be an important undeveloped source of natural gas and other hydrocarbons suitable for fuel or petrochemical feedstocks.
See also desalination, demineralization.

Brinell hardness. See hardness.

brisance. The shattering power of an explosive measured by the ratio of the weight of graded sand shattered when a charge of the test explosive is detonated in a standard manner to the weight of sand shattered by TNT detonated in the same manner.

britannia metal. See pewter, white metal.

"Britecarb."[155] TM for carbonized nickel with the carbon completely removed from one side.

"BriteSil."[531] TM for a series of solution clarified sodium silicates with controlled sodium oxide and silica contents and densities. Available in nine grades.

"BriteSorb."[531] TM for a synthetic precipitated magnesium silicate having the empirical formula $MgO \cdot 2.5SiO_2 \cdot 1.5H_2O$.
Properties: A white, finely divided, free flowing powder with large surface area;
insoluble in water and most organic solvents; pH (10% slurry) 9.5. Available in various particle sizes.
Use: Purification, clarification, adsorption, catalyst base.

British Anti-Lewisite. See 2,3-dimercaptopropanol.

British thermal unit. See Btu.

"Britone Reds."[141] TM for resinated type lithol reds.
See "Graphic Red.".

Brix degree. A measure of the density or concentration of a sugar solution. The degrees Brix equals percent by weight of sucrose in the solution and is related empirically to the density.

Broenner acid. (2-naphthylamine-6-sulfonic acid). $C_{10}H_6(NH)_2SO_3H$.
Properties: Colorless needles, soluble in boiling water.
Derivation: Heating sodium-2-naphthol-6-sulfonate with concentrated ammonia at 180C in an autoclave.
Grade: Technical. Available as the sodium salt, an odorless gray-to-pink powder.
Use: Azo dye intermediate.

bromacil. (5-bromo-3-sec-butyl-6-methyluracil). CAS: 314-40-9. Substitute approved by EPA for some uses of 2,4,5-T.
Hazard: TLV: 10 mg/m³, 1 ppm in air.
Use: Herbicide.

bromal. See tribromoacetaldehyde.

"Bromat."[430] TM for cetyl trimethyl ammonium bromide, a quaternary ammonium compound with high germicidal activity.

bromcresol green. (tetrabromo-m-cresolsulfonphthalein an acid-base). An indicator showing color change from yellow to blue over the pH range 3.8–5.4.
Properties: Yellow crystals, mp 218C, slightly soluble in water, soluble in alcohol.
See also indicator.

bromcresol purple. (dibromo-o-cresolsulfonphthalein an acid-base). An indicator which changes from yellow to purple between pH 5.2 and 6.8.
Properties: Yellow crystals, mp 241C, insoluble in water, soluble in alcohol.
See also indicator.

bromelin. (bromelain). A milk-clotting proteolytic enzyme. It is precipitated from pineapple juice with acetone or ammonium sulfide.
Use: Biochemical research, meat-tenderizing formulations, texturizer in baking, medicine.

bromeosin. See eosin.

bromic acid. $HBrO_3$.
Properties: Colorless or slightly yellow liquid; turns yellow on exposure; unstable except in very dilute solution; d 3.28; bp decomposes at 100C. Exists only in water solution.
Derivation: Sulfuric acid is added to a solution of barium bromate and the product recovered by subsequent distillation and absorption in water.
Hazard: By ingestion and inhalation. Strong irritant to tissue.
Use: Dyes, intermediates, pharmaceuticals, oxidizing agent.

brominated camphor. See camphor bromate.

bromine. CAS: 7726-95-6. Br Nonmetallic halogen element of atomic number 35, group VIIA of the Periodic Table. Aw 79.904. Valences 1, 3, 5 (valence of 7 also reported). There are two stable isotopes.
Properties: Dark, reddish-brown liquid, irritating fumes. Soluble in common organic solvents, very slightly soluble in water. Attacks most metals including platinum and palladium, aluminum reacts vigorously and potassium explosively. Dry bromine does not attack lead, nickel, magnesium, tantalum, iron, zinc, and sodium (the latter below 300C). Bp 58.8C, fp −7.3C, d 3.11 (20/4C), vapor d vs air at 15C=5.51, wt per gal 25.7 lb, specific heat 0.107 cal/g, refr index 1.647, dielectric constant 3.2.
Derivation: From seawater and natural brines by oxidation of bromine salts with chlorine; solar evaporation (Great Salt Lake); from salt beds at Stassfurt, Dead Sea.
Method of purification: Distillation.
Grade: Technical, CP, 99.8%, 99.95%.
Hazard: Toxic by ingestion and inhalation, severe skin irritant. TLV: 0.1 ppm in air, 0.7 mg/m³. Strong oxidizing agent may ignite combustible materials on contact.
Use: Manufacture of ethylene dibromide (antiknock gasoline), organic synthesis, bleaching, water purification, solvent, intermediate for fumigants (methyl bromide), analytical reagent, fire-retardant for plastics, dyes, pharmaceuticals, photography, shrink-proofing wool.

bromine azide. (bromoazide). CAS: 13973-87-0. BrN_3.
Properties: Crystals or red liquid. Mp approximately 45C, bp explodes. A strong oxidizing agent.
Hazard: Explosive when heated or shocked. Will ignite combustible materials on contact.
Use: Detonators and other explosive devices.

bromine chloride. CAS: 13863-41-7. BrCl.

Properties: Reddish-yellow, mobile liquid; fp
−66C; decomposes with evolution of chlorine
at 10C. Soluble in water, carbon disulfide, ether.
Readily hydrolyzes. Reacts with ammonia to
form bromamines.
Hazard: Irritant. Oxidizing agent.
Use: Industrial disinfectant, especially for waste-
waters.

bromine cyanide.   See cyanogen bromide.

bromine iodide.   See iodine monobromide.

bromine pentafluoride.   CAS: 7789-30-2.
BrF_5.
Properties: Colorless, fuming liquid; d 2.466
(25C); fp −61C; bp 40.5C; vap press (21.1C) 7
psia. Reacts with every known element except
inert gases, nitrogen, and oxygen.
Derivation: By reacting bromine, diluted with nitrogen and fluorine in a copper vessel at 200C.
Grade: 98% min.
Hazard: Corrosive to skin and tissue. TLV: 0.1
ppm in air. Explodes on contact with water.
Use: Synthesis, oxidizer in liquid rocket propellants.

bromine trifluoride.   CAS: 7787-71-5. BrF_3.
Properties: Colorless liquid, d 2.80, mp 9C, bp
125C, vap press (21.1C) 0.15 psia, decomposed
violently by water.
Derivation: See bromine pentafluoride.
Grade: 98% min.
Hazard: Corrosive to skin. Very reactive and dan-
gerous.
Use: Fluorinating agent, electrolytic solvent.

bromine water.   A mixture of 3.2 grams bromine
in 100 grams water.
Use: A laboratory reagent.

bromlost.   (blister gas).   See dibromodiethylsulfide.

N-bromoacetamide.   (NBA). $CH_3CONHBr$.
Properties: White powder with bromine odor. Mp
105–108C. Contains approximately 57% active
bromine, decomposes appreciably above 26.6C.
Hazard: Emits very toxic fumes of bromine on
heating.
Use: Brominating and oxidizing agent in organic
synthesis.

bromoacetic acid.   CAS: 79-08-3.
$CH_2BrCOOH$.
Properties: Colorless, deliquescent crystals. Keep
from air and moisture. Soluble in water, alcohol,
and ether. Mp 51C, bp 208C, d 1.93.

Derivation: By heating acetic acid and bromine.
Hazard: Strong irritant to skin and tissue.
Use: Organic synthesis, abscission of citrus fruit
in harvesting.

bromoacetone.   CAS: 598-31-2.
$CH_2BrCOCH_3$.
Properties: Colorless liquid when pure, rapidly becomes violet even in absence of air. Soluble in
acetone, alcohol, benzene, and ether; slightly
soluble in water, d 1.631, bp 136C (partial decomposition), fp −54C, vap d 4.75, vap press 9
mm (20C).
Derivation: By treating aqueous acetone with bromine and sodium chlorate at 30–40C.
Grade: Technical.
Hazard: Toxic by inhalation and skin contact. A
lachrymator gas, strong irritant.
Use: Organic synthesis, tear gas.

bromoacetone cyanohydrin.
$CH_2BrC(OH)(CN)CH_3$.
Properties: Colorless liquid. Soluble in alcohol,
ether, and water; d 1.584 at 13C; bp 94.5C (5
mm).
Derivation: Interaction of bromoacetone and hydrogen cyanide at approximately 0C.
Use: Organic synthesis.

bromoallylene.   See allyl bromide.

4-bromoaniline.   (p-bromoanaline; 4-bromoben-
zeneamine).   CAS: 106-40-1. C_6H_6BrN.
Properties: Colorless, rhombic crystals; mp 66C;
soluble in alcohol and ether; insoluble in cold
water.
Derivation: Steam distillation of p-bromoacetani-
lide and sodium hydroxide or bromination of
aniline.
Use: Azo dye manufacturing, preparation of dihy-
droquinazolines (with formaldehyde).

5-bromoanthranilic acid.   (2-amino-5-bromoben-
zoic acid). $C_7H_6BrNO_2$.
Properties: Colorless crystals, soluble in acetone,
partially soluble in alcohol, benzene, and acetic
acid, mp 217C.
Use: Analytical reagent for metal determination
(cobalt, copper, nickel, zinc).

bromoauric acid.   (gold tribromide acid).
$HAuBr_4 \cdot 5HOH$.
Properties: Dark, red-brown, needle crystals or
granular masses; odorless; metallic and acid
taste. Stable in air if pure but deliquescent if
chloride is present. Soluble in water and alcohol.
Mp 27C.
Derivation: By dissolving auric bromide in hydrobromic acid, concentration, and crystallization.

p-bromobenzaldehyde. BrC_6H_4CHO.
Properties: Solid, mp 58C. Used as a chemical intermediate.

bromobenzene. (phenyl bromide).
CAS: 108-86-1. C_6H_5Br.
Properties: Heavy, mobile, colorless liquid. Pungent odor, d 1.499 wt/gal 12.51 lb, bp 156.6C, fp −30.5C, flash p 124F (51.1C), refr index 1.5625. Miscible with most organic solvents, insoluble in water. Autoign temperature 1051F (566C). Combustible.
Derivation: Bromination of benzene in presence of iron.
Grade: Technical, pure.
Hazard: Skin irritant. Moderate fire risk.
Use: Solvent, top-cylinder compounds, crystallizing solvent, organic synthesis, lube oil additive.

p-bromobenzenesulfonic acid. $BrC_6H_4SO_3H$.
Properties: Crystallizes in needles. Mp 102–103C, bp 155C (25 mm), soluble in hot water and hot alcohol.

p-bromobenzoic acid. $C_6H_4BrCOOH$.
Properties: Colorless or reddish crystals. Soluble in alcohol and ether, very slightly soluble in water. Mp 254C.
Derivation: From p-bromotoluene by oxidation.
Use: Organic synthesis, detection of strontium.

o-bromobenzyl cyanide. (o-bromophenylacetonitrile; 2-bromo-α-cyanotoluene).
$BrC_6H_4CH_2CN$.
Properties: Colorless solid or liquid, mp 29C, d 1.519, bp 242C (decomposes). Nonflammable. Soluble in organic solvents.
Hazard: Strong lachrymator, irritant to tissue.

1-bromobutane. See n-butyl bromide.

2-bromobutane. See sec-butyl bromide.

5-bromo-3-sec-butyl-6-methyluracil. See bromacil.

α-bromobutyric acid. $CH_3CH_2CHBrCOOH$.
Properties: Colorless, oily liquid. Soluble in alcohol and ether; sparingly soluble in water; d 1.54; bp 181C at 258 mm, 214–217C at 760 mm with decomposition; fp −4C. Combustible.
Derivation: By heating bromine and butyric acid.
Hazard: Toxic by ingestion.
Use: Organic synthesis.

bromocarnallite. An artificial carnallite in which chlorine is replaced by bromine.

3-bromo-1-chloro-5,5-dimethylhydantoin.
$BrCl(CH_3)_2C_3N_2O_2$.
Properties: Free-flowing, white powder; faint halogen odor; mp 163–164C; soluble in benzene, methylene dichloride, chloroform. Active bromine 33% min, active chlorine 14% min.
Hazard: See bromine, chlorine.
Use: Germicide and fungicide in treatment of water, disinfectant, halogenating agent, catalyst of ionic type, selective oxidant.

sym-bromochloroethane. (ethylene chlorobromide). CH_2BrCH_2Cl.
Properties: Colorless, volatile liquid. Chloroform-like odor, soluble in alcohol and ether, insoluble in water, d 1.70, bp 107–108C, wt/gal 14.9 lb (0C), fp −16.6C. Nonflammable.
Derivation: By action of bromine and chlorine on ethylene gas.
Hazard: By ingestion and inhalation, skin irritant.
Use: Solvent especially for cellulose esters and ethers, organic synthesis, fumigant for fruits and vegetables.

bromochloromethane. (methylene chlorobromide; chlorobromomethane). CAS: 74-97-5. $BrCH_2Cl$.
Properties: Clear, colorless, volatile liquid with chloroform-like odor; d 1.93 (25C); bp 67C; fp −86.5C; refr index 1.48 (25C). Nonflammable. Soluble in organic solvents, insoluble in water.
Hazard: By inhalation. TLV: 200 ppm in air.
Use: Fire extinguishers, organic synthesis.

1-bromo-3-chloropropane. (trimethylene chlorobromide). $BrCH_2CH_2CH_2Cl$.
Properties: Colorless liquid. Fp <−50C, bp 143–145C, d 1.594 (25/25C), lb/gal 13.27 (25C), refr index 1.484 (25C). Insoluble in water. Soluble in methanol and ether. Nonflammable.
Hazard: Toxic. Avoid inhalation of fumes.
Use: Organic synthesis, pharmaceuticals.

2-bromo-2-chloro-1,1,1-trifluoroethane. See halothane.

2-bromo-α-cyanotoluene. See bromobenzyl cyanide.

bromocyclopentane. See cyclopentyl bromide.

1-bromododecane. See lauryl bromide.

bromoethane. See ethyl bromide.

2-bromoethyl alcohol. See ethylene bromohydrin.

2-bromoethylamine hydrobromide.
$BrCH_2CH_2NH_2 \cdot HBr$.
Use: Intermediate, suggested as a soldering flux.

bromoethyl chlorosulfonate.
$BrCH_2CH_2OSO_2Cl$.
Properties: Liquid, bp 100–105C (18 mm).
Derivation: Interaction of sulfuryl chloride and glycol bromohydrin.
Hazard: Irritant to skin and tissue.

p-bromofluorobenzene.
C_6H_4BrF.
Properties: Colorless liquid, bp: 151–152C, fp −17.4C, d (15C) 1.593, refr index 1.5245 (25C), insoluble in water.
Use: Intermediate, production of p-fluorophenol.

bromoform.   (tribromomethane; methyl tribromide).   CAS: 75-25-2. $CHBr_3$.
Properties: Colorless, heavy liquid; odor and taste similar to those of chloroform; soluble in alcohol, ether, chloroform, benzene, solvent naphtha, fixed and volatile oils; slightly soluble in water; d 2.887; bp 151.2C; wt/gal 24 lb; boiling range 150.3–151.2C; fp 9C; surface tension 41.53 dynes/cm (20C); dielectric constant 4.5 (20C); refr index 1.6005. Nonflammable.
Derivation: By heating acetone or ethanol with bromine and alkali hydroxide and recovery by distillation (similar to acetone process of chloroform).
Grade: Technical, pharmaceutical, spectrophotometric.
Hazard: By ingestion, inhalation and skin absorption. TLV: 0.5 ppm in air.
Use: Intermediate in organic synthesis, geological assaying, solvent for waxes, greases, and oils, medicine (sedative).

"Bromofume."[88]   TM for a soil fumigant composition, consisting of ethylene dibromide in volatile solvent.
Use: Control of wireworms and rootknot nematodes.

1-bromohexane.   See n-hexyl bromide.

bromol.   (2,4,6-tribromophenol).
CAS: 118-79-6. $C_6H_2Br_3OH$.
Properties: Soft, white needles; sweet taste; penetrating bromine odor. Soluble in alcohol, chloroform, ether, and caustic alkaline solution, almost insoluble in water, mp 96C, sublimes, d 2.55 (20/20C), bp 244C.
Derivation: Action of bromine on phenol.
Hazard: By ingestion, inhalation, skin absorption. Strong skin irritant.

bromomethane.   See methyl bromide.

bromomethylethyl ketone. $BrCH_2COC_2H_5$.
Properties: Colorless to pale-yellowish liquid. Affected by light. Soluble in alcohol, benzene, ether, insoluble in water, d 1.43, bp 145–146C (decomposes).
Derivation: Reaction of sodium bromide and methyl ethyl ketone in the presence of sodium chlorate.
Hazard: Strong irritant to skin and eyes.
Use: Organic synthesis.

α-bromonaphthalene. $C_{10}H_7Br$.
Properties: Colorless, thick liquid, pungent odor, d 1.4870, solidifies at 6.2C, bp 279C, refr index 1.6601, soluble in alcohol, ether, and benzene, slightly soluble in water.
Derivation: Bromination of naphthalene.
Use: Organic synthesis, microscopy, refractometry of fats.

1-bromo-2-naphthol. $BrC_{10}H_6OH$.   Solid, mp 121–125C. Used as a dye intermediate.

2-bromopentane. $CH_3CH_2CHBrCH_3$.
Properties: Colorless to yellow liquid, strong odor, d 1.1850 (25/25C).

bromopheniramine maleate.   (2-[p-bromo-α(2-dimethylaminoethyl)benzyl]pyridine bimaleate).
$C_{16}H_{19}BrN_2 \cdot C_4H_4O_4$.
Properties: Crystals, soluble in water, less soluble in alcohol, mp 130–135C.
Grade: NF.
Use: Medicine (antihistamine).

p-bromophenol.   CAS: 106-41-2.
$HO(C_6H_4)Br$.
Properties: Crystals, d 1.840 (15C), 1.5875 (80C), mp 64C, bp 238C, slightly soluble in water, soluble in alcohol, chloroform, ether, and glacial acetic acid.
Used as a disinfectant.

bromophenol blue.   (tetrabromophenolsulfonaphthalein).
Properties: An acid-base indicator, showing color change from yellow to purple over the range pH 3.0 to 4.6.

o-bromophenylacetonitrile.   See bromobenzyl cyanide.

2-bromo-4-phenylphenol. $C_6H_5C_6H_3BrOH$.
Properties: Light-colored solid with faint characteristic odor, d (25/4C)1.536, mp 93.6–95.6C, flash p 405F (207C), bp decomposes (18 mm) 195–200C. Soluble in alkalies, most organic solvents, insoluble in water. Combustible.
Hazard: Toxic by ingestion and inhalation.
Use: Germicide.

bromophosgene.   (carbonyl bromide; carbon oxybromide). $COBr_2$.

Properties: Heavy, colorless liquid. Strong odor. Hydrolyzed by water and is decomposed by light and heat, d 2.5 (approximately 15C), bp 64–65C.
Derivation: Action of sulfuric acid on carbon tetrabromide.
Hazard: Toxic by ingestion and inhalation.
Use: Military poison (toxic suffocant), making crystal violet-type coloring agents.

bromopicrin. (nitrobromoform; tribromonitromethane). CBr_3NO_2.
Properties: Prismatic crystals, decomposes with explosive violence if heated rapidly. Soluble in alcohol, benzene, and ether, slightly soluble in water, d 2.79 (18C), bp 127C (118 mm), mp 103C.
Derivation: Action of picric acid on an aqueous solution of bromine and calcium oxide followed by distillation under reduced pressure.
Hazard: Powerful irritant. Severe explosion hazard when heated.
Use: Organic synthesis, military poison.

3-bromopropene. See allyl bromide.

α-bromopropionic acid. (2-bromopropionic acid). $CH_3CHBrCOOH$.
Properties: Colorless liquid, d 1.69, mp 24.5C, bp 203C (decomposes), soluble in water, alcohol, and ether.
Derivation: By heating propionic acid with bromine.
Method of purification: Distillation.

3-bromo-1-propyne. See propargyl bromide.

2-bromopyridine. C_5H_4NBe.
Properties: Liquid, bp 195C, d 1.627 (20C), refr index 1.5714 (20C), solubility in 100g water 2.08 (20C).
Use: Synthesis of pyridine compounds.

3-bromopyridine. C_5H_4NBr.
Properties: Needles, bp 174.4C, d 1.628 (20/20C), refr index 1.5710 (20C). Slightly soluble in water, readily soluble in common organic solvents.

5-bromosalicylhydroxamic acid. $BrC_6H_3(OH)CONH)OH$.
Properties: Crystals, decomposes at 232C, very slightly soluble in water, forms a water-soluble sodium salt.
Used in medicine.

β-bromostyrene. (bromostyrol).
CAS: 103-64-0. $C_6H_5CHCHBr$.
Properties: Yellowish liquid, strong floral odor. Soluble in 4 volumes of 90% alcohol, d 1.395–1.424, refr index 1.602–1.608, mp min −2C.
Use: Perfumery.

bromosuccinic acid. $HOOCCH_2CHBrCOOH$.
Properties: Colorless crystals, d 2.073, mp 159–161C, soluble in water and alcohol, insoluble in ether.
Derivation: By heating bromine and succinic acid.
Use: Organic synthesis.
Note: The above are properties of the *dl* form. Optically active forms are also known.

N-bromosuccinimide. (NBS). $(CH_2CO)_2NBr$.
Properties: Fine crystals, white to cream in color, melting range 172–178C (decomposes). Soluble in carbon tetrachloride, 44.5% min active bromine.
Hazard: Use a respirator to handle dry materials which evolve toxic fumes of bromine. Strong irritant to eyes and skin.
Use: For controlled low-energy bromination.

4-bromothiophenol. BrC_6H_4SH.
Properties: White solid, fp 73C, bp 239C, almost insoluble in water, soluble in methanol, alkaline solution (with which it reacts).
Hazard: Toxic by ingestion.
Use: Intermediate.

α-bromotoluene. See benzyl bromide.

p-bromotoluene. (p-tolyl bromide). $BrC_6H_4CH_3$.
Properties: Crystals, mp 28.5C, bp 184–185C, d 1.3898 (20C), refr index 1.5490, flash p 185F (85C). Combustible. Insoluble in water, soluble in alcohol, ether, and benzene.
Hazard: Toxic by ingestion.
Use: Intermediate.

bromotrichloromethane. (trichlorobromomethane). CCl_3Br.
Properties: Colorless, heavy liquid with chloroform-like odor; miscible with many organic liquids; d 2.0; bp 104C; refr index 1.5051 (20C).
Hazard: Toxic by ingestion and inhalation of fumes.
Use: Organic synthesis.

bromotrifluoroethylene. (BFE). $BrFC:CF_2$.
Gas (monomer) or liquid (polymer). The latter are usually clear oils at room temperature and solids at −55C. Viscosities and densities vary widely. Monomer: High purity gas (97%). Shipped in cylinders.
Hazard: Flammable (gas or liquid). Dangerous fire risk.
Use: (BFE polymers). Flotation fluids for gyros or accelerometers used in inertial guidance sys-

tems. BFE polymers can also be used like CFE polymers.

bromotrifluoromethane. $CBrF_3$.
Properties: Colorless gas, noncorrosive, nonflammable, fp $-168C$, bp $-58C$, d at bp 8.71 g/L.
Derivation: Bromination of fluoroform or perfluoropropane in nonmetallic reactor.
Hazard: Toxic by inhalation.
Use: Chemical intermediate, refrigerant, metal hardening, fire extinguishment.

bromphenol blue. $C_{19}H_{10}Br_4O_5S$.
Properties: Prisms or crystals, decomposes at 280C, soluble in sodium hydroxide solution, partially soluble in alcohol and benzene, almost insoluble in water.
Use: Indicator (yellow = pH 3.0, purple = pH 4.6).

bromthymol blue. $C_{27}H_{28}Br_2O_5S$.
Properties: Yellow-white crystals, soluble in alcohol and alkaline solution, slightly soluble in water.
Use: Indicator (yellow = pH 6.0, blue pH = 7.6).

"Bromvegol"[342]. TM for brominated vegetable oils used for weighting soft-drink emulsions.

bronze. An alloy of copper and tin usually containing from 1 to 10% tin. Special types contain from 5 to 10% aluminum (Al bronze), fractional percentages of phosphorus (phosphor bronze) as deoxidizer, or low percentages of silicon (Si bronze).
Hazard: Powder is flammable.
Use: Spark-resistant tools, springs, fourdrinier wire, paint, cosmetics (as powder), electrical hardware, vacuum dryers, blenders, water gauges, flow indicators, valves, drain cocks, fine arts.
See also brass, phosphor bronze.

bronze blue. Any of a number of varieties of iron-blue pigments.

bronze orange. See red lake C.

bronzing liquid. (1) A solution of pyroxylin in amyl acetate together with a bronze powder, usually aluminum bronze. (2) Gloss oils and aluminum bronze. (3) Spirit varnishes and aluminum bronze.

brosylate ester. An ester of p-bromobenzenesulfonic acid.

broth. A liquid nutrient medium, usually containing agar, used to promote the growth of bacteria in the fermentation industry and to prepare cultures for microbiological research.

Brown, Herbert C. (1912-) An English born chemist who was the recipient of the Nobel prize for chemistry with Wittig in 1979. Via his work in organic synthesis, he discovered new routes to add substituents to olefins selectively. His early education was irregular and disjointed as a result of family circumstances and the economic depression of the 1930s. He eventually received his PhD from the University of Chicago. The reduction of carbonyl compounds with diborane was the topic of his thesis. The bulk of his career has been spent at Purdue University.

Brownian movement. The continuous zigzag motion of the particles in a colloidal suspension, e.g., rubber latex particles. The motion is caused by impact of the molecules of the liquid upon the colloidal particles. Named after the British botanist Robert Brown who first noted this phenomenon.

browning reaction. (Maillard reaction).
A complicated and not completely evaluated sequence of chemical changes occurring without the involvement of enzymes during heat exposure of foods containing carbohydrates (usually sugars) and proteins, as well as during storage. It is responsible for the surface color change of bakery products and meats. It begins with an aldol condensation reaction involving the carbonyl groups of the proteins and ends with formation of furfural, which produces the dark brown coloration. Besides color change, the reaction is accompanied by alterations in flavor and texture as well as in nutritive value. It was first noted by the French chemist Maillard.

brucine. (dimethoxy strychnine).
CAS: 357-57-3. $C_{23}H_{26}O_4N_2 \cdot 2HOH$ or 4HOH.
Properties: White, crystalline alkaloid; very bitter taste; loses water at 100C; mp 178C; soluble in alcohol, chloroform, and benzene; slightly soluble in water, ether, glycerol, and ethyl acetate. Forms brucine sulfate, hydrochloride, and nitrate (mp 230C). Also available as the sulfate.
Derivation: By extraction and subsequent crystallization from nux vomica or ignatia seeds.
Hazard: A poison by ingestion and inhalation.
Use: Denaturing alcohol, lubricant additive, separation of racemic mixtures.

brucite. (nemalite). CAS: 1317-43-7.
$Mg(OH)_2$. Natural magnesium hydroxide.
Properties: Colorless, white, gray, greenish; luster pearly or waxy; d 2.39; Mohs hardness 2.5.
Occurrence: Nevada, Washington, Canada.
Use: Refractories.

"Brush-Rhap"[266]. TM for butyl and 2-ethylhexyl esters or amine salts of 2,4,5-trichlorophenoxyacetic acid. Available in various concentrations of active ingredient and in combination with esters of 2,4-dichlorophenoxyacetic acid.
Used as a herbicide.

"BRV.[50]**"** TM for a heavy, high-boiling coaltar distillate.
Properties: Dark, coal-tar oil, d 1.14–1.18 (25/25C). Engler specific viscosity 5–10 (50C), distillation 26% max at 355C. Combustible.
Use: Rubber plasticizer, softener, and reclaiming oil; dispersing agent.

"Brymul"[51]. TM for an emulsifiable grade of cleaner for general use on metals, etc. Contains Stoddard-type solvent.
Hazard: Moderate fire risk.

"Bryton"[544]. TM for a series of oil-soluble petroleum sulfonates
Use: Detergent dispersants, rust-inhibiting agents, and alkaline carriers and as additives to motor oils and diesel fuels.

B-stage resin. (resitol). A thermosetting phenolformaldehyde type resin which has been thermally reactive beyond the A-stage so that the product has only partial solubility in common solvents (alcohols, ketones) and is not fully fusible even at 150–180C. The B-stage resin has limited commercial use.

BT. (*Bacillus thuringiensis*). A species of bacteria used as a pesticide for agricultural crops. It is of the stomach-poison type and has been approved for commercial use.

"BTC"[328]. TM for a series of cationic quaternary ammonium chlorides generally alkyldimethylbenzylammonium chloride.
Use: Disinfectant, deodorant, germicide, fungicide, algacide, slimicide.

BTDA. See 3,3',4,4'-benzophenone tetracarboxylic dianhydride.

Btu. (British thermal unit). The quantity of heat required to raise the temperature of one pound of water one degree Fahrenheit (usually from 39 to 40F). This is the accepted standard for the comparison of heating values of fuels. For example, fuel gases range from 100 (low producer gas) to 3200 (pure butane) Btu per cu ft. The usual standard for a city gas is approximately 500 Btu.

BTX. Commercial abbreviation for benzene, toluene, xylene, the three major aromatic compounds.

Bu. Informal abbreviation for butyl.

bubble cap column. See tower, distillation.

Bucherer reaction. A procedure for preparation of β-naphthylamine by heating β-naphthol with a water solution of ammonium sulfite. "A sulfite solution is prepared by saturating concentrated ammonia solution with sulfur dioxide and adding an equal volume of concentrated ammonia solution, β-naphthol is added and the charge is heated in an autoclave provided with a stirrer or a shaking mechanism." (L.F. Fieser) This reaction is also involved in the preparation of several azo dye intermediates, e.g., Tobias acid.

Bucherer-Bergs reaction. Preparation of hydantoin from carbonyl compound by reaction with potassium cyanide and ammonium carbonate, or from the corresponding cyanohydrin and ammonium carbonate.

Bucherer carbazole synthesis. Formation of carbazoles from naphthols, or naphthylamines, aryl hydrazines, and sodium bisulfite.

Buchner-Curtius-Schlotterbeck reaction. Formation of keto compounds from aldehydes and aliphatic diazo compounds; ethylene oxides may also be formed.

Buchner, Eduard. (1860-1917) A German chemist who was awarded the Nobel prize for chemistry in 1907. His works included the synthesis of diiodoacetamid, alcoholic fermentation caused by enzymes, as well as the discovery of zymase, the first enzyme to be isolated. He received his PhD at the University of Munich, where he became a lecturer. Later, he taught and performed research at Tubingen, Berlin, and Wurzburg.

Buchner method of ring enlargement. Diazoacetic acid ester reacts with benzene and homologs to give the corresponding esters of noncaradienic acid, transformed at high. temperatures to derivatives of cycloheptatriene, phenylacetic acid and β-phenylpropionic acid (when one or more methyl groups are present in the initial hydrocarbon).

buclizine hydrochloride. $C_{28}H_{33}ClN_2 \cdot 2HCl$. 1-p-chlorobenzhydryl-4-(p-(tert)-butylbenzyl)piperazinedihydrochloride.
Use: Medicine (antihistamine).

bucket elevator. See conveyor (5).

"Budene."[265] TM for a cis-1,4-polybutadiene synthetic dry rubber manufactured by solution polymerization utilizing a stereospecific catalyst. This elastomer has a relatively high cis content and is protected with a non-staining, non-discoloring antioxidant.

"Budium"[28]. TM for a polybutadiene finish for application to tin plate.

buffer. A solution containing both a weak acid and its conjugate weak base whose pH changes only slightly on addition of acid or alkali. The weak acid becomes a buffer when alkali is added and the weak base becomes a buffer on addition of acid. This action is explained by the reaction.
$$A + HOH \rightarrow B + H_3O_n.$$
in which the base B is formed by the loss of a proton from the corresponding acid A. The acid may be a cation such as NH_4+, a neutral molecule such as CH_3COOH, or an anion such as H_2PO_4-. When alkali is added hydrogen ions are removed to form water, but as long as the added alkali is not in excess of the buffer acid many of the hydrogen ions are replaced by further ionization of A to maintain the equilibrium. When acid is added this reaction is reversed as hydrogen ions combine with B to form A. The pH of a buffer solution may be calculated by the mass law equation, $pH = pK' + \log C_b/C_a$ in which pK' is the negative logarithm of the apparent ionization constant of the buffer acid and the concentration are those of the buffer base and its conjugate acid.

bufotenine. [3-(2-dimethylaminoethyl)-5-indolol]. $C_{12}H_{16}N_2O$.
Properties: Colorless prisms, insoluble in water, soluble in alcohol, slightly soluble in ether, soluble in dilute acids and alkalies.
Derivation: From toads and toadstools, also made synthetically.
Hazard: A hallucinogenic agent.
Use: Medicine (experimental).   See also hallucinogen.

builder detergent. A substance that increases the effectiveness of a soap or synthetic detergent by acting as a softener and as a sequestering and buffering agent. Phosphate-silicate formulations once widely used have been restricted for environmental reasons. They have largely been replaced by EDTA or by zeolites, sometimes combined with nitrolotriacetic acid. Certain starch derivatives can be used as builders.
See also zeolite.

bulan. (2-nitro-1,1-bis(p-chlorophenyl)butane).
CAS: 76-20-0. $C_{16}H_{15}Cl_2NO_2$.

Hazard: A toxic chlorinated nitrogenous compound used as an insecticide. When mixed with Prolan, the product is called Dilan.   See also Prolan, Dilan.

bulk density. See density.

bunamiodyl.
$C_3H_7CONHC_6HI_3CH:C(C_2H_5)COONa$.
[3(3-Butyrylamino-2,4,6-tri-iodophenyl)-2-ethyl sodium acrylate].
Used in medicine (radiopaque contrast medium, diagnostic aid).

buna rubbers. German vulcanizable synthetic rubbers from butadiene with sodium as a catalyst.
See also rubber.

bunker fuel oil. A heavy residual oil used as fuel by ships, industry, and for large-scale heating installations.

Bunsen, Robert Wilhelm. (1811–1899) Born in Germany, Bunsen is remembered chiefly for his invention of the laboratory burner named after him. He engaged in a wide range of industrial and chemical research, including blast-furnace firing, electrolytic cells, separation of metals by electric current, spectroscopic techniques (with Kirchhoff), and production of light metals by electric decomposition of their molten chlorides. He also discovered two elements, rubidium and cesium.

burette. A liquid-measuring device used extensively in chemical laboratories. A vertical glass tube, open at the top and supported on a bracket and equipped with scale graduation marks and a hand-operated stopcock at or near the bottom. The liquid to be dispensed is flowed in at the open end and can then be withdrawn in measured amounts by operating the stopcock.

Burgundy pitch. A resin obtained from Norway spruce or European silver fir. Other types, e.g., that from various species of pines, are also offered under this name. Characterized by extreme tackiness, soluble in acetone and alcohol. Used to some extent in surgeon's tape and various special adhesive compositions.

burlap. A coarse loose-woven fabric made from jute or similar fiber used in low-cost laminated composites; as liner or backing in upholstery, carpets, etc.; and as a bagging material. It is often impregnated with hot-melt adhesive.

burnable poison. A neutron absorber (poison) such as boron which when purposely incorporated in the fuel or fuel cladding of a nuclear reactor, gradually "burns up" (is changed into nonabsorbing material) under neutron irradiation. This process compensates for the loss of reactivity that occurs as fuel is consumed and fission-product poisons accumulate and keeps the overall characteristics of the reactor nearly constant during use.

burnt lime. See calcium oxide.

burnt sienna. See iron oxide red.

burnt umber. See umber.

"Buromin"[108]. TM for sodium hexa-metaphosphate in glass plate form for boiler water conditioning.

"Burosil"[108]. TM for a granular, alkaline, phosphate-silicate compound used in boiler water conditioning to precipitate calcium and magnesium as loose sludge.

burr mill. See attrition mill.

"Burundum"[326]. TM for wear- and corrosion-resistant high-alumina ceramic bodies used in cylindrical form as grinding media for particle size reduction.

bushy stunt virus. A viral protein present in tomato plant infections.
Properties: Mw 7,600,000, pH 4.1
See virus.

"Butacite"[28]. TM for polyvinyl butyral resin, available as soft pliable sheeting in 750 ft rolls 10–84 inches wide.
See polyvinyl acetal resins.

1,3-butadiene. (vinylethylene; erythrene; bivinyl; divinyl). CAS: 106-99-0.
$H_2C:CHHC:CH_2$. 36th highest-volume chemical produced in US (1985).

$$HC \overset{CH_2}{\underset{\underset{CH_2}{\|}}{\diagup}}$$

Properties: Colorless gas with mild aromatic odor, easily liquefied, bp −4.41C, d 0.6211 (liquid at 20C), fp −108.9C, flash p −105F (−76C), specific volume 6.9 cu ft/lb (700F), autoign tem-

perature 780F (414C), vap press 17.65 psia (0C). Soluble in alcohol and ether, insoluble in water. The material polymerizes readily, particularly if oxygen is present and the commercial material contains an inhibitor to prevent spontaneous polymerization during shipment or storage.
Derivation: (a) Catalytic dehydrogenation of butenes or butane; (b) oxidative dehydrogenation of butenes.
Method of Purification: Extractive distillation in the presence of furfural, absorption in aqueous cuprous ammonium acetate, or use of acetonitrile.
Grade: Technical, 98.0%, CP, 99.0%, instrument 99.4%, research 99.8%.
Hazard: Irritant in high concentration. TLV: 10 ppm in air. A suspected human carcinogen. Highly flammable gas or liquid, explosive limits in air 2–11%. May form explosive peroxides on exposure to air. Must be kept inhibited during storage and shipment. Inhibitors often used are di-n-butylamine or phenyl-β-naphthylamine. Storage is usually under pressure or in insulated tanks <35F (1.67C).
Use: Synthetic elastomers (styrene-butadiene, polybutadiene, neoprene, nitriles), ABS resins, chemical intermediate.

butadiene-acrylonitrile copolymer. See nitrile rubber.

butadiyne. See diacetylene.

butaldehyde. See butyraldehyde.

butanal. See butyraldehyde.

butane. (n-butane). CAS: 106-97-8.
$CH_3CH_2CH_2CH_3$.
Properties: Colorless gas, natural-gas odor, extremely stable, has no corrosive action on metals, does not react with moisture, very soluble in water, soluble in alcohol and chloroform, bp −0.5C, fp −138.3C, condensing pressure approximately 30 lb at 32.5C, d (liquid at 0C) 0.599, d (vapor at 0C (air = 1) 2.07, critical temperature 153.2C, critical pressure (absolute) 525 psi, heating value (25C) 3266 Btu/cu ft, specific volume (21.1C), 6.4 cu ft/lb, flash p −76F (−60C), autoign temperature 761F (405C). An asphyxiant gas.
Derivation: A by-product in petroleum refining or natural gasoline manufacture.
Grade: Research 99.99 mole %, pure 99 mole %, technical 95 mole %, also available in various mixtures with isobutane, propane, pentanes, etc.
Hazard: Highly flammable, dangerous fire and explosion risk. Explosive limits in air 1.9 to 8.5%. TLV: 800 ppm in air. Narcotic in high concentration.

Use: Organic synthesis, raw material for synthetic rubber and high octane liquid fuels, fuel for household and for many industrial purposes, manufacture of ethylene, solvent, refrigerant, standby and enricher gas, propellant in aerosols, pure grades used in calibrating instruments, food additive.   *Note*: Butane in liquid form may be stored both above and below ground. Besides storage in liquefied form under its vapor pressure at normal atmospheric temperatures, refrigerated liquid storage at atmospheric pressure may be used. Such systems are closed and insulated, and the liquid petroleum gas vapor is circulated through pumps and compressors to serve as the refrigerant for the system. Butane may be stored in pits in the earth capped by metal domes, and in underground chambers. (Compressed Gas Association). The foregoing also applies to propane.

butanedial.   See succinaldehyde.

1,4-butanedicarboxylic acid.   See adipic acid.

butanedioic anhydride.   See succinic anhydride.

1,3-butanediol.   See 1,3-butylene glycol.

1,4-butanediol.   See 1,4-butylene glycol.

2,3-butanediol.   See 2,3-butylene glycol.

butanediolamine.   See 2-amino-2-methyl-1,3-propanediol.

butanedione.   See diacetyl.

2,3-butanedione oxime thiosemicarbazone. $CH_3C(NOH)C(CH_3)N_2HCSNH_2$.   A test reagent for manganese in very dilute solution made from dimethylglyoxime and thiosemicarbazide.

butane dioxime.   See dimethylglyoxime.

butanenitrile.   See butyronitrile.

1-butanethiol.   (n-butyl mercaptan).
   CAS: 109-79-5. C_4H_9SH.
   Properties: Colorless liquid, d 0.8412 (20/4C), refr index 1.4427 (20C), flash p 35F (1.67C), bp 97.2–101.7C, strong, obnoxious odor. Slightly soluble in water, very soluble in alcohol and ether.
   Grade: 95%.
   Hazard: Toxic by inhalation. Flammable, dangerous fire risk. TLV: 0.5 ppm in air.
   Use: Intermediate, solvent.

2-butanethiol.   (sec-butyl mercaptan).
   $C_2H_5CH(SH)CH_3$.
   Properties: Colorless liquid, obnoxious odor. Boil-ing range 73–89C, d (20/4C) 0.8288, refr index 1.4363 (20C), flash p −10F (−23.3C).
   Grade: 95%.
   Hazard: Toxic by inhalation. Flammable, dangerous fire risk. TLV: 0.5 ppm in air.

1,2,4-butanetriol. $HOCH_2CHOHCH_2CH_2OH$.
   Properties: Almost colorless, odorless liquid, miscible in water and ethanol, hygroscopic, bp 312C (extrapolated), d 1.184, refr index 1.473, flash p 332F (166C). Combustible.
   Derivation: Reaction of 2-butyne-1,4-diol with water, followed by reduction.
   Grade: Technical, nitration.
   Use: Intermediate for alkyd resins and explosives, cellulose plasticizer, emulsifier for cosmetics, inks, finishes, paper, cork, textiles.

butanoic acid.   See butyric acid.

1-butanol.   See n-butyl alcohol.

2-butanol acetate.   See sec-butyl acetate.

2-butanone.   See methyl ethyl ketone.

butanoyl chloride.   See butyroyl chloride.

"Butaprene."[5]   TM for synthetic rubbers, latexes, and resins comprising copolymers of butadiene with other monomers except those copolymers of butadiene with styrene which are classified as general purpose synthetic rubbers. Latexes used in water-based paints, paper coatings and grease-resistant coatings. Resins used as rubber reinforcing agents (floor tiles, shoe soles, etc.). See "FR-S".

"Butazate."[248]   TM for zinc dibutyldithiocarbamate.

"Butazolidin"[219].   TM for phenylbutazone.

2-butenal.   See crotonaldehyde.

Butenandt, Adolf.   (1903-) A German biochemist who won a Nobel prize with Ruzicka in 1939. His work involved insecticides for plants and hormones. He received his doctorate at University of Marburg, Germany. He received a multitide of honorary degrees and awards.

butene-1.   (ethylethylene; α-butylene).
   CAS: 25167-67-3. $CH_2{:}CHCH_2CH_3$.
   A liquefied petroleum gas.
   Properties: Colorless gas, bp −6.3C, d 0.5951 (20/4C), fp approximately −185C. Specific volume

6.7 cu ft/lb (70F), flash p −110F (−79C), soluble in most organic solvents, insoluble in water. Autoign temperature approximately 700F (371C).

Derivation: (a) Gases containing appreciable content of butene-1 along with other butene and butane hydrocarbons are obtained by fractional distillation of refinery gas. (b) Can be produced directly from ethylene.

Grade: Technical 95%, CP 99.0%, research 99.4%.

Hazard: Asphyxiant gas. Highly flammable, flammable limits in air 1.6–9.3% by volume. Dangerous fire and explosion risk.

Use: Polymer and alkylate gasoline; polybutenes; butadiene; intermediate for C_4 and C_5 aldehydes, alcohols, and other derivatives; production of maleic anhydride by catalytic oxidation.

cis-**butene-2.**   (dimethylethylene; β-butylene; also called the "high-boiling" butene-2).
$CH_3HC:CHCH_3$.
A liquefied petroleum gas.

$$CH_3\diagdown \diagup CH_3$$
$$C=C$$
$$H\diagup \diagdown H$$

Properties: Colorless gas, bp 3.7C, d 0.6213 (10/4C), fp −139C, specific volume 6.7 cu ft/lb (21.1C), flash p −100F (−72C), soluble in most organic solvents. Insoluble in water. Autoign temperature 615F (323C).

Derivation: Gases containing appreciable content of cis-butene-2, along with other butene and butane hydrocarbons are obtained by fractional distillation of refinery gas.

Grade: Technical 95%, CP 99%, research 99.8%.

Hazard: Asphyxiant gas. Highly flammable. Flammable limits in air 1.8–9.7% by volume. Dangerous fire and explosion risk.

Use: Solvent; cross-linking agent; polymerize gasoline; butadiene synthesis; synthesis of C_4 and C_5 derivatives.

trans-**butene-2.**   (dimethylethylene; β-butylene; also called the "low-boiling" butene-2).
$CH_3HC:CHCH_3$.   A liquefied petroleum gas.

$$CH_3\diagdown \diagup H$$
$$C=C$$
$$H\diagup \diagdown CH_3$$

Properties: Colorless gas, bp 0.88C, fp −105.8C, d 0.6042 (20/4C), specific volume 6.7 cu ft/lb (21.1C), flash p −100F (−72C), soluble in organic solvents. Insoluble in water. Autoign temperature 615F (324C).

Derivation: Gases containing appreciable content of trans-butene-2, along with other butene and butane hydrocarbons, are obtained by fractional distillation of refinery gas.

Grade: Technical 95%, CP 99.0%, research 99.9%.

Hazard: Asphyxiant gas. Highly flammable. Flammable limits in air 1.8–9% by volume. Dangerous fire and explosion risk.

Use: See cis-butene-2.

trans-**butenedioic acid.**   See fumaric acid.

2-butene-1,3-diol. $HOCH_2CH:CHCH_2OH$.

Properties: Almost colorless, odorless liquid; miscible in water, ethanol and acetone; sparingly soluble in benzene. Technical butenediol is predominantly the cis isomer. Fp range 4.0–7.0C, bp range 232–235C, d 1.067–1.074, refr index 1.476–1.478 (25C), flash p 263F (128C). Combustible.

Derivation: reduction of 2-butyne-1,4-diol, by high pressure synthesis from acetylene and formaldehyde.

Hazard: Primary skin irritant.

Use: Intermediate for alkyd resins, plasticizers, nylon, pharmaceuticals, cross-linking agent for synthetic resins, fungicides.

3-butenenitrile.   See allyl cyanide.

butenoic acid.   See crotonic acid.

3-buten-2-one.   See vinyl methyl ketone.

butesin.   See n-butyl-p-amino-benzoate.

butethal.   (5-butyl-5-ethylbarbituric acid).
$C_{10}H_{16}N_2O_3$.

Properties: White crystals or powder, odorless, bitter taste, mp 124–127C, fairly soluble in alcohol or ether, practically insoluble in water.

Hazard: may cause addiction.

Use: Medicine (hypnotic).   See also barbiturate.

butethamine hydrochloride.   [2-(Isobutylamino)-ethyl-p-aminobenzoate hydrochloride].
$NH_2C_6H_4COOCH_2CH_2NHCH_2CH(CH_3)_2 \cdot HCl$.

Properties: White, odorless, crystals or crystalline powder with bitter taste and local anesthetizing effects on tongue. Mp 192–196C, soluble in water, slightly soluble in alcohol and chloroform, very slightly soluble in benzene, practically insoluble in ether, pH (1% solution) approximately 4.7, stable in air.

Grade: NF.

Use: Medicine (local anesthetic).

butonate. $(CH_3O)_2P(O)CH(CCl_3)OOCC_3H_7$.
(Generic name for o,o-dimethyl-2,2,2,-trichloro-1-n-butyryloxyethyl phosphonate).
Properties: Colorless, somewhat oily liquid with slight ester odor, miscible with most organic solvents, stable in neutral or acid aqueous solution, unstable in aqueous alkali, d 1.3742, refr index 1.4707, bp 129C, wt/gal 11.4 lb.
Hazard: Restrict use, toxic, cholinesterase inhibitor.
Use: Insecticide.

butopyronoxyl. (butyl mesityl oxide oxalate; n-butyl-3,4-dihydro-2,2-dimethyl-4-oxo-1,2H-pyran-6-carboxylate). CAS: 532-34-3. $C_{12}H_{18}O_4$
Properties: Yellow to pale reddish-brown liquid with aromatic odor; reasonably stable in air; slowly affected by light; insoluble in water; miscible with alcohol, chloroform, ether, glacial acetic acid; d 1.052–1.060 (25/25C); refr index 1.4745–1.4755 (25C); distilling range 256–270C.
Derivation: Condensation of mesityl oxide and dibutyl oxalate in the presence of sodium ethoxide.
Grade: Technical.
Hazard: Toxic by ingestion. May cause liver damage.
Use: Insect repellent.

2-butoxyethanol. See ethylene glycol monobutyl ether.

2-(2-butoxyethyoxy)ethyl thiocyanate. See β-butoxy-β′-thiocyanodiethyl ether.

1-butoxyethoxy-2-propanol.
$CH_3CHOHCH_2OC_2H_4OC_4H_9$.
Properties: Colorless liquid, d 0.9310 (20/20C), bp 230.3C, fp −90C, soluble in water, wt/gal 7.8 lb, flash p 250F (121C). Combustible.
Use: Solvent, hydraulic fluid components, anti-stall additive for automotive fuels, plasticizer, intermediate.

butoxyethyl laurate. See ethylene glycol monobutyl ether laurate.

butoxyethyl oleate. See ethylene glycol monobutyl ether oleate.

butoxyethyl stearate. See ethylene glycol monobutyl ether stearate.

"Butoxyne."[30] TM for a series of specialty chemicals.
Properties: Nonionic, ashless, 100% active materials differing in alkyl substituent. Effective at high temperatures and in concentrated electrolyte solution, stable to acids and alkalies.

Use: Emulsifiers for water-in systems, corrosion and rust inhibitors, gelling and thickening agents for a wide variety of organic liquids.

p-butoxyphenol. $HOC_6H_4OC_4H_9$.
Properties: White to faint yellow, crystalline powder; mp 61–65C. Soluble in alcohol, acetone, ether, benzene, aqueous alkali, insoluble in water. Combustible.
Grade: 93% pure.
Use: Synthesis.

butoxy polypropylene glycol. (Generic name for polypropylene glycol monobutyl ether). $CH_3CHOH(CH_2OCHCH_3)_n CH_2OC_4H_9$.
Colorless liquid.
Use: An insect repellent.

n-butoxypropanol. CAS: 63716-40-5.
Properties: Colorless liquid, d 0.8801 (20/20C), bp 170.2C, fp −80C (sets to glass), soluble in water, flash p 154F (67.7C). Combustible.
Use: Solvent for water-based enamels.

β-butoxy-β′-thiocyanodiethyl ether. [2-(2-butoxyethoxy)ethyl thiocyanate].
$CH_3(CH_2)_3OCH_2OCH_2CH_2SCN$.
Hazard: Toxic by ingestion and skin absorption. Skin irritant.
Use: Insecticide.

butoxytriglycol. (triethylene glycol monobutyl ether). CAS: 143-22-6. $C_4H_9O(C_2H_4O)_3H$.
Properties: Liquid, d 1.0021 (20/20C), bp decomposes, fp −47.6C, miscible in water, flash p 290F (143C). Combustible.
Use: Plasticizer, intermediate.

buttercup yellow. See zinc yellow.

butter. (1) A colloidal system (emulsion) in which the continuous phase is composed of liquid fat from fat globules disintegrated by mechanical agitation and the dispersed phase is comprised of finely divided water droplets and undamaged fat globules. (2) Outmoded term for hygroscopic metallic chlorides of viscous consistency, e.g., butter of zinc, etc.

butterfat. The oily portions of the milk of mammals.
Properties: Composition is largely glycerides of oleic, stearic and palmitic acids with smaller amounts of the glycerides of butyric, caproic, caprylic and capric acids, d range 0.910–0.914. Cow's milk contains approximately 4% butterfat.
See also milk.

butter yellow.   See dimethylaminoazobenzene.

"Butvar"[57].   TM for polyvinyl butyral resins.
Use: Coatings (metal, textile, wood, etc), film, adhesives, sealers, molded materials, insulation, and safety glass.
See polyvinyl acetal resin.

butyl.   (1) The group C_4H_9; (2) butyl rubber.

n-butyl acetate.   CAS: 123-86-4.
$CH_3COOCH_2CH_2CH_2CH_2$.
Properties: Colorless liquid, fruity odor. Soluble in alcohol, ether and hydrocarbons; slightly soluble in water. Vapor is heavier than air, d 0.8826 at 20/20C, bp 126.3C, vap press 8.7 mm (20C), fp −75C, refr index 1.2951 (20C), wt/gal 7.35 lb (20C), flash p 98F (36.6C) (TOC). Autoign temperature 790F (421C).
Derivation: Esterification and then distillation after contact of butyl alcohol with acetic acid in the presence of a catalyst such as sulfuric acid.
Hazard: Skin irritant, toxic. Flammable, moderate fire risk. TLV: 150 ppm in air.

sec-butyl acetate.   CAS: 105-46-4.
$CH_3COOCH(CH_3)(C_2H_5)$.
Properties: Colorless liquid, bp 112.2C, d 0.8905 at 0/4C, 0.870 at 20/4C, refr index 1.389 (20C), wt/gal 7.21 lb, flash p 88F (31C) (OC). Miscible with alcohol and ether, insoluble in water.
Derivation: Esterification of sec-butyl alcohol.
Hazard: Flammable, dangerous fire risk. TLV: 200 ppm in air.
Use: Solvent for nitrocellulose lacquers, thinners, nail enamels, leather finishes.

tert-butyl acetate.   CAS: 540-88-5.
$CH_3COOC(CH_3)_2$.
Properties: Colorless liquid, bp 96C, d 0.896 (20C). Insoluble in water, soluble in alcohol and ether.
Hazard: Flammable, moderate fire risk. TLV: 200 ppm in air.
Use: Solvent, gasoline additive.

butyl acetoacetate.   CAS: 591-60-6.
$CH_3COCH_2COOCH_2CH_2CH_2CH_3$.
Properties: Colorless liquid, insoluble in water, soluble in alcohol and ether. d 0.9694 (20/20C), bp 213.9C, vap press 0.19 mm (20C), flash p 185F (85C), wt/gal 8.1lb (20C). Combustible.
Grade: Technical.
Use: Intermediate in synthesis of metal derivatives, dyestuffs, pharmaceuticals, flavoring.

butyl acetoxystearate.
$CH_3(CH_2)_5CH(CH_3COO)(CH_2)_{10}COOC_4H_9$.
Properties: See butyl acetyl ricinoleate.

Derivation: From castor oil, butylalcohol and acetic anhydride with hydrogenation.
Use: Plasticizer, textile oils, adhesives.

butyl acetylene.   See 1-hexyne.

butyl acetyl ricinoleate. $C_{24}H_{44}O_4$.
Properties: Yellow, oily liquid; mild odor; miscible with most organic solvents; d 0.940 (20/20C); sapon number 125; fp indefinite; becomes cloudy at −32C; solidifies at −65C. Flash p 230F (110C) (OC), refr index 1.4614 (20C). Saybolt viscosity 123 secs at 100F, wt/gal 7.8lb (20C), almost insoluble in water. Combustible. Autoign temperature 725F (385C).
Derivation: From castor oil, butanol and acetic anhydride.
Grade: Technical.
Use: Plasticizer, emulsifier, lubricant, detergent, protective coatings, special cleansing compounds, quick breaking emulsions.

n-butyl acid phosphate.   (n-butylphosphoric acid; acid butyl phosphate).
CAS: 12788-93-1.
Properties: Water-white liquid, d 1.120–1.125 (25/4C), refr index 1.429 (25C), flash p 230F (110C) (COC). Soluble in alcohol, acetone, and toluene. Insoluble in water and petroleum naphtha. Combustible.
Hazard: Strong irritant to skin and tissue.
Use: Esterification catalyst and polymerizing agent, curing catalyst and accelerator in resins and coatings, special detergents.

N-tert-butylacrylamide.
$H_2C:CHCONHC(CH_3)_3$.
Properties: White, crystalline solid; mp 128–130C; d 1.015 (30C). Soluble in methanol, ethanol, chloroform, and acetone. Combustible.
Hazard: Toxic by ingestion and inhalation. Irritant to skin.
Use: Monomer, organic intermediate.

n-butyl acrylate.   CAS: 141-32-2.
$CH_2:CHCOOC_4H_9$.
Properties: Colorless liquid, fp −64C, boiling range 145.7–148.0C, polymerizes readily on heating, vap press (20C) 3.2 mm, d 0.9015 (20/20C), wt/gal 7.5 lb (20C), flash p 120F (49C) (OC) nearly insoluble in water. Flammable.
Derivation: Reaction of acrylic acid or methyl acrylate with butanol.
Grade: Technical (inhibited).
Hazard: Moderate fire risk. TLV: 10 ppm in air.
Use: Intermediate in organic synthesis, polymers

and copolymers for solvent coatings, adhesives, paints, binders, emulsifiers.
See also acrylic resin.

tert-**butyl-acrylate.** $CH_2:CHCOOC(CH_3)_3$.
Properties: Liquid, bp 120C, d 0.879 (25C), refr index 1.4080 (25C), flash p 66F (18.8C) (TOC). Commercial grade contains 100 ppm hydroquinone monomethyl ether as stabilizer.
Hazard: Toxic by ingestion and inhalation. Flammable, dangerous fire risk. TLV: 10 ppm in air.
Use: Monomer for acrylic resins.

n-**butyl alcohol.** (1-butanol; butyric alcohol). CAS: 71-36-3. $CH_3(CH_2)_2CH_2OH$.
Properties: Colorless liquid, vinous odor. Bp 117.7C, fp −89.0C, d (20/20C) 0.8109, wt/gal (20C) 6.76lb, refr index 1.3993 (20C), flash p 95F (35C). Soluble in water 7.7 wt % (20C), solution of water in n-butanol 20.1%. Miscible with alcohol and ether. Autoign temperature 689F (365C).
Derivation: (a) Hydrogenation of butyraldehyde, obtained in the Oxo process; (b) condensation of acetaldehyde to form crotonaldehyde, which is then hydrogenated (aldol condensation).
Hazard: Toxic on prolonged inhalation, irritant to eyes, absorbed by skin. Flammable, moderate fire risk. TLV: CL 50 ppm in air.
Use: Preparation of esters, especially butyl acetate, solvent for resins and coatings, plasticizers, dyeing assistant, hydraulic fluids, detergent formulations, dehydrating agent (by azeotropic distillation), intermediate, "butylated" melamine resins, glycol ethers, butyl acrylate.

bsec-**butyl alcohol.** (SBA; 2-butanol; methylethylcarbinol). CAS: 78-92-2. $CH_3CH_2CHOHCH_3$.
Properties: Colorless liquid, strong odor, bp 99.5C, fp −114C, d (20/4C) 0.808, wt/gal (20C) 66.74, refr index 1.3949 (25C), flash p 75F (23.8C) (CC), autoign temperature 763F (406C). Moderately soluble in water, miscible with alcohol and ether.
Derivation: Absorption of butene from cracking petroleum or natural gas in sulfuric acid with subsequent hydrolysis by steam.
Grade: Technical.
Hazard: Toxic on prolonged inhalation, irritating to eyes and skin. Flammable, dangerous fire risk. TLV: 100 ppm in air.
Use: Preparation of methyl ethyl ketone, solvent, organic synthesis, paint removers, industrial cleaners.

tert-**butyl alcohol.** (2-methyl-2-propanol; trimethyl carbinol). CAS: 75-65-0. $(CH_3)_3COH$.
Properties: Colorless liquid or crystals, camphor odor, fp 25.5C, bp 82.9C, d (liquid 26C) 0.779, refr index 1.3878 (20C), flash p 52F (11.1C) (CC), autoign temperature 892F (477C). Miscible with water, alcohol, and ether.
Derivation: Absorption of isobutene from cracking petroleum or natural gas in sulfuric acid with subsequent hydrolysis by steam.
Grade: Technical.
Hazard: Irritant to eyes and skin. Flammable, dangerous fire risk. TLV: 100 ppm in air.
Use: Alcohol denaturant, solvent for pharmaceuticals, dehydration agent, perfumery, chemical intermediate, paint removers, manufacture of methyl methacrylate, octane booster in unleaded gasoline (EPA approved).

n-**butyl aldehyde.** See butyraldehyde.

n-**butylamine.** (1-aminobutane). CAS: 109-73-9. $C_4H_9NH_2$.
Properties: Colorless, volatile liquid with amine odor; bp 77.1C; fp −49.1C; d 0.7385 (20/20C), wt/gal 6.2lb (20C); refr index 1.401 (20C); flash p 30F (1.1C) (OC), miscible with water, alcohol, ether.
Derivation: Reaction of butanol or butyl chloride with ammonia.
Grade: Technical.
Hazard: Skin irritant. Flammable, dangerous fire risk. TLV: CL 5 ppm in air.
Use: Intermediate for emulsifying agents, pharmaceuticals, insecticides, rubber chemicals, dyes, tanning agents.

sec-**butylamine.** (2-aminobutane). CAS: 13952-84-6. $CH_3CHNH_2C_2H_5$.
Properties: Colorless liquid, d 0.725 (20C), boiling range 63–68C, refr index 1.395 (20C), solidification point −104C, odor amine, flash p 15F (−9.4C), wt/gal 6.0lb (20C).
Hazard: Flammable, dangerous fire risk.
Use: Fungicide.

tert-**butylamine.** CAS: 75-64-9. $(CH_3)_3CNH_2$.
Properties: Colorless liquid, bp 44–46C, fp −72C, d 0.700 (15C), refr index 1.3794 (18C), flash p approximately 50F (10C). Miscible with water, soluble in common organic solvents.
Grade: Technical.
Hazard: Skin irritant. Flammable, dangerous fire risk.
Use: Intermediate for rubber accelerators, insecticides, fungicides, dyestuffs, pharmaceuticals.

butyl-o-aminobenzoate. See butyl anthranilate.

n-**butyl-p-aminobenzoate.** $H_2NC_6H_4COOC_4H_9$.
Properties: White powder, odorless, tasteless, mp

57–59C, bp 174C (8 mm). Soluble in dilute acids, alcohol, chloroform, ether, and fatty oils; almost insoluble in water.
Grade: NF.
Hazard: Toxic by ingestion.
Use: Medicine (local anesthetic), treatment of burns, ointments, UV absorber in suntan preparations.

N-n-butylaminoethanol. $C_4H_9NHC_2H_4OH$.
Properties: Liquid, d 0.88–0.99 (20/20C), distillation range 192–210C, wt/gal 7.4 lb, flash p 170F (76.6C). Combustible.

tert-butylaminoethyl methacrylate.
$CH_2:C(CH_3)COOCH_2CH_2NHC(CH_3)_3$.
Properties: Liquid, bp 100–105C (12 mm), d 0.914 (25C), 7.61 lb/gal, refr index 1.4440 (25C), flash p 205F (96.1C) (COC). Combustible.
Use: Coatings, textile chemicals, dispersing agent for nonaqueous systems, antistatic agent, stabilizer for chlorinated polymers, ion exchange resins, emulsifying agent, cationic precipitating agent.

N′-n-butyl-3-amino-4-methoxybenzenesulfonamide. $CH_3OC_6H_3(NH_2)SO_2NHC_4H_9$.
Properties: Pink powder, mp 96–100C, insoluble in water, partially soluble in alcohol and acetone. Used as an intermediate.

N-n-butylaniline. $C_6H_5NHC_4H_9$.
Properties: Amber liquid, d 0.932 (20C), boiling range 236–242C, refr index 1.534 (20C), odor aniline, very soluble in alcohol and ether, insoluble in water, flash p 225F (107C). Combustible.
Use: Organic synthesis, dyes.

butyl anthranilate.   (butyl-o-aminobenzoate).
$C_4H_9OOCC_6H_4NH_2$.
Used in flavoring.

2-tert-butylanthraquinone. $C_{18}H_{16}O_2$.
Properties: Yellow powder, mp 102–104C, soluble in alcohol and acetone. Combustible.
Grade: Technical (98%).
Use: Organic synthesis.

butylated hydroxyanisole.   (BHA).
CAS: 25013-16-5. $(CH_3)_3CC_6H_3OH(OCH_3)$.
A mixture of 2- and 3-tert-4-methoxyphenol.
Properties: White or slightly yellow, waxy solid having a faint characteristic odor; melting range 48–63C; not naturally water-soluble but can be made so by special treatment. Freely soluble in alcohol and propylene glycol. Combustible.
Grade: FCC, water-soluble.
Hazard: Toxic by ingestion. Use in foods restricted, consult FDA regulations.
Use: Antioxidant for fats and oils food packaging.

butylated hydroxytoluene.   See 2,6-di-tert-butyl-p-cresol.

n-butylbenzene.   (1-phenylbutane).
CAS: 104-51-8. $C_6H_5C_4H_9$.
Properties: Colorless liquid, bp 183.2C, fp −87.9C, d 0.860 (20C), refr index 1.489 (20C), flash p 160F (71.1C) (OC). Combustible, autoign temperature 774F (412C).
Grade: Technical, pure, research.
Hazard: Toxic by ingestion.
Use: Organic synthesis especially of insecticides.

sec-butylbenzene.   (2-phenylbutane).
CAS: 135-98-8. $C_6H_5C(CH_3)C_2H_5$.
Properties: Colorless liquid, bp 170.65C, vap press 15 mm (60C), fp −75.68C, d 0.8618 (204C), wt/gal 7.2 lb (20C), refr index 1.4901 (20C), flash p 145F (62.7C) (OC), combustible, autoign temperature 784F (417C).
Grade: Technical 95%, pure, research.
Hazard: Toxic by ingestion.
Use: Solvent for coating compositions, organic synthesis, plasticizer, surface-active agents.

tert-butylbenzene.   (2-methyl-2-phenylpropane).
CAS: 98-06-6. $C_6H_5C(CH_3)_3$.
Properties: Colorless liquid, insoluble in water, soluble in alcohol, bp 169.1C, fp −57.8C, d 0.866 (20C), refr index 1.492 (20C), flash p 140F (60C) (OC). Combustible. Autoign temperature 842F (450C).
Grade: Technical, pure, research.
Hazard: Toxic by ingestion.
Use: Organic synthesis, polymerization solvent, polymer linking agent.

butylbenzenesulfonamide.   (n,n-butyl benzenesulfonamide). $C_6H_5SO_2NHC_3H_9$.
Properties: Liquid, pleasant odor, amber to straw color, d 1.148 (25/25C), refr index 1.5235 (25C), bp 189–190C (4.5 mm).
Hazard: Toxic by ingestion.
Use: Synthesis of dyes, pharmaceuticals, other organic chemicals, in resin manufacturing, as plasticizer for some synthetic polymers.

butyl benzoate.   (n-butyl benzoate).
$C_6H_5COOC_4H_9$.
Properties: Colorless, oily liquid; insoluble in water; miscible with alcohol or ether; d 1.00 (20C); bp 247.3C; fp −22C; flash p 225F (107C) (OC). Combustible.
Grade: Technical.
Use: Solvent for cellulose ether, plasticizer, perfume ingredient, dyeing of textiles.

N-tert-butyl-2-benzothiazolesulfenamide.
$C_6H_4NCS(SNH)C_4H_9$.

Properties: Light buff powder or flakes (sometimes colored blue), mp 104C min, d 1.29 at 25C. Soluble in most organic solvents. Combustible.
Use: Rubber accelerator.

butyl benzyl phthalate. (BBP).
$C_4H_9OOCC_6H_4COOC_7H_7$.
Properties: Clear, oily liquid; slight odor; d 1.113–1.121 (25/25C); flash p 390F (198C). Combustible.
Grade: Technical.
Use: Plasticizer for polyvinyl and cellulosic resins, organic intermediate.

butyl benzyl sebacate.
$C_4H_9OOC(CH_2)_8COOC_7H_7$.
Properties: Light straw-colored liquid, bp 245–285C (10 mm), d 1.023 (25C), wt/gal 8.6 lb, flash p 395F (201C). Combustible.
Use: Plasticizer for resins.

butyl borate. See tributyl borate.

n-butyl bromide. (1-bromobutane).
CAS: 109-65-9. C_4H_9Br.
Properties: Colorless liquid, d 1.279 (20/20C), bp 101.6C, fp −112.4C, flash p 75F (23.9C) (OC). Autoign temperature 509F (265C), insoluble in water, soluble in alcohol and ether.
Grade: 99%.
Hazard: Flammable, dangerous fire risk.
Use: Alkylating agent.

sec-butyl bromide. (2-bromobutane).
CAS: 78-76-2. $CH_3CHBrCH_2CH_3$.
Properties: Clear, colorless liquid with pleasant odor; bp 91.2C; fp −112C; d 1.2425 (25/25C); refr index 1.4320–1.4344 (25C); flash p 70F (21.1C) (OC); soluble in alcohol and ether; insoluble in water.
Hazard: Narcotic in high concentration. Flammable, dangerous fire risk.
Use: Synthesis, alkylating agent.

butyl butanoate. See n-butyl butyrate.

n-butyl butyrate. (butyl butanoate).
CAS: 109-21-7. $CH_3(CH_2)_2COOC_4H_9$.
Properties: Colorless liquid, d 0.8721 (20/20C), refr index 1.4059 (20C), fp −91.5C, bp 165.7C (736 mm), flash p 128F (53.5C) (OC), insoluble in water, soluble in alcohol and ether. Flammable.
Grade: FCC.
Hazard: Irritant and narcotic. Moderate fire risk.
Use: Flavoring.

n-butyl carbinol. See n-amyl alcohol, primary.

sec-butyl carbinol. See 2-methyl-1-butanol.

butyl "Carbitol."[214] TM for diethylene glycol monobutyl ether.

butyl "Carbitol" acetate.[214]. TM for diethylene glycol monobutyl ether acetate.

p-tert-butylcatechol. (4-tert-butyl-1,2-dihydroxybenzene). $(CH_3)_3CC_6H_3(OH)_2$.
Properties: Colorless crystals, mp 56–57C, d 1.049 (60/25C), bp 285C, flash p 265F (129C). Combustible. Soluble in ether, alcohol, acetone, slightly soluble in water at 80C.
Hazard: Toxic by ingestion and skin absorption.
Use: Polymerization inhibitor for styrene-butadiene and other olefins.

butyl "Cellosolve."[214] TM for ethylene glycol monobutyl ether.

butyl "Cellosolve" acetate.[214]. TM for ethylene glycol monobutyl ether acetate.

butyl chloral hydrate. (trichlorobutyraldehyde hydrate). $CH_3CHClCCl_2CH(OH)_2$.
Properties: Colorless leaflets, d 1.693 (20/4C), mp 78C. Slightly soluble in water, soluble in alcohol and ether.
Derivation: Action of chlorine on paraldehyde.
Use: Medicine (hypnotic, anticonvulsant).

n-butyl chloride. (1-chlorobutane).
CAS: 109-69-3. $CH_3CH_2CH_2CH_2Cl$ or C_4H_9Cl.
Properties: Colorless liquid, insoluble in water, miscible with alcohol and ether, d 0.8875 (20/20C), bp 78.6C, wt/gal 7.35lb (20C), refr index 2.4015 (20C), vap press 80.1 mm (20C), fp −122.8C, viscosity 0.0045 poise (20C), flash p 15F (−9.4C) (OC), autoign temperature 860F (460C).
Grade: NF, technical.
Hazard: Toxic on prolonged inhalation. Flammable, dangerous fire risk.
Use: Organic synthesis (alkylating agent), solvent, anthelmintic.

sec-butyl chloride. CAS: 78-86-4.
$CH_3CHClCH_2CH$.
Properties: Colorless liquid, bp 68C, d 0.875 (20/4C), flash p 32F (0C), refr index 1.39. Miscible with alcohol and ether, sparingly soluble in water.
Hazard: Flammable, dangerous fire risk.
Use: Organic synthesis.

tert-butyl chromate. (chromic acid di-t-butyl ester). CAS: 1189-85-1.
$[(CH_3)_3CO]_2CrO_3$.
Properties: Liquid, mp −5 to 0C.
Hazards: TLV: ceiling 0.1 mg/m³. Toxic by skin

absorption. A very powerful oxidizer and a dangerous fire hazard.

butyl citrate. See tributyl citrate.

6-tert-butyl-m-cresol. (MBMC; 6-tert-butyl-3-methylphenol). $(CH_3)_3CC_6H_3(OH)CH_3$.
Properties: Clear liquid, solidifies slightly below room temperature, fp 23.1C, bp 244C, d 0.922 (80C). Soluble in organic solvents and aqueous potassium hydroxide. Flash p 116F (47C). Flammable.
Hazard: Irritant to skin. Moderate fire risk.
Use: Germicide, disinfectant, synthesis of antioxidants and rubber-processing chemicals, additives to lubricating oils, synthetic resins, perfumes (fixative).

butyl crotonate. $CH_3CH:CHCOOC_4H_9$.
Properties: Water-white liquid, persistent odor, d 0.9037 (20/20C), bp 180.5C, wt/gal 7.52lb (20C), soluble in alcohol and ether insoluble in water. Combustible.

butyl cyclohexyl phthalate.
$C_4H_9OOCC_6H_4COOC_6H_{11}$.
Properties: Clear liquid, very mild odor, d 1.078, saponification number 369, acidity (as phthalic acid) 0.01 max, miscible with most organic solvents. Combustible.
Use: Plasticizer for polymers and elastomers, nitrocellulose lacquers.

butyl decyl phthalate.
$C_4H_9OOCC_6H_4COOC_{10}H_{21}$.
Properties: Clear, oily liquid; d 0.977–0.987 (25/25C). Combustible.
Use: Primary plasticizer for polyvinyl chloride and copolymer resins.

n-butyldiamylamine. $C_4H_9N(C_5H_{11})_2$.
Properties: Straw-colored liquid, d 0.788 (20C), boiling range 229–241C, odor amine, flash p 200F (93.3C). Combustible.

n-butyldichlorarsine. $C_4H_9AsCl_2$.
Properties: Oily liquid. Somewhat agreeable odor, decomposed by water, bp 192–194C.
Hazard: Toxic by inhalation and ingestion.
Use: Military posion.

butyl dichlorophenoxyacetate. See 2,4-D.

1-n-butyl-3-(3,4-dichlorophenyl)-1-methylurea.
(neburon). $Cl_2C_6H_3NHCONCH_3(C_4H_9)$.
Properties: White, crystalline solid; mp 102C; very low solubility in water and hydrocarbon solvents. Stable toward oxidation and moisture.
Use: Weed killer.

n-butyl diethanolamine.
$C_4H_9N(CH_2CH_2OH)_2$.
Properties: Liquid, d 0.97 (20C), bp 272C, color very light straw, odor faint amine, wt/gal 18.08 lb (20C), flash p 245F (118C). Combustible.
Use: Organic synthesis.

tert-butyldiethanolamine.
$C_4H_9N(CH_2CH_2OH)_2$.
Properties: Liquid, similar to normal compound.
Use: Organic synthesis, epoxy curing agent, catalyst for polyester resins, inhibitor for printing inks.

n-butyl diethyl malonate.
$C_4H_9CH(COOC_2H_5)_2$.
Properties: Colorless liquid with a fruity odor, d 0.972–0.974 (25/25C), refr index 1.420–1.422 (25C), soluble in alcohols, ketones, esters. Combustible.
Use: Intermediate.

butyl diglycol carbonate. (diethylene glycol bis-n-butyl-carbonate). $(C_4H_9OCO_2CH_2)_2O$.
Properties: Colorless liquid of low volatility, insoluble in water (very stable to hydrolysis), widely soluble in organic solvents, compatible with many resins and plastics, d 1.07 (20/4C), boiling range 164–166C (2 mm), flash p 372F (188C), Saybolt viscosity 21 cps (20C), refr index 1.435 (20C), evaporation rate 0.59 mg/sq cm/hr (100C). Combustible.
Use: Plasticizer, high-boiling-point solvent and softening agent, manufacture of pharmaceuticals and lubricant compositions.

butyl diglyme. See diethylene glycol dibutyl ether.

4-tert-butyl-1,2-dihydroxybenzene. See p-tert-butylcatechol.

butyl "Dioxitol"[125]. TM for diethylene glycol monobutyl ether.

4-butyl-1,2-diphenyl-3,5-pyrazolidinedione.
See phenylbutazone.

butyl dodecanoate. See butyl laurate.

"Butyl Eight."[69] TM for an ultra-accelerator for rubber of the dithiocarbamate type.
Properties: Dark red liquid, odor distinct, d 1.01, partly soluble in water, soluble in acetone, benzene, carbon disulfide, chloroform, gasoline.

butylene. (butene). One of the liquefied petroleum gases butene-1, cis-butene-2, trans-butene-2, and isobutene.
See specific entry for details.

butylene dimethacrylate. $C_{12}H_{18}O_4$.
Properties: Liquid, boiling range 110C (3 mm),
d 1.011, (25/15.6C), refr index 1.4502 (25C),
flash p greater than 150F (65C). Combustible.
Use: Monomer for resins.

1,3-butylene glycol. (1,3-butanediol).
$HOCH_2CH_2CH(OH)CH_3$. Can exist in optical
isomeric forms.
Properties: Practically colorless, viscous liquid;
hygroscopic; d 1.0059 (20/20C); 8.4lb/gal (20C);
bp 207.5C; vap press 0.06 mm (20C); refr index
1.4401 (20C); flash p 250F (121C) (COC); completely soluble in water and alcohol; slightly soluble in ether. Combustible. Autoign temperature
741F (393C).
Derivation: Reduction of aldol.
Use: Polyesters, polyurethanes, surface active
agents, plasticizers, humectant, coupling agent,
solvent, food additive, flavoring.

1,4-butylene glycol. (1,4-butanediol; tetramethylene glycol).   CAS: 110-63-4.
$HOCH_2CH_2CH_2CH_2OH$.
Properties: Colorless, oily liquid. Bp 230C, mp
16C, d 1.020 (20/4C). Flash p greater than 250F
(121C). Miscible with water, soluble in alcohol,
slightly soluble in ether. Combustible.
Derivation: From acetylene and formaldehyde by
high-pressure synthesis.
Grade: Technical.
Hazard: Toxic by ingestion.
Use: Solvents, humectant, intermediate for plasticizers, pharmaceuticals, cross-linking agent in
polyurethane elastomers, manufacture of tetrahydrofuran, terephthalate plastics.

2,3-butylene glycol. (2,3-dihydroxybutane; 2,3-
butanediol; pseudobutylene glycol; sym-dimeth-
ylethylene glycol). $CH_3CHOHCHOHCH_3$.

Can exist in optical isomeric forms.
Properties: Nearly colorless, crystalline solid or
liquid; hygroscopic; d 1.045 (20/20C); mp 23–
27C; bp 179–182C; refr index 1.438 (20C); flash
p 185F (85C) (OC); soluble in alcohol and ether;
miscible with water. Combustible.
Derivation: From corn sugar by acid hydrolysis,
also from fermentation of sugar by acid hydrolysis, also from fermentation of sugar beet molasses.
Grade: 99%.
Use: Resins, solvent for dyes, intermediate, blending agent.

1,2-butylene oxide. (1,2-epoxybutane).

CAS: 106-88-7. $H_2\overset{\frown}{CO}CHCH_2CH_3$.

Properties: Colorless liquid, d 0.8312 (20/20C),
bp 63C, sets to a glass below −150C, flash p
approximately 0F (−17C) (CC). Soluble in water
and miscible with most organic solvents.
Grade: Approximately 97.5% purity.
Hazard: Toxic concentration of vapors occurs at
room temperature. Highly flammable, dangerous
fire risk.
Use: Intermediate for various polymers, stabilizer
for chlorinated solvents.

2,3-butylene oxide. (2,3-epoxybutane).

$CH_3H\overset{\frown}{CO}CHCH_3$.   Two forms, cis and trans,
are known.
Properties: Cis: fp −80C, bp 59.7C (742 mm), d
0.8266 (25/4C). Trans: fp −85C, bp 53.5C (742
mm), d 0.8010 (25/4C). Flash p approximately
0F (−17C). Very soluble in ether, benzene, organic solvents; decomposes in water.
Hazard: Toxic concentration of vapors occurs at
room temperature. Highly flammable, dangerous
fire risk.
Use: Intermediate.

butyl epoxystearate. (butyl-9,10-epoxyoctade-
canoate).

$CH_3(CH_2)_7\overset{\frown}{CHOCH}(CH_2)_7COOC_4H_9$.
Properties: Colorless liquid with mild, slightly
fatty, slightly fruity odor, d (20C) 0.910, wt/gal
7.59 lb. Combustible.
Use: Plasticizer for low-temperature flexibility improvement of vinyl resins.

n-butylethanolamine. $C_4H_9NHCH_2CH_2OH$.
Properties: Colorless liquid, d (20C) 0.892, boiling
range 194–204C, odor very faint amine type,
flash p 170F (76.6C). Combustible.

butyl ether. (n-dibutyl ether).
CAS: 142-96-1. $C_4H_9OC_4H_9$.
Properties: Colorless liquid, stable, mild ethereal
odor, d 0.7694 (20/20C), bp 142.2C, vap press
4.8 mm (20C), flash p 77F (25C), fp −95.2C,
latent heat of vaporization 67.8 cal/g at 140.9C,
refr index 1.3992 (20C), wt/gal 6.4 lb (20C), viscosity 0.0069 poise (20C). Autoign temperature
382F (194C). Miscible with most common organic solvents, immiscible in water.
Grade: Technical, spectrophotometric.
Hazard: Toxic on prolonged inhalation. Flammable, moderate fire risk. May form explosive peroxides especially in anhydrous form.
Use: Solvent for hydrocarbons, fatty materials; extracting agent used especially for separating metals, solvent purification, organic synthesis (reaction medium).

butylethylacetaldehyde.   See 2-ethylhexaldehyde.

5-butyl-5-ethylbarbituric acid.   See butethal.

n-butyl ethyl ether.   See ethyl-n-butyl ether.

butyl ethyl ketone.   (3-heptanone).
   CAS: 106-35-4. $(C_4H_9)(C_2H_5)CCO$.
   Properties: Clear liquid, d 0.8198 (20/20C), bp
   148C, fp -36.7C, refr index 1.4224 (20C), flash
   p 115F, vap d 3.93.
   Grade: Available as 20% solution in hexane.
   Hazard: Fumes are irritating. Flammable, danger-
   ous fire risk. Store solution under nitrogen.
   Use: Reactive chemical intermediate.

2-butyl-2-ethylpropanediol-1,3.   See 2-ethyl-2-bu-
   tylpropanediol-1,3.

tert-butylformamide. $(CH_3)_3CNHCOH$.
   Properties: Colorless, high-boiling liquid, soluble
   in water and common hydrocarbon solvents.
   Use: A solvent, and in petroleum additives.

butyl formate.   CAS: 592-84-7. $HCOOC_4H_9$.
   Properties: Colorless liquid, d 0.885–0.9108, bp
   107C, fp -90C, flash p 64F (17.7C) (CC), autoign temperature 612F (322.5C), miscible with
   alcohols, ethers, oils, hydrocarbons, slightly soluble in water.
   Grade: Technical.
   Hazard: Narcotic and irritating in high concentration. Flammable, dangerous fire risk.
   Use: Solvent for nitrocellulose, some types of cellulose acetate, many cellulose ethers, many natural
   and synthetic resins, lacquers, perfumes, organic
   synthesis (intermediate), flavoring.

n-butyl furfuryl ether. $C_4H_9OCH_2C_4H_3O$.
   Properties: Colorless liquid turning dark on exposure to air, extremely hygroscopic, unstable with
   moisture, d 0.955 (10/0C), bp 189–190C (765
   mm), refr index 1.4522 (20C).

n-butyl furoate. $C_4H_3OCO_2C_4H_9$.
   Properties: Colorless oil, decomposes on standing,
   d 1.055 (20/5C), bp 83–84C (1 mm), 118–120C
   at 25 mm, insoluble in water, soluble in alcohol
   and ether.

n-butyl glycidyl ether.   (glycidylbutylether;
   BGE).   CAS: 2426-08-6.
   $C_4H_9OH_2CHOCH_2$.
   Properties: Clear, colorless liquid with an irritating odor. Bp 164C, vapor p 3.2 mm at 25C,
   vapor d 3.78, d 0.908 at 25/4C. Soluble in water.
   Hazard: TLV: 25 ppm. A mild skin and eye irritant.

n-butyl glycol phthalate.   See dibutoxyethyl
   phthalate.

tert-butyl hydroperoxide.   CAS: 75-91-2.
   $(CH_3)_3COOH$.   A highly reactive peroxy
   compound.
   Properties: Water-white liquid, fp -8C, decomposes at 75C, d 0.896 (20/4C), refr index 1.396
   (25C) (90% pure), flash p (90%)130F (54.4C).
   Moderately soluble in water, very soluble in organic solvents and alkali metal hydroxide solutions. Combustible.
   Grade: 70%, 90% pure.
   Hazard: Moderate fire risk. Oxidizer.
   Use: Polymerization, oxidation, sulfonation catalyst, bleaching, deodorizing.

tert-butylhydroquinone. $C_6H_3(OH)_2C(CH_3)_3$
   Properties: Intermediate, mp 125C, insoluble in
   water, soluble in alcohol, acetone, ethyl acetate.

tert-butyl hypochlorite. $(CH_3)_3COCl$.
   Properties: Yellowish liquid.
   Hazard: Toxic by ingestion and inhalation. May
   explode at room temperature.
   Use: Organic chlorinations, oxidation of alcohols
   to ketones and sulfides to sulfoxides.

4,4′-butylidenebis(6-tert-butyl-m-cresol).
   $[(CH_3)_3CC_6H_2(OH)(CH_3)]_2CHC_3H_7$.
   Properties: White powder, mp 209C (min), d 1.03
   (25C).
   Use: Antioxidant for rubber, dry or latex.

n-butyl isocyanate.   CAS: 111-36-4.
   C_4H_9NCO.
   Properties: Colorless liquid, bp 115C, d 0.88 (20/
   4C).
   Use: Intermediate for pesticides, herbicides, phar-
   maceuticals.
   Hazard: Strong irritant to eyes and skin.

butyl isodecyl phthalate.
   Properties: Clear liquid, mild odor, color (Hazen)
   50 max, d 0.9 (20/20C), saponification number
   310, acidity (as phthalic acid) 0.01 max.
   Use: Plasticizer for polyvinyls.

tert-butylisopropylbenzene hydroperoxide.
   Properties: White crystals.
   Hazard: Dangerous fire risk. Reacts strongly with
   reducing materials. Oxidizing agent.

"Butyl Kamate."[69]   TM for an aqueous solution
   of potassium di-n-butyl dithiocarbamate, 50%
   minimum assay.
   Use: Ultra-accelerator for natural and synthetic
   latexes.

n-**butyl lactate.** CAS: 138-22-7.
$CH_3CHOHCOOC_4H_9$.
Properties: Water-white, stable liquid. Mild odor, miscible with many lacquer solvents, diluents, oils, slightly soluble in water, hydrolyzed in acids and alkalies. D 0.974–0.984 (20/20C), flash p 168F (75.5C) (TOC), mp −43C, wt per US gal 8.15 lb, bp 188C, refr index 1.4216 (20C), vap press 0.4 mm (20C), latent heat of vaporization 77.4cal/g (20C). Autoign temperature 720F (382C). Combustible.
Grade: Technical, 95% min.
Hazard: Toxic. TLV: 5 ppm in air.
Use: Solvent for nitrocellulose, ethyl cellulose, oils, dyes, natural gums and many synthetic polymers, lacquers, varnishes, inks, stencil pastes, antiskinning agent, chemical (intermediate), perfumes, dry-cleaning fluids, adhesives.

N-n-butyl lauramide. $C_{11}H_{23}CONHC_4H_9$.
Properties: White solid, boiling range 200–225C at 2 mm, odor lauric acid, flash p 375F (190C). Combustible.

butyl laurate. (butyl dodecanoate). $C_{11}H_{23}COOC_4H_9$.
Properties: Liquid, d 0.855 (25C), bp (5 mm) 130–180C, fp −10C, insoluble in water. Combustible.
Derivation: Alcoholysis of coconut oil with butyl alcohol followed by fractional distillation.
Use: Plasticizer, flavoring.

butyllithium. (n-butyllithium; sec-butyllithium; tert-butyllithium). $CH_3CH_2CH_2CH_2Li$; $CH_3CHLiCH_2CH_3$; $(CH_3)_3CLi$.
Properties: Available usually in solution in one of the C_5 to C_7 hydrocarbons in which they are quite stable. sec-Butyllithium solution must be kept at or below 15.5C.
Derivation: Sold by percent butyllithium in the solution.
Hazard: Irritant. Solid and solution highly flammable, ignites in moist air.
Use: Polymerization of isoprene and butadiene, intermediate in preparation of lithium hydride, rocket fuel component, metalating agent.

n-**butylmagnesium chloride.** C_4H_9MgCl.
Properties: Liquid, d 0.88.
Derivation: From magnesium and butyl chloride.
Grade: Available in solution in ethyl ether or in tetrahydrofuran.
Hazard: Flammable, dangerous fire risk.
Use: Grignard reagent, as an alkylating agent.

n-**and** sec-**butyl mercaptan.** See 1- and 2-butanethiol.

tert-**butylmercaptan.** Legal label name for 2-methyl-2-butanethiol.

butyl mesityl oxide. See butopyronoxyl.

n-**butyl methacrylate.** (methacrylic acid, butyl ester). CAS: 97-88-1.
$H_2C:C(CH_3)COOC_4H_9$.
Properties: Colorless liquid, bp 163.5–170.5C, fp below −75C, d 0.895 (25/25C), flash p 130F (54.4C) (OC), refr index 1.4220, readily polymerized, insoluble in water. Combustible.
Derivation: Reaction of methacrylic acid or methyl methacrylate with butanol.
Grade: Technical (inhibited).
Hazard: Toxic by ingestion. Moderate fire risk.
Use: Monomer for resins, solvent coatings, adhesives, oil additives, emulsions for textiles, leather and paper finishing. See also acrylic resin.

tert-**butyl methacrylate.**
$H_2C:C(CH_3)COOC(CH_3)_3$.
Properties: Colorless liquid, bp 66C (57 mm), d 0.877 (25C), refr index 1.4124 (24C), flash p 92F (33.3C) (TOC).
Grade: Technical containing 100 ppm hydroquinone monomethyl ether as inhibitor.
Hazard: Toxic by ingestion. Flammable, dangerous fire risk.
Use: Monomer for resins. See also acrylic resin.

tert-**butyl-4-methoxyphenol.** See butylated hydroxyanisole.

sec-**butyl-6-methyl-3-cyclohexene-1-carboxylate.** See siglure.

p-tert-**butyl-α-methylhydrocinnamaldehyde.** (α-methyl-β-(p-tert-butylphenyl)propionaldehyde. $(CH_3)_3CC_6H_4CH_2CH(CH_3)CHO$.
Properties: Light-yellow liquid, strong floral odor, d 0.942–0.949 (25/25C), refr index 1.503–1.510 (20C), flash p 204F (95.5C) (TCC), soluble in 1 part 90% alcohol. Stable, non-discoloring. Combustible.
Grade: 93%, 85% purity.
Use: Perfume.

2-tert-**butyl-4-methylphenol.** See 2-tert-butyl-p-cresol.

6-tert-**butyl-3-methylphenol.** See 6-tert-butyl-m-cresol.

n-**butyl myristate.** $CH_3(CH_2)_{12}COOC_4H_9$. The butyl ester of myristic acid.
Properties: Water-white oily liquid, saponification number 193–203, fp 1–7C, boiling range 167–197C at 5 mm, d 0.850–0.858 (25C), insoluble in water, soluble in acetone, castor oil, chloroform, methanol, mineral oil, toluene. Combustible.

Derivation: Alcoholysis of coconut oil with butyl alcohol followed by fractional distillation.

Use: Plasticizer, lubricant for textiles, paper stencils, cosmetic preparations.

"Butyl Namate"[69]. TM for an aqueous solution of sodium di-n-butyldithiocarbamate, 47% min assay.

Use: Ultra-accelerator for natural and synthetic lattices.

n-butyl nitrate. $C_4H_9NO_3$.

Properties: Liquid, d 1.103 (20C), bp 123C, water-white, odor ethereal. Insoluble in water, soluble in alcohol and ether, flash p 97F (36C).

Hazard: Flammable, moderate fire risk in contact with reducing materials. Oxidizing agent, may explode from shock or heating.

tert-butyl nitrite. (nitrous acid tert-butyl ester). $(CH_3)_3CONO$.

Properties: Yellowish liquid with pleasant odor, d 0.867, bp 63C, refr index 1.36. Soluble in alcohol, carbon disulfide, chloroform, slightly soluble in water.

Derivation: Reaction of tert-butyl alcohol sodium nitrite and sulfuric acid.

Use: Jet fuel.

butyl nonanoate. See butyl pelargonate.

butyl octadecanoate. See butyl stearate.

butyl octyl phthalate.
$C_4H_9OOCC_6H_4COOC_8H_{17}$.

Properties: Water-white liquid, mild characteristic odor, d 0.991–0.997 (20/20C), saponification number 298–308. Miscible with most organic solvents. Combustible.

Use: Plasticizer for vinyl resins.

butyl octadecanoate. See butyl stearate.

butyl oleate. CAS: 142-77-8.
$CH_3(CH_2)_7CH:CH(CH_2)_7COOC_4H_9$.

Properties: Light-colored, oleaginous liquid, mild odor, insoluble in water, miscible with alcohol, ether, vegetable and mineral oils, d 0.873 (20/20C), iodine value 76.8, fp opaque at 12C, solid at −26.4C, wt/gal 7.26 lb (20C), boiling range 173–227C (2 mm), flash p 356F (180C). Combustible.

Derivation: Alcoholysis of olein or esterification of oleic acid with butanol.

Use: Plasticizer particularly for PVC, solvent, lubricant, water-resisting agent, coating compositions, polishes, water-proofing compounds.

butyl "Oxitol."[125] TM for ethylene glycol monobutyl ether.

Hazard: Slightly irritant to skin and eyes.

Use: Solvent in various types of surface coating formulations to improve gloss and leveling, component of hydraulic fluids, coupling agent in various types of cleaning and cutting oils.

butylparaben. (n-butyl hydroxybenzoate).
$C_{11}H_{14}O_3$.

Properties: Finely divided solid, mp 68C, soluble in acetone, alcohol, and propylene glycol. Calcium and magnesium salts are available.

Use: Pharmaceutical preservative, fungistat. See also "Parabens."

n-butyl pelargonate. (n-butyl nonanoate).
$C_4H_9OOCC_8H_{17}$ Liquid with fruity odor. Combustible. D 0.865 (15.5/15.5C), bp 270C.

Use: In flavors and perfumes, as chemical intermediate.

sec-butyl pelargonate.
$C_2H_5CH(CH_3)OOC_8H_{17}$.

Properties: Liquid, d 0.8608 (20/4C), bp 123C (15 mm), combustible, refr index 1.4220. Shows optical activity.

Used as chemical intermediate.

tert-butyl peracetate. See tert-butyl peroxyacetate.

tert-butyl perbenzoate. See tert-butyl peroxybenzoate.

tert-butyl perisobutyrate. See tert-buty peroxyisobutyrate.

tert-butyl permaleic acid. See tert-butyl peroxymaleic acid.

tert-butyl peroctoate. See tert-butyl peroxy(2-ethylhexanoate).

tert-butyl peroxide. See di-tert-butyl peroxide.

tert-butyl peroxyacetate. (tert-butyl peracetate). CAS: 107-71-1. $(CH_3)_3COOOCCH_3$.

Flash p less than 80F (26.6C) (COC), d 0.923.

Grade: Available as a 72–76% solution in benzene.

Hazard: Flammable, dangerous fire risk. Oxidizer.

Use: Polymerization initiator for vinyl monomers, in manufacture of polyethylene and polystyrene.

tert-butyl peroxybenzoate. (tert-butyl perbenzoate). CAS: 614-45-9.
$(CH_3)_3COOOCC_6H_5$.

Properties: Colorless liquid with mild aromatic odor, d 1.04 (25/25C), fp 8.5C, vap press 0.33 mm (50C), flash p 200F (93.3C). Soluble in alcohols, esters, ethers, ketones, insoluble in water.

Grade: 98% min.

Hazard: Oxidizing material, do not store near combustible materials.

Use: Polymerization initiator for polyethylene, polystyrene, polyacrylates and polyesters, chemical intermediate.

tert-butyl peroxy-2-ethylhexanoate. (tert-butyl peroctoate). CAS: 62695-55-0.
$(CH_3)_3COOOCCH(C_2H_5)C_4H_9$.

Properties: Colorless liquid with a faint odor, d 0.895 (25/25C), fp below −30C, refr index 1.426 (25C), decomposes at 89C, flash p 190F (87.7C), insoluble in water, miscible with most organic solvents.

Hazard: Oxidizing material. Do not store near combustibles.

tert-butyl peroxyisobutyrate. (tert-butyl perisobutyrate). CAS: 109-13-7.
$(CH_3)_3COOOCCH(CH_3)CH_3$.

Flash p below 80F (26.6C).

Grade: Available as a 72–75% solution in benzene.

Hazard: Flammable, dangerous fire risk. Oxidizing agent.

Use: Polymerization catalyst.

tert-butylperoxy isopropyl carbonate. (BPIC).
CAS: 2372-21-6. $(CH_3)_3COOOCOCH(CH_3)_2$.

Properties: Liquid, fp −3C, d 0.945, refr index 1.4050 (20C), flash p 112–118F (44–47C) (TOC). Almost insoluble in water, miscible with hydrocarbons, esters and ethers. Relatively stable under ordinary conditions. Flammable.

Grade: Technical (8.6% active oxygen).

Hazard: Moderate fire risk. Oxidizing agent.

Use: Polymerization initiator, cross-linking agent.

tert-butyl peroxymaleic acid. (tert-butyl permaleic acid). CAS: 1931-62-0.
$(CH_3)_3COOOCCH:CHCOOH$. An unsaturated peroxide.

Properties: White crystals; mp 114–116C (decomposes); slightly soluble in water, cool 5% alkaline solution, and alcohols; moderately soluble in oxygenated organic solvents, polyester monomers; slightly soluble in naphtha, carbon tetrachloride, and chloroform; insoluble in benzene.

Grade: 95% pure.

Hazard: Oxidizing agent. Do not store near combustible materials.

Use: Polymerization catalyst, bleaching, pharmaceuticals.

tert-butyl peroxyphthalic acid. (tert-butyl perphthalic acid). $(CH_3)_3COOOCC_6H_4COOH$.

Properties: White crystals; mp 96–99C; insoluble in water; soluble in cool 5% alkaline solutions

and in alcohols; moderately soluble in oxygenated organic solvents, chlorinated hydrocarbons, polyester monomers; slightly soluble in petroleum hydrocarbons.

Grade: 95% pure.

Hazard: Oxidizing agent. Do not store near combustible materials.

Use: Polymerization catalyst and oxidizing agent.

tert-butyl peroxypivalate. CAS: 927-07-1.
$(CH_3)_3COOOCC(CH_3)_3$.

Properties: Colorless liquid, d 0.854 (25/25C), solidifies below -19C, refr index 1.410 (25C), decomposes at 70C, flash p 155–160F (68–71C). Insoluble in water and ethylene glycol, soluble in most organic solvents.

Grade: Available as 75% soluble in mineral spirits.

Hazard: (Solution) Flammable, dangerous fire risk. Oxidizing agent. May explode on heating.

Use: Polymerization initiator.

tert-butyl perphthalic acid. See tert-butyl peroxyphthalic acid.

o-sec-butylphenol. CAS: 89-72-5.
$C_2H_5(CH_3)CHC_6H_4OH$.

Properties: A slightly volatile liquid. Mw 150.22, bp 226–228C, flash p 225F, d 0.891. Insoluble in water, slightly soluble in alcohol, ether, and alkalies. Combustible.

Hazard: TLV: 5 ppm. Skin and eye irritant.

Use: Chemical intermediate in preparation of resins, plasticizers, surface active agents.

o-tert-butylphenol. $(CH_3)_3CC_6H_4OH$.

Properties: Light yellow liquid, fp −7C, d 0.982 (20C), bp 224C, flash p 230F (110C) (OC). Soluble in isopentane, toluene, and ethanol; insoluble in water. Combustible.

Hazard: Toxic by ingestion, moderate irritant to eyes and skin.

Use: Chemical intermediate for synthetic resins, plasticizers, surface-active agents, perfumes and other products, a permissible antioxidant for aviation gasoline (ASTM D910-64T).

p-tert-butylphenol. CAS: 98-54-4.
$(CH_3)_3CC_6H_4OH$.

Properties: White crystals with a distinctive odor, d (crystals) 1.03, d (molten) 0.908 (114/4C), bp 239C, mp 100C. Combustible. Insoluble in water, soluble in alcohol and ether.

Derivation: Catalytic alkylation of phenol with olefins.

Hazard: Irritant to eyes and skin.

Use: Plasticizer for cellulose acetate, intermediate for antioxidants, special starches, oil soluble phenolic resins, pour-point depressors and emulsion breakers for petroleum oils and some plastics, synthetic lubricants, insecticides, industrial odorants, motor oil additives.

n-butyl phenylacetate. $C_4H_9OOCCH_2C_6H_5$.
Properties: Colorless liquid with rose-honey odor, d 0.991–0.994 (25/25C), bp 135–141C (18 mm), refr index 1.488–1.490 (20C). Soluble in 2 volumes 80% alcohol. Combustible. Made synthetically.
Grade: 98% min.
Use: Perfumes, flavoring.

n-butyl phenyl ether. $C_4H_9OC_6H_5$.
Properties: Liquid, d 0.929 (20C), boiling range 202–212C, water-white, aromatic odor, flash p 180F (82C). Combustible.
Hazard: Toxic by ingestion.

4-tert-butylphenyl salicylate.
$(CH_3)_3CC_6H_4OOCC_6H_4OH$.
Properties: Off-white, odorless crystals, mp 62–64C, soluble in alcohol, ethyl acetate, toluene, insoluble in water.
Use: Light absorber best at 2900–3300 Å.

n-butylphosphoric acid. See n-butyl acid phosphate.

n-butyl phthalate. See dibutyl phthalate.

butyl phthalylbutyl glycolate.
$C_4H_9OOCC_6H_4COOCH_2COOC_4H_9$.
Properties: Colorless, odorless liquid; d 1.093–1.103 (25/25C); bp 219C (5 mm); solidifies below −35C; darkens on heating above 290C; flash p 390F (199C) (OC). Combustible. Insoluble in water, extremely light-stable.
Use: Plasticizer for polyvinyl chloride. FDA-approved for use in vinyl food wrappings.

n-butyl propionate. CAS: 590-01-2.
$C_2H_5CO_2C_4H_9$.
Properties: Water-white liquid, apple-like odor, soluble in alcohol and ether, miscible with all coaltar and petroleum distillates, very slightly soluble in water. D 0.875 (20C), 0.874 (15.5C), wt/gal 7.3 lb, bp 146C (commercial grades boil over a range of 130–150C due to presence of butyl alcohol and esters), fp −89C, flash p 90F (32.2C), autoign temperature 800F (426C).
Derivation: Esterification of propionic acid with butyl alcohol and sulfuric acid catalyst.
Grade: Technical (85–90% to 95% ester content).
Hazard: Skin and eye irritant. Flammable, moderate fire risk.
Use: Solvent for nitrocellulose, retarder in lacquer thinner, ingredient of perfumes, flavors.

butyl ricinoleate. $C_{17}H_{32}(OH)COOC_4H_9$.
Properties: Yellow to colorless oleaginous liquid, soluble in alcohol and ether, insoluble in water,

d 0.916 (20/20C), bp approximately 275C (13 mm), flash p 220F (104.4C). Saybolt viscosity 112 (100F), fp indefinite, slightly opaque at −30C, and very viscous at −50C, wt/gal 7.62 lb (20C). Combustible.
Derivation: Castor oil and butyl alcohol.
Use: Plasticizer, lubricant.

butyl rubber. A copolymer of isobutylene (97%) and isoprene (3%) Polymerized below −95C with aluminum chloride catalyst.
Properties: D 0.92. Vulcanizates have tensile strength up to 2000 psi (unreinforced) and 3000 psi (reinforced). Service temperature range −55 to +204C. Good abrasion resistance, excellent impermeability to gases, high dielectric constant, excellent resistance to aging and sunlight, superior shock absorbing and vibration damping qualities. Resistance to oils and greases only fair. Will support combustion
Grade: Stabilized, latex, chlorine-containing elastomer, low mw (liquid).
Use: Tire carcases and linings especially for tractors and other out-size vehicles, electric wire insulation, encapsulating compounds, steam hose and other mechanical rubber goods, pond and reservoir sealant. Latex is used for paper coating, textile and leather finishing, adhesive formulations, air bags, for tire vulcanization, self-curing cements, pressure sensitive adhesives, tire cord dips, sealants.

butyl sebacate. See dibutyl sebacate.

butyl sorbate. $C_4H_9OOCC_5H_7$.   An insect attractant for the European chafer.

n-butylstannic acid. $[(C_4H_9)Sn(OH)_3]$.
Properties: White, infusible, and insoluble free-flowing powder. Exists as a polymer of undetermined chemical structure. Further dehydration results in polymers having mw's of 1000–5000 which are soluble in organic solvents.
Hazard: Toxic by ingestion and skin absorption. TLV (as Sn): 0.1 mg/m³ of air.
Use: Polymerization catalyst, antioxidant and heat stabilizer for PVC (not approved by FDA for food containers), electrically conducting tin oxide coatings alkaline earth metal phosphates in fluorescent light bulbs, intermediate for silicones.

butyl stearamide. $C_{17}H_{35}CONHC_4H_9$.
Properties: Light straw-colored liquid, d 0.869 (20/20C), boiling range 195–200C (2 mm), flash p 430F (221C), amide odor. Combustible.
Use: Plasticizer and intermediate for the synthesis of insecticides surface-active agents, pharmaceuticals and textile assistants.

butyl stearate. (butyl octadecanoate). $C_{17}H_{35}COOC_4H_9$.
Properties: Colorless, stable, oleaginous liquid solidifying at approximately 19C. Practically odorless, sometimes with faint fatty odor, d 0.855–0.860 (25/25C), mp 19.5–20C, flash p 320F (160C) (CC). Combustible. Wt/gal 7.14lb (20C), refr index 1.4430 (20C), bp 350C. Miscible with mineral and vegetable oils, soluble in alcohol and ether, insoluble in water.
Derivation: Alcoholysis of stearin or esterification of stearic acid with butanol.
Grade: Technical, cosmetic, chemically pure.
Use: Ingredient of polishes, special lubricants and coatings, lubricants for metals and in textile and molding industries, in wax polishes as dye solvent, plasticizer for laminated fiber products, rubber hydrochloride, chlorinated rubber and cable lacquers, carbon paper and inks, emollient in cosmetic and pharmaceutical products, lipsticks, dampproofer for concrete, flavoring.

butyl sulfide. See dibutyl sulfide.

4-tert-butyl-o-thiocresol. (2-methyl-4-tert-butyl-thio-phenol). $(CH_3)_3CC_6H_3(CH_3)SH$.
Properties: Colorless liquid with mild (non-mercaptan) odor, d 0.983 (25C), fp −4C, bp 250C, refr index 1.546 (25C). Insoluble in water, soluble in hydrocarbons. Combustible.
Grade: Available as 98% pure, supplied under nitrogen atmosphere, as 55% solution in hydrocarbon.
Use: Peptizer for rubbers, polymer modifier, lube oil additive.

4-tert-butylthiophenol. $(CH_3)_3CC_6H_4SH$.
Properties: Colorless liquid with mild (non-mercaptan) odor, d 0.986 (25C), fp −11C, bp 238C, refr index 1.546 (25C). Insoluble in water, soluble in hydrocarbons.
Grade: 98%, supplied under nitrogen atmosphere.
Use: Lube oil additives, polymer modifiers, antioxidants, dyes.

n-butyltin trichloride. $C_4H_9SnCl_3$.
Properties: Colorless liquid, fumes in contact with air, d 1.71 (25/4C), bp 102C (12 mm), refr index 1.5190 (25C). Soluble in organic solvents, sparingly soluble in water with partial hydrolysis.
Hazard: Toxic by ingestion, strong irritant to skin. Avoid exposure to liquids or vapors. TLV (as Sn): 0.1 mg/m³ of air.
Use: Intermediate, catalyst, stabilizer.

butyl titanate. See tetrabutyl titanate.

p-tert-butyltoluene. (1-methyl-4-tert-butyl-benzene). CAS: 98-51-1. $(CH_3)_3CC_6H_4CH_3$.
Properties: Colorless liquid, d 0.857–0.863 (20/20C), bp 192.8C. Combustible. Insoluble in water.
Grade: Technical.
Hazard: Toxic by inhalation, ingestion, and skin absorption. TLV: 10 ppm in air.
Use: Solvent, intermediate.

n-butyltrichlorosilane. $C_4H_9SiCl_3$.
Properties: Colorless liquid, bp 142C, d 1.1608 (25/25C), refr index 1.4363 (25C), flash p 126F (52C) (COC). Readily hydrolyzed with liberation of hydrogen chloride. Soluble in benzene, ether, heptane.
Derivation: Grignard reaction of silicon tetrachloride.
Grade: Technical, 95%.
Hazard: Corrosive to skin and tissue. Moderate fire risk.
Use: Intermediate for silicones.

tert-butyltrimethylmethane. See hexamethyl-ethane.

N-n-butylurea. CAS: 592-31-4. $C_4H_9HNCONH_2$.
Properties: White solid, decomposes on heating, odorless, mp 96C, soluble in water, alcohol, and ether.

n-butyl vinyl ether. See vinyl-n-butyl ether.

"Butyl Zimate"[69]. TM for zinc dibutyldithiocarbamate.

1-butyne. See ethylacetylene.

2-butyne. See crotonylene.

butynediol. CAS: 110-65-6. $HOCH_2C:CCH_2OH$.
Properties: White, orthorhombic crystals; mp 58C; bp 238C; refr index 1.450 (25C); soluble in water, aqueous acids, alcohol, and acetone. Insoluble in ether and benzene. Combustible.
Derivation: High-pressure synthesis from acetylene and formaldehyde.
Grade: Crystalline solid, 97%, aqueous soluble 35%.
Hazard: Toxic by ingestion. May explode on contamination with mercury salts, strong acids and alkaline earth hydroxides and halides at high temperatures.
Use: Intermediate, corrosion inhibitor, electroplating brightener, defoliant, polymerization accelerator, stabilizer for chlorinated hydrocarbons, cosolvent for paint and varnish removal.

3-butyl-1-ol. (β-ethynyl ethanol).
HC:CCH₂CH₂OH.
Properties: Water-white liquid with characteristic odor, d 0.9257 (20/4C), refr index 1.4409 (20C), bp 128.9C, fp −63.6C. Combustible.
Use: Preparation of perfume bases, acetylenic esters, plastics, plasticizers, pharmaceuticals, wetting agents, medicinals, and organic synthesis.

butyraldehyde. (butaldehyde; n-butanal; n-butyl-aldehyde; butyric aldehyde).
CAS: 123-72-8.   CH₃(CH₂)₂CHO.
Properties: Water-white liquid, characteristic pungent aldehyde odor, d 0.8048 (20/20C), bp 75.7C, vap press 91.5 mm (20C), flash p 20F (−6.6C), wt/gal 6.7 lb (20C), coefficient of expansion 0.00114 (20C), fp −99C, viscosity 0.043 poise (20C), autoign temperature 446F (230C). Slightly soluble in water, soluble in alcohol and ether.
Derivation: (a) Oxo process (b) dehydrogenating butanol vapors over a catalyst the butyraldehyde being separated by distillation; (c) partial reduction of crotonaldehyde.
Grade: Technical (93% min).
Hazard: Flammable, dangerous fire risk.
Use: Plasticizers, rubber accelerators, solvents, high polymers.

butyric acid. (n-butyric acid; butanoic acid; ethylacetic acid; propylformic acid).
CAS: 107-92-6.   CH₃CH₂CH₂COOH.
Properties: Colorless liquid, penetrating and obnoxious odor, refr index 1.3981 (20C), d 0.9583 (20/4C), fp −5.0 to −8C, bp 163.5C (757 mm) and 75C (25 mm), vap press 0.84 mm (20C), autoign temperature 846F (452C). Miscible in water, alcohol, and ether. Combustible.
Derivation: Occurs as glyceride in animal milk fats. Produced as a by-product in hydrocarbon synthesis, by oxidation of butyraldehyde, and by butyric fermentation of molasses or starch.
Grade: 90%, 95%, 99%, edible, synthetic, reagent, technical, FCC.
Hazard: Strong irritant to skin and tissue.
Use: Synthesis of butyrate ester perfume and flavor ingredients, pharmaceuticals, deliming agent, disinfectants, emulsifying agents, sweetening gasolines.

butyric alcohol.   See n-butyl alcohol.

butyric aldehyde.   See butyraldehyde.

butyric anhydride.   CAS: 106-31-0.
(CH₃CH₂CH₂CO)₂O.
Properties: Water-white liquid, hydrolyzes to butyric acid, d 0.9681 (20/20C), fp −75C, bp

199.5C, vap press 0.3 mm (20C), flash p 190F (87.7C), wt/gal 8.1 lb (20C). Combustible.
Grade: Technical, 98%.
Use: Manufacture of butyrates, drugs and tanning agents.

butyrin.   See glyceryl tributyrate.

butyrolactam.   See 2-pyrrolidone.

butyrolactone. (γ-butyrolactone).
CAS: 96-48-0.   OCH₂CH₂CH₂CO.

$$
\begin{array}{ccc}
CH_2 \cdot CH_2 \cdot CH_2 \cdot \\
| \qquad\qquad | \\
O \text{————} CO
\end{array}
$$

Properties: Colorless liquid with pleasant odor, bp 204C, fp −44C, d 1.144, flash p 209F (98.3C) (OC). Miscible with water, alcohol, and ether. Combustible.
Derivation: High-pressure synthesis from acetylene and formaldehyde.
Grade: Technical.
Hazard: Toxic by ingestion.
Use: Intermediate for synthesis of butyric acid compounds, polyvinylpyrrolidone, methionine, solvent for acrylate and styrene polymers, ingredient of paint removers and textile assistants.

butyrone.   See dipropyl ketone.

butyronitrile. (propyl cyanide; butanenitrile).
CAS: 109-74-0.   CH₃(CH₂)₂CN.
Properties: Colorless liquid, d 0.796 (15C), fp −112.6C, bp 116–117C, flash p 79F (26.1C) (OC). Slightly soluble in water. Soluble in alcohol and ether.
Hazard: Flammable, dangerous fire risk.
Use: Basic material in industrial, chemical, and pharmaceutical intermediates and products, poultry medicines.

butyroyl chloride. (butyryl chloride; butanoyl chloride).   CAS: 141-75-3.   C₃H₇COCl.
Properties: Colorless liquid with pungent acid chloride odor. Reacts with alcohol and water, miscible with ether, fp −89C, distillation range 100–110C, d 1.028 (15C), refr index 1.4121 (20C).
Hazard: Toxic by ingestion and inhalation, strong irritant to tissue.
Use: Organic synthesis.

butyryl chloride.   See butyroyl chloride.

"Buxine"[227].   TM for amylcinnamic aldehyde.

"Buxinol."[227]   TM for n-amylcinnamic alcohol.

BVE. Abbreviation for butyl vinyl ether. See vinyl-n-butyl ether.

"B-X-A."[248] TM for a diarylamine-ketone-aldehyde reaction product.

Properties: Brown powder, d 1.10, melting range 85–95C, store in a cool place. Soluble in acetone, benzene, and ethylene dichloride; insoluble in water and gasoline.

Use: Antioxidant for rubber and nylon.

BZ. A nonlethal gas which causes temporary disability. It is a derivative of lysergic acid.

C

C. Symbol for carbon.

^{14}C. (carbon-14). The naturally occurring radioactive isotope of carbon used in chemical dating, tracer studies, etc.

Ca. Symbol for calcium.

CA. Abbreviation for cellulose acetate and cortisone acetate, also for controlled atmosphere.

C_3A. Abbreviation for tricalcium aluminate as used in cement.
See cement, Portland.

CAB. Abbreviation for cellulose acetate butyrate.

"Cab-O-Sil."[275] TM for colloidal silica particles sintered together in chain-like formations. Surface area ranges from 50 to 400 m^2/g, depending on grade.
Grade: Standard M-5, L-5, SD-20.
Use: Thickening and emulsifying agent for oil/water systems, drilling muds, cattle-feed supplements, tile cleaners, disperson of oil slicks on sea water, plastics, solar-heated ceiling tiles.

"Cab-O-Sperse."[275] TM for aqueous dispersions of pyrogenic silica for use in the paper and textile industries.

cacao butter. (cocoa butter). See theobroma oil.

C acid. (2-naphthylamine-4,8-disulfonic acid). $C_{10}H_5(NH)_2(SO_3H)_2$.
Properties: White, crystalline solid; slightly soluble in water.
Derivation: Reduction of 2-nitronaphthalene-4,8-disulfonic acid. The sodium salt is recrystallized from water.
Use: Azo dye intermediate.

cacodylic acid. (dimethylarsinic acid). $(CH_3)_2AsOOH$.
Properties: Colorless, odorless, deliquescent crystals; mp 200C; soluble in water, alcohol, and acetic acid; insoluble in ether.
Derivation: By distilling a mixture of arsenic trioxide and potassium acetate, and oxidizing the resulting product with mercuric oxide.
Hazard: Toxic by ingestion.
Use: Herbicide especially for control of Johnson grass on cotton, soil sterilant, chemical warfare, timber thinning.

cactinomycin. USAN name for an antibiotic produced from *Streptomyces* which is 10% dactinomycin and 90% two kinds of actinomycin C.

"Cadalume" L.[238] TM for a bright cadmium electroplating process for a protective coating on iron and steel. Materials used are CdO, NaCN, and addition agents.

"Cadalyte."[28] TM for a series of compounds for cadmium electroplating.

cadaverine. (1,5-diaminopentane; pentamethylenediamine). $NH_2(CH_2)_5NH_2$. A ptomaine formed in the decay of animal proteins after death, also made synthetically.
Properties: Syrupy, colorless, fuming, odorous liquid; mp 9C; bp 178–179C; soluble in water and alcohol, slightly soluble in ether.
Hazard: Toxic by ingestion, absorbed by skin, a skin and eye irritant.
Use: Preparation of high polymers, intermediate, biological research.

"caddy."[48] TM for a liquid cadmium fungicide used on turf grass.
Hazard: See cadmium.

Cade oil. See juniper tar oil.

cadiene. See sesquiterpene.

"Cadmate."[69] TM for cadmium diethyldithiocarbamate.
Hazard: See cadmium.

"Cadminate."[329] TM for a turf fungicide containing 60% cadmium succinate and 40% inert matter.
Hazard: See cadmium.

cadmium. CAS: 7440-43-9. Cd.
Metallic element of atomic number 48, group IIB of the periodic table. Aw 112.40. Valence 2. There are eight stable isotopes.
Properties: Soft, blue-white, malleable metal or grayish-white powder. Tarnishes in moist air, corrosion resistance poor in industrial atmospheres, becomes brittle at 80C. Resistant to alkalies, high neutron absorber. D 8.642, mp

320.9C, bp 767C, refr index 1.13, Mohs hardness 2.0. Soluble in acids, especially nitric, and in ammonium nitrate solution. Lowers melting point of certain alloys when used in low percentages. Combustible.

Occurrence: A greenockite (cadmium sulfide) ore containing zinc sulfide also with lead and copper ores containing zinc. Canada, central and western US, Peru, Australia, Mexico, Zaire.

Derivations: (1) Dust or fume from roasting zinc ores is collected, mixed with coal or coke and sodium or zinc chloride, and sintered. The cadmium fume is collected in an electrostatic precipitator, leached, fractionally precipitated, and distilled. (2) By direct distillation from cadmium-bearing zinc (3) By recovery from electrolytic zinc process (approximately 40%).

Hazard: Flammable in powder form. Toxic by inhalation of dust or fume. A carcinogen. Cadmium plating of food and beverage containers has resulted in a number of outbreaks of gastroenteritis (food poisoning). Soluble compounds of cadmium are highly toxic; however, ingestion usually induces a strong emetic action which minimizes the risk of fatal poisoning. Use as fungicide may be restricted. TLV: (Dust and soluble compounds) 0.05 mg/m³ of air; (oxide fume as Cd) ceiling 0.05 mg/m³ of air.

Grade: Technical, powder, pure sticks, ingots, slabs, high-purity crystals (less than 10 ppm impurities).

Use: Electrodeposited and dipped coatings on metals, bearing and low-melting alloys, brazing alloys, fire protection systems, nickel-cadmium storage batteries, power transmission wire, TV phosphors, basis of pigments used in ceramic glazes, machinery enamels, baking enamels, Weston standard cell control of atomic fission in nuclear reactors, fungicide, photography and lithography, selenium rectifiers, electrodes for cadmium-vapor lamps and photoelectric cells.

cadmium acetate. (a) $Cd(OOCCH_3)_2 \cdot 3HOH$; (b) $Cd(OOCCH_3)_2$.
Properties: Colorless crystals, soluble in water and alcohols. (a) d 2.01, mp loses water at 130C; (b) d 2.341, mp 256C.
Derivation: Interaction of acetic acid and cadmium oxide.
Hazard: See cadmium.
Use: Ceramics (iridescent glazes), manufacture of acetates, assistant in dyeing and printing textiles, electroplating baths, laboratory reagent.

cadmium ammonium bromide. (ammonium-cadmium bromide). $CdBr_2 \cdot 4NH_4Br$.
Properties: Colorless crystals. Soluble in alcohol and water.
Hazard: See cadmium.

cadmium antimonide. A semiconductor used in thermoelectric devices.
Hazard: See cadmium and antimony.

cadmium-base Babbitt. See Babbitt metal.

cadmium borotungstate.
$Cd_5(BW_{12}O_{40}) \cdot 18HOH$.
Properties: Yellow, heavy crystals; mp 75C; soluble in water. The solution is yellow to light brown.
Grade: Technical.
Hazard: See cadmium.
Use: Separating minerals.

cadmium bromate. $Cd(BrO_3)_2 \cdot HOH$.
Properties: White crystals or crystalline powder, d 3.758, mp decomposes. Soluble in water, insoluble in alcohol.
Derivation: By adding cadmium sulfate to a solution of barium bromate.
Hazard: Strong oxidizer, dangerous in contact with organics. Highly toxic, irritant, oxidizer.
Use: Analytical reagent.

cadmium bromide. CAS: 7789-42-6. $CdBr_2$ or $CdBr_2 \cdot 4HOH$.
Properties: White to yellowish efflorescent crystalline, d 5.192, mp (anhydrous) 568C, bp 863C, soluble in water, acetone, alcohol, and acids.
Derivation: By heating cadmium in bromine vapor.
Grade: Technical, reagent.
Hazard: See cadmium.
Use: Photography, process engraving, lithography.

cadmium carbonate. CAS: 513-78-0. $CdCO_3$.
Properties: White, amorphous powder; d 4.258; decomposes at less than 500C; soluble in dilute acids and in concentrated solution of ammonium salts; insoluble in water.
Hazard: See cadmium.
Grade: Reagent.

cadmium chlorate. $Cd(ClO_3)_2 \cdot 2HOH$.
Properties: Colorless, prismatic crystals. Hygroscopic, d 2.28 (18C), mp 80C, soluble in alcohol, water, and acetone.
Grade: Technical.
Hazard: Dangerous in contact with organic materials.

cadmium chloride. CAS: 10108-64-2.
(a) $CdCl_2$; (b) $CdCl_2 \cdot 2.5HOH$.
Properties: Small, white, odorless crystals; d (a) 4.05, (b) 3.327; mp (a) 568C; bp (a) 960C; soluble in water and acetone.

Derivation: Action of hydrochloric acid on cadmium with subsequent crystallization.

Grade: Technical, reagent.

Use: Preparation of cadmium sulfide, analytical chemistry, photography, dyeing and calico printing ingredient of electroplating baths, addition to tinning solution, manufacture of special mirrors, manufacture of cadmium yellow.

cadmium cyanide. $Cd(CN)_2$.   Obtained as a white precipitate when potassium or sodium cyanide is added to a concentrated solution of a cadmium salt. A complex ion is formed when it is dissolved in an excess of the precipitating agent, and a solution of this complex is used as an electrolyte for electrodeposition of cadmium.

Hazard: See cadmium and cyanide.

Use: Electroplating copper.

cadmium diethyldithiocarbamate.
$Cd[SC(S)N(C_2H_5)_2]_2$.
Properties: White to cream colored rods, d 1.39, melting range 68–76C, mostly soluble in benzene, carbon disulfide, chloroform, insoluble in water and gasoline.

Hazard: See cadmium.

Use: Accelerator for butyl rubber.

cadmium fluoride.   CAS: 7790-79-6. CdF_2.
Available as pure crystals, 99.89%, d 6.6, mp approximately 1110C, soluble in water and acids, insoluble in alkalies.

Hazard: See cadmium.

Use: Electronic and optical applications, high-temperature dry-film lubricants, starting material for crystals for lasers, phosphors.

cadmium hydroxide.   (cadmium hydrate).
$Cd(OH)_2$.
Properties: White, amorphous powder; d 4.79; mp loses HOH (300C); soluble in ammonium hydroxide and in dilute acids; insoluble in water and alkalies; absorbs carbon dioxide from air.

Derivation: By the action of sodium hydroxide on a cadmium salt solution.

Grade: Technical, CP.

Hazard: See cadmium.

Use: Cadmium salts, cadmium plating, storage battery electrodes.

cadmium iodate. $Cd(IO_3)_2$.
Properties: Fine, white powder; d 6.48; mp decomposes; slightly soluble in water; soluble in nitric acid or ammonium hydroxide.

Grade: Technical.

Hazard: Fire risk in contact with organic materials.

Use: Oxidizing agent.

cadmium iodide. CdI_2.
Properties: White flakes or crystals, odorless, becomes yellow on exposure to air and light. Occurs in two allotropic forms, (alpha) d 5.67, and (beta) 5.30; mp (alpha) 388C, (beta) 404C; bp (alpha) 796C; soluble in water, alcohol, ether, acetone, ammonia, and acids.

Derivation: By the action of hydriodic acid on cadmium oxide.

Hazard: See cadmium.

Use: Photography, process engraving and lithography, analytical chemistry, electroplating, lubricants, phosphors, nematocide.

cadmium molybdate. $CdMoO_4$.
Properties: Yellow crystals, d 5.347, mp approximately 1250C, slightly soluble in water, soluble in acids.

Grade: Technical, crystals, 99.98% pure.

Hazard: See cadmium.

Use: Electronic and optical applications.

cadmium nitrate.   CAS: 10325-94-7.
(a) $Cd(NO_3)_2 \cdot 4HOH$; (b) $Cd(NO_3)_2$.
Properties: White, amorphous pieces or hygroscopic needles; soluble in water, ammonia, and alcohol; (a) d 2.455, mp 59.5C, bp 132C; (b) mp 350C.

Derivation: Action of nitric acid on cadmium or cadmium oxide and crystallization.

Grade: Technical, reagent.

Hazard: Dangerous fire and explosion hazard.

Use: Cadmium salts, photographic emulsions, coloring glass and porcelain, laboratory reagent, cadmium salts.

cadmium oxalate. $Cd(COO)_2 \cdot 3HOH$.
Properties: White, amorphous powder; soluble in dilute acids, ammonium hydroxide; insoluble in alcohol and water; d 3.32 (dehydrated); mp decomposes at 340C.

cadmium oxide.   CAS: 1306-19-0.   CdO
(forms a and b).
Properties: (a) Colorless, amorphous powder; d 6.95; (b) brown or red crystals, d 8.15. Both decompose on heating at 900C. The crystals are soluble in acids and alkalies, and insoluble in water.

Derivation: Cadmium metal is distilled in a retort, the vapor reacted with air and the oxide collected in a baghouse.

Hazard: Inhalation of vapor or fume may be fatal. A carcinogen (OSHA). TLV: (fume as cadmium) CL 0.05 mg/m³ of air.

Use: Cadmium plating baths, electrodes for storage batteries, cadmium salts, catalyst, ceramic glazes, phosphors, nematocide.

cadmium pigment.   A family of pigments based on cadmium sulfide or cadmium selenide used chiefly where high color retention is required. They are light-fast and have good alkali resistance. Red shades are obtained with cadmium selenide, yellow with cadmium sulfide. Used in paints and high-gloss baking enamels, often extended with barium sulfate and are then called cadmium lithopone.

cadmium potassium iodide. $CdI_2 \cdot 2KI \cdot 2HOH$.
Properties: White powder, becomes yellowish with age; deliquescent; soluble in water, alcohol, ether, and acid; d 3.359; mp 76C (decomposes).
Derivation: By combining cadmium iodide and potassium iodide in solution in proportion of their combining weights, and subsequent crystallization.
Hazard: See cadmium.
Use: Analytical chemistry, medicine.

cadmium propionate. $Cd(OOCC_2H_5)_2$.
A solid used in scintillation counters.

cadmium ricinoleate.
$Cd[CH_3(CH_2)_5CHOHCH_2CH:CH(CH_2)_7CO_2]_2$.
Properties: Odorless, fine, white powder derived from castor oil. Mp 104C, d 1.11.
Hazard: See cadmium.
Use: Soluble used to stabilize polyvinyl chloride and copolymers against light and heat.

cadmium selenide. CdSe.
Properties: Red powder, d 5.81 (15/4C), mp above 1350C, insoluble in water, stable at high temperature.
Use: Red pigment, semiconductors, phosphors, photoelectric cells.

cadmium selenide lithopone. See cadmium pigment.

cadmium stearate. Used as a lubricant and stabilizer in plastics.
Hazard: See cadmium.

cadmium succinate. $Cd(OOCCH_2)_2$.
A white powder, slightly soluble in water, insoluble in alcohol.
Hazard: See cadmium.
Use: Fungicide.

cadmium sulfate. CAS: 10124-36-4.
(a) $CdSO_4$; (b) $3CdSO_4 \cdot 8HOH$;
(c) $CdSO_4 \cdot 4HOH$.
Properties: Colorless, odorless crystals;, d (a) 4.69, (b) 3.09, (c) 3.05; mp (a) 1000C. Soluble in water, insoluble in alcohol.
Derivation: By the action of dilute sulfuric acid on cadmium or cadmium oxide.

Grade: Technical, CP.
Hazard: A carcinogen (OSHA).
Use: Pigments, fluorescent screens, electrolyte in Weston standard cell, electroplating.
See cadmium.

cadmium sulfide. (orange cadmium; see also cadmium pigment).   CAS: 1306-23-6.   CdS.
Properties: Yellow or brown powder, d 4.82, mp 1750C (100 atm), sublimes (in nitrogen) 980C. Insoluble in cold water, forms a colloid in hot water, soluble in acids and ammonia. Can be polished like a metal. It is an n-type semiconductor.
Derivation: (a) By passing hydrogen sulfide gas into a solution of a cadmium salt acidified with hydrochloric acid. The precipitate is filtered and dried. (b) Occurs naturally as greenockite.
Grade: Technical, ND, high purity (single crystals).
Hazard: A carcinogen (OSHA), highly toxic. See cadmium.
Use: Pigments and inks, ceramic glazes, pyrotechnics, phosphors, fluorescent screens, scintillation counters, rectifiers, photoconductor in xerography, transistors, photovoltaic cells, solar cells, catalyst in photodecomposition of hydrogen sulfide.

cadmium telluride. CAS: 1306-25-8.   CdTe.
Properties: Brownish-black, cubic crystals. Oxidizes on prolonged exposure to moist air. Insoluble in water and mineral acids except nitric, in which it is soluble with decomposition. Mp 1090C, d 6.2 (25/4C).
Derivation: Fusion of the elements, reaction of hydrogen telluride and cadmium chloride.
Grade: High purity crystals, 99.99+%.
Hazard: Toxic by inhalation.
Use: Semiconductors, phosphors.

cadmium tungstate. $CdWO_4$.
Properties: White or yellow crystals or powder. Soluble in ammonium hydroxide, alkali cyanides, almost insoluble in water.
Derivation: By the interaction of cadmium nitrate and ammonium tungstate.
Forms: Single crystal rods, broken crystals (crackle).
Hazard: Toxic by inhalation.
Use: Fluorescent paint, x-ray screens, scintillation counters, catalyst, phosphors.

"Cadmolith."[296]   TM for a series of yellow and red cadmium-lithopone pigments.
Hazard: See cadmium.
Use: Automotive finishes, textile coatings, printing inks, lacquer and rubber.

"Cadmopur."[470]   TM for pure red, orange, and yellow cadmium pigments.
Hazard: See cadmium.

"Cadox."[419]   TM for a series of organic peroxide catalysts.
Hazard: May be fire risk in contact with organic materials.

"Cadoxen."[50]   Brand name for cadmium ethylenediamine hydroxide complex, a colorless, stable cellulose solvent.
Hazard: See cadmium.

caesium.   See cesium.

C₄AF.   Abbreviation for tetracalcium aluminoferrate as used in cement.
See cement, Portland.

caffeine.   (theine; methyltheobromine; 1,3,7-trimethylxanthine).   CAS: 58-08-2.
$C_8H_{10}N_4O_2 \cdot HOH$

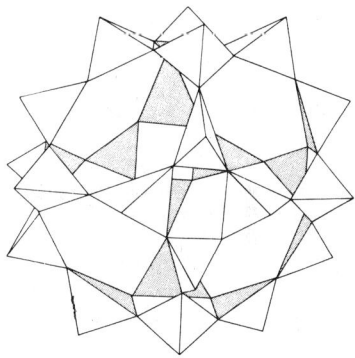

Properties: White, fleecy masses or long, flexible, silky crystals; an alkaloid; loses water at 80C. Efflorescent in air. Mp 236.8C, soluble in chloroform, slightly soluble in water and alcohol, very slightly soluble in ether. Odorless, bitter taste, solution neutral to litmus.
Derivation: By extraction of coffee beans, tea leaves, or kola nuts; also synthetically. Much of the caffeine of commerce is a by-product of decaffeinized coffee manufacture.
Method of purification: Recrystallization.
Grade: Technical, USP, FCC.
Hazard: One grain or more is toxic, 200 micrograms per mL has been found to inhibit activity of the enzyme DNA polymerase. Use in soft drinks not to exceed 0.02%.
Use: Beverages, medicine.

cage compound.   See inclusion complex, clathrate compound.

cage mill.   (bar disintegrator, squirrel-cage disintegrator).   A comminuting device which may consist of two rotating structures similar to water wheels, one fitting inside the other. They are provided with horizontal crossbars or breaker plates. The assembly is covered with a close-fitting housing. The two wheels or cages rotate at high speed in opposite directions on a horizontal axis. The material to be reduced is fed into the smaller cage from a hopper. It is ejected at speeds up to 12,000 ft/minute and is fragmented by contact with the bars. As the pieces are thrown back and forth within the cages they are disintegrated further by mutual impact. Some types of cage mills have only a single cage but others have more than two. They are used for size reduction of niter cake, fertilizers, coal, and other friable materials. Particle size can be varied by adjusting the space between the crossbars.

cage zeolite.   (sodalite).   A structure of sodium aluminosilicates often arranged in combined tetrahedra at the intersections of which are sodium atoms with oxygen atoms at the midpoints. Such sodalite units often combine to form supercages. They are common and extremely effective catalysts. Many zeolites occur naturally but synthetic types are tailor-made for special purposes and have particular catalytic functions.
See also zeolite.

"Cairox."[459]   TM for potassium permanganate.

cajeputene.   (dipentene;   dl-p-mentha-1,8-diene, inactive; limonene).   CAS: 138-86-3.
$C_{10}H_{16}$.
Properties: Bp 178C, flash p 109F (42C), refr index 1.4739 (20C), d 0.840.

Cake alum.   See aluminum sulfate.

calabarine.   See physostigmine.

"CAL."[323]   TM for a decolorizing carbon designed for use in fixed or moving beds. Total surface area (N₂, BET Method) 1000–1100 m²/g, particle size 12x40 mesh, apparent d 27.5 lb/cu ft. Used especially in beet sugar refining.

"Calade."[108]   TM for a powdered alkaline sodium hexametaphosphate compound containing wetting agents and an aluminum corrosion inhibitor.
Use: Detergent for dishwashing and general cleaning.

Caladryl.   Proprietary name for diphenhydramine, camphor, and calamine.
Use: Dermatological.

calamine.   (1) A hydrated zinc silicate containing 67.5% zinc oxide; d 3.5, Mohs hardness 4.5 to 5. Pyroelectric. Occurs in US and Europe. Source of metallic zinc. (2) Zinc oxide with low percentage of ferric oxide; zinc oxide must be 98%. Soluble in mineral acids, insoluble in water. Pharmaceutical preparation (USP).

calamus oil.
Properties: Yellow to brownish-yellow essential oil. D 0.959–0.970 (15C), refr index 1.503–1.510, saponification value 6–20. Slightly soluble in water.
Derivation: By steam distillation of calamus, the stem or root of the sweet flag. Chief known constituents, asarone (see 2,4,5-trimethoxy-1-propenylbenzene) and eugenol.
Use: Perfumery, flavoring agent.

calandria.   The heating unit of a vacuum-evaporating system.

calaverite. $AuTe_2$.   One of the gold telluride group of minerals. Corresponds to the same general formula as sylvanite and krennerite. Pale bronze-yellow color or tin-white, tarnishing to bronze yellow on exposure. Metallic luster. Contains 40–43% gold, 1–3% silver; d 9.0; Mohs hardness 2.5.
Occurrence: US (California, Colorado), Australia, Canada.
Use: Important source of gold.

"Calcene."[177]   TM for a specially prepared precipitated calcium carbonate for use in compounding rubber, paints, and plastics. The particles of TM-grade are coated with stearic acid to aid in dispersion. Grade NC is not coated.

calciferol.   See ergocalciferol.

calcimine.   (kalsomine).   Essentially chalk and glue in powdered form ready to mix with water. Used as temporary decoration for interior plaster walls. Will not withstand washing.

calcination.   Heating of a solid to a temperature below its melting point to bring about a state of thermal decomposition or a phase transition other than melting. Included are the following types of reactions: (1) thermal dissociation, including destructive distillation of organic compounds, e.g., concentration of aluminum by heating of bauxite; (2) polymorphic phase transitions, e.g., conversion of anatase to rutile form of TiO_2; (3) thermal recrystallization, e.g., devitrification of glass. Calcination is often used in the beneficiation of ores.
See also destructive distillation, pyrolysis.

calcine.   Radioactive waste prepared for disposal in the form of granular powder made by drying liquid wastes at high temperature (approximately 800C).

calcite. $CaCO_3$.   The most common form of natural calcium carbonate. Dogtooth spar, Iceland spar, nailhead spar, and satin spar are varieties of calcite. Essential ingredient of limestone, marble, and chalk.
Properties: Colorless, white, and various colored crystals, vitreous to earthy luster, good cleavage in three directions. May contain small amounts of magnesium, iron, manganese, and zinc. Reacts with acids to evolve carbon dioxide. D 2.72, Mohs hardness 2.
Use: Phosphor. Iceland spar is used in optical instruments.

calcitonin.   A thyroid hormone controlling the proportion of calcium in circulating blood. May be used in calcium balance control and possibly also in treatment of bone fractures, hypervitaminosis, and other calcium-related diseases. It is obtained in purified form from pig thyroid; has been made synthetically.

calcium.   CAS: 7440-70-2.   Ca.   Alkaline-earth element of atomic number 20, group IIA of the periodic system. Aw 40.08. Valence 2. Six stable isotopes.
Properties: Moderately soft, silver-white, crystalline metal, oxidizes in air to form adherent protective film, can be machined, extruded, or drawn. Soluble in acid. Decomposes water to liberate hydrogen. D 1.57; mp 845C; sublimes above mp in vacuum; bp 1480C; Brinell hardness 17.
Derivation: Electrolysis of fused calcium chloride, by thermal process under high vacuum from lime reduced with aluminum. Does not occur free in nature.
Forms: Crowns, nodules, ingots, crystals up to 99.9% pure.
Hazard: Evolves hydrogen with moisture. Flammable in finely divided state. Fire and explosion

hazard when heated or on contact with strong oxidizing agents.

Use: Alloying agent for aluminum, copper and lead; reducing agent for beryllium; deoxidizer for alloys. Dehydrating oils. Decarburizaton and desulfurization of iron and its alloys, getter in vacuum tubes. Separation of nitrogen from argon. Reducing agent in preparation of chromium metal powder, thorium, zirconium, and uranium. Fertilizer ingredient.

Note: Calcium is an essential component of bones, teeth, shells, and plant structures. It occurs in milk in trace amounts and is necessary in animal and human nutrition. Vitamin D aids in the deposition of calcium in bones.

calcium 45. Radioactive calcium of mass number 45.

Properties: Half-life 164 days; emits beta radiation.

Derivation: By reactor irradiation of calcium carbonate, by neutron bombardment of scandium or as a by-product of the irradiation of calcium nitrate for the preparation of C-14.

Forms available: Calcium chloride in hydrochloric acid solution and solid calcium carbonate.

Hazard: Dangerous radiation hazard, MPC in air 3×10^{-8} μCi/mL. This isotope is a boneseeker and may cause damage to the blood-forming organs.

Use: Research aid for studying water purification, calcium exchange in clays, detergency, surface wetting and other surface phenomena, calcium uptake and deposition in bone, soil characteristics as related to soil utilization of fertilizer and crop yield, diffusion of calcium in glass, etc.

calcium abietate. $(C_{20}H_{29}O_2)_2Ca$. Product of the action of lime on rosin or resin acids. See calcium resinate.

calcium acetate. (vinegar salts; gray acetate; lime acetate; calcium diacetate).
$Ca(CH_3COO)_2 \cdot HOH$.

Properties: Brown, gray, or white (when pure) powder, amorphous or crystalline; slightly bitter taste, slight odor of acetic acid; decomposes on heating. Soluble in water, slightly soluble in alcohol. Combustible.

Derivation: Action of pyroligneous acid on calcium hydroxide; the solution being filtered and evaporated to dryness yields gray acetate of lime.

Grade: Technical (80% basis), reagent, CP, pure, brown, gray, FCC.

Use: Manufacture of acetone, acetic acid, acetates, mordant in dyeing and printing of textiles, stabilizer in resins, additive to calcium soap lubricants, food additive (as antimold agent in bakery goods, in sausage casings), corrosion inhibitor.

calcium acetylsalicylate. (aspirin, soluble).
$(CH_3COOC_6H_4COO)_2 \cdot 2HOH$.

Properties: White powder. Aqueous solutions are unstable. Soluble in water.

Derivation: (a) Action of acetylsalicylic acid upon calcium carbonate in the presence of a small amount of water; (b) by passing carbon dioxide into an aqueous solution of calcium carbonate and acetylsalicyclic acid.

Use: Medicine (antipyretic).

calcium acid sulfite. See calcium hydrogen sulfite.

calcium acrylate. $(H_2C:CHCOO)_2Ca$.

Properties: Free-flowing, white powder; soluble in water; deliquescent. Solution polymerizes to form hydrophilic resin.

Use: Ion-exchange clay soil stabilizer, binder, and sealer (laboratory scale only).

calcium alginate.

Properties: White or cream-colored powder or filaments, grains or granules; slight odor and taste. Insoluble in water, insoluble in acids but soluble in alkaline solution.

Grade: FCC.

Hazard: Flammable but self-extinguishing.

Use: Pharmaceutical products; food additive; thickening agent and stabilizer in ice cream, cheese products, canned fruits and sausage casings; synthetic fibers. See also algin, alginic acid.

calcium aluminate. (tricalcium aluminate).
$3CaO \cdot Al_2O_3$.

Crystals or powder, d 3.038 (25C), mp decomposes 1535C, soluble in acids. A refractory and an important ingredient of cements, especially of aluminous cement. Fused calcium aluminate (a glass) can be used for infrared transmission and detection. Noncombustible.

calcium aluminum hydride. See aluminum calcium hydride.

calcium ammonium nitrate. A uniform mixture of approximately 60% ammonium nitrate and 40% limestone and/or dolomite. A fertilizer containing approximately 20% N.

Hazard: Oxidizer, fire risk in contact with organic materials.

calcium arsenate. (tricalcium orthoarsenate).
$Ca_3(AsO_4)_2$.

Properties: White powder, slightly soluble in water, soluble in dilute acids. Decomposes on heating.

Derivation: Interaction of calcium chloride and sodium arsenate.

Grade: Technical, CP.

Hazard: Toxic by ingestion and inhalation. TLV: 0.2 mg/m³ of air (as arsenic).

Use: Insecticide, germicide.

calcium arsenite. $CaAsO_3H$.

Properties: White, granular powder; insoluble in water; soluble in acids.

Grade: Technical.

Hazard: Toxic by inhalation and ingestion.

Use: Germicides, insecticides.

calcium ascorbate. $Ca(C_6H_7O_6)_2 \cdot 2HOH$.

Properties: A white to slightly yellow, odorless, crystalline powder; soluble in water; slightly soluble in alcohol; insoluble in ether. The pH of a 10% solution is between 6.8 and 7.4.

Grade: FCC.

Use: Food preservative.

calcium biphosphate. See calcium phosphate, monobasic.

calcium bisulfide. See calcium hydrosulfide.

calcium bisulfite. See calcium hydrogen sulfite.

calcium borate. CaB_4O_7.

Properties: White powder, soluble in dilute acids, partially soluble in water.

Use: Fire-retardant compositions, antifreeze compounds, metallurgical flux, porcelain.

calcium bromate. $Ca(BrO_3)_2 \cdot HOH$.

Properties: White, crystalline powder; d 3.329; loses its water at 180C; very soluble in water.

Grade: FCC.

Hazard: Oxidizing agent, fire risk in contact with organic materials.

Use: Maturing agent, dough conditioner.

calcium bromide. CAS: 7789-41-5.

(a) $CaBr_2 \cdot 6H_2O$; (b) $CaBr_2$.

Properties: White powder or crystals, odorless, sharp saline taste, very deliquescent, very soluble in water. Also soluble in alcohol and acetone. (a) D 2.295 (25C), mp 38C, bp 149C (decomposes); (b) d 3.353 (25C), mp 730C (slight decomposition), bp 806–812C.

Derivation: By the action of hydrobromic acid on calcium oxide; carbonate or hydroxide and subsequent crystallization.

Grade: Technical, CP.

Use: Photography, medicine, dehydrating agent, food preservative, road treatment, freezing mixtures, sizing compounds, wood preservative, fire retardant.

calcium carbide. CAS: 75-20-7. CaC_2.

Properties: Grayish-black, irregular, hard solid; must be kept dry. d 2.22, mp approximately 2300C. Garlic-like odor. Decomposes in water with formation of acetylene and calcium hydroxide and evolution of heat.

Derivation: Interaction of pulverized limestone or quicklime with crushed coke or anthracite in an electric furnace.

Grade: Technical, lumps, powder.

Hazard: Forms flammable and explosive gas and corrosive solid with moisture.

Use: Generation of acetylene gas for welding, chloroethylenes, vinyl acetate monomer, acetylene chemicals, reducing agent.   *Note:* Rapidly being replaced by hydrocarbons for most chemical uses.

calcium carbonate. CAS: 1317-65-3. $CaCO_3$.

Properties: White powder or colorless crystals, odorless, tasteless, d 2.7–2.95, decomposes at 825C, noncombustible, very slightly soluble in water (a few ppm), soluble in acids with evolution of carbon dioxide.

Occurrence: Calcium carbonate is one of the most stable, common, and widely dispersed materials. It occurs in nature as aragonite, oyster shells, calcite, chalk, limestone, marble, marl, and travertine, especially in Indiana (structural limestone), Vermont (marble), Italy (travertine), and England (chalk).

Derivation: (a) Mined from natural surface deposits. (b) Precipitated (synthetic) by reaction of calcium chloride and sodium carbonate in water solution or by passing carbon dioxide through a suspension of hydrated lime $(Ca(OH)_2)$ in water.

Hazard: A nuisance particulate dust. TLV: 5 mg/m³ of air.

Use: Source of lime; neutralizing agent, filler, and extender in rubber, plastics, paints; opacifying agent in paper; fortification of bread; putty; tooth powders; antacid; whitewash; Portland cement; sulfur dioxide removal from stack gases; metallurgical flux; analytical chemistry; carbon dioxide generation (laboratory). See also chalk, calcite, marble, limestone, whiting.   *Note:* Calcium carbonate is a major cause of boiler scale when hard water is used in heating systems.

calcium caseinate.

Properties: White or slightly yellow, nearly odorless powder. Insoluble in cold water; forms a milky solution when suspended in water, stirred, and heated.

Use: Medicine, special foods.   See also casein.

calcium chlorate. CAS: 10137-74-3.

$Ca(ClO_3)_2 \cdot 2HOH$.

Properties: White to yellowish crystals. Keep well stoppered. Melts when rapidly heated at 100C, mp (anhydrous) 340C; soluble in water and alcohol, hygroscopic; d 2.711.

Derivation: By the action of chlorine on hot calcium hydroxide slurry.

Hazard: Oxidizer, dangerous fire risk, forms explosive mixtures with combustible materials.

Use: Photography, pyrotechnics, dusting powder to kill poison ivy, herbicide.

calcium chloride. CAS: 10043-52-4.
(a) $CaCl_2$; (b) $CaCl_2 \cdot HOH$;
(c) $CaCl_2 \cdot 2HOH$; (d) $CaCl_2 \cdot 6HOH$.
46th highest-volume chemical produced in US (1985).

Properties: White, deliquescent crystals, granules, lumps, or flakes. (a) D 2.15 (25C), mp 772C, bp above 1600C; (b) mp 260C; (c) USP grade, d 0.835 (25C); (d) d 1.71 (25C), bp −4HOH at 30C and -6HOH at 200C. All forms soluble in water and alcohol. Water solution is neutral or slightly alkaline.

Derivation: (1) Action of hydrochloric acid on calcium carbonate and subsequent crystallization. (2) Commercially obtained as a by-product in the Solvay soda and other processes. (3) Recovery from brines.

Grade: Technical, CP, FCC, USP (dihydrate), various forms and purities, solution.

Use: De-icing and dust control of roads, drilling muds, dustproofing, freezeproofing and thawing coal, coke, stone, sand, ore, concrete conditioning, paper and pulp industry, fungicides, refrigeration brines, drying and desiccating agent, sequestrant in foods, firming agent in tomato canning, tire weighting, pharmaceuticals, electrolytic cells. The hexahydrate (d) is under development for solar heat storage.

calcium chlorite. CAS: 14674-72-7.
$Ca(ClO_2)_2$.
Properties: White crystals, d 2.71, decomposes in water.

Hazard: Strong oxidizer, fire risk in contact with organic materials.

calcium chromate. CAS: 13765-19-0.
$CaCrO_4$; $CaCrO_4 \cdot 2HOH$.
Properties: Bright yellow powder, hydrate loses water at 200C. Anhydrous: d 2.89, soluble in dilute acids, slightly soluble in water.

Hazard: A carcinogen.

Use: Pigment, corrosion inhibitor, oxidizing agent, depolarizer for batteries, coating for light metal alloys.

calcium citrate. (lime citrate; tricalcium citrate).
$Ca_3(C_6H_5O_7)_2 \cdot 4HOH$.
Properties: White, odorless powder; loses most of its water at 100C and all of it at 120C. Al-

most insoluble in water, insoluble in alcohol.

Grade: Reagent, technical, FCC.

Use: Dietary supplement, sequestrant, buffer, and firming agent in foods.

calcium cyanamide. (lime nitrogen; calcium carbimide). CAS: 156-62-7. $CaCN_2$.
Properties: Colorless crystals or powder, d 1.083, mp 1300C, sublimes above 1150C. decomposes in water liberating ammonia and acetylene.

Derivation: Calcium carbide powder is heated in an electric oven into which nitrogen is passed (24 to 26 hours). Any uncombined calcium carbide is leached out after removal.

Grade: Fertilizer, 21%N, industrial.

Hazard: Fire risk with moisture or combined with calcium carbide. A skin irritant. TLV: 0.5 mg/m^3 of air.

Use: Fertilizer, nitrogen products, pesticide, hardening iron and steel.

calcium cyanide. CAS: 592-01-8. $Ca(CN)_2$.
Properties: Colorless crystals or white powder, gray-black (technical), decomposes in moist air liberating hydrogen cyanide. Soluble in water and very weak acid with evolution of hydrogen cyanide. Decomposes above 350C.

Hazard: Toxic. Absorbed via skin. TLV: 5 mg/m^3 of air (as CN).

Use: Rodenticide, fumigant for greenhouses, flour mills, grain, seed, citrus trees under tents for control of scale insects, leaching of gold and silver ores, manufacture of other cyanides.

calcium cyclamate. (calcium cyclohexylsulfamate; calcium cyclohexanesulfamate).
CAS: 139-06-0. $(C_6H_{11}NHSO_3)_2Ca \cdot 2HOH$.
Properties: White crystals, nearly odorless powder with very sweet taste. Freely soluble in water (solutions are neutral to litmus); almost insoluble in alcohol, benzene, chloroform, and ether; pH (10% solution) 5.5–7.5. Sweetening power approximately 30 times that of sucrose.

Grade: NF, FCC.

Hazard: Not permitted for use in foods and soft drinks due to suspected carcinogenicity.

Use: Nonnutritive sweetener.   See also cyclamate.

calcium dehydroacetate. $(C_8H_7O_4)_2$.
Properties: White to cream powder. Almost insoluble in water and organic solvents.

Grade: 96% min.

Use: Fungicide.   See also dehydroacetic acid.

calcium diacetate. See calcium acetate.

calcium dichromate. $CaCr_2O_7 \cdot 3HOH$
(or $\cdot 4.5$ HOH).
Properties: Brownish-red crystals, deliquescent, d
(4.5HOH variety) 2.136. Soluble in water.
Grade: Technical, CP.
Use: Corrosion inhibitor, catalyst, manufacture of
chromium compounds.

calcium dihydrogen sulfite. See calcium hydrogen
sulfite.

calcium dioxide. See calcium peroxide.

calcium disodium edetate. (USAN) (calcium disodium EDTA; edathamil calcium disodium;
calcium disodium ethylenediaminetetraacetate).
$CaNa_2C_{10}H_{12}N_2O_8 \cdot xHOH$.   For other variations of the name, see ethylenediaminetetraacetic
acid. The calcium disodium salt is a mix of the
dihydrate and trihydrate.
Properties: White, odorless powder or flakes;
slightly hygroscopic; faint saline taste; stable in
air; soluble in water; insoluble in organic solvents. It acts as a chelating agent for heavy metals.
Grade: USP, FCC.
Use: Medicine (antidote in heavy-metal poisoning), in foods to "complex" trace heavy metals,
as a preservative and to retain color and flavor,
antiguwing agent in fermented malt beverages.
(For restrictions on food uses, see FDA regulations).

calcium ethylhexoate. See soaps (2).

calcium ferrocyanide. $Ca_2Fe(CN)_6 \cdot 12HOH$.
Properties: Yellow crystals, decomposes on heating, soluble in water, insoluble in alcohol, d 1.68.
Use: Removal of metallic impurities in the manufacture of citric, tartaric, and other acids.

calcium fluoride. CAS: 7789-75-5. CaF_2.
Properties: White powder occurring in nature as
fluorite (pure form) or fluorspar (mineral), mp
1402C, bp approximately 2500C, d 3.18. Reacts
with hot concentrated sulfuric acid to liberate
hydrogen fluoride. Insoluble in water, soluble in
ammonium salts, Mohs hardness 4.
Derivation: (a) By powdering pure fluorite or
fluorspar; (b) by the interaction of soluble calcium salt and sodium fluoride.
Hazard: An irritant. TLV: (asF) 2.5 mg/m³ of
air.
Grade: See fluorspar.
Use: See fluorspar. Single pure (99.93%) crystals
of calcium fluoride are also produced for use
in spectroscopy, electronics, lasers, and high-
temperature dry-film lubricants.

calcium fluorophosphate. (fluoroapatite; FAP).
$CaPO_3F$. A laser crystal said to have lowest energy threshold of any room-temperature crystal.

calcium fluosilicate. See calcium silicofluoride.

calcium formate. $Ca(OOCH)_2$.
Properties: White powder, mp above 300C, d
2.015. Soluble in water, insoluble in alcohol.
Use: Briquet binder, drilling fluids, lubricants,
chrome tanning.

calcium gluconate. CAS: 299-28-5.
$Ca(C_6H_{11}O_7)_2 \cdot HOH$.
Properties: White, odorless, practically tasteless,
fluffy powder or granules. Stable in air. Loses
water at 120C. Soluble in hot water, less soluble
in cold water, insoluble in alcohol, acetic acid
and other organic solvents, specific rotation (20/
D) approximately +6 degrees. Solution neutral
to litmus.
Derivation: Neutralization of gluconic acid with
lime or calcium carbonate.
Grade: Technical, USP, FCC, special for ampules.
Use: Food additive, buffer and sequestering agent,
vitamin tablets.

calcium glutamate. Similar to sodium glutamate.

calcium glycerophosphate. (calcium glycerinophosphate). $CaC_3H_7O_2PO_4$.
Properties: White, crystalline powder; odorless; almost tasteless; slightly hygroscopic; decomposes
above 170C; slightly soluble in water; insoluble
in alcohol.
Derivation: By esterification of phosphoric acid
with glycerol and conversion of glycerophosphoric acid to the calcium salt.
Grade: Technical, pure, FCC.
Use: Stabilizer for plastics, nutrient and dietary
supplement.

calcium glycolate. $(CH_3OHCOO)_2Ca$.
Properties: White solid.
Grade: Technical.
Use: Source of glycolic acid and of the glycolic
acid radical in chemical synthesis.

calcium hexasilicofluorate. See calcium silicofluoride.

calcium hydrate. See calcium hydroxide.

calcium hydride. CAS: 57308-10-8. CaH_2.
Properties: Grayish-white lumps or crystals.
Forms calcium hydroxide by moist air and
evolves hydrogen. D 1.7, decomposes at 675C,
decomposed by water, organic acids, and lower
alcohols.
Grade: Technical, 94% pure.

Hazard: Evolves highly flammable hydrogen when wet; solid product is slaked lime. Irritating to skin.

Use: Reducing agent, drying agent, analytical reagent in organic chemistry, easily portable source of hydrogen, cleaner for blocked-up oil wells.

calcium hydrogen sulfite. (calcium bisulfite; calcium dihydrogen sulfite; calcium acid sulfite). $Ca(HSO_3)_2$. A solution of calcium sulfite in aqueous sulfur dioxide.
Properties: Yellowish liquid with strong sulfur dioxide odor, d 1.06, corrosive to metals.
Derivation: Action of sulfur dioxide on calcium hydroxide (solution).
Hazard: Irritating and corrosive to skin and tissue.
Use: Antichlor in bleaching textiles, paper pulp (dissolving lignin) preservative, bleaching sponges, hydroxylamine salts, germicide, disinfectant.

calcium hydrosulfide. (calcium bisulfide; calcium sulfhydrate). $Ca(HS)_2 \cdot 6HOH$.
Properties: Colorless, transparent crystals; soluble in alcohol and water. Decomposes in air (15–18C).
Use: Leather industry.

calcium hydroxide. (calcium hydrate; hydrated lime; caustic lime; slaked lime).
CAS: 1305-62-0. $Ca(OH)_2$.
Properties: Soft, white, crystalline powder with alkaline, slightly bitter taste. D 2.34, mp loses its water at 580C, pH of water solution (25C) 12.4. Slightly soluble in water; soluble in glycerol, syrup, and acids; insoluble in alcohol. Absorbs carbon dioxide from air.
Derivation: Action of water on calcium oxide.
Impurities: Calcium carbonate, magnesium salts, iron.
Grade: Technical, chemical lime (insoluble matter less than 2%, magnesium less than 3%), building lime, USP, CP, FCC.
Hazard: Skin irritant, avoid inhalation. TLV: 5 mg/m³ of air.
Use: Mortar, plasters, cements, calcium salts, causticizing soda, depilatory, unhairing of hides, whitewash, soil conditioner, ammonia recovery in gas manufacture, disinfectant, water softening, purification of sugar juices, accelerator for low-grade rubber compounds, petrochemicals, food additive as buffer and neutralizing agent, shell-forming agent (poultry).

calcium hypochlorite. (calcium oxychloride).
CAS: 7778-54-3. $Ca(OCl)_2$.
Properties: White, crystalline solid; d 2.35; decomposes at 100C; decomposes in water and alcohol; not hygroscopic; practically clear in water solution. Stable chlorine carrier. An oxidizer.
Derivation: Chlorination of a slurry of lime and caustic soda with subsequent precipitation of calcium hypochlorite dihydrate, dried under vacuum.
Grade: Commercial (70%), high purity (99.2% available chlorine as calcium hypochlorite).
Hazard: Dangerous fire risk in contact with organic materials.
Use: Algicide, bactericide, deodorant, potable water purification, disinfectant for swimming pools, fungicide, bleaching agent (paper, textiles). See also lime, chlorinated.

calcium hypophosphite. $Ca(H_2PO_2)_2$.
Properties: White powder. Evolves phosphine at 300C. Soluble in water, insoluble in alcohol.
Use: Medicine, corrosion inhibitor.

calcium iodate. $Ca(IO_3)_2$.
Properties: White crystals or powder, odorless, d 4.5 (15C), decomposes 540C, soluble in water and nitric acid, insoluble in alcohol. Oxidizer.
Grade: Technical, CP, FCC.
Hazard: Fire risk in contact with organic materials.
Use: Deodorant, mouth washes, food additive, dough conditioner (up to 0.0075 part per 100 lb flour used).

calcium iodide. $CaI_2 \cdot 6H_2O$.
Properties: Yellowish-white crystals, deliquescent, decomposes in air by absorption of carbon dioxide. Soluble in water, ethanol, and pentanol, d 2.55 (anhydrous 4.0 at 25C), -6HOH at 42C, mp 783C, bp approximately 1100C.
Derivation: Action of hydriodic acid on calcium carbonate.
Use: Photography, medicine.

calcium iodobehenate. $Ca(OOCC_{21}H_{42}I)_2$.
Properties: White, odorless powder. Insoluble in alcohol, slightly soluble in cold water.
Use: Medicine and pharmaceuticals.

calcium metasilicate. $CaSiO_3$.
Properties: White powder, d 2.9. Insoluble in water.
Hazard: Irritating dust. Use in foods restricted to 5% in baking powder, 2% in table salt.
Use: Absorbent, antacid, filler for paper and paper coatings, cosmetics, food additive (anticaking agent), manufacture of glass and Portland cement.   See also dicalcium silicate.

calcium molybdate. CAS: 7789-82-4.
$CaMoO_4$.
Properties: White, crystalline powder; mp approximately 1250C; d 4.35; soluble in mineral acids;

insoluble in alcohol, ether, or water. Noncombustible.

Derivation: Fusion of calcium oxide and a molybdenum ore.

Grade: Technical, single crystals, 99.97%.

Use: Molybdic acid, alloying agent in production of iron and steel, crystals in optical and electronic applications, phosphors.

calcium naphthenate.

Properties: Sticky, tenacious mass; insoluble in water; soluble in ethyl acetate, carbon tetrachloride, gasoline, benzene, and ether. Combustible.

Derivation: Precipitation from aqueous solutions of calcium salts and sodium naphthenate.

Use: Waterproofing compositions, adhesives, driers, wood fillers, grafting waxes, cements, varnishes, color lakes. See also soap (2).

calcium nitrate. (lime nitrate; nitrocalcite; lime saltpeter; Norwegian saltpeter).

CAS: 10124-37-5. (a) $Ca(NO_3)_2 \cdot 4HOH$; (b) $Ca(NO_3)_2$.

Properties: White, deliquescent mass; soluble in water, alcohol, and acetone; d (a) 1.82, (b) 2.36; mp (a) 42C, (b) 561C.

Grade: Technical, pure, CP, reagent.

Hazard: Strong oxidizer, dangerous fire risk in contact with organic materials, may explode if shocked or heated.

Use: Pyrotechnics, explosives, matches, fertilizers, other nitrates, source of C-14 by nuclear irradiation.

calcium nitride. Ca_3N_2.

Properties: Brown crystals, d 2.63 (17C), mp 1195C. Soluble in water with evolution of ammonia (irritating gas), soluble in dilute acids, insoluble in absolute alcohol.

calcium nitrite. $Ca(NO_2)_2 \cdot H_2O$.

Properties: Colorless or yellowish crystals, hygroscopic. Soluble in water, slightly soluble in alcohol; d 2.23 (34C) (anhydrous); mp loses its water at 100C.

Grade: Technical.

Use: Corrosion inhibitor in lubricants and steel reinforced concrete.

calcium novobiocin. See novobiocin.

calcium octoate. See soap (2).

calcium oleate. (oleic acid calcium salt). $Ca(C_{18}H_{33}O_2)_2$.

Properties: Yellowish crystals; soluble in benzene and chloroform; almost insoluble in water, alcohol, and acetone; decomposes approximately 140C.

Use: Grease-thickening agent, emulsifying agent, waterproofing concrete.

calcium orthophosphate. See calcium phosphate, tribasic.

calcium orthotungstate. See calcium tungstate.

calcium oxalate. CaC_2O_4.

Properties: White, crystalline powder; soluble in dilute hydrochloric and nitric acids; insoluble in acetic acid and water. D 2.2.

Grade: Technical, CP.

Hazard: An irritant.

Use: Making oxalic acid and organic oxalates, glazes, rare-earth metal separations.

calcium oxide. (lime; quicklime; burnt lime; calx; unslaked lime; fluxing lime).

CAS: 1305-78-8. CaO.

Properties: White or grayish-white hard lumps; sometimes with a yellowish or brownish tint due to iron; odorless. Crumbles on exposure to moist air, d 3.40, mp 2570C, bp 2850C, soluble in acid, reacts with water to form calcium hydroxide with evolution of heat.

Derivation: Calcium carbonate (limestone) is roasted in kilns until all the carbon dioxide is driven off.

Impurities: Calcium carbonate; magnesium, iron, and aluminum oxides.

Grade: Technical, refractory, agricultural, FCC.

Hazard: Evolves heat on exposure to water. Dangerous near organic materials. Strong irritant. TLV: 2 mg/m³ of air.

Use: Refractory, flux in steel manufacture, pulp and paper, manufacture of calcium carbide, sulfur dioxide removal from stack gases, sewage treatment (phosphate removal, pH control), poultry feeds, neutralization of acid waste effluents, insecticides and fungicides, dehairing of hides, sugar refining, food additive, glass manufacture, sodium carbonate by Solvay process, carbon dioxide absorbent

Note: Like other high-melting solids (tungsten, zirconia, carbon), lime (CaO) becomes incandescent when heated to near its mp (2500C). Of both historic and semantic interest is the use of lime as an illuminant in stage lighting for some years before the advent of electricity (1850–1880). Invented in 1816, this technique involved an oxyhydrogen flame impinging on a cylinder of lime, causing it to emit a brilliant white light which was concentrated to a beam by a lens. The light was powerful enough to spotlight actors or simulate sunshine effects. It became known in theatrical circles as limelight and was the origin of the familiar phrase, "in the limelight."

See also lime.

calcium oxychloride. See calcium hypochlorite.

calcium palmitate.　$Ca(C_{15}H_{31}CO_2)_2$.
White or pale yellow powder produced by reacting
a soluble palmitate with a soluble calcium salt.
Insoluble in water, slightly soluble in alcohol or
ether. Combustible.
Use: Waterproofing agent, thickener for lubricating oils, manufacture of solidified oils. Available
only as technical grade.　　See also soap (2).

calcium pantothenate.　$(C_5H_{16}NO_5)_2Ca$.
The calcium salt of pantothenic acid, available
in either the dextro- or racemic forms. Only the
dextro- form has vitamin activity.
Properties: (Both forms identical) white, slightly
hygroscopic, odorless powder; sweetish taste; stable in air; solutions have a pH of 7–9; soluble
in water and glycerol; insoluble in alcohol, chloroform; and ether; mp 170–172C; decomposes
195–196C; specific rotation (5% aqueous solution) + 28.2 (25C).
Source: Same as pantothenic acid.
Grade: USP (both forms), FCC.
Use: Medicine, animal feeds, dietary supplement.
See also pantothenic acid.

calcium pectate.　A material developed in plants
such as beans, peas, potatoes, which enables them
to seal off fungus-infected areas.
See pectic acid, pectins.

calcium perborate.　$Ca(BO_3)_2 \cdot 7H_2O$.
Properties: Gray-white lumps or powder. Soluble
in acids, also in water with evolution of oxygen.
Use: Medicine, bleach, tooth powders.

calcium perchlorate.　CAS: 13477-36-6.
$Ca(ClO_4)_2$
Properties: White crystals, d 2.651, decomposes
at 270C, soluble in water and alcohol.
Hazard: Strong oxidizer, dangerous fire risk in
contact with organic materials.

calcium permanganate.　CAS: 10118-76-0.
$Ca(MnO_4)_2 \cdot 4HOH$.
Properties: Violet, deliquescent crystals; d 2.4;
soluble in water and ammonia; decomposed by
alcohol.
Grade: Technical, pure.
Hazard: Strong oxidizer, dangerous fire risk in
contact with organic materials.
Use: Textile industry, sterilizing water, dentistry,
disinfectant, deodorizer, an additive (with hydrogen peroxide) in liquid rocket propellants, in
binders for welding electrode coatings.

calcium peroxide.　(calcium superoxide; calcium
dioxide).　　CAS: 1305-79-9.　　CaO_2.
Properties: White or yellowish, odorless, almost
tasteless powder, decomposes approximately

200C, almost insoluble in water, soluble in acids
with formation of hydrogen peroxide. Available
oxygen 22.2% (min 13.3% in technical grade).
Derivation: Interaction of solution of a calcium
salt and sodium peroxide with subsequent crystallization.
Grade: 60–75%, FCC.
Hazard: Strong oxidizing agent. Dangerous fire
risk in contact with organic materials. Irritating
in concentrated form.
Use: Seed disinfectant, dentrifices, dough conditioners, antiseptic, bleaching of oils, modification
of starches, high-temperature oxidations.

calcium phenylate.　(calcium phenoxide; calcium
phenate).　　$Ca(OC_6H_5)_2$.
Properties: Finely divided, red crystals; slightly
soluble in water and alcohol.
Use: Lubricating-oil detergent, emulsifier.

calcium phosphate.　See calcium phosphate dibasic, calcium phosphate. monobasic, or calcium
phosphate tribasic.

calcium phosphate, dibasic.　(dicalcium orthophosphate; bicalcium phosphate; secondary calcium phosphate).　　$CaHPO_4 \cdot 2H_2O$ and
$CaHPO_4$.
Properties: White, tasteless, crystalline powder;
odorless; soluble in dilute hydrochloric, nitric,
and acetic acids; insoluble in alcohol; slightly
soluble in water. (Hydrate) d 2.306, loses its water at 109C. Nonflammable.
Derivation: Interaction of fluorine-free phosphoric
acid with milk of lime.
Grade: USP, FCC, dentrifice grade, feed grade,
18.5 or 21% P.
Use: Animal feed supplement, food supplement,
dentrifice, medicine, glass, fertilizer, stabilizer
for plastics, dough conditioner, yeast food.

calcium phosphate, monobasic.　(calcium biphosphate; acid calcium phosphate; calcium phosphate, primary; monocalcium phosphate).
rm $CaH_4(PO_4)_2 \cdot H_2O$.
Properties: Colorless, pearly scales or powder;
deliquescent in air. Soluble in water and acids.
Aqueous solutions are acid. Mp loses water at
100C, decomposes at 200C, d 2.20. Nonflammable.
Derivation: By dissolving either dicalcium or tricalcium phosphates in phosphoric acid and allowing the solution to evaporate spontaneously.
Grade: FCC, ceramic, anhydrous, hydrated.
Use: Baking powders, fertilizers, mineral supplement, stabilizer for plastics, to control pH in
malt, glass manufacture, buffer in foods, firming
agent.　　See also superphosphate.

calcium phosphate, precipitated.　See calcium
phosphate tribasic.

calcium phosphate, primary. See calcium phosphate, monobasic.

calcium phosphate, secondary. See calcium phosphate, dibasic.

calcium phosphate, tertiary. See calcium phosphate, tribasic.

calcium phosphate, tribasic. (calcium orthophosphate; tricalcium phosphate; precipitated calcium phosphate; tricalcium orthophosphate; tertiary calcium phosphate). $Ca_3(PO_4)_2$.
True $Ca_3(PO_4)_2$ can be prepared thermally, but is rare. Precipitated "tricalcium phosphate" is a hydroxyapatite with the approximate formula $Ca_5OH(PO_4)_3$.
Properties: White, odorless, tasteless, crystalline powder. D 3.18, mp 1670C, refr index 1.63, soluble in acids, insoluble in water, alcohol, and acetic acid. Nonflammable.
Derivation: (a) Phosphate rock, apatite, and phosphorite; (b) by the interaction of soluble of calcium chloride and sodium triphosphate with excess of ammonia; (c) by interaction of hydrated lime and phosphoric acid.
Grade: Granular, technical, CP, NF, pure precipitated, FCC.
Use: Ceramics, calcium acid phosphate, phosphorus and phosphoric acid, polishing powder, cattle foods, clarifying sugar syrups, medicine, mordant (dyeing textiles with Turkey red), fertilizers, dentifrices, stabilizer for plastics, in meat tenderizers, in foods as anticaking agent, buffer, nutrient supplement, can remove Sr-90 from milk. See also bone ash.

calcium phosphide. (photophor).
CAS: 1305-99-3. Ca_3P_2 (or Ca_2P_2).
Properties: Red-brown crystals or gray granular masses, d 2.51 (15C), mp approximately 1600C, insoluble in alcohol and ether.
Derivation: By heating calcium phosphate with aluminum or carbon, by passing phosphorus vapors over metallic calcium.
Grade: Technical.
Hazard: Dangerous fire risk, decomposed by water to phosphine which is highly toxic and flammable. See phosphine.
Use: Signal fires, torpedoes, pyrotechnics, rodenticide.

calcium phosphite. (dicalcium orthophosphite). $CaHPO_4 \cdot 2HOH$.
Properties: White powder, loses its water at 200–300C (decomposes) forming phosphine. Slightly soluble in water, insoluble in alcohol.
Use: Catalyst for polymerization reactions.

calcium phytate. (hexacalcium phytate). $C_6H_6(CaPO_4)_6$.
Properties: Free-flowing, white powder; slightly soluble in water; pH of saturated solution is neutral.
Derivation: Corn steep liquor.
Use: Sequestering agent to remove excess metals from wine and vinegar, source of calcium in pharmaceuticals and nutrition, source of phytic acid and its salts.

calcium plumbate. Ca_2PbO_4.
Properties: Orange to brown crystalline powder, decomposed by hot water or carbon dioxide, d 5.71, soluble in acids (decomposes), insoluble in cold water.
Hazard: Fire risk in contact with organic materials. Toxic by ingestion.
Use: Oxidizing agent, pyrotechnics and safety matches, glass, storage batteries.

calcium polysilicate. $CaO \cdot 12SiO_2$.   A powder used as an anticaking agent.

calcium propionate. $Ca(OOCCH_2CH_3)_2$.
(Occurs also with one water). White powder, soluble in water, slightly soluble in alcohol.
Grade: FCC.
Use: Mold-inhibiting additive in bread, other foods, tobacco, pharmaceuticals, medicine (antifungal agent).

calcium propyl arsenate. $C_3H_7AsO_3Ca$.
Crystals, soluble in water. Used for preemergence control of crabgrass.
Hazard: Toxic by ingestion.

calcium pyrophosphate. $Ca_2P_2O_7$.
Properties: White powder, soluble in dilute hydrochloric and nitric acids, insoluble in water, d 3.09, mp 1230C.
Grade: FCC.
Use: Polishing agent in dentifrices, mild abrasive for metal polishing, nutrient and dietary supplement.

calcium resinate. CAS: 9007-13-0.
Properties: Yellowish-white, amorphous powder or lumps, rosin odor. Soluble in acid, insoluble in water, soluble in amyl acetate, butyl acetate, ether, amyl alcohol.
Derivation: By boiling calcium hydroxide with rosin and filtering; fusion of hydrated lime and melted rosin.
Grade: Technical, fused.
Hazard: Flammable, dangerous fire risk, spontaneous heating.
Use: Waterproofing; manufacturing paint driers, porcelains, perfumes, cosmetics, enamels; coat-

ing for fabrics, wood, paper; tanning leather. See also soap (2).

calcium ricinoleate.
$Ca[CH_3(CH_2)_5CHOHCH_2CHCH(CH_2)_7CO_2]_2$.
Properties: White powder with a slight odor of fatty acids. Derived from castor oil. Mp 84C, d 1.04. Combustible.

calcium d-saccharate. $CaC_6H_8O_8 \cdot 4HOH$.
Properties: White, crystalline powder, odorless, tasteless, insoluble in water and alcohol, soluble in calcium gluconate solution.
Derivation: Oxidation of d-gluconic acid and neutralization with lime.
Use: Plasticizer for cement, mortar, etc.

calcium selenate. $CaSeO_4$.
Properties: Water-soluble powder, d 2.7 (dihydrate).
Use: General pesticide.

calcium silicate. See calcium metasilicate; wollastonite; Portland cement.

calcium silicide. $CaSi_2$.
Properties: Solid, d 2.5. Insoluble in cold water; decomposes in hot water; soluble in acids and alkalies.
Hazard: Flammable; may ignite spontaneously in air.

calcium silicofluoride. (calcium hexafluorosilicate). $CaSiF_6$.
Properties: (dihydrate) Finely divided solid, d 2.25, insoluble in cold water and acetone, decomposed by hot water.
Hazard: Toxic to experimental animals.
Use: Flotation agent, insecticide, rubber compounding, ceramic glazes.

calcium-silicon alloy. Contains 30% calcium.
Hazard: Flammable, may ignite spontaneously in air.

calcium sorbate. $Ca(OOCC_5H_7)_2$.
Used as a chemical preservative in foods.

calcium stannate. $CaSnO_3 \cdot 3HOH$.
Properties: White, crystalline powder; insoluble in water; dehydrates at approximately 350C.
Hazard: Toxic by ingestion and inhalation. TLV: 2 mg/m³ of air.
Use: Additive to ceramic capacitors, production of ceramic colors.

calcium stearate. $Ca(C_{18}H_{35}O_2)_2$.
Properties: White powder, mp 179C, insoluble in water, slightly soluble in hot alcohol. Decomposed by many acids and alkalies.

Derivation: Interaction of sodium stearate and calcium chloride, then filtration.
Grade: Technical, FCC.
Use: Water repellent, flatting agent in paints, lubricant in making tablets, emulsions, cements, wax crayons, stabilizer for vinyl resins, food additive, mold release agent, cosmetics.

calcium stearyl lactylate. $C_{48}H_{86}CaO_{12}$.
Properties: Finely divided, nonhygroscopic solid; almost insoluble in water.
Use: Food additive (flour mixing, egg whites, etc.).

calcium strontium sulfide. $CaSrS_2$ or $CaS \cdot SrS$.
Used as a phosphorescent pigment, a phosphor.

calcium sulfamate. $Ca(SO_3NH_2)_2 \cdot 4HOH$.
Properties: White crystals, soluble in water. Aqueous solution is stable on boiling. Nonflammable.
Grade: Technical.
Use: Flameproofing agent for textiles and certain grades of paper.

calcium sulfate. CAS: 10101-41-4. $CaSO_4$ or $CaSO_4 \cdot 2HOH$. Occurs in nature as anhydrite and in hydrated form as gypsum (plaster of paris).
Properties: (Pure anhydrous) White, odorless powder or crystals; d 2.964; mp 1450C; slightly soluble in water. (Dihydrate, pure precipitated) d 2.32, loses 1.5 waters at 128C, becomes anhydrous at 163C. Noncombustible. Neither the anhydrous nor the dihydrate form can set with water.
Derivation: From natural sources and as a by-product in many chemical operations.
Grade: Technical, pure precipitated (as the dihydrate), FCC.
Use: Portland cement retarder; tile and plaster; source of sulfur and sulfuric acid; polishing powders; paints (white pigment, filler, drier); paper (size filler, surface-coating); dyeing and calico printing; metallurgy (reduction of zinc minerals); drying industrial gases, solids and many organic liquids; in granulated form as soil conditioner; quick-setting cements, molds, and surgical casts; wallboard; food additive; desiccant.

calcium sulfhydrate. See calcium hydrosulfide.

calcium sulfide. CaS.
Properties: Yellow to light-gray powder with odor of hydrogen sulfide in moist air, unpleasant alkaline taste, gradually decomposes in moist air or in weak acids, decomposed by acids, slightly soluble in water with partial decomposition, insoluble in alcohol, d 2.6.
Derivation: Strong heating of pulverized calcium sulfate and charcoal.

Hazard: Irritating to skin and mucous membranes.

Use: Luminous paint, depilatory, preparation of arsenic-free hydrogen sulfide, lubricant additive, ore dressing and flotation agent, phosphors.

calcium sulfite. $CaSO_3 \cdot 2HOH$.
Properties: White powder, loses its water at 100C, soluble in sulfurous acid, slightly soluble in water.
Derivation: Action of sulfurous acid on calcium carbonate.
Use: Textiles (antichlor), disinfectant in sugar industry, brewing, biological cleansing, food preservative and discoloration retarder, paper manufacture.

calcium sulfocyanate. See calcium thiocyanate.

calcium superoxide. See calcium peroxide.

calcium tannate.
Properties: Yellowish-gray powder, soluble in dilute acids, slightly soluble in water.
Use: Pharmaceuticals, adhesives.

calcium tartrate. $CaC_4H_4O_6 \cdot 4HOH$.
Properties: White crystals, soluble in dilute acids, slightly soluble in water or alcohol.
Derivation: Interaction of calcium salt and crude cream of tartar.
Grade: Technical, CP.
Use: Tartaric acid, food preservative, antacid.

calcium thiocyanate. (calcium sulfocyanate).
$Ca(SCN)_2 \cdot 3HOH$.
Properties: White, hygroscopic crystals; soluble in water and alcohol.
Use: Solvent for cellulose and polyacrylate, acrylonitrile polymers, stiffening and swelling of textiles.

calcium thioglycollate. $CaCH_2COOC \cdot 3HOH$.
Properties: White powder, loses water at 100C, decomposes 250C, slightly soluble in water, insoluble in alcohol.
Use: In depilatories and hair-waving preparations.

calcium titanate. $CaTiO_3$.
Properties: Powder, d 3.98, mp 1800C.
Use: Electronics.

calcium trisodium pentetate (USAN). [calcium trisodium[(carboxymethyl)imino]bis(ethylenenitrolo)tetraacetate].
$CaNa_3C_{14}H_{18}N_3O_{10}$. A chelating agent, antidote for lead poisoning.

calcium tungstate. (calcium orthotungstate; calcium wolframate normal). $CaWO_4$.
Properties: White crystals, d 6.062, soluble in ammonium chloride, insoluble in water, decomposed by hot acids.
Derivation: (a) Interaction of calcium chloride and sodium tungstate. (b) Occurs in nature as scheelite, Nevada, California, Arizona, Utah, Colorado, New Zealand, Europe.
Method of purification: A slurry of powdered scheelite is treated with soda ash to form the soluble sodium tungstate. Insoluble impurities are filtered off and calcium tungstate is precipitated with lime.
Use: Luminous paints, fluorescent lamps, photography, medicine (radiopaque agent). *Note:* Synthetic crystals are available for use as scintillation counters and possible application in lasers.

calcium undecylenate.
$[CH_2CH(CH_2)_8COO]_2Ca$.
Properties: A fine, white powder of limited solubility, mp 155C.
Use: Bacteriostat and fungistat in cosmetics and pharmaceuticals.

calcium zirconate. $CaZrO_3$.
Properties: Solid, mp 2550C, d 4.78, soluble in nitric and other acids. Noncombustible.

calcium zirconium silicate. $CaZrSiO_5$ or $CaO \cdot ZrO_2 \cdot SiO_2$.
Properties: White solid, mp 1593C, insoluble in water, alkalies, slightly soluble in acids, soluble in hydrofluoric acid. Noncombustible.
Use: Electrical resistor ceramics, glaze opacifier.

"Calcocid."[57] TM for a series of acid dyestuffs used in the dyeing of wool and worsted goods, natural silk, jute and in coloring diversified materials.

"Calcofast."[57] TM for a series of metallized dyes containing chemically combined chromium used for dyeing wool. They can also be applied to leather, nylon, etc.

"Calcofluor."[57] TM for a series of direct dyeing dyes which possess fluorescent properties. Used for dyeing cotton, linen, viscose, acetate, nylon, wool, and certain synthetics. Used also in soaps as a brightener for textiles.

calspar. See calcite.

calender. A machine in which material is passed between heated steel rolls for any of several purposes: (1) to convert it into a sheet of uniform thickness; (2) to cause it to impregnate a textile fabric; or (3) to increase its surface gloss and hardness. Calenders, (i.e., cylinders) are composed of from three to as many as 10 or 12 hollow cast-iron rolls up to 84 in. in width, set vertically in a frame. The standard rubber or plastics calender has three steam-heated rolls which turn in opposing directions, either at the same speed or at different speeds. When moving at the same speed they deliver the mixture fed between the top and center rolls in a smooth sheet which can be as thin as 0.005 in. from between the center and bottom rolls. If fabric impregnation is desired, the center roll runs faster than the other two, thus pressing the soft and tacky mixture, called friction, into the textile material (tire carcasses, electrical tape, etc.). In paper manufacturing a high-speed calender "stack" imparts a smooth finish to the sheet as it leaves the drying unit. Smaller calenders are used to apply coating compositions. Laboratory sizes of all types are available.

See also supercalender.

"Calgolac."[108] TM for a powdered alkaline sodium hexametaphosphate detergent.
Use: Cleaning bars, fountains, and laboratory glassware and equipment.

"Calgon."[108] TM for a sodium phosphate glass cleaner commonly called sodium hexametaphosphate, it has a molecular ratio of 1:1 $Na_2O:P_2O_5$ with a guaranteed min of 67% P_2O_5. Several specialized compositions are available.
Derivation: From food-grade phosphoric acid and commercial soda ash by a thermal process.
Forms: Powder, agglomerated particles, and broken glassy plates, either pure or adjusted with mild alkalies.
Properties: Miscible in water, but is insoluble in organic solvents. It possesses sequestering, dispersing, and deflocculating properties and precipitates proteins. In very low concentration, it inhibits corrosion of steel and prevents the precipitation of slightly soluble, scale-forming compounds such as calcium carbonate and calcium sulfate.
Use: Softening water without precipitate formation as in dyeing, laundering, textile processing, and washing operations; corrosion inhibitor in deicing salt preparations; frozen desserts; pretanning hides in the manufacture of leather; dispersing clays and pigments; threshold treatment for scale; and corrosion prevention.

"Calgosil."[108] TM for a metaphosphate silicate compound, used to provide corrosion reduction in low hardness waters.

caliche. See sodium nitrate.

"Califlux."[500] TM for a series of oils composed principally of nitrogen bases and acidaffins. Used in plasticizers, extenders, as reclaiming agents for dark colored compounds.

californium. Cf. A synthetic radioactive element of the actinide group with atomic number 98 and aw 252. It has several isotopes, two of which (Cf-252 and Cf-249) are available in milligram amounts. The pure metal has not yet been obtained. Several compounds are known: the trioxide, the trifluoride, the trichloride, and the sesquioxide. The 252 isotope has potential uses in neutron activation analysis for continuous materials testing, mineral prospecting, oil-well logging, etc. Biologically, it is a bone-seeking element and has specialized applications in medicine.

"Calktite."[326] TM for an acid- and alkali-proof caulking compound, used in protective coating for masonry, acid tanks, and floors.

"Calo-Clor."[329] TM for a mercurial turf fungicide containing 73% mercury in chemical combination (principally mercuric and mercurous chlorides).
Hazard: Toxic by ingestion or inhalation.

"Calocure."[329] TM for a mercurial turf fungicide having a mercury content of 36%, mainly mercuric and mercurous chlorides.
Hazard: A poison. See mercury.

"Calogran."[329] TM for turf fungicide composed of extremely finely divided form of mercurous chloride. Contains 85% mercury (insoluble in water).
Hazard: A poison. See mercury.

calomel. See mercurous chloride.

calorie. The amount of heat needed to raise 1 g of water 1C at 1 atm. A kilogram calorie is the amount of heat required to raise 1 kg of water 1C. In the latter case, the word Calorie is capitalized when used alone, or the abbreviation kcal may be used. In connection with foods and beverages, kilogram calories are referred to.

calorizing. The process by which steel is coated with aluminum by heating it in aluminum powder. The aluminum forms an alloy with the steel surface and produces a thin, tightly adherent coating.
See also cementation.

Calsintering. A method for recovering alumina from fly ash developed in 1978 by Oak Ridge

National Laboratories. The essential steps are pelletizing and sintering a mixture of fly ash and lime (from either limestone or gypsum) at 1000 to 1200C for 20 minutes. The product is ground to 40-mesh and leached with dilute sulfuric acid. The leached solution is then solvent-extracted, crystallized, and calcined to alumina which then can be converted to aluminum by standard methods. The high recovery of alumina suggests that fly ash is a viable alternative raw material for aluminum.

calutron. A type of electromagnetic separator used for radioisotope manufacture.
See electromagnetic separation.

Calvin, Melvin. (1911-) An American chemist who won the Nobel prize for chemistry in 1961. Much of his work involved study on photosynthesis, biophysics, and the application of physics and chemistry of molecules to some of the basic problems of biology. His doctorate was from the University of Minnesota. He did post-graduate work in England and at Northwestern University and at the University of Notre Dame.

"Cambrelle."[206] TM for a nonwoven, melded fabric used for carpet backing, road reinforcement, upholstery, interlinings, tablecloths and other household applications. It is said to be composed of two different polymers, one lying within the other. Upon heating to the melting point of the external polymer, the fibers soften and unite to form a fabric. The term "melded" refers to this type of fusion.
See also nonwoven fabric.

Camp. Biochemical designation for cyclic adenosine-3′,5′- monophosphate, an activator of hormones and initiator of prostaglandin synthesis.

2-camphanol. See borneol.

2-camphanone. See camphor.

camphene. CAS: 79-92-5. $C_{10}H_{16}$.
A terpene.
Properties: Colorless crystals, soluble in ether, slightly soluble in alcohol, insoluble in water. Mp 48–52C, bp 159–162C.
Derivation: (a) By heating pinene hydrochloride with alkalies, aniline or alkali salts such as sodium acetate. (b) A constituent of certain essential oils.
Grade: Technical, (46C mp).
Hazard: Toxic by ingestion.
Use: Manufacture of synthetic camphor, camphor substitute.

camphor. (gum camphor; 2-camphanone).
CAS: 76-22-2. $C_{10}H_{16}O$. A ketone occurring naturally in the wood of the camphor tree (*Cinnamomum camphora*).

Properties: Colorless or white crystals, granules, or easily broken masses; penetrating aromatic odor, d 0.99, mp 174–179C, sublimes slowly at room temperature, slightly soluble in water, soluble in alcohol, ether, chloroform, carbon disulfide, solvent naphtha and fixed and volatile oils. Flash p 150F (65.5C) (CC), autoign temperature 871F (466C). Combustible.
Derivation: Steam distillation of the camphor-tree wood and crystallization. This product is called natural camphor and is dextrortatory. Synthetic camphor, most of which is optically inactive, may be made from pinene, which is converted into camphene which by treatment with acetic acid and nitrobenzene becomes camphor, turpentine oil is also used.
Grade: Technical (synthetic, mp 163–168C), USP (mp 174–179C).
Hazard: Evolves flammable and explosive vapors when heated. TLV: (synthetic) 12 mg/m³ of air.
Use: Medicine (internal and external), plasticizer for cellulose nitrate, other explosives and lacquers, insecticides, moth and mildew proofings, tooth powders, flavoring, embalming, pyrotechnics, intermediate.
Note: A liquid form (camphor oil) is produced almost exclusively in Taiwan; formerly used in the manufacture of sassafras oil; the available supply is used chiefly as a fragrance or flavoring material and to some extent as a pharmaceutical product.

camphor bromate. (α-bromo-1-camphor; brominated camphor). $C_{10}H_{15}BrO$.
Properties: Colorless crystals with slight camphor odor and taste. Also available as powder. Discolors in light and should be stored in cool, dark place. Mp 76C, bp 274C, d 1.449. Soluble in alcohol, ether, chloroform, and oils; insoluble in water.
Derivation: By heating camphor with bromine.
Use: Medicine, manufacture of camphor derivatives.

camphoric acid. $C_8H_{14}(COOH)_2$.
Properties: Colorless, odorless needles or scales,

soluble in alcohol, ether, fatty oils, chloroform. Partially soluble in water. D 1.0 to 1.86 (20/ 4C), mp 186–188C.
Derivation: By oxidizing camphor with nitric acid.
Use: Pharmaceuticals, medicine.

camphor, Malayan. See borneol.

dl-camphoroquinone. (2,3-bornanedione).
CAS: 10373-78-1.
Properties: Mw 166.22, mp 198–200.
Use: Synthetic intermediate.

camphor peppermint. See menthol.

Campillit. See cyanogen bromide.

Camps quinoline synthesis. Formation of hydroxyquinolines from o-acylaminoacetophenones in alcoholic sodium hydroxide. The relative proportions of the isomeric products are mainly determined by the acyl residue on the amino nitrogen.

Canada balsam. See balsam.

cananga oil. An essential oil similar in odor to ylangylang oil, strongly levorotatory. Used for floral odors in perfumery and as flavoring agent.

canavanine.
$NH_2CNHNHOCH_2CH_2CHNH_2COOH.$
A nonprotein amino acid obtained from jackbean meal. It is found naturally in the l(+) form.
Properties: Crystals, from dilute alcohol, mp 184C (decomposes), soluble in water, nearly insoluble in alcohol. Sulfate: Crystals from dilute alcohol, mp 172C (decomposes), soluble in water.
Use: Biochemical research.

candelilla wax.
Properties: Yellowish-brown, opaque to translucent solid. Soluble in chloroform, turpentine, carbon tetrachloride, trichloroethylene, toluene, hot petroleum ether, and alkalies; insoluble in water. D 0.983, mp 67–68C, saponification value 65, iodine number 37, refr index 1.4555. Combustible.
Occurrence: Mexico, Texas.
Grade: Crude, refined, powdered.
Use: Leather dressing, polishes, cements, varnishes, candles, electric insulating composition, sealing wax, waterproofing and insect-proofing containers, paint removers, dentistry, paper sizes, stiffener for soft waxes.

candicidin (USAN). An antifungal antibiotic produced by Streptomyces griseus.

candidin. An antifungal antibiotic produced by Streptomyces viridoflavus.

cane sugar. See sucrose.

cannabis. (marijuana). CAS: 8063-14-7.
Its principle, tetrahydrocannabinol, can be made synthetically.
Derivation: Dried flowering tops of pistillate plants of Cannabis sativa.
Habitat: Iran, India; cultivated in Mexico and Europe.
Hazard: A mild hallucinogen. Sale is illegal in US.
Use: Medicine, opthalmology (treatment of glaucoma).

cannel coal. A variety of bituminous or subbituminous coal of uniform and compact fine-grained texture. Dark gray to black; has a greasy luster. It is noncaking, yields a high percentage of volatile matter, and burns with a luminous, smoky flame. (ASTM definition, ASTM D493-39.) Combustible.
Hazard: Explosion risk in form of dust.

Cannizzaro reaction. Base catalyzed dismutation of aromatic aldehydes or aliphatic aldehydes with no alpha-hydrogen into the corresponding acids and alcohols. When the aldehydes are not identical, the reaction is called "crossed Cannizzaro reaction."

Cannizzaro, Stanislao. (1826–1910) Born in Italy, he extended the research of Avogadro on the molecular concentration of gases and thus was able to prove the distinction between atoms and molecules. His investigations of atomic weights helped to make possible the discovery of the Periodic Law by Mendeleev. His research in organic chemistry led to the establishment of the Cannizzaro reaction involving the oxidation-reduction of an aldehyde in the presence of concentrated alkali.

canthaxanthin. $C_{40}H_{52}O_2$. A carotenoid colorant occurring in many natural products, it has been isolated from a variety of edible mushroom. It has also been synthesized. The trans form is a violet, crystalline solid that is soluble in chloroform and various oils. It is a permissible colorant for foods, drugs, etc. It is used to produce artifical tanning of the skin, but is not approved for this use.
Hazard: Oral intake may cause loss of night vision.

"CAO-1 and CAO-3."[266] TM for technical and food grade BHT (butylated hydroxytoluene) also known as 2,6-di-tertiary butyl para cresol.

"CAO-5."[266] TM for the hindered phenolic antioxidant 2,2′-methylene-bis(4-methyl-6- tertiary-6-butyl phenol).

Caoutchouc. See rubber.

CAP. Abbreviation for chloramphenicol.

capillarity. The attraction between molecules, similar to surface tension, which results in the rise of a liquid in small tubes or fibers or in the wetting of a solid by a liquid. It also accounts for the rise of sap in plant fibers and of blood in capillary (hair-like) vessels.

"Capoten."[412] TM for captopril.

"Capracyl."[28] TM for a group of neutral-dyeing, premetalized acid colors that produce the highest possible degree of light fastness on nylon. Also suitable for dyeing wool, particularly in blends with cellulosic fibers.

capraldehyde. See decanal.

"Capran."[50] TM for transparent nylon 6 thermoplastic film used for food packaging.

capreomycin (USAN). Antibiotic produced by streptomyces capreolus. Used in medicine.

capric acid. (decanoic acid; decoic acid; decylic acid). CAS: 334-48-5.
$CH_3(CH_2)_8COOH.$
Occurs as a glyceride in natural oils.
Properties: White crystals, unpleasant odor, soluble in most organic solvents and dilute nitric acid, insoluble in water. D 0.8858 (40C), bp 270C, 172.6C (30 mm), mp 31.5C, refr index 1.4288 (40C), acid number 308–315. Combustible.
Derivation: Fractional distillation of coconut oil fatty acids.
Grade: Technical, 90%, FCC.
Use: Esters for perfumes and fruit flavors, base for wetting agents, intermediates, plasticizer, resins, intermediate for food-grade additives.

caproic acid. (hexanoic acid; hexylic acid; hexoic acid). CAS: 142-62-1.
$(CH_3(CH_2)_8COOH.$ Present in milk fats to extent of approximately 2%.
Properties: Oily, colorless or slightly yellow liquid, odor of limburger cheese. Soluble in alcohol and ether, slightly soluble in water. D 0.9276 (20/4C), fp −4.0C, bp 205C, refr index 1.4168 (20C), wt/gal 7.7lb, viscosity 0.031 poise (20C), flash p 215F (101C) (OC). Combustible.
Derivation: From crude fermentation of butyric acid, fractional distillation of natural fatty acids.
Grade: Technical, reagent to 99.8%, FCC.
Hazard: Strong irritant to tissue.

Use: Analytical chemistry, flavors, manufacture of rubber chemicals, varnish driers, resins and pharmaceuticals.

caproic aldehyde. See n-hexaldehyde.

caprolactam. (aminocaproic lactam; 2-oxo-hexamethyleneimine).   CAS: 105-60-2.

$$CH_2 \begin{array}{c} CH_2-CH_2-C{=}O \\ \\ \\ CH_2-CH_2-NH \end{array}$$

Properties: White flakes or fused; mp 68–69C; bp 180C (50 mm); d (70% solution) 1.05; refr index 1.4935 (40C); 1.4965 (31C); soluble in water, chlorinated solvents, petroleum distillated, and cyclohexene; heat of fusion 29 cal/g, heat of vaporization 116 cal/g; viscosity 9 cps at 78C; vapor press 3 mm at 100C, 50 mm at 180C.
Derivation: (1) Catalytic oxidation of cyclohexane to cyclohexanol, reacting with peracetic acid to form caprolactone and further reaction with ammonia; (2) catalytic hydrogenation of phenol to cyclohexanone, reaction with ammonia to cyclohexanone oxime with Beckmann rearrangement with sulfuric acid catalyst; (3) catalytic oxidation of cyclohexane to cyclohexanone, reaction with hydroxylamine sulfate and ammonia to cyclohexanone oxime followed by sulfuric acid catalyzed Beckmann rearrangement; (4) UV-catalyzed reaction of cyclohexane with nitrosyl chloride to cyclohexanone oxime hydrochloride, followed by Beckmann rearrangement. Method (1) was never used commercially. Method (3) has been modified to minimize formation of by-product ammonium sulfate.
Forms: Flake, molten.
Hazard: Toxic by inhalation. TLV: (vapor) 5 ppm in air, (dust) 1 mg/m³ of air.
Use: Manufacture of synthetic fibers (especially nylon 6), plastics, bristles, film, coatings, synthetic leather, plasticizers and paint vehicles, cross-linking agent for polyurethanes, synthesis of amino acid lysine.

caprolactone.
Derivation: Reaction product of peracetic acid and cyclohexanone an intermediate product in manufacture of caprolactam.
Use: Intermediate in adhesives, urethane coatings and elastomers, solvent, diluent for epoxy resins, synthetic fibers, organic synthesis.

"Caprolan."[523] TM for a polyamide fiber made from polymerized caprolactam. Has excellent dye-ability and a wide variety of end-uses, main-

tains a superior level of dimensional stability after heat setting and has outstanding mechanical qualities.
See also caprolactam, nylon.

capryl compounds. The term "approximately capryl" is generally but erroneously used in the trade to refer to octyl compounds. Thus, the definition of capryl and caprylic compounds will be found under the corresponding octyl entry, e.g., for capryl alcohol.
See 2-n-octanol; for caprylic halides, see corresponding octyl halide; for caprylic acid see octanoic acid.

caprylyl peroxide. Legal label name for octyl peroxide.

captafol. See: cis-N(1,1,2,2-tetrachloroethyl)-thio]-4-cyclohexene-1,2-dicarboximide.

captan. (N-trichloromethylmercaptotetrahydrophthalimide). CAS: 133-06-2. $C_9H_8Cl_3NO_2S$.
Properties: White to cream powder, mp 158–164C, d 1.5. Practically insoluble in water, partially soluble in acetone, benzene and toluene, slightly soluble in ethylene dichloride and chloroform.
Derivation: Reaction product of tetrahydrophthalamide and trichloromethylmercaptan.
Hazard: Irritant. Avoid contamination of feed and foodstuffs. Avoid inhalation of dust or spray mist. TLV: 5 mg/m³ of air.
Use: Seed treatment, fungicide in paints, plastics, leather, fabrics and fruit preservation, bacteriostat.

"Captax."[265] TM for 2-mercaptobenzothiazole.

captopril. (2,d-methyl-3-mercaptopropanoyl-l-proline). An orally active hypertensive drug. Its use is limited by FDA. Said to be free from usual side effects of commonly used hypertensive drugs.

capture. The process in which an atomic or nuclear system acquires an additional particle, i.e., the capture of electrons by positive ions, or capture of electrons or neutrons by nuclei.
See also cross-section (1), neutron, fission.

caramel. (1) A sugar-based food colorant made from liquid corn syrup by heating in the presence of catalysts to approximately 250F (121C) for several hours, cooling to 200F (93C), and filtering. The brown color results from either Maillard reactions, true caramelization, or oxidative reactions. Caramels are colloidal in nature, the particles being held in solution by either positive or negative electric charges.
See also caramelization,
browning reaction. (2) A low-enriched uranium reactor fuel containing 6.8% of U-235 (approximately 10 times as much as natural uranium). It has sufficient neutron density to yield plutonium.

caramelization. A type of nonenzymic browning reaction occurring during exposure of food products to heat when the products contain no nitrogen compounds, e.g., sugars.
See also browning reaction.

carbamate. A compound based on carbamic acid NH_2COOH which is used only in the form of its numerous derivatives and salts.

carbamide. See urea.

carbamide peroxide. See urea peroxide.

carbamide phosphoric acid. (urea phosphoric acid). $CO(NH_2)_2 \cdot H_3PO_4$.
Properties: White, rhombic crystals; very soluble in water and alcohol.
Hazard: Evolves toxic fumes when heated.
Use: Catalyst for acid-setting resins, flameproofing compositions, cleaning compounds, acidulant.

carbamidine. See guanidine.

carbamite. See sym-diethyldiphenylurea.

carbamylguanidine sulfate. See guanylurea sulfate.

carbamylguanidine sulfate. See guanylurea sulfate.

carbamylhydrazine hydrochloride. See semicarbazide hydrochloride.

carbamylurea. See biuret.

carbanil. See phenyl isocyanate.

carbanilide. See diphenylurea.

carbanion. A negatively charged organic ion such as H_3C^-, RC^-, having one more electron than the corresponding free radical. Carbanions are short-lived but important intermediates in base-catalyzed polymerization and alkylation reactions.
See carbonium ion, carbene, free radical.

carbaryl. (Generic name for 1-naphthyl-N-methylcarbamate). CAS: 63-25-2. $C_{10}H_7OOCNHCH_3$.
Properties: Solid, mp 142C, d 1.23. Insoluble in water.

Derivation: Synthesized directly from 1-naphthol and methyl isocyanate or from naphthyl chloroformate (1-naphthol and phosgene) plus methylamine.

Hazard: Toxic by ingestion, inhalation, and skin absorption; irritant. A reversible cholinesterase inhibitor. Use may be restricted. TLV: 5 mg/m³ of air.

Use: Insecticide.

carbazide. See carbodihydrazide.

carbazole. (dibenzopyrrole; diphenylenimine). CAS: 86-74-8. $(C_6H_4)_2NH$ (tricyclic).

Properties: White crystals with characteristic odor, mp 244–246C, bp 352–354C, partially soluble in alcohol and ether, insoluble in water.

Derivation: (a) From crude anthracene cake by selective solution of the phenanthrene with crude solvent naphtha, removal of the anthracene by conversion into a sulfonic derivative and extraction by means of water. (b) Synthetically from o-aminobiphenyl.

Grade: Technical, 97%. Use: Manufacture of dyes, reagents, explosives, insecticides, lubricants, rubber antioxidants, odor inhibitor in detergents, UV sensitizer for photographic plates.

carbazotic acid. See picric acid.

carbene. (methylene).   An organic radical containing divalent carbon. Some divalent carbon derivatives where the carbon is multiplebonded to oxygen or nitrogen are stable compounds (carbon monoxide); most are highly reactive units that are known only as reaction intermediates. Carbenes, carbonium ions, carbanions, and free radicals are the four most important classes of organic reaction intermediates containing carbon in an unstable valence state. A typical synthesis involving a carbene is that of cyclopropanes by the addition of carbenes to olefins.

2-carbethoxycyclohexanone.
$OC_6H_9COOC_2H_5$.

Properties: Colorless liquid with a characteristic ester odor, bp 106–107C (11 mm), d 1.074 (25C), refr index 1.4750 (17.5C), soluble in dilute alkali, insoluble in water. Combustible.

Use: Intermediate.

2-carbethoxycyclopentanone. (ethyl cyclopentanone-2-carboxylate; ethyl-2-oxocyclopentanecarboxylate). $OC_5H_7COOC_2H_5$.

Properties: Colorless liquid with characteristic ester odor, bp 122–124C (25 mm), flash p 191F (88.3C), refr index 1.451 (25C), d 1.0976 (0C), soluble in equimolar amounts of dilute alcohol, insoluble in water. Combustible.

Hazard: Vapors are toxic as is skin contact.
Use: Pharmaceutical intermediate.

β-carbethoxyethyltriethoxysilane.
$C_2H_5OOC(CH_2)_2Si(OC_2H_5)_3$.

Properties: Colorless liquid, bp 246C. Combustible.

Hazard: An irritant.
Use: Intermediate.

N-carbethoxypiperazine. $C_7H_{14}N_2O_2$.

Properties: Colorless, somewhat viscous liquid, slight odor, bp 116–117C (12 mm), 237C, refr index 1.4756 (25C). Miscible with water and common organic solvents.

Use: Intermediate.

β-carbethoxypropylmethyldiethoxysilane.
$C_2H_5OOC(C_3H_6)CH_3Si(OC_2H_5)_2$.

Properties: Colorless liquid, bp 228C. Combustible.

Hazard: An irritant.
Use: Intermediate.

carbide.   A binary solid compound of carbon and another element. The most familiar carbides are those of calcium, tungsten, silicon, boron, and iron (cementite). Two factors have an important bearing on the properties of carbides: (1) the difference in electronegativity between carbon and the second element, and (2) whether or not the second element is a transition metal. Salt-like carbides of alkali metals are obtained by reaction with acetylene. Those obtained from silver, copper, and mercury salts are explosive.

See acetylide; carbide, refractory; and carbide, cemented.

carbide, cemented.   A powdered form of refractory carbide united by compression with a bonding material (usually iron, nickel, or cobalt), followed by sintering. Tungsten carbide is bonded with cobalt at 1400C; from 3 to 25% of cobalt is used depending on the properties desired. Used chiefly in metal cutting tools which are hard enough to permit cutting speeds in rock or metal up to 100 times that obtained with alloy steel tools.

carbide, refractory.   A carbide characterized by great hardness, thermal stability, high melting point, and chemical resistance. Decomposed by fusion with alkali and attacked by mixtures of nitric and hydrofluoric acids. The best known refractory carbides are those of silicon, boron, tungsten, and tantalum. Used as abrasives, furnace linings, and in other high-temperature applications. Some types are bonded.

carbinol. (1) Synonym for methanol, CH_3OH; (2) hence any compound of similar structure retaining the COH radical and in which hydrocarbon radicals may be substituted for the hydrogen originally attached to the carbon. Thus, isopropanol, $(CH_3)_2CHOH$, and benzyl alcohol, $C_6H_5CH_2OH$, may be named dimethylcarbinol and phenylcarbinol respectively.

carbinoxamine maleate. (2-[p-chloro-α-(2-dimethyl-amino-ethoxy)-benzyl]pyridinemaleate). $ClC_6H_4CH(C_5H_4N)OCH_2CH_2N(CH_3)_2 \cdot C_4H_4O_4$.
Properties: White, odorless, bitter, crystalline powder; mp 116–121C. Very soluble in water, freely soluble in alcohol and chloroform, very slightly soluble in ether, pH (1% solution) 4.6–5.1.
Grade: NF.
Use: Medicine.

"Carbitol."[214] TM for a group of mono- and dialkyl ethers of diethylene glycol and their derivatives, specialized solvents with a wide variety of properties and uses. Specific types are as follows:
butyl "Carbitol"
See diethylene glycol monobutyl ether.
butyl"Carbitol" acetate
See diethylene glycol monobutyl ether acetate.
"Carbitol" acetate
See diethylene glycol monoethyl ether acetate.
"Carbitol" solvent
See diethylene glycol monoethyl ether.
dibutyl "Carbitol"
See diethylene glycol dibutyl ether.
diethyl "Carbitol"
See diethylene glycol diethyl ether.
N-hexyl "Carbitol"
See diethylene glycol monohexyl ether.
methyl "Carbitol"
See diethylene glycol monomethyl ether.
methyl "Carbitol" acetate
See diethylene glycol
monomethyletheracetate.

carbobenzyloxy-l-alanine. CAS: 1142-20-7.
$CH_3CH(NHCO_2CH_2C_6H_5)CO_2H$.
Properties: Mw 223.23, mp 82–84C, optical rotation −14.2 degrees (23C).
Grade: 98+% pure.

Carbocaine. Proprietary name for mepivacaine hydrochloride.
Use: In dentistry as a local anesthetic.

carbocyclic. Any organic compound whose "skeleton" is in the form of a closed ring of carbon atoms. This includes both alicyclic and aromatic structures.

carbodihydrazide. (carbazide).
$CO(NHNH_2)_2$.
Properties: Colorless crystals, mp 154C, d 1.1616 (−5C). Very soluble in water and alcohol.
Use: Organic intermediate and photographic chemical.

carbodiimide. See cyanamide (1).

"Carbo-Dur."[184] TM for an adsorbent of granular activated carbon used in taste, color, and odor removal.

"Carbofrax."[280] TM for bonded refractory bricks for furnace linings, muffle walls, etc.; also the cement or mortar used to install them. Contains 85% or more silicon carbide. Porosity approximately 13%.
See also refractory, silicon carbide.

carbofuran. See "Furadan."

carbohydrase. An enzyme whose catalytic activity is directed toward the breaking down of complex carbohydrates to simpler units. Illustrations are amylase, invertase, maltase.

carbohydrate. A compound of carbon, hydrogen, and oxygen that contains the saccharose unit or its first reaction product and in which the ratio

$$H-\underset{\underset{OH}{|}}{C}-\underset{\underset{O}{\|}}{C}-$$

of hydrogen to oxygen is the same as in water. Carbohydrates are the most abundant class of organic compounds, constituting three-fourths of the dry weight of all vegetation. They are also widely distributed in animals and lower forms of life. They comprise: (1) monosaccharides, simple sugars such as fructose (levulose) and its isomer glucose (dextrose), both having the formula $C_6H_{12}O_6$; (2) disaccharides, sucrose $(C_{12}H_{22}O_{11})$, maltose, cellobiose and lactose; and (3) polysaccharides (high polymeric substances). The last group includes the entire starch and cellulose families as well as pectin, the seaweed products agar and carrageenan, and natural gums. The simple sugars are crystalline and water-soluble, with a sweet taste; starches are water-soluble, tasteless, and amorphous; cellulose is insoluble in water and organic solvents and is only partially crystalline. Galactose, sorbose, xylose, arabinose, and mannose are constituents of more complex sugars. The natural gums are water-soluble plant products composed of monosaccharide units joined by glycosidic bonds (arabic, tragacanth).

Carbohydrates are an important natural source of ethanol now in extensive use in gasohol and other energy applications.
See also energy sources, gasohol, fermentation.

Carbohydrogen. See pintsch gas, oil gas.

"Carbo-Korez."[41] TM for a carbon-filled synthetic-resin, acid-proof cement of the phenol-formaldehyde type used as a mortar cement where temperatures do not exceed 370F. Especially good for high concentrations of sulfuric acid.

"Carbolac."[275] TM for high color channel blacks, used for paint, varnish, and lacquer.

carbolfuchsin. (Ziehl's stain). A staining solution of fuchsin in alcohol and aqueous phenol used in the study of microorganisms.

carbolic acid. Legal label name for phenol.

carbolic oil. (middle oil). Comprises the fraction having a boiling range of about 190–250C obtained from distillation of coal tar and containing naphthalene, phenol, and cresols.

3-carbomethoxy-1-methyl-4-piperidone hydrochloride. $C_8H_{13}NO_3 \cdot HCl$.
Properties: White, crystalline solid; mp 165C. Soluble in water, alcohols; insoluble in ether, hydrocarbons.
Use: Pharmaceutical intermediate.

2-carbomethoxy-1-methylvinyl dimethyl phosphate. The beta isomer is also a pesticide.
See mevinphos for the alpha isomer.

carbomycin. CAS: 4564-87-8. $C_{42}H_{67}NO_{16}$.
An antibiotic isolated from products of *Streptomyces halstedii* when grown in suitable media by deep culture method. It inhibits growth of certain gram-positive bacteria such as staphylococci, pneumococci and hemolytic streptococci. Mp 214C.

carbon. CAS: 7440-44-0. C. A nonmetallic element, atomic number 6, aw 12.011, Group IVA of the Periodic Table, normal valence 4, but divalent forms are known (carbenes). Carbon has two stable and four radioactive isotopes. The C^{12} isotope which comprises 99% of the element is the standard to which atomic weights of all other elements are referred (i.e., carbon = 12.00 exactly). One mole of carbon atoms (6.02×10^{23}) is contained in 12 grams of C^{12}. Carbon has two crystalline allotropes (diamond and graphite)

and several amorphous allotropes (coal, coke, carbon black, charcoal). Carbon is present in all organic and in a few inorganic compounds (carbon oxides, carbon disulfide, and metallic carbonates such as calcium carbonate). It is the active element in photosynthesis and thus occurs in all plant and animal life. The radioisotope C^{14} used in tracer research and chemical dating. Carbon is a strong reducing agent and is used as such in purifying metals. It is the only element capable of forming four covalent bonds. Its strong electrical conductivity is used to advantage in electrodes and other electrical devices. Its presence in small proportions in steel has a pronounced effect on the properties of the metal.

Carbon forms binary compounds called carbides with many metals and some nonmetals. A few compounds are known which contain divalent carbon (carbenes or methylenes).

Since its major properties and uses vary widely with its form the following entries should be consulted: diamond, graphite, activated carbon, carbon black, industrial carbon, charcoal, coke, steel, carbon cycle.

carbon-11. (C^{11}). A short-lived radioactive isotope of carbon which emits positrons which in turn become a source of gamma rays when they collide with an electron within the body. It is used experimentally in nuclear medicine for labeling pharmaceuticals.

carbon-13. (C^{13}). Stable, nonradioactive carbon isotope used for special analytical research. Commercially available in gram quantities.

carbon-14. (C^{14}). (radiocarbon). Naturally occurring, radioactive carbon isotope of mass number 14, a special case of radioactivity induced by cosmic rays in the upper atmosphere. Neutrons produced by cosmic radiation impact nitrogen atoms to yield C^{14} and a proton. Half-life 5580 years; beta radiation. Can be made by reactor irradiation of calcium nitrate.
Use: Radiation source in thickness gauges and other instruments, elucidation of mechanisms in organic chemistry, metallurgy and biochemical reactions, radiocarbon dating in geology and archaeometry.
See also chemical dating, archaeometry.

Carbona. An obsolete proprietary mixture of petroleum ether (bp 70-72C) and carbon tetrachloride.
Use: Cleaning fluid.

carbon, activated. (active carbon; activated charcoal). An amorphous form of carbon characterized by high adsorptivity for many gases, vapors,

and colloidal solids. The carbon is obtained by the destructive distillation of wood, nut shells, animal bones, or other carbonaceous material. It is "activated" by heating to 800–900C with steam or carbon dioxide which results in a porous internal structure (honeycomb-like). The internal surface area of activated carbon averages approximately 10,000 square feet per gram. The density is from 0.08 to 0.5. It is not effective in removing ethylene.
Grade: Technical, USP, as activated charcoal.
Hazard: Flammable. Toxic by inhalation of dust.
Use: Decolorizing of sugar, water and air purification, solvent recovery, waste treatment, removal of sulfur dioxide from stack gases and "clean" rooms, deodorant, removal of jet fumes from airports, catalyst natural gas purification, brewing, chromium electroplating, air conditioning.

carbonado.   See diamond, industrial.

carbon, amorphous.   See carbon, activated, carbon black.

carbonate.   A compound resulting from the reaction of either a metal or an organic compound with carbonic acid. The reaction with a metal yields a salt (calcium carbonate) and that with an aliphatic or aromatic compound forms an ester, e.g., diethyl carbonate, diphenyl carbonate. The latter are liquids used as solvents and in synthesizing polycarbonate resins.
See also carbonic acid.

carbonate mineral.   A mineral in which the basic bulding block is a carbon atom linked to three oxygen atoms. Calcite, aragonite, and dolomite are the most abundant examples found in sediments and sedimentary rocks.

carbonate rock.   A sedimentary rock that consists primarily of carbonate minerals. The dominant mineral is nearly always either calcite, in which case the rock is limestone, or dolomite, in which case the rock is dolomite.

carbonate sediment.   Unconsolidated sediment that consists primarily of carbonate minerals, usually aragonite or calcite.

carbon bisulfide.   See carbon disulfide.

carbon black.   CAS: 1333-86-4.   35th highest-volume chemical produced in US (1985). Finely divided form of carbon, practically all of which is made by burning vaporized heavy oil fractions in a furnace with 50% of the air required for complete combustion (partial oxidation). This type is also called furnace black. Carbon black can also be made from methane or natural gas by cracking (thermal black) or direct combustion (channel black), but these methods are virtually obsolete. All types are characterized by extremely fine particle size, which accounts for their reinforcing and pigmenting effectiveness.
Grade: (Furnace black) conducting (CF), fine (FF), high modulus (HMF), high elongation (HEF), reinforcing (RF), semi-reinforcing (SRF), high abrasion (HAF), super abrasion (SAF), fast extruding (FEF), general purpose (GPF), intermediate super abrasion (ISAF), channel replacement (CRF), easy processing furnace black (EPF).
Hazard: TLV 3.5 mg/m³ of air.
Use: Tire treads, belt covers, and other abrasion-resistant rubber products; plastics as reinforcing agent, opacifier, electrical conductivity, UV light absorber; colorant for printing inks; carbon paper; typewriter ribbons; paint pigment; nucleating agent in weather modification; expanders in battery plates; solar energy absorber (see note).
Note: A suspension of finely divided carbon particles in compressed air has been researched as a solar energy absorber. The heat is absorbed until the particles vaporize, yielding energy that can be used directly or for power production. The suspension is placed in a transparent container located in a solar concentrator.

carbon black oil.   A heavy refinery fraction similar to fuel oil, used as a feedstock for furnace black.

carbon, combined.   A metallurgical term for carbon which has combined chemically with iron to form cementite, as distinct from graphitic carbon in iron or steel.
See also pearlite, ferrite.

carbon cycle.   (1) The progress of carbon from air (carbon dioxide) to plants by photosynthesis (sugar and starches), then through the metabolism of animals to decomposition products which ultimately return it to the atmosphere in the form of carbon dioxide; (2) One of the processes by which the sun and other self-luminous astronomical bodies are thought to derive their energy. The net process is the combination (fusion) of four hydrogen atoms to form helium. One mechanism, called the carbon cycle, involves successive additions of hydrogen atoms followed by beta decay, to an initial carbon-12 atom until a final step is reached in which the new nucleus breaks down to a helium atom and a carbon-12 is regenerated. The carbon thus functions as a catalyst for the process. At the temperatures prevailing in the sun, all atoms are stripped of their electrons and the reaction is between the nuclei of the atoms (thermonuclear reaction). Symbolically the set of reactions is written.

$C^{12} + H^1 \rightarrow N^{13}$, $N^{13} \rightarrow C^{13} + e$; $C^{13} + H^1$ $\rightarrow N^{14}$, $N^{14} + H^1 \rightarrow O^{15}$. $O^{15} \rightarrow N^{15} +e$, N^{15} $\rightarrow C^{12} + He^4$.
See also fusion.

carbon dating. Estimation of the age of geologic structures and events by measuring the amount of radioactive decay products in existing samples. The age of a uranium-containing material can be determined by measuring the percentage of lead (or helium) formed as a result of disintegration of the uranium. Uranium decays to both helium and the 206-lead isotope, but measurement of helium content is inaccurate because of its strong tendency to escape. By determining the ratio of the percentage of lead in a sample to the percentage of uranium, the age can be calculated. Radiocarbon dating is a method of determining quite accurately the age of a carbon-bearing material derived from living plants or animals within the last 70,000 years. It is based on the ratio of carbon-14 in the material to that in a modern reference sample by measuring the radioactivity of the carbon-14 in the material. Since the half-life of carbon-14 is 5730 +/-30 years and the living precursor utilized carbon dioxide from the atmosphere or some other part of the earth's dynamic carbon reservoir, a process that ceased when the original plant or animal died, the amount of carbon-14 now present gives directly the age of the material. The carbon-14 in the reservoir is constantly being replaced by the sequence $N^{14} \rightarrow C^{14} + O \rightarrow C^{14}O_2$. This has maintained the ratio of carbon isomers constant during the ages; but burning of fossil fuels since the Industrial Revolution has lowered somewhat the quantity of carbon-14 in the atmosphere during the last few centuries, an effect that does not affect measurements on older objects. The sample to be tested must be carefully prepared to prevent contamination by younger carbon.

The radiocarbon technique was discovered by Willard F. Libby (1908–1980), Nobel Prize (1960), and has been applied with great success in the fields of archeology, geology, geochemistry, and geophysics. Its accuracy has been checked and verified by use of tree-ring counts (dendrochronology) and with the known ages of objects from ancient cultures, such as Egyptian and Chinese. The former shows that for the 2400–6000 BP (before present) age of bristlecone pine tree-rings C^{14} years equal 6,000 calendar years.

carbon dichloride. See perchloroethylene.

carbon dioxide. CAS: 124-38-9. CO_2.
17th highest-volume chemical produced in the US (1985).
Properties: (a) Gas: colorless, odorless, d 1.97g/L

(0C, 1 atm); d 1.53 (air = 1.00); (b) liquid: volatile, colorless, odorless, d (−37C) 1.101, sp volume 8.76 cu ft/lb (70F); (c) solid (dry ice): white, snow-like flakes or cubes; d 1.56 (−79C); mp −78.5C (sublimes). All forms are noncombustible. Miscible with water (1.7 volumes per volume at 0C and 0.76 volume per volume at 25C and 760 mm partial pressure of CO_2). Also miscible with hydrocarbons and most organic liquids. An asphyxiant gas in concentrations of 10% or more; low concentrations (1–3%) increase lung ventilation and are used admixed with oxygen in resuscitation equipment.
Derivation: (a) Gas: for industrial use, carbon dioxide is recovered from synthesis gas in ammonia production, from substitute natural gas production, from cracking of hydrocarbons, and from natural springs or wells. For laboratory purposes it is obtained by the action of an acid on a carbonate. It is also a by-product of the fermentation of carbohydrates and an end product of combustion and respiration. Air contains 0.033% of carbon dioxide (see greenhouse effect). (b) Liquid: by compressing and cooling the gas to approximately −37C. (c) Solid (dry ice): by expanding the liquid to vapor and snow in presses that compact the product into blocks. The vapor is recycled.
Grade: Technical, USP, commercial and welding 99.5%, bone dry (99.95%).
Hazard: Solid damaging to skin and tissue, keep away from mouth and eyes. TLV (gas): 5000 ppm in air.
Use: Refrigeration, carbonated beverages, aerosol propellant, chemical intermediate (carbonates, synthetic fibers, p-xylene, etc.), low-temperature testing, fire extinguishing, inert atmospheres, municipal water treatment, medicine, enrichment of air in greenhouses, fracturing and acidizing of oil wells, mining (Cardox method), miscible pressure source, hardening of foundry molds and cores, shielding gas for welding, cloud seeding, moderator in some types of nuclear reactors, immobilization for humane animal killing, special lasers, blowing agent, as demulsifier in tertiary oil recovery, possible source of methane, (liquid) carrier for powdered coal slurry.
Note: Carbon dioxide is the source of the carbon utilized by plants to form organic compounds in the photosynthetic reaction, catalyzed by chlorophyll.
See photosynthesis, carbon cycle (1).

carbon disulfide. (carbon bisulfide).
CAS: 75-15-0. CS_2.
Properties: Clear, colorless or faintly yellow liquid; almost odorless when pure; usually strong disagreeable odor. D 1.260 at 25/25C, bp 46.3C, fp −111C, wt/gal 10.48 lb (25C), refr index 1.6232 (25C), flash p −22F (−30C), autoign tem-

perature 212F (100C). Soluble in alcohol, benzene, and ether; slightly soluble in water. Classed as an inorganic compound.
Derivation: (a) Reaction of natural gas or petroleum fractions with sulfur. (b) From natural gas and hydrogen sulfide at very high temperature (plasma process). (c) By heating sulfur and charcoal and condensing the carbon disulfide vapors.
Method of purification: Distillation. Impurities: Sulfur compounds.
Grade: 99.9%, spectrophotometric.
Hazard: A poison. Toxic by skin absorption. TLV: 10 ppm in air. Highly flammable, dangerous fire and explosion risk, can be ignited by friction. Explosive limits in air 1–50%
Use: Viscose rayon, cellophane, manufacture of carbon tetrachloride and flotation agents, solvent.

carbon, divalent. See carbene.

carbon fiber. See graphite fiber.

carbon fluoride. $(CF)_x$, C_4F. A solid, nonconductive material formed on carbon anodes during electrolysis of molten potassium fluoride-hydrogen fluoride mixtures to yield elemental fluorine. C_4F is unstable at above 60C, $(CF)_x$ forms only at high temperatures.

carbon, graphitic. A metallurgical term referring to practically pure carbon which forms in pig iron during cooling because the absorbing power of iron for carbon decreases as its temperature falls. It exists in the iron in the form of tiny flakes distributed throughout the mass. The tendency of graphitic carbon is to weaken the metal while combined carbon up to the limit of approximately 0.90% strengthens it.
See also pearlite, cementite, ferrite.

carbon hexachloride. See hexachloroethane.

carbonic acid. H_2CO_3.

$$\begin{array}{c} HO \\ \diagdown \\ \diagup \quad C{=}O. \\ HO \end{array}$$

A weak acid formed by reaction of carbon dioxide with water. Both organic and inorganic carbonates are formed from it by reaction with organic compounds or metals, respectively. Thus inorganic carbonates, ($CaCO_3$, K_2CO_3, Na_2CO_3, etc.) are salts of carbonic acid and organic carbonates are esters of carbonic acid.

carbonic anhydrase. An enzyme in red blood cells which catalyzes the production of carbon dioxide and water from carbonic acid.
Use: Biochemical research.

carbon, industrial. Any form of pure carbon used for industrial purposes, exclusive of fuel. Coke is one of the most important. Besides its use (combined with coal-tar pitch) for refractories, furnace linings, electrodes, fibers, etc., it has tremendous volume consumption for reduction of iron in blast furnaces (see coke). Graphite in its many applications is another form, activated carbon for decolorizing and solvent recovery, carbon black for rubber and printing inks, industrial diamonds as abrasives and drilling bits, compressed carbon for electrodes and other electrical uses, and carbon fibers and whiskers are all included in this term.

carbonium ion. A positively charged organic ion such as H_3C^+, H_2RC^+, $R_3C^+{=}O$, etc., having one less electron than the corresponding free radical and acting in subsequent chemical reactions as though the positive charge was localized on the carbon atom. Such ions can exist only when corresponding negative ions are also present. An electron-deficient carbon atom is extremely reactive and has only a transitory existence in most cases, but many organic rearrangement and replacement reactions are effectively explained in terms of a carbonium ion intermediate, including acid-catalyzed polymerization of propylene and other olefins. In this case, propylene and hydrogen ion form a carbonium ion as follows:
$$H_3C{-}HC{=}CH_2 + H^+ \rightarrow H_3C{-}HC^+{-}CH_3.$$
The latter then combines with another molecule of $HC_3{-}HC{=}CH_2$ to start chain growth.
The difference between a carbonium ion, a free radical, and a carbanion may be illustrated as follows:

$$\left[\begin{array}{c} R \\ \ddot{} \\ R : C \\ \ddot{R} \end{array} \right]^+ \quad \left[\begin{array}{c} R \\ R : \overset{\cdot}{C}{\cdot} \\ R \end{array} \right]^0 \quad \left[\begin{array}{c} R \\ R : \overset{\cdot}{\underset{\cdot}{C}} : \\ R \end{array} \right]^-$$

carbonium   free   carbanion
ion   radical

carbonization. See destructive distillation.

carbon monoxide. CAS: 630-08-0. CO.
Discovered by Priestly in America in 1799.
Properties: Colorless gas or liquid, practically odorless. Burns with a violet flame, slightly soluble in water, soluble in alcohol and benzene. D 0.96716 (air = 1.0), bp −190C, fp −207C, specific volume 13.8 cu ft/lb (21.1C). Autoign temperature (liquid) 1128F (609C). Classed as an inorganic compound.
Derivation: (a) Obtained almost pure by placing a mixture of oxygen and carbon dioxide in con-

tact with incandescent graphite, coke, or anthracite. (b) Action of steam on hot coke or coal (water gas) or on natural gas (synthesis gas). In the latter case, carbon dioxide is removed by absorption in amine solution, the hydrogen and carbon monoxide separated in a low-temperature unit. (c) By-product in chemical reactions. (d) Combustion of organic compound with limited amount of oxygen, as in automobile cylinders. (e) Dehydration of formic acid.

Grade: Commercial (98%), CP (99.5%).

Hazard: Highly flammable, dangerous fire and explosion risk. Flammable limits in air 12–75% by volume. Toxic by inhalation. TLV: 50 ppm (industrial workrooms), USSR standard 35 ppm. Note: Carbon monoxide has an affinity for blood hemoglobin over 200 times that of oxygen. A major air pollutant.

Use: Organic synthesis (methanol, ethylene, isocyanates, aldehydes, acrylates, phosgene), fuels (gaseous), metallurgy (special steels, reducing oxides, nickel refining), zinc white pigments.

"Carbonox."[236] An organic humic acid material used for treatment of drilling mud to reduce viscosity and gel strength. Also used to prepare emulsion muds characterized by low filtration rates, stability, and easy maintenance.

carbon oxybromide. See bromophosgene.

carbon oxychloride. See phosgene.

carbon oxycyanide. See carbonyl cyanide.

carbon oxyfluoride. See carbonyl fluoride.

carbon oxysulfide. See carbonyl sulfide.

carbon steel. See steel.

carbon suboxide. C_3O_2. Molecular structure: $O=C=C=C=O$.

Properties: Colorless gas or liquid, strong pungent odor, bp 7C, fp −110C, refr index 1.45, d 1.12. Forms malonic acid with water. Polymerizes on storage even under pressure of 7 atm.

Derivation: From malonic acid by destructive distillation.

Hazard: Explosive limits in air 6–30%. Strong irritant to eyes and mucous membranes, causes lachrymation and impaired breathing.

Use: Dyeing auxiliary, chemical intermediate.

carbon tetrabromide. (tetrabromomethane). CAS: 558-13-4. CBr_4. A brominated hydrocarbon.

Properties: Colorless crystals, d 3.42, mp 90.1C, bp 189.5C, insoluble in water, soluble in alcohol, ether, and chloroform. Noncombustible.

Hazard: A poison; narcotic in high concentration. TLV: 0.1 ppm in air.

Use: Organic synthesis.

carbon tetrachloride. (tetrachloromethane; perchloromethane). CAS: 56-23-5. CCl_4. A chlorinated hydrocarbon.

Properties: Colorless liquid; vapor 5.3 times heavier than air; sweetish, distinctive odor; d 1.585 (25/4C); bp 76.74C; fp −23.0C; refr index 1.4607 (20C); vap press 91.3 mm (20C); wt/gal 13.22lb (25C); flash p none. Miscible with alcohol, ether, chloroform, benzene, solvent naphtha and most of the fixed and volatile oils; insoluble in water. Noncombustible.

Derivation: (a) Interaction of carbon disulfide and chlorine in presence of iron; (b) chlorination of methane or higher hydrocarbons at 250–400C.

Method of purification: Treatment with caustic alkali solution to remove sulfur chloride, followed by rectification.

Grade: Technical, CP, electronic.

Hazard: Toxic by ingestion, inhalation, and skin absorption; Do not use to extinguish fire. narcotic. A carcinogen (OSHA). TLV: 5 ppm in air. Decomposes to phosgene at high temperatures.

Use: Refrigerants. Metal degreasing, agricultural fumigant, chlorinating organic compounds, production of semiconductors, solvent (fats, oils rubber, etc).

Note: Not permitted in products intended for home use.

carbon tetrafluoride. See tetrafluoromethane.

carbon trichloride. See hexachloroethane.

N,N'-carbonyl bis(4-methoxymetanilic acid)disodium salt. (sodium methoxymetanilate urea). $[C_6H_3(OCH_3)(SO_3Na)NH]_2CO$.

Properties: Gray paste, solids approximately 70%.

Grade: Technical.

Use: Intermediate.

carbonyl bromide. See bromophosgene.

carbonyl chloride. See phosgene.

carbonyl cyanide. (carbon oxycyanide). $CO(CN)_2$.

Properties: Colorless liquid. Unstable in the presence of water, bp −83C, d 1.139 (−114C), mp 114C.

Hazard: Toxic. TLV: (as cyanide) 5 mg/m³ in air.

Use: Organic synthesis.

1,1'-carbonyldiimidazole. (N,N-carbonyldiimidazole; 1,1'-carbonylbis-1H-imidazole).

$C_7H_6N_4O$. Should be handled in absence of atmospheric moisture to avoid release of carbon dioxide.

Properties: Off-white powder or crystals. Mw 162.15, mp 118–120C.

Use: Enzyeme cross-linking agent, condensing agent for nucleoside triphosphate synthesis.

carbonyl fluoride. (fluoroformyl fluoride; carbon oxyfluoride). CAS: 353-50-4. COF_2.

Properties: Colorless, hygroscopic gas. Unstable in the presence of water. Bp −83C, d 1.139 (−114C), fp −114C. Min purity 97 mole %. Nonflammable.

Derivation: Action of silver fluoride on carbon monoxide.

Grade: Technical.

Hazard: Toxic by inhalation, strong irritant to skin. TLV: 2 ppm in air.

Use: Organic synthesis.

carbonyl group. The divalent group $=C=O$ which occurs in a wide range of chemical compounds. It is present in aldehydes, ketones, organic acids, and sugars and in the carboxyl group, i.e.,

$$-C\!\!\nwarrow_{OH}^{O}$$

In combination with transition metals, it forms coordination compounds which are highly toxic, as they decompose to release carbon monoxide when absorbed by the body, e.g., nickel carbonyl. Several metal carbonyls have antiknock properties. The carbonyl group is also found in combination with nonmetals, as in phosgene (carbonyl chloride); these compounds are also poisonous.

carbonyl sulfide. (carbon oxysulfide). CAS: 463-58-1. COS.

Properties: Colorless gas with typical sulfide odor except when pure. D gas (air = 1) 2.1, fp −138.8C, bp −50.2C (1 atm). Soluble in water and alcohol.

Derivation: Hydrolysis of ammonium or potassium thiocyanate.

Hazard: Narcotic in high concentrations. Flammable, explosive limits in air 12–28.5%.

carbophenothion. (Generic name for (S-[[p-chlorophenyl)thio]methyl]-O,O-diethyl phosphorodithioate; O-O-diethyl-S-(p-chlorophenylthiomethyl) phosphorodithioate).

$(C_2H_5O)_2P(S)SCH_2S(C_6H_4)Cl$.

Properties: Amber liquid, bp 82C (0.1 mm), d 1.29 (20C). Essentially insoluble in water, miscible in common solvents.

Hazard: Use may be restricted. A cholinesterase inhibitor.

Use: Insecticide, acaricide.

"Carbopol."[119] TM for a group of water-soluble vinyl polymers having excellent suspending, thickening, and gel-forming properties.

Use: Suspensions of glass fibers, graphite, powdered metals; gel formation in hydrocarbons; emulsifier for creosote, tars, and asphalts; lubricants; printing inks; coatings.

carborane. A crystalline compound comprised of boron, carbon, and hydrogen. It can be synthesized in various ways, chiefly by the reaction of a borane (penta- or deca-) with acetylene, either at high-temperature in the gas phase or in the presence of a Lewis base. Alkylated derivatives have been prepared. Carboranes have different structural and chemical characteristics and should not be confused with hydrocarbon derivatives of boron hydrides. The predominant structures are the cage type, the nest type, and the web type, these terms being descriptive of the arrangement of atoms in the crystals. Active research on carborane chemistry has been conducted under sponsorship of the US Office of Naval Research.

"Carbortam."[337] TM for a metallurgical processing alloy containing 15–20% titanium, 6–8% carbon, 2.4–4.0% silicon, 1.75–2.25% boron, and the balance iron except with traces of phosphorus and sulfur. Used to deoxidize and harden steel.

"Carborundum."[280] TM for abrasives and refractories of silicon carbide, fused alumina and other materials.

Properties: For silicon carbide, crystalline form ranges from small to massive crystals in the hexagonal system, the crystals varying from transparent to opaque, with colors from pale green to deep blue or black; hardness 9.17 (Mohs); d 3.06–3.20. Noncombustible, not affected by acids, slowly oxidizes at temperature above 1000C, good heat dissipator, highly refractory. For fused alumina, see properties under the TM "Aloxite."

Use: Abrasive grains and powders for cutting, grinding, and polishing; valve-grinding compounds; grinding wheels; coated abrasive products; antislip tiles and treads; refractory grains.

carbosand. Fine sand that has been treated with an organic solution and roasted to produce a material that can be sprayed onto oil slicks to aid in sinking or dispersing them.

See also oil spill treatment.

"Carboseal."[214]   TM for hygroscopic liquid compositions for joint sealing.
Use: Swelling and moistening agent for jute and other packing in cast-iron gas mains, to correct joint leakage and lay dust.

"Carbowax."[214]   TM for polyethylene glycols and methoxypolyethylene glycols.
Grade: Available in various numbered grades, i.e., 200, 400, 1000, 4000, 6000. Usually designated by approximate molecular weight of polymer.
Use: Water-soluble lubricants; solvents for dyes, resins, proteins; plasticizers for casein and gelatin compositions, glues, zein, cork, and special printing inks; solvent and ointment bases for cosmetics and pharmaceuticals; intermediates for nonionic surfactants and alkyd resins.

"Carboxide."[214]   TM for a fumigant-sterilant mixture of ethylene oxide and carbon dioxide. Ethylene oxide is the active agent. Nonflammable, odorless.
Use: Fumigant and sterilizing agent to eliminate insects such as beetles, moths, cockroaches, insect larvae and insect eggs. There is definite reduction in thermophilic bacteria and destruction of mold and fungi, including spores.

carboxybenzene.   See benzoic acid.

2-carboxy-2′-hydroxy-5′-sulfoformazylbenzene.
(o-[α-(2-hydroxy-5-sulfophenyl)-azobenzylidene]-hydrazino benzoic acid).
$HO_3SC_6H_3(OH)N{:}NC(C_6H_5){:}NNC_6H_4COOH$.
Use: Reagent used for the colorimetric determination of zinc and copper.

carboxylase.   A decarboxylase enzyme found in plant tissues which acts upon pyruvic acid, producing acetaldehyde and carbon dioxide.
Use: Biochemical research.

carboxyl group.   The chemical group characteristic of carboxylic acids which include fatty acids and amino acids. It usually occupies the terminal position in the molecule and is capable of assuming a negative charge which makes the end of the molecule water-soluble. Though it is customarily shown as either COOH or CO_2H, the structure of the group is:

$$-C\overset{\displaystyle O}{\underset{\displaystyle OH}{\big<}}$$

Thus it is composed of a carbonyl group and a hydroxyl group bonded to a carbon atom. The carbon-oxygen unsaturation within the carboxyl group is of a different order from the carbon-to-carbon unsaturation in the alkyl chain in unsaturated fatty acids. For this reason, fatty acids in which no double bond is present except that in the carboxyl group are called saturated.
See also fatty acid.

carboxylic acid.   Any of a broad array of organic acids comprised chiefly of alkyl (hydrocarbon) groups (CH_2, CH_3), usually in a straight chain (aliphatic), terminating in a carboxyl group (COOH). Exceptions to this structure are formic acid (HCOOH) and oxalic acid (HOOCCOOH). The number of carbon atoms ranges from one (formic) to 26 (cerotic), the carbon of the terminal group being counted as part of the chain. Carboxylic acids include the large and important class of fatty acids and may be either saturated or unsaturated. A few contain halogen atoms (chloracetic). There are also some natural aromatic carboxylic acids, (benzoic, salicylic) as well as alicyclic types (abietic, chaulmoogric).
See also amino acid.

carboxymethoxylamine hemihydrochloride.
$(H_2NOCH_2CO_2H)_2 \cdot HCl$.   Demonstrates anti-convulsant activity by inhibiting glutamic acid decarboxylase and γ-aminobutyric-α-keto-glutaric aminotransaminase, increasing the brain γ-aminobutyric acid concentrations, increases the sugar content of sugar cane, sugar beets, and sorghum. It forms digitoxin derivatives.
Properties: Off-white crystals. Mw 218.59, mp 156C (decomposes), hygroscopic.
Grades: 98% research.

carboxymethylcellulose.   (CMC; sodium carboxymethylcellulose; CM cellulose).
CAS: 9004-32-4.   A semisynthetic, water-soluble polymer in which CH_2COOH groups are substituted on the glucose units of the cellulose chain through an ether linkage. Mw ranges from 21,000 to 500,000. Since the reaction occurs in an alkaline medium, the product is the sodium salt of the carboxylic acid R—O—CH_2COONa.
Properties: Colorless, odorless, nontoxic, water-soluble powder or granules; pH (1% solution) 6.5–8.0; stable in pH range 2–10. D 1.59, refr index 1.51, tensile strength 8000–15,000 psi. Viscosity of 1% solution varies from 5 to 2000 centipoises, depending on the extent of etherification. Insoluble in organic liquids. Reacts with heavy-metal salts to form films that are insoluble in water, transparent, relatively tough, and unaffected by organic materials. Many of its colloidal properties are superior to those of natural hydrophilic colloids. It also has thixotropic properties and functions as a polyelectrolyte.
Derivation: By reaction of alkali cellulose and sodium chloroacetate.

Grade: Crude, technical (approximately 75% pure), high viscosity, low viscosity, semirefined, refined (99.5+% pure), USP, FCC.

Use: Detergents; soaps; food products (dietetic foods and ice cream) where it acts as water binder, thickener, suspending agent, and emulsion stabilizer; textile manufacturing (sizing); coating paper and paper board to lower porosity; drilling muds; emulsion paints; protective colloid; pharmaceuticals; cosmetics.

See also cellulose, modified.

carboxymethylmercaptosuccinic acid.
HOOCCH$_2$SCH(COOH)CH$_2$(COOH).
Properties: White powder, melting range 135–138C, water-soluble 137 g/100g at 25C, ethanol-soluble 76 g/100g at 25C.
Use: Heavy-metal chelator and deactivator.

carboxymethylpyridinium chloride hydrazide.
See Girard's "P" reagent.

carboxymethyltrimethyl ammonium chloride hydrazide. See Girard's "T" reagent.

carboxypeptidase. A proteolytic enzyme found in the pancreas which catalyzes the hydrolysis of native food proteins. It acts upon polypeptides producing simpler peptides and amino acids.
Use: Biochemical research.

carboxypolymethylene. See "Carbopol."

4-carboxyresorcinol. See β-resorcylic acid.

6-carboxyuracil. See orotic acid.

carburet of sulfur. See carbon disulfide.

carburetted hydrogen. See ethylene.

"Carbyne."[533] TM for an herbicide containing 4-chloro-3-butynyl-m-chlorocarbanilate (barban).

carcinogen. Any substance that causes the development of cancerous growths in living tissue. Such substances are usually grouped in two classifications: (1) Those that are known to induce cancer in man or animals either by operational exposure in industry or by ingestion in feedstuffs, e.g., asbestos particulates, nickel carbonyl, trichloroethylene, benzidine and compounds, vinyl monomer, benzopyrene, aflatoxin, chloromethyl ether, β-naphthylamine, as well as anthracene, phenanthrene, chrysene, and other polynuclear hydrocarbons of coal-tar origin.; (2) Experimental carcinogens are those that have been found to cause cancer in animals under experimental conditions, namely, by external applications, feeding, or injection of the substance. Among these are dimethyl sulfate, cyclamate compounds, ethyleneimine, and 4-dimethylaminoazobenzene. The Delany amendment to the Food, Drug and Cosmetic Act forbids the use in human foods of any substance falling in this group.

The substances mentioned above by no means comprise a complete list of carcinogens. There are approximately 3500 known and suspected carcinogenic compounds and new ones are constantly being discovered.

cardamom oil.
Colorless or pale-yellow essential oil, strongly aromatic, camphoraceous odor and taste, strongly dextrorotatory.
Used as flavoring for foods, confectionery, liqueurs, pharmaceuticals.

"Cardio-Green."[348] TM for indocyanine green, a diagnostic dye used in medicine.

"Cardosol" Brand Resin.[158] TM for a water-soluble ketone formaldehyde condensate which can be gelled and cured by alkali or heat.
Use: Water-resistant adhesives for box board, coatings, for glass fibers, and as a ceramic binder.

"Cadura" E Ester.[125] TM for glycidyl ester of "Versatic" 911 Acid.
Hazard: Irritant to skin.
Use: Modifier for alkyd resins and thermosetting acrylic systems, reactive diluent for epoxy resins.

δ-3-carene. 3,7,7-trimethylbicyclo[4.1.0]-hept-3-ene). CAS: 13466-78-9. C$_{10}$H$_{16}$. A terpene hydrocarbon having both a 6-member and a 3-member ring.
Properties: Clear, colorless liquid, d 0.8668 (15C), bp 170C, refr index 1.4723 (20C). Stable to approximately 250C, resinifies with oxygen. Insoluble in water, miscible with organic solutions. Combustible.
Derivation: From wood turpentine.
Use: Solvent, intermediate.

Carius (wet combustion) method. Decomposition of organic compounds containing halogen or sulfur in a sealed tube in the presence of red, fuming nitric acid at 250-300C in such a manner that halogen is converted to ionic halide and sulfur to sulfate.

"Carmethose."[305] TM for sodium carboxymethylcellulose.

carmine. An aluminum lake of the pigment from cochineal. Bright red pieces; easily powdered;

soluble in alkali solution, borax; insoluble in dilute acids; slightly soluble in hot water.
Grade: Technical.
Use: Dyes, inks, indicator in chemical analysis, coloring food materials, medicines, etc.

carminic acid. $C_{22}H_{20}O_{13}$. A tricyclic compound. The essential constituent of carmine.
Properties: Dark purplish-brown mass or bright-red powder, mp decomposes at 136C, pH 4.8 yellow, pH 6.2 violet. Soluble in water, alcohol, concentrated sulfuric acid; insoluble in ether, benzene, chloroform. Combustible.
Derivation: By extraction from the insects, Coccus cacti (cochineal).
Use: Stain in microscopy, indicator in analytical chemistry, coloring proprietary medicines, pigment for fine oil colors, color photography, dyeing.

carnallite. $KCl \cdot MgCl_2 \cdot 6HOH$ or $KMgCl_3 \cdot 6HOH$.
Properties: A natural hydrated double chloride of potassium and magnesium, white, brownish, and reddish; streak white; shining, greasy luster; strongly phosphorescent; bitter taste; deliquescent; d 1.62; Mohs hardness 1.
Occurrence: West Germany, Alsace, New Mexico.
Use: A commercial source of manufactured potash salts.

carnauba wax. (Brazil wax). The hardest and most expensive commercial wax.
Properties: Hard solid in form of yellow to greenish brown lumps, slight odor, d 0.995 (15/15C), mp 84–86C, acid number 2–9, iodine number 13.5. Soluble in ether, boiling alcohol, and alkalies; insoluble in water. Combustible.
Derivation: Exudation from leaves of the wax palm, Copernica cerifera (Brazil).
Grade: By numbers and sources, crude and refined, powdered, FCC.
Use: Shoe polishes, leather finishes, varnishes, electric insulating compositions, furniture and floor polishes, carbon paper, waterproofing, to prevent sunchecking of rubber and plastic products, confectionery, cosmetics.

carnosine. (β-alanylhistidine; ignotine). $C_9H_{14}N_4O_3$. An amino acid occurring in muscle of many animals and man. Occurs naturally in the l(+)-form.
Properties: Mp 245–250C (decomposes), soluble in water. Nitrate: crystals, mp 222C (decomposes) soluble in water. Hydrochloride: crystals, mp 245C (decomposes), soluble in water. D(-)carnosine: crystals, mp 260C.
Use: Biochemical research.

carnotite. $K_2(UO_2)_2(VO_4)_2 \cdot 3HOH$. A natural hydrated vanadate of uranium and potassium usually found in sandstones and other sedimentary rocks
Properties: Bright lemon yellow, luster dull or earthy, pearly or silky when coarsely crystalline. Soluble in acids. Radioactive. Usually occurs as a powder or in fine-grained aggregates.
Occurrence: Colorado, Utah, Arizona, New Mexico, South Dakota, Australia, Zaire, USSR.
Hazard: A radioactive poison.
Use: Ore of uranium, source of radium.

Carnot's reagent. A reagent for the determination of potassium, an alcoholic solution of sodium bismuth thiosulfate, made from sodium thiosulfate and bismuth subnitrate.

carob-seed gum. (locust bean gum). A polysaccharide plant mucilage which is essentially galactomannan (carbohydrate). Mw approximately 310,000. Swells in cold water, but viscosity increases when heated. Insoluble in organic solvents. Combustible.
Derivation: Extracted from carob seeds, from the tree Ceratonia siliqua.
Grade: Technical, FCC (as locust-bean gum).
Use: In foods as stabilizer, thickener, emulsifier, and packaging material; cosmetics; sizing and finishes for textiles; pharmaceuticals; paints bonding agent in paper manufacture; drilling fluids.

Caro's acid. (peroxysulfuric acid; persulfuric acid). H_2SO_5 or $HOSO_2OOH$.
Properties: White crystals, mp 45C (decomposes).
Derivation: Action of hydrogen peroxide on concentrated sulfuric acid; action of 40% sulfuric acid on potassium persulfate.
Hazard: Strong irritant to eyes, skin, and mucous membranes. Strong oxidizer, may explode in contact with organic materials.
Use: Caro's reagent, a pasty mass of great oxidizing power for testing aniline, pyridine, and alkaloids; dye manufacture; oxidizing agent; bleaching.

carotene. (provitamin A). $C_{40}H_{56}$.
A precursor of vitamin A occurring naturally in plants. It consists of three isomers, approximately 15% alpha, 85% beta, and 0.1% gamma. Carotene is a member of a large class of pigments called carotenoids. It has the same basic molecular structure as vitamin A and is transformed to the vitamin in the liver.
Properties: Ruby-red crystals, easily oxidized on contact with air; mp (alpha) 188C, (beta) 184C, (gamma) 178C; insoluble in water, slightly solu-

ble in alcohol, soluble in chloroform, carbon disulfide, ether, and benzene.

Source: Orange-yellow pigment in plants, algae, and some marine animals, especially in leaves, vegetation, and root crops, in trace concentrations. Notably present in butter and carrots.

Derivation: By extraction from carrots and palm oil concentration by a chromatographic process from alfalfa. β-Carotene is also made by a microbial fermentation process from corn and soybean oil.

Grade: According to USP, units of vitamin A, sold as pure crystals, as solutions in various oils, as colloidal dispersions. Also FCC.

Use: Pharmaceuticals, coloring margarine and butter, feed and food additive.

carotenoid. A class of pigments occurring in the tissues of higher plants, algae, and bacteria, as well as in fungi. Also present in some animals, as squalene in shark liver oil. They include the carotenes and xanthophylls.

Properties: (General) Yellow to deep red, crystalline solids; soluble in fats and oils; insoluble in water; high-melting; stable to alkali but unstable to acids and to oxidizing agents; color easily destroyed by hydrogenation or by oxidiation; some are optically active.

Carothers, Wallace H. (1896–1937) Born in Iowa, Carothers obtained his doctorate in chemistry at the University of Illinois. He joined the research staff of DuPont in 1928 where he undertook the development of polychloroprene (later called neoprene) that had been initiated by Nieuland's research on acetylene polymers. Carothers' crowning achievement was the synthesis of nylon, the reaction product of hexamethylenetetramine and adipic acid. Carothers' work in the polymerization mechanisms of fiber-like synthetics of cyclic organic structures was brilliant and productive, and he is regarded as one of the most original and creative American chemists of the early 20th Century.

carrageenan. (3,6-anhydro-d-galactan; carrageen). CAS: 9000-07-1. A sulfur phycocolloid. The aqueous, usually gel-forming, cell-wall polysaccharide mucilage found in the red algae (Chondrus crispus and several other species). Water-extracted from a seaweed called carragen or Irish moss (east coast of southern Canada, New England, and south to New Jersey). It is a mix of polysaccharide fractions: (1) The lambda fraction is cold-water soluble, contains D-galactose and 35% esterified sulfate, and does not gel. (2) The kappa fraction contains D-galactose and 3,6-anhydro-D-galactose (1.4 to 1 ratio) and 25% esterified sulfate. Kappa form

does not gel without addition of a solute; the properties of the gel depend on the amount and nature of the added solute. Another species of seaweed produces 100% kappa from (North Carolina to tropics). Carrageenan is a hydrophilic colloid which absorbs water readily and complexes with milk proteins.

Forms: Dehydrated, purified powder.

Grade: Technical, FCC.

Use: Emulsifier in food products, i.e., chocolate milk, toothpastes, cosmetics, pharmaceuticals, protection colloid, stabilizing aid in ice cream (0.02%).

"Carrene 16."[54]
TM for a 54% aqueous solution of lithium bromide containing an additive for corrosion inhibition in absorption refrigeration systems.

"Carrene 500."[54] TM for an azeotropic mixture of 73.8% dichlorodifluoromethane and 26.2% unsymmetrical difluoroethane boiling at −33.3C and used as a refrigerant. Nonflammable.

carrier. (1) A neutral material such as diatomaceous earth used to support a catalyst in a large-scale reaction system. (2) A gas used in chromatography to convey the volatilized mixture to be analyzed over the bed of packing which separates the components. (3) An atomic tracer carrier: a stable isotope or a normal element of which radioactive atoms of the same element have been added for purposes of chemical or biological research.

See also tracer.

Carroll reaction. Preparation of γ,δ-unsaturated ketones by base-catalyzed reaction of allylic alcohols with β-ketoesters or thermal rearrangement of allyl acetoacetates.

carthamin. (carthamic acid; safflor carmine; safflor red). $C_{21}H_{22}O_{11}$.

Properties: Dark-red powder with green luster. Slightly soluble in water, soluble in alcohol, insoluble in ether, solutions rapidly decompose.

Derivation: A glucoside from Carthamus tinctorius.

Use: Colorant in food products, cosmetics.

"Carum."[51] TM for grease-type lubricants prepared for use in chemical processing industries where insolubility in the material being processed is essential. Intended for valves and pumps handling such materials. One grade is permissible under FDA Regulations for lubrication where incidental contact with food might occur.

carvacrol. (isopropyl-o-cresol; 2-methyl-5-isopropylphenol; 2-hydroxy-p-cymene). CAS: 499-75-2.

Properties: Thick, colorless oil; thymol odor; d 0.976 (20/4C); bp 237C; fp 0C; refr index 1.523 (20C); insoluble in water; soluble in alcohol, ether, and alkalies. Combustible.
Derivation: From p-cymene by sulfonation followed by alkali fusion.
Use: Perfumes, fungicides, disinfectant, flavoring, organic synthesis.

carvone.

CH$_3$C:CHCH$_2$CH[C(CH$_3$):CH$_2$]CH$_2$CO.
CAS: 99-49-0.   A ketone derived from the terpene dipentene. It is optically active, occurring naturally in both d- and l-forms.
Properties: Pale-yellowish or colorless liquid with a strong characteristic odor. D 0.960 (20C), bp 227–230C, refr index 1.4999 (18C). Soluble in alcohol, ether, chloroform, propylene glycol, and mineral oils; insoluble in glycerol and in water. Combustible.
Derivation: The d-form is the main constituent of caraway and dill oils; the l-form occurs principally in spearmint oil and may be synthesized from d-limonene.
Method of purification: Rectification.
Grade: FCC (both d- and l-forms), technical.
Use: Flavoring, liqueurs, perfumery, soaps.

"Carwinate."[520]   TM for isocyanates used to make urethane elastomers, coatings, foam adhesives, rigid plastics, sealants, and one-shot flexible, semi-flexible, and semi-rigid foams.

caryophyllene. C$_{15}$H$_{24}$.   A mix of sesquiterpenes occurring in many essential oils. It forms the chief hydrocarbon component of clove oil.

caryophyllic acid. See eugenol.

carzinophillin A. CAS: 1403-29-8.   A natural antibiotic produced by a Streptomyces strain. Discovered in Japan in the 1950s, it has strong antitumor potential though its toxicity may preclude medical use. It is a member of the so-called bis-intercalator group of antibiotics that act by interlocking between the nucleotide bases of DNA, thus it may have mutagenic activity. The molecule contains carbon, oxygen, hydrogen, and nitrogen. The structure was elucidated in 1982.

cascade.   (1) A series of operational units or stages so arranged that the heat produced in the first unit serves as the heat source for the second unit and so on. An example of this is a triple-effect evaporator in which the latent heat of condensation is passed from one unit to another. This principle is also used in distillation each column plate representing one stage in the cascade. In flash-distillation used in the desalination of seawater the heat of condensation is used to warm the incoming water. (2) Coined term used to describe a large number of compounds derived from a common source, e.g., the arachidonic acid cascade.

"Cascamite."[65]   TM for a powdered ureaformaldehyde resin glue, water-resistant, moldproof, stainfree.

cascarilla oil.   The volatile oil obtained by steam distillation of the dried bark of Croton eleuteria Bennet. Light yellow to brown amber liquid having a pleasant spicy odor. Soluble in most fixed oils and in mineral oil, almost insoluble in glycerol and propylene glycol.
Use: Flavoring agent in foods, medicine, tobacco.

case-hardening.   A process which imparts a hard surface to steel while the interior remains soft and tough. This is accomplished by heating the steel out of contact with air while packed in carbonaceous material, cooling it to black heat, reheating to a high temperature, and quenching. The materials are usually wood charcoal with sodium, potassium, or barium carbonates, cyanides, etc.

casein. CAS: 9005-46-3.   Though commonly regarded as the principal protein in milk (approximately 3%), casein is actually a colloidal aggregate composed of several identifiable proteins together with phosphorus and calcium. It occurs in milk as a heterogeneous complex called calcium caseinate which can be fractionated by a number of methods. It can be precipitated with acid at pH 4.7 or with the enzyme rennet (rennin). The product of the latter method is called paracasein, the term being applied to any of the casein fractions involved, i.e., alpha, beta, kappa, etc.
Properties: White, tasteless, odorless, amorphous solid, d 1.25–1.31, hygroscopic, stable when kept dry but deteriorates rapidly when damp. Soluble in dilute alkalies and concentrated acids, almost insoluble in water, precipitates from weak acid solutions.

Derivation: Acid casein: Warm skim milk is acidified with dilute sulfuric, hydrochloric, or lactic acid, the whey drawn off, the curd washed, pressed, ground and dried. Rennet casein: Warm skim milk is treated with an extract of the enzyme rennin (rennet). The curd contains combined calcium and calcium phosphate.

Sources: Midwestern US, Australia, Argentina, New Zealand, Poland.

Grade: Acid-precipitated (domestic edible, imported inedible), paracasein.

Use: Cheesemaking, plastic items, paper coatings, water-dispersed paints for interior use, adhesives, especially for wood laminates, textile sizing, foods and feeds, textile fibers, dietetic preparations, binder in foundry sands.

casein-sodium.   See sodium caseinate.

cashew gum. (anacardium gum).   The exudation from the bark of the cashew-nut tree, Anacardium occidentale. Hard, yellowish-brown gum; partly soluble in water. Used for inks, insecticides, pharmaceuticals, mucilage, tanning agent, natural varnishes, bookbinders' gum.

cashew nutshell oil. (cashew nutshell liquid). The oil obtained from the spongy layer between the inner and outer shells of cashew nuts. The raw liquid contains approximately 90% anacardic acid, $C_{22}H_{32}O_3$, and a blistering compound containing sulfur. Most of the liquid used in commerce has been heated or treated with chemicals to make it safe to handle. The liquid is non-drying, but can be made drying by proper treatment. It polymerizes on heating and forms condensation products with aldehydes.

Hazard (untreated): Strong irritant.

Use: Varnishes and impregnating materials, modifier for phenol-based resins, plasticizers, germicides and insecticides, coloring materials and indelible inks, lubricants, and preservatives.

casinghead gasoline.   See gasoline.

cassava.   See tapioca.

Cassel brown.   See Van Dyke brown.

Cassel green.   See barium manganate.

Cassella's acid. (2-naphthol-7-sulfonic acid; β-naphtholsulfonic acid F; F acid). $C_{10}H_6(OH)SO_3H$.

Properties: White crystals, soluble in water and alcohol, mp 89C.

Derivation: Fusion of naphthalene-2,7-disulfonic acid with caustic soda.

Grade: Technical. Also available as the sodium salt (F salt).

Use: Intermediate for azo dyes.

Casella's Acid F. (2-naphthylamine-7-sulfonic acid; delta acid). $C_{10}H_6(NH)_2SO_3H$.

Properties: Colorless crystals, soluble in water, alcohol, and ether.

Derivation: Heating sodium 2-naphthol-7-sulfonate with aqueous ammonia and ammonium acid sulfate in an autoclave.

Use: Azo dye intermediate.

cassia oil. (Chinese cinnamon oil; cinnamon; cassia oil; cinnamon oil; USP).

CAS: 8007-80-5.

Properties: See cinnamic aldehyde, its chief constituent.

Derivation: Distilled from leaves and twigs of Cinnamomum cassia.

Grade: USP (as cinnamon oil), redistilled, technical, lead-free.

Use: Flavoring, perfumery, medicine, soaps.

cassiterite. (tinstone; wood tin; stream tin). SnO_2.   Natural tin dioxide, usually in igneous rocks.

Properties: Color brown, black, yellow, white; luster adamantine or dull submetallic; streak white; Mohs hardness 6–7; d 6.8–7.1.

Occurrence: Malaya, Bolivia, Indonesia, Africa.

Use: Principal ore of tin.

casting. (1) In the metal industries, conversion of a molten metal (iron, steel) either into bars (pigs) by pouring it into open troughs or channels made of foundry sand (sand casting) or into products of specific shape by forcing it into steel dies or molds under pressure (die casting).

See also foundry sand, cast iron, die casting, investment casting.

(2) In the plastics industry, formation of a product either by filling an open mold with liquid monomer and allowing it to polymerize in situ or (for film and sheet) by pouring the liquid mixture onto a moving flat surface.

See also molding.

cast iron.   Generic term for a group of metals that basically are alloys of carbon and silicon with iron. Relative to steel, cast irons are high in carbon and silicon, carbon ranging from 0.5 to 4.2% and silicon from 0.2 to 3.5%. All these metals may contain other alloys added to modify their properties.

Iron castings are produced in an exceptionally wide range of sizes and weights, from piston rings a fraction of an inch in diameter and weighing less than 1 oz to steam-turbine bases 20 ft long and weighing 180,000 lb.

Most cast iron is manufactured by melting a mixture of steel scrap, cast-iron scrap, pig iron, and alloys in a cupola using coke as a fuel. A small percentage is melted in electric furnaces. It is poured into molds of silica sand bonded with bentonite, fireclay, and water. A small percentage is cast into metal molds or into molds of baked or fired ceramics. Internal cavities are formed by hard but collapsible cores of sand bonded with drying oils or synthetic resins. Small molds and cores usually are made by machine using patterns of wood or metal.

"Castolat H-W."[446]   TM for a 93% high-alumina castable cement bonded with low iron calcium aluminate cement. Shipped dry with water addition, develops and maintains high strength through 3200F. Resistant to abrasion and impact.
Use: Petrochemical unit liner, burner blocks and other high temperature applications. Can be cast, trowelled, or applied with an air placement gun.

castor.   Perfume fixative obtained from secretions of the beaver. Synthetic types are available.

"Castordag."[46]   TM for a dispersion of colloidal graphite in castor oil. Used as an assembly and maintenance lubricant.

castor oil.   (ricinus oil).   CAS: 8001-79-4.
Properties: Pale-yellowish or almost colorless, transparent, viscous liquid, faint, mild odor, nauseating taste. A non-drying oil. D 0.945–0.965 (25/25C), saponification value 178, iodine value 85, fp −10C. Flash p 445F (229C), autoign temperature 840F (448C). Combustible. Soluble in alcohol, benzene, chloroform, and carbon disulfide; dextrorotatory.
Derivation: From the seeds of the castor bean, Ricinus communis (Brazil, India, USSR, US). They are cold pressed for the first grade of medicinal oil and hot pressed for the common qualities, approximately 40% of the oil content of the bean being obtained. Residual oil in the cake is obtained by solvent extraction. Chief constituent: Ricinolein (glyceride of ricinoleic acid).
Grade: USP No. 1; No. 3, refined, FCC.
Hazard: Undergoes spontaneous heating.
Use: Plasticizer in lacquers and nitrocellulose, production of dibasic acids, lipsticks, polyurethane coatings, elastomers and adhesives, fatty acids, surface-active agents, hydraulic fluids, pharmaceuticals, industrial lubricants, electrical insulating compounds, manufacture of Turkey Red oil, source of sebacic acid and of ricinoleates, medicine (laxative).   See also castor oil, dehydrated; blown oil.

castor oil, acetylated.   See glyceryl triacetylricinoleate.

castor oil acid.   See ricinoleic acid.

castor oil, blown.   See blown oil.

castor oil, dehydrated.   (DCO).   A castor oil from which approximately 5% of the chemically combined water has been removed and which as a result has drying properties similar to those of tung oil. Dehydration is carried out commercially by heating the oil in the presence of catalysts such as sulfuric and phosphoric acids, clays, and metallic oxides. The commercial product is offered in a wide range of viscosities and analytical constants. Used in protective coatings and alkyd resins.

castor oil, hydrogenated.   Principally glyceryl-tri-12-hydroxystearate. A hard, waxy product. It is insoluble in water and organic solvents. Mp 85C.
Use: Hydroxystearic acid, waterproofing fabrics, cosmetics, lubricant, mold release agent, candles, carbon paper.

castor oil, polymerized.   A rubber-like polymer results from combination of castor oil with sulfur or diisocyanates; this can be blended with polystyrene to give a tough, impact-resistant product.

castor oil, sulfonated.   See Turkey red oil.

castor seed oil meal.   (castor cake; castor meal). The residue from extraction of oil from the castor seed (ricinus). The normal product contains 29.5% crude protein, 35.8% crude fiber, 13.2% N-free extract and 1.0% crude fat. The total digestible nutrients approximately 25%. The ash content of 7.5% is high in potash and phosphate.
Hazard: Contains ricin which must be removed before internal use.
Use: Animal feeds (after removal of toxic ingredients), fertilizer.

"Castorwax."[202]   TM for hydrogenated castor oil, the triglyceride of 12-hydroxystearic acid.

catalase.   An oxidizing enzyme occurring in both plain and animal cells. It decomposes hydrogen peroxide. It can be isolated and is used in food preservation (removing oxygen in packaged foods) and in decomposing residual hydrogen peroxide in bleaching and oxidizing processes.

catalysis.   One of the most important phenomena in nature, catalysis is the "loosening" of the chemical bonds of two (or more) reactants by another substance, in such a way that a fractionally small percentage of the latter can greatly accelerate the rate of the reaction, while remaining unconsumed. (See catalyst). Thus one part by volume of catalyst can activate thousands of parts of reactants. Though the mechanism of their action is not completely known, the electronic configuration of the surface molecules of the catalyst is often the critical factor. The surface irregularities give rise to so-called "active points" at which intermediate compounds can form. Most industrial catalysis is performed by finely divided transition metals or their oxides.

Solid catalysts may combine chemically (bond) at the surface with one or more of the reactants. This is known as chemisorption and occurs on only a small portion of the catalyst surface, (i.e., at the active points), it results in changing the chemical nature of the chemisorbed molecules. Catalysis of chemical reactions by surfaces must proceed by chemisorption of at least one of the reactants.

Catalysis and catalytic mechanisms permeate almost every aspect of chemistry and are of such wide-ranging importance that they have long been the subject of continuing research. Many interrelated disciplines are involved, among which are organometallic reactions (stereospecific catalysts), electrochemistry, colloid and surface chemistry, coordination chemistry, and biochemistry. One example of comparatively recent investigation is so-called cluster catalysis in which a metal is bonded to carbon monoxide to form a metallic cluster anion. Such aggregations or crystallites may be up to 12 Å in diameter. Metal cluster research is motivated by the need to develop selective catalysts for C_1 compounds, in view of the liklihood that the basic raw materials for chemistry in the future will be derived from coal or biomass rather than oil. Another research development is the creation of a uniquely shaped catalyst particle which permits greater efficiency due to increase in surface-to-volume ratio. Such particles are known as amphora catalysts. See also following entries.

catalysis, heterogeneous. A catalytic reaction in which the reactants and the catalyst comprise two separate phases, e.g., gases over solids, or liquids containing finely divided solids as a disperse phase.

catalysis, homogeneous. A catalytic reaction in which the reactants and the catalyst comprise only one phase, e.g., an acid soluble catalyzing other liquid components.

catalyst. Any substance of which a fractional percentage notably affects the rate of a chemical reaction without itself being consumed or undergoing a chemical change. Most catalysts accelerate reactions but a few retard them (negative catalysts or inhibitors). Catalysts may be inor-

ganic, organic, or a complex of organic groups and metal halides (see catalyst, stereospecific). They may be either gases, liquids, or solids. In some cases their action is destructive and undesirable, as in the oxidation of iron to its oxide which is catalyzed by water vapor and similar types of corrosion. The life of an industrial catalyst varies from 1000 to 10,000 hours, after which it must be replaced or regenerated.

Though it is not a "substance," light in both the visible and ultrashort wavelengths can act as a catalyst, as in photosynthesis and other photochemical reactions, e.g., as polymerization initiator and crosslinking agent.

Catalysts are highly specific in their application, they are essential in virtually all industrial chemical reactions, especially in petroleum refining and synthetic organic chemical manufacturing. For details of application, see the following list. Since the activity of a solid catalyst is often centered on a small fraction of its surface, the number of active points can be increased by adding promoters which increase the surface area in one way or another, e.g., by increasing porosity. Catalytic activity is decreased by substances that act as poisons which clog and weaken the catalyst surface, e.g., lead in the catalytic converters used to control exhaust emissions.

Besides inorganic substances, there are many organic catalysts that are vital in the life processes of plants and animals. These are called enzymes and are essential in metabolic mechanisms, e.g., pepsin in digestion. Synthetic organic catalysts have been developed which imitate the action of enzymes such as chymotrypsin. Such model catalysts are examples of biomimetic chemistry. They approach the catalytic activity of natural enzymes.

Following is a partial list of catalysts, the asterisk indicates a destructive effect.

Substance	Reaction type
aluminum chloride	condensation (Friedel-Crafts)
aluminum alkyl + titanium chloride	Ziegler catalyst for stereospecific polymers
aluminum oxide	hydration dehydration
ammonia	condensation (polymers)
chromic oxide	methanol synthesis, aromatization, polymerization
cobalt	hydrocarbon synthesis Oxo process.
copper salts	oxidation (of rubber)*
ferric chloride	Friedel-Crafts
hydrogen fluoride	alkylation, condensation dehydration, isomerization

iodine	condensation, alkylation
iron	ammonia synthesis hydrocarbon synthesis
iron oxide	dehydrogenation (oxidation)
manganese dioxide	oxidation
molybdenum oxide	dehydrogenation polymerization aromatization partial oxidation
nickel	hydrogenation (oils to fats) methanation
phosphoric acid	polymerization isomerization
platinum metals	hydrogenation aromatization oxidation
silica-alumina	cracking hydrocarbons
silver	hydration oxidation
sulfuric acid	isomerization corrosion*
triethylaluminum	polymerization (stereospecific)
vanadium pentoxide	oxidation (sulfuric acid)
water (esp + NaCl)	oxidation (corrosion)*
zeolites	cracking hydrocarbons

*See also catalysis enzyme and following entries. For further information consult Center for Catalytic Science and Technology, University of Delaware.

catalyst, amphora. Catalyst particles made from a slurry of critical viscosity by compressing it into spheres or droplets followed by unidirectional heating and air-drying on a moving belt, the spheres being supported by a powder bed material. This process results in particles from 1.5 to 6 mm in diameter. The unique feature is the formation of an internal cavity having an orifice at one point so that the particle roughly resembles a doughnut from which a bite has been taken. Since the shape of the cavity suggests an amphora (Greek vase), the catalyst was so named. This shape affords a higher surface-to-volume ratio than is possible with solid spheres and other conventional forms with consequent greater efficiency. A variety of materials can be used, e.g., alumina, zeolites, metallic oxides, etc. Amphora catalysts are effective in a wide range of chemical processing applications (oxidation and reforming of hydrocarbons, hydrotreating).

catalyst, negative. See inhibitor.

catalyst, organic. See enzyme.

catalyst, shape-selective. A catalytic solid (transition metal) introduced into a crystalline aluminosilicate (zeolite) having pores or openings of approximately 5 Å. This permits the catalyst to exert a selective effect between molecules that differ in shape rather than in the reactivity of their chemical groups. The cage structure of the zeolite effectively prevents contact of the catalyst with all molecules whose shape and size exclude their entry. The ability of a catalyst to discriminate among molecules on the basis of their shape is of great value in the cracking of straight-chain hydrocarbons and has attractive possibilities in other types of catalytic reactions.
See also zeolite, cage zeolite.

catalyst, stereospecific. An organometallic catalyst which permits control of the molecular geometry of polymeric molecules. Examples are Ziegler and Natta catalysts derived from a transition metal halide and a metal alkyl or similar substances. There are many patented catalysts of this general type, most of them developed in connection with the production of polypropylene, polyethylene, or other polyolefins.
See also polymer, stereospecific, Natta catalyst, Ziegler catalyst.

catalyst, thermonuclear. See carbon cycle (2).

catechol. See pyrocatechol.

catecholborane. (1,3,2-benzodioxaborole).
CAS: 274-07-7.   A monofunctional hydroborating agent.
Properties: A liquid with mw 119.92, mp 12C, bp 50C at 50 mm, optical rotation 1.5070 degrees (20C).
Use: Preparation of alkaneboronic acide and esters from olefins.

catenane. A compound with interlocking rings which are not chemically bonded but which cannot be separated without breaking at least one valence bond. The model would resemble the links of a chain.

"Cat-Floc."[108] (diallyldimethylammonium chloride).   TM for a quaternary ammonium polymer.
Derivation: Monomer in water solution is mixed with catalytic amount of butylhydroperoxide and kept at 50–75C for 48 hours. The solid formed is taken up in water, precipitated, and washed with acetone.
Use: Flocculating agent, textile spinning aid, antistatic agent, wet-strength improvers in paper, rubber accelerators, curing epoxy resins, surfactants, bacteriostatic and fungistatic agents.

cathode. The negative electrode of an electrolytic cell to which positively charged ions migrate when a current is passed as in electroplating

baths. The cathode is the source of free electrons (cathode rays) in a vacuum tube. In a primary cell (battery), the cathode is the positive electrode.
See also anode, electrode.

cathode sputtering. See sputtered coating.

cation. An ion having a positive charge. Cations in a liquid subjected to electric potential collect at the negative pole or cathode.

cation exchange. See ion exchange.

cationic reagent. One of several surface-active substances in which the active constituent is the positive ion. Used to flocculate and collect minerals that are not flocculated by oleic acid or soaps (in which the surface-active ingredient is the negative ion). Reagents used are chiefly quaternary ammonium compounds, e.g., cetyl trimethyl ammonium bromide.

catlinite. (pipestone). A fine-grained silicate mineral related to pyrophyllite which is easily compressible, has high surface friction, and is used for gaskets in very high-pressure equipment.

"Cato."[53] TM for a cationic derivative of starch, available in ungelatinized or gelatinized (cold water solution) form. Used in manufacture of paper, warp sizing, etc.

caulking compound. See sealant.

caustic. (1) Unqualified, this term usually refers to caustic soda (NaOH). (2) As an adjective, it refers to any compound chemically similar to NaOH, e.g., caustic alcohol (C_2H_5ONa). (3) Any strongly alkaline material which has a corrosive or irritating effect on living tissue.

caustic baryta. See barium hydroxide.

causticized ash. Combinations of soda ash and caustic soda in definite proportions marketed for purposes where an alkali is needed ranging in causticity between the two materials. Causticized ash is usually designated by its caustic soda content and the range of standard marketed products embraces 7%, 10%, 15%, 25%, 36%, 45% and 67% of caustic soda.

caustic lime. See calcium hydroxide.

caustic potash. See potassium hydroxide.

caustic soda. See sodium hydroxide.

cavitation. Formation of air bubbles in a liquid such as saltwater when subjected to intense vibrations causing severe mechanical damage to the surfaces of metals exposed to it, e.g., ship propellers, steam condensers, pumps, and piping systems. The erosive effect is due to the shock waves created by collapse of the bubbles. The pressures exerted by cavitation have been calculated to be in the range of 30,000 psi. This phenomenon plays a part in corrosion of metals as well as in emulsion formation.
See also corrosive, homogenization.

Cb. Symbol for columbium, an obsolete name for niobium.

CBM. Abbreviation for chlorobromomethane (see bromochloromethane), also for constant boiling mixture.
See azeotropic mixture.

cc. Abbreviation for cubic centimeter.
See also milliliter.

c-cetyl betamine. An amphoteric product showing cationic properties in acid solutions, anionic properties in alkaline solutions.
Use: Alkalone peroxide bleaching systems.

Cd. Symbol for cadmium.

CDA. Abbreviation for completely denatured alcohol.
See alcohol, denatured.

CDAA. See α-chloro-N,N-diallylacetamide.

CDEC. See 2-chloroallyl diethyldithiocarbamate.

CDP. (1) Abbreviation for cytidine diphosphate. See cytidine phosphates. (2) Abbreviation for cresyl diphenyl phosphate.

"CDP-GA."[1] TM for a phosphorus-containing deposit modifier for automotive gasolines.

CDTA. See trans-1,2-diaminocyclohexanetetraacetic acid.

Ce. Symbol for cerium.

cedar leaf oil. An essential oil distilled from the leaves of Juniperus virginiana. Strongly dextrorotatory. Used in microscopy, perfumery, flavoring.

cedrol. $C_{15}H_{26}O$. A tertiary terpene alcohol.
Properties: Colorless crystals, cedarwood odor. Mp 86C, soluble in 11 parts of 95% alcohol. Combustible.

Use: Perfumery, for woody and spicy notes, odorant for disinfectants.

cedryl acetate. $CH_3COOC_{15}H_{25}$.
Properties: Colorless liquid having a light cedar odor. D 0.975–0.995, refr index 1.496–1.510. Soluble in one volume of 90% alcohol. Combustible.
Use: Perfumery.

CEELS. Abbreviation for characteristic electron energy-loss spectroscopy.

"Celanar."[352] TM for a polyester film made from polyethylene terephthalate.
Properties: Transparent, biaxially oriented, d 1.395, tensile strength 30,000 psi, dielectric strength 7000 volts/mil, mp 260C, service temperature −60–150C, outstanding dimensional stability and chemical resistance.
Use: Magnetic recording tape, drafting and engineering reproduction materials, metallic yarn, roll leaf, pressure-sensitive tapes, packaging, dielectric material in capacitors, wire and cable, motors, generators, transformers, and oils.

"Celanese CL."[352] TM for a series of polyvinyl acetate emulsions. Available as:. 102: Fine particle size, water resistant homopolymer emulsion. 202: Fine particle size, water resistant copolymer emulsion. 203: Vinyl-acrylic copolymer emulsion. 204: Vinyl copolymer emulsion.
Use: Paints, adhesives, and paper-coating specialties.

"Celanese Solvent."[352] TM for a series of special solvents. Available as: 203: Replacement for normal butyl alcohol in nitrocellulose lacquers, alkyd resin formulations and thinners, distillation range 115–120C, flash p 100F (37.7C) (OC). 601: Replacement for methyl ethyl ketone in vinyl and nitrocellulose applications, distillation range 74–84C, flash p 10F (−12.2C) (OC).
Hazard: Flammable, dangerous fire risk. Replacement for butanol and methylisobutyl carbinol in lacquers and brake fluids, distillation range 125–155C, flash p 120F (48.9C) (OC).

"Celanthrene."[28] TM for a group of anthraquinone disperse dyes designed especially for acetate, also suitable for application to nylon.

"Celatom."[468] TM for a group of diatomaceous silicas (diatomite) of high quality and uniformity.
Use: Filter aid, including foods and beverages, absorbents, as in insecticides and fertilizers, catalyst supports, fillers for paper, paints, explosives, concrete and asphalt, chromatography.

"Celcon."[352] TM for a highly crystalline acetal copolymer based on trioxane.
See acetal resin.

celestine blue. (CI 51050). $C_{17}H_{18}ClN_3O_4$.
Properties: Dark-green powder. Slightly soluble in water, ethylene glycol, and alcohol. Water solution gives purple color, alcoholic solution blue.
Use: Biological stain, mordant in dyeing.

celestite. $SrSO_4$. Natural strontium sulfate, usually found in sedimentary rocks.
Properties: Colorless, white, pale blue or red, luster vitreous to pearly. Resembles barite. D: 3.95, Mohs hardness 3–3.5.
Occurrence: US, Canada, Europe, Mexico.
Use: Strontium chemicals, oil-well drilling mud, sugar refining, ceramics, production of very pure sodium hydroxide.

"Celite."[247] TM for diatomaceous earth and related products.

"Celkate."[246] TM for finely divided, hydrated, synthetic magnesium silicates having high absorption properties, light tan in color, density 10–18 lb/cu ft, surface area 150–250 m^2/gram.
Grade: Available in various grades for purifying petroleum-base solvents, chemical and drug solutions, and for decolorizing of animal, fish, and vegetable oils.
Use: Filter agent to remove solids and free fatty acids, absorptive carrier of liquids, conditioning agent to improve flow properties of dry powders.

cell. (1) The fundamental unit of biological structure, comprised of (a) an outer membrane, or wall, approximately 100 Å thick, which being semipermeable maintains by osmosis the biochemical equilibrium of the intracellular fluids; (b) the cytoplasm, containing mitochondria, ribosomes, and other structures; and (c) the nucleus, in which lie the chromosomes and genes. An extremely complex biochemical organization, the cell is the dynamic unit of all life. Its ability to reproduce itself and to control its functions systematically is of basic importance to maintenance of life and growth. All organic matter is either found in cells or is produced by cellular activity and was ultimately derived from photosynthesis. Cells are comparatively large units which can be resolved in an optical microscope, the largest single cells are represented by the eggs of oviparous animals.
See also mitosis.

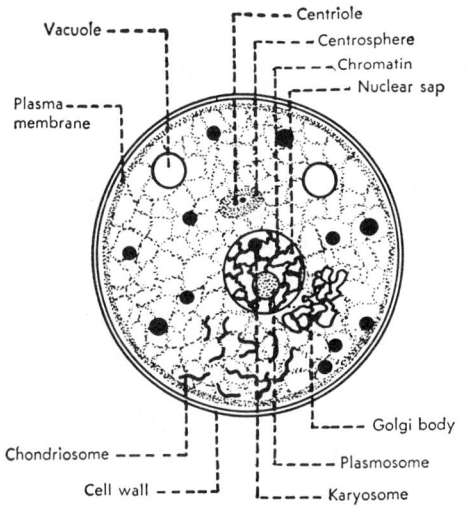

Vacuole - - - -
Centriole
Centrosphere
Chromatin
Nuclear sap
Plasma - - - -
membrane
Golgi body
Chondriosome - - - - -
Plasmosome
Cell wall - - - -
Karyosome

(2) Any self-contained unit having a specific functional purpose as follows: (a) voltaic cell (battery) to generate electric current (b) electrolytic cell to effect electrolysis, (c) fuel cell to convert chemicals into electricity, and (d) solar cell to capture heat from sunlight. All except the last involve use of electrodes and electrolytes. See also specific entries.

(3) Any completely enclosed hollow unit as in a honeycomb or cellular plastic.

cell, dry. See dry cell.

cell, electrolytic. See electrolytic cell.

"Cellex."[236] TM for sodium carboxymethylcellulose. Used as an organic colloidal compound for providing low filtrate loss in drilling mud.

cell, fuel. See fuel cell.

"Cellidor."[460] TM for cellulose acetate and acetate butyrate plastics.

"Cellitazol."[203] TM for a series of developed acetate dyestuffs.

"Celliton."[203] TM for a series of disperse dyestuffs characterized by good fastness to light, washing, etc. Used for dyeing and printing acetate fibers.

cellobiose. $C_{12}H_{22}O_{11}$. The product of the partial hydrolysis of cellulose, composed of two d-glucose molecules.
Properties: Colorless crystals, mp 225C (decomposes), soluble in water, slightly soluble in alcohol, nearly insoluble in ether, insoluble in acetone.
Use: Bacteriology.

celloidin. A pure form of pyroxylin, used for imbedding sections in microscopy.

"Cellolyn."[266] TM for a series of pale, low-melting hydroabietyl ester resins. Used for lacquers, inks, and adhesives.

cellophane. (regenerated cellulose). Film produced from wood pulp by the viscose process.
Properties: Transparent, strong, flexible, and highly resistant to grease, oil, and air. The base cellulose film is modified by softeners, flame-resisting materials and dyes, also by coating with other materials. On exposure to heat the untreated film loses strength at 149C, decomposes at 176–204C, does not melt, burns readily, and is not self-extinguishing.
Hazard: Flammable, moderate fire risk.
Use: Wrapper or protective package for fabricated articles and industrial applications.
See also rayon, viscose.

"Cellosize."[214] TM for hydroxyethyl cellulose and carboxymethylcellulose.
See also cellulose, modified.

"Cellosolve."[214] TM for mono- and dialkyl ethers of ethylene glycol and their derivatives widely used as industrial solvents.
  butyl "Cellosolve"
    See ethylene glycol monobutyl ether.
  butyl "Cellosolve" acetate
    See ethylene glycol monobutyl ether acetate.
  "Cellosolve" acetate
    See ethylene glycol monoethyl ether acetate.
  "Cellosolve" solvent
    See ethylene glycol monoethyl ether.
  dibutyl "Cellosolve"
    See ethylene glycol dibutyl ether.
  n-hexyl "Cellosolve"
    See ethylene glycol monohexyl ether.
  methyl "Cellosolve"
    See ethylene glycol monomethyl ether.
  methyl "Cellosolve" acetate
    See ethylene glycol monomethyl ether acetate.
  phenyl "Cellosolve"
    See ethylene glycol monophenyl ether.

cell, photovoltaic. See solar cell.

cell, solar. See solar cell.

"Celluflux" TPP.[1] TM for triphenyl phosphate. Used as plasticizer in cellulose acetate and phenolic plastics.

"Celluguard."[1]   TM for a water-glycol, fire-resistant hydraulic fluid.

cellular plastic.   A thermosetting or thermoplastic foam composed of cellular cores with integral skins having high strength and stiffness. The cells result from the action of a blowing agent, either at room temperature or during heat treatment of the plastic mixture. The resulting product may be either flexible or rigid, the latter being machinable. The foaming action in some cases may occur *in situ* (foamed-in-place plastics). Cellular plastics are combustible. For details, see foam, plastic.
Use (flexible): Furniture, automobile interiors, matresses, etc., where softness and resiliency are desired. (Rigid) Insulating material, boat building and similar light construction, salvage of water-logged ships.
See also foam, plastic; rubber sponge. For further information, refer to Cellular Plastics Division, Society of the Plastics Industry, 355 Lexington Ave., New York, NY 10017.

cellulase.   An enzyme complex produced by the fungi *Aspergillus niger* and *Trichoderma viride* which is capable of decomposing cellulosic polysaccharides into smaller fragments, primarily glucose. It has been used as a digestive aid in medicine and in the brewing industry. Research has been devoted to experimental application of cellulase to disposal of cellulosic solid wastes. The resulting glucose can be fermented to ethanol, used to grow yeast for animal feed proteins, or as a chemical feedstock. Note: Cellulase derived from the thermophilic soil fungus *Thielatia terrestris* functions at a much higher temperature than other types and is thus much more effective in decomposing cellulose. This indicates its possible use in conversion of biomass to energy. Commercial development of this product is expected.
See also biomass.

"Celluloid."[352]   CAS: 8050-88-2.   TM for a plastic consisting essentially of a solid solution of cellulose nitrate and camphor or other plasticizer plus a flame-retardant such as ammonium phosphate to minimize flammability; available in sheets, rods, tubes, films. Also called pyroxylin.
Hazard: Flammable, dangerous fire risk.
See also nitrocellulose.

cellulose.   CAS: 9004-34-6. $(C_6H_{10}O_5)_n$.
A natural carbohydrate high polymer (polysaccharide) consisting of anhydroglucose units joined by an oxygen linkage to form long molecular chains that are essentially linear. It can be hydrolyzed to glucose. The degree of polymerization is from 1000 for wood pulp to 3500 for cotton fiber, giving a molecular weight from 160,000 to 560,000.

Cellulose is a colorless solid, d approximately 1.50, insoluble in water and organic solvents. It will swell in sodium hydroxide solution soluble and is soluble in Schweitzer's reagent. It is the fundamental constituent of all vegetable tissues (wood, grass, cotton, etc) and is the most abundant organic material in the world. Cotton fibers are almost pure cellulose, wood contains approximately 50%.
The physical structure of cellulose is unusual in that it is not a single crystal, but consists of crystalline areas embedded in amorphous areas. Chemical reagents penetrate the latter more easily than the former. Cellulose is virtually odorless and tasteless and is combustible, with an ignition point of approximately 450F. In some forms it is flammable. For example, railroad shipping regulations require a Flammable label on such items as "burnt fiber," "burnt cotton," "wet waste paper," and "wet textiles." Fires have been known to occur in warehouses in which telephone books were stored. These were undoubtedly due to heat build-up in the paper caused by microbial activity and self-sustaining oxidation.
See also flammable material.
The most important uses of cellulose are bulk woods of many kinds; paper, most of which is made from wood pulp; cotton products (clothing, sheeting, industrial fabrics); packaging; ranging from wooden barrels to candy pats; and as a source of ethanol (enzymatic hydrolysis) and methanol (destructive distillation of wood). Specialized uses include nonwoven fabrics, medical equipment (artificial kidney), insulation and soundproofing, sausage casings, etc. Cellulose has approximately 60% of the energy content of bituminous coal; its use as a fuel has increased especially in rural locations.
See also biomass.
There are many chemical modifications of cellulose, including its esters (cellulose acetate), ethers (methylcellulose), the nitrated product (nitrocellulose), as well as rayon and cellophane (from cellulose xanthate). Thus it is the basis of many plastics, fibers, coatings, lacquers, explosives, and emulsion stabilizers. Alkali cellulose is an inter-

mediate made by the action of sodium hydroxide solution on cellulose and is used for making cellulose ethers and viscose.

See also cellulose, modified.

Cellulose exists in three forms--alpha, beta and gamma. Alpha-cellulose has the highest degree of polymerization (DP), and is the chief constituent of paper pulp. It is insoluble in strong sodium hydroxide solution. The beta and gamma forms have much lower DP and are known as hemicellulose. Methods of determining the alpha content of pulps are detailed in TAPPI Method T203 and AS TM D-588-42.

See also pulp, paper.

Cellulose has been prepared in microcrystalline form. See "Avicel."

Cellulose can be decomposed to glucose by the enzyme cellulase or by hydrolysis.

Hazard: Cellulosic materials (paper, cotton, and textile wastes) when wet with water are a dangerous fire hazard and subject to shipping regulations.

See also rayon, cellophane, nitrocellulose, carboxymethylcellulose, and following entries.

cellulose acetate. (CA). A cellulose ester in which the cellulose is not completely esterified by acetic acid.

Properties: White flakes or powder. A thermoplastic resin, softening at approximately 60–97C and melting at approximately 260C. D: 1.27–1.34, insoluble in acetone, ethyl acetate, cyclohexanol, nitropropane, ethylene dichloride. Notable for toughness, high impact strength, and ease of fabrication. Subject to dimensional change due to cold flow, heat, or moisture absorption (1–7%). Fibers weaken above 80C and are difficult to dye uniformly; not attacked by microorganisms.

Derivation: Reacting cellulose (wood pulp or cotton linters) with acetic acid or acetic anhydride, with H_2SO_4 catalyst. The cellulose is fully acetylated (three acetate groups per glucose unit) and at the same time the sulfuric acid causes appreciable degradation of the cellulose polymer so that the product contains only 200–300 glucose units per polymer chain. At this point in the process the cellulose acetate ordinarily is partially hydrolyzed by the addition of water until an average of 2–2.5 acetate groups per glucose unit remain. This product is thermoplastic and soluble in acetone. Fibers are produced by forcing an acetone solution through orifices of the spinneret into a stream of warm air which evaporates the solvent. Fibers are also produced in a similar manner from cellulose triacetate, which is insoluble in acetone but soluble in methylene chloride.

See acetate fiber, acetate film.

Forms: Sheet, film or fiber, molded items.

Grade: Filtered and unfiltered. Also graded by percent combined acetic acid content: plastic 52–54%, lacquer 54–56%, film 55.5–56.6%, water resisting 56.5–59%, triacetate 60.0–62.5%.

Hazard: Flammable, not self-extinguishing, moderate fire risk.

Use: Acetate fiber, lacquers, protective coating solutions, photographic film, transparent sheeting, thermoplastic molding composition, cigarette filters, magnetic tapes, osmotic cell membrane.

cellulose acetate butyrate. (CAB; cellulose acetobutyrate).

Properties: White pellets or granules. A thermoplastic resin. D 1.2, soluble in ketones; organic acetates; lactates; methylene, ethylene, and propylene chlorides; and high-boiling solvents. Excellent weathering properties, low water absorption, low heat conductivity and high dielectric strength, high resistance to oil and grease. Combustible.

Derivation: Reaction of purified cellulose with acetic and butyric anhydrides with sulfuric acid as catalyst and glacial acetic acid as solvent. The ratio of acetic and butyric components may be varied over a wide range.

Grade: According to percent butyryl content (17, 27, 38, 50%).

Use: Molding compositions, film and sheet, photographic film, lacquers, protective coating solutions, tail-light lenses, water desalination membrane, piping and tubing, covering for aluminum fibers, toys, packaging, brush handles, hydrometers, miscellaneous consumer products.

cellulose acetate phthalate. A reaction product of phthalic anhydride and cellulose acetate, used for coating of tablets and capsules.

cellulose acetate propionate. (cellulose propionate). Similar to cellulose acetate butyrate but made with propionic anhydride instead of butyric anhydride. Unusually stable, requires less plasticizer and is compatible with more plasticizers than the butyrate.

cellulose acetobutyrate. See cellulose acetate butyrate.

cellulose ether. See ethylcellulose, methylcellulose.

cellulose gum. A purified grade of carboxymethylcellulose.

cellulose, hydrated. (hydrocellulose). Cellulose that has been caused to react with water (approximately 8–12%) forming a gelatinous mass. Combustible.

Derivation: By mechanical pulverization and agitation with water, by the action of strong salt solutions, alkalies, or acids.

Use: In the manufacture of paper, vulcanized fiber, mercerized cotton, viscose rayon.

See also hydration.

cellulose methyl ether.   See methylcellulose.

cellulose, modified.   One of many derivatives of cellulose, formed by substitution of appropriate radicals (carboxyl, alkyl, acetate, nitrate, etc.) for hydroxyl groups along the carbon chain. Such reactions are usually not stoichiometric. Some of these products (carboxymethyl and hydroxyalkyl cellulose) are water-soluble ethers, others are organosol esters (cellulose acetate), nitrates (nitrocellulose), or xanthates (viscose).

See also modification.

Biodegraded cellulose has been used as a microbial growth medium for protein formation, a tobacco substitute based on a cellulose modification is trademarked "Cytrel."

cellulose nitrate.   See nitrocellulose.

cellulose, oxidized.   (cellulosic acid).   Derivative of cotton, cellulose produced by treatment with nitrogen dioxide. Is soluble in alkali but may be made to retain original form of the cellulose and much of its tensile strength.

Properties: Slightly off-white gauze, lint, or powder. Slight charred odor, acid taste, soluble in aqueous organic bases, in dilute alkali, and in ammonium hydroxide, forming salts and esters. It is insoluble in water, acids, and common organic solvents. It slowly degrades at room temperatures and should be kept cool. Combustible.

Grade: USP, technical.

Use: Surgery and medicine, ion-exchange medium, thickening agent.

cellulose propionate.   See cellulose acetate propionate, "Forticel."

cellulose, regenerated.   See cellophane, rayon.

cellulose sponge.   A sponge of regenerated cellulose; highly absorbent, soft, and resilient when wet; long-lasting. It will not scratch, can be sterilized by boiling, and is not affected by ordinary cleaning compounds. The pores vary in size from coarse pore (the size of a pea) to fine pore (the size of a pinhead). Yarn made of cellulose sponge consists of cotton fiber covered with the sponge product. Combustible.

Use: Washing automobiles and trucks, walls and painted surfaces, windows, etc., general cleaning;

sponge used in photographic laboratories; wet mops, cleaning pads, etc.

cellulose triacetate.   A cellulose resin in which the cellulose is almost completely esterified by acetic acid.

Properties: White flakes, d 1.2, soluble in chloroform, methylene chloride, tetrachloroethane. Combustible.

Derivation: Reaction of purified cellulose with acetic anhydride in the presence of sulfuric acid as catalyst and glacial acetic acid as solvent, followed by very slight hydrolysis.

Grade: Flake.

Use: Protective coatings resistant to most solvents, textile fibers, base for magnetic tape.

cellulose xanthate.   See viscose process.

cellulosic plastic.   One of a number of semisynthetic polymers based on cellulose.

See cellophane, cellulose acetate, cellulose, modified, nitrocellulose, rayon, viscose, carboxymethylcellulose.

cellulosic thiocarbonate.   A reactive intermediate in the graft polymerization of certain synthetic polymers to cellulosic fibers. The latter are treated with sodium hydroxide and sulfur-containing compound and the cellulose thus activated is placed in an emulsion or solution of monomers. Polymerization at 50C occurs with a peroxide catalyst to form the graft.

See also graft polymer.

"Cellulube."[1]   TM for a series of functional fluids (phosphate esters) combining fire-resistance and lubricating qualities. Available in controlled viscosities for industrial hydraulic and lubricant applications.

"Celluphos 4."[1]   TM for tributyl phosphate.

Use: Plasticizer for nitrile elastomers, foam depressant for emulsion systems, solvent, extractant and complexing agent for inorganics.

"Cellutherm."[1]   TM for a series of synthetic lubricants based on trimethylolpropane esters.

cell, voltaic.   See voltaic cell.

"Celogen."[248]   TM for a series of blowing agents, i.e.:

AZ Azodicarbonamide.

OT Phosphorus-p-oxybis-benzenesulfonylhydrazide.

RA p-Toluene sulfonyl semicarbazide.

"Celoron."[281]   TM for macerated canvas or paper-based industrial laminated or molded. plastics.

Properties: Mottled brown or black; d 1.35; high impact strength; unaffected by rapid temperature changes; resistant to heat, oil, water, and many chemicals; may be used continuously at 107-121C. Combustible.

Forms: Sheets; cut pieces; blanks; rings; molded parts.

Use: Timing gears for automobile industry; electrical insulation; structural parts.

"Celotex."[351] TM for structural building and insulation board produced in large sheets. Made from bagasse or wood fiber and treated to be resistant to fungi, termites, and water penetration. The name also includes roofing products, gypsum, wallboard, lath, plasters, mineral wool, and hard board.

Celsius, A. (1701–1744) A Swedish physicist who proposed the use of the centigrade temperature scale. His name is now generally applied to this scale (degrees centigrade = degrees Celsius).
See also centigrade.

cement aluminous. (high alumina cement). A hydraulic cement which contains at least 30–35% alumina (in contrast to Portland cement, which contains less than 5%). The alumina is usually supplied by inclusion of bauxite. Aluminous cement attains its maximum strength more rapidly than Portland cement. It is also more resistant to solutions of sulfates. It exists in two modifications, sintered and fused.

cementation. A process in which steel or iron objects are coated with another metal by immersing them in a powder of the second metal and heating to a temperature below melting point of any of the metals concerned. Zinc, chromium, aluminum, copper, and other metals are applied to iron or steel in this fashion. The process is basically diffusion of one metal into the other so that intermetallic alloy layers are formed at the interface of the basis and coating metals.
See also sherardizing.

cemented carbide. See carbide, cemented.

cement, hydraulic. Any mixture of fine-ground lime, alumina, and silica that will set to a hard product by admixture of water which combines chemically with other ingredients to form a hydrate.
See also cement, aluminous; and cement, Portland.

cementite. Fe_3C. A carbide of iron formed in the manufacture of pig iron and steel. Composed of 93.33% iron and 6.67% carbon, it is very hard and brittle and will scratch glass and feldspar but not quartz. It is about two-thirds as magnetic as pure iron under an exciting current. It occurs in ordinary steels of more than 0.85% carbon and takes its name from cement steel, made by the cementation process, which contains a great deal of this carbide.
See also carbide.

cement, organic. Any of various types of rubber cement, silicone adhesives, deKhotinsky cement.
See also adhesive, rubber-based, silicone, deKhotinsky cement.

cement, Portland. A type of hydraulic cement in the form of finely divided, gray powder composed of lime, alumina, silica, and iron oxide as tetracalcium aluminoferrate. $(4CaO \cdot Al_2O_3 \cdot Fe_2O_3)$, tricalcium aluminate $(3CaO.Al_2O_3)$, tricalcium silicate $(3CaO \cdot SiO_2)$, and dicalcium silicate $(2CaO \cdot SiO_2)$. These are abbreviated, respectively, as C_4AF, C_3A, C_3S, and C_2S. Small amounts of magnesia, sodium, potassium, and sulfur are also present. Hardening does not require air and will occur under water. Cement may be modified with various plastic latexes in proportions up to 0.2 part latex solids to one part cement to improve adhesion, strength, flexibility, and curing properties. Water evaporation can be retarded by adding such resins as methylcellulose and hydroxyethyl cellulose. For further information, refer to the Portland Cement Association, Chicago, IL.
See also concrete (1).

cement, rubber. See adhesive, rubber-based.

"CE Methyl Esters."[487] TM for a series of methyl esters of straight chain (normal) even numbered fatty acids ranging from C_8 (octanoate) to C_{18} (octadecanoate) and including mixtures of these.

Properties: Colorless to light yellow liquids or white solids. D approximately 0.87.

Use: Chemical intermediate, lubricants, cosmetic ingredients, formulating aids (rubber, wax, etc).

Center for History of Chemistry. See Appendix II.

centigrade. (Celsius). The internationally used scale for measuring temperature in which 100 degrees is the boiling point of water at sea level (1 atmosphere) and 0 degrees is the freezing point. A temperature given in centigrade degrees may be converted to the corresponding Fahrenheit temperature by multiplying it by 9/5 (or 1.8) and adding 32 to product. A temperature given in Fahrenheit degrees is converted to the corresponding centigrade temperature by subtracting 32 and multiplying the remainder by

5/9. The centigrade scale was devised by the Swedish scientist Celsius; his name is officially used to designate it even though centigrade is more meaningful.

centigrade heat unit. See chu.

centipoise. (cp). 1/100th of a poise. The poise is the metric system unit of viscosity and has the dimensions of dyne-second per square centimeter or grams per centimeter-second.

centistoke. (cs). 1/100th stoke, the kinematic unit of viscosity, it is equal to the viscosity in poises divided by the density of the fluid in grams per cubic centimeter, both measured at the same temperature.

centrifugation. A separation technique based on the application of centrifugal force to a mixture or suspension of materials of closely similar densities. The smaller the difference in density, the greater is the force required. The equipment used (centrifuge) is a chamber revolving at high speed (10,000 rpm or more) to impart a force up to 17,000 times gravity. The materials of higher density are thrown toward the outer portion of the chamber while those of lower density are concentrated at or near the inner portion. This technique is used effectively in a number of biological and industrial operations, such as separation of the components of blood, concentration of rubber latex, and of fat particles from other milk components. Separation of isotopes, i.e., those of uranium, by this method is now practicable for producing enriched uranium. This method is economically superior to the gaseous diffusion process.
See also ultracentrifuge.

centrifuge. See centrifugation, ultracentrifuge.

"Cenwax."[189] TM for a series of hydrogenated castor oil products. Used as lubricants, coatings for leather, paper and textiles, candles, crayons, sealing compounds, polishes, and wax extenders.

cephalin. (kephalin; phosphatidyl ethanol amine; phosphatidyl serine).
$CH_2OR_1CHOR_2CH_2OP(O)(OH)OR_3$. A group of phospholipids in which two fatty acids (R_1 and R_2) form ester linkages with the two hydroxyl groups of glycerophosphoric acid and either ethanolamine or serine (R_3) forms an ester linkage with the phosphate group. Cephalins are therefore either phosphatidyl ethanolamine or phosphatidyl serine. They are associated with lecithins found in brain tissue, nerve tissue, and egg yolk.
Properties: Yellowish, amorphous substance; characteristic odor and taste; insoluble in water

and acetone; soluble in chloroform and ether; slightly soluble in alcohol.
Use: Medicine, biochemical research.

cephalosporin. Any of a family of antibiotics related to penicillin discovered in 1953; an important member of this group was synthesized by Woodward in 1966. Several cephalosporins are used clinically (cephalothin, cephaloridine, and cephalexin). The molecule contains a fused β-lactam-dihydrothiazine ring system with an N-acyl side chain and an acetoxy group attached to the dihydrothiazine ring. The formula for cephalosporin (C) is $C_{16}H_{21}N_3O_8S$. Cephalosporins are reported to be free from the allergic reactions common with penicillin. Development of new cephalosporin derivatives is being actively pursued.
See also penicillin, antibiotic.

cephamycin. Any of a group of antibiotics related to cephalosporins and produced by several species of *Streptomyces*.

ceramic. A product manufactured by the action of heat on earthy raw materials in which silicon with its oxide and complex compounds known as silicates occupy a predominant position (American Ceramic Society). The chief major groups of the ceramics industry are as follows: (1) structural clay products (brick, tile, terra cotta, glazed architectural brick); (2) whitewares (dinnerware, chemical and electrical porcelain, e.g., spark plugs, sanitary ware, floor tile); (3) glass products of all types; (4) porcelain enamels; (5) refractories (materials that withstand high temperatures); (6) Portland cement, lime, plaster, and gypsum products; (7) abrasive materials such as fused alumina, silicon carbide, and related products; (8) aluminum silicate fibers. A wide range of ceramics are available as ultra-fine particles (10–150 microns), and ceramic foams are offered commercially.
See also specific entries. For further information, refer to the American Ceramic Society, 4055 North High St., Columbus, OH.

ceramic, ferroelectric. A unique type of polycrystalline ceramic having properties that make possible the production of reliable, high-density optical memories for computers that are more efficient than conventional types. Lead zirconate titanate, heated and pressed into thin plates, is one of the compounds used. As a result of its ferroelectric properties, an applied voltage aligns the electric charges in the molecules of ceramic in the direction of the field and the polarization so induced remains indefinitely until changed. Thus, the material accommodates itself to the

requirements of the digital system, namely, binary 0 and binary 1.
See also ferroelectric.

ceramic, glass. See glass ceramic.

"Ceramix."[177] TM for a technical grade of barium carbonate used in ceramic industry.
Hazard: See also barium.

"Ceramol."[400] TM for a blend of cetyl and stearyl alcohols and higher alcohol sulfates. Melts from 50 to 60C; acid number 1.0 max; saponification value 3 max; iodine number 5 max; acetyl value 185–195. Not alkaline to phenolphthalein. Combustible.
Use: Emulsifier for cosmetic creams, ointments, and lipsticks.

"Ceramvar."[155] TM for an iron–nickel–cobalt alloy designed for ceramic-to-metal sealing.

"Ceraphyl."[10] TM for a series of lactate emollients designed for use in cosmetics and pharmaceutical applications.

"Ceratak."[128] TM for a grade of petroleum microcrystalline wax, min mp 73.8C.

"Cerathane" 63-L.[128] TM for an emulsifiable microcrystalline wax, min mp 93.3C.

"Ceraweld."[128] TM for a grade of petroleum microcrystalline wax, min mp 73.8C.

"Cercor."[20] TM for thin-walled, cellular ceramic structures which can be used for a wide range of high temperature applications.
Use: Gaseous heat exhangers, burner plates, acoustics, flame arresting, filtering, and insulation.

cerebrosides. Derivatives of sphingosine in which the amino group is connected in an amide linkage to a fatty acid and the terminal hydroxyl group is connected to a molecule of sugar, usually galactose, in glycosidic linkage. They are found in brain and nervous tissue, usually in association with sphingomyelin.

"Cerelose."[30] TM for a white, crystalline, refined dextrose (pure monosaccharide), 100% fermentable. Available in hydrate, anhydrous, and liquid forms.
Use: Adhesives, chemicals, drugs and pharmaceuticals, foundry processes, and plastics.

"Ceresan."[28] TM for a series of mercury compounds used as seed disinfectants.

ceresin wax. (purified ozocerite; earth wax; mineral wax; cerosin; cerin).
Properties: White or yellow, waxy cake; white is odorless, yellow has a slight odor. D 0.92–0.94, mp 68–72C. Soluble in alcohol, benzene, chloroform, naphtha; insoluble in water. Combustible.
Derivation: Purification of ozocerite by treatment with concentrated sulfuric acid and filtration through animal charcoal.
Grade: White, yellow.
Use: Candles, sizing, bottles for hydrofluoric acid, electrical insulation, shoe and leather polishes, impregnating and preserving agent, lubricating compounds, wood filler, floor polishes, antifouling paints, waxed papers, cosmetics, ointments, matrix compositions, waterproofing textile fabrics.

ceria. See ceric oxide, rare earth.

ceric ammonium nitrate. (cerium-ammonium nitrate; ammonium hexanitratocerate).
$Ce(NO_3)_4 \cdot 2NH_4NO_3$.
Properties: Small, prismatic, yellow crystals. Soluble in water and alcohol, almost insoluble in concentrated nitric acid, soluble in other concentrated acids.
Derivation: By electrolytic oxidation of cerous nitrate in nitric acid solution, and subsequently mixing solutions of cerium nitrate and ammonium nitrate followed by crystallization.
Hazard: Strong oxidizer, dangerous fire risk in contact with organic materials.
Use: Analytical chemistry, oxidant for organic compounds, polymerization catalyst for olefins, scavenger in the manufacture of azides.

ceric hydroxide. (ceric oxide, hydrated; cerium hydrate). $CeO_2 \cdot xHOH$.
Properties: Whitish powder when pure, a hydrated oxide containing 85–90% ceric oxide. Soluble in concentrated mineral acids, insoluble in water.
Derivation: By treating a solution of a ceric salt with strong alkali. Reagent grade is prepared by adding a saturated solution of ceric ammonium nitrate to an excess of ammonium hydroxide.
Grade: Commercial, high purity, reagent.
Use: Production of cerium salts and ceric oxide, opacifier in glasses and enamels (imparts yellow color), shielding glass.

ceric oxide. (cerium dioxide; cerium oxide; ceria). CeO_2.
Properties: Pale yellow, heavy powder (white when pure). Commercial product is brown. D 7.65, mp 2600C. Soluble in sulfuric acid, insoluble in water and dilute acid, requires reducing agent with acid to dissolve the anhydrous oxide. Noncombustible.

Derivation: By decomposition of cerium oxalate by heat. Hardness depends on firing temperature.
Grade: Technical, high purity (99.8%).
Use: Ceramics, abrasive for glass polishing, opacifier in photochromic glasses, retarder of discoloration in glass, especially radiation shielding and color TV tubes, catalyst, enamels and ceramic coatings, phosphors, cathodes, capacitors, semiconductors, refractory oxides; diluent in nuclear fuels, heat stabilizer (alumina catalyst).

ceric sulfate. (cerium sulfate).
$Ce(SO_4)_2 \cdot 4H_2O$.
Properties: White or reddish-yellow crystals, d 3.91, soluble in water (decomposes), soluble in dilute sulfuric acid. Strong oxidizer.
Derivation: Action of sulfuric acid on cerium carbonate.
Hazard: Fire risk in contact with organic materials.
Use: Dyeing and printing textiles, analytical reagent, waterproofing, mildewproofing.

ceric sulfide. (cerium sulfide). A high-temperature thermoelectric material that is stable and efficient up to 1100C.

cerin. See ceresin wax.

ceritamic acid. See ascorbic acid.

cerite. A rare-earth ore found chiefly in Sweden. A minor source of cerium.

cerium. CAS: 7440-45-1. Ce. A rare-earth element of the lanthanide group of the Periodic Table. Atomic number 58, aw 140.12, valences 3,4. Four stable isotopes.
Properties: Gray, ductile, highly reactive metal, d 6.78, mp 795C, bp 3257C. Attacked by dilute and concentrated mineral acids and by alkalies. Readily oxidizes in moist air at room temperature. It has four allotropic forms. It is the second most reactive rare-earth metal. Cerium forms alloys with other lanthanides (see misch metal); it also forms a nonmetal with hydrogen, as well as carbides and intermetallic compounds. Decomposes water.
Ores: Cerite (Sweden), bastnasite (California, New Mexico), monazite (beach sands in Florida, Brazil, India, South Africa).
Derivation: Chemical processing and separation of ores.
Grade: Granules, ingots, rods (99.9% pure).
Hazard: May ignite on heating to 300F (148.9C). Strong reducing agent.
Use: Cerium salts, cerium-iron pyrophoric alloys, ignition devices, military signalling, illuminant in photography, reducing agent (scavenger), catalyst, alloys for jet engines, solid state devices, rocket propellants, getter in vacuum tubes, diluent in plutonium nuclear fuels.
See also misch metal.
Note: Cerium compounds have been found to have antiknock properties, e.g., cerium (2,2,6,6-tetramethyl-3,5-heptanedionate)$_4$.

cerium-141. Radioactive cerium of mass number 141.
Properties: Half-life 21.5 days; radiation: beta and gamma.
Derivation: From cerium-140 by capture of a neutron and emission of a gamma photon.
Form available: $CeCl_3$ in hydrochloric acid solution.
Hazard: Radioactive poison.
Use: Biological and medical research.

cerium-ammonium nitrate. See ceric ammonium nitrate.

cerium carbonate. See cerous carbonate.

cerium chloride. See cerous chloride.

cerium dioxide. See ceric oxide.

cerium hydrate. See ceric hydroxide or cerous hydroxide.

cerium naphthenate.
Properties: A rubbery material, very difficult to dry. Almost insoluble without small quantities of organic stabilizers.
Derivation: By saponifying naphthenic acids and treating the sodium naphthenate formed with a suitable cerium salt. The commercial product is a mixture of rare earth soaps.
Use: See soap (2).

cerium nitrate. See cerous nitrate.

cerium oxalate. See cerous oxalate.

cerium oxide. See ceric oxide.

cerium sulfate. See ceric sulfate, cerous sulfate.

cermet. (ceramic + metal). A semisynthetic product consisting of a mixture of ceramic and metallic components having physical properties not found solely in either one alone, e.g., metal carbides, borides, oxides, and silicides. They combine the strength and toughness of the metal with the heat and oxidation resistance of the ceramic material. The composition may range from predominantly metallic to predominantly ceramic; e.g., SAP sintered aluminum contains

85% aluminum and 15% Al_2O_3. The most important industrial cermets are titanium carbide-based, aluminum oxide-based, and special uranium dioxide types. Cermets are made by powder metallurgy techniques involving use of bonding agents such as tantalum, titanium, and zirconium. High stress-to-rupture properties, operate continuously at 982C, for short periods at 2200C.

Use: Gas turbines, rocket motor parts, turbojet engine components, nuclear fuel elements, coatings for high-temperature resistance, sensing elements in instruments, seals, bearings, etc. in special pumps and other equipment.

"Cer-O-Cillin."[327] TM for an antibacterial substance which differs from penicillin G in that the benzyl group is replaced by an allylmercaptomethyl group. Soluble in concentrations up to 500,000 units per cc in sterile water for injection, sterile sodium chloride for injection, or sterile 5% dextrose.
Use: Medicine.

cerosin. See ceresin wax.

cerotic acid. (hexacosanoic acid; cerinic acid).
$CH_3(CH_2)_{24}COOH$. A fatty acid obtained from beeswax, carnauba wax, or Chinese wax.
Properties: White, odorless crystals or powder, d 0.8198 (100/4C), mp 87.7C, refr index 1.4301 (100C). Insoluble in water; soluble in alcohol, benzene, ether, acetone. Combustible.

cerous carbonate. (cerium carbonate).
$Ce_2(CO_3)_3 \cdot 5H_2O$.
Properties: White powder, soluble in mineral acids (dilute), insoluble in water.
Derivation: By adding an alkali carbonate to a solution of a cerous salt.

cerous chloride. (cerium chloride).
CAS: 7790-86-5. $CeCl_3 \cdot xH_2O$.
Properties: White crystals, deliquescent, d 3.88 (anhydrous), mp 848C (anhydrous), bp 1727C. Soluble in water, alcohol, and acids.
Derivation: Action of hydrochloric acid on cerium carbonate or hydroxide.
Use: Incandescent gas mantles, spectrography, preparation of cerium metal, polymerization catalyst.

cerous fluoride. (cerium fluoride).
CAS: 7758-88-5. $CeF_3 \cdot xH_2O$.
Properties: Off-white powder, insoluble in water and acids. D (anhydrous) 6.16, mp 1460C, bp 2300C.
Derivation: By treating cerous oxalate with hydrofluoric acid.

Hazard: An irritant. TLV: (as F) 2.5 mg/m³ of air.
Use: In arc carbons to increase their brilliance; preparation of cerium metal.

cerous hydroxide. (cerium hydrate).
Approximate formula $Ce(OH)_3$.
Properties: White, gelatinous precipitate; yellow, brown, or pink when impurities are present. Soluble in acids, insoluble in water and alkali.
Derivation: Chief source is monazite sand.
Grade: Pure, crude.
Use: Pure form: to produce cerium salts, impart yellow color to glass, opacifying agent in glazes and enamels. Crude form: flaming arc lamp.

cerous nitrate. (cerium nitrate).
$Ce(NO_3)_3 \cdot 6HOH$.
Properties: Colorless crystals, deliquescent. Soluble in water, alcohol, and acetone; mp loses 3HOH at 150C; bp decomposes at 200C.
Derivation: Action of nitric acid on cerous carbonate.
Hazard: Strong oxidizer, fire risk in contact with organic materials.
Use: Separation of cerium from other rare earths, catalyst for hydrolysis of phosphoric acid esters.

cerous oxalate. (cerium oxalate).
$Ce_2(C_2O_4)_3 \cdot 9HOH$.
Properties: Yellowish-white, odorless, tasteless, crystalline powder; decomposes upon heating; soluble in dilute sulfuric and hydrochloric acids; insoluble in water, oxalic acid solution, alkalies, alcohol, and ether.
Derivation: By extraction from monazite sand with oxalic or hydrochloric acid and conversion into the oxalate, followed by crystallization.
Grade: Pure; the commercial product is a complex mixture of oxalates of cerium, lanthanum, and didymium.
Hazard: Toxic by ingestion.
Use: Medicine, isolation of cerium metals.

cerous sulfate. (cerium sulfate).
$Ce_2(SO_4)_3 \cdot 8H_2O$.
Properties: White crystals or powder, soluble in water and in acids, mp 630C (dehydrated), d 2.886.
Derivation: Reagent grade is prepared by reducing a solution of ceric sulfate in sulfuric acid with hydrogen peroxide.
Grade: Technical and purified (reagent).
Use: Developing agent for aniline black.

"Cerox."[455] TM for a series of high alumina refractories with resistance to thermal shock, corrosion, abrasion, erosion, and reducing atmospheres. Fusion temperature up to 1982C. Available in prefired shapes. Used in production of steel, of electronic ceramic components, of abrasives, in solid state ore reduction, metal powder sintering, in high-temperature gas reactors.

certified color. See food color, FD&C colors.

cerulean blue. A light blue pigment essentially cobaltous stannate, $CoO \cdot n(SnO_2)$.

cerussite. $PbCO_3$. Natural lead carbonate, found in the upper zone of lead deposits.
Properties: Colorless, white, gray, luster adamantine, Mohs hardness 3–3.5, d 6.55. Effervesces in nitric acid.
Occurrence: Colorado, Arizona, New Mexico, Idaho, Australia, Europe.
Use: An ore of lead.

"Cer-Vit."[191] TM for a glass ceramic having linear expansion coefficient near zero. Used for specialty products such as telescope mirrors, where minimum distortion is essential.

ceryl alcohol. $C_{26}H_{53}OH$. An alcohol obtained from Chinese insect wax.
Properties: Colorless crystals, mp 79C, insoluble in water, soluble in alcohol and ether. Combustible.

ceryl cerotate. $C_{26}H_{53}OOCC_{25}H_{51}$. The chief constituent of Chinese insect wax and typical of natural waxes. Colorless crystals with mp 84C.

CES. Abbreviation for cyanoethyl sucrose.

cesium. (caesium). CAS: 7440-46-2.
Cs. An alkali-metal element of Group IA of the Periodic Table, atomic number 55, aw 132.9054, valence 1. No stable isotopes.
Properties: Liquid at slightly greater than room temperature, soft solid below the melting point, highly reactive. Decomposes water with evolution of hydrogen, which ignites instantly. Also reacts violently with oxygen, the halogens, sulfur, and phosphorus with spontaneous ignition and/or explosion. D 1.90, mp 28C, bp 705C, Mohs hardness 0.2. Cesium has highest position in the electromotive series; it also has the lowest melting point of any alkali metal and the lowest ionization potential of any element. Soluble in acids and alcohol.
Derivation: By thermochemical reduction of cesium chloride with calcium or by electrolysis of the fused cyanide. Its chief ore is pollucite, found in Maine, South Dakota, Manitoba, Elba, South Africa.
Grade: Technical, 99.9%.

Hazard: Dangerous fire and explosion risk, ignites spontaneously in moist air, may explode in contact with sulfur or phosphorus, reacts violently with oxidizing materials, causes burns in contact with skin.
Use: Photoelectric cells, getter in vacuum tubes, hydrogenation catalyst, ion propulsion systems, plasma for thermoelectric conversion, atomic clocks, rocket propellant, heat transfer fluid in power generators, thermochemical reactions, seeding combustion gases for magnetohydrodynamic generators.

cesium 137. Radioactive cesium of mass number 137.
Properties: Half-life 33 years; radiation: beta.
Hazard: Radioactive poison. The beta decay of Cs-137 produces Ba-137, which in turn is radioactive, emitting a 0.662 Mev gamma ray with a 2.6 minute half-life.
Use: Most applications of Cs-137 depend on the fact that any Cs-137 preparation has an equivalent amount of the gamma-emitting Ba-137 daughter. Approved for sterilization of wheat, flour, potatoes, and sewage sludge.

cesium alum. See cesium aluminum sulfate.

cesium aluminum sulfate. (cesium alum). $CsAl(SO_4)_2 \cdot 12HOH$.
Properties: Colorless crystals, d 2.0215, mp 117C, soluble in water, insoluble in alcohol.
Derivation: By adding a solution of cesium sulfate to a solution of potassium alum, concentrating, and crystallizing.
Grade: Pure.
Use: Mineral waters, purification of cesium by fractional crystallizing, preparation of cesium salts.

cesium antimonide. Used as a high-purity binary semiconductor.
Hazard: Toxic by ingestion.

cesium arsenide. Used as a high-purity binary semiconductor.
Hazard: See arsenic and cesium.

cesium bromide. CsBr.
Properties: Colorless, crystalline powder, d 4.44, mp 636C, bp 1300C, soluble in water, slightly soluble in alcohol.
Grade: Technical, single crystals.
Use: Crystals for infrared spectroscopy, scintillation counters, fluorescent screens.

cesium carbonate. Cs_2CO_3.
Properties: White, hygroscopic, crystalline powder; very stable; can be heated to high tempera-

ture without loss of CO_2. Soluble in water, alcohol, and ether.
Derivation: By passing carbon dioxide into a solution of cesium hydroxide and subsequent crystallization.
Use: Brewing, mineral waters, ingredient of specialty glasses, polymerization catalyst for ethylene oxide.

cesium chloride. CsCl.
Properties: Colorless crystals, d 3.972, mp 646C, sublimes at 1290C. Soluble in water and alcohol, insoluble in acetone.
Derivation: Action of hydrochloric acid on cesium oxide and crystallization.
Grade: Pure, low optical density, single crystals.
Use: Brewing, preparation of other cesium compounds, mineral waters, evacuation of radio tubes (positive ions supplied at surface of filament), for a density gradient in ultracentrifuge separations, fluorescent screens, contrast medium.

cesium dioxide. Cs_2O_2.
Properties: Yellow needles, d 4.25, mp 400C, decomposes at 650C, soluble in water and acids.
Grade: Technical, pure.
Use: Cesium salts.

cesium fluoride. CAS: 13400-13-0. CsF.
Properties: Deliquescent crystals, d 4.115, mp 682C, bp 1251C. Very soluble in water, soluble in methanol, insoluble in dioxane and pyridine.
Grade: 99% min, single crystals.
Hazard: A poison. TLV: (as F) 2.5 mg/m³ of air.
Use: Optics, catalysis, specialty glasses.

cesium hexafluorogermanate. Cs_2GeF_6.
Properties: White, crystalline solid; mp approximately 675C; density 4.10; slightly soluble in cold water and acids; very soluble in hot water.
Hazard: An irritant, see fluorides.

cesium hydrate. See cesium hydroxide.

cesium hydroxide. (cesium hydrate).
CAS: 21351-79-1. CsOH.
Properties: Colorless or yellowish, fused, crystalline mass. Strong alkaline reaction. Hygroscopic. Keep well stoppered. D 3.675, mp 272.3C. Very soluble in water, the strongest base known.
Derivation: By adding barium hydroxide to an aqueous solution of cesium sulfate.
Grade: Technical, 50% aqueous solution.
Hazard: A poison. TLV: 2 mg/m³ of air.
Use: Recommended as electrolyte in alkaline storage batteries at subzero temperatures, polymerization catalyst for siloxanes.

cesium iodide. CsI.
Properties: Colorless, crystalline powder; deliquescent; d 4.510; mp 621C; bp 1280C; soluble in alcohol and water.
Grade: Technical, single crystals.
Use: Crystals for infrared spectroscopy, scintillation counters, fluorescent screens.

cesium nitrate. CAS: 7789-18-6. $CsNO_3$.
Properties: Crystalline powder, saltpeter taste, d 3.687, mp 414C, bp (decomposes), soluble in water and acetone, slightly soluble in alcohol.
Derivation: Action of nitric acid on cesium oxide and crystallization.
Grade: Pure, 99.0% min.
Hazard: Dangerous, may ignite organic materials on contact.
Use: Cesium salts.

cesium oxide. Cs_2O.
Properties: Orange-red crystals, d 4.36, mp decomposes 360–400C, very soluble in water, soluble in acids.
Grade: Technical, pure.
Use: Cesium salts.

cesium pentachlorocarbonylosmium(III).
Properties: Orange-red crystal. Air-stable, slightly soluble in water, hydrochloric acid, and insoluble in organic solvents.

cesium pentachloronitrosylosmium(II).
Properties: Dark-red crystals. Air-stable, sparingly soluble in water.
Use: Starting material for other cesium nitrosyl complexes.

cesium perchlorate. $CsClO_4$.
Properties: Crystals, d 3.327 (4C), mp (decomposes) 250C, soluble in water (much more in hot than cold), slightly soluble in alcohol and acetone.
Grade: 99% min.
Hazard: Dangerous fire risk, strong oxidizing agent, may ignite organic materials on contact.
Use: Optics, catalysis, specialty glasses, power generation.

cesium peroxide. See cesium tetroxide.

cesium phosphide. Used as a high-purity binary semiconductor.
Hazard: Fire risk by decomposition to phosphine which is very toxic.

cesium sulfate. Cs_2SO_4.
Properties: Colorless crystals, soluble in water, insoluble in alcohol, d 4.2434, mp 1010C.
Derivation: Action of sulfuric acid on cesium carbonate.

Grade: Pure, low optical density.
Use: Brewing, mineral waters, for density gradient in ultracentrifuge separation.

cesium tetroxide. (cesium peroxide). Cs_2O_4.
Properties: Yellow crystals, d 3.77, mp 600C, decomposes violently in water, soluble in acids. Strong oxidizing agent.
Hazard: Fire risk in contact with organic materials.

cesium trioxide. Cs_2O_3.
Properties: Chocolate-brown crystals, d 4.25, mp 400C, decomposes in water, soluble in acids.

cetalkonium chloride (USAN). (cetyldimethyl benzylammonium chloride; benzylhexadecyldimethylammonium chloride).
$C_6H_5CH_2N(CH_3)_2(C_{16}H_{33})Cl$. A quaternary ammonium germicide.
Properties: Colorless, odorless, crystalline powder; mp 58–60C. Soluble in water to form colorless, odorless solution having pH 7.2. Compatible with alkalies and antihistamines. Soluble in alcohol, acetone, esters, carbon tetrachloride.
Use: Germicide, fungicide, surface-active agent, mildew preventive.

cetane. See hexadecane.

cetane number. A rating for Diesel fuel comparable to the octane number rating for gasoline. It is the percentage of cetane ($C_{16}H_{34}$) which must be mixed with heptamethylnonane to give the same ignition performance under standard conditions as the fuel in question.

"Cetats."[430] TM for cetyltrimethylammonium tosylate.

cetene. See 1-hexadecene.

cetin. (cetyl palmitate; palmitic acid cetyl ester).
$C_{15}H_{31}COOC_{16}H_{33}$.
Properties: White, crystalline, wax-like substance. Chief constituent of commercial purified spermaceti. Mp 50C, bp 360C, d 0.832, refr index 1.4398 (70C). Soluble in alcohol and ether, insoluble in water. Combustible.
Derivation: By solution from spermaceti.
Use: Base for ointments, cerates and emulsions, manufacture of candles, soaps, etc.

cetrimonium pentachlorophenoxide.
$C_{25}H_{42}Cl_5NO$. A crystalline solid melting at 68C, used as a fungicide and as preservative for agricultural products.

cetyl alcohol. (alcohol C-16; cetylic alcohol; 1-hexadecanol, normal; primary hexadecyl alcohol; palmityl alcohol). CAS: 36653-82-4.
$C_{16}H_{33}OH$. A fatty alcohol.
Properties: White, waxy solid; faint odor; d 0.8176 (49.5C); mp 49.3C; bp 344C; refr index 1.4283 (79C); partially soluble in alcohol and ether; insoluble in water. Combustible.
Derivation: By saponifying spermaceti with caustic alkali, reduction of palmitic acid.
Method of purification: Crystals, distillation.
Grade: Technical, cosmetic, NF.
Use: Perfumery, emulsifier, emollient, foam stabilizer in detergents, face creams, lotions, lipsticks, toilet preparations, chemical intermediate, detergents, pharmaceuticals, cosmetics, base for making sulfonated fatty alcohols, to retard evaporation of water when spread as a film on reservoirs, or sprayed on growing plants.

cetyl bromide. $C_{16}H_{33}Br$.
Properties: Dark yellow liquid, fp 15C, bp 186–197C (10 mm), d 0.991 (25/25C), lb/gal 8.25 (25C), refr index 1.460 (25C), flash p 350F (176C). Soluble in ether, very slightly soluble in water, methanol. Combustible.
Use: Synthesis.

cetyldimethylbenzylammonium chloride. See cetalkonium chloride.

cetyldimethylethylammonium bromide.
$C_{16}H_{33}(CH_3)_2H_5NBr$. A quaternary ammonium salt.
Properties: Paste.
Use: Disinfectant, deodorant, germicide, fungicide, detergents.

cetyldimethylethylammonium chloride.
$C_{16}H_{33}(CH_3)_2C_2H_5NCl$. A quaternary ammonium salt.

cetylic acid. See palmitic acid.

cetylic alcohol. See cetyl alcohol.

cetyl lactate. See "Ceraphyl."

cetyl mercaptan. (hexadecyl mercaptan).
$C_{16}H_{33}SH$.
Properties: Liquid, fp 18C, bp 185–190C (7 mm), strong odor, d 0.8474 (20/4C), refr index 1.474 (20C), flash p 275F (135C). Combustible.
Grade: 95% (min) purity.
Use: Intermediate, synthetic rubber processing, surface-active agent, corrosion inhibitor.

cetyl palmitate. See cetin.

cetyl pyridinium bromide. $C_{16}H_{33}C_5H_5NBr$.
A quaternary ammonium compound.
Properties: Cream-colored, waxy solid. Soluble in acetone, ethanol, and chloroform.

Use: Germicide, deodorant, laboratory reagent, surfactant.

cetylpyridinium chloride. (Ceepryn; Cepacol; Cetamium; Dobendan; 1-hexadecylpyridinium chloride; Medilave; Pristacin; Pyrisept).
CAS: 123-03-5. $C_{21}H_8ClN \cdot H_2O$.   A quaternary ammonium salt.
Properties: (Monohydrate) White powder, mp 77–83C, surface tension (25C) 43 dynes/cm (0.1% aqueous solution). Soluble in water, alcohol, chloroform. Very slightly soluble in benzene and ether.
Use: Antibacterial in cough lozenges and syrups; emulsifier.

cetyltrimethylammonium bromide. (hexadecyltrimethylammonium bromide).
CAS: 57-09-0. $C_{16}H_{33}(CH_3)_3NBr$.
A quaternary ammonium salt.
Properties: White powder, soluble in water, alcohol and chloroform.
Grade: Technical.
Use: Surface-active agent, germicide.
Properties: White powder; soluble in water, alcohol, and chloroform.

cetyltrimethylammonium chloride.
$C_{16}H_{33}(CH_3)_3NCl$.   A quaternary ammonium salt.

cetyltrimethylammonium tosylate.
$[C_{16}H_{33}(CH_3)_3N]OSO_2C_6H_4CH_3$.   A high-temperature stable quaternary ammonium compound.
Use: Germicide, surfactant.

cetyl vinyl ether. (vinyl cetyl ether).
$C_{16}H_{33}OCH{:}CH_2$.
Properties: Colorless liquid, d 0.822 (27C), mp 16C, bp 142C (1 mm), 173C (5 mm), flash p 325F (162C) (OC), refr index 1.444 (25C). Combustible.
Grade: 97%.
Hazard: Toxic by inhalation, skin irritant. Reacts strongly with organic materials.
Use: Reactive monomer which may be copolymerized with a variety of unsaturated monomeric materials, including acrylonitrile, vinyl chloride, vinylidene chloride and vinyl acetate to yield internally plasticized resins.

Cf. Symbol for californium.

CF. Abbreviation for citrovorum factor.
See folinic acid.

"C- Fatty Acids."[487] TM for fatty acids derived from coconut oil. The major component acids are lauric and myristic. Differ primarily in amount of unsaturated acid components and color. Light-yellow solids which liquefy at approximately 25C, obtained from naturally occurring triglycerides. Combustible.
Use: Intermediate, rubber compounding, cosmetic ingredients, buffing compounds, alkyd resins, emulsifiers, grease manufacture, and candles.

CFE. Abbreviation for chlorotrifluoroethylene. Also used for polychlorotrifluoroethylene resins.

cgs. Abbreviation of centimeter gram second, the system of measurement used internationally by scientists.

chabazite. $CaAl_2Si_4O_{12} \cdot 6H_2O$.   Essentially a natural hydrated calcium aluminum silicate, usually containing some sodium and potassium. A zeolite.
Properties: Color white, reddish, yellow, brown, luster vitreous, d 2.1. Mohs hardness 4–5.
Occurrence: New Jersey, Colorado, Oregon, Europe.
Use: Water treatment.

Chadwick, Sir James. (1891–1974) A British physicist who was awarded the Nobel Prize in 1935 for his discovery of the neutron (1932), the existence of which had been predicted by Rutherford. See also neutron.

chain. A series of atoms of a particular element directly connected by chemical bonds which constitutes the structural configuration of a compound. Such chains are usually composed of carbon atoms, often shown without their accompanying hydrogen. Carbon chains may be of the following types:
(1) Open or straight chain: a sequence of carbon atoms extending in a direct line, characteristic of paraffins and olefins, the former being saturated:

H—C—C—C—C—H (with H atoms above and below each carbon)

and the latter unsaturated:

C=C—C=C (with H atoms attached)

(2) Branched chain: a linear series of carbon atoms occurring in paraffinic hydrocarbons and some alcohols that are isomeric with its straight

chain counterpart and has a subordinate chain of one or more carbon atoms:

$$H-\underset{\underset{H}{|}}{\overset{\overset{H}{|}}{C}}-\underset{\underset{\underset{H}{|}}{\overset{H}{\underset{|}{C}}-H}}{\overset{\overset{H}{|}}{C}}-\underset{\underset{H}{|}}{\overset{\overset{H}{|}}{C}}-H.$$

Such compounds are designated by the prefix iso-; the iso-paraffins are much more efficient than their straight-chain isomers, especially in gasoline. For example, octane has a low antiknock rating, whereas that of isooctane is high.

(3) Closed chain or ring: a cyclic arrangement of carbon atoms giving a closed geometric structure, i.e., a ring, pentagon or other form, characteristics of alicyclic, aromatic, and heterocyclic compounds.
See cyclic compound.

(4) Side-chain: a group of atoms attached to one or more of the locations in a cyclic or heterocyclic compound, e.g., tryptophan:

$$-CH_2CH(NH_2)COOH$$

chain mechanism. See free radical.

chain reaction. See fission, nuclear, nuclear energy.

chalcocite. (copper glance). Cu_2S. Natural cuprous sulfide, occurring with other copper minerals.
Properties: Color lead gray, tarnishing dull black, luster metallic, d 5.5–5.8, Mohs hardness 2.5–3.
Occurrence: Montana, Arizona, Utah, Nevada, Alaska, Chile, Mexico, Europe.
Use: Important ore of copper.

chalcopyrite. (copper pyrites; yellow copper). $CuFeS_2$. Natural copper-iron sulfide, found in metallic veins and igneous rocks. Also made synthetically.
Properties: Yellow or bronze iridescent crystals with metallic luster and greenish streak, d 4.1–4.3, Mohs hardness 3.5–4. May carry gold or silver or mechanically intermixed pyrite.
Occurrence: Montana, Utah, Arizona, Tennessee, Wisconsin, Europe, Chile, Canada.
Use: Important ore of copper, semiconductor research.

chalk. A natural calcium carbonate composed of the calcareous remains of minute marine organisms. Decomposed by acids and heat. Odorless, tasteless.
See also calcite, calcium carbonate, whiting, chalk, prepared.

chalk, drop. See chalk, prepared.

chalk, French. A variety of soapstone or steatite. See also talc.

chalking. A natural process by which paints develop a loose, powdery surface formed from the film. Chalking results from decomposition of the binder, due principally to the action of UV rays.

chalk, precipitated. See calcium carbonate, precipitated.

chalk, prepared. (drop chalk; calcium carbonate, prepared).
Properties: Fine, white to grayish-white impalpable powder; often formed in "conical drops." Odorless, tasteless and stable in air. Mp decomposes at 825C with evolution of carbon dioxide, decomposed by acids, practically insoluble in water, insoluble in alcohol. Noncombustible.
Derivation: By grinding native calcium carbonate to a fine powder, agitating with water, allowing the coarser particles to settle, decanting the suspension and allowing the fine particles to settle slowly.
Use: Medicine (antacid), tooth powders, calcimine, polishing powders, silicate cements.
See also whiting and calcium carbonate.

chamber process. An obsolete method for manufacturing sulfuric acid in lead-lined chambers from sulfur dioxide, air, and steam in the presence of NO_x as catalysts. It is no longer used in the US.

chamois. A very soft, flexible leather made from the flesh layer of a split sheepskin by treating with fish oils, piling in contact with similarly treated skins, and allowing the fish oils to oxidize.

channel black. See carbon black.

Chapman rearrangement. Thermal rearrangement of aryl imidates to N,N-diaryl amides.

charcoal, activated. See carbon, activated.

charcoal, animal. See bone black, ivory black.

charcoal, bone. See bone black.

charcoal, vegetable. See vegetable black, carbon, activated.

charcoal, wood. A highly porous form of amorphous carbon.
Derivation: Destructive distillation of wood.
Grade: Technical, in lumps, powdered, briquettes.
Hazard: Dangerous fire risk in briquette form or when wet, may ignite spontaneously in air.
Use: Chemical (precipitant in the cyanide process, precipitant of iodine and lead salts from their solutions, catalyst, calcium carbide); decolorizing and filtering medium; gas adsorbent; component of black powder and other explosives; fuel; arc light electrodes; decolorizing and purifying oils; solvent recovery; deodorant.

Chardonnet, H. (1839–1924) A native of France, he has been called the father of rayon because of his successful research in producing what was then called artificial silk from nitrocellulose. He was able to extrude fine threads of this semisynthetic material through a spinneret-like nozzle and the textile product was made on a commercial scale in several European countries. He was awarded the Perkin medal for this work.

Charles' Law. At constant volume, the pressure of a confined gas is proportional to its absolute temperature.
See also Gay-Lussac's Law.

Charpy. A standard testing device for impact strength.

chaulmoogra oil. (gynocardia oil; hydnocarpus oil).
Properties: Brownish-yellow oil or soft fat, characteristic odor, somewhat acrid taste. Soluble in ether, chloroform, benzene, solvent naphtha; sparingly soluble in cold alcohol; almost entirely soluble in hot alcohol, carbon disulfide. D 0.940, iodine value 85–105 (based on type) optically active.
Chief constituents: Glycerides of chaulmoogric and hydnocarpic acids.
Use: Medicine (treatment of leprosy and other infective skin diseases).

chaulmoogric acid. (hydnocarpyl acetic acid).
$CH_2CH_2CHCHCH(CH_2)_{12}COOH.$ A cyclic fatty acid.
Properties: Colorless, shiny leaflets; mp 68.5C; soluble in ether, chloroform, and ethyl acetate.
Source: Chaulmoogra oil.
Use: Medicine, biochemical research.

chavicol. (p-allylphenol; 1-allyl-4-hydroxybenzene). $C_3H_5C_6H_4OH.$
Properties: Liquid, mp 16C, bp 230C, d 1.033 (18/4C), soluble in water and alcohol. Occurs in many essential oils.

CHDM. See 1,4-cyclohexanedimethanol.

checking. Development of small cracks in the surface of a material such as rubber, paint or ceramic glaze.

"Cheelox."[307] TM for a series of organic chelating and sequestering agents, consisting of polycarboxylic acid derivatives of amines or polyamines or their salts, i.e., ethylenediaminetetraacetic acid.

"Chel."[443] TM for chelating agents based on polyaminocarboxyllic acids.
Use: To reduce the harmful effects of trace metals.

chelate. The type of coordination compound in which a central metal ion such as Co^{2+}, Ni^{2+}, Cu^{2+}, or Zn^{2+}, is attached by coordinate links to two or more nonmetal atoms in the same molecule, called ligands. Heterocyclic rings are formed with the central (metal) atom as part of each ring. Ligands offering two groups for attachment to the metal are termed bidentate (two-toothed); three groups, tridentate; etc.
A common chelating agent is ethylenediaminetetraacetic acid (EDTA). Nitrilotriacetic acid $N(CH_2COOH)_3$; and ethyleneglycolbis(β-aminoethyl ether)-N,N-tetraacetic acid $(HOOCCH_2)_2NCH_2CH_2OCH_2CH_2OCH_2CH_2N(CH_2COOH)_2$ are used in analytical chemical titrations and to remove ions from solutions and soils. Metal chelates are found in biological systems, e.g., the iron-binding porphyrin group of hemoglobin and the magnesium-binding chlorophyll of plants. Medicinally, metal chelates are used against Gram-positive bacteria, fungi, viruses, etc.
See also ammine, sequestration, complex, cobaltammine.

"Chelon."[428] TM for a series of chelating agents.

"Chem-Hoe."[177] TM for isopropyl-N-phenylcarbamate (IPC). A selective herbicide.

Chemical Abstracts. A weekly publication of the American Chemical Society which consists of research articles and patents in all major fields of chemistry throughout the world. It is the most indispensable publication in chemical literature and is the largest scientific abstract journal in the world. For further information, see Appendix II.

Chemical Abstracts Service. See chemical data processing.

chemical bond. See bond, chemical.

chemical change. Rearrangement of the atoms, ions, or radicals of one or more substances resulting in the formation of new substances often having entirely different properties. Such a change is called a chemical reaction. In some cases, energy in the form of heat, light, or electricity is required to initiate the change; this is known as an endothermic reaction. When energy is given off as a result of rupture of chemical bonds, the change is said to be exothermic. Chemical changes should be distinguished from physical changes in which only the state or condition of a substance is modified its chemical nature remaining the same. A physicochemical change has some of the characteristics of both. Examples of the three types are:

Chemical Changes

fuel + oxygen →
$$CO_2 + water + heat \text{ (exothermic)}$$
water + CO_2 + energy →
$$sugar + oxygen \text{ (endothermic)}$$

Physical Changes.

water to ice or steam;   crystallization;
coagulation of latex.   distillation processes.

Physiochemical Changes.

cooking of food,
vulcanization of rubber,
tanning of leather,
drying of oil- or plastic-based paints.

chemical dating. Estimation of the age of geologic structures and events by measuring the amount of radioactive decay products in existing samples. The age of a uranium-containing material can be determined by measuring the percentage of lead (or helium) formed as a result of disintegration of the uranium. Uranium decays to both helium and the 206 lead isotope, but measurement of helium content is inaccurate because of its strong tendency to escape. By determining the ratio of the percentage of lead in a sample to the percentage of uranium, the age can be calculated. A more recent method, applicable to events within about 10,000 years, involves the use of the natural radioactive carbon isotope (C^{14}). The percentage of this isotope determined in a carefully prepared sample is an index of its age based on the half-life of C^{14} (5700 years), which was present in the atmospheric carbon dioxide absorbed by plants centuries ago. This method has yielded valuable results in the study of archaeological specimens, deep-sea sedimentation, and dates of volcanic and glacier activity.

chemical data storage. The recording, coding, storage, and retrieval of chemical information. Early methods involved the use of punched card systems, of which there were several types. The advent of digital computers made it possible to store and retrieve such data as molecular structure, physical and chemical properties, and other relevant facts with much greater efficiency. The ultimate in chemical documentation is the sophisticated and extensive information bank assembled by Chemical Abstracts Service, a division of ACS. It features online call-up and display of the most complex organic formulas and property data of over six million compounds. To have made this knowledge accessible so rapidly is a tremendous achievement of the utmost value to chemical research. Among its pioneers were Dyson, Perry, Crane, and Tate. Chemical Abstracts Service is located in Columbus, Ohio.

chemical economics. The principles and practical application of industrial economics in particular reference to the manufacture of chemicals and chemically derived products. Economic factors that apply mainly to the chemical industries are: (1) heavy capital investment; (2) vast range of raw materials and products; (3) complex and varied production methods; (4) high-level R & D for new products, plus extensive testing for safety and acceptability; (5) market research to develop new product uses, etc.; (6) substitution of synthetics for natural products or (in some cases) the reverse; (7) utilization of by-product materials; (8) waste reclamation and pollution control. See also chemical process industry.

chemical education. The instruction and training of students at secondary, college, and graduate levels in both the theoretical and practical aspects of chemistry. Well-balanced courses include a substantial amount of laboratory experimentation in addition to lectures and textbook study. Comprehensive one-year courses are offered by most colleges for non-science majors. The more advanced curricula emphasize quantum mechanical and thermodynamic considerations, stoichiometry, spectroscopy. Mechanical models of molecular structure are useful in teaching organic chemistry. A few schools provide courses in industrial chemistry. Student employment in a chemical process plant for part of the semester has proved successful. Tape recordings of symposia, analytical methods and lectures in general chemistry, as well as a broad group of high-level short courses in special subjects are offered by ACS, which also sponsors radio broadcasts on topics of popular interest in chemistry. The field is well served by two publications--the *Journal of Chemical Education & Chemistry*. Among the leaders in the development of chemical education may be mentioned Ira Remsen, James B. Conant, Gilbert N. Lewis, Louis Fieser, Roger Adams, and Joel Hildebrand. In recent years, the pervasive influence of computer technology has penetrated almost all of the physical sciences. Thus basic training in its applications to chemistry has become essential adjunct to chemical education. See also chemical literature, model, computational chemistry.

chemical engineering. That branch of engineering concerned with the production of bulk materials from basic raw materials by large-scale application of chemical reactions worked out in a laboratory. The unit operations of chemical engineering are those common to all chemical processing: fluid flow, heat transfer, filtration, evaporation, distillation, drying, mixing, adsorption, solvent extraction, and gas absorption, based on the fundamentals involved in every technical problem in chemical engineering: conservation, equilibrium, kinetics and control. Conservation involves the laws of the conservation of matter and energy. The limits of any chemical process are established by its equilibrium conditions. The principles underlying the equilibrium concept (the Law of Mass Action) are of fundamental importance (see chemical laws). Kinetics means time-dependent processes or rate processes (momentum, thermal, mass transfer, and chemical kinetics). Control involves either the systems approach or feedback or closed-loop methods, as well as considerations of the stability of the system.
See also equilibrium constant; stoichiometry; kinetics, chemical.

chemical flooding. A method of enhanced oil recovery in which a mixture of detergents, alcohols, and other solvents is pumped into the porous oil-bearing strata, followed by much brine which furnishes the pressure necessary to drive the mixture through the rocky structure. Field tests show this process to be no more than 50% efficient and will probably not be important in secondary oil recovery for some years. Another technique, called micellar flooding, involves the use of surfactants and a mixture of water and a high polymer. It is a two-step process: the surfactants (petroleum sulfonate and alcohol) are injected into the oil-bearing strata, followed by the water-polymer mixture. Volumes of these materials are required.
See also hydraulic fracturing.

Chemical Industry Institute of Toxicology.
An industry-supported organization whose aim is investigation of environmental and occupa-

tional health concerns of widely used basic chemicals. It tests the toxicity of commodity chemicals and develops testing methods. It is located at Research Triangle Park, Raleigh, NC.

chemical laws. A group of basic principles governing the combining power and reaction characteristics of elements. Among the more important are:

(1) Law of Mass Action: The rate of a homogeneous (uniform) chemical reaction at constant temperature is proportional to the concentration of the reacting substances.

(2) Law of Definite or Constant Composition: Any chemical compound contains the same elements in the same fixed proportions by weight. Exceptions to this law occur in solid compounds such as silicates which are known as nonstoichiometric compounds.

(3) Law of Multiple Proportions: When two elements unite to form two or more compounds, (e.g., nitrogen and oxygen can form five different oxides) the weight of one element that combines with a given weight of the other is in the ratio of small whole numbers. Hydrogen and oxygen unite in the ratio of 1 to 8 in water and of 1 to 16 in hydrogen peroxide. Thus, the weights of oxygen that unite with one gram of hydrogen are in the ratio of 1 to 8.

(4) Law of Conservation of Mass: Any chemical reaction between two or more elements or compounds leaves the total mass unchanged, the reaction products having exactly the same mass as present in the reactants, regardless of the extent to which other properties are changed.

(5) Law of Avogadro: Equal volumes of gases at constant temperature and pressure contain the same number of molecules if the gases are the same or different. 22.4 liters of any gas contain 6.02×10^{23} molecules.
See also Avogadro's law, mole.

chemical literature. World-wide chemical information available in the form of journals, patents and books (handbooks, dictionaries, encyclopedias, reports of conferences and symposia, review volumes, research treatises, and textbooks). The output of journal articles has increased almost exponentially in the last 70 years. Of major importance in making this information explosion available to scientists are the abstract journals, particularly *Chemical Abstracts*, described in Appendix II.

Outstanding among periodicals is the *Journal of the American Chemical Society* in which over 2000 articles appear yearly. Hundreds of other journals throughout the world add to the vast stream of chemical literature.

In book form, there are four monumental multivolume compendia (the dates are for the origi-

nal edition, but in most cases there have been several revisions): Abegg's and Gmelin's "Handbuch der Anorganischen Chemie" (1870), Beilstein's "Handbuch der Organischen Chemie" (1881), Mellor's "Inorganic and Theoretical Chemistry" (1922), and Heilbron's "Dictionary of Organic Compounds" (approximately 1920, new edition, 7 volumes, 1982). Other noteworthy reference sources are "Merck Index" (1889), "Handbook of Chemistry and Physics" (1918), Lange's "Handbook of Chemistry" (1934), "Chemical Engineers Handbook" (1934), "Ring Index" (1940), Hackh's "Chemical Dictionary" (1929) and Kirk-Othmer's "Encyclopedia of Chemical Technology" (1947), the third edition of which comprises 25 volumes published from 1978 to 1983. The ACS Monograph Series established in 1919, is devoted to high-level treatises in developing areas of chemistry. Many useful series of "Advances" reviewing progress in specific fields did appear from time to time.
See also chemical education.

"Chemical Mace." See "Mace."

Chemical Manufacturers Association.
(CMA). Founded in 1863, this nonprofit trade association of chemical manufacturers represents more than 90% of the chemical industry of the US and Canada. It is located at 2501 M Street, N.W., Washington, DC 20037. It is particularly active in the fields of packaging, transportation, safe handling of hazardous chemicals, and wording of precautionary labels, as well as in general chemical education and all aspects of pollution control. It has instituted an emergency telephone information service called "Chem-trec" to provide instant information for safety precautions in accidents involving chemicals.

Chemical Marketing Research Association.
(CMRA). A professional organization whose primary functions concern market research in the chemical and process industries. Techniques for this comparatively new research are still being evaluated. Major aims of the CMRA are opportunities for members to keep in touch with the latest consumer and industrial research concepts, to encourage the general welfare of the chemical industries, to improve the service of these industries to the public, and to cooperate with government officials in furthering the national welfare. The offices of the organization are at 100 Church St., New York, NY 10007.

chemical microscopy. See microscopy, chemical.

chemical milling. The process of producing metal parts of predetermined dimensions by removing metal from the surface with chemicals. Acid or alkaline pickling or etching baths are used for this purpose. Immersion of a metal part will result in uniform removal of metal from all surfaces exposed to the solution. This is used by the aircraft industry for weight reduction of large parts. It is also used in the manufacturing of instruments and other components where exact tolerances are required.

chemical nomenclature. The origin and use of the names of elements, compounds, and other chemical entities, as individuals and as groups, as well as various proposals for systematizing them. It may be considered to include three major aspects: (1) the gradual sporadic development of these names, which may go back to the alchemists of the Middle Ages; (2) the proliferation of terminology due to rapid extension of organic chemical in the mid-nineteenth century, which led to the recommendations of the Geneva System in 1892; (3) the additional reforms adopted by the International Union of Pure and Applied Chemistry in 1930. There is yet no clear-cut elimination of the older names in spite of changes introduced in these reformed systems. Thus, present nomenclature is in some respects a hybrid; for example, the earlier terms paraffin and olefin are still widely used instead of more modern alkane and alkene, and methyl alcohol is still in common use for methanol.

A comparatively recent development in the nomenclature of inorganic and complex compounds is use of the Stock system in which Roman numerals indicate the oxidation state or coordination value. For example, iron II chloride stands for ferrous chloride ($FeCl_2$), and iron III chloride for ferric chloride ($FeCl_3$).
See also Geneva System, benzene.

chemical oxygen demand. See oxygen consumed.

chemical planetology. Application of various branches of chemistry (analytical, physical, and geochemistry) to study the composition of the surface and atmosphere of the planets, mainly Venus, Mars, and Jupiter. Much information has been obtained by spectrographic methods and valuable additional data have resulted from space probes.
See also astrochemistry.

chemical process industry. An industry whose product(s) results from (a) one or more chemical or physicochemical changes; (b) extraction, separation, or purification of a natural product with or without chemical reactions; (c) the preparation of specifically formulated mixtures of materials, either natural or synthetic. Examples are as follows (with allowance for some overlapping): (a) the plastics, rubber, leather, food, dye, and synthetic organic industries; (b) the petroleum, paper, textile, and perfume industries. Many of these involve one or more unit operations of chemical engineering as well as basic processes as polymerization, oxidation, reduction, hydrogenation, etc., usually with the aid of a catalyst. This definition may be interpreted to include ore processing, separation, and refinement, as well as the manufacture of metal products; however, these are usually considered to comprise the metal and metallurgical industries.

chemical reaction.   A chemical change that may occur in several ways, e.g., by combination, by replacement, by decomposition or by some modification of these. Reactions are endothermic when heat is needed to maintain them and exothermic when they evolve heat. All chemical reactions are in balance, i.e., the number of atoms in the reacting substances is always equal to the number of atoms in the reaction products. Common types of reactions are oxidation, reduction, ionization, combustion, polymerization, hydrolysis, condensation, enolization, saponification, rearrangement, etc. Chemical reactions involve rupture of only the bonds which hold the molecules together, and should not be confused with nuclear reactions where the atomic nucleus is involved. A reversible reaction is one in which the reaction product is unstable and thus changes back into the original substance spontaneously. In a complete reaction, the activity goes to the right and is indicated by an arrow $\rightarrow$; if heat or a catalyst is used, it is indicated by a symbol or word usually placed in small type above the arrow as: $\xrightarrow{\Delta}$, $\xrightarrow{catalyst}$; a reversible reaction is shown by either $\rightleftharpoons$ or $\leftrightarrow$.

chemical research.   See applied research, fundamental research.

chemical sediment.   A sediment created by precipitation of one or more minerals from natural waters.

chemical smoke.   Chemically generated aerosols used primarily for military purposes. They are of four types: (a) FS, a mixture of sulfuric anhydride and chlorosulfonic acid, used in shells and bombs and sprayed from airplanes; (b) FM, titanium tetrachloride, same as FS but brilliant white and will drop like a curtain when sprayed; (c) HC, a mixture of hexachloroethane, aluminum, and zinc oxide, burns to yield a white cloud;

(d) WP, a white phosphorus, burns to form white cloud of phosphoric acid, an excellent smoke producer.
See also fog, smoke, chemical warfare.

chemical stoneware.   (brick, chemical).
A clay pottery product widely employed to resist acids and alkalies. It is used for utensils, pipes, stopcocks, ball mills, laboratory sinks, etc.

chemical technology.   A general term covering a broad spectrum of physicochemical knowledge of the materials, processes, and operations used in the chemical process industries. It includes (1) basic phenomena such as activation, adsorption, oxidation, catalysis, corrosion, surface activity, polymerization, etc.; (2) the properties, behavior, and handling of industrial materials and products (plastics, textiles, coatings, soap, foods, metals, pharmaceuticals, etc.); and (3) their formulation, fabrication, and testing (compounding, extruding, molding, assembly, and the like).
See also chemical process industry.

chemical thermodynamics.   That aspect of thermodynamics concerned with the relationship of heat work and other forms of energy to equilibrium in chemical reactions and changes of state.
See also thermodynamics; thermochemistry; kinetics, chemical; equilibrium constant.

chemical warfare.   The employment of a chemical agent directly for military purposes, i.e., to cause casualties by irritating, burning, asphyxiation or poisoning, to contaminate ground, to screen action by smoke, or to cause incendiary damage. Includes use of all forms of toxic or irritating gases including nerve gases, smoke-inducing agents, flammable gels, such as napalm, and such incendiary materials as magnesium and thermite. Biological warfare involves the use of bacteria and other infective agents against enemy forces or populations.

chemical waste.   Unusable by-products from many chemical and metal-processing operations which often contain toxic or polluting materials that become environmental threats if improperly disposed of, such as digester streams from paper manufacture, fluorides from aluminum manufacturing, mercury from chlor-alkali cells, tailings from asbestos production, and insecticidal wastes. Disposal techniques approved by EPA are: (1) landfill in which an impermeable barrier is placed between the waste and ground water, such as a layer of hard-packed clay; (2) incineration of organic materials plus use of specially equipped ships in which liquid wastes are inciner-

ated at sea; (3) such chemical and biological methods as neutralization of acidic and basic wastes, oxidation, and activated sludge treatment of organic wastes. Dumping of wastes into lakes and watercourses is strictly forbidden. Storage in metal drums is inadequate due to corrosion and subsequent leakage. Federal authority to provide for control and safe disposal of hazardous wastes is provided by the Resource Conservation and Recovery Act (RCRA), passed by Congress in 1976.
See also radioactive waste.

Chemico process. A technique used for extracting sulfur from low-grade ores (25 to 50% sulfur) by means of hot water.

"Chemigum."[265] TM for synthetic acrylonitrile/ butadiene dry elastomers and latices produced by emulsion polymerization. They have great resistance to oils and aromatic fuels. They range from high to low nitrile content polymers.
Use: (Dry) Auto gaskets and hoses, refrigerator hoses, fuel cell interliners, oil-resistant shoe soles and heels, belt covers, hard packing compounds, adhesives, kitchen and drug sundries, tubing, textile coatings, frictioning compounds, oil-resistant auto parts, rubber rolls, printers' supplies, flooring. (Latex): Saturated and beater-impregnated paper, nonwoven fabric binders, carpet backings, textile and paper coating, foam, dipped goods, textile inks, leather finishes.

chemiluminescence. The emission of absorbed energy (as light) due to a chemical reaction of the components of the system. It includes the subclasses bioluminescence and oxyluminescence in which light is produced by chemical reactions involving organisms and oxygen, respectively. Chemiluminescence occurs in thousands of chemical reactions covering a wide variety of compounds, both organic and inorganic. Emission of light by fireflies is a common example of bioluminescence.
See also luminescence.

chemisorption. The formation of bonds between the surface molecules of a metal (or other material of high surface energy) and another substance (gas or liquid) in contact with it. These bonds are comparable in strength to ordinary chemical bonds and much stronger than the Van der Waals type characterizing physical adsorption. Chemisorbed molecules are often altered. Hydrogen is chemisorbed on metal surfaces as hydrogen atoms. Chemisorption of hydrocarbons may result in formation of chemisorbed hydrogen atoms and hydrocarbon fragments. Even when dissociation does not occur, the properties of the molecules are changed by the surface in important ways. This mechanism is the activating force of catalysis.
An example of chemisorption is the boundary lubrication of moving metal parts in machinery. An oil film forms a chemisorbed layer at the interface and averts the high frictional forces that would otherwise exist. Solids with high surface energies are necessary for chemisorption to occur, e.g., nickel, silver, platinum, iron.
See also catalysis.

chemistry. A basic science whose central concerns are (1) the structure and behavior of atoms (elements); (2) the composition and properties of compounds; (3) the reactions between substances with their accompanying energy exchange; and (4) the laws that unite these phenomena into a comprehensive system. Chemistry is not an isolated discipline, for it merges into physics and biology. The origin of the term is obscure. Chemistry evolved from the medieval practice of alchemy. Its bases were laid by such men as Boyle, Lavoisier, Priestly, Berzelius, Avogadro, Dalton, and Pasteur.
See inorganic chemistry, organic chemistry, physical chemistry, and Appendix II A-E..
Note: Chemistry has been variously defined to the point where definition has become a semantic exercise of questionable, if not negative, value. "Chemistry is the science of matter" and "Chemistry is a branch of physics" are two instances of such definitions. The first relegates physics to the background while the second accords it supremacy.

chemistry history. See Appendix II A-E.

chemistry in space. See space, chemistry in.

"Chemlok."[547] TM for a series of adhesives, primarily for bonding rubber to metal. "Chemlok" 205. A mixture of polymers, organic compounds, and mineral fillers in a methyl isobutyl ketone and "Cellosolve" solvent system. Used as a primer with "Chemlok" 220 when the bond must have exceptional resistance to adverse environmental conditions or when surface preparation must be minimized. Also used as a single coat adhesive for bonding nitrile elastomers. "Chemlok" 220. Dissolved organic polymers and dispersed fillers in a xylene and perchloroethylene solvent system. Used as a versatile one-coat adhesive for bonding uncured elastomers to metals during vulcanization.

chemodynamics. A comprehensive, interdisciplinary study of chemicals in the environment with special reference to pesticides.

chemometrics. Application of computer data-analysis techniques to the classification, assimilation, and interpretation of chemical information. Its major purpose is to correlate data in such a way that trends or patterns are indicated. Molecular spectra, thermodynamic functions, and distribution of chemicals in the atmosphere in relation to rainfall are some fields to which this science was utilized. Formal programs in chemometrics are under way at the University of Washington and at Umea University in Sweden.

chemonite. (copper arsenite, ammonical). A wood-preservative solution prescribed by Federal Specification TT-W-549 to contain copper hydroxide ($Cu(OH)_2$) 1.84%, arsenic trioxide (As_2O_3) 1.3%, ammonia (NH_3) 2.8%, acetic acid 0.05%, water as necessary to 100.0%.
Hazard: Toxic by ingestion.

chemonuclear production. Manufacture of chemicals using the energy of a nuclear reactor. Feasibility studies on making hydrogen cyanide from nitrogen and methane using fission fragments as the energy source for the heat of reaction indicated that this process cannot yet compete economically with standard methods. In the case of hydrazine, the economic aspect is more favorable because of the high cost of conventional methods.

chemosterilant. A term coined by the USDA for materials or processes which sterilize insects, usually the males, thus preventing their reproduction. Gamma radiation is one method used. The males are brought to the radiation by sex attractants.

chemotaxis. The tendency of certain bacteria, especially *E. coli* and *Salmonella*, to move toward nutrients and other attractants and away from repellents and toxic chemicals. Such bacteria live in a liquid medium and propel themselves through it by means of thread-like appendages called flagellae. They apparently can determine differences in the concentration of chemicals in the medium as they move through it, thus exhibiting a "memory" in respect to their environment.

chemotherapy. The treatment or prevention of a disease by administration of a chemical. The term was first used by Paul Ehrlich, discoverer of the arsphenamine treatment for syphilis (1910). He said that chemotherapy results from the interaction of chemically reactive groups on drugs and of chemically active receptor groups on parasitic cells and that an effective drug must be of quite low molecular weight. His achievement was one of the great triumphs of biomedical science. More recent important achievements were the development of antimalarials, the synthesis of sulfa drugs, the discovery and proliferative development of antibiotics (penicillin, streptomycin, etc.), and the synthesis of cortisone. Much research effort has been devoted to the chemotherapeutic investigation of cancer. An antibiotic, adriamycin, is said to be effective against certain types of cancer. Recently, treatment with Co-60, either by irradiation or as a tissue implant, was successfully used. A broad spectrum of antidepressant drugs to treat acute mental depression is another outstanding development of chemotherapy.

CHEMRAWN. Abbreviation of Chemical Research Applied to World Needs, a program established by the International Union of Pure and Applied Chemistry and co-sponsored by the American Chemical Society and the Chemical Institute of Canada to discuss and study the problems of long-range conversion to a post-petroleum economy. The subject range includes all forms of natural raw materials, particularly those used for fuel, food, and the manufacturing of organic chemicals. The conference's first meeting was in Toronto in 1978, and a second is scheduled for The Hague. The results were published in book form in 1986.

CHEMTREC. Abbreviation of Chemical Transportation Emergency Center. It is a division of the Chemical Manufacturers Association established as an emergency information source for transportation accidents involving flammable, toxic, or explosive materials.

"Chemtree."[461] TM for a group of metallic mortars used for various kinds of nuclear shielding where a combination of formability and high attenuation values is required. They are dry powders which are mixed with water before use.

chemurgy. Development of nonfood uses for farm products, especially such waste materials as corn cobs, nut shells, etc. This term was introduced in 1935, but has since become obsolescent.
See also biomass, biotechnology.

chenopodium oil. (wormseed oil; American goose-foot oil).   CAS: 8006-99-3.
Properties: Colorless or yellowish oil, characteristic penetrating odor, bitterish burning taste. Soluble in 3–10 volumes of 70% alcohol (inferior and adulterated oils do not yield a clear solution). D 0.965–0.990 (15C), optical rotation −4 to −8 degrees, refr index 1.4740–1.4790 (20C). Chief known constituents: Ascaridole ($C_{10}H_{16}O_2$); o-cymene; l-limonene.

Derivation: Distilled from the seeds and leaves of Chenopodium ambrosioides anthelminticum.
Use: Medicine (antihelminthic).

chert. An impure rock, often gray in color, that consists primarily of extremely small quartz crystals precipitated from water solutions.

chi acid. See anthraquinone-1,8-disulfonic acid.

Chicago acid. (1-amino-8-naphthol-2,4-disulfonic acid; 8-amino-1-naphthol-5,7-disulfonic acid; SS acid; 1,8,2,4-acid; 2 sulfur acid).

Properties: Gray paste, white when pure. Soluble in water and sodium hydroxide solution.
Derivation: 1-Naphthylamine-8-sulfonic acid is reacted with sulfuric acid to yield 1-naphthylamine-4,8-disulfonic acid; further reaction with 25% oleum gives 1,8-naphthosulfam-2,4-disulfonic acid which melts at 155C with 40% NaOH.
Use: Azo dye intermediate.

Chichibabin pyridine synthesis. Condensation of carbonyl compounds with ammonia or amines under pressure to form pyridine derivatives; the reaction is reversible and produces different pyridine derivatives along with byproducts.

Chichibabin reaction. Amination of pyridines and other heterocyclic nitrogen compounds with alkali-metal amides.

chicle. A thermoplastic, gum-like substance obtained from the latex of the sapodilla tree native to Mexico and Central America. Softens at 32.3C. Insoluble in water, soluble in most organic solvents. Chief use is as a chewing gum after incorporation of sugar and specific flavoring.
Hazard: Ingestion should be avoided.

Chilean nitrate. See sodium nitrate.

Chilean saltpeter. See sodium nitrate.

China clay. See kaolin.

China-wood oil. See tung oil.

Chinese insect wax. A wax secreted by a louse-like insect upon the leaves of plants in China. Its chief ingredient is ceryl cerotate.

chiral. In chemistry this term describes asymmetric molecules that are mirror-images of each other, i.e., they are related optically as right and left hands. Such molecules are also called enantiomers and are characterized by optical activity. See also optical isomer, enantiomer.

chitin. $(C_8H_{13}NO_5)_n$. A glucosamine polysaccharide. Contains approximately 7% nitrogen and is structurally similar to cellulose. Principal constituent of the shells of crabs, lobsters, and beetles. It is also found in some fungi, algae, and yeasts.
Properties: White, amorphous, semitransparent mass; insoluble in the common solvents; soluble in concentrated hydrochloric, sulfuric, and nitric acids.
Use: Biological research, source of chitosan.

chitosan. Deacylated derivative of chitin. Absorbs heavy metals from water and industrial waste streams, also used as dyeing assistant and in photographic emulsions.

chlophedianol hydrochloride (USAN). (α-(2-dimethylaminoethyl)-o-chlorobenzyhydrol hydrochloride). $C_{17}H_{20}ClNO \cdot HCl$. Used in medicine.

chloracetyl chloride. See chloroacetyl chloride.

chloral. (trichloracetaldehyde).
CAS: 75-87-6. CCl_3CHO.
Properties: Colorless, mobile, oily liquid; penetrating odor; d 1.505 (25/4C); fp −57.5C; bp 97.7C; vap press 35 mm (20C); index of refr 1.4557 (20C); latent heat of vaporization 97.1 Btu/lb. Soluble in water, alcohol, ether, and chloroform; combines with water forming chloral hydrate.
Derivations: (a) By chlorination of ethyl alcohol, addition of sulfuric acid, and subsequent distillation; (b) by the chlorination of acetaldehyde.
Grade: Technical, 94% min.
Hazard: Toxic by ingestion.
Use: Manufacture of chloral hydrate and DDT.

chloral hydrate. ("knockout drops"; trichloro-acetaldehyde, hydrated; trichloroethylidene glycol). CAS: 302-17-0. $CCl_3CH(OH)_2$.
Properties: Transparent, colorless crystals; aromatic, penetrating, slightly acrid odor and slightly bitter, sharp taste. Slowly volatilizes when exposed to air. Soluble in water, alcohol, chloroform and ether, also soluble in olive oil and turpentine oil. D 1.901, mp 52C, bp 97.5C.
Derivation: Action of one-fifth of its volume of water on chloral.
Grade: Technical, USP.

Hazard: Overdose toxic, hypnotic drug, dangerous to eyes.
Use: Medicine (sedative), manufacture of DDT, liniments.

chloral hydrate antipyrine. (antipyrine chloral hydrate). $C_{11}H_{12}N_2OCl_3CH(OH)_2$.
Properties: Colorless crystals, moderately soluble in water, soluble in alcohol, mp 67C.
Use: Medicine (sedative).

chlor-alkali cell. See electrolytic cell.

α-chloralose. CAS: 15879-93-3. $C_8H_{11}Cl_3O_6$.
Properties: Acicular solid, mp 186C, soluble in water, ether, and glacial acetic acid.
Derivation: By heating a mixture of chloral and glucose.
Hazard: May cause addiction, highly toxic.
Use: Coating seeds to protect them from birds.

Chloramben. (3-amino-2,5-dichlorobenzoic acid).
CAS: 133-90-4.
Use: Herbicide.

chlorambucil. (4-(p[bis(2-chloroethyl)amino]-phenyl)butyric acid).
$(ClC_2H_4)_2NC_6H_4(CH_2)_3COOH$, a nitrogen mustard derivative.
Properties: Off-white powder, mp 65–69C, slightly soluble in water, soluble in acetone and ether.
Grade: USP.
Hazard: A carcinogen.
Use: Medicine (leukemia treatment), insect sterilant.

chloramine. NH_2Cl. A colorless, unstable, pungent liquid; soluble in water; decomposes (slowly in dilute solution) to form nitrogen plus hydrochloric acid and ammonium chloride. Mp −66C, soluble in alcohol and ether. (Do not confuse with chloramine-T.) Chloramine is an intermediate in the manufacturing of hydrazine.

chloramine-B. (sodium benzenesulfochloramide). $C_6H_5SO_2NClNa$.
Properties: White powder with faint chlorine odor, soluble in water.
Use: Medicine (antiseptic).

chloramine-T. (sodium p-toluenesulfochlor-amine). $CH_3C_6H_4SO_2NNaCl \cdot 3HOH$.
Properties: White or slightly yellow crystals or crystalline powder, containing more than 11.5% and less than 13% active chlorine. Slight odor of chlorine. Decomposes slowly in air, liberating chlorine. (Not to be confused with NH_2Cl which is also termed chloramine.) Soluble in water; in-

soluble in benzene, chloroform, ether; decomposed by alcohol.
Derivation: Reaction of ammonia and p-toluene-sulfochloride under pressure. The latter is reacted with sodium hypochlorite in the presence of an alkali and the chloramine produced by crystallization.
Use: Medicine (antiseptic), reagent.
See also dichloramine-T.

chloramphenicol. (d(-)threo-1-(p-nitrophenyl)-2-dichloroacetamido-1,3-propandiol).
CAS: 56-75-7. $C_{11}H_{12}Cl_2N_2O_5$.

An antibiotic derived from *Streptomyces venezuelae* or by organic synthesis, it was the first substance of natural origin shown to contain an aromatic nitro group.
Properties: Fine, white to grayish-white or yellowish-white, needle-like crystals or elongated plates. Bitter taste, neutral to litmus and reasonably stable in neutral or slightly acid solutions. Mp 149–153C, alcohol solution is dextrorotatory while ethyl acetate solution is levorotatory. Very slightly soluble in water, freely soluble in alcohol, propylene glycol, acetone and ethyl acetate.
Grade: USP.
Hazard: Has deleterious and dangerous side effects. Must conform to FDA labeling requirements. Use is closely restricted.
Use: Medicine (antibiotic), antifungal agent.

chloranil. (tetrachloroquinone; tetrachloro-p-benzo-quinone). CAS: 118-75-2.
$C_6Cl_4O_2$.
Properties: Yellow crystals, mp 290C, d 1.97, soluble in ether, insoluble in water, good storage stability.
Derivation: From phenol, p-chlorophenol or p-phenylenediamine by treatment with potassium chlorate and hydrochloric acid.
Hazard: Skin irritant.
Use: Agricultural fungicide, dye intermediate, electrodes for pH measurements, reagent.

chloranthrene yellow. See flavanthrene.

chlorapatite. See appatite.

"Chlorasol."[214] TM for a fumigant composition, 70.3% ethylene dichloride, 29.7% carbon tetrachloride by weight. Boiling range 75–78C. Non-flammable.

Use: Fumigant for meal, grain and clothes moths; grain weevils; grain, flour and carpet beetles; the rice weevil; book lice; penetrates stored grain, rolled rugs, upholstered furniture, cartons, sacks, and stacked material.

chlorauric acid. See gold trichloride.

chlorazine. (Generic name for 2-chloro-4,6-bis-(diethylamino)-s-triazine). CAS: 580-48-3.

$[(C_2H_5)_2N]_2CNC(Cl)NCN.$

Properties: Solid, mp 15–18C, d 1.096 (20C), soluble in hydrocarbons, alcohols, ketones, insoluble in water.

Use: Herbicide.

chlorbenside. (p-chlorobenzyl-p-chlorophenyl sulfide). CAS: 103-17-3.

$ClC_6H_4CH_2SC_6H_4Cl.$ Generic name for an agricultural toxicant.

Properties: Crystals, almond-like odor (technical grade), mp 75–76C, insoluble in water, soluble in aromatic hydrocarbons, acetone, resistant to acid and alkaline hydrolysis.

Grade: Technical.

Hazard: Skin irritant.

Use: Acaricide.

chlor- compounds. Most organic compounds of chlorine retain the letter "o" in accepted chemical terminology, i.e., chlorobenzene, chloroacetic, etc., although the form without the "o" is sometimes used. Therefore, for chlor- compounds see also chloro-.

chlorcyclizine hydrochloride.

$ClC_6H_4CH(C_6H_5)C_4H_8N_2CH_3 \cdot HCl.$

Solutions acid to litmus, pH (1 in 100 solution) 4.8–5.5.

Grade: USP.

Use: Medicine (antihistamine).

chlordane. (1,2,4,5,6,7,8,8-octachloro-4,7-methano-3a,4,7,7a-tetrahydroindane).

CAS: 57-74-9. $C_{10}H_6Cl_8.$

Properties: Colorless, viscous liquid; d 1.57–1.67 (60/60F); viscosity SSU 100 seconds (38C); organic chlorine 64–67% by weight; purity 98%; bp 175C (2 mm); refr index 1.56–1.57 (25C); soluble in many organic solvents; insoluble in water; miscible in deodorized kerosene; decomposes in weak alkalies.

Grade: Technical and pure.

Hazard: Toxic by ingestion, inhalation, and skin absorption. Only for termite control. TLV: 0.5 mg/m³ of air.

Use: Insecticide, fumigant.

chlordiazepoxide hydrochloride (USAN).

(7-chloro-2-methylamino-5-phenyl-3,H,1,4-benzodiazepine-4-oxide hydrochloride).

$C_{16}H_{14}ClN_3O \cdot HCl.$

Properties: Crystals, mp 212–218C, soluble in water, sparingly soluble in alcohol, insoluble in ether and chloroform.

Hazard: CNS depressant. Manufacture and dosage controlled by law.

Use: Medicine (tranquilizer).

chlordimeform. (TM "Galecron"; [N'-(4-chloro-o-tolyl)-N,N-dimethylformamine]).

Ovicide, insecticide, and miticide designed for use on cotton and vegetable crops. Available in a concentrated emulsion form, it is stated to be less toxic than organophosphates and to be biodegradable.

chlorendic anhydride. (hexachloroendomethylenetetrahydrophthalic anhydride; 1,4,5,6,7,7-hexachlorobicyclo-(2,2,1)-5-heptene-2,3-dicarboxylic anhydride). $C_9H_2Cl_6O_3$

Properties: Fine, white, free-flowing crystals; mp 239–240C; d 1.73. Readily soluble in acetone, benzene, toluene; slightly soluble in water, n-hexane, and carbon tetrachloride. Nonflammable.

Derivation: By Diels-Alder reaction of maleic anhydride and hexachlorocyclopentadiene.

Grade: Technical, pure.

Use: Flame-resistant polyester resins, hardening epoxy resins, chemical intermediate, source of chlorendic acid.

"Chlorex."[214] TM for 2,2'-dichloroethyl ether.

chlorfenvinphos. (O,O-diethyl-O-[2-chloro-1-(2,4-dichlorophenyl)vinyl]phosphate).

CAS: 470-90-6. $C_{12}H_{14}Cl_3O_4P.$

Properties: Yellowish liquid, soluble in most organic solvents, slightly soluble in water.

Hazard: A cholinesterase inhibitor.

Use: Insecticide, nematocide, parasiticide.

"Chlorhydrol."[15a] TM for an aluminum chlorohydroxide antiperspirant, offered in liquid or powdered form for use in lotions, creams, and gels; noncorrosive and nonirritating to skin. Does not damage fabrics.

chloric acid. CAS: 7790-93-4.

$HClO_3 \cdot 7HOH.$

Occurs only in aqueous solution.

Derivation: Reaction of barium chlorate and sulfuric acid.

Hazard: Toxic by ingestion and inhalation. Strong oxidizer, ignites organic materials on contact.

Use: Catalyst in polymerization of acrylonitrile

chloridizing. Heating in the presence of chlorine as a step in the recovery of certain metals from their oxides or other compounds.

"Chlorimets."[47] TM for a series of nickel-base cast alloys. "Chlorimet" 2 contains 32% molybdenum, 3% iron (max), 1% silicon and 0.10% carbon. "Chlorimet" 3 contains 18% molybdenum, 18% chromium, 3% iron (max), 1.0% silicon and 0.07% carbon.

chlorinated acetone. See chloroacetone.

chlorinated camphene. See toxaphene.

chlorinated diphenyl. See chlorodiphenyl.

chlorinated hydrocarbon. See hydrocarbon, halogenated.

chlorinated isocyanuric acid. Used as dry bleach. See dichloro- or trichloroisocyanuric acid, and potassium or sodium dichloroisocyanurate.

chlorinated lime. See lime, chlorinated.

chlorinated naphthalene. (chloronaphthalene). $C_{10}H_7Cl$. From the chlorination of naphthalene. Physical state varies from oily liquids to crystalline solids,depending on the extent of chlorination.
Hazard: (Tri- and higher) Toxic by ingestion, inhalation, and skin absorption. Strong irritants. TLV: (tetra compound CAS: 1335-88-2) 2 mg/m^3 of air.
Use: Solvent, immersion liquid in microscopy.
See also oil, chloronaphthalene, wax, chloronaphthalene.

chlorinated paraffin. See paraffin, chlorinated.

chlorinated para red. A modification of para red that contains some chlorine. Much lighter than para or toluidine red, has excellent brilliance but poorer heat resistance.

chlorinated polyether. See "Penton."

chlorinated polyolefin. See rubber, chlorinated; polypropylene, chlorinated.

chlorinated rubber. See rubber, chlorinated.

chlorinated trisodium phosphate. See trisodium phosphate, chlorinated.

chlorine. CAS: 7782-50-5. Cl. Ninth highest-volume chemical produced in US (1985).

Nonmetallic halogen element of atomic number 17, Group VIIA of the Periodic Table. Aw 35.453, valences = 1,3,4,5,7. Two stable isotopes Cl-35 (75.4%) and Cl-37 (24.6%).
Properties: (1) Dense, greenish-yellow, diatomic gas. Noncombustible, but supports combustion (oxidizing agent); pungent, very irritating odor. Liquefaction press 7.86 atm (25C), 1 atm at −35C. Water solubility 0.64 g Cl_2 per 100 g water. D 3.21 g/L (0C, 1 atm) (air = 1.29). Thermodynamic properties: (a) critical temperature 144.0C; (b) critical press 78.525 atm absolute; (c) critical volume 1.763 L/kg. Strongly electronegative.
(2) Liquid: Clear amber, very irritating odor, d 1.56 (−35C), fp −101C, 1 L of liquid = 456.8 L of gas at 0C and 1 atm. Very low electrical conductivity. Soluble in chlorides and alcohols. Extremely strong oxidizing agent. Slightly soluble in cold water.
Occurrence: Not free in nature, component of minerals halite (rock salt), sylvite, and carnallite, chloride ion in seawater.
Derivation: (1) Electrolysis of sodium chloride brine in either diaphragm or mercury cathode cells; chlorine is released at the anode. (2) Fused salt electrolysis of sodium or magnesium chloride; (3) Electrolysis of hydrochloric acid. (4) Oxidation of hydrogen chloride with nitrogen oxide as catalyst and absorption of steam with sulfuric acid ("KeloChlor" process). No byproduct caustic is produced.
Grade: Technical (gas and liquid), pure (99.9%).
Hazard: A military poison; TLV: 1 ppm in air. Dangerous in contact with turpentine, ether, ammonia, hydrocarbons, hydrogen, powdered metals and other reducing materials.
Use: Manufacture of carbon tetrchloride, trichloroethylene, chlorinated hydrocarbons, polychloroprene (neoprene), polyvinyl chloride, hydrogen chloride, ethylene dichloride, hypochlorous acid, metallic chlorides, chloracetic acid, chlorobenzene, chlorinated lime, water purification, shrinkproofing wool, in flame-retardant compounds, in special batteries (with lithium or zinc), processing of meat, fish, vegetables and fruit. For information, refer to the Chlorine Institute, 342 Madison Ave., New York, NY.

chlorine-36. Radioactive chlorine of mass number 36. Half-life approximately 440,000 years; radiation: beta.
Derivation: Separated from various isotopes produced during irradiation of potassium chloride.
Forms available: As hydrochloric acid solution and as solid potassium chloride.
Hazards: MPC 4×10^{-7} mCi/mL of air.
Use: Tracer in studying the salt water corrosion

of metals, especially steel, reaction mechanism of chlorinated hydrocarbons, location and flow of salt waters in porous media, etc.

chlorine-bromide. See bromine-chloride.

chlorine dioxide. CAS: 10049-04-4. ClO_2.
Properties: Red-yellow gas, fp −59.5C, bp 10C, very reactive, unstable, strong oxidizer, decomposes in water. Dissolves in alkalies forming a mixture of chlorite and chlorate.
Derivation: Usually made at point of consumption from sodium chlorate, sulfuric acid, and methanol, or from sodium chlorate and sulfur dioxide. Concentration of gas is limited to 10% to reduce explosion hazard.
Grade: Sold as hydrate in frozen form.
Hazard: Explodes when heated or by reaction with organic materials. Very irritating to skin and mucous membranes. TLV: 0.1 ppm in air.
Use: Bleaching wood pulp, fats and oils; controversial maturing agent for flour; water treatment (purification and taste removal); swimming pools; odor control; biocide.

chlorine heptoxide. Cl_2O_7.
Properties: Colorless, viscous liquid; fp −91C; bp 82C; d 1.86; hydrolyzes to form perchloric acid.
Hazard: Explodes on contact with iodine or flame, or by shock. Strong irritant to tissue and very toxic.
Use: Cellulose esterification catalyst.

chlorine monofluoride. (chlorine fluoride). ClF.
Properties: Colorless gas, slightly yellow when liquid. Mp −155.6C, bp −100.1C, critical temperature −14C.
Hazard: Extremely reactive. Destroys glass instantly, attacks quartz readily in presence of moisture. Organic matter bursts into flame on contact. Violent reaction with water. Extremely corrosive to skin, eyes, mucous membranes, and respiratory tissues.
Use: Fluorinating reagent.

chlorine monoxide. Cl_2O.
Properties: Yellow gas; strong, unpleasant odor; fp −120C; bp 2.2C; soluble in water and carbon tetrachloride.
Derivation: Reaction of mercuric oxide and chlorine.
Hazard: Explodes on contact with organic materials. Strong irritant to eyes, skin, and mucous membranes.
Use: Chlorination.

chlorine trifluoride. CAS: 7790-91-2. ClF_3.
Properties: Nearly colorless gas, pale green liquid or white solid, bp 11.4C, fp −76.3C, d of gas (air = 1.29) 3.14g/L. Very reactive, comparable to fluorine.
Derivation: By reaction of chlorine and fluorine at 280C and condensation of the product at −80C. Obtained 99.0% pure.
Hazard: Explodes in contact with organic materials or with water. Dangerous fire risk. A poison, very toxic, corrosive to skin. TLV: ceiling of 0.1 ppm in air.
Use: Fluorination, cutting oil well tubes, reprocessing reactor fuels, oxidizer in propellants.

chlorine water. Clear, yellowish liquid; deteriorates on exposure to air and light. Made by saturating water with approximately 0.4% chlorine.
Use: Deodorizer, disinfectant, antiseptic.

chloriodized oil. Chlorinated and iodinated vegetable oil. Contains 26.0–28.0% iodine in organic combination.
Properties: Pale yellow, viscous, oily liquid with faint, bland taste. Practically insoluble in water; slightly soluble in alcohol; freely soluble in benzene, chloroform, and ether.
Derivation: Formed by chemical addition of iodine monochloride to a vegetable oil.
Hazard: A poison. Very toxic by ingestion.
Use: Medicine (radiopaque medium).

chlorisondamine chloride. (4,5,6,7-tetrachloro-2(2-dimethylaminoethyl)isoindoline dimethylchloride). $C_{14}H_{20}Cl_6N_2$, a quaternary ammonium compound.
Properties: Crystals, decomposes 258–265C, soluble in water and alcohol.
Use: Medicine (blood-pressure control).

chlormequat. See 2-chloroethyl-trimethyl ammonium chloride.

chlormethazanone. (2-(4-chlorophenyl-3-methyl-4-meta-thiazanone-1-dioxide). $C_{11}H_{12}ClNO_3S$.
Properties: Crystals, mp 117C, insoluble in water, slightly soluble in alcohol.
Use: Medicine (tranquilizer and muscle relaxant).

chloroacetaldehyde. CAS: 107-20-0. $ClCH_2CHO$.
Properties: (of 40% aqueous solution): Clear, colorless liquid; pungent odor; bp 85C; fp −16.3C; d 1.19 (25/25C); refr index 1.397 (25C); wt/gal 9.9lb (25C). Soluble in water, acetone, methanol; at greater than 50% concentration in water, it forms an insoluble hemihydrate. Pure substance has flash p 190F (87.7C).
Hazard: Corrosive to skin and mucous membranes. TLV: CL of 1 ppm in air.
Use: Intermediate, fungicide.

chloroacetaldehyde dimethyl acetal. See dimethyl chloroacetal.

chloroacetamide. (α-chloroacetamide; 2-chloroethanamide). $ClCH_2CONH_2$.
Properties: Colorless to pale yellow crystals, characteristic odor, mp 117–119C, bp 220C (decomposes). Soluble in water and alcohol, insoluble in ether.
Hazard: Strong irritant to skin and tissue.
Use: Intermediate.

chloroacetic acid. (chloracetic acid; MCA; monochloroacetic acid).   CAS: 79-11-8. $CH_2ClCOOH$.
Properties: Colorless to light-brownish crystals, deliquescent, d 1.58, crystallizing point alpha form 61.0–61.7C, beta form 55.5–56.5C, gamma form 50C. The commercial material melts at 61–63C, boiling range 186–191C. Soluble in water, alcohol, ether, chloroform, carbon disulfide.
Derivation: Action of chlorine on acetic acid in the presence of acetic anhydride, phosphorus, or sulfur.
Grade: Technical, medicinal, 99.5% pure.
Hazard: Use in foods prohibited by FDA. Irritating and corrosive to skin.
Use: Herbicide, preservative, bacteriostat, intermediate in production of carboxymethylcellulose; ethyl chloroacetate; glycine; synthetic caffeine; sarcosine; thioglycolic acid; EDTA; 2,4-D; 2,4,5-T.

chloroacetic anhydride. (chloroethanoic anhydride; symmetrical dichloroacetic anhydride). $(ClCH_2CO)_2O$.
Properties: Colorless to slightly yellow crystals with pungent odor, mp 51–55C, bp 203C, d 1.55, soluble in chloroform and ether, hydrolyzes to chloroacetic acid.
Hazard: Irritating to skin and eyes, moderately toxic by inhalation.
Use: Intermediate for acetylation of amino acids, cellulose chloroacetates.

o-chloroacetoacetanilide.
$CH_3COCH_2CONHC_6H_4Cl$.
Properties: White crystals resemble ethyl acetoacetate in chemical reactivity. Mp 107C, vap press 0.1 mm (20C). Nonflammable. Insoluble in water.
Use: Organic synthesis, dyestuffs.

chloroacetone. (monochloroacetone; 1-chloro-2-propanone; chloracetone; chlorinated acetone).   CAS: 78-95-5. CH_3COCH_2Cl.
A lachrymator.
Properties: Colorless liquid, pungent, irritating odor. D 1.162 (16C), bp 119C, fp −44.5C. Soluble in alcohol, ether, chloroform, and water.

Derivation: Chlorination of acetone.
Hazard: Strong irritant to tissue, eyes, and mucous membranes; toxic by ingestion.
Use: Couplers for color photography, enzyme inactivator, insecticides, perfumes, intermediate, organic synthesis, tear gas, polymerization of vinyl monomers.

chloroacetonitrile. (chloroethane nitrile; chloromethyl cyanide).   CAS: 107-14-2.
$ClCH_2CN$.
Properties: Colorless liquid with pungent odor, d 1.202–1.2035 (25/25C), refr index 1.4210–1.4240 (25C), 5–95% distils between 124–129C. Soluble in hydrocarbons, alcohols; insoluble in water.
Hazard: Irritant.
Use: Fumigant, intermediate.

α-chloroacetophenone. (chloracetophenone; phenacylchloride; phenyl chloromethyl ketone).
CAS: 532-27-4. $C_6H_5COCH_2Cl$.
A strong lachrymator.
Properties: (omega isomer) White crystals, floral odor, mp 56C, bp 247C. The para isomer has mp 20C, bp 237C. Insoluble in water; soluble in acetone, benzene, carbon disulfide.
Derivation: From chloroacetylchloride, benzene, and aluminum chloride.
Hazard: Strong irritant to eyes and tissue as gas or liquid. TLV: (alpha) 0.05 ppm in air.
Use: Pharmaceutical intermediate, riot control gas.
See also "Mace."

chloroacetyl chloride. (chloracetyl chloride).
CAS:79-04-9. $ClCH_2COCl$.   A lachrymator.
Properties: Water-white liquid, pungent odor. D 1.495 (0C), bp 105–110C, decomposes in water.
Derivation: (a) Action of chlorine on acetyl chloride in sunlight. (b) Dropping phosphorus trichloride on chloroacetic acid.
Hazard: Irritant to eyes, corrosive to skin. TLV: 0.05 ppm in air.
Use: Preparation of chloroacetophenone, intermediate, tear gas.

chloroacetylurethane.
$ClCH_2CONHCOOC_2H_5$.
Properties: Crystals, soluble in alcohol, sparingly soluble in water, mp 129C.
Derivation: By interaction of a urethane derivative and ethyl chloracetate.

chloroacrolein. $H_2C:CClCHO$.
Properties: Colorless liquid, d 1.205 at 15C, bp 29–31C (17 mm).
Derivation: Chlorination of acrolein.

Hazard: Irritant to eyes and skin.
Use: Tear gas.

α-chloroacrylonitrile.
Properties: Readily polymerizes and copolymerizes with other unsaturated monomers. High cross-linking ability.
Derivation: Chlorination of acrylonitrile and dehydro halogenation by cracking.
Use: Synthetic fibers, coatings and films, acrylic polymers, treatment of cotton fiber, intermediate.

2-chloroallyl diethyldithiocarbamate.
(CDEC). $(C_2H_5)_2NCSSCH_2CCl:CH_2$.
Properties: Amber liquid, bp 128–130C (1 mm), very slightly soluble in water, soluble in benzene, alcohol, acetone, chloroform, and ether.
Forms available: Liquid and granular.
Hazard: Dry preparations are irritating to eyes and skin.
Use: Herbicide, pesticide.

1-(3-chloroallyl)-3,5,7-triaza-1-azoniaadamantane chloride. $C_6H_{12}N_4(CH_2CHCHCl)Cl$.
White to cream-colored powder, soluble in water and methanol, almost insoluble in acetone.
Use: Bactericide used as preservative in latexes, paints, floor polishes, joint cements, adhesives, inks, starches, etc.

chloroaluminum diisopropoxide.
$[(CH_3)_2CHO]_2AlCl$. White crystals, mp 160C (decomposes), soluble in most organic solvents, hydrolyzes.
Use: Catalyst and intermediate.
Hazard: An irritant.

chloroamino-. See aminochloro-.

2-chloro-5-aminobenzoic acid.
Grades: Technical.
Use: Intermediate in manufacture of azo dyes for textiles and plastics and as coupling agent of color pigments in color photography.

p-chloro-o-aminophenol. See 2-amino-4-chlorophenol.

2-chloro-4-tert-amylphenol. $C_5H_{11}C_6H_3ClOH$.
Properties: Water-white liquid with aromatic odor, d 1.11 (20C), boiling range 253–265C, flash p 225F (107C). Combustible.

m-chloroaniline. (m-aminochlorobenzene).
CAS: 108-42-9. $ClC_6H_4NH_2$.
Properties: Colorless to light amber liquid, tends to darken during storage, boiling range 228–231C, fp −10.6C. Insoluble in water, soluble in organic solvents.
Grade: Technical.
Use: Intermediate for azo dyes and pigments, pharmaceuticals, insecticides, agricultural chemicals.

o-chloroaniline. (o-aminochlorobenzene).
CAS: 95-51-2. $ClC_6H_4NH_2$.
Properties: Amber liquid, amine odor, darkens on exposure to air, distillation range 208–210C, fp −2.3C, d 1.213 (20/4C), refr index 1.5896 (20C). Miscible with alcohol and ether, insoluble in water.
Grade: Technical.
Hazard: Toxic by ingestion.
Use: Dye intermediate, standards for colorimetric apparatus, manufacture of petroleum solvents and fungicides.

p-chloroaniline. (p-aminochlorobenzene).
CAS: 106-47-8. $ClC_6H_4NH_2$.
Properties: White or pale yellow solid, mp 69.5C, distilling range 229–233C, d 1.17, soluble in hot water and organic solvents.
Grade: Technical.
Hazard: Toxic by inhalation and ingestion.
Use: Dye intermediate, pharmaceuticals, agricultural chemicals.

4-chloroaniline-3-sulfonic acid.
$HSO_3C_6H_3ClNH_2$.
Properties: White to light-gray powder.
Use: Intermediate for dyes.

2-chloroanthraquinone. $C_{14}H_7ClO_2$.
Properties: Mp 208–211C, insoluble in water, soluble in hot benzene.
Derivation: Condensing phthalic anhydride and chlorobenzene in the presence of anhydrous aluminum chloride to form p-chlorobenzoylbenzoic acid. Ring closure of the intermediate acid is brought about by heating in sulfuric acid solution.
Use: Starting material for certain vat dyes. See anthraquinone and 2-methylanthraquinone.

chloroauric acid. See gold trichloride.

chloroazotic acid. See aqua regia.

chlorobenzal. See benzyl dichloride.

chlorobenzaldehyde. C_6H_4CHOCl.
Properties: Colorless to yellowish-liquid (o-) or powder (p-), boiling range 209–214C, fp 8.0C (min), d 1.240–1.245 (25/25C). Soluble in alcohol, ether, and acetone; insoluble in water. Combustible.

Use: Intermediate in the preparation of triphenyl methane and related dyes, organic intermediate.

3-chloro-4-benzamido-6-methylaniline.
$ClC_6H_2NH_2CH_3(NHCOC_6H_5)$.
Properties: White solid, mp 198–199C.
Use: Azoic dyes, pigments.

chlorobenzanthrone. $C_{17}H_9ClO$.
Properties: All isomers: yellow needles; soluble in alcohol, benzene, toluene, acetic acid.
Derivation: From benzanthrone by treatment with chlorine.

chlorobenzene. (monochlorobenzene; phenyl chloride). CAS: 108-90-7. C_6H_5Cl.

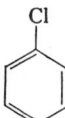

Properties: Clear, volatile liquid. Almond-like odor, d 1.105 (25/25C), bp 131.6C, fp −45C, wt/gal 9.19 lb (25C), refr index 1.5216 (25C), flash p 85F (29.4C) (CC), autoign temperature 1180F (637C). Miscible with most organic solvents, insoluble in water.
Derivation: By passing dry chlorine into benzene with a catalyst.
Grade: Technical.
Hazard: Moderate fire risk. Explosive limits 1.8 to 9.6% in air. Avoid inhalation and skin contact. TLV: 75 ppm in air.
Use: Phenol, chloronitrobenzene, aniline, solvent carrier for methylene diisocyanate, solvent, pesticide intermediate, heat transfer.

o-chlorobenzylidene malononitrile. (CS; OCBM). CAS: 2698-41-1. $ClC_6H_4CH{=}C(CN)_2$.
Properties: White crystals with the odor of pepper. Mw 189, mp 93–95C, bp 310–315C. Insoluble in water.
Available forms: Available both unground and ground with 5% silica aerogel or treated "Cab-O-Sil".
Hazard: Strong irritant to eyes and mucous membranes. TLV: ceiling 0.05 ppm.
Use: An incapacitating agent used by the military and law enforcement officers.

p-chlorobenzenesulfonamide.
$ClC_6H_4SO_2NH_2$.
Properties: White, odorless powder; mp 145–148C; soluble in alcohol.
Grade: 98–99% purity.
Use: Intermediate for pharmaceuticals and resins.

o-**chlorobenzenethiol.** See p-chlorothiophenol.

p-**chlorobenzhydrol.** (p-chlorobenzohydrol).
$ClC_6H_4C(C_6H_5)HOH$.
Properties: White to off-white, crystalline powder; mp 57–61C; insoluble in water; soluble in ether, alcohol, and benzene.
Use: Organic synthesis.

"Chlorobenzilate."[219] TM for ethyl-4,4′-dichlorobenzilate.
$(C_6H_4Cl)_2C(OH)COOC_2H_5$.
Properties: Viscous, yellow liquid; bp 141–142C (0.06 mm); d (technical 90%) 1.2816 (20/4C). Slightly soluble in water; soluble in acetone, benzene, methanol.
Hazard: Carcinogen. Caused testicular damage in farm workers. Use has been restricted.
Use: Pesticide, acaricide.

p-**chlorobenzohydrol.** See p-chlorobenzhydrol.

chlorobenzoic acid. ClC_6H_4COOH.
Properties: Nearly white, coarse powder, mp (o-) 142C, (m-) 158C, (p-) 243C. Soluble in methanol, alcohol, ether, and hot water.
Use: Intermediate for the preparation of dyes, fungicides, pharmaceuticals, and other organic chemicals; preservative for adhesives and paints.

p-**chlorobenzophenone.** $ClC_6H_4COC_6H_5$.
Properties: White to off-white, crystalline powder; mp 73–78C; bp 332C. Soluble in acetone, benzene, carbon tetrachloride, ether, and hot alcohol. Insoluble in water.
Use: Intermediate.

chlorobenzotriazole. $Cl\overline{C_6H_4NHN}{:}N$.
Properties: White solid, mp 157–159C.
Use: Intermediate and photographic chemical.

o-**chlorobenzotrichloride.** $ClC_6H_4CCl_3$.
Properties: Colorless liquid or solid, mp 29.37C, bp 264.3C, d 1.5131 (25/4C), refr index 1.5836 (20C). Soluble in alcohol, ether, and acetone; decomposed by water.
Use: Intermediate for pharmaceuticals, dyes, and other organic chemicals.

p-**chlorobenzotrichloride.** $ClC_6H_4CCl_3$.
Properties: Water-white liquid, boiling range 248–257C, fp approximately 3.8C, d 1.480–1.490 (25/25C). Soluble in alcohol, ether, and acetone; insoluble in water.
Use: Same as o-chlorobenzotrichloride.

m-**chlorobenzotrifluoride.** (m-chlorotrifluoromethylbenzene; m-chloro-α,α,α-trifluoroto-

luene).   CAS: 98-15-7. $ClC_6H_4CF_3$.
Properties: Water-white, aromatic liquid; bp 138C; fp −56C; refr index 1.446 (20C); flash p 122F (50C) (CC); d 1.351 (15.5/15.5C). Combustible.
Hazard: Moderate fire risk. TLV: (as F) 2.5 mg/m³ of air.
Use: Intermediate in manufacturing of dyes and pharmaceuticals, dielectrics, insecticides.

o-**chlorobenzotrifluoride.** (o-chlorotrifluoromethylbenzene; o-chloro-α,α,α-trifluorotoluene).   CAS: 88-16-4. $ClC_6H_4CF_3$.
Properties: Colorless liquid with aromatic odor, d 1.379 (15.5/15.5C), refr index 1.456 (20C), bp 152C, fp −7.4C, flash p 138F (58.8C) (CC), wt/gal 11.50 lb (15.5C).
Hazard: See m-form.
Use: Dye intermediate, chemical intermediate, solvent, and dielectric fluid.

p-**chlorobenzotrifluoride.** (p-chlorotrifluoromethylbenzene; p-chloro-α,α,α-trifluorotoluene).   CAS: 98-56-6. $ClC_6H_4CF_3$.
Properties: Water-white liquid, aromatic odor, bp 139.3C, fp −36C, refr index 1.446 (20C), flash p 116F (46.6C) (CC), d 1.3533 (15.5/15.5C), wt/gal 11.28 lb (15.5C). Flammable.
Hazard: See m-form.
Use: See o-chlorobenzotrifluoride.

chlorobenzoyl chloride. ClC_6H_4COCl.
Properties: Colorless liquid, boiling range 227–239C, fp −4 to −6C, fp (o-) −4 to −6C, (p-) 10–12C, d 1.374–1.376 (25/15C). Soluble in alcohol, ether, and acetone. Insoluble in water.
Use: Intermediate for pharmaceuticals, dyes, and other organic chemicals.

p-**chlorobenzoyl peroxide.**   CAS: 94-17-7. $(ClC_6H_4CO)_2OO$.
Properties: White, odorless powder; decomposes violently on heating or contamination. Insoluble in water, soluble in organic solvents.
Hazard: Dangerous fire and explosion risk, explodes when heated to 38C, strong oxidizer, will ignite on contact with organic materials. Store in dark, cool locality. Toxic.
Use: Bleaching agent, polymerization catalyst.

chlorobenzyl chloride.   CAS: 104-83-6. $ClC_6H_4CH_2Cl$.
Properties: Colorless liquid, boiling range 216–222C, mp (o-) −17C, (p-) 31C, d 1.270–1.280 (25/15C). Soluble in alcohol, ether, and acetone; insoluble in water.
Hazard: Irritating to skin, eyes.

Use: Intermediate for organic chemicals, pharmaceuticals and dyes.

p-**chlorobenzyl-p-chlorophenyl sulfide.**   See chlorobenside.

p-**chlorobenzyl cyanide.** $ClC_6H_4CH_2CN$.
Properties: Colorless to pale yellow solid, mp 27C, soluble in acetone and alcohol.
Hazard: A poison. · TLV: (as CN) 5 mg/m³ of air.
Use: Organic synthesis.

p-**chlorobenzyl-p-fluorophenyl sulfide.**   See fluorbenside.

2-(p-chlorobenzyl)pyridine. (2-(4-chlorobenzyl)pyridine). $ClC_6H_4CH_2C_5H_4N$.
Properties: Liquid, bp 310.5C, fp 8.4C, d 1.168 (25C), refr index 1.5865 (20C), insoluble in water.
Use: Organic synthesis.

chlorobromo-.   See bromochloro-.

2-chlorobutadiene-1,3.   See chloroprene.

1-chlorobutane.   See n-butyl chloride.

chlorobutanol. (trichloro-tert-butyl alcohol; 1,1-trichloro-2-methyl-2-propanol; acetone chloroform).   CAS: 57-15-8. $Cl_3CC(CH_3)_2OH$.
Properties: Colorless to white crystals with characteristic odor and taste. Soluble in alcohol and glycerol, hot water, ether, chloroform, and volatile oils. Mp (anhydrous form) 97C, mp (hemihydrate) 78C, bp 167C, sublimes easily.
Derivation: By action of potassium hydroxide on a solution of chloroform and acetone.
Grade: USP.
Hazard: Action similar to chloral hydrate. Combustible.
Use: Plasticizer for cellulose esters and ethers, preservative for biological fluids and solutions, antimicrobial agent, anesthetic in dentistry.

4-chloro-2-butynyl-m-chlorocarbanilate.   See barban.

chlorocarbon. A compound of carbon and chlorine or carbon, hydrogen, and chlorine, such as carbon tetrachloride, chloroform, tetrachloroethylene, etc.

chlorocarbonyl ferrocene. (ferrocenoyl chloride). $C_5H_5FeC_5H_4COCl$.
Properties: Orange-red solid, mp 48–49C.
Use: Intermediate.

chlorochromic anhydride. See chromyl chloride.

chlorocosane. See paraffin, chlorinated.

3-chlorocoumarin. $C_9H_5O_2Cl$.
Properties: Slightly yellow, crystalline solid; mp 118C.
Grade: Technical.
Use: Tin-plating solutions.

p-chloro-m-cresol. See 4-chloro-3-methylphenol.

α-chloro-N,N-diallylacetamide. (CDAA).
$ClCH_2CON(CH_2CH:CH_2)_2$.
Properties: Amber liquid or granules, bp 74C (0.3 mm), slightly soluble in water, soluble in alcohol, hexane, and xylene.
Hazard: Toxic by ingestion. Dry formulations are irritating to eyes and skin.
Use: Herbicide.

1-chloro-2-dichloroarsinoethene. See β-chlorovinyl dichloroarsine.

Chlorodecone [1,2,3,4,5,5,6,7,8,9,10,10-dodeca-chlorooctahydro-1,3,4-metheno-2-cyclobuta-(c,d)-pentalone]. CAS: 143-50-0.
Properties: Mp 349C (decomposes), solubility in water 0.4% at 100C.
Use: Insecticide.

2-chloro-1-(2,4-dichlorophenyl)vinyl diethyl phosphate. See diethyl-1-(2,4-dichlorophenyl)-2-chlorovinyl phosphate.

chlorodifluoroacetic acid. $CClF_2 \cdot COOH$.
Properties: Colorless, pungent liquid, bp 122C, fp 23C, miscible in water and most organic solvents, strong acid, dissolves cellulose and proteins.
Use: Catalyst, particularly for esterification and condensation reactions, herbicides, intermediate.

1,1,1-chlorodifluoroethane. (1,1,1-difluorochloroethane; difluoromonochloroethane).
CAS: 75-68-3. CH_3CClF_2.
Properties: Colorless, nearly odorless gas; bp −9.2C; fp −130.8C; d 1.194 (−9C). Insoluble in water.
Hazard: Flammable gas. Explosive limits in air 6.2–18%.
Derivation: Chlorinating 1,1-difluoroethane in UV light.
Grade: Technical.
Use: Refrigerant, solvent, intermediate.

chlorodifluoromethane. (monochlorodifluoromethane; difluorochloromethane; difluoromono-chloromethane; refrigerant 22).
CAS: 74-97-5. $CHClF_2$.
Properties: Colorless, nearly odorless gas; d (gas at its bp) 4.82g/L; bp −40.8C; fp −160C; partly soluble in water.
Derivation: Reaction of chloroform with anhydrous hydrogen fluoride with antimony chloride catalyst.
Grade: Technical, 99.9% pure.
Hazard: Asphyxiant. TLV: 1000 ppm in air.
Use: Refrigerant, low-temperature solvent, fluorocarbon resins, especially tetrafluoroethylene polymers. See also chlorofluorocarbon.

1-chloro-2,4-dinitrobenzene. (dinitrochlorobenzene). $C_6H_3(NO_2)_2Cl$.
Properties: Pale yellow needles; almond odor; soluble in hot alcohol, ether, benzene, carbon disulfide; insoluble in water; d 1.69; mp 53C; bp 315C; flash p 382F (194C).
Hazard: Toxic by ingestion, inhalation, and skin absorption. Combustible. Upper explosive limit 22%. A skin irritant.
Derivation: Chlorination of dinitrobenzene.
Grade: Technical, fused.
Use: Dyes, organic synthesis.

chlorodiphenyl. CAS: (42% chlorine) 53469-21-9; (54% chlorine) 11097-21-9.
Properties: Colorless, mobile liquid; bp 340–375C; flash p 383F (195C) (OC). Resistant to acids and alkalies.
Grade: 54% chlorine, 42% chlorine.
Hazard: Toxic by ingestion, inhalation, and skin absorption; strong irritant. TLV: (54% Cl) 0.5 mg/m³ of air. (42% chlorine) 1 mg/m³ of air.
Use: Plasticizer for cellulosics, vinyl resins, and chlorinated rubbers.
See also diphenyl.

4-chlorodiphenyl sulfone. See p-chlorophenyl phenyl sulfone.

1-chloro-2,3-epoxypropane. See epichlorohydrin.

2-chloroethanamide. See chloroacetamide.

chloroethane. See ethyl chloride.

chloroethane nitrile. See chloroacetonitrile.

chloroethanoic anhydride. See chloroacetic anhydride.

2-chloroethanol. See ethylene chlorohydrin.

chloroethene. See vinyl chloride.

chloroethyl alcohol. See ethylene chlorohydrin.

2-chloro-4-ethyl-amino-s-triazine.   See atrazine.

β-chloroethylchloroformate.
CH_2ClCH_2OOCCl.
Properties: Colorless liquid, decomposed by alkaline solutions and hot water, insoluble in cold water, d 1.3825 (20C), bp 152.5C (752 mm).
Derivation: By bubbling gaseous phosgene into ethylene chlorohydrin at 0C.
Hazard: Very irritating to eyes and skin. Toxic by ingestion and inhalation.

β-chloroethyl chlorosulfonate.
$ClCH_2CH_2OSO_2Cl$.
Properties: Colorless liquid. Chloropicrin-like odor. Darkens on long storage and decomposes with evolution of hydrogen chloride, bp 101C (23 mm).
Hazard: Toxic by ingestion and inhalation. Strong irritant to tissue.
Derivation: Interaction of sulfuryl chloride and ethylene chlorohydrin. Also from action of sulfur trioxide on ethylene chloride at less than 45C.

chloroethylene.   See vinyl chloride.

2-chloroethylphosphonic acid.   (ethephon; "Ethrel"). $ClCH_2CH_2PO_3H_2$.   A plant growth regulator that increases rate of ripening of plants by promoting release of ethylene, stimulates rubber latex formation, ripening of sugarcane, flowering agent for pineapple, color enhancer.

2-chloroethyltrimethyl ammonium chloride.
(chlormequat). $ClCH_2CH_2N(CH_3)_3Cl$.   A plant growth regulator stated to be effective in shortening the height of wheat and as ripening agent in sugarcane.

2-chloroethyl vinyl ether.   (2-chloroethoxy ethane).   CAS: 110-75-8.
$CH_2ClCH_2OCHCH_2$.
Properties: Colorless liquid, bp 110C, d 1.052, stable in caustic solution, hydrolyzes in acid solutions.
Derivation: Reaction of a mixture of sodium hydroxide and triethanolamine with 2,2'-dichlorodiethyl ether.
Hazard: To experimental animals. Moderate fire risk. Combustible.
Use: Manufacture of cellulose ethers.

chlorofenethol.   (di(p-chlorophenyl)methylcarbinol; 1,2-bis(p-chlorophenyl)ethaneol).
CAS: 80-08-6.
Properties: White solid, mp 69.5–70C. Insoluble in water; soluble in petroleum ether, ethanol, toluene.
Use: Miticide.

Chlorofenvinphos.   (Birlane; 2-chloro-1-(2,4-dichlorophenyl)-vinyl diethyl phosphate).
CAS: 470-90-6.
Properties: Bp 168–170C (67C at 0.5 mm Hg),
Use: General purpose stored product insecticide.

chlorofluorocarbon.   Any of several compounds comprised of carbon, fluorine, chlorine, and hydrogen, the best known of which are trichlorofluoromethane and dichlorodifluoromethane. Their use was prohibited in 1979 except for a few specialized items because of their depleting effect on stratospheric ozone.

chloroform.   (trichloromethane).
CAS: 67-66-3. $CHCl_3$.
Properties: Colorless, highly refractive, heavy, volatile liquid; characteristic odor; sweet taste. Keep from light. Miscible with alcohol, ether, benzene, carbon disulfide, carbon tetrachloride, fixed and volatile oils; slightly soluble in water; d 1.485 (20/20C); bp 61.2C; fp −63.5C; wt/gal 12.29 lb (25C); refr index 1.4422 (25C).
Derivation: (a) Reaction of chlorinated lime with acetone, acetaldehyde, or ethanol; (b) by-product from the chlorination of methane.
Method of purification: Extraction with concentrated sulfuric acid and rectification.
Grade: Technical, CP, ACS, NF, reagent.
Hazard: Toxic by inhalation; anesthetic, prolonged inhalation or ingestion may be fatal. TLV: 10 ppm in air, 50 mg/m³ of air. OSHA PEL: 50 ppm for 10 minutes. A carcinogen. It has been prohibited by FDA from use in drugs, cosmetics and food packaging, including cough medicines, toothpastes, etc. Nonflammable. Will burn on prolonged exposure to flame or high temperature.
Use: Fluorocarbon plastics, solvent, analytical chemistry, fumigant, insecticides.

chloroformoxime.   ClHCNOH.
Properties: Needles. Odor resembles that of hydrocyanic acid. Stable at 0C, unstable at normal temperature. Small quantities volatilize, large quantities decompose. Aqueous solutions slowly decompose. Soluble in water, alcohol, ether, benzene; slightly soluble in carbon disulfide.
Derivation: Interaction of hydrochloric acid and sodium cyanate.
Hazard: Toxic by inhalation; strong irritant to tissue.
Use: Organic synthesis, tear gas, and vesicant.

chloroformyl chloride.   See phosgene.

chlorogenic acid.   (3-(3,4-dihydroxycinnamoyl)quinic acid).
$(HO)_2C_6H_3CHCHCOOC_6H_7(OH)_3COOH$.
Important metabolic factor in many plant tissues.

Properties: Crystals; mp 208C; slightly soluble in cold water; soluble in hot water, alcohol, acetone.

chlorohydrin. (α-chlorohydrin; 1-chloropane-2,3-diol; glyceryl α-chlorohydrin).
CAS: 96-24-2. $CH_2OHCHOHCH_2Cl$.
Properties: Colorless, heavy liquid; unstable; hygroscopic. The commercial grade is a mixture of the two isomers, alpha and beta, of which alpha is in a greater proportion. D 1.326 (18C), bp 213C (decomposes), wt/gal 11.102 lb, fp −40C, viscosity 2.388 poise (20C). Soluble in water, alcohol, and ether; immiscible with oils. Nonflammable.
Derivation: By passing hydrogen chloride gas into glycerol containing 2% acetic acid.
Grade: Technical.
Hazard : Toxic by ingestion, inhalation.
Use: Solvent for cellulose acetate, glyceryl phthalate resins, partial solvent for gums, intermediate in organic synthesis, antifreeze agent for dynamite.

chlorohydrin rubber. An elastomer made from epichlorohydrin. Both a homopolymer and a copolymer with ethylene oxide are available.
See also "Hydrin."

chlorohydroquinone. (2-chloro-1,4-dihydroxybenzene; 2,5-dihydroxychlorobenzene [Cl = 1]).
$ClC_6H_3(OH)_2$.
Properties: White to light-tan fine crystals, mp 100C, bp 263C. Very soluble in water and alcohol, slightly soluble in ether.
Grade: Photographic, commercial.
Use: Photographic developer, organic intermediate, dyestuffs, bactericide.

chlorohydroxybenzene. See chlorophenol.

5-chloro-2-hydroxybenzophenone.
$C_6H_5COC_6H_3OHCl$.
Properties: Yellow crystals; nearly odorless; mp 93–95C; soluble in alcohol, ethyl acetate, methyl ethyl ketone; insoluble in water.
Use: Light absorber, best at 320–380 mμ.

2-chloro-4-(hydroxymercuri)phenol. See hydroxymercurichlorophenol.

4-chloro-1-hydroxy-3-methylbenzene. See 4-chloro-3-methylphenol.

6-chloro-3-hydroxytoluene. See 4-chloro-3-methylphenol.

chloro-IPC. (isopropyl-N-(3-chlorophenyl) carbamate; isopropyl-3-chlorocarbanilate;

CIPC; chlorpropham).   CAS: 101-21-3.
$C_6H_4ClNHCOOC_3H_7$.
Properties: Light tan powder, mp 41.4C, vap press 2 mm (149C), d 1.18 (30C), very slightly soluble in water.
Hazard: Toxic by ingestion.
Use: Pre-emergence herbicide, prevents sprouting of potatoes.

2-chloro-N-isopropylacetanilide. (N-isopropyl-α-chloroacetanilide).   CAS: 1918-16-7.
$C_6H_5N[CH(CH_3)_2]COCH_2Cl$.
Properties: Light tan powder or granules; mp 67–76C; bp 110C (0.03 mm); very slightly soluble in water; soluble in acetone, alcohol, benzene, xylene, and carbon tetrachloride.
Hazard: Toxic by ingestion and skin absorption.
Use: Herbicide.

chloroisopropyl alcohol. See propylene chlorohydrin.

6-chloro-4-isopropyl-1-methyl-3-phenol. See chlorothymol.

chloromadinone acetate. A nonestrogenic sex hormone used in oral contraception.
See also antifertility agent.

chloromaleic anhydride. $CH{:}CClC(O)OC(O)$.
Properties: Yellow liquid, d 1.5, mp 10–15C, bp 192C.
Use: Catalyst for epoxy resins, intermediate.

chloromercuriferrocene. $C_5H_5FeC_5H_4HgCl$.
Orange, crystalline solid; mp 193–194. Used as an intermediate and for inorganic polymers.
Hazard: Toxic by ingestion and inhalation.

o-chloromercuriphenol. See o-hydroxyphenylmercuric chloride.

chloromethane. See methyl chloride.

chloromethapyrilene citrate. See chlorothen citrate.

3-chloro-2-methylaniline. See 2-amino-6-chlorotoluene.

3-chloro-4-methylaniline. See 4-amino-2-chlorotoluene.

5-chloro-2-methylaniline. See 2-amino-4-chlorotoluene.

chloromethylated diphenyl oxide.
$C_6H_5OC_6H_5$.   Up to three —CH_2Cl radicals can be substituted for hydrogens.
Straw-colored liquids or white solids, mp up to 55C, d 1.19–1.30 (25/25C), flash p 307F (152C).

Insoluble in water, very soluble in ether. Combustible.
Use: Intermediate, resins, plasticizers.

chloromethylbenzene. See chlorotoluene.

1-chloro-3-methylbutane. See isoamyl chloride.

chloromethylchloroformate. CAS: 22128-62-7.
$ClCOOCH_2Cl$.
Properties: Mobile, colorless liquid; penetrating, irritating odor. Hydrolyzed by water. Decomposed by alkalies. D 1.465 at 15C, bp 106.5–107C, vap d 4.5, vapor press 5.6 mm (20C). Soluble in most organic solvents.
Hazard: Toxic by ingestion and inhalation.

chloromethylchlorosulfonate. $ClCH_2OClSO_2$.
Properties: Colorless liquid, d 1.63, bp 49–50C (14 mm).
Derivation: By protracted boiling of chlorosulfonic acid with chloromethylchloroformate; also from paraformaldehyde and chlorosulfonic acid.
Hazard: Toxic by ingestion and inhalation; strong irritant to tissue.

chloromethyl cyanide. See chloroacetonitrile.

1-chloromethylethylbenzene. See ethylbenzyl chloride.

1-chloromethylnaphthalene. (α-naphthyl-methyl chloride). $C_{10}H_7CH_2Cl$.
Properties: Colorless to greenish-yellow liquid with sharp pungent odor, d 1.182 (25/25C), coagulation point 23C, refr index 1.6354–1.6360 (25C), insoluble in water, soluble in usual organic solvents. Very reactive.
Hazard: Vapor irritating to eyes.
Use: A lachrymator, intermediate.

4-chloro-3-methylphenol. (4-chloro-1-hydroxy-3-methylbenzene; 6-chloro-3-hydroxytoluene; 4-chloro-m-cresol; so-called p-chloro-m-cresol). $C_6H_3CH_3OHCl$.
Properties: White or slightly pink crystals with phenolic odor; mp 64–66C; bp 235C; volatile with steam; soluble 1:250 in water at 25C; soluble in alkalies, organic solvents, fats, and oils.
Hazard: Irritant to skin.
Use: External germicide; preservative for glues, gums, paints, inks, textile, and leather goods.

4-chloro-2-methylphenoxyacetic acid.
See MCP.

chloromethylphosphonic acid.
$ClCH_2PO(OH)_2$.
Properties: White, hygroscopic solid; mp 85–95C.

Use: Intermediate for flameproofing agents, resins, lubricants, additives, plasticizers.

chloromethylphosphonic dichloride.
$ClCH_2POCl_2$.
Properties: Water-white to light straw liquid, highly reactive, d 1.638 (25C), refr index 1.4960–1.4970 (25C).
Hazard: Toxic by inhalation; irritant to eyes, lungs, and mucous membranes.
Use: Intermediate for flameproofing agents, resins, lubricants, additives and plasticizers.

3-chlor-2-methyl-1-propene. See methylallyl chloride.

"Chloromycetin."[330] TM for a proprietary form of chloramphenicol, an antibiotic. Must comply with FDA requirements.
See chloramphenicol.

chloronaphthalene. See chlorinated naphthalene.

α-chloro-m-nitroacetophenone.
$NO_2C_6H_4COCH_2Cl$.
Properties: Off-white, free-flowing granules; mp approximately 95–100C; soluble in chlorinated solvents; insoluble in water.
Use: Bacteriostat and fungistat in cutting oils, water systems, paint, plastics, textiles; chemical intermediate.

2-chloro-4-nitroaniline. (o-chloro-p-nitro-aniline). CAS: 121-87-9.
$C_6H_3ClNO_2NH_2$.
Properties: Yellow needles; soluble in alcohol, benzene, ether; slightly soluble in water and strong acids; mp 107C.
Derivation: (a) From 1,2-dichloro-4-nitrobenzene by heating with alcoholic ammonia. (b) From the chlorination of p-nitroaniline in acid solution.
Hazard: Toxic by ingestion and inhalation.
Use: Intermediate in manufacture of dyes.

4-chloro-2-nitroaniline. (p-chloro-o-nitro-aniline). CAS: 89-63-4.
$C_6H_3ClNO_2NH_2$.
Properties: Orange crystals, mp 163C, insoluble in water, soluble in methanol and ether.
Hazard: Toxic by ingestion and inhalation.
Use: Dye and pigment intermediate.

4-chloro-3-nitroaniline. CAS: 635-22-3.
$C_6H_3ClNO_2NH_2$.
Properties: Yellow to tan powder, mp 95–97C, soluble in alcohol and acetone, partially soluble in hot water.
Hazard: Toxic by ingestion and inhalation.

Use: Intermediate in the manufacture of azo dyes, pharmaceuticals and other organic compounds.

m-chloronitrobenzene. CAS: 121-73-3.
$C_6H_4ClNO_2$.
Properties: Yellowish crystals, d 1.534, mp 44C, bp 236C, soluble in most organic solvents, insoluble in water.
Derivation: By chlorinating nitrobenzene in the presence of iodine and recrystallizing.
Hazard: Toxic by inhalation and ingestion. Combustible.
Use: Intermediate for dyes.

o-chloronitrobenzene. CAS: 88-73-3.
$C_6H_4ClNO_2$.
Properties: Yellow crystals, d 1.368, bp 245.5C, mp 32C, soluble in alcohol and benzene, insoluble in water, flash p 261F (127C).
Derivation: By nitrating chlorobenzene and purifying by rectification.
Hazard: Toxic by inhalation and ingestion. Combustible.
Use: Intermediate, especially for dyes.

p-chloronitrobenzene. CAS: 100-00-5.
$C_6H_4ClNO_2$.
Properties: Yellowish crystals, d 1.520, mp 83C, bp 242C, soluble in organic solvents, insoluble in water.
Derivation: Nitration of chlorobenzene and recrystallization.
Hazard: Very toxic by inhalation and ingestion. Absorbed via skin. Combustible.
Use: Intermediate, especially for dyes, manufacture of p-nitrophenol from which parathion is made, agricultural chemicals, rubber chemicals.

2-chloro-5-nitrobenzenesulfonamide.
$ClNO_2C_6H_3SO_2NH_2$.
Properties: Grayish-white solid, insoluble in water, soluble in benzene.
Hazard: Toxic by ingestion.
Use: Dye and pharmaceutical intermediates.

6-chloro-3-nitrobenzenesulfonic acid, sodium salt.
$NaSO_3C_6H_3NO_2Cl$.
Properties: Off-white, moist crystals.
Hazard: Toxic by ingestion.
Use: Intermediate for dyes and pharmaceuticals.

4-chloro-3-nitrobenzoic acid.
$ClC_6H_3NO_2COOH$.
Properties: Light gray or white powder, mp 170–174C.
Hazard: Toxic by ingestion and inhalation.
Use: Intermediate for dyes, perfumes, flavors, pharmaceuticals.

4-chloro-3-nitrobenzotrifluoride. (p-chloro-m-nitrotrifluorotoluene). $C_6H_3CF_3NO_2Cl$.
Properties: Thin, oily liquid; d 1.542 (15.5/15.5C); fp −7.5C; flash p 275F (135C). Combustible. Refr index 1.491 (20C), bp 222C, soluble in organic solvents, insoluble in water.
Grade: 97.5%.
Hazard: Very toxic by inhalation and ingestion. TLV: (as F) 2.5 mg/m³ of air.
Use: Intermediate for dyestuffs, agricultural chemicals, pharmaceuticals.

4-chloro-2-nitrophenol, sodium salt.
$ClC_6H_3NO_2ONa$.
Properties: Red needles with one molecule of water of crystallization. Soluble in hot water.
Derivation: Nitration of p-dichlorobenzene followed by hydrolysis.
Grade: 90% anhydrous sodium salt, containing 80% base.
Use: Dye intermediate, manufacture of 2-amino-4-chlorophenol.

1-chloro-1-nitropropane. See korax.

2-chloro-6-nitrotoluene. $ClNO_2C_6H_3CH_3$.
Solid, mp 36.5–40C, insoluble in water. Intermediate.
Hazard: Dangerous fire risk.

4-chloro-2-nitrotoluene. $ClNO_2C_6H_3CH_3$.
Solid, mp 35–37C, insoluble in water, soluble in alcohol and ether, intermediate.
Hazard: Dangerous fire risk.

p-chloro-m-nitrotrifluorotoluene. See 4-chloro-3-nitrobenzotrifluoride.

chloronitrous acid. See aqua regia.

chloropentafluoroacetone. ClF_2CCOCF_3.
Properties: Colorless, nonflammable, hygroscopic, highly reactive gas; bp 7.8C; fp −133C; d 1.43 g/mL (25C).
Hazard: Evolves heat on contact with water.
Use: Intermediate.

chloropentafluoroethane. (monochloropentafluorethane; fluorocarbon 115).
CAS: 76-15-3. $CClF_2CF_3$.
Properties: Colorless gas, bp −37.7F, fp −106C, insoluble in water, soluble in alcohol and ether. Has good thermal stability. Nonflammable.
Hazard: TLV: 1,000 ppm.
Use: Dielectric gas.

1-chloropentane. See n-amyl chloride.

chlorophacinone. $C_{23}H_{15}ClO_3$.
Properties: Crystals, mp 139C, almost insoluble in water, soluble in common organic solvents.

Hazard: Inhibits blood coagulation.
Use: Rodenticide.

chlorophene. USAN for o-benzyl-p-chlorophenol.

m-chlorophenol. (3-chloro-1-hydroxybenzene).
CAS: 108-43-0. C_6H_4OHCl.
Properties: White crystals with odor similar to phenol; discolors on exposure to air; d 1.245; mp 33C; bp 214C; soluble in alcohol, ether, and aqueous alkali; slightly soluble in water.
Derivation: From m-chloraniline through the diazonium salt.
Hazard: Toxic by skin absorption, inhalation, or ingestion.
Use: Intermediate in organic synthesis.

o-chlorophenol. (2-chloro-1-hydroxybenzene).
CAS: 95-57-8. C_6H_4OHCl.
Properties: Colorless to yellow brown liquid with unpleasant penetrating odor; very soluble in water; soluble in alcohol, ether, and aqueous sodium hydroxide;, volatile with steam; bp 175C; fp 9.3C; d 1.265 (15.5C); flash p 225F (107C). Combustible.
Derivation: Chlorination of phenol.
Hazard: Toxic by skin absorption, inhalation, or ingestion. Strong irritant to tissue.
Use: Organic synthesis (dyes).

p-chlorophenol. (4-chloro-1-hydroxybenzene).
CAS: 106-48-9. C_6H_4OHCl.
Properties: White crystals (yellow or pink when impure) with unpleasant penetrating odor. Slightly soluble in water; soluble in benzene, alcohol, and ether. Volatile with steam. Bp 217C, mp 42–43C, d 1.306, refr index 1.5579 (40C). A 1% solution is acid to litmus. Flash p 250C (121C). Combustible.
Derivation: Chlorination of phenol, from chloraniline through the diazonium salt.
Grade: NF, technical.
Hazard: Toxic by skin absorption, inhalation, or ingestion; strong irritant to tissue.
Use: Intermediate in synthesis of dyes and drugs, denaturant for alcohol, selective solvent in refining mineral oils, antiseptic.

p-chlorophenyl benzenesulfonate. See fenson.

p-chlorophenyl-p-chlorobenzenesulfonate.
See ovex.

4-chloro-α-phenyl-o-cresol. See o-benzyl-p-chlorophenol.

3-p-chlorophenyl-1,1-dimethylurea. See monuron.

m-chlorophenyl isocyanate. ClC_6H_4NCO.
Properties: Clear water-white to light yellow liquid, fp −4.4C, flash p 215F (101.6C) (COC). Combustible.
Hazard: Toxic via strong irritant to skin, eyes, and mucous membranes.
Use: Intermediate for pharmaceuticals, herbicides, and pesticides.

p-chlorophenyl isocyanate. ClC_6H_4NCO.
Properties: Colorless to slightly yellow liquid or white crystals. Fp 29.9C, flash p 230F (110C) (COC). Combustible.
Hazard: Strong irritant to skin, eyes, and mucous membranes.
Use: Intermediate for pharmaceuticals, herbicides, and pesticides.

3-(p-chlorophenyl)-5-methylrhodanine.
$C_{10}H_8ClNOS_2$.
Yellow crystals, mp 106–110C, insoluble in water, soluble in acetone.
Hazard: Toxic by ingestion.
Use: Fungicide and for nematode control.

chloro-o-phenylphenol. (chloro-2-phenylphenol).
$C_6H_3(OH)ClC_6H_5$.
Properties: Clear, colorless to straw colored, viscous liquid with faint characteristic odor; d 1.228 (20/4C); fp less than −20C; boiling range 5–95% 146–158.7C (5 mm); flash p 273F (134C); readily soluble in most organic solvents. Combustible.
Composition: (80%) 4-chloro-2-phenylphenol, (20%) 6-chloro-2-phenylphenol.
Hazard: Toxic by ingestion and inhalation.
Use: Fungicide.

p-chlorophenyl phenylsulfone. (4-chlorodiphenyl sulfone; sulphenone). $ClC_6H_4SO_2C_6H_5$.
Properties: Dimorphic crystals, slight aromatic odor, tasteless, insoluble in water, soluble in most organic solvents. Relatively stable in acids and alkalies.
Hazard: Toxic by ingestion.
Use: Insecticide and acaricide (for pests, harmful to most grapes and pears).

chlorophenyltrichlorosilane. CAS: 26571-79-9.
$ClC_6H_4SiCl_3$. A mixture of isomers.
Properties: Colorless to pale yellow liquid, bp 230C, d 1.439 (25/25C), refr index 1.5414 (20C), flash p 255F (124C) (COC), readily hydrolyzed with liberation of hydrogen chloride. Combustible.
Derivation: By Grignard reaction of silicon tetrachloride and chlorophenylmagnesium chloride.
Grade: Technical.
Hazard: Corrosive to wet skin.
Use: Intermediate for silicones.

4-chlorophthalic acid. $C_6H_3Cl(COOH)_2$.
Properties: Colorless crystals, mp 150C, decomposes on further heating, soluble in alcohol and ether, insoluble in water.
Derivation: Chlorination of phthalic acid.

chlorophyll. The green pigment essential to photosynthesis. It is present in all plants except fungi and bacteria. It occurs in three forms (a, b, and c) all of which are magnesium-centered porphyrins containing a hydrophilic carbocyclic ring with a lipophilic phytyl tail. Chlorophyll is a photoreceptor for wavelengths up to 700 mμ. It can readily transfer radiant energy to its chemical environment and thus acts as a transducer in photosynthesis. It is structurally analogous to the red blood pigment hemin. Chlorophyll has been synthesized by two different routes. Its derivatives are relatively unstable to light, oxidizing agents, and chemical reagents.
Properties: Chlorophyll a: $C_{55}H_{72}MgN_4O_5$.

Blue-green, microcrystalline wax; approximately three times as plentiful as chlorophyll b; mp 117–120C; freely soluble in ether, ethanol, acetone, chloroform, carbon disulfide, benzene; sparingly soluble in cold methanol; insoluble in petroleum ether. The alcoholic solution is blue-green with a deep-red fluorescence.
   Chlorophyll b: $C_{55}H_{70}MgN_4O_6$. Yellow-green, microcrystalline wax; sparingly soluble in absolute alcohol, ether. Ether solution has a brilliant green color. Solutions with other organic solvents are usually green to yellow-green with red fluorescence.
   Chlorophyll c: Occurs in marine organisms and may be as important as chlorophyll b.
Derivation: Alcoholic extraction of green plants; isolation by chromatography.
Grade: Aqueous, alcoholic, or oil solutions; water solutions are prepared by saponifiction of oil-soluble chlorophyll.
Use: Colorant for soaps, oils, fats, waxes, liquors, confectionery, preserves, cosmetics, perfumes,

dentistry, source of phytol, sensitizer for color film, toothpaste additive, deodorant.
See also photosynthesis, porphyrin, chelate.
Note: An experimental use of chlorophyll to act as energy converter in a synthetic photovoltaic cell has aroused interest in connection with solar energy research. The cell is termed a "synthetic leaf," as it is an attempt to approximate the energy-trapping function of a natural leaf.

chlorophyllin. Reaction product of alcoholic potassium or sodium hydroxide and alcoholic leaf extracts. The methyl and phytyl groups are replaced by alkali but the magnesium is not replaced. Used in food coloring, dyes, deodorants, and medicine.

chloropicrin. (chlorpicrin; nitrotrichloromethane; trichloronitromethane; nitrochloroform).
CAS: 76-06-2. CCl_3NO_2.
Properties: Pure product slightly oily, colorless, refractive liquid. Relatively stable, no decomposition by water or mineral acids, d 1.692 (0C), bp 112C, fp −69.2C. Soluble in alcohol, benzene, ether, carbon disulfide; slightly soluble in water. Nonflammable.
Derivation: (a) Action of picric acid on calcium hypochlorite; (b) nitrification of chlorinated hydrocarbons.
Hazard: Very toxic by ingestion and inhalation; strong irritant. TLV: 0.1 ppm in air.
Use: Organic synthesis, dye-stuffs (crystal violet), fumigants, fungicides, insecticides, rat exterminator, tear gas.

chloroplatinic acid. CAS: 16941-12-1.
$H_2PtCl_6 \cdot 6H_2O$.
Properties: Red-brown crystals; soluble in water, alcohol, and ether; d 1.431; mp 60C.
Derivation: By solution of platinum in aqua regia, evaporation, and crystallization.
Hazard: See platinum chloride.
Use: Electroplating, platinizing pumice and the like for catalysts, etching zinc for printing, platinum mirrors, indelible ink, ceramics, (producing fine color effects on high-grade porcelain), microscopy.

β-chloroprene. (2-chlorobutadiene-1,3).
CAS: 126-99-8. $H_2C{:}CHCCl{:}CH_2$.

$$H_2C{=}CH{-}\overset{\overset{\textstyle CH_2}{\|}}{C}Cl$$

Properties: Colorless liquid, bp 59.4C, d 0.9583 (20/20C), soluble in alcohol, slightly soluble in water, flash p −4F (−20C).

Derivation: Addition of hydrochloric acid to vinylacetylene; chlorination of butadiene.

Grade: Pure, 95% min.

Hazard: Flammable, dangerous fire risk, explosive limits in air 4.0 to 20%. Toxic by ingestion, inhalation, and skin absorption. TLV: 10 ppm in air. NIOSH has recommended 1 ppm.

Use: Manufacture of neoprene.

1-chloropropane. See propyl chloride.

3-chloropropane-1,2-diol. See chlorohydrin.

1-chloro-2-propanol. See propylene chlorohydrin.

1-chloro-2-propanone. See chloroacetone.

1- or 3-chloropropene. See allyl chloride.

2-chloropropene. (isopropenyl chloride).
CAS: 557-98-2. $CH_3CCl:CH_2$.

Properties: Colorless gas or liquid, d 0.918 (9C), fp −137.4C, bp 22.65C (1 atm).

Derivation: By treating propylene dichloride with alcoholic potassium hydroxide and fractionating from the simultaneously formed 1-chloropropene.

Grade: 95% purity.

Hazard: Toxic by ingestion and inhalation. Flammable, dangerous fire risk.

Use: Intermediate in organic synthesis, formulation of copolymers.

2-chloropropionic acid. (α-chloropropionic acid).
CAS: 598-78-7. $CH_3CHClCOOH$.

Properties: Crystals, d 1.260–1.268 (20C), bp 183–187C, soluble in water. Combustible.

Use: Intermediate for weed killers.

3-chloropropionic acid. (β-chloropropionic acid).
CAS: 107-94-8. CH_2ClCH_2COOH.

Properties: Crystals, mp 41C, bp 200C, soluble in water, alcohol, chloroform. Combustible.

Use: Intermediate.

3-chloropropionitrile. CAS: 542-76-7.
$ClCH_2CH_2CN$.

Properties: Colorless liquid, fp −51C, flash p 168F (75.5C) (CC), refr index 1.4341 (25C), d 1.1363 (25C), bp 176C (decomposes). Miscible with acetone, benzene, carbon tetrachloride, alcohol, and ether.

Hazard: Toxic by ingestion, inhalation, and skin contact. Combustible.

Use: Intermediate in polymer synthesis.

α-chloropropylene. See allyl chloride.

chloropropylene oxide. See epichlorohydrin.

3-chloropropyl mercaptan. $ClCH_2CH_2CH_2SH$.

Properties: Liquid, distillation range 141.1–145.6C, d 1.131 (15.5C), refr index 1.492 (20C), flash p approximately 110F (43C).

Hazard: Moderate fire risk. Combustible.

Use: Intermediate.

3-chloro-1-propyne. See propargyl chloride.

2-chloropyridine. CAS: 109-09-1. ClC_5H_4N.

Properties: Oily liquid, d 1.205 (15C), bp 170C, slightly soluble in water, soluble in alcohol and ether.

Hazard: Toxic by ingestion.

Use: Production of antihistamines, germicides, pesticides, and agricultural chemicals.

6-chloroquinaldine. $C_9H_5N(CH_3)Cl$.

Properties: Brownish-black, oily, crystalline mass.

Grade: Technical.

Use: Intermediate.

chloroquine.
$C_9H_5NClNHCH(CH_3)(CH_2)_3N(C_2H_5)_2$.
(7-chloro-4-(4-diethylamino-1-methylbutyl-amino)-quinoline).

Properties: Colorless crystals. Bitter taste; insoluble in alcohol, benzene, chloroform, ether.

Derivation: Condensation of 4,7-dichloroquinoline with 1-diethylamino-4-aminopentane.

Hazard: Toxic by ingestion.

Use: Medicine (antimalarial). Usually dispensed as the phosphate.

5-chlorosalicylanilide.
$ClC_6H_3OHCONHC_6H_5$.

Properties: White crystals, mp 209–211C. Slightly soluble in water; soluble in alcohol, ether, chloroform, and benzene.

Hazard: Toxic via ingestion.

Use: Fungicide, antimildew agent, intermediate for pharmaceuticals, dyes, pesticides.

5-chlorosalicyclic acid. CAS: 321-14-2.
$ClC_6H_3OHCOOH$.

Properties: White crystals, mp 174–176C. Slightly soluble in water; soluble in alcohol, ether, chloroform, and benzene.

Hazard: Toxic by ingestion.

Use: Mothproofing, insecticide, intermediate for pharmaceuticals, dyes, pesticides.

o-chlorostyrene. CAS: 1331-28-8. C_8H_7Cl.

Properties: Mp 138.6, d 1.1 at 20C, mp −63.15C, bp 188.7C. Soluble in alcohol, ether, and acetone.

Hazard: TLV: 50 ppm.

N-chlorosuccinimide. (NCS).
COCH$_2$CH$_2$CONCl.
Properties: White crystals, mp 148–149C, soluble in water, sparingly soluble in chloroform and carbon tetrachloride.
Use: Chlorinating agent, disinfectant for swimming pools, bactericide.

chlorosulfonic acid. (sulfuric chlorohydrin).
CAS: 7790-94-5.   ClSO$_2$OH.
Properties: Colorless to light yellow, fuming, slightly cloudy liquid, pungent odor, d 1.76–1.77 (20/20C, fp −80C, bp 158C, decomposes in water to sulfuric and hydrochloric acids. Decomposed by alcohol and acids.
Derivation: By treating sulfur trioxide or fuming sulfuric acid with hydrochloric acid.
Grade: Technical.
Hazard: Toxic by inhalation; strong irritant to eyes and skin; causes severe burns. Can ignite combustible materials. Evolves hydrogen on contact with most metals.
Use: Synthetic detergents, pharmaceuticals, sulfonating agent for dyes, pesticides, intermediates, ion-exchange resins, anhydrous hydrogen chloride, and smoke-producing chemicals.

4-chlorosulfonylbenzoic acid.
ClSO$_2$C$_6$H$_4$COOH.
Properties: Light tan powder, soluble in benzene, slightly soluble in ether.
Use: Intermediate.

chlorosulfuric acid.   See sulfuryl chloride.

chlorotetracycline.   See chlortetracycline.

chlorotetrafluoroethane. (monochlorotetrafluoroethane).   C$_2$CF$_4$Cl.
Properties: Nonflammable, odorless and colorless gas, much heavier than air.

chlorothen citrate. (chloromethapyrilene citrate).
C$_{14}$H$_{18}$ClN$_3$S·C$_6$H$_8$O$_7$.
Properties: White, practically odorless, crystalline powder; slightly soluble in alcohol and water; practically insoluble in chloroform, ether, and benzene. 1% solution is clear and colorless; pH (1% solution) 3.9–4.1, mp 112–116C, on further heating solidifies and remelts, 125–140C decomposes.
Grade: NF.
Use: Medicine (antihistamine).

"Chlorothene."[233]   TM for a series of chlorinated organic solvents.
Grade: NU and Industrial are both inhibited 1,1,1-trichloroethane.

Use: For cold cleansing of metal parts and as industrial solvents.

p-chlorothiophenol. (p-chlorobenzenethiol).
CAS: 106-54-7.   ClC$_6$H$_4$SH.
Properties: Moist white to cream crystals, mp 52–55C, bp 205–207C, soluble in most organic solvents.
Hazard: Toxic by ingestion.
Use: Oil additives, agricultural chemicals, plasticizers, rubber chemical, dyes, wetting agents and stabilizers.

chlorothymol. (6-chloro-4-isopropyl-1-methyl-3-phenol).   CH$_3$C$_6$H$_2$(OH)(C$_3$H$_7$)Cl.
Properties: White crystals or granular powder, characteristic odor, aromatic, pungent taste, becomes discolored with age, affected by light, mp 59–61C. Soluble in benzene, chloroform, dilute caustic soda, alcohol; slightly soluble in water.
Derivation: Action of sulfuryl chloride on thymol in a solution of carbon tetrachloride.
Grade: NF.
Hazard: Irritant to skin and mucous membranes in concentrated solution.
Use: Bactericide, component of antiseptic solutions.

α-chlorotoluene.   See benzyl chloride.

m-chlorotoluene. (3-chloro-1-methylbenzene).
CAS: 108-41-8.   CH$_3$C$_6$H$_4$Cl.
Properties: Colorless liquid, d 1.07218 (20/4C), bp 161.6C, fp −48.0C, refr index 1.52 (20C).
Derivation: Diazotization of m-toluidine followed by treating with cuprous chloride.
Hazard: Narcotic in high concentration. Avoid inhalation.
Use: Solvent, intermediate.

o-chlorotoluene. (2-chloro-1-methylbenzene).
CAS: 95-49-8.   CH$_3$C$_6$H$_4$Cl.
Properties: Colorless liquid, bp 159.2C, fp −35.1C, d 1.0776 (25/4C), refr index 1.5268 (20C). Miscible with alcohol, acetone, ether, benzene, carbon tetrachloride, and n-heptane; slightly soluble in water.
Derivation: By catalytic chlorination of toluene.
Hazard: Toxic by inhalation. TLV: 50 ppm in air.
Use: Solvent and intermediate for organic chemicals and dyes.

p-chlorotoluene. (4-chloro-1-methylbenzene).
CAS: 106-43-4.   CH$_3$C$_6$H$_4$Cl.
Properties: Colorless liquid, boiling range 162–166C, fp approximately 6.5C, d 1.065–1.067 (25/15C), refr index 1.5184 (22C). Soluble in alcohol, ether, acetone, benzene, and chloroform; slightly soluble in water.

Hazard: Avoid inhalation; strong irritant.

Use: Solvent and intermediate for organic chemicals and dyes.

2-chlorotoluene-4-sulfonic acid. (o-chlorotoluene-p-sulfonic acid). $CH_3C_6H_3(SO_3H)Cl$.
Properties: White, glistening plates; soluble in hot water.
Derivation: Chlorination of toluene-p-sulfonic acid.
Method of purification: Recrystallization from water.
Use: Dye intermediate.

2-chloro-p-toluidine.   See 4-amino-2-chlorotoluene.

4-chloro-o-toluidine.   See 2-amino-4-chlorotoluene.

6-chloro-o-toluidine.   See 2-amino-6-chlorotoluene.

4-chloro-o-toluidine hydrochloride.
$CH_3C_6H_3(Cl)NH_2 \cdot HCl$.
Hazard: Toxic by ingestion, inhalation, and skin absorption.

2-chloro-5-toluidine-4-sulfonic acid. (6-chloro-m-toluidine-4-sulfonic acid).
$CH_3C_6H_2(NH_2)(SO_3H)Cl$.
Properties: Fine, white crystals; soluble in dilute caustic solution.
Derivation: From o-chlorotoluene-p-sulfonic acid by nitration and subsequent reduction.
Use: Intermediate.

chlorotriazinyl dye. A fiber-reactive dye, both mono- and di-derivatives have been developed. They react readily and permanently with cellulose under alkaline conditions with 70–80% efficiency.

chlorotrifluoroethylene. (CFE; CTFE; trifluorochloroethylene).   CAS: 79-38-9.
$ClFC:CH_2$.
Properties: Colorless gas with faint ethereal odor, bp −27.9C, fp −157.5C, d (liquid) 1.305 (20C), decomposes in water.
Derivation: From trichlorotrifluoroethane and zinc.
Grade: Technical 99.0%.
Hazard: Dangerous fire risk. Flammable limits in air 8.4–38.7%.
Use: Intermediate, monomer for chlorotrifluoroethylene resins.

chlorotrifluoroethylene polymer. (polytrifluorochloroethylene resin; fluorothene). A polymer of chlorotrifluoroethylene, usually including vinylidine fluoride, characterized by the repeating structure $(CF_2—CCl)$.
Properties: Colorless, d 2.10–2.15, refr index 1.43. Impervious to corrosive chemicals, resistant to most organic solvents, heat-resistant, nonflammable, thermoplastic. Zero moisture absorption, high impact strength, transparent films and thin sheets.
Use: Chemical piping, gaskets tank linings, connectors, valve diaphragms, wire and cable insulation, electronic components.
See also 'Kel-F," fluorocarbon polymer.

chlorotrifluoromethane. (monochlorotrifluoromethane; trifluorochloromethane).
CAS: 75-72-9. $CClF_3$.
Properties: Colorless, nonflammable gas, ethereal odor, bp −81.4C, fp −181C, heavier than air.
Derivation: From dichlorodifluoromethane in vapor phase with aluminum chloride catalyst.
Grade: 99.0% min purity.
Hazard: Toxic by inhalation; slightly irritant.
Use: Dielectric and aerospace chemical, hardening of metals, pharmaceutical processing.
See also chlorofluorocarbon.

2-chloro-5-trifluoromethylaniline.
$C_6H_3ClCF_3NH_2$.
Properties: Amber colored oil, fp 6–8C, insoluble in water, soluble in alcohol and acetone. Forms water-soluble salts with mineral acids.
Hazard: Irritant to eyes and skin.
Use: Intermediate for dyes and other organic chemicals.

chlorotrifluoromethylbenzene.   See chlorobenzotrifluoride.

chloro-α,α,α-trifluorotoluene.   See chlorobenzotrifluoride.

β-chlorovinyldichloroarsine. (1-chloro-2-dichloroarsinoethane; dichloro[2-chlorovinyl]arsine; chlorovinylarsinedichloride; Lewisite). Two isomers, probably cis and trans, are known. $ClCH:CHAsCl_2$.
Properties: Colorless liquid when pure. Impurities influence a color change ranging from violet to brown. Geranium-like odor, decomposed by water and alkalies. Inactivated by bleaching powder. Antidote is dimercaptopropanol.
Derivation: Condensation of arsenic trichloride with acetylene in the presence of aluminum or copper or mercury chloride. The mixed arsines are separated by fractionating.
Hazard: Vesicant gas, a poison. See arsenic.
Use: Poison gas, skin blistering agent.

β-chlorovinyl ethyl ethynyl carbinol.
See ethchlorvynol.

β-chlorovinylmethylchloroarsine.
ClCH:CHAsClCH₃.
Properties: Liquid, decomposed by water, bp 112–115C (10 mm).
Derivation: Interaction of acetylene and methyldichloroarsine in the presence of aluminum chloride.
Hazard: Strong irritant poison, absorbed by skin.

chloroxine. (5,7-dichloro-8-quinolinol).
Properties: Colorless crystals; mp 180C; soluble in benzene, acetone, acids, and sodium hydroxide solution.
Use: Analytical reagent, antibacterial.

Chloroxuron. (3-(p-(p-chlorophenoxy)phenyl)-1,1-dimethylurea). CAS: 1982-47-4.
Use: Herbicide.

p-chloro-m-xylenol. (4-chloro-3,5-dimethylphenol). C₆H₂(CH₃)₂OHCl. Crystals with phenolic odor.
Hazard: Toxic by ingestion; strong irritant, absorbed by skin.
Use: Active ingredient in germicides, antiseptics, etc., fungistat; mildew preventive; preservative; chemical intermediate.

6-chloro-3,4-xylyl methylcarbamate. (6-chloro-3,4-dimethylphenyl-N-methylcarbamate).
ClC₆H₂(CH₃)₂OCONHCH₃.
Properties: Solid, mp 120–133C.
Hazard: Toxic by ingestion and inhalation.
Use: Pesticide.

chlorpheniramine maleate. (chlorprophenpyridamine maleate). C₁₆H₁₉ClN₂·C₄H₄O₄.
(1-(p-chlorophenyl)-1-(2-pyridyl)3-dimethylaminopropane maleate).
Properties: White, odorless crystals; mp 130–135C; slightly soluble in ether; soluble in alcohol, chloroform, and water; pH (1% solution) approximately 4.8.
Grade: USP.
Use: Medicine (antihistamine).

chlorphenol red. (dichlorosulfonphthalein).
C₆H₄SO₂OC(C₆H₃ClOH)₂. An acid-base indicator, showing color change from yellow to red over the pH range 5.2 to 6.8.

chlorpromazine. (2-chloro-10-(3-dimethylaminopropyl)-phenothiazine). CAS: 50-53-3.
C₁₇H₁₉ClN₂S.

Properties: Oily liquid, amine odor, alkaline reaction, bp 200–205C (0.8 mm).
Hazard: Toxic by ingestion.
Use: Medicine (antipsychotic drug).

chloropropham. See chloro-IPC.

chlorpyrifos. See "Dursban."

chlorquinaldol. (5,7-dichloro-8-hydroxyquinaldine; 5,7-dichloro-2-methyl-8-quinolinol).
CH₃C₉H₃N(OH)Cl₂.
Properties: Yellow, crystalline, tasteless powder with a pleasant medicinal odor; mp 114C. Soluble in benzene, alcohol, chloroform; insoluble in water.
Use: Medicine (bactericide and fungicide).

chlortetracycline. (CTC; chlorotetracycline).
C₂₂H₂₃ClN₂O₈. An antibiotic produced by growth of Streptomyces aureofaciens in submerged cultures. It has a wide antimicrobial spectrum including many Gram-positive and Gram-negative bacteria, rickettsiae and several viruses. Its chemical structure is that of a modified naphthacene molecule. Also available as the hydrochloride.
Properties: Golden-yellow crystals, mp 168–169C. Slightly soluble in water; very soluble in aqueous solutions above pH 8.5; freely soluble in the "Cellosolves" dioxane and "Carbitol," slightly soluble in methanol, ethanol, butanol, acetone, ethyl acetate, and benzene; insoluble in ether and naphtha.
Derivation: By submerged aerobic fermentation, filtration, solvent extraction, and crystallization.
Use: Medicine (antibiotic), feed supplement, preservative for raw fish.

chlorthion. Generic name for O,O-dimethyl-O-(3-chloro-4-nitrophenyl)thiophosphate.
A phosphoric acid ester containing chlorine.
Hazard: Toxic by inhalation and ingestion, cholinesterase inhibitor, absorbed via intact skin. Use may be restricted.
Use: Insecticide.

CHOC. Abbreviation for Center of History of Chemistry.
See Appendix II-D.

cholaic acid. See taurocholic acid.

cholecalciferol. (5,7-cholestadien-3-β-ol; 7-dehydrocholesterol; activated vitamin D_3).
$C_{27}H_{44}O$. A free vitamin D_3, isolated in crystalline state from the 3,5-dinitrobenzoate, produced by irradiation, and equivalent in activity to the vitamin D_3 of tuna liver oil.
Properties: Colorless crystals, unstable in light and air. Insoluble in water, soluble in alcohol, chloroform, and fatty oils. Melting range 84–88C, specific rotation +105–112 degrees.
Grade: USP, FCC.
Available forms: Hermetically sealed under nitrogen.
Use: Medicine (antirachitic vitamin).

choleic acid. A general term applied to the coordination complexes formed by deoxycholic acid (a bile acid) with fatty acids or other lipids and with a variety of other compounds including such aromatics as phenol and naphthalene. These complexes are similar to those used in separation processes such as the urea adducts for large-scale purification.
See also cholic acid.

cholesteric. A molecular structure found in some liquid crystals, so called because it was first noted in cholesteryl alcohol in 1888. It occurs in some optically active compounds or in mixtures of chiral compounds and nematic liquid crystals.

cholesterol. (cholesterin; 5-cholesten-3-β-ol).
CAS: 57-88-5. $C_{27}H_{45}OH$. The most common animal sterol, a monohydric secondary alcohol of the cyclopentenophenanthrene (4-ring fused) system, containing one double bond. It occurs in part as the free sterol and in part esterified with higher fatty acids as a lipid in human blood serum. The primary precursor in biosynthesis appears to be acetic acid or sodium acetate. Cholesterol itself in the animal system is the precursor of bile acids, steroid hormones and provitamin D_3.

Properties: White or faintly yellow, almost odorless, pearly granules or crystals; affected by light; mp 148.5C; bp 360C (decomposes); d 1.067 (20/4C); levorotatory; specific rotation (25C) −34–

38 degrees. Sparingly soluble in water; moderately soluble in hot alcohol; soluble in benzene, oils, fats and aqueous solutions of bile salts.
Occurrence: Egg yolk, liver, kidneys, saturated fats and oils. All body cells contain cholesterol produced by the liver (approximately 1000 mg a day).
Source: Prepared from beef spinal cord by petroleum ether extraction of the nonsaponifiable matter; purification by repeated bromination.
Grade: Technical, USP, SCW (standard for clinical work).
Use: Emulsifying agent in cosmetic and pharmaceutical products, source of estradiol.

cholestyramine. A synthetic anion-exchange polymer in which quaternary ammonium groups are attached to a copolymer of styrene and divinylbenzene. A white to buff-colored powder having a particle size from 50–100 mesh, it is stable to 150C. Insoluble in water and organic solvents. It is effective in binding bile salts such as cholesterol. Research on test animals and in clinical trials on humans indicates that it is effective in eliminating from the body such toxic organochlorine compounds as "Kepone."

cholic acid. CAS: 81-25-4. $C_{23}H_{49}O_3COOH$.

The most abundant bile acid. In bile, it is conjugated with the amino acids glycine and taurine as glycocholic acid and taurocholic acid, respectively, and does not occur free.
Properties: The monohydrate crystallizes in plates from dilute acetic acid. Bitter taste with sweetish aftertaste. Anhydrous form, mp 198C. Not precipitated by digitonin. Soluble in glacial acetic acid, acetone, and alcohol; slightly soluble in chloroform; practically insoluble in water and benzene.
Derivation: From glycocholic and taurocholic acids in bile, organic synthesis.
Grade: FCC.
Use: Biochemical research, pharmaceutical intermediate, emulsifying agent in foods, up to 0.1%.

choline. (choline base; β-hydroxyethyltrimethylammonium hydroxide). CAS: 62-49-7.
$(CH_3)_3N(OH)CH_2CH_2OH$. Member of the

vitamin B complex. Essential in the diet of rats, rabbits, chickens, and dogs. In man it is required for lecithin formation and can replace methionine in the diet. There is no evidence of disease in man due to choline deficiency. It is a dietary factor important in furnishing free methyl groups for transmethylation; has a lipotropic function.

Source: Egg yolk, kidney, liver, heart, seeds, vegetables, and legumes; synthetic preparation from trimethylamine and ethylene chlorohydrin or ethylene oxide.

Properties: Viscous, alkaline liquid; soluble in water and alcohol. Amounts are expressed in milligram of choline.

Use: Medicine, nutrition, feed supplement, catalyst, curing agent, control of pH, neutralizing agent, solubilizer.

choline bicarbonate. See 2-hydroxyethyltrimethylammonium bicarbonate.

choline bitartrate. $(C_5H_{14}NO)C_4H_5O_6$.
Properties: White, crystalline powder; odorless or faint trimethylamine-like odor; acid taste; hygroscopic; soluble in water and alcohol; insoluble in ether, chloroform, and benzene.
Grade: FCC.
Use: Medicine, dietary supplement, nutrient.

choline chloride. CAS: 67-48-1.
$(CH_3)_3N(Cl)CH_2CH_2OH$. Animal feed additive derived from agricultural waste or made synthetically. Available as 50% dry feed-grade and 70% solution.

cholinesterase. (1) (acetylcholinesterase) Enzyme specific for the hydrolysis of acetylcholine to acetic acid and choline in the body. It is found in the brain, nerve cells, and red blood cells, and is important in the mechanism of nerve action.
See nerve gases; parathion; insecticide.
Derivation: From bovine erythrocytes.
Use: Biochemical research, determination of phosphorus in insecticides and poisons.
(2) "Pseudo" or nonspecific cholinesterase, prepared from horse serum. This esterase hydrolyzes other esters as well as choline esters. It occurs in blood serum, pancreas and liver.

cholinesterase inhibitor. A chemical compound which deactivates the enzyme cholinesterase thus preventing or retarding hydrolytic breakdown of the highly toxic acetylcholine formed in the body by the nervous system. Nerve gases act in this way and so do a number of insecticides, usually organic esters of phosphoric acid derivatives. Serious poisoning and death may occur on inges-

tion or prolonged inhalation of such compounds. Cholinesterase can be reactivated by administration of atropine sulfate or pralidoxime iodide.
See also parathion, nerve gas.

cholytaurine. See taurocholic acid.

chondroitin sulfate. A major constituent of the cartilaginous tissue in the body.

chondrus. See carrageenan.

chorionic gonadotropin. (HCG). A hormone isolated from blood and urine of pregnant women, is secreted by the placenta. It is a glycoprotein containing approximately 11% galactose and having a molecular weight of approximately 100,000.
Properties: Rods or needle-like crystals. Soluble in water and glycols. Relatively unstable in aqueous solution, stable in dry form. It enhances estrone and progesterone production. Units: One international unit equals the activity of 0.1 mg of a standard preparation.
Use: Medicine, veterinary medicine.

"Chromacyl."[28] TM for a group of dyes that contain chromium in the molecule. Suitable for wool and nylon.

chromated zinc chloride. See zinc chloride chromated.

chromatin. A deoxyribonucleoprotein complex consisting of (1) double-stranded DNA molecules, (2) a basic protein called histone, and (3) other proteins. The latter protect the DNA from attack by enzymes. Chromatin occurs in the cell nucleus, where it forms chromosomes, the carriers of genes. Its name is derived from its sensitivity to biological stains.
See also deoxyribonucleic acid, chromosome, gene.

chromatography. A group of laboratory separation techniques based on selective adsorption by which components of complex mixtures (vapors, liquids, solutions) can be identified. Its discoverer, Tswett (1906), named the procedure chromatography because the plant pigments used in his early experiments produced bands of characteristic color. Since the 1930s, the method has been widely applied in many variations to the analysis of colorless mixtures such as hydrocarbons, metallic salts, etc. Separation is due to redistribution of the molecules of the mixture between the thin phase (adsorption layer) and the bulk phase (adsorbent) with which it is in contact. As the thin phase sometimes approaches molecular dimensions, the size and shape of the molecules of the mixture are of great significance.

Basically, chromatography involves the flow of a mobile (gas or liquid) phase over a stationary phase (which may be a solid or a liquid). Liquid chromatography is used for soluble substances and gas (vapor-phase) chromatography for volatile substances. As the mobile phase moves past the stationary phase, repeated adsorption and desorption of the solute occurs at a rate determined chiefly by its ratio of distribution between the two phases. If the ratio is large enough, the components of the mixture will move at different rates, producing a series of bands (chromatographs) by which their identity can be determined.
See also liquid chromatography, gas chromatography, paper chromatrography, thin-layer chromatography, ion-exchange chromatography, gel filtration.

"Chromax" Castings.[350] TM for ferrous alloys containing 20–30% Nickel and 15–20% chromium.

chrome alum. See chromium potassium sulfate.

chrome ammonium alum. See chromium ammonium sulfate.

chrome cake. A green form of salt cake (sodium sulfate) containing a low percentage of chromium. A by-product of sodium dichromate manufacture used in the paper industry.

chrome dye. A mordant dye, most frequently one in which sodium dichromate is used as the mordant.

chrome green. See chrome pigment.

"Chromekill 4A."[142] TM for a nearly neutral material composed of organic and inorganic reducing agents designed for the reduction of hexavalent chromium to trivalent chromium when it is present in low percentage in alkaline cleaning solutions and electroplating baths.

"Chromel."[166] TM for a series of nickel-chromium alloys. There are a number of different compositions ranging from 35–80% nickel and 16–20% chromium, some containing iron. The 80Ni-20Cr types are useful up to 1150–1200C.

chrome-molybdenum steel. Steel made by any accepted method of quality steel-making containing both chromium and molybdenum, usually in the ranges of chromium 0.35–1.10% and molybdenum 0.08–0.35%.

chrome-nickel steel. See steel, stainless.

chrome pigment. An inorganic pigment containing chromium. The most important types are: (1) chrome oxide green, one of the most permanent and stable pigments known, the pure grade consisting of 99% Cr_2O_3, used in paints applied to cement and lime-containing surfaces; (2) chrome green, chrome yellow, and chrome red, consisting chiefly of lead chromate and used in paints, rubber, and plastic products; (3) miscellaneous pigments such as molybdate orange and zinc yellow, based on lead and zinc compounds of chromium, respectively. All these are more stable to sunlight, weathering and chemical action than the brighter organic dyes.
Hazard: Toxic by ingestion.

chrome potash alum. See chromium potassium sulfate.

chrome red. See chrome pigment.

chrome steel. A steel made by any accepted method of quality steel-making containing chromium as alloying element usually in the range 0.20–1.60%, although the content may be as high as 25% in specialized heat-resistant and wear-resistant steels and in stainless steels.
See also steel, stainless.

chrome tanning. See tanning.

chrome-vanadium steel. A steel made by any accepted method of quality steel-making containing both chromium and vanadium usually in the ranges of chromium 0.50–1.10% and vanadium 0.10–0.20%.

chrome yellow. See lead chromate, chrome pigment.

chromia. See chromic oxide.

chromic acetate. (chromium acetate).
CAS: 1066-30-4. $Cr(C_2H_3O_2)_3 \cdot H_2O$.
Properties: Grayish-green powder or bluish-green, pasty mass. Soluble in water, insoluble in alcohol.
Derivation: Action of acetic acid on chromium hydroxide. The solution is evaporated and crystallized.
Hazard: Toxic by ingestion.
Use: Textile mordant, tanning, polymerization and oxidation catalyst, emulsion hardener.

chromic acid. (chromium trioxide; chromic anhydride).   CAS: 7738-94-5. CrO_3.
The name is in common use although the true chromic acid, H_2CrO_4, exists only in solution.
Properties: Dark purplish-red crystals; soluble in water, alcohol, and mineral acids. Deliquescent; d 1.67–2.82; mp 196C.

Derivation: (a) Sulfuric acid is added to a solution of sodium dichromate and the product is crystallized out; (b) chromite is fused with soda ash and limestone and then treated with sulfuric acid; (c) electrolysis.

Grade: Technical, CP.

Hazard: Powerful oxidizing agent, may explode on contact with reducing agents, may ignite on contact with organic materials. A poison. Corrosive to skin. TLV: 0.05 mg/m³ of air. A human carcinogen.

Use: Chemicals (chromates, oxidizing agents, catalysts), chromium plating intermediate, medicine (caustic), process engraving, anodizing, ceramic glazes, colored glass, metal cleaning, inks, tanning, paints, textile mordant, etchant for plastics.

chromic bromide. $CrBr_3$.
Properties: Black crystals, soluble in boiling water, insoluble in cold water unless chromous salts are added.
Derivation: Passage of bromine vapor over pulverized chromium at 1000C.
Use: Olefin polymerization catalyst.

chromic chloride. (chromium chloride; chromium trichloride; chromium sesquichloride).
CAS: 10025-73-7.   (a) $CrCl_3$ or
(b) $CrCl_3 \cdot 6H_2O$.
Properties: (a) Violet crystals, d 1.76, mp 1150C, sublimes at approximately 1300C. Insoluble in water and alcohol. (b) Greenish-black or violet deliquescent crystals, depending on whether or not chlorine is coordinated with the chromium, d 1.76, mp 83C. Soluble in water and alcohol, insoluble in ether.
Derivation: (a) By passing chlorine over a mixture of chromic oxide and carbon. (b) By the action of hydrochloric acid on chromium hydroxide.
Hazard: A poison. TLV: 0.5 mg/m³ in air.
Use: Chromium salts, intermediates, textile mordant, chromium plating including vapor plating, preparation of sponge chromium, catalyst for polymerizing olefins, waterproofing.

chromic fluoride. (chromium fluoride; chromium trifluoride).   CAS: 7788-97-8.
$CrF_3 \cdot 4H_2O$ or $CrF_3 \cdot 9H_2O$.
Properties: Fine, green crystals; d (anhydrous): 3.8; mp greater than 1000C; bp (sublimes) 1100–1200C. Insoluble in water and alcohol, soluble in hydrochloric acid.
Derivation: Interaction of chromium hydroxide and hydrofluoric acid.
Grade: Technical, high purity (CrF_3).
Hazard: Irritant to skin and eyes, especially in solution. TLV: 0.5 mg/m³ of air.
Use: Printing and dyeing woolens, mothproofing, halogenation catalyst.

chromic hydroxide. (chromic hydrate; chromium hydroxide; chromium hydrate). $Cr(OH)_3$.
Properties: Green, gelatinous precipitate; decomposes to chromic oxide by heat. Insoluble in water, soluble in acids and strong alkalies.
Derivation: By adding a solution of ammonium hydroxide to the solution of a chromium salt.
Use: Guignet's green, catalyst, tanning agent, mordant.

chromic nitrate. (chromium nitrate).
CAS: 13548-38-4. $Cr(NO_3)_3 \cdot 9H_2O$.
Properties: Purple crystals, soluble in alcohol and water, mp 60C, decomposes 100C.
Derivation: By the action of nitric acid on chromium hydroxide.
Hazard: May ignite organic materials on contact, may be explosive when shocked or heated, powerful oxidizer. Very toxic.
Use: Catalyst, corrosion inhibitor.

chromic oxide. (chromium(III) oxide; chromia; chromium sequioxide; green cinnabar).
CAS: 1308-38-9. Cr_2O_3.
Properties: Bright-green, extremely hard crystals; d 5.2; mp 2435C; bp 4000C; insoluble in water, acids, and alkalies.
Derivation: (a) By heating chromium hydroxide, (b) by heating dry ammonium dichromate, (c) by heating sodium dichromate with sulfur and washing out the sodium sulfate.
Hazard: Toxic by ingestion and inhalation. TLV: 0.5 mg/m³ of air.
Use: Metallurgy, green paint pigment, ceramics, catalyst in organic synthesis, green granules in asphalt roofing, component of refractory brick, abrasive.

chromic phosphate. (chromium phosphate).
CAS: 7789-04-0.   (a) $CrPO_4 \cdot 6H_2O$;
(b) $CrPO_4 \cdot 4H_2O$.
Properties: (a) Violet crystals, d 2.12 (14C); (b) green crystals, soluble in acids, insoluble in water.
Derivation: (a) Interaction of solutions of chromium chloride and sodium phosphate; (b) by mixing chrome alum and disodium hydrogen phosphate. Violet, amorphous powder (not the hexahydrate) is formed which becomes crystalline on contact with water. On boiling, it is converted into green crystalline hydrate.
Use: Paint pigment, catalyst.

chromic sulfate. (chromium sulfate).
CAS: 10101-53-8.   (a) $Cr_2(SO_4)_3$;
(b) $Cr_2(SO_4)_3 \cdot 15H_2O$; (c) $Cr_2(SO_4)_3 \cdot 18H_2O$.
Properties: (a) Violet or red powder; (b) dark-green amorphous scales; (c) violet cubes. D

(a) 3.012; (b) 1.867; (c) 1.70. (a) Insoluble in water and acids; (b) soluble in water, insoluble in alcohol; (c) soluble in water and alcohol.

Derivation: Action of sulfuric acid on chromium hydroxide with subsequent crystallization.

Use: Chrome plating, chromium alloys, mordant, catalyst, green paints and varnishes, green ink, ceramics (glazes). The basic form (reduction of sodium dichromate) is used in tanning.

chromite. (chrome iron ore).
CAS: 1308-31-2. $FeCr_2O_4$. A natural oxide of ferrous iron and chromium, sometimes with magnesium and aluminum present. Usually occurs in magnesium and iron rich igneous rocks.
Properties: Color iron-black to brownish-black, streak dark brown, luster metallic to submetallic, d 3.6, Mohs hardness 5.5.
Grade: Metallurgical, refractory, chemical.
Occurrence: USSR, South Africa, Zimbabwe, Philippines, Cuba, Turkey.
Hazard: A carcinogen. TLV: 0.05 mg/m³ of air.
Use: Only commercial source of chromium and its compounds.

chromium. CAS: 7440-47-3. Cr. Metallic element of atomic number 24, group VIB of the Periodic Table, aw 51.996, valences=2,3,6; 4 stable isotopes. Name derived from Greek for color.
Properties: Hard, brittle, semi-gray metal; d 7.1; mp 1900C; bp 2200C. Compounds have strong and varied colors. Cr ion forms many coordination compounds. Exists in active and passive forms, the latter giving rise to its corrosion resistance due to a thin surface oxide layer that passivates the metal when treated with oxidizing agents. Active form reacts readily with dilute acids to form chromous salts. Soluble in acids (except nitric) and strong alkalies, insoluble in water.
Occurrence: USSR, South Africa, Turkey, Philippines, Zimbabwe, Cuba.
Derivation: From chromite by direct reduction (ferrochrome), by reducing the oxide with finely divided aluminum or carbon, and by electrolysis of chromium solutions.
Grade: (ore): Chromium ores are classified as (1) metallurgical, (2) refractory, and (3) chemical and their consumption in the US is in that order. (1) Must contain a minimum of 48% Cr_2O_3 and have Cr-Fe ratio of 3:1; (2) must be high in Cr_2O_3, and Al_2O_3 and low in iron; (3) must be low in SiO_2 and Al_2O_3 and high in Cr_2O_3.
Forms available: (1) Chromium metal as lumps, granules or powder; (2) high- or low-carbon ferrochromium; (3) single crystals, high-purity crystals, or powder run 99.97% pure.
Hazard: Hexavalent chromium compounds are carcinogenic (OSHA) and corrosive on tissue, resulting in ulcers and dermatitis on prolonged contact. TLV: For chromium dust and fume is 0.5 mg/m³ of air.
Use: Alloying and plating element on metal and plastic substrates for corrosion resistance, chromium-containing and stainless steels, protective coating for automotive and equipment accessories, nuclear and high-temperature research, constituent of inorganic pigments.

chromium-51. Radioactive Chromium of mass number 51.
Properties: Half-life 26.5D, radiation: gamma (0.32 MeV).
Grade: USP (as sodium chromate Cr-51 injection).
Hazard: Radioactive poison.
Use: Diagnosis of blood volume (as tracer).

chromium acetate. See chromic acetate.

chromium acetylacetonate.
$[CH_3COCHC(CH_3)O]_3Cr$.
Properties: Purple powder or red-violet crystals, mp 216C, bp 340C, insoluble in water, soluble in acetone and alcohol.
Derivation: Reaction of chromium chloride, acetylacetone and sodium carbonate.
Use: Reduction of detonation of nitromethane.

chromium ammonium sulfate. (ammonium chromium sulfate; chrome ammonium alum).
$CrNH_4(SO_4)_2 \cdot 12HOH$.
Properties: Green powder or deep violet crystals, d 1.72, mp 94C. Soluble in water, slightly soluble in alcohol. The aqueous solution is violet when cold, green when hot.
Grade: Technical.
Use: Mordant, tanning.

chromium boride. One of several compounds of chromium and boron, e.g., CrB, CrB_2, and Cr_3B_2. They have high melting points, are very hard and corrosion-resistant and may be suitable for use in jet and rocket engines.
Properties: CrB may be crystals, d 6.2, mp 1550C, Mohs hardness 8.5, resistivity 67 μ-ohm cm (20C). CrB_2: d 5.15, mp 1850C, hardness 2010 (Knoop), resists oxidation up to 1100C. Cr_3B_2: may be crystals, d 6.1, Mohs hardness 9+.
Use: Metallurgical additives, high-temperature electrical conductors, cermets, refractories, coatings resistant to attack by molten metals.

chromium bromide. See chromous bromide.

chromium carbide. Cr_3C_2.
Properties: Ortho-rhombic crystals, d 6.65, microhardness 2700 kg/sq mm (load 50 g), mp 1890C,

bp 3800C, resistivity 95 μ-ohm cm (room temperature). Highest oxidation resistance at high temperature of all metal carbides, also resistant to acids and alkalies.
Use: Gage blocks and hot extrusion dies, in powder form as spray coating material, components for pumps and valves.

chromium carbonate.   See chromous carbonate.

chromium carbonyl.   See chromium hexacarbonyl.

chromium chloride.   See chromic chloride and chromous chloride.

chromium-copper.   A copper-chromium alloy containing 8–11% chromium. Used in the manufacture of hard steels for increased elasticity.

chromium dioxide. CrO_2.
Properties: Black, semiconducting material; d 4.9. Acicular crystals having strong magnetic properties. Also in powder form.
Derivation: By heating chromic acid.
Use: Magnetic component in recording tapes, catalyst.

chromium fluoride.   See chromic fluoride.

chromium hexacarbonyl.   (chromium carbonyl).   CAS: 13007-92-6. $Cr(CO)_6$.
Properties: White, crystalline solid; d 1.77; mp 150–151C (inert atmosphere); 210C (explodes). Slightly soluble in iodoform and carbon tetrachloride; insoluble in water, alcohol, ether, or acetic acid. More stable than most metal carbonyls, decomposed by chlorine and fuming nitric acid but resists bromine, iodine, water, and cold concentrated nitric acid. Decomposes photochemically when solutions are exposed to light.
Grade: 98% pure.
Hazard: Explodes at 400F (204C). Toxic by inhalation and ingestion.
Use: Isomerization and polymerization catalyst, gasoline additive, intermediate.

chromium hydrate.   See chromic hydroxide.

chromium hydroxide.   See chromic hydroxide.

chromium manganese antimonide.   Brittle gray solid having magnetic properties when above a definite temperature. If the composition of the compound is intentionally changed the temperature required also changes.

chromium naphthenate.
Properties: Dark-green liquid or violet powder.

Derivation: By addition of chromium salts to solution of sodium naphthenate and recovery of the precipitate.
Grade: 6% chromium.
Hazard: Toxic by ingestion.
Use: Paints (anti-chalking agent).

chromium nitrate.   See chromic nitrate.

chromium oxide.   See chromic oxide.

chromium oxychloride.   See chromyl chloride.

chromium oxyfluoride.   See chromyl fluoride.

chromium phosphate.   See chromic phosphate.

chromium potassium sulfate.   (chrome alum; potassium chromium sulfate; chrome potash alum). CAS: 10141-00-1. $CrK(SO_4)_2 \cdot 12H_2O$.
Properties: Dark, violet-red crystals; efflorescent; d 1.813; mp 89C; loses 10HOH at 100C. Soluble in water.
Derivation: By reducing potassium dichromate in dilute sulfuric acid with sulfurous acid.
Hazard: Toxic by ingestion.
Use: Tanning (chrome-tan liquors), textile dye (mordant), photography (fixing bath), ceramics.

chromium sesquichloride.   See chromic chloride.

chromium sesquioxide.   See chromic oxide.

chromium steel.   See steel stainless, iron stainless.

chromium sulfate.   See chromic sulfate.

chromium trichloride.   See chromic chloride.

chromium trifluoride.   See chromic fluoride.

chromium trioxide.   See chromic acid.

chromogen.   See chromophore.

"Chromegene."[203]   TM for mordant dyestuffs used on wool and leather. Characterized by very good fastness to light, fulling, etc.

"Chromolan."[243]   TM for metallized acid dyes.

chromophore.   A chemical grouping which when present in an aromatic compound (the chromogen) gives color to the compound by causing a displacement of, or appearance of, absorbent bands in the visible spectrum. Dyes are sometimes classified on the basis of their chief chromophores, e.g., —NO, nitroso dyes; —NO$_2$, nitro dyes; —N=N—, azo dyes; etc.

chromosome. The heredity-bearing gene carrier of the living cell derived from chromatin and consisting largely of nucleoproteins (DNA) together with other protein components (histones). See also deoxyribonucleic acid, gene.

"Chromosorb."[247] TM for a series of screened calcined and flux-calcined diatomite aggregates. Available in non-acid washed, acid-washed, and acid-washed dimethyldichlorosilane treatments.
Grade: Chromosorb phosphorus, W, G for analytical use; Chromosorb A for preparative chromatography.
Use: Supports in gas or liquid chromatography.

chromotropic acid. (1,8-dihydroxynaphthalene-3,6-disulfonic acid).

HO$_3$S ⟨6⟩ ⟨3⟩ SO$_3$H
         ⟨8⟩   ⟨1⟩
        HO   OH

Properties: White needles, soluble in water, insoluble in alcohol and ether.
Derivation: Reaction of H-acid with 10% NaOH at 280C.
Use: Azo dye intermediate, analytical reagent.

chromous bromide. (chromium bromide). CrBr$_2$.
Properties: White crystals, changes to yellow on heating. Oxidizes in moist air but stable in dry air, d 4.356, mp 842C, soluble in water (blue color).
Hazard: Toxic by ingestion, irritant to skin and tissue. TLV: (as Cr) 0.5 mg/m^3 in air.

chromous carbonate. (chromium carbonate). CrCO$_3$.
Properties: Grayish-blue, amorphous mass; d 2.75. Soluble in mineral acids; slightly soluble in water containing carbon dioxide; insoluble in alcohol.
Hazard: Toxic by ingestion; irritant to skin and tissue. TLV: (as Cr) 0.5 mg/m^3 in air.

chromous chloride. (chromium chloride). CrCl$_2$.
Properties: White, deliquescent needles; d 2.878 (25C); mp 824C; active reducing agent; very soluble in water; insoluble in alcohol and ether.
Derivation: Reaction of the metal with anhydrous hydrogen chloride.
Hazard: A poison. TLV: (as Cr) 0.5 mg/m^3 in air.

Use: Reducing agent, catalyst, reagent, chromizing.

chromous fluoride. CrF$_2$.
Properties: Greenish, shiny crystals; mp 890C; d 3.80; slightly soluble in water; insoluble in alcohol; soluble in boiling hydrochloric acid.
Derivation: Reaction of chromous chloride with hydrofluoric acid.
Hazard: A poison. Strong irritant to eyes and skin.
Use: Catalyst for alkylation and hydrocarbon cracking.

chromous formate. Cr(HCOO)$_2$.
Properties: (monohydrate): Reddish, acicular crystals; soluble in water.
Derivation: Reaction of chromous chloride with sodium formate.
Use: Chromium electroplating solutions, catalyst for organic reactions.

chromous oxalate. CrC$_2$O$_4$·HOH.
Properties: Yellow, crystalline powder; d 2.468; soluble in water; active reducing agent.
Hazard: Toxic by ingestion; irritant to skin and tissue; a poison. TLV: (as Cr) 0.5 mg/m^3 in air.

chromous sulfate. CrSO$_4$ · 5HOH.
Properties: (pentahydrate): Blue crystals, soluble in water, insoluble in acetone, slightly soluble in alcohol. Solutions are subject to atmospheric oxidation.
Use: Oxygen scavenger, reducing agent, analytical reagent.

"Chromoxane."[203] TM for mordant dyestuffs. Used on wool, characterized by fairly good fastness to light, very good fastness to fulling, etc., and by relatively bright shade.

"Chromspun"[256]. TM for solution-dyed acetate fiber.

chromyl chloride. (chromium oxychloride; chlorochromic anhydride).   CAS: 14977-61-8. CrO$_2$Cl$_2$.
Properties: Mobile, dark-red liquid; bp 116C; fp —96.5C; d 1.911. Fumes in air, reacts vigorously with water to form chromic acid, chromic chloride, hydrochloric acid and chlorine. Miscible with carbon tetrachloride, tetrachloroethane, carbon disulfide. Nonconductor of electricity. Protect from light and moisture.
Derivation: By heating sodium dichromate and sodium chloride with sulfuric acid.
Hazard: TLV: 0.025 ppm. Corrosive to tissue. Strong oxidizing agent.
Use: Organic oxidations and chlorinations, solvent for chromic anhydride, chromium complexes and dyes, catalyst.

chromyl fluoride. (chromium oxyfluoride).
CrF_2O_2.
Properties: Black crystals, sublimes at 29.6C, polymerizes on exposure to light.
Hazard: Highly reactive, ignites hydrocarbon gases at high temperatures.
Use: Fluorination catalyst, dyeing polyolefins.

chrysamine G. $C_{26}H_{16}N_4O_6Na_2$.
Properties: Yellowish-brown powder. Very sparingly soluble in water.
Derivation: By coupling diazotized benzidine with salicylic acid.
Use: Yellow direct cotton dyestuff.

chrysanthemummonocarboxylic acid ethyl ester.
$(CH_3)_2CCH(COOC_2H_5)CHCH:C(CH_3)_2$.
Properties: Colorless to pale yellow with pleasant ester odor, d 0.924–0.927 (25/25C), insoluble in water, soluble in alcohols and ketones.
Hazard: Toxic by ingestion.
Use: Synthesis of perfumes, pharmaceuticals, insecticides, other organic chemicals.
See also cinerin I, pyrethrin I.

chrysazin. See 1,8-dihydroxyanthraquinone.

chrysene. (1,2-benzphenanthrene).
CAS: 218-01-9. $C_{18}H_{12}$ (a tetracyclic hydrocarbon).
Properties: Crystals, d 1.274 (20/4C), mp 254C, bp 448C, sublimes easily in a vacuum. Slightly soluble in alcohol, ether, glacial acetic acid; insoluble in water.
Derivation: Distillation of coal tar.
Hazard: Human carcinogen.
Use: Organic synthesis.

chrysoidine hydrochloride. (m-diaminoazobenzene hydrochloride).
$(C_6H_5NNC_6H_3(NH_2)_2 \cdot HCl$.
Properties: Red-brown powder or large black shiny crystals with a green luster. Soluble in alcohol and water giving orange-brown solutions, insoluble in ether, mp 117C.
Use: Orange dye for cotton and silk.

chrysolite. See olivine.

chrysophanic acid. (1,8-dihydroxy-3-methyl-anthraquinone). CAS: 481-74-3.
$C_{15}H_{10}O_4$.
Properties: Mw 254.25.

chrysotile. See asbestos.

chu. Abbreviation for centigrade heat unit. It is the amount of heat required to raise the temperature of one pound of water one centigrade degree from 15C to 16C. It is sometimes called a pcu (pound centigrade unit).

Chugaev reaction. Formation of olefins from alcohols without rearrangement through pyrolysis of the corresponding xanthates via cis-elimination.

chymosin. See rennin.

chymotrypsin. A proteolytic enzyme found in the intestine which catalyzes the hydrolysis of various proteins (especially casein) and protein digestion products to form polypeptides and amino acids.
Properties: White to yellowish-white, odorless crystals or amorphous powder. Soluble in water or saline solution.
Derivation: Crystallized extract of the pancreas gland of the ox.
Use: Biochemical research, medicine.

chymotrypsinogen. A crystalline enzyme occurring in the pancreas which gives rise to chymotrypsin.
Use: Biochemical research.

Ci. Abbreviation for curie.

Ciamician-Dennstedt rearrangement. Expansion of the pyrrole ring by heating with chloroform or other halogeno compounds in the alkaline solution. The intermediate dichlorocarbene, by addition to the pyrrole, forms an unstable dihalogenocyclopropane which rearranges to a 3-halogenopyridine.

ciaphos. See cyanophos.

"Cibacet."[443] TM for disperse dyes for polyamide and cellulose acetate fibers.

"Cibacron."[443] TM for reactive dyes for cellulosic fibers.

"Cibanone."[443] TM for anthraquinone vat dyes.

"Cibaphasol"[443]. TM for coacervate-forming colloidal product used in textile dyeing.

cigarette tar. The comparatively non-volatile residue from the burning of cigarette tobacco which appears in finely divided form in the smoke. Cigarette tar is known to contain various aromatic ring compounds (especially benzo[a]pyrene) found in coal tar which are carcinogens. Various forms of activated carbon are used in filters to try to adsorb the toxic components of the tars that find their way into the smoke.
See also smoke (5).

ciguatoxin. $C_{28}H_{52}NO_5Cl$. A complex toxic principle in bony fishes, has both fat- and water-soluble fractions. Ninhydrin test positive. It is a type of quaternary ammonium compound and one fraction is said to be an irreversible anticholinesterase. The pharmacology is unknown.

CIIT. Abbreviation for Chemical Industry Institute of Toxicology.

cincholepidine. See lepidine.

cinchona bark. The bark of one of several species of Cinchona trees native to South America and cultivated in Indonesia, Peru, Ecuador, and western Africa. The best-known types are calisaya, loxa, and succirubra. It is used primarily as the source of natural quinine, quinidine, chinchonidine, cinchonine, and related alkaloids formerly used as antimalarials.

cinder. See slag.

cinene. See dipentene.

cineol. See eucalyptol.

cinerin I. CAS: 8003-34-7. $C_{20}H_{28}O_3$.
One of the four primary active insecticidal principles of pyrethrum flowers. It is the 3-(2-butenyl)-4-methyl-2-oxo-3-cyclopenten-1-yl ester of chrysanthemummonocarboxylic acid.
Properties: A viscous liquid, quickly oxidized in air, bp 200C (0.1 mm). Insoluble in water, soluble in organic solvents, incompatible with alkalies.
Hazard: Toxic by ingestion.
Use: Household insecticide.
See also cinerin II, pyrethrin I and II.

cinerin II. CAS: 8003-34-7. $C_{21}H_{28}O_5$.
One of the four primary active insecticidal principles of pyrethrum flowers. It is the 3-(2-butenyl)-4-methyl-2-oxo-3-cyclopenten-1-yl ester of chrysanthemumdicarboxylic acid monomethyl ester.
Properties: A viscous liquid, quickly oxidized in air, bp 200C (0.1 mm). Insoluble in water, soluble in organic solvents.
Hazard: Toxic by ingestion.
Use: Household insecticide.
See also cinerin I, pyrethrin I and II.

cinnabar. (natural vermilion; liver ore).
HgS.   Natural mercuric sulfide occurring in veins near recent volcanic rocks and hot springs.
Properties: Red, scarlet, reddish-brown to blackish solid, streak scarlet; luster adamantine to dull earthy when impure; d 8.10; hardness 2.5. Soluble in aqua regia. Has greater optical rotation than any other substance (+325 degrees).

Occurrence: California, Nevada, Spain, Italy, Mexico, Yugoslavia.
Hazard: See mercuric sulfide.
Use: The only important ore of mercury.

cinnamaldehyde. See cinnamic aldehyde.

cinnamein. See benzyl cinnamate.

cinnamene. See styrene.

cinnamic acid. (β-phenylacrylic acid; 3-phenylpropenoic acid; cinnamylic acid).
CAS: 621-82-9. $C_6H_5CH:CHCOOH$.
Properties: White, crystalline scales; soluble in benzene, ether, acetone, glacial acetic acid, carbon disulfide, oils; insoluble in water. Congealing point 133C (min), bp 300C. Combustible.
Derivation: By heating benzaldehyde with sodium acetate in presence of a dehydrating agent (acetic anhydride) or by heating benzyl chloride with sodium acetate in an autoclave. Occurs naturally in tobacco and some balsams.
Grade: Technical, refined.
Use: Medicine (anthelmintic), perfumes, intermediate.

cinnamic alcohol. (cinnamyl alcohol; phenylallylic alcohol; 3-phenyl-1-propen-1-ol; styryl carbinol).   CAS: 104-54-1.
$C_6H_5CH:CHCH_2OH$.

Properties: White needles or crystals, hyacinth-like odor. Soluble in water, alcohol, glycerol and organic solvents. Congealing point 33C (min): (pure) as low as 24C (technical), bp 257C. Combustible.
Derivation: (a) From oil of cassia or oil of cinnamon. Occurs as an ester. (b) Reduction of cinnamic aldehyde.
Method of purification: Recrystallization.
Grade: Technical, FCC.
Use: Perfumery, particularly for lilac and other floral scents; flavoring agent; soaps; cosmetics.

cinnamic aldehyde. (cinnamaldehyde; 3-phenylpropenal; cinnamyl aldehyde).
CAS: 104-55-2. $C_6H_5CH:CHCHO$.
Properties: Yellowish oil, cinnamon odor, sweet taste. Thickens on exposure to air. Soluble in 5 volumes of 60% alcohol, very slightly soluble in water, d 1.048–1.052, refr index 1.618–1.623, fp −8C, bp 248C. Combustible.
Derivation: (a) From Ceylon and Chinese cinnamon oils; (b) by condensation of benzaldehyde and acetaldehyde.

Method of purification: Rectification.
Use: Flavors, perfumery.
See also cassia oil.

cinnamon oil.   See cassia oil.

cinnamoyl chloride.   (phenylacrylyl chloride).
$C_6H_5CH:CHCOCl$.
Properties: Yellow crystals, mp 35C, bp 170C (58 mm), d 1.1617 (45/4C), decomposes in water, soluble in hydrocarbons and esters.
Hazard: Skin irritant.
Use: Reagent for determination of water, chemical intermediate.

cinnamyl acetate. $C_6H_5CH:CHCH_2OOCCH_3$.
Properties: Colorless liquid having a floral-spicy odor. Soluble in 4 volumes of 70% alcohol. D 1.048–1.052, refr index 1.539–1.542. Combustible.
Use: Perfumery (fixative), flavoring.

cinnamyl alcohol.   See cinnamic alcohol.

cinnamyl aldehyde.   See cinnamic aldehyde.

cinnamyl cinnamate.   (styracin).
CAS: 122-69-0. $C_9H_7O_2C_9H_9$.
Properties: Rectangular, prismic crystals; mp 40C (min). Soluble in alcohol, ether, benzene.
Derivation: Esterification of cinnamic acid with cinnamic alcohol.
Use: Perfumery, flavoring.

cinnamylic acid.   See cinnamic acid.

CIPC.   See chloro-IPC.

"Circosol."[494]   TM for various grades of naphthenic oils which provide overall balance of nonstaining and processing properties for most polymers. Aid processing and breakdown of rubber or polymer.

cis-.   (Latin: "on this side".) A prefix used in designating geometrical isomers in which there is a double bond between two carbon atoms. This prevents free rotation and thus two different spatial arrangements of substituent groups or atoms are possible. When a given atom or radical is positioned on one side of the carbon axis or backbone, the isomer is called *cis-*; when it is in the opposite location relative to the carbon axis, the isomer is called *trans-* (Latin: on the other side). This is indicated in the following formulas:

cis   trans

This is often called *cis-trans* isomerism. These prefixes are disregarded in alphabetizing chemical names.

citiolone.   (N-acetylhomocysteinethiolactone).
CAS: 1195-16-0. $C_6H_9NO_2S$.
Properties: Acicular crystals, mp 112C, maximum UV absorption 238 nanometers.
Use: Fogging preventive in photography.

citraconic anhydride.   (methylmaleic anhydride).   CAS: 616-02-4. $C_5H_4O_3$.
Properties: Colorless liquid. Mp 7–8C, bp 213–214C, d 1.25 (15/4C). Soluble in ether.
Grade: Reagent.
Use: Reagent for alkalies, alcohols, and amines.

citral.   (geranial; geranialdehyde; 3,7-dimethyl-2,6-octadienal).   CAS: 5392-40-5.
$(CH_3)_2CCHC_2H_4C(CH_3)CHCHO$.   Commercial material is a mixture of alpha and beta isomers.
Properties: Mobile, pale yellow liquid; strong lemon odor; d 0.891–0.897 (15C); refr index 1.4860–1.4900 (20C); not optically active. Soluble in 5 volumes of 60% alcohol; soluble in all proportions of benzyl benzoate, diethyl phthalate, glycerol, propylene glycol, mineral oil, fixed oils, and 95% alcohol; insoluble in water. Combustible.
Derivation: Principal component of lemon grass oil and can be isolated by fractional distillation. Obtained synthetically by oxidation of geraniol, nerol or linalool by chromic acid.
Grade: Technical, pure, FCC.
Use: Perfumes, flavoring agent, intermediate for other fragrances, vitamin A synthesis.

citric acid.   (2-hydroxy-1,2,3-propanetricarboxylic acid).
$HOOCCH_2C(OH)(COOH)CH_2COOH \cdot H_2O$.
Properties: Colorless, translucent crystals or powder; odorless; strongly acid taste; hydrated form is efflorescent in dry air. D 1.542, mp 153C (anhydrous form), decomposes before boiling. Very soluble in water and alcohol, soluble in ether. Combustible.
Occurrence: In living cells, both animal and plant.
Derivation: By mold fermentation of carbohydrates, including deep fermentation, from lemon, lime, pineapple juice, molasses.
Grade: Both hydrous and anhydrous, technical, CP, USP, FCC.
Use: Preparation of citrates, flavoring extracts, confections, soft drinks, effervescent salts, acidifier, dispersing agent, medicines, acidulant and antioxidant in foods (for details see regulations of Meat Inspection Division of USDA), seques-

tering agent, water-conditioning agent and detergent builder, cleaning and polishing stainless steel and other metals, alkyd resins, mordant, removal of sulfur dioxide from smelter waste gases, abscission of citrus fruit in harvesting, cultured dairy products.
See TCA cycle.

citric acid cycle. See TCA cycle.

"Citroflex."[299] TM for a series of organic citrates.

"Citronel B' and C'."[188] TM for lemon oil substitutes for technical use.

citronellal. (3,7-dimethyl-6(or 7)-octenal).
CAS: 8000-29-1. Has both d- and l-isomers. $C_9H_{17}CHO$.
Derivation: From citronella oil, of which it is the main component; also from lemongrass oil.
Use: Soap perfumery, manufacture of hydroxycitronellal, insect repellent.

citronellal hydrate. See hydroxycitronellal.

citronella oil. CAS: 8000-29-1.
Properties: Light yellowish, essential oil with rather pungent, citrus-like odor. Soluble in 80% alcohol. D 0.887–0.906, refr index 1.468–1.483, solutions are levorotatory. Combustible.
Hazard: May not be used on food crops.
Use: Insect repellent, perfumes for soaps and disinfectants, manufacture of cironellal, geraniol, denaturant for alcohol.

citronellol. (3,7-dimethyl-6(or 7)-octen-1-ol).
CAS: 106-22-9.
$CH_2:C(CH_3)(CH_2)_3CH(CH_3)CH_2CH_2OH$.
The 7-octene form, which predominates.
Occurrence: Citronella, geranium, rose, savin, and other essential oils.
Use: Perfumery (floral odors, mainly rose types).
See also rhodinol.

citronellyl acetate. $C_{10}H_{10}OOCCH_3$.
CAS: 150-84-5.
Properties: Colorless liquid. Odor somewhat like that of bergamot oil. D 0.884–0.891, bp 119–121 (15 mm), optical rotation usually slightly dextro, up to +1 degree, refr index 1.450–1.452. Soluble in nine volumes of 70% alcohol. Combustible.
Derivation: Action of acetic anhydride on citronellol.
Use: Perfumery, flavoring.

citronellyl formate. $C_{10}H_{19}OOCH$.
CAS: 105-85-1.
Properties: Colorless liquid, floral odor. D 0.890–0.903 (25/25C), refr index 1.4430–1.4490 (20C).

One volume dissolves in three volumes 80% alcohol, soluble in most oils. Combustible.
Grade: FCC.
Use: Flavoring.

citron yellow. See zinc yellow.

citrovorum factor. See folinic acid.

citrulline. $NH_2CONH(CH_2)_3CHNH_2COOH$.
An arginine derivative. An amino acid found in watermelon juice in the l(+) form.
Properties: Crystals from methanol-water mixture, mp 222C, soluble in water, insoluble in methanol and ethanol.
Use: Biochemical research.

citrus peel oils. Edible oils expressed from the peel or rind of grapefruit, lemon, lime, orange, and tangerine.
Properties: Color, odor, and taste characteristic of source. D 0.84–0.89, refr index 1.473–1.478, optically active, unsaponifiable. Soluble in alcohol, vegetable oils, mineral oil; orange and lemon oils are soluble in glacial acetic acid. Combustible.
Constituents: Limonene, citral, and terpenes in varying percentages.
Use: Flavoring agents in desserts, soft drinks, ice cream, odorants in perfumery and cosmetics, furniture polish (lemon oil).
Note: Terpene-free grades are available at much higher concentration (from 15–30 times) the original oil. These grades have much lower optical rotation and less tendency to spoil on storage since terpenes tend to oxidize to undesirable components such as carvone and p-cymene.
See terpeneless oil.

citrus seed oils. Edible oils expressed from seeds of grapefruit, orange, lemon, lime, and tangerine reclaimed from cannery processing.
Properties: Nondrying; odor, color, and taste characteristic of source; bitter taste removed by alkali refining. D 0.91–0.92, saponification value 190–195, iodine number 100–110, optically inactive. Combustible. May be bleached, deodorized, or hydrogenated.
Constituents: Chiefly palmitic, oleic, and linoleic acids.
Use: Flavoring, food products, cosmetics, odorants in special soaps.

civet. (zibeth). Unctuous secretion from the civet cat used as a perfume fixative. Synthetic types are available.

Civex process. See reprocessing.

civettal. See 1,2,3,4-tetrahydro-6-methylquinoline.

Cl. Symbol for chlorine; the molecular formula is Cl_2.

"Cladkote."[41] TM for a modified polyester composite of resins and siliceous reinforcing material which cures to a tough, chemical-resistant topping, suggested for use in food plants under acid conditions.

cladding. The process in which two metals are bonded by being rolled together at suitable pressure and temperature. Controlled explosion is also used. At the interface, each metal diffuses sufficiently into the other to form an alloy. Cladding is generally from 5–20% of total thickness, but may be heavier depending on the properties desired. The base metal is usually carbon or low-alloy steel clad with stainless steel, nickel or other protective metal. Nonferrous metals are also clad; copper with cupronickel is used for coinage.

Claisen condensation. A condensation reaction discovered in 1890 in which aldehydes react with esters of the type RCH_2COOR' in the presence of metallic sodium or sodium ethylate to form unsaturated esters $R''CH{=}C(R)COOR'$. The mechanism of the reaction parallels that of acetoacetic ester condensation. An addition compound of the ester and sodium ethylate is first formed, which is transformed into an unsaturated sodio ether; an addition compound between this ether and the aldehyde forms next; this changes to the unsaturated ester through a molecular rearrangement and cleavage of sodium hydroxide. Amides of alkali metals are excellent catalysts for the reaction,but they tend to form some amide byproducts. Union with the carbonyl carbon takes place at the alpha-carbon in the acid.

Claisen rearrangement. Thermal rearrangement of allyl ethers of enols or phenols to gamma, delta-unsaturated ketones or omicron-allylphenols, respectively, by a 3,3-sigmatropic shift.

Claisen-Schmidt condensation. Condensation of an aromatic aldehyde with an aliphatic aldehyde or ketone in the presence of a relatively strong base (hydroxide or alkoxide ion) to form an alpha-,beta-unsaturated aldehyde or ketone.

"Clalite."[159] TM for a buffered sodium hydrosulfite used to improve whiteness of coating clay in paper industry.

Clapeyron equation. The equation $dp/dT = \Delta H/T\Delta V$. It states that the change of vapor pressure of a liquid (expressed in ergs) with absolute temperature (°K) equals the heat of vaporization (in calories) per gram divided by the product of the absolute temperature and the increase in volume (V, in cubic centimeters) when a specified quantity (a gram) of the liquid changes to vapor. Other consistent units may be used. The approximate Clapeyron equation $d \ln p/dT = \Delta H/RT^2$, expresses the same relation in a less exact form because in its derivation, it is assumed that ΔV is equal to the volume of vapor. This assumption (that the volume of liquid is negligibly small) is usually true within a few percent under ordinary conditions of temperature and pressure.

"Clarase."[212] TM for a fungal alpha-amylase with strong liquefying and dextrinogenic activity together with considerable saccharogenic action. Use: Fruit juices and pectin, chocolate syrup, brewing.

clarification. Removal of bulk water from a dilute suspension of solids by gravity sedimentation, aided by chemical flocculating agents. Large circular tanks equipped with revolving plows or rakes are used for this purpose. Important applications of clarification are for separation of wood pulp waste from paper mill effluents and excess water from activated sludge.
See also dewatering.

Clark-Lubs indicators. Phenol red, cresol red, bromophenol blue, bromocresol purple, thymol blue, bromthymol blue, and methyl red.

clary sage oil. See sage oil.

classification. In chemical engineering parlance, the mechanical process of separating subdivided solids (crushed stone, cement, mineral aggregate, and the like) into two (or more) "classes" each containing a specific size range. The latter may vary from an inch or more in diameter to powders of considerable fineness. Customary classifying equipment includes so-called grizzlies, perforated metal and vibrating screens, sifters, sieves, and similar devices. A magnetic separator is often used in conjunction with the screen to remove tramp metal. A similar method is used in the food industry for size grading of certain fruits and vegetables, e.g., peas, berries, etc.

clathrate compound. An inclusion complex in which molecules of one substance are completely enclosed within the other, as argon within hydroquinone crystals. Urea adducts are inclusion complexes of the channel or canal type. In these, the complexing urea crystals wrap around the molecule of the other substance, usually a straight-chain unbranched aliphatic hydrocarbon. Similar complexes are formed with thiourea. See also inclusion complex.

Claude system. A process for the production of liquid air in which the compressed gas is made to perform work in an expansion engine and thus cool itself.

clay. A hydrated aluminum silicate. Generalized formula $Al_2O_3SiO_2 \cdot xHOH$. Component of soils in varying percentages.
Properties: Fine, irregularly-shaped crystals ranging from 150 microns to less than 1 micron (colloidal); color reddish-brown to pale buff, depending on iron oxide content; d approximately 2.50; insoluble in water and organic solvents; odorless; absorbs water to form a plastic, moldable mass and in some cases forms a thixotropic gel (bentonite); refractory material; strong ion-exchange capability; important in soil chemistry and construction engineering.
Derivation: Weathering of rocks.
Occurrence: Southeastern US, Wyoming, Texas, Canada, England, France, USSR.
Types: Kaolinite, montmorillonite, atapulgite, illite, bentonite, halloysite.
Grade: Natural, refined, air-floated.
Hazard: Dusts may be irritant to nose and throat. Suspensions of dust are a fire hazard.
Use: Ceramic products, refractories, colloidal suspensions, oil-well drilling fluids, filler for rubber and plastic products, films, paper coating, decolorizing oils, temporary molds, filtration, carrier in insecticidal sprays, catalyst support.
See also fullers earth, bentonite, ceramic, refractory, kaolin, slip clay, polyorganosilicate graft polymer.

"Clearate."[48] TM for a high grade soya lecithin. Used in foods, inks, cosmetics, and paints.

cleave. (1) Of a crystal, to break or separate along definite planes defined by the crystalline structure. It may cleave in one direction, as in mica, or in several. (2) Of an alkene molecule, to divide into two compounds (aldehydes or ketones) at the double bond. This is usually done by ozone followed by hydrolysis in the presence of powdered zinc.

Cleland's reagent. See dithiothreitol.

Clemmensen reaction. The Clemmensen method of reduction (1913) consists in refluxing a ketone with amalgamated zinc and hydrochloric acid. Acetophenone, for example, is reduced to ethylbenzene. The method is applicable to the reduction of most aromatic-aliphatic ketones to at least some aliphatic and alicyclic ketones and to the

γ-keto acids obtainable by Friedel-Crafts condensations with succinic anhydride (succinolylation) and to the cyclic ketones formed by intramolecular condensation.

Cleveland Open Cup. See COC.

Cleve's acid. (1-naphthylamine-6-sulfonic acid). Properties: Colorless needles, slightly soluble in water, mp greater than 330C.
Derivation: Nitration of naphthalene-β-sulfonic acid. On reduction with iron, this yields a mixture of 1-naphthylamine-6-sulfonic acid (Cleve's acid) and 1-naphthylamine-7-sulfonic acid (or Cleve's acid-1,7). The latter is separated by crystallization as the sodium salt, the 1,6-acid precipitates on acidification.
Use: Azo dye intermediate.

Cleve's acid-1,7 (1-naphthylamine-7-sulfonic acid). See Cleve's acid.

cliffstone Paris white. A special grade of whiting made from a hard grade of English chalk.

clinical chemistry. A subdivision of chemistry that deals with the behavior and composition of all types of body fluids including the blood, urine, perspiration, glandular secretions, etc. It involves analysis and testing of these for content of numerous metabolic constituents as well as foreign materials; thus it also includes toxicological factors.

clinoptilolite. A natural, inorganic zeolite used as a selective ion-exchange medium for removal of ammonia from plant waste water.

clomiphene. See 1-chloro-2-(4-diethylamino-ethoxy)phenyl)-1,2-diphenylethylene.

clopidol. (3,5-dichloro-2,6-dimethyl-4-pyridinol; Coyden). CAS: 2971-90-6. $C_7H_7Cl_2NO$. A penta-substituted pyridine derivative.
Properties: A solid with mw 192.06, mp greater than 320C. Insoluble in water.
Hazard: Toxic. TLV: 10 mg/m³.

"Clorafin."[266] TM for a series of chlorinated, nonflammable paraffins. Used as plasticizers and lubricant additives.

"Cloran."[505] TM for a reactive difunctional chlorinated carboxylic acid anhydride.
Use: Reactive intermediate to produce flame retardant polymers.

cloud point. In petroleum technology, the temperature at which a waxy solid material appears as a diesel fuel is cooled. This material is harmful to engine performance.

cloud seeding.   See nucleation.

clove oil.   (caryophyllus oil).   An essential oil distilled from cloves. Optically active.
Use: Medicine (local), flavoring, dentistry, perfumery, confectionery, soaps.

clupanodonic acid. $C_{21}H_{35}COOH$.   Derived from herring oil.

clupeine.   A protamine (simple protein) from herring. Contains no sulfur.
Properties: Water soluble.

cluster catalysis.   See under catalysis.

Cm.   Symbol for curium.

"CM."[28]   TM for a flame-retardant composition based on ammonium sulfamate and modified to prevent afterglow and to improve penetration.
Properties: Fine, white granules; soluble in water; insoluble in dry cleaning solvents.
Use: Treatment of fabrics, paper, paper products, and other cellulosic materials.

CMA.   Abbreviation for Chemical Manufacturers Association.

CMC.   See carboxymethylcellulose.

CM-cellulose.   Abbreviation for carboxymethylcellulose, used especially by biochemists.

CMHEC.   Abbreviation for carboxymethyl hydroxyethyl cellulose.

CMP.   Abbreviation for cytidine monophosphate.
See cytidylic acid.

CMPP.   Abbreviation for 2-(4-chloro-2-methylphenoxyl) propionic acid.
See mecoprop.

CMRA.   Abbreviation for Chemical Marketing Research Association.

CMU.   Abbreviation for chlorophenyldimethylurea.
See monuron.

CNS.   Abbreviation for central nervous system (as applied to the action of certain drugs).

Co.   Symbol for cobalt.

coacervation.   An important equilibrium state of colloidal or macromolecular systems. It may be defined as the partial miscibility of two or more optically isotropic liquids, at least one of which is in the colloidal state. For example, gum arabic shows the phenomenon of coacervation when mixed with gelatin.

coagulant.   A substance that induces coagulation. Coagulants are used in precipitating solids or semisolids from solution, as casein from milk, rubber particles from latex, or impurities from water. Compounds that dissociate into strongly charged ions are normally used for this purpose. Blood contains the natural coagulant thrombin.

coagulation.   Irreversible combination or aggregation of semisolid particles, such as fats or proteins, to form a clot or mass. This can often be brought about by addition of appropriate electrolytes, for example, by the addition of an acid to milk or of aluminum sulfate to turbid water. Mechanical agitation and removal of stabilizing ions, as in dialysis, also cause coagulation. The clotting of blood by thrombin, of rubber particles in latex by acetic acid, and of egg-white by heat are additional instances.
See also agglomeration (1); flocculation.

Coahran process.   Recovery of acetic acid from pyroligneous acid by extracting with ether. It is an improved version of the Brewster process, but is basically the same.

coal.   A natural, solid, combustible material formed from prehistoric plant life, it occurs in layers or veins in sedimentary rocks. It is far more plentiful in the US than petroleum, and is an important source of heat and energy. It occurs chiefly in West Virginia and Kentucky, as well as in Wyoming and other western states. Much of it is too high in sulfur content to meet desirable pollution standards unless sulfur is removed by scrubbing.

Chemically, coal is a macromolecular network comprised of groups of polynuclear aromatic rings, to which are attached subordinate rings connected by oxygen, sulfur, and aliphatic bridges. This extended open structure is conducive to catalytic reactions, which in effect subdivide it into smaller molecules that can be defined readily.

Coal is an important source of chemical raw materials: pyrolysis (destructive distillation) yields coal tar and hydrocarbon gases which can be upgraded by hydrogenation or methanation to synthetic crude oil and fuel gas respectively. Catalytic hydrogenation yields hydrocarbon oils and gasoline. Gasification produces carbon monoxide and hydrogen (synthesis gas), from which ammonia and other products can be made.

Numerous processes for adapting these reactions to large-scale production of fuel oil and gasoline have been developed in the US, but none has yet proved economically successful. The process is being used abroad (Lurgi and Sasol methods). See also gasification, hydrogenolysis, hydrosolvation, peat, lignite.

coal, conversion.   See gasification.

coal gas.   (bench gas; coke-oven gas).   A mixture of gases produced by the destructive distillation of bituminous coal in highly heated fireclay or silica retorts or in byproduct coke ovens.
Hazard: Flammable.
Use: Directly on open hearth furnaces.

coal, gasification.   See gasification.

coal, hydrogenation.   See hydrogenolysis, gasification.

coal oil.   The crude oil obtained by the destructive distillation of bituminous coal, or the distillate obtained from this oil.
Hazard: Flammable, moderate fire risk.

coal tar.   (coal tar pitch volatiles).
CAS: 8007-45-2.   Coal tar may be hydrogenated under pressure to form a petroleum-like fuel suitable for residual use. Coal tar fractions obtained by distillation and the chemicals found in each are as follows: (1) light oil (up to 200C), benzene, toluene, xylenes, cumenes, coumarone, indene; (2) middle oil (200–250C); and (3) heavy oil (250–300C): naphthalene, acenaphthene, methylnaphthalenes, fluorene, phenol, cresols, pyridine, picolines; (4) anthracene oil (300–350C): phenanthrene, anthracene, carbazole, quinolines; and (5) pitch. Typical yields: 5% light oil, 17% middle oil, 7% heavy oil, 9% anthracene oil, 62% pitch. Treatment with alkalies, acids, and solvents is necessary to separate the individual chemicals.
Properties: Black, viscous liquid (or semi-solid); naphthalene-like odor; sharp, burning taste; obtained by destructive distillation of bituminous coal as in coke ovens, One ton of coal yields 8.8 gal of coal tar. Combustible. D 1.18–1.23. Soluble in ether, benzene, carbon disulfide, chloroform; partially soluble in alcohol, acetone, methanol; only slightly soluble in water.
Grade: Crude, refined, USP.
Hazard: A human carcinogen. Toxic by inhalation. TLV: (volatiles) 0.2 mg/m^3 in air.
Use: Raw material for plastics, solvents, dyes, drugs, and other organic chemicals. The crude or refined product or fractions thereof are also used for waterproofing, paints, pipe coating, roads, roofing, insulation, as pesticides and sealants.

coal tar creosote.   Low-cost, water-insoluble, low volatility wood preservative.

coal-tar distillate.   The lighter fractions of coal tar. The terms coal tar light oil, coal tar naphta, and coal tar oil are loosely defined and are sometimes regarded as synonymous.
Hazard: Flammable, dangerous fire risk. Toxic by inhalation and skin absorption.
See also under naphtha.

coal-tar dye.   A dye produced from the coal-tar hydrocarbons or their derivatives such as benzene, toluene, xylene, naphthalene, anthracene, aniline, etc.
See also dye, synthetic.

coal-tar light oil.   See coal-tar distillate, light oil.

coal-tar naphtha.   See coal-tar distillate, naphtha, solvent.

coal-tar oil.   See coal-tar distillate.

coal-tar pitch.   Dark brown to black amorphous residue left after coal tar is redistilled. It is composed almost entirely of polynuclear aromatic compounds and constitutes 48–65% of the usual grades of coal tar. Different grades have different softening points: roofing pitch softens at 65C, electrode pitch at 110–115C. Combustible.
Hazard: Volatile components (anthracene, phenanthrene, acridine) are carcinogens. TLV: 0.2 mg/m^3 in air.
Use: Binder for carbon electrodes, base for paints and coatings. Impregnation of fiber pipe for electrical conduits and drainage, foundry core compounds, briquetting coal, tar-bonded refractory brick, paving and roofing, plasticizers for elastomers and polymers, extenders, saturants, impregnants, sealants.

coating.   A film or thin layer applied to a base material called a substrate. The coatings most commonly used in industry are metals, alloys, resin solutions, and solid/liquid suspensions on various substrates (metals, plastics, wood, paper, leather, etc.). They may be applied by electrolysis, vapor deposition, vacuum, or mechanical means such as brushing, spraying, calendering, and roller coating. Products such as cables and power cord are coated by extrusion. Such thermosetting resins as acrylics, epoxies, and polyesters appropriately compounded are available in powder form and are applied to auto bodies, machinery, and other industrial products by electrostatic spraying techniques. This process is claimed to minimize the pollution problems and waste encountered with solvent-based sprayed coatings.

See also protective coating, film, paint, vacuum deposition.

cobalamin. Imprecise name for vitamin B_{12}, it refers to the molecule cyanocobalamin without the cyano- group (CN).

See cyanocobalamin.

cobalt. CAS: 7440-48-4. Co. Metallic element of atomic number 27, group VIII of the Periodic Table. Aw 58.9332. Valences 2,3; no stable isotopes. There are several artificial radioactive isotopes, the most important being Co-60. Coordination number: 4 (divalent), 6 (trivalent).

Properties: Steel-gray, shining, hard, ductile, somewhat malleable metal; ferromagnetic, with permeability two-thirds that of iron, has exceptional magnetic properties in alloys. D 8.9, mp 1493C, bp 3100C. Attacked by dilute hydrochloric and sulfuric acids, soluble in nitric acid. Corrodes readily in air. Hardness: cast 124 Brinell, electrodeposited 300 Brinell. An important trace element in soils and necessary for animal nutrition. Cobalt has unusual coordinating properties, especially the trivalent ion. Noncombustible except as powder.

Occurrence: Principal ores are smaltite, cobaltite, chloanthite, linnaeite (Canada, Zaire, Zambia).

Derivation: From ore concentration by roasting followed (a) by thermal reduction by aluminum, (b) by electrolytic reduction of solutions of metal, or (c) by leaching with either ammonia or acid in an autoclave under elevated temperatures and pressures and subsequent reduction by hydrogen.

Forms available: Rondels (1 by 3/4 inch), shot, anodes, 150 and finer mesh powder, to 99.6% purity, ductile strips (95% cobalt, 5% iron) and high purity strips, 99.9% pure, single crystals.

Hazard: Dust is flammable; toxic by inhalation. TLV: 0.05 mg/m³ of air.

Use: Chemical (cobalt salts, oxidizing agent), electroplating ceramics, lamp filaments, catalyst, (sulfur removal from petroleum, Oxo process, organic synthesis), trace element in fertilizers, glass, drier in printing inks, paints and varnishes, colors, cermets. Principal use in alloys, especially cobalt steels for permanent and soft magnets and cobalt-chromium high-speed tool steels, cemented carbides, jet engines, coordination and complexing agent

See cobaltammine.

cobalt-57. Radioactive cobalt of mass number 57. Properties: Half-life 267 days, radiation: gamma, K x-ray.

Derivation: Bombardment of a nickel target with protons.

Forms available: Cobaltous chloride in hydrochloric acid solution, cyanocobalamin cobalt-57 solution (USP).

Hazard: Radioactive poison.

Use: Biological research.

cobalt-58. Radioactive cobalt of mass number 58. Properties: Half-life 72 days, radiation: positron, gamma, K x-ray.

Forms available: Cobaltous chloride in hydrochloric acid solution.

Hazard: Radioactive poison.

Use: Biological and medical research.

cobalt-60. Radioactive cobalt of mass number 60. One of the most common radioisotopes. Properties: Half-life 5.3 years, radiation: beta and gamma. Radiocobalt is available in larger quantities and is cheaper than radium.

Derivation: Irradiation of cobalt oxide, Co_2O_3, or of cobalt metal.

Forms available: Encapsulated pellets or wire needles, cobaltous chloride in hydrochloric acid solution, solid cobaltic oxides, labeled compounds such as cyanocobalamin (USP).

Hazard: Radioactive poison.

Use: Radiation therapy (cancer), radiographic testing of welds and castings, as a source of ions in gas-discharge devices, as the radiation source in liquid-level gauges, for locating buried telephone and electrical conduits, portable radiation units, gamma irradiation for wheat and potatoes, as a research aid in studying the permeability of porous media to flow of oil, wearing quality of floor wax, oil consumption in internal combustion engines, wool dyeing, etc.

cobalt acetate. See cobaltous acetate.

cobaltammine. A coordination or complexing compound containing the group $[Co(NH_3)_6]_3{}^+$ or its derivatives in which some of the ammonia has been replaced by other groups or ions. The names hexammine, pentammine, etc., are used to indicate the number of ammonia groups present in any case. Prepared by adding excess ammonia to a cobaltous salt, exposing to air so that oxygen is absorbed, and boiling to oxidize the cobalt.

These compounds show none of the ordinary properties of cobalt. Different types of salts with various acid radicals are known. The ammonia in the ammines may be replaced, molecule for molecule, by other nitrogen compounds such as hydroxylamine or ethylene diamine, by water, by ions such as hydroxyl, chloride, nitrate, etc., or by groups such as nitro (NO_2).

See coordination compound.

cobalt ammonium sulfate. See cobaltous ammonium sulfate.

cobalt arsenate. See cobaltous arsenate.

cobalt black. See cobaltic oxide.

cobalt bloom. See erythrite (2).

cobalt blue. (Thenard's blue; cobalt ultramarine; azure blue).
  Properties: Blue to green pigment of variable composition, consisting essentially of mixtures of cobalt oxide and alumina, approximating cobaltous aluminate, $Co(AlO_2)_2$. Cobalt blue is said to be the most durable of all blue pigments, being resistant to both weathering and chemicals.
  Derivation: By heating alumina with (a) cobaltous oxide, or a material yielding this oxide on calcination; (b) cobalt phosphate; (c) cobalt arsenate. Greenish shades may be made by incorporating zinc oxide.
  Grade: Technical (called genuine to distinguish it from the imitation, which is ultramarine blue).
  Use: Pigments in oil or water, cosmetics (eyeshadows, grease paints).

cobalt bromide. See cobaltous bromide.

cobalt carbonate. See cobaltous carbonate, basic.

cobalt chloride. See cobaltous chloride.

cobalt chromate. See cobaltous chromate.

cobalt chromate, basic. See cobaltous chromate.

cobalt difluoride. See cobaltous fluoride.

cobalt-2-ethylhexoate. (cobalt octoate).
  $C_4H_9CH(C_2H_5)COOH$.   Probably the cobaltous salt of 2-ethylhexoic acid.
  Properties: Blue liquid, d 1.013 (25C).
  Use: Paint drier, whitener, catalyst.

cobalt-gold alloy. Made by a vapor deposition technique into magnetic films. Compositions range from 25–60% gold.

cobalt hydrate. See cobaltic hydroxide, cobaltous hydroxide.

cobalt hydroxide. See cobaltic hydroxide, cobaltous hydroxide.

cobaltic acetylacetonate.
  $Co[CH_3COCHC(CH_3)O]_3$.
  Properties: Dark green or black crystals, d 1.43, mp 241C.

Derivation: Reaction of cobaltous carbonate with acetylacetone and peroxide.
  Use: Vapor plating of cobalt.

cobaltic boride. CoB.
  Properties: Crystal prisms, d 7.25 (18C), mp greater than 1400C. Decomposes in water, soluble in nitric acid.
  Use: Ceramics.

cobaltic fluoride. See cobalt trifluoride.

cobaltic hydroxide. (cobalt hydroxide; cobalt hydrate).   CAS: 1307-86-4. $Co(OH)_3$, actually considered to be $Co_2O_3 \cdot 3H_2O$.
  Properties: Dark-brown powder, soluble in cold concentrated acids, insoluble in water and alcohol. D 4.46, loses water at 100C.
  Derivation: Addition of sodium hydroxide to a solution of cobaltic salt, action of chlorine on a suspension of cobaltous hydroxide, action of sodium hypochlorite on a cobaltous salt.
  Use: Cobalt salts, catalyst.

cobaltic oxide. (cobalt oxide; cobalt black).
  CAS: 1308-04-9. Co_2O_3.   Sometimes incorrectly called cobalt peroxide.
  Properties: Steel-gray or black powder. Soluble in concentrated acids, insoluble in water, d 4.81–5.60, mp decomposes at 895C.
  Derivation: By heating cobalt compounds at low temperature with excess of air.
  Use: Pigment, coloring enamels, glazing pottery.

cobaltic potassium nitrite. (cobalt yellow).
  $CoK_3N_6O_{12}$ or $CoK_3(NO_2)_6$.
  Properties: Yellow, crystalline solid; insoluble in alcohol; slightly soluble in water and acetic acid; decomposed by sulfuric acid.
  Derivation: By adding potassium nitrite to a cobalt salt solution.
  Use: Colorant for rubber, glass and ceramic products, separation of cobalt from nickel, analytical chemistry.

cobalt iodide. See cobaltous iodide.

cobaltite. (cobalt glance).   CoAsS.   Silver-white to gray mineral, metallic luster. Contains 35.5% cobalt, d 6–6.3, Mohs hardness 5.5.
  Occurrence: Canada, Zaire, Sweden.
  Hazard: Toxic dust or powder when inhaled.
  Use: An important cobalt ore, ceramics.

cobalt linoleate. See cobaltous linoleate.

cobalt molybdate. CAS: 13762-14-6.
  $CoMoO_4$.   A molybdenum catalyst (a gray-

green powder) used in petroleum technology, in reforming and desulfurization.

cobalt monoxide. See cobaltous oxide.

cobalt naphthenate. See cobaltous naphthenate.

cobalt neodecanoate. Deep blue paste containing 12% cobalt. Used as a drying additive for printing inks.

cobalt nitrate. See cobaltous nitrate.

cobaltocene. (dicyclopentadienylcobalt).
CAS: 1277-43-6. $(C_5H_5)_2Co$.   A metallocene.
Properties: Purple crystals; mp 172–173C; soluble in hydrocarbons; highly reactive compound which is readily oxidized by air, water, and dilute acids.
Hazard: Toxic by ingestion.
Use: Polymerization inhibitor of olefins up to 200C, Diels-Alder reaction, catalyst, paint drier, oxygen stripping agent.

cobalto-cobaltic oxide. (tricobalt tetraoxide).
Co_3O_4.
Properties: Steel-gray to black in anhydrous form. Insoluble in water, hydrochloric and nitric acids, soluble in sulfuric acid and fused sodium hydroxide, hygroscopic. D 6.07, changes to cobaltous oxide at 900–950C.
Derivation: By heating strongly other cobalt oxides in air. Thus, the commercial oxides contain a substantial quantity of Co_3O_4.
Use: Ceramics, pigments, catalyst, preparation of cobalt metal, semiconductors.

cobalt octoate. See cobalt-2-ethylhexoate. See also soap (2).

cobalt oleate. See cobaltous oleate.

cobaltous acetate. (cobalt acetate).
CAS: 71-48-7. $Co(C_2H_3O_2)_2 \cdot 4HOH$.
Properties: Reddish-violet, deliquescent crystals. Soluble in water, acids, and alcohol; d 1.7043, mp $-H_2O$ at 140C.
Derivation: Action of acetic acid on cobalt carbonate with subsequent crystallization.
Grade: Technical, pure crystals, CP.
Hazard: May not be used in food products (FDA).
Use: Sympathetic inks, paint and varnish driers, catalyst, anodizing, mineral supplement in feed additives, foam stabilizer.

cobaltous aluminate. See cobalt blue.

cobaltous ammonium phosphate. (ammonium cobaltous phosphate; cobalt violet).
CoH_4NO_4P.
Properties: Violet to dark red crystals, insoluble in water, soluble in common acids.

Use: Plant nutrient, colorant in glass, glazes, enamels, analytical chemistry.

cobaltous ammonium sulfate. (cobalt ammonium sulfate). $CoSO_4 \cdot (NH_4)_2SO_4 \cdot 6H_2O$.
Properties: Ruby-red crystals, soluble in water, insoluble in alcohol, d 1.902.
Derivation: Crystals of cobaltous sulfate with ammonium sulfate.
Use: Ceramics, cobalt plating, catalyst.

cobaltous arsenate. (cobalt arsenate).
$Co_3(AsO_4)_2 \cdot 8H_2O$.
Properties: Violet-red powder. Soluble in acids, insoluble in water. D 3.178 (15C).
Derivation: Interaction of solutions of sodium arsenate and of a cobalt salt.
Grade: Technical.
Use: Painting on glass and porcelain in light blue colors, coloring glass.   See also erythrite.

cobaltous bromide. (cobalt bromide).
CAS: 7789-43-7. $CoBr_2 \cdot 6H_2O$.
Properties: Red-violet crystals. Soluble in water, alcohol, and ether. Anhydrous crystals, bright green. D 2.46, mp 47–48C, loses $6H_2O$ at 130C.
Derivation: By the action of bromine on cobalt or of hydrobromic acid on cobaltous hydroxide or carbonate followed by crystallization.
Grade: Technical, CP.
Use: In hygrometers, catalyst.

cobaltous carbonate. $CoCO_3$.
Properties: Red crystals, insoluble in water and ammonia, soluble in acids, d 4.13, mp decomposes. The cobalt carbonate of commerce is usually the basic salt (see following entry).
Derivation: By heating cobaltous sulfate with a solution of sodium bicarbonate.
Use: Ceramics, trace element added to soils and animal feed, temperature indicator, catalyst, pigments.

cobaltous carbonate, basic.
$2CoCO_3 \cdot 3Co(OH)_2 \cdot H_2O$.   The cobalt carbonate of commerce.
Properties: Red-violet crystals, soluble in acids, insoluble in cold water, decomposes in hot water, mp decomposes.
Derivation: By adding sodium carbonate to a solution of cobaltous acetate followed by filtration and drying.
Use: Manufacturing cobaltous oxide, cobalt pigments, cobalt salts, intermediate.

cobaltous chloride. (cobalt chloride).
CAS: 7646-79-9.   (a) $CoCl_2$, (b) $CoCl_2 \cdot 6H_2O$.
Properties: (a) Blue, (b) ruby-red crystals. Soluble in water and alcohol, also soluble in acetone. D (a) 3.348, (b) 1.924; mp (a) sublimes, (b) 86.75C.

Derivation: By the action of hydrochloric acid on cobalt, its oxide, hydroxide, or carbonate. Concentration gives (b) and dehydration (a).

Hazard: May not be used in food products (FDA). Can cause blood damage.

Use: Absorbent for ammonia, gas-masks, electroplating, sympathetic inks, hygrometers, manufacture of vitamin B_{12}, flux for magnesium refining, solid lubricant, dye mordant, catalyst, barometers, laboratory reagent, fertilizer additive.

cobaltous chromate.   (basic cobalt chromate; cobalt chromate). $CoCrO_4$.

Properties: Brown or yellowish-brown powder. Variable composition. (Pure cobaltous chromate is $CoCrO_4$, gray-black crystals.) Soluble in mineral acids, in solutions of chromium trioxide; insoluble in water.

Hazard: A carcinogen.

Use: Ceramics (tinting).

cobaltous citrate. $Co_3(C_6H_5O_7)_2 \cdot 2H_2O$.

Properties: Rose-red, amorphous powder; mp 150C (loses 2HOH). Slightly soluble in water, soluble in dilute acids.

Use: Vitamin preparations, therapeutic agents.

cobaltous cyanide.   (a) $Co(CN)_2 \cdot 2H_2O$, (b) $Co(CN)_2$.

Properties: (a) Buff crystals, (b) blue-violet powder; d (b) 1.872; mp (b) 280C. Insoluble in water, soluble in potassium cyanide, hydrochloric acid, ammonium hydroxide.

Hazard: Highly toxic. TLV: (as CN) 5 mg/m³ of air.

cobaltous ferrite. $CoOFe_2O_3$.   A constituent of magnetically soft ferrites which have high permeability, low coercive force, low magnetic saturation and high resistivity.

See also ferrite.

cobaltous fluoride.   (cobalt difluoride).

CAS: 10026-17-2. CoF_2.

Properties: Rose-red crystals or powder, d 4.46, mp approximately 1200C, bp 1400C. Soluble in cold water, hydrofluoric acid. Decomposes in hot water. Ammine complexes can be prepared from the hydrate.

Hazard: Highly toxic. TLV: (as F) 2.5 mg/m³ of air.

Use: Catalyst.

cobaltous formate. $Co(CHO_2)_2 \cdot 2H_2O$.

CAS: 544-18-3.

Properties: Red crystals, d 2.129 (22C), mp 140C (loses 2HOH), decomposes 175C. Soluble in cold water, insoluble in alcohol.

Use: Catalyst.

cobaltous hydroxide.   (cobalt hydroxide; cobalt hydrate). $Co(OH)_2$.

Properties: Rose-red powder. Soluble in acids and ammonium salt solutions, insoluble in water and alkalies, soluble in ammonia and acids. D 3.597.

Derivation: Addition of sodium hydroxide to a solution of cobaltous salt.

Use: Cobalt salts, paint and varnish driers, catalyst, storage battery electrodes.

cobaltous iodide.   (cobalt iodide).

$CoI_2 \cdot 6H_2O$.

Properties: Brownish-red crystals, loses iodine on exposure to air, d 2.90, loses $6H_2O$ at 27C. Soluble in water and alcohol. Anhydrous cobaltous io-

dide, CoI_2, in the form of black crystals, d 5.68, or in a yellow beta-modification which gives a colorless aqueous solution.

Derivation: Heating cobalt powder with hydriodic acid; anhydrous cobaltous iodide is prepared by heating cobalt in iodine vapor.

Use: In hygrometers, determination of water in organic solvents, as a catalyst.

cobaltous linoleate.   (cobalt linoleate).

$Co(C_{18}H_{31}O_2)_2$.

Properties: Brown, amorphous powder. Soluble in alcohol, ether, and acids; insoluble in water. Combustible.

Derivation: By boiling a cobalt salt and sodium linoleate.

Use: Paint and varnish drier, especially enamels and white paints.

See also soap (2).

cobaltous naphthenate.

Properties: Brown, amorphous powder or bluish-red solid. Insoluble in water. Soluble in alcohol, ether, oils. Composition indefinite. Combustible.

Derivation: By treating cobaltous hydroxide or cobaltous acetate with naphthenic acid.

Use: Paint and varnish drier, bonding rubber to steel and other metals.

cobaltous nitrate.   (cobalt nitrate).

CAS: 10141-05-6. $Co(NO_3)_2 \cdot 6H_2O$.

Properties: Red crystals, deliquescent in moist air. Soluble in most organic solvents. D 1.88 (25C), mp 56C, decomposes 75C.

Derivation: Action of nitric acid on metallic cobalt, cobalt oxide, hydroxide, or carbonate, with subsequent crystallization.

Use: Sympathetic inks, cobalt pigments, catalysts, additive to soils and animal feeds, vitamin preparations, hair dyes, porcelain decoration.

Hazard: Oxidizing agent, dangerous fire risk in contact with organic materials.

cobaltous oleate. (cobalt oleate).
$Co(C_{18}H_{33}O_2)_2$.
Properties: Brown, amorphous powder. Soluble in alcohol and ether, insoluble in water. Mp 235C. Combustible.
Derivation: By heating cobaltous chloride and sodium oleate followed by filtration and drying.
Use: Paint and varnish driers.
See also soaps (2).

cobaltous oxide. (cobalt oxide; cobalt monoxide).
CAS: 1307-96-6. CoO.
Properties: Grayish powder under most conditions, can form green-brown crystals. Soluble in acids and alkali hydroxides, insoluble in water and ammonium hydroxide. D 6.45, mp 1935C.
Derivation: Calcination of cobalt carbonate or its oxides at high temperature in a neutral or slightly reducing atmosphere.
Grade: Technical, ceramic.
Use: Pigment in paints and ceramics, preparation of cobalt salts, catalyst, porcelain enamels, coloring glass, feed additive, cobalt metal powder.

cobaltous perchlorate. $Co(ClO_4)_2$.
Properties: Red needles; d 3.327; soluble in water, alcohol, acetone.
Hazard: Fire and explosion risk in contact with organic materials. Strong oxidizing agent.
Use: Chemical reagent.

cobaltous phosphate. (cobalt phosphate).
CAS: 13455-36-2. $Co_3(PO_4)_2 \cdot 8H_2O$.
Properties: Reddish powder, d 2.769 (25C), loses 8HOH at 200C, slightly soluble in cold water, soluble in mineral acids.
Derivation: Interaction of solutions of cobalt salts and sodium phosphate.
Use: Cobalt pigments, coloring glass, painting porcelain, animal feed supplement.

cobaltous resinate. (cobalt resinate). Principally cobalt abietate.
Properties: Brown-red powder, insoluble in water, soluble in oils.
Derivation: See soap (2). The precipitated grade is higher in cobalt, more expensive, and more effective.
Grade: Fused, precipitated.
Use: Varnish drier.
Hazard: Spontaneously flammable in air, reacts strongly with oxidizing materials.
See abietic acid.

cobaltous silicofluoride. $CoSiF_6 \cdot 6H_2O$.
Properties: Pale red crystals, d 2.087, soluble in water.
Hazard: Irritating to tissue.
Use: Ceramics.

cobaltous succinate. $Co(C_4H_4O_4) \cdot 4H_2O$.
Properties: Violet crystals, slightly soluble in cold water, soluble in alkalies, insoluble in alcohol.
Use: Vitamin preparations, therapeutic agents.

cobaltous sulfate. (cobalt sulfate).
CAS: 10124-43-3. (a) $CoSO_4$,
(b) $CoSO_4 \cdot 7H_2O$.
Properties: Red powder, soluble in water. D (a) 3.472, (b) 1.948; mp (a) 735C, (b) loses 7HOH at 420C.
Derivation: Action of sulfuric acid on cobaltous oxide.
Hazard: May not be used in food products (FDA).
Use: Ceramics, pigments, glazes, in plating baths for cobalt, additive to soils, catalyst, paint and ink drier, storage batteries.

cobaltous tungstate. (cobalt tungstate; cobalt wolframate). $CoWO_4$.
Properties: Reddish-orange powder, insoluble in water, soluble in hot concentrated acids, d 8.42.
Derivation: By adding a sodium tungstate solution to a solution of a cobalt salt.
Use: Pigment, drier for enamels, inks, paints, electronic devices, antiknock agents.

cobalt oxide. The commercial cobalt oxides are not usually definite chemical compounds but are mixtures of two or more cobalt oxides.
See cobaltic oxide, cobaltous oxide, cobalto-cobaltic oxide.

cobalt phosphate. See cobaltous phosphate.

cobalt potassium cyanide. $K_3Co(CN)_6$.
Properties: Yellow crystals, d 1.878 (25C), mp decomposes, soluble in water, slightly soluble in alcohol.
Grade: Pure, electronic.
Hazard: A poison. TLV: (as CN) 5 mg/m³ of air.
Use: Electronic research.

cobalt potassium nitrite. (cobalt yellow; potassium cobaltinitrate; Fischer's salt; potassium hexanitrocobaltate III). $K_3Co(NO_2)_6$.
Properties: Yellow, microcrystalline powder. Slightly soluble in water, insoluble in alcohol, mp decomposes at 200C.
Derivation: By adding potassium nitrate and acetic acid to a solution of a cobalt salt.
Use: Medicine, yellow pigment, painting on glass or porcelain.

cobalt resinate. See cobaltous resinate.

cobalt selenite. $CoSe_2O_3 \cdot 2H_2O$. A blue-red powder, insoluble in water.

cobalt silicide. A semiconductor reported to have as much as 15% efficiency in converting heat to electricity in the temperature range 20–800C.

cobalt soap. See cobaltous linoleate, cobaltous naphthenate, cobaltous oleate, cobaltous resinate, cobalt tallate.

cobalt sulfate. See cobaltous sulfate.

cobalt tallate. Cobalt derivative of refined tall oil, of varying composition. Used as a drier in paints and varnishes. Combustible.
See soap (2).

cobalt tetracarbonyl. (dicobalt octacarbonyl). $Co_2(CO)_8$.
Properties: Orange or dark brown crystals. White when pure. D 1.78, mp 51C (decomposes above this temperature). Insoluble in water; soluble in alcohol, ether, and carbon disulfide.
Derivation: Combination of finely divided cobalt with carbon monoxide under pressure.
Hazard: Toxic by ingestion and inhalation.
Use: Catalyst in Oxo process, high-purity cobalt salts.

cobalt titanate. Co_2TiO_4.
Properties: Greenish-black crystals, d 5.07–5.12. Soluble in concentrated hydrochloric acid.

cobalt trifluoride. (cobaltic fluoride). CoF_3.
Properties: Light brown free-flowing powder, d 3.88 (25C), no odor except hydrogen fluoride odor developed in moist air, stable in sealed containers, reacts readily with moisture in the atmosphere to form a dark, almost black powder, reacts with water to form a black, finely divided precipitate (cobaltic hydroxide). Insoluble in alcohol and benzene. As a fluorinating agent yields one atom of fluorine and reverts to the difluoride. The spent cobalt difluoride may be regenerated with elemental fluorine.
Hazard: Strong irritant to tissue. TLV: (as F) 2.5 mg/m³ of air.
Use: Fluorination of hydrocarbons.

cobalt tungstate. See cobaltous tungstate.

cobalt ultramarine. See cobalt blue.

cobalt violet. See cobaltous ammonium phosphate.

cobalt wolframate. See cobaltous tungstate.

cobalt yellow. See cobalt potassium nitrite.

"Cobenium."[155] TM for a heat-treatable, high-cobalt alloy. Comprising cobalt 40%, chromium 20%, nickel 15%, molybdenum 7%, manganese 2%, beryllium 0.04%, carbon 0.15%, iron balance.
Properties: Corrosion-resistant, non-magnetic, resistant to set and fatigue, heat-treatable, high strength, high elasticity.
Use: Springs, corrosion-resistant products.

"Coblac."[133] TM for a series of carbon black-nitrocellulose dispersions. which make it possible to produce black lacquers without milling or grinding. Available in several types for pigmenting automotive lacquers, industrial lacquers, leather finishes, etc.

"Cobon."[169] TM for 2-nitroso-1-naphthol used for the colorimetric determination of cobalt. Sensitivity: 0.005 ppm cobalt.

"Cobrate."[266] TM for corrosion inhibitors primarily for copper and copper alloys.

COC. Abbreviation for Cleveland Open Cup, a standard technique of flash point determination.

cocaine. (methylbenzoylecgonine).
CAS: 50-36-2. $C_{17}H_{21}NO_4$.   An alkaloid.

Properties: Colorless to white crystals or white powder; soluble in alcohol, chloroform, and ether; slightly soluble in water (solution is alkaline to litmus). The hydrochloric acid solution is levorotatory. Mp 98C.
Derivation: By extraction of the leaves of Erythroxylon coca with sodium carbonate solution, treatment of the latter with dilute acid and extraction with ether, evaporation of the solvent, re-solution of the alkaloid and subsequent crystallization. Also synthetically from the alkaloid ecgonine.
Method of purification: Recrystallization.
Grade: Technical, NF.
Hazard: Central nervous system stimulant, a poison. Possession is illegal in US.
Use: Local anesthetic (medicine, dentistry), usually as the hydrochloride.

cocarboxylase. (TPP; thiamine pyrophosphate chloride). $C_{12}H_{19}ClN_4O_7P_2S \cdot H_2O$. The coenzyme of the yeast enzyme carboxylase. The key substance in decarboxylation, an energy-producing reaction in the body.
Properties: Crystals from alcohol containing some hydrochloric acid, mp 240–244C (decomposes) soluble in water, dry substance very stable.
Use: Biochemical research.

coccus. (1) A type of bacteria characterized by a spherical shape, e.g., pneumococcus. (2) Synonym for cochineal.

cocculin. See picrotoxin.

cocculus, solid. Legal label name (Air) for picrotoxin.

cochineal. (coccus). A red coloring matter consisting of the dried bodies of the female insects of Coccus cacti. The coloring principle is carminic acid $C_{22}H_{20}O_{13}$.
Grade: Technical, NF, silver grain, black grain.
Use: Coloring food, medicinal products, toilet preparations, manufacture of red and pink lakes.

cocoa. (cacao). Powder prepared from roasted, cured kernels of ripe seed of Theobroma cacao.
Grade: Commercial, USP.
Use: Flavoring agent for pharmaceuticals, food products, beverage.
See also cacao bean.

cocoa butter. See theobroma oil.

cocoa oil. See theobroma oil.

"Cocoloid."[322] TM for an algin-carrageenan composition, a hydrophilic colloid.
Use: Stabilizer for chocolate-milk products, sterilized cream, and other milk products.

coconut acid. Mixture of fatty acids derived from hydrolysis of coconut oil. Acid chain lengths vary from 6 to 18 carbons, but are mostly 10, 12, and 14. Combustible.
Grade: Distilled, double-distilled.
Use: Soaps, detergents, source of long-chain alkyl groups.

coconut cake. (coconut palm cake; copra cake). The residual product from expression of oil from the seed of the coconut.
See coconut oil meal.

coconut oil. CAS: 8001-31-8.
Properties: White, semi-solid fat containing C_{12} to C_{15}, slight odor. D 0.92, saponification value 250–264, iodine value 7–10, sharp mp at 25C. Soluble in alcohol, ether, chloroform, carbon disulfide; immiscible with water. Highly digestible, resists oxidative rancidity but is susceptible to that induced by molds and other microorganisms. Nondrying. Combustible.
Chief constituents: Glycerides of lauric acid and of capric, myristic, palmitic and oleic acids.
Derivation: Hydraulic press or expeller extraction from coconut meat followed by alkali-refining, bleaching, and deodorizing.
Grade: Crude, refined, Ceylon, Manila.
Use: Food products (margarine, hydrogenated shortenings), synthetic cocoa butter, soaps, cosmetics, emulsions, cotton dyeing, synthetic detergents, source of fatty acids, fatty alcohols and methyl esters, base for laundering and cleaning preparations for soft leathers.

coconut oil meal. The dried and crushed form of coconut cake recovered from the hydraulic or expeller process of extraction of oil from the meat. The usual product of commerce contains 24.2% crude protein, 13.3% crude fiber, 35.7% nitrogen-free extract, 7.4% ether soluble (fat) and 6.0% ash. The toal digestible nutrients approximate 72%.
Use: Animal feeds, fertilizer ingredient.

cocoonase. A proteolytic enzyme derived from silk moths, closely similar to trypsin, isoelectric point approximately 9.5, slightly less than for trypsin, pH approximately 8.0, mw approximately 25,000. As secreted by the insect, it is 80% pure. Identical with trypsin in catalytic specificity and mechanism of action.

cocoyl sarcosine. $C_{11}H_{23}CON(CH_3)CH_2COOH$. Yellow liquid, d 0.970, mp 22–28C. Combustible. Used as a detergent emulsifier.

"C-O-C-S."[55] TM for insecticides containing copper oxychloride sulfate.
Use: Insecticide for fruits and vegetables.

COD. Abbreviation for chemical oxygen demand. See oxygen consumed, biochemical oxygen demand (BOD), dissolved oxygen (DO).

codehydrogenase 1. See nicotinamide adenine dinucleotide.

codeine. (methylmorphine). CAS: 76-57-3. $C_{18}H_{21}NO_3 \cdot H_2O$. A narcotic alkaloid.
Properties: Colorless or white crystals or powder. Effloresces slowly in dry air, affected by light, mp 154.9C. Slightly soluble in water, soluble in

alcohol and chloroform, levorotatory in acid and alcohol solutions.

Derivation: From opium by extraction and subsequent crystals, also by the methylation of morphine.

Grade: Technical, NF.

Hazard: Habit-forming narcotic, sale legally restricted.

Use: Medicine (analgesic).

cod-liver oil. (morrhua oil). CAS: 8001-69-2.
Chief constituents: Glycerides of palmitic, stearic acids, cholesterol, butyl alcohol esters, etc.

Properties: Pale yellow, viscous liquid; fixed, nondrying oil; slightly fishy odor and taste. Soluble in ether, chloroform, ethyl acetate, and carbon disulfide; slightly soluble in alcohol. D 0.918–0.927, saponification value 180–192, iodine value 145–180, maumene test 102–113, acid value 204–207. Combustible.

Derivation: From the livers of codfish (Gadus morrhua) and other species of Galidae. These are rendered by steam heat and the oil separated and chilled until the stearin solidifies, then it is pressed and the clear oil collected.

Method of purification: Filtration.

Grade: Pale, light-brown, dark-brown, NF.

Use: Medicine (for its vitamin A and D content, now largely replaced by synthetic products), chamois-leather tanning.

coenzyme. A comparatively low molecular weight organic substance which can attach itself to, and supplement, a specific protein to form an active enzyme system. Generally synonymous with the term prosthetic group.

See also following entries.

coenzyme I. See nicotinamide adenine dinucleotide.

coenzyme A. (CoA). $C_{21}H_{36}O_{16}N_7P_3S$.
Essential for the formation of acetylcholine and for acetylation reactions in the body. It has been synthesized and is built up from pantothenic acid, cysteamine, adenosine, and phosphoric acid.

Properties: White powder, water-soluble, insoluble in alcohol and ether.

coenzyme Q. CAS: 303-98-0.
$CH_3C_6(O)_2(OCH_3)_2[CH_2CH:C(CH_3)CH_2]_n H$.
Found in animal organs and yeast. Active in the citric acid cycle in carbohydrate metabolism. The n in the formula varies according to the source.

coffearine. See trigonelline.

coffinite. $U(SiO_4)_{l-x}(OH)_{4x}$, (or $USiO_4$, with appreciable $(OH)_4$ in place of some SiO_4). A naturally occurring uranium mineral. Color

black, d 5.1, luster adamantine, commonly fine-grained, mixed with organic matter and other minerals.

Occurrence: Colorado, Utah, Wyoming, Arizona.

Use: Ore of uranium (Colorado).

cogeneration. Simultaneous production of electricity and steam from the same energy source. Research has indicated a potential fuel saving of 30% by use of this process.

coherent light. A high-energy beam of light having a single wavelength and frequency generated by vibration of certain crystals.

See also laser.

cohune oil. An edible, nondrying oil with properties similar to coconut and babassu oils. Its composition is 46% lauric acid, 16% myristic acid, and 10% oleic acid, balance mixed acids. Obtained from a palm native to Mexico and Central America. Combustible.

"Coilife."[308] TM for special epoxy resin encapsulation of random wound stators using solventless epoxy resin formulations and rotational seasoning process.

coke. The carbonaceous residue of the destructive distillation (carbonization) of bituminous coal, petroleum, and coal-tar pitch. The principal type is that produced by heating bituminous coal in chemical recovery or beehive coke ovens (metallurgical coke), one ton of coal yielding approximately 0.7 ton of coke. It is used chiefly for reduction of iron ore in blast furnaces, and as a source of synthesis gas. Petroleum yields coke during the cracking process. Coke from petroleum residues and coal-tar pitch is used for refractory furnace linings in the electrorefining of aluminum and other high-temperature service, and for electrodes in electrolytic reduction of Al_2O_3 to aluminum, as well as in electrothermal production of phosphorus, silicon carbide, and calcium carbide.

coke oven gas. 53% hydrogen, 26% methane, 11% nitrogen, 7% carbon monoxide, 3% heavier hydrocarbons.

Use: To produce hydrogen.

cola. (kola; kola nuts; kola seeds; Soudan coffee; guru). Contains caffeine, theobromine.

Derivation: Seeds of Cola nitida or other species of Cola.

Habitat: West Africa, West Indies, India.

Use: Soft drinks.

colchicine. CAS: 64-86-8. $C_{22}H_{25}NO_6$.
An alkaloid plant hormone.

Properties: Yellow crystals or powder; odorless, or nearly so. Soluble in water, alcohol, and chloroform; affected by light; mp 135–150C. Solutions are levorotatory.

Derivation: From Colchicum autumnale by extraction and subsequent crystallization. Has been synthesized.

Grade: Technical, USP.

Hazard: As little as 20 mg may be fatal if ingested.

Use: To induce chromosome doubling in plants, phytopathology.

cold flow. The permanent deformation of a material that occurs as a result of prolonged compression or extension at or near room temperature. Some plastics and vulcanized rubber exhibit this behavior. In metals it is known as creep.

cold rubber. Synthetic rubber produced by polymerization at relatively low temperature, specifically SBR or butadiene-styrene elastomers produced by polymerization at approximately 4.4C compared with usual temperature of approximately 49C. A special catalyst system is required.

cold-short. See short.

colemanite. ($Ca_2B_6O_{11} \cdot 5H_2O$). The ore of calcium borate. Used to replace boric acid in the manufacture of glass fibers. Mined in Turkey, it began to be imported into the US in large volume in 1965, and is competitive with domestically produced B_2O_3.

Properties: D 2.26–2.48.

Derivation: From kernite.

colestipol. An anion exchange resin, highly crosslinked and insoluble. A specific formula has not been ascertained. Said to be a copolymer of diethylene triamine and chloroepoxypropane. It is used as a cholesterol-sequestering agent in medicine.

colistin. $C_{45}H_{85}N_{13}O_{10}$. Antibiotic produced by a soil microorganism. Probably identical to polymyxin E and closely related chemically to polymyxin B since it is a polypeptide composed of amino acids and a fatty acid.

See polymyxin.

collagen. A fibrous protein comprising most of the white fiber in the connective tissues of animals and man, especially in the skin, muscles, and tendons. The most abundant protein in the animal kingdom, it is rich in proline and hydroxyproline. The molecule is analogous to a three-strand rope, in which each strand is a polypeptide chain. It has a molecular weight of approximately 100,000. Glue made from the collagen of animal hides and skins is still widely used as an adhesive. So-called "soluble" collagen is that first formed in the skin; upon aging it becomes increasingly crosslinked and less hygroscopic. "Soluble" collagen is being used in the cosmetic industry as the basis for face creams, lotions, and hair preparations. Special forms of collagen were developed for dialysis membranes. Microcrystalline collagen is being used in prosthetic devices and other medical and surgical applications. Regenerated collagen, used in sausage casings, is made by neutralizing with acid collagen that has been purified by alkaline treatment. Collagen is converted to gelatin by boiling water, which causes hydrolytic cleavage of the protein to a mixture of degradation products.

See also gelatin.

2,4,6-collidine. (2,4,6-trimethylpyridine). $(CH_3)_3C_5H_2N$.

Properties: Colorless liquid, bp 170.4C, fp −44.5C, d 0.913 (20/20C), refr index 1.4981 (20C). Soluble in alcohol, slightly soluble in water. Combustible.

Grade: Technical, (97.5% purity).

Use: Chemical intermediate, dehydrohalogenating agent.

collodion. A solution of pyroxylin (nitrocellulose) in ether and alcohol. USP specifications are pyroxylin 40 g, ether 750 ml, and alcohol 250 ml.

Properties: Pale yellow, syrupy liquid; odor of ether; immiscible with water; flash p approximately 0F (−17.7C).

Grade: Technical, USP.

Hazard: Flammable, dangerous fire risk.

Use: Cements, coating wounds and abrasions, solvent for drugs, corn removers, process engraving, lithography, photography.

See also nitrocellulose.

colloid chemistry. A subdivision of physical chemistry comprising the study of phenomena characteristic of matter when one or more of its dimensions lie in the range between 1 millimicron (nanometer) and 1 micron (micrometer). It thus includes not only finely divided particles but also films, fibers, foams, pores, and surface irregularities. It is dimension that is critical, rather than the nature of the material. Colloidal particles may be gaseous, liquid, or solid, and occur in various types of suspensions (imprecisely called solutions), e.g., solid/gas (aerosol), solid/solid, liquid/liquid (emulsion), gas/liquid (foam). In this size range, the surface area of the particle is so much greater than its volume that unusual phenomena occur, i.e., the particles do not settle out of the suspension by gravity and are small enough to pass through filter membranes. Macromolecules (proteins and other high polymers) are at the lower limit of this range; the upper limit is usually taken to be the point at which the particles can be resolved in an optical microscope. The first specific observations were made by Thomas Graham approximately 1860, and were extended by Ostwald, Hatchek, and Freundlich. Though the term is often used synonymously with surface chemistry, in a strict sense it is limited to the size range noted in at least one dimension, whereas surface chemistry is not. Natural colloid systems include rubber latex, milk, blood, egg-white, etc.

See also surface chemistry, colloid, protective, emulsion.

colloid, association.   See association.

colloid mill.   See homogenization.

colloid, protective.   A hydrophilic high polymer whose particles (molecules) are of colloidal size, such as protein or gum. It may be either naturally present in such systems as milk and rubber latex, or intentionally added to mixes to stop coagulation or coalescence of the particles of fat or other dispersed material. Protective colloids are also called stabilizing, suspending, or thickening agents; they also act as emulsifiers. Examples are (a) hydrocarbon particles of latex that are covered with a layer of protein which keeps them from cohering as a result of the impact due to their Brownian motion; (b) gelatin, sodium alginate, or gum arabic, which are added to ice cream to inhibit formation of ice particles, and to confectionery and other food products to obtain a smooth, creamy texture. They are readily adsorbed by the suspended particles and reinforce the protective effect of proteins that may be naturally present.

See also thickening agent, gum, natural, gelatin.

"Colloisol."[440]   TM for a series of vat dyes for dyeing and printing textiles of cellulosic fibers.

cologne.   (toilet water).   A scented, alcohol-based liquid used as a perfume, after-shave lotion, or deodorant. Combustible.

Cologne brown.   See Van Dyke brown.

colophony.   A rosin residue which remains after the volatiles have been removed by distillation of crude turpentine from the Pinus species.

colorant.   Any substance that imparts color to another material or mixture. Colorants are either dyes or pigments and may either be (1) naturally present in a material (chlorophyll in vegetation), (2) admixed with it mechanically (dry pigments in paints), or (3) applied to it in a solution (organic dyes to fibers). *Note:* A valid distinction between dyes and pigments is almost impossible to draw. Some have established it on the basis of solubility, or on physical form and method of application. Most pigments are insoluble, inorganic powders, the coloring effect being a result of their dispersion in a solid or liquid medium. Most dyes, on the other hand, are soluble synthetic organic products which are chemically bound to and actually become part of the applied material. Organic dyes are usually brighter and more varied than pigments. but tend to be less stable to heat, sunlight, and chemical effects. The term colorant applies to black and white as well as to actual colors. Instruments for measuring, comparing, and matching hue, tone, and depth of colors are called colorimeters.

See also dye, pigment, colorimetry, food color, FD&C color.

"Colorex."[1]   TM for titanium trichloride in aqueous solution with zinc chloride. Dark violet to black liquid.

Use: Powerful reducing agent, dye stripper for textiles.

colorimetry.   An analytical method based on measuring the color intensity of a substance or a colored derivative of it. For example, the yellow carotene content of butter is determined by saponifying a sample of butter in an alkaline solution, extracting the carotene with ether, and measuring the intensity of yellow color in the ether extract. Colorimetric methods are used to determine very minute amounts. They are used in hospital laboratories for blood and urine analysis; in food laboratories for determination of vitamins, preservatives, coloring matter, etc; and in metallurgical laboratories for traces of metals in raw materials and finished products.

colorless dye.   Synonym for optical brightener.

"Colorundum."[205]   TM for a balanced mixture of abrasive aggregates, mineral oxide color, and stearate for surfacing concrete floors, in a choice of colors.

Columbite.   An ore of tantalum.

See tantalite.

columbium. Cb. Alternate name for the element niobium. The latter name became official in 1949; columbium is still used by metallurgists.

column, distillation. See tower, distillation.

"Coly-Mycin."[546] TM for colistin in form of sulfate or methane sulfonate (colistimethate sodium).

combination. A chemical reaction in which two substances unite to form a third, the reacting substances may be elements or compounds, but the product is always a compound. Examples: $Cu + Cl_2 \rightarrow CuCl_2$ and $2NH_3 + H_2SO_4 \rightarrow (NH_4)_2SO_4$. Polymerization. is a special case of combination, where complex organic molecules of the same kind unite to form chains or clusters of high molecular weight.
See also polymerization.

combining number. See valence.

combining weight. See equivalent weight.

combustible material. Any substance that will burn, regardless of its autoignition temperature or whether it is a solid, liquid, or gas. Although this definition necessarily includes all flammable materials as well, this fact is disregarded in official classifications. Usually defined, the term "combustible" refers to solids that are relatively difficult to ignite and that burn relatively slowly, and to liquids having a flash point greater than 100F (37.7C). It is difficult to generalize about the combustibility of solids. The rate and ease of combustion may depend as much on their state of subdivision as on their chemical nature. Many metals in powder or flake form will ignite and burn rapidly, whereas most are noncombustible as bulk solids. Cellulose is combustible as a textile fabric or as paper, and is flammable as fine fibers (cotton linters). A plastic that burns at flame temperature will be a greater fire hazard as a foam than as a bulk solid because of the large surface area exposed to air and thinness of the cell walls. Some polymers, such as nylon and polyvinylidene chloride, will melt and burn but will not propagate flame; others, e.g., polyvinyl chloride and polyurethane, ignite at high temperature and evolve toxic fumes. Acrylics and cellulose-derived plastics, such as rayon and cellulose acetate, are readily combustible. This may be partially offset by use of fire-retardant chemicals. Glass is noncombustible in all forms.
See also flammable material, hazardous material, combustion.

combustion. An exothermic oxidation reaction which may occur with any organic compound as well as with certain elements, e.g., hydrogen, sulfur, phosphorus, magnesium. The end products of elemental combustion are oxides; of organic compounds are carbon dioxide and water. Examples are (a) for an element: $2H_2 + O_2 \rightarrow 2H_2O$, (b) for an organic compound (carbohydrate): $C_6H_{12}O_6 + 6O_2 \rightarrow 6CO_2 + 6H_2O$. Here, combustion is the reverse of photosynthesis. The heat of combustion is due to rupture of chemical bonds and formation of new compounds. Substances differ greatly in their combustibility, that is, in their ignition points (solids and gases) or their flash points (liquids). Carbon disulfide burns almost explosively at 100C, while rubber hydrocarbon and nylon are difficult to ignite at any temperature. Oxygen is not combustible but actively supports combustion; no oxygen is needed if another oxidizing agent is present, as in the combustion of a mixture of hydrogen and chlorine to form hydrogen chloride.

Spontaneous combustion may occur at or even below room temperature (1) by exposure to air of substances that are highly sensitive to oxidation, (e.g., phosphorus); (2) by heat build-up from bacterial activity (compost, sewage sludge) or oxidation catalyzed by moisture, as in wet waste materials (paper, cotton, wool); (3) by internal heat accumulation due to autoxidation (fish oils, linseed oil).
See also oxidation, pyrophoric, autoignition temperature, flash point.

Comirin. Exact formula undetermined. An antifungal agent produced by *Bacterium antimyceticum* comprised mostly of amino acids. A gray to white powder which resists heat and acids, but decomposes in presence of alkalies; soluble in organic acids and bases; slightly soluble in water. A fungicide in paints, textile products, plant growth.

comminution. Size reduction of materials by any of several means, e.g., grinding. cutting, shredding, chopping, etc. Solids can be reduced to a particle size approaching 1 micron in special fine-grinding equipment. Comminution of coal by chemical means is possible via a low molecular weight compound such as sodium hydroxide or anhydrous ethanol, which penetrates the natural fault system of the coal causing fragmentation without mechanical crushing. This permits removal of sulfur from coal without burning or grinding. Approximately 100 pounds of chemical are needed per ton of coal.

compatibility. The ability of two or more materials to exist in close and permanent association indefinitely. Liquids (solvents) are compatible if they are miscible and do not undergo phase separation on standing. Thus, water is compatible with alcohol but not with gasoline. Liquids and solids are compatible if the solid is soluble in the liquid only. Solids are compatible if they can exist in intimate contact for long periods with no adverse effect of one on the other.

"Compazine."[71] TM for a brand of prochlorperazine, as the maleate or the edisylate.

complement. A term used in immunochemistry to refer to a number of blood proteins that act in conjunction with antibodies to cause disintegration of invading cells. It is an essential component of immune serum.

complex compound. See coordination compound.

complexing agent. See ligand, chelate, ethylenediaminetetraacetic acid.

complex ion. An ion which has a molecular structure consisting of a central atom bonded to other atoms by coordinate covalent bonds.
See coordination compound.

component. One of the minimum number of substances required to state the composition of all phases of a system in the absence of chemical reaction of any substances in a mixture.
See also constituent.

composite. A mixture or mechanical combination on a macro scale of two or more materials that are solid in the finished state, are mutually insoluble, and differ in chemical nature. The major types are: (1) laminates of paper, fabric or wood (veneer), and a thermosetting material (resin, rubber or adhesive); examples are tire carcases, plywood, and electrical insulating structures. (2) Reinforced plastics, principally of glass fiber and a thermosetting resin, other types of fibers such as boron, aluminum silicate, and silicon carbide may be used (see also whiskers). (3) Cermets, which are mixtures of ceramic and metal powders, heat-treated and compressed. (4) Fabrics, i.e., woven combinations of wool or cotton and a synthetic fiber. (5) Filled composites where a bonding material, i.e., linseed oil, resin, or asphalt, is loaded with a filler in the form of flakes or small particles; examples are linoleum, glass/flake-plastic mixtures for battery cases and asphalt/gravel road-surfacing mixtures.

composting. Aerobic bacterial decomposition of solid organic wastes, both agricultural and urban, including sewage sludge. As much as 500 tons

a day can be handled in the larger installations, the waste degrading quickly at low temperature. Decomposition is accelerated by adding ammonium bicarbonate. This can be used as a soil conditioner and for landfill. The waste is piled, turned frequently to provide aeration, and maintaining a high temperature in the pile to destroy pathogenic organisms. Volume of composted waste is from 20–60% of original volume.

compound. (1) A substance composed of atoms or ions of two or more elements in chemical combination. The constituents are united by bonds or valence forces. A compound, a homogeneous entity where the elements have definite proportions by weight and are represented by a chemical formula. A compound has characteristic properties quite different from those of its constituent elements. It is decomposed by energy in the form of a chemical reaction, of heat, or of an electric current. Example: Water is a *liquid* formed by chemical combination of two *gases*; it can be separated into hydrogen and oxygen by an electric current (electrolysis); in certain reactions it is split into its constituent ions (H, OH) (hydrolysis); it is not chemically changed by heat or cold.
See also mixture, homogeneous, reaction.
   (2) Loosely, a product formula (often proprietary) of various types, e.g., pharmaceuticals (a vegetable compound), rubber (a fast-curing compound), etc.
   (3) Having two sets of lenses (compound microscope).

Combes quinoline synthesis. Formation of quinolines by condensation of beta-diketones with primary arylamines followed by acid catalyzed ring closure of the intermediate Schiff base.

compound 1080. Use may be restricted.
See sodium fluoroacetate.

compreg. A hardwood impregnated with a phenolformaldehyde resin under heat and pressure.

compressed gas. Any material or mixture that, when enclosed in a container, has an absolute pressure exceeding 40 psi at 21.1C or, regardless of the pressure at 21.1C, has an absolute pressure greater than 140 psi at 54.4C, or any flammable material having a vap pressure greater than 40 psi abs at 37.7C. (Vapor pressure determined by Reid method (ASTM)). Compressed gases include liquefied petroleum gases and oxygen, nitrogen, anhydrous ammonia, acetylene, nitrous oxide, and fluorocarbon gases. Some of these are shipped in tonnage volume. For details on properties, containers, and shipping regulations, see specific gas. For additional information, see Compressed Gas Association, 500 Fifth Avenue, NY, NY.

compression molding. Formation of a rubber or plastic article to a desired shape, either by placing the raw mixture in a specially designed cavity or bringing it into contact with a contoured metal surface. After the material is in place, heat and pressure are supplied by a hydraulic press, the time and temperature varying with the nature of the material. For rubber products, vulcanization occurs simultaneously. Most plastic molding is now done by the injection method, which is more economically efficient.
See also injection molding.

"Compresto."[84] TM for an electrical conductor consisting of layers of shaped pure aluminum wires concentrically stranded about one round core wire.

Compton effect. One of the principal processes by which high energy electromagnetic radiation or gamma rays interact with or are absorbed by matter. In the Compton process the gamma ray frees an electron in matter as if the electron was unbound dividing the momentum of the gamma ray between the ejected electron and a new gamma ray of lower energy going off in a new direction.

computational chemistry. Use of computers in organic synthesis and in chemical engineering as a more efficient means of research than conventional laboratory experimentation. The phenomenal capacity of sophisticated computers for almost instantaneous mathematical calculations has made them an invaluable aid in exploring and evaluating the more likely pathways for a given organic synthesis, for which there may be innumerable possible sequences. The term "heuristic" is applied to such procedures. Computers can also handle the vast complexity of quantum mechanical calculations and aid in the elucidation of the complicated molecular structures that occur in pharmaceutical compounds and recombinant DNA research. The Quantum Chemistry Program Exchange at Indiana University offers many programs in this field, from subroutines to major computational systems. Chemical engineers utilize computers to develop more thermodynamically efficient procedures and to consolidate overall plant operations, especially in the areas of energy consumption, reaction rates and hazardous waste problems.
See also retrosynthesis.

Note: Notwithstanding the immense capability of computers to point the way to solutions of chemical and engineering problems, experimentation will remain the ultimate means of achievement. It is interesting to speculate how much time and effort such empirical scientists as Goodyear and Edison could have saved had computers been available to them.

Conant, James Bryant. (1893–1978). An American chemist and educator, born in Boston, who received his doctorate in chemistry from Harvard in 1916 and was President of Harvard for 20 years (1933–53). His major scientific activities included pioneering research on chlorophyll and important contributions to the Manhattan Project. Perhaps his greatest achievements lay in the educational field in which he exerted a strong liberalizing influence at both the collegiate and secondary school levels. He also was ambassador to postwar Germany and educational advisor to Berlin. He wrote many books on science and education including basic chemical texts, and received a number of scientific and educational awards.

concave. See gyratory crusher.

concentration. The amount of a given substance in a stated unit of a mixture, solution, or ore. Common methods of stating concentration are percent by weight or by volume, normality, or weight per unit volume as grams per cubic centimeter or pounds per gallon. The concentration of an atom, ion, or molecule in a solution may be indicated by square brackets, as $[Cl^-]$. For radioactivity, the concentration is usually expressed as millicuries per milliliter (mCi/mL) or millicuries per millimole (mCi/mM).

conchiolin. $C_{32}H_{98}N_2O_{11}$. A natural bonding agent in the calcium carbonate structure of pearls.

conchoidal. A term adopted from mineralogy by chemists to describe a type of surface formed by fracturing a hard solid by impact. Certain materials present involutely curved fracture surfaces suggestive of the shape of the shells of bivalves (conch), from which the term is derived. Examples are glass, blown asphalt, and numerous minerals.

concrete. (1) A conglomerate of gravel, pebbles, sand, broken stone, blast-furnace slag or cinders, termed the aggregate, embedded in a matrix of either mortar or cement, usually standard Portland cement in the US. Reinforced concrete and ferro-concrete contain steel in various forms.

Further information can be obtained from the American Concrete Institute, Detroit, MI.
See cement, Portland.
Use: Building and road construction, radiation shielding.
(2) A waxy solid obtained from roses by extraction with non-polar solvents (benzene) after trace quantities of solvent have been removed. When the concrete is dewaxed by a properly chosen second solvent (alcohol), the desired essential oil remains. This is called an absolute.
See also perfume.

concrete, cellular.   A light-weight, concrete foam which may be made in several ways: (a) by addition of aluminum powder to the concrete mix and applying heat, which releases hydrogen; (b) by whipping air into the mix containing an entraining agent; (c) by adding preformed foam to the mix. Such foams are made from a foaming agent such as dried blood, a stabilizer, organic solvents, and a germicide.
See also foam.

concrete, reinforced.   See concrete.

condensation.   (1) A type of chemical reaction in which two or more molecules combine with the separation of water, alcohol, or other simple substance. The distinction from an addition reaction is not sharp, as some reactions can be either stopped at the addition stage or carried a step beyond it. If a polymer is formed by condensation the process is called polycondensation, e.g., phenolformaldehyde resin.
See also Claisen condensation, aldol. (2) The change of state of a substance from the vapor to the liquid (or solid) form.

conductance.   The conductivity of a solution, defined as the reciprocal of the resistance. It is usually used in connection with electrolytic solutions.
See conduction (2).

conduction.   (1) Transference of heat through a substance or from one substance to another when the two substances are in physical contact (thermal conduction). Crystalline solids (especially metals and alloys) are good thermal conductors because of their high density; liquids such as water and glass and high polymers such as rubber and cellulose usually are not.
(2) Transference of an electric current through a solid or liquid (electrical conduction). In metallic or electronic conductors the current is carried by a flow of electrons from atom to atom; the atomic nuclei remaining stationary. This type of conduction is common to all metals and alloys,

carbon and graphite, and certain solid compounds (manganese dioxide, lead sulfide). In electrolytic or ionic conductors the current is carried by ions, as in solutions of acids, bases, and salts and in many fused compounds. In electrolytic conduction as in metallic conduction heat is generated and a magnetic field is formed around the conductor; a transfer of matter also occurs. In a few materials, such as solutions of alkali and alkaline earth metals in anhydrous liquid ammonia, both types of conduction take place simultaneously; such conductors are called mixed conductors.
See also semiconductor, transference number.

conductivity.   The property of a substance or mixture that describes its ability to transfer heat or electricity. It is the reverse of resistivity.
See also conduction.

configuration.   In an organic molecule, the location or disposition of substituent atoms or groups around asymmetric carbon atoms. This can be changed only by severing single covalent bonds.

conformation.   The shapes or arrangements in three-dimensional space that an organic molecule can assume by rotating carbon atoms or their substituents around single covalent bonds. The conformation of a molecule is not fixed, though one or another shape may be more likely to occur. The number of conformational isomers is infinite. Conformational analysis involves the study of the preferred (or most likely) conformations of a molecule in the ground, transition, and excited states. Research on the conformations of cyclohexane and various sugars has contributed much to this aspect of stereochemistry.

Congo red.   (Sodium diphenyl-bis-α-naphthylamine sulfonate; C.I. 22120).
CAS: 573-58-0. $C_{32}H_{22}O_6N_6S_2Na_2$.
Properties: Brownish-red powder, soluble in water and alcohol, insoluble in ether, odorless, decomposes on exposure to acid fumes.
Derivation: Combination of tetraazotized benzidine and naphthionic acid.
Use: Dye, medicine (diagnostic aid), indicator, biological stain.

Congo resin.   A variety of copal fossil resin. The natural product is insoluble in organic solvents, but forms transparent gels with some alcohols and hot solvents. When thermally processed (cracked), it is soluble in all organic solvents, fatty acids, and vegetable oils. Its high acid number prevents its use in paints containing reactive pigments.
Use: High-gloss varnishes, paints for metal surfaces, wrinkle finishes.

coniferin. $C_{16}H_{22}O_8$. A glucoside contained in pine bark and other conifers. When decomposed, it yields coniferyl alcohol which can be oxidized to vanillin. Used as a raw material for manufacture of synthetic vanillin.

coning oil. Usually an emulsified mineral oil used as lubricant for textile fibers in processing to the finished yarn. Fatty acid esters are often used for the oil-in-water emulsions.

conjugated double bonds. Two or more double bonds which alternate with single bonds in an unsaturated compound, as in the formula for butadiene-1,3 ($H_2C{=}CH{-}CH{=}CH_2$) or maleic acid (the $O{=}C{-}C{=}C{-}C{=}O$ skeleton).

conjugate layers. Two layers of a liquid system each composed of a different ternary mixture and in equilibrium with one another.

Conrad-Limpach reaction. Thermal condensation of arylamines with β-ketoesters followed by cyclization of the intermediate Schiff bases to 4-hydroxyquinolines.

conservation of energy, law. (Also known as the first law of thermodynamics). See energy.

consistency. Resistance to flow of a material, usually a liquid. For Newtonian liquids, consistency and viscosity are synonymous. For non-Newtonian liquids, it qualitatively represents plastic flow. The term is used by food technologists. See also body (1); viscosity.

"Consol C."[1] TM for a chemically modified protein colloid derived from bones and hides. Used in the paper industry to improve interfiber bonding and filler retention.

constantan. Generic name for an alloy containing from 40 to 45% nickel and from 55 to 60% copper. Used in thermocouples and specialized heat-measuring devices.

constant-boiling mixture. See azeotropic mixture.

constant composition, law. See chemical laws (2).

constituent. Any of the elements or subgroups in a molecule of a compound. For example, nitrogen is a constituent of proteins; the carboxyl group is a constituent of fatty acids. See also component.

contact acid. Sulfuric acid made by the contact process.

contact process. A process for manufacture of sulfuric acid and oleum in which the sulfur dioxide from combustion of sulfur, pyrites, or other sulfur sources is oxidized with air to sulfur trioxide by contact with a vanadium pentoxide catalyst. Most of the sulfuric acid produced in the US is made by this process. See also sulfuric acid.

contact resin. (impression resin; low-pressure resin). A synthetic thermosetting resin characterized by cure at relatively low pressure. The usual components are an unsaturated, high molecular weight monomer, such as an allyl ester, or a mixture of styrene or other vinyl monomer with an unsaturated polyester or alkyd. Cure requires heat and a catalyst as well as some pressure. The curing does not result in water formation as with phenol-formaldehyde resins.

container. This term refers not only to units used to pack chemical products for shipment, but also to transport them. The factors that dictate their selection include size of shipment, compatibility of the product with container material, ease of storage and handling, and cost. Common containers are:

    tank cars, tank trucks, hopper cars (bulk chemicals)

    barges (petroleum products, petrochemicals, other chemicals)

    tankers (crude oil, refined products)

    pipelines (gases, liquid chemicals, etc.)

    55-gal drums of steel, plastic, or fiberboard, with or without polyethylene lining.

    5-, 15-, 30-gal metal cans and pails

    5-gal plastic carboys

    1-gal metal cans

    steel cylinders (compressed gases)

    multiwall paper bags (dry powders)

    glass and plastic bottles, vials, etc.

    wooden kegs, barrels, boxes, bales; burlap bags (bulk imports, such as gums, etc.)

    76-lb flasks (mercury)

contaminant. Any substance accidentally or unwillingly introduced into air, water, or food products which has the effect of rendering them toxic or otherwise harmful. Examples are sulfur dioxide resulting from combustion of high-sulfur fuels; pesticide residues in vegetables, fish, or other food products; industrial dusts; and radioactive materials resulting from nuclear explosions. See also fallout; decontamination.

Note: Pesticide residues in foods are often referred to as unintentional additives though they are actually contaminants.

"Continental."[104] TM for a series of channel blacks used in natural and synthetic rubber, paints, inks, and plastics.

"Continex."[104] TM for a furnace blacks. Used in rubber, plastics, paints, paper.

continuous distillation. Distillation in which a feed, usually of nearly constant composition, is supplied continuously to a fractionating column and the product is continuously withdrawn at the top, bottom, and sometimes at intermediate points.

continuous phase. See phase (2).

control. (1) In any chemical or other scientific experiment, the reference base with which the results are compared. This base is invariably a sample of identical constitution and prepared under the same conditions from which all experimental variations are omitted. Thus, the control represents known values as far as any specific experiment is concerned. Such a sample is often called a "blank." The use of a control is vital to significant interpretation of experimental data. (2) Automatic.
See automatic control.

"Control GP."[266] TM for corrosion inhibitor and stabilizer for glycol natural gas dehydrators.

controlled atmosphere. See atmosphere, controlled.

controlled-release. Descriptive of a compound manufactured in such a way that its effect will be kept uniform over an extended time period; this not only provides more effective control, but also reduces the waste involved in using unnecessarily high concentrations. This principle has been applied successfully to fertilizers, pesticides, pharmaceuticals, flavors, and fragrances. It may involve (1) encapsulation of the agent, (2) incorporating it into a neutral matrix such as a rubber or plastic, (3) coating the particles with sulfur (fertilizers), (4) absorbing the agent into substrates of various types.

convection. The transfer of heat from one place to another by a moving gas or liquid. Natural convection results from differences in density caused by temperature differences. Thus, warm air is less dense than cool air; the warm air rises relative to the cool air, and vice versa. Forced convection involves motion caused by pumps, blowers, or other mechnical devices.

converting. A term used specifically in the paper industry to refer to (1) modification of raw paper

by coating, impregnating, laminating, and corrugating (wet converting); and (2) fabrication of a multitude of finished products such as bags, cartons, packaging materials, napkins, facial tissues, cups, plates, and the like (dry converting). Specialized equipment is required for these operations, e.g., rollcoaters, extrusion coaters, embossers, crepers, carton and bag-making machines, etc.

conveyor. Any device for continuous transport of raw materials, assembly units, or finished products. The more important types are as follows, all except (4), (6), and (7) being mechanically driven.
(1) Chain:
    (a) plain chain: four or more parallel chains whose tops are a little higher than the tracks on which they move (used for flat objects, cartons, metal bars, etc.).
    (b) in-floor: two parallel chains operating in metal channels about 16 inches apart and set into the floor (used for cartons of finished goods to storage or loading dock).
    (c) enclosed chain: steel sprockets or drags, wholly enclosed; horizontal, vertical, or inclined movements are possible (used for powders, grain, vegetable products.)
(2) Belt:
    (a) light: a flat, rubber/fabric structure for carrying units through a packaging or assembly line; also for feeding rubber or plastic mixes to extruder.
    (b) heavy: rubber/fabric structure with thick cover, for long- distance conveying of bulk solids (coal, coke, gravel, ores).
(3) Screw: an auger rotating in a metal trough or channel; used for feeding particulates to mixing equipment; handling light solids (wood chips, sawdust, etc.).
(4) Gravity roller: cylindrical metal rollers attached to metal track or frame; used for conveying cartons or boxes where plant layout permits uniform gravity movement; the angle of incline is critical.
(5) Bucket: metal scoops mounted on belts or steel frames (for vertical carriage of gravel, crushed stone, ores, etc.).
(6) Pneumatic: metal tubes of varying diameters operated by positive or negative air pressure; used for light particulates (unloading bins and hopper cars and intraplant transport).
(7) Pipeline: plastic or welded steel tubes for long-distance movement of crude oil, refined products, ammonia, natural gas, water, steam. Positive pressure required.

cooking. (1) Conversion of a foodstuff from the raw, inedible state to a palatable and more readily digestible condition by application of heat, as in boiling, frying, roasting, or baking. One or more chemical and physical changes occur: (1) hydrolysis of collagen in the connective tissue of meats; (2) softening and partial hydrolysis of cellulose and starches to sugars in fruits and vegetables; (3) denaturation and coagulation of proteins in meat resulting in a less tightly ordered structure and improved texture; (4) coagulation of egg albumin and wheat gluten (bakery products); (5) modification or "shortening" of wheat flour by added fats which coat the particles to form a laminar structure, thus preventing formation of chains of gluten; (6) evolution of carbon dioxide by the action of leavening agents (yeast, baking powder), which cause "rising" or gas bubbles that are retained by the gluten to form a stable solid foam; (7) browning of meats and bakery products due to reaction of sugars and amino acids (nonenzymatic); (8) deactivation of enzymes and destruction of bacteria; (9) loss or deactivation of water-soluble vitamins in meat and vegetables. (2) In the paper industry, digestion of wood pulp with such compounds as sodium sulfate, sodium hydroxide, and sodium sulfide to separate the lignin content from the cellulose.

See also pulp, paper.

coolant. (heat-transfer medium; thermofor). Any liquid or gas having the property of absorbing heat from its environment and transferring it effectively away from its source. Coolants are used in all types of automobiles, as well as in chemical processing and nuclear engineering equipment. One of the most effective and cheapest coolants is water, which is almost universally used in automotive and ordinary reaction equipment. Air is also used. Where intense heating requires a more efficient medium, special coolants are used: liquid sodium in nuclear reactors, liquid hydrogen in high thrust nuclear rocket engines; carbon dioxide, propylene glycol, and "Dowtherm" in chemical processing reactors. Methoxy propanol has been introduced for diesel engines. Some coolants provide antifreeze protection.

See also antifreeze.

Coolidge, William D. (1873-1975) An American physical chemist born in Massachusetts. He received a degree in electrical engineering at M.I.T. (1896) and doctorate in physics at Leipzig (1899). In 1905, he joined the General Electric Research Laboratory which had been established five years earlier. Here he invented the ductile tungsten filament and developed the use of tungsten in electrical switches and medical x-ray tubes. He also did pioneer research in experimental metallurgy and powder metallurgy. He also had a prominent part in evaluating uranium research

(1941) and in setting up the Manhattan project. He was the recipient of many honors and awards, not the least of which was induction into the National Inventors Hall of Fame in 1975.

cooperage. The manufacture of barrels, formerly of wood but now including drums made of various materials such as resin-bonded fiber, metal, and plastic.

coordination compound. (complex compound). A compound formed by the union of a metal ion (usually a transition metal) with a nonmetallic ion or molecule called a ligand or complexing agent. The ligand may be either positively or negatively charged (such ions as Cl^- or $NH_2NH_3^+$) or it may be a molecule of water or ammonia. The most common metal ions are those of cobalt, platnium, iron, copper, and nickel which form highly stable compounds. When ammonia is the ligand, the compounds are called ammines. The total number of bonds linking the metal to the ligand is called its coordination number. It is usually 2, 4, or 6, and often depends on the type of ligand involved. All ligands have electron pairs on the coordinating atom, e.g., nitrogen that can be either donated to or shared with the metal ions. The metal ion acts as a Lewis acid (electron acceptor) and the ligand as a Lewis base (electron donor). The bonding is neither covalent nor electrostatic but may be considered intermediate between the two types. The charge on the complex ion is the sum of the charges on the metal ion and the ligands; for example, $4NH_3 + 2Cl^- + Co^{+3}$ forms the complex $[Co(NH_3)_4Cl_2]^+$. The brackets enclose the metal ion and the coordinated ligands.

See also chelate, sequestration, metallocene.

coordination number. (CN). The number of points at which ligands are attached to the metal ion in a complex. Common coordination numbers are 2, 4, and 6, exemplified by the ions $[Ag(NH_3)_2]^+$, $[Ni(CN)_4]^{-2}$, and $[PdCl_6]^{-2}$.

copaiba oil. A levorotatory, essential oil obtained from a tree native to Brazil; a component of copaiba resin (a type of balsam). The sap of this tree is a turpentine-like liquid which can be used as an automotive fuel; intensive cultivation of copaiba trees for this purpose is under experimentation in Brazil.

copaiba resin. See balsam (3), copaiba oil.

copaivic acid. $C_{20}H_{30}O_2$. A monobasic acid derived from copaiba balsam.

copal. A group of fossil resins still used to some extent in varnishes and lacquers. Insoluble in oils and water. Most important types are Congo, kauri, and manila.

Cope elimination reaction.   Formation of an olefin and a hydroxylamine by pyrolysis of an amine oxide.

Cope rearrangement.   Thermal isomerization of 1,5-dienes by 1 3,3-sigmatropic shift.

"Copel."[166]   TM for 55–45 copper-nickel alloy used as a resistor material in the construction of electrical instruments where temperature coefficient of resistance must be very low.

Cope's rule.   The tendency for body size to increase during the evolution of a group of animals.

copolymer.   An elastomer produced by the simultaneous polymerization of two or more dissimilar monomers, as SBR synthetic rubber from styrene and butadiene.

copper.   CAS: 7440-50-8.   Cu.   Metallic element of atomic number 29, of group IB of the periodic system, aw 63.546, valences 1, 2; two stable isotopes.
Properties: Distinctive reddish color, d 8.96, mp 1083C, bp 2595C, ductile, excellent conductor of electricity. Complexing agent, coordination numbers 2 and 4. Dissolves readily in nitric and hot concentrated sulfuric acids; in hydrochloric and dilute sulfuric acids slowly, but only when exposed to the atmosphere. More resistant to atmospheric corrosion than iron, forming a green layer of hydrated basic carbonate. Readily attacked by alkalies. A necessary trace element in human diet, a factor in plant metabolism. Essentially nontoxic in elemental form. Noncombustible, except as powder.
Ores: Azurite, azurmalachite, chalococite, chalcopyrite (copper pyrites), covellite, cuprite malachite.
Sources: Michigan, Arizona, Utah, Montana, New Mexico, Nevada, Tennessee, Chile, Canada, USSR, Zambia, Zaire.
Derivation: Varies with the type of ore. With sulfide ores, the steps may be (1) concentration (of low grade ores) by flotation and leaching, (2) roasting, (3) formation of copper "matte" (40–50% Cu), (4) reduction of matte to "blister" copper (96–98%), (5) electrolytic refining to 99.9+% copper.
Forms available: Ingot, sheet, rod, wire, tubing, shot, powder, high purity (impurities less than 10 ppm) as single crystals or whiskers.
Hazard: Flammable in finely divided form. TLV: (fume) 0.2 mg/m³, (dusts and mists) 1 mg/m³.
Use: Electric wiring; switches, plumbing, heating,

roofing and building construction; chemical and pharmaceutical machinery; alloys (brass, bronze, Monel metal, beryllium-copper); electroplated protective coatings and undercoats for nickel, chromium, zinc, etc., cooking utensils; corrosion-resistant piping; insecticides; catalyst; antifouling paints. Flakes used as insulation for liquid fuels. Whiskers used in thermal and electrical composites.

copper abietate.   (cupric abietate).
$Cu(C_{20}H_{29}O_2)_2$.
Properties: Green scales, soluble in alcohol and in oils with fine green color, insoluble in water.
Derivation: Heating copper hydroxide with abietic acid.
Use: Preservative metal paint, fungicide.

copper acetate.   (cupric acetate; crystals of Venus; verdigris, crystallized).   CAS: 142-71-2.
$Cu(C_2H_3O_2)_2 \cdot H_2O$.
Properties: Greenish-blue, fine powder. Soluble in water, alcohol and ether. D 1.9, mp 115C, decomposes at 240C.
Derivation: Action of acetic acid on copper oxide and subsequent crystallization.
Use: Pesticide, catalyst, fungicide, pigments, manufacture of Paris green.

copper acetate, basic.   (copper subacetate; verdigris; verdigris blue; verdigris green).
Properties: Masses of minute, silky crystals either pale green or bright blue in color. Blue variety approximate formula $(C_2H_3O_2)_2Cu_2O$. Green variety approximate formula $CuO \cdot 2Cu(C_2H_3O_2)_2$. Coppery taste. The green rust with which uncleaned copper vessels become coated and which is commonly termed verdigris is a copper carbonate and must not be confused with true verdigris. Apart from its impurities, verdigris is a variable mixture of the basic copper acetates. Soluble in acids, very slightly soluble in water and alcohol.
Derivation: Action of acetic acid on copper in the presence of air.
Use: Paint pigment, insecticide, fungicide, mildew preventive, mordant in dyeing and printing.

copper acetoarsenite.   (cupric acetoarsenite; Paris green; king's green; Schweinfurt green; imperial green). $(CuO)_3As_2O_3 \cdot Cu(Cu_2H_3O_2)_2$.
Properties: Emerald-green powder. Soluble in acids, insoluble in alcohol and water. Phytotoxic.
Derivation: By reacting sodium arsenite with copper sulfate and acetic acid.
Hazard: Toxic by ingestion.
Use: Wood preservative, larvicide, marine antifouling paints.

copper acetylacetonate.   (copper-2,4-pentanedione). $Cu(C_5H_7O_2)_2$.   Crystalline powder, slightly soluble in water and alcohol, soluble in chloroform, mp greater than 230C. Resistant to hydrolysis. A chelating nonionizing compound.

copper amalgam.
Properties: Hard, brown leaflets. Contains approximately 74% mercury and approximately 24% copper. Soluble in nitric acid.
Use: Dental cement.

copper aminoacetate. See copper glycinate.

copper aminosulfate. See copper sulfate, ammoniated.

copper ammonium acetate. (cuprous acetate, ammoniacal). Used in an absorption process for separating butadiene from C_4 streams at refineries or from byproducts of ethylene production.

copper arsenate. Probably the basic cupric arsenate.
Properties: Light blue, blue, or bluish-green powder. Variable composition. Contains approximately 33% copper and approximately 29% arsenic. Soluble in dilute acids, ammonium hydroxide; insoluble in alcohol, water.
Use: Insecticide, fungicide.

copper arsenite. (cupric arsenite; copper orthoarsenite; Scheele's green).
CAS: 10290-12-7. $CuHAsO_3$, or $Cu_3(AsO_3)_2 \cdot 3H_2O$, variable.
Properties: Fine, light-green powder. Soluble in acids, insoluble in water and alcohol, mp decomposes.
Use: Insecticide, fungicide, pigment, wood preservative, rodenticide.

copper arsenite ammoniacal. See chemonite.

copperas. See ferrous sulfate.

copperas, blue. See copper sulfate.

copperas, green. See ferrous sulfate.

copperas, white. See zinc sulfate.

copper benzoate. $(C_6H_5COO)_2Cu \cdot 2HOH$.
Properties: Blue, crystalline, odorless powder; slightly soluble in cold water, acids, and alcohol; mp (loses water) 110C.
Derivation: Interaction of solutions of a benzoate and copper salt.

copper-beryllium. See beryllium-copper.

copper, blister. See blister copper.

copper blue. See mountain blue.

copper borate. See copper metaborate.

copper bromide. (cupric bromide). $CuBr_2$.
Properties: Black powder or crystals, deliquescent. Soluble in acetone, alcohol, water; mp 498C; d 4.77 (25C).
Use: Photography (intensifier), organic synthesis (brominating agent), battery electrolyte, wood preservative.

copper carbonate. (cupric carbonate; copper carbonate, basic; artificial malachite; mineral green. For the native mineral see malachite).
CAS: 12069-69-1. $Cu_2(OH)_2CO_3$.
Properties: Green powder. Soluble in acids, insoluble in water, d 3.7–4.0, decomposes at 200C.
Derivation: By adding sodium carbonate to a solution of copper sulfate, filtering, and drying.
Grade: Technical, CP.
Hazard: Toxic by ingestion.
Use: Pigments, pyrotechnics, insecticides, copper salts, coloring brass black, astringent in pomade preparations, antidote for phosphorus poisoning, smut preventive, fungicide for seed treatment, feed additive (in small amounts).

copper chloride. (cupric chloride).
CAS: 1344-67-8. (a) $CuCl_2$, (b) $CuCl_2 \cdot 2H_2O$.
Properties: (a) Brownish-yellow powder, hygroscopic, (b) green, deliquescent crystals, soluble in water and alcohol. D (a) 3.386 (25C), (b) 2.54 (25C); (a) mp 620C at 993C decomposes to cuprous chloride; (b) loses 2HOH at 100C.
Derivation: (a) By the union of copper and chlorine. (b) Copper carbonate is treated with hydrochloric acid and the product crystallized.
Grade: Technical, CP, reagent.
Hazard: Toxic by ingestion and inhalation.
Use: Isomerization and cracking catalyst, mordant in dyeing and printing fabrics, sympathetic ink, disinfectant, pyrotechnics, wood preservation, fungicides, metallurgy, preservation of pulpwood, deodorizing and desulfurizing petroleum distillates, photography, water purification, feed additive, electroplating baths, pigment for glass and ceramics, acrylonitrile manufacturing. See also cuprous chloride.

copper chromate. (cupric chromate, basic). $CuCrO_4 \cdot 2CuO \cdot 2H_2O$.
Properties: Light chocolate-brown powder. Loses water at 260C. Soluble in nitric acid insoluble in water.
Derivation: Action of chromic acid on copper hydroxide.
Hazard: A carcinogen. TLV: (as Cr) 0.5 mg/m³ of air.

Use: Mordant in dyeing, wood preservative, seed treatment fungicides.

copper cyanide. (cupric cyanide).
CAS: 544-92-3.　　Cu(CN)$_2$.
Properties: Green powder. Keep well stoppered. Soluble in acids and alkalies, insoluble in water.
Derivation: Addition of potassium cyanide to a solution of copper sulfate; cupric cyanide is precipitated. This can be dried but is not stable.
Grade: Technical.
Hazard: A poison. TLV: (as CN) 5 mg/m^3 of air.
Use: Electroplating copper on iron, intermediate (introduction of the cyanide group in place of the amino radical in aromatic organic compounds).
See also cuprous cyanide.

copper, deoxidized. (copper, oxygen-free).
Copper metal specially treated (as by addition of phosphorus) to remove all or a part of the 0.05% oxygen, normally present. It is more ductile than ordinary copper.

copper dihydrazinium sulfate.
(CuSO$_4$(N$_2$H$_4$)$_2$·H$_2$SO$_4$.
Properties: Bluish powder, mp greater than 300C, starts to decomposes at 140C, very slightly soluble in water (250 ppm at 80C).
Hazard: Toxic by ingestion or inhalation. Skin and eye irritant.
Use: Foliage fungicide.

copper dimethyldithiocarbamate. See "Cumate."

copper, electrolytic. Copper, refined by electrolysis. The purest form of copper available commercially.

copper ethylacetoacetate. Cu(C$_6$H$_9$O$_3$)$_2$.
Properties: Blue-green powder, mp 192–193C. Insoluble in water, soluble in most organic solvents.
Use: Fungicide, intermediate.

copper ferrocyanide. (cupric ferrocyanide).
Cu$_2$Fe(CN)$_6$·7H$_2$O.
Properties: Reddish-brown powder, insoluble in water and acids, soluble in ammonium hydroxide and potassium cyanide solutions.
Use: Pigment in paints and enamels, analytical test for traces of copper, inorganic osmotic membranes.

copper fluoride. (cupric fluoride).
CuF$_2$·2H$_2$O.　　The anhydrous form, CuF$_2$, is also available.
Properties: Blue crystals. Slightly soluble in water, soluble in acids and alcohol, d 2.93 (25C).

Derivation: (1) By decomposing copper carbonate with hydrofluoric acid and subsequent crystallization; (2) fluorination of copper hydroxyfluoride at 525C.
Grade: Technical.
Hazard: A poison. TLV: (as F) 2.5 mg/m^3 of air.
Use: Ceramics, enamels, flux in metallurgy, high-energy batteries, fluorinating agent.

copper fluosilicate. (copper silicofluoride; cupric fluosilicate; cupric silicofluoride).
CuSiF$_6$·4H$_2$O.
Properties: Blue, hygroscopic crystals. Soluble in water, slightly soluble in alcohol, d 2.158, decomposed by heat.
Derivation: Interaction of copper hydroxide and hydrofluosilicic acid.
Method of purification: Crystallization.
Grade: Technical.
Hazard: A poison. TLV: (as F) 2.5 mg/m^3 of air.
Use: Dyeing and hardening white marble, treating grape vines for "white disease."

copper glance. See chalcocite.

copper gluconate. (cupric gluconate).
[Cu$_2$OH(CHOH)$_4$COO]$_2$Cu.
Properties: Odorless, light blue, fine, crystalline powder. Soluble in water; insoluble in acetone, alcohol, and ether.
Grade: Pharmaceutical, FCC.
Use: Feed additive, dietary supplement, mouth deodorant.

copper glycinate. (copper aminoacetate).
(NH$_2$CH$_2$COO)$_2$Cu.
Properties: Blue, triboluminescent crystals; mp 130C. Slightly soluble in water and alcohol; insoluble in hydrocarbons, ethers, and ketones.
Grade: Anhydrous, hydrate.
Use: Catalyst for rapid biochemical assimilation of iron, electroplating baths, photometric analysis, feed additive.

copper hemioxide. See copper oxide, red.

copper hydrate. See copper hydroxide.

copper hydroxide. (cupric hydroxide; copper oxide hydrated; copper hydrate).
CAS: 20427-59-2.　　Cu(OH)$_2$.
Properties: Blue powder. Soluble in acids and ammonium hydroxide. Insoluble in water, d 3.368, mp decomposes.
Derivation: Interaction of a solution of a copper salt with an alkali.
Grade: Technical.
Hazard: Toxic by ingestion and inhalation.
Use: Copper salts, mordant, cuprammonium

rayon, pigment, staining paper, feed additive, pesticide and fungicide, catalyst.

copper-8-hydroxyquinoline. See copper-8-quinolinolate.

"Copper Inhibitor 50."[28] TM for 50% disalicylalpropylenediamine and 50% aromatic solvent. Used to prevent catalytic action of copper on oxidation of natural and synthetic rubbers.
Hazard: Flammable, moderate fire risk.

copper iodide. See cuprous iodide.

copper lactate. (cupric lactate).
$Cu(C_3H_5O_3)_2 \cdot 2H_2O$.
Properties: Greenish-blue crystals or granular powder, soluble in water and ammonium hydroxide.
Use: Source of copper in copper plating fungicides.

copper mercury iodide. See mercuric cuprous iodide.

copper metaborate. (copper borate; cupric borate). $Cu(BO_2)_2$.
Properties: Bluish-green, crystalline powder; insoluble in water; soluble in acids; d 3.859.
Derivation: Interaction of copper sulfate and sodium borate.
Grade: Technical, CP.
Hazard: Toxic by ingestion.
Use: Dehydrogenation catalyst, wood preservative, fire-retardant, paint and ceramic pigment, insecticides (especially wheat-rust compounds).

copper methane arsenate. CH_3AsO_3Cu.
Properties: Greenish solid.
Derivation: Reaction of disodium methyl arsenate with copper salts.
Hazard: Toxic by ingestion.
Use: Algicide.

copper molybdate. $CuMoO_4$.
Properties: Crystals, density 3.4g/cc, mp approximately 500C. Insoluble in water.
Grade: 99.98% pure. Electronic and optical equipment.
Use: Paint pigment and protective coatings, corrosion inhibitor.

copper monoxide. See copper oxide, black.

copper naphthenate. CAS: 1338-02-9.
Properties: Green-blue solid. High germicidal power. Soluble in gasoline, benzene, and mineral oil distillates.
Derivation: Addition of solution of cupric sulfate to aqueous solution of sodium naphthenate.

Grade: 6, 8, 11.5% copper.
Hazard: Flammable, moderate fire risk (solution). Toxic by ingestion and inhalation.
Use: Wood, canvas and rope preservative, insecticide, fungicide, antifouling paints.
See also soap (2).

copper nitrate. (cupric nitrate).
CAS: 3251-23-8.   (a) $Cu(NO_3)_2 \cdot 3H_2O$,
(b) $Cu(NO_3)_2 \cdot 6H_2O$.
Properties: Blue, deliquescent crystals. Soluble in water and alcohol. D (a) 2.32 (25C), (b) 2.074; mp (a) 114.5C, (b) loses 3HOH at 26.4C; (a) decomposes 170C.
Derivation: By treating copper or copper oxide with nitric acid. The solution is evaporated and product recovered by crystallization.
Grade: Technical, CP.
Hazard: Oxidizer, with organic materials causes violent combustion or explosion.
Use: Light-sensitive papers; analytical reagent; mordant in textile dyeing; nitrating agent; insecticide for vines; coloring copper black; electroplating; production of burnished effect on iron; paints; varnishes, enamels; pharmaceutical preparations; catalyst.

copper nitrite. (copper nitrite basic; cupric nitrite). $Cu(NO_2)_2 \cdot 3Cu(OH)_2$, variable.
Properties: Green powder. Decomposes at 120C. Soluble (with decomposition) in dilute acids, ammonium hydroxide, slightly soluble in water.

copper nitrite basic. See copper nitrite.

copper octoate. See soap (2).

copper oleate. (cupric oleate).
$Cu(C_{18}H_{33}O_2)_2$.
Properties: Brown powder or greenish-blue mass. Soluble in ether, insoluble in water. Combustible.
Derivation: Interaction of copper sulfate and sodium oleate.
Use: Preserving fish nets and marine lines, fungicide, insecticide, ore flotation, lube oil antioxidant, emulsifying agent, fuel oil ignition improver, catalyst.
See also soap (2).

copper oxalate. (cupric oxalate). CuC_2O_4.
Properties: Bluish-green powder, decomposes at approximately 300C to copper oxide. Insoluble in water, alcohol, and acetic acid; soluble in ammonium hydroxide.
Derivation: Reaction of oxalic acid with copper sulfate.
Hazard: Toxic by ingestion; tissue irritant.
Use: Catalyst in organic synthesis, rodent repellent in seed coatings.

copper oxide black.   (cupric oxide; copper monoxide).   CAS: 1317-39-1.   CuO.
For native black copper oxide see tenorite.
Properties: Brownish-black powder. Soluble in acids, difficultly soluble in water, d 6.32, decomposes at 1026C.
Derivation: Ignition of copper carbonate or copper nitrate.
Hazard: Toxic by ingestion.
Use: Ceramic colorant, reagent in analytical chemistry, insecticide for potato plants, catalyst, purification of hydrogen, batteries and electrodes, aromatic acids from cresols, electroplating, solvent for chromic iron ores, desulfurizing oils, rayon, metallurgical and welding fluxes, antifouling paints, phosphors.

copper oxide hydrated.   See copper hydroxide.

copper oxide red.   (cuprous oxide; copper protoxide; copper hemioxide; copper suboxide).
Cu_2O.   For the native ore see cuprite.
Properties: Reddish-brown, octahedral crystals. Soluble in acids and ammonium hydroxide, insoluble in water, d 5.75–6.09, mp 1235C, bp 1800C.
Derivation: (a) Oxidation of finely divided copper. (b) Addition of bases to cuprous chloride. (c) Action of glucose on cupric hydroxide.
Grade: Technical, CP, 97% min (for pigments), also USN Type I (97%), USN Type II (90%).
Hazard: Toxic by ingestion.
Use: Copper salts, ceramics, porcelain red glaze, red glass, electroplating, antifouling paints, fungicide, catalyst, brazing preparations, photocells.

copper oxinate.   See copper-8-quinolinolate.

copper oxychloride.   (cupric oxychloride).
Composition variable, possibly $3CuO \cdot CuCl_2 \cdot 3.5HOH$.
Properties: Bluish-green powder. Soluble in acids, ammonia; insoluble in water.
Hazard: Toxic by ingestion and inhalation.
Use: Pigment, pesticide, fungus control in grapevines.

copper-2,4-pentanedione.   See copper acetylacetonate.

copper phenolsulfonate.   (copper sulfocarbolate).
$[C_6H_4(OH)SO_3]_2Cu \cdot 6H_2O$.
Properties: Green, prismatic crystals. Soluble in water and alcohol. Combustible.
Derivation: Interaction of barium phenolsulfonate and copper sulfate.
Use: Esterification catalyst, electroplating.

copper phosphate.   (copper orthophosphate).
$Cu_3(PO_4)_2 \cdot 3H_2O$.
Properties: Light-blue powder. Soluble in acids, ammonium hydroxide; insoluble in water.

Use: Analysis, fungicide, catalyst, oxidation inhibitor for metals.

copper phosphide. Cu_3P_2.
Properties: Grayish-black, metallic powder. Insoluble in water, soluble in nitric acid; insoluble in hydrochloric acid. D 6.67.
Derivation: By heating copper and phosphorus.
Hazard: Dangerous, spontaneously flammable and toxic phosphine evolved on reaction with water. May explode when mixed with potassium nitrate.
Use: Manufacturing phosphor-bronze.

copper phthalate. $C_8H_4O_4Cu$.
Properties: Fine, blue powder; assay min 95%; very slightly soluble in common organic solvents or water. Combustible.
Hazard: Toxic by ingestion.
Use: Fungicide.

copper phthalocyanine blue.   See Pigment Blue 15.

copper phthalocyanine green.   See Pigment Green 7.

copper potassium ferrocyanide.   (potassium copper ferrocyanide). $K_2CuFe(CN)_6 \cdot H_2O$.
Properties: Brownish-red powder. Insoluble in water.
Use: Pigment.

copper protoxide.   See copper oxide red.

copper pyrites.   See chalcopyrite.

copper-8-quinolinolate.   (copper-8; copper oxinate; copper-8-hydroxyquinoline).
$Cu(C_9H_6ON)_2$.
Properties: Yellow-green, nonhygroscopic, odorless powder. Insoluble in water. Somewhat soluble in weak acids, soluble in strong acids. Insoluble in most organic solvents, but somewhat soluble in pyridine and quinoline. Soluble copper-8 refers to the product formed by heating copper-8-quinolinolate with certain organic acids (naphthenic, lactic, stearic, etc.) or their salts. In such products, the copper-8-quinolinolate does not settle out on standing even after dilution with various solvents.
Derivation: From 8-quinolinol and copper salt such as copper acetate.
Grade: 10% active salt (1.8% Cu) solution.
Hazard: Toxic by ingestion.
Use: Fungicide and mildew-proofing of fabrics, analysis for copper.

copper resinate.   (cupric resinate).
Properties: Green powder. Soluble in ether and oils, insoluble in water. Combustible.

Derivation: By heating copper sulfate and rosin oil and filtering and drying the precipitate.
Use: Antifouling paints, insecticide.
See also soaps (2).

copper ricinoleate. $Cu(C_{17}H_{32}OHCOO)_2$.
Properties: Green plastic solid. Soluble in water and aliphatic hydrocarbons, partially soluble in alcohols and glycols, soluble in ketones and aromatic hydrocarbons. Softening p 64C.
Hazard: Toxic by ingestion. Combustible.
Use: Fungicides, insecticides.

copper scale. A coating formed on copper after heating, composed of cupric and cuprous oxides.

copper selenate. (cupric selenate).
$CuSeO_4 \cdot 5H_2O$.
Properties: Light-blue crystals. Soluble in acids, ammonium hydroxide, water; insoluble in alcohol; d 2.559; loses 4HOH at 50–100C.
Use: Black colorant for copper.

copper silicate. A complex mixture precipitated by solutions of copper salts from sodium silicate solutions. Used in pigments, catalysts, and insecticides.

copper silicide. See silicon-copper.

copper silicofluoride. See copper fluosilicate.

copper sodium chloride. (sodium copper chloride). $CuCl_2 \cdot 2NaCl \cdot 2H_2O$.
Properties: Light-green crystals. Soluble in water.

copper sodium cyanide. See sodium copper cyanide.

copper stearate. (cupric stearate).
$Cu(C_{18}H_{35}O_2)_2$.
Properties: Light blue powder. Soluble in ether, chloroform, benzene, and turpentine. Insoluble in water and alcohol; soluble in hot benzene, toluene, carbon tetrachloride. Mp 125C. Combustible.
Derivation: By the interaction of copper sulfate and sodium stearate.
Use: Preservative for cellulosic materials, antifouling paints, catalyst.

copper subacetate. See copper acetate basic.

copper suboxide. See copper oxide red.

copper sulfate. (cupric sulfate; blue vitriol; blue stone; blue copperas). $CuSO_4 \cdot 5H_2O$.
Properties: Blue crystals or blue, crystalline granules or powder; slowly efflorescing in air; white when dehydrated; nauseous metallic taste. Soluble in water, methanol; slightly soluble in alcohol and glycerol; d 2.284.
Derivation: Action of dilute sulfuric acid on copper or copper oxide (often as oxide ores) in large quantities with evaporation and crystallization.
Method of purification: Recrystallization.
Grade: Technical, CP, NF, also sold as monohydrate. Available as crystals or powder.
Hazard: Toxic by ingestion, strong irritant.
Use: Agriculture (soil additive, pesticides, Bordeaux mixture), feed additive, germicides, textile mordant, leather industry, pigments, electric batteries, electroplated coatings, copper salts, reagent in analytical chemistry, medicine, wood preservative, preservation of pulp wood and ground pulp, process engraving and lithography, ore flotation, petroleum industry, synthetic rubber, steel manufacture, treatment of natural asphalts. The anhydrous salt is used as a dehydrating agent.

copper sulfate ammoniated. (cupric ammonia sulfate; ammonio-cupric sulfate; copper aminosulfate). $Cu(NH_3)_4SO_4 \cdot H_2O$.
Properties: Dark blue, crystalline powder; decomposes in air; soluble in water; insoluble in alcohol.
Derivation: By dissolving copper sulfate in ammonium hydroxide and precipitating with alcohol.
Hazard: Toxic by ingestion.
Use: Calico printing, manufacturing copper arsenate, insecticide, treating fiber products.

copper sulfate, tribasic. CAS: 7758-98-7.
$CuSO_4 \cdot 3Cu(OH)_2 \cdot H_2O$.
Properties: Aqua colored powder of extremely fine particle size, water-soluble, stable in storage, forms essentially neutral water dispersion.
Hazard: Toxic by ingestion.
Use: A fixed copper fungicide. Also micronutrient for plants. Compatible with arsenicals, organic insecticides, sulfur and cryolite. Used as spray or dust. Does not inhibit photosynthesis.

copper sulfide. (cupric sulfide). CuS.
Properties: Black powder or lumps. Soluble in nitric acid, insoluble in water. Occurs as the mineral covellite. D 3.9–4.6, decomposes 220C.
Derivation: By passing hydrogen sulfide gas into a solution of a copper salt.
Hazard: Toxic by ingestion.
Use: Antifouling paints, dyeing with aniline black, catalyst preparation.
See also cuprous sulfide.

copper sulfocarbolate. See copper phenolsulfonate.

copper sulfocyanide. See cuprous thiocyanate.

copper tallate. See soaps (2).

copper trifluoroacetylacetonate.
Cu[OC(CH₃):CHCO(CF₃)]₂.
Properties: Solid, mp 188–190C.
Use: Metal analysis standards, vapor phase deposition of metals, laser studies.

copper tungstate. (cupric tungstate).
CuWO₄·2H₂O.
Properties: Light-green powder, soluble in ammonium hydroxide, slightly soluble in acetic acid, insoluble in alcohol and water.
Use: Polyester catalyst, semiconductors, nuclear reactors.

copper yellow. See chalcopyrite.

copper zinc chromate. Variable in composition. Used as a fungicide.
Hazard: Toxic by ingestion and inhalation. A carcinogen. TLV: (as Cr) 0.5 mg/m³.

"Coppralyte."²⁸ TM for a group of products for electroplating copper.

"Coprantine."⁴⁴³ TM for dyes requiring aftertreatment with copper compounds.

copra oil. The name applied to lower grades of coconut oil.

"Coral."²⁷⁸ TM for a stereospecific, polyisoprene rubber consisting essentially of cis-1,4-polyisoprene.
Properties: Similar to those of natural rubber, both unvulcanized and vulcanized.
Use: Replacement for natural rubber.

coral. Skeletons of the coral polyps found in the warmer oceans and consisting mainly of calcium carbonate colored with ferric oxide. Coral rock is porous. It occurs in the form of reefs east of Australia and at other locations in the southwest Pacific. It has been found to be the habitat of organisms that are rich in prostaglandins.
Hazard: Will infect open wounds.
Use: Building stone, cement, road and air field construction, jewelry.

"Coray."⁵¹ TM for a series of general-purpose, naphthenic-base oils that serve as lubricants, in process applications, and as components in proprietary formulations.

"Corcel" 46.¹⁹ᵃ TM for cellular glass microspheres in the form of fine, white powder, obtained by benefication of a natural siliceous earth;

has high water absorption value, low thermal conductivity, bulk density of 3–4.5 lb/cubic foot. Used as filler for plastics, thermal insulating material, admixture for concrete, surface coatings, and catalyst carrier.

cord (wood). Unit volume of cut wood stacked in a pile measuring 4 feet by 4 feet by 8 feet and containing 128 cubic feet.

cordite. A smokeless powder which is a mixture of nitrocellulose and nitroglycerin with approximately 5% petrolatum added to thicken and stabilize the mixture. Materials are dissolved in acetone and mixed. Evaporation of the excess acetone leaves a gelatinous mass which is extruded into cords.

core of the earth. The central part of the earth below a depth of 2900 kilometers. It is thought to be composed largely of iron and to be molten on the outside with a solid central region.

"Corephen 10."⁴⁷⁰ TM for phenol formaldehyde condensation product; used as vehicle in highly acid-resistant paints.

Corey-Winter olefin synthesis. Synthesis of olefins from 1,2-diols and thiocarbonyldiimidazole. Treatment of the intermediate cyclic thionocarbonate with trimethylphosphite yields the olefin by cis-elimination.

"CoRezyn."⁴⁶³ TM for a variety of polymer resins tailored for specific uses and for materials used with the polymers in making finished products.

"Corgard."²⁰ TM for a borosilicate glass armored with an opaque, filament-wound laminate of glass fiber and a modified polyester resin.

coriander oil. An essential oil with strong spicy taste; may be irritating when ingested. Dextrorotatory. Used chiefly to flavor gin.

coriandrol. See d-linalool.

Cori ester. See glucose-1-phosphate.

coriolis effect. The tendency of a current of air or water flowing over the surface of the earth to bend to the right in the Northern Hemisphere and to the left in the Southern Hemisphere.

cork. A form of cellulose comprising the light outer bark of the oak tree known as Quercus suber. It grows naturally in southern Europe and northern Africa, and has been cultivated in southwestern US. Its special properties are extreme lightness, relatively impervious to water, resilient structure, and low rate of heat transfer. These account for its usefulness as bottle stoppers, insulation, wallboard, life preservers, gaskets, and sound-deadening insertions. D 0.1–0.25 g/cc. Combustible.

corkboard. A mixture of ground cork and paper pulp formed into thick sheets for insulating purposes.

"Colar."[28] TM for chemical resistant finishes having a base of polyamide catalyzed epoxy resins.

Cornforth, John. (1917-) An Australian born chemist who won the Nobel prize for chemistry in 1975 with V. Prelog for work on the chemical synthesis of organic compounds. Although deaf since childhood, he attained his doctorate from Oxford and held prestigious posts all over the world, as well as authoring many papers on organic and biochemical subjects.

corn oil. (maize oil).
Properties: Pale yellow liquid, characteristic taste and odor. Insoluble in water; soluble in ether, chloroform, amyl acetate, benzene, and carbon disulfide; slightly soluble in alcohol. D 0.914–0.921, saponification value 188–193, iodine value 102–128, flash p 490F (254C), combustible, nontoxic and nondrying, moderate tendency to spontaneous heating. Chief constituents: Linoleic and oleic acids (unsaturated), palmitic and stearic (saturated).
Derivation: The germ of common corn (Indian corn, Zea mays) is removed from the grain and pressed.
Grade: Crude, refined, USP, technical.
Use: Foodstuffs, soap, lubricants, leather dressing, factice, margarine, salad oil, hair dressing, solvent.

cornstarch. A carbohydrate polymer derived from corn of various types, composed of 25% amylose and 75% amylopectin. A white powder which swells in water, it is the most widely used starch in the US. The so-called waxy variety (made from waxy corn) contains only branched amylopectin molecules. Its chief uses are as a source of glucose, in the food industry as a filler in baking powder and a thickening agent in various food products, and in adhesives and coatings. It has been proposed as an additive to plastics to promote rapid degradation in such products as bottles and waste containers.
See also starch.

corn steep liquor. The dilute aqueous solution obtained by soaking corn kernels in warm 0.2%

sulfur dioxide solution for 48 hours as the first step in the recovery of corn starch, corn oil, and gluten from corn. The solution contains mineral matter as well as soluble organic material extracted from the corn. It is used as a growth medium for penicillin and other antibiotics, and it is also concentrated and used as an ingredient of cattle feeds.

corn sugar. See dextrose.

corn syrup. See glucose syrup.

"Corobex."[159] TM for a series of organotin salt compounds, phenylmercuric salt compounds, and quaternary compounds used as bacteriostatic and fungistatic finish in the textile, plastics, and rubber industries.

corona. An electrical discharge effect which causes ionization of oxygen and the formation of ozone. It is particularly evident near high-tension wires and in spark-ignited automotive engines. The ozone formed can have a drastic oxidizing effect on wire insulation, cable covers, and hose connections. For this reason, such accessories are made of oxidation-resistant materials such as nylon, neoprene, and other synthetics.

coroxon. (O,O-diethyl-O-(3-chloro-4-methylcoumarin-7-yl)phosphate).
Hazard: A cholinesterase inhibitor.
Use: Insecticide, fungicide.

corresponding states. (reduced states).
Two substances are in corresponding states when their pressures, volumes (or densities), and temperatures are proportional respectively to their critical pressures, volumes (or densities), and temperatures. If any two of these ratios are equal, the third must also be equal. This principle has been useful in the development of physical and thermal properties of substances.

corrosion. (1) The electrochemical degradation of metals or alloys due to reaction with their environment, which is accelerated by the presence of acids or bases. In general, the corrodability of a metal or alloy depends upon its position in the activity series. Corrosion products often take the form of metallic oxides. This is actually beneficial in the case of aluminum and stainless steel, for the oxide forms a strongly adherent coating which effectively prevents further degradation. Hence, these metals are widely used for structural purposes. The rusting of iron is a familiar example of corrosion which is catalyzed by moisture. Acidic soils are highly corrosive. Sulfur is a corrosive agent in automotive fuels and in the atmosphere (as SO_2). sodium chloride in the air at locations near the sea is also strongly corrosive, especially at temperatures above 21C. Copper, nickel, chromium and zinc are among the more corrosion-resistant metals, and are widely used as protective coatings for other metals. Excellent corrosion-resistant alloys are stainless steel (18 Ni–8Cr), monel metal (66 Ni–34Cu), and duralumin.

(2) The destruction of body tissues by strong acids and bases.
See corrosive material, protective coatings, paints. See also tarnish.

"Corrosion Inhibitor CS."[108]   TM for a synergistic combination of sodium nitrate-borax and organic inhibitors, used to prevent corrosion of ferrous and nonferrous metal and alloy surfaces in low-makeup closed cooling and heating systems.

corrosive material.   Any solid, liquid, or gaseous substance that attacks building materials, metals, burns, irritates, or destructively attacks organic tissues, most notably the skin, and when taken internally, the lungs and stomach. Among the more widely used chemicals that have corrosive properties are the following:

acetic acid, glacial   hydrofluoric acid
acetic anhydride   nitric acid
bromine   potassium hydroxide
chlorine   sodium hydroxide
fluorine   sulfuric acid
hydrochloric acid

See also toxic materials.

corrosive sublimate.   Obsolete term for mercuric chloride.

"Corrosol."[526]   TM for phosphoric acid-type metal conditioners and rust removers used to remove rust from steel and prepare it for further processing or prepare normally nonreceptive zinc aluminum surfaces for paint.

corticoid hormone.   A hormone produced or isolated from the cortex (external layer) of the adrenal gland. Corticoid hormones now used in medicine include cortisone, hydrocortisone, deoxycorticosterone, fluorcortisone, prednisone, prednislone, methyl prednisolone, triamcinolone, dexamethasone, corticotropin (ACTH),and aldosterone. Some occur naturally in adrenal extract, others are modifications of the natural hormones. All are now made synthetically. They are derivatives of cyclopentanophenanthrene.
See also cortisone, ACTH.

corticosterone.   CAS: 50-22-6. $C_{21}H_{30}O_4$.
One of the less active adrenal cortical steroid hormones.
Properties: Crystalline plates, mp 180–182C. Soluble in organic solvents, insoluble in water.
Derivation: Isolation from adrenal cortex extract, synthesis from deoxycholic acid.
Use: Biochemical research, medicine.

corticotropin.   See ACTH.

cortisol.   See hydrocortisone.

cortisone.   (11-dehydro-17-hydroxycorticoster one).   CAS: 53-06-5. $C_{21}H_{28}O_5$.
An adrenal, cortical, steroid hormone. It affects carbohydrate and protein metabolism.
Properties: White, crystalline solid; mp 220–224 (decomposes). Dextrorotatory in solutions. Slightly soluble in water; sparingly soluble in ether, benzene, and chloroform; fairly soluble in methanol, ethanol, and acetone.
Derivation: From adrenal gland extract (usually from cattle) (historical method), synthetically from bile acids, from other steroids or sapogenins.
Hazard: Damaging side effects, e.g., sodium retention from ingestion.
Use: Medicine, an anti-inflammatory drug used in treatment of acute arthritis, Addison's disease, inflammable diseases of eyes and skin and others.

corundum.   (emery).   CAS: 1302-74-5.
Al_2O_3.   Natural aluminum oxide sometimes with small amounts of iron, magnesium, silica, etc.
Occurrence: New York, Greece, Asia Minor.
Use: Various polishing and abrasive operations, grinding wheels.
See also aluminum oxide, diaspore, sapphire.

corynine.   See yohimbine.

cosmetic.   Any preparation in the form of a liquid, semi-liquid, paste, or powder applied to the skin to improve its appearance, and for cleaning, softening, or protecting the skin or its adjuncts but without specific medicinal or curative effects. They include hairsprays, shampoos, nail polish, deodorants, shaving creams, facial creams, dusting powders, rouge, etc. Detergents, common soap, and bactericidal agents are not themselves classed as cosmetics though they may be components of cosmetic mixtures. A partial list of cosmetic ingredients follows:
   animal fats (lanolin)
   vegetable oils, waxes
   alcohols (glycerol, glycols)
   surfactants (alkyl sulfonates)
   UV blocking agents (PABA)

phenylene diamine
aluminum chlorohydrate
FDC organic dyes
talc (magnesium silicate)
essential oils
inorganic pigments
chlorophyllins
nitrocellulose lacquers.
steroid hormones.

Knowledge of the structure and function of the skin is essential for proper cosmetic formulation (cosmeticology). All ingredients must be tested for possible toxic effects since the skin is an important means of access for poisons. Addition of proteins to cosmetic preparations is of questionable value.

cosmochemistry. See astrochemistry, chemical planetology, elements, origin.

"Cosol."[21] TM for high-boiling coal-tar solvents for use in alkyd resin enamels and synthetic lacquers.

cotton. Staple fibers, surrounding the seeds of various species of *Gossypium*. Both Egyptian and Sea Island cotton have unusually long staple (approximately 2 inches). Cotton is the major textile fiber and an important source of cellulose which constitutes 88–96% of the fiber. So-called "absorbent cotton" is almost pure cellulose.
Properties: Tenacity, 3–6 g per denier (dry), 4–8 g per denier (wet); elongation 3–7%; d 1.54; moisture regain 7% (21C, 65% relative humidity); yellows slowly at 121C; decomposes at approximately 148C; low permanent set; decomposed by acids; swells in caustic but is undamaged. Soluble in cuprammonium hydroxide. Subject to mildew. May be dyed by direct, vat, azoic, sulfur, and basic dyes. Combustible.
Source: US, Brazil, Egypt, India.
Hazard: Toxic by inhalation. TLV: (dust) 0.2 mg/m³ of air. Moderately flammable in the form of dust or linters, fiber ignites readily. In the form of dust or linters, exposure of workers in textile mills may cause "brown lung."
Use: Apparel, industrial and household fabrics, upholstery, medicine, thread.
See also cellulose.

cotton, acetylated. Cotton fibers, threads, or fabrics treated with acetic anhydride, acetic acid, perchloric acid, and catalyst to improve the heat, rot, and mildew resistance by forming a surface coating of cellulose acetate.
Hazard: Ignites readily, not self-extinguishing.

cotton, aminized. A cotton fabric produced by reacting 2-aminoethylsulfuric acid with the cellulose of the fabric in a strongly alkaline solution. The treated cotton can take acid wool dyes and can be made rot-resistant and water-repellent.

cotton, cyanoethylated. Cotton treated with acrylonitrile. It is passed through a caustic bath which induces mild swelling of the fiber and catalyzes the subsequent reaction with acrylonitrile. The fabric is then neutralized with acetic acid, washed, and dried. The treatment leaves 3–5% nitrogen attached to the cellulose polymer. The cyanoethylated fiber is claimed to have permanent rot- and mildew-resistance, greater retention of strength after exposure to heat, improved receptiveness to dyes, and higher abrasion- and stretch-resistance.

cotton linters. Short, fleecy fibers which adhere to cottonseed after it has been passed once through a cotton-gin. They are removed from the seed by a second ginning.
Hazard: Flammable, dangerous fire risk.
Use: Rayon manufacture, cellulosic plastics, nitrocellulose lacquers, soil-cement binder in road construction, explosives.

cotton, mercerized. Cotton which has been strengthened by passing through 25–30% solution of sodium hydroxide under tension and then washed with water, while under tension. This causes the fibers to shrink, increases their strength and attraction for colors as well as imparting luster. A process using liquid ammonia for this purpose has been introduced in the UK.

cotton oil. (cotton spraying oil). A compounded oil sprayed (in the form of a fine-mist) onto cotton to condition the fibers for yarn-making operations. Used to lubricate the fibers, to reduce static, "fly," and dust and generally improve the suppleness and strength of the fibers.

cottonseed. The seed of the cotton plant, *Gossypium hirsutum*. It contains about 22% crude fiber, 20% protein, 20% oil, 10% moisture and 24% N-free extract; also contains from 1–2% of toxic pigment gossypol; specially processed kernels are free of this.
Hazard: Generates dangerous amounts of heat if piled or stored wet or hot. Powerful allergen, may cause asthma and other respiratory difficulties on inhalation.
Use: Source of cottonseed oil and meal, cotton linters, source of nutritional protein after removal of gossypol by centrifugation.

cottonseed meal. (cottonseed cake). The pulverized cottonseed press cake. Depending on the extractive process, varying percentages of

protein will remain in the meal and it is normally sold with 36–45% protein content. The 42% product contains approximately 42% crude protein, 6% crude fiber, 25% N-free extract, 10% other extract (fat), and 7% ash. The total digestible nutrient averages 79%. The ash is high in potash and phosphate; some types contain gossypol.
Use: Animal feeds, fertilizer ingredient, filler for plastics.

cottonseed oil.
Properties: Pale yellow or yellowish-brown to dark ruby-red or black-red, fixed, semi-drying oil depending on the nature and condition of the seed. The pure oil is odorless and has a bland taste. Soluble in ether, benzene, chloroform, and carbon disulfide; slightly soluble in alcohol, d 0.915–0.921, iodine value 109–116, flash p 486F (252C). Combustible. Chief constituents: Glycerides of palmitic, oleic and linoleic acids.
Derivation: From cotton seeds by hot-pressing or solvent extraction.
Grade: Crude, refined, prime summer yellow, bleachable, USP.
Use: Leather dressing, soap stock, lubricant, glycerol, base for cosmetic creams, hydrogenated to semisolid for use in food products, waterproofing compositions, dietary supplement.

Cottrell, Frederick G. (1877–1948). A native of California, Cottrell obtained his doctorate from Liebig in 1902. His major contribution to industrial chemistry was his discovery of a practical method of dust elimination by electrical precipitation. Used in factory stacks and other large units, this process has contributed greatly to purifying the atmosphere of industrial areas. The principle involves charging a suspended wire with electricity. This creates a field which ionizes the surrounding air, the particles assuming the charge on contact, and then moving to the wall of the stack where they are electrically discharged and precipitated.

couch roll. (pronounced cooch). A hollow suction roll on a Fourdrinier paper machine over which the formed sheet or web passes as it leaves the wire. The suction is provided by a vacuum or suction box inside the roll whose face is perforated to offer as large a vacuum area as possible. The chief feature of the couch roll is its great water-removing capacity; this gives the sheet enough strength to enable it to hold together as it passes to the pick-up felts.

coumaphos. Generic name for O,O-diethyl-O-(3-chloro-4-methyl-2-oxo-2H-1-benzopyran-7-yl)-phosphorothioate). CAS: 56-72-4. $C_9H_3O_2(CH_3)ClOPS(OC_2H_5)_2$.
Properties: Crystals, mp 91C. Insoluble in water, soluble in aromatic solvents.
Hazard: Use may be restricted; cholinesterase inhibitor.
Use: Insecticide, anthelmintic.

coumarin. (cumarin; benzopyrone; tonka bean camphor). CAS: 91-64-5. $C_9H_6O_2$. A lactone.
Properties: Colorless crystals, flakes, or powder; fragrant odor similar to vanilla; bitter, aromatic burning taste; mp 69C, bp 290C. Soluble in 10 vols of 95% alcohol, in ether, chloroform, and fixed volatile oils; slightly soluble in water. Combustible.
Derivation: (a) By heating salicylic aldehyde, sodium lactate, and acetic anhydride; (b) fine grades are isolated from Tonka beans.
Hazard: Toxic by ingestion; carcinogenic. Use in food products prohibited (FDA).
Use: Deodorizing and odor-enhancing agent, pharmaceutical preparations.

coumarone. (benzofuran). CAS: 271-89-6. C_8H_6O. A bicyclic ring compound derived from coal-tar naphtha, the parent substance of coumarone-indene resins.
Properties: Colorless, oily liquid; d 1.09, bp 165 to 175C, insoluble in water.

coumarone-indene resin. A thermosetting resin derived by heating a mixture of coumarone and indene with sulfuric acid, which induces polymerization. It is soft and sticky at room temperature; it hardens on heating to a resinous solid. Soluble in hydrocarbon solvents, pyridine, acetone, carbon disulfide, and carbon tetrachloride; insoluble in water, alcohol. Combustible. Said to have been the first synthetic polymer.
Use: Adhesives, printing inks, floor tile binder, friction tape.
See also "Cumar," "Nevidene," "Paradene."

countercurrent. Descriptive of a process in which a liquid and a vapor stream or two streams of immiscible liquids or a liquid and a solid are caused to flow in opposite directions and past or through one another with more or less intimate contact so that the individual substances present are more or less completely transferred to that stream in which they are more soluble or stable under the conditions existing. The streams leaving such a process are usually of higher purity that can be attained otherwise at equal cost. Distillation with a fractionating column is also a typical countercurrent process in which rising vapor is purified by contact with

descending liquid (reflux). Leaching, washing, and chemical reaction are frequently carried out in a countercurrent manner.
See extraction, liquid-liquid.

count. (1) The external indication given by a radiation detector, such as a Geiger counter, of the amount of radioactivity to which the detector is exposed. The background counts are those which come from a source external to that being measured. (2) The number of warp and filler threads in a linear inch of a textile fabric, e.g., the count of a sheeting may be 80 x 60.

coupling. (1) The combination of an amine or phenol with a diazonium compound to give an azo compound, the reaction by which azo dyes are prepared. Thus m-phenylenediamine $C_6H_4(NH_2)_2$ couples with benzene diazonium chloride $C_6H_5N_2Cl$ to produce the dye chrysoidine $C_6H_5N_2C_6H_3(NH_2)_2$. See also azo dye intermediate. (2) Oxidative coupling. (3) An agent, e.g., a vinyl silane, used to protect fiberglass laminates from effects of water absorption. (4) A condensation polymerization of amino acids to form proteins; it can be done synthetically only by suppressing certain active sites on the amino acid molecules.

covalent bond. See bond, chemical.

covering power. See opacity.

Cox chart. A special semilogarithmic plot of vapor pressure versus temperature especially useful for the petroleum hydrocarbons. The graph corresponding to each separate hydrocarbon is a straight line. All the lines appear to intersect at a point outside the chart.

CP. Abbreviation for chemically pure, an accepted grade of drugs and fine chemicals that contain a minimum of impurities.

"CP-40."[62] TM for chlorination derivatives of paraffin wax.
Use: Rendering fabrics waterproof and fire retardant, fire-retardant paints, plasticizer and extender for certain plastic materials, etc.

"C-P-B-."[248] TM for dibutyl xanthogen disulfide.
Properties: Amber-colored, free-flowing liquid; d 1.15; soluble in acetone, benzene, gasoline, and ethylene dichloride; insoluble in water.
Use: Accelerator for pure gum hand-made druggists sundries and medical supplies, bathing shoes, bathing caps, novelties, and cold-cure cements.

"CP" Bond.[30] TM for custom-tailored dry polyvinyl alcohol-based adhesive blends.

CPR. Abbreviation for cyclonene-pyrethrin-rotenone. Applied to various insecticide formulations containing as active ingredients approximately 10 parts piperonyl cyclonene, 5 parts rotenone, and 1 part pyrethrin.

Cr. Symbol for chromium.

"CR-39."[177] TM for allyl diglycol carbonate, an optical plastic.
Properties: Clear, optical plastic highly resistant to impact and abrasion. Furnished as a clear liquid. Thermosetting.
Use: Ophthalmic lenses, shields, instrument panels, marine and aircraft glazing.

cracking. A refining process involving decomposition and molecular recombination of organic compounds, especially hydrocarbons obtained by means of heat, to form molecules suitable for motor fuels, monomers, petrochemicals, etc. A series of condensation reactions takes place accompanied by transfer of hydrogen atoms between molecules which brings about fundamental changes in their structure. Thermal cracking, the older method, exposes the distillate to temperatures of approximately 540–650C (1000–1200F) for varying periods of time; it is no longer used for gasoline, but is still of value in producing hydrocarbon gases for plastics monomers. The development of premium fuels for airplanes and automobiles resulted from the use of catalysts in cracking. In this process, hydrocarbon vapors are passed at approximately 400C (750F) over a metallic catalyst (e.g., silica-alumina or platinum); the complex recombinations (alkylation, polymerization isomerization, etc.), occur within seconds to yield high-octane gasoline. Among the chemical changes induced are conversion of alicyclic compounds (cyclohexane) to aromatic compounds and of straight-chain to branched-chain structures (isomerization). Cracking reactions are exothermic. Free radicals, carbonium ions, and other chain-initiating agents are involved in these rearrangements.

Catalytic cracking is carried out by either the moving-bed or fluid-bed technique. In the former the catalyst is pelleted, while in the latter it is finely divided. Instances in which cracking does not involve production of gasoline are the steam-cracking of methane or naphtha to form synthesis gas, thermal cracking of naphtha to ethylene, and thermal decomposition of methane to carbon black and hydrogen.
See also catalysis, synthesis gas, fluidization, pyrolysis.

"Crag."[214] (Sesone). CAS: 136-78-7.
TM for agricultural chemicals including: (1) Fly Repellent (active ingredient, butoxy polypropylene glycol). Colorless liquid, 100% active material. (2) Fungicide 974 (active ingredient, 3,5-dimethyl-tetrahydro-1,3,5-2H-thiadiazine-2-thione). Wettable powder, 85% active material. Irritant. (3) Glyodin Solution (active ingredient, 2-heptadecyl glyoxaldine acetate) 34% active solution. (4) Herbicide-1 (SES) (active ingredient, sodium-2,4-dichlorophenoxyethyl sulfate). Water-soluble powder, 90% active material.
Hazard: Toxic by inhalation. TLV: 10 mg/m³ of air.

Craig method. Introduction of a halogen into the alpha-position of pyridine by treatment of a solution of the alpha-aminopyridine with sodium nitrite in hydrogen halide, followed by warming.

crambe seed oil. A vegetable oil obtained from the seeds of the crambe, a plant related to mustard and rape. Growth and processing techniques have been studied by the Agricultural Research Service of USDA. The plant can be grown on marginal and strip-mined land. Its high content of erucic acid and the high protein content of its meal make it economically attractive. The oil itself is useful as an industrial lubricant, especially for molds for continuous steel casting.

Cram's rule of asymmetric induction.
Rule governing the stereochemistry of addition to carbonyl compounds containing an asymmetric center. The diastereomer which is formed by approach of the reagent from the less hindered side of the carbonyl group predominates in the product. The alpha-substituents are schematically represented as S (small), M (medium), and L (large).

crazing. Development of minute cracks in the surface of a material, such as ceramic glaze, varnish, paint, etc., often as a result of exposure to sunlight or weathering.

cream of tartar. See potassium bitartrate.

creatine. (N-methyl-N-guanylglycine; (α-methylguanido)acetic acid).
$HN:C(NH_2)N(CH_3)CH_2COOH$. A nitrogenous acid widely distributed in the muscular tissue of the body.
Properties: (monohydrate) Prisms from water, anhydrous at 100C, decomposes 303C, slightly soluble in water, insoluble in ether.
Source: Commercially isolated from meat extracts.

Grade: Technical, CP.
Use: Biochemical research.

creatinine. $C_4H_7N_3O$. The anhydride of creatine, a metabolic waste product.
Properties: Colorless to yellow liquid; d 1.092 (25C); mp 5.5C; bp 220C; slightly soluble in water, alcohol, benzene, chloroform, ether, acetic acid.

Creighton process. Electrochemical reduction of sugars, e.g., reduction of glucose, xylose, and galactose to sorbitol, xylitol, and dulcitol.

creosote, coal-tar. (creosote oil; liquid pitch oil; tar oil).
Properties: Yellowish to dark green-brown, oily liquid; clear at 38C or higher; naphthenic odor; frequently contains substantial amounts of naphthalene and anthracene; distilling range 200–400C; flash p 165F (74C) (CC). Soluble in alcohol, benzene, and toluene; immiscible with water. Autoign temperature 637F (335C), d 1.06–1.10.
Derivation: Fractional distillation of coal-tar.
Method of purification: Rectification.
Grade: Technical, crude, refined.
Hazard: Toxic by inhalation of fumes, skin and eye irritant. Use may be restricted.
Use: Wood preservative (ties, telephone poles, marine piling, etc.), disinfectants, fungicide, biocide.

p-cresidine. See 5-methyl-o-anisidine.

"Creslan."[57] TM for an acrylic fiber.

cresol. (methyl phenol; hydroxymethylbenzene; cresylic acid). CAS: 1319-77-3.
$CH_3C_6H_4OH$. A mixture of isomers obtained from coal tar or petroleum.
Properties: Colorless, yellowish, or pinkish liquid; phenolic odor; d 1.030–1.047; wt/gal 8.66–8.68 lb; flash p approximately 180F (82C); mp 11–35C; bp 191–203C. Soluble in alcohol, glycol, dilute alkalies, and water.
Derivation: Coal tar (from coke and gas works), also from toluene by sulfonation or oxidation.
Grade: Various, depending on phenol content or other properties. NF grade contains not more than 5% phenol.
Hazard: Irritant, corrosive to skin and mucous membranes, absorbed via skin. TLV: 5 ppm in air.
Use: Disinfectant, phenolic resins, tricresyl phosphate, ore flotation, textile scouring agent, organic intermediate, manufacture of salicylaldehyde, coumarin, and herbicides, surfactant, synthetic food flavors (para isomer only).
See also cresylic acids.

m-cresol. (m-cresylic acid; 3-methylphenol).
CAS: 108-39-4. $CH_3C_6H_4OH$.
Properties: Colorless to yellowish liquid, phenol-like odor. Soluble in alcohol ether, and chloroform; soluble in water. D 1.034, mp 12C, bp 203C, wt/gal 8.66 lb, flash p 187F (86C), autoign temperature 1038F (558C).
Derivation: By fractional distillation of crude cresol (from coal tar), also synthetically.
Method of purification: Rectification.
Grade: Technical (95–98%).
See cresol.

o-cresol. (o-cresylic acid; 2-methylphenol).
CAS: 95-48-7. $CH_3C_6H_4OH$.
Properties: White crystals, phenol-like odor. Soluble in alcohol, ether, chloroform, and hot water; d 1.047; mp 30.9C; flash p 178F (81C); autoign temperature 1110F (598C); bp 191C; lb/gal 8.68.
Derivation: (a) By fractional distillation of crude cresol from coal tar. (b) Interaction of methanol and phenol.
Method of purification: Crystals.
Grade: According to fp: 25, 29, 30, 30.5C, etc.
See cresol.

p-cresol. (p-cresylic acid; 4-methylphenol).
CAS: 106-44-5. $CH_3C_6H_4OH$.

Properties: Crystalline mass, phenol-like odor. Soluble in alcohol, ether, chloroform and hot water. Wt/gal 8.67 lb, d 1.039, bp 202C, mp 35.25C, flash p 187F (86C), autoign temperature 1038F (558C).
Derivation: (a) By fractional distillation of crude cresol. (b) From benzene by the cumene process (see phenol).
Method of purification: Crystallization.
Grade: Technical, 98%, 99.0% min purity or 34C min fp.
See cresol.

cresolphthalein. $C_6H_4COOC(C_6H_3(OH)CH_3)_2$.
An acid-base indicator, changes from colorless to red between pH 8.2 and 9.8, reagent.
See also indicator.

cresol purple. $C_6H_4SO_2OC(C_6H_3(OH)CH_3)_2$.
M-cresolsulfonphthalein, an acid-base indicator, showing color change from red to yellow over

the range pH 1.2–2.8 and from yellow to purple over the range pH 7.4–9.0.
See also indicator.

cresol red. $C_6H_4SO_2OC(C_6H_3(OH)CH_3)_2$.
O-cresol-sulfonphthalein, an acid-base indicator, changes from yellow to red between pH 7.0 and 8.8.
See also indicator.

cresotic acid. (cresotinic acid; hydroxytoluic acid). $CH_3C_6H_3(OH)COOH$. Ten possible isomers, most common is 2-hydroxy-3-methylbenzoic acid, also known as o-cresotic acid or o-homosalicylic acid. The description which follows is of this isomer.
Properties: White crystals or powder, mp 166C, bp approximately 250C, partially soluble in hot water, soluble in alcohol and ether, combustible.
Derivation: Treatment of o-cresol with caustic and carbon dioxide under pressure.
Use: Dye intermediate, research on plant growth inhibition.

"Crestalkyd."[263] TM for a group of oil-modified alkyd resins. Oil length from 30–80%, acid value less than 15%.
Use: High-quality paints, lacquers, enamels and varnishes. Plasticizing resin for nitrocellulose and similar finishes.

"Crestapol."[263] TM for a series of plasticizers for polyvinyl chloride; also a designation for specialty polyesters used in dispersing media.

m-cresyl acetate. (m-tolyl acetate).
CAS: 140-39-6. $CH_3C_6H_4OCOCH_3$.
Properties: Colorless, oily liquid; odor similar to phenol; bp approximately 112C; distillation with steam; insoluble in water; soluble in common organic solvents; combustible.
Use: Medicine (antiseptic, fungicide).

o-cresyl acetate. (o-tolyl acetate).
$CH_3COOC_6H_4CH_3$.
Properties: Liquid, bp 208C. Nearly insoluble in cold water, soluble in hot water, soluble in organic solvents, combustible.
Use: Flavoring.

p-cresyl acetate. (p-tolyl acetate).
$CH_3C_6H_4COCH_3$.
Properties: Colorless liquid, floral odor, soluble in 2.5 volumes of 70% alcohol, in most fixed oils, insoluble in glycerol, d 1.0532 (15C), optical rotation (100 mm) 0 degrees, refr index 1.500–1.504, acid value 0.7, ester value 341.6, combustible.

Grade: Technical, FCC.
Use: Perfumery, flavoring.

cresyl diglycol carbonate. (diethylene glycol bis(cresylcarbonate). $C_{20}H_{22}O_7$.
Properties: Colorless liquid of low volatility, d 1.19 (20/4C), bp approximately 250C (2 mm), flash p 475F (246C), refr index 1.523 (20C). Insoluble in water (very stable to hydrolysis). Widely soluble in organic solvents. Compatible with many resins and plastics. Combustible.
Use: Plasticizer.

cresyl diphenyl phosphate. (cresyl phenyl phosphate). $(CH_3C_6H_4)(C_6H_5)_2PO_4$. Probably seldom a pure compound, but a mixture of, o- m-, and p-cresyl and phenyl phosphates.
Properties: Colorless, transparent liquid; very slight odor; insoluble in water; soluble in most organic solvents except glycerol; d 1.20 (20/20C); fp −38C; boiling range 235–255C (4 mm); flash p 450F (232C); combustible.
Use: Plasticizer, extreme-pressure lubricant, hydraulic fluids, gasoline additive, food packaging.

cresylic acids. Commercial mixtures of phenolic materials boiling above the cresol range. An arbitrary standard in use for cresylic acids is that 50% must boil above 204C. If the boiling point is less than 204C, the material is called cresol. Cresylic acid varies widely according to its source and boiling range.

A typical commercial cut, bp 220–250C, has the composition m- and p-cresols 0–1%; 2,4- and 2,5-xylenols 0–3%; 2,3- and 3,5-xylenols 10–20%; 3,4-xylenol 20–30%; and C_9 phenols 50–60%. Excellent electrical insulators.
Derivation: Petroleum, coal tar. Imported cresylic acid is derived from coal tar (gas works), also made synthetically.
Hazard: Corrosive to skin, absorbed via skin.
Use: Phosphate esters, phenolic resins, wire enamel solvent, plasticizers, gasoline additives, laminates, coating for magnet wire for small electric motors. Disinfectants, metal-cleaning compounds, phenolic resins, flotation agents, surfactants, chemical intermediates, oil additives, solvent refining of lubricating oils, scouring compounds, pesticides.

p-cresyl isobutyrate. (p-tolyl isobutyrate). $CH_3H_6H_4OCOCH(CH_3)_2$.
Use: Flavoring.

cresyl phenyl phosphate. See cresyl diphenyl phosphate.

cresyl silicate. $(CH_3C_6H_4O)_4Si$.
Properties: Colorless liquid, bp 450C.

Derivation: Reaction of cresol and silicon tetrachloride.
Use: Heat-transfer fluid.

cresyl-p-toluene sulfonate. (tolyl-p-toluene sulfonate). $CH_3C_6H_4SO_3C_6H_4CH_3$.
Properties: Brown, oily liquid; faint odor; d 1.207; flash p 365F (185C); mp 68.70C. Combustible.
Derivation: From reaction of p-toluenesulfonyl chloride with p-cresol.
Use: Plasticizer.

"CRI."[239] TM for a concentrated, rust-inhibiting germicide containing 12.5% (by weight) alkyl (mostly C_{12}, C_{14}, C_{16}, but ranging from C_8 to C_{18}) dimethylbenzylammonium chloride as active ingredient. Inert ingredients are 40% 2-hydroxypropylamine nitrite, water (47.5%), and Fastusol Turquoise Blue.

Crick-Watson. See deoxyribonucleic acid.

Criegee reaction. Oxidative cleavage of vicinal glycols by lead tetraacetate.

critical assembly. A system of fissionable material (enriched uranium) and moderator sufficient to sustain a chain reaction at a low and controllable power level, as in a nuclear reactor.
See also fission, nuclear, nuclear reactor.

critical constant. A maximum or minimum value for a physical constant which is characteristic of a substance, e.g., the critical temperature of a gas is the temperature above which it cannot be liquefied by an increase in pressure.

criticality. The state of a nuclear reactor when it is sustaining a chain reaction.

critical mass. The minimum mass of a fissionable material (^{235}U or ^{239}Pu) that will initiate an uncontrolled chain reaction as in an atomic bomb. The critical mass of pure ^{239}Pu is about 10 lbs and of ^{235}U about 33 lbs. This phenomenon was unknown before 1940.

critical solution temperature. The temperature above or below which two liquids are miscible in all proportions. Some pairs of liquids have both an upper and a lower critical soluble temperature, that is they can exist in two phases only in a medium temperature range.

crocein acid. (croceic acid; Bayer's acid; 2-naphthol-8-sulfonic acid).

Derivation: Sulfonation of β-naphthol with 94% sulfuric acid at 95C. and recrystallization from a salt.

Use: Azo dye intermediate.

Crocein Scarlet MOO.   See Brilliant Crocein.

crocetin. $C_{20}H_{24}O_4$.   A dicarboxylic carotenoid derived from saffron.

Properties: Red, rhomboid crystals; soluble in pyridine and dilute sodium hydroxide; slightly soluble in water and organic solvents; mp 285C; combustible.

Use: Experimental treatment of arteriosclerosis by increasing oxygen diffusion through arterial walls, thus decreasing build-up of cholesterol.

crocidolite.   A type of asbestos.
See asbestos.

crocking.   Removal of a dye or pigment from the surface of a paint or textile by rubbing or attrition.

Cross-Bevan (viscose) process.   Production of rayon by treatment of cellulose with alkali and carbon disulfide to yield cellulose xanthate, solution in dilute caustic and extrusion of the viscous "Viscose" into a coagulating bath, a 7-10% sulfuric acid containing 1-5% zinc sulfate and an active surface agent.

crosshead.   A device attached to the head of an extrusion machine which permits the material to be extruded in opposite directions simultaneously, i.e., at right angles to the barrel. It is applicable chiefly to coating of wire, cable, and small-diameter hose.

cross-linking.   Attachment of two chains of polymer molecules by bridges composed of either an element, a group, or a compound which join certain carbon atoms of the chains by primary chemical bonds, as indicated in the schematic diagram.

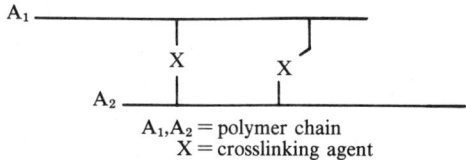

$$A_1, A_2 = \text{polymer chain}$$
$$X = \text{crosslinking agent}$$

Cross-linking occurs in nature in substances made up of polypeptide. chains which are joined by the disulfide bonds of the cystine residue, as in keratins, insulin, and other proteins. Polysaccharide molecules can also cross-link to form

stable gel structures (dextran). Cross-linking can be effected artificially, either by addition of a chemical substance (cross-linking agent) and exposing the mixture to heat, or by subjecting the polymer to high-energy radiation. Examples are (1) vulcanization of rubber with sulfur or organic peroxides, (2) cross-linking of polystyrene with divinylbenzene, (3) cross-linking of polyethylene by means of high-energy radiation or with an organic peroxide, (4) cross-linking of cellulose with dimethylol carbamate (10% solution) in durable-press cotton textiles. Cross-linking has the effect of changing a plastic from thermoplastic to thermosetting. Thus, it also increases strength, heat- and electrical resistance, and especially resistance to solvents and other chemicals.
See also vulcanization, polyethylene (cross-linked), keratin.

cross-section.   (1) A measure of the probability that a nuclear reaction will occur. Usually measured in barns, it is the apparent (or effective) area presented by a target nucleus (or particle) to an oncoming particle or other nuclear radiation, such as a photon or gamma radiation. (Atomic Energy Commission). Also called capture cross-section.

(2) A section made by a plane cutting through a solid, usually at right angles. Tissue cross-sections are widely used for microscopic observation.

crotonaldehyde.   (2-butenal; crotonic aldehyde; β-methyl acrolein).   CAS: 123-73-9.
$CH_3CH:CHCHO$.   Commercial crotonaldehyde is the trans isomer.

Properties: Water-white, mobile liquid; pungent, suffocating odor; turns to a pale yellow color in contact with light and air. A lachrymator. Very soluble in water; miscible in all proportions with alcohol, ether, benzene, toluene, kerosene, gasoline, solvent naphtha. D 0.8531 (20/20C), bp 102C, flash p 55F (12.7C), fp −69C, vap press 30 mm (20C).

Derivation: Aldol condensation of two molecules of acetaldehyde.

Grade: Technical, 87% water-wet form.

Hazard: Irritating to eyes and skin. TLV: 2 ppm in air. Flammable, dangerous fire risk. Explosive limits in air 2.9–15.5% by volume.

Use: Intermediate for n-butyl alcohol and 2-ethylhexyl alcohol, solvent, preparation of rubber accelerators, purification of lubricating oils, insecticides, tear gas, fuel-gas warning agent, organic synthesis, leather tanning, alcohol denaturant.

crotonic acid.   (2-butenoic acid; β-methacrylic acid).   CAS: 3724-65-0.
$CH_3CH:CHCOOH$.   Exists in cis and trans

isomeric forms, the latter being the stable isomer used commercially. The cis form melts at 15C and is sometimes called isocrotonic acid.

Properties: White, crystalline solid; d 0.9730, mp 72C; bp 185C; soluble in water, ethanol, toluene, acetone; flash p 190F (87.7C) (COC). Combustible.

Derivation: Oxidation of crotonaldehyde.

Grade: 97%.

Hazard: Strong irritant to tissue.

Use: Synthesis of resins, polymers, plasticizers, drugs.

crotonic aldehyde.   See crotonaldehyde.

croton oil.   (tiglium oil).   CAS: 8001-28-3.

Properties: Brownish-yellow liquid, d 0.935–0.950 (25C), refr index (40C) 1.470–1.473. Soluble in ether, chloroform, and fixed or volatile oils; slightly soluble in alcohol. Chief constituents: Glycerides of stearic, palmitic, myristic, lauric and oleic acids and croton resin, a vesicant.

Derivation: By expression from the seeds of Croton tiglium.

Hazard: Strong skin irritant, ingestion of small amounts may be fatal.

Use: Medicine (counterirritant, cathartic).

crotonylene.   (2-butyne; dimethylacetylene).
$CH_3C{:}CCH_3$.

Properties: Liquid, bp 27C.

Hazard: Flammable, dangerous fire risk. Moderate explosion hazard.

crotoxin.   Exact formula undetermined. A toxic principle of rattlesnake venom having a polypeptide structure, one component being basic and the other acidic. Classed as a neurotoxin. Injection may be fatal.

crotoxyphos.   (dimethyl-2-(α-methylbenzyloxy-carbonyl)-1-methylvinyl phosphate; O,O-dimethyl-O-(1-methyl-2-(1-phenylcarbethoxy) vinyl)ophosphate; α,α-methyl-benzyl 3-(dimethoxy-phosphinyloxy)-isocrotonate).
CAS: 7700-17-6. $C_{14}H_{19}O_6P$.

Properties: Yellowish liquid; d 1.19; bp 135C at 4.0 (0.03 mm Hg); refr index 1.50; soluble in acetone, alcohol, chlorinated hydrocarbons; slightly soluble in saturated hydrocarbons.

Hazard: A cholinesterase inhibitor.

Use: Insecticide used externally on farm animals.

crotyl alcohol.   (2-buten-1-ol; 3-methylallyl alcohol).   CAS: 6117-91-5.
$CH_3CH{:}CHCH_2OH$.

Properties: Clear, stable liquid; d 0.8550 (20/20C); boiling range 121–126C; flash p (TOC) 113F

(45C). Partially soluble in water (17%), wholly soluble in alcohol and ether. Combustible.

Hazard: Toxic by ingestion, strong eye and skin irritant. Moderate fire risk.

Use: Chemical intermediate, source of monomers, herbicide and soil fumigant.

crown ether.   A cyclic molecule in which ether groups (polyethers) connected by dimethylene linkages are coordinated to a centrally located metal atom via the oxygen atoms of the ethers, which function as ligands (electron donors). Such compounds have strong complexing or chelating capabilities. In some types, silicon replaces the dimethylene linkages. They were so named because their molecular models resemble a crown. See also porphyrin, chelate.

crown filler.   A mineral filler, usually calcium sulfate or carbonate or a mixture thereof used in paper manufacture.

crown glass.   See glass, optical.

crucible.   (1) A cone-shaped container having a curved base and made of a refractory material, used for laboratory calcination and combustibles. Some types are equipped with a cover. A gooch cruciable has openings in its base to permit filtration with suction; named after its inventor, an American chemist. (2) In the steel industry, a special type of furnace provided with a cavity for collecting the molten metal.

crufomate.   (O-methyl-O-(4-tert-butyl-2-chloro-phenyl)methylphosphoramidate; 4-t-butyl-2-chlorophenyl-methyl methyl phosphoramidate; ruelene).

Properties: Crystals with mw 291.71, mp 61C, insoluble in water. Soluble in alcohol, benzene, and carbon tetrachloride. Commerical product is a yellow oil, bp 117C.

Hazard: TLV: 5 mg/m³.

Use: Systemic insecticide and antihelminthic.

crusher, gyratory.   See gyratory crusher.

crust.   The outermost layer of the lithosphere, consisting of felsic and mafic rocks less dense than the rocks of the mantle below.

cryochemistry.   That branch of chemistry devoted to the study of reactions occurring at extremely low temperature (−200C and lower). It permits synthesis of compounds that are too unstable or too reactive to exist at normal temperature.

cryogenics.   Study of the behavior of matter at temperatures below −200C. The use of the liquefied gases, oxygen, nitrogen, and hydrogen at approximately −260C is standard industrial practice. Examples: Use of liquid nitrogen for quick-freezing of foods and of liquid oxygen in steel production. Some electronic devices and specialized instruments, such as the cryogenic gyro, operate at liquid helium temperature (approximately 4K). Many lasers and computer circuits require low temperature. Original research in this field was carried out by W. F. Giauque in the US and by Kamerlingh-Onnes in Holland. See also superconductivity.

cryolite. (Greenland spar; icestone).
Na_3AlF_6. A natural fluoride of sodium and aluminum or made synthetically from fluorspar, sulfuric acid, hydrated alumina and sodium carbonate.
Properties: Colorless to white, sometimes red, brown, or black; luster vitreous to greasy; hardness 2.5; d 2.95–3.0. Refr index 1.338, mp 1000C, soluble in concentrated sulfuric acid and in fused aluminum and ferric salts.
Occurrence: Colorado, USSR; Greenland (only commercial source).
Derivation: Synthetic product is made by fusing NaF and aluminum fluoride.
Use: Electrolyte in the reduction of alumina to aluminum, ceramics, insecticide, binder for abrasives, electric insulation, explosives, polishes.

"Cryovac."[311] TM for a light, shrink-film, transparent packaging material based on polyvinylidene chloride. Used especially for meats and other perishables.

cryptocyanine. (1,1'-diethyl-4,4'-carbocyanine iodide). $C_{25}H_{25}N_2I$.
Properties: Solid, mp 250.5C.
Use: Organic dye soluble used as a chemical shutter in laser operation.
See also cyanine dye.

cryptostegia rubber. Rubber from leaves of *Cryptostegia grandiflora* and *C. madagascariensis*.

cryptoxanthin. (provitamin A; hydroxy-β-carotene). $C_{40}H_{56}O$. A carotenoid pigment with vitamin A activity.
Properties: Garnet-red prisms with metallic luster; mp 170C; soluble in chloroform, benzene, and pyridine; slightly soluble in alcohol and methanol.
Occurrence: In many plants, egg yolk, butter, blood serum. Can be made synthetically.
Use: Nutrition, medicine.

crystal. The normal form of the solid state of matter. Crystals have characteristic shapes and cleavage planes due to the arrangement of their atoms, ions, or molecules, which comprise a definite pattern called a lattice. Crystals may be face-centered, body-centered, cubic, ortho-rhombic, monoclinic, prismatic, etc. They have flat surfaces, sharp edges, and a definite angle between a given pair of surfaces. The form of a crystal is called its "habit." One of the most important features of a crystal is its optical properties, chief of which is its index of refraction, i.e., the extent to which a beam of light is deflected on passing through the crystal. Depending on the manner of light transmission, a crystal may be isotropic or anisotropic. Anisotropic crystals can polarize light (see also optical isomerism, optical rotation). Crystals also have electrical and magnetic properties now being used in computers and other electronic devices. Crystals are almost always imperfect and contain impurities (atoms of other elements). These are utilized in semiconductors. For methods of growing crystals, see nucleation.

Single crystals are used in masers, lasers, semiconductors, miniaturized components, computer memory systems, and as "whiskers." Many metals are now available in large, single crystalline form and such natural crystals as ruby, garnet, sapphire, etc., are used in these applications.
See also crystallization, nucleation, liquid crystals, hole, vacancy.

crystalline rocks.
Igneous or metamorphic rocks.

crystal liquid. See liquid crystal.

crystallite. That portion of a crystal whose constituent atoms, ions, or molecules form a perfect lattice, without strains or other imperfections. Single crystals may be quite large, but crystallites are usually in the microscopic range.
See also crystal.

crystallization. The phenomenon of crystalline formation by nucleation and accretion. The freezing of water into ice is one of the commonest examples of crystallization in nature. Industrially, it is used as a means of purifying materials by evaporation and solidification. The sugar of commerce is made in this way. Similarly, salt cake is derived from crystallization of natural brines (Searles Lake). Nucleated crystallization is also used to form polycrystalline ceramic structures.
See also crystal.

crystals of Venus. See copper acetate.

crystallography. The study of the crystal formation of solids, including x-ray determination of lattice structures, crystal habit, and the shape, form, and defects of crystals. When applied to metals, this science is called metallography.

crystal violet. See methyl violet.

"Crystamet."[18] TM for sodium metasilicate pentahydrate.

"Crystarose."[188] TM for a highly purified grade of trichloromethylphenylcarbinol acetate.

"Crystex."[1] TM for a rubber-insoluble sulfur used as vulcanizing agent in natural and synthetic rubbers; 85% of the sulfur is insoluble at the usual milling temperature. This metastable form is converted to the stable soluble sulfur at the usual vulcanizing temperature.

"Crystic."[263] TM for a series of unsaturated polyester resins. Available in several grades as well as self-extinguishing and thixotropic types.
Use: Low-pressure laminating for glass fiber reinforced plastics, potting, casting, embedding, and coating.

cs. Abbreviation for centistoke.

Cs. Symbol for cesium.

"CS-137."[304] TM for a barium-sodium organic-complex vinyl stabilizer.
Properties: Creamy-white paste, d 1.54, refr index 1.48.
Use: Stabilizer for transparent organosols and soluble coatings to impart superior light and weathering resistance.

C₂S. Abbreviation for dicalcium silicate as used in cement.
See cement, Portland.

C₃S. Abbreviation for tricalcium silicate as used in cement.
See cement, Portland.

"CSC."[319] TM for choline chloride aqueous solutions and choline chloride feed supplements.

CS gas. See o-chlorobenzylidine malononitrile.

C-stage resin. (resite). The fully cross-linked phenolformaldehyde type resin which is infusible and insoluble in all solvents.
See A-stage resin, phenolformaldehyde resin.

CTFE. See chlorotrifluoroethylene.

CTP. Abbreviation for cytidine triphosphate.

Cu. Symbol for copper.

cuam. Abbreviation for cuprammonium, the copper ammonium radical.

cube root. A powdered insecticidal preparation containing 5% rotenone.

"Cubex."[308] TM for an oriented silicon-iron alloy in rolled sheet form for use as cores for transformers and other inductive devices. The alloy sheet comprises cubic grains with faces parallel to the sheet surface. In one form, the sheet is doubly oriented with two directions of easy magnetization parallel to the surface to the sheet. One direction of easy magnetization is parallel to the rolling direction and the second is perpendicular to the rolling direction.

cubic centimeter. (cc). Unit of volume for liquids and finely divided solids.
See also milliliter (mL).

"Cubidow."[233] TM for compacted salt comprising either or both calcium and sodium chloride.

"Cubond."[296] TM for a copper brazing paste. Consists of metallic copper powder on cuprous oxide pigments of high purity in organic or petroleum vehicles which impart satisfactory suspension properties.

"Cubor Dusts."[147] TM for insecticides containing 0.75–1.0% pure rotenone.

cue-lure. Generic name for 4-(p-hydroxphenyl)-2-butanone acetate.
$CH_3COCH_2CH_2C_6H_4OOCH_3$.
Properties: Liquid, boiling range 117–124C. Insoluble in water, soluble in most organic solvents.
Use: Insect attractant.

cuen. Abbreviation for cupriethylenediamine.

cullet. In ancient glass manufacturing, chunks of glass of varying sizes and colors furnished to artisans for shaping and finishing. In modern practice, the term refers to fragments of scrap glass from production operations which are collected and recycled to the furnace for remelting.

culm. Anthracite tailings, especially prevalent in eastern Pennsylvania. They represent a considerable source of energy which could be used for example in fluidized bed boilers.

"Culofix."[300] TM for dye fixatives of fatty or resin cationic type for application to dyed textiles.

cumaldehyde.   See cuminic aldehyde.

"Cumar."[21]   TM for a series of neutral, stable, synthetic resins of the coumarone-indene type, manufactured from selected distillates of tar.
Use: Softener and tackifier in varnishes, floor tile, rubber products, printing ink, adhesives and waterproofing materials, leather, electrical, radio, paper, and other industries.
See also coumarone-indene resin.

"Cumate."[69]   TM for preparation of copper dimethyldithiocarbamate.
$[(CH_3)_2NC(S)S]_2Cu$.
Properties: Dark brown powder; d 1.75; melts at over 325C; moderately soluble in acetone, benzene, chloroform; insoluble in water, alcohol, gasoline.
Use: In SBR, primary accelerator, secondary accelerator with thiazoles. In butyl rubber, primary accelerator. For molded and extruded goods.

cumene.   (isopropylbenzene).   CAS: 98-82-8.
$C_6H_5CH(CH_3)_2$.   31st highest-volume chemical produced in US (1985).

Properties: Colorless liquid; soluble in alcohol, carbon tetrachloride, ether, and benzene; insoluble in water. D 0.8620, bp 152.7C, wt/gal 7.19 lb (25C), fp −96C, refr index 1.489 (25C), flash p 115F (46C), autoign temperature 795F (424C). Combustible.
Derivation: (a) Alkylation of benzene with propylene (phosphoric acid catalyst), (b) distillation from coal tar naptha fractions or from petroleum.
Grade: Technical, research, pure.
Hazard: Toxic by ingestion, inhalation, and skin absorption; a narcotic. TLV: 50 ppm in air. Moderate fire risk.
Use: Production of phenol, acetone, and alpha-methylstyrene; solvent.

cumene hydroperoxide.   (α,α-dimethylbenzyl hydroperoxide).   CAS: 80-15-9.
$C_6H_5C(CH_3)_2OOH$.

Properties: Colorless to pale yellow liquid; slightly soluble in water; readily soluble in alcohol, acetone, esters, hydrocarbons, chlorinated hydrocarbons; flash p 175F (79.4C). Combustible.
Derivation: A solution or emulsion of cumene is oxidized with air at approximately 130C.
Hazard: Toxic by inhalation and skin absorption. Strong oxidizing agent; may ignite organic materials.
Use: Production of acetone and phenol; polymerization catalyst, particularly in redox systems, used for rapid polymerization.

cumerone. C_8H_{60}.
Properties: Colorless oily liq, d 1.09.6, bp 165–175C, insoluble in water.

cumic alcohol.   See cuminic alcohol.

cumic aldehyde.   See cuminic aldehyde.

cumidine.   (o-isopropylaniline).
CAS: 643-28-7. $(CH_3)_2CHC_6H_4NH_2$.
Properties: Colorless liquid, d 0.957 (20/4C), fp −63C, bp 225C. Insoluble in water.
Use: Reagent in determination of tungsten.
See also pseudocumidine.

cuminic alcohol.   (p-isopropylbenzyl alcohol; cuminyl alcohol; cumic alcohol).
CAS: 536-60-7. $CH_2OH(C_6H_4)CH(CH_3)_2$.
Found in caraway seed.
Properties: Colorless liquid, caraway-like odor, aromatic taste, d 0.981 (15C), bp 248C, refr index (24C) 1.522. Insoluble in water; miscible with alcohol, ether. Combustible.
Use: Flavoring.

cuminic aldehyde.   (cumic aldehyde; cumaldehyde; p-isopropylbenzaldehyde).
CAS: 122-03-2. $(CH_3)_2CHC_6H_4CHO$.
Properties: Colorless to yellow liquid with a cumin odor, d 0.986 (22C), bp 235C. Insoluble in water, soluble in alcohol and ether. Combustible.
Use: Perfumery, flavoring, synthesis.

cumin oil.   A dextrorotatory essential oil used in perfumery and flavoring.

cumulene.   A chain of up to 6 double-bonded carbon atoms derived from acetylene.

cumyl phenol. $C_6H_5C(CH_3)_2C_6H_4OH$.
Properties: White to tan crystals with characteristic phenol odor. Fp 72.0C, d 1.115 g/mL (25C), distillation range 188.9–190.9C (10 mm), flash p 320F (160C). Combustible.
Use: Intermediate for resins, insecticides, lubricants.

"Cunife."[166] TM for a ductile permanent magnet material composed of 60% copper, 20% nickel, and 20% iron. Forms available: wire, strip, finished magnets.

"Cunilate."[8] TM for "solubilized copper-8-quinolinolate." One type contains 10% copper-8-quinolinolate with 2-ethylhexoic acid and an aromatic solvent as carrier with 40% total solids.
Properties: D 0.9542–0.9545 at 77F, flash p 110F (43.3C), color greenish yellow, pH 5.5–6.0. Combustible.
Hazard: Moderate fire risk.
Use: Fungicidal treatment of fabrics, wood (by immersion or pressure), cotton rope and other materials as well as in combination with water repellents and auxiliary chemicals.

"Cunimene."[8] TM for a series of metal complexes of dehydroabietyl amine 8-hydroxyquinolinium 2-ethylhexoate.
Use: Fungicide and bactericide used in vinyl plasticizers, rubber, cordage fibers, textiles, etc.

"Cunisil-647."[324] TM for an alloy containing 97.50% copper, 1.90% nickel, and 0.60% silicon. This high-strength, corrosion-resistant alloy is available in round rod, with or without the final precipitation-hardening heat treatment. Used in electrical equipment.

cupellation process. A process for freeing silver, gold, or other nonoxidizing metals from base metals which can be oxidized. The metallic mixture is placed in a cupel, which is a shallow, porous cup, and roasted in a blast of air. The base metal oxides are absorbed in the cupel, leaving the pure metal to be decanted.

cupferron. (ammonium nitroso-β-phenylhydroxylamine). CAS: 135-20-6.
$C_6H_5N(NO)ONH_4$.
Properties: Creamy-white crystals, mp 163–164C. Soluble in water and alcohol.
Derivation: By treating an ether solution of β-phenylhydroxylamine with dry ammonia gas and amyl nitrite.
Use: Analytical reagent, especially for separation and precipitation of metals, e.g., copper, iron, vanadium.

cupola. A vertical furnace similar to a blast-furnace used for melting iron or other metals for casting.

cuprammonium process. A minor process for making rayon by dissolving cellulose in an ammoniacal copper solution and reconverting it to cellulose by treatment with acid.

cupreine. (hydroxycinchonine).
$C_{19}H_{22}O_2N_2 \cdot 2H_2O$. One of the cinchona alkaloids.
Properties: Colorless crystals, mp (anhydrous) 198C. Soluble in alcohol; slightly soluble in water, chloroform, ether, benzene.
Derivation: From cuprea bark Remijia pedunculata.
Use: Medicine (antimalarial).

cupric. Form of the word copper used in naming copper compounds in which the copper has a valence of 2.
See the corresponding compound under copper.

cupric chromate basic. See copper chromate.

cupriethylene diamine. Purple liquid, ammoniacal odor. Dissolves cellulose products.
Hazard: Strong irritant to tissue.

cuprinol. A copper naphthenate or sodium pentachlorophenate preparation used as a wood and fabric preservative.

cuprite. (copper ore ruby; red oxide of copper).
Cu_2O. Crimson, scarlet, vermilion, deep or brownish-red; secondary mineral; adamantine or dull luster; brownish-red streak. Soluble in nitric and concentrated hydrochloric acids, d 5.85–6.15, Mohs hardness 3.5–4.
Occurrence: US, England, Germany, France, Siberia, Australia, China, Peru, Bolivia.
Use: Source of copper.

cupronickel. An alloy of copper and nickel used in coinage, condenser and heat-exchanger tubes. Most types contain from 10–30% nickel. Strongly corrosion-resistant, especially to seawater.

"Cuprophenyl."[443] TM for after-coppering direct dyes for cellulosic fibers.

cuprotungsten. An alloy of copper and tungsten.

cuprous. Form of the word copper used in naming copper compounds in which the copper has a valence of 1.

cuprous acetate ammoniacal. See copper ammonium acetate.

cuprous acetylide. Cu_2C_2.
Properties: Amorphous, red powder. A salt of acetylene.
Derivation: Reaction of acetylene with aqueous soluble of cuprous salts.

Hazard: Severe explosion risk when shocked or heated.
Use: Detonators and other explosive devices.

cuprous bromide. CuBr or Cu_2Br_2.
Properties: White, crystalline solid becoming green in light; mp 500C; d 4.71. Soluble in hydrochloric acid and ammonium hydroxide, decomposes in hot water, insoluble in acetone and sulfuric acid. Protect from light when stored.
Use: Catalyst in organic reactions.

cuprous chloride. CuCl or Cu_2Cl_2.
Properties: White, cubical crystals; d 4.14; mp 430C; bp 1490C; becomes greenish on exposure to air and brown on exposure to light. Slightly soluble in water; soluble in acids, ammonia, ether; insoluble in alcohol and acetone.
Derivation: Copper and cupric chloride solution or copper and hydrochloric acid in air.
Grade: Technical, reagent, single crystals.
Use: Catalyst, preservative and fungicide, desulfurizing and decolorizing agent in petroleum industry, absorbent for carbon monoxide.

cuprous cyanide. $Cu_2(CN)_2$ or CuCN.
Properties: Cream-colored powder. Insoluble in water; soluble in sodium and potassium cyanides, in hydrochloric acid and ammonium hydroxide. D 1.9, mp 475C.
TLV: (as CN) 5 mg/m^3.
Use: Electroplating, antifouling paints, insecticide, catalyst.

cuprous iodide. CuI.
Properties: White to brownish-yellow powder. Soluble in ammonia and potassium iodide solutions, insoluble in water, d 5.653 at 15C, bp 1290C, mp 606C.
Derivation: Interaction of solutions of potassium iodide and copper sulfate.
Hazard: Toxic.
Use: Feed additive, in table salt as source of dietary iodine (up to 0.01%), catalyst, cloud seeding.

cuprous mercuric iodide. Cu_2HgI_4.
Properties: Dark-red crystals which become dark brown to black when heated to about 65C, insoluble in water and alcohol.
Hazard: Toxic by ingestion.
Use: Temperature indicator for bearings and other moving machinery parts.

cuprous oxide. See copper oxide red.

cuprous potassium cyanide. $KCu(CN)_2$.
Properties: Crystalline solid, d 2.38, insoluble in water, soluble in dimethylsulfoxide, decomposed by heating in water.

Derivation: By evaporating a water soluble of cuprous cyanide and potassium cyanide.
Hazard: A poison by ingestion.
Use: Copper electroplating vats.

cuprous selenide. Cu_2Se.
Properties: Dark blue to black crystalline solid, mp 1100C, d 6.84. , Soluble in sulfuric acid (sulfur dioxide evolved), also in potassium cyanide solution.
Use: Semiconductor research.

cuprous sulfide. Cu_2S.
Properties: Black powder or lumps, soluble in nitric acid and ammonium hydroxide, insoluble in water. Occurs as the mineral chalcocite, d 5.52–5.82, mp approximately 1100C.
Derivation: By heating cupric sulfide in a stream of hydrogen.
Grade: Technical, single crystals.
Use: Antifouling paints, solar cells, electrodes, solid lubricants, luminous paints, catalyst.

cuprous sulfite. $Cu_2SO_3 \cdot H_2O$.
Properties: White, crystalline powder; soluble in ammonium hydroxide, hydrochloric acid (decomposes); insoluble in water; d 3.83.
Use: Catalyst, fungicide, textile dyeing.

cuprous thiocyanate. (copper sulfocyanide). CuSCN.
Properties: Yellow-white powder. Insoluble in water, soluble in ammonia, d 2.843, mp 1084C.
Use: Manufacture of organic chemicals, antifouling paints, printing textiles, primers of explosives.

"Curacid."[586] TM. *400* (cis-3-methyl Δ_4-tetrahydrophthalic anhydride).
Properties: White crystals or powder, d 1.20 g/cc, mp 55C.
Use: Air-drying unsaturated polyester and alkyd resins. *600* (3-methylhexahydrophthalic anhydride).
Properties: Colorless, transparent liquid; d 1.17 g/cc; viscosity 65 cps (25C).
Use: Hardener for epoxy and alkyd resins.

"Curafos."[108] TM for food grade sodium hexametaphosphate and sodium tripolyphosphate used for curing, preventing undesirable color change, and loss of moisture in meat.

"Curalon."[248] TM for a series of urethane curatives. "Curalon L," a mixture of hindered aromatic primary diamines. "Curalon M," p,p'-methylene bis(o-chloroaniline).

curare. CAS: 8063-06-7. A highly toxic mixture of approximately 40 alkaloids occurring in several species of South American trees. It acts on the central nervous system, and derivatives are used to some extent in medicine as a muscle relaxant.

See also snake venom.

curative. (1) Any substance or agent that effects a fundamental and desirable change in a material to make it suitable for practical use, e.g., meats, rubber, tobacco. (2) Any substance that combats disease by killing bacteria or that restores health by chemical means.

See curing.

"Curavis."[108] TM for a dry, pulverized, food grade, polymeric phosphate composition exhibiting high viscosity in water solution. For use in the meat packing industry.

curcumin. (tumeric yellow; 1,7-bis(4-hydroxy-methoxyphenyl)-1,6-heptadiene-3,5-dione). $[CH_3OC_6H_3(OH)CH:CHCO]_2CH_2$.

Properties: Orange-yellow needles, mp 183C, soluble in water and ether, soluble in alcohol. CI 75300.

Derivation: The coloring principle from curcuma.

Use: Analytical reagent, food dye, biological stain. As an acid-base indicator it is brownish-red with alkalies, yellow with acids (pH range 7.4–8.6), also an indicator for boron.

curie. (abbreviation Ci). The official unit of radioactivity, defined as exactly 3.70×10^{10} disintegrations per second. This decay rate is nearly equivalent to that exhibited by 1 g of radium in equilibrium with its disintegration products. A millicurie (mCi) is 0.001 curie. A microcurie (μ) is one millionth curie.

Curie, Marie S. (1867–1934). Born in Warsaw Poland, she and her husband Pierre made an intensive study of the radioactive properties of uranium. They isolated polonium in 1898 from pitchblende ore. By devising a tedious and painstaking separation method, they obtained a salt of radium in 1912, receiving the Nobel prize in physics for this achievement in 1903 jointly with Becquerel. In 1911, Mme Curie alone received the Nobel prize in chemistry. Her work laid the foundation of radioactive elements which culminated in control of nuclear fission.

See also Rutherford.

curie point. Transition temperature above which magnetism ceases to exist.

curing. Conversion of a raw product to a finished and useful condition, usually by application of heat and/or chemicals which induce physico-chemical changes. Many food products require aging under specified temperature conditions. The more common types of curing are as follows:.

(a) Meats: Use of sodium chloride, sugars, sodium nitrite, sodium nitrate, ascorbic acid. These not only act as preservatives, but also aid in color retention. Some types are smoked subsequently. Conversion of collagen to gelatin occurs as a result of "hanging" meat for several days.

(b) Leather: Treatment of hides and skins with tanning agents of vegetable or mineral origin. This converts the protein structure into a firm and durable product as a result of complexing reactions.

See also tanning.

(c) Tobacco: Exposure for 3–5 days to temperatures from 37–65C to reduce moisture content, convert starches to reducing sugars, and discharge the chlorophyll, followed by aging from 1–5 years to remove odors and improve smoking quality.

(d) Cheese: Aging for 9–12 months at 4.5–10C to develop sharp flavor; the process is also called ripening.

(e) Rubber: Addition of sulfur and accelerator, followed by exposure to heat which effects crosslinking. This converts the material from a thermoplastic to a thermosetting product. High-energy radiation can also be used.

See also vulcanization.

curium. Cm. Synthetic radioactive element of atomic number 96, aw 244, valences 3,4. Isotopes available: 244 and 242 (gram quantities).

Properties: Silvery-white metal, d 13.5, mp 1340C, chemically reactive, more electropositive than aluminum. An alpha emitter. Biologically it is a bone-seeking element. Forms compounds such as CmO_2, Cm_2O_3, CmF_3, CmF_4, $Cm(OH)_3$, $CmCl_3$, $CmBr_3$, $CmI_3 \cdot Cm_2(C_2O_4)_3$.

Use: Thermoelectric power generation for instrument operation in remote locations on earth or in space vehicles.

See also actinide elements.

"Curona."[173] TM for sodium isoascorbate as a curing aid in cured and comminuted meat products. It is used as an anti-oxidant in the meat packing industry for preserving the natural color of meat products, to shorten processing time and reduce shrinkage.

current density. In an electroplating bath or solution the electric current per unit area of the object or surface being plated. Expressed in amperes per square centimeter or more usually amperes per square decimeter.

Curtius rearrangement. Formation of isocyanates by thermal decomposition of acyl azides.

cutting fluid. A liquid applied to a cutting tool to assist in the machining operation by washing away the chips or serving as a lubricant or coolant. Commonly used cutting fluids are water, water solutions or emulsions of detergents and oils, mineral oils, fatty oils, chlorinated mineral oils, sulfurized mineral oils, mixtures of the foregoing oils. Transparent grades are available.

"Cyana."[57] TM used in connection with the textile finishes obtained by applying "Aerotex" Resins and similar products.

"Cyanamer."[57] TM for an acrylic polymer.

cyanamide. (1) (cyanogenamide; carbodiimide).
CAS: 420-04-2.   HN:C:NH or N:CNH$_2$.
Properties: Deliquescent crystals, mp 43C, d 1.08. Very soluble in water, alcohol, ether, phenols, ketones.
Hazard: Strong irritant to skin and mucous membranes; avoid inhalation or ingestion. TLV: 2 mg/m^3 of air.
    (2) See calcium cyanamide.

cyanamide process. See nitrogen fixation, ammonia anhydrous.

cyanic acid. See isocyanic acid.

cyanide pulp. The mixture obtained by grinding crude gold and silver ore and dissolving the precious metal content in sodium cyanide solution.

cyanine dye. One of a series of dyes consisting of two heterocyclic groups (usually quinoline nuclei) connected by a chain of conjugated double bonds containing an odd number of carbon atoms. Example: cyanine blue C$_2$H$_5$NC$_9$H$_6$:CHC$_9$H$_6$NC$_2$H$_5$. They include the isocyanines, merocyanines, cryptocyanines, and dicyanines.
Use: Sensitizers for photographic emulsions.

cyanoacetamide. (malonamide nitrile; propionamide nitrile).   CNCH$_2$CONH$_2$.
Properties: White crystals, bp decomposes, mp 119C, soluble in water and alcohol. Combustible.
Derivation: Ammonolysis of cyanoacetic ester or dehydration of ammonium cyanoacetate.
Hazard: Toxic by ingestion.
Use: Organic pharmaceutical synthesis, plastics.

cyanoacetic acid. (malonic nitrile).
CNCH$_2$COOH.
Properties: White crystals, hygroscopic. Soluble in water, alcohol, and ether; mp 66.1–66.4C; decomposes at 160C.

Derivation: Interaction of sodium chloroacetate and potassium cyanide solution.
Hazard: Toxic by ingestion.
Use: Organic synthesis.

cyanoacrylate adhesive. An adhesive based on the alkyl 2-cyanoacrylates (see for example methyl-2-cyanoacrylate). The latter are prepared by pyrolyzing the poly(alkyl)-2-cyanoacrylates produced when formaldehyde is condensed with the corresponding alkyl cyanoacetates. These adhesives have excellent polymerizing and bonding properties. To prevent premature polymerization, inhibitors are added. Supplied commercially as "Eastman 910."

"Cyanobrik."[28] TM for 98% sodium cyanide in 1-oz, pillow-shaped, briquette form.

cyanocarbon. Any of a class of compounds in which the cyanide radical (—CN) replaces hydrogen in organic compounds as in tetracyanoethylene, (CN)$_2$C:C(CN)$_2$. The compounds are quite reactive and form colored complexes with aromatic hydrocarbons.
See also nitrile.

"Cyanocel."[57] TM for chemically modified (cyanoethylated) cellulose.

cyanocobalamin. (vitamin B$_{12}$).
CAS: 68-19-9.   C$_{63}$H$_{88}$CoN$_{14}$O$_{14}$P.

The antipernicious anemia vitamin. All vitamin B$_{12}$ compounds contain the cobalt atom in its trivalent state. There are at least three active forms: cyanocobalamin, hydroxocobalmin, nitrocobalamin. Vitamin B$_{12}$ is a component of a coenzyme which takes part in the shift of carboxyl groups within molecules. As such it has an influence on nucleic acid synthesis, fat metabolism, conversion of carbohydrate to fat, and metabolism of glycine, serine, methionine and choline.

Source: (food) Liver, eggs, milk, meats and fish. Commercial source: Produced by microbial action on various nutrients (spent antibiotic liquors, sugar beet molasses, whey, also from sewage sludge).

Properties: Dark red crystals or red powder. Very hygroscopic, odorless and tasteless. Slightly soluble in water, soluble in alcohol, insoluble in acetone and ether.

Grade: USP, radioactive.

Use: Medicine (blood and nerve treatment), nutrition, animal feed supplements.

2-cyanoethyl acrylate.

CH_2:$CHCOOCH_2CH_2CN$.

Properties: Liquid; d 2.0690, bp polymerizes when heated, fp −16.9C., lb/gal 8.9, flash p 255F (124C) (COC), soluble in water, combustible.

Hazard: Toxic by ingestion and inhalation.

Use: Forms polymers, copolymers for viscosity index improvers, adhesives, textile finishes and sizes.

See also cyanocarylate adhesive.

cyanoethylation. Process for introducing the group —OCH_2CH_2CH into an organic molecule by reaction of acrylonitrile with a reactive hydrogen such as that on a hydroxyl or amino group.

cyanoethyl sucrose. (CES).

$C_{12}H_{14}O_3(OH)_{0.7}(OC_2H_4CN)_{7.3}$.

Properties: Clear, very viscous, pale yellow liquid; d 1.20 (20/20C); fp sets to a glass at −10C; bp above 300C; flash p over 375F (190C); combustible; refr index (25C) 1.615. Viscosity decreases very rapidly when heated. Has high volume resistivity, low power factor, unusually high dielectric constant.

Use: Capacitor impregnation, phosphor binding in electroluminescent panels, modification of electrical properties in coatings.

cyanoformic chloride. $CNCOCl$.

Properties: Oily liquid, bp 126–18C (750 mm).

Derivation: Reaction between phthalocyl chloride and the amide of ethyl oxalate.

"Cyanogas."[57] TM for a pesticide containing not less than 42% calcium cyanide; evolves hydrogen cyanide gas on exposure to atmospheric moisture.

Hazard: Poisonous.

cyanogen. (dicyan; oxalonitrile).

CAS: 460-19-5. $N\equiv C—C\equiv N$

Properties: Colorless gas, pungent penetrating odor, burns with a purple-tinged flame. Soluble

in water, alcohol, and ether. Specific gravity 1.8064 (air = 1), fp −28C, bp −20.7C.

Derivation: (a) Potassium cyanide solution is slowly dropped into copper sulfate solution; (b) mercury cyanide is heated.

Grade: Technical, pure.

Hazard: Flammable limits in air 6–32%. Store away from light and heat. A very toxic material. TLV: 10 ppm in air.

Use: Organic synthesis, welding and cutting metals, fumigant, rocket propellant.

cyanogenamide. See cyanamide (1).

cyanogen azide. (carbonpernitride).

$N\equiv N+\equiv N—C\equiv N$.

Properties: Colorless, oily liquid; unstable at room temperature.

Hazard: Explodes when shocked or heated; store and handle in solvents, e.g., acetonitrile.

Use: Organic synthesis; has wide range of reactivity.

cyanogen bromide. (bromine cyanide).

CAS: 506-68-3. $BrC\equiv N$.

Properties: Crystals, penetrating odor, slowly decomposed by cold water. Corrodes most metals. Soluble in water, alcohol, benzene, ether. D 2.02, bp 61.2C, mp 52C, vap d 3.6.

Derivation: (a) Action of bromine on potassium cyanide. (b) Interaction of sodium bromide, sodium cyanide, sodium chlorate, and sulfuric acid.

Hazard: A poison. Strong irritant to skin and eyes.

Use: Organic synthesis, parasiticide, fumigating compositions, rat exterminants, cyaniding reagent in gold extraction processes.

cyanogen chloride. CAS: 506-77-4. $CNCl$.

Properties: Colorless gas or liquid. Soluble in water, alcohol, and ether. D 1.2, bp 12.5C, fp −6C, vap d 2.1. Min purity 97 mole %.

Derivation: Action of chlorine on moist sodium cyanide suspended in carbon tetrachloride and kept cooled to −3C, followed by distillation.

Hazard: Toxic by ingestion and inhalation, strong irritant to eyes and skin. TLV: CL 0.3 ppm.

Use: Organic synthesis, tear gas, warning agent in fumigant gases.

cyanogen fluoride. (fluorine cyanide). CNF.

Properties: Colorless gas. Forms a white, pulverulent mass if cooled strongly and sublimes at −72C. Insoluble in water.

Derivation: Interaction of silver fluoride and cyanogen iodide.

Hazard: Inhalation or ingestion poison, strong irritant to eyes and skin. TLV: (as F) $2.5mg/m_3$.

Use: Organic synthesis, tear gas.

cyanogen iodide. (iodine cyanide). CNI.
Properties: Colorless needles, very pungent odor, acrid taste. Soluble in water, alcohol, and ether, mp 146.5C, d 1.84.
Derivation: By heating a metal cyanide with iodine.
Hazard: Strong irritant to eyes and skin. A poison.
Use: Taxidermists' preservatives.

"Cyanogran."[28] TM for a 98% sodium cyanide in granular form. White, crystalline solid crushed to pass 100% through 10 mesh, retained on 50 mesh.
Hazard: A deadly poison by inhalation and ingestion, strong irritant to eyes and skin.

cyanoguanidine. See dicyandiamide.

cyanomethyl acetate. (methyl cyanoethanoate). $CNCH_2COOCH_3$.
Properties: Colorless liquid, fp −22C, bp 200C, d 1.12.
Hazard: Highly toxic.
Use: Organic synthesis, pesticide.

cyano(methylmercuri)guanidine. (methylmercury dicyandiamide). $CH_3Hg(NHC(:NH)NHCN$.
Properties: Crystals, mp 156C, soluble in water.
Hazard: A poison by inhalation or ingestion, strong skin irritant.
Use: Seed fungicide and disinfectant.

cyanophenfos. $C_{15}H_{14}NO_2PS$.
Properties: Colorless crystals, mp 80C, refr index 1.58, partially soluble in aromatic solvents and ketones, almost insoluble in water.
Hazard: A poison.
Use: Insecticide.

cyanophos. (ciafos; O,O-dimethyl O-(4-cyanophenyl)phosphorothioate).
CAS: 2636-26-2. $C_9H_{10}NO_3PS$.
Properties: Yellowish liquid, bp (0.1 mm) 120C, fp 14C, refr index 1.54. Soluble in alcohol, methanol, acetone; slightly soluble in water. Decomposed by light and in alkaline environment.
Hazard: Poison. A cholinesterase inhibitor.
Use: Pesticide.

3-cyanopyridine. (3-azabenzonitrile; nicotinonitrile). CAS: 100-54-9. C_5H_4NCN.
Properties: Colorless liquid, bp 206.2C, mp 49.6C, soluble in water.
Use: Organic synthesis.

4-cyanopyridine. C_5H_4NCN.
Properties: Colorless liquid, bp 195.4C, mp 78.5C, partially soluble in water, soluble in most organic solvents.
Use: Organic synthesis.

cyanuramide. See ammelide.

cyanurdiamide. See ammeline.

cyanuric acid. (tricarbimide; tricyanide).
CAS: 108-80-5.
$HOCHC(OH)NC(OH)N \cdot 2H_2O$.
Properties: White crystals, odorless, slight bitter taste. Soluble in hot water and concentrated mineral acids, insoluble in alcohol and acetone, d 1.768, decomposes to cyanic acid at 320C.
Use: Intermediate for chlorinated bleaches, selective herbicide, whitening agents.
See also isocyanuric acid, the ketone isomer.

cyanuric chloride. (2,4,6-trichloro-1,3,5-triazine). CAS: 108-77-0. $C_3N_3Cl_3$ (cyclic).
Properties: Crystals with pungent odor, d 1.32, mp 146C, bp 194C (764 mm). Soluble in chloroform, carbon tetrachloride, hot ether, dioxane, ketones. Very slightly soluble in water (hydrolyzes in cold water).
Hazard: Toxic by ingestion and inhalation.
Use: Chemical synthesis, dyestuffs, herbicides, optical brighteners.

cyanurtriamide. See melamine.

cyclamate. Group name for synthetic nonnutritive sweetening agents derived from cyclohexylamine or cyclamic acid. The series includes sodium, potassium, and calcium cyclamates. As a result of a study made on laboratory animals in 1970, which indicated that these compounds cause incidence of genetic damage in chick embryos and cancer in rats from high dosage of cyclamates, their use in beverages and food products was banned in the US. More recent research has failed to confirm the carcinogenicity of these compounds in laboratory animals even at levels up to 240 times human intake. Notwithstanding these results, FDA has not yet withdrawn its ban on use of cyclamates as food additives or as table-top sweeteners, in view of the continuing uncertainty about its safety.
See also sweetener, nonnutritive.

cyclamen alcohol. The alcohol corresponding to cyclamen aldehyde, used as a stabilizer of cyclamen aldehyde.

cyclamen aldehyde. (methyl-p-isopropylphenylpropyl aldehyde). CAS: 103-95-7.
$(CH_3)_2CHC_6H_4CH(CH_3)CH_2CHO$.
Properties: Colorless liquid, floral odor, d 0.949–0.959, refr index 1.507–1.520. Soluble in 1 volume of 80% alcohol, in most oils.
Grade: FCC.
Use: Perfumery, soap perfumes, flavoring.

cyclamic acid. (USAN name for cyclohexanesulfamic acid; cyclohexylsufamic acid.
CAS: 100-88-9. $C_6H_{11}NHSO_3H$.
Properties: Odorless, white, crystalline solid with a sweet-sour taste; mp 170C. Strong, stable acid; soluble in water and alcohol; insoluble in oils.
Hazard: Suspected carcinogen.
Use: Nonnutritive sweetener, acidulant.
See also cyclamate.

cyclethrin. (3-(2-cyclopentenyl)-2-methyl-4-oxo-2-cyclopentenyl ester of chrysanthemum-monocarboxylic acid).
Properties: Viscous, brown liquid; soluble in petroleum solvents and other common organic solvents. Formulated principally as liquid for spray applications corresponding to natural pyrethrins.
Hazard: Toxic by inhalation and ingestion.
Use: Insecticide with applications similar to allethrin and other analogs.
See also furethrin, barthrin, ethyrthrin.

cyclic compound. An organic compound whose structure is characterized by one or more closed rings, it may be mono-, bi-, tri-, or polycyclic depending on the number of rings present. There are three major groups of cyclic compounds: (1) alicyclic, (2) aromatic (also called arene), and (3) heterocyclic. For more detailed information, consult specific entries.
See also organic compounds.

cyclizine hydrochloride. (1-diphenylmethyl-4-methylpiperazine hydrochloride).
CAS: 303-25-3.
$(C_6H_5)_2CHC_4H_8N_2CH_3 \cdot HCl$.
Properties: White, crystalline powder or small colorless crystals. Odorless or nearly so; bitter taste; mp 285C (decomposes); slightly soluble in water, alcohol, chloroform; insoluble in ether; pH (2% solution) 4.5–5.5.
Grade: USP.
Use: Medicine (antiemetic).

cycloaliphatic epoxy resin. (cycloalkenyl epoxides). A polymer prepared by epoxidation of multicycloalkenyls (polycyclic aliphatic compounds containing carbon-carbon double bonds) with organic peracids such as peracetic acid. Resistant to high temperatures.
Use: Space vehicles, outdoor electrical installations in polluted and humid atmospheres, high-temperature adhesives.

cyclobarbital. [5-(1-cyclohexenyl)-5-ethylbarbituric acid; tetrahydrophenobarbital].
CAS: 52-31-3. $C_{12}H_{16}N_2O_3$.
Properties: White crystals or crystalline powder, odorless, bitter taste, mp 170–174C, soluble in alcohol or ether, very slightly soluble in cold water or benzene.

Derivation: Hydrogenation of phenobarbital with colloidal palladium in alcohol as a catalyst.
Hazard: See barbiturates.
Use: Medicine (hypnotic, sedative).

cyclobutane. (tetramethylene).
CAS: 287-23-0. C_4H_8.

$$\begin{array}{ccc} CH_2 & - & CH_2 \\ | & & | \\ CH_2 & - & CH_2 \end{array}$$

Properties: Colorless gas, d 0.7083 (11C), bp 13C, fp −80C, flash p less than 50F. Insoluble in water, soluble in alcohol and acetone.
Derivation: Catalytic hydrogenation of cyclobutene.
Hazard: Flammable, dangerous fire risk.

cyclobutene. (cyclobutylene). C_4H_6,
Properties: Gas, d 0.733, bp 2.0C.
Derivation: From petroleum.
Hazard: Flammable, dangerous fire risk.

cyclocitrylideneacetone. See ionone.

cyclocumarol. $C_{20}H_{18}O_4$.   CAS: 518-20-7.
A synthetic blood anticoagulant.
Properties: White, crystalline powder with slight odor; mp 164–168C. Soluble in water, slightly soluble in alcohol.

"Cyclodex."[74] TM for water-dispersible driers with certified metal content of cobalt, lead, or manganese.

cycloheptane. (heptamethylene; suberane).
CAS: 291-64-5. C_7H_{14}.
Properties: Colorless liquid. Soluble in alcohol, insoluble in water, d 0.809, bp 117, fp −12C, aniline equivalent −6, flash p less than 70F (20C).
Grade: Technical.
Hazard: Flammable, dangerous fire risk. Narcotic by inhalation.
Use: Organic synthesis.

cycloheptanone. (suberone).   CAS: 502-42-1.
$C_7H_{12}O$.
Properties: Colorless liquid, peppermint odor, bp 179C, d 0.95, insoluble in water, soluble in ether and alcohol. Combustible.
Use: Research, intermediate.

cyclohexane. (hexamethylene; hexanaphthene; hexalhydrobenzene).   CAS: 110-87-7.
C_6H_{12}.   43rd highest-volume chemical produced in US (1985).

Structure: A typical alicyclic hydrocarbon. It may exist in two modifications called the "boat" and the "chair," as shown. This is due to slight distortion of the bond angles in accordance with the modified version of Baeyer's strain theory. Cyclohexane has been studied extensively on a theoretical basis in a branch of advanced chemistry called conformational analysis.
See conformation.

chair   boat

Properties: Colorless, mobile liquid; pungent odor; d 0.779 (20/4C); bp 807C; fp 6.3C; refr index 1.4263; aniline equivalent 7. Insoluble in water; soluble in alcohol, acetone, benzene; flash p (98% grade) −1F (−18.3C) (CC); autoign temperature 473F (245C). Flammable limits in air 1.3–8.4%.
Derivation: (a) Catalytic hydrogenation of benzene. (b) Constituent of crude petroleum.
Grade: 85, 98, 99.86%, spectrophotometric.
Hazard: Flammable, dangerous fire risk. Moderately toxic by inhalation and skin contact. TLV: 300 ppm in air.
Use: Manufacture of nylon; solvent for cellulose ethers, fats, oils, waxes, bitumens, resins, crude rubber; extracting essential oils; chemicals (organic synthesis, recrystallizing medium); paint and varnish remover; glass substitutes; solid fuels; fungicides; analytical chemistry.

1,4-cyclohexanebis(methylamine).
$C_6H_{10}(CH_2NH_2)_2$. The commercial product is about 40% cis and 60% trans.
Properties: Clear liquid, d 0.9419 (20/4C), bp 239–244C. Miscible in water, alcohol, and most other organic solvents. Combustible.
Use: Intermediate, resins.

cyclohexanecarboxylic acid. See hexahydrobenzoic acid.

1,2-cyclohexanedicarboxylic anhydride.
See hexahydrophthalic anhydride.

1,4-cyclohexanedimethanol. (CHDM).
$C_6H_{10}(CH_2OH)_2$. Cis and trans isomers are known and are present in the commercial product in about 30–70%.
Properties: Liquid; bp 286.0C (735 mm cis-isomer); mp 41–61C; d (super-cooled) 1.0381 (25/4C); flash p 330F (165C) (COC); refr index (20C) 1.4893; soluble in water, ethanol. Combustible.
Use: Polyester films and protective coatings, reduction of reaction time in esterification.

cyclohexanesulfamic acid. See cyclamic acid.

cyclohexanol. (hexahydrophenol).
CAS: 108-93-0. $C_6H_{11}OH$.
Properties: Colorless oily liquid, camphor-like odor, hygroscopic. Sparingly soluble in water, miscible with most organic solvents and oils. D 0.937 (37.4C), mp 23C, bp 160.9C, wt/gal/approximately 8 lb, flash p 154C (67.7C), refr index 1.465 (22C). Combustible, autoign temperature 572F (300C).
Derivation: Phenol is reduced with hydrogen over active nickel at 71–76.6C. The cyclohexanone is removed by condensing with benzaldehyde in the presence of alkali.
Grade: Technical (contains freezing inhibitor).
Hazard: Toxic by skin absorption and inhalation; narcotic. TLV: 50 ppm in air.
Use: Soap making to incorporate solvents and phenolic insecticides, source of adipic acid for nylon, textile finishing, solvent for alkyd and phenolic resins, cellulosics, blending agent, lacquers, paints and varnishes, finish removers, emulsified products, leather degreasing, polishes, plasticizers, plastics, germicides.

cyclohexanol acetate. (cyclohexanyl acetate).
CAS: 622-45-7. $CH_3COOC_6H_{11}$.
Properties: Colorless liquid, odor resembling that of amyl acetate. Miscible with most lacquer solvents and dilutions and with halogenated and hydrogenated hydrocarbons. Soluble in alcohol, insoluble in water. D 0.966, bp 177C, flash p 136F (57C), autoign temperature 633F (333C). Combustible.
Hazard: Narcotic.
Use: Solvent for nitrocellulose, cellulose ether, bitumens, metallic soaps, basic dyes, blown oils, crude rubber, many natural and synthetic resins and gums, lacquers.

cyclohexanone. (pimelic ketone; ketohexamethylene).   CAS: 108-94-1. $C_6H_{10}O$.

Properties: Water-white to pale yellow liquid with acetone- and peppermint-like odor. Slightly soluble in water, miscible with most solvents, bp 156.7C, fp −32C, d 0.948, flash p 111F (44C), refr index (20C) 1.4507, vap press (100C)135 mm, autoign temperature 788F (420C). Combustible.

Derivation: By passing cyclohexanol over copper with air at 280F, also by oxidation of cyclohexanol with chromic acid or oxide.

Hazard: Moderate fire risk. Toxic via inhalation and skin contact. TLV: 25 ppm in air.

Use: Organic synthesis, particularly of adipic acid and caprolactam (about 95%), polyvinyl chloride and its copolymers, and methacrylate ester polymers, wood stains, paint and varnish removers, spot removers, degreasing of metals, polishes, leveling agent dyeing and delustering silk, lube oil additive, solvent for cellulosics, natural and synthetic resins, waxes, fats, etc.

cyclohexanone peroxide. (1-hydroperoxycyclohexyl 1-hydroxycyclohexyl peroxide).
CAS: 12262-58-7.
$C_6H_{10}(OOH)OOC_6H_{10}OH$.
Properties: Grayish paste, insoluble in water, soluble in most organic solvents.
Hazard: Dangerous fire risk, powerful oxidizer.

cyclohexanyl acetate.   See cyclohexanol acetate.

cyclohexene.   (1,2,3,4-tetrahydrobenzene).
CAS: 110-83-8. C_6H_{10}.

Properties: Colorless liquid, d 0.811 (20/4C), bp 83C, fp −1037C, refr index. 1.445 (25C), flash p 11F (−11.6C), aniline equivalent 10, wt/gal 6.7 lb (25C), autoign temperature 590F (310C), soluble in alcohol, insoluble in water.

Grade: Technical 95%, 99%, research 99.9 mole %.

Hazard: Flammable, dangerous fire risk. Toxic by inhalation. TLV: 300 ppm in air.

Use: Organic synthesis, catalyst solvent, oil extraction, manufacturing of adipic and maleic acids.

3-cyclohexene-1-carboxaldehyde.   (1,2,3,6-tetrahydrobenzaldehyde).

$CH_2CH:CHCH_2CH_2CHCHO$.
Properties: Liquid, d 0.9721, bp 164.2C, fp −100C, lb/gal 8.1 (20C), flash p 135F (57.2C), slightly soluble in water, combustible.

Hazard: Strong irritant to tissue.
Use: Intermediates, improves water resistance of textiles.

cyclohexene oxide. $C_6H_{10}O$.
Properties: Colorless liquid with a strong odor, bp 129–130C, d 0.967 (25/4C), refr index (25C) 1.4503, flash p 81F (27.2C). Soluble in alcohol, ether, acetone; insoluble in water.
Grade: 98% pure.
Hazard: Flammable, dangerous fire risk.
Use: Chemical intermediate.

cyclohexenylethylbarbituric acid.   See cyclobarbital.

cyclohexenylethylene.   See vinylcyclohexene.

cyclohexenyltrichlorosilane. $C_6H_9SiCl_3$.
Properties: Colorless liquid, bp 202C, d 1.263 (25/25C), refr index (25C) 1.488, flash p 200F (93.3C) (COC). Readily hydrolyzed by moisture, with liberation of hydrogen chloride. Combustible.
Grade: Technical.
Hazard: Strong irritant, corrosive to skin.
Use: Intermediate for silicones.

cycloheximide.   (Generic name for 3-[2-(3,5-dimethyl-2-oxocyclohexyl)-2-hydroxy-ethyl]glutarimide).   CAS: 66-81-9.

$CH_2CH(CH_3)CH_2CH(CH_3)COCHCH(OH)$
$CH_2CHCH_2CONHCOCH_2$.   A plant growth regulator. Byproduct in the manufacture of streptomycin.
Properties: Crystals, mp 115.5–117C. Slightly soluble in acetone, alcohol, and chlorinated solvents.
Hazard: Toxic by ingestion.
Use: Fungicide, antibiotic, abscission of citrus fruit in harvesting, turf disease control.

cyclohexylamine.   (hexahydroaniline; aminocyclohexane).   CAS: 108-91-8.
$C_6H_{11}NH_2$.
Properties: Colorless liquid with unpleasant odor, bp 134.5C, d 0.8647 (25/25C), fp −18C. Strong organic base, pH of 0.01% aqueous solution 10.5, forms an azeotrope with water, bp 96.4C. Miscible with most solvents, flash p 90F (32.2C) (OC), autoign temperature 560F (293C).
Grade: Technical (98%).
Hazard: Flammable, moderate fire risk. Toxic by ingestion, inhalation, skin absorption. TLV: 10 ppm in air.
Use: Boiler-water treatment, rubber accelerator, intermediate in organic synthesis.

cyclohexylbenzene. See phenylcyclohexane.

N-cyclohexyl-2-benzothiazolesulfenamide.
(benzothiazyl-2-cyclohexylsulfenamide).
$C_6H_4SNCSNHC_6H_{11}$.
Properties: Cream-colored powder, d 1.27, melting range 93–100C, insoluble in water, soluble in benzene.
Use: Rubber accelerator.

cyclohexyl bromide. CAS: 108-85-0.
$C_6H_{11}Br$.
Properties: A liquid, not more than faintly yellow, having a penetrating odor, d 1.32–1.34 (25/25C), refr index 1.4926–1.4936 (25C).

cyclohexyl chloride. CAS: 542-18-7.
$C_6H_{11}Cl$.
Properties: Colorless liquid, fp −43C, bp 142C, flash p 89F (31.6C), d 0.992.
Hazard: Flammable, moderate fire risk.

2-cyclohexylcyclohexanol. $C_6H_{11}H_{10}OH$.
Properties: Colorless liquid, fp 29C, bp 271–277C, d 0.977 (25/25C), lb/gal 8.13, refr index 1.495 (25C), flash p 255F (124C), soluble in methanol and ether, slightly soluble in water. Combustible.

2-cyclohexyl-4,6-dinitrophenol. See dinitrocyclohexylphenol.

cyclohexyl isocyanate. CAS: 3173-53-3.
$C_6H_{11}NCO$.   Used to form cyclohexyl carbamates or ureas for agricultural chemicals or pharmaceutical use.

cyclohexyl methacrylate.
$H_2C:C(CH_3)COOC_6H_{11}$.
Properties: Colorless monomeric liquid with pleasant odor, bp 210C, refr index (20C) 1.4578, d 0.9626 (20/20C), viscosity (25C) 5.0 centipoises, insoluble in water. Combustible.
Use: Optical lens systems, dental resins, encapsulation of electronic assemblies.

p-cyclohexylphenol. $C_6H_{11}C_6H_4OH$.
Properties: Crystals, mp 120C (min), combustible.
Grade: Technical.
Use: Intermediate for resins and organic synthesis.

N-cyclohexylpiperdine. $C_6H_{11}NC_5H_{10}$.
Properties: Yellow liquid, refr index (20C)1.4856. Combustible.
Use: Intermediate.

cyclohexyl stearate. $C_6H_{11}OOCC_{17}H_{35}$.
Properties: Pale yellow powder; d 0.882 at 30/15.5C; mp 26–28C; soluble in benzene, toluene, and acetone; insoluble in water.
Use: Plasticizer for natural and synthetic resins.

cyclohexylsulfamic acid. See cyclamic acid.

N-cyclohexyl-p-toluenesulfonamide.
$C_6H_{11}NHSO_2C_6H_4CH_3$.
Properties: Yellow-brown fused mass, relatively light-stable, mp 86C, bp 350C. Combustible. Soluble in alcohol, esters, ketones, aromatic hydrocarbons, and vegetable oils; insoluble in water. Compatible with a wide variety of resins including most of the cellulosic and vinyl resins.
Use: Resin plasticizer.

cyclohexyl trichlorosilane. CAS: 98-12-4.
$C_6H_{11}SiCl_3$.
Properties: Colorless to pale yellow liquid, bp 206C, d 1.226 (25/25C), refr index (25C) 1.4759, flash p 185F (85C) (COC). Readily hydrolyzed by moisture with liberation of hydrogen chloride. Combustible.
Derivation: By Grignard reaction of silicon tetrachloride and cyclohexylmagnesium chloride.
Grade: Technical.
Hazard: Toxic by ingestion and inhalation, strong irritant to tissue.
Use: Intermediate for silicones.

"Cyclolube."[500]   TM for a series of oils composed principally of cycloparaffins. Used as plasticizers for nonpolar polymers, lubricants for polar polymers and extenders for relatively saturated polymers.

cyclomethylcaine. (3-(2-methylpiperidine)propyl-p-cyclohexyloxybenzoate). $C_{22}H_{33}NO_3$.
Properties: White, odorless crystalline powder. Sparingly soluble in water, alcohol, and chloroform; and very slightly soluble in acetone, ether, and dilute acids.
Use: Medicine (topical anesthetic).

cyclone. A dust-collecting device consisting of a cylindrical chamber the lower portion of which is tapered to fit into a cone-shaped receptacle placed below it. The dust-laden air enters through a vertical slot-like duct on the upper wall of the chamber at the rate of at least 100 feet per second. Since the particles enter at a tanget, they whirl in a circular or cyclonic path within the chamber. The centrifugal force exerted on the particles is proportional to their weight and to the square of their velocity. The particles slide along the walls of the chamber and gradually circulate down into the conical receptor while the clean air escapes through a central pipe at the bottom. The dust accumulates in the cone and is discharged at intervals or continuously. The larger the particles the more efficient is the removal; in simple cyclones those less than 50 microns in diameter are not retained but improved models retain particles as small as 20 microns. Cyclones are also used in cleaning and firing pulverized coal.

cyclonite. (sym-trimethylene trinitramine; hexahydro-1,2,5-trintro-sym-triazine; trinitro-trimethylenetriamine; cyclotrimethylenetrini-tramine; RDX).   CAS: 121-82-4. $N(NO_2)CH_2N(NO_2)CH_2N(NO_2CH_2$.
Properties: White, crystalline solid; d 1.82; mp 203.5C;, soluble in acetone; insoluble in water, alcohol, carbon tetrachloride, and carbon disulfide; slightly soluble in methanol and ether.
Derivation: Reaction of hexamethylenetetramine with concentrated nitric acid.
Hazard: High explosive, easily initiated by mercury fulminate. Toxic by inhalation and skin contact. TLV: 1.5 mg/m³ of air.
Use: Explosive 1.5 times as powerful as TNT.

3-cyclooctadiene.   CAS: 29965-97-7. C_8H_{12}.
Intermediate for such compounds as suberic acid

1,5-cyclooctadiene.

$HC:CH(CH_2)_2CH:CHCH_2CH_2$. A butadiene dimer.
Properties: Liquid, fp −56.39C, distillation range 301–303F (technical), bp 149.34C (pure), d 0.88328 (20/4C), lb/gal 7.38, vap press 0.50 psia (37.7C), refr index 1.4933 (20C), flash p 100F (37.7C), combustible.
Derivation: Catalytic dimerization of butadiene.
Grade: Technical 95%, 99%, 99.8 mole %.
Hazard: Moderate fire risk.
Use: Resin intermediate, third monomer in EPT rubber.

cyclooctane. C_8H_{16}.
Properties: Colorless liquid, d 0.835, bp 148C, mp 14C, combustible.

1,3,5,7-cyclooctatetraene.   CAS: 629-20-9. C_8H_8.
Properties: Colorless liquid, fp −7C, bp 140C, d 0.943 (0/4C), refr index 1.5394 (20C). It behaves like an aliphatic hydrocarbon, is relatively reactive, and resinifies on standing in air. Combustible.

Derivation: Nickel-catalyzed polymerization of acetylene, a reaction discovered by Reppe in Ger-

many about 1940. The mechanism has several possible pathways.
Use: Organic research
See also polyacetylene.

cycloolefin.   An alicyclic hydrocarbon having two or more double bonds, e.g., the very reactive and widely used cyclopentadiene derived from coal tar as well as cyclohexadiene and cyclooctatetraene containing six and eight carbon atoms respectively. The latter has four double bonds and is a polymer of acetylene.

cycloparaffin.   An alicyclic hydrocarbon in which three or more of the carbon atoms in each molecule are united in a ring structure and each of these ring carbon atoms is joined to two hydrogen atoms or alkyl groups. The simplest members are cyclopropane (C_3H_6), cyclobutane (C_4H_8), cyclopentane (C_5H_{10}), cyclohexane (C_6H_{12}), and derivatives of these such as methylcyclohexane ($C_6H_{11}CH_3$).
Hazard: All members of the cycloparaffin series are narcotic and may cause death through respiratory paralysis. For most of the members there appears to be a narrow range between the concentration causing deep narcosis and those causing death.

1,3-cyclopentadiene.   CAS: 542-92-7. C_5H_6.

Properties: Colorless liquid with d 0.805 and bp 425C. Insoluble in water; soluble in alcohol, ether, and benzene.
Derivation: From coal tar and cracked petroleum oils.
Hazard: Decomposes violently at high temperature. Toxic. TLV: 75 ppm in air.
Use: Chemical intermediate, organic synthesis (Diels-Alder reaction), starting material for synthetic prostaglandin, chlorinated insecticides, and formation of sandwich compounds by chelation, e.g., cyclopentadienyl iron dicarbonyl dimer $[C_5H_5Fe(CO)_2]_2$.

cyclopentamine hydrochloride.   (1-cyclopentyl-2-methyl-aminopropane hydrochloride). $C_5H_9CH_2(CH_3)NHCH_3:HCl$.
Properties: White, crystalline powder. Mild characteristic odor and bitter taste. Mp 113.0–116.0C. Freely soluble in water, alcohol, and chloroform; soluble in benzene; slightly soluble in ether; pH (1% solution) approximately 6.2.

Grade: NF.
Use: Medicine (vasoconstrictor).

cyclopentane. (pentamethylene).
CAS: 287-92-3. C_5H_{10}.

$$
\begin{array}{c}
CH_2 \\
CH_2 \quad CH_2 \\
| \qquad | \\
CH_2 \!\!-\!\! CH_2
\end{array}
$$

Properties: Colorless liquid; soluble in alcohol; insoluble in water; d 0.7445 (20/4C); bp 49.27C; fp −94C; refr index 1.406 (20/D); flash p −35F (−37.2C).
Derivation: Catalytic cracking of cyclohexane.
Grade: Technical, 95%, 99%, research.
Hazard: Flammable; dangerous fire risk. Moderately toxic by ingestion and inhalation. TLV: 600 ppm in air.
Use: Solvent for cellulose ethers, motor fuel, azeotropic distillation agent.

1,2,3,4-cyclopentanetetracarboxylic acid.
$C_5H_6(COOH)_4$.
Properties: Crystalline powder, mp 195–196C, soluble in water but insoluble in most organic solvents.
Use: Curing agent for resins; imparts thermal stability and high-temperature properties.

1,2,3,4-cyclopentanetetracarboxylic acid dianhydride. $C_5H_6(C_2O_3)_2$.
Properties: White solid, mp 220–221C. Moderately soluble in dimethyl formamide, butyrolactone, and dimethylsulfoxide, but only slightly soluble in acetone; insoluble in hydrocarbons.
Use: Curing agent for epoxy resins.

cyclopentanol. (cyclopentyl alcohol).

CAS: 96-41-3. $\overline{CH_2CH_2CH_2}CH_2CHOH$.
Properties: Colorless, viscous liquid; pleasant odor; d 0.946 (20/4C); refr index 1.4575 (20C); fp −19C; bp 139–140C; flash p 124F (51C). Slightly soluble in water; soluble in alcohol. Combustible.
Hazard: Moderate fire risk.
Use: Perfume and pharmaceutical solvent, intermediate for dyes, pharmaceuticals and other organics.

cyclopentanone.   CAS: 120-92-3. C_5H_8O.

$$
\begin{array}{c}
CH_2\!\!-\!\!CH_2 \\
| \qquad\quad \rangle CO \\
CH_2\!\!-\!\!CH_2
\end{array}
$$

Properties: Water-white, mobile liquid; distinctive ethereal odor somewhat like peppermint; bp

131C; d 0.943; refr index 1.437; flash p 87F (30.5C) (CC); insoluble in water; soluble in alcohol and ether.
Hazard: Flammable, moderate fire risk. Narcotic in high concentration.
Use: Intermediate for pharmaceuticals, biologicals, insecticides and rubber chemicals.

cyclopentanone oxime. C_5H_8NOH.   Nearly colorless and odorless, crystalline solid; mp 56C; bp 196C; soluble in water, alcohol. Used as intermediate in synthesis of amino acids, proline, and ornithine.

cyclopentene.   CAS: 142-29-0.

$\overline{CH{:}CHCH_2CH_2}CH_2$.
Properties: Colorless liquid, d 0.772, bp 44C, fp −135.21C, refr index (20/D) 1.4225, flash p −20F (−29C).
Grade: Technical, research (99.89 mole %).
Hazard: Flammable, dangerous fire risk. Moderate narcotic action.
Use: Organic synthesis, polyolefins, epoxies, crosslinking agent.

cyclopentenylacetone.   [1-(1-cyclopentenyl)-2-propanone]. $C_5H_7CH_2COCH_3$.
Properties: Clear, colorless liquid; ketone odor; bp 170C; refr index 1.4545–1.4550 at 25C. Combustible.
Use: Organic synthesis.

cyclopentylacetone. (1-cyclopentyl-2-propanone). $C_5H_9CH_2COCH_3$.
Properties: Liquid, d 0.893 (25/25C), bp 180–184C, refr index 1.4420 (25C). Combustible.
Use: Organic synthesis.

cyclopentyl alcohol.   See cyclopentanol.

cyclopentyl bromide. (bromocyclopentane). C_5H_9Br.
Properties: Clear, mobile liquid with sweet aromatic odor; bp 137–138C; d 1.3866 (20/4C); wt/gal 11.6 lb (20C); refr index (n 20/D) 1.4885; flash p 108F (42.2C) (CC); insoluble in water. Combustible.
Hazard: Moderate fire risk and moderately toxic.
Use: Organic synthesis (pharmaceuticals).

1-cyclopentyl-2-methylaminopropane hydrochloride.   See cyclopentamine hydrochloride.

cyclopentyl phenyl ketone. $C_5H_9COC_6H_5$.
Properties: Colorless to light yellow liquid, bp 145–146C (15 mm), soluble in most common organic solvents, insoluble in water. Combustible.
Use: Pharmaceutical intermediate.

1-cyclopentyl-2-propanone. See cyclopentyl acetone.

cyclopentylpropionic acid.
$C_5H_9CH_2CH_2COOH$.
Properties: Liquid, bp 130–132C (12 mm), flash p 116F (46.6C), insoluble in water. Combustible.
Hazard: Moderate fire risk.
Use: Intermediate, wood preservatives.

cyclopentylpropionyl chloride. (cyclopentylpropionic acid chloride). $C_5H_9CH_2CH_2COCl$.
Properties: Liquid, bp 81–82C (10 mm), flash p 104F (40C), soluble in water. Combustible.
Hazard: Moderate fire risk.
Use: Intermediate.

cyclophane. A term applied in organic research to two molecules, e.g., of benzene, that are connected by covalent bonds involving all their carbon atoms (superphane). Such molecules exhibit a high degree of strain. In other types some of the carbon atoms are not involved. Cyclophanes can be made by a reaction mechanism in which dimerization of benzocyclobutenes plays a part.

"Cyclophos."[1] TM for colorless crystalline hydrated sodium polyphosphate. Cyclic compound. FCC grade. Coagulates albumin and certain other proteins. Soluble in water.
Use: Precipitation of proteins.

cyclophosphamide. CAS: 50-18-0.
$C_7H_{15}Cl_2N_2O_2P$.   A nitrogen mustard.
Properties: Crystalline solid; mp 41C; slightly soluble in benzene, alcohol, carbon tetrachloride; very slightly soluble in ether and acetone.
Hazard: An experimental carcinogen by injection.
Use: Antineoplastic for treatment of leukemia, etc., tested for use in chemical shearing of sheep and insect chemosterilant.

cyclopropane. (trimethylene).   CAS: 75-19-4.
C_3H_6.

CH₂
/ \
CH₂—CH₂

Properties: Colorless gas of characteristic odor resembling that of solvent naphtha and having a pungent taste. D 0.72–079, bp −32.9C, fp −126.6C. Soluble in alcohol and ether, partially soluble in water. Autoign temperature 928F (497C).
Derivation: Reduction of dibromocyclopropane with zinc dust.
Grade: Technical, USP, 99.5% min.

Hazard: Highly flammable. Forms flammable and explosive mixtures with air or oxygen. Explosive limits in air 2.4–10.3% by volume. Moderately toxic by inhalation. Narcotic in high concentration.
Use: Organic synthesis, anesthetic.

cyclopropanespirocyclopropane. See spiropentane.

cyclosilane. See silane.

"Cyclo Sol."[125] TM for a series of hydrocarbon solvents composed of 50–99% aromatic hydrocarbons.
Hazard: Flammable; dangerous fire risk.

methylcyclopentenolone. Used in soap, cosmetic perfumery, and flavor compounding.

cyclotrimethylenetrinitramine. See cyclonite.

cyclotron. A circular electromagnetic device for accelerating positively charged particles (protons, deuterons, alpha particles). It was invented in 1929 by E. O. Lawrence (1901–1958) at the University of California at Berkeley. It is now used chiefly for basic nuclear research. The acceleration is achieved by successive applications of small accelerations at low voltage synchronized with the rotational period of the particles in a magnetic field. Energies of over 700 MeV for protons and over 900 MeV for alpha particles have been attained. The energized particles emerging from the cyclotron impinge upon a target nucleus, resulting in formation of radioactive isotopes, neutrons, and ionizing radiation. The first plutonium was made in a cyclotron in 1939. Powerful cyclotrons with huge electromagnets are in use in physical research laboratories throughout the world.
See also bombardment.

cycloversion. A process using bauxite as a catalyst for (1) desulfurization, (2) reforming, and (3) cracking of petroleum to form high-octane gasoline.

cyclyd. A coined term referring to the cyclic alkyd coatings prepared from "Polycyclol 1222," used in baking metal primers and air-drying maintenance paints.

"Cycocel."[57] TM for a plant growth regulator (2-chloroethyltrimethyl ammonium chloride) said to be effective for cereal grains, tomatoes, and peppers.

"Cycolac."[525] TM for a series of acrylonitrile-butadiene-styrene polymers. Especially suitable for metal plating.
See ABS resin.

"Cycolon."[525] TM for medium-impact material possessing most of the properties of ABS resins.
Use: Automotive parts, housewares, toys, refrigeration parts.

"Cycoloy."[525] TM for a family of alloys of ABS and other thermoplastic resins, providing a high-performance balance of properties.
Use: Automotive, sporting goods, electrical/electronics markets.

"Cycopol."[11] TM for a group of copolymer resins including styrenated alkyds, acrylonitrile-modified styrenated alkyd, vinyl, toluene type, special acrylic type resins.
Use: Vehicle for solvents, hammer finishes, coil coatings, metal decorating.

"Cycovin."[525] TM for alloys of ABS and PVC, providing a family of materials with flame-retardant properties in addition to the balanced features of ABS.
Use: Instructional and electrical applications.

cymene. (cymol; isopropyltoluene; methylpropylbenzene). CAS: m- 535-77-3; o- 527-84-4; p- 99-87-6; mixed 25155-15-1.
$CH_3C_6H_4CH(CH_3)_2$. The o- m- and p-isomers are known. The p-isomer is:

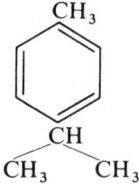

Properties: Colorless, transparent liquids with aromatic odor. Combustible. D: o- 0.8748, m- 0.862, p- 0.8551. Fp: o- −71C, m- −64C, p- −68C. Bp: o- 177C, m- 175.6C, p- 176.5C. Refr index: p- 1.489 (20C). Soluble in alcohol, ether, and chloroform. Insoluble in water. Flash p (p-) 127F (51C) (CC).
Derivation: Mixed cymenes are produced from toluene by alkylation. P-cymene occurs in several essential oils and is made from monocyclic terpenes by dehydrogenation. These terpenes can be made from turpentine or obtained as a byproduct from the sulfite digestion of spruce pulp in paper manufacture.
Method of purification: Washing with sulfuric acid, water, and alkali.
Grade: Technical.
Hazard: Moderate fire risk. Moderately toxic by ingestion.
Use: Solvents, synthetic resin manufacture, metal

polishes, organic synthesis (oxidation to hydroperoxides used as catalysts for synthetic rubber manufacture; cymene alcohols are made by hydrogenating the hydroperoxides). Pure p-cresol and carvacrol are made from p-cymene.

cypermethrin. (cyano(3-phenoxyphenyl) methyl; 3-(2,2-dichlorovinyl-2,2-dimethylcyclopropanecarboxylate). CAS: 52315-07-8.
Use: Insecticide.

Cys. Abbreviation for cysteine.

cysteamine. See 2-aminoethanethiol.

cysteine. (α-amino-β-thiolpropionic acid; β-mercaptoalanine). CAS: 52-90-4.
$HSCH_2CH(NH_2)COOH$. A nonessential amino acid derived from cystine occurring naturally in the L(+) form.
Properties: Colorless crystals, soluble in water, ammonium hydroxide, and acetic acid. Insoluble in ether, acetone, benzene, carbon disulfide, and carbon tetrachloride.
Derivation: Hydrolysis of protein; degradation of cystine. Found in urinary calculi.
Available forms: Available commercially as L(+)-cysteine hydrochloride.
Use: Biochemical and nutrition research, reducing agent in bread doughs (up to 90 ppm).

cystine. (β,β'-dithiobisalanine; di[α-amino-β-thiolpropionic acid]). CAS: 56-89-3.
$HOOCCH(NH_2)CH_2SSCH_2CH(NH_2)COOH$.
A nonessential amino acid.
Properties: White, crystalline plates; soluble in water; insoluble in alcohol. Optically active. DL-cystine, mp 260C. D(+)-cystine, mp 247–249C. L(-)-cystine, mp 258–261C with decomposition; the naturally occurring form.
Derivation: Hydrolysis of protein (keratin), organic synthesis. Occurs as small hexagonal crystals in urine.
Grade: FCC.
Use: Biochemical and nutrition research, nutrient and dietary supplement.

"Cystokon."[329] TM for a 30% solution of sodium acetrizoate.

cythion. Proprietary malathion.
Use: Insecticide.

cytidine. CAS: 65-46-3. $C_9H_{13}N_3O_5$.
The nucleoside consisting of D-ribose and cytosine.
Properties: White, crystalline powder; soluble in water, acid, alkali; insoluble in alcohol.

Derivation: From yeast ribonucleic acid. Also available as the hemisulfate, $(C_9H_{13}N_3O_5)_2 \cdot H_2SO_4$.
Use: Biochemical research.

cytidine phosphates. Nucleotides used by the body in growth processes. important in biochemical and physiological research. Those isolated and commercially available (as sodium salts) are the monophsophate (CMP; see cytidylic acid), the diphosphate (CDP), and the triphosphate (CTP).

cytidylic acid. (cytosylic acid; cytidinephosphoric acid; cytidine monophosphate; CMP). $C_9H_{14}N_3O_8P$. The monophosphoric ester of cytosine, i.e., the nucleotide containing cytosine D-ribose and phosphoric acid. The phosphate may be esterified to either the 2,3, or 5 carbon of ribose yielding cytidine-2'-phosphate, cytidine-3'-phosphate, and cytidine-5'-phosphate, respectively.
Properties: (cytidine-3'-phosphate): White, crystalline powder; odorless; mild sour taste; mp: crystals from 50% alcohol 230–233C (with decomposition). Crystals from water 227C (with decomposition). Slightly soluble in water and dilute alkalies, insoluble in alcohol and other organic solvents.
Derivation of commercial product: From yeast nucleic acid by hydrolysis. The 5'-monophosphate is made synthetically by phosphorylation and hydrolysis of isopropylidene cytidine.
Use: Biochemical research.

cytochemistry. The branch of biochemistry devoted to study of the chemical composition of cells and cell membranes, including chromosomes, genes, and the complex reactions involved in cell growth and replication, as well as the mechanism of enzyme activity.
See also molecular biology.

cytochrome. A class of iron-porphyrin proteins Of great importance in cell metabolism. They are pigments occurring in the cells of nearly all animals and plants. Several types have been identified: Cytochrome carbon is the most abundant and has been obtained in pure forms. The cytochromes and cytochrome oxidase have important functions in cell respiration. The latter is an iron-prophyrin-containing protein which is an important enzyme in cell respiration. It catalyzes the oxidation of cytochrome carbon and is reduced itself in the reaction. It is then reoxidized by oxygen.
See prophyrin

cytokinins. See kinins.

cytoplasm. The extra-nuclear components of the living cell, containing mitochondria, plastids, spherosomes, etc. This, together with the nucleus, constitutes the protoplasm. The chemical constituents are chiefly proteins, plus a high percentage of water.

"Cytosar."[327] TM for cytosine arabinoside.

cytosine. (2-oxo-4-aminopyrimidine). CAS: 71-30-7. $C_4H_5N_3O$. A pyrimidine found in both ribonucleic and deoxyribonucleic acids and certain coenzymes.
Properties: (monohydrate): Lustrous platelets. Decomposes at 320–325C. Slightly soluble in water and alcohol, insoluble in ether.
Derivation: Isolation following hydrolysis of nucleic acids; organic synthesis.
Use: Biochemical research.

cytosine arabinoside. (Ara-C; 1-β-D- arabinofuranosyl cytosine). CAS: 147-94-4. $C_9H_{13}N_3O_5$. A drug synthesized in 1969 at Salk Institute. It is useful in combating myelocytic leukemia in adults and has been approved by FDA as a prescription drug.

cytosine monophosphate. See cytidylic acid.

cytosylic acid. See cytidylic acid.

"Cytrel."[352] TM for a tobacco substitute derived from cellulose. It is free from nicotine and has 15–30% of the tar content of tobacco.

"Cytrol" Amitrol-T[57]. TM for 3-amino-1,2,4-triazole liquid formulation herbicide.

"Cyzine."[57] TM for 10% 2-acetylamino-5-nitrothiazole feed supplement.

D

d-. Prefix indicating that a substance is dextrorotatory. A plus sign (+) is now preferred.

D. Symbol for deuterium.

D-. Prefix indicating the right-handed enantiomer of an optical isomer.
See also glyceraldehyde, asymmetry, enantiomer.

2,4-D. (2,4-dichlorophenoxyacetic acid).
CAS: 94-75-7. $Cl_2C_6H_3OCH_2COOH$.
Properties: White to yellow crystalline powder, difficultly soluble in water or oils, soluble in alcohols. Stable, mp 138C, bp 160C (0.4 mm).
Derivation: Reaction of 2,4-dichlorophenol and chloroacetic acid in aqueous sodium hydroxide.
Forms available: Sodium salt (60–85% acid), amine salts (10–60% acid), esters (10–45% acid). These forms are dispersible in water or oils (esters) and can be applied as sprays.
Grade: Technical.
Hazard: Irritant. Use may be restricted. TLV: 10 mg/m³.
Use: Selective weed killer and defoliant, fruit drop control.

DAA. Abbreviation for diacetone acrylamide.

"DABCO."[472] TM for triethylenediamine.

"Dacron."[28] TM for a polyester fiber made from polyethylene terephthalate. Available as filament yarn, staple, tow, and fiber fill.
Properties: D 1.38, tensile strength (psi) 4–5 grams/denier (about 50–60 thousand psi), break elongation 10–36%, moisture regain 0.4%, high elastic recovery, good insect resistance, difficult to ignite, self-extinguishing, soluble in m-cresol (hot), trifluoroacetic acid, and o-chlorophenol, mp 250C.
Derivation: Reaction of dimethyl terephthalate and ethylene glycol. The resulting polymer is meltextruded through a spinneret and stretched.
Use: Textile fabrics and suitings, often combined with wool and other fibers; cordage; fire hose; etc.
See also dimethyl terephthalate, terephthalic acid.

"Dacthal."[309] TM for a composition of dimethyl ester 2,3,5,6-tetrachloroterephthalic acid. Commonly known as DCPA. Used in pre-emergence weed control.

dactinomycin. USAN for actinomycin D.
$C_{62}H_{86}N_{12}O_{16}$.

An antibiotic produced from *Streptomyces*. Used in medicine.

"dag" Dispersions.[46] TM for a series of dispersions useful as lubricants. Most are colloidal. The base types are: graphite-water, graphite-oil, graphite-solvent, molybdenum disulfide, resin-bonded solid film (with alkyd, epoxy, or phenolic resin solutions) and some special types.

DAHQ. See 2,5-di-tert-amylhydroquinone.

Dakin reaction. Replacement of the aldehyde or acetyl groups in phenolic aldehydes or ketones by a hydroxyl group by means of hydrogen peroxide.

Dakin's solution. An aqueous solution containing 0.5% sodium hypochlorite, used as an antiseptic, especially for wound treatment.

Dakin-West reaction. Reaction of alpha-amino acids with acetic anhydride in the presence of base to give alpha-acetamido ketones. The reaction occurs via the intermediate azlactone.

"Dalamar."[28] TM for an azo yellow pigment.

dalapon. CAS: 75-99-0. CH_3CCl_2COOH.
Generic name for 2,2-dichloropropionic acid. The sodium salt (sodium 2,2-dichloropropionate) is commonly used.
Properties: (free acid): Liquid, bp 185–190C, 90–92C (14 mm), d 1.389 (22.8 4C). Very soluble in water and alcohol, soluble in ether. (sodium salt): Crystals, decomposes 174–176C, salty taste. Corrosive to iron. Soluble in water, aqueous solutions hydrolyze above 70C.
Hazard: Strong irritant to eyes and skin. TLV: 1 ppm in air.
Use: Herbicide.

"Dalpad."[233] TM for a coalescing agent, a stable, low-odor, low-temperature, film-forming aid for polyvinyl acetate and acrylic latex paints.

dalton. A unit of mass introduced in comparatively recent years. It designates 1/16th the mass of oxygen⁻¹⁶, the lightest and most abundant isotope of oxygen. Since this is 15.9949, the dalton is equivalent to 0.9997 mass unit.

Dalton, John. (1766–1844). The first theorist since the Greek philosopher Democritus to con-

ceive of matter in terms of small particles. The founder of the atomic theory on which all succeeding chemical investigation has been based (1807). His essential concept of the indivisibility of the atom was not called into question until 1910 when radioactive decay was established by Rutherford. Dalton's theories relating to pressures of gases and atomic combinations led to the basic generalizations stated in the law of multiple proportions, the law of constant composition, and the law of conservation of matter. Dalton's law of partial pressures states that in any mixture of gases each constituent exerts its pressure independently as if the other constituents were absent and that the solubility of mixed gases in liquid is proportional to the partial pressure of each.
See Priestley, chemical laws.

"Dama."[313] TM for fatty dialkyl methylamines.

dammar. A group of tree-derived resins soluble in hydrocarbon and chlorinated hydrocarbon solvents, partially soluble in alcohols, insoluble in water. Used in colorless and overprint varnishes, cellulosic lacquers, alkyd baking enamels, and paper and textile coatings.

dandy roll. A light roller covered with wire cloth which rides on the formed sheet near the dry end of the Fourdrinier wire. Its purpose is to provide a closer finish to the sheet as well as to impress a screen pattern on the upper surface of the sheet similar to that made by the wire on the under surface. Paper so marked is called "wove" in the trade. The dandy roll may also impress a pattern of parallel lines, in which case the paper is called "laid". It is also used to apply watermarks.

danthron. See 1,8-dihydroxyanthraquinone.

DAP. (1) Abbreviation for diallyl phthalate. (2) Abbreviation for diammonium phosphate. See ammonium phosphate dibasic.

"Dapon."[55] TM for a series of diallyl phthalate resins. Used for molding compounds, prepregs, and coatings.

dapsone. USAN and USP name for 4,4'-sulfonyldianiline.

"Daran."[311] TM for polyvinylidene chloride latexes, used as barrier coatings for packaging papers, paperboards, plastic films, and specialty saturants.

"Daratak."[311] TM for polyvinyl acetate homopolymer emulsions; used in adhesives and specialty coatings.

"Darco."[89] TM for activated carbon. Available in various grades for use in sugar refining, removal of impurities from electroplating solutions, purification of drycleaning solvents, drug and chemical purification, and purification and decolorization of animal and vegetable oils, fats, and waxes.

"Darex."[311] TM for styrene-butadiene latexes and related vehicles.
Use: Textile coatings, rug backings, saturants, shoe products, coatings, paints, and adhesives.

"Darvan."[69] TM for a series of anionic surfactants used as dispersants, emulsion stabilizers and latex stabilizers.

Darzens condensation. Formation of α-,β-epoxy esters (glycidic esters) by the condensation of aldehydes or ketones with esters of α-haloacids; the corresponding thermally unstable glycidic acids yield aldehydes or ketones on decarboxylation.

Darzens-Nenitzescu synthesis of ketones. Acylation of olefins with acid chlorides or anhydrides catalyzed by Lewis acids. When performed in the presence of a saturated hydrocarbon, the product is the saturated ketone.

Darzens synthesis of tetralin derivatives. Cyclization of α-benzyl-α-allylacetic acid type compounds by moderate heating in concentrated sulfuric acid to yield tetralin derivatives.

DAS. Abbreviation for 4,4'-diamino-2,2'-stibenedisulfonic acid.

dating chemical. See chemical dating.

daturic acid. See n-heptadecanoic acid.

daughter element. The element formed when another element undergoes radioactive decay. The latter is called the parent. The daughter may or may not be radioactive.

daunomycin. (daunorubicin).
CAS: 20830-81-3. $C_{27}H_{29}NO_{10}$. An antibiotic.
Properties: Reddish needles, bp (decomposes) 190C, soluble in water and methyl alcohol, insoluble in chloroform and benzene.
Hazard: A carcinogen (OSHA).

Davy, Sir Humphry. (1778–1829). Born in Cornwall, Davy was the first to isolate the alkali metals and recognize the identity of chemical and electrical energy. A pioneer in the science of electrochemistry, he carried out basic studies of electrolysis of salts and water, and his application of electricity to the decomposition of molten caustic potash led to the isolation of metallic potassium.

"Daxad."[311] TM for anionic, polymer-type dispersing agents. Supplied as light-colored powders or aqueous solutions. Effective dispersant for aqueous suspensions of insoluble dyestuffs, polymers, clays, tanning agents, and pigments.
Use: Manufacture of dyestuff pastes, textile backings, latex paints and paper coatings, retanning and bleaching of leather, dye resist in leather dyeing, disperson of pitch in paper manufacture, pre-floc prevention in the manufacture of synthetic rubber.

2,4-DB. Abbreviation for 2,4-dichlorophenoxybutyric acid.

DBCP. Abbreviation for 1,2-dibromo-3-chloropropane.

DBM. Abbreviation for dibutyl maleate.

DBMC. Abbreviation for 4,6-di-tert-butyl-m-cresol.

"DB" Oil.[202] TM for a castor oil specially refined to minimize acidity and moisture content for dielectric and sonar applications and for urethane polymers.

DBP. Abbreviation for dibutyl phthalate.

"DBPC."[11] (Also "dbpc"). TM for a polyalkyl hydroxybenzene, i.e., 2,6-di-tert-butyl-p-cresol.

DBS. Abbreviation for dibutyl sebacate.

DCA. Abbreviation for deoxycorticosterone acetate.

DBC. Abbreviation for 1,4-dichlorobutane.

DCHP. Abbreviation for dicyclohexyl phthalate.

DCO. Abbreviation for dehydrated castor oil.
See castor oil, dehydrated.

DCP. Abbreviation for dicapryl phthalate.

DCPA. Abbreviation for dimethyl-2,3,5,6-tetrachloroterephthalate.

DCPC. Abbreviation for dichlorophenyl methyl carbinol.
See di(p-chlorophenyl)ethanol.

DDB. Abbreviation for dodecylbenzene.

DDBSA. Abbreviation for dodecylbenzenesulfonic acid.
See sodium dodecylbenzenesulfonate.

DDD. Abbreviation for dichlorodiphenyldichloroethane.
See TDE.

DDDA. Abbreviation for dodecanedioic acid.

DDDM. Abbreviation for 2,2'-dihydroxy-5,5'-dichlorodiphenylmethane.
See dichlorophene.

DDE. Abbreviation for dichlorodiphenyldichloroethylene, $(ClC_6H_4)_2C:CCl_2$. It is a degradation product of DDT found as an impurity in DDT residues.

DDH. Abbreviation for dichlorodimethylhydantoin.

"DDI Diisocyanate."[590] TM for an intermediate made from a 36-carbon dimer aliphatic dibasic acid; exact structure is a complex mixture of isomers. May be used for reaction with diamines to give polyureas with highly desirable properties; also a source of polyurethanes.

DDM. (1) Abbreviation for diaminodiphenylmethane. (2) Abbreviation for n-dodecyl mercaptan.

DDNP. Abbreviation for diazodinitrophenol.

DDP. Abbreviation for dodecyl phthalte.

DDQ. See 2,3-dichloro-5,6-dicyanobenzoquinone.

DDS. Abbreviation for diaminodiphenylsulfone.
See sulfonyldianiline.

DDT. (dichlorodiphenyltrichloroethane; dicophane; chlorophenothane; 1,1,1-trichloro-2,2-bis(chlorophenyl)ethane). CAS: 50-29-3. $(ClC_6H_4)_2CHCCl_3$.

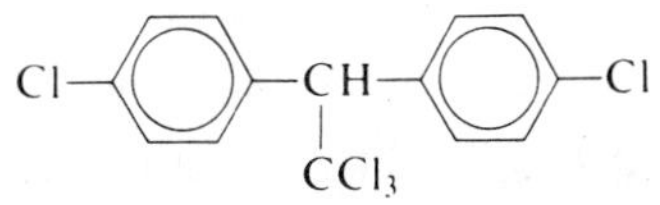

Cl—⟨⟩—CH—⟨⟩—Cl
      |
     CCl₃

Properties: Colorless crystals or white to slightly off-white powder, odorless or with slight aromatic odor. Insoluble in water; soluble in acetone, ether, benzene, carbon tetrachloride, kerosene, dioxane, and pyridine. Not compatible with alkaline materials.

Derivation: Condensing chloral or chloral hydrate with chlorobenzene in presence of sulfuric acid.

Grade: Technical, purified, aerosol, USP, as chlorophenothane.

Hazard: Toxic by ingestion, inhalation, and skin absorption, especially in solution. Lethal dosage for humans estimated to be 500 mg/kg of body weight (solid material). Since DDT is not biodegradable and is ecologically damaging, its agricultural use in the US was prohibited in 1973 (though its manufacturing for export is permitted). Claims of human carcinogenicity have not been proved. DDT can be used for a few specialized purposes, e.g., to combat the tussock moth. TLV: 1 mg/m³ of air. FDA tolerance: 5 ppm in foods.

Use: Insecticide, for tobacco and cotton, pesticide (tussock moth).

DDVP. Abbreviation for dimethyl dichlorovinyl phosphate.
See dichlorovos.

DE. Abbreviation for dextrose equivalent.

DEA. Abbreviation for diethanolamine, also abbreviation for Drug Enforcement Administration, a government agency which replaces the Bureau of Narcotics.

DEAC. Abbreviation for diethylaluminum chloride.

"Deacidite."[184] TM for a commercial grade anion exchanger consisting of a porous aliphatic polyamine weak base. Used in streptomycin conversion and strong acid adsorption.

"Deacon process." A method of converting hydrogen chloride to chlorine by oxidation of hydrogen chloride with oxygen at 400–500C over a copper salt catalyst, $2HCl + O \rightarrow Cl_2 + H_2O$. It is a means of producing chlorine without caustic and of utilizing the large amounts of byproduct hydrogen chloride from the chlorination of organic compounds. When conducted in the presence of an organic compound which reacts with the chlorine formed it is known as oxychlorination, e.g., $CH_2{=}CH_2 + 2HCl + O \rightarrow CH_2ClCH_2Cl + H_2O$.

DEAE. Abbreviation for diethylaminoethyl.

DEAE-cellulose. (diethylaminoethyl cellulose). A cellulose ether containing the group $(C_2H_5)_2NCH_2CH_2$ bound to the cellulose in an ether linkage. An anionic ion-exchange material.
Use: Chromatography.

DEAE-dextran. A diethylaminoethyl ether of dextran, an electropositively charged polymer.

deanol. See 2-dimethylaminoethanol.

deblooming agent. A substance added to mineral oils to mask fluorescence. Nitronaphthalene and yellow coal-tar dyes are among such products.

Debye, Peter J. M. (1884–1966). A Dutch chemist and physicist who received the Nobel Prize in 1936 for his pioneer studies of molecular structure by x-ray diffraction methods. The interference patterns are still called Debye-Sherrer rings. He also made outstanding contributions to knowledge of polar molecules and to fundamental electrochemical theory.
See also Debye-Huckel theory.

Debye-Huckel theory. A theory advanced in 1923 for quantitatively predicting the deviations from ideality of dilute electrolytic solutions. It involves the assumption that every ion in a solution is surrounded by an ion atmosphere of opposite charge. Results deduced from this theory have been verified for dilute solutions of strong electrolytes and it provides a means of extrapolating the thermodynamic properties of electrolytic solutions to infinite dilution.

DEC. Abbreviation for β-diethylaminoethyl chloride hydrochloride.

decaborane. CAS: 17702-41-9. $B_{10}H_{14}$.
Properties: Colorless crystals, stable indefinitely at room temperature, decomposes slowly into boron and hydrogen at 300C, d (25/4C) 0.94, mp 99.7C, d 0.78 (100C), bp 213C. Slightly soluble in cold water; hydrolyzes in hot water; soluble in benzene, hexane, alcohol, carbon tetrachloride toluene.
Derivation: Byproduct of the pyrolysis of diborane.
Grade: Technical 95%, high purity 99%.
Hazard: May explode in contact with heat or flame or with oxygenated and halogenated solvents. Ignites in contact with oxygen. Absorbed by skin. TLV: 0.05 ppm in air.
Use: Catalyst, corrosion inhibitor, fuel additive, stabilizer, rayon delustrant, mothproofing agent, dye stripping agent, reducing agent, fluxing agent, oxygen scavenger, propellant.

decachloro-1,1'-bis-2,4-cyclopentadienyl.
See bis-(pentachloro-2,4-cyclopenta dien-1-yl).

decachloro-octahydro-1,2,4-metheno-2H-cyclobuta[cd]-pentalen-2-one. See kepone.

decaglycerol. See polyglycerol.

decahydronaphthalene. $C_{10}H_{18}$.
 CAS: 91-17-8. Cis- and trans- forms are known.
 Properties: Colorless liquid, aromatic odor. Insoluble in water, soluble in alcohol and ether. Cis: d (20/4C)0.8927, fp −43.2C, bp 194.6C, refr index (20C) 1.48113. Trans: d (20/4C)0.8700, fp −31.5C, bp 185.5C, refr index (20C) 1.46968. Flash p 136F (57.7C) (closed cup), autoign temperature 482F (250C). Combustible.
 Derivation: By treatment of naphthalene in a fused state (above 100C) with hydrogen in the presence of a copper or nickel catalyst.
 Grade: Technical.
 Hazard: Moderate fire risk. Irritant to eyes and skin.
 Use: Solvent for oils, fats, waxes, resins, rubber, etc. Substitute for turpentine, cleaning machinery, stain-remover, shoe creams, floor waxes, etc., cleaning fluids, lubricants, motor fuel additive.

Δ-decalactone. Artificial flavoring for margarine. Approved by FDA.

"Decalin."[28] TM for decahydronaphthalene.

"Decalso."[184] TM for a commercial grade cation exchanger consisting of a synthetic alumino-silicate gel (inorganic).
 Use: Water softening, separation of amino acids, radioactive waste treatment, milk treatment.

decamethrin. (1R-(1-alpha(S*),3-alpha))-cyano-(3-phenoxyphenyl)methyl-3-(2,2-dibromovinyl)-2,2-dimethylcyclopropanecarboxylate).
 CAS: 52918-63-5.
 Use: Insecticide.

decamethyltetrasiloxane. $C_{10}H_{30}O_3Si_4$.
 Properties: Colorless liquid, bp 195C, fp −70C, d 0.853, refr index 1.34. Soluble in light hydrocarbons and benzene, slightly soluble in alcohol. Stable over wide temperature range.
 Use: Silicone oils, antifoam agent in lube oils.

n-decanal. (capraldehyde; capric aldehyde; n-decyl aldehyde; aldehyde C-10).
 $CH_3(CH_2)_8CHO$.
 Properties: Colorless to light yellow liquid, floral-fatty odor, d 0.831–0.838 (15C), refr index (20C) 1.427–1.431. Soluble in 80% alcohol, fixed oils, volatile oils, mineral oil; insoluble in water and glycerol. Combustible.
 Derivation: Occurs in lemongrass, citronella, orange, and many other oils. Synthetically by oxidation of the corresponding alcohol or reduction of the acid.
 Grade: Technical, FCC.
 Use: Perfumery, flavoring.

n-decane. (decyl hydride). CAS: 124-18-5.
 $CH_3(CH_2)_8CH_3$.
 Properties: Colorless liquid, d 0.7298, bp 174C, fp −30C, refr index (20C) 1.4114, flash p 111F (44C) (closed cup), autoign temperature 482F (250C). Soluble in alcohol, insoluble in water. Combustible.
 Grade: Technical, 95%, 99%, research.
 Hazard: Moderate fire risk. Narcotic.
 Use: Organic synthesis, solvent, standardized hydrocarbon, jet fuel research.

decanedioic acid. See sebacic acid.

decanoic acid. See capric acid.

l-decanol. (n-decyl alcohol; alcohol C-10).
 $CH_3(CH_2)_8CH_2OH$.
 Properties: Colorless, water-white liquid. Sweet odor, d 0.829, bp 232.9C, mp 6C, flash p (OC) 180F (82.2C), refr index (20C) 1.4372. Insoluble in water (25C), soluble in alcohol and ether. Combustible.
 Derivation: Reduction of coconut oil fatty acids, from C_9 olefin and synthesis gas, by the Oxo process.
 Grade: Technical, high purity.
 Use: Plasticizers, detergents, synthetic lubricants, solvents, perfumes, flavorings, antifoam agent.

decanoyl chloride. (sometimes called caproyl chloride). $CH_3(CH_2)_8COCl$. Available in bottles; carboys and drums. Intermediate, polymerization initiator.

decanoyl peroxide,.
 $CH_3(CH_2)_8C(O)OOC(O)(CH_2)_8CH_3$.
 Properties: Soft, white granules; mp 38–42C (decomposes); insoluble in water and alcohol; soluble in ether and benzene.
 Hazard: Strong oxidizer, fire risk in contact with organic materials.
 Use: Polymerization catalyst.

"Decanox."[154] TM for decanoyl peroxide.

decarboxylase. One of a group of enzymes in the living cell that removes carbon dioxide from various carboxylic acids without oxidation.

decay. Spontaneous disintegration of an unstable atomic nucleus, e.g., uranium, radium, with emission of alpha, beta, and gamma radiation, and eventual formation of another element of lower atomic weight.
 See radioactivity, half-life.

1-decene. See decylene.

"Dechlorane."[597] TM for a series of chlorinated hydrocarbon flame retardants, e.g., perchloropentacyclodecane.

"Decholin."[272] TM for dehydrocholic acid.

deck. The platen of a compression molding press.

deckle. A strip or bar placed along the sides of a Fourdrinier wire to equalize the flow of the pulp slurry and give the sheet a straight edge. When a ragged or "deckle" edge is desired on certain specialty papers the strip is removed.

"Declomycin."[315] TM for demthylchlortetracycline hydrochloride.

decoction. Pharmaceutical term for a liquid produced by boiling one or more drugs in water and filtering.

decoic acid. See capric acid.

decolorizing agent. Any material that removes color by a physical or chemical reaction. Charcoals, blacks, clays, earths, or other materials of highly adsorbent character used to remove undesirable color as from sugar, vegetable and animal fats and oils, etc. Also refers to bleaches involving a chemical reaction for removing color.

decomposition. A fundamental type of chemical change. In simple decomposition, one substance breaks down into two simpler substances, e.g., water yields hydrogen and oxygen. In double decomposition, two compounds break down and recombine to form two different compounds, e.g., $2HCl + CaCO_3 \rightarrow CaCl_2 + H_2CO_3$. In some cases, heat is absorbed and in others it is evolved.

Decomposition may occur as a result of: (1) reaction at room temperature ($NaOH + HCl \rightarrow NaCl + H_2O$), (2) heating in air ($C + H_2O \rightarrow CO + H_2$), (3) electrolysis (inorganic compounds), ($NaCl_{aq} + e \rightarrow Na^+ + Cl^-$), (4) bacterial or enzymic action (fermentation, $C_6H_{12}O_6 \rightarrow 2C_2H_5OH + 2CO_2$), (5) radiation (photodecomposition) as in the breakdown of chlorofluorocarbons in the upper atmosphere and of biodegradable polymers exposed to sunlight, (6) heating in absence of air (thermal decomposition) in which carbonaceous raw materials such as coal and natural gas are converted into carbon and volatile organic compounds without undergoing combustion (coal → coke, coal tar and coal gas). The term thermal decomposition is virtually synonymous with pyrolysis and destructive distillation.

See also degradation, pyrolysis, destructive distillation.

decontamination. Removal of radioactive poisons from skin, clothing, equipment, etc. Skin can often be decontaminated by washing with soap and water, application of titanium dioxide paste or a saturated solution of potassium permanganate followed by a rinse of 5% sodium bisulfite is approved procedure. Contaminated clothing should not be sent to commercial laundries nor burned in open incinerators. Water, steam, and detergents are effective on painted or metal surfaces.

decortication. Removal of the hard coating (cortex) from certain vegetables, nuts, fruits, etc, by either mechanical or manual means.

See hydraulic barking.

"Decroline."[307] TM for a series of stripping agents consisting of zinc sulfoxylate formaldehyde.

decyl acetate. (acetate C-10).
$CH_3(CH_2)_9OOCCH_3$.
Properties: Liquid with floral orange-rose odor, bp 187–190C, d 0.862–0.864, refr index 1.426. Soluble in 80% alcohol, ether, benzene, glacial acetic acid; insoluble in water. Combustible.
Grade: Technical.
Use: Perfumery.

n-decyl alcohol. See 1-decanol.

n-decyl aldehyde. See n-decanal.

n-decylamine. $CH_3(CH_2)_9NH_2$.
Properties: Water-white liquid, amine odor, boiling range 215–221C, d 0.797 (20/20C), refr index 1.437 (20C), flash p 210F (99C). Combustible.
Hazard: An irritant.

decyl carbinol. See 1-undecanol.

decylene. (1-decene). $C_{10}H_{20}$ or $H_2C{:}CH(CH_2)_7CH_3$.
Properties: Colorless liquid, d 0.7396 (20/4C), bp 172C, fp −66.3C, refr index (20C) 1.4220, flash p 130F (55C), autoign temperature 455F (235C). Soluble in alcohol, insoluble in water. Combustible.
Grade: Technical, high purity.
Use: Organic synthesis of flavors, perfumes, pharmaceuticals, dyes, oils, resins.

decyl hydride. See n-decane.

decylic acid. See capric acid.

decyl mercaptan. $C_{10}H_{21}SH$.
Properties: Liquid, fp −26C, bp 114C (13 mm), d 0.8410 (20/4C), refr index 1.4536 (20C). Combustible, strong odor.

Grade: 95% (min) purity.
Use: Intermediate, synthetic rubber processing.

decyl-octyl methacrylate.
$H_2C:C(CH_3)COO(CH_2CH_3)$.
Use: Monomer for plastics, molding powders, solvent coatings, adhesives, oil additives, emulsions for textile, leather and paper finishing.

"DeeGee."[212]   See "Dee O."

"Dee O."[212]   TM for a glucose oxidase enzyme system catalyzing the reaction of glucose to gluconic acid with the uptake of oxygen.
Use: Beverages; salad dressing and other oxygen sensitive foods; analytical; stabilizing egg solids by desugaring whites, yolks, or whole eggs.

deet.   See N,N-diethyl-m-toluamide.

"DEF."[181]   TM for S,S,S-tributyl phosphorotrithioate.

defecation.   Purification, used specifically in the industrial clarification of sugar solutions.

deflagration.   Very rapid autocombustion of particles of explosive as a surface phenomenon. Initiated by contact of a flame or spark but may be caused by impact or friction. Deflagration is a characteristics of low explosives.
See also detonation.

defoaming agent.   (antifoaming agent).   A substance used to reduce foaming due to proteins, gases or nitrogenous materials which may interfere with processing. Examples are 2-octanol, sulfonated oils, organic phosphates, silicone fluids, dimethylpolysiloxane, etc. For restrictions on their use in foods, see FDA regulations.

defoliant.   An herbicide that removes leaves from trees and growing plants. They may be either organic or inorganic. Some examples: (organic) phenoxyacetic acids, trichloropicolinic acid, carbamates, and nitro compounds; (inorganic) arsenic compounds, cyanides, thiocyanates, and chlorates. Several of the more persistent types have been used in military operations. Many defoliants are toxic.
See 2,4-D, 2,4,5-T, Agent Orange.

DEG.   (1) Abbreviation for diethylene glycol, (2) Abbreviation for diethanolglycine.

DEGN.   Abbreviation for diethylene glycol dinitrate.

degradation.   A type of decomposition characteristic of high molecular weights substances such as proteins, polymers, branched-chain sulfonates, etc. It may result from oxidation, heat, sunlight, solvents, bacterial action, or, in the case of body proteins, from infectious microorganisms.
See also biodegradability, decomposition.

degras.   Crude wool grease obtained by solvent-washing of wool. It is a dark brown semisolid with strong unpleasant odor and high water-absorbing capacity. A type known as moellen degras is a byproduct of tanning chamois leather with various fish oils. The chief use of degras is as the source of lanolin; minor uses are in leather dressing and printing inks. Available in several grades (neutral, common, and technical).

degree of polymerization.   (DP).   The number of monomer units in an average polymer molecule in a given sample. For natural cellulose it is about 3000, but in most polymers it is still higher. It can be controlled by appropriate processing techniques. DP is an important factor in plastics technology as it directly affects the viscosity of solutions and properties of the end product.
See also polymerization, shortstopping agent.

deguelin. $C_{23}H_{22}O_8$.
Properties: Greenish powder, mp 170C, soluble in alcohol, insoluble in water.
Hazard: Toxic by inhalation; skin irritant.
Use: Insecticide.

"DEH."[233]   TM for a variety of polyamines and polyamides suitable for curing epoxy resins.

dehairing.   See unhairing.

dehumidification.   The removal of moisture (water vapor) from air. Also sometimes. extended to analogous processes of removing a vapor from a gas mixture.

"Dehybor."[441]   TM for anhydrous sodium tetraborate.

"Dehydratine."[205]   TM for bituminous water barrier coatings.

dehydration.   (1) Removal of 95% or more of the water from a material, usually a foodstuff, by exposure to high temperature by various means. Its primary purpose is to reduce the volume of the product, increase its shelf-life, and lower transportation costs. Special equipment for dehydration includes tunnel dryers, vacuum (shelf) dryers, drum dryers, etc., in which the bulk product is exposed to a hot-air environment. Another method is spray-drying, in which a liquid product is ejected from a nozzle into hot air; dried milk and egg-white are prepared in this way. The term dehydration is not applied to loss of water by evaporation or sun-drying. See also drying. (2) Removal of one or more molecules of H_2O from a chemical compound, e.g., of ethanol to ethylene.

"Dehydrite."[16] TM for anhydrous granular magnesium perchlorate.

dehydroabietic acid. $C_{20}H_{28}O_2$. Solid, used as a basis for thermoplastic resins.

dehydroacetic acid. (DHA, methylacetopyranone). CAS: 520-45-6.

$CH_3C:CHC(O)CH(COCH_3)C(O)$.

Properties: Colorless, odorless, tasteless crystals; mp 108.5C; bp 270C. Partially soluble in acetone and benzene, insoluble in water, highly reactive. Combustible.
Derivation: (a) By action of N-bromosuccinimide on ketene dimer; (b) by strong heating of acetoacetic ester.
Grade: Technical, FCC.
Hazard: Toxic by ingestion.
Use: Fungicide and bactericide, plasticizer, chemical intermediate, medicated toothpastes.

dehydroascorbic acid.

$OCOCOCOCHCHOHCH_2OH$. The oxidized form of ascorbic acid with the same vitamin activity.
Properties: Needles, mp 225C (decomposes) soluble in water at 60C.
Derivation: Synthesized from ascorbic acid.
Use: Nutrition, medicine.

7-dehydrocholesterol. (provitamin D_3).
CAS: 434-16-2. $C_{27}H_{44}O \cdot H_2O$. A sterol found in the skin of man and animals which forms vitamin D_3 upon UV irradiation.

Properties: Slender platelets from ether-methanol, mp 150C, insoluble in water, soluble in organic solvents.
Use: Nutrition, medicine, biochemical research. See also cholecalciferol.

dehydrocholic acid. $C_{24}H_{34}O_5$. A polycyclic compound.
Properties: White, fluffy, odorless powder with bitter taste; mp 231–240C. Almost insoluble in water; slightly soluble in ether and alcohol; soluble in chloroform, glacial acetic acid and solutions of alkali hydroxides and carbonates.
Derivation: Oxidation of cholic acid.
Grade: NF.
Use: Medicine, pharmaceutical intermediate.

dehydrocyclodimerization. A method of converting paraffin (straight-chain) hydrocarbons containing three to five carbon atoms into aromatic (ring-type) hydrocarbons. Its main steps are (a) removal of hydrogen from the paraffins, (b) dimerization of the resulting olefins, (c) aromatization of the dimerized olefins and diolefins, (d) isomerization or transalkylation to C_8 to C_{10} alkyl benzene isomers. Metallic catalysts are essential in some or all of these steps. The process is not in large-scale use.

dehydroepiandrosterone. See dehydroisoandrosterone.

dehydrogenase. An enzyme which catalyzes oxidation by the removal of hydrogen. See oxidase.

dehydrogenation. The process whereby hydrogen is removed from compounds by chemical means. Dehydrogenation of primary alcohols yields the group of compounds called aldehydes. It is considered to be a form of oxidation as two hydrogen atoms, each of which contains an electron, have been removed, as in the reaction $CH_3CH_2OH \rightarrow CH_3CHO + H_2$

11-dehydro-17-hydroxycorticosterone. See cortisone.

dehydroisoandrosterone. (dehydroepiandrosterone). $C_{19}H_{28}O_2$. An androgenic steroid, a metabolic product of the adrenal steroid hormones with about one-third of the androgenic activity of androsterone.
Properties: (dimorphous) Needles with mp 140–141C, leaflets with mp 152–153C, precipitated by digitonin. Soluble in benzene, alcohol, and ether; sparingly soluble in chloroform and petroleum ether. Also available as the acetate salt.
Derivation: Isolated from male urine, synthesis from cholesterol or sitosterol.
Use: Medicine, biochemical research.

dehydrothio-p-toluidine.

$CH_3C_6H_3SC(C_6H_4NH_2)N$.
Properties: Long, yellowish, iridescent needles. Solutions have a violet-blue fluorescence, mp 191C,

bp 434C, soluble in alcohol, very slightly soluble in water.
Derivation: By heating p-toluidine and primuline base with sulfur and separation from the primuline base by distillation in vacuo.
Use: Dyestuffs, intermediate.

deicing compound.   See calcium chloride, sodium chloride, alcohol.

de-inking.   The removal of printing inks from paper by use of strong alkaline solutions such as soda-ash liquor, caustic soda or lime which dissolve varnish and free the ink carbon. Removal of the carbon is accomplished by use of colloidal agents such as talc or bentonite and by mechanical agitation with water.

deKhotinsky cement.   A thermoplastic adhesive mixture of shellac and pine tar. It is not attacked by water, sulfuric acid, nitric acid, hydrochloric acid, carbon disulfide, benzene, gasoline, or turpentine; very little affected by ether, chloroform, alkalies, but readily dissolved by ethanol.

"Delac."[248]   TM for a series of delayed action rubber accelerators.

Delepine reaction.   Preparation of primary amines by reaction of alkyl halides with hexamethylenetetramine followed by acid hydrolysis of the formed quaternary salts.

delhi hard.   A ferrous alloy (d: 7.75, mp 500C) containing in addition to iron, 16.5–18% chromium, 1.1% carbon, 0.75–1% silicon, 0.35–0.5% manganese. It is resistant to cold ammonium hydroxide in all concentrations and to mine and sea waters and moist sulfurous atmospheres.

deliquescent.   Tending to absorb atmospheric water vapor and become liquid. The term refers specifically to water-soluble chemical salts in form of powders which dissolve in the water absorbed from the air. Such salts should be kept closely stoppered or otherwise enclosed.
See also hygroscopic.

"Delrin."[28]   TM for a type of acetal resin. White and colors available. Also supplied as pipe and fittings. Thermoplastic.
Use: Injection-molded and extruded parts, door handles, bushings, other mechanical items. Underground pipe, automotive parts.

"Delsan."[28]   TM for fungicide-insecticide seed treatment containing 60% thiram and 15% dieldrin.
Hazard: Toxic by ingestion and inhalation.

delta acid.   See Casella's Acid F.

"Deltyl."[227]   TM for a mixture of isopropyl esters of lauric, myristic, and palmitic acids. "Deltyl Extra" is predominantly isopropyl myristate, "Deltyl Prime" isopropyl palmitate.
Use: Replaces vegetable or mineral oils in cosmetics; emollient and auxiliary emulsifying agent.

delustrant.   A substance used to produce dull surfaces on a textile fabric. Chiefly used are barium sulfate, clays, chalk, etc. They are applied in the finishing coat.

De Mayo reaction.   Synthesis of 1,5-diketones by photoaddition of enol derivatives of 1,3-diketones. to olefins, followed by a retro-aldol reaction.

"Demerol" Hydrochloride.[162]   TM for meperidine hydrochloride.

demethylchlortetracycline hydrochloride.
$C_{21}H_{21}ClN_2O_8 \cdot HCl$.
Properties: Yellow, crystalline powder; odorless and has a bitter taste. Partially soluble in water and slightly soluble in alcohol.
Grade: NF.
Use: Medicine (antibiotic)

demeton.   (Systox).   CAS: 8065-48-3.
$C_8H_{19}O_3PS_2$.   A mixture of O,O-diethyl-O-2-(ethylthio)-ethyl phosphorothioate (demeton-O) and O,O-diethyl-S-2-(ethylthio)ethyl phosphorothioate (demeton-S).
Properties: (of mixture): Pale yellow liquid, bp 134C (2 mm), d 1.118, slightly soluble in water, soluble in most organic solvents.
Hazard: Toxic by skin absorbtion; cholinesterase inhibitor. Use may be restricted. TLV: 0.01 ppm.
Use: Systemic insecticide (absorbed by plant which then becomes toxic to sucking and chewing insects)

demeton methyl.   (O,O-dimethyl-S,2-(ethylthio) ethyl phosphorothiolate).   CAS: 802-00-2.
Use: Systemtic insecticide.

demineralization.   Removal from water of mineral contaminants, usually present in ionized form. The methods used include ion-exchange techniques, flash distillation, or electrodialysis. Acid mine wastes may be purified in this way, thus aiding the pollution problem.
See also desalination, water conditioning.

Demjanov rearrangement.   Deamination of primary amines by diazotization to give rearranged alcohols.

Democritus. (approximately 465 BC). A Greek philosopher, the first thinker of record to conceive of matter as existing in the form of small indivisible particles, which he called atoms. However, this concept was overshadowed by Aristotle's theories and it was not until some 2000 years later that it was developed by John Dalton in England--an astonishing length of dormancy for one of the most creative ideas in the history of science.
See also Dalton.

demulsification. The process of destroying or "breaking" an unwanted emulsion, especially water-in-oil types occurring in crude petroleum. Both chemical and physical means are used. Chemical means include addition of polyvalent ions to neutralize electrical charges or of a strong acid; physical means include heating, centrifuging, or use of high-potential alternating current.
See also emulsion, nonylphenol.

demurrage. A fee imposed on shippers of chemicals and other products by the railroads for retaining freight cars at loading docks for more than a given period of time (usually 24 hr).

"DEN."[233] TM for a series of epoxy novolacs for multifunctional resins for all uses where maximum chemical or heat resistance is required.

denatonium benzoate. USAN for benzyldiethyl-[(2,6-xylylcarbamoyl)methyl]ammonium benzoate (Bitrex), a bitter-tasting compound approved as a denaturant for alcohol, mp 165C, soluble in water and alcohol, insoluble in ether.

denaturant. See alcohol denatured.

denaturation. A change in the molecular structure of globular proteins that may be induced by bringing a protein solution to its boiling point or by exposing it to acids or alkalies or to various detergents. Denaturation reduces the solubility of proteins and prevents crystallization. It involves rupture of hydrogen bonds so that the highly ordered structure of the native protein is replaced by a looser and more random structure. It is usually irreversible but in some cases is reversible, depending on the protein and the treatment involved.
See also degradation

denatured alcohol. See alcohol denatured.

denier. A unit used in the textile industry to indicate the fineness of a filament. If 9000 meters of a filament weighs 1 gram, the filament is 1 denier; if 10,000 meters weighs 1 gram, the fila-

ment is 1 grex. Sheer women's hosiery usually runs from 15 to 10 denier.

density. Weight/unit volume (W/V) expressed as grams/cubic centimeter for solids and liquids and usually as grams/liter for gases. Densities of some common substances follow:

	grams/cm^3	grams/liter
sulfur	2.06	
aluminum	3.7	
sodium	0.967	
glycerol	1.27	
water*	1.0	
chlorine		3.214
carbon dioxide		1.977
air**		1.293
oxygen		1.429
hydrogen		0.0899

* Basis of comparison for solids and liquids.
** Basis of comparison for gases.

For discussion of density vs specific gravity, see specific gravity. Apparent density is the weight of a unit volume of powder, usually expressed as grams per cubic centimeter, determined by a specified method. (MPA definition, MPA Standard 9–50T). Bulk density is an alternate term for apparent density.
See also current density.

"Deo-Base."[45] TM for light petroleum distillate, superfine grade of kerosene without its objectionable odor.

deodorant. A substance used to remove or mask an unpleasant odor. It may or may not have a distinctive odor of its own. Deodorants act (1) by adsorption (activated carbon, charcoal, chlorophyllin), (2) by replacement (pine oil or other perfume), (3) by neutralization (aluminum chlorohydrate), and (4) by oxidation or hydrogenation, e.g., of fish oils. The cosmetic industry supplies a wide variety of deodorants and antiperspirants chiefly based on neutralization. Mouthwashes and breath "sweeteners" often contain calcium iodate, thymol, peppermint, or similar substance to mask or replace odors.
See also odor, cosmetic.

deoxidizer. An agent which removes oxygen from a compound or from a molten metal.

deoxy-. Preferred prefix indicating replacement of hydroxyl by hydrogen in the parent compound. The meaning is the same as that of desoxy- and the two prefixes are used interchangeably.

deoxyanisoin. (4'-methoxy-2-(p-methoxyphenyl) acetophenone).
$CH_3OC_6H_4COCH_2C_6H_4OCH_3$.
Properties: Off-white to buff, crystalline powder with a sweet, faint, cinnamon-like odor; mp 110–112C.
Use: Intermediate.

deoxybenzoin. (α-phenylacetophenone, benzyl phenyl ketone). $C_6H_5CH_2COC_6H_5$.
Properties: Colorless crystals, mp 53–60C, slightly soluble in hot water, soluble in alcohols and ketones.
Use: Intermediate.

deoxycholic acid. (desocycholic acid).
CAS: 83-44-3. $C_{24}H_{40}O_4$.   A bile acid, contains one less hydroxyl group than cholic acid.
Properties: Crystals, mp 172–173C. Not precipitated by digitonin. Practically insoluble in water and benzene, slightly soluble in chloroform and ether, soluble in acetone and solutions of alkali hydroxides and carbonates, freely soluble in alcohol. Also available as sodium salt. Forms coordination compounds with fatty acids.
Derivation: Isolation from bile, organic synthesis.
Grade: Technical, FCC (as desoxycholic acid).
Use: Medicine, precursor for organic synthesis of cortisone, emulsifying agent in foods (up to 0.1%).

deoxycorticosterone. (4-pregnen-21-ol-3,20-dione; 11-deoxycorticosteroid).
CAS: 64-85-7. $C_{21}H_{30}O_3$.   An adrenal cortical steroid hormone. Active in causing the retention of salt and water by the kidney.
Properties: Crystalline plates, mp 141–142C. Freely soluble in alcohol and acetone.
Derivation: From adrenal cortex extract, synthesis from other steroids.
Use: Medicine (usually as acetate).

deoxyribonuclease. One of a group of enzymes which cause the splitting of deoxyribonucleic acids. Pancreatic deoxyribonuclease, the most widely studied, cleaves the acid at the 3'-phosphate bond. Other deoxyribonucleases cleave the 5'-phosphate bond.

deoxyribonucleic acid. (DNA).   A complex sugar-protein polymer of nucleoprotein which contains the complete genetic code for every enzyme in the cell. It occurs as a major component of the genes which are located on the chromosomes in the cell nucleus. The DNA molecule is a unique and vastly intricate structure first elucidated by the English chemists Crick and Watson in 1953. It is comprised of from 3000 to several million nucleotide units arranged in a double helix containing phosphoric acid, 2-deoxyribose, and the nitrogenous bases adenine,

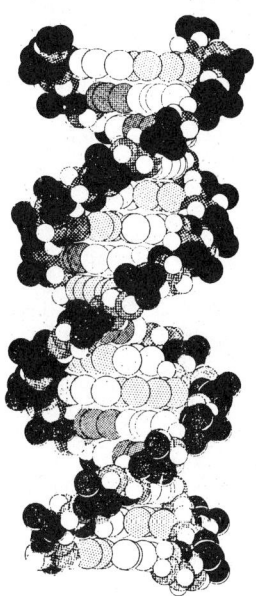

guanine, cytosine, and thymine. The spiral consists of two chains of alternating phosphate and deoxyribose units in continuous linkages. The nitrogenous bases project toward the axis of the spiral and are joined to the chains by hydrogen bonds. Adenine always unites with thymine, and cytosine with guanine. The complementarity of the bases on the joined chains allows each chain to act as a template for replication of the other when the chains are separated thus producing two new strands of DNA. The sequence of the bases on the chains varies with the individual and it is this sequence that governs the genetic code. DNA works in conjunction with ribonucleic acid (RNA). Synthesis of self-replicating DNA was reported in 1967. Elucidation of the structure of the DNA molecule is under continuing research. Recent studies on synthetic DNA indicate that the helix may have a left-handed rather than a right-handed form.
See also ribonucleic acid, gene, nucleic acid, genetic code, replication, recombinant DNA.

D-deoxyribose.
$CH_2OHCHOHCHOHCH_2CHO$.   A five carbon atom sugar that is unusual because there is no oxygen atom attached to the second carbon

atom. It is a constituent of deoxyribonucleic acid.

"Deoxy-Sol."[552] TM for an aqueous solution of hydrazine, used for boiler feed treatment.

2,4-DEP. (Tris-(2,4-dichlorophenoxy) ethylphos phite). An herbicide.

DEP. Abbreviation for diethylphthalate.

"Depakene."[3] TM for sodium valporate.

Department of Transportation. (DOT). The Federal agency which has been responsible since 1967 for the regulation and control of transportation of hazardous materials.

DEPC. Abbreviation for (1) diethylpyrocarbonate; (2) γ-diethylaminopropyl chloride hydrochloride.

"Dependip."[200] TM for a petroleum solvent prepared by straight-run overhead distillation.
Properties: Water-white, d 0.758 (15.5C), wt/gal 6.31 lb (15.5C), flash p (Tagliabue open cup) 52F (11.1C)
Hazard: Flammable, dangerous fire risk.
Use: Solvent in rubber-dipping cement.

dephlegmation. Partial condensation of vapor from a distillation operation to produce a liquid richer in higher-boiling constituents than the original vapor. The residual vapor is richer in the lower-boiling constituents.

depilatory. A substance used to remove hair from skin. Sulfides are largely used for this purpose. The leather industry uses large amounts of sodium sulfide for unhairing hides. The cosmetic industry also offers various sulfide preparations for removing unwanted body hair.

dequalinium chloride. $C_{30}H_{40}Cl_2N_4$.
Properties: Crystals, mp 326C, soluble in water.
Use: Bacteriostat, antiseptic.

"DER."[233] TM for a series of epoxies, including liquid resins, solid resins and solutions, flexible resins, and flame-retardant resins.

"Deraspan."[233] TM for a group of epoxy resins and curing agents.

"Dergon."[300] TM for a series of liquid detergents used for textile scouring.

"Dergopal."[300] TM for a series of fluorescent whitening agents for natural and synthetic textile fibers. Different members applied to specific fibers. Includes both coumarin and stilbene types.

"Deriphat."[590] TM for a series of amphoteric surfactants.
Use: Cosmetic and detergent formulations.

derris root. The root of the shrubs Derris elliptica and D. malaccensis. Chief active constituent is rotenone. Used as an insecticide.

DES. Abbreviation for diethylstilbestrol.

desalination. (desalting). Any of several processes for removing dissolved mineral salts from ocean water and other brines. The most important are: (1) Distillation with reuse of vapors by compressive distillation or multiple-effect evaporation. Solar distillation has been in use on the Greek islands for some years. (2) Electrodialysis, an ion-exchange method more efficient for purification of brackish water than seawater (see demineralization). (3) Reverse osmosis which uses pressure applied to the surface of a saline solution which is separated from pure water by a semipermeable membrane which ions cannot easily penetrate (see also osmosis). The pressure forces the water component of the solution through the membrane, thus effectively separating the components of the solution. Membranes used are cellulose acetate or graphitic oxide. This method is planned for use in a desalination plant proposed for the brackish waters of the lower Colorado River and is said to be the world's largest. It is also used in a Potomac River installation. (4) Flash distillation appears to be the most effective method so far developed for seawater desalination accounting for about 90% of world production capacity.

There are approximately 350 desalination plants in the US producing over 65 million gallons of fresh water a day. Development is under control of the Office of Saline Water, Dept. of Interior.

desert. A terrestrial environment that receives less than about 25 centimeters (10 inches) of rain per year, and consequently supports only a few kinds of plants.

"Desiccite."[217] TM of adsorbent used for static dehumidification in protective packaging of metal equipment, food and pharmaceuticals.

desiccant. A hygroscopic substance such as activated alumina, calcium chloride, silica gel, or zinc chloride. Such substances adsorb water vapor from the air and are used to maintain a dry atmosphere in containers for food packaging, chemical reagents, etc.
See also molecular sieve.

desiccator. A tightly closed vessel containing a desiccant. Used in the laboratory for drying test materials. Some types have partial vacuum.

"Desicote."[274] TM for a mixture of hydrophobic monomers stabilized in chlorinated hydrocarbon and aromatic solvents. Rapidly decomposes on contact with sorbed water on glass surfaces, leaving surface water-repellent.

desiodothyroxine. See thyronine.

"Desmocoll."[470] TM for a series of isocyanate-modified polyesters for adhesive application.

"Desmodur."[470] TM for a group of isocyanates and isocyanate prepolymers for urethane coatings, foams, adhesives, etc.

"Desmophen."[470] TM for a group of polyesters and polyethers for crosslinking with isocyanates.

desorption. The process of removing an adsorbed material from the solid on which it is adsorbed. See adsorption. Desorption may be accomplished by heating, by reduction of pressure, by the presence of another more strongly adsorbed substance, or by a combination of these means.

desoxy-. See deoxy-

desoxycholic acid. FCC name for deoxycholic acid.

destructive distillation. An operation in which a highly carbonaceous material, such as coal, oil shale or tar sands is subjected to high temperature in the absence of air or oxygen resulting in decomposition to solids, liquids, and gases. As the solid end product is carbon, the term carbonization is often used. Other terms with the same general meaning as destructive distillation are pyrolysis and thermal decomposition. Destructive distillation of coal is carried out in the temperature range of 350–1000C yielding coal tar, coal gas, and char (coke, carbon).

detergent. Any substance that reduces the surface tension of water, specifically a surface-active agent which concentrates at oil-water interfaces, exerts emulsifying action, and thus aids in removing soils. The older and still widely used types are the common sodium soaps of fatty acids which are relatively weak. The much stronger synthetic detergents are classed as anionic, cationic, or nonionic, depending on their mode of chemical action. The latter functions by a hydrogen bonding mechanism. The most widely used group comprises linear alkyl sulfonates (LAS), often aided by "builders." LAS are preferable to alkyl benzene sulfonates (ABS) because they are readily decomposed by microorganisms (biodegradable). LAS are straight-chain compounds having 10 or more carbon atoms in the chain. The branched-chains characteristic of ABS resist decomposition; these have been largely replaced by LAS because of water pollution.
See surface tension, emulsion, wetting agent, soap (1), alkylate (3), biodegradability, eutrophication, builder.

detonation. The extremely rapid, self-propagating decomposition of an explosive accompanied by a high pressure-temperature wave that moves at from 1000–9000 meters/second. May be initiated by mechanical impact, friction or heat. Detonation is a characteristic of high explosives which vary considerably in their sensitivity to shock, nitroglycerin being one of the most dangerous in this respect.
See also explosive, high; deflagration.

deuterium. (heavy hydrogen). Symbol D. An isotope of hydrogen whose nucleus contains one neutron and one proton and is therefore twice as heavy (aw 2.014) as the nucleus of normal hydrogen. The ratio in nature is 1 part deuterium to 6500 parts normal hydrogen.
See deuteron.
Properties: Almost identical with hydrogen, d (H=1) 2.0, fp −254.4C (121 mm), bp −249.5C, autoign temperature 1085F, noncorrosive.
Derivation: Electrolysis of high purity heavy water, fractional distillation of liquid hydrogen.
Grade: 98, 99.5 atom %.
Hazard: Highly flammable and explosive. Explosive range 4–74%.
Use: Bombardment of atomic nuclei, tracer element, thermonuclear reactions.
See also deutero-, heavy water.

deuterium oxide. See heavy water.

deutero-. (deuterated). Prefix indicating that one or more of the hydrogens in a compound is the deuterium isotope. Example: deuteroborane solution, used for labeling olefinic unsaturation. The adjective form, deuterated, has the same meaning. Deuterated ethylene, sometimes written ethylene-1,1-D_2, has the formula $CH_2{:}CD_2$.

deuteron. (deuton). A nuclear particle having mass two and a positive charge of 1, identical with the nucleus of the deuterium atom.

Devarda's metal. (Devarda's alloy).
Properties: Gray powder. Contains copper, aluminum, and zinc in the proportion of 50:45:5. Slightly soluble in hydrochloric acid.

Grade: Reagent (20-mesh and finer).
Use: Analysis (testing for nitrogen).

developer. (1) A term applied in the dyeing industry to certain organic compounds which in combination with some other organic compound already deposited upon the fiber will develop a colored compound or if united with a dye already upon the fiber will form a new coloring matter possessing a more desirable or a faster color.

(2) A substance used in photography to convert a latent image to a visible one by chemical reduction of a silver compound to metallic silver more rapidly in the portions exposed to light than in those not exposed. Such reducing agents as hydroquinone, pyrogallol, and p-phenylenediamine are used.
See also photographic chemistry.

devitrification. Unwanted crystals of silica on heating or cooling. The term is used largely in the glass industry. The tendency to devitrify results from the unstable nature of glasses. It usually occurs if the melt is cooled too slowly.

De Vrys reagent. Contains molybdate ion.
Use: Test reagent for alkaloids.

devulcanization. Technically a misnomer since vulcanization is irreversible. The term is used to describe the softening of a vulcanizate caused by heat and chemical additives during reclaiming.

dewatering. Removal of gross water from a suspension or sludge by filtration, expression, centrifugation or clarification. Paper pulp is dewatered by the Fourdrinier wire, rubber latex may be concentrated in a centrifuge in which half or more of the water is removed. Sludges and organic wastes are also dewatered centrifugally.
See drying, dehydration, centrifuge, clarification.

dew of death. See beta-chlorovinyl dichloroarsine.

dew point. Temperature at which air is saturated with moisture or in general the temperature at which a gas is saturated with respect to a condensable component.

"Dexedrine."[71] TM for dextroamphetamine sulfate.

"Dexon."[181] TM for p-dimethylaminobenzenediazo sodium sulfonate.

dextran. (macrose). Certain polymers of glucose which have chain-like structures and molecular weights up to 200,000. Produced from sucrose by Leuconostoc bacteria, occurs as slimes in sugar refineries, on fermenting vegetables or in dairy products. Clinical dextran is standardized to a low molecular weight (75,000), is made by partial hydrolysis and fractional precipitation of the high molecular weight particles.
Properties: Stable to heat and storage. Soluble in water making very viscous solutions. Solutions can be sterilized. Combustible.
Use: Blood plasma substitute or expander, confections, lacquers, oil-well drilling muds, filtration gel, food additive.

dextranase. An enzyme reported to be effective in reducing dental caries.

dextran sulfate. See sodium dextran sulfate.

"Dextrid."[236] TM for an organic polymer used for control of filtration mud rheology and solids in drilling muds. Stabilized against microbiological degradation. A low percentage controls filtration without appreciable viscosity increase.

dextrin. (starch gum). A group of colloidal products formed by the hydrolysis of starches. Industrially, it is made by treatment of various starches with dilute acids or by heating dry starch. The yellow or white powder or granules are soluble in boiling water and insoluble in alcohol and ether.
Use: Adhesives, thickening agent, sizing paper and textiles, substitute for natural gums, food industry, glass-silvering compositions, printing inks, felt manufacture, substitute for lactose in penicillin manufacture, fuel in pyrotechnic devices.

"Dextrinase."[A212] TM for a fungal amylase which converts starches and dextrins to maltose and dextrose.
Use: Syrups and other products high in reducing sugars.

dextrorotatory. Having the property when in solution of rotating the plane of polarized light to the right or clockwise. Dextrorotatory compounds are given the prefix d or (+) to distinguish them from their levorotatory, l or (-) isomers. The plus sign (+) is preferred.
See optical rotation.

dextrose. See glucose which is the preferred term.

dextrose equivalent. (DE). The total amount of reducing sugars expressed as dextrose that is present in a corn syrup, calculated as a percentage of the total dry substance. The usual technique for determining DE in the corn products industry is the volumetric alkaline copper method.
See also glucose syrup.

DFDD. (difluorodiphenyldichloroethane). $C_{14}H_{10}Cl_2F_2$.
Properties: Colorless crystals, mp 75C.
Hazard: Toxic by ingestion and skin contact.
Use: Contact insecticide.

DFDT. (difluorodipheynltrichloroethane). $FC_6H_4)_2CHCCl_3$. Fluorine analog of DDT.
Properties: A low-melting white solid, mp 45.5C. Odor resembling ripe apples. Insoluble in water, soluble in organic solvents.
Derivation: By condensing chloral and fluoroben-zene in the presence of sulfuric acid or chlorosul-fonic acid.
Hazard: Toxic by ingestion, inhalation, and skin absorption. Use may be restricted.
Use: Contact insecticide.

DFL No. 3.[28] TM for a soluble of buffered phos-phate esters, used as a lubricant release agent and corrosive inhibitor for synthetic rubber dri-ers.

DFP. Abbreviation for diisopropyl fluorophos-phate.

DHA. Abbreviation for dihydroxyacetone.

D-homo rearrangement of steroids. Originally discovered in 17β-hydroxy-20-ketosteroids, but thoroughly studied in the 17α-hydroxy-20-keto series, this reaction involves an acid- or base-catalyzed acyloin rearrangement which yields a 6-membered D-ring.

"DHP-MP."[203] TM for 1,4-bis(2-hydroxpropyl)-2-methylpiperazine.

DHS. Abbreviation for dihydrostreptomycin.

Di. Symbol for didymium.

di-. Prefix meaning two.
See also bi-.

diacetic acid. See acetoacetic acid.

diacetin. (glyceryl diacetate). $CH_2O(OCCH_3)CHOHCH_2O(OCCH_3)$.
Properties: Hygroscopic liquid. It is a mixture of isomers, d 1.18, bp approximately 259C, refr in-dex 1.44, miscible with water, benzene, and alco-hol. The commercial mixture gells at approxi-mately −30C. Combustible.
Derivation: Heating one mole of glycerol with two moles of glacial acetic acid.
Grade: Technical.
Use: Plasticizer and softening agent, solvent for cellulose derivatives, "Glyptal" resins, shellac.

diacetone acrylamide. (DAA). $C_9H_{15}NO_2$.
A vinyl monomer.
Properties: White, crystalline solid; purity 99+%; highly soluble in water and most organic sol-vents; polymerizes readily. (The DAA homo-polymer is insoluble in water.) Mp 57C, bp (8 mm) 120C.
Use: Imparts water tolerance and vapor permea-bility to copolymer films; latex and water-based coating compositions; adhesion improver for cel-lulosics, concrete, glass; cross-linking agent in polyester resins; color photography.

diacetone alcohol. (diacetone; 4-hydroxy-4-methylpentanone-2; 4-hyroxy-2-keto-4-methyl-pentane). CAS: 123-42-2. $CH_3COCH_2C(CH_3)_2OH$.
Properties: Colorless liquid, pleasant odor, d 0.9406 at 20/20C, bp 169.1C, flash p varies from less than 73F (23C) to 100F (38C) or higher depending on grade, wt/gal 7.8 lb (20C), viscos-ity 0.032 poise (20C), fp −42.8C, refr index 1.42416 (20C), autoign temperature 1118F (603C). Miscible with alcohols, aromatic and hal-ogenated hydrocarbons, esters, and water. A constant-boiling mixture with water has bp 99.6C and contains approximately 13% diacetone alco-hol.
Derivation: Condensation of acetone.
Grade: Technical, acetone-free, reagent.
Hazard: Flammable, dangerous fire risk, explosive limits in air 1.8–6.9%. An irritant. TLV: 50 ppm in air.
Use: Solvent for nitrocellulose, cellulose acetate, various oils, resins, waxes, fats, dyes, tars, lac-quers, dopes, coating compositions, wood preser-vatives, stains, rayon and artificial leather, imita-tion gold leaf, dyeing mixtures, antifreeze mixtures, extraction of resins and waxes, preser-vative for animal tissue, metal-cleaning com-pounds, hydraulic compression fluids, stripping agent (textiles), laboratory reagent. The technical grade containing acetone has greater solvent power.

diacetonyl sulfide. $(CH_3COCH_2)_2S$.
Properties: Crystals, bp 136–137C (15 mm), mp 47C.
Derivation: Interaction of chloroacetone and hy-drogen sulfide gas.

diacetyl. (biacetyl; butanedione; diketobutane; di-methyl diketone; dimethylglyoxal). CAS: 431-03-8. $CH_3COCOCH_3$.
Properties: Yellow liquid, strong odor. Soluble in water, alcohol, and ether; d 0.990 (15/15C); mp greater than 3 to greater than 4C; bp 88–91C; refr index (18C) 1.3933; flash p less than 80F (26C).
Derivation: Special fermentation of glucose, syn-thesis from methyl ethyl ketone.
Grade: Technical, flavor grade, FCC.
Hazard: Flammable, dangerous fire risk.
Use: Aroma carrier in food products.

diacetylaminoazotoluene. (4-o-tolylazo-o-diacetotoluide).
[$CH_3C_6H_4NNC_6H_3(CH)N(CH_3CO)_2$].
Properties: Crystalline powder. Color varies from yellowish-red through rose to red. Acted upon by atmospheric water vapor. Soluble in alcohol, chloroform, and ether; also in fats, oils, and greases; insoluble in water; mp 74–76C.
Use: Medicine (external).

diacetylene. $HC \equiv CC \equiv CH$. An unsaturated hydrocarbon containing two triple bonds with the type formula C_nH_{2n-6}. The simplest is butadiyne or biacetylene, a gas which boils at 10C. Combustible.
Hazard: Ignites spontaneously in contact with moist silver salts, may explode at −25C.

1,2-diacetylethane. See acetonylacetone.

1,1′-diacetyl ferrocene. $(C_5H_4COCH_3)_2Fe$.
Red crystalline solid, mp 122–124C. Used as an intermediate.
See ferrocene.

diacetylmethane. See acetylacetone.

diacetylmorphone. (diamorphone; heroin).
CAS: 561-27-3. $C_{17}H_{17}NO(C_2H_3O_2)_2$.
Properties: White, odorless, bitter crystals or crystalline powder. Soluble in alcohol, mp 173C.
Derivation: By acetylization of morphine.
Hazard: Addictive narcotic; ingestion of less than 1 grain may be fatal. Cannot be legally sold in US.

diacetyl peroxide. See acetyl peroxide.

"Diadem Chrome."[232] TM for a series of chrome dyestuffs suitable for application by the after-chrome method.

diagenesis. The set of processes, including solution, that alter sediments at low temperatures after burial.

"Diak."[28] TM for a series of rubber accelerators used to vulcanize "Viton" fluoroelastomer and polyacrylate elastomers.

dialdehyde starch. See starch dialdehyde.

dialifor. (S-(2-chloro-1-phthalmidoethyl)-O,O-diethylphosphorothionate).
CAS: 10311-84-9.
Properties: White crystals, mp 167-169C, insoluble in water, soluble in common organic solvents.
Hazard: Toxic by ingestion, cholinesterase inhibitor.

Use: Acaricide, and pesticide against codling moth, red spider mite, etc., of deciduous fruit.

dialkylchloroalkylamine hydrochloride. A group of amine salts having the formula RCl.HCl when R represents such groups as $(CH_3)_2NCH_2CH_2$— (β-dimethylaminoethyl chloride hydrochloride), $(CH_3)_2NCH_2CH(CH_3)$—(β-dimethylaminoisopropyl chloride hydrochloride), etc. Used in organic synthesis.

"Diall."[175] TM for a series of diallyl phthalate thermosetting molding compounds.
Properties: Excellent electrical resistance, high physical strength, flame resistance, good dimensional stability, resistance to virtually all solvents and most chemicals, outstanding colorfastness in sunlight and heat, fungus resistant.
Grade: Mineral-filled, synthetic fiber-filled and glass fiber-filled.

di-allate. See 2,3-dichloroallyl diisopropylthiocarbamate.

diallyl adipate. $C_3H_5OOC(CH_2)_4COOC_3H_5$.
Properties: Liquid, color-maximum #100 Pt-Co, characteristic odor, d (20C)1.025. Combustible.
Use: Monomer.

diallylamine. (di-2-propenylamine).
CAS: 124-02-7. $(CH_2{:}CHCH_2)_2NH$.
Properties: Liquid, d (20C) 0.7889, bp 112C, fp −100C, refr index (20C) 1.4404, soluble in water, combustible.
Derivation: From allylamine and allylbromide.
Hazard: Toxic by inhalation and skin absorption.
Use: Intermediate.

diallybarbituric acid. (5,5-diallylbarbituric acid).
CAS: 52-43-7. $C_{10}H_{12}N_2O_3$.
Properties: White, odorless crystals or crystalline powder; slightly bitter taste; soluble in alcohol or ether; slightly soluble in water; mp 171–173C.
Hazard: See barbiturates.
Use: Medicine (sedative).

diallyl chlorendate. $(C_3H_5OOC)_2C_7H_2Cl_6$.
Solid, fp 29.5C, viscosity (20C)4.0 cp, d (20C)1.47. Used as a monomer for allyl resins especially in flame-retardant compositions.

diallyl cyanamide. $(H_2C{:}CHCH_2)_2NCN$.
Properties: Liquid, fp less than −70C, bp 222C, d 0.90, insoluble in water, soluble in organic solvents.
Derivation: Reaction of allylbromide and disodium cyanamide.
Hazard: Yields very toxic cyanide fumes on heating.
Use: Organic intermediate, polymers.

diallyl diglycollate. $(C_3H_5OOCCH_2)_2O$.
Properties: Liquid, color-maximum #100 Pt-Co, characteristic odor, d (20C)1.1113.
Use: Monomer.

diallyldimethylammonium chloride. See "Catfloc."

diallyl isophthalate.
$C_6H_4(COOH_2C:CHCH_2)_2$.
Properties: Monomer is liquid, color-maximum #175 Pt-Co, mild characteristic odor, d (20C)1.124, prepolymer is solid, d (25C)1.256.
Use: Molding and laminating, cross-linker for polyesters.

diallyl maleate. $C_3H_5OOCCH:CHCOOC_3H_5$.
CAS: 999-21-3.
Properties: Colorless or straw-colored liquid, bp 109–110C (3 mm), d 1.077 (20C), refr index (20C) 1.4699. Polymerizes readily when exposed to light or temperature above approximately 50C. Combustible.
Hazard: Toxic by ingestion, irritating to skin.
Use: Polymers and copolymers, insecticide formulations.

diallylmelamine.

$(C_3H_5)_2NCNC(NH_2)NC(NH_2)N$.
Properties: White, crystalline solid; mp 142C; d 1.24 (30C); combustible.
Hazard: Toxic by ingestion, irritating to skin, evolves cyanide on heating.
Use: Monomer for resins.

diallyl phosphite. $(CH_2:CHCH_2O)_2PHO$.
Properties: Water-white liquid, fp 0C, bp 62C (1 mm), refr index (25C) 1.444, d 1.080 (25/15C), combustible.
Use: Synthesis of organophosphorus compounds.

diallyl phthalate. (DAP). CAS: 131-17-9.
$C_6H_4(COOCH_2CH:CH_2)_2$. The name is also used for the polymer.
Properties: Nearly colorless, oily liquid. Limited solubility in gasoline, mineral oil, glycerol, glycols, and certain amines. Soluble in most other organic liquids. Insoluble in water. D 1.120 (20/20C), fp −70C (viscous liquid), boiling range 158–165C (4 mm), odor mild lachrymatory, flash p 330F (165.5C), viscosity 13 cp (20C). Combustible.
Hazard: Toxic by ingestion.
Use: Primary plasticizer which will polymerize if not inhibited, a monomer which will polymerize with heat and catalyst. Forms low-pressure laminates with various fillers such as glass cloth, paper, etc., for electrical insulation.

diallyl sulfide. See allyl sulfide.

dialysis. The separation of small molecules from macromolecules in a solution by means of a semipermeable membrane such as parchment or collodion. The rates of diffusion of the small and the large molecules are so widely different that the former will readily pass through the membrane, whereas the latter will penetrate with extreme difficulty. For example the diffusion rates are about 2.3 for sodium chloride, 7 for cane sugar, and from 50–100 for proteins and other macromolecules. This differential led Thomas Graham to define substances that would pass through the membrane easily as crystalloids and those having a tendency to be retained by the membrane as colloids.
See also colloid chemistry, electrodialysis, Graham.

"Diam."[590] TM for a series of fatty diamines, $RNH(CH_2)_3NH_2$.
Use: Corrosion inhibitors, petroleum additives, asphalt emulsifiers and chemical intermediates.

diamide hydrate. See hydrazine hydrate.

diamine. See hydrazine.

3,6-diaminoacridine. See acriflavine.

m-diaminoazobenzene hydrochloride.
See chrysoidine.

diaminoazoxytoluene. (azoxytoluidine).
$C_6H_3(CH_3)(NH_2)N_2OC_6H_3(NH_2)(CH_3)$.
Properties: Yellow or orange crystals, mp 168C, soluble in alcohol, insoluble in water. Combustible.
Derivation: By alkaline reduction of p-nitro-o-toluidine.
Use: Dye intermediate.

diaminobenzene. See phenylenediamine.

3,3′-diaminobenzidine. (3,3′4,4′-biphenyltetramine). $(H_2N)_2C_6H_3C_6H_3(NH_2)_2$. Solid, mp 178–180C.
Use: Copolymerized with diphenyl isophthalate to make high-temperature-resistant polybenzimidazoles.

1,3-diaminobutane. CAS: 590-88-5.
$NH_2CH_2CH_2CHNH_2CH_3$.
Properties: Water-white liquid, amine odor, boiling range 143–150C, d 0.858 (20/20C), refr index 1.450 (20C), flash p 125F (51.6C). Combustible.

Hazard: Toxic by ingestion and skin absorption. Moderate fire risk.

alpha,epsilon-diaminocaproic acid.   See lysine.

diaminochrysazin. $(NH_2)_2(OH)_2C_{14}H_4O_2$. (1,8-diamino-4,5-dihydroxyanthraquinone). Use: Colorimetric determination of boron.

trans-1,2-diaminocyclohexanetetraacetic acid monohydrate.   (CDTA). $C_6H_{10}[N(CH_2COOH)_2]_2 \cdot H_2O$. Properties: White, crystalline solid; mp 200–220C. Very slightly soluble in water and insoluble in most common organic solvents. Partially soluble in dimethylformamide and dimethylsulfoxide upon heating. Forms stable complexes. Use: Chelating agent similar to ethylenediaminetetraacetic acid.

diaminodiethyl sulfide. $S(CH_2CH_2NH_2)_2$. Properties: Mobile, colorless liquid with amine-like odor. Miscible with water and benzene, insoluble in aliphatic hydrocarbons, bp 230–240C, d 1.054 (25C). Combustible.

1,8-diamino-4,5-dihydroxyanthraquinone. See diaminochrysazin.

diaminodihydroxyarsenobenzene dihydrochloride. See arsphenamine.

di-p-aminodimethoxydiphenyl.   See dianisidine.

diaminodiphenic acid.   (benzidinedicarboxylic acid). $C_6H_3(CO_2H)NH_2C_6H_3(CO_2H)NH_2$. Properties: White crystals, soluble in alcohol and ether, insoluble in water. Derivation: By boiling m-nitrobenzaldehyde with caustic soda, reducing with zinc dust and acidifying. Hazard: See benzidine. Use: Dyestuff.

p-diaminodiphenyl.   See benzidine.

diaminodiphenylamine. $HN(C_6H_4NH_2)_2$. Properties: Yellowish crystals, soluble in alcohol and ether, insoluble in water, mp 158C. Use: Dye intermediate, detection of hydrogen cyanide.

diaminodiphenylethylene.   See p-diaminostilbene.

p,p'-diaminodiphenylmethane.   (4,4'-methylenedianiline; MDA).   CAS: 107-77-9. $H_2NC_6H_4CH_2C_6H_4NH_2$. Properties: Crystals from water or benzene; mp 92–93C; bp 398–399C; slightly soluble in cold water; very soluble in alcohol, benzene, ether; flash p 440F. Combustible. Hazard: Toxic by inhalation. TLV: 0.1 ppm in air, suspected human carcinogen. Use: Determination of tungsten and sulfates, polymer and dye intermediate, corrosion inhibitor, epoxy resin hardening agent, isocyanate resins, polyamides.

4,4'diaminodiphenyl sulfone.   See sulfonyldianiline.

diaminodiphenylthiourea.   (diaminothiocarbanilide). $(NH_2C_6H_4NH)_2CS$. Properties: Colorless plates or crystalline solid, soluble in alcohol and ether, sparingly soluble in water, mp 195C. Derivation: By boiling p-phenylenediamine with carbon disulfide.

diaminodiphenylureadisulfonic acid. $CO(NHC_6H_3NH_2SO_3H)_2$. Properties: Colorless, needle-like crystals; slightly soluble in water. Derivation: Action of phosgene upon either p-phenylenediaminesulfonic acid or 4-nitroaniline-3-sulfonic acid. Use: Dye manufacture.

3,3'-diaminodipropylamine.   See 3,3'-imino-bis-propylamine.

diaminoditolyl.   See tolidine.

p,p'-diaminoditolylmethane. $NH_2C_7H_6CH_2C_7H_6NH_2$. Properties: Glistening, crystalline plates; soluble in alcohol and ether; mp 149C. Derivation: By heating formaldehyde and o-toluidine.

1,2-diaminoethane.   See ethylenediamine.

6,9-diamino-2-ethoxyacridine lactate monohydrate. See ethodin.

di-p-aminoethoxydiphenyl.   See ethoxybenzidine.

diaminoethyl ether tetraacetic acid. $(HOOCCH_2)_2NCH_2CH_2OCH_2CH_2N$ $(CH_2COOH)_2$. Properties: Slightly soluble in water, purity 98% min. Use: A chelating agent.

1,6-diaminohexane.   See hexamethylenediamine.

3,6-diamino-10-methylacridinium chloride.   See acriflavine.

diaminoaphthalene. See naphthylenediamine.

1,5-diaminopentane. See cadaverine.

2,3-diaminophenazine. $C_{12}H_{10}N_4$.
Properties: Brownish, needle-like crystals; mp 265C; soluble in alcohol and benzene; sublimes on heating.
Use: Analytical reagent for detection of metals.

2,4-diaminophenol dihydrochloride.
$C_6H_{10}ClN_2O$.
Properties: Colorless crystals, mp 205C, soluble in water, slightly soluble in alcohol.
Use: Photographic developer, hair and fur dyeing, reagent.

2,5-diaminophenol. $C_6H_3OH(NH_2)_2$.
Properties: Colorless crystals, mp 68C, soluble in water.
Derivation: By reduction of 2,5-dinitrophenol.
Hazard: May be skin irritant.
Use: Organic synthesis.

2,4-diaminophenol hydrochloride. (amidol).
$C_6H_3(NH_2)_2OH \cdot 2HCl$.
Properties: Grayish-white crystals, soluble in water, slightly soluble in alcohol.
Derivation: By interaction of dinitrophenol with iron and hydrochloric acid.
Use: Photographic developer, dyeing furs and hair, analytical reagent.

2,4-diamino-6-phenyl-s-triazine. See benzoguanamine.

1,2-diaminopropane. (propylenediamine; 1,2-propanediamine). $NH_2CH_2CH(NH_2)CH_3$.
Properties: Colorless, very hygroscopic, strongly alkaline liquid. Very soluble in water. Ammoniacal odor, d 0.8732 (20/20C), refr index 1.4460 (20C), flash p 92F (33C), bp 117C.
Grade: Technical, 75%, 90%, 98% solution.
Hazard: Dangerous fire risk. Toxic by ingestion and skin absorption.
Use: Synthesis of medicinals, dyes, rubber accelerators, electroplating, analytical reagent.

1,3-diaminopropane. (1,3-propanediamine).
$NH_2CH_2CH_2CH_2NH_2$.
Properties: Water-white mobile liquid, amine odor, d 0.8881 (20/20)C, bp 139.7C, fp −12C, refr index 1.459 (20C), flash p 120F (49C) (Tagliabue open cup). Completely soluble in water, methanol, and ether. Combustible.
Hazard: Moderate fire risk. Strong irritant to eyes and skin.
Use: Intermediate.

2,6-diaminopyridine. $NC_5H_3(NH_2)_2$.
Properties: Crystals, mp 120.8C, bp 285C, soluble in water. Combustible.
Derivation: From 2-aminopyridine.

p-diaminostilbene. (diaminodiphenylethylene).
$C_6H_4(NH_2)CHCHC_6H_4(NH_2)$.
Properties: Colorless needles or plates, soluble in alcohol and ether, insoluble in water, mp 227C. Combustible.
Derivation: Reduction of dinitrostilbene.

4,4′-diamino-2,2′-stilbenedisulfonic acid. (DAS).
$C_6H_3(NH_2)(SO_3H)CHCHC_6H_3(SO_3H)(NH_2)$.
Properties: Yellowish, microscopic needles; soluble in alcohol and ether; insoluble in water.
Derivation: Boiling sodium salt of p-nitrotoluene-o-sulfonate in water and caustic soda and reduction with zinc dust.
Hazard: Toxic by ingestion.
Use: Dyestuffs.

diaminothiocarbanilide. See diaminodiphenylthiourea.

di-α-amino-β-thiolpropionic acid. See cystine.

diaminotoluene. See toluene-2,4-diamine.

4,6-diamino-m-toluenesulfonic acid. See m-tolylenediaminesulfonic acid.

4,6-diamino-s-triazine-2-ol. See ammeline.

2,5-diaminovaleric acid. See ornithine.

diammonium ethylenebisdithiocarbamate.
$NH_4S_2CNH(CH_2)_2NHCS_2NH_4$.
Properties: Mp 72.5C, very soluble in water.
Grade: 42% solution in water.
Use: Fungicide, intermediate, corrosion inhibitor. See also nabam.

diammonium hydrogen phosphate. See ammonium phosphate dibasic.

diammonium phosphate. See ammonium phosphate dibasic.

diamond. An allotropic form of carbon, crystallizes isometrically, consists of carbon atoms covalently bound by single bonds only in a predominantly octahedral structure.

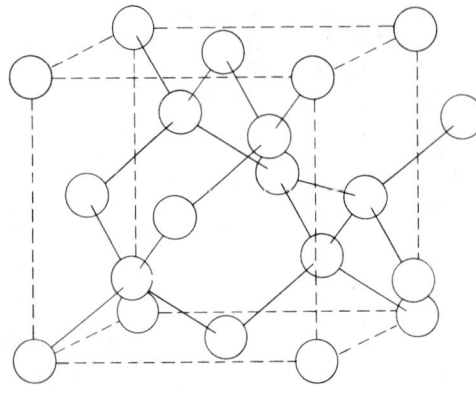

This accounts for its extreme hardness (Mohs 10) and great stability. It has a high refractive index (2.42), d 3.50, coefficient of friction 0.05, has highest thermal conductivity of any substance, transparent to infrared, melting point of 3700C, boiling point 4200C. The purest diamonds used for gems are mined in South Africa, lower grades in Brazil, Venezuela, India, Borneo, Arkansas, also made synthetically by heating carbon and a metal catalyst in an electric furnace at about 3000F under high pressure.
See also diamond, industrial.
Use: Special surgical knives, windows in space probes, high-capacity transmitters.

diamond, industrial. Low-grade diamonds (bort and carbonado) as well as those made synthetically in an electric furnace (3000F, 1.3 million psi).
Use: Oil-well drill bits, primary grinding of steel, wire-drawing dies, glass and metal cutting, grinding wheels.
See also abrasive.

diamond pyramid hardness. See hardness.

diamthazole dihydrochloride.
$C_{15}H_{23}N_3OS \cdot 2HCl$.   (6-(2-Diethylamino-ethoxy)-2-dimethylaminobenzothiazole dihydro-chloride).
Properties: Crystals, decomposes 269C. Soluble in water, ethanol, and methanol.
Use: Topical therapy (medicine), antifungal agent.

di-n-amylamine. (di-n-pentylamine).
CAS: 2050-92-2. $(C_5H_{11})_2NH$.
Properties: Colorless liquid, bp 202–3C (745 mm), very slightly soluble in water, soluble in alcohol and ether, d 0.77–0.78 (20C), refr index 1.430 (20C), flash p 124F (51.1C). Combustible.
Derivation: From reaction of amyl chloride and ammonia.

Hazard: Moderate fire risk. Toxic by ingestion and inhalation.
Use: Rubber accelerators, flotation reagents, dye-stuffs and corrosion inhibitors, solvent for oils, resins, and some cellulose esters.

N,N-diamylaniline. (mixed isomers).
$C_5H_5N(C_5H_{11})_2$.
Properties: Dark amber liquid, d 0.898 (20C), boiling range 276–292C, faint aniline odor, flash p 260F (126C). Combustible.
Use: Organic dyes.

di-tert-amyl disulfide.
$CH_3CH_2C(CH_3)_2SSC(CH_3)_2CH_2CH_3$.
Properties: Liquid, d 0.931 (15.5/15.5C), vacuum distillation range 86–102C, refr index 1.495 (20C), flash p 220F (104.4C). Combustible.

diamylene. $C_{10}H_{20}$.
Properties: Colorless liquid, fp less than −50C, bp 150C, flash p 118F (47.7C) (open cup), d 0.77. Combustible.
Hazard: Moderate fire risk.
Use: Solvent, organic synthesis.

2,5-di(tert-amyl)hydroquinone. (DAHQ, 2,5-di-[tert-pentyl]hydroquinone).
$(C_5H_{11})_2C_6H_2OH)_2$.
Properties: Buff powder, mp 176C, d 1.05 (25C). Slightly soluble in water, soluble in alcohol and benzene.
Use: Antioxidant for uncured rubber and for unsaturated resins and oils, food packaging, polymerization inhibitor.

diamyl maleate. $(CHCOOC_5H_{11})_2$.
Properties: Water-white liquid, d 0.981 (20C), boiling range 263–300C, odor faintly alcoholic, flash p 270F (132C). Combustible.

diamyl phenol. (1-hydroxy-2,4-diamylbenzene).
$(C_5H_{11})_2C_6H_3OH$.   Commercial form is a mixture of isomers including both secondary amyl and tertiary amyl groups mainly in 2, 4 positions.
Properties: Light straw-colored liquid with mild phenolic odor, miscible with both aliphatic and aromatic hydrocarbons, insoluble in water and 10% aqueous alkalies. Boiling range (ASTM 5–95%)280–295C, d 0.930 (20C), wt.gal 7.8 lb (20C), flash p (Tagliabue open cup) 260F (126C). Combustible.
Hazard: Irritant to skin.
Use: Synthetic resins, lubricating oil additives, rust preventives, plasticizers, synthetic detergents, antioxidants and antiskinning agents, rubber chemicals, rodenticide, fungicide.

diamyl phthalate. CAS: 131-18-0.
$C_6H_4(COOC_5H_{11})_2$.
Properties: Colorless, nearly odorless, oily liquid; d 1.022 (20C); wt/gal 8.52 lb (20C); refr index

1.488 (25C); bp 342C; fp less than −55C; flash p 357F (180C) (closed cup). Combustible.
Derivation: By esterification of phthalic anhydride with amyl alcohol in the presence of approximately 1% concentration sulfuric acid as catalyst.
Use: Plasticizer.

diamyl sodium sulfosuccinate. $C_{14}H_{25}NaO_7S$.
Properties: White powder; soluble in water, acetone, carbon tetrachloride, glycerol.
Derivation: Action of alcohol on maleic anhydride, followed by addition of sodium bisulfite.
Use: Wetting agent, emulsion polymerization.

diamyl sulfide. (amyl sulfide). $(C_5H_{11})_2S$.
A mixture of isomers.
Properties: Yellow liquid, d 0.85–0.91 (20/20C), distillation range 170–180C, flash p (185F) (85C) (open cup), refr index 1.477 (19C). Combustible. Obnoxious odor.
Hazard: Irritating by inhalation, ingestion, and skin absorption.
Use: Organic sulfur compounds by addition reactions, flotation agent in metallurgical processes, odorant.

dianhydrosorbitol. See sorbide.

1,4,3,6-dianhydrosorbitol. See isosorbide.

1,2-dianilinoethane. $C_{14}H_{16}N_2$.
Properties: Colorless crystals, mp 67C, bp 228C (12 mm), soluble in alcohol and ether.
Derivation: Heating aniline with dichloro- or dibromoethane.
Use: Manufacture of antihistamines, resin and rubber stabilizer, reagent for aldehydes.

dianisidine. (di-p-aminodi-m-methoxydiphenyl; 3,3′-dimethoxybenzidine).   CAS: 119-90-4. $[C_6H_3(OCH_3)NH_2]_2$.
Properties: Colorless crystals, soluble in alcohol and ether, insoluble in water, mp 137C, flash p 403F (206C). Combustible.
Derivation: The methyl ether of o-nitrophenol is reduced by zinc dust and caustic soda to the hydrazo compound which is then rearranged with hydrochloric acid.
Hazard: See anisidine, benzidine.
Use: Azo dye intermediate.

dianisidine diisocyanate. (3,3′-dimethoxybenzidine 4,4′-diisocyanate).   CAS: 91-93-0. $[OCN(CH_3O)C_6H_3]_2$.
Properties: Grey to brown powder, mp 112C min, soluble in ketones and esters.
Hazard: Toxic by inhalation and ingestion, skin irritant.

Use: Polymers and adhesive systems, high-strength backbone or cross-linking intermediate.

diaphragm cell. A type of electrolytic cell for the production of sodium hydroxide and chlorine from sodium chloride brine. The cell contains anode and cathode compartments separated by a porous diaphragm or membrane to prevent mixing of the solutions. Asbestos fibers are usually used for this diaphragm though a recent development is a plastic material made from perfluorosulfonic acid (see "Nafion"). The brine is fed continuously to the anode compartment where chlorine is released at the graphite anode and flows through the diaphragm to the steel cathode where hydrogen is liberated. Sodium hydroxide accumulates in the liquid and is continuously drained from the cathode compartment. The Hooker cell and the Vorce cell are two widely used types of diaphragm cell.

"Diaron."[36] TM for powdered melamine adhesives.

"Diasan."[570] TM for a quaternary ammonium bacteriostat for textile sanitizing.

diaspore. $Al_2O_3 \cdot H_2O$.   A natural hydrous aluminum oxide occurring in bauxite and with corundum and dolomite.
Properties: White, gray, yellowish and greenish; luster vitreous to pearly; d 3.35–3.45; Mohs hardness 6.5–7.
Occurrence: Arkansas, Missouri, Pennsylvania, Switzerland, USSR, Czechoslovakia.
Use: Refractory, abrasive.

diastase malt. A commercial mixture containing amylolytic enzymes.
Properties: Yellowish-white, amorphous powder or syrupy liquid. Soluble in water, almost insoluble in alcohol.
Derivation: The filtrate from the mash of malted grain is concentrated at low temperature in vacuum. The sugar acts as preservative. The diastase hydrolyzes starch to malt sugar.
Use: Desizing of textiles, bread-making, malted foods, converting starch to sugar.
See also amylase.

diastereoisomer. (diastereomer).   In any group of four optical isomers occurring in compounds containing two asymmetric carbon atoms, such as 4-carbon sugars, there are two pairs of enantiomers (structures which are mirror images of each other) indicated by the letters D and L. The two D and the two L isomers are not mirror images, and these are called diastereoisomers. For example, in the structures below (1) and (2) are enantiomers and so are (3) and (4), (1) and (3) and (2) and (4) are diastereoisomers:

$$\begin{array}{c} \text{CHO} \\ | \\ \text{H—C—OH} \\ (1) \quad \text{HO—C—H} \\ | \\ \text{CH}_2\text{OH} \\ \text{L} \end{array} \qquad \begin{array}{c} \text{CHO} \\ | \\ \text{HO—C—H} \\ \text{H—-COH} \quad (2) \\ | \\ \text{CH}_2\text{OH} \\ \text{D} \end{array}$$

$$\begin{array}{c} \text{CHO} \\ | \\ \text{HO—C—H} \\ (3) \quad \text{HO—C—H} \\ | \\ \text{CH}_2\text{OH} \end{array} \qquad \begin{array}{c} \text{CHO} \\ | \\ \text{H—C—OH} \\ \text{H—C—OH} \quad (4) \\ | \\ \text{CH}_2\text{OH} \\ \text{D} \end{array}$$

See also optical isomer, enantiomer, epimer, anomer.

diatomaceous earth. (diatomite; kieselguhr; infusorial earth).
Properties: Soft, bulky, solid material (88% silica) composed of skeletons of small prehistoric aquatic plants related to algae (diatoms). They have intricate geometric forms. Available as light-colored blocks, bricks, powder, etc. True d 1.9–2.35, bulk density from 5–15 lb/cubic ft. Insoluble in acids except hydrogen fluoride, soluble in strong alkalies. Absorbs 1.5–4 times its weight of water, also has high oil absorption capacity. Poor conductor of sound, heat, and electricity. Noncombustible.
Occurrence: Western US, Europe, Algeria, USSR.
Grade: Natural, chemical, airfloated.
Hazard: Inhalation of dust.
Use: Filtration, clarifying and decolorizing; insulation; absorbent; mild abrasive; drilling mud thickener; extender in paints, rubber, and plastic products; ceramics; paper coating; anticaking agent in fertilizers; asphalt compositions; chromatography; refractories; acid-proof liners; catalyst carrier.

diatomic. Descriptive of a gas whose molecules are composed of two atoms, e.g., O_2, N_2, Cl_2, H_2. Gases in which the element is present as single atoms are called monatomic, e.g., Ar, Ne.

diatomite. See diatomaceous earth.

diatrizoate sodium. See sodium diatrizoate.

1,4-diazabicyclo[2.22]octane.

$$\overline{\text{CH}_2\text{CH}_2\text{NCH}_2\text{CH}_2\text{NCH}_2\text{CH}_2}.$$

Properties: Crystals, hygroscopic, mp 158C, bp 174C, forms crystalline hydrate, sublimes easily, soluble in water and organic solvents.

Use: Possible catalyst for urethane foams and coatings, chemical intermediate.

diazepam. USAN for 7-chloro-1,3-dihydro-1-methyl-5-phenyl-2H-1,4-benzodiazepin-2-one.
CAS: 439-14-5. $C_{16}H_{13}ClN_2O$.
Properties: Slightly yellowish, crystalline powder; practically no odor; mp 131.5–134.5C. One gram of diazepam dissolves in about 350 mL of water, in approximately 15 mL of 95% ethanol, or in approximately 2 mL of chloroform.
Hazard: Central nervous system depressant. Addictive. Manufacture and usage restricted.
Use: Medicine (tranquilizer).

"Diazine."[243] TM for a group of direct dyes, applied to cotton, diazotized, and then coupled onto phenols or amines.

1,3-diazine. (pyrimidine; miazine).

$$\overline{\text{CHN(CH)}_3\text{N}}.$$

Properties: Liquid or crystalline mass with a penetrating odor, mp 20–22C, bp 123–124C. Soluble in water, alcohol, and ether. See also pyrimidine.

"Diazinon."[593] CAS: 333-41-5.
$[\text{C(CH}_3)_2\text{CHC}_4\text{N}_2\text{H(CH}_3)\text{O}]\text{PS(OC}_2\text{H}_5)_2$.
TM for O,O-diethyl- O-(2-isopropyl-4-methyl-6-pyrimidinyl)phosphorothioate.
Properties: Colorless liquid; bp 83–84C (0.002 mm); slightly soluble in water; freely soluble in petroleum solvents, alcohol, and ketones. More stable in alkaline than neutral or acid solutions.
Hazard: Toxic by ingestion, inhalation, and skin absorption. Cholinesterase inhibitor. Use may be restricted. TLV: 0.1 mg/m³ of air.
Use: Insecticide (use against fire ants permitted by EPA).

diazoaminobenzene. (1,3-diphenyltriazene; benzeneazoanilide). CAS: 136-35-6.
$C_6H_5NNNHC_6H_5$.
Properties: Golden-yellow scales. Soluble in alcohol, ether, and benzene; insoluble in water; mp 96C.
Derivation: Interactions of nitrous acid and an alcoholic solution of aniline.
Hazard: Explodes on heating to 150C. Dangerous.
Use: Organic synthesis, dyes, insecticide.

p-diazobenzenesulfonic acid. $C_6H_4SO_3N_2$.
Properties: White or slightly red crystals or white paste. Soluble in water and ether, insoluble in alcohol.
Derivation: From sulfanilic acid, sodium nitrite, and sulfuric acid.

Hazard: Explodes when shocked or heated.
Use: Dyestuffs, reagent.

p-diazodimethylaniline zinc chloride double salt.
(p-dimethylaminobenzene diazonium chloride,
zinc chloride double salt; p-diazotized aminodi-
methylaniline zinc chloride double salt).
$(CH_3)_2NC_6H_4N_2Cl \cdot ZnCl_2$.
Properties: Yellow to orange (light-sensitive) crystals. Specification: Moisture content 5–20%, zinc
17–23%, chloride 31–35%.
Hazard: Irritant.
Use: Rapid diazotype coupler used in coatings for
light-sensitive paper.

2-diazo-4,6-dinitrobenzene-1-oxide.   See diazodinitrophenol.

diazodinitrophenol.   (2-diazo-4,6-dinitrobenzene-
1-oxide; 5,7-dinitro-1,2,3-benzoxadiazole;
DDNP).   CAS: 87-31-0.
$(NO_2)_2C_6H_2ON_2$ (bicyclic).
Properties: Yellow, crystalline compound; darkens
rapidly by exposure to sunlight; d 1.6; soluble
in nitrobenzene, acetone, acetic acid, and nitroglycerin. Desensitized by water.
Derivation: Diazotization of picramic acid in
aqueous solution with sodium nitrite and hydrochloric acid.
Hazard: Explodes when shocked or heated to
180C; dangerous, an initiating explosive.
Use: Primary charge in blasting caps.

p-diazodiphenylamine sulfate.
$(C_6H_5NHC_6H_4N_2)_2SO_4$.
Properties: Yellow-green solid with unpleasant
odor. Sensitive to light, soluble in water.
Use: Used as a light sensitive diazo compound
for coating on reproduction paper, giving direct
positive prints of various colors with different
developers or coupling agents.

1,2-diazole.   See pyrazole.

diazomethane.   (azimethylene).
CAS: 334-88-3. $H_2C{=}N^+{=}N^-$.
Properties: Yellow gas at room temperature, fp
−145C, bp −23C, d 1.45.
Hazard: Severe explosion risk when shocked, may
explode on contact with alkali metals, rough surfaces, or heat (100C). Toxic by inhalation, a carcinogen (OSHA). TLV: 0.2 ppm in air.
Use: Organic synthesis (methylating agent).

1-diazo-2-naphthol-4-sulfonic acid.
$C_{10}H_5N_2OSO_3H$.
Properties: Yellow needles in paste or dry form.
Slightly soluble in water. Also available as so-
dium salt, mp 168C. Combustible.

Derivation: Diazotization of 1-amino-2-naphthol-
4-sulfonic acid and filtering of the diazo compound.
Hazard: May explode if heated above 100C.
Use: Azo dyes, valuable chrome dyestuff component.

"Diazopon AN."[307]   TM for a nonionic dispersing
and stabilizing agent consisting of polyoxyethylated fatty alcohol.
Properties: Clear, oily, yellow liquid; d 1.03–1.04;
soluble in water; stable to strong acids, alkalies,
and metallic ions.
Use: Dispersing and stabilizing agent in the
naphthol dyeing process to improve fastness to
rubbing. Used in dissolving fast color salts, it
has a stabilizing effect on the diazonium compound.

diazotization.   The reaction of a primary aromatic
amine with nitrous acid in the presence of excess
mineral acid to produce a diazo (—N≡N—)
compound. Widely used in organic synthesis, especially production of dyes.

diazotizing salt.   See sodium nitrite.

"Diazyme."[212]   TM for an amyloglucosidase
which splits starch almost completely to glucose.

DIBA.   See diisobutyl adipate.

DIBAC.   Abbreviation for diisobutylaluminum
chloride.

"Dibactol."[430]   TM for myristyldimethylbenzylammonium chloride.

DIBAL-H.   Abbreviation for diisobutylaluminum
hydride.

dibasic.   See monobasic.

dibenalacetone.   (dibenzylidene acetone).
$C_6H_5HC{=}CHCOHC{=}CHC_6H_5$.
Properties: (cis-trans) Yellow crystals, mp 60C.
Derivation: Reaction of benzaldehyde and ace-
tone. There are three geometric isomers (trans-
trans, cis-trans, cis-cis).
Use: Sun-tan lotions, cosmetics.

dibenzanthrone.   (volanthrone). $C_{34}H_{16}O_2$.
Violet-blue vat dye.
Properties: Bluish-black powder. Soluble in nitro-
benzene and concentrated sulfuric acid.
Derivation: From benzanthrone.
Use: Intermediate.

3,3′-dibenzanthronyl.   (also called the
13,13′-compound). $C_{34}H_{18}O_2$.

Properties: Dark yellow needles. Soluble in concentrated sulfuric acid, mp 412C.
Use: Intermediate.

4,4′-dibenzanthronyl. (also called the 2,2′-compound). $C_{34}H_{18}O_2$.
Properties: Yellow needles. Soluble in nitrobenzene; slightly soluble in benzene, alcohol, ether; mp 320C.
Use: Intermediate.

2,3,6,7-dibenzoanthracene. See pentacene.

dibenzocycloheptadienone. $C_{15}H_{12}O$. A tricyclic compound, light yellow to amber solid, mp 28.5C, bp 203–204C (7 mm), d 1.1635 (20C), refr index (20C) 1.6324. Soluble in alcohol and most organic solvents, insoluble in water. Combustible.
Use: Intermediate.

dibenzofuran. See diphenylene oxide.

"Dibenzo G-M-F."[248] TM for dibenzoyl-p-quinonedioxime. $(C_6H_5COON)_2C_6H_4$.
Properties: Brownish-gray powder, d 1.37, starts to decompose above 200C, good storage stability. Insoluble in acetone, benzene, gasoline, ethylene dichloride, and water.
Grade: Available in superdispersing grades.
Use: Non-sulfur vulcanizing agent, in tire-curing bags, gaskets and wire insulation to impart heat resistance.
See also benzoyl peroxide.

dibenzopyran. See xanthene.

dibenzopyrone. See xanthone.

dibenzopyrrole. See carbazole.

dibenzothiophene. $\overline{C_6H_4C_6H_4}S$.
Properties: Colorless crystals, mp 97–98C. Combustible.
Use: Cosmetics and pharmaceuticals, intermediate.

dibenzoyl. See benzil.

trans-1,2-dibenzoylethylene.
$C_6H_5COCHCHCOC_6H_5$.
Properties: Yellow-orange crystals; mp 111C; soluble in glacial acetic acid, ethyl acetate, benzene, and chloroform; sparingly soluble in alcohol; insoluble in water and petroleum ether. Combustible.
Use: Enzyme inhibitor, bactericide, and intermediate.

dibenzoylmethane. (1,3-diphenyl-1,3-propanedione). $C_6H_5COCH_2COC_6H_5$.
Properties: Crystals, mp 80C, bp 219–221C (18

mm). Partially soluble in alcohol; soluble in ether, chloroform, and aqueous sodium hydroxide. Combustible.
Use: Colorimetric determination of uranium. .

dibenzoyl peroxide. See benzoyl peroxide.

dibenzoyl-p-quinonedioxime. See "Dibenzo G-M-F."

2,4-dibenzoylresorcinol.
$C_6H_5COC_6H_2(OH)_2COC_6H_5$.
Properties: Light yellow crystals; nearly odorless; mp 125–128C; soluble in alcohol, ethyl acetate, methyl ethyl ketone; insoluble in water.
Use: Light absorber, best at 280–370 microns.

dibenzyl. See sym-diphenylethane.

N,N-dibenzylamine. $HN(CH_2C_6H_5)_2$.
Properties: Colorless to light yellow, oily liquid; bp 300C (partial decomposition); mp −26C; ammonia-like odor. Combustible, d 1.017 (20C), refr index 1.5730–1.5740 (25C), distilling range 168–172C (10 mm). Insoluble in water, soluble in alcohol and ether.
Hazard: Toxic by ingestion and inhalation.
Use: Intermediate, rubber activator, reagent for metals.

N,N-dibenzyl-p-aminophenol.
$(C_6H_5CH_2)_2NC_6H_4OH$.
Properties: Brown powder; soluble in acetone, benzene, anhydrous methanol; mp greater than 110C.

N,N-dibenzylaniline. $C_6H_5N(CH_2C_6H_5)_2$.
Properties: Yellowish-white crystals, soluble in alcohol and ether, insoluble in water, mp 70C, bp above 300C.
Hazard: See aniline.
Use: Manufacture of dyes.

dibenzyl disulfide. (benzyl disulfide).
$C_6H_5CH_2SSCH_2C_6H_5$.
Properties: Pink solid, odor similar to benzaldehyde, mp 70–72C, dissolves in most organic solvents, very slightly soluble in water. Combustible.
Use: Antioxidant and antisludging agent for petroleum oils, extreme-pressure lube oils and greases, silicone oils.

dibenzyl ether. (benzyl ether).
CAS: 103-50-4. $C_6H_5CH_2OCH_2C_6H_5$.
Properties: Colorless, unstable liquid. Faint, almond odor. Insoluble in water, soluble in most organic solvents, d 1.035, bp 298–300C, flash p 275F (135C). Combustible.
Grade: Technical.

Hazard: Irritant and narcotic.
Use: Plasticizer for various resins, perfumery (solvent for nitro-musks), flavoring.

dibenzylidene acetone.   See dibenalacetone.

N,N-dibenzylmethylamine. $CH_3N(CH_2C_6H_5)_2$.
Properties: Colorless to light yellow liquid, d 0.99 (25C), refr index 1.5560–1.5590 (25C), distilling range 152–158C (11 mm). Partially soluble in water, soluble in organic solvents. Combustible.
Use: Intermediate, oil-soluble rust inhibitor, cutting oils, hydraulic fluids, lubricants.

dibenzyl sebacate.
$C_6H_5CH_2OOC(CH_2)_8COOCH_2C_6H_5$.
Properties: Light straw-colored liquid, mp 28C, bp 265C (4 mm), d 1.055 (30/20C), complete nonvolatility, gives excellent low-temperature flexibility, flash p 450F (232C). Combustible.
Use: Plasticizer, especially for plastic linings for containers.

dibenzyl succinate.   See benzyl succinate.

2,5-dibiphenylyloxazole.   (BBO). $C_{27}H_{21}NO$.
Properties: Crystalline solid, mp 237–239C.
Grade: Purified.
Use: Scintillation counter or a wavelength shifter in soluble scintillators.

diborane.   (diboron hexahydride; boroethane).
CAS: 19287-45-7. B_2H_6.
Properties: Colorless gas with repulsive odor, bp −92.5C, fp −165C, density 0.18 g/mL (17C). Soluble in carbon disulfide, decomposes in water. Highly reactive, flash p −130F (−90C), autoign temperature 100–125F (38–51C)
Derivation: Reaction of lithium hydride and boron trifluoride catalyzed by ether at 25C.
Grade: Technical 95%, high purity 99+%.
Hazard: Highly flammable, dangerous fire risk, reacts violently with oxidizing materials including chlorine. Toxic by inhalation, strong irritant. TLV: 0.1 ppm in air.
Use: Synthesis of organic boron compounds and metal borohydrides, polymerization catalyst for ethylene, fuel for air-breathing engines and rockets, reducing agent, doping agent for p-type semiconductors.

dibromoacetylene.   (dibromomethyne).
CAS: 624-61-3. $BrC{\equiv}CBr$.
Properties: Heavy, colorless liquid. Disagreeable odor. Soluble in most organic solvents, d approximately 2, mp 76–76.5C.
Derivation: (a) Interaction of magnesium dibromoacetylene and an ethereal solution of cyanogen bromide. (b) Interaction of tribromoethylene and alcoholic potash.

Hazard: Explodes on contact with oxygen or on heating. Dangerous! Toxic by inhalation and injection, strong irritant.
Use: Organic synthesis (halogenated ethylene).

9,10-dibromoanthracene. $C_6H_4C_2Br_2C_6H_4$ (tricyclic).
Properties: Yellow crystals. Soluble in chloroform, slightly soluble in alcohol and ether, insoluble in water, mp 226C, sublimes.
Derivation: Bromination of anthracene.
Use: Organic synthesis.

o-dibromobenzene.   (benzene dibromide).
CAS: 26249-12-7. $C_6H_4Br_2$.
Properties: Heavy liquid with pleasant, aromatic odor; bp 225.5C; fp 7.13C; d 1.9767 (25/4C); refr index (20C) 1.6155. Combustible. Miscible with alcohol, acetone, ether, benzene, carbon tetrachloride, and n-heptane; insoluble in water.
Derivation: Interaction of benzene with an excess of bromine in presence of iron.
Use: Solvent for oils, motor fuels, top-cylinder compounds, organic synthesis, ore flotations.

p-dibromobenzene.   (benzene dibromide).
CAS: 26249-12-7. $C_6H_4Br_2$.
Properties: Colorless crystals, soluble in alcohol and ether, mp 89C, bp 219C, d 2.261, refr index (99C) 1.5743.
Use: Organic synthesis of dyestuffs and drugs, manufacture of intermediates, fumigant.

N,N-dibromobenzenesulfonamide.
$C_6H_5SO_2NBr_2$.
Properties: Solid, mp 109–111C. Active bromine 50.4%.
Use: Halogenating agent.

dibromochloromethane.   CAS: 124-48-1.
$CHBr_2Cl$.
Properties: Clear, colorless, heavy liquid; d 2.38; bp 116C.
Hazard: Irritant and narcotic.
Use: Organic synthesis.

1,2-dibromo-3-chloropropane.   (DBCP).
CAS: 96-12-8. $CH_2BrCHBrCH_2Cl$.
Properties: Colorless (when pure) liquid, d 2.05 (20C), bp 195.5C, fp 6.7C, refr index 1.5530 (20C), flash p (TOC) 170F (76.6C). Combustible, slightly soluble in water, miscible with oils.
Hazard: A carcinogen, reported to cause sterility. Use regulated by EPA. OSHA standard of 1 ppb. No longer made in the US.
Use: Pesticide, nematocide, soil fumigant.

dibromodiethyl sulfide. (CH$_2$CH$_2$Br)$_2$S. The bromine analog of mustard gas.
Properties: White crystals; soluble in alcohol, benzene, ether; insoluble in water. D 2.05 (15C), bp 240C (decomposes), mp 31–34C.
Derivation: Action of hydrobromic acid on an aqueous solution of thiodiglycol.
Hazard: Toxic by inhalation, strongly irritant poison gas.
Use: Organic synthesis.

dibromodiethyl sulfone. (BrCH$_2$CH$_2$)$_2$SO$_2$.
Properties: Plates; soluble in alcohol, benzene, ether; mp 111–112C.
Derivation: Interaction of dibromodiethyl sulfide, chromic anhydride, and dilute sulfuric acid.

dibromodiethyl sulfoxide. (BrCH$_2$CH$_2$)$_2$SO.
Properties: Glittering crystals; soluble in alcohol, benzene, ether; mp 100–101C.
Derivation: Interaction of benzoyl peroxide and a hot solution of dibromodiethyl sulfide in chloroform.

dibromodifluoromethane. (difluorodibromomethane). CAS: 75-61-6. CF$_2$Br$_2$.
Properties: Colorless, heavy liquid; fp −141C; bp 24.5C; d 2.288 (15/4C); refr index 1.399 (12C); insoluble in water; soluble in methanol and ether; nonflammable.
Derivation: Vapor phase bromination of difluoromethane.
Grade: Pure (95.0% min).
Hazard: Irritant. TLV: 100 ppm in air.
Use: Synthesis of dyes, pharmaceuticals, quaternary ammonium compounds, fire-extinguishing agent.

1,3-dibromo-5,5-dimethylhydantoin.

BrNCONBrCOC(CH$_3$)$_2$.
Properties: Free-flowing, cream-colored powder with slight bromine odor; mp 187–191C (decomposes); quite stable at 75C; soluble in benzene, chloroform, glacial acetic acid; slightly soluble in water and carbon tetrachloride; insoluble in hexane. Contains 55% active bromine which is slowly released in aqueous solution.
Derivation: Bromination of dimethylhydantoin.
Use: Controlled bromination and oxidation of organic compounds, water treatment, polymerization catalyst, potential germicide and sanitizer.

1,2-dibromoethane. See ethylene dibromide.

4,5′-dibromofluorescein. C$_{20}$H$_{10}$Br$_2$O$_5$.
Properties: Yellow powder (clean shade), mp 265–267C.

Grade: 99+%.
Use: Lipstick dye.

2,4-dibromofluorobenzene. C$_6$H$_3$Br$_2$F.
Properties: Colorless liquid, d 2.047 (20C), bp 214C, refr index (25C) 1.5790. Insoluble in water; soluble in alcohol, acetone, ether, benzene, chloroform, ethyl acetate, and glacial acetic acid.
Hazard: Irritant to skin and eyes.
Use: Intermediate for agricultural and pharmaceutical chemicals.

dibromoformoxime. CBr$_2$NOH.
Properties: Crystals, mp 70–71C, distills between 75 and 85C (3 mm).
Hazard: Evolves highly toxic fumes on heating. A military poison.

dibromoiodoethylene. Br$_2$CCHI.
Properties: Liquid, d 2.952 (24C), bp 91C (15 mm).
Derivation: Reaction of iodine and dibromoacetylene.

dibromomalonic acid. HOOC·CBr$_2$COOH.
Properties: Light yellow needles or prisms, mp 147C (decomposes).
Use: Intermediate for drugs and fine chemicals.

dibromomalonyl chloride. ClOCCBr$_2$COCl.
Properties: Yellowish, oily liquid; bp 75–77C (15 mm).
Use: Chemical intermediate.

dibromomethane. See methylene bromide.

dibromomethyl ether. (CH$_2$Br)$_2$O.
Properties: Colorless liquid; decomposed by water; soluble in acetone, benzene, ether; d 2.2; bp 154–155C; fp −34C.
Derivation: (a) The reaction product of paraformaldehyde and sulfuric acid is treated with ammonium bromide. (b) Interaction of hydrobromic acid and paraformaldehyde.
Hazard: Evolves highly toxic fumes on heating, strong irritant to eyes. Fire risk in contact with oxidizers.
Use: Military poison (lachrymator).

9,10-dibromooctadecanoic acid. (9,10-dibromostearic acid).
CH$_3$(CH$_2$)$_7$CHBrCHBr(CH$_2$)CO$_2$H.
Properties: Yellow solid or liquid, d 1.2458 (30/4C), mp 29–30C, refr index 1.4893 (42C). Insoluble in water; soluble in alcohols, ketones, aromatic and chlorinated hydrocarbons. Also available as methyl ester.
Grade: Technical (amber liquid).
Use: Chemical intermediate.

1,5-dibromopentane. See pentamethylene dibromide.

1,3-dibromopropane. See trimethylene bromide.

dibromopropanol. (2,3-dibromo-1-propanol). CAS: 96-13-9. CH₂BrCHBrCH₂OH.
Properties: Colorless liquid; d 2.120 (20/4C); bp 219C; soluble in acetone, alcohol, ether, and benzene.
Hazard: A carcinogen.
Use: Intermediate in preparation of flame retardants, insecticides, and pharmaceuticals.

dibromoquinonechlorimide. $C_6H_2Br_2ClNO$.
A reagent used for spot visualizations in chromatographic systems.
Properties: Yellow powder, slightly soluble in water, moderately soluble in hot alcohol.
Hazard: Highly sensitive to heat. Explodes at 120C, decomposes with rapid heat evolution at 60C.

3,5-dibromosalicylaldehyde. (3,5-dibromo-2-hydroxybenzadehyde). $Br_2(OH)C_6H_2CHO$.
Properties: Pale yellow crystals; mp 86C; slightly soluble in water; soluble in ether, benzene, chloroform, alcohol, and acetic acid.
Use: Medicine (external), fungicide.

4′,5-dibromosalicylanilide. See dibromsalan.

9,10-dibromostearic acid. See 9,10-dibromooctadecanoic acid.

2,5-dibromoterephthalic acid.
$C_6H_2Br_2(COOH)_2$.   A flame-retardant monomer for production of polyester fibers which are made by reacting this acid with dimethyl terephthalate and ethylene glycol. Permanent lowering of flammability is said to be gained by this method of incorporating bromine in molecular combination.

sym-dibromotetrafluoroethane. $CBrF_2CBrF_2$.
Properties: Liquid, bp 47.3C, d 2.18 (21.1C), nonflammable.
Use: Refrigerant, fire-extinguishing agent, control fluid.

dibromsalan. USAN for 4′,5-dibromosalicylanilide.
$BrC_6H_3(OH)CONHC_6H_4Br$.
Hazard: A suspected carcinogen. Use in cosmetics prohibited (FDA)
Use: Disinfectant.

dibucaine. CAS: 85-79-0. $C_{20}H_{29}N_3O_2$.
(2-n-butoxy-N-(2-diethylaminoethyl)cinchoninamide).
Properties: Colorless or almost colorless powder, mp 62–65C, odorless, somewhat hygroscopic, affected by light, soluble in alcohol and acetone, slightly soluble in water. Also available as the hydrochloride.
Grade: NF.
Hazard: Possibly allergenic.
Use: Medicine (anesthetic).

dibutoline sulfate. $(C_{15}H_{33}N_2O_2)_2SO_4$.
Bisdibutylcarbamate of ethyl(2-hydroxyethyl)dimethylammonium sulfate.
Properties: Hygroscopic powder, decomposes 166C, soluble in water and benzene.
Use: Surface-active agent in medicine.

2,5-dibutoxyaniline. $C_6H_3(OC_4H_9)_2NH_2$.
Properties: Mp 18C, insoluble in water, soluble in organic solvents.
Hazard: See aniline.
Use: Dyes, synthesis.

1,4-dibutoxy benzene. See hydroquinone di-n-butyl ether.

dibutoxyethyl adipate.
$(C_2H_4COOC_2H_4OC_4H_9)_2$.
Properties: Colorless, oily liquid; d 0.997 (20/20C); fp −34C; boiling range 205–215 (4 mm); mild, butyl type odor; flash p 370F (187C); refr index 1.442 (25C); wt/gal 8 lb. Soluble or only slightly soluble in mineral oil, glycerol, glycols, and some amines; soluble in most other organic liquids. Combustible.
Use: Primary plasticizer for most resins, imparting flexibility at very low temperature as well as stability to UV light.

dibutoxyethyl phthalate. (n-butyl glycol phthalate).   CAS: 117-83-9.
$C_6H_4(COOC_2H_4OC_4H_9)_2$.
Properties: Colorless liquid, fp −55C, d 1.06 (20C), bp 270C, wt/gal 8.86 lb, fast to light, water-resistant, flash p 407F (208C). Combustible. Soluble in organic solvents.
Use: Plasticizer for polyvinyl chloride, polyvinyl acetate, and other resins.

dibutoxymethane. $CH_2(OC_4H_9)_2$.
Properties: Colorless liquid, wt/gal 6.97 lb (20C), refr index 1.40615 (20C), d 0.838 (20/20C), flash p 140F (60C) (closed cup), boiling range 164–186C, insoluble in water. Combustible.
Hazard: Moderate fire risk.

dibutoxytetraglycol. (tetraethylene glycol dibutyl ether). $(C_4H_9OC_2H_4OC_2H_4)_2O$.
Properties: Practically colorless liquid with characteristic odor, slightly soluble in water (1.3% by weight), d 0.9436 (20/20C), lb/gal 7.85 (20C), bp 237C (50 mm), 330C, vap press less than

0.01 mm (20C), fp −20C, flash p 355F (180C), solubility of water in product 4.8% by wt (20C), refr index 1.4357. Combustible.
Use: Solvent, especially for DDT.

N,N-di-n-butyl acetamide. $CH_3CON(C_4H_9)_2$.
Properties: Colorless liquid, d 0.890 (20C), boiling range 245–250C, faint odor, flash p 225F (107C). Combustible.

di-n-butylamine. CAS: 111-92-2.
$(C_4H_9)_2NH$.
Properties: Colorless liquid with amine odor, bp 159.6C, fp −62C, d 0.7613 (20/20C), wt/gal 6.33 lb (20C), refr index (20C) 1.4175, flash p (open cup) 125F (51.6C). Partially soluble in water, soluble in alcohol and ether, miscible with hydrocarbons. Combustible.
Derivation: By reaction of butanol or butyl chloride with ammonia.
Grade: Technical.
Hazard: Moderate fire risk. Toxic by ingestion and inhalation.
Use: Corrosion inhibitor; intermediate for emulsifiers, rubber accelerators, dyes, insecticides, and flotation agents; inhibitor for butadiene.

di-sec-butylamine. $(CH_3CHCH_2CH_3)_2NH$.
Properties: Water-white liquid, amine odor, boiling range 132–135C, d 0.754 (20/20C), refr index 1.412 (20C), flash p 75F (23.9C).
Hazard: Flammable, dangerous fire risk. Toxic by ingestion and inhalation.

dibutylamine pyrophosphate.
Grade: Available as 40% solution in ethanol-acetone.
Hazard: Flammable, dangerous fire risk.
Use: Anticorrosion agent in lacquers and cotton solutions, for light and heat stabilization of vinyl chloride and vinyl copolymer resins.

N,N-di-n-butylaminoethanol. (2-N-dibutylaminoethanol. CAS: 102-81-8.
$(C_4H_9)_2NCH_2CH_2OH$.
Properties: Colorless liquid, d 0.859 (20C), boiling range 224–232C, faint odor, amine-like, flash p 200F (93.3C). Combustible.
Hazard: Toxic by ingestion and skin absorption. TLV: 2 ppm in air.
Use: Synthesis.

dibutylammonium oleate. See "Barak."

di-n-butylammonium tetrafluoroborate.
$(C_4H_9)_2NH_2BF_4$.
Properties: Solid, mp 266C.
Use: Lubricating, surface treating, fluxing of aluminum.

N,N-di-n-butylaniline. $C_6H_5N(C_4H_9)_2$.
Properties: Amber liquid, faint aniline odor, d 0.904 (20C), boiling range 267–275C, refr index 1.519 (20C), soluble in alcohol and ether, insoluble in water, flash p 230F (110C). Combustible.
Hazard: See aniline.

2,5-di-tert-butyl benzoquinone.
$[C(CH_3)_3]_2C_6H_2O_2$.
Properties: Yellow crystals; insoluble in water; soluble in ethyl acetate, acetone, benzene; slightly soluble in ethyl alcohol; mp 149–151C.
Hazard: See quinone.
Use: Oxidant, polymerization catalyst.

dibutyl butyl phosphonate.
$C_4H_9P(O)(OC_4H_9)_2$.
Properties: Colorless liquid with mild odor. Stable, insoluble in water, miscible with most common organic solvents, d 0.948 (20/4C), bp 127–128C (2.5 mm), flash p 310F (154.4C) (COC). Combustible.
Use: Heavy-metal extraction and solvent separation, gasoline additives, antifoam agent, plasticizer, textile conditioner, and antistatic agent.

dibutyl "Carbitol."[214] TM for diethylene glycol dibutyl ether.

dibutyl "Cellosolve."[214] TM for ethylene glycol dibutyl ether.

dibutyl chlorophosphate. $(C_4H_9O)_2P(O)Cl$.
Properties: Water-white liquid, bp 103–106C (1.5 mm), d 1.0742 (25C), refr index 1.4289 (25C), soluble in alcohols, hydrolyzes slowly.
Use: Intermediate in organic synthesis.

4,6-di-tert-butyl-m-cresol. (DBMC; 4,6-tert-butyl-3-methylphenol).
$[C(CH_3)_3]_2CH_3C_6H_2OH$.
Properties: Crystalline solid, mp 62.1C, bp 282C, d 0.912 (80/4C), viscosity 9.9 centistokes (80C), 1.42 centistokes (160C), flash p 262F (127.7C) (open cup), Combustible. Very soluble in ethanol, benzene, carbon tetrachloride, ethyl ether, and acetone; essentially insoluble in water, ethylene glycol, and 10% aqueous sodium hydroxide.
Use: Rubber reclaiming, surface-active agents, resins and plasticizers, antioxidants and perfumes.

2,6-di-tert-butyl-p-cresol. (2,6-tert-butyl-4-methylphenol; butylated hydroxytoluene; BHT).
CAS: 128-37-0. $[C(CH_3)_3]_2CH_3C_6H_2OH$.
Properties: White, crystalline solid; fp 70C; bp 265C; d 1.048 (20/4C); viscosity 3.47 centistokes

0C; 1.54 centistokes (120C); refr index 1.4859 (75C); soluble in methanol, ethanol, isopropanol, "Cellosolve" (12C), naphtha, benzene, methyl ethyl ketone, and linseed oil; insoluble in water and 10% sodium hydroxide; flash p (COC) 275F (135C). Combustible.

Grade: Technical, feed, FCC.

Hazard: Use in foods restricted. TLV: 10 mg/m³ of air.

Use: Antioxidant for petroleum products, jet fuels, rubber, plastics and food products, food packaging, animal feeds. Satisfies ASTM D910–64T for use in aviation gasoline.

2,6-di-tert-butyl-α-dimethylamino-p-cresol.
("Ethyl" Antioxidant 703).
$(C_4H_9)_2C_6H_2OH[CH_2N(CH_3)_2]$.
Properties: Light yellow, crystalline solid; mp 93.9C; flash p 280F (137.7C) (open cup). Insoluble in water and 10% sodium hydroxide, soluble in organic solvents. Combustible.
Use: Antioxidant in gasoline and oils including jet-engine oils.

di-tert-butyl diperphthalate.
$(CH_3)_3COOCOC_6H_4COOOC(CH_3)_3$.
Available as a 50% solution in dibutyl phthalate.
Properties: Clear liquid, insoluble in water, soluble in most organic solvents, flash p 145F (62.7C) (open cup). Combustible, d 1.056
Hazard: May ignite organic materials, strong oxidizer, may explode when shocked or in contact with reducing materials.
Use: High-temperature polymerization catalyst for vinyl and polyester resins.

dibutyl diphenyl tin.
$(CH_3CH_2CH_2CH_2)_2Sn(C_6H_5)_2$.
Properties: Clear, slightly greenish liquid; contains 30.7% Sn; bp 175C (2 mm); refr index 1.563 (17.5C); d 1.19. Combustible.
Hazard: Toxic. TLV: (as Sn) 0.1 mg/m³ of air, toxic by skin absorption.

di-tert-butyl disulfide. $C(CH_3)_3SSC(CH_3)_3$.
Properties: Liquid, d 0.9291, boiling range 190–207C, refr index 1.491 (20C), flash p approximately 170F (76.6C). Combustible.
Use: Intermediate, diesel and jet fuel additive, lubricant additive.

n-dibutyl ether. See butyl ether.

dibutyl fumarate.
$(C_4H_9)OOCCH:CHCOO(C_4H_9$.
Properties: Colorless liquid, d 0.9873 (20C), bp 285.2C, fp −15.6C, refr index 1.4466 (20C), insoluble in water, flash p 300F (149C). Combustible.

Use: Monomeric plasticizers, copolymers, intermediate.

dibutyl hexahydrophthalate.
$C_6H_{10}(COOC_4H_9)_2$.
Properties: Liquid, d 1.005, bp 185–190C, flash p 305F (151.6C). Combustible.
Use: Plasticizer for nitrocellulose.

2,5-di-tert-butyl hydroquinone.
$[C(CH_3)_3]_2C_6H_2(OH)_2$.
Properties: White powder; soluble in acetone, alcohol, benzene; insoluble in water, aqueous alkali; mp 210–212C.
Use: Polymerization inhibitor, antioxidant, stabilizer against UV deterioration of rubber.

di-n-butyl itaconate.
$CH_2:C(COOC_4H_9)CH_2(COOC_4H_9$.
Properties: Clear, colorless liquid with slight odor; bp 145C (10 mm); d 0.9833 (22C); refr index 1.442 (25C); insoluble in water. Combustible.
Use: Resins, lube oil additives, plasticizers.

N,N-di-n-butyl lauramide.
$C_{11}H_{23}CON(C_4H_9)_2$.
Properties: Straw-colored liquid, d 0.861 (20C), boiling range 200–230C (3 mm), odor of lauric acid, flash p 375F (190C). Combustible.

dibutyl maleate. (DBM). CAS: 105-76-0.
$C_4H_9OOCCH:CHCOOC_4H_9$.
Properties: Colorless, oily liquid; bp 280.6C; sets to a glass below −85C; d 0.9964 (20/20C); wt/gal 8.3 lb (20C); flash p 285F (140.5C) (OC); insoluble in water. Combustible.
Use: Copolymers, plasticizers, intermediate.

2,6-di-tert-butyl-4-methylphenol. See 2,6-di-tert-butyl-p-cresol.

4,6-di-tert-butyl-3-methylphenol. See 4,6-di-tert-butyl-m-cresol.

dibutyl oxalate. $(COOC_4H_9)_2$.
Properties: Water-white, high-boiling liquid; mild odor; bp 240–250C; refr index 1.425; fp −30C; wt/gal approximately 8.24 lb (20C); flash p 220F (104.4C) (closed cup). Miscible with most alcohols, ketones, esters, oils, hydrocarbons. Combustible.
Derivation: By the standard esterification process using normal butyl alcohol and oxalic acid.
Grade: According to ester content 90%, 95%, 99–100%.
Hazard: Toxic by ingestion and inhalation, strong skin irritant.
Use: Organic synthesis, solvent.

di-tert-butyl peroxide. (tert-butyl peroxide; DTBP). CAS: 110-05-4.
$(CH_3)_3COOC(CH_3)_3$.
Properties: Clear, water-white liquid; d 0.791 (25/25C); fp −40C;, bp 111C; refr index 1.389 (20C); flash p 65F (18.3C) (closed cup). Soluble in styrene, ketones, most aliphatic and aromatic hydrocarbons; insoluble in water.
Hazard: Flammable, dangerous fire hazard. Strong oxidizer, may ignite organic materials or explode when shocked or in contact with reducing materials.
Use: Polymerization catalyst for resins, including olefins, styrene, styrenated alkyds, and silicones; ignition accelerator for diesel fuel; organic synthesis; intermediate.

2,4-di-tert-butylphenol. CAS: 96-76-4.
$[(CH_3)_3C]_2C_6H_3OH$.
Properties: Tan, crystalline solid; mp 52C; bp 152–157C (25 mm); d 0.907 (60/4C); lb/gal 7.57 (60C); flash p 265F (129C). Soluble in methanol; ether, very slightly soluble in water. Combustible.
Hazard: See phenol.
Use: Intermediate, antioxidant, stabilizer, germicide.

2,6-di-tert-butylphenol. CAS: 128-39-2.
$[(CH_3)_3C]_2C_6H_3OH$.
Properties: Light straw, crystalline solid; mp 37C; d 0.914 (20C); bp 253C; flash p 245F (118C). Combustible. Soluble in alcohol and benzene, insoluble in water.
Hazard: See phenol.
Use: Intermediate, antioxidant, satisfies ASTM D910–64T for use as antioxidant in aviation gasoline.

N,N′-di-sec-butyl-p-phenylenediamine.
$C_{14}H_{24}N_2$.
Properties: Amber to red liquid (normally supercooled below 18C), d 0.94 (15.5/15.5C), fp 20C, flash p 290F (COC). Soluble in gasoline, absolute alcohol, and benzene; insoluble in water or caustic solutions. Combustible.
Hazard: Toxic by ingestion, inhalation, and skin absorption; strong irritant to tissue, causes skin burns.
Use: Oxidation inhibitor and stabilizer in gasoline (satisfies ASTM D910–64T as antioxidant in aviation gasoline), prevents decomposition of tetraethyl lead in gasoline.

dibutyl phosphate. (di-n-butyl phosphate).
CAS: 107-66-4.
Properties: Pale amber liquid, mw 210.21, mp decomposes abpve 100C, vap p less than 1 torr at 20C.

Hazard: TLV: 1 ppm. Respiratory tract irritant.
Use: Organic catalyst and antifoaming agent.

dibutyl phosphite. CAS: 1809-19-4.
$(C_4H_9O)_2PHO$.
Properties: Water-white liquid, bp 95C (1 mm), d 0.9860 (25C), refr index 1.4228 (25C), flash p 120F (49C), soluble in common organic solvents. Combustible.
Hazard: Moderate fire risk.
Use: Solvent, antioxidant, intermediate.

dibutyl phthalate. CAS: 84-74-2.
$C_6H_4(COOC_4H_9)_2$.
Properties: Colorless, odorless, stable, oily liquid; d 1.0484 (20/20C); fp −35C; viscosity 0.203 poise (20C); distillation range 227–235 (37 mm); flash p (COC) 340F (171C); wt/gal 8.72 lb (20C); refr index 1.4915 (25C); bp 340.0C; vap press 1.11 mm (150C); miscible with common organic solvents; insoluble in water. Combustible, autoign temperature approximately 750F (398.8C).
Derivation: By treating n-butyl alcohol with phthalic anhydride followed by purification which results in a product unusually free from odor and color.
Grade: Technical, 99–100% dibutyl phthalate.
Hazard: Toxic. TLV: 5 mg/m³ of air.
Use: Plasticizer in nitrocellulose lacquers, elastomers, explosives, nail polish and solid rocket propellants; solvent for perfume oils; perfume fixative; textile lubricating agent; safety glass; insecticides; printing inks; resin solvent; paper coatings; adhesives; insect repellent for textiles.

2,5-di-tert-butylquinone. $[C(CH_3)_3]_2C_6H_2O_2$.
Properties: Yellow powder; mp 149–151C; insoluble in water; soluble in alcohol, acetone, ethyl acetate, and benzene.
Hazard: Fire risk in contact with organic materials.
Use: Oxidizing agent.

dibutylsebacate. (DBS).
$C_4H_9OCO(CH_2)_8OCOC_4H_9$.
Properties: Clear, colorless, odorless liquid; bp 349C (760 mm), 180C (3 mm); fp −11C; d 0.936 (20/20C); wt/gal 7.81 lb (20C); refr index 1.4395 (25C); flash p 350F (176C). Insoluble in water. Combustible.
Grade: Technical.
Use: Plasticizer, rubber softener, dielectric liquid, cosmetics and perfumes, sealing food containers, flavoring.

N,N-dibutylstearamide. $C_{17}H_{35}CON(C_4H_9)_2$.
Properties: Yellow liquid, d 0.860 (20/20C), boiling range 173–175C (0.4 mm), flash p 420F (215C), fatty-acid odor. Combustible.

di-n-butyl succinate. $C_{12}H_{22}O_4$.
Properties: Colorless liquid, bp 120C, fp −29C, d 0.977, refr index 1.43
Derivation: Reaction of butyl alcohol with succinic acid.
Use: Insect repellent.

di-tert-butyl sulfide. (butyl sulfide).
$[(CH_3)_3C]_2S$.
Properties: Liquid, fp −11C, boiling range 297–303F, d 0.8316, wt/gal 6.93 lb, refr index 1.451 (20C), flash p 125F (51.6C). Combustible.
Hazard: Moderate fire risk.
Use: Intermediate, flavoring.

dibutyl tartrate.
$C_4H_9OOCCHOHCHOHCOOC_4H_9$.
Properties: Light tan liquid, mp 21C, bp approximately 204C (26 mm), refr index 1.4463 (20C), flash p 195F (90.5C) (closed cup), combustible, autoign temperature 544F (284.4C), wt/gal 9.07 lb (20C). Miscible with the common organic solvents, oils, hydrocarbons.
Use: Solvent and plasticizer for cellulose esters and ethers, elastomers, lubricant; rubberized fabrics; lacquers; dopes; transfer inks.

dibutylthiourea. $C_4H_9NHCSNHC_4H_9$.
Properties: White to light tan powder; mp 59–69C; slightly soluble in water; soluble in methanol, ether, acetone, benzene, ethyl acetate; insoluble in gasoline.
Use: Corrosion inhibitor, for pickling cast iron or carbon steel, reducing corrosion of ferrous metals and aluminum alloys in brine, intermediate.

dibutyltin-bis-(lauryl) mercaptide.
$(C_4H_9)_2Sn(SC_{12}H_{25})_2$.
Yellow liquid, tin content 18.5%, soluble in toluene and heptane.
Hazard: TLV: (as Sn) 0.1 mg/m³ of air, toxic by skin absorption.
Use: Antioxidant and metal cleaning (or protective) agent.

dibutyltin diacetate. CAS: 1067-33-0.
$(C_4H_9)_2Sn(C_2H_3O_2)_2$.
Properties: Clear, yellow liquid; bp 130C (2 mm); fp less than 12C; soluble in water and most organic solvents; flash p 290F (143C). Combustible.
Derivation: Reaction of acetic acid with dibutyltin oxide.
Hazard: TLV: (as Sn) 0.1 mg/m³ of air, toxic by skin absorption.
Use: Stabilizer for chlorinated organics, catalyst for condensation reactions.

dibutyltin dichloride. CAS: 683-18-1.
$(C_4H_9)_2SnCl_2$.
Properties: White, crystalline solid; mp 43C; bp 135C (10 mm); d 1.36 (24/4C); refr index 1.4991 (51C); insoluble in cold water; hydrolyzed by hot water; soluble in many organic solvents; flash p 335F (168C). Combustible.
Derivation: Reaction of butylmagnesium chloride with tin tetrachloride.
Hazard: TLV: (as Sn) 0.1 mg/m³ of air, toxic by skin absorption.
Use: Organotin intermediate.

dibutyltin di-2-ethylhexoate. CAS: 2781-10-4.
$(C_4H_9)_2Sn(O_2CC_7H_{15})_2$.
Properties: Waxy, white solid; mp 52.54C; insoluble in water; soluble in most organic solvents.
Derivation: Reaction of dibutyltin oxide with 2-ethyl-hexoic acid.
Hazard: TLV: (as Sn) 0.1 mg/m³ of air, toxic by skin absorption.
Use: Catalyst for silicone curing, polyether foams.

dibutyltin dilaurate. CAS: 77-58-7.
$(C_4H_9)_2Sn(OCOC_{10}H_{20}CH_3)_2$.
Properties: Clear, pale yellow liquid; d 1.066; mp 23C; fp 8.0C; lb/gal 8.84 (20C); flash p 440F (226C); soluble in acetone and benzene; insoluble in water. Combustible.
Hazard: TLV: (as Sn) 0.1 mg/m³ of air, toxic by skin absorption.
Use: Stabilizer for vinyl resins, lacquers, elastomers, catalyst for urethane and silicones.

dibutyltin maleate. CAS: 15535-69-0.
$[(C_4H_9)_2Sn(OOCCH)_2]_x$.
Properties: White, amorphous powder; mp 110C; insoluble in water; soluble in benzene and organic esters; flash p 400F (204C). Combustible.
Derivation: Reaction of maleic acid with di-butyltin oxide.
Hazard: TLV: (as Sn) 0.1 mg/m³ of air, toxic by skin absorption.
Use: Stabilizer for polyvinyl chloride resins, condensation catalyst.

dibutyltin oxide. CAS: 818-08-6.
$[(C_4H_9)_2SnO^-]_x$.
Properties: White powder, mp (decomposes). Insoluble in water. Combustible.
Derivation: Hydrolysis of dibutyltin dichloride with caustic.
Hazard: TLV: (as Sn) 0.1 mg/m³ of air, toxic by skin absorption.
Use: Condensation catalyst, intermediate for other organotins.

dibutyltin sulfide. CAS: 4253-22-9.
$[(C_4H_9)_2SnS]_3$.
Properties: Colorless, oily liquid. Combustible.

Derivation: Reaction of dibutyltin oxide with hydrogen sulfide.

Hazard: TLV: (as Sn) 0.1 mg/m^3 of air, toxic by skin absorption.

Use: Vinyl stabilizer, antioxidant, lubricating additive.

2,6-di-tert-butyl-p-tolyl-N-methylcarbamate.
$[(CH_3)_3C]_2CH_3C_6H_2OOCNHCH_3$.   White solid, mp 200C, insoluble in water, soluble in alcohol.
Use: Herbicide.

1,1-dibutylurea.   (n,n-dibutylurea).
$NH_2CON(C_4H_9)_2$.
Liquid, mp 22–25C, boiling range 118–119C (2–3 mm), soluble in alcohol and ether, flash p 279F (137C). Combustible.
Use: Thermoplastic resins by copolymerization with urea with formaldehyde catalyst.

dibutyl xanthogen disulfide.   See "C-P-B."

DIC.   Abbreviation for β-diisopropylaminoethyl chloride hydrochloride.

dicalcium magnesium aconitate.   (calcium magnesium aconitate). $[C_3H_3(COO)_3]_2Ca_2Mg$.
Properties: White, crystalline powder or lumps.
Derivation: Precipitation from molasses with lime.
Use: Conversion to aconitic acid, tributyl aconitate, and similar ester plasticizers.

dicalcium orthophosphate.   See calcium phosphate, dibasic.

dicalcium orthophosphite.   See calcium phosphite.

dicalcium phosphate dihydrate.   See calcium phosphate, dibasic.

dicalcium silicate. $2CaO \cdot SiO_2$.   One of the components of cement.
See cement, Portland.
Also obtained as a byproduct in electric furnace operation used to neutralize acid soils, noncombustible.

"Dicalite."[218]   TM for a group of products made from either diatomite or perlite, used in filters and filteraids.
See also diatomaceous earth.

dicamba.   Generic name for 3-6-dichloro-o-anisic acid (2-methoxy-3,6-dichlorobenzoic acid).
$HOOC(Cl)C_6H_2Cl(OCH_3)$.
Properties: Crystals, mp 114–116C, slightly soluble in water, moderately soluble in xylene, very soluble in alcohol.
Use: Herbicide, pest control.

dicapryl adipate.
$C_8H_{17}OOC(CH_2)_4COOC_8H_{17}$.
Properties: Almost water-white liquid, bp 213–216.5C (4 mm), flash p 352F (177C). Combustible.
Use: Plasticizer for vinyl resins and cellulose esters. See also capryl compounds.

dicapryl phthalate.   (DCP, di-(2-octyl)phthalate).
$(C_8H_{17}OOC)_2C_6H_4$.
Properties: Colorless, viscous liquid; bp 227–234C (4.5 mm); fp −60C; refr index 1.480 (20C); d 0.965 (25C); flash p 395F (201C); insoluble in water; compatible with vinyl chloride resins and some cellulosic resins. Combustible.
Use: Monomeric plasticizer for vinyl and cellulosic resins. See also capryl compounds.

dicapryl sebacate.
$C_8H_{17}OOC(CH_2)_8COOC_8H_{17}$.
Properties: Light straw-colored liquid, bp 231.5–239C (4 mm), nonvolatile, gives excellent low-temperature flexibility, flash p 445F (229C). Combustible.
Use: Plasticizer for vinyl resins and acrylonitrile rubber.   See also capryl compounds.

dicapthon.   Generic name for O-(2-chloro-4-nitrophenyl) O,O-dimethyl phosphorothioate).
$(CH_3O)_2P(:S)OC_6H_3(Cl)NO_2$.
Properties: White solid; mp 51–52C; insoluble in water; soluble in acetone, cyclohexanone, ethylacetate, toluene, and xylene.
Hazard: Cholinesterase inhibitor. Use may be restricted.
Use: Insecticide.

dicarboxylic acid.   A carboxylic acid containing two —COOH groups, e.g., adipic, oxalic, phthalic, sebacic, and maleic acids.

dicetyl.   See dotriacontane.

dicetyl ether.   (dihexadecyl ether).
Properties: Crystals, mp 54C, bp decomposes at 300C, d 0.8117 (54/4C)
Use: Electrical insulators, water repellents, lubricants in plastic molding and processing, antistatic substances, chemical intermediate.

dicetyl sulfide.   (dihexadecyl thioether; dihexadecyl sulfide). $(C_{16}H_{33})_2S$.
Properties: Solid, mp 57–58C, bp decomposes, d 0.8253 (60/4C).
Use: Organic synthesis (formation of sulfonium compounds).

dichlobenil.   (Generic name for 2,6-dichlorobenzonitrile). $Cl(Cl)C_6H_3CN$.
Properties: White solid, mp 144C. Almost insoluble in water, soluble in organic solvents.

Hazard: Toxic by ingestion and inhalation.
Use: Herbicide.

dichlofenthion. (O,O-diethyl-O-2,4-dichloro-phenyl phosphorothioate).   CAS: 97-17-6. $C_{10}H_{13}Cl_2O_3PS$.
Properties: Colorless liquid, bp 165C (0.3 mm), d 1.30, refr index 1.52, slightly soluble in water, soluble in organic solvents.
Hazard: Cholinesterase inhibitor, use may be restricted.
Use: Pesticide.

dichlone. (Generic name for 2,3-dichloro-1,4-naphthoquinone).   CAS: 117-80-6. $C_{10}H_4Cl_2O_2$.
Properties: Yellow needles; mp 193C; soluble in xylene and o-dichlorobenzene; slightly soluble in ethyl alcohol, glacial acetic acid, and carbon tetrachloride; almost insoluble in water.
Hazard: Toxic by ingestion and inhalation, irritant to skin and eyes.
Use: Seed disinfectant, fungicide for foilage and textiles, insecticide, organic catalyst.

dichloramine-T. (p-toluenesulfondichloramide). $CH_3C_6H_4SO_2NCl_2$.
Properties: Pale yellow crystals containing not less than 28% nor more than 30% active chlorine. Chlorine odor, stable when pure, decomposed slowly by air, rapidly by impurities, petrolatums, kerosene, olive oil, and alcohol. Soluble in glacial acetic acid, chlorinated paraffin hydrocarbons, eucalyptol, benzene, chloroform, and carbon tetrachloride; almost insoluble in water; mp 80C.
Derivation: Reaction between toluene-p-sulfonamide and calcium hypochlorite solution is acidified with acetic acid and extracted with chloroform. The chloroform solution is dried chemically, filtered, and evaporated.
Use: Germicide, antibacterial.
See also chloramine-T.

dichloroacetaldehyde. $CHCl_2CHO$.
Properties: Colorless liquid with a penetrating pungent odor, d 1.436 (25C), wt/gal 12.1lb, flash p 140F (60C) (closed cup). Combustible.
Hazard: Moderate fire hazard. Toxic by ingestion and inhalation, strong skin irritant.
Use: Manufacture of insecticides.

dichloroacetic acid.   CAS: 79-43-6. $CHCl_2COOH$.
Properties: Colorless liquid, d 1.5724 (13C), fp −4C, bp 193–194C. Soluble in water, alcohol, and ether.
Derivation: Chlorination of acetic acid in presence of iodine.

Hazard: Toxic by ingestion and inhalation, strong irritant to eyes and skin.
Use: Intermediate, pharmaceuticals, medicine.

α,α-dichloroacetophenone. $C_6H_5COCHCl_2$.
Properties: Crystals, d 1.34 (15C), bp 247C (decomposes) mp 20–21.5C.
Hazard: Toxic by ingestion and inhalation, strong irritant to eyes and skin.

dichloroacetyl chloride. $Cl_2CHCOCl$.
Properties: Fuming liquid, acrid odor, d 1.5315 (16/4C), bp 107–108C, refr index 1.4638 (16C), soluble in ether, decomposes in water and alcohol.
Hazard: Strong irritant to skin and eyes.
Use: Intermediate.

dichloroacetylene.   CAS: 7572-29-4. $ClC{\equiv}CCl$.
Properties: A liquid with mw 94.93, mp −66C, bp explodes. Soluble in alcohol, acetone.
Ocurrence: Thermal decomposition of trichloroethylene, closed circuit anesthesia using trichloroethylene and soda-lime absorption of carbon dioxide. May be synthesized by passing trichloroethylene over alkaline materials at 70F.
Hazard: TLV: 0.1 ppm.

2,3-dichloroallyl diisopropylthiolcarbamate. (diallate). $[(CH_3)_2CH]_2NCOSCH_2C(Cl){:}CHCl$.
Properties: Brown liquid, bp 159C (9 mm). Almost insoluble in water; soluble in acetone, benzene, chloroform, kerosene, and xylene.
Hazard: Toxic. Absorbed by skin. Use may be restricted.
Use: Herbicide.

2,4-dichloro-6-aminophenol. (2-amino-4,6-dichlorophenol). $C_6H_5Cl_2NO$.
Properties: Acicular crystals, mp 95C, soluble in benzene, less so in carbon disulfide. The hydrochloride is soluble in water and alcohol.
Use: Azo dye intermediate.

2,5-dichloroaniline. $C_6H_3NH_2Cl_2$.
Properties: Light brown or amber-colored crystalline mass. Slightly soluble in water; soluble in alcohol, benzene, and dilute hydrochloric acid; mp 47–50C; bp 251–252C. Combustible.
Derivation: Nitration of p-dichlorobenzene with subsequent reduction.
Hazard: See aniline.
Use: Dye intermediate.

3,4-dichloroaniline.   CAS: 95-76-1. $C_6H_3NH_2Cl_2$.
Properties: Crystals, mp 68–72C, bp 272C, slightly soluble in water, soluble in most organic solvents, flash p 331F (166C). Combustible.

Derivation: Nitration of o-dichlorobenzene with subsequent reduction.

Hazard: See aniline.

Use: Intermediate for manufacture of dyes and pesticides.

dichlorobenzaldehyde. $C_6H_3CHOCl_2$.
(2,4-, 2,5-, and 3,4-isomers).

Properties: White, crystalline solid; bp 233C; mp 65–67C (2,4-), 35C (3,4-); 2,4-isomer soluble in methanol, ethanol, ether, and acetone, insoluble in water. 2,5-isomer soluble in ethanol and ether. 3,4-isomer soluble in ethanol, ether, and acetone; slightly soluble in methanol and amyl ether; insoluble in water.

Use: Intermediate in the manufacture of pharmaceuticals, dyes, and other organic chemicals.

dichlorobenzalkonium chloride. $C_{21}H_{36}Cl_3N$.

Properties: Colorless crystals, very bitter taste, soluble in alcohol and water.

Derivation: Reaction of N-dimethyldodecylamine with 3,4-dichlorobenzyl chloride.

Hazard: Skin irritant.

Use: Algicide, antiseptic, sterilant.

m-dichlorobenzene. (1,3-dichlorobenzene).
CAS: 541-73-1. $C_6H_4Cl_2$.

Properties: Colorless liquid, d 1.288 (20/4C), bp 172C, fp −24C, refr index 1.547 (20.9C), soluble in alcohol and ether, insoluble in water. Combustible.

Derivation: Chlorination of monochlorobenzene.

Use: Fumigant and insecticide.

o-dichlorobenzene. (1,2-dichlorobenzene).
CAS: 95-50-1. $C_6H_4Cl_2$.

Properties: Colorless liquid, pleasant odor, a mixture of isomers containing at least 85% o- and varying percentages of p- and m-. Flash p 150F (65.5C). Combustible, miscible with most organic solvents, insoluble in water, autoign temperature 1198F (647C), d 1.284, wt/gal 10.7 lb, bp 172–179C, fp −17C.

Derivation: Chlorination of monochlorobenzene.

Method of purification: Rectification.

Grade: Purified, technical.

Hazard: Toxic by inhalation and ingestion. TLV: ceiling of 50 ppm in air.

Use: Manufacture of 3,4-dichloroaniline, solvent for a wide range of organic materials and for oxides of nonferrous metals, solvent carrier in production of toluene diisocyanate, dye manufacture, fumigant and insecticide, degreasing hides and wool, metal polishes, industrial odor control, heat transfer.

p-dichlorobenzene. (1,4-dichlorobenzene; PDB). CAS: 106-46-7. $C_6H_4Cl_2$.

Properties: White crystals, volatile (sublimes readily), penetrating odor, d 1.458, bp 173.7C, mp 53C, flash p 150F (65.5C) (closed cup). Soluble in alcohol, benzene, and ether; insoluble in water. Combustible.

Derivation: Chlorination of monochlorobenzene.

Grade: Technical.

Hazard: Toxic by ingestion, irritant to eyes. TLV: 75 ppm in air.

Use: Moth repellent, general insecticide, germicide, space odorant manufacture of 2,5-dichloroaniline, dyes, intermediates, pharmacy, agriculture (fumigating soil).

N,N-dichlorobenzenesulfonamide.
$C_6H_5SO_2NCl_2$.

Properties: Solid, fine, white crystals; mp 68–71C.

Use: A source of positive chlorine.

3,3′-dichlorobenzidine. CAS: 91-94-1.
$C_6H_3ClNH_2C_6H_3ClNH_2$.

Properties: Gray to purple, crystalline solid; insoluble in water; soluble in alcohol and ether; mp 165C.

Hazard: A tumorigen and carcinogen; absorbed by skin. TLV: Suspected human carcinogen.

Use: Intermediate for dyes and pigments, curing agent for isocyanate-terminated resins for urethane plastics.

2,4-dichlorobenzoic acid. $Cl_2C_6H_3COOH$.

Properties: White to slightly yellowish powder; mp 158–162C; soluble in alcohol, ether, acetone, 5% caustic; insoluble in water and heptane.

Use: Intermediate for antimalarials, dyes, fungicides, pharmaceuticals and other organic chemicals.

3,4-dichlorobenzoic acid. $Cl_2C_6H_3COOH$.

Properties: White to slightly yellowish powder; mp 202–204C; soluble in alkali, alcohol, ether, and acetone; slightly soluble in diacetone; insoluble in water ethylene dichloride, and toluene.

Use: Intermediate for pharmaceuticals, dyes, etc.

2,6-dichlorobenzonitrile. See dichlobenil.

3,4-dichlorobenzotrichloride. $Cl_2C_6H_3CCl_3$.

Properties: Water-white liquid, boiling range 276–285C, fp approximately 24.0C, d 1.585–1.590 (25/15C). Soluble in alcohol, ether, and acetone; insoluble in water. Combustible.

Use: Intermediate for pharmaceuticals, dyes, etc.

dichlorobenzoyl chloride. $Cl_2C_6H_3COCl$.
Properties: Colorless liquid, boiling range 250–260C, fp 15–16C, d 1.500–1.510 (25/15C). Soluble in alcohol, ether, and acetone; slightly soluble in heptane; insoluble in water. Combustible.
Use: Intermediate in the manufacture of pharmaceuticals, dyes, and other organic chemicals.

dichlorobenzyl chloride. $Cl_2C_6H_3CH_2Cl$.
Properties: Colorless liquid; boiling range 245–252C; d 1.415–1.420 (25/15C); soluble in alcohol, ether, and acetone; insoluble in water. Combustible.
Use: Intermediate for organic chemicals, pharmaceuticals and dyes, insecticide.

1,1-dichloro-2,2-bis(p-chlorophenyl) ethane. See TDE.

1,1-dichloro-2,2-bis(p-ethylphenyl) ethane. (diethyldiphenyldichloroethane; "Perthane"). $C_{18}H_{20}Cl_2$.
Properties: Crystals; mp 56C; insoluble in water; soluble in acetone, kerosene, diesel fuel.
Hazard: Toxic. Absorbed by skin.
Use: Insecticide, formulated as emulsifiable concentrate or wettable powder. Used especially against insects, including moths.

1,4-dichlorobutane. (tetramethylene dichloride; DCB). $ClCH_2(CH_2)_2CH_2Cl$.
Properties: Colorless, mobile liquid; pleasant odor; d 1.141(20/4C); boiling point 155C; flash p 104F (40C) (Tagliabue open cup); refr index 1.4542 (20C). Insoluble in water, soluble in most common organic solvents. Combustible.
Hazard: Moderate fire risk.
Use: Organic synthesis, including adiponitrile.

1,3-dichlorobutene-2. $ClH_2CCH:CClCH_3$.
Properties: Clear to straw-colored liquid, bp 125–130C, insoluble in water, soluble in organic solvents, flash p 80F (26.6C) (COC).
Hazard: Flammable, moderate fire risk. Toxic by inhalation.

1,4-dichlorobutene-2. CAS: 764-41-0. $ClH_2CCH:CHCH_2Cl$.
Properties: Colorless liquid; distinct odor; miscible with benzene, alcohol, carbon tetrachloride; immiscible with ethylene glycol, glycerol, and water; bp 158C (60C) (20 mm); mp 3.5C; d 1.1858 (25/4C); refr index 1.4863 (25C). Combustible.
Grade: Available as 95–98% trans-isomer, 2–5% cis-isomer. Above constants are for the pure trans-isomer.
Hazard: Irritant to skin and eyes, causes blisters. Moderate fire hazard.
Use: Intermediate.

dichlorocarbene. CCl_2.
Exists only at low temperature and pressure, fp −114C, bp −20C, decomposes on distillation at normal pressure to hexachloroethane and hexachlorobenzene.
Derivation: Reaction of carbon tetrachloride vapor with carbon at 1300C and 10^{-3} mm.
Hazard: Explosive reaction with carbon, forms phosgene on reaction with oxygen.
Use: Research.

1,1′-dichlorocarbonyl ferrocene. (ferrocenoyl dichloride). $[C_5H_4(COCl)]_2Fe$.
Properties: Red, crystalline solid; mp 93–95C.
Use: Intermediate.

4,6-dichloro-N-(2-chlorophenyl)-1,3,5-triazin-2-amine. $ClC_6H_4NHC_3N_3Cl_2$.
Properties: Tan, crystalline solid; mp 159–160C; insoluble in water.
Hazard: See aniline.
Use: Foliage fungicide.

dichloro(2-chlorovinyl)arsine. See chlorovinyldichloroarsine.

3,3′-dichloro-4,4′-diaminodiphenylmethane. See 4,4′-methylenebis(2-chloroaniline).

2,3-dichloro-5,6-dicyanobenzoquinone.

(DDQ). $OCC(Cl):C(Cl)COC(CN):C(CN)$.
Properties: Bright, yellow-orange solid; mp 213–215C.
Use: Highly selective oxidizing agent for organic compounds.

2,2′-dichlorodiethyl ether. See dichloroethyl ether.

dichlorodiethyl formal. See dichloroethyl formal.

dichlorodiethyl sulfide. (mustard gas; dichloroethyl sulfide). $S(CH_2CH_2Cl)_2$.
Properties: Yellow liquid, bp 228C, fp 14C, d 1.27, flash p 220F (104C).
Derivation: Bubbling ethylene through sulfur chloride, also from thiodiglycol and hydrogen chloride.
Grade: Pure, technical (containing excess sulfur as a polysulfide).
Hazard: Vesicant war gas, causes conjunctivitis and blindness. Can be decontaminated by chloramines or bleaching powder. Vapor is extremely poisonous can can be absorbed via skin.
Use: Organic synthesis, poison gas, medicine.

dichlorodiethyl sulfone. $(ClCH_2CH_2)_2SO_2$.
Properties: Colorless crystals; bp 179–181C (14–15 mm); mp 52C; soluble in alcohol, chloroform, and ether; slightly soluble in water.
Hazard: Strong irritant to eyes and skin.

2,2-dichloro-1,1-difluoroethyl methyl ether.
(methoxyflurane). $HCCl_2CF_2OCH_3$.
Properties: Clear, colorless liquid; fruity odor; bp 104.65C; fp −35; d 1.4223 (25C); completely stable in the presence of alkali, air, light, or moisture. Slightly soluble in water. Combustible.
Grade: ND.
Use: Anesthetic.

dichlorodifluoromethane. (difluorodichloromethane; fluorocarbon-12). CAS: 75-71-8. CCl_2F_2.
Properties: Colorless, odorless, noncorrosive gas; bp −29.8C; fp −158C; critical pressure 43.2 atm; insoluble in water; soluble in most organic solvents. Nonflammable.
Derivation: (a) Reaction of carbon tetrachloride and anhydrous hydrogen fluoride, in the presence of an antimony halide catalyst, (b) high temperature chlorination of vinylidene fluoride (vinylidene fluorides made by addition of hydrogen fluoride to acetylene).
Grade: 99.9% min purity.
Hazard: Narcotic in high concentration. TLV: 1000 ppm in air.
Use: Refrigerant in air conditioners, plastics, blowing agent, low-temperature solvent, leak-detecting agent, freezing of foods by direct contact, chilling cocktail glasses. See also chlorofluorocarbon.

1,3-dichloro-5,5-dimethylhydantoin. (DDH).
CAS: 118-52-5. $ClNCONClOC(CH_3)_2$.
Properties: White powder with mild chlorine odor, mp approximately 130C, sublimes approximately 100C without decomposition. Contains approximately 36% active chlorine. Slightly soluble in water with gradual liberation of hypochlorous acid; soluble in benzene, chloroform, ethylene dichloride, alcohol. Combustible with evolution of chlorine at 210C.
Derivation: Chlorination of dimethylhydantoin.
Grade: Technical.
Hazard: Toxic by inhalation, skin irritant. TLV: 0.2 mg/m³ of air.
Use: Household laundry bleach, water treatment, mild chlorinating agent, pharmaceutical intermediate, catalyst.

dichlorodimethylsilane. See dimethyldichlorosilane.

dichlorodiphenyldichloroethane. See TDE.

dichlorodiphenyldichloroethylene. See DDE.

dichlorodiphenyltrichloroethane. See DDT.

1,1-dichloroethane. See ethylidene chloride.

1,2-dichloroethane. See ethylene dichloride.

dichloroether. See dichloroethyl ether.

dichloroethoxymethane. See dichloroethylformal.

1,2-dichloroethyl acetate.
$CH_3COOCHClCH_2Cl$.
Properties: Water-white liquid, d 1.296 (20C), boiling range, 58–65C (13 mm), fp less than −32C, refr index 1.444 (20C), bp decomposes, flash p 307F (152C), combustible, miscible with alcohol and ethyl ether, immiscible with water.
Hazard: Toxic by inhalation.
Use: Organic synthesis.

p-di(2-chloroethyl)aminophenylalamine.
See melphalan.

dichloroethylarsine. See ethyldichloroarsine.

dichloroethyl carbonate. $(ClH_2CCH_2O)_2CO$.
Properties: Colorless liquid, slowly hydrolyzed by alkalies, volatile in steam, d 1.3506 (20C), bp 240C (partial decomposition), insoluble in water.
Derivation: By heating ethylene chlorohydrin and trichloromethylchloroformate together (under reflux).

sym-**dichloroethylene.** (1,2-dichloroethylene; acetylene dichloride; dichloracetylene).
CAS: 540-59-0.   ClHC:CHCl.   Exists as cis and trans isomers.
Properties: Colorless, low-boiling liquid; pleasant odor; decomposes slowly on exposure to air, light, and moisture; soluble in most organic solvents; slightly soluble in water. Trans-isomer: d 1.257, bp 47–49C. Cis-isomer: d 1.282, bp 58–60C, flash p 39F (3.9C), fp −80C.
Derivation: Two stereoisomeric compounds made by the partial chlorination of acetylene.
Grade: Technical, as cis, trans, and mixture of both.
Hazard: Flammable, dangerous fire hazard. Toxic by ingestion, inhalation, and skin contact; irritant and narcotic in high concentration. TLV: 200 ppm in air.
Use: General solvent for organic materials, dye extraction, perfumes, lacquers, thermoplastics, organic synthesis.

sym-**dichloroethyl ether.** (dichloroether; dichloroethyl oxide; bis(2-chloroethyl)ether; 2,2′-dichlorodiethyl ether). CAS: 111-44-4. $ClCH_2CH_2OCH_2CH_2Cl$.
Properties: Colorless liquid, odor like that of ethylene dichloride, bp 178.5C, d 1.2220 (20/20C),

wt/gal 10.2 lb (20C), refr index 1.457 (20C), flash p 131F (55C) (CC), fp −51.8C, autoign temperature 696F (368C), miscible with most organic solvents, insoluble in water. Combustible.
Derivation: Chlorination of ethyl ether.
Grade: Technical.
Hazard: Moderate fire hazard. Toxic by inhalation and ingestion, strong irritant. TLV: 5 ppm in air, toxic by skin absorption.
Use: General solvent, selective solvent for production of high-grade lubricating oils, textile scouring and cleansing, fulling compounds, wetting and penetrating compounds, paints, varnishes, lacquers, finish removers, spotting and dry cleaning, soil fumigant, intermediate and cross-linker in organic synthesis.

dichloroethyl formal. (dichlorodiethyl formal). $CH_2(OCH_2CH_2Cl)_2$.
Properties: Colorless liquid, bp 218.1C, fp −32.8C, d 1.2339 (20/20C), wt/gal 10.3 lb (20C), flash p 230F (110C) (OC), slightly soluble in water, decomposed by mineral acids. Combustible.
Hazard: Toxic by inhalation and ingestion, strong irritant.
Use: Solvent, intermediate for polysulfide rubber.

dichloroethyl oxide. See dichloroethyl ether.

dichloroethyl sulfide. See dichlorodiethyl sulfide.

dichloromonofluoromethane. (fluorodichloromethane; fluorocarbon 21). CAS: 75-43-4. $CHCl_2F$.
Properties: Colorless, nearly odorless, heavy gas; bp 8.9C; fp −135C; d 1.426 (0C); critical pressure 51.0 atm; soluble in alcohol and ether; insoluble in water; nonflammable.
Derivation: Reaction of chloroform and hydrogen fluoride.
Grade: Technical.
Hazard: Toxic by inhalation. TLV: 10 ppm in air.
Use: Solvent, refrigerant.
See chlorofluorocarbon.

dichloroformoxime. CCl_2NOH.
Properties: Colorless, prismatic crystals; disagreeable, penetrating odor; high vap press; slowly decomposes at normal temperature, the rate depending on temperature and humidity; bp 53–54C (18 mm); mp 39–40C; soluble in water, alcohol, ether, and benzene.
Derivation: (a) Action of chlorine on fulminic acid, HONC, (b) reduction of trichloronitrosomethane with either aluminum amalgam or hydrogen sulfide.
Hazard: Strong irritant to skin and eyes.
Use: A military poison.

α-dichlorohydrin. (α-propenyldichlorohydrin; 1,3-dichloro-2-propanol; sym-dichloroisopropyl alcohol). $CH_2ClCHOHCH_2Cl$.
Properties: Colorless, slightly viscous, unstable liquid; faint chloroform-like odor. The commercial product is a mixture of two isomers. Miscible with most organic solvents, vegetable oils; slightly soluble in water; d 1.36–1.39; fp −4C; bp 174C; refr index 1.47–1.48; flash p 165F (73.9C); combustible; vap press 7 mm.
Derivation: Interaction of glycerol and dry hydrogen chloride gas and subsequent distillation.
Hazard: Moderate fire risk. Toxic by inhalation and ingestion.
Use: General solvent, intermediate in organic synthesis, paints, varnishes, lacquers, watercolor binder, photographic lacquers.

5,7-dichloro-8-hydroxyquinaldine. See chlorquinaldol.

dichloroisocyanuric acid. (dichloro-s-triazine-2,4,6-trione). $\overline{OCNClCONClCONH}$.
Properties: White, slightly hygroscopic, crystalline powder, granules. D (loose bulk, approximately) powder 34 lb/cu ft, granular 53 lb/cu ft. Active ingredient approximately 70% available chlorine, decomposes 225C.
Hazard: Oxidizer, may ignite organic materials on contact. Irritant to eyes.
Use: Household dry bleaches, dishwashing compounds, scouring powders and detergent sanitizers, replacement for calcium hypochlorite.

sym-dichloroisopropyl alcohol. See α-dichlorohydrin.

dichloroisopropyl ether. CAS: 108-60-1. $[ClCH_2C(CH_3)H]_2O$.
Properties: Colorless liquid, d 1.1135 (20/20C), bp 187.4C (760 mm), vap press 0.10 mm (20C), flash p 185F (85C), wt/gal 9.3 lb (20C), coefficient of expansion 0.00096 (20C), viscosity 0.0230 poise (20C). Miscible with most oils and organic solvents, immiscible with water. Combustible.
Hazard: Toxic by ingestion and inhalation, strong irritant.
Use: Solvent for fats, waxes, greases, extractant, paint and varnish removers, spotting agents and cleaning solutions.

dichloromethane. Legal label name for methylene chloride.

3′,4′-dichloro-2-methylacrylanilide. See dicryl.

dichloromethylchloroformate. $ClCOOCHCl_2$.
Properties: Colorless liquid; decomposed by water and alkalies; soluble in alcohol, benzene, and

ether; d 1.56 (15C); bp 110–111C; vap d 5.7 (air=1.29).

Derivation: (a) By chlorinating methyl formate, (b) by chlorinating methylchloroformate. In both methods, the mixture of chloro-derivatives is then separated by fractionation.

Hazard: Strong irritant to eyes and skin.

sym-dichloromethyl ether. $O(CH_2Cl)_2$.

Properties: Colorless, volatile liquid; suffocating odor; decomposed by heat and water; soluble in acetone, benzene, ethyl alcohol, and methyl alcohol; insoluble in water; d 1.315 (20C); bp 105C.

Derivation: (a) Action of chlorine on methyl ether, (b) interaction of hydrochloric acid and formaldehyde with subsequent dehydration of the chloromethyl alcohol formed.

Hazard: A carcinogen by ingestion and inhalation. Strong irritant to eyes and mucous membranes.

5,7-dichloro-2-methyl-8-quinolinol. See chlorquinaldol.

dichloromethylsilane. CH_3SiHCl_2.

Colorless liquid, bp 41C, d 1.113 (25/25C), refr index 1.3983 (20C). Combustible.

Hazard: Corrosive to tissue.

Use: Intermediate.

dichloromethyl sulfate. $(ClCH_2O)_2SO_2$.

Properties: Colorless, odorless liquid; soluble in alcohol, benzene, and ether; d 1.60 (20C), bp 96–97C (14 mm). Combustible.

Derivation: (a) By bubbling sulfur trioxide through (cooled) dichloromethyl ether, (b) by heating chlorosulfonic acid with formaldehyde.

dichloronaphthalene. See chlorinated naphthalenes.

2,3-dichloro-1,4-naphthoquinone. See dichlone.

1,2-dichloro-4-nitrobenzene. $Cl_2C_6H_3NO_2$.

Properties: Solid; mp 43C; bp 255–256C; d 1.4266 (100/4C); insoluble in water; soluble in hot alcohol, ether.

Hazard: Fire risk by spontaneous reaction.

Use: Intermediate.

2,5-dichloronitrobenzene. $Cl_2C_6H_3NO_2$.

Properties: Pale yellow crystals, d 1.669 (22C), mp 55C, bp 266C, insoluble in water, soluble in chloroform and hot alcohol.

Hazard: Fire risk by spontaneous reaction.

Use: Intermediate.

1,1-dichloro-1-nitroethane. CAS: 594-72-9.

$H_3CC(Cl)_2NO_2$.

Properties: Bp 124C, d 1.4153 (20/20C), flash p 168F (75.5C) (OC). Combustible.

Hazard: Strong irritant. TLV: 2 ppm in air.

Use: Grain fumigant, solvent.

dichloropentane. CAS: 30586-10-8.

$C_6H_{10}Cl_2$. A mixture of the dichloro-derivatives of both normal and isopentane. About 40% are amylene dichlorides having two chlorine atoms attached to adjacent carbon atoms.

Properties: Light yellow liquid, d 1.06–1.08 (20C), acidity as hydrochloric acid not over 0.025% (distillation = 95%) 130–200C, flash p less than 80F (26C), wt/gal 8.94 lb. Water solubility negligible; water azeotrope at 80–97C = approximately 66% $C_6H_{10}Cl_2$.

Hazard: Flammable, dangerous fire risk.

Use: Solvent for oils, greases, rubber resins, and bituminous materials; removal of tar; reclaiming rubber; paint and varnish removers; degreasing of metals; insecticide; soil fumigant; removal of wax deposits on oil-well equipment.

dichlorophenarsine hydrochloride. (3-amino-4-hydroxyphenyldichloroarsine hydrochloride).

$C_6H_3(AsCl_2)(OH)NH_2 \cdot HCl$.

Properties: White, hygroscopic powder; mp 200C; soluble in water and solutions of alkali hydroxides and carbonates and in dilute mineral acids.

Hazard: Toxic by ingestion.

Use: Medicine (syphilis treatment).

dichlorophene. (2,2'-dihydroxy-5,5'-dichlorodiphenylmethane; DDDM). CAS: 97-23-4.

$(C_6H_3ClOH)_2CH_2$.

Properties: Light tan, free-flowing powder; weakly phenolic odor; mp 177C; vap press 10^{-4} mm (100C) and about 10^{-10} mm (25C) (extrapolated value). Soluble in acetone and alcohols; slightly soluble in benzene, toluene, carbon tetrachloride; insoluble in water.

Derivation: Condensation of p-chlorophenol with formaldehyde in the presence of sulfuric acid.

Grade: Pure and technical.

Hazard: Toxic by ingestion.

Use: Fungicide and bactericide, textile preservative, some dermatological and cosmetic applications, veterinary medicine.

2,4-dichlorophenol. CAS: 120-83-2.

$Cl_2C_6H_3OH$.

Properties: White, low-melting solid; bp 210C; mp 45C; flash p 237F (113C). Combustible. Soluble in alcohol and carbon tetrachloride, slightly soluble in water.

Derivation: By chlorination of phenol.

Hazard: Toxic by ingestion, strong irritant to tissue.

Use: Organic synthesis.

2,4-dichlorophenoxyacetic acid. See 2,4-D.

2,4-dichlorophenoxybutyric acid. (2,4-DB). $Cl_2C_6H_3O(CH_2)_3COOH$.   Mp 118–120C, used as a herbicide.

di-(4-chlorophenoxy)methane. $(ClC_6H_4O)_2CH_2$.
Properties: Solid, mp 65C, insoluble in water and oils, soluble in ether and acetone.
Use: Acaricide.

2-(2,4-dichlorophenoxy)propionic acid. (dichloroprop). $Cl_2C_6H_3OCH(CH_3)COOH$.
Properties: Solid; mp 117–118C; soluble in acetone, alcohol, and ether; insoluble in water.
Use: Herbicide.

2,4-dichlorophenylbenzenesulfonate. $Cl_2C_6H_3OSO_2C_6H_5$.
Properties: Waxy solid, vap press 2.7×10^{-4} mm at 30C, insoluble in water, soluble in most organic solvents.
Use: Acaricide, insecticide.

O-(2,4-dichlorophenyl)-O,O-diethyl phosphorothioate. $Cl_2C_6H_3OP(S)(OC_2H_5)_2$.
Properties: (pure compound) Liquid, bp 120–123C (0.2 mm), slightly soluble in water, miscible with most organic solvents.
Hazard: Cholinesterase inhibitor. Absorbed by skin, use may be restricted.

3-(3,4-dichlorophenyl)-1,1-dimethylurea. See diuron.

di(p-chlorophenyl)ethanol. (1,1-bis(p-chlorophenyl)ethanol; di(p-chlorophenyl)methyl carbinol; DMC; DCPC). $CH_3C(C_6H_4Cl)_2OH$.
Properties: Colorless crystals, mp 70C, insoluble in water, soluble in common organic solvents.
Derivation: Reaction of 4,4′-dichlorobenzophenone with methyl magnesium bromide followed by treatment with water.
Use: Insecticide.

3,4-dichlorophenyl isocyanate. $Cl_2C_6H_3NCO$.
White to yellow crystals. Strong irritant to tissue, especially eyes and mucous membranes.
Use: Chemical intermediate, organic synthesis.

3-(3,4-dichlorophenyl)-1-methoxy-1-methylurea. See linuron.

di(p-chlorophenyl)methyl carbinol. See di(p-chlorophenyl)ethanol.

o-(2,4-dichlorophenyl)-o-methyl isopropylphosphoramidothioate. (DMPA). $Cl_2C_6H_3OP(S)(OCH_3)NHCH(CH_3)_2$.
Properties: Solid; mp 51.4C; vap press 2 mm at 150C; slightly soluble in water (5 ppm); freely soluble in acetone, benzene, and carbon tetrachloride.
Hazard: Cholinesterase inhibitor, use may be restricted.
Use: Herbicide, insecticide.

2,4-dichlorophenyl-4-nitrophenyl ether. $Cl_2C_6H_3OC_6H_4NO_2$.
Properties: Dark brown solid; setting point 62.5C; soluble in acetone, methanol, and xylene.
Hazard: Toxic by ingestion.
Use: Herbicide.

dichlorophenyltrichlorosilane. $Cl_2C_6H_3SiCl_3$.
A mixture of isomers.
Properties: Straw-colored liquid, d 1.562, bp 260C, refr index 1.5638 (20C), flash p (COC) 286F (141C); readily hydrolyzed with liberation of hydrogen chloride; soluble in benzene, perchloroethylene. Combustible.
Derivation: Chlorination of phenyltrichlorosilane.
Grade: Technical.
Hazard: Strong irritant to skin and eyes.
Use: Intermediate for silicones.

3,6-dichlorophthalic acid. $C_6H_2Cl_2(COOH)_2$.
Properties: Colorless, thick crystals; soluble in hot water.
Derivation: By oxidizing dichloronaphthalene tetrachloride (see chloronaphthalene) with nitric acid.

dichloroprop. See 2-(2,4-dichlorophenoxy)propionic acid.

1,2-dichloropropane. See propylene dichloride.

1,3-dichloro-2-propanol. See α-dichlorohydrin.

1,3-dichloropropene. (1,3-dichloropropylene).
CAS: 542-75-6. $CHCl:CHCH_2Cl$.
Properties: Exists in cis and trans isomeric forms, both colorless liquids; d 1.225 (10/4C); flash p 95F (35C) (OC); insoluble in water; soluble in acetone, toluene, octane. Bp cis isomer 104C, trans 112C; refr index (20C) cis 1.469, trans 1.475.
Derivation: Chlorination of propylene.
Hazard: Flammable, moderate fire risk. Strong irritant. TLV: 1 ppm in air, toxic by skin absorption.
Use: Organic synthesis, soil fumigants.

dichloropropene-dichloropropane mixture.
Hazard: A poison via ingestion, inhalation, or absorbed through the skin; corrosive to tissue (skin and eyes).
Use: Pesticide, insecticide.

3,4-dichloropropionanilide. See propanil.

2,2-dichloropropionic acid.   See dalapon.

2,6-dichlorostyrene.   CAS: 6607-45-0.
  $C_6H_3(CH:CH_2)Cl_2$.
  Properties: Colorless liquid, bp 92–94C(5 mm), flash p 225F (107C) (open cup), insoluble in water, soluble in most organic solvents. Polymerizes slowly on standing unless inhibited. Combustible.
  Use: Monomer and co-monomer in plastic research.

p-N,N-dichlorosulfamylbenzoic acid.   See halazone.

dichlorosulfonphthalein.   See chlorphenol red.

2,6-dichloroquinonechlorimide.   Reagent used for spot visualizations in chromatographic systems.
  Hazard: Explodes readily on slight heating.

dichlorotetrafluoroacetone. $CClF_2COCClF_2$.
  Properties: Colorless liquid, bp 45.2C, fp less than $-100C$, soluble in water and most organic solvents, stable to acids but not alkalies.
  Hazard: Toxic by ingestion and inhalation.
  Use: Solvent in acidic media, complexing agent for active hydrogen compound separation.

sym-dichlorotetrafluorethane.   (fluorocarbon-114; tetrafluorodichloroethane).
  CAS: 76-14-2. $CClF_2CClF_2$.
  Properties: Colorless, nearly odorless, nonflammable gas; bp 3.55C; fp $-94C$; critical pressure 32.3 atm; insoluble in water.
  Derivation: By treating perchloroethylene with hydrogen fluoride.
  Grade: Technical 95%.
  Hazard: Toxic by inhalation. TLV: 1000 ppm in air.
  Use: Solvent, fire extinguishers, refrigerant and air conditioner, blowing agent, dielectric fluid. See also chlorofluorocarbon.

2,5-dichlorothiophene. $C_4H_2Cl_2S$ (cyclic).
  Properties: Colorless to light-yellow liquid, bp 161C. Combustible.
  Use: Intermediate.

α, α-dichlorotoluene.   See benzyldichloride.

dichlorotoluene.   (chlorobenzyl chloride).
  $C_6H_3CH_3Cl_2$.
  Properties: Colorless liquid, boiling range 200–300C, fp approximately $-13C$, d 1.245–1.247 (25/15C), refr index 1.5480 (22C). Soluble in alcohol, ether, and acetone; insoluble in water. Exists as o- and p- isomers. Combustible.
  Hazard: Irritant, see benzyl chloride.

Use: High-boiling solvent, intermediate for organic synthesis.

dichloro-sym-triazine-2,4,6-trione.   See dichloroisocyanuric acid.

β,β'-dichlorovinylchloroarsine.
  $(ClCH:CH)_2AsCl$.
  Properties: Yellowish liquid, bp 230C (decomposes), d 1.70.
  Hazard: Toxic by inhalation and ingestion, strong irritant to skin and mucous membranes.
  Use: Military poison.

2,2-dichlorovinyl dimethyl phosphate.   See dichlorovos.

β,β'-dichlorovinylmethylarsine.
  $(ClCH:CH)_2AsCH_3$.
  Properties: Liquid, bp 140–145C (10 mm).
  Derivation: Interaction of acetylene and methyldichloroarsine in the presence of aluminum chloride.
  Hazard: Toxic by inhalation and ingestion, strong irritant to skin and mucous membranes.
  Use: Military poison.

dichlorovos.   (2,2-dichlorovinyl dimethyl phosphate; DDVP).   CAS: 62-73-7.
  $(CH_3O)_2P(O))CH:CCl_2$.
  Properties: Liquid, bp 120C (14 mm), slightly soluble in water and glycerol, miscible with aromatic and chlorinated hydrocarbon solvents and alcohols.
  Hazard: Cholinesterase inhibitor, a poison. TLV: 1 mg/m³ of air, toxic by skin absorption.
  Use: Insecticide, fumigant.

dichloroxylenol.   (2,4-dichloro-3,5-dimethyl-phenol). $C_8H_8Cl_2O$.
  Properties: Crystals; mp 95C; partially soluble in acetone, diethyl ketone, and chloroform; almost insoluble in water.
  Use: Mold inhibitor, antibacterial agent, preservative.

dichroic.   A term used in crystallography to denote crystals which refract incident light in two directions thus displaying two colors when observed from different angles, e.g., calcite.
  See also anisotropic, birefringent.

dichromatic.   Characterizing certain dyes and indicators for which different colors may be seen depending on the thickness of the solution viewed.

dicobalt octacarbonyl.   See cobalt tetracarbonyl.

"Dicodic."[9]   TM for dihydrocodeinone, used in medicine as bitartrate and hydrochloride salts.

dicofol. (Generic name for 4,4'-dichloro-α-trichloromethylbenzhydrol). See 1,1'-bis-(chlorophenyl)-2,2,2-trichloroethanol.

dicomarol. See bishydroxycoumarin.

dicresyl glyceryl ether. (glyceryl ditolyl ether). This may be a mix of o-, m- and p-isomers. CH₃C₆H₄OCH₂CHOHCH₂OC₆H₄CH₃.
Properties: Similar to cresyl glyceryl ether (see mephenesin). D 1.136, refr index 1.549, boiling range 328–340C.

dicresyl glyceryl ether acetate.
CH₃C₆H₄OCH₂CHOOCCH₃OC₆H₄CH₃.
Properties: Fairly stable liquid, d 1.115, bp 360C, refr index 1.53. Combustible.

dicrotophos. See "Bidrin."

dicryl. (Generic name for 3',4'-dichloro-2-methylacrylanilide). CAS: 2164-09-2. Cl₂C₆H₃NHCOC(CH₃):CH₂.
Properties: Solid; mp 128C; insoluble in water; soluble in acetone, alcohol, isophorone, and dimethyl sulfoxide.
Use: Herbicide, pest control.

"Dicrylan."[443] TM for a group of products which includes silicone elastomers and other polymeric chemicals used for coating textile fabrics.

dicumyl peroxide. [C₆H₅C(CH₃)₂O]₂.
Hazard: Strong oxidizer, may ignite organic materials on contact.
Use: Polymerization catalyst and vulcanizing agent.

"Di-cup."[266] TM for a series of vulcanizing and polymerization agents containing dicumyl peroxide.

dicyan. See cyanogen.

dicyandiamide. (cyanoguanidine). NH₂C(NH)(NHCN).
Properties: Pure white crystals, d 1.400 (25C), stable when dry, melting range 207–209C, soluble in liquid ammonia, partly soluble in hot water. Nonflammable.
Derivation: Polymerization of cyanamide in the presence of bases.
Grade: 99% pure, technical.
Use: Fertilizers; nitrocellulose stabilizer; organic synthesis, especially of melamine, barbituric acid, and guanidine salts; pharmaceutical products; dyestuffs; explosives; retarding rancidity in fats and oils; fire-proofing compounds; case-hardening preparations; cleaning compounds; soldering compounds; accelerator; thinner for oil-well drilling muds; stabilizer in detergent compositions; modifier for starch products; catalyst for epoxy resins.

dicyanine. See cyanine dye.

o-dicyanobenzene. See phthalonitrile.

1,4-dicyanobutane. See adiponitrile.

2,4-dicyanobutene-1. See methylene·glutaronitrile.

dicyclohexyl. See bicyclohexyl.

dicyclohexyl adipate. (CH₂CH₂COOC₆H₁₁)₂.
Properties: White crystals, odorless, compatible with most natural and synthetic resins, soluble in most organic solvents, insoluble in water, d 0.913–0.919 (15/15C), bp 256C, fp −0.1C, density 45 lb/cu ft, acidity (as adipic acid) less than 0.05%
Use: Plasticizer.

dicyclohexylamine. CAS: 101-83-7. (C₆H₁₁)₂NH.
Properties: Colorless liquid with faint amine odor, d 0.913–0.919 (15/15C), bp 256C, fp −0.1C, refr index 1.4823 (25C), flash p 210F (98.9C). Combustible, slightly soluble in water, miscible with organic solvents, strongly basic.
Hazard: Toxic by ingestion, strong irritant to skin and mucous membranes.
Use: Intermediate, insecticides, plasticizer, corrosion inhibitors, antioxidants in rubber, lubricating oils, fuels, catalyst for paint, varnishes and inks, detergents, extractant.

dicyclohexyl carbodiimide. C₆H₁₁NCNC₆H₁₁.
Properties: White crystals with a heavy sweet odor. Set point 29–30C, bp 138–140C (2 mm), soluble in organic solvents.
Use: Chemical intermediate, coupling agent in peptide synthesis.

dicyclohexyl phthalate. (DCHP). C₆H₄(COOC₆H₁₁)₂.
Properties: White, granular solid; nonvolatile; mildly aromatic odor; d 1.20 (25/25C); mp 62–65C; soluble in most organic solvents; insoluble in water; compatible with a large number of polymers; flash p 405F (207C). Combustible.
Use: Plasticizer for nitrocellulose, ethyl cellulose, chlorinated rubber, polyvinyl acetate, polyvinyl chloride, and other polymers.

dicyclopentadiene. CAS: 77-73-6. C₁₀H₁₂.
Properties: Liquid, d 0.979 (20/20C), bp 172C, mp 33.6C, bulk d 8.2 lb/gal (60F), refr index

1.5073 (31C), flash p 90F (32.2C), soluble in alcohol, insoluble in water.
Derivation: Olefin manufacture.
Hazard: Flammable, moderate fire risk. Toxic by ingestion. TLV: 5 ppm in air.
Use: Chemical intermediate for insecticides, EPDM elastomers, metallocenes, paints and varnishes, flame retardants for plastics.

dicyclopentadiene dioxide. $C_{10}H_{12}O_2$.
Properties: White, crystalline powder; mp 180–184C; d 1.331 (25C); slightly soluble in water; soluble in acetone and benzene.
Use: Intermediate for epoxy resins, plasticizers, protective coatings.

dicyclopentadienylcobalt. See cobaltocene.

dicyclopentadienyl iron. (bis-cyclopentadienyl iron). CAS: 102-54-5. The first organometallic "sandwich" compound (synthesized in 1951) which served as a prototype for metallocenes. Such compounds are based on cyclic unsaturates combined with a transition metal or its halide.
Hazard: Toxic. TLV: 10 mg/m³ of air.
See also ferrocene.

dicyclopentadienylnickel. See nickelocene.

dicyclopentadienyl osmium. See osmocene.

dicyclopentadienyltitanium dichloride. See titanocene dichloride.

dicyclopentadienylzirconium dichloride. See zirconocene dichloride.

DIDA. See diisodecyl adipate.

didecyl adipate. $(CH_2CH_2COOC_{10}H_{21})_2$.
Properties: Light-colored liquid, d 0.9181 (20/20C), 7.7 lb/gal (20C), bp 245C (5 mm), vap press (0.58 mm (100C), insoluble in water, viscosity 26.3 cp (20C), flash p 425F (218C). Combustible.
Use: Plasticizer.

didecylamine. $[CH_3(CH_2)_9]_2NH$.
Properties: Light straw-colored liquid, faint amine odor, boiling range 195–215C (12 mm), d 0.840 (20/20C) (solid). Combustible.

didecyl ether. $(C_{10}H_{21})_2O$.
Properties: Liquid, mp 16C, bp 170–180C (6 mm), d 0.819 (20/4C), refr index 1.4418 (20C). Combustible.
Grade: 95% (min) purity.

Use: Electrical insulators, water repellent, lubricant in plastic molding and processing, antistatic agent, intermediate.

didecyl phthalate. (DDP).   CAS: 84-77-5. $C_6H_4(COOC_{10}H_{21})_2$.
Properties: Light-colored liquid, d 0.9675 (20/20C) 8.05 lb/gal (20C), bp 261C (5 mm), insoluble in water, vap press 0.3 mm (200C), viscosity 113.2 cp (20C), flash p 445F (229C). Combustible.
Use: Plasticizer, especially for vinyl resins.

didecyl sulfide. (didecyl thioether). $(C_{10}H_{21})_2S$.
Properties: Liquid, solidifies at 22.2C, bp 205–206C (4 mm), d 0.831 (20/4C), refr index 1.4569 (33.5C). Combustible.
Grade: 95% (min) purity.
Use: Organic synthesis (formation of sulfonium compounds).

didecyl thioether. See didecyl sulfide.

didodecylamine. See dilaurylamine.

didodecyl ether. See dilauryl ether.

didodecyl-3,3'-thiodipropionate. See dilauryl thiodipropionate.

didodecyl thioether. See dilauryl sulfide.

DIDP. See diisodecyl phthalate.

didymium. Di.   Commercial mix of rare earth elements obtained from monazite sand by extraction followed by the elimination of cerium and thorium. The name is used like that of an element in naming mixed oxides and salts. The approximate composition of didymium from monazite, expressed as rare earth oxides, is 46% La_2O_3, 10% Pr_6O_{11}, 32% Nd_2O_3, 5% Sm_2O_3, 0.4% yttrium earth oxides, 1% CeO_2, 3% Gd_2O_3, 2% others. Commercially used didymium salts are acetate, carbonate, chloride, fluoride, nitrate, etc.

didymium oxide. Di_2O_3.
Brown powder, insoluble in water, soluble in acids.
Use: (as salts) Coloring and decolorizing glass, in temperature-compensating capacitors for radio, television, and radar. In carbon arc cores (fluoride), in stainless steel (oxide), metallurgical research, textile treatment.

die. A device, usually of steel, having a specific shape or design which it imparts to such materials as metals and plastics either by impact

(stamping), by the contour of a negative cavity (casting), or by passing the material through it (extrusion). Diamond dies may be used for wire-drawing. The terms "die" and "mold" are virtually synonymous in the sense of a negative cavity into which a molten metal or plastic is introduced under pressure, the former being used in reference to metals and the latter for plastics, rubber, etc.
See also die casting, investment casting, injection molding, extrusion.

die casting. Shaping of metal products by forcing a molten metal or alloy under high pressure into a negative-cavity die by means of a hydraulic ram. The die is usually made of an alloy steel. Metals commonly used for die casting are zinc, aluminum, copper, lead and their alloys, some of which also include silicon. Die castings can be held to tolerances as low as 0.001–0.0015 inches and sizes from 75–100 lb (Al) are possible. The largest end-use area for die castings is automobile and airplane parts. They are also used in washing and drying machines, electrical equipment and appliances, and for various military applications.

Dieckmann reaction. Base-catalyzed intramolecular cyclization of esters of dicarboxylic acids to give β-ketoesters.

dieldrin. (HEOD)   CAS: 60-57-1.
$C_{12}H_{10}OCl_6$.   (not less than 85% of 1,2,3,-4,10,10-hexachloro-6,7-epoxy-1,4,4a,5,6,7,8,8a-octahydro-1,4-endo,exo-51,2,3,4,10,10-hexa,8-dimethanonaphthalene and not less than 15% active related compounds.)
Properties: Light-tan, flaked solid; mp 175C. Insoluble in water, methanol, aliphatic hydrocarbons; soluble in acetone, benzene. Compatible with most fertilizers, herbicides, fungicides, and insecticides.
Derivation: By oxidation of aldrin with peracids.
Hazard: Toxic by ingestion, inhalation, and skin absorption. Use restricted to nonagricultural applications. Carcinogenic. TLV: 0.25 mg/m³ of air.
Use: Insecticide.
See also endrin, a stereoisomer of dieldrin.

dielectric. A substance with very low electrical conductivity, i.e., an insulator. Such substances have electrical conductivity of less than 1,000,-000th mho/cm. Those with a somewhat higher conductivity (10⁻⁶ to 10⁻³ mho/cm) are called semiconductors. Among the more common solid dielectrics are glass, rubber and similar elastomers, wood and other cellulosics. Liquid dielec-trics are hydrocarbon oils, askarel, and silicone oils.
See also transformer oil.

dielectric constant. A value that serves as an index of the ability of a substance to resist the transmission of an electrostatic force from one charged body to another as in a condenser. The lower the value the greater the resistance. The standard apparatus utilizes a vacuum whose dielectric constant is 1. In reference to this, various materials interposed between the charged terminal have the following values at 20C: air 1.00058, glass 3, benzene 2.3, acetic acid 6.2, ammonia 15.5, ethanol 25, glycerol 56, and water 81. The exceptionally high value for water accounts for its unique behavior as a solvent and in electrolytic solutions. Most hydrocarbons have high resistance (low conductivity). Dielectric constant values decrease as the temperature rises.

dielectric strength. The maximum electric field that an insulator or dielectric can withstand without breakdown, usually measured in kV/cm. At breakdown, a considerable current passes as an arc, usually with more or less decomposition of the material along the path of the current.

Diels, Otto P. H. (1876-1954) A German chemist who won a Nobel prize in chemistry with Alder in 1950. His prize was awarded for diene synthesis work which led to improved methods of analyzing and synthesizing organic compounds. His research resulted in the discovery of carbon suboxide, methods of dehydrating cyclical hydrocarbons using selenium, structure of steroids. A student of Fischer, he graduated from the University of Berlin.

Diels-Alder reaction. An important organic reaction for the synthesis of 6-membered rings discovered in 1928. It involves the addition of an ethylenic double bond to a conjugated diene, i.e., a compound containing two double bonds separated by one single bond, as in 1,3-butadiene $(CH_2{=}CH{-}CH{=}CH_2)$ or cyclopentadiene. The ease of addition of the ethylenic compound is greatly enhanced by adjacent carbonyl groups, hence maleic anhydride reacts quantitatively with hexachlorocyclopentadiene to form chlorendic anhydride.

dien. Abbreviation of diethylenetriamine as used in formulas for coordination compounds.
See also en; pn; py.

diene. See diolefin.

"Diene."[278]   TM for a commercial stereospecific polybutadiene rubber.
Properties: (Unvulcanized) Narrow molecular weight distribution, soluble in aliphatic or aro-

matic hydrocarbons, free from gel, low "tackiness," no elastic memory ("nerve"). Vulcanized: High resilience, excellent hysteretic properties, excellent resistance to abrasion and cold (brittle point below −87C).
Use: To enhance the low temperature and hysteretic properties of other elastomers.

dienestrol.
$HOC_6H_4C(CHCH_3)C(CHCH_3)C_6H_4OH$.
(3,4-bis(p-hydroxyphenyl)-2,4-hexadiene).
A synthetic with estrogenic activity.
Properties: Colorless, odorless needles or powder; mp 227C; soluble in alcohol; practically insoluble in water; sensitive to light.
Grade: NF.
Use: Medicine (estrogenic hormone).

dienone-phenol rearrangement. Transformation of a 4,4-disubstituted cyclohexadienone into a 3,4-disubstituted phenol upon acid treatment.

diesel ignition improver. A substance such as amyl nitrate which is added to diesel fuels to improve fuel ignition and to raise the cetane number.

diesel oil. (Fuel oil No. 2). Fuel for diesel engines obtained from distillation of petroleum. Its efficiency is measured by the cetane number. It is composed chiefly of unbranched paraffins; its volatility is similar to that of gas oil. Flash p 110–190F, d less than 1. Combustible.
Hazard: Moderate fire risk. Environmental hazard.
Use: Fuel for trucks, ships and other automotive equipment, drilling muds, mosquito control (coating on breeding waters).
See also fuel oil.

"Diesel-Treat."[108] TM for dry, granular, orange sodium dichromate used as a corrosion inhibitor. Sold in 50-lb drums.
Use: Closed cooling systems, particularly diesel engines, cooling tower systems.

dietary food supplement. Any food product to which enough vitamins and minerals have been added to furnish more than 50% of the recommended daily allowance in a single serving (FDA). Such foods must have added ingredients identified on labels.

diethanolamine. (DEA; di(2-hydroxyethyl)-amine). CAS: 111-42-2.
$(HOCH_2CH_2)_2NH$.
Properties: Colorless crystals or liquid; active base; mp 28.0; bp 269C; d 1.092 (30/20C); flash p 306F (152C) (open cup); very soluble in water and alcohol; insoluble in ether, benzene. Combustible.
Derivation: Ethylene oxide and ammonia.
Hazard: Toxic. TLV: 3 ppm in air.
Use: Liquid detergents for emulsion paints, cutting oils, shampoos, cleaners, and polishes, textile specialties; absorbent for acid gases; chemical intermediate for resins, plasticizers, etc.; solubilizing 2,4-D; humectant; dispersing agent.

N,N-diethanolglycine. (DEG).
$(HOCH_2CH_2)_2NCH_2COOH$.
Used as a chelating agent. Also available as the sodium salt.

2,5-diethoxyaniline. $NH_2C_6H_3(OC_2H_5)_2$.
Properties: White to gray powder, mp 83–85C, insoluble in water, soluble in organic solvents.
Hazard: See aniline.
Use: Intermediate.

1,4-diethoxybenzene. See hydroquinone diethyl ether.

1,1-diethoxyethane. See acetal.

diethoxyethyl phthalate. See diethyl glycol phthalate.

diethylacetal. See acetal.

diethyl acetaldehyde. See 2-ethylbutyraldehyde.

N,N-diethylacetamide. $CH_3CON(C_2H_5)_2$.
Properties: Colorless liquid, d 0.920 (20C), boiling range 182–186C, faint odor, flash p 170F (76.6C). Combustible.
Hazard: Toxic by ingestion.

diethylacetic acid. See 2-ethylbutyric acid.

N,N-diethylacetoacetamide.
$CH_3COCH_2CON(C_2H_5)_2$.
Properties: Liquid, d 0.9950 (20/20C), bp decomposes, fp −70C (sets to glass below this temperature). Miscible in water, flash p 250F. Combustible.
Use: Intermediate for pigments.

diethyl adipate. $C_2H_5OCO(CH_2)_4OCOC_2H_5$.
Properties: Colorless liquid, d 1.002 (25C), refr index 1.426 (25C), bp 245C, fp −21C. Insoluble in water. Combustible.
Use: Plasticizer.

diethylaluminum chloride. (aluminum diethyl monochloride; DEAC). $(C_2H_5)_2AlCl$.
Properties: Colorless pyrophoric liquid, bp 208C, fp −50C.

Derivation: Reaction of triethylaluminum with ethylaluminum sesquichloride.

Hazard: Highly flammable in air, reacts violently with water. Dangerous fire and explosion hazard.

Use: Polyolefin catalyst, intermediate in production of organometallics.

diethylaluminum hydride. $(C_2H_5)_2AlH$. A pyrophoric mix with triethylaluminum.

Derivation: Action of ethylene and hydrogen on aluminum.

Hazard: Highly flammable in air, reacts violently with water. Dangerous fire and explosion hazard.

Use: Catalyst, reducing agent.

diethylamine. CAS: 109-89-7. $(C_2H_5)_2NH$.

Properties: Colorless liquid; ammoniacal odor; alkaline reaction; bp 55.5C; fp −49.8C; d 0.7062 (20/20C); wt/gal 5.91 lb (20C); autoign temperature 594F (312C); flash p less than −15F (−26C); miscible with water, alcohol, most organic solvents.

Derivation: From ethyl chloride and ammonia under heat and pressure.

Grade: Technical.

Hazard: Highly flammable, dangerous fire risk. Flammable limits in air 1.8–10.1%. Toxic by ingestion, strong irritant. TLV: 10 ppm in air.

Use: Rubber chemicals, textile specialties, selective solvent, dyes, flotation agents, resins, pesticides, polymerization inhibitors, pharmaceuticals, petroleum chemicals, electroplating, corrosion inhibitors.

α-diethylaminoaceto-2,6-xylidide. See lidocaine.

5-diethylamino-2-aminopentane. (1-diethylamino4-aminopentane).
$CH_3CH(NH_2)(CH_2)_2CH2N(C_2H_5)_2$.

Properties: Liquid with an amine odor, d 0.82, bp 142–144C. Soluble in water, alcohol, and ether. Combustible.

Use: Pharmaceuticals.

diethylaminoaniline. See p-aminodiethylaniline.

diethylaminocellulose. (DEAE cotton).

A cellulose derivative containing a tertiary amine group which acts as a catalyst for epoxide reactions. It is also used in ion-exchange fractionations. It is made by adding β-chloroethyldiethylamine hydrochloride to cellulose in sodium hydroxide. Repeated treatments increase the nitrogen content of the cotton (cellulose) to over 1% with beneficial effect on crease resistance. Combustible.

See also epoxide.

2-diethylaminoethanethiol hydrochloride.
$(C_2H_5)_2NCH_2CH_2SH \cdot HCl$.

Solid, mp 170C, soluble in water and alcohol, insoluble in benzene.

Use: Pharmaceutical intermediate, pesticides, polymerization promoter.

diethylaminoethanol. (N,N-diethylethanolamine). CAS: 100-37-8.
$(C_2H_5)_2NCH_2CH_2OH$.

Properties: Colorless, hygroscopic liquid base combining the properties of amines and alcohols. Bp 161C, d 0.88–0.89 (20/20C), vap press 21 mm (20C), flash p 140F (60C) (open cup), wt/gal 7.14 lb (20C), fp −70C, combustible. Soluble in water, alcohol, benzene.

Grade: Technical.

Hazard: Moderate fire risk. Toxic by ingestion and skin absorption. TLV: 10 ppm in air.

Use: Water-soluble salts, fatty acid derivatives, textile softeners, pharmaceuticals, antirust compositions, emulsifying agents in acid media, derivatives containing tertiary amine groups, curing agent for resins.

diethylaminoethoxyethanol.
$(C_2H_5)_2NC_2H_4OC_2H_4OH$.

Properties: Colorless liquid, d 0.930–0.950 (20/20C), boiling range (95%) 215.0–228.0C, combustible.

Use: Intermediate.

β-diethylaminoethyl chloride hydrochloride. (DEC). CAS: 869-24-9.
$(C_2H_5)_2NCH_2CH_2Cl \cdot HCl$. Intermediate in the manufacture of pharmaceuticals and as an organic intermediate for attaching the diethylaminoethyl radical.

N,N′-diethyl-3-amino-4-methoxy-benzene-sulfonamide. $NH_2(CH_3O)C_6H_3SO_2N(C_2H_5)_2$.

White to pink crystals, mp 100–103C, insoluble in water and ether, partially soluble in alcohol.

Use: Intermediate.

1-diethylamino-2-methylbenzene. See N,N-diethyl-o-toluidine.

7-diethylamino-4-methylcoumarin. (MDAC; 4-methyl-7-[diethylamino]coumarin).
$CH_3C_9H_4O(O)N(C_2H_5)_2$.

Properties: Granular, light-tan color, mp 68–72C, gives a bright blue-white fluorescence in very dilute solutions. Soluble in aqueous acid solutions, resins, varnishes, vinyls, and nearly all common organic solvents; slightly soluble in aliphatic hydrocarbons.

Use: Optical bleach in textile industry, in coatings for paper, labels, book covers, etc; to lighten plastics, resins, varnishes, and lacquers; invisible marking agent.

5-diethylamino-2-pentanone. (1-diethylamino-4-pentanone). $CH_3CO(CH_2)_3N(C_2H_5)_2$.
Liquid with an amine odor; combustible.
Use: Pharmaceuticals.

m-diethylaminophenol. $C_6H_4OHN(C_2H_5)_2$.
Properties: White, crystalline solid; mp 78C; bp 276–280C; soluble in alcohol, caustic soda, ether.
Derivation: Diethylaniline is sulfonated with oleum and resulting diethylaniline-m-sulfonic acid fused with caustic soda.
Hazard: See phenol.
Use: Dyes.

3-diethylaminopropylamine.
$(C_2H_5)_2NCH_2CH_2CH_2NH_2$.
Properties: Water-white liquid, amine odor, bp 169C, fp −100C (sets to a glass), d 0.82 (20/20C), refr index 1.442 (10C), flash p 138F (58.9C) (open cup), Combustible.
Hazard: Moderate fire risk. Irritant to skin.
Use: Curing agent for epoxy resins, intermediate.

γ-diethylaminopropyl chloride hydrochloride.
(DEPC). $(C_2H_5)_2NCH_2CH_2CH_2Cl \cdot HCl$.
Use: Manufacture of pharmaceuticals, intermediate for attaching the diethylaminopropyl radical.

N,N-diethylaniline. CAS: 91-66-7.
$(C_2H_5)_2NC_6H_5$.
Properties: Colorless to yellow liquid, d 0.9351, fp −38 to −39C, bp 215–216C, flash p 185F (85C), slightly soluble in alcohol and ether, soluble in water. Combustible.
Derivation: By heating aniline hydrochloride with alcohol at 180C under pressure.
Grade: Technical.
Hazard: See aniline.
Use: Organic synthesis, dyestuff intermediate.

diethylbarbituric acid. See barbital.

diethylbenzene. CAS: 25340-17-4.
$(C_6H_4(C_2H_5)_2$. The commercial product is either a mixture of isomers or the p-isomer, the latter being available in both pure and technical grades.
Properties: Colorless liquid, boiling range 179.8–184.8C, d 0.865 (25/25C), flash p 132F (55.5C). Soluble in alcohol, benzene, carbon tetrachloride, ether; insoluble in water; wt/gal 7.22lb; refr index 1.49; autoign temperature 806F (430C), Combustible.
Hazard: Moderate fire risk. Moderately toxic.
Use: Intermediate, solvent.

di(2-ethylbutyl)azelate.
$C_6H_{13}OOC(CH_2)_7COOC_6H_{13}$.
Properties: Pale yellow to water-white liquid, d 0.9340 (20/20C), viscosity 56 sec Saybolt (100F), flash p 385F (196C), fp less than −40C, acid number less than 1.0, faint odor. Stable to heat, light, and hydrolysis. Combustible.
Use: As plasticizer for polyvinyl chloride and its copolymers and for cellulose esters.

di(2-ethylbutyl)phthalate. (dihexyl phthalate).
$C_6H_4(COOC_6H_{13})_2$.
Properties: Oily, slightly aromatic liquid; d 1.010–1.016 (20/20C); bp 350C (735 mm); acidity (as phthalic acid) 0.01% max; fp −50C; ester content 98% max; flash p 381F (193C). Combustible.
Grade: Technical.
Use: Plasticizer for cellulose ester and vinyl plastics.

diethyl cadmium. $(C_2H_5)_2Cd$.
Properties: Colorless, pyrophoric, oily liquid; bp 64C (19 mm); fp −21C.
Derivation: Reaction of cadmium acetate with triethyl aluminum.
Hazard: Ignites spontaneously in air, dangerous fire hazard. See cadmium.
Use: TEL production, synthesis of ketones from acid chlorides.

diethylcarbamazine citrate. CAS: 1642-54-2.
$C_{10}H_{21}N_3O \cdot C_6H_8O_7$. (1-diethylcarbamyl-4-methylpiperazine dihydrogen citrate).
Properties: White, crystalline powder; odorless or slight odor; slightly hygroscopic; very soluble in water; sparingly soluble in alcohol; practically insoluble in acetone, chloroform, and ether. Mp 135–138C.
Grade: USP.
Use: Medicine (anthelmintic).

N,N'-diethylcarbanilide. See sym-diethyldiphenylurea.

diethyl carbinol. See 3-pentanol.

diethyl "Carbitol."[14] TM for diethylene glycol diethyl ether.

1,1'-diethyl-4,4'-carbocyanine iodide. See cryptocyanine.

diethyl carbonate. (ethyl carbonate).
CAS: 105-58-8. $(C_2H_5)_2CO_3$.
Properties: Colorless liquid, mild odor, stable, d 0.975 (20/4C), bp 125C, fp −43C, flash p 77F (25C) (OC). Miscible with alcohols, ketones, esters, aromatic hydrocarbons, some aliphatic solvents; insoluble in water. Combustible.

Derivation: The steps in its manufacture are: (a) reacting chlorine and carbon monoxide to produce phosgene ($COCl_2$), (b) reacting phosgene with ethanol to make ethyl chlorocarbonate ($ClCO_2C_2H_5$), (c) reacting ethyl chlorocarbonate with anhydrous ethanol to produce diethyl carbonate.

Grade: Technical.

Hazard: Flammable, dangerous fire risk.

Use: Solvent for nitrocellulose, cellulose ethers, many synthetic and natural resins; organic synthesis; adhering rare earths to cathodes.

diethyl chlorophosphate. $(C_2H_5O)_2P(O)Cl$.
Properties: Water-white liquid, bp 60C (2 mm), d 1.1915 (25C), refr index 1.4153 (25C), soluble in alcohols. Combustible.

Grade: Technical.

Hazard: Toxic by ingestion, inhalation, and skin absorption; cholinesterase inhibitor.

Use: Intermediate in organic synthesis.

diethylcyclohexane. $(C_2H_5)_2C_6H_{10}$.
Properties: Liquid, d 0.8037 (20/20C), bp 174C, fp −100C, insoluble in water, flash p 125F (51.6C) (open cup), autoign temperature 465F (240.5C). Combustible.

Hazard: Moderate fire risk.

N,N-diethylcyclohexylamine. $C_6H_{11}N(C_2H_5)_2$.
Properties: Colorless liquid, bp 194.5C, soluble in ether and benzene, slightly soluble in water. Combustible.

Grade: Technical.

Use: Solvent, intermediate.

diethyl-1-(2,4-dichlorophenyl)2-chlorovinyl phosphate. $(C_2H_5)_2PO \cdot OC(C_6H_3Cl_2)$:CHCl.
Properties: Liquid; bp 167–170C (5 mm); insoluble in water; miscible with acetone, ethanol, kerosene, and xylene.

Hazard: Cholinesterase inhibitor.

Use: Insecticide.

diethyldichlorosilane. CAS: 1719-53-5.
$(C_2H_5)_2SiCl_2$.
Properties: Colorless liquid, bp 130.4C, d 1.053 (25/25C), refr index 1.4309 (25C), flash p (COC) 77F (25C); readily hydrolyzed with liberation of hydrogen chloride.

Derivation: Reaction of powdered silicon and ethyl chloride at 300C, in presence of copper powder.

Grade: Technical.

Hazard: Flammable, dangerous fire risk. Corrosive to tissue.

Use: Intermediate for silicones.

O,O-diethyl-S,2-diethylaminoethyl phosphorothioate hydrogen oxalate. See tetram.

diethyl diethylmalonate.
$(C_2H_5)_2C(COOC_2H_5)_2$.
Colorless liquid, sweet odor, d 0.984 (25/25C). Used as an intermediate.

diethyldiphenyldichloroethane. See 1,1-dichloro-2,2-bis(p-ethylphenyl)ethane.

sym-**diethyldiphenylurea.** (N,N′-diethylcarbanilide; ethyl centralite; carbamite).
$C_2H_5(C_6H_5)NCON(C_6H_5)C_2H_5$.
Properties: White crystalline, solid; peppery odor; mp 79C; bp 325–330C; d 1.12 (20C); insoluble in water; soluble in organic solvents; flash p 302F (150C).
Use: Stabilizer for nitrocellulose-based smokeless powder and in solid rocket propellants.

diethylenediamine. See piperazine.

1,4-diethylene dioxide. See 1,4-dioxane.

diethylene disulfide. See 1,4-dithiane.

diethylene ether. See 1,4-dioxane.

diethylene glycol. (dihydroxydiethyl ether; diglycol; DEG).   CAS: 111-46-6.
$CH_2OHCH_2OCH_2CH_2OH$.
Properties: Colorless, practically odorless, syrupy liquid; sweetish taste; extremely hygroscopic; noncorrosive; lowers freezing point of water; miscible with water, ethanol, acetone, ether, ethylene glycol; immiscible with benzene, toluene, carbon tetrachloride; bp 245.0C; fp −80C; d 1.1184 (20/20C); wt/gal 9.35 (15C); refr index 1.446 (25C); surface tension 48.5 dynes/cm (25C); viscosity 0.30 poise (25C); vap press 0.01 mm (30C); flash p 255F (123.9C). Combustible. Autoign temperature 444F (228.9C).

Derivation: Byproduct of manufacture of ethylene glycol.

Grade: Technical.

Hazardous for household use in concentration of 10% or more (FDA).

Use: Production of polyurethane and unsaturated polyester resins, triethylene glycol; textile softener; petroleum solvent extraction; dehydration of natural gas, plasticizers, and surfactants; solvent for nitrocellulose and many dyes and oils; humectant for tobacco, casein, synthetic sponges, paper products; cork compositions, book-binding adhesives, dyeing assistant, cosmetics, antifreeze solutions.

diethylene glycol acetate. See diethylene glycol monoacetate.

diethylene glycol bis(allyl carbonate). (allyl diglycol carbonate). CAS: 132-22-3. $O[CH_2CH_2OCOO(C_3H_5)]_2$.
Properties: Liquid, fp −4C, bp 160C (4 mm), d 1.143 (20C), viscosity 9 cp (20C), monomer for allyl resins, particularly in optically clear castings.

diethylene glycol bis(n-butylcarbonate). See butyl diglycol carbonate.

diethylene glycol bis(chloroformate). See diglycol chloroformate.

diethylene glycol bis(cresyl carbonate). See cresyl diglycol carbonate.

diethylene glycol bis(2,2-dichloropropionate).
A herbicide.
See "Garlon."

diethylene glycol bis(phenyl carbonate).
See phenyl diglycol carbonate.

diethylene glycol diacetate. (diglycol acetate). $(CH_3COOCH_2CH_2)_2O$.
Properties: Colorless liquid, miscible with water, d 1.1159, bp 250C, mp 19.1C, flash p 275F (135C), vap press 0.02 mm. Combustible.
Grade: Technical.
Use: Solvent for cellulose esters, printing inks, lacquers.

diethylene glycol dibenzoate.
$C_6H_5COO(CH_2)_2O(CH_2)_2OOCC_6H_5$.
Properties: Liquid, bp 225–227C (3 mm), flash p 450F (232C). Combustible.
Use: Plasticizer.

diethylene glycol dibutyl ether. (dibutyl "Carbitol"). CAS: 112-73-2.
$C_4H_9O(C_2H_4O)_2C_4H_9$.
Properties: Almost colorless liquid, slightly soluble in water, d 0.8853 (20/20C), 7.36 lb/gal (20C), bp 256C, vap press 0.02 mm (20C), fp −60.2C, viscosity 2.39 cp (20C), flash p 245F (118C). Combustible.
Use: High-boiling, inert solvent with application in extraction processes and in coatings and inks; diluent in vinyl chloride dispersions; extractant for uranium ores.

diethylene glycol dicarbamate. See diglycol carbamate.

diethylene glycol diethyl ether. (diethyl "Carbitol"; ethyl diglyme). CAS: 112-36-7.
$C_2H_5O(C_2H_4O)_2C_2H_5$.
Properties: Colorless liquid, very stable, d 0.9082 (20/20C), bp 189C, flash p 180F (82.2C), wt/

gal 7.56 lb (20C), fp −44.3C, soluble in hydrocarbons and water. Combustible.
Use: Solvent for nitrocellulose, resins, lacquers; high-boiling medium and solvent for organic synthesis.

diethylene glycol dimethyl ether. (diglyme; diglycol methyl ether). $CH_3(OCH_2CH_2)_2OCH_3$.
Properties: Colorless liquid with mild odor, bp 162.0C, fp −68.0C, d 0.9451 (20/20C), flash p 153F (67.2C), viscosity 1.089 cp (20C), miscible with water and hydrocarbons. Combustible.
Grade: Technical.
Use: Solvent, anhydrous reaction medium for organo-metallic synthesis.

diethylene glycol dinitrate. (DEGN; diglycol nitrate; dinitroglycol). CAS: 693-21-0.
$(O_2NOCH_2CH_2)_2O$.
Properties: Liquid, d 1.377 (25/4C), fp −11.3C, bp 161C, slightly soluble in water and alcohol, soluble in ether.
Hazard: Severe explosion hazard when shocked or heated. Toxic by ingestion.
Use: Plasticizer in solid rocket propellants.

diethylene glycol dipelargonate.
$(C_8H_{17}COOCH_2CH_2)_2O$.   A simple ester of pelargonic acid. Acid number 2.0, bp 229C (5 mm), pour point 10F, viscosity (SUV at 110C) 36 seconds, flash p 410F (210C). Combustible.
Use: Secondary plasticizer for vinyl resins.

diethylene glycol distearate. See diglycol stearate.

diethylene glycol monoacetate. (diethylene glycol acetate). $HO(CH_2)_2(CH_2)_2OOCCH_3$.
Miscible with water and aromatic hydrocarbons. Solvent for nitrocellulose, cellulose acetate, camphor, and rosin.

diethylene glycol monobutyl ether. (butyl "Carbitol"). $C_4H_9OCH_2CH_2OCH_2CH_2OH$.
Properties: Colorless liquid, faint butyl odor, bp 230.6C, d 0.9536 (20/20C), wt/gal 7.94 lb (20C), refr index 1.4316 (20C), viscosity 0.0649 poise (20C), vap press 0.01 mm (20C), specific heat 0.546 cal/g (20–25C), flash p 172F (77.7C), autoign temperature 442F (227.7C), coefficient of expansion 0.00088 (per C) to 20C, fp −68.1C, soluble in oils and water. Combustible.
Grade: Technical.
Use: Solvent for nitrocellulose, oils, dyes, gums, soaps, polymers; plasticizer intermediate.

diethylene glycol monobutyl ether acetate. (butyl "Carbitol" acetate). CAS: 124-17-4.
$CH_3CO(OC_2H_4)_2OC_4H_9$.
Properties: Colorless liquid, miscible with most organic liquids, d 0.9810 (20/20C), bp 246.7C,

vap press <0.01 mm (20C), flash p 240F (115.5C), wt/gal 8.16 lb (20C), nitrocellulosexylene dilute ratio 1:8, coefficient of expansion 0.0010 (20C), fp −32.3C, viscosity 0.0356 poise (20C), autoign temperature 570F (298.8C). Combustible.
Grade: Technical.
Use: Solvent for oils, resins, gums, also for cellulose nitrate and polymeric coatings; plasticizer in lacquers and coatings.

diethylene glycol monoethyl ether. ("Carbitol" solvent).   CAS: 111-90-0.
$CH_2OHCH_2OCH_2CH_2OC_2H_5$.
Properties: Colorless, hygroscopic liquid; mild, pleasant odor; slightly viscous; stable; bp 195–202C; d 1.0272 (20/20C); refr index 1.425 (25C); flash p 205F (96.1C); wt/gal 8.55 lb (20C); miscible with water and the common organic solvents. Combustible.
Grade: Technical.
Use: Solvent for dyes, nitrocellulose, and resins, mutual solvent for mineral oil-soap and mineral oil-sulfonated oil mixtures, nonaqueous stains for wood, for setting the twist and conditioning yarns and cloth; textile printing, textile soaps, lacquers, organic synthesis, brake fluid diluent.

diethylene glycol monoethyl ether acetate. ("Carbitol" acetate).   CAS: 112-15-2.
$CH_3COOCH_2OCH_2CH_2OC_2H_5$.
Properties: Colorless liquid, d 1.0114 (20/20C), bp 217.4C, vap press 0.05 mm (20C), flash p 230F (110C), wt/gal 8.4 lb (20C), coefficient of expansion 0.00105, fp −25C, refr index 1.418 (30C), viscosity 0.0279 poise (20C), soluble in water, miscible with most organic solvents. Combustible.
Grade: Technical.
Use: Solvent for cellulose esters, gums, resins; coatings and lacquers; printing inks.

diethylene glycol monohexyl ether. (n-hexyl "Carbitol").   CAS: 112-59-4.
$C_6H_{13}OC_2H_4OC_2H_4OH$.
Properties: Water-white liquid, d 0.9346 (20/20C), 7.8 lb/gal (20C), bp 259.1C, vap press less than 0.01 mm (20C), fp −33C, viscosity 8.6 cp (20C), flash p 285F (140.5C). Combustible.
Use: High-boiling solvent.

diethylene glycol monolaurate.   See diglycol laurate.

diethylene glycol monomethyl ether. [(2-β-methyl "Carbitol"), methoxyethoxy ethanol]. CAS: 111-77-3.
$CH_3OCH_2CH_2OCH_2CH_2OH$.
Properties: Colorless liquid, refr index 1.4264 (27C), d 1.0211 (20/4C), bp 194C, soluble in water, flash p 200F (93.3C), wt/gal 8.51 lb (20C). Combustible.
Use: Solvent, brake fluid component, intermediate.

diethylene glycol monomethyl ether acetate. (methyl "Carbitol" acetate).   CAS: 629-38-9.
$CH_3COOC_2H_4OC_2H_4OCH_3$.
Properties: Colorless liquid, bp 209.1C, flash p 180F (82.2C) (open cup), d 1.04 (20/20C), vap press 0.12 mm (20C). Combustible.
Use: Solvent.

diethylene glycol monooleate.   See diglycol oleate.

diethylene glycol monoricinoleate.   See diglycol ricinoleate.

diethylene glycol monostearate.   See diglycol monostearate.

diethylene glycol phthalate.   See diglycol phthalate.

diethylene glycol stearate.   See diglycol stearate.

diethylene oxide.   See 1,4-dioxane.

diethylenetriamine.   CAS: 111-40-0.
$NH_2C_2H_4NHC_2H_4NH_2$.
Properties: Yellow liquid, ammoniacal odor, strongly alkaline, hygroscopic, somewhat viscous, soluble in water and hydrocarbons, corrosive to copper and its alloys, bp 206.7C, fp −39C, d 0.9542 (20/20C), vap press 0.37 mm (20C), flash p 215F (101.6C), wt/gal 7.9 lb (20C), viscosity 0.0714 poise (20C), coefficient of expansion 0.00088.
Grade: Technical.
Hazard: Toxic by ingestion, inhalation, and skin absorption. Strong irritant to eyes and skin. TLV: 1 ppm in air.
Use: Solvent for sulfur, acid gases, various resins, dyes; saponification agent for acidic materials; fuel component.
See hydyne.

diethylenetriamine pentaacetic acid.
CAS: 67-43-6.
$HOOCCH_2N[CH_2CH_2N(CH_2COOH)_2]_2$.
Properties: White, crystalline solid; mp 230C (decomposes); slightly soluble in cold water; soluble in hot water.
Grade: Technical.
Use: Chelating agent.

N,N-diethylethanolamine.   See diethylaminoethanol.

diethyl ether.   See ethyl ether.

diethyl ethoxymethylenemalonate.
$C_2H_5OCH:C(COOC_2H_5)_2$.
Properties: Liquid, d 1.0855 (15/15C), refr index 1.4625 (20C), bp 279–281C (decomposes) flash p 190F (87.7C), insoluble in water.
Grade: 98% (min purity). Combustible.
Use: Synthesis.

uns-diethylethylene.   See 2-ethyl-1-butene.

N,N-diethylethylenediamine.   CAS: 100-36-7.
$(C_2H_5)_2NC_2H_4NH_2$.
Properties: Colorless liquid, bp 145.2C, sets to a glass below −100C, d 0.8211 (20/20C), wt/gal 6.8 lb (20C), flash p 115F (46.1C) (OC), miscible with water. Combustible.
Hazard: Moderate fire risk. Moderately toxic.
Use: Intermediate.

p,p'-(1,2-diethylethylene)diphenol.   See hexestrol.

diethyl ethylmalonate. $C_2H_5CH(COOC_2H_5)_2$.
Properties: Colorless liquid, ester odor, d 0.9994 (25/25C). Combustible.
Use: Intermediate.

diethyl ethylphosphonate. $C_2H_5P(O)(OC_2H_5)_2$.
Properties: Colorless liquid with mild odor, stable, miscible with most common organic solvents, slightly soluble in water, soluble in alcohol, d 1.025 (20/4C), bp 82–83C (11 mm), flash p 220F (104.4C) (COC). Combustible.
Use: Heavy metal extraction and solvent separation, gasoline additives, antifoam agent, plasticizer, textile conditioner and antistatic agent.

1,1'-diethyl ferrocenoate.   (1,1'-ferrocene dicarboxylic acid diethyl ester).
$(C_2H_5OOCC_5H_4)_2Fe$.
Orange crystals, mp 38–40C.
Use: Intermediate, high-temperature plasticizer.

diethylgermanium dichloride. $(C_2H_5)_2GeCl_2$.
Properties: Colorless liquid, fp −38C, bp 175C, decomposed by water.
Use: Biocide, intermediate.

diethylglycol phthalate.   (diethoxyethyl phthalate). $(C_2H_5OCH_2CH_2OOC)_2C_6H_4$.
Properties: Water-white to pale straw liquid, d 1.115–1.120 (20/20C), wt/gal 9.31 lb, flash p 343F (172C). Combustible.
Use: Plasticizer.

di(2-ethylhexyl) adipate.   (DOA; dioctyl adipate).
CAS: 103-23-1.
$[CH_2CH_2COOCH_2CH(C_2H_5)C_4H_9]_2$.
Properties: Light-colored, oily liquid; d 0.9268 (20/20C); refr index 1.4472; flash p 385F (196C); pour point −75C; bp 417C (214 Cat 5 mm); vap press 2.60 mm (200C); insoluble in water; viscosity 13.7 cp (20C); 7.7 lb/gal (20C). Combustible.
Assay: 99% min.
Use: Plasticizer, commonly blended with general purpose plasticizers, such as DOP and DIOP in processing polyvinyl and other polymers, solvent, aircraft lubes.

di(2-ethylhexyl)amine.   (dioctylamine).
CAS: 20830-75-5.
$[C_4H_9CH(C_2H_5)CH_2]_2NH$.
Properties: Water-white liquid with slightly ammoniacal odor, d 0.8062 (20/20C), 6.7 lb/gal (20C), bp 281.1C, vap press 0.01 mm (20C), viscosity 3.70 cp (20C), flash p 270F (132C), high solubility in hydrocarbons and low solubility in water, refr index 1.4420 (20C). Combustible.
Hazard: Moderately toxic.
Use: Synthesis of dyestuffs, insecticides, emulsifying agents, etc.

di(2-ethylhexyl)aminoethanol.   See di(2-ethylhexyl)ethanolamine.

di(2-ethylhexyl) azelate.   (DOZ; dioctyl azelate).
$(CH_2)_7[COOCH_2CH(C_2H_5)C_4H_9]_2$.
Properties: Colorless, odorless liquid; d 0.919 (20/20C); refr index 1.4472; bp 376C; flash p 430F (221C). Combustible.
Grade: 99% pure.
Use: Plasticizer for vinyls, especially used as low-temperature plasticizer, base for synthetic lubricants.

di(2-ethylhexyl)ethanolamine.   (di(2-ethylhexyl)aminoethanol; dioctylaminoethanol).
$[C_4H_9CH(C_2H_5)CH_2]_2N(CH_2)_2OH$.
Properties: Colorless liquid, insoluble in water, wt/gal 7.2 lb, flash p 280F (137C). Combustible.
Use: Emulsifier, acid-stable wetting agent.

di(2-ethylhexyl) ether.   CAS: 10143-60-9.
$[C_4H_9CH(C_2H_5)CH_2]_2O$.
Properties: Colorless, stable liquid; mild odor; almost insoluble in water; d 0.8121 (20/20C); 6.6 lb/gal (20C); bp 269.4C; vap press less than 0.01 mm (20C); sets to glass below −95C; viscosity 2.89 cp (20C); refr index 1.4525 (20C). Combustible.
Use: High-boiling, inert reaction medium, component of certain foam breakers.

di(2-ethylhexyl)2-ethylhexylphosphonate.   (bis(2-ethylhexyl)2-ethylhexylphosphonate).
$C_8H_{17}PO(OC_8H_{17})_2$.
Properties: Colorless liquid with a mold odor, d 0.908 (20/4C), bp 160–161C (0.26 mm), flash p 420F (215C), insoluble in water, miscible with most common organic solvents. Combustible.

Use: Heavy-metal extraction, solvent separation, gasoline additive, anti-foam agent, plasticizer, stabilizer, textile conditioner and antistatic agent.

di(2-ethylhexyl)fumarate.   (dioctyl fumarate; DOF).   CAS: 141-02-6.
$C_8H_{17}OOCCH:CHCOOC_8H_{17}$.
Properties: Clear, mobile liquid; d 0.937–0.940 (25/25C); bp 211–220C; flash p 365F (185C). Combustible.
Use: Monomer for polymerization and copolymerization.

di(2-ethylhexyl) hexahydrophthalate.   (dioctyl hexahydrophthalate).
$C_6H_{10}[COOCH_2CH(C_2H_5)C_4H_9]_2$.
Properties: Light-colored liquid, d 0.9586 (20/20C), 8.0 lb/gal (20C), bp 216C (5 mm), vap press 2.2 mm (200C), insoluble in water, viscosity 42.1 cp (20C), flash p 425F (218C). Combustible.
Use: Plasticizer.

di(2-ethylhexyl) hydrogen phosphate.   (bis(2-ethylhexyl)hydrogen phosphate).
$(C_8H_{17})_2HPO_4$.
Properties: Solid, d 0.972 (20/4C), flash p (COC) 340F (171C); insoluble in water.
Use: Heavy-metal extraction.

di(2-ethylhexyl) isophthalate.   (dioctyl isophthalate). $C_6H_4[COOCH_2CH(C_2H_5)C_4H_9]_2$.
Properties: Colorless liquid, bp 258C at 10 mm, d 0.984 (20/20C), 8.2 lb/gal, pour point −46C, insoluble in water, viscosity 86.5 cp (20C). Combustible.
Use: Plasticizer.

di(2-ethylhexyl) maleate.   (dioctyl maleate; DOM).   CAS: 142-16-5.
$C_8H_{17}OOCCH:CHCOOC_8H_{17}$.
Properties: Liquid, bp 209C (10 mm), freezing point sets to glass below −60C, d 0.9436 (20/20C), wt/gal 7.9 lb (20C), flash p 365F (185C) (OC), insoluble in water. Combustible.
Use: Copolymers, intermediate.

di(2-ethylhexyl) phosphite.   (bis(2-ethylhexyl) phosphite). $(C_8H_{17}O)_2PHO$.
Properties: Mobile, colorless liquid with mild odor and a high degree of thermal stability; insoluble in water (hydrolyzes very slowly); miscible with most common organic solvents; d 0.937 (20/4C); bp 163–164C (3 mm); refr index 1.444 (25C); flash p 330F (165C). Combustible.
Use: Lubricant additive, intermediate, adhesive.

di(2-ethylhexyl) phosphoric acid.   (di-n-octyl phosphoric acid; di[2-ethylhexyl]hydrogen phosphate).   CAS: 298-07-7.

$[C_4H_9CH(C_2H_5)CH_2]_2HPO_4$.
Properties: Strongly acidic liquid, d 0.973 (25/25C), fp −60C, refr index 1.4420 (25C), flash p 385F (196C), wt/gal 8.2 lb, insoluble in water, soluble in organic solvents. Combustible.
Use: Metal extraction and separation, intermediate for wetting agents and detergents.

di(2-ethylhexyl) phthalate.   (di-sec-octyl phthalate; dioctyl phthalate; DOP).
CAS: 117-81-7.
$C_6H_4[COOCH_2CH(C_2H_5)C_4H_9]_2$.
Properties: Light-colored, odorless liquid; d 0.9861 (20/20C); pour point −46C; refr index 1.4836; flash p 425F (218C); 8.20 lb/gal (20C); bp 231C (5 mm) vap press 1.32 mm (200C); viscosity 81.4 cp (20C); insoluble in water; miscible with mineral oil. Combustible.
Derivation: Reaction of 2-ethylhexanol and phthalic anhydride.
Hazard: TLV: 5 mg/m³.
Use: Plasticizer for many resins and elastomers.

di(2-ethylhexyl) sebacate.   (dioctyl sebacate).
CAS: 122-62-3. $C_4H_8COOC_8H_{17})_2$.
Properties: Pale straw-colored liquid, d 0.91 (25C), refr index 1.447 (28C), bp 248C (4 mm), fp −55C, flash p 410F (210C) (COC). Insoluble in water; partially compatible with cellulose acetate and cellulose acetate butyrate; compatible with ethyl cellulose, polystyrene, polyethylene, vinyl chloride, and vinyl chloride acetate. Combustible.
Use: Plasticizer.

di(2-ethylhexyl) sodium sulfosuccinate.   See dioctyl sodium sulfosuccinate.

di(2-ethylhexyl) succinate.   (dioctyl succinate).
$C_8H_{17}OCOCH_2CH_2COOC_8H_{17}$.
Properties: Liquid, bp 257C (50 mm), fp sets to glass below −60C, d 0.9346 (20/20C), wt/gal 7.8lb (20C), flash p 315F (157C) (OC), vap press less than 0.01 mm (20C), solubility in water less than 0.01% by wt (20C). Combustible.
Use: Plasticizer, intermediate.

diethylhydroxylamine.   CAS: 3710-84-7.
$(C_2H_5)_2NOH$.
Properties: Liquid, refr index 1.4238 (20C). Combustible.
Grade: 85%.
Use: Photographic developer, antioxidant, corrosion inhibitor.

diethyl isoamylethylmalonate.
$(C_2H_5)(C_5H_{11})C(COOC_2H_5)_2$.
Properties: Colorless liquid, sweet odor, d 0.950 (25/25C). Combustible;
Use: Intermediate.

diethyl ketone. (metacetone; propione; 3-penta-none; ethyl propionyl). CAS: 96-22-0. $C_2H_5COC_2H_5$.

Properties: Colorless, mobile liquid; acetone-like odor; soluble in alcohol and ether; slightly soluble in water; autoign temperature 846F (452C); bp 101C; d 0.816; fp −42C; flash p 55F (12.78C) (OC).

Derivation: By distilling sugar with an excess of lime.

Method of purification: Rectification.

Grade: Technical.

Hazard: Flammable, dangerous fire hazard. TLV: 200 ppm.

Use: Medicine, organic synthesis.

diethyl maleate. CAS: 141-05-9. $C_2H_5OOCCH:CHCOOC_2H_5$.

Properties: Water-white liquid, d 1.0687, 8.92 lb/gal (20C), refr index 1.4400 (20C), bp 225C, fp approximately −115C, viscosity 3.567 cp (20C), flash p 200F (93.3C) (COC), dielectric constant 2.18 (calc) (25C), surface tension 3.70 dynes/cm (20C). Readily soluble in alcohol, diethyl ether, paraffinic hydrocarbons and common organic solvents; soluble in water, readily hydrolyzed by alkaline solutions. Combustible.

Derivation: Reaction of maleic anhydride with ethanol in the presence of a catalyst.

Hazard: Irritant to eyes and skin.

Use: Organic synthesis, flavoring.

diethyl malonate. See ethyl malonate.

diethylmalonylurea. See barbital.

diethyl(1-methylbutyl)malonate.
$[C_3H_7CH(CH_3)]CH(COOC_2H_5)_2$.
Colorless liquid, ester odor, d 0.969 (25/25C).
Use: Intermediate, organic synthesis.

diethylmethylmethane. See 3-methylpentane.

O,O-diethyl-O-p-nitrophenyl phosphorothioate. See parathion.

diethyl oxalate. See ethyl oxalate.

diethyl oxide. See ethyl ether.

di(p-ethylphenyl)dichloroethane. See 1,1-dichloro-2,2-bis(p-ethylphenyl)ethane.

N,N-diethyl-p-phenylenediamine. See p-amino-diethylaniline.

diethyl phosphite. CAS: 762-04-9.
$(C_2H_5O)_2HPO$.
Properties: Water white liquid, bp 138C, d 1.069

(25C), refr index 1.4061 (25C), flash p 195F (90.5C) (COC), soluble in water, common organic solvents. Combustible.

Use: Paint solvent, lubricant additive, antioxidant, reducing agent, intermediate for flame retardants and insecticides, phosphorylating agent.

O,O-diethyl phosphorochloridothioate. (ethyl PCT). CAS: 2524-04-1. $(C_2H_5O)P(S)Cl$.

Properties: Colorless to light amber liquid, d 1.196 (25/25C), fp less than −75C, bp 49C (below 1 mm), refr index 1.4705 (25C), insoluble in water, soluble in most organic solvents, stable at room temperature, slowly isomerizes at 100C.

Hazard: Cholinesterase inhibitor, irritant to eyes and lungs.

Use: Intermediate for pesticides, oil and gasoline additives, flame retardants, flotation agents.

diethyl phthalate. (ethyl phthalate; DEP). CAS: 84-66-2. $C_6H_4(CO_2C_2H_5)_2$.

Properties: Water-white, stable, odorless, liquid; bitter taste; fp −40.5C; refr index 1.5002 (25C); surface tension 37.5 dynes/cm (20C); viscosity 31.3 centistokes (0C); vap press 14 mm (163C); 30 mm (182C); 734 mm (295C); bp 298C; flash p 325F (162.7C) (open cup); wt/gal approximately 9.31 lb (20C); d 1.120 (25/25C); miscible with alcohols, ketones, esters, aromatic hydrocarbons; partly miscible with aliphatic solvents; insoluble in water. Combustible.

Derivation: By reacting phthalic anhydride with ethanol followed by careful purification.

Grade: Technical.

Hazard: Toxic by ingestion and inhalation, strong irritant to eyes and mucous membranes, narcotic. TLV: 5 mg/m³ of air.

Use: Solvent for nitrocellulose and cellulose acetate, plasticizer, wetting agent, insecticidal sprays, camphor substitute, plastics, perfumery as fixative and solvent, alcohol denaturant, mosquito repellents, plasticizer in solid rocket propellants.

2,2-diethyl-1,3-propanediol.
$HOCH_2C(C_2H_5)_2CH_2OH$.
Properties: Colorless liquid, mp 61.3C, bp 160C (50 mm), d 0.949 (at melting point), wt/gal 8.2 lb (60C), flash p 215C (101.6C) (OC), soluble in water. Combustible.

Grade: Technical, pharmaceutical.

Hazard: Toxic by ingestion.

Use: Emulsifying agent, intermediate, medicine.

O,O-diethyl-O-2-pyrazinyl phosphorothioate. See thionazin.

diethylpyrocarbonate. (DEPC).

$$C_2H_5O-\overset{\overset{\displaystyle O}{\|}}{C}-O-\overset{\overset{\displaystyle O}{\|}}{C}-OC_2H_5$$

Properties: Colorless liquid, sweet ester-like odor, miscible with ethanol and methanol, refr index 1.395–1.398 (25C).
Grade: FCC.
Hazard: Toxic. Use in food products prohibited (FDA), irritant to eyes and skin.
Use: Fermentation inhibitor.

diethylstilbestrol. (stilbestrol; DES; 3,4-bis(p-hydroxyphenyl)-3-hexene).   CAS: 56-53-1.
A non-steroid, synthetic estrogen, always in the trans form. It is the most active of the commonly used stilbene compounds.
Properties: White, odorless, crystalline powder; mp 169–172C; almost insoluble in water; soluble in alcohol, chloroform, ether, fatty oils, and dilute alkali hydroxide.
Derivation: From anethole hydrobromide, from anisole, from anisoin.
Grade: USP.
Hazard: A carcinogen. Under USDA regulations, no residues are permitted in tissues of slaughtered animals; not permitted in cattle feeds (FDA).
Use: Biochemical research, medicine (prevents spontaneous abortion), veterinary medicine.

diethyl succinate. CAS: 123-25-1.
$(-CH_2COOC_2H_5)_2$.
Properties: Colorless liquid with faint pleasant odor, bp 216.2C, fp −21C, d 1.0418 (20/20C), wt/gal 8.7 lb (20C), refr index 1.4201 (20C), flash p 230F (110C) (OC), miscible with alcohol and ether, slightly soluble in water, Combustible.
Use: Plasticizer, intermediate, flavoring.

diethyl sulfate. (ethyl sulfate).   CAS: 64-67-5.
$(C_2H_5)_2SO_4$.
Properties: Colorless liquid, faint ethereal odor, irritating after-effects, noncorrosive, soluble in alcohol and ether, insoluble in water, d 1.1803,

bp 208C (decomposes) vap press 0.19 mm (20C), flash p 220F (104.4C), autoign temperature 817F (436C), wt/gal 9.8 lb (20C), fp −24.4C, viscosity 1.79 cp (20C). Combustible.
Derivation: Action of fuming sulfuric acid on ethanol.
Method of purification: Rectification in vacuo.
Grade: Technical.
Hazard: Toxic by ingestion and inhalation, strong irritant.
Use: Ethylating agent in organic synthesis.

diethyl sulfide. See ethyl sulfide.

diethyl tartrate. $C_4H_4O_6(C_2H_5)_2$.
Properties: Colorless, thick, oily liquid; bp 280C; mp 17C; soluble in water and alcohol; d 1.204 (20/4C). Combustible.
Use: Plasticizer for automobile lacquers; solvent for nitrocellulose, gums, and resins.

1,3-diethylthiourea. CAS: 105-55-5.
$C_2H_5NHCSNHC_2H_5$.
Properties: Buff solid; mp 68–71C; slightly soluble in water; soluble in methanol, ether, acetone, benzene, and ethyl acetate; insoluble in gasoline.
Use: Inhibitor of corrosion in metal pickling solutions, accelerator, activator in elastomers.

N,N-diethyl-m-toluamide. (deet).
CAS: 134-62-3. $CH_3C_6H_4CON(C_2H_5)_2$.
Properties: Colorless liquid, mild bland odor, bp 160C (19 mm), d 0.996–1.002 (25/25C), refr index 1.5200–1.5235 (25C). Soluble in water, alcohol, ether, and benzene. Combustible.
Grade: USP.
Hazard: Toxic by ingestion, irritant to eyes and mucous membranes.
Use: Insect repellents, resin solvent, film formers.

N,N-diethyl-m-toluidine. $CH_3C_6H_4N(C_2H_5)_2$.
Properties: Light amber oil, bp 231C, refr index (20C) 1.5361.
Hazard: See toluidine.
Use: Dye intermediate.

N,N-diethyl-o-toluidine. (1-diethylamino-2-methyl-benzene). $CH_3C_6H_4N(C_2H_5)_2$.
Properties: Prisms from water; mp 72.3C; bp 209C; soluble in water, alcohol, and ether.
Derivation: From o-toluidine.
Hazard: See toluidine.
Use: Dye intermediate.

O,O-diethyl-O-3,5,6-trichloro-2-pyridyl phosphorothioate. $Cl_3C_5NHOP(S)(OC_2H_5)_2$.
Properties: Solid; mp 41.5–43C; soluble in acetone, benzene, ether; almost insoluble in water.

Hazard: Cholinesterase inhibitor.
Use: Insecticide.

3,9-diethyl-6-tridecanol. (heptadecanol).
$C_4H_9CH(C_2H_5)C_2H_4CH(OH)C_2H_4CH(C_2H_5)_2$.
Properties: White solid, d 0.8475, bp 309C, flash p 310F (154C) (open cup), refr index 1.4531 (20C), insoluble in water. Combustible.
Use: Intermediate for synthetic lubricants, defoamers and surfactants.

1,1-diethylurea. $NH_2CON(C_2H_5)_2$. White solid; mp 75C; soluble in water, alcohol, and benzene. When copolymerized with simple urea by the use of formaldehyde, it yields modified resins that differ in nature from those made from monosubstituted ureas. These resins tend to be permanently thermoplastic.

diethylzinc. (ethylzinc; zinc ethyl; zinc diethyl). CAS: 557-20-0. $Zn(C_2H_5)_2$. The first known organometallic compound.
Properties: Colorless, pyrophoric liquid; d 1.207 (20C); fp −28C; bp 118C; soluble in most hydrocarbons.
Derivation: Action of ethyl iodide on zinc and sodium-zinc, or by interaction of zinc chloride with triethyl aluminum.
Grade: Technical.
Hazard: Ignites spontaneously on contact with air, dangerous fire hazard, decomposes violently in water.
Use: Organic synthesis, catalyst for polymerization of olefins, high energy aircraft and missile fuel, production of ethyl mercuric chloride.

differential gravimetric analysis. (DGA). A variation of differential thermal analysis in which additional information is obtained by determining the rate of change in weight during the heating process.

differential thermal analysis. (DTA). The method of precisely measuring the temperature and the rate of temperature change as heat is added to or abstracted from a sample of material that is in a controlled constant environment. The method determines whether the sample is a pure substance or a mix and yields information about its composition and thermal properties.

diffraction, neutron. An analytical technique analogous to x-ray diffraction in which an incident beam of neutrons is scattered by the atoms of a crystal. Because elements that are close to each other in the Periodic Table differ considerably in their neutron-scattering ability, neutron diffraction is capable of distinguishing between them. For example, carbon, nitrogen, and oxygen atoms can be readily identified by neutron diffraction whereas they appear almost identical by the x-ray method. More accurate determination of the bond lengths of light atoms and distribution of molecular bonding electrons is also possible. Details of molecular structure that can only be inferred by other techniques can often be observed directly by neutron diffraction. Investigations using this method include hydrogen bonding, so-called metal cluster compounds (C—H—M relationships) and electronic charge distributions.

diffraction, x-ray. A method of spectroscopic analysis involving the reflection or scattering of x-radiation by the atoms of a substance (lattice) as the rays pass through it. The rays are reflected by the atoms at an angle that is characteristic of the substance, yielding a spectrum that indicates its atomic or molecular structure. The spectra thus obtained are well-defined and specific; from them the properties of elements and the structure of both crystalline and amorphous materials can be obtained. For example, unvulcanized rubber gives an amorphous pattern while vulcanized rubber is crystalline; the cellulose macromolecule has alternating crystalline and amorphous areas. X-ray diffraction was one of the earliest and most successful methods of instrumental analysis; developed by Bragg and van Laue early in this century it was used with dramatic effect by Moseley (1912) in establishing the location of several elements in the Periodic System.
See also lattice, crystal, x-radiation.

diffusion. The spontaneous mixing of one substance with another when in contact or separated by a permeable membrane or microporous barrier. The rate of diffusion is proportional to the concentration of the substances and increases with temperature. Diffusion occurs most readily in gases, less so in liquids, and least in solids. The theoretical principles are stated in Fick's laws. In gases diffusion takes place counter to gravity and the rate at which different gases diffuse into a particular gas, (e.g., air) is inversely proportional to the square roots of the densities. Carbon dioxide and chlorine vapor will diffuse in air until a uniform mixture results. Diffusion occurs in the cell walls of plants and animals (see osmosis). Many substances diffuse through a parchment membrane.
See also dialysis; diffusion, gaseous.

diffusion, gaseous. A technique used for separating the light isotope of uranium (U-235) from the heavy isotope (U-238). The uranium is allowed to diffuse through a series of microporous barriers whose apertures are of molecular dimensions in the form of the gas uranium hexafluoride; this is a mix of $U^{238}F_6$ and $U^{235}F_6$ in a ratio of 140:1. Because of the vastly greater number of the heavier molecules and the extremely small difference in their masses, the mix must pass through a barrier a great many times to obtain a high concentration of the 235 isotope. Assuming that the diffusion rate of two gases through a porous barrier is inversely proportional to the square root of their molecular weights, the ideal separation factor is the square root of the product of M_1 times M_2, where M_2 is the molecular weight of $U^{238}F_6$ and M_1 that of $U^{235}F_6$. This method is still in use for uranium enrichment for nuclear fuel.

diffusion length. A property of materials used in reactors for moderators or reflectors. It is a measure of the distance a thermal neutron diffuses after it is thermalized until it is captured. It is related to the density of the material and to the scattering and absorption cross sections.

diflubenzuron. (N-((4-chlorophenyl)aminocarbonyl)-2,6-difluorobenzamide; 1-(4-chlorophenyl)-3-(2,6-difluorobenzoyl)urea).
CAS: 35367-38-5.
Use: Insecticide.

difluophosphoric acid. See difluorophosphoric acid.

2,4-difluoroaniline. $C_6H_5F_2N$.
Properties: Liquid, bp 170C (753 mm), fp −7.5C, density 10.7lb/gal, flash p 158F (70C).
Use: Organic synthesis.

1,1,1-difluorochlorethane. See 1,1,1-chlorodifluoroethane.

difluorochloromethane. See chlorodifluoromethane.

difluorodiazine. FN:NF.
Properties: Gas, can exist as cis and trans isomers.
Grade: All trans isomer, 95–99.8%.
Hazard: See fluorine.
Use: Trans form--preparation of ionic fluorine compounds; cis form--polymerization initiator.

difluorodibromomethane. See dibromodifluoromethane.

difluorodichloromethane. See dichlorodifluoromethane.

difluorodiphenyltrichloroethane. See DFDT.

1,1-difluoroethane. (ethylidene fluoride).
CAS: 75-37-6. CH_3CHF_2.
Properties: Colorless, odorless gas; bp −24.7C; fp −117C; d 1.004 (−25C); refr index 1.255 (20C); insoluble in water.
Derivation: By adding hydrogen fluoride to acetylene.
Grade: Technical, 98%.
Hazard: Flammable, dangerous fire risk. Flammable limits in air 3.7–18%. Narcotic in high concentration.
Use: Intermediate.

1,1-difluorethylene. (Air)Legal label name for vinylidene fluoride.

difluoromethane. CH_2F_2.
Properties: Gas, bp −51.6C, soluble in alcohol, insoluble in water, high thermal stability.
Use: Refrigeration, organic synthesis.

difluoromonochloroethane. Legal label name for 1,1,1-chlorodifluoroethane.

difluoromonochloromethane. See chlorodifluoromethane.

difluorophosphoric acid. (difluophosphoric acid).
HPO_2F_2.
Properties: Mobile, strongly fuming, colorless liquid; d 1.583 (25/4C); fp −75C; bp 116C; corrosive to glass and fabric. Noncombustible.
Hazard: When heated, corrosive to tissue.
Use: Chemical polishing agent, protective coatings for metal surfaces, catalyst.

difluron. CAS: 35367-38-5. $C_{14}H_9ClF_2N_2O_2$.
Properties: Solid, mp 239C, sparingly soluble in water.
Hazard: Toxic by ingestion.
Use: Larvicide.

difolaton. See: cis-N(1,1,2,2-tetrachloroethyl)-thio]-4-cyclohexene-1,2-dicarboximide.

"Digesta". TM for a series of fine calcium phosphates with varying proportions of calcium and phosphate.
Grade: FCC.
Use: Feed supplement for livestock and poultry.

digester. A cylindrical metal vessel used chiefly in the preparation of wood pulp for papermaking in which lignin is separated from cellulose by chemical means. It operates at approximately 150 psi and 170C. The wood is fed to the digester in the form of chips to which cooking liquor is added. Standard digesters are 12 ft in diameter and 45 ft high with a wall thickness of 2 inches

These hold about 20 cords of wood and some are even larger. The cooking cycle varies from 2.5 hr for board stock to 5 hr for bleached paper. Heat supply is by circulating steam and heat exchanger though some types have direct steam injection. Digesters are designed for both batch and continuous operation. They are also used in reclaiming fabricated rubber products (tires, boot and shoe, etc.).
See also pulp, paper; digestion (2).

digestion. (1) The physiological processes involved in the assimilation of nutrients from ingested foods by the animal organism. Hydrochloric acid in the gastric juice plays a prominent part aided by the saliva which initiates carbohydrate breakdown and by bile and pancreatic secretions in the intestine. Numerous types of enzymes catalyze these processes. (2) In chemical engineering the term refers to several processes: (a) the preferential dissolution of certain mineral constituents in some ore concentrates, (b) the liquefaction of organic waste materials by microbiological action as in activated sludge, (c) removal of lignin from wood by hot chemical solutions in the manufacture of chemical cellulose and paper pulp, (d) separation of fabric from scrap tires by hot sodium hydroxide solution in the reclaiming of rubber. The equipment for (c) and (d) is called a digester.
See also metabolism, nutrition.

digitalis. A drug obtained from dried leaves of the purple foxglove, native to southern Europe but grown in the US. Used in treatment of cardiac diseases both human and animal. Allergic reactions are infrequent. Contains both digitonin and digitoxin.
Hazard: Be aware that an overdose can be fatal.

digitonin. CAS: 11024-24-1. $C_{56}H_{92}O_{29}$.
A saponin derived from digitalis seeds.
Use: Determination of cholesterol (an insoluble addition compound is formed), analytical reagent.

digitoxin. CAS: 71-63-6. $C_{41}H_{64}O_{13}$.
Most active glycoside of Digitalis purpurea.
Properties: White, odorless, bitter leaflets or powder; mp 255–256C; slightly soluble in water or ether; soluble in alcohol.
Derivation: From digitalis leaves, usually digitalis purpurea (foxglove).
Grade: USP.
Hazard: Toxic by ingestion, overdose can be fatal.
Use: Medicine (cardiac treatment).

diglycerol. See polyglycerols.

diglycidyl ether. (DGE; di(2,3-epoxypropyl) ether). CAS: 2238-07-5. $C_6H_{10}O_3$.
Properties: A colorless liquid with a strong, irritating odor. Mw 130.14, d 1.262 at 25C, bp 260C, vap d 3.78 at 25C, vap p 0.09 Torr.
Hazard: TLV 0.1 ppm. Severe eye respiratory tract irritant.
Use: Intermediate.

1,3-diglycidyloxybenzene. See resorcinol diglycidyl ether.

diglycol. See diethylene glycol.

diglycol acetate. See diethylene glycol diacetate.

diglycol carbamate. (diethylene glycol dicarbamate). $O(CH_2CH_2OCONH_2)_2$.
Properties: White, crystalline solid; relatively stable to acid hydrolysis, but less stable to basic conditions.
Use: Manufacture of resins.

diglycol chloroformate. [diethylene glycol bis-(chloroformate)]. $O(CH_2CH_2OCOCl)_2$.
Properties: Liquid, bp 125–127C (5 mm), flash p 295F (146C). Combustible. Soluble in acetone, alcohol, ether, chloroform, and benzene.
Use: Preparation of nonvolatile plasticizers or modifying agent.

diglycol chlorohydrin. CAS: 628-89-7. $ClCH_2CH_2OCH_2CH_2OH$.
Properties: Colorless liquid, miscible with water, d 1.1698, bp 196.8C, flash p 225F (107C), vap press 0.17 mm. Combustible.
Hazard: Toxic by ingestion and inhalation, an irritant.

diglycolic acid. $(O(CH_2COOH)_2$.
Properties: White, crystalline solid; mp 148C; soluble in water and alcohol; pH of 10% aqueous solution 1.4. Forms a nonhygroscopic monohydrate at relative humidities above 72% at 25C.
Use: Manufacture of resins and placticizers, organic synthesis, sequestering agent, emulsion breaker in petroleum.

diglycol laurate. (diethylene glycol monolaurate). $C_{11}H_{23}COOC_2H_4OC_2H_4OH$.
Properties: Light straw-colored, oily liquid; practically odorless and edible; d 0.96. Insoluble in water; soluble in methanol, ethanol, toluene, naphtha and mineral oil; miscible in certain proportions in cottonseed oil, acetone, and ethyl acetate. Flash p 290F (143C). Combustible.
Derivation: Lauric acid ester of diethylene glycol.
Use: Emulsifying agent for oils and hydrocarbon solvents; emulsions for lubrication, sizing, finish-

ing, of textiles, paper, and leather; fluid emulsions of oils for hand lotions, hair dressings etc.; cutting and spraying oils; dry-cleaning soap base; antifoaming agent.

diglycol methyl ether. See diethylene glycol dimethyl ether.

diglycol monostearate. (diethylene glycol monostearate). CAS: 106-11-6. $C_{17}H_{35}COOC_2H_4OC_2H_4OH$.
Properties: Small, white flakes available in regular or water-dispersible types.
Use: Emulsifier and thickener in cosmetics, mold release lubricant for die casting, temporary binder for ceramics and grinding wheels.

diglycol nitrate. See diethylene glycol dinitrate.

diglycol oleate. (diethylene glycol monooleate). $C_{17}H_{33}COOC_2H_4OC_2H_4OH$.
Properties: Light red, oily liquid; fatty odor. Soluble in ethanol, naphtha, ethyl acetate, methanol; partly soluble in cottonseed oil; insoluble in water. D 0.93, iodine value 65–75, titer below 0C, pH (25C) 7.7–8.2 (5% aqueous dispersion). Combustible.
Derivation: Oleic acid ester of diethylene glycol.
Use: Emulsifying agent for fluid water-in-oil emulsions for the manufacture of furniture and automobile polish, water-emulsion paints, agricultural sprays.

diglycol phthalate. (diethylene glycol phthalate). $C_6H_4(COOC_2H_4OH)_2$.
Properties: Pale yellow, liquid resin. Soluble in methanol, ethanol, acetone, ethyl acetate; partly soluble in toluene, naphtha, mineral oil, cottonseed oil; insoluble in water. D 1.29, saponification value 430–450, acid value 170–175. Combustible.
Use: Plasticizer, emulsifier.

diglycol ricinoleate. (diethylene glycol monoricinoleate). $C_{17}H_{32}(OH)COOC_2H_4OC_2H_4OH$.
Properties: Light yellow liquid, fp less than −60C, d 0.980 (25C). Soluble in alcohol, acetone, and ethyl acetate; insoluble in water. Combustible.
Use: Plasticizer for high polymers and elastomers.

diglycol stearate. (diethylene glycol distearate). $C_{17}H_{35}COOC_2H_4)_2O$.
Properties: White, wax-like solid; faint fatty odor; disperses in hot water; soluble in hot alcohol, oils, and hydrocarbons; mp 54–55C; d 0.9333 (20/4C). Combustible.
Derivation: Stearic acid ester of diethylene glycol.
Grade: Technical, cosmetic.
Use: Emulsifying agent for oils, solvents, and waxes; lubricating agent for paper and cardboard; suspending medium for powders in the manufacture of polishes, cleaners, and textile delusterants; temporary binder for abrasive powders and clays for ceramic insulation; protective coating for hygroscopic powders; thickening agent; pharmaceuticals.

diglyme. See diethylene glycol dimethyl ether.

digoxin. CAS: 20830-75-5. $C_{41}H_{64}O_{14}$.
A cardiotonic digitalis glycoside.
Properties: Colorless to white crystals or white, crystalline powder; odorless; melts at approximately 235C (decomposes); insoluble in water, chloroform, and ether; soluble in pyridine and dilute alcohol.
Derivation: From the leaves of Digitalis lanata.
Grade: USP.
Hazard: See digitalis.
Use: Medicine (cardiac diseases).

diheptyl-p-phenylenediamine. See N,N'-bis(1,4-dimethylpentyl)-p-phenylenediamine.

dihexadecylamine. See dipalmitylamine.

dihexadecyl ether. See dicetyl ether.

dihexadecyl sulfide. See dicetyl sulfide.

dihexadecyl thioether. See dicetyl sulfide.

dihexyl. See n-dodecane.

di-n-hexyl adipate. $(—CH_2CH_2COOC_6H_{13})_2$.
Properties: Liquid, colorless, water-white to maximum 100 Pt-Co, d 0.939 (20C), refr index 1.438 (25C), surface tension 32.7 dynes/cm (20C), viscosity 8.8 cp (20C), bp 183–192C (4 mm) (midpoint 191C), water soluble 0.1% (25C), gasoline and oil solubility complete, flash p 325F (162.7C). Combustible.
Use: Low-temperature plasticizer for SBR elastomers.

di-n-hexylamine. $[CH_3(CH_2)_5]_2NH$.
Properties: Water-white liquid, bp 233–243C, d 0.788 (20/20C), refr index 1.434 (20C), flash p 220F (104.4C). Combustible.
Hazard: Toxic by ingestion and skin absorption.

di-n-hexyl maleate. CAS: 105-52-2. $C_6H_{13}OOCCH:CHCOOC_6H_{13}$.
Properties: Liquid, d 0.9602 (20/20C), bp 179C (10 mm), refr index 1.449 (20C), vap press <0.01 mm (20C), fp −70C, viscosity 10.2 cp (20C), soluble in water <0.01% by wt (20C). Combustible.
Use: Preparation of resins.

dihexyl phthalate.   See di(2-ethylbutyl) phthalate.

dihexyl sebacate.
(—CH$_2$CH$_2$CH$_2$CH$_2$COOC$_6$H$_{13}$)$_2$.
Properties: Light straw-colored liquid, bp 203C (4 mm), flash p 415F (212C). Combustible.
Derivation: By reacting dodecyl alcohol with sebacic acid.
Use: Plasticizer for vinyl resins.

dihydrazine sulfate.   (N$_2$H$_4$)$_2$·H$_2$SO$_4$.
Properties: White, crystalline flakes; mp approximately 104C; decomposes at approximately 180C; soluble in water; insoluble in most organic solvents.
Grade: 95% is available commercially.
Hazard: An irritant.
Use: Reducing agent.

dihydric.   Containing two hydroxyl groups connected to different carbon atoms, e.g., a dihydric alcohol. Dihydric alcohols are collectively called glycols (diols).
See also ethylene glycol.

dihydroabietyl alcohol.   (hydroabietyl alcohol).
C$_{19}$H$_{31}$CH$_2$OH.
Properties: Solid, d 1.007–1.008, refr index 1.5280, vap press 1.5 × 10^{-5} mm (25C), mp 32–33C, flash p 375F (190C), insoluble in water. Combustible.
Use: Plasticizer.

1,8-dihydroacenaphthylene.   See acenaphthene.

dihydrochalcone.   Any of a group of disaccharide sugars having a sweet taste, they are derived from such flavanones as naringin and hesperidin. Basic research on the conversion of these compounds into acceptable synthetic sweeteners has been in progress for some years.
See also flavanone.

dihydrocholesterol.   (β-cholestanol; 3-β-hydroxycholestane).   CAS: 360-68-9.   C$_{27}$H$_{47}$OH.
A sterol found in the feces.
Properties: White crystals, mp (monohydrate) 142C, optical rotation alpha (25C) + 23 degrees, soluble in fat solvents, insoluble in water.
Derivation: By a series of oxidation and reduction reactions from cholesterol.
Use: Biochemical research, pharmaceutical preparations.

dihydrocoumarin.   See benzodihydropyrone.

6,7-dihydrodipyrido(1,2-a:2′,1′-c)pyrazidinium salt.   See diquat.

dihydrofolic reductase.   An enzyme acting as an essential catalyst of DNA synthesis.

dihyrogen ferrous EDTA.   See ethylenediaminetetraacetic acid (note).

dihydro-2(3)-imidazolone.   See ethylene urea.

2,3-dihydroindene.   See indan.

9,20-dihydro-9-oxoanthracene.   See anthrone.

2,3-dihydropyran.   C$_5$H$_8$O.
Properties: Colorless, mobile liquid; ether-like odor; bp 84.3C; fp −70C; d 0.927 (20/4C); refr index 1.4180 (25C); flash p 0F (−17.7C); soluble in water and alcohol.
Hazard: Highly flammable, dangerous fire risk. Toxic by ingestion and inhalation.
Use: Intermediate.

1,2-dihydro-3,6-pyridazinedione.   See maleic hydrazide.

dihydrostreptomycin.   (DHS).   C$_{21}$H$_{41}$N$_7$O$_{12}$.
A derivative of streptomycin in which the carbonyl group of the streptose portion has been reduced by the addition of two hydrogen atoms. It has antibiotic properties similar to streptomycin, and is mainly used in the treatment of tuberculosis.

2,5-dihydrothiophene-1,1-dioxide.   (sulfolene).

O$_2$SCH$_2$CH:CHCH$_2$.
Properties: Solid; bp decomposes; d 1.314 (70C); flash p 235F (112C); partially soluble in water, acetone, and toluene. Combustible.

1,2-dihydro-2,2,4-trimethylquinoline.
("Agerite" Resin D).   C$_{12}$H$_{15}$N.
Properties: Brown pellets or flakes, d 1.03 (25C), melting range 75–100C (initial melt).
Use: Rubber antioxidant.

dihydroxyacetone.   (DHA; dihydroxypropanone).
HOCH$_2$COCH$_2$OH.
Properties: Colorless, crystalline solid; mp 80C; hygroscopic; soluble in water and alcohol.
Derivation: Action of sorbose bacterium on glycerol.
Use: Intermediate, emulsifier, humectant, plasticizers, fungicides, cosmetics (creates synthetic suntan).

2,4-dihydroxyacetophenone.   See 4-acetylresorcinol.

dihydroxyaluminum sodium carbonate.
Al(OH)$_2$OOCONa.   White powder.
Derivation: Reaction of aluminum isopropoxide and an aqueous soluble of sodium bicarbonate.
Use: Medicine (antacid).

1,3-dihydroxy-2-amino-4-octadecene.
See sphingosine.

1,8-dihydroxyanthranol. See anthralin.

1,2-dihydroxyanthraquinone. See alizarin.

1,4-dihydroxyanthraquinone. See quinizarin.

1,5-dihydroxyanthraquinone. (anthrarufin).
$C_{14}H_6O_2(OH)_2$.
Properties: Yellow crystals, soluble in alcohol, sparingly soluble in water, mp 280C.
Derivation: By heating anthraquinone with boric acid and sulfuric anhydride.
Use: Intermediate for alizarin and indanthrene dyes.

1,8-dihyroxyanthraquinone. (chrysazin; danthron). $C_{14}H_6O_2(OH)_2$.
Properties: Orange-colored powder or reddish-brown needles, mp 191C, soluble in alcohol, sparingly soluble in water.
Derivation: From 1,8-anthraquinone potassium disulfonate.
Grade: Technical, NF.
Use: Dyes, medicine.

m-dihydroxybenzene. See resorcinol.

o-dihyroxybenzene. See pyrocatechol.

p-dihydroxybenzene. See hydroquinone.

2,4-dihydroxybenzenecarboxylic acid. See β-resorcylic acid.

2,4-dihydroxybenzoic acid. See β-resorcylic acid.

2,5-dihydroxybenzoic acid. See gentisic acid.

3,5-dihydroxybenzoic acid. See α-resorcylic acid.

2,4-dihydroxybenzophenone.
$C_6H_5COC_6H_3(OH)_2$.
Properties: Light-yellow, crystalline solid; mp 142C; bp 194C (1 mm); density 5.8 lb/gal (20C); insoluble in water; soluble in ethanol, methanol, methyl ethyl ketone, and ethyl acetate.
Use: UV absorber in polymers.

2,5-dihydroxybenzoquinone. $C_6H_2(OH)_2O_2$.
Properties: Yellow-orange solid; mp 216C (decomposes); soluble in concentrated sulfuric acid; slightly soluble in ethanol, acetone, water, benzene.
Derivation: From hydroquinone.
Hazard: Irritant to eyes and skin.

Use: Metal chelating, insecticides, polymerization inhibitor, tanning agent, dyestuff manufacture.

2,3-dihydroxybutane. See 2,3-butylene glycol.

2,5-dihydroxycholorobenzene. See chlorohydroquinone.

3-(3,4-dihydroxycinnamoyl)quinic acid. See chlorogenic acid.

dihydroxydiaminomercurobenzene.
$OHNH_2C_6H_3HgC_6H_3OHNH_2$. A mercury compound analogous to arsphenamine.
Use: In medicine as a source of mercury.
Hazard: See mercury.

2,2′-dihydroxy-5,5′dichlorophenylmethane.
See dichlorophene.

dihydroxydiethyl ether. See diethylene glycol.

2,2′-dihydroxy-5,5′-difluorodiphenyl sulfide.
$FC_6H_3(OH)S(OH)C_6H_3F$.
Properties: White, amorphous solid; mp 119–121C; soluble in acetone, ether, chloroform, ethanol, ethyl acetate, and glacial acetic acid; moderately soluble in benzene; insoluble in water.
Use: Fungicide (textile, agricultural chemical).

2,4-dihydroxy-3,3-dimethylbutyric acid, γ lactone.
See pantolactone.

n-(2,4-dihydroxy-3,3-dimethylbutyryl)-β-alanine.
See pantothenic acid.

5,7-dihydroxydimethylcoumarin. $C_{11}H_{10}O_4$.
Properties and uses closely resemble those of 5,7-dihydroxy-4-methylcoumarin.

dihydroxydiphenyl sulfone. (sulfonyl bisphenol).
$(C_6H_4OH)_2SO_2$. The commercial product is a mixture of isomers.
Properties: White, free-flowing, odorless crystals; mp 215–240C; soluble in alcohol and acetone; insoluble in water.
Grade: Technical.
Use: Electroplating, phenolic resins, polyvinyl chloride, intermediate.

5,5′-dihydroxy-7,7′-disulfonic-2,2′-dinaphthylurea.
(6,6′-ureylenebis-1-naphthol-3-sulfonic acid).
$HOC_{10}H_5(SO_3H)NHCONHC_{10}H_5(SO_3H)OH$.
Properties: (Crude) Light gray paste, soluble in water, very soluble in alkaline solutions.
Derivation: Phosgenation of J acid.
Use: Dye intermediate.

p-di-(2-hydroxyethoxy)benzene. See hydroquinone di-(β-hydroxyethyl)ether.

di(2-hydroxyethyl)amine. See diethanolamine.

N,N-dihydroxyethyl ethylenediamine.
$(CH_2NHC_2H_4OH)_2$.
Properties: Solid crystals, mp 98C, bp 196C (10 mm), flash p 355F (179C). Combustible.
Use: Manufacture of textile-finishing assistants.

dihydroxyethyl sulfide. See thiodiglycol.

N,N-dihydroxyethyl-m-toluidine.
$CH_3C_6H_4N(C_2H_4OH)_2$.
Properties: Light-gray solid, mp 62C, distillation range 175–185C (2 mm).

2,2′-dihydroxy-3,5,6,3′,5′,6′-hexachlorodiphenyl methane. See hexachlorophene.

1,3-dihydroxy-4-hexylbenzene. See hexyl resorcinol.

3′,4′dihydroxy-2-isopropylaminoacetophenone hydrochloride.
$C_6H_3(OH)_2COCH_2NH(C_3H_7)·HCl$.   Light-colored, crystalline powder with a faint odor.
Use: Intermediate.

3,4-dihydroxy-α-(methylaminomethyl)benzyl alcohol. See epinephrine.

5,7-dihydroxy-4-methylcoumarin.
$C_{10}H_8O_4·H_2O$.
Properties: Yellow to white solid, fluoresces blue, absorbs UV light, melting range 270–285C. Insoluble in water, benzene, ether; soluble in alcohol and sodium hydroxide.
Derivation: From phloroglucinol.
Use: In suntan oils as a sun screen, in wall paints as a whitening agent.

1,1′-dihydroxymethylferrocene.
$[C_5H_4(CH_2OH)]_2Fe$.
Yellow, crystalline solid; mp 85–86C.
Use: Intermediate.
See also ferrocene.

1,3-dihydroxynaphthalene. (naphthoresorcinol).
$C_{10}H_6(OH)_2$.
Properties: Transparent, crystalline plates; mp 124–125C; soluble in alcohol, ether, and water. Combustible.
Derivation: By heating naphthalene-1,3-disulfonic acid with alkali at 230C under pressure.
Grade: Technical, reagent.
Use: Dyes; pharmaceuticals; analytical reagent for sugars, oils, glucuronic acid.
Note: There are several other isomeric forms of dihydroxynaphthalene, (1,5-; 1,6-; 1,7-; 1,8-; 2,3-; 2,6-; 2,7-). They are derived by heating a naphthalene disulfonic acid isomer with caustic soda, are soluble in alcohol and ether and sparingly soluble in water, mp ranges from 136C (1,6-) to 260C (2,6-).
Use: Organic dyes.

2,8-dihydroxy-3-naphthoic acid.
$(C_{10}H_5(OH)_2COOH)$.
Properties: Light-green powder, slightly soluble in hot water, soluble in alcohol and acetone, mp 235–240C.
Use: Intermediate.

L-dihydroxyphenylalanine. (L-dopa).
An amino acid used in treating Parkinson's disease, manganese poisoning and muscular dystonia. Derived from several types of beans including vanilla, also made synthetically.

1,2-dihydroxypropane. See 1,2-propylene glycol

dihydroxypropanone. See dihydroxyacetone.

dihydroxystearic acid. $C_{17}H_{33}(OH)_2COOH$.
Properties: White crystals, odorless, tasteless, soluble in alcohol and ether, insoluble in water, mp 135C.
Derivation: By heating dibromide of isooleic acid with silver oxide.
Use: Cosmetics, lotions.

dihydroxysuccinic acid. See tartaric acid.

3,5-dihydroxytoluene. See orcin.

DII. Abbreviation for diesel ignition improver.

diiodacetylene. (diiodethyne).
CAS: 624-74-8. $IC≡CI$
Properties: White crystals, unpleasant odor, light acts upon it causing a. gradual change in color to red and a separation of iodine. Soluble in alcohol, ether, benzene; insoluble in water. Mp 78.5C (decomposes).
Derivation: By dissolving iodine in liquid ammonia and passing acetylene into the solution.
Hazard: Highly volatile. Toxic by inhalation, vapors irritating to eyes and mucous membranes.
Use: Organic synthesis, military poison.

sym-diiododibromoethylene. $BrIC:CIBr$.
Properties: Crystals, mp 95–96C.
Derivation: Reaction of iodine and dibromoacetylene.
Use: Organic synthesis.

diiododiethyl sulfide. $(ICH_2CH_2)_2S$.
Properties: Bright yellow prisms. Slowly decomposes, the rate being accelerated by light and

by heat. Hydrolyzed by alkali solutions. Soluble in alcohol, benzene, ether; insoluble in water. Mp 62C.

Derivation: Interaction of dichlorodiethyl sulfide with an acetic acid solution of sodium iodide.

Hazard: Toxic by ingestion and inhalation, strong irritant.

4',5'-diiodofluorescein. $C_{20}H_{10}I_2O_5$.

Properties: Orange powder, soluble in alcohol, slightly soluble in water, CI# 45425A, Solvent #73.

Use: Dyeing textile, etc., analytical reagent.

diiodomethane. See methylene iodide.

3,5-diiodosalicylic acid. $I_2C_6H_2(OH)COOH$.

Properties: White to pale pink, crystalline powder; slightly soluble in water.

Use: Source of iodine for animal nutrition.

diiodothyronine. (3,5-Diiodothyronine). $HOC_6H_4OC_6H_2I_2CH_2CH(NH_2)COOH$.

A thyronine derivative which is an intermediate obtained in the manufacture of synthetic thyroxine, also probably an intermediate in the synthesis of thyroxine by the thyroid gland.

diisobutyl adipate. (DIBA).   CAS: 141-04-8. $[C_2H_4COOCH_2CH(CH_3)_2]_2$.

Properties: Colorless liquid, odorless, compatible with most natural and synthetic polymers, soluble in most organic solvents, insoluble in water, d 0.950 (25C), bp 278–280C, fp −20C, wt/gal 7.95 lb, acidity (as adipic acid) less than 0.05%. Combustible.

Use: Plasticizer.

diisobutyl aluminum chloride. (DIBAC). CAS: 1779-25-5. $[(CH_3)_2CHCH_2]_2AlCl$.

Properties: Colorless liquid, bulk d 0.905, fp −39.5C.

Derivation: Reaction of isobutylene and hydrogen with aluminum chloride.

Hazard: Strong irritant to tissue.

Use: Polyolefin catalyst.

diisobutyl aluminum hydride. (DIBAL-H). CAS: 1191-15-7. $[(CH_3)_2CHCH_2]_2AlH$.

Properties: Colorless, pyrophoric liquid; fp −80C; bulk d 0.798; bp 105C (0.2 mm); miscible in hydrocarbon solvents; dilute solutions nonpyrophoric.

Derivation: Reaction of isobutylene and hydrogen with aluminum.

Hazard: Ignites spontaneously in air, dangerous fire risk.

Use: Reducing agent in pharmaceuticals.

diisobutylamine. CAS: 110-96-3. $[(CH_3)_2CHCH_2]_2NH$.

Properties: Water-white liquid, d 0.745 (20C), boiling range 136–140C, amine odor, flash p 85F (29.4C).

Hazard: Flammable, moderate fire risk. Toxic by ingestion.

Use: Intermediate.

diisobutylcarbinol. See 2,6-dimethyl-4-heptanol.

diisobutyl carbinyl acetate. See nonyl acetate.

diisobutylene. A group of isomers of the formula C_8H_{16} of which 2,4,4-trimethylpentene-1 and 2,4,4-trimethylpentene-2 are the most important since they are formed in appreciable amounts when isobutene (isobutylene) is polymerized.

Properties: Colorless liquids, d 0.7227 (15.5C), boiling range 101–104C, flash p 20F (−6.6C).

Hazard: Fire risk. Narcotic in high concentrations.

Use: Alkylation, intermediate, antioxidants, surfactants, lube additive, plasticizers, rubber chemicals.

α-diisobutylene. See 2,4,4-trimethylpentene-1.

β-diisobutylene. See 2,4,4-trimethylpentene-2.

diisobutyl ketone. (2,6-dimethyl-4-heptanone). CAS: 108-83-8. $(CH_3)_2CHCH_2COCH_2CH(CH_3)_2$.

Properties: Colorless liquid, stable, mild odor, miscible with most organic liquids, immiscible with water, d 0.8089 (20/20C), bp 168.1C, vap press 1.7 mm (20C), flash p 140F (60C), bulk density 6.7 lb/gal (20C), fp −41.5C, coefficient of expansion 0.00101 (20C). Combustible.

Grade: Technical.

Hazard: Toxic. TLV: 25 ppm in air.

Use: Solvent for nitrocellulose, rubber, synthetic resins; lacquers, coating compositions, organic synthesis, roll-coating inks, stains.

diisobutyl phenol. See octyl phenol.

diisobutyl phthalate. CAS: 84-69-5. $C_6H_4[COOCH_2CH(CH_3)_2]_2$.

Properties: Liquid, refr index 1.4900 (25C), d 1.040 (20/20C), flash p 385F (196C), bp 327C. Combustible.

Use: Plasticizer.

diisobutyl sodium sulfosuccinate. $C_{12}H_{21}NaO_7S$.

Properties: The commercial product is a mix of three esters in the form of a fine, white powder; quite soluble in water and glycerol; insoluble in acetone, benzene, and carbon tetrachloride; surface tension of 1% water solution is 50 dynes/cm.

Hazard: Irritant to eyes, skin, and mucous membranes.

Use: Surfactant, emulsifier.

diisocyanate. An organic compound with two isocyanate groups (—NCO) formed by treating diamines, (e.g., toluene-2,4-diamine, hexamethylenediamine, p, p'-diaminodiphenylmethane) with phosgene. Combustible.

Use: Production of polyurethane foams and elastomers, in phenol-formaldehyde resins to improve water and alkali resistance, bonding rubber to rayon or nylon.

See also polyurethane.

diisodecyl adipate. (DIDA).
$C_{10}H_{21}OOC(CH_2)_4COOC_{10}H_{21}$.
Properties: Light-colored, oily liquid; mild odor; d 0.918 (20/20C); fp −71C; boiling range 239–246C (4 mm); refr index 1.450 (25C); bulk d 7.5 lb/gal; flash p 225F (107.2C); insoluble in glycerol, glycols, and some amines; soluble in most other organics. Combustible.

Use: Primary plasticizer for polymers.

diisodecyl-4,5-epoxytetrahydrophthalate.
("Flexol" PEP). $OC_6H_8(COOC_{10}H_{21})_2$.
Properties: Liquid, d 0.9867 (20/20C), bulk density 8.2 lb/gal, pour point 38C, oxirane oxygen 3%. Combustible.

Use: Plasticizer-stabilizer resistant to fungi, in vinyl plastics for outdoor use.

diisodecyl phthalate. (DIDP).
$C_6H_4(COOC_{10}H_{21})_2$.
Properties: Clear liquid with a mild odor, d 0.966 (20/20C), fp −50C, bp 250–257C (4 mm), refr index 1.483 (25C), viscosity 108 cp (20C), bulk density 8 lb/gal, flash p 450F (232C). Insoluble in glycerol, glycols, and some amines; soluble in most other organics. Combustible.

Grade: Technical.

Hazard: An irritant.

Use: Plasticizer.

diisononyl adipate.
$[C_9H_{19}OOC(CH_2)_4COOC_9H_{19}]$. A low-volatility plasticizer based on isononyl alcohol. Combustible.

diisooctyl acid phosphate. CAS: 27215-10-7.
$(C_8H_{17})_2HPO_4$.
Hazard: Irritant to skin and eyes.

diisooctyl adipate. (DIOA).
$C_8H_{17}OOC(CH_2)_4COOC_8H_{17}$.
Properties: A light straw-colored liquid, mild odor, d 0.924 (25C), bp 214–226C (4 mm), fp

above 75C, flash p 370F (187C). Combustible.

Use: Plasticizer, especially at low temperatures.

diisooctyl azelate. (DIOZ).
$C_8H_{17}OOC(CH_2)_7COOC_8H_{17}$. Liquid diester of azelaic acid, bp 237C (5 mm), pour point −85F, acid number 1.0, viscosity (min) 3.2 cs (210F), d 0.92 at 20C, flash p 415F (212C). Combustible.

Use: Plasticizer for vinyl resins, base for synthetic lubricants.

See also di-2-ethylhexyl azelate.

diisooctyl phthalate. (DIOP).
CAS: 27554-26-3. $(C_8H_{17}COO)_2C_6H_4$.
Isomeric esters obtained from phthalic anhydride and the mixed octyl alcohols made by the Oxo process.

Properties: Nearly colorless, viscous liquid; mild odor; bp 370C; d 0.980–0.983 (20/20C); bulk d 8.20 lb/gal (20C); flash p 450F (232C); insoluble in water; compatible with vinyl chloride resins and some cellulosic resins. Combustible.

Grade: Technical.

Use: Plasticizer for vinyl, cellulosic and acrylate resins and synthetic rubber.

See isooctyl alcohol.

diisooctyl sebacate. (DIOS).
$C_8H_{17}OOC(CH_2)_8COOC_8H_{17}$.
Properties: Liquid, d 0.915 (25/25C), flash p 440F (226C), pour point −40C, viscosity 24 cp (20C), bulk d 7.65 lb/gal. Combustible.

Use: Plasticizer.

diisopropanolamine. (DIPA).
$(CH_3CHOHCH_2)_2NH$.
Properties: White, crystalline solid; d 0.9890 (45/20C); bp 248.7C; bulk d 8.2 lb/gal (45C); vap press 0.02 mm (42C); mp 42C; viscosity 1.98 poise (45C); miscible with water; flash p 260F (126C). Combustible.

Use: Emulsifying agents for polishes, textile specialties, leather compounds, insecticides, cutting oils, and water paints.

diisopropyl. See 2,3-dimethylbutane.

diisopropylamine. CAS: 108-18-9.
$[(CH_3)_2CH]_2NH$.
Properties: Colorless, volatile liquid with amine odor; bp 84.1C; fp −96.3C; d 0.7178 (20/20C); bulk d 6.0 lb/gal (20C); refr index 1.3924 (20C); flash p 30F (−1.11C) (OC); slightly soluble in water; soluble in most organic solvents.

Derivation: From isopropyl chloride and ammonia.

Grade: Technical.

Hazard: Flammable, dangerous fire risk. Toxic

by ingestion, inhalation, and skin absorption. TLV: 5 ppm in air.
Use: Intermediate, catalyst.

diisopropylaminoethanol. See N,N-diisopropyl-ethanolamine.

β-diisopropylaminoethyl chloride hydrochoride. (DIC). [(CH₃)₂CH]₂NCH₂CH₂Cl·HCl.
Use: Organic synthesis especially for introduction of the β-diisopropylaminoethyl radical.

m-diisopropylbenzene.
Properties: Colorless liquid, d 0.8559 (20/4C), fp −63C, bp 231C, refr index 1.4883 (20C), flash p 170F (76.6C). Insoluble in water; miscible with alcohol, ether, acetone, benzene, and carbon tetrachloride; autoign temperature 840F. Combustible.
Use: Solvent, intermediate.

p-diisopropylbenzene.

Properties: White solid, mp 64C, d 0.8568 (20/4C), bp 210C, refr index 1.4898 (20C). Insoluble in water; miscible with alcohol, ether, acetone, benzene, and carbon tetrachloride. Combustible.
Use: Solvent, intermediate.

diisopropylbenzene hydroperoxide. Available as a 52% solution in a nonvolatile solvent. Colorless to pale yellow liquid, strong oxidizing agent.
Hazard: Dangerous fire risk in contact with organic materials.

N,N-diisopropylbenzothiazyl-2-sulfenamide. C₆H₄NC[SN(C₃H₇)₂]S. Rubber accelerator.

diisopropyl carbinol. (2,4-dimethylpentanol-3). [(CH₃)₂CH]₂CHOH.
Properties: Colorless liquid, bp 140C, fp approximately −70C, bulk density 6.9 lb/gal, flash p 120F (48.9C). Combustible.
Hazard: Moderate fire risk.
Use: Solvent, organic synthesis (intermediate), denaturant.

diisopropyl cresol. Antioxidant or stabilizer in medicine (external).
See isopropyl cresol.

N,N'-diisopropyldiamidophosphoryl fluoride. (mipafox).
(CH₃)₂CHNHPO(F)NHCH(CH₃)₂.
Hazard: Cholinesterase inhibitor. TLV: (as F) 2.5 mg/m³ of air.
Use: Insecticide.

diisopropyl dixanthogen. (C₃H₇OCS₂)₂.
Properties: Yellow to greenish pellets, d 1.28, mp 52C (min), purity 98% (min). Insoluble in water, soluble in ethyl alcohol, acetone, benzene, and gasoline.
Hazard: Toxic by ingestion and inhalation, strong irritant.
Use: Modifier in polymerization reactions, additive for lubricants, flotation reagent, fungicide, weed killer.

N,N-diisopropylethanolamine. (diisopropylaminoethanol). [(CH₃)₂CH]₂NCH₂CH₂OH.
Properties: Colorless liquid, d 0.8742 (20C), vap press 0.08 mm (20C), fp −39.3C, bp 191C, flash p 175F (79.4C), slightly soluble in water. Combustible.
Hazard: Strong irritant to tissue.
Use: Organic synthesis.

diisopropyl ether. See isopropyl ether.

diisopropyl fluorophosphate. (DFP; isofluorophate). CAS: 55-91-4.
[(CH₃)₂CHO]₂POF. Oily liquid, forms hydrogen fluoride in the presence of moisture. One member of a series of compounds, the fluorophosphate alkyl esters, characterized by extremely high toxicity, marked miotic effects noted even in concentration that are chemically undetectable.
Hazard: Cholinesterase inhibitor.
Use: Insecticide.

diisopropyl ketone. (2,4-dimethylpentanone-3).
[(CH₃)₂CH]₂CO.
Properties: Colorless liquid, bp 123.7C, bulk d 6.9 lb/gal.
Use: Solvent.

2,6-diisopropylnaphthalene. C₁₆H₂₀.
Properties: Clear, yellowish-brown liquid; faint, sweet odor; boiling range 290–295C; flash p (COC) 284F (140C); d 0.95 (30C); bulk density 7.9 lb/gal; insoluble in water. Combustible.
Hazard: Avoid inhalation of vapors and prolonged skin contact.
Use: Organic intermediate.

diisopropyl oxide. See isopropyl ether.

diisopropyl peroxydicarbonate. (isopropyl percarbonate; isopropyl peroxydicarbonate; IPP). CAS: 105-64-6.

$(CH_3)_2CHOC(O)OOC(O)OCH(CH_3)_2$.

Properties: Colorless, crystalline solid; d 1.080 (15.5/4C); mp 8–10C; refr index 1.4034 (20C); almost insoluble in water; miscible with aliphatic and aromatic hydrocarbons, esters, ethers, chlorinated hydrocarbons.

Derivation: By reaction of sodium peroxide with isopropyl chloroformate.

Grade: Stabilized and unstabilized.

Hazard: Spontaneous decomposition at room temperature releases flammable and corrosive products, dangerous fire hazard, explodes on heating. Store in open containers at low temperature with adequate ventilation.

Use: Low temperature polymerization catalyst.

2,6-diisopropylphenol. $C_6H_3OH[CH(CH_3)_2]_2$.

Properties: Light straw-colored liquid, fp 18C, d 0.955 (20C), bp 242C, flash p 240F (115.5C), soluble in toluene and alcohol, insoluble in water. Combustible.

Use: Intermediate for synthetic polymers, plasticizers, surface-active agents.

diisopropyl-p-phenylenediamine.

$H_7C_3NHC_6H_4NHC_3H_7$.

Properties: Dark red liquid, d 0.88, soluble in alcohol.

Use: Gasoline antioxidant and sweetener (permissible for aviation gasoline, ASTM D910–64T).

N,N'-diisopropylthiourea.

$(CH_3(_2CHNHCSNHCH(CH_3)_2$.

Properties: Grayish-white solid, mp 138.5–142.5C. Slightly soluble in water, soluble in methanol, acetone and ethyl acetate, insoluble in ether, benzene and gasoline.

Use: Corrosion inhibitor, metal pickling with hydrochloric acid or sulfuric acid, for reducing corrosion of ferrous metals and aluminum alloys in brine, intermediate.

diketene. (acetyl ketene).   CAS: 674-82-8.

$CH_2:\overline{CCH_2C(O)O}$.

Properties: Colorless, non-hygroscopic liquid; pungent odor; readily polymerizes on standing; d 1.096 (20/20C); fp −7.5C; bp 127.4C; soluble in common organic solvents; flash p 93F (33.9C) (TOC).

Derivation: By spontaneous polymerization of ketene obtained by thermal decomposition of acetone or from bromoacetylbromide and zinc.

Hazard: Flammable, moderate fire risk. An irritant.

Use: Production of acetoarylamides, pigments and toners, pesticides, food preservatives, pharmaceutical intermediates.

diketobutane. See diacetyl.

2,5-diketohexane. See acetonylacetone.

2,5-diketopyrrolidone. See succinimide.

2,5-diketotetrahydrofurane. See succinic anhydride.

Dilan. A 1:2 mix of Prolan and Bulan used as an insecticide.

Hazard: Toxic. See also Prolan, Bulan.

dilatancy. A system is said to be dilatant if its rate of increase of strain decreases with increased shear. Among the better known systems exhibiting dilatant behavior are pastry doughs, highly pigmented paints, and many other industrially important materials. Dilatancy is usually associated with suspension, especially those containing high concentration of suspended matter which is often of colloidal dimensions.

dilaudid. Proprietary dihyromorphine hydrochloride.

dilaurylamine. (didodecylamine).

$(C_{12}H_{25})_2NH$.

Properties: Liquid, mp 45C, d 0.89, almost insoluble in water.

Use: Chemical intermediate.

dilauryl ether. (didodecyl ether).

$(C_{12}H_{25})_2O$.

Properties: Liquid, mp 33C, bp 190–195C (1 mm), d 0.8147 (33/4C). Combustible.

Grade: 95% (min) purity.

Use: Electrical insulators, water repellents, lubricants for plastic molding and processing, antistatic substances, chemical intermediates.

dilauryl phosphite. $(C_{12}H_{25}O)_2PHO$.

Water-white liquid. Combustible.

Use: Synthesis of organophosphorus compounds for extreme pressure lubricants, adhesives, textile finishing agents, pesticides; catalyst in polymerization of unsaturated compounds.

dilauryl sulfide. (didodecyl thioether).

$(C_{12}H_{25})_2S$.

Properties: Liquid, mp 40–40.5C, bp 260–263C (4 mm), d 0.8275 (40/4C). Combustible.

Grade: 95% (min) purity.

Use: Organic synthesis (formation of sulfonium compounds).

dilauryl thiodipropionate. (didodecyl-3,3'-thiodipropionate; thiodipropionic acid, dilauryl ester).

$(C_{12}H_{25}OOCCH_2CH_2)_2S$.

Properties: White flakes having sweetish odor, mp 40C, d 0.975 (solid 25C), insoluble in water, soluble in most organic solvents, extremely resistant to heat and hydrolysis. Combustible.

Grade: FCC.

Use: Antioxidant, additive for high-pressure lubricants and greases, plasticizer and softening agent, antioxidant for edible fats and oils (up to 0.02% oil content).

dilinoleic acid. $C_{34}H_{62}(COOH)_2$.

Properties: Light yellow, viscous liquid with slight odor; d 0.921 (100C); refr index 1.4851 (40C); iodine value 80. Combustible.

Use: Modifier in alkyd and polyamide resins, polyester or metallic soap for petroleum additive, emulsifying agent, adhesives, shellac substitute, to upgrade drying oils.

dilithium sodium phosphate. Li_2NaPO_4.

A commercial source of lithium found in Searles Lake brine.

dilituric acid. See 5-nitrobarbituric acid.

diluent. (1) An ingredient used to reduce the concentration of an active material to achieve a desirable and beneficial effect. Examples are combination of diatomaceous earth with nitroglycerin to form the much less shock-sensitive dynamite; addition of sand to cement mixes to improve workability with no serious loss of strength; addition of an organic liquid having no solvent power to a paint or lacquer to reduce viscosity and achieve suitable application properties (see also thinner). (2) Low-gravity materials used primarily to reduce cost, i.e., blown asphalt, wood floc, etc., in rubber and plastic mixes. In this sense, there is no clear distinction between a diluent and an extender. (3) An ingredient of rocket fuels such as helium, hydrazine, or hydrogen.

dilution ratio. (hydrocarbon tolerance). The maximum number of unit volumes of hydrocarbon that can be added per unit volume of active solvent to cause the first trace of gelation to occur when the concentration of nitrocellulose in the solution is 8 g/100 mL. This may be used to evaluate the solvent power of active solvents by comparing them with a standard hydrocarbon or to evaluate hydrocarbon solvents by comparing them with an active solvent.

See solvent, lacquer.

dimagnesium orthophosphate. See magnesium phosphate dibasic.

dimagnesium phosphate. See magnesium phosphate dibasic

dimedone. (1,1-dimethyl-3,5-diketocyclohexane). $(CH_3)_2C_6H_6O_2$.

Properties: Greenish-yellow needles or prisms, mp 148–149C. Slightly soluble in cold water and naphtha; soluble in alcohol, chloroform, benzene.

Use: Reagent for the detection of ethyl alcohol and the identification of aldehydes.

dimefox. (bis(dimethylamino)fluorophosphate; tetramethyldiamidophosphoric fluoride); (BFPO). CAS: 115-26-4. $[(CH_3)_2N]_2POF$.

Properties: Liquid, fishy odor, d 1.1151 (20/4C), bp 67C (4.0 mm) 86C (15 mm), refr index 1.4267 (20C). Soluble in water, ether, benzene.

Hazard: Cholinesterase inhibitor, use may be restricted. TLV: (as F) 2.5 mg/m³ of air.

Use: A systemic pesticide primarily for ornamental and non-food plants.

dimenhydrinate. (2-(Benzohydryloxy)-N,N-dimethylethylamine-8-chlorotheophyllinate; "Dramamine"). CAS: 523-87-5. $C_{17}H_{22}NO \cdot C_7H_6ClN_4O_2$.

Properties: White, crystalline, odorless powder; freely soluble in alcohol and chloroform; soluble in benzene; sparingly soluble in ether; slightly soluble in water; mp 102–107C; pH (saturated solution) 6.8–7.3.

Grade: USP.

Use: Medicine (antiemetic).

dimer. An oligomer whose molecule is composed of two molecules of the same chemical composition. For example, 1,4-dioxane is a dimer of ethylene oxide.

A dimer acid is a high molecular weightw dibasic acid which is liquid (viscous), stable, resistant to high temperatures, and which polymerizes with alcohols and polyols to yield a variety of products such as plasticizers, lube oils, hydraulic fluids. It is produced by dimerization of unsaturated fatty acids at mid-molecule and usually contains 36 carbons. Trimer acid which contains three carboxyl groups and 54 carbons is similar. See also polymer.

dimercaprol. USP name for 2,3-dimercaptopropanol.

2,3-dimercaptopropanol. (BAL; British Anti-Lewisite; dimercaprol; 1,2-diethioglycerol). CAS: 59-52-9. $CH_2(SH)CH(SH)CH_2OH$.

Properties: Colorless, oily, viscous liquid with strong odor of mercaptans; bp 80C (1.9 mm) 140C (40 mm); mp 77C; d 1.2385 (25/4C); refr index 1.5720 (25C); soluble in vegetable oils; moderately soluble in water with decomposition; soluble in alcohol.

Derivation: Bromination of allyl alcohol followed by reaction with NaSH.

Grade: USP, as dimercaprol.

Use: Antidote to Lewisite, organic arsenicals, and heavy metals.

dimetan. (5,5-dimethyldihydroresorcinol dimethylcarbamate). $C_{11}H_{17}NO_3$.

Properties: Yellow crystals, mp 43–45C, slightly soluble in water and oils but readily soluble in organic solvents.

Hazard: Toxic by ingestion and inhalation.

Use: Insecticide, especially for aphids.

dimethyallyl. See 2,5-dimethylhexadiene-1,5.

dimethicone. $CH_3[Si(CH_3)_2O]Si(CH_3)_3$.

Properties: Colorless silicone oil consisting of dimethylsiloxane polymers (range in viscosities from 0.65 to 1,000,000 centistokes at room temperature). Viscosity grades higher than 50 centistokes are immiscible in water; miscible with chloroform, ether.

Use: Ointments and topical drug ingredient, skin protectant.

dimethisoquin hydrochloride. 3-butyl-1-(2-dimethylaminoethoxy)isoquinoline hydrochloride).

$CH_3(CH_2)_2CH_2C_9H_5NOCH_2CH_2N(CH_3)_2 \cdot HCl$.

Properties: White powder, odorless with bitter numbing taste, mp 144–147C, freely soluble in alcohol, soluble in water, pH (1% solution) 3.5–5.0.

Grade: NF.

Use: Medicine (topical anesthetic).

dimethoate. (O,O-dimethyl-S-(N-methylcarbamoylmethyl)phosphorodithioate).

CAS: 60-51-5. $(CH_3O)_2PSSCH_2CONHCH_3$.

Properties: White solid, mp 51–52C, moderately soluble in water, soluble in most organic solvents except hydrocarbons

Hazard: A cholinesterase inhibitor, use has been restricted.

Use: Insecticide (dry formulations not permitted).

dimethoxane. (6-acetoxy-2,4-dimethyl-m-dioxane).   CAS: 828-00-2.

$CH_3COOC_4H_5O_2(CH_3)_2$.

Properties: Clear yellow to light-amber liquid, d 1.069–1.076 (25/25C), refr index 1.431–1.438 (20C), bp 66–68C (3 mm), fp less than −25C, soluble in or miscible with water and organic solvents. Combustible.

Use: Preservative for cosmetics, inks, emulsions; gasoline additive.

1,2-dimethoxy-4-allylbenzene. See methyl eugenol.

2,5-dimethoxyaniline. $NH_2C_6H_3(OCH_3)_2$.

Properties: Gray flakes, mp 69–73C, insoluble in cold water, soluble in organic solvents and hot water.

Hazard: See aniline.

Use: Intermediate for dyes, pharmaceuticals and insecticides; antioxidant.

2,5-dimethoxybenzaldehyde.

$(CH_3O)_2C_6H_3CHO$.

Properties: Flaked solid, mp 46–49C, soluble in organic solvents, insoluble in water. Combustible.

Use: Organic synthesis.

1,2-dimethoxybenzene. See veratrole.

1,3-dimethoxybenzene. See resorcinol dimethyl ether.

1,4-dimethoxybenzene. See hydroquinone dimethyl ether.

dimethoxybenzidine. See dianisidine.

3,3′-dimethoxybenzidine-4,4′-diisocyanate. See dianisidine diisocyanate.

3,4-dimethoxybenzyl alcohol.

$C_6H_3(OCH_3)_2CH_2OH$.

Properties: Viscous brown liquid or low-melting solid. Combustible.

Use: Organic synthesis.

2,4-dimethoxy-5-chloroaniline.

$NH_2(Cl)C_6H_2(OCH_3)_2$.

Properties: Violet-gray crystals, mp 70–71C. Insoluble in water; soluble in alcohol, benzene, and other organic solvents.

Grade: 98.5%.

Hazard: See aniline.

Use: Intermediate for dyes and other organics.

p,p′-dimethoxydiphenylamine.

$(CH_3OC_6H_4)_2NH$.   A rubber antioxidant.

1,2-dimethoxyethane. See ethylene glycol dimethyl ether.

(2-dimethoxyethyl) adipate.

$CH_3OC_2H_4OOC(CH_2)_4COOC_2H_4OCH_3$.

Properties: Liquid, d 1.075 (25C), refr index 1.439 (25C), bp 185–190C (11 mm), fp −16C, slightly soluble in water. Combustible.

Use: Plasticizer.

di(2-methoxyethyl)phthalate. (DMEP).

CAS: 117-82-8. $C_6H_4(COOCH_2CH_2OCH_3)_2$.

Properties: Oily liquid with mild odor, d 1.172

(20/20C), bp 340C, fp —45C, flash p 381F (194C) (open cup). Combustible.

Use: Plasticizer, especially for cellulose acetate; solvent.

dimethoxymethane. See methylal.

2,5-dimethoxy-4-methylamphetamine.
(STP; DOM). A hallucinogenic, habit-forming drug, used in medicine, manufacture and use controlled by law in the US.
See also amphetamine.

3,4-dimethoxyphenethylamine. See homoveratrylamine.

3,4-dimethoxyphenylacetic acid. See homoveratric acid.

2,6-dimethoxyphenyllithium. $(CH_3O)_2C_6H_3Li$.
Properties: White to tan, free-flowing, pyrophoric powder; moderately soluble in toluene and benzene; slightly soluble in ethyl ether; stable indefinitely in sealed containers.
Hazard: Ignites spontaneously in air.

1-(3,4-dimethoxyphenyl)-2-nitro-1-propene.
$(CH_3O)_2C_6H_3CH:C(NO_2)CH_3$.
Properties: Yellow crystals, mp 68–75C.
Use: Intermediate.

3-(dimethoxyphosphinyloxy)-N,N-dimethyl-cis-crotonamide.
$(CH_3O)_2P(O)OC)CH_3):CHC(O)N(CH_3)_2$.
Properties: Brown liquid; bp 400C; miscible with water, ethanol, xylene; very slightly soluble in kerosene.
Hazard: Cholinesterase inhibitor, use may be restricted.
Use: Insecticide.

dimethoxystrychnine. See brucine.

dimethoxytetraglycol. (tetraethylene glycol dimethyl ether; bis(2-methoxyethoxyethyl ether; tetraglyme). CAS: 143-24-8.
$CH_3(OCH_2CH_2)_4OCH_3$.
Properties: Water-white, practically odorless liquid; stable; soluble in hydrocarbons, water. D 1.0132 (20/20C), bp 275.8C, 189C (100 mm), vap press less than 0.01 mm (20C), flash p 285F (141C), bulk density 8.4 lb/gal (20C), fp —29.7C, viscosity 0.0405 poise (20C), coefficient of expansion 0.00091 (20C). Combustible.
Grade: Technical.
Use: Solvent.

dimethrin. Generic name for 2,4-dimethylbenzyl-2,2-dimethyl-3-(2-methylpropenyl) cyclopropane carboxylate (2,4-dimethylbenzyl chrysanthemumate). CAS: 70-38-2. $C_{19}H_{26}O_2$.
Properties: Amber liquid, d 0.986 (20C), bp 175C (3.8 mm). Insoluble in water; soluble in petroleum hydrocarbons, aromatic petroleum derivatives, alcohols, and methylene chloride; decomposed by strong alkali.
Hazard: Toxic by ingestion.
Use: Insecticide.

dimethyl. See ethane.

dimethylacetal. (ethylidenedimethyl ether).
CAS: 534-15-6. $CH_3CH(OCH_3)_2$.
Properties: Colorless liquid, strongly aromatic odor. Soluble in water, alcohol, ether, and chloroform. D 0.848 (25C), bp 62–63C, flash p approximately 80F (26.6C).
Derivation: By heating acetaldehyde with methyl alcohol and glacial acetic acid and distilling.
Hazard: Flammable, dangerous fire risk. Toxic by ingestion and inhalation.
Use: Medicine, organic synthesis.

N,N-dimethyl acetamide. (DMAC).
CAS: 127-19-5. $CH_3CON(CH_3)_2$.
Properties: Colorless liquid, bp 166C, d 0.9366 (25C), refr index 1.4351 (25C). Miscible with water, aromatics, esters, ketones, and ethers. Flash p 171F (77C). Combustible.
Derivation: From acetic anhydride and dimethylformamide.
Grade: Technical, high purity certified.
Hazard: Toxic by inhalation, absorbed by skin, strong irritant. TLV: 10 ppm in air.
Use: Solvent for plastics, resins, gums and electrolytes; intermediate; catalyst; paint remover; high purity solvent for crystallization and purification.

N,N-dimethylacetoacetamide.
$CH_3COCH_2CON(CH_3)_2$.
Properties: Liquid, bp 220C, d 1.048–1.053 (20/20C), refr index 1.4379 (20C), miscible in water and organic solvents, flash p 252F (122C) (COC). Combustible.
Use: Chemical intermediate.

2,4-dimethyl acetophenone.
$CH_3COC_6H_3(CH_3)_2$.
Properties: Colorless liquid, odor of mimosa, d 0.994–0.997, refr index 1.532–1.534, soluble in four volumes of 60% alcohol. Combustible.
Use: Perfumery, flavoring.

dimethylacetylene. See crotonylene.

dimethylamine. (DMA). CAS: 124-40-3.
$(CH_3)_2NH$.
Properties: (anhydrous) Gas with ammoniacal odor; (25% water solution) d 0.6865 at —6C,

bp 6.88C, fp −92.2C, flash p 0F (−17.7C), bulk density approximately 7.8 lb/gal, soluble in alcohol and ether, autoign temperature (anhydrous) 806F (430C).

Derivation: Interaction of methanol and ammonia over a catalyst at high temperatures. The mono-, di- and trimethylamines are all produced.

Method of separation: Azeotropic or extractive distillation.

Grade: Technical (anhydrous, 25% and 40% aqueous solutions), 99%.

Hazard: Flammable, dangerous fire risk, explosive limits in air 2.8–14% Irritant. TLV: (Anhydrous) 10 ppm in air.

Use: Acid gas absorbent, solvent antioxidants, manufacture of dimethylformamide and dimethylacetamide, dyes, flotation agent, gasoline stabilizers, pharmaceuticals, textile chemicals, rubber accelerators, electroplating, dehairing agent, missile fuels, pesticide propellant, rocket propellants, surfactants, reagent for magnesium.

dimethylaminoaniline.   See p-aminodimethylaniline.

dimethylaminoantipyrine.   See aminopyrine.

dimethylaminoazobenzene. (methyl yellow; butter yellow).   CAS: 60-11-7.
$C_6H_5NNC_6H_4N(CH_3)_2$.

Properties: Yellow, crystalline leaflets; mp 116C; soluble in alcohol, ether, strong mineral acids, and oils; insoluble in water.

Derivation: Action of benzenediazonium chloride on dimethyl aniline.

Hazard: Toxic by inhalation and skin absorption, a carcinogen, may not be used in foods or beverages (FDA).

Use: Organic research, indicator, dyes.

p-**dimethylaminobenzaldehyde.**   CAS: 100-10-7.
$C_6H_4[N(CH_3)_2]CHO$.

Properties: Yellow, crystalline plates; soluble in alcohol and ether; mp 73C; bp 176–177C(17 mm). Combustible.

Derivation: By mixing dimethylaniline, anhydrous chloral, and phenol, and allowing the mix to stand. The phenol is removed by shaking with dilute caustic soda and the residue dissolved in water and hydrochloric acid and crystallized.

Grade: Technical, reagent.

Hazard: Toxic by ingestion.

Use: Dyes, medicine, reagent.

dimethylaminobenzene.   See xylidene.

p-**dimethylaminobenzene diazonium chloride, zinc chloride double salt.**   See p-diazodimethylaniline, zinc chloride double salt.

p-**dimethylaminobenzenediazo sodium sulfonate.** ("Dexon").   CAS: 140-56-7.
$(CH_3)_2NC_6H_4NNSO_3Na$.

Properties: Solid, melts with decomposition above 200C, soluble in water.

Hazard: Toxic by ingestion.

Use: Fungicide for protection of germinating seed and seedlings.

3-dimethylaminobenzoic acid.
$(CH_3)_2NC_6H_4COOH$.

Properties: Pale yellow crystals, mp 147–153C.

Use: Intermediate.

dl-**dimethylamino-4,4-diphenyl-3-heptanone hydrochloride.**   See methadone hydrochloride.

2-dimethylaminoethanol. (deanol; dimethylethanolamine).   CAS: 108-01-0.
$(CH_3)_2NCH_2CH_2OH$.

Properties: Colorless liquid with amine odor, bp 134.6C, fp −59.0C, d 0.8879 (20/20C), bulk density 7.4 lb/gal (20C), refr index 1.4300 (20C), flash p 105F (40.5C) (OC). Miscible with water, acetone, ether, and benzene. Combustible.

Preparation: From ethylene oxide and dimethylamine.

Grade: Anhydrous and 70% aqueous solution.

Hazard: Moderate fire risk.

Use: Intermediate in the synthesis of dyestuffs, textile auxiliaries, pharmaceuticals, and corrosion inhibitors; curing epoxy, amine, and polyamide resins; emulsifier in paints and coatings.

β-**dimethylaminoethyl chloride hydrochloride.** (DMC). $(CH_3)_2NCH_2CH_2Cl:HCl$.

Use: Manufacture of antihistamines and other pharmaceuticals. Organic intermediate for introduction of β-dimethylaminoethyl radical.

dimethylaminoethyl methacrylate.
$CH_2:C(CH_3)COOCH_2N(CH_3)_3$.

Properties: Liquid, d 0.933 (25C), bp 182–190C, refr index 1.4376 (25C), flash p 165F (73.9C) (Tagliabue open cup). Combustible.

Hazard: Irritant to skin, eyes, and mucous membranes; strong lachrymator.

Use: Binders for coatings, textile chemicals, dispersing agents for nonaqueous systems, antistatic agents stabilizers for chlorinated polymers, ion exchange resins, emulsifying agents, cationic precipitating agents.

dimethylaminomethyl phenol.
$C_6H_4OHCH_2N(CH_3)_2$.   Exists as o-, m- and p- isomers, the commercial material is a mix of o- and p-.

Properties: Dark red liquid, odor phenolic, free of methylamine, d 1.010 (25/25C), refr index 1.530 (25C), distillation range 80–130C (2 mm),

water content (Karl Fischer) 0.5%, soluble in organic solvents, moderately soluble in water.
Hazard: See phenol.
Use: Antioxidants, stabilizers, catalysts, intermediates.

4-dimethylamino-3-methylphenolmethyl carbamate (ester). (CH₃)₂NC₆H₃(CH₃)OOCNHCH₃.
Properties: Tan, crystalline solid; mp 93–94C; slightly soluble in water; moderately soluble in aromatic solvents; unstable in highly alkaline media.
Hazard: Toxic by ingestion, inhalation, and skin absorption.

1-dimethylamino-2-propanol.
(CH₃)₂NCH₂CHOHCH₃.
Properties: Water-white liquid, amine odor, bp 125.6C, d 0.850 (20/20C), refr index 1.421 (20C), flash p 90F (32C), soluble in water and most organic solvents.
Hazard: Flammable, moderate fire risk.
Use: Organic synthesis.

3-dimethylaminopropylamine.
(CH₃)₂NCH₂CH₂CH₂NH₂.
Properties: Colorless liquid, bp 123C, fp −70C, sets to a glass below this temperature, d 0.8100 (30C), refr index 1.4328 (25C), flash p 95F (35C) (Tagliabue closed cup), soluble in water and organic solvents.
Hazard: Flammable, moderate fire risk. Toxic by ingestion and inhalation, strong irritant.
Use: Curing agent for epoxy resins, organic intermediate.

1-dimethylamino-2-propyl chloride. (β-dimethylaminoisopropyl chloride).
(CH₃)₂NCH₂CHClCH₃.
Yellow liquid which darkens with age, distillation range 113–120C, refr index 1.422–1.423 (25C).
Use: Intermediate.

1-dimethylamino-3-propyl chloride. (DMPC).
(CH₃)₂NCH₂CH₂CH₂Cl.
Use: Intermediate for pharmaceutical and organic synthesis.

dimethylaminopropylmethacrylamide.
(DMAPMA.).
H₂C=C(CH₂)COOCH₂N(CH₃)₃
Properties: Pale yellow liquid, flash p 140C (CC), refr index 1.4763, d 0.94, bulk density 7.8 lb/gal.
Use: Reactive cationic monomer, incorporation of amine group into polymers.

4-dimethylamino-3,5-xylyl-N-methylcarbamate.
(CH₃)₂N(CH₃)₂C₆H₂OOCNHCH₃.
Properties: White, crystalline solid; mp 85C; al-

most insoluble in water; soluble in benzene, alcohol, and chloroform.
Hazard: Toxic by ingestion.
Use: Systemic insecticide.

N,N-dimethylaniline. (aniline N,N-dimethyl).
CAS: 121-69-7.   C₆H₅N(CH₃)₂.
Properties: Yellowish to brownish, oily liquid; soluble in alcohol and ether; insoluble in water; d 0.954; mp 2.5C; bp 192.5–193.5C; flash p 145F (62.7C) (closed cup); refr index 1.5582; autoign temperature 700F (370C). Combustible.
Derivation: By heating a mix of aniline, aniline hydrochloride, and methyl alcohol (free from acetone) in an autoclave and distilling.
Grade: Technical, reagent, 99.9% pure.
Hazard: Toxic material absorbed by skin. TLV: 5 ppm in air.
Use: Dyes, intermediates, solvent, manufacture of vanillin, stabilizer (acid acceptor), reagent.

dimethyl anthranilate. (N-methyl methyl anthranilate).   CAS: 85-91-6.
CH₃OOCC₆H₄NHCH₃.
Properties: Colorless or pale yellow liquid, grape-like odor, d 1.132–1.138 (15C), refr index 1.578–1.581 (20C). Soluble in three volumes or more of 80% alcohol, in benzyl benzoate, diethyl phthalate, fixed oils, mineral oils and volatile oils; insoluble in glycerol; slightly soluble in propylene glycol. Congealing point 18C (4% impurity) to 10C (20% impurity). Combustible.
Derivation: Methylation of methyl anthranilate or esterification of N-methyl anthranilic acid.
Use: Perfumes, flavorings, and drugs.

dimethylarsinic acid. See cacodylic acid.

1,2-dimethylbenzene. See o-xylene.

1,3-dimethylbenzene. See m-xylene.

1,4-dimethylbenzene. See p-xylene.

3,3-dimethylbenzidine. See o-tolidine.

dimethylbenzylcarbinyl acetate. See α,α-dimethylphenethyl acetate.

2,5-dimethylbenzyl chloride. (2,5-dimethyl-α-chlorotoluene; α-chloro-p-xylene).
C₆H₃(CH₃)₂CH₂Cl.
Properties: Colorless to pale-yellow liquid, sharp pungent odor, d 1.035–1.045 (25/25C), bp 221–226C, refr index 1.5350–1.5360 (25C), soluble in alcohols and ethers, insoluble in water. Combustible.
Hazard: Irritant to eyes and mucous membranes, a lachrymator.
Use: Intermediate for pharmaceuticals, dyes, perfumes, plasticizers, resins, wetting agents, germicides, etc.

2,4-dimethylbenzyl chrysanthemumate. See dimethrin.

α,α-dimethylbenzyl hydroperoxide. See cumene hydroperoxide.

dimethyl benzylphosphonate.
$(CH_3O)_2P(O)CH_2C_6H_5$.
Water-white liquid, bp 114C (0.75 mm).
Use: Extraction of mineral salts from special solutions, fire retardant, low-temperature plasticizer.

3,3'-dimethyl-4,4'-biphenylene diisocyanate.
(4,4'-bi-o-tolylene diisocyanate).
Properties: Flaked, white, crystalline solid; assay 98% min; mp 69C min.
Use: High-strength elastomers, coatings, rigid plastics.

1,1'-dimethyl-4,4'-bipyridinium salt. See paraquat.

dimethyl brassylate.
$CH_3OOC(CH_2)_{11}COOCH_3$.
An ester of a 13-carbon saturated aliphatic dibasic acid. Waxy, low-melting solid, made by an ozone oxidation process.
Use: Preparation of ethylene brassylate, a synthetic musk; chemical intermediate.

2,3-dimethyl-1,3-butadiene. C_6H_{10}.
Properties: Dark liquid, bp 70C, fp −76C, d 0.727, refr index 1.43.
Derivation: Shale oil fractions, by distillation of pinacol.
Use: Manufacture of elastomers.

2,2-dimethylbutane. See neohexane.

2,3-dimethylbutane. (diisopropyl).
  CAS: 79-29-8. $(CH_3)_2CHCH(CH_3)_2$.

$$CH_3-CH-CH-CH_3$$
$$\qquad\ \ |\qquad\ |$$
$$\qquad\ \ CH_3\ \ CH_3$$

Properties: Colorless liquid bp 57.9C, d 0.66164 (20C), fp −128.41C, refr index 1.37495 (20C), flash p −20F (−28.9C), autoign temperature 788F (420C).
Derivation: Alkylation of ethylene with isobutane using aluminum chloride catalyst.
Grade: Technical 95%, 99%, 99.8 mole %.
Hazard: Flammable, dangerous fire and explosion risk.
Use: High-octane fuel, organic synthesis.

2,2-dimethyl-1,3-butanediol.
$CH_3CH(OH)C(CH_3)CH_2OH$.

Properties: Liquid, d 0.9700, bp 202.4C, fp −12.8C, bulk density 8.1 lb/gal, very soluble in water. Combustible.

2,3-dimethylbutene-1.
$CH_3CH(CH_3)C(CH_3):CH_2$.
Properties: Liquid, fp −157.27C, bp 55.6C, d 0.678 (20/4C); bulk density 5.68 lb/gal (60F), refr index 1.390 (20C), flash p −20F (−28.9C).
Hazard: Flammable, dangerous fire and explosion risk.
Use: Perfume synthesis, isomerization reactions.

2,3-dimethylbutene-2.
$CH_3C(CH_3):C(CH_3)CH_3$.
Properties: Liquid, fp −74.3C, bp 73.2C, d 0.708 (20.4C), bulk density 5.94 lb/gal (60F), refr index 1.412 (20C), flash p 0F (−17.7C).
Hazard: Flammable, dangerous fire and explosion risk.
Use: Isomerization reactions.

2,4-dimethyl-6-tert-butyl phenol.
$(CH_3)_2C_6H_2OH[C(CH_3)_3]$.
Properties: A low-viscosity, straw-colored liquid; readily soluble in gasoline and oils; d 0.961. Combustible.
Use: Antiskinner, antioxidant mainly for gasoline (ASTM D91064T).

dimethyl cadmium. C_2H_6Cd.
Properties: Liquid with unpleasant odor, fp −4.0C, bp 105C, d 1.98, refr index 1.54, soluble in hydrocarbon solvents.
Hazard: Explodes on heating to 150C, pyrophoric.
Use: Polymerization catalyst, organic synthesis.

dimethylcarbamoyl chloride. (dimethyl carbamyl chloride).   CAS: 79-44-7. $(CH_3)_2NCOCl$.
Properties: A colorless liquid that hydrolyzes rapidly; bp 165C, fp −33C, d 1.67.
Hazard: TLV: A suspected human carcinogen and a strong lachrymator.
Use: Chemical intermediate in production of dyes, pharmaceuticals and pesticides.

dimethyl carbate. (cis-dimethyl ester-5-norbornene-2,3-dicarboxylic acid).
  CAS: 5826-73-3. $C_{11}H_{14}O_4$.   Generic name for bicyclo(2,2,1)-5-heptene-2,3-dicarboxylic acid dimethyl ester.
Properties: Clear, oily liquid or crystalline solid; d 1.165 (35/4C); insoluble in water.
Derivation: Esterification of Diels-Alder condensation product of maleic anhydride and cyclopentadiene.
Hazard: Toxic by ingestion and inhalation.
Use: Insect repellent.

dimethylcarbinol. See isopropyl alcohol.

dimethyl carbonate. Legal label name for methyl carbonate.

dimethyl chloroacetal. (chloroacetaldehyde dimethyl acetal). $ClCH_2CH(OCH_3)_2$.
Properties: Colorless liquid with a pleasant odor, boiling range 126–132C, flash p 110F (43.3C), d 1.082–1.092 (25/4C), refr index 1.4110–1.4130 (25C), purity 97% (min), bulk density 9.07 lb/gal. Combustible.
Grade: Technical.
Hazard: Moderate fire risk.
Use: Organic synthesis, pharmaceuticals, solvent.

2,5-dimethyl-α-chlorotoluene. See 2,5-dimethylbenzyl chloride.

dimethyl cyanamide. $(CH_3)_2NCN$.
Properties: Colorless, mobile liquid; fp −41C; d 0.876; bp 160C; flash p 160F (71.1C) (Tagliabue closed cup). Combustible.
Hazard: Toxic by ingestion and inhalation.
Use: Chemical intermediate, solvent.

dimethylcyclohexane. (hexahydroxylene). Mix of o-, m- and p-isomers. $C_6H_{10}(CH_3)_2$.
Properties: Water-white liquid of mild odor, d 0.776 (15/15C), boiling range 120–129C, fp less than −65C, flash p approximately 50F (10C) (closed cup), soluble in most common solvents, almost insoluble in water.
Hazard: Flammable, dangerous fire risk.
Use: Synthesis, special solvent.
Note: There are several isomers of dimethylcyclohexane, i.e., 1,3-; 1,4-; cis-1,2-; trans-1,2-. They have properties and uses closely similar to those given above. All are flammable.

dimethyl-1,4-cyclohexanedicarboxylate. $CH_3OOCC_6H_{10}COOCH_3$.
Properties: Partially crystalline solid, d 1.102 (35/4C), consists of approximately 60% cis and 40% trans isomers, bp (mixed isomer) 265C, bulk density 9.18 lb/gal (20C), soluble in all proportions in most organic solvents.
Use: Plasticizers, polymers.

N,N-dimethylcyclohexylamine. $(CH_3)_2NC_6H_{11}$.
Properties: Water-white liquid, distilling range 157–160C, d 0.8490 (20/20C), fp approximately −77C, flash p 110F (43.3C) (COC), partly soluble in water, miscible with alcohol, benzene, acetone. Combustible.
Hazard: Moderate fire risk.
Use: Catalyst for polyurethane foams, intermediate for rubber accelerators, treatment of textiles.

1,2-dimethylcyclopentane. $C_5H_8(CH_3)_2$.
Properties: Colorless liquid, Cis: bp 99.5C, d 0.772 (20C); trans: bp 91.8C, d 0.751 (20C). Combustible.
Grade: Technical.
Use: Organic synthesis.

2,2′-dimethyl-1,1′-dianthraquinone. $C_{30}H_{18}O_4$.
Properties: Yellow crystals, soluble in hot nitrobenzene, aniline, and chlorobenzene, mp 365–367C.
Use: Dye intermediate.

2,5-dimethyl-2,5-di(tert-butylperoxy)hexane. $C_4H_9OOC(CH_3)_2CH_2CH_2C(CH_3)_2OOC_4H_9$.
Properties: Stable, colorless, liquid; bp 50–52C (0.1 mm); active oxygen 10.5% min; fp 8C; flash p 185F (85C); soluble in alcohol; insoluble in water. Combustible.
Hazard: Strong oxidizer, may ignite organic materials. Irritant to eyes and skin.
Use: Catalyst in polyethylene cross-linking, styrene polymerization, polyester resins.

dimethyldichlorosilane. (dichlorodimethylsilane). CAS: 75-78-5. $(CH_3)_2SiCl_2$.
Properties: Colorless liquid, bp 70C, fp −86C, d 1.062 (20C), refr index 1.4023 (25C), flash p (COC) 16F (−8.9C). Reacts with water to form complex mixture of dimethylsiloxanes and liberates hydrogen chloride. Soluble in benzene and ether.
Derivation: Action of silicon on methyl chloride in presence of a copper catalyst or by Grignard reaction from methyl chloride and silicon tetrachloride.
Grade: Technical.
Hazard: Flammable, dangerous fire and explosion risk.
Use: Intermediate for silicone products.

dimethyl dichlorovinyl phosphate. See dichlorovos.

5,5-dimethyldihydroresorcinol dimethylcarbamate. See dimetan.

1,1-dimethyl-3,5-diketocyclohexane. See dimedone.

dimethyldiketone. See diacetyl.

N,N′-dimethyl-N,N′-di(1-methylpropyl)-p-phenylene diamine. Forms a continuous protective film.
Properties: Volatile, reddish-brown liquid.
Use: Antiozonant in rubber.

2,2-dimethyl-1,3-dioxolane-4-methanol.
$C_6H_{12}O_3$.
Properties: Colorless, odorless liquid; bp 82C (10 mm); d 1.065; refr index 1.43; flash p 194F (90C); soluble in water and most organic solvents; optically active. Combustible.
Use: General solvent, plasticizer.

N,N-dimethyl-N,N-dintrosterephthalamide.
$C_6H_4[CON(CH_3)NO]_2$.
Use: Blowing agent liberating nitrogen at 100C with a residue of dimethyl terephthalate.

dimethyl dioxane.   CAS: 25136-55-4.

$\overline{OCH(CH_3)CH_2OCH_2}CH(CH_3)$.
Properties: Water-white liquid, soluble in water, d 0.9268, bp 117.5C, flash p 75F (23.9), vap press 154.4 mm at 20C.
Hazard: Flammable, dangerous fire risk.

N,N-dimethyl-2,2-diphenylacetamide.
See diphenamid.

dimethylenemethane.   See allene.

dimethylethanolamine.   See dimethylaminoethanol.

dimethyl ether.   (methyl ether; methyl oxide; wood ether).   CAS: 115-10-6. CH_3OCH_3.
Properties: Colorless compressed gas or liquid, soluble in water and alcohol, d 0.661, bp −24.5C, fp −141.4C, flash p −42F (−41C), autoign temperature 662F (350C), soluble in organic solvents.
Derivation: By dehydration of methanol.
Grade: Technical, 99.5%.
Hazard: Highly flammable, dangerous fire and explosion hazard.
Use: Refrigerant, solvent, extraction agent, propellant for sprays, chemical (reaction medium), catalyst and stabilizer in polymerization.

dimethyl ethyl carbinol.   See tert-amyl alcohol.

dimethylethylene.   See butene-2.

sym-**dimethylethylene glycol.**   See 2,3-butylene glycol.

O,O-dimethyl-S-2-(ethylsulfinyl)ethyl phosphorothioate.   (oxydemetonmethyl).
$(CH_3O)_2P(O)SC_2H_4SOC_2H_5$.
Properties: Amber liquid, bp 106C (0.01 mm), d 1.28 (20/4C), miscible in water.
Hazard: Cholinesterase inhibitor, use may be restricted.
Use: Systemic insecticide.

dimethyl ferrocenoate.   (1,1′-ferrocene dicarboxylic acid dimethyl ester).
$(C_5H_4COOCH_3)_2Fe$.
Properties: Orange, crystalline solid; mp 114–115C.
See also ferrocene.

N,N-dimethylformamide.   (DMF).
CAS: 68-12-2. $HCON(CH_3)_2$.
Properties: Water-white liquid, a dipolar aprotic solvent, bp 152.8C, fp −61C, refr index 1.4269 (25C), d 0.953–0.954 (15.6/15.6C), flash p 136F (57.7C), autoign temperature 833F (445C), miscible with water and most organic solvents (except halogenated hydrocarbons). Combustible.
Derivation: Reaction of methyl formate with dimethylamine.
Hazard: Moderate fire risk. Toxic by skin absorption. Strong irritant to skin and tissue. TLV: 10 ppm in air.
Use: Solvent for vinyl resins and acetylene, butadiene, acid gases; polyacrylic fibers; catalyst in carboxylation reactions; organic synthesis; carrier for gases.

dimethylfuran. $\overline{OC(CH_3)CHCHC}(CH_3)$.
Properties: Colorless liquid, insoluble in water, d 0.8900, bp 94C, flash p 45F (7.2C).
Hazard: Flammable, dangerous fire risk.

dimethyl glycol phthalate.
$C_6H_4(COOCH_2CH_2OCH_3)_2$.
Properties: Colorless liquid, d 1.17, bp 230C. Combustible.
Use: Solvent mix for cellulose esters, plasticizing mix for cellulose esters.

dimethylglyoxal.   See diacetyl.

dimethylglyoxime.   (butane dioxime).
$CH_3C(NOH)C(NOH)CH_3$.
Properties: White crystals or powder, mp 242C, soluble in alcohol and ether, insoluble in water.
Use: Analytical chemistry, especially as a reagent for nickel, biochemical research.

2,6-dimethyl-4-heptanol.   (diisobutylcarbinol).
$[(CH_3)_2CHCH_2]_2CHOH$.
Properties: Colorless liquid, refr index 1.423 (21C), d 0.8121 (20C), fp sets to glass approximately −65C, bp 178C (750 mm), insoluble in water, soluble in alcohol and ether, flash p 162F (72.2C) (TOC). Combustible.
Use: Surface-active agents, lubricant additives, rubber chemicals, flotation agents, antifoam agent.

2,6-dimethyl-4-heptanone.   See diisobutyl ketone.

2,6-dimethyl-5-hepten-1-al.
$(CH_3)_2C:CH(CH_2)_2CH(CH_3)CHO$.
Properties: Yellow liquid, moderately stable but not likely to cause discoloration, d 0.845–0.855 (25/25C), refr index 1.441–1.447 (20C), flash p 144F (62.2C) (TOC), soluble in two parts of 70% alcohol. Combustible.
Use: Perfumery.

2,6-diemthylheptene-3.
$(CH_3)_2CHCH:CHCH_2CH(CH_3)_2$.   Mixed cis and trans isomers.
Properties: Liquid, distillation range 128–129C, d 0.722 (60/60F), refr index 1.412 (20C), flash p 70F (21.1C) (Tagliabue open cup).
Grade: 95%.
Hazard: Flammable, dangerous fire risk.

2,5-dimethylhexadiene-1,5. (dimethyallyl).
$CH_2:C(CH_3)CH_2CH_2C(CH_3):CH_2$.
Properties: Water-white liquid with hydrocarbon odor, d 0.740–0.760 (25/25C), refr index 1.426–1.429 (25C), ASTM distillation 90% between 114–123C, soluble in hydrocarbons, insoluble in water, flash p 56F (13.3C).
Hazard: Flammable, dangerous fire risk.
Use: Solvent.

2,5-dimethylhexane-2,5-dihydroperoxide.
$(CH_3)_2C(OOH)CH_2CH_2C(OOH)(CH_3)_2$.
Properties: Fine powder, 90% peroxide, mp 102–104C, insoluble in hydrocarbons, slightly soluble in water, soluble in alcohols.
Hazard: Dangerous fire risk, strong oxidizer, store away from organic materials.
Use: High-temperature catalyst for polyester premix compounds and silicone resins.

dimethylhexanediol. (2,5-dimethylhexane-2,5-diol). $(CH_3)_2COH(CH_2)_2COH(CH_3)_2$.
Properties: White crystals, mp 88.5–89C, bp 204–215C, d 0.898 (20/20C); soluble in water, acetone and alcohol; insoluble in benzene, carbon tetrachloride, and kerosene. Combustible.
Use: Chemical intermediate.

2,5-dimethylhexane-2,5-diperoxybenzoate.
$C_{22}H_{26}O_6$.
Properties: Fine, white granules; mp 114C; insoluble in alcohols and hydrocarbons; soluble in acetone and chlorinated hydrocarbons.
Hazard: Strong oxidant, fire risk in contact with organic materials.
Use: Oxidizing agent, polymerization agent.

dimethylhexynediol. (2,5-dimethyl-3-hexyne-2,5-diol). $(CH_3)_2CHOC\equiv CCOH(CH_3)_2$.
Properties: White crystals, mp 94–95C, bp 205–206C, d 0.949 (20/20C). Soluble in water,

slightly soluble in benzene, carbon tetrachloride, naphtha; very soluble in acetone, alcohol, and ethyl acetate.
Use: Wire-drawing lubricant, antifoaming agent, coupling agent in resin coatings, chemical intermediate.

dimethyl hexynol. (3,5-dimethyl-1-hexyne-3-ol).
$HC\equiv CCOH(CH_3)CH_2CH(CH_3)_2$.
Properties: Colorless liquid with camphor-like odor, bp 150–151C, sets to a glass approximately −68C, d 0.8545 (20/20C), slightly soluble in water, flash p 134F (56.6C) (TOC). Combustible.
Hazard: Moderate fire risk.
Use: Stabilizer for chlorinated organic compounds, surface active agent, intermediate, solvent lubricant.

5,5-dimethylhydantoin. (DMH).
CAS: 77-71-4. $HNCONHCOC(CH_3)_2$.
Properties: White, crystalline solid; mp 178C; soluble in water, alcohol, and ether.
Derivation: (a) From acetone, urea, and ammonium carbonate; (b) from acetone, potassium cyanate and hydrogen cyanide.
Hazard: Central nervous system depressant.
Use: Synthesis, preparation of water-soluble polymers.

dimethylhydantoin-formaldehyde polymer.
Properties: Light-colored, brittle resin containing 0.3% max of formaldehyde; d 1.30; softening point 59–80C; dissolves readily in cold and hot water, methanol, ethylacetate, methyl ethyl ketone, chloroform, methylene chloride, and hot glycerol; insoluble in benzene, xylene, diethyl ether, trichloroethylene, and carbon tetrachloride.
Use: Sizing, adhesives, blending agent, aerosol hair sprays.

1,1-dimethylhydrazine. CAS: 57-14-7.
$(CH_3)_2NNH_2$.   An unsymmetrical compound.
Properties: Colorless, hygroscopic liquid with ammonia-like odor; fumes in air; evolves heat on contact with water; fp −58C; bp 63C; d 0.782 (25C); soluble in water and alcohol; flash p approximately 5F (−15C); autoign temperature 480F (249C).
Derivation: (a) Reaction of dimethylamine and chloramine, (b) catalytic oxidation of dimethylamine and ammonia.
Hazard: Flammable, dangerous fire risk. Suspected human carcinogen; toxic by skin absorption. Corrosive to skin. TLV: 0.5 ppm in air.
Use: Component of jet and rocket fuels, chemical synthesis, stabilizer for organic peroxide fuel additives, absorbent for acid gases, photography, plant growth control agent.

dimethylhydroquinone. See hydroquinone dimethyl ether.

dimethylhydroxybenzene. See xylenol.

dimethyl-3-hydroxglutaconate dimethyl phosphate. See "Bomyl."

3,7-dimethyl-7-hydroxy-octenal. See hydroxycitronellal.

dimethylisopropanolamine.
$(CH_3)_2NCH_2CH(OH)CH_3$.
Properties: Colorless liquid, d 0.8645 (25/20C), bulk density 7.4 lb/gal (20C), bp 125.8C, miscible in water, viscosity 1.51 cp (20C), vap press 9 mm (20C), fp sets to glass approximately −85C, refr index 1.4189 (20C), flash p 95F (35C) (open cup), solubility of water in compound complete at 20C.
Hazard: Flammable, moderate fire risk.
Use: Synthesis of methadone, other chemical syntheses. Combining the properties of tertiary amine and secondary alcohol.

dimethyl itaconate.
$CH_2C(COOCH_3)CH_2(COOCH_3)$.
Properties: White crystals with slight odor, mp 36C, bp 91.5C (10 mm), d 1.27 (24C), refr index 1.441 (20C), slightly soluble in water. Combustible.
Use: Polymers and copolymers, plasticizers, intermediate.

dimethylketol. See acetylmethylcarbinol.

dimethylketone. See acetone.

dimethyl maleate. $CH_3OOCCH:CHCOOCH_3$.
Properties: Colorless liquid, d 1.153, bulk density 9.62 lb/gal, bp 200.4C, flash p 235F (112C). Combustible.

dimethyl malonate. CAS: 108-59-8.
$CH_2(COOCH_3)_2$.
Properties: Colorless liquid, bp 180–181C, fp −62C, refr index 1.4140, flash p 194F (90C), very slightly soluble in water, soluble in alcohol and ether. Combustible.
Use: Chemical intermediate.

dimethylmethane. See propane.

2,6-dimethylmorpholine. CAS: 141-91-3.

$\overline{OCH(CH_3)CH_2NHCH_2CH}(CH_3)$.
Properties: Liquid, d 0.99346, bp 146.6C, fp −85C, soluble in water, bulk density 7.8 lb/gal, flash p 112F (44.4C). Combustible.

Hazard: Moderate fire risk.
Use: Corrosive inhibitors, stabilizers for chlorinated solvents, rubless floor polishes, rubber accelerators, germicides, and textile finishing agents.

2-(2,6-dimethyl-4-morpholinothio)benzothiazole.
$C_{13}H_{16}N_2OS_2$.
Properties: Cream to light-yellow powder, mp 88C, d 1.26 (25C).
Use: Delayed-action vulcanization accelerator.

dimethyl-α-naphthylamine. $C_{10}H_7N(CH_3)_2$.
Properties: Colorless liquid, soluble in alcohol and ether, insoluble in water, d 1.045, bp 275C.
Derivation: Action of methylsulfate on α-naphthylamine.
Grade: CP, analytical.
Use: Determination of nitrates.

dimethyl-β-naphthylamine. $C_{10}H_7N(CH_3)_2$.
Properties: Crystalline solid, soluble in alcohol and ether, insoluble in water, d 1.039 (70/70C), mp 46C, bp 305C.
Derivation: Interaction of dimethylamine and β-naphthol.

dimethylnitrobenzene. See nitroxylene.

O,O-dimethyl-O-p-nitrophenyl phosphorothioate. See methyl parathion.

dimethyl nitrosamine. See N-nitrosodimethylamine.

N,N-dimethyl-p-nitrosoaniline. (p-nitrosodimethylaniline). CAS: 138-89-6.
$(CH_3)_2NC_6H_4NO$.
Properties: Green leaflets, soluble in alcohol and ether, insoluble in water, mp 93C.
Derivation: Nitrous acid and N-dimethylaniline.
Hazard: Flammable. Toxic by ingestion.
Use: Production of methylene blue, vulcanization accelerator.

3,6-dimethyl-3,6-octanediol.
$C_2H_5(CH_3)COH(CH_2)_2COH(CH_3)C_2H_5$.
Properties: White, waxy solid; mp 44C; bp 241–242C; d 0.919 (20/20C); soluble in water, acetone, alcohol, benzene, carbon tetrachloride.
Use: Non-foaming surface-active agent; chemical intermediate.

dimethyloctanoic acid. See isodecanoic acid.

3,6-dimethyl-3-octanol.
$(C_2H_5)CHCHCH_3(CH_2)_2COHCH_3(C_2H_5)$.
Properties: Colorless liquid, sweet odor, d 0.8366 (20/20C), refr index 1.4370 (20C), bp 202–203C, fp −67.5C. Combustible.
Use: Perfumery (floral odors), flavoring.

3,7-dimethyl-1-octanol. (tetrahydrogeraniol).
$(CH_3)_2CH(CH_2)_3CH(CH_3)CH_2CH_2OH$.
Properties: Colorless liquid, sweet odor, soluble in mineral oil and in propylene glycol, insoluble in glycerol, refr index 1.4350–1.4450 (20C), d 0.826–0.842 (25C). Combustible.
Grade: FCC.
Use: Flavoring agent, perfumery.

3,7-dimethyl-3-octanol. See tetrahydrolinalool.

2,6-dimethyl-1,5,7-octatriene. See ocimene.

2,6-dimethyl-2,4,6-octatriene. See alloocimene.

3,7-dimethyl-6-octenal. See citronellal.

3,7-dimethyl-6(or 7)-octen-1-ol. See citronellol.

dimethyloctynediol. (3,6-dimethyl-4-octyne-3,6-diol).
$C_2H_5(CH_3)COHC{\equiv}CCOH(CH_3)C_2H_5$
Properties: White crystals, mp 55–56C, bp 222C, d 0.923 (solid 20C) 0.908 (liquid 60C). Moderately soluble in water; slightly soluble in kerosene; very soluble in acetone, alcohol, benzene, and carbon tetrachloride. Combustible.
Use: Surface-active agent, intermediate.

dimethylol ethylene urea.

$OCN(CH_2OH)CH_2CH_2N(CH_2OH)$.
Use: Wrinkle-resistant textile finishes.

dimethylolpropionic acid. (DMPA; 2,2-bis(hydroxymethyl)propionic acid).
$CH_3C(CH_2OH)_2COOH$.
Properties: Off-white, crystalline solid; mp 192–194C; soluble in water and methanol; slightly soluble in acetone; insoluble in benzene.
Use: Water-soluble alkyd resins, textile finishing, cosmetics, plasticizers.

dimethylolurea. (DMU; 1,3-bishydroxymethylurea).

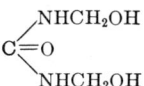

Properties: Colorless crystals, mp 126C (technical 85–90C), d 1.34 (20C), slight formaldehyde odor, soluble in water and methanol, insoluble in ether, capable of polymerization.
Derivation: Combination of urea and formaldehyde in the presence of salts or alkaline catalysts.
Hazard: Irritant to skin.
Use: First stage of ureaformaldehyde resins; impregnating wood to increase hardness and fire

resistance and to form self-binding laminations for plywood manufacture; permanent-press fabrics.

2,3-dimethylpentaldehyde.
$CH_3CH_2CH(CH_3)CH(CH_3)CHO$.
A branched-chain heptanal.
Properties: Liquid, d 0.8293, bp 140.5C, fp −110C, bulk density 6.91 lb/gal (20C), flash p 94F (34.4C), slightly soluble in water.
Hazard: Flammable, moderate fire risk.
Use: Intermediate.

2,4-dimethylpentane.
$(CH_3)_2CHCH_2CH(CH_3)_2$.
Properties: Colorless liquid, d 0.6684 (25C), bp 80.5C, refr index 1.382 (20C), flash p 10F (−12.1C), fp −119C, soluble in alcohol, insoluble in water.
Grade: 95%, 99%, research, 99.7 mole %.
Hazard: Flammable, dangerous fire hazard.
Use: Organic synthesis. *Note*: Other isomers (2,2-; 2,3-; 3,3-) of closely similar properties are available.

2,4-dimethylpentanol-3. See diisopropyl carbinol.

2,4-dimethylpentanone-3. See diisopropyl ketone.

α,α-dimethylphenethyl acetate. (dimethylbenzylcarbinyl acetate).
$C_6H_5CH_2C(CH_3)_2OOCCH_3$.
Properties: Colorless liquid; floral, fruity odor; solidifies at room temperature; soluble in mineral oil; insoluble in glycerol; refr index 1.4910–1.4950 (20C) in supercooled liquid form; d 0.995–1.002 in supercooled liquid form; mp 29–30C. Combustible.
Grade: FCC.
Use: Flavoring agent.

dimethylphenol. See xylenol.

dimethyl-p-phenylenediamine. See p-aminodimethylaniline.

N,β-dimethylphenylethylamine. See phenylpropylmethylamine.

2,3-dimethyl-1-phenyl-3-pyrazolin-5-one. See antipyrine.

dimethyl phosphite. $(CH_3O)_2P(O)H$.
Properties: Mobile, colorless liquid; mild odor; d 1.200 (20/4C); refr index 1.400 (25C); bp 72–73C (25 mm); flash p 205F (96C); soluble in water; miscible with most common organic solvents. Combustible.
Use: Lubricant additive, intermediate, adhesive.

O,O-dimethyl phosphorochloridothioate.
(methyl PCT).   CAS: 2524-03-0.
$(CH_3O)_2P(S)Cl$.
Properties: Colorless to light amber liquid, bp 66–67C (16 mm) d 1.320 (25C), refr index 1.4795 (25C). Soluble in alcohol, benzene, acetone, carbon tetrachloride, chloroform, ethyl acetate; slightly soluble in hexane; insoluble in water.
Grade: 96–100% purity.
Hazard: Strong irritant to eyes, skin, and mucous membranes; cholinesterase inhibitor; use may be restricted.
Use: Intermediate for insecticides, pesticides, fungicides, oil and gasoline additives, plasticizers, corrosive inhibitors, flame retardants, flotation agents.

dimethyl phthalate.   CAS: 131-11-3.
$C_6H_4(COOCH_3)_2$.
Properties: Colorless, oily liquid; refr index 1.5138 (25C); heat of combustion 5769 cal/g; d 1.189 (25/25C); bp 282C; flash p 300F (149C); bulk density 9.93 lb/gal (68F); vap press less than 0.1 mm (20C); miscible with alcohol and ether; insoluble in water and paraffinic hydrocarbons; slightly soluble in mineral oil; autoign temperature 1032F (555C). Combustible.
Grade: Technical.
Hazard: Irritant to eyes and mucous membranes, not absorbed by skin. TLV: 5 mg/m³ of air.
Use: Plasticizer for nitrocellulose and cellulose acetate, resins, rubber, and in solid rocket propellants; lacquers; plastics; rubber; coating agents; safety glass; molding powders; insect repellent.

N,N'-dimethylpiperazine.   (1,4-dimethylpiperazine). $(CH_3)_2C_4H_8N_2$.
Properties: Colorless, mobile liquid; d 0.8565 (20/4C); bp 131C; fp −1C; flash p 176F (80C) (TOC). Combustible.
Use: Curing agent for polyurethane foams, intermediate for cationic surface-active agents. *Note*: Two other isomers are available, *cis*-2,5-, and *trans*-2,5-; neither is flammable, but both are combustible.

2,6-dimethylpiperidine.   (2,6-lupetidine).
$(CH_3)_2C_5H_8NH$.
Properties: Liquid, bp 127.9C, d 0.8199 (20/20C), refr index 1.4383 (20C), miscible in water at 20C. Combustible.
Use: Intermediate.

dimethylpolysiloxane.   A liquid defoaming agent, refr index 1.40, viscosity 300 centistokes.
Hazard: Toxic material. Use in foods limited to 10 ppm (0 ppm in milk).

dimethyl-POPOP.   See 1,4-bis[2-(4-methyl-5-phenyloxazolyl)]benzene.

2,2-dimethylpropane.   See neopentane.

2,2-dimethyl-1,3-propanediol.   See neopentyl glycol.

dimethylpyridine.   See lutidine.

2,7-dimethylquinoline. $(CH_3)_2C_9H_5N$.
Properties: Liquid, fp approximately −40C, distillation range 140–150C (20 mm), soluble in benzene and diethyl ether. Combustible.
Use: Organic synthesis, dye intermediate.

dimethyl resorcinol.   See resorcinol dimethyl ether.

dimethyl sebacate. $[(CH_2)_4COOCH_3]_2$.
Properties: Liquid, water-white, d 0.9896 (20/20C), mp 24.5C, flash p 293F (145C), bp approximately 294C, refr index 1.4376 (20C). Combustible.
Grade: Technical.
Use: Solvent or plasticizer for nitrocellulose, vinyl resins; intermediate.

dimethyl silicone.   General term for a family of silicones of composition $[(CH_3)_2SiO]_x$, the more volatile materials formed on hydrolysis of dimethyldichlorosilane. Colorless oils with boiling point ranging from 134C (for x = 3) to 188C (20 mm) (for x = 9).
Use: Transformer liquid, brake fluids.
See also polydimethylsiloxane.

dimethyl sulfate.   (methyl sulfate).
CAS: 77-78-1. $(CH_3)_2SO_4$.
Properties: Colorless liquid; soluble in alcohol, ether, and water; d 1.3516; fp −26.8C; bp 188C (decomposes); flash p 182F (83.3C) (closed cup). Combustible.
Derivation: By adding fuming sulfuric acid to methanol and distilling in vacuo.
Grade: Technical.
Hazard: Strong irritant, absorbed by skin, a carcinogen, induces tumors in animals, protective clothing required. TLV: 0.1 ppm in air.
Use: Methylating agent for amines and phenols, polyurethane-based adhesives.

dimethyl sulfide.   (methyl sulfide).
CAS: 75-18-3. $(CH_3)_2S$.
Properties: Colorless, volatile liquid; disagreeable odor; soluble in alcohol and ether; insoluble in water; d 0.845 (20C); fp −83C; bp 37.5C; evolves sulfur dioxide when heated; autoign temperature 403F (206C); flash p 0F (−17.7C).
Derivation: (a) From kraft pulping black liquor by heating with inorganic sulfur compounds, (b)

by interaction of a solution of potassium sulfide and methyl chloride in methanol.

Hazard: Flammable, dangerous fire risk, moderate explosion risk. Flammable limits in air 2.2–19.7%.

Use: Gas odorant, solvent for many inorganic substances, catalyst impregnator.

2,4-dimethylsulfolane. $C_6H_{12}O_2S$.

Properties: Slightly yellow liquid, bp 280C, d 1.13, flash p 290F (143C), refr index 1.47, slightly soluble in water, partially soluble in alkanes and alkenes, miscible with most aromatics.

Hazard: Toxic by ingestion.

Use: Solvent extraction (liquid-liquid and vapor-liquid).

dimethylsulfone. (methyl sulfone; methylsulfonylmethane). $C_2H_6O_2S$.

Properties: Colorless crystals, mp 110C, bp 237C. Soluble in water, alcohol, and acetone.

Derivation: Oxidation of dimethyl sulfide.

Use: Solvent.

dimethyl sulfoxide. (methyl sulfoxide; DMSO). CAS: 67-68-5. $(CH_3)_2SO$.

Properties: Colorless, hygroscopic liquid; bp 189C; mp 18.5C; d 1.10 (20/20C); specific heat 0.7; flash p 203F (95C) (open cup); dielectric constant 48.9 (20C); nearly odorless; slightly bitter taste; extremely powerful aprotic solvent; soluble in water, alcohol, benzene, acetone, chloroform. Combustible.

Derivation: Oxidation of dimethyl sulfide with nitrogen tetroxide under anhydrous conditions; sulfide waste liquors.

Hazard: Readily penetrates skin and other tissues; approved by FDA for humans but, must comply with FDA regulations.

Use: Solvent for polymerization and cyanide reactions; analytical reagent; spinning polyacrylonitrile and other synthetic fibers; industrial cleaners, pesticides, paint stripping; hydraulic fluids; preservation of cells at low temperatures; diffusion of drugs, etc., into blood stream by topical application; medicine (anti-inflammatory); veterinary medicine; plant pathology and nutrition; pharmaceutical products; metal-complexing agent.

dimethyl terephthalate. (DMT). CAS: 120-61-6. $C_6H_4(COOCH_3)_2$.

Properties: Colorless crystals, mp 140C, sublimes above 300C, insoluble in water, soluble in ether and hot alcohol.

Derivation: Oxidation of p-xylene or mixed xylene isomers with concurrent esterification.

Use: Polyester resins for film and fiber production, especially polyethylene terephthalate; intermediate.

dimethyl 2,3,5,6-tetrachloroterephthalate. $C_6Cl_4(COOCH_3)_2$.

Properties: Crystals, mp 156C, insoluble in water, slightly soluble in acetone and benzene.

Use: Herbicide.

3,5-dimethyltetrahydro-1,3,5(2H)thiadiazine-2-thione. CAS: 533-74-4. $C_5H_{10}N_2S_2$.

Properties: Crystals, mp 100C, d 1.30 (20C), slightly soluble in water and alcohol, soluble in acetone.

Hazard: Toxic by ingestion and inhalation. TLV: 10 mg/m³ of air.

Use: Herbicide, nematocide, preservative for adhesives and proteinaceous additives, fungicide. See also sesone, "Crag."

dimethyltin dichloride. CAS: 753-73-1. $(CH_3)_2SnCl_2$.

Properties: White crystals; mp 106–108C; soluble in water, alcohol, and hydrocarbons.

Hazard: Toxic material by skin absorption. TLV: (as Sn) 0.1 mg/m³.

Use: Electroluminescence, PVC stabilizer, catalyst.

dimethyltin oxide. CAS: 2273-45-2. $(CH_3)_2SnO$.

Properties: White powder, 98.5% min purity.

Hazard: Toxic material by skin absorption. TLV: (as Sn) 0.1 mg/m³.

Use: Intermediate, PVC stabilizer.

N,N'-dimethylurea. (DMU; sym-dimethylurea; 1,3-dimethylurea). CAS: 96-31-1. $(CH_3NH)_2CO$.

Properties: Colorless prisms, d 1.14, mp 106C, bp 270C, soluble in water and alcohol, insoluble in ether.

Use: Intermediate in synthesis of drugs.

1,3-dimethylxanthine. See theophylline.

3,7-dimethylxanthine. See theobromine.

dimetilan. Generic name for 1-dimethylcarbamoyl-5-methyl-3-pyrazolyl dimethylcarbamate.

Properties: Colorless to red-brown solid, mp 68–71C, bp 200C (13 mm), soluble in water and most organic solvents.

Hazard: Toxic by ingestion and inhalation.

Use: Insecticide.

dimolybdenum trioxide. See molybdenum sesquioxide.

"Di-MoN."[93] TM for modified diammonium phosphate.
Use: Low cost source of phosphorus and nitrogen as nutrients in biological waste treatment.

dimpylate. See diazinon.

Dimroth rearrangement. Rearranagement of N-alkylated or arylated iminoheterocycles to the corresponding alkylamino or arylamino heterocycles.

dimyristyl amine. (ditetradecylamine). $(C_{14}H_{29})_2NH$.
Solid, mp 52C, d 0.89, almost insoluble in water.
Use: Intermediate.

dimyristyl ether. (ditetradecyl ether). $(C_{14}H_{29})_2O$.
Properties: Liquid, mp 38–40C, bp 238–248C (4 mm) d 0.8127 (45/4C). Combustible.
Grade: 95% (min) purity.
Use: Electrical insulators, water repellents, lubricants in plastic molding, antistatic substances, chemical intermediates.

dimyristyl sulfide. (ditetradecyl sulfide, dimyristyl thioether). $(C_{14}H_{29})_2S$.
Properties: Solid, mp 49–50C, bp decomposes, d 0.8258 (50/4C).
Grade: 95% (min) purity.
Use: Organic synthesis (formation of sulfonium compounds).

N,N'-di-β-2-naphthyl-m-phenylenediamine. $C_6H_4(NHC_{10}H_7)_2$.
Properties: Colorless needles, mp 191C, sparingly soluble in alcohol, insoluble in water and ether.
Derivation: By heating m-phenylenediamine with β-naphthol and extraction with alcohol.
Use: Organic synthesis.

N,N'-di-β-naphthyl-p-phenylenediamine. (sym-di-β-naphthyl-p-phenylamine; DNPD). $C_6H_4(NHC_{10}H_7)_2$.
Properties: Gray powder, d 1.25, set point 225C (min), purity 98% (min), insoluble in water, slightly soluble in acetone and chlorobenzene.
Use: Antioxidant, stabilizer, polymerization inhibitor, intermediate in organic synthesis.

dinitolmide. See zoalene.

dinitraniline orange. (Permanent Orange). A pigment made from dinitroaniline and β-naphthol. It is a reddish shade of orange that has excellent light-fastness.
Hazard: Toxic by ingestion.

dinitroaminophenol. See picramic acid.

2,4-dinitroaniline. (2,4-dinitraniline).
CAS: 92-02-9. $C_6H_3NH_2(NO_2)_2$.
Properties: Yellow crystals, slightly soluble in alcohol, insoluble in water, d 1.615, mp 188F, flash p 435F (223C). Combustible.
Derivation: Nitration of p-nitroaniline with hot mixed acid.
Grade: Technical, pure.
Hazard: Toxic by ingestion and inhalation, strong irritant.
Use: Intermediate for azo pigments, toner pigment in printing inks, corrosion inhibitor.

2,4-dinitroanisole. (2,4-dintriophenyl methyl ether). $CH_3OC_6H_3(NO)_2$.
Properties: Colorless to yellow, monoclinic needles from water or alcohol; mp 88C; d 1.341 (20/4C) sublimes; slightly soluble in hot water; soluble in alcohol and ether.
Hazard: Toxic by ingestion.
Use: Ovicide, effective against moths, furniture and carpet beetles, cockroaches, and body lice.

dinitrobenzene. CAS: o- 528-29-0; m- 99-65-0; p- 100-25-4. $C_6H_4(NO_2)_2$. M-, o-, and p-isomers.
Properties: Yellow crystals, soluble in chloroform and ethyl acetate, sparingly soluble in benzene, slightly soluble in water; d m- 1.546, o- 1.565, p- 1.6; mp m- 89.9C, o- 117.9C, p- 172–173C; bp m- 302.8C, o- 319C, p- 299C.
Derivation: Nitration of nitrobenzene with hot mixed acid.
Hazard: o-isomer explodes when shocked or heated. Toxic By inhalation and ingestion, absorbed by skin. TLV: 1 mg/m³ of air.
Use: Organic synthesis, dyes, camphor substitute in cellulose nitrate.

5,7-dinitro-1,2,3-benzoxadiazole. See diazodinitrophenol.

3,5-dinitrobenzoyl chloride. $(NO_2(_2C_6H_3COCl)$.

Properties: Yellow crystals, mp 66–68C, bp 196C (12 mm), decomposed by water and. alcohol.
Hazard: An irritant.
Use: Reagent for amino acids and presence of alcohols in acetals and ketals.

2-(2,4-dinitrobenzyl)pyridine.
$(C_5H_4NCH_2C_6H_3(NO_2)_2)$.   Behaves photo-
chromically.
Use: To make plastics sensitive to light.
See photochromism.

2,4-dinitro-6-sec-butylphenol.   (2-sec-butyl-4,6-
dinitrophenol; dinoseb; DNBP).
CAS: 88-85-7.
$CH_3(C_2H_5)CHC_6H_2(NO_2)_2OH$.
A plant growth regulator.
Properties: Reddish-brown liquid, slightly soluble
in water, soluble in alcohol and other organic
solvents. Forms salts with metals and organic
bases.
Hazard: Possible fire risk. Absorbed by skin,
strong irritant.
Use: Insecticide and ovicide but must be used in
the dormant growth season or as a salt form
to reduce toxicity. Herbicide for preemergence
treatment, increases yield of corn 5–10%.

dinitrochlorbenzene.   See 1-chloro-2,4-dinitroben-
zene.

4,6-dinitro-o-cresol.   (4,6-dinitro-2-methylphe-
nol).   CAS: 121-14-2.
$CH_3C_6H_2(NO_2)_2OH$.
Properties: Yellow solid; mp 85.8C; slightly soluble in water; soluble in alcohol, acetone, ether.
Hazard: Toxic, absorbed by skin, use may be restricted. TLV: 0.2 mg/m³ of air.
Use: Dormant ovicidal spray for fruit trees (highly
phytotoxic and cannot be used successfully on
actively growing plants), herbicide, insecticide.

2,6-dinitro-p-cresol.   (DNPC).
$(NO_2)_2CH_3C_6H_2OH$.
Light yellow, crystalline solid.
Grade: Presscake (36–43% active 2,6-DNCP).
Use: Parent compound for intermediates, dyes,
and pharmaceuticals.

dinitrocyclohexylphenol.   (2-cyclohexyl-4,6-dini-
trophenol; dinitro-o-cyclohexylphenol;
DNOCHP).   CAS: 131-89-5.
$C_6H_{11}C_6H_2(NO_2)_2OH$.
Properties: Crystalline solid.
Use: Control of mites on citrus fruits.

2,4-dinitrofluorobenzene.   (1-fluoro-2,4-dinitro-
benzene; DNFB). $C_6H_3F(NO_2)_2$.
Properties: Crystals; soluble in ether, benzene,
propylene glycol; mp 26C; bp 137C.
Hazard: Toxic by ingestion, mutagenic and carcinogenic.
Use: Alkylating agent, reagent in elucidating
amino acid sequence in proteins.

dinitrogen tetroxide.   See nitrogen dioxide, nitrogen oxides, NO_x.

dinitrogen trioxide.   See nitrogen trioxide, nitrogen oxides, NO_x.

dinitroglycol.   See diethylene glycol dinitrate.

2,4-dinitro-4-hydroxydiphenylamine.
$(NO_2)_2C_6H_3NHC_6H_4OH$.
Properties: Yellow solid, mp 190C, insoluble in
water.
Derivation: Condensation of 2,4-dinitro-1-chloro-
benzene and p-aminophenol.

dinitro(1-methylheptyl)phenyl crotonate.   See dinocap.

4,6-dinitro-2-methylphenol.   See 4,6-dinitro-o-cre-
sol.

dinitronaphthalene. $C_{10}H_6(NO_2)_2$.   Isomers:
(a) 1,5-, (b) 1,8-.
Properties: (a) Yellowish-white needles, (b) yellowish-white, thick, crystalline tablets; mp (a)
217C, (b) 172C); (a) sparingly soluble in pyridine,
(b) soluble in pyridine; bp (a) sublimes, (b) decomposes.
Derivation: By dissolving α-nitronaphthalene in
sulfuric acid and adding nitric acid.
Method of purification: Crystallization.
Hazard: Moderate fire and explosion risk.
Use: Dyes, especially sulfur colors, intermediates.

2,4-dinitro-1-naphthol-7-sulfonic acid.   (flavianic
acid). $C_{10}H_6O_8N_2S$.
Properties: Yellow needles, mp 151C, very soluble
in water.
Use: Intermediate, precipitant for organic bases,
reagent for amino acids.

dinitrophenol. $C_6H_3OH(NO_2)_2$.
CAS: 25550-58-7.   Commercial product is
usually a mixture of 2,3-, 2,4-, and 2,6-isomers.
Properties: Yellow crystals, (2,3) d 1.681, mp
144C; (2,4) d 1.683, mp 114–115C; (2,6) mp 63C.
Soluble in alcohol, ether, benzene, and chloroform; slightly soluble in water.
Derivation: (a) By heating phenol with dilute sulfuric acid, cooling the product, and then nitrating, keeping the temperature approximately 50C;
(b) by nitration with mixed acid with careful
temperature control.
Method of purification: Crystallization.
Hazard: Severe explosion hazard when dry. Absorbed by skin; dust inhalation may be fatal.
Use: Dyes, especially sulfur colors, picric acid,
picramic acid, preservation of lumber, manufacture of the photographic developer diaminophe-
nol hydrochloride, explosives manufacture, indicator, reagent for potassium and ammonium
ions.

dinitrophenylhydrazine. $(NO_2)_2C_6H_3NHNH_2$.

Properties: Red, crystalline powder; mp about 200C; slightly soluble in water and alcohol; soluble in moderately dilute inorganic acids; readily soluble in diglyme.
Hazard: Severe explosion and fire risk.
Use: Explosive, reagent for aldehydes and ketones.

2,4-dinitrophenyl methyl ether. See 2,4-dinitroanisole.

3,5-dinitrosalicylic acid.
$C_6H_2(OH)(NO_2)_2COOH$.
Properties: Yellow crystals, slightly soluble in water, soluble in alcohol and benzene, mp 174C.
Derivation: Nitration of salicylic acid.
Use: Determination of glucose.

dinitrosopentamethylenetetramine. (DNPT).
$(NO)_2C_5H_{10}N_4$. A bicyclic compound.
Properties: Light cream-colored powder; decomposes in air at 190–200C; soluble in dimethyl formamide; somewhat soluble in pyridine, methyl ethyl ketone, and acetonitrile. Combustible.
Hazard: May explode at 200C (390F).
Use: Blowing agent for rubber and plastics.

2,4-dinitrosoresorcinol. CAS: 35860-51-6.
$C_6H_2(OH)_2(NO)_2 \cdot H_2O$.
Properties: Light brown powder, mp 162–163C, decomposes sometimes violently, soluble in water and most organic solvents.
Grade: Technical (13.7% N).
Hazard: Severe explosion risk when shocked or heated. An irritant.
Use: Chelation of heavy metals, cross-linking agent, blasting caps and explosive primers.

3,5-dinitro-o-toluamide. See zoalene.

dinitrotoluene. (DNT). CAS: 121-14-2.
$C_6H_3CH_3(NO_2)_2$. (a) 2,4-; (b) 3,4-; (c) 3,5-.
Properties: Yellow crystals, soluble in alcohol and ether, very slightly soluble in water, d (a) 1.3208, (b) 1.32, (c) 1.277; mp (a)70.5C, (b) 61C, (c) 92.3C. A commercial grade consisting of a mixture of the three isomers is an oily liquid. Combustible.
Derivation: Nitration of nitrotoluene with mixed acid.

Method of purification: Crystallization.
Hazard: Absorbed by skin. TLV: 1.5 mg/m³ of air.
Use: Organic synthesis, toluidines, dyes, explosives.

dinker. A machine for cutting forms from flat sheets of plastic, rubber, metal, paper, etc., by impact of a metal die.

dinocap. CAS: 6119-92-2. Generic name for 2-(2-methylheptyl)-4,6-dinitrophenyl crotonate. $CH_3CH:CHCOOC_6H_2(NO_2)_2CH(CH_3)C_6H_{13}$.
Properties: Brown liquid, bp 138–140C (0.05 mm), insoluble in water, soluble in most organic solvents.
Hazard: Toxic by ingestion and inhalation, strong irritant.
Use: Acaricide, fungicide.

dinonyl adipate. Ester of nonyl alcohol.
Properties: Colorless liquid, bp 201–210C (1 mm), d 0.926 (25C), refr index 1.4523 (20C), viscosity 14.9 centistokes (100F). Combustible.
Use: Plasticizer where special low-temperature properties are desired.

dinonyl carbonate. $(C_9H_{19})_2CO_3$. Ester of nonyl alcohol, colorless liquid, bp 135–140C (0.3 mm), d 0.894 (25C), refr index 1.4427 (20C). Combustible.

dinonyl ether. $C_9H_{19}OC_9H_{19}$.
Properties: Colorless liquid, bp 148–153C (5 mm), d 0.817 (25C), refr index 1.4405 (20C). Can be made from nonyl alcohol plus nonyl halide by the Williamson synthesis. Combustible.

dinonylmaleate.
$C_9H_{19}OOCCH:CHCOOC_9H_{19}$.
Ester of nonyl alcohol, colorless liquid, bp 157–167C (0.1 mm), d 0.941 (25C), refr index 1.4586 (20C), viscosity 6900 centistokes (−40F), 17.47 centistokes (100F), 3.50 centistokes (210F). Combustible.

dinonyl phenol. $(C_9H_{19})_2C_6H_3OH$.
Properties: Colorless liquid, insoluble in water, soluble in common organic solvents. Combustible.
Hazard: See phenol.
Use: Solvent.

dinonyl phthalate. (DNP).
$C_6H_4(COOC_9H_{19})$. Ester of nonyl alcohol.
Properties: Colorless liquid, bp 205–220C (1 mm), d 0.979 (25C), refr index 1.4871 (20C), viscosity 55.3 centistokes (37.7C), flash p 420F (215C). Combustible.

Use: General-purpose, low-volatility plasticizer for vinyl resins; pure grade as stationary liquid phase in chromatography.

dinoseb. Legal label name (air) for 2,4-dinitro-6-sec-butyl-phenol.

DIOA. Abbreviation for diisooctyl adipate.

dioctadecylamine. See distearylamine.

2,6-dioctadecyl-p-cresol. See 2,6-distearyl-p-cresol.

dioctadecyl ether. See distearyl ether.

dioctadecyl sulfide. See distearyl sulfide.

3,3'-dioctadecyl thiodipropionate. See distearyl thiodipropionate.

dioctyl adipate. See di(2-ethylhexyl) adipate.

dioctylamine. See di-2-ethylhexylamine.

dioctylaminoethanol. See di(2-ethylhexyl)ethanolamine.

dioctyl azelate. See di(2-ethylhexyl) azelate.

dioctyl chlorophosphate. (dioctyl phosphorochloridate). $(C_8H_{17}O)_2P(O)Cl$.
Properties: Water-white liquid, d 0.991 (25C), refr index 1.445 (25C), decomposes on distillation, soluble in common inert organic solvents, insoluble in water. Combustible.
Use: Intermediate, insecticide.

di(n-octyl-n-decyl)adipate. (DNODA).
Properties: Clear, oily liquid; d 0.912–0.920 (25/25C); refr index 1.443–1.447 (25C). Combustible.
Use: Low-temperature plasticizer.

di(n-octyl-n-decyl) phthalate. (DNODP).
Properties: Clear, oily liquid; slight odor; acidity (as phthalic acid) 0.01% max; d 0.968–0.977 (25/25C); crystallizing point −30C; bp 232–267C (4 mm); flash p 426F (219C). Combustible.
Use: Plasticizer for polyvinyl chloride and other vinyls.

di-n-octyl diphenylamine.
$C_8H_{17}C_6H_4NHC_6H_4C_8H_{17}$.
Properties: Light tan powder; d 0.99; mp 80–90C; soluble in benzene, gasoline, acetone, and ethylene dichloride; insoluble in water.
Use: Antioxidant for petroleum-based and synthetic lubricants and plastics.

dioctyl ether. $(C_8H_{17})_2O$.
Properties: Liquid, fp −7C, bp 291.7C, d 0.805 (17/4C) refr index 1.4329 (24C). Combustible.
Grade: 95% (min) purity.
Use: Electrical insulator, water repellent, mold lubricant, antistatic agent, intermediate.

dioctyl fumarate. See di(2-ethylhexyl) fumarate.

dioctyl hexahydrophthalate. See di(2-ethylhexyl) hexahydrophthalate.

N,N'-di-n-octyl-p-phenylenediamine.
$C_6H_4(NHC_8H_{17})_2$.
Properties: Colorless liquid, bp approximately 390C, d 0.912 (15C), pour point −4C, flash p 395F (201C) (Pensky-Martin), refr index 1.5129 (20C). Completely miscible in methanol, pentane, and benzene; vap press (absolute) 0.33 mm (150C). Combustible.
Use: Antioxidant, antiozonant for gasoline, mercaptans, synthetic rubber.

dioctyl phosphite. (dioctyl phosphonate).
$(C_8H_{17}O)_2P(O)H$.
Properties: Water-white liquid, bp 150–155C (2–3 mm), d 0.929 (25C), refr index 1.4418 (25C), soluble in common organic solvents. Combustible.
Use: Solvent, antioxidant, intermediate.

dioctyl phosphoric acid. See di(2-ethylhexyl) phosphoric acid.

dioctyl phosphorochloridate. See dioctyl chlorophosphate.

dioctyl phthalate. See di(2-ethylhexyl) phthalate.

di(2-octyl) phthalate. See dicapryl phthalate.

dioctyl sebacate. See di(2-ethylhexyl) sebacate.

dioctyl sodium sulfosuccinate. (di(2-ethylhexyl) sodium sulfosuccinate; sodium dioctyl sulfosuccinate). CAS: 577-11-7.
$C_8H_{17}OOCCH_2CH(SO_3Na)COOC_8H_{17}$.
An anionic surface-active agent.
Properties: White, wax-like solid with characteristic odor. Slowly soluble in water; freely soluble in alcohol, glycerol, carbon tetrachloride, acetone, xylene; saponification value 240–253; stable in acid and neutral solutions; hydrolyzes in alkaline solutions.
Derivation: By esterification of maleic anhydride with 2-ethylhexyl alcohol followed by addition of sodium bisulfite.
Hazard: Use in food products restricted.
Grade: NF, FCC.
Use: Food additive (processing aid in sugar indus-

try, stabilizer for hydrophilic colloids), wetting agent, dispersant, emulsifier.

dioctyl succinate. See di(2-ethylhexyl) succinate.

dioctyl sulfide. (dioctyl thioether).
$(C_8H_{17})_2S$.
Properties: Liquid, mp 0.5C, bp 180C (10 mm), d 0.8419 (17/17C), refr index 1.4606 (20C). Combustible.
Grade: 95% (min) purity.
Use: Organic synthesis (formation of sulfonium compounds).

dioctyl thioether. See dioctyl sulfide.

dioctyl thiopropionate. See 3,3'-(2-ethylhexyl) thiodipropionate.

di(n-octyl)tin-S,S'-bis-(isooctylmercaptoacetate).
CAS: 26401-97-8. A heat stabilizer for PVC food packaging materials, especially for clear plastic bottles. Approved by FDA for all foods except malt beverages, carbonated soft drinks, milk, and other dairy products.

"Diofan."[440] TM for dispersions of vinylidene copolymers for paper coating.

diol. Synonym for glycol or dihydric alcohol.

diolefin. (diene; alkadiene). An aliphatic compound (olefin) containing two double bonds, e.g., butadiene.

"Dionin."[123] $Cl_9H_{23}NO_3$. TM for ethylmorphine hydrochloride.
Hazard: Toxic by ingestion, narcotic and habituating.
Use: Medicinally as a substitute for morphine.

DIOP. Abbreviation for diisooctyl phthalate.

DIOS. Abbreviation for diisooctyl sebacate.

dioxacarb. (O-(1,3-dioxolan-2-yl)-phenyl-N-methylcarbamate). CAS: 6988-21-2.
Properties: Mp 114–15C, vap press 0.3 mm at 20C, soluble in water to 0.6%.
Use: Control for cockroaches and stored product pests.

1,4-dioxane. (diethylene ether; 1,4-diethylene dioxide; diethylene oxide; dioxyethylene ether).

CAS: 123-91-1. $\overline{OCH_2CH_2OCH_2CH_2}$

Properties: Colorless liquid, ethereal odor, stable, miscible with water and most organic solvents, bp 101.3C, fp 10–12C, d 1.0356 (20/20C), 8.61 lb/gal (20C), refr index 1.4221 (20C), flash p 65F (18.3C) (ASTM OC), autoign temperature 356F (180C).
Derivation: (a) Ethylene glycol by treatment with acid, (b) from β,β-dichloroethyl ether by treatment with alkali.
Grade: Reagent, technical, spectrophotometric, scintillation.
Hazard: Flammable, dangerous fire risk, may form explosive peroxides. Toxic by inhalation, absorbed by skin, a carcinogen. TLV: (technical grade) 25 ppm in air. OSHA standard: 100 ppm in air.
Use: Solvent for cellulosics and wide range of organic products; lacquers; paints; varnishes; paint and varnish removers; wetting and dispersing agent in textile processing, dye baths, stain and printing compositions; cleaning and detergent preparations; cements; cosmetics; deodorants; fumigants; emulsions; polishing compositions; stabilizer for chlorinated solvents, scintillation counter.

dioxathion. ("Delnav"; p-dioxane-2,3-diyl ethyl phosphorodithioate). CAS: 78-34-2.
$C_4H_6O_2[SPS(OC_2H_5)_2]_2$.
Properties: Viscous, brown liquid; fp −20C; bulk d 1.257 (26C); insoluble in water; soluble in hexane. The technical material is a mixture of cis and trans isomers.
Hazard: Toxic by inhalation, ingestion, and skin absorption; cholinesterase inhibitor; use may be restricted. TLV: 0.2 mg/m³ of air.
Use: Insecticide, miticide.

dioxin. The commonly accepted, though chemically imprecise, name for the compound 2,3,7,8-tetrachlorodibenzo-p-dioxin (TCDD), which is only one of more than 70 members of the family of chlorinated dioxins. It was found to be a contaminant of the herbicide 2,4,5-T (trichlorophenoxyacetic acid) some ten years after the latter was approved for use; it was then banned by FDA for most purposes. Synthesized in 1957, dioxin is a white, crystalline solid in pure form. It was present as a contaminant in defoliants used in Vietnam (agent orange), and its toxicity was widely publicized. Though it is undoubtedly harmful to humans, no deaths have occurred. It is a carcinogen, a teratogen, and a mutagen. Its toxicity to laboratory animals varies widely with the species; it is lethally toxic to guinea pigs, but hamsters appear relatively unaffected. Where soil contamination occurs, the concentration is no greater than several parts per million at most. Wastes contaminated with dioxin must be disposed of in officially approved landfills. Human toxicology is under continuing investigation. *Note*: A detailed study of the whole dioxin situation appeared in C&E News of June 6, 1983.

"Dioxitol."[125] TM for diethylene glycol monoethyl ether, bp 202C, d 0.990 (20/20C).

3,5-dioxo-1,2-diphenyl-4-n-butylpyrazolidine. See phenylbutazone.

1,3-dioxolane. (ethylene glycol formal).

CAS: 646-06-0. $\overline{OCH_2CH_2OCH_2}$. A cyclic acetal.
Properties: Water-white liquid, soluble in water, stable under neutral or slightly alkaline conditions, d 1.065, bp 74C, flash p 35F (1.67C) (open cup), vap press 70 mm (20C) bulk density 8.2 lb/gal (20C).
Derivation: Reaction of formaldehyde with ethylene glycol.
Hazard: Flammable, dangerous fire risk. Toxic by inhalation and ingestion.
Use: Low-boiling solvent and extractant for oils, fats, waxes, dyes, and cellulose derivatives.

dioxolone-2. See ethylene carbonate.

dioxopurine. See xanthine.

dioxyanthraquinone. See dihydroxyanthraquinone.

dioxybenzene. See dihydroxybenzene.

dioxyethylene ether. See 1,4-dioxane.

DIOZ. Abbreviation for diisoctyl azelate.

DIPA. Abbreviation for diisopropanolamine.

dipalmitylamine. (dihexadecylamine).
$(C_{16}H_{33})_2NH$.
Properties: Solid, mp 65C, d 0.83, slightly soluble in water.
Use: Intermediate.

dipentaerythritol.
$(CH_2OH)_3CCH_2OCH_2C(CH_2OH)_3$. Occurs in technical pentaerythritol, an off-white, free-flowing powder. The molecule contains six primary hydroxyl groups, all esterifiable, mp 212–220C, d 1.33 (25/4C).
Use: Paints and coatings.

dipentamethylenethiuram tetrasulfide.

$[\overline{CH_2(CH_2)_4NCS}]_2S_4$.
Properties: Light gray powder; d 1.53; mp 110C min; soluble in chloroform, benzene, acetone; insoluble in water.
Use: Ultra-accelerator for rubber.

dipentene. (cinene; limonene, inactive; *dl*-p-mentha-1,8-diene; cajeputene). CAS: 138-86-3. $C_{10}H_{16}$. Commercial form is high in dipentene content, but also contains other terpenes and related compounds in varying amounts.
Properties: Colorless liquid, lemon-like odor, d 0.847 (15.5/15.5C), bp 175–176C, refr index 1.473 (20C), flash p 113F (45C) (closed cup), fp −97C, 7.15 lb/gal (15.5C), autoign temperature 458F (236C), miscible with alcohol, insoluble in water.
Derivation: (a) From various essential oils, (b) by close fractionation of wood turpentine, (c) by-product in making synthetic camphor.
Grade: Steam-distilled, destructively distilled.
Hazard: Moderate fire risk. Combustible.
Use: Solvent for oleoresinous products, rosin, ester gum, alkyd resins, waxes, metallic soap driers, rubber, etc.; rubber compounding and reclaiming; dispersing agent for oils, resins, resin-oil combinations, pigments and driers; paints, enamels, lacquers, and varnishes; general wetting and dispersing agent; printing inks; perfumes; flavors; floor waxes and furniture polishes; synthetic resins, polyterpenes.

dipentene dioxide. (limonene dioxide).
CAS: 96-08-2. $C_{10}H_{16}O_2$.
Properties: Liquid, d 1.0287 (20C), bp 242C, fp −100C, soluble in water.
Use: Intermediate for plasticizers, epoxy resins; pharmaceuticals.

dipentene glycol. See terpin hydrate.

dipentene monoxide. (limonene monoxide).
$C_{10}H_{14}O$.
Properties: Liquid, d 0.929 (20C), fp −6C, bp 75C, flash p 152F (66C). Combustible.
Use: Organic intermediate, epoxy resins.

"Dipentite."[62] TM for diphenylpentaerythritol diphosphite.

di-n-pentylamine. See di-n-amylamine.

2,5-di(tert-pentyl)hydroquinone. See 2,5-di(tert-amyl)-hydroquinone.

diphacinone. Generic name for 2-diphenylacetyl-1,3-indandione.
Use: Rodenticide.
See diphenadione.

diphemanil methyl sulfate. (4-diphenylmethylene-1,1-dimethylpiperidinium methyl sulfate).

$(C_6H_5)_2C\overline{CCH_2CH_2N(CH_3)_2CH_2CH_2}\cdot CH_3SO_4$.
Properties: White, bitter, crystalline solid with faint characteristic odor; mp 189–196C; very

slightly soluble in ether; slightly soluble in alcohol, chloroform, and water; stable to heat and light; somewhat hygroscopic; pH (1% solution) 4.0–6.0.
Grade: NF.
Use: Medicine (anticholinergic).

diphenadione. (2-diphenylacetyl-1,3-indandione; diphacinone). CAS: 82-66-6. $C_{23}H_{16}O_3$.
Properties: Yellow, odorless crystals or crystalline powder; mp 145–147C; practically insoluble in water; soluble in acetone and acetic acid.
Hazard: Prevents blood clotting.
Use: Medicine (anticoagulant), rodenticide.

diphenamid. CAS: 957-51-7. Generic name for N,N-dimethyl-2,2-diphenylacetamide.
$(C_6H_5)_2CHCON(CH_3)_2$.
Properties: White solid, mp 134.5–135.5C, soluble in water, acetone, dimethyl formamide, and phenyl "Cellosolve."
Hazard: Toxic by ingestion.
Use: Herbicide, plant growth regulator.

diphenatrile. See diphenylacetonitrile.

diphenhydramine hydrochloride. [-(2-benzhydryloxy)-N,N-dimethylethylamine hydrochloride].
$(C_6H_5)_2CHOCH_2CH_2N(CH_3)_2 \cdot HCl$.
Properties: White, odorless, crystalline powder; darkens slowly on exposure to light; mp 166–170C; solutions practically neutral to litmus paper; soluble in water, alcohol, and chloroform; very slightly soluble in benzene and ether.
Grade: USP.
Hazard: Toxic. Prescription only, do not use with alcohol or other central nervous system depressants.
Use: Medicine (antihistamine).

diphenic acid. (2,2′-biphenyldicarboxylic acid).
$HOOCC_6H_4C_6H_4COOH$.
Properties: White needles, mp 228–229C, soluble in hot water.
Use: Synthesis of dyes, detergents, pharmaceuticals.

diphenolic acid. See 4,4-bis(4-hydroxyphenyl) pentanoic acid.

diphenyl. (biphenyl). CAS: 92-52-4.
$C_6H_5C_6H_5$. Several crystalline forms are known.

Properties: White scales, pleasant odor, soluble in alcohol and ether, insoluble in water, d ap-

proximately 1, mp 70C, bp 256C, flash p 235F (112.7C).
Derivation: (a) By slowly passing benzene through a red hot iron tube, (b) by heating bromobenzene and sodium with subsequent distillation.
Hazard: Toxic. TLV: 0.2 ppm in air.
Use: Organic synthesis, heat-transfer agent, fungistat in packaging of citrus fruit, plant disease control, manufacture of benzidine, dyeing assistant for polyesters.
See also chloridiphenyl.

diphenylacetic acid. $(C_6H_5)_2CHCOOH$.
Properties: Colorless, odorless crystals; bp sublimes; mp 147.8–148.2C. Soluble in hot water, alcohol, ether, chloroform.

diphenylacetonitrile. (diphenatrile).
$(C_6H_5)_2CHCN$.
Properties: Yellow, crystalline powder; mp 73–73.5C; insoluble in water but very soluble in alcohol.
Use: Preparation of diphenylacetic acid, synthesis of antispasmodics, herbicide.

diphenylacetylene. See tolan.

2-diphenylacetyl-1,3-indandione. See diphenadione.

diphenylamine. (DPA; N-phenylaniline).
CAS: 122-39-4. $(C_6H_5)_2NH$.
Properties: Colorless to grayish crystals; soluble in carbon disulfide, benzene, alcohol, and ether; insoluble in water. D 1.159, mp 52.85C, bp 302C, flash p 307F (152.7C), autoign temperature 1173F (633C). Combustible.
Derivation: By heating equal formula weights of aniline and aniline hydrochloride in an autoclave. The product is boiled with dilute hydrochloric acid to remove the unaltered aniline and the residue is distilled.
Grade: Technical; refined, flake, and fused.
Hazard: Toxic by ingestion. TLV: 10 mg/m³ of air.
Use: Rubber antioxidants and accelerators, solid rocket propellants, pesticides, dyes, pharmaceuticals, veterinary medicine, storage preservation of apples, stabilizer for nitrocellulose, analytical chemistry.

diphenylamine chloroarsine. (adamsite; phenarsazine chloride; DM). CAS: 578-94-9.
$C_6H_4(AsCl)(NH)C_6H_4$.
Properties: Canary-yellow crystals, sublimes readily, d 1.65, mp 195C, bp 410C (decomposes). Insoluble in water; soluble in benzene, xylene, carbon tetrachloride.
Derivation: By heating diphenylamine with arsenic trichloride.

Hazard: Toxic by inhalation and ingestion, strong irritant.
Use: Military poison gas, wood treating.

9,10-diphenylanthracene. $C_{14}H_8(C_6H_5)_2$.
Properties: Crystals, mp 248–250C, insoluble in water and alcohol, slightly soluble in toluene.
Grade: Purified.
Use: Primary fluor or wavelength shifter in soluble scintillators.

1,4-diphenylbenzene. See terphenyl.

diphenylbenzidine.
$C_6H_5HNC_6H_4C_6H_4NHC_6H_5$.
Properties: White powder, insoluble in water, slightly soluble in alcohol and acetone, soluble in boiling toluene, sensitive to light, mp 242C.
Derivation: Diphenylamine and fuming sulfuric acid.
Hazard: May be carcinogenic; see benzidine.
Use: Determination of zinc and nitrites.

2,5-diphenyl-p-benzoquinone.
$C_6H_5C_6H_2O_2C_6H_5$.
Properties: Greenish-yellow solid; mp 210–214C; slightly soluble in styrene, benzene, acetone, and ethyl acetate.
Use: Polymerization inhibitor.

diphenylbromoarsine. $(C_6H_5)_2AsBr$.
Properties: White crystals, mp 54–56C.
Derivation: (a) Hydrobromic acid and diphenylarsenious oxide are heated together for approximately 4 hours at 115–120C; (b) by action of arsenic tribromide on triphenyl arsine at 300–350C.
Hazard: Strong irritant. A poison.

1,3-diphenyl-2-buten-1-one. See dypnone.

diphenylcarbazide. $(C_6H_5NHNH)_2CO$.
Properties: White crystals or flakes, insoluble in water, soluble in alcohol and acetone, mp 173C, decomposes in light.
Derivation: Phenylhydrazine and urea.
Use: Determination of chromium and other metals, indicator for iron.

diphenylcarbinol. See benzhydrol.

diphenyl carbonate. CAS: 102-09-0.
$(C_6H_5O)_2CO_3$.
Properties: White, crystalline solid; can be halogenated and nitrated in characteristic manner; readily undergoes hydrolysis and ammonolysis; soluble in acetone, hot alcohol, benzene, carbon tetrachloride, ether, glacial acetic acid, and other organic solvents; insoluble in water; bp 302C; mp 78C; d 1.1215 (87/4C).

Grade: Technical.
Use: Plasticizer and solvent, synthesis of polycarbonate resins.

diphenyl, chlorinated. See chlorodiphenyl.

diphenylchloroarsine. CAS: 712-48-1.
$(C_6H_5)_2AsCl$.
Properties: Colorless crystals or dark-brown liquid which slowly becomes semi-solid, decomposed by water (slowly). Soluble in carbon tetrachloride, chloropicrin, phenyldichloroarsine; almost insoluble in water. D 1.363 (40C) (solid), or 1.356 (45C) (liquid), bp 333C (in CO_2 atm), mp 41C.
Derivation: Benzene and arsenic trichloride are heated in presence of aluminum chloride.
Hazard: A poison. Toxic by inhalation, strong irritant to tissue.
Use: Military poison gas.

diphenyldecyl phosphite. $(C_6H_5O)_2POC_{10}H_{21}$.
Properties: Nearly water-white liquid, d 1.023 (25/15.5C), mp 18C, refr index 1.5160 (25C). Combustible.
Use: Chemical intermediate, stabilizer for polyvinyl and polyolefin resins.

2,4′-diphenyldiamine. (2,4′-biphenyldiamine; 2,4′-diaminodiphenyl). $C_{12}H_{12}N_2$.
Properties: Crystalline needles, slightly soluble in alcohol and ether, mp 45C, bp 360C.
Derivation: Reduction of azobenzene with tin and hydrochloric acid.
Use: Azo dye manufacture, tungsten determination.

diphenyldichlorosilane. CAS: 80-10-4.
$(C_6H_5)_2SiCl_2$.
Properties: Colorless liquid, bp 305C, fp −22C, d 1.19 (20C), refr index 1.5773 (25C), flash p (COC) 288F (142C), readily hydrolyzed by moisture with liberation of hydrogen chloride. Combustible.
Derivation: (a) Reaction of powdered silicon and chlorobenzene in the presence of copper powder as catalyst, (b) reaction of phenylmagnesium chloride with silicon tetrachloride.
Grade: Technical.
Hazard: Strong irritant to tissue.
Use: Intermediate for silicone lubricants.

diphenyldi-n-dodecylsilane.
$(C_6H_5)_2Si(C_{12}H_{25})_2$.
Properties: Colorless oil.
Derivation: Reaction of didodecyldichlorosilane with phenyl lithium.
Use: High-temperature lubricant.

diphenyldiimide. See azobenzene.

diphenyldimethoxysilane. $(C_6H_5)_2Si(OCH_3)_2$.
Properties: Liquid, d 1.080 (25C), bp 191C (53 mm), refr index 1.5404 (25C). Soluble in acetone, benzene, methyl alcohol. Combustible.
Use: Treatment of powders, glass, paper and fabrics.

diphenyleneimine. See carbazole.

α-diphenylenemethane. See fluorene.

diphenylene oxide. (dibenzofuran).
CAS: 38178-38-0. $C_{12}H_8O$ (tricyclic).
Properties: Crystalline solid, mp 87C, bp 288C. Insoluble in water; slightly soluble in alcohol, ether, and benzene.
Derivation: From coal-tar.
Use: Insecticide.

uns-diphenylethane. (1,1-diphenylethane).
$(C_6H_5)_2CHCH_3$.
Properties: Colorless liquid; soluble in chloroform, ether, carbon dilsulfide; bp 286C; d 1.004 (20C); fp −21.5C; flash p 264F (129C). Combustible.
Derivation: Action of acetaldehyde on benzene in presence of concentrated sulfuric acid.
Use: Solvent for nitrocellulose, organic synthesis.

sym-diphenylethane. (bibenzyl; dibenzyl; 1,2-diphenyl-ethane). $C_6H_5CH_2CH_2C_6H_5$.
Properties: White, crystalline needles or small plates; soluble in alcohol, chloroform, ether, carbon disulfide; insoluble in water; d 0.9782; bp 284C; mp 52C.
Derivation: (a) By treating benzyl chloride with metallic sodium, (b) action of benzyl chloride on benzylmagnesium chloride.
Use: Organic synthesis.

diphenyl ether. See diphenyl oxide.

diphenylethylene. See stilbene.

N,N-diphenylethylenediamine. (ethyl diphenyldiamine). $C_6H_5NHCH_2CH_2NHC_6H_5$.
Properties: Cream-colored solid, d 1.14, softening point 64C. Insoluble in water; soluble in acetone, ethylene dichloride, benzene, and gasoline.
Use: Antioxidant in rubber compounding.

diphenylglycolic acid. See benzilic acid.

N,N'-diphenylguanidine. (DPG; melaniline).
CAS: 102-06-7. $HN:C(NHC_6H_5)_2$.
Properties: White powder, bitter taste, slight odor, d 1.13, mp 147C, decomposes above 170C. Soluble in ethanol, carbon tetrachloride, chloroform, hot benzene, and toluene; slightly soluble in water.

Derivation: Treatment of aniline with cyanogen chloride.
Hazard: Toxic by ingestion.
Use: Basic rubber accelerator, primary standard for acids.

1,6-diphenylhexatriene. (DPH).
$C_6H_5HC:CHCH:CHCCH:CHC_6H_5$.
Use: Wavelength shifter in soluble scintillation counting.

diphenyl isophthalate. (DPIP).
$C_6H_5OOCC_6H_4COOC_6H_5$.
Properties: White solid, mp 138–139C. Combustible.
Use: Manufacture of polybenzimidazoles, high-temperature-resistant polymers.

diphenylketone. See benzophenone.

diphenylmethane. (benzylbenzene).
$(C_6H_5)_2CH_2$.

Properties: Long, colorless needles; soluble in alcohol, chloroform, hexane, benzene, ether; insoluble in liquid ammonia; d 1.0056; mp 26.5C; bp 264.7C; flash p 266F (130C). Combustible.
Derivation: Condensation of benzyl chloride or methylene chloride and benzene in presence of aluminum chloride.
Hazard: Toxic by ingestion.
Use: Organic synthesis, dyes, perfumery.

diphenylmethane-4,4'-diisocyanate. (MDI; methylene-di-p-phenylene isocyanate; methylene bisphenyl isocyanate). CAS: 101-68-8.
$CH_2(C_6H_4NCO)_2$.
Properties: Light-yellow, fused solid; solidification point 37C; d 1.197 (70C); soluble in acetone, benzene, kerosene, and nitrobenzene. Combustible.
Derivation: p,p'-Diaminodiphenylmethane and phosgene.
Hazard: Toxic by inhalation of fumes; strong irritant. TLV: CL 0.02 ppm in air.
Use: Preparation of polyurethane resin and spandex fibers, bonding rubber to rayon and nylon.

diphenylmethanol. See benzhydrol.

diphenylmethyl bromide. (benzhydryl bromide).
$BrCH(C_6H_5)_2$.
Properties: Solid, mp 45C, bp 193C (26 mm), decomposes in hot water, soluble in alcohol, very soluble in benzene.

Hazard: Strong irritant to eyes and skin.
Use: Organic synthesis.

diphenylmethylchlorosilane.
$(C_6H_5)_2(CH_3)SiCl$.
Properties: Colorless liquid, d 1.107 (25C), bp 295C, flash p 135F (57.2C). Combustible.
Derivation: Grignard reaction of diphenyldichlorosilane with methylmagnesium chloride.
Hazard: Moderate fire risk.
Use: Intermediate, end stopper for silicone oils.

diphenylnaphthylenediamine.
$C_{10}H_6(NHC_6H_5)_2$.
Properties: Silvery, crystalline plates; slightly soluble in alcohol; insoluble in water; mp 164C.
Derivation: By heating 2,7-dihydroxynaphthalene with aniline and aniline hydrochloride.
Use: Organic synthesis.

diphenylnitrosamine.   See N-nitrosodiphenylamine.

2,5-diphenyloxazole.   (DPO).

$\overline{OOC_6H_5}{:}CC_6H_5$.
Properties: White, fluffy solid; mp 70–72C.
Grade: Scintillation.
Use: Primary fluor used as scintillation counter or wavelength shifter.

diphenyl oxide.   (phenyl ether; diphenyl ether).
CAS: 101-84-8. $(C_6H_5)_2O$.
Properties: Colorless crystals or liquid, geranium-like odor, soluble in alcohol and ether, insoluble in water, d 1.072–1.075, mp 27C, bp 259C, flash p 239F (115C), autoign temperature 1144F (618C). Combustible.
Derivation: Reaction of bromobenzene and sodium phenate heated under pressure.
Grade: Technical, perfume, industrial.
Hazard: TLV: 1 ppm. Toxic by inhalation of vapor.
Use: Perfumery, particularly soaps; heat-transfer medium resins for laminated electrical insulation; chemical intermediate for such reactions as halogenation, acylation, alkylation, etc.

diphenylpentaerythritol diphosphite.
$C_6H_5OP(OCH_2)_2C(CH_2O)_2POC_6H_5$ (spiro).
Properties: White powder, bp 190–200C (0.1 mm).
Use: Stabilizer for resins.

N,N′-diphenyl-m-phenylenediamine.
$C_6H_4(NHC_6H_5)_2$.
Properties: Crystalline needles, soluble in hot alcohol, insoluble in water, mp 95C. Combustible.
Derivation: By heating resorcinol with aniline in presence of calcium chloride and zinc chloride.
Use: Organic synthesis.

N,N′-diphenyl-p-phenylenediamine.   (DPPD).
CAS: 74-31-7. $(C_6H_5NH)_2C_6H_4$.
Properties: Gray powder, d 1.28, mp 145–152C, purity 9.2% (min). Insoluble in water; soluble in acetone, benzene, monochlorobenzene, and isopropyl acetate. Combustible.
Use: Flex-resistant antioxidant in rubbers; stabilizer; polymerization inhibitor; retards copper degradation; intermediate for dyes, drugs, plastics and detergents.

diphenyl phosphite. $(C_6H_5O)_2PHO$.
Properties: Clear, straw-colored liquid; mp 12C; refr index 1.557 (25C); d 1.221 (25/15C); flash p 350F (176C). Combustible.
Use: Synthesis of organophosphorus compounds.

diphenyl phthalate. $C_6H_4(COOC_6H_5)_2$.
Properties: Yellow-white powder, mp 68–70C, d 1.28 (20C), flash p 435F (224C), bp 405C, bulk d 10.68 lb/gal, refr index 1.572 (74C). Combustible.
Use: Plasticizer for ethylcellulose, nitrocellulose, and various polymers.

diphenyl-4-piperidylmethane.
$(C_6H_5)_2(C_5H_{10}H)CH$.
Properties: White solid, difficultly soluble in water but readily soluble in dilute acids, moderately soluble in organic solvents, fp 99.7C min.
Use: Intermediate.

1,3-diphenyl-1,3-propanedione.   See dibenzoylmethane.

N,N′-diphenylpropylenediamine.
$C_6H_5NHCH_2CH(CH_3)NHC_6H_5$.
Properties: Clear, deep reddish-brown, thick liquid; d 1.07; stable in storage; insoluble in water; soluble in acetone, ethylene dichloride, benzene, and gasoline; readily disperses. Combustible.
Use: Antioxidant for rubber latexes.

diphenyl-4-pyridylcarbinol.
$(C_6H_5)_2(C_5H_4N)COH$.
Properties: White solid, very weak base, slightly soluble in most organic solvents, soluble in hot glacial acetic acid, mp 236–241C.
Use: Intermediate.

diphenyl-4-pyridylmethane.
$(C_6H_5)_2(C_5H_4N)CH$.
Properties: White to pale-yellow, crystalline solid; moderately soluble in common organic solvents; bp 234C (20 min); fp 123C min.
Use: Intermediate.

diphenylsilanediol. $(C_6H_5)_2Si(OH)_2$.
Properties: White solid, mp 130–150.

Derivation: Hydrolysis of diphenyldichlorosilane.
Use: Silicone chemical.

p,p'-diphenylstilbene.
$C_6H_5C_6H_4CH:CHC_6H_4C_6H_5$.
Properties: Crystals, mp 308–310C.
Use: In purified form as fluor in plastic scintillators.

diphenylsulfone. $C_{12}H_{10}SO_2$.
Properties: White, crystalline solid; mp 128C; bp 376C; soluble in benzene and hot alcohol; insoluble in water.
Derivation: Reaction of benzene with sulfuric acid.
Hazard: Toxic by ingestion.
Use: Larvicide and ovicide.

N,N-diphenylthiourea. See thiocarbanilide.

1,3-diphenyltriazene. See diazoaminobenzene.

diphenylurea. (carbanilide).
$(NHC_6H_5)CO(NHC_6H_5)$.
Properties: Colorless prisms, soluble in alcohol and ether, very slightly soluble in water, d 1.239, mp 235C, bp 260C.
Use: Organic synthesis.

diphenyl-o-xenyl phosphate.
$(C_6H_5O)_2(C_6H_5C_6H_4O)PO$.
Properties: d 1.20 (20C), refr index 1.582–1.590 (60C), boiling range 250–285C (5 mm), flash p 225F, insoluble in water. Combustible.
Use: Plasticizer.

diphosgene. Legal label name (air) for trichloromethylchloroformate.

diphosphopyridine nucleotide. See nicotinamide adenine dinucleotide.

diphosphoric acid. See pyrophosphoric acid.

dipicrylamine. See hexanitrodiphenylamine.

dipicryl sulfide. Legal label name (air) for hexanitrodiphenyl sulfide.

1,3-di-4-piperidylpropane. (4-di-pip).
$(HNC_5H_9)CH_2CH_2CH_2C_5H_9NH$.
Properties: Solid, fp 67C, bp 329C, soluble in water, a stable high-boiling strong organic base.
Use: Intermediate.

dipole. An assemblage of atoms or subatomic particles having equal electric charges of opposite sign separated by a finite distance; for instance the proton and orbital electron of a hydrogen atom or the hydrogen and chlorine atoms of a hydrogen chloride molecule.
See also polar.

dipole moment. In many molecules, the atoms and their electrons and nuclei are so arranged that one part of the molecule has a positive electrical charge while the other part is negatively charged. Such a molecule, therefore, becomes a small magnet or dipole. Changing electrical or magnetic fields causes the molecule to turn, or rotate, in one direction or another, depending on the charge of the field. The dipole moment (v) is the distance between the charges multiplied by the quantity of charge in electrostatic units.
See also polar.

dippel's oil. (Bone oil). Dark brown oil obtained from distillation of bones. Contains pyridine and substituted pyridines.
Use: To denature alcohol.

dipotassium orthophosphate. See potassium phosphate dibasic.

dipropargyl. (bipropargyl; 1,5-hexadiyne).
$HC:CCH_2CH_2C:CH$.
Properties: Colorless liquid, soluble in alcohol, insoluble in water, d 0.805, bp 85C, fp −6.0C.
Hazard: Moderate fire and explosion risk when exposed to heat.

dipropenyl. See 2,4-hexadiene.

di-2-propenylamine. See diallylamine.

di-n-propylamine. CAS: 142-84-7.
$(C_3H_7)_2NH$.
Properties: Water-white liquid, fp −63C, d 0.741 (20C), bp 109C, color water-white, amine odor, bulk density 6.1 lb/gal, flash p 63F (17.2C) (Tagliabue open cup), soluble in water.
Hazard: Flammable, dangerous fire risk. Skin irritant.
Use: Intermediate.

dipropylene. See 2,4-hexadiene.

dipropylene glycol. (2,2'-dihydroxydopropyl ether). CAS: 110-98-5.
$(CH_3CHOHCH_2)_2O$.
Properties: Colorless, slightly viscous liquid; soluble in toluene and water; d 1.0252 (20/20C), bp 233C, vap press 0.01 mm (20C), flash p 280F (137.7C) (open cup), bulk d 8.5 lb/gal (20C), coefficient of expansion 0.00073 (20C), viscosity 1.07 poise (20C). Combustible.
Grade: Technical.
Use: Polyester and alkyd resins, reinforced plastics, plasticizers, solvent.

dipropylene glycol dibenzoate. (Benzoflex; 3,3'-oxydyl-1-propanol dibenzoate).
CAS: 94-51-9. $C_{20}H_{22}O_5$
Properties: Light-colored liquid, d 1.1271 (20/20C), bulk d 9.4 lb/gal (20C), bp 250C (10 mm), mp 200C, insoluble in water, viscosity 227 cps (20C). Combustible.
Use: Plasticizer.

dipropylene glycol dipelargonate.
$[CH_3(CH_2)_7COOCH(CH_3)CH_2]_2O$.   A synthetic lubricant.

dipropylene glycol monomethyl ether.
(dipropylene glycol methyl ether).
CAS: 34590-94-8. $CH_3OC_3H_6OC_3H_6OH$.
Properties: Colorless liquid, d 0.950 (25/4C), bp 189C, 74.5C (10 mm), fp −80C, viscosity 3.5 cp (25C), refr index 1.419 (25C), flash p 185F (85C) (open cup). Completely miscible with water, VM&P naphtha, acetone, ethanol, benzene, carbon tetrachloride, ether, methanol, monochlorobenzene, and petroleum ether. Combustible.
Hazard: Toxic. Absorbed by skin. TLV: 100 ppm in air.
Use: Solvent, hydraulic brake fluids.

dipropylene glycol monosalicylate. (dipropylene glycol monoester; salicylic acid dipropylene glycol monoester).
$C_3H_6(OOCC_6H_4OH)OC_3H_6OH$.
Properties: Light-colored oil, fragrant odor, d 1.16 (40C), refr index approximately 1.52, soluble in alcohol, insoluble in water.
Use: UV light-screening agents, protective coatings, plasticizers.

dipropylene triamine.   See 3,3'-iminobispropylamine.

dipropyl ketone. (butyrone; 4-heptanone).
CAS: 123-19-3. $(CH_3CH_2CH_2)_2CO$.
Properties: Stable, colorless liquid; pleasant odor; insoluble in water; miscible with many organic solvents. Bp 143.7C, fp −32.1C, d 0.8162 (20/20C), bulk d 6.79 lb/gal (20C), refr index 1.4068 (20C), surface tension 25.2 dynes/cm (25C) viscosity 0.0074 poise (20C), vap press 5.2 mm (20C), flash p 120F (40C) (CC).
Grade: Technical.
Hazard: Moderate fire risk. Toxic by inhalation, skin irritant. TLV: 50 ppm in air.
Use: Solvent for nitrocellulose, raw and blown oils, resins and polymers; lacquers, flavoring.

dipropylmethane.   See heptane.

dipropyl phthalate. $C_6H_4(COOC_3H_7)_2$.
Properties: Colorless liquid, d 1.071 (25C), refr index 1.494 (25C), bp 129–132C (1 mm), soluble in water 0.015% by weight. Combustible.
Use: Plasticizer.

"Dipterex."[181]   TM for trichlorfon.

α,α'-dipyridyl. (2,2'-bipyridine).
CAS: 366-18-7. $(C_5H_4N)_2$.
Properties: White crystals, mp 69–70C, bp 272–273C. Slightly soluble in water; soluble in alcohol, ether, benzene, chloroform, petroleum ether.
Grade: Reagent.
Use: Reagent for iron determination.

2,2-dipyridylamine. $(C_5H_4N)_2NH$.
Properties: Solid, fp 92.3C (min), bp 222C (50 mm), very slightly soluble in water. Combustible.
Derivation: From 2-aminopyridine.
Use: Intermediate.

dipyridylethyl sulfide. $[C_5H_4N(CH_2)_2]_2S$.
Properties: Liquid, d 1.113 (25C), refr index 1.5841 (20C), mp 1.5C, soluble in water and common organic solvents. Combustible.
Grade: Technical (95% purity).
Use: Synthesis of pharmaceuticals, dyestuffs, rubber chemicals, flotation agents, insecticides, fungicides, plasticizers, textile assistants, herbicides, oil additives, rust preventives and pickling inhibitors.

diquat. (6,7-dihydrodipyrido(1,2-a:2',1'-c)pyrazidinium salt; 1,1'-ethylene-2,2'-di-pyridinium dibromide).   CAS: 85-00-7.
$(C_5H_4NCH_2^-)_2Br_2$.
Yellow crystals, mp 335C, soluble in water.
Hazard: Toxic by ingestion. TLV: 0.5 mg/m³ of air.
Use: Herbicide and plant growth regulator, sugarcane flowering suppressant.

direct dye.   See dye direct.

diresorcinol. (tetrahydroxydiphenyl).
$(OH)_2C_6H_3C_6H_3(OH)_2$.
Properties: White to slight yellowish, crystalline powder; mp 310C; slightly soluble in cold water.
Derivation: By fusing resorcinol and phenol with caustic soda.
Use: Organic synthesis.

diresorcinolphthalein.   See fluorescein.

N,N'-disalicylidene-1,2-diaminopropane.
(N,N'-disalicylidene propylenediamine; disalicylalaminopropane disalicylalpropylenediimine).
$HOC_6H_4CH:NCH_2CH(CH_3)N:CHC_6H_4OH$.
Properties: (80% active compound): Liquid, d 1.08 (15.5/15.5C), bulk d 9.0 lb/gal, pour point

0F (−17.7C), flash p approximately 70F (21C) (Tagliabue open cup), viscosity 25 cs (37.7C), insoluble in water, miscible with benzene and xylene.
Hazard: Flammable, dangerous fire risk.
Use: Metal deactivator in motor fuels.

disc. (disk). A small thin circular section or platelet of a material, especially of a biological specimen. "Disc" is the spelling preferred by scientists.

"Discaloy."[308] TM for an austenitic iron-base alloy containing nickel, chromium and relatively small proportions of molybdenum, titanium, silicon, and manganese. This alloy is precipitation-hardened and was developed primarily to meet the need for improved gas turbine disks, one of the most critical components of jet engines.

discharging agent. A substance capable of destroying a dye or mordant present within the fibers of a fabric. There are various methods of utilizing this property so that it is possible to produce a colorless figure upon a colored ground or a colored figure upon a different-colored ground. Some examples are titanous sulfate, sodium hydrosulfide, zinc formaldehyde sulfoxylate.
See also stripping (2).

Dische reaction. The reaction between the Dische reagent (diphenylamine, acetic acid, and sulfuric acid) and 2-deoxypentoses resulting in the development of a characteristic blue color.

"Discolite."[159] TM for sodium sulfoxylate formaldehyde.
Use: Dye discharge, vat printing, stripping of fibers; accelerator in emulsion polymerization systems.

disilanyl. See silane.

disiloxane. See siloxane.

disinfectant. A substance used on inanimate objects which destroys harmful microorganisms or inhibits their activity. Disinfectants are either complete or incomplete. Complete disinfectants destroy spores as well as vegetative forms of microorganisms; incomplete disinfectants destroy vegetative forms of the organisms, but do not injure spores.
Some representative disinfectants are: (1) mercury compounds (mercuric chloride, phenylmercuric borate); (2) halogens and halogen compounds (chlorine, iodine, fluorine, bromine, calcium and sodium hypochlorite); (3) phenols including cresol from coal tar, o-phenylphenol; (4) synthetic detergents (anionic such as sodium alkyl benzene sulfonates and cationic such as quaternary ammonium compounds); (5) alcohols (of low molecular weight except methanol); (6) natural products (pine oil); (7) gases (sulfur dioxide, formaldehyde, ethylene oxide). Heat and electromagnetic waves are used as disinfectants.
A number of compounds (mercurous and mercuric chlorides, copper sulfate or carbonate, and a mixture of zinc oxide and zinc hydroxide) have been employed as seed disinfectants.
Effectiveness of disinfectants is rated by the phenol coefficient.
See antiseptic.

dislocation. Any variation from perfect order and symmetry in a crystalline lattice. In some cases the imperfection is due to missing atoms resulting in "holes" or vacancies in the lattice, or one or more atoms of another element may be present effecting important changes in the conductivity, hardness, and other properties of the crystals. Disordered arrangement of the constituent atoms may also occur.
See also impurity, semiconductor, hole.

disodium acetarsenate. (aricyl).
$NaOOCCH_2As(OH)O(ONa) \cdot 2H_2O$.
Properties: White, crystalline powder; soluble in water.
Derivation: By reacting sodium arsenite with sodium monochloracetate.

2,7-disodium dibromo-4-hydroxymercurifluorescein. See merbromin.

disodium dibutyl-o-phenylphenoldisulfonate.
Properties: Light-brown paste; soluble in alcohol, acetone, dibutyl tartrate, ethylene glycol.
Use: Wetting, penetrating, and spreading agent used in kier-boiling; scouring and dyeing textiles; industrial cleaners; deodorant preparations; insecticidal formulations; metal cleaning; stabilizer and wetting agent in latex used to treat cord or other fabrics.

disodium dihydrogen pyrophosphate. See sodium pyrophosphate acid.

disodium 1,2-dihydroxybenzene-3,5-disulfonate.
(4,5-dihydroxy-m-benzenedisulfonic acid disodium salt, sodium catechol disulfonate).
$C_6H_2(OH)_2(SO_3Na)_2$.
Properties: Non-hygroscopic crystals, freely soluble in water, produces water-soluble colored compounds with metal salts.
Use: Colorimetric reagent for iron, manganese, titanium, molybdenum.

disodium diphosphate. See sodium pyrophosphate acid.

disodium EDTA. (ethylenediaminetetraacetic acid, disodium salt).
CAS: 139-33-3. $C_{10}H_{14}N_2Na_2O_8.2H_2O$.
Properties: White, crystalline powder; freely soluble in water; bulk density 6.5 lb/gal; pH (5% solution) between 4 and 6.
Grade: USP, FCC.
Use: Food preservative, chelating and sequestering agent.

disodium endothal. See endothal.

disodium ethylenebisdithiocarbamate. See nabam.

disodium guanylate. See sodium guanylate.

disodium hydrogen phosphate. See sodium phosphate dibasic.

disodium inosinate. See sodium inosinate.

disodium methylarsonate. (DMA; disodium methanearsonate; methanearsonic acid, disodium salt). CAS: 144-21-8. $CH_3AsO(ONa)_2$, sometimes with $6H_2O$.
Properties: Colorless, crystalline solid; hygroscopic; mp above 355C (hexahydrate mp 132–139C); soluble in water and methanol.
Derivation: Reaction of methyl chloride with sodium arsenate.
Grade: 55–65% powder concentration; 31.5% blend.
Hazard: Toxic by ingestion and inhalation.
Use: Pharmaceuticals, herbicide (crabgrass killer).

disodium orthophosphate. See sodium phosphate dibasic.

disodium phenyl phosphate. $C_6H_5Na_2PO_4$.
Properties: White powder, soluble in water, insoluble in acetone and ether.
Use: Reagent for milk pasteurization.

disodium phosphate. See sodium phosphate dibasic.

disodium pyrophosphate. See sodium pyrophosphate acid.

disodium tartrate. See sodium tartrate.

disperse dye. See dye, disperse.

disperse phase. See phase (2); collid chemistry.

dispersing agent. A surface-active agent added to a suspending medium to promote uniform and maximum separation of extremely fine solid particles, often of colloidal size. True dispersing agents are polymeric electrolytes (condensed sodium silicates, polyphosphates, lignin derivatives); in nonaqueous media sterols, lecithin and fatty acids are effective.
Use: Wet-grinding of pigments and sulfur; preparation of ceramic glazes, oil well drilling muds, insecticidal mix, carbon black in rubber, and water-insoluble dyes.
See also emulsion; detergent.

dispersion. (1) A two-phase system where one phase consists of finely divided particles (often in the colloidal size range) distributed throughout a bulk substance, the particles being the disperse or internal phase and the bulk substance the continuous or external phase. Under natural conditions, the distribution is seldom uniform; but under controlled conditions, the uniformity can be increased by addition of wetting or dispersing agents (surfactants) such as a fatty acid. The various possible systems are: gas/liquid (foam), solid/gas (aerosol), gas/solid (foamed plastic), liquid/gas (fog), liquid/liquid (emulsions), solid/liquid (paint), and solid/solid (carbon black in rubber). Some types such as milk and rubber latex are stabilized by a protective colloid which prevents agglomeration of the dispersed particles by an abherent coating. Solid-in-liquid colloidal dispersions (loosely called solutions) can be precipitated by adding electrolytes which neutralize the electrical charges on the particles. Larger particles will either gradually coalesce and rise to the top or settle out, depending upon their specific gravity.
See also suspension; colloid chemistry.
(2) In the field of optics, dispersion denotes the retardation of a light ray, usually resulting in a change of direction as it passes into or through a substance, the extent of this effect depending on the frequency. Dispersion is a critically important property of optical glass.
See also refractive index.

"Dispersite."[248] TM for water dispersions of natural, synthetic, and reclaimed rubbers and resins.
Use: Adhesives for textiles, paper, shoes, leather, tapes; coatings for metal, paper, fabrics, carpets; protective (strippable) for saturating paper, felt, book covers, tape, jute pads; for dipping tire cords. Can be applied by spraying, spreading, impregnation, saturation.

"Disperson."[104] TM for wettable grades of zinc, calcium and other metallic stearates.
Use: Where easy dispersion in water is desired.

"Disperson OS."[206] TM for an oil-soluble emulsifying agent comprised of an 8% solution of a polyethenoxy compound in isopropanol. Designed especially for dispersion of oil spills in seawater. Claimed to be biodegradable and to have low toxicity for fish and other marine organisms. Amount needed said to be from 20–25% of the oil volume.

"Dispersol."VL[325]   TM for an ethylene oxide condensate. Nonionic dispersant and retardant for vat dyes and dispersant for acetate dyes. 20% aqueous solution.

displacement series.   See activity series.

disposal, waste.   See waste control, chemical waste, radioactive waste.

disproportionation.   A chemical reaction in which a single compound serves as both oxidizing and reducing agent and is thereby converted into a more oxidized and a more reduced derivative. Thus, a hypochlorite upon appropriate heating yields a chlorate and a chloride and an ethyl radical formed as an intermediate is converted into ethane and ethylene.
See also transalkylation.

dissociation.   The process by which a chemical combination breaks up into simpler constituents as a result of either (1) added energy as in the case of gaseous molecules dissociated by heat, or (2) the effect of a solvent on a dissolved polar compound (electrolytic dissociation), e.g., water on hydrogen chloride. It may occur in the gaseous, solid, or liquid state or in solution. All electrolytes dissociate to a greater or less extent in polar solvents. The degree of dissociation can be used to determine the equilibrium constant for dissociation, an important factor in ascertaining the extent of a chemical process.
See also ionization.

dissolved oxygen (DO).   One of the most important indicators of the condition of a water supply for biological, chemical, and sanitary investigations. Adequate dissolved oxygen is necessary for the life of fish and other aquatic organisms, and is an indicator of corrosivity of water, photosynthetic activity, septicity, etc.
See also biochemical oxygen demand.

distearylamine.   (dioctadecylamine).
$(C_{18}H_{37})_2NH$.
Properties: Solid, d 0.85, mp 69C, almost insoluble in water.
Use: Intermediate.

2,6-distearyl-p-cresol.   (2,6-dioctadecyl-p-cresol).
$(C_{18}H_{37})_2CH_3C_6H_2OH$.
A viscous, pale yellow liquid; soluble in most nonpolar solvents; refr index 1.4825–1.4855 (25C). Combustible.

Use: Antioxidant, heat stabilizer for polypropylene.

distearyl ether.   (dioctadecyl ether).
$(C_{19}H_{37})_2O$.
Properties: Solid, mp 58–60C, bp decomposes.
Grade: 95% (min) purity.
Use: Electrical insulators, water repellents, lubricants in plastic molding and processing, antistatic agent, intermediate.

distearyl sulfide.   (dioctadecyl sulfide; distearyl thioether). $(C_{18}H_{37})_2S$.
Properties: Solid, mp 68–69C, bp decomposes, d 0.8148 (70/4C).
Grade: 95% (min) purity.
Use: Organic synthesis (formation of sulfonium compounds).

distearyl thiodipropionate.   (e,e'-dioctadecyl thiodipropionate; thiodipropionic acid, distearyl ester). $(C_{18}H_{37}OOCCH_2CH_2)_2S$.
Properties: White flakes, mp 58–62C, mp 55C, insoluble in water, very soluble in benzene and olefin polymers, resistant to heat and hydrolysis.
Use: Antioxidant, plasticizer, softening agent.

distearyl thioether.   See distearyl sulfide.

distillate.   Any product of distillation, especially of petroleum.

distillation.   A separation process in which a liquid is converted to vapor and the vapor then condensed to a liquid. The latter is referred to as the distillate, while the liquid material being vaporized is the charge or distilland. Distillation is thus a combination of evaporation, or vaporization, and condensation. Simple examples are the natural cycle of evaporation and condensation of steam from a tea kettle on a cold surface.
   The usual purpose of distillation is purification or separation of the components of a mix. This is possible because the composition of the vapor is usually different from that of the liquid mix from which it is obtained. Alcohol has been so purified for generations to separate it from water, fusel oil, and aldehydes produced in the fermentation process. Gasoline, kerosene, fuel oil, and lubricating oil are produced from petroleum by distillation. It is the key operation in removing salt from seawater.
Use: Chemical analysis, in laboratory research and for manufacture of many chemical products.
See also destructive distillation, batch distillation, extractive distillation, rectification, dephlegmation, flash distillation, continuous distillation, simple distillation, reflux, fractional distillation, azeotropic distillation, vacuum distillation, molecular distillation, and hydrodistillation.

disulfiram. See tetraethylthiuram disulfide.

3,5-disulfobenzoic acid. $C_6H_3(HSO_3)_2COOH$.
Properties: White powder, soluble in water.
Grade: CP.
Use: Intermediate for detergents, dyes, and pharmaceuticals.

disulfoton. (Di-Syston; O,O-diethyl S-[2-(ethylthio)ethyl]phosphorodithioate).
CAS: 298-04-4.
$(C_2H_5O)_2P(S)SCH_2CH_2SCH_2CH_3$.
Properties: Pure compound yellow liquid, technical compound brown liquid, bp 108C (0.01 mm), d 1.144 (20C), insoluble in water.
Hazard: Toxic by ingestion and inhalation, cholinesterase inhibitor, absorbed by skin, use may be restricted. TLV: 0.1 mg/m³ of air.
Use: Systemic insecticide, acaricide.

disulfuryl chloride. See pyrosulfuryl chloride.

"Di-Syston."[181] TM for disulfoton.
See disulfoton.

ditetradecylamine. See dimyristylamine.

ditetradecyl ether. See dimyristyl ether.

ditetradecyl sulfide. See dimyristyl sulfide.

"Dithane."[23] TM for agricultural fungicides based on salts of ethylene bisdithiocarbamate. Supplied in zinc, manganese, and sodium forms as wettable powder or liquid concentration.

1,4-dithiane. (diethylene disulfide).

$\overline{SCH_2CH_2SCH_2}CH_2$.
Properties: White crystals, volatile in steam, soluble in alcohol and ether, slightly soluble in water, bp 115.6C (60 mm), mp 108C.
Derivation: Interaction of dichloroethyl sulfide with sodium or potassium sulfide.
Use: Organic synthesis.

dithiane methiodide. $C_4H_8S_2 \cdot CH_3I$.
Properties: Crystals, soluble in hot water, slightly soluble in alcohol, insoluble in ether, mp 174C.
Derivation: Interaction of dichloroethyl sulfide and methyliodide.

dithianone. $C_{14}H_4N_2O_2S_2$.
Properties: Brownish solid, mp 220C. Insoluble in water, soluble in acetone, chloroform, dioxane.
Hazard: Toxic by ingestion.
Use: Fungicide.

β-β'-dithiobisalanine. See cystine.

2,2'-dithiobis(benzothiazole). (benzothiazolyl disulfide; benzothiazyl disulfide; 2-mercapto-

benzothiazyl disulfide; MBTS).
CAS: 120-78-5. $(C_6H_4SCN)_2S_2$.
Properties: Pale yellow, free-flowing, odorless powder; d 1.54; mp 168C; insoluble in water; sparingly soluble in organic solvents.
Use: Primary accelerator in natural and nitrile rubber and SBR, plasticizer and vulcanization retarder in neoprene Type G, cure modifier in neoprene Type W, oxidation cure activator in butyl. For extruded and molded goods, tire and tubes, wire and cable, sponge.

2,4-dithiobiuret. $C_2H_5N_3S_2$.
Properties: Colorless crystals, bulk d 1.52 g/mL, decomposes 181C, partially soluble in boiling water, acetone and "Cellosolve."
Derivation: Reaction of dicyandiamide and hydrogen sulfide.
Hazard: Toxic by ingestion and inhalation, causes respiratory paralysis.
Use: Intermediate in manufacture of insecticides, synthetic resins; plasticizer, rubber accelerator.

dithiocarbamic acid. (aminodithioformic acid).
NH_2CS_2H.
Properties: Colorless needles, soluble in alcohol.
Use: The metal salts of the acid are important as strong (ultra) rubber accelerators, as are the thiuram disulfide derivatives. Seed disinfectant. See thiuram, selenium diethyldithiocarbamate, zinc dibutyldithiocarbamate, zinc diethyldithiocarbamate; ziram.

2,2'-dithiodibenzoic acid. $(C_6H_4COOH)_2S_2$.
Properties: Tan to gray powder, mp 280C (min).
Use: Intermediate for pharmaceuticals.

4,4'-dithiodimorpholine. CAS: 103-34-4.
$C_4H_8ONSSNOC_4H_8$.
Properties: Gray to tan powder, mp 122C min, d 1.36 (25C).
Hazard: Toxic by ingestion, inhalation, and skin absorption.
Use: Rubber accelerator, fungicide.

1,2-dithioglycerol. See 2,3-dimercaptopropanol.

6,8-dithiooctanoic acid. See dl-α-lipoic acid.

dithiooxamide. (rubeanic acid).
$SC(NH_2)C(NH_2)S$.
Properties: Stable, orange-red powder; decomposes at 140C; insoluble in water; soluble in acetone and chloroform. Forms highly colored stable complexes with many metal ions; can form a series of N,N'-derivatives.
Use: Chemical intermediate.

dithiothreitol. (Cleland's reagent).
Properties: Solid, mp 42–43.5C, 99% pure, available in 1-, 5-, 25-g quantities.

Use: Reducing agent for proteins and enzymes, biochemical research.

dithymol diiodide.   See thymol iodide.

1,4-di-p-toluidinoanthraquinone.   (FD&C Green No. 6). $C_{14}H_6O_2(NHC_6H_4CH_3)_2$.
Properties: Solid, mp 213C.
Use: Dye, approved for restricted use in drugs and cosmetics.

di-o-tolyl carbodiimide.
$(CH_3)C_6H_4N:C:NC_6H_4(CH_3)$.
Properties: Yellow to red-brown liquid with faint acrid odor, d 1.063 (25/4C), bp 140C (0.9 mm), refr index 1.6248 (20C), soluble in organic solvents such as chloroform carbon tetrachloride, benzene, and dioxane. Combustible.
Hazard: Toxic by inhalation and ingestion, irritant to skin and eyes.
Use: Surface coatings, textile processing, stabilizers of polyesters and urethanes, scavenger compounds for materials sensitive to active hydrogen.

N,N-di-o-tolylethylenediamine.
$CH_3C_6H_4NHCH_2CH_2NHC_6H_4CH_3$.
Properties: Light-brown to purple granular solid, d 1.13, softens at approximately 57C. Insoluble in water; soluble in acetone, ethylene dichloride, benzene, and gasoline.
Use: Antioxidant for light-colored rubber goods.

di-o-tolylguanidine.   (DOTG Accelerator).
CAS: 97-39-2. $(CH_3C_6H_4NH)_2CNH$.
Properties: White powder, non-hygroscopic, very slightly soluble in water, soluble in warm alcohol from which it crystallizes on cooling, d 1.10, mp 179C.
Derivation: Desulfurization of di-o-tolylthiourea with a lead compound in the presence of ammonia.
Hazard: Toxic by ingestion.
Use: Basic rubber accelerator.

2,7-di-p-tolylnaphthylenediamine.
$C_{10}H_6(NHC_6H_4CH_3)_2$.
Properties: Fine needles, sparingly soluble in alcohol, insoluble in water, mp 237C.
Derivation: By heating 2,7-dihydroxynaphthalene with p-toluidine and p-toluidine hydrochloride.

1,3-di-p-tolylphenylenediamine.
$C_6H_4(NHC_6H_4CH_3)_2$.
Properties: Long needles, soluble in alcohol and ether, insoluble in water, mp 137C.
Derivation: By heating resorcinol and p-toluidine in presence of zinc chloride.

di-o-tolythiourea.   (DOTT).
$SC(NHC_6H_4CH_3)_2$.
Properties: Colorless, crystalline leaflets; pungent odor; not hygroscopic; soluble in alcohol, ether,

and benzene; insoluble in water; mp 144–148C.
Derivation: By the interaction of o-toluidine and carbon disulfide.
Hazard: Toxic by ingestion.
Use: Metal pickling inhibitor.

ditridecyl phthalate.   (DTDP; Jayflex).
CAS: 119-06-2. $C_6H_4(COOC_{13}H_{27})_2$.
Properties: Colorless liquid, d 0.951 (20/20C), bp greater than 285C at 5 mm, pour point −37C, viscosity 190 cps (25C), flash p 460F (237C). Combustible.
Use: Plasticizer.

ditridecyl thiodipropionate.   (3,3′-tetramethylnonyl thiodipropionate; thiodipropionic acid ditridecyl ester). $(C_{13}H_{27}OOCCH_2CH_2)_2S$.
Properties: Colorless liquid, d 0.932 (25C), insoluble in water, soluble in most organic solvents. Combustible.
Use: Stabilizer, plasticizer and softening agent for plastics, lubricant additive.

diuretic.   A drug which promotes water elimination from the body via kidney function.

diuron.   (3-(3,4-dichlorophenyl)-1,1-dimethylurea).   CAS: 330-54-1.
$C_6H_3Cl_2NHCON(CH_3)_2$.
Properties: White, crystalline solid; mp 159C; vap press (30C) 2×10^{-7} mm; very low solubility in hydrocarbon solvents; approximately 42 ppm at 25C in distilled water; stable toward oxidation and moisture; decomposes at 180C.
Hazard: Toxic. TLV: 10 mg/m³ of air.
Use: Pre-emergence herbicide, sugar cane flowering suppressant.

divalent carbon.   See carbene.

divanadyl tetrachloride.   See vanadyl chloride.

divinyl acetylene. $H_2C:CHC:CCH:CH_2$.
Trimer of acetylene formed by passing it into a hydrochloric acid solution containing a metallic catalyst.
Use: Intermediate in manufacture of neoprene.

divinylbenzene.   (DVB; vinylstyrene).
CAS: 108-57-6. $C_6H_4(CH:CH_2)_2$.   Exists as o-, m- and p-isomers. The commercial form contains the three isomeric forms together with ethylvinylbenzene and diethylbenzene.
Properties: (pure m-isomer) Water-white liquid, easily polymerized, bp 199.5C, fp −66.90C, d 0.9289 (20C), viscosity 1.09 centipoise (20C), refr index 1.5772 (20C), flash p 165F (73.9C). (Divinylbenzene 55%) Pale straw-colored liquid, fp −87C, bp 195C, d 0.918 (25/25C), insoluble in

water, soluble in methanol and ether. Combustible.
Grade: 50–60%, 20–25%.
Hazard: Velocity of polymerization involves an explosion risk. If uninhibited, store at below 90F (32.3C). Toxic by inhalation. TLV: 10 ppm in air.
Use: Polymerization monomer for special synthetic rubbers, drying oils, ion-exchange resins, casting resins and polyesters. *Note:* Should contain inhibitor when stored or shipped.

divinyl ether. Legal label name for vinyl ether.

divinyl oxide. See vinyl ether.

3,9-divinylspirobi-m-dioxane. (3,9-divinyl-2,4,8, 10-tetraoxaspiroundecane).
$[CH_2:CHCH(OCH_2)_2]_2C$. Bicyclic.
Properties: Liquid, mp 42C, 120C (2 mm), d 1.251 (20/20C), flash p 290F (143C) (COC), slightly soluble in water. Combustible.
Use: Intermediate and monomer.

divinyl sulfide. $(CH_2:CH)_2S$.
Properties: Mobile liquid, unpleasant odor, polymerizes readily, d 0.9174 (15C), bp 85–86C. Combustible.
Derivation: Interaction of dichlorodiethyl sulfide and an alcoholic solution of potassium hydroxide.
Grade: Technical.

divinyl sulfone. $CH_2:CHSO_2CH:CH_2$.
Properties: Liquid, d 1.1788 (20/20C), bp 234C, fp −26C, soluble in water, flash p 255F (124C). Combustible.
Use: Monomer used in manufacture of polymers with diols, urea, and malonic esters; shrinkage control agent (textiles).

dixanthogen. (bis(ethylxanthogen); Sulfasan).
CAS: 502-55-6. $(C_2H_5OC=S)_2S_2$.
Properties: Yellow crystals, mp 30C, strong odor, insoluble in water, soluble in benzene and ether.
Use: Herbicide, insecticide, parasiticide.

di-o-xenyl phenyl phosphate.
$(C_6H_5C_6H_4O)_2(C_6H_5O)PO$.
Properties: Liquid, d 1.20 (60C), refr index 1.603–5 (60C), boiling range 285–330C (5 mm), flash p 250F (121C), insoluble in water. Combustible.
Use: Plasticizer.

"Dixie" Clay.[69] TM for a hard kaolin.
Properties: White to cream powder, d 2.62, fineness (through 325 mesh) 99.8%.
Use: Filler and reinforcing agent for rubber and plastics, paper coating.

di-p-xylylene. $(—CH_2C_6H_4CH_2-)_2$. Stable, white crystals; mp 280C.
Derivation: Pyrolysis of p-xylene at 398C in presence of steam. An organic quench (benzene or toluene) gives the dimer in yields of 10–15%.
Use: Parylene resins.

dixylylethane. (DXE).
$[C_6H_3(CH_3)_2]_2HCCH_3$.
Properties: Colorless liquid, bp 315C, pour point −34C, flash p 323F (162C), d 0.97.
Use: Dielectric fluid, heat-transfer medium, solvent, chemical intermediate.

DKP. Abbreviation for dipotassium phosphate. See potassium phosphate dibasic.

DL-. A prefix indicating that a compound contains equal parts of D and L stereoisomers. Small capitals are conventionally used to indicate the right- and left-handed structure of such molecules, they do not indicate rotation.
See also *dl*, racemic, mesoenantiomer.

dl. A prefix denoting a crystal that rotates the plane of polarized light equally to both left and right, resulting in optical inactivity. The symbol ± is preferably used.

DM. See diphenylamine chloroarsine.

DMA. Abbreviation for dimethylamine or disodium methyl arsonate.

DMAC. Abbreviation for dimethylacetamide.

DMAPMA. Abbreviation of dimethylaminopropylmethacrylamide.

DMB. Abbreviation for dimethoxybenzene. See hydroquinone dimethyl ether.

DMC. (1) Abbreviation for dichlorophenyl methyl carbinol.
See di(p-chlorophenyl) ethanol.
(2) Abbreviation for β-dimethylaminoethyl chloride hydrochloride.

DMDT. Abbreviation for dimethoxydiphenyltrichloroethane. See methoxychlor.

DMF. Abbreviation for dimethyl formamide.

DMH. Abbreviation for dimethyl hydantoin.

DMHF. Abbreviation for dimethylhydantoin formaldehyde.
See methylol dimethylhydantoin.

"DMP."[23] TM for dimethylaminomethyl-substituted phenols.
Use: Chemical intermediate, curing agents for epoxy resins, antioxidants.

DMPA. See O-(2,4-dichlorophenyl)-O-methyl isopropylphosphoramidothioate dimethylol propionic acid.

DMPC. Abbreviation for 1-dimethylamino-3-propyl chloride.

DMSO. Abbreviation for dimethyl sulfoxide.

DMT. Abbreviation for dimethyl terephthalate.

DMU. Abbreviation for dimethylurea and dimethylolurea.

DNA. Abbreviation for deoxyribonucleic acid.

DNBP. Abbreviation for dinitro-o-sec-butylphenol.

DNC. Abbreviation for dinitrocresol.

DNFB. See 2,4-dinitrofluorobenzene.

DNOC. Abbreviation for 4,6-dinitro-o-cresol.

DNOCHP. Abbreviation for dinitro-o-cyclohexylphenol.

DNODA. Abbreviation for di(n-octyl,n-decyl)-adipate.

DNODP. Abbreviation for di(n-octyl,n-decyl)-phthalate.

DNP. Abbreviation for dinonyl phthalate.

DNPC. See 2,6-dinitro-p-cresol.

DNPD. Abbreviation for N,N'-di-β-naphthyl-p-phenylenediamine.

DNPT. Abbreviation for dinitrosopentamethyl-enetetramine.

DNT. Abbreviation for dinitrotoluene.

DO. Abbreviation for dissolved oxygen.

DOA. Abbreviation for dioctyl adipate.
See the more exact name di(2-ethylhexyl) adipate.

DOC. Abbreviation for dichromate oxygen consumed.
See oxygen consumed.

n-docosane. $C_{22}H_{46}$.
Properties: Solid, mp 45.7C, bp 230C (15 mm), d 0.778 (45/5C), refr index 1.4400 (45C). Combustible.
Grade: 95% (min) purity.

Use: Organic synthesis, calibration, temperature-sensing devices.

docosanoic acid. See behenic acid.

1-docosanol. See behenyl alcohol.

cis-13-docosenoic acid. See erucic acid.

doctor knife. A metal straight-edge (or equivalent) which detaches adhering product from a rotating drum or roll, or removes excess coating from a web. No cutting is involved.
Use: To scrape dried films, such as soap flakes, from drum dryers; to remove ink from printing rolls; to clean residual paper stock from rolls and to provide a crimped paper for toweling, etc.; and to remove soft or sticky rubber and adhesive compositions from mixing rolls. A unique application is the so-called air doctor (air knife); it is actually a stream of air delivered by a turbine blower through a tube with a discharge slot. Located at the take-off end of a paper-coating machine, it efficiently levels off the coating suspension across the full width of the web.

doctor treatment. A method of improving or "sweetening" the odor of gasoline, petroleum solvents, or kerosene. A doctor solution of sodium plumbite, Na_2PbO_2, is made by dissolving litharge in sodium hydroxide solution, the feed to be sweetened is passed through the doctor solution. The action of the sodium plumbite and the lead sulfide formed from it, in conjunction with free sulfur (either naturally present or added), converts the malodorous mercaptans to the pleasanter disulfides.

dodecarbonium chloride. $C_{23}H_{41}ClN_2O$.
Properties: Crystalline solid, acidic taste, mp 145C, insoluble in benzene and acetone, soluble in water and alcohol.
Hazard: Toxic by ingestion.
Use: Biocide, disinfectant.

dodecahydrosqualane. See squalane.

dodecamethylpentasiloxane. $C_{12}H_{36}O_4Si_5$.
Properties: Inert, colorless liquid; fp about −75C; bp 230C; refr index 1.40, soluble in benzene and low molecular weight hydrocarbons; little change in viscosity with temperature.
Use: Silicone fluids, antifoam agent in lube oils.

dodecanal. See lauryl aldehyde.

n-dodecane. (dihexyl). CAS: 112-40-3.
$CH_3(CH_2)_{10}CH_3$.
Properties: Colorless liquid, d 0.749 (20/4C), fp −10C, bp 213C, refr index 1.4221 (20C), flash

p 160F (71.1C). Soluble in alcohol, acetone, ether; insoluble in water; autoign temperature 400F. Combustible.

Grade: 95%, 99%, 99.7 mole %.

Use: Solvent, organic synthesis, distillation chaser, jet fuel research.

1,12-dodecanedioic acid.
$HOOCC_{10}H_{20}COOH$.
A twelve-carbon straight-chain saturated aliphatic dibasic acid.

Properties: White, crystalline powder; mp 130–132C; soluble in hot toluene, alcohols, hot acetic acid; slightly soluble in hot water.

Use: Intermediate for plasticizers, lubricants, adhesives, polyesters, etc.

dodecanoic acid. See lauric acid.

n-dodecanol. See lauryl alcohol.

dodecanoyl peroxide. See lauryl peroxide.

dodecene. $C_{12}H_{24}$. Many possible isomers.
See 1-dodecene; tetrapropylene; sodium dodecylbenzenesulfonate.

1-dodecene. (α-dodecylene; tetrapropylene).
CAS: 6842-15-5. $H_2C:CH(CH_2)_9CH_3$.
Properties: Colorless liquid, d 0.7600 (20/4C), fp −33.6C, bp 213C, refr index 1.4327 (20C). Soluble in alcohol, acetone, ether, petroleum, coal tar solvents; insoluble in water. Combustible.

Hazard: Irritant and narcotic in high concentration.

Use: Flavors, perfumes, medicine, oils, dyes, resins.

dodecenylsuccinic acid.
$HOOCCH(C_{12}H_{23})CH_2COOH$.
Properties: Extremely viscous liquid, practically insoluble in water, miscible in oil. Combustible.

Use: Synthesis, corrosion inhibitor in oils, waterproofing.

dodecenylsuccinic anhydride. (DDS).

CAS: 25377-73-5. $C_{12}H_{23}CHCOOOCCH_2$.
The normal and at least two branched-chain dodecenyls are used commercially. The following properties are those of a branched chain compound.

Properties: Light-yellow, clear, viscous oil; bp 180–182C (5 mm); d 1.002 (25C), flash p 352F (177C) (COC), viscosity 400 entipoises (20C); 15.5 centipoises (70C). Combustible.

Use: Alkyd, epoxy and other resins, anticorrosive agents, plasticizers, wetting agents for bituminous compounds.

dodecyl acetate. (acetate C-12; lauryl acetate).
CAS: 112-66-3. $C_{12}H_{25}OOCCH_3$.
Properties: Colorless liquid, fruity odor, d 0.860–0.862, refr index 1.430–1.433, bp 150.5–151.5C, soluble in three volumes of 80% alcohol. Combustible.

Use: Perfumery, flavoring.

n-dodecyl alcohol. See lauryl alcohol.

dodecyl aldehyde. See lauryl aldehyde.

dodecylaniline. $C_{12}H_{25}C_6H_4NH_2$ (Probably the p-isomer).
Properties: Colorless liquid, d 0.907–0.912 (25/25C), bp 340–350C, oil-soluble aromatic amine, insoluble in water, soluble in most organic solvents. Combustible.

Hazard: See aniline.

Use: Intermediate.

dodecylbenzene. (detergent alkylate).
$C_{12}H_{25}C_6H_5$. A commercial blend of isomeric, predominantly monoalkyl benzenes. The side chains are saturated, averaging 12 carbon atoms, flash p 285F (140C). Combustible.

Derivation: Alkylation of benzene with isomeric dodecenes, obtained usually by polymerization of propylene.

See tetrapropylene. For linear (normal) dodecylbenzene, see sodium dodecylbenzenesulfonate.

Hazard: Toxic by ingestion.

Use: Detergents of ABS or LAS type.

dodecylbenzenesulfonic acid. (DDBSA). See sodium dodecylbenzenesulfonate.

dodecylbenzyl mercaptan.
$C_{12}H_{25}C_6H_4CH_2SH$. Offered as branched-chain isomers.
Properties: Light-yellow oil, unpleasant odor, bp approximately 150C (0.5 mm). Soluble in acetone, benzene, and heptane; insoluble in water and alcohol. Combustible.

Hazard: Irritant.

Use: Polymerization modifier, intermediate, metal-cleaning and polishing compounds.
See also thiol.

n-dodecyl bromide. See lauryl bromide.

6-dodecyl-1,2-dihydro-2,2,4-trimethylquinoline.
$C_{12}H_{25}C_9H_5N(CH_3)_3$.
Properties: Dark viscous liquid, d 0.93 (45C). Combustible.

Use: Rubber antioxidant (flex-cracking).

dodecyldimethyl(2-phenoxyethyl)ammonium bromide. See domiphen bromide.

n-dodecylguanidine acetate. See dodine.

n-dodecyl mercaptan. (DDM). $C_{12}H_{25}SH$.
Many isomers of dodecyl mercaptan are possible and a variety of these occur in the technical material known as tert-dodecyl mercaptan or lauryl mercaptan or simply dodecyl mercaptan.
See lauryl mercaptan.
Properties of n-isomer: Colorless liquid, bp 143C, refr index 1.4589, soluble in ether and alcohol, insoluble in water, flash p 262F (127C). Combustible.
See also thiol.

4-dodecyloxy-2-hydroxybenzophenone.
$C_{12}H_{25}OC_6H_3(OH)C(O)C_6H_5$.
Properties: Pale-yellow flakes, setting point 43C, soluble in polar and nonpolar organic solvents.
Use: UV light inhibitor in plastics.

dodecylphenol. CAS: 27193-86-8.
$C_{12}H_{25}C_6H_4OH$. A mix of isomers.
Properties: Straw-colored liquid, d 0.94 (20/20C), flash p 325F (162.7C), boiling range 310–335F (154–168C), phenolic odor, soluble in water. Combustible.
Hazard: See phenol.
Use: Solvent, intermediate for surface-active agents, oil additives, resins, fungicides, bactericides, dyes, pharmaceuticals, adhesives, rubber chemicals.

dodecyltrichlorosilane. CAS: 4484-72-4.
$C_{12}H_{25}SiCl_3$.
Properties: Colorless to yellow liquid, bp 288C, d 1.026 (25/25C), refr index 1.4521 (25C), readily hydrolyzed with liberation of hydrogen chloride.
Derivation: By Grignard reaction of silicon tetrachloride and dodecylmagnesium chloride.
Hazard: Strong irritant to tissue.
Use: Intermediate for silicones.

dodecyltrimethylammonium chloride.
$C_{12}H_{25}N(CH_3)_3Cl$.
Properties: White liquid, surface tension 0.1% in water, 33 dynes/cm (25C); soluble in water and alcohol.
Use: Germicides, fungicides, textile fiber softeners, cationic emulsifiers, flotation reagents, mildew proofing.

dodine. (N-dodecylguanidine acetate).
CAS: 2439-10-3.
$C_{12}H_{25}NHC(:NH)NH_2 \cdot CH_3COOH$.
Properties: Crystals, mp 136C, soluble in hot water and alcohol.
Hazard: Strong irritant to eyes and skin at greater than 50% concentration.
Use: Fungicide.

DOE. Abbreviation of Department of Energy.

Doebner-Miller reaction (Beyer method for quinolines). Synthesis of quinolines from primary aromatic amines and α,β-unsaturated carbonyl compounds in acid catalysis. The carbonyl compounds may be prepared in situ from two molecules of aldehyde or an aldehyde and methyl ketone. The latter is known as the Beyer method.

Doebner reaction. Formation of substituted cinchoninic acids from aromatic amines on heating with aldehydes and pyruvic acid.

Doering-LaFlamme Allene synthesis. Treatment of an olefin with bromoform in the presence of an alkoxide and reaction of the 1,1-dibromocyclopropane with an active metal to produce an allene in which a carbon atom is inserted between the carbon atoms of the original olefinic double bond.

DOF. See di(2-ethylhexyl) fumarate.

"Dolcoseal."[528] TM for spray-dried flavoring products such as essential oils, aromatic chemicals, flavor and perfume compounds.

Dollo's law. The rule that any substantial evolutionary change is virtually irreversible because genetic changes are not likely to be reversed in an order exactly opposite to the order in which they originally developed.

"Dolocotone."[528] TM for purified benzodihydropyrone.

dolomite. $CaMg(CO_3)_2$. A carbonate of calcium and magnesium.
Properties: Color gray, pink, white; vitreous luster; d 2.85; Mohs hardness 3.5-4; good cleavage in three directions; similar to calcite but less soluble in acids (reacts with acid when powdered or with hot acid). Noncombustible.
Use: Refractory for furnaces, manufacture of magnesium compounds and magnesium metals, as building material, in fertilizers, stock feeds, paper making, ceramics, mineral wool, removal of sulfur dioxide from stack gases.

dolphin oil. See porpoise oil.

DOM. A hallucinogenic drug. Use: Restricted by FDA. See di(2-ethylhexyl) maleate.

domiphen bromide. (dodecyldimethyl[2-phenoxyethyl]-ammonium bromide).
$(C_6H_5OC_2H_4N(C_{12}H_{25})(CH_3)_2Br$. A quaternary ammonium salt.
Properties: Crystals, mp 112C, soluble in water and organic solvents.
Use: Medicine (anti-infective).

donor. An atom which furnishes a pair of electrons to form a covalent bond or linkage with another atom called the acceptor.
See also bond, chemical; Lewis electron theory.

DOP. Abbreviation for dioctyl phthalate. See the more exact name di(2-ethylhexyl) phthalate.

L-dopa. See L-dihydroxyphenylalanine.

dope. (1) Sizing formulation consisting of solutions of nitrocellulose, cellulose acetate, or other cellulose derivations applied to crepe yard to set the twist and assist creping, to leather to form a high-gloss finish. (2) A combustible, such as wood pulp, starch, sulfur, etc., used in "straight" dynamites. (3) A trace impurity introduced into ultrapure crystals to obtain desired physical properties, especially electrical properties. Examples: erbium oxide doped with thulium for use as laser crystals; germanium or silicon doped with boron or arsenic for use as semiconductors.

Doppler effect. A shift toward longer wavelengths for waves reaching an observer when the source of the waves is moving away from the observer.

"Dortan."[51] TM for light-colored, sulfurized cutting oils which permit work visibility.

dosimeter. (radiation). Measurement of the amount of radiation delivered to the body of an individual. The permissible dose is the quantity of radiation which may be received by an individual over a given period with no detectable harmful effects. For X- or gamma ray exposure the permissible dose is 0.3 roentgen/week, measured in air. All workers with radioactive materials are expected to wear some device for detecting incident radiation. A dosimeter based on fiber optics has been developed for possible application in radiation therapy.
See also rad; rem.

DOT. Abbreviation for Department of Transportation, the agency responsible for shipping regulations for hazardous products in the US.

DOTG. Abbreviation for di-o-tolylguanidine.

dotriacontane. (dicetyl). $CH_3(CH_2)_{30}CH_3$.
Properties: Crystals, d 0.823, bp 310C, mp 70C.
Use: Research.

DOTT. Abbreviation for di-o-tolylthiourea.

double bond. See unsaturation.

double salt. A hydrated compound resulting from crystallization of a mixture of ions in aqueous solution. Common examples are the alums, made by crystallizing from solution either potassium or ammonium sulfate and aluminum sulfate; Rochelle salt (potassium sodium tartrate), made from a water solution of potassium acid tartrate treated with sodium carbonate; and Mohr's salt (ferrous ammonium sulfate), crystallized from mixed solutions of ferrous sulfate and ammonium sulfate.
See also nickel ammonium sulfate.

"Dowanol."[233] TM for a series of glycol monoethers.
Use: Solvents, intermediates for plasticizers, bactericidal agents, and fixatives for soap and perfumes.

"Dowclene."[233] TM for a series of solvents for specialized cleaning. A stabilized emulsion of caustic soda, a detergent, and a sequestering agent. EC A colorless liquid, fp −56.6, bp 77–122C, d 1.381. WR Inhibited 1,1,1-trichloroethane.

"Dow Corning."[149] TM for a wide range of silicone and polysiloxane products including emulsions, lubricants, greases, mold-release agents, laminating polymers, electrical varnishes, and heat-resistant coatings.

"Dowetch."[233] TM for magnesium photoengraving sheet, plate and extruded tube. Also applies to chemicals used in the one-step engraving process.

"Dowex."[233] TM for a series of synthetic ion-exchange resins made from styrene-divinylbenzene copolymers having a large number of ionizable or functional groups attached to this hydrocarbon matrix. These functional groups determine the chemical behavior and types of ion-exchange resin. The strong acid cation resins are capable of exchanging cations--for example, sodium for calcium and magnesium as in softening water. The strong base anion resins are capable of exchanging anions.

"Dowfax 9N."[233] TM for a series of nonylphenolethylene oxide adducts.
Use: Surfactants; textile, pulp and paper, leather, latex paint.

"Dowflake."[233] TM for calcium chloride, 77–80% in a special flake form.

"Dowfrost."[233] TM for a heat-transfer medium consisting of inhibited propylene glycol.
Use: Immersion freezing of poultry and other foods.

"Dowfume."[233]   TM for a series of proprietary products used as fumigants and pesticides.

"Dowlex."[233]   TM for a hybrid polyethylene of both low and high density, i.e., it has both linear and branched molecular structure. It can be used for both containers and bags. Forms available are film, injection molding, and extrusion.

"Dowmetal."[233]   TM for magnesium alloys containing more than 85% magnesium.

downcomer.   A pipe or flue which conveys gases, vapors or condensate downward in blast furnaces, distillation towers, refineries, etc.

"Dowtherm."[233]   TM for a group of liquid heat-transfer media.

doxorubicin.   See adriamycin.

DOZ.   Abbreviation for dioctyl azelate.
See the more exact name di(2-ethylhexyl) azelate.

DP.   Abbreviation for degree of polymerization.

DPA.   Abbreviation for diphenylamine, also for diphenolic acid.

DPG.   Abbreviation for diphenylguanidine.

DPH.   Abbreviation for 1,6-diphenylhexatriene.

DPIP.   See diphenyl isophthalate.

DPN.   Abbreviation for diphosphopyridine dinucleotide.
See nicotinamide adenine dinucleotide.

DPO.   Abbreviation for 2,5-diphenyloxazole.

DPPD.   Abbreviation for N,N'-diphenyl-p-phenylenediamine.

"DPS."[342]   TM for bacitracin methylene disalicylate.

Dragendorfs reagent.   (Kraut's reagent; potassium iodobismuthate).
Use: Testing alkaloids.

Dragonic acid.   See anisic acid.

dragon's blood.
Properties: Deep red, amorphous lumps; mp 120C; soluble in alcohol, ether, and volatile and fixed oils; insoluble in water. Chief constituents: Dracoalban, dracoresene, draconine, and esters.
Derivation: the resin from the surface of the fruit

of several species of Daemonoraps; habitat is Indonesia and Borneo.
Use: Pigment for coloring paints, polishes, lacquers, etc.; photoengraving, to protect zinc plates from acid.

"Drakeol."[25]   TM for a series of colorless, odorless, tasteless, white mineral oils.

dram.   Unit of weight used in pharmacy, equivalent to 0.125 ounce, 60 grain or 3.888 gram.

"Dramamine."[70]   TM for dimenhydrinate.

"Drawcote."[204]   TM for a class of compounds used as dry film lubricants in the cold working of metals.

"Draw-ex."[51]   TM for a group of drawing compounds suitable for hot and cold metal working. Oil- and water-soluble grades are available, some pigmented.

draw.   In metalworking technology, to gradually reduce the diameter of a metal rod or plastic cylinder by pulling it through perforations of successively diminishing size in a series of plates. Wire and various specialty filaments are made in this way; they may be either cold-drawn (without preheating) or hot-drawn (heated to softening point). Dies made of low-grade diamonds are often used for precision work. An analogous method is used in sheet-metal forming.

drazoxolon.   (3-methyl-4-(o-chlorophenylhydrazono)-5-isoxazolone).   CAS: 5707-69-7. $C_{10}H_8ClN_3O_2$.
Properties: Yellowish crystals, mp 167C. Insoluble in water, acids, and straight-chain hydrocarbons; partially soluble in chloroform, alkalies, ketones, and aromatic solvents.
Hazard: Toxic by ingestion.
Use: Antifungal agent.

"Dresinate."[266]   TM for liquid, paste, and powder forms of sodium and potassium soaps of rosins and modified rosins used as emulsifiers, detergents and dispersants in soluble oils, cleaning compounds, and other compositions.

"Dresinol."[236]   TM for 40–45% solids dispersions of rosins, modified rosins, or ester resins using aqueous ammonia as a dispersant plus a protective colloid stabilizing agent.

"Drewmulse."[555]   TM for a series of glycerol and glycol esters and sorbitan and polyoxyethylene sorbitan esters of fatty acids.
Use: Emulsifiers and opacifiers.

"Dri-Clor."[204] TM for a powdered laundry bleach with more than 38% available chlorine.

drier. A substance used to accelerate the drying of paints, varnishes, printing inks, and the like by catalyzing the oxidation of drying oils or synthetic resin varnishes such as alkyds. The usual driers are salts of metals with a valence of two or greater and unsaturated organic acids. The approximately order of effectiveness of the more common metals is cobalt, magnesium, cerium, lead, chromium, iron, nickel, uranium, and zinc. These are usually prepared as the linoleates, naphthenates, and resinates of the metals. Paste driers are commonly the metal salts as acetates, borates, or oxalates dispersed in a dry oil. See also soap (2).
Note: The spelling "dryer" refers to equipment used for drying (of paper, textiles, food products, etc.).

"Drierite."[345] TM for a special form of anhydrous calcium sulfate having a highly porous granular structure and a high affinity for water. Absorbs water vapor both by hydration and capillary action.
Grade: Regular, indicating (turns blue to red in use), "Du-Cal" (for drying air and gases).
Use: Drying of solids, liquids, gases.

"Dri-Film."[245] TM for a group of silicone resins designed to impart durable moisture and weather resistance to surfaces.
Use: Dri-Film 88 is used as a protective coating for electric motors, transformers, field coils, etc.; Dri-Film 144 a masonary water repellent; Dri-Film 1040 and 1042 imparts durable water repellency and other properties including water-borne spot and stain resistance, improved hand and drape, increased flex abrasion resistance, and improved tear strength and wrinkle recovery.

drilling fluid. (drilling mud). A suspension of barytes and bentonite or attapulgite clay in either water or a petroleum oil. When circulated through oil-well drilling pipes, it acts as a coolant and lubricant and keeps the hole free from bore cuttings. To be effective, it must have a specific gravity of at least 2.0 and should be thixotropic, with appreciable gel strength. Lignosulfonates are used as thinners in the water-based type. Oil-based muds require thickening additives such as blown asphalt and metallic soaps of tall oil and rosin acids.

"Driltreat."[236] TM for a phosphatide liquid used as a stabilizer in oil emulsion drilling muds to offset effects of water contamination. Also effectively controls filtration.

"Drimix."[267] TM for a form of dry dispersion of liquid or solid materials in a siliceous carrier. Also called dry liquid concentrates.
Use: To dry up liquids to convert them to free-flowing powders for easier handling in rubber and plastics industries.

"Drinox."[401] TM for a series of insecticides. H-34 = A liquid containing 24.5% heptachlor. PX = A planter-box seed treatment containing 25% heptachlor.
Hazard: Toxic by ingestion, inhalation, and skin absorption.
Use: Seed treatment.

"Driocel."[99] TM for a solid granular desiccant for drying process liquids and gases. Manufactured from a selected grade of natural bauxite, reduced to the required particle size and thermally activated to its maximum absorbing activity. Mesh grades 4/8, 4/10, 8/14.

"Dri-Pax."[24] TM for a group of silica gel products used in packaging pharmaceuticals and similar packaged products. Extends shelf life and deodorizes products having unpleasant odors.

"Drisoy."[64] TM for a group of chemically treated soybean oils for replacement of linseed oil.

"Dritomic."[50] TM for a micron-fine wettable sulfur agricultural fungicide.

"Dri-Tri."[1] TM for sodium phosphate, tribasic, Na_3PO_4.

"Driwall."[448] TM for transparent coatings (silicone).
Use: Prevents water penetrating exterior walls of brick, stone, masonry. dross.
See slag.

"Drucomine" 9650.[555] TM for a cationic fatty amino amide.
Use: Softener and finisher.

drug. (1) A substance that acts on the central nervous system, e.g., a narcotic, hallucinogen, barbiturate, or psychotropic drug. (2) A substance that kills or inactivates disease-causing infectious organisms. (3) A substance that affects the activity of a specific bodily organ or function. (4) According to the FDA, a drug is a substance "intended for use in the diagnosis, cure, mitigation, treatment or prevention of disease, or to affect the structure or function of the body." Under this definition, high-potency preparations of vitamins A and D are classified as drugs, though other vitamins are not; antiperspirants and suntan lotions are also considered to be drugs.

The Drug Enforcement Administration is the federal agency responsible for supervision and control of the manufacture and use of drugs, especially those in class (1). The term "ethical drug" is equivalent to "prescription drug,", i.e., a drug sold only on a physician's prescription. See also pharmaceutical; see Appendix II for history of the industry.

"Drupene."[555]   TM for nonionic ethoxylated fatty alcohols, available as liquid or paste.
Use: Detergents and penetrants.

"Drustate."[555]   TM for a series of cationic fatty amino amide condensates, alkyl sulfates.
Use: Antistatic agents.

dry cell.   (Lechanche cell). A primary battery having a zinc anode, a carbon or graphite cathode surrounded by manganese dioxide and a paste containing ammonium chloride as electrolyte. Such batteries are not reversible and therefore have a limited operating life. Their chief use is in flashlights and similar devices requiring low voltage.
See also battery.

dry chemical.   A mixture of inorganic substances containing sodium bicarbonate (or frequently potassium bicarbonate) with small percentages of added ingredients to render it free-flowing and water repellent.
Use: Fire extinguisher on fires in electric equipment, oils, greases, gasoline, paints, and flammable gases.

dry deposition.   Deposition of materials from the atmosphere without the aid of rain or snow, e.g., particulates in the range of 2.5 microns as well as pollutant gases (SO_2, NO_2). See also acid precipitation.

dryer.   Any of numerous types of equipment used in the chemical industries to remove water from a product during processing. Space does not permit description of even a few of the great variety and multiplicity of choices available. The major types include the following:

belt	fluid-bed	screw
centrifugal	freeze	spray
convection	pan	tubular
conveyor	rotary drum	tunnel
flash	rotary tray	truck tray
	rotary vacuum	vibrating

See also drying.

"Dry-Flo."[53]   TM for starch ester derivative containing hydrophobic groups. Especially free-flowing in dry state. Cannot be wetted with water, yet when moistened with a water-miscible solvent it can be gelatinized and used to produce films with water-repellent properties.
Use: Dusting powder, no-offset spray for printing, approved by FDA for use in foods.

"Drygas."[580]   TM for a proprietary preparation used as a gasoline line antifreeze.

dry ice.   See carbon dioxide (solid).

drying.   (1) Polymerization of the glycerides of unsaturated vegetable oils induced by exposure to air or oxygen. See drying oil, drier.
(2) Removal of from 90 to 95% of the water from a material, usually by exposure to heat. Industrial drying is performed by both continuous and batch methods. The type of equipment and the temperatures used depend on the physical state of the material, i.e., whether liquid (solution or slurry), semiliquid (paste), solid units, or sheet. *Continuous drying*: The rotating-drum dryer is used for flaked or powdered products (soap flakes); a heated metal drum revolves slowly in contact with a solution of the material, the dry product being removed with a doctor knife. In paper manufacturing, drying is performed by a battery of staggered steam-heated revolving drums located at the dry end of the fourdrinier machine, the paper passing around the drums at high speed; the moisture content is thus reduced from 60% to about 5%. In spray drying, milk, egg-white, and other liquid food products are passed through an atomizing device into a stream of hot air. In tunnel drying, the product travels on a conveyor belt through a heated chamber of considerable length. *Batch drying*: Steam-jacketed pans are used if the material is in paste or slurry form, or in removable trays placed in an oven, if in solid units (fruits, vegetables, meats, etc.). The revolving-tube dryer, used for granular solids and coarse powders, is a long, horizontal cylinder in which a current of warm air runs countercurrent to the movement of the material. Freeze-drying is a specialized technique utilizing high vacuum and low temperatures.
See also dehydration, evaporation, freeze-drying.

drying oil.   An organic liquid which, when applied as a thin film, readily absorbs oxygen from the air and polymerizes to form a relatively tough, elastic film. The oxidation is catalyzed by such metals as cobalt and manganese.
See drier.

Drying oils are usually natural products such as linseed, tung, perilla, soybean, fish, and dehydrated castor oils, but are also prepared by combination of natural oils or their fatty acids with various synthetic resins. The drying ability is due to the presence of unsaturated fatty acids, especially linoleic and linolenic, usually in the form of glycerides. The degree of unsaturation of an oil, and hence its drying ability, is expressed by its iodine number. The drying oils have the greatest capacity for iodine and the nondrying oils the least.

"Dryolene."[200]   TM for a petroleum solvent prepared by straightrun overhead distillation.
Properties: Water-white, boiling range 205–287F (96.1–141C), d 0.747 (15.5C), flash p 25F (−3.89C) (TOC), bulk d 6.22 lb/gal (15.5C).
Hazard: Flammable, dangerous fire risk.
Use: Paint and rubber cement for fast setting and relatively slow final drying.

"Dryspersion."[267]   TM for dry dispersion of rubber compounding chemicals in powder form. Deagglomerated and treated with non-staining oil.

"DS-207."[304]   TM for a mix of dibasic lead salts of C_{16}-C_{18} fatty acids.
Properties: Soft, white, unctuous powder; d 2.0; refr index 1.6; non-melting at high temperatures; apparent lubricity of normal lead stearate.
Use: Stabilizer-lubricants for vinyls.

DSP.   Abbreviation for disodium phosphate. See sodium phosphate dibasic.

DSMA.   (disodium methanearsonate).
CAS: 144-21-8.
Use: Herbicide.

D-stoff   See phosgene, carbonyl chloride, $COCl_2$.

DTA.   See differential thermal analysis.

DTBP.   Abbreviation for di-tert-butyl peroxide.

DTDP.   Abbreviation for ditridecyl phthalate.

"Duclean."[28]   TM for acids containing pickling inhibitors.
Use: Pickling iron and steel. No. 1 sulfuric acid 60 degrees Bé, d 1.706, fp −10.8C. No. 2 hydrochloric acid technical, d 1.142, fp 40C
Hazard: Corrosive liquids.

Duff reaction.   ortho-Formylation of phenols or para-formylation of aromatic amines with hexamethylenetetramine in the presence of an acidic catalyst.

Duhring's rule.   Relates the vap press of similar substances at different temps. A straight or nearly straight line results if the temperature at which a liquid exerts a particular pressure is plotted on a graph against the temperature at which some similar reference liquid exerts the same vap press. Water is most frequently used as a reference liquid since its vap press at various temperatures is well known.

dulcin.   (4-ethoxyphenylurea).   CAS: 150-69-6. $H_2NOCNHC_6H_4OC_2H_5$.   This substance should not be confused with dulcitol.
Properties: White needle crystals or powder with taste approximately 200X as sweet as sugar, mp 173–174C, soluble in alcohol and ether, moderately soluble in hot water.
Derivation: From p-aminophenol.
Use: Artificial sweetening agent, prohibited by FDA for use in foods.

dulcitol.   (dulcite; dulcose). $C_6H_8(OH)_6$ A sugar.
Properties: White crystalline powder, slightly sweet taste, soluble in hot water, slightly soluble in cold water, very slightly soluble in alcohol, d 1.466, mp 188.5C.
Derivation: By hydrogenation of lactose, occurs naturally in Melampyrum nemorosum.
Grade: Technical, reagent.
Use: Bacteriology, medicine.

Dulong and Petit's law.   The atomic heat capacity (atomic weight times specific heat) of elementary substances is a constant whose average value at room temperature is 6.2. A few elements, notably boron, carbon, and silicon obey the law only at high temperatures.

Dumas method.   Condensation of nitrogenous organic compounds with copper oxide in a stream of carbon dioxide, collection of the elementary nitrogen over aqueous potassium hydroxide, and volumetric estimation in an azotometer. Any nitrogen oxides formed are reduced by passage over a red-hot copper spiral.

dumortierite. $8Al_2O_3 \cdot 6SiO_2 \cdot B_2O_3 \cdot H_2O$ (Perhaps).   A natural basic aluminum silicate, bright smalt-blue to greenish-blue in color, vitreous luster, d 3.26–3.36, Mohs hardness 7.
Occurrence: US (New York, Arizona, Nevada), France, Norway.
Use: Spark-plug porcelain, special refractories.

"Duocarb."[155]   TM for carbonized nickel from which all excess carbon has been removed and both surfaces polished.

"Duo-Sol" process.   A proprietary process for refining lubricating oils by extraction with a solvent consisting of liquid propane and a cresol base.

"Duponol."[28] TM for a series of surface-active agents based on lauryl sulfate. These have detergent, emulsifying, dispersing, and wetting properties and are used in the textile, paper, leather, cosmetic, electroplating industries; in dental and medical preparations, and in agricultural products.

"Duradene."[278] TM for a stereoregular copolymer of butadiene and styrene with exceptional purity, soluble in aromatics and aliphatics.
Use: Rubber products having superior abrasion resistance, good wet-skin resistance, flex-cracking resistance, and good resilience. Automotive tire treads, belt conveyors, shoe soles, and resilient products.

duralumin. A high-strength aluminum alloy, containing 4% copper, 0.5% magnesium, 0.25–1.0% manganese, and low percentages of iron and silicon. Resistant to corrosion by acids and seawater.
Use: Aircraft parts, railroad cars, boats, machinery.

"Duran."[573] TM for a series of polyvinylidine chloride packaging materials.

"Durana."[197] TM for a nitrogen fertilizer solution used in the manufacture of solid fertilizer compositions.

"Duranickel" 301.[283] TM for an alloy of 94% nickel and 4.5% aluminum.

"Duraphos."[55] TM for a complex sodium-calcium-aluminum polyphosphate containing 67% phosphate.
Use: Slow-dissolving glassy phosphate.

"Duraplex."[23] TM for drying and non-drying oil-modified alkyd resins derived from phthalic anhydride, polyhydric alcohols, and vegetable oils. Air drying, baking, and non-drying grades, in solvent solution or viscous 100% resins. Produce tough, glossy, light-colored coatings with excellent durability.
Use: Primers, lacquers and enamels; metal decorating, automotive coatings, furniture finishes, architectural enamels, inks.

"Duratone."[236] TM for a synthetic organophilic colloid used to control filtration and aid in suspending drilled solid particles in oil-well muds. It is effective at high temperature and is affected little by dissolved salts in the emulsified water phase.

"Durco" D-10.[47] TM for a nickel-base alloy. Typical analysis: nickel 65%, chromium 23%, iron 5%, copper 3.5%, molybdenum 2%, manganese 1%.
Properties: Wt 0.310 lb/in^3, tensile strength 60,000 psi, hardness (Brinell) 180, corrosion-resistant, machinable.

"Durcon." 6[47] TM for an epoxy resin, corrosion resistant to a variety of materials while retaining high impact resistance and low moisture absorption.
Properties: Wt 0.069 lb/in^3, tensile strength 13,000 psi, hardness (Rockwell M) 114.

durene. (sym-1,2,4,5-tetramethylbenzene). $C_6H_2(CH_3)_4$.

Properties: Colorless crystals; camphor-like odor; soluble in alcohol, ether, and benzene; insoluble in water; sublimes and is volatile with steam. D 0.838, mp 77C, bp 196–197C, flash p 165F (74C) (COC). Combustible.
Derivation: By heating o-xylene and methyl chloride in presence of aluminum chloride. Occurs in coal tar.
Use: Organic synthesis, plasticizers, polymers, fibers.

"Durez."[62] TM for synthetic resins, plastics, molding compositions, phenolformaldehyde resins, furfuryl alcohol resins, polyester resins, and polyurethane resins.
Use: Industrial chemicals; curing agents, rubber accelerators, plasticizers, and binders.

"Durichlor" 51.[47] TM for a high silicon iron alloy. Typical analysis: silicon 14.5%, chromium 4.5, carbon 0.9% manganese 0.65%.
Properties: Wt 0.255 lb/in^3, hardness (Brinell) 520, tensile strength 16,000 psi; resistant to moist chlorine, ferric chloride, chlorinated solutions and hydrochloric acid.

"Durimet 20."[47] TM for an austenitic stainless steel. Typical analysis: nickel 29%, chromium 20%, copper 3%, molybdenum 2%, silicon 1%, carbon 0.07% max.
Properties: Wt 0.287 lb/in^3, tensile strength 62,500 psi, hardness (Brinell) 120–150, machinable, sulfuric acid-resistant, weldable.

"Duriron."[47] TM for a high silicon iron alloy. Typical analysis: silicon 14.5%, carbon 0.9%, manganese 0.65%.
Properties: Wt 0.255 lb/in^3, hardness (Brinell) 520, tensile strength 16,000 psi, resistant to abra-

sion, erosion, and corrosion; machined by grinding.

"Durisite."[326] TM for a furan acid- and alkali-resistant resin cement mortar.

"Durite."[65] TM for a series of phenolformaldehyde resins used in the manufacture of grinding wheels, brake linings, clutch facings, lamp-basing cements.

"Durobrite."[28] TM for zinc cyanide plating brightening agents; amber liquids.

Durometer hardness. See hardness.

"Dursban."[233] (chlorpyrifos).
  CAS: 2921-88-2. TM for insecticides containing O,O-diethyl-O-3,5,6-trichloro-2-pyridyl phosphorothioate. $C_9H_{11}Cl_3NO_3PS$.
  Hazard: Toxic. Use may be restricted. TLV: 0.2 mg/m³ of air, toxic by skin absorption.
  Use: Control of chinch bugs in Gulf Coast states, tick control on cattle and sheep (Australia).

dust, industrial. Finely divided solid particles that may have damaging effects on personnel by inhalation or that constitute a fire hazard. They are air suspensions of particles 10 microns or less in diameter, though sizes up to 50 microns may be present. Such dusts include (1) metallic particles of all types, some being more harmful than others; (2) silica, mica, talc, quartz, graphite, clays, calcium carbonate, asbestos; (3) organic materials such as chemicals, pesticides, flour or other cereals, cellulose, coal, etc. The size range of mineral dust most damaging to the lungs is between 0.5 and 2 microns. Silicosis is caused by chronic exposure to uncombined silica (quartz, cristobalite) in mines and quarries. Baghouses, dust collectors, or electrostatic precipitators can be used for dust control.
  Hazard: Dust suspensions in enclosed industrial areas are an imminent fire hazard regardless of the chemical nature of the dust; an explosion may be initiated by static sparks or any open flame. Threshold Limit Values (TLV) for industrial dusts are given by the ACGIH.
  See threshold limit value, ACGIH.

dusting agent. A powdery solid used as an abherent and mold-release agent in the plastics and rubber industries. Typical materials in general use are talc (soapstone), mica, slate, flour, and clay. Graphite and mica have a flat, crystalline structure which causes them to act as lubricants and thus are especially effective in preventing sheets or slabs of hot solid mix from sticking together when stacked.
  See also antiblock agent.

Dutch oil. See ethylene dichloride.

Dutch process. Process for making white lead. See lead carbonate basic.

"Dutrex."[125] TM for a series of aromatic hydrocarbon concentrates derived from petroleum.
  Properties: Colors ranging from light amber to black, viscosity 3–10,000 cp at 210F, d approximately 1.0, odor slight to none, very low volatilities.
  Hazard: Flammable.
  Use: Rubber processing and extending oils, plasticizer extenders for polyvinyl chloride, resin solvents, tackifier for nitrile-butadiene rubber.

Dutt-Wormall reaction. Preparation of diazoaminosulfinates by reaction of diazonium salts with aryl- or alkylsulfonamides followed by alkaline hydrolysis to the corresponding sulfinic acid of the sulfonamide and the azide.

Du Vigneaud, Vincent (1901–1978) An American biochemist who won the Nobel prize for chemistry in 1955. His work involved the study of the metabolism of biologically significant sulfur-compounds which led to the finding of transmethylation in mammalian metabolism. He isolated and proved the structure of the vitamin Biotin, synthesized penicillin, oxytocin, and the vasopressin hormone of the posterior pituitary. His education was at the Universities of Rochester, Yale, St. Louis, and George Washington.

DVB. Abbreviation for divinylbenzene.

DXE. See dixylylethane.

Dy. Symbol for dysprosium.

dyclonine hydrochloride. (4'-butoxy-3-piperidinopropiophenone hydrochloride). $C_{18}H_{27}NO_2 \cdot HCl$.
  Properties: Crystals, mp 175–176C. Soluble in water, alcohol, acetone, phenol; coefficient 3.6.
  Grade: ND.
  Use: Medicine (topical anesthetic).

"Dycril."[28] TM for a photosensitive plastic bonded to steel, aluminum, or "Cronar" base supports. Exposure to a UV light source renders exposed areas insoluble to a subsequent mild alkaline washout solution so that a relief image of the exposed pattern remains. Depth of image is dependent on the thickness of the photosensitive plastic layer. There are eight types available.
  Use: Printing plates.

dye, azo.   See azo dye.

dye, azoic.   See azoic dye.

dye, certified.   See food color, FD&C colors.

dye, direct.   A water-soluble dye taken up directly by fibers from an aqueous solution containing an electrolyte, presumably due to selective adsorption. Usually applied to cellulosic fibers. Dyeing assistants such as sodium chloride or sodium sulfate are used to obtain a high concentration of dye on the fibers.
See also dye, synthetic.

dye, disperse.   A dye that may be in any of three clearly defined chemical classes: (a) nitroarylamine, (b) azo, and (3) anthraquinone; almost all contain amino or substituted amino groups but no solubilizing sulfonic acid groups. They are water-insoluble dyes introduced as a dispersion or colloidal suspension in water and are absorbed by the fiber, after which they may remain untreated or be after-treated (diazotized) to produce the final color. Their use is primarily for cellulose acetate, nylon, polyester and other synthetic fibers, and for thermoplastics.

dye, fiber-reactive.   A synthetic dye containing reactive groups capable of forming covalent linkages with certain portions of the molecules of natural or synthetic fibers, e.g., covalently bound azo dye on cellulose. This problem was not solved until 1953, when cholortriazinyl dyes were introduced. Since then, hundreds of fiber-reactive dyes for cellulose have been patented.

dyeing assistant.   Any material added to a dye bath to promote or control dyeing. The action of assistants differs with the classes of dyes but in most cases they aid in level deposition of the dye, either by delaying its absorption, increasing its solubility, or assisting the dye solution to penetrate the material. Examples: sodium sulfate decahydrate, pine oils, alkylaryl sulfonates.

dye intermediate.   See azo dye intermediate.

dye, metal.   An organic dye suitable for use with a metal such as aluminum or steel. Alizarin Cyanin RR, Alizarin Green sulfur, Nigrosine 2Y, and Naphthalene Blue RS are used for this purpose.

dye, natural.   An organic colorant obtained from an animal or plant source. Few of these are now in major use. Among the best known are madder, cochineal, logwood, and indigo. The distinction between natural dyes and natural pigments is often arbitrary.
See colorant, pigment.

dye retarding agent.   An additive to dye baths to prevent, by decreasing absorption of the dyes, the rapid exhaustion of the bath. Examples: sodium lauryl sulfate, sulfonated oils.

dye, solvent.   Any of several organosoluble dyes used in plastics, printing inks, cosmetics, etc., as well as for polyester and other synthetic fibers. They are chemically related to disperse dyes. The solvents used are such chlorinated hydrocarbons as 1,1,1-trichloroethane, trichloroethylene, and perchloroethylene. Use of solvent dyes avoids the necessity of fine-grinding and gives ideal dispersion. (The term "solvent dye" is somewhat misleading as it would apply equally well to dyes that are soluble in water.)

dye, synthetic.   An organic colorant derived from coal tar- and petroleum-based intermediates and applied by a variety of methods to impart bright, permanent colors to textile fibers. Organic dyes were first synthesized by Perkin in England (1856), and were later developed by Hofmann in Germany. Some, termed "fugitive," are unstable to sunlight, heat, and acids or bases; others, called "fast," are not. Direct (or substantive) dyes can be used effectively without "assistants;" indirect dyes require either chemical reduction (vat type) or a third substance (mordant), usually a metal salt or tannic acid, to bind the dye to the fiber. A noteworthy development is the fiber-reactive group wherein the dye reacts chemically with cellulose. Dyes may be either acidic or basic, and their effectiveness on a given fiber depends on this factor. Some types are soluble in water, others are not, but can be made so by specific chemical treatment.

The central problem is the affinity of a dye for a fiber and this involves both the chemical nature and the physical state of the dye, i.e., whether acidic or basic and whether colloidal, molecular, or ionic. Neither colloidal particles nor dissociated ions are accepted by a fiber; the dye must be in molecular dispersion to be effective. Further details will be found in the following entries: acetate dye, anthraquinone dye, acid dye, azo dye, alizarin, aniline, fiber-reactive dye, eosin, intermediate, resist, stilbene dye, sulfide dye. The chemical classes of coloring matters and their arrangement according to chemical structures have been designated numerically in the "Colour Index," a British publication.
See also colorant, pigment.

dye, vat. See vat dye.

dyfonate. (fonofos).   CAS: 944-22-9. $C_6H_5SPS(C_2H_5O)_2$.
Hazard: Cholinesterase inhibitor. TLV: 0.1 mg/ m^3 of air, toxic by skin absorption.
Use: Soil insecticide.

"Dylan."[11]   TM for polyethylene made by the high pressure process.
Use: Film packaging, flexible pipe, squeeze bottles, wire insulation, filaments, paper and cloth laminates, housewares.
See "Super Dylan."

"Dylene."[11]   TM for polystyrene, available in regular, medium, and high-impact grades.

"Dylex" Latices.[11]   TM for acrylic, styrene-butadiene latices.
Use: Exterior and interior paints.

"Dylite."[11]   TM for expandable polystyrene beads and pellets. They expand on application of heat to a foam with completely closed cell structure.
Use: Thermal insulator, protective packaging, flotation equipment, and containers.

"Dylopon."[300]   TM for a low-temperature dyeing auxiliary for wool.

"Dylox."[181]   TM for trichlorfon.

"Dylux."[28]   TM for a photosensitive paper free from silver compounds and coated with organic dyes that are activated by UV radiation. Gives good continuous tones and high resolution. Used in lithography, photoproofing, and information-handling systems. Permits instantaneous direct recording without processing.

dynamite.   An industrial explosive detonated by blasting caps. The principal explosive ingredient is nitroglycerin (straight dynamite) or especially sensitized ammonium nitrate (ammonia gelatin dynamite) dispersed in carbonaceous materials. Diethyleneglycol dinitrate, which is also explosive, is often added as a freezing point depressant. A dope such as wood pulp and an antacid as calcium carbonate are also essential.
Hazard: Moderately sensitive to shock or heat, dynamite is a rather serious explosion hazard, also may ignite or explode as a result of contact with powerful oxidizing agents.
Note: Use of dynamite is decreasing; it may be entirely replaced by safer and more efficient explosives.
See explosive, high.

"Dyphene."[141]   TM for pure phenolic resins characterized by extreme hardness, excellent chemical resistance, and fast-drying properties.

"Dyphos."[304]   TM for lead phosphite, dibasic. Used as a stabilizer for vinyls and other chlorinated resins.

dyphylline.   [7(2,3-dihydroxypropyl)theophylline; hyphylline].   CAS: 479-18-5. $C_{10}H_{14}N_4O_4$.
Properties: Crystals, bitter taste, mp 158C, soluble in water, alcohol, and chloroform.
Use: Medicine.

dypnone.   (phenyl α-methylstyryl ketone; 1,3-diphenyl-2-butene-1-one).   CAS: 1322-90-3. $C_6H_5COCHC(CH_3)C_6H_5$.
Properties: Stable, light-colored liquid with a mild, fruity odor; bp 340C; d 1.093 (20/20C); fp sets to a glass approximately −30C; insoluble in water, flash p 350F (176C). Combustible.
Use: Softening agent, plasticizer and perfume base. High absorption of UV light, low water-solubility, low evaporation rate and good solvent action make it useful in light-stable coatings and sunscreen preparations.

dysprosia.   See dysprosium oxide.

dysprosium.   Dy.   Rare earth or lanthanide element having atomic number 66; aw 162.50; valence = 3; six stable isotopes.
Properties: Non-corroding metal, does not react with moist air to form hydroxide, mp 1407C, bp 2330C, d 8.54, reacts slowly with water and halogen gases, soluble in dilute acids, high cross-section for thermal neutrons.
Derivation: Reduction of the fluoride with calcium.
Forms: High-purity lumps, ingots, sponge, powder.
Use: Measurement of neutron flux, reactor fuels, fluorescence activator in phosphors.
See also rare earth metal.

dysprosium nitrate. $Dy(NO_3)_3 \cdot 5H_2O$.
Properties: Yellow crystals, mp 88.6C (in its water of crystallization), soluble in water.
Derivation: treatment of oxides, carbonates, or hydroxide with nitric acid.
Grade: Up to 99.9% pure.
Hazard: Strong oxidizing agent, may ignite organic materials.

dysprosium oxide.   (dysprosia)). Dy_2O_3.
Properties: White powder, much more magnetic than ferric oxide, slightly hygroscopic, absorbing

moisture and carbon dioxide from the air, d 7.81 (27/4C) mp 2340C, soluble in acids and alcohol.
Derivation: Ignition of hydroxides and oxyacids (carbonates, oxalates, sulfates, etc.).
Use: With nickel, in cermets used as nuclear reactor control rods that do not require water-cooling.

dysprosium salts. Salts available commercially are the chloride $DyCl_3 \cdot xH_2O$, fluoride $DyF_3 \cdot 2H_2O$, arsenide, antimonide, and phosphide.
Use: The last three are used as high-purity binary semiconductors.

dysprosium sulfate. $Dy_2(SO_4)_3 \cdot 8H_2O$.
Properties: Brilliant yellow crystals, stable at 110C and completely dehydrated at 360C, soluble in water.

Derivation: Dissolving the hydroxide, carbonate, or oxide in dilute sulfuric acid.
Use: Atomic weight determination.

"Dyterge."[300] TM for a one-bath scouring and dyeing assistant for wool and wool blends.

"Dythal."[304] TM for lead phthalate, dibasic.
Use: Stabilizer for vinyl resins.

"Dytol."[23] TM for aliphatic primary alcohols derived from natural fats and oils (C_{10} to C_{18}).
Use: Additives for cosmetic creams, polymerization regulators for elastomers and plastics, detergents and viscosity index improvers for lubricating oils, finishing and softening agents for textiles, preparation of quaternary ammonium compounds, surfactants, water evaporation control, and anti-foam.

E

EADC. Abbreviation for ethylaluminum dichloride.

EAK. Abbreviation for ethyl amyl ketone.

earth. (1) Any siliceous or clay-like compound or mixture, e.g., fullers earth, diatomaceous earth. (2) A natural metallic oxide, sometimes used as a pigment, e.g., red and yellow iron oxide, ocher, or umber. (3) A series of chemically related metals that are difficult to separate from their oxides or other combined forms, specifically, rare earths, alkaline earths.
See also specific entry.

earth wax. General name for ozocerite, ceresin, and montan waxes.
See also wax.

EASC. Abbreviation for ethylaluminum sesquichloride.

East India. A type of fossil or semirecent resin similar to dammar. Varieties are batu, black and pale.
Use: In spirit and oleoresinous varnishes.

"Eastobond."[256] TM for a series of hot-melt adhesives used in the packaging industry for bonding paper, board, film, foil, and glassine.

"Eastone."[256] TM for disperse dyestuffs for use with acetate and nylon fibers.

"Eastozone."[256] TM for rubber antiozonants based on p-phenylenediamine.

ebonite. See rubber, hard.

"Ebonol."[142] TM for various blackening agents for metals. Materials are supplied as powders that are added to water at various temperatures to accomplish blackening.

echinomycin. CAS: 512-64-1.
$C_{51}H_{64}N_{12}O_{12}S_2$.
Properties: Colorless crystals, slightly hygroscopic, mp 218C. Soluble in fats, chloroform, dioxane; insoluble in water.
Use: Medicine (antibiotic).

ecology. The study of the interactions between plant and animal organisms and their environ-

ment; the latter is conceived to include everything that is not an intrinsic part of the organism and thus includes both living and non-living components. Though primarily a biological phenomenon, ecology does involve chemistry in respect to plant and animal nutrients, metabolism, photosynthesis, etc., especially interferences that may occur in connection with these. Thus, insecticides, chemical waste disposal, air and water pollution, oil spills, and radioactive contaminants have direct bearing on the ecology of a given area.
See also environmental chemistry.

economic poison. See pesticide.

economics, chemical. See chemical economics.

economizer. A device that acts like a heat exchanger whereby the heat produced by an operation is used to warm incoming air or water. It is widely used in the paper industry, in boilers, and in chemical processing. Heat recovery up to 30% is possible.

ecosystem. The organisms of an ecologic community together with the physical environment that they occupy.

ECTEOLA-cellulose. (epichlorohydrin triethanolamine cellulose). A dry, powdered cellulose derivative containing tertiary amine groups.
Use: As an anion exchanger in chromatography. It is less basic than DEAE-cellulose and serves to separate viruses, nucleic acids, and nucleoproteins.

EDB. Abbreviation for ethylene dibromide.

Edeleanu process. A solvent extraction process using liquid sulfur dioxide for the removal of undesirable aromatics from heavy lubrication oils.

edestin. A protein having a molecular weight of 310,000 obtained from hempseed. Closely similar proteins occur in seeds of pumpkin, squash, etc.

edible oil. As commonly used, the term refers to any fatty oil obtained from the flesh or seeds of plants that is used primarily in foodstuffs (margarine, salad dressing, shortening, etc.). Among these are olive, safflower, cottonseed, coconut, peanut, soybean, and corn oils, some of which may be hydrogenated to solid form. They vary in degree of unsaturation, ranging from 78% for safflower to about 10% for coconut. Castor oil, though technically edible, is not usually considered in this classification, nor are medicinal oils derived from animal sources (cod liver, mineral oil, etc.).

Edman degradation. Sequential degradation of peptides beginning at the N-terminal residue based on the reaction of phenylisocyanate with the alpha-amino group of the terminal amino acid of the peptide chain.

EDTA. Abbreviation for ethylenediaminetetraacetic acid.

EDTAN. Abbreviation for ethylene diamine tetraacetonitrile.

EDTA Na$_4$. Abbreviation for ethylene diamine tetraacetic acid tetrasodium salt.
See tetrasodium EDTA.

edetate. See ethylene diamine tetraacetic acid (note).

effect. An evaporation-condensation unit.
See evaporation.

efflorescence. Loss of combined water molecules by a hydrate when exposed to air, resulting in partial decomposition indicated by presence of a powdery coating on the material. This commonly occurs with washing soda ($Na_2CO_3 \cdot 10H_2O$) which loses almost all its water constituent spontaneously.

effluent. Any gas or liquid emerging from a pipe or similar outlet; usually refers to waste products from chemical or industrial plants as stack gases or liquid mix.

egg oil. Fatty oil obtained from egg yolk by extraction with ethylene dichloride; insoluble in water, but readily forms emulsions on strong agitation.
Use: Ointments, cosmetic creams.

egg yolk.
Properties: Yellow, semi-solid mass; d 0.95; mp 22C; high cholesterol content.
Grade: Technical, edible.
Use: Baking, dairy products, mayonnaise, pharmaceuticals, soap, perfumery.
See also albumin, egg.

EHEC. Abbreviation for ethyl hydroxyethyl cellulose.

Ehrlich, Paul. (1854–1915) A native of Silesia, Ehrlich is considered the founder of the science of chemotherapy or the treatment of diseases by chemical agents. He did fundamental work on immunity which gave him the Nobel prize in medicine in 1908, and also developed the famous neoarsphenamine (salvarsan or 606) treatment for syphilis (1910) which was not improved upon until the discovery of penicillin.

Ehrlich-Sachs reaction. Formation of anils by the base-catalyzed condensation of compounds containing active methylene groups with aromatic nitroso compounds.

eicosamethyl nonasiloxane. $C_{20}H_{60}O_8Si_9$.
Properties: Inert liquid, bp 173C (5 mm), d 0.918, soluble in benzene and light hydrocarbons, slightly soluble in alcohol.
Use: Silicone fluids, foam suppressor in lube oils.

eicosane. $C_{20}H_{42}$. Most technical eicosane is a mixture of predominantly straight-chain hydrocarbons averaging 20 carbon atoms to the molecule.
Properties: (pure n-eicosane): White, crystalline solid; fp 36.7C; bp 205C (15 mm); flash p 212F (100C); refr index 1.4348 (20C); d 0.778 (at melting point); insoluble in water; soluble in ether; can be readily chlorinated. Combustible.
Grade: Pure normal (99+%), technical.
Use: Cosmetic, lubricants, plasticizers.

eicosanoic acid. See arachidic acid.

eicosanoid. Any of a number of biochemically active compounds resulting from enzymic oxidation of arachidonic acid, e.g., prostaglandins, thromboxanes, prostacyclin, and leukotrienes. As a group, they comprise what is known as the arachidonic acid cascade. They have many pharmacological and medical possibilities.
See also prostaglandin, arachidonic acid.

1-eicosanol. See arachidyl alcohol.

5,8,11,14-eicostetraenoic acid. See arachidonic acid.

Eigen, Manfred. (1927-) A German physicist who won the Nobel prize for chemistry in 1967. His research concerned the rate of hydrogen ion formation through disassociation of water. He also was concerned with enzyme control. He received his degree at the University of Gottingen.

Einhorn-Brunner reaction. Formation of substituted 1,2,4-triazoles by condensation of hydrazines or semicarbazides with diacylamines in the presence of acid catalysts.

einsteinium. Es.   A synthetic radioactive element with atomic number 99 and aw 253, discovered in the debris from the 1952 hydrogen bomb explosion. Eisteinium has since been prepared in a cyclotron by bombarding uranium with accelerated nitrogen ions, in a nuclear reactor by irradiating plutonium or californium with neutrons, and by other nuclear reactions. The element is named for Albert Einstein. It has chemical properties similar to those of the rare earth holmium. Isotopes are known with mass numbers ranging from 246 to 253, valence = 2, and the lowest heat of vaporization of any divalent element.
See also actinide element.

ekahafnium.   One of the recently discovered transuranic elements; atomic number 104. It has two alpha-emitting isotopes (257 and 259) and possibly a third (258). The former, made by bombardment of californium-249 with carbon-12 nuclei, has a half-life of five seconds and decays into nobelium-253. The 259 isotope, made by merging a carbon-13 nucleus with californium-249, has a half-life of three seconds and decays to nobelium-255.

"Ekonol."[280]   TM for an engineering plastic composed of poly-p-oxybenzoate. Resistant to temperatures above 600C, self-lubricating surface.
Use: In pumps handling corrosive liquids, protective coating for titanium skins on supersonic transports, disk brakes, etc.

"ELA."[28]   Brand name for elastomer lubricating agent, a mixture of phosphate esters. Light amber liquid.
Use: Unvulcanized rubbers.

elaboration.   A term used in biochemistry to describe chemical transformations within an organism resulting in formation of specific types of substances; for example, plants elaborate proteins and fats, and poisonous snakes elaborate their venom. It also refers to formation of metabolic end products such as purines and uric acid.

elaidic acid.   (trans-9-octadecenoic acid). $CH_3[CH_2]_7HC{:}CH[CH_2]_7COOH$.   The trans form of an unsaturated fatty acid of which the cis-form is oleic acid

$$CH_3(CH_2)_7\diagdown \qquad \diagup H$$
$$C{=}C$$
$$H\diagup \qquad \diagdown (CH_2)_7CO_2H$$

Properties: White solid, d 0.8505 (79/4C), mp 43.7C, bp 288C (100 mm), 234C (15 mm), refr

index 1.4358 (79C). Insoluble in water; soluble in alcohol, ether, benzene, and chloroform. Combustible.
Derivation: Synthesized from oleic acid by elaidinization.
Grade: Purified, 99+%.
Use: Medical research, reference standard in chromatography.

elaidinization.   Originally the reaction by which oleic acid is converted into elaidic acid, but now used in a more general sense to indicate the conversion of any unsaturated fatty acid or related compound from the geometric cis to the corresponding trans form. Nitrous acid and selenium compounds are commonly used as catalysts for this reaction. The resulting trans acids are more stable to oxidation.

elastic modulus.   See modulus of elasticity.

elasticity.   The ability of a material to recover its original shape partially or completely after the deforming force has been removed. The small amount of deformation that is not recovered is called permanent set or permanent elongation. Among common materials glass and some metals are virtually 100% elastic, whereas vulcanized rubber and other elastomeric substances are in the range of 90% elastic after extension to rupture. So-called perfect elasticity is a property of atoms which show no energy loss on collision.
See also modulus of elasticity, stress, strain, plasticity.

elastin.   A scleroprotein which occurs in connective tissue.
Properties: Yellow, fibrous mass; insoluble in water, dilute acids, alkalies, and salt solutions, and alcohol. Is partially digested by pepsin solution and wholly by trypsin.

elastomer.   As originally defined by Fisher (1940), this term referred to synthetic thermosetting high polymers having properties similar to those of vulcanized natural rubber, namely the ability to be stretched to at least twice their original length and to retract very rapidly to approximately their original length when released. Among the better-known elastomers introduced since the 1930s are styrene-butadiene copolymer, polychloroprene (neoprene), nitrile rubber, butyl rubber, polysulfide rubber ("Thiokol"), cis-1,4-polyisoprene, ethylene-propylene terpolymers (EPDM rubber), silicone rubber, and polyurethane rubber. These can be crosslinked with sulfur, peroxides, or similar agents. The term was later extended to include uncrosslinked polyolefins that are thermoplastic; these are generally known as TPO rubbers. Their extension and retraction properties are notably different from those of thermosetting elastomers but they are well adapted to such specific uses as wire and cable coating, automobile bumpers, vibration dampers, and specialized mechanical products.

Elbs persulfate oxidation.   Hydroxylation of phenols to p-diphenols by potassium persulfate in alkaline solution.

Elbs reaction.   Formation of anthracenes by intramolecular condensation of diaryl ketones containing a methyl or methylene substituent adjacent to the carbonyl group.

electric double layer.   A diffuse aggregation of positive and negative electric charges surrounding a suspended colloidal particle which aids in maintaining its stability. According to the Gouy-Freundlich theory, advanced about 1920, a close-packed array of charges is attached to the surface of the particle while a diffuse double layer of charges of opposite sign extends into the liquid. The particle is electrically neutral. There is an electrokinetic potential gradient across the double layer that is called the zeta potential. The diagram is an approximation of this phenomenon. Modifications of this theory have been introduced in recent years, notably by Derjaguin and Landau and Verwey and Overbeek (DLVO theory).
See also zeta potential.

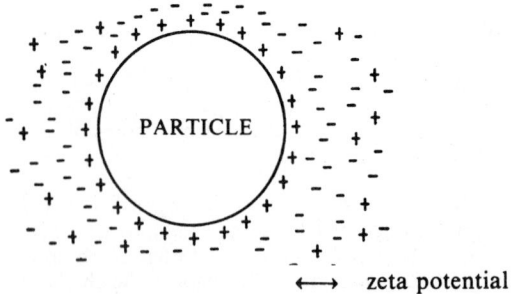

⟷   zeta potential

electric furnace.   See furnace.

electric steel.   Steel made in an electric furnace.

electric vehicle propulsion.   See storage battery.

electride.   An experimental compound composed of an alkali metal cation and an electron in which the electron functions as a chemical element, (e.g., a halogen) in salt formation. Several such compounds have been made in the US and

abroad. The phenomenon is reported to be one which challenges accepted concepts of compound formation.

electrochemical equivalent.   The number of grams of an element or group of elements liberated by the passage of one coulomb of electricity (one ampere for one second).

Electrochemical Society.   Established in 1902, this society was organized to promote the advance of the science of electrochemistry and related fields. It is comprised of 11 divisions, each devoted to a special branch of electrochemistry, e.g., corrosion, batteries, rare metals, electrodeposition, etc. It publishes a journal and sponsors books relating to its major interests. Its office is in Pennington, NJ.

electrochemistry.   That branch of chemistry concerned primarily with the relationship between electrical forces and chemical reactions. This relationship is fundamental and far-reaching as the structure of matter is basically electrical. Electrochemistry is directly involved in chemical bonding, ionization, electrolysis, metallurgy, battery science, fuel cells, and corrosion--in short, in any situation in which a chemical change is caused by, or associated with, electrical phenomena, e.g., production of aluminum. It has certain applications in biochemistry (in nerve reactions and the electric organs of fish) as well as in organic chemistry. Michael Faraday (1791–1867) is generally regarded as the founder of electrochemistry.
See also following entries.

electrocoating.   A process of applying primer paint to household appliances, automobiles, etc., in which the metal piece to be coated becomes the anode in a tank of water-based paint. The coating deposited on the metal is uniform regardless of the shape of the article. Large-scale use of electrocoating is on automobile bodies.
See also electrostatic coating.

electroconductive polymer.   See polymer, electroconductive.

electrocratic.   Descriptive of a liquid colloidal dispersion of insoluble solid particles whose stability is maintained by either positive or negative charges on the particles. As the like charges are mutually repellent, they offset the attraction of gravity and prevent the particles from sinking or coalescing. A colloidal gold suspension is a well-known example. Electrocratic dispersions can be readily precipitated by addition of an oppositely charged electrolyte, as in the purification of water with Al^{+++} ions from aluminum sulfate.

"Electrocure." TM for a process for hardening or curing paint films on plastics by use of low voltage electron beams. Paints so treated cure in four seconds or less at room temperature; the finishes produced are said to be superior in resistance to abrasion and chemicals. The process can also be applied to other substrates such as wood, glass, metals, etc.

electrode. Either of two substances having different electromotive activity which enables an electric current to flow in the presence of an electrolyte. Electrodes are sometimes called plates or terminals. Commercial electrodes are made of a number of materials which vary widely in electrical conductivity, i.e., lead, lead dioxide, zinc, aluminum, copper, iron, manganese dioxide, nickel, cadmium, mercury, titanium, and graphite; research electrodes may be calomel (mercurous chloride), platinum, glass, or hydrogen. Electrodes are essential components of both batteries and electrolytic cells; in batteries the negative plate is the anode and the positive plate the cathode, whereas in electrolytic cells the reverse is the case.
Use: Electrodes are also used in welding devices.
See also anode, cathode, battery, electrolytic cell.

electrode, glass. A thin glass membrane which when immersed in a suitable liquid medium develops a measurable electrical potential that can be readily related to the activity of ionic species present in the solution. By appropriate manipulation of the glass composition, careful pretreatment of the glass surface, and reproducible experimental conditions, electrodes can be devised which not only yield information about the concentration of ions in solution but also have the ability to discriminate in terms of a selective response between a number of different ions of similar chemical characteristics. Because of their ability to give both qualitative and quantitative information about ions in solution, glass electrodes are widely used for purposes of chemical measurement, especially in electrochemical research.

electrode, hydrogen. A platinum surface coated with platinum black immersed in a solution and bathed with a stream of pure hydrogen gas. The potential developed depends on the equilibrium between the hydrogen gas and the hydrogen ions in solution.
Use: Standard reference electrode.

electrodeposition. The precipitation of a material at an electrode as the result of the passage of an electric current through a solution or suspen-

sion of the material, for example, alkaline-earth carbonates, rubber from latex, paint films on metal. A technique for electrodepositing refractory carbide coatings on metal has been reported. The electrode is in the shape of the desired article. An important advantage of electrodeposition is its ability to coat complex shapes having small and irregular cavities with exact thickness control.
See also electrophoresis.

electrodialysis. A form of dialysis in which an electric current aids the separation of substances that ionize in solution. Seawater can be desalted by this method on a large scale by placing it in the center chamber of a three-compartment container having two semipermeable membranes and a positive electrode in one end chamber and a negative electrode in the other. The ions migrate to their respective electrodes under a difference of potential, leaving the water salt-free.
See also dialysis, desalination, demineralization.

electroforming. An electrolytic plating process for manufacturing metal parts. A mold of the object to be reproduced is made in a soft metal or in wax (by impression). The mold surface is made conducting by coating with graphite. Some suitable metal is then deposited electrolytically on the mold surface. This mold is then (in most cases) a negative of the object to be produced. Other industrial applications are phonograph records, plastic tile, ducting, tubing, etc.

"Electroless" coating. A protective coating of copper, cobalt, nickel, gold, or palladium deposited in a bath without application of an electric current, i.e., by chemical reduction.

electroluminescence. Luminescence generated in crystals by electric fields or currents in the absence of bombardment or other means of excitation. It is a solid-state phenomenon involving p- and n-type semiconductors, and is observed in many crystalline substances, especially silicon carbide, zinc sulfide, gallium arsenide, as well as in silicon, germanium and diamond.
See also phosphor.

electrolysis. Decomposition of water and other inorganic compounds in aqueous solution by means of an electric current, the extent being proportional to the quantity of electricity passing through the solution. The positive and negative ions formed are carried by the current to the oppositely charged electrodes, where they are collected (if wanted) or released (if unwanted). Metallic ions deposited on the electrode form a

coating. A simple electrolysis is the separation of water into oxygen and hydrogen; this is one method of producing hydrogen. Somewhat more complicated is electrolysis of brine to chlorine and sodium hydroxide; this is carried out in electrolytic cells of the diaphragm or mercury type with water taking part in the reaction. In electroplating, metal salts dissociate into their constituent ions, the positively charged metal ions coating the cathode. There are a number of variations of this process (electrodeposition, electrocoating, electroforming).
See also electrolytic cell, ionization.

electrolyte. A substance that will provide ionic conductivity when dissolved in water or when in contact with it; such compounds may be either solid or liquid. Familiar types are sulfuric acid and sodium chloride which ionize in solution. A solid electrolyte used originally in fuel cells is a polymer of perfluorinated sulfonic acid used as the core of a water electrolysis cell for production of hydrogen and oxygen. When saturated with water it has high conductivity. Another solid type is a ceramic mixture of sodium, aluminum, lithium, and magnesium used as a separating medium in the liquid sodium/sulfur (beta) battery now under development. The most common application of electrolytes is in electroplating of metals in which dissolved (ionized) metal salts are the electrolytes.
See also electrolysis, electroplating.

electrolyte acid. Legal label name for battery acid.

electrolytic cell. An electrochemical device in which electrolysis occurs when an electric current is passed through it. Ionizable compounds dissociate in the aqueous solution with which the electrodes are in contact. Such cells are of two types: (1) the diaphragm cell, which has two compartments separated by a porous membrane; and (2) the mercury cell, in which mercury is the cathode. The anodes of both types have long been made of graphite; because this decomposes rapidly as electrolysis progresses, they are being replaced with dimensionally stable types consisting of titanium coated with oxides of ruthenium and other rare metals which are also much more efficient. In electrolysis of sodium chloride, the current causes chloride ion to migrate to the anode, where it is collected as chlorine gas; sodium hydroxide and hydrogen are also formed, the hydrogen being discharged. The overall cell reaction is: $2NaCl + 2H_2O \rightarrow H_2 + Cl_2 + 2NaOH$. This principle is applied in the electroplating of metals, electrodeposition of colloids, and similar processes.
See also diaphragm cell, mercury cell, electroplating.

electromagnetic separation. Separation of isotopes, especially those of uranium, by first accelerating them by means of an electrostatic field and then passing them through a magnetic field. The effect of this is to cause all the particles to take a curved path; the heavier ones having higher kinetic energy describe a wider arc than the lighter ones. Thus, two isotopes of closely similar masses can be separated and collected.
See also mass spectrometry, magnetic separation.

electromagnetic spectrum. See radiation.

electrometallurgy. Application of the principles and technqiues of electrochemistry to the production of such metals as aluminum and titanium.

electromotive series. See activity series.

electron. Discovered by J. J.Thompson in 1896, the electron is a fundamental particle of matter that can exist either as a constituent of an atom or in the free state. It has a negative electric charge (4.8×10^{-10} esu) and a mass 1/1837 that of a proton, equivalent to 9.1×10^{-28}gram. The number of electrons in an atom of any element is the same as the number of protons in the nucleus, i.e., the atomic number. Thus, the range is from one electron in hydrogen to 103 in lawrencium. As the negative charge of the electrons equals the positive charge of the protons, all atoms are electrically neutral. Electrons are arranged in from 1 to 7 shells around the nucleus; the maximum number of electrons in each shell is strictly limited by the laws of physics. The tendency of electrons to form complete outer shells accounts for the valence of an element, and they play an essential part in chemical bonding. The outer shells are not always filled: sodium has two electrons in the first shell, eight in the second, and only one in the third. A single electron in the outer shell may be attracted into an incomplete shell of another element, leaving the original atom with a net positive charge. The latter then is called an ion. Valence electrons are those which can be captured by or shared with another atom.

Electrons can be removed from the atoms of metals and some other elements by heat, light, electric energy, or bombardment with high-energy particles (see radiation, ionizing). In such cases, they are totally free from the atomic orbit and their energy can be utilized by means of a conductor (electricity) or a vacuum tube or semiconductor. Current is generated by detaching the electrons of a metallic conductor (silver, copper) by means of an electric or magnetic field; the electrons then flow along the conductor to a positively charged terminal. The entire science of electronics is made possible by the tendency of a heated metal cathode to emit a continuous stream of electrons in a vacuum tube.

Free electrons, called beta particles, are spontaneously emitted by decaying radioactive nuclei; they have comparatively low energy but can be accelerated to velocities approaching that of light. The basic nature of the electron has been the subject of much research of the highest order of mathematical rigor. In simplest terms, the electron has the properties of both a particle and a wave, i.e., a standing wave is associated with an electron moving in its orbit. The energy state of any electron is described by four quantum numbers.
See also shell, atom, orbital theory, Lewis electron theory.

electron-beam welding.   See welding.

electronegativity.   All atoms (except those of helium) that have fewer than eight electrons in their highest principal quantum level have low energy orbital vacancies capable of accommodating electrons from outside the atom. The existence of these vacancies is evidence that within these regions the nuclear charge can exert a significant attraction for such electrons, even though as a whole the atom is electrically neutral. This attraction is called "electronegativity." To the extent that the initially neutral atom may be able to acquire electrons from outside, it will acquire also their negative charge. The word "electronegativity" means "tendency to become negatively charged." The concept of electronegativity is an extremely useful one in chemistry since the attractive force exerted by the nuclei of atoms that have vacancies in their outer shells makes possible the formation of both covalent and ionic bonds and is thus a fundamental factor in the formation of chemical compounds. The most highly electronegative elements are the halogens, sulfur, and oxygen.
See also bond, chemical.

electron microscope.   A microscope in which the source of illumination is a stream of electrons emanating from a tungsten cathode in a high vacuum and accelerated by a strong electric impulse (300 kilovolts). The electrons are focused by a series of magnetic fields which function as lenses in the same way as glass affects waves of visible light, i.e., the electron stream curves as it passes through the magnetic field. Such lenses were developed in Germany by Busch, Knoll, and Tuska in the 1930s, and were adapted to microscopy by Zworykin and Hillier at the RCA Laboratories in the early 1940s when the first commercial instruments were produced. The electron microscope is characterized by ex-

tremely high resolving power due to the ultrashort wavelength of electronic radiation--a small fraction of an Angstrom unit (Å). Resolution of 2 Å is possible which permits determination of the structure of macromolecules (DNA) and even observation of large atoms (uranium). A new instrument of the scanning transmission type expected to have a resolving power of 0.05 Å is in use at the University of Chicago.

Two kinds of electron microscopies are in general use: the transmission type, in which the electrons penetrate the specimen, and the scanning type introduced in 1970, in which the electrons, condensed to a fine beam, repeatedly traverse the surface of the specimen producing a 3-dimensional contour effect by means of secondary electrons emanating from the specimen itself. Pictures of astonishing accuracy have been obtained, especially of surface structures, a matter of great importance in the study of catalysis and other critical phenomena in both industry and the biological sciences. A unique combination of these techniques is the scanning transmission electron microscope (STEM), by means of which colored motion pictures of uranium atoms on a thin-film carbon substrate have been obtained.
See also optical microscope, resolving power, ultramicroscope, field-ion microscope.

electron paramagnetic resonance.   (EPR).
A method of spectroscopic analysis similar to nuclear magnetic resonance except that microwave radiation is employed instead of radiofrequencies. It is used for studying free radicals, crystalline centers, transition elements, and structures involving unpaired electrons.

electron-volt.   (eV).   An extremely small unit used in measuring the energy of electrons and other nuclear constituents. It is the energy developed by an electron in falling through a potential difference of one volt, equivalent to 1.6×10^{-19} joule. The rupture of a carbon-to-carbon bond has been calculated to yield approximately 5 eV.

electrophile.   See nucleophile, Lewis acid.

electrophoresis.   Migration of suspended or colloidal particles in a liquid such as rubber latex due to the effect of potential difference across immersed electrodes. The migration is toward electrodes of charge opposite to that of the particles. Most solids, being negatively charged, migrate to the anode, the exception being basic dyes, hydroxide solutions, and colloids which have adsorbed positive ions, all of which are positively charged and migrate to the cathode. Migrating particles lose their charge at the electrode and generally agglomerate around it. Clay suspensions can be filtered by means of forced flow electrophoresis.

Electrophoresis is important in the study of proteins because the molecules of such materials act like colloidal particles and their charge is positive or negative according to whether the surrounding solution is acidic or basic. Thus, the acidity of the solution can be used to control the direction in which a protein moves upon electrophoresis. It has been found that electrophoresis can be carried out more efficiently under zero gravity conditions in outer space than on Earth. See also electrodeposition.

electroplating.   The deposition of a thin layer or coating of metal, (e.g., chromium, nickel, copper, silver, etc.) on an object by passing an electric current through an aqueous solution of a salt containing ions of the element being deposited, for example, Cu^{++}. The material being plated (usually a metal but often a plastic) constitutes the cathode. The anode is often composed of the metal being deposited; ideally it dissolves as the process proceeds. The thin layer deposited is sometimes composed of two or more metals, in which case it is an alloy. The solution or plating bath contains dissolved salts of all the metals being deposited. Electrolytic cells are used for this process.

The anode must be an electrical conductor but may or may not be of the same chemical composition as the material being deposited, and may or may not dissolve during the process. The purpose of electroplating is usually protection of the base metal from corrosion. Silver is electroplated on copper for economy reasons; plastics may be electroplated for decorative effects.
See also electrophoresis, protective coating, electroless coating, throwing power, current density.

electropolishing.   A nonmechanical method of polishing metal surfaces by a method that is actually the reverse of electroplating. This is achieved by making the object to be polished the anode in an electrolytic circuit, the cathode usually being carbon. The electrolytes used are phosphoric, hydrofluoric, nitric, or sulfuric acids (sometimes called polishing acids).

electrostatic bond.   Alternative name for an ionic bond.
See bond, chemical.

electrostatic coating.   A metal painting technique in which electrostatically charged pigment particles are sprayed onto a substrate metal followed by baking. The electric charge attracts the particles to the metal and holds them in place until heat treatment is applied. Maintenance of the charge is thus essential; factors affecting this are relative humidity (the lower the better) and the chemical nature of the pigment, e.g., phthalocyanine blue retains the charge much longer than titanium dioxide.

electrostatic precipitator.   See Cottrell.

electrovalent bond.   Alternative name for an ionic bond.
See bond, chemical.

electrowinning.   The technique of extracting a metal from its soluble salt by an electrolytic cell. It is used in recovery of zinc, cobalt, chromium, and manganese, and has recently been applied to copper when in the form of a silicate ore. For any specific metal, the salt in solution is subjected to electrolysis and is electrodeposited on a cathode made of the metal being extracted.

element.   One of the 109 presently known kinds of substances that comprise all matter at and above the atomic level. According to a theory that has gained acceptance, the lightest elements were formed in less than half an hour from a primordial complex called ylem, a mixture of neutrons and electromagnetic radiation. The smallest unit of any element is the atom. All the atoms of a given element are identical in nuclear charge and number of electrons and protons, but they may differ in mass, e.g., hydrogen has mass numbers of 1, 2, and 3, called hydrogen, deuterium, and tritium, respectively. These are the isotopes of hydrgen; most elements have isotopic forms which are due to the presence of one or more extra neutrons in the nucleus. The atomic number of an element indicates its position in the Periodic Table and represents the number of protons present, which is the same as the number of electrons.

All elements heavier than lead are unstable and radioactive. About 90% of the earth's crust is made up of elements with even numbers of protons and neutrons. No stable elements heavier than nitrogen have an odd number of both protons and neutrons. Elements of even atomic number normally have several isotopes while those of odd atomic number never have more than two stable isotopes. All elements beyond uranium (transuranic) were nonexistent in 1940. They are artificially created by bombardment of other elements with neutrons or other heavy particles. Research on new elements is actively carried on at the Lawrence Livermore Laboratories which reported discovery of Element 106 in 1974. Creation of Element 109 was announced in 1982. A single atom of it was made by West Germany physicists by bombarding Bi-209 with Fe-58 nuclei. Many more (possibly up to 150) are theoretically possible according to Dr. G. T. Seaborg. See also Periodic Table, isotope, radioactivity, abundance. *Note*: For origin of elements, see nucleogenesis.

elemi. A soft, balsam-like resin obtained from a tree in the Philippines, soluble in coal-tar hydrocarbons, but not in petroleum solvents, alcohols, and ketones.
Use: Plasticizer, adhesion of lacquers to metals, cements and adhesives, wax compositions, printing inks, textile and paper coatings, perfumery, waterproofing, engraving.

"Elprene."[41] TM for a self-curing, synthetic-rubber coating of the neoprene type used as a general maintenance coating.

Eltekoff reaction. Production of highly-branched hydrocarbons by methylation of olefins with methyl choride or methyl iodide in the presence of lead oxide or calcium oxide at high temperatures.

elutriation. A process of washing, decantation, and settling which separates a suspension of a finely divided solid into parts according to their weight. It is especially useful for very fine particles below the usual screen sizes, and is used for pigments, clay dressing, and ore flotation.

"Elvace."[36] TM for a series of vinyl acetate ethylene emulsions.

"Elvanol."[28] TM for various grades of polyvinyl alcohol.

"EMA" Resins.[58] TM for ethylene-maleic anhydride copolymers. Water-soluble resins which serve as dispersing agents in cold water detergents, thickeners, binders, stabilizers, and emulsifiers.

embosser. See fiber roll.

embrittlement. Hardening of a metal (especially steel) or of an ABS resin resulting in loss of strength and impairment of other physical properties. In metals, the primary cause is exposure to hydrogen, though other factors such as corrosion also are involved. In copolymer plastics, such as ABS resins, embrittlement is due to formation of a vitreous matrix as well as to oxidation of the butadiene particles in the matrix. Embrittlement due to thermal shock occurs in pressurized-water reactors. This may result in rupture of reactor walls and is a constant source of trouble in reactors of this type.

Emde degradation. Modification of the Hofmann degradation method for reductive cleavage of the carbon-nitrogen bond by treatment of an alcoholic or aqueous solution of a quaternary ammonium halide with sodium amalgam. Also used as a catalytic method with palladium and platinum catalysts. The method succeeds with ring compounds not degraded by the Hofmann procedure.

emerald green. A pigment consisting of copper acetoarsenite.
Hazard: Toxic by ingestion.

emeralds, synthetic. Artificial crystals produced by a high pressure-high temperature process from beryllium aluminum silicate containing a small amount of chromium.
Use: Lasers, masers, semiconductors.

"Emerest."[242] TM for a series of fatty esters.
Use: Wetting agents, detergents, softeners, antistats, defoamers, lubricants, stabilizers, emulsifiers, corrosion inhibitors, and other applications not requiring food-grade products.

"Emerez."[242] TM for a series of dimer acid-based polyamide resins.
Use: For the manufacture of printing inks, hot melt adhesives, and thixotropic paints.

"Emersol."[242] TM for a series of fatty acids including linoleic, oleic, palmitic, and stearic acid of improved stability and color.

emery. See corundum, abrasive.

emetine. (cephaeline methyl ether; 6',7',10,11-tetramethoxyemetan). CAS: 483-18-1. $C_{29}H_{40}O_4N_2$. An alkaloid from ipecac.
Properties: White powder, mp 74C, very bitter taste, darkens on exposure to light, soluble in alcohol and ether, slightly soluble in water.
Derivation: By extraction from root of Cephalis ipecacuanha (ipecac) or synthetically.
Hazard: Toxic by ingestion.
Use: Medicine (antiamebic).

"Emisaloy."[155] TM for an alloy composed of 80% nickel and 20% cobalt.
Properties: Mp 1500C, d 8.84, resistivity at 20C ll.0 microohm cm, tensile strength at 20C 85,000 psi.
Use: Electron tubes.

emission spectroscopy. Study of the composition of substances and identification of elements by observation of the wavelengths of radiation they emit as they return to a normal state after excita-

tion by an external energy source. When atoms or molecules are excited by energy input from an arc, spark, or flame, they respond in a characteristic manner; their identity and composition are signaled by the wavelengths of incident light they emit. The spectra of elements are in the form of lines of distinctive color, such as the yellow sodium D line of sodium, those of molecules are groups of lines called bands. The number of lines present in an emission spectrum depends on the number and position of the outermost electrons and the degree of excitation of the atoms. The first application of emission spectra was identification of sodium in the solar spectrum (1814).
See also spectroscopy.

emmenagogue. A drug used to induce menstruation.

Emmert reaction. Formation of 2-pyridyldialkylcarbinols by condensation of ketones with pyridine or its homologs in the presence of aluminum or magnesium amalgam.

emodin. (frangula emodin; frangulic acid; 1,3,8-trihydroxy-6-methylanthraquinone).
CAS: 518-82-1. $C_{14}H_4O_2(OH)_3CH_3$.
Occurs either free or combined with a sugar in a glucoside, in rhubarb, cascara sagrada, and other plants. A synthetic product is also available.
Properties: Orange crystals, mp 256C, soluble in alcohol, insoluble in water.
Use: Medicine (cathartic).

"Emolein."[242] TM for synthetic lubricant esters.
Use: Compounding of synthetic jet engine lubricants to meet both military and civilian specifications. They include diisooctyl azelate, di-2-ethylhexyl azelate, dipropylene glycol dipelargonate, isodecyl pelargonate.

empirical formula. See formula, chemical.

"Empol."[242] TM for dimer acids, 36-carbon aliphatic dibasic acids.
Use: Modifiers of synthetic polymers, soaps, corrosive inhibitors, etc. Six grades available.

"Emralon."[46] TM for series of resin-bonded, tetrafluoroethylene solid film lubricants. Colloidal TFE resin is dispersed in phenolic resin, acrylic resin latex, epoxy resin, or thermoplastic resin solutions.
Properties: Flash p varies from 4.4 to 14.4C.
Hazard: Flammable, dangerous fire hazard.

"Emsal."[242] TM for sulfated fatty alcohols.

"Emsorb."[242] TM for sorbitan and ethoxylated sorbitan esters.

"Emtall."[242] TM for a series of fractionated tall-oil fatty acids and distilled tall oils.

EMTS. Abbreviation for ethylmercury-p-toluenesulfonanilide.

"Emulphogene BC."[307] TM for a series of tridecyloxopoly(ethyleneoxy)ethanols.

"Emulphor."[307] TM for a series of nonionic emulsifying agents and dispersants. Some are polyoxyethylated vegetable oils, alcohols, and fatty acids.

emulsifier. A surface-active agent.
See emulsion.

emulsifying oil. See soluble oil.

emulsion. (synaptase; amygdalase; β-glucosidase). An enzyme catalyzing the production of glucose from β-glucosides.
Properties: White powder, odorless and tasteless, capable of hydrolyzing glucosides such as amygdalin to glucose and the other component substances. Soluble in water, insoluble in ether and alcohol.
Source: Sweet almonds.
Derivation: By extracting an emulsion of almonds with ether filtering the clear solution and precipitating the emulsion with alcohol.

emulsion. A stable mixture of two or more immiscible liquids held in suspension by small percentages of substances called emulsifiers. These are of two types: (1) Proteins or carbohydrate polymers which act by coating the surfaces of the dispersed fat or oil particles thus preventing them from coalescing; these are sometimes called protective colloids. (2) Long-chain alcohols and fatty acids which are able to reduce the surface tension at the interface of the suspended particles because of the solubility properties of their molecules. Soaps behave in this manner; they exert cleaning action by emulsifying the oily components of soils. All such substances, both natural and synthetic, are known collectively as detergents.

Polymerization reactions are often carried out in emulsion form; a wide variety of food and industrial products are emulsions of one kind or another, e.g., floor and glass waxes, drugs, paints, shortenings, textile and leather dressings, etc.

All emulsions are comprised of a continuous phase and a disperse phase; in an oil-in-water (o/w) emulsion, such as milk, water is the continuous phase and butterfat (oil) the disperse phase; in a water-in-oil (w/o) emulsion, such as butter, free fat (from crushed fat globules) is the continuous phase and unbroken fat globules plus water droplets are the disperse phase.

See also colloid, protective; phase (2); detergent; surface-active agent; wetting agent.

emulsion breaker. See demulsification.

emulsion paint. See paint, emulsion.

emulsion polymerization. Polymerization reaction carried out with the reactants in emulsified form. Performed at normal pressure -20 to $+60C$. Many copolymers (synthetic rubbers) are made in this way.

"Emulvin."[470] TM for non-ionic emulsifying, stabilizing, and wetting agent for latex processing.

"Emulvis."[94] TM for polyoxyethylene stearate. Use: Viscosity builder and solubility retarder for cosmetics, soaps, and shampoos.

en. Abbreviation for ethylenediamine used in formulas for coordination compounds, e.g., the cobalt complex $Co[en]_3(NO_3)_3$.
See also dien, pn, py.

enamel. (1) A type of paint consisting of an intimate dispersion of pigments in a varnish or resin vehicle. The vehicle may be an oil-resin mix or entirely synthetic resin. Those containing drying oils are converted to films by oxidation; those comprised wholly of synthetic resins may be converted by either heat or oxidation, or both.
See also baking finish. (2) Porcelain enamel.

enamine. A group of amino olefins; the name refers especially to unsaturated tertiary amines of the general formula

$$R_2N-C=C-$$

where R is any alkyl group. Though of little use as end products, enamines are valuable intermediates for many organic syntheses.

enanthaldehyde. See heptanal.

enanthic acid. See n-heptanoic acid.

enanthyl alcohol. See 1-heptanol.

enantiomer. (enantiomorph). One of a pair of optical isomers containing one or more asymmetric carbon atoms C^* whose molecular configurations have left- and right-hand (chiral) forms. These forms are conventionally designated dextro (d) and levo (l) because they compare to each other structurally as do the right and left hands when the carbon atoms are lined

up vertically. This is apparent in the enantiomorphic forms of glyceraldehyde; the two structures are mirror images of each other and cannot be made to coincide:

$$\begin{array}{cc} CHO & CHO \\ | & | \\ HO-C^*-H \quad (L) & H-C^*-OH \quad (D) \\ | & | \\ CH_2OH & CH_2OH \end{array}$$

Several pairs of enantiomers are possible depending on the number of asymmetric carbon atoms in the molecule. Compounds in which an asymmetric carbon is present display optical rotation. See also asymmetry, optical isomer, optical rotation.

enantiomorph. See enantiomer.

"Enbond."[142] TM for alkaline compounds for the cleaning of metal surfaces prior to electroplating, painting or other processes.

encapsulation. The process in which a material or an assembly of small, discrete units is coated with or imbedded in a molten film, sheath, or foam, usually of an elastomer. A foam-forming plastic may be used to fill the spaces between various electrical or electronic components so that they are imbedded in and supported by the foam. Plastics and other materials used for this purpose are often called potting compounds. A specialized use of this technique is in growing crystals for semiconductors in which a coating of liquid boric oxide is the encapsulating agent. Use of a glassy silicate coating to encapsulate nuclear waste for permanent disposal is under investigation.
See also microencapsulation.

"Endic" Anhydride.[316] TM for endo-cis-bi-cyclo-(2.21)-5-heptene-2,3-dicarboxylic anhydride. $(C_9H_8O_3)$.
Properties: White crystals; mp 163C; soluble in aromatic hydrocarbons, acetone, ethanol.
Use: Elastomers, plasticizers, fire retardant chemicals, resins, and epoxy curing systems.

endo-. A prefix used in chemical names to indicate an inner position, specifically (a) in a ring rather than a side chain or (b) attached as a bridge within a ring.
See also exo-.

endomycin. An antifungal antibiotic complex produced by streptomyces.

"Endor."[28] TM for a rubber peptizing agent containing activated zinc salt of pentachlorothiophenol.

$(C_6Cl_5S)_2Zn$ and 80% inert filler.
Properties: Grayish-green powder, d 2.39.

endorphin. Any of a group of polypeptides formed in the brain tissue and pituitary gland of higher animals which are thought to control the transfer of signals at nerve junctions, thus insuring that behavior patterns in the individual remain normal. Imbalance or malfunction of these polypeptides has been reported to be a factor in irrational and violent actions and other emotional disorders as well as in epilepsy and memory processes. This is a developing field of medicinal chemistry called neuropharmacology.

endosulfan. (6,7,8,9,10,10-hexachloro-1,5,5a,6,9, 9a-hexahydro-6,9-methano-2,4,3-benzodioxa-thiepin-3-oxide). CAS: 115-29-7. $C_6H_6Cl_6O_3S$.
Properties: (commercial product): Brown crystals, mp 70–100C (pure mp 106C), mixture of two isomers, mp 108–109C and 206–208C.
Hazard: Toxic by ingestion, inhalation, and skin absorption; use may be restricted. TLV: 0.1 mg/m³ of air.
Use: Insecticide.

endothall. (7-oxalobicyclo-[2.2.1]-heptane-2,3-dicarboxylic acid disodium salt).
CAS: 145-73-3. $C_8H_8Na_2O_5$.
Hazard: Strong irritant to eyes and skin.
Use: Defoliant, herbicide.

endothermic. A process or change that takes place with absorption of heat and requires high temperature for initiation and maintenance. An example is production of carbon monoxide and hydrogen by passing steam over hot coke.

endothion. (S-[5-methoxy-4-oxo-4H-pyran-2-yl)-methyl] O,O-dimethyl phosphorothioate).
CAS: 2778-04-3.

$(CH_3O)_2P(O)SCH_2C:CHC(O)C(OCH_3):CHO.$
Properties: Crystals, mp 90–91C, very soluble in water, soluble in chloroform and ethanol.
Hazard: Toxic by ingestion, cholinesterase inhibitor, use may be restricted.
Use: Insecticide.

"Endox."[142] TM for alkaline rust-removal and descaling products supplied in powder form. These products are added to water and the solutions are used both electrolytically and non-electrolytically for removal of rust, etc., from iron and steel.

end point. (1) In chemical analysis, the point during a titration at which a marked color change is observed indicating that no more titrating solution is to be added.

See indicator.
(2) The highest temperature reached during an assay distillation of hydrocarbon liquids indicating the overall volatility of the liquid (ASTM).

endrin. (1,2,3,4,10,10-hexachloro-6,7-epoxy-1,4, 4a,5,6,7,8,8a-octahydro-1,4-endo,endo-5,8-dimethanonaphthalene). CAS: 72-20-8. $C_{12}H_8OCl_6$. A stereoisomer of dieldrin which is the endo, exo isomer.
Properties: White, crystalline powder; mp approximately 200C (rearranges above this point); insoluble in water and methanol; moderately soluble in other common organic solvents; compatible with non-acidic fertilizers, herbicides, fungicides, and insecticides.
Hazard: Toxic by inhalation and skin absorption, a carcinogen, use may be restricted. TLV: 0.1 mg/m³ in air.
Use: Insecticide.
See also aldrin.

"Enduro."[251] TM for a series of stainless and heat-resisting steels: low nickel group (200 Series), chromium-nickel group (300 Series), straight-chrome group (400 Series).

enercology. A coined word defined as the balanced relationship between energy and ecology. A foundation devoted to practical applications of this relationship has been established at Alma College, Michigan.

Ene reaction. Addition of an olefin with an allylic hydrogen (ene) to a compound with a double bond (enophile), involving the allylic shift of one double bond, transfer of the allylic hydrogen to the enophile, and bonding between the two unsaturated termini.

energy. The fundamental active entity in the universe defined by physicists as either $E = mc^2$ (Einstein) or $E = h\nu$ (Planck). More simply, it is considered as the capacity for doing work. These equations show that energy cannot be completely divorced from mass as the two are to some extent interconvertible. The law of conservation of energy, simply stated, is that the sum total of energy in the universe is constant, therefore, energy cannot be either created or destroyed but only converted from one form to another.
    Radiant energy (light) comprises the electromagnetic spectrum; all wavelengths of light are composed of photons or packets of energy traveling at the speed of light. Theoretically, they have no mass except that associated with their speed. Protons, electrons, and neutrons are forms of highly condensed energy which possess determinable mass but move at lower speeds.

Energy is directly related to chemical phenomena in the formation and decomposition of compounds, the many important reactions that occur in electrochemistry and in the release of energy in nuclear fission and fusion. Free energy is a thermodynamic function; in chemical reactions, it is a measure of the extent to which a substance can react. Kinetic energy (the energy of motion) is most clearly exhibited in gases, in which molecules have much greater freedom of motion than in liquids and solids.

See also radiation, matter, thermodynamics, free energy.

Note: "One of the most difficult challenges we face is to find ways to ensure that all peoples of the world share more equitably the vast human benefits that energy can bring. The foundation of worldwide energy policy must be based on energy conservation and the development of additional sources through a judicious application of science and technology." (Glenn T. Seaborg, ACS meeting, April 1976).

energy converter. Any element or compound having the ability to convert the radiant energy of sunlight into electrical, thermal, or chemical energy. Prominent among them are silicon, selenium, and tellurium, as well as the chlorophyll of plants in photosynthesis.

See also solar cell, magnetohydrodynamics.

energy sources. Non-renewable energy sources are materials of geologic origin, i.e., petroleum, natural gas, coal, shale oil, and uranium which cannot be replaced once their supply is exhausted. Renewable sources, on the other hand, are those that can be replenished on a predictable time basis known collectively as biomass; these include such cellulosic products as wood, bagasse, agricultural waste, and residue from the forest products industries (shavings, bark, sawdust, etc.) which yield the same heating value per dry ton as one barrel of crude oil. Several of these sources have been in use for some years as on-site fuels in the sugarcane, plywood, and paper pulp industries. Another instance is methane obtained from animal manures (biogas) which is being developed on a large scale in the west and southwest. Interest in the mass cultivation of algae and hydrocarbon-producing plants such as guayule and copaiba for fuel has been reported.

There are also a number of mechanical energy sources whose development involves engineering rather than chemistry, namely, solar radiation, wind, water flow, tides, and thermal gradients in ocean water. Several of these are already in limited operation but in most cases their development will be slow and costly. Hydroelectric power has long been an accomplished fact, though it accounts for only about 1% of electrical needs. Solar energy is being actively researched, but is unlikely to be a factor of consequence for at least another decade.

The following entries discuss the chemical sources of energy more specifically:

battery	carbohydrate
biogas	cellulose
biomass	coal
breeder	copaiba
ethyl alcohol (note)	ocean thermal energy
fission	conversion
fusion	oil sands
fuel cell	peat
fuel oil	petroleum
gasoline	plutonium
gasohol	radiation
gasification	shale oil
geothermal energy	solar cell
guayule	solar energy
hydrocarbon (note)	solar pond
hydrogen	storage (4)
methane	thermoelectricity
natural gas	uranium
nuclear energy	wood

energy storage. See storage (4).

enfleurage. Extraction of odoriferous components of flowers by means of fats or mixtures of fat and tallow, the process being carried out at room temperature to avoid decomposition of the desired perfumes. The latter are separated from the fat by washing with alcohol.

See also essential oil, perfume.

engineering material. A metal, alloy, plastic, or ceramic used in the fabrication of machinery and its components, structural shapes, chemical process equipment, castings, tools, instruments, drums, tanks, piping, ductwork, and auxiliary items (excluded are materials used chiefly as protective coatings or as components of alloys). Engineering materials are characterized by hardness, strength, machinability, dimensional stability, nonflammability, and resistance to corrosion, most acids, solvents, and heat. The more important of these are listed below; for specific uses, see individual entries.

Metals and alloys	Plastics
aluminum	ABS resin
beryllium	acetal resin
brass	acrylic resin
bronze	fluorocarbon polymer
cast iron	nylon

Metals and alloys	Plastics
copper	phenolformaldehyde
lead	resin
magnesium	polybutylene terephthal-
steel	ate
tantalum	polycarbonate
titanium	polyethylene
zinc	polyphenylene oxide
zirconium	polypropylene
various trademarked	polystyrene
alloys	polyvinyl chloride
	reinforced plastics (FRP)
Ceramics	ureaformaldehyde resin
glass	
porcelain	

enhanced oil recovery. Any of several methods for increasing the productivity of oil wells (after pumping is no longer effective) by emulsifying as much as possible of the oil trapped in the rock structure. Techniques that have been researched and used to some extent involve pumping into the formations pressurized carbon dioxide, water, steam, detergent solutions, brine, and a mixture of various high polymers, guar gum, xanthan biopolymer, bauxite, or sand. Considerable success has been achieved, but such methods have proved to be both slow and inefficient.
See also chemical flooding, hydraulic fracturing.

enhancer. A food additive that brings out the taste of a food product without contributing any taste of its own. Sodium glutamate is the most widely used substance of this class; its effective concentration is in parts per thousand.
See also potentiator.

enol. A chemical grouping containing both a double bond (ene) and a hydroxyl group (OH) forming an intermediate and reversible product. Enols are characteristic of racemic compounds.

enolase. An enzyme active in glycolysis which catalyzes the conversion of 2-phosphoglyceric acid to the phosphorylated enol of pyruvic acid.

enology. (oenology). The art and science of wine making.

"Enovid."[70] TM for norethynodrel with mestranol. Oral contraceptive approved by FDA.

enrichment. (1) In food technology, the addition to a foodstuff of various nutrient substances during manufacture to increase the dietary value of the food, e.g., addition to wheat flour of vitamins B_1, B_2, niacin, and iron. In this way, the food is brought up to a specific nutritional standard. (2) Increase in the abundance of certain isotopes of an element by any of several methods: (a) By a chemical reaction accompanied by irradiation from a laser beam; enrichment of boron, chlorine, and sulfur isotopes has been achieved in this way in a number of research laboratories. Uranium enrichment is also possible, either by adding previously prepared U-235 to natural uranium or by the laser technique. (b) Uranium can also be enriched by gas centrifugation and gaseous diffusion. The latter is the usual procedure. (3) Addition of oxygen to air to increase its combustion-supporting ability.

enrobe. To coat candy centers, fruits, nuts, etc., with a liquid confection such as chocolate or sugar solution by mechanical dipping.

ensilage. A feed for livestock prepared by long storage of corn husks, stalks, etc., in an air-tight vertical cylindrical structure (silo) in which the material undergoes anaerobic fermentation. It is a component of biomass.

enterokinase. An enzyme found in the small intestine which converts trypsinogen into trypsin.

entrainer. An additive for liquid mixtures difficult to separate by ordinary distillation. The entrainer usually forms an azeotrope with one of the compounds of the mixture and thereby aids in the separation of such a compound from the remainder of the mixture.

entrainment. The presence of minute drops of water or other liquid in the vapor produced by evaporation or distillation; also small bubbles of air or other gas in a liquid as a result of turbulence induced by agitation. Entrained water or gas is often undesirable and care should be taken to eliminate or prevent it; in certain cases it is beneficial.
See venturi, entrainer.

environmental chemistry. That aspect of chemistry concerned with air and water pollution, pesticides, and chemical and radioactive waste disposal. A random selection of specific areas of research includes lead and other toxic chemicals in the air, effects of increased burning of coal, biological modification of wastes, detoxification methods, pesticide content of fish, environmental analytical and monitoring techniques, utilization of biomass, drinking water quality, organic contaminants in lakes and rivers, effect of deforestation on carbon dioxide and oxygen content of air.

Environmental Protection Agency. (EPA). A federal agency established in 1970. Under the Toxic Substances Control Act of 1976, it is required to ensure the safe manufacture, use, and transportation of hazardous chemicals. It includes an Office of Hazardous Materials Control which administers Congressional legislation pertaining to this field. The EPA may require manufacturers to conduct tests on materials or products which adversely affect the environment or public health and safety. One of the most important aspects of its activities is the establishment and supervision of automotive emission standards. It is also concerned with pesticides, fungicides, and other potentially detrimental materials, as well as industrial waste disposal. It operates in close conjunction with the US Department of Agriculture and the Food and Drug Administration. The construction of new plants for manufacturing products must conform to EPA standards, especially as regards effluents that contribute to water pollution.
See also toxic material; environmental chemistry.

enzyme. Any of a unique class of proteins which catalyze a broad spectrum of biochemical reactions. Enzymes are formed in living cells; they are comprised of one or more polypeptide chains with molecular weight from 10,000 to a million or more. An important characteristic of enzymes is their specificity, i.e., a given enzyme can catalyze one particular reaction and no others. Six types are recognized which catalyze the following reactions: (1) redox (oxidoreductases), (2) transference of specific radicals or groups (transferases), (3) hydrolysis (proteolytic), (4) removal from or addition to the substrate of specific chemical groups (lyases), (5) isomerizaton (isomerases), (6) combination or binding together substrate units (ligases). The names of enzymes invariably terminate in either -ase or -in. The following partial list indicates some of the more important functions performed by enzymes; among these are the ability to cleave the peptide bonds of proteins (hydrolysis) with simultaneous formation of water and to decompose sugars and starches to ethyl alcohol and carbon dioxide (fermentation). Enzymes are essential to many biochemical processes, especially in the food, beverage, and pharmaceutical industries.

amylase	starch hydrolysis
carboxylase	decomposes pyruvic acid
cellulase	converts cellulose to glucose
cholinesterase	inactivates acetylcholine
chymotrypsin	hydrolysis of proteins
invertase	converts sucrose to glucose and fructose
lipase	hydrolysis of fats
maltase	converts maltose to glucose
pepsin	hydrolysis of proteins
protease	hydrolysis of peptide linkages
rennin	hydrolysis of proteins
ribonuclease	decomposes RNA
trypsin	splits proteins to amino acids
urease	decomposes urea to NH_4 and CO_2.
zymase	converts sugars to alcohol and CO_2 (fermentation)

Recent research in biomimetic chemistry has succeeded in creating synthetic enzymes that imitate the behavior of natural enzymes, e.g., chymotrypsin, and are almost as effective.
See also catalyst, biomimetic chemistry, fermentation, hydrolysis.

eosin. (bromeosin; CI No.45380; tetrabromofluorescein). CAS: 15086-94-9. $C_{20}H_8Br_4O_5$.
Properties: Red, crystalline powder; soluble in alcohol and acetic acid; insoluble in water; the potassium and sodium salts are soluble in water.
Derivation: Bromination of fluorescein.
Use: Dyeing silk, cotton, and wool; red writing ink; cosmetic products; biological stain; coloring motor fuel.

EP. (1) Abbreviation for extreme pressure as applied to lubricants. (2) Abbreviation for ethylene-propylene.

"EPAL."[313] TM for linear primary alcohols.

EPA. Abbreviation for Environmental Protection Agency.

EPC black. Abbreviation for easy processing channel black.
See carbon black.

EPDM. Abbreviation for a terpolymer elastomer made from ethylene-propylene diene monomer.
See ethylene-propylene terpolymer.

"Epex."[267] TM for a series of epoxy resin extenders.
Use: Extension and dilute of epoxy resin systems for all applications.

ephedrine. (1-phenyl-2-methylaminopropanol). CAS: 299-42-3. $C_6H_5CH(OH)CH(NHCH_3)CH_3$. Optically active (levorotatory) form.

Properties: White to colorless granules, pieces, or crystals; unctuous to touch, hygroscopic, gradually decomposes on exposure to light. Soluble in water, alcohol, ether, chloroform, and oils; mp 33–40C, bp 255C (decomposes).
Derivation: Isolation from stems or leaves of Ephedra, especially Ma huang (China and India).
Grade: Technical, NF.
Hazard: Toxic by ingestion.
Use: Medicine (bronchodilator).
See racephedrine for the inactive mix of isomers.

epi. (1) A prefix denoting a bridge or intramolecular connection, e.g., epoxide. (2) An abbreviation for epichlorohydrin.

"Epiall."[175] TM for a series of epoxy molding compounds.
Properties: Excellent electrical resistivity, high physical strength, flame resistant, good dimensional stability, outstanding resistance to high temperature (500F). Resistant to virtually all solvents and most chemicals, good color fastness in sunlight and heat, fungus resistance, water absorption 0.17% in 48 hours. Available in mineral glass-filled and long-glass-fiber-filled grades.

epichlorohydrin. (chloropropylene oxide).
CAS: 106-89-8. An epoxide.

$\overline{CH_2OCHCH_2Cl}$.

$$\begin{array}{c} CH_2 \\ | \quad\quad O \\ CH \\ | \\ CH_2Cl \end{array}$$

Properties: Highly volatile, unstable liquid; chloroform-like odor; miscible with most organic solvents; slightly soluble in water; d 1.1761 (20/20C); bp 115.2C; bulk d 9.78 lb/gal; vap press 12.5 mm (20C); fp −25C; viscosity 1.12 cp (20C) refr index 1.4358 (25C); flash p 93F (33.9C) (TOC).
Derivation: By removing hydrogen chloride from dichlorohydrin.
Hazard: Toxic by inhalation, ingestion, and skin absorption; strong irritant, a carcinogen. Flammable, moderate fire risk. TLV: 2 ppm in air.
Use: Major raw material for epoxy and phenoxy resins, manufacture of glycerol, curing propylene-based rubbers, solvent for cellulose esters and ethers, high wet-strength resins for paper industry.

epichlorohydrin triethanolamine cellulose.
See ECTEOLA-cellulose.

epimer. An isomer which differs from the compound with which it is being compared only in the relative positions of an attached hydrogen and hydroxyl. The isomerism may be represented as -HCOH- and -HOCH-. It is common in sugars.
See also diastereoisomer.

epinephrine. (l-methylaminoethanolcatechol; "Adrenalin"). CAS: 51-43-4.

A hormone of the adrenal glands.
Properties: (l-form): Light brown or nearly white, odorless, crystalline powder; affected by light; mp 211–212C; specific rotation −50 to −53.5 (25C); sparingly soluble in water; insoluble in alcohol, chloroform, ether, acetone, oils; readily soluble in aqueous solutions of mineral acids, sodium hydroxide, and potassium hydroxide.
Derivation: From the adrenal glands of sheep and cattle or synthetically from pyrocatechol.
Grade: USP.
Use: Medicine (vasoconstrictor).

"Epiphen."[65] TM for an epoxy resin in liquid form. "Epiphen" ER-823 is used in adhesives for rubber, steel, aluminum, or glass. Catalyst is supplied for specific end uses.

epitaxis. An oriented crystalline growth between two crystalline solid surfaces of different chemical composition in which the surface of one crystal offers suitable positions for deposition of a second crystal. This behavior is characteristic of some types of high polymers.

"E phosphorus Mudlube."[236] TM for a solution of modified higher organic acids which imparts extreme pressure lubricating properties to drilling muds.

EPN. (o-ethyl-o,p-nitrophenyl phenylphosphorothioate). CAS: 2104-64-5.
$C_6H_5P(C_2H_5O)(S)OC_6H_4NO_2$.
Properties: Light-yellow crystals, mp 36C, d 1.5978 (30C), insoluble in water, soluble in most organic solvents, decomposes in alkaline solutions.
Grade: Wettable powders and dusts.
Hazard: A cholinesterase inhibitor, absorbed by skin, use may be restricted. TLV: 0.5 mg/m³ of air.
Use: Cotton insect pest control, acaricide.

"Epolene."[256]   TM for a series of low molecular weight polyethylene resins. Available in both emulsifiable and nonemulsifiable types.

"Eponol" Resins.   TM for high molecular weight linear copolymers of bisphenol A and epichlorohydrin; produce outstanding surface coatings by solvent evaporation alone.

"Epon" Resins.[125]   TM for a series of condensation products of epichlorohydrin and bisphenol-A having excellent adhesion, strength, chemical resistance, and electrical properties when formulated into protective coatings, adhesives, and structural plastics.

"Epotuf."[36]   TM for epoxy resins, epoxy hardeners, and epoxy esters used as coating vehicles.

epoxide.   An organic compound containing a reactive group resulting from the union of an oxygen atom with two other atoms (usually carbon) that are joined in some other way as indicated:

$$=C\underset{\diagdown O \diagup}{\underline{\qquad}}C=$$

This group, commonly called "epoxy," characterizes the epoxy resins. Epichlorohydrin and ethylene oxide are well-known epoxides. The compounds are also being used in certain types of cellulose derivatives and fluorocarbons.

"Epoxols."[152]   TM for epoxidized oils and esters. 7-4 High purity epoxidized soybean oil with a minimum of 7% oxirane oxygen. 9-5 A high oxirane (9% minimum) epoxidized triglyceride. 5-2E An epoxidized, higher alkyl ester. 7-4 and 9-5 are approved for use in food packaging materials.
Use: Vinyl films, sheetings, extrusions and coatings, polyvinylidene chloride, chlorinated rubber, chlorinated paraffins, nitrocellulose, and other compounds.

"Epoxybond."[41]   TM for an epoxy adhesive putty in stick form.

1,2-epoxybutane.   See 1,2-butylene oxide.

3,4-epoxycyclohexane carbonitrile.
$O(C_6H_9)CN$.
Properties: Liquid, d 1.0929 (20/20C), bp 244.5C, fp −33C, soluble in water.
Hazard: Toxic by skin absorption, ingestion, and inhalation.
Use: Intermediate, stabilizer.

epoxyethane.   See ethylene oxide.

2,3-epoxy-2-ethylhexanol.
$C_3H_7\overline{CHOC}(C_2H_5)CH_2OH$.
Properties: Liquid, d 0.9517 (20/20C), bp decomposes, fp −65C, slightly soluble in water. Combustible.
Hazard: Skin irritant.
Use: Stabilizer, intermediate.

epoxy novolak.   Epoxy resin made by the reaction of epichlorohydrin with a novolak resin (phenol-formaldehyde; see novolak). These have a repeating epoxide structure which offers better resistance to high temperatures than epichlorohydrin-bisphenol A type, and are especially useful as adhesives.

2,3-epoxy-1-propanol.   See glycidol.

epoxy resin.   A thermosetting resin based on the reactivity of the epoxide group. One type is made from epichlorohydrin and bisphenol A. Aliphatic polyols such as glycerol may be used instead of the aromatic bisphenol A. Molecules of this type have glycidyl ether structures, $—O\overline{CH_2CHOCH_2}$, in the terminal positions, have many hydroxyl groups, and cure readily with amines.
Another type is made from polyolefins oxidized with peracetic acid. These have more epoxide groups, within the molecule as well as in terminal positions, and can be cured with anhydrides, but require high temperatures. Many modifications of both types are made commercially. Halogenated bisphenols can be used to add flame-retardant properties.
See also epoxy novolak.
The reactive epoxies form a tight crosslinked polymer network and are characterized by toughness, good adhesion, corrosive-chemical resistance, and good dielectric properties.
Most epoxy resins are the two-part type which harden when blended. A one-component liquid type for filament winding and a pelletized type for injection molding are available under the TM "Arnox."
Hazard: Strong skin irritant in uncured state.
Use: Surface coatings as on household appliances and gas storage vessels; adhesive for composites and for metals, glass, and ceramics; casting metal-forming tools and dies; encapsulation of electrical parts; filament-wound pipe and pressure vessels; floor surfacing and wall panels; neutron-shielding materials; cements and mortars; nonskid road surfacing; rigid foams; oil wells (to solidify sandy formations); matrix for stained glass windows; low-temperature mortars.

EPR. Abbreviation for ethylene propylene rubber, also for electron paramagnetic resonance.

epsilon acid. (1-naphthylamine-3,8-disulfonic acid). $C_{10}H_5(NH)_2(SO_3H)_2$.
Properties: White, crystalline scales; soluble in hot water.
Derivation: Naphthalene-1,5- and 1,6-sulfonic acids are nitrated and reduced, giving 1-naphthylamine-3,8- and 4,8-disulfonic acids. Separation is effected by crystallizing out the acid sodium salts of 1-naphthylamine-3,8-disulfonic acid.
Use: Azo dye intermediate.

Epsom salts. See magnesium sulfate.

EPT. Abbreviation for ethylene-propylene terpolymer.

"Eptac No.1."[28] TM for zinc dimethyldithiocarbamate, an ultra-accelerator for rubber.

"Eptam."[1] TM for a selective herbicide containing ethyl-N,N-di-n-propylthiolcarbamate.

EPTC. (S-ethyl di-N,N-propylthiocarbamate).
CAS: 759-94-4. $C_2H_5SC(O)N(C_3H_7)_2$.
Available forms: Liquid and granular formulations.
Use: A pre-emergence herbicide.

eq. Abbreviation for gram equivalent weight, i.e., the equivalent weight in grams. Recommended as an international unit.

Equanil. Proprietary meprobamate.
Use: Sedative.

equilibrium. (1) Chemical equilibrium is a condition in which a reaction and its opposite or reverse reaction occur at the same rate, resulting in a constant concentration of reactants; for example, ammonia synthesis is at equilibrium when ammonia molecules form and decomposes at equal velocities ($N_2 + 3H_2 \leftrightarrows 2NH_3$). (2) Physical equilibrium is exhibited when two or more phases of a system are changing at the same rate so that the net change in the system is zero. An example is the liquid-to-vapor/vapor-to-liquid interchange in an enclosed system which reaches equilibrium when the number of molecules leaving the liquid is equal to the number entering it.

equilibrium constant. A number that relates the concentration of starting materials and products of a reversible chemical reaction to one another. For example, for a chemical reaction represented by the equation $aAB + bCD \leftrightarrows cAD + dBC$ the equilibrium constant would be $K = [(AD)^c(BD)^d]/[(AB)^a(CD)^b]$ where a, b, c, and d are the numbers of molecules of AB, CD, AD and BC that occur in the balanced equation and (AD), (BC), (AB), and (CD) are the molecular concentration of AD, BC, AB, and CD in any mixture that is at equilibrium. At any one temperature, K is usually at least approximately constant regardless of the relative quantities of the several substances, so that when K is known it is often possible to predict concentration of products when those of the starting materials are known. The constant changes markedly with temperature. The constant can often be calculated from the relations of thermodynamics if the free energy for the chemical reaction is known or can be calculated by measuring all concentrations in one or more carefully conducted experiments.

equivalent weight. (combining weight).
The weight of an element that combines chemically with 8 g of oxygen or its equivalent. Since 8 g of oxygen combines with 1.008 g of hydrogen, the latter is considered equivalent to 8 g of oxygen. When 8 g is selected for the combining weight of oxygen, no element has a combining weight value of greater than one. The equivalent weight of an acid is the weight that contains one atomic weight of acidic hydrogen, i.e., the hydrogen that reacts during neutralization of acid with base. The equivalent weight of a base or hydroxide is the weight that will react with an equivalent weight of acid. Equivalent weights of other substances are defined in a similar manner.

Er. Symbol for erbium.

erbia. See erbium oxide.

erbium. Er. Element with atomic number 68, aw 167.26, valence = 3, one of the rare earth elements of the yttrium subgroup.
See rare earth metals.
Properties: Soft, malleable solid with metallic luster; insoluble in water; soluble in acids; salts are pink to red; d 9.16 (15C); mp 1522C; bp approximately 2500C; high electrical resistivity.
Derivation: Reduction of the fluoride with calcium, electrolysis of the fused chloride.
Source: Rare earth minerals.
Forms: Lumps, ingots of high purity, sponge, powder.
Hazard: Flammable in finely divided form.

Use: Nuclear controls, special alloys, room-temperature laser.

erbium nitrate. $Er(NO_3)_3 \cdot 5H_2O$.
Properties: Large, reddish crystals; soluble in water; alcohol, ether, and acetone; loses $4H_2O$ at 130C.
Derivation: Treatment of oxides, carbonates or hydroxide with nitric acid.
Grade: 99%.
Hazard: May explode if shocked or heated.

erbium oxalate. $Er_2(C_2O_4)_3 \cdot 10H_2O$.
Properties: Reddish, microcrystalline powder; decomposes at 575C; d 2.64; soluble in water and dilute acids.
Hazard: Corrosive.
Use: Oxalates of the rare earth metals are used to separate the latter from common metals.

erbium oxide.   (erbia). Er_2O_3.
Properties: Pink powder which readily absorbs moisture and carbon dioxide from the atmosphere, d 8.64, specific heat 0.065, infusible, insoluble in water, slightly soluble in mineral acids.
Derivation: By heating the oxalate or other oxyacid salts.
Grade: 98–99%.
Use: Phosphor activator, infrared-absorbing glass.
See also rare earth.

erbium sulfate. $Er_2(SO_4)_3 \cdot 8H_2O$.
Properties: Pink, monoclinic crystals; soluble in water; d 3.68; dehydrated at 400C.
Derivation: Dissolving hydroxides, carbonates, or oxides in dilute sulfuric acid.
Grade: 99.9%.
Use: To determine atomic weight of the rare-earth element.

erbon.   (Generic name for 2-(2,4,5-trichlorophenoxy)-ethyl-2,2-dichloropropionate).
CAS: 136-25-4.
$(Cl)_3C_6H_2OC_2H_4OC(O)C(Cl)_2CH_3$.
Properties: Solid, mp 49–50C (technical), bp 161C (0.5 mm), insoluble in water, soluble in most organic solvents.
Hazard: Toxic by inhalation and ingestion, strong irritant to eyes and skin.
Use: Herbicide.

"Ercusol."[470]   TM for an aqueous dispersion of an acrylonitrile-based copolymer and other monomers.

erepsin.   A mixture of peptidase enzymes formerly thought to be a single enzyme which catalyzes the hydrolysis of peptides in the small intestine.

ergamine.   See histamine.

ergocalciferol.   (calciferol; vitamin D_2).
CAS: 50-14-6. $C_{28}H_{44}O$.

Properties: White, odorless crystals; affected by air and light; insoluble in water; soluble in alcohol, chloroform, ether, and fatty oils; mp 115–118C; specific rotation +103 to +106.
Derivation: From ergosterol by irradiation with UV light.
Grade: USP, FCC, as vitamin D_2.
Use: Nutrition, dietary supplement.

ergosterol.   (provitamine D_2). $C_{28}H_{43}OH$.
A plant sterol widely distributed in nature.

Properties: Colorless crystals, mp 166C (with 1.5 H_2O), bp 250C (0.01 mm), d 1.04, specific rotation −135 (in chloroform). Insoluble in water; soluble in alcohol, benzene, ether.
Derivation: Synthesized by yeast from simple sugars, obtained from fungus ergot.
Hazard: Due to its ability to catalyze calcium deposition in the bony structure (thus preventing rickets), overdosage of vitamin D may be harmful.
Use: Antirachitic vitamin; when irradiated with UV light, it has vitamin D activity; source of estradiol.

ergot.   (secale cornutum; rye ergot).
A fungus growth, Claviceps purpurea, on rye.
Habitat: Europe, cultivated in Spain and Russia.
Grade: Spanish, Russian.

Hazard: Toxic by ingestion.
Use: Source of many alkaloids, medicine (vasoconstrictor).

ergotamine. CAS: 113-15-5. Alkaloid present in rye ergot.
Properties: Mw 581.73.

"Erionyl."[443] TM for a group of condensation products of aromatic sulfonic acids used as fixative/reserving agents in dyeing nylon and nylon/cellulosic blends.

Erlenmeyer flask. A useful type of laboratory glassware, it is an open container whose dimensions are, for example, about 8 inches tall with a relatively narrow neck section about 1.5 inches in diameter and 2 inches long, below which the contour becomes cone-shaped. The bottom is flat. It was named after its inventor, a German chemist.
Use: Numerous experiments involving liquids, especially titrations and extractive testing.

Erlenmeyer-Plochl azlactone and amino acid synthesis. Formation of azlactones by intramolecular condensation of acylglycines in the presence of acetic anhydride. The reaction of azlactones with carbonyl compounds followed by hydrolysis to the unsaturated alpha-acylamino acid and by reduction yields the amino acid; drastic hydrolysis gives the alpha-oxo acid.

erucamide. (erucylamide). $C_{21}H_{41}CONH_2$.
Properties: Solid, d 0.888, mp 75–80C, iodine value 70–80, soluble in isopropanol, slightly soluble in alcohol and acetone. Combustible.
Use: Foam stabilizer, solvent for waxes and resins, emulsions, antiblock agent for polyethylene.

erucic acid. (cis-13-docosenoic acid). $C_8H_{17}CH:CH(CH_2)_{11}COOH$. A C_{22} (solid) fatty acid with one double bond; a homolog of oleic acid with four more carbons.
Properties: Mp 33–34C, bp 264C (15 mm), iodine value 75. Combustible.
Derivation: Fats and oils from mustard seed, rapeseed, and crambe seed.
Use: Preparation of dibasic acids and other chemicals, polyethylene film additive, water-resistant nylon.

erucyl alcohol. $C_{22}H_{43}OH$. A C_{22} (solid) fatty alcohol having one double bond.
Properties: White, soft solid; almost odorless, d 0.8486, cloud point 27.2C, boiling range 334–376C, iodine value 83, flash p 395F (201C), soluble in alcohol and most organic solvents. Combustible.

Derivation: Sodium reduction of erucic acid.
Use: Lubricants, surfactants, petrochemicals, plastics, textiles, rubber.

"Erusticator."[204] TM for a rust remover which dissolves rust rapidly from fabrics.

"Erusto."[204] TM for a series of laundry and drycleaning products.

erythorbic acid. (trivial name adopted officially for d-erythro-ascorbic acid, formerly called isoascorbic acid). $C_6H_8O_6$.
Properties: Shiny, granular crystals; decomposes 164–169C; soluble in water, alcohol, pyridine; moderately soluble in acetone; slightly soluble in glycerol.
Grade: FCC.
Use: Antioxidant (industrial and food) especially in brewing industry, reducing agent in photography.

erythrite. [(1) Synonym for erythritol. (2) cobalt bloom]. $Co_3(As)_4)_2 \cdot 8H_2O$. A natural hydrated cobalt arsenate.
Properties: Crimson, peach, red, pink, or pearl gray; contains 37.5% cobalt oxide; soluble in hydrochloric acid; d 2.91–2.95; Mohs hardness 1.5–2.5; mp decomposes.
Occurrence: US (California, Colorado, Idaho, Nevada), Ontario.
Hazard: Toxic by ingestion and inhalation.
Use: Coloring glass and ceramics.

erythritol. (tetrahydroxybutane.)
CAS: 149-32-6.
$CH_2OHCHOHCHOHCH_2OH$.
A tetrahydric alcohol found in Protococcus vulgaris and other lichens of Rocella species. Can be made synthetically.

$$CH_2OH$$
$$|$$
$$HCOH$$
$$|$$
$$HCOH$$
$$|$$
$$CH_2OH$$

Properties: White, sweet crystals; mp 121–122C; bp 329–331C; d 1.45; soluble in water; slightly soluble in alcohol; insoluble in ether.
Use: Manufacture of erythrityl tetranitrate.

erythritol anhydride. (butadiene dioxide).
CAS: 564-00-1. $C_4H_6O_2$.
Properties: Colorless liquid, bp 138C, fp −18C, d 1.11, hydrolyzed to erythritol when dissolved in water.

Hazard: Quite toxic.
Use: Cross-linking agent, biocide, bacteriostat.

erythrityl tetranitrate. (erythrol tetranitrate). $CH_2ONO_2(CHONO_2)_2CH_2ONO_2$.
Properties: Crystals; mp 61C; soluble in alcohol, ether, and glycerol; insoluble in water.
Derivation: By nitration of erythritol.
Hazard: Severe explosion risk when shocked or heated.
Use: Medicine (diluted with lactose in non-explosive tablets).

erythro-ascorbic acid. See erythorbic acid.

erythrocyte. A red blood cell containing hemoglobin, the iron-carrying protein of the blood. See also leucocyte.

erythrogenic acid. See isanic acid.

erythromycin. CAS: 114-07-8. $C_{37}H_{67}NO_{13}$.
An antibiotic produced by growth of Streptomyces erythreus Waksman. It is effective against infections caused by Gram-positive bacteria, including some beta-hemolytic streptococci, pneumococci, and staphylococci.
Properties: White or slightly yellow, odorless, bitter, crystalline powder; mp 133–138C; freely soluble in alcohol, chloroform, and ether; very slightly soluble in water; slightly hygroscopic; pH (saturated solution) 8–10.5; pH less than 4 is destructive; alcoholic solution is levorotatory.
Grade: USP.
Use: Medicine (antibiotic); various salts are available.
Note: The erythromycin molecule was synthesized by the Harvard chemist Robert Woodward. The synthesis was almost completed at the time of his death in 1979, and was finished by his associates in 1981. It is an extremely complex structure containing a lactone ring of 14 members with 10 asymmetric centers; it also has two specialized sugar molecules, L-cladinose and D-desosamine. The reported molecular configuration is:

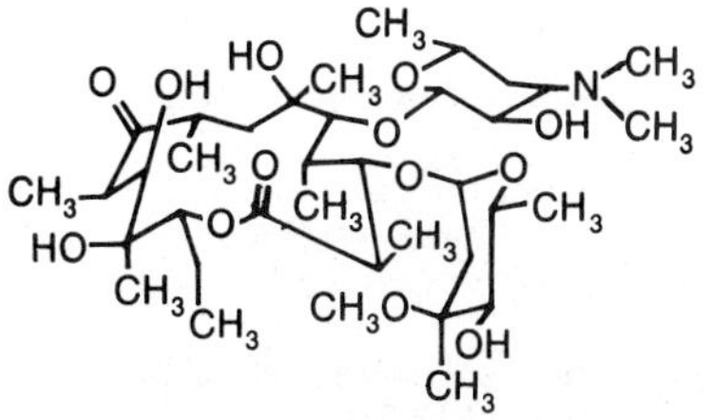

erythrosine. CAS: 16423-68-0.
$C_{20}H_6I_4Na_2O_5$.   (Sodium (or potassium) salt of iodeosin; CI No.45430; FD&C Red No.3).
Properties: Brown powder, forms cherry red solution in water, soluble in alcohol.
Use: Biological stain, certified food color.

Es. See einsteinium.

ESCA. Electron spectroscopy for chemical analysis.

"Escalol 106."[10] TM for glyceryl-p-amino-benzoate
$CH_2OHCH(OH)CH_2OOCC_6H_4NH_2$.
Use: Sun-screening compound.

"Escalol 506."[10] TM for amyl-p-dimethylamino-benzoate,
$C_5H_{11}OOCC_6H_4N(CH_3)_2$.
Use: UV screen.

Eschweiler-Clarke reaction. Reductive methylation of rimary or secondary amines with formaldehyde and formic acid (special form of the Leuckart-Wallach reaction).

ESD. Electron-stimulated desorption.

eserine. See physostigmine.

esparto. The leaves of a desert plant (Stipa tenacissima and Lygeum spartum) of the Mediterranean area. Relatively high content of cellulose and alpha-cellulose make it usable for high-quality paper for intaglio color printing. A wax derived from this plant is used as a substitute for carnauba, especially in carbon paper.

essential. (1) Containing the characteristic odor or flavor, (i.e., the essence) of the original flower or fruit; an essential oil, usually obtained by steam distillation of the flowers or leaves or cold-pressing of the skin. (2) As applied to certain amino acids, fatty acids, and vitamins, this term is used by biochemists to mean that the compound in question is a necessary nutritional factor that is not synthesized within the body of the animal and thus must be obtained from external sources. Eight amino acids are classified as essential on this basis.
See also amino acid.

essential oil. A volatile oil derived from the leaves, stem, flower, or twigs of plants, and usually carrying the odor or flavor of the plant. Chemically, they are often principally terpenes (hydrocarbons), but many other types also occur. Essential oils (except for those containing esters) are unsaponifiable. Some are nearly pure single compounds, as oil of wintergreen, which is methyl salicylate. Others are mixtures, as turpentine oil (pinene, dipentene), and oil of bitter almond (benzaldehyde, hydrocyanic acid). Some contain resins in solution and are called oleoresins or balsams.

Properties: Pungent taste and odor, usually nearly colorless when fresh but becoming darker and thick on exposure to the air, optically active, d 0.850–1.100; soluble in alcohol, carbon disulfide, carbon tetrachloride, chloroform, petroleum ether, and fatty oils; insoluble in water except for individual constituents of some oils which may be partially water-soluble, resulting in a loss of these constituents during steam distillation.

Derivation: (a) By steam distillation, (b) by pressing (fruit rinds), (c) by solvent extraction, (c) by maceration of the flowers and leaves in fat and treating the fat with a solvent, (e) by enfleurage.

Use: Perfumery, flavors, thinning precious metal preparations used in decorating ceramic ware.

See also terpeneless oil and specific entries. Further information can be obtained from the Essential Oil Association of US.

Note: Many essential oils are now made synthetically for a wide variety of fragrances and flavoring agents. Use of these synthetics is increasing because of a shortage of natural products.

"Esskol."[64]   TM for modified linseed oils used in redwood finishes, varnishes, and enamels.

"Estan."[51]   TM for light colored, general purpose, lime-base greases. Available in wide range of consistencies and suitable for all methods of application. Made with an oil having a minimum of internal frication and bearing drag.

"Estane."[119]   TM for thermoplastic polyester and polyether urethane elastomers which provide good physical and chemical properties without curing. Extremely tough and abrasion resistant with high tensile strength at high ultimate elongation; good solvent resistance particularly to gasoline; low air permeability and exceptional low-temperature flexibility.

Use: Wire and cable jacketing, fuel hose and tanks, belting, shoe heels, coated fabrics, free film, adhesives.

ester.   An organic compound corresponding in structure to a salt in inorganic chemistry. Esters are considered as derived from acids by the exchange of the replaceable hydrogen of the latter for an organic radical. The usual reaction is that of an acid (organic or inorganic) with an alcohol or other organic compound rich in OH groups. Esters of acetic acid are called acetates and esters of carbonic acid carbonates.

See also fatty ester.

ester gum.   Hard, semisynthetic resin produced by esterification of natural resins (especially rosin) with polyhydric alcohols (principally glycerol but also pentaerythritol). Flash p 375F (190C). Combustible.

Grade: By color, also as gum rosin or wood rosin.

Use: Paints, varnishes, and cellulosic lacquers.

"Esteron."[233]   TM for a series of weed and brush control products; they are formulated esters of 2,4-D and 2,4,5-T.

"Estonmite."[88]   TM for p-chlorophenyl-p-chlorobenzene sulfonate, miticide, available as a dust base, wettable powder, and emulsifiable solution.

Use: Ovicide, specifically against the eggs of spider mites.

"Estonox."[88]   TM for toxaphene in a dust base, wettable powder, and in a stabilized emulsifiable carrier.

Use: Control of insects on cotton, seed alfalfa, sugar beets, beans, and potatoes.

Hazard: See toxaphene.

estradiol.   CAS: 50-28-2. $C_{18}H_{24}O_2$.   A female sex hormone. It occurs in two isomeric forms, alpha and beta. Beta-estradiol has the greatest physiological activity of any naturally occurring estrogen. The alpha form is relatively inactive. Commonly used preparations are the benzoate, dipropionate, and valerate, as well as ethinylestradiol.

Properties: (of beta form): White or slightly yellow, small crystals or crystalline powder, odorless, mp 173–179C, stable in air. Almost insoluble in water; soluble in alcohol, acetone, dioxane, and in solutions of alkali hydroxides; sparingly soluble in vegetable oils.

Derivation: Isolated from human and mare pregnancy urine, commercial synthesis from cholesterol or ergosterol.

Grade: NF (beta form).

Hazard: A carcinogen (OSHA).

Use: Medicine (estrogenic hormone).

estragole.   (chavicol methyl ether; methyl chavicol).   CAS: 140-67-0. $C_6H_4(C_3H_5)(OCH_3)$.

Properties: Colorless liquid, anise odor, d 0.965–0.975 (20/4C), 1.5230 (17.5C), bp 216C, soluble in alcohol and chloroform.

Occurrence: In tarragon oil, basil oils, anise bark oil, and others.
Use: Perfumes, flavors.

"Estrex."[152]   TM for a series of methyl, butyl, propyl, glyceryl, and polyethylene glycol esters of fatty acids.
Use: Adhesives, agricultural sprays, aluminum rolling, antioxidants, cosmetics, cutting oils, detergents, leather tanning, lubricants and lubricating additives, plasticizers, rubber and textile finishes.

estriol.   CAS: 50-27-1. $C_{18}H_{24}O_3$.
Properties: White, odorless, microcrystalline powder; mp 282C; exhibits reddish fluorescence under filtered UV light. Undergoes phase change at 270–275C; practically insoluble in water; soluble in alcohol, dioxane, and oils.
Derivation: Isolation from pregnant human urine, isolated from human placenta, organic synthesis.
Hazard: A carcinogen (OSHA).
Use: Medicine (estrogenic hormone).

estrogen.   A general term for female sex hormones. They are responsible for the development of the female secondary sex characteristics, such as the deposition of fat and the development of the breasts. The naturally occurring estrogens, such as estradiol, estrone, and estriol, are steroids. Estrogens are produced by the ovary and, to a lesser degree, by the adrenal cortex and testis. Some synthetic nonsteroid compounds such as diethylstilbestrol and hexestrol have estrogenic activity.
Hazard: Carcinogenic, damaging side effects (thromboembolism)
Use: Medicine, biochemical research, oral contraceptives.
See also antifertility agent.

"Estron."[236]   TM for synthetic yarn and staple fiber acetate used in cigarette filter tips, tobacco smoke filters, and tobacco smoke filter tip rods.

estrone.   CAS: 53-16-7. $C_{18}H_{22}O_2$.   A steroid with some estrogenic activity.

Properties: Small, white crystals or white, crystalline powder; mp 258–262C; odorless; stable in air; insoluble in water; soluble in alcohol, acetone, dioxane, and solutions of fixed alkali hydroxides.
Derivation: Isolated from pregnant human urine, synthesis from ergosterol.
Hazard: A carcinogen (OSHA).

"Estynox."[202]   TM for epoxidized esters of castor or other fatty acids produced by reacting the double bonds in these oils to produce oxirane or epoxy groups.
Use: Stabilizing plasticizers for polyvinyl chloride and nitrocellulose resins. Acid scavengers in cutting fluids, lubricants, textile processing, alkyd manufacture, corrosion inhibitors.

Et.   Informal abbreviation for ethyl. Example: EtOH, ethyl alcohol.

Etard reaction.   Oxidation of an arylmethyl group to an aldehyde by means of chromyl chloride.

"Ethacure."[313]   TM for curing agents for thermosetting plastics.

"Ethafoam."[233]   TM for a lightweight, low-density polyethylene foam.

"Ethanox."[313]   TM for hindered phenolic antioxidants.

ethanal.   See acetaldehyde.

ethanamide.   See acetamide.

ethane.   (dimethyl; methylmethane).
CAS: 74-84-0. C_2H_6.
Properties: Colorless, odorless gas, insoluble in water, soluble in alcohol, relatively inactive chemically, bp −88.63C, fp −183.23C (triple point), d of liquid 0.446 (0C), d of vapor (air = 1) 1.04 (0C), critical temperature 32.1C, critical pressure (absolute) 718 psi, specific heat at constant pressure 0.897, specific heat at constant volume 0.325, ratio of specific heats (cp/cv) 1.224, heat of combustion approximately 22,300 Btu/lb or 1800 Btu/cu ft, flash p −211F (−135C), autoign temperature 959F (515C), an asphyxiant gas.
Derivation: Fractionation of natural gas.
Grade: 95%, 99%, research, 99.98%.
Hazard: Severe fire risk if exposed to sparks or open flame. Flammable limits in air 3–12%.
Use: Petrochemicals (source of ethylene, halogenated ethanes), refrigerant, fuel.

ethanedioyl chloride.   See oxalyl chloride.

1,2-ethanedithiol.   (dithioethyleneglycol; ethylenedimercaptan).   CAS: 540-63-6.
$HSCH_2CH_2SH$.
Properties: Liquid, mw 94.20, bp 144–146C, d 1.123. Soluble in alcohol and alkalies.
Hazard: Vapors cause severe headache and nausea.

Use: Metal complexing agent. Reverses the inhibition by α-keto aldehydes on mitosis in *E. coli*.

ethane hydrate. See gas hydrate.

ethanethiol. (ethyl sulfhydrate; ethyl mercaptan).
CAS: 75-08-1. C_2H_5SH.
Properties: Colorless liquid, has one of the most penetrating and persistent odors known (skunk). Slightly soluble in water; soluble in alcohol, ether, petroleum naphtha. D 0.83907 (20/4C), bp 36C, fp −121C, refr index 1.4305 (20C), flammable limits in air 2.8–18.2% by volume, flash p approximately 80F (26.6C) (CC), autoign temperature 570F (298C).
Derivation: By saturating potassium hydroxide solution with hydrogen sulfide, mixing with calcium ethylsulfate solution and distilling on a water bath.
Hazard: Toxic by ingestion and inhalation. Flammable, dangerous fire risk. TLV: 0.5 ppm in air.
Use: LPG odorant, adhesive, stabilizer, chemical intermediate. *Note*: Tomato juice is reported to deodorize materials contaminated with this compound.

ethanethiolic acid. See thiacetic acid.

ethanoic acid. See acetic acid.

ethanol. See ethyl alcohol.

ethanolamine. (MEA; monoethanolamine; colamine; 2-aminoethanol; 2-hydroxyethylamine).
CAS: 141-43-5. $HOCH_2CH_2NH_2$.
Properties: Colorless, hygroscopic, viscous liquid; ammoniacal odor; strong base; miscible with water, methanol, acetone; d 1.0179 (20/20C); bp 170.5C; mp 10.5C; vap press 0.48 mm (20C); flash p 200F (93.3C) (OC); bulk d 8.5lb/gal (20C). Combustible.
Derivation: Reaction of ethylene oxide and ammonia gives a mix of mono-, di-, and triethanolamines.
Grade: Technical, NF.
Hazard: Skin irritant. TLV: 3 ppm in air.
Use: Scrubbing acid gases (H_2S, CO_2), especially in synthesis of ammonia, from gas streams; nonionic detergents used in drycleaning, wool treatment, emulsion paints, polishes, agricultural sprays; chemical intermediates, pharmaceuticals, corrosion inhibitor, rubber accelerator.

ethanol formamide. $HOCH_2NHOCH$.
Properties: Somewhat viscous liquid; miscible with water, alcohol, and glycerol; compatible with polyvinyl alcohol, many cellulosic and natural resins; bp 143C (2.5 mm); fp approximately −72C; d 1.170 (25/4C); flash p 347F (175C). Combustible.

ethanol hydrazine. See β-hydroxyethylhydrazine.

ethanolurea. $NH_2CONHCH_2CH_2OH$.
Properties: Liquid, solidification point 71–74C. Formaldehyde condensation products are permanently thermoplastic and water-soluble. As increasing amounts of simple urea are mixed with ethanolurea, the condensation products gradually change from pliable film-forming resins into the brittle types. Thus, almost any degree of water-solubility and flexibility may be obtained in the final resin. The modified resins formed with ethanolurea are compatible with polyvinyl alcohol, methyl cellulose, cooked starch, and other water-dispersible materials.

ethaverine. (1-[(3,4-diethoxyphenyl)methyl]-6,7-diethoxyisoquinoline; isoquinoline).
$C_{24}H_{29}NO_4$. CAS: 486-47-5.
Properties: Crystals with mw 395.48, 99–101C. Insoluble in water, very soluble in hot alcohol, slightly soluble in ether and chloroform.
Use: Antispasmodic drug.

ethchlorvynol. (1-chloro-3-ethyl-1-penten-4-yn-3-ol; β-chlorovinyl ethyl ethynyl carbinol).
CAS: 113-18-8.
$HC{\equiv}CCOH(C_2H_5)CH{=}CHCl$.
Properties: Colorless to yellow liquid, pungent aromatic odor, darkens on exposure to light and to air, d 1.068–1.071, refr index 1.4765–1.4800 (25C), bp 173–181C, immiscible with water, miscible with most organic solvents.
Grade: NF.
Hazard: Abuse may cause addiction.
Use: Medicine (sedative).

ethene. See ethylene.

ethenol. See vinyl alcohol.

ethephon. See 2-chloroethylphosphonic acid.

ether. A class of organic compounds in which an oxygen atom is interposed between two carbon atoms (organic groups) in the molecular structure giving the generic formula ROR. They may be derived from alcohols by elimination of water, but the major method is catalytic hydration of olefins. Only the lowest member of the series, methyl ether, is gaseous; most are liquid and the highest members are solid (cellulose ethers). The term "ether" is often used synonymously with "ethyl ether" and is the legal label name for it.
Hazard: The lower molecular weight ethers are dangerous fire and explosion hazards; when con-

taining peroxides they can detonate on heating.
Use: See ethyl ether; polymer; water-soluble; ethylene oxide; propylene oxide; diethylene glycol; ethyl cellulose; polyether.
Note: An illogical and archaic use of the term "ether" survives in such names as petroleum ether.
See also crown ether.

ethereal.   Descriptive of a liquid characterized by high volatility, often a mixture of ethyl ether and an essential oil.

ethical drug.   A prescription drug.
See also drug.

ethinylestradiol.   (ethynylestradiol; 19-nor-17-α-pregna-1,3,5(1)-trien-20-yne-3,17-diol).
CAS: 57-63-6. $C_{20}H_{24}O_2$.
Properties: Fine, white to creamy white, odorless, crystalline powder; sensitive to light; mp 142–146C; may also exist in a polymorphic modification; mp 180–186C; soluble in acetone, alcohol, chloroform, dioxane, ether, and vegetable oils; practically insoluble in water; soluble in solutions of sodium hydroxide or potassium hydroxide; slightly dextrorotatory in dioxane solution.
Derivation: Preparation from estrone.
Grade: USP.
Use: Medicine (estrogenic hormone).

ethion.   (Generic name for O,O,O',O'-tetraethyl-S,S'-methylenediphosphorodithioate).
CAS: 563-12-2. $[(C_2H_5O)_2P(S)S]_2CH_2$.
Properties: Liquid, d 1.220 (20C), fp −13C. Slightly soluble in water; soluble in acetone, xylene, chloroform, and methylated naphthalene.
Hazard: Cholinesterase inhibitor, use may be restricted. TLV: 0.4 mg/m³ of air.
Use: Insecticide and miticide.

ethisterone.   (preneninolone; anhydrohydroxyprogesterone; ethynyltestosterone).
CAS: 434-03-7. $C_{21}H_{28}O_2$.
Properties: White or slightly yellow crystals or as crystalline powder; odorless; stable in air; affected by light; mp 267–275 (decomposes); almost insoluble in water; slightly soluble in alcohol, chloroform, ether, and vegetable oils.
Derivation: From progesterone and other steroids.
Grade: NF.
Use: Medicine (estrogenic hormone).

"Ethocel."[233]   TM for ethylcellulose resins which are able to withstand shock and maintain toughness over a temperature range of +93 to −40C. Available in transparent, translucent, and opaque colors. Insoluble in water, soluble in most organic solvents.
Use: Household articles, automotive parts, tools for aircraft industry.

ethodin.   (6,9-diamino-2-ethoxyacridine lactate monohydrate).   CAS: 1837-57-6.
$C_{15}H_{15}N_3O \cdot C_3H_6O_3 \cdot H_2O$.
Properties: Pale yellow crystals, darkens at 200C, mp 235C, slowly soluble in 15 parts water, soluble in 9 parts boiling water, soluble in 110 parts alcohol (22C), solutions are yellow, fluorescent and stable to boiling. Purity: 97% (dry basis).
Use: Bactericide, surgical antisepsis, preparation of pure gamma globulin.

ethofumesate.   (2-ethoxy-2,3-dihydro-3,3-dimethyl-5-benzofuranol methanesulfonate).
CAS: 67293-74-7. $C_{13}H_{18}O_5S$.
Properties: Whitish crystals, mp 71C, does not hydrolyze with water at neutral pH.
Use: Herbicide, silvicide.

ethoheptazine.   (ethyl heptazine; 1-methyl-4-carbethoxy-4-phenylhexamethyleneimine).
$C_{16}H_{23}NO_2$.
Properties: Liquid, d 1.038 (26/4C), bp 133–134C (1.0 mm), refr index 1.5210 (26C).
Use: Medicine (analgesic).

ethohexadiol.   USP name for 2-ethylhexanediol-1,3).

"Ethomid."[15]   TM for polyethoxylated high-molecular weight amides.
Use: Emulsifiers, wetting agents.

"Ethonic."[313]   TM for ethoxylated alcohol surfactants.

"Ethosperse."[73]   TM for a series of surface-active compounds, the reaction products of fatty alcohols, glycols, glycerol and sorbitol and ethylene oxide.
Use: Food, cosmetics, pharmaceuticals, and industrial applications.

*p***-ethoxyacetanilide.**   See acetophenetidin.

ethoxycarbonyl isothiocyanate.   (ethyl isothiocyanateformate).   CAS: 16182-04-0.
$C_2H_5O_2CNCS$.
Properties: Moisture senstive solid with mw 131.15, bp 56C at 18 mm, d 1.112.
Hazard: Lachrymator.
Use: Versatile reagent for organic synthesis.

2-ethoxy-3,4-dihydro-2H-pyran.

$\overline{OCH}$:CHCH₂CH₂CHOC₂H₅.
Properties: Liquid, d 0.970 (20/20C), bp 143C, fp −100C, sets to glass below this temperature, flash p 111F (open cup), very slightly soluble in water. Combustible.

Hazard: Moderate fire risk.
Use: Stabilizer, intermediate.

6-ethoxy-1,2-dihydro-2,2,4-trimethylquinoline.
(ethoxyquin).   CAS: 91-53-2. $C_{14}H_{19}NO$.
Properties: Yellow liquid, bp 125C (2 mm), mp
approximately 0C, refr index 1.569–1.572 (25C),
d 1.029–1.031 (25C), discolors and stains badly.
Hazard: Toxic by ingestion.
Use: Insecticide, antioxidant, flex-cracking inhibitor, post-harvest preservation of apples (scald
inhibitor).

2-ethoxyethanol.   See: ethylene glycol monoethyl
ether.

2-ethoxyethyl acetate.   See ethylene glycol monoethyl ether acetate.

2-ethoxyethyl-p-methoxycinnamate.
$CH_3OC_6H_4CH:CHCOOC_2H_4OC_2H_5$.
Properties: Slightly yellow, viscous liquid; practically odorless; d 1.1000–1.1035 (25/25C); refr
index 1.5650–1.5675 (20C); flash p greater than
212F (100C) (Tagliabue closed cup); miscible
with alcohol and isopropyl alcohol; almost insoluble in water. Combustible.
Use: UV absorber in suntan preparations.

3-ethoxy-4-hydroxybenzaldehyde.   See ethyl vanillin.

1-ethoxy-2-hydroxy-4-propenylbenzene.
See propenyl guaethol.

4-ethoxy-3-methoxybenzaldehyde.
$C_6H_3(OC_2H_5)(OCH_3)CHO$.
Properties: White to light brown crystals having
a slight vanillin odor, mp 62–64C. Combustible.
Use: Intermediate.

4-ethoxy-3-methoxyphenylacetic acid.
$C_6H_3(OC_2H_5)(OCH_3)CH_2COOH$.
Properties: An off-white powder, mp 119–122C.
Use: Intermediate.

1-ethoxy-2-methoxy-4-propenylbenzene.
See isoeugenyl ethyl ether.

ethoxymethylenemalononitrile.
$C_2H_5OCH:C(CN)_2$.
Properties: Colorless liquid, bp 160C, mp 64C,
flash p 311F (155C). Combustible.
Use: Chemical intermediate.

4-ethoxyphenol.   See hydroquinone monoethyl
ether.

ethoxyquin.   Coined name for 6-ethoxy-1,2-dihy-
dro-2,2,4-trimethylquinoline.

ethoxytriglycol.   CAS: 112-50-5.
$C_2H_5O(C_2H_4O)_3H$.
Properties: Colorless liquid, d 1.0208 (20/20C),
bulk density 8.5 lb/gal (20C), bp 255.4C, vap
press less than 0.01 (20C), fp −18.7C, viscosity
7.80 cp (20C), completely soluble in water. Flash
p 275F (135C) (closed cup). Combustible.
Use: Chemical intermediate.

"Ethrel."[214]   TM for a plant-growth regulator,
2-chloroethylphosphonic acid.

Ethyl.   TM for anti-knock fuel additive, tetraethyl
lead.

ethyl abietate. $C_{19}H_{29}COOC_2H_5$.
Properties: Amber-colored, viscous liquid; hardens upon oxidation. Soluble in ether and most
varnish solvents, insoluble in water, bp 350C,
flash p 352F (177C), mp 45C, refr index 1.4980,
d 1.02. Combustible.
Derivation: (a) By heating together ethyl chloride
and an alcoholic solution of rosin and caustic
soda. (b) By reacting ethyl iodide with silver abietate.
Hazard: Irritant.
Use: Varnishes, lacquers, and coating compositions.

"Ethylac."[204]   TM   for   2-benzothiazyl-N,N-di-
ethylthiocarbamylsulfide.

n-ethylacetamide. $CH_3CONHC_2H_5$.
Properties: Colorless liquid, d 0.920 (20/20C),
boiling range 206–208.5C, flash p 230F (110C),
faint odor. Combustible.

ethyl acetamidocyanoacetate.   (acetamidocyanoacetic ester; ethyl n-acetyl-α-cyanoglycine).
$NCCH(NHCOCH_3)COOC_2H_5$.
Properties: Solid, mp 129C.
Use: Synthesis of amino acids and related compounds.

n-ethylacetanilide.   (ethylphenylacetamide).
$C_6H_5NC_2H_5COCH_3$.
Properties: White, crystalline solid; faint odor;
soluble in most organic solvents; insoluble in wa-
ter; d 0.994; bp 258C; flash p 126F (52.2C); mp
54C. Combustible.
Grade: Technical.
Hazard: Toxic by ingestion. Moderate fire risk.
Use: Substitute for camphor in nitrocellulose.

ethyl acetate.   (acetic ether; acetic ester; vinegar
naphtha).   CAS: 141-78-6.
$CH_3COOC_2H_5$.
Properties: Colorless, fragrant liquid; soluble in
chloroform, alcohol, and ether; slightly soluble

in water; bp 77C; vap press 73 mm (20C); fp −83.6C; bulk density 0.8945 g/ml (25C); flash p 24F (−4.4C); autoign temperature 800F (426C).

Derivation: By heating acetic acid and ethyl alcohol in presence of sulfuric acid and distilling.

Grade: Commercial 85–88%, 95–98%, 99%, NF (99%), FCC.

Hazard: Toxic by inhalation and skin absorption; irritant to eyes and skin. Flammable; dangerous fire and explosion risk, flammable limits in air 2.2–9%. TLV: 400 ppm in air.

Use: General solvent in coatings and plastics, organic synthesis, smokeless powders, pharmaceuticals, synthetic fruit essences.

ethyl acetate, anhydrous. ethyl acetate, grade 99%.

ethyl-o-acetate. $CH_3C(OC_2H_5)_3$.
Properties: Colorless liquid, bp 144–148C, refr index 1.395 (25C), insoluble in water, soluble in alcohol and ether, flash p 131F (55C). Combustible.
Hazard: Moderate fire risk.
Use: Intermediate.

ethyl acetic acid. See butyric acid.

ethyl acetoacetate. (diacetic ester; acetoacetic ester). $CH_3COCH_2COOC_2H_5$ (keto form), $CH_3C(OH):CHCOOC_2H_5$ (enol form).
This compound is a tautomer at room temperature consisting of about 93% keto form and 7% enol form.
Properties: Colorless liquid, fruity odor, soluble in water and common organic solvents, d 1.0250 (20/4C), fp (enol) -80C (keto) -39C, bp 180–181C, bulk d 8.5 lb/gal, vap press 0.8 mm (20C), flash p 185F (85C) (COC), coefficient of expansion 0.00101/C. Combustible.
Derivation: Action of metallic sodium on ethyl acetate with subsequent distillation.
Grade: Technical, 98%.
Hazard: Toxic by ingestion and inhalation; irritant to skin and eyes.
Use: Organic synthesis, antipyrine, lacquers, dopes, plastics, manufacture of dyes, pharmaceuticals antimalarials, vitamin B; flavoring.

ethyl acetone. See methyl propyl ketone.

ethylacetylene. (1-butyne). CAS: 107-00-6. $C_2H_5C{\equiv}CH$.
Properties: Available as liquefied gas, bp 8.3C, d 0.669 (0/0C), fp −130C, flash p less than 20F (−6.6C) (TOC), specific volume 7.2 cu/ft/lb (21.2C), insoluble in water.

Hazard: Flammable, dangerous fire risk.
Use: Specialty fuel, chemical intermediate.

ethyl-n-acetyl-α-cyanoglycine. See ethyl acetamidocyanoacetate.

ethyl acrylate. CAS: 140-88-5. $CH_2:CHCOOC_2H_5$.
Properties: Colorless liquid, bp 99.4C, fp −72.0C, d 0.9230 (20/20C), refr index 1.4037 (25C), bulk d 7.6 lb/gal (20C), soluble in alcohol and ether, readily polymerized, flash p 60F (15.5C) (open cup).
Derivation: (a) Ethylene cyanohydrin, ethyl alcohol, and dilute sulfuric acid; (b) Oxo reaction of acetylene, carbon monoxide, and ethyl alcohol in the presence of nickel or cobalt catalyst.
Grade: Technical (inhibited, usually with hydroquinone or its monomethyl ether), pure uninhibited.
Hazard: Toxic by ingestion, inhalation, skin absorption; irritant to skin and eyes. Flammable, dangerous fire and explosion hazard. TLV: 5 ppm in air.
Use: Monomer for acrylic resins.
See also acrylate and acrylic resin.

ethyl alcohol. (alcohol; grain alcohol; ethanol; EtOH). CAS: 64-17-5. C_2H_5OH.
Properties of pure 100% absolute alcohol (dehydrated): Colorless, limpid, volatile liquid. Bp 78.3C, fp −117.3C, ethereal vinous odor, pungent taste. Miscible with water, methanol, ether, chloroform, and acetone. Properties: (95%) Refr index 1.3651 (15C), surface tension 22.3 dynes/cm (20C), viscosity 0.0141 poise (20C), vap press 43 mm (20C), specific heat 0.618 cal/g (23C), flash p 55F (12.7C), d 0.816 (15.56C), bp 78C, fp −114C, autoign temperature 793F (422C).
Derivation: (a) From ethylene by direct catalytic hydration or with ethyl sulfate as intermediate; (b) fermentation of biomass, especially agricultural wastes; (c) enzymatic hydrolysis of cellulose (see also cellulase).
Grade: USP (95% by volume), absolute, pure, completely denatured, specially denatured, industrial, various proofs (one-half the proof number is the percentage of alcohol by volume).
Hazard: Is classified as a depressant drug. Though it is rapidly oxidized in the body and is therefore noncumulative, ingestion of even moderate amounts causes lowering of inhibitions, often succeeded by dizziness, headache, or nausea. Larger intake causes loss of motor nerve control, shallow respiration, and in extreme cases unconsciousness and even death. Degree of intoxication is determined by concentration of alcohol in the brain. Of primary importance is the fact that intake of even moderate amounts together with

barbiturates or similar drugs is extremely dangerous and may even be fatal. Flammable, dangerous fire risk; flammable limits in air 3.3–19%. TLV: 1000 ppm in air.

Use: Solvent for resins, fats, fatty acids, oils, hydrocarbons; extraction medium; manufacture of acetaldehyde, acetic acid, ethylene, butadiene, 2-ethyl hexanol, dyes, pharmaceuticals, elastomers, detergents, cleaning preparations, surface coatings, cosmetics, explosives; antifreeze, beverages, antisepsis, gasohol, yeast-growth medium, octane booster in gasoline.

See also alcohol, denatured; industrial alcohol, biomass.

Note: Ethanol from fermentation of biomass and hydrolysis of cellulose is a significant alternate energy source, especially as an automotive fuel. Its use in gasoline will continue to increase. Further information can be obtained from the National Alcohol Fuels Information Center, 1617 Cole Blvd., Golden, Colorado, 80401.

ethyl α-allylacetoacetate.
$CH_3COCH(CH_2CH:CH_2)COOC_2H_5$.
Properties: Water-white liquid, d 0.989 (20C), bulk density 8.24 lb/gal (20C). Combustible.
Use: Intermediate for pharmaceuticals, perfumes, fungicides, insecticides, fine chemicals.

ethyl aluminum dichloride. (EADC).
$C_2H_5AlCl_2$.
Properties: Clear, yellow, pyrophoric liquid. Bp (extrapolated) 194C, fp 32C, d 1.222, bulk d 10.28 lb/gal (25C).
Derivation: Reaction of aluminum chloride with ethyl aluminum sesquichloride.
Hazard: Ignites on contact with air, dangerous fire risk, reacts violently with water. Skin irritant.
Use: Catalyst for olefin polymerization, aromatic hydrogenation; intermediate.

ethyl aluminum sesquichloride. (EASC).
$(C_2H_5)_3Al_2Cl_3$.
Properties: Clear, yellow, pyrophoric liquid. Bp 204C, fp −50C, d 1.08.
Derivation: Reaction of ethyl chloride and aluminum.
Grade: Commercial.
Hazard: Ignites on contact with air, dangerous fire risk, reacts violently with water.
Use: Catalyst for olefin polymerization, aromatic hydrogenation; intermediate.

ethylamine. (monoethylamine; aminoethane).
CAS: 75-04-7. $CH_3CH_2NH_2$.
Properties: Colorless, volatile liquid (or gas). Ammonia odor, strong alkaline reaction, bp 16.6C, fp −81.2C, d 0.689 (liquid 15/15C), bulk d 5.7 lb/gal (20C), flash p approximately 0F (−17.7C)

(OC), autoign temperature 723F (383C). Miscible with water, alcohol, and ether.
Derivation: From ethyl chloride and alcoholic ammonia under heat and pressure.
Grade: Technical (anhydrous and 70% aqueous solution), pure 98.5% min.
Hazard: Strong irritant. Flammable, dangerous fire risk, flammable limits in air 3.5–14%. TLV: 10 ppm in air.
Use: Dye intermediate, solvent extraction, petroleum refining, stabilizer for rubber latex, detergents, organic synthesis.

ethylamine hydrobromide. $C_2H_5NH_2 \cdot HBr$.
Properties: White, almost odorless granules; mp 158–161C; very soluble in water.
Use: Intermediate (where liquid ethylamine or liquid hydrobromic acid cannot be used).

ethyl-o-aminobenzoate. See ethyl anthranilate.

ethyl-p-aminobenzoate hydrochloride. (anesthesol; benzocaine; procaine hydrochloride).
CAS: 51-05-8. $C_6H_4NH_2CO_2C_2H_5 \cdot HCl$.
Properties: White, crystalline, odorless, tasteless powder; stable in air; mp 88–92C; soluble in dilute acids; less soluble in chloroform, ether, and alcohol; very slightly soluble in water.
Derivation: Ethylation of p-nitrobenzoic acid followed by reduction.
Grade: Technical, pure, NF (as benzocaine).
Hazard: Toxic by ingestion.
Use: Medicine (local anesthetic), suntan preparations.

ethylaminoethanol. See ethylethanolamine.
Mixed ethylaminoethanols (sold in up to tank car lots) may also contain diethylaminoethanol.

2-ethylamino-4-isopropylamino-6-methylthio-s-triazine. $C_2H_5HNC_3N_3(SCH_3)NHCH(CH_3)_2$.
Properties: White, crystalline powder; mp 84–85C; slightly soluble in water; soluble in organic solvents.
Hazard: Toxic by ingestion.
Use: Weed-killing agent in pineapple and sugar cane.

ethyl-1-(p-aminophenyl)-4-phenylisonipecotate.
See anileridine.

ethyl amyl ketone. (EAK; 5-methyl-3-heptanone).   CAS: 541-85-5.
$CH_3CH_2CO(CH_2)_4CH_3$.
Properties: Colorless liquid, pungent odor, insoluble in water, soluble in 4 volumes of 60% alcohol, bp 157C, bulk d 83 lb/gal, d 0.819–0.824, refr index 1.416, flash p 138F (58C). Combustible.

Hazard: Narcotic in high concentration. Moderate fire risk. TLV: 25 ppm in air.

Use: Perfumery, solvent for nitrocellulose and vinyl resins.

n-ethylaniline.   CAS: 103-69-5.
$C_2H_5NHC_6H_5$.

Properties: Colorless liquid becoming brown on exposure to light, soluble in alcohol, insoluble in water, d 0.9631, fp −63.5C, bp 206C, refr index 1.5559 (20C), flash p 185F (85C) (open cup). Combustible.

Derivation: By heating aniline and ethyl alcohol in presence of sulfuric acid with subsequent distillation.

Hazard: Toxic by ingestion, inhalation, and skin absorption.

Use: Organic synthesis.

o-ethylaniline.   CAS: 578-54-1.
$C_6H_4(NH_2)C_2H_5$.

Properties: Brown liquid, fp −44C, d 0.982 (20C), bp 214C, flash p (open cup) 185F (85C), soluble in alcohol and toluene, insoluble in water. Combustible.

Hazard: Toxic by ingestion, inhalation, and skin absorption.

Use: Intermediate for pharmaceuticals, dyestuffs, pesticides and other products.

ethyl anthranilate.   (ethyl-o-aminobenzoate).
CAS: 87-25-2. $C_6H_4(NH_2)COOCH_2CH_3$.

Properties: Colorless liquid, fruity odor, d 1.117, refr index 1.564, bp 260C, soluble in alcohol and propylene glycol. Combustible.

Grade: Technical, FCC.

Use: Perfumery and flavors, similar to methyl anthranilate.

2-ethylanthraquinone. $C_{14}H_8O_2C_2H_5$.

Properties: Buff to light yellow paste, mp 108C.

Use: Synthesis, especially of hydrogen peroxide.

"Ethyl" Antiknock Compounds.[313]   TM for a series of fuel additives containing various percentages of tetraethyl lead, ethylene dibromide, ethylene dichloride, dye, kerosene and antioxidant. All are used to improve the octane rating of motor fuels.

"Ethyl" Antioxidants.[313]   TM for a series of gasoline antioxidants based on phenols (chiefly ditertbutyl phenol). They inhibit formation of gum and peroxides in gasoline and the formation of decomposed products of jet fuels in storage. Also used for steam-turbine and industrial oils and to retard decomposition of antiknock compounds in gasoline.

ethylarsenious oxide. C_2H_5AsO.

Properties: Colorless oil; garlic-like, nauseating odor; oxidizes in air and forms colorless crystals; soluble in acetone, benzene, ether; d 1.802 (11C); bp 158C (10 mm).

Hazard: Highly toxic.

Use: Organic synthesis.

"Ethyl" Automate Liquid Dyes.[313]   TM for single-phase organic dyes in aromatic solvent. Full range of liquid dyes in various concentrations, colors, and combinations.

2-ethylaziridine.   See ethylethyleneimine.

ethylbenzene.   (phenylethane).
CAS: 100-41-4. $C_6H_5C_2H_5$.   18th highest-volume chemical produced in US (1985).

Properties: Colorless liquid, aromatic odor, vapor heavier than air, bp 136.187C, refr index 1.49594 (20C), d 0.867 (20C), fp −95C, bulk d 7.21 lb/gal (25C), flash p 59F (15C), autoign temperature 810F (432C), specific heat 0.41/cal/gal/C, viscosity 0.64 centipoise (25C). Soluble in alcohol, benzene, carbon tetrachloride, and ether; almost insoluble in water.

Derivation: (a) By heating benzene and ethylene in presence of aluminum chloride with subsequent distillation, (b) by fractionation directly from the mixed xylene stream in petroleum refining.

Grade: Technical, pure, research.

Hazard: Toxic by ingestion, inhalation and skin absorption; irritant to skin and eyes. Flammable, dangerous fire risk. TLV: 100 ppm in air.

Use: Intermediate in production of styrene, solvent.

4-ethylbenzenesulfonic acid.   (p-ethylbenzenesulfonic acid).   CAS: 98-69-1.
$C_2H_5C_6H_4SO_3H$.

Properties: Solid with mw 186.23, d 1.229, fp above 110C.

Hazard: Corrosive.

ethyl benzoate.   CAS: 93-89-0.
$C_6H_5CO_2C_2H_5$.

Properties: Colorless aromatic liquid, soluble in alcohol and ether, insoluble in water, d 1.043–1.046, fp −32.7C, bp 212.9C, refr index 1.505, flash p 200F (93C). Combustible.

Derivation: By heating ethanol and benzoic acid in presence of sulfuric acid.

Grade: Technical, FCC.
Use: Flavoring, perfumery, solvent mixture, lacquers, solvent for many cellulose derivatives and natural and synthetic resins.

ethyl benzoylacetate.   CAS: 94-02-0.
$C_6H_5COCH_2COOC_2H_5$.
Properties: Light yellow oil, bp 265C (decomposes), d 1.111–1.117 (20C), flash p 275F (135C), soluble in most organic solvents, insoluble in water. Combustible.
Derivation: Reaction of ethyl acetate and ethyl benzoate with metallic sodium.
Method of purification: Vacuum distillation.
Grade: 95% pure.
Use: Dye and pharmaceutical intermediate.

ethyl-o-benzoylbenzoate.
$C_6H_5COC_6H_4COOC_2H_5$.
Properties: Yellowish-white solid; odorless; insoluble in water; soluble in alcohol, acetone, ethyl acetate, and benzene; mp 56–58C; bp 325C. Combustible.
Use: Plasticizer for nitrocellulose and synthetic resins.

1-ethylbenzyl alcohol.   (1-phenyl-1-propanol; ethyl phenyl carbinol).   CAS: 93-54-9.
$C_9H_{11}OH$.
Properties: Oily liquid; bp 220C; d 0.99; refr index 1.51; soluble in ethanol and methanol, benzene, and toluene.
Derivation: Benzaldehyde or phenyl ethyl ketone.
Use: Heat-transfer fluid.

ethylbenzylaniline.   CAS: 92-59-1.
$C_6H_5N(C_2H_5)CH_2C_6H_5$.
Properties: Light yellow liquid, soluble in alcohol and ether, insoluble in water, d 1.034, bp 286C. Combustible.
Derivation: By heating ethylaniline, benzyl chloride, and aqueous caustic soda with subsequent distillation.
Hazard: Toxic by ingestion and inhalation.
Use: Dyestuffs, organic synthesis.

ethylbenzyl chloride.   (1-chloromethylethylbenzene). $ClCH_2C_6H_4C_2H_5$.   Consists of 70% p- and 30% o-ethylbenzyl chloride.
Properties: Colorless liquid, d 1.0460–1.0475 (25/25C), refr index 1.5290–1.5305 (25C), soluble in alcohols, insoluble in water.
Hazard: Irritant to eyes, a lachrymator.
Use: Intermediate.

ethyl biscoumacetate.   (ethyl bis(4-hydroxycoumarinyl)-acetate).   CAS: 548-00-5.
$C_{22}H_{16}O_8$.   A synthetic derivative of bishydroxycoumarin.

Properties: White, odorless, bitter, crystalline solid; mp 177–182C; another form melts 154–157C; soluble in acetone and benzene; slightly soluble in alcohol and ether; insoluble in water.
Grade: NF.
Use: Medicine (anticoagulant).

ethyl borate.   Legal label name for triethyl borate.

ethyl bromide.   (bromoethane).   CAS: 74-96-4.
C_2H_5Br.
Properties: Colorless liquid, soluble in alcohol and ether, sparingly soluble in water, d 1.431 (20/4C), bp 38.4C, bulk d 12–12.1/lb/gal, vap press 386 mm (20C), autoign temperature 952F (511C), fp −119C, flash p approximately 80F (26C).
Derivation: From ethanol or ethylene and hydrobromic acid. One process uses gamma radiation to initiate the combination.
Grade: Technical (98%).
Hazard: Toxic by ingestion, inhalation, and skin absorption; strong irritant. TLV: 200 ppm in air. Flammable, dangerous fire hazard, explosion limits in air 6–11%.
Use: Organic synthesis, medicine (anesthetic), refrigerant, solvent, grain and fruit fumigant.

ethyl bromoacetate.   CAS: 105-36-2.
$CH_2BrCOOC_2H_5$.
Properties: Clear, colorless liquid; partially decomposed by water; soluble in alcohol, benzene, ether; insoluble in water; d 1.53 (4C); bp 168C; fp −13.8C; vap d 5.8.
Derivation: Interaction of bromine and acetic acid in the presence of red phosphorus.
Grade: Technical.
Hazard: Toxic by ingestion, inhalation, and skin absorption; strong irritant.

ethylbromopyruvate.   CAS: 70-23-5.
$BrCH_2COCO_2C_2H_5$.
Properties: Light yellow liquid with mw 195.02, bp 90–100C 10 mm, d 1.554, fp 98C.
Hazard: Severe poison, lachrymator, irritant.
Use: Pharmaceutical intermediate.

ethyl butanoate.   See ethyl butyrate.

2-ethylbutanol.   See 2-ethylbutyl alcohol.

2-ethyl-1-butene.   (uns-diethylethylene).
$CH_3CH_2(C_2H_5)C:CH_2$.
Properties: Colorless liquid; d 0.6894 (20/4C); bp 64.95C; refr index 1.3969 (20C); soluble in alcohol, acetone, ether, and benzene; insoluble in water. Combustible.
Grade: 95% pure.
Use: Organic synthesis of flavors, perfumes, medicines, dyes, resins.

3-(2-ethylbutoxy)propionic acid.
$CH_3CH_2CH(C_2H_5)CH_2OCH_2CH_2COOH$.
Properties: Water-white liquid, d 0.9600 (20/20C), bp 200C (100 mm), vap press less than 0.1 mm (20C), fp glass at approximately −90C, insoluble in water, flash p 280F (137C). Combustible.
Use: Preparation of metallic salts for paint driers and gelling agents.

2-ethylbutyl acetate. CAS: 123-66-0.
$C_2H_5CH(C_2H_5)CH_2OOCCH_3$
Properties: Colorless liquid, mild odor, d 0.875–0.881 (20/20C), boiling range 155–164C, purity not less than 90% ethylbutyl acetate, bulk d 7.33 lb/gal (20C), flash p 130F (54.4C) (open cup).
Hazard: Moderate fire risk.
Use: Solvent for nitrocellulose lacquers, flavoring.

2-ethylbutyl alcohol. (2-ethylbutanol; pseudo-hexyl alcohol). CAS: 97-95-0.
$CH_3CH_2CH(C_2H_5)CH_2OH$.
Properties: Colorless liquid, stable, miscible with most organic solvents, slightly soluble in water, bp 148.9C, fp −114C, d 0.8328 (20/20C), bulk d 6.93 lb/gal (20C), refr index 1.4229 (20C), flash p 137F (58.3C) (ASTM OC), vap press 0.9 mm (20C). Combustible.
Hazard: Moderate fire risk.
Use: Solvent for oils, resins, waxes, dyes; diluent; synthesis of perfumes, drugs; flavoring.

n-ethylbutylamine. CAS: 617-79-8.
$C_2H_5NHCH_2CH_2CH_2CH_3$.
Properties: Water-white liquid, bp 108C, fp −78C, d 0.7401 (20/20C), refr index 1.407 (20C), flash p 65F (18.3C), partly soluble in water.
Hazard: Flammable, dangerous fire risk. Toxic by ingestion.
Use: Intermediate.

ethylbutyl carbonate. $C_2H_5CO_3C_4H_9$.
Properties: Colorless liquid used as solvent for many natural and synthetic resins, in mixtures for nitrocellulose, d 0.92–0.93 (20C), bp 135–175C, flash p 122F (50C) (closed cup). Combustible.
Hazard: Moderate fire risk.

ethyl-n-butyl ether. (n-butyl ethyl ether).
CAS: 628-81-9. $C_2H_5OC_4H_9$.
Properties: Liquid, d 0.7528 (20C), fp −103C, bp 92.2C, flash p 40F (4.4C), vap press 43 mm (20C), slightly soluble in water.
Hazard: Flammable, dangerous fire risk.
Use: Extraction solvent, inert reaction medium.

ethyl butyl ketone. (3-heptanone). CAS: 106-35-4.
$CH_3CH_2CH_2CH_2COCH_2CH_3$.
Properties: Clear liquid, d 0.8191 (20/20C), fp −39C, boiling range 142.8–147.8C, 95% purity,

bulk d 6.8 lb/gal, flash p 115F (46C) (open cup), insoluble in water, soluble in alcohol. Combustible.
Hazard: Moderate fire risk. TLV: 50 ppm in air.
Use: Solvent mix for air-dried and baked finishes, for polyvinyl and nitrocellulose resins.

2-ethyl-2-butylpropanediol-1,3. (2-butyl-2-ethyl-propanediol-1,3).
$HOCH_2C(C_2H_5)(C_4H_9)CH_2OH$.
Properties: White crystals or liquid, d 0.931 (50/20C), bp 178C (50 mm), fp 41.4C, solubility in water 0.8% by wt (20C), flash p 280F (137C). Combustible.
Use: Synthesis of lubricants, emulsifying agents; insect repellents, plastics.

2-ethylbutyl silicate.
$[CH_3CH_2CH(C_2H_5)CH_2O]_4Si$.
Properties: Colorless liquid, bp 164C (1 mm).
Derivation: Reaction of silicon tetrachloride with 2-ethylbutanol.
Use: Hydraulic fluid, heat-transfer liquid.

2-ethylbutyraldehyde. (diethylacetaldehyde).
CAS: 97-96-1. $(C_2H_5)_2CHCHO$.
Properties: Colorless liquid, insoluble in water, d 0.8164 (20/20C), bp 116.8C, vap press 13.7 mm (20C), flash p 70F (21.1C) (open cup), bulk d 6.8 lb/gal (20C), fp −89C.
Grade: Technical.
Hazard: Irritant to eyes and skin. Flammable, dangerous fire risk.
Use: Organic synthesis, pharmaceuticals, rubber accelerators, synthetic resins.

ethyl butyrate. (ethyl butanoate).
CAS: 105-54-4. $C_3H_7CO_2C_2H_5$.
Properties: Colorless liquid, pineapple-like odor, soluble in alcohol and ether, almost insoluble in water and glycerol, d 0.8788, fp −93.3C, bp 120.6C, refr index 1.400 (20C), flash p 78F (25.5C) (closed cup), autoign temperature 865F (462C).
Derivation: Ethyl alcohol and butyric acid heated in presence of sulfuric acid, with subsequent distillation.
Grade: Technical, FCC.
Hazard: Irritant to eyes and mucous membranes, narcotic in high concentration. Flammable, dangerous fire risk.
Use: Flavoring extracts, perfumery, solvent mixture for cellulose esters and ethers.

2-ethylbutyric acid. (diethyl acetic acid).
CAS: 88-09-5. $(C_2H_5)_2CHCOOH$.
Properties: Water-white liquid, resembles butyric acid in most properties except that its odor is less pronounced and its water solubility limited, d 0.9225 (20/20C), bp 190C, vap press 0.08 mm

(20C), flash p 210F (98C), bulk d 7.7 lb/gal, fp −15C. Combustible.
Use: Ester formation; intermediate for drugs, dyestuffs, chemicals; flavoring.

ethyl caffeate. (ethyl-3,4-dihydroxycinnamate). $C_6H_3(OH)_2CH:CHCOOC_2H_5$.
Properties: Yellow to tan crystals, characteristic aromatic odor, insoluble in water, very soluble in alcohol.
Grade: CP.
Use: Food antioxidant.

ethyl caprate. (ethyl decanoate).
CAS: 110-38-3. $C_9H_{19}COOC_2H_5$.
Properties: Colorless liquid, fragrant odor, soluble in alcohol and ether, insoluble in water, d 0.862, bp 243C. Combustible.
Derivation: By heating capric acid, absolute alcohol, and sulfuric acid with subsequent distillation.
Use: Organic synthesis, flavoring agent.

ethyl caproate. (ethyl hexoate; ethyl hexanoate).
CAS: 123-66-0. $C_5H_{11}COOC_2H_5$.
Properties: Colorless to yellowish liquid, pleasant odor, soluble in alcohol and ether, insoluble in water and glycerol, d 0.873, bp 167. Combustible.
Derivation: Heating absolute alcohol and n-caproic acid in presence of sulfuric acid with subsequent distillation.
Grade: Technical, FCC.
Use: Organic synthesis, artificial fruit essences.

ethyl caprylate. (ethyl octoate; ethyl octanoate).
CAS: 106-32-1. $CH_3(CH_2)_6COOC_2H_5$.
Properties: Colorless liquid, pineapple odor, soluble in alcohol and ether, insoluble in water and glycerol, d 0.865–0.869 (25C), fp −48C, bp 207–209C. Combustible.
Derivation: Heating caprylic acid, alcohol, and sulfuric acid with subsequent distillation.
Grade: Technical, FCC.
Use: Flavoring, fruit essences.

ethyl carbamate. See urethane.

ethyl carbazate. CAS: 4114-31-2.
$H_2NNHCO_2C_2H_5$.
Properties: Off-white crystals with mw 104.11, mp 44–47C, bp 108–110C at 22 mm, fp 86C.
Hazard: Irritant.
Use: Synthetic intermediate.

N-ethylcarbazole. (9-ethylcarbazole).
$(C_6H_4)_2NC_2H_5$ (tricyclic).
Properties: Leaflets, soluble in ether and hot alcohol, mp 69–70C, bp 175C (5 mm).
Derivation: Action of ethyl chloride on the potassium salt of carbazole.

Use: Intermediate for dyes, pharmaceuticals; agricultural chemicals.

ethyl carbonate. See diethyl carbonate.

ethylcellulose. An ethyl ether of cellulose.
Properties: White granular thermoplastic solid, lowest density of the commercial cellulose plastics, properties vary with extent to which hydroxyl radicals of cellulose have been replaced by ethoxy groups. Standard commercial product has 47–48% ethoxy content, d 1.07–1.18, refr index 1.47, high dielectric strength, softening point 100–130C. Soluble in most organic liquids; compatible with resins, waxes, oils, and plasticizers; inert to alkalies and dilute acids; insoluble in water and glycerol. Combustible.
Derivation: From alkali cellulose and ethyl chloride or sulfate; from cellulose and ethanol in presence of dehydrating agents.
Grade: Technical, NF, FCC.
Use: Hot-melt adhesives and coatings for cables, paper, textiles, etc.; extrusion wire insulation; protective coatings, pigment-grinding bases, toughening agent for plastics, printing inks, molding powders, proximity fuses, vitamin preparations, casing for rocket propellants, food and feed additive.

ethyl centralite. See sym-diethyldiphenylurea.

ethyl chloride. (chloroethane).   CAS: 75-00-3.
C_2H_5Cl.
Properties: Gas at room temperature, when compressed, a colorless volatile liquid, ether-like odor, burning taste, stable and noncorrosive when dry but will hydrolyze in the presence of water or alkalies, miscible with most of the commonly used solvents, slightly soluble in water, d 0.9214, fp −140.85C, bp 12.5C, critical point 187.2C (52 atmospheres, d 0.33), vap press 1000 mm (20C), flash p −58F (−50C) (CC), autoign temperature 966F (518C).
Derivation: (a) From ethylene and hydrogen chloride, (b) by passing hydrogen chloride into a solution of zinc chloride and ethanol.
Grade: Technical, USP.
Hazard: Highly flammable, severe fire and explosion risk; flammable limits in air 3.8–15.4%. Irritant to eyes. TLV: 1000 ppm in air.
Use: Manufacture of tetraethyl lead and ethylcellulose, anesthetic, organic synthesis, alkylating agent, refrigeration, analytical reagent; solvent for phosphorus, sulfur, fats, oils, resins, and waxes; insecticides.

ethyl chloroacetal. $ClCH_2CH(OC_2H_5)_2$.
Properties: Water-white liquid with pleasant odor, d 1.022 (20C), boiling range 54–61C (20 mm), 149–153C, fp −32C, flash p 117F (47.2C), refr

index 1.418 (20C), soluble in alcohol and ethyl ether, insoluble in water. Combustible.
Hazard: Moderate fire risk.

ethyl chloroacetate.   CAS: 105-39-5.
$CCH_2ClCO_2C_2H_5$.
Properties: Water-white, mobile liquid, pungent fruity odor, decomposed by hot water and alkalies. Soluble in alcohol, benzene and ether; insoluble in water. D 1.1585 (20C), bp 144.2C, vap d 4.23–4.46, flash p 131F (55C), refr index 1.4227 (20C). Combustible.
Derivation: (a) Action of chloroacetyl chloride on alcohol, (b) by treating chloroacetic acid with alcohol and sulfuric acid.
Hazard: Strong irritant to eyes.
Use: Solvent, organic synthesis, military poison, vat dyestuffs.

ethylchlorocarbonate.   (ethyl chloroformate). CAS: 541-41-3. $ClCOOC_2H_5$.
Properties: Water-white liquid with irritating odor, d 1.135–1.139 (20/20C), bp 93–95C, refr index 1.3974 (20C), flash p 61F (16.1C) (closed cup); decomposes in water and alcohol; soluble in benzene, chloroform, and ether.
Derivation: Reaction of carbon monoxide with gaseous chlorine, producing phosgene $(COCl_2)$ which is then treated with anhydrous ethanol giving ethyl chlorocarbonate and splitting off hydrogen chloride.
Grade: Technical.
Hazard: Flammable, dangerous fire risk. Strong irritant to eyes and skin.
Use: Organic synthesis, intermediate in making diethyl carbonate, flotation agents, polymers, isocyanates.

ethyl chloroformate.   See ethyl chlorocarbonate.

ethylchlorosulfonate. $C_2H_5OClSO_2$.
Properties: Colorless, oily liquid; pungent odor; fumes in moist air; decomposed by water; attacks lead and tin but copper mildly; iron and steel not affected; soluble in chloroform and ether; insoluble in water; d 1.379 (0C); bp 153C; vap d 5 (air = 1.29); volatility 18,000 mg/m³ (20C).
Derivation: (a) Action of fuming sulfuric acid on ethylchloroformate, (b) interaction of ethylene and chlorosulfonic acid.
Grade: Technical.
Hazard: Strong irritant to eyes and skin, evolves phosgene when heated.
Use: Organic synthesis, military poison.

ethyl cinnamate.   (ethyl phenylacrylate; cinnamylic ether). $C_6H_5CH:CHCOOC_2H_5$.
Properties: Limpid, oily liquid; strawberry-like odor. Soluble in alcohol and ether, insoluble in water, d 1.045–1.048, refr index 1.560 (20C),

congealing point 7C (min), bp 271C. Combustible.
Derivation: Heating ethyl alcohol and cinnamic acid in presence of sulfuric acid.
Use: Perfumery, flavoring extracts.

ethyl citrate.   See triethyl citrate.

ethyl crotonate.   CAS: 623-70-1.
$CH_3CH:CHCOOC_2H_5$.
Properties: Water-white solid or liquid; characteristic, pungent, persistent odor. Soluble in alcohol and ether, insoluble in water, d 0.9207 (20/20C), mp (solid) 45C, bp (solid) 209C, liquid 139C, flash p 36F (2.2C), refr index 1.4242 (20C), 7.65 lb/gal (20C).
Hazard: Flammable, dangerous fire risk. Strong irritant.
Use: Solvent and softening agent, lacquers, organic synthesis.

ethyl cyanide.   (propionitrile; propanenitrile). CAS: 107-12-0. C_2H_5CN.
Properties: Mobile, colorless liquid; ethereal odor. Soluble in alcohol and water, d 0.7829 (20/20C), refr index 1.3664 (20C), bp 97.4C, fp −92.9C, flash p 61F (16.1C) (open cup).
Derivation: Heating barium ethyl sulfate and KCN with subsequent distillation.
Hazard: Toxic by ingestion and inhalation. TLV (as CN): 5 mg/m³ of air. Flammable, dangerous fire risk.
Use: Solvent, dielectric fluid, intermediate.

ethyl cyanoacetate.   (malonic ethyl ester nitrile). CAS: 105-56-6. $CNCH_2COOC_2H_5$.
Properties: Colorless liquid, bp 206–208C, fp −22.5C, refr index 1.41751 (20C), soluble in alcohol and ether, soluble in alkaline solutions, flash p 230F (110C). Combustible.
Derivation: Esterification of cyanoacetic acid with ethanol, reaction of an alkali cyanide and chloroacetic ethyl ester.
Method of purification: Vacuum distillation.
Grade: Reagent, technical.
Hazard: Toxic by ingestion and inhalation.
Use: Organic synthesis, pharmaceuticals, dyes.

5-ethyl-5-cycloheptenylbarbituric acid.   See heptabarbital.

ethylcyclohexane. $C_2H_5C_6H_{11}$.
Properties: Colorless liquid, d 0.787, bp 131.8C, refr index 1.4330 (20C), flash p 95F (35C), autoign temperature 504F (262C).
Hazard: Flammable, moderate fire risk; flammable limits in air 0.9–6.6%.
Use: Organic synthesis.

N-ethylcyclohexylamine.   CAS: 5459-93-8.
$C_6H_{11}NHC_2H_5$.
Properties: Liquid with mw 127.23, bp 165C, d 0.844, freezing p 43C.

Hazard: Corrosive and toxic.
Use: Intermediate in production of herbicides and pharmaceuticals.

ethylcyclopentane. $C_2H_5C_5H_9$.
Properties: Colorless liquid, d 0.766, bp 103.5C, refr index 1.4198 (20C), autoign temperature 504F (262C).
Hazard: Flammable, moderate fire risk; flammable limits in air 1.1–6.7%.

ethyl cyclopentanone-2-carboxylate. See 2-carbethoxy cyclopentanone.

ethyl decanoate. See ethyl caprate.

ethyl diazoacetate. CAS: 623-73-4.
$N_2CHCO_2C_2H_5$.
Properties: Liquid with mw 114.10, mp 22C, bp 140–141 at 720 mm, d 1.085.
Available forms: Research quantities.
Hazard: Flammable liquid which explodes when heated.

ethyldichloroarsine. (dichloroethylarsine).
CAS: 598-14-1. $C_2H_5AsCl_2$.
Properties: Colorless, mobile liquid; becomes yellowish under the action of light and air. Fruit-like odor (high dilution), decomposed by water, attacks brass but not iron (dry). Soluble in alcohol, benzene, ether, and water. D 1.742 (14C), bp 156C (decomposes), fp −65C, coefficient of thermal expansion 0.0011, vap d 6 (air = 1.29), volatility 20,000 mg/m³ (20C), vap press 2.29 mm (21.5C).
Derivation: Chlorination of ethyl arsenious oxide.
Hazard: Toxic by ingestion, inhalation, and skin absorption; strong irritant.
Use: Military poison.

ethyl-4,4′-dichlorobenzilate. See chlorobenzilate.

ethyl dichlorophenoxyacetate. See 2,4-D.

ethyldichlorosilane. CAS: 1789-58-8.
$C_2H_5SiHCl_2$.
Properties: Colorless liquid, bp 75.5C, d 1.088 (25/25C), flash p 30F (−1.1C) (COC), readily hydrolyzed by moisture with the liberation of hydrogen and hydrogen chloride.
Derivation: By Grignard reaction of trichlorosilane and ethylmagnesium chloride.
Hazard: Flammable, dangerous fire risk. Strong irritant to eyes and skin.
Use: Intermediate for silicones.

ethyldiethanolamine. $C_2H_5N(CH_2CH_2OH)_2$.
Properties: Water-white liquid, d 1.015 (20C), boiling range 246–252C, odor amine, flash p 255F (123C). Combustible.
Use: Solvent, detergents.

ethyl diglyme. See diethylene glycol diethyl ether.

ethyldimethylmethane. See isopentane.

ethyl dimethyl-9-octadecenylammonium bromide. $C_{22}H_{46}NBr$. Cationic surfactant and algicide, soluble in propylene glycol and isopropyl alcohol.

ethyldipropylmethane. See 4-ethylheptane.

S-ethyl di-n,n-propylthiocarbamate. See EPTC.

ethyl enanthate. (ethyl heptanoate; cognac oil).
$CH_3(CH_2)_5COOC_2H_5$.
Properties: Clear, colorless oil with fruity odor and taste; d 0.87; bp 187C; soluble in alcohol, chloroform, and ether; insoluble in water. Combustible.
Derivation: By heating enanthic acid and ethanol in presence of sulfuric acid and subsequent recovery by distillation.
Grade: Technical.
Use: Artificial cognac flavor, flavor for liquors and fruity-type soft drinks.

ethylene. (bicarburetted hydrogen; ethene).
CAS: 74-85-1. $H_2C:CH_2$. Sixth highest-volume chemical produced in US (1985).
Properties: Colorless gas with sweet odor and taste, fp −169C, bp −103.9C, flash p −213F (−135C), d of liquid 0.610 (0C), vap d 0.975 (air = 1.29), critical temperature 9.5C, autoign temperature 1009F (543C), critical pressure (absolute) 744 psi, purity not less than 96% ethylene by gas volume, not more than 0.5% acetylene, not more than 4% methane and ethane, 13.4 cu ft/lb (15.6C), slightly soluble in water, alcohol and ethyl ether, an asphyxiant gas.
Derivation: (1) Thermal cracking of hydrocarbon gases (800–900C), (2) dehydration of ethanol, (3) from synthesis gas with Ru as catalyst.
Hazard: Highly flammable, dangerous fire and explosion risk; explosive limits in air 3–36% by volume.
Grade: Technical (95% min), 99.5% min, 99.9 mole percentage.
Use: Polyethylene, polypropylene, ethylene oxide, ethylene dichloride, ethylene glycols, aluminum alkyls, vinyl chloride, vinyl acetate, ethyl chloride, ethylene chlorohydrin, acetaldehyde, ethyl alcohol, polystyrene, styrene, polyvinyl chloride, SBR, polyester resins, trichloroethylene, etc.; refrigerant, welding and cutting of metals, anesthetic, in orchard sprays to accelerate fruit ripening.
Note: Under development is a method of producing ethylene directly from petroleum by so-called flame-cracking. The advanced cracking reactor

(ACR) process involves combustion at approximately 2000C of a mixture of crude oil and high-temperature gases. Combustion is supported by pure oxygen. If the prototype unit is successful, large-scale application of this method will be possible within the decade.

ethylene bis(iminodiacetic acid). See ethylene-diaminetetraacetic acid.

ethylene bis(oxyethylenenitrilo)tetraacetic acid. See ethylene glycol-bis(β-aminoethyl ether)-N,N-tetraacetic acid.

ethylene bromide. See ethylene dibromide.

ethylene bromohydrin. (glycol bromohydrin; 2-bromoethyl alcohol).   CAS: 540-51-2. $BrCH_2CH_2OH$.
Properties: Hygroscopic liquid, d 1.7629 (20C), bp 149–150C (750 mm), refr index 1.4915 (20C), soluble in most organic solvents and completely miscible with water, aqueous solutions have a sweet burning taste. Hydrolysis of aqueous solutions is accelerated by heat, acids, and alkalies. Combustible.
Derivation: Action of hydrogen bromide on ethylene oxide.
Hazard: Irritant to eyes and mucous membranes.
Use: Organic synthesis.

ethylene carbonate. (glycol carbonate; dioxo-lone-2).   CAS: 96-49-1. $(—CH_2O)_2CO$.
Properties: Colorless, odorless solid or liquid. Mp 36.4C, bp 248C, d 1.3218 (39/4C), refr index 1.4158 (50C), flash p 290F (143C) (open cup). Miscible (40%) with water, alcohol, ethyl acetate, benzene, and chloroform; soluble in ether, n-butanol, and carbon tetrachloride. Combustible.
Derivation: Interaction of ethylene glycol and phosgene.
Use: Solvent for many polymers and resins, plasticizer, intermediate for pharmaceuticals, rubber chemicals, textile finishing agents, hydroxyethylation reactions.

ethylene chloride. See ethylene dichloride.

ethylene chlorobromide. See sym-bromochloro-ethylene.

ethylene chlorohydrin. (2-chloroethyl alcohol; glycol chlorohydrin).   CAS: 107-07-3. $ClCH_2CH_2OH$.
Properties: Colorless liquid, faint ethereal odor, soluble in most organic liquids and completely miscible with water, d 1.2045 (20/20C), bp 128.7C, refr index 1.4419 (20C), vap press 4.9 mm (20C), flash p 140F (60C) (open cup), bulk

d 10.0 lb/gal (20C), coefficient of expansion 0.00089 (20C), fp −62.6C, viscosity 0.0343 poise (20C), autoign temperature 797F (425C). Combustible.
Derivation: Action of hypochlorous acid on ethylene.
Grade: Anhydrous, 38%.
Hazard: Deadly via ingestion, inhalation, and skin absorption, strong irritant, penetrates ordinary rubber gloves and protective clothing. Moderate fire hazard. TLV: Ceiling of 1 ppm in air.
Use: Solvent for cellulose acetate, ethylcellulose; introduction of hydroxyethyl group in organic synthesis, to activate sprouting of dormant potatoes, manufacture of ethylene oxide and ethylene glycol, insecticides.

ethylene cyanide. (ethylene dicyanide; succinonitrile).   CAS: 110-61-2. $C_2H_4(CN)_2$.
Properties: Colorless, waxy solid; soluble in alcohol, water, and chloroform; mp 57–57.5C; bp 265.7C; flash p 270F (132C). Combustible.
Derivation: Interaction of ethylene dibromide and potassium cyanide in presence of alcohol.
Hazard: TLV (as CN): 5 mg/m³ of air.

ethylene cyanohydrin. (β-hydroxypropionitrile). CAS: 109-78-4. $HOCH_2CH_2CN$.
Properties: Straw-colored liquid, fp −46C, bp 227–228C (decomposes) d 1.0404 (25/4C), vap press 0.08 mm (25C), 20 mm (117C). Miscible with water, acetone, methyl ethyl ketone, ethanol, chloroform, and diethyl ether; insoluble in benzene, carbon tetrachloride, and naphtha. Combustible.
Derivation: Ethylene oxide and hydrogen cyanide.
Hazard: Toxic by ingestion.
Use: Solvent for certain cellulose esters and inorganic salts; organic intermediate for acrylates.

ethylenediamine. (1,2-diaminoethane).
CAS: 107-15-3. $NH_2CH_2CH_2NH_2$.
Properties: Colorless, alkaline liquid; ammonia odor. Strong base, soluble in water and alcohol, slightly soluble in ether, insoluble in benzene, readily absorbs carbon dioxide from air, d 0.8995 (20/20C), bulk d 7.50 lb/gal (20C), bp 116–117C, vap press 10.7 mm (20C), mp 8.5C, viscosity 0.0154 poise (25C), flash p 93F (33.9C) (closed cup), refr index 1.4540 (26C), pH of 25% soluble 11.9 (25C).
Derivation: Catalytic reaction of ethylene glycol or ethylene dichloride and ammonia (nickel or copper catalysts are used).
Method of purification: Redistillation.
Grade: Technical, USP 97%, solutions of various strengths.
Hazard: Toxic by inhalation and skin absorption,

strong irritant to skin and eyes. Flammable, moderate fire risk. TLV: 10 ppm in air.
Use: Fungicide, manufacture of chelating agents (EDTA), dimethylolethylene-urea resins, chemical intermediate, solvent, emulsifying agent, textile lubricants, antifreeze inhibitor.

ethylenediamine carbamate. See "Diak."

ethylenediamine dihydroiodide.
$(-CH_2NH_2)_2 \cdot 2HI$.
Use: Feed additive.

ethylenediamine-di-o-hydroxyphenylacetic acid.
$[-CH_2NHCH(COOH)C_6H_4OH]_2$.   A phenolic analog of ethylenediaminetetraacetic acid.
Properties: Mp 218C (decomposes), insoluble in water and most organic solvents, soluble in sulfuric acid, also forms water-soluble alkali metal and ammonium salts.
Use: Chelating iron in mildly alkaline solutions.

ethylenediamine tartrate.
Use: Make piezoelectric crystals for control of electric frequencies, etc., as in television.

ethylenediaminetetraacetic acid.   (EDTA; ethylenebisiminodiacetic acid; ethylenedinitrilotetraacetic acid).   CAS: 60-00-4.
$(HOOCCH_2)_2NCH_2CH_2N(CH_2COOH)_2$.
An organic chelating agent.

$$\begin{array}{ccc} NaOOCH_2C & & CH_2COONa \\ & N-H_2C-H_2C-N & \\ NaOOCH_2C & & CH_2COONa \end{array}$$

Properties: Colorless crystals, decomposes at 240C, slightly soluble in water, insoluble in common organic solvents, neutralized by alkali metal hydroxides to form a series of water-soluble salts containing from one to four alkali metal cations.
Derivation: (a) Addition of sodium cyanide and formaldehyde to a basic solution of ethylenediamine (forms the tetrasodium salt), (b) heating tetrahydroxyethylethylenediamine with sodium hydroxide or potassium hydroxide with cadmium oxide catalyst.
Use: Detergents, liquid soaps, shampoos, agricultural chemical sprays; metal cleaning and plating, metal chelating agent, treatment of chlorosis, decontamination of radioactive surfaces, metal deactivator in vegetable oils, oil emulsions, pharmaceutical products, etc.; anticoagulant of blood, eluting agent in ion exchange, to remove insoluble deposits of calcium and magnesium soaps; in textiles to improve dyeing, scouring, and deter-

gent operations; antioxidant, clarification of liquids, analytical chemistry, spectrophotometric titration, aid in reducing blood cholesterol, in medicine to treat lead poisoning and calcinosis, food additive (preservative).
Note: A number of salts of EDTA are available with uses identical or similar to the acid. The USP salts are called edetates (calcium disodium, disodium edetates), others are usually abbreviation to EDTA (tetrasodium, trisodium EDTA). Other salts, known chiefly under trademark names, are the sodium ferric, dihydrogen ferrous and a range of disodium salts with magnesium, divalent cobalt, manganese, copper, zinc, and nickel.

ethylenediaminetetraacetonitrile.   (EDTAN).
$[-CH_2NCH_2CN_2]_2$.
Properties: White, crystalline solid. Melting range 126–132C, bulk d 48.4 lb/cu ft, slightly soluble in water, soluble in acetone.
Hazard: Toxic by ingestion and inhalation.
Use: Chelating agent and intermediate.

ethylene dibromide.   (EDB; 1,2-dibromoethane; ethylene bromide).   CAS: 106-93-4.
$BrCH_2CH_2Br$.
Properties: Colorless, nonflammable liquid. Sweetish odor, emulsifiable, miscible with most solvents and thinners, slightly soluble in water, d 2.17–2.18 (20C), bulk d 18.1 lb/gal, bp 131C, vap press 17.4 mm (30C), fp 9C, refr index 1.5337 (25C), flash p none.
Derivation: Action of bromine on ethylene.
Hazard: A carcinogen. Toxic by inhalation, ingestion, and skin absorption, strong irritant to eyes and skin.
Use: Scavenger for lead in gasoline, grain fumigant, general solvent, waterproofing preparations, organic synthesis, fumigant for tree crops.
Note: May poison platinum catalysts.

ethylene dichloride.   (sym-dichloroethane; 1,2-dichloroethane; ethylene chloride; Dutch oil).
CAS: 107-06-2. $ClCH_2CH_2Cl$.   14th highest-volume chemical produced in US (1985).
Properties: Colorless, oily liquid. Chloroform-like odor, sweet taste. Stable to water, alkalies, acids, or active chemicals; resistant to oxidation, will not corrode metals, miscible with most common solvents, slightly soluble in water, bp 83.5C, fp $-35.5C$, d 1.2554 (20/4C), 10.4 lb/gal, refr index 1.444, flash p 56C (13.3C).
Derivation: Action of chlorine on ethylene with subsequent distillation with metallic catalyst, also by reaction of acetylene and hydrochloric acid.
Grade: Technical, spectrophotometric.
Hazard: Toxic by ingestion, inhalation, and skin

absorption; strong irritant to eyes and skin, a carcinogen. Flammable, dangerous fire risk, explosive limits in air 6–16%. TLV: 10 ppm in air.

Use: Production of vinyl chloride, trichloroethylene, vinylidene chloride, and trichloroethane; lead scavenger in antiknock gasoline; paint, varnish, and finish removers; metal degreasing, soaps and scouring compounds, wetting and penetrating agents, organic synthesis, ore flotation, solvent, fumigant.

ethylene dicyanide.   See ethylene cyanide.

ethylenedinitrilotetraacetic acid.   See ethylenediaminetetraacetic acid.

ethylenedinitrilotetra-2-propanol.   See N,N,N′,N′-tetrakis(2-hydroxypropyl) ethylenediamine.

ethylene diphenyldiamine.   See N,N-diphenylethylenediamine.

1,1′-ethylene-2,2′-dipyridinium dibromide. See diquat.

ethylene glycol.   (ethylene alcohol; glycol; 1,2-ethanediol).   CAS: 107-21-1. CH_2OHCH_2OH.   The simplest glycol; 29th highest-volume chemical produced in US (1985).

Properties: Clear, colorless, syrupy liquid. Sweet taste, hygroscopic, lowers fp of water, relatively non-volatile, odorless. Soluble in water, alcohol, and acetone. D 1.1155 (20C), bp 197.2C, fp −13.5C, bulk d 9.31 lb/gal (15/15C), refr index 1.430 (25C), flash p 240.8F (116C), autoign temperature 775F (412C). Combustible.

Derivation: (1) Air oxidation of ethylene followed by hydration of the ethylene oxide formed, (2) acetoxylation, (3) from carbon monoxide and hydrogen (synthesis gas) from coal gasification, (4) Oxirane process.

Grade: Technical.

Hazard: TLV: (vapor) ceiling 50 ppm. Toxic by ingestion and inhalation. Lethal dose reported to be 100 cc.

Use: Coolant and antifreeze, asphalt-emulsion paints, heat-transfer agent, low-pressure laminates, brake fluids, glycol diacetate, polyester fibers and films, low-freezing dynamite, solvent, extractant for various purposes, solvent mixture for cellulose esters and ethers especially cellophane, cosmetics (up to 5%), lacquers, alkyd resins, printing inks, wood stains, adhesives, leather dyeing, textile processing, tobacco, ingredient of deicing fluid for airport runways, humectant, ball-point pen inks, foam stabilizer.

ethylene glycol-bis(β-aminoethyl ether)-N,N-tetraacetic acid.   (ethylene bis(oxyethylenenitrilo) tetraacetic acid).
$[—CH_2OC_2H_4N(CH_2COOH)_2]_4$.
Properties: Crystals, mp 241 (decomposes), soluble in water.
Use: Chelating agent.

ethylene glycol bis(mercaptopropionate).   See glycol dimercaptopropionate.

ethylene glycol bisthioglycolate.   See glycol dimercaptoacetate.

ethylene glycol diacetate.   (glycol diacetate). CAS: 111-55-7.
$CH_3COOCH_2CH_2OOCCH_3$.
Properties: Colorless liquid; faint odor; soluble in alcohol, ether, benzene; slightly soluble in water (10%). D 1.1063 (20/20C), bp 190.5C, vap press 0.3 mm (20C), flash p 205F (96C) (open cup), bulk d 9.2 lb/gal (20C), fp −31C, refr index 1.415 (20C). Combustible.
Derivation: (a) Ethylene glycol and acetic acid, (b) ethylene dichloride and sodium acetate.
Use: Solvent for cellulose esters and ethers, resins, lacquers, printing inks, perfume fixative, nondiscoloring plasticizer for ethyl and benzyl cellulose.

ethylene glycol dibutyl ether.
(ethylene glycol dibutyl).   CAS: 112-48-1.
$C_4H_9OC_2H_4OC_4H_9$.
Properties: Almost colorless liquid, slight odor, slightly soluble in water, d 0.8374 (20/20C), bulk d 7.0 lb/gal (20C), bp 203.1C, vap press 0.09 mm (20C), fp −69.1C, flash p 185F (85C). Combustible.
Use: High-boiling inert solvent, specialized solvent and extraction applications.

ethylene glycol dibutyrate.   (glycol dibutyrate). $(—CH_2OCOC_3H_7)_2$.
Properties: Colorless liquid, d 1.024 (0C), refr index 1.424 (25C), bp 240C, fp less than −80C, soluble in water 0.050% by weight. Combustible.
Use: Plasticizer.

ethylene glycol diethyl ether.   (ethyl glyme).
$C_2H_5OCH_2CH_2OC_2H_5$.
Properties: Colorless liquid, slight odor, stable, d 0.8417 (20/20C), bp 121.4C, vap press 9.4 mm (20C), flash p 95F (35C), bulk d 7 lb/gal (20C), fp −74C, immiscible with water.
Grade: Technical.
Hazard: Flammable, moderate fire risk.
Use: Organic synthesis (reaction medium), solvent and diluente for detergents.

ethylene glycol diformate. (glycol diformate). HCOOCH$_2$CH$_2$OOCH.
Properties: Water-white liquid, soluble in water, alcohol and ether, d 1.2277 (20/20C), bulk d 10.2 lb/gal (20C), bp 177.1C, flash p 200F (93C), vap press 0.5 mm (20C), fp −10C, hydrolyzes slowly liberating formic acid. Combustible.
Hazard: Toxic by ingestion.
Use: Embalming fluids.

ethylene glycol dimethyl ether. (GDME; glycol dimethyl ether; 1,2-dimethoxyethane; monoglyme). CH$_3$OCH$_2$CH$_2$OCH$_3$.
Properties: Water-white liquid with a mild odor, d 0.8683 (20C), bp 85.2C, fp −69C, refr index 1.3792 (20C), flash p 104F (40C), soluble in water and hydrocarbons, pH 8.2.
Hazard: Moderate fire risk.
Use: Solvent.

ethylene glycol dinitrate. CAS: 628-96-6.
A freezing point-depressant for nitroglycerine.
Use: Low-freezing dynamites.
Hazard: Toxic by skin absorption. TLV: 0.05 ppm in air.

ethylene glycol dipropionate. (glycol propionate; glycol dipropionate). (−CH$_2$OCOC$_2$H$_5$)$_2$.
Properties: Liquid, d 1.054 (15C), refr index 1.419 (25C), bp 211C, fp less than −80C, soluble in water 0.16% by weight. Combustible.
Use: Plasticizer.

ethylene glycol monoacetate. (glycol monoacetate). CAS: 542-59-6. HOCH$_2$CH$_2$OOCCH$_3$.
Properties: Colorless liquid; almost odorless; soluble in water, alcohol, ether, benzene, and toluene. Bp 181–182C, d 1.108, flash p 215F (101C). Combustible.
Derivation: (a) Heating ethylene glycol with acetic acid (glacial) or acetic anhydride, (b) passing ethylene oxide into hot acetic acid containing sodium acetate or sulfuric acid.
Use: Solvent for nitrocellulose, cellulose acetate, camphor.

ethylene glycol monobenzyl ether. (benzyl "Cellosolve"). C$_6$H$_5$CH$_2$OC$_2$H$_4$OH.
Properties: Water-white liquid, faint rose-like odor, d 1.070 (20/20C), bp 255.9C, vap press 0.02 mm (20C), flash p 265F (129C), 8.9 lb/gal (20C), autoign temperature 665F (351C). Combustible.
Hazard: Toxic by ingestion.
Use: Solvent for cellulose acetate, dyes, inks, resins, perfume fixative; organic synthesis (selective hydroxyethylating agent); coating compositions for leather, paper and cloth; lacquers.

ethylene glycol monobutyl ether. (2-butoxyethanol; butyl"Cellosolve"). CAS: 111-76-2. HOCH$_2$CH$_2$OC$_4$H$_9$.
Properties: Colorless liquid, mild odor, high dilution ratio with petroleum hydrocarbons, soluble in alcohol and water, bp 171.2C, d 0.9019 (20/20C) 7.51 lb/gal (20C), refr index 1.4190 (25C), vap press 0.76 mm (20C), flash p 142F (61C), autoign temperature 472F (244C). Combustible.
Grade: Technical.
Hazard: A toxic material. TLV: 25 ppm in air.
Use: Solvent for nitrocellulose resins, spray lacquers, quick-drying lacquers, varnishes, enamels, drycleaning compounds, varnish removers, textile (preventing spotting in printing or dyeing), mutual solvent for "soluble." mineral oils to hold soap in solution and to improve the emulsifying properties.

ethylene glycol monobutyl ether acetate. (butyl"Cellosolve" acetate). C$_4$H$_9$OCH$_2$CH$_2$OOCCH$_3$.
Properties: Colorless liquid, fruity odor, soluble in hydrocarbons and organic solvents, insoluble in water, bp 192.3C, d 0.9424 (20/20C), fp −63.5C, flash p 190F (87.7C). Combustible.
Grade: Technical.
Use: High-boiling solvent for nitrocellulose lacquers, epoxy resins, multicolor lacquers; film coalescing aid for polyvinyl acetate latex.

ethylene glycol monobutyl ether laurate. (butoxyethyl laurate). C$_{11}$H$_{23}$COO(CH$_2$)$_2$OC$_4$H$_9$.
Properties: Liquid, d 0.985 (25C), fp −10 to −15C, insoluble in water. Combustible.
Use: Plasticizer.

ethylene glycol monobutyl ether oleate. (butoxyethyl oleate). C$_{17}$COOCH$_2$CH$_2$OC$_4$H$_9$.
Properties: Liquid, d 0.892 (25C), fp less than −45C, insoluble in water. Combustible.
Use: Plasticizer.

ethylene glycol monobutyl ether stearate. (butoxyethyl stearate). C$_{17}$H$_{35}$COOC$_2$H$_4$OC$_4$H$_9$.
Properties: Colorless liquid, d 0.882 (20C) vap press less than 0.01 mm (20C), bp 210–233C (4 mm), mp 16.5C, insoluble in water. Combustible.
Use: Plasticizer and solvent.

ethylene glycol monoethyl ether. (2-ethoxyethanol; "Cellosolve" solvent). CAS: 110-80-5. HOCH$_2$CH$_2$OC$_2$H$_5$.
Properties: Colorless liquid, practically odorless, bp 135.6C, d 0.9311 (20/20C), 7.74 lb/gal (20C),

refr index 1.4060 (25C), flash p 120F (48.9C), pour point less than 100F (37.7C), autoign temperature 460F (237C), miscible with hydrocarbons and water. Combustible.
Grade: Technical.
Hazard: TLV: 5 ppm, toxic by skin absorption. Moderate fire risk.
Use: Solvent for nitrocellulose, natural and synthetic resins; mutual solvent for formulation of soluble oils; lacquers and lacquer thinners, dyeing and printing textiles, varnish removers, cleaning solutions, leather, anti-icing additive for aviation fuels.

ethylene glycol monoethyl ether acetate. ("Cellosolve" acetate; 2-ethoxyethyl acetate). CAS: 111-15-9. $CH_3COOCH_2CH_2OC_2H_5$.
Properties: Colorless liquid, pleasant ester-like odor, bp 156.3C, d 0.9748 (20/20C), 8.1 lb/gal (20C), refr index 1.4030 (25C), viscosity 1.32 cp (20C), flash p 120F (48.9C), fp −61.7C, vap press 2 mm (20C), miscible with aromatic hydrocarbons, slightly miscible with water. Combustible.
Grade: Technical.
Hazard: Toxic by ingestion and skin absorption. TLV: 5 ppm in air.
Use: Solvent for nitrocellulose, oils, and resins; retards "blushing" in lacquers, varnish removers, wood stains, textiles, leather.

ethylene glycol monoethyl ether laurate. $C_{11}H_{23}COO(CH_2)_2OC_2H_5$.
Properties: Liquid, d 0.89 (25C), fp −7 to −11C, insoluble in water. Combustible.
Use: Plasticizer.

ethylene glycol monoethyl ether ricinoleate. $C_{17}H_{32}(OH)COO(CH_2)_2OC_2H_5$.
Properties: Liquid, d 0.929 (25C), fp less than −10C, insoluble in water. Combustible.
Use: Plasticizer.

ethylene glycol monohexyl ether. (n-hexyl "Cellosolve"). $C_6H_{13}OCH_2CH_2OH$.
Properties: Water-white liquid, d 0.8887 (20/20C), bulk d 7.4 lb/gal (20C), bp 208.1C, vap press 0.05 mm (20C), fp −50.1C, flash p 195F (90.5C). Combustible.
Use: High-boiling solvent.

ethylene glycol monomethyl ether. (2-methoxyethanol; methyl "Cellosolve").
CAS: 109-86-4. $CH_3OCH_2CH_2OH$.
Properties: Colorless liquid, mild agreeable odor, stable. Miscible with hydrocarbons, alcohols, ketones, glycols and water. Bp 124.5C, d 0.9663 (20/20C), bulk d 8.0 lb/gal (20C), refr index 1.4021 (20C), flash p 110F (43.3C), fp −85.1C, autoign temperature 551F (288C). Combustible.

Derivation: From ethylene oxide.
Grade: Technical.
Hazard: Toxic by ingestion and inhalation. Moderate fire risk. TLV: 5 ppm in air.
Use: Solvent for nitrocellulose, cellulose acetate, alcohol-soluble dyes, natural and synthetic resins, solvent mixtures, lacquers, enamels, varnishes, leather, perfume fixative, wood stains, sealing moisture-proof cellophane, jet fuel deicing additive.

ethylene glycol monomethyl ether acetate. (methyl "Cellosolve" acetate).   CAS: 110-49-6. $CH_3COOCH_2CH_2OCH_3$.
Properties: Colorless liquid, pleasant odor, stable, miscible with common organic solvents, soluble in water, d 1.0067 (20/20C), bp 145C, vap press 2 mm (20C), flash p 120F (48.9C), bulk d 8.4 lb/gal (20C), fp −65.1C. Combustible.
Hazard: Toxic by ingestion, inhalation, and skin absorption. Moderate fire risk. TLV: 5 ppm in air.
Use: Solvent for nitrocellulose, cellulose acetate, various gums, resins, waxes, oils; textile printing, photographic film, lacquers, dopes.

ethylene glycol monomethyl ether acetyl ricinoleate. $C_{17}H_{32}(OCOCH_3)COOCH_2CH_2OCH_3$.
Properties: Liquid, d 0.966, refr index 1.460, boiling range 220–260C, fp less than −60C, flash p 425F (218C), insoluble in water. Combustible.
Use: Plasticizer.

ethylene glycol monomethyl ether ricinoleate. $C_{17}H_{32}(OH)COOCH_2CH_2OCH_3$.
Properties: Liquid, d 0.935 (25C), fp less than −60C, insoluble in water. Combustible.
Use: Plasticizer.

ethylene glycol monomethyl ether stearate. $C_{17}H_{35}COOCH_2CH_2OCH_3$.
Properties: Liquid, d 0.890, mp 21C, insoluble in water. Combustible.
Use: Plasticizer.

ethylene glycol monooctyl ether. $C_4H_9CHC_2H_5CH_{23}OCH_2CH_2OH$.
Properties: Colorless, odorless liquid. Bp 228.3C, d 0.8859, flash p 230F (110C), vap press 0.02 mm (20C). Combustible.
Use: Solvent for cellulose esters, plasticizers.

ethylene glycol monophenyl ether. (2-phenoxyethanol; phenyl "Cellosolve").
CAS: 122-99-6. $C_6H_5OCH_2CH_2OH$.
Properties: Colorless liquid, faint aromatic odor, stable in presence of acids and alkalies, partially soluble in water, d 1.1094 (20/20C), bp 244.9C,

fp 14C, vap press less than 0.01 mm (20C), flash p 250F (121C), phenol 0.3% (max), bulk d 9.2 lb/gal. Combustible.

Grade: Technical.

Use: Solvent for cellulose acetate, dyes; inks, resins; perfume fixative, bactericidal agent, organic synthesis of plasticizers, germicides, pharmaceuticals; insect repellent.

ethylene glycol monoricinoleate.
$C_{17}H_{32}(OH)COO(CH_2)_2OH$.

Properties: Clear, moderately viscous, pale yellow liquid. Mild odor, miscible with most organic solvents, d 0.965 (25/25C), saponification value 170, hydroxyl value 270, solidifies at −20C, insoluble in water. Combustible.

Derivation: Castor oil and ethylene glycol.

Grade: Technical.

Use: Plasticizer, greases, urethane polymers.

ethylene glycol monostearate. (glycol stearate).
CAS: 111-60-4. $C_{17}H_{35}COO(CH_2)_2OH$.

Properties: Yellow, waxy solid; mp 57–60C; d 0.96 (25C); soluble in alcohol, hot ether, and acetone; insoluble in water. Combustible.

ethylene glycol silicate. $(HOCH_2CH_2)_4SiO_4$.

Properties: Colorless liquid, slowly hydrolyzed by acids, miscible with water.

Use: Nonvolatile bonding agent for pigments, weather-proofing paints for protecting concrete, stone, brick, and plastic surfaces.

ethylene hexene-1-copolymer. Product developed especially for use in blow-molded items. Other possible uses are coatings for pipe, wire, and cables; sheeting; and monofilament.

ethylene hydrate. See gas hydrate.

ethyleneimine. (aziridine; ethylenimine).
CAS: 151-56-4.

$$CH_2{-}CH_2$$
$$\diagdown \diagup$$
$$NH$$

Properties: Clear, colorless liquid with amine odor. Bp 57C, d 0.832 (20/4C), refr index 1.4123 (25C), fp −78C, flash p 12F (−11.1C), miscible with water and most organic solvents, autoign temperature 612F (322C).

Derivation: From ethylene dichloride and ammonia by use of an acid acceptor.

Hazard: Corrosive, absorbed by skin, causes tumors, exposure should be minimized, a carcinogen. Dangerous fire and explosion hazard, flammable limits in air 3.6–46%. TLV: 0.5 ppm in air.

Use: Intermediate and monomer for fuel oil and lubricant refining, ion exchange, protective coatings, pharmaceuticals, adhesives, polymer stabilizers, surfactants. Alkyl substituted forms, called alkyl aziranes, are used as intermediates and for microbial control, aziridinyl compounds are also used as polymers and intermediates.

ethylene-maleic anhydride copolymer.

Properties: Fine, water-soluble powder available as both straight-chain and crosslinked polymers in a variety of molecular weights. May be in form of free acid, sodium, or amide-ammonium salts. Reacts readily with alcohols and amines.

Use: Oil-well drilling muds, stabilizers and thickeners in liquid detergents, cosmetics, paints; textile sizes, printing inks, suspending agents, ceramic binders.

ethylenenaphthalene.   See acenaphthene.

ethylene oxide.   (epoxyethane; oxirane).
CAS: 75-21-8.   23rd highest-volume chemical produced in US (1985).

$$CH_2{-}CH_2$$
$$\diagdown \diagup$$
$$O$$

Properties: Colorless gas at room temperature, liquid at approximately 12C, soluble in organic solvents, miscible with water, fp −111.3C, bp 10.73C, d 0.8711 (20/20C), bulk d 7.25 lb/gal (20C), viscosity 0.32 cp (0C), flash p approximately 0F (−17.7C) (TOC), autoign temperature 805F (429C).

Derivation: (a) oxidation of ethylene in air or oxygen with silver catalyst, (b) action of an alkali on ethylene chlorohydrin.

Grade: Technical, pure (99.7%).

Hazard: Irritant to eyes and skin. TLV: 1 ppm in air, a suspected human carcinogen. Highly flammable, dangerous fire and explosion risk, flammable limits in air 3–100%.

Use: Manufacture of ethylene glycol and higher glycols, surfactants, acrylonitrile, ethanolamines, petroleum demulsifier, fumigant, rocket propellant, industrial sterilant, e.g., medical plastic tubing, fungicide.

ethylene-propylene-diene monomer.   See ethylene-propylene terpolymer.

ethylene-propylene rubber. (EPR).   An elastomer made by the stereospecific copolymerization of ethylene and propylene. Has no unsaturation, cannot be vulcanized with sulfur but can be cured with peroxides.

ethylene-propylene terpolymer. (EPT; EPDM). An elastomer based on stereospecific linear terpolymers of ethylene, propylene, and small amounts of a nonconjugated diene, e.g., a cyclic or aliphatic diene (hexadiene, dicyclopentadiene, or ethylidene norbornene). The unsaturated part of the polymer molecule is pendant from the main chain which is completely saturated. Can be vulcanized with sulfur.

Properties of vulcanizate: Light cream to white, excellent resistance to ozone, to high and low temperatures (from -51 to $+148C$) and to acids and alkalies, good electrical resistance, susceptible to attack by oils, pelletized forms available.

Use: Automotive parts, gaskets, cable coating, mechanical rubber products, cover strips for tire sidewalls, tire tubes, safety bumpers, coated fabrics, footwear, wire and cable coating, thermoplastic resin modifier.

See also "Nordel."

ethylene thiourea. CAS: 96-45-7.

$\overline{NHCH_2CH_2NHCS}$ (2-imidazolidinethione).
Properties: White to pale green crystals, faint amine odor, mp 199–204C. Slightly soluble in cold water; very soluble in hot water; slightly soluble at room temperature in methanol, ethanol, acetic acid, naphtha.

Use: Electroplating baths, intermediate for antioxidants, insecticides, fungicides, vulcanization accelerators, dyes, pharmaceuticals, synthetic resins.

ethylene urea. (2-imidazolidinone; dihydro-2(3)-imidazolone; 2-imidazolidone).

CAS: 120-93-4. $\overline{CH_2CH_2NHCONH}$.
Properties: White, lumpy powder; odorless; mp 125–128C; soluble in water.
Derivation: From ethylenediamine and urea.
Use: Drip-dry textile products, ingredient of plasticizers and adhesives, insecticide.

ethylene-vinyl acetate copolymer. (EVA).
An elastomer used to improve adhesion properties of hot-melt and pressure-sensitive adhesives, also for conversion coatings and thermoplastics.
See also "Ultrathene."

ethylenimine. See ethyleneimine.

N-ethylethanolamine. (ethylaminoethanol).
$C_2H_5NHCH_2CH_2OH$.
Properties: Colorless liquid, d 0.914 (20C), boiling range 167–169C, odor amine, soluble in water, alcohol and ether, flash p 160F (71.1C) (OC). Combustible.
Use: Solvent, intermediate.

ethyl ether. (ether; diethyl ether; sulfuric ether; ethyl oxide; diethyl oxide). CAS: 60-29-7. $(C_2H_5)_2O$.
Properties: Colorless, volatile, mobile liquid. Hygroscopic, aromatic odor, burning and sweet taste, bp 34.5C, fp $-116.2C$, d 0.7147 (20/20C), surface tension 17.0 dynes/cm (20C), refr index 1.3526 (20C), viscosity 0.00233 poise (20C), vap press 442 mm (20C), specific heat 0.5476 cal/g (30C), flash p $-49F$ ($-45C$), autoign temperature 356F (180C), latent heat of evaporation 83.96 cal/g at bp, electric conductivity 4×10^{-3} mhos (25C), bulk d 6 lb/gal (20C). Soluble in alcohol, chloroform, benzene, solvent naphtha, and oils; slightly soluble in water.
Derivation: By the action of sulfuric acid on ethanol or ethylene followed by distillation.
Method of purification: Rectification, dehydration, treatment with alkali and charcoal.
Grade: USP (for anesthesia), ACS Reagent, ACS Absolute, CP, concentrated, USP 1880, washed, motor, electronic.
Hazard: Central nervous system depressant by inhalation and skin absorption. Very flammable, severe fire and explosion hazard when exposed to heat or flame. TLV: 400 ppm in air. Forms explosive peroxides. Explosive limits in air 1.85–48%.
Use: Organic synthesis, smokeless powder, industrial solvent (nitrocellulose, alkaloids, fats, waxes, etc.); analytical chemistry, anesthetic, extractant.

ethyl-3-ethoxypropionate.
$C_2H_5OOCCH_2CH_2OC_2H_5$.
Properties: Liquid, d 0.9496 (20C), bp 170.1C, vap press 0.9 mm (20C), sets to glass at $-100C$, slightly soluble in water, flash p 180F (82.2C) (open cup). Combustible.
Use: Intermediate for vitamin B_1, other chemicals.

ethylethylene. See butene-1.

ethylethyleneimine. (2-ethylaziridine).

$C_2H_5HCNHCH_2$.
Properties: Water-white liquid, bp 87–90C, d 0.812–0.816 (25/25C).
Use: Organic intermediate whose derivatives are used in the textile, paper, rubber and pharmaceutical industries.

ethyl ferrocenoate. (ferrocenecarboxylic acid ethyl ester). $(C_5H_4COOC_2H_5)_2Fe$.
Properties: Orange, crystalline solid; mp 63–64C.
Use: Intermediate.

ethylfluoroformate. $FCOOC_2H_5$.
Properties: Liquid, d 1.11 (33C), bp 57C.

Derivation: Interaction of ethylchloroformate and a reactive fluoride.

Hazard: Strong irritant.

ethylfluorosulfonate. $C_2H_5OFSO_2$.
Properties: Liquid, ethereal odor.
Hazard: Strong irritant.

ethyl formate. CAS: 109-94-4. $HCOOC_2H_5$.
Properties: Water-white, unstable liquid; pleasant aromatic odor; miscible with benzene, ether, alcohol; slightly soluble in water; gradual decomposition in water. Flash p −4F (−20C) (closed cup), explosive limits in air 2.8–13.5%, autoign temperature 851F (455C), d 0.9236 (20/20C), fp −80.5C, bp 54.3C, vap press 200 mm (20.6C), 300 mm (30.2C) 7.61 lb/gal (68F), refr index 1.35975 (20C).
Derivation: Heating ethanol with formic acid in presence of sulfuric acid.
Grade: Technical, FCC.
Hazard: Narcotic and irritant to skin and eyes, use in foods restricted to 0.0015%. Highly flammable, dangerous fire and explosion risk. TLV: 100 ppm in air.
Use: Solvent for cellulose nitrate and acetate, acetone substitute, fumigant, larvicide, synthetic flavors, organic synthesis.

ethyl-o-formate. $CH(OC_2H_5)_3$.
Properties: Colorless liquid, bp 145.9C, fp less than −18C, refr index 1.392 (20C), slightly soluble in water, soluble in alcohol and ethers, flash p 85F (29.4C).
Hazard: Flammable, moderate fire risk.
Use: Intermediate.

ethyl-3-formylpropionate.
$C_2H_5OOCC_2H_4C(O)H$.
Properties: Liquid, d 1.0625 (20/20C), bp 190.9C, fp less than −80C, bulk d 8.9 lb/gal, flash p 200F (93.3C), somewhat soluble in water. Combustible.
Hazard: Toxic by ingestion and inhalation.
Use: Solvent for lacquers, antibiotic extraction, acetic acid separation, coalescing acids for emulsion paints.

2-ethylfuran. CAS: 3208-16-0.
Properties: Liquid with mw 96.13, bp 92–93C at 768 mm, d 0.912.
Grades: Food.
Hazard: Flammable liquid.

ethyl furoate. $C_4H_3OCO_2C_2H_5$.
Properties: White leaflets or prisms, insoluble in water, soluble in alcohol and ether, d 1.1174 (20.8/4C), mp 34C, bp 195C (706 mm).

ethyl glyme. See ethylene glycol diethyl ether.

4-ethylheptane. (ethyldipropylmethane).
$CH_3(CH_2)_2CHC_2H_5(CH_2)_2CH_3$.
Properties: Colorless liquid, d 0.730, bp 141.2C, refr index 1.4109 (20C). Combustible.
Grade: Technical.
Use: Organic synthesis.

ethyl heptanoate. See ethyl enanthate.

ethyl heptazine. See ethoheptazine.

2-ethylhexaldehyde. (butylethylacetaldehyde; octyl aldehyde; 2-ethylhexanal).
$C_4H_9CH(C_2H_5)CHO$.
Properties: Colorless high-boiling liquid, mild odor, miscible with most organic solvents, slightly soluble in water, d 0.8205 (20C), bp 163.4C, vap press 1.8 mm (20C), flash p 125F (51.6C) (open cup), 6.8 lb/gal. Combustible.
Grade: Technical.
Hazard: Ignites in air.
Use: Organic synthesis, perfumes.

2,2′-(2-ethylhexamido)diethyl di(2-ethylhexoate). ("Flexol" 8N8).
$(C_7H_{15}OCOC_2H_4)_2NCOC_7H_{15}$.
Properties: Light-colored liquid, d 0.9564 (20/20C), bulk d 8.0 lb/gal (20C), bp 256C (5 mm), flash p 420F (215C), vap press 0.60 mm (200C), insoluble in water, viscosity 139.2 cp (20C). Combustible.
Use: Plasticizer.

2-ethylhexanal. See 2-ethylhexaldehyde.

2-ethylhexanediol-1,3. (ethohexadiol; 2-ethyl-3-propyl-1,3-propanediol). CAS: 94-96-2.
$C_3H_7CH(OH)CH(C_2H_5)CH_2OH$.

$$CH_3CH_2CH_2\overset{\displaystyle OH}{\underset{\displaystyle CH_2CH_3}{\overset{|}{\underset{|}{C}}HCHCH_2OH}}$$

Properties: Colorless, slightly viscous, odorless liquid. Hygroscopic, d 0.9422 (20/20C), bulk d 7.8 lb/gal (20C), bp 244C, vap press less than 0.01 mm (20C), flash p 260F (126C), fp approximately −40C, refr index 1.4465–1.4515, viscosity 323 cp (20C), soluble in alcohol and ether, partially soluble in water. Combustible.
Grade: USP (as ethohexadiol), industrial.
Use: Insect repellent, cosmetics, vehicle and solvent in printing inks, medicine, chelating agent for boric acid.

ethyl hexanoate. See ethyl caproate.

2-ethylhexanol. See 2-ethylhexyl alcohol.

2-ethylhexenal. See 2-ethyl-3-propylacrolein.

2-ethyl-1-hexene. $CH_3(CH_2)_3(C_2H_5)C:CH_2$.
Properties: Colorless liquid, d 0.7270 (20/4C), bp
120C, refr index 1.4157 (20C). Soluble in alcohol,
acetone, ether, petroleum, coal-tar solvents; insoluble in water. Combustible.
Grade: 95% min purity.
Hazard: Toxic by ingestion and inhalation.
Use: Organic synthesis of flavors, perfumes, medicines, dyes, resins.

ethyl hexoate. See ethyl caproate.

2-ethylhexoic acid. (butylethylacetic acid).
CAS: 78-42-2. $C_4H_9CH(C_2H_5)COOH$.
Properties: Mild-odored liquid, slightly soluble in
water, d 0.9077 (20/20C), bulk d 7.6 lb/gal
(20C), bp 226.9C, vap press 0.03 mm (20C), fp
−83C, viscosity 7.73 cp (20C), acid number 370,
flash p 260F (126C). Combustible.
Grade: 99%.
Use: Paint and varnish driers (metallic salts).
Ethylhexoates of light metals are used to convert
some mineral oils to greases. Its esters are used
as plasticizers.

2-ethylhexyl. An 8-carbon radical of the formula,
$CH_3(CH_2)_3CH(C_2H_5)CH_2—$. Many of its compounds were formerly called octyl.

2-ethylhexyl acetate. CAS: 103-09-0.
$CH_3COOCH_2CHC_2H_5C_4H_9$.
Properties: Water-white stable liquid, very slightly
soluble in water, miscible with alcohol, d 0.8733
(20C), bp 198.6C, fp −93C, vap press 0.4 mm
(20C), flash p 180F (82.2C), bulk d 7.3 lb/gal
(20C). Combustible.
Grade: Technical (approximately 95%).
Use: Solvent for nitrocellulose, resins, lacquers,
baking finishes.

2-ethylhexyl acrylate. CAS: 103-11-7.
$CH_2:CHCOOCH_2CH(C_2H_5)C_4H_9$
Properties: Liquid, pleasant odor, d 0.8869, bp
214–218C, vap press 0.1 mm (20C), sets to glass
at −90C, flash p 180F (82.2C) (open cup), insoluble in water. Combustible.
Use: Monomer for plastics, protective coatings,
paper treatment, water-based paints.

2-ethylhexyl alcohol. (2-ethylhexanol; octyl
alcohol).   CAS: 104-76-7.
$CH_3(CH_2)_3CHC_2H_5CH_2OH$.
Properties: Colorless liquid, miscible with most
organic solvents, slightly soluble in water, d 0.83

(20C), bp 183.5C, fp −76C, vap press 0.36 (20C),
refr index 1.4300 (20C), bulk d 6.9 lb/gal (20C),
flash p 178F (81.1C). Combustible.
Derivation: (a) Oxo process from propylene and
synthesis gas, (b) aldolization of acetaldehyde
or butyraldehyde followed by hydrogenation, (c)
from fermentation alcohol.
Grade: Technical.
Use: Plasticizer for PVC resins, defoaming agent,
wetting agent, organic synthesis, solvent mix for
nitrocellulose, paints, lacquers, baking finishes;
penetrant for mercerizing cotton, textile finishing
compounds, plasticizers, inks, rubber, paper, lubricants, photography, dry cleaning.

2-ethylhexylamine. CAS: 104-75-6.
$C_4H_9CH(C_2H_5)CH_2NH_2$.
Properties: Colorless liquid, d 0.7894 (20/20C),
bulk d 6.56 lb/gal (20C), bp 169.2C, vap press
1.2 (20C), viscosity 1.11 cp (20C), flash p 140F
(60C) (open cup), soluble in water, soluble of
water in (20C) 25.3%. Combustible.
Hazard: Moderate fire risk. Toxic by ingestion
and inhalation.
Use: Synthesis of detergents, rubber chemicals,
oil additives and insecticides.

n-2-ethylhexylaniline.
$C_6H_5NHCH_2CH(C_2H_5)C_4H_9$.
Properties: Light yellow liquid with mild odor,
d 0.9119 (20/20C), bp 194C (50 mm), vap press
less than 0.01 mm (20C), fp sets to a glass approximately −70C, viscosity 7.4 cp (20C), soluble
in water less than 0.01% (20C), flash p 325F
(162C) (COC). Combustible.
Hazard: Chronic exposure toxic.
Use: Solvent, organic synthesis.

2-ethylhexyl bromide. $C_4H_9CH(C_2H_5)CH_2Br$.
Properties: Water-white liquid, bp 56–58C
(6 mm), insoluble in water.
Use: Introduction of the 2-ethylhexyl group in
organic synthesis. Preparation of disinfectants,
pharmaceuticals.

2-ethylhexyl chloride. CAS: 123-04-6.
$C_4H_9CH(C_2H_5)CH_2Cl$.
Properties: Colorless liquid, d 0.8833 (20C), bp
172.9C, refr index 1.4310, bulk d 7.33 lb/gal,
flash p 140F (60C) (open cup), fp −135C, insoluble in water. Combustible.
Hazard: Moderate fire risk.
Use: Synthesis of cellulose derivatives, dyestuffs,
pharmaceuticals, textile auxiliaries, insecticides,
resins.

2-ethylhexyl cyanoacetate.
$CNCH_2COOCH_2CH(C_2H_5)C_4H_9$.
Properties: Liquid, bp 150C, refr index 1.4389

(20C). Insoluble in water; soluble in alcohol, benzene, and ether. Combustible.
Use: Organic intermediate.

2-ethylhexyl isodecyl phthalate. $C_{26}H_{43}O_4$
Properties: Colorless, high-boiling liquid. Good dielectric properties, refr index 1.478–1.488 (25C), d 0.969–0.977 (25/25C), miscible with most common solvents, thinners and oils. Combustible.
Use: Plasticizer.

2-ethylhexylmagnesium chloride.
$C_4H_9CH(C_2H_5)CH_2MgCl$.   Grignard reagent available commercially in tetrahydrofuran solution.

2-ethylhexyl octylphenyl phosphite.
$(C_8H_{17}O)_2(C_8H_{17}C_6H_{17}C_6H_4O)P$.
Properties: Colorless to light-yellow liquid with characteristic odor, d 0.935–0.950 (20/4C), flash p 385F (196C), insoluble in water. Combustible.
Use: Antioxidant, plasticizer, flame retardant, lubricating oil additive.

3,3'-(2-ethylhexyl) thiodipropionate.   (dioctyl thiopropionate). $(C_8H_{17}OOCCH_2CH_2)_2S$.
Properties: Colorless liquid, d 0.952 (25C), insoluble in water, soluble in most organic solvents. Combustible.
Use: Antioxidant, stabilizer and lubricant.

ethyl-2-hydroxy-2,2-bis(4-chlorophenyl)acetate.
See chlorobenzilate.

ethyl-3-hydroxybutyrate.   CAS: 5405-41-4.
$CH_3CH(OH)CH_2CO_2C_2H_5$.
Properties: A solid with mw 132.16, bp 170C, d 1.017, freezing p 64C.
Grade: Food.

ethyl hydroxyethyl cellulose.   (EHEC).   A cellulose ether.
Properties: White, granular solid; available in extra low and high-viscosity types. Soluble in mixture of aliphatic hydrocarbons containing alcohol; soluble in water. Combustible.
Use: Stabilizer, thickener, binder, film former in silk screen and gravure printing inks, protective coatings, aqueous, aqueous-organic, and organic solvent systems.

ethyl-α-hydroxyisobutyrate.
$(CH_3)_2COHCOOC_2H_5$.
Properties: Water-white liquid, d 0.978–0.986 (20C), bp 149–150C. Soluble in water, alcohol, and ether. Combustible.
Use: Solvent for nitrocellulose and cellulose ace-

tate, solvent mixture for cellulose ethers, organic synthesis, pharmaceuticals.

ethylidene chloride.   (1,1-dichloroethane).
CAS: 75-34-3. CH_3CHCl_2.
Properties: Colorless, neutral, mobile liquid. An aromatic ethereal odor, saccharin taste. Soluble in alcohol, ether, fixed and volatile oils; very sparingly soluble in water. D 1.174 (17C), bp 57–59C, fp −98C, refr index 1.4166 (20C). Combustible.
Hazard: Toxic. TLV: 200 ppm in air.
Use: Extraction solvent, fumigant.

ethylidenediethyl ether.   See acetal.

ethylidenedimethyl ether.   See dimethylacetal.

ethylidene fluoride.   See 1,1-difluoroethane.

5-ethylidene-2-norbornene.   (ENB).   A diene used as the third monomer in EPDM elastomers.
Hazard: Toxic.

ethyl iodide.   (iodoethane).   CAS: 75-03-6.
C_2H_5I.
Properties: Colorless liquid, turns brown on exposure to light, soluble in alcohol and ether, slightly soluble in water, d 1.90–1.93 (25/25C), fp −108C, bp 72C, refr index 1.5168 (15C). Combustible.
Derivation: By digesting red phosphorus with absolute ethanol after which iodine is added and the mix distilled.
Hazard: Toxic by inhalation and skin absorption, narcotic in high concentration.
Use: Medicine, organic synthesis.

ethyl iodoacetate. $CH_2ICOOC_2H_5$.
Properties: Dense, colorless liquid. Decomposed by light and air, also (very slowly) by alkaline solutions and water, d 1.8, bp 179C, vapor d 7.4, vap press 0.54 mm (20C).
Derivation: Interaction of potassium iodide with ethyl bromo- or chloroacetate.
Hazard: Strong irritant to eyes and skin.

5-ethyl-5-isoamylbarbituric acid.   See amobarbital.

ethylisobutylmethane.   See 2-methylhexane.

ethyl isobutyrate.   CAS: 97-62-1.
$(CH_3)_2CHCOOC_2H_5$.
Properties: Colorless, volatile liquid. Soluble in alcohol and ether, slightly soluble in water, d 0.870, bp 110–111C, fp −88C, refr index 1.3903 (20C). Combustible.
Derivation: Heating isobutyric acid and ethyl alcohol with subsequent distillation.
Use: Organic synthesis, flavoring extracts.

ethyl isocyanate. CAS: 109-90-0. C_2H_5NCO.
Properties: Liquid, d 0.898, bp 60C, soluble in chlorinated and aromatic hydrocarbons.
Hazard: Strong irritant to tissue.
Use: Pharmaceutical and pesticide intermediate.

2-ethylisohexanol.
$(CH_3)_2CHCH_2CH(C_2H_5)CH_2OH$.
Properties: Liquid, d 0.825–8.035 (20/20C), boiling range 173.0–181.0C, refr index 1.4235 (20C), bulk d 6.89 lb/gal (20C), flash p 170F (76.6C) (COC). Combustible.
Use: Chemical intermediate.

ethyl isothiocyanate. See ethyl thiocarbimide.

ethyl isovalerate. (ethyl valerate; ethyl 2-methylbutyrate). CAS: 108-64-5. $(CH_3)_2CHCH_2COOC_2H_5$.
Properties: Colorless, oily liquid with fruity odor. Bp 135C, fp −99C, d 0.864 (20/20C), refr index 1.3950–1.3990 (20C). Slightly soluble in water; miscible with alcohol, ether, and benzene. Combustible.
Derivation: Heating sodium valerate and ethanol in presence of sulfuric acid or hydrochloric acid with subsequent distillation.
Grade: Technical, FCC.
Use: Essential oils, perfumery, artificial fruit essences, flavoring.

ethyl lactate. CAS: 97-64-3.
$CH_3CHOHCOOC_2H_5$.
Properties: Colorless liquid, mild odor. Miscible with water, alcohols, ketones, esters, hydrocarbons, oil. D 1.020–1.036 (20/20C), bp 154C, flash p 115F (46.1C) (closed cup), bulk d 8.55 lb/gal (20C), autoign temperature 752F (400C). Combustible.
Derivation: (a) By the esterification of lactic acid with ethanol, (b) by combining acetaldehyde with hydrogen cyanide to form acetaldehyde cyanohydrin which is converted into ethyl lactate by treatment with ethanol and an inorganic acid. Combustible.
Grade: Technical (96%).
Hazard: Moderate fire risk.
Use: Solvent for nitrocellulose, cellulose acetate, many cellulose ethers, resins; lacquers, paints, enamels, varnishes, stencil sheets, safety glass, flavoring.

ethyl levulinate. $CH_3CO(CH_2)_2COOC_2H_5$.
Properties: Colorless liquid, d 1.012, bp 205–206C, soluble in water, miscible with alcohol, refr index 1.4229 (20C). Combustible.
Use: Solvent for cellulose acetate and starch ethers, flavoring.

ethyllithium. (lithium ethyl).
Properties: Transparent crystals, decomposed by water; the commercial product is a 2 M suspension of C_2H_5Li in benzene.
Hazard: Ignites in air, reacts with oxidizing materials.
Use: Grignard-type reactions.

ethylmagnesium bromide. C_2H_5MgBr.
Dissolved in ether.
Properties: Liquid, d 1.01.
Hazard: Flammable, dangerous fire risk.
Use: Grignard-type reactions.

ethylmagnesium chloride. C_2H_5MgCl.
Dissolved in ether (also offered commercially in tetrahydrofuran).
Properties: Liquid, d 0.85.
Hazard: Flammable, dangerous fire risk.
Use: Grignard-type reactions.

N-ethylmaleimide. (1-ethyl-1-H-pyrrole-2,5-dione). CAS: 128-53-0. $C_6H_7NO_2$.
Properties: Crystals with mw 125.13, mp 45C.
Hazard: Lacrymator when liquid, a strong irritant.
Use: Cancer research.

ethyl malonate. (malonic ester; diethyl malonate). CAS: 105-53-3. $CH_2(COOC_2H_5)_2$.
Properties: Colorless liquid, sweet ester odor. Insoluble in water; soluble in alcohol, ether, chloroform, and benzene. Bp 198C, fp −50C, d 1.055 (25/25C), flash p 200F (93.3C). Combustible.
Derivation: By passing hydrogen chloride into cyanoacetic acid dissolved in absolute alcohol with subsequent distillation.
Use: Intermediate for barbiturates and certain pigments, flavoring.

ethyl mercaptan. Legal label name for ethanethiol.

ethylmercuric acetate. CAS: 109-62-6.
$C_2H_5HgOOCCH_3$.
Properties: White, crystalline powder. Mp 178C, slightly soluble in water, soluble in many organic solvents, may be steam-distilled.
Hazard: Strong irritant; see mercury.
Use: Seed fungicide as dust or slurry with water.

ethylmercuric chloride. CAS: 107-27-7.
C_2H_5HgCl.
Properties: Crystals, d 3.482, mp 193C, insoluble in water, slightly soluble in ether, soluble in hot alcohol, sublimes readily.
Derivation: Reaction of zinc diethyl and mercuric chloride.
Hazard: Strong irritant. See mercury.
Use: Fungicide for seed or bulb treatment either along or with other organic mercury compounds.

ethylmercuric phosphate. CAS: 2440-45-1. $(C_2H_5HgO)_3PO$.
Properties: White powder, soluble in water, garlic-like odor.
Derivation: Reaction of ethylmercuric acetate with phosphoric acid.
Hazard: Strong irritant. See mercury.
Use: Seed fungicide, timber preservative.

ethylmercurithiosalicylic acid, sodium salt. See thimerosal.

n-ethylmercuri-p-toluenesulfonanilide.
CAS: 517-16-8. $C_{15}H_{17}SO_2NHg$.
Properties: Crystalline solid with strong, garlic-like odor; insoluble in water.
Hazard: Ingestion may prove fatal.
Use: Agricultural chemical (grain smut control).

ethylmercury-2,3-dihydroxypropyl mercaptide.
$C_2H_5HgSCH_2CHOHCH_2OH$. Organic mercurial.
Hazard: Very toxic. See mercury.
Use: Fungicidal dust or in slurry treatment for control of seedborne disease and to reduce losses from seed decay and damping-off of wheat, oats, rye, etc.

ethylmercury-p-toluenesulfonanilide.
(EMTS; "Ceresan" M).
$C_6H_5N(HgC_2H_5)SO_2C_6H_4CH_3$.
Properties: Crystals, pungent odor, mp 154–157C, nearly insoluble in water, soluble in acetone and chloroform.
Hazard: Very toxic. See mercury.
Use: Dust or slurry for control of seed-borne diseases of fungi by treatment of seeds or bulbs.

"Ethyl" Metal Deactivator.[313] TM for a product containing 80% N,N'-disalicylidene-1,2-diaminopropane and 20% toluene solvent.
Properties: Amber liquid, bulk d 1.0672 at 20C, flash p 84F (29C) (OC), soluble in gasoline, insoluble in water.
Use: Neutralize the catalytic effect of copper in promoting fuel oxidation.
Hazard: Flammable, moderate fire risk. Irritant to eyes and skin.

ethyl methacrylate. CAS: 97-63-2.
$H_2C:CCH_3COOC_2H_5$.
Properties: Colorless liquid, bp 119C, fp approximately −75C, d 0.911, refr index 1.4116 (25C), flash p 70F (21.1C) (OC), insoluble in water, readily polymerized.
Derivation: Reaction of methacrylic acid or methyl methacrylate with ethanol.
Grade: Technical (inhibited).

Hazard: Flammable, dangerous fire and explosion hazard. An irritant.
Use: Polymers, chemical intermediates.
See also acrylic resin.

N-ethyl-3-methylaniline. See n-ethyl-m-toluidine.

ethyl-2-methylbutyrate. See ethyl isovalerate.

ethyl methylcellulose. A water-soluble cellulose ether used for thickening, sizing, emulsifying, and dispersing.
See also polymer, water-soluble.

ethyl methyl ether. CAS: 540-67-0.
$C_2H_5OCH_3$.
Properties: Colorless liquid, d 0.725, bp 10.8C, soluble in water, miscible with alcohol and ether, flash p −35F (−37.2C), autoign temperature 374F (190C).
Hazard: Highly flammable, dangerous fire and explosion risk.
Use: Medicine (anesthetic).

2-ethyl-4-methylimidazole. $C_2H_5C_3N_2H_2CH_3$.
Properties: A supercooled amber liquid which if crystallized is a low-melting solid, mp 45C, bp 154C (10 mm).
Use: Curing epoxy resin systems.

ethyl methyl ketone. See methyl ethyl ketone.

ethyl methylphenylglycidate. (so-called aldehyde C-16; "Strawberry aldehyde").

$$CH_3(C_6H_5)COCHCOOC_2H_5$$

Properties: Colorless to yellowish liquid having a strong odor suggestive of strawberry, soluble in three volumes of 60% alcohol, d 1.104–1.123, refr index 1.509–1.511. Combustible.
Grade: Technical, FCC (as aldehyde C-16).
Use: Perfumery, flavors.

5-ethyl-2-methylpyridine. See 2-methyl-5-ethylpyridine.

7-ethyl-2-methyl-4-undecanol. (tetradecanol).
$C_4H_9CH(C_2H_5)C_2H_4CH(OH)CH_2CH(CH_3)_2$
Properties: Liquid, d 0.8355 (20/20C), bp 264C, flash p 285F (140C), insoluble in water. Combustible.
Use: Intermediate for synthetic lubricants, defoamers, and surfactants.

ethylmorphine hydrochloride. CAS: 125-30-4.
$C_{19}H_{23}NO_3 \cdot HCl \cdot 2H_2O$.
Properties: White, crystalline powder; odorless. soluble in water and alcohol, slightly soluble in

ether and chloroform, mp approximately 123C (decomposes).

Derivation: Action of hydrochloric acid on ethylmorphine which is made by action of ethyl iodide on morphine in alkaline solution.

Grade: Technical, NF.

Hazard: Toxic by ingestion; habit-forming narcotic.

Use: Medicine (analgesic).

N-ethylmorpholine. CAS: 100-74-3.

$CH_2CH_2OCH_2CH_2NCH_2CH_3$.

Properties: Colorless liquid, ammoniacal odor, miscible with water, d 0.916 (20/20C), fp −63C, bp 138C, bulk d 7.6 lb/gal (20C), flash p 90F (32.3C) (open cup).

Grade: Technical.

Hazard: Irritant to skin and eyes, absorbed by skin. Flammable, moderate fire risk. TLV: 20 ppm in air.

Use: Intermediate for dyestuffs, pharmaceuticals, rubber accelerators and emulsifying agents, solvent for dyes, resins, oils; catalyst in making polyurethane foams.

"ethyl" Multi-Purpose Additive.[313] TM for a product containing 52% mixed substitute oleamides, 37% isopropyl alcohol, 7% aromatic solvent, and 4% water.

Properties: Clear amber liquid, bulk d 0.888 at 20C, flash p 75F (23.8C) (closed cup), pour point 50F (10C).

Hazard: Flammable, dangerous fire risk.

Use: Remove and prevent deposits on the throat walls of carburetors, to prevent carburetor icing and as a corrosion preventive.

ethyl mustard oil. See ethyl thiocarbimide.

ethyl myristate. (ethyl tetradecanoate). $CH_3(CH_2)_{12}COOC_2H_5$.

Properties: Liquid, d 0.856, mp 12C, bp 295C, insoluble in water, soluble in alcohol, slightly soluble in ether. Combustible.

Use: Flavoring.

n-ethyl-α-naphthylamine. (n-ethyl-1-naphthylamine). $C_{10}H_7NCH_2H_5$.

Properties: Colorless liquid, refr index 1.6475, bp 305C, insoluble in water, soluble in alcohol and ether. Combustible.

Use: Intermediate.

ethyl nitrate. CAS: 625-58-1. $C_2H_5NO_3$.

Properties: Colorless liquid, pleasant odor, sweet taste, soluble in alcohol and ether, insoluble in water, d 1.116, fp −112C, bp 87.6C, vapor three times heavier than air, flash p 50F (10C) (closed cup), lower explosive limit 3.8%.

Derivation: By heating alcohol, urea nitrate and nitric acid with subsequent distillation.

Hazard: Flammable, dangerous fire and explosion risk.

Use: Organic synthesis, drugs, perfumes, dyes, rocket propellant.

ethyl nitrite. CAS: 109-95-5. $C_2H_5NO_2$.

Properties: Yellowish volatile liquid, soluble in alcohol and ether, decomposes in water, d 0.90, bp 16.4C, flash p −31F (−35C), decomposes spontaneously at 194F (90C). Explosive limits in air 3–50%, narcotic in high concentration.

Hazard: Highly flammable, dangerous, explodes!

Use: Organic reactions, synthetic flavoring.

ethyl nonanoate. See ethyl pelargonate.

ethyl octanoate. See ethyl caprylate.

ethyl octoate. See ethyl caprylate.

ethyl oenanthate. See ethyl enanthate.

ethyl oleate. $C_{17}H_{33}COOC_2H_5$.

Properties: Light-colored, yellowish, oily liquid. Insoluble in water, soluble in alcohol and ether, soluble of water in product 1.0 cc/100 cc (202C), bulk d 7.27 lb/gal (20C), refr index 1.45189 (20C), bp 205C, fp approximately −32C, d 0.867, flash p 348F (175C). Combustible.

Use: Solvent, plasticizer, lubricant, water-resisting agent, flavoring.

ethyl orthopropionate. $CH_3CH_2C(OC_2H_5)_3$.

Properties: Colorless liquid, bp 155 to 160C, refr index 1.401 (20C), insoluble in water, soluble in alcohol and ether, flash p 140F (60C). Combustible.

Hazard: Moderate fire risk.

Use: Intermediate.

ethyl oxalate. (diethyl oxalate). CAS: 95-92-1. $(COOC_2H_5)_2$.

Properties: Colorless, unstable, oily, aromatic liquid; miscible in alcohol, ether, ethyl acetate, and other common organic solvents; slightly soluble in water and gradually decomposed by it. D 1.09 (20/20C), bp 186C, fp −40.6C, bulk d approximately 8.96 lb/gal (20C), flash p 168F (75.5C) (closed cup). Combustible.

Derivation: By standard esterification procedure using ethanol and oxalic acid. The final purification calls for unusual technique and equipment. The last traces of water are most difficult to remove and this is accomplished by a special step in the rectification.

Hazard: Toxic by ingestion, strong irritant to skin and mucous membranes.

Use: Solvent for cellulose esters and ethers, many

natural and synthetic resins; radio tube cathode fixing lacquers, dye intermediate, pharmaceuticals, perfume preparations, organic synthesis.

ethyl-2-oxocyclopentanecarboxylate. See 2-carbethoxycyclopentanone.

ethylparaben. (ethyl-p-hydroxybenzoate).
CAS: 120-47-8. $C_9H_{10}O_3$.
Properties: Colorless crystals, mp 115C, decomposes at 297C, soluble in alcohol and ether, almost insoluble in water.
Derivation: Esterification of p-hydroxybenzoic acid.
Use: Pharmaceutical preservative.
See also "Parabens."

ethyl parathion. See parathion.

ethyl PCT. See O,O-diethylphosphorochlorlodothioate.

ethyl pelargonate. (ethyl nonanoate; wine-ether). $CH_3(CH_2)_7COOC_2H_5$.
Properties: Colorless liquid, fruity odor, refr index 1.4220 (20C), d 0.866 (18/4C), bp approximately 220C, fp −44C, insoluble in water, soluble in alcohol and ether. Combustible.
Grade: Technical, FCC.
Use: Flavoring alcoholic beverages, perfumes, chemical intermediate.

3-ethylpentane. (triethylmethane).
$(C_2H_5)_3CH$.
Properties: Colorless liquid, bp 93.5C, fp −118.6C, d 0.69818(20C), refr index 1.3934 (20C), soluble in alcohol, insoluble in water. Combustible.
Grade: Technical.
Use: Organic synthesis.

m-ethylphenol. (3-ethylphenol).
$HOC_6H_4C_2H_5$.
Properties: Colorless liquid, fp −4C, bp 214C, d 1.001, very slightly soluble in water, miscible with alcohol and ether. Combustible.
Hazard: Toxic material.
See phenol.

p-ethylphenol. (4-ethylphenol).
$HOC_6H_4C_2H_5$.
Properties: Colorless needles, mp 46C, bp 219C, flash p 219F, soluble in alcohol or ether, slightly soluble in water. Combustible.
Hazard: Toxic material.
See phenol.

ethylphenylacetamide. See ethylacetanilide.

ethyl phenylacetate. CAS: 101-97-3.
$C_6H_5CH_2COOC_2H_5$.
Properties: Colorless liquid, honey-like odor, d 1.027–1.032, refr index 1.498, bp 226C, soluble

in 8 parts of 60% alcohol, insoluble in water. Combustible.
Grade: Technical, FCC.
Use: Perfumery, flavors, intermediate especially for phenobarbital, herbicide.

ethyl phenylacrylate. See ethyl cinnamate.

5-ethyl-5-phenylbarbituric acid. See phenobarbital.

ethyl phenylcarbamate. (phenylurethane; ethyl phenylurethane). $C_6H_5NHCOOC_2H_5$.
Properties: White, crystalline solid; aromatic odor; clove-like taste; mp 51C; soluble in alcohol, ether, and boiling water; insoluble in cold water.
Derivation: Action of ethanol on phenyl isocyanate.

ethyl phenyl dichlorosilane. $C_2H_5(C_6H_5)SiCl_2$.
Properties: Colorless liquid which fumes strongly in moist air.
Hazard: Toxic by inhalation and ingestion, strong irritant to eyes and skin.

N,N-ethylphenylethanolamine.
$C_6H_5NC_2H_5CH_2CH_2OH$.
Properties: Liquid, d 1.04 (20/20C), bp 268C (740 mm), bulk d 8.7 lb/gal. Combustible.
Use: Organic synthesis, dyestuffs.

2-ethyl-2-phenylglutarimide. (glutethimide).
$C_{13}H_{15}NO_2$.
Properties: White, crystalline powder; a saturated solution is slightly acid; freely soluble in acetone, ethyl acetate, and chloroform; soluble in ethanol and methanol; practically insoluble in water; melting range 85–87C.
Grade: NF (as glutethimide).
Hazard: Manufacture and use controlled by law.
Use: Medicine (sedative).

ethyl phenyl ketone. See propiophenone.

ethyl phenylurethane. See ethyl phenylcarbamate.

ethylphosphoric acid. $C_2H_5H_2PO_4$.
Properties: Pale straw-colored liquid, d 1.33 (25C), can be neutralized with alkalies or amines to give water-soluble salts. Combustible. Purity: 97% remainder being orthophosphoric acid and ethanol.
Hazard: An irritant.
Use: Catalyst, rust remover, soldering flux, intermediate.

ethyl phthalate. See diethyl phthalate.

ethyl phthalyl ethyl glycolate.
$C_2H_5OCOC_6H_4COOCH_2COOC_2H_5$.

Properties: Liquid with slight odor, d 1.180 (25C), refr index 1.498 (25C), bp 190C (5 mm), soluble in water 0.018% by weight, miscible with most organic solvents. Combustible.
Use: Plasticizer.

5-ethyl-2-picoline. See 2-methyl-5-ethylpyridine.

2-ethylpiperidine. CAS: 1484-89-6.
Properties: Liquid with mw 113.20, bp 143C, d 0.850.
Hazard: Flammable and an irritant.

1-ethyl-1-propanol. See 3-pentanol.

ethyl propiolate. CAS: 623-47-2.
$HC{\equiv}CCO_2C_2H_5$.
Properties: Liquid with mw 98.10, bp 120C, d 0.968.
Hazard: Flammable and a lacrymator.

ethyl propionate. CAS: 105-37-3.
$C_2H_5COOC_2H_5$.
Properties: Water-white liquid, odor of pineapples, soluble in alcohol and ether, soluble in water 2.5% at 15C, d 0.888 (25C), bp 99C, fp −73C, flash p 54F (12.2C) (closed cup), autoign temperature 890F (476C), refr index 1.3844 (20C).
Derivation: Treating ethanol with propionic acid.
Grade: Commercial 85–90% ester content, FCC.
Hazard: Flammable, dangerous fire risk.
Use: Solvent for cellulose ethers and esters, various natural and synthetic resins; flavoring agent, fruit syrups, cutting agent for pyroxylin.

ethylpropionyl. See diethyl ketone.

2-ethyl-3-propylacrolein. (2-ethylhexenal).
CAS: 645-62-5. $C_3H_7CH{:}C(C_2H_5)CHO$.
Properties: Yellow liquid, powerful odor, d 0.8518 (20/20C), bp 175.0C, vap press 10 mm (20C), flash p 155F (68.3) (open cup), bulk d 7.1 lb/gal (20C), viscosity 0.113 poise (20C). Combustible.
Hazard: Toxic by inhalation and ingestion; strong irritant.
Use: Insecticide, organic synthesis (intermediate), warning agents, and leak detectors.

α-ethyl-β-propylacrylanilide.
$C_6H_5NHCOC(C_2H_5){:}CH(CH_2)_2CH_3$.
Use: Rubber accelerator.

2-ethyl-3-propylacrylic acid.
$C_3H_7CH{:}C(C_2H_5)COOH$.
Properties: Liquid, d 0.9484 (20C), fp −7.8C, bp 232.1C, vap press less than 0.1 mm (20C), flash p 330F (165C), insoluble in water. Combustible.
Use: Pharmaceuticals, resins and plastics, lubricants.

ethyl propyl ketone. See 3-hexanone.

3-ethyl-3-propyl-1,3-propanediol. See 2-ethylhexanediol-1,3.

2-(5-ethyl-2-pyridyl)ethyl acrylate.
$CH_2{:}CHCOOC_2H_4C_5H_3NC_2H_5$.
Properties: Liquid, d 1.0458 (20C), bp 181C (50 mm), fp −75C, very slightly soluble in water. Combustible.
Use: Manufacture of plastics and fibers, adhesives, textile finishes and sizes.

ethyl pyrophosphate. See tepp.

ethyl salicylate. CAS: 118-61-6.
$C_6H_4(OH)COOC_2H_5$.
Properties: Colorless liquid with a faint odor of methyl salicylate, soluble in ether and alcohol, slightly soluble in water, d 1.127–1.130, refr index 1.523, bp 231–234C. Combustible.
Use: Perfumery, flavors.

ethyl silicate. (tetraethyl orthosilicate).
CAS: 78-10-4. $(C_2H_5)_4SiO_4$.
Properties: Colorless liquid, faint odor, miscible with alcohol, hydrolyzed to an adhesive form of silica, d 0.9356 (20/20C), bp 168.1C, mp 110C (sublimes), vap press 1.0 mm (20C), flash p 125F (51.6C) (open cup), bulk d 7.8 lb/gal (20C), fp −77C, viscosity 0.0179 poise (20C). Combustible.
Grade: 29% silicon, 40% silicon.
Hazard: Moderate fire risk. Strong irritant to eyes, nose, throat. TLV: 10 ppm in air.
Use: Weatherproof and acid proof mortar and cements, refractory bricks, other molded objects, heat-resistant paints, chemical-resistant paints, protective coatings for industrial buildings and castings, lacquers, bonding agent, intermediate.

ethyl silicate, condensed. Light-yellow liquid with mild odor consisting of 85% by weight tetraethyl orthosilicate and 15% polyethoxysiloxanes.
Yields high purity, refractory silica on hydrolysis or ignition.
Hazard: Toxic.
See ethyl silicate. Use: Intermediate for siloxane compounds, for precision casting of high-melting alloys, pigments binder for paints, surface hardener for sandstones.

ethyl sodium oxalacetate.
$C_2H_5OOCC(ONa){:}CHCOOC_2H_5$.
Properties: Light yellow powder, 92% pure.
Derivation: Reaction of pure ethyl acetate and diethyl oxalate with metallic sodium.
Use: Dyes, synthesis.

ethyl sulfate. See diethyl sulfate.

ethyl sulfhydrate. See ethanethiol.

ethyl sulfide. (diethyl sulfide). CAS: 352-93-2. $(C_2H_5)_2S$.
Properties: Colorless, oily liquid; garlic-like odor. Soluble in alcohol and ether, slightly soluble in water, d 0.837, fp −102C, bp 92–93C, refr index 1.4423 (20C). Combustible.
Derivation: Heating sodium ethyl sulfate and potassium sulfide with subsequent distillation.
Use: Organic synthesis, solvent for mineral salts, electroplating baths.

ethylsulfonylethyl alcohol. (2-(ethylsulfonyl)ethanol). $CH_3CH_2SO_2CH_2CH_2OH$.
Properties: Colorless crystals, mp 45C, bp 153C (2.5 mm), soluble in water and common organic solvents, hygroscopic.
Derivation: Reaction of 2-ethylthioethanol with hydrogen peroxide.
Use: Solvent and organic intermediate, antistatic agent, humectant.

ethylsulfuric acid. (acid ethylsulfate).
CAS: 540-82-9. $C_2H_5HSO_4$.
Properties: Colorless, oily liquid;, soluble in water, alcohol, and ether; d 1.316; bp 280C (decomposes). Combustible.
Derivation: Action of sulfuric acid on ethanol.
Hazard: Toxic by ingestion and inhalation; strong irritant to eyes and skin.
Use: Precipitant for casein, organic synthesis.

4-ethyl-1,4-thiazane. $C_4H_8NSC_2H_5$.
Properties: Colorless, mobile liquid; soluble in water; d 0.9929 (15C); bp 184C (763 mm). Combustible.
Derivation: Interaction of dichlorodiethyl sulfide and an aliphatic amine in the presence of alcohol and sodium carbonate.

ethyl thiocarbimide. (ethyl mustard oil; ethyl isothiocyanate). C_2H_5NCS.
Properties: Colorless liquid, pungent odor, soluble in alcohol, insoluble in water, d 1.004 (15/4C), fp −5.9C, bp 131–132C.
Hazard: Toxic by inhalation; strong irritant to skin and mucous membranes.
Use: Military poison gas.

ethyl thioethanol. $C_2H_5SC_2H_4OH$.
Properties: Pale straw-colored liquid, d 1.015–1.025 (20/20C), distillation range 180–184C, refr index 1.486 (20C). Combustible.
Grade: 95% min.
Use: Organic intermediate (pesticides, plasticizers, etc.).

ethylthiopyrophosphate. See sulfotepp.

n-ethyl-p-toluenesulfonamide.
$C_2H_5NHSO_2C_6H_4CH_3$.
Properties: Colorless crystals, mp 64C, soluble in alcohol, flash p 260F (126C). Combustible.
Grade: A mixture of the o- and p-isomers is available commercially.
Use: Plasticizer.

ethyl-p-toluenesulfonate. CAS: 80-40-0.
$CH_3C_6H_4SO_3C_2H_5$.
Properties: Unstable solid, mp 33C, bp 221.3C, d 1.17, soluble in many organic solvents, insoluble in water, flash p 316F (157C). Combustible.
Use: Plasticizer for cellulose acetate, ethylating agent.

N-ethyl-m-toluidine. (N-ethyl-3-methylaniline). CAS: 102-27-2. $CH_3C_6H_4NHC_2H_5$.
Properties: Light amber liquid, refr index 1.5451 (20C), bp 221C. Combustible.
Use: Chemical intermediate.

N-ethyl-o-toluidine. CAS: 94-68-8.
$CH_3C_6H_4NHC_2H_5$.
Properties: Colorless to yellowish oil; soluble in alcohol, ether, and hydrochloric acid; insoluble in water; d 0.9534; bp 214C. Combustible.
Derivation: Heating ethanol with o-toluidine and hydrochloric acid.

ethyl triacetylgallate.
$C_2H_5OOCC_6H_2(OOCCH_3)_3$.
Properties: Colorless crystals or white crystalline powder, insipid taste, odorless, soluble in warm alcohol and acetone, slightly soluble in ether and alcohol, insoluble in water, mp 134–136C.

ethyltrichlorosilane. CAS: 115-21-9.
$C_2H_5SiCl_3$.
Properties: Colorless liquid, bp 99.5C, d 1.236 (25/25C), refr index 1.4257 (25C), flash p 72F (22.2C) (open cup). Soluble in benzene, ether, heptane, perchloroethylene; readily hydrolyzed with liberation of hydrogen chloride.
Derivation: By reaction of ethylene and trichlorosilane in the presence of a peroxide catalyst.
Hazard: Flammable, dangerous fire risk, may form explosive mixture with air. A strong irritant.
Use: Intermediate for silicones.

ethyl urethane. See urethane.

ethyl valerate. See ethyl isovalerate.

ethyl vanillate. $(CH_3O)HOC_6H_3COOC_2H_5$.
Properties: Solid, mp 44C, bp 291–293C, insoluble in water, soluble in alcohol and ether.
Use: Food preservative, medicine, sunburn preventive.

ethyl vanillin. (3-ethoxy-4-hydroxybenzaldehyde). CAS: 121-32-4.
$OHC_6H_3(OC_2H_5)CHO$.
Properties: Fine, white crystals; intense odor of vanillin;, affected by light; mp 76.5C; soluble in alcohol, chloroform, and ether; slightly soluble in water. Combustible.
Grade: NF, FCC.
Use: Flavors, to replace or fortify vanillin.

ethyl vinyl ether. See vinyl ethyl ether.

ethylxanthic acid. See xanthic acid.

ethylzinc. See diethylzinc.

ethyne. See acetylene.

ethynylation. Condensation of acetylene with a reagent such as an aldehyde to yield an acetylenic derivative. The best example is the union of formaldehyde and acetylene to produce butynediol.

1-ethynylcyclohexanol. CAS: 78-27-3.
$HC{\equiv}CC_6H_{10}OH$.
Properties: Colorless, low-melting solid with sweet odor; mp 30–31C; bp 180C; d 0.967 (20/20C); slightly soluble in water; soluble in most organic solvents.
Use: Stabilization of chlorinated organic compounds, intermediate, corrosion inhibitor for mineral acids, medicine (sedative).

ethynylestradiol. See ethinylestradiol.

ethythrin. Ethyl analog of allethrin.
See also barthrin, cyclethrin, and furethrin.
Use: Insecticide with applications similar to allethrin.

ETU. See ethylene thiourea.

Eu. Symbol for europium.

eucalyptol. (cineol; cajeputol).
CAS: 470-82-6. $C_{10}H_{18}O$.
Properties: Colorless, essential oil; a terpene ether having a camphor-like odor and pungent, cooling, spicy taste; slightly soluble in water; miscible with alcohol, chloroform, ether, glacial acetic acid, and fixed or volatile oils; d 0.921–0.923 (25C); bp 174–177C; congealing point not less than about 0C, refr index 1.4550–1.4600 (20C). Combustible.
Derivation: By fractionally distilling eucalyptus oil followed by freezing. The oil is imported from Spain, Portugal, and Australia.
Grade: Technical, NF.
Use: Pharmaceuticals (cough syrups, expectorants), flavoring, perfumery.

eucryptite. A lithium silicate mineral ($LiAlSiO_4$) containing up to 4.8% lithia.
Use: Glass manufacture and as a source of lithium.

eugenic acid. See eugenol.

eugenol. (4-allyl-2-methoxyphenol; caryophyllic acid; eugenic acid). CAS: 97-53-0.
$C_3H_5C_6H_3(OH)OCH_3$.
Properties: Colorless or yellowish liquid, oily, becomes brown in air, spicy odor and taste. Soluble in alcohol, chloroform, ether, and volatile oils; very slightly soluble in water. D 1.064–0.070, bp 253.5C, fp −9C, refr index 1.5400–1.5420 (20C), optically inactive. Combustible.
Derivation: By extraction of clove oil with aqueous potash, liberation with acid, and rectification in a stream of carbon dioxide.
Grade: Technical, USP, FCC.
Use: Perfumes, essential oils, medicine (analgesic), production of isoeugenol for the manufacture of vanillin, flavoring.

eugenyl methyl ether. See methyl eugenol.

"Eureka."[28] TM for a soldering flux crystalline composition based on zinc chloride and ammonium chloride.

europia. See europium oxide.

europium. Eu. Atomic number 63, one of the lanthanide or rare-earth elements of the cerium subgroup, aw 151.96, valences = 2,3; two stable isotopes.
Properties: Steel-gray metal, difficult to prepare, quite soft and malleable (DPH = 20), mp 826C, bp approximately 1489C, d 5.24, oxidizes rapidly in air, may burn spontaneously, most reactive of the rare earth metals, liberates hydrogen from water, reduces metallic oxides.
Derivation: Reduction of the oxide with lanthanum or misch metal.
Source: Rare-earth minerals.
Grade: High purity (ingots, lumps).
Hazard: Highly reactive, may ignite spontaneously in powder form.
Use: Neutron absorber in nuclear control, color TV phosphors to activate yttrium, phosphors in postage stamp glues to permit electronic recognition of first-class mail.

europium chloride. CAS: 10025-76-0.
$EuCl_3 \cdot 6H_2O$.
Properties: Yellow needles, soluble in water, d 4.89 (20C), mp 850C,
Derivation: Obtained by treating the oxide with hydrochloric acid.

europium fluoride. $EuF_3 \cdot 3H_2O$.
Properties: Crystals, mp 1390C, bp 2280C, insoluble in water and dilute acids.

europium nitrate. $Eu(NO_3)_3 \cdot 6H_2O$. Colorless to pale pink crystals, mp 85C (sealed tube), soluble in water, obtained by treating the oxide with nitric acid.

europium oxalate. $Eu_2(C_2O_4)_3 \cdot 10H_2O$.
White powder, insoluble in water, slightly soluble in acids.
Grade: 25–50% and 99.8% Eu salt. Impure grade may be colored.

europium oxide. (europia). Eu_2O_3.
Properties: Pale rose powder, d 7.42, insoluble in water but soluble in acids to give the corresponding salt.
Derivation: Calcination of the oxalate, solvent extraction or liquid ion-exchange processes
Grade: 25–50%, 99.9%.
Use: Nuclear-reactor control rods, especially in red- and infrared-sensitive phosphors.

europium sulfate. $Eu_2(SO_4)_3$ anhydrous and $8H_2O$.
Properties: Colorless to pale pink crystals, d (anhydrous) 4.99 (20C), hydrate loses $8H_2O$ at 375C and is slightly soluble in water.

eutectic. The lowest melting point of an alloy or solution of two or more substances (usually metals) that is obtainable by varying the percentage of the components. Eutectic alloys are relatively few; they are the particular alloys that have definite and minimum melting points compared with other combinations of the same metals.
See also alloy, fusible.

eutrophication. The unintentional enrichment or "fertilization" of either fresh- or saltwater by chemical elements or compounds present in various types of industrial wastes. Phosphates and nitrogenous compounds in detergent and chemical processing wastes are particularly effective eutrophying agents. They supply nutrients to algae which proliferate so abundantly that a large proportion die for lack of light; their decomposition products deplete the water of its dissolved oxygen and thus cause the death of fish and other marine life. One process for removing phosphates involves addition of a metal ion source to waste effluents to insolubilize dissolved phosphates, after which the particles are agglomerated by anionic polymers.
See also algae, nitrilotriacetic acid (note).

eV. Abbreviation for electron-volt.

EVA. Abbreviation for ethylene-vinyl acetate copolymer.

"Evanacid 3CS."[312] TM for carboxymethylmercaptosuccinic acid.

"Evanohm."[155] TM for an alloy of 75% nickel, 20% chromium, 2.5% aluminum, and 2.5% copper.
Forms: Wire, insulated wire.
Use: Precision-wound resistors.

evaporation. The change of a substance from the solid or liquid phase to the gaseous or vapor phase. In the case of such solids as ice, snow, dry ice, and a few others, the substances do not go through a liquid phase; the phenomenon is called sublimation. The rate of evaporation of liquids varies with their chemical nature and with the temperature; in general, organic liquids (benzene, gasoline) evaporate at lower temperatures and higher rates than water. The thermal energy required to vaporize a given volume of a liquid is known as its latent heat of vaporization; it remains in the vapor (steam, in the case of water heated to its boiling point) and is released when the vapor condenses. For water, this latent heat value is 540 cal/g. In chemical processing installations requiring a series of evaporations and condensations, the units are set up in series and the latent heat of vaporization from one unit is utilized to supply energy for the next. Such units are called "effects" in engineering parlance as, e.g., a triple-effect evaporator.
See also distillation.

evaporite. A mineral or rock formed by precipitation of crystals from evaporating water.

EVE. Abbreviation for ethyl vinyl ether.
See vinyl ethyl ether.

"Evenglo."[11] TM for a durable lightweight polystyrene, varying in its properties of light transmission, degree of diffusion, and color.
Use: Lighting applications.

"Everdur."[324] TM for a group of 5 copper-silicon alloys with compositions adjusted to hot and cold working, hot forging, welding, free machining and for ingots for remelting and casting.

"Everflex."[311] Brand name for polyvinyl acetate copolymer emulsions.
Use: Paints, paper coating, and acoustic tile coatings.

evolution. Emission of a gas as a result of a chemical reaction as hydrogen from acids reacted with a metal or carbon dioxide from the action of an acid on a carbonate. In formulas it may be indicated by an upward vertical arrow

↑

exchange reaction. A chemical reaction in which atoms of the same element in two different molecules or in two different positions in the same molecule transfer places. Exchange reactions are usually studied with the aid of a tracer or tagged atom.

excipient. A natural, inert, and somewhat tacky material used in the pharmaceutical industry as a binder in tablets, etc. Commonly used materials are gum arabic, honey, and beet pulp.

excited state. A higher than normal energy level (vibrational frequency) of the electrons of an atom, radical, or molecule resulting from absorption of photons (quanta) from a radiation source (arc, flame, spark, etc.) in any wavelength of the electromagnetic spectrum. X-ray, UV, visible, infrared, microwave, and radiofrequencies are used for excitation in various types of spectroscopy. When the energizing source is removed or discontinued, the atom or molecule returns to its normal or stable state either by emitting the absorbed photons or by transferring the energy to other atoms or molecules. The increased vibrational activity of the atom or molecule yields line or band spectra characteristic of its structure, thus permitting identification. Photochemical reactions are induced by excited chemical entities which are also responsible for the phenomena of luminescence (phosphorescence and fluorescence).
See also spectroscopy, photochemistry, absorption (2).

exciton. An energetic entity induced in semiconductor crystals by incident radiation and occurring in the field area between a hole and its displaced electron. It is conceived as behaving like an uncharged particle having quantum-mechanical properties.

exhaust emission control. See air pollution.

"Exkin 1,2,3." TM for a series of antiskinning agents of the volatile oxime type.
Use: Paints.

exo-. A prefix used in chemical names to indicate attachment to a side chain rather than to a ring. See also endo-.

"Exon."[35] TM for a series of polyvinyl resins, compounds, and latexes composed of polymers and copolymers based on vinyl chloride.

Use: Molding, sheeting film, strip coatings, protective coatings, extrusions, paints, ink, adhesives, plastisols, flooring, phonograph records.

exothermic. A process or chemical reaction which is accompanied by evolution of heat, for example, combustion reactions.

exotic. A term applied to materials of various functional types that have extra power and that are often derived from unusual sources. Examples are rocket propellants derived from boron hydrides and certain special-purpose solvents.

expander. (1) A mixture of lampblack, barium sulfate, and an organic material usually derived from lignin that increases the capacity of storage batteries, presumably by coating the anode and thus preventing the deposit of lead sulfate on the underlying lead metal.
(2) A substance used in medicine as a substitute for blood plasma, e.g., dextran, gelatin, or polyvinylpyrollidone.

expeller. See extractor; expression.

explosive, high. A chemical compound, usually containing nitrogen, that detonates as a result of shock or heat. (See detonation.) Dynamite was the most widely used explosive for blasting and other industrial purposes until 1955 when it was largely replaced by prills-and-oil and slurry types. The former consists of 94% ammonium nitrate prills and 6% fuel oil (ANFO). Slurry blasting agents (SBA) are based on thickened or gelatinized ammonium nitrate slurries sensitized with TNT, other solid explosives or aluminum. An unusual type of explosive is represented by acetylides of copper and silver, which are examples of commercially used explosives that contain neither oxygen or nitrogen.

High explosives vary greatly in their shock sensitivity; most sensitive are mercury fulminate and nitroglycerin, while TNT and ammonium nitrate are comparatively difficult to detonate, requiring the use of blasting caps or similar activating device. For further information, refer to Institute of Makers of Explosives, 1575 Eye St., N.W., Washington, D.C.

explosive, initiating. An explosive composition used as a component of blasting caps, detonators, and primers. They are highly sensitive to flame, heat, impact, or friction. Examples are lead azide, silver acetylide, mercury fulminate, diazodinitrophenol, nitrosoguanidine, lead styphnate, and pentaerythritol tetranitrate.

explosive limits. The range of concentration of a flammable gas or vapor (% by volume in air) in which explosion can occur upon ignition in a confined area. Explosive limits for some common substances are:

Substance	Lower(%)	Upper(%)
carbon disulfide	1	50
benzene	1.5	8
methane	5	15
butadiene	2	11.5
butane	1.9	8.5
propane	2.4	9.5
natural gas	3.8	17
hydrogen	4	75
acetylene	2.5	80

explosive, low. An explosive which deflagrates rather than detonates, such as black powder. See deflagration.

explosive, permissible. Explosives approved for use in coal mines. Usually they are modified dynamites.

exposure testing. Determination of the degradation of a material by exposing samples to an environment selected for its adverse effect. Materials most frequently tested are paints, metals and alloys, rubber and plastics. An area frequently chosen is the coast of southern Florida where the combination of high temperature, strong sunlight, salt air, and moisture is particularly severe, especially as regards metal corrosion. Burial of metals in acid soils for long periods of time and immersion of impregnated wood samples in seawater are other exposure testing techniques.
See also testing.

expression. Removal of a liquid from a solid by hydraulic pressure, as in manufacture of vegetable oils from seeds, rinds, or meal cake. Worm devices similar to extrusion machines are also used for this purpose; they are called expellers or extractors.

extender. A low-gravity material used in paint, ink, plastic, and rubber formulations chiefly to reduce cost per unit volume by increasing bulk. Extenders include diatomaceous earth, wood flock, mineral rubber, liquid asphalt, etc. Microscopic droplets of water fixed permanently in a plastic matrix are an efficient extender for polyester resins. In the food industry, the term refers to certain extruded proteins, especially those derived from soybeans which are used in meat products to provide equivalent nutrient values

at lower cost. Made from defatted soy flour, they are often called textured proteins.
See also diluent (2), filler.

extinguishing agent. See fire extinguishment.

extraction, solvent. See solvent extraction.

extractive distillation. A variety of distillation that always involves the use of a fractionating column and is characterized by use of a purposely added substance which modifies the vaporization characteristics of the materials undergoing separation, to make them easier to separate. The additive is often called a solvent and is usually chosen to be much less volatile than any of the substances being separated. It is added to the downflowing liquid reflux stream near the top of the column and is removed from the still pot or reboiler at the base. The addition of furfural to mixtures of butadiene and butene hydrocarbons to separate the butadiene more easily is an example of extractive distillation.

extractive metallurgy. That portion of metallurgical science devoted to the technology of mining and processing of metals and their ores.

extractor. A machine designed for expression of the oil from seeds such as cottonseed, flaxseed, etc., as well as for extracting the juice from fruits. It is similar in principle to an extrusion machine having a tapering screw rotating in a cylindrical barrel; the oil or juice is delivered to a container while the residual pulp and fiber are extruded as waste material. Fruit is fed into the machine after being crushed and preheated. Juices can also be separated centrifugally. Machines of this type are also called expellers.

extreme-pressure additive. (EP additive).
(1) Material added to cutting fluids to impart high film strength. They are mainly sulfur, chlorine, and occasionally phosphorus compounds. Actual conditions, amounts, etc., are proprietary.
(2) Lubricating oil and grease additives that prevent metal-to-metal contact in highly loaded gears. Some react with the metal gears to form a protective coating. Saponified lead salts are often used.
See "Aroclor."

extrusion. A fundamental processing operation in many industries in which a material is forced through a metal forming die, followed by cooling or chemical hardening. The material may be liquid (molten glass or a polymer dispersion); a viscous polymer, as in injection molding; a semisolid mass, such as a rubber or plastic mix; or a hot metal billet. High-viscosity materials are fed into a rotating screw of variable pitch which forces the materials through the die with considerable pressure; a ram is used for metals at temperatures from 537–1093C. Film is made by passing a low-viscosity mixture through a narrow slit. Molten glass and polymer suspensions are forced through a nozzle having a tiny orifice (spinneret); the latter are hardened after extrusion by immersion in a bath of formaldehyde or similar agent. Food items (spaghetti, etc.) are also extruded. Extrusion involves rheological principles of some complexity, critical factors being viscosity, temperature, flow rate, and die design.

See also injection molding.

"Eypel."[313]   TM for polymers and elastomers containing a nitrogen-phosphorous backbone.

"Eza."[313]   TM for zeolite A.

F

F. Symbol for fluorine; the molecular formula is F_2.

"FA."[224] TM for furfuryl alcohol.

FAB. Abbreviation of fast atom bombardment.

fabric. A textile structure composed of mechanically interlocked fibers or filaments. It may be randomly integrated (nonwoven) or closely oriented by warp and filler strands at right angles to each other (woven). While the word usually refers to wool, cotton, or synthetic fibers, fabrics can also be made of glass fiber and graphite.

fabrication. The molding, forming, machining, assembly, and finishing of metals, rubber, and plastics into end-use products. In the paper industry, the term "converting" is used in this sense.

"Fabrikoid."[28] TM for pyroxylin-coated fabrics which are water-resistant, soap- and water-washable.
Use: Book binding, luggage.

"Fabrilite."[28] TM for vinyl coated fabrics and selected vinyl compounds without fabric backing.
Use: Pocketbooks, bags, upholstery, etc.

F acid. See Casella's acid.

factice. (vulcanized oil). A soft, mealy material made by reaction of sulfur or sulfur chloride with a vegetable oil.
Use: Erasers, rubber goods (bath spray tubing, etc.) to give soft "hand."

factor. A term used chiefly by biochemists to indicate any member of a biologically active complex, especially if its exact chemical nature is unknown or if its function in cellular metabolism has not been elucidated. Several of the B complex vitamins were originally referred to as "factors" until their identity had been established by research. There are a number of blood-coagulation factors.
See also Rh factor.

facultative. See bacteria.

Fahrenheit. The scale of temperature in which 212 degrees is the boiling point of water at 760 mm Hg and 32 degrees is the freezing point of water. The scale was invented by a German physicist, G. D. Fahrenheit (1686-1736), who introduced the use of mercury instead of alcohol in thermometers.
See centigrade for method of converting from Fahrenheit to centigrade. See absolute temperature for converting Fahrenheit to absolute Rankine.

fallout. Deposition upon the earth of the radioactive particles resulting from a nuclear explosion, e.g., strontium-90.

"Falone."[248] TM for tris(2,4-dichlorophenoxyethyl) phosphite.
Properties: Viscous, amber liquid; d 1.434; mp 70–72C. Soluble in benzene, xylene and aromatic hydrocarbons; insoluble in water; available as an emulsifiable concentration and a granular solid.
Use: A pre-emergence herbicide.

famphur. (famophos). CAS: 52-85-7. (O,O-dimethyl-O-[p-(dimethylsulfamoyl) phenyl]phosphorothioate (generic)). $(CH_3O)_2P(S)OC_6H_4SO_2N(CH_3)_2$.
Properties: Crystalline powder, mp 55C, very soluble in chloroform and carbon tetrachloride, slightly soluble in water.
Hazard: Cholinesterase inhibitor, use may be restricted.
Use: Insecticide.

"Fanal."[203] TM for phosphotungstic lakes. Characterized by brilliancy of shade and good fastness to light.
Use: Printing inks.

Faraday, Michael. (1791-1867) A native of England, Faraday did more to advance the science of electrochemistry than any other scientist. A profound thinker and accurate experimentalist and observer, he was the first to propound correct ideas as to the nature of electrical phenomena, not only in chemistry but in other fields. His contributions to chemistry include the basic laws of electrolysis, electrochemical decomposition (the basis of corrosion of metals) as well as of battery science and electrometallurgy. His work in physics led to the invention of the dynamo. Faraday was in many respects the exemplar of a true scientist, combining meticulous effort and interpretive genius.

faraday. The quantity of electricity that can deposit (or dissolve) one gram-equivalent weight of a substance during electrolysis (approximately 96,500 coulombs).

farnesol. (Generic name for 3,7,11-trimethyl-2,6,10-dodecatrienol). CAS: 4602-84-0. $C_{15}H_{25}OH$.
Properties: Colorless liquid, delicate floral odor, soluble in three volumes of 70% alcohol, d 0.885 (15C), bp 145–146C (3 mm). Combustible.
Derivation: Found in nature in many flowers and essential oils such as cassia, neroli, cananga, rose, balsams, ambrette seed.
Use: Perfumery, flavoring, insect hormone.

fast. (1) Descriptive of a dye or pigment whose color is not impaired by prolonged exposure to light, steam, high temperature, or other environmental conditions. Inorganic pigments are normally superior in this respect to organic dyes. (2) In nuclear technology, the term refers to neutrons moving at the speed at which they emerge from a ruptured nucleus as opposed to "slow" or thermal neutrons whose speed has been reduced by impinging on a neutral substance called a moderator. Fast neutrons are used in breeder reactors.

fast atom bombardment. (FAB). One of several techniques for ionizing solids from solutions. In FAB, a thin film of the dissolved solid to be analyzed is bombarded with fast atoms. These dislocate ions by impact, which are then analyzed by mass spectroscopy. Peptide ions of with molecular weight of approximately 6000 have been produced and analyzed by this method.

fat. A glyceryl ester of higher fatty acids such as stearic and palmitic. Such esters and their mixtures are solids at room temperatures and exhibit crystalline structure. Lard and tallow are examples. There is no chemical difference between a fat and an oil, the only distinction being that fats are solid at room temperature and oils are liquid. The term "fat" usually refers to triglycerides specifically, whereas "lipid" is all-inclusive.
See also lipid.

fat dyes. Oil-soluble dyes for candles, wax, etc.

fatigue. Incremental weakening of a material as a result of repeated cycles of stresses that are far lower than its breaking load, ending in failure. For metals, to which the term usually refers, the number of low-stress cycles may be of the order of 10^7. Failure is due to development of cumulative imperfections in the crystal structure, with consequent minute interior cracks. Gear failure is often caused by fatigue. It has been reported in experimental windmills for power generation in which steel blades have failed after a few hundred hours of operation due to centrifugal stress. In elastomeric materials, fatigue involves complete dissipation of their resilient energy by repeated cycles of low-order stresses.

fat liquoring agent. An oil-in-water emulsion usually made from raw oils such as neatsfoot, cod, etc., made soluble by dispersing agents such as sulfonated oils.
Use: Leather processing to replace natural oils removed from hides by tanning operations.
See also neatsfoot oil, emulsion.

fat splitting. See hydrolysis.

fatty acid. A carboxylic acid derived from or contained in an animal or vegetable fat or oil. All fatty acids are composed of a chain of alkyl groups containing from 4 to 22 carbon atoms (usually even-numbered) and characterized by a terminal carboxyl group —COOH. The generic formula for all above acetic is $CH_3(CH_2)_x COOH$ (the carbon atom count includes the carboxyl group). Fatty acids may be saturated or unsaturated (olefinic), and either solid, semisolid, or liquid. They are classed among the lipids together with soap and waxes.
Saturated: A fatty acid in which the carbon atoms of the alkyl chain are connected by single bonds. The most important of these are butyric (C_4), lauric (C_{12}), palmitic (C_{16}), and stearic (C_{18}). They have a variety of special uses (see specific entry). Stearic acid leads all other fatty acids in industrial use, primarily as a dispersing agent and accelerator activator in rubber products and in soaps.
Unsaturated: A fatty acid in which there are one or more double bonds between the carbon atoms in the alkyl chain. These acids are usually vegetable-derived and consist of alkyl chains containing 18 or more carbon atoms with the characteristic end group —COOH. Most vegetable oils are mixtures of several fatty acids or their glycerides; the unsaturation accounts for the broad chemical utility of these substances, especially of drying oils. The most common unsaturated acids are oleic, linoleic, and linolenic (all C_{18}). Safflower oil is high in linoleic acid, peanut oil contains 21%, olive oil is 38% oleic acid, palmitoleic acid is abundant in fish oils. Aromatic fatty acids are now available.
See phenylstearic acid.
Note: Linoleic, linolenic, and arachidonic acids are called essential fatty acids by biochemists because they are necessary nutrients that are not synthesized in the animal body.
Use: Special soaps, heavy-metal soap, lubricants,

paints and lacquers (drying oils), candles, salad oil, shortening, synthetic detergents, cosmetics, emulsifiers. For further details, refer to Fatty Acid Producer's Council, 485 Madison Ave., New York.

fatty acid enol ester. A fatty acid reacted with enolic form of acetone for the purpose of increasing the chemical reactivity of the acid. Stearic acid (18-carbon) combined with acetone (3-carbon) gives isopropenyl stearate (21-carbon). This is effective in making the fatty stearoyl group available for synthesis of polymers, medicinals, and the like.
See also fatty ester.

fatty acid pitch. A byproduct residue from
(a) soap stock and candle stock manufacture, (b) refining of vegetable oils, (c) refining of refuse greases, (d) refining of wool grease.
Properties: Dark brown to black, properties analgous to complex hydrocarbons, contains fixed carbon (5–35%), soluble in naphtha and carbon disulfide.
Use: Manufacture of black paints and varnishes, tarred papers, printers' rolls, rubber filing agent, impregnating agent, electrical insulations, marine caulking, waterproofing, sealant.

fatty alcohol. A primary alcohol (from C_8 to C_{20}), usually straight-chain. High molecular weight alcohols are produced synthetically by the Oxo and Ziegler processes. Those from C_8 to C_{11} are oily liquids; those greater than C_{11} are solids. Other methods of production are (a) reduction of vegetable seed oils and their fatty acids with sodium, (b) catalytic hydrogenation at elevated temperatures and pressures, and (c) hydrolysis of spermaceti and sperm oil by saponification and vacuum fractional distillation. The more important commercial saturated alcohols are octyl, decyl, lauryl, myristyl, cetyl, and stearyl. The commercially important unsaturated alcohols, such as oleyl, linoleyl, and linolenyl, are also normally included in this group. The odor tends to disappear as the chain length increases.
Use: Solvent for fats, waxes, gums and resins; pharmaceutical salves and lotions, lube oil additives, detergents and emulsifiers, textile antistatic and finishing agents, plasticizers, nonionic surfactants, cosmetics.

fatty amine. A normal aliphatic amine derived from fats and oils. May be saturated or unsaturated, primary, secondary or tertiary, but the alkyl groups are straight-chain and have an even number of carbons in each. The length varies from 8 to 22 carbon atoms.
Derivation: Fatty acids are treated with ammonia and heated to form fatty acid amides which are

converted to nitriles and reduced to the amine.
Use: Organic bases, soaps, plasticizers, tire cords, fabric softeners, water-resistant asphalt, hair conditioners, cosmetics, medicinals.

fatty ester. A fatty acid with the active hydrogen replaced by the alkyl group of a monohydric alcohol. The esterification of a fatty acid, RCOOH, by an alcohol, R'OH yields the fatty ester RCOOR'. The most common alcohol used is methanol, yielding the methyl ester $RCOOCH_3$. The methyl esters of fatty acids have higher vapor pressures than the corresponding acids and are distilled more easily.

fatty nitrile. (RCN). An organic cyanide derived from a fatty acid.
Derivation: Fatty acids are treated with ammonia and heated to form fatty acid amides which are converted to nitriles.
Use: Intermediates for fatty amines, lube oil additives, plasticizers.

faujasite. $Na_2CaO \cdot Al_2O_3 \cdot 5SiO_2 \cdot 10HOH$, Mineral.
Use: A zeolite or molecular sieve.

Favorskii-Babayan synthesis. Synthesis of acetylenic alcohols from ketones and terminal acetylenes in the presence of anhydrous alkali.

Favorskii rearrangement. Base-catalyzed rearrangement of α-haloketones to acids or esters. The rearrangement of α,α'-dibromocyclohexanones to 1-hydroxycyclopentanecarboxylic acids, followed by oxidation to the ketones, is known as the Wallach degradation.

"Faxam."[51] TM for a series of general-purpose paraffin-base oils that serve as lubricants in process applications and as components in proprietary formulations.

FBR. Abbreviation for fast breeder reactor.
See breeder.

FCC. (1) Abbreviation for Food Chemicals Codex, a publication giving specifications and test methods for chemicals used in foods. (2) Abbreviation for fluid-cracking catalyst as used in the petroleum refining industry. Examples are powdered silica-alumina in which alumina is impregnated with dry synthetic silica gel and various natural clays impregnated with alumina.

FDA. Abbreviation for Food and Drug Administration.

FD&C color. A series of colorants permitted in food products, marking inks, etc., certified by the FDA. Among the more important are the following:

Blue No. 1: disodium salt of 4-((4-(N-ethyl-p-sulfobenzylamino)-phenyl)-(2-sulfoniumphenyl)-methylene)-(1-(N-ethyl-N-p-sulfobenzyl)-sup2,5-cyclo- hexadienimine).

Blue No. 2: Disodium salt of 5,5'-indigotin disulfonic acid.

Green No. 3: Disodium salt of 4-((4-(N-ethyl-p-sulfobenzylamino)-phenyl-(4-hydroxy-2-sulfonium phenyl)-methylene)-(1-(N-ethyl-N-p-sulfobenzyl)- sup2,3-cyclohexadienimine).

Green No. 6: 1,4-di-p-toluidinoanthraquinone.

Red No. 2: Trisodium salt of 1-(4-sulfo-1-naphthylazo)-2-naphthol-3,6-disulfonic acid. Formerly the largest volume food color in commercial use. A carcinogen. Use prohibited by FDA. Red No. 40 is presently a permissible substitute.

Red No. 3: Disodium salt of 9-o-carboxyphenyl-6-hydroxy-2,4,5,7-tetraiodo-3-isoxanthone (erythrosin).

Red No. 4: Disodium salt of 2-(5-sulfo-2,4-xylylazo)-naphthyl-4-sulfonic acid. Use in foods prohibited by FDA.

Violet No. 1: Monosodium salt of 4-((N-ethyl-p-sulfobenzylamino)-phenyl)-(4(N-ethyl-p-sulfoniumbenzylamino)-phenyl)-methylene)-(N,N=dimethyl-$\Delta^{2,5}$-cyclohexadienimine). Use prohibited by FDA in 1973 and by USDA in 1976.

Yellow No. 5: Trisodium salt of 3-carboxy-5-hydroxy-1-p-sulfophenyl-4-sulfophenylazopyrazole.

Yellow No. 6: Disodium salt of 1-p-sulfophenylazo-2-naphthol-6-sulfonic acid.

See also food color, food additive.

feathers. See keratin.

Fe. Symbol for iron.

Federal Trade Commission. (FTC). A consumer protection agency.

feeder. An accessory equipment unit which provides controlled flow of materials of a wide range of particulate sizes to or from processing operations. Major types include the following: (1) Vibratory: an enclosed bowl or open trough activated electromagnetically which vibrates at a constant rate of 3600 oscillations a minute (electromechanical, hydraulic, and pneumatic types are also used). Capacities are up to 2000 lb/hr. The bowl type is applicable to large-size units of materials up to several inches in diameter (wood, plastics, ceramics, etc.). (2) Volumetric: an enclosed device which meters a particulate by volume; there are a number of types, including the rotary lock, the helix, and the roll. Bulk density, particle size, and moisture content of the material handled are important factors. (3) Gravimetric: a belt conveyor provided with a scale which continuously weighs the material passing over it. These are used in operations that are not suitable for volumetric feed.

feedstock. Gaseous or liquid petroleum-derived hydrocarbons or mixture of hydrocarbons from which gasoline, fuel oil, and petrochemicals are produced by thermal or catalytic cracking. It is also called charging stock. Feedstocks commonly used include ethane, propane, butane, butene, benzene, toluene, xylene, naphtha, and gas oils.

Fehling's solution. A reagent used as a test for sugars, aldehydes, etc. It consists of two solutions, one of copper sulfate, the other of alkaline tartrate, which are mixed just before use. Benedict's modification is a one-solution preparation. For details See Book of Methods, Association of Official Analytical Chemists.

Feist-Benary synthesis. Formation of furans from α-halogeno ketones or ethers and 1,3-dicarbonyl compounds in the presence of pyridine. With ammonia as condensing agent pyrrole derivatives are always formed as secondary products.

feldspar. (potassium aluminosilicate). General name for a group of sodium, potassium, calcium, and barium aluminum silicates. Commercially, feldspar usually refers to the potassium feldspars with the formula $KAlSi_3O_8$, usually with a little sodium. Noncombustible.
Grade: Usually based on silicon dioxide content, potassium-sodium ratio, iron content, and fineness of grinding.
Occurrence: North Carolina, Colorado, New Hampshire, South Dakota, California, Arizona, Wyoming, Virginia, Texas.
Hazard: Toxic as fine-ground powder.
Use: Pottery, enamel, and ceramic ware; glass; soaps; abrasive; bond for abrasive wheels; cements and concretes; insulating compositions; fertilizer; poultry grit; tarred roofing materials.

felsic rock. A silicon-rich igneous rock that contains only a small percentage of iron and magnesium. Granite is the most abundant example. Felsic rocks dominate the crust of continents.

felt. A compressed, porous, nonwoven fabric usually made of wool and used as a vibration damper and caulking agent. Its moisture-absorbing property is utilized in the drying section of fourdrinier machines.

FEMA. Acronym of Flavor Extract Manufacturers Association. It makes recommendations to FDA on safety aspects of flavoring materials.

femto-. Prefix meaning one-quadrillionth (10^{-15}). Laser pulses as short as 30 femtoseconds have been produced.

fenac. (2,3,6-trichlorophenylacetic acid). CAS: 85-34-7. Use: Herbicide.

fenchone. CAS: 1195-79-5. $C_{10}H_{16}O$.
Properties: Oil with camphor-like odor, d 0.9465 (19C), bp 193C, soluble in ether, insoluble in water. Combustible.
Derivation: A ketone found (a) as dextrofenchone in oil of fennel, (b) as levofenchone in oil of thuja.
Use: Flavoring.

fenchyl alcohol. (fenchol; 2-fenchanol; 1-hydroxyfenchane). CAS: 512-13-0. $C_{10}H_{18}O$.

Properties: Colorless, oily liquid (*d* and *l* forms) or solid (*dl* form). D approximately 0.96, bp 201C, mp 39C, refr index 1.473.
Derivation: Pine oil, fennel oil, also made synthetically.
Use: Solvent, organic intermediate, odorant, flavoring.

fenitrothion. [O,O-diemthyl-O-(3-methyl-4-nitrophenyl)phosphorothioate].
CAS: 122-14-5. $C_9H_{12}NO_5PS$.
Properties: Yellow, oily liquid. Bp 118C (0.05 mm), d 1.322, refr index 1.552, insoluble in water, soluble in most organic solvents except aliphatics.
Hazard: Cholinesterase inhibitor, use may be restricted.
Use: Insecticide.

fenoprop. See silvex.

fenson. (p-chlorophenyl benzene sulfonate; murvesco). CAS: 80-38-6.
$ClC_6H_4OSO_2C_6H_5$.
Properties: Colorless crystals, mp 61–62C, soluble in organic solvents, insoluble in water.
Use: Acaricide.

fensulfothion. (O,O-diethyl-O-[p-(methylsulfinyl)phenyl]phosphorothioate).
CAS: 115-90-2. $C_{11}H_{17}O_4PS_2$.
Properties: Liquid, bp 138C (0.01 mm).
Hazard: Cholinesterase inhibitor, TLV: 0.1 mg/m^3 of air. Toxic by skin absorption.
Use: Insecticide, especially for nematocide control.

fenthion. (O,O-diemthyl-O-[4-(methylthio)-m-tolyl]phosphorothioate (generic)).
CAS: 55-38-9.
$(CH_3O)_2P(S)OC_6H_3(CH_3)SCH_3$.
Properties: Brown liquid, bp 105C (0.01 mm), insoluble in water, soluble in most organic solvents.
Hazard: Toxic by ingestion, inhalation, and skin absorption; use may be restricted, cholinesterase inhibitor; TLV: 0.1 mg/m^3 of air.
Use: Insecticide, acaricide.

fenticlor. (2,2'-thiobis(4-chlorophenol); novex).
CAS: 97-24-5. $C_{12}H_8Cl_2O_2S$.
Properties: Acicular crystals, mp 170C. Soluble in alcohol, hot benzene, and sodium hydroxide solution.
Derivation: Cholorination of bis(2-hydroxyphenyl)sulfide.
Hazard: Toxic by ingestion.
Use: Fungicide.

Fenton reaction. Oxidation of α-hydroxy acids with hydrogen peroxide and ferrous salts (Fenton's reagent) to α-keto acids or of 1,2-glycols to hydroxy aldehydes.

Fenton's reagent. A solution of sulfuric acid and a ferrous salt.
Use: Oxidation of polyhydric alcohols.

fenuron. (Generic name for 3-phenyl-1,1-dimethylurea). CAS: 101-42-8.
$C_6H_5NHCON(CH_3)_2$.
Properties: White, crystalline solid. Almost insoluble in water (0.3% at 25C), sparingly soluble in hydrocarbon solvents, stable toward oxidation and moisture, mp 127–129C.
Use: Weed and brush killer.

fenvalerate. (cyano(3-phenoxyphenyl)methyl-4-chloro-alpha-(1-methylethyl)phenylacetate).
CAS: 51630-58-1.
Use: Insecticide.

FEP resin. Abbreviation for fluorinated ethylene-propylene resin.

"Feran."[197] TM for a nitrogen fertilizer solution containing ammonium nitrate and water.
Use: Manufacturing solid fertilizers and for direct application.

ferbam. (Generic name for ferric dimethyl-dithiocarbamate). CAS: 14484-64-1. [(CH₃)₂NCSS]₃Fe.

Properties: Black or dark colored, fluffy powder; decomposes above 180C; usually readily dispersible but very slightly soluble in water; pH of saturated solution 5.0.

Derivation: By addition of carbon disulfide to an alcoholic solution of dimethylamine and precipitation with a ferric salt.

Grade: 76% wettable powder, 87% technical powder.

Hazard: Irritant to eyes and mucous membranes. TLV: 10 mg/m³.

Use: Fungicide.

"Fergon."[508] TM for ferrous gluconate.

"Fermate."[28] (ferric diethyl dithiocarbamate.) TM for a wettable powder containing 76% ferbam.

Use: Fruit fungicide.

"Fermentase."[114] TM for a diastatic malt syrup.
Use: In baking.

fermentation. A chemical change induced by a living organism or enzyme, specifically bacteria or the microorganisms occurring in unicellular plants such as yeast, molds, or fungi. The reaction usually involves the decomposition of sugars and starches to ethanol and carbon dioxide, the acidulation of milk or the oxidation of nitrogenous organic compounds. The basic reaction is

$$C_6H_{12}O_6 \xrightarrow{\text{catalyst}} 2C_2H_5OH + 2CO_2$$

Enzymes are usually involved in such reactions; with yeast, the effective enzyme is zymase. Fermentation is essential in the preparation of breads and other food products, and in the manufacture of beer, wine, and other alcoholic beverages, as well as of citric acid, gluconic acid, sodium gluconate, and synthetic biopolymers. Much of the industrial alcohol used in the US is made by fermentation of blackstrap molasses, a byproduct of sugar manufacture. Antibiotics are produced by various forms of microorganisms active in molds, especially bacteria and Actinomycetes. The activated sludge process for sewage digestion is a form of fermentation. A continuous fermentation process for deriving edible protein from petroleum has been introduced. Fermentation is also used in making synthetic amino acids. Research in this field is being directed toward conversion of agricultural, urban, and animal wastes to fuels by fermentation processes.

See also yeast, sewage sludge, antibiotics, bacteria, biotechnology.

fermentation alcohol. See ethanol.

"Fermex."[173] TM for a diastatic (proteaseamylase) enzyme supplement for baking fermentation used to improve quality and uniformity of bread and other yeast-raised bakery products.

Fermi, Enrico. (1901-1954) An Italian physicist who later became a US citizen. He developed a statistical approach to fundamental problems of physical chemistry based on Pauli's exclusion principle. He discovered induced or artificial radioactivity resulting from neutron impingement as well as slow or thermal neutrons. He was Professor of Physics at Columbia (1939) and was awarded the Nobel Prize in Physics in 1938. He was the first to achieve a controlled nuclear chain reaction, directed the construction of the first nuclear reactor at University of Chicago (1942), and worked on the atomic bomb at Los Alamos. He also carried on fundamental research on subatomic particles using sophisticated statistical techniques. Element 100 (fermium) is named after him.

fermium. (element 100). Fm. Aw 254, valence = 3, half-life three hours, a synthetic radioactive element with atomic number 100 discovered in 1952. Fermium has since been prepared in a nuclear reactor by irradiating californium, plutonium, or einsteinium with neutrons in a cyclotron by bombarding uranium with accelerated oxygen ions, and by other nuclear reactions. The element is named for Enrico Fermi. It has chemical properties similar to those of the rare earth erbium. Isotopes are known with mass numbers 254, 255, and 256. ·

Use: Tracer studies.
See also actinide element.

Ferrario reaction. Formation of phenoxathiins by cyclization of diphenyl ethers with sulfur in the presence of aluminum chloride.

ferrate. See ferrite (2).

ferredoxin. An iron-containing protein thought to be involved in photosynthesis as an acceptor of energy-rich electrons from chlorophyll. It occurs in green plants and in bacteria that metabolize elemental hydrogen.

ferric acetate, basic. (iron acetate, basic). CAS: 10450-55-2. Fe(C₂H₃O₂)₂OH.

Properties: Red powder, soluble in alcohol and acids, insoluble in water. Combustible.

Derivation: Action of acetic acid on ferric hydroxide with subsequent crystallization.

Use: Mordant in dyeing textiles, wood preservative, medicine.

ferric acetylacetonate.
$Fe[OC(CH_3):CHC(O)CH_3]_3$.
Properties: Crystalline powder, mp 179–182C, slightly soluble in water, soluble in most organic solvents, resistant to hydrolysis, a chelating nonionizing compound. Combustible.
Use: Moderating and combustion catalyst, solid fuel catalyst, bonding agent, curing accelerator, intermediate.

ferric ammonium alum. See ferric ammonium sulfate.

ferric ammonium citrate. (iron ammonium citrate).
Properties: Thin, transparent, garnet red scales or granules or as a brownish yellow powder; odorless (or slight ammonia odor); saline, mildly ferruginous taste; deliquescent; affected by light; soluble in water; insoluble in alcohol. Combustible.
Derivation: Addition of citric acid to ferric hydroxide, then adding ammonium hydroxide, followed by filtration.
Grade: Technical.
Use: Medicine, blueprint photography, feed additive.

ferric ammonium oxalate. (iron ammonium oxalate; ammonioferric oxalate).
CAS: 14221-47-7. $(NH_4)_3Fe(C_2O_4)_3 \cdot 3H_2O$.
Properties: Green crystals, soluble in water and alcohol, sensitive to light.
Derivation: Interaction of ammonium binoxalate and ferric hydroxide.
Hazard: Irritant to skin and mucous membranes.
Use: Blueprint photography.

ferric ammonium sulfate. (iron ammonium sulfate; ferric ammonium alum; ammonio ferric sulfate). $FeNH_4(SO_4)_2 \cdot 12H_2O$.
Properties: Lilac to violet, efflorescent crystals. D 1.71, mp 39–41C, bp loses 12HOH at 230C, soluble in water, insoluble in alcohol.
Derivation: By mixing solutions of ferric sulfate and ammonium sulfate followed by evaporation and crystallization.
Use: Medicine, analytical chemistry, textile dyeing (mordant).

ferric arsenate. CAS: 10102-49-5.
$FeAsO_4 \cdot 2H_2O$.
Properties: A green or brown powder, d 3.18, decomposes on heating, insoluble in water, soluble in dilute mineral acids. Nonflammable.
Hazard: Toxic by ingestion and inhalation, strong irritant.
Use: Insecticide.

ferric arsenite. CAS: 63989-69-5.
$2FeAsO_3 \cdot Fe_2O_3 \cdot 5H_2O$. A basic salt of variable composition.
Properties: Brownish-yellow powder, soluble in acids, insoluble in water. Nonflammable.
Hazard: Toxic by ingestion and inhalation, strong irritant.
Use: Combined with ammonium citrate (ferric ammonium citrate) and used in medicine.

ferric bromide. (ferric tribromide; iron bromide).
CAS: 10031-26-2. $FeBr_3$.
Properties: Dark red, deliquescent crystals; soluble in water, alcohol, and ether; mp sublimes.
Derivation: By the action of bromine on iron filings.
Use: Bromination catalyst.

ferric chloride, anhydrous. (ferric trichloride; ferric perchloride; iron chloride, iron trichloride, iron perchloride). CAS: 7705-08-0.
$FeCl_3$.
Properties: Black-brown solid, d 2.898 (25C), mp 306C (decomposes), bp 319C. Soluble in water, alcohol, glycerol, methanol, and ether. Noncombustible.
Derivation: Action of chlorine on ferrous sulfate or chloride.
Grade: Anhydrous 96%; 42 degrees Bé solution, photographic and sewage grades.
Hazard: Toxic by ingestion, strong irritant to skin and tissue.
Use: Treatment of sewage and industrial wastes; etching agent for engraving, photography, and printed circuitry; condensation catalyst in Friedel-Crafts reactions; mordant; oxidizing, chlorinating, and condensing agent; disinfectant; pigment; feed additive; water purification.

ferric chromate. (iron chromate).
CAS: 10294-52-7. $Fe_2(CrO_4)_3$.
Properties: Yellow powder, soluble in acids, insoluble in water and alcohol, soluble in hydrochloric acid.
Derivation: By adding sodium chromate to a solution of a ferric salt.
Hazard: Carcinogenic. Toxic by ingestion and inhalation. Strong irritant. Moderate fire risk by reaction with reducing agents.
Use: Metallurgy, ceramics (color), paint pigment.

ferric citrate. (iron citrate). CAS: 2338-05-8.
$FeC_6H_5O_7 \cdot 5H_2O$.
Properties: Reddish-brown scales, keep away from light, soluble in water, insoluble in alcohol.
Derivation: By the action of citric acid on ferric hydroxide and crystallization.
Use: Medicine, blueprint paper.

ferric dichromate. (iron dichromate; ferric bichromate). $Fe_2(Cr_2O_7)_3$.
Properties: Reddish-brown granules, soluble in water and acids.
Derivation: By heating aqueous chromic acid and moist ferric hydroxide.
Hazard: Toxic by inhalation and ingestion, strong irritant. Moderate fire risk by reaction with reducing agents.
Use: Preparation of pigments.

ferric dimethyldithiocarbamate. See ferbam.

ferric ferrocyanide. (iron ferrocyanide; Prussian Blue). Blue pigment described under iron blue.

ferric fluoride. (iron fluoride).
CAS: 7783-50-8. FeF_3.
Properties: Green crystals, d 3.52, soluble in dilute hydrogen fluoride, insoluble in alcohol and ether.
Hazard: Strong irritant. TLV (as F): 2.5 mg/m^3 of air.
Use: Ceramics (porcelain, pottery), catalyst.

ferric fluoroborate.
Use: Rebuilding of worn iron parts, such as cylinders, stereotypes, and electrotypes; plating of solder iron tips.

ferric glycerophosphate. (iron glycerophosphate). $Fe_2[C_3H_5(OH)_2PO_4]_3 \cdot xH_2O$.
Properties: Yellowish scales, odorless and nearly tasteless, soluble in water, insoluble in alcohol.
Use: Pharmaceuticals.

ferrichrome. $C_{27}H_{42}FeN_9O_{12}$. A cyclic iron chelate compound.
Properties: Yellow needles; soluble in water and methanol; slightly soluble in alcohol, acetone, and chloroform.
Derivation: Isolated from rust fungus in 1952, synthesized in 1969.
Use: Growth-promoting factor in medicine.

ferric hydroxide. (ferric hydrate; iron hydroxide; iron hydrate; iron oxide, hydrated; ferric oxide, hydrated). CAS: 20344-49-4. $Fe(OH)_3$.
Properties: Brown flocculant precipitate which dries as the oxide; d 3.4–3.9; mp loses water at approximately 500C; soluble in acids; insoluble in water, alcohol, and ether. Noncombustible.
Derivation: Addition of ferrous sulfate solution to ammonia solution.
Use: Water purification, manufacturing pigments, rubber pigment, catalyst.

ferric hypophosphite. (iron hypophosphite).
CAS: 7783-84-8. $Fe(H_2PO_2)_3$.
Properties: White or grayish-white powder, odorless, tasteless. Slightly soluble in water, more soluble in boiling water.
Hazard: Explosion may occur if triturated or heated with nitrates, chlorates, or other oxidizing agents.

"Ferriclear."[1] TM for a grade of ferric sulfate largely used as a conditioner of alkaline soils.

ferric naphthenate.
Properties: A heavy-metal soap. Combustible.
Derivation: Fusion method by heating naphthenic acids with the metallic oxide.
Use: Conditioning and waterproofing agent, sludge preventive, fungicide and paint drier. See also soap (2).

ferric nitrate. (iron nitrate).
CAS: 10421-48-4. $Fe(NO_3)_3 \cdot 9H_2O$.
Properties: Violet crystals, d 1.684, mp 47.2C, decomposes at 125C, soluble in water and alcohol.
Derivation: Action of concentrated nitric acid on scrap iron or iron oxide and crystallizing.
Hazard: Dangerous fire risk in contact with organic materials. Strong oxidant and irritant.
Use: Dyeing (mordant for buffs and blacks), tanning, analytical chemistry.

ferric octoate. See soap (2).

ferric oleate. (iron oleate). $Fe(C_{18}H_{33}O_2)_3$.
Properties: Brownish-red lumps; soluble in alcohol, ether, and acids; insoluble in water. Combustible. See also soap (2).

ferric oxalate. $Fe_2(C_2O_4)_3$.
Properties: Pale-yellow amorphous scales or powder, odorless, decomposes on heating to 100C, soluble in water and acids, insoluble in alkali. Combustible.
Hazard: Toxic by ingestion and inhalation.
Use: Catalyst in making O_2, silvertone photographic printing papers.

ferric oxide. (ferric oxide, red; iron oxide; red iron trioxide; ferric trioxide). CAS: 1309-37-1. Fe_2O_3.
Properties: Dense, dark-red powder or lumps; d 5.12–5.24; mp 1565C; soluble in acids; insoluble in water.
Grade: Technical, 99.5% pure, electronic.
Use: Metallurgy, gas purification, paint and rubber pigment, component of thermite, polishing compounds, mordant, laboratory reagent, catalyst (p-hydrogen), feed additive, electronic pigments for TV, permanent magnets, mem-

ory cores for computers, magnetic tapes. See also iron oxide reds.

ferric perchloride. See ferric chloride.

ferric phosphate. (iron phosphate).
CAS: 10045-86-0. $FePO_4 \cdot 2H_2O$.
Properties: Yellowish-white powder, insoluble in water, soluble in acids, d 2.87.
Derivation: By adding a solution of sodium phosphate to a solution of ferric chloride. The product is filtered and then dried.
Use: Fertilizers, feed and food additive.

ferric pyrophosphate. (iron pyrophosphate).
CAS: 10058-44-3. $Fe_4(P_2O_7)_3 \cdot xH_2O$.
Properties: Yellowish-white powder, insoluble in water, soluble in dilute acid, 24% iron min, not to be confused with ferric pyrophosphate, soluble.
Use: Source of nutritional iron, catalyst, pigments, flame retardant.

ferric pyrophosphate, soluble. A combination of ferric pyrophosphate and sodium citrate.
Properties: Apple-green crystals, very soluble in water, insoluble in alcohol, protect from light, 11% iron.
Use: Feed additive.

ferric resinate. (iron resinate).
Properties: Reddish-brown powder; soluble in ligroin, carbon disulfide, ether, oil of turpentine; slightly soluble in alcohol; insoluble in water.
Use: Drier (paints, varnish).
See also soap (2).

ferric sodium oxalate. (iron sodium oxalate).
$Na_3Fe(C_2O_4)_3 \cdot 5.5H_2O$.
Properties: Emerald-green crystals, decomposed by heat or light, soluble in water and alcohol, d 1.973 (18C), decomposes at 300C, protect from light.
Derivation: By the interaction of sodium acid oxalate and ferric hydroxide.
Use: Photography, blueprinting.

ferric stearate. (iron stearate).
$Fe(C_{18}H_{35}O_2)_3$.
Properties: Light brown powder, soluble in alcohol and ether, insoluble in water. Combustible.
Derivation: Interaction of solutions of ferric sulfate and sodium stearate.
Use: Varnish driers, photocopying.
See also soap (2).

ferric sulfate. (iron sulfate; ferric trisulfate; iron tersulfate; iron persulfate).
CAS: 10028-22-5.   (a) $Fe_2(SO_4)_3$,
(b) $Fe_2(SO_4)_3 \cdot 9H_2O$.
Properties: Yellow crystals or grayish-white pow-

der, d (a) 3.097, (b) 2.0–2.1, mp decomposes at 480C, (a) slightly soluble in water, (b) very soluble in water, keep well closed and protected from light. Noncombustible.
Derivation: By adding sulfuric acid to ferric hydroxide.
Grade: Technical, CP, partly hydrated.
Use: Pigments, reagent, etching aluminum, disinfectant, textiles (dyeing and calico printing), flocculant in water and sewage purification, soil conditioner, polymerization catalyst, metal pickling, chelated iron products, intermediate.

ferric tallate. See soap (2).

ferric tannate. (iron tannate; iron gallotannate).
$Fe_2(C_{14}H_7O_9)(OH)_3$.
Properties: Dark brown or bluish black powder; variable composition; soluble in alkalies; insoluble in water, alcohol, and ether; soluble in dilute acids. Combustible.
Derivation: Interaction of ferric acetate and tannic acid solutions.
Use: Medicine.

ferric tribromide. See ferric bromide.

ferric trichloride. See ferric chloride.

ferric trioxide. See ferric oxide.

ferric trisulfate. See ferric sulfate.

ferric vanadate. (iron metavanadate).
$Fe(VO_3)_3$.
Properties: Grayish-brown powder, soluble in acids, insoluble in water and alcohol. Noncombustible.
Derivation: By adding a solution of a ferric salt to the liquor obtained by leaching vanadium ores with caustic soda solution or by lixivating the slags obtained when vanadium ores are fused with soda ash, etc.
Grade: Technical.
Use: Metallurgy.

ferrite. (1) Iron in the body-centered cubic form, commonly occurs in steels, cast iron, and pig iron approximately at 910C. Alpha and beta iron are the common varieties of ferrite and the name is also applied to delta iron. (2) A compound, a multiple oxide, of ferric oxide with another oxide, as sodium ferrite, $NaFeO_2$, but more commonly a multiple oxide crystal.

Ferrites are made by dissolving hydrated ferric oxide in concentrated alkali solution; by fusing ferric oxide with alkali metal chloride, carbonate, or hydroxide; or by simply heating metal oxides with ferric oxide. Ceramic ferrites are made by pressforming powdered ingredients (with a binder) into a sheet then sintering or firing.

Use: The oxide ferrites in rectifiers on memory and record tapes, for permanent magnets, semiconductors, insulating materials, dielectrics, high frequency components, and various related uses in radio, television, radar, computers, and automatic control systems.

ferro-alloy. An alloy of iron with some element other than carbon used as a vehicle for introducing such an element into steel during its manufacture. The element may alloy with the steel by solution or as the carbide, neutralize the harmful impurities by combining with them and separating from the steel as flux or slag before solidification.

ferroboron. A ferro-alloy used as hardening agent in special steels. It also is an efficient deoxidizer. Boron steel is used in controlling the operating rate of nuclear reactors. Two grades are available, 10 and 17% boron.

"Ferrocarbo."[280] TM for briquetted or granular silicon carbide.
Use: Cupola addition in the production of gray iron or as a ladle addition to steel. Decomposes into its component elements and acts as a powerful deoxidizer and graphitizer. Machinability and strength of the iron or steel are increased with no loss of hardness.

ferrocene. (dicyclopentadienyliron).
$(C_5H_5)_2Fe.$ A coordination compound of ferrous iron and two molecules of cyclopentadiene in which the organic portions have typically aromatic chemical properties. Its activity is intermediate between phenol and anisole. The first compound shown to have the "sandwich" structure found in certain types of metallocene molecules. Two structures of ferrocene are shown below:

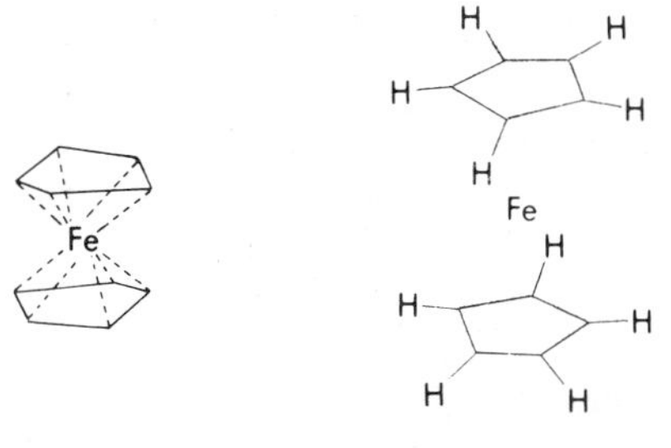

Properties: Orange, crystalline solid; camphor-like odor; mp 173–174C; resists pyrolysis at 400C; insoluble in water; soluble in benzene, ether, and alcohol; iron content 29.4–30.6%.
Derivation: From ferrous chloride and cyclopentadiene sodium.
Hazard: Moderate fire risk. Evolves toxic products on decomposition and heating.

Use: Additive to fuel oils to improve efficiency of combustion and eliminate smoke, antiknock agent, catalyst, coating for missiles and satellites, high-temperature lubricant, intermediate for high-temperature polymers, UV absorber.
See also metallocene.

ferrocenecarboxylic acid ethyl ester.
See ethyl ferrocenoate.

1,1'-ferrocenedicarboxylic acid diethyl ester.
See 1,1'-diethyl ferrocenoate.

1,1'-ferrocenedicarboxylic acid dimethyl ester.
See dimethyl ferrocenoate.

1,1'-ferrocenediyl dichlorosilane.
$(C_5H_5)_2FeSiCl_2.$ An experimental ferrocene derivative which prevents oxidative deterioration of the surfaces of photoelectrodes with which it is in contact. It increases the stability of light-sensitive electrodes in energy conversion reactions occurring in liquid media.

ferrocenoyl chloride. See chlorocarbonyl ferrocene.

ferrocenoyl dichloride. See 1,1'-dichlorocarbonyl ferrocene.

ferrocenylborane polymer.
Properties: Long-term heat resistance at 315C, short-term stability approximately 815C, good resistance to oxidation and hydrolysis, contains up to 30% iron directly combined.
Use: Specialty plastics, coatings, fibers; ablative material for space vehicles.

ferrocenyl methyl ketone. See acetylferrocene.

ferrocerium. A pyrophoric alloy of iron and misch metal.

ferrochromium. (ferrochrome). An alloy composed principally of iron and chromium used as a means of adding chromium to steels (low-, medium-, and high-carbon) and to cast iron. Available in several classifications and grades, generally containing between 60–70% chromium, in crushed sizes and lumps up to 75 pounds which readily dissolve in molten steel.

ferroconcrete. See concrete.

ferroelectric. A crystalline material such as barium titanate, monobasic potassium phosphate, or potassium-sodium tartrate (Rochelle salts) that, over certain limited temperature ranges, has a natural or inherent deformation (polarization)

of the electrical fields or electrons associated with the atoms and groups in the crystal lattice. This results in the development of positive and negative poles and a consequent "direction" of polarization, which can be reversed when the crystal is exposed to an external electric field. Ferroelectric crystals are internally strained and, as a consequence, show unusual piezoelectric and elastic properties.
Use: Capacitors, transducers, computer technology.
See also ceramic, ferroelectric.

ferroin chelation group.   A functional group characteristic of heterocyclic ring nitrogen compounds:

$$=N-C-C-N=$$

Among such compounds are 2,2'-bipyridine; 1,10-phenanthroline; and the 2-pyridyl triazines. These provide a large number of terminal ($\equiv C-H$) groups in which the hydrogen can be replaced by many chemical groupings (carboxyl, hydroxyl, halogen, etc.). Thus, synthesis of an almost endless number of substituted ferroin reactants is possible. About 200 such chelation reagents have been synthesized. Ferroin chelation chemicals in general form complex undissociated cations with divalent metal ions, e.g., $[(C_{12}H_8N_2)_3Fe]^{+2}$.

ferromagnesite.   An iron-bearing variety of magnesite.
Use: Refractory owing to its ability to bond under heat.

ferromagnetic oxide.   See ferrite (2).

ferromanganese.   An alloy consisting of manganese (approximately 48%) plus iron and carbon. Available in standard, low-carbon, and medium carbon grades in ground, crushed, and lump sizes ranging from 80 mesh to 75-lb lumps, suitable for ladle or furnace addition.
Use: Vehicle for adding manganese to steel.
See also manganese steel, nodules.

ferromolybdenum.   An alloy composed largely of iron and molybdenum used as a means of adding molybdenum to steel. Engineering steels rarely contain more than 1% molybdenum, stainless steels may contain 3%, and tool steels as much as 10%. Ferromolybdenum is available in several grades in which molybdenum ranges from 55 to 75% and the maximum carbon content is either 1.10%, 0.60%, or 2.50%. It is generally

added in the furnace since it does not oxidize under steel-making conditions. Mp approximately 1630C. Available in crushed sizes up to one inch.

ferroniobium.   An alloy of iron and niobium made by reducing the ore columbite with silicon.
Use: Stainless steels and other alloys for welding rods.

ferrophosphorus.   An alloy of iron and phosphorus used in the steel industry for adjustments of phosphorus content of special steels.
Grade: (a) 18% phosphorus, (b) 25% phosphorus.
Use: In preventing thin sheets from sticking together when rolled and annealed in bundles.

ferrosilicon.   An alloy of iron and silicon used to add silicon to steel and iron. D 5.4, insoluble in water. Small quantities of silicon deoxidize the iron and larger amounts impart special properties. Available in six grades containing from 20 to 95% silicon. The 20% grade is made in a blast furnace, but grades of higher silicon content are made in electric furnaces.
Hazard: Ferrosilicon containing from 30 to 90% silicon is flammable and evolves gases in presence of moisture.
Use: Pidgeon process for producing metallic magnesium.

ferrosoferric oxide.   See iron oxide, black.

ferrotitanium.   An alloy composed principally of iron and titanium used to add titanium to steel. It is often made from titanium scrap. Three classifications are available: low, high, and medium carbon content. Furnished in various lump, crushed, and ground sizes.

ferrotungsten.   An alloy of iron and tungsten used as a means of adding tungsten to steel. Contains 70 to 80% tungsten and no more than 0.6% carbon. Melting range 1648–2750C, dissolves readily in molten steel. Furnished in ground and crushed sizes up to one inch.
See tungsten steels.

ferrous acetate.   (iron acetate).
   $Fe(C_2H_3O_2)_2 \cdot 4H_2O$.
Properties: Greenish crystals when pure and unexposed to air; usually partly brown from action of air, soluble in water and alcohol, oxidizes to basic ferric acetate in air. Combustible.
Derivation: Action of acetic acid or pyroligneous acid on iron with subsequent crystallization.
Use: Textile dyeing, medicine, dyeing leather, wood preservative.

ferrous ammonium sulfate. (Mohr's salt; iron ammonium sulfate). CAS: 10045-89-3.
$Fe(SO_4) \cdot (NH_4)_2SO_4 \cdot 6H_2O$.
Properties: Light green crystals, soluble in water, insoluble in alcohol, d 1.865, decomposes at 100–110C, deliquescent, affected by light.
Derivation: By mixing solutions of ferrous sulfate and ammonium sulfate followed by evaporation and subsequent crystallization.
Use: Analytical chemistry, metallurgy.

ferrous arsenate. (iron arsenate).
CAS: 10102-50-8. $Fe(AsO_4)_2 \cdot 6H_2O$.
Properties: Green, amorphous powder; insoluble in water; soluble in acids.
Derivation: Interaction of solutions of sodium arsenate and ferrous sulfate.
Use: Insecticide.

ferrous bromide. (iron bromide).
$FeBr_2 \cdot 6H_2O$.
Properties: Green, crystalline powder; very deliquescent. Readily oxidized in moist air, soluble in water and alcohol, d 4.636, mp 27C.
Derivation: Action of bromine on iron filings.
Use: Polymerization catalyst.

ferrous chloride. (iron chloride; iron dichloride; iron protochloride). CAS: 7758-95-3.
(a) $FeCl_2$, (b) $FeCl_2 \cdot 4H_2O$.
Properties: Greenish-white crystals, soluble in alcohol and water, d (a) 3.16 (25C), (b) 1.93, mp (a) 670–674C, deliquescent, readily oxidized.
Derivation: Action of hydrochloric acid on an excess of iron with subsequent crystallization.
Use: Mordant in dyeing, metallurgy, pharmaceutical preparations, manufacture of ferric chloride, sewage treatment.

ferrous-2-ethylhexoate. A paint drier.
See soap (2).

ferrous fluoride. (iron fluoride). FeF_2.
Properties: Green crystals, soluble in acids, slightly soluble in water, insoluble in alcohol and ether, d 4.09.
Hazard: Strong irritant. TLV (as F): 2.6 mg/m³ of air.
Use: Ceramics, catalyst.

ferrous fumarate. CAS: 141-01-5.
$FeC_4H_2O_4$. Anhydrous salt of a combination of ferrous iron and fumaric acid, stable, odorless, substantially tasteless. Reddish-brown, anhydrous powder, contains 33% iron by weight, does not melt at temperatures up to 280C, insoluble in alcohol, very slightly soluble in water. Combustible.
Grade: USP.
Use: Dietary supplement.

ferrous gluconate. (iron gluconate).
CAS: 299-29-6. $Fe(C_6H_{11}O_7)_2 \cdot 2H_2O$.
Properties: Yellowish-gray or pale greenish-yellow fine powder or granules with slight odor, solution (1 in 20) is acid to litmus, soluble in water and glycerol, insoluble in alcohol. Combustible.
Grade: Pharmaceutical, NF.
Use: Feed and food additive, vitamin tablets.

ferrous iodide. (iron iodide; iron protoiodide).
CAS: 7783-86-0. $FeI_2 \cdot 4H_2O$.
Properties: Dark violet to black hygroscopic leaflets, soluble in water and alcohol, d 2.873, decomposes at 90–98C, mp (anhydrous) 177C, deliquescent, affected by light.
Derivation: By the action of iodine on iron filings.
Use: Manufacture of alkali metal iodides, pharmaceutical preparations, catalyst.

ferrous lactate. (iron lactate).
CAS: 5905-52-2. $Fe(C_3H_5O_3) \cdot 3H_2O$.
Properties: Greenish-white crystals, slight peculiar odor, soluble in water, insoluble in alcohol, deliquescent, affected by light. Combustible.
Derivation: By interaction of calcium lactate with ferrous sulfate or direct action of lactic acid on iron filings.
Use: Food additive and dietary supplement.

ferrous naphthenate. A soap based on mixed naphthenic acids. Available commercially as a liquid containing 6% iron.
See soap (2).

ferrous octoate. A paint drier.
See soap (2).

ferrous oxalate. (iron oxalate).
CAS: 516-03-0. $FeC_2O_4 \cdot 2H_2O$.
Properties: Pale yellow, odorless, crystalline powder. Soluble in acids, insoluble in water, d 2.28, decomposes at 160C releasing carbon monoxide.
Derivation: By the interaction of solutions of ferrous sulfate and sodium oxalate.
Hazard: Toxic. Evolves carbon monoxide on heating.
Use: Photographic developer, pigment in glass, plastics, paints.

ferrous oxide. (iron monoxide).
CAS: 1345-25-1. FeO.
Properties: Black powder, d 5.7, mp 1420C, insoluble in water, soluble in acid.
Derivation: Prepared from the oxalate by heating but the product contains some ferric oxide.
Use: Catalyst, glass colorant, steel manufacture.

ferrous phosphate. CAS: 14940-41-1.
Fe$_3$(PO$_4$)$_2$·8H$_2$O.
Properties: Bluish-gray powder, d 2.58, soluble in inorganic acids, insoluble in water, hygroscopic.
Use: Catalyst, ceramics.

ferrous phosphide. (iron phosphide).
CAS: 1310-43-6. Fe$_2$P.
Properties: Bluish-gray powder, d 6.56, mp 1290C, ferromagnetic, insoluble in water.
Grade: 24–25% phosphorus.
Hazard: Evolves toxic and flammable products on exposure to moisture or acids.
Use: Iron and steel manufacture.

ferrous selenide. CAS: 1310-32-3. FeSe.
Properties: Black, shiny solid; d 6.8. Almost insoluble in water, soluble in hydrochloric acid evolving selenium hydride.
Use: Semiconductor technology.

ferrous sulfate. (iron sulfate; iron vitriol; copperas; green vitriol; sal chalybis).
CAS: 7720-78-7. FeSO$_4$·7H$_2$O.
Properties: Greenish or yellow-brown crystals or granules, odorless, soluble in water with saline taste, insoluble in alcohol, d 1.89, mp 64C, loses 7H$_2$O by 300C, pH 3.7 (10% solution), hygroscopic.
Derivation: (a) Byproduct from the pickling of steel and many chemical operations, (b) by action of dilute sulfuric acid and iron, (c) oxidation of pyrites in air followed by leaching and treatment with scrap iron.
Method of purification: Recrystallization.
Grade: Technical, anhydrous, CP, USP.
Hazard: Ingestion causes intestinal disorders.
Use: Iron oxide pigment, other iron salts, ferrites, water and sewage treatment, catalyst especially for synthetic ammonia, fertilizer, feed additive, flour enrichment, reducing agent, herbicide, wood preservative, process engraving.

ferrous sulfide. (iron sulfide; iron protosulfide).
CAS: 1317-37-9. FeS.
Properties: Dark brown or black metallic pieces, sticks, or granules; soluble in acids, insoluble in water, d 4.75, mp 1195, bp decomposes.
Derivation: By fusing iron and sulfur.
Use: Generating hydrogen sulfide, ceramics, other sulfides, pigment.
See also pyrite.

ferrovanadium. CAS: 12604-58-9. An iron-vanadium alloy used to add vanadium to steel. Vanadium is used in engineering steels to the extent of 0.1–0.25% and in high speed steels to the extent of 1–2.5% or higher. Melting range 1482–1521C. Furnished in a variety of lump, crushed, and ground sizes.
Derivation: By reduction of the oxide with aluminum or silicon in presence of iron in an electric furnace.
Grade: Available containing from 50 to 80% vanadium.
Hazard: Moderate fire risk. TLV: (dust) 1 mg/m^3 of air.

ferrozirconium. Alloys used in the manufacture of steel, 12–15% zirconium alloy. Approximate analysis: zirconium 12–15%, silicon 39–43%, iron 40–45%. Application: steel of high silicon content. 35–40% zirconium alloy: approximate analysis: zirconium 35–40%, silicon 47–52%, iron 8–12%. Application: steel of low silicon content.

ferrum. Latin name for iron, hence the symbol Fe.

fertile material. In nuclear technology, any substance not capable of fission but which can be converted into a fissionable material in a nuclear reactor. Uranium 238 (converted to plutonium 239) and thorium 232 (converted to uranium 233) are the most important fertile materials.

fertilizer. A substance or mixture that contains one or more of the primary plant nutrients and sometimes also secondary and/or trace nutrients. The primary nutrients are nitrogen (supplied as anhydrous ammonia or solutions containing nitrogen derived from ammonia, ammonium nitrate, or urea), phosphorus (as superphosphates derived from phosphate rock), and potassium (in the form of KCl from sylvite ore or natural brines). Secondary nutrients are calcium, magnesium, and sulfur. Trace elements (iron, copper, boron, manganese, zinc, and molybdenum) are also among the 12 elements considered essential for plant growth. Nitrogen solutions and anhydrous ammonia are used both in fertilizer manufacture and for direct application to the soil. Substantial amounts of both separate materials and mixtures are used in liquid form. Controlled-release fertilizers are those whose particles are coated with polymeric sulfur by a proprietary process. Their advantages include more uniform supply of nutrient, lower labor costs, and reduced leaching losses in areas of irrigation and high rainfall.
See also superphosphate, nutrient solution. For further information, refer to National Fertilizer Solutions Association, Peoria, IL.

FFA. Abbreviation for free fatty acid.
Use: Describing specifications for fatty esters, glycerides, oils, etc.

FFPA.   Abbreviation for "free from prussic acid."

FGAN.   Fertilizer grade ammonium nitrate.
Use: In blasting agents as well as fertilizers because its coating of kieselguhr and its prilled form, making it safer to handle than the usual grades.

fiber.   A fundamental form of solid (usually crystalline) characterized by relatively high tenacity and an extremely high ratio of length to diameter (several hundred to one). Natural fibers are animal, e.g., wool and silk (proteins), vegetable, e.g., cotton (cellulose), and mineral (asbestos). Cotton fiber is called staple and rarely exceeds 2 inches in length.

Semisynthetic fibers include rayon and inorganic substances extruded in fibrous form, such as glass, boron, boron carbide, boron nitride, carbon, graphite, aluminum silicate, fused silica, and some metals (steel). Synthetic fibers are made from high polymers (polyamides, polyesters, acrylics, and polyolefins) by extruding from spinnerets (nylon, "Orlon," etc.). Some are being used in specialty papers, though the primary use is in textile fabrics. For ceramic fibers see "Fiberfax."

Metal fibers are used in several ways: (1) As "whiskers," which are single-crystal fibers up to 2 inches long having extremely high tensile strength; they are made from tungsten, cobalt, tantalum, and other metals, and are used largely in composite structures for specialized functions. (2) As filaments, which are alloys drawn through diamond dies to diameters as small as 0.002 cm; steel for tire cord and antistatic devices has been developed for such applications. (3) In biconstituent structures composed of a metal and a polymeric material; for example, aluminum filament covered with cellulose acetate butyrate.

Hollow fibers of cellulose acetate and nylon are used as membranes in the reverse osmosis method of water purification.
See also filament, denier, whiskers, glass fiber, and following entries.

fiber, biconstituent.   A composite fiber comprising a dispersion of fibrils of one synthetic material within, and parallel to the axis of another, also a fiber made up of polymeric material and a metal or alloy filament.

fiberfill.   A fiber designed specifically for use as a filling material in such products as pillows, comforters, quilted linings, furniture battings, e.g., sisal, jute.

"Fiberfrax."[280]   TM for ceramic fiber made from alumina and silica. Available in bulk as blown, chopped and washed, long staple, paper, rope, roving, blocks.
Properties: Retains properties to 1260C and under some conditions used to 1648C, light weight, inert to most acids and unaffected by hydrogen atmosphere, resilient.
Use: High-temperature insulation of kilns and furnaces, packing expansion joints, heating elements, burner blocks; rolls for roller hearth furnaces and piping, fine filtration, insulating electrical wire and motors, insulating jet motors, sound deadening.

fiber gear.   A driver gear made of a material of somewhat lower strength than the driven gear (cast iron); for example, a composite such as fiberglass-reinforced plastic or an engineering plastic, e.g., nylon. It is intended to fail under overload, thus protecting the driven master gear from destructive stress.

"Fiberglas."[191]   TM for a variety of products made of or with glass fibers or glass flakes including insulating wools, mats and rovings, coarse fibers, acoustical products, yarns, electrical insulation and reinforced plastics.
See also glass fiber, reinforced plastic.

fiber glass.   See glass fiber.

fiber, graphite.   See graphite fiber.

fiber, optical.   A fine-drawn silica (glass) fiber or filament of exceptionally high purity and specific optical properties (refractive index) that transmits laser light impulses almost instantaneously with high fidelity. Such fibers are made from quartz coated with germanium-doped silica by vapor deposition; 100 or more filaments are assembled into a cable which has extremely high data-carrying capacity. These are applicable not only to telephonic communication systems, for which they are now being used, but also to remote-sensing devices which permit analysis of samples at widely separated locations. Thus, one of the most important developing uses of optical fibers is in analytical instrumentation. As they are nonelectrical and noncorrosive, optical fiber cables are safe to use in highly toxic or explosive environments, e.g., radioactive separations and hazardous waste analyses. The laser beam is coupled to the end of the cable (which may be up to 1000 meters long) by a device called an optrode; the light traverses the cable and interacts with the sample, eliciting a signal that is reflected back through the same cable to a spectrometer. Fiber optics are also used in other forms of instrumentation, e.g., radiation dosimeters and high-temperature thermometers. In the latter case, the fibers are made from single crystals of alumina.

See also glass, optical; laser; thermometer (5).

fiber-reactive dye. See dye, fiber-reactive.

fiber roll. A calender roll constructed of specially prepared papers or fabrics on a steel base. The fibrous material is cut into circular sheets with a hole at the center; these are stacked on a steel core and then compressed under high pressure, producing a dense, hard material with a smooth surface. As such, it is used in supercalenders for paper finishing. An intaglio design can be impressed upon it by an engraved steel roll; this operation requires several days and is facilitated by application of water, soap, or other softener. So prepared, it is used in embossing calenders for applying decorative patterns on special paper or plastic products.

See also supercalender.

fibrid. Generic name for a fibrous form of synthetic polymeric material used for example as a binder material in the manufacture of textryl.

fibrin. An insoluble blood protein resulting from the hydrolysis of fibrinogen by the action of thrombin; it polymerizes to form blood clots. Recent research has found that it forms a protective coating on tumors which inhibits antigenic activity, thus protecting the tumor and neutralizing the immune system of the organism.

fibrinogen. A sterile fraction of normal human blood plasma, dried from the frozen state. In solution, it has the property of being converted into insoluble fibrin when thrombin is added. It is an essential factor in blood-clotting mechanism.

Properties: White or grayish amorphous substance.

Grade: USP.

Use: Medicine (coagulant).

fibrinolysin. (plasmin). A proteolytic enzyme which dissolves fibrin and hastens the solution of clots that may form in the bloodstream. It is prepared by activating a fraction of normal human plasma with highly purified streptokinase.

fibroin. The fibrous material in silk, a scleroprotein containing glycine and alanine, light yellow silk-like mass, insoluble in water, soluble in concentrated alkalies and concentrated acids.

fibrolite. See sillimanite.

ficin. A proteolytic enzyme hydrolyzing casein, collagen, edestin, fibrin, liver and other protein-like material.

Properties: Buff to cream-colored powder with an acrid odor, very hygroscopic, partially soluble in water, insoluble in organic solvents.

Source: Fig latex or sap, commercially prepared by filtering and drying the latex.

Use: Food industry, bating leather, tenderizing meat, shrinkproofing wool, coagulation of milk, chillproofing beer, Rh factor determination.

"Ficoll."[485] TM for an inert, non-ionized, synthetic high polymer made by the crosslinking reaction of epichlorohydrin and sucrose. Contains no ionized groups, and its extreme solubility is due to the high content of hydroxyl groups (23%).

Use: Density gradients for centrifugation, electrophoresis, and specific gravity determinations; component in the isotonic solutions for preparation of cells and cell fragments; concentrating solutions of sensitive materials by means of dialysis.

fictile. Descriptive of certain molecules which have no permanent structure, but are constantly changing their shapes and arrangements. An example is the metal carbonyl $Fe_3(CO)_{12}$ in which, according to Dr. F. Albert Cotton, originator of the term, "carbonyl groups readily move from one iron atom to another through the rapid formation and dissolution of carbonyl bridges between iron atoms."

field-ion microscope. A type of microscope whose unique feature is that it has no lens system. Invented by Muller in 1951, it is capable of resolving metal atoms 2–3 Å in diameter. Its essential components are an evacuated glass chamber through which runs a wire carrying an electric impulse of 30,000 volts which establishes a field strength of 500 million volts/cm. A specimen of the metal to be observed which is machined to an extremely fine tip and is positively charged is connected to the wire. An inert gas such as helium or neon is then admitted. The positively charged tip of the specimen attracts electrons from the helium atoms creating positive helium ions. These are strongly repelled by the metal atoms and stream to the negatively charged fluorescent screen producing an image of the individual atoms of the metal. Magnifications of one million times or more have been obtained of atoms of indum, tungsten, and others.

See also electron microscope.

Fieser, Louis F. (1899-1977) A distinguished American chemist, Fieser became Professor of

Organic Chemistry at Harvard in 1930 after teaching for several years at Bryn Mawr. He achieved the synthesis of vitamin K_1 and did fundamental research on cortisone, the chemistry of steroids and aromatic carcinogens. His achievements as a chemist and educator were recognized throughout the world. Unique in his facility in laboratory demonstration and as a lecturer and author, he exemplified that rare combination of a great teacher and a profound scholar.

filament. A continuous fiber usually made by extrusion from a spinneret (nylon, rayon, glass, polyethylene). It also may be a drawn metal (tungsten, gold) or a metal carbide.
See fiber.

filament winding. The process of winding fibers under tension on to a prepared core. Before or during the winding operation, the assembly is impregnated with a thermosetting resin. Structures of considerable size and strength can be made in this way. The fibers used are chiefly glass, boron, or silicon carbide.
See filament.

filler. (1) An inert mineral powder of rather high specific gravity (2.00–4.50) used in plastic products and rubber mix to provide a certain degree of stiffness and hardness and to decrease cost. Examples are calcium carbonate (whiting), barytes, blanc fixe, silicates, glass spheres and bubbles, slate flour, soft clays, etc. Fillers have neither reinforcing nor coloring properties, and the term should not be applied to materials that do, i.e., reinforcing agents or pigments. Fillers are similar to extenders and diluents in their cost-reducing function; exact lines of distinction between these terms are difficult, if not impossible, to draw. Use of fillers and extenders in plastics has increased in recent years due to shortages of basic materials.
(2) The cross or transverse thread in a fabric or other textile structure.
(3) A metal or alloy used in brazing and soldering to effect union of the metals being joined.
See also diluent, extender, reinforcing agent.

film. An extremely thin continuous sheet of a substance which may or may not be in contact with a substrate. There is no precise upper limit of thickness, but a reasonable assumption is 0.010 inch. The protective value of any film depends on its being 100% continuous, i.e., without holes or cracks, since it must form an efficient barrier to molecules of atmospheric water vapor, oxygen, etc. A long-chain fatty acid or alcohol on water produces a film whose "thickness" is the length of one molecule (approximately 200 Å). The fatty

acid molecules are oriented with the radical end in the water. Such films are good evaporation barriers and have been successfully imposed on glass. Soap bubbles are elastic films about one micron thick and have considerable strength.

Film-forming agents (drying oils) are essential in paints and lacquers. Oxide films formed automatically on the surface of aluminum protect it from corrosion. Thin metallic oxide films are widely used in electronic and semiconducting devices. Electro-deposited metals (chromium, copper, nickel) are conventionally (and perhaps illogically) called coatings.

The term film is also applied to sheets of cellophane, polyethylene, polyvinylidene chloride, etc., used for wrapping and packaging of food products, meats, and poultry (especially shrink films which are stretched before application). These function as a moisture vapor barrier. Plastic films are also used as slip surfaces in concrete structures such as air strips, ice rinks, and highways. Photographic film is made from cellulose acetate.

filter. See filtration; leaf, filter; baghouse.

filter aid. See filter media, filtration.

filter alum. See aluminum sulfate.

filter media. Almost any water-insoluble porous material having a reasonable degree of rigidity can serve as a filter. Sand is used in simple large-scale water filtration, the voids between the grains providing the porosity. In industrial operations, cotton duck, woven wire cloth, nylon cloth, and glass cloth are used. For laboratory work, Whatman filter paper, diatomaceous earth, and closely packed glass fibers are standard materials. Plastics membranes containing over a million pores per square inch are used in bacteriological filtration.
See also filtration, screen.

filter sand. Sand used to separate sediment and suspended matter from water.

"Filtrasorb Carbon."[323] TM for specially engineered activated carbon granules having mechanical strength great enough to withstand repeated filter backwashings and regeneration at high temp.
Use: In conjunction with polymeric water treatment chemicals for coagulating solids in raw sewage with subsequent filtration over the carbon granules.

filtration. The operation of separating suspended solids from a liquid (or gas) by forcing the mix-

ture through a porous barrier (see filter media). The construction and operation of the many kinds of industrial filtration equipment are too detailed to permit description. The most widely used types may be classified as follows: (1) gravity filters, used largely for water purification and consisting of thick beds of sand and gravel which retain the flocculated impurities as the water passes through, (2) pressure filters of plate-and-frame or shell-and-leaf construction which utilize filter cloths of coarse fabric as a separating medium, (3) vacuum or suction filters of the rotating drum or disk type, used on thick sludges and slurries, (4) edge filters, (5) clarification filters, (6) bag filters (dust collectors). Gel filtration is a chromatographic technique involving separation at the molecular level. For bacteriological filtration, membranes having over a million pores per square inch are used, e.g., collodion or synthetic film. Some types of viruses will pass through such membranes and are thus known as filterable viruses.
See baghouse.

"Filtrol."[217] TM for acid-activated clays used as decolorizing adsorbents and catalysts.

fine chemical. A chemical produced in comparatively small quantities and in a relatively pure state. Examples are pharmaceutical and biological products, perfumes, photographic chemicals, and reagent chemicals.

fines. The portion of a powder composed of particles which are smaller than a specified size (MPA definition, MPA Standard 9–50T).

finishing compounds. Materials that impart softness, flexibility, stiffness, color, water and fire resistance, etc.
Use: In the final or finishing stages of manufacture of a product, usually textiles and leather, to make them suitable for specific purposes.

Finkelstein reaction. Reaction of alkyl halides with sodium iodide in acetone.

fireclay. See refractory.

fire extinguishment. Fires are divided into 4 classes, each requiring special treatment. The essential point in extinguishing all types is exclusion of air from the fire by an effective means.
Class A includes fires in combustible materials, such as wood, paper, and cloth where the quenching and cooling effect of quantities of water or of solutions containing a high percentage of water is of first importance. Fire extinguishers utilizing the pressure of carbon dioxide to throw

a stream of water onto the fire are the most widely used for this class. In the soda-acid extinguisher, the carbon dioxide is generated within the cylinder at the time of use. In another type, carbon dioxide gas is stored in the cylinder under pressure and is released by opening a valve.
Class B includes fires in flammable liquids where a blanketing or smothering effect is essential. Carbon dioxide gas, dry chemical, or foam are suitable. Water should not be used.
Class C includes fires in electrical equipment. The use of carbon dioxide gas or dry chemical extinguishers is recommended. Water should not be used.
Class D fires are burning metals. A powder formulation such as "Met-L-X" powdered graphite or trimethoxyboroxine will extinguish a metal fire; water should not be used.
In general, for small fires, salt (sodium chloride) and sodium bicarbonate are effective, either dry or in concentrated solution. Carbon tetrachloride and methyl bromide are to be avoided as extinguishing agents because of the toxicity of their decomposition products, for example, phosgene.
See also foam, fire-extinguishing.

"Firefrax."[442] TM for a group of refractory cements made from kaolin or fireclay base materials for applications where aluminum silicate cements are best suited.
Use: Laying and repairing fireclay and silica brick work, bond for crushed firebrick or ganister for patching furnace linings and for making rammed-up or monolithic linings, patching materials for byproduct coke ovens, and as a wash for small pouring ladles in non-ferrous foundry.

fire point. The lowest temperature at which a liquid evolves vapors fast enough to support continuous combustion. It is usually close to the flash point.
See also autoignition temperature.

fire-retarding agent. See flame retarding agent.

fire sand. See furnace sand.

"Fi-Retard."[300] TM for a group of flame retardants for cotton, rayon, nylon, and paper.

fir needle oil. (fir oil). An essential oil obtained by the steam distillation of needles and twigs of several varieties of coniferous trees (Abies) native to both Canada and Siberia.
Use: Odorant in perfumery, flavoring agent.

Fischer, Emil. (1852-1919) A German organic chemist, recipient of Nobel Prize in chemistry

(1902) for his original research in the chemistry of purines and sugars. He was Professor of Chemistry at the University of Berlin (1882), succeeding Hofmann. He synthesized fructose and glucose, and elucidated their stereochemical configurations; he also established the nature of uric acid and its derivatives. Additional work included enzyme chemistry, proteins, synthetic nitric acid, and ammonia production.

Fischer, Ernst Otto. (1918-) A German inorganic chemist who won this Nobel prize for chemistry in 1973 with Wilkinson for their independent work on the chemistry of organometallic "sandwich compounds." He was the contributor to many publications on organometallic chemistry. His education and work was primarily in Munich.

Fischer, Hans. (1881-1945) A German biochemist who studied under Emil Fischer. He was awarded the Nobel Prize in chemistry in 1930 for his synthesis of the blood pigment hemin. He also did important fundamental research on chlorophyll, the porphyrins, and carotene.

Fischer-Hepp rearrangement. Rearrangement of secondary aromatic nitrosamines to p-nitrosoarylamines.

Fischer indole synthesis. Formation of indoles on heating aryl hydrazones of aldehydes or ketones in the presence of catalysts such as zinc chloride or other Lewis acids, or proton acids.

Fischer oxazole synthesis. Condensation of equimolar amounts of aldehyde cyanohydrins and aromatic aldehydes in dry ether in the presence of dry hydrochloric acid.

Fischer peptide synthesis. Formation of polypeptides by treatment of an α-chloro or α-bromo acyl chloride with an amino acid ester, hydrolysis to the acid, and conversion to a new acid chloride which is again condensed with a second amino acid ester, and so on. The terminal chloride is finally converted to an amino group with ammonia.

Fischer phenylhydrazine synthesis. Formation of arylhydrazines by reduction of diazo compounds with excess sodium sulfite and hydrolysis of the substituted hydrazine sulfonic acid salt with hydrochloric acid. The process is a standard industrial method for production of arylhydrazines.

Fischer phenylhydrazone and osazone reaction. Formation of phenylhydrazones and osazones by

heating sugars with phenylhydrazine in dilute acid.

Fischer's reagent. A reagent used as a test for sugars.
Preparation: Three parts of sodium acetate and two parts of phenylhydrazine hydrochloride in 20 parts of water.
Note: Do not confuse with Karl Fischer reagent.

Fischer's salt. See cobalt potassium nitrite.

Fischer-Speier esterification method. Esterification of acids by refluxing with excess alcohol in the presence of hydrochloric acid or other acid catalysts.

Fischer-Tropsch Process. Synthesis of liquid or gaseous hydrocarbons or their oxygenated derivatives from the carbon monoxide and hydrogen mixture (synthesis gas) obtained by passing steam over hot coal. The synthesis is carried out with metallic catalysts such as iron, cobalt, or nickel at high temperature and pressure. The process was developed in Germany in 1923 by F. Fischer and H. Tropsch, and was used there for making synthetic fuels before and during World War II. It has never been used for this purpose in the US; the only coal to gasoline conversion plant using this process is Sasol in South Africa, though the closely related Lurgi process is being used rather extensively in a number of locations. Easing of the petroleum crisis has tended to diminish conversion activity in the US.

Fischer-Tropsch syntheses. (Synthol process; oxo synthesis). Synthesis of hydrocarbons, aliphatic alcohols, aldehydes, and ketones by the catalytic hydrogenation of carbon monoxide using enriched synthesis gas from passage of steam over heated coke. The ratio of products varies with conditions. The high-pressure Synthol process gives mainly oxygenated products and addition of olefins in the presence of cobalt catalyst (oxo synthesis) produces aldehydes. Normal-pressure synthesis leads mainly to petroleum-like hydrocarbons.

fisetin. (3,7,3'-Tetrahydroxyflavone). $C_{15}H_{10}O_6$. See flavanol.

Fisher's solution. See physiological salt solution.

fish glue. An adhesive derived from the skins of commercial fish (chiefly cod). A ton of skins yields about 50 gals of liquid glue. Bond strength on wood is approximately 2500 psi, pH approximately 6.5–7.2. Compatible with animal glues, some dextrins, some polyvinyl acetate emulsions and with rubber latex. Chief applications are in

gummed tape, cartons, blue print paper and letterpress printing plates. Fish glue can be made light-sensitive by adding ammonium bichromate and water-insoluble by UV radiation, hence its usefulness in the photoengraving process.
See also adhesive.

fish-liver oil. An oil containing a high percentage of vitamin A, high-potency livers as from cod, shark, and halibut, contain from 100,000 to 1,500,000 A units/g. The oil is extracted by cooking the livers under low pressure steam and removing the oil which floats on the condensate. Livers of low oil content are processed with a weak solution of sodium hydroxide or sodium carbonate which extracts the oil in emulsified form.
Use: Medicine and dietary supplement.
See also fish oil.

fish meal. A fishery byproduct consisting essentially of processed scrap from the filleting operation or from whole fish. In the dry process, the waste from cod, halibut, and haddock heads is disintegrated and dried. The oil and proteins are largely retained. In the wet process, the whole fish (chiefly menhaden and pilchard) are used. These are steam-cooked and run through a screw press to remove the oil. The resulting meal is then dried and packed. Its chief use is now for animal feeds and as a raw material for fish protein concentrate.
Hazard: Flammable, strong tendency to spontaneous heating.

fish oil. A drying oil obtained chiefly from menhaden, pilchard, sardine and herring. Extracted from the entire body of the fish by cooking and compressing. Should not be confused with fish-liver oil. It contains approximately 20% polyunsaturated fatty acids, which enables it to lower cholesterol content of the human diet. Chemically modified fish oil is used in soaps, detergents, protective coatings, and alkyd resins. The hydrogenated product is used as a base for margarines and shortenings and as an industrial dispersing agent.
Hazard: Subject to spontaneous heating.

fish protein concentrate. (FPC). A flour or paste-like product prepared from whole fish including bones and viscera, of a size and type not acceptable for sale as such. Both biological (enzymatic) and chemical (solvent extraction) methods are used to obtain the proteins.

fissile. Synonymous with fissionable.
See fission.

fissiochemistry. The process by which a chemical change or reaction is brought about by nuclear energy, for example, the production of anhydrous hydrazine from liquid ammonia in a nuclear reactor.

fission. The splitting of an atomic nucleus induced by bombardment with neutrons from an external source and propagated by the neutrons so released. When a fissionable (unstable) nucleus, such as ^{235}U or ^{239}Pu, is struck by a neutron in a critical area, the following events occur: (1) the nucleus disintegrates to form several other elements, called fission products or fragments, all of which are radioactive and have high kinetic energy; (2) the disrupted nucleus emits an average of 2.5 neutrons (^{235}U) or three neutrons (^{239}Pu), which in turn split other nuclei of the fissionable material in a chain reaction which is self-perpetuating, (3) it also emits the energy equivalent of the mass defect of the nucleus, usually approximately 200 MeV per nucleus, some of which is in the form of gamma rays.

A nuclear explosion will not occur until a critical mass of fissionable material is attained, that is, the smallest amount capable of sustaining a chain reaction. Similarly, a nuclear reactor will not produce power until the assembly achieves a critical activity. This occurs as follows. The neutrons introduced to the system are continually escaping, some are lost through the walls, some are captured by structural materials, and some are absorbed by the fissionable atoms themselves without fission taking place. When the neutrons entering the system are very slightly in excess of those lost to it, the assembly is said to be critical and measurable power generation takes place. The ratio is carefully controlled, the rate of energy production rising exponentially. Control rods made of cadmium absorb neutrons so readily that the reactor can function at precisely predetermined levels of activity. Nuclear fission is used for electric power generation and for making radioactive isotopes.
See also nuclear reactor.

Fittig's synthesis. The preparation of aromatic hydrocarbons by condensation of aryl halides with alkyl halides in the presence of metallic sodium.

fix. (1) To cause an unreactive element, e.g., nitrogen to combine into a chemical compound, as in ammonia synthesis. (2) To hold a dye permanently on a fiber or fabric by chemical or mechanical action or a combination of both. (3) To retard the evaporation rate of the volatile components of essential oils, as in perfumes.
See also nitrogen fixation. See also following entries.

fixation. See nitrogen fixation.

fixative. (1) See fixing agent, perfume; (2) a substance applied as a spray or solution to harden and preserve objects for microscopic examination, or to pencil and ink drawings to prevent blurring, e.g., a sodium silicate solution.

fixed oil. A nonvolatile, fatty oil characteristic of vegetables as opposed to the volatile essential oils of flowers.

fixing agent, chemical. (1) A substance which aids fixation of mordants upon textiles by uniting chemically with them and holding them on the fiber until the dyes can react with them. (2) A substance which causes actual precipitation of mordant on the fiber by double decomposition.

fixing agent, mechanical. (1) A substance (e.g., albumin) capable of holding pigments permanently upon textile fibers. (2) Certain gums and starches which hold dyes and other substances upon textile fibers long enough to permit a desirable reaction to take place.

fixing agent, perfume. (fixative). A substance which prevents too rapid volatilization of the components of a perfume and tends to equalize their rates of volatilization. It thus increases the odor life of a perfume and keeps the odor unchanged. For many years, the chief fixatives were animal products (ambergris, civet, musk, castoreum), but these have been largely replaced by synthetics.
See also perfume.

flake lead. See lead carbonate, basic.

flame cracking. See ethylene (note).

"Flamenol."[245] TM for electrical conductors insulated with a vinyl halide resin such as plasticized polyvinyl chloride.

flame-retarding agent. A substance applied to or incorporated in a combustible material to reduce or eliminate its tendency to ignite when exposed to a low-energy flame such as a match or cigarette. There are three methods of application: (1) as a coating or surface finish (non-durable, readily removed), (2) in solution form to penetrate the fibers (semi-durable, reasonably stable), and (3) as an integral part of the polymer structure of a synthetic fiber (durable, not removable). The latter method provides permanent protection as it not only makes the material self-extinguishing but cannot be leached out by laundering or drycleaning. Substances commonly used in methods (1) and (2) include such inorganic salts as ammonium sulfamate, zinc borate, and anti-

mony oxychloride, chlorinated organic compounds such as chlorendic anhydride, alumina trihydrate, and certain organic phosphates and phosphonates. Method (3) is exemplified by a polyester fiber TM "Trevira 271" composed of polyethylene terephthalate and an undisclosed flame-retardant chemically linked to the polymer molecule. A copolymer of styrene and phosphazene has also been researched. Certain types of fibers (polyamides and aramids) are inherently flame-retardant, e.g., nylon, "Nomex," "Kevlar."
Use: Carpets, rugs, upholstery, plastics used in construction and miscellaneous wearing apparel.

flammability. The ease with which a material (gas, liquid, or solid) will ignite either spontaneously (pyrophoric), from exposure to a high-temperature environment (autoignition) or to a spark or open flame. It also involves the rate of spreading of a flame once it has started. The more readily ignition occurs, the more flammable the material; less easily ignited materials are said to be combustible, but the line of demarcation is often indefinite and depends on the state of subdivision of the material as well as on its chemical nature. The Flammable Fabrics Act establishes standards of flammability to which all textile manufacturers must conform.
See also flammable material, combustible material.

flammable material. Any solid, liquid, vapor, or gas that will ignite easily and burn rapidly. Flammable solids are of several types: (1) dusts or fine powders (metals or organic substances such as cellulose, flour, etc.); (2) those that ignite spontaneously at low temperatures (white phosphorus); (3) those in which internal heat is built up by microbial or other degradation activity (fish meal, wet cellulosic materials); (4) films, fibers, and fabrics of low-ignition point materials.

Flammable liquids are defined by the National Fire Protection Association and the Department of Transportation (DOT) as those having a flash point (flash p) less than 100F (37.7C) (CC) and a vapor pressure of not over 40 psia at 100F.

Flammable gases are ignited very easily; the flame and heat propagation rate is so great as to resemble an explosion, especially if the gas is confined. The most common flammable gases are hydrogen, carbon monoxide, acetylene, and other hydrocarbon gases. Oxygen, though essential for the occurrence of combustion, is not itself either flammable or combustible; neither are the halogen gases, sulfur dioxide, or nitrogen. Flammable gases are extremely dangerous fire hazards and require precisely regulated storage conditions. *Note:* The terms "flammable," "nonflam-

mable," and "combustible" are difficult to delimit. Since any material that will burn at any temperature is combustible by definition, it follows that this word covers all such materials, irrespective of their ease of ignition. Thus, the term "flammable" actually applies to a special group of combustible materials that ignite easily and burn rapidly. Some materials (usually gases) classified in shipping and safety regulations as nonflammable are actually noncombustible. The distinction between these terms should not be overlooked. For example, sodium chloride, carbon tetrachloride, and carbon dioxide are noncombustible; sugar, cellulose, and ammonia are combustible but nonflammable.
See also combustible material.

flash. The overflow of rubber or plastic at the parting line of a mold when subjected to full pressure. It is removed in the finishing operation. See also following entries.

flash distillation. Distillation in which an appreciable proportion of a liquid is quickly converted to vapor in such a way that the final vapor is in equilibrium with the final liquid. This method is now widely used for desalination of seawater.

flash photolysis. A method of investigating the mechanism of extremely rapid photochemical reactions involving the formation of free radicals (both inorganic and organic) by irradiating a given reaction mixture with a flash of high-intensity light, thus producing the short-lived radicals which activate photochemical reactions. These products are instantaneously analyzed spectroscopically which permits identification of the radical species from the spectra obtained. The time lapses involved in this technique are approximately 1/100,000 second. It has also been applied to study of the exceedingly fast reactions occurring in flames and explosions.
See also photochemistry, free radical, photolysis.

flash point. The temperature at which a liquid or volatile solid gives off vapor sufficient to form an ignitable mixture with the air near the surface of the liquid or within the test vessel (NFPA). For the purposes of the official shipping regulations, the flash point is determined by the Tagliabue open cup method (ASTM D1310–63), usually abbreviated TOC. (IATA also permits the Abel or Abel-Pensky closed-cup tester.) Other methods used, generally for the higher flash points, are the Tag closed cup (Tagliabue closed cup, TCC) and Cleveland open cup (COC). The open cup method more nearly approximates actual conditions.
See also flammable material.

flatting agent. A substance ground into minute particles of irregular shape and used in paints and varnishes to disperse incident light rays so that a dull or "flat" effect is produced. Standard flatting agents are heavy-metal soaps, finely divided silica, and diatomaceous earth.

flavanol. (3-hydroxyflavone). A derivative of flavanone; yellow needles melting at 169C, has violet fluorescence in concentrated sulfuric acid. It is a flavonoid pigment. Dyes cotton a bright yellow when mordanted with aluminum hydroxide. Other hydroxyflavones are chrysin, fisetin, and quercitin. Eleven different flavonols are known. Not identical with flavonol.

flavanone. (2,3-dihydroflavone). A group of colorless derivatives of flavone distributed in higher plants either in free form or as glucosides. About 25 different types have been isolated. It comprises one of the major groups of flavonoids. Examples are hesperidin and naringin.

flavanthrene. (indanthrene yellow; chloranthrene yellow). $C_{28}H_{12}O_2N_2$.
Properties: Brownish-yellow needles, soluble in dilute alkaline solutions.
Derivation: Action of antimony pentachloride on β-aminoanthraquinone in boiling nitrobenzene.
Use: Vat dye for textiles, etc.

flavianic acid. See 2,4-dinitro-1-naphthol-7-sulfonic acid.

flavin. (1) isoalloxazine. $C_{10}H_6N_4O_2$.
The nucleus of various natural yellow pigments. See riboflavin and flavin enzymes.
(2) tetrahydroxyflavanol. $C_{15}H_{10}O_7 \cdot 2H_2O$.
A yellow dye derived from oak bark.

flavine. See acriflavine.

flavin enzyme. (flavoprotein). An enzyme composed of protein linked to coenzymes which are mono- or di-nucleotides containing riboflavin. Because of their distinctive color, they are also called "yellow enzymes." The flavin enzymes function in tissue respiration as dehydrogenases, the hydrogen atoms being taken up by the riboflavin group.

flavin mononucleotide. See riboflavin phosphate.

flavone. (2-phenylchromone). One of a group of flavonoid plant pigments existing as colorless needles, insoluble in water and melting at 100C. It fluoresces violet in concentrated sulfuric acid. It can be synthesized. Treatment with alcoholic alkali yields flavanone. The flavones produce ivory and yellow colors in plants and flowers. See also flavonoid.

flavonoid. A group of aromatic, oxygen-containing, heterocyclic pigments widely distributed among higher plants. They constitute most of the yellow, red, and blue colors in flowers and fruits. Exceptions are the carotenoids. The flavonoids include the following subgroups: (1) catechins; (2) leucoanthocyanidins and flavanones; (3) flavanins, flavones, and anthocyanins; and (4) flavonols.
For details consult specific entries.

flavonol. (flavon-3-ol). A flavonoid plant pigment giving ivory and yellow colors to flowers. Not identical with flavanol.

flavoprotein. See flavin enzyme.

flavor. (1) The simultaneous physiological and psychological response obtained from a substance in the mouth that includes the senses of taste (salty, sour, bitter, sweet), smell (fruity, pungent), and feel. The sense of feel as related to flavor encompasses only the effect of chemical action on the mouth membranes such as heat from pepper, coolness from peppermint, and the like (Institute of Food Technologists). No reliable correlation of taste with chemical structure has yet been possible. Flavor is a critical factor in the acceptability of foods, medicines, confectionery, and beverages. Flavors are used in insect and animal baits to induce ingestion of the bait, also to prevent rodent attack on organic materials, e.g., tributyl tin in cable covers. Substances that affect flavor often have a synergistic effect (for example, monosodium glutamate and certain nucleotides). Sodium chloride is classed as a seasoning agent.
See also potentiator, enhancer.
(2) Any substance or mixture of substances that contributes a positive taste to a food product, such as vanillin, cacao, and fruit extracts among natural products, together with numerous synthetic compounds that imitate or duplicate these tastes. Undesirable or "off-flavors" occur in milk, meat, and other food products as a result of improper preparation, oxidation, and incipient rancidity. There are over 1500 flavoring materials listed as food additives under provisions of the Food, Drug, and Cosmetics Act.
See also odor.

flax. Bast fibers, approximately 20 inches long, obtained from the stems of the linseed plant, Linum usitatissimum. Stronger and more durable than cotton. Combustible.
Use: Apparel fabrics (linens), thread, rope, twine, cigarette paper, duplicating papers.

flaxseed oil. See linseed oil.

"Flectol." H[58] TM for polymerized 1,2-dihydro-2,2,4-trimethylquinoline.
Use: Rubber antioxidant especially for belting and tire carcasses.

Fleming, Sir Alexander. (1881-1955) A Scottish biochemist and bacteriologist who discovered (1928) the bactericidal properties of molds produced from the plant Penicillium notatum. A broad spectrum of antibiotics developed from this discovery.
See also antibiotic.

"Flexamine G."[248] TM for a mixture of N,N'-diphenyl-p-phenylenediamine and a complex diarylamine-ketone reaction product.
Properties: Brownish-gray granules, d 1.20, melting range 75–90C. S, soluble in acetone, benzene, and ethylene dichloride; insoluble in water and gasoline.
Use: Antioxidant used in tires, camelback, wire insulation, neoprene belting, and soles.

"Flexane."[445] TM for room-temperature-curing urethane.

"Flexbond."[472] TM for polyvinyl acetate co-polymer emulsions.
Properties: 48–57% solids, 60–3800 cp, 4.0–6.5 pH, 0.14μ–1.5μ average particle size, nonionic, borax stable except for 803 grade, mechanically stable.
Use: Adhesive bases for paper, cloth, plastic, and leather; vehicles for house paints or industrial finishes; binders for pigments in paper coatings.

"Flexichem."[152] TM for a series of soaps composed of sodium stearate, sodium tallowate, calcium stearate, zinc stearate, potassium stearate, and sodium oleate. Food grade materials available.
Use: Gelling and thickening agents, lubricant, concrete water absorption reducing agents, plasticizer and processing aids for rubber and metal products.

"Fleximet."[152] TM for a series of soaps used as rod and wire processing lubricants.

"Flexol."[214] TM for a series of plasticizers and stabilizers including phthalates, adipates, polyalkylene glycol derivatives, polymeric epoxies, decanoates, octoates, hexoates, tri(2-ethylhexyl) phosphate and dibutyltin dilaurate.
Use: Film and sheeting, flooring, coated fabrics, wire and cable and other extrusions, organosols, plastisols and plastigels, lacquers, and rubbers.

"Flexo Wax C."[73]   TM for a synthetic, noncrystalline, hydrocarbon wax. Mp 66–74C, has cold flow of 63–65C, high adhesive and waterproofing properties.
Use: Adhesives, coatings for paper, greaseproofing, lacquers and enamels, nickel alloy stamping, plating, water-repellent coatings.

"Flexricin."[202]   TM for a group of castor oil derivatives including (1) alkyl ricinoleates, compatible with cellulosic resins, polyvinyl butyral, polyamide, rosin and shellac and used as plasticizers; (2) acetyl ricinoleates, the alkyl esters of acetylated ricinoleic acid used as plasticizers for nitrocellulose, ethylcellulose, polyvinyl chloride resins and copolymers, natural and synthetic rubbers and other polymers and as textile lubricants, and (3) polyol ricinoleates used for synthesis, cosmetics, brake fluids, in waxes and greases and as anti-foam agents.

"Flexzone."[248]   TM for a series of antiozonants, antioxidants and stabilizers based on p-phenylenediamine.

flint.   A crystalline form of native silica or quartz.
Properties: Color smoky gray, brownish, blackish, dull yellowish; luster waxy to greasy. Mohs hardness 6.5–7, d 2.60–2.65, more easily soluble in hot caustic alkali than crystallized quartz.
Occurrence: Europe, US.
Use: Abrasive; balls for ball mills; paint extender; filler for fertilizer, insecticides, rubber, plastics, and road asphalt; ceramics, chemical tower packing.

"Flintflex."[28]   TM for air- or force-dried organic coating system.
Use: Linings in interiors of containers that haul dry bulk ladings of edibles or chemicals. Complies with FDA Regulations. Accepted by the Meat Inspection Division of the USDA for interiors of freight cars, motor trucks, and trailers and in federally inspected meat-processing plants.

flint glass.   See glass, optical.

flit.   Proprietary insecticide containing coal-tar oil and refined petroleum. May contain chlorinated benzenes.

floatation.   Purification and/or classification of finely divided solids, e.g., clays by passing them through an airblast. Do not confuse with flotation.

flocculant.   A substance that induces flocculation. Flocculants are used in water purification, liquid waste treatment, and other special applications. Inorganic flocculants are lime, alum, and ferric chloride, polyelectrolytes are examples of organic flocculants.

flocculation.   The combination or aggregation of suspended colloidal particles in such a way that they form small clumps or tufts. The word is derived from this appearance. Carbon black displays a tendency to flocculate in rubber when improperly dispersed and some clays have the same property. Oil-well drilling muds are made alkaline to prevent flocculation of their components. Flocculation can often be reversed by agitation, as the cohesive forces are relatively weak. This is not true of other forms of aggregation (coalescence and coagulation) which are irreversible.
See also agglomeration, aggregation.

flock.   A light powder comprised of ground wood or cotton fibers used as an extender or filler in plastics, low-grade rubber, and flooring compositions.

"Flo-Gard."[177]   TM for amorphous calcium polysilicate used as an anticaking agent for salt.

"Flomet-Z."[329]   TM for a fine, white, grit-free powder containing 12.5–14.0% zinc oxide.
Use: Lubricant in powdered iron metallurgy.

flooding, chemical.   See chemical flooding.

Flood reaction.   Formation of trialkylsilyl halides from hexaalkyldisiloxanes using concentrated sulfuric acid in the presence of ammonium chloride or fluoride, or by treatment of the intermediate silane sulfates with hydrochloric acid in the presence of ammonium sulfate.

"Floropyrl."[123]   TM for diisopropyl fluorophosphate.

Flory, Paul J.   (1910-1986) An American chemist who won the Nobel prize in 1974 for his work in polymer chemistry. He published extensive work on the physical chemistry of polymers and macromolecules. He held many medals and awards. Flory received his doctorate from Ohio State University in 1934. His most recent post was the C. J. Wood professor in chemistry at Stanford.

flotation.   A method of separating minerals from waste rock or solids of different kinds from one another by agitating the pulverized mixture of solids with water, oil, and special chemicals which cause preferential wetting of solid particles of certain types by the oil, while other kinds are not wet. The unwetted particles are carried to the surface by the air bubbles and thus separated from the wetted particles. A frothing agent is also used to stabilize the bubbles in the form of a froth which can be easily separated from the body of the liquid (froth flotation). Do not confuse with floatation.

"Flovis."[73] TM for a modified polyoxyethylene fatty-acid ester.
Properties: Cream to tan solid, mp 39–42C, d 1.02 (25C), pH 5% aqueous dispersion 3–5 (25C).
Use: Textile and adhesive industries for stabilizing starch solutions, fluid or heavy paste, against "setting-up."

flow diagram. (flow sheet). A chart or line drawing used by chemical engineers to indicate successive steps in the production of a chemical, materials input and output, byproducts, waste, and other relevant data.

flowers. A fine powder usually resulting from sublimation of a substance, e.g., flowers of sulfur. The term is now obsolete.

flowers of Benjamin. See Benzoic acid.

flox. A mixture of liquid fluorine (30%) and liquid oxygen (70%), designed for use as a space vehicle propellant.
Hazard: Explosively flammable.

fluid. Any material or substance that changes shape or direction uniformly in response to an external force imposed upon it. The term applies not only to liquids, but to gases and to finely divided solids. Fluids are broadly classified as Newtonian and non-Newtonian depending on their obedience to the laws of classical mechanics.
See also liquid, Newtonian; rheology; fluidization; hydraulic fluid.

fluid bed. See fluidization.

fluidization. A technique in which a finely divided solid is caused to behave like a fluid by suspending it in a moving gas or liquid. The solids so treated are frequently catalysts; hence, the term fluid catalysis. The fluidized catalyst, e.g., alumina-silica gel, is brought into intimate contact with the suspending liquid or gas mix, usually a petroleum fraction. Local overheating of the catalyst is greatly reduced and portions of catalyst can be easily removed for regeneration without shutting down the unit. There are also non-catalytic applications of fluidization, e.g., reduction of iron ore. Important uses of the fluidized bed process are (1) cracking of petroleum fractions, (2) gasification of coal, (3) application of organic coatings to metals (fusion bond method), (4) coal combustion, in which sulfur-bearing coal (1–1/3 inches diameter) is fed into a fluidized bed of limestone. Combustion occurs at 1600F, at which temperature the limestone is reduced to lime which reacts with the sulfur in the coal to form gypsum. This technique makes possible the use of high-sulfur coal without necessity of scrubbers. The bed material is approximately 5% coal and 95% limestone products.

fluid, supercritical. See supercritical fluid.

fluoboric acid. (fluoroboric acid; hydrogen tetrafluoroborate). CAS: 16872-11-0. HBF_4.
Properties: Colorless, strongly acid liquid; d approximately 1.84; bp 130C (decomposes); miscible with water and alcohol.
Derivation: Action of boric and sulfuric acids on fluorspar.
Grade: Technical (approximately 48%), pure.
Hazard: Highly toxic; corrosive, irritant.
Use: Production of fluoborates, electrolytic brightening of aluminum, throwing power aid in electrolytic plating baths, esterification catalyst, metal cleaning, making stabilized diazo salt.

"Fluokarb."[184] TM for an activated bonechar.
Use: Fluoride removal.

fluometuron. (N-(3-trifluoromethylphenyl)-N',N'-dimethylurea). CAS: 2164-17-2. $C_{10}H_{11}F_3N_2O$.
Properties: Crystalline solid, mp 163C, partially soluble in water, soluble in alcohol and acetone.
Derivation: Reaction of dimethylamine with 3-trifluoromethylphenyl isocyanate.
Hazard: Toxic by ingestion.
Use: Herbicide.

fluor. See phosphor.

fluorol. See sodium fluoride.

fluophosphate alkyl ester. See diisopropyl fluorophosphate.

fluoranthene. (idryl). CAS: 206-44-0. $C_{16}H_{10}$. A tetracyclic hydrocarbon.
Properties: Colored needles, fp 107C, bp 250C (60 mm), insoluble in water, soluble in ether and benzene. Combustible.
Derivation: From coal tar.

fluorapatite. See apatite.

fluorbenside. $ClC_6H_4CH_2SC_6H_4F$. (Generic name accepted for p-chlorobenzyl-p-fluorophenyl sulfide). Crystals, mp 36C, insoluble in water, soluble in acetone and oils.
Use: Acaricide.

"Fluorel."[158] TM for a fully-saturated fluorinated polymer containing more than 60% fluorine by weight.
Use: O-rings, gaskets; hoses, wire and fabric coatings, diaphragms, fuel cells, expellent bladders, sealants, insulation, containers.

fluorene. (α-diphenylenemethane).
CAS: 86-73-7.

Properties: Small, white, crystalline plates; fluorescent when impure; soluble in alcohol, ether, benzene, and carbon disulfide; insoluble in water; mp 116C; bp 295C (decomposes). Combustible.
Derivation: By reduction of diphenylene ketone with zinc, from coal tar.
Grade: Technical, 98% pure.
Use: Resinous products, dyestuffs.

9-fluorenone. CAS: 486-25-9. $C_{13}H_8O$.
Properties: A solid with mw 180.21, mp 82-85C, bp 342C.
Use: Intermediate.

fluorescein. (resorcinolphthalein; diresorcinolphthalein). CAS: 2321-07-5. $C_{20}H_{12}O_5$
Properties: Orange-red, crystalline powder; very dilute alkaline solutions exhibit intense greenish-yellow fluorescence by reflected light while the solution is reddish-orange by transmitted light; mp decomposes at 290C; soluble in dilute alkalies, boiling alcohol, ether, and dilute acids, glacial acetic acid; insoluble in water, benzene, chloroform. Combustible.
Derivation: By heating phthalic anhydride and resorcinol.
Grade: The sodium salt (uranine) and potassium salt are marketed.
Use: Dyeing seawater for spotting purposes, tracer to locate impurities in wells, dyeing silk and wool, diagnostic aid in ophthalmology, indicator and reagent for bromine.
See also uranine CI 45350.

fluorescence. A type of luminescence in which an atom or molecule emits visible radiation in passing from a higher to a lower electronic state. The term is restricted to phenomena in which the time interval between absorption and emission of energy is extremely short (10^{-8} to 10^{-3} second). This distinguishes fluorescence from phosphorescence, in which the time interval may extend to several hours. Fluorescent materials may be liquid or solid, organic or inorganic. Fluorescent crystals such as zinc or cadmium sulfide are used in lamp tubes, television screens, scintillation counters, and similar devices. Fluorescent dyes are used for labeling molecules in biochemical research.
See also phosphorescence, phosphor, resonance (2).

fluoridation. Addition to public drinking water supplies of 1 ppm of a fluoride salt for the purpose of reducing the incidence of dental caries. The chemicals most commonly used for city fluoridation programs are fluosilicic acid, sodium silicofluoride, and sodium fluoride. The concentration used has been established to be far below the permissible level of toxicity of fluorine-containing compounds in the human body. The program was successfully tested for over 20 years on local populations, and since then has been widely adopted in large cities in the US. The protection is especially effective for children, whose teeth are usually more susceptible to caries than those of adults. Fluorine is a bone-seeking element; tooth protection is due to the ability of fluoride ion to replace other ions in hydroxyapatite, the chief mineral component of bones and teeth. Fluorides are used in toothpastes and other dentifrices.

fluorinated ethylene-propylene resin. (FEP resin). A copolymer of tetrafluoroethylene and hexafluoropropylene with properties similar to polytetrafluoroethylene resin. The repeating structure of the molecule is $[-CF_2-CF_2-CF_2CF(CF_3)-]_n$.
See also "Teflon."
Properties: Similar to polytetrafluoroethylene, but has higher coefficient of friction. Available as extrusion and molding powder, aqueous dispersion, film, monofilament fiber, and nonsticking finish.
Use: Wire and cable insulation, pipe linings, lining for processing equipment. Fibers are used for filtration screening and mist separators.

fluorine. F CAS: 7782-41-4. Nonmetallic halogen element in group VIIA of the periodic classification. Atomic number 9, aw 18.99840, valence = 1, no stable isotopes, the most electronegative element and powerful oxidizing agent known.
Properties: Pale yellow diatomic gas or liquid, pungent odor, bp $-188C$, fp $-219C$, d of gas

1.695 (air = 1.29), d of liquid 1.108 (−188C), sp volume 10.2 cu ft/lb (21C). Reacts vigorously with most oxidizable substances at room temperature, frequently with ignition; forms fluorides with all elements except helium, neon, and argon.

Occurrence: Widely distributed to the extent of 0.03% of the earth's crust. The chief minerals are fluorapatite, cryolite, and fluorspar (Spain, Mexico, South Africa).

Derivation: Electrolysis of molten anhydrous hydrofluoric acid-potassium fluoride melts with special copper-bearing carbon anodes, steel cathodes and containers, and monel screens.

Hazard: Powerful oxidizing agent; though nonflammable, it reacts violently with a wide range of both organic and inorganic compounds and thus is a dangerous fire and explosion risk in contact with such materials. Toxic by inhalation, extremely strong irritant to tissue. TLV: 1 ppm in air.

Use: Production of metallic and other fluorides, production of fluorocarbons, active constituent of fluoridating compounds used in drinking water, toothpastes, etc.

fluorine cyanide. See cyanogen fluoride.

fluorine nitrate. CAS: 7789-26-6. FNO_3.
Properties: Gas or liquid; ignites in contact with alcohol, aniline, and ether.
Hazard: Strong oxidizing agent, liquid explodes on shock or friction.
Use: Rocket propellants.

fluoroacetic acid. CAS: 144-49-0.
CH_2FCOOH.
Properties: Colorless crystals mp 33C, bp 165C, soluble in water and alcohol.
Hazard: Toxic by ingestion.
Use: Rodenticide.

fluoroacetophenone. (phenacyl fluoride; phenyl fluoromethylketone). $C_6H_5COCH_2F$.
Properties: Brown liquid, pungent odor, bp 98C (8 mm).
Derivation: By Friedel-Crafts synthesis.
Hazard: Irritant. TLV (as F): 2.5 mg/m³.

p-fluoroaniline. CAS: 371-40-4.
$FC_6H_4NH_2$.
Properties: Liquid, d 1.1524 (25C), bp 187.4C, fp −2C, refr index 1.5395 (20C).
Hazard: Toxic material. See aniline.
Use: Intermediate (herbicides), preparation of p-fluorophenol.

fluorobenzene. (phenyl fluoride).
CAS: 462-06-6. C_6H_5F.
Properties: Colorless liquid with benzene odor, d 1.0252 (20C), refr index 1.4646 (25C), bp 84.9C,

fp −40C, insoluble in water, miscible with alcohol, ether.
Hazard: Flammable, dangerous fire risk. Irritant. TLV (as fluorine): 2.5 mg/m³ of air.
Use: Insecticide and larvicide intermediate, identification reagent for plastic or resin polymers.

fluoroboric acid. Legal label name (Air) for fluoboric acid.

fluorocarbon. Any of a number of organic compounds analogous to hydrocarbons in which the hydrogen atoms have been replaced by fluorine. The term is loosely used to include fluorocarbons that contain chlorine. These should properly be called chlorofluorocarbons or fluorocarbon chloride since it is these which deplete the ozone layer of the upper atmosphere.
See chlorofluorocarbon.
Properties: Fluorocarbons are chemically inert, nonflammable, and stable to heat up to 260–315C. They are denser and more volatile than the corresponding hydrocarbons and have low refractive indices, low dielectric constants, low solubilities, low surface tensions, and viscosities comparable to hydrocarbons. Some are compressed gases, others are liquids.
Hazard: Nonflammable, react violently with reactive substances, e.g., barium, sodium, and potassium.
Use: Refrigerants, solvents, blowing agents, fire extinguishment, lubricants and hydraulic fluids, floatation and damping fluids, dielectric, plastics, electrical insulation, wax coatings for alkali cleaning tanks, air conditioning.
Note: The rather considerable number of these compounds is designated by a number system preceded by the word "refrigerant," "propellant," "fluorocarbon" or by a TM ("Freon," "Ucon," "Genetron."). They are cross-referenced in this book as follows:

11. See trichlorofluoromethane.
12. See dichlorodifluoromethane.
13. See chlorotrifluoromethane.
14. See tetrafluoromethane.
21. See dichlorofluoromethane.
22. See chlorofluoromethane.
23. See fluoroform.
113. See 1,1,2-trichloro-1,2,2-trifluoroethane.
114. See 1,2-dichloro-1,1,2,2-tetrafluoroethane.
115. See chloropentafluoroethane.
116. See hexafluoroethane.

fluorocarbon polymer. This term includes polytetrafluoroethylene, polymers of chlorotrifluoroethylene, fluorinated ethylene-propylene polymers, polyvinylidene fluoride, hexafluoropropylene, etc.
Properties: Thermoplastic, resistant to chemicals and oxidation, broad useful temperature range

(up to 287C), high dielectric constant, resistant to moisture, weathering, ozone, and UV radiation. Their structure comprises a straight backbone of carbon atoms symmetrically surrounded by fluorine atoms. Noncombustible.

Forms: Powders, dispersions, film, sheet, tubes, rods, tapes, and fibers.

Use: High-temperature wire and cable insulation; electrical equipment, drug and chemical equipment, coating of cooking utensils, piping gaskets, continuous sheet, bonding industrial diamonds to metal (grinding wheels).

See also fluoroelastomer.

fluorochemical. Organic compounds, not necessarily hydrocarbons, in which a large percentage of the hydrogen directly attached to carbon has been replaced by fluorine. The presence of two or more fluorine atoms on a carbon atom usually imparts stability and inertness to the compound, and fluorine usually increases the acidity of organic acids.

Derivation: (a) Electrolysis of solutions in hydrogen fluoride (Simons process), (b) replacement of chlorine or bromine by fluorine with hydrogen fluoride in the presence of a catalyst (antimony trifluoride or pentafluoride), (c) addition of hydrogen fluoride to olefins or acetylene.

Use: Dielectric and heat-transfer liquids, pump sealants, surfactants, metering devices, special solvents.

See also fluorocarbon, fluroelastomer.

5-fluorocytosine. (flucytosine).
CAS: 2022-85-7. $C_4H_4FN_3O$.
Properties: Crystalline solid with mw 129.09, mp 295C (decomposes), light sensitive.

fluorodichloromethane. See dichlorofluoromethane.

fluoroelastomer. Any elastomeric high polymer containing fluorine; they may be homopolymers or copolymers. Fluorocarbon polymers include a large group of fluoroelastomers including a copolymer in which the molecular skeleton is a —P≡N— chain containing approximately equal numbers of tri- and heptafluoroethyoxy side groups. Such polymers are amorphous, thermally stable, noncombustible, have low glass transition temperature (−77C) and are generally resistant to attack by solvents and chemicals.

fluoroform. (trifluoromethane; propellant 23; refrigerant 23).   CAS: 75-46-7. CHF_3.
Properties: Colorless gas, bp −84C, mp 160C.
Grade: 98% min purity. Nonflammable.
Use: Refrigerant, intermediate in organic synthesis, direct coolant for infrared detector cells, blowing agent for urethane foams.

fluoroformyl fluoride. See carbonyl fluoride.

"Fluoroinert"[158]. TM for a series of perfluorinated liquids used for cleaning electronic components after testing, bp range 31–173C, high dielectric strength, colorless, nonflammable.

"Fluorolubes"[62]. TM for polymers of trifluorovinyl chloride (—CF_2—CFCl—)$_x$ containing 49% fluorine and 31% chlorine. Products are light oils, heavy oils, and grease-like materials.
Use: Lubricant and sealant for plug cocks, valves, and vacuum pumps; impregnant for gaskets and packings; fluid for hydraulic equipment, heat exchange, and instrument damping.

fluoromethane. (methyl fluoride).
CAS: 593-53-3. CH_3F.
Properties: Colorless gas, bp −78.2C, fp −142C, d 1.19 (Air = 1.29), soluble in alcohol and ether.
Hazard: Flammable. Narcotic in high concentrations. TLV (as fluorine): 2.5 mg/m³.

p-fluorophenol. (4-fluorophenol).
CAS: 371-41-5. FC_6H_4OH.
Properties: White, crystalline solid. D 1.1889 (56C), mp 48.2C (stable form), 28.5C (unstable form), bp 185.6C, soluble in water.
Hazard: Irritant.
Use: Fungicide, intermediate for pharmaceuticals.

fluorophosphoric acid, anhydrous.
CAS: 13537-32-1. H_2PO_3F.
Properties: Colorless, viscous liquid; d 1.1818 (25C); miscible with water.
Hazard: Strong irritant to tissue.
Use: Metal cleaners, electrolytic or chemical polishing agents, formation of protective coatings for metal surfaces, catalyst.
See also difluorophosphoric acid and hexafluorophosphoric acid.

fluorosilicic acid. See fluosilicic acid.

fluorosulfonic acid. See fluosulfonic acid.

fluorosulfuric acid. See fluosulfonic acid.

fluorothane. (ethyl fluoride). CH_3CH_2F.
Use: Replaces ether in surgery.

fluorothene. See chlorotrifluoroethylene polymer.

fluorometholone. (9-fluoro-11,17-dihydroxy-6-methylpregna-1,4-diene-3,20-dione; fluormetholone).   CAS: 426-13-1.
$C_{22}H_{29}FO_4$.
Properties: Crystalline solid with mp 292-303C.
Use: A steroid, glucocorticoid, anti-inflammatory.

fluorotrichloromethane. See trichlorofluoromethane.

5-fluorouracul. [5-fluoro-2,4(1H,3H)-pyrimidi-nedione]. CAS: 51-21-8. $C_4H_3FN_2O_2$.
Properties: Crystalline solid with mw 130.08, mp 282C (decomposes). Soluble in water or methanol-water mixtures.
Use: Antineoplastic agent.

fluorspar. (fluorite; florspar). CaF_2.
Natural calcium fluoride; color yellow, green, or purple crystals; Mohs hardness 4, d 3.2, mp 1350C.
Grade: Metallurgical, ceramic and acid, containing more than 85 and 98% CaF_2, respectively.
Occurrence: US, Canada, Europe, Mexico.
Use: Principal source of fluorine and its compounds by way of hydrogen fluoride, flux in open hearth steel furnaces and in metal smelting, in ceramics, for synthetic cryolite, in carbon electrodes, emery wheels, electric arc welders, certain cements, dentifrices, phosphors, paint pigment, catalyst in wood preservatives, optical equipment.

fluosilicate. A salt of fluosilicic acid, H_2SiF_6. For possible synonyms see fluosilicic acid.

fluosilicic acid. (hydrofluosilicic acid; fluorosilicic acid; hexafluorosilicic acid; hydrogen hexafluorosilicate; hydrosilicofluoric acid; hydrofluorosilicic acid). CAS: 16961-83-4. H_2SiF_6.
Properties: (aqueous solution): Colorless fuming liquid, attacks glass and stoneware.
Derivation: Byproduct of the action of sulfuric acid on phosphate rock containing fluorides and silica or silicates. The hydrogen fluoride acts on the silica to produce silicon tetrafluoride, SiF_4, which reacts with water to form fluosilicic acid, H_2SiF_6.
Grade: Technical, CP.
Hazard: Extremely corrosive by skin contact and inhalation.
Use: Water fluoridation, ceramics (to increase hardness), disinfecting copper and brass vessels, hardening cement, etc., wood preservative and impregnating compounds, electroplating, manufacture of aluminum fluoride, synthetic cryolite and hydrogen fluoride, sterilizing bottling and brewing equipment (1–2% solution).

fluosulfonic acid. (fluorosulfuric acid; fluorosulfonic acid). CAS: 7789-21-1. HSO_3F.
Properties: Colorless, fuming liquid; d 1.745 (15C); fp −87C; bp 165C. Soluble in nitrobenzene, reacts violently with water, does not attack glass.
Derivation: Reaction of anhydrous hydrogen fluoride with sulfuric acid or sulfuric acid anhydride.
Hazard: Extremely irritating to eyes and tissue.

Use: Catalyst in organic synthesis, electropolishing, fluorinating agent.

flushed color. A pigment dispersed in oil, varnish, etc., the transfer from the water phase to the oil phase having been effected without the usual drying and subsequent grinding of the dry pigment. It is claimed that flushed colors are ready for use without grinding.

flux. (1) A substance that promotes the fusing of minerals or metals or prevents the formation of oxides. For example, in metal refining lime is added to the furnace charge to absorb mineral impurities in the metal. A slag is formed which floats on the bath and is run off. (2) A substance applied to metals that are to be united. On application of heat, it aids the flow of solder and prevents formation of oxides. (3) Any readily fusible glass or enamel used as a base or ground in ceramic processing. (4) The rate of flow or transfer of electricity, magnetism, water, heat, energy, etc., the term being used to denote the quantity that crosses a unit area of a given surface in a unit of time. (5) The intensity of neutron radiation, expressed as the number of neutrons passing through one square centimeter in one second. (6) A mixture of sodium nitrate and sodium nitrite; oxidizing agent used as a low explosive.

fluxing lime. See calcium oxide.

fly ash. The very fine ash produced by combustion of powdered coal with forced draft and often carried off with the flue gases. Special equipment is required for effective recovery, e.g., electrostatic precipitators. Fly ash is a mixture of alumina, silica, unburned carbon, and various metallic oxides. It is reported to have mutagenic properties after passing through stack precipitators. The alumina is recoverable by calsintering, which makes it a possible alternative source of aluminum.
Use: Cement additive for oil-well casings, absorbent for oil spills (silicone-coated), to replace lime in scrubbing sulfur dioxide from flue gas, as a filler in plastics, source of germanium (England), proposed as catalyst for coal liquefaction, removal of heavy metals from industrial waste waters, separation of oil-sand tailings.
See also calsintering.

Fm. Symbol for fermium.

FMN. Abbreviation for flavin mononucleotide. See riboflavin phosphate.

foam. A dispersion of a gas in a liquid or solid. The gas globules may be of any size, from colloidal to macroscopic, as in soap bubbles. Bak-ers' bread and sponge rubber are examples of solid foams. Typical liquid foams are those used in fire-fighting, shaving creams, etc. In such foams, the liquid must have sufficient cohesion to form an elastic film, e.g., soap, oil, protein, fatty acids, etc. Surfactant-induced foams have been developed to increase the efficiency of fuel cells.

Foams made by mechanical incorporation of air are widely used in the food industry, e.g., whipped cream, eggwhite, ice cream, etc. Useful foams for automobile seats, mattresses, and similar uses are made from natural and synthetic latexes, e.g., polystyrene, polyurethane.

A glass foam is based on sodium silicate and rock wool, and vitreous ceramic foams are also available. Metals can be caused to foam. Concrete foams are also in general use.

Foams designed for fire extinguishment are agglomerations of small bubbles of gas produced by two methods: (1) by chemical reaction between aluminum sulfate and sodium bicarbonate to generate carbon dioxide (chemical foams), and (2) by mixing or agitation of air with water containing the foaming ingredients (mechanical foams). The two types are equally efficient in fire extinguishing ability.

Besides the foaming ingredients, the foams contain stabilizing agents to assure permanence; there are many of these, for example, soaps, proteins, extract of licorice root, fatty acids and sulfite liquors. The ingredients of chemical foams are assembled in two separate units, which generate the foam on blending.

Fire foams are used primarily on fires in hydrocarbon liquids (Class B fires). There are many special types tailored for specific uses. Recently developed fire protection systems for aircraft include an instant-generating foam for cabin interiors, using a 2.5% aqueous solution of alkyl sulfonate and rigid polyurethane foams for use in fuel tanks.

See also fire extinguishment.

An unusually stable foam that remains intact much longer than fire-fighting foams has been developed at Sandia National Laboratories. It results from a synergistic action caused when a water-soluble polymer and a fatty alcohol are blended with a solvent and a surfactant. A possible agricultural use would be in insecticide application.

"Foamex"[73]. TM for a mixture of aliphatic esters.
Properties: Very pale yellow liquid, insoluble in water, d 0.96–0.97.
Use: Foam retardation and prevention in water solutions of glue, casein, shellac, gelatin, etc.

foam, metal. A cellular metallic structure, usually of aluminum or zinc alloys, made by incorporat-

ing titanium or zirconium hydride in the base metal. This subsequently evolves hydrogen to produce a uniform, foam-like material. Its density is approximately that of seawater so that it is weightless when submerged. The principal use of foamed metals is in absorption of shock impact without elastic rebound. Fiber-reinforced light-metal foams have potential application in reducing the weight of automobile bodies.

foam, plastic. A cellular plastic which may be either flexible or rigid. Flexible foams may be polyurethane, rubber latex, polyethylene or vinyl polymers, rigid foams are chiefly polystyrene, polyurethane, epoxy, and polyvinyl chloride. The blowing agents used are sodium bicarbonate, halocarbons such as CCl_3F, and hydrazine. Flexible polystyrene foam is available in extruded sheets and also in the form of "beads" made by treating a polystyrene suspension with pentane; these expand from 30–50 times on heating and are used as automobile radiator sealants. Rigid foams are widely used for boat construction, filtration, fillers in packing cases, absorption of oil spills, and building insulation. The latter application involves a fire risk described below.
Hazard: The most widely used types of organic foam plastics (polystyrene, polyurethane, polyisocyanurate) are combustible; even when fire-retardant agents are incorporated, such foams will burn. The extent of burning or fire severity will vary with surface treatment, end-use location, recipe, and degree of protection. Thin coatings of fire-retardant paint, metal, or automatic sprinkler systems may not adequately protect against rapid fire spread. Organic foamed plastic surfaces should not be left exposed. Multiple adjacent surfaces like walls and ceiling create a most severe hazard because of the chemical kinetics associated with radiative, conductive, and convective currents developed during a building fire.

New methods of making such plastic foams as polyurethane that are reported to reduce their combustibility have been developed, for example, use of trichlorobutylene oxide instead of propylene oxide.

Use of urea-formaldehyde foams for building insulation has been restricted in some areas due to potential toxic effect of the formaldehyde.

fob. Abbreviation for "freight on board," a designation used in shipping a material to indicate that freight charges are to be paid by the purchaser. This is in contrast to "freight prepaid and allowed," indicating that freight charges are paid by the manufacturer.

fog. A suspension of liquid droplets in air; an aerosol. The size of the droplets ranges from col-

loidal to macroscopic. "Synthetic" fogs can be produced on a laboratory scale by ultrasonic vibrations and natural fogs can be precipitated by the same means. Mists or fogs comprised of atomized particles of oil are used as military concealment screens and for insecticidal purposes in orchards and truck gardens.
See also smog, chemical smoke, aerosol.

folacin. See folic acid.

"Foliafume"[342]. TM for pyrethrin-rotenone plant spray concentrate.

folic acid. (pteroylglutamic acid; folacin; PGA). $C_{19}H_{19}N_7O_6$. Considered a member of the vitamin B complex. At least three substances with folic acid activity occur in nature, one of which, pteroylglutamic acid, is made synthetically.
Properties: (pteroylglutamic acid): Orange-yellow needles or platelets, tasteless, odorless; slightly soluble in methanol and springly soluble in water; insoluble in acetone, ether, benzene; moderately soluble in dilute alkali hydroxide and carbonate solutions; stable in heat in neutral and alkaline solution; destroyed by heating with acid; inactivated by light.
Sources: Green plant tissue, fresh fruit, liver, and yeast. Synthetic pteroylglutamic acid made by the reaction of 2,3-di-bromopropanol, 2,4,5-triamino-6-hydroxypyrimidine and p-aminobenzoyl glutamic acid.
Grade: 10% feed grade, USP.
Use: Medicine, nutrition, food additive (maximum daily ingestion not to exceed 0.01 mg).
See also folinic acid.

folinic acid. (5-formyl-5,6,7,8-tetrahydropteroyl-L-glutamic acid; citrovorum factor; leucoverin). CAS: 58-05-9. $C_{20}H_{23}N_7O_7$. A member of the folic acid group of vitamins and a growth factor for the bacterium Leuconostoc citrovorum. Folinic acid is an important metabolite of folic acid and may be the active form in cellular metabolism. It is an effective hematopoietic factor. Ascorbic acid and vitamin B_{12} are essential for the conversion of folic acid to folinic acid.
Properties: dl-L-form: Crystals, decompose 240–250C, sparingly soluble in water.
Derivation: (a) Prepared by catalytic reduction of folic acid, (b) produced microbially.
Use: Medicine, nutrition, biochemical research.

folpet. (phaltan; (N-(trichloromethylthio)-phthalimide)). CAS: 133-07-3. $C_6H_4(CO)_2NSCCl_3$.
Properties: Light colored powder, insoluble in water, slightly soluble in organic solvents.
Use: Fungicide-bactericide for vinyls, paints, and enamels.

"Fomade"[69]. TM for a combination of foam former, foam stabilizer, and vinyl stabilizer.
Properties: Light tan, semi-fluid paste; d 1.02.
Use: Foaming agent for polyvinyl chloride.

"Fomerez"[104]. TM for a series of polyester and polyether resins, stannous octoate catalysts and coupling agents used in the manufacture of urethane foams.

fonofos. See dyfonate.

"Fonoline"[45]. TM for petrolatum of soft consistency and low melting point with color range of white to yellow and meeting USP or NF purity requirements for petrolatum.

fonophos. (ethylphosphonodithioic acid-O-ethyl-S-phenyl ester). CAS: 944-22-9. $C_{10}H_{15}OPS_2$.
Properties: Yellow liquid, insoluble in water, miscible with organic solvents.
Hazard: Cholinesterase inhibitor.
Use: Insecticide, soil fumigant.

food. Any substance or mixture which, when ingested by man or animals, contributes to the maintenance of vital processes. With the exception of sodium chloride and water, all foods are derived from plants, either by direct consumption or by ingestion of animal tissue or such animal products as eggs, milk, etc., which are derived metabolically from vegetable sources. Basic foods are composed of proteins, fats (lipids), and carbohydrates, together with vitamins and minerals. Ancillary items that are associated with foods, though with little or no nutritive value, are collectively called food additives, e.g., flavorings, spices, preservatives, and colorants. Many of these are also plant-derived, though some are now made synthetically.
See also plant (1), nutrient, and following entries.

food additive. (a) Intentional: The Food Protection Committee of the National Research council states that a food additive is "a substance or mixture other than a basic foodstuff that is present in food as a result of any aspect of production, processing, storage, or packaging." (b): Unintentional: Substances that may become part of a food product as a result of chance contamination, such as insecticide residues, fertilizers and the like. The permissible content of insecticide residues has been established by the FDA.
The Food Additives Amendment to the Food, Drug, and Cosmetic Act empowers the FDA to disapprove or discontinue any food additive that it determines to be unsafe at the level of intended use, based on data supplied by the manufacturer. *Note*: Unintentional additives should more logically be called contaminants.

food chain. The sequence of nutritional steps in an ecosystem, with producers at the bottom and consumers at the top.

food web. The nutritional structure of an ecosystem in which more than one species occupies each level. Thus, there are usually several producer species and several consumer species in a food web.

food color. (certified color). A colorant which may be either dye (soluble) or a lake (insoluble) permissible for use in foods, drugs, or cosmetics by FDA. The dyes color by solution and the lakes by dispersion. All must satisfy strict regulations as to toxicity.
See also FD&C colors.

Food and Drug Administration. (FDA). The Federal agency responsible for administering and enforcing the Food, Drug, and Cosmetic Act, including the Food Additives Amendment which went into effect in March 1960. It has the authority to require proof of the efficacy and safety of drugs, foods, and pharmaceuticals, to conduct and evaluate screening tests, and to compel withdrawal from the market of any such product that it finds ineffective or hazardous. It establishes tolerances on food and animal feed additives of all types, including pesticides, as well as on cosmetic products, flammable fabrics, and packaging and labeling materials. It can also require specific statement on labels of the components or ingredients of a product, as well as precautionary warnings.
See also Environmental Protection Agency, food additive.

food engineering. Application of engineering principles to the design of equipment for large-scale food processing, e.g., automatic harvesting devices, dryers of various types, crystallizers, ovens and heat-exchangers, comminuting and mixing equipment, distillation units, packaging machines. Food engineering requires an understanding of thermodynamics, conditions of state and equilibrium, rate processes, and transport phenomena. The unit operations of chemical engineering and basic physics and mechanics are also involved.

food technology. Practice of the techniques used in the preparation of foods for large-scale human use. Among others, these include harvesting, post-harvest treatment, all forms of cooking, tenderizing, preservation by chemicals, heating, dehydration, drying and freezing, distillation and solvent extraction, milling, refining, hydrogenation, emulsification, packaging materials and storage, labeling and transportation. Other aspects of food technology are bacteriology, sanitation, quality control and formulation of ingredients for a wide variety of end products. A recent development of importance is the growth of convenience and quick-service foods.

foots. (soapstock). The mixture of soap, oil, and impurities that precipitates when natural fatty oils are refined by treatment with caustic soda or soda ash. Usually contains 30–50% of free and combined fatty acids. A related meaning is the suspended solid matter in crude oils.
Use: Manufacture of relatively low-grade soaps, as a source of free fatty acids.

"Foray"[548]. TM for a monoammonium phosphate-based formulation used to extinguish fires in flammable liquids (Class B fires) and in combustible materials such as wood and paper (Class A fires).

forensic chemistry. See legal chemistry.

forge. A furnace used for heating and softening a metal in preparation for hot-working, e.g., wrought iron.

forging. Shaping a heated metal by means of repeated impact, thus improving its strength.

formal. See methylal.

formaldehyde. (oxymethylene; formic aldehyde; methanal). CAS: 50-00-0. HCHO.
24th highest-volume chemical produced in US (1985). A readily polymerizable gas. It is commercially offered as a 37–50% aqueous solution which may contain up to 15% methanol to inhibit polymerization. These commercial grades are called formalin. It is one of the few organic compounds known to exist in outer space.

$$H-\overset{\displaystyle}{\underset{\displaystyle O}{\overset{\|}{C}}}-H$$

Properties: (gas): Strong, pungent odor, vap d 1.067 (air = 1.000), at −20/4C vap d 0.815, bp −19C, fp −118C, autoign temperature 806F (430C), soluble in water and alcohol. (Aqueous 37% solution with 15% methanol): bp 96C, flash p 122F (50C). (Methanol-free): bp 101C, flash p 185F (85C).
Derivation: Oxidation of synthetic methanol or low-boiling petroleum gases such as propane and butane. Silver, copper, or an iron-molybdenum oxide are the most common catalysts.
Grade: Aqueous solutions: 37%, 44%, 50% inhib-

ited (with varying percentages of methanol) or stabilized or unstabilized (methanol-free), also available in solution in n-butanol, ethanol, or urea; USP (37% aqueous solution containing methanol).

See also paraformaldehyde, the polymerized, solid form.

Hazard: Moderate fire risk. Explosive limits in air 7–73%. Toxic by inhalation, strong irritant, a carcinogen. (Solution): Avoid breathing vapor and avoid skin contact. TLV: 1 ppm in air.

Use: Urea and melamine resins, polyacetal resins, phenolic resins, ethylene glycol, pentaerythritol, hexamethylenetetramine, fertilizer, disinfectant, biocide, embalming fluids, preservative, reducing agent as in recovery of gold and silver, corrosion corrosive inhibitor in oil wells, durable-press treatment of textile fabrics, industrial sterilant, treatment of grain smut, foam insulation, particle board, plywood, a versatile chemical intermediate.

formaldehyde aniline. (formaniline).
$C_6H_5NCH_2$.
Properties: Colorless to yellowish crystals, initial mp 133C, bp 271C, d 1.14, but these vary somewhat from sample to sample. Soluble in water, ether, and alcohol.
Derivation: Condensation of formaldehyde and aniline.
Hazard: Toxic by ingestion.
Use: Rubber accelerator, intermediate.

formaldehyde cyanohydrin. See glycolonitrile.

formaldehyde-p-toluidine. (methylene-p-toluidine). $(CH_3C_6H_4NCH_2)_x$.
Properties: White powder with grayish-yellow cast, aromatic odor, soluble in acetone, d 1.11.
Derivation: Reaction between formaldehyde and p-toluidine.
Use: Rubber accelerator, dyes.

formalin. An aqueous 37–50% solution of formaldehyde which may contain 15% methyl alcohol.
See formaldehyde.

formamide. (methanamide).   CAS: 75-12-7.
$HCONH_2$.
Properties: Colorless, hygroscopic, oily liquid. D 1.146, bp 200–212C with partial decomposition beginning approximately 180C, mp 2.5C, soluble in water and alcohol, flash p 310F (154C). Combustible.
Derivation: Interaction of ethyl formate and ammonia with subsequent distillation.
Hazard: Toxic material. TLV: 20 ppm in air.

Use: Solvent, softener, intermediate in organic synthesis.

formaniline.   See formaldehyde aniline.

"Formaset" LC-1[42]. TM for a syrup-type urea formaldehyde resin.
Use: Stabilizing and bodying agent for textiles.

"Formcel"[352]. TM for a series of water-free formaldehyde solutions in alcohols.
Use: Alcoholated urea and melamine resins, embalming fluids.

formetanate. (3-dimethylaminoethyleneiminophenyl-N-methylcarbamate hydrochloride)
CAS: 222-30-9.
Properties: Water soluble.
Use: Acaricide for deciduous fruits.

"Formica"[13]. TM for high-pressure laminated sheets of melamine and phenolic plastics for decorative applications as surfacing, adhesives for bonding laminated plastic to other surfaces.

formic acid. (hydrogen carboxylic acid; methanoic acid).   CAS: 64-18-6.   HCOOH.
Properties: Colorless, fuming liquid; penetrating odor; soluble in water, alcohol, and ether; d 1.2201 (20/4C), mp 8.3C, bp 100.8C, flash p 156F (69C) (OC), bulk d 10.16 lb/gal (20C), refr index 1.3719 (20C), autoign temperature 1114F (600C), strong reducing agent. Combustible.
Derivation: (a) By treatment of sodium formate and sodium acid formate with sulfuric acid at low temperatures and distilling in vacuo, (b) by acid hydrolysis of methyl formate, (c) as a byproduct in the manufacture of acetaldehyde and formaldehyde.
Method of purification: Rectification.
Grade: Technical, 85%, 90%, CP, FCC.
Hazard: Corrosive to skin and tissue. TLV: 5 ppm in air.
Use: Dyeing and finishing of textile, leather treatment, chemicals (formates, oxalic acid, organic esters), manufacture of fumigants, insecticides, refrigerants, solvents for perfumes, lacquers; electroplating, brewing (antiseptic), silvering glass, cellulose formate, natural latex coagulant, ore flotation, vinyl resin plasticizers.

formic aldehyde.   See formaldehyde.

formonitrile.   See hydrocyanic acid.

"Formopon"[23]. TM for sodium formaldehyde hydrosulfite. "Formopon" Extra is the basic zinc salt.

formothion. [S-(N-formyl-N-methylcarbamoyl-methyl)-dimethyl phosphorodithioate].
CAS: 2540-82-1. $C_6H_{12}NO_4PS_2$.
Properties: Yellow liquid, fp 25C, insoluble in water, miscible with common organic solvents.
Hazard: Cholinesterase inhibitor.
Use: Systemic insecticide.

formula, chemical. A written representation using symbols of a chemical entity or relationship. There are several kinds of formulas as follows: (1) Empirical: Expresses in simplest form the *relative* number and the kind of atoms in a molecule of one or more compounds; it indicates composition only, not structure. Example: CH is the empirical formula for both acetylene and benzene. (2) Molecular: shows the actual number and kind of atoms in a chemical entity (i.e. a molecule, group or ion). Examples: H_2 (one molecule of hydrogen), $2H_2SO_4$ (two molecules of sulfuric acid), CH_3 (a methyl group), $Co(NH_3)_6^{++}$ (an ion). (3) Structural: Indicates the location of the atoms, groups or ions relative to one another in a molecule as well as the number and location of chemical bonds: Examples:

$$CH_2{=}\underset{\underset{CH_3}{|}}{C}{-}CH{=}CH_2 \text{ (isoprene)}$$

benzene

Since all molecules are three-dimensional, they cannot properly be shown in the plane of the paper. This is sometimes indicated by extra-heavy lines or three-dimensional artwork (configurational formula) as in this representation of an ethanol molecule:

(4) Generic: Expresses a generalized type of organic compound where the variables stand for the number of atoms or for the kind of radical in a homologous series. Examples:

C_nH_{2n+2}	C_nH_{2n}
a paraffin	an olefin
ROR	ROH
an ether	an alcohol
(R = a hydrocarbon radical)	

(5) Electronic: A structural formula in which the bonds are replaced by dots indicating electron pairs, a single bond being equivalent to one pair of electrons shared by two atoms. Example: the electronic formula for methane is:

$$H{:}\overset{\overset{H}{\displaystyle\cdot\cdot}}{\underset{\underset{H}{\displaystyle\cdot\cdot}}{C}}{:}H$$

formula, product. A list of the ingredients and their amounts or percentages required in an industrial product. Such formulas (or recipes) are mixtures, not compounds; they are generally used in such industries as adhesives, food, paint, rubber, and plastics.
See also formulation.

formulation. Selection of components of a product formula or mixture to provide optimum specific properties for the end use desired. Formulation by experienced technologists are essential for products intended to meet specifications or special service conditions.

formula weight. The sum of the atomic weights represented in a chemical formula. Thus, since the atomic weight of hydrogen is 1 and that of oxygen is 16, the formula weight of water (H_2O) is 18 (approximate atomic weights used).

"Formvar."[58] TM for polyvinyl formal resins.
Use: Wire enamels, electrical insulation, coatings, adhesives, films, and molded materials.

2-formyl-3,4-dihydro-2H-pyran. See acrolein dimer.

formyl fluoride. CAS: 1493-02-3. HCOF.
Properties: Gas at normal temperature and pressure, decomposes slowly with formation of hydrogen fluoride and carbon monoxide, soluble in water (decomposes), bp −26C, fp −142C.
Derivation: Interaction of benzoyl chloride and a formic acid solution of potassium fluoride.
Grade: Technical.
Hazard: Toxic by inhalation, strong irritant to tissue. TLV (as fluorine): 2.5 mg/m³ of air.
Use: Organic synthesis (acetylating agent).

1-formylpiperidine. $C_5H_5NCH{=}O$.
Properties: Colorless liquid, liquid from -30 to 222C, aprotic, low-volatility. Miscible with alcohols, esters, ketones, amines, amides, inorganic acids, organometallics; soluble in water and hexane.
Use: Solvent for polar and nonpolar compounds as well as many high polymers, gas absorption, plastics modifiers.
See also N,N-dimethylformamide.

Forster reaction. Formation of secondary amines by condensation of a primary amine with an aldehyde, addition of alkyl halide to the Schiff base, and subsequent hydrolysis.

"Forticel"[352]. TM for a cellulosic thermoplastic for use in injection molding, extrusion, rotational casting, and blow molding.
Properties: Pellets (crystals, translucent, metallic, and opaque colors), d 1.20, highest use temperature 80C, soluble in organic solvents, insoluble in mineral oils. Combustible.
Use: Pen and pencil barrels, telephone bases, spectacle frames, tool handles, sheeting, steering wheels, etc.

fortification. In food technology, addition to a food ingredient or product of nutrients that are not normally present, for example, addition of vitamin D to milk or of vitamin C to cake-fillings. Nutritionists apply this term to foods especially designed for school children and elderly persons.
See also nutrification, enrichment.

"Fortiflex"[352]. TM for a high-density polyethylene consisting mainly of long molecules with occasional short side branches. Thermoplastic.
Properties: Milk-white, translucent pellets (colors are also available); density 0.95; melt index 0.2–8; tensile strength 3,100–3,700 psi; highest use temperature 225F. Combustible.
See polyethylene.

"Fortisan"[352]. TM for a cellulosic fiber manufactured by partial saponification of stretched cellulose acetate. A semisynthetic product. It resembles cellulose (cotton) in many respects. The high-tenacity product has a dry strength of 5–7 lbs/denier (100,000–130,000 psi), wet strength is 85% of dry strength, it has relatively low elongation under stress, elastic recovery is approximately 70% after extension to break, immediate elastic recovery is 46%, delayed recovery 30% at 5% strain; Young's modulus 1650; can be dyed in the same way as cotton. The monofilament can be produced to a fineness of one denier; d 1.50, resists stretching both dry and wet. Combustible.

"Fortracin"[342]. TM for animal feed supplements of bacitracin methylene disalicylate.

"Fortrel"[352]. TM for a polyester type synthetic fiber.

"Fosbond"[204]. TM for a group of chemicals used to provide a corrosion-resistant bond between zinc or ferrous metals and a paint film.

"Foscoat"[204]. TM for a class of chemicals designed to provide a phosphate coating prior to cold working.

"Fosfodril"[55]. TM for a glassy phosphate of high molecular weight (sodium hexametaphosphate).
Use: Thickener for drilling muds, water treatment in oil-well flooding operations.

fosfomycin. CAS: 23155-02-4. $C_3H_7O_4P$.
Properties: Water-soluble crystals, mp 95C.
Derivation: Produced by *Streptomyces*, also made synthetically.
Use: General antibiotic.

"Foslube"[204]. TM for a class of lubricants used to impregnate a phosphate coating prior to cold working.

"Fosrinse"[204]. TM for a class of chemicals used to render insoluble the acid salts remaining after phosphate treatment of ferrous metals.

fossil. Any material that results from an animal or vegetable source in past geologic ages and has been buried (compressed) in the earth. Examples are fossil fuels (petroleum, natural gas, coal, lignite), fossil waxes (ozocerite, montan), fossil resins (amber), and fossil woods partially preserved (petrified) by the action of silica.

"Fosterite"[308]. TM for a family of resins. Largest application is as "solventless" varnishes for electric insulation, also as a photoelastic resin and as a bond for impregnating and laminating asbestos sheets. Rods made of this plastic will carry a beam of light without the dispersion which occurs in air, making it possible to bend the beam.

fostion. (O,O-diethyl-S-(N-isopropyl carbamoylmethyl) phosphorodithioate).
CAS: 2275-18-5.
$(C_2H_5O)_2SPSCH_2SCH(CH_3)_2$.
Properties: White, crystalline solid; soluble in most organic solvents; mp 24C.
Hazard: Toxic by ingestion and skin absorption, cholinesterase inhibitor.
Use: Fungicide, ovicide.

"Fotoceram"[20]. TM for a photosensitive, crystalline ceramic.
Use: Electronics and industrial arts.

"Foundrez"[36]. TM for a group of water-soluble phenol-formaldehyde and urea-formaldehyde resins for foundry applications.

foundry sand. (greensand; molding sand).
Sand containing zirconium, titanium, and other metals and mixed with suitable binders used in making molds for casting metals. The sand/binder mixture is either rammed into place around the mold or baked into a core at 204–260C (dry-sand molding). The binders used are resins of various types, casein, etc.

Fourdrinier. The machine most widely used for papermaking, named for its English inventors who introduced it in the mid-19th Century. It provides a wide range of papers from heavy board to light tissue and is of impressive size and complexity. Its unique feature is the traveling wire mesh belt onto which the slurry of fiber and water is run from the headbox. The wire is a screen made of specially annealed bronze and brass, and is usually 55–85 mesh. The sheet is formed on this wire almost instantly, most of the water draining through the interstices of the wire. After leaving the wire, the sheet (called the web) passes through the press section of the machine where a number of rollers express enough of the remaining water to enable the sheet to hold together. It then moves into the multi-roller drying section. The dried sheet (4–6% moisture content) is then fed to a high-speed calender for compaction and finishing. The entire process is continuous and rapid, the machine often operating for several days without shutdown.
See also calender, drying (2), paper, couch roll, dandy roll, supercalender.

"FP" Acids.[484] TM for a series of fluorophosphoric acids.

FPC. (1) Abbreviation for fish protein concentrate. (2) Abbreviation for Federal Power Commission.

Fr. Symbol for francium.

fraction. Any portion of a mixture characterized by closely similar properties. The most important fractions of petroleum are naphtha, gasoline, fuel oil, kerosene, and tarry or waxy residues. These are obtained by fractional distillation.
See also separation.

fractional distillation. Distillation in which rectification is used to obtain product as nearly pure as possible. A part of the vapor is condensed and the resulting liquid contacted with more vapor, usually in a column with plates or packing.

The term is also applied to any distillation in which the product is collected in a series of separate components of similar boiling range.
See also reflux.

fractionation. In general, the separation or isolation of components of a mixture or a micromolecular complex. In distillation, this is done by means of a tower or column in which rising vapor and descending liquid are brought into contact (counter-current flow). Macromolecular components (proteins and other high polymers) can be separated by a number of methods, including electrophoresis, gel filtration, chromatography, centrifugation, foam fractionation, and partition.
See also reflux.

fracturing, hydraulic. See hydraulic fracturing.

fraissite. (benzyl iodide). $C_6H_5CH_2I$.
Use: A tear gas.

Franchimont reaction. Carboxylic acid dimerization to 1,2-dicarboxylic acids by treating α-bromocarboxylic acids with potassium cyanide followed by hydrolysis and decarboxylation.

francium. Fr. Element of atomic number 87, group IA of the periodic system, aw 223, valence = 1; it appears to exist only as radioactive isotopes. One isotope is actinium K (Fr[223]). Other isotopes have been made artificially: Fr[223] is the longest-lived isotope having a half-life of 21 minutes and is the only natural isotope. Francium is the heaviest of the alkali-metal family.

fragrance. An odorant used to impart a pleasant smell to shaving lotions, toothpastes, men's accessories, etc.; balsamic and piny odors are typical.

frankincense. (olibanum). A gum resin.

Frankland-Duppa reaction. Formation of α-hydroxycarboxylic esters by reaction of dialkyl oxalates with alkyl halides in the presence of zinc, or amalgamated zinc, and acid.

Frankland synthesis. Synthesis of zinc dialkyls from alkyl halides and zinc.

franklinite. (iron, manganese, zinc) $(FeMn)_2O_4$.
Black mineral resembling magnetite.

Frary metal. A lead-based bearing metal containing 97–98% lead alloyed with 1–2% each of barium and calcium, excellent for low-pressure bearings at moderate temperatures.

Frasch process. A process by which much of the worlds sulfur is obtained. Developed about 1900 by Herman Frasch, the process involves melting sulfur underground by introducing superheated water through a pipe under pressure and forcing the molten sulfur to the surface by compressed air.

Fraunhofer lines. See spectroscopy.

free energy. An exact thermodynamic quantity used to predict the maximum work obtainable from the spontaneous transformation of a given system. It also provides a criterion for the spontaneity of a transformation or reaction and predicts the greatest extent to which the reaction can occur, i.e., its maximum yield. Transformation of a system can be brought about by either heat or mechanical work. Free energy is derived from the internal energy and entropy of a system in accordance with the laws of thermodynamics.

free radical. A molecular fragment having one or more unpaired electrons, usually short-lived and highly reactive. In formulas, a free radical is conventionally indicated by a dot as $Cl^{\bullet}$, $(C_2H_5)^{\bullet}$. In spite of their transitory existence, they are capable of initiating many kinds of chemical reactions by means of a chain mechanism. Free radicals are formed only by the splitting of a molecular bond. A chain can result only if (1) radicals attack the substrate and (2) the radicals lost by this reaction are regenerated. Chain mechanisms for the thermal decomposition of many substances have been established. Free radicals are known to be formed by ionizing radiation and thus play a part in deleterious degradation effects that occur in irradiated tissue. They also act as initiators or intermediates in such basic phenomena as oxidation, combustion, photolysis, and polymerization.
See also carbonium ion.

free sulfur. Sulfur which is left chemically uncombined after vulcanization of a rubber compound. When this exceeds 1% the upper limit of solubility of sulfur in rubber, blooming will occur. Most rubber products are vulcanized with as low a sulfur content as possible so that the free sulfur content of the product is seldom over 0.5%.
See also bloom, vulcanization.

freeze-drying. (lyophilization). A method of dehydration or of separating water from biological materials. The material is first frozen and then placed in a high vacuum so that the water (ice) vaporizes in the vacuum (sublimes) without melting and the non-water components are left behind in an undamaged state.
Use: Blood plasma, certain antibiotics, vaccines, hormone preparations, food products such as

coffee and vegetables. One technique prepares freeze-dried ceramic pellets from water solutions of metal salts.

"Freezene"[45]. TM for a series of refrigeration white mineral oils.
Use: Low-temperature lubrication.

freezing point. See melting point.

"Freon"[28]. TM for a series of fluorocarbon products used in refrigeration and air-conditioning equipment, as blowing agents, fire extinguishing agents, and cleaning fluids and solvents.
Properties: Clear, water-white liquids; vapors have a mild somewhat ethereal odor and are not irritating; essentially stable and inert. Nonflammable, nonexplosive. Noncorrosive. For listing of specific types, see fluorocarbon.
Note: Many types contain chlorine as well as fluorine and should be called chlorofluorocarbons.

"Freon" E.[28] TM for a series of hydrogen end-capped tetrafluoroethylene epoxide polymers having a DP up to 10, boiling range 39–490C, high dielectric constant.
Use: Coolants in electronic devices.

"Freon" C-51-12. See perfluorodimethylcyclobutane.

Freund synthesis. Formation of alicyclic hydrocarbons by the action of sodium (Freund) or zinc (Gustavson) on open chain dihalo compounds; 1,3-dichloropropane derived from the chlorination of propane obtained from natural gas is cyclized in the Hass process by treating with zinc dust in aqueous alcohol in the presence of sodium iodide as catalyst.

"Frianite"[118]. TM for a processed anhydrous potassium aluminum silicate. Typical chemical analysis: Silicon dioxide 74.7%, aluminum oxide 13.7%, potassium oxide 5.6%, other oxides 4.3%.
Properties: Fine, pinkish, inert powder containing no soluble salts. D 2.37, pH 5.4–6.5, bulk d 47 lb/cu ft, free-flowing liquid holding capacity is approximately 1-1/2% max.
Use: Diluent for dry blending or formulating dusting pesticides.

friction. A soft and extremely tacky mixture of rubber and softener applied to a fabric by means of a three-roll calender. The differential speed of the calender rolls drives the material into the interstices of the fabric, forming a strongly adherent coating. Uncured friction on a light sheeting is used for electrical insulating or friction tape.

Heavy-weave fabric coated with high-grade friction (rubber plus softener and curing agents) is used as piles in tire carcases, transmission belts, and other laminated products which are vulcanized.
See also calender.

friction welding. See welding.

Friedel-Crafts reaction. A type of reaction involving anhydrous aluminum chloride and similar metallic halides as catalysts, discovered in 1877 by Charles Friedel, a French chemist (1832–1899), and James Mason Crafts, an American chemist (1830–1917), during joint research in France; it has been developed since then for many important industrial uses, exemplified by the condensation of ethyl chloride and benzene to form ethylbenzene and the manufacture of acetophenone from acetyl chloride and benzene. The name is now applied to a wide variety of acid-catalyzed organic reactions.
Use: Alkylation and acylation in general. Some examples are the production of high-octane gasoline, cumene, detergent alkylate, and various plastics and elastomers.

Friedlaender synthesis. Base-catalyzed condensation of 2-aminobenzaldehydes with ketones to form quinoline derivatives.

Fries rearrangement. Rearrangement of phenolic esters to o- and/or p-phenolic ketones on heating with aluminum chloride or other Lewis acid catalysts.

Fries rule. The most stable form of a polynuclear hydrocarbon is that in which the maximum number of rings has the benzenoid arrangement of three double bonds. A benzenoid electronic configuration is energetically favored and, therefore, particularly stable.

frit. A ground glass used in making glazes and enamels and also for making so-called frit seals. Finely powdered glass may be called a frit. The term is also used for finely ground inorganic minerals, mixed with fluxes and coloring agents which turn into a glass or enamel on heating.

Fritsch-Buttenberg-Wiechell rearrangement.
The rearrangement of 1,1-diaryl-2-haloethylenes to diaryl acetylenes with strong bases.

"FR-N"[278]. TM for a series of butadieneacrylonitrile elastomeric polymers and latices.
Properties: Oil and solvent resistance combined with good flexibility and resistance to low temperatures, water absorption, and permanent set.
Use: Oil-resistant seals, shoe soles, gasoline hose, belt conveyors, plasticizers, paper saturation, adhesives, leather finishes, and carpet backing.

froth flotation. See flotation.

FRP. Abbreviation for glass fiber-reinforced plastic.
See reinforced plastic.

"FR-S"[278]. TM for general-purpose rubbers and latexes, composed of copolymers of butadiene and styrene.
Use: Rubber: Tires, hose, belting and packing; molded and extruded automotive and industrial products; soles and heels; hard rubber. Latex: Adhesives, foamed rubber, textile and rug backing, paper coating and impregnation, modification of plastics to produce high impact strength, asphalt additive.

"FRTP"[539]. TM for glass fiber-reinforced thermoplastic.

fructose. (fruit sugar; D(-)-fructose; levulose). $C_6H_{12}O_6$. A sugar occurring naturally in a large number of fruits and in honey. It is the sweetest of the common sugars.
Properties: White crystals; soluble in water, alcohol, and ether; mp 103–105C (decomposes); specific rotation −89 to −91 degrees. Combustible.
Derivation: Hydrolysis of inulin, hydrolysis of beet sugar followed by lime separation, from cornstarch by enzymic or microbial action.
Grade: Technical, NF, food, parenteral.
Use: Foodstuffs, medicine, preservative.

fructose-1,6-diphosphate. (FDP; fructosediphosphoric acid; Harden-Young ester). $H_2PO_4(C_6H_{10}O_4)H_2PO_4$. Can be prepared from fructose and certain other sugars by the use of yeasts. It is known to take part in cell metabolism; an intermediate in carbohydrate metabolism. Usually handled in the form of its barium or calcium salts, white amorphous powders, soluble in ice water and dilute acid solutions, insoluble in hot water and alcohol.
Use: Organic synthesis, research in cell metabolism.

FT black. Abbreviation for fine thermal black.
See thermal black.

FTC. Abbreviation for Federal Trade Commission.

fuchsin. (basic fuchsin; magenta). A synthetic rosaniline dyestuff, a mixture of rosaniline and p-rosaniline hydrochlorides.
Properties: Dark green powder or greenish crystals with a bronze luster, faint odor, soluble in water and alcohol.

Grade: NF.
Use: Textiles and leather industries, as a red dye, pharmaceutical.

fuel.   Any substance that evolves energy in a controlled chemical or nuclear reaction. The most common type of chemical reaction is combustion, the type of oxidation occurring with petroleum products, natural gas, coal, and wood; more rapid oxidation takes place in rocket fuels (hydrogen, hydrogen peroxide, hydrazine) which approaches the rate of an explosion. The nuclear fuels used for power generation release their energy by fission of the atomic nucleus (uranium, plutonium, thorium).
See also combustion, fission.

fuel cell.   (1) An electrochemical device for continuously converting chemicals--a fuel and an oxidant--into direct-current electricity. It consists of two electronic-conductor electrodes separated by an ionic-conducting electrolyte with provision for the continuous movement of fuel, oxidant, and reaction product into and out of the cell. The fuel can be gaseous, liquid, or solid; the electrolyte liquid or solid; the oxidant gaseous or liquid. The electrodes are solid, but may be porous and may contain a catalyst. Fuel cells differ from batteries in that electricity is produced from chemical fuels fed to them as needed, so that their operating life is theoretically unlimited. The cell products can be regenerated externally into fuel for return to the cell, e.g., carbon dioxide from the cell can be reacted with coal to form carbon monoxide for feed to the cell. Fuel is oxidized at the anode (negative electrode), giving electrons to an external circuit; the oxidant accepts electrons from the anode and is reduced at the cathode. Simultaneously with the electron transfer, an ionic current in the electrolyte completes the circuit. One type of electrolyte is a solid polymer of perfluorinated sulfonic acid. The fuels range from hydrogen, carbon monoxide, and carbonaceous materials to redox compounds, alkali metals, and biochemical materials. Fuel cells based on hydrogen and oxygen have a significant future as a primary energy source. Cells of this type are under development for use as a power source for electric automobiles, the hydrogen being derived from methanol. Large-scale development of fuel cells for on-site power generation for housing units is well advanced and research has been completed for construction of a 26-megawatt cell capable of serving the needs of a small community.   C. A. Hampel
(2) An aircraft fuel tank or container made of or lined with an oil-resistant synthetic rubber.

fuel element.   A fabricated rod, form, or other shape which consists of or contains the fissiona-

ble fuel for a nuclear reactor. The term does not refer to a chemical element, but rather to a device from which power is derived.

fuel oil.   Any liquid petroleum product that is burned in a furnace for the generation of heat or used in an engine for the generation of power, except oils having a flash point of approximately 100F (37.7C) and oils burned in cotton or wool-wick burners. The oil may be a distilled fraction of petroleum, a residuum from refinery operations, a crude petroleum, or a blend of two or more of these.
Because fuel oils are used with burners of various types and capacities, different grades are required. ASTM has developed specifications for six grades of fuel oil. No. 1 is a straight-run distillate, a little heavier than kerosene, used almost exclusively for domestic heating. No. 2 (diesel oil) is a straight-run or cracked distillate used as a general purpose domestic or commercial fuel in atomizing-type burners. No. 4 is made up of heavier straight-run or cracked distillates and is used in commercial or industrial burner installations not equipped with preheating facilities. The viscous residuum fuel oils, Nos. 5 and 6, sometimes referred to as bunker fuels, usually must be preheated before being burned. ASTM specifications list two grades of No. 5 oil, one of which is lighter and under some climatic conditions may be handled and burned without preheating. These fuels are used in furnaces and boilers of utility power plants, ships, locomotives, metallurgical operations, and industrial power plants.
See also diesel oil, fuel oil, synthetic.
Use: Domestic and industrial heating, power for heavy units (ships, trucks, trains), source of synthesis gas.

Fujimoto-Belleau reaction.   Synthesis of cyclic α-substituted α,β-unsaturated ketones from enol lactones and Grignard reagents prepared from primary halides.

Fukui, Kenichi.   (1918-) A Japanese professor who was co-recipient of Nobel prize for chemistry along with Hoffmann in 1981. His work involved the quantum mechanical studies of chemical reactivity. Fukui's entire career has been at Kyoto University.

fuller's earth.   A porous colloidal aluminum silicate (clay) which has high natural adsorptive power. Gray to yellow color, noncombustible.
See also bentonite, diatomite.
Occurrence: Florida, England, Canada.
Use: Decolorizing of oils and other liquids, oil-well drilling muds, insecticide carrier, floor-sweeping compounds, cosmetics, rubber filler, carrier for catalysts, filtering medium.

fulminates. Materials with carbon-nitrogen-oxygen groups.
Use: Sensitive explosives.

fumagillin. CAS: 23110-15-8. $C_{27}H_{36}O_7$.
An antibiotic substance produced by Aspergillus fumigatus.
Properties: Light yellow crystals from dilute methanol, mp 189–194C, insoluble in water, dilute acids, saturated hydrocarbons; soluble in most other organic solvents.
Use: Medicine.

fumaric acid. (boletic acid; lichenic acid; allomaleic acid; trans-butenedioic acid).
CAS: 110-17-8. The trans-isomer of maleic acid.

$$HO-\overset{\overset{O}{\|}}{C}-\overset{}{C}-H$$
$$H-\overset{}{C}-\overset{\underset{\|}{O}}{C}-OH$$

Properties: Colorless, odorless crystals; fruit acid taste. Stable in air, d 1.635, sublimes at 290C, mp 287C (sealed tube), soluble in water 0.63g/100 g (25C), soluble in alcohol 5.76 g/100 g (30C), insoluble in chloroform and benzene. Combustible.
Derivation: (a) Isomerization of maleic acid, (b) catalytic oxidation of benzene.
Grade: Technical, crystals, FCC.
Use: Modifier for polyester, alkyd, and phenolic resins; paper-size resins; plasticizers, rosin esters and adducts, alkyd resin coatings, upgrading natural drying oils (especially tall oil) to improve drying characteristics, in foods as substitute for tartaric acid, as acidulant and flavoring agent (FDA approved), mordant, organic synthesis, printing inks.

fumaryl chloride. CAS: 627-63-4.
ClCOCH:CHCOCl.
Properties: Clear, straw-colored liquid; bp 158–160C, 62–64C (13 mm); d 1.408 (20C).
Hazard: Corrosive to eyes and skin.
Use: Chemical intermediate for pharmaceuticals, dyestuffs, and insecticides.

fume. The particulate, smoke-like emanation from the surface of heated metals. Also the vapor evolved from concentrated acids (sulfuric, nitric), from evaporating solvents, or as a result of combustion or other decomposition reactions (exhaust fume). Many of these fumes are toxic.

fumigant. A toxic agent in vapor form that destroys rodents, insects and infectious organisms,

a type of pesticide. The most effective temperature for their use is approximately 70F. They are used chiefly in enclosed or limited areas (barns, greenhouses, ships' holds, and the like), and also are applied locally to soils, grains, fruits, and garments. Some commonly used fumigants are formaldehyde, sulfur dioxide and other sulfur compounds, hydrogen cyanide, methyl bromide, carbon tetrachloride, p-dichlorobenzene, and ethylene oxide. Care must be used when handling and applying fumigants in view of their toxicity.
See also repellent, pesticide.

fuming. A characteristic of some highly active liquids which evolve visible smoke-like emanations on contact with air. Most familiar are the forms of nitric and sulfuric acids designated as "fuming." These are not pure, concentrated acids; low percentage of nitrogen dioxide and water are present in fuming nitric acid and fuming sulfuric acid contains sulfur trioxide. Hydrofluoric acid (a mixture of hydrogen fluoride and water) also fumes. Pure compounds in which fuming occurs are fluosilicic acid and hydrazine.

fundamental particle. (elementary particle).
One of the many constituents of atoms in either the normal or the excited state, i.e., protons, electrons, and neutrons that are directly involved in chemical reactions. Other subatomic species are mesons, positrons, neutrinos, hyperons, etc. Photons are light quanta, i.e., "particles" of energy which have no rest mass.
See also particle, photon.

fundamental research. (basic research).
Scientific investigations undertaken primarily to increase knowledge of a given field on a long-range basis. It is free from the time factor usually present in applied research and is comparatively unlimited by economic restrictions. In general, it seeks basic causes for phenomena rather than immediate results. It has no predetermined goal or purpose. None the less, tremendous achievements in chemistry and other sciences have resulted and it will always remain the essential cornerstone of science.
"Fundamental research is essentially a matter of inquiring into nature. The motivation for this activity is only imperfectly understood but it is primarily an intellectual pursuit. Unlike applied research, the reward being sought is attained through the ability to understand and explain natural phenomena. The interest centers on elucidating the laws of nature, not on manipulating or exploiting them" (Howard McMahon).

fungal protease enzyme. See acid fungal protease.

fungicide. Any substance which kills or inhibits the growth of fungi. Older types include a mixture of lime and sulfur, copper oxychloride, and Bordeaux mixture. Copper naphthenate has been used to impregnate textile fabrics such as tenting and military clothing. Dithiocarbamate and quinone types were introduced about 1940. Mercury compounds are also effective but have been discontinued because of their toxicity to humans. Hypochlorite solutions are used in swimming pools and water-cooled heat exchangers. Some types of fungi that infect the human body are extremely hard to eradicate and require highly specific medical treatment.

fungus. Any of a plant-like group of organisms that does not produce chlorophyll; they derive their food either by decomposing organic matter from dead plants and animals or by parasitic attachment to living organisms, thus often causing infections and disease. Examples of fungi are molds, mildews, mushrooms, and the rusts and smuts that infect grain and other plants. They grow best in a moist environment at temperatures of about 25C, little or no light being required. See also mycotoxin.

Funk, Casimir. (1884–1948) Born in Poland and later becoming an American citizen, Funk in 1911 isolated a food factor extracted from rice hulls which he found to be a cure for a disease due to malnutrition (beri beri). Believing this to be an amine compound essential to life, he coined the name "vitamine," from which the final "e" was later dropped. The various types and functions of vitamins were not differentiated till some years later as a result of the work of McCollum, Szent-Gyorgi, R. J. Williams and others.

furacrylic acid. See furylacrylic acid.

"Furadan". (2,3-dihydro-2,2-dimethyl-7-benzofuranylmethylcarbamate).
CAS: 1563-66-2.   TM for a pesticide designed to combat corn rootworm and rice water weevil. Approved by USDA. Also effective on alfalfa, sugar cane, rice, peanuts, and potatoes.
Hazard: Toxic material. TLV: 0.1 mg/m³ of air.

"Furafil"[224]. TM for lignocellulose produced by the pressure digestion of the acidified residue remaining after extraction of furfural from agricultural raw materials.
Use: Additive for bulk, absorbency, or conditioning action; extender for phenolic glues for plywood, foundry facings; oil well drilling muds, fertilizer.

furamide. See furoamide.

furan. (furfuran; tetrol).   CAS: 110-00-9.

HC:CHCH:CHO.   A   heterocyclic compound. Its basic structure is:

Properties: Colorless liquid turning brown on standing, this color change is retarded if a small amount of water is added, d 0.938 (20/4C), fp −86C, bp 31.4C, flash p less than 32F (0C) (TOC), refr index 1.4216 (20C), insoluble in water, soluble in alcohol and ether.
Derivation: Dry distillation of furoic acid from furfural.
Hazard: Flammable, dangerous fire risk, flammable limits 2–24%, forms peroxides on exposure to air. Absorbed by skin.
Use: Organic synthesis, especially for pyrrole, tetrahydrofuran, thiophene.
See also furan polymer.

furancarboxylic acid.   See furoic acid.

2,5-furandione.   See maleic anhydride.

2-furanmethanethiol.   OCH:CHCH:CCH₂SH.
Properties: Liquid, d 1.1319 (20/4C), bp 155C, refr index 1.5324 (20C), insoluble in water.
Use: Ingredient for synthetic coffee compositions and fortifier for natural coffee blends and flavor adjunct. Also inhibits the corrosive power of nitric acid.

furan polymer. A plastic derived (1) from furfuryl alcohol or (2) from furfural or reaction products of furfural and a ketone. The materials are dark-colored and are resistant to solvents, most nonoxidizing acids, and alkalies, and to specific corrosives such as dinitrogen pentoxide.
Properties: Physical properties of a typical asbestos-reinforced furan polymer are d 1.7, tensile strength 5,000 psi, flexural strength 7,500 psi, impact strength 0.5 ft lb/in. notch, water absorption 0.2% and coefficient of thermal expansion $3.0–10^{-5}$ in/in/F.
Use: Coating asphaltic pavements, foundry sand cores, shell molding and corrosion resistant materials of construction. Since furfural is readily obtainable by heating pentosan-containing products such as corn cobs with mineral acid, these resins are inexpensive and have great potential use where products with their characteristics are required.

furazolidone. (N-(5-nitro-2-furfurylidene-3-amino)-2-oxazolidinone).   CAS: 67-45-8.
$C_8H_7N_3O_5$
Properties: Yellow powder, odorless, mp 255C; slightly soluble in polyethylene glycol; insoluble in water, alcohol, and peanut oil.

Derivation: Synthetically from furfural, hydroxy-ethylhydrazine, and diethyl carbonate.
Grade: NF.
Hazard: A carcinogen, use has been restricted. See also nitrofuran.

furcellaran.
Properties: Seaweed gum (a natural phycocolloid) available as an odorless white powder, soluble in warm water. Form gels at low concentrations. Reputed to be more stable to heat and acids than those of other vegetable gums. Available in the form of salts.
Derivation: From the seaweed *Furcellaria fastigiata.*
Use: Gelling agent, viscosity control agent, puddings, jams, tooth pastes; bacterial cultural media, pharmaceuticals.

furethrin. (3-furfuryl-2-methyl-4-oxo-2-cyclopenten-1-ylchrysanthemumate (generic)).
$C_{21}H_{27}O_4$. A synthetic analog of allethrin substituting the 2-furfuryl for allyl in the side chain.
Properties: Yellow liquid, bp 187–188C (0.4 mm), insoluble in water, soluble in light oils.
Use: Insecticide, use like allethrin.

furfural. (ant oil, artificial; pyromucic aldehyde; furfuraldehyde; bran oil). CAS: 98-01-1.
C_4H_3OCHO

$$
\begin{array}{c}
CH{-}CH \\
\| \quad \| \\
CH \quad CCHO \\
\diagdown O \diagup
\end{array}
$$

Properties: Colorless liquid when very pure, becomes reddish-brown upon exposure to light and air. Odor somewhat similar to benzaldehyde. Forms condensation products with many types of compounds, phenol, amines, urea, etc. Soluble in alcohol, ether, and benzene; 8.3% soluble in water at 20C. D 1.1598 (20/4C), fp −36.5C, bp 161.7C, heat of vaporization 107.5 cal, refr index 1.5260 (20C), flash p 140F (60C) (CC), autoign temperature 392C (797F). Combustible.
Derivation: From oat hulls, rice hulls, corn cobs, bagasse, and other cellulosic waste materials by steam-acid digestion.
Grade: Technical, refined.
Hazard: Absorbed by skin, irritant to eyes, skin, and mucous membranes. TLV: 2 ppm in air.
Use: Solvent refining of lubricating oils, butadiene, rosin, and other organic materials; solvent for nitrocellulose, cellulose acetate, shoe dyes; intermediate for tetrahydrofuran and furfuryl alcohol, phenolic and furan polymers; wetting agent in manufacture of abrasive wheels and brake lin-

ings, weed killer, fungicide, adipic acid and adiponitrile, road construction, production of lysine, refining of rare earths and metals, flavoring, analytical reagent.

furfuralacetic acid. See furylacrylic acid.

furfuraldehyde. See furfural.

furfuramide. See hydrofuramide.

furfuran. See furan.

furfuryl acetate. $C_4H_3OCH_2OOCCH_3$.
Properties: Colorless liquid turning brown upon exposure to light and air, pungent odor, insoluble in water, soluble in alcohol and ether, d 1.1175 (20/4C), bp 175–177C, refr index 1.4627. Combustible.
Derivation: By treatment of furfuryl alcohol with acetic anhydride.
Grade: Refined.
Use: Flavor.

furfuryl alcohol. (furyl carbinol).
CAS: 98-00-0. $C_4H_3OCH_2OH$.

$$
\begin{array}{c}
CH{-}CH \\
\| \quad \| \\
CH \quad CCH_2OH \\
\diagdown O \diagup
\end{array}
$$

Properties: Colorless, mobile liquid becoming brown to dark-red upon exposure to light and air. It autopolymerizes with acid catalysts, often with explosive violence, to form a thermosetting resin that cures to an insoluble, infusible solid, highly resistant to chemical attack. Soluble in alcohol, chloroform, benzene, and water; insoluble in paraffin hydrocarbons; d 1.1285 (20/4C), bp 170C, refr index 1.4850 (25C), flash p (OC) 167F (75C), autoign temperature 915F (490C). Combustible.
Derivation: Continuous vapor-phase hydrogenation of furfural.
Grade: Technical, refined.
Hazard: May react explosively with mineral and some organic acids. Toxic by inhalation and skin absorption. Approved for food products. TLV: 10 ppm in air.
Use: Wetting agent, furan polymers, corrosion-resistant sealants and cements, foundry cores,; modified urea-formaldehyde polymers, penetrant, solvent for dyes and resins, flavoring. The polymer is used as a mortar for bonding acid-proof brick and chemical masonry.
See also furan polymer.

α-furfuryl amine. $C_4H_3OCH_2NH_2$.
Properties: Colorless liquid; soluble in water, alcohol, and ether; d 1.0550 (17C); bp 145C (757

mm); refr index 1.4900 (17C); flash p 99F (37.2C) (OC).
Derivation: Furfural and ammonia.
Hazard: Flammable, moderate fire risk.
Use: Corrosion inhibitor, component of soldering flux, chemical intermediate.

6-furfurylaminopurine. See kinetin.

furfuryl mercaptan. See 2-furanmethanethiol.

α-furildioxime. (di-2-furanylethanedione dioxime). $C_{10}H_8N_2O_4$. CAS: 522-27-0.
Properties: Needle-like crystals with mp 167C, mw 220.18. Very soluble in alcohol, ether; slightly soluble in benzene, petroleum ether.
Use: Reagent in alcohol solution for nickel, gives an orange-red compound.

furnace. An enclosed chamber or structure lined with firebrick or similar refractory and containing a heat source (coal, coke, gas, or electric elements). It may have various designs depending on its function. Furnaces are used for steel production (open-hearth and basic oxygen types), for smelting iron and other ores (blast furnace), for manufacture of furnace carbon black, etc. Electric furnaces are used for special types of steel as well as for high-temperature reactions such as the manufacture of pyrolytic graphite, synthetic diamonds, silicon, silicon carbide, and salt cake. Temperatures obtained range up to 3000C. Laboratory electric furnaces are used for high-temperature experiments and product testing.
See: Muffle furnace, reverberatory furnace, Mannheim furnace, kiln, forge, cupola.

furnace black. See carbon black.

furnace oil. Usually No. 1 fuel oil.
See fuel oil.

furnace sand. (fire sand). Sand used to line furnace bottoms or walls, particularly in open hearth steel furnaces.

"Furnex"[133]. TM for a series of semi-reinforcing furnace carbon blacks (SRF).

furnish. Term used by papermakers for the mixture containing the constituents of paper as supplied to the fourdrinier wire on which the sheet is formed.

furoamide. (pyromucamide; furamide). C_4H_3OCON.
Properties: Crystals, sublimes partly at 100C, mp 142C.

Derivation: Treatment of furoyl chloride with ammonia.

furoic acid. (pyromucic acid; furancarboxylic acid). CAS: 88-14-2. C_4H_3OCOOH or OCH:CHCH:CCOOH.
Properties: Colorless crystals, mp 133–134C, sublimes at 130C (50–60 mm). Slightly soluble in cold water; very soluble in hot water, alcohol, and ether; insoluble in paraffin hydrocarbons. Combustible.
Derivation: Cannizzaro reaction from furfural, oxidation of furfural.
Purification: Sublimation, fractional crystals from hot water.
Grade: Technical.
Use: Preservative, bactericide, furoates for perfume and flavoring, fumigant, textile processing, chemical intermediate.

Furol viscosity. The efflux time in seconds (SFS) of 60 ml of sample flowing through a calibrated Furol orifice in a Saybolt viscometer under specified conditions. Furol viscosity is approximately 1/10 of Saybolt Universal viscosity and is used for fuel oil and residual materials of relatively high viscosity. Furol is derived from the words "fuel and road oils."

furoyl chloride. C_4H_3OCOCl.
Properties: Colorless liquid, powerful lachrymator, soluble in ether, decomposes in water, mp −2C, bp 176C. Combustible.
Derivation: Treatment of furoic acid with phosphorus pentachloride.
Hazard: Strong irritant to eyes and skin.
Use: Substitute for chloropicrin in disinfecting grain elevators.

furylacrylic acid. (furfural acetic acid; furacrylic acid) $C_4H_3OCH:CHCOOH$.
Properties: White powder; slightly soluble in cold water; easily soluble in hot water; soluble in alcohol, ether, and glacial acetic acid; mp 141C; bp 117C (8 mm), 286C.
Derivation: From furfural.
Use: Derivatives used in perfumes.

furyl carbinol. See furfuryl alcohol.

fuse. Of a solid, to melt, e.g., fused salt. An electric fuse acts as a circuit breaker by the melting of a thin strip of metal. The term has the connotation of uniting or joining, as in welding. The union of hydrogen nuclei to yield energy is called fusion.

fused ring. A ring having one or more of its sides in common with another ring, as shown:

fused salt. A salt (i.e., ionic compound) in the molten state. Halides and nitrates are the salts most commonly used. High temperatures (500–1000C for the alkali halides) are usually required. Most fused salts are liquids with viscosities, diffusion coefficients, thermal conductivities and surface tensions in the same range as water. They conduct electricity exceptionally well.

Use: Production of sodium by electrolysis, heat-transfer agents, reaction medium in chemical synthesis, heat-treatment of metals (from 350–2400F), solvents for the metals corresponding to their cations, nuclear power reactors.
See also salt bath.

fused silica. See silica.

fusel oil. (amyl alcohol; fermentation; grain oil; potato oil). A volatile, oily mixture consisting largely of amyl alcohols. Isoamyl alcohol (isobutyl carbinol) and active amyl alcohol (2-methyl-1-butanol) are chief constituents. Ethyl, propyl, butyl, hexyl, and heptyl alcohols as well as other alcohols have been separated. Acids, esters, and aldehdyes are also present. Normal primary amyl alcohol (1-pentanol) is not found in fusel oil. Combustible.

Hazard: Moderate fire risk. Toxic by ingestion and inhalation.

Use: Chemicals (amyl ether, amyl acetate, pure amyl alcohols, nitrous ether, various esters), explosives (gelatinizing agent), solvent for fats and oils, intermediate, pharmaceuticals, nitrocellulose plastics, synthetic rubber, varnishes, lacquers, solvent for resins and waxes, perfumery.

fusible alloy. See alloy, fusible.

fusion. (thermonuclear reaction). An endothermic nuclear reaction yielding large amounts of energy in which the nuclei of light atoms (chiefly the hydrogen isotopes D (deuterium) and T (tritium) unite or fuse to form helium. Uncontrolled fusion was achieved some years ago in the hydrogen bomb, in which the initiating temperature was supplied by a fission reaction. Research efforts are now being devoted to developing a controlled and sustained fusion reaction which would utilize the deuterium and tritium in water. Several reactions are possible but the most efficient is, for each fusion event, $D + T + e \rightarrow He^{-4} + n + 17.5$ MeV. An energy input (e) equivalent to at least 44 million degrees C is necessary.

One approach utilizes powerful laser beams impinging on a mixture of deuterium and tritium in glass microspheres coated with Teflon and beryllium, which have an ablative effect. Fusion reactions have been successfully attained by this method at the Lawrence Livermore Laboratory, where a new and more powerful magnetic fusion laser is expected to confine a plasma in the range of 5×10^8 C. The other approach is a magnetic fusion device called a tokamak, located in Princeton, NJ. The Joint European Torus (JET) is under construction in Cambridge, England.

Fusion has two great advantages over fission as an energy source: (1) it utilizes water and readily available lithium as its raw materials instead of scarce and costly uranium; (2) it produces only tritium as a radioactive byproduct. As indicated above, the $D + T$ reaction yields He-4 nuclei, as well as 24-MeV neutrons, which carry off 80% of the energy. Though active research is continuing, no power production from fusion can be expected till the 1990s. A center for Fusion Engineering has been established at the University of Texas, Austin.
See also JET, tokamak.

"Fybrene"[45]. TM for petrolatum, USP, of medium melting point and medium consistency.
Use: Paper industry.

"Fybrol"[309]. TM for a series of wool lubricants based on petroleum sulfonates, fatty esters, and mineral oil.

"Fyrex."[1] TM for a substantially neutral ammonium phosphate, fine crystals, soluble in water.
Use: Flameproofing textiles, wood and fibers, manufacture of matches to prevent afterglow.

G

γ. See gamma.

g. (1) Abbreviation of gram, (2) acceleration due to gravity.

μg. Abbreviation of microgram.

"G-942."[28] TM for an aqueous solution of the partial sodium salt of a polymeric carboxylic acid.
Use: Tanning agent.

Ga. Symbol for gallium.

GABA. Abbreviation for γ-aminobutyric acid. See under aminobutyric acid.

Gabriel-Colman rearrangement. Formation of isoquinoline derivatives by the action of sodium ethoxide on phthalimidoacetic esters.

Gabriel ethylenimine method. Formation of ethylenimines (aziridines) by elimination of hydrogen halides from aliphatic vicinal haloamines with alkali. The method can be extended to the preparation of five- and six-membered-ring amines.

Gabriel synthesis. Conversion of alkyl halides to primary amines by treatment with potassium phthalimide and hydrolysis of the N-alkylphthalimides formed.

G-acid. (2-naphthol-6,8-disulfonic acid).

Properties: White needles, soluble in water.
Derivation: Sulfonation of β-naphthol.
See under Schaeffer acid.
Use: Azo dye intermediate.

gadolinium. CAS: 7440-54-2. Gd. A rare-earth element of the lanthanide series, atomic number 64, group IIIB of the periodic table, aw 157.25, valence = 3; seven natural isotopes.
Properties: Lustrous metal, d 7.87, mp 1312C, bp 3000C; reacts slowly with water, soluble in dilute acid, insoluble in water; exhibits a high degree of magnetism, especially at low temperatures; salts are colorless; has highest neutron absorption cross-section of any known element; has superconductive properties, burns in air to form the oxide. Combustible.
Derivation: (a) Reduction of the fluoride with calcium, (b) electrolysis of the chloride with sodium chloride or potassium chloride in an iron pot which serves as an anode and graphite cathode.
Grade: Ingots, lumps, turnings, powder up to 99.9+% pure.
Use: Neutron shielding, garnets in microwave filters, phosphor activator, catalyst, scavenger for oxygen in titanium production.

gadolinium chloride. $GdCl_3 \cdot xH_2O$. Colorless crystals, soluble in water, purities up to 99.9%.
Use: Gadolinium sponge metal, by contact with a reducing metal vapor; source of gadolinium.

gadolinium fluoride. $GdF_3 \cdot 2H_2O$. Available up to 99.9% purity.
Properties: Semisolid mass.
Hazard: Toxic material. TLV (as F): 2.5 mg/m³ of air.
Use: Source of gadolinium.

gadolinium nitrate. $Gd(NO_3)_3 \cdot xH_2O$.
Colorless crystals, soluble in water, decomposes 110C, purities up to 99.9% gadolinium salt.
Hazard: Fire risk in contact with organic materials.

gadolinium oxalate. $Gd_2(C_2O_4)_3 \cdot 10H_2O$.
Properties: White powder, insoluble in water, slightly soluble in acids, loses $6H_2O$ at 110C, purities up to 99.9% gadolinium salt.

gadolinium oxide. Gd_2O_3. Properties: White to cream-colored powder, d 7.41, mp 2330C, insoluble in water, soluble in acids to form the corresponding salts, hygroscopic, absorbs carbon dioxide from the air, purities up to 99.8% gadolinium oxide.
Derivation: Calcination of gadolinium salts, liquid/liquid ion-exchange separation.
Use: Neutron shields; catalysts; dielectric ceramics; filament coatings; special glasses; TV phosphor activator; lasers, masers, and telecommunication; laboratory reagent.

gadolinium sulfate. $Gd_2(SO_4)_3 \cdot 8H_2O$.
Properties: Colorless crystals, slightly soluble in hot water, more soluble in cold, d 3.01 (15C), purities up to 99.9% gadolinium salt.

Use: Cryogenic research; the selenide is used in thermoelectric generating devices.

"Gafac."[307]   TM for a series of anionic complex organic phosphate esters with good emulsifying and detergent properties.
Use: Surfactants in emulsion polymerization (FDA approved).

"Gafamide."[307]   TM for a series of fatty acid ethanolamides used as foam boosters, stabilizers, thickeners, detergents, and emulsifiers.

"Gafanol."[307]   TM for a series of polyethylene glycol polymers ranging from free-flowing liquids to waxes.
Use: Surfactant intermediates, solvents for pharmaceuticals, plasticizers, etc.

"GAF" Carbonyl Iron Powders.[307]   TM for microscopic spheres of extremely pure iron. Produced in 11 carefully controlled grades ranging in particle size from 3–20 microns in diameter. The iron content of some types is as high as 99.5%.
Use: High-frequency cores for radio, telephone, television, short-wave transmitters, radar receivers, direction finders; alloying agents; catalysts; powder metallurgy; magnetic fluids.

"Gafcol."[307]   TM for a series of mild-odored glycol ethers used as solvents for nitrocellulose lacquers, dyes, insecticides and intermediates in the manufacture of plasticizers. Miscible with most liquids.

"Gafstate."[307]   TM for a series of free acids of complex phosphate esters. Antistatic agents for plastic packaging films (FDA approved).

Gal.   Abbreviation for galactose.

galactose.   CAS: 59-23-4. $C_6H_{12}O_6$.
A monosaccharide commonly occurring in milk sugar or lactose.
Properties: White crystals, soluble in hot water and pyridine, slightly soluble in glycerol, mp 165–168C.
Derivation: By acid hydrolysis of lactose.
Use: Organic synthesis, medicine (diagnostic aid).

D-galactosamine hydrochloride.   (2-amino-2-dioxy-D-galactose).   CAS: 1772-03-8. $C_6H_{14}ClNO_5$.
Properties: Crystalline solid, mw 215.64, mp 182-185C (decomposes),
Derivation: Amino sugar isolated from chondroitin sulfate.

D(+)-galacturonic acid.   CAS: 685-73-4.
$COOH(CHOH)_4CHO$.   A major constituent

of plant pectins. It exhibits mutarotation, having both an alpha and a beta form.
Properties: The alpha form melts with decomposition at 159–160C. Soluble in water, slightly soluble in hot alcohol, insoluble in ether.
Derivation: Hydrolysis of pectins.
Use: Biochemical research.

galena.   (galenite; lead glance).
CAS: 1314-87-0.   PbS.   Natural lead sulfide.
Properties: Color lead gray, streak lead gray, metallic luster, good cubic cleavage, d 7.4–7.6, Mohs hardness 2.5; soluble in strong nitric acid, also in excess of hot hydrochloric acid.
Occurrence: Western US, Canada, Africa, South America.
Use: Chief ore of lead, frequently recovered for the silver it sometimes contains.

gallic acid.   (3,4,5-trihydroxybenzoic acid).
CAS: 149-91-7. $C_6H_2(OH)_3CO_2H$.

Properties: Colorless or slightly yellow crystalline needles or prisms; soluble in alcohol and glycerol, sparingly soluble in water and ether; d 1.694, mp 222–240C.
Derivation: Action of mold on solutions of tannin or by boiling the latter with strong acid or caustic soda.
Use: Photography, writing ink, dyeing, manufacture of pyrogallol, tanning agent and manufacture of tannins, paper manufacture, pharmaceuticals, engraving and lithography, analytical reagent.

gallium.   CAS: 7440-55-3.   Ga.   Metallic element of atomic number 31; group IIIA of the periodic system; aw 69.72; valences = 2,3; two stable isotopes.
Properties: Silvery-white liquid at room temperature; mp 29.7C; bp 2403C; may be undercooled to almost 0C without solidifying; d 5.9 (25C); more dense as a liquid than as a solid; soluble in acid, alkali, and slightly soluble in mercury; reacts with most metals at high temperatures; grades up to 99.9999% purity are available.
Occurrence: Prepared commercially from zinc ores and bauxite.
Derivation: Extraction of gallium as gallium chloride by ethyl ether or isopropyl ether and subsequent electrodeposition from an alkaline gallium oxide solution.

Use: The metal has no significant commercial uses. Its compounds are used as semiconductors.

gallium antimonide. GaSb. Available in an electronic grade.
Use: Semiconducting devices.

gallium arsenide. CAS: 1303-00-0. GaAs.
Properties: Crystals, mp 1238C, electroluminescent in infrared light.
Grade: Ingots, polycrystalline form in high purity electronic grade, single crystals. Often alloyed with gallium phosphide or indium arsenide.
Hazard: Toxic metal. See arsenic.
Use: Semiconductor in light-emitting diodes for telephone dials, injection lasers, solar cells, magneto-resistance devices, thermistors, microwave generation.

gallium oxides. The sesquioxide, Ga_2O_3, and suboxide, Ga_2O, are known. Both are stable at ordinary temperatures.
See gallium sesquioxide.

gallium phosphide. CAS: 12063-98-8. GaP.
Properties: Pale orange, transparent crystals or whiskers up to 2 cm long, made by vapor phase reaction at relatively low temperatures between phosphorus and gallium suboxide. These crystals are intermediate between normal semiconductors and insulators or phosphors. They operate over a temperature range of -55 to 500C. Gallium phosphide is electroluminescent in visible light.
Grade: Polycrystalline form in high-purity electronic grade, single crystals, whiskers.
Use: Semiconducting devices.

gallium sesquioxide. Ga_2O_3.
Properties (alpha and beta forms): Alpha form: white crystals, d 6.44, mp 1900C, changes to beta form at 600C. Beta form: white crystals, d 5.88. Both forms are insoluble in hot acid.
Grade: High-purity electronic.
Use: Spectroscopic analysis.

gallocyanine. CAS: 1562-85-2. $C_{15}H_{13}ClN_2O_5$.
Properties: Greenish solid; insoluble in water; soluble in alcohol, glacial acetic acid, alkali carbonate, and concentrated hydrochloric acid. CI 51030.
Derivation: Reaction of gallic acid with nitrosodimethylaniline.
Use: Analytical reagent (for lead), dye, and biological stain.

gallotannic acid. See tannic acid.

galls. (nutgalls). Excrescences on various kinds of oak trees resulting from the deposition of insect eggs.
Grade: The best grades (55–60% tannic acid) come from Iran, Syria, Turkey and Tripoli.
Use: Source of gallic and gallotanic acids, leather tanning, writing inks, medicine, textile printing, pharmaceuticals.

"Galvan."[275] TM for battery and electrically conductive carbon black.

galvanizing. Coating of a ferrous metal by passing it through a bath of molten zinc or by electrodeposition of zinc. In the former process, the iron and zinc combine to form an intermetallic compound at the interface, the outer surface being relatively pure zinc, which crystallizes as it cools to form the characteristic "spangle." The electrodeposition method gives a uniform surface which may be either dull or bright. Duration of corrosion protection is directly related to the thickness of the zinc coating.
See also sacrificial protection.

"Galvaseal."[204] TM for a light chromate-type conversion coating for all types of zinc or galvanized steel surfaces, producing salt spray resistance and white rust prevention as well as base for paint. Application by spray, immersion, or roller coat.

"Galvoline."[233] TM for a cored magnesium ribbon used as a continuous anode for the cathode protection of buried pipe lines and other metal structures. Combustible.

"Galvomag."[233] TM for magnesium alloy composition used in anodes in cathodic protection.

"Galvorod."[233] TM for a cored magnesium rod used as an anode in the cathodic protection of water-heater tanks. Combustible.

gamete. A sex cell (egg or sperm) which carries half the normal complement of chromosomes and combines with another sex cell to produce a new individual possessing the normal complement.

gametocide. A substance which can control pollinization of plants by selectively killing plant sex cells (gametes). Some suggested gametocides are maleic hydrazide and sodium-α,β-dichloroisobutyrate.

gamma. (γ). A prefix having meanings analogous to those of alpha, namely, to designate locations of substituents in a compound or a particu-

lar form or modification of an organic substance (gamma-globulin) or a metal crystal. It also identifies the most intense form of short-wave radiation.
See gamma ray.

gamma acid. (2-amino-8-naphthol-6-sulfonic acid; aminonaphtholsulfonic acid-2,8,6).

Properties: White crystals, soluble in alcohol and ether, slightly soluble in water.
Derivation: β-naphthylamine sulfate is sulfonated with 66% oleum; the precipitate, 2-naphthylamine-6,8-disulfonic acid, is heated with 65% sodium hydroxide solution at 210C and 210 psi.
Use: Azo dye intermediate.

gamma-globulin. See globulin.

gamma ray. Electromagnetic radiation of extremely short wavelength and intensely high energy. Gamma rays originate in the atomic nucleus; they usually accompany alpha and beta emission as in the decay of radium, and always accompany fission. Gamma rays are extremely penetrating and are best absorbed by dense materials like lead and depleted uranium.
Hazard: Exposure to gamma radiation may be lethal. Complete protection is essential.
Use: To initiate chemical syntheses (ethyl bromide) and crosslinking of polyethylene and other polymers; biochemical research; food preservation; analytical chemistry.
See also x-ray, radiation.

gamma ray spectroscopy. An analytical technique involving the use of gamma radiation which is emitted from radioactive nuclei in discrete energies. The spectrum of energies and the relative intensities of the gamma rays often characterize the radionuclide that emits them; it is these that are determined by gamma ray spectroscopy. It is possible to identify quantitatively the elements present by their characteristic gamma ray spectra as well as to determine radioactive decay rates.

"Gamtox."[253] TM for pesticide formulation containing benzene hexachloride.
See benzene hexachloride.

"Ganex."[307] TM for a series of modified polyvinylpyrrolidone-based products. Available in a wide range of physical forms from liquid to waxy solid to granular solid. Used as emollients and lubricity additives, dispersing and suspending agents, pour point depressants, sizing additives.

gangue. The minerals and rock mined with a metallic ore but valueless in themselves or used only as a byproduct. They are separated from the ore in the milling and extraction processes, often as slag. Common gangue materials are quartz, calcite, limonite, feldspar, pyrite, etc.

"Gantrez" AN.[307] TM for an interpolymer of methyl vinyl ether and maleic anhydride.
Properties: Soluble in water and many organic solvents. Compatible with a wide range of gums, resins, and plasticizers and with most metallic salts. Stable in acid and alkaline solutions. Available in a range of molecular weights.
Use: Ammonium nitrate slurry explosives, for its suspending action and crosslinking ability; rust-preventive films; antistatic and finishing agent in natural and synthetic textiles and glass fibers; adhesives and coatings, thickening agent and protective colloid; flocculant and foam stabilizer in paper-making; photoreproduction, pharmaceutical preparations; cosmetics; etc.

"Gantrez" M.[307] TM for polyvinyl methyl ether.

"Gardenol."[227] TM for methyl phenyl carbinyl acetate.

"Garlon."[233] TM for herbicides containing water-soluble salts or oil-soluble esters of 3,5,6-trichloro-2-pyridyloxyacetate acid.

garnet, natural. A group of silicate minerals. Garnets in nature are usually composed of mix of various garnet subspecies.
Use: Abrasive, blast cleaning of buildings.

garnet, synthetic. Single-crystal garnets designed, by the introduction of bismuth, calcium and vanadium, for microwave devices such as low-frequency ferri-magnetic resonators. Special yttrium, iron, and aluminum garnets are used in lasers and electronic devices.

garnierite. $(Ni, Mg)_6(OH)_6Si_4O_{11} \cdot H_2O$. A natural nickel-magnesium silicate, occurring as a natural alteration of magnesium silicate rocks. A nickel ore.

gas. ($\uparrow$). A state of matter characterized by very low density and viscosity (relative to liquids and solids), comparatively great expansion and contraction with changes in pressure and temperature, ability to diffuse readily into other gases, and ability to occupy with almost complete uniformity the whole of any container. Gases may be either elements (argon) or compounds (carbon dioxide); elemental gases may be monatomic (helium), diatomic (chlorine), or triatomic (ozone). All exist in the gaseous state at standard temperature and pressure, but can be liquefied by pressure. The most abundant gases are oxygen, hydrogen, nitrogen (diatomic), and carbon dioxide. Gases are used for fundamental research on the behavior of matter largely because the low concentration permits isolation of the phenomena far better than is possible in liquids or solids.

A "perfect" or ideal gas is one which closely conforms to the simple gas laws for expansion and contraction (Boyle's Law, Charles' Law). *Note*: Use of the word "gas" for gasoline, natural gas, or the anesthetic nitrous oxide is acceptable only in informal communication.

See also kinetic theory, compressed gas, noxious gas, noble gas, vapor.

gas, asphyxiant. See asphyxiant gas.

gas black. See carbon black.

"Gas-Chrom."[425] TM for a series of diatomaceous earths designed for use as supports in gas chromatography. They have been treated in various ways, such as flux-calcined and screened, washed with acids and bases, and/or silanized for inertness. "Gas-Chrom" Q, which has been treated with dichlorodimethylsilane, is specially recommended for the analytical separation of steroids, pesticides, and other high molecular weight materials.

gas chromatography. (GC; gas-liquid chromatography; GLC; vapor-phase chromatography; VPC). The process in which the components of a mix are separated from one another by volatilizing the sample into a carrier gas stream which is passing through and over a bed of packing consisting of a 20–200 mesh solid support. The surface of the latter is usually coated with a relatively nonvolatile liquid (the stationary phase). This gives rise to the term gas-liquid chromatography. If the liquid is not present, the process is gas-solid chromatography, which is also widely useful for analysis. Different components move through the bed of packing at different rates and so appear separately at the effluent end, where they are detected and measured by thermal conductivity changes, density differences, or ionization detectors.

Gas chromatography is advantageous as a means of analysis of minute quantities of complex mixtures from industrial, biological, and chemical sources and is also of potential value in actually preparing moderate quantities of highly purified compounds otherwise difficult to separate from the mixture in which they occur.

gas, compressed. See compressed gas, liquefied petroleum gas.

gaseous diffusion. See diffusion, gaseous.

gas hydrate. A clathrate compound formed by a gas (either noble or reactive) and water. The compounds are crystalline solids and are insoluble in water. They usually form (only at relatively low temperatures and high pressures) directly by contact of gas and liquid water. From 6–18 molecules of water may combine with each molecule of gas, depending upon the nature of the gas.

The best known gas hydrates are those of ethane, ethylene, propane, and isobutane. Others include: methane and 1-butene, most of the fluorocarbon refrigerant gases, nitrous oxide, acetylene, vinyl chloride, carbon dioxide, methyl and ethyl chloride, methyl and ethyl bromide, cyclopropane, hydrogen sulfide, methyl mercaptan, and sulfur dioxide.

Interest in the gas hydrates originated mainly because of the nuisance of such compound formation in gas pipelines. In recent years, propane has been successfully used to precipitate water from salt solution (or seawater), thus yielding potable water.

gasification. Production of gaseous or liquid hydrocarbon fuels from coal (1) by direct addition of hydrogen to form methane (hydrogasification), (2) by reacting steam with hot coal (800C) in the presence of air or oxygen to form carbon monoxide and hydrogen (synthesis gas) followed by a methanation reaction (Fischer-Tropsch method), (3) by underground aeration. In methods (1) and (2), the coal enters the reaction sequence in finely divided form; hydrocarbon gases, naphtha, and fuel oils of various grades can be produced by several modifications. In method (3), air is pumped down to previously ignited coal seams and the combustible products evolved are collected.

Many variations of methods (1) and (2) have been researched in the US, but none is economically competitive with crude oil. Several have advanced to the pilot stage, and projections of cost, investment, and installation for large-scale operation have been made. Environmental and geographic considerations have also been investigated. The controlling factor for future development is the price of crude.

All three methods mentioned above are technically feasible. During World War II hydrocarbon fuels were produced from coal on large scale in Germany by both hydrogasification and Fischer-Tropsch methods. More recently, large-scale coal conversion has been successfully achieved by the Lurgi and Sasol methods; the latter, in South Africa, is the only large-scale installation utilizing the Fischer-Tropsch technology. The Lurgi technique is in production at a number of locations.

Catalytic gasification of peat for production of methanol is now utilized on a commercial scale, use of lignite and wood for this purpose is under development.
See also Fischer-Tropsch process, synthesis gas.

gas, inert. See noble gas, inert.

gas laws. See Boyle's law, Charles' law, Gay-Lussac's law, ideal gas.

gas, liquefied petroleum. See liquefied petroleum gas.

gas liquid. See light hydrocarbon.

gas, natural. See natural gas.

gas, noble. See noble.

gas oil. A liquid petroleum distillate with viscosity and boiling range between kerosene and lubricating oil. Boiling range 232–426C, flash p 150F (65.5C), autoign temperature 640F (337C). Combustible.
Use: Absorption oil, manufacture of ethylene.

gasoline. CAS: 8006-61-9. A mixture of volatile hydrocarbons suitable for use in a spark-ignited internal combustion engine and having an octane number of at least 60. The major components are branched-chain paraffins, cycloparaffins, and aromatics. There are several methods of production: distillation or fractionation which yields straight-run product of relatively low octane number, used primarily for blending; thermal and catalytic cracking; reforming; polymerization; isomerization; and dehydrocyclodimerization. All but the first are various means of converting hydrocarbon gases into motor fuels by modifications of chemical structure, usually involving catalysis. The present source of gasoline is petroleum, but it may also be produced from shale oil and Athabasca tar sands as well as by hydrogenation or gasification of coal.

antinock gasoline A gasoline to which a low percentage of methyl-tert-butyl ether (MBTE) has been added to eliminate knocking and increase octane number. This compound has almost completely replaced tetraethyllead. Gasolines of octane number 100 or more are used chiefly as aviation fuel; those having a research octane number of approximately 90 are in general automotive use.
See also antiknock agent, octane number.

casinghead gasoline, see natural gasoline (below).

cracked gasoline Gasolines produced by the catalytic decomposition of high-boiling components of petroleum. In general, such gasolines have much higher octane ratings (80–100) than that produced by fractional distillation. The difference is due to the prevalence of unsaturated, aromatic and branched-chain hydrocarbons in the cracked gasoline. The actual properties vary widely with the nature of the starting material and the temperature, time, pressure, and catalyst used in cracking.

high-octane gasoline A gasoline with an octane number of 90–100.
See antiknock gasoline, octane number.

lead-free gasoline An automotive fuel containing no more than 0.05 g of lead per gallon, designed for use in engines equipped with catalytic converters.

natural gasoline A gasoline obtained by recovering the butane, pentane, and hexane hydrocarbons present in small proportion in certain natural gases. Used in blending to produce a finished gasoline with adjusted volatility but low octane number. Do not confuse with natural gas.

polymer gasoline A gasoline produced by polymerization of low molecular weight hydrocarbons such as ethylene, propene, and butenes. Used in small amounts for blending with other gasolines to improve their octane number.

pyrolysis gasoline Gasoline produced by thermal cracking as a byproduct of ethylene manufacture. It is used as a source of benzene by the hydrodealkylation process.

reformed gasoline A high octane gasoline obtained from low octane gasoline by heating the vapors to a high temperature or by passing the vapors through a suitable catalyst.

straight-run gasoline Gasoline produced from petroleum by distillation without use of cracking or other chemical conversion processes. Its octane number is low.

white gasoline An unleaded gasoline especially designed for use in motorboats; it is uncracked and strongly inhibited against oxidation to avoid gum formation and is usually not colored to distinguish it from other grades. It also serves as a fuel for camp lanterns and portable stoves.
Hazard: Highly flammable, dangerous fire and explosion risk. TLV 300 ppm.

gas, perfect. See ideal gas.

Gastaldi synthesis. Formation of dicyanopyrazines by cyclization of two molecules of an aminocyanomethyl ketone, produced by treatment of an isonitrosomethyl ketone bisulfite compound with potassium cyanide, heating in hydrochloric acid, and oxidation.

gastric juice. A mixture of hydrochloric acid and pepsin secreted by glands in the stomach in response to a conditioned nerve reflex. Its pH is about 2.0. Its action in the metabolic breakdown of food components is essential to the digestive process, though carbohydrate decomposition is initiated by the saliva.
See also digestion (1).

gate. The opening in an injection mold leading from the sprue or runner to the mold cavity; the term also refers to molded material removed from the aperture when the product is ejected.

Gattermann aldehyde synthesis. Preparation of aldehydes of phenols, of phenol ethers or of heterocyclic compounds by treatment of the aromatic substrate with hydrogen cyanide and hydrochloric acid in the presence of Lewis acid catalysts.

Gattermann-Koch reaction. Formylation of benzene, alkylbenzenes, or polycyclic aromatic hydrocarbons with carbon monoxide and hydrochloric acid in the presence of aluminum chloride at high pressure. Addition of cuprous chloride allows the reaction to proceed at atmospheric pressure.

gauge. An instrument for measuring and indicating such process variables as pressure (hydraulic, steam, air), liquid level, thickness, vacuum, etc. The many types of gauges are activated by mechanical, ultrasonic, electronic, magnetic, and pneumatic means. Some operate on the principle of automatic control. In materials technology, the term "gauge" is often synonymous with thickness, especially in the metals, rubber, and plastics fields. Light-gauge refers to thicknesses from about 0.005–0.05 inch, and heavy gauge to thicknesses from about 0.05–0.150 inch.
See also mil, meter (2).

Gay-Lussac, Joseph Louis. (1778–1850) A French chemist and physicist noted for the brilliance and accuracy of his reasoning and experimental work. He contributed greatly to the knowledge of gases in his discovery (1808) of the law of combining volumes and his independent discovery (1802) of the law of Charles, the relationship of temperature to the volume of gases. He graduated from and taught at the Ecole Polytechnique, becoming a full professor in 1810. His work in chemistry was extensive, resulting

in the discovery of boron with Louis-Jacques Thenard which he named, and a variety of compounds such as boron trifluoride, chloric acid, and dithionic acid ($H_2S_2O_6$). He identified iodine as an element, named it, and studied its properties. He investigated the relationship of acids and bases and introduced many analytical techniques (such as the use of litmus as an indicator). Among his many contributions to industrial chemistry were improvements in the production of sulfuric acid. Much of the progress of chemistry in the early 19th Century is associated with his career.

Gay-Lussac's law. A modification of Charles' Law to state the following: At constant pressure the volume of a confined gas is proportional to its absolute temperature. The volumes of gases involved in a chemical change can always be represented by the ratio of small whole numbers.

GC. Abbreviation for gas chromatography.

Gd. Symbol for gadolinium.

GDME. Abbreviation for glycol dimethyl ether. See ethylene glycol dimethyl ether.

GDP. Abbreviation for guanosine diphosphate. See guanosine phosphates.

Ge. Symbol for germanium.

gel. A colloid in which the disperse phase has combined with the continuous phase to produce a viscous jelly-like product. Only 2% gelatin in water forms a stiff gel. A gel is made by cooling a solution whereupon certain kinds of solutes (gelatin) form submicroscopic crystalline particle groups which retain much solvent in the interstices (so-called "brush-heap" structure). Gels are usually transparent, but may become opalescent.
See also pectin.

gelatin. A mixture of proteins obtained by hydrolysis of collagen by boiling skin, ligaments, tendons, etc. Its production differs from that of animal glue in that the raw materials are selected, cleaned, and treated with special care so that the product is cleaner and purer than glue. Type A gelatin is obtained from acid-treated raw materials, and Type B from alkali-treated raw materials. Gelatin is strongly hydrophilic, absorbing up to 10 times its weight of water and forming reversible gels of high strength and viscosity. It can be chemically modified to make it insoluble in water for such special applications as microencapsulation of fish nutrients for fish culture.
Properties: Flakes or powder, odorless, tasteless, soluble in warm water and glycerol; insoluble in organic solvents.

Grade: Edible, photographic, technical, USP.

Use: Photographic film; sizing; textile and paper adhesives; cements; capsules for medicinals; matches; light filters; clarifying agent; desserts, jellies, etc. culture medium for bacteria; blood plasma volume expander; microencapsulation; printing inks; nutrient; protective colloid in ice cream.

gelatin dynamite. A high explosive which contains nitrocellulose in addition to nitroglycerin. The product is a gelatinized mass, less sensitive to shock and friction than straight dynamite.

"Gelcarin."[124] TM for carrageenan extratives from Irish moss. Dry free-flowing powder packed in fiber drums.

Grade: Food grade, meeting FDA requirements.

Use: Gelling, suspending, binding,and viscosity-building.

gel filtration. A type of fractionation procedure in which molecules are separated from each other according to differences in size and shape; the action is similar to that of molecular sieves. Dextran gels (3-dimensional networks of polysaccharide chains) are usually used in this method known as gel filtration chromatography.

See also fractionation, molecular sieve.

"Gelgard."[233] TM for a synthetic polymeric water gelling material.

Use: Fire control.

"Gel-Kote."[448] TM for pigmented polyester resin coatings for polyester products.

gelled hydrogen. Liquid hydrogen thickened with silica powder.

Use: Rocket fuel.

"Gelloid."[309] TM for a series of purified extracts of various types of Irish moss seaweed, rich in mucilagous content. Manufactured into dry, odorless, edible powders used in food and pharmaceutical applications.

gel paint. (thixotropic paint). A paint formulation which has a semi-solid or gel consistency when undisturbed,but which flows readily under the brush or when stirred or shaken. After removal of the stress,it becomes stiff again and has little tendency to spill, drip,or run. The thixotropic quality is obtained by the carefully controlled reaction of a relatively small proportion of a polyamide resin with an alkyd resin vehicle.

See thixotropy.

gelsemine. CAS: 509-15-9. $C_{20}H_{22}N_2O_2$.

A toxic plant alkaloid. Used in medicine as a central nervous system stimulant.

"Geltone."[236] TM for a synthetic organophilic colloid that imparts gel to any oil/mud system.

"Gelva."[58] TM for vinyl acetate polymers.

Use: Adhesives, binders, chewing gum bases, coatings, hot melt adhesives, paints, paper treatment, permanent starches, slush molding, textile sizes and finishes, and thickeners.

"Gelvatol."[58] TM for polyvinyl alcohol resins.

Use: Adhesive, emulsifier, hydraulic cement additive, textile sizes, paper coating.

"Gelwhite."[471] TM for a series of white grades of water-refined montmorillonite clays.

Grade: GP and L, high sodium content in exchange position and low sodium content.

gem-. Prefix. Abbreviation of geminate, meaning two identical groups attached to the same carbon atom.

"Gemon-3010."[245] TM for a glass-reinforced thermosetting polyimide molding compound with excellent high-temperature properties, mechanical performance, and processing ease.

Use: Space applications, electric and electronic industries.

"Genacron."[307] TM for a series of disperse dyes for polyester fibers.

"Genacryl."[307] TM for a series of basic dyes for acrylic fibers.

"Genamid."[590] TM for resinous amine-based coreactants for epoxy resins.

Use: Coatings, castings, potting adhesives, laminates, tooling.

gene. A complex of nucleoproteins (chiefly DNA) that is the active transmitter of genetic information. Genes occur on the chromosomes of every living cell where they are arranged in a linear order. Genetic mechanism is the same in all organisms ranging from the lowest forms of life, both plant and animal, to man. Every organism has a large number of different genes; there are over 10,000 in each cell of the human body. These control the intricate and well-balanced system of biochemical reactions in the cell. The first synthesis of a gene was reported in 1970. In 1976, it was announced that a synthetic gene was successfully introduced into a microorganism in which it functioned in the same manner as would a normal gene. Because of the ability of DNA to store and transfer "coded" genetic instructions, genes determine the sequence of amino acids in specific polypeptides (see proteins); thus they prescribe the structure of the proteins synthesized. In viruses, the genes consist of ribonucleic acid (RNA). Mutation of genes is a change in the basic sequence of amino acids in DNA and may be induced by ionizing radiation. Mutations can occur in any living cell.

See also genetic code; radiation, ionizing; recombinant DNA; molecular biology.

"GenEpoxy."[590]   TM for liquid and solid resins having terminal epoxide groups.

Use: Coatings, structural adhesives, laminates, casting, encapsulation, sealants, concrete topping.

generic formula.   See formula, chemical.

"Genesolv."[50]   TM for ultrapure halogenated hydrocarbon solvents of the methane and ethane series.

Properties: High densities and low viscosities, combined residue (soluble plus insoluble matter) is less than 1 ppm.

Grade: Standard and electronic, Genesolv A, B, C, D.

Hazard: Nonflammable, nonexplosive, thermally stable, and relatively nonhydrolyzable.

Use: Removal of greases and oils from metals, plastic, elastomer, and paint or varnish surfaces. Used with all cleaning techniques on assembled motors and parts, electronic devices, precision components, motion picture film, refrigeration systems, etc. Also used for isolation of viruses, for fire-extinguishing, and as dielectric coolants.

gene-splicing.   See genetic engineering.

genetic code.   Information stored in the genes which "program" the linear sequence of amino acids within the protein polypeptide chain synthesized during cell development. The agencies involved are DNA (deoxyribonucleic acid) and RNA (ribonucleic acid). It is this information code (analogous to a computer language) that determines (1) the nature and type of an organism, i.e., its heredity, and (2) the kind of cell structure to be formed (muscle, bone, organ, etc.).

See also protein, amino acid, ribonuclease.

genetic engineering.   Transference of genetic material from the genes of one species to those of another by uniting a portion of the DNA of one organism with extranuclear sections of DNA from another organism (also called gene-splicing). Such "recombinant" molecules will replicate in the same way as in normal DNA behavior. This technique is being utilized in basic genetic research sponsored by the National Institutes of Health.

There have been a number of outstanding achievements in genetic engineering in which

bacteria are made to form chains of amino acids programmed by nucleic acids supplied by the experimenters. (1) Production of human insulin from *E. coli* by implanting synthetic genes in the bacterium; commercial production of such synthetic insulin has been going on for some time and the product has been approved by FDA. (2) Formation and secretion by *E. coli* of the protein ovalbumin. (3) Synthesis of rat growth hormone in bacteria. (4) Synthesis of human growth hormone in bacteria; it will probably be developed on a commercial scale. (5) Production of a biologically active protein closely similar to interferon by exposing to the action of viruses bacteria which have been programmed to form it by introduction of the appropriate genetic code. (6) Creation of a wholly synthetic bacterium characterized by unusual ability to consume crude oil.

Various patents are pending on some of these techniques for commercial exploitation. The Supreme Court has ruled that these and other results of genetic engineering are patentable inventions.

See also biotechnology.

"Genetron."[50]   TM for a group of fluorinated hydrocarbons.

Properties: Low moisture content, noncorrosive, nonflammable, stable, low power requirement per unit of refrigeration, high dielectric strength. For specific identification, see fluorocarbon.

Geneva System.   A system of nomenclature for organic compounds recommended in 1892. It is based on compounds derived from hydrocarbons as a starting point, the names corresponding to the longest straight carbon chain present. The position is indicated by numbers applied to the carbons of the straight chain, beginning with the end carbon nearest the substituent element or group.

Normal compounds, in which the carbons are linked in a straight chain, are named ethane, propane, pentane, etc. Normal propane is:

$$
\begin{array}{ccccc}
 & H & H & H \\
 & | & | & | \\
H- & C- & C- & C-H \\
 & | & | & | \\
 & H & H & H \\
\end{array}
$$

Branched compounds are so named as to show the proper deviation, e.g., 2-methyl propane

$$
\begin{array}{ccc}
1 & 2 & 3 \\
H_3C- & CH- & CH_3 \\
 & | & \\
 & CH_3 & \\
\end{array}
$$

is a derivation of propane, not of methane. Similarly, isopentane

$$H_3C - \overset{\overset{\displaystyle CH_3}{|}}{\underset{\underset{\displaystyle CH_3}{|}}{C}} - CH_3$$

is named 2-methylbutane. The four carbons in a straight line give the name butane and the side group is attached to the second carbon. Another pentane has the structure:

$$\overset{1}{H_3C} - \overset{2}{\underset{\underset{\displaystyle CH_3}{|}}{CH}} - \overset{3}{CH} - \overset{4}{CH}$$

and was once called tetramethyl pentane. Under the Geneva system, it is named in terms of two methyl groups joined to carbon number two in a propane chain—2,2-dimethyl propane.

Some special recommendations of the Geneva system define the use of certain suffixes: open-chain hydrocarbons with one double bond end in -ene; those with two double bonds end in -diene (alkenes and alkadienes rather than olefins and diolefins). Triple-bonded compounds end in -yne.

Alcohols are named by use of the hydrocarbon name with -ol as the characteristic suffix, i.e., methanol, ethanol, etc. In order to extend this plan to polyhydric alcohols, a syllable such as di, tri, or tetra is inserted between the name of the parent hydrocarbon and the suffix ol.

CH₂OHCH₂OH   1,2-ethanediol

$\overset{1}{C}H_2OH\overset{2}{C}HOH\overset{3}{C}H_2OH$ 1,2,3-propanetriol

Sulfides, disulfides, sulfoxides, and sulfones are named like ethers, the oxy-term being replaced with thio-, dithio-, sulfinyl, and sulfonyl. The acids are also named in accordance with the Geneva System.
See also chemical nomenclature.

"Gen-Flo Latices."[179]   TM for styrene-butadiene emulsion polymers with varying amounts of styrene and butadiene.
Use: Latex paints, paper coatings, adhesives, textile printing inks, sizes.

genin.   The steroid portion linked to a sugar residue in certain glycosides. Important genins are found in the digitalis glycosides used as heart stimulants.

"Genite" 923.[50]   TM for 2,4-dichlorophenyl ester benzenesulfonic acid; available as 50% emulsifiable or 50% wettable powder. Miticide specific for European red mite and clover mite.

"Gensol No. 6."[79]   TM for terpene solvent.
Use: Odorant for masking other odors, solvent and softener in rubber reclaiming, solvent in printing ink manufacture.

"Gen-Tac Latex."[179]   TM for a latex containing vinyl pyridine, butadiene, and styrene.
Use: Adhesive for rubber-to-fabric constructions, as in tires and mechanical products.

gentamicin. (gentamycin).   CAS: 1403-66-3. $C_{21}H_{43}N_5O_7$.
Properties: Amorphous solid with mp 102-108C. Freely soluble in water, pyridine, in acid solutions; moderately soluble in methanol, ethanol, and acetone; practically insoluble in benzene and halogenated hydrocarbons.
Use: Antibacterial.

"Genthane-S."[179]   TM for a polyurethane elastomer.
Use: Mechanical goods, grommets, packings, and extrusions.

gentian violet.   USP name for methyl violet.

gentisic acid.   (2,5-dihydroxybenzoic acid). CAS: 490-79-9. $C_6H_3(OH)_2COOH$.
Properties: Crystals; mp 199–200C; soluble in water, alcohol, and ether; insoluble in carbon disulfide, chloroform, and benzene.
Use: Medicine as sodium gentisate (analgesic).

"Gentro."[179]   TM for a series of vulcanizable polymers containing butadiene and styrene manufactured by the cold process.
Properties: Contains approximately 23.5% styrene and 76.5% butadiene. Polymers mixed with reinforcing pigments give a variety of properties, abrasion resistance, flexibility, color, etc.
Use: Tires, mechanical goods, proofed goods, extruded goods, shoe soles, heels, sponge, etc.

"Gentro-Jet."[179]   TM for a series of carbon blacks co-precipitated with SBR polymer.
Use: Tires, tread rubber, mechanical goods, conveyor belts, V-belts, etc.

geochemistry.   The study of the chemical composition of the earth in terms of the physicochemical and geological processes and principles that produce and modify minerals and rocks. Of practical importance in discovering and establishing the limits of ore deposits, petroleum, tar sands, salt, sulfur, and other valuable resources.

geometric isomer.   A type of stereoisomer in which a chemical group or atom occupies different spatial positions in relation to the double bonds. If the double bond is between two carbon atoms the isomers are called *cis* and *trans*, as in crotonic acid, where the H and COOH reverse locations:

cis   trans

If the double bond is between a nitrogen and a carbon atom the isomers are named *anti* and *syn*, as in benzaldoxime, where the OH group shifts location:

anti   syn

This phenomenon also occurs in saturated ring compounds having three or more members in the ring as well as in certain coordination compounds.

See also stereochemistry.

"Geon."[119] TM for a group of polyvinyl chloride polymers, available as general-purpose, rigid, insulation, compounded, latex, paste, polyblend, or soluble types. Also a high-temperature type (polyvinyl dichloride) that will withstand temperatures 15.5C higher than other vinyls and can be processed in standard equipment.

See also polyvinyl chloride.

geothermal energy. Superheated water and steam trapped in rock strata in areas characterized by volcanic activity or by intrusions of molten magma. Its temperatures range from 150–300C. It escapes either from natural surface vents (geysers, fumaroles, hot springs) or from bore holes drilled through the strata. A contributing source of heat is the natural radioactivity of rocks in the earth's upper mantle. Power was produced in Italy from geothermal sources as long ago as 1913; since then geothermal power plants have been installed in Iceland, New Zealand, France, Hungary, Japan, Mexico, and El Salvador. Many nonelectrical uses have been developed for home heating and industrial purposes, especially in Iceland.

The chief source of geothermal energy in the continental US is the California-Nevada area. The geysers in central California have been generating electric power from steam in substantial amounts for some years. The same is true of Hawaii. Geological formations appropriate for geothermal heat are so few that this form of energy will always have limited potential.

geranial. See citral.

geranialdehyde. See citral.

geraniol. (trans-3,7-diemthyl-2,6-octadien-1-ol). $(CH_3)_2C:CH(CH_2)_2C(CH_3):CHCH_2OH$.
A terpene alcohol.
Properties: Colorless to pale yellow, liquid oil; pleasant geranium-like odor; d 0.870–0.890 (15C); fp −15C; bp 230C; refr index 1.4710–1.4780 (20C); optical rotation −2 to +2 degrees. Soluble in alcohol, ether, mineral oil, fixed oils; insoluble in water and glycerol. Combustible.
Derivation: From citronella oil (Java), citronellol-free grades from palmerosa oil and (synthetically) from pinene. These are of higher quality.
Grade: Standard, soap, synthetic, FCC, EOA.
Use: Perfumery, constituent of synthetic fragrances and synthetic linalool.

geranyl acetate. (geraniol acetate).
CAS: 105-87-3. $CH_3COOC_{10}H_{17}$.
Properties: Clear, colorless liquid. Odor of lavender, d 0.907–0.918 (15C), bp 242C, optical rotation −2 to + 2 degrees, refr index 1.4580–1.4640 (20C), soluble in alcohol and ether, insoluble in water and glycerol. Combustible.
Derivation: (a) Constituent of several essential oils. (b) By heating geraniol and sodium acetate with acetic anhydride.
Grade: Technical, FCC.
Use: Perfumery, flavoring.

geranyl butyrate. (geraniol butyrate).
$C_3H_7COOC_{10}H_{17}$.
Properties: Colorless liquid, bp 151C (18 mm), rose-like odor, d 0.9008 (17/4C), insoluble in water and glycerol, soluble in alcohol, ether. Occurs in several essential oils. Combustible.
Grade: FCC.
Use: Perfumes and soaps, flavoring, synthetic attar of rose.

geranyl formate. (geraniol formate).
CAS: 105-86-2. $HCOOC_{10}H_{17}$.
Properties: Colorless liquid, bp 113 (15 mm), d 0.927 (20/4C), rose-like odor, insoluble in alcohol and ether. Occurs in several essential oils. Combustible.
Grade: FCC.
Use: Perfumes and soaps, flavoring, synthetic neroli oil.

geranyl propionate. (geraniol propionate).
$C_2H_5COOC_{10}H_{17}$.
Properties: Colorless liquid, rose-like odor, d 0.896–0.913 (25C), refr index 1.4570–1.4650 (20C), soluble in most oils, insoluble in glycerol. Combustible.
Use: Perfumery, flavoring.

germane. A germanium hydride of the general formula $Ge_n H_{2n} H_{2n+2}$.
See germanium tetrahydride.

germanium. Ge. Nonmetallic element of atomic number 32, aw 72.59, valences = 2,4; group IVA of the periodic system.
Properties: Grayish-white solid, a p-type semiconductor, conductivity depends largely on added impurities, d 5.323, mp 937.4C, bp 2830C, oxidizes readily at 600–700C, does not volatilize approximately 1350C, hardness 6 on Mohs scale. Attacked by nitric acid and aqua regia, stable to water, acids and alkalies in absence of dissolved oxygen.
Derivation: Recovered from residues from refining of zinc and other sources, by heating in the presence of air and chlorine. It is also present in some coals and can be recovered from their ash.
Occurrence: Missouri, Kansas, Oklahoma.
Method of purification: The chloride is distilled and then hydrolyzed to the oxide which is reduced by hydrogen to the metal. Zone-melting is used for final purification. Single crystals are made by vaporization of germanium diiodide. The impurities in germanium are of controllig importance in its use in transistors. These are added to high-purity germanium in trace amounts during growth of single crystals.
Grade: Transistor, i.e., impurities 1 part in 10^{10}.
Forms: Ingots, single crystals, pure or doped, powder.
Use: Solid state electronic devices (transistors, diodes), semiconducting applications, brazing alloys, phosphors, gold and beryllium alloys, infrared-transmitting glass.

germanium dichloride. CAS: 10060-11-4.
$GeCl_2$ White powder, mp decomposes, decomposes in water, soluble in germanium tetrachloride, insoluble in alcohol and chloroform.

germanium dioxide. (germanium oxide).
CAS: 1310-53-8. GeO_2.
Properties: White powder, hexagonal, tetragonal, and amorphous.
Grade: Technical, semiconductor, 99.999% pure.
Use: Phosphors, transistors and diodes, infrared-transmitting glass.

germanium monoxide. GeO.
Properties: Black solid, mp 710C (sublimes), insoluble in water, soluble in oxidizing agents. Available commercially 99.999% pure.

germanium potassium fluoride. (potassium germanium fluoride). K_2GeF_6.
Properties: White crystals, soluble in water (hot), insoluble in alcohol.

Grade: Technical.
Hazard: Toxic material. TLV (as F): 2.5 mg/m³.

germanium telluride. GeTe.
Properties: An efficient semiconductor, mp 725C.

germanium tetrachloride. CAS: 10038-98-9.
$GeCl_4$.
Properties: Colorless liquid, d 1.874 (25/25C), fp −49.5C (1 atm), bp 83.1C, refr index 1.464, decomposes in water. Insoluble in concentrated hydrochloric acid; soluble in carbon disulfide, chloroform, benzene, alcohol, and ether.
Derivation: Reaction of chlorine with elemental germanium.

germanium tetrahydride. CAS: 7782-65-2.
GeH_4. Colorless gas, d 3.43g/L, fp −165C, bp −88C, decomposes at 350C, insoluble in water, soluble in liquid ammonia, slightly soluble in hot hydrochloric acid.
Hazard: Toxic material. TLV: 0.2 ppm in air.

germicide. See bactericide.

"Germ-I-Tol."[430] TM for higher alkyl dimethyl benzyl ammonium chloride. Meets the requirements for benzalkonium chloride, USP.

getter. See scavenger.

ghatti gum. Exudation from the stem of *Anogeissus latifolia*.
Properties: Colorless to pale yellow tears, rounded or vermiform. Almost tasteless and odorless, partially soluble in water. Can be solubilized by autoclaving.
Use: Thickener and protective colloid; emulsifier for oils, fats, waxes.

Giauque, William F. (1895–1982) An American chemist who achieved distinction for his studies of the properties of matter at temperatures approaching absolute zero (−272C). This research established the science of cryogenics. Giauque received the Nobel Prize in chemistry in 1949. He was professor and research director at University of California at Berkeley. One of his most significant contributions was the invention of a magnetic cooling device which made it possible to attain cryogenic temperatures. An important property of matter discovered as a result of his work is superconductivity.
See also cryogenics, superconductivity.

gibberellic acid. CAS: 77-06-5. $C_{19}H_{22}O_6$.
A plant growth-promoting hormone. It is a tetracyclic dihydroxylactonic acid. It was synthesized in 1978.
Properties: Crystals; mp 233–235C; slightly soluble in water; soluble in methanol, ethanol, ace-

tone; soluble in aqueous solutions of sodium bicarbonate and sodium acetate.
Use: Agriculture and horticulture, malting of barley with improved enzymatic characteristics.

gibberellin. A group of plant growth regulators (hormones) isolated in 1938 and widely distributed in flowering plants which promotes elongation of shoots and coleoptiles. They differ from auxins in not stimulating the growth of roots and in various other properties. The presence of auxin appears to be necessary for gibberellins to function. All gibberellins have closely related structures, being weak acids with a ring system containing double bonds and 8 asymmetric carbons.

Gibberellin A₃

See also gibberellic acid, plant growth regulator.

Gibbs-Duhem equation. (GDE). An exact thermodynamic relation that permits computation of the changes of chemical potential for one component of a uniform mixture over a range of compositions, provided the changes of potential for each of the other components have been measured over the same range.

Gibbs, Josiah Willard. (1839–1903) The father of modern thermodynamics. During his lifelong post as professor of mathematical physics at Yale, he stated the fundamental concepts embraced by the three laws of thermodynamics, especially the nature of entropy. A theorist rather than an experimenter, Gibbs was the first to expound with mathematical rigor the "relation between chemical, electrical, and thermal energy and capacity for work." It has been said that throughout his adult life Gibbs did nothing but think. The results established him as a great creative scientist.
See also thermodynamics.

Gibbs phthalic anhydride process. Oxidation of naphthalene to phthalic anhydride with air at 360 degrees over vanadium pentoxide and other catalysts.

"Gibrel."[123] TM for compound for promoting plant growth, the potassium salt of gibberellic acid.

giga-. Prefix meaning 10^9 units (symbol = G). 1 Gg = 1 gigagram = 10^9 grams.

Gilbert, Walter. (1932-) An American molecular biochemist who won the Nobel prize for chemistry in 1980 along with Berg and Sanger for their studies of the chemical structure of nucleic acid. Author of many papers on theoretical physics and molecular biology. He has been at Harvard since 1972.

gilsonite. An asphaltic material or solidified hydrocarbon found only in Utah and Colorado. One of the purest (9.9%) natural bitumens. Said to be the first solid hydrocarbon to be converted to gasoline.
Hazard: Irritant, skin sensitizer.
Use: Acid, alkali, and waterproof coatings; black varnishes, lacquers, baking enamels, and japans; wire-insulation compounds; linoleum and floor tile; paving; insulation; diluent in low-grade rubber compounds; a possible source of gasoline, fuel oil, and metallurgical coke.
See also asphalt, bitumen.

gin. An alcoholic beverage made by distilling alcohol through a mixture of herbs and berries (juniper, coriander, etc.), and adjusting to 80–100 proof.
Properties: Flash p 90F (32.3C).
Hazard: Flammable, moderate fire risk. Slight irritant, intoxicant.

Girard's reagent. (Girard's "P:" carboxymethylpyridinium chloride hydrazide; acethydrazide-pyridinium chloride).
$C_5H_5NClCH_2CONHNH_2$. (Girard's "T:" carboxymethyltrimethyl ammonium chloride hydrazide; trimethylacethydrazide ammonium chloride).
$(CH_3)_3NClCH_2CONHNH_2$.
Properties: White to faintly pinkish crystals with little or no odor, mp 190–200C, soluble in water, insoluble in oils. "T" is hygroscopic.
Use: Separation of aldehydes and ketones from natural oily or fatty materials: extraction of hormones.

Girbotol absorption. (amine absorption). A process for the removal of hydrogen sulfide or carbon dioxide from a gaseous mixture. An organic amine (ethanolamine or diethanolamine, which are basic) is allowed to flow down a tortuous path through a tower where it is contacted by and absorbs (acidic) hydrogen sulfide or carbon dioxide from the gas to be purified as it moves up the tower. The amine, contaminated with these products, is then sent from the bottom of the tower to a steam stripper where it flows countercurrent to steam, which strips the hydrogen sulfide or carbon dioxide from it. The amine is then returned to the top of the tower. The process is widely used in the petroleum industry for purifying refinery and natural gases and for recovery of hydrogen sulfide for sulfur manufacture. Removal of carbon dioxide from gases is usually done with monoethanolamine.

glacial.   A term applied to a number of acids, e.g., acetic and phosphoric which have a freezing point slightly below room temperature when in a highly pure state. For example, glacial acetic acid is 99.8% pure and crystallizes at 16.6C.

glance.   A mineralogical term meaning brilliant or lustrous, used to describe hard, brittle materials that exhibit a bright reflecting surface when fractured. Examples of such materials are hard asphalts (glance pitch) and ores of certain metals such as lead glance (galena).

Glaser coupling.   Coupling of terminal acetylenes by shaking an aqueous solution of cuprous chloride-ammonium chloride and the alkyne in an atmosphere of air or oxygen.

glass.   A ceramic material consisting of a uniformly dispersed mixture of silica (sand) (75%), soda ash (20%), and lime (5%), often combined with such metallic oxides as those of calcium, lead, lithium, cerium, etc., depending on the specific properties desired. The blend (or "melt") is heated to fusion temperature (approximately 700–800C) and then gradually cooled (annealed) to a rigid, friable state, often referred to as vitreous. Technically, glass is an amorphous, undercooled liquid of extremely high viscosity which has all the appearances of a solid. It has almost 100% elastic recovery.
See also glass, optical.
Properties: (soda-lime glass) Lowest electrical conductivity of any common material (below 10^{-6} mho/cm). Low thermal conductivity. High tensile and structural strength. Relatively impermeable to gases. Inert to all chemicals except hydrofluoric, fluosilicic, and phosphoric acids and hot, strong alkaline solutions. Continuous upper use temperature about 121C but may be higher, depending on composition. Good thermal insulator in fibrous form. Molten glass is extrudable into extremely fine filaments. Glass is almost opaque to UV radiation; in the absence of added colorant it transmits 95–98% of light to which it is exposed. Noncombustible.
Occurrence: Natural glass is rare but exists in the form of obsidian in areas of volcanic activity and meteor strikes. Excellent sand for glass-making

occurs in Virginia (James River), Pennsylvania, Massachusetts, New Jersey, West Virginia, Illinois, and Maryland; also in southern Germany and Czechoslovakia.
Forms available: Plate, sheet, fiber, filament, fabric, rods, tubing, pipe, powder, beads, flakes, hollow spheres.
See also sodium silicate.
Hazard: TLV: fibers or dust 10 mg/m^3 of air.
Use: Windows, structural building blocks, chemical reaction equipment, pumps and piping, vacuum tubes, light bulbs, glass fibers, yarns and fabrics, containers, optical equipment. Minute glass spheres with partial vacuum interior and treated exterior are available for compounding with resins for use in deep-sea floats, potting compounds, and other composites.
See also following entries.

glass, borosilicate.   See glass, heat-resistant.

glass-ceramic.   A devitrified or crystallized form of glass whose properties can be made to vary over a wide range.
Properties: Rupture modulus up to 50,000 psi, d 2.5, thermal shock resistance 900C, upper continuous use temperature 700C. Glass-ceramics lie between borosilicate glasses and fused silica in high-temperature capability.
Derivation: A standard glass formula to which a nucleating agent, such as titania, has been added is melted, rolled into sheet, and cooled. It is then heated to a temperature at which nucleation occurs, causing formation of crystals.
See also nucleation; "Pyroceram," "Cer-vit."
Use: Range and stove tops, laboratory bench tops, architectural panels, restaurant heating and warming equipment, telescope mirrors.

glass electrode.   See electrode, glass.

glass enamel.   A finely ground flux, basically lead borosilicate, intimately blended with colored ceramic pigments. Different grades give characteristics of acid resistance, alkali resistance, sulfide resistance, or low lead release to meet requirements for various uses. Firing range 540–760C.
Use: For fired-on labels and decorations on glass ware, tumblers, milk bottles, beverage bottles, glass containers, illuminating ware, architectural glass, and signs.
See also porcelain enamel.

glass fiber.   Generic name for a manufactured fiber in which the fiber-forming substance is glass (Federal Trade Commission). Noncombustible.
Properties: Tensile strength 15 g/denier, elongation 3–4%, d 2.54, moisture regain - none, loses

strength above 315C, softens approximately 815C.

See also under glass.

Derivation: Molten glass is extruded at high speed through extremely small orifices.

Use: Thermal, acoustic, and electrical insulation (coarse fibers in bats or sheets); decorative and utility fabrics such as drapes, curtains, table linen, carpet backing, tenting, etc.; tire cord as belt between tread and carcass; filter medium; reinforced plastics; light transmission for communication signals; reinforcement of cement products for construction use.

See also fiber, optical.

Note: Glass fibers made by heating equal parts of slate and limestone to 1400C, first made in 1978, have exceptionally high resistance to alkalies and are a suitable substitute for asbestos for cement reinforcement. Large-scale production for this purpose is under development.

glass, heat-resistant. (1) A soda-lime glass containing approximately 5% boric oxide which lowers the viscosity of the silica without increasing its thermal expansion. Such glasses (known as borosilicates) have a very low expansion coefficient and high softening point (about 593C). Tensile strength is about 10,000 psi. Continuous use temperature 482C. Transmits UV light in higher wavelengths and is used in "sunlight" lamps and similar equipment for this reason.

See also "Pyrex".

Note: Use of borosilicate glass as a storage-disposal medium for high-level radioactive wastes has been under research for some time. Tentative conclusions based on high-temperature and high-pressure autoclave tests indicate that this method would be suitable for geologic storage of such wastes.

See radioactive waste.

(2) A pure silica glass trademarked "Vycor" which softens at about 1482C.

See also silica, fused.

glassine. A thin, transparent, and very flexible paper obtained by excessive beating of the pulp. It may contain an admixture of urea-formaldehyde to improve strength.

Use: Packaging, dust covers for books; general household purposes.

glassmaker's soap. Term for manganese dioxide (MnO_2).

glass, metallic. Metal alloys having an amorphous atomic structure similar to that of silica glass, achieved by cooling of the molten alloy so rapidly that no crystalline structure is formed. Such alloys are said to be harder than their crystalline

counterparts and are more resistant to corrosion. Those containing iron have unusual ferromagnetic properties which make them suitable for use as transformer coils.

See also "Metglas."

glass, optical Glasses intended for vision-correcting and such applications as lenses for cameras, microscopes and other instruments; must be of extremely high quality and uniformity to meet requirements for refractive index and light dispersion. Optical glass may be either crown (lime) or flint (lead). Lead oxide is a major ingredient of flint glass, imparting high refractive index and dispersion as well as surface brilliance. Flint glasses are also used in vacuum tubes and electrical equipment. Many special ingredients are used in both crown and flint glasses for specific refractivity and dispersion properties.

glass, photochromic. A glass that changes color on exposure to light and returns to original color when the light has been removed. One type is a silicate glass containing dispersed crystals of colloidal silver halide which is precipitated within the melt during cooling. Alkali borosilicates are the most suitable types of glasses for this purpose.

Use: Variable-tint prescription lenses which darken in sunlight and return to original clearness indoors (85% light transmission when clear, 45% in sunlight).

glass, photosensitive. A glass containing a small amount of a photosensitive substance such as a gold, silver, or copper compound. When UV light is passed through a photographic negative onto the surface of this glass a latent image is formed within the glass which is converted to a visible image made up of tiny metal particles when the glass is heated. In a special type of photosensitive glass (photosensitive opal), the metal particles of the photographic image within the glass serve as nuclei for the growth of nonmetallic crystals; crystalline growth is confined to the area of the image. These crystalline areas are dissolved much more rapidly than the adjacent glass by hydrogen fluoride. Thus, the glass can be formed into intricate shapes without the use of mechanical tools.

glass, plate. Plate glass has the same composition as window glass (soda-lime-silica), differing from it only in method of manufacture. These differences are primarily (1) the longer time of annealing (3 or 4 days), which eliminates the distortion and strain effects of rapid cooling, and (2) intensive grinding and polishing, which removes local imperfections and produces a bright, highly reflecting finish.

glass, ruby. A deep-red glass made by incorporating colloidal gold in the silicate mixture. It is used chiefly in the decorative arts.

glass, safety. See safety glass.

glass transition temperature. (T_g). The temperature at which an amorphous material (such as glass or a high polymer) changes from a brittle vitreous state to a plastic state. Many high polymers such as acrylics and their derivatives have this transition point which is related to the number of carbon atoms in the ester group. The T_g of glass depends on its composition and extent of annealing.

glass, water. See sodium silicate.

"Glasteel."[522] TM for an engineering laminate formed by spraying a slurry of powdered glass onto a base of mild steel followed by high-temperature firing. A chemical reaction takes place during the firing between the glass and steel, forming a chemical bond which locks the pieces firmly together.
Use: Corrosion-resistant process equipment, piping, heat exchangers, pumps and valves.

Glauber's salt. See sodium sulfate decahydrate.

glauconite. $K_2(Mg, Fe)_2Al_6(Si_4O_{10})_3(OH)_{12}$.
A natural silicate of potassium, aluminum, iron, and magnesium found in greensands and other sedimentary rocks.
Properties: Color green, luster earthy, d 2.3.
Occurrence: New Jersey, Virginia.
Use: Water softener, foundry molds, fertilizer.

glaze. A mixture similar to procelain enamel, applied to a ceramic substrate. It may refer to (1) a vitreous coating on pottery or enamelware, (2) the mixed dry powders of the batch to be used for the coating, or (3) a water suspension of these materials (wet glaze). Glazes must be low in sodium and are usually mixtures of silicates and flint, lead compounds, boric acid, calcium carbonate, etc.
See also frit, procelain enamel.

glaze stain. Finely ground calcined oxide of cobalt, copper, iron, and manganese.
Used for coloring ceramic glazes.

GLC. Abbreviation for gas-liquid chromatography.
See gas chromatography.

gliadin. A prolamin occurring in gluten, the protein of wheat, rye, and other grains. Wheat gliadin has the composition: 52.7% carbon, 17.7% nitrogen, 21.7% oxygen, 6.9% hydrogen, 1.0% sulfur. It is composed of 18 amino acids, 40% being glutamic acid. Insoluble in water, soluble in 70–90% alcohol, soluble in dilute acid and in alkali.
Use: Chemical synthesis of spinal anesthetics, pharmaceutical preparations.

"Glidco."[296] TM for a series of products containing or derived from terpene hydrocarbons. They include tars used for impregnating paper, twine, cordage, etc.; dipentene is used in paints and rubber reclaiming and terpineol is used in soap perfumes and the preparation of essential oils.

"Globar."[280] TM for silicon carbide heating elements and resistors and accessories.
Properties: Elements have a working temperature up to 1510C which can be extended to 1648C for short periods, low coefficient of expansion, structure not affected by rapid heating and quick cooling, resistance remains practically constant above 482C.
Use: Electric resistors and heating elements, terminals, and other accessories for electric heating elements, electric heating appliances, electric furnaces.

globulin. (1) Any of a group of proteins synthesized by the body when invaded by infective organism. They are coagulated by heat, insoluble in water, soluble in dilute solutions of salts, strong acids, and strong alkalies. Enzymes and acids cause hydrolysis to amino acids as the only products. Examples are immune serum or gamma globulins in blood, myosin in muscle. The blood globulins are used in immunizing against specific diseases and in medical research. The gamma globulin molecule is reported to consist of 19,996 atoms associated in 1320 amino acid units. (2) A protein occurring naturally in wheat and other cereal grains.
See also immunochemistry.

Glu. Abbreviation for glutamic acid.

glucagon. (hyperglycemic-glycogenolytic factor; HG-factor; HGF). Produced by the alpha cells of the islands of Langerhans and also by the gastric mucosa. It is opposite in effect to insulin. It appears to be a straight-chain polypeptide with a molecular weight of approximately 3500. Small amounts have been detected in commercial insulin preparations.
Grade: USP.
Use: Medicine, biochemical research.

"Glucarine B."[73] TM for glycol carbohydrate complex.

Properties: Water-white, syrupy, clear fluid; soluble in water, alcohols, glycols, and glycerol.

Use: Replaces glycerin where a cheaper colorless product is desired, especially for cosmetic and technical purposes.

glucase. See maltase.

gluconic acid. (glyconic acid; glycogenic acid). CAS: 526-95-4. $CH_2OH(CHOH)_4COOH$.

Properties: Pure product is crystalline, mp 131C, commercial grade is 50% aqueous solution, d 1.24, light brown color, insoluble in organic solvents.

Derivation: Bacterial, chemical, or electrochemical oxidation of glucose. It is the chief acid in honey.

Grade: Technical, 50% solution.

Use: Pharmaceutical and food products, cleaning and pickling metals, sequestrant, cleansers for bottle washing, paint strippers, alkaline derusters, catalyst in textile printing (ammonium salt).

glucono-δ-lactone. (D-gluconic acid, δ-lactone). CAS: 90-80-2. $CH_2OHCH(CHOH)_3C(O)O$.

Properties: White crystals, mp 155C, bp (decomposes), readily soluble in water, slightly soluble in alcohol.

Derivation: Oxidation of glucose.

See gluconic acid.

D(+)-glucosamine.   CAS: 3416-24-8. $CH_2OH(CHOH)_3CHNH_2CHO$.

Properties: (beta-form): Colorless needles, mp 110C (decomposes), very soluble in water, slightly soluble in methanol and ethanol, insoluble in ether and chloroform.

Use: Biochemical research.

glucose. (dextrose; grape sugar; corn sugar). CAS: 60-99-7.

$$
\begin{array}{c}
CH{=}O \\
|\\
H{-}C{-}OH \\
|\\
HO{-}C{-}H \\
|\\
H{-}C{-}OH \\
|\\
H{-}C{-}OH \\
|\\
CH_2OH
\end{array}
$$

Properties: Colorless crystals or white granular powder, odorless, sweet taste, d 1.544, mp 146C, soluble in water, slightly soluble in alcohol, it has the D (right-handed) configuration and is dextrorotatory. Combustible.

Occurrence: Formed in plants by photosynthesis, also in the blood.

Derivation: Hydrolysis of corn starch with acids or enzymes, hydrolysis of cellulosic wastes.

Grade: Technical, USP, anhydrous, hydrated.

Use: Confectionery, infant foods, medicine, brewing and wine-making, intermediate, caramel coloring, baking and canning, source of methane by anaerobic fermentation, source of certain amino acids, e.g., lysine, by fermentation.

See also glucose syrup, invert sugar.

glucose oxidase. An enzyme commercially available under various trademarks; catalyzes oxidation of glucose to gluconic acid. Water-soluble amorphous powder. Removes excess oxygen from canned foods, beer, etc., and from stored food.

Use: Food preservative, analytical reagent for glucose, stabilizer for vitamins C and B_{12}.

glucose syrup. (corn syrup).   A mixture of D-glucose, maltose, and maltodextrins made by hydrolysis of cornstarch by the action of acids or enzymes. The degree of conversion of the starch varies with consequent effect on the dextrose equivalent (DE) or reducing power of the syrup.

Use: Food industry as a sweetener (high DE), thickener, or bodying agent in soft drinks (low DE).

See also glucose.

α-glucosidase.   See maltase.

β-glucosidase.   See emulsin.

glucoside.   See glycoside.

D(+)-glucuronic acid.   CAS: 6556-12-3. $COOH(CHOH)_4CHO$.   A widely distributed substance in both plants and animals, usually occurs as part of a larger molecule as in various gums or combined with phenols or alcohols.

Properties: Needle-like crystals; exhibits mutarotation; the beta form has mp 165C; soluble in water and alcohol.

Derivation: From gum acacia.

Use: Biochemical research, medicine.

D-glucuronolactone.   CAS: 32449-92-6. $C_6H_8O_6$.   The gamma lactone of glucuronic acid. Found in plant gums and animal connective tissues.

Properties: Colorless, odorless, white powder, d 1.76 (30/4C), mp 172–178C, soluble in water.

Derivation: From glucuronic acid or by synthesis.

Use: Growth factor, medicine, pharmaceutical intermediate.

glue. A colloidal suspension of various proteinaceous materials in water. Most familiar are

those derived by boiling animal hides, tendons, or bones, which are high in collagen. Chief sources are slaughterhouse wastes and fish scraps. Other animal-derived glues are made from casein (milk) and blood. The most important vegetable glue is made from soybean protein. Combustible in solid form.
See adhesive, fish glue, collagen, gelatin.

"Gluflex."[157] TM for a clear, noncrystallizing product consisting of sucrose, levulose, dextrose, sodium bisulfite, and water, and having a high solids content.
Use: Animal glues, gummed tapes.

"Glueglis." TM for a composition containing chiefly glue and glycerol used in printing press rollers. Has soft, rubbery consistency but is easily decomposed by heat.

gluside. See saccharin.

glutamic acid. (α-aminoglutaric acid; 2-amino pentanedioic acid). CAS: 56-86-0. $COOH(CH_2)_2CH(NH_2)COOH$. A nonessential amino acid. The naturally occurring form is L(+)-glutamic acid.
Properties: *dl*-glutamic acid (synthetic racemic mix): crystals; mp 224–227C (decomposes); slightly soluble in ether, alcohol, and petroleum ether; d 1.4601 (20/4C); moderately soluble in water. *d*(-)-glutamic acid, mp 247–249C (decomposes), d 1.538 (20/4C). *l*(+)-glutamic acid: crystals; sublimes at 200C; decomposes at 247–249C; slightly soluble in ether, acetone, cold glacial acetic acid; insoluble in alcohol; d 1.538 (20/4C); specific rotation +37 to +38.9 degrees (25C); available commercially.
Derivation: Hydrolysis of vegetable protein (e.g., beet sugar waste, wheat gluten), organic synthesis based on acrylonitrile. It comprises 40% of the gliadin in wheat gluten.
Grade: FCC (l-form).
Use: Medicine, biochemical research, salt substitute, flavor enhancer (l-form only).
See also sodium glutamate.

glutamine. (2-amino-4-carbamoylbutanoic acid). CAS: 56-85-9. $H_2NC(O)(CH_2)_2CH(NH_2)COOH$ A nonessential amino acid. Both the l- and dl-forms are available.
Properties: White, crystalline powder; soluble in water; insoluble in most organic solvents. Should be kept dry and refrigerated, mp (l-form) 184–185C (decomposes) (dl-form) 176C. Its presence in wheat gluten contributes to the elastic properties of flour by hydrogen bonding and disulfide crosslinking.

Derivation: Action of enzymes on gluten, from beet roots, constituent of many proteins.
Use: Medicine, culture media, biochemical research, feed additive.

γ-glutamylcysteinylglycine. See glutathion.

glutaraldehyde. CAS: 111-30-8. $OCH(CH_2)_3CHO$.
Properties: Liquid, d 0.72, bp 188C (decomposes), fp −14C, vap press 17 mm (20C), soluble in water and alcohol, no flash point, nonflammable.
Grade: 99%, 50% biological solution, 25% solution.
Hazard: Irritant. TLV: ceiling 0.2 ppm in air.
Use: Intermediate, fixative for tissues, for cross-linking protein and polyhydroxy materials, tanning of soft leathers.

glutaric acid. (n-pyrotartaric acid; pentanedioic acid). CAS: 110-94-1. $HOOC(CH_2)_3COOH$.
Properties: Colorless crystals; mp 97C; refr index 1.419 (106C); soluble in water, alcohol, ether, benzene, and chloroform.
Derivation: From cyclopentanone.
Use: Organic synthesis, biochemical research.

glutaric anhydride. (pentanedioic acid anhydride). $H_2C(CH_2CO)_2O$.
Properties: Solid, mp 56.5C, bp 303C, soluble in benzene and toluene, soluble in water on complete hydrolysis.
Hazard: Irritant.
Use: Plasticizers, resin, lubricant, adhesive synthesis, dyes and pharmaceuticals.

glutaronitrile. (trimethylenedicyanide; pentanedinitrile). CAS: 544-13-8. $NC(CH_2)_3CN$.
Properties: Colorless to straw-colored, viscous liquid; bp 286C; d 0.989; soluble in water and alcohol; insoluble in ether and carbon disulfide. Combustible.
Use: Chemical intermediate.

glutathione. (γ-glutamylcysteinylglycine). CAS: 70-18-8. $C_{10}H_{17}O_6N_3S$. A universal component of the living cell. Contains glutamic acid, cysteine, and glycine. These are chemically bound but can be separated by hydrolysis.
Properties: White, crystalline powder; odorless; mp 190-192C; mild, sour taste; soluble in water and dilute alcohol.
Use: Nutritional and metabolic research.

gluten. A mixture of many proteins in which gliadin, glutenin, globulin, and albumin predomi-

nate; it occurs in highest percentage in wheat (Manitoba wheat contains approximately 12%) and also to some extent in other cereal grains, usually associated with starch. It comprises 18 amino acids. Gluten is insoluble in water and is hydrophilic. Its specific adaptability to bread making is due to its elastic, cohesive nature that enables it to retain the bubbles of carbon dioxide evolved by leavening agents; this also imparts to doughs their characteristic dilatant properties. This behavior is due to disulfide crosslinks and hydrogen bonding between the proteins or their constituent amino acids.
Use: Special breakfast foods and other cereals and foods, cattle food, adhesives, production of certain amino acids.

glutenin. One of the proteins present in wheat flour in substantial percentage. It is composed of 18 amino acids.

glutethimide. See 2-ethyl-2-phenylglutarimide.

Gly. Abbreviation for glycine.

glycarbylamide. (4,5-imidazoledeicarboxamide). CAS: 83-39-6. $C_3H_2N_2(CONH_2)_2$.
Properties: White powder, melting above 360C, insoluble in water, a coccidiostat for chickens.

glyceraldehyde. (glyceric aldehyde). CAS: 367-47-5. $HOCH_2CHOHCHO$.
Isomeric with dihydroxyacetone. It is produced by the oxidation of sugars in the body. As the simplest aldose, the conformation of d- and l-glyceraldehydes has been designated the reference standard for d- and l-carbohydrates and derivatives.

L   D

CHO   CHO

HO—C*—H   H—C*—OH

CH₂OH   CH₂OH

In these isomers, the central carbon atom (C*) is asymmetric.
Properties: (dl-glyceraldehyde) Tasteless crystals from alcohol-ether mixture, mp 145C, insoluble in benzene, petroleum ether, pentane.
Grade: 40% aqueous solution.
Use: Biochemical research, intermediate, nutrition, prepraration of polyesters, adhesives; cellulose modifier, leather tanning.

glyceride. An ester of glycerol and fatty acids in which one or more of the hydroxyl groups of the glycerol have been replaced by acid radicals.

The latter may be identical or different so that the glyceride may contain up to three different acid groups. Glycerides can be made synthetically. The most common are based on fatty acids which occur naturally in oils and fats.
See also mono-, di-, and triglyceride.

glycerin. See glycerol.

glycerin carbonate. (hydroxymethylethylene carbonate). $CH_2O(CO)OCHCH_2OH$.
Properties: Pale yellow, odorless, hygroscopic liquid; boiling range 125–130C (0.1–0.2 mm); fp supercools to a glass; d 1.4000 (20/4C); refr index 1.4580 (20C); flash p 415F (212C); miscible with water, alcohol, ether; soluble in ethylene dichloride; insoluble in carbon tetrachloride, benzene, and aliphatic hydrocarbons. Combustible.
Grade: Technical.
Use: Solvent, intermediate.

glycerol. (glycerin; glycyl alcohol; 1,2,3-propanetriol). CAS: 56-81-5.

$$CH_2—OH$$
$$|$$
$$CH—OH$$
$$|$$
$$CH_2—OH$$

A trihydric (polyhydric) alcohol.
Properties: Clear, colorless, odorless, syrupy liquid; sweet taste; hygroscopic; d anhydrous 1.2653; USP greater than 1.249 (25/25C); dynamite 1.2620; mp 18C; bp 290C; soluble in water and alcohol (aqueous solutions are neutral); insoluble in ether, benzene, and chloroform and in fixed and volatile oils; flash p 320F (160C); autoign temperature 739F (392C). Combustible.
Derivation: (1) Byproduct of soap manufacture; (2) from propylene and chlorine to form allyl chloride which is converted to the dichlorohydrin with hypochlorous acid; this is then saponified to glycerol with caustic solution; (3) isomerization of propylene oxide to allyl alcohol, which is reacted with peracetic acid; the resulting glycidol is hydrolyzed to glycerol; (4) hydrogenation of carbohydrates with nickel catalyst; (5) from acrolein and hydrogen peroxide.
Method of purification: Redistillation, ion-exchange techniques.
Grade: USP, CP (pharmaceutical and commercial, where highest grade is required), saponification soap lye, crude yellow distilled (for commerical purposes where color and extreme purity are not factors), high gravity or dynamite (dehydrated to 99.8–99.9% purity), natural, synthetic, FCC.

Use: Alkyd resins, dynamite, ester gums, pharmaceuticals, perfumery, plasticizer for regenerated cellulose, cosmetics, foodstuffs, conditioning tobacco, liquors, solvent, printer's ink rolls, polyurethane polyols, emulsifying agent, rubber stamp and copying inks, binder for cements and mixes, special soaps, lubricant and softener, bacteriostat, penetrant, hydraulic fluid, humectant, fermentation nutrients, antifreeze mixtures.

glycerol boriborate.
Properties: Pale yellow liquid obtained by heating glycerol, sodium borate and boric acid, composition varies, soluble in cold water, absolute alcohol, other alcohols, glycerol.
Use: Adhesive, binder, fabric softener, fire retardant on fabrics.

glycerol dichlorohydrin. See α-dichlorohydrin.

glycerol-1,3-distearate. (glyceryl-1,3-distearate). $C_{39}H_{76}O_5$.
Properties: Solid, mp 29.1, very slightly soluble in cold alcohol and ether, soluble in hot organic solvents.

glycerol monolaurate. (glyceryl monolaurate). $C_{11}H_{23}COOCH_2CHOHCH_2OH$.
Properties: Cream-colored paste; faint odor; dispersible in water; soluble in methanol and ethanol, toluene, naphtha, mineral oil, cottonseed oil, ethyl acetate; mp 23–27C; d 0.98; FFA less than 2.5%; iodine value 5–8; pH 8.0–8.6 (25C) (5% aqueous dispersion). Combustible.
See monoglyceride.
Grade: Edible, technical.
Use: Emulsifying and dispersing agent for food products, oils, waxes, and solvents; antifoaming agent; dry-cleaning soap base.

glycerol monooleate. (glyceryl monooleate). $C_{17}H_{33}COOCH_2CHOHCH_2OH$.
Properties: Yellow oil or soft solid, d 0.95, mp 14–19C depending on purity, iodine value 65–80, insoluble in water, somewhat soluble in alcohol and most organic solvents. Combustible.
See monoglyceride.
Grade: Edible, technical.
Use: Foods, pharmaceuticals and cosmetics; rust-preventive oils, textile finishing; vinyl light stabilizers, odorless base paints, flavoring.

glycerol monoricinoleate. (glyceryl monoricinoleate).
$C_6H_{13}CHOHC_{10}H_{18}COOCH_2CHOHCH_2OH$.
Properties: Yellow liquid, d 1.10, mp less than −5C, iodine value 65–70, dispersible in water, soluble in most organic solvents. Combustible.
See monoglyceride.

Use: Non-drying emulsifying agent, solvent, plasticizer, in polishes, in cosmetics, in textile, paper, and leather processing; low-temperature lubricant. Stabilizes latex paints against breakdown due to repeated freeze-thaws.

glycerol monostearate. (GMS; glyceryl monostearate; monostearin). $(C_{17}H_{35})COOCH_2CHOHCH_2OH$.
Properties: Pure white or cream-colored, wax-like solid with faint odor and fatty agreeable taste. Affected by light, mp 58–59C (capillary tube), d 0.97, FFA less than 5%, iodine value 3–4, dispersible in hot water, soluble (hot) in alcohol, oils, and hydrocarbons. Combustible.
See monoglyceride.
Grade: Edible, cosmetic, NF.
Use: Thickening and emulsifying agent for margarine, shortenings, and other food products; flavoring; emulsifying agent for oils, waxes, and solvents; protective coating for hygroscopic powders; cosmetics; pharmaceuticals; opacifier; detackifier; resin lubricant.

glycerol phthalate. See glyceryl phthalate.

glycerol tributyrate. See glyceryl tributyrate.

glycerol tripropionate. See glyceryl tripropionate.

glycerol tristearate. See stearin.

glycerophosphoric acid. $C_3H_5(OH)_2H_2PO_4$.
Properties: Colorless, odorless, liquid; soluble in water and alcohol; d 1.60; fp −25C. Combustible.
Derivation: Interaction of glycerol and phosphoric acid.
Use: Manufacture of glycerophosphates.

glyceryl abietate. An ester gum.
Use: Additive in citrus-flavored beverages.

glyceryl-p-aminobenzoate. See "Escalol 106."

glyceryl α-chlorohydrin. See chlorohydrin.

glyceryl diacetate. See diacetin.

glyceryl-1,3-distearate. See glycerol-1,3-distearate.

glyceryl ditolyl ether. See dicresyl glyceryl ether.

glyceryl monoacetate. See acetin.

glyceryl monolaurate. See glycerol monolaurate.

glyceryl monooleate. See glycerol monooleate.

glyceryl monoricinoleate. See glycerol monoricinoleate.

glyceryl monostearate. See glycerol monostearate.

glyceryl phthalate. (glycerol phthalate).
Properties: Water-white, solid resin; insoluble in water; soluble (hot) in methanol and ethanol, acetone, ethyl acetate; partly soluble in toluene, naphtha; d 1.29; saponifiction value 605–615; acid value 300–315; softening point approximately 67C.
Grade: Technical.
Use: Varnishes, lacquers, etc.
See also alkyd resin.

glyceryl ricinoleate. See glyceryl triricinoleate.

glyceryl triacetate. See triacetin.

glyceryl tri-(12-acetoxystearate). (castor oil, acetylated and hydrogenated).
$C_3H_5(OOCC_{17}H_{34}OCOCH_3)_3$.
Properties: Clear, pale yellow, oily liquid; mild odor; soluble in most organic solvents; insoluble in water; d 0.955 (25/25C); saponification value 298; iodine value 3; solidifies at 4C. Combustible.
Derivation: Hydrogenation of acetylated castor oil.
Grade: Technical.
Use: Plasticizer for nitrocellulose, ethylcellulose, and polyvinyl chloride; lubricants; protective coatings.

glyceryl tri-(12-acetylricinoleate). (castor oil, acetylated). $C_3H_5(OOCC_{17}H_{32}OCOCH_3)_3$.
Properties: Clear, pale yellow, oily liquid; mild odor; soluble in most organic liquids; insoluble in water; d 0.967 (25/25C); saponification value 300; iodine value 76; solidifies at −40C. Combustible.
Grade: Technical.
Derivation: Acetylation of castor oil.
Use: Plasticizer for nitrocellulose, ethylcellulose, and polyvinyl chloride; lubricants; protective coatings.

glyceryl tributyrate. (tributyrin; butyrin; glycerol tributyrate). $C_3H_5(OCOC_3H_7)_3$.
Properties: Colorless, oily liquid; d 1.035 (20C); refr index 1.4359 (20C); bp 315C; soluble in water 0.010%; soluble in alcohol and ether. Combustible.
Grade: Technical, FCC.
Use: Plasticizer, flavoring.

glyceryl tri(12-hydroxystearate). (castor oil, hydrogenated). $C_3H_5(OOCC_{17}H_{34}OH)_3$.
Glyceryl triricinoleate in which hydrogen has saturated the ricinoleic groups.
Properties: Hard, brittle, wax-like solid; yellowish to milky-white in color; mp 86–88C; d 0.899 (100/25C).

Use: Lubricants, heavy-metal soaps, waxes, plasticizers, cosmetics, chemical intermediate. The lithium compound is used in high-temperature greases.

glyceryl trinitrate. See nitroglycerin.

glyceryl trioleate. See olein.

glyceryl tripalmitate. See tripalmitin.

glyceryl tripropionate. (glycerol tripropionate; tripropionin). $C_3H_5(OCOC_2H_5)_3$.
Properties: Solid, d 1.078 (20C), refr index 1.431 (20C), bp 177–182C (20 mm), fp less than −50C, solubility in water 0.313% of weight, soluble in alcohol.
Use: Plasticizer.

glyceryl triricinoleate. (glyceryl ricinoleate). $C_3H_5(OOCC_{17}H_{32}OH)_3$. The triglyceride of ricinoleic acid. It constitutes approximately 80% of castor oil.
Properties: Light amber, oily liquid.
Use: Emulsifying agent.

glyceryl tristearate. See stearin.

glycidol. (2,3-epoxy-1-propanol).
CAS: 556-52-5.

$$CH_2OHCHCH_2 \overset{O}{\diagup\diagdown}$$

An epoxide.
Properties: Colorless liquid; bp 162C; soluble in water, alcohol, and ether; d 1.12. Combustible.
Derivation: Treatment of monochlorohydrin with bases; reaction product of allyl alcohol and perbenzoic acid.
Hazard: Toxic material. TLV: 25 ppm in air.
Use: Stabilizer for natural oils, demulsifier, dye-leveling agent, stabilizer for vinyl polymers.

γ-glycidoxypropyltrimethoxysilane.
$\overline{OCH_2CHCH_2}O(CH_2)_3Si(OCH_3)_3$
Properties: Liquid, d 1.070 (25C), bp approximately 120C (2 mm), refr index 1.4280 (25C). Soluble in acetone, benzene, ether; reacts with water.
Derivation: Addition of hypochlorite to allyl alcohol and reaction with soda lime.
Use: Coupling agent for glass- and mineral-filled plastics.

glycidyl acrylate. CAS: 106-90-1.
$H_2C:CHCOOCH_2\overline{CHCH_2}O$.
Properties: Liquid, d 1.1074 (20/20C), bp 57C (2 mm) with polymerization, fp −41.5C, flash

p 141F (60.5C) (TOC), insoluble in water. Combustible.

Hazard: Irritant to skin and eyes.   See acrylate.

glycidyl ether. OCH_2CHOCH_2.   Appears as the terminal group of epoxy resin structures resulting from reaction of epichlorohydrin and bisphenol A with alkaline catalyst. Also reacts with novolac resins.

Properties: Combustible liquid.

glycine. (1) (aminoacetic acid).
CAS: 56-40-6. NH_2CH_2COOH.
A nonessential amino acid.

Properties: White, sweet, odorless crystals; mp 232–236C with decomposition; d 1.1607; combines with hydrochloric acid to form the hydrochloride; soluble in water; insoluble in alcohol and ether.

Derivation: Action of ammonia on chloroacetic acid; occurs in many proteins and is especially abundant in silk fibroin, gelatin, and sugar cane.

Grade: Technical, NF, FCC.

Hazard: Use in fats restricted to 0.01%.

Use: Organic synthesis, nutrient, biochemical research, buffering agent, chicken-feed additive, reduces bitter taste of saccharin, retards rancidity in animal and vegetable fats.

(2) The extreme dilution of methylacetophenone gives a perfume resembling the odor of the climbing plant glycine (Wisteria sinensis), native to China and cultivated elsewhere. The name is also given to bouquets made from violet, lilac, and jasmin ottos.

(3) p-hydroxyphenylglycine. A photographic developer.

glycocholic acid. (cholylglycine). $C_{26}H_{43}NO_6$
The sodium salt occurs in bile where it is formed by the combination of glycine with cholic acid. It aids in the digestion and absorption of fats.

Properties: Crystallizes from water with 1.5 moles H_2O, becomes anhydrous at 100C, anhydrous form decomposes at 165C, practically insoluble in water, the sodium salt is soluble in water and alcohol.

Derivation: Precipitation from bile.

Use: Biochemical research, food emulsifying agent (up to 0.1%).

glycocoll. See glycine (1).

glycocoll-p-phenetidine hydrochloride.
See phenocoll hydrochloride.

glycogen. (animal starch; liver starch).
CAS: 9005-79-2. $(C_6H_{10}O_5)$.   A glycose polysaccharide, the storage carbohydrate of the

animal organism, found especially in the liver and rested muscle.

Properties: White powder, forms a dextrorotatory colloidal solution; insoluble in alcohol, soluble in water, sweet taste.

Derivation: Isolated from liver by treatment with 30% sodium hydroxide solution.

Use: Biochemical research.

glycogenic acid. See gluconic acid.

glycol. See ethylene glycol, it is also a general term for dihydric alcohols which are physically and chemically similar to glycerol.

glycol bromohydrin. See ethylene bromohydrin.

glycol carbonate. See ethylene carbonate.

glycol chlorohydrin. See ethylene chlorohydrin.

glycol diacetate. See ethylene glycol diacetate.

glycol dibutyrate. See ethylene glycol dibutyrate.

glycol diformate. See ethylene glycol diformate.

glycol dimercaptoacetate. (ethylene glycol bisthioglycolate).
$HSCH_2COOCH_2CH_2OOCCH_2SH$.

Properties: Liquid, d 1.313, bp 137–139C (2 mm), refr index 1.519 (25C). Insoluble in water; soluble in alcohol, acetone, and benzene. Combustible.

Use: Crosslinking agent for rubbers, accelerator in curing epoxy resins.

glycol dimercaptopropionate. [ethylene glycol bi-(mercaptopropionate)].
$(HSCH_2CH_2COOCH_2)_2$.

Properties: Liquid, d 1.219 (25C), bp 175–195C, refr index 1.5150 (25C). Insoluble in water and hexane; soluble in alcohol, acetone, and benzene. Combustible.

Use: Crosslinking agent for polymers, especially epoxy resins, chemical intermediate.

glycol dimethyl ether. See ethylene glycol dimethyl ether.

glycol dipropionate. See ethylene glycol dipropionate.

glycolic acid. See hydroxyacetic acid.

glycol monoacetate. See ethylene glycol monoacetate.

glycolonitrile. (glyconitrile; formaldehyde cyanohydrin). $HOCH_2CN$.

Properties: Mobile, colorless, odorless oil. Supplied commercially as a 70% aqueous solution

stabilized with phosphoric acid, bp 183C (slight decomposition), mp does not solidify when cooled to −72C, d 1.1039 (19C), refr index 1.4090 (25C), electrolytic dissociation constant $K = 0.843 \times 10^{-5}$ (25C).
Derivation: Formaldehyde and hydrogen cyanide.
Hazard: Toxic by ingestion, inhalation, and skin absorption.
Use: Solvent and organic intermediate.

glycol propionate. See ethylene glycol dipropionate.

glycol stearate. See ethylene glycol monostearate.

glycothiourea. See 2-thiohydantoin.

glycoylurea. See hydantoin.

glycolysis. Enzymatic (anaerobic) decomposition of sugars, starches, and other carbohydrates with release of energy, a type of reaction occurring in yeast fermentation and in certain metabolic processes. Lactic acid is one of the products formed.

"Glycomuls."[73] TM for a series of sorbitol fatty acid esters ranging from liquids to relatively high-melting, wax-like solids and with varying surface-active characteristics.
Use: Foods, cosmetics, pharmaceuticals, chemical specialties.

glyconic acid. See gluconic acid.

glyconitrile. See glycolonitrile.

glycoprotein. A composite molecule made up of a carbohydrate group and a simple protein. An example is the taste-modifying sweetener occurring in so-called "miracle fruit."

glycoside. One of a group of organic compounds, of abundant occurrence in plants which can be resolved by hydrolysis into sugars and other organic substances known as aglycones. Specifically glycosides are acetals which are derived from a combination of various hydroxy compounds with various sugars. They are designated individually as glucosides, mannosides, galactosides, etc. Glycosides were formerly called glucosides, but the latter term now refers to any glycoside having glucose as its sugar constituent.

glycyl alcohol. See glycerol.

glycyrrhizin. A glycoside of the triterpene group, the active principle of licorice root from which it is extracted. It has an intensely sweet taste,

and is used as a humectant in tobacco and a flavoring in confectionery and pharmaceutical products. The ammoniated derivative, which is 50 times as sweet as sucrose, is used as a foaming agent in root beer and mouthwashes; as a sweetener in chocolate, cocoa, and chewing gum; and as a taste-masking agent in pharmaceuticals such as aspirin. Its ability to exert strong synergistic action with sucrose makes it useful in low-calorie foods (from 30–100 ppm are effective).
See also sweetener, nonnutritive.

"Glydag" B.[46] TM for a dispersion of colloidal graphite in 1,3-butylene glycol.
Use: Rubber lubricant and soluble-oil additive.

"Glyecine" A.[307] TM for a dyeing assistant comprising thiodiethylene glycol, 100% active.
Use: Hygroscopic agent in textile printing, solvent for basic colors.

glyme. Trivial name for a series of glycol ethers used as aprotic solvents. The group includes monoglyme (bp 85C), ethyl glyme (bp 121C), diglyme (bp 162C), ethyl diglyme (bp 190C), triglyme (bp 216C), butyl diglyme (bp 256C), and tetraglyme (bp 276C). Each is separately listed and referred to its conventional name.

glyodin. (Generic name for 2-heptadecyl-2-imidazoline acetate; 2-heptadecylglyoxalidine acetate). CAS: 556-22-9.
$C_{17}H_{35}C_3H_5N_2 \cdot CH_3COOH$.
Properties: Light orange crystals, mp 62C, d 1.03, insoluble in water.
Derivation: Ethylenediamine and stearic acid.
Use: Fungicide (fruits and vegetables).

glyoxal. CAS: 107-22-2.   OHCCHO.
Properties: Yellow crystals or light yellow liquid, mild odor, mp 15C, bp 51C, d 1.14 (20/20C), bulk d 10.0 lb/gal (20C), vapor has a green color and burns with a violet flame, refr index 1.3826 (20C), polymerizes on standing or in presence of a trace of water. An aqueous solution contains monomolecular glyoxal and reacts weakly to acid. Undergoes many addition and condensation reactions with amines, amides, aldehydes and hydroxyl-containing materials. Glyoxal VP resists discoloration.
Derivation: Oxidation of acetaldehyde.
Grade: 40% solution; pure, solid; VP.
Hazard: Mixture of vapor and air may explode.
Use: Permanent-press fabrics; dimensional stabilization of rayon and other fibers. Insolubilizing agent for compounds containing polyhydroxyl groups (polyvinyl alcohol, starch, and cellulosic materials); insolubilizing of proteins (casein, gelatin, and animal glue); embalming fluids; leather

tanning; paper coatings with hydroxyethylcellulose; reducing agent in dyeing textiles.

glyoxaline. See imidazole.

glyoxyldiureide. See allantoin.

glyphosate. (N-(phosphonomethyl)glycine).
CAS: 1071-83-6.
Use: Herbicide.

glyphosine. $C_4H_{11}NO_8P_2$. A ripening agent for sugar cane.

"Glyptal."[245] TM for a group of alkyd-type polymers and plasticizers.

"G-M-F."[248] TM for p-quinonedioxime.
$HONC_6H_4NOH$. A rubber accelerator.

GMP. Abbreviation for guanosine monophosphate.
See guanosine phosphates, guanylic acid, sodium guanylate.

GMS. Abbreviation for glycerol monostearate.

goa powder. (Araroba). See chrysarobin.

Gogte synthesis. Formation of α-pyrone derivatives by rearrangement of acyl-substituted glutaconic anhydrides.

gold. CAS: 7440-57-5. Au. Metallic element of atomic number 79, Group IB of the Periodic Table, aw 196.9665, valence = 1,3; no stable isotopes.
Properties: Yellow, ductile metal; relatively soft; does not corrode in air but is tarnished by sulfur. Chemically nonreactive; attacked by chlorine and cyanide solutions in presence of oxygen. Soluble in aqua regia, insoluble in acids; d 19.3, mp 1063C, bp 2800C; excellent reflector of infrared and heat; electrical resistivity (0C) 2.06 microhm-cm; extremely high light reflectivity.
Occurrence: South Africa, USSR, Northwest Canada, US (South Dakota, Nevada, Utah, Alaska, California), Australia. Oceans are estimated to contain 70 million tons in solution with 10 billion additional tons on the ocean floor. There is no present economical method of exploiting these resources.
Derivation: Ore is treated with cyanide solution, the dissolved gold cyanate is recovered by precipitation with zinc dust, aluminum, or by hydrolysis. Placer methods are also used.
Forms available: Ingots, sheet, wire, tubing, powder, leaf; alloys with copper or other metals, the gold content being expressed in carats (the number of parts of gold in 24 parts of alloy). Single

crystals and aqueous colloidal suspensions. Leaf may be made in near-colloidal thickness; one troy ounce covers 68 sq ft at 0.0001 inch thickness.
Use: Infrared reflectors; electrical contact alloys; brazing alloys; polarographic electrodes; spinnarets; laboratory ware; decorative arts (ceramics); in electronics for bonding transistors and diodes to wires, for metallizing ceramic and mica capacitors, and for printed circuits; space vehicle instruments; dental alloys; jewelry. Colloidal dispersions are used in coloring glass, as nucleating agent in photosensitive glasses, and for specialized medical treatments; gold leaf is used in surgery. Further details can be obtained from Gold Information Center, Box 934, Madison Sq. Station, New York, 10159.

gold 198. Radioactive gold of mass number 198.
Properties: Half-life, 2.7 days.
Derivation: Neutron irradiation of the element.
Forms available: Gold metal, colloidal gold.
See radio-gold and gold sodium thiosulfate. The NF solution is colloidal Au-198.
Hazard: Toxic from beta and gamma radiation.
Use: Internal radiation therapy, to detect leaks in bacterial filters, to locate solidification boundary in continuously cast aluminum, to determine metallic silver in photographic materials.

gold, artificial. See stannic sulfide.

gold bromide. (aurous bromide). AuBr.
Properties: Yellowish-gray mass, decomposes at approximately 165C, insoluble in water.
Grade: Technical.
See also gold tribromide.

gold bronze. See aluminum bronze.

gold chloride. See gold trichloride.

gold-cobalt alloys. See cobalt-gold alloys.

gold cyanide. (auric cyanide; cyanoauric acid).
$Au(CN)_3 \cdot 3H_2O$ or $HAu(CN)_4 \cdot 3H_2O$.
Properties: Colorless, hygroscopic crystals, mp 50C (decomposes) very soluble in water, soluble in alcohol and ether.
Hazard: Toxic material. TLV (as CN): 5 mg/m³ of air.
Use: Electrolyte in the electroplating industry.

gold, filled. A thin gold alloy bonded to a base metal, also called clad stock.
Use: Inexpensive jewelry.

gold hydroxide. (auric hydroxide). $Au(OH)_3$.
Properties: Brown powder; sensitive to light, keep in amber bottle; probably a hydrated gold oxide (Au_2O_3) and loses water easily; soluble in hydro-

chloric acid, solutions of sodium cyanide and alkali hydroxides; insoluble in water.
Use: Gilding liquids, decorating porcelain, gold plating.

gold, liquid bright. See "Liquid Bright Gold."

gold oxide. (auric oxide; auric trioxide; gold trioxide). Au_2O_3.
Properties: Brownish-black powder, decomposed by heat, keep in dark bottle, soluble in hydrochloric acid, insoluble in water.
Use: Gold plating.

gold potassium chloride. See potassium gold chloride.

gold potassium cyanide. See potassium gold cyanide.

gold potassium iodide. See potassium gold iodide.

gold salts. See sodium gold chloride.

Goldschmidt, Hans. See thermite.

Goldschmidt process. Formation of sodium formate by absorption of carbon monoxide in caustic soda at increased pressures and temperatures around 200 degrees. At temperatures above 400 degrees, alkali formate liberates hydrogen and yields alkali oxalate.

gold-silicon alloy. (silicon-gold alloy).
Formed in amorphous foils 10 microns thick by cooling molten gold and silicon almost instantaneously by spreading on a moving wheel. The atoms are "frozen" before crystals can form.
Use: Electronics.

gold sodium chloride. See sodium gold chloride.

gold sodium cyanide. See sodium gold cyanide.

gold sodium thiomalate. CAS: 12244-57-4.
$NaOOCCH(SAu)CH_2COONa \cdot H_2O$.
Properties: White to yellowish-white odorless powder with metallic taste, affected by light, very soluble in water, practically insoluble in alcohol and ether, aqueous solutions are colorless to pale yellow, pH (5% solution) 5.8–6.5.
Derivation: Reaction of sodium thiomalate with a gold halide.
Grade: USP.
Use: Medicine (antirheumatic).

gold solder. A solder usually composed of gold, silver, copper, zinc, or brass.
Use: Principally by jewelers.

gold tin purple. (purple of Cassius; gold stannate; aurous stannate; gold-tin precipitate; CI 77482).
Properties: Brown powder, insoluble in water, soluble in ammonia).
Derivation: By the reaction of a neutral solution of gold trichloride with stannous and stannic chlorides, yielding a mixture of colloidal gold and tin oxide in varying proportions.
Grade: Technical.
Use: Manufacture of ruby glass, coloring enamels and porcelain.

gold tribromide. (auric bromide; gold bromide).
CAS: 10294-28-7. $AuBr_3$.
Properties: Brownish-black powder, mp 160C with decomposition, soluble in water and alcohol.
Use: Analysis (testing for alkaloids, spermatic fluid), medicine.

gold trichloride. (a) (auric chloride; gold chloride). CAS: 13453-07-1. $AuCl_3$.
(b) $AuCl_3 \cdot 2HOH$. (c) (chlorauric acid; chloroauric acid; gold trichloride, acid).
$AuCl_2 \cdot HCl \cdot 4HOH$ or $HAuCl_4 \cdot 4HOH$.
Properties: Yellow to red crystals, decomposed by heat, soluble in water, alcohol, and ether.
Derivation: Action of aqua regia on gold.
Grade: Technical, CP, usually as chlorauric acid.
Use: (c): Photography, gold plating, special inks, medicine, ceramics (enamels, gilding and painting porcelain), glass (gilding, ruby glass), manufacture of finely divided gold and purple of Cassius.

gold trioxide. See gold oxide.

gold, white. A jeweler's alloy consisting of approximately 58% gold, 17% nickel, 7% zinc, and 17% copper.

Gomberg-Bachmann reaction. Formation of diaryl compounds from aryl diazonium salts and aromatic compounds in the presence of alkali.

Gomberg free radical reaction. Formation of free radicals by abstraction of the halogen from triarylmethyl halides with metals.

Gooch. See crucible.

Goodyear, Charles. (1800–1860) Born in Woburn, MA, Goodyear was the first to realize the potentialities of natural rubber. Frustrated by its lack of stability to temperature and other weaknesses in the uncured state, he experimented with additives such as magnesium and sulfur. The discovery of vulcanization was not accidental, as

is often stated, but the result of intelligent trials and correct evaluation of their results. Though Goodyear's patents were contested by Hancock in England, he well merits the credit for making rubber usable in countless ways, and helping to make the automobile possible.
See also vulcanization.

Gordon Research Conferences. See American Association for Advancement of Science.

gossyplure. (7,11-hexadecadien-1-ol acetate). CAS: 50933-33-0. $C_{18}H_{32}O_2$.
Properties: A yellow liquid with mw 280.46, bp 130-132C. Soluble in most organic solvents.
Hazard: Extremely flammable.
Use: Sex attractant for pink bollworm.

gossypol. CAS: 303-45-7. $C_{30}H_{30}O_8$.
A natural polyphenol.
Properties: Yellow, crystalline pigment having three modifications; insoluble in water; soluble in alcohol.
Occurrence: Cottonseed kernels.
Hazard: Toxic by ingestion but is inactivated by heat, 0.04% max allowed in foods.
Use: Stabilizer for vinyl polymers, has possibilities as a biodegradable insecticide and male contraceptive, actively investigated in China.

Gould-Jacobs reaction. Synthesis of 4-hydroxyquinolines from anilines and diethyl ethoxymalonate via cyclization of the intermediate anilinomethylenemalonate followed by hydrolysis and decarboxylation.

GPF black. Abbreviation for general purpose furnace black.
See carbon black.

grade. Any of a number of purity standards for chemicals and chemical products established by various specifications. Some of these grades are as follows:
ACS (American Chemical Society specifications) reagent (analytical reagent quality)
CP (chemically pure)
USP (conforms to US Pharmacopeia specifications)
NF (conforms to National Formulary specifications)
purified
FCC (Food Chemicals Codex specifications)
technical (industrial chemicals)

food	spectro
feed	commercial
semiconductor	chemical
radio	injectable
research	nitration

graduate. A cylindrical glass container with etched volumetric gradations usually ranging from 5–100 or more milliliters.
Use: Measuring liquids in chemical and biological laboratories.

Graebe-Ullmann synthesis. Formation of carbazoles by the action of nitrous acid on 2-aminodiphenylamines, followed by decomposition of the resulting benzotriazoles.

"Grafoil."[521] TM for pure, flexible, graphite tape with highly directional properties similar to pyrolytic graphite. Thermal insulating properties up to 3650C.

grafting. A deposition technique whereby organic polymers can be bonded to a wide variety of other materials, both organic and inorganic, in the form of fibers, films, chips, particles, or other shapes. Grafting occurs at specific catalyst sites on the "host" materials, which must have some capacity for ion exchange, methathesis, or complex formation. Ionizable groups may be added artificially.
One proprietary application is polymerization of acrylonitrile with wood pulp fibers to make synthetic soil blocks, the polymer imparts high water-holding capacity to the pulp. Plant nutrient materials are added and the mixture pressed into blocks to be used for starting seedlings.

graft polymer. A copolymer molecule comprised of a main backbone chain to which side chains containing different atomic constituents are attached at various points. The main chain may be either a homopolymer or a copolymer. This process may be applied to the union of cellulosic molecules (cotton, rayon) with synthetic polymers (except polyesters, acrylics, and polypropylene) to form modified fibers having improved flame resistance, dimensional stability, resilience, and bacterial resistance. An intermediate called cellulose thiocarbonate is formed in this proprietary process.

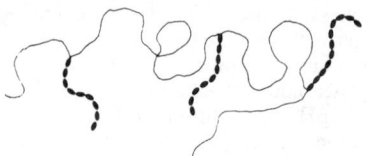

See also polyorganosilicate graft polymer.

Graham, Thomas. (1805–1869) Born in Scotland, Graham is famous for his basic studies in diffusion which led to the development of colloid chemistry. He was the first to observe a marked

difference in the rate of passage of certain types of substances through a parchment membrane. Those which readily crystallize, like sugar, pass rapidly through the membrane, but gelatinous types are "slow in the extreme." The latter, which comprise albumin, starch, gums, etc., he designated as colloids and their solutions as colloidal solutions. The former, which he called crystalloids, form "true" or molecularly dispersed solutions.
See also colloid chemistry.

Graham's salt.   See sodium metaphosphate.

grain.   (1) The smallest unit of weight in the avoirdupois system; 1 grain = 0.0648 gram; one ounce contains 437.5 grains. (2) Any cereal plant, as wheat, corn, barley, etc. (3) Crystalline particles of metals. (4) The dehaired side of a skin or hide.

grain alcohol.   See ethanol.

grain oil.   See fusel oil.

gram.   (g).   One one-thousandth kilogram. It is the weight of one milliliter (approximately 1 cubic centimeter) of water at 4C. One pound contains 453.5 grams.

gram atomic weight.   The atomic weight of an element in grams, i.e., the gram atomic weight of oxygen is 15.994 grams.
See mole.

gramicidin.   An antibiotic produced by the metabolic processes of the bacteria *Bacillus brevis*. It is a polypeptide which is active against most Gram-positive pathogenic bacteria. It is one of the two antibiotic components of tyrothricin but has been isolated and used alone.
Properties: White, crystalline platelets; mp 229–230C; soluble in lower alcohols, acetic acid, and pyridine; moderately soluble in dry acetone and dioxane; almost insoluble in water, ether, and hydrocarbons. Depresses surface tension, forms a fairly stable colloidal emulsion in distilled water.
Derivation: From tyrothricin by extraction with a mixture of equal volumes of acetone and ether, followed by concentration in vacuo and dissolving in hot acetone.
Grade: NF.
Use: Medicine (antibacterial).

gramine.   [3-(dimethylaminomethyl)indole].
CAS: 87-52-5. $C_{11}H_{14}N_2$.
Properties: Shiny, flat needles; mw 174.24; mp 138-139C. Soluble in alcohol, ether, chloroform; slightly soluble in cold acetone; practically insoluble in petroleum ether, water.

gram molecular weight.   The molecular weight of a compound in grams, i.e., the gram molecular weight of carbon dioxide is 44.01 grams.
See mole.

Gram-positive, -negative.   A characteristic property of bacteria in reacting to a staining method developed by Gram around 1880. The bacteria are stained with crystalline violet, treated with Gram's solution and again stained with safranine. If the dye is retained the bacteria are called Gram-positive; and vice versa.
See also bacteria.

granule.   A piece of gravel of small size (between two and four millimeters).

grapefruit oil.   See citrus peel oil.

grape sugar.   See glucose.

"Graphallast."[82]   TM for a group of graphite and hydrocarbon oilless materials. Resistant to scuffing or abrasion.
Use: Low friction bushings, bearings, and seals in submerged applications in many corrosive chemicals at normal temperatures less than 65.5C.

"Graphalloy."[82]   TM for a series of oilless, self-lubricating, long life, low-friction materials consisting of graphite and a metal or alloy such as Babbitt, bronze, cadmium, copper, gold, iron, nickel, or silver. Many grades are available to meet most chemical applications.
Use: Bearings, bushings, seals, electric brushes, brush assemblies, and nonfreezing electric contacts.

"Graphic Red."[141]   TM for lithol red pigments composed of sodium, barium, and calcium salts of diazotized Tobias acid coupled with β-naphthol.
Use: Printing inks, paints, rubber, floor coverings, crayons.

"Graphilm."[82]   TM for a superior liquid lubricant containing graphite and a carrying agent which dries to a film approximately 0.001 inch thick. The resulting film is insoluble in water and adheres tenaciously to the surface upon which it has been baked. Suitable for high temperature applications to 537C and may be applied by brushing, dipping, or spraying.
Use: Provides lubrication for hard vacuum, high temperature and cryogenic applications.

graphite. (black lead; plumbago).

CAS: 7782-42-5. The crystalline allotropic form of carbon. Occurs naturally in Madagascar, Ceylon, Mexico, Korea, Austria, USSR, and China. Also produced synthetically by heating petroleum coke to approximately 3000C in an electric resistance furnace. Approximately 70% used in US is synthetic.

Properties: Relatively soft, greasy feel; steel gray to black color with a metallic sheen; d 2.0–2.25 depending upon origin; apparent d artificial graphite 1.5–1.8. High electrical and thermal conductivity, specific heat 0.17 at room temperature, 0.48 at 1500C, tensile strength 400–2000 psi, compressive strength usually approximately 2000–8000 psi. Coefficient of friction 0.1μ. Resistant to oxidation and thermal shock. Sublimes at 3650C.

Grade: Powdered, flake, crystals, rods, plates, fibers.

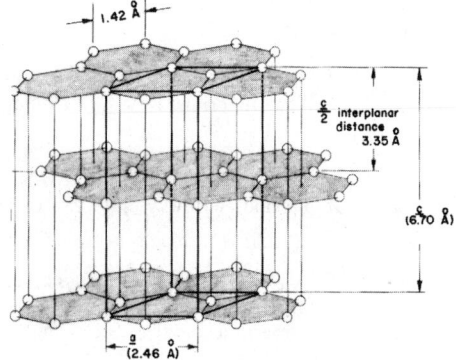

Hexagonal form of graphite.

Hazard: (powder, natural) fire risk. TLV: 2.5 mg/m³, respirable dust

Use: Crucibles, retorts, foundry facings, molds, lubricants, paints and coatings, boiler compounds, powder glazing, electrotyping, monochromator in x-ray diffraction analysis, fluorinated graphite polymers with fluorine to carbon ratios of 0.1–1.25, electrodes, bricks, chemical equipment, motor and generator brushes, seal rings, rocket nozzles, moderator in nuclear reactors, cathodes in electrolytic cells, pencils, fibers, self-lubricating bearings, intercalation compounds.

See also carbon, graphite fiber, carbon industrial.

graphite fiber. High tensile fibers or whiskers made from (1) rayon, (2) polyacrylonitrile, or (3) petroleum pitch.

Properties: (1) Amorphous structure, oxidizes readily, requiring silicon carbide coating. Resistant to acids and bases, including hydrogen fluoride; tensile strength 50,000–150,000 psi, elastic modulus 4–9 million psi, upper temperature limit in oxidizing atm 500C, self-lubricating, resistant to electricity, lightweight. (2) Polycrystalline structure, tensile strength up to 350,000 psi, elastic modulus up to 70 million psi, smooth surface, lightweight. (3) Properties similar to (2).

Derivation: (1) Heating rayon in air at 1400–1700C. (2) Heating polyacrylonitrile in air at 220C (20 hours) to oxygenate, then in hydrogen to 1000C (24 hours) to carbonize, finally in argon at 2500C (2 hours). (3) Heating pitch materials (petroleum residues, asphalt, etc.) to carbonization temperature.

Form: Filament, yarn, fabric, whiskers. Fibers may be 7–8 microns in diameter.

Use: Heating pads (combined with glass fiber), protection clothing, polyester and epoxy composites for jet engine components, space craft, compressor blades, airframe structure, electrodes for spark-hardening metals, flame-proof textile products, engineering thermoplastics.

See also fiber; whiskers; composite; carbon, industrial.

graphite, pyrolytic. (pyrographite). A dense, nonporous graphite stronger and more resistant to heat and corrosion than ordinary graphite, intended for rocket nozzles, missiles in general, and nuclear reactors; exhibits tensile strengths 5–10 times higher than commercial graphite and maintains its strength above 3300C. It is ultrapure, impermeable to all fluids, and is an anisotropic thermal and electrical conductor.

graphitic oxide. (GO). $C_7O_2(H)_2$.

Properties: Light-yellow flakes or plates which deposit in layers to form a membrane, usually supported on a cellulose ester.

Derivation: Slow oxidation of graphite with potassium nitrate in nitric and sulfuric acids.

Use: Reverse-osmosis membrane for desalination of sea water.

"Graphlon."[82] TM for a group of graphite and resin materials that exhibit extremely high chemical inertness for submerged applications. "Graphlon" bearings, bushings and seals withstand highly corrosive chemicals Operating range from −272 to +232C in air.

"Graphmetex."[82] TM for a group of special insulating resins employed as insulating spacers and holders for electrical components.

GRAS. Abbreviation for "Generally recognized as safe," applied to food additives approved by FDA.

gravel. Sediment particles larger than sand (larger than two millimeters).

gravimetric analysis. A type of quantitative analysis involving precipitation of a compound which can be weighed and analyzed after drying. It is also used in determining specific gravity.

"Gredag."[46] TM for specialized dispersions based on graphite, aluminum, and molybdenum disulfide.
Use: Die casting compounds, release agents, plunger lubricants, die pretreatments, toggle assembly, and machine and ejector pin lubricants.

Greek letters. See alpha, beta, and gamma, respectively.

green. Used in the process industries in a number of senses in addition to that of a dye or pigment. For example, it often is applied to any material that is uncured or untreated, e.g., leather, rubber, cement, etc.
See green liquor, greensand, greensalt, green soap, Paris green.

green cross. (superpalite). See trichloromethyl chloroformate.

greenhouse effect. Gradual rise in average global temperature due to absorption of infrared radiation by increasing amounts of carbon dioxide in the air which retards dissipation of heat from the earths surface. This phenomenon, which has been well documented, is ascribed to the burning of fossil fuels, especially coal, aided by aerosols and other contaminants. Definitive studies by NRC (1979) and NASA (1981) project a substantial increase in average temperature during the next century, perhaps as much as 3–4C. The increase of 1C reported to have occurred from 1900–1950 indicates that this has been going on for the last 100 years or more, coinciding with the industrial revolution.
Since the CO_2 concentration tends to be greatest at the poles (perhaps due to the absence of vegetation), it is probable that the ice caps there will deteriorate, with a concomitant rise in sea levels through the world. Other deleterious effects are also predicted, e.g., droughts in North America and central Asia caused by basic changes in rainfall distribution. As combustion of organic fuels is the underlying cause of this situation, reduction in their use is desirable wherever possible, combined with accelerated development of energy sources which do not produce carbon dioxide.

Greenland spar. See cryolite, natural.

green liquor. See liquor (b).

greenockite. CdS. A native cadmium sulfide containing 77.7% cadmium. Ore of cadmium.

greensalt. A wood preservative containing chromated copper arsenate.
Hazard: Highly toxic.

greensand. See foundry sand.

green soap. A liquid soap made with potassium hydroxide and a vegetable oil (except coconut and palm kernel oil).
See also soap.

greenstone.
Available forms: Mineral granules, various screen gradings and colors.
Use: In roofing materials where outdoor weathering resistance is required, as a filler for asbestos shingles, tennis court surfaces.

green-verditer. A paint pigment consisting of the hydroxycarbonate of copper.
See copper carbonate, malachite.

"Greenz."[48] TM for an ammonium lignin sulfonate containing 4.5% iron.
Use: Agricultural spray or soil additive in the treatment of iron chlorosis.

grex. See denier.

Griess diazo reaction. Formation of aromatic diazonium salts from primary aromatic amines and nitrous acid or other nitrosating agents.

"Griffco."[309] TM for a polyvinyl acetate emulsion useful as a base for adhesives and paint.

Grignard degradation. Stepwise dehalogenation of a polyhalo compound through its Grignard reagent which on treatment with water yields a product containing one halogen atom less.

Grignard reaction. Addition of organomagnesium compounds (Grignard reagents) to carbonyl groups or other unsaturated groups to give alcohols or ketones.

Grignard reagent. An important class of reagents used in synthetic organic chemistry, made by union of metallic magnesium with an organic chloride, bromide, or iodide, usually in the presence of an ether and in the complete absence of water. General formula RMgX, where R is an alkyl or aryl or other organic group and X a halogen. The value of the reagents lies in their ease of reaction with water, carbon dioxide, alcohols, aldehydes, ketones, amines, etc., to produce a great variety of organic compounds, usually with good yields. Examples are ethyl magnesium chloride (C_2H_5MgCl), methyl magnesium bromide (CH_3MgBr), etc.

Hazard: Because the heat of decomposition of Grignard reagents with water is great and the ether they are dissolved in is highly volatile and flammable, they must be handled with extreme care. Some, especially the solution of MeMgBr in ethyl ether, may ignite spontaneously on contact with water or even damp floors; nearly all will ignite on a wet rag or similar material. Since all Grignard reagents react rapidly with both water and oxygen, contact must be avoided. Ordinary materials of construction are satisfactory for use with Grignard reagents.

Grignard, Victor. (1871-1935) A French chemist who worked with Sabatier to win Nobel prize in 1912. He authored "Theses sur les Combinaisons Organomagnesiennes Mixtes et Leurs Applications a des Syntheses." As a student of Berbier, he went from Lyons, France to Besancon to Nancy, France. He discovered organomagnesium compounds which are used for organic synthesis by the named "Grignard" reaction.

grinding media. Balls ranging in diameter from 1/4 to 2 inches (in special cases even larger) made of porcelain, flint, bronze-alumina or alloy steels of various compositions. These perform an efficient size-reduction action on solid materials that are softer than the balls themselves. Steel rods are also used.
See ball mill, rod mill.

griseofulvin. $C_{17}H_{17}ClO_6$.

Properties: White to creamy white, odorless powder. Insoluble in water; slightly soluble in alcohol, benzene, acetone, and chloroform.
Derivation: By growth of Penicillium griseofulvum or synthetically.
Grade: USP.
Use: Medicine (antifungal), feed additive.

Grob fragmentation reaction. Carbon-carbon bond cleavage involving a system of four atoms.

grizzly. A coarse size-grading device used for ores, coal, and mineral aggregate. It consists of a grid of parallel steel bars separated by a distance determined by the size separation desired. It is set at an angle of approximately 45 degrees so that lumps that are too large to pass through it can roll down the bars while the smaller sizes fall into a container beneath.

"Groco."[410] TM for a series of fatty acids which include stearic acid; tallow and hydrogenated tallow; oleic acid; white oleine; and coconut, soya bean, corn, cottonseed, and palm oils.

grog. Crushed refractory materials added to ceramic mixes to reduce lamination in plastic clays and also to reduce shrinkage on drying. Materials crushed for this purpose are pottery, fire brick, quartz, quartzite, burned ware, saggers.

ground-nut oil. See peanut oil.

groundwood. A type of wood pulp produced by direct friction of a rotating stone on the end of a log. In the paper trade this is called mechanical pulp.
Use: Newsprint and other low-grade papers.
See also pulp, paper.

group. (1) One of the major classes or divisions into which elements are arranged in the Periodic Table (vertical columns). The classification is made according to the properties of the elements, those whose properties are similar occupying one group. Groups I through VII are divided into subgroups (A and B), but Group VIII is not so divided. The noble gas group has no number, though it was formerly referred to as Group 0.
See also Periodic Table.
Note: Though often loosely so used, the word "group" should not be applied to a number of elements of similar properties that are not actual groups in the Periodic Table; the proper term for these is "series," i.e., lanthanide series, rare-earth series, etc.
  (2) A combination of two or more closely associated elements which tend to remain together in reactions, usually behaving chemically as if they were individual entities, i.e., in respect to valence, ionization and related properties. Among the more familiar are OH (hydroxyl), COOH (carboxyl), CO_3 (carbonate), NH_4 (ammonium), SO_4 (sulfate), CH_3 and homologs (methyl, etc.), SH (sulfhydryl), and CO (carbonyl). When a group acquires an electric charge it is called a radical.
  (3) Any combination of elements that has a specific functional property, for example, a chromophore group in dyes.

growth. (1) In biochemistry, the continuous process of cell division and reproduction characteristic of all living organisms. The basic phenomenon is considered to be osmosis by which nutrients are transferred through cell walls and tissue structures; it is thus essential to the metabolic functioning of the organism.
  (2) In crystallography, the process of crystalline formation and development by nucleation

and accretion. Crystals of many kinds are artificially produced for a variety of uses, (e.g., lasers) by vapor condensation, electrodeposition, or rapid cooling of a saturated solution.

growth hormone. See somatotropic hormone.

growth regulator. See plant growth regulator.

Grundmann aldehyde synthesis. Transformation of an acid into an aldehyde of the same chain length by conversion of the acid chloride via the diazo ketone to the acetoxy ketone, reduction with aluminum isopropoxide and hydrolysis to the glycol, and cleavage with lead tetraacetate.

G salt. The sodium or potassium salt of 2-naphthol-6,8-disulfonic acid (G acid).
Use: Dye intermediate.

GTP. Abbreviation for guanosine triphosphate. See guanosine phosphates.

guaiac. (guaiac gum; guaiac resin).   A resin from certain Mexican and West Indian trees, especially *Guaiacum santum* and *G. officinale*.
Properties: Brownish lumps; mp 85C; insoluble in water; soluble in alcohol, ether, acetone, chloroform, and caustic soda.
Use: Flavoring agent in foods; medicine (diagnostic aid for blood or hemoglobin).

guaiacol. (methylcatechol; pyrocatechol methyl ether; pyrocatechol methyl ester; o-methoxyphenol; o-hydroxyanisole).   CAS: 90-05-1. $OHC_6H_4OCH_3$.

Properties: Faintly yellowish, limpid, oily liquid or yellow crystals; aromatic odor; constitutes 60–90% of beechwood creosote; soluble in alcohol, ether, chloroform, and glacial acetic acid; moderately soluble in water; d 1.1395; mp 27.9C; bp 205C; flash p 180F (82.2C) (OC). Combustible.
Derivation: (a) By extracting beechwood creosote with alcoholic potash, washing with ether, crystallizing the potash compound from alcohol and decomposing it with dilute sulfuric acid. (b) Also from o-anisidine by diazotization and subsequent action of dilute sulfuric acid.
Hazard: Toxic by ingestion and skin absorption.
Use: Synthetic flavors, medicine (expectorant).

guaiacwood oil.
Properties: Yellow to amber semisolid mass with floral odor; soluble in alcohol, ether and chloroform; insoluble in water; d 0.965–0.975; optical rotation −6 to −7 degrees.
Derivation: Steam distillation of guaiacwood (Paraguay). Chief constituent: Guaiol.
Use: Perfume fixative and modifier, soap odorant, fragrances, production of guaiacwood acetate.

"Guai-A-Phene."[296]   TM for a phenolic type antiskinning agent used for the prevention of gelling, skinning, and oxidation in paint, varnishes, printing inks, linoleum, etc.

guaiol.   CAS: 489-86-1. $C_{15}H_{26}O$. A bicyclic sesquiterpene alcohol found in guaiac wood oil.
Properties: Crystalline solid, mp 91C, insoluble in water, soluble in alcohol.

guanethidine sulfate. [2-(hexahydro-1(2H)-azocinyl)ethyl]guanidine sulfate).   CAS: 60-02-6. $(C_{10}H_{22}N_4)_2 \cdot H_2SO_4$.
Properties: White crystalline powder, strong odor, very soluble in water, slightly soluble in alcohol, practically insoluble in chloroform.
Grade: USP.
Use: Medicine (antihypertensive agent).

guanidine.   (carbamidine; iminourea).
CAS: 113-00-8. $HN{=}C(NH_2)_2$.
Properties: Colorless crystals, mp 50C, decomposes at 160C, soluble in water and alcohol.
Derivation: (a) By heating calcium cyanamide with ammonium iodide. (b) By treating urea with ammonia under pressure. Combustible.
Use: Organic synthesis.

guanidine-aminovaleric acid.   See arginine.

guanidine carbonate.
$(H_2NCNHNH_2)_2 \cdot H_2CO_3$.
Properties: White granules, soluble in water, slightly soluble in alcohol and acetone. Decomposes without melting at 197–199C, d 1.25. Combustible.
Derivation: From dicyandiamide.
Grade: Technical, over 95% pure.
Hazard: Toxic by ingestion.
Use: As a strong organic alkali, organic intermediate, soap and cosmetic products.

guanidine hydrochloride. $HNC(NH_2)_2 \cdot HCl$.
Properties: White powder, mp approximately 183C, soluble in water and alcohol, pH of aqueous solution 6.2 for 10% solution.
Grade: 88% and 95% pure.
Hazard: Toxic by ingestion, evolves hydrogen chloride fumes on heating.

guanidine nitrate. $H_2NC(NH)NH_2 \cdot HNO_3$.
Properties: White granules, soluble in water and alcohol, slightly soluble in acetone, melting range 206–212C.

Derivation: From cyanamide or dicyandiamide.

Grade: Technical, greater than 95% pure.

Hazard: Strong oxidant, may ignite organic materials on contact, may explode by shock or heat.

Use: Manufacture of explosives, disinfectants, photographic chemicals.

guanidine thiocyanate.   CAS: 593-84-0.
$H_2NC(=NH)NH_2 \cdot HSCN$

Properties: Solid with mw 118.16, mp 114-119C. Contains 2% hydrochloric acid.

Hazard: Irritant.

Use: Potent protein denaturant used in isolation of intact DNA, RNA.

guanine.   (2-amino-6-oxypurine).
CAS: 73-40-5. $C_5H_5N_5O$.   A purine constituent of ribonucleic acid and deoxyribonucleic acid. Usual sources are guano, sugar beets, yeast, clover seed, and fish scales.

Properties: Colorless, rhombic crystals; mp 360C (decomposes); insoluble in water; sparingly soluble in alcohol and ether; freely soluble in ammonium hydroxide, alkali hydroxides, and dilute acids. Available as hydrochloride or hemisulfate.

Derivation: Isolation following hydrolysis of nucleic acids (usually from yeast), organic synthesis.

Use: Biochemical research, cosmetics.

See also pearl essence.

guano.   Excrement of sea birds found chiefly on islands off coast of Peru, Chile. Contains 10% nitrogen, 6% phosphorus, 2% potassium. Formerly the main source of nitrates.

guanosine.   (guanine riboside).
CAS: 118-00-3. $C_{10}H_{13}N_5O_5$.   The nucleoside containing guanine and D-ribose.

Properties: White, crystalline, odorless powder with mild saline taste; mp 237–240C (decomposes); very slightly soluble in cold water; soluble in boiling water, dilute mineral acids, hot acetic acid, and dilute bases; insoluble in alcohol, ether, chloroform and benzene.

Derivation: Found in pancreas, clover, coffee plant, and pollen of pines; prepared from yeast nucleic acid.

Use: Biochemical research.

guanosine monophosphate.   See guanylic acid.

guanosine phosphates.   Nucleotides used by the body in growth processes; important in biochemical and physiological research. Those isolated are the monophosphate (GMP), the diphosphate (GDP), and the triphosphate (GTP).

guanosine phosphoric acid.   See guanylic acid.

"Guantal."[58]   TM for diphenylguanidine phthalate, a rubber accelerator.

guanylic acid.   (GMP; guanosine monophosphate; guanosine phosphoric acid).
CAS: 3′ 117-68-0; 5′ 85-32-5. $C_{10}H_{14}N_5O_8P$.
The monophosphoric ester of guanine, i.e., the nucleotide containing guanine, D-ribose, and phosphoric acid. The phosphate may be esterified to either the 2,3, or 5 carbon of ribose, yielding guanosine-2′-phosphate, guanosine-3′-phosphate, and guanosine-5′-phosphate, respectively. It is important in growth processes of the body.

Derivation of commercial product: Isolation from nucleic acid of yeast or pancreas; also made synthetically.

Use: Biochemical research, flavor potentiator (disodium salt).

guanyl nitrosaminoguanylidene hydrazine.
A high explosive.

guanyl nitrosaminoguanyl tetrazene.
See tetrazene.

guanylurea sulfate.   (carbamylguanidine sulfate).
$(C_2H_6ON_4)_2 \cdot H_2SO_4 \cdot 2H_2O$.

Properties: White powder, greater than 97% pure, soluble in water and alcohol.

Derivation: From cyanamide or dicyandiamide.

Use: Analytical reagent for nickel, manufacture of dyes, organic synthesis.

"Guardkote."[125]   TM for fast-setting two-component liquid systems based on "Epon" resins and designed specifically for highway resurfacing and repair. They are used in combination with sharp aggregates to waterproof and deslick portland cement or bituminous concrete pavements.

guar gum.   (guar flour).   CAS: 9000-30-0.
A water-soluble plant mucilage obtained from the ground endosperms of *Cyanopsis tetragonoloba*, cultivated in India and Pakistan as livestock feed as well as in southwestern US. The water-soluble portion of the flour (85%) is called guaran and consists of 35% galactose, 63% mannose, probably combined in a polysaccharide, and 5–7% protein.

Properties: Yellowish-white free-flowing powder, completly soluble in hot or cold water. Practically insoluble in oils, greases, hydrocarbons, ketones, esters. Water solutions are tasteless, odorless, non-toxic. Has 5–8 times the thickening power of starch. Reduces the friction drag of water on metals.

Grade: Industrial, technical, FCC.

Use: Paper coating, cosmetics, pharmaceuticals, binder in tablet mixtures, interior coating of fire-hose nozzles, fracturing aid in oil wells, textiles, printing, polishing, thickener and emulsifier in food products, e.g., cheese spreads, ice cream, and frozen desserts.

Guareschi-Thorpe condensation. Synthesis of pyridine derivatives by condensation of cyanoacetic ester with acetoacetic ester in the presence of ammonia. In a second type of synthesis, a mixture of cyanoacetic ester and a ketone is treated with alcoholic ammonia.

guayule. A rubber-like hydrocarbon, almost identical with cis-polyisoprene, obtained from a shrub which is a member of the Compositae family. It is grown extensively in Mexico and the southwestern US. It contains approximately 15% of rubber hydrocarbon in the form of latex in cells in the roots and stem. Conventional practice involves crushing and parboiling the entire plant; this coagulates the latex, which is removed by milling. An improved process utilizes steam from 200–240C for 6 minutes instead of milling. Resins which comprise 10–15% of the plant are then removed with acetone followed by treatment with hexane to extract the hydrocarbon. This method can increase the yield of hydrocarbon by up to 50%. The yield is 1 lb of rubber/6 lb of comminuted shrub.

Guerbet reaction. Condensation of alcohols at high temperature and pressure in the presence of sodium alkoxide or copper by a dehydrogenation, aldol condensation, and hydrogenation sequence.

"Guidon."[446] TM for a magnesia-chrome refractory made from electrically fused grains. Has high temperature strength approximately 10 times that of conventional magnesia-chrome refractories, which makes it well suited to abrasive environments.
Use: Electric furnace sidewalls, copper converter tuyere lines, open hearth front walls and roof.

Guignet's green. A chrome green pigment made by fusing potassium chromate and boric acid.

gum arabic. See arabic, gum.

gum benzoin. See benzoin resin.

gum camphor. See camphor.

gum, gasoline. A viscous oxidation product occurring after long standing in gasoline that is not stabilized with an antioxidant.

gum, natural. A carbohydrate high polymer that is insoluble in alcohol and other organic solvents, but generally soluble or dispersible in water. Natural gums are hydrophilic polysaccharides composed of monosaccharide units joined by glycosidic bonds. They occur as exudations from various trees and shrubs in tropical areas or as phycocolloids (algae) and differ from natural resins in both chemical composition and solubility properties. Some contain acidic components and others are neutral.
Use: Protective colloids and emulsifying agents in food products and pharmaceuticals, as sizing for textiles and in electrolytic deposition of metals.
See arabic, tragacanth, guar, karaya.
Note: The terminology of natural gums and resins is inconsistent and often confusing. The word "gum," often used as an adjective, seems to acquire a different meaning from the noun. For example, the resinous products obtained from pine pitch (produced by the parenchyma cells of softwoods) are conventionally called "gum turpentine," and "gum rosin." There are also such "gum resins" as gum benzoin, gum camphor, and others. The so-called ester gum is a semisynthetic reaction product of rosin and a polyhydric alcohol. All these are actually resinous products having properties quite different from those of natural gums. Furthermore, resins are complex *mixtures*, whereas gums are *compounds* that can be represented by a formula. Still further complicating the matter is the common application of the word "gum," to such plant latices as chicle and natural rubber, which are different from both carbohydrate gums and resins. It is probable that the confusion originated in the casual use of "gum" to refer to any soft sticky product derived from trees. In view of this situation, any specific definition of these terms is likely to be controversial.
See also resin, natural; polymer, water-soluble.

gum rosin. See rosin; gum, natural (note).

gum sugar. See arabinose.

gum, synthetic. See ester gum.

gum tragacanth. See tragacanth gum.

gum turpentine. See turpentine; gum, natural (note).

guncotton. See nitrocellulose.

gun metal. An alloy of copper with 10% tin.

gunpowder. See black powder.

"Guthion."[181] TM for O,O-dimethyl-S-4-oxo-1,2,3-benzotriazin-3(4H)-ylmethyl phosphorodithioate.
See azinphos methyl.
Hazard: A cholinesterase inhibitor.

Gutknecht pyrazine synthesis. Cyclization of α-amino ketones produced by reduction of isonitroso ketones; the dihydropyrazines formed are dehydrogenated with Hg_2O or $CuSO_4$ or sometimes with atmospheric oxygen.

gutta percha. (trans-polyisoprene).   A geometric isomer of natural rubber obtained from trees native to Malaya. It is stiff, hard, and inelastic when cold; softens at 60C to moldable condition and melts at 100C, insoluble in water, essentially soluble in carbon disulfide, petroleum ether, and chloroform. It can be vulcanized with sulfur.
Use: Formerly used widely for golf ball covers, its present uses are in dentistry, surgical accessories, and as an insulating medium in electrical devices.

glyocardia oil.   See chaulmoogra oil.

gyplure. (Generic name for cis-9-octadecen-1,12-diol-12-acetate).   It is a synthetic product used as a sex attractant for the male gypsy moth. The natural product, found in female moths, is said to be d-10-acetoxy-1-hydroxy-cis-7-hexadecene.

gyratory crusher. A machine for large-scale size reduction of rocks, ore, or mineral aggragate. Essentially, it comprises a vertical cone-shaped crushing head or mantle which moves in a shell shaped like an inverted cone which is open at the top where the feed enters. The space between the mantle and the shell decreases gradually, being smallest at the bottom where the crushed material is discharged. The lower third of the mantle and shell have concave surfaces to minimize packing of the crushed product. The crushing head functions by executing a gyratory or circular motion, i.e., a rolling action in which the upper portion describes a circle while the lower portion is almost stationary. From 300–500 gyrations per minute are possible. The mechanism is activated by a shaft attached to the crushing head, the bottom of which rests in a gear-driven eccentric. Heavy-duty gyratories have feed openings up to 72 inches; fine-reduction sizes are as small as 12 inches.

gypsum. CAS: 1010-14-4. $CaSO_4 \cdot 2H_2O$.
A mineral that consists of calcium sulfate with water molecules attached, or the rock that consists primarily of this mineral.
See calcium sulfate.

gypsum cement. (plaster of Paris; Keene's cement; Parian cement; Martin's cement; Mack's cement).   A group of cements which consist essentially of calcium sulfate and are produced by the partial dehydration of gypsum to the hemihydrate, $CaSO_4 \cdot 1/2HOH$. They usually contain additions of various sorts. For example, Keene's cement contains alum or aluminum sulfate; Mack's cement contains sodium or potassium sulfate; Martin's cement contains potassium carbonate; and Parian cement contains borax.

H

H. Symbol for hydrogen, the molecular formula is H_2.

Haber, Fritz. (1868–1934) Born in Breslau, Germany, Haber's great contribution to chemistry, for which he was awarded the Nobel prize in 1918, was his development (with Bosch) of a workable method for synthesizing ammonia by the water gas reaction from hot coke, air, and steam; the gas mixture obtained includes nitrogen from the air as well as hydrogen from the steam. It was the first successful attempt to "fix" atmospheric nitrogen in an industrial process. This discovery was developed to production scale in approximately 1912; it enabled Germany to manufacture an independent supply of explosives for World War I.

habit. The type of geometric structure that a given crystalline material invariably forms, e.g., cubic, orthorhombic, monoclinic, tetragonal, hexagonal, etc. Each of these types has several subclasses. Thus, crystals may have the form of thin sheets or plates, cubes, rhomboids, and even more complicated geometric structures. For example, the crystalline habit of mica is monoclinic with formation of extremely thin sheets. See also crystal.

H-acid. (1-amino-8-napthol-3,6-disulfonic acid). Properties: Gray powder, soluble in water, alcohol and ether.

HO_3S [ring structure, positions 6, 8, 3, 1] SO_3H / HO NH_2

Derivation: Fusion of 1-naphthylamine-3,6,8-trisulfonic acid (Koch acid) with 30% sodium hydroxide solution at 180C. Use: Azo dye intermediate.

hafnia. See hafnium oxide.

hafnium. CAS: 7440-58-6. Hf. Metallic element of atomic number 72, Group IVB of the Periodic Table, atomic weight 178.49, valences = 2,3,4; 6 stable isotopes. Properties: Generally similar to zirconium. Gray crystals, d 13.1, mp approximately 2150C, bp above 5400C, high thermal neutron cross-section (115 barns). Good corrosive resistance and high strength.

See also zirconium.
Occurrence: Zirconium ores.
Derivation: Extremely difficult to separate from zirconium. Most important methods are: (1) solvent extraction of the thiocyanates by hexone, (2) solvent extraction of the nitrates by tributyl phosphate, (3) fractional crystallization of the double fluorides.
Forms available: Powder, rods, single crystals.
Hazard: Toxic by inhalation; compounds are also toxic. Powder form explosive in air either dry or wet with less than 25% water. TLV: 0.5 mg/m^3 of air.
Use: Control rods in water-cooled nuclear reactors; light-bulb filaments, electrodes, special glasses, getter in vacuum tubes.

hafnium boride. HfB_2. Crystalline solid, mp 3100C.
Use: Refractory material.

hafnium carbide. HfC.
Properties: High thermal neutron absorption cross-section, very high mp 3890C (7030F), most refractory binary substance known.
Use: Control rods in nuclear reactors.

hafnium disulfide. HfS_2. Available in a particle size of 40 microns as a solid lubricant. The diselenide and ditelluride are also available.

hafnium nitride. HfN. Yellow-brown crystals, mp 3305C, the most refractory of known metal nitrides.

hafnium oxide. (hafnia). HfO_2.
Properties: White solid, d 9.68 (20C), mp 2812C, bp approximately 5400C, insoluble in water.
Use: Refractory metal oxide.

Hahn, Otto. (1879-1968) A German physical chemist who won the Nobel prize for chemistry in 1944 for atom splitting and the principle of the chain reaction. Well known for work on nuclear fission. He discovered protactinium and transuranium elements with atomic numbers 94, 95, and 96. After receiving his doctorate at the University of Munich, he worked in Canada before returning to Europe.

hair. See keratin.

583

halazone. (p-N,N-dichloro sulfamylbenzoic acid; p-sulfondichloraminobenzoic acid).
CAS: 80-13-7. $HOOCC_6H_4SO_2NCl_2$.
Properties: White crystalline powder, strong chlorine odor, affected by light. Soluble in glacial acetic acid, benzene; slightly soluble in water, chloroform; insoluble in petroleum ether; mp 195C with decomposition.
Grade: NF.
Use: Water disinfectant.

half-life. The time required for an unstable element or nuclide to lose one-half of its radioactive intensity in the form of alpha, beta, and gamma radiation. It is a constant for each radioactive element or nuclide. Half-lives vary from fractions of a second for some artificially produced radioactive elements to millions of years. The half-life of U-235 is 7.1×10^8 years; that of Pu-239 is 24,360 years.

halibut liver oil. (haliver oil).
Properties: Pale yellow to dark red liquid, fishy odor and taste. Soluble in alcohol, ether, chloroform, and carbon disulfide; insoluble in water. D 0.920–0.930, saponification number 160–180, iodine number 120–136, refr index about 1.47.
Derivation: By expressing and boiling halibut livers.
Grade: Crude, refined.
Use: Source of vitamins A and D, leather dressing.

halides. Binary compounds of the halogens.

halite. The mineral that consists of sodium chloride (NaCl), popularly known as rock salt, or the rock that consists primarily of this mineral.

Hall, Charles Martin. (1863–1914) A native of Ohio, Hall invented a method of reducing aluminum oxide in molten cryolite by electrochemical means. This discovery made possible the large-scale production of metallic aluminum and resulted in formation of the Aluminum Co. of America. The process requires high electric power input. Hall is generally considered as the founder of the aluminum industry.
See also Hall process.

"Hallco."[94] TM for specialty esters based on acetic, lauric, and sebacic acids.

"Hallcomid."[94] TM for a series of N,N'-dimethyl amides and esters of fatty acids.

"Hallcote."[94] TM for clay containing anti-blocking coatings (slab dips) used in rubber processing.

Haller-Bauer reaction. Cleavage of non-enolizable ketones with sodium amide, most often applied to ketones $ArCOCR_3$ to yield trisubstituted acetic acids.

halloysite. $Al_2O_3 \cdot 3SiO_2 \cdot 2HOH$. A clay used in refactories and as a catalyst support.

Hall process. The electrolytic recovery of aluminum from bauxite or more specifically, from the alumina extracted from it (See Bayer process). A typical cell for this process consists of a rectangular steel shell, lined with insulating brick and block carbon. The cell holds a molten cryolite-alumina electrolyte, commonly called the "bath." The carbon bottom is covered by a pad of molten aluminum and serves as the cathode. The anodes are prebacked carbon blocks, suspended in the electrolyte. The cathodic current is collected from the carbon bottom by imbedded steel bars that protude through the shell to connect with the cathode bus. During electrolysis, aluminum is deposited on the metal pad and the oxygen, liberated at the anode, reacts with the carbon to form carbon dioxide, some of which is reduced to carbon monoxide by secondary reactions. At 24–48 hour intervals, aluminum is tapped from the cell by a siphon. The process requires large amounts of electric power (from 4–5% of total US production). Disposition of the toxic fluoride waste is a problem.
See also Toth process.

hallucinogen. Any of a number of drugs acting on the central nervous system in such a way as to cause mental disturbance, imaginary experiences, coma, and even death. Many of these are narcotics and/or alkaloids, some are derived from plants and others made synthetically. They differ in degree of addiction and hallucinatory effect. Their sale and possession (other than by physicians) is illegal in the US. Most common hallucinogens are cannabis (marijuana, hashish), lysergic acid (LSD), amphetamine, and numerous morphine derivatives.

halocarbon. A compound containing carbon, one or more halogens, and sometimes hydrogen. The lower members of the various homologous series are used as refrigerants, propellant gases, fire-extinguishing agents, and blowing agents for urethane foams. When polymerized, they yield plastics characterized by extreme chemical resistance, high electrical resistivity, and good heat resistance.
See also fluorocarbon.

halogenation. Incorporation of one of the halogen elements, usually chlorine or bromine, into a chemical compound. Thus benzene (C_6H_6) is treated with chlorine to form chlorobenzene

(C_6H_5Cl) and ethylene (C_2H_4) is treated with bromine to form ethylene dibromide ($C_2H_4Br_2$). Compounds of chlorine and bromine are sometimes used as the source of the halogen, e.g., phosphorus pentachloride.

halogen. One of the electronegative elements of Group VII A of the Periodic Table (fluorine, chlorine, bromine, iodine, astatine, listed in order of their activity, fluorine being the most active of all chemical elements).

"Halon."[175] TM for tetrafluoroethylene polymer. Properties: Inertness to almost all chemicals, resistance to high and low temperatures, zero moisture absorption, high impact strength, excellent dielectric properties, nonstick surface with low coefficient of friction, self-extinguishing (ASTM D-635).
Grade: Varies with particle size as 600, 300, 20 micron average size.
See also polytetrafluoroethylene, fluorocarbon polymer.

halothane. (2-bromo-2-chloro-1,1,1-trifluoroethane). CAS: 151-67-7. $CF_3CHBrCl$.
Properties: Colorless, volatile liquid. Sweetish odor, d 1.872–1.877 (20/4C), bp 50.2C, 20C (243 mm), light sensitive, may be stabilized with 0.01% thymol, slightly soluble in water, miscible with many organic solvents.
Grade: USP.
Use: Medicine (anesthetic).

"Halowax."[11] TM for chlorinated hydrocarbons of varying chlorine content, including chlorinated naphthalenes, ranging from a low viscosity oil to hard microcrystalline wax.
Hazard: Toxic by ingestion.
Use: Electrical insulating and fire-resisting materials, impregnants, sealing compounds, crankcase additive, ingredient in penetrating oils, plasticizer, protective coatings.

Hamilton, Alice. (1869-1970) The first American physician to devote her life to the practice of industrial medicine. In studying the lead industries in Illinois, she discovered and ameliorated lead poisoning among bathtub enamelers in Chicago. She wrote about phossy jaw, which occurred among American matchmakers using white or yellow phosphorus. She studied the effects of carbon monoxide among steelworkers, the toxicity of nitroglycerin among munitions makers during World War I, the symptoms of hatters exposed to mercury in Danbury, Connecticut, and the 'dead fingers' syndrome of workers utilizing the early jackhammers. She

also described the toxic effects to the blood forming cells from benzol, and the neurologic and psychological responses of workers in the viscose rayon industry. In 1919, Dr. Hamilton was appointed assistant professor of industrial medicine at Harvard Medical School. The first woman on the Harvard faculty, she gave occupational medicine respectability as an academic pursuit.

hammer mill. A crushing or shredding device consisting of four or more rectangular metal hammers or sledges mounted on a rotating shaft, the hammers being free to swing on a pin. As the shaft rotates the hammers impact the material introduced from above, crushing it against a stationary breaker plate. The fragments are carried downward and are sorted by a grid located beneath the shaft; those that are too large to pass through it are returned for another cycle. A complete mill is made up of several such units working on a single shaft, with suitable housing. Products such as coal, limestone, and other mineral aggregate, as well as wood, sugarcane, and similar materials can be effectively disintegrated with equipment of this type.

Hammick reaction. Decarboxylation of α-picolinic or related acids in the presence of carbonyl compounds accompanied by the formation of a new carbon-carbon bond.

"Hamp-ene."[517] TM for a series of chelating agents including ethylenediaminetetraacetic acid; the disodium salt, dihydrate; the trisodium salt, monohydrate; the tetrasodium salt, monohydrate; the tetrasodium salt; and the tetrasodium salt, dihydrate.

hand. (handle). A term used chiefly in the textile industry to describe a fabric in a qualitative manner, as determined by the sensory perception of its feeling. It is also used to some extent in the leather industry. Such pragmatic physical-perception tests are of great practical value, although the properties being ascertained are often not objectively definable.
See also texture.

handedness. See asymmetry; enantiomer; chiral.

"Hansa."[203] TM for a group of yellow to orange, insoluble azo pigments based on toluidine and β-naphthol. Have good lightfastness in deep shades but tend to fade in pastels. Notably poor resistance to bleeding, good weather resistance, and are relatively unaffected by acids and alkalies. Used chiefly in emulsion paints, toy enamels, and other applications.

Hantzch pyridine synthesis. Synthesis of alkylpyridines by condensation of two moles of a β-dicarbonyl compound with one mole of an aldehyde in the presence of ammonia. The resulting dihydropyridine is dehydrogenated with an oxidizing agent.

Hantach pyrrole synthesis. Formation of pyrrole derivatives from α-chloromethyl ketone, β-keto esters, and ammonia or amines.

"Harbide."[446] TM for a silicon carbide brick, formed by impact pressing, low permeability, dense impervious surfaces, high resistance to oxidation.
Use: Refractory for ceramic kilns, furnace linings, recuperator tubes, radiant tubes, retorts, and in applications subject to mechanical abrasion.

"Harchemex."[189] TM for a compound of mainly C_{14} and C_{16} straight-chain primary alcohols in the approximate ratio of 2 to 1.
Use: Wetting agents; germicidal quaternary ammonium compounds; lubricating oil additive.

hard. A nontechnical word used by chemists with a variety of meanings. It describes the following:
(1) Water in which calcium carbonate or calcium sulfate is present.
See water, hard.
(2) An acid having high positive oxidation and whose valence electrons are not readily excited.
See acid.
(3) Extremely short-wave radiation, e.g., hard x-rays. (4) The resistance of a pesticide or detergent to biodegradation. (5) Rubber cured with 30% or more of sulfur.
See rubber, hard.
(6) Wood from deciduous trees.
See hardwood, hardness.

Harden, Sir Arthur. (1865-1940) An English chemist who won Nobel prize in chemistry in 1929 along with Hans von-Euler-Chelpin. He discovered fermentation enzymes and demonstrated the structure of zymase. His fermentation work proved how inorganic phosphates speeded process. Born in England, he received his doctorate in Germany.

hardness. The resistance of a material to deformation of an indenter of specific size and shape under a known load. This definition applies to all types of hardness scales except the Mohs scale, which is based on the concept of scratch hardness and is used chiefly for minerals. The most generally used hardness scales are Brinell (for cast iron), Rockwell (for sheet metal and heat-treated steel), diamond pyramid, Knoop, and sclero-scope (for metals). Durometer hardness is used for materials such as rubber and plastics.

hardwood. In papermaking terminology, the wood from deciduous trees (maple, oak, birch), regardless of whether it is actually hard or soft.
Use: Components of paper pulp, but in much less volume than softwoods.

"Harflex."[189] TM for a series of polymeric plasticizers for vinyl resins, synthetic rubbers and cellulose esters. Characterized by permanence and resistance to extraction.

Hargreaves process. The manufacture of sodium sulfate (salt cake) from sodium chloride and sulfur dioxide. A mixture of sulfur dioxide and air is passed over briquettes of sodium chloride in a countercurrent manner to produce sodium sulfate and hydrogen chloride. This process accounts for only a small amount of the salt cake produced in the US.

Harries ozonide reaction. Treatment of olefins with ozone as a method of cleaving olefinic linkages. On hydrolysis or catalytic hydrogenation, the initially formed ozonide yields two molecules of carbonyl compounds.

hartshorn. See ammonium carbonate.

hartshorn oil. See bone oil.

HAS. Abbreviation for hydroxylamine acid sulfate.

hashish. Extract of cannabis (marijuana) more concentrated and thus more powerful than the base drug.
See cannabis.

Hass chlorination rules. A series of rules pertaining to the chlorination of saturated hydrocarbons.

Hassel, Odd. (1897-) A Norwegian chemist who won Nobel prize for chemistry in 1969 with Barton. A great deal of his work was concerned with using x-ray and electron differentiation methods of crystal and molecular structures. He also researched stereochemistry and conformational analysis. His education and teaching career were in his homeland.

Hass vapor phase nitrogen. Production of nitroparaffins by vapor phase reaction of aliphatic hydrocarbons with gaseous nitric acid at 420C; a series of thirteen rules governing these reactions.

"Hastelloy."[275] TM for a series of high-strength, nickel-base, corrosion-resistant alloys. With the exception of "Hastelloy alloy D," which is cast only, all are produced in the forms of sheet, plate, bars, rods, welding electrodes, and wire and can be fabricated into all types of process equipment.

Hauptman, Herbert A. (1917-) An American biophysicist who won the Nobel prize for chemistry in 1985 along with Karle. Work involved developing equations which allow determination of phase information from x-ray crystallogrphy intensity patterns. The use of computers permitted use of the equations to determine the conformation of thousands of chemicals. Hauptman has been director of research and vice president of the Medical Foundation of Buffalo since 1972. He is also a professor of biophysics in Buffalo at the State University of New York.

"Haveg."[349]   TM for a series of molding compounds fabricated into corrosive resistant chemical process equipment. Composites of various synthetic polymers with silicate fillers, graphite, etc.
Use: Tanks, towers, scrubbers, agitators, piping, etc., in the handling of sulfuric, hydrochloric, and phosphoric acids and chlorinated solvents.

"Havelast."[349]   TM for an elastomeric binder or impregnant for various reinforcing materials such as "Sil-Temp," as fabric or rovings, asbestos, glass or graphite.
Use: Rocket and missile industry when resiliency is desired.

Haworth methylation.   Formation of methylated methyl glycosides from monosaccharideswith dimethyl sulfate and 30% sodium hydroxide. The glycosidic methyl group is hydrolyzed with acid to yield the free methylated sugar.

Haworth phenanthrene synthesis.   Acylation of aromatic compounds with aliphatic dibasic acid anhydrides to β-aroylpropionic acids, reduction of the carbonyl group according to Clemmensen or Wolff-Kishner procedures, cyclization of the γ-arylbutyric acid with 85% sulfuric acid, and conversion of the cyclic ketone to polycyclic hydroaromatic and subsequently to aromatic compounds.

Haworth, Sir Walter N.   (1883-1950) An English chemist who received the Nobel prize in chemistry in 1937 along with Paul Karrer. He recommended the name ascorbic acid and synthesized vitamin C. He accomplished much work on carbohydrate structure and developed a substitute for blood plasma using carbohydrates. During World War II, he developed gaseous diffusion separation on uranium isotopes. He received his PhD in Manchester.

Hayashi rearrangement.   Rearrangement of α-benzoylbenzoic acids in the presence of sulfuric acid or phosphorous pentoxide.

"Haystellite."[275]   TM for tungsten carbide products, principally in the form of hard-surfacing rods for protecting parts from severe abrasion.

hazardous material.   Any material or substance which in normal use can be damaging to the health and well-being of man. Such materials cover a broad range of types which may be classified as follows: (1) Toxic agents (see poison; toxicity) including drugs, chemicals, and natural or synthetic products that in normal use are in any way harmful, ranging from poisons to skin irritants and allergens. When improperly used, all materials can be hazardous to man. (2) Corrosive chemicals such as sodium hydroxide or sulfuric acid that destroy or otherwise damage the skin and mucous membranes on external contact or inhalation. (3) Flammable materials including (a) organic solvents, (b) finely divided metals or powders, (c) some classes of fibers, textiles or plastics, and (d) chemicals that either evolve or absorb oxygen during storage, thus constituting a fire risk in contact with organic materials. (4) Explosives and strong oxidizers such as peroxides and nitrates. (5) Materials in which dangerous heat build-up occurs on storage, either by oxidation or microbiological action. Examples are fish meal, wet cellulosics, and other organic waste materials. (6) Radioactive chemicals that emit ionizing radiation. Packaging, labeling and shipping of hazardous materials by rail, highway, water and air is regulated in the US by the Department of Transportation (DOT), 1920 L St., N.W., Washington, D.C. In general, its regulations for air shipment follow those of the International Air Transport Association (IATA), though there are numerous points of difference.
   An authoritative guide to the safety labeling of hazardous materials and products is issued by the American National Standards Institute, 1430 Broadway, NY, NY 10018.
See also toxicity, flammable material, label.

hazardous waste.   See chemical waste, radioactive waste.

Hb.   Abbreviation for hemoglobin.

"HCA."[50]   TM for a highly potent herbicide whose active ingredient is hexachloroacetone.
Use: Control of all weeds in non-crop areas. It increases the initial and residual effectiveness of oil sprays. Available in mixtures or concentrated.
Hazard: Irritant.

HCCH.   Abbreviation for hexachlorocyclohexane.

HCG. Abbreviation for human chorionic gonadotropin.

"HCR."[233] TM applied to ion exchange resin used in water treating and chemical process applications; strong acid cation exchange resin, 8% divinylbenzene cross-linked.

HDPE. Abbreviation for high-density polyethylene.

He. Symbol for helium.

HEA. Abbreviation for 2-hydroxyethyl acrylate.

head box. A large container holding the prepared slurry of paper pulp and additives which is located at the head of the fourdrinier machine; the slurry is fed continuously from slots in the bottom of the box to the wire on which the sheet is formed.

head space. A term used in the canning industry to refer to the space intentionally left between the top of the filled liquid and the cover of the can to allow for expansion of the contents during heat-processing.

heat. A form of energy associated with and proportional to molecular motion. It can be transferred from one body to another by radiation, conduction or convection; sensible heat is accompanied by a change in temperature, but latent heat is not.—*of combustion*: The heat evolved when a definite quantity of a substance is completely oxidized (burned).—*of crystallization*: The heat evolved or absorbed when a crystal forms from a saturated solution of a substance; —*of dilution*: The heat evolved per mole of solute when a solution is diluted from one specific concentration to another. —*of formation*: The heat evolved or absorbed when a compound is formed in its standard state from elements in their standard states at a specified temperature and pressure. —*of fusion*: The heat required to convert a substance from the solid to the liquid state with no temperature change (also called latent heat of fusion or melting). —*of hydration*: The heat associated with the hydration or solvation of ions in solution; also the heat evolved or absorbed when a hydrate of a compound is formed. —*of reaction*: The heat evolved or absorbed when a chemical reaction occurs in which the final state of the system is brought to the same temperature and pressure as that of the initial state of the reacting system. —*of solution*: The heat evolved or absorbed when a susbtance is dissolved in a solvent. —*of sublimation*: The heat required to convert a unit mass of a substance from the solid to the vapor state (sublimation) at a specified temperature and pressure without the appearance of the liquid state. —*of transition*: The heat evolved or absorbed when a unit mass of a given substance is convereted from one crystalline form to another. —*of vaporization*: The heat required to convert a substance from the liquid to the gaseous state with no temperature change (also called latent heat of vaporization).

heat exchanger. A vessel in which an outgoing hot liquid or vapor transfers a large part of its heat to an incoming cool liquid; in the case of vapors, the latent heat of condensation is thus utilized to heat the entering liquid. The shell-and-tube type is widely used, in which the hot liquid or vapor is contained in the shell while the cool liquid passes through the tubes, which are usually arranged in coils for maximum contact with the heat source. Heat exchangers are used in many chemical operations, e.g., evaporation and pulp manufacture, as well as to produce steam from the heat developed in nuclear reactors for power generation.
See also evaporation; heat transfer.

heat transfer. Transmission of thermal energy from one location to another by means of a temperature gradient existing between the two locations. It may take place by conduction, convection, or radiation. Heat transfer is involved in many types of industrial operations, including distillation, evaporation, canning of foods, baking, curing, etc. In some cases, a heat exchanger is utilized. Fluids of high heat capacity are widely used to remove unwanted heat or to transfer it from one place to another within a system; examples are air, water, ethylene glycol.
See also coolant.

heavy. A nontechnical word used in a number of scientific senses: (1) referring to atomic weight (heavy water, heavy metal), (2) referring to production volume (heavy chemical), (3) referring to physical weight (heavy spar), (4) referring to thickness (heavy-gauge wire), (5) referring to distillation range (heavy oil).
See following entries.

heavy chemical. A chemical produced in tonnage quantities, often in a relatively impure state. Examples are sodium chloride, sulfuric acid, soda ash, salt cake, sodium hydroxide, etc.
See also fine chemical.

heavy hydrogen. See deuterium and tritium.

heavy metal. A metal of atomic weight greater than sodium (22.9) that forms soaps on reaction with fatty acids, e.g., aluminum, lead, cobalt.
See soap (2).

heavy oil. An oil distilled from coal-tar between 230C and 330C, the exact range not at all definite.
See coal tar.

heavy oxygen. See oxygen 18.

heavy spar. See barite.

heavy water. (deuterium oxide; DOD; tritium oxide; TOT). Water composed of two atoms of deuterium and one atom of oxygen; or one or two atoms of tritium and one of oxygen. In lower states of purity, the proportion of heavy water molecules decreases. Deuterium oxide is present to the extent of approximately one part in 6,500 of ordinary water. Deuterium oxide freezes at 3.8C, boils at 101.4C and has density of 1.1056 at 25C.
Derivation: There are several methods of separating or concentrating DOD: (a) Fractional distillation, (b) Girdler-Spevack process, (c) hydrogen sulfide exchange process, (d) electrolysis, (e) cryogenic methane distillation.
Available forms: DOD 99.75% pure, in up to 5000g lots.
Use: Moderator in some types of nuclear reactors.
See also tritium (hydrogen of atomic weight 3), which combines with oxygen to give another variety of heavy water, TOT, i.e., tritium oxide.

hecto-. (h). Prefix meaning 10^2 units, e.g., 1 hg = 1 hectogram = 100 grams.

hectorite. (hector clay). CAS: 12173-47-6. $Na_{0.67}(Mg,Li)_6Si_8O_{20}(OH,F)_4$. One of the montmorillonite group of minerals that are principal constituents of bentonite.
Hazard: Respiratory irritant.
Use: Chill-proofing of beer.

hedeoma oil. See pennyroyal oil.

hedonal. (methylpropylcarbinolurethane).
CAS: 120-36-5.
$CH_3CH_2CH(CH_3)OCONH_2$.
Properties: White, crystalline powder; faint aromatic odor and taste; soluble in alcohol, ether, organic solvents; sparingly soluble in cold water; more soluble in hot water; fusing point 76C; bp 215C.
Use: Medicine (sedative).

HEDTA. See hydroxyethylethylenediaminetriacetic acid.

HEED. Abbreviation for high energy electron diffraction.

Hehner number. The percentage of weight of water-insoluble fatty acids in oils and fats.

Heisenberg, Werner P. (1901–1976) A native of Germany, Heisenberg received his doctorate from the University of Munich in 1923, after which he was closely associated for several years with Niels Bohr in Copenhagen. He was awarded the Nobel prize in physics in 1932 for his brilliant work in quantum mechanics. In 1946, he became Director of the Max Planck Institute. His notable contributions to theoretical physics, best known of which was the Uncertainty Principle, imparted new impetus to nuclear physics and made possible a better understanding of atomic structure and chemical bonding.
See also uncertainty principle.

helenine. A nucleoprotein derived from the mold Penicillium funiculosum.
Use: An antiviral drug.

Helferich method. Glycosidation of an acetylated sugar by heating with a phenol in the presence of zinc chloride or p-toluenesulfonic acid as catalyst.

Helianthine B. See methyl orange.

"Helindon."[203] TM for vat dyestuffs used for dyeing wool.

"Heliogen."[203] TM for phthalocyanine dyestuffs.
Use: Paint, lacquers, printing inks, wallpaper, coated paper, rubber, and plastics.

heliotropyl acetate. See piperonyl acetate.

"He-Li-Ox" 500 Series.[329] TM for a series of PVC heat stabilizers for extruding, injection molding, calendering, and blow molding. All ingredients conform to FDA requirements for food packaging.

"Heliozone."[28] TM for a blend of waxy material.
Use: Retard sun checking and cracking of rubbers.

helium. CAS: 7440-59-7. He. Noble element of atomic number 2; first element in the noble gas group of the periodic table, atomic weight 4.00260, valence = 0. Helium nuclei are alpha particles. Most important isotope is helium-3.
Properties: Colorless, noncombustible gas; odorless and tasteless; liquefies at 4.2K to form helium I; at 2.2K there is a transiton (lambda) point at which helium II is formed. See Note below.
Properties: Bp −268.9C (1 atm), fp −272.2C (25 atm) lowest of any substance, bulk d 0.1785 g/L at 0C, very slightly soluble in water, insoluble in alcohol; rate of diffusion through solids is three times that of air; an asphyxiant gas.

See also noble gas.

Occurrence: Texas, Oklahoma, Kansas, New Mexico, Arizona, Canada. Originally discovered in the sun's atmosphere (1868) and recently confirmed in the atmosphere of Jupiter.

Derivation: From natural gas, by liquefaction of all other components, followed by purification over activated charcoal.

Grade: USP, technical, 99.995% pure.

Use: To pressurize rocket fuels, welding, inert atmosphere (growing germanium and silicon crystals), inflation of weather and research balloons, heat-transfer medium, leak detection, chromatography, cryogenic research, magnetohydrodynamics, luminous signs, geological dating, aerodynamic research, lasers, diving and space vehicle breathing equipment. Possible future uses include coolant for nuclear fusion power plants and in superconducting electric systems.

Note: Liquid helium has unique thermodynamic properties too complex to be adequately described here. Liquid He I has refr index 1.026, d 0.125, and is called a "quantum fluid" because it exhibits atomic properties on a macroscopic scale. Its bp is near absolute zero and viscosity is 25 micropoises (water = 10,000). He II, formed on cooling He I below its transition point, has the unusual property of superfluidity, extremely high thermal conductivity, and viscosity approaching zero.

See also superconductivity.

Hell-Volhard-Zelinsky reaction. α-Halogenation of carboxylic acids with halogen and phosphorus, presumably involving the enol form of the intermediate acyl halide.

"Heloplex."[416]   TM for the sodium bisulfite complex of heliotropin.

$[(CH_2O_2)C_6H_3CH(OH)SO_3Na$ (bicyclic)].

Properties: Dry white to gray powder, slightly soluble in water.

Use: zinc plating industry for compounding of brighteners.

HEMA.   See hydroxyethyl methacrylate.

hematein.   CAS: 475-25-2. $C_{16}H_{12}O_6$.

A tetracyclic compound. An oxidation product of hematoxylin, the coloring principle of logwood. Not to be confused with hematin.

Properties: Dark purple solid, mp 200+C, almost insoluble in water, slightly soluble in alcohol and ether, soluble in dilute sodium hydroxide giving a bright red color, soluble in ammonia with brownish-violet color.

Derivation: By adding ammonia to logwood extract and exposing to air.

Use: Indicator, biological stain.

hematin.   CAS: 15489-90-4.

$C_{34}H_{32}N_4O_4FeOH$.   The hydroxide of heme. Not to be confused with hematein.

Properties: Blue to brown-black powder; decomposes at 200C without melting; soluble in alkalies, hot alcohol, or ammonia; slightly soluble in hot pyridine; insoluble in water, ether, and chloroform.

Derivation: By dissolving hemin in dilute potassium hydroxide, precipitating with acetic acid, and recrystallizing from pyridine.

Use: Biochemical research.

hematite, red.   (red iron ore; bloodstone; iron oxide).   CAS: 1317-60-8. Fe_2O_3 with impurities.

Properties: Brilliant black to blackish red or brick red mineral with brown to cherry red streak and metallic to dull luster, d 4.9–5.3, Mohs hardness 6. Noncombustible.

Hazard: A carcinogen (OSHA).

Use: The most important ore of iron. Also certain varieties are used as paint pigments and for rouge.   See also iron oxide reds and ferric oxide.

hematoporphyrin.   CAS: 14459-29-1.

$C_{34}H_{38}O_6N_4$.   Deep red crystals, soluble in alcohol, sparingly soluble in ether, insoluble in water.

Obtained from hemin or hematin by the action of strong acids.

Hazard: Toxic. Reported to be preferentially absorbed by cancerous tissues, making them fluoresce under UV light.

Use: Medicine (antidepressant).

hematoxylin.   CAS: 517-28-2.

$C_{16}H_{14}O_6 \cdot 3HOH$.

Properties: Slightly yellow crystals, turning red in light, mp 100–120C, soluble in hot water and alcohol, in glycerol, and in alkali hydroxides.

Hazard: May be carcinogenic.

Use: Colorant in inks, biological stains.

heme.   (hem).   CAS: 14875-96-8.

$C_{34}H_{32}FeN_4O_4$.   The nonprotein portion of hemoglobin and myoglobin, consisting of reduced (ferrous) iron bound to protoporphyrin.

See porphyrin hemin.

Use: Medical and biochemical research.

hemel.   (Generic name for hexamethylmelamine; HMM). $C_3N_3[N(CH_3)_2]_3$ (cyclic).

Properties: Solid, insoluble in water, soluble in acetone.

Use: Chemosterilant for insects.

hemellitic acid.   (2,3-xylic acid.)

$C_6H_3(CH_3)_2COOH$.

Properties: Mp 144C.

hemicellulose. Cellulose having a degree of polymerization of 150 or less. A collective term for alpha and gamma cellulose. It is that portion of holocellulose that is soluble in mild caustic solution. The pure form is obtained from corn grain hulls. It is not an important component of cellulosic products and is of chiefly theoretical interest. Hemicellulose obtained by treating a mixture of hard- and soft-woods with steam has been used as an animal feed supplement.

hemimellitene. (1,2,3-trimethylbenzene). $C_6H_3(CH_3)_3$.
Properties: Liquid, d 0.8944 (20/4C), fp −25.5C, bp 176C, insoluble in water, soluble in alcohol. Occurs in some petroleums. Combustible.

hemin. (Teichmann's crystals; the chloride of heme). $C_{34}H_{32}N_4O_4FeCl$.

Properties: Crystals which are brown by transmitted light and steel blue by reflected light; sinters at 240C; freely soluble in ammonia water; soluble in strong organic bases; insoluble in carbonate solutions, dilute acid solutions; insoluble but stable in water.
Derivation: By heating hemoglobin with acetic acid and sodium chloride.
Use: Identification of blood stains, biochemical research, complexing agent.
See also chelate.

hemoglobin. (Hb). The respiratory protein of the red blood cells, it transfers oxygen from the lungs to the tissues and carbon dioxide from the tissues to the lungs. Its affinity for carbon monoxide is greater than 200 times that for oxygen.
Hemoglobin is a conjugated protein of molecular weight 65,000, consisting of approximately 94% globin (protein portion), and 6% heme. Each molecule can combine with one molecule of oxygen to form oxyhemoglobin (HbO_2). The

iron (in the heme portion) must be in the reduced (ferrous) state to enable the hemoglobin to combine with oxygen.
Oxyhemoglobin is available commercially as a brownish red powder or crystals, soluble in water.
Use: Medicine, usually called hemoglobin.

hemp. Soft white fibers 3–6 ft long. It is coarser than flax but stronger, more glossy, and more durable than cotton. Combustible.
Obtained from the stems of Cannabis sativa.
Sources: Central Asia, Italy, USSR, India, US.
Hazard: May ignite spontaneously when wet.
Use: Blended with cotton or flax in toweling and heavy fabrics, twine, cordage, packing.
See also cannabis.

hempa. (Generic name for hexamethylphosphoric triamide; hexamethylphosphoramide; HMPA; hexametapol). CAS: 680-31-9.
$[(N(CH_3)_2]_3PO$.
Properties: Water-white liquid, mild amine odor, soluble in water and polar and nonpolar solvents, bp 230–232C (739.4 mm), d 1.021 (15.5/15.5C). Combustible.
Hazard: A carcinogen.
Use: UV inhibitor in polyvinyl chloride, chemosterilant for insects, promoting stereospecific reactions, specialty solvent.

hempseed oil.
Properties: A drying oil similar in properties and uses to linseed, edible, iodine value approximately 160, d 0.923, refr index 1.470–1.472. Contains approximately 10% saturated fatty acids (palmitic and stearic), unsaturated acids present are linoleic, linolenic, and oleic. Saponification value 190-193. Combustible.
Produced chiefly in USSR, southern Europe, Japan, Chile, little in US.

henda- compounds. See corresponding undecompound.

n-heneicosanoic acid. $CH_3(CH_2)_{19}COOH$.
A saturated fatty acid not normally found in natural fats or waxes.
Properties: White, crystalline solid; mp 74.3C. Synthetic product available for organic synthesis, 99% purity. Combustible.

henequen.
Properties: Hard, strong, reddish fibers obtained from the leaves of *Agave fourcroydes*. It is similar to sisal but coarser and stiffer. Denier ranges from 300–500. Combustible.
Source: Mexico, Cuba.
Use: Binder twine, cordage.

Henkel reaction. Industrial scale thermal rearrangement or disproportionation of alkaline salts of aromatic açids to symmetrical diacids in the presence of cadmium or other metallic salts.

henna. A coloring principle obtained from dried leaves of certain tropical plants (North Africa, India).
Use: Commercial hair-dyeing preparations to give a yellow red color, medicine as an antifungal agent.

"Hennig Purifier."[177] TM for a preparation having a soda-ash base and other materials. Produced as walnut-sized briquettes. Packed in 100-lb paper bags.
Use: Ladle addition to produce cleaner steel by aiding in removal of dissolved oxides and silicates and fluxing non-metallic inclusions to slag.

Henry reaction. Formation of nitroalcohols by an aldol-type condensation of nitroparaffins with aldehydes in the presence of base (Henry) or by the condensation of sodium salts of acinitroparaffins with the sodium bisulfite addition products of aldehydes in the presence of a trace of alkali or weak acid (Kamlet). Widely used in sugar chemistry.

Henry's law. When a liquid and a gas are in contact, the weight of the gas that dissolves in a given quantity of liquid is proportional to the pressure of the gas above the liquid. Thus, if air is kept in contact with water at standard atmosphere pressure, each kg of water dissolves 0.017 g of oxygen at 20C; if this pressure is halved (by doing the experiment at high altitude where the pressure is only 0.5 atm) the water dissolves only 0.0085 g of oxygen. The law holds true only for equilibrium conditions, i.e., when enough time has elapsed so that the quantity of gas dissolved is no longer changing.

hentriacontane. $C_{31}H_{64}$ or $CH_3(CH_2)_{29}CH_3$.
Properties: Crystals, d 0.781 (68C), bp 302C (15 mm), mp 68C. Combustible.

HEOD. See dieldrin.

heparin. CAS: 9005-49-6. A complex organic acid (mucopolysaccharide) present in mammalian tissues, a strong inhibitor of blood coagulation, a dextrorotatory polysaccharide built up from hexosamine and hexuronic acid units containing sulfuric acid ester groups. Precise chemical formula and structure uncertain; a formula of $(C_{12}H_{16}NS_2Na_3)_{20}$ and molecular weight of 12,000 have been suggested for sodium heparinate.
Properties: White or pale colored amorphous powder; nearly odorless; hygroscopic; soluble in water; insoluble in alcohol, benzene, acetone, chloroform, and ether; pH in 17% solution between 5.0–7.5.
Derivation: Animal livers or lungs.
Grade: USP.
Hazard: May cause internal bleeding.
Use: Medicine (anticoagulant), biochemical research, rodenticides.

heptabarbital. (5-[1-cyclohepten-1-yl]-5-ethylbarbituric acid; 5-ethyl-5-cycloheptenylbarbituric acid). CAS: 509-86-4. $C_{13}H_{18}N_2O_3$.
Properties: White, odorless, crystalline powder; slightly bitter taste; mp 174C. Very sparingly soluble in water, slightly soluble in alcohol, soluble in alkaline solutions. Forms water-soluble sodium, magnesium, and calcium salts.
Use: Medicine (sedative).
See also barbiturate.

heptachlor. CAS: 76-44-8. $C_{10}H_7Cl_7$.
(1,4,5,6,7,8,8-heptachloro-3a,4,7,7a-tetrahydro-4,7-methanoindene. Generic).
Properties: White to light tan waxy solid, mp 95–96C, d 1.57–1.59, insoluble in water, soluble in xylene and alcohol.
Hazard: Toxic by ingestion, inhalation, and skin absorption; use has been restricted and discontinued except for termite control. TLV: 0.5 mg/m^3 of air.
Use: Insecticide.

heptachlorepoxide. $C_{10}H_9Cl_7O$. A degradation product of heptachlor which also acts as an insecticide.

heptachlorotetrahydromethanoindene.
See heptachlor.

n-heptacosane. $CH_3(CH_2)_{25}CH_3$.
Properties: Crystals, soluble in alcohol, insoluble in water, d 0.804, bp 270C (15 mm), mp 59.5C. Combustible.

n-heptadecane. $C_{17}H_{36}$ or $CH_3(CH_2)_{15}CH_3$.
Properties: Leaflets, soluble in alcohol, insoluble in water, d 0.778, bp 303C, mp 22.5C. Combustible.

n-heptadecanoic acid. (margaric acid). CAS: 506-12-7. $CH_3(CH_2)_{15}COOH$.
A saturated fatty acid not normally found in natural fats or waxes.
Properties: Colorless crystals, mp 61C, d 0.8355 (90.6/4C), bp 363.8C, 230.7C (16 mm), refr index 1.4324 (70C), soluble in alcohol and ether, insoluble in water, available as a 99% pure synthetic product.
Use: Organic synthesis.

heptadecanol. Any saturated C_{17} alcohol. Combustible.
See: (for example) n-heptadecanol and 3,9-diethyl-6-tridecanol.

n-heptadecanol. $C_{17}H_{35}OH$.
Properties: Colorless liquid, slightly soluble in water, d 0.8475 (20/20C), bp 308.5C, vap press less than 0.01 mm (20C), flash p 310F (154C), bulk d 7.1lb/gal (20C). Combustible.
Grade: Technical.
Use: Organic synthesis, plasticizer, intermediates, perfume fixatives, soaps and cosmetics, manufacture of wetting agents and detergents.

2-(8-heptadecenyl)-2-imidazoline-1-ethanol. See amine 220.

2-heptadecylglyoxalidine. (2-heptadecylimidazoline). $C_{20}H_{40}N_2$ or $C_{17}H_{35}C_3H_5N_2$.
Properties: Waxy solid; mp 85C; bp 200C (2 mm); slightly soluble in water; soluble in alcohol, benzene; hydrolyzes on standing to form N-2-aminoethyl stearamide. Combustible.
Derivation: By reacting stearic acid with ethylene diamine.
Use: Fungicide.

2-heptadecylglyoxalidine acetate. See glyodin.

2-heptadecylimidazoline. See 2-heptadecylglyoxalidine.

2-heptadecyl-2-imidazoline acetate. See glyodin.

heptafluorobutyric acid. (perfluorobutyric acid). C_3F_7COOH.
Properties: Colorless, hygroscopic liquid with sharp odor. Bp 120C (735 mm), fp −17.5C, d 1.641 (25C), refr index 1.290 (25C), surface tension 15.8 dynes/cm (30C). Miscible with water, acetone, ether, and petroleum ether; soluble in benzene and carbon tetrachloride; insoluble in carbon disulfide and mineral oil.
Derivation: By electrolysis of a solution of butyric acid in hydrogen fluoride.
Hazard: Irritant to tissue.
Use: Intermediate, surfactant, acidulant.

heptaldehyde. See heptanal.

heptalin acetate. See methylcyclohexanol acetate.

heptamethylene. See cycloheptane.

heptamethylnonane. $C_{16}H_{34}$. Isomer of cetane (hexadecane). In 1964 it replaced α-methylnaphthalene as ignition standard for diesel fuels.
Properties: Ignition value 15 on cetane-α-methylnaphthalene scale.
Hazard: Flammable, moderate fire risk.

heptanal. (heptaldehyde; enanthaldehyde; aldehyde C-7). CAS: 111-71-7. $C_6H_{13}CHO$.
Properties: Oily, colorless liquid; penetrating, fruity odor; hygroscopic. Soluble in three volumes of 60% alcohol, slightly soluble in water, soluble in ether, d 0.814–0.819, refr index 1.42, mp 43C, bp 153C. Combustible.
Derivation: Castor oil, from decomposition of the ricinoleic acid glyceride.
Use: Manufacture of 1-heptanol, organic synthesis, perfumery, pharmaceuticals, flavoring.

n-heptane. (dipropylmethane).
CAS: 142-82-5. $CH_3(CH_2)_5CH_3$.
Properties: Volatile, colorless liquid. Fp −90.595C, bp 98.428C, refr index 1.38764 (20C), d 0.68368 (20C), flash p 25F (−3.89C) (CC). Soluble in alcohol, ether, chloroform; insoluble in water; distillation range 93.3–98.9C; vap press 2.0 psi absolute (37.7C) (max). Color Saybolt +30 (min), maximum sulfur content 0.01 wt%, corrosive passes ASTM D 130–30 test, autoign temperature 433F (222C).
Derivation: Fractional distillation of petroleum, purified by rectification.
Grade: Commercial, 99%, spectro, ASTM reference fuel, research, 99.92 mole %.
Hazard: Toxic by inhalation. Flammable, dangerous fire risk. TLV: 400 ppm in air.
Use: Standard for octane rating determinations (pure normal heptane has zero octane number), anesthetic, solvent, organic synthesis, preparation of laboratory reagents.
See also octane number.

1,7-heptanedicarboxylic acid. See azelaic acid.

1,7-heptanedioic acid. See pimelic acid.

n-heptanoic acid. (enanthic acid; n-heptylic acid; heptoic acid). CAS: 111-14-8. $CH_3(CH_2)_5COOH$.
Properties: Clear, oily liquid; unpleasant odor. Soluble in alcohol and ether, insoluble in water, d 0.9181 (20/4C), fp 8C, bp 221.9C, refr index 1.4229. Combustible.
Derivation: By oxidizing heptanal with potassium permanganate in dilute sulfuric acid.
Use: Organic synthesis, production of special lubricants for aircraft and brake fluids.

1-heptanol. See heptyl alcohol.

2-heptanol. See methyl amyl carbinol.

3-heptanol. CAS: 589-82-2. $CH_3CH_2CH(OH)C_4H_9$.
Properties: Liquid, d 0.8224 (20C), fp −70C, bp 156.2C, flash p 140F (60C) (CC), slightly soluble in water.

Hazard: Toxic by ingestion. Moderate fire risk.
Use: Flotation frother, solvent and diluent in organic coatings, intermediates.

2-heptanone. See methyl-n-amyl ketone.

3-heptanone. See ethyl butyl ketone.

4-heptanone. See dipropyl ketone.

heptanoyl chloride. CAS: 2528-61-2.
$CH_3(CH_2)_5COCl$.
Properties: Solid, mw 148.63, bp 173C, d 0.960, fp 58C.
Hazard: Corrosive and a lachrymator.

1-heptene. (1-heptylene).
$CH_3(CH_2)_4CH:CH_2$.
Properties: Colorless liquid, d 0.6968 (20/4C), bp 93.3C, fp −10C, flash p 32F (0C), refr index 1.3994 (20C). Soluble in alcohol, acetone, ether, petroleum and coal-tar solvents; insoluble in water.
Hazard: Flammable, dangerous fire risk.
Grade: 95% purity.
Use: Organic synthesis.

2-heptene. (2-heptylene).
$CH_3(CH_2)_3CH:CHCH_3$ (cis and trans isomers).
Properties: Colorless liquid; d (cis) 0.708, (trans) 0.704, (commercial) 0.7010–0.7050 (20/4C); bp (trans) 98C, (cis) 98.5C, (commercial) 97–99C, refr index 1.406 (20C); flash p (commercial) 28F (−2.2C); soluble in alcohol, acetone, ether, petroleum and coal-tar solvents; insoluble in water.
Hazard: Flammable, dangerous fire risk.
Use: Suggested as plant growth retardant.

3-heptene. (3-heptylene). $C_3H_7CH:CHC_2H_5$.
Properties: (mixed cis and trans isomers) Colorless liquid, bp 95C, d 0.705 (15.5/15.5C), refr index 1.405 (20C), flash p 21F (−6.1C).
Hazard: Flammable, dangerous fire risk.
Use: Suggested as plant growth retardant.

heptoic acid. See heptanoic acid.

1-heptyl acetate. CAS: 112-06-1.
$C_7H_{15}OOCCH_3$. Liquid with fruity odor.
Use: Artifical fruit essences.

heptyl alcohol. (1-heptanol; enanthyl alcohol).
CAS: 111-70-6. $C_7H_{15}OH$.
Properties: Colorless, fragrant liquid. Fp −34.6C, bp 175C, d 0.824 (20/4C), refr index 1.4233 (20C), flash p 170F (76.6C), slightly soluble in water, miscible with alcohol and ether. Combustible.
Derivation: From heptaldehyde by reduction.

Use: Organic intermediate, solvent, cosmetic formulations.

heptylamine. $C_7H_{15}NH_2$.
Properties: Colorless liquid, d 0.777 (20/4C), fp −23C, bp 155C, flash p 140F (60C) (OC), slightly soluble in water, soluble in alcohol or ether. Combustible.

heptyl formate. $HCOOC_7H_{15}$.
Properties: Colorless liquid with fruity odor, bp 176.7C, d 0.894 (0C). Combustible.
Use: Artificial fruit essences.

heptyl heptoate. $C_7H_5OOCC_6H_{13}$.
Properties: Colorless liquid with fruity odor, d 0.865 (19C), bp 273–274C (754 mm). Combustible.
Use: Artificial fruit essences.

n-heptylic acid. See heptanoic acid.

heptyl pelargonate. $C_7H_{15}OOCC_8H_{17}$.
Properties: Liquid with pleasant odor, d 0.866 (15.5/15.5C), bp 300C, refr index 1.4360. Combustible.
Use: Flavors and perfumes.

herbicide. (weed killer). A pesticide, either organic or inorganic, used to destroy unwanted vegetation, especially various types of weeds, grasses, and woody plants. Until 1924 inorganics such as sodium chlorate, sodium chloride, ammonium sulfamate, arsenic, and boron compounds were used. At that time the more specific organics were introduced, typified by 2,4-dichlorophenoxyacetic acid (2,4-D). Herbicides may be of two major types: (1) selective, such as 2,4-D, 2,4,5-T, phenols, carbamates and urea derivatives, permitting elimination of weeds without injury to the crop, and (2) nonselective, comprising soil sterilants (sodium compounds) and silvicides (ammonium sulfamate). The latter kill woody plants and trees. Some types act as overstimulating growth hormones. Many herbicides are highly toxic and should be handled and applied with care; use of chlorinated types may be restricted.
See also defoliant.

"Herclor."[266] TM for a group of specialty elastomers based on epichlorohydrin, claimed to have unique service performance properties.
Use: Automotive and aircraft parts, wire and cable coating, seals and gaskets, packings, hose, belting, and coated fabrics. "Herclor H" is a homopolymer and "Herclor C" a copolymer with ethylene oxide; both have high resistance to ozone, heat, solvents, and chemical attack.

"Hercoflex."[266]   TM for a series of plasticizers. 150 Di(n-octyl, n-decyl) phthalate. 290 Di(n-octyl, n-decyl) adipate. 600 High-boiling ester of pentaerythritol and a saturated aliphatic acid. 707 High mw polyol ester.
Use: High-temperature vinyl electrical insulation. 900 High mw polyester plasticizer for polyvinyl acetate. J15 Saturated aliphatic ester of pentaerythritol for plasticizing vinylidene chloride.

"Hercolube."[266]   TM for synthetic lubricant base stocks.
Derived from pentaerythritol esters of saturated fatty acids.

"Hercolyn." D[266]   TM for a pale, viscous liquid, the hydrogenated methyl ester of rosin.
Use: Plasticizing resin.

"Herculoid."[266]   TM for nitrocellulose containing 10.9–11.2% nitrogen.
Hazard: Like nitrocellulose.
Use: Pyroxylin plastics.

"Herculon."[266]   TM for polypropylene olefin fibers. Available in bulked continuous and continuous multifilament yarns, staple, and uncut tow.
Use: Apparel, home furnishings, and industrial applications.

"Heresite."[17]   TM for a series of pure phenolformaldehyde resinous coatings and related products of the thermosetting type. Applied by spraying, dipping, or roller-coating, followed by curing at temperatures of approximately 200C.
Use: Anticorrosive lining for shipping containers, machinery, and equipment for chemicals, food, drugs, and petroleum industries. Lining of tank cars for sulfuric acid and other corrosive chemicals. Anticorrosive coatings for heat exchangers, fin tube coils, and air handling equipment.

heroin.   See diacetylmorphine.

herring oil.   See fish oil.

Herzberg, Gerhard. (1904-) A German-born physicist who won the Nobel prize for chemistry in 1971 for his work on the composition of molecules. His research involved the spectroscopy of atoms and molecules and their excitation behavior. He became a Canadian citizen and was the director of the Division of Pure Physics of the National Research Council of Canada.

Herz reaction; Herz compounds. Formation of o-aminothiophenols by heating aromatic amines with excess sulfur monochloride. The first products formed are thiazothionium halides known as Herz compounds. If the position para to the amino group is unoccupied, chloride is substituted at this position during the reaction.

Herzig-Meyer determination of N-alkyl groups. N-Alkylamines are refluxed with hydriodic acid, the quaternary alkyl ammonium iodides are pyrolyzed to split off alkyl iodide, which is determined gravimetrically by conversion to silver iodide or titrated as iodate.

hesperidin.   CAS: 520-26-3. $C_{28}H_{34}O_{15}$.
A natural bioflavonoid of the flavanone group.
Properties: Fine needles, mp 258–262C, soluble in dilute alkalies, pyridine.
Derivation: Extraction from citrus fruit peel.
Use: Synthetic sweetener research.

Hess's law.   The heat evolved or absorbed in a chemical process is the same whether the process takes place in one or in several steps; also known as the law of constant heat summation.

hetastarch.   A starch derivative containing 90% amylopectin.
Use: Blood plasma volume expander.

heteroaromatic.   See heterocyclic.

heterocyclic.   Designating a closed-ring structure, usually of either 5 or 6 members, in which one or more of the atoms in the ring is an element other than carbon, e.g., sulfur, nitrogen, etc. Examples are pyridine, pyrole, furan, thiophene and purine.

thiophene   pyridine

heterogeneous. (Latin "different kinds") Any mixture or solution comprised of two or more substances, whether or not they are uniformly dispersed. Common examples are such diverse materials as air (a mixture of 20% oxygen and 80% nitrogen), milk, marble, paint, gasoline, blood, mayonnaise. In all such cases, the mixture can be separated mechanically into their components. "Homogenized" milk is as heterogeneous as regular milk and the term is, strictly speaking, a misnomer.
See also homogeneous; mixture.

heterogeneous catalysis.   See catalysis, heterogeneous.

heteromolybdates. (heteropolymolybdates). A large group of complex molybdenum salts and acids in which the anion contains oxygen atoms and from 2–18 hexavalent molybdenum atoms, as well as one or more other metal or nonmetal atoms (phosphorus, arsenic, iron, and tellurium). The latter are referred to as hetero atoms and any of approximately 35 elements may be present in this manner. Example: $Na_3PMo_{12}O_{40}$, sodium phospho-12-molybdate. The molecular weights of these compounds range up to 3000. The acids and most of the salts are very soluble in water and the acids and some salts are soluble in organic solvents.

Use: Phosphomolybdates and phosphotungstates are used as precipitants for basic dyes to form lakes and toners. The phospho- and silicomolybdate groups are of key importance in the functioning of certain enzymes. There are many uses in analytical chemistry.

HETP. (1) Abbreviation for hexaethyl tetraphosphate; (2) abbreviation for height equivalent to a theoretical plate.
See theoretical plate.

Heumann-Pfleger indigo synthesis. Cyclization of phenylglycine to indoxyl followed by oxidation by air or oxidizing agents, such as ferric chloride, to yield indigo.

heuristic. See computational chemistry.

Hevesy, Georg de. (1885-1966) A Hungarian chemist who won Nobel prize in chemistry in 1943. He discovered the element hafnium in 1923. One of his interesting projects involved the calculation of the percentages of chemical elements in the universe. He also was involved in research using radioactive lead and phosphorus traces. His work included the separation of isotopes by physical means. His PhD was granted at Freiburg in 1908.

hexabromoethane. C_2Br_6.
Properties: Yellowish-white, rhombic needles; slightly soluble in water, alcohol; mp 149C (decomposes with separation of bromine).
Derivation: Action of bromine on diiodoacetylene.
Use: Organic synthesis.

hexacalcium phytate. See calcium phytate.

hexachloroacetone. (hexachloro-2-propanone). CAS: 116-16-5. $Cl_3CCOCCl_3$.
Properties: Yellow liquid, bp 204C, fp −3C, d 1.744 (12/12C), slightly soluble in water, soluble in acetone. Combustible.
Hazard: Toxic by ingestion and inhalation, strong irritant, evolves phosgene when heated.
Use: Desiccant, herbicide.

hexachlorobenzene. (perchlorobenzene).
CAS: 118-74-1. C_6Cl_6.
Properties: White needles, d 2.04, soluble in benzene and boiling alcohol, insoluble in water, mp 229C, bp 326C, flash p 468F (242C). Combustible.
Hazard: Toxic by ingestion.
Use: Organic synthesis, fungicide for seeds, wood preservative.

hexachlorobutadiene. CAS: 87-68-3.
$Cl_2C:CClCCl:CCl_2$.
Properties: Clear, colorless liquid with mild odor. Freezing range −19 to −22C, boiling range 210–220C, refr index 1.552 (20C), flash p none, d 1.675 (15.5/15.5C), bulk d 13.97 lb/gal (15.5C), purity 98% (min). Vap press 22 mm (100C), 500 mm (200C). Viscosity (37.7C) 2.447 cp, 1.479 centistokes; (98.9C) 1.131 cp , 0.724 centistokes. Insoluble in water, compatible with numerous resins, soluble in alcohol and ether. Nonflammable.
Hazard: Toxic by ingestion and inhalation, a suspected carcinogen. TLV: 0.02 ppm.
Use: Solvent for elastomers, heat-transfer liquid, transformer and hydraulic fluid, wash liquor for removing C_4 and higher hydrocarbons.

1,2,3,4,5,6-hexachlorocyclohexane. (BHC; HCCH; HCH; TBH; benzene hexachloride). CAS: 608-73-1. $C_6H_6Cl_6$. A systemic insecticide. The gamma isomer is known as lindane.
Properties: White or yellowish powder or flakes; musty odor; color, odor, melting point vary with isomeric composition; d 1.87; vap press approximately 0.5 mm (60C); stable toward moderate heat but decomposed by alkaline substances. Mp of the pure isomers are: (alpha-trans) 157–158C, (beta-cis) 297C (sublimes), (gamma) 112.5C, (delta) 138–139C, (epsilon) 217–219C. Insoluble in water; soluble in 100% alcohol, chloroform, and ether.
Derivation: Chlorination of benzene in actinic light.
Method of purification: Fractional crystallization. The technical grade may run 10–15% gamma isomer, but can be brought up to 99% (lindane).
Grade: Technical (mixture of isomers), 25% gamma isomer, and 99% gamma isomer (lindane).
Hazard: Toxic by ingestion and inhalation, absorbed by skin, strong irritant to skin and eyes. Central nervous system depressant. Use may be restricted. TLV (lindane): 0.5 mg/m³ of air.
Use: Component of insecticides; toxic to flies, cockroaches, aphids, grasshoppers, wire worms, and boll weevils.

hexachlorocyclopentadiene. (perchlorocyclopentadiene). CAS: 77-47-4. C_5Cl_6.
Properties: Pale yellow liquid, pungent odor, bp 239C, fp 9.6C, d 1.717 (15/15C), bulk d 14.30 lb/gal (15.5C), refr index 1.563 (25C), flash p none. Nonflammable.
Grade: Technical.
Hazard: Toxic by ingestion, inhalation, and skin absorption. TLV: 0.01 ppm in air.
Use: Intermediate for resins, dyes, pesticides, fungicides, pharmaceuticals.

hexachlorodiphenyl oxide. CAS: 55720-99-5. $C_{12}H_4Cl_6O$.
Properties: Light yellow, very viscous liquid. Bp 230–260C (8 mm), d 1.60 (20/20C), bulk d 13.12 lb/gal at 25C, refr index 1.621 (25C), flash p none. Soluble in methanol, ether; very slightly soluble in water. Nomflammable.
Hazard: Toxic by ingestion.
Use: Solvent, intermediate.

1,1,1,4,4,4-hexachloro-1,4-disilabutane.
See bis(tricholorsilyl)ethane.

hexachloro endomethylene tetrahydrophthalic acid. See chlorendic acid.

hexachloro endomethylene tetrahydrophthalic anhydride. See chlorendic anhydride.

hexachloroethane. (perchloroethane; carbon trichloride; carbon hexachloride). CAS: 67-72-1. Cl_3CCCl_3.
Properties: Colorless crystals, camphor-like odor, d 2.091, mp 185C, bp sublimes at 185C, soluble in alcohol and ether, insoluble in water.
Hazard: Toxic by ingestion and inhalation, strong irritant, absorbed by skin. TLV: 10 ppm in air.
Use: Organic synthesis, retarding agent in fermentation, camphor substitute in nitrocellulose, pyrotechnics and smoke devices, solvent, explosives.

hexachloromethylcarbonate. (triphosgene). $(OCCl_3)_2CO$.
Properties: White crystals, odor similar to that of phosgene, decomposed by hot water and alkali hydroxides. Only slowly aceted upon by cold water. Soluble in alcohol, benzene, ether; d approximately 2, bp 205–206C (partial decomposition), mp 78–79C.
Derivation: Chlorination of dimethyl carbonate exposed to direct sunlight.
Hazard: Strong irritant to eyes and skin.
Use: Lachrymator.

hexachloromethyl ether. $O(CCl_3)_2$.
Properties: Liquid, phosgene-like odor, d 1.538 (18C), bp 98C (partial decomposition).
Derivation: Chlorination of dichloromethyl ether.
Hazard: Strong irritant to eyes and skin.

hexachloronaphthalene. CAS: 1335-87-1. $C_{10}H_2Cl_6$.
Properties: White solid.
Hazard: Toxic by inhalation, strong irritant, absorbed by skin. TLV: 0.2 mg/m³ of air.

hexachlorophene. (2,2'-methylene-bis-(3,4,6-trichlorophenol; bis-(3,5,6-trichloro-2-hydroxyphenyl) methane). CAS: 70-30-4. $(C_6HCl_3OH)_2CH_2$.
Properties: White, free-flowing powder; odorless; mp 161–167C; soluble in acetone, alcohol, ether, chloroform; insoluble in water.
Derivation: Condensation of 3,4,5-trichlorophenol with formaldehyde in the presence of sulfuric acid.
Hazard: FDA prohibits use unless prescribed by a physician.
Use: Topical antiinfective (restricted), germicidal soaps, veterinary medicine.

hexachloro-2-propane. See hexachloroacetone.

hexachloropropylene. (hexachloropropene; perchloropropylene). $CCl_3CCl:CCl_2$.
Properties: Water-white liquid, bp 210C, insoluble in water, miscible with alcohol, ether, chlorinated compounds.
Use: Solvent, plasticizer, hydraulic fluid.

hexacontane. $C_{60}H_{122}$. High molecular weight hydrocarbon.
Properties: Waxy solid, mp 101C. Combustible.

hexacosanoic acid. See cerotic acid.

n-hexadecane. (cetane). $C_{16}H_{34}$.
Properties: Colorless liquid, d 0.77335 (20/4C), flash p 200F (93C), bp 286.5C, mp 18.14C, refr index 1.43435 (20C), soluble in alcohol, acetone, ether; insoluble in water, autoign temperature 401F (205C). Combustible.
Grade: Technical, ASTM.
Use: Solvent, organic intermediate, ignition standard for diesel fuels.
See also cetane number.

hexadecanoic acid. See palmitic acid.

1-hexadecanol. See cetyl alcohol.

hexadecanoyl chloride. See palmitoyl chloride.

1-hexadecene. (cetene; α-hexadecylene). $CH_3(CH_2)_{13}CH:CH_2$.
Properties: Colorless liquid, mp 4C, bp 274C, flash p 200F (93C), d 0.784 (15/4C), refr index 1.441 (20C). Insoluble in water; soluble in alcohol, ether, petroleum, and coal-tar solvents. Combustible.
Derivation: Treatment of cetyl alcohol with phosphorus pentoxide.
Grade: 95% purity.
Use: Organic synthesis.

cis-9-hexadecenoic acid. See palmitoleic acid.

6-hexadecenolide. See ambrettolide.

hexadecyl mercaptan. See cetyl mercaptan.

tert-hexadecyl mercaptan. $C_{16}H_{33}SH$.
Properties: Colorless liquid, unpleasant odor, boiling range 121–149C (5 mm), d 0.874 (60/60F), refr index 1.474 (20C), flash p 265F (129.4C). Combustible.
Use: Polymer modification.
See also thiol.

hexadecyltrichlorosilane. CAS: 5894-60-0. $C_{16}H_{33}SiCl_3$.
Properties: Colorless to yellow liquid, bp 269C, d 0.996 (25/25C), refr index 1.4568 (25C), flash p 295F (146C). Combustible.
Derivation: By Grignard reaction of silicon tetrachloride and hexadecylmagnesium chloride.
Grade: Technical.
Hazard: Strong irritant, evolves hydrogen chloride in presence of moisture.
Use: Intermediate for silicones.

hexadecyltrimethylammonium bromide.
See cetyl trimethylammonium bromide.

1,4-hexadiene. CAS: 42296-74-2. $H_2C:CHCH_2HC:CHCH_3$.
Properties: Colorless liquid, d 0.6996 (20/4C), bp 64C (745 mm), refr index 1.4162 (20C), flash p −6F (−21.1C), insoluble in water.
Derivation: Reaction between ethylene and butadiene with a special catalyst.
Hazard: Highly flammable, explosive limits in air 2–6.1%.
Use: As third monomer in EPDM synthetic elastomers.

2,4-hexadienoic acid. See sorbic acid.

1,5-hexadiyne. See dipropargyl.

hexa-2-ethylbutoxydisiloxane.
$[(CH_3CH(C_2H_5)CH_2CH_2O)_3Si]_2O$.
Properties: Colorless oil, bp 195C (0.2 mm). Combustible.

Derivation: Reaction of silicon tetrachloride, 2-ethylbutanol, and water.
Use: Aircraft hydraulic fluid.

hexaethyl tetraphosphate. (HETP).
CAS: 757-58-4. A mixture of ethyl phosphates and ethyl pyrophosphate (TEPP).
Properties: Yellow liquid, d 1.26–1.28 (25/4C), fp −90C, refr index 1.427, decomposes at high temperatures, soluble or miscible in water and many organic solvents except kerosene, hydrolyzes in low concentration, hygroscopic.
Hazard: Toxic by ingestion, inhalation, and skin absorption; cholinesterase inhibitor.
Use: Contact insecticide.
Note: Hexethyl tetraphosphate and compressed gas mixture not accepted by air or by rail passenger. For details consult regulations.

hexafluoroacetone. CAS: 648-16-2. CF_3COCF_3.
Properties: Colorless, hygroscopic, highly reactive gas. Bp −27C, fp −122C, liquid density 1.33 (25C), minimum purity 95%.
Hazard: Toxic by inhalation and skin absorption. Reacts vigorously with water and other substances, releasing considerable heat. Nonflammable. TLV: 0.1 ppm in air.
Use: Intermediate in organic synthesis.

hexafluorobenzene. C_6F_6.
Properties: Liquid, bp 80.26C, mp 5.2C, d 1.613. Combustible.
Hazard: Toxic by inhalation.
Use: Chemical intermediate, solvent in NMR spectroscopy.

hexafluoroethane. (fluorocarbon 116).
CAS: 76-16-4. CF_3CH_3.
Properties: A gas, bp −78.2C, d 1.59, insoluble in water, slightly soluble in alcohol. One of the most stable of all organic compounds.
Grade: 99.6% pure.
Use: Dielectric and coolant, aerosol propellant, refrigerant.

hexafluorophosphoric acid. CAS: 16940-81-1. HPF_6.
Properties: (65% solution): Colorless, fuming liquid, d 1.81, mp 31C ($6H_2O$), stable in neutral and alkaline solutions.
Hazard: Strong irritant to tissue.
Use: Metal cleaners, electrolytic or chemical polishing agents for the formation of protective coatings for metal surfaces, and as a catalyst.

hexafluoropropylene. (perfluoropropene). $CF_3CF:CF_2$.
Properties: Gas, fp −156C, bp −29C, d 1.583 (−40/4C).

hexafluoropropylene epoxide. (HFPO).
$CF_2CF_2CF_2O$.
Derivation: Oxidation of hexafluoropropylene with alkaline hydrogen peroxide at approximately $-30C$.
Use: Monomer for HFPO polymers that are heat-resistant to 410C. Noncombustible.
See also "Freon E" and "Krytox."

hexafluorosilicic acid. See fluosilicic acid.

hexaglycerol. See trimethylolpropane, polyglycerol.

hexahydric alcohol. See mannitol, sorbitol, and dulcitol.

hexahydroaniline. See cyclohexylamine.

hexahydrobenzene. See cyclohexane.

hexahydrobenzoic acid. (cyclohexanecarboxylic acid [a naphthenic acid]). $C_6H_{11}COOH$.
Properties: Colorless monoclinic prisms, mp 31C, bp 233C, d 1.048 (15/4C), refr index 1.4561 (33.8C), slightly soluble in water, soluble in alcohol and ether.
Use: Paint and varnish driers, drycleaning soaps, lubricating oils; stabilizer for rubber.

hexahydrocresol. See methylcyclohexanol.

hexahydromethylphenol. See methylcyclohexanol.

hexahydrophenol. See cyclohexanol.

hexahydrophthalic anhydride. (1,2-cyclohexandicarboxylic anhydride). $C_6H_{10}(CO)_2O$.
Properties: Clear, colorless, viscous liquid which becomes a glassy solid at 35–36C; bp 158C (17 mm); d 1.19 (40C); miscible with benzene, toluene, acetone, carbon tetrachloride, chloroform, ethanol, and ethyl acetate; slightly soluble in petroleum ether.
Hazard: Toxic by inhalation, strong irritant to eyes and skin.
Use: Intermediate for alkyds, plasticizers, insect repellents, and rust inhibitors; hardener in epoxy resins.

hexahydropyridine. See piperidine.

hexahydrotoluene. See methylcyclohexane.

hexahydro-1,3,5-trinitro-sym-triazine.
See cyclonite.

hexahydroxycyclohexane. See inositol.

hexahydroxylene. See dimethylcyclohexane.

hexakis(methoxymethyl)melamine.

NC(NR)NC(NR)NC(NR) where
R = (CH$_2$OCH$_3$)$_2$.
Use: Crosslinking agent for alkyds, epoxies, cellulosics and vinyls.

n-hexaldehyde. (caproic aldehyde).
CAS: 66-25-1. $CH_3(CH_2)_4CHO$.
Properties: Colorless liquid, sharp aldehyde odor, d 0.8156 (20/20C), bp 128.6C, vap press 10.5 mm (20C), flash p 90F (32.2C) (OC), bulk d 6.9 wt/gal (20C), fp $-56.3C$, immiscible with water.
Grade: Technical.
Hazard: Flammable, moderate fire risk.
Use: Organic synthesis of plasticizers, rubber chemicals, dyes, synthetic resins, insecticides.

"Hexalin."[28]   TM for cyclohexanol (usually shipped with 2.25% methanol as antifreeze).
Hazard: Toxic by ingestion.

hexametapol. See hempa.

hexamethonium chloride.   CAS: 60-25-3.
$(CH_3)_3NCl(CH_2)_6NCl(CH_3)_3$.   [hexamethylenebis-(trimethylammonium chloride].
Properties: White, crystalline, hygroscopic powder with faint odor; mp 289–292C (decomposes); very soluble in water; soluble in alcohol, methanol, and n-propanol; insoluble in chloroform and ether. Available commercially as unhydrated form or as dihydrate.
Use: Medicine (antihypertensive).

hexamethylbenzene. $C_{12}H_{18}$ or $C_6(CH_3)_6$.
Properties: Colorless plates, soluble in alcohol, insoluble in water, bp 265C, mp 165.5C. Combustible.

hexamethyldiaminoisopropanol diiodide.
See propiodal.

hexamethyldisilazane.   (HMDS).
CAS: 999-97-3. $(CH_3)_3SiNHSi(CH_3)_3$.
Properties: Liquid, d 0.77 (25C), refr index 1.4057 (25C), bp 125C, flash p 77F (25C). Soluble in acetone, benzene, ethyl ether, heptane, perchloroethylene; reactive with methanol and water.
Purity: 99% min.
Hazard: Flammable, moderate fire risk.
Use: Chemical intermediate, chromatographic packings.

hexamethylene. See cyclohexane.

hexamethylenediamine. (1,6-diaminohexane; 1,6-hexanediamine).   CAS: 124-09-4. $H_2N(CH_2)_6NH_2$.
Properties: Colorless leaflets, mp 39–42C, bp 205C, soluble in water, slightly soluble in alcohol and benzene. Combustible.
Derivation: (1) Reaction of adipic acid and ammonia (catalytic vapor-phase) to yield adiponitrile, followed by liquid-phase catalytic hydrogenation. (2) Chlorination of butadiene followed by reaction with sodium cyanide (cuprous chloride catalyst) to 1,4-dicyanobutylene and hydrogenation.
Hazard: Toxic by ingestion, strong irritant to tissue.
Use: Formation of high polymers, e.g., nylon 66.

hexamethylenediamine carbamate. See "Diak."

hexamethylene diisocyanate.
$OCN(CH_2)_6NCO$.
Properties: Liquid, d 1.04 (25/15.5C), flash p 284F (140C). Combustible.
Use: Chemical intermediate.

hexamethylene glycol. (1,6-hexanediol).
CAS: 629-11-8. $CH_2OH(CH_2)_4CH_2OH$.
Properties: Crystalline needles, mp 42C, bp 210C, refr index 1.457, d 0.953 (50C), flash p 130C (266F).
Derivation: Reduction of adipic acid ester with copper chromite catalyst.
Hazard: Toxic by ingestion.
Use: Solvent, intermediate for high polymers (nylon, polyesters), coupling agent, coil coating.

hexamethyleneimine. CAS: 111-49-9.
$C_6H_{12}NH$ (cyclic).
Properties: Clear, colorless liquid with an ammonia-like odor; bp 138C; fp −37C; d 0.8799 (20/4C).
Hazard: Toxic by ingestion, strong irritant to tissue.
Use: Intermediate for pharmaceutical, agricultural, and rubber chemicals.

hexamethylenetetramine. (methenamine; HMTA; aminoform; hexamine, erroneously "Hexamethyleneamine").   CAS: 100-97-0.
$(CH_2)_6N_4$.   A heterocyclic fused ring structure

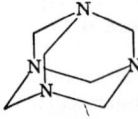

Properties: White, crystalline powder or colorless lustrous crystals. Practically odorless, d 1.27

(25C), soluble in water, alcohol and chloroform, insoluble in ether, sublimes approximately 200C, partly decomposes.
Derivation: Action of ammonia on formaldehyde.
Grade: Technical, NF (as methenamine).
Hazard: Skin irritant. Flammable, dangerous fire risk.
Use: Curing of phenolformaldehyde and resorcinolformaldehyde resins, rubber-to-textile adhesives, protein modifier, organic synthesis, pharmaceuticals, ingredient of high explosive cyclonite, fuel tablets, rubber accelerator, fungicide, corrosion inhibitor, shrink-proofing textiles, antibacterial.

hexamethylmelamine. See hemel.

hexamethylpararosaniline chloride. See methyl violet.

hexamethylphosphoramide. See hempa.

hexamethylphosphoric triamide. See hempa.

hexamethyltetracosahexaene. See squalene.

hexamethyltetracosane. See squalane.

hexamine. See hexamethylenetetramine.

hexanaphthene. See cyclohexane.

n-hexane. CAS: 100-54-3. $CH_3(CH_2)_4CH_3$.
Properties: Colorless, volatile liquid. Faint odor, d 0.65937 (20/4), bp 68.742C, fp −95C, refr index 1.37486 (20C), flash p −9F (−22.7C), autoign temperature 500F (260C). Soluble in alcohol, acetone, and ether; insoluble in water.
Derivation: By fractional distillation from petroleum (molecular sieve process).
Grade: 85%, 95%, 99%, spectro, research, and nanograde.
Hazard: Flammable, dangerous fire risk. TLV: 50 ppm in air.
Use: Solvent, especially for vegetable oils, low temperature thermometers, calibrations, polymerization reaction medium, paint diluent, alcohol denaturant.

hexanedioic acid. See adipic acid.

1,6-hexanediol. See hexamethylene glycol.

hexanedione-2,5. See acetonyl acetone.

1,2,6-hexanetriol. CAS: 106-69-4.
$HOCH_2CH(OH)CH_2CH_2CH_2CH_2OH$.
Properties: Water-white liquid, d 1.1063, sets to glass approximately −20C (fp under controlled

conditions 32.8C), bp at 5 mm (178C), flash p 380F (193C), miscible with water. Combustible.
Use: Alkyd and polyester resin intermediate, softener, moistening agent, and solvent.

hexanitrodiphenylamine. (hexil; hexyl; hexite; dipicrylamine).
$(NO_2)_3C_6H_2NHC_6H_2(NO_2)_3$.
Properties: Yellow solid, mp 238–244C, decomposes violently at higher temperatures, insoluble in water and alcohol, soluble in alkalies and warm acetic or nitric acid.
Derivation: Nitration of diphenylamine, also from dinitrochlorobenzene.
Hazard: Explodes on shock or exposure to heat, dangerous.
Use: Booster explosive, analysis for potassium.

hexanitrodiphenyl sulfide. (dipicryl sulfide).
$[(NO_2)_3C_6H_2]_2S$.
Properties: Golden-yellow leaflets, mp 234C, sparingly soluble in alcohol and ether, more soluble in glacial acetic acid and acetone.
Derivation: Interaction of picryl chloride and sodium thiosulfate in alcohol solution in the presence of magnesium carbonate.
Hazard: Explodes on shock or exposure to heat, dangerous.
Use: High explosive.

hexanitromannite. See mannitol hexanitrate.

hexanoic acid. Legal label name for caproic acid.

1-hexanol. See hexyl alcohol.

2-hexanone. See methyl-n-butyl ketone.

3-hexanone. (ethyl propyl ketone).
CAS: 589-38-8. $C_2H_5CO(CH_2)_2CH_3$.
Properties: Colorless liquid, bp 124C, d 0.813 at 22C, flash p 95F, (35C) (OC).
Hazard: Toxic by ingestion and inhalation, strong irritant. Flammable, moderate fire risk.
Use: Solvent.

hexanoyl chloride. $CH_3(CH_2)_4COCl$.
Properties: Colorless liquid, bp 151–153C, refr index 1.4867 (20C), decomposed by water and alcohol, soluble in ether and chloroform. Combustible.
Use: Chemical intermediate.

hexaphenyldisilane. $[(C_6H_5)_3Si]_2$.
Properties: White powder, mp 352C.
Derivation: Sodium condensation of triphenylchlorosilane.
Use: High-temperature applications.

"Hexaphos."[55] TM for a glassy phosphate of high molecular weight having superior water-softening properties.
Use: Water-softening, boiler-scale control, component of cleansers, laundry mixes, dishwashing compounds, pitch control in pulp industry, prevention of lime soap deposits in textile operations.

hexatriacontane. $C_{36}H_{74}$.
Properties: Waxy solid, d 0.797, mp 75C. Combustible.
See also paraffin wax.

1-hexene. (hexylene). CAS: 592-41-6.
$CH_3CH_2CH_2CH_2CH:CH_2$.
Properties: Colorless liquid, d 0.6734 (20/4C), bp 63.55C, fp −139.8C, refr index 1.3876 (20C), flash p −15F (−26.1C), insoluble in water, soluble in alcohol.
Grade: 95%, 99%, research.
Hazard: Irritant. Highly flammable, dangerous fire risk.
Use: Synthesis of flavors, perfumes, dyes, resins; polymer modifier.

2-hexene. $CH_3CH_2CH_2CH:CHCH_3$.
Properties: (mixed cis and trans isomers): Colorless liquid, bp 68C, fp −146C, refr index 1.3948 (20C), d 0.686 (15.5/15.5C), flash p −5F (20.5C), insoluble in water, soluble in alcohol.
Grade: 95%, 99%.
Hazard: Highly flammable, dangerous fire risk.
Use: Chemical intermediate.

5-hexene-2-one. See allylacetone.

hexenol. (3-hexen-1-ol; leaf alcohol).
$C_6H_{11}OH$.
Properties: Liquid, odor of green leaves, bp 156C, refr index 1.438, d 0.85. Combustible.
Occurrence: Grasses, leaves, herbs, tea, etc.
Use: Odorant in perfumery.

hexestrol. (p,p′-(1,2-diethylethylene)diphenol).
CAS: 84-16-2.
$HOC_6H_4CH(C_2H_5)CH(C_2H_5)C_6H_4OH$.
A nonsteroid, synthetic estrogen.
Properties: Odorless, white, crystalline powder; mp 185–188C; soluble in ether, acetone, alcohol, and methanol; practically insoluble in water; sensitive to light.
Derivation: From anethole, by reaction of diacetyl peroxide on p-methoxy-n-propylbenzene.
Use: Medicine (estrogenic hormone).

hexetidine. (amino-1,3-bis[β-ethylhexyl]-5-methylhexahydropyrimidine).
CAS: 141-94-6. $C_{21}H_{45}N_3$.
Properties: Liquid, d 0.860–0.875 (25/25C), bp 172–176C (1 mm), refr index 1.460–1.466 (25C).

Soluble in methanol, benzene, petroleum ether; insoluble in water. Combustible.

Grade: Technical, NF.

Use: Fungicide, bactericide, algicide; antistatic agent for synthetics, insect repellent, medicine (antifungal agent).

hexil. See hexanitrodiphenyl amine.

hexite. See hexanitrodiphenyl amine.

hexobarbital. (n-methyl-5-cyclohexenyl-5-methylbarbituric acid). CAS: 56-29-1. $C_{12}H_{16}N_2O_3$.
Properties: White crystals, mp 145–147C.
Use: Medicine (sedative).
See also barbiturate.

"Hexogen."[230] TM for a series of paint driers made with odorless solvents, essentially solutions of metallic salts of 2-ethylhexoic acid. Supplied in a variety of high metal concentrations including calcium 4%, calcium 5%, cobalt 6%, lead 24%, manganese 6%, iron 6%, and zinc 8%.

hexoic acid. See caproic acid.

hexokinase. An enzyme which catalyzes the formation of adenosine diphosphate and hexose-6-phosphate from adenosine triphosphate and glucose or fructose.
Use: Biochemical research.

hexone. See methyl isobutyl ketone.

hexyl. (1) The straight-chain group C_6H_{13}. (2) Hexanitrodiphenylamine.

hexyl acetate. CAS: 142-92-7.
$CH_3COOC_6H_{13}$.
Properties: Colorless liquid, sweet ester odor, insoluble in water, very soluble in alcohol or ether, bp 169.2C, d 0.890. Combustible.
Derivation: From primary and sec-hexyl alcohols.
Use: Solvent for cellulose esters and other resins, spray base.

sec-hexyl acetate. See methyl amyl acetate.

hexyl alcohol. (1-hexanol; amyl carbinol).
CAS: 111-27-3. $CH_3(CH_2)_4CH_2OH$.
Properties: Colorless liquid, d 0.8186, fp −51.6C, bp 157.2C, bulk d 6.8 lb/gal (20C), refr index 1.1469 (25C), flash p 149F (65C) (TOC), autoign temperature 559F (292C), slightly soluble in water, soluble in alcohol and ether. Combustible.
Derivation: (a) By reduction of ethyl caproate, (b) from olefins by the Oxo process.
Grade: Technical (90–99%), purified (99.8%).
Use: Pharmaceuticals (introduction of hexyl group

into hyponotics, antiseptics, perfume esters, etc.), solvent, plasticizer, intermediate for textile and leather finishing agents.

n-hexylamine. CAS: 111-26-2.
$CH_3(CH_2)_5NH_2$.
Properties: Water-white liquid, amine odor, boiling range 126–132C, fp −21C, d 0.767 (20/20C), refr index 1.419 (20C), flash p 85F (29.4C) (OC), slightly soluble in water.
Hazard: Toxic by ingestion, inhalation, and skin absorption. Flammable, moderate fire risk.

n-hexyl bromide. (1-bromohexane).
$CH_3(CH_2)_5Br$.
Properties: Colorless to slightly yellow liquid, d 1.165 (20/20C), bp 155.5C, soluble in alcohol, esters, ethers; insoluble in water.
Grade: 96–98% pure.
Use: Intermediate, for introduction of hexyl group.

n-hexyl "Carbitol."[214] TM for diethylene glycol monohexyl ether.

n-hexyl "Cellosolve."[214] TM for ethylene glycol monohexyl ether.

hexyl cinnamaldehyde. CAS: 101-86-0.
$C_6H_{13}C(CHO):CHC_6H_5$.
Properties: Pale yellow liquid; jasmin-like odor, particularly on dilution, d 0.953–0.959 (25C), refr index 1.5480–1.5520 (20C), soluble in most fixed oils and in mineral oil, insoluble in glycerol and in propylene glycol. Combustible.
Grade: FCC.
Use: Flavoring agent.

hexylene. See 1-hexene.

hexylene glycol. (4-methyl-2,4-pentanediol).
CAS: 107-41-5. $(CH_3)_2COHCH_2CHOHCH_3$.
Properties: Colorless, nearly odorless liquid. D 0.9216 (20/4C), bp 198.3C, refr index 1.4276 (20C), flash p 200F (93C) (OC) bulk d 7.69 lb/gal. Miscible with water, hydrocarbons, and fatty acids. Combustible.
Hazard: Toxic by ingestion and inhalation; irritant to skin, eyes, and mucous membranes. TLV: CL 25 ppm in air.
Use: Hydraulic brake fluids, printing inks, coupling agent and penetrant for textiles, fuel and lubricant additive, emulsifying agent, inhibitor of ice formation in carburetors, cosmetics.
See also 1,6-hexanediol.

n-hexyl ether. CAS: 112-58-3.
$C_6H_{13}OC_6H_{13}$.
Properties: Colorless liquid with characteristic odor, very slightly soluble in water, d 0.7942 (20/20C), bulk d 6.6 lb/gal (20C), fp −43.0C, viscosity 1.68 cp (20C), flash p 170F (76.6C),

autoign temperature 369F (187C). Combustible.
Use: Extraction processes, manufacture of collodion, photographic film and smokeless powder.

hexylic acid. See caproic acid.

hexyl mercaptan. $C_6H_{13}SH$.
Properties: Colorless liquid, bp 149–150C (768 mm), d 0.8450 (20/4C), refr index 1.4492 (20C), unpleasant odor. Combustible.
Grade: 95% min purity.
Use: Intermediate, synthetic rubber processing. See also thiol.

hexyl methacrylate. $C_6H_{13}OOCC(CH_3):CH_2$.
Properties: Liquid, d 0.88, bp 67–85C (8 mm). Combustible.
Use: Monomer for plastics, molding powder, solvent coatings, adhesives, oil additives; emulsions for textile, leather, and paper finishing.

p-tert-hexylphenol. $C_6H_{13}C_6H_4OH$.
Properties: Water-white liquid, faint phenol odor, d 0.986 (20/20C), boiling range 155–165C, refr index 1.520 (20C), flash p 285F (140C). Combustible.
Use: Organic synthesis, preparation of resinous condensation products.

hexylresorcinol. (1,3-dihydroxy-4-hexylbenzene).   CAS: 136-77-6.
$C_6H_{13}C_6H_{13}(OH)_2$.

Properties: Yellow, viscous liquid which solidifies on standing, or needle-shaped crystals with a faint fatty odor and astringent taste; slightly soluble in water; freely soluble in alcohol, glycerol, and vegetable oils; mp 62–67C; bp 333C.
Grade: NF.
Hazard: Irritant to respiratory tract and skin, concentrated solutions are vesicant.
Use: Medicine (topical antiseptic).

hexyltrichlorosilane. $C_6H_{13}SiCl_3$.
Properties: Colorless liquid with a sharp penetrating odor, fumes strongly in moist air. Combustible.
Hazard: Toxic by ingestion and inhalation, strong irritant.
Use: Chemical intermediate.

1-hexyne. (butyl acetylene). $C_4H_9C{\equiv}CH$.
Properties: Water-white liquid with characteristic odor, d 0.7152 (20/4C), refr index 1.3990 (20C), bp 71.4C, fp −132C.
Hazard: Probably flammable.

hexynol. (1-hexyn-3-ol).
$CH_3(CH_2)_2CHOHC{\equiv}CH$.
Properties: Light yellow liquid with strong odor; bp 142C; d 0.882 (20/20C); slightly soluble in water and miscible with most hydrocarbons, chlorinated solvents, ketones, alcohols, and glycols.
Hazard: Toxic by ingestion and inhalation, absorbed by skin. Probably flammable.
Use: Corrosion inhibitor against mineral acids, high-temperature oil-well-acidizing inhibitor.

Heyrovsky, Jaroslav. (1890-1967) A Czechoslovakian physiochemist who won the Nobel prize for chemistry in 1959. He is known for work in electrochemistry. He developed polarographic and oscillo-polarographic methods. Although his PhD was from the University of Prague, he later studied in London.

Hf. Symbol for hafnium.

HFPO. See hexafluoropropylene epoxide.

Hg. Symbol for mercury (Latin: hydrargyrum).

HFG. See glucagon.

"HGR."[233] TM for an ion exchnage resin used in water treating and chemical process applications, strong acid cation exchange resin, 10% divinylbenzene crosslinked.

HHDN. Abbreviation for hexachlorohexahydrodimethanonaphthalene.
See aldrin.

hiding power. See opacity.

"Hi-Eff."[425] TM for a series of polyester stationary phases for use in gas chromatography. Examples are "Hi-Eff" 4B, butane-1,4-diol succinate; "Hi-Eff" 1AP, a pretested diethylene glycol adipate.

hi-flash naphtha. See naphtha (2a).

high vacuum distillation. See molecular distillation.

Hilbert-Johnson reaction. Reaction of 2,4-dialkoxypyrimidines with halogenoses to yield pyrimidine nucleosides.

Hildebrand, Joel. (1891–1983) One of the most distinguished American chemists and teachers. Born in New Jersey, be obtained his doctorate in chemistry and physics from the University of Pennsylvania. After studying abroad under

Nernst and van't Hoff, he became Professor of Chemistry at the University of California, Berkeley, in 1913 where he remained until retirement. He made many important contributions to physical chemistry, particularly in the area of nonelectrolyte solutions; his treatise on this subject is a recognized classic and his textbook on the principles of chemistry established a new standard of excellence. He also made important contributions to the thermodynamics of vaporization of liquids. He proposed the use of helium in deep-sea diving equipment, which has become accepted practice. A gifted teacher and lecturer, he continued his constructive research to the end of his long life. He was unusually active in outdoor sports such as swimming, skiing, and hiking. Among his numerous awards were the Nichols and William Gibbs Medals and the Priestley Medal.

hindered isocyanate. See isocyanate generator.

hindrance. See steric hindrance.

Hinsberg oxindole synthesis. Formation of oxindoles from secondary aryl amines and sodium bisulfite addition compound of glyoxal; primary aryl amines give glycine or glycinamide derivatives.

Hinsberg reaction. Reaction of primary and secondary amines with sulfonyl halides to give sulfonamides; because the products from primary amines are soluble in alkali and those from secondary amines are not, and since tertiary amines do not react, this method is useful for the separation and identification of amines.

Hinsberg sulfone synthesis. Formation of sulfonylquinol derivatives by addition of quinones to cold dilute aqueous solutions of sulfinic acids.

Hinsberg synthesis of thiophene derivatives. Formation of thiophene carboxylic acids from α-diketones and dialkyl thiodiacetate.

Hinshelwood, Sir Cyril N. (1897-1968) An English chemist who won the Nobel prize for chemistry in 1956 along with Semenov, a Russian. He authored "The Kinetics of Chemical Change," "The Structure of Physical Chemistry," and many other journal articles. His work clarified inorganic and organic reactions. His education was at Oxford before he began lecturing and research.

"Hippuran."[329] TM for brand of iodohippurate sodium, a water-soluble x-ray contrast medium.

hippuric acid. (benzaminoacetic acid; benzoylaminoacetic acid; benzoylglycocoll; benzoylglycin). CAS: 495-69-2. $C_6H_5CONHCH_2COOH$.
Properties: Colorless crystals; d 1.371 (20C); mp 188C; decomposes on further heating; soluble in hot water, alcohol, and ether.
Use: Organic synthesis and medicine.

His. Abbreviation for histidine.

"Hi-Sil."[177] TM for a group of hydrated, amorphous silicas used as reinforcing pigments in elastomers as fillers and brightening agents in paper and paints and as flow-conditioners.

histaminase. An enzyme occurring in the animal digestive system, it converts histidine to histamine.

histamine. (4-aminoethylglyoxaline; 4-(2-aminoethyl)imidazole; 4-imidazole ethylamine). CAS: 51-45-6. $NH_2CH_2CH_2C_3H_3N_2$.

$$HC = CCH_2CH_2NH_2$$

Properties: White crystals, mp 83–84C, bp 209–210C (18 mm), soluble in water, slightly soluble in alcohol. A product of the degradation of histidine, histamine occurs in animal and human body tissues and is liberated by injury to the tissue or whenever a protein is decomposed by putrefactive bacteria.
Use: (as hydrochloride or phosphate) medicine (diagnostic aid).
See also antihistamine.

histidine. (α-amino-β-imidazolepropionic acid). CAS: 71-00-1. $HOOCCH(NH_2)CH_2C_3H_3N_2$.
An amino acid essential for rats. It is found naturally in the L(-) form.

$$H-C = CCH_2CH(NH_2)COOH$$

Properties: Colorless crystals, soluble in water, insoluble in alcohol and ether, shows optical activity. DL-histidine, mp 285–6C with decomposition; D(+)-histidine, mp 287–8C; L(-)-histidine, mp 277C with decomposition. Available commercially as L(+)-histidine hydrochloride and as the free base.
Derivation: From blood corpuscles, organic synthesis.

Use: Medicine, feed additive, biochemical resarch, dietary supplement.

histochemistry. A branch of biochemistry devoted to the study of the chemical composition and structure of animal and plant tissues. It involves the use of microscopic, x-ray diffraction and radioactive tracer techniques in examining the cellular composition and structure of bones, blood, muscle, and other animal and vegetable tissues. It is also applied to study of the action of herbicides, defoliants, etc.
See also cytochemistry.

histone. Any of several proteins of low molecular weight (12,000–20,000) occurring in cell nuclei attached to DNA by ionic bonds. They contain varying amounts of the amino acids lysine, arginine, cysteine and glycine. They have been classified into four basic types and can be obtained from both plant and animal sources.

history, chemistry. See Appendix II A-E.

"Hitec."[28] TM for a eutectic mixture comprised of sodium nitrite, sodium nitrate, and potassium nitrate.
Use: Heat transfer medium for both heating and cooling operations in the range of 149–537C, such as maintaining reactor temperature, high temperature distillation, and preheating of reactants.

HMDS. Abbreviation for hexamethyldisilazane.

HMF black. Abbreviation for high-modulus furnace black.
See carbon black.

HMM. Abbreviation for hexamethylmelamine.
See hemel.

HMPA. Abbreviation for hexamethylphosphoramide.
See hempa.

HMTA. Abbreviation for hexamethylenetetramine.

HNM. Abbreviation for hexanitromannite.
See mannitol hexanitrate.

Ho. Symbol for holmium.

hob. A hardened steel master die used to make multiple mold cavities by forcing it into soft steel or beryllium-copper blanks.

Hoch-Campbell aziridine synthesis. Formation of aziridines by treatment of ketoximes with Grig-

nard reagents and subsequent hydrolysis of the organometallic complex.

"Hodag."[512] TM for a series of nonionic, surface-active chemicals, including sorbitan and polyoxyethylene sorbitan esters; polyhydric alcohol, glycol, polyglycol and glycerol esters; polyoxyethylene ethers, alkanolamine condensates. Some are FDA-cleared for food applications.
Use: Emulsifiers, antifoam agents, solubilizers, lubricants.

Hodgkin, Dorothy C. (1910-) An Egyptian born chemist who was recipient of the Nobel prize for chemistry in 1964. Her work involved determining the structure of vitamin B_{12}, cholesterol iodide, and the antibiotic penicillin by using x-ray crystallographic analysis. She was educated at Oxford and Cambridge.

Hofmann, August Whilhelm. (1818–1892) A German organic chemist who studied under Liebig. While professor of chemistry at the Royal College of Chemistry in London, he did original research on coal-tar derivatives which later led him into a study of organic dyes. Perkin, who first synthesized the dye mauveine in England, was a student of Hofmann. When the latter returned to Germany he continued his work in the field of dyes, which became the basis of German leadership in synthetic dye manufacture which continued until World War I.

Hofmann degradation. Formation of an olefin and a tertiary amine by pyrolysis of a quaternary ammonium hydroxide; useful for the preparation of some cyclic olefins and for opening nitrogen-containing ring compounds.

Hofmann isonitrile synthesis. Formation of isonitriles by the reaction of primary amines with chloroform in the presence of an alkali; the odor of the isocyanide is a test for a primary amine.

Hofmann-Loffler-Freytag reaction. Formation of pyrrolidines or piperidines by thermal or photochemical decomposition of protonated N-halo-amines.

Hofmann-Martius rearrangement. Thermal conversion of N-alkylaniline hydrohalides to o- and p-alkylanilines.

Hofmann's reaction. Reaction used for preparation of a primary amine from an amide by treatment with a halogen (bromine, usually) and caustic soda. The resulting amine has one less carbon atom than the amide used.

Hofmann rule. When a quaternary ammonium hydroxide containing different primary alkyl radicals is decomposed, the least substituted olefin is formed preferentially.

Hofmann-Sand reaction. Olefin mercuration with mercuric salts (halides, acetates, nitrates, or sulfates) in aqueous solution. In alcoholic solutions, the accelerated reaction produces alkoxylakyl compounds.

Hofmann's Violet. (triethylrosaniline hydrochloride; CI42530). $C_{26}H_{32}N_3HCl$. Water-soluble green powder.
Use: Dye for inks and textiles, biological stain.

Hoffman, Roald. (1937-) A Polish born chemist who won the Nobel prize for chemistry with Fukui in 1981. His work involved applying the theories of quantum mechanics to predict the course of chemical reactions.

hog. A large enclosed chamber equipped either with rotating knives or heavy hammers in which wood is disintegrated to a uniform degree of fineness.

hole. In semiconductor terminology, a hole is an energy deficit in a crystalline lattice due to (1) electrons ejected from unsatisfied covalent bonds at sites where an atom is missing, i.e., a vacancy, or to (2) electrons supplied by atoms of impurities in the crystal, e.g., arsenic or boron. The free electrons from these sources move through the crystal, leaving positively charged energy deficits which are considered to move as they become alternately filled and vacated by electrons, creating a flow of positive electricity.
See also semiconductor.

holmium. Ho. Metallic element of atomic number 67, group IIIB of the periodic table, one of the rare-earth elements of the yttrium subgroup, atomic weight 164.9303, valence = 3, no stable isotopes.
See rare-earth metals.
Properties: Crystalline solid with metallic luster, d 8.803, mp 1470C, bp 2720C, reacts slowly with water, soluble in dilute acids, has one of the highest nuclear moments of any rare earth, important magnetic and electrical properties.
Occurrence: In gadolinite and monazite.
Derivation: Reduction of the fluoride by calcium.
Grade: Lumps, ingots, bulk sponge, powder. Highest purity is nuclear grade 99.9+%.
Use: Getter in vacuum tubes, research in electrochemistry, spectroscopy.

holmium chloride. $HoCl_3$.
Properties: Bright yellow solid, mp 718C, bp 1500C, soluble in water.

holmium fluoride. HoF_3.
Properties: Bright yellow solid, mp 1143C, bp above 2200C, insoluble in water.

holmium oxide. (holmia). Ho_2O_3.
Properties: Light yellow solid, slightly hygroscopic, soluble in inorganic acids.
Grade: 98–99%.
Use: Refractories, special catalyst.

holocellulose. The entire water-insoluble carbohydrate fraction of wood (60–80%). It is composed of alpha-cellulose and hemicellulose; it contains hexosan and pentosan polymers and varies widely in degree of crystallinity.

holography. Production of a unique 3-dimensional image on photographic film by means of an interference pattern created by a laser beam that is split by a mirror-like device. The beam is divided in such a way that one portion is reflected from the subject while the other forms a direct image; the resulting superimposition results in an unusual 3-D effect.
Use: The technique has a number of practical applications in airplane flight control, missile guidance systems; other uses will undoubtedly develop.

holopulping. A method for making paper pulp without use of sulfur compounds which may eventually replace the kraft process (sodium sulfate). More selective delignification of the wood fibers is obtained by alkaline oxidation of extremely thin (0.03 inches) wood chips at low temperature and pressures, followed by solubilization of the lignin fraction. Holopulping has other advantages over the kraft process: (a) A 65–80% carbohydrate yield compared to 45–50% for kraft. (b) Holopulp may be used for a dense paper such as glassine or for a bulky board. Its use in tissue and printing grades offers improved strength. (c) Low temperatures and atmospheric pressure. (Kraft pulping is carried out at 170C and under pressure.) Readily adaptable to continuous operation and automatic control. (d) Air pollution is greatly reduced because the organic materials are burned and few odorous compounds are formed. Stream pollution is minimized by countercurrent washing. The remaining calcium sludge waste is easily disposed of without harmful effects.

holothurin. Steroid glucoside (saponin) having antibiotic properties, extracted from the sea cucumber. It is reported to have suppressed growth of tumors in mice.

homatropine. $C_{16}H_{21}NO_3$. An alkaloid.
Properties: White crystals, slightly soluble in water, mp 95.5C.

Derivation: Condensation of tropine and mandelic acid.

Hazard: Toxic by ingestion and inhalation.

Use: Medicine (usually in the form of its salts).

homo-. A prefix meaning the same or similar; usually designating a homolog of a compound, differing in formula from the latter by an increase of CH_2.

See homologous series.

homocyclic. A ring compound containing only one kind of atom in the ring structure, e.g., benzene.

See also heterocyclic.

homogeneous. (Latin, "the same kind") This term, in its strict sense, describes the chemical constitution of a compound or element. A compound is homogeneous since it is composed of one and only one group of atoms represented by a formula. For example, pure water is homogeneous as it contains no other substance than is indicated by its formula, H_2O. Homogeneity is a characteristic property of compounds and elements (collectively called substances) as opposed to mixtures. The term is often loosely used to describe a mixture or solution comprised of two or more compounds or elements that are uniformly dispersed in each other. Actually, no solution or mixture can be homogeneous; the situation is more accurately described by the phrase "uniformly dispersed." Thus so-called "homogenized" milk is not truly homogeneous; it is a mixture in which the fat particles have been mechanically reduced to a size that permits uniform dispersion and consequent stability.

See also mixture; compound; heterogeneous; substance.

homogeneous catalysis. See catalysis, homogeneous.

homogeneous reaction. A chemical reaction in which the reacting substances are in the same phase of matter, i.e., solid, liquid, or gaseous.

See also catalysis, homogeneous.

homogenization. A mechanical process for reducing the size of the fat particles of an emulsion (usually milk) to uniform size, thus creating a colloidal system that is unaffected by gravity. The original diameter of the fat particles (6–10 microns) is reduced to 1–2 microns, with an increase in total surface area of 4–6 times. This is done by passing the milk through a homogenizer (or colloid mill), a machine having small channels, under a pressure of 2000–2500 psi at a speed of approximately 700 ft/sec. This operation not only brings about a permanently stable system, but also changes the properties of the milk in respect to taste, color, and the chemical nature of the protective coating on the fat particles. It also increases its sensitivity to light and its tendency to foam. The forces involved are shear, impingement, distention, and cavitation.

See also homogeneous, colloid mill.

homologous series. A series of organic compounds in which each successive member has one more CH_2 group in its molecule than the preceding member. For instance CH_3OH (methanol), C_2H_5OH (ethanol), C_3H_7OH (propanol), etc., form a homologous series.

homomenthyl salicylate. (3,3,5-trimethylcyclohexyl salicylate). $(CH_3)_3C_6H_8OOCC_6H_4OH$. A homolog of menthyl salicylate.

Properties: Light yellow almost odorless oil, neutral and nonirritating to the skin. Absorbs UV radiation in sunlight (2940–3200Å). Insoluble in water; soluble in alcohol, chloroform, and ether.

Use: UV filter for antisunburn creams.

homomorphs. Molecules similar in size and shape. They need have no other characteristics in common. Many properties of several homomorphs can be predicted by knowing properties of one.

homophthalic acid. $C_6H_4(CH_2COOH)COOH$.

Properties: Light tan powder.

Use: Intermediate.

homopolymer. A natural or synthetic high polymer derived from a single monomer. An example of a natural homopolymer is rubber hydrocarbon, whose monomer is isoprene; a synthetic homopolymer is typified by polychloroprene or polystyrene whose monomers are, respectively, chloroprene and styrene.

See also polyblend.

homosalate. $C_{16}H_{22}O_3$.

Properties: Liquid, bp 162C (4 mm), d 1.05, refr index 1.51.

Use: Sunscreening agent.

o-homosalicylic acid. See cresotic acid.

homoveratric acid. (3,4-dimethoxyphenylacetic acid). $(CH_3O)_2C_6H_3CH_2COOH$.

Properties: Crystals, mp 94–101C, very slightly soluble in water, soluble in most organic solvents.

homoveratrylamine. (3,4-dimethoxyphenylethylamine). $(CH_3O)_2C_6H_3(CH_2)_2NH_2$.

Properties: Colorless to pale yellow liquid with slight vanilla odor, d 1.09 (25/25C), solidifies

15C, bp 295C (decomposes), refr index 1.5442–1.5452 (25C).

honey. A unique mixture of a number of low molecular weight sugars (except sucrose) but including invert sugar. It is considerably sweeter than glucose.
Use: A food and sweetener since the beginning of civilization, also has applications in medicine and tobacco processing.

Hooke's law. When a load is applied to any elastic body so that the body is deformed or strained, then the resulting stress (the tendency of the body to resume its normal condition) is proportional to the strain. Stress is measured in units of force per unit area, strain is the extent of the deformation. For example, when a bar of metal is subjected to a stretching load, the extent of the increase in length of the bar is directly proportional to the force per unit area, i.e., to the stretching load or stress. In general Hooke's law applies only up to a certain stress called the yield strength.

Hooker reaction. Oxidation of 2-hydroxy-3-alkyl-1,4-quinones with dilute alkaline permanganate with shortening of the alkyl side chain by a methylene group and simultaneous exchange of hydroxyl and alkyl or alkenyl group positions.

hopcalite. A mixture of oxides of copper, cobalt, manganese, and silver.
Use: Gas masks as a catalyst converting carbon monoxide to carbon dioxide.
Hazard: Not safe for a respirator when nitroparaffin vapors are present.

hops. A plant of the genus *Humulus* widely grown in temperate climates.
Use: Beer making,
See also brewing.

horizon. See soil.

"Hormodin."[123] TM for a formulation of indolebutyric acid.

hormone, animal. An organic compound secreted by an endocrine (ductless) gland, whose products are released into the circulating fluid. Hormones regulate such physiological processes as metabolism, growth, reproduction, molting, pigmentation, and osmotic balance. They are sometimes called "chemical messengers." Hormones produced by one species usually show similar action in other species. They vary widely in chemical nature, some are steroids--estrogen, progesterone, cortisone--while others are amino acids (thyroxine), polypeptides (vasopressin), low molecular weight proteins (insulin), and conjugated proteins. Amino acid and steroid hormones have been isolated and many (including insulin) have been synthesized and are manufactured for medical purposes. Other types are made directly from the endocrine organs of animals.

hormone, plant. See plant growth regulator.

"Horne's Dry Lead."[50] TM for a laboratory chemical used for clarification in sugar analysis. Adopted by the Association of Official Agricultural Chemists. Available in 1-lb, 5-lb and 10-lb bottles and 25-lb cartons.

"Hornstone."[205] TM for zinc fluosilicate concrete hardener.

hot. Slang for highly radioactive, e.g., hot laboratory.

hot-melt composition. See adhesive, hot-melt; sealant, asphalt, bitumen.

hot-short. See short.

Houben-Fischer synthesis. Formation of aromatic nitriles by basic hydrolysis of trichloromethyl aryl ketimines obtained by Hoesch synthesis (but not from dichloro- or monochloromethyl aryl ketimines). Acidic hydrolysis yields ketones.

Houben-Hoesch reaction. Synthesis of acylphenols from phenols or phenolic ethers by the action of organic nitriles in the presence of hydrochloric acid and aluminum chloride as catalyst.

Houdriflow process. A moving-bed type of catalytic cracking in which the catalyst pellets move downward through a reactor concurrently with the feed. The catalyst is then separated and regenerated for further use.
See also Thermofor process.

Houdry cracking process. Decomposition of petroleum or heavy petroleum fractions into more useful lower boiling materials by heating at 500C and 30 psi over a silica-alumina-manganese oxide catalyst.

HPA. Abbreviation for hydroxypropyl acrylate.

HPC black. Abbreviation for hard-processing channel black.
See carbon black.

HPLC. Abbreviation of high-performance liquid chromatography.

HS. Abbreviation for hydroxylamine sulfate.

"HT-44."[212] TM for an extremely heat-stable liquefying enzyme (an amylase) from a bacterial source high in α-amylase activity.
Use: Textiles, starch adhesives, paper, brewing, industrial grain alcohol.

"HTH."[84] TM for a high-test calcium hypochlorite product commercially available as a stable, water-soluble material in both granular and tablet form, containing a minimum of 70% available chlorine as calcium hypochlorite.
Use: Bleaching, sterilizing, oxidizing.

"HTH-15."[84] TM for an all-purpose germicide, disinfectant and stain remover. Contains 15% of available chlorine and yields sodium hypochlorite solutions directly when added to water.
Use: Dairy and poultry farm sanitation, for sterilizing glasses and food utensils and general sanitation.

HTST. Abbreviation for high-temperature short-time, refers to processes such as pasteurization, sterilization, etc.

HTU. Abbreviation for height of a transfer unit: the height of a distillation column or fractionating tower in which unit separation is achieved by transfer from liquid to vapor or vice versa, of the materials being separated. Unit separation is defined by the differential equation that takes into account the varying concentrations along the column. HTU is also applied to extraction and other countercurrent separation processes.

Huber's reagent. An aqueous solution of ammonium molybdate and potassium ferrocyanide used for detecting free mineral acid. With the exception of boric acid and arsenic trioxide, free mineral acids produce a reddish-brown precipitate, or a turbidity with the reagent.

Hubl's reagent. (a) 50 g iodine dissolved in one L of 95% alcohol. (b) 60 g mercuric chloride dissolved in one L of alcohol. (c) Make up an iodine monochloride solution from (a) and (b). Add an excess to a known weight of the fat or oil dissolved in chloroform. The excess of iodine chloride can be estimated by the potassium iodide and thiosulfate method. By running a blank test, the amount of iodine absorbed can be estimated.
Use: Determination of iodine values of oils and fats.

Hudson isorotation rules. For anomeric (alpha and beta) sugars, Hudson's isorotation rule states that (a) the rotation of carbon 1 in many sugar derivatives is affected in only a minor degree by changes in the structure of the rest of the molecule and (b) changes in the structure of carbon 1 affect in only a minor degree the rotation of the remainder of the molecule. Another way of stating the rule is to say that the rotation of any aldose derivative is the algebraic sum of A and B where A is the contribution of the anomeric center and B is the contribution of the rest of the molecule.

Hudson lactone rule. The value of the rotation of aldonic acid lactones is decisively affected by the configuration of that carbon atom whose hydroxyl group is engaged in the cyclization. If, in the normal Fischer projection formula, the lactone ring is written on the right, the lactone is dextrorotatory; if it is written on the left, the lactone is levoratatory.

humectant. A substance having affinity for water with stabilizing action on the water content of a material. A humectant keeps within a narrow range the moisture content caused by humidity fluctuations. Example, glycerol.
Use: Tobacco, baked products, dentifrices.

humic acid. A brown, polymeric constituent of soils, lignite and peat; it contains the brownish-black pigment melanin. It is soluble in bases, but insoluble in mineral acids and alcohols. It is not a well-defined compound, but a mixture of polymers containing aromatic and heterocyclic structures, carboxyl groups and nitrogen. An excellent chelating agent, important in the exchange of cations in soils. It is a natural stream pollutant and is thought to be capable of triggering the "red tide" phenomenon due to microorganisms in seawater. Detectable to 0.1 ppm in water.
Use: Drilling fluids, printing inks, plant growth.

humidity, absolute. The pounds of water vapor per pound of dry air in an air-water vapor mixture.

humidity indicator. A cobalt salt (e.g., cobaltous chloride) that changes color as the humidity of the environment changes. Cobaltous compounds are pink when hydrated and greenish-blue when anhydrous.

humidity, relative. The percentage relation between the actual amount of water vapor in a given volume of air at a definite temperature and the maximum amount of water vapor that would be present if the air were saturated with water vapor at that temperature.

"Humulin."[100] TM for synthetic insulin. It is the first recombinant DNA product to be made commercially; approved by FDA.

humus. The organic component of soils containing humic acid, fulvic acid, and humin. It is formed by the decay of leaves, wood, and other vegetable matter.
Use: Top soil additive in horticulture, golf courses, and truck gardens; available in 100-lb paper bags and bulk shipments.
See soil (1).

Hunsdiecker reaction. Synthesis of organic halides by thermal decarboxylation of silver salts of carboxylic acids in the presence of halogens.

Hunter process. Production of titanium by reduction with sodium in an atmosphere of argon or helium.

hyaluronic acid. A polymer of acetylglycosamine, $C_8H_{15}NO_6$, and glucuronic acid occurring as alternate units with a high molecular weight. Found in vitreous humor (of the eye), synovial fluid, pathologic joints, group A and C hemolytic streptococci, and skin. It appears to bind water in the interstitial spaces, forming a gel-like substance which holds the cells together. Its solutions are highly viscous. The polymeric structure is broken down by the enzyme hyaluronidase.

"Hyamine."[23] TM for quaternary-ammonium-type bactericides, algicides, and fungicides, supplied as water-soluble crystals or aqueous solutions.

Hyatt, John Wesley. (1837–1920) Hyatt is generally credited as being the father of the plastics industry. In 1869, he and his brother patented a mixture of cellulose nitrate and camphor which could be molded and hardened. Its first commercial use was for billard balls. The TM "Celluloid" was the first ever applied to a synthetic plastic product; its flammability hazard limits its use.

"Hybase."[544] TM for overbased petroleum sulfonates used as additives in motor oils to reduce corrosion; overbased sulfonates have base numbers ranging from 65–400.

"Hycar."[119] TM for various types of synthetic elastomers. Nitrile Rubbers: Copolymers of butadiene and acrylonitrile. Polyacrylic Rubbers: Polymers of an acrylic acid ester. Useful in applications where oil and solvent-resistance at high temperatures (to 218C) is required.
Use: Oil and gasoline hose, packings, automotive transmission gaskets and conveyor belts. Styrene Rubber: Copolymer of styrene and butadiene; oil solution. Latexes: Nitrile and acrylic polymers, available as water emulsions with total solids ranging from 40–52%. Use: Adhesives, abrasion-resistant coatings.

"Hydan."[28] TM for methionine hydroxy analogue *ca* 90%.
Use: A source of methionine (an essential amino acid) for poultry, dog, and livestock feeds.

hydantoin. (glycolylurea). CAS: 461-72-3. $NHCONHCOCH_2$.
Properties: White, odorless solid; crystallizing in needles; slightly soluble in water, ether; soluble in alcohols and solutions of alkali hydroxides; mp 220C.
Use: Intermediate in the synthesis of pharmaceuticals, textile lubricants, and certain high polymers, including epoxy resins.

hydnocarpic acid. CAS: 459-67-6. $C_{16}H_{28}O_2$.
A component of chaulmoogra oil.

hydrabamine pencillin V. (hydrabamine phenoxymethylpenicillin).
Properties: A water-insoluble mixture of crystalline phenoxymethylpenicillin salts consisting chiefly of the salt of N,N'-bis(dehydroabietyl)-ethylenediamine with smaller amounts of the salts of the dihydro and tetrahydro derivatives.
Use: Medicine (antibacterial).

"Hydraid."[108] TM for a family of water soluble organic polymers, some cationic, anionic and nonionic of various molecular weights and coagulation properties.
Use: Paper and pulp mill retention, drainage and clarification aids.

hydralazine hydrochloride. (1-hydrazinophthalazine hydrochloride). $C_8H_5N_2NHNH_2HCl$.
Properties: White, odorless, crystalline powder. Mp 270–280C (decomposes), very slightly soluble in ether and alcohol, soluble in water, pH (2% solution) 3.5–4.5.
Grade: NF.
Use: Medicine (antihypertensive agent)

"Hydral"700 Series.[226] TM for several grades of hydrated aluminum oxides. $Al_2O_3 \cdot 3H_2O$ or $Al(OH)_3$, of extremely fine, uniform particle size.
Properties: Fluffy, snow-white powders.
Use: As fillers in rubber, paper, plastics, adhesives, polishes, inks, paints, cosmetics.

"Hydraphthal."[28] TM for a combination solvent and detergent for textile scouring.

hydrargaphen. (phenylmercury methylenedinaphthalenesulfonate). CAS: 14235-86-0. $C_{33}H_{24}Hg_2O_6S_2$.
Properties: Extremely fine powder, insoluble in water, forms colloidal dispersons with strong adsorptive power in sodium or potassium dinaphthylmethane disulfonates.

Hazard: A poison.

Use: Biocide for protection of wool, leather, paints, and wood products.

hydrase. See hydrolase.

hydrate. See hydration.

hydrated aluminum oxide. See alumina trihydrate.

hydrated silica. See silicic acid.

hydration. (1) The reaction of molecules of water with a substance in which the H—OH bond is not split. The products of hydration are called hydrates, i.e.,

$$CuSO_4 + 5HOH \rightarrow CuSO_4 \cdot 5HOH$$

A given compound often forms more than one hydrate; the hydration of sodium sulfate can give $Na_2SO_4 \cdot 10HOH$ (decahydrate), $Na_2SO_4 \cdot 7HOH$ (heptahydrate), and $Na_2SO_4 \cdot HOH$ (monohydrate). In formulas of hydrates, the addition of the water molecules is conventionally indicated by a centered dot. The water is usually split off by heat, yielding the anhydrous compound.
See also water of crystallization, gas hydrate.

(2) The strong affinity of water molecules for particles of dissolved or suspended substances that is the fundamental cause of electrolytic dissociation. Ions and other charged particles thus acquire a tightly held film of water, an effect that is important in the stabilization of colloidal solutions. The phenomenon is also called solvation. The term hydration is used in the paper industry to describe the combination of water with wood pulp in the beater, as a result of which fiber-to-fiber adhesion is increased by hydrogen bonding.
See also solvation.

hydraulic. (1) Descriptive of a machine or operation in which a liquid is used to exert or transfer pressure, e.g., hydraulic press, hydraulic fracturing. The liquid is usually water, but it may also be of higher viscosity such as a heavy oil or glycol-type lubricant, as in brake fluid. (2) Descriptive of a material that hardens on addition of water, e.g., hydraulic cement.
See also following entries.

hydraulic barking. Removal of bark from logs by impingement of a stream of water delivered from one or more nozzles at a pressure of 1200–1400 psi. Several types of machines are used, the best known being the Hansel barker.

hydraulic cement. See cement, hydraulic.

hydraulic fluid. A liquid or mixture of liquids designed to transfer pressure from one point to another in a system on the basis of Pascal's Law, i.e., pressure on a confined liquid is transmitted equally in all directions. For industrial use, such fluids are based on paraffinic and cycloparaffinic petroleum fractions, usually with added antioxidant and viscosity index improvers. Flame-resistant types include additives such as phosphate esters or emulsions of water and ethylene glycol. The brake fluids used in autos are composed of (1) a lubricant (polypropylene glycol of 1000–2000 mw, a castor oil derivative, or a synthetic polymeric mixture of monobutyl ethers of oxyethylene and oxypropylene glycols); (2) a solvent blend (mixture of glycol ethers); and (3) additives for corrosive resistance, buffering, etc.; bp 375–550F. The composition and performance characteristics are specified by the Society of Automotive Engineers.

hydraulic fracturing. A method of enhanced recovery of natural gas and petroleum. An aqueous solution of a water-soluble gum (e.g., guar), in which coarse sand or sintered bauxite is suspended, is introduced through a well bore under extremely high pressure into the rock structure in which the gas or oil is entrained. This creates minute fissures (fractures) in the rock which are held open by the suspended particles after the liquid has drained off. The hydrocarbon flows through these fissures to the well bore, and is evacuated to a pipeline. The sand and bauxite are called "proppants" by petroleum engineers as they prevent the fissures from closing. Sand is used in shallower wells and bauxite in formations over 10,000 ft deep.
See also chemical flooding.

hydraulic lime. See lime, hydraulic.

hydraulic press. A simple machine (the only one discovered since prehistoric times) that operates on Pascal's principle (1650): pressure applied to a unit area of a confined liquid is transmitted equally in all directions throughout the liquid. A hydraulic press is comprised of a large piston in an enclosed chamber; its top is attached to a platen that rests on the members of a metal frame when the press is open. Water (or oil) is pumped into the chamber through a valve; once it has been filled, whatever pressure per square inch is applied at the valve will be transmitted to *every* square inch of the piston and of the walls of the chamber as well. Thus, for a piston whose cross-sectional area is 100 sq in., 10 psi at the valve will exert 1000 lb pressure on the bottom of the piston, causing it to rise and the press to close. The pressure on the object being pressed varies inversely with its area. Hydraulic presses exerting pressures up to 15 *tons* are used for shaping steel products. Less dramatic are those for molding rubber and plastics, compressing laminates, dewatering solids, and expressing vegetable oils. Some have up to a dozen platens (decks) for multiple-product work. The same principle is used to activate plungers on injection-molding presses.

hydrazine. (hydrazine base; hydrazine, anhydrous; diamine). CAS: 302-01-2. H_2NNH_2.
Properties: Colorless, fuming, hygroscopic liquid; mp 2.0C, bp 113.5C, mp 1.4C, d 1.004 (25/4C), bulk density 8.38 lbs/gal, flash p 126F (52.2C) (OC), miscible with water and alcohol, insoluble in chloroform and ether, strong reducing agent and diacidic but weak base, autoign temperature 518F (270C). Combustion of hydrazine is highly exothermic, yielding 148.6 kcal/mole, nitrogen and water are products.
Derivation: The preferred method is a 2-step process: (1) reaction of sodium hypochlorite and ammonia to yield chloramine (NH_2Cl) and sodium hydroxide; (2) reaction of chloramine, ammonia, and sodium hydroxide to yield hydrazine, sodium chloride, and water. Noteworthy is the need to carry out the reactions in the presence of such colloidal materials as gelatin, glue, or starch to prevent unwanted side reactions that would reduce the yield of hydrazine. An older method utilized the reaction of sodium hypochlorite or calcium hypochlorite with urea.
Grade: to 99% pure.
Hazard: Severe explosion hazard when exposed to heat or by reaction with oxidizers. Toxic by ingestion, inhalation, and skin absorption; strong irritant to skin and eyes; a carcinogen (OSHA). TLV: 0.1 ppm in air.
Use: Reducing agent for many transition metals and some nonmetals (arsenic, selenium, tellurium), as well as uranium and plutonium; corrosion inhibitor in boiler feedwater and reactor cooling water; wastewater treatment; electrolytic plating of metals on glass and plastics; nuclear fuel reprocessing; redox reactions; polymerization catalyst; shortstopping agent; fuel cells; blowing agent; scavenger for gases; drugs and agricultural chemicals (maleic hydrazide); component of high-energy fuels; rocket propellent.

hydrazine acid tartrate. (hydrazine tartrate). $N_2H_4 \cdot C_4H_6O_6$.
Properties: Colorless crystals, mp 182–183C, soluble in water.
Hazard: See hydrazine.

hydrazine dihydrochloride. CAS: 5341-61-7. $N_2H_4 \cdot 2HCl$.
Properties: Colorless crystals, d 1.42, mp 198C (loses hydrogen chloride), bp 200C (decom-

poses), soluble in water, slightly soluble in alcohol.

hydrazine hydrate. (diamide hydrate). CAS: 7803-57-8. $H_2NNH_2 \cdot HOH$.
Properties: Colorless fuming liquid, fp −51.7C, bp 119.4C, d 1.032, bulk d 8.61 lb/gal, flash p 163F (OC), miscible with water and alcohol, insoluble in chloroform and ether, strong reducing agent, weak base. Combustible.
Hazard: See hydrazine.
Use: Chemical intermediate, catalyst, solvent for inorganic materials.

hydrazine monobromide. $N_2H_4 \cdot HBr$.
Properties: White, crystalline flakes. Mp 81–87C, decomposes at approximately 190C, soluble in water and lower alcohols, insoluble in most organic solvents.
Grade: 95%.
Use: Soldering flux.

hydrazine monochloride. $N_2H_4 \cdot HCl$.
Properties: White, crystalline flakes. Mp 87–92C, decomposes at approximately 240C, soluble in water (37g/100g HOH at 20C), somewhat soluble in lower alcohols, insoluble in most organic solvents.

hydrazine nitrate. $N_2H_4NO_3$.
Hazard: Severe explosion risk. Poison.

hydrazine perchlorate.
$N_2H_4 \cdot HClO_4 \cdot 1/2HOH$.
Properties: Solid; d 1.939; mp 137C; bp 145C; decomposes in water; soluble in alcohol; insoluble in ether, benzene, chloroform, and carbon disulfide.
Hazard: Severe explosion risk.
Use: Rocket propellant.

hydrazine sulfate. (diamine sulfate; diamidogen sulfate). CAS: 10034-93-2. $NH_2NH_2 \cdot H_2SO_4$.
Properties: White, crystalline powder. Very soluble in hot water, soluble at 1 part in 33 cold water, insoluble in alcohol, stable in storage, strong reducing agent, d 1.37, mp 85C.
Hazard: A carcinogen (OSHA).
Use: Manufacture of chemicals, condensation reactions, catalyst in making acetate fibers. Analysis of minerals, slags, and fluxes; determination of arsenic in metals; separation of polonium from tellurium; fungicide, germicide.

hydrazine tartrate. CAS: 634-62-8. $C_4H_{10}N_2O_6$.
Properties: Colorless crystals, mp 182C.
Use: Deposition of metals as in silvering mirrors.

hydrazobenzene. (1,2-diphenylhydrazine).
CAS: 12-2-66-7. $C_6H_5NHNHC_6H_5$.
Properties: Mw 184.24, mp 123–126C, d 1.158(16/4C).

hydrazoic acid. (hydrogen azide).
CAS: 7782-79-8. HN_3.
Properties: Colorless, volatile liquid. Soluble in water, obnoxious odor, fp −80C, bp 37C.
Derivation: Reaction of hydrazine and nitrous acid, or of nitrous oxide and sodium amide (with heat).
Hazard: Dangerous explosion risk when shocked or heated. Strong irritant to eyes and mucous membranes.

hydride. An inorganic compound of hydrogen with another element. Some are ionic and others are covalent. Hydrides may be either binary or complex; the latter are transition-metal complexes, e.g., carbonyl hydrides and cyclopentadienyl hydrides. Most common are hydrides of sodium, lithium, aluminum, boron, etc.
Hazard: Flammable, dangerous fire risk, react violently with water and oxidizing agents. Irritant.
See Lithium aluminum hydride, sodium borohydride.

"Hydroholac."[23] TM for plasticized nitrocellulose lacquer emulsions, including clear finishes, binders and colors. Produce flexible, lacquer-type, cleanable leather finishes from aqueous systems.
Use: Finishes on glove, garment, handbag and shoe leather.

"Hydricin."[202] TM for a castor-derived base oil for industrial functional fluids and SAE hydraulic brake fluids.

"Hydrin."[119] TM for a series of synthetic rubbers based on epichlorohydrin.
Use: Industrial and automotive molded goods.

hydriodic acid. CAS: 10034-85-2.
Properties: Colorless or pale yellow liquid (an aqueous solution of hydrogen iodide, which is a gas at room temperature), a constant-boiling solution is formed (bp 127, d 1.7) containing 57% hydrogen iodide; strong acid and an active reducing agent.
Derivation: (a) By passing hydrogen with iodine vapor over warm platinum sponge and absorption in water. (b) By the action of iodine on a solution of hydrogen sulfide.
Grade: Technical, 47%; NF, diluted, 10%.
Hazard: Strong irritant to eyes and skin.
Use: Preparation of iodine salts, organic preparations, analytical reagent, disinfectant, pharmaceuticals.
See also hydrogen iodide.

hydroabietyl alcohol. See dihydroabietyl alcohol.

hydrobiotite. A natural ore of magnesium, iron, and aluminum silicate occurring in Montana. Source of verxite.

hydroboration. The reaction of diboranes either with alkenes (olefins) to form trialkylboron compounds or with acetylene to yield alkenylboranes. Much research has been devoted to developing these reactions, the products of which are called organoboranes. They are useful in many complex organic syntheses, including prostaglandins and insect pheronomes.
See also borane, organoborane, carborane.

hydrobromic acid. CAS: 10035-10-6.
Hydrogen bromide in aqueous solution.
Properties: Colorless or faintly yellow liquid consisting of an aqueous solution of hydrogen bromide, which is a gas at room temperature. Soluble in water and alcohol, a constant-boiling solution is formed of d 1.49, containing 48% hydrogen bromide; bp at 700 mm (122C), saturated solution contains 68.8% hydrogen bromide at 0C. Hydrobromic acid is a strong acid and sensitive to light. Noncombustible.
Derivation: By dissolving hydrogen bromide in water or by distilling from a mixture of sodium bromide and 50% sulfuric acid.
Grade: Technical 40%; medicinal 48%, 62%.
Hazard: Strong irritant to eyes and skin.
Use: Analytical chemistry, solvent for ore minerals, manufacture of inorganic and some alkyl bromides, alkylation catalyst.
See also hydrogen bromide.

hydrocarbon. An organic compound consisting exclusively of the elements carbon and hydrogen. Derived principally from petroleum, coal tar, and plant sources. Following is a resume of the principal types.
 I. Aliphatic (straight-chain)
   (1) Paraffins (alkanes): generic formula C_nH_{2n+2}. Saturated, single bonds only.
   (2) Olefins: generic formula C_nH_{2n}.
     (a) alkenes: unsaturated (one double bond).
     (b) alkadienes: unsaturated (two double bonds) (butadiene).
   (3) Acetylenes: generic formula C_nH_{2n-2}. Unsaturated (triple bond).
   (4) Acyclic terpenes. Unsaturated (polymers of isoprene C_5H_8). *Note:* Some aliphatic compounds have branched chains in which the subchain also contains carbon atoms (isobutane); both chains are essentially straight.

II. Cyclic (closed ring)
  (1) Alicyclic: three or more carbon atoms
      in a ring structure with properties simi-
      lar to those of aliphatics.
      (a) Cycloparaffins (naphthenes): satu-
          rated compounds often having a
          boat or chair structure, e.g., cyclo-
          hexane, cyclopentane.
      (b) Cycloolefins: unsaturated, two or
          more double bonds, e.g., cyclopen-
          tadiene (2), cyclooctatetraene (4).
      (c) Cycloacetylenes (cyclynes): unsatu-
          rated (triple bond).
  (2) Aromatic: unsaturated, hexagonal ring
      structure (three double bonds), single
      rings and double or triple fused rings.
      (a) benzene group (1 ring).
      (b) naphthalene group (2 rings).
      (c) anthracene group (3 rings).
  (3) Cyclic terpenes: monocyclic (dipentene)
      dicyclic (pinene).
Note: Olefinic (isoprenoid) hydrocarbons are produced by a number of plants. Notably *Hevea
braziliensis* (rubber), guayule, and various members of the Euphorbiaceae family. Current research on the latter group indicates that they
could be used as a source of liquid fuels and
chemical feedstocks by genetic modification of
the plants and control of their molecular constitution. It is estimated that oil obtained by large-
scale cultivation of such plants, which grow well
in semi-arid environments, could become economically competitive with petroleum within a
few years.
See also guayule, biomass, copaiba.

hydrocarbon, halogenated. A hydrocarbon in
which one or more of the hydrogen atoms has
been replaced by fluorine, chlorine, bromine, or
iodine. Examples: carbon tetrachloride, chlorobenzene, chloroform, trifluoromethane. This
greatly increases the anesthetic and narcotic action of aliphatic hydrocarbons. Many halogenated hydrocarbons are highly toxic; some may
detonate on contact with barium. A number of
the chlorinated types are used as insecticides.
See also fluorocarbon, chlorofluorocarbon.

hydrocellulose. See cellulose, hydrated.

hydrochloric acid. (HCl).   CAS: 7647-01-0.
25th highest-volume chemical produced in US
(1985). Hydrogen chloride in aqueous solution.
Properties: Colorless or slightly yellow, fuming,
pungent liquid; flash p none; a constant-boiling
acid containing 20% hydrochloric acid is
formed. Hydrochloric acid is a strong, highly
corrosive acid. The commercial "concentrated"
or fuming acid contains 38% hydrochloric acid
and has a d 1.19; soluble in water, alcohol, and
benzene. Noncombustible.

Derivation: Dissolving hydrogen chloride in water
at various concentrations.
Grade: USP (35–38%), NF dilution (10%), technical (usually 18, 20, 22, 23 degrees Bé, corresponding to 28, 31, 35, 37% hydrogen chloride),
FCC.
Hazard: Toxic by ingestion and inhalation, strong
irritant to eyes and skin.
Use: Acidizing (activation) of petroleum wells,
boiler scale removal, chemical intermediate, ore
reduction, food processing (corn syrup, sodium
glutamate), pickling and metal cleaning, industrial acidizing, general cleaning, e.g., of membrane in desalination plants, alcohol denaturant,
laboratory reagent.
See also hydrogen chloride.

hydrocinnamic acid. (3-phenylpropionic acid).
CAS: 501-52-0. $C_6H_5CH_2CH_2COOH$.
Properties: Crystals with hyacinth-rose odor; mp
46C; bp 280C; soluble in hot water, alcohol, ben-
zene, ether.
Derivaton: Reduction of cinnamic acid with so-
dium amalgam.
Use: Fixative for perfumes, flavoring.

hydrocinnamic alcohol. See phenylpropyl alcohol.

hydrocinnamic aldehyde. See phenylpropyl aldehyde.

hydrocinnamyl acetate. See phenylpropyl acetate.

hydrocolloid. A hydrophilic colloidal material
used largely in food products as emulsifying,
thickening, and gelling agents. They readily absorb water, thus increasing viscosity and imparting smoothness body texture to the product, even
in concentrations of less than 1%. Natural types
are plant exudates (gum arabic), seaweed extracts
(agar), plant seed gums or mucilages (guar gum),
cereal gums (starches), fermentation gums (dextran), and animal products (gelatin). Semisynthetic types are modified celluloses and modified
starches. Completely synthetic types are also
available, e.g., polyvinylpyrolidone. Most are
carbohydrate polymers but a few such as gelatin
and casein are proteins.

hydrocortisone. (17-hydroxycorticosterone; cortisol; hydrocortisone alcohol).   CAS: 50-23-7.
$C_{21}H_{30}O_5$.   An adrenal cortical steroid hormone.
Properties: White, odorless, crystalline powder;
sensitive to light; bitter taste. Mp 212–220C with
some decomposition, soluble in sulfuric acid, partially soluble in alcohol and propylene glycol.
Derivaton: Isolation from extracts of adrenal
glands, synthesis from other steroids.

Grade: USP.
Use: Medicine (anti-inflammatory agent), also used as the acetate and sodium succinate salts. See also cortisone.

hydrocracking. The cracking of petroleum or its products in the presence of hydrogen. Special catalysts are used, for example, platinum on a solid base of mixed silica and alumina or zinc chloride.
See also hydrogenation, hydrogenolysis, hydroforming.

hydrocyanic acid. (prussic acid; hydrogen cyanide; formonitrile).   CAS: 74-90-8.   HCN.
Properties: Water-white liquid at temperatures below 26.5C; faint odor of bitter almonds; usual commercial material is 96–99% pure; d (liquid) 0.688 (20/4C), (gas) 0.938 g/L, bp 25.6C; fp −13.3C; flash p 0F (−17.7C); soluble in water. The solution is weakly acidic, sensitive to light. When not absolutely pure or stabilized, hydrogen cyanide polymerizes spontaneously with explosive violence. Miscible with water, alcohol, soluble in ether, autoign temperature 1000F (537C).
Derivation: (a) By catalytically reacting ammonia and air with methane or natural gas. (b) By recovery from coke oven gases. (c) From bituminous coal and ammonia at 1250C. Hydrogen cyanide occurs naturally in some plants (almond, oleander).
Grade: Technical (96–98%), 2, 5, and 10% solutions. All grades usually contain a stabilizer, usually 0.05% phosphoric acid.
Hazard: Flammable, dangerous fire risk, explosive limits in air 6–41%. Toxic by ingestion inhalation, and skin absorption. TLV: 10 ppm in air.
Use: Manufacture of acrylonitrile, acrylates, adiponitrile, cyanide salts, dyes, chelates, rodenticides, pesticides.

hydrodealkylation. (HDA).   A type of hydrogenation used in petroleum refining in which heat and pressure in the presence of hydrogen are used to remove methyl or larger alkyl groups from hydrocarbon molecules, or to change the position of such groups. The process is used to upgrade products of low value, such as heavy reformate fractions, naphthenic crudes, or recycle stocks from catalytic cracking. Also toluene and pyrolysis gasoline are converted to benzene and methyl naphthalenes to naphthalene by this process.
See also transalkylation.

hydrodistillation. (steam distillation).   Removal of essential oils from plant components (flowers, leaves, bark, etc.) by the use of high-temperature steam. The process is used chiefly in the perfume and fragrance industry.

hydrofining. A petroleum refining process in which a limited amount of hydrogenation converts the sulfur and nitrogen in a petroleum fraction to forms in which they can be easily removed. Hydrofining is generally a separate treatment prior to more extensive hydrogenation. The usual catalysts are oxides of cobalt and molybdenum. Desulfurization, ultrafining, and catfining have a similar meaning.

hydroflumethiazide. (trifluoromethylhydrothiazide).   CAS: 135-09-1.
$C_8H_8F_3N_3O_4S_2$.
Properties: White, crystalline, odorless solid. Mp 260–275C, insoluble in water and acid, soluble in dilute alkali but unstable in alkaline solutions.
Grade: NF.
Use: Medicine (antihypertensive).

hydrofluoric acid. CAS: 7664-39-3.
Hydrogen fluoride in aqueous solution.
Properties: Colorless, fuming, mobile liquid; bp (38% solution) 112C. Will attack glass and any silicon-containing material.
Derivation: Dissolving hydrogen fluoride in water to various concentrations.
Grade: CP, technical, 38%, 47%, 53%, 70%.
Hazard: Toxic by ingestion and inhalation, highly corrosive to skin and mucous membranes. TLV: ceiling 3 ppm.
Use: Aluminum production, fluorocarbons, pickling stainless steel, etching glass, acidizing oil wells, fluorides, gasoline production (alkylation), processing uranium.
See also hydrogen fluoride.

hydrofluorosilicic acid. Legal label name (Rail) for fluosilicic acid.

hydrofluosilicic acid. See fluosilicic acid.

hydroforming. The use of hydrogen in the presence of heat, pressure, and catalysts (usually platinum) to convert olefinic hydrocarbons to branched chain paraffins (isomerization) to yield high-octane gasoline. Catforming and similar terms are often used in the same sense.

hydrofuramide. (furfuramide).
CAS: 494-47-3. $OC_4H_3CH(NCHC_4H_3O)_2$.
Properties: Light brown to white powder, mp 117C, boils about 250C with decomposition, insoluble in cold water, soluble in alcohol and ether.
Derivation: Treatment of furfural with ammonia.
Use: Rubber accelerator, hardening agent for resins, rodenticides, fungicides.

hydrogasification. Production of gaseous or liquid fuels by direct addition of hydrogen to coal.
See also gasification.

hydrogen. CAS: 1333-74-0. H_2. Nonmetallic element of atomic number 1, group IA of periodic table, atomic weight 1.0079, valence = 1. Molecular formula is H_2. Isotopes: deuterium (D_2), tritium (T_3). Hydrogen discovered by Cavendish in 1766, named by Lavoisier in 1783.

Properties: A diatomic gas, density 0.8999 g/L, d 0.0694 (air = 1.0), specific volume 193 cu ft/lb (21.1C), fp −259C, bp −252C, autoign temperature 1075F (580C). Very slightly soluble in water, alcohol, and ether; Noncorrosive; can exist in crystalline state at from 4–1 degrees Kelvin; classed as an asphyxiant gas; rate of permeation through solids is approximately four times that of air.

Occurrence: Chiefly in combined form (water, hydrocarbons, and other organic compounds), traces in earth's atmosphere. Unlimited quantities in sun and stars. It is the most abundant element in the universe.

Derivation: (1) Reaction of steam with natural gas (steam reforming) and subsequent purification; (2) partial oxidation of hydrocarbons to carbon monoxide and interaction of carbon monoxide and steam; (3) gasification of coal (See Note 1); (4) dissociation of ammonia; (5) thermal or catalytic decomposition of hydrocarbon gases; (6) catalytic reforming of naphtha; (7) reaction of iron and steam; (8) catalytic reaction of methanol and steam; (9) electrolysis of water (See Note 2). In view of the importance of hydrogen as a major energy source of the future, development of the most promising of these methods may be expected.

See also gasification, and "Hypo" process.

Note 1: The projected cost of producing hydrogen from coal by proven gasification techniques has been estimated to be competitive with gasoline.

Note 2: More efficient methods than electrolysis for obtaining hydrogen from water are under investigation. One of these is thermochemical decomposition. Another is photochemical decomposition by solar radiation, either directly or via a solar power generator. Photolytic decomposition of water with platinum catalyst has been achieved. Hydrogen can also be obtained by photolytic decomposition of hydrogen sulfide with cadmium sulfide catalyst.

See also photolysis, thermochemistry.

Method of purification: By scrubbing with various solutions (see especially the Girbitol absorption process). For very pure hydrogen, by diffusion through palladium.

Grade: Technical, pure, from an electrolytic grade of 99.8% to ultra-pure with less than 10 ppm impurities.

See also para-hydrogen.

Hazard: Highly flammable and explosive, dangerous when exposed to heat or flame, explosive limits in air 4–75% by volume.

Use: Production of ammonia, ethanol, and aniline; hydrocracking, hydroforming, and hydrofining of petroleum; hydrogenation of vegetable oils; hydrogenolysis of coal; reducing agent for organic synthesis and metallic ores; reducing atmosphere to prevent oxidation; as oxyhydrogen flame for high temperatures; atomic-hydrogen welding; instrument-carrying balloons; making hydrogen chloride and hydrogen bromide; production of high-purity metals; fuel for nuclear rocket engines for hypersonic transport; missile fuel; cryogenic research. A treatise on the physical properties of hydrogen compiled by the National Bureau of Standards is available from the US Government Printing Office.

Note: A safe storage method for hydrogen for possible use as automotive fuel involves the use of metal hydrides from which the hydrogen is released at specified temperatures. Iron titanium hydride has been found the most satisfactory.

ortho-hydrogen. See para-hydrogen.

para-hydrogen. Type of molecular hydrogen preferred for rocket fuels. Molecular hydrogen (H_2) exists in two varieties, ortho- and para-, named according to their nuclear spin types. Ortho-hydrogen molecules have a parallel spin, para- an antiparallel spin. By cooling to liquid air temperature and use of a ferric oxide gel catalyst, the normal equilibrium of 3 ortho- to 1 para- is displaced and para-hydrogen may be isolated. It is being produced with less than 5 ppm impurities.

hydrogenated terphenyls. CAS: 61788-32-7. Complex mixtures of ortho- meta- and para- terphenyls in various stages of hydrogenation.

Hazard: TLV: 0.5 ppm.

Use: A heat-transfer medium and plasticizer.

hydrogenation. Any reaction of hydrogen with an organic compound. It may occur either as direct addition of hydrogen to the double bonds of unsaturated molecules, resulting in a saturated product, or it may cause rupture of the bonds of organic compounds, with subsequent reaction of hydrogen with the molecular fragments. Examples of the first type (called addition hydrogenation) are the conversion of aromatics to cycloparaffins and the hydrogenation of unsaturated vegetable oils to solid fats by addition of hydrogen to their double bonds. Examples of the second type (called hydrogenolysis) are hydrocracking of petroleum and hydrogenolysis of coal to hydrocarbon fuels.

See also hydrogenolysis, hydrocracking, hydroforming.

hydrogen azide. See hydrazoic acid.

hydrogen bond. An attractive force, or bridge, occurring in polar compounds such as water in which a hydrogen atom of one molecule is attracted to two unshared electrons of another. The hydrogen atom is the positive end of one polar molecule and forms a linkage with the electronegative end of another such molecule. In the formula below, the hydrogen atom in the center is the "bridge."

$$H:\overset{..}{\underset{..}{O}}:H:\overset{..}{\underset{..}{O}}:H:\overset{..}{\underset{..}{O}}:$$
$$\quad H \qquad H \qquad H$$

hydrogen bonds are only one-tenth to one-thirteenth as strong as covalent bonds but they have pronounced effects on the properties of substances in which they occur, especially as regards melting point, boiling point, and crystalline structure. They are found in compounds containing such strongly electronegative atoms as nitrogen, oxygen, and fluorine. They play an important part in the bonding of cellulosic compounds, e.g., in the paper industry, and occur also in many complex structures of biochemical importance, e.g., adenine-uracil linkage in DNA.

hydrogen bromide. CAS: 10035-10-6.   HBr.
Properties: Colorless gas, d 2.71 (air = 1.00), fp −86C, bp −66.4C, specific volume 4.8 cu ft/lb (70F, 1 atm), soluble in water and alcohol, nonflammable.
Derivation: (a) By passing hydrogen with bromine vapor over warm platinum sponge which acts as a catalyst. (b) As a byproduct in the bromination of organic compounds.
Grade: Up to 99.8% min purity.
Hazard: Toxic by inhalation, strong irritant to eyes and skin. TLV: CL 3 ppm in air.
Use: Organic synthesis, makes bromides by direct reaction with alcohols, pharmaceutical intermediate; alkylation and oxidation catalyst, reducing agent.
See also hydrobromic acid.

hydrogen chloride. CAS: 7647-01-0.   HCl.
Properties: Colorless, fuming, with a suffocating odor, D 1.268 (air = 1.00), fp −114C, bp −85C, specific volume 10.9 ft/lb (21.1C, 1 atm), very soluble in water, soluble in alcohol and ether. Nonflammable.
Derivation: (1) Byproduct of organic chlorination reactions (approximately 90%), (2) reaction of sodium chloride and sulfuric acid, (3) burning hydrogen in an atmosphere of chlorine in absence of air.
Hazard: Toxic by inhalation, strong irritant to eyes and skin. TLV: CL 5 ppm in air.
Use: Production of vinyl chloride from acetylene and alkyl chlorides from olefins, hydrochlorination (see rubber hydrochloride), polymerization,

isomerization, alkylation, and nitration reactions.
See also hydrochloric acid.

hydrogen cyanide. See hydrocyanic acid.

hydrogen dioxide. See hydrogen peroxide.

hydrogen electrode. See electrode, hydrogen.

hydrogen fluoride. CAS: 7664-39-3.   HF.
Properties: Colorless, fuming gas or liquid; very soluble in water; the liquid and gas consist of associated molecules; the vapor density corresponds to hydrogen fluoride only at high temperatures; fp −83C, bp 19.5C, d (liquid) 0.988 (14C), sp vol 17 cu ft/lb (21.1C, 1 atm). Nonflammable.
Derivation: Distillation from the reaction product of calcium fluoride and sulfuric acid, also from fluosilicic acid.
Grade: To 99.9% min purity.
Hazard: Toxic by ingestion and inhalation, strong irritant to eyes, skin and mucous membranes. TLV: 3 ppm in air.
Use: Catalyst in alkylation, isomerization, condensation, dehydration, and polymerization reactions; fluorinating agent in organic and inorganic reactions; production of fluorine and aluminum fluoride; additive in liquid rocket propellants; refining of uranium.
See also hydrofluoric acid.

hydrogen hexafluorosilicate. See fluosilicic acid.

hydrogen iodide. CAS: 10034-85-2.   HI.
Properties: Colorless gas, bp −35C, fp −51C, fumes in moist air, d 5.2 (25C), freely soluble in water. Nonflammable.
Hazard: Strong irritant. Poison.
Use: Making hydriodic acid.

hydrogen ion concentration. See pH.

hydrogenolysis. (destructive hydrogenation). A type of hydrogenation reaction in which molecular cleavage of an organic compound occurs with addition of hydrogen to each portion. An important application is hydrocracking (hydrogenative splitting) of large organic molecules, with formation of fragments that react with hydrogen by use of catalysts and high temperatures. Hydrogenolysis of coal to gaseous and liquid fuels was used in Germany in the 1940s, a similar method (Oil/Gas Process) is under development in the US. The German process used pulverized coal made into a paste with heavy oil and a metallic catalyst. The mixture plus the necessary hydrogen was subjected to 300–700 atm at approximately 500C. The coal was converted into heavy oil, distillable oil, gasoline, and hydrocarbon gases. Large quantities of hydrogen are necessary.

See also gasification, hydrogenation.

hydrogen peroxide. CAS: 7722-84-1. H_2O_2 (molecular formula), H—O—O—H (structural formula).

Properties: (pure anhydrous) Density of solid, 1.71 g/cc, density of liquid 1.450 g/cc at 20C, viscosity, liquid 1.245 centipoise, surface tension 80.4 dynes/cm at 20C, fp −0.41C, bp 150.2C, soluble in water and alcohol. (Solutions): Pure hydrogen peroxide solutions, completely free from contamination, are highly stable; a low percentage of an inhibitor such as acetanilide or sodium stannate, is usually added to counteract the catalytic effect of traces of impurities such as iron, copper, and other heavy metals. A relatively stable sample of hydrogen peroxide typically, decomposes at the rate of approximately 0.5% per year at room temperature.

Derivation: (a) autoxidation of an alkyl anthrahydroquinone such as the 2-ethyl derivative in a cyclic continuous process in which the quinone formed in the oxidation step is reduced to the starting material by hydrogen in the presence of a supported palladium catalyst; (b) by electrolytic processes in which aqueous sulfuric acid or acidic ammonium bisulfate is converted electrolytically to the peroxydisulfate, which is then hydrolyzed to form hydrogen peroxide; (c) by autoxidation of isopropyl alcohol. Method (a) is most widely used.

Grade: USP (3%), technical (3, 6, 27.5, 30, 35, 50, and 90%), FCC. Most common commercial strengths are 27.5, 35, 50, and 70%.

Hazard: Dangerous fire and explosion risk, strong oxidizing agent. Concentrated solutions are highly toxic and strongly irritating. TLV: 1 ppm in air.

Use: Bleaching and deodorizing of textiles, wood pulp, hair, fur, etc.; source of organic and inorganic peroxides; pulp and paper industry; plasticizers; rocket fuel; foam rubber; manufacture of glycerol; antichlor; dyeing; electroplating; antiseptic; laboratory reagent; epoxidation, hydroxylation, oxidation, and reduction; viscosity control for starch and cellulose derivatives; refining and cleaning metals; bleaching and oxidizing agent in foods; neutralizing agent in wine distillation; seed disinfectant; substitute for chlorine in water and sewage treatment.

hydrogen phosphide. See phosphine.

hydrogen, phosphoretted. See phosphine.

hydrogen selenide. CAS: 7783-07-5. H_2Se.
Properties: Colorless gas; soluble in water, carbon disulfide, phosgene; bp −42C; fp −64C; d 2.00 (air = 1).

Grade: 98% pure.

Hazard: Dangerous fire and explosion risk; reacts violently with oxidizing materials. Toxic by inhalation, strong irritant to skin, damaging to lungs and liver. TLV: 0.05 ppm (as selenium) in air.

Use: Prepartion of metallic selenides and organoselenium compounds; in doping as mix for preparation of semiconductor materials containing controlled amounts of significant impurities.

hydrogen slush. A mixture of solid and liquid hydrogen at the hydrogen triple point 13.8K and 1.02 psia. It is denser and less hazardous than liquid hydrogen.

hydrogen sulfide. (sulfuretted hydrogen).
CAS: 7783-06-4. H_2S.
Properties: Colorless gas, offensive odor, soluble in water and alcohol, d 1.189 (air = 1.00), fp −83.8C, bp −60.2C, sp vol 11.23 cu ft/lb (21.1C, 1 atm), autoign temperature 500F (260C).

Derivation: (a) By the action of dilute sulfuric acid on a sulfide, usually iron sulfide; (b) by direct union of hydrogen and sulfur vapor at a definite temperature and pressure; (c) as a byproduct of petroleum refining.

Grade: Technical 98.5%, purified 99.5% min, CP.

Hazard: Highly flammable, dangerous fire risk, explosive limits in air 4.3–46%. Toxic by inhalation, strong irritant to eyes and mucous membranes. TLV: 10 ppm in air.

Use: Purification of hydrochloric acid and sulfuric acid, precipitating sulfides of metals, analytical reagent, source of sulfur and hydrogen.

hydrogen tellurate. See telluric acid.

hydrogen telluride. CAS: 7783-09-7. H_2Te.
Properties: Colorless gas, d (liquid): 2.57 (−20C), fp −49C, bp −2C, soluble in water but unstable, soluble in alcohol and alkalies.

Hazard: See hydrogen selenide.

hydrol. See tetramethyldiaminobenzhydrol.

hydrolase. (hydrase). An enzyme which catalyzes the removal of water from the substrate. See enzyme.

"Hydrolin."[84] TM for an ammonium nitrate base blasting agent which requires specially constructed primers for detonation.

Use: For seismic prospecting at sea.

Hazard: See ammonium nitrate.

hydroliquefaction. Production of liquid hydrocarbon fuels by hydrogenation of coal.

See gasification; Oil/Gas Process.

hydrolube. A water-glycol base noncombustible hydraulic fluid.

hydrolysis. A chemical reaction in which water reacts with another substance to form two or more new substances. This involves ionization of the water molecule as well as splitting of the compound hydrolyzed, e.g., $CH_3COOC_2H_5 + HOH \rightarrow CH_3COOH + C_2H_5OH$. Examples are conversion of starch to glucose by water in the presence of suitable catalysts; conversion of sucrose (cane sugar) to glucose and fructose by reaction with water in the presence of an enzyme or acid catalyst; conversion of natural fats into fatty acids and glycerol by reaction with water in one process of soap manufacture; and reaction of the ions of a dissolved salt to form various products, such as acids, complex ions, etc.

hydrometer. Device for measuring the density of liquids.
See also Baume'

hydronium ion. An ion (H_3O^+) formed by the transfer of a proton (hydrogen nucleus) from one molecule of water to another, a companion ion (OH^-) is also formed, the reaction is $2HOH \rightarrow H_3O^+ + OH^-$. Formation of such ions is statistically rare, resulting from the interaction of water molecules in a ratio of $1:556$ million.

hydroperoxide. An organic peroxide having the generalized formula ROOH. An example is ethyl hydroperoxide (C_2H_5OOH). Methyl and ethyl hydroperoxides are unstable and thus are strong oxidizing agents and explosion hazards; those of higher molecular weight are more stable. Hydroperoxides can be derived by oxidation of saturated hydrocarbons, or by alkylating hydrogen peroxide in a strongly acidic environment. They are used as polymerization initiators.

hydrophilic. Having a strong tendency to bind or absorb water, which results in swelling and formation of reversible gels. This property is characteristic of carbohydrates, such as algin, vegetable gums, pectins and starches, and of complex proteins such as gelatin and collagen.

hydrophobic. Antagonistic to water, incapable of dissolving in water. This property is characteristic of all oils, fats, waxes, and many resins, as well as of finely divided powders like carbon black and magnesium carbonate.

hydroponics. See nutrient solution.

"Hydro-Pruf."[300] TM for a silicone water repellent for fabrics. Applied with a catalyst at high curing temperatures.

hydroquinol. See hydroquinone.

hydroquinone. (quinol; hydroquinol; p-dihydroxybenzene).   CAS: 123-31-9.
$C_6H_4(OH)_2$.
Properties: White crystals; soluble in water, alcohol, and ether; d 1.330; mp 170C; bp 285C; flash p 329F (165C); autoign temperature 960F (515.5C). Combustible.
Derivation: Aniline is oxidized to quinone by manganese dioxide and is then reduced to hydroquinone.
Grade: Technical, photographic.
Hazard: Toxic by ingestion and inhalation, irritant. TLV: 2 mg/m³ of air.
Use: Photographic developer (except color film); dye intermediate; inhibitor; stabilizer in paints and varnishes, motor fuels, and oils; antioxidant for fats and oils; inhibitor of polymerization.

hydroquinone benzyl ether. See p-benzyloxyphenol.

hydroquinone dibenzyl ether. CAS: 103-16-2.
$C_6H_5CH_2OC_6H_4OCH_2C_6H_5$.
Properties: Tan powder; mp 119C (min); purity 90% (min); insoluble in water; soluble in acetone, benzene, and chlorobenzene. Combustible.
Use: Solvent; perfumes, soap, plastics, and pharmaceuticals.

hydroquinone di-n-butyl ether. (1,4-dibutoxybenzene). $C_6H_4[O(CH_2)_3CH_3]_2$.
Properties: White flakes with no appreciable odor; mp 45–46C; bp 124C (1.3 mm), 158C (15.0 mm); insoluble in water; soluble in benzene, acetone, ethyl acetate, and alcohol. Combustible.

hydroquinone diethyl ether. (1,4-diethoxybenzene). $C_6H_4(OC_2H_5)_2$.
Properties: White granular solid with anise-like odor, mp 71–72C, bp 246C. Neither boiling caustic nor acid solution cause any hydrolysis. Absorbs UV light. Insoluble in water; soluble in benzene, acetone, ethyl acetate, and alcohol. Combustible.

hydroquinone di(β-hydroxyethyl) ether. (p-di-[2-hydroxyethoxy]benzene).
$C_6H_4(OC_2H_4OH)_2$.
Properties: White solid, mp 99C, bp 185–200C (0.3 mm), slightly soluble in water and most organic solvents, miscible with water at 80C. Combustible.
Use: Preparation of polyester, polyolefins, polyurethanes and hard waxy resins, organic synthesis.

hydroquinone dimethyl ether. (1,4-dimethoxybenzene; DMB; dimethyl hydroquinone).
CAS: 654-42-2. $C_6H_4(OCH_3)_2$.
Properties: White flakes with sweet clover odor, bp 213C, mp 56C, d 1.0293 (65C), viscosity 1.04

cp (65C), dielectric constant 2.8, absorbs UV light in range 2800–3000Å, soluble in benzene and alcohol, insoluble in water. Combustible.
Use: Weathering agent in paints and plastics, fixative in perfumes, dyes, resin intermediate, cosmetics, especially suntan preparations, flavoring.

hydroquinine mono-n-butyl ether.
$CH_3(CH_2)_3OC_6H_4OH$.
Properties: White flakes; mp 64–65C; bp 115C (1.4 mm); insoluble in water; soluble in benzene, acetone, ethyl acetate, and alcohol. Combustible.

hydroquinone monoethyl ether.   (4-ethoxyphenol). $C_2H_5OC_6H_4OH$.
White solid; mp 63–65C; bp 246–247C; slightly soluble in water; soluble in benzene, acetone, ethyl acetate, and alcohol. Combustible.
Use: See hydroquinone monomethyl ether.

hydroquinone monomethyl ether.   (4-methoxyphenol; p-hydroxyanisole).   CAS: 150-76-5. $CH_3OC_6H_4OH$.
Properties: White, waxy solid; mp 52.5C; bp 243C; d 1.55 (20/20C); slightly soluble in water; readily soluble in benzene, acetone, ethanol, ethyl acetate. Combustible.
Use: Manufacture of antioxidants, pharmaceuticals, plasticizers, dyestuffs; stabilizer for chlorinated hydrocarbons and ethyl cellulose, inhibitor for acrylic monomers and acrylonitriles, UV inhibitor.

hydrosilicofluoric acid.   See fluosilicic acid.

hydrosolvation.   Solvent extraction of coal (containing up to 5% sulfur) under hydrogen pressure with the use of a catalyst such as zinc chloride; pressures from 1000–2000 psi are necessary for suitable conversion. This process offers a means of deriving fuel oil and petrochemical feedstocks directly from coal.

hydrosulfite.   See sodium hydrosulfite.

hydrosulfite-formaldehyde.   One of several mixtures of sodium formaldehyde hydrosulfite and sodium formaldehyde bisulfite used as discharges and stripping or reducing agents in dyeing and other textile operations. In some cases the zinc derivatives are used. Derivation is by the action of formaldehyde on aqueous sodium hydrosulfite or from zinc, formaldehyde, sulfur dioxide, and sodium hydroxide.
Hazard: Toxic by ingestion.

hydrothermal energy.   See geothermal energy.

hydrotrope.   A chemical which has the property of increasing the aqueous solubility of various slightly soluble organic chemicals.
Use: Formulation of liquid detergents.

hydroxocobalamin.   (USAN) (α[5,6-dimethylbenzimidazolyl]hydroxocobalamide).
$C_{62}H_{89}CoN_{13}O_{15}P$.   A form of vitamin B_{12}.
Use: Medicine.
See vitamin B_{12}.

hydroxyacetal.   See hydroxycitronellal dimethyl acetal.

p-**hydroxyacetanilide.**   See p-acetylaminophenol.

hydroxyacetic acid.   (glycolic acid).
$CH_2OHCOOH$.
Properties: Colorless crystals, deliquescent, mp 78–79C, bp decomposes. Soluble in water, alcohol, and ether; available commercially as a 70% solution, light straw-colored liquid, odor like burnt sugar, d 1.27, mp 10C. Combustible.
Derivation: From chloroacetic acid by reaction with sodium hydroxide, or by reduction of oxalic acid. Occurs naturally in sugar cane syrup.
Grade: Technical 70% solution, pure crystals.
Use: Leather dyeing and tanning; textile dyeing; cleaning, polishing, and soldering compounds; copper pickling; adhesives; electroplating; breaking of petroleum emulsions; chelating agent for iron; chemical milling; pH control.

hydroxyacetone.   See hydroxy-2-propanone.

o-**hydroxyacetophenone.** $C_6H_4(OH)COCH_3$.
Properties: Greenish-yellow liquid with minty odor, d 1.1307 (20.8C), bp 213C (717 min), refr index 1.5580 (20C), slightly soluble in water. Combustible.

2-hydroxyadipaldehyde.
$OHCCH_2CH_2CH_2CHOHCHO$.
Properties: (25% aqueous solution) D 1.066 (20C), bp 37C (50 mm), vap press 17 mm (20C), fp −3.5C, pH approximately 3.0. Combustible.
Hazard: Toxic by inhalation.
Use: Intermediate, insolubilizing agent for proteins and polyhydroxy materials, crosslinking agent for polyvinyl compounds, shrinkage control agent (textiles).

β-**hydroxyalanine.**   See serine.

5-hydroxy-3-(β-aminoethyl)indole.   See serotonin.

hydroxyaniline.   See aminophenol.

o-**hydroxyanisole.**   See guaiacol.

p-**hydroxyanisole.**   See hydroquinone monomethyl ether.

9-hydroxyanthracene.   See anthranol.

hydroxyapatite. $Ca_{10}(PO_4)_6(OH)_2$. The major constituent of bone and tooth mineral. It is finely divided, crystalline, non-stoichiometric material rich in surface ions (carbonate, magnesium, citrate) which are readily replaced by fluoride ion, thus affording protection to the teeth. See also fluoridation; calcium phosphate tribasic.

p-hydroxyazobenzene-p-sulfonic acid. $HOC_6N_4NNC_6H_4SO_3H$.
Properties: Orange-red crystals, very soluble in water.
Use: Analytical reagent, precipitant for numerous organic bases.

m-hydroxybenzaldehyde. CAS: 90-2-8. HOC_6H_4CHO.
Properties: Orange-pink crystals, mp 101.5C, slightly soluble in cold water, very soluble in hot water and aromatic hydrocarbons.
Hazard: Toxic by ingestion.
Use: Intermediate for dyes, plastics, pharmaceuticals and bactericides; color reagent for Schiff's reagent, sensitizing agent in photographic emulsions.

o-hydroxybenzaldehyde. See salicylaldehyde.

p-hydroxybenzaldehyde. CAS: 123-08-0. HOC_6H_4CHO.

$$HO-\bigcirc-CHO$$

Properties: Colorless needles, soluble in alcohol, ether or hot water, d 1.129, mp 116C (sublimes).
Use: Pharmaceuticals.

o-hydroxybenzamide. See salicylamide.

hydroxybenzene. See phenol.

2-hydroxy-1′,2′-benzocarbazole-3-carboxylic acid. $C_{17}H_{11}NO_3$. A four-ring structure, light green powder, mp 315–320C, soluble in ethanol and acetone, insoluble in water.
Use: Manufacture of dye intermediates and other organic chemicals.

m-hydroxybenzoic acid. CAS: 99-06-9. $C_6H_4(OH)COOH$.
Properties: White powder, mp 200C, soluble in water and hot alcohol.
Use: Intermediate for plasticizers, resins, light stabiliziers, petroleum additives, pharmaceuticals.

o-hydroxybenzoic acid. See salicylic acid.

p-hydroxybenzoic acid. CAS: 99-96-7. $C_6H_4(OH)COOH \cdot HOH$.
Properties: Colorless crystals, soluble in alcohol and ether, partially soluble in water, d 1.46, mp 210C.
Derivation: Interaction of p-aminobenzoic acid and nitrous acid.
Use: Intermediate, synthetic drugs, food preservative (up to 0.1%) (approved by FDA). Its methyl, propyl, and butyl esters are preservatives for cosmetics and pharmaceuticals.

2-hydroxybenzophenone. $C_6H_5COC_6H_4OH$.
Properties: A solid, mp 41C, bp 210C (27 mm), insoluble in water, soluble in alcohol.
Use: UV absorber in plastics.

o-hydroxybenzyl alcohol. See salicyl alcohol.

β-hydroxybutyraldehyde. See aldol.

p-hydroxybutyranilide. (4′-hydroxybutyranilide). CAS: 101-91-7. $C_{10}H_{13}NO_2$.
Properties: Acicular crystals, mp 138C, soluble in alcohol, partially soluble in hot water.
Use: Antioxidant for petroleum products.

β-hydroxybutyric acid. CAS: 300-85-6. $CH_3CH(OH)CH_2COOH$.
Properties: Viscid yellow mass, mp 48–50C, bp 130C (12 mm), very soluble in water, alcohol and ether.
Derivation: Interaction of acetoacetic acid and sodium amalgam.
Grade: Technical, reagent.
Use: Intermediate.

2-hydroxycamphane. See borneol.

hydroxy-β-carotene. See cryptoxanthin.

3-β-hydroxycholestane. See dihydrocholesterol.

hydroxycitronellal. (citronellal hydrate; 3,7-dimethyl-7-hydroxyoctenal). CAS: 107-75-5. $(CH_3)_2C(OH)(CH_2)_3CH(CH_3)CH_2CHO$.
Properties: Viscous, colorless or faintly yellow liquid, sweet lily-type odor, d 0.925–0.930 (15C), refr index 1.448–1.450 (20C), optical rotation (Java type) +9 to +10.5 degrees, boiling range 94–96C (1 mm), soluble in alcohol (50%), fixed oils, slightly soluble in water, glycerol and mineral oil. Combustible.
Derivation: From citronellal (*Java citronella* or *Eucalyptus citriodora*).
Grade: Perfume, FCC.
Use: Perfumery (fixative, muguet odor), flavoring, soap and cosmetic fragrances.

hydroxycitronellal dimethyl acetal. (hydroxy-acetal).

$(CH_3)_2C(OH)(CH_2)CH(CH_3)CH_2CH(OCH_3)_2$.

Properties: Colorless liquid, light floral odor, d 0.925–0.930 (25/25C), refr index 1.4410–1.4440 (20C), soluble in most fixed oils, mineral oil, propylene glycol, insoluble in glycerol. Combustible.

Grade: Perfume, FCC.

Use: Flavoring agent in foods, perfumery.

hydroxycitronellal-methyl anthranilate Schiff base.

$C_{18}H_{27}O_3N$.

Properties: Linden-orange flower odor; yellow, honeylike, viscous liquid. Stable, refr index 1.5350–1.5460 (20C), flash p 206F (96.6C), (TCC) soluble in 2 parts of 70% alcohol, 1 part of 80% alcohol. Combustible.

Use: Perfumery.

17-hydroxycorticosterone.   See hydrocortisone.

2-hydroxy-p-cymene.   See caracrol.

3-hydroxy-p-cymene.   See thymol.

1-hydroxy-2,4-diamylbenzene.   See diamyl phenol.

hydroxydiphenyl.   See phenylphenol.

p-hydroxydiphenylamine. (anilinophenol).

CAS: 122-37-2. $C_6H_5NHC_6H_4OH$.

Gray solid leaflets, mp approximately 70C, purity 98% (min), bp 330C. Insoluble in water; soluble in alcohol, ether, acetone, chloroform, alkali, and benzene.

Hazard: Irritant.

2-hydroxyethanesulfonic acid.   See isethionic acid.

hydroxyethylacetamide.   See n-acetylethanolamine.

2-hydroxyethyl acrylate. (HEA).   A functional monomer for the manufacture of thermosetting acrylic resins.

2-hydroxyethylamine.   See ethanolamine.

2-hydroxyethyl carbamate.   CAS: 589-41-3.

$H_2NCOOCH_2CH_2OH$.

Properties: Crystalline, deliquescent solid. Mp 43C, bp 130–135C (1 mm), d 1.2852 g/cc (20C), flash p 370F (187.7C), miscible with water, soluble in alcohol and acetone, insoluble in benzene and chloroform. Combustible.

Use: Chemical intermediate, especially for wash-and-wear cotton finishing agents.

hydroxyethylcellulose. ("Cellosize").

Properties: Nonionic, water-soluble cellulose ether; white, free-flowing powder; insoluble in organic solvents; soluble in hot or cold water; stable in concentrated salt solutions; grease and oil resistant. Combustible.

Use: Thickening and suspending agent; stabilizer for vinyl polymerization; retards evaporation of water in cements, mortars, and concrete; binder in ceramic glazes, films, and sheeting; protective colloid; paper and textile sizing; secondary petroleum recovery.

hydroxyethylethylenediamine. (aminoethylethanol amine).   CAS: 111-41-1.

$NH_2CH_2CH_2NHCH_2CH_2OH$.

Properties: Hygroscopic liquid, mild ammoniacal odor, soluble in water, d 1.0304 (20/20C), bp 243.7C, vap press 0.01 mm (20C), flash p 275F (135C), bulk d 8.6 lb/gal (20C). Combustible.

Use: Textile finishing compounds (antifuming agents, dyestuffs, cationic surfactants), resins, rubber products, insecticides and certain medicinals.

hydroxyethylethylenediaminetriacetic acid. (HEDTA).   CAS:150-39-0.

$C_{10}H_{18}O_7N_2$.

Properties: Solid, mp 159C, soluble in water and methanol.

Use: Chelating compound.

n-(2-hydroxyethyl)ethyleneimine.   See 1-aziridineethanol.

1-(2-hydroxyethyl)-2-n-heptadecenyl-2-imidazoline. $C_{22}H_{42}N_2O$.   A liquid cationic detergent.

Use: Corrosion inhibitor, acid-stable emulsifier.

β-hydroxyethylhydrazine. (ethanolhydrazine).

CAS: 109-84-2. $HOCH_2CH_2NHNH_2$.

Properties: Colorless, slightly viscous liquid. D 1.11 (20C), fp −70C, boiling range 145–153C (25 mm), flash p 224F (106.6C), miscible with water, soluble in lower alcohols, slightly soluble in ether. Combustible.

Grade: 70%.

Hazard: Moderate fire risk. Irritant.

Use: Intermediate, plant growth regulator, flowering inducer for pineapples.

2-hydroxyethyl methacrylate. (HEMA).

CAS: 868-77-9. $C_6H_{10}O_3$.

Properties: Clear, mobile liquid. D 1.064 (77/60F), fp −12C, refr index 1.4505 (25C), miscible with water and soluble in common organic solvents. An inhibitor is usually added to solutions to prolong shelf life. The 30% solution is made with xylene.

Hazard: 30% grade (with xylene): Flammable, moderate fire risk.

Grade: 30%, 96%.

Use: Acrylic resins, binder for nonwoven fabrics, enamels.

N-hydroxyethylmorpholine.   See n-morpholine ethanol.

N-hydroxyethyl piperazine.   CAS: 103-76-4.

HOCH₂CH₂NCH₂CH₂NHCH₂CH₂.

Properties: Liquid, d 1.0614 (20/20C), bp 246.3C, fp −10C, flash p 255F (124C), miscible with water. Combustible.

Use: Intermediate for pharmaceuticals, anthelmintics, surface-active agents, and synthetic fibers.

N-2-hydroxyethylpiperidine.   See 2-piperidinoethanol.

hydroxyethyltrimethylammonium bicarbonate.
   (choline bicarbonate).
   (CH₃)₃NCH₂CH₂OH·HCO₃.

Properties: Colorless liquid, d 1.0965 (25/4C), fp −21.3C, refr index 1.3967 (25C), miscible with water.

Grade: Industrial grade is a 44–46% aqueous solution of a mixture of carbonate and bicarbonate.

Use: Alkaline catalyst, intermediate for choline salts and surfactants.

β-hydroxyethyltrimethylammonium hydroxide.
See choline.

1-hydroxyfenchane.   See fenchyl alcohol.

4-hydroxy-2-keto-4-methylpentane.   See diacetone alcohol.

hydroxylamine.   (oxammonium).
   CAS: 7803-49-8.   NH₂OH.

Properties: Colorless crystals. Soluble in alcohol, acids, and cold water. D 1.227, mp 33C, bp 70C (60 mm), the free base is unstable.

Derivation: By decomposing hydroxylamine hydrochloride or sulfate with a base and distilling in vacuo.

Hazard: Decomposes rapidly at room temperature, violently when heated, detonates in flame-heated test-tube. Irritant to tissue.

Use: Reducing agent, organic synthesis.

hydroxylamine acid sulfate.   (hydroxylammonium acid sulfate; hydroxylamine hemisulfate; HAS).   NH₂OH·H₂SO₄.

Properties: White to brown crystalline solid, very hygroscopic, soluble in water and methanol, slightly soluble in alcohol, bulk d 15–16 lb/gal

(20C), mp indefinite (decomposes), pH of 0.1 M aqueous solution 1.6.

Use: Reducing agent, photographic developer, purification agent for aldehydes and ketone, synthesis of dyes and pharmaceuticals; rubber chemicals, reagent.

hydroxylamine hydrochloride.   (hydroxylammonium chloride).   CAS: 5470-11-1.
   NH₂OH·HCl.

Properties: Colorless, hygroscopic crystals; d 1.67 (17C). Soluble in water, glycerol, and alcohol; insoluble in ether; mp 152C (decomposes); pH of 0.1 M aqueous solution 3.4.

Derivation: Reduction of ammonium chloride, frequently by electrolysis.

Hazard: Toxic by ingestion, strong irritant to tissue.

Use: Organic synthesis, photographic developer, medicine, controlled reduction reactions, nondiscoloring short-stopper for synthetic rubbers, antioxidant for fatty acids.

hydroxylamine sulfate.   (HS; hydroxylammonium sulfate).   CAS: 10039-54-0.
   (NH₂OH)₂·H₂SO₄.

Properties: Colorless crystals, solution has a corrosive action on the skin, mp 177C (decomposes), soluble in water, slightly soluble in alcohol.

Hazard: Irritant to tissue.

Use: Reducing agent, photographic developer, purification agent for aldehydes and ketones, chemical synthesis, textile chemical, oxidation inhibitor for fatty acids, catalyst, biological and biochemical research, making oximes for paints and varnishes, rustproofing, nondiscoloring short-stopper for synthetic rubbers, unhairing hides.

hydroxylammonium acid sulfate.   See hydroxylamine acid sulfate.

hydroxylammonium chloride.   See hydroxylamine hydrochloride.

hydroxylammonium sulfate.   See hydroxylamine sulfate.

hydroxyl group.   The univalent group —OH, occurring in many inorganic compounds that ionize in solution to yield OH⁻ radicals. Also the characteristic group of alcohols.
See also -ol.

hydroxymercurichlorophenol.   (2-chloro-4-[hydroxymercuri]phenol).
   C₆H₃Cl(HgOH)(OH).   Insoluble in water and common organic solvents, soluble in solu-

tions of acids and alkalies with the formation of salts.

Hazard: Toxic by ingestion and inhalation, strong irritant to eyes and skin, use may be restricted. TLV: 0.01 (Hg) mg/m³ of air.

Use: Seed disinfectant and fungicide.

hydroxymercuricresol.
$C_6H_3CH_3(HgOH)(OH)$.
Hazard: Toxic by ingestion and inhalation, strong irritant to eyes and skin, use may be restricted. TLV: 0.01 (Hg) mg/m³ of air.
Use: Pesticide.

hydroxymercurinitrophenol.
$C_6H_3NO_2(HgOH)(OH)$.
Hazard: Toxic by ingestion and inhalation, strong irritant to eyes and skin, use may be restricted. TLV: 0.01 (Hg) mg/m³ of air.
Use: Pesticide.

2-hydroxy-3-methylbenzoic acid. See cresotic acid.

3-hydroxy-3-methylbutanone-2.
$CH_3COH(CH_3)C(O)CH_3$. Colorless liquid, d 0.95 (25/25C). Combustible.
Use: Intermediate.

7-hydroxy-4-methylcoumarin. See β-methyl umbelliferone.

4-hydroxymethyl-2,6-di-tert-butylphenol.
$[(CH_3)_3C]_2C_6H_2(OH)(CH_2OH)$.
Properties: White, crystalline powder. Mp 140–141C, bp 162C (2.6 mm), flash p 375F (190.5C), vap press 0.03 mm (100C), bulk d 26 lb/ft³. No odor. Partially soluble in methanol, ethanol, and acetone; insoluble in water. Combustible.
Hazard: Toxic by ingestion.
Use: Antioxidant for gasoline and other hydrocarbons.

hydroxymethylethylene carbonate. See glycerin carbonate.

5-hydroxymethyl-2-furaldehyde. (5-hydroxymethyl-2-formylfuran). $C_6H_6O_3$.
Properties: Crystalline solid (needles), d 1.20, refr index 1.56, UV absorption max: 283 nm, soluble in water and common organic solvents, protect from light and oxygen.
Derivation: From glucose, cornstarch, sucrose, molasses.
Use: Organic synthesis for glycols, acetals, aldehydes, etc.

Dl-α-hydroxy-γ-methylmercaptobutyric acid, calcium salt. See methionine hydroxy analog calcium.

2-hydroxymethyl-5-norbornene. See 5-norbornene-2-methanol.

4-hydroxy-4-methylpentanone-2. See diacetone alcohol.

2-(2′-hydroxy-5′-methylphenyl)benzotriazole.
$HOC_6H_3(CH_3)N_3C_6H_4$.
Properties: Off-white, crystalline powder; mp 129–130C; bp 225C (10 mm); insoluble in water; soluble in methylethyl ketone, methyl methacrylate.
Use: Protector against UV radiation, thermal stabilizer, antioxidant, chelation of metals.

3-hydroxy-2-methyl-1,4-pyrone. See maltol.

hydroxynaphthalene. See naphthol.

3-hydroxy-2-naphthoic acid. (β-hydroxynaphthoic acid; 3-naphthol-2-carboxylic acid; β-oxynaphthoic acid). CAS: 92-70-6.
$C_{10}H_6OHCOOH$.
Properties: Yellow, rhombic leaflets. Soluble in alcohol and ether, mp 217–219C, insoluble in water.
Derivation: By treating sodium-2-naphtholate with carbon dioxide under pressure.
Hazard: Irritant.
Use: Dyes and pigments.

β-hydroxynaphthoic anilide. (naphthol AS; azoic coupling component 2).
$C_{10}H_6OHCONHC_6H_5$.
Properties: Cream-colored crystals, sodium salt is soluble in water, mp 246.0C.
Derivation: Condensation of β-hydroxynaphthoic acid and aniline.
Use: Dyes.

2-hydroxy-1,4-naphthoquinone. $C_{10}H_5O_2(OH)$.
Properties: Yellow to orange-yellow needles or powder, mp 192–195C (decomposes). Redox potential 0.362 volt; soluble in cold water, benzene, carbon tetrachloride, and petroleum ether.
Use: Intermediate for pharmaceuticals, henna hair and wool dyes, bactericides; seed disinfectant.

5-hydroxy-1,4-naphthoquinone. (juglone).
CAS: 481-39-0. This compound derived from walnuts is the starting point in the 16-step synthesis of oxytetracycline. It has also been found to be a natural herbicide with allelopathic properties.

N-(7-hydroxyl-1-naphthyl)acetamide.
(1-acetylamino-7-naphthol).
$C_{10}H_6(OH)NHCOOH_3$. A granular paste.
Use: Chemical intermediate.

4-hydroxy-3-nitrobenzenearsonic acid. (3-nitro-4-hydroxyphenylarsonic acid). $HOC_6H_3(NO_2)AsO(OH)_2$.
Properties: Pale yellow crystals.
Derivation: Heating phenol with arsenic and treating the p-hydroxyphenyl arsonate with nitric and sulfuric acids.
Hazard: Toxic by ingestion.

4-hydroxynonanoic acid, γ-lactone. See γ-nonyl lactone.

cis-**12-hydroxyoctadec-9-enoic acid.** See ricinoleic acid.

12-hydroxyoleic acid. See ricinoleic acid.

15-hydroxypentadecanoic acid lactone. See pentadecanolide.

3-hydroxyphenol. See resorcinol.

2-hydroxy-2-phenylacetophenone. See benzoin.

β-p-hydroxyphenylalanine. See tyrosine.

2-(2′-hydroxyphenyl)benzotriazole. $HOC_6H_4N_3C_6H_4$. An UV light absorber.
Use: Plastics.

p-hydroxyphenyl benzyl ether. See p-benzyloxyphenol.

4-(p-hydroxyphenyl)-2-butanone acetate. See Cuelure.

N-(p-**hydroxyphenyl)glycine.** (glycine[photographic]; photo-glycin). CAS: 122-87-2. $HOC_6H_4NHCH_2COOH$.
Properties: White to buff crystals or powder, mp 240C (with decomposition), slightly soluble in water, soluble in alkaline solutions.
Derivation: By condensation of p-aminophenol with chloroacetic acid.
Use: Photographic developer, cellulose and nitrocellulose acetate lacquers and varnishes, analytical reagent.

2-hydroxyphenylmercuric chloride. (chloromercuriphenol). CAS: 90-03-9. HOC_6H_4HgCl.
Properties: White to faint pink fine crystals; 0.1 part in 100 soluble in water (25C); soluble in hot water, alkali, and alcohol; mp 152C.
Hazard: Skin irritant. TLV: 0.01 (Hg) mg/m^3 of air.
Use: Antiseptic, fungicide.

hydroxyphenylstearic acid. A derivative of phenylstearic acid potentially useful as oxidation and corrosion inhibitor.

1-hydroxypiperidine. $C_5H_{10}NOH$.
Properties: Soluble in water, organic solvents, aqueous acids, and bases.
Use: Intermediate, reducing agent, polymerization inhibitor for vinyl monomers, antioxidant, metal ion reduction.

11-α-hydroxyprogesterone. $C_{21}H_{30}O_3$.
Properties: White, crystalline powder. Mp 163, specific rotation +179 degrees, insoluble in water, soluble in alcohol.
Derivation: From progesterone by microbiological oxidation.
Use: A steroid intermediate, biochemical research.

4-hydroxyproline. (Hyp; 4-hydroxy-2-pyrrolidinecarboxylic acid). CAS: 51-35-4. HOC_4H_7NCOOH.

Properties: Colorless crystals, very soluble in water, slightly soluble in alcohol, insoluble in ether, optically active. DL-hydroxyproline, mp 261–262C with decomposition. L-hydroxyproline, mp 270C (natural). D-hydroxyproline, mp 274C.
Derivation: Hydrolysis of protein (gelatin), organic synthesis.
Use: Biochemical research. Available commercially as L-hydroxyproline.

2-hydroxy-1,2,3-propanetricarboxylic acid. See citric acid.

hydroxy-2-propanone. (acetol; acetonyl alcohol; acetylcarbinol; hydroxyacetone; pyruvic alcohol). CH_3COCH_2OH.
Properties: Colorless liquid, d 1.0824 at 20/20C, bp 146C, fp −17C, soluble in water, alcohol and ether. Combustible.
Derivation: (a) By action of potassium acetate or potassium formate on a solution of bromo- or chloro-acetone in dry methanol, (b) by bacterial fermentation of propylene glycol.
Grade: Technical.
Use: Solvent for nitrocellulose.

α-hydroxypropionic acid. See lactic acid.

α-hydroxypropionitrile. See lactonitrile.

β-hydroxypropionitrile. See ethylene cyanohydrin.

2-hydroxypropyl acrylate. (HPA). CAS: 999-61-1. $CH_2CHCOOCH_2CHOHCH_3$. A functional

monomer used in manufacture of thermosetting acrylic resins for surface coatings.
Properties: A liquid with mw 130.14, bp 77C at 5 torr.
Hazard: TLV: 0.5 ppm in air. Corrosive to skin and eyes. Toxic by skin absorption.
Use: In manufacture of thermosetting resins for surface coatings.

2-hydroxypropylamine. See isopropanolamine.

hydroxypropyl cellulose. CAS: 9004-64-2.
A cellulose ether with hydroxypropyl substitution.
Properties: White powder; soluble in water, methyl and ethyl alcohols, and other organic solvents. Thermoplastic; can be extruded and molded. Insoluble in water above 37.7C. Combustible.
Grade: FCC.
Use: Emulsifier, film former, protective colloid, stabilizer, suspending agent, thickener, food additive.

hydroxypropylglycerin.
Properties: Pale straw-colored liquid, d 1.084 (25/25C), refr index 1.459 (25C), flash p 380F (193C), pour p −23C, soluble in water and methanol. Combustible.
Use: Intermediate for alkyd resins and polyesters, plasticizer for cellulosics, glue, starch, etc.

hydroxypropyl methacrylate.
$CH_3CHOHCH_2OOCC(CH_3):CH_2$.
Properties: Clear, mobile liquid. D 1.066 (25/16C), refr index 1.446 (25C), flash p 206F (96.6C), limited solubility in water, soluble in common organic solvents. Combustible.
Use: Monomer for acrylic resins, nonwoven fabric binders, detergent lube oil additives.

hydroxypropyl methylcellulose. (methylcellulose; propylene glycol ether).
Properties: White powder, swells in water producing clear to opalescent, viscous, colloidal solution; insoluble in anhydrous alcohol, ether, and chloroform. Combustible.
Grade: NF, FCC.
Use: Food products (except confectionery), as thickening agent, stabilizer, emulsifier; thickener in paint-stripping preparations.

N-β-hydroxypropyl-o-toluidine.
$CH_3C_6H_4NHCH_2CH(OH)CH_3$.
Properties: Amber color, distillation range 170–180C (20 mm), d 1.035–1.045 (20/20C), refr index 1.540–1.550 (20C).
Use: Dye intermediate.

4-hydroxy-2H-pyran-3,3,5,5(4H,6H)tetramethanol. See anhydroenneaheptitol.

2-hydroxypyridine-N-oxide. Bactericidal agent related to aspergillic acid, made from pyridine-N-oxide.

1-hydroxy-2-pyridine thione. (2-pyridinethiol-1-oxide). $C_5H_4NOH(S)$. Apparently exists in equilibrium with the -SH form. Forms chelates with iron, manganese, zinc, etc.
Use: fungicide, bactericide.

4-hydroxy-2-pyrrolidinecarboxylic acid. See hydroxyproline.

8-hydroxyquinoline. (8-quinolinol; oxyquinoline; oxine). CAS: 148-24-3. C_9H_6NOH.
Properties: White crystals or powder, darkens when exposed to light, technical grade usually tan; almost insoluble in water; soluble in alcohol, acetone, chloroform, benzene, also in formic, acetic hydrochloric, and sulfuric acids and in alkalies; phenolic odor; mp 73–75C; bp 267C.
Grade: CP, technical.
Hazard: Toxic by ingestion.
Use: Precipitating and separating metals, preparation of fungicides, chelating agent, disinfectant.

8-hydroxyquinoline benzoate. CAS: 86-75-9. $C_9H_6NOH:C_6H_5COOH$.
Properties: Yellowish-white crystals with a saffron odor, mp 56–61C, almost insoluble in water, soluble in alcohol and glycerol.
Use: Antiseptics, fungicide, recommended against Dutch elm disease.

8-hydroxyquinoline sulfate. CAS: 134-31-6. $(C_6H_7NO)_2 \cdot H_2SO_4$.
Properties: Pale yellow powder, slight saffron odor, burning taste, melting range 167–182C, soluble in water, slightly soluble in alcohol, insoluble in ether.
Use: Antiseptic, antiperspirant, deodorant, fungicide.

4-hydroxysalicylic acid. See β-resorcylic acid.

12-hydroxystearic acid. CAS: 106-14-9.
$CH_3(CH_2)_5(CHOH)(CH_2)_{10}COOH$. A C_{18} straight chain fatty acid with an —OH group attached to the carbon chain, mp 79–82C. It is produced by hydrogenation of ricinoleic acid. Combustible.
Use: Lithium greases, chemical intermediates.

1,12-hydroxystearyl alcohol. (1,12-octadecanediol). A long-chain fatty alcohol made by re-

duction of 12-hydroxystearic acid by replacing the —COOH group with a —CH$_2$OH. Combustible.
Properties: Boiling range 315–335C, mp 69C.
Use: Chemical intermediate, synthetic fibers, organic synthesis, pharmaceuticals, surface-active agents, plastics and resins, protective coatings.

hydroxysuccinic acid.   See malic acid.

α-hydroxytoluene.   See benzyl alcohol.

hydroxytoluene.   See cresol.

hydroxytoluic acid.   See cresotic acid.

1-hydroxytriacontane.   See 1-triacontanol.

5-hydroxytryptamine.   See serotonin.

4-hydroxyundecanoic acid, γ-lactone.
See γ-undecalactone.

hydrozincite.   (zinc bloom).   Zn$_5$(OH)$_6$(CO$_3$)$_2$.
A natural basic carbonate of zinc found in the upper zones of zinc deposits.
Properties: Color white to gray or yellowish, luster dull to silky, fluorescent in UV light, d 3.5–4.0, hardness 2.0–2.5.
Occurrence: Missouri, Pennsylvania, Utah, California, Nevada, Europe.
Use: An ore of zinc.

hydyne.   Mixture of 60% (by weight) of uns-dimethylhydrazine and 50% diethylenetriamine.
Use: High-energy fuel.

"Hyfac."[242]   TM for saturated or hydrogenated fatty acids and glycerides and castor oil derivatives.

"Hyform."[57]   TM for water emulsions of pure paraffin wax, microcrystalline wax or a modification of one of these waxes.
Use: Binders for pressed ceramic pieces, lubricants for die or mold release, and plasticizers during mold forming.

"Hygromix."[530]   TM for hygromycin B, a feed and food additive.

hygromycin.   CAS: 6379-56-2.   C$_{23}$H$_{29}$NO$_{12}$.
Properties: White powder, weakly acidic, freely soluble in water and alcohol. Produced by Streptomyces hygroscopicus.
Use: Medicine (broad-spectrum antibiotic).

hygroscopic.   Descriptive of a substance that has the property of adsorbing moisture from the air,

such as silica gel, calcium chloride or zinc chloride. The water-vapor molecules are held by the molecules of the agent which is called a desiccant when used primarily for this purpose. Paper and cotton fabrics are hygroscopic, normally containing 5–8% water after standing in an atmosphere of normal humidity; they are usually kept in constant-humidity rooms before use. Many dry chemicals are hygroscopic and should be kept in well-stoppered bottles or tightly closed containers.
See also deliquescent.

"Hylene."[28]   TM for a series of organic isocyanates. M-50, 50% methylenebis(4-phenyl isocyanate) (C$_6$H$_4$NCO)$_2$, in monochlorobenzene. Used in adhesives.
See diphenylmethane diisocyanate. MP. Bisphenol adduct of methylenebis(4-phenyl)isocyanate. Used as bonding agent for adhering "Dacron" fiber to rubber compositions. T Toluene-2,4-diisocyanate, used for urethane products. TM 80% toluene-2,4-diisocyanate and 20% toluene-2,6-diisocyanate. TM-65 65% 2,4-compound and 35% 2,6-compound.

"Hylox."[214]   TM for a polysulfone matrix used as the base for stereotype printing plates. The plate is produced by extruding polypropylene onto the matrix and pressure-forming.

"Hyonic."[309]   TM for a series of products used in liquid detergents, shampoos, textile scouring, metal cleaners, dairy detergents, rug and upholstery shampoos, hard surface cleaners.

Hyp.   Abbreviation for hydroxyproline.

"Hypalon."[28]   TM for chlorosulfonated polyethylene, a synthetic rubber.
Properties: White chips, d 1.10–1.28, resistant to ozone as well as the weather, oil, solvents, chemicals, and abrasion.
Use: Insulation for wire and cable, shoe soles and heels, automotive components, building products, coatings, flexible tubes and hoses, seals, gaskets, diaphragms. "Hypalon" 45 can accept large amounts of filler and is used as a binder for powdered metal to produce magnetic gaskets for doors and sheet goods for x-ray barriers.

hyperglycemic-glycogenolytic factor.
See glucagon.

hypergolic fuel.   A liquid rocket fuel or propellant which consists of combinations of fuels and oxidizers which ignite spontaneously on contact.

hypersorption.   Process in which activated carbon selectively adsorbs less volatile components from a gaseous mix, while the more volatile components pass on unaffected. Particularly applicable to separations of low-boiling mixtures such as hydrogen and methane, ethane from natural gas, ethylene from refinery gas, etc.

hypertensin. See angiotensin.

hypnone. See acetophenone.

hypo-. (1) A prefix used in chemical terminology to indicate a compound (usually an acid) in its lowest oxidation state or containing the lowest proportion of oxygen in a series of compounds, e.g., nitric acid (HNO_3), nitrous acid (HNO_2), hyponitrous acid ($H_2N_2O_2$). (2) Common term for a photographic chemical.
See sodium thiosulfate.

hypoallergenic. A term recently introduced in the cosmetics industry and defined by FDA as describing a cosmetic product that is less likely to cause adverse allergenic reactions than competing products. Claims of hypoallergenicity must be substantiated by specific dermatological tests made by the manufacturer of the product.

hypobromous acid. HBrO. An unstable compound resulting from hydrolysis of bromine chloride.
Use: Bactericide and wastewater disinfectant.

hypochlorite solution. An aqueous solution of a metallic salt of hypochlorous acid. Strong oxidizing agent.
Hazard: Irritant to skin and eyes.
Use: Bleaching of textiles, antiseptic agent.

hypochlorous acid. CAS: 7790-92-3. HOCl.
Properties: Greenish-yellow aqueous solution, highly unstable, weak acid, decomposes to hydrogen chloride and oxygen. Can exist only in dilute solutions.
Derivation: Water solution of chloride of lime (bleaching powder).
Hazard: Irritant to skin and eyes.
Use: Textile and fiber bleaching, water purification, antiseptic.
See also hypochlorite solution, calcium hypochlorite, sodium hypochlorite.

"Hypol."[311] TM for a family of polyurethane prepolymers derived from. toluene diisocyanate (TDI). Foams with addition of water. Good fire retardancey, high additive loading.
Use: Consumer, medical, biomedical, and industrial applications.

"Hypol Plus."[311] TM for a family of polyurethane prepolymers derived from. methylenedi-

phenyl diisocyanate (MDI). Foams with addition of water. High purity, high additive loading.
Use: Medical and surgical aids, health and personal car products.

α-hypophamine. See oxytocin.

β-hypophamine. See vasopressin.

hypophosphoric acid. CAS: 7803-60-3. $H_4O_6P_2$.
Properties: Platelike crystals, commercially available as water solution, readily forms hydrates, mp 70C, mp (hydrated) 55C, the solution decomposes when concentrated.
Use: Baking powder (sodium salt).

hypophosphorous acid. CAS: 6303-21-5. H_3PO_2.
Properties: Colorless, oily liquid or deliquescent crystals. Sour odor, soluble in water, d 1.439, mp 26.5C, a strong monobasic acid and reducing agent, sold in solution.
Derivation: Heating concentrated baryta water with white phosphorus and decomposing the barium hypophosphite with sulfuric acid, filtering the liquid, and concentrating under reduced pressure.
Grade: Technical, NF (30–32% solution, d 1.13), 50% purified.
Hazard: Fire and explosion risk in contact with oxidizing agents.
Use: Preparation of hypophosphites, electroplating baths.

hypoxanthine. CAS: 68-94-0. $C_5H_4N_4O$.
An intermediate in the metabolism of animal purines, also widely distributed in the vegetable kingdom.
Properties: White to cream powder, decomposes at 150C, almost insoluble in cold water, slightly soluble in boiling water, soluble in dilute acids and alkalies.
Derivation: Deamination of adenine, reduction of uric acid.
Use: Biochemical research.

hypoxanthine riboside. See inosine.

hypoxanthine riboside-5-phosphoric acid.
See inosinic acid.

hydrocarbon gas streams. A hydrocarbon such as methane is contacted with a catalyst under moderate conditions of temperature and pressure and decomposed into carbon, which remains on the catalyst, and hydrogen, which is mechanically removed. The hydrogen produced is about 94% pure when the charge stock is methane.

hysteresis. (Derived from the Greek word meaning "to lag behind".) A retardation of the effect, as if from viscosity, when the forces acting upon the body are changing. A common illustration is the retentivity of induction in ferromagnetic materials such as iron and its alloys when the magnetizing force is changed. When such a substance is placed in a magnetizing coil and the magnetizing field is gradually increased to a given value, and then decreased, the magnetic induction in decreasing does not follow the same relation to the magnetizing field that it did when the field was increasing but lags behind the decreasing field. Hysteresis is analogous to mechanical inertia and the energy lost is analogous to that lost in mechanical friction. It presents a major problem in the design of electrical machines with iron cores such as transformers and rotating armatures. In instruments designed for very high frequencies, the retardation and losses are so great as to render iron cores useless.

The stress-strain curves of vulcanized rubber also display hysteresis, in that strain (elongation, crystallization) persists when the deforming stress is removed, thus producing a hysteresis *loop* instead of a reversible pathway of the curves. This loop indicates a *loss* of resilient energy (Norman E. Gilbert). *Note:* The simple diagram shown below is a generalized representation of a hysteresis loop.

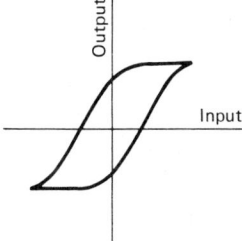

I

I. Symbol for iodine.

I-131. See iodine-131.

IAA. Abbreviation for 3-indoleacetic acid.

IATA. Abbreviation for International Air Transport Association (Geneva) which publishes annual regulations for air shipment of hazardous materials.

IBIB. Abbreviation for isobutyl isobutyrate.

ibuprofen. (p-Isobutyl-hydratropic acid; 2-(4-isobutylphenyl)propionic acid).
CAS: 15687-27-1.
Properties: Mw 206.31. $C_{13}H_{18}O_2$.
Use: Ingredient in over-the-counter pain relievers.

-ic. A suffix used in naming inorganic compounds which indicates that the central element is present in its highest oxidation state; thus in ferric chloride ($FeCl_3$) the iron has an oxidation number of +3, equivalent to its valence: in an ionized state it would have three positive charges (Fe^{+++}). (A recommended change in this system of nomenclature is to use the common name of the element (iron) together with a Roman numeral showing the oxidation number; thus, ferric chloride would be iron (III) chloride).

ICC. Abbreviation for ignition control compound, also Interstate Commerce Commission.

ice. (H_2O).   An allotropic, crystalline form of water, mp 0C (32F), latent heat of melting 80 calories/g, d 0.91; its property of melting under pressure accounts for "slipperiness." Occurs in nature as ice I but several other forms are known.
Use: Preservation of fish at sea, medicine (reduction of swelling).
See also water.

Iceland moss.
Properties: A water-soluble gum which gels on cooling.
Derivation: A lichen growing in Scandinavia and Iceland.
Use: Flavoring alcoholic beverages, food additive, cosmetics.

Iceland spar.   A form of calcite having unique optical properties.
Use: In polarizing light (Nicol prism).

ichthammol.   (ammonium ichthosulfonate).
Properties: Brownish-black, syrupy liquid, burning taste, tarry odor. Incompatible with acids, alkaloids, carbonate, hydroxides, mercuric chloride; soluble in water, alcohol-ether, or alcohol-ether-water mixture; partially soluble in alcohol and ether; miscible with glycerol.
Derivation: Aqueous solution of sulfonated ammonium compounds derived from the action of sulfuric acid upon distillates from bituminous shales.
Grade: NF.
Use: Pharmaceutical products such as skin ointments, cosmetic preparations, special dermatological soaps.

"Ichthymall."[329]   TM for ichthammol.

"Ichthyol."   TM for ichthammol.

-ide. A suffix used in naming compounds comprised of two elements; in such names the first (electropositive) element retains its name without change, while the second (electronegative) bears the suffix -ide as a modification of the elemental name. Examples: sodium hydroxide, magnesium chloride, hydrogen sulfide, etc. Similarly, oxygen is modified to oxide, fluorine to fluoride, phosphorus to phosphide, and carbon to carbide.

ideal gas.   (perfect gas).   A gas in which there is complete absence of cohesive forces between the component molecules; the behavior of such a gas can be predicted accurately by the ideal gas equation through all ranges of temperature and pressure. The concept is theoretical as no actual gas meets the ideal requirement; carbon dioxide especially lacks conformity. The generalized ideal gas law is derived from a combination of the laws of Boyle and Charles, namely $pv = RT$, where p is pressure, v is volume, T is absolute temperature and R is the gas constant ($p_0 v_0 / 273.2C$).

ideal solution.   A solution which exhibits no change of internal energy on mixing and complete uniformity of cohesive forces. Its behavior is described by Raoult's Law over all ranges of temperature and concentration.

identity period.   The repeating unit or monomer which occurs n times in a natural or synthetic polymer molecule; for example, the anhydroglucose unit in cellulose is enclosed in brackets:

Ig. Abbreviation for immunoglobulin.

"Igenal."[203] TM for a series of dyestuffs for chrome-tanned leather. Characterized by unusual tinctorial power.

"Igepal."[307] TM for a series of biodegradeable nonionic surfactants used as detergents, dispersants, emulsifiers and wetting agents. CA, CO, DM, RC: A series of alkylphenoxypoly(oxyethylene)ethanols resulting from the combination of an alkylphenol with ethylene oxide. The general formula is $RC_6H_4O(CH_2CH_2O)_n\,CH_2CH_2OH$, in which R may be C_8H_{17} or a higher homolog. LO and A: A series of linear alkylphenol ethoxylates (LO series) and a series of linear aliphatic ethoxylates (A series).

"Igepon."[307] TM for a series of anionic surfactants used as detergents, wetting agents, emulsifiers, dispersants and foaming agents. T and C types are sulfo-amides derived from N-methyltaurine or N-cyclohexyltaurine and fatty acids and have the general formula: $RCON(R')CH_2CH_2SO_3Na$. A types are sulfoesters derived from isethionic acid and a fatty acid and have the general formula: $RCOOCH_2CH_2SO_3Na$ (R and R′ represent alkyl groups).

ignition control compound. A substance such as methyl diphenyl phosphate or trimethyl phosphate which is added to gasoline to control spark plug fouling, surface ignition, and engine rumble.

ignition point. See autoignition temperature.

ignotine. See carnosine.

IILE. Ion-induced light emission.

"Illium."[491] TM for a series of superstainless steel alloys with high corrosive resistance.

ilmenite. (titanic iron ore). $FeO \cdot TiO_2$. Iron-black mineral, black to brownish-red streak, submetallic luster, resembles magnetite in appearance but is readily distinguished by feeble magnetic character, d 4.5–5, Mohs hardness 5–6.
Occurrence: Widely in US, Canada, Sweden, USSR, India, also made synthetically.

Use: Titanium paints and enamel, source of titanium metal, welding rods, titanium alloys, ceramics.

Imhoff tank. A reinforced concrete structure of considerable size (approximately 35 ft high) designed especially for sewage clarification. Its principal features are (1) an upper or sedimentation compartment in which in-flowing sewage deposits its suspended solids by gravity (residence time 2–3 hours), the free water being drawn off through an outlet, and (2) a separate lower compartment in which digestion of the accumulated sediment (sludge) takes place. The sludge is passed from the upper compartment to the digestion chamber through an inclined slot or channel. The gases generated by digestion are released through suitably located vents. The digested sludge is removed through outlet pipes at intervals of about 6 months. The dried sludge contains 2–3% ammonia and 1% phosphoric acid, which make it suitable as a soil conditioner. See also sewage sludge.

imidazole. (glyoxalin).   CAS: 288-32-4.

HNCHNCHCH.   A dinitrogen ring compound. An antimetabolite and inhibitor of histamine. Colorless crystals; mp 90C; bp 257C; soluble in water, alcohol, and ether.
Use: Biological control of pests, especially fabric-feeding insects, often in combination with *dl*-p-fluorophenylalanine, an amino-acid inhibitor; also as a contact insecticide in an oil spray. The mechanism is that of structural antagonism rather than active toxicity.
See also antihistamine; antagonist, structural.

4,5-imidazoledicarboxamide. See glycarbylamide.

4-imidazole ethylamine. See histamine.

2-imidazolidinone. See ethylene urea.

2-imidazolidone. See ethylene urea.

imidazo(4,5-d)pyrimidine. See purine.

imide. A nitrogen-containing acid having two double bonds.
See succinimide, phthalimide.

imine. A nitrogen-containing organic substance having a carbon-to-nitrogen double bond

$$R \text{---} CH$$
$$\|$$
$$NH$$

Such compounds are highly reactive, even more so than the carbon-nitrogen triple bond characteristic of nitriles.

3,3'-iminobispropylamine. (dipropylene triamine; 3,3'-diaminodipropylamine). CAS: 56-18-8. $H_2NC_3H_6NHC_3H_6NH_2$.
Properties: Colorless liquid, d 0.9307 (20/20C), bp 240.6C, fp −6.1C, flash p 175F (79.4C) (CC), soluble in water and polar organic solvents. Combustible.
Hazard: Toxic by ingestion and inhalation; irritant.
Use: Intermediate for soaps, dyestuffs, rubber chemicals, emulsifying agents, petroleum specialties, insecticides, and pharmaceuticals.

iminodiacetic acid disodium salt hydrate.
(iminodiethanoic acid disodium salt hydrate). CAS: 142-73-4. $HN(CH_2CO_2NA)_2 \cdot xH_2O$.
Properties: Crystalline solid.
Hazard: Irritant.
Use: Intermediate for surface active agents, complex salts, chelating agents, and aminocarboxylic acid synthesis.

iminodiacetonitrile. $HN(CH_2CN)_2$.
Properties: Light tan, crystalline solid; mp 77–78C; soluble in water and acetone.
Use: Chemical intermediate.

iminourea. See guanidine.

"Imlar."[28] TM for vinyl resin-base finish used where extreme resistance to abnormal chemical exposure is required.

"Immedial."[203] TM for a series of sulfur dyestuffs. Characterized by very good fastness to light and good fastness to washing and perspiration.
Use: Dyeing of cotton and rayon.

immiscible. Descriptive of substances of the same phase or state of matter that cannot be uniformly mixed or blended. Though usually applied to liquids such as oil and water, the term also may refer to powders that differ widely in some physical property, e.g., specific gravity, such as magnesium carbonate and barium sulfate.
See also miscibility.

immune serum globulin. A sterile solution of globulins which contains those antibodies normally present in adult blood. Over 90% of the total protein is globulin. It is a transparent, nearly colorless, nearly odorless liquid. Must be kept refrigerated.
Derivation: From a plasma or serum pool of venous or placental blood from 1000 or more individuals.

Grade: USP.
Use: Medicine (immunology).
See also antigen, globulin.

immunochemistry. That branch of chemistry concerned with the various defense mechanisms of the animal organism against infective agents, particularly the response between the body and foreign macromolecules (antigens) and the interaction between the products of the response (antibodies) and the agents that have elicited them. This involves study of the many proteins (serum globulins, enzymes, bacteria, and viruses) involved in these responses. It developed from the original work of Jenner (1775) and Pasteur (1880).
See also antigen-antibody, complement.

immunoglobulin. See globulin, immune serum globulin

IMP. (1) Abbreviation for inosine monophosphate.
See inosinic acid or sodium inosinate.
(2) Abbreviation for insoluble metaphosphate (Maddrell salt).
See sodium meta-phosphate.

impact strength. The ability of a material to accept a sudden blow or shock without fracture or other substantial damage, measured by standard impact-testing equipment (Izod, Charpy). It is a property of hard, friable materials such as metals, hard rubber, engineering plastics, Portland cement, glass, etc.

impalpable. Descriptive of a state of subdivision of particles so fine that the individual particles cannot be distinguished as such by pressing a powder between the thumb and index finger.

impeller. A type of agitator used in mixing or blending fluids of low viscosity, usually in a cylindrical chamber, either open or closed. The motion induced by an impeller is a combination of flow and turbulence, the proportion of each depending on the size, speed and position of the impeller. In common use are the marine propeller, the turbine and the helical ribbon types. Propeller and turbine impellers are attached to a power-driven rotating shaft which enters the container either vertically (top-entering) or at an angle (side-entering), they may be centered in a liquid or placed off-center, depending on the flow pattern desired. The propeller type has from two to four elliptical blades, whereas the turbine has a number of rectangular blades set vertically or at an angle. A wide range of flow-turbulence patterns can be obtained with either type. Helical

ribbon impellers are used for viscous liquids and dry powders. Many variations of impellers are available for liquids up to medium viscosity for a multitude of special mixing techniques.
See also agitator, mixing.

imperial green. See copper acetoarsenite.

"Impermex."[236] TM for a water-dispersible organic colloid, developed for the purpose of decreasing the water loss of drilling muds, even in those contaminated with salt, salt water, cement, or other water-soluble electrolyte.

"Implex."[23] TM for thermoplastic, high-impact acrylic molding powder, supplied in natural and colored forms. Maximum toughness, gloss, stain-, and heat-resistant grades.
Use: Shoe heels, business-machine and musical instrument keys, housings, automotive parts, knobs, metalized parts, etc.

"impregnole."[42] TM for a series of water-repellents. "FH" A zirconium-synthetic wax complex. "367" A solvent solution of a mixture of silicone polymers and other ingredients.

impurity. The presence of one substance in another, often in such low concentration that it cannot be measured quantitatively by ordinary analytical methods. It is impossible to prepare an ideally pure substance. In certain metal crystal lattices, foreign substances can exist in as low a concentration as one millionth of an atomic percent. For example, arsenic atoms are present in germanium crystals in this percentage; this fact is largely responsible for the semiconducting properties of germanium. Here the impurity is beneficial, but often it is detrimental, for example, in graphite used as a moderator in nuclear reactors, and in many metallic catalysts. In the air, trace amounts of sulfur dioxide and carbon monoxide are potentially dangerous impurities in concentration of 5 ppm of sulfur dioxide and 50 ppm for carbon monoxide.
See also purity, chemical; trace element, air pollution, semiconductor, purification.

incendiary gel. (1) Mixture of thermite suspended in oil set to a jelly with a small amount of soap which undergoes spontaneous ignition on contact with air. Another type may contain magnesium in jellied oil. (2) Jellied gasoline combined with thickening agents such as napalm or finely divided magnesium.

incineration. Disposal of solid and liquid organic waste materials by burning at temperatures from 1200–1500C. This method is approved by EPA for use on very toxic organic chemicals and chemical wastes. Use of specially equipped incinerator ships for burning chemical wastes at sea has become a common practice.
See also waste control.

inclusion complex. (adduct). An unbonded association of two molecules in which a molecule of one component is either wholly or partly locked within the crystal lattice of the other. There are several types of such complexes, the most familiar being the so-called clathrates (from Latin, crossbars of a grating). The clathrate compound $3C_6H_4(OH)_2 \cdot SO_2$ may be depicted as:

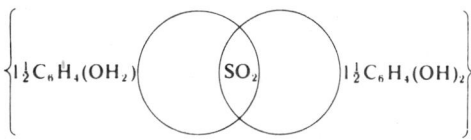

where the interlocked rings denote mutual enclosure of two identical cages. The formula for any clathrate compound is determined by the ratio of available cavities to the amount of cage material. Inclusion compounds can be used to separate molecules of different shapes, e.g., straight-chain hydrocarbons from those containing side chains as well as structural isomers. They can also be used as templates for directing chemical reactions.
See also clathrate compound, gas hydrate.

"Incoloy."[283] TM for a group of corrosion-resistant alloys of nickel, iron, and chromium.

"Inconel."[283] TM for a group of corrosion-resistant alloys of nickel and chromium.

"Indalone."[55] TM for n-butyl mesityl oxide oxalate.
See butopyronoxyl.
Use: Insect repellents which can be applied directly on the skin.

indan. (hydrindene; 2,3-dihydroindene).
CAS: 496-11-7.

CHCHCHCHCCCH₂CH₂CH₂.   Bicyclic

Properties: Colorless liquid, bp 176.5C, fp −51.4C, refr index 1.5388 (16.4C), d 0.965 (20/4C) insoluble in water, soluble in alcohol and ether. Combustible.
Derivation: From coal tar.
Hazard: Irritant to skin and eyes.
Use: Organic synthesis.

indanthrene. See indanthrone.

indanthrene yellow. See flavanthrene.

indanthrone. (Indanthrene Blue R; 6,15-dihydro-5,14,18-anthrazinetetrone). CAS: 81-77-6. $C_{28}H_{14}N_2O_4$. A blue vat dye or pigment. The molecule consists of two anthraquinone nuclei linked through two —NH groups.
Properties: Heat- and light-stable blue powder, decomposes 470C, soluble in concentrated sulfuric acid and dilute alkaline solutions.
Derivation: By fusion of β-aminoanthroquinone with caustic potash in the presence of potassium nitrate.
Use: Dyeing unmordanted cotton, pigment in quality paints and enamels.

indene. CAS: 95-13-6.

CHCHCHCHCCCHCHCH₂

Properties: Colorless liquid, d 0.996 (20/4C), fp −3.5C, bp 182C, refr index 1.5726 (25C), flash p 173F (78.3C), insoluble in water, soluble in most organic solvents, oxidizes readily in air, forms polymers on exposure to air and sunlight. Combustible.
Derivation: Contained in the fraction of crude coal tar distillates which boils from 176–182C.
Hazard: Toxic by inhalation. TLV: 10 ppm in air.
Use: Preparation of coumarone-indene resins, intermediate.
See also coumarone.

Indian red. (iron saffron). A red (maroon) pigment made by calcining copperas to obtain red ferric oxide. Fine particle size.
Use: Pigment in paint, rubber, and plastics: polishing agent.
See also iron oxide red, rouge.

Indian Yellow. (1) (aureolin) A yellow pigment distinguished by being unaffected by hydrogen sulfide. It is durable, without action upon other pigments and is permanent in oils and water color. It consists of a double nitrite of cobalt and potassium and is prepared by adding excess of potassium nitrite solution to a solution of cobalt nitrate acidified with acetic acid. (2) Also sometimes used for the yellow synthetic dye primuline.
See cobalt potassium nitrite.

indicator. An organic substance (usually a dye or intermediate) which indicates by a change in its color the presence or absence or concentration of some other substance, or the degree of reaction between two or more other substances. The most common example is the use of acid-base indicators such as litmus, phenolphthalein, and methyl orange to indicate the presence or absence of acids and bases, or the approximate concentration of hydrogen ion in a solution.
Use: Analytical chemistry.
The pH ranges of several typical indicators are as follows:

alizarin yellow R	10.1-12.0	yellow to red
methyl orange	3.1-4.4	red to yellow
phenolphthalein	8.3-10.0	colorless to red
phenol red	6.8-8.4	yellow to red
litmus	4.4-8.3	red to blue
Congo red	3.0-5.2	blue to red
bromthymol blue	6.0-7.6	yellow to blue
chlorphenol red	5.2-6.8	yellow to red
cresol purple	7.4-9.0	yellow to purple

See also titration, pH.

indigo. (indigotin; synthetic indigo blue; CI No.73000). CAS: 482-89-3. $C_{16}H_{10}N_2O_2$. A double indole derivative.

Properties: Dark blue, crystalline powder. Bronze luster, d 1.35, sublimes at 300C (decomposes), soluble in aniline, nitrobenzene, chloroform, glacial acetic acid, and concentrated sulfuric acid; insoluble in water, ether, and alcohol.
Derivation: From aniline and chloroacetic acid, and fusing the resulting phenylglycine with alkali and sodium amide. Formerly from plants of genus Indigofera.
Grade: Technical, pure.
Use: Textile dyeing and printing inks, manufacture of indigo derivatives, paints, analytical reagents.

"Indigolite."[309] TM for a combination of sodium formaldehyde sulfoxylate and a sulfonated quaternary base used to give a discharge on indigo-dyed grounds and discharge printing of vat dyestuffs.

indirect dye. A mordant dye.

indium. CAS: 7440-74-6. In. Metallic element of atomic number 49, of Group IIIA of the periodic system, aw 114.82, valences = 1,3; 2 stable isotopes.
Properties: Ductile shiny silver-white metal, softer than lead, soluble in acids, insoluble in alkalies, d 7.31 (20C), mp 156C, bp 2075C, corrosion-

resistant at room temperature, oxidizes readily at higher temperatures, Mohs hardness 1.2.

Occurrence: Not found native but in a variety of zinc and other ores. The indium content is generally very low, rarely exceeding 0.001%. Indium-bearing ores occur in western US, Canada, Peru, Japan, Europe, USSR.

Forms available: Small ingots or bars, shot, pencils, wire, sheets, powder, single crystals.

Purity: Technical, high purity (below 10 ppm impurities).

Hazard: Metal and its compounds toxic by inhalation. TLV: 0.1 (as In) mg/m³ of air .

Use: Automobile bearings, electronic and semiconductor devices, low-melting brazing and soldering alloys, reactor control rods, electroplated coatings on silver-plated steel aircraft bearings which are tarnish-resistant, radiation detector.

indium acetylacetonate. $In(C_5H_7O_2)_3$.
Properties: Mp 186C.
Hazard: See indium.
Use: catalyst.

indium antimonide. InSb.
Properties: Crystalline solid, mp 535C.
Hazard: See indium; antimony.
Use: Electronic grade used for semiconductor devices and infrared detector, computer technology (Hall effect).

indium arsenide. InAs.
Properties: Crystals, mp 943C, insoluble in acids.
Hazard: See indium; arsenic.
Use: Electronic grade for semiconductor devices, in injection lasers.

indium chloride. (indium trichloride).
CAS: 10025-82-8. $InCl_3$.
Properties: White powder, deliquescent, soluble in alcohol and water, d 3.46 (25C), mp 586C, sublimes at 300C.
Derivation: Direct union of the elements or by the action of hydrochloric acid on the metal.
Hazard: See indium.

indium oxide. (indium sesquioxide; indium trioxide). CAS: 1312-43-2. In_2O_3.
Properties: White to light-yellow powder in both amorphous and crystalline forms, depending on temperature, soluble in hot acid (amorphous), insoluble (crystals), d 7.179.
Derivation: By burning the metal in air or heating the hydroxide, nitrate, or carbonate.
Hazard: See indium.
Use: Manufacture of special glasses.

indium phosphide. CAS: 22398-80-7. InP.
A brittle metallic mass, mp 1070C, slightly soluble in mineral acids.

Hazard: See indium.
Use: Electronic grade in semiconductor devices, injection lasers, and experimental solar cells.

indium sulfate. CAS: 13464-82-9. $In_2(SO_4)_3$.
Properties: Grayish powder, deliquescent, soluble in water, d 3.438, decomposed by heat.
Hazard: See indium.

indium telluride. CAS: 1312-45-4. In_2Te_3.
Properties: Black, friable crystals, d 5.8, mp 665C.
Hazard: See indium; tellurium.
Use: Semiconductor technology.

indium trichloride. CAS: 10025-82-8. $InCl_3$.
Properties: Tan to yellow deliquescent crystals, d 4.0, mp 585C (sublimes), soluble in water, keep in closed containers.
Hazard: See indium.
Use: Electroplating baths.

"Indo Carbon."[203] TM for sulfur dyestuffs.
Use: Dyeing and printing of cotton and rayon.

"Indocin." TM for the anti-inflammatory drug indomethacin.
Use: Treatment of arthritis.

"Indofast."[438] TM for vat dyestuff pigments, including carbazole dioxazine violet.
Use: Paints, printing inks, and plastics.

indole. (2,3-benzopyrrole). CAS: 120-72-9.

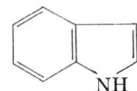

Properties: White to yellowish scales, turning red on exposure to light and air; unpleasant odor in high concentration but pleasant in dilute solutions, soluble in alcohol, ether, hot water, and fixed oils; insoluble in mineral oil and glycerol; mp 52C; bp 254C.
Derivation: From indigo and by numerous syntheses. Also can be produced from 220–260 degree fraction from coal tar.
Grade: Technical, CP, FCC.
Hazard: A carcinogen.
Use: Chemical reagent, perfumery.

3-indoleacetic acid. (IA; IAA; β-indoleacetic acid). CAS: 87-51-4. $C_8H_6NCH_2COOH$.
A plant hormone.
Properties: Crystals, mp 168–170C, the natural material is levorotatory, specific rotation −3.8 degrees in alcohol (20C), insoluble in water and chloroform, soluble in alcohol and ether.

Use: Agriculture and horticulture, growth-promoting hormone, plant cell enlarger.
See also auxin, plant growth regulator.

indole-α-aminopropionic acid.   See tryptophan.

3-indolebutyric acid.   (hormodin).
   CAS: 133-32-4. $C_8H_6N(CH_2)_3COOH$.
Properties: White or off-white powder, essentially odorless, mp 123C, insoluble in water, soluble in alcohols and ketones.
Use: Plant hormone, especially used in rooting plants.
See also auxin, plant growth regulator.

indole-2,3-dione.   See isatin.

indomethacin.   (USAN name for 1-(p-chlorobenzoyl)-5-methoxy-2-methylindole-3-acetic acid).
   CAS: 53-86-1. $C_{19}H_{16}ClNO_4$.
Use: Medicine for treatment of arthritis.

"Indopol."[216]   TM for polybutenes (mono-olefin polymers). Viscous, non-drying liquids of medium to high molecular weight.
Use: Lubricants and greases, sealants, caulks, adhesives, electrical insulation, chemical intermediates.

indurite.   Explosive containing 40% guncotton and 60% nitrobenzene.

"Indusoil."[228]   TM for distilled or fractionated tall oils and tall oil products.
Use: For coatings and resins, drying oils, heavy-metal soaps, detergents, cleaners, polishes, core oils, lubricants, floor coverings and flotation.

industrial alcohol.   See alcohol, industrial.

industrial carbon.   See carbon, industrial.

industrial chemistry.   See chemical technology, chemical process industry.

industrial diamonds.   See diamonds, industrial.

industrial dust.   See dust, industrial.

industrial waste.   See waste control, chemical waste.

inert.   A term used to indicate chemical inactivity in an element or compound. Helium, neon, and argon are practically inert gaseous elements; carbon dioxide is a gaseous compound of low activity. Ingredients added to mixtures chiefly for bulk and weight purposes are said to be inert.
See also noble, extender.

infrared.   The region of the electromagnetic spectrum including wavelengths from 0.78 micron to approximately 300 microns, (i.e., longer than visible light and shorter than microwave).
See also radiation.
Use: Spectroscopic analysis, medicine, baking of enamels, drying, photography.

infrared spectroscopy.   An analytical technique which may measure either (1) the range of wavelengths in the infrared that are absorbed by a specimen, which characterize its molecular constitution (absorption spectroscopy), or (2) the infrared waves emitted by excited atoms or molecules (emission spectroscopy). Extremely hot bodies (stars) emit spectra in which the atomic composition can be determined by characteristic lines such as the sodium D line in the sun's spectrum. Infrared absorption bands identify molecular components and structures, some of which are:

absorption band (μ)	structure indicated
2.3–3.2	OH and NH groups; H_2
3.2–3.3	aromatics, olefins
3.33–3.55	aliphatics
5.7–6.1	aldehydes, ketones, acids, amides

See also microwave spectroscopy, absorption (2).

"Infrax."[280]   TM for a refractory insulation used as primary linings of fuel-fired and electric furnaces only when protected by a cement facing. Available in brick form.

infusion.   An aqueous solution obtained by treating drugs with hot or cold water, without boiling. Generally prepared by pouring boiling water upon the vegetable substance and macerating the mixture in a tightly closed vessel until the liquid cools. When not otherwise specified, they are of 5% strength by weight.

infusorial earth.   See diatomaceous earth.

ingot iron.   Highly refined steel with a maximum of 0.15% impurity. Due to high purity it has excellent ductility and resistance to rusting.

ingrain dye.   An insoluble dye developed by impregnating a fabric with one or more intermediates and then producing the dye by reaction with a different intermediate.

inhibitor. (1) A compound (usually organic) that retards or stops an undesired chemical reaction, such as corrosion, oxidation or polymerization. Examples are acetanilide which retards decomposition of hydrogen peroxide and salicylic acid, used to prevent prevulcanization of rubber. Such substances are sometimes called negative catalysts. (2) A biological antagonist used to retard growth of pests and insects and in medicine.
See antagonist, structural; antioxidant.

"Inhibitor NPH."[329] TM for a synthetic organic chemical which provides an effective means of controlling hard polymer formation in synthetic rubber production.
Properties: Fine white to yellow-white platelets, bulk d 5 lb/gal, mp 160–164C (with decomposition), ammoniacal odor.

initiating explosive. See explosive, initiating.

initiator. An agent used to start the polymerization of a monomer. Its action is similar to that of a catalyst, except that it is usually consumed in the reaction. Organic peroxides and similar compounds are often used and short-wave radiation has a similar initiating effect. Free radicals usually play a part.
See also activator, free radical.

injection molding. A plastic-molding operation, introduced approximately 1935, performed in a single machine capable of producing both small articles of complex geometry (combs) and large units that could not be made economically in any other way (auto body parts, tubs, etc.). Though used primarily for thermoplastics, the injection method can be modified to handle thermosets. A simplified description is as follows: (1) A molding powder is fed into the heating chamber of the machine, which holds several times as much material as is necessary to fill the mold. The powder is heated to a viscous liquid. (2) An amount of molding powder that is just sufficient to fill the mold cavity is then forced into the rear of the heating chamber by a plunger, thus injecting an equal amount of liquid plastic from the front of the heating chamber into the mold. (3) The material remains in the mold under high pressure until it cools, and is then ejected. The rheological properties of the fluid plastic are of critical importance, as it must flow readily through the sprue and mold gate and fill the mold uniformly. The amount of material injected into a mold can range from less than 1 ounce to 25 pounds or more. Modern injection-molding machines have many specialized mechanical features and are of impressive dimensions.

Of comparatively recent development is the technique of reaction injection molding, in which a two-part semi-liquid resin blend is flowed through a nozzle into the mold cavity, where it polymerizes as a result of chemical reaction. One of the components contains the activating agent or curative; the two parts must be mixed with the greatest care for satisfactory results. Isocyanate resins (urethanes) and epoxy resins are well adapted to this fast, low-energy process.

ink. See printing ink, writing ink.

inorganic chemistry. A major branch of chemistry that is generally considered to embrace all substances except hydrocarbons and their derivatives, or all substances that are not compounds of carbon, with the exception of carbon oxides and carbon disulfide. It covers a broad range of subjects, among which are atomic structure, crystallography, chemical bonding, coordination compounds, acid-base reactions, ceramics, and the various subdivisions of electrochemistry (electrolysis, battery science, corrosion, semiconduction, etc.). It is important to state that inorganic and organic chemistry often overlap. For example, chemical bonding applies to both disciplines, electrochemistry and acid-base reactions have their organic counterparts; catalysts and coordination compounds may be either organic or inorganic.
Regarding the importance of inorganic chemistry, R. T. Sanderson has written: "All chemistry is the science of atoms, involving an understanding of why they possess certain characteristic qualities and why these qualities dictate the behavior of atoms when they come together. All properties of material substances are the inevitable result of the kind of atoms and the manner in which they are attached and assembled. All chemical change involves a rearrangement of atoms. Inorganic chemistry [is] the only discipline within chemistry that . . . examines specifically the differences among all the different kinds of atoms."

inosine. (hypoxanthine riboside).
CAS: 58-63-9. $C_{10}H_{12}N_4O_5$.   An important intermediate in animal purine metabolism. Also available as its barium salt.
Properties: (dihydrate) Crystallizes in needles from water, mp 90C (anhydrous), 218C (decomposes), levorotatory in solution, slightly soluble in water, soluble in alcohol.
Derivation: By deamination of adenosine.
Use: Biochemical research.

inosinic acid. (IMP; hypoxanthine riboside-5-phosphoric acid).   CAS: 131-99-7.

$C_{10}H_{13}N_4O_8P$.   An important intermediate in the synthesis and metabolism of animal purines.

Properties: Syrup with agreeable sour taste, freely soluble in water and formic acid, very sparingly soluble in alcohol and ether.

Derivation: From meat extract or by enzymatic deamination of muscle adenylic acid.

Use: Biochemical research, flavor enhancer.

See sodium inosinate.

inositol.   (hexahydroxycyclohexane).

CAS: 87-89-8. $C_6H_6(OH)_6 \cdot 2HOH$.   A constituent of body tissue. There are nine isomeric forms of inositol (myo-inositol or meso-inositol or specifically the cis-1,2,3,5-trans-4,6-hexahydroxycyclohexane has vitamin activity).

Properties: (I-inositol) White crystals, odorless, mp 224–227C, sweet taste, soluble in water, insoluble in absolute alcohol and ether. Stable to heat, strong acid, and alkali. D 1.524 (dihydrate), 1.752 when anhydrous, dihydrate melts at 215–216C.

Source: Food source: vegetables, citrus fruits, cereal grains, liver, kidney, heart, and other meats. Commercial source: corn steep liquor by precipitation and hydrolysis of crude phytate.

Units: Amounts are expressed in milligrams of inositol. Grade: NF, FCC.

Use: Medicine, nutrition, intermediate.

inositol hexaphosphoric acid.   See phytic acid.

inositol hexaphosphoric acid ester, sodium salt. See sodium phytate.

INPC.   Abbreviation for isopropyl-N-phenylcarbamate.

See IPC.

INS.   Ion neutralization spectroscopy.

insecticide.   A type of pesticide designed to control insect life that is harmful to man, either directly as disease vectors or indirectly as destroyers of crops, food products or textile fabrics. General types are as follows: (1) Inorganic: arsenic, lead and copper (inorganic compounds and mixtures); the use of these has diminished sharply in recent years because of the development of more effective types less toxic to man. (2) Natural organic compounds, such as rotenone and pyrethrins (relatively harmless to man since they quickly decompose to nontoxic compounds), nicotine, copper naphthenate, and petroleum derivatives. (3) Synthetic organic compounds: (a) chlorinated hydrocarbons such as DDT, dieldrin, endrin, chlordane, lindane, p-dichlorobenzene; (b) the organic esters of phosphorus (the parathions and related substances). (4) Of comparatively recent development are pyreth-

roids, or insect growth regulators, which act as neurotoxins, preventing larvae from becoming adult forms (juvenile hormones); and metabolic inhibitors, e.g., imidazole, which function as structural antagonists.

Hazard: Insecticides are toxic to man in varying degrees. Among the safest are the pyrethrins, rotenone, and methoxychlor. Most of the organophosphorus types (parathion and related compounds) are highly toxic, but are reasonably biodegradable. The chlorinated hydrocarbons resist biodegradation; their use has been restricted and in some cases (DDT) banned, for agricultural application, due to their harmful ecological effects. EPA is constantly monitoring new insecticides and issues warnings or restrictions on their use; its approval is required before they can be registered.

See also pesticide, fumigant, herbicide, rodenticide, repellent.

"Insecti-sol."[25]   TM for a highly purified odorless solvent exceeding CSMA minimum requirements for an insecticide base. The deodorized hydrocarbon distillate is water white and has a 170F (76.6C) (min) flash p, a 465–480F (240–248C) distillation end point, and a 98% unsulfonatable residue. Combustible.

Use: Food processing and cosmetic manufacture.

insect wax.   See shellac, Chinese insect wax.

"Insidol."[300]   TM for a wetting agent used in textile processing. It is a sulfodicarboxylic acid ester composition.

instantizing.   See agglomeration (2).

"Instantreat."[323]   TM for a high purity, specially processed, powdered complex phosphate. Contains enough chlorinated ingredient to disinfect feed solution and feed equipment.

Use: Controls corrosion, lime scale, and red water trouble in homes, hotels, restaurants, and small industrial water systems.

instrument.   Any of a wide variety of devices used for one of the following purposes: (1) observation (microscope), (2) measurement (thermometer, thermocouple, flowmeter, balance), (3) chemical analysis (spectrometer).

See also analytical chemistry, instrumentation.

instrumentation.   (1) Plant. An inclusive term for sensing devices that measure, record, and control temperature, flow rate, thickness, pH, liquid level, and other process variables on a continuous basis. Some types, e.g., thickness gauges, utilize radioisotopes. Particularly sophisticated computerized instrumentation is required in petroleum refining and nuclear reactor control. (2) Labora-

tory. A broad range of analytical techniques and devices utilized in the many types of chromatography and spectroscopy. Significant advances in analytical instrumentation have been made in recent years, e.g., in liquid chromatography (HPLC). A notable event in this field is the annual Pittsburgh Conference on Analytical Chemistry and Applied Spectroscopy. (3) General. Of special significance for both laboratory and process control is the introduction of fiber optical devices which can transmit signals from many remote locations. They are particularly useful in analyzing samples of radioactive chemicals and other hazardous materials. They are also used in high-temperature thermometry and dosimetry.
See fiber, optical.

insulating oil.   See transformer oil.

insulator.   Any substance or mixture that has an extremely low dielectric constant, low thermal conductivity, or both. Electrical insulators are either solid or liquid, the latter being used in transformers (askarel, mineral oils, silicone oils). A wide variety of solid types includes porcelains, glass, mica, alumina, various high polymers (epoxies, polyethylene, polystyrene, phenolics), cellulosic materials, nylon, and silicone resins. All these may be used alone or combined with other insulators as composites.
See also dielectric, transformer oil.

  Thermal insulators comprise an equally broad range of materials. Such inorganics as mineral fibers, magnesia, aluminum silicate, cellulose, and glass fibers are widely used for steam and hot-water pipes, furnaces, and blown-in home insulation. Organic products that are effective include plastic foams (polyurethane, polyvinyl chloride, polystyrene) and cellular rubber. There are a number of materials that may be called double insulators, since they have both electrical and thermal insulating properties, e.g., polystyrene, PVC, cellulose, glass, magnesia, and aluminum silicate.

  Air is a unique case, as it is the only gaseous material in actual use as an insulator. Its dielectric constant is 1.0058, far less than that of any other dielectric material, and it has low thermal conductivity as well. It is particularly effective when trapped within a solid network, as in wool, cellular plastics, or glass fibers, or as an interlayer between wall panels.

"Insulgrease."[245]   TM for a series of grease-like silicone dielectric compounds.
Use: Electrical insulation in connectors, electrical and electronic equipment, as heat transfer media, moisture and corrosion resistant coatings, antiseize compounds, and insulator coatings.

insulin.   A polypeptide hormone having a molecular weight of 5733. It is formed in the isles of Langerhans located in the pancreas and was so named for this reason. Insulin is composed of 16 amino acids arranged in a coiled chain and crosslinked in several places by the disulfide bonds of cystine residues. The sequence of amino acids has been elucidated. The insulin molecule was synthesized in 1963. In 1977, rat insulin was produced in the bacterium *E. coli* by recombinant DNA techniques. A year later human insulin was generated after chemically synthesized genes were added to *E. coli*. This synthetic insulin is now in commercial production and has been approved by FDA. Insulin regulates carbohydrate metabolism in the body by decreasing the blood glucose level. A systemic deficiency leads to diabetes.
Properties: White powder or hexagonal crystals, readily soluble in dilute acids, soluble in water.
Derivation: Extraction of minced pancreas with acidified dilute alcohol, followed by precipitation with absolute alcohol; also by gene-splicing methods.
Grade: USP, in various solutions or suspensions which include insulin injection; isophane insulin suspension; protamine zinc insulin suspension; NF, as globin zinc insulin injection.
Hazard: Overdosage can be fatal.
Use: Medicine (for diabetes control).
See also Banting, recombinant DNA.

"Insuloxide."[337]   TM for a mixture containing a minimum of 94% zirconium oxide and a maximum of 5% silicon dioxide, mp 2485C.
Use: Thermal insulation, super-refractories.

"Insurok."[63]   TM for a series of laminated and molding plastics characterized by durability, light weight, resistance to many chemicals, and high dielectric strength.
Use: Replacement for cast aluminum in airplane parts, switches, distributors, commutators, etc.

"Intalox."[249]   TM for a particular shape of tower filling materials available in porcelain, chemical stoneware, and carbon.

intercalation compound.   A compound comprised of a crystalline lattice which acts as an electron donor and "foreign" electron acceptor atoms interspersed or diffused between the planes of the lattice. An important group of intercalated compounds are those of graphite, where bromine, for example, can act as electron acceptor. Graphite is particularly susceptible to this phenomenon because of its orderly stacked layers of crystals. Anhydrous metal nitrates such as copper and zinc nitrates also form intercalated compounds with graphite. A further example is trilithium nitride, whose structure consists of a series of layers of dilithium nitride, between which is a layer of lithium atoms. This markedly increases the conductivity, so that the material becomes an effective solid electrolyte in batteries. Other substances having this property are sodium beta alumina, titanium disulfide and some metal dioxides. The phenomenon does not impair the crystalline structure and is reversible. Intercalated compounds are used for superconductors, synthetic lubricants, catalysts, and storage batteries. They are used in biochemical research; an acridine-based compound that can intercalate between stacked pairs of bases in a DNA helix is used in cancer research.

interface. The area of contact between two immiscible phases of a dispersion which may involve either the same or different states of matter. Five types are possible: (1) solid/solid (carbon black/rubber), (2) liquid/liquid (water/oil), (3) solid/gas (smoke/air), (4) solid/liquid (clay/water), (5) liquid/gas (water/air). At a fresh surface of either liquid or solid the molecular attraction exerts a net inward pull. Hence the characteristic property of a liquid is surface tension and that of a solid surface is adsorption. Both have the same cause, namely, the inward cohesive forces acting on the molecules at the surface. These phenomena provide to some degree the fundamental mechanism for many industrially important processes (catalysis, emulsification, mixing, alloying) and products (detergents, adhesives, lubricants, paints). Such properties as wettability of solid powders, spreading coefficients of liquids, and protective action of colloidal substances are intimately associated with interfacial behavior.
See also surface, surface tension, catalysis, emulsion, detergent, wetting agent.

interferon. An antiviral protein produced by vertebrate cells in response to virus infection. Discovered in 1957, it is a product of the infected cell, rather than of the disease-inducing virus. It can be formed by any type of infected cell and is effective against many viruses, but only in the cells of the organism that produced it. It is quite different from an antibody which is produced only by specialized cells and acts by combining directly with a specific virus. Interferon does not inactivate viruses directly, but reacts with susceptible cells which then resist virus multiplication. Prostaglandins are believed to be effective in maintaining normal interferon production in the body. Four types of interferon are presently known. As a result of advanced recombinant DNA techniques, a biologically active protein almost identical with interferon has been made outside the body by exposing to virus action bacteria into which the genetic code controlling interferon formation in the body has been introduced. The "synthetic" interferon is available in quantities large enough to permit its use in treatment of various virus diseases. The interferon molecule itself has not yet been synthesized.
See also recombinant DNA.

intermediate. An organic compound, either cyclic (derived from coal tar or petroleum products such as benzene, toluene, naphthalene, etc.) or acyclic, (e.g., ethyl and methyl alcohol). These compounds may be considered as chemical stepping stones between the parent substance and the final product. The cyclic type (e.g., aniline, β-naphthol, and benzoylbenzoic acid) still predominate as intermediates for synthetic dyes and have few other uses; the acyclic type in general have many independent uses. Exceptions are hexamethylenetetramine, an acyclic intermediate for phenolformaldehyde resin and butadiene for synthetic elastomers. Intermediates are the foundation of the modern approach to organic technology. The distinction between an intermediate and an end product is not always precise.
See also azo dye intermediate.

intermetallic compound. (1) A compound or alloy formed by two metals that have been placed in intimate contact during the process of brazing or coating, the compound occurring at the interface between the metal surfaces. In some cases, as in galvanizing, the metals form bimetallic compounds; in others they form alloys or solid solutions of varying composition. (2) A two- or three-component metal system prepared with special metals having semiconducting properties, e.g., gallium, for use in lasers, diodes, transistors, etc.
See also galvanizing, semiconductor.

International Union of Pure and Applied Chemistry. (IUPAC). A voluntary nonprofit association of national organizations representing chemists in 45 member countries. It was formed in 1919 with the object of facilitating international agreement and uniform practice in both academic and industrial aspects of chemistry. Examples are nomenclature, atomic weights, symbols and terminology, physicochemical constants, and certain methods of analysis and assay. In addition to standardization, IUPAC carries on several hundred research projects relating to food technology, water and air quality, single-cell proteins, etc. Its offices are in Basel, Switzerland.

interstellar space. See astrochemistry.

interstitial. (1) Descriptive of a nonstoichiometric compound of a metal and a nonmetal whose structure conforms to a simple chemical formula, but exists over a limited range of chemical composition. Interstitial compounds are represented by borides, nitrides, and carbides of the transition metals. (2) Descriptive of an atom of an impurity that causes a defect or dislocation in a crystalline lattice, e.g., an atom of carbon or nitrogen in an iron crystal, or of arsenic in a semiconductor. (3) In a biological sense, the term describes cells located between or within layers of tissue.

introns. Intervening sequences in the genes of higher organisms; they vary in number and may be as high as 17 (in the protein conalbumin).

intumescence. The foaming and swelling of a plastic or other material when exposed to high surface temperatures or flames.
Use: Polyurethane base coating materials for rocket reentry.

inulin. (alantin; alant atarch).
CAS: 9005-80-5.
Properties: Spherical crystals with mp 158–165C. Hygroscopic in moist air. Soluble in hot water.
Source: From chicory root.
Use: Diagnostic aid for kidney function.

"Invar." TM for an iron-nickel alloy containing 40–50% nickel and characterized by an extremely low coefficient of thermal expansion.
Use: Precision instruments, measuring tapes, weights, etc.

invention. The chief requirement of an invention is that it be unobvious to a person having ordinary skill in the art to which the claim pertains and knowing everything that has gone before, as shown by publication anywhere or public use in the US. When reliance for patentability is placed on a new mixture of components that have been used separately, it must be shown that there is some unexpected coaction between the ingredients and not just the additive effects of the several materials. The inventor is the one who contributes the inventive concept in workable detail, not the one who demonstrates it or tests it.

"Invermul."[236] TM for a basic emulsifier in the preparation of water in oil-type emulsion drilling muds for oil wells.

invertase. (sucrase; invertin). Enzyme produced by yeast and by the lining of the intestines.
It is a white powder, soluble in water. It catalyzes the conversion of sucrose (ordinary sugar) to glucose and levulose (fructose) during fermentation of sugars.
Use: Production of invert sugar for syrups and candy; analytical reagent for sucrose.

invert soaps. Cationic detergents with disinfectant properties.

invert sugar. A mixture of 50% glucose and 50% fructose obtained by the hydrolysis of sucrose. It absorbs water readily and is usually only handled as a syrup. Because of its fructose content, invert sugar is levorotatory in solution and sweeter than sucrose. Invert sugar is often incorporated in products where loss of water must be avoided. Commercially it is obtained from the inversion of a 96% cane sugar solution.
Use: Food industry, brewing industry, confectionery, humectant.

investment casting. A ceramic or metal casting method originally used to make reproductions of sculptured pieces (lost wax process) and adapted to industry for manufacture of precision metal parts. It is generally known as precision investment casting. The sequence of operations is (1) a wax prototype is made in a metal mold; (2) several of these are attached to a central member, also made of wax to form a "tree;" (3) the tree is dipped and dried eight times in a ceramic glaze, thus building up a coating or "investment;" (4) the assembly is baked in an oven, thus melting out the wax, leaving a cavity which is then used as a mold for liquid metal. Great accuracy can be obtained with this method. The exact nature of the waxes and coatings are not disclosed.

in vitro. A condition in which a reaction is carried out in a laboratory experiment (i.e., a glass container, test tube, or beaker), as opposed to a reaction occurring in a living organism (in vivo).

iodeosin. (tetraiodofluorescein). $C_{20}H_8I_4O_5$.
Properties: Red powder, soluble in dilute alkalies, slightly soluble in alcohol and ether, insoluble in water.
Derivation: Interaction of fluorescein and iodine in presence of iodic acid.
Use: Indicator in analytical chemistry.

iodic acid. CAS: 7782-68-5. HIO_3.
Properties: Colorless, rhombic crystals or white, crystalline powder. A moderately strong acid, soluble in cold and hot water, d 4.629, mp 110C (decomposes).
Derivation: By adding sulfuric acid to a solution

of barium iodate and subsequent filtration and crystallization.

Hazard: Toxic by ingestion, strong irritant to eyes and skin.

Use: Analytical chemistry, medicine (1–3% solution).

iodic acid anhydride. (iodine pentoxide).
I_2O_5.

Properties: White, crystalline powder; soluble in water and nitric acid; insoluble in absolute alcohol, chloroform, ether, carbon disulfide; d 4.799; mp 300C (decomposes).

Hazard: Toxic by ingestion, strong irritant to eyes and skin, strong oxidizing agent.

Use: Oxidizing agent, organic synthesis.

iodine. CAS: 7553-56-2. I. Nonmetallic halogen element of atomic number 53; group VIIA of the periodic table; the least reactive of the halogens, aw 126.9045; valences = 1,3,5,7; no stable isotopes but many artificial radioactive isotopes.

Properties: Heavy, grayish-black plates or granules having a metallic luster and characteristic odor; readily sublimed having a violet vapor; d 4.98; mp 113.5C; bp 184C; soluble in alcohol, carbon disulfide, chloroform, ether, carbon tetrachloride, glycerol, and alkaline iodide solutions; insoluble in water; a semiconductor.

Derivation: (a) From brine wells in Michigan, Oklahoma, Japan, Indonesia; (b) from mother liquors of Chilean nitrate. It can also be extracted from kelp.

Method of purification: Sublimation.

Grade: Crude, CP, USP.

Hazard: Toxic by ingestion and inhalation, strong irritant to eyes and skin. TLV: ceiling 0.1 ppm in air.

Use: Dyes (aniline dyes, phthalein dyes), alkylation and condensation catalyst, iodides, iodates, antiseptics and germicides, x-ray contrast media, food and feed additive, stabilizers, photographic film, water treatment, pharmaceuticals, medicinal soaps, unsaturation indicator.

iodine 131. (I-131). Radioactive iodine of mass number 131.

Properties: Half-life 8 days; radiation, beta and gamma.

Derivation: By pile irradiation of tellurium and from the fission products of nuclear reactor fuels.

Forms available: As elemental iodine and in a weakly basic solution of sodium iodide in sodium sulfite; iodine 131 is also available in tagged compounds such as dithymol diiodide, potassium iodate, diiodofluorescein, insulin, ACTH, etc.

See also grades.

Grade: USP lists iodinated I-131 serum albumin, rose bengal sodium I-131 injection, sodium iodide I-131 capsules and solution, sodium iodohippurate I-131 injection.

Use: Diagnosis and treatment of goiter, hyperthyroidism, and other thyroid disorders; internal radiation therapy; in film gauges to measure film thicknesses of the order of one micron; for detecting leaks in water lines; as a source of radiation in oil field tests; as a tracer in chemical analysis; as a tracer in studying diet iodine for cattle, the functions of the thyroid gland, the efficiency of mixing pulp fibers, the thermal stability of potassium iodate in bread dough, chemical reaction mechanisms, etc.

iodine bromide. See iodine monobromide.

iodine chloride. See iodine monochloride and iodine trichloride.

iodine cyanide. See cyanogen iodide.

iodine monobromide. (bromine iodide).
CAS: 7789-33-5. IBr.

Properties: Crystals, purplish-black mass. Soluble in water with decomposition, alcohol, and ether. Mp 42C, bp 116C (decomposes), d 4.41.

Derivation: By the interaction of iodine and bromine.

Hazard: Toxic by ingestion and inhalation, vapors corrosive to tissue.

Use: Organic synthesis.

iodine monochloride. CAS: 7790-99-0.
ICl.

Properties: Black crystals (alpha and beta forms) or reddish-brown, oily liquid. Soluble in alcohol, water (with decomposition), and dilute hydrochloric acid. Mp (alpha) 27C, (beta) 14C, bp 101C (decomposes), d (alpha) 3.18 (0C), (beta) 3.24 (liquid at 34C).

Derivation: By the action of dry chlorine on iodine.

Hazard: Toxic by ingestion and inhalation, strong irritant to eyes and skin.

Use: Analytical chemistry, organic synthesis.

iodine number. (iodine value). The percentage of iodine that will be absorbed by a chemically unsaturated substance (vegetable oils, rubber, etc.) in a given time under arbitrary conditions. A measure of unsaturation.

iodine pentafluoride. CAS: 7783-66-6.
IF_5.

Properties: Fuming liquid, bp 98C, mp 9.4C, d 3.189 (25C), attacks glass.

Derivation: Passing fluorine over iodine. Available in cylinders at 98.0% min purity.

Hazard: Dangerous fire risk, reacts violently with water. Toxic by ingestion and inhalation, corrosive to skin and mucous membranes. TLV (as F): 2.5 mg/m^3 of air.
Use: Fluorinating and incendiary agent.

iodine pentoxide. See iodic acid anhydride.

iodine tincture. A solution of iodine and potassium iodide or sodium iodide in alcohol, a reddish-brown liquid having the odors of iodine and alcohol, contains 44–50% by volume alcohol, 2 g of iodine and 2.4 g sodium iodide per 100 cc.
Grade: USP
Hazard: Toxic by ingestion, avoid using on open cuts.
Use: Antiseptic (use on skin surface only).

iodine trichloride. CAS: 865-44-1. ICl$_3$.
Properties: Orange-yellow, deliquescent, crystalline powder; pungent irritating odor; soluble in water (with decomposition), alcohol, and benzene; mp 33C; d 3.11.
Derivation: By interaction of iodine and chlorine.
Hazard: Toxic by ingestion and inhalation, corrosive to tissue.
Use: Agent for introducing iodine and chlorine in organic synthesis; topical antiseptic.

iodine value. See iodine number.

iodipamide. CAS: 606-17-7.
(CH$_2$)$_4$(CONHC$_6$HI$_3$COOH)$_2$.
Properties: White, nearly odorless, crystalline powder; very slightly soluble in alcohol, chloroform, ether; insoluble in water; pH of saturated solution is 3.5–3.9.
Hazard: Toxic by ingestion.
Use: Medicine (x-ray contrast medium).

iodisan. See propiodal.

iodized oil. An iodine addition product of vegetable oil or oils containing 38–42% organically combined iodine.
Properties: Thick, viscous, oily liquid; affected by air and light; soluble in solvent naphtha.
Hazard: Toxic by ingestion.
Use: Medicine (radiopaque medium).

iodoacetic acid, sodium salt. (sodium iodoacetate). CAS: 305-53-3. ICH$_2$CO$_2$Na.
Properties: Colorless or white crystals with mw 207.93, mp 210C. Soluble in water, alcohol, and very slightly soluble in ether. Hygroscopic in moist air.
Hazard: Toxic.
Use: Analytical reagent.

iodoethane. See ethyl iodide.

iodoethylene. See tetraiodoethylene.

iodoform. (triiodomethane). CAS: 75-47-8. CHI$_3$.
Properties: Small greenish yellow or lustrous crystals or powder, penetrating odor. Soluble in benzene and acetone; partially soluble in alcohol, glycerol, chloroform, carbon disulfide, and ether; insoluble in water. D 4.08, mp 115C.
Derivation: (a) By heating acetone or methanol with iodine in presence of an alkali or alkaline carbonates. (b) Electrolytically, by passing a current through a solution containing potassium iodide, alcohol, and sodium carbonate.
Grade: Technical, NF.
Hazard: Irritant, decomposes violently at 400F (204C). TLV: 0.6 ppm in air.
Use: Medicine (antiseptic for external use only).

iodomethane. See methyl iodide.

1-iodooctadecane. (octadecyl iodide). CAS: 629-93-6. CH$_3$(CH$_2$)$_{17}$I.
Properties: Solid with mw 380.40, mp 33-35C, bp 194–197C/2 mm, fp 110C. Light sensitive.
Hazard: Irritant.
Use: Reagent for introduction of C$_{18}$ chain.

iodopanoic acid. See iopanoic acid.

iodophor. ("tamed iodine"). (1) A complex of iodine with certain types of surface-active agents that have detergent properties. (2) More generally, any carrier of iodine.

iodophosphonium. See phosphonium iodide.

2-iodopropane. See isopropyl iodide.

N-iodosuccinimide. (succiniodimide). CAS: 516-12-1. (—CH$_2$CO)$_2$NI).
Properties: Colorless crystals, mp 200–201C, soluble in acetone and methanol; insoluble in carbon tetrachloride and ether, decomposes in water.
Hazard: Skin irritant.
Use: Iodinizing agent in synthetic organic chemistry.

"Imag."[329] TM for a potassium iodide mixture containing 90% potassium iodide and made free-flowing with 8% magnesium carbonate and 2% potassium hydroxide.

ion. An atom or radical that has lost or gained one or more electrons and has thus acquired an electric charge. Positively charged ions are cations and those having a negative charge are anions. An ion often has entirely different properties from the element (atom) from which it was formed.

In sodium chloride solution, sodium exists as sodium ion (Na^+), i.e., sodium atoms that have lost one electron. The chlorine is present as chloride ion (Cl^-), i.e., chlorine atoms that have gained one electron. Copper sulfate solution contains copper ion (Cu^{++}), i.e., copper atoms that have lost two electrons, and sulfate ion (SO_4^{--}), i.e., sulfate radicals that have gained two electrons.

Ions occur in water solution or in the fused state (except in the case of gases). Compounds that form ions are called electrolytes because they enable the solution to conduct electricity. Ion formation causes an abnormal increase in the boiling point of water and also lowers the freezing point, the extent depending on the concentration of the solution. Ions are also formed in gases as a result of electrical discharge.
See also ionization, electrolysis, ion exchange.

ion exchange. A reversible chemical reaction between a solid (ion exchanger) and a fluid (usually a water solution) by means of which ions may be interchanged from one substance to another. The superficial physical structure of the solid is not affected. The customary procedure is to pass the fluid through a bed of the solid, which is granular and porous and has only a limited capacity for exchange. The process is essentially a batch type in which the ion exchanger, upon nearing depletion, is regenerated by inexpensive brines, carbonate solutions, etc. Ion exchange occurs extensively in soils.

Ion exchange resins are synthetic resins containing active groups (usually sulfonic, carboxylic, phenol, or substituted amino groups) that give the resin the property of combining with or exchanging ions between the resin and a solution. Thus, a resin with active sulfonic groups can be converted to the sodium form and will then exchange its sodium ions with the calcium ions present in hard water. Some specific applications of ion exchange: Water softening, milk softening (substitution of sodium ions for calcium ions in milk), removal of iron from wine (substitution of hydrogen ions), recovery of chromate from plating solutions, uranium from acid solutions, streptomycin from broths, removal of formic acid from formaldehyde solutions, demineralization of sugar solutions, recovery of valuable metals from wastes, recovery of nicotine from tobacco-dryer gases, catalysis of reaction between butyl alcohol and fatty acids, recovery and separation of radioactive isotopes from atomic fission, chromatography, establishment of mass micro standards, in cigarette filters to remove polonium from smoke.
See also zeolite.

ion-exchange chromatography. A chromatographic method based on the ability of polymers to sorb ionized solutes reversibly, e.g., crosslinked resins with exchangeable hydrogen or hydroxyl ions. It can be carried out both in columns and on sheets.

ion-exchange resin. See ion exchange.

ion exclusion. The process in which a synthetic resin of the ion exchange type absorbs nonionized solutes such as glycerine or sugar while it does not absorb ionized solutes that are also present in a solution in contact with the resin. Thus, sodium chloride and glycerine can be separated by passage of their aqueous solution through a bed of particles of an ion exclusion resin.

ionic bond. See bond, chemical.

ionic detergent. See detergent.

"Ionite."[271] TM for a lignite-type material.
Use: Oil-well drilling muds, as low-density filler in dark-colored rubber, as an organic base filler for fertilizers, and as a source of humic acid.

ionization. A chemical change by which ions are formed from a neutral molecule of an inorganic solid, liquid, or gas. The most common type of ionization occurs when an ionically bonded inorganic compound such as sodium chloride or sulfuric acid is dissolved in water (or other solvent), the molecule separates or dissociates into two ions, the metallic ion being positively charged by loss of an electron and the nonmetallic ion being negatively charged by gaining an electron. The degree of dissociation varies with the type of compound, the solvent, and the temperature. Molecules or atoms of gases are ionized by passage of an electric current through the gas; this removes electrons and leaves a positive charge.

Compounds that ionize in solution greatly increase the conductivity of the solvent. Ionization is most effective in water because its high dielectric constant lowers the ionic bonding forces in the solute molecules enough to cause separation of their constituent atoms. Ion formation produces a notable rise in the boiling point and a depression of the freezing point of water. An electric current passed through a solution containing ions causes them to move to the oppositely charged electrode; this effect is the basis of many industrial electrochemical operations, such as electroplating and the manufacture of sodium hydroxide and chlorine.

See also ion, electroplating, electrolysis, electrolyte, dissociation.

ionizing radiation. See radiation, ionizing.

ionomer resin. A copolymer of ethylene and a vinyl monomer with an acid group, such as methacrylic acid. They are crosslinked polymers in which the linkages are ionic as well as covalent bonds.
See bond, chemical.
There are positively and negatively charged groups which are not associated with each other and this polar character makes these resins unique.
Properties: Transparent, electrically conductive, resilient, and thermoplastic. Cannot be completely dissolved in any commercial solvents. High resistance to abrasion, cracking, corona attack; high tensile strength. Working temperature approximately -108 to $+65.5C$.
Use: Break-resistant transparent bottles, packaging films, mercury flasks, protective equipment, pipe and tubing, electric distribution elements. As foam, insulation of fresh concrete.

ionone. (α- or β-cyclocitrylideneacetone).
CAS: 8013-90-9. $C_{13}H_{20}O$.
Properties: Light yellow to colorless liquid, violet odor, d 0.927–0.933 (25C), bp 126–128C (12 mm), refr index 1.4970–1.5020 (20C). Soluble in alcohol, ether, mineral oil, and propylene glycol; insoluble in water and glycerol.
Grade: 95%, 99%, mixed isomers, FCC.
Derivation: (1) Condensing citral with acetone followed by ring closure with an acid, (2) reaction of acetylene with acetone followed by hydrogenation, condensation with diketone, and further reaction with acetylene.
Use: Perfumery, chemical synthesis, flavoring, vitamin A production (beta isomer).

ion retardation. A process based on amphoteric (bifunctional) ion-exchange resins containing both anion and cation adsorption sites. These sites will associate with mobile anions and cations in solution and thus remove both kinds of ions from solutions. These ions may be eluted by rinsing with water. Process can make clean separations of ionic-nonionic mixtures. Has also been suggested for demineralization of salt solutions.
See also ion.

iopanoic acid. (iodopanoic acid; 3-amino-α-ethyl-2,4,6-triiodohydrocinnamic acid).
CAS: 96-83-3.
$C_6HI_3NH_2CH_2CH(COOH)C_2H_5$.
Properties: Cream-colored, tasteless powder with faintly aromatic odor, mp 152–158C (decomposes), darkens on exposure to light. Soluble in acetone, ether, alcohol, chloroform, and dilute alkalies; insoluble in water.

Grade: USP.
Use: Medicine (radiopaque medium).

IPA. (1) Abbreviation for isophthalic acid. (2) Abbreviation for isopropyl alcohol.

IPAE. See isopropylaminoethanol.

Ipatieff, Vladimir N. (1890–1952) Born in Russia, Ipatieff was an army officer as well as a chemist. He was a member of the Academy of Science and carried out organic research at the Institute of Chemistry in Leningrad. He left the USSR under the Stalin regime and at the invitation of Gustav Egloff joined the Universal Oil Products Co. He and his close associate, Herman Pines, did basic development on catalytic alkylation and isomerization of hydrocarbons of the greatest importance for high-octane aviation gasoline.

IPC. (INPC; isopropyl-N-phenylcarbamate).
CAS: 122-42-9. $C_6H_5NHCOOCH(CH_3)_2$.
Properties: White to gray crystalline needles, odorless when pure, mp 84C (technical grade). Soluble in alcohol, acetone, isopropyl alcohol; insoluble in water.
Hazard: Toxic by ingestion.
Use: Herbicide.

IPC, chloro-. See chloro-IPC.

ipecac. Dried root of Cephaelis ipecacuanha.
Habitat: Brazil and Bolivia, cultivated in India.
Grade: Technical, USP.
Hazard: Contains emetine, a toxic alkaloid.
Use: Medicine (emetic), production of emetine.

Ir. Symbol for iridium

"Irco" Aluminum Coatings.[526] TM for chromate conversion coatings for aluminum which provide corrosion resistance by themselves and also increase the corrosion resistance and adhesion of subsequent paint films. Application of spray, dip, or roller coat.

"Irco" Bond.[526] TM for crystalline zinc phosphates.
Use: Base for oil and paint, as an aid to drawing and forming of steel, as a nonembrittling etchant for steel, and as a wearing surface on certain types of steel parts. Application by dip, automatic, or barrel methods.

"Ircolene."[526] TM for special polar-type rustproofing oils for indoor-outdoor protection of ferrous or nonferrous metals. Application by spray, dip, or dip-spin centrifugal methods.

"Irco" Lube.[526] TM for oil-absorptive manganese phosphate coatings for mating moving steel surfaces. Application by dip, automatic, or barrel methods.

"Irco" Iron Phosphates.[526] TM for specially blended compounds of acidic salts which produce an iridescent iron phosphate coating on cold-rolled steel surfaces, providing paint base by increasing adhesion and corrosive resistance. Application by spray or immersion.

"Irugalan."[443] TM for 1:2 metal complex dyes for wool and polyamide fibers.

"Irganox."[219] TM for a series of complex, high-molecular weight stabilizers that inhibit oxidation and thermal degradation of many organic materials. They contain multifunctional chemical groupings. Several are hindered polyphenols. All are white, crystalline, free-flowing powders; non-staining, non-volatile, and odorless.

"Irgasan."[443] TM for broad spectrum antibacterial agent.
Use: Deodorant soaps and underarm deodorants.

iridic chloride. (iridium chloride; iridium tetrachloride). $IrCl_4$.
Properties: Brownish-black mass, hygroscopic. Soluble in water, alcohol, and dilute hydrochloric acid.
Derivation: Action of chlorine or aqua regia on the ammonium salt $(NH_4)_2IrCl_6$.
Use: Analysis (testing for nitric acid in the presence of nitrous acid), microscopy, plating solution.

iridium. CAS: 7439-88-5. Ir. Metallic element of atomic number 77, one of the platinum metals, group VIII of the periodic system, aw 192.22, valences = 1,2,3,4,6, two stable isotopes.
Properties: Silver-white, low ductility, does not tarnish in air, on heating strongly a slightly volatile oxide is formed, insoluble in acids, slowly soluble in aqua regia and in fused alkalies, bulk d 22.65 (20C) (calculated) (the heaviest element known), mp 2443C, bp approximately 4500C, most corrosion-resistant element, modulus of elasticity is one of highest (75,000,000 psi), Brinell hardness (cast) 218, highly resistant to chemical attack.
Occurrence: Canada, South Africa, USSR, Alaska.
Derivation: Occurs with platinum, remains insoluble when the crude platinum is treated with aqua regia, occurs as iridosmine. The powder is obtained by hydrogen reduction of ammonium chloroiridate.
Forms: Powder, single crystals.
Use: Alloy with platinum for ammonia fuel-cell catalyst, electric contacts and thermocouples,

commercial electrodes and resistance wires, laboratory ware, extrusion dies for glass fibers, jewelry. Primary standards of weight and length.

iridium 192. Radioactive iridium of mass number 192.
Properties: Half-life, 74 days; radiation, beta and gamma.
Derivation: Pile irradiation of iridium.
Forms available: Iridium metal or potassium or sodium chloroiridate in hydrochloric acid solution.
Use: Radiography of light castings, treatment of cancer.

iridium bromide. See iridium tribromide.

iridium chloride. See iridic chloride, iridium trichloride.

iridium potassium chloride. (potassium chloroiridate; potassium iridium chloride). K_2IrCl_6.
Properties: Dark-red crystals, soluble in water (hot), d 3.549.
Use: Black pigment (porcelain decoration). The iridium (III) salt is also known: K_3IrCl_6, greenish yellow.

iridium sesquioxide. CAS: 1312-46-5. Ir_2O_3.
Properties: Black powder, slightly soluble in hydrochloric acid (concentrated), insoluble in water, decomposes at 400C.
Derivation: By heating the chloroiridate K_2IrCl_6 with sodium carbonate.
Use: Ceramics (porcelain decoration).

iridium tetrachloride. See iridic chloride.

iridium tribromide. (iridium bromide). $IrBr_3 \cdot 4HOH$. Olive-green, brown, or black crystals; soluble in water; insoluble in alcohol. Prepared by action of hydrobromic acid on iridium trihydroxide.

iridium trichloride. (iridium chloride). CAS: 10025-83-9. $IrCl_3$. Dull green to blue-black particles, mp 763C (decomposes), insoluble in water and alcohol. Prepared by action of chlorine on iridium powder at 600C.

iridomyrmecin. CAS: 485-43-8. $C_{10}H_{16}O_2$.
Properties: Colorless crystals, mp 60C, bp 104C (1.5 mm), characteristic odor, refr index 1.46, optically active, soluble in ether and fatty alcohols, almost insoluble in water.
Use: Insecticide, biocide.

iridosmine. (osmiridium). A natural alloy of iridium and osmium containing some platinum, rhenium, ruthenium, iron, copper, palladium. Tin-white to light steel gray in color; streak,

same; metallic luster. Composition is variable ranging from 10.0–77.2% iridium, 17.2–80.0% osmium, 0–10.1% platinum, 0–17.2% rhenium, 0–8.9% ruthenium, 0–1.5% iron, 0.0.9% copper, trace, palladium. Unattacked by aqua regia, d 18.8–21.12, Mohs hardness 6–7.
Occurrence: Alaska, South Africa.
Use: Fountain-pen point tips, surgical needles, watch pivots, compass bearings, hardening platinum (standard weights, jewelry), source of iridium and osmium.

Irish moss. See carrageenan.

"IRN."[299] TM for a group of magnetic iron oxides, black ferroso-ferric oxides (Fe_3O_4) and brown gamma ferric oxides (Fe_2O_3), with application in ferric cores and electronic parts, magnetic inks, data-handling accounting systems, recording and instrumentation tapes.

iron. Fe. CAS: 7439-89-6. Metallic element of atomic number 26, Group VIII of the Periodic Table, aw 55.847, valences = 2,3; 4 stable isotopes, 4 artificially radioactive isotopes.
Properties: Silver-white malleable metal, tensile strength 30,000 psi, Brinell hardness 60, mp 1536C, bp 3000C, d 7.87 (20C), magnetic permeability 88,400 gauss (25C), the only metal that can be tempered, mechanical properties are altered by impurities, especially carbon.
    Iron is highly reactive chemically, a strong reducing agent, oxidizes readily in moist air, reacts with steam when hot, to yield hydrogen and iron oxides. Dissolves in nonoxidizing acids (sulfuric and hydrochloric acid) and in cold dilute nitric acid. For biochemical properties, see below.
Major ores: Hematite, limonite, magnetite, siderite, also taconite (low-grade 25–30% iron).
Occurrence: Minnesota (Mesabi), Alabama, Labrador, Yukon, Europe, South America.
Major types: (1) Molten (or pig) iron. Derived (a) by smelting ore with limestone and coke in blast furnaces (purity 91–92%), (b) by continuous direct reduction in which iron ore and limestone are preheated in a fluidized bed, followed by heating to 926C, by melting at 1926C, and reduction to iron at 1648C with powdered coal (purity 99%) (proprietary process).
Use: Steels of various types, other alloys (cast and wrought iron), source of hydrogen by reaction with steam.
See also steel.
    (2) Powdered iron. Derived (a) by treatment of ore or scrap with hydrochloric acid to give ferrous chloride solution, which is then purified by filtration, vacuum-crystallized, and dehydrated to ferrous chloride dehydrate powder; this is reduced at 800C to metallic iron (briquettes or powder) of 99.5% purity; (b) by thermal decomposition of iron carbonyl [$Fe(CO_5)$] at 250C (99.6–99.9% pure); (c) by hydrogen reduction of high-purity ferric oxide or oxalate; (d) by electrolytic deposition from solutions of a ferrous salt (99.9% pure).
Use: Powder metallurgy products, magnets, high-frequency cores, auto parts, catalyst in ammonia synthesis.
    (3) Cast iron and wrought iron are mixtures of iron and other materials.
See specific entry.
    (4) Single crystals and whiskers are also available.
    (5) Iron sponge.
Hazard: Dust and fine particles suspended in air are flammable and an explosion risk. TLV (as oxide fume): 5 mg/m³; for soluble salts (as iron) 1 mg/m³. Biochemistry: Iron is a constituent of hemoglobin and is essential to plant and animal life, an important factor in cellular oxidation mechanism.
Use: Medicine and dietary supplements. For further information refer to the American Iron and Steel Institute, 150 East 42 St., New York, NY.

iron 55. Radioactive iron, mass number 55.
Properties: Half life 2.91 years, decays through K capture.
See iron 59.
Hazard: Toxic material.

iron 59. Radioactive iron of mass number 59.
Properties: Half-life 46.3 days; radiation beta and gamma.
Derivation: Pile irradiation of iron metal, giving a product which contains iron 55 impurity. Both iron 55 and iron 59 are produced pure in the cyclotron. Enriched samples of each are also available.
Hazard: Toxic material.
Use: Medicine, tracer element in biochemical and metallurgical research.

iron acetate liquor. (iron liquor; black liquor; black mordant; iron pyrolignite).
Properties: Intensely black liquor, sometimes containing copperas or tannin, absorbs oxygen from the air, d 1.09–1.115, containing 5–5.5% iron.
Derivation: (a) By the action of pyroligneous acid on iron filings, (b) double decomposition of ferrous sulfate with calcium pyrolignite (calcium acetate).
Use: Mordant, especially for alizarine and nitroso dyes.

iron alum. See ferric potassium sulfate.

iron black. (contains no iron).
Properties: Fine black powder.
Derivation: Action of zinc upon an acid solution

of an antimony salt, a black antimony being precipitated as a fine powder.
Use: Imparting the appearance of polished steel to papier mache and plaster of Paris.

iron blue. A pigment (of which there are several varieties) prepared by precipitating ferrous ferrocyanide from a solution of ferrocyanide and ferrous sulfate. Subsequent oxidation produces a complex ferriferrocyanide whose shade and pigment properties are dependent upon the oxidizing agent, reactant concentrations, pH, temperature, size of batch, and other conditions of manufacture. Common oxidants are nitric acid, sulfuric acid, potassium dichromate with sulfuric acid, perchlorates, and peroxides.
Properties: Insoluble in water, oils, alcohol, hot paraffin, organic solvents and unaffected by dilute acids. Unstable to alkalies of all concentrations or reducing media. Resistant to light and ordinary baking temperatures.
Use: Paints, printing inks, plastics, cosmetics (eye shadow), artist colors, laundry blue, paper dyeing, fertilizer ingredient, baked enamel finishes for autos and appliances, industrial finishes.

iron carbonate, precipitated. See iron oxide, brown.

iron carbonyl. See iron pentacarbonyl.

iron, cast. See cast iron.

iron compounds. See corresponding ferric or ferrous compounds.

irone. (6-methylionone).
CAS: (alpha) 79-69-6; (beta) 79-70-9; (gamma) 79-68-5. $C_{14}H_{22}O$.
Properties: Yellowish liquid with mw 206.32, mp 89–90C, refractive index 1.5017 (20C), d 0.9434 (21/4C). Soluble in alcohol. A mixture of three isomers (alpha-, beta- and gamma-irone).
Use: Perfumery, violet odor. The alpha isomer is also used as a flavoring agent.

iron formation. Complex sedimentary or weakly metamorphosed rock that usually consists of oxides, sulfides, or carbonates of iron, often in association with chert.

iron mass. See iron sponge (2).

iron octoate. Same as ferric octoate.
Use: Catalyst for curing silicone resins and rubbers.

iron-ore cement. Cements in which ferric oxide replaces a large part of the alumina. There must be some alumina present, however. Iron-ore cement is rather slow setting and hardening, but is more resistant to sea water than is Portland cement. It is light to chocolate brown in color, d about 3.31, higher than Portland cement.

iron ore, chrome. See chromite.

iron ore, magnetic. See magnetite.

iron ore, red. See hematite, red.

iron oxide, black. (ferrosoferric oxide; ferroferric oxide; iron oxide, magnetic; black rouge). CAS: 1309-37-1. $FeO \cdot Fe_2O_3$ or Fe_3O_4.
See also the mineral form, magnetite.
Properties: Reddish or bluish black amorphous powder, d 5.18, mp 1538C (decomposes). Soluble in acids; insoluble in water, alcohol, and ether.
Derivation: (a) Action of air, steam, or carbon dioxide on iron; (b) specially pure grade by precipitating hydrated ferric oxide from a solution of iron salts, dehydrating, and reducing with hydrogen, (c) occurs in nature as the mineral magnetite.
Grade: Technical, pure (96% min).
Use: Pigment, polishing compound, metallurgy, magnetic inks, and in ferrites for electronic industry, coatings for magnetic tape, catalyst.

iron oxide, brown. (iron subcarbonate; iron carbonate, precipitated).
Properties: Reddish-brown powder containing ferric carbonate with ferric hydroxide $Fe(OH)_3$ and ferrous hydroxide $Fe(OH)_2$ in varying quantities; not a true oxide; soluble in acids, insoluble in water and alcohol.
Derivation: By the interaction of solution of ferrous sulfate and sodium carbonate.
Grade: Technical.
Use: Paint pigment.

iron oxide, hydrated. See ferric hydroxide.

iron oxide, metallic brown. A naturally occurring earth, principally ferric oxide to which extenders have been added.
Use: Paint pigment.

iron oxide process. A process for the removal of sulfides from a gas by passing the gas through a mixture of iron oxide, Fe_2O_3 (called iron sponge or iron mass), and wood shavings. The iron oxide is converted to iron sulfide and can be regenerated by allowing the iron sulfide to contact air. See iron sponge, spent.

iron oxide red. (burnt sienna; Indian red; red iron oxide; red oxide; rouge (1); Turkey red).

CAS: 1332-37-2.   Pigments composed mainly of ferric oxide (Fe_2O_3).
See also ferric oxide for a description of the pure material.
Use: Marine paints, metal primers, polishing compounds, pigment in rubber and plastic products, theatrical rouge, grease paints.

iron oxide yellow.   (hydrated ferric oxide).
$Fe_2O_3 \cdot HOH$.   A precipitated pigment of finer particle size and greater tinctorial strength than the naturally occurring oxides such as ocher; excellent lightfastness and resistance to alkali.
Use: Paints, rubber products, plastics.

iron pentacarbonyl.   (iron carbonyl).
CAS: 13463-40-6. $Fe(CO)_5$.
Properties: Mobile, yellow liquid. Evolves carbon monoxide on exposure to air or to light, soluble in nickel tetracarbonyl and most organic solvents, soluble with decomposition in acids and alkalies, insoluble in water, d 1.466 (18C), bp 102.8C (749 mm), decomposes at 200C, mp −21C, flash p 5F (−15C).
Derivation: Finely divided iron is treated with carbon monoxide, in the presence of a catalyst such as ammonia.
Hazard: Flammable, dangerous fire risk. Toxic by ingestion, inhalation, and skin absorption. TLV: 0.01 ppm in air.
Use: Catalyst in organic reactions, carbonyl iron for high-frequency coils.

iron powder.   See iron.

iron protochloride.   See ferrous chloride.

iron protoiodide.   See ferrous iodide.

iron protosulfide.   See ferrous sulfide.

iron pyrite.   See pyrite.

iron pyrolignite.   See iron acetate liquor.

iron pyrophosphate.   See ferric pyrophosphate.

iron red.   Red varieties of ferric oxide that are used as pigments.
See iron oxide red.

iron sponge.   (1) Sponge iron, a finely divided porous form of iron made by reducing an iron oxide at such low temperatures that melting does not occur, usually by mixing iron oxide and coke and applying limited increase in temperature.
Use: For precipitating copper or lead from solutions of their salts, and in electric furnace steel

operations. (2) Iron mass (iron oxide). Finely divided iron oxide, distributed on a support so as to give a large surface area. One form is a mixture of wood shavings covered with a hydrated iron oxide. This may be made by mixing wet wood shavings with iron borings or similar material and allowing rusting to occur to produce finely divided iron oxide. In another method, wood shavings are mixed with a slurry of the hydrated ferric oxide produced in purifying alum and then dried. The iron sponge or iron mass is used for removing sulfur from coal gas or similar materials.
See iron oxide process; see also iron sponge, spent.

iron sponge, spent.   (iron mass, spent; spent oxide).   Iron sponge after saturation with sulfur.
It is liable to spontaneous heating.
See iron oxide process.

iron, stainless.   Alloys containing 3–38% chromium, with or without traces of nickel, essentially magnetic and ferritic in character. High chromium irons are brittle after welding. Most popular composition for fabrication is 15–18% chromium, 0.1% C (max).
See ferritic stainless steel, under steel, stainless.

irradiation.   Exposure to radiation of wavelengths shorter than those of visible light (gamma, x-ray, or UV) either for medical purposes (cancer therapy, removal of skin blemishes), for destruction of bacteria in milk and other foodstuffs or for inducing polymerization of monomers or vulcanization of rubber. UV irradiation was formerly used to induce activation of vitamin D in milk and has been used for some time to sterilize the air in operating rooms, etc.

"Irrathene."[245]   TM for a thermosetting form of polyethylene formed by irradiation of polyethylene with high energy cathode rays (electrons). Does not melt (up to 250C) but oxidizes rapidly at elevated temperatures unless protected by an inhibitor. Its resistance to acids, alkalies, and solvents is superior to that of polyethylene and it has excellent electrical resistance even at 200C. It is available in films and tapes used for packaging and electrical insulation.

irreversible.   A permanent chemical or physico-chemical change. Examples are: (1) A chemical reaction that can proceed only to the right, giving a product that is stable and that cannot revert to the original constituents; most reactions are of this type. (2) A colloidal system that cannot be restored to its original form after coagulation or precipitation, for example, the hardening of egg white or milk protein by heat, the formation

of butter from milk (mechanical action) and the coagulation of rubber latex by acid (chemical action).
See also reversible.

"Irron."[169] TM for 8-quinolinol-7-iodo-5-sulfonic acid.
Use: Colorimetric determination of iron.

"Irtran" 4.[115] TM for an optical material made from zinc selenide.

ISAF black. Abbreviation for intermediate super abrasion furnace black.

isanic acid. (erythrogenic acid).
CAS: 506-25-2. $C_{18}H_{26}O_2$.   A fatty acid with acetylenic triple bonds at the 9 and 11 positions.
Properties: Crystals, mp 39C, d 0.93, refr index 1.49, readily polymerizes and may explode on rapid heating to 250C, soluble in acetone and alcohol, turns red on exposure to air and heat.
Derivation: Vegetable oil from isano nuts grown in Zaire.

isatin. (indole-2,3-dione).   CAS: 91-56-5.
$C_6H_4COC(OH)N$.

Properties: Yellowish-red or orange crystals, bitter taste. Soluble in water, alcohol, and ether; mp 200–203C.
Derivation: From indigo by oxidation.
Use: Dyestuffs, pharmaceuticals, analytical reagent.

isatoic anhydride. $C_8NO_3H_5$.   A bicyclic molecule composed of a benzene ring attached at the o- and m- positions to a heterocyclic ring. It forms useful anthranilic acid derivatives by reaction with hydrogen.
Properties: Tan powder.
Grade: Technical, 96% min.
Use: Intermediate for polymer curing agents, anthranilic esters, heterocyclic compounds, and benzyne.

isethionic acid. (2-hydroxyethanesulfonic acid).
CAS: 107-36-8. $HOCH_2CH_2SO_3H$.
Properties: Liquid, bp 100C (decomposes), very soluble in water, insoluble in alcohol.
Use: Detergents, surfactants, synthesis.

iso-. A prefix meaning "the same" as in such terms as isomer (the same part), isotope (the

same place), isometric (the same measure), isobar (the same pressure), etc. In organic chemistry, it denotes an isomer of an alkyl hydrocarbon, alcohol, etc., having a subordinate chain of one or more carbon atoms attached to a carbon of the straight chain.
See isobutane, isooctane, branched chain.

isoalloxazine. (flavin). $C_{10}H_6N_4O_2$.
A derivative of isoalloxazine widely distributed in plants and animals, usually as yellow pigments.
See also riboflavin.

isoamyl acetate. CAS: 123-92-2.
$CH_3COOCH_2CH_2CH(CH_3)_2$.
Properties: Colorless liquid, bp 142C, fp −78.5C, d 0.876 (15/4C), bulk d 7.30 lb/gal, (15C), flash p 80F (26.6C), slightly soluble in water, miscible with alcohol and ether, banana-like odor.
Derivation: Rectification of commercial amyl acetate.
Grade: Reagent, technical.
Hazard: Flammable, moderate fire risk. Irritant. TLV: 100 ppm in air; explosive limits in air 1–7.5%.
Use: Flavoring, perfumes, solvent for nitrocellulose, masking undesirable odors.   See also amyl acetate.

isoamyl alcohol, primary. (3-methyl-1-butanol; isopentyl alcohol; isobutyl carbinol).
CAS: 123-51-3. $(CH_3)_2CHCH_2CH_2OH$.
Properties: Colorless liquid, pungent taste, disagreeable odor, bp 132.0, fp −117.2C, d 0.813 (15/4C), bulk d 6.79 lb/gal, refr index 1.4075 (20C), flash p 109F (42.7C) (CC), autoign temperature 657F (347C), slightly soluble in water, miscible with alcohol and ether. Combustible.
Derivation: Distillation of fusel oil or the mixed alcohols resulting from the chlorination and hydrolysis of pentane.
Grade: Technical.
Hazard: Moderate fire risk. Vapor is toxic and irritant. TLV: 100 ppm in air, explosive limits in air 1.2–9%.
Use: Photographic chemicals, organic synthesis, pharmaceutical products, solvent, determination of fat in milk, microscopy.
See also fusel oil.

isoamyl alcohol, secondary. Similar to the primary form except bp is 113C, flash p 103F (39.4C), d 0.819. Combustible.

isoamyl benzoate. (amyl benzoate).
CAS: 94-46-2. $C_6H_5COOC_5H_{11}$.
Properties: Colorless liquid with fruity odor, d 0.986–0.989, refr index 1.493, bp 260C, insoluble in water, soluble in alcohol. Combustible.
Use: Perfumery, flavors, cosmetics.

isoamyl benzyl ether. (benzyl isoamyl ether). CAS: 122-73-6. $C_5H_{11}OCH_2C_6H_5$.
Properties: Colorless liquid, fruity odor, d 0.904–0.908, refr index 1.481–1.485, soluble in four parts of 80% alcohol. Combustible.
Grade: Technical.
Use: Soap perfumes.

isoamyl butyrate. CAS: 106-27-4. $C_5H_{11}OOCC_3H_7$.
Properties: Practically water-white, d 0.866–0.868 (15.5C), bp 189C, flash p 138F (58C), soluble in alcohol and ether, slightly soluble in water. Combustible.
Derivation: By treating isoamyl alcohol with butyric acid.
Method of purification: Distillation.
Grade: Commercial 95–100% ester content, FCC (as amyl butyrate).
Use: Flavoring extracts, solvent and plasticizer for cellulose acetate.

isoamyl chloride. CAS: 107-84-6. $(CH_3)_2CHCH_2CH_2Cl$. Any of several compounds or mixtures thereof may be referred to by this name, since numerous isomers are possible, the most common of which is 1-chloro-3-methylbutane. Combustible.
Properties: Colorless or slightly yellow liquid, bp 99.7C (758 mm), d 0.893, refr index 1.410, slightly soluble in water, soluble in alcohol and ether.
Derivation: Isoamyl alcohol and hydrochloric acid or chlorination of isopentane.
Use: (mixture, usually also containing normal amyl chloride): solvent (nitrocellulose, varnishes, lacquers, neoprene), rotogravure inks, soil fumigation, organic compounds.

isoamyldichloroarsine. CAS: 64049-23-6. $C_5H_{11}AsCl_2$.
Properties: Oily liquid, sweetish odor, decomposed by water, bp 88.5–91.5C (15 mm). Combustible.
Derivation: Interaction of phosphorus trichloride and isoamylarsenic acid.
Hazard: Strong irritant.

α-isoamylene. See 3-methyl-1-butene.

β-isoamylene. See 3-methyl-2-butene.

isoamyl ether. CAS: 544-01-4. $[(CH_3)_2CHCH_2CH_2]_2O$.
Properties: Colorless liquid with pleasant odor, bp 172C, d 0.783 (12/4C), refr index 1.40, insoluble in water, soluble in alcohol and ether. Combustible.
Use: Grignard reaction solvent, lacquer solvent.

isoamyl formate. CAS: 110-45-2. $HCOOCH_2CH_2CH(CH_3)_2$.
Properties: Colorless liquid with fruity odor, bp 122–124C, d 0.877, partially soluble in water, soluble in alcohol and ether. Combustible.
Hazard: Strong irritant.
Use: Artificial fruit essences.

isoamyl furoate. $C_4H_3OCO_2C_5H_{11}$.
Properties: Colorless liquid, becoming brown in light, insoluble in water, soluble in alcohol and ether, d 1.0335 (20/4C), bp 232–234C, 135–137C (25 mm), refr index 1.4720.

isoamyl isovalerate. See isoamyl valerate.

sec-isoamyl mercaptan. (2-methylbutyl-3-thiol). $(CH_3)_2CHCH(SH)CH_3$.
Properties: Colorless liquid, offensive odor, distillation range 101–127C, d 0.841 (20/4C), insoluble in water, soluble in organic solvents, refr index 1.445 (20C), flash p 46F (7.7C).
Hazard: Flammable, dangerous fire risk.
Use: Polymerization modifier, insecticide intermediate, vulcanization accelerator intermediate, nonionic surface-active agent.

isoamyl nitrite. See amyl nitrite.

isoamyl pelargonate. $(CH_3)_2CHCH_2CH_2OOCC_8H_{17}$.
Properties: Liquid with fruity odor, d 0.860 (15.5/15.5C), bp 260C, refr index 1.4300 (20C). Combustible.
Use: Flavors and perfumes, chemical intermediate.

isoamyl phthalate. CAS: 605-50-5. $C_{18}H_{26}O_4$.
Properties: Colorless liquid, no odor, bp 225C, d 1.02, insoluble in water, soluble in organic solvents.
Use: Plasticizer for nitrocellulose, rubber cements, foam suppressant.

isoamyl propionate. See amyl propionate.

isoamyl salicylate. (amyl salicylate). CAS: 87-20-7. $C_6H_4OHCOOC_5H_{11}$.
Properties: Water-white liquid sometimes having a faint yellow tinge which should not be pink or red, orchid-like odor, d 1.053 (15C), refr index 1.5050–1.5080 (20C), optical rotation 0 to +2.30 degrees, bp 280C. Soluble in alcohol, ether; insoluble in water and glycerol. Flash p 132C; combustible.
Derivation: By esterifying salicylic acid with amyl alcohol. The usual article of commerce is the isoamyl ester.
Method of purification: Distillation.

Grade: A pure grade of at least 99% ester content which should not exceed 100% on analysis (indicating lower esters).
Use: Soap perfumes.

isoamyl valerate. ("apple essence," "apple oil," isoamyl isovalerate, amyl valerianate, amyl valerate). $C_4H_9CO_2C_5H_{11}$.
Properties: Clear liquid, odor of apples when diluted with alcohol, d 0.8812, bp 203.7C, soluble in alcohol and ether, slightly soluble in water. Combustible.
Derivation: By adding sulfuric acid to a mixture of amyl alcohol and valeric acid. Subsequent recovery by distillation.
Grade: Technical, FCC (as isoamyl isovalerate).
Use: Fruit essences, flavoring agent.

isoascorbic acid. See erthorbic acid.

isobars. Nuclides having the same mass number but different atomic numbers, in contrast to isotopes, which have the same atomic number but different mass number. C-14 and N-14 are isobars.

isobenzan. (Generic name for 1,3,4,5,6,7,8,8-octachloro-1,3,3a,4,7,7a-hexahydro-4,7-methano-isobenzofuran). CAS: 297-78-9. $C_9H_6OCl_8$.
Properties: Solid, mp 248–257C, soluble in acetone and ether, slightly soluble in alcohols and kerosene, insoluble in water.
Use: Insecticide.

isoborneol. CAS: 124-76-5. $C_{10}H_{17}OH$.
A geometrical isomer of borneol.
Properties: White solid with camphor odor, mp 216C (sublimes), more soluble in most solvents than borneol.
Derivation: By reduction of camphor.
Use: Perfumery, chemical esters.

isobornyl acetate. $C_{10}H_{17}OOCCH_3$.
Properties: Colorless liquid, pine-needle odor, d 0.978 (20C), bp 220–224C, soluble in most fixed oils and in mineral oil, insoluble in glycerol and water. Combustible.
Derivation: By heating camphene (50–60C) with glacial acetic acid and sulfuric acid and separating by adding water.
Grade: Technical, FCC.
Use: Compounding pine-needle odors, toilet waters, bath preparations, antiseptics, theater sprays, soaps, making synthetic camphor, flavoring agent.

isobornyl salicylate.
Properties: Viscous, colorless oil; sweet odor; ester content 96%. Combustible.

Grade: Technical.
Use: Perfumery (fixative), cosmetics (filter for suntan preparations).

isobornyl thiocyanoacetate. CAS: 115-31-1. $C_{10}H_{17}OOCCH_2SCN$. The technical grade contains 82% or more of isobornyl thiocyanoacetate, also other terpenes and derivatives.
Properties: Yellow oily liquid, terpene-like odor, d 1.1465 (25/4C), refr index 1.512 (25C), acid number 1.19. Very soluble in alcohol, benzene, chloroform, and ether; practically insoluble in water. Combustible.
Derivation: By treating isoborneol with chloroacetyl chloride and potassium thiocyanate.
Hazard: Toxic by ingestion, strong irritant.
Use: Insecticide chiefly in cattle spray, medicine.

isobornyl valerate.
$(CH_3)_2CHCH_2COOC_{10}H_{17}$.
Properties: Colorless, neutral liquid; oily taste; peculiar, aromatic odor; does not irritate the stomach; soluble in alcohol, ether; sparingly soluble in water; bp 132–138C (12 mm); d 0.954.

isobutane. (2-methylpropane; trimethylmethane). CAS: 75-28-5. $(CH_3)_2CHCH_3$.

A liquefied petroleum gas.
Properties: Colorless gas, slight odor, stable, does not react with water, has no corrosive action on metals, bp −11.73C, fp −159C, d 0.5572 (20C at saturation pressure), d (air = 1) 2.01, soluble in water, slightly soluble in alcohol, soluble in ether, flash p −117F (−83C), autoign temperature 864F (462C).
Derivation: An important component of natural gasoline, refinery gases, wet natural gas; also obtained by isomerization of butane.
Grade: Technical, 99 mole % (pure grade), 99.96 mole % (research grade), and other high-purity grades.
Hazard: Highly flammable, dangerous fire and explosive risk; explosive limits in air 1.9–8.5%.
Use: Organic synthesis, refrigerant, motor fuels, aerosol propellant, synthetic rubber, instrument calibration fluid.

isobutane hydrate. See gas hydrates.

isobutanol. See isobutyl alcohol.

isobutanolamine. See 2-amino-2-methyl-1-propanol.

isobutene. (2-methylpropene; isobutylene). CAS: 115-11-7. (CH₃)₂C:CH₂. A liquefied petroleum gas.

$$CH_3-\underset{\underset{}{\overset{\displaystyle CH_3}{|}}}{C}=CH_2$$

Properties: Colorless, volatile liquid or easily liquefied gas. Coal gas odor, bp −6.9C, fp −139C, flash p −105F (−76C), d 0.6 (20C), soluble in organic solvents. Polymerizes easily and also reacts easily with numerous materials, autoign temperature 869F (465C).
Derivation: Fractionation of refinery gases, catalytic cracking of MTBE.
Hazard: Highly flammable, dangerous fire and explosion risk, explosive limits in air 1.8–8.8%.
Use: Production of isooctane, high-octane aviation gasoline, butyl rubber, polyisobutene resins, tert-butyl chloride, tert-butanol methacrylates; copolymer resins with butadiene, acrylonitrile, etc.; methyl-tert-butyl ether.

isobutyl acetate. CAS: 110-19-0. C₄H₉OOCCH₃.
Properties: Colorless, neutral liquid; fruit-like odor; soluble in alcohols, ether, and hydrocarbons; partially soluble in water. Bp 116–117C, flash p 64F (17.7C) (CC), d 0.8685 (15C), refr index approximately 1.392, bulk d 7.23 lb/gal, fp −99C, autoign temperature 793F (422C).
Derivation: Treating isobutanol with acetic acid in the presence of catalysts.
Grade: Technical, solvent, perfume, FCC.
Hazard: Flammable, dangerous fire risk. TLV: 150 ppm in air.
Use: Solvent for nitrocellulose; in thinners, sealants, and topcoat lacquers; perfumery; flavoring agent.

isobutyl acrylate. CAS: 106-63-8. (CH₃)₂CHCH₂OOCCH:CH₂.
Properties: Liquid, bp 61–63C (51 mm), d 0.884 (25C), refr index 1.4124 (25C), flash p 86F (30C) (TOC), contains 100 ppm monomethyl ether hydroquinone as inhibitor.
Hazard: Flammable, moderate fire risk.
Use: Monomer for acrylate resins.

isobutyl alcohol. (isopropylcarbinol; 2-methyl-1-propanol). CAS: 78-83-1. (CH₃)₂CHCH₂OH.
Properties: Colorless liquid, partially soluble in water, soluble in alcohol and ether, d 0.806 (15C), bp 107C, flash p 100F (37.7C) (OC), fp −108C, refr index 1.397 (15C), autoign temperature 800F (426C).
Derivation: Byproduct of synthetic methanol production, purified by rectification.

Hazard: Flammable, moderate fire risk. Strong irritant. TLV: 50 ppm in air.
Use: Organic synthesis, latent solvent in paints and lacquers, intermediate for amino coating resins, substitute for n-butanol. Paint removers, fluorometric determinations, liquid chromatography, fruit flavor concentrates.

$$CH_3-\underset{\underset{\displaystyle CH_3}{|}}{\overset{\overset{\displaystyle CH_2OH}{|}}{CH}}$$

isobutyl aldehyde. See isobutyraldehyde.

isobutylamine. CAS: 78-81-9. (CH₃)₂CHCH₂NH₂.
Properties: Colorless liquid, amine odor, strongly caustic, soluble in water, alcohol, ether and hydrocarbons, d 0.731 (20C), boiling range 66–69C, fp −85C, flash p 15F (−9.4C), autoign temperature 712F (377C).
Hazard: Flammable, dangerous fire risk. Strong irritant to skin and mucous membranes.
Use: Organic synthesis, insecticides.

isobutyl-p-aminobenzoate. CAS: 94-14-4. NH₂C₆H₄COOCH₂CH(CH₃)₂.
Properties: White, crystalline scales; mp 64–65C; almost insoluble in water; soluble in alcohol, benzene, acetone.
Use: Medicine (topical anesthetic), sunscreen preparations.

isobutylbenzene. CAS: 538-93-2. (CH₃)₂CHCH₂C₆H₅.
Properties: Liquid, d 0.8532 (20/4C), fp −51.6C, bp 171.1, refr index 1.486 (20C), flash p 140F (60C), autoign temperature 802F (427C). Combustible.
Hazard: Moderate fire risk. Toxic in high concentration.
Use: Pharmaceutical intermediate, perfume synthesis, flavoring.

isobutyl benzoate. (eglantine). C₆H₅CO₂CH₂CH(CH₃)₂.
Properties: Colorless liquid, characteristic odor, d 1.002, bp 237C, insoluble in water, miscible with alcohol and ether. Combustible.
Use: Perfumes, flavors.

isobutyl carbinol. See isoamyl alcohol, primary.

isobutyl chloroformate. CAS: 543-27-1. ClCO₂CH₂CH(CH₃)₂.
Properties: Liquid with mw 136.58, bp 128.8C, d 1.053.

Hazard: Flammable with flash p 27C. Corrosive.
Use: Peptide reagent.

isobutyl cinnamate. CAS: 122-67-8.
$C_4H_9OOCCH:CHC_6H_5$. Colorless oil, amber fragrance, d 1.001–1.004, refr index 1.541, soluble in two volumes of 70% alcohol. Combustible.
Use: Perfumery.

isobutyl cyanoacrylate. A tissue adhesive effective in surgery and medicine to retard bleeding from internal organs. Also applicable to the mounting of pearls and other jewels. Available as an aerosol spray.

isobutylene. See isobutene.

isobutylene-isoprene rubber. See butyl rubber.

isobutyl furoate. $C_4H_3OCO_3C_4H_9$.
Properties: Colorless liquid becoming brown in light, insoluble in water, soluble in alcohol and ether, d 1.0383 (26.5/4C), bp 221–223C (corrosive), refr index 1.4676 (26.5C). Combustible.

N-isobutylhendecenamide. See N-isobutylundecylenamide.

isobutyl heptyl ketone. See 2,6,8-trimethyl-4-nonanone.

isobutyl isobutyrate. (IBIB). CAS: 97-85-8.
$(CH_3)_2CHCOOCH_2CH(CH_3)_2$.
Properties: Colorless liquid, fruity odor, d 0.853–0.857 (20/20C), fp −80.7C, bp 148.7C, refr index 1.3999 (20C), insoluble in water, soluble in alcohol and ether. Combustible.
Hazard: Toxic by inhalation.
Use: Flavoring, insect repellent, nitrocellulose lacquers and thinners, substitute for methyl amyl acetate.

isobutyl mercaptan. See 2-methyl-1-propanethiol.

isobutyl methacrylate. CAS: 97-86-9.
$(CH_3)_2CHCH_2OOCC(CH_3):CH_2$.
Properties: Liquid, bulk d 0.882 g/mL (25/25C), boiling range 155C, refr index 1.4172 (25C), flash p 120F (49C) (TOC). Contains 25 ppm hydroquinone monomethyl ether as inhibitor. Combustible.
Hazard: Moderate fire risk.
Use: Monomer for acrylic resins.

isobutyl phenylacetate.
$(CH_3)_2CHCH_2OOCCH_2C_6H_5$.
Properties: Colorless liquid, honey-like odor. Soluble in most fixed oils; insoluble in glycerol, mineral oil, and propylene glycol, d 0.984–0.988 (25C), refr index 1.4860–1.4880 (20C). Combustible.
Grade: Technical, FCC.
Use: Flavoring agent, perfumes.

isobutyl propionate. CAS: 540-42-1.
$CH_3CH_2COOCH_2CH(CH_3)_2$.
Properties: Water-white liquid, d 0.86–0.8635 (20/20C), bp 138C, fp −71.4C, refr index 1.3975 (20C), insoluble in water, very soluble in alcohol and ether. Combustible.
Use: Solvent, fruit flavor concentration.

isobutyl salicylate. CAS: 87-19-4.
$HOC_6H_4COOCH_2CH(CH_3)_2$.
Properties: Colorless liquid, d 1.064–1.065 (25C), may have slightly yellowish tinge, bp 259C, soluble in alcohol and mineral oil, insoluble in water and glycerol. Combustible.
Grade: Technical, FCC.
Use: Perfumery, flavoring.

isobutyl stearate. CAS: 646-13-9. $C_{22}H_{44}O_2$.
Properties: Waxy, crystalline solid; mp 20C.
Use: Cosmetics, inks, coatings, polishes.

N-isobutylundecylenamide. (N-isobutylhendecenamide). $CH_3(CH_2)_7CH:CHCONHC_4H_9$.
A synergist for pyrethrum.
Use: Insecticides.

isobutyl valerate. $C_9H_{19}O_2$.
Properties: Colorless liquid, pleasant odor, d 0.85, bp 170C, refr index 1.40, insoluble in water, soluble in ether and alcohol.
Use: Food industry as flavoring, fruit extracts, and similar products.

isobutyl vinyl ether. See vinyl isobutyl ether.

isobutyraldehyde. (isobutyl aldehyde). CAS: 78-84-2. $(CH_3)_2CHCHO$.
Properties: Transparent, colorless, highly refractive liquid, pungent odor, d 0.794 (20/4C), bp 65C, fp −66C, refr index 1.3730 (20C), flash p (CC) −40F (−40C), (OC) −11F (−23.9C), explosive limits in air 1.6–10.6%, autoign temperature 490F (254C), soluble in alcohol, insoluble in water.
Derivation: (a) Oxo process reaction of propylene with carbon monoxide and hydrogen, (b) dehydrogenation of isobutanol.
Hazard: Highly flammable, dangerous fire and explosion risk. Irritant to skin and eyes.
Use: Intermediate for rubber antioxidants and accelerators, for neopentyl glycol; synthesis of amino acids, cellulose esters, flavors, etc.

isobutyric acid. (2-methylpropanoic acid). CAS: 79-31-2.   (CH₃)₂CHCOOH.

$$CH_3CH-C-OH$$
$$|\qquad||$$
$$CH_3\quad O$$

Properties: Colorless liquid, soluble in water, alcohol and ether, d 0.946–0.950 (20/20C), bp 154.4C, fp −47C, refr index 1.3930 (20C), flash p 170F (76.6C) (TOC), autoign temperature 935F (501C). Combustible.
Grade: Technical.
Hazard: Toxic by ingestion, strong irritant to tissue.
Use: Manufacture of esters for solvents, flavors and perfume bases, disinfecting agent, varnish, deliming hides, tanning agent.

isobutyric anhydride. CAS: 97-72-3. [(CH₃)₂CHCO]₂O.
Properties: Liquid with boiling range 180–187C, d 0.951–0.956 (20/20C), flash p 139F (59.4C), autoign temperature 665F (351C). Combustible.
Hazard: Strong irritant to tissue.
Use: Chemical intermediate.

isobutyronitrile. (2-methylpropanenitrile; isopropyl cyanide).   CAS: 78-82-0. (CH₃)₂CHCN.
Properties: Colorless liquid, d 0.773 (20/20C), bp 107C, fp −75C, slightly soluble in water, very soluble in alcohol and ether. Combustible.
Hazard: Toxic by ingestion, inhalation, and skin absorption.
Use: Intermediate for insecticides, etc.

isobutyroyl chloride. (isobutyryl chloride; 2-methylpropanoyl chloride).   CAS: 79-30-1. (CH₃)₂CHCOCl.
Properties: Colorless liquid, refr index 1.4079, d 1.017 (20/4C), fp −90C, bp 92C, soluble in ether, reacts with water and alcohol.
Use: Chemical intermediate.

isocetyl laurate. C₁₁H₂₃COOC₁₆H₃₃.
Properties: Oily liquid, almost odorless, d 0.858, fp approximately −65C, viscosity 19.6 cp at 25C, insoluble in water, soluble in most organic solvents. Combustible.
Use: Cosmetics and pharmaceuticals (lubricant, fixative and solvent), plasticizer, mold release agent, textile softener.

isocetyl myristate. C₁₃H₂₇COOC₁₆H₃₃.
Properties: Oily liquid with practically no odor, d 0.857, fp −39C, viscosity 25.6 cp at 25C, insoluble in water, soluble in most organic solvents. Combustible.
See isocetyl laurate.

isocetyl oleate. C₁₇H₃₃COOC₁₆H₃₃.
Properties: Oily liquid with practically no odor, d 0.862, fp −57C, viscosity 29.0 at 25C, insoluble in water, soluble in most organic solvents. Combustible.
See isocetyl laurate.

isocetyl stearate. C₁₇H₃₅COOC₁₆H₃₃.
Properties: Oily liquid with practically no odor, d 0.862, fp +57C, viscosity 29.0 cp at 25C, insoluble in water, soluble in most organic solvents. Combustible.
See isocetyl laurate.

isocil. (Generic name for 5-bromo-3-isopropyl-6-methyluracil).   CAS: 314-42-1.

OCCBrC(CH₃)NHC(O)NCH(CH₃)₂.
Crystals, mp 158C, soluble in absolute alcohol.
Hazard: Toxic by ingestion.
Use: Herbicide.

isocinchomeronic acid. (2,5-pyridinedicarboxylic acid).   CAS: 100-26-5. HOOC(C₅H₃N)COOH.
Properties: Light-tan powder, leaflets, or prisms. No odor, mp 254C, sublimes as nicotinic acid above this temperature. Insoluble in cold water, alcohol, ether, benzene; slightly soluble in boiling water, boiling alcohol; soluble in hot dilute mineral acids.
Use: Intermediate for nicotinic acid, insecticides, polymers, dyestuffs.

isocrotonic acid. See crotonic acid.

isocyanate. A compound containing the isocyanate radical —NCO. Monoisocyanates are in use, as in the treatment of cellulose to obtain a cellulose tricarbamate, but the term isocyanate usually refers to a diisocyanate.

isocyanate generator. (hindered isocyanate). An isocyanate derivative that will decompose to an isocyanate upon heating. In one type phenol is combined with an isocyanate and the resulting urethane is stable at room temperature, but dissociates at 160C to the original phenol and isocyanate. These generators are used commercially in a mixture with a polyester which can be stored indefinitely, but which upon heating produces a polyurethane resin.

isocyanate resin. See polyurethane resin.

isocyanic acid. (cyanic acid).   CAS: 75-13-8. HN=C=O.   A gas resulting from depolymerization of cyanuric acid at 300–400C in a stream of carbon dioxide. An intermediate product in the formation of urethanes and allophanates. Direct synthesis from nitric oxide, carbon monoxide, and hydrogen with iridium or palladium catalyst at 300C was reported by Bell Laboratories in 1978.

Hazard: Severe explosion risk. Strong irritant to eyes, skin, and mucous membranes.

isocyanurate. A compound closely related to isocyanate but containing three NCO groups. Its products are similar to polyurethane resins and are particularly useful as rigid foams for insulation in the building and construction industry. For combustibility, see foam, plastic.

See also isocyanuric acid.

isocyanuric acid. (s-triazine-2,4,6-trione).

CAS: 108-80-5. $\overline{OCNHCONHCONH}$.

Properties: White, crystalline powder.

The ketone isomer of cyanuric acid. Derivatives of isocyanuric acid, such as dichloro- and trichloroisocyanuric acid and potassium and sodium dichloroisocyanurate, are bleaches and sanitizers.

isodecaldehyde. $C_9H_{19}CHO$. Mixed isomers.

Properties: Liquid, d 0.8290, bp 197.0C, fp −80C, insoluble in water, bulk d 6.9 lb/gal, flash p 185F (85C). Combustible.

Use: Intermediate for pharmaceuticals, dyes, resins.

isodecane. See 2-methylnonane.

isodecanoic acid. $C_{10}H_{20}O_2$. Mixture of branched chain acids, primarily trimethylheptanoic and dimethyloctanoic.

Properties: Liquid, bp 254C, 137C (10 mm), d 0.9019 (20/20C), fp glass at approximately −60C, very slightly soluble in water, viscosity 12.9 cp at 20C, refr index 1.4358 (20C). Combustible.

Use: Intermediate for metal salts, ester type lubricants, plasticizers.

isodecanol. $C_{10}H_{21}OH$. Mixed isomers.

Properties: Colorless liquid, d 0.8395, insoluble in water, flash p 220F (104C), bp 220C. Combustible.

Use: Antifoaming agent in textile processing.

isodecyl chloride. $C_{10}H_{21}Cl$. Mixed isomers.

Properties: Colorless liquid, d 0.8767, bp 210.6C, fp glass at −180C, insoluble in water, flash p 200F (93.3C). Combustible.

Use: Solvent for oils, fats, greases, resins, gums; extractants, cleaning compounds; intermediate for insecticides, pharmaceuticals, plasticizers, polysulfide rubbers, resins, and cationic surfactants.

isodecyl octyl adipate.

Properties: Light-colored oily liquid, d 0.924 (20/20C), mid-bp 227C (4 mm), refr index 1.448 (25C), viscosity at 20C 20 cp, flash p 400F (204C). Combustible.

Use: Plasticizer.

isodecyl pelargonate.

$(CH_3)_2CH(CH_2)_6CH_2OOC(CH_2)CH_3$.

A synthetic lubricant.

isodrin. (Generic name for 1,2,3,4,10,10-hexachloro-1,4,4a,5,8,8a-hexahydro-1,4-endo,endo-5,8-dimethanonaphthalene). CAS: 465-73-6. $C_{12}H_8Cl_6$. An isomer of aldrin.

Properties: Crystals, decomposes above 100C.

Hazard: Toxic, use may be restricted.

Use: Insecticide.

isodurene. (1,2,3,5-tetramethylbenzene).

$(CH_3)_4C_6H_2$.

Properties: Liquid, soluble in alcohol and ether, insoluble in water, d 0.896, bp 197C, fp −24C. Combustible.

Derivation: From coal tar.

Grade: Technical.

Use: Organic synthesis.

isoelectric point. The pH at which the net charge on a molecule in solution is zero. At this pH, amino acids exist almost entirely in the zwitterion state, that is, the positive and negative groups are equally ionized. A solution of proteins or amino acids at the isoelectric point exhibits minimum conductivity, osmotic pressure, and viscosity. Proteins coagulate best at this point. Typical isoelectric points (pH) are: glycine 6.6; gelatin 4.7; serum albumin 5.4.

isoeugenol. (1-hydroxy-2-methoxy-4-propenylbenzene). CAS: 97-54-1.

$(CH_3CHCH)C_6H_3OHOCH_3$.

Properties: Pale yellow oil, spice-clove type odor, soluble in alcohol, ether and other organic solvents, slightly soluble in water, d 1.081–1.084, mp 19C, bp 268C, refr index 1.5739 (19C). Combustible.

Derivation: From eugenol by isomerization with caustic potash.

Grade: Perfumer's grade, FCC.

Use: Perfumes, vanillin, flavoring agent.

isoeugenol acetate. See acetylisoeugenol.

isoeugenol ethyl ether. (1-ethoxy-2-methoxy-4-propenyl-benzene).

$C_3H_5(CH_3O)C_6H_3OC_2H_5$.

Properties: Synthetic, white, crystalline powder; mp 64C; insoluble in water; soluble in alcohol, ether, benzene. Combustible.

Use: Sweetening agent and odorant fixative.

isoflurophate. See diisopropyl fluorophosphate.

"Isoforming." Proprietary process for fixed-bed hydroisomerization, requiring a non-noble metal catalyst. Claimed to give high yields of C_8 (xylene) isomers with low hydrogen consumption and minimal catalyst regeneration.

isoheptane. See 2-methylhexane.

isohexane. CAS: 107-83-5. C_6H_{14}.
A mixture of branched-chain isomers.
Properties: Colorless liquid, boiling range 54–61C, d 0.671 (15.5/15.5C), flash p −26F (−32C) (CC).
Grade: Commercial.
Hazard: Highly flammable, dangerous fire and explosion risk, explosive limits in air 1–7%.
Use: Solvent, freezing-point depressant.

isolan. See 1-isopropyl-3-methyl-5-pyrazolyl dimethylcarbamate.

isolation. Identification and separation of a pure substance which is present in trace amounts in a complex mixture. A famous instance of this was the isolation of polonium (1898) and radium (1912) from pitch-blende by the Curies by coprecipitation techniques followed by repeated fractional crystallization.

isoleucine. (2-amino-3-methylpentanoic acid; Ile). CAS: 73-32-5.
$CH_3CH_2CH(CH_3)CH(NH_2)COOH$.　An essential amino acid, found naturally in the l(+) form.
Properties: Crystals, slightly soluble in water, nearly insoluble in alcohol, insoluble in ether.
Derivation: Hydrolysis of protein (zein, edestin), amination of α-bromo-β-methylvaleric acid.
Use: Medicine, nutrition, biochemical research.

"Isome."[342] TM for di-n-propyl-6,7-methylenedioxy-3-methyl-1,2,3,4-tetrahydronaphthalene-1,2-dicarboxylate.
Use: Insecticides.
See propyl isome.

isomer. (1) One of two or more molecules having the same number and kind of atoms and hence the same molecular weight, but differing in respect to the arrangement or configuration of the atoms. Butanol (C_4H_9OH or $C_4H_{10}O$) and ethyl ether ($C_2H_5OC_2H_5$ or $C_4H_{10}O$) have the same empirical formulas but are entirely different kinds of substances; normal butanol ($CH_3CH_2CH_2CH_2OH$) and isobutanol ($[CH_3]_2CHCH_2OH$) are the same kinds of substances, differing chiefly in the shape of the mole-

cules; sec-butanol ($CH_3CHOHCH_2CH_3$) exists in two forms, one a mirror image of the other (enantiomer). Isomers often result from location of an atom or group of a compound at various positions on a benzene ring, e.g., xylene, dichlorobenzene. (2) Nuclides, (i.e., kinds of atomic nuclei) having the same atomic and mass numbers, but existing in different energy states. One is always unstable with respect to the other, or both may be unstable with respect to a third. In the latter instance the energy of transformation in the two cases will differ.
See also geometric isomer, optical isomer.

isomerization. A method used in petroleum refining to convert straight-chain to branched-chain hydrocarbons or alicyclic to aromatic hydrocarbons, to increase their suitability for high-octane motor fuels. For examples, butane (a gaseous paraffin hydrocarbon, $CH_3CH_2CH_2CH_3$) can be slightly modified in structure by catalytic reactions to give the isomeric isobutane ($CH_3CH_3CHCH_3$) used as a component of aviation fuel. Similarly, methylcyclopentane can be isomerized to cyclohexane, which is then dehydrogenated to benzene. Isomerization techniques were introduced on large scale during World War II.
See also isomer, chain.

α-isomethylionone. (γ-methylionone).
$C_{14}H_{22}O$.
Properties: Slightly yellow liquid, d 0.925–0.929 (25/25C), refr index 1.5000–1.5010 (20C), flash p 217F (102.7C) (TCC), soluble in 5 parts of 70% alcohol. a synthetic product. Combustible.
Use: Floral perfumes, particularly of a violet character; flavoring.

isomorphism. The state in which two or more compounds that form crystals of similar shape have similar chemical properties and can usually be represented by analogous formulas, e.g., Ag_2S and Cu_2S.

"Isomate."[520] TM for isocyanate foam systems. Available as non-burning, pour-in-place froth, or spray foams.

isonipecaine hydrochloride. See meperidine hydrochloride.

"Isonol C100."[520] $C_6H_5N[CH_2CH(CH_3)OH]_2$.
An aromatic reinforcing polyol.
Properties: Amber liquid, viscosity (50C) 1000 cp (max), d 1.055 (23C), water content 0.05%. Combustible.
Use: Ingredient of polyurethane foams, coatings,

sealants, and elastomers; intermediate in organic synthesis.

isononyl alcohol. $C_6H_{17}CH_2OH$. A higher alcohol developed in early 1968. Combustible.
Use: Basis of plasticizers such as diisononyl adipate.

isooctane. (2,2,4-trimethylpentane).
CAS: 540-84-1. $(CH_3)_3CCH_2CH(CH_3)_2$.

$$CH_3-\overset{\overset{\displaystyle CH_3}{|}}{\underset{\underset{\displaystyle CH_3}{|}}{C}}-CH_2-\overset{\overset{}{}}{\underset{\underset{\displaystyle CH_3}{|}}{CH}}-CH_3$$

A branched-chain hydrocarbon.
Properties: Colorless liquid, d 0.6919 (20/4C), fp −107.4C, bp 99.2C, refr index 1.3914 (20C), flash p 10F (−12.2C), insoluble in water, slightly soluble in alcohol and ether, autoign temperature 784F (417C).
Grade: Technical, pure, research, spectrophotometric.
Hazard: Flammable, dangerous fire risk, explosive limits in air 1.1–6%. Toxic by ingestion and inhalation.
Use: Organic synthesis, solvent, motor fuel; used with normal heptane to prepare standard mix to determine anti-knock property of gasoline. See octane number.

isooctene. CAS: 11071-47-9. C_8H_{16}.
Mixture of isomers.
Properties: Colorless liquid, boiling range 87.7–93.3C bromine number 137, d 0.726 (60/60F), flash p approximately 20F (−6.6C).
Hazard: Flammable, dangerous fire risk.

isooctyl adipate. $(C_8H_{17}OOCCH_2CH_2-)_2$.
Plasticizer providing low-temperature stability.
Use: In calendering film, sheeting, vinyl dispersions, extrusions.

isooctyl alcohol. (isooctanol).
CAS: 26952-21-6. General term applied to any isomer of the formula $C_7H_{15}CH_2OH$ in which the eight carbon atoms form a branched chain. Usually refers to a mixture of isomers made by the Oxo process. A selected C_7 hydrocarbon fraction is reacted with hydrogen and carbon monoxide in the presence of a catalyst at pressures up to 3000 psi. The crude alcohol is recovered and purified.
Properties: Clear liquid, distillation range 182–195C, bulk d 6.95 lb/gal, d 0.832 (20/20C), flash p 180F (82.2C) (TOC). Combustible.
Hazards: TLV: 50 ppm. Toxic by skin absorption.
Use: Ingredient of plasticizers, intermediate for nonionic detergents and surfactants, synthetic drying oils, cutting and lubricating oils, hydraulic fluids; resin solvent, emulsifier, antifoaming agent, intermediate for introducing the isooctyl group into other compounds.

isooctyl isodecyl phthalate.
$C_8H_{17}OOCC_6H_4COOC_{10}H_{21}$.
Properties: Clear liquid, d 0.976 (20/20C), flash p 445F (229C), mild odor. Combustible.
Grade: Technical.
Use: Plasticizer.

isooctylmercaptoacetate. See isooctyl thioglycolate.

isooctyl palmitate. $C_8H_{17}OOCC_{15}H_{31}$.
Properties: Clear liquid, d 0.863 (20C), acidity 0.2% max (palmitic), moisture 0.05% max, mp 6–9C, bp 228C (5 mm), soluble in most organic solvents. Combustible.
Use: Secondary plasticizer for synthetic resins, extrusion aid and plasticizer.

isooctylphenoxypolyoxyethylene ethanol.
(isooctylphenylpolyethylene glycol ether).
$C_{32}H_{55}O_{10}$.
Properties: Slightly viscous, pale amber-colored liquid, oily musty odor, mp 2–5C, bp 150C (initial) at one micron, bulk d 1.06 (20C), flash p 227F (108C). Combustible.
Use: Surface-active agent.

isooctyl thioglycolate. (isooctylmercaptoacetate).
CAS: 25103-09-7. $HSCH_2COOCH_2C_7H_{15}$.
Properties: Water-white liquid with faint fruity odor, bp 125C (17 mm) d 0.9736 (25C), refr index 1.4606 (21C), acid number less than 1. Combustible.
Grade: 99% (min purity).
Hazard: Toxic by ingestion.
Use: Antioxidants, fungicides, oil additives, plasticizers, insecticides, stabilizers, polymerization modifiers, stabilizer in tin-sulfur compounds, stripping agent for polysulfide rubber.

"Isopar."[592] TM for a group of high-purity isoparaffinic materials.
Hazard: Skin irritant; little tendency to migrate through polyethylene containers and are permitted under a number of FDA regulations for food-related applications.
Use: As odorless solvents, reaction diluents, in proprietary formulations, etc.

isopentaldehyde. CAS: 590-86-3. C_4H_9CHO.
A mix of isomeric 5-carbon aldehydes.
Properties: Water-white liquid with sharp odor, d 0.8089 (20/20C), bp 98.6C, fp −95.4C, water

dissolves 0.85% aldehyde at 20C, water-soluble to 2.2% in the aldehyde. Combustible.

Use: Possible intermediate for bis-phenols, epoxy and polycarbonate resins, and modified formaldehyde resins.

isopentane. (2-methylbutane; ethyldimethylmethane). CAS: 78-78-4. $(CH_3)_2CHCH_2CH_3$.
Properties: Colorless liquid, pleasant odor, fp −159.890C, bp 27.854C, d 0.61967 (20C), flash p −70F (−57C), soluble in hydrocarbons, oils, ether; very slightly soluble in alcohol, insoluble in water, autoign temperature 788F (420C).
Derivation: Fractional distillation from petroleum, purified by rectification.
Grade: Research (99.99%), pure (99%), technical (95%), commercial.
Use: Solvent, manufacture of chlorinated derivatives, blowing agent for polystyrene.
Hazard: Highly flammable, dangerous fire risk.

isopentanoic acid. CAS: 503-74-2. C_4H_9COOH. A mixture of isomeric 5-carbon acids.
Properties: Water-white liquid, penetrating odor, d 0.9388 (20/20C), bp 183.2C, vap press 0.14 mm at 20C, fp −44C, water dissolves 3.24 wt% of acid at 20C, acid dissolves 10.4% water at 20C. Combustible.
Hazard: Strong irritant to tissue.
Use: Intermediate for plasticizers, synthetic lubricants, pharmaceuticals, metallic salts, vinyl stabilizers; extractant for mercaptans from hydrocarbons.

isopentyl alcohol. See isoamyl alcohol, primary.

"Isoplast."[327] TM for an engineering grade thermoplastic polyurethane for extrusion and injection molding. It is an impact-resistant glass based on methylenediphenyl isocyanate.

isophane insulin suspension. See insulin.

isophorone. (3,5,5-trimethyl-2-cyclohexen-1-one). CAS: 78-59-1.

$\overline{C(O)CHC(CH_3)CH_2C(CH_3)_2}CH_2$.
Properties: Water-white liquid, d 0.9229 (20/20C), bulk d 7.7 lb/gal (20C), bp 215.2C, vap press 0.2 mm (20C), fp −8.1C, viscosity 2.62 cp (20C), flash p 205F (96C), autoign temperature 864F (462C), has high solvent power for vinyl resins, cellulose esters and ether and many substances soluble with difficulty in other solvents, slightly soluble in water. Combustible.
Hazard: Irritant to skin and eyes. TLV: CL 5 ppm in air.

Use: In solvent mixtures for finishes, for polyvinyl and nitrocellulose resins, pesticides, stoving lacquers.

isophorone diisocyanate. (3-isocyanatomethyl-3,5,5-trimethylcyclohexyl-isocyanate; IPDI). CAS: 4098-71-9. $C_{12}H_{18}N_2O_2$.
Properties: Colorless to slightly yellow liquid with mw 222.3, mp about −60C, bp 158C at 10 torr, vap press 0.0003 torr at 20C, d 1.056. Completely miscible with esters, ketones, ethers, and aromatic and aliphatic hydrocarbons.
Hazard: TLV: 0.01 ppm. A severe irritant which is toxic by skin absorption.
Use: Yields polyurethanes with high stability, resistance to light discoloration, and chemical resistance.

isophthalic acid. (m-phthalic acid; IPA). CAS: 121-91-5. $C_6H_4(COOH)_2$.
Properties: Colorless crystals, mp 345–348C, sublimes, slightly soluble in water, soluble in alcohol and acetic acid, insoluble in benzene and petroleum ether. Combustible.
Derivation: (a) Oxidation of m-xylene, (b) liquid phase oxidation of mixed xylenes, (c) direct oxidation of mixed alkyl aromatics with heavy metal salts and bromine as catalysts.
Grade: Technical.
Use: Polyester, alkyd; polyurethane and other high polymers, plasticizers.

isophthaloyl chloride. (m-phthalyl dichloride). CAS: 99-63-8. $C_6H_4(COCl)_2$.

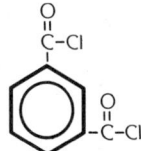

Properties: Crystalline solid, mp 41C, bp 276C, flash p 356F (180C), (COC) soluble in ether and other organic solvents, reactive with water and alcohol. Combustible.
Use: Intermediate, dyes, synthetic fibers, resins, films, protective coatings, laboratory reagent.

isopolyester. A polyester resin based on isophthalic acid.

isoprene. (3-methyl-1,3-butadiene; 2-methyl-1,3-butadiene). CAS: 78-79-5. $CH_2{=}C(CH_3)CH{=}CH_2$. The molecular unit of natural rubber.
Properties: Colorless volatile liquid, fp −146C, bp 34.08C, refr index 1.4216 (20C), d 0.6808 (20/4C), flash p −55F (−48C), insoluble in water,

soluble in alcohol, ether and hydrocarbon solvents, autoign temperature 802F (427C).

Derivation: (1) From cracked products of heavy petroleum oils, (2) dehydrogenation of isopentene, (3) pyrolysis of methyl pentene or of isobutyleneformaldehyde condensation products, (4) dehydration of methyl butenol.

Grade: Polymerization (min purity 99%), research (99.99%).

Hazard: Highly flammable, dangerous fire and explosion risk. Irritant.

Use: Monomer for manufacture of polyisoprene, chemical intermediate, component of butyl rubber.

See also polyisoprene, rubber, natural.

isoprenoid. A compound based on the isoprene structure. These include many naturally occurring materials such as terpenes, rubber, cholesterol, and other steroids.

isopropanol. Legal label name for isopropyl alcohol.

isopropanolamine. (MIPA; 2-hydroxypropylamine; 1-amino-2-propanol). CAS: 78-96-6. $CH_3CH(OH)CH_2NH_2$.

Properties: Liquid, slight ammonia odor, d 0.9619, mp 1.4C, bp 159.9C, refr index 1.4462 (20C), flash p 170F (76.6C), soluble in water. Combustible.

Use: Emulsifying agent, drycleaning soaps, soluble textile oils, wax removers, metal cutting oils, cosmetics, emulsion paints, plasticizers, insecticides.

isopropenyl acetate. CAS: 108-22-5. $CH_3COOC(CH_3):CH_2$.

Properties: Water white liquid, d 0.9226, bp 97.4C, fp −92.9C, solubility in water 3.25% by wt (20C), refr index 1.4020 (20C), flash p 60F (15.5C) (OC).

Hazard: Flammable, dangerous fire risk.

Use: Acylation reagent.

isopropenylacetylene. (2-methyl-1-buten-3-yne). $H_2C=C(CH_3)C\equiv CH$.

Properties: Colorless liquid, bp 33–34C, fp −113C, d 0.695 (20/20C, refr index 1.4168 (20C), flash p below 20F (−6.6C) (TOC). Very slightly soluble in water and miscible with acetone, alcohol, benzene, carbon tetrachloride, and kerosene.

Hazard: Flammable, dangerous fire risk.

isopropenyl chloride. See chloropropene.

isopropenylchloroformate. $ClCOOC(CH_3):CH_2$.

Properties: Liquid, d 1.103 (20C), bp 93C (746 mm).

Derivation: Distillation of the reaction products of acetone and phosgene.

Hazard: Strong irritant to eyes and skin.

p-isopropoxydiphenylamine. $C_6H_5NHC_6H_4OCH(CH_3)_2$.

Properties: Dark gray flakes, d 1.10, set point 80–86C, purity 92% (min), ash 0.10% (max). Insoluble in water; soluble in ethanol, acetone, benzene, and gasoline.

Use: Rubber antioxidant.

2-isopropoxyethanol. (IPE; isopropyl glycol; isopropyl 'Cellosolve'). CAS: 109-59-1. $(CH_3)_2CHOCH_2CH_2OH$.

Properties: A mobile liquid with mw 104.15, d 0.91, bp 139.5–144.5, vap press 2.6 torr at 20C.

Hazard: Combustible liquid with flash p 49C. TLV: 25 ppm. Toxic by skin absorption.

Use: As a component of lacquers and other coatings, and as a solvent.

o-isopropoxyphenyl-N-methylcarbamate. (propoxur). CAS: 114-26-1. $(CH_3)_2CHOC_6H_4OOCNHCH_3$.

Properties: White, crystalline powder; odorless; mp 91C. Soluble in most polar solvents, very slightly soluble in water, unstable in highly alkaline media, stable under normal use conditions.

Hazard: Toxic by ingestion and inhalation. TLV: 0.5 mg/m³ of air.

Use: Insecticide, molluscicide.

β-isopropoxypropionitrile. $(CH_3)_2CHO(CH_2)_2CN$.

Properties: Colorless to straw-colored liquid, combines the chemical and physical properties of ethers and nitriles, fp −67C, bp 82–86C (25 mm) 65–65.5C (10 mm), d 0.9058 (25C), slightly soluble in water, soluble in organic solvents, flash p 155F (68.3C). Combustible.

isopropyl acetate. CAS: 108-21-4. $CH_3COOCH(CH_3)_2$.

Properties: Colorless liquid, aromatic odor, bp 89.4C, d 0.8690 (25/4C), refr index 1.378 (20C), sp ht 0.46 cal/g, fp −73.4C, heat of vaporization 135 Btu/lb, viscosity (25C) 0.49 cp, solubility in water 2.9 wt%, flash p 40F (4.4C), bulk d 7.17 lb/gal (20C), miscible with most organic solvents, autoign temperature 860F (460C).

Derivation: By reacting isopropyl alcohol with acetic acid in the presence of catalysts.

Grade: 95%, 85–88%.

Hazard: Flammable, dangerous fire risk. TLV: 250 ppm in air.

Use: Solvent for nitrocellulose, resin gums, etc.; paints, lacquers, and printing inks; organic synthesis, perfumery.

isopropylacetone. See methyl isobutyl ketone.

N-isopropylacrylamide. (NIPAM). Crystalline solid; homopolymers and copolymers prepared with this material show inverse solubility in water.

Use: Binders in textiles, paper, adhesives, detergents, cosmetics.

See also acrylic resin.

isopropyl alcohol. (IPA; dimethylcarbinol; sec propyl alcohol; isopropanol; 2-propanol). CAS: 67-63-0. $(CH_3)_2CHOH$. 50th highest-volume chemical produced in US (1985).

Properties: Colorless liquid, pleasant odor, bp 82.4C, d 0.7863 (20/20C), refr index 1.3756 (20C), sp ht 0.65 cal/g, fp −86C, critical temperature 235C, critical pressure 53 atm, vap press 33 mm at 20C, flash p 53F 11.7C (TOC), heat of combustion 14,346 Btu/lb, heat of vaporization 288 Btu/lb, viscosity 2.1 cp (25C), autoign temperature 850F (453C). Soluble in water, alcohol, and ether.

Derivation: By treatment of propylene with sulfuric acid and hydrolyzing.

Method of purification: Rectification.

Grade: 91%, 95%, 99%, NF (99%), nanograde.

Hazard: Flammable, dangerous fire risk, explosive limits in air 2–12%. Toxic by ingestion and inhalation. TLV: 400 ppm in air.

Use: Manufacture of acetone and its derivatives, manufacture of glycerol and isopropyl acetate, solvent for essential and other oils, alkaloids, gums, resins, etc.; latent solvent for cellulose derivatives, coatings solvent, deicing agent for liquid fuels, lacquers, extraction processes, dehydrating agent, preservative, lotions, denaturant.

isopropylamine. (2-aminopropane). CAS: 75-31-0. $(CH_3)_2CHNH_2$.

Properties: Colorless, volatile liquid. Amine odor, strong alkaline reaction, bp 32.4C, fp −101C, d 0.6881 (20/20C), bulk d 5.7 lb/gal (20C), refr index 1.3770 (15C), flash p −35F (−37.2C) (OC), autoign temperature 756F (402C). Miscible with water, alcohol, and ether.

Derivation: From acetone and ammonia under pressure.

Hazard: Highly flammable, dangerous fire risk. Strong irritant to tissue. TLV: 5 ppm in air.

Use: Solvent, intermediate in synthesis of rubber accelerators, pharmaceuticals, dyes, insecticides, bactericides, textile specialties, and surface-active agents, dehairing agent, solubilizer for 2,4-D acid.

p-isopropylaminodiphenylamine. See N-isopropyl-N-phenyl-p-phenylenediamine.

isopropylaminoethanol. (IPAE). CAS: 109-56-8. A commercial mixture of approximately 60% isopropylethanolamine

$(CH_3)_2CHNHCH_2CH_2OH$ and 40% isopropyldiethanolamine $(CH_3)_2CHN(CH_2CH_2OH)_2$.

Properties: Amber to straw-colored liquid, distillation range 110–165C, fp approximately −50C, d 0.91–0.94 (20/20C), flash p 145–155F (62.7–68.3C) (OC). Combustible.

Use: Synthesis of emulsifiers.

N-isopropylaniline. CAS: 768-52-5. $C_6H_5NHCH(CH_3)_2$.

Properties: Yellowish liquid, bp 206C, pour point below −67C, refr index 1.5365 (20C), flash p 190F (87.7C) (COC). Combustible.

Hazard: Toxic by inhalation and skin absorption. TLV: 2 ppm in air.

Use: Dyeing acrylic fibers, chemical intermediate.

p-isopropylaniline. See cumidine.

isopropyl antimonite. $[(CH_3)_2CHO]_3Sb$.

Properties: Colorless liquid, bp 82C at 7 mm Hg pressure.

Derivation: Reaction of antimony trichloride with isopropanol.

Use: Crosslinking agent, flameproofing agent.

p-isopropylbenzaldehyde. See cuminic aldehyde.

isopropylbenzene. See cumene.

p-isopropylbenzyl alcohol. See cuminic alcohol.

isopropyl bromide. CAS: 75-26-3. $CH_3CHBrCH_3$.

Properties: Colorless liquid, d 1.304 (25/25C), bp 58.5–60.5C, fp −90C, refr index 1.422 (25C), flash p none, slightly soluble in water, soluble in ethanol and ether. Nonflammable.

Use: Synthesis of pharmaceuticals, dyes, other organics.

isopropyl butyrate. CAS: 638-11-9. $(CH_3)_2CHOOCC_3H_7$.

Properties: Colorless liquid, d 0.8652 (13C), bp 128C.

Use: Solvent for cellulose ethers, flavoring.

isopropylcarbinol. See isobutyl alcohol.

isopropyl chloride. CAS: 75-29-6. $CH_3CHClCH_3$.

Properties: Colorless liquid, d 0.858 (25/25C), bp 34.8C, fp −117.6C, refr index 1.374 (25C), flash p −26F (−32.3C), autoign temperature 1100F (593C), slightly soluble in water, soluble in ethanol and ether.

Hazard: Highly flammable, fire and explosion risk, explosive limits in air 2.8–10.7%.

Use: Solvent, intermediate.

N-isopropyl-α-chloroacetanilide. See 2-chloro-N-isopropylacetanilide.

isopropyl-3-chlorocarbanilate. See chloro-IPC.

isopropyl chloroformate. CAS: 108-23-6. $(CH_3)_2CHOOCCl$.
Properties: Colorless liquid, a phosgene derivative.
Hazard: Toxic by inhalation.
Use: Chemical intermediate for free-radical polymerization initiators, organic synthesis.

isopropyl-N-(3-chlorophenyl)carbamate. See chloro-IPC.

isopropyl-m-cresol. See thymol.

isopropyl-o-cresol. See carvacrol.

isopropyl cresol. A mix of di- and monoisopropyl cresols.
Use: Antioxidant.
See thymol and carvacrol.

isopropyl cyanide. See isobutyronitrile.

isopropyl dichlorophenoxyacetate. See 2,4-D.

isopropyldiethanolamine. See isopropylaminoethanol.

2-isopropyl-4-dimethylamino-5-methylphenyl-1-piperidinecarboxylate methyl chloride.
$C_3H_7C_6H_2(CH_3)N(CH_3)_2OOCNC_5H_{10} \cdot CH_3Cl$
Properties: White solid; mp 151–153C; insoluble in ether; soluble in methanol.
Use: A plant tranquilizer or antigibberellin, which causes some plants to become dwarfs without otherwise affecting their growth or health.

isopropylethanolamine. See isopropylaminoethanol.

isopropyl ether. (diisopropyl ether; diisopropyl oxide). CAS: 108-20-3.
$(CH_3)_2CHOCH(CH_3)_2$.
Properties: Colorless, volatile liquid; ethereal odor. Somewhat similar to ethyl ether in properties but does tend to form peroxides more readily than ethyl ether. Consequently, the presence or absence of peroxides should be determined; if present, they should be destroyed with sodium sulfite before distillation; bp 67.5C, d 0.723 (15.5/4C), refr index 1.368, fp −60C, soluble in water 0.65% wt (25C), flash p approx 0F (−17.7C), autoign temperature 830F (443C), bulk d 6.05 lb/gal (15.5C), miscible with most organic solvents.
Grade: Technical.
See ether.

Hazard: Flammable, dangerous fire risk, explosive limits in air 1.4–21%. Toxic by inhalation, strong irritant. TLV: 250 ppm in air.
Use: Solvent for animal, vegetable, mineral oils, waxes, and resins; extraction of acetic acid from aqueous solutions; solvent for dyes in presence of small amount of alcohol; paint and varnish removers; spotting compositions; rubber cements.

isopropylethylene. See 3-methyl-1-butene.

isopropyl furoate. $C_4H_3OCO_2C_3H_7$.
Properties: Colorless liquid becoming brown in light, insoluble in water, soluble in alcohol and ether, d 1.0655 (23.7/4C), bp 198.6C (corrosive), refr index 1.4682 (23.7C).

isopropyl glycidyl ether. (IGE).
CAS: 4016-14-2. $C_6H_{12}O_2$.
Properties: A liquid with mw 116.16, d 0.9186, bp 127C, vap press about 9.4 torr at 25C.
Hazard: Flammable liquid with flash p 92F (33.33C). TLV: 50 ppm. Skin, eye, and respiratory irritant.
Use: Stabalizer of chlorinated solvents, and viscosity reducer of epoxy resins.

isopropylideneacetone. See mesityl oxide.

p,p-isopropylidenediphenol. See bisphenol A.

isopropyl iodide. (2-iodopropane).
CAS: 75-30-9. CH_3CHICH_3.
Properties: Colorless liquid that discolors in air and light; miscible with chloroform, ether, alcohol, and benzene; slightly soluble in water. D 1.703, fp −90C, bp about 90C, refr index 1.5026 (20C).
Use: Organic synthesis, pharmaceuticals.

isopropyl mercaptan. CAS: 75-33-2.
$(CH_3)_2CH(HS)$.
Properties: Liquid, extremely unpleasant odor, d 0.814 (15.5/15.5C), boiling range 51–55C, flash p −30F (−34.4C).
Derivation: Propylene and hydrogen sulfide.
Hazard: Highly flammable, dangerous fire hazard.
Use: Standard for petroleum analysis, intermediate.

2-isopropyl-5-methylbenzoquinone. See p-thymoquinone.

5-isopropyl-2-methyl-1,3-cyclohexadiene. See α-phellandrene.

3-isopropyl-6-methylene-1-cyclohexene. See β-phaellandrene.

1-isopropyl-2-methylethylene. See 4-methyl-2-pentene.

1-isopropyl-3-methyl-5-pyrazolyl dimethylcarbamate. (isolan).   CAS: 119-38-0.
$C_{10}H_{17}N_3O_2$.
Properties: Liquid, d 1.07 (20C), bp 103C (0.7 mm), miscible with water.
Derivation: By treating 1-isopropyl-3-methyl-5-pyrazolone with dimethylcarbamoyl chloride.
Hazard: Cholinesterase inhibitor.
Use: Insecticide.

2-isopropylnaphthalene. $C_{13}H_{14}$.
Properties: Clear, yellowish-brown liquid. Faint sweet odor, bp 268C, flash p 252F (122C) (COC), autoign temperature 475C, bulk d 8.1 lb/gal, d 0.973, insoluble in water. Combustible.
Hazard: Avoid inhalation of vapors and prolonged skin contact.
Use: Organic intermediate.

isopropylmyristate.   CAS: 110-27-0.
$CH_3(CH_2)_{12}CO_2CH(CH_3)_2$.
Properties: Colorless oil, practically odorless, d 0.850–0.860, fp 3C, refr index 1.435–1.438 (20C), soluble in most organic solvents, insoluble in water, also soluble in vegetable oils, dissolves waxes, lanolin, and similar products. Combustible.
Grade: Double-distilled from coconut oil.
Use: Cosmetic creams, topical medicinals.

isopropyl nitrate.   (2-propanol nitrate).
CAS: 1712-64-7. $(CH_3)_2CHNO_3$.
Properties: Colorless liquid, bp 102C.
Hazard: Oxidizing material, fire risk in contact with organic materials.

isopropyl palmitate. $(CH_3)_2CHOOCC_{15}H_{31}$.
Properties: Colorless liquid, d 0.850–0.855 (25/25C), refr index 1.4350–1.4390 (20C), mp 14C, soluble in 4 parts 90% alcohol, soluble in mineral and fixed oils, insoluble in water. Combustible.
Use: Emollient and emulsifier in lotions, creams, and similar cosmetic products.

isopropyl percarbonate.   Legal label name for diisopropyl peroxydicarbonate.

isopropyl peroxydicarbonate.   See diisopropyl peroxydicarbonate.

m,p-isopropylphenol. $(CH_3)_2CHC_6H_4OH$.
Properties: A solid mixture of the m- and p- isomers, completely soluble in 10% sodium hydroxide, fp (m-) 25.9C, (p-) 63.2C, bp (m-) 228.6C, (p-) 228.5C. Combustible.

o-isopropylphenol.   CAS: 88-69-7.
$(CH_3)_2CHC_6H_4OH$.
Properties: Light yellow liquid, bp 214C, fp 17C, d 0.995 at 20C, flash p 220F (104C) (OC). Insoluble in water; soluble in isopentane, toluene, ethyl alcohol, 10% sodium hydroxide. Combustible.
Use: Intermediate for synthetic resins, plasticizers, surface active agents, perfumes.

isopropyl-N-phenylcarbamate.   See IPC.

N-isopropyl-N'-phenyl-p-phenylenediamine.
(p-isopropylaminodiphenylamine).
CAS: 101-72-4. $C_3H_7NHC_6H_4NHC_6H_5$.
Properties: Dark gray to black flakes, fp range 72–76C, d 1.04 (25C), soluble in benzene and gasoline; insoluble in water.
Use: Protection of rubbers against oxidation, ozone, flexcracking, and poisoning by copper and manganese.

4-isopropylpyridine. $C_5NH_4C_3H_7$.
Properties: Liquid, bp 182.2, d 0.9282 at 20C, refr index 1.4960 (20C); solubility in 100 g of water at 20C, 1.17 g; solubility of water in 100 g, 19.4 g at 20C.

isopropyl titanate.   See tetraisopropyl titanate.

isopropyltoluene.   See cymene.

isopropyl-2,4,5-trichlorophenoxyacetate.
See 2,4,5-trichlorophenoxyacetic acid.

isopropyltrimethylmethane.   See 2,2,3-trimethylbutane.

isopulegol.   (1-methyl-4-isopropenylcyclohexan-3-ol).   CAS: 7786-67-6. $C_{10}H_{17}OH$.
A terpene derivative.
Properties: Water-white liquid. Mint-like odor, d 0.904–0.911, refr index 1.471–1.474. Combustible.
Use: Perfumery (geranium and rose compounds), flavoring. Also available as the acetate.

isoquinoline.   CAS: 119-65-3.

CHCHCHCHCCCHCHNCH

Properties: Colorless plates or liquid, d 1.099 (20C), mp 23C, bp 243C, insoluble in water, soluble in dilute mineral acids and most organic solvents, refr index 1.6223 (25C). Combustible.
Derivation: From coal-tar, also synthetic.
Hazard: Toxic by ingestion.
Use: Manufacture of pharmaceuticals (such as nicotinic acid), dyes, insecticides, rubber accelerators and in organic synthesis.

1,3-isoquinolinediol. $C_9H_5N(OH)_2$
Properties: Cream colored paste, solids approx 80%.
Use: Intermediate.

isosafrole. CAS: 120-58-1. $C_{10}H_{10}O_2$.
Bicyclic.
Properties: Colorless, fragrant liquid; odor of anise; d 1.117–1.120; ref index 1.576; bp 253C; soluble in alcohol, ether, and benzene. Combustible.
Derivation: Treatment of safrole with alcoholic potash.
Use: Manufacture of heliotropin, perfumes, flavors, pesticide synergists.

isosorbide. (1,4,3,6-dianhydrosorbitol).
CAS: 625-67-5.

$\overline{OCH_2CHOHCHCHCHOHCH_2O}$.   A polyol

with a hydroxyl group attached to each of two cis-oriented saturated furan rings. Intermediate for pharmaceuticals. Combustible.
Use: Medicine (diuretic).

isostearic acid. A coined name for a C_{18} saturated fatty acid of the formula $C_{17}H_{35}COOH$. It is a complex mixture of isomers, primarily of the methyl-branched series that are mutually soluble and virtually inseparable.
Use: Similar to stearic or oleic acids.

isosterism. Similarity in physical properties of elements, ions, or compounds, due to similar or identical outer shell electron arrangements.

isostilbene. See stilbene.

isotactic polymer. See polymer, isotactic.

"Isothan."[328] TM for a series of liquid cationic detergents.
Use: Emulsifying, mothproofing, bacteriostat, fungicide, and germicide.

isotherm. Constant temperature line used on climatic maps or in graphs of thermodynamic relations, particularly the graph of pressure-volume relations at constant temperature.

isotone. A nuclide which has the same excess of neutrons over protons as another nuclide.

isotonic. A solution having the same osmotic pressure as another solution, for example, human blood and physiological salt solution.

isotope. One of two or more forms or species of an element that have the same atomic number, i.e., the same position in the Periodic Table, but different masses. The difference in mass is due to the presence of one or more extra neutrons in the nucleus. Thus, "regular" hydrogen, with atomic number 1 and a mass of 1 (proton), is one of the three isotopes of hydrogen. The other two are the naturally occurring deuterium, which has a neutron in its nucleus as well as a proton, giving a mass of 2; and the artificially produced tritium (1 proton and 2 neutrons) with a mass of 3 (approximately). The atomic weight of an element is the average weight percent of all its natural isotopes. The heavier isotopes usually occur very rarely in the atomic population (1 part in 4500 for hydrogen-2, and 1 part in 140 for U-235; in the exceptional case of chlorine, the ratio of isotopes 35 and 37 is about 3 to 1).

The occurrence of isotopes among the 83 most abundant elements is widespread, but separation methods are complicated and costly. 21 elements have no isotopes, each consisting of only one kind of atom (see note below). The remaining 62 natural elements have from 2–10 isotopes each. There are 287 different isotopic species in nature; noteworthy among them are oxygen-17, carbon-14, uranium-235, cobalt-60, and strontium-90, all but the first being radioactive.

There are three kinds of isotopes: (1) natural nonradioactive, (2) natural radioactive, and (3) artificial radioactive (made by neutron bombardment).
Use: (Nonradioactive) Preparation of heavy water to moderate nuclear reactors. (Radioactive): Tracers in biochemical, metallurgical, and medical research; in geochemical and archeological research (C-14); irradiation source for polymerization, sterilization, etc.; therapeutic agents in various diseases (iodine, sodium, gold, etc.); electric power generation.
See also Aston, chemical dating, radioactivity, heavy water, decay, tracer, nuclide.
Note: According to this definition, it is strictly improper to refer to elements that exist in only one atomic form as having "one isotope"; actually such elements as beryllium, aluminum, arsenic, iodine, and others have *no* isotopes, that is, they have no other atomic form that is like them in all respects except mass. The term "isotope" requires the existence of at least *two* elemental forms, in the same sense that the word "twin" requires the existence of a pair.

"Isotron."[204] TM for fluorinated hydrocarbons.
Use: Refrigerants, blowing agents and solvents.
11. Trichlorofluoromethane.
12. Dichlorodifluoromethane.
13. Chlorotrifluoromethane.
113. Trichlorotrifluoroethane.
114. Dichlorotetrafluoroethane.
M Solvent. Trichlorofluoromethane.
T Solvent. Trichlorotrifluoroethane.
Precision Solvent cleaner. Trichlorotrifluoroethane. An ultrapure solvent.

isotropic. Descriptive of the property of transmitting light equally in all directions, cubic (isometric) crystals have this property, as well as liquids, gases, and most glasses.
See also anisotropic.

isovaleral. See isovaleraldehyde.

isovaleraldehyde. (isovaleral; isovaleric aldehyde; 3-methylbutyraldehyde).   CAS: 590-86-3. $(CH_3)_2CHCH_2CHO.$ Occurs in orange, lemon, peppermint, and other essential oils.
Properties: Colorless liquid, apple-like odor, d 0.785, fp −51C, bp 92C, refr index 1.390 (20C), soluble in alcohol and ether, slightly soluble in water. Combustible.
Derivation: Oxidation of isoamyl alcohol, also by Oxo process from petroleum.
Use: Flavoring, perfumes, pharmaceuticals, synthetic resins.

isovaleric acid. (isopropylacetic acid).
CAS: 503-74-2. $(CH_3)_2CHCH_2COOH.$
Occurs in valerian, hop oil, tobacco, and other plants.
Properties: Colorless liquid, disagreeable taste and odor, d 0.931 (20/20C), bp 176C, refr index 1.4043 (20C), fp −37C, slightly soluble in water, soluble in alcohol and ether. Combustible.
Derivation: With other valeric acids, by distillation from valerian, by oxidation of isoamyl alcohol.
Use: Medicine, flavors, perfumes.

isovaleric aldehyde. See isovaleraldehyde.

isovaleryl chloride. (3-methylbutanoyl chloride). CAS: 108-12-3. $(CH_3)_2CHCH_2COCl.$
Properties: Colorless liquid, refr index 1.4136 (24C), d 0.9854 (24/4C), bp 113C, soluble in ether, decomposed by water and alcohols.
Use: Intermediate in synthesis.

isovaleryl-p-phenetidine.
$C_2H_5OC_6H_4NHCOCH_2CH(CH_3)_2.$
Properties: White, glistening needles; almost insoluble in water and ether; soluble in alcohol and chloroform.
Derivation: By heating isovaleric acid with p-phenetidine.

itaconic acid. (methylene succinic acid).
CAS: 97-65-4. $CH_2{:}C(COOH)CH_2COOH.$
Properties: White, odorless, hygroscopic crystals; mp 167–168C; soluble in water, alcohols, and acetone; sparingly soluble in other organic solvents.
Derivation: Submerged fermentation by mold of various carbohydrates.
Use: Copolymerizations, resins, plasticizers, lube oil additive, intermediate.

-ite. A suffix indicating an intermediate oxidation state of a metallic salt, analogous to -ous for acids, e.g., sodium sulfite ($NaSO_3$), containing one less oxygen atom than the sulfate.

IUPAC. Abbreviation for International Union of Pure and Applied Chemistry.

Ivanov reagent. Reaction product of aryl acetic acids and excess Grignard reagent. The reagent may be used in condensations with carbonyl compounds and other Grignard-type reactions.

IVE. Abbreviation for isobutyl vinyl ether.
See vinyl isobutyl ether.

Izod. An impact-testing device of the notched-bar type.

J

J acid. (2-amino-5-naphthol-7-sulfonic acid).

Properties: Gray needles, white when pure, soluble in hot water, sparingly soluble in cold water.
Derivation: β-naphthylamine sulfate is sulfonated with 66% oleum, the filtrate yields 2-naphthyl-amine-1,5,7-trisulfonic acid. On reaction with dilute sulfuric acid at 125C, this yields 2-naphthylamine-5,7-disulfonic acid; from this J acid is obtained on reaction with 58% sodium hydroxide solution at 200C and 210 psi.
Use: Azo dye intermediate.

J acid urea. See 5,5'-dihydroxy-7,7'-disulfonic-2,2'-disulfonic-2,2'-dinaphthylurea.

Jacobsen rearrangement. Reaction of polymethylbenzenes with concentrated sulfuric acid to give rearranged polymethylbenzenesulfonic acids. Under identical conditions, halogenated polymethylbenzenes undergo disproportionation.

Janovsky reaction. Reaction of aldehydes and ketones containing α-methylene groups with m-dinitrobenzenes in the presence of a strong base, resulting in the formation of an intense purple coloration, used for the detection of carbonyl compounds. The color is due to the formation of a Meisenheimer complex.

Japan. A varnish yielding a hard, glossy, dark-colored film. Japans are usually dried by baking at relatively high temperatures (ASTM D 16–52). True Japan varnishes contain a strongly irritating chemical, more recent types contain kauri or copal resin, linseed oil, lead oxide, pigments, and solvents such as kerosene or turpentine.
Hazard: Flammable, irritant to eyes and skin.
Use: Coatings for miscellaneous wood and metal products.

Japan wax. (Japan tallow; sumac wax).
Properties: Pale, yellow solid; tallow-like rancid odor. Contains 10–15% palmatin and other glycerides. Soluble in benzene and naphtha, insoluble in water and in cold alcohol, d 0.970–0.980, mp 53C, saponification number 220, iodine number 10-15. Combustible.

Derivation: From a species of Rhus by boiling the fruit in water.
Use: Candles, floor waxes, polishes, substitute for beeswax, food packaging.

Japp-Klingemann reaction. Formation of hydrazones by coupling of aryldiazonium salts with active methylene compounds in which at least one of the activating groups is acyl or carboxyl. This group usually cleaves during the process.

jar mill. An assembly of small ceramic jars containing porcelain or flint pebbles; the jars are often arranged in parallel tiers, each containing four or more jars, each tier being mounted on a rotating shaft. Such mills are used for production of small quantities of pulverized material. See also ball mill, pebble mill.

jasmine aldehyde. See α-amylcinnamic aldehyde.

jasmine oil. An essential oil in perfumery and flavoring. It is dextrorotatory.

jamolin. $C_{21}H_{30}O_3$. Insecticidal principle of pyrethrum.

jasmone. (3-methyl-2-(2-pentenyl)-2-cyclopenten-1-one). CAS: 488-10-8. $C_{11}H_{16}O$.
A ketone found in jasmine oil and other flower oils.
Properties: Odor of jasmine, d 0.944 (22/0C).
Use: Perfumery.

"Jasmonyl."[227] TM for isomeric mixture of nonane-1,3-diol monoacetates.

jatrophone. A diterpenoid growth inhibitor isolated from an alcohol extract of the plant *Jatropha gossypiifolia*. Its unique structure includes a 12-membered ring. It is readily attacked by nucleophiles. Useful in study of tumor growth inhibition and other biochemical research.

Javel water. NaHClO.
Use: A bleach.

"Javollal."[188] TM for an aromatic concentration used as a substitute for oil of citronella.

"Jaysol."[29] TM for an ethyl alcohol composition containing an aliphatic solvent, methyl isobutyl ketone and ethyl acetate.

"Jellitic."[103]   TM for a prepared dry wheat paste which is pre-gelatinized over hot rolls to make the starch water-soluble.

jelly.   A modified form of the word "gel" widely used in popular language, but also used in chemical literature to refer to the mechanical strength of the gel structures occurring with pectins, gelatin, and various natural gums. "Jelly strength" is frequently specified in the food industry. Other uses of the word are found in "petroleum jelly" obtained as a distillation product of petroleum residues (petrolatum) and in the so-called "royal jelly," a natural nutrient mixture of proteins and carbohydrates produced by bees as food for the queen bee.
See also gel.

"Jel-O-Mer."[36]   TM for a thixotropic additive used in coatings.

Jenner, Edward.   (1749–1823) An English physician, Jenner studied medicine in London and established his practice in the rural area of Gloucestershire. Here he discovered the technique of vaccination as preventive of smallpox (1776). The idea of utilizing cowpox, a disease of cattle, as a protective medium was suggested by his observation that personnel working in dairies developed immunity to smallpox after contracting the much milder cowpox. Jenner's work not only led to almost complete elimination of smallpox in Europe, but also anticipated the development of immune reactions by Pasteur a century later. His success was no accident, but rather the result of detailed observations from which he drew correct conclusions. He was a scientist of the highest caliber and a noteworthy benefactor of mankind.

JET.   Abbreviation for Joint European Torus, an experimental nuclear fusion device under construction in England. It is a project jointly undertaken by several European countries.
See also tokamak; fusion.

jet fuel.   A fuel for jet (turbine) engines, usually a petroleum distillate similar to kerosine. A number of types with somewhat different compositions and properties have been used.

Jetset.   A fast-setting cement developed by the Portland Cement Association. Reported to harden in 20 minutes after pouring. Accelerating agent has not been disclosed.

JH.   (methyl-cis-10,11-epoxy-7-ethyl-3,11-dimethyl-trans, trans-2,6-tridecadienoate).
A synthetic hormone containing a 13-carbon chain; said to have possibilities as an insecticide.

It acts by preventing insects from maturing. Its future depends on the possibility of large-scale production.
See also juvenile hormone.

"JHR Compound."[554]   TM for a thermoplastic compound impervious to mineral acid; does not decompose hydrogen peroxide.
Use: To coat interiors of tanks and containers for shipment of hydrogen peroxide, acids, etc.

"Jiffix."[329]   TM for an acid-hardening, ammonium thiosulfate fixing bath. Ready-mixed and rapid-acting.

Joint European Torus.   See JET.

jojoba oil.
Properties: Colorless, odorless, waxy liquid; chemically similar to sperm oil.
Derivation: By crushing seeds of an evergreen desert shrub found in southwestern US and northern Mexico. Experimental cultivation in California and Israel. Yield of oil from seeds approaches 50%.
Use: Substitute for sperm oil, especially in transmission lubricants, high-pressure lubricant, antifoam agent (antibiotic fermentation), substitute for carnauba wax and beeswax, cosmetic preparations.

Joliot-Curie, Irene.   (1897-1956) A French nuclear scientist who won a Nobel prize for chemistry with her husband Frederic Joliet-Curie. Their joint work involved production of artificial radioactive elements by using alpha rays to bombard boron. They discovered that hydrogen-containing material when exposed to what they considered gamma rays would emit protons. They were involved in many firsts: they gave the first chemical proof of artificial transmutation and of capture of alpha particles, and were the first to prepare positron emitters. Her career started with a ScD, at the University of Paris, and included scores of honors and awards.

Joliot-Curie, Frederick.   (1900-1958) A French physicist who, along with his wife Irene Joliot-Curie, won the Nobel prize in chemistry in 1935. His important discoveries included artificial radioactivity. He did much work on atom structure, dematerialization of electrons, and inverse transformation. Work on hormone synthesis and thyroid substances containing radioactively labeled elements was significant. ScD from the University of Paris was followed by a distinguished career filled with honors and appointments.

Jones oxidation.   The oxidation of primary and secondary alcohols to acids and ketones by the addition of the calculated amount of chromic anhydride in dilute sulfuric acid to a solution of the alcohol in acetone. This procedure does not attack triple bonds nor shift double bonds into conjugation with the ketone formed in the oxidation.

jordan. See beater.

Joule-Thomas coefficient. The change in temperature per atmosphere change of pressure on a gas or other fluid when the enthalpy remains constant. It is found by measuring the temperature change from T' to T when the pressure P' of a gas on one side of a porous plug changes to P on the other side. The change of temperature (T' - T) and of (P' - P) are measured under conditions such that no heat is gained or lost and the pressure of the plug is great enough to insure a nearly constant pressure in the incoming and outgoing gas. The ratio of (T' - T)/(P' - P) at several pressure ranges is extrapolated to the limiting case as (P' - P) approaches zero. This limiting value is the Joule-Thomson coefficient, μ. Thus,

$$\mu = \left(\frac{\alpha T}{\alpha P}\right)_H$$

The subscript H indicates that the enthalpy, H, remains constant during the expansion ($\Delta H = 0$).

Jourdan-Ullmann-Goldberg synthesis.
Synthesis of substituted diphenylamines. The re-action products can be used as intermediates in the synthesis of acridones.

juglone. See 5-hydroxy-1,4-naphthoquinone.

juniper tar oil. (oil of cade, cade oil, oleum cadium). Chief constitutent is cadinene, a sesquiterpene.
Properties: Yellow oil, d 0.980–1.055. Soluble in alcohol.
Derivation: By product of distillation of *Juniperus oxycedrus*

jute. Bast fibers, 4–10 ft long, obtained from the stems of several species of Corchorus, especially C. capularis. Contains a higher proportion of lignin and less cellulose than any other commercial vegetable fiber and has relatively poor strength and durability. The fibers are soft and lustrous but lose strength when wet. Combustible. Not self-extinguishing.
Hazard: Flammable in form of dust, may ignite spontaneously when wet.
Use: Burlap, sacking, linoleum, twine, carpet backing, packing, coarse paper.

juvenile hormone. One of several hormones which retard the development of insects in the larval stage; so called because they prevent the insect from maturing by maintaining its juvenile characteristics. Obtained naturally from silk moths; various syntheses indicate possible use as insecticides, especially for fire ants. Composition of one type is $C_{18}H_{30}O_3$.
See also JH.

K

k. Abbreviation for kilo-, as in kcal.

K. (1) Symbol for potassium (from Latin kalium); (2) symbol for Kelvin scale.

K acid. (1-amino-8-naphthol-4,6-disulfonic acid). $C_{10}H_4NH_2OH(SO_3H)_2$.
Derivation: Fusion of a naphthylamine trisulfonic acid with sodium hydroxide.
Use: Azo dye intermediate.

kainite. CAS: 1318-72-5.
$MgSO_4 \cdot KCl \cdot 3H_2O$. A natural hydrated double salt of potassium and magnesium; color white, gray, reddish, or colorless; streak, colorless; vitreous luster; contains 30% potassium chloride; d 2.05–2.13; hardness 2.5–3.
Occurrence: Germany; one of the Stassfurt minerals.
See potash.
Use: Chemicals (potassium salts), fertilizer (as such).

kalsomine. See calcimine.

kanamycin sulfate. CAS: 25389-94-0.
$C_{18}H_{36}N_4O_{11} \cdot H_2SO_4$. A wide-spectrum antibiotic.
Properties: White, odorless, crystalline powder. Decomposes over a wide range above 250C, soluble in water, practically insoluble in methanol and ethanol.
Grade: USP.
Use: Medicine (antibacterial).

"Kaocast."[455] TM for an alumina-silica refractory which can withstand temperatures up to 3000F.

"Kaolex."[285] TM for a series of hydrous aluminum silicates (sedimentary kaolins) from Georgia and South Carolina.
Properties: PCE 33-35, d 2.60, DMR to 350 psi for casting, jiggering, pressing and extruding. Air-floated or water-washed (lump or pulverized).
Use: Ceramics, refractories, rubber and plastic products.

kaolin. (China clay). A white-burning aluminum silicate which, due to its great purity, has a high fusion point and is the most refractory of all clays. Composition: Mainly kaolinite (50% alumina, 55% silica, plus impurities and water).
Properties: White to yellowish or grayish fine powder; d 1.8–2.6; darkens and develops clay odor when moistened; insoluble in water, dilute acids, and alkali hydroxides; has high lubricity (slipperiness). Noncombustible.
Occurrence: Southeastern US, England, France.
Grade: Technical, NF, also graded on basis of color and particle size.
Use: Filler and coatings for paper and rubber, refractories, ceramics, cements, fertilizers, chemicals (especially aluminum sulfate), catalyst carrier, anticaking preparations, cosmetics, insecticides, paint, source of alumina, adsorbent for clarification of liquids, electrical insulators.
See Toth process, kaolinite, aluminum silicate, clay.

kaolinite. $Al_2O_3 \cdot 2SiO_2 \cdot 2H_2O$. A clay mineral rarely found pure, the main constituent of kaolin and some other clays.

"Kaomul."[455] TM for a mullite firebrick that combines the properties of alumina-diaspore and mullite refractories, mp 3280F, chemical analysis 61% alumina and 35% silica.

"Kaowool."[455] TM for a stable, high-temperature alumina-silica ceramic fiber. Can be used up to 2300F, mp 3200F, diameter of fibers 2.8 microns, length up to 10 inches. Available in bulk, strip or blanket forms. Noncombustible.
Use: Insulating material, high temperature filter, pipe and joint protection, for sound absorption.

kapok. Cotton-like fibers obtained from the seed pods of various species of Ceiba and Bombax. Extremely light and resilient but too brittle for spinning. Combustible; not self-extinguishing.
Sources: Indonesia, Philippines, Ecuador.
Use: Life jackets, insulation, pillows, upholstery.

"Karathane."[23] TM for an agricultural fungicide-miticide based on dinitro(1-methylheptyl)phenyl crotonate, supplied as a wettable powder or liquid concentration. May be combined with most other insecticides and fungicides, except oil-based products.
Use: Controls powdery mildew and various species of mites on plants.
See dinocap.

karaya gum. (sterculia gum; India tragacanth; kadaya gum). A hydrophilic polysaccharide which exudes from certain Indian trees of the genus Sterculia. Color varies from white to dark brown or black.

Properties: A carbohydrate polymer of varying chemical composition. The properties depend on freshness and time of storage. Viscosity greatly decreases over 6 months storage. Forms a translucent colloidal gel in water.

Grade: Technical, FCC.

Use: Pharmaceuticals, textile coatings, ice cream and other food products, adhesives, protective colloids, stabilizers, thickeners, emulsifiers.

See also gum, natural.

"Karbate."[214] TM for carbon and graphite made impervious to fluids under pressure by impregnation with chemically resistant materials. Strength is increased but thermal conductivity is not lowered by this impregnation nor are the other properties modified to any extent. Resistant to thermal shock and to attack by most nonoxidizing chemicals.

Forms: Supplied as complete equipment items; also in blocks, cylinders, tubes.

Use: Pipe, fittings; valves, pumps, heat exchangers, towers, and absorbers for chemical process equipment.

See also carbon, industrial.

"Kardel."[214] TM for biaxially oriented polystyrene film.

Use: Window envelopes and packaging products.

Karl Fischer reagent. A solution of iodine, sulfur dioxide, and pyridine in methanol or methyl "Cellosolve."

Use: Determination of water.

Note: Do not confuse with Fischer's reagent.

Karle, Jerome. (1918-) An American physical chemist who won the Nobel prize for chemistry along with Hauptman in 1985. He developed a series of math equations which allow determination of phase information from x-ray crystallography intensity patterns. The advent of computers allowed the use of the equations to determine the conformation of thousands of chemicals. The work was done at the Naval Research Laboratory in Washington, D. C. Karle heads the laboratory for the Structure of Matter.

Karrer, Paul. (1889-1971) A recipient of the Nobel prize for chemistry in 1937 with Haworth. Although born in Moscow, he attended European universities and received his Doctorate in Zurich. He initiated work on flavins, carotenoids, and vitamins A and B, and accomplished work on structure and synthesis of vitamin B2 as well as vitamins A and E.

"Katadyne" process. A proprietary method of sterilizing water and other potable liquids with a specially prepared form of silver.

kauri. A fossil (hard) copal resin, derived from the kauri pine (Agathis australis) of New Zealand. Soluble in alcohols and ketones, acid value 60–80, must be heat-treated (cracked) before use in varnishes. Combustible.

Use: Varnishes and lacquers, paints, organic cements, to evaluate the solvent power of hydrocarbons.

kauri-butanol value. A measure of the aromatic content and hence the solvent power of a hydrocarbon liquid. Kauri gum is readily soluble in butanol but insoluble in hydrocarbons. The k.b. value is the measure of the volume of solvent required to produce turbidity in a standard solution containing kauri gum dissolved in butanol. Naphtha fractions have a k.b. value of about 30, and toluene about 105.

kcal. Abbreviation for kilogram calorie; Cal has the same meaning.

K-capture. (K-radiation). A type of radioactive decay in which an electron is captured by an atomic nucleus and immediately combines with a proton to form a neutron. The product of this radioactivity has the same mass number as the parent but the atomic number is one unit less. Thus iron-55 with atomic number 26 decays by K-capture to form manganese-55, with atomic number 25. Terms synonymous with K-capture are K-electron capture and orbital electron capture.

Keene's cement. See gypsum cement.

Kekule, August. (1829–1896) Born in Darmstadt, Germany, Kekule laid the basis for the ensuing development of aromatic chemistry. His idea of a hexagonal structure for benzene in 1865 was a monumental contribution to theoretical organic chemistry. "This had been preceded in 1858 by the remarkable notion that carbon was tetravalent and that carbon atoms could be joined to each other in molecules. The theory of the benzene ring has been called the 'most brilliant piece of scientific prediction to be found in the whole field of organic chemistry,' for besides promulgating the idea, he had predicted the number and types of isomers which might be expected in various substitutions on the ring." (L. B. Clapp)

"Kelco-gel."[232] TM for refined sodium alginate. Available as HV and LV grades which vary in viscosity. A hydrophilic colloid.

Use: Thickening, suspending, stabilizing, binding and gelling agent for foods, pharmaceuticals, cosmetics, welding rods, ceramics, latex paints, industrial gels, pastes, coatings, films.

"Kelecin."[64]   TM for a group of soybean lecithins.
Use: Emulsifying and dispersing agents for protective coatings, mastics, animal feeds, automotive specialty chemicals, textiles, rubber processing, plastics, cosmetics, and food products.

"Kel-F."[158]   TM for a series of fluorocarbon products including polymers of chlorotrifluoroethylene and certain copolymers available as extrusion and molding powders, resins, dispersions, gums, oils, waxes, and greases that are characterized by high termal stability, resistance to chemical corrosion, high dielectric strength, and high impact, tensile, and compressive strength.
Use: Corrosion control, contamination prevention, insulation, electrical equipment, molded and fabricated industrial equipment, lubricants, gyro and damping fluids. Especially useful under extreme conditions, including jet and space technology.
See also fluorocarbon polymer.

kelp.   A large coarse seaweed occurring chiefly off the coast of California. It is a type of algae and is mechanically harvested by specially equipped barges. Dried kelp contains 2–4% ammonia, 1–2% phosphoric acid, 15–20% potash, and traces of iodine.
Use: Fertilizers, plastics and conversion to methane by microorganisms; permissible chewing-gum base.

Kelvin scale.   See absolute temperature.

"Kenamine."[473]   TM for a series of straight-chain amines from primary through quaternary; available in chain lengths from C_{12} through C_{22}. They are strongly hydrophobic and are biodegradable.
Use: Cationic intermediate, removal of moisture from surfaces.

Kendall-Mattox reaction.   Formation of a conjugated ketone from an α-bromoketone via a phenylhydrazone or semicarbozone.

Kendrew, John C.   (1917-) An English molecular biologist who won the Nobel prize for chemistry in 1962 with Perutz. His work verified Pauling's earlier thesis concerning the alpha-helix structure of the polypeptide chain. After receiving his PhD from Cambridge, he was science advisor to the allied air commander-in-chief during World War II. He was also editor of the journal "Molecular Biology."

"Kenflex."[267]   TM for a synthetic polymer of aromatic hydrocarbons.
Use: Processing and compounding aid for neoprene, "Hypalon," SBR, vinyl compounds and other plastics, potting compounds, protective coatings, paper and textile coatings; insecticides, inks, chemical synthesis.

"Kennametal."[347]   TM for a group of hard cemented carbides whose principal ingredient is tungsten carbide plus cobalt with additions of carbides of tantalum, titanium, niobium. Available in varied degrees of hardness from 82.0–94.0 Rockwell A and of shock resistance by variation of composition, manufacturing techniques, and structure. There are six major series.

"Kennertium."[347]   TM for a group of high-density tungsten alloys. Grades W-2 and W-10 are 18.5 and 17.0 g/cc, respectively. Easily machinable.
Use: High-inertia applications, balancing, and radioactive shielding.

"Kentanium."[347]   TM for a group of hard cemented carbides whose principal ingredient is titanium carbide.

kepone.   (chlordecone; decachlorooctahydro-1,3,4-metheno-2H-cyclobuta(cd)pentalene-2-one).   CAS: 143-50-0. $C_{10}Cl_{10}O$.
Properties: Crystalline solid, mp 350C (decomposes), soluble in acetic acid, alcohols, ketones, acetone; slightly soluble in water.
Hazard: A carcinogen. Toxic by ingestion, inhalation, and skin absorption. Manufacture and use have been restricted.
Use: Insecticide, fungicide.

"Keranol."[300]   TM for a modified cationic softener for fabrics. Also used on synthetic fibers as antistatic agent.

keratin.   A class of natural fibrous proteins occurring in vertebrate animals and man, they are characterized by their high content of several amino acids, especially cystine, arginine, and serine. They are generally harder than the fibrous collagen group of proteins. The softer keratins are components of the external layers of skin, wool, hair, and feathers, while the harder types predominate in such structures as nails, claws, and hoofs. The hardness is largely due to the extent of crosslinking by the disulfide bonds of cystine by the mechanism shown below:

$$\cdots R_1CH-CO-HN \cdot CH \cdot CO-NH-CHR_2 \cdots$$
$$|$$
$$CH_2$$
$$|$$
$$S$$
$$|$$
$$S$$
$$|$$
$$CH_2$$
$$|$$
$$\cdots R_1CH \cdot OC-HN \cdot CH \cdot CO-NH-CHR_4 \cdots$$

Keratins are insoluble in organic solvents but do absorb and hold water. The molecules contain both acidic and basic groups and are thus amphoteric.

Use: Tablet coatings that dissolve only in the intestines, foam extinguishers, protein hydrolyzates.

keratinase. A water-soluble proteolytic enzyme having the ability to digest the keratin in wool and other forms of hair, converting a portion of it to a water-soluble form. It thus acts as a depilatory and is used both in removing hair from pelts and hides, as well as from human skin. It is inactivated by heating to 100C.

kernite. CAS: 9025-41-6. $Na_2B_4O_7 \cdot 4H_2O$. A natural sodium borate found in Kern County, California.

Properties: Colorless to white, two good cleavages, luster vitreous to pearly, Mohs hardness 3, d 1.95. Noncombustible.

Use: Major source of borax and boron compounds.

kerogen. CAS: 8032-30-2. The organic component of oil shale, it is a bitumen-like solid whose approximate composition is 75–80% carbon, 10% hydrogen, 2.5% nitrogen, 1% sulfur, and the balance oxygen. It is a mixture of aliphatic and aromatic compounds of humic and algal origin and comprises a substantial proportion of the shale; after fractionating and refining, the oil is reported to yield 18% gasoline, 30% kerosene, 27% gas oil, 15% light lube oil and 10% heavy lube oil.

kerosene. (kerosine). CAS: 8008-20-6.
Properties: Water-white, oily liquid; strong odor. D 0.81, boiling range 180–300C, flash p 100–150F (37.7–65.5C), autoign temperature 444F (228C). Combustion properties can be greatly improved by a proprietary hydrotreating process involving a selective catalyst.

Derivation: Distilled from petroleum.

Hazard: Moderate fire risk, explosive limits in air 0.7–5.0%. Toxic by inhalation.

Use: Rocket and jet engine fuel, domestic heating, solvent, insecticidal sprays, diesel and tractor fuels.

ketene. CAS: 463-51-4. $H_2C\!\!=\!\!C\!\!=\!\!O$.
Properties: Colorless gas, disagreeable taste, readily polymerizes, cannot be shipped or stored in a gaseous state, mp −151C, bp −56C.

Derivation: Pyrolysis of acetone or acetic acid by passing its vapor through a tube at 500–600C.

Hazard: Toxic by inhalation, strong irritant to skin and mucous membranes. TLV: 0.5 ppm in air.

Use: Acetylating agent, generally reacting with compounds having an active hydrogen atom; reacts with ammonia to give acetamide. Starting point for making various commercially important products, especially acetic anhydride and acetate esters.

ketimine. A type or class of curing agent for epoxy resins which makes it possible to use very high solids content coatings in spray equipment. Reacts with epoxies very slowly and thus delays curing time which prevents setting up of the resin during spraying operation. In presence of water or water vapor, ketimine breaks down to a polyamine and a ketone. Epoxy coatings cured with ketimine should not exceed thickness of 10 mils.

4-ketobenzotriazine. (benzazimide; 4-keto-(3H)-1,2,3-benzotriazine). $C_7H_5N_3O$ (bicyclic).
Properties: Tan powder, mp 210C (decomposes), soluble in alkaline solutions and organic bases.

Use: Organic synthesis.

α-ketoglutaric acid. (2-oxopentanedioic acid). CAS: 328-50-7. $HOOCCH_2CH_2COCOOH$. Soluble in water and alcohol, mp 113.5C, important in amino-acid metabolism.

β-ketoglutaric acid. (ADA, acetonedicarboxylic acid). $HOOCCH_2CH_2COCOOH$
Properties: Colorless needles, mp 135C (decomposes), soluble in water and alcohol, insoluble in benzene and chloroform.

Derivation: By heating dehydrated citric acid and concentrated sulfuric acid together.

Use: Organic synthesis.

ketohexamethylene. See cyclohexanone.

ketone. A class of liquid organic compounds in which the carbonyl group, $C\!\!=\!\!O$, is attached to two alkyl groups; they are derived by oxidation of secondary alcohols. The simplest member of the series is acetone, $CH_3C(O)CH_3$, but many more complex ketones are known.

Use: Solvents, especially for cellulose derivatives, in lacquers, paints, explosives and textile processing.

See also acetone, diethyl ketone, methyl ethyl ketone.

ketone, Michler's. See tetramethyldiaminobenzophenone.

ketonimine dye. A dye whose molecules contain the $-NH\!\!=\!\!C\!\!=$ chromophore group. There are only two members in the class: auramine and a closely related homolog, methyl aurin in which a methyl group replaces one of the hydrogen atoms of aurin. These are basic dyes used on cotton with tannin or tartar emetic as mordant.

α-ketopropionic acid.   See pyruvic acid.

γ-ketovaleric acid.   See levulinic acid.

"Kevlar."[28]   TM for an aromatic polyamide fiber of extremely high tensile strength and greater resistance of elongation than steel. Its high energy-absorption property makes it particularly suitable for use in belting radial tires for which it was specifically developed; it is also used as a reinforcing material for plastic composites in bullet-proof vests, and in cordage products.
See also aramid.

Keyes process.   A distillation process involving the addition of benzene to a constant-boiling 95% alcohol-water solution to obtain absolute (100%) alcohol. On distillation a ternary azeotropic mixture containing all three components leaves the top of the column while anhydrous alcohol leaves the bottom. The azeotrope (which separates into two layers) is redistilled separately for recovery and reuse of the benzene and alcohol.

"Keyval TN."[496]   TM for a modified EDTA complex.
Use: High-capacity iron sequestrant.

kg.   Abbreviation for kilogram = 1000 grams.

Kick's law.   The amount of energy required to crush a given quantity of material to a specified fraction of its original size is the same, regardless of the original size.

kier.   A large metal tank or vessel in which wool or cotton fibers or fabrics are scoured, bleached, or dyed, usually in an alkaline solution (kier boiling).

kieselguhr.   See diatomaceous earth.

kieserite.   CAS: 14567-64-7. $MgSO_4 \cdot H_2O$.
A natural magnesium sulfate occurring in enormous quantities in the Stassfurt salt beds (Germany), Austria, and India.
See also epsomite and magnesium sulfate.

Kiliani-Fischer synthesis.   Extension of the carbon atom chain of aldoses by treatment with cyanide. Hydrolysis of the cyanohydrins followed by reduction of the lactone yields the homologous aldose.

"killed" steel.   Steel deoxidized by the addition of aluminum, ferrosilicon, etc., while the mixture is maintained at melting temperature until all bubbling ceases. The steel is quiet and begins to solidify at once without any evolution of gas when poured into the ingot molds.

"Kilmag."[50]   TM for a formulation of calcium arsenate for the control of certain fly maggots under poultry cages.
Hazard: Very toxic.

kiln.   (1) A refactory-lined cylinder, either stationary or rotary.
Use: Calcination of lime, magnesia, cement, ores, etc., and for incinerating gaseous, liquid, and solid wastes. (2) A furnace for firing ceramic products.

kilo-.   Prefix meaning 10^3 units (symbol k), e.g., 1 kg = 1 kilogram = 1,000 grams.

kilogram.   (1) A mass identical with that of the international kilogram at the International Bureau of Weights and Measures in France. It is the mass of a liter of water at 4C. (2) A force equal to the weight of one kilogram mass, measured at sea level.

kinematic viscosity.   See viscosity.

kinetics, chemical.   Chemical phenomena can be studied from two fundamental approaches: (1) *thermodynamics*, a rigorous and exact method concerned with equilibrium conditions of initial and final states of chemical changes; (2) *kinetics*, which is less rigorous and deals with the rate of change from initial to final states under nonequilibrium conditions. The two methods are related. Thermodynamics, which yields the driving potential--a measure of the tendency of a system to change from one state to another--is the foundation upon which kinetics is built. The rate at which a change will occur depends upon two factors: (a) directly with driving force or potential, and (b) inversely with a resistance. A measure of the tendency of a system to resist chemical change is the so-called activation energy, which is independent of the driving force or so-called free energy of the reaction.
   The diagram is a mechanical analogy illustrating the difference between activation energy and driving potential. The chemical system is represented by a sphere resting in a valley. The initial equilibrium state A is at a higher elevation than the final state B. The difference in elevation between A and B is a measure of the free energy change of the reaction, that is, the driving force which will take the system from A to B. This quantity ΔG is determined by the classical methods of thermodynamics. Now A and B are equi-

librium states represented by the valleys. For the system to go from A to B it must first overcome the hill separating the valleys. The elevation of this hill from the valley of the initial state is a measure of the resistance to change in the system in going from A to B. The quantity ΔG^*, known as free energy of activation, is determined by the methods of kinetics.

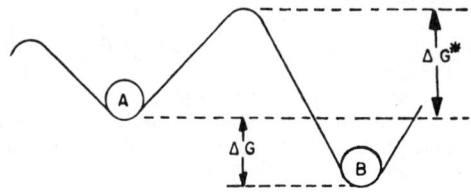

The system of molecules which is undergoing reaction consists of these molecules in different energy states. If the temperature of a gas is raised, there is an increase in the energy of these states and hence an increase in the collisions of molecules which have the necessary activation energy; as a result, the rate of the reaction will increase. Also, if by means of a catalyst, the activation energy is decreased, more colliding molecules will react and again the rate of reaction will increase. Note: Adapted from article by Roger Gilmont in "Encyclopedia of Chemistry" Hampel and Hawley, editors, 3rd ed., 1973.
See thermodynamics, chemical. See also thermodynamics.

kinetic theory. A theory of matter based on the mathematical description of the relationship between the volumes, temperatures, and pressures of gases (P-V-T phenomena). This relationship is summarized in the so-called gas laws as follows: (1) Boyle's law (at constant temperature the volume of a gas is inversely proportional to its pressure), (2) Charles' law (at constant volume the pressure exerted by a gas is proportional to its absolute temperature), (3) Avogadro's law (equal volumes of the same or different gases under the same conditions of temperature and pressure contain the same number of molecules).

The theory involves the basic concept of matter as comprised of atoms and/or molecules which move more rapidly (gases) or vibrate more energetically (solids) as temperature increaes. Thus, crystals melt at a point where the heat or energy input exceeds the bond energy of the solid state.
See also kinetics, chemical; gas; thermodynamics.

kinetin. (6-furfurylaminopurine).
CAS: 9001-54-1. $C_{10}H_9N_5O$. A plant growth regulator used to promote growth in bacterial cultures and as dormancy breaker.

king's green. See copper acetoarsenite.

kinin. (cytokinin). One of a group of plant growth regulators which promote cell division and differentiation.
See kinetin.

Kishner cyclopropane synthesis. Formation of cyclopropane derivatives by decomposition of pyrazolines formed by reacting α,β-unsaturated ketones or aldehydes with hydrazine.

Kistiakowsky, George B. (1900–1982) Born in Kiev, Russia, where he fought in the White Russian Army, he studied in Germany under Maxwell Bodenheim, where he obtained his doctorate in chemistry. In 1926, he came to the US and became an American citizen in 1933. For 41 years, he was Professor of Chemistry at Harvard; in addition to many chemical awards, he was the recipient of the Priestley Medal as well as distinguished Medals of Honor from three Presidents of the US. President Eisenhower appointed him as his assistant for science and technology and he was Chairman of the Science Advisory Committee from 1957–1963. Among his many achievements in both chemistry and physics, he was a world-famous authority on explosives. A key member of the Manhattan Project, he devised the detonating mechanism for the first experimental atomic bomb in New Mexico, at which time he was head of the Los Alamos Laboratory. Though he ranks high among those who developed the bomb, he perceived the awesome destructive potential of nuclear weapons and became an ardent opponent of their future use. Resigning from the Pentagon in 1967, he returned to teaching at Harvard. Among other distinguished organizations, he was a member of the Royal Society of London, the AAAS and the ACS. His opinions on nuclear warfare are clearly stated in the following two comments: "The last 20 years I have been trying mostly to undo the nuclear weapons--and this has been my least successful effort; " and again, shortly before his death: "There simply is not time enough left before the world explodes. Concentrate a mass movement for peace such as there has never been before. The threat of annihilation is unprecedented."

Kjeldahl test. An analytical method for determination of nitrogen in certain organic compounds. It involves addition of a small amount of anhydrous potassium sulfate to the test compound, followed by heating the mixture with concentrated sulfuric acid, often with a catalyst such as copper sulfate. As a result ammonia is formed. After alkalyzing the mixture with sodium hydroxide, the ammonia is separated by distillation, collected in standard acid and the nitrogen determined by back-titration.

"Kleanrol."[28]   TM for a soldering flux crystal based on zinc chloride and ammonium chloride.

"Klearol."[45]   TM for a white mineral oil, technical grade, d 0.828–0.838.
Use: Cosmetic preparations, shell egg preservation.

Klein's reagent.   A saturated solution of cadmium borotungstate, formula variously given, possibly $2CdO \cdot B_2O_3 \cdot 18H_2O$, d 3.28.
Use: Separation of minerals by specific gravity.

Klug, Aaron S.   (1926-) A South African born chemist who won the Nobel prize for chemistry in 1982 for his work with the electron microscope and research into the structure of nucleic and protein complexes. His use of crystallographic electron microscopy to analyze the structures of biologically important complex chemicals. He was cited in particular for his establishment of Fourier microscopy.

kneading.   Blending of soft plastic semi-solid materials into a uniform mixture by subjecting them to a rolling pressure exerted by agitators of specific shape rotating in through-like containers. The action is a combination of turning, folding and pressing. The operation is used in processing bakery doughs, printing inks, clays, and pastes of various types.
See also blend, sigma blade.

knife.   See doctor knife.

knock.   Ignition of a portion of the gasoline in the cylinder head due to spontaneous oxidation reactions rather than to the spark. It causes serious power loss, especially in high-compression engines.
See also octane number.

Knoevenagel condensation; Doebner modification.
Condensation of aldehydes or ketones with active methylene compounds (specifically malonic esters) in the presence of ammonia or amines; the use of malonic acid and pyridine is known as the Doebner modification.

Knoop hardness.   See hardness.

Knoop-Oesterlin amino acid synthesis.
Preparation of α-amino acids by catalytic hydrogenation of α-oxo acids in aqueous ammonia in the presence of platinum, palladium, or Raney nickel catalysts, probably via an unstable iminocarboxylate ion intermediate.

Knorr pyrazole synthesis.   Formation of pyrazole derivatives from hydrazines, hydrazides, semicarbazides, and aminoguanidines by condensation with 1,3-dicarbonyl compounds; with substituted hydrazines, two structurally isomeric pyrazoles are formed.

Knorr pyrrole synthesis.   Formation of pyrrole derivatives by condensation of α-amino ketones as such or generated in situ from isonitrosoketones with carbonyl compounds containing active α-methylene groups.

Knorr quinoline synthesis.   Formation of α-hydroxyquinolines from β-ketoesters and arylamines above 100C. The intermediate anilide undergoes cyclization by dehydration with concentrated sulfuric acid.

Koch acid.   (1-naphthylamine-3,6,8,-trisulfonic acid).

Properties: White solid, slightly soluble in water.
Derivation: Naphthalene-β-sufonic acid is sulfonated with 60% oleum at 165C, the resulting naphthalene-1,3,6-trisulfonic acid is nitrated and the product reduced with iron.
Use: Azo dye intermediate.

Koch-Haaf carboxylations.   Formation of tertiary carboxylic acids by treating alcohols with carbon monoxide in strong acid.

Kochi reaction.   Synthesis of organic chlorides by decarboxylation of carboxylic acids in the presence of lead tetraacetate and lithium chloride.

"Kodaflex AD-2."[256]   TM for adhesion-promoting plasticizer for vinyl coatings.

"Kodar."[256]   TM for a thermoplastic polyester resin (1,4-cyclohexylenedimethylene terephthalate/isophthalate copolymer). It is made from terephthalic acid, isophthalic acid, and the glycol cyclohexylene-dimethanol. Its main field of use is in so-called blister packaging for which it is claimed to be superior to polyvinylchloride and polyethylene terephthalate. Available in clear-colored amorphous pellets. Complies with food packaging safety regulations, including meats and poultry.

"Kodel."[256] TM for a polyester-type fiber.

Koenigs-Knorr synthesis. Formation of glycosides from acetylated glycosyl halides and alcohols or phenols in the presence of silver carbonate or silver oxide. The reaction proceeds with inversion of configuration.

kojic acid. [5-hydroxy-2-(hydroxymethyl)-4-pyrone]. CAS: 5-1-30-4. $C_6H_6O_4$.
Properties: Crystals, mp 152–154C, soluble in water, acetone, alcohol; slightly soluble in ether, insoluble in benzene, mildly antibiotic.
Derivation: Fermentation of starches and sugars by certain molds.
Use: Chemical intermediate, metal chelates, insecticide, antifungal and antimicrobial agent.

kola. See cola.

Kolbe electrolytic synthesis. Formation of hydrocarbons by the electrolysis of alkali salts of carboxylic acids (decarboxylative dimerization).

Kolbe-Schmidt reaction. The preparation of salicyclic acid or its derivatives from carbon dioxide and sodium or potassium phenolate.

"Kolene" DGS Salt[557]. TM for an anhydrous molten oxidizing salt bath using a sodium hydroxide base with additives necessary to provide controlled chemical oxidizing and dissolving properties.
Use: Descaling of heat treated and hot work oxides and scales, deglassing (removal of glass drawing lubricants), investment, and silica removal; removal of burned-in carbon deposits, cleaning of oils, greases, and organic materials from the surface of metals.

"Kollercast."[263] TM for molded synthetic resin products for industrial purposes.

Knoop scale. Comparative hardness scale, ranges from glass (300–600) to diamond (6000–6500).

Komarowsky reaction. The reaction between certain alcohols and p-hydroxybenzaldehyde in dilute sulfuric acid solutions to give soluble colored complexes. 1,2-Propylene glycol gives a colored product while ethylene glycol does not: The reaction has also been employed to determine cyclohexanol in cyclohexanone.

Kondakov rule. Olefins which add mineral acids readily react with chlorine or bromine to give unsaturated monohalides; those which do not add mineral acids readily form dihalides.

korax. CAS: 600-25-9. Generic name for 1-chloro-1-nitropropane. $ClCH_2CH(CH_3)NO_2$.
Properties: Liquid, bp 170.6C (745 mm), miscible with most organic solvents, slightly soluble in water, flash p 144F (62.2C). Combustible.

Hazard: Moderate fire risk. Strong irritant. TLV: 2 ppm in air.
Use: Fungicide.

"Koreon."[292] TM for a gruop of one-bath crome-tanning compounds, usually of the basic chromic sulfate type. Range of chromic oxide equivalence 23.5–25.5%, basicity range 33.0–59.0%, soluble in water.

Kostanecki acylation. Formation of chromones or coumarines by acylation of o-hydroxyaryl ketones with aliphatic acid anhydrides followed by cyclization.

Kr. Symbol for krypton.

Krafft degradation. Conversion of carboxylic acids, especially of high molecular weight, into the next lower homolog by dry distillation of the alkaline earth salt with the corresponding acetate, followed by chromic acid oxidation of the methyl ketone.

kraft paper. A strong and relatively cheap paper made cheifly from pine by digestion with a mixture of caustic soda, sodium sulfate, sodium carbonate, and sodium sulfide. It is by far the largest volume paper made in US.
See also holopulping process.

"Kralastic."[248] TM for a series of ABS resins.
Properties: Granular rubber-plasticized resins, rigid and tough, dimensionally stable, light in weight, chemically resistant, good electrical resistance.
Use: Injection and extrusion applications, chemical pipe, cathode edge strips, cams, gears, cable floats, wheels, etc.

"Kraton" 101[125]. TM for a styrene-butadiene elastomer which requires no vulcanization while displaying most of the properties of conventional vulcanized polymers. White free-flowing crumb, readily soluble in a large number of commercially used solvents.
Use: In pressure sensitive wet lay-up and hot-melt adhesives, dip coating, spraying, and spreading applications.

Krebs cycle. See TCA cycle.

"Krene."[214] TM for flexible vinyl film and sheeting supplied in continuous roll form for packaging, laminations, seat covers, and shower curtains.

"Krenite" 10[28]. TM for a solution of urea and ammonium nitrate in aqueous ammonia; contains 43.5% nitrogen.
Use: Manufacture of mixed fertilizers.

Krohnke aldehyde synthesis. Transformation of benzyl halides into aldehydes via their pyridinium salts which, on treatment with p-nitrosodimethylaniline, give nitrones. Hydrolysis of the nitrones yields aldehydes.

Kroll process. A widely used process for obtaining titanium metal. Titanium tetrachloride is reduced with magnesium metal at red heat and atmospheric pressure, in the presence of an inert gas blanket of helium or argon. Magnesium chloride and titanium metal are produced. The reaction is $TiCl_4 + 2Mg \rightarrow Ti + 2MgCl_2$. Essentially the same process is also used for obtaining zirconium.

"Kromatherm."[299] TM for high-temperature pigments designed for silicone- and fluorocarbon resin-based paint vehicles.

"Kronitex."[55] TM for a series of synthetic phosphate esters to replace such natural products as tricresyl and cresyl diphenyl phosphates.
Use: Flame-retardant platicizers for vinyls, dust filter medium, gas additives, wood-treating chemical, foam control.

krypton. CAS: 7439-90-9. Kr. Element of atomic number 36, noble gas group of the periodic system, aw 83.80, valence = 2 (possibly others), has six stable isotopes and a number of artificially radioactive forms.
Properties: Colorless, odorless gas. Bp 152.9C (1 atmosphere), fp −157.1C, d 2.818 (air = 1), sp vol 4.61 cu ft/lb (21C, 1 atmosphere), only slightly soluble in water. Known to combine with fluorine at liquid nitrogen temperature by means of electric discharges or ionizing radiation to make KrF_2 and KrF_4. These compounds decompose at room temperature. Noncombustible.
See noble gas.
Derivation: By fractional distillation of liquid air. Air contains 0.000108% of krypton by volume.
Use: Incandescent bulbs and fluorescent light tubes, lasers, high-speed photography.
Note: Solid krypton exists at cryogenic temperatures as a white, crystalline substance; mp 116K.

krypton 85. Radioactive krypton of mass number 85.
Properties: Half-life 10.3 Y; radiations, beta with a small component of gamma. Low radiotoxicity.
Derivation: A fission product extracted from irradiated nuclear fuel.
Forms available: Gas of high chemical purity, but mixed with other isotopes of krypton in sealed glass flasks.
See also kryptonates.
Use: Activation of phosphors for self-luminous markers, detecting leaks, medicine to trace blood flow.

krypton 86. Isotope of krypton used in measurement of standard meter.

kryptonates. Materials impregnated with krypton-85 in such a way that the radioactive atoms are held within the crystalline lattice structure. Elements, alloys, glasses, inorganic compounds, rubbers, and plastics have been so impregnated with tracer atoms.

"Krytox."[28] TM for a series of hexafluroropropylene epoxide polymers of medium molecular weight, used as lubricating oils and greases; in high temperature or corrosive conditions, good chemical inertness, even with boiling sulfuric acid; low solubility in most solvents; good lubricity under load; nonflammable; have thermal stability up to 260C.

K-Selectride.[541] TM for potassium tri-sec-butylborohydride, 1.0 molar solution in tetrahydrofuran. CAS: 54575-49-4.
$KB[CH(CH_3)C_2H_5]_3H$.
Properties: Moisture sensitive liquid with mw 222.27, d 0.913, fp −17C.
Hazard: Flammable liquid, handle under nitrogen.
Use: Stereoselective reduction of ketones; conjugate reduction and alkylation of alpha,beta-unsaturated ketones.

KTPP. Abbreviation for potassium tripolyphosphate.

Kucherov reaction. Hydration of acetylenic hydrocarbons with dilute sulfuric acid in the presence of mercuric sulfate or boron trifluoride as catalyst.

Kuhn, Richard. (1900-1967) A German chemist who won the Nobel prize in 1938. He worked on cartinoids and synthetic vitamins, and discovered chemical formula for vitamin B6. He also discovered method for dissolving symplexes from plants using invert-soaps. He received his PhD in Munich, and went on to teach in Switzerland.

Kuhn-Roth method for C-methyl determination. Oxidation of organic compounds with chromic and sulfuric acids in such a manner that the C-methyl groups are converted to acetic acid which can be assayed volumetrically. The method has been modified and extended (1) to saturated fatty acids and alcohols containing up to about 20 carbon atoms, and (2) to aliphatic long-chain compounds of very high molecular weight.

Kuhn-Winterstein reaction. Conversion of 1,2-glycols into trans olefins by reaction with diphosphotetraiodide (P_2I_4) or other halogenated reagents. This reaction is useful in the preparation of polyenes.

"Kuron."[233] TM for a hormone-type weed and brush killer.

Kurroll's salt. $NaPO_3$ (IV) A high-temperature crystalline form of sodium metaphosphate.

kyanite. (cyanite; disthene; rhoetizite). $Al_2O_3 \cdot SiO_2$.   A mineral.

"Kydex."[23] TM for a thermoformed acrylic polyvinyl chloride alloy plastic sheet.
Use: Housings, trays, covers, containers, protective guards, and decorative parts exposed to severe service conditions.

"Kynol."[280] TM for a flame-resistant fiber available as 1.5 inch staple, 1.7 denier. It is a cross-linked amorphous phenolic polymer, inert to all solvents and with fair resistance to oxidizing acids and strong alkalies. Will not ignite up to 2500C, but will char slowly at 260C. Potential uses are as ablative agent in spacecraft, flameproof apparel, protective clothing, etc.
See ablation.

kynurenine. CAS: 343-65-7. $C_{10}H_{12}N_2O_3$.
An amino acid that is a metabolic product of tryptophan.
Use: Biochemical and nutritional research, especially in connection with B complex deficiencies.

L

l-.   Prefix indicating that a compound is levorotatory. A minus sign (-) is now preferred.

L.   Prefix indicating the left-handed enantiomer of an optical isomer.
See also D.

L.   (sometimes l).   Abbreviation for liter.

"L-26."[304]   TM for the normal lead salt of 2-ethylhexoic acid. Straw-colored, viscous liquid; soluble in many common organic solvents; stable at comparatively high temperatures.
Use: Curing agent for silicone paints and insulating varnishes.

La.   Symbol for lanthanum.

labile.   Descriptive of a substance that is inactivated by high temperature or radiation, e.g., a heat-labile vitamin, unstable.

laboratory conditions.   An ideal set of conditions in which all variant factors except the one under test can be held constant, as for example, rooms provided with constant temperature and humidity control, clean rooms, and the like. Less specifically, the term refers to experimental or small batch conditions as opposed to large-scale production.

laboratory machinery.   Small-scale working models of basic types of equipment used for experimental purposes in the laboratories of many chemical process industries. Commercially available are such items as calenders, mills of various kinds, mixers, autoclaves, extruders, electric furnaces, distillation columns, etc. A machine as complex as a fourdrinier can be duplicated in laboratory size (about 10 ft by 18 in.).

labèl.   (1) A warning notice required by DOT and IATA to be placed on a shipping container of a hazardous material transported by air, highway, rail, or water. Names of labels are as follows:

Corrosive	Flammable Gas
Irritant	Oxidizer Flammable
Dangerous when Wet	Liquid
Nonflammable Gas	Poison
Explosive	Flammable Solid
Organic Peroxide	Radioactive

(2) A notice required to appear on a food product indicating its composition and nutritional value (RDA), or on a pharmaceutical or household product stating its hazardous properties.
(3) A radioactive isotope or fluorescent dye added to a chemical compound to trace its course and behavior through a series of reactions, usually in living organisms. This technique has also been used to measure frictional wear of moving parts of automotive engines.
See also tracer; tagged compound.

laccase.   An enzyme that oxidizes phenols or o- and p-quinones.
Occurrence: It occurs in the latex of the lac tree, in potatoes, sugar beets, apples, cabbages, and other plants.

lac dye.   (C.I. Natural Red 25; lacchaid acid). CAS: 60687-93-6.   A brilliant red dye obtained by maceration of crude lac.
See shellac.

lachrymator.   (lacrimator).   A material (gas) that is strongly irritant to the eyes; tear gas.

L acid.   (1-naphthol-5-sulfonic acid). $C_{10}H_6(OH)SO_3H$.
Properties: White solid, soluble in water.
Derivation: From naphthalene-1,5-disulfonic acid by fusion with sodium hydroxide.
Use: Azo dye intermediate.

lacmoid.   (resorcinol blue). $(HO)_2C_6H_3N[C_6H_2(OH)_3]_2$.
Properties: Lustrous, dark-violet, crystalline scales. Soluble in alcohol, ether, acetone, phenol, and glacial acetic acid; slightly soluble in water; pH 4.4–6.2.
Derivation: From resorcinol by treatment with sodium nitrite.
Use: Indicator in analytical chemistry (more sensitive than litmus).

lacquer.   A protective or decorative coating that dries primarily by evaporation of solvent, rather than by oxidation or polymerization. Lacquers were originally comprised of high-viscosity nitrocellulose, a plasticizer (dibutyl phthalate or blown castor oil), and a solvent. Later, low-viscosity nitrocellulose became available; this was frequently modified with resins such as ester gum

or rosin. The solvents used are ethanol, toluene, xylene, and butyl acetate. Together with nitrocellulose, alkyd resins are used to improve durability. The nitrocellulose used for lacquers has a nitrogen content of 11–13.5% and is available in a wide range of viscosities, compatibilities, and solvencies. Chief uses of nitrocellulose-alkyd lacquers are for coatings for metal, paper products, textiles, plastics, furniture, and nail polish. Various types of modified cellulose are also used as lacquer bases, combined with resins, and plasticizers. Many non-cellulosic materials, such as vinyl and acrylic resins are also used, as are bitumens, with or without drying oils, resins, etc. See also nitrocellulose.
Hazard: Flammable, dangerous fire and explosion risk.

lactalbumin. See albumin, milk.

lactam. A cyclic amide produced from amino acids by the removal of one molecule of water. An example is caprolactam, $CH_2(CH_2)_4CONH$, derived from epsilon-aminocaproic acid, $NH_2(CH_2)_5COOH$.

lactase. An enzyme present in intestinal juices and mucosa which catalyzes the production of glucose and galactose from lactose.
Use: Biochemical research.

lactic acid. (α-hydroxypropionic acid; milk acid). CAS: 50-21-5. $CH_3CHOHCOOH$.

$$CH_3-\overset{\displaystyle H}{\underset{\displaystyle OH}{C}}-COOH$$

Properties: Colorless or yellowish, odorless, hygroscopic, syrupy liquid; bp 122C (15 mm); mp 18C; d 1.2; miscible with water, alcohol, glycerol, and furfural; insoluble in chloroform, petroleum ether, carbon disulfide. Cannot be distilled at atmospheric pressure without decomposition; when concentrated above 50% it is partially converted to lactic anhydride. It has one asymmetric carbon and two enantiomorphic isomers. The commercial form is a racemic mixture.
Derivation: (a) By fermenting starch, milk whey, molasses, potatoes, etc and neutralizing the acid as soon as formed with calcium or zinc carbonate. The solution of lactates is concentrated and decomposed with sulfuric acid; (b) synthetically by hydrolysis of lactonitrile.
Grade: Technical 22% and 44%, food 50–80%, plastic 50–80%, USP (85–90%), CP, FCC.
Use: Cultured dairy products, as acidulant, chemi-

cals (salts, plasticizers, adhesives, pharmaceuticals), mordant in dyeing wool, general-purpose food additive, manufacture of lactates.

lactic acid dehydrogenase. An enzyme found in animal tissues and yeast which takes part in controlling carbohydrate metabolism in the cell.
Use: Biochemical research.

lactogenic hormone. See luteotropin.

lactoglobulin. A protein found in milk. It comprises from 7–12% of the skim milk protein and is closely associated with casein.

lactone. An inner ester of a carboxylic acid formed by intramolecular reaction of hydroxylated or halogenated carboxylic acids with elimination of water. They occur in nature as odor-bearing components of various plant products, also made synthetically.
See butyrolactone, propiolactone.

lactonitrile. (α-hydroxypropionitrile; acetaldehyde cyanohydrin).   CAS: 78-97-7. $CH_3CHOHCN$.
Properties: Straw-colored liquid, acid to methyl red, fp −40C, bp 182–184C (slight decomposition), d 0.9919 (18.4C), refr index 1.4058 (18.4C), soluble in water and alcohol, insoluble in petroleum ether and carbon disulfide, flash p 170F (76.6C). Combustible.
Derivation: Acetaldehyde and hydrocyanic acid.
Grade: Technical, 95–97% purity.
Hazard: Toxic by inhalation, ingestion, and skin absorption; evolves hydrocyanic acid in presence of alkali.
Use: Solvent, intermediate in production of ethyl lactate and lactic acid.

lactose. (milk sugar; saccharum lactis). CAS: 63-42-3. $C_{12}H_{22}O_{11} \cdot H_2O$.

Properties: White, hard, crystalline mass or white powder; sweet taste; odorless; stable in air. Soluble in water, insoluble in ether and chloroform, very slightly soluble in alcohol, d 1.525, mp decomposes at 203.5C.
Derivation: From whey, by concentration and crystallization. Cows' milk contains about 5% lactose.

Grade: Crude, fermentation, spray-dried, edible, USP.

Use: Pharmacy; infant foods; bacteriology; baking and confectionery; margarine and butter manufacture; manufacture of penicillin, yeast, edible protein, and riboflavin; culture media; adsorbent in chromatography.

lac wax. A wax obtained from lac consisting of myricyl and ceryl alcohols, free and combined with various fatty acids. Combustible.
See shellac.

LAD. Abbreviation for lithium aluminum deuteride.

ladder polymer. An ordered molecular network of double-stranded chains connected by hydrogen or chemical bonds located at regular intervals along the chains. Many complex proteins, including DNA, are of this nature.

Ladenburg rearrangement. Thermal rearrangement of an alkyl- or benzylpyridinium halide to an alkyl- or benzylpyridine.

ladle. A refractory-lined pot or vessel equipped with a spout and with lugs for handling by a crane used to transport molten metal from the furnace to ingot molds, into which the metal is poured from the ladle.

lagoon. A scientifically constructed pond 3–5 ft deep in which sewage and other organic wastes are decomposed by the action of algae, sunlight, and oxygen thus restoring water to a purity equal to that obtained with other types of treatment. These so-called oxidation ponds are often used following activated sludge treatment. The waste may be retained in the lagoon for as long as 30 days and is then chlorinated and passed through a trickle filter.

LAH. Abbreviation for lithium aluminum hydride.

laid (papermaking). See dandy roll.

lake. An organic pigment produced by the interaction of an oil-soluble organic dye, a precipitant, and an absorptive inorganic substrate. Insoluble in water; poor light-fastness makes lakes unsuitable for use in exterior paints.
Use: Printing inks, wallpaper inks, metal decorative coatings, coated fabrics, rubber, plastics, food colorants.
See also toner, alizarin, madder lake.

Lake Red carbon. Red pigments made by coupling 2-chloro-5-aminotoluene-4-sulfonic acid with β-naphthol and forming various metal salts.
Properties: Good resistance to bleeding, reasonable light resistance, good transparency, produces inks with good flow.
Grade: Resinated and non-resinated.
Use: General purpose colors for letterpress, gravure, flexographic, moisture set, heat set inks, specially for offset printing inks.

"Laktane."[51] TM for a solvent especially prepared for use as lacquer diluent and in rotogravure printing inks. Its boiling range is typically 103–109C.

"Lambast."[58] TM for a contact herbicide containing 25.5% 2,4-bis(3-methoxypropylamino)-6-methylthio-s-triazine $[CH_3O(CH_2)_3NH]_2C_3N_3SCH_3$.
Properties: White solid, mp 55C, insoluble in water, slightly soluble in ethanol, very soluble in acetone and benzene.
Hazard: Toxic by ingestion.

lamepon. An acetylated peptide used as a surface-active agent.

laminate. A composite made of any one of several types of thermosetting plastic (phenolic, polyester, epoxy, or silicone) bonded to paper, cloth, asbestos, wood, or glass fiber. High tensile and dielectric strength and low moisture absorption are characteristic of these products. Available as sheet, rod, or tubing in mechanical, electrical, and general-purpose grades (National Electrical Manufacturers Association). Plywood is composed of a veneer with grain oriented at a 90 degree angle on successive layers and bonded with a thermosetting adhesive of the urea or phenol-formaldehyde type to give a high strength, dimensionally stable, weather-resistant construction material. It can be made non-flammable by treatment with salt solution. Polyvinyl butyral sheet is used in safety glass.
See also reinforced plastic.

lampblack. A black or gray pigment made by burning low-grade heavy oils or similar carbonaceous materials with insufficient air and in a closed system such that the soot can be collected in settling chambers.
Properties: Markedly different from carbon black. Strongly hydrophobic. Nonflammable.
Use: Black pigment for cements, ceramic ware, mortar, inks, linoleum, surface coatings, crayons, polishes, carbon paper, soap, etc.; ingredient of insulating compositions, liquid-air explosives, matches, fertilizer, furnace lutes, lubricating compositions, carbon brushes; reagent in cementation of steel.
See also carbon black.

"Lanaset."[57]   TM for wool dyes with optimal fastness and minimal fiber damage.

land.   (1) The portion of a mold which provides the separation or cutoff of the flash from the molded article. (2) The surface along the top of the flights or ribs of the screw of an extruder. (3) In an extrusion die, the surface parallel to the flow of material.

langbeinite. $K_2Mg_2(SO_4)_3$.   A natural sulfate of potassium and magnesium found in salt deposits.
Properties: Colorless to yellowish, reddish greenish; luster vitreous; Mohs hardness 3.5–4; d 2.83.
Occurrence: New Mexico, Germany, India.
Use: Source of potash.

Langmuir, I.   (1881–1957) An American physical chemist who was awarded the Nobel Prize in 1932 for his fundamental research in surface chemistry, especially monomolecular films. This led to development of modern knowledge of emulsification and detergency. Langmuir also investigated electrical discharges in gases and did pioneer work on cloud-seeding techniques.

"Lanitol."[300]   TM for a group of alkylarylsulfonate type detergents. A: Liquid biodegradable type. F: Flake, sodium salt. CW: Powder, same as flake plus alkaline builders.

"Lannate."[28]   TM for methomyl insecticides.
See methomyl.

"Lanogene."[493]   TM for liquid lanolin consisting of low melting point esters. A crystal-clear, amber, viscous liquid; anhydrous; soluble in mineral and vegetable oils.
Use: Nonionic with out emulsifier; pigment dispersant; plasticizer; emollient.

lanolin.
Properties: (Hydrous): Yellowish to gray semisolid containing 25–30% water, slight odor. (Anhydrous): Brownish-yellow semisolid containing no more than 0.25% water but can be mixed with about twice its weight of water without separation.
Derivation: Purification of degras, a crude grease obtained by solvent-treatment of wool. Contains cholesterol esters of higher fatty acids. Hydrogenated, ethoxylated and acetylated derivatives are available.
Grade: Technical, cosmetic, USP.
Use: Ointments, leather finishing, soaps, face creams, facial tissues, hair-set and suntan preparations.
See also degras.

lanosterol.   (isocholesterol). $C_{30}H_{50}O$.
An unsaturated sterol closely related to cholesterol.
Properties: mp 139–140C, optically active crystals

lanthana.   See lanthanum oxide.

lanthanide series.   (lanthanoid series).   The rare earth series of elements, atomic numbers 58 through 71.
See also rare earth.

lanthanum.   CAS: 7439-91-0.   La.   Metallic element of atomic number 57, group IIIB of the periodic table, a rare earth of the cerium group, aw 138.9055, valence = 3; two stable isotopes.
Properties: White malleable ductile metal, oxidizes rapidly in air, d 6.18–6.19, mp 920C, bp 3454C, corrodes in moist air, soluble in acids, decomposes water to lanthanum hydroxide and hydrogen, superconducting at approximately 6K.
Derivation: By cracking of monazite or bastnasite ores with concentrated sulfuric acid and subsequent separation.
Forms available: Ingots, rods, 20-mil sheets, powdered, 99.9% pure.
Hazard: Ignites spontaneously in powdered form. May delay blood clotting and cause liver injury upon ingestion.
Use: Lanthanum salts, electronic devices, pyrophoric alloys, rocket propellants, reducing agent catalyst for conversion of nitrogen oxides to nitrogen in exhaust gases (usually in combination with cobalt, lead, or other metals), phosphors in x-ray screens.
See also rare-earth metal.

lanthanum acetate. $La(C_2H_4O_2)_3 \cdot xH_2O$.
Properties: White powder, soluble in water, purities up to 99.9+%, soluble in acids.

lanthanum ammonium nitrate.
Properties: Colorless crystals, soluble in water.
Grade: Purities to 99.9+%.
Hazard: Oxidizer; explosion and fire risk.

lanthanum antimonide.   LaSb.   A binary semiconductor.

lanthanum arsenide.   LaAs.   Made in high purity for use as a binary semiconductor.
Hazard: Highly toxic.

lanthanum carbonate. $La_2(CO_3)_3 \cdot 8H_2O$.
Properties: White powder, insoluble in water, soluble in acids, d 2.6.
Grade: Up to 99.9+% lanthanum salts.

lanthanum chloranilate.
$La_2(O:C_6Cl_2O_2:O)_3 \cdot xH_2O$.
Use: Reagent for fluoride determination.

lanthanum chloride.   CAS: 10099-58-8.
$LaCl_3 \cdot 7H_2O$.
Properties: White crystals, transparent, hygroscopic, mp (hydrate) 91C (decomposes); (for anhydrous) d 3.842 (25C), mp 872C; soluble in alcohol, water, acids.
Derivation: Treatment of lanthanum carbonates or oxides with hydrochloric acid in an atmosphere of dry hydrogen chloride.
Grade: Purities to 99.9+%, single crystals available.
Use: Anhydrous trichloride of rare-earth metal is often used to prepare the metal.

lanthanum fluoride. LaF_3.
Properties: White powder, insoluble in water, acids.
Grade: Purities up to 99.9+%, single crystals available.
Hazard: Toxic by ingestion. TLV (as F): 2.5 mg/m^3 of air.
Use: Phosphor lamp coating (gallium arsenide solid-state lamp), carbon arc electrodes, lasers.

lanthanum nitrate.   CAS: 10099-59-9.
$La(NO_3)_3 \cdot 6H_2O$.
Properties: White crystals, hygroscopic, bp 126C, mp 40C, soluble in alcohol, water, acids.
Grade: Purities to 99.9+%.
Hazard: Explosion and fire risk.
Use: Antiseptic, gas mantles.

lanthanum oxalate. $La_2(C_2O_4)_3 \cdot 9H_2O$.
Properties: White powder, insoluble in water; soluble in acids.
Grade: Purities to 99.9+%.

lanthanum oxide.   (lanthana; lanthanum trioxide; lanthanum sesquioxide). La_2O_3.
Properties: White or buff amorphous powder, d 6.51 (15C), mp 2315C, bp 4200C, soluble in acids, insoluble in water.
Derivaton: Ignition of hydroxide or oxyacid (oxalate, sulfate nitrate, etc.), direct combustion of free metal (burns with brilliant, white light).
Grade: Purities to 99.9+%.
Use: Calcium lights, optical glass, technical ceramics, cores for carbon-arc electrodes, fluorescent phosphors, refractories.

lanthanum phosphide.   LaP.   Made in high purity for use as binary semiconductor.

lanthanum sesquioxide.   See lanthanum oxide.

lanthanum sulfate. $La_2(SO_4)_3 \cdot 9H_2O$.
Properties: White crystals, d 2.821, refr index 1.564 (20C), insoluble in alcohol, slightly soluble in water, acids.
Derivation: By dissolving hydroxide, carbonate, or oxide in dilute sulfuric acid.
Grade: Purities to 99.9+%.
Use: The sulfates of the rare-earth elements are often used for atomic weight determination of the element.

lanthanum trioxide.   See lanthanum oxide.

lanthionine. $S(CH_2CHNH_2COOH)_2$.
A non-essential amino acid first obtained from deaminated wool.

lard.   Purified internal fat of the hog.
Properties: Soft, white, unctuous mass; faint odor; bland taste. Soluble in ether, chloroform, light petroleum hydrocarbons, carbon disulfide; insoluble in water. Mp 36–42C, high in saturated fats. Combustible.
Chief constituents: Stearin, palmitin, olein.
Use: Cooking, pharmacy (ointments, cerates), perfumery (pomades), source of lard oil.

lard oil.
Properties: Colorless or yellowish liquid; peculiar odor and bland taste; soluble in benzene, ether, chloroform, and carbon disulfide; slightly soluble in alcohol. Mp −2C, refr index 1.470 (20C), d 0.915, saponification value 195–196, iodine value 56–74, subject to spontaneous heating, flash p 420F (215C), autoign temperature 883F (472C). Combustible.
Chief constituents: Olein with a small percentage of the glycerides of solid fatty acids. Derivation: By cold-pressing lard.
Grade: Prime winter edible, prime winter inedible, antibiotic, off prime, extra #1, #1, #2.
Use: Lubricant, metal cutting compounds, oiling wool, soap manufacture, antibiotic fermentation.

"Larex."[152]   TM for a series of nine grades of lard oil.
Use: Sulfonating, sulfurizing, cutting oils, textile lubricant, nutrient and defoamer for antibiotic fermentations.

larvicide.   A chemical agent that kills the eggs of insects. Examples are p-dichlorobenzene, chloropicrin, and copper acetoarsenite.

LAS.   See alkyl sulfonate, linear.

laser.   A device which produces a beam of coherent or monochromatic light as a result of photon-stimulated emission. Such beams have extremely high energy as they consist of a single wavelength and frequency. Materials capable of producing this effect are certain high-purity crystals (ruby, yttrium garnet, and metallic tungstates or molybdates doped with rare-earth ions); semiconductors such as gallium arsenide, neodymium-doped glass; various gases, including carbon dioxide, helium, argon, and neon; and plasmas. A chemical laser is one in which the excitation energy is furnished by a chemical reaction, e.g., H + $Cl_2 \rightarrow$ HCl (active) + Cl; or combustion of carbon monoxide to form excited carbon dioxide.

Hazard: Laser radiation can irreparably damage the eyes. Proper shielding is essential at all times.

Use: Laser beams are used in industry for cutting diamonds for wire-drawing dies, in flash photolysis, spectroscopy, and photography. They also have developing applications in medicine and surgery. They are being used for controlled fusion reactions, for biomedical investigations, for organic chemical research, for sophisticated analytical techniques, and in three-dimensional photography (holography). It is possible to increase the abundance of certain isotopes of such elements as uranium, chlorine, and boron by use of laser irradiation. Research on uranium enrichment by this method has been under way for several years.

See also fusion (2), enrichment, holography.

LATB. See lithium aluminum tri-tert-butoxyhydride.

latent heat. The quantity of energy in calories per gram absorbed or given off as a substance undergoes a change of state, that is, as it changes from liquid to solid (freezes), from solid to liquid (melts), from liquid to vapor (boils), or from vapor to liquid (condenses). No change in temperature occurs. Water has unusually high latent heat values; the latent heat of fusion (melting) of ice is 80 cal/g and the latent heat of condensation of steam (latent heat of vaporization of water) is 540 cal/g. The considerable energy delivered by steam condensation is utilized for power generation and for heating a variety of chemical plant equipment (dryers, evaporators, reactors and distillation columns).

See also evaporation; heat.

latent solvent. See solvent, latent.

laterite. A low-grade ore similar to bauxite, but containing only half as much aluminum oxide. Possible substitute for bauxite.

latex. A white, tacky, aqueous suspension of a hydrocarbon polymer occurring naturally in some species of trees, shrubs, or plants, or made synthetically. the most important natural latex is that of the tropical tree *Hevea braziliensis*, which was the only source of rubber up to 1945. It is comprised of globules of rubber hydrocarbon coated with protein; the particles are of irregular shape, varying from 0.5–3 microns in diameter; the suspension is stabilized by electric charges. The composition is about 60% water, 35% hydrocarbon, 2% protein, and low percentages of sugars and inorganic salts. For commercial purposes, rubber latex can be concentrated by evaporation or centrifugation. Ammonia is added as a preservative. Coagulation is induced by addition of acetic or formic acid. A vulcanized form is available. Natural latex is used in the manufacture of thin articles (surgeons' gloves and other medical equipment), as an adhesive, in foamed products and for coating various products such as tire cord. Conversion of latex to gasoline via zeolite catalysts has been reported.

Other sources of rubber-containing latex are guayule, a shrub grown in Mexico and the southwestern US, and several types of dandelions and related species. The botanical function of latex in the plant is unknown.

Synthetic latexes are made by emulsion polymerization techniques from styrene-butadiene copolymer, acrylate resins, polyvinyl acetate, and similar materials. Their particle size is much smaller than in natural latex, ranging from 0.05–0.15 micron; thus, they are truly colloidal suspensions. Their chief use is as a binder in exterior and interior paints, replacing drying oils; they are also used for foams and coatings.

See also guayule, gutta percha, electrophoresis, paint, emulsion, "Vultex."

latex paint. See paint, emulsion.

"Lanthanol" LAL[243]. TM for a highly refined sodium "lauryl" sulfoacetate, biodegradable organic detergent possessing wetting, scouring, emulsifying, and dispersing properties; a foaming agent.

Properties: White dry powder, pH 6.9–7.1 in 0.25% water soluble, stable to hard water, stable to acid and alkali in a pH range of 5.0–8.5, soluble in water solution 1% at 25C, 25% at 100C, hygroscopic, d 0.55, pleasant odor, tasteless.

Use: Tooth pastes, tooth powders, liquid dentrifices, foaming bath salts, shampoos, synthetic detergents.

"Laticrete."[596] TM for a series of latex admixtures designed to replace the water in Portland cement mortars, providing tremendous improvement to many of the key physical properties, i.e., adhesion, shock resistance, frost resistance, etc.

lattice. (1) The structural arrangement of atoms in a crystal. Accurate information is obtained by x-rays, which are diffracted by the lattice at various angles. As the atoms are from 1.5–3 Å apart in most crystals, the lattice acts as a diffraction grating. See also crystals, dislocation. (2) The array of nuclear fuel elements and moderator in a nuclear reactor.

"Latyl."[28] TM for a group of disperse dyes developed particularly for coloration of "Dacron" polyester fiber, on which they have exceptionally good light and wet fastness properties.

laudanidine. (levo-laudanine; tritopine). CAS: 301-21-3. $C_{20}H_{25}O_4N$.   An alkaloid. Properties: White crystals, mp 182–185C. Insoluble in water; soluble in alcohol, benzene, chloroform, and slightly soluble in ether.
Derivation: From opium.
Hazard: Toxic by ingestion.
Use: Medicine (analgesic).

laudanum. Tincture of opium (tinctura opii BP).

laughing gas. See nitrous oxide.

laundry sour. See sour.

lauraldehyde. See lauryl aldehyde.

Laurent's acid. (1-naphthylamine-5-sulfonic acid).

Properties: White or pink needles, gives greenish fluorescence in dilute aqueous solution.
Derivation: From naphthalene-α-sulfonic acid by nitration, reduction with iron, and separation from the 1-naphthylamine-8-sulfonic acid also formed.
Use: Azo dye intermediae.

Laurent's α acid. (1-nitronaphthalene-5-sulfonic acid). $C_{10}H_6(NO)_2(SO_3H)$.
Properties: Pale yellow needles, soluble in water, alcohol and ether. Combustible.
Derivation: By sulfonating nitronaphthalene with a mixture of chlorohydrin and sulfuric acid.
Use: Dye intermediate.

"Laurex."[248] TM for the zinc salts of a mixture of fatty acids in which lauric acid predominates.
Properties: Yellowish white granulated waxy powder, d 1.15, mp 95–105C. Soluble in benzene; insoluble in acetone, gasoline, ethylene dichloride, and water. Combustible.
Use: Accelerator, activator, and plasticizer for rubber.

lauric acid. (dodecanoic acid).
CAS: 143-07-7. $CH_3(CH_2)_{10}COOH$.
A fatty acid occurring in many vegetable fats as the glyceride, especially in coconut oil and laurel oil. Combustible.
Properties: Colorless needles, d 0.833, mp 44C, bp 225C (100 mm), refr index 1.4323 (45C), insoluble in water, soluble in benzene and ether.
Derivaton: Fractional distillation of mixed coconut or other acids.
Grade: 99.8% pure, technical, FCC.
Use: Alkyd resins, wetting agents, soaps, detergents, cosmetics, insecticides, food additives.

lauric aldehyde. See lauryl aldehyde.

laurone. An aliphatic ketone, insoluble in water, stable to high temperatures, acids, alkalies. Compatible with high-melting vegetable waxes, fatty acids, paraffins, etc. Incompatible with resins, polymers and organic solvents at room temperature, but compatible with them at high temperature.
Use: As antiblock agent.

N-lauroyl-p-aminophenol.
$HO(C_6H_4)NHCOCH_2(CH_2)_9CH_3$.
Properties: White to off-white powder; mp 123–126C; insoluble in water; soluble in polar organic solvents (especially when heated) including alcohol, acetone, and dimethylformamide.
Use: Rubber antioxidant.

lauroyl chloride. $C_{11}H_{23}COCl$.
Properties: Water-white liquid, refr index 1.445 (20C), fp −17C, bp 145C (18 mm), decomposes in water and alcohol, soluble in ether.
Use: Surfactant, polymerization initiator, antienzyme agent, foamer; synthesis of lauroyl peroxide, sodium-N-lauroyl sarcosinate, and other sarcosinates.

lauroyl peroxide. (dodecanoyl peroxide).
CAS: 105-74-8. $(C_{11}H_{23}CO)O_2$.
Properties: White, coarse powder; tasteless; faint odor. Soluble in oils and in most organic solvents, slightly soluble in alcohols, insoluble in water, mp 49C.
Grade: Technical (about 95%).
Hazard: Dangerous fire and explosion risk, will ignite organic materials, strong oxidizer. Toxic by ingestion and inhalation, strong irritant to skin.
Use: Bleaching agent, intermediate and drying

agent for fats, oils, and waxes; polymerization catalyst.

N-laurolsarcosine.
$CH_3(CH_2)_{10}CON(CH_3)CH_2COOH$.
Properties: White solid, mp 31–35C, d 0.970.
Use: Surfactant, antienzyme, in cosmetics and pharmaceuticals. Also used in form of sodium-N-lauroyl sarcosinate.

lauryl acetate. See dodecyl acetate.

lauryl alcohol. (alcohol C12; n-dodecanol; dodecyl alcohol). CAS: 112-53-8.
$CH_3(CH_2)_{10}CH_2OH$.
Properties: Colorless solid, floral odor, d 0.830–0.836, refr index 1.428, mp 24C, bp 259C, insoluble in water, soluble in two parts of 70% alcohol, flash p above 212F (100C) (CC). Combustible.
Derivation: Reduction of coconut oil fatty acids.
Grade: Technical, FCC.
Use: Synthetic detergents, lube additives, pharmaceuticals, rubber, textiles, perfumes; flavoring agent.

lauryl aldehyde. (lauric aldehyde; dodecyl aldehyde; aldehyde C-12 lauric; dodecanal; lauraldehyde). CAS: 112-54-9.
$CH_3(CH_2)_{10}CHO$.
Properties: Colorless solid or liquid, strong fatty floral odor, d 0.828–0.836, refr index 1.433–1.440, mp 44C, soluble in 90% alcohol, insoluble in water. Combustible.
Grade: Technical, FCC.
Use: Perfumery, flavoring agent.

lauryl bromide. (n-dodecyl bromide; 1-bromododecane). CAS: 143-15-7. $C_{12}H_{25}Br$.
Properties: Amber liquid with coconut odor and low volatility, d 1.026 (25/25C), bp 175C (45 mm), insoluble in water, soluble in alcohol, ether, fp −15.5C, flash p 291F (144C). Combustible.
Grade: Technical, approx 60% pure.
Derivation: Coconut oil.
Use: Intermediate for quaternary ammonium compounds, organometallics, and vinyl stabilizers.

lauryl chloride. Commercially a mixture of n-alkyl chlorides, with $C_{12}H_{25}Cl$ dominant. A clear, water-white, oily liquid with a faint fatty odor. Completely miscible with most organic solvents, slightly miscible with alcohol, immiscible with water.
Properties: Crystallization point −19C, d 0.863 (15.5/15.5C), distillation range 112–160C (5 mm), flash p 235F (112C). Combustible.
Grade: Refined, technical.
Use: Synthesis of esters, sulfides, lauryl mercaptan

(used in styrene-butadiene polymerization), other organics.

lauryldimethylamine. CAS: 112-18-5.
$CH_3(CH_2)_{11}N(CH_3)_2$. A liquid cationic detergent.
Use: Corrosion inhibitor, acid-stable emulsifier.

lauryldimethylamine oxide. CAS: 1643-20-5.
$CH_3(CH_2)_{11}N(CH_3)_2 \cdot H_2O$. A nonionic detergent.
Use: As a foam stabilizer; stable at high concentration of electrolytes and over a wide pH range.

lauryl lactate. See "Ceraphyl."

lauryl mercaptan. (n-dodecyl mercaptan).
CAS: 112-55-0. $C_{12}H_{25}SH$ (approximately).
Properties: (technical material, mixture of isomers): Water-white or pale-yellow liquid, mild odor, d 0.85 (20/20C), fp −7.5C, distillation range 200–235C, refr index 1.45–1.47. Insoluble in water; soluble in methanol, ether, acetone, benzene, gasoline, and ethyl acetate. Flash p 210F (99C). Combustible.
Grade: 95% min.
Hazard: May be injurious to eyes.
Use: Manufacture of synthetic rubber and plastics, also in the synthesis of pharmaceuticals and in insecticides and fungicides; nonionic detergent.
See also thiol.

lauryl methacrylate. CAS: 142-90-5.
$CH_2{:}C(CH_3)COO(CH_2)_{11}CH_3$. The commercial material is a mixture containing also lower and higher fatty derivatives.
Properties: Boiling range 272–344C, bulk d 0.868 g/mL, flash p 270F (132C) (COC). Combustible.
Use: Polymerizable monomer for plastics, molding powders, solvent coatings, adhesives, oil additives; emulsions for textile, leather, and paper finishing.
See also acrylic resin.

lauryl pyridinium chloride. $C_5H_5NClC_{12}H_{25}$.
Properties: Mottled tan semisolid, soluble in water and organic solvents, flash p 347F (175C). Combustible.
Grade: Technical, contains higher and lower fatty acid derivatives.
Use: Cationic detergent, dispersing and wetting agent, ingredient of fungicides and bactericides.

lauryl pyridinium-5-chloro-2-benzothiazyl sulfide. See "Vancide 26EC."

lautal. A hard aluminum alloy containing 4–5% copper, 1.5–2% silicon and fractional percentages of other metals such as iron, manganese, or magnesium.

lava. Molten rock (magma) that has reached the surface of the earth.

lavandin oil. See lavender oil.

lavender oil. CAS: 8000-28-0. An essential oil used in perfumery, 35% ester content as linalyl acetate required. Terpeneless grade has about twice the concentration of the natural oil.

"Lavenol."[188] TM for a series of synthetic lavender oil substitute of various types.

Lavoisier, Antoine Laurent. (1743–1794) A French chemist generally regarded as the "father" of chemistry. His "Traite Elementaire de Chimie" (1789) listed 30 elements, clarified the nomenclature of acids, bases and salts, and described the composition of numerous organic substances. He erroneously believed that oxygen is the characteristic element of acids. However, his fundamental work on combustion, as a result of which he identified and named nitrogen (azote), and on the separation of hydrogen from water by a unique reduction experiment carried out in a heated gun barrel, earned him a leading position among early chemists.
See also Mendeleef.

Lawrence, Ernest O. (1901–1958) An American physicist who invented the cyclotron in 1929. Both the element lawrencium and the Lawrence Livermore Research Laboratory at the University of California were named after him.
See also cyclotron, bombardment.

lawrencium. CAS: 22537-19-5. Lr. A synthetic radioactive element with atomic number 103, discovered in 1961, aw 257, only one other isotope is known (256); the 257 isotope has a half-life of 8 seconds. It has been made by bombarding californium with boron ions. It exhibits alpha radiation.
See actinide series.

lay-up. In the reinforced plastics industry, a term used to refer to placement of the reinforcing material in the mold.

LC_{50}. (lethal concentration 50%). That quantity of a substance administered by inhalation that is necessary to kill 50% of test animals exposed to it within a specified time. The test applies not only to gases and vapors but to fumes, dusts, and other particulates, suspended in air.

LCL. Abbreviation for "less than carload lot," used by shippers, traffic managers, railroads, etc.

LD_{50}. (lethal dose 50%). That quantity of a substance necessary to kill 50% of exposed animals in laboratory tests within a specified time. A substance having an oral LD_{50} of less than 400 mg/kg of body weight is considered to be highly toxic.

LDPE. Abbreviation for low-density polyethylene.

leaching. See solvent extraction.

lead. CAS: 7439-92-1. Pb. (from Latin plumbum). Metallic element of atomic number 82, Group IVA of the periodic table, aw 207.2, valences = 2,4, four stable isotopes. The isotopes are the end products of the disintegration of three series of natural radioactive elements uranium (206), thorium (208), and actinium (207).
Properties: Heavy, ductile, soft, gray solid, d 11.35, mp 327.4C, bp 1755C, soluble in dilute nitric acid, insoluble in water but dissolves slowly in water containing a weak acid, resists corrosion, relatively impenetrable to radiation. Poor electrical conductor, good sound and vibration absorber. Noncombustible.
Occurrence: US, Mexico, Canada, South America, Australia, Africa, Europe.
Derivation: Roasting and reduction of galena (lead sulfide), anglesite (lead sulfate), and cerussite (lead carbonate). Also from scrap.
Purification method: Desilvering (Parkes process), electrolytic refining (Betts process), pyrometallurgical refining (Harris process). Bismuth is removed by Betterton-Kroll process.
Grade: High purity (less than 10 ppm impurity), pure (99.9+%), powdered (99% pure), pig lead, paste.
Forms available: Ingots, sheet, pipe, shot, buckles or straps, grids, rod, wire, etc.; paste; powder; single crystals.
Hazard: Toxic by ingestion and inhalation of dust or fume. TLV (as Pb): (fumes and dusts and inorganic compounds) 0.15 mg/m^3 of air. For ambient air the EPA standard is 1.5 micrograms/m^3. A cumulative poison. FDA regulations require zero lead content in foods and less than 0.05% in house paints.
Use: Storage batteries, tetraethyllead (gasoline additive), radiation shielding, cable covering, ammunition, chemical reaction equipment (piping, tank linings, etc.), solder and fusible alloys, type metal, vibration damping in heavy construction, foil, babbit and other bearing alloys. For further information, refer to Lead Industries Association, 292 Madison Ave., New York.

lead acetate. (sugar of lead). CAS: 301-04-2. $Pb(C_2H_3O_2)_2 \cdot 3H_2O$.
Properties: White crystals or flakes (commercial grades are frequently brown or gray lumps),

sweetish taste, absorbs carbon dioxide when exposed to air, becoming insoluble in water. Soluble in water, slightly soluble in alcohol, freely soluble in glycerol, d 2.50, mp loses water at 75C, at 200C decomposes, bp (anhydrous) 280C. Combustible.
Derivation: By the action of acetic acid on litharge or thin lead plates.
Grade: Powdered, granular, crystals, flakes, CP.
Hazard: Toxic by ingestion, inhalation, and skin absorption; use may be restricted.
Use: Dyeing of textiles, waterproofing, varnishes, lead driers, chrome pigments, gold cyanidation process, insecticide, antifouling paints, analytical reagent, hair dye.

lead alkyl, mixed. A mixture containing various methyl and ethyl derivatives of tetraethyl lead and tetramethyl lead. Thus, methyl triethyl lead, dimethyl diethyl lead, and ethyl trimethyl lead may all be present with or without tetraethyl and tetramethyl lead.
Hazard: Toxic by ingestion and skin absorption.
Use: Antiknock agents in aviation gasoline.

lead antimonate. (Naples yellow; antimony yellow). CAS: 13510-89-9. $Pb_3(SbO_4)_2$.
Properties: Orange-yellow powder, insoluble in water, d 6.58 (20C). Noncombustible.
Derivation: Interaction of solutions of lead nitrate and potassium antimonate, concentration, and crystallization.
Hazard: Toxic by inhalation. TLV (as Pb): 0.15 mg/m³ of air.
Use: Staining glass, crockery, and porcelain.

lead arsenate. (lead orthoarsenate).
CAS: 7784-40-9. $Pb_3(AsO_4)_2$.
Properties: White crystals, soluble in nitric acid, insoluble in water, d 5.8, mp 1042 C (decomposes).
Derivation: By the action of a soluble lead salt on a solution of sodium arsenate, concentration, and crystallization.
Hazard: Highly toxic. TLV: 0.15 mg/m³ of air, a carcinogen.
Use: Insecticide, herbicide.

lead arsenite. CAS: 10031-13-7. $Pb(AsO_2)_2$.
Properties: White powder, soluble in nitric acid, insoluble in water, d 5.85.
Hazard: Highly toxic.
Use: Insecticide.

lead azide. CAS: 13424-46-9. $Pb(N_3)_2$.
Properties: Colorless, very sensitive needles; an initiating explosive. Should always be handled submerged in water.
Derivation: Reaction of sodium azide with a lead salt.

Hazard: Severe explosion risk, detonates at 350C (660F). Highly toxic. TLV (as Pb): 0.15 mg/m³ of air.
Use: Primary detonating compound for high explosives. *Note*: Explosions have occurred in cases where azide compounds have reacted with the lead in plumbing after being washed down sinks.

lead-base Babbitt. See Babbitt metal.

lead biorthophosphate. See lead phosphate, dibasic.

lead, blue. A term applied to galena to distinguish it from white lead ore. It is also applied to blue basic lead sulfate.

lead borate. CAS: 10214-39-8.
$Pb(BO_2)_2 \cdot H_2O$.
Properties: White powder, soluble in dilute nitric acid, insoluble in water, d 5.598, mp 160C (loses water). Noncombustible.
Derivation: Interaction of solutions of lead hydroxide and boric acid with subsequent crystallization.
Hazard: Toxic by inhalation. TLV: 0.15 mg/m³ of air.
Use: Varnish and paint drier, waterproofing paints, lead glass, electrically conductive ceramic coatings.

lead borosilicate. A constituent of optical glass, composed of a mixture of the borate and silicate of lead.

lead bromate. CAS: 34018-28-5.
$Pb(BrO_3)_2 \cdot H_2O$.
Properties: Colorless crystals, soluble in hot water, d 5.53, decomposes at about 180C.
Hazard: Toxic by inhalation or ingestion.

lead bromide. CAS: 10031-22-8. $PbBr_2$.
Properties: White powder, slightly soluble in hot water, insoluble in alcohol, d 6.66, bp 916C, mp 373C.
Hazard: Toxic by inhalation. TLV (as Pb): 0.15 mg/m³ of air.

lead carbolate. See lead phenate.

lead carbonate. $PbCO_3$. See lead carbonate, basic.

lead carbonate, basic. (lead subcarbonate; white lead; lead flake). $2PbCO_3 \cdot Pb(OH)_2$.
Properties: White amorphous powder, soluble in acids, insoluble in water, decomposes at 400C, d 6.86. Noncombustible.
Derivation: (a) Dutch process. By the corrosion

of lead buckles in pots by acetic acid and carbon dioxide generated by the fermentation of waste tanbark. (b) Carter process. By treating very finely divided lead in revolving wooden cylinders with dilute acetic acid and carbon dioxide.

Hazard: Toxic by inhalation. TLV (as Pb): 0.15 mg/m³ of air.

Use: Exterior paint pigment, ceramic glazes.

lead chloride.   CAS: 7758-95-4. $PbCl_2$.

Properties: White crystals, slightly soluble in hot water, insoluble in alcohol and cold water, d 5.88, mp 498C, bp 950C. Noncombustible.

Derivation: By the addition of hydrochloric acid or sodium chloride to a solution of a lead salt with subsequent crystallization.

Hazard: Toxic by inhalation. TLV (as Pb): 0.15 mg/m³ of air.

Use: Preparation of lead salts, lead chromate pigments, analytical reagent.

lead cholorsilicate complex.   See "Lectro" 60.

lead chromate.   (chrome yellow). $PbCrO_4$.

Properties: Yellow crystals, soluble in strong acids and alkalies, insoluble in water, d 6.123, mp 844C.

Derivation: Reaction of sodium chromate and lead nitrate in solution.

Hazard: Toxic by ingestion and inhalation; human suspected carcinogen. TLV (as chromium): 0.05 mg/m³ of air.

Use: Pigment in industrial paints, rubber, plastics, ceramic coatings; organic analysis.

See also chrome pigment.

lead coating.   Coatings of lead or lead-rich alloys are (1) deposited by dipping into the molten metal after applying a layer of tin to secure good adhesion of the lead coating, (2) by electroplating from a fluosilicate or fluoborate bath, or (3) by spraying.

lead cyanide.   CAS: 592-05-2. $Pb(CN)_2$.

Properties: White to yellowish powder, slightly soluble in water, decomposes in acid.

Derivation: Interaction of solutions of potassium cyanide and lead acetate.

Hazard: Toxic by ingestion, inhalation, and skin absorption. TLV (as Pb): 0.15 mg/m³ of air.

Use: Metallurgy.

lead dimethyldithiocarbamate.   CAS: 19010-66-3. $Pb[SCSN(CH_3)_2]_2$.

Properties: White powder, d 2.43, melting range 310C, insoluble in all common organic solvents, slightly soluble in cyclohexanone.

Use: Vulcanization accelerator with litharge.

lead dioxide.   (lead oxide, brown; plumbic acid, anhydrous; lead peroxide; lead superoxide). CAS: 1309-60-0. PbO_2.

Properties: Brown, hexagonal crystals, soluble in glacial acetic acid, insoluble in water and alcohol, d 9.375, mp 290C (decomp), an oxidizing agent.

Derivation: By adding bleaching powder to an alkaline solution of lead hydroxide.

Hazard: Dangerous fire risk in contact with organic materials. TLV (as Pb): 0.15 mg/m³ of air.

Use: Oxidizing agent, electrodes, lead-acid storage batteries, curing agent for polysulfide elastomers, textiles (mordant, discharge in dyeing with indigo), matches, explosives, analytical reagent.

lead dross.   (lead scrap).   Consists of the scrap, dross, or waste from sulfuric acid tanks; a mixture of metallic lead, lead sulfate, and free sulfuric acid.

lead, electrolytic.   Pure lead obtained by electrolytic deposition.

See Betts process.

lead ethylhexoate.   See "L-26."

lead flake.   See lead carbonate, basic.

lead fluoborate.   CAS: 13814-96-5 $B_2F_8 \cdot Pb$.

Properties: Liquid with mw 380.81.

Grades: Technical 51%.

Hazard: TLV (as Pb): 0.15 mg/m³ of air.

Use: Salt for electroplating lead; can be mixed with stannous fluoborate to electroplate any composition of tin and lead as an alloy.

lead fluoride.   CAS: 7783-46-2. PbF_2.

Properties: Colorless crystals, density 8.2 g/cc, mp (approximately) 824C, very slightly soluble in water. Noncombustible.

Grade: Crystals, 99.93%.

Hazard: Toxic. Strong irritant. TLV (as Pb): 0.15 mg/m³ of air.

Use: Electronic and optical applications, starting materials for growing single crystal solid-state lasers, high-temperature dry film lubricants in the form of ceramic-bonded coatings.

lead fluosilicate.   (lead silicofluoride). CAS: 25808-74-6. $PbSiF_6 \cdot 2H_2O$.

Properties: Colorless crystals, soluble in water, decomposes when heated.

Hazard: Toxic. Strong irritant. TLV (as Pb): 0.15 mg/m³ of air.

Use: Solution for electrorefining lead.

lead formate.   CAS: 811-54-1. $Pb(CHO_2)_2$.

Properties: Brownish-white lustrous needles, d 4.63, soluble in water, decomposes at 190C. Noncombustible.

Hazard: Toxic by ingestion.
Use: Reagent in analytical determinations.

lead fumarate, tetrabasic. See "Lectro" 78.

lead glass. See glass.

lead hydroxide. (lead hydrate; hydrated lead oxide). CAS: 12362-20-8. $Pb(OH)_2$.
Properties: White, bulky powder. Soluble in alkalies, slightly soluble in water, soluble in nitric and acetic acids, d 7.592, mp decomposes at 145C, absorbs carbon dioxide from air. Noncombustible.
Derivation: By the addition of sodium or ammonium hydroxide to a solution of a lead salt with subsequent filtration and drying.
Hazard: Toxic material. TLV (as Pb): 0.15 mg/m³ of air.
Use: Lead salts, lead dioxide.

lead hyposulfite. See lead thiosulfate.

lead iodide. CAS: 10101-63-0. PbI_2.
Properties: Golden-yellow crystals or powder, odorless, soluble in potassium iodide and concentrated sodium acetate solutions, soluble in boiling water, d 6.16, mp 402C, bp 954C. Noncombustible.
Derivation: Interaction of lead acetate and potassium iodide.
Hazard: Toxic material. TLV (as Pb): 0.15 mg/m³ of air.
Use: Bronzing, printing, photography, cloud seeding.

lead linoleate. (lead plaster). $Pb(C_{18}H_{31}O_2)_2$.
Properties: Yellowish-white paste, soluble in oils, insoluble in water. Combustible.
Derivation: By heating a solution of lead nitrate with sodium linoleate.
Grade: Technical, fused (contains 26.5% lead).
Hazard: Toxic material. Absorbed by skin.
Use: Medicine, drier in paints and varnishes.

lead malate. CAS: 816-68-2. $C_4H_6O_5 \cdot Pb$.
Properties: Soft, yellowish-white, crystalline powder; d 6.3; refr index 2.08.
Hazard: Toxic material. Absorbed by skin.
Use: Vulcanizing agent for chlorosulfonated polyethylene. Highly basic stabilizer with high heat stability in vinyls.

lead metavanadate. See lead vanadate.

lead molybdate. CAS: 10190-55-3. $PbMoO_4$.
Properties: Yellow powder, soluble in nitric acid, insoluble in water and alcohol, d 5.9, mp 1060–1070C. Noncombustible.
Derivation: By adding a solution of lead nitrate to a solution of ammonium molybdate, concentration, and crystallization.
Hazard: Toxic material. TLV (as Pb): 0.15 mg/m³ of air.
Use: Analytical chemistry, pigments. See molybdate oranges. Single crystals available for electronic and optical uses.

lead monohydrogen phosphate. See lead phosphate, di-basic.

lead mononitroresorcinate. CAS: 51317-24-9. $PbO_2C_6H_3NO_2$.
Hazard: An initiating explosive, dangerous. Forbidden for transport.

lead monoxide. See litharge, massicot.

lead β-naphthalenesulfonate. $Pb(C_{10}H_7SO_3)_2$.
Properties: White, crystalline powder; soluble in alcohol; insoluble in water. Combustible.
Derivation: By the action of lead acetate on β-naphthalenesulfonic acid.
Hazard: Toxic material. Absorbed by skin.
Use: Organic preparations.

lead naphthenate. CAS: 61790-14-5. $C_7H_{12}O_2 \cdot xPb$.
Properties: Soft, yellow, resinous, semi-transparent. Gives deposits in highly acid oils but not when mixed with suitable quantities of cobalt or manganese, soluble in alcohol, mp approximately 0C. Combustible.
Derivation: Addition of lead salt to aqueous sodium naphthenate solution.
Grade: liquid 16%, 24% lead, solid 37% lead.
Hazard: Toxic material. A known carcinogen (OSHA), absorbed by skin.
Use: Paint and varnish drier, wood preservative, insecticide, catalyst for reaction between unsaturated fatty acids and sulfates in the presence of air, lube oil additive.
See also soap (2).

lead nitrate. CAS: 10099-74-8. $Pb(NO_3)_2$.
Properties: White crystals, soluble in water and alcohol, d 4.53, decomposes at 470C.
Derivation: By the action of nitric acid on lead.
Hazard: Strong oxidizing material, dangerous fire risk in contact with organic materials. TLV (as Pb): 0.15 mg/m³ of air.
Use: Lead salts, mordant in dyeing and printing calico, matches, mordant for staining mother-of-pearl, oxidizer in the dye industry, sensitizer in photography, explosives, tanning, process engraving, and lithography.

lead nitrite, basic. (basic lead nitrite; lead subnitrite).
Properties: Light-yellow powder, variable composition, essentially $3PbO \cdot N_2O \cdot H_2O$, soluble in dilute nitric acid, easily decomposed.

Hazard: Toxic material. TLV (as Pb): 0.15 mg/m^3 of air.

lead ocher.   See massicot (1).

lead octoate.   See soap (2).

lead oleate.   CAS: 1120-46-3.
[CH$_3$(CH$_2$)$_7$CH:CH(CH$_2$)$_7$COO]$_2$Pb.
Properties: White powder or ointment-like granules or mass; soluble in alcohol, ether, turpentine, and benzene; insoluble in water. Combustible.
Derivation: Reaction of oleic acid with lead hydrate or carbonate, or of lead acetate and sodium oleate.
Hazard: Toxic material. Absorbed by skin.
Use: Varnishes, lacquers, paint drier, high-pressure lubricants.
See also soap (2).

lead orthoarsenate.   See lead arsenate.

lead orthophosphate, normal.   See lead phosphate.

lead orthosilicate.   See lead silicate.

lead oxide, black.   See lead suboxide.

lead oxide, brown.   See lead dioxide.

lead oxide, hydrated.   See lead hydroxide.

lead oxide, red.   (red lead; minium; lead
tetroxide).   CAS: 1314-41-6.   Pb$_3$O$_4$.
Properties: Bright red powder, partly soluble in acids, insoluble in water, d reported variously 8.32–9.16, decomposes between 500 and 530C. An oxidizing agent, may react with reducing agents.
Derivation: By carefully heating litharge in a furnace in a current of air.
Grade: Technical, 95%, 97%, 98%.
Hazard: Toxic as dust. TLV (as Pb): 0.15 mg/m^3 of air.
Use: Storage batteries, glass, pottery and enameling, varnish, purification of alcohol, packing pipe joints, metal-protective paints, fluxes and, ceramic glazes.

lead oxide, yellow.   See litharge.

lead perchlorate.   CAS: 13637-76-8.
Pb(ClO$_4$)$_2$·3H$_2$O.
Properties: White crystals, d 2.6, mp 100C (decomposes), very soluble in cold water, soluble in alcohol.
Hazard: Dangerous in contact with organic materials, strong oxidizing agent. Very toxic material.

lead peroxide.   See lead dioxide.

lead phenate.   (lead phenolate; lead carbolate).
Pb(OH)OC$_6$H$_5$.
Properties: Yellowish to grayish-white powder, soluble in nitric acid, insoluble in water and alcohol.
Derivation: By boiling phenol with litharge.
Hazard: Toxic material. Absorbed by skin.

lead phenolsulfonate.   (lead sulfocarbolate).
Pb(C$_6$H$_4$OHSO$_3$)$_2$·5H$_2$O.
Properties: White crystals or powder, soluble in water and alcohol.
Hazard: Toxic material. Absorbed by skin.

lead phosphate.   (normal lead orthophosphate).
CAS: 7446-27-7.   Pb$_3$(PO$_4$)$_2$.
Properties: White powder, d 6.9–7.3, mp 1014C, insoluble in water, soluble in acids and alkalies.
Hazard: Toxic material. TLV (as Pb): 0.15 mg/m^3 of air.
Use: Stabilizing agent in plastics.

lead phosphate, dibasic.   (lead monohydrogen phosphate; lead biorthophosphate).
PbHPO$_4$.
Properties: Soft, white powder or fine plate-like crystals; d 5.66 (15C); mp decomposes.
Hazard: Toxic material. TLV (as Pb): 0.15 mg/m^3 of air.
Use: Imparting heat resistance and pearlescence to polystyrene and casein plastics.

lead phosphite, dibasic.   CAS: 1344-40-7.
2PbO·PbHPO$_3$·1/2H$_2$O.
Properties: Fine, white, acicular crystals; d 6.94; refr index 2.25; insoluble in water.
Hazard: Toxic material. TLV (as Pb): 0.15 mg/m^3 of air. Store in closed containers away from open flame or sparks and at temperatures not to exceed 400F.
Use: Heat and light stabilizer for vinyl plastics and chlorinated paraffins. As an UV screening and antioxidizing stabilizer for vinyl and other chlorinated resins in paints and plastics.

lead phthalate, dibasic.   C$_6$H$_4$(COO)$_2$Pb·PbO.
Properties: Fluffy white crystalline powder, insoluble in water, d 4.5, refr index 1.99 (avg).
Derivation: By boiling litharge with phthalic acid.
Hazard: Toxic by inhalation and skin absorption.
Use: Heat and light stabilizer for general vinyl use.

lead plaster.   See lead linoleate.

lead protoxide.   See litharge.

lead, red.   See lead oxide, red.

lead resinate.   Pb(C$_{20}$H$_{29}$O$_2$)$_2$.
Properties: Brown, lustrous, translucent lumps or yellow-white powder, or yellowish-white paste, insoluble in most solvents. Combustible.

Derivation: By heating a solution of lead acetate and rosin oil.

Grade: Precipitated 23% lead.

Hazard: Toxic material. Absorbed by skin.

Use: Paint and varnish drier, textile water-proofing agent.

lead salicylate. $Pb(OOCC_6H_4OH)_2 \cdot H_2O$.

Properties: Soft, creamy-white, crystalline powder; d 2.3; refr index 1.78; soluble in hot water and alcohol. Combustible.

Use: Stabilizer or costabilizer for flooring and other vinyl compounds requiring good light stability.

lead selenide. PbSe. Gray crystals, d 8.10 (15C), mp 1065C, insoluble in water, soluble in nitric acid. A semiconductor used in infrared detectors and thermoelectric devices.

Hazard: Moderate fire heazard as dust or in presence of moisture. TLV: (as lead) 0.15 mg/m³ of air.

lead sesquioxide. Pb_2O_3.

Properties: Reddish-yellow powder, soluble in alkalies and acids, insoluble in water, decomposes at 370C.

Derivation: By gently heating metallic lead.

Hazard: Toxic material. TLV (as Pb): 0.15 mg/m³ of air.

Use: Ceramics, ceramic cements, metallurgy, varnishes.

lead silicate. (lead metasilicate).
CAS: 10099-76-0. $PbSiO_3$.

Properties: White, crystalline powder; insoluble in most solvents. Noncombustible.

Derivations: Interaction of lead acetate and sodium silicate.

Hazard: Toxic material. TLV (as Pb): 0.15 mg/m³.

Use: Ceramics, fireproofing fabrics.

lead silicate, basic. (white lead silicate; lead silicate sulfate). A pigment made up of an adherent surface layer of basic lead silicate and basic lead sulfate cemented to silica.

Properties: Excellent film-forming properties with drying oils combined with low density.

Derivation: Fine silica is mixed with litharge and sulfuric acid. The mixture is then furnaced in a rotary kiln and ground to break up agglomerates.

Hazard: Toxic material. TLV (as Pb): 0.15 mg/m³ of air.

Use: Pigment in industrial paints.

lead silicochromate. A yellow lead-silicon pigment. Normal lead silicon chromate is used as a yellow prime pigment for traffic marking paints. Basic lead silicon chromate is used as a corrosive inhibitive pigment for metal protective coatings, primers and finishers. Also for industrial enamels requiring a high gloss.

Hazard: Toxic material. TLV (as Pb): 0.15 mg/m³ of air.

lead silicofluoride. See lead fluosilicate.

lead-soap lubricants. Lead salts saponified with fats, hard at low temperatures, viscous at medium temperatures, but become somewhat fluid on heating by friction.

Use: "Extreme-pressure lubricants," but are not suited for high speeds.

See lead naphthenate, lead oleate, lead stearate, also soap (2).

lead sodium hyposulfite. See lead sodium thiosulfate.

lead sodium thiosulfate. (lead sodium hyposulfite; sodium lead hyposulfite; sodium lead thiosulfate). CAS: 10101-94-7.
$PbS_2O_3 \cdot 2Na_2S_2O_3$.

Properties: Heavy, small, white crystals; soluble in solutions of thiosulfates.

Hazard: Toxic material. TLV (as Pb): 0.15 mg/m³ of air.

Use: Matches.

lead stannate. $PbSnO_3 \cdot 2H_2O$.

Properties: Light-colored powder, insoluble in water. Approximate temperature of dehydration 170C.

Hazard: Toxic material. TLV (as Pb): 0.15 mg/m³ of air.

Use: Additive in ceramic capacitors, pyrotechnics.

lead stearate. CAS: 1072-35-1.
$Pb(C_{18}H_{35}O_2)_2$.

Properties: White powder, mp 100–115C, d 1.4, soluble in hot alcohol, insoluble in water. Combustible.

Derivation: By heating a solution of lead acetate with sodium stearate.

Hazard: Toxic material. Absorbed by skin.

Use: Varnish and lacquer drier, high-pressure lubricants, lubricant in extrusion processes, stabilizer for vinyl polymers, corrosion inhibitor for petroleum, component of greases, waxes, and paints.

lead styphnate. Legal label name for lead trinitroresorcinate.

lead subacetate. CAS: 1335-32-6.
$(2Pb(OH)_2Pb(C_2H_3O_2)_2$.

Properties: White, heavy powder with mw 807.75. Partially soluble in cold water, more soluble in

hot water, absorbs atmospheric carbon dioxide becoming almost insoluble.
Hazard: Suspected carcinogen and poisonous. Toxic by ingestion.
Use: Decolorizing agent (sugar solutions, etc.).

lead subcarbonate.   See lead carbonate, basic.

lead subnitrite.   See lead nitrite.

lead suboxide.   (lead oxide, black; litharge, leaded). Pb_2O.
Properties: Black amorphous material, d 8.342, decomposes on heating, insoluble in water, soluble in acids and bases.
Hazard: Toxic material. TLV (as Pb): 0.15 mg/m^3 of air.
Use: In storage batteries.

lead, sugar of.   See lead acetate.

lead sulfate.   CAS: 7446-14-2. $PbSO_4$.
Properties: White, rhombic crystals. Slightly soluble in hot water, insoluble in alcohol, soluble in sodium hydroxide solution and concentrated hydriodic acid, d 6.12–6.39, mp 1170C. Noncombustible.
Derivation: Interaction of solutions of lead nitrate and sodium sulfate.
Hazard: Strong irritant to tissue. TLV (as Pb): 0.15 mg/m^3 of air.
Use: Storage batteries, paint pigments.

lead sulfate, basic.   (white lead, sublimed; white lead sulfate).   Approximate formula $PbSO_4 \cdot PbO$.
Properties: White, monoclinic crystals, d 6.92, mp 977C, slightly soluble in hot water or acids. Noncombustible.
Grade: Vary from 72–85% lead sulfate and remainder lead oxide. Sold dry or ground in oil.
Derivation: Three methods are used: (a) Lead sulfide ore (galena) is subjected to high temperatures in an oxidizing atmosphere, (b) molten lead is sprayed into a jet of ignited fuel gas and air in the presence of sulfur dioxide gas, (c) atomized metallic lead is mixed with water and sulfuric acid is added under controlled conditions.
Hazard: Toxic material. TLV (as Pb): 0.15 mg/m^3 of air.
Use: Paints, ceramics, pigments.

lead sulfate, blue basic.   (sublimed blue lead; blue lead).   Composition: Lead sulfate (min) 45%, lead oxide (min) 30%, lead sulfide (max) 12%, lead sulfite (max) 5%, zinc oxide 5%, carbon and undetermined matter (max) 5%.
Properties: Blue-gray corrosion-inhibiting pigment, insoluble in water or alcohol, d 6.2, Noncombustible.

Derivation: By heating lead ores in special furnaces.
Hazard: Toxic material. TLV (as Pb): 0.15 mg/m^3 of air.
Use: Component of structural-metal priming coat paints, rust-inhibitor in paints, lubricants, vinyl plastics, and rubber products.

lead sulfate, tribasic. $3PbO \cdot PbSO_4 \cdot H_2O$.
Properties: Fine, white powder; d 6.4; refr index 2.1.
Hazard: Toxic material. TLV (as Pb): 0.15 mg/m^3.
Use: Electrical and other vinyl compounds requiring high heat stability.

lead sulfide.   (plumbous sulfide).
CAS: 1314-87-0.   PbS.
Properties: Silvery, metallic crystals or black powder. Soluble in acids, insoluble in water and alkalies, d 7.13–7.7, mp 1114C, sublimes at 1281C.
Derivation: (a) Found in nature as the mineral galena, (b) by passing hydrogen sulfide gas into an acid solution of lead nitrate.
Grade: Technical, CP, electronic.
Hazard: Toxic by ingestion and inhalation. TLV (as Pb): 0.15 mg/m^3 of air.
Use: Ceramics, infrared radiation detector, semiconductor, ceramic glaze, source of lead.

lead sulfite. $PbSO_3$.
Properties: White powder, decomposes on heating, insoluble in water, soluble in nitric acid.
Hazard: Toxic by ingestion and inhalation. TLV (as Pb): 0.15 mg/m^3 of air.

lead sulfocarbolate.   See lead phenolsulfonate.

lead sulfocyanide.   See lead thiocyanate.

lead superoxide.   See lead dioxide.

lead tallate.   A mixture of lead and tall oil. Combustible.
Grade: liquid 16% lead, 24% lead, solid 30% lead.
Derivation: By the fusion process.
Hazard: Toxic material. Absorbed by skin.
See soap (2).

lead telluride.   CAS: 1314-91-6.   PbTe.
Properties: Crystalline solid, d 8.2, mp 905C, insoluble in water and most acids.
Hazard: Toxic by ingestion and inhalation. TLV (as Pb): 0.15 mg/m^3 of air.
Use: Single crystals used as photoconductor and semiconductor in thermocouples.

lead tetraacetate.   CAS: 546-67-8.
$Pb(CH_3COO)_4$.
Properties: Colorless or faintly pink crystals, sometimes moist with glacial acetic acid, mp

175C, d 2.228 (17C). Soluble in benzene, chloroform, nitrobenzene, hot glacial acetic acid. Combustible.
Derivation: From red lead (Pb_3O_4) and glacial acetic acid in the presence of acetic anhydride.
Hazard: Toxic material. Absorbed by skin.
Use: Oxidizing agent in organic synthesis, laboratory reagent.

lead tetraethyl. See tetraethyl lead.

lead tetroxide. See lead oxide, red.

lead thiocyanate. (lead sulfocyanate).
CAS: 592-87-0. $Pb(SCN)_2$.
Properties: White or light-yellow crystalline powder, soluble in potassium thiocyanate, nitric acid and slightly soluble in cold water, decomposes in hot water, d about 3.8.
Hazard: Toxic by ingestion and inhalation. TLV (as Pb): 0.15 mg/m^3 of air.
Use: Ingredient of priming mix for small-arms cartridges, safety matches, dyeing.

lead thiosulfate. (lead hyposulfite). PbS_2O_3.
Properties: White crystals, d 5.18, soluble in acids and sodium thiosulfate solution, insoluble in water, mp decomposes.
Derivation: By the interaction of solutions of lead nitrate and sodium thiosulfate, concentration, and crystallization.
Hazard: Toxic by ingestion and inhalation. TLV (as Pb): 0.15 mg/m^3 of air.

lead titanate. $PbTiO_3$.
Properties: Pale-yellow solid, insoluble in water, d 7.52.
Derivation: Interaction of oxides of lead and titanium at a high temperature. Contains lead sulfate and lead oxide as impurities.
Hazard: Toxic by ingestion and inhalation. TLV (as Pb): 0.15 mg/m^3 of air.
Use: Industrial paint pigment.

lead trinitroresorcinate. (lead styphnate).
CAS: 63918-97-8. $C_6H(NO_2)_3(O_2Pb)$.
Properties: Monohydrate is monoclinic orange-yellow crystals, practically insoluble in water, d 3.1 (monohydrate), 2.9 (anhydrous).
Derivation: Prepared by adding a solution of magnesium styphnate (from magnesium oxide and styphnic acid) to a lead salt solution.
Hazard: Detonates at 500F (260C), dangerous explosion risk, an initiating explosive.

lead tungstate. (lead wolframate).
CAS: 7759-01-5. $PbWO_4$.
Properties: White powder, soluble in sodium hydroxide solutions, insoluble in water, d 8.235, mp 1130C.

Derivation: By mixing solutions of lead nitrate and sodium tungstate, concentrating, and crystallizing.
Hazard: Toxic material. TLV (as Pb): 0.15 mg/m^3 of air.
Use: Pigment.

lead vanadate. (lead metavanadate).
CAS: 10099-79-3. $Pb(VO_3)_2$.
Properties: Yellow powder, insoluble in water, decomposes in nitric acid.
Hazard: Toxic material. TLV (as Pb): 0.15 mg/m^3 of air.
Use: Preparation of other vanadium compounds, pigment.

lead water. A 1% solution of basic lead acetate.
Hazard: Toxic. See lead acetate.

lead, white. See lead carbonate, basic; also lead silicate, basic; and lead sulfate, basic.

lead wolframate. See lead tungstate.

lead wool. Fine filaments or threads of metallic lead, prepared and used for packing pipe joints.

lead zirconate titanate. (LZT). $PbTiZrO_3$.
Forms piezoelectric crystals.
Use: Element in hi-fi sets and as a transducer for ultrasonic cleaners, ferroelectric materials in computer memory units.

leaf, filter. A unit of a shell-and-leaf filter press on which the cake is formed. In general, a leaf consists of a circular or rectangular metal frame in which is fastened a coarse wire screen. This is covered on both sides with a fine-mesh wire cloth, over which is placed the filter medium proper, e.g., nylon fabric. The filtrate passes through the fabric and into an escape pipe to the discharge port. Each shell may contain as few as six or as many as 50 leaves of varying dimensions; the entire assembly can be pulled out of the shell for cake removal. In some models the leaves rotate.

"Leafseal."[563] TM for a formulation of decenylsuccinic acid and its esters.
Use: Direct application to plants to enable them to resist frost and drought.

leather. An animal skin or hide that has been permanently combined with a tanning agent which causes a physicochemical change in the protein components of the skin. This change renders it resistant to putrefactive bacteria, enzymes, and hot water, increases its strength and abrasion resistance, and makes it serviceable for long periods of time. Tanning agents are either vegetable, mineral, or synthetic. Hides from cows or steers are chiefly used for men's shoes, transmission belting, and other heavy-duty service. These are usually vegetable-tanned. Lighter grades made from the skins of sheep, calves, or reptiles are used for shoe uppers, luggage, gloves, and similar end products (chrome-tanned).

Leather is a naturally poromeric material which retains the microporosity of the original skin; this property makes it uniquely applicable to footwear; to a limited extent it is able to conform to the contour of the individual foot. Leather is made in many colors, weights and finishes. However, it has been replaced to an increasing extent by plastics for many minor uses and by synthetics for shoe uppers and soling. For further information refer to American Leather Chemists' Association, University of Cincinnati, Cincinnati, Ohio.

See poromeric, and see tanning.

leavening agent. See yeast, baking powder.

Lebedev process. Formation of butadiene from ethanol by catalytic pyrolysis. The catalysts used are mixtures of silicates and aluminum and zinc oxides.

Le Blanc. (1742–1806) A French inventor of the first successful process for making soda ash. His patent was confiscated by the Revolutionist government and the process was used widely for years without either acknowledgment or remuneration. His original formula was 100 parts salt cake, 100 parts limestone and 50 parts coal.

Le Chatelier. (1850–1936) A French physical chemist, famous chiefly for his statement of the equilibrium principle (often known as Le Chatelier's Law). His work included investigations of cements, alloys, and gaseous combustion. The principle may be stated: every system in equilibrium is conservative and tends to resist changes upon it by reacting in such a way as to help nullify the imposed change.

Lechanche cell. See dry cell.

lecithin. $C_8H_{17}O_5NRR'$, R and R' being fatty acid groups. Pure lecithin is a phosphatidyl choline. The lecithins are mixtures of diglycerides of fatty acids linked to the choline ester of phosphoric acid. The lecithins are classed as phosphoglycerides or phosphatides (phospholipids). Commercial lecithin is a mixture of acetone-insoluble phosphatides. FCC specifies not less than 50% acetone-insoluble matter (phosphatides).

Properties: Light brown to brown, viscous semi-liquid with a characteristic odor, partly soluble in water and acetone, soluble in chloroform and benzene.

Derivation: Usually from soybean oil, also from corn, other vegetable seeds, egg yolk and other animal sources.

Grade: Technical, unbleached, bleached; fluid, plastic, edible, FCC, 96+% for biochemical or chromatographic standards.

Use: Emulsifying, dispersing, wetting, penetrating agent and antioxidant; in margarine, mayonnaise, chocolate and candies, baked goods, animal feeds, paints, petroleum industry (drilling, leaded gasoline), printing inks, soaps and cosmetics, mold release for plastics, blending agent in oils and resins, rubber processing, lubricant for textile fibers.

lectin. A type of protein occurring in the seeds of certain plants, especially legumes, characterized by unusual binding specificity; their precise function in the plant is being researched. Studies have been made on the molecular, structure, and carbohydrate content of the lectin found in the European herb sainfoin.

"Lectro."[304] TM for a series of lead vinyl stabilizers.

60 A lead chlorosilicate complex.

Properties: A fine white powder, d 3.9, refr index 2.1.

Use: For vinyl electrical insulation and tapes.

78 Tetrabasic lead fumarate.

Properties: Creamy white powder, d 6.5, refr index, 2.1.

Use: For heat stabilizer for electrical grade plastisols, phonograph records, and electrical insulation.

"Ledate."[69] TM for lead dimethyldithiocarbamate.

LEED. Low energy electron diffraction.

Leeuwenhoek, van. See van Leeuwenhoek.

legal chemistry. (forensic chemistry). The application of chemical knowledge and procedures to matters involving civil or criminal law and to all questions where control of chemical compounds, products, or processes is vested in agencies of Federal or state governments. Legal chemistry applies to the following areas:

(1) Crime detection: primarily identification of poisons, of bloodstains, writing and typewriter inks, and a host of miscible materials such as textile fibers from clothing, hair, skin, etc. A variety of analytical methods are used in police

laboratories, including microscopes, spot tests, color reactions, and spectrophotometry.

(2) Food, drugs, and cosmetics are under the control of the US Food and Drug Administration. New products and proposed additives must be submitted by the manufacturer and approved before being placed on public sale. Control of the manufacture of illicit drugs is an important phase of legal chemistry.

(3) Pesticides are subject to Federal regulation. New products must be registered, labeling must be specific as to chemical composition, active and inert ingredients, and directions for use.

(4) Marketing and competitive pricing of chemical products' fair trade agreements and discriminatory practices are also under Federal supervision (Robinson-Patman Act). This includes mergers, tie-in sales, and other merchandising practices.

(5) Interstate shipment and labeling of hazardous chemicals is regulated by the Department of Transportation and the Federal Aviation Agency as well as by state and local laws.
See labeling, and toxicology.

(6) Patent law comprises a vast body of legal practice and court decisions. The patent system is designed to protect inventions and new discoveries, and most chemical companies retain legal counsel in this field.

(7) Water pollution is subject to Federal regulations (Federal Water Pollution Control Act, 1956). This covers the discharge of contaminating industrial waste, sewage, oil, etc., into navigable streams and their tributaries as well as into coastal waters.

(8) Flammability of fabrics.

(9) Use of volatile, toxic solvents on an industrial scale.

(10) Air pollution including gases and particulates from industrial stacks and auto exhaust emissions.

Lehmstedt-Tanasescu reaction. Preparation of acridones (and 10-hydroxyacridones) from o-nitrobenzaldehyde and a halobenzene in the presence of concentrated sulfuric acid containing nitrous acid as catalyst.

lehr. A long oven designed for controlled slow cooling of glass (annealing). The hot glass is carried through the lehr on a conveyor at a predetermined rate, traversing areas of gradually decreasing temperature. The entire process may require several days.
See annealing.

Leloir, Luis F. (1906-) A French born biochemist who won the Nobel prize for chemistry in 1970 for work in biosynthesis of carbohydrates. He discovered chemical compounds that affect the storage of chemical energy in humans and animals. He has been the head of the Department of Biochemistry at the University of Buenos Aires since 1962.

"Lemol."[65] TM for a series of polyvinyl alcohols in partially and fully hydrolyzed form at various molecular weights. Supplied as nondusting white granules with d 1.2–1.3.
Use: Adhesives, emulsions, polymerization, film coatings, polyester release agents, textile printing, finishing, and sizing.

lemon chrome. See barium chromate.

lemongrass oil. (verbena oil, Indian).
CAS: 8007-02-1 (West Indian).   An essential oil obtained by steam distillation of a grass (Cymbopogon).
Properties: Dark yellow to light brown-red, pronounced heavy lemon-like odor, d 0.900–0.910 (15/15C), optical rotation −3 to +1 degrees, refr index 1.4830–1.4890 (20C), soluble in alcohol, slightly soluble in glycerol. Combustible.
Source: India, East and West Indies, Guatemala.
Chief constituents: Citral (75–85%), geraniol, ethylheptenone.
Use: Perfumes, flavoring, isolates and ionones, source of citral.

lemon oil. See citrus peel oils.

lenacil. CAS: 2164-08-1. $C_{13}H_{18}N_2O_2$.
Properties: Colorless solid, mp 315C, d 1.32, almost insoluble in water, soluble in pyridine.
Use: Herbicide.

lepidine. (γ-methylquinoline; cincholepidine).
CAS: 491-35-0. $C_9H_6NCH_3$.   An alkaloid.
Properties: Oily liquid, quinoline-like odor, turns red-brown on exposure to light, d 1.086, bp 266C, solidifies at about 0C, soluble in alcohol, ether, and benzene, slightly soluble in water.
Derivation: From cinchonine.
Use: Organic preparations, medicine.

lepidolite. (lithia mica).
$K_2Li_3Al_4Si_7O_{21}(OH,F)_3$.   A fluosilicate of potassium, lithium, and aluminum, found in pegmatites. Rubidium occurs as an impurity. A variety of mica.
Properties: Color pink and lilac to gray, luster pearly, perfect micaceous cleavage, Mohs hardness 2.5–4, d 2.8–3.0.
Occurrence: California, South Dakota, New Mexico, South Africa.

Use: Source of lithium and rubidium, flux in glass and ceramics production.

"Lethane."[23] TM for a group of thiocyanate insecticides. 60: 2-thiocyanoethyl laurate. 384: β-butoxy-β'-thiocyanodiethyl ether. A-70: diethylene glycol dithiocyanate.
Hazard: Toxic by ingestion.

Letts nitrile synthesis. Formation of nitriles by heating aromatic carboxylic acids with metal thiocyanates.

Leu. Abbreviation for leucine.

leucine. (α-amino-γ-methylvaleric acid; α-aminoisocaproic acid). CAS: 61-90-5. $(CH_3)_2CHCH_2CH(NH_2)COOH$. An essential amino acid. Found naturally in the $l(-)$ form.
Properties: White crystals, soluble in water, slightly soluble in alcohol, insoluble in ether, optically active (natural form). Dl-leucine mp 332C with decomposition; $l(-)$-leucine mp 295C, d 1.239 (18/4C).
Derivation: Hydrolysis of protein (edestin, hemoglobin, zein), organic synthesis from the α-bromo acid.
Grade: Commercial (dl-), FCC (l-).
Use: Nutrient and dietary supplement, biochemical research.

Leuckart thiophenol reaction. Decomposition of diazoxanthates by warming gently in faintly acidic cuprous media to the corresponding aryl xanthates which afford aryl thiols on alkaline hydrolysis and aryl thioethers on warming.

Leuckart-Wallach reaction. Reductive alkylation of ammonia or of primary or secondary amines with carbonyl compounds and formic acid or formamides as reducing agents.

leuco-compound. See vat dye.

leucocyte. A white blood cell.

leucovorin. Preferred name for folinic acid.

"Leukanol."[23] TM for synthetic tanning assistants of the sulfonic type supplied in liquid and solid grades. Powerful dispersants for vegetable tannins and bleaches for chrome-tanned leather.

leukotriene. One of a group of physiologically active compounds derived directly from arachidonic acid. They are chemically related to the prostaglandins and occur in white blood cells (leucocytes).

"Levapren 450."[470] TM for ethylene-vinyl acetate copolymer containing approx 45% vinyl acetate.

leveling. (1) A term used in the paint industry to describe the application properties of a paint, i.e., its ability to cover a dry surface easily and to hold its level without sagging or running. (2) The ability of a nickel-plated coating to cover surface irregularities of the substrate, achieved by the incorporation of one or more brighteners in the plating formulation. (3) Aiding the uniform dispersion of a dye in a dye bath or solution by addition of a suitable material, e.g., lignin.

"Levelume."[288] TM for bright high-leveling nickel process. Prepared from nickel sulfate, nickel chloride, boric acid and organic addition agents.
Use: Electrical appliance, automotive trim, plumbing fixtures.

Levene-Hudson phenylhydrazide rule. "The direction of rotation of the phenylhydrazides of the sugar acids indicates the configuration of the hydroxyl on the alpha-carbon atom. If the phenylhydrazide rotates to the right, the hydroxyl on the alpha-carbon is to the right, and vice-versa." The rule was shown to be valid for salts, amides, and corresponding acylated nitriles. In connection with the rule, Hudson mentioned that "the sugar benzylphenylhydrazones rotate to the left when the asymmetric alpha-carbon atom of the configuration has its hydroxyl to the right, and vice-versa."

levorotatory. Having the property when in solution of rotating the plane of polarized light to the left or counterclockwise. Levorotatory compounds may have the prefix l- to distinguish them from their dextrorotatory or d- isomers, but the minus sign (-) is preferred.

levulinic acid. (γ-ketovaleric acid; acetylpropionic acid; 4-oxopentanoic acid; levulic acid). CAS: 123-76-2. $CH_3CO(CH_2)_2COOH$.
Properties: Crystals, bp 245–246C, mp 33–35C, d 1.1447 (25/4C), refr index 1.442 (16C). Miscible in water, alcohol, esters, ethers, ketones, aromatic hydrocarbons; insoluble in aliphatic hydrocarbons. Combustible.
Use: Intermediate for plasticizers, solvents, resins, flavors, pharmaceuticals, acidulant and preservative; chrome plating; solder flux; stabilizer for calcium greases; control of lime deposits.

levulose. See fructose.

Lewis acid. Any molecule or ion (called an electrophile) that can combine with another molecule or ion by forming a covalent bond with two electrons from the second molecule or ion. An acid is thus an electron acceptor. Hydrogen ion (proton) is the simplest substance that will do this, but many compounds such as boron trifluoride, BF_3, and aluminum chloride, $AlCl_3$, exhibit the same behavior and are therefore properly called acids. Such substances show acid effects on indicator colors and when dissolved in the proper solvents.

Lewis base. A substance that forms a covalent bond by donating a pair of electrons, neutralization resulting from reaction between the base and the acid with formation of a coordinate covalent bond. It is also called a nucleophile.
See also Lewis electron theory.

Lewis electron theory. A theory involving acid and base formation, neutralization, and related phenomena on the basis of exchange of electrons between substances and the formation of coordinate bonds. It represented an important advance in chemical theory, largely replacing earlier concepts. Advanced in 1923 by Gilbert N. Lewis, it contributed much to the development of coordination chemistry in which the base is represented by the ligand and the acid by the metal ion.

Lewis, Gilbert N. (1875–1946) An American chemist, native of Massachusetts, Professor of chemistry at MIT from 1905–1912 after which he became dean of chemistry at University of California at Berkeley. His most creative contribution was the electron-pair theory of acids and bases which laid the groundwork for coordination chemistry. He was also a leading authority on thermodynamics.

lewisite. Legal label name for β-chlorovinyl-dichloroarsine.

Lewis metal. Alloy of one part tin and one part bismuth.
Properties: Expands when cooling, mp 138C.
Use: Sealing and holding die parts.

"Lewisol" 28.[266] TM for a pale hard resin, a maleic-modified glycerol ester of rosin. Acid number 36, softening point 141C, USDA color WG.

Lewis, Warren P. (1882–1974) Born in Laurel, Maryland, graduated from MIT in 1905, PhD from University of Breslau, Germany in 1908. He became professor of chemical engineering at MIT in 1910. He is often regarded as the father of chemical engineering in the US, as his outstanding books and other publications did much to establish the fundamental principles of this field.

"Lexan."[245] TM for thermoplastic carbonate-linked polymers produced by reacting bisphenol A and phosgene.
Use: Molding applications and other industrial arts.
See also polycarbonate resin.

"L-310 Fatty Acid."[487] TM for a fatty acid derived from linseed oil.
Properties: The major component acids are oleic, linoleic, and linolenic. Light yellow liquid at ambient temperature, obtained from naturally occurring triglycerides.
Use: Chemical intermediate, paint and varnishes, alkyd resins and soaps.

Li. Symbol for lithium.

Libby, Willard F. (1908-1980) An American chemist who won the Nobel prize for chemistry in 1960 and in 1980. His first Nobel work involved chemistry in the space program such as radiochemistry and isotope tracer work. He is known for the 'atomic time clock' which is a way of estimating the age of ancient materials by measuring the amount of radioactive carbon 14 in organic or carbon containing objects. His Doctorate was awarded at Wesleyan University. He also worked at Syracuse, Carnegie Institute of Technology, and Georgetown among others.

"Librium" Hydrochloride.[190] TM for chlordiazepoxide hydrochloride. Manufacture and use restricted.

licanic acid. (4-keto-9,11,13-octadecatrienoic acid).
$CH_3(CH_2)_3(CH:CH)_3(CH_2)_4COCH_2COOH$.
Properties: White crystals, α-licanic acid (naturally occurring isomer) melts at 74–75C, readily isomerizes to the beta-form, mp 99.5C, soluble in organic solvents.
Derivation: Occurs in oiticica and other oils as glycerides.

lichenic acid. See fumaric acid.

licorice. See glycyrrhizin.

lidocaine. (α-diethylaminoaceto-2,6-xylidide).
CAS: 137-58-6.
$C_6H_3(CH_3)_2NHCOCH_2N(C_2H_5)_2$.
Properties: White or slightly yellow crystalline powder, characteristic odor, mp 66–69C, bp 180–182C (at 4 mm). Soluble in alcohol, ether, or chloroform; insoluble in water.
Derivation: By action of diethylamine on chloroacetylxylidide.
Grade: USP.
Use: Medicine (local anesthetic).

Lieben iodoform reaction. Cleavage of methyl ketones with halogens (mostly iodine) and base to carboxylic acids and haloform.

Liebig, Justus Von. (1803–1873) A German chemist who founded the *Annalen*, a world-famous chemical journal. He was a great teacher of chemistry, training such men as Hofmann, who did basic work on organic dyes. Liebig contributed original research in the fields of human physiology, plant life, soil chemistry and was the discoverer of chloroform, chloral, and cyanogen compounds. He was the first to recommend addition of nutrients to soils and thus may be considered the originator of the fertilizer industry.

life, origin. (biogenesis). The succession of chemical events that led up to the appearance of living organisms on earth about 3.3 billion years ago. According to one theory, substantiated by experimental evidence, this occurred as follows. The inorganic compounds originally present were carbides, water, ammonia, and carbon dioxide. The carbides reacted with water to form methane, which in turn reacted with ammonia and water vapor as a result of an electric impulse to form amino acids, porphyrins, and nucleotides (or their precursors). All these compounds have been created artificially in the laboratory. It has further been shown that amino acids and nucleotides can be concentrated into proteins (and probably nucleic acids) by the action of zinc-bearing clays, which were present along the shores of the primeval oceans. Little or no free oxygen existed in the primordial atmosphere, which consisted chiefly of reducing gases. The complex chemical reactions which eventually resulted in the formation of DNA took place in an anaerobic aqueous environment and the earliest living organisms developed in a nutrient solution in which free oxygen finally appeared by the photosynthesis of algae. Another theory advances the idea that essential life chemicals such as purines and amino acids were formed under primitive conditions from aqueous solutions of hydrogen cyanide. Both these theories are based on research carried out by highly competent biochemists.

ligand. A molecule, ion or atom that is attached to the central atom of a coordination compound, a chelate, or other complex. Thus, the ammonia molecules in $[Co(NH3)_6]^{+++}$ and the chlorine atoms in $[PtCl_6]^{--}$ are ligands. Ligands are also called complexing agents, for example EDTA, ammonia, etc.
See also chelate, coordination compound.

light hydrocarbon. One of a group of hydrocarbon products derived from natural gas or petroleum; ethane, propane, iso- and normal butane and natural gasoline (C_5 and heavier). Produced largely in southwest Texas and Louisiana, these are used as feedstocks for a wide vareity of organics.
See also liquefied petroleum gas.

light metal. In engineering terminology, a metal of specific gravity less than three that is strong enough for construction use (aluminum, magnesium, beryllium).

light microscope. See optical microscope.

light oil. (coal tar light oil). A fractional distillate from coal-tar with bp range from 110–210C, consisting of a mixture of benzene, pyridine, toluene, phenol, and cresols. The term is also sometimes used for oils of about the same bp range, but from other sources.
Grade: Technical.
Hazard: Highly flammable, dangerous fire risk.
Use: Source of benzene, solvent naphthas, toluene, phenol, and cresols.

light water. (1) A fire-fighting agent consisting of a water solution of perfluorocarbon compounds mixed with a water-soluble thickener of the polyoxyethylene compound type. It can be used simultaneously with dry chemical to smother gasoline or similar fires. (2) Ordinary water (as distinct from heavy water) used to both cool and moderate nuclear reactors.

lignin. A phenylpropane polymer of amorphous structure comprising 17–30% of wood. It is so closely associated with the holocellulose which makes up the balance of woody material that it can be separated from it only by chemical reaction at high temperature. It is believed to function as a plastic binder for the holocellulose fibers. It is recovered from wood-processing wastes in limited amounts.
Use: Stabilization of asphalt emulsions, ceramic binder and deflocculant, dye leveler and dispersant, drilling fluid additive, precipitation of proteins, extender for phenolic plastics, special molded products, source of vanillin, phenol, and of a component of battery expanders.

lignin sulfonate. (lignosulfonate). A metallic sulfonate salt made from the lignin of sulfite pulp-mill liquors, mw range 1000–20,000.
Properties: Light-tan to dark-brown powder, no pronounced odor, stable in dry form and relatively stable in aqueous solution, nonhygroscopic, no definite mp, decomposes above 200C,

d about 1.5, forms colloidal solutions or dispersions in water, practically insoluble in all organic solvents.
Use: Dispersing agent in concrete and carbon black-rubber mixes, extender for tanning agents, oil-well drilling mud additives, ore flotation agents, production of vanillin, industrial cleaners, gypsum slurried, dyestuffs, pesticide formulations. Commercially available as the salts of most metals and of ammonium.

lignite. (brown coal). A low rank of coal between peat and sub-bituminous, it contains 35–40% water. It occurs in the continental US, Alaska, Germany and the Netherlands. Its Btu value is low. Drying, crushing and pelletizing lignite with an asphaltic binder for direct use as fuel has been successfully demonstrated. Polymer resins (polyesters and polyamides) can be derived from lignite by oxidation with nitric acid, followed by extraction of the nitro-coal acids, which are the basis of the polymer molecules. Peat can also be used. A process for gasification of lignite to produce methanol is approaching commercial development in Sweden.
See also peat, gasification.

lignoceric acid. (n-tetracosanoic acid).
CAS: 557-59-5. $CH_3(CH_2)_{22}COOH$.
A long-chain saturated fatty acid found in minor quantities in most natural fats.
Properties: Crystals, mp 84.2C, bp 272C (10 mm), d 0.8207 (100/rC), refr index 1.4287 (100C), nearly insoluble in ethanol.
Source: Lignite and beechwood tar, peanut oil, sphingomyelin.
Grade: Technical, 99%.
Use: Biochemical research.

"Lignosol."[476] TM for a series of calcium, sodium, and ammonium lignosulfonates. These compounds are mixtures of lignosulfonates and wood sugars. Special grades are available which have low wood sugar content.
Use: Binders, dispersing agents and tanning agents.

lignosulfonate. See lignin sulfonate.

"Lignox."[236] TM for a soluble calcium lignosulfonate in dry powder form.
Use: Treatment of drilling mud containing calcium ions and in brine emulsion muds.

ligroin. A saturated volatile fraction of petroleum boiling in the range 60–110C. There is a special grade of ligroin known as petroleum benzin.
Hazard: Highly flammable, dangerous fire risk. Toxic by ingestion and inhalation.
Use: Solvent for resins, paints, varnishes, etc.

"Lilial."[227] TM for p-tert-butyl-α-methylhydrocinnamaldehyde. $C_{14}H_{20}O$.

lime. Fourth highest-volume chemical produced in US (1985). Specifically, calcium oxide (CaO), more generally, any of the various chemical and physical forms of quicklime, hydrated lime, and hydraulic lime (adapted from ASTM definition C41-47). Noncombustible. For further information, see National Lime Association, 925 16th St. N.W., Washington, D.C.
Hazard: Unslaked lime (quicklime) yields heat on mixing with water and is a caustic irritant.
Use: See calcium oxide, calcium hydroxide.
See also following entries and note under calcium oxide.

lime acetate. See calcium acetate.

lime, agricultural. Lime slaked with a minimum amount of water to form calcium hydroxide.

lime, air-slaked. Lime which has absorbed carbon dioxide and moisture from the atmosphere. It consists of a powder composed of calcium carbonate and calcium hydroxide.

lime citrate. See calcium citrate.

lime, chlorinated. (chloride of lime; bleaching powder). $CaCl(ClO) \cdot 4H_2O$
Properties: White powder, chlorine odor, mp (decomposes), decomposes in water, acids.
Derivation: By conducting chlorine into a box-like structure containing slaked lime spread upon perforated shelves.
Grade: 35–37% active chlorine, technical.
Hazard: Evolves chlorine and at higher temperatures oxygen. With acids or moisture evolves chlorine freely at ordinary temperatures.
Use: Textile and other bleaching applications, organic synthesis, deodorizer, disinfectant.
See also calcium hypochlorite, bleach.

lime, fat. A pure lime which combines readily with water to form a fine white powder, free from grit, and makes a smooth stiff paste with excess of water. Must not be loaded hot.
See also lime, lean.

lime, hydrated. See calcium hydroxide.

lime, hydraulic. A variety of calcined limestone which when pulverized absorbs water without swelling or heating and gives a cement that hardens under water. The limestone burned for this purpose usually contains 10–17% silica, alumina, and iron and 40–45% lime, magnesia sometimes replacing lime. Must not be loaded hot.

lime hypophosphite. See calcium hypophosphite.

lime, lean. A lime which does not lake freely with water because it has been prepared from limestone containing a high percentage of impurities, e.g., silica, iron, alumina, etc. Must not be loaded hot.

lime-nitrogen. See calcium cyanamide.

lime oil, distilled. CAS: 8008-26-2. Colorless to greenish yellow volatile oil obtained by distillation from the juice or whole crushed fruit of *Citrus aurantifolia Swingle*.
Properties: Refr index 1.4745–1.4770 (20C), d 0.855–0.863 (25C), angular rotation +34 to +47 degrees, soluble in most fixed oils and mineral oil, insoluble in glycerol and propylene glycol. Combustible.
Chief constituents: Terpineol, citral.
Grade: FCC (contains between 0.5 and 2.5% of aldehydes, calculated as citral).
Use: Extracts, flavoring, perfumery, toilet soaps, cosmetics.

lime oil, expressed. See citrus peel oil.

lime saltpeter. See calcium nitrate.

lime, slaked. See calcium hydroxide.

limestone. CAS: 1317-65-3. $CaCO_3$. A noncombustible solid characteristic of sedimentary rocks and composed mainly of calcium carbonate in the form of the mineral calcite, Mohs hardness about 3. Limestones are sometimes classed according to the impurities contained, e.g.: Dolomitic limestone: Usually a limestone containing more than 5% magnesium carbonate. Magnesium limestone: Dolomitic limestone. Used as a solid diluent and carrier in pesticides. Argillaceous limestone: Contains clays, used in cement manufacture as "cement rock." Siliceous limestone: A limestone containing sand or quartz. Limestones are also named according to the formation in which they occur.
See also marble and dolomite.
Use: Building stone, metallurgy (flux), manufacture of lime, source of carbon dioxide, agriculture, road ballast, cement (Portland and natural), alkali manufacture, removal of sulfur dioxide from stack gases and sulfur from coal.

lime, sulfurated. (calcium sulfide, crude).
A mix of calcium sulfide and calcium sulfate.
Properties: Yellowish-gray or grayish-white powder, odor of hydrogen sulfide, soluble in acids, insoluble in water and alcohol. Noncombustible.
Derivation: By roasting calcium sulfate with coke.
Use: Medicine, depilatory, luminous paint.

lime-sulfur solution. A solution made by boiling together lime (50 lbs), sulfur (100 lbs) and water (100 gals) and diluting to one-tenth strength. Contains calcium polysulfide, free sulfur, and calcium thiosulfate.
Use: Fungicidal spray on fruit trees, sheep dip.

lime, unslaked. See calcium oxide.

lime water. (calcium hydroxide solution).
CAS: 1305-62-0.
Properties: Clear, colorless, odorless, alkaline aqueous solution of calcium hydroxide containing more than 0.14 g of $Ca(OH)_2$ in each 100 mL at 25C (the strength varies with the temperature at which the solution is stored), d about 1.00 (25C), absorbs carbon dioxide from air.
Grade: USP (as calcium hydroxide solution).
Use: Medicine (external).

limonene. CAS: 138-86-3. $C_{10}H_{16}$.
A widely distributed optically active terpene, closely related to isoprene. It occurs naturally in both d- and l-forms. The racemic mixture of the two isomers is known as dipentene.

Properties: Colorless liquid, (a) d 0.8411 (20C), bp 176–176.4C; (b) d 0.8422 (20C), bp 176–176.4C, oxidizes to film in air, oxidation behavior similar to that of rubber or drying oils.
Derivation: (a) Lemon, bergamot, caraway, orange, and other oils, (b) peppermint and spearmint oils.
Use: Flavoring, fragrance and perfume materials, solvent, wetting agent, resin manufacture.

limonene dioxide. See dipentene dioxide.

limonene, inactive. (or racemic or *dl*).
See dipentene.

limonene monoxide. See dipentene monoxide.

limonite. See hematite, brown.

"Lin-All."[480] TM for a liquid paste or flake form of tallates of calcium, cobalt, copper, iron, lead, manganese, and zinc.
Use: Drier for paint and printing inks.

linalool. (linalol; 3,7-dimethyl-1,6-octadien-3-ol). CAS: 78-70-6.
$(CH_3)_2C:CHCH_2CH_2C(CH_3)OHCH:CH_2$.
Linalool is the l-isomer, coridandrol is the d-isomer.
Properties: Colorless liquid, odor similar to that of bergamot oil and French lavender, soluble in alcohol and ether, d 0.858–0.868 (25C), bp 195–199C, angular rotation −2 to +2 degrees, soluble in fixed oils. Combustible.
Derivation: Citrus peel oils, especially from oranges. Made synthetically from geraniol.
Method of purification: Rectification.
Grade: Ex bois de rose oil, synthetic, FCC.
Use: perfumery, flavoring agent.

linalool oxide. (tetrahydro-α,α-5-trimethyl-5-vinylfurfuryl alcohol). $C_{10}H_{17}O_2$.
Properties: liquid, refr index 1.4523 (20C).
Derivation: Synthetically from acetone.
Use: Perfuming and flavoring agent.

linalyl acetate. CAS: 115-95-7.
$C_{10}H_{17}C_2H_3O_2$.
Properties: Clear, colorless, oily liquid. Odor of bergamot, bp 108–110C, d 0.908–0.920, refr index 1.450–1.458 (20C), angular rotation −1 to +1 degrees. Soluble in alcohol, ether, diethyl phthalate, benzyl benzoate, mineral oil, fixed oils, alcohol; slightly soluble in propylene glycol; insoluble in water, glycerol. Combustible.
Derivation: Action of acetic anhydride on linalool in presence of sulfuric acid. May also be obtained from bergamot and other oils.
Method of purification: Rectification.
Grade: Ex bois de rose oil 92%, 96–98%, FCC (natural and synthetic).
Use: Extracts, perfumery, flavoring agent, substitute for petitgrain oil.

linalyl formate. $C_{10}H_{17}OOCH$.
Properties: D 0.915 (25/4C), bp 100–103C (10C), refr index 1.456–1.457 (20C), insoluble in water, soluble in alcohol. Combustible.
Derivation: Synthetically from acetone.
Use: Perfume, flavoring.

linalyl isobutyrate. CAS: 78-35-3.
$C_{10}H_{17}OOCCH(CH_3)_2$.
Properties: liquid, d 0.890, refr index 1.4490. Combustible.
Derivation: Synthetically from acetone.
Use: Perfume, flavoring.

linalyl propionate. $C_{10}H_{17}OOCC_2H_5$.
Properties: Colorless liquid, floral odor similar to bergamot oil, d 0.895–0.902 (25C), refr index 1.4500–1.4550 (20C), soluble in most fixed oils and in mineral oil, slightly soluble in propylene glycol, insoluble in glycerol. Combustible.

Derivation: Synthetically, starting with acetone.
Grade: FCC.
Use: Perfumery, flavoring agent.

lincomycin. CAS: 154-21-2. $C_{18}H_{34}N_2O_6S$.
An antibacterial substance produced by *Streptomyces lincolnensis*. The hydrochloride is stable in the dry state and in aqueous solution for at least 24 months, soluble in water at room temperature in concentrations up to 500 mg/mL.

lindane. (γ-benzene hexachloride).
CAS: 58-89-9. $C_6H_6Cl_6$. Legal label name for gamma isomer of 1,2,3,4,5,6-hexachlorocyclohexane.

Hazard: Toxic by inhalation, ingestion, and skin absorption. TLV: 0.5 mg/m^3 of air.
Use: Pesticide. Use may be restricted.

"Lindiste."[88] TM for a rare earth oxide glass polishing agent. Contains cerium and other rare earth oxides in the proportion in which they occur naturally.

linear molecule. See alkyl sulfonate, linear; polyethylene, high-density.

linoleic acid. (linolic acid). CAS: 60-33-3.

A polyunsaturated fatty acid (2 double bonds) existing in both conjugated and unconjugated forms. A plant glyceride essential to human diet.
Properties: Colorless to straw-colored liquid, d 0.905 (15/4C), fp −5C, bp 228C (14 mm), refr index 1.4710 (15C), insoluble in water, soluble in alcohol and ether. Combustible.
Commercial sources: Linseed oil, safflower oil, tall oil.
Grade: Technical, purified (99+%), edible.
Use: Soaps, special driers for protective coatings, emulsifying agents, feeds, biochemical research, dietary supplement, margarine.
Note: Do not confuse with linolenic acid.

linolein. A glyceride of linoleic acid. It is one of the constituents of linseed oil which induces drying.

linolenic acid. (9,12,15-octadecatrienoic acid). CAS: 463-40-1.

A polyunsaturated fatty acid (3 double bonds) which occurs as the glyceride in many seed fats. It is an essential fatty acid in the diet. It is also a constituent of drying oils.
Properties: Colorless liquid, soluble in most organic solvents, insoluble in water, d 0.916 (20/4C), fp −11C, bp 230C (17 mm). Combustible.
Grade: Purified 99+%.
Use: Nutrient, biochemical research, drying oils.
Note: Do not confuse with linoleic acid.

linolenin. A glyceride of linolenic acid. Like linolein it is a constituent of drying oils.
See linseed oil.

linolenyl alcohol. (octadecatrienol).
$C_{18}H_{32}O$.   The fatty alcohol derived from linolenic acid. Available commercially as 50% pure.
Properties: Colorless solid, iodine value 190, cloud point 50.0F (10C), d 0.864. Combustible.
Derivation: Reduction of acid made from linseed oil.
Impurities: Oleyl and linoleyl alcohols with some saturated alcohols.
Use: Paints, flotation, lubricants; surface-active agents, resins, synthetic fibers.

linoleyl alcohol.
$CH_3(CH_2)_4CH:CHCH_2CH:CH(CH_2)_7CH_2OH$.
The fatty alcohol derived from linoleic acid. Available commercially as 50–60% pure alcohol.
Properties: Colorless solid, iodine value 137, d 0.855, cloud point 59F (15C). Combustible.
Derivation: Reduction of linoleic acid.
Impurities: Mostly oleyl alcohol with some linolenyl and saturated alcohols.
Use: Paints, flotation, paper, surface-active agents, resins, leather.

linoleyltrimethylammonium bromide. An amorphous solid from yellow to very light brown in color, soluble in water, alcohol.
Use: Germicide, deodorant, algicide, slime control.

linolic acid. See linoleic acid.

linseed cake. The press cake formed when the seeds are crushed and the oil is extracted.
See linseed oil meal.

linseed oil. (flaxseed oil).   CAS: 8001-26-1.
Properties: Golden-yellow, amber, or brown drying oil with peculiar odor and bland taste. D 0.921–0.936, iodine value 177, saponification value 189–195, acid number (max) 4 (ASTM D 234-48), flash p 432F (222C), autoign temperature approximately 650C (343C), polymerizes on exposure to air. Soluble in ether, chloroform, carbon disulfide, and turpentine; slightly soluble in alcohol; spontaneous heating. Combustible.
Chief constituents: Glycerides of linolenic, oleic, linoleic, and saturated fatty acids. The drying property is due to the linoleic and linolenic groups. Derivation: From seeds of the flax plant Linum usitatissimum by expression or solvent extraction. Various refining and bleaching methods are used.
Grade: Raw, boiled, doubled-boiled, blown, varnish makers, refined.
See linseed oil, boiled.
Use: Paints, varnishes, oilcloth, putty, printing inks, core oils, linings and packings, alkyd resins, soap, pharmaceuticals.
Note: Use of linseed oil in paints has decreased sharply since the introduction of emulsion paints.

linseed oil, blown. Linseed oil whose viscosity is increased by air bubbled through it at 93C. The reaction is mainly oxidation followed by polymerization. The resulting product dries to a harder film than heat-bodied oils and is used in interior paints and enamels.

linseed oil, boiled. The term is a misnomer since the oil does not boil. Small amounts of driers, (e.g., oxides of manganese, lead, or cobalt or their naphthenates, resinates, or linoleates) are added to hot linseed oil to accelerate drying. The "boiled oil" becomes thicker and darker.

linseed oil, heat-bodied. Linseed oil which has been polymerized by heating at 287–315C. This increases viscosity and acid content and reduces iodine value. Bodied oil dries much faster than unprocessed oil.

linseed oil meal. (linseed cake).   The crushed and extracted residue from flaxseed (linseed), generally prepared by crushing the seeds, cooking with steam, and hydraulic expression of the oil. The resulting cake is sold by its protein content.

linters, cotton. See cotton linters.

linuron. (Generic name for 3-(3,4-dichlorophenyl)-1-methoxy-1-methylurea).
CAS: 330-55-2.
$C_6H_3Cl_2NHC(O)N(OCH_3)CH_3$.
Properties: Solid, mp 93–94C, slightly soluble in water, partially soluble in acetone and alcohol.
Use: Herbicide.

"Lipal."[555] TM for a series of polyoxyethylene esters and ethers of fatty acids.
Use: Detergents, emulsifiers, solubilizers, wetting agents, and coupling agents.

lipase. CAS: 9001-62-1. Any of a class of enzymes that hydrolyze fats to glycerol and fatty acids. Lipase is abundant in the pancreas but also occurs in gastric mucosa, in the small intestine, and in fatty tissue. It is found in milk, wheat germ, and various fungi. Commercial pancreatin and most trypsin preparations contain lipase.
Derivation: From pork pancreas or calf glands.
Use: Manufacture of cheese and similar foods, for removal of fat spots in dry cleaning or grease accumulations, in analytical chemistry of fats because it selectively hydrolyzes only fatty acids on the ends of triglycerides.

lipid. (lipide). An inclusive term for fats and fat-derived materials. It includes all substances which are (1) relatively insoluble in water but soluble in organic solvents (benzene, chloroform, acetone, ether, etc.); (2) related either actually or potentially to fatty acid esters, fatty alcohols, sterols, waxes, etc.; and (3) utilizable by the animal organism. One of the chief structural components of living cells.
See also phospholipid, fat, fatty acid.

Lipmann, Fritz. (1899-1986) A German-born biochemist who won the Nobel Prize in 1953 for the discovery of coenzyme A (CoA). He earned doctorates at the University of Berlin in both chemistry and medicine. He worked at Cornell and Harvard Universities. He founded the biochemistry department of Brandeis University and later joined the factulty of Rockfeller University.

dl-α-lipoic acid. (6,8-dithiooctanoic acid; thioctic acid; POF). CAS: 62-46-4.

$\overline{SSCH_2CH_2CH(CH_2)_4COOH}$. A pyruvate oxidation factor. Pyruvate is a normal intermediate in carbohydrate metabolism.
Properties: Crystals, mp 60–61C, bp 160–165C, practically insoluble in water, soluble in fat solvents, forms a water-soluble sodium salt.
Sources: Yeast and liver.
Use: Nutrition, biochemical research.

"Lipopure."[425] Brand name for a group of solvents purified to remove contaminants which might interfere with lipid procedures.

lipotropic agent. An agent which, because of its affinity for fats and oils, helps to regulate the metabolism of fat and cholesterol in the animal body. Inositol is an example.

Lipowitz's metal. A fusible alloy.
See alloy, fusible.

lipoxidase. An enzyme which catalyzes the addition of oxygen to the double bonds of unsaturated fatty acids of plant origin.
Use: Biochemical research, whitening bread.

Lipscomb, William N. (1919-) An American chemist who won the Nobel prize for chemistry in 1976 for his studies on the structure and bonding mechanisms of boranes. Much of the research concerned structure and function of enzymes and natural products in organic and theoretical chemistry. He studied at the Universities of Kentucky, California, and Minnesota.

liquation. The separation of two or more components of a mixture by heating to a temperature at which one component melts, leaving the others as solids.
Use: Separation of alloy components.

liquefaction. (coal). See gasification; Fischer-Tropsch process.

liquefied natural gas. See natural gas.

liquefied petroleum gas. (compressed petroleum gas; liquefied hydrocarbon gas; LPG). A compressed or liquefied gas obtained as a by-product in petroleum refining or natural gasoline manufacture, e.g., butane, isobutane, propane, propylene, butylenes and their mixtures.
Properties: Colorless, noncorrosive, flash p −100F (−74C), autoign temperature 800–1000F (426–537C).
Hazard: Highly flammable, dangerous fire and explosion risk. TLV: 1000 ppm in air.
Use: Domestic and industrial fuel, automotive fuel, welding, brazing, and metal-cutting.
See also compressed gas, natural gas.

liquid. An amorphous (noncrystalline) form of matter intermediate between gases and solids in which the molecules are much more highly concentrated than in gases but much less concentrated than in solids. The molecules of liquids are free to move within the limits set by intermolecular attractive forces. At the air/liquid interface the vibration of the molecules causes some to be ejected from the liquid at a rate depending on the surface tension. The tendency of molecules to escape from a liquid surface is called fugacity and is largely responsible for evaporation, which occurs when the air space above the liquid is unrestricted. In a closed system, where the air space is restricted, the escaping molecules eventually saturate the air and thus the number of molecules leaving the liquid will be equal to those returning to it as a result of molecular attraction. In these circumstances the liquid/air system is said to be in equilibrium.

Liquids vary greatly in viscosity, melting point, vapor pressure, and surface tension. Mercury has a density of 13.6 and the highest surface tension of all liquids. Glass has the highest viscosity. Polar liquids are those whose molecules have opposite electrical charges on their terminal atoms or groups, which impart a force called dipole moment. Water is a polar liquid with high dielectric constant. Pure hydrocarbon liquids are generally nonpolar.

See also liquid, Newtonian; glass; amorphous; solid; liquid crystals; kinetic theory.

liquid chromatography. An analytical method based on separation of the components of a mixture in solution by selective adsorption. All systems include a moving solvent, a means of producing solvent motion, such as gravity or (in more recently developed equipment) a pump; a means of sample introduction, a fractionating column, and a detector. Innovations in functional systems provide the analytical capability for operating in three separation modes: (1) Liquid/liquid: partition in which separations depend on relative solubilities of sample components in two immiscible solvents (one of which is usually water). (2) Liquid/solid: adsorption where the differences in polarities of sample components and their relative adsorption on an active surface determine the degree of separation. (3) Molecular size separations which depend on the effective molecular size of sample components in solution.

Solvents, often referred to as carriers, include isooctane, methyl ethyl ketone, acetone/chloroform, tetrahydrofuran, hexane, and toluene.

Packing materials in columns of various lengths include silica gel, alumina, glass beads, polystyrene gel and ion exchange resins.

High-performance liquid chromatography (HPLC) is the term applied to new and more effective instrumental techniques developed in recent years which have greatly increased the scope of this analytical method. It can now be applied to biological as well as chemical research. Among the separations possible are peptides (by reverse phase chromatography), proteins and enzymes (hydrophobic and size exclusion modes of chromatography), amino acids, and inorganic and organometallic compounds (neutral species, including clusters, by adsorption and size exclusion, and ionic species including coordination compounds). A comparatively recent develop-

ment is the use of supercritical fluids as solvents, e.g., carbon dioxide and sulfur hexafluoride.

See also supercritical fluid. See also gas chromatography, paper chromatography, thin-layer chromatography, instrumentation.

liquid crystal. An organic compound in an intermediate or mesomorphic state between solid and liquid. This phenomenon was first noted in 1888 in cholesteryl benzoate, a crystalline solid. It becomes a turbid liquid, or liquid crystals, when heated to 145C; on further heating to 179C the liquid becomes isotropic. This sequence is reversed when the substance is cooled. Color changes occur on both heating and cooling. Many organic compounds, e.g., sodium benzoate, exhibiting this behavior, are known and used extensively in electric and electronic displays, thermometers, color TV tubes, electronic clocks and calculators, and similar devices dependent on temperature determination. Liquid crystals have several varieties of molecular order: nematic, smectic (nine types), and cholesteric. They indicate small temperature differences by changing color when applied to the skin and are used in medicine for this purpose. They are available in microencapsulated form. They can align with dichroic dye molecules in a thin-layer cell to produce color changes.

See also nematic, smectic, cholesteric.

liquid dioxide. See nitrogen dioxide.

liquid, Newtonian. Characteristic of liquids is their ability to flow, a property depending largely on their viscosity, and sometimes also on the rate of shear. A Newtonian liquid is one that flows immediately on application of force and for which the rate of flow is directly proportional to the force applied. Water, gasoline, and motor oils at high temperatures are examples. Some liquids have abnormal flow response when force is applied, that is, their viscosity is dependent on the rate of shear. Such liquids are said to exhibit non-Newtonian flow properties. Some will not flow until the force exerted is greater than a definite value called the yield point. [W. A. Gruse].

liquid pitch oil. See creosote, coal tar.

liquid rosin. See tall oil.

liquid rubber. See rubber, liquid.

liquor. In chemical technology, any aqueous solution of one or more chemical compounds. In sugar manufacturing, it refers to the sirups obtained from various refining steps (mother liquors). The paper industry uses this term extensively as follows: (a) black liquor is liquid digester waste (also called spent sulfate liquor) containing sulfonated lignin, rosin acids, and other waste wood components from which tall oil is made; (b) green liquor is a solution made by dissolving chemicals recovered in the alkaline pulping process in water; (c) white liquor is made by adding caustic soda to sodium sulfide solution. In dyeing technology, red liquor is an alternate name for mordant rouge.

liter. (L, l). The volume of one kilogram of water at its temperature of maximum density (4C) at standard atmospheric pressure. A liter is 1.05 quart, or 0.26 gallon.

"Lithaflux."[250] TM for lepidolite.
Use: Batch material for glass, porcelain enamel ground coats; ceramic body flux and glaze constituents.

"Lithafrax."[280] TM for a ceramic material made from β-spodumene.

litharge. (lead oxide, yellow; plumbous oxide; lead monoxide). CAS: 1317-36-8. PbO. An oxide of lead made by controlled heating of metallic lead.
Properties: Yellow crystals, d 9.53, mp 888C, insoluble in water, soluble in acids and alkalies, a strong base, commercial grades are yellow to reddish, depending on treatment and purity.
See also massicot.
Derivation: By oxidizing metallic lead in air. Various forms of lead and various temperatures from 500–1000C are used.
Grade: CP, fused, powdered.
Hazard: Toxic by ingestion and inhalation. TLV (as Pb): 0.15 mg/m³ of air.
Use: Storage batteries, ceramic cements and fluxes, pottery and glazes, glass, chromium pigments, oil refining, varnishes, paints, enamels; assay of precious metal ores, manufacture of red lead, cement (with glycerol), acid-resisting compositions, match-head compositions, other lead compounds, rubber accelerator.

litharge-glycerol cement. Made by mixing glycerol with 1/6 to 1/2 portion of water and mixing with enough litharge to give a paste of desired consistency. Must be used as soon as mixed. Fillers retard the setting and avoid cracking. The product is somewhat resistant to acids.
Hazard: Toxic by ingestion.

litharge, leaded. See lead suboxide.

"Lithocote."[145] TM for a series of protective coatings available as LC-19, LC-24, LC-25, LC-34, LC-73 baked phenolic or modified epoxy linings or as LC-82, LC-610 catalyzed epoxy coatings.
Use: Tank cars, storage tanks, pipe, etc., to prevent corrosion and product contamination.

lithia. See lithium oxide.

lithic acid. See uric acid.

lithium. CAS: 7439-93-2. Li. Metallic element of atomic number 3, group IA of Periodic Table, aw 6.941, valence = 1; two isotopes. It is the lightest and least reactive of the alkali metals and the lightest solid element.
Properties: Very soft silvery metal, d 0.534 (20C), mp 179C, bp 1317C, Mohs hardness 0.6, viscosity of liquid lithium less than water, heat capacity about the same as water. Reacts exothermally with nitrogen in moist air at high temperatures; high electrical conductivity; soluble in liquid ammonia.
Sources: Spodumene, lepidolite, ambylgonite occurring in US, Canada, central Africa, Brazil, Argentina, Australia, Europe. Also desert lake brines.
Derivation: By electrolysis of a mixture of lithium chloride and potassium chloride, high-temperature extraction from spodumene by sodium carbonate, solar evaporation of lake brines.
Grade: 99.86% to 99.9999%. Available as ingots, rods, wire, ribbon, and pellets.
Hazard: Ignites in air near its melting point; dangerous fire and explosion risk when exposed to water, acids, or oxidizing agents. Extinguish lithium fires only with chemicals. Lithium in solution is toxic to the central nervous system.
Use: Production of tritium, reducing and hydrogenating agents, alloy hardeners, pharmaceuticals, Grignard reagents. Scavenger and degasifier for stainless and mild steels in molten state, modular iron, soaps and greases, deoxidizer in copper and copper alloys, catalyst, heat-transfer liquid, storage batteries (with sulfur, selenium, tellurium, and chlorine). Rocket propellants, vitamin A synthesis, silver solders, underwater buoyancy devices, nuclear reactor coolant.
See "Lith-Ex."

lithium acetate, dihydrate. (acetic acid, lithium salt; Quilonum). CAS: 6108-17-4. $CH_3CO_2Li \cdot 2H_2O$.
Properties: White, crystalline powder with mw 102.02; mp 53–56C. Soluble in water and alcohol.

lithium aluminate. $LiAlO_2$.
Properties: White powder, mp 1900–2000C, d 2.55 (25C), insoluble in water.
Grade: Ceramic.

Use: As a flux in high-refractory porcelain enamels.

lithium aluminum deuteride. (LAD).
LiAlD$_4$.
Properties: White to gray crystals, d 1.02, stable in dry air at room temperature, but very sensitive to moisture, decomposes above 140C liberating deuterium. Soluble in diethyl ether, tetrahydrofuran; slightly soluble in other low molecular weight ethers. Preparation: By reacting aluminum chloride with lithium deuteride.
Hazard: Flammable, dangerous fire risk; requires special handling; ignites in air.
Use: Introduction of deuterium atoms into molecules by reduction of same groups attacked by lithium aluminum hydride.

lithium aluminum hydride. (LAH).
CAS: 16853-85-3.   LiAlH$_4$.
Properties: White powder; d 0.917; sometimes turns gray on standing; stable in dry air at room temperature, but highly sensitive to moisture, including atmospheric; decomposes to lithium hydride, aluminum metal, and hydrogen above 125C without melting; soluble in diethyl ether, tetrahydrofuran, dimethyl "Cellosolve;" slightly soluble in dibutyl ether; insoluble or very slightly soluble in hydrocarbons and dioxane. Preparation: Reaction of aluminum chloride with lithium hydride.
Hazard: Flammable, dangerous fire risk; may ignite spontaneously on grinding or rubbing, or from static sparks. Reacts violently with air, water, and many organic materials. Fires must be extinguished with powdered limestone or dry chemical.
Use: Reducing agent for over 60 different functional groups, especially for pharmaceutical, perfume, and fine organic chemicals; converts esters, aldehydes, and ketones to alcohols and nitriles to amines; source of hydrogen; propellant; catalyst in polymerizations.

lithium aluminum hydride, ethereal. LiAlH$_4$, plus ether.
Properties: Colorless solution in ether, very reactive to water.
Derivation: From lithium hydride and ether solution of aluminum chloride.
Hazard: Flammable, dangerous fire risk.
Use: See lithium aluminum hydride.

lithium aluminum-tri-tert-butoxyhydride.
(LATB; lithium-tri-tert-butoxyaluminohydride).
CAS: 17476-04-9.   LiAl[OC(CH$_3$)$_3$]$_3$H.
Properties: White powder, d 1.03, stable in dry air but sensitive to moisture, soluble in the dimethyl ether of diethylene glycol, tetrahydrofuran, diethyl ether; slightly soluble in other ethers.
Hazard: Evolves flammable hydrogen at 400C; dangerous.
Use: For stereospecific reductions of steroid ketones and for reductions of acid chlorides to aldehydes.

lithium amide. CAS: 7782-89-0.   LiNH$_2$.
Properties: White, crystalline solid; ammonia-like odor; d 1.18; mp 380–400C; decomposes in water to form ammonia.
Derivation: Reaction of lithium hydride with ammonia.
Grade: 92–95% lithium amide.
Hazard: Flammable, dangerous fire risk; do not use water to extinguish.
Use: Organic synthesis, including antihistamines and other pharmaceuticals.
See "Lith-Ex."

lithium arsenate. Li$_3$AsO$_4$.
Properties: White powder, d 3.07 (15C), slightly soluble in water, soluble in dilute acetic acid.
Hazard: A poison.

lithium benzoate. CAS: 553-54-8.   LiC$_7$H$_5$O$_2$.
Properties: White crystals or powder, soluble in water and alcohol.
Derivation: Reaction of benzoic acid with lithium carbonate.

lithium bicarbonate. LiHCO$_3$.   A lithium salt formed by dissolving lithium carbonate in water with excess carbon dioxide. The solution, called Lithia Water, is used in medicine and prepared mineral waters.

lithium borate. See lithium tetraborate.

lithium borohydride. CAS: 16949-15-8.
LiBH$_4$.
Properties: White to gray crystalline powder, decomposes in vacuum above 200C, in air at 275C. Soluble in water, lower primary amines, and ethers; d 0.66; extremely hygroscopic.
Derivation: Reaction of sodium borohydride and lithium chloride.
Hazard: Flammable, dangerous fire and explosion risk.
Use: Source of hydrogen and borohydrides; reducing agent for aldehydes, ketones, and esters.

lithium bromide. CAS: 7550-35-8.   LiBr.
Properties: White, cubic, deliquescent crystals, or as a white to pinkish white granular powder; odorless; sharp bitter taste. D 3.464, mp 547C, bp 1265C. Very soluble in water, alcohol, and ether; slightly soluble in pyridine; soluble in methanol, acetone, glycol; a hot concentrated solution dissolves cellulose; forms addition com-

pounds with ammonia and amines; forms double salts with $CuBr_2$, $HgBr_2$, HgI_2, $Hg(CN)_2$, and $SrBr_2$; greatly depresses vapor pressure over its solutions.

Derivation: Reaction of hydrobromic acid with lithium carbonate.

Grade: 53% (min) LiBr brine; solid, single crystals.

Use: Pharmaceuticals, air conditioning, humectant, drying agent, batteries, low temperature heat-exchange medium, medicine (sedative).

lithium butyl. See butyllithium.

lithium carbide. Li_2C_2. Crystalline, white powder; d 1.65 (18C); decomposes in water; soluble in acid with evolution of acetylene.

Hazard: Fire risk in contact with acids.

lithium carbonate. CAS: 554-13-2. Li_2CO_3.
Properties: White powder, d 2.111, mp 735C, bp decomposes at 1200C, slightly soluble in water, insoluble in alcohol, soluble in dilute acid.

Derivation: (a) Finely ground ore is roasted with sulfuric acid at 250C, lithium sulfate is leached from the mass and converted to the carbonate by precipitation with soda ash. (b) Reaction of lithium oxide with carbon dioxide or ammonium carbonate solution.

Grade: Technical, CP.

Hazard: Water solution is strong irritant.

Use: Ceramics and porcelain glazes, pharmaceuticals, catalyst, other lithium compounds, coating of arc-welding electrodes, nucleonics, luminescent paints, varnishes and dyes, glass ceramics, aluminum production.

lithium chlorate. $LiClO_3$.
Properties: Needlelike crystals, deliquescent, d 1.119 (18C); mp 128C, decomposes at 270C, more soluble in water than any other inorganic salt (313 g per 100 mL water at 18C), very soluble in alcohol.

Hazard: Dangerous explosion hazard when shocked or combined with organic materials. Strong oxidant.

Use: Air conditioning, inorganic and organic chemicals, propellants.

See "Lith-X."

lithium chloride. CAS: 7447-41-8. LiCl.
Properties: White, deliquescent crystals; d 2.068; mp 614C; bp 1360C; very soluble in water, alcohols, ether, pyridine, nitrobenzene. One of the most hygroscopic salts known. Not to be used as dietary salt substitute.

Derivation: Reaction of lithium ores with chlorides, natural brines.

Grade: Technical 99% (min) assay, 35–40% brine, inhibited; single crystals.

Use: Air conditioning, welding and soldering flux, dry batteries, heat-exchange media, salt baths, desiccant, production of lithium metal, soft drinks and mineral water to reduce escape of carbon dioxide.

lithium chromate. CAS: 14307-35-8.
$Li_2CrO_4 \cdot 2HOH$.
Properties: Yellow, crystalline, deliquescent powder; soluble in water and forms a eutectic at −60C; soluble in alcohols.

Hazard: Toxic by ingestion.

Use: Corrosion inhibitor in alcohol-base antifreezes and water-cooled reactors. Oxidizing agent for organic material, especially in the presence of light; heat-transfer medium.

lithium citrate. CAS: 919-16-4.
$Li_3C_6H_5O_7 \cdot 4HOH$.
Properties: White powder or granules, loses four waters at 105C, mp decomposes, soluble in water, slightly soluble in alcohol.

Derivation: Reaction of citric acid with lithium carbonate.

Use: Beverages, pharmaceuticals, dispersion stabilizer (clay deflocculant).

lithium cobaltite. $LiCoO_2$. Dark blue powder, insoluble in water, the compound exhibits both the fluxing property of lithium oxide and the adherence-promoting property of cobalt oxide.

Use: Ceramics.

lithium deuteride. LiD.
Properties: Gray crystals, d 0.906, reacts slowly with moist air, thermally stable to its mp of 680C.

Use: thermonuclear fusion.

See also deuterium.

lithium dichromate. CAS: 13843-81-7.
$Li_2Cr_2O_7 \cdot 2HOH$.
Properties: Yellowish-red, crystalline powder, d 2.34 (30C), mp 130C, deliquescent, soluble in water, forms eutectic at −70C.

Use: Dehumidifying, refrigeration.

lithium ferrosilicon. CAS: 64082-35-5.
Properties: Dark, crystalline, brittle, metallic lumps or powder; evolves flammable gas in contact with moisture; must be kept cool and dry.

Hazard: Flammable, dangerous fire risk.

lithium fluoride. CAS: 7789-24-4. LiF.
Properties: Fine white powder, d 2.635 (20C), mp 842C, bp 1670C, slightly soluble in water, does not react with water at red heat, soluble in acids, insoluble in alcohol.

Derivation: Reaction of hydrogen fluoride with lithium carbonate.

Grade: Guaranteed 98% (min) lithium fluoride, CP, single pure crystals.

Hazard: Strong irritant to eyes and skin. TLV (as F): 2.5 mg/m³ of air.

Use: Welding and soldering flux, ceramics, heat-exchange media, synthetic crystals in infrared and ultraviolet instruments, rocket fuel component, radiation dosimetry. Component of fuel for molten salt reactors, x-ray diffraction.

lithium fluorophosphate. $LiF \cdot Li_3PO_4 \cdot HOH$.
Properties: White crystals.
Derivation: Interaction of lithium fluoride and lithium phosphate.
Use: Ceramics.

lithium grease. A grease using a lithium soap of the higher fatty acids as a base. Water-resistant, stable when heated above its melting point and cooled again.
Use: Aircraft and low temperature service.
See lithium stearate, lubricating grease. Lithium hydroxy-stearate from hydrogenated castor oil is also widely used.

lithium hydride. CAS: 7580-67-8. LiH.
Properties: White, translucent, crystalline mass or powder. Commercial product is light bluish-gray due to minute amount of colloidally dispersed lithium, d 0.82 (20C), mp 680C, decomposition pressure nil at 25C, 0.7 mm at 500C, 760 mm at approximately 850C; decomposed by water forming hydrogen and lithium hydroxide. Insoluble in benzene and toluene, soluble in ether.
Derivation: Reaction of molten lithium with hydrogen.
Grade: 93–95%, based on hydrogen evolution.
Hazard: Flammable, dangerous fire risk, ignites spontaneously in moist air, use dry chemical to extinguish. TLV: 0.025 mg/m³ of air.
Use: Desiccant, source of hydrogen, condensing agent in organic synthesis, preparation of lithium amide and double hydrides, nuclear shielding material, reducing agent.
See "Lith-X."

lithium hydroxide. CAS: 1310-65-2. LiOH.
Properties: Colorless crystals, d 2.54, mp 470C, bp 924C (decomposes) slightly soluble in alcohol, soluble in water, absorbs carbon dioxide and water from air.
Derivation: Causticizing of lithium carbonate, action of water on metallic lithium, or by addition of Li_2O to water.
Hazard: Water solutions are strongly irritant.
Use: Storage battery electrolyte, carbon dioxide absorbent in space vehicles, lubricating greases, ceramics, catalyst, photographic developers, lithium soaps.

lithium hydroxystearate.
$LiOOC(CH_2)_{10}CHOH(CH_2)_5CH_3$.
Properties: White powder, mp 205C, dissolves in hot petroleum oil to form greases. Combustible.
Derivation: From hydrogenated castor oil.
Use: Lubricating greases.

lithium hypochlorite. CAS: 13840-33-0.
LiOCl.
Hazard: Strong oxidant, may ignite organic materials.
Use: Bleach, sanitizing agent.

lithium iodate. $LiIO_3$.
Properties: White powder, d 4.487 (25C), mp 50–60C (transition point from alpha to beta form), soluble in water, insoluble in alcohol, an oxidizing material.

lithium iodide trihydrate. $LiI \cdot 3HOH$.
Properties: White crystals, soluble in water and in alcohol, d 3.48 (25C), mp 72C, loses water (anhydrous) 450C. bp 1171C, extremely hygroscopic.
Derivation: Action of hydriodic acid on lithium hydroxide with subsequent crystallization.
Use: Air conditioning, catalyst in acetal formation, solubilizes lithium in propylamine, nucleonics.

lithium lactate. $LiC_3H_5O_3$.
Properties: White, odorless powder; very soluble in water with practically neutral reaction. Lithium lactate is nonhygroscopic and stable, whereas sodium lactate can be prepared in solution only.
Use: Wherever a dry alkali lactate is required.

lithium manganite. Li_2MnO_3.
Properties: Reddish-brown powder, insoluble in water, extremely stable.
Use: Smelter addition in the manufacturing of frit, as a mill addition, ceramic-bonded grinding wheels.

lithium metaborate dihydrate. $LiBO_2 \cdot 2HOH$.
Properties: White, crystalline powder; soluble in water; mp 840C (anhydrous).
Use: Ceramics (flux in enamel cover coats, increases resistance to torsion), welding and brazing (anhydrous form), nucleonics.

lithium metasilicate. (lithium silicate).
Li_2SiO_3.
Properties: White powder, mp 1201C, d 2.52 (25C), insoluble in water.
Use: Flux in glazes and ceramic enamels, welding rod coating.

lithium methoxide. (lithium methylate).
$LiOCH_3$.
Properties: White, free-flowing powder; soluble in methanol; must be protected from moisture.

Grade: Commercially available as a 10% solution in methanol.

Hazard: Solution very toxic by ingestion.

Use: Strong base and chemical intermediate where water is undesirable.

lithium methylate. See lithium methoxide.

lithium molybdate. Li_2MoO_4.

Properties: White, crystalline compound soluble in water, mp 705C, d 2.66.

Use: Steel coating, petroleum cracking catalyst.

lithium niobate. (lithium metaniobate). $LiNbO_3$.

Properties: Single crystals available, mp 1250C, sizes up to 6 inches long and 1/2 inch in diameter.

Use: Infrared detectors, transducer in laser technology.

lithium nitrate. CAS: 7790-69-4. $LiNO_3$.

Properties: Colorless powder, d 2.38, mp 261C, soluble in water and alcohol.

Derivation: Reaction of nitric acid with lithium carbonate.

Grade: Technical, commercially pure, reagent.

Hazard: Dangerous explosion risk when shocked or heated, strong oxidizing agent.

Use: Ceramics, pyrotechnics, salt baths, heat-exchange media, refrigeration systems, rocket propellant.

lithium nitride. CAS: 26134-62-3. Li_3N.

Properties: Brownish-red crystals of hexagonal structure or fine, free-flowing red powder. Water vapor in moist air causes slow decomposition. Reacts with water giving LiOH and ammonia, insoluble in polyethers, density approximately 1.3 (25C), mp 845C, decomposition pressure not measurable below 1250C.

Hazard: Ignites in air, use dry chemicals to extinguish ("Lith-X").

Use: Nitriding agent in metallurgy, reducing and nucleophilic reagent in organic reactions, source of bound nitrogen in organic reactions.

lithium orthophosphate. See lithium phosphate.

lithium oxide. (lithia). CAS: 12057-24-8. Li_2O.

Properties: White powder, mp 1427C, d 2.023 (25C), a strong alkali, absorbs carbon dioxide and water from air.

Hazard: Water solutions of high concentration are strongly irritant.

Use: Ceramics and special glass formulations, carbon dioxide absorbent (mineral water).

lithium perchlorate. CAS: 7791-03-9. $LiClO_4$.

Properties: Colorless, deliquescent crystals; d 2.429; mp 236C; decomposes at 430C; has more available oxygen than liquid oxygen (on a volume basis); reacts with $4NH_3$ to form an ammoniate. Forms $LiClO_4 \cdot 3HOH$, colorless crystals with d 1.84, mp 75C, soluble in water and alcohol.

Hazard: Oxidizing agent, dangerous fire and explosion risk in contact with organic materials. Irritant to skin and mucous membranes.

Use: Solid rocket propellants.

lithium peroxide. CAS: 12031-80-0. Li_2O_2.

Properties: Fine white powder, mp decomposes, d 2.14 (20C), in closed container no detectable loss of available oxygen. Soluble in water, 8% (20C); anhydrous acetic acid 5.6% (20C); insoluble in absolute alcohol (20C).

Derivation: Addition of hydrogen peroxide to lithium hydroxide.

Hazard: Dangerous fire and explosion risk in contact with organic materials. Strong oxidizing agent.

Use: As supplier of active oxygen, commercial samples have 32.5–34% available oxygen content.

lithium phosphate. (lithium orthophosphate). CAS: 13762-75-9. $2Li_3PO_4 \cdot HOH$.

Properties: White, crystalline powder; soluble in dilute acids; slightly soluble in water; d 2.41.

lithium ricinoleate. $LiOOC_{17}H_{32}OH$.

Properties: White powder, mp 174C, insoluble or limited solubility in most organic solvents. Combustible.

Derivation: Castor oil.

Use: Alcoholysis and ester interchange catalyst.

lithium silicate. See lithium metasilicate.

lithium silicon. CAS: 68848-64-6. An alloy of lithium and silicon.

Properties: Black, shiny lumps or powder with sharp irritant odor; keep cool and dry.

Hazard: Dangerous fire risk, reacts with water to form flammable gases.

lithium stearate. $LiC_{18}H_{35}O_2$.

Properties: White crystals, d 1.025, mp 220C, insoluble in cold and hot water, alcohol and ethyl acetate; forms gels and mineral oils.

Derivation: Reaction of stearic acid with lithium carbonate.

Grade: Grease, cosmetic.

Use: Cosmetics, plastics, waxes, greases, lubricant in powder metallurgy, corrosive inhibitor in petroleum, flatting agent in varnishes and lacquers, high-temperature lubricant.

lithium sulfate.   CAS: 10377-48-7.
  $Li_2SO_4 \cdot HOH$.
Properties: Colorless crystals, d 2.06, mp 130C, soluble in water, insoluble in 80% alcohol. Does not form alums.
Derivation: Reaction of sulfuric acid with lithium carbonate or with spodumene ore.
Grade: Technical and pharmaceutical.
Use: Pharmaceutical products, ceramics.

lithium tetraborate. $Li_2B_4O_7 \cdot 5HOH$.
Properties: White, crystalline powder; mp loses water at 200C; very soluble in water; insoluble in alcohol.
Derivation: Reaction of boric acid with lithium carbonate.
Use: Ceramics, vacuum spectroscopy, metal refining and degassing.

lithium titanate. Li_2TiO_3.   White powder, insoluble in water, has strong fluxing properties in small percentage in titanium-bearing enamels. The insolubility permits its use in vitreous and semi-vitreous glazes.

lithium tri-tert-butoxyaluminohydride.   See lithium aluminum tri-tert-butoxyhydride.

L-Selectride.[541]   TM for lithium tri-sec-butylborohydride, 1.0 molar solution in tetrahydrofuran.
CAS: 38721-52-7. $LiB[CH(CH_3)C_2H_5]_3H$.
Properties: Liquid with mw 190.11, d 0.890, freezing p −17C. Moisture sensitive. Packaged under nitrogen.
Hazard: Pyrophoric, must be handled under inert atmosphere.
Use: Reagent for the stereoselective reduction of ketones. Has been used in prostaglandin synthesis.

lithium tungstate. Li_2WO_4.   White crystals, d 3.71, soluble in water.

lithium vanadate.   (lithium metavanadate).
  $LiVO_3 \cdot 2HOH$.
Properties: Yellowish powder, soluble in water.

lithium zirconate. Li_2ZrO_3.   White powder, insoluble in water, efficient flux in glasses containing zirconium dioxide, recommended as a flux in zirconium-opacified enamels, glazes, and porcelains.

lithium-zirconium silicate. $Li_2O \cdot ZrO_2 \cdot SiO_2$.
White powder, a strong flux in enamels, glazes, and porcelains. It can be used in place of lithium zirconate.

lithocholic acid.   CAS: 434-13-9.
  $C_{24}H_{39}O_2OH$.   A bile acid.
Properties: Crystals in leaflets from alcohol, mp 184–189C, not precipitated by digitonin, freely

soluble in hot alcohol, soluble in ethyl acetate, slightly soluble in glacial acetic acid, insoluble in water, ligroin.
Derivation: From bile and gallstones, from deoxycholic acid or cholic acid.
Use: Biochemical research.

Lithol Red.[293]   TM for one of a group of pigments made by combining Tobias acid and β-naphthol. Available as sodium, barium, and calcium lithols. Poor resistance to sunlight and weathering, generally good resistance to bleeding and to chemicals.
Use: Industrial enamels, toys and dipping enamels, rubber, plastics, etc.

Lithol Rubine.[293]

$OOCC_{10}H_5(OH)N{:}NC_6H_3(CH_3)SO_2OCa$.
TM for the calcium salt of an azo pigment made by diazotizing p-toluidine-m-sulfonic acid and coupling with 3-hydroxy-2-naphthoic acid. Has poor hiding power, good resistance to bleeding and baking, fair lightfastness and alkali resistance.
Use: Paints, plastics, printing inks, cosmetics.

lithopone.   CAS: 1345-05-7.   A white pigment consisting of zinc sulfide, barium sulfate, and zinc oxide; formerly used widely in paints, white rubber goods, paper, white leather etc. It has been largely replaced by titanium dioxide.

"Lith-X."[548]   TM for a graphite-base dry chemical extinguishing agent suitable for fires of lithium, titanium, and zirconium.

litmus.   (lichen blue).
Properties: A blue, amorphous powder (frequently compressed into small cakes or strips ("paper"). Soluble in water, changes color with acidity of solution, red at pH 4.5, blue at pH 8.3.
Derivation: By treating lichens (particularly Variolaria lecanora and V. rocella) with ammonia and potash and then fermenting the mass.
Use: Indicator in analytical chemistry where precision is not required, soil testing.

Little, Arthur D.   (1863–1935) Born in Boston, Little was a pioneer in the field of industrial research and chemical consulting. Originally an authority on paper technology, he established a consulting industrial chemical laboratory in 1886, which has since become a large institution of world-wide reputation, located in Cambridge, MA. It has served as a prototype of many industrially-oriented consulting firms that have become a significant factor in the growth of research in the last half century. It has made significant contributions in such fields as flavors, food chemistry and acceptability, paper chemistry, and rubber chemistry, as well as in corporate management.

liver. A term formerly used to characterize compounds containing sulfides or sulfates, possibly because of the dark brown color, e.g., liver of sulfur, liver of lime, etc. The word is outmoded and should be avoided.
See also livering.

livering. A pronounced increase in the viscosity of a paint or printing ink as a result of which it becomes a coagulated and unusable mass. This may be due to oxidation or to partial polymerization of the components.

lixiviation. See leaching.

LNG. Abbreviation for liquefied natural gas.
See natural gas.

loam. A soil comprised of 25–50% sand, 30–50% silt, and 5–25% clay.
See also soil.

"Lo-Bax."[84] TM for a chlorine sterilizer designed especially for handlers of milk and other dairy and food products who require clear, fast-killing bactericidal solutions. It is a stable, quick-dissolving powder containing 50% available chlorine.

Lobry de Bruyn-van Ekenstein transformation. Isomerization of carbohydrates in alkaline media, considered to embrace both epimerization of aldoses and ketoses and aldose-ketose interconversion.

"Loctite."[477] TM for anaerobic polymers which retain liquid while exposed to air and which automatically harden without heat or catalysts when confined between closely fitted metal parts. Applied to mating surfaces before, during, and after assembly. Available in range of strengths and viscosities for a wide variety of applications.

locust-bean gum. See carob-seed gum.

"Lodex."[245] TM for fine particle permenent magnet material and permanent magnets.

logwood. (hematoxylin). $C_{10}H_{14}O_6 \cdot 3HOH$.
See hematin; dye, natural.

"Lomar."[309] TM for condensed naphthalene sulfonate.
Use: Dispersing agents in paper-making, cement and concrete, gypsum wallboard, emulsion polymerization, dyeing, etc.

London purple. CAS: 8012-74-6. An insecticide containing arsenic trioxide, aniline, lime, and ferrous oxide; insoluble in water.
Hazard: Toxic by ingestion and inhalation.

"Lorox."[23] TM for either a powder or solution containing linuron.

Lossen rearrangement. Conversion of a hydroxamic acid, via a Hofmann-like arrangement, to an amine containing one less carbon than the original hydroxamic acid.

"Lotol."[248] TM for a series of compounded latices, based on all synthetic and natural types.

"Lovibond."[581] TM for a color analysis technique for petroleum products used chiefly in England but never adopted by ASTM.

low-alloy steel. See alloy steel.

low explosive. See explosive, low.

low-melting alloy. See alloy, fusible.

low-pressure resin. See contact resin.

low-soda alumina. Aluminum oxide (Al_2O_3) with less than 0.15% sodium oxide content.
Use: High-grade electric insulator and other ceramic bodies.

LOX. Abbreviation for liquid oxygen, especially when used as a rocket fuel.

LPG. Abbreviation for liquefied petroleum gas.

Lr. Symbol for element lawrencium.

LSD. Abbreviation for lysergic acid diethylamide.

"L-70 Series Surfactants."[214] TM for a class of organo-silicones.
Use: Wetting agents, emulsifiers and coemulsifiers, penetrants, foamers and foam stabilizers, defoamers, lubricants, leveling agents, antistats, and textile softeners.

LTH. Abbreviation for luteotropic hormone.
See luteotropin.

Lu. Symbol for lutetium.

lube oil additive. A chemical added in small amounts to lubricating oils to impart special qualities, such as low pour point when chlorinated hydrocarbons are added. Other special properties are:

low viscosity index . . . . . . .butene polymers
detergent and
  suspensoid properties . . . .metallic stearate soaps
oxidation stability . . . . . . .calcium stearate
reduced foaming tendency .silicone compounds
resistance to high operating
  temperatures . . . . .phosphorus penta-
                    sulfide, zinc dithi-
                    ophosphate

lubricant, solid.   A material having a characteristic crystalline habit which causes it to shear into thin, flat plates, which readily slide over one another and thus produce an antifriction or lubricating effect, for example, mica, graphite, molybdenum disulfide, talc, boron nitride.

lubricant, synthetic.   Any of a number of organic fluids having specialized and effective properties that are required in cases where petroleum-derived lubricants are inadequate. Each type has at least one property not found in conventional lubricants. Though their cost is much higher, they can be used over a wide range of temperatures and are stable to heat and oxidation. The major types are polyglycols (hydraulic and brake fluids), phosphate esters (fire-resistant), dibasic acid esters (aircraft turbine engines), chlorofluorocarbons (aerospace), silicone oils and greases (electric motors, antifriction bearings), silicate esters (heat-transfer agents and hydraulic fluids), neopentyl polyol esters (turbine engines), and polyphenyl ethers (excellent heat and oxidation resistance, but poor performance at low temperatures). An unusual property of synthetic lubricants is their exceptional resistance to ionizing radiation.

lubricating grease.   A mixture of a mineral oil or oils with one or more soaps. The most common soaps are those of sodium, calcium, barium, aluminum, lead, lithium, potassium, and zinc. Oils thickened with residuum, petrolatum, or wax may be called greases. Some form of graphite may be added. Greases range in consistency from thin liquids to solid blocks and in color from transparent to black. The specifications for a grease are determined by the speed, load, temperature, environment, and metals in the desired application. Texture of grease may be smooth, buttery, ropy or stringy, fibrous, spongy, or rubbery. The texture does not necessarily indicate the viscosity of the grease, but is related to the formulation and methods of manufacture.
See also lubricating oil.

lubricating oil.   (lube oil).   A selected fraction of refined mineral oil used for lubrication of moving surfaces, usually metallic and ranging from small precision machinery (watches) to the heaviest equipment. Lubricating oils usually have small amounts of additives to impart special properties such as viscosity index and detergency. They range in consistency from thin liquids to grease-like substances. In contrast to lubricating greases, lube oils do not contain solid or fibrous materials.
See also porpoise oil; lubricant, synthetic; extreme-pressure additive (2); lube oil additive.

lubrication.   The introduction of a substance of low viscosity between two adjacent solid surfaces, one of which is in motion (bearing). From an engineering point of view, the chemical nature of the substance is not of critical importance. Thus, materials as diverse as air, water, and molasses could theoretically be used as lubricants under appropriate conditions. Air and water have been used, as well as some solids such as graphite, but in general oils, fats, and waxes are utilized. The ability of a substance to act as a lubricant is sometimes called lubricity.

"Lubricin" N-1[202].   TM for a low viscosity ricinoleate derivative that increases the oiliness and wetting power of mineral oils, decreases corrosion, and exhibits detergent action on tar, varnish, and carbon deposits.
Use: Additive for lubricating oils, motor fuels and cutting oils.

"Lubricin" V-1[202].   TM for a lubricant which facilitates the processing of rigid vinyl plastics.

lubricity.   See lubrication.

"Lucalox."[245]   TM for a transparent, polycrystalline alumina, used in special lamp bulbs to produce a golden white light considered superior for industrial, commercial, outdoor, and consumer lighting applications.

luciferase.   An enzyme occurring only in fireflies. See luciferin.

luciferin.   An albumin present in some life forms (notably fireflies) which, under the influence of the enzyme luciferase, exhibits bioluminescence. When luciferin (along with luciferase) comes in contact with adenosine triphosphate, present in all living cells, a chemical reaction occurs which causes luminescence. By correlating its intensity with the amount of adenosine triphosphate present, a means of measuring bacterial growth is obtained. Applications in bio-medical technology are envisaged for this so-called "luminescence biometer."

Lucifer Yellow CH.[541]   CAS: 67769-47-5.
$C_{13}H_9Li_2N_5O_9S$.   The dye spreads rapidly
through the injected cell and binds effectively
to tissue by a variety of fixatives. The movement
from cell to cell is termed dye-coupling.
Properties: Fluffy orange hygroscopic powder
with mw 457.25.
Use: Highly fluorescent dye for marking nerve
cells.

"Lucite."[28]   TM for acrylic resins consisting of a
series of polymeric esters of methacrylic acid,
$CH_2C(CH_3)COOR$, in which R is methyl, ethyl,
n-butyl, isobutyl, or combinations of these alkyl
groups. These are water-white, transparent, thermoplastic resins in granular or solution forms,
used in lacquers, coatings, adhesives, modifiers
for other resins. "Lucite" is also a TM for an
acrylic monomer, methyl methacrylate, for
acrylic resins in the form of injection molding
and extrusion powders; for acrylic resin dental
materials; and for acrylic syrup, a liquid methyl
methacrylate for use in translucent and decorative panels.
See acrylic resin.

"Ludox."[28]   TM for series of aqueous colloidal silica solutions.
Use: Floor polishes, paints, adhesives, paper coatings, catalyst supports, latex products, textile
cleaning treatments, photosensitized paper, binder for inorganic fibrous materials, reactant for
synthetic silicates.

"Lugatol."[440]   TM for anionic dyes having good
tinctorial and fastness properties for all kinds
of leather.

Lugol's solution.   An aqueous solution containing
5 g iodine and 10 g potassium iodide per 100
mL water. Grade USP.
Use: Medicine and pharmaceuticals.

"Lumatex."[440]   TM for a series of mineral and
organic pigment dyestuffs which can be fixed
on all types of textile fibers with suitable binders.

"Luminal."[162]   TM for phenobarbital.
See also barbiturate.

luminescence.   The emission of visible or invisible
radiation unaccompanied by high temperature
by any substance as a result of absorption of
exciting energy in the form of photons, charged
particles or chemical change. It is a general term
which includes both fluorescence and phosphorescence. Special types of chemiluminescence, bioluminescence, electroluminescence,
photoluminescence, and triboluminescence.

Common examples are light from the firefly,
fluorescent lamp tubes, and television screens.
See also fluorescence, phosphorescence, phosphor.

LUMO.   Acronym for lowest unoccupied molecular orbital.

"Luperco."[154]   TM for a series of organic peroxides including aromatic diacyl, ketone, and alkyl
peroxides.
Hazard: See peroxide (1).
Use: Crosslinking polyethylene, curing synthetic
and natural rubber, vinyl polymerization initiator, catalyst in polyester and vinyl monomers,
mild organic oxidizing agents.

"Luperox."[154]   TM for a group of organic peroxides.
(2,5-2,5; 2,5-dimethylhexane-2,5-dihydroperox-
ide).
6 Technical bis(1-hydroxycyclohexyl)peroxide.
118 2,5-dimethylhexane-2,5-diperoxybenzoate.
Hazard: See peroxide (1).
Use: Catalyst for polymerization of polyesters and
vinyls, elastomer vulcanization.

2,6-lupetidine.   See 2,6-dimethylpiperidine.

"Luran S."[440]   (ASA polymer).   TM for
acrylic ester-modified styrene-acrylonitrile terpolymer. A thermoplastic material claimed to
be excellent for outdoor use. Highly resistant to
sunlight and weathering.

Lurgi process.   See gasification.

luster.   The appearance of the surface of a substance in reflected light. The term is used particularly in describing minerals. Types of luster are:
(a) metallic, like metals or the mineral pyrite;
(b) vitreous, like glass or quartz; (c) adamantine,
exceedingly brilliant, like diamond; (d) resinous,
like resin, or sphalerite; (e) dull, not bright or
shiny, like chalk.
See also delustrant.

"Lustralite."[36]   TM for sulfonamide formaldehyde resins used as coating vehicles.

"Lustran."[58]   TM for ABS and SAN molding and
extruding compounds.

"Lustrasol."[36]   TM for acrylic and acrylic-modi-
fied alkyd solutions.

"Lustrex."[58]   TM for styrene molding compound
and extruding and calendering material.

"Lustrone."[23]   TM for plasticized nitrocellulose
lacquers tinted with dyes.
Use: Transparent-color effects on leather.

lutein. A yellow pigment isolated from the corpus luteum and found in body fats and egg yolks. It is a carotenoid and is similar to or identical with xanthophyll.

luteotropin. (adenohypophyseal luteotropin; prolactin; lactogenic hormone; LTH). One of the hormones secreted by the anterior lobe of the pituitary gland. It aids in causing growth of the mammary gland and initiates milk secretion by the mammary gland, it also influences the activity of the corpus luteum, including the secretion of progesterone.
Properties: Crystalline protein, mw approximately 33,300, almost insoluble in water, soluble in dilute acids and acidified methanol and ethanol.

lutes. Cements and adhesives composed of oxides, clays, and silicas.

lutetia. See lutetium oxide. See also rare earths.

lutetium. CAS: 7439-94-3. Lu. Metallic element, atomic number 71, group IIIB of the periodic table, a lanthanide rare-earth, aw 174.97, valence = 3, it has one natural radioactive isotope (Lu-176) with half-life of 2.2×10^{10} years.
See rare-earth metals.
Properties: Metallic luster, soft and ductile, d 0.849, mp 1652C, bp 3327C, reacts slowly with water, soluble in dilute acids, difficult to isolate.
Occurrence: Monazite (approximately 0.003%) (India, Brazil, Africa, Australia, US).
Derivation: Reduction of the fluoride or chloride with calcium.
Grade: Regular, high purity (ingots, lumps).
Use: Nuclear technology.

lutetium chloride. $LuCl_3 \cdot xHOH$ (anhydrous).
Properties: D 3.98, mp 905C, water-soluble. Purity: Up to 99.9% lutetium salts.

lutetium fluoride. $LuF_3 \cdot 2HOH$ (anhydrous).
Properties: Mp 1182C, bp 2200C, insoluble in water. Purity: Up to 99.9% lutetium salts.
Hazard: Irritant. TLV (as F): 2.5 mg/m³ of air.

lutetium nitrate. $Lu(NO_3)_3 \cdot 6HOH$.
Purity: Up to 99.9% lutetium salts.
Hazard: Fire and explosion risk.

lutetium oxide. (lutetia). Lu_2O_3.
Properties: White solid, slightly hygroscopic, mp approximately 2500C, d 9.42, absorbs water and carbon dioxide.
Derivation: Monazite sand.
Grade: Up to 99.9% purity.

lutetium sulfate. $Lu_2(SO_4)_3 \cdot 8HOH$.
Properties: Soluble in water. Purity: Up to 99.9% lutetium salts.

2,6-lutidine. (2,6-dimethylpyridine).
$NC(CH_3)CHCHCHCCH_3$.
Properties: Colorless, oily liquid; peppermint odor. D 0.932, bp 143C, fp −6.6C, refr index 1.4973 (20C), derived from coal tar.
Hazard: Flammable.
Use: Pharmaceuticals, resins, dyestuffs, rubber accelerators, insecticides. Also available as 2,3-, 2,4-, 2,5-, 3,4-, 3,5-isomers.

"Lutonal."[440] TM for vinyl ether polymers, solid and in solution.

"Luxol."[28] TM for a series of spirit and lacquer-soluble dyes.
Use: Lacquers, wood stains, stamp inks, ball point pen inks.

"LX-685."[21] TM for a heat-reactive resin used in the manufacture of ready-mixed aluminum paints, grease- and gasoline-resistant coatings, floor and deck enamels and concrete curing compounds.

"Lyamine."[123] TM for lysine for use as an animal feed supplement.

"Lycedan."[91] TM for a brand of adenosine-5-phosphoric acid for medicinal use.

lycopene. (carotene). CAS: 502-65-8.
$C_{40}H_{56}$. A long-chain hydrocarbon with 13 double bonds of which 11 are conjugated. Combustible.
Properties: Red crystals, mp 175C, soluble in chloroform and carbon disulfide, slightly soluble in alcohol, insoluble in water.
Source: The main pigment of tomato, paprika, rose hips, etc.

lycopodium. Spores from *Lycopodium clavatum*, a fungus native to North America and Europe.
Properties: A flammable yellowish powder.
Hazard: Flammable, protect from open flame.
Use: In the pharmaceutical industry as a coating, in flashlight powders and pyrotechnics.

"Lycra."[28] TM for a spandex fiber in the form of continuous monofilaments. Available in yarn with denier from 40–2240.
Properties: Tensile strength 0.8 gpd, d 1.21, break elongation 550%, moisture regain 1.3%, mp 230C, tensile recovery 95% from 50% elongation, soluble in dimethyl formamide (hot).
Hazard: Combustible, but self-extinguishing.
Use: Foundation garments, swim wear, surgical hose and other elastic products.

lye. See sodium hydroxide and potassium hydroxide.

"Lykopon."[23]   TM for sodium hydrosulfite.

lyocratic.   See lyophilic.

lyophilic.   Characterizing a material which readily goes into colloidal suspension in a liquid; if into water, it is called hydrophilic. The colloid is stabilized by the formation of an adsorbed layer of molecules of the dispersing medium about the suspended particles. Systems of this type are said to be lyocratic. Examples: glue; gelatin; milk fat particles.

lyophilization.   See freeze-drying.

lyophobic.   Characterizing a material which exists in the colloidal state but with a tendency to repel liquids; if the liquid is water, the material is called hydrophobic. Such colloids are generally stabilized by the adsorption of ions and coagulate when the charge is neutralized. Examples: colloidal gold, colloidal arsenic sulfide.

Lys.   Abbreviation for lysine.

lysergic acid.   CAS: 82-58-6. $C_{16}H_{16}N_2O_2$.
Structure based on a condensed four-ring nucleus.
Properties: Crystals (with 1–2 molecules of water) in plates from water, mp 240C (decomposes). It is amphoteric, moderately soluble in pyridine, slightly soluble in water and neutral organic solvents, soluble in alkaline and acid solutions.
Derivation: Alkaline hydrolysis of ergot alkaloids, organic synthesis.
Use: Medical research, synthesis of ergonovine.

D-lysergic acid diethylamide.   (LSD).
CAS: 50-37-3. $C_{20}H_{25}N_3O$.   Synthetic derivative of lysergic acid. Only the D-form is active. Tasteless, colorless, odorless.
Hazard: A strong hallucinatory and habit-forming drug with possible mutagenic effects. Manufacture and sale are under legal restraint.

lysine.   (α,ϵ-diaminocaproic acid; Lys).
CAS: 56-87-1. $NH_2(CH_2)_4CH(NH_2)COOH$.
An essential amino acid.
Properties: Colorless crystals, soluble in water, slightly soluble in alcohol, insoluble in ether. Optically active: d(-)-lysine, mp 224C with decomposition; l(+)-lysine mp 224C with decomposition.
Derivation: Extraction of natural proteins, synthetically by fermentation of glucose or other carbohydrates and by synthesis from caprolactam.
Use: Biochemical and nutritional research, pharmaceuticals, culture media, fortification of foods and feeds (wheat flour), nutrient and dietary supplement, animal feed additive. Commercially available as dl- and l-lysine monohydrochloride.

lysozyme.   CAS: 9001-63-2.   An antibiotic enzyme found in egg white. By hydrolyzing certain sugar linkages in glycoproteins, it can dissolve the mucopolysaccharides found in the walls of certain bacteria and hence acts as a mild antiseptic. Its 3-dimensional structure has been determined by x-ray crystallography. It is a molecule containing approximately 2200 atoms composing 129 amino acid units strung together and intricately folded (mw 14,500).

lyxoflavine.   CAS: 13123-37-0. $C_{17}H_{20}N_4O_6$.
Properties: Yellowish crystals, mp 284C (decomposes), optically active, almost insoluble in water and alcohol, soluble in alkaline solutions. It is a chemical analog of riboflavin.
Use: Growth promoter for agricultural products.

M

m. Abbreviation for meter.

M. Abbreviation for molar, used to characterize the concentration of a solution. A molar solution contains one mole of a substance in one liter of solution.

m-. Abbreviation for meta-.

μm. Abbreviation for micrometer.

mμ. Abbreviation for millimicron.

MAC. Abbreviation for methyl allyl chloride.

"Mace." ("Chemical Mace"). TM for a riot control gas dispersed as an aerosol.
See chloroacetophenone.

macerate. To soften or break up a fibrous substance by long soaking in water at or near room temperature, often accompanied by mechanical action, as in the preparation of paper stock in the beater. In the plastics industry, to comminute a fabric so that it can be used as a filler in a plastics composition. The term is also used in pharmacies to describe a method of preparing medicinal compositions.

M acid. (1-amino-5-naphthol-7-sulfonic acid). $C_{10}H_5NH_2OHSO_3H$.
Properties: Gray needles, slightly soluble in cold water, soluble in hot water and alcohol.
Use: Azo dye intermediate.

macromolecule. A molecule, usually organic, comprised of an aggregation of hundreds or thousands of atoms. Such giant molecules are generally of two types. (1) Individual entities (compounds) that cannot be subdivided without losing their chemical identity. Typically these are proteins, many of which have molecular weights running into the millions. (2) Combinations of repeating chemical units (monomers) linked together into chain or network structures called polymers; each monomer has the same chemical constitution as the polymer, e.g., isoprene (C_5H_8) and polyisoprene (C_5H_8)x. Synthetic elastomers (plastics) are typical of this kind of macromolecule; cellulose is the most common example found in nature. Most macromolecules are in the colloidal size range.
See also polymer, high; protein; colloid chemistry.

macrose. See dextran.

madder. A natural dyestuff.
See alizarin, lake, dye, natural.

Maddrell's salts. (IMP). Insoluble sodium metaphosphate, $NaPO_3$-II and $NaPO_3$-III.
See sodium metaphosphate.

Madelung synthesis. Formation of indole derivatives by intramolecular cyclization of an N- (2-alkylphenyl)alkanamide by a strong base at high temperature.

magenta. See fuchsin.

mag-lith. A magnesium-lithium alloy used as a structural metal in space vehicles.

magma. (1) In medicine, a class of preparations in which finely divided, freshly precipitated, insoluble, inorganic hydroxides are suspended in water to form a viscous, opaque mixture which may settle out on standing. Magmas of bismuth, magnesium, and iron are used, commonly called milk of bismuth, milk of magnesia, etc. (2) In geology, a molten mass within the earth's crust (e.g., lava). The source of igneous rock.

magnalium. An alloy of aluminum and magnesium.

magnesia. Magnesium oxide that has been specially processed.
See magnesium oxide.

magnesia alba. See magnesium carbonate; magnesium carbonate, basic.

magnesia-alumina. $MgO \cdot Al_2O_3$. A synthetic spinel.

magnesia, burnt. See magnesite, dead-burned.

magnesia, calcined. See magnesite, caustic-calcined.

magnesia, caustic-calcined. See magnesite, caustic-calcined.

magnesia-chromia. $MgO \cdot Cr_2O_3$. A synthetic spinel.

magnesia, dead-burned. See magnesite, dead-burned.

magnesia, fused.   Used as a refractory and to handle electricity at high temperatures.
See "Magnorite."

magnesia, lightburned.   A special high-purity magnesium oxide.

magnesite.   (natural magnesium carbonate).
($MgCO_3$).   The term magnesite is loosely used as a synonym for magnesia as are also the terms caustic-calcined magnesite, dead-burned magnesite, and synthetic magnesite.
Hazard: A nuisance particulate.
Properties: White, yellowish, grayish-white, or brown crystalline solid; d 3–3.12; Mohs hardness 3.5–4.5.
Occurrence: US (California, Washington, Nevada), Austria, Greece.
Use: To make the various grades of magnesium oxide, to produce carbon dioxide, refractory.
See also magnesium carbonate and following entries.

magnesite, burnt.   See magnesite, dead-burned.

magnesite, caustic-calcined.   (caustic-calcined magnesia; calcined magnesite; calcined magnesia). Principally magnesia (magnesium oxide) MgO).   The product obtained by firing magnesite or other substances convertible to magnesia upon heating at some temperature below 1450C so that some carbon dioxide is retained (2–10%) and the magnesium oxide displays adsorptive capacity or activity.
Grade: Technical, chemical, synthetic rubber, USP (light, medium light, heavy).
Use: Magnesium oxychloride and oxysulfate cements, 85% magnesia insulation, rubber (reinforcing agent, accelerator), uranium processing, chemical processing, rayon, refractories, paper pulp, acid-neutralizing fertilizers, welding rod coatings, fillers, glass constituents, abrasives.
See also magnesium oxide.

magnesite, dead-burned.   (burnt magnesia; dead-burned magnesia; refractory magnesia; burnt magnesite; (magnesium oxide), MgO). The granular product obtained by burning (firing) magnesite or other substances convertible to magnesia upon heating above 1450C long enough to form granules suitable for use as a refractory (ASTM). Synthetic magnesium hydroxide or chloride is sometimes used instead of magnesite as a source.
Grade: 85–87% (from magnesite ores); 97–99% (from sea water and brines).
Use: Refractories, as grains or basic brick, the latter especially in open hearth furnaces for steel, furnaces for nonferrous metal smelting, and in cement and other kilns.
See also magnesium oxide.

magnesite, synthetic.   Magnesium oxide, MgO, as obtained from sea water, sea water bitterns, or well brines. The preliminary product is usually magnesium hydroxide or chloride, which is then heated, or sometimes treated with steam and heated in the case of the chloride, to obtain the oxide. Synthetic magnesite constitutes the purer grades of dead-burned magnesite.

magnesium.   CAS: 7439-95-4.   Mg.
Metallic element of atomic number 12; Group IIA of the Periodic Table; aw 24.305; valence = 2; 3 isotopes.
Properties: Silvery, moderately hard, alkaline-earth metal; readily fabricated by all standard methods; lightest of the structural metals; strong reducing agent; electrical conductivity similar to aluminum. D 1.74, mp 650C, bp 1107C, soluble in acids, insoluble in water. Magnesium is the central element of the chlorophyll molecule; it is also an important component of red blood corpuscles.
Sources: Magnesite and dolomite; sea water and brines.
Derivation: (a) Electrolysis of fused magnesium chloride (Dow sea water process), (b) reduction of magnesium oxide with ferrosilicon (Pidgeon process).
Forms available: Ingots, bars, fine powder (up to 99.6% pure), sheet and plate, rods, tubing, ribbon, flakes.
Hazard: (solid metal) Combustible at 650C. (powder, flakes, etc.): Flammable, dangerous fire hazard. Use dry sand or talc to extinguish.
Use: Aluminum alloys for structural parts, die-cast auto parts, missiles, space vehicles; powder for pyrotechnics and flash photography, production of iron, nickel, zinc, titanium, zirconium; antiknock gasoline additives; magnesium compounds and Grignard syntheses; cathodic protection; reducing agent; desulfurizing iron in steel manufacture; precision instruments; optical mirrors; dry and wet batteries.

magnesium acetate.   CAS: 142-72-3.   (a) $Mg(OOCCH_3)_2$ or (b) $Mg(OOCCH_3)_2 \cdot 4HOH$.
Properties: Colorless, crystalline aggregate or monoclinic crystals; acetic acid odor; (a) mp 323C, d 1.42; (b) mp 80C, d 1.45; soluble in water and dilute alcohol.
Derivation: Interaction of magnesium carbonate and acetic acid.
Use: Dye fixative in textile printing, deodorant, disinfectant, and antiseptic.

magnesium acetylacetonate. $Mg(C_5H_7O_2)_2$.
Crystalline powder, slightly soluble in water, resistant to hydrolysis, a chelating nonionizing compound.

magnesium amide.   CAS: 7803-54-5.
  $Mg(NH_2)_2$.
  Properties: Whitish to gray crystals, d 1.40, decomposes when heated.
  Derivation: Reaction of magnesium with ammonia under elevated pressure.
  Hazard: A pyrophoric material igniting in air at room temperature. Evolves ammonia on vigorous reaction with water.
  Use: Catalyst for polymerization.

magnesium ammonium orthophosphate.
  See magnesium ammonium phosphate.

magnesium ammonium phosphate.   (magnesium ammonium orthophosphate).
  $MgNH_4PO_4 \cdot 6HOH$.
  Properties: White powder, d 1.71, mp decomposes to magnesium pyrophosphate, $Mg_2P_2O_7$, soluble in acids, insoluble in alcohol and water.
  Derivation: By the interaction of solutions of a magnesium salt and ammonium phosphate.
  Use: Fire retardant for fabrics, fertilizer.

magnesium arsenate.   (arsenic acid, magnesium salt).   CAS: 10103-50-1.
  $Mg_3(AsO_4)_2 \cdot xH_2O$.
  Properties: White powder which, when pure, is insoluble in water. Technical material is highly hydrated and made from magnesium carbonate and arsenic acid.
  Hazard: Toxic by ingestion and inhalation.
  Use: Insecticide.

magnesium benzoate.   CAS: 553-70-8.
  $Mg(C_7H_5O_2)_2 \cdot 3HOH$.
  Properties: White, crystalline powder; loses 3 HOH at 110C, mp approximately 200C, soluble in water and alcohol.

magnesium biphosphate.   See magnesium phosphate, monobasic.

magnesium borate.   CAS: 13703-82-7.
  $3MgO \cdot B_2O_3$ (orthoborate) or $Mg(BO_2)_2 \cdot 8HOH$ (metaborate).
  Properties: Transparent, colorless crystals or white powder; soluble in alcohol, acetic acid, and inorganic acids; slightly soluble in water.
  Derivation: By heating magnesium oxide, boric anhydride.
  Use: Preservative, antiseptic, fungicide.

magnesium borocitrate.
  $Mg(BO_2)_2 \cdot Mg_3(C_6H_5O_7)_2 \cdot 14HOH$.
  Properties: White powder or small, white, lustrous scales; soluble in water.
  Derivation: By mixing magnesium borate and magnesium citrate.

magnesium boron fluoride.
  Grade: Technical.
  Hazard: Strong irritant. TLV (as F): 2.5 mg/m³ of air.
  Use: Metal flux.

magnesium bromate.   CAS: 7789-36-8.
  $Mg(BrO_3)_2 \cdot 6HOH$.
  Properties: White crystals or crystalline powder, soluble in water, insoluble in alcohol, d 2.29, mp loses 6HOH at 200C, bp decomposes.
  Derivation: By adding magnesium sulfate to a solution of barium bromate.
  Hazard: Dangerous fire risk in contact with organic materials.
  Use: Analytical reagent, oxidizing agent.

magnesium bromide.   CAS: 7789-48-2.
  $MgBr_2 \cdot 6HOH$.
  Properties: Colorless, very deliquescent crystals; bitter taste. Soluble in water, slightly soluble in alcohol, d 2.00, mp 172C, mp (anhydrous) 700C.
  Derivation: Reaction of hydrobromic acid with magnesium oxide and subsequent crystallization.
  Use: Organic syntheses, medicine (sedative).

magnesium calcium chloride.   See calcium magnesium chloride.

magnesium carbonate. $MgCO_3$.   The term magnesium carbonate is generally reserved for the synthetic pure variety. The naturally occurring material is called magnesite.
  Properties: Light, bulky, white powder. Bulk d approximately 4 lb/ft³, d about 3.0, decomposes 350C, refr index about 1.52, soluble in acids, very slightly soluble in water, insoluble in alcohol. Noncombustible.
  Derivation: (1) Mined as natural material, (2) carbonation of magnesium oxide or $Mg(OH)_2$ with CO_2, (3) reaction of a soluble magnesium salt solution with sodium carbonate or bicarbonate.
  Grade: Technical, NF, FCC.
  Use: Magnesium salts, heat insulation and refractory, rubber reinforcing agent, inks, glass, pharmaceuticals, dentifrices and cosmetics, free-running table salts, antacid, making magnesium citrate, filtering medium. Used in foods as drying agent, color retention agent, anticaking agent, carrier.

magnesium carbonate, basic.   (magnesia alba).
  Various formulas are given and may all be possible because of the method of derivation.
  A typical formula is $Mg(OH)_2 \cdot 3MgCO_3 \cdot 3HOH$.
  Properties and uses are almost identical with those listed above.

Derivation: Precipitation from magnesium salt solution.
See magnesium carbonate.

magnesium chlorate.   CAS: 10326-21-3.
$Mg(ClO_3)_2 \cdot 6HOH$.
Properties: White powder, bitter taste, very hygroscopic, soluble in water, slightly soluble in alcohol, d 1.8, mp 35C (decomposes at 120C).
Hazard: Dangerous fire risk in contact with organic materials, strong oxidizing agent.
Use: Defoilant, desiccant.

magnesium chloride.   CAS: 7786-30-3.
(a) $MgCl_2$, (b) $MgCl_2 \cdot 6HOH$.
Properties: Colorless or white crystals, deliquescent, d (a) 2.32, (b) 1.56, mp (a) 708C, (b) loses 2HOH at 100C, if heated rapidly melts at 116–118C, bp (a) 1412C, (b) decomposes to oxychloride, soluble in water and alcohol.
Derivation: Action of hydrochloric acid on magnesium oxide or hydroxide, especially the latter when precipitated from sea water or brines (Great Salt Lake).
Method of purification: Recrystallization.
Grade: Technical (crystals, fused, flakes, granulated), CP.
Hazard: Toxic by ingestion.
Use: Source of magnesium metal, disinfectants, fire extinguishers, fireproofing wood, magnesium oxychloride cement, refrigerating brines, ceramics, cooling drilling tools, textiles (size, dressing and filling of cotton and woolen fabrics, thread lubricant, carbonization of wool), paper manufacture, road dust-laying compounds, floor sweeping compounds, flocculating agent, catalyst.

magnesium chromate. $MgCrO_4 \cdot 5HOH$.
Properties: Small, readily soluble, yellow crystals.
Use: Since it does not produce a fusible alkaline residue when thermally decomposed, it is used as a corrosion inhibitor in the water coolant of gas turbine engines. Insoluble basic magnesium chromates also are available. Their potential applications are in the treatment of light metal surfaces.
Hazard: Toxic by ingestion.

magnesium citrate, dibasic.   (acid magnesium citrate).   CAS: 144-23-0.
$MgHC_6H_5O_7 \cdot 5HOH$.
Properties: White or slightly yellow, odorless granules or powder, soluble in water, insoluble in alcohol.
Derivation: Reaction of citric acid and magnesium hydroxide or carbonate.
Use: Laxative, dietary supplement.

magnesium dichromate. $MgCr_2 \cdot 6HOH$.
Properties: Characterized by high solubility in water. It is an orange-red, deliquescent, crystalline hydrate.
Use: Potential applications are in formulations for corrosion prevention and metal treatment. Noncombustible.

magnesium dioxide.   See magnesium peroxide.

magnesium fluoride.   (magnesium flux).
CAS: 7783-40-6. MgF_2.
Properties: White crystals, exhibits fluorescence by electric light, soluble in nitric acid, insoluble in alcohol and water, d 3.15, mp 1263C, bp 2239C. Noncombustible.
Derivation: By adding sodium fluoride or hydrofluoric acid to a solution of magnesium salt.
Grade: Technical, CP, single crystals.
Hazard: Strong irritant. TLV (as F): 2.5 mg/m³ of air.
Use: Ceramics, glass, single crystals for polarizing prisms, lenses and windows.

magnesium fluosilicate.   (magnesium silicofluoride).   CAS: 18972-56-0.
$MgSiF_6 \cdot 6HOH$.
Properties: White, efflorescent, crystalline powder; d 1.788; decomposes 120C; soluble in water.
Derivation: By treating magnesium hydroxide or carbonate with hydrofluosilicic acid.
Grade: Technical (crystals, solution).
Hazard: Strong irritant. TLV (as F): 2.5 mg/m³ of air.
Use: Ceramics, concrete hardeners, waterproofing, mothproofing, laundry sour, magnesium casting.

magnesium flux.   See magnesium fluoride.

magnesium formate.   CAS: 557-39-1.
$Mg(CHO_2)_2 \cdot 2HOH$.
Properties: Colorless crystals, soluble in water, insoluble in alcohol and ether. Combustible.
Derivation: Action of formic acid on magnesium oxide.
Use: Analytical chemistry.

magnesium gluconate. $Mg(C_6H_{11}O_7)_2 \cdot 2HOH$.
Properties: Odorless, almost tasteless, white powder or fine needles; soluble in water. Combustible.
Derivation: Magnesia or magnesium carbonate dissolved in gluconic acid.
Grade: Pharmaceutical.
Use: Medicine, vitamin tablets.

magnesium glycerophosphate.   (magnesium glycerinophosphate).
$CH_2(OH)CH(OH)CH_2OP(O)O_2Mg$.
Properties: Colorless powder, soluble in water, insoluble in alcohol.

Derivation: Action of glycerophosphoric acid on magnesium hydroxide.
Use: Stabilizer for plastics.

magnesium hydrogen phosphate.   See magnesium phosphate, dibasic.

magnesium hydroxide.   (magnesium hydrate in aqueous suspension; milk of magnesia; magnesia magma).   CAS: 1309-42-8. $Mg(OH)_2$.
Properties: White powder, odorless, soluble in solution of ammonium salts and dilute acids, almost insoluble in water and alcohol, d 2.36, mp decomposes at 350C. Noncombustible.
Derivation: Precipitation from a solution of a magnesium salt by sodium hydroxide, precipitation from sea water with lime. It occurs naturally as brucite.
Grade: Technical, NF, FCC.
Use: Intermediate for obtaining magnesium metal, sugar refining, medicine (antacid, laxative), residual fuel oil additive, sulfite pulp, uranium processing, dentrifices, in foods as drying agent, color retention agent, frozen desserts.

magnesium lauryl sulfate. $Mg(OSO_3C_{12}H_{25})_2$.
Properties: Pale yellow liquid with a mild odor; soluble in methanol, acetone, and water; insoluble in kerosene. Combustible.
Derivation: Sulfonation of lauryl alcohol and interaction with a magnesium salt.
Use: Surfactant and anionic detergent, foaming, wetting, and emulsifying agent.

magnesium lime.   Same as magnesium limestone. See limestone.

magnesium limestone.   See limestone.

magnesium methoxide.   (magnesium methylate). $(CH_3O)_2Mg$.
Properties: Colorless, crystalline solid; decomposes on warming.
Derivation: Reaction of magnesium and methanol.
Use: Dielectric coating, crosslinking agent to form stable gels, catalyst.

magnesium methylate.   See magnesium methoxide.

magnesium molybdate. $MgMoO_4$.   Crystalline powder, absolute density 2.8, mp approximately 1060C, soluble in water.
Use: Electronic and optical applications.

magnesium nitrate.   CAS: 10377-60-3. $Mg(NO_3)_2 \cdot 2HOH$.
Properties: White crystals, soluble in water and alcohol, deliquescent, d 1.45, mp 95–100C, decomposes at 330C.

Derivation: Action of nitric acid on magnesium oxide with subsequent crystallization.
Hazard: Dangerous fire and explosion risk in contact with organic materials, strong oxidizing agent.
Use: Pyrotechnics, concentrated nitric acid.

magnesium oleate.   CAS: 1555-53-9. $Mg(C_{18}H_{33}O_2)_2$.
Properties: Yellowish mass; soluble in linseed oil, hydrocarbons, alcohol, and ether; insoluble in water. Combustible.
Derivation: Interaction of magnesium chloride and sodium oleate.
Use: Varnish driers, in dry-cleaning solvents (to prevent spontaneous ignition), emulsifying agent, lubricant for plasticizers.

magnesium orthophosphate.   See magnesium phosphate.

magnesium oxide.   (magnesia).
CAS: 1309-48-4.   MgO.   Two forms are produced, one a light fluffy material prepared by a relatively low-temperature dehydration of the hydroxide, the other a dense material made by high-temperature furnacing of the oxide after it has been formed from the carbonate or hydroxide.
See also periclase.
Properties: White powder, either light or heavy depending upon whether it is prepared by heating magnesium carbonate or the basic magnesium carbonate, d approximately 0.36 (varies), mp 2800C, bp 3600C, slightly soluble in water, soluble in acids and ammonium salt solutions. Noncombustible.
Derivation: (a) By calcining magnesium carbonate or magnesium hydroxide, (b) by treating magnesium chloride with lime and heating or by heating it in air, (c) from sea water via the hydroxide.
Grade: Technical, CP, USS, FCC, 99.5%, fused, low boron, rubber, semiconductor, single crystals.
Hazard: Toxic by inhalation of fume. TLV (as magnesium): 10 mg/m^3.
Use: Refractories, especially for steel furnace linings, polycrystalline ceramic for aircraft windshields, electrical insulation, pharmaceuticals and cosmetics, inorganic rubber accelerator, oxychloride and oxysulfate cements, paper manufacture, fertilizers, removal of sulfur dioxide from stack gases, adsorption and catalysis, semiconductors, pharmaceuticals, food and feed additive.

magnesium oxychloride cement.   (Sorel cement).
A mixture of magnesium chloride and magnesium oxide that reacts with water to form a solid mass, presumed to be magnesium oxychloride. Fillers such as wood flour, sawdust, sand, pow-

dered stone, talc, etc., are usually present. A variety of proprietary mixtures are available. Strength ranges from 7000 to 10,000 psi. Copper powder minimizes water solubility.

magnesium palmitate. $Mg(C_{16}H_{31}O_2)_2$.
Properties: Crystalline needles or white lumps, mp 121.5C, insoluble in water and alcohol. Combustible.
Use: Varnish drier, lubricant for plastics.

magnesium perborate. CAS: 17097-11-9. $Mg(BO_3)_2 \cdot 7HOH$.
Properties: White powder, sparingly soluble in water, decomposes with evolution of oxygen.
Derivation: Action of peroxide or electrolytic oxidation of borate solutions.
Hazard: Moderate fire risk in contact with organic materials.
Use: Driers, bleaching, antiseptic (tooth powders).

magnesium perchlorate. CAS: 10034-81-8.
(a) $Mg(ClO_4)_2$; (b) $Mg(ClO_4)_2 \cdot 6HOH$.
Properties: White crystals, deliquescent, very soluble in water and alcohol, (a) d 2.21 (18C), decomposes at 251C (b) d 1.98, decomposes at 185–190C.
Derivation: Reaction of magnesium hydroxide and perchloric acid.
Hazard: Dangerous fire and explosion risk in contact with organic materials.
Use: (a) As a regenerable drying agent for gases and (b) oxidizing agent.

magnesium permanganate. CAS: 10377-62-5. $Mg(MnO_4)_2 \cdot 6HOH$.
Properties: Bluish-black, friable, deliquescent crystals, d 2.18, mp decomposes, soluble in water.
Hazard: Fire hazard in contact with organic materials. Powerful oxidizer.
Use: Polymerization catalyst, antiseptic.

magnesium peroxide. (magnesium dioxide).
CAS: 1335-26-8. MgO_2.
Properties: White, tasteless, odorless powder. Decomposes above 100C, insoluble in water, soluble in dilute acids with formation of hydrogen peroxide, available oxygen 28.4%, keep cool and dry, a powerful oxidizing material.
Derivation: From sodium or barium peroxide with magnesium sulfate in a concentrated solution.
Grade: Technical, 15, 25, and 50%.
Hazard: Powerful oxidizer and dangerous fire risk, reacts with acidic materials and moisture.
Use: Bleaching and oxidizing agent, medicine (antacid).

magnesium phosphate. (magnesium orthophosphate). See magnesium phosphate, dibasic; magnesium phosphate, monobasic; or magnesium phosphate, tribasic.

magnesium phosphate, dibasic. (dimagnesium orthophosphate; dimagnesium phosphate; magnesium phosphate, secondary; magnesium hydrogen phosphate). CAS: 7757-86-0. $MgHPO_4 \cdot 3HOH$.
Properties: White, crystalline powder; decomposes to pyrophosphate on heating. Soluble in dilute acids, slightly soluble in water, d 2.13, loses water at 205C, decomposes at 550–650C. Nonflammable.
Derivation: Action of orthophosphoric acid on magnesium oxide.
Grade: Technical, FCC.
Use: Stabilizer for plastics, food additive, medicine (laxative).

magnesium phosphate, monobasic. (magnesium biphosphate; acid magnesium phosphate; magnesium tetrahydrogen phosphate).
CAS: 13092-66-5. $MgH_4(PO_4)_2 \cdot 2HOH$.
Properties: White, hygroscopic, crystalline powder. Decomposes to metaphosphate on heating, soluble in water and acids, insoluble in alcohol. Nonflammable.
Derivation: Action of orthophosphoric acid on magnesium hydroxide.
Use: Fireproofing wood, stabilizer for plastics.

magnesium phosphate, neutral. See magnesium phosphate, tribasic.

magnesium phosphate, secondary. See magnesium phosphate, dibasic.

magnesium phosphate, tribasic. (Magnesium phosphate, neutral; trimagnesium phosphate).
CAS: 7757-86-0. $Mg_3(PO_4)_2 \cdot 8HOH$ or 4HOH.
Properties: Soft, bulky, white powder; odorless and tasteless. Loses all water at 400C, soluble in acids, insoluble in water. Nonflammable.
Derivation: Reaction of magnesium oxide and phosphoric acid at high temperatures.
Grade: Technical, reagent, NF (five water variety), FCC (4, 5, or 8 waters).
Use: Dentifrice polishing agent, pharmaceutical antacid, adsorbent, stabilizer for plastics, food additive and dietary supplement.

magnesium pyrophosphate. CAS: 13446-24-7. $Mg_2P_2O_7 \cdot 3HOH$.
Properties: White powder, soluble in acids, insoluble in alcohol and water, d 2.56, loses water at 100C, mp (anhydrous) 1383C.

magnesium ricinoleate. $Mg(OOCC_{17}H_{32}OH)_2$.
Properties: Coarse, yellow granules with faint fatty acid odor; mp 98C; d 1.03 (25/25C). Combustible.
Use: Cosmetics.

magnesium salicylate. CAS: 18917-89-0.
$Mg(C_7H_5O_3)_2 \cdot 4HOH$.
Properties: Colorless, efflorescent, crystalline powder; soluble in water and alcohol.
Derivation: Action of salicylic acid on magnesium hydroxide.
Use: Medicine (anti-infective).

magnesium silicate. $3MgSiO_3 \cdot 5HOH$ (variable). The FCC specifies a ratio of $2MgO:5SiO_2$.
See also magnesium trisilicate, serpentine, and talc.
Properties: Fine white powder, insoluble in water or alcohol, an absorbent. Noncombustible.
Derivation: Interaction of a magnesium salt and a soluble silicate.
Grade: Technical, FCC.
Hazard: Toxic by inhalation, use in foods restricted to 2%.
Use: Rubber filler; ceramics, glass, refractories; absorbent for crude oil spills; manufacture of permanently dry resins and resinous compositions; paints, varnishes, and paper (filler); animal and vegetable oils (bleaching agent); odor absorbent; filter medium; catalyst and catalyst carrier; anticaking agent in foods.
See also asbestos.

magnesium silicide. CAS: 22831-39-6.
Mg_2Si.
Properties: Bluish crystals, mp 1085C, d 1.9, decomposes on heating above 500C, also by water and hydrochloric acid.
Derivation: By heating magnesium powder with silicon in ratio of 20:6.
Use: Semiconductor technology, electrical equipment.

magnesium silicofluoride. See magnesium fluosilicate.

magnesium stannate. $MgSnO_3 \cdot 3HOH$.
Properties: White, crystalline powder; soluble in water; approximately decomposes at 340C.
Hazard: Toxic by inhalation. TLV (as Sn): 2 mg/m^3 of air.
Use: Additive in ceramic capacitors.

magnesium stannide. CAS: 1313-08-2.
Mg_2Sn.
Properties: Blue-white crystals, mp 775C, soluble in water and dilute hydrochloric acid, has electrical and magnetic properties.
Use: Semiconductor technology, magnetochemistry, thermoelectric research.

magnesium stearate. CAS: 557-04-0.
$Mg(C_{18}H_{35}O_2)_2$ or with one H_2O. Technical grade contains small amounts of the oleate and 7% magnesium oxide, MgO.

Properties: Soft, white, light powder. D 1.028, mp 88.5C (pure), 132C (technical), tasteless, odorless, insoluble in water and alcohol. Nonflammable.
Grade: Technical, USP FCC.
Use: Dusting powder, lubricant in making tablets, drier in paints and varnishes, flatting agent, in medicines, stabilizer and lubricant for plastics, emulsifying agent in cosmetics, dietary supplement.

magnesium sulfate. CAS: 7587-88-9.
(a) $MgSO_4$, (b) (epsom salts) $MgSO_4 \cdot 7H_2O$.
Properties: Colorless crystals; saline bitter taste; neutral to litmus; d (a) 2.65, (b) 1.678; (a) decomposes at 1124C, (b) loses 6HOH at 150C, loses water at 200C, soluble in glycerol, very soluble in water, sparingly soluble in alcohol. Noncombustible.
Derivation: (a, b) Action of sulfuric acid on magnesium oxide, hydroxide, or carbonate, (b) mined in a high degree of purity.
Grade: Technical, CP, USP, FCC.
Use: Fireproofing, textiles (warp-sizing and loading cotton goods, weighting silk, dyeing and calico printing), mineral waters, catalyst carrier, ceramics, fertilizers, paper (sizing), cosmetic lotions, dietary supplement.

magnesium sulfide. MgS.
Properties: Red-brown, crystalline solid; d 2.84; decomposes above 2000C; decomposes in water.
Use: Source of hydrogen sulfide, laboratory reagent.

magnesium sulfite. CAS: 77570-88-2.
$MgSO_3 \cdot 6HOH$.
Properties: White, crystalline powder; slightly soluble in water; insoluble in alcohol. D 1.725, mp loses 6HOH at 200C, bp decomposes.
Derivation: Action of sulfurous acid on magnesium hydroxide.
Use: Manufacture of paper pulp (as bisulfite).

magnesium tetrahydrogen phosphate.
See magnesium phosphate, monobasic.

magnesium trisilicate. USP specifies not less than 20% magnesium oxide and 45% SiO_2, similar to the FCC requirements for magnesium silicate.
See also talc.
Properties: Fine, white, odorless, tasteless powder; free from grittiness. Insoluble in water and alcohol, readily decomposed by mineral acids. Noncombustible.
Derivation: By reaction of soluble magnesium salts with soluble silicates.
Grade: Technical, USP.

Use: Industrial odor absorbent, decolorizing agent, antioxidant, medicine (antacid).

magnesium tungstate. (magnesium wolframate). CAS: 13573-11-0. $MgWO_4$.
Properties: White crystals, d 5.66, soluble in acids, insoluble in water and alcohol. Noncombustible.
Derivation: Interaction of solutions of magnesium sulfate and ammonium tungstate.
Use: Fluorescent screens for x-rays, luminescent paint.

magnesium zirconate. MgO,ZrO_2.
Properties: Powder, d 4.23, mp 2060C.
Use: Electronics.

magnesium zirconium silicate. $MgZrSiO_5$, or $MgO \cdot ZrO_2 \cdot SiO_2$.
Properties: White solid, mp 1760C, d 80 lb/ft³, insoluble in water and alkalies, slightly soluble in acids. Noncombustible.
Use: Electrical resistor, ceramics, glaze opacifier.

magnetic separation. Use of a magnetic field to remove unwanted magnetic particulates from solid or liquid mixtures of nonmagnetic materials, e.g., removal of impurities from clays, bauxite, glass sands, and mineral processing. Low-gradient fields are suitable for separation of strongly magnetic materials, whereas high-gradient fields can separate particles of materials that are weakly magnetic, such as coliform bacteria from municipal wastes and sulfur from coal. Removal of magnetic impurities from industrial waste water is called magnetic filtration, for example, reconditioning of boiler water and regeneration of condensate in power plants.
See also electromagnetic separation, mass spectrometry.

magnetite. (lodestone; iron ore, magnetic).
Fe_3O_4 often with titanium or magnesium. A component of taconite.
Properties: Black mineral, black streak, submetallic or dull to metallic luster. Contains 72.4% iron. Readily recognized by strong attraction by magnet. Soluble in powder form in hydrochloric acid. Decomposes at 1538C to ferric oxide Fe_2O_3; d 4.9–5.2; hardness 5.5–6.5.
See also iron oxide, black.

magnetochemistry. A subdivision of chemistry concerned with the effect of magnetic fields on chemical compounds; analysis and measurement of these effects, (e.g., magnetic moment and magnetic susceptibility) are important tools in crystallographic research and determination of molecular structures. Substances that are repelled

by a magnetic field are diamagnetic (water, benzene); those that are attracted are paramagnetic (oxygen, transition element compounds). Diamagnetic materials have only induced magnetic moment; paramagnetic materials have permanent magnetic moment. Magnetochemistry has been useful in detection of free radicals, elucidation of molecular configurations of highly complex compounds, and in its application to catalytic and chemisorption phenomena.
See also nuclear magnetic resonance.

magnetohydrodynamics. (MHD). The behavior of high-temperature ionized gases passed through a magnetic field. A power-generating method using MHD involves an open cycle in which hot combustible gases from coal, seeded with cerium or potassium to increase electrical conductivity, constitute the working fluid. These are sent through a nozzle surrounded by a magnet; the electricity induced by movement of the ionized gas through the magnetic field is passed to electrodes and the gas sent to a steam generator. Efficiency is rated at 50–60% compared with 40% for conventional fossil fuel plants and 33% for plants using nuclear fuels. Two-phase liquid-metal systems are being studied as auxiliary units for a number of energy converters. MHD is an important field of expansion of research activity on new sources of energy; its high efficiency and low pollution factor indicate that it may have a significant future in electric power supply.
See also plasma (2).

Maillard reaction. See browning reaction.

"Maintain."[266] TM for a line of products for cleaning, inhibiting corrosion, and coating architectural copper, brass, and bronze.

Malachite Green. (Benzaldehyde Green; Victoria Green). CAS: 569-64-2. $C_{23}H_{25}ClN_2$.
Properties: Water-soluble green crystals, CI 42,000, soluble in ethyl and methyl alcohol, amyl alcohol.
Derivation: Condensation of benzaldehyde with N,N-dimethylaniline, oxidation of the phenylmethane product, and reaction with hydrochloric acid. It may be formed as a double salt of zinc chloride.
Hazard: Toxic by ingestion.
Use: Dyeing textiles, either directly or with mordant; plant fungicide, staining bacteria, antiseptic.

Malachite Green G. See Brilliant Green.

Malaprade reaction. Compounds containing two hydroxyl groups or a hydroxyl and an amino group attached to adjacent carbon atoms undergo cleavage of the carbon-to-carbon bond when treated with periodic acid.

malathion. (S-[1,2-bis(ethoxycarbonyl)ethyl]O,O-dimethyl phosphorodithioate).
CAS: 121-75-5.
$(CH_3O)_2P(S)SCH(COOC_2H_5)CH_2COOC_2H_5$.
Properties: Yellow, high-boiling liquid (bp 156–157C under 0.7 mm with slight decomposition). Mp 3.0C, refr index 1.4985 (25C), d 1.2315 (25C), vap press (20C) approximately 0.00004 mm, miscible with most polar organic solvents, slightly soluble in water. Combustible. Purity: Technical grade is 95+% pure.
Derivation: From diethyl maleate and dimethyldithiophosphoric acid.
Hazard: Absorbed by skin, cholinesterase inhibitor. TLV: 10 mg/m³ of air.
Use: Insecticide; has been used effectively on the Mediterranean fruit fly.

malaxate. To soften and mix dry materials in the presence of water or other liquid by rubbing, kneading, or rolling, thus producing a soft plastic mass. This term is used by one manufacturer, the Fitzpatrick Co., 832 Industrial Drive, Elmhurst, IL, 60126, to describe a machine designed for this purpose, it provides continuous mixing of dry solids with one or more liquids by means of single or double screw-type agitators rotating in a channel.

maleic acid. (maleinic acid; cis-butenedioic acid).
CAS: 110-16-7.

$$
\begin{array}{c}
\quad\quad O \\
\quad\quad \| \\
HC\!-\!COH \\
\| \\
HC\!-\!COH \\
\quad\quad \| \\
\quad\quad O
\end{array}
$$

Properties: Colorless crystals, repulsive astringent taste, faint odor, soluble in water, alcohol and acetone, very slightly soluble in benzene, d 1.59, mp 130–131C. At room temperatures slightly higher than its melting point, it is converted partly to fumaric acid. Combustible.
Derivation: Same as for maleic anhydride with special recovery conditions.
Grade: Technical, reagent.
Hazard: Toxic by ingestion.
Use: Organic synthesis (malic, succinic, aspartic, tartaric, propionic, lactic, malonic and acrylic acids); dyeing and finishing of cotton, wool and silk; preservative for oils and fats.

maleic anhydride. (2,5-furandione).
CAS: 108-31-6.

$$
\begin{array}{c}
\quad O \\
\quad \| \\
\quad C \\
HC\diagup \quad\diagdown \\
\| \quad\quad O \\
HC\diagdown \quad\diagup \\
\quad C \\
\quad \| \\
\quad O
\end{array}
$$

Properties: Colorless needles, d 0.934 (20/4C), mp 53C, bp 200C, flash p 218F (103C). Soluble in water, acetone, alcohol, and dioxane; partially soluble in chloroform and benzene. Autoign temperature 890F (476C).
Derivation: (1) Vapor-phase oxidation of benzene with atmospheric oxygen with V_2O_5 catalyst at 400C. (2) Under development is a fixed-bed process involving oxidation of butane with undisclosed catalyst.
Grade: Technical; rods, flakes, lumps, briquettes, and molten.
Hazard: Irritant to tissue. TLV: 0.25 ppm in air.
Use: Polyester resins, alkyd coating resins, fumaric and tartaric acid manufacture, pesticides, preservative for oils and fats, permanent-press resins (textiles), Diels-Alder reactions.

maleic hydrazide. (1,2-dihydro-3,6-pyridazinedione). CAS: 108-31-6.

$\overline{HC\!:\!CHC(O)}NHNHC(O)$.   A plant growth regulator.
Properties: Crystals, mp 297C, slightly soluble in hot alcohol, more soluble in hot water.
Derivation: By treating maleic anhydride with hydrazine hydrate.
Hazard: Toxic by ingestion.
Use: Systemic herbicide, treatment of tobacco plants, post-harvest sprouting inhibitor, weed control, sugar content stabilizer in beets.

maleinic acid. See maleic acid.

maleo-pimaric acid. Reaction product of maleic anhydride and l-pimaric acid; derived from pine gum.
Properties: Crystalline solid, mp approximately 225C, soluble in most organic solvents, insoluble in water or aliphatic hydrocarbons.
Use: Resins.

malic acid. (hydroxysuccinic acid; apple acid).
CAS: 6915-15-7.
$COOHCH_2CH(OH)COOH$ (do not confuse with maleic acid).
Properties: Colorless crystals; sour taste; d (dl-form) 1.601, (d or l form) 1.595 (20/4C); mp

(*dl*) 128C, (*d* or *l*) 100C; bp (*dl*) 150C (decomposes), (*d* or *l*) 140C (decomposes); very soluble in water and alcohol, slightly soluble in ether. Combustible.

Derivation: Occurs naturally in unripe apples and other fruits. Made synthetically by catalytic oxidation of benzene to maleic acid, which is converted to malic acid by heating with steam under pressure.

Grade: Technical, active and inactive; FCC. The natural material is levorotatory but the synthetic material is inactive.

Use: Manufacture of various esters and salts, wine manufacture, chelating agent, food acidulant, flavoring.

malonamide nitrile.   See cyanoacetamide.

malonic acid.   (methanedicarbonic acid).
     CAS: 141-82-2. $CH_2(COOH)_2$.
Properties: White crystals, soluble in water, alcohol and ether, mp 132–134C, bp decomposes, d 1.63.
Derivation: From monochloroacetic acid by reaction with potassium cyanide followed by hydrolysis.
Hazard: Strong irritant.
Use: Intermediate for barbiturates and pharmaceuticals.

malonic dinitrile.   (malononitrile).
     CAS: 109-77-3. $CH_2(CN)_2$.
Properties: Colored crystals, mp 32.1C, bp 220C.
Hazard: Toxic by ingestion and inhalation.
Use: Organic synthesis, leaching agent for gold.

malonic ester.   See ethyl malonate.

malonic ester synthesis.   Syntheses based on the strongly activated methylene group of malonic esters which on reaction with sodium ethoxide form a resonance-stabilized ion that can be alkylated and acylated. After hydrolysis, the free alkylmalonic acids readily decarboxylate to mono- or disubstituted monocarboxylic acids.

malonic ethyl ester nitrile.   See ethyl cyanoacetate.

malonic methyl ester nitrile.   See methyl cyanoacetate.

malonic mononitrile.   See cyanoacetic acid.

malononitrile.   See malonic dinitrile.

malonylurea.   See barbituric acid.

malt.   Yellowish or amber-colored grains of barley which have been partially germinated by artificial means. It contains dextrin, maltose, and amylase; characteristic odor and taste. Black malt is grain which has been scorched in the drying process.
Use: Brewing, malted milk and similar food products, extract of malt (with 10% glycerol).

maltase.   (glucase; α-glucosidase).   An enzyme that hydrolyzes maltose to glucose. Occurs in the small intestine, in yeast, molds, and malt; usually associated with the enzyme amylase.

malt extract.   (maltine).
Properties: Light brown, sweet, viscous liquid; contains dextrin, maltose, a little glucose, and an amylolytic enzyme. It is capable of converting not less than five times its weight of starch into water-soluble sugars; soluble in cold water but more readily soluble in warm water, d greater than 1.350 and less than 1.430 (25C).
Derivation: By infusing malt with water at 60C, concentrating the expressed liquid below 60C, and adding 10% by weight of glycerol.
Use: Nutrient, emulsifying agent.

maltha.   A black, viscous, natural bitumen consisting of a complex mixture of hydrocarbons. Its viscosity and rheological properties lie between those of crude oil and semisolid asphalt. It is the chief component of Athabaska oil sands.

maltol.   (3-hydroxy-2-methyl-4-pyrone).
     CAS: 118-71-8. $CH_3C_5H_2O(O)(OH)$.
Properties: White, crystalline powder having a characteristic caramel- butterscotch odor and suggestive of a fruity-strawberry aroma in dilute solution. Slightly soluble in water, more soluble in alcohol and propylene glycol, melting range 160–164C.
Grade: FCC.
Use: Flavoring agent in bakery products.

maltose.   (malt sugar; maltobiose).
     CAS: 69-79-4. $C_{12}H_{22}O_{11} \cdot HOH$.
The most common reducing disaccharide, composed of two molecules of glucose. Found in starch and glycogen.
Properties: Colorless crystals, mp 102–103C, soluble in water, insoluble in ether, slightly soluble in alcohol. Combustible.
Derivation: By the enzymatic action of diastase (usually obtained from malt extract) on starch.
Use: Nutrient, sweetener, culture media, stabilizer for polysulfides, brewing.

Man.   Abbreviation for mannose, also for methyacrylonitrile.

mandarin oil.   See citrus peel oils.

mandelic acid.   (phenylglycolic acid; α-phenylhydroxyacetic acid; benzoglycolic acid, known also

as amygdalic acid).   CAS: 90-64-2.
$C_6H_5CHOHCOOH$.   Exists in stereoisomeric forms. The properties are those of the *dl*-form.

$$C_6H_5-\overset{\overset{\displaystyle H}{|}}{\underset{\underset{\displaystyle OH}{|}}{C^*}}-\overset{\overset{\displaystyle}{}}{\underset{\underset{\displaystyle O}{\|}}{C}}-OH$$

Properties: Large, white crystals or powder with a faint odor. D 1.30, mp 117–119C, darkens on exposure to light, soluble in ether, slightly soluble in water and alcohol.
Derivation: Hydrolysis of the cyanohydrin formed from benzaldehyde, sodium bisulfite, and sodium cyanide. Can be obtained from amygdalin.
Hazard: Toxic by ingestion.
Use: Organic synthesis, medicine (urinary antiseptic).

mandelonitrile.   (benzaldehyde cyanohydrin; "Laetrile").   CAS: 532-28-5.
$C_6H_5CH(OH)CN$.
Properties: Oily, yellow liquid; d 1.1165 (20/4C); fp −10C; bp 170C (decomposes). Soluble in alcohol, chloroform, ether; nearly insoluble in water.

maneb.   (Generic name for manganese ethylenebisdithiocarbamate).   CAS: 12427-38-2.
$(SSCNCH_2CH_2NHCSS)Mn$.
Properties: Brown powder, decomposes on heating, partially soluble in water, soluble in chloroform.
See also zineb.
Derivation: Reaction of disodium ethylenebisdithiocarbamate and a manganese salt.
Use: Fungicide for foliage.

manganese.   CAS: 7439-96-5.   Mn.
Metallic element of atomic number 25; Group VIIB of Periodic Table; aw 64.9380; valences = 2, 3, 4, 6, 7; no stable isotopes; four artificial radioisotopes.
Properties: There are four allotropic forms of which alpha is most important. Brittle silvery metal, d 7.44, Mohs hardness 5, mp 1245C, bp 2097C, decomposes water, readily dissolves in dilute mineral acids. Pure manganese cannot be fabricated. Manganese is considered essential for plant and animal life.
Occurrence: Usually associated with iron ores in submarginal concentration. Important ores of manganese are pyrolusite, manganite, psilomelane, rhodochrosite. An important source of manganese is open-hearth slags. Ores occur chiefly in Brazil, India, South Africa, Gabon, Ghana, Zaire, Montana; 90% of US consumption is imported. So-called nodules rich in manganese and containing also cobalt, nickel, and copper have been found in huge quantities (estimated at 1.5 trillion tons) on the floor of the Pacific south of Hawaii. Such nodules have been located in other areas, as well as in Lake Michigan.
Derivation: Reduction of the oxide with aluminum or carbon. Pure manganese is obtained electrolytically from sulfate or chloride solution.
Grade: Technical, pure or electrolytic, powdered.
Hazard: Dust or powder is flammable. Use dry chemical to extinguish. Toxic. TLV (fume): 1 mg/m³ of air; (metal and most compounds): CL of 5 mg/m³ of air.
Use: Ferroalloys (steel manufacture), nonferrous alloys (improved corrosion resistance and hardness), high-purity salt for various chemical uses, purifying and scavenging agent in metal production, manufacture of aluminum by Toth process.

manganese acetate.   (acetic acid manganese(2+) salt).   CAS: 638-38-0.
$Mn(C_2H_3O_2)_2 \cdot 4HOH$.
Properties: Pale red crystals, soluble in water and alcohol, d 1.59, mp 80C. Combustible.
Derivation: Action of acetic acid on manganese hydroxide.
Use: Textile dyeing, oxidation catalyst, paint and varnish (drier, boiled oil manufacture), fertilizers, food packaging, feed additive.

manganese ammonium sulfate.   (manganous ammonium sulfate).
$MnSO_4 \cdot (NH_4)_2SO_4 \cdot 6HOH$.
Properties: Light-red crystals, d 1.83, soluble in water.

manganese arsenate.   See manganous arsenate.

manganese binoxide.   See manganese dioxide.

manganese black.   See manganese dioxide.

manganese borate.   CAS: 1303-95-3.
MnB_4O_7.
Properties: Reddish-white powder, insoluble in water.
Derivation: By the action of boric acid on manganese hydroxide.
Grade: Technical.
Use: Varnish and oil drier.

manganese-boron.   An alloy of manganese and boron used in making brass, bronze, and other alloys.

manganese bromide.   See manganous bromide.

manganese bronze.   Alloy of 55–60% copper, 38–42% zinc, up to 3.5% manganese with or without small amounts of iron, aluminum, tin or lead.

manganese carbonate. (manganous carbonate; rhodocrosite). CAS: 598-62-9. $MnCO_3$.
Properties: Rose-colored crystals, almost white when precipitated, soluble in dilute inorganic acids, almost insoluble in common organic acids, both concentrated and dilute, insoluble in water, d 3.125, mp decomposes.
Derivation: (1) A precipitate from the addition of sodium carbonate to a solution of a manganese salt. (2) Hydrometallurgical treatment of manganiferous iron ore.
Grade: Chemical (46% manganese).

manganese carbonyl. CAS: 10170-69-1. $Mn_2(CO)_{10}$.
Properties: Yellow crystals, mp 154C, decomposition begins at 110C in absence of CO, d 1.75, insoluble in water, soluble in most organic solvents.
Hazard: Toxic material. TLV (as manganese): CL of 5 mg/m³ of air.
Use: Antiknock gasoline, catalyst.

manganese chloride. See manganous chloride.

manganese chromate. See manganous chromate.

manganese citrate. (manganous citrate). $Mn_3(C_6H_5O_7)_2$.
Properties: White powder, soluble in water in presence of sodium citrate. Combustible.
Derivation: Action of citric acid on manganese hydroxide.
Use: Feed additive, food additive and dietary supplement.

manganese cyclopentadienyl tricarbonyl. CAS: 12079-65-1. $C_5H_4Mn(CO)_3$.
Hazard: Toxic material absorbed by skin. TLV (as manganese): 0.1 ppm in air.
Use: Antiknock agent.

manganese dioxide. (manganese binoxide; manganese black; battery manganese; manganese peroxide). CAS: 1313-13-9. MnO_2.
Properties: Black crystals or powder, soluble in hydrochloric acid, insoluble in water, d 5.026, mp decomposes.
Derivation: (a) Natural as pyrolusite and as a special African ore of different atomic structure used exclusively for the battery grade; (b) by electrolysis; (c) by heating manganese oxide in presence of oxygen; (d) by decomposition of manganese nitrate.
Grade: Technical, CP, Battery.
Hazard: Oxidizing agent, may ignite organic materials.
Use: Oxidizing agent, depolarizer in dry cell batteries (African and synthetic types only), pyrotechnics, matches, etc., catalyst, laboratory reagent, scavenger and decolorizer, textile dyeing, source of metallic manganese (as pyrolusite).

manganese dithionate. MnS_2O_6.
Properties: Crystals, d 1.76, soluble in water.

manganese ethylenebisdithiocarbamate. See maneb.

manganese ethylhexoate. See manganese octoate.

manganese fluoride. See manganous fluoride.

manganese gluconate. $Mn(C_6H_{11}O_7)_2 \cdot 2HOH$.
Properties: Light pinkish powder or coarse pink granules, soluble in water, insoluble in alcohol and benzene. Combustible.
Grade: Pharmaceutical, FCC.
Use: Feed additive, food additive and dietary supplement, vitamin tablets.

manganese glycerophosphate. $CH_2OHCHOHCH_2OP(O)O_2Mn$.
Properties: Yellow-white or pinkish powder, odorless, nearly tasteless, soluble in water in presence of citric acid, insoluble in alcohol.
Derivation: Action of glycerophosphoric acid on manganese hydroxide.
Grade: Technical, FCC.
Use: Food additive and dietary supplement.

manganese green. See barium manganate.

manganese hydrate. See manganic hydroxide, manganous hydroxide.

manganese hydrogen phosphate. See manganous phosphate, acid.

manganese hydroxide. See manganic hydroxide, manganous hydroxide.

manganese hypophosphite. CAS: 10043-84-2. $Mn(H_2PO_2)_2 \cdot HOH$.
Properties: Pink crystals or powder, odorless, tasteless, soluble in water, insoluble in alcohol.
Derivation: Interaction of manganese sulfate and calcium hypophosphite.
Grade: Technical, FCC.
Hazard: Dangerous fire and explosion risk when heated (evolves phosphine) or in presence of strong oxidizing agents.
Use: Food additive and dietary supplement.

manganese iodide. See manganous iodide.

manganese linoleate. $Mn(C_{18}H_{31}O_2)_2$.
Properties: Dark brown, plaster-like mass; soluble in linseed oil. Combustible.

Derivation: By boiling a manganese salt, sodium linoleate, and water.
Use: Paint and varnish drier, pharmaceutical preparations.

manganese monoxide.   See manganous oxide.

manganese naphthenate.
Properties: Hard, brown, resinous mass. It is a pale buff in color when precipitated in the cold, but darkens immediately in solution. Soluble in mineral spirits, hardens on exposure to air, mp approximately 130–140C, commercial solution contains 6% manganese. Combustible.
Derivation: Precipitation from a mixture of soluble manganese salts and aqueous sodium naphthenate solution.
Hazard: Solution is flammable.
Use: Paint and varnish drier.

manganese nitrate.   See manganous nitrate.

manganese octoate.   (manganese ethylhexoate). $Mn(OOCC_6H_{13}[C_2H_5])_2$.   Commercially formed from 2-ethylhexoic acid and manganous hydroxide. Sold as a clear brown solution in a petroleum solvent containing 6% manganese.
Hazard: Solution is flammable.
Use: Primarily as drier for paints, enamels, varnishes and printing inks.

manganese oleate.   CAS: 23250-73-9.
$Mn(C_{18}H_{33}O_2)_2$.
Properties: Brown, granular mass; soluble in oleic acid and ether; insoluble in water.
Derivation: By boiling manganese chloride, sodium oleate, and water.
Hazard: Solution is flammable.
Use: Paint and varnish drier.

manganese oxalate.   CAS: 640-67-5.
$MnC_2O_4 \cdot 2HOH$.
Properties: White, crystalline powder. Soluble in dilute acids, very slightly soluble in water, d 2.453, loses 2HOH at 100C. Combustible.
Derivation: By adding sodium oxalate to manganese chloride.
Use: Paint and varnish drier.

manganese peroxide.   See manganese dioxide.

manganese phosphate.   See manganous phosphate. See also manganous phosphate, acid.

manganese phosphate, dibasic.   See manganous phosphate, acid.

manganese protoxide.   See manganous oxide.

manganese pyrophosphate.   See manganous pyrophosphate.

manganese resinate.   CAS: 9008-34-8.
$Mn(C_{20}H_{29}O_2)_2$.
Properties: Dark, brownish-black mass or flesh-colored powder; soluble in hot linseed oil; insoluble in water.
Derivation: By boiling manganese hydroxide, rosin oil, and water.
Hazard: Flammable, dangerous fire risk.
Use: Varnish and oil drier.

manganese sesquioxide.   See manganic oxide.

manganese silicate.   See manganous silicate.

manganese sulfate.   See manganous sulfate.

manganese sulfide.   See manganous sulfide.

manganese sulfite.   See manganous sulfite.

manganese tallate.   Manganese salts of tall oil fatty acids. Marketed as a solution containing 6% manganese. Combustible.
Use: Drier.

manganese tetroxide.   (manganese oxide; trimanganese tetroxide).   CAS: 1317-35-7.
Mn_3O_4.
Properties: A brownish-powder with mw 228.79, d 4.876, mp 1564C. Insoluble in water, soluble in hydrochloric acid with evolution of chlorine.
Ocurrence: Generated in the pouring and casting of molten ferromanganese.
Hazard: Chronic manganese poisoning and pulminary effects. TLV: 1 mg/m³.

manganese-titanium.   An alloy containing manganese, titanium, aluminum, iron, silicon.
Properties: (regular) Mp 1454C, (special) mp 1332C.
Use: (regular) Deoxidizer in high grade steel; (special) non-ferrous alloys deoxidizer.

manganic acetylacetonate.
$Mn[OC(CH_3):CHCO(CH_3)]_3$.
Properties: Brown, crystalline solid; mp 172C. Combustible.
Derivation: Reaction of a manganese salt with acetylacetone and sodium carbonate.

manganic fluoride. MnF_3.
Properties: Red, crystalline solid; d 3.54; decomposed by water and by heat; soluble in acids; attacks glass when hot.
Hazard: Toxic material. TLV (as F): 2.5 mg/m³.
Use: Fluorinating agent.

manganic hydroxide.   (manganese hydroxide; hydrated manganic oxide). $Mn(OH)_3$; rapidly loses water to form MnO(OH).
Properties: Brown powder, d 4.2–4.4, mp decomposes, decomposes in acids, insoluble in water.

Derivation: By the action of oxygen on precipitated manganous hydroxide.
Use: Pigment for fabrics, ceramics.

manganic oxide. (manganese sesquioxide). Mn_2O_3. In nature as manganite, a manganese ore.
Properties: Black lustrous powder, sometimes tinged brown, very hard, d 4.5, decomposes at 1080C, soluble in cold hydrochloric acid, not soluble in nitric acid (decomposes), hot sulfuric acid, insoluble in water. Noncombustible.
Hazard: Flammable in finely divided form. Toxic by inhalation of dust.

manganic oxide, hydrated. See manganic hydroxide.

"Manganin."[155] TM for a resistance alloy of copper 87% and manganese 13%.
Use: Thermocouples and electrical instruments.

manganous ammonium sulfate. See manganese ammonium sulfate.

manganous arsenate. (manganese arsenate; manganous arsenate, acid). $MnHAsO_4$.
Properties: Reddish-white powder, hygroscopic, soluble in acids, slightly soluble in water.
Hazard: Highly toxic.

manganous bromide. (manganese bromide). $MnBr_2 \cdot 4HOH$.
Properties: Red crystals, loses water at 64C, very soluble in water, deliquescent. Noncombustible.
Derivation: Action of hydrobromic acid with manganous carbonate or manganous hydroxide.
Hazard: Irritant.

manganous carbonate. See manganese carbonate.

manganous chloride. (manganese chloride).
CAS: 7773-01-5. (a) $MnCl_2$, (b) $MnCl_2 \cdot 4HOH$.
Properties: Rose-colored crystals; deliquescent; d (a) 2.98, (b) 1.913; mp (a) 650C, (b) 87.5C; bp (a) 1190C. Very soluble in water, soluble in alcohol, insoluble in ether. Noncombustible.
Grade: CP, anhydrous.
Use: Catalyst in the chlorination of organic compounds, paint drier, dyeing, pharmaceutical preparations, fertilizer compositions, feed additive, dietary supplement.

manganous chromate. (manganese chromate; manganous chromate, basic). $2MnO \cdot CrO_3 \cdot 2HOH$.
Properties: Brown powder, slightly soluble in water with hydrolysis.
Hazard: Toxic by inhalation.

manganous citrate. See manganese citrate.

manganous fluoride. (manganese fluoride). MnF_2.
Properties: Reddish powder; soluble in acids; insoluble in water, alcohol, and ether; d 3.98; mp 856C. Noncombustible.
Derivation: Action of hydrogen fluoride on manganous hydroxide.
Grade: Technical.
Hazard: TLV (as F): 2.5 mg/m³ of air.

manganous hydroxide. (manganese hydroxide). $Mn(OH)_2$. Occurs naturally as pyrochroite.
Properties: White to pink crystals, d 3.258, Mohs hardness 2.5, decomposes with heat, insoluble in water and alkali, soluble in acids and ammonium salts.

manganous iodide. (manganese iodide).
(a) MnI_2, (b) $MnI_2 \cdot 4HOH$.
Properties: (a) White, deliquescent, crystalline mass; (b) rose crystals, d (a) 5.01, mp (a) 638C (in vacuum), bp (a) 1061C, soluble in water with gradual decompostion, soluble in alcohol.
Derivation: Action of hydriodic acid on manganous hydroxide.

manganous nitrate. (manganese nitrate).
CAS: 10377-66-9. $Mn(NO_3)_2 \cdot 6HOH$.
Properties: Pink crystals, very soluble in water, deliquescent, soluble in alcohol, d 1.82, bp 129C, mp 26C.
Hazard: Fire and explosion risk in contact with organic materials.
Use: Ceramics, intermediates, catalyst, manganese dioxide.

manganous orthophosphate. See manganous phosphate.

manganous oxide. (manganese protoxide; manganese monoxide). CAS: 1317-35-7. MnO.
Properties: Grass-green powder, soluble in acids, insoluble in water, d 5.45, mp 1650C, but converted to Mn_3O_4 if heated in air. Noncombustible.
Derivation: (a) By reduction of the dioxide in hydrogen, (b) by heating the carbonate with exclusion of air.
Grade: Technical.
Use: Textile printing, analytical chemistry, catalyst in manufacture of allyl alcohol, ceramics, paints, colored glass, bleaching tallow, animal feeds, fertilizers, food additive and dietary supplement.

manganous phosphate. (manganese phosphate; manganous orthophosphate).
$Mn_3(PO_4)_2 \cdot 7HOH$.
Properties: Reddish-white powder, soluble in mineral acids, insoluble in water.
Derivation: By the action of orthophosphoric acid on manganous hydroxide.
Use: Conversion coating of steels, aluminum, and other metals.

manganous phosphate, acid. (manganese hydrogen phosphate; manganous phosphate, secondary; manganese phosphate, dibasic).
$MnHPO_4 \cdot 3HOH$.
Properties: Pink powder, contains some tribasic phosphate, soluble in acids, slightly soluble in water.
Use: Feed additive.

manganous pyrophosphate. (manganese pyrophosphate).   CAS: 53731-35-4.
(a) $Mn_2P_2O_7$, (b) $Mn_2P_2O_7 \cdot 3HOH$.
Properties: White powder, d (a) 3.71, mp 1196C, soluble in solutions of potassium or sodium pyrophosphate, insoluble in water.

manganous silicate. (manganese silicate).
CAS: 7759-00-4. $MnSiO_3$.   Occurs naturally as rhodonite.
Properties: Red crystals or yellowish-red powder, insoluble in water, d 3.72, mp 1323C. Noncombustible.
Derivation: Interaction of manganous salts with sodium silicate.
Use: Colorant for glass and ceramic glazes.

manganous sulfate. (manganese sulfate).
CAS: 7785-87-7. $MnSO_4 \cdot 4HOH$.
Properties: Translucent, pale rose-red, efflorescent prisms. Soluble in water, insoluble in alcohol, d 2.107, mp 30C, anhydrous mp 700C, decomposes at 850C.
Derivation: Byproduct of production of hydroquinone, or by the action of sulfuric acid on manganous hydroxide or carbonate.
Grade: Technical, CP, fertilizer, feed.
Use: Fertilizers, feed additive, paints and varnishes, ceramics, textile dyes, medicines, fungicides, ore flotation, catalyst in viscose process, synthetic manganese dioxide.

manganous sulfide. (manganese sulfide).
CAS: 18820-29-6.   MnS.   Green crystals, d 3.99, decomposes on melting, almost insoluble in water.
Use: Additive in steel making.

manganous sulfite. (manganese sulfite; manganous sulfite, normal). $MnSO_3$.
Properties: Grayish-black or brownish-red powder, soluble in solution of sulfur dioxide, insoluble in water.

Manila fiber.   See abaca.

Manila resin.   A type of copal resin similar to Congo and kauri.
Properties: Soluble in ether, methanol, and ethanol; partially soluble in amyl alcohol; insoluble in water; d 1.062; mp 230–250C. Combustible.
Occurrence: Philippine Islands and East Indies.
Use: Varnishes, paints, lacquers, printing inks.

manna.   The water-soluble exudate of a plant occurring in the Mediterranean area, used in medicine as a laxative. It has a high carbohydrate content, especially of mannitol.

Mannheim furnace.   A muffle furnace used for the manufacture of salt cake and hydrogen chloride. It consists of a firebrick compartment in which is a circular pan some 10–15 ft in diameter, usually made of cast iron. The charge is fed through a hopper at the top and plows maintain continuous movement of the materials being heated. The temperature of the hydrogen chloride gas evolved is approximately 145C. Salt cake is removed through an opening at the side. The furnace is run continuously. The Mannheim furnace is no longer widely used, since most salt cake is now obtained from natural sources.

Mannich reaction.   Reaction of active methylene compounds with formaldehyde and ammonia or primary or secondary amines to give β-aminocarbonyl compounds.

mannitol.   (manna sugar; mannite).
CAS: 69-65-8. $C_6H_8(OH)_6$.   A straight-chain hexahydric alcohol.

$$
\begin{array}{c}
CH_2OH \\
| \\
HCOH \\
| \\
HCOH \\
| \\
HOCH \\
| \\
HOCH \\
| \\
CH_2OH
\end{array}
$$

Properties: White, crystalline powder or granules; odorless; faint sweet taste. Soluble in water, slightly soluble in lower alcohols and amines, almost insoluble in other organic solvents, d 1.52, mp 165–167C, specific rotation (20C) between +23 and +24 degrees, bp 290–295C (3–3.5 mm). Combustible.

Derivation: (1) A natural plant exudate; (2) by hydrogenation of corn sugar or glucose.

Grade: Reagent, commercial, NF, FCC.

Use: Base for dietetic foods, diluent, determination of boron, pharmaceutical products, medicine, thickener and stabilizer in food products.

mannitol hexanitrate. (hexanitromannite; HNM; nitromannite; nitromannitol).
CAS: 15825-70-4. $C_6H_8(ONO_2)_6$.
Properties: Colorless crystals; mp 112–113C; soluble in alcohol, acetone, ether; insoluble in water.
Derivation: By nitrating mannitol with mixed acid, purifying the precipitate from organic solvents and stabilizing.
Grade: Technical, pharmaceutical.
Hazard: Dangerous fire and explosion risk; an initiating explosive.
Use: Explosive cap ingredient, medicine (admixed with at least 10 parts of carbohydrate).

d(+)-mannose. CAS: 3458-28-4. $C_6H_{12}O_6$.
A carbohydrate occurring in some plant polysaccharides.
Properties: Crystals from alcohol or acetic acid, sweet taste with bitter after-taste, mp 132C (decomposes).
Use: Biochemical research.

"Maprofix."[328]   TM for a series of sodium, ammonium, and magnesium lauryl sulfates and lauryl ether sulfates, sodium myristyl sulfate, and several di- and triethanolamine lauryl sulfates. These are anionic detergents used for foaming, wetting, and emulsifying in cosmetic, household, and industrial uses.

"Marabond."[121]   TM for a partially purified calcium lignosulfonate.
Use: Oil well cement retarders, foundry supplies, ceramic products.

"Maracarb."[121]   TM for a series of chelating humectant and dispersing agents which are complex mixtures of the salts of lower molecular weight lignosulfonic acids and the salts of the alkaline reversion products of hexoses and pentoses which are produced from wood in the sulfite pulping process. Available in liquid and powder form.
Use: Fertilizers, agricultural chemicals, dyestuff pastes.

maraging steel. An alloy steel containing nickel, cobalt, and titanium.
Use: Solid rocket cases.

"Marasperse."[121]   TM for a series of lignosulfonates used as dispersants or emulsion-stabilizing agents. The basic lignin monomer unit is a substituted phenyl propane. Available in various types for specific uses.
Properties: Brown powder, completely soluble in water, insoluble in oils and most organic solvents.
Use: Dyestuffs, oil-well drilling fluids, gypsum board, agricultural chemical formulations, industrial cleaners, carbon black dispersions, ceramics.

Marathon-Howard process.   A treatment of waste sulfite liquor from sulfite pulp manufacture to recover chemicals and reduce stream pollution. The waste sulfite is treated with lime and precipitates (1) calcium sulfite for use in preparing fresh cooking acid for the sulfite pulp process and (2) a basic calcium salt of lignin sulfonic acid (lignin sulfonates) which can be pressed and used as a fuel or used as raw material for vanillin, lignin plastics, and other chemicals. The remaining liquor with its BOD reduced 80% is the effluent.

marble.   A metamorphic form of calcium carbonate, usually containing admixtures of iron and other minerals which impart variegated color patterns. Marble chips are often used as source of carbon dioxide in laboratory experiments.

"Marbon."[525]   TM for a series of high-styrene rubber-reinforcing resins. 800A is an easy-processing resin used with natural and synthetic rubbers for smoother calendering and extrusion, reduced shrinkage, faster mixing with lowered power consumption. 8000 AE is an electrical grade resin.
Properties: Light color, d 1.04.

margaric acid.   See n-heptadecanoic acid.

marijuana.   See cannabis.

marjoram oil.   A yellowish essential oil, optically active.
Use: Soap perfume and in toilet preparations. The Spanish grade is used as a flavoring ingredient.

Markownikoff rule.   When a halogen acid adds to an unsymmetrical ethylenic compound, the halogen usually appears on the carbon atom carrying the smaller number of hydrogen atoms; this order of addition is frequently reversed with hydrogen bromide if peroxides are present (peroxide effect).

"Marlate."[28]   TM for technical grade methoxychlor insecticides, supplied in 50% wettable powder and an emulsifiable formulation containing 2 lb/gal.

marsh gas. See methane.

martensite. The chief constituent of hardened carbon tool steels. It is a solution of carbon or Fe_3C in beta-iron, or an exceedingly fine-grained alpha-iron with carbon or Fe_3C in atomic or molecular dispersion. Carbon content up to 1%; easily obtained by quenching small bodies of hypereutectoid steel in cold water; more difficult to obtain in low-carbon steels.

Martin, Archer J. P. (1910-) An English chemist and engineer who won the Nobel prize for chemistry in 1952 along with Synge. His work involved partition chromatography in analysis which led to development of new antibiotics and amino acids. He received his Doctorate from Cambridge University.

Martinet dioxindole synthesis. Formation of derivatives of dioxindole from esters of mesoxalic acid and aromatic amines or amino quinolines.

martonite. A lachrymator gas containing 80% bromoacetone and 20% chloroacetone.

"Marvinol."[248] TM for polyvinyl chloride thermoplastic resins.
Properties: White powder or pellets. Actual properties vary with selection of stabilizer, plasticizer, pigment, and filler.
Grade: Thirteen types for calendering, molding, extrusion, and coating including high, medium and low molecular weight resins, plastisol resins, and copolymer resins. Also includes rigid, electrical, flexible extrusion, and injection molding compounds.

mash. A mixture of malted barley (or other grain) and water used for preparing wort in brewing operations. Also a mixture of grain, etc., prepared for fermentation in distilling, e.g., "sour mash whiskey."

"Masonex."[92] TM for a hemicellulose extract.
Properties: Liquid containing 65% solids, 55% carbohydrates, pH 5.5, water-soluble; the spray-dried product is also water-soluble with 97% solids and 84% carbohydrates (pH 3.7).
Derivation: From wood fibers after mild acid hydrolysis in steam.
Use: (liquid) Intermediate in furfuryl production, animal feeds; (dried) binder in refractory bricks, tackifier in adhesives.

"Masonite."[92] TM for composition hardboard made by treating wood chips with steam at high pressure and compressing the resulting fibers into mats from which rigid panels are made by hot-

pressing. The fiber is waterproofed with a water repellent sizing agent having a paraffin base.

masonry cement. A group of special cements more workable and plastic than Portland cement. Some are similar to waterproofed Portland cement while others are Portland cement mixed with hydrated lime, crushed limestone, diatomaceous earth or granulated slag.

mass. The quantity of matter contained in a particle or body regardless of its location in the universe. Mass is constant, whereas weight is affected by the distance of a body from the center of the earth (or of any other planet or satellite, e.g., the moon). At extremely high temperatures, for example, the sun's interior, mass is converted into energy. According to the Einstein equation $E = mc^2$, all forms of energy, such as radiant energy and energy of motion, possess a mass equivalent, even though they have no independent rest mass (photons); thus there is no absolute distinction between mass and energy.
See also energy, matter.

mass action, law. See chemical laws (1).

mass conservation law. See chemical laws (4).

mass defect. The difference between the total mass of the constituents of an atomic nucleus (protons and neutrons) taken independently and the actual mass of the nucleus as a whole. The latter is always slightly less than the sum of the masses of the constituents, the difference being the mass equivalent of the energy of formation (binding energy) of the nucleus. This accounts for the high energy release obtained from nuclear fission.
See also fission, nuclear; mass number.

massecuite. Term used in the sugar industry for the mixture of sugar and molasses prior to removal of the molasses.

massicot. PbO.   (1) (lead ocher) Natural lead monoxide, PbO. Contains 92.8% lead; found in US (Colorado, Idaho, Nevada and Virginia). (2) An oxide of lead corresponding to the same formula as litharge (PbO), but having a different physical state.
Properties: It is a yellow powder formed by the oxidation of a bath of metallic lead at approximately 345C so that the oxide formed is not melted; d 9.3, mp 600C. If the oxide is melted, it is converted into litharge.

mass number. The number of neutrons and protons in the nucleus of an atom. Thus, the mass number of normal helium is 4, of carbon 12, of oxygen 16, and of uranium 238. A given atom is characterized by its atomic number, equivalent to the number of protons, which give it its charge and thus determine the kind of element and by its mass number, which includes the neutrons that make up the remainder of its mass. Helium has 2 protons and 2 neutrons (mass number 4 and atomic number 2). Protons and neutrons each have very close to unit mass and since the mass change associated with binding the particles together in the nucleus is very small, the mass number is always within 1/10th unit of the atomic weight of the nuclide.

See also mass, mass defect, atomic weight.

mass spectrometry. A method of chemical analysis in which the substance to be analyzed is heated and placed in a vacuum. The resulting vapor is exposed to a beam of electrons which causes ionization to occur, either of the molecules or their fragments. The ions thus produced are accelerated by an electric impulse and then passed through a magnetic field, where they describe curved paths whose directions depend on the speed and mass-to-charge ratio of the ions. This has the effect of separating the ions according to their mass (electromagnetic separation). Because of their greater kinetic energy, the heavier ions describe a wider arc than the lighter ones and can be identified on this basis. The ions are collected in appropriate devices as they emerge from the magnetic field.

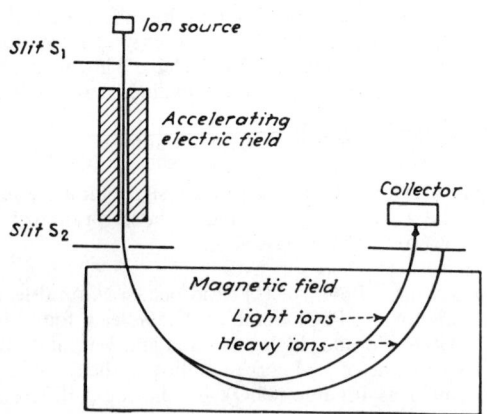

master batch. A previously prepared mixture comprised of a base material and a high percentage of an ingredient (usually a dry powder) that is critical to the product being manufactured. Aliquot parts of this mixture are added to production-size quantities (batches) during the mixing operation. This method permits uniform dis-

persion of very small amounts (less than 1%) of such additives as dry curing agents in rubber and colorants in plastics and paints. Modifying elements may be incorporated in alloys in this way. Master batches of organic dyes dispersed in rubber or plastic are prepared by manufacturers of colorants for direct use. Master-batched accelerators (mixtures of rubber, zinc oxide and accelerator) are commonly used in rubber mixes.

mastic. (1) A solid, resinous exudation of the tree Pistachia lentiscus, of recent geologic origin. Found in the Mediterranean area. Soluble in alcohol and ether, balsamic odor, turpentine taste. Used in chewing gum, varnishes, and to some extent in adhesives and dentistry. (2) A soft, putty-like sealant; often packed in cartridges and applied by a gun with a nozzle of appropriate size or in bulk for application by a knife or other spreading device. Such mastics often contain bituminous ingredients or polymerized rosin acids, used for laying floor and wall tiles and similar applications.

mastication. Permanent softening of crude natural rubber and certain other elastomers by application of mechanical energy, as on a roll mill or in a Banbury mixer. The softening is said to be due to formation of free radicals resulting from the rupture of the polymer chain and the addition of oxygen at these active points. The study of this phenomenon has been called mechanochemistry. This "breaking down" of a high-polymer substance, first practiced by Hancock in England, is essential in preparing it for the incorporation of curatives and other modifying substances and is an essential operation in the manufacture of useful products from rubber and rubber-like substances.

material. A nonspecific term used with various shades of meaning in the technical literature. It should not be used as a synonym for substance, but is generally used in the collective expressions "raw materials" and "materials handling." The term "material balance" in chemical engineering denotes the sum of all the substances entering a reaction and all those that leave it in a given time period. "Material," also loosely refers to closely associated mixture, either of natural origin (petroleum, wood, ores) or manmade (glass, cement, composites).

See also substance, engineering material, raw material.

material balance. See material.

materials handling. A general term that includes the methods used for in-plant transportation, distribution, and storage of raw materials and semi-processed products, as by fork-lift trucks, elevators, conveyor systems, pipelines, etc., as well as safe practices for storage and movement of toxic and flammable substances.

matte. A product containing a metal sulfide as obtained after roasting and fusion of sulfide minerals. Oxides or metals may also be present. Common examples are copper matte and nickel matte.

matter. Anything that has mass or occupies space.
See also mass.
(a) *States*: There are three generally accepted states (phases) in which substances can exist, namely, solid, liquid, and gas (vapor). From time to time it has been proposed that specialized forms of matter be regarded as states, such as the vitreous (glassy) state, the colloidal state, and the plasma state, but none of these suggestions has gained substantial acceptance.
See phase (1), solid, liquid, gas.
(b) *Levels*: Matter is basically composed of particles in the following levels of size and complexity: (1) subatomic (protons, neutrons, electrons); (2) atomic and molecular (below 10 angstroms), (3) colloidal (from 10 angstroms to 1 micron), (4) microscopic, (5) macroscopic, (6) space or celestial. Level (1) is invisible by any means, levels (2) and (3) can be resolved in field-ion or electron microscopes, level (4) lies in the range of the optical microscope, (5) is visible to the naked eye, (6) requires telescopes for detailed observation.
See also particle, electron microscope, field-ion microscope.

Mayer's reagent. See mercuric potassium iodide.

mayonnaise. A semiliquid salad dressing composed of vegetable oil, vinegar, and egg yolk. It is a typical colloidal emulsion in which the mutually repellent base ingredients are brought together and held in a stable form by the lecithin in egg yolk.
See also emulsion.

MBMC. Abbreviation for monobutyl-m-cresol.
See tert-butyl-m-cresol.

MBT. Abbreviation for mercaptobenzothiazole.

MBTS. Abbreviation for 2-mercaptobenzothiazyl disulfide.
See 2,2'-dithiobisbenzothiazole.

"MBX."[11] TM for 2,4-dimethyl-6-tert-butylphenol.
Use: Liquid antioxidant.

mC. Abbreviation for milliCurie.
See Curie.

MCA. Abbreviation for monochloroacetic acid.
See chloroacetic acid.

"MCC."[84] TM for caustic soda flake.
Use: Cleaning and sterilizing milk, beer, and beverage bottles.

McFadyen-Stevens reaction. Base-catalyzed thermal decomposition of aroylbenzenesulfonylhydrazines to aldehydes.

McLafferty rearrangement. Electron-impact-induced cleavage of carbonyl compounds having a hydrogen in the gamma-position, to an enolic fragment and an olefin.

McMillan, Edwin M. (1907-) An American physicist who won the Nobel prize for chemistry in 1951 along with Seaborg. His work included research in nuclear physics and particle accelerator development as well as microwave radar and sonar. He and his colleagues discovered neptunium and plutonium. He was the recipient of the Atoms for Peace prize in 1963. His PhD in physics was awarded from Princeton University.

MCPA. (2-methyl-4-chlorophenoxyacetic acid; 4-chloro-2-methylphenoxyacetic acid).
CAS: 94-74-6. $CH_3ClC_6H_3OCH_2COOH$.
Properties: White, crystalline solid; mp 118–119C; free acid insoluble in water but sodium and amine salts are soluble.
Grade: Emulsifiable concentration.
Use: Selective herbicide.

MCPB. (4-(2-methyl-4-chlorophenoxy)butyric acid). CAS: 94-81-5.
Hazard: Toxic by ingestion.
Use: Herbicide.

MCPP. (2-(2-methyl-4-chlorophenoxy)-propionic acid).
See mecoprop.

Md. Symbol for mendelevium.

MDA. (1) Abbreviation for metal deactivator, (2) Abbreviation for p,p'-methylenedianiline; p,p'-diaminodiphenylmethane.

MDAC. 4-methyl-7-diethylaminocoumarin.
See 7-diethylamino-4-methylcoumarin.

MDI. methylene di-p-phenylene isocyanate.
See diphenylmethane-4,4'-diisocyanate.

Me. Symbol for methyl.

MEA. Abbreviation for monoethanolamine. See ethanolamine.

mean free path. In a given system, the average distance that particles travel between collisons with other particles. The particles may be molecules of a gas, free electrons, neutrons, etc. The value is determined by statistical methods.

mecamylamine hydrochloride. (methylaminoisocamphane hydrochloride). $C_{11}H_{21}N \cdot HCl$.
Properties: White, crystalline powder; almost odorless; mp 245C with some decompostion; soluble in water, alcohol, chloroform; somewhat soluble in benzene, isopropyl alcohol; insoluble in ether.
Grade: USP.
Use: Medicine (antihypertensive).

mecarbam. (S-[(ethoxycarbonyl)methylcarbamoyl]methyl-O,O-diethyl phosphorodithioate). CAS: 2595-54-2.
$(C_2H_5O)_2P(S)SCH_2CON(CH_3)COOC_2H_5$.
Properties: Yellow oil, bp 144C (0.02 mm).
Hazard: Highly toxic, cholinesterase inhibitor, use may be restricted.
Use: Insecticide.

mechanochemistry. See mastication.

mechlorethamine hydrochloride. [methyl-bis-(2-chloroethyl)amine hydrochloride].
CAS: 55-86-7. $CH_3N(CH_2CH_2Cl)_2 \cdot HCl$.
A nitrogen mustard.
Properties: White, crystalline, hygroscopic powder; nasal irritant and a vesicant; soluble in water; mp 108–111C.
Grade: USP.
Hazard: Highly toxic, vesicant, and strongly irritant to mucous membranes.
Use: Medicine (antineoplastic).

meclizine hydrochloride. (1-p-chlorobenzhydryl-4-methylbenzylpiperazine dihydrochloride).
CAS: 36236-67-6. $C_{25}H_{27}ClN_2 \cdot 2HCl \cdot HOH$.
Properties: White or yellowish powder or crystals; slight odor; insoluble in water and ether, very soluble in chloroform, pyridine, and acid-alcohol-water mixture; slightly soluble in dilute acids and alcohol.
Grade: USP.
Use: Medicine (antihistamine).

meclofenoxate hydrochloride. CAS: 3685-84-5. $C_{12}H_{17}Cl_2NO_3$.
Properties: Colorless crystals, mp 135C, soluble in water, insoluble in benzene.
Use: Plant growth hormone.

"Mecopex."[401] TM for a broad-leaf herbicide containing potassium salt of mecoprop (31.5%).
Hazard: See mecoprop.

mecoprop. (2-(4-chloro-2-methylphenoxy)propionic acid; MCPP; CMPP).   CAS: 93-65-2.
$ClC_6H_3(CH_3)OCH(CH_3)COOH$.
Properties: Solid; mp 93–94C; insoluble in water; soluble in alcohol, acetone, and ether.
Hazard: Toxic by ingestion and inhalation, irritant to skin and eyes.
Use: Herbicide.

medicinal chemistry. A subdivision of chemistry which deals with the effects of drugs and pharmaceuticals on the human body and on various infective organisms and with the synthesis of compounds specifically for certain diseases, such as antimalarials and antihypertensive agents. It also is concerned with immunology, hormone activity, etc.
See also: clinical chemistry.

medlure. (sec-butyl-4-(or 5)-chloro-2-methylcyclohexanecarboxylate). $C_{12}H_{21}ClO_2$.
Properties: Liquid, bp 78–79C (0.25 mm), insoluble in water, soluble in most organic solvents.
Use: Insect attractant.

medroxyprogesterone acetate. (17-hydroxy-6a-methyl-preg-4-ene-3,20-dione-17-acetate).
CAS: 71-58-9. $C_{24}H_{34}O_4$.   A hormone derivative.
Properties: White to off-white, odorless, crystalline powder; stable in air. Insoluble in water, freely soluble in chloroform, sparingly soluble in alcohol, melting range 200–208C.
Use: Medicine (injectable contraceptive).

Meerwein-Ponndorf-Verley reduction.
Reduction of aldehydes or ketones to the corresponding alcohols with aluminum alkoxides.

mega-. Prefix meaning 10^6 units (symbol M), e.g., 1 megaton = 1,000,000 tons.

megatomoic acid. (trans-3,cis-5-tetradecadienoic acid). The active ingredient of the sex attractant in the female carpet beetle. Research is underway on synthesis of this substance to be used as a lure in control of this pest.

megaton. One million tons, usually used in defining the blast effect of a nuclear explosion. A 1-megaton bomb is equivalent in destructive potential to 1 million tons of TNT.

megestrol acetate. (17-hydroxy-6-methylpregna-4,6-diene-3,20-dione acetate).
CAS: 595-33-5. $C_{24}H_{32}O_4$.
Properties: Crystals with mw 384.50, mp 214–216C. Soluble in water.
Use: Antineoplastic and oral contraceptive.

meglumine diatrizoate. (methylglucamine diatrizoate; diatrizoate methylglucamine).
CAS: 131-49-7.
$(CH_3CONH)_2C_6I_3COOH$
$\cdot CH_3NHCH_2(CHOH)_4 \cdot CH_2OH.$
Properties: Available in solution for injection, pH between 6.0–7.6.
Grade: USP (as injection).
Use: Medicine (radiopaque medium).
Note: The iodipamide and iothalamate are also available.

Meisenheimer complexes. Sigma-complexes obtained as brightly colored solutions on interaction of 1,3,5-trinitrobenzene to similar compounds.

Meisenheimer rearrangement. Rearrangement of tertiary amine oxide to O,N,N-trisubstituted hydroxylamines.

MEK. Abbreviation for methyl ethyl ketone.

melamine. (cyanurtriamide; 2,4,6-triamino sym-triazine). CAS: 108-78-1.

Properties: White, monoclinic crystals; sparingly soluble in water, glycol, glycerol, pyridine; very slightly soluble in ethanol; insoluble in ether, benzene, carbon tetrachloride; d 1.573 (14C); mp 354C; bp sublimes. Nonflammable.
Derivation: (1) By heating urea and ammonia. The resulting mixture of isocyanic acid and ammonia reacts over a solid catalyst at approximately 400C to form melamine. (2) From cyanamide, dicyanamide or cyanuric chloride.
Method of purification: Recrystallization from water.
Grade: 99% min.
Hazard: Toxic by ingestion, skin, and eye irritant.
Use: Melamine resins, organic syntheses, leather tanning.

melamine resin. A type of amino resin made from melamine and formaldehyde. The first step in resin formation is the production of trimethylol melamine, $C_3N_3(NHCH_2OH)_3$, the molecules of which contain a ring with 3 carbon and 3 nitrogen atoms, the —NHCH₂OH groups being attached to the carbon atoms. This molecule can combine further with others of the same kind by a condensation reaction. Excess formaldehyde or melamine can also react with trimethylol melamine or its polymers, providing many possibilities of chain growth and crosslinking. The nature and degree of polymerization depend upon pH, but heat is always needed for curing. Melamine resins are more water- and heat-resistant than urea resins. They may be water-soluble syrups (low molecular weight) or insoluble powders (high molecular weight) dispersible in water. Widely used as molding compounds with alpha-cellulose, wood flour, or mineral powders as fillers and with coloring materials; also for laminating, boilproof adhesives, increasing wet strength of paper, textile treatment, leather processing, and for dinnerware and decorative plastic items.

Butylated melamine resins are formed by incorporating butyl or other alcohols during resin formation, whereupon the —NHCH₂OH groups convert to —NHCH₂OC₄H₉. These resins are soluble in paint and enamel solvents and in surface coatings, often in combination with alkyds. They give exceptional curing speed, hardness, wear resistance, and resistance to solvents, soaps, and foods.

Melamine-acrylic resins are water-soluble and are used for formation of water-base industrial and automotive finishes.
See also urea-formaldehyde resin.

melaniline. See diphenylguanidine.

melanin. A brownish-black pigment that occurs normally in the retina, skin, and hair of higher animals with the exception of albinos. Formed from tyrosine by the action of tyrosinase.

melissic acid. (triacontanoic acid).
$CH_3(CH_2)_{28}COOH.$ A long-chain fatty acid.
Properties: Crystalline solid, mp 94C, soluble in benzene and hot alcohol, insoluble in water. Combustible.
Derivation: By oxidation of 1-triacontanol, occurs in minor amounts in many plant and insect waxes and in montan wax.
Use: Biochemical research.

melissyl alcohol. See 1-triacontanol and 1-hentriacontanol.

melittin. CAS: 20449-79-0. $C_{131}H_{229}N_{39}O_{31}.$
A polypeptide derived from bee venom that has strong antibacterial activity, especially against Staphylococcus aureus 80 which is resistant to penicillin. It inhibits growth of many Gram-positive and Gram-negative bacteria.
Properties: White water-soluble powder.
Use: Antirheumatic drug.

"Melmac."[57]   TM for products molded from melamine-formaldehyde resins.

"Melonal."[227]   TM for 2,6-dimethyl-5-hepten-l-al.

melphalan.   (p-di(2-chloroethyl)aminophenylala-nine) formerly called sarcolysin.
CAS: 148-82-3.
$(ClCH_2CH_2)_2NC_6H_4CH_2CH(NH_2)COOH$.
Melphalan is both the USAN name for the acid and the generic name for the hydrochloride.
Properties: A nitrogen mustard, crystals, mp 180C.
Grade: ND (in medicine, for the acid).
Hazard: Strong irritant to eyes and mucous membranes.
Use: Medicine, insect chemosterilant.

melt index.   The viscosity of a thermoplastic polymer at a specified temperature and pressure, it is a function of the molecular weight. Specifically, the number of grams of such a polymer that can be forced through a 0.0825 inch orifice in 10 minutes at 190C by a pressure of 2160 g.

melting point.   The melting point or freezing point of a pure substance is the temperature at which its crystals are in equilibrium with the liquid phase at atmospheric pressure. The term "melting point" (mp) is used when the equilibrium temperature is approached by heating the solid. Ordinarily mp refers to temperatures above 0C, the melting point of ice. The terms "melting point" and "freezing point" are often used interchangeably, depending on whether the substance is being heated or cooled. The number of calories required to convert one mole of pure crystals to the liquid state is called the molar heat of fusion.

"Meltopax."[337]   TM for a double silicate containing 55.5–57.5% zirconium oxide, 27–29% silicon dioxide, and 13.5–15.5% sodium oxide; mp 1426C.
Use: Ingredient for enamel frits and special glasses.

"Melurac."[57]   TM for urea-melamine-formaldehyde condensation products used mainly as adhesives for bonding of veneers for the production of exterior grade plywood.

membrane cell.   See diaphragm cell.

membrane, semipermeable.   A microporous structure, either natural or synthetic, which acts as a highly efficient filter in the range of molecular dimensions, allowing passage of ions, water, and other solvents and very small molecules, but al-most impermeable to macromolecules (proteins) and colloidal particles. The thickness is about 100Å, the pore diameter is from 8Å for the walls of tissue cells to 100Å or more for manufactured membranes. Plant cell wall membranes are proteinaceous substances which function in natural osmosis. Membranes of cellophane, collodion, asbestos fiber, etc., are used in such industrial operations as waste liquor recovery, desalination, and electrolysis; processing whey proteins, biomedical research, gas separations, e.g., adjusting carbon monoxide/hydrogen ratios for ammonia production from synthesis gas; metal extraction and recovery.
See also osmosis, dialysis.

memtetrahydrophthalic anhydride.   (methyl norbornene dicarboxylic anhydride).
CAS: 85-43-8. $C_8H_8O_3$.
Properties: Clear, transparent, slightly viscous liquid; colorless to light yellow.
Hazard: Strong irritant to eyes and skin.
Use: Curing epoxy resins, electrical laminating and filament winding, intermediate for polyesters, alkyd resins, and plasticizers.

MENA.   Abbreviation for the methyl ester of naphthaleneacetic acid.
See α-naphthaleneactic acid methyl ester.

menadione.   (2-methyl-1,4-naphthoquinone; menaphthone; vitamin K_3).   CAS: 58-27-5.
$C_{10}H_5CH_3O_2$.
Properties: Yellow, crystalline powder; nearly odorless; mp 105–107C; affected by sunlight. Soluble in alcohol, benzene, and vegetable oils; insoluble in water.
Derivation: Oxidation of β-methylnaphthalene.
Grade: USP.
Hazard: Irritant to skin and mucous membranes, especially the alcoholic solution.
Use: Medicine, fungicides, animal feed additives.

menazon.   (Generic name for S-[(4,6-diamino-S-triazin-2-yl)methyl]-O,O-dimethyl phosphorodithioate).   CAS: 78-57-9.
$(CH_3O)_2P(S)SCH_2C{:}NC(NH_2){:}NC(NH_2){:}N$.
Properties: Off-white solid, mp 160–162C, slightly soluble in water and organic solvents.
Hazard: Highly toxic, cholinesterase inhibitor.
Use: Acaricide, insecticide.

Mendeleef, D. I.   (1834–1907) Born in Siberia, Mendeleef made a fundamental contribution to chemistry in 1869 by establishing the principle of periodicity of the elements. His first Periodic Table was compiled on the basis of arranging the elements in ascending order of atomic weight and grouping them by similarity of properties. So accurate was Mendeleef's thinking that he predicted the existence and atomic weights of several elements that were not actually discovered until years later. The original table has been modified and corrected several times, notably by Moseley, but it has accommodated the discovery of isotopes, rare gases, etc. Its importance in the development of chemical theory can hardly be overestimated.

See Becquerel, Moseley, periodic table.

mendelevium. CAS: 7440-11-1. Md.
Synthetic radioactive element produced in a cyclotron by bombarding einsteinium with alpha particles; atomic number 101, aw 256, 4 isotopes, valence = 3. Mendelevium decays by spontaneous fission with a half-life of 1.5 hour. The heaviest isotope, Md-258, has half-life of 60D. Mendelevium is thought to have chemical properties similar to those of the rare earth thulium. It is made in research quantities only and no uses are reported.

See actinide series.

menhaden oil.
Properties: Yellowish-brown or reddish-brown, drying oil; characteristic odor; soluble in ether, benzene, naphtha, and carbon disulfide; d 0.927–0.933; saponification value 191–196; iodine value 139–180; refr index 1.480. Combustible.
Derivation: By cooking or pressing the body of the menhaden fish.
Method of purification: Filtration and bleaching with Fuller's earth.
Grade: Prime crude, brown strained, strained, bleached, winter oil, bleached winter white oil.
Hazard: Subject to spontaneous heating.
Use: Hydrogenated fats for cooking and industrial use (soap, rubber compounding), printing inks, animal feed, leather dressing lubricants, paint drier, cleansers, lipstick.

meniscus. The concave curve of a liquid surface in a graduate or narrow tube. Caused by surface tension. In reading a value (e.g., 5 cc), it is conventional to ignore the higher liquid around the perimeter. In the case of mercury, which does not wet the tube because of its extremely high surface tension, the meniscus is convex.

Menschutkin reaction. Reaction of tertiary amines with alkyl halides to form quaternary salts.

menthanediamine. (p-menthane-1,8-diamine).
CAS: 80-52-4.

$(CH_3)_2C(NH_2)CHCH_2CH_2C(CH_3)(NH_2)CH_2CH_2$.

A primary alicyclic diamine, also a tert-alkyl-amine.
Properties: Clear liquid, terpene odor, boiling range 107–126C (10 mm), fp −45C, refr index 1.4794 (25C), miscible with water and most organic solvents.
Hazard: Strong irritant to eyes and skin, calls for eye protection.
Use: Curing agent for epoxy resins, chemical intermediate.

p-menthane-8-hydroperoxide. CAS: 80-47-7.
$C_{10}H_{20}O_2$.
Properties: Clear, pale yellow liquid; d 0.910–0.925 (15.5/4C); refr index 1.460–1.475 (20C).
Hazard: Strong oxidizing agent, dangerous in contact with organic materials. Strong irritant to skin and eyes.
Use: Catalyst for rubber and polymerization reactions, coatings.

p-menthan-3-one. See menthone.

menthol. (hexahydrothymol; methylhydroxyiso-propyl-cyclohexane; p-menthan-3-ol; peppermint camphor). CAS: 89-78-1.
$CH_3C_6H_9(C_3H_7)OH$. It may be l- (from natural sources) or dl- (natural or synthetic).
Properties: White crystals with cooling odor and taste; mp 41–43C (l-form); congealing temperature 27–28C (dl-form), specific rotation (25C) (l-menthol) −45 to −51 degrees, (dl-menthol) −2 to +2 degrees; soluble in alcohol, light petroleum solvents, glacial acetic acid, and fixed or volatile oils; slightly soluble in water. Combustible.
Occurrence: Brazil (natural product), Japan.
Derivation: By freezing from peppermint oil, by hydrogenation of thymol.
Grade: Technical, USP, FCC.
Hazard: Irritant to mucous membranes on inhalation.
Use: Perfumery, cigarettes, liqueurs, flavoring agent, chewing gum, chest rubs, cough drops.

menthol acetic ester. See menthyl acetate.

menthol valerate. (menthyl isovalerate).
CAS: 89-47-4. $(CH_3)_2CHCH_2COOC_{10}H_{19}$.
Properties: Colorless liquid; mild odor; cooling, faintly bitter taste; d 0.907 (15.4C); insoluble in water; soluble in alcohol, chloroform, ether, and oils.
Derivation: By action of valeric acid on menthol.
Use: Medicine, flavoring.

menthone. (p-menthan-3-one).
CAS: 1074-95-9. $C_{10}H_{18}O$.
Properties: Colorless, oily, mobile liquid; slight peppermint odor. Slightly soluble in water, solu-

ble in organic solvents, d 0.897 (15C), bp 207C. Combustible.
Derivation: A ketone found in oil of peppermint.
Use: Flavoring.

menthyl acetate. (menthol acetic ester).
CAS: 89-48-5. $C_{10}H_{19}OOCCH_3$.
Properties: Colorless liquid, menthol-like odor, slightly soluble in water, miscible with alcohol and ether, bp 227–228C, d 0.922–0.927, optical rotation −72 degrees 47 minutes to −73 degrees 18 minutes, refr index 1.447. Combustible.
Derivation: (a) By boiling menthol with acetic anhydride in the presence of sodium acetate. (b) Peppermint oil.
Use: Perfumery, flavoring.

menthyl isovalerate. See menthol valerate.

menthyl salicylate. CAS: 89-46-3. $C_6H_4(OH)COOC_{10}H_{19}$.
Properties: Colorless liquid; miscible with alcohol, ether, chloroform, and fatty oils; insoluble in water; soluble in organic solvents. Combustible.
Use: Sunscreen preparations.
See also homomenthyl salicylate.

MEP. Abbreviation for methyl ethyl pyridine.

meperidine hydrochloride. (Demerol).
CAS: 50-13-5. $C_{15}H_{21}NO_2HCl$. An addictive drug, use by prescription only.
Use: Analgesic.

mepivacaine. (Carbocaine; 1-methyl-2-(2,6-xylylcarbamoyl)piperidine).
Use: Local anesthetic.

meprobamate. (2-methyl-2-n-propyl-1,3-propanediol dicarbamate). CAS: 57-53-4.
$H_2NCOOCH_2C(CH_3)(C_3H_7)CH_2OOCNH_2$.
Properties: White powder, mp 103–107C, characteristic odor and bitter taste, slightly soluble in water and ether, soluble in alcohol and acetone.
Grade: NF.
Hazard: Central nervous system depressant, use restricted by law.
Use: Medicine (antianxiety agent).

meq. Abbreviation for milliequivalent. See eq.

merbromin. (dibromohydroxymercurifluorescein disodium salt; 2,7-disodiumdibromo-4-hydroxymercurifluorescein). CAS: 129-16-8.
$C_{20}H_8Br_2HgNa_2O_6$.
Properties: Iridescent green scales or granules; odorless; soluble in water; insoluble in alcohol, acetone, chloroform, or ether; stable in air.
Derivation: From dibromofluorescein and mercuric acetate.

Grade: Technical, NF.
Hazard: Toxic by ingestion. TLV (as mercury): 0.05 mg/m³ of air.
Use: Medicine (antiseptic).

mercaptamine. See 2-aminoethanethiol.

mercaptan. See thiol.

mercaptoacetic acid. See thioglycolic acid.

2-mercaptobenzoic acid. See thiosalicylic acid.

2-mercaptobenzothiazole. (MBT).

CAS: 149-30-4. CHCHCHCHCCSC(SN)N.

Properties: Yellowish powder; distinctive odor (depends on degree of purification); soluble in dilute caustic, alcohol, acetone, benzene, chloroform; insoluble in water and gasoline; d 1.52; melting range 164–175C. Combustible.
Grade: Technical, 97%.
Use: Vulcanization accelerator for rubber (requires use of stearic acid for full activation), tire treads and carcasses, mechanical specialties, etc., fungicide, corrosion inhibitor in cutting oils and petroleum products, extreme-pressure additive in greases.

2-mercaptobenzothiazyl disulfide. See 2,2'-dithiobisbenzothiazole.

2-mercaptoethanol. $HSCH_2CH_2OH$.
CAS: 60-24-2.
Properties: Water-white, mobile liquid with disagreeable odor. D 1.1168 (20/20C), 9.29 lb/gal (20C), bp 157.1C, vap press 1.0 mm (20C), viscosity 3.43 cp (20C), miscible in water, and most organic solvents, flash p 165F (74C), fp sets to a glass approximately −100C, refr index 1.5011 (20C). Combustible.
Hazard: Toxic by inhalation and ingestion.
Use: Solvent for dyestuffs, intermediate for producing dyestuffs, pharmaceuticals, rubber chemicals, flotation agents, insecticides, plasticizers, water-soluble reducing agent, biochemical reagent, PVC stabilizers, agricultural chemicals, textile auxiliary.

β-mercaptoethylamine hydrochloride.
CAS: 156-57-0. $HS(CH_2)_2NH_2 \cdot HCl$.
Properties: Hygroscopic, white powder; mp 71C; very soluble in water and ethanol.
Hazard: Toxic by inhalation and ingestion.

N-(2-mercaptoethyl)benzenesulfonamide.
(S-(O,O-diisopropyl phosphorodithioate; N-β-diisopropyldithiophosphorylethyl)benzenesul-

fonamide).   CAS: 741-58-2.
$C_6H_5SO_2NH(CH_2)_2SP(S)[OCH(CH_3)_2]_2$.
Properties: Colorless liquid, slight odor, d 1.25 (22C), insoluble in water, soluble in kerosene and xylene.
Hazard: Highly toxic, cholinesterase inhibitor, use may be restricted.
Use: Herbicide.

2-mercapto-4-hydroxypyrimidine. See thiouracil.

2-mercaptoimidazoline. See "NA-22."

β-mercaptopropionic acid. $HSCH_2CH_2COOH$.
Properties: Clear liquid, d 1.218 (21C), mp 16.8C, bp 111C (15 mm), soluble in water and alcohol.
Use: Stabilizer, antioxidant, catalyst, chemical intermediate.

6-mercaptopurine. (6-MP; 6-purinethiol).
CAS: 50-44-2. $C_5H_4N_4S \cdot HOH$.   A sulfur-containing purine base not found in animal nucleoproteins.
Properties: Yellow, crystalline powder; mp 308C (decomposes); nearly odorless. Insoluble in water, soluble in hot alcohol and dilute alkali solutions, slightly soluble in dilute sulfuric acid.
Grade: USP.
Use: Medicine (to prevent rejection of kidney transplants).

mercaptosuccinic acid. See thiomalic acid.

2-mercaptothiazoline. C_2H_4NSSH.   White crystals.
Use: Synthesis of pharmaceuticals.

D-3-mercaptovaline. See penicillamine.

mercerized cotton. See cotton, mercerized.

mercerizing assistant. A compound used to increase the penetration of mercerizing baths. Cresylic acid derivatives, special sulfonated oils and other wetting agents are typical.

mercuric acetate. (mercury acetate).
CAS: 1600-27-7. $Hg(C_2H_3O_2)_2$.
Properties: White, crystalline powder; soluble in alcohol and water; sensitive to light; d 3.2544.
Hazard: Toxic by ingestion, inhalation, and skin absorption; strong irritant. TLV (as mercury): 0.01 mg/m³ of air.
Use: Catalyst in organic synthesis, pharmaceuticals.

mercuric arsenate. (mercury arsenate).
CAS: 7784-37-4. $HgHAsO_4$.
Properties: Yellow powder, soluble in hydrochloric acid, slightly soluble in nitric acid, insoluble in water.

Hazard: Toxic by ingestion, inhalation, and skin absorption; strong irritant. TLV (as mercury): 0.05 mg/m³ of air.
Use: Waterproof paints, antifouling paints.

mercuric barium bromide. (barium mercury bromide; mercury barium bromide).
$HgBr_2 \cdot BaBr_2$.
Properties: Colorless, crystalline mass; very hygroscopic; soluble in water.
Hazard: Toxic by ingestion, inhalation, and skin absorption; strong irritant. TLV (as mercury): 0.05 mg/m³ of air.

mercuric barium iodide. (barium mercury iodide; mercury barium iodide). $HgI_2 \cdot BaI_2 \cdot 5HOH$.
Properties: Reddish or yellow crystalline mass, unstable, deliquescent, soluble in alcohol and water.
Hazard: Toxic by ingestion, inhalation, and skin absorption; strong irritant. TLV (as mercury): 0.05 mg/m³ of air.
Use: Microanalysis (testing for alkaloids), preparing Rohrbach solution.

mercuric benzoate. (mercury benzoate).
CAS: 583-15-3. $Hg(C_7H_5O_2)_2 \cdot H_2O$.
Properties: White crystals, mp 165C, sensitive to light, soluble in solutions of sodium chloride and ammonium benzoate, slightly soluble in alcohol and water.
Derivation: By the interaction of a mercuric salt and sodium benzoate.
Hazard: Toxic by ingestion, inhalation, and skin absorption; strong irritant. TLV (as mercury): 0.01 mg/m³ of air.
Use: Medicine (antisyphilitic).

mercuric bromide. (mercury bromide).
CAS: 7789-47-1. $HgBr_2$.
Properties: White, rhombic crystals; sensitive to light. Soluble in alcohol and ether, sparingly soluble in water, d 6.11, mp 235C, bp 322C.
Derivation: By adding potassium bromide to a solution of a mercuric salt and crystallizing.
Hazard: Toxic by inhalation, ingestion, and skin absorption; strong irritant. TLV (as mercury): 0.05 mg/m³ of air.
Use: Medicine.

mercuric chloride. (mercury bichloride; mercury chloride; corrosive sublimate).
CAS: 7487-94-7. $HgCl_2$.
Properties: White crystals or powder, odorless. Soluble in water, alcohol, ether, pyridine, glycerol, and acetic acid; d 5.44 (25C); bp 303C; mp 276C.
Derivation: (a) Direct combination of chlorine with mercury heated to volatilizing point, (b)

by subliming mercuric sulfate with common salt. Method of purification: Recrystallization and sublimation. Impurities: Mercurous chloride. Grade: Technical, crystals, granular, powder, CP, NF.

Hazard: Toxic by ingestion, inhalation, and skin absorption; a poison. TLV (as mercury): 0.05 mg/m³ of air.

Use: Manufacture of calomel and other mercury compounds, disinfectant, organic synthesis, analytical reagent, metallurgy, tanning, catalyst for vinyl chloride, sterilant for seed potatoes; fungicide, insecticide, and wood preservative; embalming fluids, textile printing, dry batteries, photography, process engraving and lithography.

mercuric chloride, ammoniated. See mercury, ammoniated.

mercuric cuprous iodide. (copper mercury iodide; mercury copper iodide). $HgI_2 \cdot 2CuI$.

Properties: Dark red, crystalline powder; d 6.12. Insoluble in alcohol and water.

Hazard: Highly toxic. TLV (as mercury): 0.05 mg/m³ of air.

Use: Thermoscopy (detecting overheating of machine bearings) by reversible color change.

mercuric cyanate. See mercury fulminate.

mercuric cyanide. (mercury cyanide).
CAS: 592-04-1. $Hg(CN)_2$.

Properties: Colorless, transparent prisms darkened by light. Soluble in water and alcohol, d 4.018, mp decomposes.

Derivation: Interaction of mercuric oxide and an aqueous solution of hydrogen cyanide.

Grade: Technical, CP.

Hazard: Toxic by ingestion, inhalation, and skin absorption. TLV (as mercury): 0.05 mg/m³ of air.

Use: Medicine (antiseptic), germicidal soaps, manufacturing cyanogen gas, photography.

mercuric dichromate. See mercury dichromate.

mercuric dimethyldithiocarbamate.
$C_6H_{12}N_2S_4Hg$.

Hazard: Highly toxic by ingestion, inhalation, and skin absorption. TLV (as mercury): 0.01 mg/m³ of air.

Use: Turf fungicides.

mercuric fluoride. (mercury fluoride).
CAS: 7783-39-3. HgF_2.

Properties: Transparent crystals, d 8.95, mp 645C decomposes, moderately soluble in water and alcohol.

Derivation: Mercuric oxide and hydrogen fluoride.

Hazard: Highly toxic by ingestion, inhalation, and skin absorption; strong irritant. TLV (as mercury): 0.05 mg/m³ of air.

Use: Synthesis of organic fluorine compounds.

mercuric iodide. (mercury iodide, red).
CAS: 7774-29-0. HgI_2.

Properties: Red, tetragonal crystals; turn yellow when heated to 130C, returning to red on cooling. Soluble in boiling alcohol and in solutions of sodium thiosulfate or potassium iodide or other hot alkali chloride solutions, almost completely insoluble in water, d 6.28, mp 259C, bp 349C.

Derivation: By the direct union of mercury and iodine.

Grade: Technical, reagent.

Hazard: Highly toxic by ingestion, inhalation, and skin absorption; strong irritant. TLV (as mercury): 0.05 mg/m³ of air.

Use: Medicine (antiseptic), analytical reagents (Nessler's reagent, Mayer's reagent).

See also mercuric potassium iodide.

mercuric lactate. $Hg(C_3H_5O_3)_2$.

Properties: White, crystalline powder; soluble in water; decomposed by heat.

Hazard: Toxic by ingestion, inhalation, and skin absorption. TLV (as mercury): 0.01 mg/m³ of air.

Use: Fungicide.

mercuric naphthenate. (mercury naphthenate).

Properties: Dark amber liquid, soluble in lubricating oils and mineral spirits, bulk d 10.4 lb/gal.

Grade: 25% mercury.

Hazard: Toxic by ingestion, inhalation, and skin absorption. TLV (as mercury): 0.05 mg/m³ of air.

Use: Mildew-resistance promoter in paints.

mercuric nitrate. (mercury nitrate; mercury pernitrate).   CAS: 10045-94-0.
$Hg(NO_3)_2$.

Properties: Colorless crystals or white, deliquescent powder; d 4.3; mp 79C. Decomposed by heat, soluble in water and nitric acid, insoluble in alcohol.

Derivation: By action of hot nitric acid on mercury.

Hazard: Dangerous fire risk in contact with organic materials. Very toxic. TLV (as mercury): 0.05 mg/m³ of air.

Use: Nitration of aromatic organic compounds, felt manufacture, mercury fulminate manufacturing.

mercuric oleate. (mercury oleate).
CAS: 1191-80-6C010. $C_{36}H_{66}HgO_4$.

Properties: Yellowish to red semi-solid or solid mass, soluble in fixed oils, insoluble in water, slightly soluble in alcohol or ether.

Derivation: By mixing yellow mercuric oxide with oleic acid.

Grade: Technical, pharmaceutical.

Hazard: Highly toxic. TLV (as mercury): 0.05 mg/m³ of air.

Use: Antiseptic, antifouling paints.

mercuric oxide, red. (red precipitate; mercury oxide, red).   CAS: 21908-53-2.   HgO.

Properties: Heavy, bright orange-red powder; soluble in dilute hydrochloric acid and nitric acid;, insoluble in water, alcohol, and ether; d 11.00–11.29; mp decomposes.

Derivation: By heating mercurous nitrate.

Grade: Technical, paint, ACS reagent.

Hazard: Fire risk in contact with organic materials. Highly toxic. TLV (as mercury): 0.05 mg/m³ of air.

Use: Chemicals, paint pigment, perfumery and cosmetics, pharmaceuticals for topical disinfection, ceramics (pigment), dry batteries, especially for miniaturized equipment, polishing compounds, analytical reagent, antifouling paints, fungicide, antiseptic.

mercuric oxide, yellow. (mercury oxide, yellow; yellow precipitate).   CAS: 21908-53-2. HgO.

Properties: Light, orange-yellow powder; odorless; stable in air but turns dark on exposure to light; finer powder than the red form; d 11.03 (27.5C); mp decomposes; slightly soluble in cold water, more so after boiling; soluble in dilute hydrochloric acid and nitric acid, potassium iodide solution, concentrated solutions of alkaline-earth chloride, magnesium chloride; insoluble in alcohol.

Derivation: (a) By the action of either potassium hydroxide or sodium hydroxide on mercuric chloride. (b) By the action of sodium carbonate upon mercuric nitrate solution.

Grade: CP, technical, NF.

Hazard: Fire risk in contact with organic materials. Highly toxic. TLV (as mercury): 0.05 mg/m³ of air.

Use: Antiseptic, mercury compounds.

See also mercuric oxide, red.

mercuric oxycyanide. CAS: 1335-31-5. HgO·Hg(CN)₂.

Properties: White, crystalline powder; d 4.44; moderately soluble in water.

Hazard: Detonates on heating, dangerous explosion risk. Highly toxic. TLV (as mercury): 0.05 mg/m³ of air.

Use: Medicine (topical antiseptic).

mercuric phosphate. (normal mercuric phosphate; neutral mercuric phosphate; trimercuric orthophosphate; mercuric phosphate, tertiary; mercury phosphate).   Hg₃(PO₄)₂.

Properties: Heavy, white or yellowish powder; soluble in acids; insoluble in alcohol, water.

Hazard: Highly toxic. TLV (as mercury): 0.05 mg/m³ of air.

mercuric potassium cyanide. (mercury potassium cyanide).   CAS: 591-89-9. Hg(CN)₂·2KCN.

Properties: Colorless crystals, soluble in water and alcohol.

Derivation: By mixing mercuric and potassium cyanides and crystallizing.

Hazard: Highly toxic by ingestion, inhalation, and skin absorption. TLV (as mercury): 0.05 mg/m³ of air.

Use: Silvering glass in mirror manufacture.

mercuric-potassium iodide. (Mayer's reagent; potassium mercuric iodide).   K₂HgI₄ or 2KI·HgI₂.

Properties: Odorless, yellow crystals; deliquescent in air; crystallizes with either 1, 2, or 3 molecules of water; the commercial product is the anhydrous form containing approximately 25.5% mercury; d 4.29; neutral or alkaline to litmus; very soluble in water; soluble in alcohol, ether, and acetone.

Derivation: (a) Action of hydrochloric acid and potassium iodide on mercuric cyanide or mercuric chloride. (b) Action of potassium iodide on mercuric oxide.

Hazard: Highly toxic by ingestion, inhalation, and skin absorption. TLV (as mercury): 0.05 mg/m³ of air.

Use: Analytical chemistry.

See also Nesslers reagent.

mercuric salicylate. (salicylated mercury). CAS: 5970-32-1.

Properties: White powder, yellow or pink tinge, a compound of mercury and salicylic acid of somewhat varying composition, mercury replacing both phenolic and carboxylic hydrogen. Contains more than 54% but less than 59.5% mercury, odorless, tasteless, soluble in solutions of the fixed alkalies or their carbonates, and in warm solutions of the alkali halides; insoluble in water and alcohol.

Derivation: By gently heating freshly precipitated yellow mercuric oxide and salicylic acid in presence of water.

Hazard: Highly toxic by ingestion, inhalation, and skin absorption. TLV (as mercury): 0.05 mg/m³ of air.

Use: Medicine (topical antiseptic).

mercuric silver iodide. (mercury silver iodide; silver mercury iodide).   HgI₂·2AgI.

Properties: Yellow powder, becomes red at 40–50C, d 6.08, soluble in solutions of potassium cyanide or potassium iodide, insoluble in dilute acids and water.

Hazard: Highly toxic by ingestion, inhalation, and skin absorption. TLV (as mercury): 0.05 mg/m^3 of air.

Use: Thermoscopy (detecting overheating in journal bearings).

mercuric sodium phenolsulfonate.
CAS: 535-55-7. $C_{12}H_8HgNa_2O_8S_2$.
Properties: Colorless, water-soluble powder.
Derivation: Reaction of sodium salt of p-phenolsulfonic acid and mercuric oxide.
Hazard: Toxic by ingestion and skin absorption.
Use: Antiseptic and germicide, in soaps, ointments, etc. (1% concentration).

mercuric stearate.   (mercury stearate).
CAS: 645-99-8. $(C_{17}H_{35}CO_2)_2Hg$.
Properties: Yellow, granular powder; soluble in fatty oils; insoluble in water and alcohol.
Hazard: Highly toxic by ingestion, inhalation, and skin absorption. TLV (as mercury): 0.05 mg/m^3 of air.
Use: Germicide.

mercuric sulfate.   (mercury persulfate; mercury sulfate).   CAS: 7783-35-9. $HgSO_4$.
Properties: White, crystalline powder. Soluble in acids, insoluble in alcohol, decomposes in water, d 6.466, decomposes at red heat.
Derivation: By the action of sulfuric acid on mercury with subsequent crystallization.
Hazard: Highly toxic by ingestion, inhalation, and skin absorption. TLV (as mercury): 0.05 mg/m^3 of air.
Use: Calomel and corrosive sublimate, catalyst in the conversion of acetylene to acetaldehyde, extracting gold and silver from roasted pyrites, battery electrolyte.

mercuric sulfide, black.   (mercury sulfide, black).
CAS: 1344-48-5.   HgS.
Properties: Black powder; soluble in sodium sulfide solution; insoluble in water, alcohol and nitric acid; d 7.55–7.70; mp sublimes at 446C; mp 583C.
Derivation: By passing hydrogen sulfide into a solution of mercury salt, reaction of mercury with sulfur.
Hazard: Highly toxic by ingestion, inhalation, and skin absorption. TLV (as mercury): 0.05 mg/m^3 of air.
Use: Pigment.

mercuric sulfide, red.   (vermilion; quicksilver vermilion; chinese vermilion; red mercury sulfide;

artificial cinnabar; red mercury sulfuret).
CAS: 1344-48-5.   HgS.
Properties: Fine, bright scarlet powder. Insoluble in water and alcohol, d 8.06–8.12, mp sublimes at 583C.
Derivation: By heating mercury and sulfur with subsequent recovery by sublimation. A precipitated form is known as English vermilion.
Hazard: Highly toxic by ingestion, inhalation, and skin absorption. TLV (as mercury): 0.05 mg/m^3 of air.
Use: Pigment.

mercuric thiocyanate.   (mercuric sulfocyanate; mercuric sulfocyanide; mercury sulfocyanate; mercury thiocyanate).   CAS: 592-85-8.
$Hg(SCN)_2$.
Properties: White powder, slightly soluble in alcohol, insoluble in water, mp decomposes.
Derivation: By precipitation of mercuric nitrate with ammonium sulfocyanate and subsequent solution in a large amount of hot water and crystallizing.
Hazard: Highly toxic by ingestion, inhalation, and skin absorption. TLV (as mercury): 0.05 mg/m^3 of air.
Use: Photography, pyrotechnics.

"Mercurochrome."[348]   TM for merbromin.

mercurous acetate.   (mercury proto-acetate; mercury acetate).   CAS: 631-60-7.
$C_4H_6Hg_2O_4$.
Properties: Colorless scales or plates, decomposed by boiling water and by light into mercury and mercuric acetate, slightly soluble in water, insoluble in alcohol and ether, soluble in dilute nitric acid.
Derivation: Reaction of sodium acetate with mercurous nitrate solution acidified with nitric acid.
Grade: Technical.
Hazard: Highly toxic by ingestion, inhalation, and skin absorption. TLV (as mercury): 0.05 mg/m^3 of air.
Use: Medicine (anti-bacterial).

mercurous acetylide. Hg_2C_2.
Properties: White powder, a salt of acetylene.
Derivation: Reaction of acetylene with aqueous solution of mercurous salts.
Hazard: Severe explosion risk when shocked or heated. Highly toxic.

mercurous bromide.   (mercury bromide).
CAS: 15385-58-7.   HgBr or Hg_2Br_2.
Properties: White powder or colorless crystals, odorless, tasteless, becomes yellow on heating, returning to white on cooling. Darkens on exposure to light; soluble in fuming nitric acid (pro-

longed heating), hot concentrated sulfuric acid, hot ammonium carbonate or ammonium succinate solutions; sparingly soluble in water; insoluble in alcohol and ether; d 7.307; sublimes at 340–350C; mp 405C.
Derivation: (a) Action of potassium bromide on solution of mercurous nitrate in dilute nitric acid. (b) Sublimation from mixture of mercury and mercuric bromide.
Hazard: Highly toxic by ingestion, inhalation, and skin absorption. TLV (as mercury): 0.05 mg/m³ of air.

mercurous chlorate. (mercury chlorate).
CAS: 10294-44-7. $Hg_2(ClO_3)_2$.
Properties: White crystals, d 6.409, mp 250C (decomposes), soluble in alcohol, water, and acetic acid.
Hazard: Explodes in contact with organic or combustible materials, keep away from light. Highly toxic.

mercurous chloride. (mercury monochloride; mercury protochloride; mercury chloride, mild; calomel). CAS: 10112-91-1. Hg_2Cl_2.
Properties: White, rhombic crystals or crystalline powder; odorless; stable in air but darkens on exposure to light; insoluble in water, ether, alcohol, and cold dilute acids; d 6.993; mp 302C; bp 384C; decomposed by alkalies.
Derivation: By heating mercuric chloride and mercury with subsequent sublimation.
Grade: Technical, CP, NF.
Hazard: Toxic dose is uncertain.
Use: Fungicide, electrodes, pharmaceuticals, pyrotechnics, ceramic painting, maggot control in agriculture.

mercurous chromate. (mercury chromate).
Hg_2CrO_4.
Properties: Brick-red powder, variable composition, decomposes on heating, soluble in nitric acid (concentrated), insoluble in alcohol and water.
Hazard: Moderate fire hazard in contact with organic materials. Highly toxic. TLV (as mercury): 0.05 mg/m³ of air.
Use: Ceramics (coloring green).

mercurous iodide. (mercury protoiodide).
CAS: 15385-57-6. HgI or Hg_2I_2.
Properties: Bright yellow powder, becoming greenish on exposure to light due to decomposition into metallic mercury and mercuric iodide. Becomes dark yellow, orange, and orange-red on heating. Undergoes same color change in opposite order on cooling. Odorless and tasteless; soluble in castor oil, liquid ammonia, aqua ammonia; insoluble in water, alcohol, and ether; d

7.6445–7.75; sublimes at 140C; mp 290C (with partial decomposition).
Derivation: (a) Action of potassium iodide on a mercurous salt. (b) Boiling a solution of mercurous nitrate containing nitric acid with excess of iodine.
Grade: Technical.
Hazard: Toxic by ingestion, inhalation, and skin absorption. TLV (as mercury): 0.05 mg/m³ of air.
Use: Medicine (topical antibacterial).

mercurous nitrate, hydrated. CAS: 10415-75-5.
$HgNO_3 \cdot 2HOH$.
Properties: Short prismatic crystals, effloresces and becomes anhydrous in dry air, sensitive to light, soluble in small quantities of warm water (hydrolyzes in larger quantities)and water acidified with nitric acid, d 4.785 (3.9C), mp 70C (decomposes).
Derivation: Action of cold dilute nitric acid upon an excess of mercury and warming slightly.
Hazard: May be explosive if shocked or heated. Highly toxic. TLV (as mercury): 0.05 mg/m³ of air.
Use: Analytical agent.

mercurous oxide. CAS: 15829-53-5. Hg_2O.
Properties: Black powder, d 9.8, decomposes at 100C, soluble in acids, insoluble in water.
Derivation: Action of sodium hydroxide on mercurous nitrate.
Hazard: Highly toxic. TLV (as mercury): 0.05 mg/m³ of air.

mercurous sulfate. CAS: 7783-36-0. Hg_2SO_4.
Properties: White to yellow, crystalline powder; soluble in hot sulfuric acid, dilute nitric acid; slightly soluble in water, d 7.56, decomposes on heating.
Derivation: (a) Dissolving mercury in sulfuric acid and heating gently. (b) Adding sulfuric acid to mercurous nitrate solution.
Hazard: Highly toxic. TLV (as mercury): 0.05 mg/m³ of air.
Use: Chemical (admixed with sulfuric acid as a catalyst, in oxidation of naphthalene to phthalic acid), batteries (Clark cell, Weston cell).

mercury. (quicksilver; hydrargyrum).
CAS: 7439-97-6. Hg. Metallic element of atomic number 80, group IIB of the Periodic Table, aw 200.59, valences = 1,2; 4 stable isotopes and 12 artificially radioactive isotopes.
Properties: Silvery, extremely heavy liquid, sometimes found native; insoluble in hydrochloric acid; soluble in sulfuric acid upon boiling; readily soluble in nitric acid; insoluble in water, alcohol, and ether; soluble in lipids;, extremely high sur-

face tension (480 dynes/cm) giving it unique rheological behavior; high electric conductivity; d 13.59; fp −38.85C; bp 356.6. Noncombustible.

Chief ore: Cinnabar.

Occurrence: Spain, Yugoslavia, Mexico, Canada, Algeria.

Derivation: By heating cinnabar in air or with lime and condensing the vapor.

Method of purification: Distillation; an important proportion of used mercury is recovered by redistillation.

Grade: Technical, virgin, redistilled, ACS.

Hazard: (1) Mercury, metallic: Highly toxic by skin absorption and inhalation of fume or vapor, absorbed by respiratory and intestinal tract. FDA permits zero addition to the 20 micrograms of mercury contained in average daily diet. TLV (as mercury): 0.05 mg/m³ of air. (2) All inorganic compounds of mercury are highly toxic by ingestion, inhalation, and skin absorption. TLV: 0.05 mg/m³ of air. (3) Most organic compounds of mercury are highly toxic; inorganic mercury can be converted to methylmercury by bacteria in water. TLV (as mercury): (alkyl compounds) 0.01 mg/m³ of air; all others 0.05 mg. *Note*: Spillage may be a toxic hazard due to droplet proliferation. Clean-up requires special care.

Use: Amalgams, catalyst, electrical apparatus, cathodes for production of chlorine and caustic soda, instruments (thermometers, barometers, etc.), mercury vapor lamps, extractive metallurgy, mirror coating, arc lamps, boilers, coolant and neutron absorber in nuclear power plants.

mercury, ammoniated. (mercuric chloride, ammoniated; ammonobasic mercuric chloride; ammoniated mercury chloride, white precipitate; white precipitate, fusible; aminomercuric chloride; mercury cosmetic). $HgNH_2Cl$.

Properties: White, pulverulent lumps or powder; earthy, metallic taste. Odorless, stable in air, darkens on exposure to light, soluble in ammonium carbonate and sodium thiosulfate solutions and in warm acids, insoluble in water and alcohol.

Derivation: By precipitating mercuric chloride with ammonium hydroxide in excess.

Grade: USP, technical.

Hazard: Highly toxic. TLV (as mercury): 0.05 mg/m³ of air.

Use: Medicine (local anti-infective), pharmaceuticals.

mercury bichloride. Legal label name for mercuric chloride.

mercury cell. An electrolytic cell for the production of caustic soda and chlorine from sodium chloride brine. Continuously fed brine is decomposed in one compartment between graphite anodes where chlorine is liberated and a mercury cathode where a sodium amalgam is formed. The amalgam flows continuously or intermittently to a second compartment where it is decomposed with water, forming a caustic solution. The decomposition is usually performed electrolytically by making the amalgam anodic with respect to an iron or graphite cathode. Pure water is supplied to the decomposition compartment at such a rate as to maintain a constant concentration of caustic in the product. With respect to the diaphragm cell, the mercury cathode cell has generally a more concentrated solution (50–70%), it has the disadvantages of higher operating voltage and lower efficiency (52–55%) and a high capital investment in mercury. Examples of mercury-cathode cells are Castner cell and DeNora cell.

mercury compounds. See corresponding mercurous or mercuric compound, e.g., mercury chlorate, see mercurous chlorate; mercury oxide, see mercuric oxide.

mercury dichromate. (mercuric dichromate; mercury bichromate). $HgCr_2O_7$.

Properties: Heavy, red, crystalline powder; soluble in acids; insoluble in water.

Hazard: Highly toxic. TLV (as mercury): 0.05 mg/m³ of air.

mercury fulminate. (mercuric cyanate). $Hg(CNO)_2$.

Properties: Gray, crystalline powder. Soluble in alcohol, ammonium hydroxide, and hot water; slightly soluble in cold water; d 4.42; mp explodes.

Derivation: By treating mercury with strong nitric acid and alcohol.

Grade: Technical.

Hazard: Explodes readily when dry, keep moist till use, an initiating explosive. Highly toxic. TLV (as mercury): 0.05 mg/m³ of air.

Use: Manufacture of caps and detonators for producing explosions of military, industrial, and sporting purposes.

mercury selenide. HgSe. Sublimes in a vacuum, d 8.266, insoluble in water.

Hazard: Highly toxic. TLV (as mercury): 0.05 mg/m³ of air.

Use: Semiconductor in solar cells, thin-film transistors, infrared detectors, ultrasonic amplifiers.

mercury telluride. HgTe.

Grade: 99.99%.

Use: Semiconductor in solar cells, thin-film transistors, infrared detectors, ultrasonic amplifiers.

myricyl alcohol. See 1-triacontanol.

"Merlon."[567] TM for a polycarbonate resin.
Properties: Transparent, slightly straw-colored resin, d 1.2, refr index 1.587. Rods, tubes, pipes, sheets, film extrusions, and special types may be produced by extrusion, thermoforming, injection, and blow moldings.
Use: Electrical and electronic component industries, protective equipment, graphic arts and photographic film, cable wrapping and protective overlays for other thermoplastic sheeting, missile applications.

merocyanine. See cyanine dye.

"Merpentine."[28] TM for a surface-active compounded anionic wetting, penetrating and leveling agent used in leather and textile fields.

Merrifield, R. Bruce. (1921-) An American chemist who won the Nobel prize for chemistry in 1984. Merrifield was cited for work on the use of a solid matrix as an aid to chemical synthesis of complex peptides and proteins. His synthesis techniques have been used in the development of solid matrix-bound inorganic and organic agents. PhD awarded at U.C.L.A.

Merrifield solid-phase peptide synthesis.
Synthesis of large peptides by stepwise chain elongation on a polymeric support.

"Mersize."[58] TM for paper-sizing agents. "Mersize" TFL 70%, 77% and "Mersize" RM Dry are forms of chemically fortified rosin-based sizes. "Mersize" CD-2 and CD-2 Dry are used in conjunction with typical rosin sizes.

"Merthiolate."[100] TM for a 0.1% solution of thiomersal in alcohol.
Hazard: Highly toxic by ingestion, for external use only.
Use: General antiseptic.

mescaline. (3,4,5-trimethoxyphenethylamine).
CAS: 54-04-6. $(CH_3O)_3C_6H_2CH_2CH_2NH_2$.
An alkaloid derived from a Mexican cactus.
Properties: Crystals, mp 35–36C, bp 180C. Soluble in water, alcohol, chloroform, benzene; nearly insoluble in ether.
Hazard: Highly toxic by ingestion.
Use: Biochemical and medical research, hallucinogenic drug.

mesh. The number of apertures per square inch of a screen or sieve, it is the square of the number of strands of metal or plastic per linear inch.
See also screen.

mesitylene. (1,3,5-trimethylbenzene; sym-trimethylbenzene).   CAS: 108-67-8.
$C_6H_3(CH_3)_3$.
Properties: Liquid, d 0.863, fp −52.7C, bp 164.6C, insoluble in water, soluble in alcohol and ether, derived from coal tar. Combustible.

Hazard: Moderate fire hazard. Toxic by inhalation.
Use: Intermediate, including anthraquinone vat dyes, UV oxidation stabilizers for plastics.

2-mesitylenesulfonyl chloride. CAS: 68985-08-0.
$1,3,5-(CH_3)_3C_6H(SO_2Cl)_2$.
Properties: White crystalline solid, mw 317.21, mp 123–125C, moisture sensitive.
Hazard: Corrosive.
Use: Pharmaceutical intermediate, protective reagent for the guanidino group in peptide synthesis, selective sulfonating agent.

mesityl oxide. (isopropylideneacetone; methyl isobutenyl ketone; 4-methyl-3-penten-2-one).
CAS: 141-79-7. $(CH_3)_2{:}CHCOCH_3$.
Properties: Oily, colorless liquid; honey-like odor; soluble in water, alcohols, and ethers. D 0.8569 (20/20C), bp 130–131C, vap press 8.7 mm (20C), flash p 90F (32.3C), bulk d 7.1 lb/gal (20C), fp −46.4C, viscosity 0.0060 poise (20C), autoign temperature 652F (344C).
Derivation: Dehydration of acetone or diacetone alcohol.
Grade: Technical.
Hazard: Flammable, moderate fire risk. Toxic by ingestion, inhalation, and skin absorption. TLV: 15 ppm in air.
Use: Solvent for cellulose esters and ethers, vinyl resins, lacquers, roll-coating inks, stains, ore flotation, paint and varnish-removers, insect repellent, manufacture of methyl isobutyl ketone.

meso-. A prefix meaning middle or intermediate, specifically (1) an optically inactive stereoisomeric form resulting from the presence of an even number of dextro and levorotatory isomers in a natural substance. This cancels the optical activity and causes formation of an intermediate structure which is its own mirror image. This effect is called internal compensation. Tartaric acid is an example.
See also racemization, tartaric acid. (2) An intermediate hydrated form of an inorganic acid.

(3) Designating a middle position in certain cyclic organic compounds. (4) A ring system characterized by a middle position of certain rings.

mesomerism. See resonance (1).

mesomorphic. A molecular arrangement intermediate between crystals (solid) and amorphous (liquid) which characterizes liquid crystals.

messenger RNA. See ribonucleic acid.

mesyl chloride. See methanesulfonyl chloride.

Met. Symbol for methionine.

meta-. A prefix. For definition of meta-compounds
See ortho-.

metabolism. The chemical transformations occurring in an organism from the time a nutrient substance enters it until it has been utilized and the waste products eliminated. In animals and man, digestion and absorption are primary steps, followed by a complicated series of degradations, syntheses, hydrolyses, and oxidations, in which agents such as enzymes, bile acids, and hydrochloric acid take part. These transformations are often localized with respect to organs, tissues, and types of cells involved.

Basal metabolism is the rate of total heat production of an individual who is awake but in complete mental and physical repose, at comfortable temperature and without having had food for at least 12 hours. Under these conditions, oxidation of stored nutrients provides the sole source of energy expended and heat is measurable by calorimetry.
See also digestion (1).

metabolite. An intermediate substance produced and used in the processes of a living cell or organism. Metabolites effect replacement and growth in living tissue and are also a source of energy in the body. Examples are nucleic acids, enzymes, glucose, cholesterol, and many similar substances.

"Meta Bond."[526] TM for microcrystalline zinc phosphates.
Use: Preparation of steel, galvanized steel and aluminum for decorative and protective finishing. Application by spray or immersion.

metaboric acid. HBO_2. White crystals, d 2.486, mp 236C, slightly soluble in water.

metacetone. See diethylketone.

Metachrome Yellow. (alizarine yellow GG; sodium-m-nitrobenzeneazosalicylate; CI 14025).   CAS: 584-42-9. $C_{13}H_8N_3NaO_5$.
Properties: Finely divided yellow crystals, soluble in hot water.
Derivation: Diazotizing m-nitroaniline and coupling with salicylic acid.
Use: Acid-base indicator, biological stain.

metaformaldehyde. See sym-trioxane.

metal. An element that forms positive ions when its compounds are in solution and whose oxides form hydroxides rather than acids with water. Approximately 75% of the elements are metals which occur in every group of the Periodic Table except VIIA and the noble gas group. Most are crystalline solids with metallic luster, conductors of electricity and have rather high chemical reactivity; many are quite hard and have high physical strength. They also readily form solutions (alloys) with other metals. The presence of very low percentage of other elements (not necessarily metals) profoundly affects the properties of many metals e.g., carbon in iron. Mercury, cesium, and gallium are liquid at room temperature. Geologically, metals usually occur in the form of compounds that must be physically or chemically processed to yield the pure metal, common methods are application of heat (smelting), carbon reduction, electrolysis, and reduction with aluminum or magnesium. Metals fall into the following classifications which are not mutually exclusive:

alkali metals	rare metals
alkaline-earth metals	rare-earth metals
transition metals	actinide metals
noble metals	light metals
platinum metals	heavy metals

See specific entry for further information.

The chemistry of metals, i.e., their behavior as atoms or ions is a fundamental factor in electrochemical reactions, as well as in the metabolism of plants and animals, where many have essential nutrient and other biochemical functions. Among these are iron, copper, cobalt, potassium, and sodium, often in traces. Some metals are quite toxic, especially cadmium, mercury, lead, barium, chromium, and beryllium, both in elemental form and as compounds.
See also alloy, trace element, electroplating.

metal deactivator. A compound added to gasoline to neutralize the catalytic effect of copper in promoting fuel oxidation.

metaldehyde. CAS: 9002-91-9. $(CH_3CHO)_n$. A polymer of acetaldehyde in which n usually is 4–6.

Properties: White prisms, decomposes with partial regeneration of acetaldehyde, when heated above 80C; soluble in benzene, chloroform; slightly soluble in alcohol, ether; insoluble in water; sublimes at 112–115C; mp 246C in sealed tube.

Hazard: Flammable, dangerous fire risk. Strong irritant to skin and mucous membranes.

Use: Fuel to replace alcohol, to destroy snails and slugs.

metal dye. See dye, metal.

metal fiber. See fiber.

metal, foamed. A light metal such as aluminum or zinc to which titanium hydride has been added. The latter evolves hydrogen to give a blown structure with about the same specific gravity as water.

metal glass. See glass, metallic.

metallic bond. See bond, chemical.

metallic soap. See soap (2).

"Metalliding."[245] A proprietary surface-alloying process in which ions are electrolytically dissolved from a metal anode in a bath of molten fluoride salts and diffuse into the surface of the cathode. The latter may be any of a number of transition metals or rare earths. The alloys are intended for use in space vehicles and also have potential application in the automotive field.

metallized dye. A soluble dye including any one of a variety of metals chemically combined, applied to wool in an acid bath, by use of sodium chloride to salt out the dye onto the fiber.

metallizing. Coating a plastic or similar material with a deposit of metal (usually aluminum) by means of vacuum deposition. The thickness of such films may vary from 0.01 to as much as 3 mils. Metallized plastic films are used for yarns, packaging, stamping foil, labels, etc.

metallocene. An organometallic coordination compound obtained as a cyclopentadienyl derivative of a transition metal or metal halide. The metal is bonded to the cyclopentadienyl ring by electrons moving in orbitals extending above and below the plane of the ring (pi bond). There are three types of metallocenes: (a) dicyclopentadienyl-metals with the general formula $(C_5H_5)_2M$, (b) dicyclopentadienyl-metal halides with the general formula $(C_5H_5)_2MX_{1-3}$, (c) monocyclopentadienyl-metal compounds with the general formula $C_5H_5MR_{1-3}$, where R is CO, NO, halide group, alkyl group, etc. Types (a) and (b) are known as molecular sandwiches since the two cyclopentadiene rings lie above and below the plane on which the metal atom is situated.

Most metallocenes are crystalline and those belonging to the first transition series of the periodic table, from vanadium to nickel, have the same melting point, 173C. The metal complexes are soluble in many organic solvents while the halides are soluble primarily in polar solvents. When the central metal atom is in a stable oxidation state the metallocene is not decomposed by high temperature, air, water, dilute acids, or bases.

Some metallocenes, such as ferrocene, undergo a wide variety of aromatic ring substitution reactions, including Friedel-Crafts acylation, arylation, and sulfonation; a few such as nickelocene and cobaltocene are too unstable to be directly substituted.

Derivation: The most important industrial method is the reaction of a cyclopentadienide (for example the sodium salt) with a transition metal halide in an organic solvent.

Use: Catalysts, polymers, UV absorbers, reducing agents, free radical scavengers, antiknock agents. See also ferrocene, cobaltocene, nickelocene, titanocene dichloride, zirconocene dichloride, uranocene.

metalloid. See nonmetal, semiconductor.

metallurgical coke. See coke.

metal, powdered. Metals are produced in powdered form for a variety of uses in several industries. In this form they are the raw materials for powder metallurgy in which the powders are pressed in molds and heated (sintered) at high temperature. Metal powders range in size from −325 mesh (0.045–0.060 mm "diameter") to +100 mesh and are available in practically all industrial metals. They are produced by machining, milling, shotting, granulation, atomizing, condensation, reduction, chemical precipitation, or electrodeposition. Their properties and purities vary with the method of preparation.

Hazard: Dangerous fire risk, especially as dust. Toxic by inhalation.

Use: Electric, automotive, machinery, tool and refractory-metal industries; paint pigments, flares and incendiary bombs, brazing materials, calorizing, metal-spraying, metallurgical agents; heat-generating agents, catalysts, etc.

"Metalyn."[266] TM for a series of distilled methyl esters of tall oil fatty acids.

Metamucil. Proprietary preparation of psyllium mucilloid, dextrose, benzyl benzoate, $NaHCO_3$, KH_2PO_4, and citric acid.
Use: Laxative.

metanilic acid. (m-sulfanilic acid; m-aminobenzenesulfonic acid).   CAS: 121-47-1. $C_6H_4(NH_2)SO_3H$.
Properties: Small, colorless needles; soluble in water, alcohol, and ether.
Derivation: By the reduction of metal-nitrobenzene-sulfonic acid. Nitrobenzene is sulfonated until the product is soluble in water. The mixture is then poured into water and reduced with iron, made alkaline with lime, and the lime salt dissociated with sodium carbonate.
Hazard: See aniline.
Use: Azo dye manufacture (sodium salt), sulfa drug synthesis.

meta-phosphoric acid. See phosphoric acid, meta-.

metapon. (methyldihydromorphine).
A narcotic drug.
See also narcotic.

"Metasol."[123] TM for a series of chemicals used as fungicides, mildewicides, bactericides, slime control agents for pulp and paper mill systems, and as preservatives for all types of systems. A recently developed type (J-26) is claimed to be a broad-spectrum antimicrobial agent with a minimum ecological hazard.

"Meta-Systox R."[181] TM for O,O-dimethyl-S-2--(ethylsulfinyl)ethyl phosphorothioate.

metepa. (Generic name for tris(methyl-1-aziridinyl)phosphine oxide).   CAS: 57-39-6. $(C_3H_6N)_3PO$.
Properties: Amber liquid, amine odor, bp 188C at 1 mm, d 1.079, miscible with water and organic solvents.
Hazard: Toxic by ingestion and skin absorption, strong irritant to skin.
Use: Insect chemosteriliant, addition products for textile treatments, adhesives, paper and rubber processing, crosslinking agent in polymer systems which contain active hydrogens, monomer.

meter. (1) The basic unit of length of the metric system (39.37 in.). Originally defined as one ten-millionth of the distance from the equator to the North Pole. Now defined as 1,650,763.73 wavelengths of the orange-red line of the isotope krypton 86. (2) A device for measuring the flow rate of liquids, gases, or particulate solids, e.g., flowmeters, rotameters, proportioning equipment.
See also gauge.

"Metglas."[50] TM for an amorphous metal alloy developed for use as transformer coils.

"Methac."[65] TM for a series of blends of methyl acetate with methanol in varying proportions.
Hazard: Toxic by ingestion.
Use: Lacquer solvents, paint removers, organic synthesis.

methacrolein. (methacrylaldehyde).
CAS: 78-85-3. $CH_2{:}C(CH_3)CHO$.
Properties: Liquid, d 0.8474 (20/20C), bp 68.0C, flash p 35F (1.7C) (OC), soluble in water (20C) 5.9% by wt, shipped with 0.1% hydroquinone as polymerization inhibitor.
Hazard: Flammable, dangerous fire risk. Strong irritant.
Use: Copolymers, resins.

methacrylaldehyde. See methacrolein.

methacrylamide. (methacrylic acid amide).
CAS: 79-39-0. $CH_2{:}C(CH_3)CONH_2$.
Properties: Solid, mp 110C.
Use: A monomer for acrylic resins.

methacrylamidopropyltrimethylammonium chloride. (MAPTAC).

$$CH_2 = \overset{\overset{\textstyle CH_3}{|}}{\underset{\underset{\textstyle O}{\|}}{C}}CNHCH_2CH_2CH_2N(CH_3)_3\ Cl$$

Properties: (50% water solution): Amber liquid, d 1.059, refr index 1.427, flash p (CC) none, bulk d 8.66 lb/gal.
Use: Reactive cationic monomer for wide range of industrially useful polymers.

methacrylate ester. $CH_2{:}C(CH_3)COOR$ where R is usually methyl, ethyl, isobutyl, or n-butyl-isobutyl (50-50).   Esters of methacrylic acid; supplied commercially as the polymers.
See acrylic resin.

methacrylate resin. See acrylic resin.

methacrylatochromic chloride.

$H_2C{:}C(CH_3)\overset{\hspace{4em}\rule{5em}{0.4pt}}{C}{:}OCrCl_2OHCrCl_2O$
Properties: Water-soluble solid.
Derivation: Reaction of methacrylic acid with basic chromic chloride.
Use: Water repellent, nonadhesive, insolubilizer for vinyl polymer.

methyacrylic acid. (α-methacrylic acid) (monomer).   CAS: 79-41-4. $H_2C{:}C(CH_3)COOH$.

$$CH_3 \quad O$$
$$H_2C=C-C-OH$$

Properties: Colorless liquid, mp 15–16C, bp 161–162C, d 1.015 (20C), flash p 170F (76.6C). Soluble in water, alcohol, ether, most organic solvents. Polymerizes readily to give water-soluble polymers. Combustible.
Derivation: Reaction of acetone cyanohydrin and dilute sulfuric acid; oxidation of isobutylene.
Grade: 40% aqueous solution, bp 76–78C (25 mm), crude monomer 85% pure, glacial (99% assay).
Hazard: Toxic material. Strong irritant to skin. TLV: 20 ppm.
Use: Monomer for large-volume resins and polymers, organic synthesis. Many of the polymers are based on esters of the acid, as the methyl, butyl, or isobutyl esters.
See acrylic resin.

α-methacrylic acid.   See methacrylic acid.

β-methacrylic acid.   See crotonic acid.

methacrylonitrile.   (2-methyl-2-propenenitrile; MAN).   CAS: 126-98-7.
$CH_2=C(CH_3)C\equiv N$.
Properties: Clear, colorless liquid. Bp 90.3C, fp −38.8C, flash p 55F (12.7C) (TOC), d 0.789, slightly soluble in water, soluble in acetone, thermoplastic, resistant to acids and alkalies.
Hazard: Flammable. Toxic by ingestion, inhalation, and skin absorption. TLV: 1 ppm.
Use: Vinyl nitrile monomer, copolymer with styrene, butadiene, etc., elastomers, coatings, plastics.

γ-methacryloxypropyltrimethoxysilane.
$CH_2:C(CH_3)COOCH_2CH_2CH_2Si(OCH_3)_3$.
Properties: Liquid, d 1.045 (25C), bp approximately 80C (1 mm), refr index 1.4285 (25C), flash p 135F (57.2C). Soluble in acetone, benzene, ether, methanol, and hydrocarbons. Combustible.
Grade: 97% min purity.
Hazard: Moderate fire risk.
Use: Coupling agent for promotion of resin-glass, resin-metal, and resin-resin bonds, for formulation of adhesives having "built in" primer systems.

methacryloyl chloride.   CAS: 920-46-7.
$H_2C=C(CH_3)COCl$.
Properties: Liquid with mw 104.54, bp 95–96C, d 1.070, fp 2C.
Available forms: Technical 90% stabalized with phenothiazine.
Hazard: Flammable and corrosive liquid.

methadone hydrochloride.   (dl-6-dimethylamino-4,4-diphenyl-3-heptanone hydrochloride).
CAS: 1095-90-5.
$(C_6H_5)_2C(COC_2H_5)CH_2CH(CH_3)N(CH_3)_2\cdot HCl$.
A synthetic narcotic.
Properties: Crystalline substance with a bitter taste; no odor; mp 232–235C; soluble in water, alcohol, and chloroform; practically insoluble in ether and glycerol; pH (1% aqueous solution) 4.5–6.5.
Grade: USP.
Hazard: Toxic. Addictive narcotic. Use restricted.
Use: Medicine (sedative, treating heroin addiction).

methallenestril.   (β-ethyl-6-methoxy-α,α-dimethyl-2-naphthalenepropionic acid).
CAS: 517-18-0.
$CH_3OC_{10}H_6CH(C_2H_5)C(CH_3)_2COOH$.
Properties: Crystals, mp 132.5C, soluble in ether, vegetable oils.
Use: Medicine (estrogen).

methallyl acetate.   See methylallyl acetate.

methallyl alcohol.   See methylallyl alcohol.

β-methallyl chloride.   See β-methylallyl chloride.

methallylidene diacetate.
$CH_2:C(CH_3)CH(OCOCH_3)_2$.
Properties: Liquid, d 1.510 (20/20C), bp 191.0C, fp −15.4C, flash p 215F (101C) (COC), slightly soluble in water. Combustible.
Use: Chemical intermediate, can provide controlled release of methacrolein in acid solution.

methamidophos.   (O,S-dimethyl phosphoramidothioate).   CAS: 10265-92-6.
Properties: Mp 39–41C, water-miscibile.
Use: Insecticide for cotton, cole crops, lettuce, potatoes.

methamphetamine hydrochloride.   See amphetamine.

methanal.   See formaldehyde.

methanamide.   See formamide.

methanation.   A reaction by which methane is formed from the hydrogen and carbon monoxide derived from coal gasification. It requires a catalyst, e.g., nickel, and temperatures in the range of 500C. In one process the reaction is performed in an adiabatic fixed-bed reactor. The reaction is: $3H_2 + CO \rightarrow CH_4 + water$
See also gasification.

methane. (marsh gas; methyl hydride).
CAS: 74-82-8. CH_4. The first member of the paraffin (alkane) hydrocarbon series.
Properties: Colorless, odorless, tasteless gas; lighter than air; practically inert toward sulfuric acid, nitric acid, alkalies, and salts but reacts with chlorine and bromine in light (explosively in direct sunlight); flash p $-306F$ ($-188C$); bp $-161.6C$; fp $-182.5C$; autoign temperature $1000F$ ($537C$); vap d 0.554 ($0C$); critical temperature $-82.1C$; critical pressure 672 psia; heating value 1009 Btu/cu ft; soluble in alcohol, ether; slightly soluble in water; an asphyxiant gas.
Occurrence: Natural gas and coal gas, from decaying vegetation and other organic matter in swamps and marshes.
Derivation: (1) From natural gas by absorption or adsorption. (2) From coal mines for use as fuel gas. (3) From a mixture of carbon monoxide and hydrogen (synthesis gas) obtained by reaction of hot coal with steam; the mixed gas is passed over a nickel-based catalyst at high temperature. See methanation. Methane can also be obtained by a nickel-catalyzed reaction of carbon dioxide and hydrogen. (4) Anaerobic decomposition of manures and other agricultural wastes. (5) By horizontal drilling of coal seams.
Grade: Research 99.99%, CP 99%, technical 95%, Btu grade, must have heating value of 1000 Btu/cu ft at 15.5C and a pressure of 30 inches of mercury.
Hazard: Severe fire and explosion hazard, forms explosive mixture with air (5–15% by volume).
Use: Source of petrochemicals by conversion to hydrogen and carbon monoxide by steam cracking or partial oxidation. Important products are methanol, acetylene, hydrogen cyanide, and ammonia. Chlorination gives carbon tetrachloride, chloroform, methylene chloride, and methyl chloride. In the form of natural gas, methane is used as a fuel, as a source of carbon black, and as the starting material for manufacture of synthetic proteins.
See also natural gas, synthetic natural gas, biogas.

methanecarboxylic acid. See acetic acid.

methanedicarbonic acid. See malonic acid.

methanesulfonic acid. CAS: 75-75-2. CH_3SO_3H.
Properties: Liquid at room temperature, d 1.4812 (18/4C), mp 17–20C, bp 200C, refr index 1.4317 (16C). Soluble in water, alcohol, ether; flash p none.
Grade: 70%.
Hazard: Corrosive to tissue (eyes, skin, and mucous membranes).
Use: Catalyst in esterification, alkylation, olefin polymerization, peroxidation reactions.

methanesulfonyl chloride. (mesyl chloride).
CAS: 124-63-0. CH_3SO_2Cl.
Properties: Pale yellow liquid, d 1.485 (20/20C), bp 164C, fp $-32C$, soluble in most organic solvents, insoluble in water (hydrolyzes slowly).
Grade: 98%, 99+%.
Use: Intermediate, flame-resistant products, stabilizer for liquid sulfur trioxide, biological chemicals.

methanethiol. (methyl mercaptan).
CAS: 74-93-1. CH_3SH.
Properties: Water-white liquid when below its boiling point or colorless gas; powerful, unpleasant odor. Fp $-121C$, d 0.87 (20C), flash p 0F ($-17C$), bp 5.96C. Slightly soluble in water; soluble in alcohol, ether, and petroleum naphtha.
Derivation: Methanol and hydrogen sulfide.
Grade: 98.0% purity.
Hazard: Flammable, dangerous fire risk. Explosive limits in air 3.9–21.8%. Strong irritant. TLV: 0.5 ppm in air.
Use: Synthesis, especially of methionine, jet fuel additives, fungicides; also as catalyst.

methanethiomethane. See dimethyl sulfide.

methanoic acid. See formic acid.

methanol. See methyl alcohol.

methapyrilene. CAS: 91-80-5. $C_{14}H_{19}N_3S$.
Properties: Colorless liquid, bp 173C (3 mm), soluble in water and alcohol, insoluble in benzene.
Hazard: Toxic by ingestion, a carcinogen.
Use: Medicine (antihistamine).

methenamine. See hexamethylenetetramine.

"Methendic" Anhydride.[316] TM for a mix of bicyclic unsaturated dibasic anhydrides as a relatively nonvolatile liquid at room temperature.
Properties: Pale amber liquid, color Gardner 3–6, d 1.2–1.3 (25C), 1.5052 (27C), flash p 275–285F (135–140C) (COC). Miscible with acetone, aromatic and aliphatic hydrocarbons at room temperature. Combustible.
Use: Crosslinking or curing agent for epoxy type resin systems.

methenyl tribromide. See bromoform.

methicillin. $C_6H_3(OCH_3)C{=}O$. A semisynthetic antibiotic.
Use: Effective against resistant staphylococci.

methidathion. (an ester of dithiophosphoric acid).
CAS: 950-37-8. $C_6H_{11}N_2O_4PS_3$.
Properties: Crystalline solid, almost insoluble in water, soluble in common organic solvents.

Hazard: Toxic by ingestion, a cholinesterase inhibitor.
Use: Insecticide.

methiocarb. (4-Methylthio-3,5-xylyl-N-methylcarbamate). CAS: 2032-65-7.
Properties: Mp 121C.
Use: Insecticide for vegetables and fruits.

methiodal sodium. See sodium methiodal.

methionine. (2-amino-4-(methylthio)butyric acid). CAS: 59-51-8.
$CH_3SCH_2CH_2CH(NH_2)COOH$. An optically active essential sulfur-containing amino acid important in biological trans-methylation processes. The levo form is biologically active.
Properties: (DL racemic mix): White, crystalline platelets or powder; faint odor. Soluble in water, dilute acids and alkalies, very slightly soluble in alcohol, practically insoluble in ether, pH (1% aqueous solution) 5.6–6.1.
Derivation: Hydrolysis of protein, synthesized from hydrogen cyanide, acrolein, and methyl mercaptan.
Grade: NF, feed 98%.
Use: Pharmaceuticals, feed additive, vegetable oil enrichment, single-cell protein.

methionine hydroxy-analog calcium. (dl-α-hydroxyγ-methylmercaptobutyric acid, calcium salt; 2-hydroxy-4-methylthiobutyric acid, calcium salt). $(CH_3SCH_2CH_2CHOHCOO)_2Ca$.
Free methionine hydroxy analog is a metabolite in methionine utilization.
Properties: Free-flowing light tan powder, soluble in water, insoluble in common organic solvents.
Use: Animal feed, synthesis of pharmaceuticals.

methiotepa. Generic name for tris(2-methyl-1-aziridinyl)-phosphine sulfide.
See metepa.

"Methocel."[233] TM for methylcellulose.

"Metholene."[242] TM for a series of methyl esters of fatty acids, from methyl caproate through methyl stearate.

methomyl. (S-methyl-N-(methylcarbamoyloxy)-thioacetimidate). CAS: 16752-77-5.
$C_5H_{10}N_2O_2S$.
Properties: Crystalline solid, mp 78C, d 1.30. Partially soluble in water, alcohol, and acetone; soluble in methanol.
Hazard: Toxic by ingestion. TLV: 2.5 mg/m³
Use: Insecticide, nematocide.

methoprene. CAS: 40596-69-8. $C_{19}H_{34}O_3$.
An insecticidal preparation said to act in the manner of a juvenile hormone, which arrests development of insects in the larval stage.

methotrexate. (amethopterin; 4-amino-10-methylfolic acid). CAS: 59-05-2.
Properties: Orange-brown, crystalline powder; insoluble in water, alcohol, chloroform, ether; slightly soluble in dilute hydrochloric acid; soluble in dilute solutions of alkali hydroxides and carbonates; folic acid antagonist.
Grade: USP.
Hazard: Very toxic.
Use: Chemosterilant, cancer treatment.

methoxsalen. (8-methoxypsoralen).
CAS: 298-81-7. $C_{12}H_8O_4$, tricyclic.
Properties: White to cream-colored, odorless, crystalline solid; slightly soluble in alcohol; practically insoluble in water. Combustible.
Use: Suntan accelerator, sunburn protector.

methoxyacetaldehyde. A highly reactive aldehyde derivative available in 77% aqueous solution, clear colorless liquid with odor characteristic of lower aldehydes, miscible with water and many organic solvents. Resembles butyraldehyde in structure and some properties. Claimed as possible antimicrobial agent, preservative, and polymer modifier.

methoxyacetic acid. CAS: 625-45-6.
CH_3OCH_2COOH.
Properties: Liquid, mp (min) 7.7C, boiling range 197–198C (min at 733 mm), d 1.1738 (25/4C), refr index 1.415 (25C), flash p 260F (126C), acid number 612 (min). Combustible.
Use: Synthesis.

p-methoxyacetophenone. (p-acetoanisole; acetanisole; p-acetylanisole). CAS: 100-06-1.
$CH_3OC_6H_4COCH_3$.
Properties: Crystalline solid with pleasant odor, bp 258C, congealing p 36.5C. Soluble in alcohol, ether, fixed oils. Combustible.
Derivation: Interaction of anisole and acetyl chloride in the presence of aluminum chloride and carbon disulfide.
Grade: Technical, FCC.
Use: Perfumery (for floral odors), flavoring.

methoxyamine. (hydroxylamine methyl ether).
CAS: 67-62-9. CH_3ONH_2.
Properties: Colorless liquid, unpleasant odor, bp 50C, soluble in water and alcohol.
Derivation: From hydroxylamine disulfonic acid and methyl sulfate.
Hazard: Toxic by ingestion, strong skin irritant.
Use: Analytical reagent, mainly for ketones and aldehydes.

o-methoxyaniline. See o-anisidine.

p-methoxyaniline. See p-anisidine.

methoxybenzaldehyde. See anisaldehyde.

methoxybenzene. See anisole.

p-methoxybenzoic acid. See anisic acid.

p-methoxybenzyl acetate. See anisyl acetate.

p-methoxybenzyl alcohol. See anisic alcohol.

p-methoxybenzyl formate. See anisyl formate.

2-methoxy-4,6-bis(isopropylamino)-s-triazine.
(2,4-bis(isopropylamino)-6-methoxy-s-triazine).
$CH_3OC_3N_3[NHCH(CH_3)_2]_2$.
Properties: White solid, almost insoluble in water.
Use: Herbicide.

3-methoxybutanol.
$CH_3CH(OCH_3)CH_2CH_2OH$.
Properties: Liquid, d 0.9229, bp 161.1C, flash p
165F (74C), vap press 0.9 mm (20C), sets to
glass at −85C, soluble in water. Combustible.
Hazard: Toxic by ingestion.
Use: High-boiling lacquer solvent, coupling agent
for brake fluids, intermediate for plasticizers,
herbicides, film-forming additive in PVA emulsions, solvent for pharmaceuticals.

1-methoxycarbonyl-1-propen-1-yl
dimethylphosphate. See mevinphos.

methoxychlor. (methoxy DDT; DMDT;
2,2-bis(p-methoxyphenol)-1,1,1-trichloroeth-
ane). CAS: 72-43-5.
$Cl_3CCH(C_6H_4OCH_3)_2$.
Properties: White, crystalline solid; mp 89C; insoluble in water; soluble in alcohol. Not compatible with alkaline materials.
Derivation: Reaction of methyl phenyl ether and
chloral hydrate.
Hazard: Toxic material. TLV: 10 mg/m³ of air.
Use: Insecticide effective against mosquito larvae
and house flies; recommended for use in dairy
barns.

2-methoxy-3,6-dichlorobenzoic acid. See dicamba.

2-(β-methoxyethoxy)ethanol. See diethylene glycol monomethyl ether.

methoxyethylene. See vinyl methyl ether.

2-methoxyethylmercury acetate. CAS: 151-38-2.
$CH_3OCH_2CH_2HgOOCCH_3$. A fungicide
and disinfectant used in treating seeds.
Hazard: Highly toxic. TLV: (as mercury) 0.01
mg/m³ of air.

methoxyethyl oleate.
$CH_3OCH_2CH_2OOCC_{17}H_{33}$.
Properties: Oily liquid, mild odor, fp approximately −18C, d 0.898 (25C), boiling range 180–
206C (4 mm), flash p 385F (196C) (OC), viscosity 8 cp (25C). Combustible.
Use: Plasticizer and solvent.

methoxyethyl stearate.
$CH_3OCH_2CH_2OOCC_{17}H_{35}$.
Properties: Oily liquid, mild odor, fp 19–24C, boil-
ing range 186–205C (4 mm), flash p 378F (192C)
(OC), viscosity 9 cp at 25C. Combustible.
Use: Plasticizer and solvent.

methoxyflurane. See 2,2-dichloro-1,1-difluoro-
ethyl methyl ether.

3-methoxy-4-hydroxybenzaldehyde. See vanillin.

methoxyhydroxymercuripropylsuccinyl urea.
3-hydroxymercuri-2-methoxypropylcarbamoyl-
succinamic acid. $C_9H_{16}HgN_2O_6$.
Properties: Bitter crystals, mp 198.5C.
Derivation: Made by the mercuration of allylsuc-
cinylurea.
Hazard: Highly toxic.

4′-methoxy-2-(p-methoxyphenyl)acetophenone.
See deoxyanisoin.

2-methoxy-5-methylaniline. See 5-methyl-o-anisi-
dine.

4-methoxy-4-methylpentanol-2.
$CH_3C(CH_3)(OCH_3)CH_2CHOHCH_3$.
Properties: Liquid, boiling range 163.8–167C.
Combustible.
Use: Solvent for resin-coating formulation.

4-methoxy-4-methylpentanone-2.
$CH_3C(CH_3)(OCH_3)CH_2COCH_3$.
Properties: Water-white liquid, boiling range 147–
163C, flash p 141F (60.5C). Combustible.
Derivation: Diacetone alcohol.
Hazard: Moderate fire risk. Irritant to skin and
eyes.
Use: Solvent for a variety of resin coatings.

2-methoxy-4-methylphenol. See creosol.

2-methoxynaphthalene. See β-naphthyl methyl
ether.

1-methoxy-4-nitrobenzene. See p-nitroanisole.

4-methoxyphenol. (p-methoxyphenol).
See hydroquinone monomethyl ether.

o-methoxyphenol. See guaiacol.

p-methoxyphenylacetic acid. (p-methoxy-α-to-luic acid). $CH_3OC_6H_4CH_2COOH$.
Properties: Off-white to pale yellow flakes, mp 85C.
Use: Preparation of pharmaceuticals, other organic compounds.

p-methoxyphenylbutanone.   See anisylacetone.

methoxypolyethylene glycol.   One of a series of compounds with properties similar to the polyethylene glycols of comparable molecular weight, slightly viscous liquids to soft wax-like solids.
Use: Manufacture of detergents and emulsifying and dispersing agents through the preparation of the mono-fatty-acid derivatives.

methoxy propanol. (monopropylene glycol methyl ether). $CH_3OCH_2CH(OH)CH_3$.
Properties: Colorless liquid, bp 120C, flash p 102F (39C), d 0.92 (25C). Combustible.
Hazard: Moderate fire risk.
Use: Antifreeze and coolant for diesel engines.

p-methoxypropenylbenzene.   See anethole.

p-methoxypropiophenone.
$C_2H_5COC_6H_4OCH_3$.
Properties: Clear colorless liquid, distillation range 110–140C (3 mm), refr index 1.543–1.545 (25C).

3-methoxypropylamine. (3-MPA).
$CH_3OCH_2CH_2CH_2NH_2$.
Properties: Colorless liquid, bp 119C, d 0.873 (20/20C), refr index 1.4153 (25C), flash p 80F (26.6C) (TCC). Miscible with water, ethanol, toluene, acetone, carbon tetrachloride, hexane, and ether.
Hazard: Flammable, moderate fire risk. Toxic by ingestion and inhalation.
Use: Organic intermediate, emulsifier in anionic coatings and wax formulations.

8-methoxypsoralen.   See methoxsalen.

p-methoxytoluene.   See p-methylanisole.

p-methoxy-α-toluic acid.   See p-methoxyphenylacetic acid.

methoxytriethylene glycol acetate.   See methoxytriglycol acetate.

methoxytriglycol. $CH_3O[C_2H_4O]_3H$.
Properties: Colorless liquid, d 1.0494, bp 249C, fp −44C, flash p 245F (118C), soluble in water. Combustible.
Use: Plasticizer intermediate.

methoxytriglycol acetate. (methoxytriethyleneglycol acetate). $CH_3COO(C_2H_4O)_3CH_3$.
Properties: Colorless liquid with fruity odor, d 1.0940 (20/20C), bulk d 9.2 lb/gal (20C), bp 244.0C, flash p 260F (126C), water-soluble, low volatility. Combustible.
Use: Antidusting agent for finely powdered materials, especially for certain dyestuffs.

methyl abietate.   CAS: 127-25-3.
$C_{19}H_{29}COOCH_3$.
Properties: Colorless to yellow liquid, d 1.033–1.043 (20C), refr index 1.525–1.535, flash p 360F (182C), bp 365C, miscible with most organic solvents. Combustible.
Use: Solvent and plasticizer, lacquers, varnishes, coating compositions, adhesives.

methyl acetate.   CAS: 79-20-9. $CH_3CO_2CH_3$.
Properties: Colorless, volatile liquid; fragrant odor. Miscible with the common hydrocarbon solvents, soluble in water, d 0.92438, fp −98.05C, bp 54.05C, flash p 15F (−9.4C), refr index 1.3619 (20C), bulk d 7.76 lb/gal (20C), autoign temperature 935F (501C).
Derivation: By heating methanol and acetic acid in presence of sulfuric acid and distilling.
Grade: Technical, CP.
Hazard: Flammable, dangerous fire and explosion risk, explosive limits in air 3–16%. Irritant to respiratory tract. TLV: 200 ppm in air.
Use: Paint remover compounds, lacquer solvent, intermediate, synthetic flavoring.

methylacetic acid.   See propionic acid.

methyl acetoacetate.   CAS: 105-45-3.
$CH_3COCH_2CO_2CH_3$.
Properties: Colorless liquid, soluble in alcohol, slightly soluble in water, d 1.0785 (20/20C), bp 171.7C, vap press 0.7 mm (20C), flash p 170F (78C), bulk d 9.0 lb/gal (20C), fp −80C. Combustible.
Hazard: Toxic by ingestion and inhalation.
Use: Solvent for cellulose ethers, ingredient of solvent mixture for cellulose esters, organic synthesis.

methyl acetone.
Properties: Water-white, anhydrous liquid, consisting of various mixtures of acetone, ethyl acetate, and methanol; miscible with hydrocarbons, oils, and water; flash p near 0F (−17C).
Derivation: A byproduct in the wood-distillation industry, also synthetic.
Hazard: Flammable, dangerous fire risk. Toxic by ingestion.
Use: Solvent for nitrocellulose, cellulose acetate, rubber, gum, resins; lacquers, paint and varnish

removers, extracting perfumes, dewaxing natural gums.

methylacetophenone. (methyl-p-tolyl ketone). $CH_3C_6H_4COCH_3$.
Properties: Colorless or pale-yellow liquid, fragrant coumarin odor, soluble in seven parts of 50% alcohol and in most fixed oils, d 1.001–1.004, refr index 1.533–1.535. Combustible.
Derivation: Action of acetic anhydride on toluene.
Grade: Technical, FCC.
Use: Perfumery, flavoring.
See also glycine (2).

methylacetopyranone. See dehydroacetic acid.

methyl acetylene. (allylene; propyne).
CAS: 74-99-7. $CH_2C{\equiv}CH$.
Properties: Colorless, liquefied gas. Bp −23.1C, fp −101.5C, sp vol 9.7 cu ft/lb (70F).
Hazard: Flammable, dangerous fire risk. Toxic by inhalation. TLV: 1000 ppm in air.
Use: Specialty fuel, chemical intermediate.

methylacetylene-propadiene, stabilized. (MAPP).
Properties: Colorless, liquefied gas. D (liquid): 0.576 (15/15C), boiling range −39 to −20C, flame temperature (in oxygen) 2925C. A mixture containing 60–66.5% methylacetylene and propadiene, balance propane and butane.
Hazard: Flammable, dangerous fire risk. Toxic by inhalation. TLV: 1000 ppm in air.
Use: Industrial fuel gas for cutting, welding, brazing, heat treating, metallizing.
See also "Mapp."

methyl acetylricinoleate.
$C_{17}H_{32}(OCOCH_3)COOCH_3$.
Properties: Pale yellow, low viscosity, oily liquid. Mild odor, d 0.938 (25/25C), solidifies at −26C, soluble in most organic liquids, insoluble in water. Combustible.
Derivation: Castor oil, methanol and acetic anhydride.
Use: Plasticizer, lubricant, protective coatings, synthetic rubbers, vinyl compounds.

methyl acetylsalicylate. CAS: 580-02-9.
$CH_3COOC_6H_4COOCH_3$.
Properties: White crystals, mp 52C, 134–136C (9 mm).
Derivation: By heating methyl salicylate with a slight excess of acetic anhydride, adding alcohol, then water, and separating the resulting precipitate.
Use: Perfumery (fixative).

methyl acid phosphate. See methylphosphoric acid.

β-methylacrolein. See crotonaldehyde.

methyl acrylate. CAS: 96-33-3.
$CH_2{:}CHCOOCH_3$.
Properties: Colorless, volatile liquid. Bp 80.5C, fp −76.5C, vap press 65 mm (20C), d 0.9574 (20/20C), bulk d 8.0 lb/gal, slightly soluble in water, readily polymerized, flash p 25F (−3.8C) (TOC).
Derivation: (a) Ethylene cyanohydrin, methanol, and dilute sulfuric acid; (b) Oxo reaction of acetylene, carbon monoxide, and methanol in the presence of nickel or cobalt catalyst; (c) from β-propiolactone.
Grade: Technical (inhibited).
Hazard: Flammable, dangerous fire and explosion risk. Toxic by inhalation, ingestion, and skin absorption; irritant to skin and eyes. TLV: 10 ppm in air.
Use: Acrylic polymers, amphoteric surfactants, vitamin B_1, chemical intermediate.
See also acrylate.

β-methylacrylic acid. See crotonic acid.

methylal. (dimethoxymethane; formal).
CAS: 109-87-5. $CH_3OCH_2OCH_3$.
Properties: Colorless, volatile liquid; chloroform-like odor; pungent taste; fp −105C; d 0.86 (20/4C); bp 42.3C; soluble in water at 20C to extent of 32 wt%; miscible in alcohol, ether, and hydrocarbons; flash p approximately 0F (−17.7C) (OC); autoign temperature 459F (237C).
Hazard: Flammable, dangerous fire and explosion risk. Toxic by ingestion and inhalation. TLV: 1000 ppm in air.
Use: Solvent, organic synthesis, perfumes, adhesives, and protective coatings; special fuel.

2-methylalanine. See aminoisobutyric acid.

methyl alcohol. (methanol; wood alcohol).
CAS: 67-56-1. CH_3OH.   22nd highest-volume chemical produced in US (1985).
Properties: Clear, colorless, mobile, highly polar liquid; miscible with water, alcohols, and ether. D 0.7924, fp −97.8C, bp 64.5C, bulk d 6.59 lb/gal (20C), refr index 1.329 (20C), surface tension 22.6 dynes/dm (20C), viscosity 0.00593 poise (20C), vap press 92 mm (20C), flash p 54F (12.2C) (OC), autoign temperature 867F (464C).
Derivation: (a) By high-pressure catalytic synthesis from carbon monoxide and hydrogen; (b) partial oxidation of natural gas hydrocarbons; (c) several processes for making methanol by gasification of wood, peat, and lignite have been developed but have not yet proved out commercially; (d) from methane with molybdenum catalyst (experimental).

Method of purification: Rectification.

Grade: Technical, CP (99.85%), electronic (used to cleanse and dry components), fuel.

Hazard: Flammable, dangerous fire risk. Explosive limits in air 6–36.5% by volume. Toxic by ingestion (causes blindness). TLV: 200 ppm in air.

Use: Manufacture of formaldehyde, acetic acid, and dimethyl terephthalate; chemical synthesis (methyl amines, methyl chloride, methyl methacrylate); antifreeze; solvent for nitrocellulose, ethylcellulose, polyvinyl butyral, shellac, rosin, manila resin, dyes; denaturant for ethanol; dehydrator for natural gas; fuel for utility plants (methyl fuel); feedstock for manufacture of synthetic proteins by continuous fermentation; source of hydrogen for fuel cells; home heating oil extender.

methylallyl acetate. (methylallyl acetate).
$CH_2:C(CH_3)CH_2OOCCH_3$.

Properties: Colorless liquid, d 0.9162 (20C), bulk d 7.6 lb/gal.

Hazard: Probably flammable.

Use: Monomer, preparation of methallyl derivatives.

methylallyl alcohol. (methallyl alcohol; 2-methyl-2-propen-1-ol). $H_2C:C(CH_3)CH_2OH$.

Properties: Colorless liquid with pungent odor, d 0.8515 (20/4C), bp 115C, refr index 1.4255 (25C), flash p 92F (33.3C). Soluble in water, alcohols, esters, ketones, and hydrocarbons.

Grade: 98.5% min purity.

Hazard: Flammable, moderate fire risk. Irritant to eyes and skin.

Use: Intermediate.

β-methylallyl chloride. (methallyl chloride; 3-chloro-2-methyl-1-propene; MAC).
$CH_2:C(CH_3)CH_2Cl$.

$$H_2C{=\!=}C-\!\!\!\!\overset{|}{\underset{|}{C}}-\!\!\!Cl$$
$$CH_3\ CH_2$$

Properties: Colorless to straw-colored, volatile liquid with a sharp penetrating odor. D 0.925 (20C), bp 73C, refr index 1.427 (25C), flash p −3F (−19.4C) (TOC).

Hazard: Flammable, dangerous fire risk. Explosive limits in air 3.2–8.1%. Toxic by ingestion, irritant to eyes and skin.

Use: Intermediate for production of insecticides, plastics, pharmaceuticals, and other organic chemicals; fumigant for grains, tobacco, and soil.

methylaluminum sesquibromide.
$(CH_3)_3Al_2Br_3$.

Properties: Cloudy, yellow liquid at 25C. Fp −4C, bp (extrapolated) 166C, d 1.514 (25C).

Hazard: Ignites spontaneously in air, reacts violently with water, keep out of contact with air and moisture.

Use: Catalyst for polymerization of olefins and hydrogenation of aromatics, chemical intermediate.

methylaluminum sesquichloride.
$(CH_3)_3Al_2Cl_3$.

Properties: Colorless liquid at 25C, fp 22.8C, bp (extrapolated) 143.7C, d 1.1629 (25C), 9.705 lb/gal (25C).

Hazard: Ignites spontaneously in air, reacts violently with water, keep out of contact with air and moisture.

Use: Catalyst for polymerization of olefins, catalyst for hydrogenation of aromatics.

methylamine. (monomethylamine; aminomethane).   CAS: 74-89-5. CH_3NH_2.

Properties: Colorless gas, strong ammoniacal odor, bp −6.79C, fp −92.5C, flash p (gas) 14F (−10C), (30% solution) (TOC) 34F (1.1C); soluble in water, alcohol, ether; autoign temperature 806F (430C).

Derivation: Interaction of methanol and ammonia over a catalyst at high temperature. The mono-, di-, and trimethylamines are all produced and yields are regulated by conditions. They are separated by azeotropic or extractive distillation.

Grade: Technical (anhydrous, 30–40% solutions).

Hazard (gas and liquid): Dangerous fire risk. Explosive limits in air 5–21%. Strong irritant to tissue. TLV: 10 ppm in air.

Use: Intermediate for accelerators, dyes, pharmaceuticals, insecticides, fungicides, surface active agents, tanning, dyeing of acetate textiles, fuel additive, polymerization inhibitor, component of paint removers, solvent, photographic developer, rocket propellant.

methyl-o-aminobenzoate. See methyl anthranilate.

methylaminoacetic acid. See sarcosine.

methylaminodimethylacetal.
$(CH_3O)_2CHCH_2NHCH_3$.

Properties: Water-white to slightly yellow, clear liquid having a sharp ammoniacal odor; refr index 1.406–1.409 (20C); d 0.924 (25/25C). Combustible.

l-methylaminoethanolcatechol. See epinephrine.

2-(methylamino)glucose. See N-methylglucosamine.

2-methylaminoheptane.
$CH_3(CH_2)_4CH(NHCH_3)CH_3$.
Properties: Oily liquid with a slight amine odor, bp 155C, somewhat soluble in water.
Use: Medicine, usually as the hydrochloride.

N-methyl-p-aminophenol. $CH_3NHC_6H_4OH$.
Properties: Colorless needles; soluble in water, alcohol, and ether; mp 87C. Combustible.
Derivation: (a) Interaction of hydroquinone and methylamine. (b) Methylation of p-aminophenol hydrochloride.
Hazard: Eye and skin irritant.
Use: Organic synthesis, photographic developer.

N-methyl-p-aminophenol sulfate.
$HOC_6NHCH_3 \cdot 1/2H_2SO_4$.
Properties: Colorless needles, mp 250–260C with decomposition, soluble in water and alcohol, insoluble in ether, discolors in air. Combustible.
Derivation: By methylation of p-aminophenol and conversion of the resulting methylated base by neutralization with sulfuric acid.
Grade: CP, photographic.
Use: Photographic developer.

methylamyl acetate. (methylisobutyl carbinol acetate; sec-hexyl acetate; 4-methyl-2-pentanol acetate). CAS: 108-84-9.
$CH_3COOCH(CH_3)CH_2CH(CH_3)_2$.
Properties: Colorless liquid, mild odor, d 0.8598 (20/20C), bp 146.3C, fp −64C, vap press 3 mm (20C), flash p 110F (43.3C) (OC), bulk d 7.1 lb/gal, insoluble in water, soluble in alcohol. Combustible.
Grade: Technical.
Hazard: Moderate fire risk. Toxic by inhalation.
TLV: 50 ppm in air.
Use: Solvent for nitrocellulose and other lacquers.

methylamyl alcohol. (methylisobutyl carbinol; MIBC; 4-methylpentanol-2). CAS: 108-11-2.
$(CH_3)_2CHCH_2CH(CH_3)OH$.
Properties: Colorless, stable liquid; miscible with most common organic solvents; slightly soluble in water. Bp 131.8C, d 0.8079 (20/20C), bulk d 6.72 lb/gal (20C), sets to a glass at approximately −90C, refr index 1.4089 (25C), vap press 2.8 mm (20C), flash p 105F (40.5C) (OC). Combustible.
Derivation: From methyl isobutyl ketone.
Grade: Technical.
Hazard: Moderate fire risk, explosive limits in air 1–5.5%. TLV: 25 ppm in air.
Use: Solvent for dyestuffs, oils, gums, resins, waxes, nitrocellulose, and ethylcellulose; organic synthesis; froth flotation; brake fluids.

methyl-n-amyl carbinol. (heptanol-2,2-methyl-1-pentanol). CAS: 543-49-7.
$CH_3(CH_2)_4CHOHCH_3$.
Properties: Stable, colorless liquid; mild odor; miscible with common organic liquids. D 0.8187 (20/20C), bp 160.4C, vap press 1.0 mm (20C), flash p 130F (54.4C) (OC), bulk d 6.8 lb/gal (20C). Combustible.
Grade: Technical.
Hazard: Moderate fire risk.
Use: Solvent for synthetic resins, frothing agent in ore flotation.

methyl-n-amylketone. (2-heptanone). CAS: 110-43-0.
$CH_3CH_2CH_2CH_2CH_2COCH_3$.
Properties: Water-white liquid, almost insoluble in water, miscible with organic solvents, d 0.8166 (20/20C), bp 150.6C, vap press 2.6 mm (20C), flash p 120F (49C), autoign temperature 991F (532C), refr index 1.4110 (20C), bulk d 6.8 lb/gal (20C), nitrocellulose-toluene dilution ratio 3.9, fp −35C. Combustible.
Grade: Technical.
Hazard: Moderate fire risk. Toxic by inhalation, skin irritant, narcotic in high concentration. TLV: 50 ppm in air.
Use: Solvent for nitrocellulose lacquers, synthetic flavoring, perfumery.

N-methylaniline. CAS: 100-61-8.
$C_6H_5NH(CH_3)$.
Properties: Colorless to reddish-brown oily liquid, discolors on standing, soluble in alcohol and ether, d 0.991, fp −57C, bp 190–191C. Combustible.
Derivation: By heating aniline chloride and methanol and subsequent distillation.
Grade: Technical.
Hazard: Toxic by ingestion, inhalation, and skin absorption. TLV: 0.5 ppm.
Use: Organic synthesis, solvent, acid acceptor.

α-methylanisalacetone. (1-[p-methoxyphenyl]-1-penten-3-one).
$CH_3OC_6H_4CH:CHCOCH_2CH_3$.
Properties: White to pale yellow solid, sharp odor, stable, mp 60C min, 1 g is soluble in 5 mL 95% alcohol. Combustible.
Grade: 99% pure min.
Use: Flavoring.

5-methyl-o-anisidine. (p-cresidine; 2-methoxy-5-methylaniline). CAS: 120-71-8.
$CH_3C_6H_3(NH_2)OCH_3$.
Properties: White crystals, mp 51.5C, bp 235C, insoluble in water, soluble in organic solvents.
Derivation: 2-Nitro-p-cresol, obtained by the action of nitrous and excess nitric acids upon p-toluidine, is methylated and reduced.

Grade: Technical.
Hazard: A carcinogen.
Use: Dyes.

p-methylanisole. (p-cresyl methyl ether; p-methoxytoluene; methyl-p-cresol).
CAS: 104-93-8. $CH_3C_6H_4OCH_3$.
Properties: Colorless liquid, strong floral odor, d 0.966–0.970, refr index 1.5100–1.5130 (20C), one part dissolves in three parts of 80% alcohol. Combustible.
Grade: FCC, technical.
Use: Perfumery, flavoring.

1-methylanthracene. See α-methylanthracene.

α-methylanthracene. (1-methylanthracene).
$C_{15}H_{12}$ or $C_6H_4(CH)_2C_6H_3CH_3$.   (a tricyclic aromatic).
Properties: Colorless leaflets, soluble in alcohol, insoluble in water, d 1.101, bp 200C, mp 86C. Combustible.
Grade: Technical.
Use: Organic synthesis.

methyl anthranilate. (methyl-o-aminobenzoate; neroli oil, artificial).   CAS: 134-20-3.
$H_2NC_6H_4CO_2CH_3$.
Properties: Crystals or pale-yellow liquid with bluish fluorescence, grape-like odor, d 1.167–1.175 (15C), refr index 1.5820–1.5840 (20C), bp 135C, mp 23.8C (min); soluble in five volumes or more of 60% alcohol; soluble in fixed oils, propylene glycol, volatile oils; slightly soluble in water, mineral oil; insoluble in glycerol. Combustible.
Derivation: By heating anthranilic acid and methanol in presence of hydrochloric acid with subsequent distillation. Occurs in many flower oils.
Grade: Technical, FCC.
Use: Flavoring, perfume (cosmetics and pomades).

methylanthraquinone. CAS: 84-54-8.
$CH_3C_6H_3(CO)_2C_6H_4$.   (tricyclic).
Properties: White needles, soluble in organic solvents, insoluble in water, mp 177C, bp sublimes. Combustible.
Derivation: By heating anthraquinone and methanol in presence of sulfuric acid.
Use: Organic synthesis.

methyl apholate. 2,2,4,4,6,6-hexahydro-2,2,4,-4,6,6-hexakis(2-methyl-1-aziridinyl)-1,3,5,2,4,6-
triazatriphosphorine. $N_3P_3(NCH_2CHCH_3)_6$.
Use: Insect chemosterilant.

methyl arachidate. (methyl eicosanoate).
$CH_3(CH_2)_{18}COOCH_3$.   The methyl ester of arachidic acid.
Properties: Waxlike solid, mp 45.8C, bp 284C (100 mm), 216C (10 mm), refr index 1.4352 (50C),

insoluble in water, soluble in alcohol and ether.
Derivation: Esterification of arachidic acid with methanol and vacuum distillation.
Grade: Purified (99.8%+).
Use: Special synthesis, intermediate for pure arachidic acid, reference standard for gas chromatography, medical research.

methyl azinphos. See azinphos methyl.

methyl behenate. (methyl docosanoate).
$CH_3(CH_2)_{20}COOCH_3$.   The methyl ester of behenic acid.
Properties: Wax-like solid, mp 53.2C, bp 215.5C (3.75 mm), refr index 1.4262 (80C), insoluble in water, soluble in alcohol and ether. Combustible.
Derivation: Esterification of behenic acid with methanol followed by fractional distillation.
Grade: Purified (99.8%+).
Use: Special synthesis, intermediate for pure behenic acid, biochemical and medical research, reference standard in gas chromatography.

methylbenzaldehydes. See tolyl aldehydes.

methylbenzene. See toluene.

methylbenzethonium chloride. (benzyldimethyl (2-[2-(para-1,1,3,3-tetramethylbutylcresoxy)ethoxy]ethyl) ammonium chloride).
CAS: 25155-18-4.
$(CH_3)_3CCH_2C(CH_3)_2C_6H_3(CH_3O)(CH_2)_2O(CH_2)_2$ $N(CH_3)_2(CH_2C_65)Cl \cdot H_2O$.   A quaternary ammonium compound.
Properties: Colorless, odorless crystals with bitter taste; mp 161–163C; readily soluble in alcohol, hot benzene, "Cellosolve," chloroform, and water; insoluble in carbon tetrachloride and ether.
Grade: NF.
Use: Medicine (bactericide).

methyl benzoate. (benzoic acid, methyl ester; niobe oil).   CAS: 93-58-3.
$C_6H_5COOCH_3$.

$$CH_3O—C{=}O$$

Properties: Liquid of fragrant odor, colorless, oily, d 1.085–1.088, refr index 1.514, fp −12.3C, bp 198.6C, flash p 181F (82.7C). Soluble in three parts of 60% alcohol, in most fixed oils, in ether; insoluble in water. Combustible.
Derivation: (a) By heating methanol and benzoic acid in presence of sulfuric acid. (b) Passing dry

hydrogen chloride through a solution of benzoic acid and methanol. (c) Occurs naturally in oils of clove, ylang ylang, tuberose.
Grade: Technical, FCC.
Hazard: Toxic by ingestion.
Use: Perfumery, solvent for cellulose esters and ethers, resins, rubber; flavoring.

methylbenzoic acid. See o-, m- and p-toluic acid.

methylbenzophenone. See phenyl tolyl ketone.

methyl-o-benzoylbenzoate.
$C_6H_5COC_6H_4COOCH_3$.
Properties: Colorless liquid, d 1.190 (25C), refr index 1.587 (25C), vap press 4.0 mm (175C), bp 351C, mp 40C, flash p 350F (176C), very slightly soluble in water. Combustible.
Use: Plasticizer.

α-methylbenzyl acetate. (methylphenylcarbinyl acetate; styralyl acetate; sec-phenylethyl acetate; phenylmethylcarbinyl acetate).
$C_6H_5CH(CH_3)OOCCH_3$.
Properties: Colorless liquid, strong floral odor, d 1.023–1.026 (25/25C), refr index 1.4935–1.4960 (20C), flash p 178F (81.1C) (TCC). Soluble in 70% alcohol, glycerol, and mineral oil; insoluble in water. Combustible.
Grade: 98% min, FCC.
Use: Perfumery, flavoring.

α-methylbenzyl alcohol. (styralyl alcohol; phenylmethylcarbinol; 1-phenylethan-1-ol; sec-phenethyl alcohol; methylphenylcarbinol).
CAS: 589-18-4. $C_6H_5CH(CH_3)OH$.
Properties: Colorless liquid, mild floral odor, congeals below room temperature, d 1.009–1.014 (25C), mp 20.7C, bp 204C, refr index 1.525–1.529 (20C), flash p 205F (96.1C) (COC). Soluble in alcohol, glycerol, mineral oil; slightly soluble in water. Combustible.
Grade: FCC.
Use: Perfumery, flavoring dyes, laboratory reagent.

α-methylbenzylamine. CAS: 98-84-0.
$C_6H_5CH(CH3)NH_2$.
Properties: Water-white liquid, mild ammoniacal odor, d 0.9535 (20/20C), refr index 1.5366 (20C), bp 80C (18 mm), vap press 0.5 mm (20C), fp sets to a glass approximately −65C, flash p 175F (79.4C) (COC), soluble in most organic solvents, somewhat soluble in water. Combustible.
Hazard: Toxic by ingestion.
Use: Synthesis, emulsifying agent.

α-methylbenzyldiethanolamine.
$C_6H_5CH(CH_3)N(C_2H_4OH)_2$.
Properties: Dark amber liquid, ammonia-like odor, d 1.0812 (20C), bp 244C (50 mm), flash

p 370F (187C) (OC), vap press less than 0.01 mm (20C), sets to glass at −7C, moderately soluble in water. Combustible.
Use: Emulsifying agents, textile specialties, quaternaries.

α-methylbenzyldimethylamine.
$C_6H_5CH(CH_3)N(CH_3)_2$.
Properties: Colorless liquid, d 0.9044 (20/20C), bp 195.6C, vap press 0.6 mm (20C), fp sets to a glass approximately −70C, refr index 1.5024 (20C), flash p 175F (79.4C) (OC), viscosity 1.85 cp (20C), slightly soluble in water. Combustible.
Use: Polymerization catalyst.

α-methylbenzyl ether. CAS: 93-96-9.
$C_6H_5CH(CH_3)OCH(CH_3)C_6H_5$.
Properties: Straw yellow, mobile liquid with faint odor. D 1.0017 (20/20C), bp 286.3C, vap press less than 0.01 mm (20C), fp sets to a glass at approximately −30C, very slightly soluble in water, flash p 275F (135C) (COC), soluble in most organic solvents. Combustible.
Use: Solvent, styrenating agent, softener for synthetic rubbers.

methyl blue. (sodium triphenyl-p-rosaniline sulfonate; CI 42780). CAS: 28983-56-4.
$C_{37}H_{27}N_3O_3S \cdot 2NaSO_3$. A dark blue powder or dye.
Use: Medicine as an antiseptic and in biological and bacteriological stains.
Note: Do not confuse with methylene blue.

methyl borate. See trimethyl borate.

methyl bromide. (bromomethane).
CAS: 74-83-9. CH_3Br.
Properties: Colorless, transparent, easily liquefied gas or volatile liquid. Burning taste, chloroform-like odor, miscible with most organic solvents, forms a voluminous crystalline hydrate with cold water, d 1.732 (0C), bp 3.46C, vap press 1250 mm (20C), fp −94C, flash p none, nonflammable in air, burns in oxygen.
Derivation: Action of bromine on methanol in presence of phosphorus with subsequent distillation.
Grade: Technical, pure (99.5% min).
Hazard: Toxic by ingestion, inhalation, and skin absorption; strong irritant to skin. TLV: 5 ppm in air.
Use: Soil and space fumigant; disinfestation of potatoes, tomatoes, and other crops; organic synthesis; extraction solvent for vegetable oils.

methyl bromoacetate. CAS: 96-32-2.
$BrCH_2COOCH_3$.
Properties: Colorless to straw-colored liquid, fp approximately −50C, bp 145.0–146.7C, d 1.655

(25/25C), ref index 1.456 (25C), very slightly soluble in water, soluble in methanol, ether.
Hazard: Vapor is strong irritant to eyes.
Use: Synthesis of weed killers, dyes, vitamins, pharmaceuticals; lachrymator.

2-methyl-1,3-butadiene. See isoprene.

2-methylbutanal. See 2-methylbutyraldehyde.

2-methylbutane. See isopentane.

2-methyl-2-butanethiol. (tert-butyl mercaptan). $(CH_3)_2CSH(C_2H)_5$.
Properties: Boiling range 95–119C, d 0.828 (15.5/15.5C), refr index 1.438 (20C), flash p 30F (−1.1C), strong offensive odor.
Grade: 95%.
Hazard: Flammable, dangerous fire risk.
Use: Odorant, intermediate, bacterial nutrient.

2-methyl-1-butanol. (amyl alcohol, primary, active; sec butyl carbinol).   CAS: 137-32-6.
$CH_3CH_2CH(CH_3)CH_2OH$.   The active alcohol from fusel oil. The synthetic product is a racemic mixture of both dextro- and levorotatory compounds and therefore not optically active.
Properties: Colorless liquid, d 0.81–0.82 (20C), fp below −70C, bp 128C, refr index 1.41 (20C), slightly soluble in water, miscible with alcohol and ether, flash p 115F (46.1C) (OC). Combustible.
Derivation: Occurs in fusel oil; is made synthetically by fractional distillation of the mixed alcohols resulting from the chlorination and alkaline hydrolysis of pentane.
Hazard: Moderate fire and explosion risk. Toxic by ingestion, inhalation, and skin absorption.
Use: Solvent, organic synthesis (introduction of active amyl group), lubricants, plasticizers, additives for oils and paints.

2-methyl-2-butanol. See tert-amyl alcohol.

3-methyl-1-butanol. See isoamyl alcohol, primary.

3-methyl-2-butanone. See methyl isopropyl ketone.

methylbutanoyl chloride. See isovaleroyl chloride.

2-methyl-1-butene. CAS: 563-46-2. C_5H_{10} or $H_2C:C(CH_3)CH_2CH_3$.
Properties: Colorless, volatile liquid. Disagreeable odor, bp 31.11C, refr index 1.378 (20C), d 0.650 (20/20C), fp −137.52C, flash p approximately −20F (−28C), soluble in alcohol, insoluble in water.
Derivation: Refinery gas.

Grade: 95%, 99% and research.
Hazard: Highly flammable, dangerous fire and explosion risk.
Use: Organic synthesis, pesticide formulations.

2-methyl-2-butene. See 3-methyl-2-butene.

3-methyl-1-butene. (isopropylethylene; α-isoamylene).   CAS: 563-45-1. C_5H_{10} or $H_2C:CHCH(CH_3)_2$.
Properties: Colorless, extremely volatile liquid or gas. Disagreeable odor, bp 20.1C, refr index 1.3643 (20C), d 0.6272 (20C), fp −168.5C, flash p −70F (−57C), soluble in alcohol, insoluble in water.
Derivation: Cracking of petroleum, a component of refinery gas.
Grade: Research, 99% min, technical 95% min.
Hazard: Highly flammable, dangerous fire and explosion risk, explosive limits 1.6–9.1%.
Use: Organic synthesis, high-octane fuel manufacture.

3-methyl-2-butene. (2-methyl-2-butene; trimethylethylene; β-isoamylene). C_5H_{10} or $H_3CCH:C(CH_3)_2$.
Properties: Colorless, volatile liquid. Disagreeable odor, bp 38.51C, refr index 1.387 (20C), d 0.6623 (20/4C), fp −133.83C, flash p −50F (−45C), soluble in alcohol, insoluble in water.
Derivation: Cracking of petroleum, a component of refinery gas.
Grade: 90%, 95% (technical), 99% (pure), and research.
Hazard: Highly flammable, dangerous fire and explosion risk.
Use: Organic synthesis, hydrogenation, halogenation, alkylation, condensation reactions.

cis-2-methyl-2-butenoic acid. See angelic acid.

trans-2-methyl-2-butenoic acid. See tiglic acid.

2-methyl-1-buten-3-yne. See isopropenylacetylene.

1-methylbutyl alcohol. See 2-pentanol.

N-methylbutylamine. CAS: 110-68-9. $CH_3CH_2CH_2CH_2NHCH_3$.
Properties: Liquid, d 0.7335, bp 91.1C, fp −75.0C, miscible in water, flash p 35F (1.6C) (TOC).
Hazard: Flammable, dangerous fire risk.
Use: Intermediate.

methyl-tert-butyl ether. (MTBE).
CAS: 1634-04-4. $(CH_3)_3COCH_3$.   44th highest-volume chemical produced in US (1985).
Properties: Colorless liquid, bp 55C, fp −110C, d 0.74, bulk d 6.18 lb/gal, solubility in water 4

wt%, solution of water in 1.3 wt%, heat of vaporization (55C) 145 Btu/lb (7 kcal/mole), heat of combustion (25C) 101,000 Btu/gal (804 kcal/mole). Octane blending value: 115–125 (Research), 98–110 (Motor).
Derivation: Catalytic reaction of methanol and isobutene (38–93C at 100–200 psi). There are several variations of the process.
Hazard: Flammable, moderate fire risk.
Use: Octane booster for unleaded gasoline (up to 7% by volume), manufacture of isobutene. Approved by EPA.
See also octane number.

methyl butyl ketone. (propylacetone; 2-hexanone). CAS: 591-78-6. $CH_3COC_4H_9$.
Properties: Colorless liquid, bp 127.2C, d 0.830 (20/20C), refr index 1.4024 (20C), vap press 10 mm (20C), soluble in alcohol and ether, flash p 95F (35C) (OC).
Grade: Technical.
Hazard: Flammable, moderate fire risk, explosive limits 1.2–8% in air. Irritant to eyes and mucous membranes, narcotic in high concentration, absorbed by skin. TLV: 5 ppm in air.
Use: Solvent.

2-methyl-6-tert-butylphenol.
$C_6H_3(OH)(CH_3)tert-C_4H_9$).
Properties: Crystalline solid, light straw color, mp 28C, d 0.9618 (30C), bp 230C, flash p 220F (104C) (OC). Soluble in methyl ethyl ketone, ethanol, benzene, and isooctane; insoluble in water. Combustible.
Use: Chemical intermediate.

2-methylbutyl-3-thiol. See sec-isoamyl mercaptan.

2-methyl-4-tert-butylthiophenol. (4-tert-butyl-o-thiocresol). $(CH_3)_3CC_6H_3(CH_3)SH$.
Properties: Water-white liquid, no mercaptan odor, d 0.983 (25C), refr index 1.546 (25C), fp −4C, bp 177C (100 mm), soluble in aliphatic and aromatic hydrocarbons, insoluble in water. Combustible.
Use: Chemical intermediate.

methyl butynol. (2-methyl-3-butyn-2-ol).
$HC≡CCOH(CH_3)_2$.
Properties: Colorless liquid with fragrant odor, bp 104–105C, mp 2.6C, d 0.8672 (20/20C), refr index 1.4211 (20C), flash p 77F (25C) (TOC), miscible with water, soluble in most organic solvents.
Grade: Technical, 95% min.
Hazard: Flammable, dangerous fire risk.
Use: Stabilizer in chlorinated solvents, viscosity reducer and stabilizer, electroplating brightener, intermediate.

2-methylbutyraldehyde. (2-methylbutanal). $CH_3CH_2CH(CH_3)CHO$.
Properties: Liquid, d 0.8029 (20/4C), bp 92–93C, refr index 1.3869 (20C), soluble in alcohol and ether, insoluble in water. Combustible.
Use: Flavoring.

3-methylbutyraldehyde. See isovaleraldehyde.

methyl butyrate. CAS: 623-42-7.
$CH_3CH_2CH_2COOCH_3$.
Properties: Colorless liquid, slightly soluble in water, soluble in alcohol, d 0.898 (20C), bp 102C, fp −92C, refr index 1.3875 (20C), flash p 57F (14C) (CC).
Grade: Technical.
Hazard: Flammable, dangerous fire risk.
Use: Solvent for ethylcellulose, solvent mixture for nitrocellulose, flavoring.

2-methylbutyric acid. See isopentanoic acid.

3-methylbutyric acid. See isopentanoic acid.

methyl caprate. (methyl decanoate).
$CH_3(CH_2)_8COOCH_3$.
Properties: Colorless liquid, d 0.8733 (20/4C), fp −13.3C, bp 224C, 130.6C (30 mm), refr index 1.4237 (25C), insoluble in water, soluble in alcohol and ether. Combustible.
Derivation: Esterification of capric acid with methanol or alcoholysis of coconut oil, purified by fractional vacuum distillation.
Grade: Technical, 99.8% pure.
Use: Intermediate for detergents, emulsifiers, wetting agents, stabilizers, resins, lubricants, plasticizers.

methyl caproate. (methyl hexanoate).
CAS: 106-70-7. $CH_3(CH_2)_4COOCH_3$.
The methyl ester of caproic acid.
Properties: Colorless liquid, d 0.8850 (20/4C), fp −71C, bp 151.2C, 63.0C (30 mm), refr index 1.4054 (20C), insoluble in water, soluble in alcohol and ether. Combustible.
Derivation: Esterification of caproic acid with methanol or alcoholysis of coconut oil.
Grade: Technical, 99.8+%.
Use: Intermediate for caproic acid detergents, emulsifiers, wetting agents, stabilizers, resins, lubricants, plasticizers, flavoring.

methyl caprylate. (methyl octanoate).
$CH_3(CH_2)_6COOCH_3$. The methyl ester of caprylic acid.
Properties: Colorless liquid, d 0.8784 (20/4C), fp −37.3C, bp 192C (759 mm), 98.3 (30 mm), refr index 1.4152 (25C), insoluble in water, soluble in alcohol and ether. Combustible.

Derivation: (a) Esterification of caprylic acid with methanol, (b) alcoholysis of coconut oil.

Grade: Technical, 99.8%.

Use: Intermediate for caprylic acid detergents, emulsifiers, wetting agents, stabilizers, resins, lubricants, plasticizers, flavoring.

methyl "Carbitol."[214]　TM for diethylene glycol monomethyl ether.

methyl "Carbitol" acetate.　TM for diethylene glycol monomethyl ether acetate.

methyl carbonate.　(dimethyl carbonate).
CAS: 616-38-6.　$CO(OCH_3)_2$.
Properties: Colorless liquid, pleasant odor, miscible with acids and alkalies, stable in the presence of water, soluble in most organic solvents, insoluble in water, d 1.0718 (20C), bp 90.6C, mp 0C.
Derivation: Interaction of phosgene and methanol.
Grade: Technical.
Hazard: Flammable, dangerous fire risk. Toxic by inhalation, strong irritant.
Use: Organic synthesis, specialty solvent.

methyl "Cellosolve."[214]　TM for ethylene glycol monomethyl ether.

methyl "Cellosolve" acetate.[214]　TM for ethylene glycol monomethyl ether acetate.

methylcellulose.　(cellulose methyl ether; "Methocel").　CAS: 9004-67-5.
Properties: Grayish-white, fibrous powder; aqueous suspensions neutral to litmus; swells in water to a viscous colloidal solution; insoluble in alcohol, ether, chloroform, and in water warmer than 50.5C; soluble in glacial acetic acid; unaffected by oils and greases; stable up to approximately 300C; stable to light. Combustible.
　Molecular weights vary from 40,000 to 180,000. Specifications call for methoxy group content of narrow or wide ranges within 25–33%.
Derivation: From cellulose by conversion to alkali cellulose and then reacting this with methyl chloride, dimethyl sulfate, or methanol and dehydrating agents. The proportions of the reacting materials are varied to control the properties of the product, such as water solubility and viscosity of water solutions.
Grade: USP, technical, FCC.
Use: Protective colloid in water-based paints to prevent flocculation of pigment; film and sheeting; binder in ceramic glazes; leather tanning; dispersing, thickening, and sizing agent; adhesive; food additive.
See also cellulose, modified; carboxymethylcellulose, hydroxyethylcellulose.

methylcellulose, propylene glycol ether.
See hydroxypropyl methylcellulose.

methyl cerotate.　(methyl hexacosanoate).
$CH_3(CH_2)_{24}COOCH_3$.　The methyl ester of cerotic acid.
Properties: Wax-like solid, insoluble in water, soluble in alcohol and ether, mp 62.9C, bp 237C (1.95 mm), refr index 1.4301 (80C). Combustible.
Derivation: Esterification of cerotic acid with methanol.
Grade: Purified (99+%).
Use: Intermediate in special synthesis, medical research, reference standard for gas chromatography.

methyl chloride.　(chloromethane; monochloromethane).　CAS: 74-87-3.
CH_3Cl.
Properties: Colorless compressed gas or liquid, faintly sweet, ethereal odor, d 0.92 (20C), bp −23.7C, fp −97.6C, flash p approximately 32F (0C) (OC), refr index 1.3712 (−23.7C), critical temperature 143C, critical pressure 970 psi absolute, autoign temperature 1170F (632C), bulk d 7.68 lb/gal (20C). Slightly soluble in water, by which it is decomposed; soluble in alcohol, chloroform, benzene, carbon tetrachloride, glacial acetic acid; attacks aluminum, magnesium, and zinc.
Derivation: (a) Chlorination of methane, (b) action of hydrochloric acid on methanol either in vapor or liquid phase.
Grade: Pure (99.5% min), technical, and two refrigerator grades.
Hazard: Flammable, dangerous fire risk, explosive limits in air 10.7–17%. Narcotic. Psychic effects. TLV: 50 ppm in air.
Use: Catalyst carrier in low-temperature polymerization (butyl rubber), tetramethyl lead, silicones, refrigerant, fluid for thermometric and thermostatic equipment, methylating agent in organic synthesis, such as methylcellulose, extractant and low-temperature solvent, herbicide, topical anesthetic.

methyl chloroacetate.　CAS: 96-34-4.
$ClCH_2COOCH_3$.
Properties: Colorless liquid with sweet pungent odor, d 1.236 (20/4C), fp −32.7C, bp 131C, refr index 1.419–1.420 (25C), insoluble in water, miscible with alcohol and ether. Combustible.
Hazard: Toxic by ingestion and inhalation.
Use: Solvent, intermediate.

methyl chloroform.　See 1,1,1-trichloroethane.

methyl chloroformate.　(methyl chlorocarbonate).
CAS: 79-22-1.　$ClCOOCH_3$.

Properties: Colorless liquid, decomposed by hot water, stable to cold water. Soluble in methanol, ether, and benzene. D 1.23 (15C), bp 71.4C, vapor d 3.9 (air = 1), flash p 54F (12.2C).
Derivation: Reaction between methanol and carbonyl chloride.
Grade: Technical (95% min).
Hazard: Flammable, dangerous fire risk. Highly corrosive and irritant to skin and eyes.
Use: Military poison (lachrymator), organic synthesis, insecticides.

methylchloromethyl ether. (chloromethyl methyl ether). CAS: 107-30-2. $ClCH_2OCH_3$.
Properties: Colorless liquid, d 1.0625 (10/4C), fp −103.5C, bp 59.5C, decomposes in water, soluble in alcohol and ether.
Hazard: Flammable, dangerous fire and explosion risk. Toxic by ingestion and inhalation. Suspected human carcinogen.

2-methyl-4-chlorophenoxyacetic acid.
See MCPA.

4-(2-methyl-4-chlorophenoxy)butyric acid.
See 4-MCPB.

2-(2-methyl-4-chlorophenoxy)propionic acid.
See mecoprop.

methyl chlorosilane. CAS: 993-00-0.
CH_5ClSi. One of several intermediates in the formation of silicones or siloxanes, they react with hydroxyl groups on many types of surfaces to produce a permanent thin surface film of silicone that imparts water-repellency. Examples are methyltrichlorosilane, dimethyldichlorosilane and trimethylchlorosilane.
Hazard: Toxic by ingestion and inhalation, strong irritant to skin and eyes.

methyl chlorosulfonate. CH_3OSO_2Cl.
Properties: Colorless liquid, pungent odor, decomposed by water. Soluble in alcohol, carbon tetrachloride, chloroform; insoluble in water. D 1.492 (10C), bp 133–135C (decomposes), fp −70C, vap d 4.5 (air = 1).
Derivation: Interaction of sulfuryl chloride and methanol.
Grade: Technical.
Hazard: Highly toxic by ingestion and inhalation, strong irritant to skin and eyes.
Use: Organic synthesis, military poison.

methylcholanthrene. CAS: 56-49-5. $C_{21}H_{16}$.
A polynuclear hydrocarbon.

Properties: Yellow crystals melting at 180C, soluble in benzene, insoluble in water.
Derivation: From bile acids via 1,2-benzanthracene.
Hazard: Powerful carcinogen.
Use: Biochemical research.

methyl cinnamate. CAS: 103-26-4. $C_6H_5CH:CHCOOCH_3$.
Properties: White crystals, strawberry-like odor, d 1.0415, mp 34C, bp 259.6C. Soluble in alcohol and ether, in glycerol, most fixed oils, and mineral oil; insoluble in water, Combustible.
Derivation: By heating methanol, cinnamic acid, and sulfuric acid with subsequent distillation.
Grade: Technical, FCC.
Use: Perfumes, flavoring.

methylcoumarin. $C_{10}H_8O_2$.
Properties: White crystals with vanilla flavor, exists as α and β forms, mp (α) 90C, (β) 82C, both forms are soluble in alcohol. Combustible.
Use: Perfumes, flavoring.

methyl-p-cresol. See p-methylanisole.

cis-α-methylcrotonic acid. See angelic acid.

trans-α-methylcrotonic acid. See tiglic acid.

methyl cyanide. See acetonitrile.

methyl cyanoacetate. (malonic methyl ester nitrile). $CNCH_2COOCH_3$.
Properties: Colorless liquid, bp 203C (115C) at 6 mm, fp −22.5C, d 1.1225 (15/4C), refr index 1.419–1.420 (20C). Soluble in water, alcohol, and ether. Combustible.
Derivation: Esterification of cyanoacetic acid with methanol, reaction of an alkali cyanide and chloroacetic methyl ester.
Use: Organic synthesis, pharmaceuticals, dyes.

methyl-2-cyanoacrylate. CAS: 137-05-3. $CH_2:C(CN)COOCH_3$.
Properties: Colorless liquid, bp 48–49C (2.5–2.7 mm), d 1.1044 (27/4C), viscosity 2.2 cp (25C).
Hazard: Toxic by inhalation. TLV: 2 ppm in air.
Use: Adhesive, dentistry.
See also cyanoacrylate adhesives.

methyl cyanoethanoate.   See cyanomethyl acetate.

methyl cyanoformate. $CNCOOCH_3$.
Properties: Colorless liquid, ethereal odor, decomposed by alkalies and water. Soluble in alcohol, benzene, ether. D approximately 1.00 (20C), bp 100C.
Derivation: Methylchloroformate is dissolved in methanol and subjected to the action of (hot) sodium cyanide or potassium cyanide.
Use: Organic synthesis.

methylcyclohexane.   (hexahydrotoluene). $CH_3C_6H_{11}$.
Properties: Colorless liquid, d 0.769, bp 100.8C, fp −126.9C, refr index 1.42312, flash p 25F (−3.89C) (CC), autoign temperature 545F (285C).
Source: Petroleum.
Grade: Technical (95%), 99%, and research.
Hazard: Flammable, dangerous fire risk. Lower explosive limit 1.2% in air. TLV: 400 ppm in air.
Use: Solvent for cellulose ethers, organic synthesis.

methylcyclohexanol.   (hexahydromethyl phenol; hexahydrocresol).   CAS: 25639-42-3. $CH_3C_6H_{10}OH$.   o-, m-, and p- forms.
Properties: Colorless, viscous liquid; aromatic menthol-like odor. Bp 155–180C, d 0.924, flash p 154F (67.7C) (CC). Combustible.
Derivation: (a) A mixture of three isomeric (o-, m-, and p-) cyclic secondary alcohols made by the hydrogenation of cresol, (b) catalytic oxidation of methylcyclohexane.
Grade: Technical.
Hazard: Toxic by ingestion. TLV: 50 ppm in air.
Use: Solvent for cellulose esters and ethers and for lacquers, antioxidant for lubricants, blending agent for special textile soaps and detergents.

methylcyclohexanol acetate.   (heptalin acetate; methyl cyclohexyl acetate). $C_7H_{13}OOCCH_3$.
Properties: Colorless liquid, ester-like odor, slower rate of evaporation than amyl acetate, bp 176–193C, d 0.941, flash p 147F (64C) (CC), toluene dilution ratio 2.5. Combustible.
Derivation: Catalytic hydrogenation and esterification of cresols by means of acetic acid.
Use: Solvent.

methylcyclohexanone.   CAS: 1331-22-2. $CH_3C_5H_9CO$.
Properties: Water-white to pale-yellow liquid, acetone-like odor, a mixture of cyclic ketones. Closely resembles cyclohexanone in physical properties, miscibility, tolerance for nonsolvent and solvent action; bp 160–170C; d 0.925; flash p 138F (58.9C). Combustible.
Derivation: By high-temperature, catalytic hydro-

genation of cresols or by the dehydrogenation of methylcyclohexanol.
Hazard: Moderate fire risk. Toxic by ingestion, inhalation, and skin absorption. TLV (o-isomer): 50 ppm in air.
Use: Solvent, lacquers.

methylcyclohexanone glyceryl acetal. $CH_3C_6H_9I_2C_3H_5OH$.   spiro rings.
Properties: Colorless liquid, d 1.074 (20C), refr index 1.474 (20C), bp 130–140C (20 mm), flash p 200F (93.3C), insoluble in water. Combustible.
Use: Plasticizer.

methylcyclohexanyl oxalate. $(CH_3C_6H_{10}OOC)_2$.
Properties: Colorless, odorless, neutral, stable liquid comprising a mixture of isomers; miscible with most lacquer solvents and diluents.

4-methylcyclohexene-1. $CH_3C_6H_9$.
Properties: Colorless liquid, fp −121.1C, distillation range 110C–117C, d 0.818 (60/60F), refr index 1.450 (20C), flash p 30F (−1.1C), insoluble in water, soluble in alcohol.
Grade: Pure, 99.0 mol%; technical, 95.0 mol%.
Hazard: Flammable, dangerous fire risk. May be irritant to skin and eyes.
Use: Intermediate.

6-methyl-3-cyclohexene carboxaldehyde.

$\overline{CH_3CHCH_2CH{:}CHCH_2CHCHO}$.
Properties: Colorless liquid, d 0.9484, bp 176.4C, fp −39.0C, soluble in water 0.3% by wt at 20C.
Use: Intermediate.

N-methyl-5-cyclohexenyl-5-methylbarbituric acid.   See hexobarbital.

N-methylcyclohexylamine. $C_6H_{11}NHCH_3$.
Properties: Water-white liquid, d 0.86 (20C), soluble in alcohol and ether, slightly soluble in water, purity 99%, distillation range 5–95 cc within 2C, including 149C, corrected to 760 mm. Combustible.
Hazard: Toxic. Strong irritant to tissue.
Use: Intermediate, solvent, acid acceptor.

methylcyclopentadiene dimer.   (methyl-1,3-cyclopentadiene). $C_{12}H_{16}$.
Properties: Colorless liquid, d 0.9341 (20/4C), bp 78–183C, flash p 140F (60C) (TOC). Insoluble in water; very soluble in alcohol, benzene, and ether. Combustible.
Hazard: Moderate fire risk.
Use: High-energy fuels, curing agents, plasticizers, resins, surface coatings, pharmaceuticals, dyes.

methylcyclopentadienyl manganese tricarbonyl.
CAS: 12108-13-3. $CH_3C_5H_4Mn(CO)_3$.
Derivation: Reaction of methylcyclopentadiene with manganese carbonyl.
Hazard: Toxic by ingestion, inhalation, and skin absorption. TLV (as manganese): 0.2 ppm in air.
Use: Antiknock agent for unleaded gasoline. Its use has been prohibited by EPA because of its deleterious effect on catalytic converters.

methylcyclopentane.   CAS: 96-37-7.
$C_5H_9CH_3$.
Properties: Colorless liquid, d 0.750 (20/4C), fp −142.5C, refr index 1.40983 (20C), bp 72C (742 mm), flash p approximately 20F (−6.6C), immiscible with water.
Grade: Technical (95%), 99%, and research.
Hazard: Flammable, dangerous fire and explosion risk. May be irritant and narcotic.
Use: Organic synthesis, extractive solvent, azeotropic distillation agent.

5-methylcytosine.   (5-methyl-2-oxy-4-aminopyrimidine).   CAS: 554-01-8. $C_5H_7N_3O$.
A pyrimidine found in deoxyribonucleic acids, nucleotides, and nucleosides.
Properties: Crystals in prisms from water, mp 270C (decomposes).

methyl decanoate.   See methyl caprate.

methyl demeton.   (Generic name for O,O-dimethyl-O (and S)-2-ethylthio)ethyl phosphorothioates).   CAS: 8022-00-2.
Properties: Slightly soluble in water, soluble in most organic solvents.
Hazard: Cholinesterase inhibitor, use may be restricted, absorbed by skin. TLV: 0.5 mg/m³ of air.
Use: Systemic insecticide.

methyl dichloracetate.   CAS: 116-54-1.
$Cl_2CHCOOCH_3$.
Properties: liquid, d 1.3759–1.3839 (20/20C), refr index 1.4374–1.4474 (20C), bp 143C, slightly soluble in water, soluble in ether and alcohol, flash p 176F (80C). Combustible.
Grade: 99.0% pure.
Hazard: Strong irritant to tissue, forms corrosive products on hydrolysis, keep dry.
Use: Organic intermediate.

methyldichloroarsine.   CAS: 593-89-5.
CH_3AsCl_2.
Properties: A colorless, mobile liquid having an agreeable odor. Decomposed by water, bp 136C, mp −59C, d 1.838.
Hazard: Toxic. Highly irritant.
Use: Military poison, intermediate.

methyl-3,4-dichlorocarbanilate.   See swep.

methyl-N-3,4-dichlorophenylcarbamate.
See swep.

methyldichlorosilane.   CAS: 75-54-7.
CH_3SiHCl_2.
Properties: Colorless liquid, bp 41C, d 1.10 (27C), flash p −26F (−32.2C), soluble in benzene, ether, heptane.
Derivation: Reaction of methyl chloride with silicon and copper.
Hazard: Flammable, dangerous fire risk. Very toxic.
Use: Manufacture of siloxanes.

methyl dichlorostearate. $C_{17}H_{33}Cl_2COOCH_3$.
Properties: Light, yellow, oily liquid with a slight odor. Completely miscible with most organic solvents, freezing range +7 to −5C, bp decomposes 250C, d 0.997 (15.5/15.5C), refr index 1.4599 (25C), flash p 358F (181C). Combustible.
Use: Intermediate, plasticizer extender.

methyldiethanolamine.   CAS: 105-59-9.
$CH_3N(C_2H_4OH)_2$.
Properties: Colorless liquid with amine-like odor, miscible with benzene and water, d 1.0418 (20C), bp 247.2C, bulk d 8.7 lb/gal, vap press less than 0.01 mm (20C), fp −21.0C, refr index 1.4699, flash p 260F (126.6C) (COC). Combustible.
Grade: Technical.
Use: Intermediate, absorption of acidic gases, catalyst for polyurethane foams, pH control agent.

methyldiethylamine. $CH_3N(C_2H_5)_2$.
Properties: Water-white to straw-colored liquid, bp 62.5C, d 0.724 (20C), has inverse water solubility. Combustible.
Use: Desalination of brackish water, chemical intermediate, acid neutralizer.

4-methyl-7-(diethylamino)coumarin.   See 7-diethyl-amino-4-methylcoumarin.

3-methyl-2,5-dihydrothiophene-1,1-dioxide.
(3-methylsulfolene). $CH_3C_4H_5SO_2$ (cyclic).
Properties: Solid, mp 63C, bp decomposes. Slightly soluble in water, acetone, and toluene. Combustible.
Use: Chemical intermediate, catalyst.

methyl-3-(dimethoxyphosphinyloxy)crotonate.
See mevinphos.

methyl-N,3,7-dimethyl-7-hydroxyoctylidenanthranilate.   See hydroxycitronellal-methyl anthranilate. Schiff base.

Methyl "Dioxitol."[125]   TM for diethylene glycol monomethyl ether having an ASTM distillation range 192–196C.

methyldioxolane.   (2-methyl-1,3-dioxolane).

$\overline{OCH(CH_3)OCH_2CH_2}$.
Properties: Water-white liquid, soluble in water, d 0.982 (20/20C), bp 81C. Combustible.
Hazard: Irritant.
Use: Extractant and solvent for oils, fats, waxes, dyestuffs, and cellulose derivatives.

methyldiphenylamine.   (diphenylmethylamine).
CAS: 552-82-9. $(C_6H_5)_2NCH_3$.
Properties: Colorless liquid, d 1.048, fp −7.5C, bp 295C, refr index 1.62, insoluble in water, soluble in alcohol.
Derivation: Heating diphenylamine with methanol and hydrochloric acid.
Hazard: Toxic by ingestion.
Use: Analytical reagent, dye synthesis.

methyl diphenyl phosphate.   (methyl phenyl phosphate). $CH_3OP(O)(OC_6H_5)_2$.
Properties: Clear, oily liquid; d 1.225–1.235 (25C); refr index 1.5370 (20C); pour point −85F (−65C). Nonflammable.
Use: Ignition control compound.

methyldipropylmethane.   See 4-methylheptane.

methyl docosanoate.   See methyl behenate.

methyl dodecanoate.   See methyl laurate.

methyl eicosanoate.   See methyl arachidate.

methyl elaidate.
$(CH_3(CH_2)_7CH:CH(CH_2)_7COOCH_3$.
The methyl ester of elaidic acid (trans-octadec-9-enoic acid).
Properties: Colorless liquid, insoluble in water, soluble in most organic solvents, d 0.8702 (25C), mp less than 15C, bp 213.5C (15 mm), refr index 1.4462 (25C). Combustible.
Derivation: Prepared from oleic acid by elaidinization and esterification.
Use: Pure grade (99+%) used in biochemical research.

methylene.   (1) The divalent hydrocarbon group —CH₂—, in which the carbon atom has its normal valence of 4; it is derived from methane by dropping two hydrogen atoms. It occurs in many organic compounds, e.g., methylene chloride (dichloromethane), CH_2Cl_2. (2) See carbene.

N,N′-methylenebisacrylamide.
$CH_2(NHCOCH:CH_2)_2$.

Properties: Colorless, crystalline powder; mp 185C.
Hazard: Toxic by ingestion.
Use: Organic intermediate, crosslinking agent.

4,4′-methylenebis(2-chloroaniline).   (3,3′-dichloro-4,4′-diaminodiphenylmethane; p,p′-methylene-bis-o-chloroaniline; MOCA).
CAS: 101-14-4. $CH_2(C_6H_4ClNH_2)_2$.
Properties: Tan-colored pellets, d 1.44, melting range 99–107C. Soluble in hot methyl ethyl ketone, acetone, esters, and aromatic hydrocarbons.
Hazard: Toxic: A carcinogen, absorbed by skin.
TLV: 0.02 ppm in air.
Use: Curing agent for polyurethanes and epoxy resins.

p,p′-methylenebis(o-chloroaniline).   See 4,4′-methylenebis(2-chloroaniline).

2,2′-methylenebis(4-chlorophenol).   See dichlorophene.

4,4′-methylenebis(2,6-di-tert-butylphenol).
$[(C_4H_9)_2C_6H_2(OH)]_2CH_2$.   A sterically hindered bis-phenol.
Properties: Light yellow powder, mp 155C, bp 217C (1 mm), d 0.99 (20C), insoluble in water and 1.0N sodium hydroxide. Combustible.
Use: Oxidation inhibitor and antiwear agent for motor oils, aviation piston engine oils, industrial oils, antioxidant for rubbers, resins, adhesives.

3,3′-methylenebis(4-hydroxycoumarin).
See bishydroxycoumarin.

methylenebis(phenylisocyanate).   See diphenylmethane diisocyanate.

2,2′-methylenebis(3,4,6-trichlorophenol).
See hexachlorophene.

methylene blue.   (methylthionine chloride).
CAS: 61-73-4. $C_{16}H_{18}N_3SCl \cdot 3HOH$ (medicinal); $(C_{16}H_{18}N_3SCl)_2 \cdot ZnCl_2 \cdot HOH$ (dye).
Properties: Dark green crystals or powder with bronze-like luster, odorless or slight odor, stable in air. Soluble in water, alcohol, chloroform. Water solutions are deep blue. CI 52015.
Derivation: By oxidation of p-aminodimethylaniline with ferric chloride in the presence of hydrogen sulfide. The dye is the zinc chloride double salt of the chloride.
Grade: USP, technical.
Hazard: Toxic by ingestion.
Use: Dyeing cotton and wool, biological and bac-

teriological stains, reagent in oxidation-reduction titrations in volumetric analysis, indicator. Note: Do not confuse with methyl blue.

methylene bromide. (dibromomethane).
CAS: 74-95-3. CH_2Br_2.
Properties: Clear, colorless liquid; d 2.47; solidifies −52C; bp 97C. Slightly soluble in water, miscible with alcohol, ether, chloroform, and acetone. Nonflammable.
Use: Organic synthesis, solvent.

methylene chloride. (methylene dichloride; dichloromethane). CAS: 75-09-2.
CH_2Cl_2.
Properties: Colorless, volatile liquid; penetrating ether-like odor. Soluble in alcohol and ether, slightly soluble in water, d 1.335 (15/4C), fp −97C, bp 40.1C, bulk d 11.07 lb/gal (20C), refr index 1.4244 (20C), viscosity 0.430 cp (20C), autoign temperature 1224F (662C). Nonflammable and nonexplosive in air.
Derivation: Chlorination of methyl chloride and subsequent distillation.
Hazard: Toxic. A carcinogen, narcotic. TLV: 100 ppm in air.
Use: Paint removers, solvent degreasing, plastics processing, blowing agent in foams, solvent extraction, solvent for cellulose acetate, aerosol propellant.

methylene chlorobromide. See bromochloromethane.

4,4′-methylenedianiline. See p,p′-diaminodiphenylmethane.

methylene dichloride. See methylene chloride.

3,4-methylenedioxybenzaldehyde. See piperonal.

3,4-methylenedioxypropylbenzene.
(dihydrosafrole). $C_3H_7C_6H_3O_2CH_2$.
Properties: Colorless liquid, d 1.065 (25/25C), somewhat soluble in alcohol. Combustible.
Use: Essential oil compositions.

methylene-di-p-phenylene isocyanate.
See diphenylmethane-4,4′-diisocyanate.

5,5′-methylenedisalicylic acid. CAS: 122-25-8.
$CH_2[C_6H_3(OH)COOH]_2$.
Properties: Nonhygroscopic, light tan, coarse powder; stable in air (darkens in light); tends to decarboxylate at very high temperatures; decomposes at 238C; soluble in alcohol, ether, and acetone; insoluble in benzene, chloroform. Combustible.
Use: Alkyd resins and modified phenolic composi-

tions for paints and varnishes, intermediate for dyestuffs and printing ink.

methylene glutaronitrile. (acrylonitrile dimer; 2,4-dicyanobutene-1). $NCC_2H_4C(:CH_2)CN$.
Properties: Colorless liquid, bp 103C (5 mm), fp −9.6C, refr index 1.4504 (25C), d 0.9756 (25/4C), slightly soluble in water, insoluble in aliphatic and alicyclic hydrocarbons, soluble in aromatic hydrocarbons and polar organic solvents.
Derivation: From acrylonitrile by catalytic dimerization.
Use: Vinyl monomer for polymerization and copolymerization, intermediate in making fibers and pharmaceuticals.

methylene iodide. (diiodomethane).
CAS: 75-11-6. CH_2I_2.
Properties: Yellow liquid, soluble in alcohol and ether, insoluble in water, d 3.33, mp 6C, bp 180C (decomposes).
Hazard: May be irritating and narcotic.
Use: Separating mixture of minerals, organic synthesis, x-ray contrast media.

5-methylene-2-norbornene. A copolymer in EPDM elastomers.

methylene succinic acid. See itaconic acid.

methylene-p-toluidine. See formaldehyde-p-toluidine.

methyl ester. Any of a group of fatty esters derived from coconut and other vegetable oils, tallow, etc.; alkyl groups range from C_8 to C_{18} in varying percentages.
Use: Lubricants for metal-cutting fluids, high-temperature grinding, cold-rolling of steel.

N-methylethanolamine. CAS: 109-83-1.
$CH_3NHC_2H_4OH$.
Properties: Liquid, d 0.9414, bp 159.5C, vap press 0.7 mm (20C), fp −4.5C, flash p 165F (73.9C), soluble in water. Combustible.
Use: Textile chemicals, pharmaceuticals.

methyl ether. See dimethyl ether.

4-methyl-7-ethoxycoumarin.
$C_2H_5OC_6H_3C(CH_3):CHC(O)O$.
Properties: White solid with a walnut odor, mp 113–114C, slightly soluble in alcohol.
Use: Perfumery, flavoring.

2-(1-methylethoxy)phenol methylcarbamate.
$(CH_3)_2CHOC_6H_4OOCNHCH_3$.
Properties: White, odorless, crystalline powder; mp 91C. Soluble in most alcohols, very slightly

soluble in water, unstable in highly alkaline media, stable under normal conditions.
Hazard: Toxic by ingestion and inhalation.
Use: Insecticide.

methylethylcarbinol. See sec-butyl alcohol.

methylethylcellulose. The methyl ether of ethylcellulose in which both methyl and ethyl groups are attached to anhydroglucose units by ether linkages.
Properties: White to pale cream-colored fibrous solid or powder, practically odorless, disperses in cold water to form aqueous solutions which undergo a reversible transformation from solution to gel upon heating and cooling, respectively. Combustible.
Grade: Technical, FCC.
Use: Emulsifier, stabilizer, foaming agent.
See also cellulose, modified.

methyl ethyl diketone. See acetyl propionyl.

2-methyl-2-ethyl-1,3-dioxolane.

$(CH_3)(C_2H_5)COCH_2CH_2O$.
Properties: Colorless liquid, d 0.9392, bp 117.6C, fp −81.96C, flash p 74F (23.3C) (OC), soluble in water 2.2% by wt.
Hazard: Flammable, dangerous fire risk.
Use: Solvent.

sym-**methylethylethylene.** See 2-pentene.

methylethylglyoxal. See acetyl propionyl.

methylethylhydantoin formaldehyde resin.
A reaction product of 5,5-methylethylhydantoin,

$NHCONHCOC(CH_3)C_2H_5$, and formaldehyde.
Properties: Clear, pale, solid resin; softening point (B&R) 85C; d 1.30; pH (10% aqueous solution) 6.5–7.5.

methyl ethyl ketone. (ethyl methyl ketone; 2-butanone; MEK).   CAS: 78-93-3.
$CH_3COCH_2CH_3$.
Properties: Colorless liquid, acetone-like odor, bp 79.6C, d 0.8255 (0/4C), 0.805 (20/4C) and 0.7997 (25/4C), refr index 1.379 (20C), sp heat 0.549 cal/g, fp −86.4C, viscosity 0.40 cp (25C), soluble in water 22.6 wt%, solubility of water 9.9 wt %, flash p (TOC) 24F (−4.4C), bulk d 6.71 lb/gal (20C), autoign temperature 960F (515C). Soluble in benzene, alcohol, and ether; miscible with oils.
Derivation: (a) From mixed n-butylenes and sulfuric acid to cause hydrolysis followed by distillation to separate sec-butyl alcohol which is dehy-

drogenated, (b) by controlled oxidation of butane, (c) by fermentation.
Grade: Technical.
Use: Solvent in nitrocellulose coatings and vinyl films, "Glyptal" resins, paint removers, cements and adhesives, organic synthesis, manufacture of smokeless powder, cleaning fluids, printing, catalyst carrier, acrylic coatings. *Note*: Does not dissolve cellulose acetate and most waxes.
Hazard: Flammable, dangerous fire risk, explosive limits in air 2–10%. Toxic by inhalation. TLV: 200 ppm in air.

methyl ethyl ketone peroxide. (ethyl methyl ketone peroxide).   CAS: 1338-23-4.
$C_8H_{16}O_4$.
Properties: Colorless liquid, strong oxidizing agent.
Hazard: Fire risk in contact with organic materials. Strong irritant to skin and tissue. TLV: CL 0.2 ppm in air.
Use: Manufacture of acrylic resins, hardening agent for fiber glass reinforced platics.

2-methyl-5-ethylpyridine. (MEP; aldehydine; aldehyde collidine; 5-ethyl-2-picoline).
CAS: 104-90-5. $CH_2C_5H_3NC_2H_5$.
Properties: Colorless liquid, sharp penetrating odor, d 0.921 (20/20), bp 178.3C, fp −70.3C, flash p (COC) 165F (73.9C), refr index 1.4970 (20C). Almost insoluble in water; soluble in alcohol, ether, benzene, concentrated sulfuric acid.
Derivation: Paraldehyde is treated with ammonia under high pressure and in the presence of ammonium acetate as a catalyst. Picolines and other substituted pyridines are byproducts.
Grade: Technical.
Hazard: Toxic. Corrosive and strong irritant to tissue.
Use: Nicotinic acid and nicotinamide, vinyl pyridines for copolymers, intermediates for germicides and textile finishes, corrosion inhibitor for chlorinated solvents.

methyl eugenol. (Generic name for 4-allyl-1,2-dimethyoxybenzene; 4-allyl veratrole; 1,2-dimethoxy-4-allylbenzene; eugenyl methyl ether).
CAS: 93-15-2. $(CH_3O)_2C_6H_3CH_2CH:CH_2$.
Properties: Colorless to pale yellow liquid, bp 91–95C (0.3 mm), d 1.032–1.036 (25C), insoluble in water, soluble in most organic solvents. Combustible.
Grade: Technical, FCC.
Use: Insect attractant, flavoring.

methyl fluoride. Legal label name (Air) for fluoromethane.

methyl fluorosulfonate. CAS: 421-20-5.
CH_3OSO_2F.
Properties: Colorless liquid with ethereal odor, bp 92C, d 1.42.
Hazard: Toxic. Strong irritant to tissue, inhalation of fume must be avoided. Reacts with water, steam, and acids evolving corrosive vapor.
Use: Methylating agent, organic synthesis.

N-methylformanilide. $C_6H_5N(CH_3)CHO$.
Properties: Colorless to light yellow liquid, refr index 1.5570–1.5600 (25C), distillation range 127–131C (16 mm).
Grade: 95% min.
Use: Organic synthesis.

methyl formate. CAS: 107-31-3. $HCOOCH_3$.
Properties: Colorless liquid, agreeable odor, saponified by water or alkaline solutions. Soluble in water, alcohol, and ether. D 0.950–0.980 (20/20C), fp −99.8C, bp 31.8C, flash p −2F (−18.9C), bulk d 8.03 lbs/gal (68F), refr index 1.3431 (20C), autoign temperature 853F (456C).
Derivation: By heating methanol with sodium formate and hydrochloric acid with subsequent distillation.
Grade: Technical, refined, FCC.
Hazard: Flammable, dangerous fire and explosion risk, explosive limits in air 5.9–20%. Irritant. TLV: 100 ppm in air.
Use: Organic synthesis, cellulose acetate solvent, fumigant, larvicides.

methyl fuel. Auxiliary fuel for automotive equipment, electric power production, fuel cells, etc. A mixture of methanol and alcohols of up to four carbon atoms, it is made by catalytic treatment of synthesis gas. It can be blended with gasoline in low percentage. Its combustible products contain less polluting components than No. 5 fuel oil.
See also methanol.

2-methylfuran. CAS: 534-22-5. $C_4H_3OCH_3$.
Properties: Colorless liquid, ether-like odor, fp −88.68C, bp 63.2–65.6C, d 0.913 (20/4C), refr index 1.4320 (20C), flash p −22F (−30C), insoluble in water 0.3 g/100 g water. Miscible with most organic solvents. Forms a binary azeotrope with methanol, a ternary azeotrope with methanol-water.
Hazard: Highly flammable, dangerous fire and explosion risk. Irritant.
Use: Chemical intermediate.

N-methylfurfurylamine. $C_4H_3OCH_2NHCH_3$.
Properties: Colorless to light yellow, refr index 1.4700–1.4720 (25C), distilling range 144–153C. Combustible.
Use: Intermediate.

methyl-2-furoate. (methyl pyromucate).
CAS: 611-13-2. $C_4H_3OCO_2CH_3$.
Properties: Colorless liquid turning yellow in light, pleasant odor, insoluble in water, soluble in alcohol and ether, d 1.1739 (15/15C), bp 181.3C (corrosive), refr index 1.4860 (20C). Combustible.
Derivation: By esterification of furoic acid.
Use: Solvent, organic synthesis.

methyl gallate. (methyl-3,4,5-trihydroxybenzoate). CAS: 99-24-1.
$C_6H_2(OH)_3COOCH_3$.
Properties: White, crystalline powder.
Use: Industrial antioxidant.

N-methylglucamine. CAS: 6284-40-8.
$CH_2OH(CHOH)_4CH_2NHCH_3$.
Properties: White crystals, mp 128C, soluble in water, slightly soluble in alcohol, complexes with metals. Preparation: From glucose and methylamine.
Use: Detergents, pharmaceuticals, dyes.

methyl-α-d-glucopyranoside. See methyl glucoside.

α-methyl glucoside. (methyl-α-d-glycopyranoside). CAS: 97-30-3.

$CH_2OHCH(CHOH)_3CHOOCH_3$.
Properties: Odorless, white crystals; mp 168C; bp 200C (0.2 mm). Specific optical rotation (aqueous solution) +158.9C (20C), d (30/4C) 1.46, soluble in water, slightly soluble in 80% alcohol and methanol, insoluble in ether. Combustible.
Derivation: (a) By treating dextrose with methanol in the presence of hydrochloric acid or cation exchange resin, (b) enzymatic synthesis from yeast.
Grade: Technical.
Use: Plasticizer for phenolic, amine, and alkyd resins; nonionic surfactants, tall oil varnishes, reclaiming drying oils, polyurethane foams.

methyl glycocoll. See sarcosine.

methyl glycol. See propylene glycol.

methyl group. The simplest alkyl group, CH_3, formed by dropping a hydrogen atom from methane (CH_4). It occurs at both ends of paraffinic molecules having two or more carbon atoms in the chain, as well as in many other organic compounds.
See also alkyl.

4-methylguaiacol. See creosol.

N-methyl-N-guanylglycine. See creatine.

methyl heneicosanoate. $CH_3(CH_2)_{19}COOCH_3$.
The methyl ester of heneicosanoic acid.
Properties: White wax-like solid, insoluble in water, soluble in alcohol and ether, mp 48–9C, bp 207C (3.75 mm). Combustible.
Grade: Purified 96%, 99.5%.
Use: Intermediate in organic synthesis.

methyl heptadecanoate. (methyl margarate).
$CH_3(CH_2)_{15}COOCH_3$.   The methyl ester of heptadecanoic acid (margaric acid).
Properties: White, wax-like solid. Insoluble in water, soluble in alcohol and ether, mp 29C, bp 184–7C, 130C (1 mm). Combustible.
Grade: Purified 96%, 99.5%.
Use: Intermediate in organic synthesis.

2-methylheptane. $(CH_3)_2CH(CH_2)_4CH_3$.
See isooctane.

3-methylheptane. $C_2H_5CH(CH_3)(CH_2)_3CH_3$.
Properties: Colorless liquid, fp −120.5C, bp 118.9C, d 0.70582 (20/4C), refr index 1.39849 (20C).
Grade: 99%, 95%.
Hazard: Flammable, dangerous fire risk.
Use: Calibration, organic synthesis.

4-methylheptane. (methyldipropylmethane).
C_8H_{18} or $CH_3(CH_2)_2CHCH_3(CH_2)_2CH_3$.
Properties: Colorless liquid, soluble in alcohol and ether, insoluble in water, d 0.7161, bp 122.2C.
Grade: Technical.
Hazard: Flammable, dangerous fire risk.
Use: Organic synthesis.

methylheptenone. (6-methyl-5-heptene-2-one).
CAS: 409-02-9. $(CH_3)_2C{:}CH(CH_2)_2COCH_3$.
Constituent of lemongrass oil and many other essential oils.
Properties: Colorless liquid, insoluble in water but miscible with alcohol or ether, d 0.860 (20C), fp −67.1C, bp 173–174C. Combustible.
Derivation: From oil of lemon grass or by controlled oxidation of corresponding secondary alcohol.
Use: Organic synthesis, inexpensive perfumes, flavoring.

methyl heptyne carbonate. (methyl-2-octynoate).
CAS: 111-12-6. $CH_3(CH_2)_4C{\equiv}CCOOCH_3$
Properties: Colorless liquid, strong violet-type odor, d 0.919–0.923, refr index 1.446–1.450 (20C), soluble in most fixed oils and mineral oil, soluble in five parts of 70% alcohol. Combustible.
Derivation: From heptaldehyde.

Grade: Technical, FCC.
Use: Perfumery, flavoring.

2-(1-methylheptyl)-4,6-dinitrophenyl crotonate.
See cinocap.

methyl hexacosanoate. See methyl cerotate.

methyl hexadecanoate. See methyl palmitate.

methylhexamine. See 4-methyl-2-hexylamine.

2-methylhexane. (ethylisobutylmethane; isoheptane). $(CH_3)_2CH(CH_2)_2CH_2CH_3$.
Properties: Colorless liquid, soluble in alcohol, insoluble in water, d 0.6789, bp 90.0C, fp −118.5C, refr index 1.38498 (20C), flash p approximately 0F (−17.7C).
Grade: Technical.
Hazard: Flammable, dangerous fire risk, explosive limits in air 1–6%.
Use: Organic synthesis.

3-methylhexane.
$H_3CCH_2CH(CH_3)CH_2CH_2CH_3$.
Properties: Colorless liquid, bp 92C, d 0.692 (15.5/15.5C), refr index 1.388 (20C), flash p 25F (−3.9C).
Grade: Technical 95%.
Hazard: Flammable, dangerous fire risk.
Use: Organic synthesis, oil extender solvent.

methyl hexanoate. See methyl caproate.

5-methyl-2-hexanone. See methyl isoamyl ketone.

methyl hexyl ketone. (2-octanone).
CAS: 111-13-7. $CH_3COC_6H_{13}$.
Properties: Colorless liquid with pleasant odor, camphor taste, d 0.82 (20/4C), mp 20.9C, bp 173.5C, distillation range 166–173C, flash p 160F (71.1C), refr index 1.416 (20C), insoluble in water, soluble in alcohol, hydrocarbons, ether, esters, etc. Combustible. Preparation: By distilling sodium ricinoleate with caustic soda.
Use: Perfumes, high-boiling solvent, especially for epoxy resin coatings, leather finishes, flavoring, odorant, antiblushing agent for nitrocellulose lacquers.

methyl hydrazine. (monomethylhydrazine; MMH).   CAS: 60-34-4. CH_3NHNH_2.
Properties: Colorless, hygroscopic liquid; ammonia-like odor; d 0.874 (25C); fp −52.4C; bp 87.5C; soluble in water, hydrazine, hydrocarbons, and monohydric alcohols; flash p approximately 80F (26.6C).
Hazard: Flammable, dangerous fire risk, vapors may explode, may self-ignite in air and on contact

with oxidizing agents. Toxic by ingestion and inhalation. TLV: CL 0.2 ppm, suspected human carcinogen.
Use: Missile propellant, intermediate, solvent.

methyl hydride. See methane.

methylhydrogen sulfate. See methylsulfuric acid.

methyl-p-hydroxybenzoate. See methyl paraben.

methylhydroxybutanone. (3-methyl-3-hydroxy-2-butanone). $(CH_3)_2COHCOCH_3$.
Properties: Clear, colorless liquid with sweet camphor-like odor. Bp 140.3C, fp −86.5C, d 0.9553 (20/20C), refr index 1.4153 (20C). Miscible with water, acetone, benzene, mineral spirits. Combustible.
Use: Specialty solvent, chemical intermediate, flavor formulations.

4-methyl-7-hydroxycoumarin diethoxythiophosphate. See potasan.

methyl-3-hydroxy-α-crotonate dimethyl phosphate. See mevinphos.

methylhydroxyisopropylcyclohexane. See menthol.

methyl-12-hydroxystearate. CAS: 141-23-1. $C_{17}H_{34}OHCOOCH_3$.
Properties: White, waxy solid in the form of short flat rods. Mp 48C, acid value 4, saponification value 177, iodine value 5, insoluble in water, limited solubility in organic solvents. Combustible.
Use: Adhesives, inks, cosmetics greases.

2-methylimidazole. (2MZ).

$CHCHNC(CH_3)NH$.
Properties: Solid, mp 142–143C.
Use: Dyeing auxiliary for acrylic fibers, plastic foams.

3-methylindole. See skatole.

methyl iodide. (iodomethane). CAS: 74-88-4. CH_3I.
Properties: Colorless liquid, turns brown on exposure to light, d 2.24–2.27 (25/25C), fp −66.1C, bp 42C, refr index 1.526–1.527 (25C), soluble in alcohol and ether, partially soluble in water. Nonflammable.
Derivation: Interaction of methanol, sodium iodide, and sulfuric acid with subsequent distillation.
Hazard: Toxic by ingestion, inhalation, and skin absorption; narcotic, irritant to skin. TLV: 2 ppm in air, suspected human carcinogen.

Use: Organic synthesis, microscopy, testing for pyridine.

methylionone. (irone). $C_{14}H_{22}O$.
Properties: Colorless to amber-yellow liquid, floral odor.
Grade: Several isomers are available as α, β, δ, γ and mixtures. The constants are approximately: d 0.926–0.939, refr index 1.501–1.504, bp 144C (16 mm), soluble in alcohol, insoluble in water.
Derivation: Oil of orris.
Use: Perfumery, flavoring.
See also α-isomethylionone.

methyl isoamyl ketone. (5-methyl-2-hexanone; MIAK). CAS: 110-12-3. $CH_3COC_2H_4CH(CH_3)_2$.
Properties: Colorless, stable liquid; pleasant odor. D 0.8132 (20/20C), refr index 1.4062 (20C), bp 144C, fp −73.9C, bulk d 6.77 lb/gal, flash p (OC) 110F (43.3C), slightly soluble in water, miscible with most organic solvents. Combustible.
Grade: 97.5%.
Hazard: Moderate fire risk. TLV: 50 ppm in air.
Use: Solvent for nitrocellulose, cellulose acetate butyrate, acrylics, and vinyl copolymers.

methyl isobutenyl ketone. See mesityl oxide.

methylisobutyl carbinol. See methylamyl alcohol.

methylisobutyl carbinol acetate. See methylamyl acetate.

methyl isobutyl ketone. (hexone; 4-methyl-2-pentanone; isopropylacetone). CAS: 108-10-1. $(CH_3)_2CHCH_2COCH_3$.
Properties: Colorless, stable liquid; pleasant odor. Slightly soluble in water, miscible with most organic solvents, d 0.8042 (20/20C), bp 115.8C, fp −85C, bulk d 6.68 lbs/gal (20C), vap press 15.7 mm (20C), refr index 1.3959 (20C), flash p 73F (22.7C), autoign temperature 860F (460C).
Derivation: Mild hydrogenation of mesityl oxide.
Grade: Technical, 98.5%.
Hazard: Flammable, dangerous fire risk, explosive limits in air 1.4–7.5%. Avoid ingestion and inhalation. TLV: 50 ppm in air, absorbed by skin.
Use: Solvent for paints, varnishes, nitrocellulose, lacquers, manufacture of methyl amyl alcohol, extraction processes including extraction of uranium from fission products, organic synthesis, denaturant for alcohol.

methyl isocyanate. CAS: 624-83-9. CH_3NCO.
Properties: Colorless liquid, d 0.9599 (20/20C), bp 39.1C, reacts with water, flash p less than 20F (−6.6C).

Hazard: Flammable, dangerous fire risk. Toxic by skin absorption and a strong irritant. TLV: 0.02 ppm in air.
Use: Intermediate.

methyl isothiocyanate. (methyl mustard oil). CAS: 556-61-6. $CH_3N:CS$.
Properties: Colorless crystals, mp 35C, bp 120C, soluble in alcohol and ether, partially soluble in water.
Derivation: Reaction of methylamine and carbon disulfide.
Hazard: Toxic by ingestion, strong irritant to eyes and skin.
Use: Insecticide, a possible military poison.

methylisoeugenol. (propenyl guaiacol). CAS: 93-16-3. $CH_3CH:CHC_6H_3(OCH_3)_2$.
Properties: Colorless to light-yellowish liquid with spicy odor, d 1.050–1.053, bp 262–264C, refr index 1.566–1.569, soluble in two parts of 70% alcohol, almost insoluble in mineral oil, insoluble in glycerol. Combustible.
Grade: Technical, FCC.
Use: Perfumery, flavoring agent.

methyl isonicotinate. $C_5NH_4COOCH_3$.
Properties: Clear amber to red liquid, mild odor, d 1.15 (20/20C).
Use: Intermediate for synthesis of isonicotinic acid hydrazide.

1-methyl-4-isopropenylcyclohexan-3-ol.
See isopulegol.

methyl isopropenyl ketone. CAS: 814-78-8. $CH_3COC(CH_3):CH_2$.
Properties: Colorless liquid, d 0.854 (20C), pleasant odor and sweet taste, polymerizes readily.
Hazard: Flammable, dangerous fire risk, explosive limits in air 1.8–9.0%.
Use: Plastics.

methyl isopropyl ketone. (3-methyl-2-butanone). CAS: 563-80-4. $CH_3COCH(CH_3)_2$.
Properties: Colorless liquid, bp 93C, fp −92C, refr index 1.38788 (16C), d 0.815 (15/4C), very slightly soluble in water, soluble in alcohol and ether.
Derivation: Synthetic, also by fermentation.
Hazard: Toxic material. TLV: 200 ppm in air.
Use: Solvent for nitrocellulose lacquers.
See also ketones.

2-methyl-5-isopropylphenol. See carvacrol.

5-methyl-2-isopropylphenol. See thymol.

methyl-p-isopropylphenyl propyl aldehyde.
See cyclamen aldehyde.

methyl lactate. CAS: 547-64-8. $CH_3CHOHCOOCH_3$.
Properties: Colorless liquid, soluble in alcohol and ether, decomposed by water, bp 144.8C, fp approximately −66C, refr index 1.4156 (20C), flash p (CC) 125F (51.6C), autoign temperature 725F (385C), bulk d 9 lb/gal (68F). Combustible.
Hazard: Moderate fire risk.
Use: Solvent for cellulose acetate, nitrocellulose, cellulose acetobutyrate, cellulose acetopropionate, lacquers, stains.

methyllactonitrile. See acetone cyanohydrin.

methyl laurate. (methyl dodecanoate). $CH_3(CH_2)_{10}COOCH_3$. The methyl ester of lauric acid.
Properties: Water-white liquid, d 0.8702 (20/4C), mp 4.8C, bp 262C (766 mm), 160C (30 mm), refr index 1.4301 (25C), insoluble in water, noncorrosive. Combustible.
Derivation: From coconut oil.
Method of purification: Vacuum fractional distillation.
Grade: 69%, 74%, 90%, 96%, 99.8%.
Use: Intermediate for detergents, emulsifiers, wetting agents, stabilizers, lubricants, plasticizers, textiles, flavoring.

methyl lauroleate.
$CH_3CH_2CH:CH(CH_2)_7COOCH_3$.
The methyl ester of lauroleic acid.
Properties: Colorless liquid, insoluble in water, soluble in common organic solvents. Combustible.
Grade: Purified product, 99.5%.
Use: Organic synthesis, reference standard for gas chromatography, biochemical research.

methyl lignocerate. (methyl tetracosanoate). $CH_3(CH_2)_{22}COOCH_3$. The methyl ester of lignoceric acid.
Properties: Wax-like solid, insoluble in water, soluble in alcohol and ether, mp 57.8C, bp 232C (3.75 mm), refr index 1.4283 (80C). Combustible.
Derivation: Esterification of lignoceric acid with methanol followed by vacuum distillation.
Grade: Purified (99.8%+).
Use: Intermediate in special synthesis, biochemical research, reference standard in gas chromatography.

methyl linoleate. CAS: 112-63-0.
$CH_3(CH_2)_4CH:CHCH_2CH:(CH_2)_7COOCH_3$.
The methyl ester of linoleic acid (cis, cis-octadec-9,12-dienoic acid).
Properties: Colorless oil, insoluble in water, soluble in alcohol and ether, d 0.8886 (18/4C), fp −35C, bp 212C (16 mm), refr index 1.4593 (25C). Combustible.

Derivation: Urea fractionation and vacuum distillation of methyl esters of safflower oil.

Grade: Technical, purified (99+%).

Use: Intermediate for detergents, emulsifiers, wetting agents, stabilizers, resins, lubricants, plasticizers, textiles, reference standard in gas chromatography, biochemical research.

methyl linolenate.
$C_{19}H_{32}O_2$. The methyl ester of linolenic acid (cis, cis, cis-octadec-9,12,15-trienoic acid).

Properties: Colorless liquid, insoluble in water, soluble in alcohol and ether, d 0.892 (20/4C), mp less than 15C, bp 207C (14 mm), refr index 1.4709 (20C). Combustible.

Derivation: Esterification and vacuum fractional distillation.

Use: Organic synthesis, biochemical research.

methyllithium. CH_3Li. Commercially available as a 5% solution in ether.

Hazard: Flammable, dangerous fire and explosion risk, self-ignites in air.

Use: In Grignard-type reactions.

methylmagnesium bromide. CH_3MgBr.
Available in solution in ether.

Derivation: Reaction of magnesium and methyl-bromide.

Hazard: Flammable, dangerous fire and explosion risk.

Use: Alkylating agent in organic synthesis, Grignard reagent.

methylmagnesium chloride. CH_3MgCl.
Available as a solution in tetrahydrofuran.

Hazard: Flammable, dangerous fire and explosion risk.

Use: Alkylating agent in organic synthesis, Grignard reagent.

methylmagnesium iodide. CH_3MgI.
Available as solution in ether.

Derivation: Reaction of magnesium and methyl iodide.

Hazard: Flammable, dangerous fire and explosion risk.

Use: Alkylating agent in organic synthesis, Grignard reagent.

methylmaleic anhydride. See citraconic anhydride.

methyl margarate. See methyl heptadecanoate.

methyl mercaptan. Legal label name for methanethiol.

methylmercury acetate. $CH_3HgOOCCH_3$.
Hazard: Toxic by ingestion, absorbed by skin. TLV: 0.01 mg/m³ of air.
Use: Seed disinfectant.

methylmercury cyanide. (methylmercury nitrile).
CAS: 2597-97-9. CH_3HgCN.
Properties: Crystals, mp 95C, soluble in water and organic solvents.
Hazard: Toxic by ingestion, absorbed by skin. TLV: 0.01 mg/m³ of air.
Use: Fungicide, seed disinfectant.

methylmercury dicyandiamide. See cyano (methylmercuri) guanidine.

methylmercury-2,3-dihydroxypropylmercaptide.
$CH_3HgSCH_2CHOHCH_2OH$.
Hazard: Toxic by ingestion, absorbed by skin. TLV: 0.01 mg/m³ of air.
Use: Seed disinfectant.

methylmercury-8-hydroxyquinolate. See methylmercury quinolinolate.

methylmercury quinolinolate. (methylmercury-8-hydroxyquinolate; methylmercury oxyquinolinate). CAS: 86-85-1.
$C_9H_6NOHgCH_3$.
Properties: Yellow crystals, mp 133–137C.
Hazard: Toxic material. TLV: 0.01 mg/m³ of air.
Use: Seed disinfectant.

methyl methacrylate. CAS: 80-62-6.
$CH_2{:}C(CH_3)COOCH_3$. Acrylic resin monomer.
Properties: Colorless, volatile liquid. Bp 101C, fp −48.2C, d 0.940 (25/25C), flash p (OC) 50F (10C), autoign temperature 790F (421C), slightly soluble in water, soluble in most organic solvents, readily polymerized by light, heat, ionizing radiation and catalysts. Can be copolymerized with other methacrylate esters and many other monomers.
Derivation: (1) Acetone cyanohydrin, methanol, and dilute sulfuric acid. (2) Oxidation of tert-butyl alcohol to methacrolein and then to methacrylic acid, followed by reaction with methanol.
Grade: Technical (inhibited).
Hazard: Flammable, dangerous fire risk, explosive limits in air 2.1–12.5%. TLV: 100 ppm in air.
Use: Monomer for polymethacrylate resins, impregnation of concrete.
See also acrylic resin.

methylmethane. See ethane.

N-methyl methyl anthranilate. See dimethyl anthranilate.

7-methyl-3-methylene-1,6-octadiene. See myrcene.

methylmorphine. See codeine.

N-methyl morpholine. CAS: 109-02-4.

$\overline{CH_2CH_2OCH_2CH_2NCH_3}$.
Properties: Water-white liquid with ammonia odor, forms constant-boiling mixture with 25% water and boiling at 97C, miscible with benzene, water, d 0.921 (20/20C, bp 115.4C, fp −66C, flash p (TOC) 75F (23.9C).
Grade: Technical.
Hazard: Flammable, dangerous fire risk. Skin irritant.
Use: Catalyst in polyurethane foams, extraction solvent, stabilizing agent for chlorinated hydrocarbons, self-polishing waxes, oil emulsions, corrosion inhibitors, pharmaceuticals.

methyl myristate. (methyl tetradecanoate).
$CH_3(CH_2)_{12}COOCH_3$.   The methyl ester of myristic acid.
Properties: Colorless liquid, mp 17.8C, bp 186.8C (30 mm), 157.5C (1 mm), refr index 1.4351 (25C), insoluble in water. Combustible.
Derivation: (a) Esterification of myristic acid with methanol, (b) alcoholysis of coconut oil with methanol.
Method of purification: Vacuum fractional distillation.
Grade: Technical (93%), purified (99.8+%).
Use: Intermediate for myristic acid detergents, emulsifiers, wetting agents, stabilizers, resins, lubricants, plasticizers, textiles, animal feeds, standard for gas chromatography, flavoring.

methyl myristoleate.
$CH_3(CH_2)_3CH:CH(CH_2)_7COOCH_3$.   The methyl ester of myristoleic acid (cis-tetradec-9-enoic acid).
Properties: Colorless liquid, insoluble in water, soluble in alcohol and ether, bp 108.9C (1 mm). Combustible.
Use: Purified product used in medical research and organic synthesis.

α-methylnaphthalene. CAS: 90-12-0.
$C_{10}H_7CH_3$.
Properties: Colorless liquid, d 1.025, fp −32C, bp 240–243C, refr index 1.6140 (25C), insoluble in water, soluble in alcohol and ether, autoign temperature 984F (529C). Combustible.
Derivation: From coal tar.
Hazard: Moderate fire risk.
Use: Organic synthesis.

β-methylnaphthalene. CAS: 91-57-6.
$C_{10}H_7CH_3$.
Properties: Solid, d 0.994 (40/4C), bp 241–242C, mp 34C, refr index 1.6015 (25C), insoluble in water, soluble in alcohol and ether. Combustible.
Derivation: From coal tar.

Grade: Technical, 95% min.
Use: Organic synthesis, insecticides.

2-methyl-1,4-naphthoquinone. See menadione.

methyl naphthyl dodecyl dimethylammonium chloride. $C_{10}H_7HC_2C_{12}H_{25}(CH_3)_2NCl$.
Quaternary ammonium salt.
Properties: White to slightly yellow, crystalline powder with mild odor and taste; mp 159–160C; bulk d 5 lb/gal; soluble in cold water, lower alcohols, glycerin, and acetone; pH of 5% solution 8.9.
Grade: Min purity 97%.
Use: Germicide.

methyl naphthyl ether. See β-naphthyl methyl ether.

methyl nitrate. CAS: 598-58-3. CH_3NO_3.
Properties: Colorless liquid, bp 66C (explodes), d 1.217 (15C), slightly soluble in water, soluble in alcohol and ether.
Derivation: By reaction of nitric acid and methanol in the presence of urea.
Hazard: Explodes when heated, severe hazard when exposed to heat or shock. Narcotic, strong irritant to tissue.
Use: Rocket propellant.

methyl nitrite. CAS: 624-91-9. CH_3NO_2.
Properties: A gas, bp −12C, fp −17C, d 0.991 at 15C.
Hazard: Severe explosion risk when shocked or heated. Toxic by inhalation, narcotic.
Use: Synthesis of nitrile and nitroso esters.

methylnitrobenzene. See nitrotoluene.

4-methyl-2-nitrophenol. See 2-nitro-p-cresol.

methyl-4-nitrosoperfluorobutyrate. A monomer for a nitroso ester terpolymer, prepared by reacting methyl nitrite with perfluorosuccinic and perfluoroglutaric anhydrides. The reaction is exothermic and gives high conversion to liquid nitrile esters, which yield nitroso esters on pyrolysis or photolysis.

N-methyl-N-nitroso-p-toluenesulfonamide.
$CH_3C_6H_4SO_2N(NO)CH_3$.
Properties: Fine, yellow crystals; mp 56–59C; soluble in ether and most organic solvents; insoluble in water.
Use: Reagent for the preparation of diazomethane.

methyl nonadecanoate. $CH_3(CH_2)_{17}COOCH_3$.
The methyl ester of nonadecanoic acid.
Properties: White, waxy solid. Mp 39.5C, bp 190.5C (3.75 mm), insoluble in water, soluble in alcohol and ether. Combustible.

Grade: Purified 96%, 99.5%.
Use: Intermediate in organic synthesis, medical research.

2-methylnonane. (isodecane).
$(CH_3)_2CH(CH_2)_6CH_3$.
Properties: Colorless liquid, d 0.728, fp −74.7C, bp 167C. Combustible.

methyl nonanoate. (methyl pelargonate).
$CH_3(CH_2)_7COOCH_3$.   The methyl ester of pelargonic acid.
Properties: Colorless liquid with fruity odor, fp −35C, bp 213.5C, d 0.877 (18C), refr index 1.4302 (25C), insoluble in water, soluble in alcohol and ether. Combustible.
Derivation: Esterification of nonanoic (pelargonic) acid with methanol followed by fractional distillation.
Grade: Purified (96+%).
Use: Perfumes, flavors, reference standard for gas chromatography, intermediate in organic synthesis, medical research.

methyl-2-nonenoate.
$CH_3(CH_2)_5HC:CHCOOCH_3$.
Properties: Colorless to slightly yellow liquid with strong violet-leaf odor, d 0.893–0.898 (25/25C), refr index 1.4400–1.4440, stable, soluble in alcohol. Combustible.
Use: Perfumes.

methylnonylacetaldehyde. (aldehyde C-12; MNA; methylundecanal).
$CH_3(CH_2)_8CH(CH_3)CHO$.
Properties: Colorless liquid, fruity odor. Soluble in three volumes of 80% alcohol, in fixed oils, mineral oil; d 0.824–0.828; refr index 1.432–1.435 (20C). Combustible.
Grade: Technical, FCC (as methylundecanal).
Use: Perfumery, flavoring.

methyl nonyl ketone. (2-undecanone).
CAS: 112-12-9. $CH_3COC_9H_{19}$.
Properties: Oily liquid, strong odor, soluble in two parts of 70% alcohol, d 0.822–0.826, bp 225C, refr index 1.429–1.433, flash p 192F (88.9C). Combustible.
Derivation: Oil of rue, also made synthetically.
Use: Perfumery, flavoring, animal repellent.

methyl norbornene dicarboxylic anhydride.
Legal label name for memtetrahydrophthalic anhydride.

methyl octadecanoate. See methyl stearate.

methyl-2-octynoate. See methyl heptyne carbonate.

methylolacrylamide. Available in 60% aqueous solution.
Use: Intermediate for copolymerization of vinyl acetate and acrylic acid, polymers for coatings, varnishes, adhesives, crease-proof and wrinkle-resistant fabrics, permanent-press textiles by irradiation bonding.

methylol dimethylhydantoin. (dimethylhydantoin formaldehyde; DMHF).
$(CH_3)_2CN(CH_2OH)CONHCO$.
Properties: White, odorless, crystalline solid; mp 110–117C. Soluble in water, methanol, acetone; insoluble in hydrocarbons.
Use: Textile and paper finishing, preservative for cosmetics; source of formaldehyde.
See also dimethylhydantoinformaldehyde resin.

methyl oleate. CAS: 112-62-9.
$CH_3(CH_2)_7COOCH_3$.   The methyl ester of oleic acid (cis-octadec-9-enoic acid).
Properties: Clear to amber liquid, faint fatty odor, soluble in alcohols and most organic solvents, insoluble in water, d 0.8739 (20C), fp −19.9C, bp 218.5C (20 mm), refr index 1.4521 (20C). Combustible.
Derivation: Esterification of oleic acid, vacuum fractional distillation, solvent crystallization.
Grade: Technical, purified 99+%.
Use: Intermediate for detergents, emulsifiers, wetting agents, stabilizers, textile treatment, plasticizers for duplicating inks, rubbers, waxes, etc.; biochemical research, chromatographic reference standard.

methylol imidazolidone. $C_3H_4N_2(O)CH_2OH$.
Properties: Straw-colored to water-white, clear liquid; mild odor; bulk d 10 lb/gal (15.5C).
Use: Wash and wear fabrics.

methylol riboflavin. A mixture of methylol derivatives of riboflavin exhibiting the same activity.
Properties: Orange to yellow hygroscopic powder, nearly odorless, soluble in water, nearly insoluble in alcohol, benzene, chloroform and ether; dextrorotatory. Dry powder is unstable and on standing loses biological activity by liberation of formaldehyde.
Derivation: Action of formaldehyde on riboflavin in weakly alkaline solutions.
Use: Nutrition, vitamin source.

methylol urea. $H_2NCONHCH_2OH$.
Properties: Colorless crystals, mp 11C, soluble in water and methanol, insoluble in ether, capable of polymerization.
Derivation: Combination of urea and formaldehyde, in the presence of salts or alkaline catalysts.

Use: Urea-formaldehyde resins, molding adhesives, treating textiles and wood.
See A-stage resin, urea-formaldehyde resin.

"Methylon."[245] TM for resinous compositions made from condensation products of substituted aromatic hydrocarbons and aldehydes used in the surface coating and protective arts.

methyl orange. (p-(p-dimethylamino phenylazo)-benzene sulfonate of sodium; Helianthine B; orange III; gold orange; tropaeolin D; CI 13025). CAS: 547-58-0.
$(CH_3)_2NC_6H_4NNC_6H_4SO_3Na$.
Properties: Orange-yellow powder, soluble in hot water, insoluble in alcohol.
Use: Acid-base indicator, red in acid, yellow-orange in alkaline, pH range 3.1–4.4.
See indicator.

methyl orthophosphoric acid. See methylphosphoric acid.

methyl oxide. See dimethyl ether.

methyl "Oxitol."[125] TM for ethylene glycol monomethyl ether whose ASTM distillation range is 123.5–125.5C.

methyl palmitate. (methyl hexadecanoate). $CH_3(CH_2)_{14}COOCH_3$. The methyl ester of palmitic acid.
Properties: Colorless liquid, mp 29.5C, bp 211.5C (30 mm), 180.5C (10 mm), refr index 1.4310 (45C), insoluble in water, soluble in alcohol and ether. Combustible.
Derivation: Esterification of palmitic acid with methanol or alcoholysis of palm oil plus vacuum distillation.
Grade: 80%, pure (99.8%).
Use: Intermediate for detergents, emulsifiers, wetting agents, stabilizers, resins, lubricants, plasticizers, animal feeds, medical research.

methylparaben. (methyl-p-hydroxybenzoate). CAS: 99-76-3. $CH_3OOCC_6H_4OH$.
Properties: Colorless crystals or white, crystalline powder; mp 125–128C; odorless or faint characteristic odor; slight burning taste; soluble in alcohol, ether; slightly soluble in water, benzene, and carbon tetrachloride.
Grade: USP, FCC.
Hazard: Toxic. Use in foods restricted to 0.1%.
Use: Food additive (preservative), antimicrobial agent.
See also "Parabens."

methylparafynol. See 3-methyl-1-pentyn-3-ol.

methyl parathion. (O,O-dimethyl-O-p-nitrophenylphosphorothioate). CAS: 298-00-0.

$(CH_3O)_2P(SO)OC_6H_4NO_2$. The methyl homolog of parathion.
Properties: White, crystalline solid; mp 35–36C. D 1.358 (20/4C), refr index 1.5515 (35C), slightly soluble in water, miscible in all proportions with acids and alcohols, esters and ketones, slightly decomposed by acid solutions, rapidly in dilute alkalies. The commercial product is a tan liquid (xylene solution) with pungent odor, decomposes violently at 248F. It is not classed as a "hard" insecticide, but is relatively biodegradable.
Forms available: Emulsifiable concentrate, wettable powder, dusts, technical (80%), liquid solution, encapsulated.
Hazard: Explosion risk when heated. Toxic by skin absorption, inhalation, and ingestion; cholinesterase inhibitor; use has been restricted. TLV: 0.2 mg/m³ of air.
Use: Insecticide, especially for cotton.

methyl PCT. See O,O-dimethyl phosphorochloridothioate.

methyl pelargonate. See methyl nonanoate.

methyl pentadecanoate. $CH_3(CH_2)_{13}COOCH_3$. The methyl ester of pentadecanoic acid.
Properties: Colorless liquid, insoluble in water, soluble in alcohol and ether, d 0.8618 (25/4C), mp 18.5C, bp 199C (30 mm), refr index 1.4374 (25C). Combustible.
Grade: Reagent, 96% and 99.5%.
Use: Intermediate in organic synthesis, reagent medical research.

methylpentadiene. C_6H_{10}. Numerous isomers are possible. Commercially available mixture contains 2- and 4-methyl-1,3-pentadiene.
Properties: Colorless liquid, d 0.7184 (20/4C), bp 75–77C, flash p −30F (−34.4C), reactive with halogens, hydrohalogens, sulfur dioxide, and maleic anhydride.
Hazard: Flammable, dangerous fire risk.
Use: Organic synthesis, alkyd and other polymers.

2-methylpentaldehyde. $C_3H_7CH(CH_3)CHO$.
Properties: Colorless liquid, d 0.8092, bp 118.3C, fp −100C, soluble in water 0.42% by wt, flash p (OC) 68F (20C).
Hazard: Flammable, dangerous fire risk. Strong irritant to skin and mucous membranes.
Use: Intermediates for dyes, resins, pharmaceuticals.

2-methylpentane. (dimethylpropylmethane). CAS: 107-83-5. $CH_3(CH_2)_2CH(CH_3)_2$.
Properties: Colorless liquid, fp −153C, bp 60C, refr index 1.372 (20C), d 0.658 (60/60F), flash

p −10F (−23.3C), autoign temperature 583F (306C).

Grade: 95%, 99%, research.

Hazard: Flammable, dangerous fire risk, reacts vigorously with oxidizing materials.

Use: Organic synthesis, solvent.

3-methylpentane. (diethylmethylmethane).
CAS: 96-14-0. $CH_3CH_2CH(CH_3)CH_2CH_3$.
Properties: Colorless liquid, soluble in alcohol, insoluble in water, slightly soluble in ether, d 0.6645 (20/4C), bp 64.0C, refr index 1.37662 (20C), flash p approximately 20F (−6.6C).
Grade: Technical (95%), 99%, research.
Hazard: Flammable, dangerous fire risk.
Use: Organic synthesis, solvent.

2-methyl-1,3-pentanediol.
$C_2H_5CH(OH)CH(CH_3)CH_2OH$.
Properties: Colorless liquid, d 0.9745, bp 220.3C, fp −30C, freely soluble in water. Combustible.
Use: Solvent, coupling agent.
See also hexylene glycol, a mixture of isomers.

2-methyl-2,4-pentanediol. (4-methyl-2,4-pentanediol).
$CH_3CHOHCH_2COH(CH_3)CH_3$.
Properties: Colorless liquid, miscible with water and most organic solvents including lower aliphatic hydrocarbons, d 0.9235 (20/20C), bp 197.1C, viscosity 34 cp (20C), vap press 10.8 mm (95.2C), 334 mm (169.7C), flash p (OC) 201F (93.9C), 7.59 lb/gal (20C). Combustible.
Use: Coupling agent, chemical synthesis.
See also hexylene glycol, a mixture of isomers.

4-methyl-2,4-pentanediol. See 2-methyl-2,4-pentanediol.

2-methylpentanoic acid. CAS: 97-61-0.
$CH_3CH_2CH_2CH(CH_3)COOH$.
Properties: Water-white liquid, d 0.9242 (20/20C), bp 196.4C, vap press 0.02 mm (20C), fp sets to glass approximately −85C, flash p 225F (107C), soluble in water 1.3% by wt (20C). Combustible.
Use: Plasticizers, vinyl stabilizers, metallic salts, alkyd resins.

4-methylpentanoic acid.
$(CH_3)_2CH(CH_2)_2COOH$.
Properties: Colorless liquid, bp 197C, d 0.921 (20/4C), bulk d 7.66 lb/gal (20C). Miscible in alcohol, benzene, and acetone; low solubility in water. Combustible.
Use: Intermediate for plasticizers, pharmaceuticals and perfumes.

2-methyl-1-pentanol. $C_3H_7CH(CH_3)CH_2OH$.
Properties: Colorless liquid, d 0.8252, bp 148.0C, vap press 1.1 mm (20C), solubility in water

0.31% by wt, flash p 135F (57.2C) (OC). Combustible.
Hazard: Moderate fire risk.
Use: Solvent, intermediate.

4-methyl-2-pentanol. See methylamyl alcohol.

4-methyl-2-pentanol acetate. See methylamyl acetate.

4-methyl-2-pentanone. See methyl isobutyl ketone.

2-methyl-1-pentene. (1-methyl-1-propylethylene). $H_2C:C(CH_3)CH_2CH_2CH_3$.
Properties: Colorless liquid, d 0.6820 (20/4C), bp 62.6C, fp −135C, refr index 1.3925 (20C), flash p −15F (26.1C), soluble in alcohol, acetone, ether, petroleum, coal-tar solvents, insoluble in water.
Grade: 95%, 99%, research.
Hazard: Flammable, dangerous fire risk.
Use: Organic synthesis, flavors, perfumes, medicines, dyes, oils, resins.

4-methyl-1-pentene.
$H_2C:CHCH_2CH(CH_3)CH_3$.
Properties: Colorless liquid, fp −153C, bp 53.5C, d 0.6640 (20/4C), refr index 1.3826 (20C), flash p −25F (−31.6C).
Grade: 95%, 99%, research.
Hazard: Same as for 2-methyl-1-pentene.
Use: Organic synthesis, monomer for plastics used in automobiles, electronic components, and laboratory ware.

4-methyl-2-pentene. (1-isopropyl-2-methylethylene (cis-trans mixture)).
$CH_3CH:CHCH(CH_3)_2$.
Properties: Colorless liquid, d 0.670 (20/4C), bp (mixture) 55C, (cis) 56.1C, (trans) 58.3C, refr index 1.388 (20C), flash p −20F (−28.9C), soluble in alcohol, acetone, ether, petroleum and coal-tar solvents, insoluble in water.
Hazard: Flammable, dangerous fire risk.
Use: Organic synthesis.

4-methyl-3-penten-2-one. See mesityl oxide.

3-methyl-1-pentyn-3-ol. (meparfynol; methylparafynol; methyl pentynol).
$HC{\equiv}CCOH(CH_3)CH_2CH_3$
Properties: Colorless liquid, bp 121.4C, fp −30.6C, d 0.8721 (20/20C), refr index 1.4318 (20C), flash p (TOC) 101F (38.3C), moderately soluble in water, miscible with acetone, benzene, carbon tetrachloride, ethyl acetate. Combustible.
Grade: High purity 98.5%, technical 95.0% min, pharmaceutical.

Hazard: Moderate fire risk. Toxic.

Use: Stabilizer in chlorinated solvents, viscous reducer, electroplating brightener, intermediate in syntheses of hypnotics and isoprenoid chemicals, solvent for polyamide resins, acid inhibitor, prevention of hydrogen embrittlement, medicine (soporific and anesthetic).

methylphenethylamine. See amphetamine.

methyl phenylacetate. $C_6H_5CH_2COOCH_3$.

Properties: Colorless liquid, honey-like odor, soluble in five parts of 60% alcohol, in fixed oils; insoluble in water, d 1.062–1.066 (25C), refr index 1.506–1.509 (20C), bp 218C. Combustible.
Grade: Technical, FCC.
Use: Perfumery, flavors for tobacco, flavoring.

methylphenylcarbinol. See α-methylbenzyl alcohol.

methylphenylcarbinyl acetate. See α-methylbenzyl acetate.

methylphenyldichlorosilane. $CH_3(C_6H_5)SiCl_2$.

Properties: Colorless liquid, bp 82C (13 mm), 205C, refr index 1.5199 (25C), d 1.19, soluble in benzene, ether, methanol, flash p 83F (28.3C).
Derivation: From chlorobenzene Grignard reagent and methyltrichlorosilane or from benzene and methyldichlorosilane.
Hazard: Flammable, moderate fire risk, reacts strongly with oxidizing materials. Irritant.
Use: Manufacture of silicones.

methyl phenyl ether. See anisole.

4-methyl-1-phenyl-2-pentanone. (benzyl isobutyl ketone). $C_6H_5CH_2C(O)CH_2CH(CH_3)_2$.
Properties: Combustible.
Use: Flavoring.

methyl phenyl phosphate. See methyl diphenyl phosphate.

2-methyl-2-phenylpropane. See tert-butylbenzene.

3-methyl-1-phenyl-2-pyrazolin-5-one.
(3-methyl-1-phenyl-5-pyrazolone; 1-phenyl-3-methyl-5-pyrazolone).
$C_6H_5NN:C(CH_3)CH_2CO$.
Properties: White powder or crystals, soluble in water, slightly soluble in alcohol or benzene, insoluble in ether, bp 287C (205 mm), mp 127C, vap press less than 0.01 mm (20C).
Derivation: By condensation of phenylhydrazine with ethylacetoacetate.
Hazard: Toxic by ingestion.

Use: Intermediate for dyes and drugs, sensitive reagent for detection of cyanide.

methylphloroglucinol. (2,4,6-trihydroxytoluene). $C_6H_2(OH)_3CH_3$.
Properties: Cream to light tan fine crystals, odorless, mp 210–214C, soluble in water, alcohol and ether, insoluble in benzene. Combustible.
Use: Reactive coupling agent, dye and plastic intermediate.

methylphosphonic acid. $CH_3PO(OH)_2$.
Properties: White solid, mp 103–104C.
Use: Organic synthesis.

methylphosphoric acid. (methyl orthophosphoric acid; methyl acid phosphate). $CH_3H_2PO_4$.
Properties: Pale straw-colored liquid, d 1.42 (25C), can be neutralized with alkalies or amines to give water-soluble salts. Combustible. Purity: 97% (remainder orthophosphoric acid and methanol).
Use: Textile and paper processing compounds, catalysts in urea-resin formation, polymerizing agent for resins and oils, rust remover, soldering flux, chemical intermediate.

3′-methylphthalanilic acid. (N-m-tolylphthalamic acid). $C_{15}H_{13}NO_3$.
Properties: Crystalline solid, mp 150C, soluble in alcohol and other polar solvents with decomposition, slightly soluble in water.
Hazard: Toxic. Avoid ingestion.
Use: In agriculture as an antiscission agent for fruits and vegetables.

methyl phthalyl ethyl glycolate.
$C_2H_5OOCCH_2OOCC_6H_4COOCH_3$.
Properties: Colorless liquid with slight characteristic odor, d 1.217–1.227, flash p 375F (190C), miscible with most organic solvents, very slightly soluble in water. Combustible.
Use: Plasticizer, solvent.

methyl picrate. See trinitroanisole.

N-methylpiperazine. CAS: 109-01-3.

$\overline{CH_3NCH_2CH_2NHCH_2CH_2}$.
Properties: Colorless liquid, d 0.9038, bp 138.0C, fp −6.4C, hygroscopic, amine-like odor, flash p 108F (42.2C). Combustible.
Hazard: Moderate fire risk.
Use: Intermediate for pharmaceuticals, surface agents, synthetic fibers.

2-methylpiperidine. (2-pipecoline).
$C_5H_{10}NCH_3$.
Properties: Colorless liquid, bp 118.2C, fp −4.2C, d 0.8401 (20/20C), refr index 1.4457 (20C), miscible in water at 20C. Combustible.

1-methyl-4-piperidinol. $CH_3C_5H_9NOH.$
Properties: Colorless to light amber, oily liquid with amine-like odor, crystals may be present, or material may be solid, but will melt on warming; refr index 1.4757–1.4777 (25C), congealing temperature 27.5C min. Combustible.
Use: Organic synthesis.

3-(2-methylpiperidino)propyl-3,4-dichlorobenzoate. See piperalin.

methylprednisolone. CAS: 83-43-2.
$C_{22}H_{30}O_5.$ A steroid.
Properties: White, odorless, crystalline powder; mp 240C with decomposition. Sparingly soluble in alcohol, slightly soluble in acetone, insoluble in water.
Grade: NF.
Use: Medicine (hormone).

2-methylpropane. See isobutane.

2-methylpropanenitrile. See isobutyronitirile.

2-methyl-1-propanethiol. (isobutyl mercaptan). $(CH_3)_2CHCH_2SH.$
Properties: Liquid, bp 85–95C, unpleasant odor, d 0.8363 (15.5C), flash p 15F (−9.4C).
Grade: 95%.
Hazard: Flammable, dangerous fire risk.

2-methyl-2-propanethiol. (tert-butyl mercaptan). $(CH_3)_3CSH.$
Properties: Colorless liquid with strong skunk odor, d 0.79–0.82 (60/60F), distillation range 62–67C, refr index 1.422 (20C), flash p −15F (−26.1C), bulk d 6.71lb/gal.
Hazard: Flammable, dangerous fire risk. Very toxic.
Use: Intermediate, gas odorant for detecting leaks.

2-methylpropanoic acid. See isobutyric acid.

2-methyl-1-propanol. See isobutyl alcohol.

2-methyl-2-propanol. See tert-butyl alcohol.

2-methylpropanoyl chloride. See isobutyrol chloride.

2-methylpropene. See isobutene.

2-methyl-2-propen-1-ol. See methylallyl alcohol.

methyl propionate. CAS: 554-12-1.
$CH_3CH_2COOCH_3.$
Properties: Colorless liquid, soluble in most organic solvents, somewhat soluble in water, d 0.937 (4C), refr index 1.3769 (20C), boiling range 78.0–79.5C, autoign temperature 876F (468C), flash p 28F (−2.2C) (CC), 7.58 lb/gal.
Grade: Technical.
Hazard: Flammable, dangerous fire risk, explosive limits in air 2.5–13%.
Use: Solvent for cellulose nitrate, solvent mixture for cellulose derivative, lacquers, paints, varnishes, coating compositions, flavoring.

methylpropylbenzene. See cymene.

methyl propyl carbinol. See 2-pentanol.

methyl propyl carbinol urethane. See hedonal.

methyl propyl ketone. (ethyl acetone; 2-pentanone; MPK).   CAS: 107-87-9.
$CH_3(CH_2)_2COCH_3.$
Properties: Water-white liquid. The commercial material consists of a mixture of methyl propyl and diethyl ketones in the approximate ratio of 3:1 and contains at least 97% of these ketones, the balance being sec-amyl alcohol. Soluble in alcohol and ether, slightly soluble in water, d 0.809 (20/20C), fp −77.5C, bp 101.7C, refr index 1.3895 (20C), viscosity 0.473 centipoise (25C), flash p 45F (7.22C), autoign temperature 941F (505C).
Grade: Technical.
Hazard: Flammable, dangerous fire risk. TLV: 200 ppm in air, explosive limits in air 1.6–8.2%.
Use: Solvent, substitute for diethyl ketone, flavoring.

2-methyl-2-n-propyl-1,3-propanediol dicarbamate. See meprobamate.

methylpyridine. See picoline.

methyl pyromucate. See methyl-2-furoate.

N-methylpyrrole. $C_4H_4NCH_3.$
Properties: Colorless liquid, bp 112C, fp −57C, d 0.914 (20C) refr index 1.4898 (17D), flash p 61F (16.1C), insoluble in water, soluble in alcohol.
Grade: 98% min purity.
Hazard: Flammable, dangerous fire risk.
Use: Organic synthesis.

N-methylpyrrolidine. $CH_3NCH_2CH_2CH_2CH_2.$
Properties: Colorless liquid, ammonia-like odor, refr index 1.4200–1.4230 (25C), bp 80.5C, fp −90C, d 0.805, flash p 7F (−13.9C).
Hazard: Flammable, dangerous fire risk. Irritant to skin and eyes.

N-methyl-2-pyrrolidone.

$CH_3NCH_2CH_2CH_2CO.$
Properties: Colorless liquid with mild amine odor, d 1.027, fp −24C, bp 202C, flash p 204F (95.5C),

miscible with water, various organic solvents, castor oil. Combustible.
Derivation: High-pressure synthesis from acetylene and formaldehyde.
Use: Solvent for resins, acetylene, etc., pigment dispersant, petroleum processing, spinning agent for polyvinyl chloride, intermediate. *Note*: A proprietary adaptation of this solvent to cleanup of vinyl chloride reaction vessels is available under trademark of "M-Pyrol."

α-methylquinoline. See quinaldine.

γ-methylquinoline. See lepidine.

2-methylquinone. See toluquinone.

6-methyl-2,3-quinoxalinedithiol cyclic carbonate. (6-methyl-2-oxo-1,3-dithio-(4,5-b)quinoxaline). $C_{10}H_6N_2S_2O$.
Properties: Yellow, crystalline powder; mp 172C. Insoluble in water, slightly soluble in acetone and alcohol, soluble in organic solvents.
Hazard: Toxic by ingestion and inhalation.
Use: Acaricide, insecticide, fungicide.

methyl red. (p-dimethylaminoazobenzenecarboxylic acid; CI 13020). CAS: 493-52-7. $(CH_3)_2NC_6H_4NNC_6H_4COOH$.
Properties: Dark-red powder or violet crystals, mp 180C, insoluble in water, soluble in alcohol, ether, glacial acetic acid. Color fades quickly due to reduction.
Use: Acid-base indicator in the range pH 4.2–6.2 (red to yellow). Note: No longer widely used because of instability.

methylresorcinol. See orcin.

methyl ricinoleate. $C_{19}H_{36}O_3$. The methyl ester of ricinoleic acid.
Properties: Colorless liquid, insoluble in water, soluble in alcohol and ether, d 0.9236 (22/4C), fp −4.5C, bp 245C (10 mm), refr index 1.4628. Combustible.
Derivation: Esterification of ricinoleic acid or alcoholysis of castor oil, purification by vacuum distillation.
Grade: Technical, purified (99+%).
Use: Plasticizer, lubricant, cutting oil additive, wetting agent.

methylrosaniline chloride. See methyl violet.

methyl salicylate. (gaultheria oil; wintergreen oil; betula oil; sweet-birch oil). CAS: 119-36-8. $C_6H_4OHCOOCH_3$.
Properties: Yellow to red liquid, odor of wintergreen, refr index 1.535–1.538, d 1.180–1.185, fp −8.3C, flash p 214F (101C), autoign temperature 850F (454C), bp 222.2C; natural oil optically inactive, synthetic oil, angular rotation not more than −1.5 degrees, soluble in 7 parts of 70% alcohol, soluble in ether and in glacial acetic acid, sparingly soluble in water. Combustible.
Derivation: By heating methanol and salicylic acid in presence of sulfuric acid or by distillation from leaves of Gaultheria procumbens or bark of Betula lenta.
Method of purification: Rectification.
Grade: Technical, USP, FCC.
Hazard: Toxic by ingestion, use in foods restricted by FDA, lethal dose 30 cc in adults, 10 cc in children.
Use: Flavor in foods, beverages, pharmaceuticals, odorant, perfumery, UV-absorber in sunburn lotions.

"Methyl Selenac."[69] (TM for selenium dimethyldithiocarbamate). $[(CH_3)_2NC(S)S]_4Se$.
Properties: Yellow powder, d 1.58, melting range 140–172C, slightly soluble in carbon disulfide, benzene, chloroform; insoluble in water, dilute caustic, gasoline.
Use: vulcanizing agent.

methyl silicone. See dimethyl silicone. See also silicone and siloxane.

methyl stearate. (methyl octadecanoate). $CH_3(CH_2)_{16}COOCH_3$. The methyl ester of stearic acid.
Properties: Semisolid, mp 37.8C, bp 234.5C (30 mm), 204.5C (10 mm), flash p 307F (152C), refr index 1.4328 (50C), insoluble in water, soluble in ether and alcohol. Combustible.
Derivation: Esterification of stearic acid with methanol or alcoholysis of stearin with methanol.
Method of purification: Vacuum fraction distillation.
Impurities: Most technical methyl stearate is 55% stearate and 45% methyl palmitate.
Grade: Distilled, pressed, technical, pure (99.8+%).
Use: Intermediate for stearic acid detergents, emulsifiers, wetting agents, stabilizers, resins, lubricants, plasticizers.

α-methylstyrene. CAS: 98-83-9. $C_6H_5C(CH_3):CH_2$.
Properties: Colorless liquid, subject to polymerization by heat or catalysts, bp 165.38C, fp −23.21C, d 0.9062 (25/25C), viscosity 0.940 cp (20C), flash p 129F (53.9C), autoign temperature 1065F (573C), refr index 1.5359 (25/25C), insoluble in water. A polymerization inhibitor such as tert-butyl catechol is usually present in commercial quantities. Combustible.

Derivation: From benzene and propylene by use of aluminum chloride and hydrochloric acid to yield cumene, which is then dehydrogenated.

Hazard: Moderate fire risk. Explosive limits in air 1.9–6.1%. Avoid inhalation and skin contact. TLV: 50 ppm in air.

Use: Polymerization monomer, especially for polyesters.

methyl styryl ketone. See benzylidene acetone.

methylsuccinic acid. See pyrotartaric acid.

methyl sulfate. Legal label name (Rail) for dimethyl sulfate.

methyl sulfide. Legal label name (Rail) for dimethyl sulfide.

3-methylsulfolane. See 3-methyltetrahydrothiophene-1,1-dioxide.

3-methylsulfolene. See 3-methyl-2,5-dihydrothiophene-1,1-dioxide.

methylsulfonic acid. (methanesulfonic acid). CH_3SO_2OH.

Properties: Colorless solid, d 1.48, mp 20C, bp 167C (10 mm), slightly soluble in benzene, almost insoluble in other organic solvents. Attacks iron, copper, and steel.

Derivation: Reaction of methane and sulfur trioxide.

Hazard: Toxic by ingestion, skin irritant, corrosive to tissue.

Use: Polymerization catalyst.

methyl sulfoxide. See dimethyl sulfoxide.

methylsulfuric acid. (acid methyl sulfate; methyl hydrogen sulfate). CH_3OSO_2OH or CH_3HSO_4.

Properties: Oily liquid, soluble in anhydrous ether, slightly soluble in alcohol and water, bp 188C, d 1.352, fp −27C.

Derivation: Interaction of methanol and chlorosulfonic acid.

Use: Sulfonating agent, specialty solvent.

N-methyltaurine. (sodium-N-methyltaurate). CAS: 107-68-6. $CH_3NHCH_2CH_2SO_3Na$. Available in commercial quantities as an aqueous solution of the sodium salt.

Properties: (of solution) Clear light-colored liquid, approximately 34–36% sodium salt, d 1.21 (25/4C), at fp (−28C average) becomes a suspension of white crystals.

Use: Intermediate for detergents, dyestuffs, pharmaceuticals and other organics.

17-methyltestosterone. CAS: 58-18-4. $C_{20}H_{30}O_2$. A synthetic androgenic steroid.

Properties: White or creamy-white crystals or crystalline powder, odorless, stable in air, slightly hygroscopic, affected by light, mp 163–168C, soluble in ethanol, methanol, ether and other organic solvents, insoluble in water.

Derivation: By organic synthesis.

Grade: NF.

Use: Medicine (hormone).

methyl tetracosanoate. See methyl lignocerate.

methyl tetradecanoate. See methyl myristate.

2-methyltetrahydrofuran. CAS: 25265-68-3. $C_4H_7CH_3$.

Properties: Colorless liquid, ether-like odor, bp 80.2C, fp −136C, d 0.854 (20/4C), refr index 1.4025 (25C), flash p (TOC) 12F (−11.1C), soluble in water 15.1 g/100 g water (25C), solubility in water increases with a decrease in temperature, freely soluble in most organic solvents.

Hazard: Flammable, dangerous fire risk.

Use: Chemical intermediate, reaction solvent.

3-methyltetrahydrothiophene-1,1-dioxide. (3-methylsulfolane). $CH_3C_4H_7SO_2$.

Properties: Colorless liquid, d 1.194 (20/4C), mp 0C, bp 276C, flash p 325F (162C). Completely soluble in water, acetone, and toluene. Combustible.

Use: Solvent and extractive medium.

methyl-2-thienyl ketone. See 2-acetylthiophene.

m-(methylthio)aniline. $H_2NC_6H_4SCH_3$.

Properties: Pale yellow oil, d 1.140 (25C), bp 163–165C (16 mm), fp −3.0C. Insoluble in water; soluble in alcohol, benzene, and acetic acid. Combustible.

Hazard: See aniline.

Use: Pharmaceutical intermediate.

4-(methylthio)-3,5-dimethylphenyl-n-methylcarbamate. [(4-methylthio)-3,5-xylyl methylcarbamate]. $CH_3S(CH_3)_2C_6H_2OOCNHCH_3$.

Properties: Solid, mp 121C, insoluble in water, soluble in alcohol and acetone.

Use: Pesticide.

methylthionine chloride. See methylene blue.

methyltin. See organotin compounds.

methyl-p-toluate. CAS: 99-75-2. $CH_3C_6H_4COOCH_3$.

Properties: White, crystalline solid; fp 34C; d 1.058 (25/25.6C).

Use: Organic synthesis.

methyl-p-toluenesulfonate.   CAS: 80-48-8. $CH_3C_6H_4SO_3CH_3$.
Properties: White, damp crystals; solidification point 24C; bp 157C (8 mm); decomposes 262C. A vesicant material, insoluble in water, soluble in alcohol and benzene.
Grade: 97% min.
Hazard: Toxic by ingestion and inhalation, strong irritant to skin and eyes.
Use: Accelerator, methylating agent, catalyst for alkyd resins.

methyl-p-tolyl ketone.   See methyl acetophenone.

methyltrichlorosilane.   CAS: 75-79-6. CH_3SiCl_3.
Properties: Colorless liquid, bp 66.4C, d 1.270 (25/25C), refr index 1.4085 (25C), flash p 8F (−13.3C), readily hydrolyzed by moisture with the liberation of hydrogen chloride.
Derivation: By Grignard reaction of silicon tetrachloride and methylmagnesium chloride.
Hazard: Flammable, dangerous fire risk, may form explosive mixture with air. Strong irritant.
Use: Intermediate for silicones.

methyl tricosanoate. $CH_3(CH_2)_{21}COOCH_3$.
The methyl ester of tricosanoic acid.
Properties: White, wax-like solid. Insoluble in water, soluble in alcohol and ether, mp 55–56C. Combustible.
Grade: Purified 96% and 99.5%.
Use: Intermediate in organic synthesis, biochemical research.

methyl tridecanoate. $CH_3(CH_2)_{11}COOCH_3$.
The methylester of tridecanoic acid.
Properties: Colorless liquid, insoluble in water, soluble in alcohol and ether, mp 5.5C, bp 130–132C (4 mm) refr index 1.4327 (25C). Combustible.
Derivation: Esterification of tridecanoic acid with methanol followed by fractional distillation.
Grade: Purified 96% and 99.5%.
Use: Intermediate in organic synthesis, biochemical research, reference standard in gas chromatography.

methyl trimethylolmethane.   See trimethylol ethane.

"Methyl Trithion."[1]   TM for an insecticide-acaricide containing various percentages of S-(p-chlorophenylthio)methyl-O,O-dimethyl phosphorodithioate. Available as liquid or powder.
Hazard: Cholinesterase inhibitor.

"Methyl Tuads."[69]   TM for tetramethylthiuram disulfide, $[(CH_3)_2NC(S)S]_2$.
Use: Vulcanizing agent and primary accelerator in natural, nitrile and butyl rubbers and in SBR. As secondary accelerator in natural and nitrile rubbers and SBR. Used in coated fabrics, extruded and molded goods, steam hose. Increases heat resistance.

β-methylumbelliferone.   (7-hydroxy-4-methylcoumarin; BMU). $C_{10}H_8O_3$.
Properties: White to light tan powder, mp 186–188C, soluble in concentrated sulfuric acid, partly soluble in ethanol, isopropanol, 5% aqueous sodium carbonate solution, very slightly soluble in water, very dilute aqueous alkaline solutions give a bright blue-white fluorescence in daylight or UV light.
Grade: Technical.
Use: Optical bleach on soaps, starches and laundry products, suntan lotions.

2-methylundecanal.   See methylnonylacetaldehyde.

methylundecanoate. $(CH_3(CH_2)_9COOCH_3$.
The methyl ester of undecanoic acid.
Properties: Colorless liquid, insoluble in water, soluble in alcohol and ether, bp 123C (10 mm), refr index 1.4270 (25/4C). Combustible.
Derivation: Esterification of undecanoic acid with methanol followed by fractional distillation.
Grade: Purified 96%, 99.5%.
Use: Organic intermediate for synthesis, flavoring, biochemical research.

5-methyluracil.   See thymine.

methylvinyldichlorosilane. $(CH_3)(C_2H_3)SiCl_2$.
Properties: Colorless liquid, bp 92C, d 1.08 (25C), refr index 1.4270 (25C), flash p 40F (4.4C), soluble in benzene and ether, reacts with methanol and water.
Derivation: From methyldichlorosilane and acetylene or vinyl chloride.
Hazard: Flammable, dangerous fire risk. Irritant.
Use: Manufacture of silicones.

methyl vinyl ether.   See vinyl methyl ether.

methyl vinyl ketone.   Legal label name for vinyl methyl ketone.

2-methyl-5-vinylpyridine. $CH_3C_5H_3NCH{:}CH_2$.
Properties: Clear to faintly opalescent liquid, d 0.978–0.982 (20/20C), bp 181C, refr index 1.5400–1.5454 (20C), fp (anhydrous) −14.3C, flash p (TOC) 165F (73.9C). Combustible.
Use: Monomer for resins, oil additive, ore flotation agent, dye acceptor.

methyl violet. (Gentian Violet, USP; hexamethyl-p-rosaniline chloride; CI 42555).
CAS: 8004-87-3. $C_{25}H_{30}N_3Cl$.
Properties: Green powder, soluble in water and chloroform, partially soluble in alcohol and glycerol.
Use: Medicine (topical antibacterial and antiallergen), acid-base indicator, alcohol denaturant, biological stain, textile dye.

methyl yellow. See dimethylaminoazobenzene.

"Methyl Zimate."[69] TM for zinc dimethyldithiocarbamate, $[(CH_3)_2NC(S)S]_2Zn$.
Use: Ultra-accelerator for natural and synthetic rubbers.

methylprylon. (3,3-diethyl-5-methyl-2,4-piperidinedione). CAS: 125-64-4.
$C_5NH_4(O)_2(C_2H_5)_2(CH_3)$.
Properties: Nearly white, crystalline powder; slight characteristic odor; bitter taste; melting range 74–77C. Soluble in water; very soluble in alcohol, chloroform, ether, and benzene.
Grade: NF.
Hazard: Abuse may cause addiction.
Use: Medicine (sedative, hypnotic).

metiram. CAS: 12122-67-7. Mixture of (ethylenebis(dithiocarbamato))zinc ammoniates with ethylenebis(dithiocarbamic acid)anhydrosulfides.
Use: Fungicide.

"Met-L-KYL."[548] TM for a dry chemical which extinguishes fires caused by pyrophoric liquids such as triethylaluminum and adsorbs the spilled metal alkyl to prevent reignition.

"Met-L-X."[548] TM for a dry chemical based on sodium chloride, approved for use on sodium, potassium, sodium-potassium alloy, and magnesium fires.

metobromuron. CAS: 3060-89-7.
$C_9H_{11}BrN_2O_2$.
Properties: Colorless crystals, mp 95C, soluble in alcohols, slightly soluble in water.
Use: Preemergence herbicide.

metolachlor. (2-chloro-N-(2-ethyl-6-methylphenyl)-N-(2-methoxy-1-methylethyl)acetamide).
CAS: 51218-45-2
Use: A herbicide.

metopon hydrochloride. (6-methyldihydromorphinone hydrochloride). $C_{18}H_{21}O_3N \cdot HCl$.
A morphine derivative.
Properties: White, odorless, crystalline powder. Very soluble in water, sparingly soluble in alcohol, insoluble in benzene.

Hazard: Addictive narcotic, prescription only.
Use: Medicine (analgesic).

metoquinone. CAS: 622-91-3. $C_{20}H_{24}N_2O_4$.
Properties: Colorless crystals, mp 130C (decomposes), partially soluble in water, slightly soluble in alcohol and acetone, almost insoluble in benzene.
Use: Photographic developer.

metribuzin. CAS: 21087-64-9. $C_8H_{14}N_4OS$.
Properties: Colorless crystals, mp 125C, soluble in alcohols, slightly soluble in water.
Hazard: Toxic material. TLV: 5 mg/m³ of air.
Use: Herbicide.

"Metso."[201] TM for a series of detergents which include the pentahydrate of sodium-metasilicate, hydrated sodium sesquisilicate and technical anhydrous sodium orthosilicate. Available in lump and powder forms.

MeV. Abbreviation for million electron-volts.

mevinphos. (2-carbomethoxy-1-methylvinyl dimethyl phosphate; methyl-3-hydroxy-α-crotonate dimethyl phosphate). CAS: 7786-34-7.
$(CH_3O)_2P(O)OC(CH_3:CHCOOCH_3$.
Properties: Yellow liquid, bp 99–103C (0.03 mm), slightly soluble in oils, miscible with water and benzene.
Hazard: Toxic by ingestion, inhalation, and skin absorption; use may be restricted; cholinesterase inhibitor. TLV: 0.01 ppm.
Use: Insecticide and acaricide.

mexacarbate. CAS: 315-18-4. $C_{12}H_{18}N_2O_2$.
Properties: Colorless crystals, mp 85C. Soluble in acetone, alcohol, and benzene; insoluble in water.
Hazard: Toxic by ingestion.
Use: Insecticide.

mexicain. Proteolytic enzyme with properties and uses similar to papain.

Meyer reaction. Preparation of alkylstannonic acids by reacting alkali stannite with an alkyl iodide. When applied to alkali arsenites or plumbites, the reaction yields alkylarsonic and alkylplumbonic acids, respectively.

Meyer-Schuster rearrangement. Acid catalyzed rearrangement of secondary and tertiary α-acetylenic alcohols to α,β-unsaturated carbonyl compounds: aldehydes when the acetylenic group is terminal, ketones when it is internal.

Meyer synthesis. Formation of aliphatic nitrites and nitro derivatives by the reaction of aliphatic halides with metal nitrites.

Meyers aldehyde synthesis. Synthesis of aldehydes R-CH$_2$CHO from alkylhalides R-X and 2-lithiomethyl- tetrahydro-3-oxazine.

Mg. Symbol for magnesium.

mg. Abbreviation for milligram.

MH. See maleic hydrazide.

"MH-30."[248] TM for a 30% solution of maleic hydrazide.

"MHA."[58] TM for methionine hydroxy analog, calcium salt.

MHD. See magnetohydrodynamics.

MIAK. See methyl isoamyl ketone.

miazine. See pyrimidine.

MIBC. Abbreviation for methylisobutyl carbinol. See methylamyl alcohol.

MIBK. Abbreviation for methyl isobutyl ketone.

mica. Any of several silicates of varying chemical composition but with similar physical properties and crystalline structure. All characteristically cleave into thin sheets which are flexible and elastic. Synthetic mica is available. It has electrical and mechanical properties superior to those of natural mica; it is also water-free.
Properties: Soft, translucent solid; d 2.6–3.2. Mohs hardness 2.8–3.2, refr index 1.56–1.60, dielectric constant 6.5–8.7, noncombustible, heat-resistant to 600C, colorless to slight red (ruby), brown to greenish yellow (amber).
Derivation: From muscovite (ruby mica), phlogopite (amber mica), and pegmatite. Synthetic single crystals are "grown" electrothermally.
Occurrence: US, Canada, India, Brazil, Malagasy Republic.
Forms: Block, sheet, powder, single crystals.
Grade: Dry-ground, wet-ground.
Hazard (dust): Irritant by inhalation, may be damaging to lungs.
Use: Electrical equipment, vacuum tubes, incandescent lamps, dusting agent, lubricant, windows in high-temperature equipment, filler in exterior paints, cosmetics, glass and ceramic flux, roofing, rubber, mold-release agent, specialty paper for insulation and filtration, wallpaper and wallboard joint cement, oil-well drilling muds.

"Mica-Flex."[216] TM for a flexible electrical insulation containing a large percentage of mica.
Use: Mainly on high voltage motor and generator coils.

"Micarta."[308] TM for a group of laminated plastics composed of paper or fabric made from cellulose, glass, asbestos or synthetic fibers bonded with phenolic or melamine resins and cured at elevated temperature and pressure.
Use: Plating barrels, rayon-manufacturing equipment, pickling tanks, electrical and thermal insulation, oil-handling equipment, steel rolling-mill bearings, chemical-handling valve bodies, paper-mill suction box covers and equipment.

"Micatex."[236] TM for mica prepared for addition to drilling fluids to reduce water loss to the formation and for overcoming mild losses of circulation. An effective seal is formed over mildly permeable formations when the mud in which it is entrained forces the material against the formation. Will not disintegrate appreciably, nor will it corrode or abrade slush-pump liners or other metal or moving parts of the mud system.

micellar flooding. See chemical flooding.

micelle. An electrically charged colloidal particle, usually organic in nature, composed of aggregates of large molecules, e.g., in soaps and surfactants. The term is also applied to the casein complex in milk.
See also colloid chemistry.

Michael reaction. Base catalyzed addition of carbanions to activated unsaturated systems.

Michaelis-Arbuzov reaction. Formation of monoalkylphosphonic esters from alkyl halides and trialkyl phosphites, via the intermediate phosphonium salt.

Michler's hydrol. See tetramethyldiaminobenzhydrol.

Michler's ketone. See tetramethyldiaminobenzophenone.

micro-. Prefix meaning 10^{-6} unit (symbol μ), e.g., 1 microgram = 0.000001 g or 1 μg.
See also micron.

microanalysis. See microchemistry.

"Microballoon."[572] TM for hollow, finely divided, hole-free, low-density particles of synthetic resins or similar film-forming materials. Glass is one of the materials used.
Use: To form a protecting layer of the tiny spheres over liquid surfaces, such as oils in big tanks, to reduce evaporation; to separate helium from natural gas because of the wide difference in relative rates of diffusion through the spheres; as an extender in plastics to achieve low density.

microchemistry. A branch of analytical chemistry that involves procedures that require handling of very small quantities of materials. Specifically, it refers to carrying out various chemical operations (weighing, purification, quantitative and qualitative analysis) on samples ranging from 0.1–10 mg; this often involves use of a microscope, and still more often of chromatography. See also microscopy, chemical.

microcrystalline. A form in which a number of high-polymeric substances have been prepared. They include cellulose, chrysotile asbestos, amylose (starch), collagen, nylon, and certain mineral waxes. On the microscopic level, these substances are composed of colloidal microcrystals connected by molecular chains. The process involves breaking up the network of microcrystals (by acid hydrolysis in the case of cellulose) and separating them by mechanical agitation. The size range of the microcrystals is from 2.5–500 nanometers (millimicrons). The products form extremely stable gels which have a number of commercial use possibilities. Petroleum-derived waxes of high molecular weight have been available in microcrystalline form for many years. Chlorophyll has a naturally microcrystalline structure.
See also cellulose; wax, microcrystalline; "Avicel."

microCurie. See Curie.

microencapsulation. Enclosure of a material in hollow spheres or capsules (microspheres) in the micron size range (20–150μ), they can be made of glass, silica, various high polymers or proteins (gelatin, albumen). The silica type can be incorporated in plastics, elastomers, and metals for weight-saving purposes; they can also be bonded to one another to give extremely thin sheets of silica. Microspheres coated with layers of Teflon and beryllium are used to contain the deuterium and tritium used in laser fusion experimentation.
Polymeric or proteinaceous microspheres are used to introduce drugs to specific locations in the body. The coating material acts as a semipermeable membrane, permitting slow release and high concentration of a drug at the desired site. Enzymes, hormones, and other biochemical substances can be temporarily immobilized by this technique.

microgram (μg). One millionth (10^{-6}) gram.

"Microlith."[219] TM for organic pigment stir-in dispersions compatible with a broad range of organic solvents and polymers.

"Micromet."[323] TM for a specially formulated phosphate glass, slowly soluble in water and used to inhibit scale, corrosion, and red water in water systems and air conditioning systems.

micrometer. (μm). One millionth (10^{-6}) meter or 1 micron (10,000Å units).

micron. See micrometer.

micronutrient. See trace element.

microorganism. A living organism of microscopic size generally considered to include bacteria, molds (actimonyces), and fungi, but excluding viruses.
See also bacteria.

microscope. See optical microscope, electron microscope, field-ion microscope, ultramicroscope.

microscopy, chemical. Use of a microscope primarily for study of physical structure and identification of materials. This is especially useful in forensic chemistry and police laboratories. Many types of microscopes are used in industry; most important are the optical, ultra-, polarizing, stereoscopic, electron, and x-ray microscopes. Organic dyes of various types are used to stain samples for precise identification.

"Microsol."[443] TM for aqueous pigment dispersions for spin-coloring of regenerated cellulose fibers.

microsphere. See microencapsulation.

microwave spectroscopy. A type of absorption spectroscopy used in instrumental chemical analysis which involves use of that portion of the electromagnetic spectrum having wavelengths in the range between the far infrared and the radiofrequencies, i.e., between 1 millimeter and 30 centimeters. Substances to be analyzed are usually in the gaseous state. Klystron tubes are used as microwave source.

middle oil. A fraction distilled from coal tar.
See coal tar.

middlings. The granular part of the interior of the wheat berry obtained in the process of milling. This product, when reduced by grinding to the desired fineness, produces the finest quality of flour.

Midgley, Thomas Jr. (1889–1944) An American chemist and inventor. One of the most creative and brilliant chemists of his era, Midgley's early work was in the field of rubber chemistry and technology, especially in the development of synthetic and substitute rubbers which were being introduced in the 1930s. He worked with Kettering at General Motors and then became vice-president of Ethyl Corporation, as well as of the Ohio State University Research Foundation. His innovative genius was responsible for the development of organic lead compounds for antiknock gasoline and later for the discovery of fluorocarbon refrigerants for which he did the basic research. He was recipient of many of chemistry's highest honors including the Nichols Medal, the Perkin Medal, and the Priestly Medal.

Miescher degradation. Adaptation of the Barbier-Wieland carboxylic acid degradation to permit simultaneous elimination of three carbon atoms, as in degradation of the bile acid side chain to the methyl ketone stage. Conversion of the methyl ester of the bile acid to the tertiary alcohol, followed by dehydration, bromination, dehydrohalogenation, and oxidation of the diene yields the required degraded ketone.

Mignonac reaction. Formation of amines by catalytic hydrogenation of aldehydes and ketones in liquid ammonia and absolute ethanol in the presence of a nickel catalyst.

migration. Movement of a substance from one material to another with which it is in intimate contact. For example into the packaged produce, similarly a portion of the sulfur in a rubber mixture may migrate into a material to which it is laminated.

migration area. A term used in nuclear technology as a measure of the moderation or slowing down of neutrons. It is one-sixth of the mean square distance a neutron travels before thermal capture.

mike. A term adopted by the American Standards Association for a microinch or 10^{-6} inch.

mil. One thousandth (0.001) inch.
Use: In reference to surface coatings, metal sheet, films, cable covers, friction tape, etc.

Milas hydroxylation of olefins. Formation of cis-glycols by reaction of alkenes with hydrogen peroxide and either a catalytic amount of osmium, vanadium, or chromium oxide or UV light.

mildew preventive. A compound used to prevent the growth of parasitic fungi, usually stain-producing, on such organic materials as textiles, leather, paper, farinaceous products, etc. Compounds used widely include cresols, phenols, benzoic acid, formaldehyde, and or-ganic derivatives or salts of copper, zinc, and mercury.

mill. See specific type (ball mill, pebble mill, etc.).

Millon's reagent. Mercury dissolved in equal weight of nitric acid (d: 1.41), and solution diluted to twice the volume and decanted from the precipitate.
Use: Test for albumin.

Mills-Nixon effect. Selective reactivity of certain substituted aromatic systems suggestive of certain degree of preponderance of one of the tautomeric Kekule forms in the benzenoid nucleus.

milk. A heterogeneous liquid secreted by the mammary gland and composed (for cows' milk) of approximately 87% water, 3.8% emulsified particles of fat and fatty acids, 3% casein, 5% sugar (lactose), serum proteins, calcium, phosphorus, potassium, iron, magnesium, copper, and several vitamins. The fat particles are from 6–10 micrometers in diameter, much larger than colloidal size; they are coated with an adsorbed layer of protein (protective colloid), which maintains the emulsion. the casein is closely associated with the calcium, forming micelles of calcium caseinate which are of colloidal dimensions. The white color of milk is largely due to light scattering by these particles and to some extent by the fat particles, rather than to the presence of a pigment. The casein complex coagulates (a) when high temperature or bacteria convert the lactose to lactic acid, as in souring, or (b) when acid or certain enzymes (rennet) are intentionally added. The serum proteins lactalbumin and lactoglobulin are also colloidally dispersed. The lactose and mineral salts are in true molecular solution. Thus milk is a complex system exhibiting several levels of dispersion, from the molecular through colloidal and into the microscopic size range. Important processes applied to milk on an industrial scale include pasteurization, homogenizaton, coagulation, dehydration, and condensation.
See specific entries. The milks of animals other than cows show considerable variations in composition, especially fat and protein content. See also colloid chemistry, emulsion, casein.

milk of lime. (lime water). Calcium hydroxide suspended in water.

milk of magnesia. (magnesia magma).
A white, opaque, more or less viscous suspension of magnesium hydroxide in water from which varying proportions of water usually separate on standing.

Grade: USP.
Use: Medicine (laxative).

milk sugar. See lactose.

milli-. Prefix meaning 10^{-3} unit or 1/1000th part.

milliCurie. See Curie.

milligram. (mg). One-thousandth gram (10^{-3} gram).

milliequivalent. (meq). One-thousandth of the equivalent weight of a substance.

milliliter. (mL). One thousandth liter, the volume occupied by one gram of pure water at 4C and 760 mm pressure. One milliliter (mL) equals 1.000027 cubic centimeters (cc).

millimeter. (mm). One-thousandth meter, about 0.03937 inch.

millimicron. (mμ; mu). One-thousandth micron, 10 Å, 1 billionth meter, 1 nanometer.

millirem. One-thousandth rem.
See rem.

"Milorganite." TM for blended fertilizer containing approximately 20% of an activated sludge marketed in dry granular form by the Milwaukee sewage disposal plant. Contains 5–10% moisture, 6.5–7.5% ammonia, 2.5–3.5% available phosphoric acid, 3–4% total phosphoric acid. Source of vitamin B_{12}.

Milori blue. Any of a number of the varieties of iron blue pigments.
See iron blues.

"Milvex."[590] TM for a group of specialty nylon resins said to differ from conventional nylons in the following ways: (1) Highly water-resistant (dimensionally stable), (2) clarity, as opposed to the opacity of existing types, (3) adhesion to metals, glass, and other smooth surfaces, (4) tensile strength (up to 7500 psi) and elongation (up to 600%), (5) high flexibility and internal plasticization.

mineral. A widely used general term referring to the nonliving constituents of the earth's crust which include naturally occurring elements, compounds, and mixtures that have a definite range of chemical composition and properties. Usually inorganic, but sometimes including fossil fuels, e.g., coal, minerals are the raw materials for a wide vareity of elements (chiefly metals) and chemical compounds. Minerals can be and many are synthesized to achieve purity greater than that found in natural products. The term *mineral industry* statistically comprehends the mining and production of metals (ores) fossil fuels, clay, gemstones, cement, glass, rocks, sulfur, sand, etc. Mineralogy is the study and classification of minerals by source, chemical composition and properties, chiefly physical, such as color, hardness, and crystalline structure. This term was used by early chemists to describe a variety of substances; many of these uses are obsolescent, but a few persist, e.g.:

> mineral black: inorganic black pigments
> mineral blue: varieties of blue pigments
> mineral dust: industrial dust, nuisance dust
> mineral green: copper carbonate
> mineral oil: a liquid petroleum derivative
> mineral pitch: asphalt
> mineral red: iron oxide red
> mineral rubber: blown asphalt
> mineral spirits: a grade of naphtha
> mineral water: natural spring water containing sulfur, iron, etc.
> mineral wax: a wax found in the earth (ozocerite), or derived from petroleum
> mineral wool: fibers made by blowing air or steam through slag

As used by nutritionists the term refers to such components of foods as iron, copper, phosphorus, calcium, iodine, selenium, fluorine, and trace micronutrients.

minim. In the US, a unit of volume equal to approximately 0.06 mL.
Use: Pharmacy.

minium. Pb_3O_4. Natural red oxide of lead. Found in Colorado, Idaho, Utah, Wisconsin.
See also lead oxide, red.

MIPA. Abbreviation for monoisopropanolamine. See isopropanolamine.

mipafox. See N,N'-diisopropyldiamidophosphoryl fluoride.

miracle fruit. (miraculin). See sweetener, nonnutritive, glycoprotein.

"Mirasol."[223] TM for a series of alkyd type resins. Epoxy resin esters are also marketed under this name. Available in all modifications including drying oils, semi-drying oils, non-drying oils, natural and phenolic resins.
Use: Air-drying and baking finishes including architectural, lacquer, wrinkle, hammer, and other

industrial enamels; also printing inks and textile finishes.

mirbane oil. See nitrobenzene.

"Mirrex."[482] TM for a calendered, unplasticized PVC film. Available in film or sheeting for a wide range of packaging applications.

mirror-image molecules. See optical isomerism, enantiomorph, chiral.

misch metal. The primary commercial form of mixed rare-earth metals (95%) prepared by the electrolysis of fused rare earth chloride mixture, d approximately 6.67, mp approximately 648C. Form: waffle-like plates weighing 40–60 lb packed in oiled paper, immersed in oil, or coated with vinyl paint.
Hazard: Flammable, dangerous fire risk.
Use: Lighter flints, ferrous and non-ferrous alloys, cast iron, aluminum, nickel, magnesium and copper alloys, getter in vacuum tubes, magnetic alloys.

miscibility. The ability of a liquid or gas to dissolve uniformly in another liquid or gas. Gases mix with one another in all proportions. This may or may not be true of liquids, whose miscibility properties depend on their chemical nature. Alcohol and water are completely miscible because of their chemical similarity, but some liquids are only partially miscible in others because of their chemical difference, e.g., benzene and water. Many gases are miscible with liquids to a greater or lesser extent, e.g., formaldehyde mixes readily with water; CO_2 is partially miscible with water and oxygen only very slightly. Liquids that do not mix at all are said to be immiscible, as oil and water. The term "solubility" is often used synonymously with "miscibility" in reference to liquids, but it more properly applies to solids.

Mitchell, Peter. (1920-) A British biochemist who was the recipient of the Nobel prize for chemistry in 1978 for his work on studies of cellular energy transfer. A graduate of Cambridge and recipient of many awards, he has been Director of Research, Glynn Research Institute, since 1964.

miticide. A pesticide which kills mites, small animals of the spider class, among them the European red mite and the common red spider which infest fruit trees.

mitochondria. Particles of cytoplasm found in most respiring cells. They synthesize most of the cell's adenosine triphosphate and are the chief energy sources of living cells. They are highly plastic, mobile structures which may fragment or fuse together at random. Many enzymes, especially those involved in converting food-derived energy into a form usable by the cell, are located in the mitochondria and DNA molecules have also been found there. Yeast is a particularly rich source of mitochondria for research purposes.

mitomycin C. $C_{15}H_{18}N_4O_5$. Antibiotic derived from Streptomyces, stated to be effective against tumors.

mitosis. The division of a cell nucleus to produce two new cells, each having the same chemical and genetic constitution as the parent cell. The deoxyribose (nucleic acid) component of the chromosomes is present in duplicate in the original nucleus. The amount of nucleic acid is doubled just before cell division begins; subsequent events (called phases) permit separation of the products of replication to form the new nuclei. Each half-chromosome carries the identical nucleic acids of the original chromosome.
See also cell (1).

Mitsunobu reaction. Intermolecular dehydration reaction occurring between alcohols and acidic components on treatment with diethyl azodicarboxylate and triphenyl phosphine under mild neutral conditions. The reaction exhibits streospecificity and regional and functional selectivity.

mixed acid. (nitrating acid). A mixture of sulfuric and nitric acids used for nitrating, e.g., in the manufacture of explosives, plastics, etc. Consists of 36% nitric acid and 61% sulfuric acid.
Hazard: Spillage may cause fire or liberate dangerous gas. Causes severe burns, irritant by ingestion and inhalation, may cause NO_x poisoning.

mixing. Effecting a uniform dispersion of liquid, semi-solid or solid ingredients of a mixture by means of mechanical agitation. Low-viscosity liquids and suspensions are mixed with impellers of the turbine or propellor type. The mixing action results both from direct contact of the impeller blades with the liquid and from the turbulence induced by the impeller in the outer portions of the liquid. For this reason the diameter of the impeller need be only from one-fourth to one-half that of the container. For liquids of medium viscosity, revolving paddles of various shapes are used. Thicker mixtures involving volatile solvents are mixed in closed containers (churns) equipped with fin-like members mounted on a rotating shaft. For liquids of very high viscosity, helical rotors, sigma blades and similar devices are necessary. Because turbulence cannot be initiated in such fluids, the blades must fit closely within the walls of the container so as to make contact with every part of the material being mixed. While most industrial mixing of such pasty materials is done batchwise in kneaders, Banbury mixers, etc., continuous mixing of these can also be effectively carried out in horizontal compartments equipped with rotating screws, whose pitch and flight are contoured in such a way as to provide both rotary and axial motion. There are a number of ingeniously engineered types of these for mixing plastics, rubber, food products, and similar products. Dry solid particulates are mixed in rotating cylinders or tumbling barrels.

See also impeller, agitator, kneader, muller.

mixture. (mix). A heterogeneous association of substances which cannot be represented by a chemical formula. Its components may or may not be uniformly dispersed and can usually be separated by mechanical means. Liquids that are uniformly dispersed are called solutions. Mixtures may be natural or artificial, as indicated by the following:

Natural	Artificial
air	glass
petroleum	paint
milk	cement
blood	perfumes
marble	plastics
wood	cermets
latex	alloys
vegetable oils	
sea water	

See also compound, blend, solution, mixing.

MKP. Abbreviation for monopotassium phosphate.
See potassium phosphate, monobasic.

mL. Abbreviation for milliliter.

MLA. Abbreviation for mixed lead alkyls.

mm. Abbreviation for millimeter.

MMH. Abbreviation for monomethylhydrazine.

Mn. Symbol for manganese.

Mo. Symbol for molybdenum.

mobility. The ease with which a liquid moves or flows. Hydrocarbon liquids (nonpolar) which

have low viscosity, surface tension, and density respond more readily to an applied force than does water (a polar liquid). For this reason, fires involving hydrocarbon liquids should be extinguished with foam rather than with a direct stream of water.

MOCA. See 4,4'-methylenebis(2-chloroaniline).

modacrylic fiber. A manufactured fiber in which the fiber-forming substance is any long chain synthetic polymer composed of less than 85% but at least 35% by weight of acrylonitrile units, —$CH_2CH(CN)$— (Federal Trade Commission). Other chemicals, such as vinyl chloride, are incorporated as modifiers. Characterized by moderate tenacity, low water absorption, and resistance to combustion; self-extinguishing.
Use: Deep pile and fleece fabrics, industrial filters, carpets, underwear, blends with other fibers.
See also acrylonitrile, acrylic fiber.

model. A representation, either abstract or physical, of a system, arrangement, or structure which cannot be perceived objectively. (1) A mathematical model is one in which all or most of the parameters of a complex system such as an ocean are assigned symbolic values that can be utilized to give a theoretical approximation of actuality. Such models are useful in physical chemical analyses. (2) A space-lattice model is a 3-dimensional duplication of the shape and structure of a crystal in which the atoms comprising the lattice are plastic spheres or balls connected by rods to represent bonds. (3) A molecular model is similar, except that it represents an individual chemical compound rather than a crystal. The spheres are

made to scale based on the known diameter of the atoms represented; they are often colored to suggest the nature of the element (black for carbon, white for hydrogen, red for halogens, etc.). In one type, both single and double bonds are plastic rods which join the spheres at appropriate angles; in another the spheres are fused in clusters. The two types are illustrated by the models of isobutane shown; a clustered model of the DNA molecule is shown in the entry on deoxyribonucleic acid. Both space-lattice and molecular models are useful for classroom demonstration.

moderator. A substance of low atomic weight, such as beryllium, carbon (graphite), deuterium (in heavy water), or ordinary water, which is capable of reducing the speed of neutrons but which has little tendency toward neutron absorption. The neutrons lose speed when they collide with the atomic nuclei of the moderator. Moderators are used in nuclear reactors, since slow neutrons are most likely to produce fission. A typical graphite-moderated reactor may contain 50 tons of uranium for 472 tons of graphite. Reactors in the US are cooled and moderated with light water.

modification. A chemical reaction in which some or all of the substituent radicals of a high polymer are replaced by other chemical entities, resulting in a marked change in one or more properties of the polymer without destroying its structural identity. Cellulose, for example, can be modified by substitution of its hydroxyl groups by carboxyl or alkyl radicals along the carbon chain. These reactions are usually not stoichiometric. Their products have many properties foreign to the original cellulose, e.g., water solubility, high viscosity, gel and film forming ability. Other polymeric substances that can undergo modification are rubber, starches, polyacrylonitrile, and some other synthetic resins.
See also cellulose, modified.

modulus of elasticity. (elastic modulus).
A coefficient of elasticity representing the ratio of stress to strain as a material is deformed under dynamic load. It is a measure of the softness or stiffness of the material (Young's modulus).

moellon degras. See degras.

mohair. A natural fiber similar to wool obtained from angora goats.
Properties: Tenacity 14 g/denier. Combustible.
Use: Fabrics for outer clothing, draperies, upholstery.

Mohr's salt. See ferrous-ammonium sulfate.

Mohs scale. An empirical scale of the hardness of mineral or mineral-like materials originally consisting of 10 values, ranging from talc, with a rating of 1, to diamond, with a rating of 10, the rating being based on the ability of each material to scratch the one next below it in the series. The number of materials has been expanded from 10–15 with the addition of several synthetically produced substances (e.g., silicon carbide) between the original 9–10 positions. The scale is named after the German mineralogist, Friedrich Mohs (1773–1839).
See also hardness (1).

moiety. An indefinite portion of a sample.

Moissan, Henri. (1852–1907) A native of Paris, Moissan was a professor at the School of Pharmacy from 1886 to 1900 and at the Sorbonne from 1900 to 1907. At the former institution, he first isolated and liquefied fluorine in 1886 by the electrolysis of potassium acid fluoride in anhydrous hydrogen fluoride. His work with fluorine undoubtedly shortened his life as it did that of many other early experimenters in the field of fluorine chemistry. He won great fame by his development of the electric furnace and pioneered its use in the production of calcium carbide, making acetylene production and use commercially feasible, in the preparation of pure metals, such as magnesium, chromium, uranium, tungsten, etc,, and in the production of many new compounds, e.g., silicides and carbides and refractories. In 1906, he was awarded the Nobel Prize in chemistry.

molal. A concentration in which the amount of solute is stated in moles and the amount of solvent in kilograms. The unit of molality is moles of solute per kilogram of solvent and is designated by a small m, 1 mole of $NaCl$ in 1 kg of solvent is a 1 molal concentration. Note: Do not confuse with molar.

molar. A concentration in which 1 molecular weight in grams (1 mole) of a substance is dissolved in enough solvent to make one liter of solution. Molarity is indicated by an italic capital M. Molar quantities are proportional to the molecular weight of the substances.

molassess. The thick liquid left after sucrose has been removed from the mother liquor in sugar manufacture. Blackstrap molasses is the syrup from which no more sugar can be obtained economically. It contains approximately sucrose 20%, reducing sugars 20%, ash 10%, organic nonsugars 20%, water 20%. Combustible.

Use: Feed, food, raw material for various alcohols, acetone, citric acid and yeast propagation. Sodium glutamate is made from Steffens molasses, a waste liquor from beet sugar manufacture.
See also fermentation.

mold. See fungus.

mold preventive. See mildew preventive.

molding. Forming a plastic or rubber article in a desired shape by application of heat and pressure, either in a negative cavity, usually of metal, or in contact with a contoured metal or phenolic resin surface.
See injection molding, blow molding, compression molding.

molding powder. A mixture in a granular or pelleted form of a plastic base material together with necessary modifying ingredients (filler, plasticizer, pigment, etc.). Such mixtures are normally prepared by resin manufacturers and sold as such to processors ready for use in injection molding or extrusion operations.

molding sand. See foundry sand.

mold-release agent. See abherent.

mole. The amount of pure substance containing the same number of chemical units as there are atoms in exactly 12 grams of carbon-12 (i.e., 6.023×10^{23}). This involves the acceptance of two dictates--the scale of atomic masses and the magnitude of the gram. Both have been established by international agreement. Formerly, the connotation of "mole" was "gram molecular weight." Current usage tends to apply the term "mole" to an amount containing Avogadro's number of whatever units are being considered. Thus, it is possible to have a mole of atoms, ions, radicals, electrons, or quanta. This usage makes unnecessary such terms as "gram-atom," "gram-formula weight," etc. All stoichiometry essentially is based on the evaluation of the number of moles of substance. The most common involves the measurement of mass. Thus 25.000 grams of water will contain 25000/18.015 moles of water, 25.000 grams of sodium will contain 25.000/22.990 moles of sodium. The convenient measurements on gases are pressure, volume, and temperature. Use of the ideal gas law constant R allows direct calculation of the number of moles: $n = (PV)(RT)$. T is the absolute temperature, R must be chosen in units appropriate for P, V, and T. The acceptance of Avogadro's law is inherent in this calculation; so too are approximations of the ideal gas. William F. Kieffer.
See also Avogadro's law.

molecular biology. A subdivision of biology that approaches the subject of life at the molecular level. This applies to phenomena occurring within the cell nucleus, where the chromosomes and genes are located. These structures, which determine heredity, are in turn composed of nucleic acids, which direct the selection and assembly of amino acids in the dividing chromosomes. Much of the essential mechanism of life can be understood by study of specific protein molecules (DNA and RNA) and their determination of the amino acid composition of the genes.
See also genetic code, deoxyribonucleic acid, recombinant DNA.

molecular distillation. (high vacuum distillation). Distillation at low pressures of the order of 0.001 mm. A molecular distillation is distinguished by the fact that the distance from the surface of the liquid being vaporized to the condenser is less than the mean free path (the average distance traveled by a molecule between collisions) of the vapor at the operating pressure and temperature. This distance is usually of the order of magnitude of a few inches. This process is useful in separation of extremely high boiling and heat-sensitive materials such as glycerides and some vitamins.

molecular formula. See formula, chemical.

molecular rearrangement. See rearrangement.

molecular sandwich. (sandwich molecule).
See metallocene.

molecular sieve. A microporous structure composed of either crystalline aluminosilicates, chemically similar to clays and feldspars and belonging to a class of materials known as zeolites, or crystalline aluminophosphates derived from mixtures containing an organic amine or quaternary ammonium salt. Pore sizes range from 5–10Å. The outstanding characteristic of these materials is their ability to undergo dehydration with little or no change in crystalline structure. The dehydrated crystals are interlaced with regularly spaced channels of molecular dimensions, which comprise almost 50% of the total volume of the crystals.

The empty cavities in activated "molecular sieve" crystals have a strong tendency to recapture the water molecules that have been driven off. This tendency is so strong that if no water is present they will accept any material that can get into the cavity. However, only those molecules that are small enough to pass through the pores of the crystals can enter the cavities and be adsorbed on the interior surface. This sieving or screening action, which makes it possible to separate smaller molecules from larger ones, is

the most unusual characteristic of molecular sieves. They are used in many fields of technology, to dry gases and liquids; for selective molecular separations based on size and polar properties; as ion-exchangers, as catalysts, as chemical carriers, in gas chromatography, and in the petroleum industry to remove normal paraffins from distillates.
See also zeolite, gel filtration, pore.

molecular weight. The sum of the atomic weights of the atoms in a molecule. That of methane (CH$_4$) is 16.043, the atomic weights being carbon = 12.011, hydrogen = 1.008. The chemical formula used in such a calculation must be the true molecular formula of the substance designated. For example, the molecular formula of oxygen is O$_2$ and its molecular weight is 31.998 (atomic weight of oxygen = 15.999). For ozone the molecular formula is O$_3$ and the molecular weight is 47.997. The true molecular weight of a gas or vapor is found by measuring the volume of a given weight and then calculating the weight of 22.4 L at 0C and 760 mm. The molecular weight of many complex organic molecules runs as high as a million or more (proteins and high polymers).
See also Avogadro, atomic weight, mole.

molecule. A chemical unit composed of one or more atoms. The simplest molecules contain only one atom, for example, helium molecules (1 atom/molecule). Oxygen molecules (O$_2$) are composed of two atoms and ozone (O$_3$) of three. Molecules may contain several different sorts of atoms. Water contains two different kinds, hydrogen and oxygen, and dimethylamine [CH$_3$)$_2$NH] has three kinds. Molecules of many common gases [hydrogen (H$_2$), oxygen (O$_2$), nitrogen (N$_2$), and chlorine (Cl$_2$)] consist of two atoms each. The atoms of a molecule are held together by chemical bonds. Molecules vary in size from less than 1 to more than 500 millimicrons and in weight from 4 (He) to 40 million for tobacco mosaic virus.
See also macromolecule, bond, chemical, atom.

molten salt. See fused salt.

molybdate chrome orange. See molybdate orange.

molybdate orange. (molybdenum orange; molybdate chrome orange). CAS: 12656-85-8. A solid solution of lead chromate, lead molybdate and lead sulfate.
Properties: Fine dark orange or light red powder.
Derivation: By adding solutions of sodium chromate, sodium molybdate, and sodium sulfate to a lead nitrate solution under carefully controlled conditions and filtering off the precipitates.

Hazard: Toxic by ingestion.
Use: Pigment in printing inks, paints, plastics.

molybdenite. (molybdenum glance).
CAS: 1309-56-4.   MoS$_2$.   Natural molybdenum sulfide found in igneous rocks and metallic veins.
Properties: Color, bluish-lead gray, streak gray-black, luster metallic, one perfect cleavage, greasy feel, Mohs hardness 1–1.5, d 4.6–4.8, similar in appearance to graphite, soluble in sulfuric and strong nitric acids.
Occurrence: Colorado, Utah, New Mexico, Chile.
Use: Principal ore of molybdenum.

molybdenite concentrate. Commercial molybdenite ore after the first processing operations. Contains approximately 90% molybdenum disulfide along with quartz, feldspar, water, and processing oil.

molybdenum. CAS: 7439-98-7.   Mo.
Metallic element of atomic number 42, group VIB of the Periodic Table, aw 95.94, valences = 2, 3, 4, 5, 6. Seven stable isotopes.
Properties: Gray metal or black powder, does not occur free in nature a necessary trace element in plant nutrition. Insoluble in hydrochloric acid and hydrogen fluoride, ammonia, sodium hydroxide, or dilute sulfuric acid; soluble in hot concentrated sulfuric or nitric acids; insoluble in water. D 10.2, mp 2610C, bp 5560C, high strength at very high temperatures, oxidizes rapidly above 1000F in air at sea level, but is stable in upper atmosphere.
Derivation: By aluminothermic, hydrogen, or electric furnace reduction of molybdenum trioxide or ammonium molybdate.
Forms available: Ingots, rods, wire, powder, ingots (from powder), high ductility sheets, also as large single crystal.
Purity: Rods and wire 99.9%, powder 99.9%.
Hazard: Flammable in form of dust or powder. TLV (of molybdenum compounds as Mo): 10 mg/m^3 of air (insoluble); 5 mg/^{3} of air (soluble).
Use: Alloying agent in steels and cast iron; high-temperature alloys, tool steels; pigments for printing inks, paints, and ceramics; catalyst; solid lubricants; missile and aircraft parts; reactor vessels; cermets; die-casting copperbase alloys; special batteries.
See also ferromolybdenum, heteromolybdates.

molybdenum acetylacetonate. Mo(C$_5$H$_7$O$_2$).
Properties: Crystalline powder, slightly soluble in water, resistant to hydrolysis, a chelating nonionizing compound.
Use: Catalyst for polymerization of ethylene and formation of polyurethane foam.

molybdenum aluminide. A cermet which can be flame-sprayed.

molybdenum anhydride. See molybdenum trioxide.

molybdenum boride. CAS: 12007-97-5.
Several borides are known: Mo_2B, mp 2000C; MO_3B_2, mp 2070C; MoB (ordinary and beta-forms), mp 2180C; Mo_2B_5, mp 1600C (transforms to MoB_2).
Derivation: By heating molybdenum powder and boron to 1500–1600C in hydrogen.
Use: Brazes to join molybdenum, tungsten, tantalum, and niobium parts, especially electronic components, corrosion and abrasion-resistant parts, cutting tools, refractory cermets.

molybdenum carbonyl. See molybdenum hexacarbonyl.

molybdenum dioxide. CAS: 18868-43-4.
MoO_2.
Properties: Lead-gray, nonvolatile powder. Insoluble in hydrochloric acid and hydrogen fluoride and alkalies, sparingly soluble in sulfuric acid, d approximately 6.4.
Derivation: Reduction of molybdenum trioxide or molybdates by hydrogen, partial oxidation of metallic molybdenum.
Hazard: Toxic material. TLV (as Mo): 10 mg/m^3 of air.

molybdenum diselenide. $MoSe_2$. Available as a 40-micron powder.
Use: Solid lubricant.

molybdenum disilicide. $MoSi_2$. A cermet.
Properties: Dark gray, crystalline powder. D 6.31 (20C), mp 1870–2030C, not affected by air up to 1648C, not attacked by most inorganic acids including aqua regia, but very soluble in hydrofluoric and nitric acids. Has high stress-rupture strength.
Derivation: By fusion of hydrogen-reduced molybdenum with silicon.
Forms: Cylinders, lumps, granules, powder, whiskers. May be coated on materials by vapor deposition and by flame spraying.
Grade: 98%, 99.5%, mesh size 200 and 325.
Hazard: Toxic material. TLV (as Mo): 5 mg/m^3 of air.
Use: Electrical resistors, protective coatings at high temperatures, engine parts in space vehicles (molybdenum coated with molybdenum disilicide).

molybdenum disulfide. (molybdic sulfide; molybdenum sulfide). CAS: 1317-33-5. MoS_2.
Properties: Black, crystalline powder; d 4.80; mp 1185C. Soluble in aqua regia, sulfuric acid (concentrated); insoluble in water. Mohs hardness 1, coefficient of friction 0.02-0.06.

Hazard: Toxic material. TLV (as Mo): 5 mg/m^3 of air.
Derivations: Purification of molybdenite, reaction of sulfur or hyrodgen sulfide on molybdenus trioxide.
Use: Lubricants in greases, oil dispersions, resin-bonded films, dry powders, etc., especially at extreme pressures and high vacua; hydrogenation catalyst.
See also molybdenite.

molybdenum ditelluride. $MoTe_2$. Available as a 40-micron powder.
Use: Solid lubricant.

molybdenum glance. See molybdenite.

molybdenum hexacarbonyl. (molybdenum carbonyl). $Mo(CO)_6$.
Properties: White, shiny crystals; decomposes at 150C (sublimes); d 1.96; bp approximately 155C; vap press approximately 0.1 mm (20C), approximately 43 mm (101C); insoluble in water; soluble in ceresin, paraffin oil, benzene, aminoanthraquinone; slightly soluble in ether and other organic solvents.
Derivation: From molybdenum pentachloride by reaction with zinc dust and carbon monoxide in ether at high pressures.
Hazard: Decomposes above 150C to evolve carbon monoxide. TLV (as Mo): 5 mg/m^3 of air.
Use: Plating molybdenum, i.e., molybdenum mirrors; intermediate.

molybdenum hexafluoride. CAS: 7783-77-9.
MoF_6.
Properties: White, crystalline solid; hygroscopic. Mp 17.5C, bp 35C, d (liquid) approximately 2.5, readily hydrolyzed.
Derivation: Action of fluorine on molybdenum powder.
Hazard: Strong irritant. TLV (as Mo): 5 mg/m^3 of air.
Use: Separation of molybdenum isotopes.

molybdenum lake. See phosphomolybdic pigment.

molybdenum metaphosphate. $Mo(PO_3)_6$.
Properties: Yellow powder, d 3.28 (0C), insoluble in water and in most acids, slightly soluble in hot aqua regia.

molybdenum naphthalene. Dark purple, viscous liquid; soluble in most hydrocarbons; catalyst for commercial production of propylene oxide using hydroperoxides.
Hazard: Toxic material. TLV (as Mo): 5 mg/m^3 of air.

molybdenum orange. See molybdate orange.

molybdenum III oxide. See molybdenum sesquioxide.

molybdenum oxides. See molybdenum sesquioxide, molybdenum dioxide, molybdenum trioxide.

molybdenum pentachloride. CAS: 10241-05-1. $MoCl_5$.

Properties: Green-black solid, dark red as liquid or vapor, mp 194C, bp 268C, d 2.9, hygroscopic, reacting with water and air, soluble in dry ether, dry alcohol, and other organic solvents.

Derivation: direct action of chlorine on finely divided molybdenum metal.

Hazard: Irritant. TLV (as Mo): 5 mg/m^3 in air.

Use: Chlorination catalyst, vapor-deposited molybdenum coatings, component of fire-retardant resins, brazing and soldering flux, intermediate for organometallic compounds, e.g., molybdenum hexacarbonyl.

molybdenum sesquioxide. (dimolybdenum trioxide; molybdenum III oxide).

CAS: 1313-29-7. Mo_2O_3. Known only in the hydrated form, $Mo(OH)_3$, although commonly assigned the formula Mo_2O_3. A compound formed by a dry reaction of molybdenum and oxygen which approximates the composition of the sesquioxide is probably a mixture of molybdenum and molybdenum dioxide.

Properties: Gray-black powder, slightly soluble in acids, insoluble in alkalies and water.

Derivation: Zinc reduction of acid solutions of molybdic acids and molybdates, electrolytic deposition from acid solutions of molybdates.

Hazard: TLV (as Mo): 10 mg/m^3 of air.

Use: Catalyst in organic synthesis, decoration and protection for metal articles, feed additive.

molybdenum silicide. Alloy of 60% molybdenum, 30% silicon, and 10% iron used as means of introducing molybdenum into steel.

molybdenum sulfide. See molybdenum disulfide.

molybdenum trioxide. (molybdenum anhydride; molybdic oxide; molybdic acid hydride).

CAS: 1313-27-5. MoO_3.

Properties: White powder at room temperature, yellow at elevated temperatures, d 4.69, mp 795C, sublimes starting at 700C, bp 1150C, sparingly soluble in water, very soluble in excess alkali with formation of molybdates, soluble in concentrated mixture of nitric acid and hydrochloric acid or nitric and sulfuric acids. Two hydrates are known: $MoO_3 \cdot HOH$ and $MoO_3 \cdot 2HOH$. Readily combines with acids and bases to form a series of polymeric compounds.

Derivation: Roasting of molybdenite, by ignition of the metal, the sulfides, the lower oxides, and of molybdic acids.

Purification: Sublimation.

Grade: Technical, pure, reagent, ACS.

Hazard: Toxic material. TLV (as Mo): 5 mg/m^3 of air.

Use: Source of molybdenum compounds, agriculture, analytical chemistry, manufacture of metallic molybdenum, introduction of molybdenum in alloys, corrosive inhibitor, ceramic glazes, enamels, pigments, catalyst.

molybdic acid. CAS: 7782-91-4. Molybdic acid of commerce is either an ammonium molybdate (molybdic acid 85%) or molybdenum trioxide. The use of the term interchangeably for these compounds has caused confusion. Solutions of molybdic acid are very complex chemically since they show a great tendency to polymerize.

molybdic acid, anhydride. See molybdenum trioxide.

molybdic oxide. See molybdenum trioxide.

molybdic sulfide. See molybdenum disulfide.

molybdophosphates. See heteromolybdates.

12-molybdophosphoric acid. See phosphomolybdic acid.

molybdosilicates. See heteromolybdates.

12-molybdosilicic acid. (silicomolybdic acid). $H_4SiMo_{12}O_{40} \cdot xHOH$ where x is usually 6–8.

Properties: Yellow, crystalline powder; d 2.82; soluble in water, ethanol, acetone; insoluble in benzene and cyclohexane; decomposes in strongly basic solutions; thermally stable.

Grade: Reagent.

Hazard: Strong oxidizing agent in aqueous solution. TLV (as Mo): 5 mg/m^3 in air.

Use: Catalysts; reagents; photography; precipitants and ion exchangers in atomic energy; additives in plating processes; imparting water resistance to plastics, adhesives, and cement.

See also heteromolybdates.

MON. Abbreviation for Motor Octane Number.

monacetin. See acetin.

"Monacide."[405] TM for a series of insecticides containing 5% DDVP.

"Monamids."[405] TM for a group of dialkylolamides which include various grades of coconut fatty acid monoethanolamide, coconut fatty acid monopropanolamide, lauric acid monoethanolamide, lauric acid monoisopropanolamide, and stearic acid monoethanolamide.

"Monamines."[405] TM for a group of dialkylolamides used as detergent, detergent additives, foam booosters, wetters, emulsifiers, dispersing agents, thickeners, and conditioners.

"Monastrip."[405]   TM for a solvent-stripper for un-cured and cured epoxy, polyester, and silicone rubber casting and encapsulating compounds.

"Monaterics."[405]   TM for a special group of substituted imidazolines classified as amphoteric surfactants. Excellent wetting, emulsifying, penetrating and spreading properties in systems requiring broad pH ranges.

monatomic.   See diatomic.

"Monawets."[405]   TM for a group of surfactants of di-octyl, di-hexyl, di-isobutyl, and di-tridecyl sulfosuccinates known for their wetting, spreading, penetrating and emulsifying power.

monazite.   A natural phosphate of the rare earth metals, principally the cerium and lanthanide metals, usually with some thorium. Yttrium, calcium, iron and silica are frequently present. Monazite sand is the crude natural material and is usually purified from other minerals before entering commerce.
Properties: Color yellowish to reddish brown, luster vitreous to resinous, streak white, Mohs hardness 5–5.5, d 4.9–5.3.
Occurrence: North Carolina, South Carolina, Idaho, Colorado, Montana, Florida, Brazil, India, Australia, Canada.
Use: Source of thorium, cerium, and other rare-earth metals and compounds.

"Monazolines."[405]   TM for a series of cationic imidazolines useful as emulsifiers, antistatic agents, water displacers, and corrosion inhibitors in agricultural sprays, acid and solvent cleaners, cosmetics, and water-oil systems.

Mond process.   Mixed ores obtained from roasting crude ores are heated from 50–80C in a stream of producer gas. Oxides other than nickel are reduced to the metallic state while nickel forms nickel carbonyl [Ni(CO)$_4$] which passes off as a vapor. The vapor is subsequently resolved into carbon monoxide and free nickel.

"Mondur."[567]   TM for a series of isocyanates.
Use: Surface coatings, adhesives, chemical intermediate, hydrophobic agent to increase water repellency of textiles, leather, and paper products.

"Monel."[283]   TM for a large group of corrosion-resistant alloys of predominantly nickel and copper and a very small percentage of carbon, manganese, iron, sulfur, and silicon. Some also contain a small percentage of aluminum, titanium, and cobalt. Have good corrosion-resisting properties.
Use: Electronics industry.

"Monex."[248]   TM for tetramethylthiuram monosulfide.

mono-.   Prefix denoting one, see under specific compound, e.g., monochloroacetic acid.
See chloroacetic acid. Only a few of the many possible cross references from mono-compounds are given here. (See below.)

monoanhydrosorbitol.   See sorbitan.

monoazo dye.   See azo dye.

monobasic.   Descriptive of acids having one displaceable hydrogen atom per molecule. Acids having two, three, or more displaceable hydrogen atoms are called dibasic, tribasic, and polybasic, respectively.
See also acid.

"Monobed."[23]   TM for intimate mixture of "Amberlite" cation and anion exchange resins.
Use: Complete removal of ionizable impurities from water and other solutions in a one-step treatment.

monocalcium phosphate.   See calcium phosphate, monobasic.

monochloroacetic acid.   Legal label name (Rail) for solid chloroacetic acid.

monochloroacetone.   Legal label name (Rail) for chloroacetone.

monochlorobenzene.   See chlorobenzene.

monochlorodifluoromethane.   Legal label name (Rail) for chlorodifluoromethane.

monochloroethane.   See ethyl chloride.

monochloromethane.   See methyl chloride.

monochloropentafluoroethane.   Legal label name (Rail) for chloropentafluoroethane.

monochlorophenol.   See chlorophenol.

monochlorotetrafluoroethane.   Legal label name (Rail) for chlorotetrafluoroethane.

monochlorotriazinyl dye.   A fiber-reactive dye for cellulose fibers.
See dye, fiber-reactive.

monochlorotrifluoromethane. Legal label name (Rail) for chlorotrifluoromethane.

"Monochrome."[203] (USA). TM for a series of mordant dyestuffs. Characterized by good fastness properties.
Use: Dyeing of wool.

monocrotophos. See "Azodrin."

monoethanolamine. See ethanolamine.

monoethylamine. Legal label name (Rail) for ethylamine.

monofilament. A single continuous strand of glass or synthetic fiber as extruded from a spinneret. See filament.

monoglyceride. A glycerol ester of fatty acids in which only one acid group is attached to the glycerol group. A typical formula is $RCOOCH_2CHOHCH_2OH$. Small amounts of monoglycerides occur naturally.
Derivation: Produced synthetically by the alcoholysis of fats with glycerol, yielding a mixture of mono-, di-, and triglycerides which is predominantly monoglycerides.
Use: Emulsifiers, cosmetics, lubricants.
See glycerol monostearate, glycerol monolaurate, etc.

monoglyme. See ethylene glycol dimethyl ether.

monohydric alcohol. An alcohol in which a hydroxyl group (—OH) has replaced one of the hydrogen atoms of a hydrocarbon, for example:

C_2H_5H C_2H_5OH
ethane   *ethanol*

RH   ROH
alkane   *alkyl alcohol*

There are a number of classifications analogous to those of hydrocarbons: (1) paraffinic or simple alcohols, whose formula may be represented as C_nH_{2n+1}; (2) olefinic or fatty alcohols which contain one or more double bonds; (3) alicyclic alcohols, closed ring structures which may or may not contain a double bond, e.g., cyclohexanol; (4) aromatic alcohols in which the hydroxyl group is attached to a benzene nucleus as in phenol; (5) heterocyclic alcohols, based on the pentagonal furan ring; and (6) polycyclic alcohols of high molecular weight, known collectively as sterols. Any of these types that contain 12 or more carbon atoms are semisolid to solid and have a wax-like consistency; the others are colorless liquids.

Monohydric alcohols are also classified as primary, secondary, or tertiary on the basis of the number of alkyl (methyl) groups substituted for the hydrogen atoms on the central or methanol carbon atom.
See also primary.

monomer. (momer). A molecule or compound usually containing carbon and of relatively low molecular weight and simple structure which is capable of conversion to polymers, synthetic resins, or elastomers by combination with itself or other similar molecules or compounds. Thus, styrene is the monomer from which polystyrene resins are produced, vinyl chloride is the monomer of polyvinyl chloride. Other common monomers are methyl methacrylate, adipic acid, and hexamethylenediamine.

monomethylamine. Legal label name (Rail) for methylamine.

monomolecular film. See film.

"Monoplex."[94] TM for monomeric liquid plasticizers for polyvinyl chloride and other high polymers. Primarily esters, but also some epoxides which impart heat and light stability.
Use: Plasticizers, stabilizers, process aids.

monopropellant. A propellant which combines fuel and oxidizer in one compound or mixture. Gunpowder is an example of a solid monopropellant. Liquid monopropellants, for rockets, include: methyl nitrate, nitromethane, a mixture of hydrocarbons with tetranitromethane, a mixture of methyl nitrate and methanol.
See also rocket fuel.

monosaccharide. Any of several simple sugars having the formula $C_6H_{12}O_6$; the best-known are glucose, fructose, and galactose. Monosaccharides combine to form more complex sugars known as oligo- and polysaccharides.

monosodium glutamate. See sodium glutamate.

monostearin. See glycerol monostearate.

monoxychlor. Trichloroethane analog of DDT, less toxic to warm-blooded animals than DDT.

montan wax. (lignite wax).
Properties: White, hard earth wax; crude product is dark brown; mp 80–90C; soluble in carbon tetrachloride, benzene, and chloroform; insoluble in water. Combustible.
Derivation: By countercurrent extraction of lignite. American and German lignite are usual sources.

Method of purification: Distillation with super-heated steam.

Grade: Crude, refined.

Use: Substitute for carnauba and beeswax, shoe and furniture polishes, phonograph records, roofing paints, rendering paints waterproof, adhesive pastes, electric insulating compositions, paper-sizing compositions, carbon papers, wire coating, suncrack preventive in rubber products.

montmorillonite. $Al_2O_3 \cdot 4SiO_2 \cdot HOH$ (approximately). A type of clay. One of the major components of bentonite and fullers earth.

monuron. (3-(p-chlorophenyl)-1,1-dimethylurea; CMU). CAS: 150-68-5. $ClC_6H_4NHCON(CH_3)_2$. A plant growth regulator.

Properties: White, crystalline, odorless solid; mp 175C. Very low solubility in water and hydrocarbon solvents, slightly soluble in oils, partially soluble in alcohols, stable toward oxidation and moisture.

Use: Herbicide, sugarcane flowering suppressant.

Moore, Stanford. (1913-1982) An American biochemist who won the Nobel prize for chemistry in 1972 with Anfinsen and Stein for enzyme studies. He was involved with the analysis of the action of the complex enzyme deoxyribonuclease. His PhD was granted from the University of Wisconsin.

"Mo-Permalloy."[155] TM for a magnetic alloy.

Properties: D 8.72, tensile strength 85,000 psi.

Use: Laminated cores for high quality communication inductors, transformers and magnetic field detectors.

mordant. A substance capable of binding a dye to a textile fiber. The mordant forms an insoluble lake in the fiber, the color depending on the metal of the mordant. The most important mordants are trivalent chromium complexes, metallic hydroxides, tannic acid, etc. Mordants are used with acid dyes, basic dyes, direct dyes and sulfur dyes. Premetallized dyes contain chromium in the dye molecule. A mordant dye is a dye requiring use of a mordant to be effective.

See also dye, fiber-reactive.

"Mordantine." TM for liquid antimony lactate containing 11% available antimony oxide. Recommended as a replacement for technical tartar emetic.

See "Antilac."

mordanting assistant. A chemical such as lactic, oxalic, and sulfuric acids, tartrates, etc.

Use: In conjunction with mordants to bring about a gradual decomposition of the latter and to assist in producing a uniform deposition of the actual mordant upon and within textile materials.

mordant rouge. See aluminum diacetate.

"Morestan."[181] TM for 6-methyl-2,3-quinoxalinedithiol cyclic carbonate.

"Morflex."[299] TM for specialty plasticizers (including phthalates, adipates, azelates, sebacates, trimellitates, triacetin, and polymerics) for vinyl compounds.

morin. (2′,3,4,5,7-pentahydroxyflavone). CAS: 480-16-0. $C_{15}H_{10}O_7 \cdot 2HOH$. One of the two coloring principles of yellow Brazilwood.

Properties: Colorless needles, mp 285C (decomposes). Soluble in alcohol, alkaline solutions; slightly soluble in boiling water. Combustible.

Use: Mordant dye, spot test reagent for metal salts, luminescence indicator.

See brasilin.

"Morlex."[214] TM for corrosion inhibitors.

Use: Steam boilers and steam heating systems, for example, Corrosion Inhibitor A. A mixture of 91% morpholine in water.

"Morosodren."[401] TM for a fungicide concentration containing methylmercury dicyandiamide (2.2%).

Hazard: Toxic by ingestion.

morphine. CAS: 57-27-2. $C_{17}H_{19}NO_3 \cdot HOH$.

Properties: White, crystalline alkaloid. Slightly soluble in water, alcohol, and ether; bp 254C (decomposes); d 1.31.

Derivation: From opium by extraction and crystallization. Opium contains approximately 10% morphine.

Hazard: Narcotic, habit-forming drug, sale restricted by law in the US.

Use: Analgesic (in form of acetate, hydrochloride, tartrate, and other soluble salts).

p-morphine. See thebaine.

morpholine. (tetrahydro-1,4-oxazine).

CAS: 110-91-8. $OCH_2CH_2NHCH_2CH_2$ or C_4H_8ONH.

Properties: Colorless, hygroscopic liquid; amine-like odor. Soluble in water and organic solvents, bp 128.9C, fp −4.9C, d 1.002 (20/20C), bulk d 8.34 lb/gal (20C), vap press 6.6 mm (20C), viscosity 2.23 cp (20C), flash p (100F) (37.7C) (OC), autoign temperature 590F (310C).

Derivation: Dehydration of diethanolamine.

Grade: Technical, 98%.

Hazard: Flammable, moderate fire risk. Toxic by ingestion and inhalation, irritant to skin, absorbed by skin. TLV: 20 ppm in air.

Use: Rubber accelerator, solvent, additive to boiler water, waxes and polishes, optical brightener for detergents, corrosion inhibitor, preservation of book paper, organic intermediate (catalyst, antioxidants, pharmaceuticals, bactericides, etc.).

morpholine borane. $C_4H_8ONH \cdot BH_3$.

Properties: White, needle-shaped, crystalline compound; mp 93C; soluble in hot water and alcohol; insoluble in carbon tetrachloride.

Use: Reducing agent for aldehydes and ketones. Useful in acid media where sodium borohydride is ineffectual because of its instability in acid.

morpholine ethanol. (N-hydroxyethylmorpholine). $C_4H_8ONCH_2CH_2OH$.

Properties: Colorless liquid, miscible with water, d 1.0724, bp 225.5C, flash p 210F (98.9C). Combustible.

2-(morpholinothio)benzothiazole. (N-oxydiethylene-2-benzothiazolesulfenamide). $C_{11}H_{12}N_2OS_2$.

Properties: Buff to brown flakes with sweet odor, mp 80C min, d 1.34 (25C). Insoluble in water; soluble in benzene, acetone, methanol.

Use: Delayed action vulcanization accelerator.

morphology. A term borrowed from the biological sciences by physical chemists to denote the shape, structure or form of such substances as high polymers, crystals, reinforcing agents, and the like, e.g., the morphology of carbon black in rubber.

morphothion. Generic for O,O-dimethyl-S-(morpholinocarbonylmethyl)phosphorodithioate.

$(CH_3O)_2P(S)SCH_2C(O)NCH_2CH_2OCH_2CH_2$.

Properties: Colorless solid, mp 65C, soluble in acetone, dioxane, and acetonitrile.

Hazard: Cholinesterase inhibitor, use may be restricted.

Use: Insecticide.

mortar. (1) A type of adhesive or bonding agent which may be either inorganic or organic, soft and workable when fresh but sets to a hard, infusible solid on standing, either by hydraulic action or by chemical crosslinking. The chief ingredients of inorganic mortars are cement, lime, silica, sulfur, and sodium or potassium silicate. Organic mortars are based on various synthetic resins (epoxy, phenolic, polyester, and furan). All types are resistant to acids. Some (potassium sili-

cate) are useful up to 1600F, others are used for bonding acid-proof brick, tile, etc., for masonry construction and for lining chemical reaction equipment.

See also sealant, adhesive, cement.

(2) A ceramic receptacle used by pharmacists for preparing mixtures of medicinals and for hand-pulverizing soft solids.

mortar, metallic. A mixture of powdered metal and other ingredients that have been mixed with water. Lead, tungsten, and depleted uranium have been used as the metal component. These mortars resist weathering, mild acids and alkalies, intense radiation, and extreme temperature variation.

Use: Space technology.

morzid. (Generic for bis(1-aziridinyl)-morpholinophosphine sulfide). CAS: 2168-68-5. $CH_2CH_2OCH_2CH_2NP(NCH_2CH_2)_2$:S.

Use: Insect chemosterilant.

Moseley, Henry. (1887–1915) A British chemist who studied under Rutherford and brilliantly developed the application of X-ray spectra to study of atomic structure; his discoveries resulted in a more accurate positioning of elements in the Periodic Table by closer determination of atomic numbers. Tragically for the development of science, Moseley was killed in action at Gallipoli in 1915.

"Moskene."[227] TM for 1,1,3,3,5-pentamethyl-4,6-dinitroindane.

Mossbauer effect. A nuclear phenomenon discovered in 1957. Defined as the elastic (recoil-free) emission of a gamma particle by the nucleus of a radioactive isotope and the subsequent absorption (resonance scattering) of the particle by another atomic nucleus. Occurs in crystalline solids and glasses, but not in liquids. Examples of gamma-emitting isotopes are: iron-57, nickel-61, zinc-67, tin-119. The Mossbauer effect is used to obtain information on isomer shift, on vibrational properties and atomic motions in a solid, and on location of atoms within a complex molecule.

mossy zinc. Zinc powder formed by pouring molten zinc into water.

mother. (1) A mold of bacterial complex containing enzymes which promote fermentation, as in manufacture of vinegar from cider or of cultured dairy products from milk. (2) A substance secreted by epithelial cells of the oyster. (3) A mother liquor is a concentrated solution from

which the product is obtained by evaporation and/or crystallization, e.g., in sugar manufacture.
See nacre.

Motor Octane Number (MON).   See octane number.

mountain blue.   (copper blue).
Derivation: The mineral azurite in ground form.
Use: Paint pigment.

6-MP.   Abbreviation for 6-mercaptopurine.

MPA.   Abbreviation for multipurpose additive.

"MPA."[202]   TM for a colloidal thixotropic agent used in paints, inks, and lacquers to prevent pigment settling and excess film flow. Dispersed paste available in mineral spirits, xylene, or toluene.

3-MPA.   Abbreviation for 3-methoxypropyl-amine.

MPC black.   Abbreviation for medium processing channel black.
See carbon black.

MPK.   Abbreviation for methyl propyl ketone.

"MPS-500."[62]   TM for a stabilized chlorinated ester of a fatty acid. A viscous, light yellow liquid recommended as a low-cost plasticizer for polyvinyl chloride formulations.

"M-Pyrol."[307]   TM for an aprotic solvent, N-methyl-2-pyrrolidone. Claimed to be effective cleaner of vinyl chloride reaction equipment.

"MS."[241]   TM for miscrospherical silica alumina.
Use: Cracking catalyst.

MSG.   Abbreviation for monosodium glutamate.
See sodium glutamate.

MSP.   Abbreviation for monosodium phosphate.
See sodium phosphate, monobasic.

MTBE.   Abbreviation for methyl-tert-butyl ether.

MT black.   Abbreviation for medium thermal black.

MTD.   Abbreviation for m-tolylenediamine.
See toluene-2,4-diamine.

mucic acid.   (saccharolactic acid; galactaric acid; tetrahydroxyadipic acid).
HOOC(CHOH)$_4$COOH.
Properties: White, crystalline powder; mp approximately 210C (decomposes); soluble in water; insoluble in alcohol. Combustible.

Derivation: Oxidation of lactose or similar carbohydrates with nitric acid.
Use: Substitute for tartaric acid, sequesterant for metal ions (calcium, iron), retards hardening of concrete; intermediate for synthesis of heterocyclic compounds (pyrroles).

mucilage.   A plant product obtained from seeds, roots, or other parts of plants by extraction with either hot or cold water. Mucilages give slippery or gelatinous solutions, e.g., those from guar bean, linseed, locust bean, and related leguminous plant seeds. Generally plant mucilates are insoluble in alcohol, but some are partly soluble in water and partly soluble in alcohol. From various types of salt-water algae the so-called seaweed mucilages, such as agar, algin, and carrageenin (sometimes referred to as algal polysaccharides) may be obtained by extraction with hot water. Mucilages are closely related to gums and the distinction between them is not always clear.
See also adhesives; gum, natural.

mucopolysaccharide.   A polysaccharide composed of alternate units of uronic acids and amino sugars (in which a hydroxyl group is replaced with an amino group, which in turn may be N-substituted by other groups). The mucopolysaccharides act as structural support for connective tissue and mucous membranes of animal organisms.

mud, drilling.   See drilling fluid.

muffle furnace.   A furnace or kiln in which the materials being heated are kept out of direct contact with the heat source, the combustion being effected by heat reflected from the walls of the furnace.
See Mannheim furnace, reverberatory furnace.

muller.   A device for uniform mixing of dry and wetted solids by a combined rubbing and smearing action analogous to that of a mortar and pestle. It consists of a stationary circular pan within which two heavy wheel-like members, together with scrapers (plows) revolve. The mulling wheels have flat, wide surfaces (outer rims) which ride on the material and effect the mixing action. As the inner edges of the wheels travel less distance than the outer edges, a smearing effect is provided across the surface of the wheels. The plows continually rake the material into the path of the wheels as the unit revolves. Continuous mulling is obtained with two such machines arranged in tandem, space being provided for constant recirculation of the material between them. Mullers are used for fine dispersion and blending of a wide range of products that are dense enough to support the wheels and fluid enough to provide traction, e.g., putty, explosives and heavy pastes.

"Mullfrax."[280] TM for refractory products made from mullite produced in electric furnaces.
Use: Construction materials for furnaces and kilns.

Mulliken, Robert S. (1896-) An American chemist, physicist, and educator who won the Nobel prize for chemistry in 1961. He did research on isotope separation and on spectroscopy of electrons in molecule formation. After M.I.T. granted his PhD, he did graduate work before working in industry, government, and academia.

mullite. $3Al_2O_3 \cdot 2SiO_2$. A stable form of aluminum silicate formed by heating other aluminum silicates (such as cyanite, sillimanite and andalusite) to high temperatures; also found in nature.
Properties: Colorless crystals, d 3.15, mp 1810C, insoluble in water.
Use: Refractories, glass.
See also aluminum silicate.

"Mulram."[455] TM for a ramming mix with a fused mullite base and a use limit up to 3200F. Available in several grain sizes.
Use: Refractory where resistance to metal or slag penetration is needed.

"Multimet."[214] TM for a high-temperature alloy composed of approximately equal parts of iron, nickel, cobalt, and chromium.
Use: Applications involving high stress up to 815C and moderate stress to 1093C.

multiple proportions law. See chemical laws (3).

"Multisorb" A.R.[329] TM for a grade of manganese dioxide used as a solid absorbent for SO_2 and NO_2 in analytical chemistry.

"Multi-Sperse."[141] TM for stir-in pulps of yellows, oranges, reds, blues, greens, violets, and black compatible with all types of latex paints, including acrylic, butadiene-styrene types, and polyvinyl acetate.

"Multiwax."[45] TM for refined microcrystalline wax obtained from crude petroleum. Composed primarily of alkylated naphthenes and isoparaffins with small amounts of normal paraffins.

"Multranil 176."[567] TM for a polymer which in the solid form is similar in appearance to SBR rubber. A base resin of a two-component system which when mixed with the proper curing agents forms a versatile adhesive.

"Multrathane."[567] TM for a series of compounds used mainly in formulations for solid urethane elastomers. Some are also used in the formulations of spandex fibers, urethane coatings, and adhesives.

municipal waste. See sewage sludge.

Muntz metal. An alloy containing approximately 60% copper and 40% zinc; a low percentage of lead is sometimes added for free-cutting. It is classified as a brass, and is used primarily for condenser tube plates and other electrical applications. It is formed by hot-working, is not amenable to cold-working.
See also brass.

murexide. CAS: 3051-09-0. $C_8H_8N_6O_6$.
Properties: Dark red crystals, partially soluble in hot water, insoluble in alcohol.
Use: Indicator.

muriatic acid. Obsolete name for hydrochloric acid. The related term muriate, indicating presence of chlorine in an inorganic compound, is also obsolete.

murvesco. See fenson.

muscone. See musk.

muscovite. (white mica; potassium mica; isinglass). $KAl_2(AlSi_3O_{10})(OH)_2$. A natural hydrous potassium aluminum silicate of the mica group.

musk. CAS: 300-54-9.
Properties: An unctuous, brownish, semi-liquid when fresh; dried, in grains or lumps with color resembling dried blood. Strong characteristic odor. The odor-bearing constituent is muscone, $CH_3C_{15}H_{27}O$, a 15-carbon ring with ketone oxygen.
Derivation: Natural: Secretion from preputial follicles of the musk deer. Synthetic: (a) Ketones and lactones with 15- or 16-carbon rings, structurally resembling the odoriferous principles of natural musk, civet, and musky-type plants. Among these are ambrettolide, civetone, muscone, exaltolide. (b) Nitrated compounds, usually nitrated tert-butyl-toluenes or xylenes or related compounds. The three most commonly used in perfumery are musk ambrette, musk ketone, and musk xylene.
Use: Cosmetics and perfumery (fixative), fragrances, mothproofing agent.

mustard gas. Legal label name (Air) for dichlorodiethyl sulfide.

mustard oil, artificial. See allyl isothiocyanate.

mustard oil. Any of several organic compounds having the formula R—N=C=S, in which R is an alkyl or aryl radical —NCS an isothiocyanate group. Its best known member is allyl isothiocyanate, the characteristic ingredient of mustard oils.
See also nitrogen mustard.
Hazard: Irritant to mucous membranes.

mutagenic agent. (1) Any of a number of chemical compounds able to induce mutations in DNA and in living cells. The alkyl mustards as well as dimethyl sulfate, diethyl sulfate, and ethylmethane sulfonate, comprise a group of so-called alkylating agents, reacting with the nitrogen atoms of guanine, a constituent of both RNA and DNA. This reaction affects the guanine molecule in such a way as ultimately to induce a mutation in DNA by depurination. Nitrous oxide can deaminate both guanine and cytosine. If DNA having transforming activity is exposed to such deamination conditions, it is slowly deactivated. Nitrous oxide also produces mutants in whole cells, whole bacteriophage, some viruses, and in DNA having transforming ability. (2) Ionizing radiation.

MVE. Abbreviation for methyl vinyl ether.
See vinyl methyl ether.

mw. Abbreviation for molecular weight.

"MX."[280] TM for fiber-bonded abrasives.
Properties: High tensile strength and resistance to impact and heat shock, unusually resilient.
Use: For finishing and polishing flutes of taps, drill end mills, reamers, etc.; removing burrs from milling and drilling operations, breaking edges of cast aluminum parts, etc.; cleaning cast iron molds, removing flash from molded plastics.

"Mycoban."[299] TM for sodium and calcium propionates. These salts inhibit the growth of many fungi and of some microorganisms, particularly Bacillus mesentericus, for commercially significant periods of time.
Use: Inhibit mold and rope in bakery products.

"Mycostatin."[412] TM for nystatin.

mycotoxin. A highly toxic principle produced by molds or fungi. One type, the aflatoxins, is produced by the Aspergillus flavus fungus; another is a member of the tricothecene group produced by the fusarium fungus. This has been identified in samples of the so-called "yellow rain" in Southeast Asia, where it is said to have been the cause of many deaths among war refugees. Its presence there is subject to some conjecture, since the fusarium fungus cannot germinate in the humid environment of that area. There is substantial evidence (blood tests, autopsies, and contaminated gas masks) that the Soviets have used such lethal agents in Afghanistan also.

myelin. A unique, sheath-like structure which encloses major nerve trunks, somewhat like insulation around a wire. It is comprised of approximately 80% lipid, the balance being made up of proteins, polysaccharides, salts, and water. The lipid fraction is composed of sphingolipids and glycerophosphates, which in turn contain long-chain fatty acids. It has a low concentration of polyunsaturated lipids and high concentration of long-chain sphingolipids. Its composition is essentially constant in different species of animals, and also as between adults and infants. The breakdown of the lipid structure of myelin is a characteristic of multiple sclerosis.

"Mylar."[28] TM for a polyester film. Seven available types used for electrical, industrial, and packaging purposes.
Forms: Roll and sheet.

Mylone. See "Crag" (fungicide).

myoglobin. A protein-iron-porphyrin molecule similar to hemoglobin. The chief difference is that myoglobin complexes one heme group per molecule, whereas hemoglobin complexes four heme groups.
See also heme.

myo-inositol. See inositol.

myokinase. An enzyme found in muscle and other tissues that catalyzes the reaction 2ADP ⇌ ATP + AMP.

myosin. A protein of molecular weight above 500,000 which is an essential component of muscular tissue and strongly affects its contractile properties.

myrcene. (7-methyl-3-methylene-1,6-octadiene). $C_{10}H_{16}$. A triply unsaturated aliphatic hydrocarbon found in oil of bay, verbena, hops, and others.
Properties: Yellow, oily liquid; pleasant odor. Bp 167C, d of 80% myrcene 0.806 (15.5/15.5C), refr index of 81% myrcene 1.471 (20C). Insolu-

ble in water; soluble in alcohol, chloroform, ether, glacial acetic acid. Combustible.

Use: Preparation of perfume chemicals, flavoring.

"Myrcene 85."[296] TM for a special grade of the triply unsaturated aliphatic hydrocarbon, $C_{10}H_{16}$, 7-methyl-3-methylene-1,6-octadiene. Minimum purity 75%. Balance mainly l-limonene.

myrcia oil. (bay oil; bayleaf oil). A yellow essential oil, slightly levorotatory.

Use: Bay rum, fragrances and flavors.

myricyl alcohol. See 1-triacontanol and 1-hentriacontanol.

myricyl palmitate. $C_{30}H_{61} \cdot C_{16}H_{31}O_2$ (approximately). A wax ester found in beeswax.

myristic acid. (tetradecanoic acid).
CAS: 544-63-8. $CH_3(CH_2)_{12}COOH$.
Properties: Oily, white, crystalline solid. Soluble in alcohol and ether, soluble in water, d 0.8739 (80C), bp 326.2C, 204.3C (20 mm), mp 54.4C, refr index 1.4310 (60C). Combustible.
Derivation: Fractional distillation of coconut acid and other vegetable oils, occurs in sperm oil.
Grade: Technical, 99.8%, FCC.
Use: Soaps, cosmetics, synthesis of esters for flavors and perfumes, component of food-grade additives.

myristin. (glyceryl trimyristate).
$C_3H_5(OOCC_{13}H_{27})_3$. A triglyceride occurring, usually to a small extent, in natural fatty oils.

myristoleic acid. (cis-tetradec-9-enoic acid).
$CH_3(CH_2)_3CH{:}CH(CH_2)_7COOH$.
Properties: Colorless liquid, mp −4C, found in fat of some seeds and in fish oil.

myristoyl peroxide. $(C_{13}H_{27}CO)_2O_2$.
Properties: Soft granules, 90% peroxide.

Hazard: Oxidizing materials, dangerous fire and explosion risk.
Use: Catalyst for vinyl type monomers.

myristyl alcohol. (1-tetradecanol).
CAS: 112-72-1. $C_{14}H_{29}OH$.
Properties: White solid, d 0.8355 at 20/20C, bp 264.1C (20 mm) 171.5C, mp 38C, flash p 285F (140.5C), bulk d 7.0 lb/gal (20C), insoluble in water, soluble in ether, partially soluble in ethanol . Combustible.
Grade: Technical.
Use: Organic synthesis, plasticizers, anti-foam agent, intermediate, perfume fixative for soaps and cosmetics, wetting agents and detergents, ointments and suppositories, shampoos, toothpaste and cold creams, specialty cleaning preparations.

myristyl chloride. See tetradecyl chloride.

myristyldimethylamine. $CH_3(CH_2)_{13}N(CH_3)_2$.
A liquid cationic detergent.
Use: Corrosion inhibitor, acid-stable.

myristyldimethylbenzylammonium chloride.
$C_{14}H_{29}(CH_3)_2C_6H_5CH_2NCl$. A quaternary ammonium compound. Free-flowing powder.
Use: Surfactant and detergent.

myristyl lactate. See "Ceraphyl."

myristyl mercaptan. See tetradecyl thiol.

myrrh. Gum resin obtained from various species of Balsamodendron and Cammiphora.
Use: Perfumery, incense, and toiletries.

myxin. (6-methoxy-1-phenazinol-5,10-dioxide).
CAS: 13925-12-7. $C_{13}H_{10}N_2O_4$.
Properties: Reddish, acicular crystals; mp 120C. Evolves heat near 150C and can decompose with explosive violence at this temperature, soluble in acetone.
Use: Bacteriostat and antifungal agent, antibiotic.

N

N. (1) Symbol for nitrogen. The names of certain compounds (such as N,N-dibutyl urea) contain this symbol as an indication that the group or groups appearing next in the name (i.e., the butyl groups in the example cited) are joined to the nitrogen atoms in the molecule. The molecular formula is N_2. (2) Mathematical symbol for Avogadro's Number.

N. Abbreviation for normal solution.
See normal (2).

n. Symbol for refractive index: n20/D is refractive index under standard conditions of temperature and wavelength (sodium D line).

n-. Abbreviation for normal.
See normal (1).

Na. Symbol for sodium.

"NA-22."[28] TM for 2-mercaptoimidazoline

$(\overline{CH_2CH_2NC(SH)}NH)$.
Properties: A white powder, d 1.42, mp above 195C.
Use: To accelerate vulcanization of neoprene.

nabam. (disodium ethylenebisdithiocarbamate).
CAS: 142-59-6.
$NaSSCNHCH_2CH_2NHCSSNa$.
Properties: Colorless crystals when pure, easily soluble in water.
Derivation: Addition of carbon disulfide to an alcoholic solution of ethylenediamine followed by neutralization with sodium hydroxide, or by reaction of ethylenediamine with carbon disulfide in aqueous sodium hydroxide.
Grade: 19% aqueous solution.
Hazard: Irritant to skin and mucous membranes, narcotic in high concentrations, use may be restricted.
Use: Plant fungicide, starting material for derivatives that are also pesticides.

"Nabor."[50] TM for basic dyes intended especially for dyeing acrylic fibers.

NAC. Abbreviation for National Agricultural Chemicals Association.

"Nacconate."[50] TM for a group of diisocyanates.
Use: In preparation of flexible, semi-rigid, and rigid urethane foams; elastomers; protective coatings, adhesives; spandex-type fibers, and textile finishes.

"Nacconol."[50] TM for a group of technical biodegradable linear alkyl sulfonate detergents.
See alkyl sulfonate, linear.
Use: Industrial cleaners, specialty compounding; air entrainment; emulsion polymerization; textile scouring agent; detergent intermediate.

"Naccotan." A[50] TM for an alkyl aryl sodium sulfonate.
Use: Retanning of chrome-tanned leather, useful in the dispersion of thick slurries.

"Naccufix."[50] TM for a copper organocomplex for fixing direct dyes on cellulose fibers.

"Nacelan."[50] TM for water-insoluble dyes dispersed in water and forming solid solutions in synthetic fibers.

nacre. (mother of pearl). A form of calcium carbonate secreted by the epithelial cells in the mantle of the oyster. The crystals are bonded by conchiolin ($C_{32}H_{98}N_2O_{11}$), the layers built up by excretion form pearls.

nacreous pigment. A pigment containing guanine crystals obtained from fish scales or skin, which produces a pearly luster. May be applied as surface coatings, as in simulated pearls, or incorporated in plastics. The pigment particle is generally a very thin platelet of high index of refraction. The crystals are readily oriented into parallel layers because of their shape. Being transparent, each crystal reflects only part of the incident light reaching it and transmits the remainder to the crystal below. The nacreous effect is obtained from the simultaneous reflection of light from the many parallel microscopic layers.

"Nadic Methyl Anhydride."[175] TM for methylbicyclo[2.2.1]heptene-2,3-dicarboxylic anhydride isomers. $C_{10}H_{10}O_3$.
Properties: Colorless to light-yellow, viscous liquid. Viscosity 175–225 cp (25C), refr index 1.500–1.506 (20C), d 1.200–1.250 (20/20C). Miscible in all proportions with acetone, benzene, naphtha, and xylene.
Use: Curing agent for epoxy resins; intermediate for polyesters, alkyd resins, and plasticizers; electrical laminating and filament-winding.

NADP. See nicotinamide adenine dinucleotide phosphate.

"Naflon."[28] TM for a perfluorosulfonic acid membrane.
Use: Manufacture of chlorine and caustic soda. It is a chemically stable ion-exchange resin.

NaK. (sodium-potassium alloy).
Properties: Soft, silvery solid or liquid; must be kept away from air and moisture. The liquid forms come under the class name potassium (or sodium), metallic liquid alloy.
Grade: (a) 78%, potassium, 22% sodium; mp −11C, bp 784C, d 0.847 (100C); (b) 56% potassium, 44% sodium; mp 19C, bp 825C, d 0.886 (100C).
Hazard: Ignites in air; explodes in presence of moisture, oxygen, halogens, acids. Store under kerosene. Use *dry* salt or soda ash to extinguish, *not* water, foam.
Use: Heat exchange fluid, electric conductor, for organic synthesis and catalysis.

"Nalan."[28] TM for durable water repellents used in textile industry.

"Nalco."[182] TM for a broad class of chemicals, organic or otherwise, employed in the treatment of water and hydrocarbons; also paper-making chemicals, cleaning compounds, combustion aids, weed and brush controls, lubricating and antilubricating compositions, apparatus, pumps, and mechanisms for proportioning chemicals.

Nalcoag[182]. TM for a colloidal silica available in particle sizes from 4–100 mμ.
Use: Reinforcing agent, antiblock agent, and dispersing agent.

"Nalcolyte." 671[182] TM for a synthetic high polymer used for clarifying industrial plant water and municipal water supplies. A coagulant behaving as a polyelectrolyte. Effective at concentrations of less than 1 ppm, also used in still lower concentrations as a filter aid.

naled. (1,2-dibromo-2,2-dichloroethyl dimethyl phosphate).   CAS: 300-76-5.
$(CH_3O)_2P(O)OC(Br)HCBr(Cl)_2$.
Properties: Pure compound is a solid, mp 26C, technical compound is a moderately volatile liquid, bp 110C (0.5 mm), insoluble in water, slightly soluble in aliphatic solvents, very soluble in aromatic solvents, hydrolyzes in water.
Hazard: Cholinesterase inhibitor, use may be restricted. Toxic by skin absorption. TLV: 3 mg/m³.
Use: Insecticide, acaricide.

"Nalgene."[559] TM for plastic laboratory ware for industry, research, and education. Made of polypropylene, polyethylene, "Teflon" FEP, polyallomer, and polycarbonate.

nalidixic acid. (USAN; 1-ethyl-7-methyl-1,8-naphthyridin-4-one-3-carboxylic acid).
CAS: 389-08-2. $C_{12}H_{12}N_2O_3$.
An antibacterial compound used in medicine.

"Nalkylene."[544] TM for biodegradable detergent alkylates made from straight-chain hydrocarbons. Available in molecular weight ranges to replace branched-chain dodecylbenzene or tridecylbenzene. The linear alkylbenzenes sulfonate readily with SO_3 or oleum by batch or continuous processes to form low-viscosity "soft" sulfonic acid; neutralization gives soft LAS slurries.

nalorphine. (N-allylnormorphine).
CAS: 62-67-9. $C_{19}H_{21}NO_3$.   The allyl ($-CH_2-CH{=}CH_2$) derivative of morphine. It is able to "antagonize" or neutralize most of the effects of narcotic drugs (morphine, codeine), but not those of other types of depressants.
Use: Biochemical research tool for studying the mechanism of narcotic action; also as an antidote for acute morphine poisoning.
See also narcotic.

NaMBT. See sodium MBT.

nameplate. The officially rated capacity of a chemical plant, as opposed to effective or actual maximum; the latter is usually 85–95% of nameplate.

name reaction. A chemical reaction, usually organic, that is commonly identified by the name of its discoverer(s), for example, Friedel-Crafts, Fischer-Tropsch, Claisen, Clemmensen, Willegerodt, Diels-Alder, etc. Many have important industrial applications.
See specific name.

nano-. Prefix meaning 10^{-9} unit (symbol n); 1 ng = 1 nanogram = 0.000000001 gram; 1 nanometer = 1 millimicron.

"Nanograde."[329] TM for a grade of chemical purity. Impurities guaranteed to be less than ten parts per trillion.

nanometer. (nm).   One billionth (10^{-9}) meter, equal to 1 millimicron or 10Å units.

napalm. An aluminum soap of a mixture of oleic, naphthenic, and coconut fatty acids.
Properties: Granular powder; mixed with gasoline it forms a sticky gel that is stable from −40 to 100C.

Hazard: Flammable, dangerous fire risk.
Use: Incendiary agent.

naphtha. (benzin).   CAS: 8030-30-6.
(1) *petroleum* (petroleum ether). A general term
applied to refined, partly refined, or unrefined
petroleum products and liquid products of natural gas, not less than 10% of which distil below
347F (175C) and not less than 95% of which
distil below 464F (240C) when subjected to distillation in accordance with the Standard Method
of Test for Distillation of Gasoline, Naphtha,
Kerosene, and Similar Petroleum Products
(ASTM D86); fp −73C, bp 30–60C, flash p −
57F (−50C), autoign temperature 550F (287C),
d 0.6.
Hazard: Flammable, dangerous fire risk, explosive
limits in air 1–6%.
Use: Source (by various cracking processes) of gasoline, special naphthas, petroleum chemicals, especially ethylene. Cracking for ethylene also produces propylene, butadiene, pyrolysis gasoline,
and fuel oil, source of synthetic natural gas. (a)
VM&P (Varnish Makers and Painters) (petroleum spirits, petroleum thinner)
  Any of a number of narrow-boiling-range fractions of petroleum with bp of approximately 93–
204C, according to the specific use; distillation
range 119–143C, d 07543, bulk d 6.280 lb/gal,
pour point approximately −70F (−56C), flash
p (TCC) 20F (−6.6C), autoign temperature 450F
(232C).
Hazard: Flammable, dangerous fire risk.
Use: Thinners in paints and varnish. (b) *blending*
A petroleum fraction with volatility similar to
the higher boiling fractions of gasoline. It is used
primarily in blending with natural gasoline to
produce a finished gasoline of specified volatility.
(c) *cleaners'* A dry-cleaning fluid derived from
petroleum and similar to Stoddard solvent, but
not necessarily meeting all its specifications; flash
p 100F (37.7C).
  (2)*coal-tar* (a) *heavy* (high-flash naphtha)
Properties: Deep amber to dark red liquid, a mixture of xylene and higher homologs, d 0.885–
0.970, bp 160–220C (approximately 90% at
200C), flash p 100F (37.7C), evaporation 303
minutes.
Derivation: From coal-tar by fractional distillation.
Hazard: Moderate fire risk. Toxic by ingestion,
inhalation, and skin absorption.
Use: Coumarone resins; solvent for asphalts, road
tars, pitches, etc.; cleansing compositions; process engraving and lithography; rubber cements
(solvent); naphtha soaps; manufacture of ethylene and acetic acid. (b) *solvent* (160 degree benzol).

Properties: A mixture of a small percentage of
benzene and toluene and xylene and higher homologs from coal-tar. (a) Crude: dark straw-colored liquid, (b) refined: water-white liquid; d (a)
0.862–0.892, (b) 0.862–0.872; bp (a) approximately 160C (80%), (b) approximately 160C
(90%); flash p (a) and (b) approximately 78F
(25.5C).
Derivation: From coal-tar by fractional distillation.
Grade: Dark straw, water-white.
Hazard: Flammable, dangerous fire risk.
Use: Solvent; xylene; cumene; nitrated, for incorporation in dynamite.

naphthacene. (tetracene; rubene).
CAS: 92-24-0. $C_{18}H_{12}$.   The molecule
consists of four fused benzene rings.
Properties: Orange solid, d 1.35, mp approximately 350C, not easily soluble, slight green
fluorescence in daylight.
Occurrence: In commercial anthracene and coal
tar.
Hazard: Explodes when shocked, reacts with oxidizing materials.
Use: Organic synthesis.

naphthalene. (tar camphor).   CAS: 91-20-3.
$C_{10}H_8$.

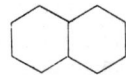

Properties: White, crystalline, volatile flakes;
strong coal-tar odor; soluble in benzene, absolute
alcohol, and ether; insoluble in water. D 1.145
(20/4C), mp 80.2C, bp 217.96C, flash p 176F
(80C), sublimes at room temperature, autoign
temperature 979F (526C). Combustible.
Derivation: (a) From coal-tar oils boiling between
200–250C (middle oil) by crystallization and distillation. (b) From petroleum fractions after various catalytic processing operations.
Grade: By melting point, 74C min (crude) to
greater than 79C (refined); scintillation (80–
81C).
Forms: Flakes, cubes, spheres, powder.
Hazard: Toxic by inhalation. TLV: 10 ppm in
air.
Use: Intermediate (phthalic anhydride, naphthol,
"Tertralin," "Decalin," chlorinated naphthalenes, naphthyl and naphthol derivatives, dyes),
moth repellent, fungicide, smokeless powder,
cutting fluid, lubricant, synthetic resins, synthetic tanning, preservative, textile chemicals,
emulsion breakers, scintillation counters, antiseptic.

α-naphthaleneacetic acid. (1-naphthylacetic acid). CAS: 86-87-3. $C_{10}H_7CH_2COOH$. A plant growth regulator.
Properties: White crystals, odorless, mp 132–135C. Soluble in acetone, ether, and chloroform; slightly soluble in water and alcohol.
Grade: Usually supplied in dilute form, either as a powder or liquid solution ready for use.
Hazard: Skin irritant.
Use: Inducing rooting of plant cuttings, spraying apple trees to prevent early drop, fruit thinner.

α-naphthaleneacetic acid, methyl ester. (MENA). $C_{10}H_7CH_2COOCH_3$. A plant growth regulator.
Use: Delaying sprouting of potatoes, weed control, thinning of peaches, etc.

naphthalene, chlorinated. See chlorinated naphthalene.

naphthalenediamine. See naphthylenediamine.

1,5-naphthalene diisocyanate. $C_{10}H_6(NCO)_2$. White to light yellow, crystalline solid; mp 127–131C.
Hazard: Irritant.
Use: Manufacture of polyurethane solid elastomers.

naphthalene-1,5-disulfonic acid. See Armstrong's acid.

naphthalene-2,7-disulfonic acid. CAS: 92-41-1. $C_{10}H_6(SO_3H)_2$.
Properties: White, crystalline solid; soluble in water. Combustible.
Derivation: Sulfonation of naphthalene at high temperature and separation from 2,6-isomer.
Use: Intermediate for dyes.

α-naphthalenesulfonic acid. (1-naphthalenesulfonic acid.). CAS: 85-47-2. $C_{10}H_7SO_3H \cdot HOH$.
Properties: Deliquescent crystals; soluble in water, alcohol, and ether; mp 90C. Combustible.
Derivation: Interaction of naphthalene and sulfuric acid.
Use: Starting point in the manufacture of α-naphthol, α-naphthalene sulfonic acid, α-naphthylaminesulfonic acid, solvent (Na salt) for phenol in the manufacture of disinfectant soaps.

β-naphthalenesulfonic acid. (2-naphthalenesulfonic acid.). CAS: 120-18-3. $C_{10}H_7SO_3H$ or $C_{10}H_7SO_3H \cdot HOH$.
Properties: Non-deliquescent, white plates; mp

124–125C; soluble in water, alcohol, and ether. Combustible.
Derivation: Sulfonation of naphthalene.
Use: Starting point in the manufacture of β-naphthol, β-naphtholsulfonic acid, β-naphthylaminesulfonic acid, etc.

2-naphthalenethiol. See "RPA."

1,3,6-naphthalenetrisulfonic acid, trisodium salt. $C_{10}H_5(SO_3Na)_3$. Fine buff crystals.
Use: Diazo type stabilizer.

1,8-naphthalic acid anhydride. $C_{12}H_6O_3$.
Properties: Light tan powder, mp 268–270C.
Use: Dyestuffs, organic synthesis in general.

"Naphthanil."[28] TM for a series of dye bases. Prior to coupling the bases must first be diazotized to form the diazo salt. Also represents a series of diazo pigments.
Use: Widely used on cotton and rayon textiles.

naphtha, petroleum. See naphtha (1).

naphtha, solvent. See naphtha (2b).

naphtha, VM&P. See naphtha (1a).

naphthene. See cycloparaffin. The term naphthene is misleading and obsolescent.

naphthenic acid. Any of a group of saturated higher fatty acids derived from the gas-oil fraction of petroleum by extraction with caustic soda solution and subsequent acidification. Gulf and West-coast crudes are relatively high in these acids. The commercial grade is a mixture, usually of dark color and unpleasant odor, corrosive to metals. The chief use of naphthenic acids is in the production of metallic naphthenates for paint driers and cellulose preservatives. Other uses are as solvents, detergents, rubber reclaiming agent, etc.

naphthionic acid. (1-naphthylamine-4-sulfonic acid; 1-aminonaphthalene-4-sulfonic acid; 4-amino-1-naphthalenesulfonic acid).

Properties: White crystals or powder, soluble in water, slightly soluble in alcohol and ether.
Derivation: Heating equimolar amounts of

α-naphthylamine and sulfuric acid at 10–15 mm (several hours).
Use: Intermediate for azo dyes, e.g., Congo Red.

α-naphthol. (1-naphthol; 1-hydroxynaphthalene). CAS: 90-15-3. $C_{10}H_7OH$.

OH

Properties: Colorless or yellow prisms or powder, disagreeable taste. Soluble in benzene, alcohol, and ether; insoluble in water. D 1.224 (4C), 1.0954 (95/4C), mp 96C, bp 278C, volatile in steam, sublimes, refr index 1.6206 (98.7C). Combustible.
Derivation: By fusing sodium α-naphthalene sulfonate and caustic soda. The melt is decomposed with hydrochloric acid and distilled.
Hazard: Toxic by ingestion and skin absorption.
Use: Dyes, organic synthesis, synthetic perfumes.

β-naphthol. (2-naphthol; 2-hydroxynaphthalene). CAS: 135-19-3. $C_{10}H_7OH$.

OH

Properties: White, lustrous, bulky leaflets or white powder. Darkens with age, faint phenol-like odor, stable in air but darkens on exposure to sunlight, d 1.217, mp 121.6C, bp 285C, flash p 307F (152.7C). Soluble in alcohol, ether, chloroform, glycerol, oils, and alkaline solutions; almost insoluble in water. Combustible.
Derivation: By fusing sodium β-naphthalene sulfonate with caustic soda. The product is distilled in vacuo and then sublimed.
Grade: Technical, sublimed, resublimed.
Hazard: See α-naphthol.
Use: Dyes; pigments; antioxidants for rubber, fats, oils; insecticides; synthesis of fungicides; pharmaceuticals; perfumes; antiseptic.

naphthol AS. See β-hydroxynaphthoic anilide.

β-naphthol benzoate. See benzonaphthol.

3-naphthol-2-carboxylic acid. See 3-hydroxy-2-naphthoic acid.

Naphthol Green B. A dye used in crystallizing solar salt, it increases the evaporation rate by added absorption of energy; 5 ppm is said to increase salt production 15–20%.

β-naphthol methyl ether. See β-naphthyl methyl ether.

β-naphthol sodium. See sodium-β-naphtholate.

naphtholsulfonic acid. Any of several sulfonated aromatic acids derived from α- or β-naphthol or naphthalene and used as azo dye intermediates.

1,2-naphthoquinone. (β-naphthoquinone). CAS: 524-42-5. $C_{10}H_6O_2$.
Properties: Yellow crystals, mp 120C (decomposes); soluble in ether, benzene, alcohol.
Hazard: Irritant.
Use: Chemical reagent and intermediate.

1,4-naphthoquinone. (α-naphthoquinone). CAS: 130-15-4. $C_{10}H_6O_2$.
Properties: Yellow powder; odor like benzoquinone; mp 123–126C; sublimes at 100C. Slightly soluble in water; soluble in ethanol, ethyl ether, chloroform, benzene, and acetic acid. Combustible.
Hazard: Irritant.
Use: Polymerization regulator for rubber and polyester resins, synthesis of dyes and pharmaceuticals, fungicide, algicide.

naphthoquinone oxime. See 1-nitroso-2-naphthol; 2-nitroso-1-naphthol.

1,2-naphthoquinone-4-sulfonic acid. (β-naphthoquinone-4-sulfonic acid). $C_{10}H_5(O)_2SO_3H$.
Derivation: Oxidation with nitric acid of 2-amino-1-naphthol-4-sulfonic acid or 1-amino-2-naphthol-4-sulfonic acid.
Use: Dye intermediate, identification of sulfonamide derivatives.

naphthoresorcinol. See 1,3-dihydroxynaphthalene.

β-naphthoxyacetic acid. (2-naphthoxyacetic acid). CAS: 120-23-0. $C_{10}H_7OCH_2COOH$. A plant growth regulator.
Properties: Crystals; mp 156C; soluble in water, alcohol, acetic acid.
Use: Rooting clippings, inhibits early fall of fruit, growth promoter.

1-naphthylacetic acid. See naphthaleneacetic acid.

α-naphthylamine. (1-naphthylamine). CAS: 134-32-7. $C_{10}H_7NH_2$.
Properties: White crystals becoming red on exposure to air, soluble in alcohol and ether, slightly soluble in water, flash p 157C, d 1.13, mp 50C, bp 301C. Combustible.

Derivation: Reduction of α-nitro-naphthalene with iron and hydrochloric acid. The mass is then mixed with milk of lime and distilled.

Method of purification: Crystallization.

Hazard: Toxic, especially if containing the beta isomer; a carcinogen (OSHA).

Use: Dyes and dye intermediates, agricultural chemicals.

β-naphthylamine. (2-naphthylamine).
CAS: 91-59-8. $C_{10}H_7NH_2$.

Properties: White to reddish, lustrous leaflets; soluble in hot water, alcohol, ether. Commercial: mp 109.5C, d 1.061 (98.4C), bp 306C. Combustible.

Derivation: From β-naphthol by heating in an autoclave with ammonium sulfite and ammonia (Bucherer reaction).

Method of purification: Distillation.

Hazard: Toxic by ingestion, inhalation, skin absorption; a carcinogen (OSHA).

α-naphthylamine hydrochloride.
$C_{10}H_7NH_2 \cdot HCl$.

Properties: White to gray, crystalline powder; soluble in water, alcohol, and ether.

Derivation: By the action of hydrochloric acid on α-naphthylamine.

Use: Dyes, organic synthesis.

naphthylaminesulfonic acid. Any of several sulfonated aromatic acids derived from α- or β-naphthylamine and used as azo dye intermediates.

o-2-naphthyl-m-N-dimethylthiocarbanilate.
See tolnaftate.

1,5-naphthylenediamine. (1,5-diaminonaphthalene). CAS: 2243-62-1. $C_{10}H_6(NH_2)_2$.

Properties: Colorless crystals, mp 190C, bp sublimes, soluble in alcohol and hot water, very sparingly soluble in cold water. Combustible.

Derivation: (a) By the reduction of α-dinitronaphthalene, (b) by heating dihydroxynaphthalene with aqueous ammonia.

Use: Organic synthesis.

1,8-naphthylenediamine. (1,8-diaminonaphthalene). $C_{10}H_6(NH_2)_2$.

Properties: Colorless crystals, mp 66C, bp 205C (12 mm), d 1.12, refr index 1.68, soluble in alcohol, slightly soluble in water.

Derivation: Reduction of 1,8-dinitronaphthalene with PI_3.

Use: Lube oil antioxidant, analytical reagent.

N-α-naphthylethylenediamine dihydrochloride.
$C_{10}N_7NHCH_2CH_2NH_2 \cdot 2HCl$.

Properties: Colorless crystals, soluble in water.

Use: Reagent for the quantitative determination of sulfa drugs, for the detection of nitrogen dioxide in air.

β-naphthyl ethyl ether. (nerolin).
$C_{10}H_7OC_2H_5$.

Properties: White crystals, orange-blossom odor, congealing p 35C, soluble in 5 parts of 95% alcohol. Combustible.

Derivation: Interaction of β-naphthol and ethanol in presence of sulfuric acid.

Use: Perfumes, soaps, flavoring.

1-naphthyl-N-methylcarbamate. See carbaryl.

α-naphthylmethyl chloride. See 1-chloromethylnaphthylene.

β-naphthyl methyl ether. (β-naphthol methyl ether; 2-methoxynaphthalene; methyl naphthyl ether). $C_{10}H_7OCH_3$.

Properties: White, crystalline scales; soluble in alcohol and ether; insoluble in water; mp 72C; bp 274C. Combustible.

Derivation: (a) By heating β-naphthol and methanol in presence of sulfuric acid. (b) By methylating b-naphthol with dimethyl sulfate.

Use: Perfumery (soaps).

α-naphthylphenyloxazole. (NPO; ANPO; 2-(1-naphthyl)-5-phenyloxazole).
$C_{19}H_{13}NO$.

Properties: Fluorescent yellow needles, mp 104–106C.

Grade: Scintillation.

Use: Scintillation counter or wave-length shifter in solution scintillators.

N-1-naphthylphthalamic acid.
$C_{10}H_7NHCOC_6H_4COOH$.

Properties: Crystalline solid, mp 185C. Almost insoluble in water; slightly soluble in acetone, benzene, and ethanol. Not stable in solutions above pH 9.5 nor at temperatures above 180C, noncorrosive, do not store near seeds or fertilizers. Combustible.

Use: Selective pre-emergence herbicide.

α-naphthylthiourea. (ANTU). CAS: 86-88-4.
$C_{10}H_7NHCSNH_2$.

Properties: Odorless, gray powder; mp 198C; insoluble in water and only very slightly soluble in most organic solvents.

Derivation: From α-naphthylthiocarbamide and alkali or ammonium thiocyanate.

Hazard: Toxic by ingestion. TLV: 0.3 mg/m³ of air.

Use: Rodenticide.

naphite. See trinitronaphthalene.

Naples yellow. See lead antimonate.

"Napon."[53] TM for a starch derivative containing both hydrophilic and hydrophobic groups.
Use: Emulsifying agent.

naptalam. (α-napthylphthalamic acid).
  CAS: 132-66-1. $C_{18}H_{13}NO_3$.
  Properties: Colorless crystals, mp 203C, d 1.40. Soluble in alkaline solutions; slightly soluble in alcohol, benzene, and acetone.
  Use: Analytical reagent (thorium, zirconium); herbicide.

narcotic. (1) A natural, semisynthetic, or synthetic nitrogen-containing heterocyclic drug that characteristically effects sleep (coma) and relief of pain, but also may result in addiction, i.e., a biochemical situation in which the body tissues become so adapted to the drug that they can no longer function normally without it. Natural narcotics are the plant products morphine and codeine (constituents of opium), both of which are alkaloids. Opium is obtained from the seed of the oriental poppy, Papaver somniferens. Semisynthetic narcotics are modifications of the morphine molecule, e.g., diacetylmorphine (heroin), ethylmorphine ("Dionin"), and methyldihydromorphine (metopon). Synthetic narcotics are meperidine, ethadone, and phenazocine (there are other addictive agents that are not narcotics). The sale of narcotics is strictly controlled by law in the US. (2) Inducing sleep or coma. Many chemicals that are not narcotics in essence (1) have this property (chloroform, barbiturates, benzene, etc.).

l-α-**narcotine.** See noscapine.

naringin. (naringenin-7-rhamnoglucoside; naringenin-7-rutinoside; aurantiin).
  CAS: 10236-47-2. $C_{27}H_{32}O_{14}$.
  Properties: A flavanone glycoside (bioflavonoid), crystals, mp 171C, bitter taste. Soluble in acetone, alcohol, warm acetic acid and warm water.
  Source: Extracted from flowers and rind of grapefruit and immature fruit.
  Use: Beverages, sweetener research.

nascent. Descriptive of the abnormally active condition of an element, for example, the atomic oxygen released from hydrogen peroxide and sulfur atoms evolved from thiuramsulfide accelerators. The term is now obsolescent.

National Academy of Sciences--National Research Council. A private, nonprofit organization of scientists devoted to the expansion of science and its use for the general welfare. The Academy was established in 1863, in part to act as advisor to the Federal Government on scientific matters; the Council was established in 1916, its members being appointed by the President of the Academy. Its headquarters are in Washington, DC.

National Agricultural Chemicals Association. (NAC). An organization of chemical manufacturers concerned with the safety aspects of pesticides. Its Pesticide Safety Team Network can be called upon to decontaminate accidental spills of the more toxic pesticides.

"National" Boronated Graphites.[214] TM for a series of nuclear grades of graphite with unique characteristics of the combination of neutron absorption with heat dissipation, accomplished by dispersing boron as boron carbide particles in graphite. Safe to handle before and after exposure.
  Use: Moderators, reflectors, thermal columns, shielding, control rods, fuel elements, crucibles, and molds.
  See also graphite.

National Fire Protection Association. (NFPA). An organization devoted to promoting knowledge of fire protection methods. For many years its publication (the NFPA Handbook) has been the accepted standard for all matters relating to combustion and flammable materials, firefighting methods, safety, and protection of property. Its headquarters are at Batterymarch Park, Quincy, MA, 02269.

National Institute for Occupational Safety and Health. A federal agency under the Department of Health and Human Services, Public Health Service. It is responsible for investigating the toxicity of workroom environments and all other matters relating to safe industrial practice.

National Research Council. See National Academy of Sciences.

National Resources Defense Council. (NRDC). A private environmental advisory group founded in 1970 whose function is to point out serious environmental hazards and to oversee the enforcement of regulations pertaining to them. It was influential in postponement of the breeder reactor and in recognition of the chlorofluorocarbon-ozone problem. It operates primarily through the courts.

natrium. The Latin name for sodium, hence the symbol Na in chemical nomenclature.

natrolite. CAS: 1318-95-2.
$Na_2Al_2Si_3O_{10} \cdot 2HOH$.   A mineral of the zeolite group.
Properties: Colorless or white to gray, yellow, greenish, or red; d 2.2–2.25; Mohs hardness 5–5.5.
See zeolite.

natron. A complex salt found in dry lake beds of Egypt. Originally an ingredient of ceramic glazes. It has the percentage composition: sodium carbonate 4.9, sodium bicarbonate 12.6, sodium chloride 30.6, sodium sulfate 20.6, silica 10, calcium carbonate 2, magnesium carbonate 1.9, alumina 0.7, iron oxide 0.3, water 4.7 and organic matter 11.7.

"Natron 86."[53]   TM for a cationic polyelectrolyte.
Use: Water purification, clarification of paper mill effluents, removal of colored waste and other solids from water.

"Natsyn."[265]   TM for a series of cis-1,4-polyisoprene synthetic rubbers essentially duplicating the chemical structure of natural rubber.

Natta catalyst. A stereospecific catalyst made from titanium chloride and aluminum alkyl or similar materials by a special process which includes grinding the materials together to produce an active catalyst surface.
See also Ziegler catalyst.

Natta, Giulio. (1903-1979) An Italian chemist born in Imperia on the Riviera, co-recipient (with Karl Ziegler) of the Nobel prize in 1963 for his fundamental work on catalytic polymerization. In 1954 he developed isotactic polypropylene at his laboratory at the Polytechnic Institute of Milan, which led to wide application of various stereospecific polymers with organometallic catalysts such as triethylaluminum. He was for many years consultant for the Montecatini chemical firm. The researchers of Natta, together with those of Ziegler, made possible the chemical manipulation of monomers to form specifically ordered 3-dimensional polymers having predetermined properties, to which the term "tailor-made" is often applied.

natural. Descriptive of a substance or mixture which occurs in nature; the opposite of synthetic or manmade. Elements from 1–92 are natural substances that may occur in either the free or combined state. The transuranium elements and all artificial isotopes of other elements are synthetic. Many mixtures occur naturally, e.g., petroleum, shale oil, wood, metallic ores, natural gas; others are manmade modifications of natural compounds, e.g., glass, cement, paper, gasoline. Such materials may be considered to be semisynthetic. The term "synthetic natural" is often applied to synthesized compounds that are identical with the natural substance; for example, synthetic natural gas and synthetic natural rubber. The term "natural product" is defined as any organic compound formed by living organisms, "natural gasoline" has a specialized meaning.
See also synthesis, natural gas, biomass, gasoline.

natural gas. A mixture of low molecular weight hydrocarbons obtained in petroleum-bearing regions throughout the world. Its composition is 85% methane, 10% ethane, the balance being made up of propane, butane, and nitrogen. In the US it occurs chiefly in the southwestern states and Alaska. An as yet unexploited source of natural gas under extremely high pressure (so-called geopressurized gas) exists in Texas and Louisiana at depths of 15–20,000 ft. The tremendous pressures involved present formidable engineering problems. Natural gas is classed as a simple asphyxiant. It should not be confused with natural gasoline. About 3% of the natural gas consumed in the US is used as feedstocks by the chemical industries.
Properties: Colorless flammable gas or liquid, almost odorless, autoign temperature 900–1100F, heating value 1000 Btu/ft³. A warning odor is added to household fuel gas as a safety precaution.
Hazard: Flammable, dangerous fire and explosion risk; explosive limits in air 3.8–17%.
Use: Fuel and cooking gas, ammonia synthesis, formaldehyde and other petrochemical feedstocks, source of synthesis gas and methanol.
See also liquefied petroleum gas, synthetic natural gas.

natural gasoline. See under gasoline.

natural product. See under natural.

"Naugapol."[248]   TM for a series of butadiene-styrene copolymers which have received special processing for minimum water-soluble salts and low ash content. Included are "Naugapol K" series which are master batches of high styrene resin ("Kralac" A-EP) and "Naugapol" elastomers.
Use: Wire and cable insulation, mechanical goods, adhesives and cements. The "K" series are used for shoe soles, floor tile, and wire and cable insulation.

"Naugawhite."[248]   TM for an alkylated phenol antioxidant.
Use: General-purpose antioxidant for rubber and latex in foam sponge, tire carcass, refrigerator

gaskets, footwear, proofing, wire insulation, and sundries.

naval stores. Historically, the pitch and rosin used on wooden ships. The term now includes all products derived from pine wood and stumps, including rosin, turpentine, pine oils, tall oil, and its derivatives.

"Navane."[299] TM for the *cis* isomer of N,N-dimethyl- 9[3-4-methyl-1-piperazinyl-propylidene]-thioxanthene-2-sulfonamide. An antipsychotic drug (thiothixene) said to be as effective as most of the phenothiazines. Approved by FDA.

Nazarov cyclization reaction. Synthesis of cyclopentenones by the acid-catalyzed electrocyclic ring closure of divinyl or allylvinyl ketones available by hydration of divinylacetylenes.

Nb. Symbol for niobium.

NBA. Abbreviation for N-bromoacetamide.

NBR. Abbreviation for nitrile-butadiene rubber.

NBS. Abbreviation for N-bromosuccinimide.

NC. Abbreviation for nitrocellulose.

NCI. Abbreviation for National Cancer Institute.

NCS. Abbreviation for N-chlorosuccinimide.

Nd. Symbol for neodymium.

NDGA. Abbreviation for nordihydroguaiaretic acid.

"N-Dure."[197] TM for a heavy liquid solution containing urea and formaldehyde.
Use: Manufacture of solid fertilizer compositions in which the nitrogen is slowly available to growing plants.

Ne. Symbol for neon.

neatsfoot oil.
Properties: A fixed pale yellow oil with a peculiar odor; soluble in alcohol, ether, chloroform, and kerosene. D 0.916, saponification value 194–199, flash p 470F (243C), autoign temperature 828F (442C), iodine value 70. Combustible.
Derivation: By boiling in water the shin bones and feet (deprived of hoofs) of cattle and separating the oil from the fat obtained.
Grade: 15, 20, 30, 40F cold test (the temperature in degrees Fahrenheit at which stearin separates).

Hazard: Subject to spontaneous heating.
Use: Leather industry for fat-liquoring and softening, lubricant, oiling wool.

Neber rearrangement. Formation of α-amino ketones by treatment of sulfonic esters of ketoximes with potassium ethoxide, followed by hydrolysis.

neburon. Generic name for 1-n-butyl-3-(3,4-dichlorophenyl)-1-methylurea.

"Neelium."[41] TM for a synthetic rubber coating of the neoprene type which can be applied in films up to 20 mils. It is supplied in a two-package system of liquid and accelerator.

Nef reaction. Formation of aldehydes and ketones from primary and secondary nitroparaffins, respectively, by treatment of their salts with sulfuric acid.

Nef synthesis. Addition of sodium acetylides to aldehydes and ketones to yield acetylenic carbinols; occasionally and erroneously referred to as the Nef reaction.

negatol. A condensation product of m-cresolsulfonic acid with formaldehyde. A polymerized dihydroxydimethyldiphenylmethanedisulfonic acid. It is dispersible in water, forming colloidal solutions which are very acidic. The pH of a 5% dispersion is approximately 1.0.
Use: Medicine (antiinfective).

"Nelio."[420] TM for a series of pine products (rosin, pinene, turpentine, etc.) characterized by extreme purity and uniformity.

nematic. A linear molecular structure occurring in some liquid crystals and characterized by a thread-like appearance under a polarizing microscope.

nematocide. An agent which is destructive to soil nematodes (roundworms or threadworms).

Nencki reaction. The ring acylation of phenols with acids in the presence of zinc chloride, or the modification of the Friedel-Crafts alkylation-acylation procedure by substitution of ferric chloride for aluminum chloride.

Nenitzescu indole synthesis. Hydrogenative acylation of cycloolefins with acid chlorides in the presence of aluminum chloride; with five- and six-membered rings no change in ring size occurs, but with seven-membered rings, rearrangement takes place with formation of a cyclohexane derivative.

neo-. (1) A prefix meaning new and designating a compound related in some way to an older one, e.g., neoprene. (2) A prefix indicating a hydrocarbon in which at least one carbon atom is connected directly to four other carbon atoms; as, neopentane, neohexane.

"Neochel."[288] TM for a wetting agent used in copper plating. A liquid replacement for Rochelle salt with proprietary additives.

"NeoCryl."[516] TM for acrylic polymers and copolymers.
Use: Protective and decorative coatings in the leather, floor polish, paper, paint, textile, and adhesive fields.

neodecanoic acid. CAS: 26896-20-8.
$C_9H_{19}COOH$. Clear, colorless liquid in 97% purity; available commercially. Its derivatives are especially effective as paint driers and are being widely used. Applications as plasticizers and lubricants are also possible.

"Neodol 25."[125] TM for a C_2—C_{15} linear, primary alcohol.
Use: Manufacture of biodegradable surfactants, dispersants, solvents, emulsifiers and chemical intermediates.

neodymia. See neodymium oxide.

neodymium. CAS: 7440-00-8. Nd.
Metallic element having atomic number 60, group IIIB of the Periodic Table, aw 144.24, valence = 3. A rare-earth element of the lanthanide (cerium) group. There are seven isotopes.
Properties: Soft, malleable, yellowish metal; tarnishes easily. D 7.0, mp 1024C, bp approximately 3030C, ignites to oxide (200–400C), liberates hydrogen from water, soluble in dilute acids. High electrical resistivity, paramagnetic. Readily cut and machined. Store under mineral oil or inert gas to prevent tarnish and corrosion.
Derivation: Monazite, bastnasite, allanite. Ores are cracked by heating with sulfuric acid.
Forms available: Ingots, rods, sheet, powder to 99.9+% purity.
Hazard (Salts): Irritant to eyes and abraded skin.
Use: Neodymium salts, electronics, alloys, colored glass, especially in astronomical lenses and lasers, to increase heat resistance of magnesium, metallurgical research, yttrium-garnet laser dope, gas scavenger in iron and steel manufacture.
See also didymium.

neodymium acetate. $Nd(C_2H_3O_2)_3 \cdot xHOH$.
Properties: Pink powder, soluble in water.

neodymium ammonium nitrate.
$Nd(NO_3)_3 \cdot 2NH_4NO_3 \cdot 4HOH$.
Properties: Pink crystals, soluble in water, technical grade contains 75% neodymium salt, principal impurities praseodymium and samarium compounds.

neodymium carbonate. $Nd(CO_3)_3 \cdot xHOH$.
Properties: Pink powder, insoluble in water, soluble in acids.
Grade: 75%, 95%, and 99% neodymium salt.

neodymium chloride. (a) $NdCl_3$,
(b) $NdCl_3 \cdot 6HOH$.
Properties: (a) Violet crystals, d 4.134 (25C), mp 784C, bp 1600C, very soluble in water, soluble in alcohol, insoluble in ether and chloroform; (b) Red crystals, mp 124C, loses 6HOH at 160C, very soluble in water and alcohol.
Grade: 75%, 95%, 99%, and 99.9%.

neodymium fluoride. NdF_3.
Properties: Pink powder, mp 1410C, bp 2300C, insoluble in water.
Grade: 65%, 75%, 99% and 99.9%.
Hazard: Irritant.

neodymium nitrate. $Nd(NO_3)_3 \cdot 6HOH$.
Properties: Pink crystals, very soluble in water, soluble in alcohol and in acetone.
Grade: 75%, 95%, 99%, and 99.9%.
Hazard: Possible explosion risk.

neodymium oxalate. $Nd_2(C_2O_4)_3 \cdot 10HOH$.
Properties: Pink powder, insoluble in water, slightly soluble in acids.
Grade: 75%, 95% and 99%.

neodymium oxide. (neodymia). Nd_2O_3.
Properties: Pure product a blue-gray powder, technical grade a brown powder, d 7.24, mp 2270C, insoluble in water, soluble in acids, hygroscopic, absorbs carbon dioxide from the air.
Grade: 65%, 75%, 85%, 95%, 99%, and 99.9% oxide.
Use: (65%): To counteract color of iron in glass. (Purified grade): Ceramic capacitors, coloring glass, refractories, carbon arc-light electrodes, color TV tubes, dehydrogenation catalyst.

neodymium sulfate. $Nd_2(SO_4)_3 \cdot 8HOH$.
Properties: Pink crystals, d 2.85, mp 800C (decomposes) soluble in cold water, sparingly soluble in hot water.
Grade: 75%, 99%, and 99.9%.

"Neoflex."[125] TM for a series of linear primary alcohols, from C_6 through C_{11}, including blends.
Properties: (C_7): Colorless liquid, d 0.8217 (25/25C), distillation range 174–182C.

Hazard: Combustible, moderate fire risk.

Use: Solvent, reaction intermediate, esterifying agent.

neohexane. (2,2-dimethylbutane).
CAS: 75-83-2. C_6H_{14}.

$$CH_3-\overset{\overset{\displaystyle CH_3}{|}}{\underset{\underset{\displaystyle CH_3}{|}}{C}}-CH_2-CH_3$$

Properties: Colorless, volatile liquid. Bp 49.7C, refr index 1.3659 (25C), d 0.6570 (25C), fp −99.7C, flash p −54F (−47C), octane rating 100+, autoign temperature 797F (425C).

Derivation: By the thermal or catalytic union (alkylation) of ethylene and isobutane, both recovered from refinery gases.

Grade: 95%, 99%, research.

Hazard: Highly flammable, dangerous fire and explosion risk, explosion limits 1.2–7%.

Use: Component of high-octane motor and aviation fuels, intermediate for agricultural chemicals.

"Neol."[594] TM for neopentyl glycol.

"Neolan."[463] TM for metal-complex dyes for wool and polyamide fibers.

"Neolith."[434] TM for pellet-type, free-flowing, dust-free litharge, designed for ease of handling.
Use: Ceramic industry.

"Neolyn."[266] TM for a series of soft-to-medium hard modifying rosin-derived alkyd resins. Available in solution and/or solid forms of various grades.
Use: Adhesives, lacquers, organosols, plastisols and floor tile.

neomycin. CAS: 1404-04-2. An antibiotic complex obtained from *Streptomyces fradiae*; it is soluble in water and methanol, but insoluble in most organic solvents. It consists of three component substances, all of which function as anti-infective agents; some derivatives have fungicidal properties. The three types are as follows: A (also called neamine): $C_{12}H_{26}N_4O_6$, B: $C_{23}H_{46}N_6O_{13}$ (also available as hydrochloride and sulfate); and C: $C_{23}H_{46}O_{13}$.

neon. CAS: 7440-01-9. Ne. Inert element of atomic number 10, noble gas group of the periodic system, aw 20.179. Three stable isotopes.

Properties: Colorless, odorless, tasteless gas; does not form compounds, but ionizes in electric discharge tubes. Liquefies at −245.92C, fp −248.6C, d 0.6964 (air = 1), sp vol 11.96 cu ft/lb (21C, 1 atm), slightly soluble in water, an asphyxiant gas. Noncombustible.

Derivation: By fractional distillation of liquid air. It constitutes 0.0012% of normal air.

Grade: Technical, highest purity.

Use: (Gas): Luminescent electric tubes and photoelectric bulbs, electronic industry, high-voltage indicators, lasers; (liquid): cryogenic research.

neonicotine. See anabasine.

neopentane. (2,2-dimethylpropane; tetramethylmethane). CAS: 463-82-1.
C_5H_{12} or $C(CH_3)_4$. Present in small amounts in natural gas.

$$CH_3-\overset{\overset{\displaystyle CH_3}{|}}{\underset{\underset{\displaystyle CH_3}{|}}{C}}-CH_3$$

Properties: Colorless gas or very volatile liquid, bp 9.5C, d 0.591 (20/4C), fp −19.5C, soluble in alcohol, insoluble in water, flash p −85F (−65C), autoign temperature 842F (450C).

Grade: Technical 95%, pure 99%, research 99.9%.

Hazard: Highly flammable, dangerous fire risk, explosive limits in air 1.4–7.5%.

Use: Research, butyl rubber.

neopentanoic acid. See trimethylacetic acid.

neopentyl glycol. (2,2-dimethyl-1,3-propanediol). CAS: 126-30-7. $HOCH_2C(CH_3)_2CH_2OH$.

Properties: White, crystalline solid. Boiling range 95% between 204–208C, mp 120–130C, d 1.066 (25/4C), partially soluble in water, miscible in alcohol and ether.

Use: Polyester foams, insect repellent alkyd modifier, plasticizers, urethanes, synthetic lubricants.

neoplasm. An abnormal growth of tissue in the body which may or may not be malignant.
See also antineoplastic.

neoprene. (polychloroprene). CAS: 126-99-8. $(CH_3ClC:CHCH_3)_n$. A synthetic elastomer available in solid form, as a latex or as a flexible foam.

Properties: Vulcanized with metallic oxides rather than sulfur; d 1.23; resistant to oils, oxygen, ozone, corona discharge, and electric current; an isocyanate-modified form has high flame resistance. Combustible, but less so than natural rubber.

Use: (Solid): mechanical rubber products, lining oil-loading hose and reaction equipment, adhesive cement, binder for rocket fuels, coatings for electric wiring, gaskets and seals; (liquid): specialty items made by dipping or electrophoresis from the latex; (foam): adhesive tape to replace metal fasteners for automotive accessories, seat cushions, carpet backing, sealant.

neopyrithiamine. See pyrithiamine.

"NeoRez."[516]   TM for polystyrene copolymer emulsions and urethane prepolymers.
Use: Floor polish, textile and paper fields, leather finishes, adhesives, and protective coatings.

Neosalvarsan. (sodium 3:3'-diamino-4:4'-dihydroxyarsenobenzene formaldehyde sulfoxylate). CAS: 457-60-3. $C_{13}H_{14}O_4N_2SAs_2Na$.
Use: Treatment against syphilis.

"Neosol."[125]   TM for ethanol based on a formulation approved by the Bureau of Internal Revenue.

Neosporin. Proprietary formulation of polymyxin B sulfate, neomycin sulfate, and gramicidin.
Use: Ocular antibiotic.

neotridecanoic acid. $C_{12}H_{25}COOH$.
Colorless liquid of 97% purity. Suggested for plasticizers, lubricants, paint driers, fungicides, cosmetics, alkyd resins.

"Neowet."[159]   TM for a complex polyethylene ether non-ionic wetting agent.
Use: General wetting purposes, particularly useful with enzymatic desizing agents, scouring all types of textile fabrics, good for use with resin finishes.

"Neozone."[28]   TM for a group of rubber antioxidants. An N-phenyl-α-naphthylamine. d N-phenyl-β-naphthylamine. C Fusion of "Neozone" A and toluene-2,4-diamine.

"Neozymes."[159]   TM for a series of pancreatic and bacterial enzymes, proteolytic and amylolytic activity for low and high temperatures, continuous or batch desizing.
Grade: Regular, HT, 3-LC and 5-LC, available in powder and liquid form.
Use: Desizing fabrics coated with starch or gelatin.

nephelite. (nepheline). $(Na,K)(Al,Si)_2O_4$.
Essentially a silicate of sodium, found in silica-poor igneous rocks.
Properties: Colorless, white, yellowish; luster vitreous to greasy; Mohs hardness 5.5–6; d 2.55–2.65.

Occurrence: USSR, Ontario, Norway, South Africa, Maine, Arkansas, New Jersey.
Use: Ceramic and glass manufacture, enamels, source of potash and aluminum (USSR).

nephelometry. Photometric analytical techniques for measuring the light scattered by finely divided particles of a substance in suspension. It is used to estimate the extent of turbidity in such products as beer and wine in which colloidally dispersed particles are present.

"Nepoxide."[41]   TM for a coating of the epoxy type which exhibits excellent adhesive properties and resistance to general chemicals and solvents. Can be deposited in high film thickness.

neptunium. CAS: 7439-99-8.   Np.
A radioactive transuranic element, having atomic number 93, first formed by bombarding uranium with high-speed deuterons; aw 237.0482, valences = 3, 4, 5, 6; d 20.45. Neptunium 237, the longest-lived of the 11 isotopes, has been found naturally in extremely small amounts in uranium ores. It is produced in weighable amounts as a byproduct in the production of plutonium.
Metallic neptunium is obtained by first preparing neptunium trifluoride, which is reduced with barium vapor at 1200C. It is a silvery white metal, mp 640C. Neptunium is similar chemically to uranium; it forms such intermetallic compounds as $NpAl_2$ and $NpBe_{13}$ as well as NpC, $NpSi_2$, NpN, NpF_3, NpF_6, NpF_4, $NpMO_2$, Np_3O_8, etc; soluble in hydrochloric acid, strong reducing agents.
Use: Np 237 is used in neutron detection instruments.
Hazard: A radioactive poison.

neptunium dioxide. NpO_2.
Properties: Dark olive, free-flowing powder.
Derivation: Neptunium oxalate is precipitated from solutions containing nitric acid and neptunium. The neptunium oxalate is calcined at 500–550C, producing neptunium dioxide.
Hazard: Toxic. A radioactive poison.
Use: Fabrication by powder metallurgy into target elements to be irradiated to produce Pu 238.

nerol. (cis-3,7-dimethyl-2,6-octadien-1-ol).
CAS: 106-25-2.
$(CH_3)_2C:CH(CH_2)_2C(CH_3):CHCH_2OH$.
The cis isomer of geraniol.
Properties: Colorless liquid, rose-neroli odor, soluble in absolute alcohol. Combustible.
Derivation: Iodization of geraniol with hydriodic acid, followed by treatment with alcoholic soda.
Use: Perfumery, flavoring.

nerolidol. (3,7,11-trimethyl-1,6,10-dodecatrien-3-ol). CAS: 7212-44-4.
$(CH_3)_2C:CH(CH_2)_2C(CH_3):$
$CH(CH_2)_2(CH_3) \cdot (OH)CH:CH_2.$
A sesquiterpene alcohol.
Properties: Straw-colored liquid with an odor similar to rose and apple, d 0.878, refr index 1.480–1.482, angular rotation (natural) +11 to +14 degrees; (synthetic) optically inactive, stable in air, soluble in alcohol and most fixed oils, insoluble in glycerol. Combustible.
Derivation: Occurs naturally in Peru balsam and oils of orange flower, neroli, sweet orange, and ylang ylang. Also made synthetically.
Grade: FCC.
Use: Perfumery, flavoring.

neolin. See β-naphthyl ethyl ether.

neroli oil. An essential oil used in perfumery and flavoring.

"Nerosol."[342] TM for a blend of sesquiterpenic alcohols used in perfumery.

Nernst, Walther. (1864-1941) A German chemist who won the Nobel prize in 1920. He was educated at Zurich and Berlin, and received his PhD at Wurzburg. He authored many works concerning theory of electric potential and conduction of electrolytic solutions. He developed the third law of thermodynamics which states that at absolute zero the entropy of every material in perfect equilibrium is zero, therefore volume, pressure, and surface tension all become independent of temperature. He also invented Nernst's lamp which required no vacuum and little current.

nerve gas. (nerve poison). One of several toxic chemical warfare agents developed in Germany during World War II. They are organic derivatives of phosphoric acid (principally alkyl phosphates, fluorophosphates, and thiophosphates). They inhibit the enzyme cholinesterase and cause acetylcholine poisoning and cessation of nerve transmission. They are colorless, odorless, tasteless liquids of low volatility and are absorbed rapidly through the eyes, lungs, or skin; they are lethally toxic to higher animals and man. Many insecticides have the same structure and properties. Antidotes are atropine sulfate and pralidoxime iodide.
See also parathion; cholinesterase inhibitor.

R   X
 \   /
  P
 /   \\
R   O

Nessler's reagent. CAS: 7783-33-7.
Solution of mercuric iodide in potassium iodide.
Hazard: High toxicity.
Use: Detecting the presence of ammonia, particularly in very small amounts.

"Neto."[492] TM for a dual acid-enzyme converted product with a dextrose equivalent of approximately 42. Has a maltose content three times that of an acid-converted corn syrup of the same degree of conversion.
Use: Confections.

Neuberg blue. A mixture of copper blue (powdered azurite) and an iron blue (Prussian blue). It can be more easily ground in oil than pure copper blue.

neurine. (trimethylvinylammonium hydroxide).
CAS: 463-88-7. $CH_2:CHN(CH_3)_3OH.$
A poisonous ptomaine formed during putrefaction by the dehydration of choline.
Properties: Syrupy liquid, fishy odor, absorbs carbon dioxide from the air, soluble in water and alcohol.
Hazard: Highly toxic.
Use: Biochemical research.

neutral. (1) Of particles, without electric charge. See neutron; atom.
(2) Of solutions, neither acidic nor basic. See pH.

neutralization. A chemical reaction in which water is formed by mutual interaction of the ions that characterize acids and bases when both are present in an aqueous solution, i.e., $H^+ + OH^- \rightarrow H_2O$, the remaining product being a salt. R. T. Sanderson states: "An aqueous solution containing an excess of hydronium ions is called acidic. It readily releases protons to electron-donating substances. An aqueous solution containing an excess of hydroxyl ions is called basic. It readily accepts protons from substances that can release them, and is in general an excellent donor. No aqueous solution can contain an excess of both hydronium and hydroxyl ions, because when these ions collide, a proton is immediately transferred from the hydronium to the hydroxyl ion, and both become water molecules."
Neutralization occurs with both (a) inorganic and (b) organic compounds: (a) $Ca(OH)_2 + H_2SO_4 \rightarrow CaSO_4 + 2HOH$; (b) $HCOOH + NaHCO_3 \rightarrow HCOONa + CO_2 + HOH$. It should be noted that neutralization can occur without formation of water, as in the reaction $CaO + CO_2 \rightarrow CaCO_3$. Neutralization does not mean the attaining of pH 7.0; rather it means the

equivalence point for an acid-base reaction. When a strong acid reacts with a weak base, the pH will be less than 7.0 and when a strong base reacts with a weak acid, the pH will be greater than 7.0.

neutral oil. A lubricating oil of medium or low viscosity obtained by distillation and dewaxing of crude petroleum or its cracking products.

neutral red. (toluylene red; 3-amino-7-(dimethylamino)-2-methylphenazine monohydrochloride; CI# 50040).
$(CH_3)_2NC_6H_3N_2C_6H_2CH_3NH_2 \cdot HCl$ (tricyclic).
Properties: Green powder, dissolves in water or alcohol to give red color.
Use: Acid-base indicator in the range pH 6.8–8.0 (red in acid, yellow brown in alkali); biological stain.

"Neutrol."[217] TM for an acid-activated clay used as decolorizing adsorbent for vegetable and animal fats and oils.
See also "Filtrol."

neutron. Discovered by Chadwick in 1932, the neutron is a fundamental particle of matter having a mass of 1.009 but no electric charge. It is a constituent of the nucleus of all elements except hydrogen, the number of neutrons present being the difference between the mass number and the atomic number of the element. Neutrons may be liberated from the nucleus by fission of Uranium-235, plutonium, and a few other elements, each nucleus yielding an average of 2.5 neutrons; they can also be produced by bombardment of other elements, e.g., beryllium with positively charged particles.

As free neutrons are uncharged they have tremendous penetrating power as a result of their electrical neutrality; hence they have a highly damaging effect on living tissue, requiring the use of shielding of all equipment in which they are produced. Neutrons directly emitted from atomic nuclei are termed "fast;" it is these that bring about the chain reaction in the atomic bomb. In a nuclear power reactor, where a less rapid reaction is desired, the energy of fast neutrons is partially absorbed by the moderator, and the neutrons so retarded are called "slow" or thermal.
See also electron, proton, fission.
Use: Nuclear fission, manufacture of plutonium and radioactive isotopes; activation analysis.

neutron activation analysis. See activation analysis.

neutron diffraction. See diffraction, neutron.

"Neutronyx."[328] TM for a group of nonionic detergents composed of alkylphenol polyglycol ethers containing ethylene oxide or of polyethylene glycol fatty acid esters. "Neutronyx." S-60 and S-30 are anionics, the ammonium and sodium salts of a sulfated alkylphenol polygycol ether.
Use: Detergents; wetting, emulsifying, dispersing agents.

"Neutroscents."[188] TM for a series of perfumes designed particularly to cover objectionable odors, available in water-soluble form for sprays, air-conditioning apparatus and other dispersion devices. Also available in a highly concentrated form.

"Nevinol."[21] TM for a series of (alkyl) hydroxy resins.
Use: Adhesives, lacquers, paper coatings, special inks and varnishes.

Neville-Winter acid. (1-naphthol-4-sulfonic acid; α-naphtholsulfonic acid; NW acid).

Properties: Transparent plates, soluble in water, mp 170C.
Derivation: From sodium salt of naphthionic acid by hydrolysis of amino group.
Use: Azo dye intermediate, e.g., Congo Corinth.

"Nevindene."[21] TM for high-melting coumarone-indene resins of extreme hardness.
Use: Dental compounds, fast-drying varnishes, rotogravure inks, aluminum paints, and insulating compounds.

NF. Abbreviation for National Formulary, a compendium of pharmaceutical formulations widely used as a standard reference.

NFPA. Abbreviation for National Fire Protection Association.

Ni. Symbol for nickel.

"Niacet."[214] TM for various metallic acetate salts, including aluminum formoacetate, copper acetate, potassium acetate, sodium acetate, sodium diacetate, zinc acetate, and "Niaproof" aluminum acetate.

"Niacide."[55] TM for fungicidal products containing dimethyl dithiocarbamates used mainly for scab control.

niacin. (nicotinic acid; pyridine-3-carboxylic acid). CAS: 59-67-6.

The antipellagra vitamin, essential to many animals for growth and health. In man, niacin is believed necessary along with other vitamins for the prevention and cure of pellagra. It functions in protein and carbohydrate metabolism. As a component of two important enzymes, coenzyme I and coenzyme II, it functions in glycolysis and tissue respiration.
Properties: Colorless needles, odorless, mp 236C, sublimes above melting point, sour taste, soluble in water and alcohol, insoluble in most lipid solvents, quite stable to heat and oxidation, d 1.473, a vasodilator in high concentration.
Units: Amounts of niacin are expressed in milligrams.
Sources: Food sources, meat, fish, milk, whole grains, yeast. Commercial sources: synthetic niacin is made by oxidation of nicotine, quinoline, or 2-methyl-5-ethylpyridine (from ammonia and formaldehyde or acetaldehyde).
Grade: NF, FCC, blended with soy flour (animal feeds).
Use: Medicine (cholesterol-lowering agent), nutrition, feeds, enriched flours, dietary supplement.
See also niacinamide.

niacinamide. (nicotinamide; nicotinic acid amide). CAS: 98-92-0. $C_5N_4NCONH_2$.
Same biological function as niacin.
Properties: Colorless needles, mp 129C, d 1.40. Soluble in water, ethanol, and glycerol; bitter taste.
Sources: Synthetic made by conversion of niacin to the amide.
Grade: USP, FCC. Also commercially available as the hydrochloride.
Use: Medicine, dietary supplement.

niacinamide ascorbate. A complex of ascorbic acid and niacinamide.
Properties: Lemon-yellow powder, odorless or with a very slight odor. May gradually darken upon exposure to air. Soluble in water and alcohol, sparingly soluble in glycerol, practically insoluble in benzene.
Grade: FCC.
Use: Dietary supplement.

"Nial."[155] TM for a nickel-base thermocouple alloy, wherein the negative thermal EMF with reference to pure platinum is obtained by adding manganese, aluminum and silicon. Low percentages of iron, cobalt, zirconium, and magnesium are added to control both the EMF and type of oxide.
Properties: Magnetic, mp 2550F, d 8.47, sp heat at 20C, 0.12 cal/g, tensile strength 83,000 psi.

nialamide. (1-(2-benzylcarbamyl)ethyl-2-isonicotinoylhydrazine). CAS: 51-12-7. $C_5H_4NCO(NH)_2(CH_2)_2CONHCH_2C_6H_5$.
Properties: White, crystalline powder of low solubility in water and good solubility in slightly acid solution. It is stable in crystalline form, suspension, and solution.
Use: In medicine as an antidepressant.
Hazard: Toxic in overdose.

"Nialk.."[62] TM for chlorine, caustic soda, caustic potash, carbonate of potash, paradichlorobenzene, and trichloroethylene.

"Niaproof."[214] TM for a water-repellent compound. Substantially a soluble basic aluminum acetate salt.
Use: Source of aluminum ion for water-repellent finishes for textile, paper, and leather products, particularly in processes using wax or soap emulsions.

"Niax."[214] TM for a polyurethane foam preparation packaged in two units: (1) the resin base and (2) the expander or catalyst, said to be composed chiefly of dimethylaminopropionitrile.

nicarbazin. Equimolar complex of 4,4'-dinitrocarbanilide and 2-hydroxy-4,6-dimethylpyrimidine.
Properties: Forms crystals, decomposes at 265–275C, insoluble in water.
Use: Coccidiostat.

niccolite. (arsenical nickel). NiAs.
An ore of nickel.
Properties: Pale copper-red mineral with dark tarnish, metallic luster. Contains 43.9% nickel, soluble in concentrated nitric acid, d 7.3–7.67, hardness 5–5.5.
Hazard: Toxic by inhalation of dust.

"Nichrome."[330] TM for an alloy containing 60% nickel, 24% iron, 16% chromium, 0.1% C.
Use: It is used principally for electric resistance purposes. It also offers good resistance to mine and sea waters and moist sulfurous atmospheres.

nickel. CAS: 7440-02-0. Ni. Metallic element of atomic number 28, group VIII of the Periodic Table, aw 58.70, valences = 2, 4. Five stable isotopes.
Properties: Malleable, silvery metal. Readily fabricated by hot- and cold-working, takes high polish, excellent resistance to corrosion, d 8.908, mp 1455C, bp 2900C, electrical resistivity (20C)

6.844 microhm-cm. Attacked slightly by hydrochloric acid and sulfuric acid, somewhat more by nitric acid, highly resistant to strong alkalies.

Occurrence: Ontario, Cuba, Norway, Dominican Republic.

Derivation: Nickel ores are of two types, sulfide and oxide, the former accounting for two-thirds of the world's consumption. Sulfide ores are refined by flotation and roasting to sintered nickel oxide, and either sold as such or reduced to metal, which is cast into anodes and refined electrolytically or by the carbonyl process (See Mond process). Oxide ores are treated by hydrometallurgical refining, e.g., leaching with ammonia. Much secondary nickel is recovered from scrap.

Grade: Electrolytic, ingot, pellets, shot, sponge, powder, high-purity strip, single crystals.

Hazard: Flammable and toxic as dust or fume. A carcinogen (OSHA). TLV: (metal) 1 mg/m^3 of air; (soluble compounds) (as nickel): 0.1 mg/m^3 of air.

Use: Alloys (low-alloy steels, stainless steel, copper and brass, permanent magnets, electrical resistance alloys), electroplated protective coatings, electroformed coatings, alkaline storage battery, fuel cell electrodes, catalyst for methanation of fuel gases and hydrogenation of vegetable oils.

See also Raney's nickel.

nickel acetate.　CAS: 373-02-4.
$Ni(OOCCH_3)_2 \cdot 4HOH$.
Properties: Green, monoclinic crystals; effloresces somewhat in air. D 1.74, decomposes on heating to 250C, soluble in water and alcohol.
Derivation: By heating nickel hydroxide with acetic acid in the presence of metallic nickel.
Hazard: Toxic by ingestion, a carcinogen (OSHA).
Use: Textiles (mordant), catalyst.

nickel acetylacetonate.　CAS: 3264-82-2.
$(CH_3COCHCOCH_3)_2Ni$.
Properties: Green, crystalline solid; mp 230C; bp 220C (11 mm); d 1.45. Soluble in water, alcohol, and benzene; insoluble in ether.
Use: Catalyst for organic reactions.

nickel aluminide.　A cermet that can be flame-sprayed.

nickel ammonium chloride.　(ammonium nickel chloride).　(a) $NiCl_2 \cdot NH_4Cl$,
(b) $NiCl_2 \cdot NH_4Cl \cdot 6HOH$.
Properties: (a) Yellow powder, (b) green crystals, d 1.65, soluble in water, deliquescent.
Hazard: See nickel.
Use: Electroplating, dyeing (mordant).

nickel ammonium sulfate.　(nickel salts, double; ammonium nickel sulfate).

CAS: 15699-18-0.
$NiSO_4 \cdot (NH_4)_2SO_4 \cdot 6HOH$.
Properties: Green crystals, decomposed by heat, soluble in water, less in ammonium sulfate solution, insoluble in alcohol, d 1.929.
Derivation: An aqueous solution of nickel sulfate is acidified with sulfuric acid, then an aqueous solution of ammonium sulfate is added. On concentrating, crystals of the double sulfate separate out.
Hazard: See nickel.
Use: Nickel electrolyte for electroplating.

nickel arsenate.　(nickelous arsenate).
$Ni_3(AsO_4)_2 \cdot 8HOH$.
Properties: Yellow-green powder, soluble in acids, insoluble in water, d 4.98.
Hazard: Highly toxic.
Use: Catalyst (hardening fats used in soap).

nickel bromide.　(nickelous bromide).
CAS: 13462-88-9.　(a) $NiBr_2$,
(b) $NiBr_2 \cdot 3HOH$.
Properties: (a) Brownish-yellow solid or yellow, lustrous scales; d 5.098; mp 963C. (b) Deliquescent, greenish scales; loses 3HOH at 300C; soluble in water, alcohol, ether, and ammonium hydroxide.
Derivation: Bromination of nickel powder or nickel carbonyl.
Hazard: See nickel.

nickel carbonate, basic.　CAS: 3333-67-3.
$NiCO_3 \cdot 2Ni(OH)_2 \cdot 4HOH$.
Properties: Light green crystals or brown powder, d 2.6, insoluble in water, soluble in ammonia and dilute acids.
Derivation: (1) In nature as the mineral zaratite. (2) Synthetically, by addition of soda ash to a solution of nickel sulfate.
Hazard: See nickel.
Use: Electroplating, preparation of nickel catalysts, ceramic colors and glazes.

nickel carbonyl.　(nickel tetracarbonyl).
CAS: 13463-39-3.　$Ni(CO)_4$.　A zero-valent compound. The four carbonyl groups form a tetrahedral arrangement and are linked covalently to the metal through the carbons.
Properties: Colorless liquid, soluble in alcohol and many organic solvents, soluble in concentrated nitric acid, insoluble in water, d 1.3185, fp −19C, bp 43C, vap press 400 mm at 25.8C.
Derivation: By passing carbon monoxide gas over finely divided nickel.
Grade: Technical.
Hazard: Flammable, dangerous fire risk, explodes at 60C (140F). A carcinogen (OSHA). TLV (as nickel): 0.05 ppm in air.

Use: Production of high-purity nickel powder by Mond process, continuous nickel coatings on steel and other metals.

nickel chloride. (nickelous chloride).
  CAS: 7718-54-9.   (a) $NiCl_2$,
  (b) $NiCl_2 \cdot 6HOH$.
Properties: (a) Brown scales, deliquescent,
  (b) green scales, deliquescent; soluble in water, alcohol, and ammonium hydroxide, (a) d 3.55, mp 1001C. Nonflammable.
Derivation: Action of hydrochloric acid on nickel.
Hazard: See nickel.
Use: Electroplated nickel coatings, reagent chemical.

nickel cyanide. CAS: 557-19-7.
  $Ni(CN)_2 \cdot 4HOH$.
Properties: Apple-green plates or powder, soluble in ammonium hydroxide and potassium cyanide solution, insoluble in water and acids, mp loses 4HOH at 200C, bp decomposes.
Derivation: By adding potassium cyanide to a solution of a nickel salt.
Hazard: Highly toxic. TLV (as CN): 5 mg/m³ of air.
Use: Metallurgy, electroplating.

nickel dibutyldithiocarbamate. CAS: 13927-77-0.
  $Ni[SC(S)N(C_4H_9)_2]_2$.
Properties: Dark green flakes, d 1.26, mp 86C min.
Use: Antioxidant for synthetic rubbers.

nickel dimethylglyoxime. $C_8H_{14}N_4NiO_4$.
Properties: Bright red powder, insoluble in water, soluble in mineral acids, sublimes at 250C.
Use: Pigment in paints, cosmetics, cellulosics.

nickel formate. CAS: 3349-06-2.
  $(HCOO)_2Ni \cdot 2HOH$.
Properties: Green crystals, d 2.15, partially soluble in water, insoluble in alcohol.
Derivation: (a) Reaction of sodium formate and nickel sulfate, (b) dissolving nickel hydroxide in formic acid.
Hazard: See nickel.
Use: Production of nickel catalysts.

nickel hydroxide. See nickelous hydroxide, nickelic hydroxide.

nickelic hydroxide. (nickel hydroxide).
  CAS: 12054-48-7. $Ni(OH)_3$.
Properties: Black powder, mp decomposes.
Derivation: By adding a hypochlorite to a solution of a nickel salt.
Use: Nickel salts.

nickelic oxide. (nickel peroxide; nickel(III) oxide.; nickel sesquioxide; black nickel oxide).
  CAS: 1314-06-3. Ni_2O_3.
Properties: Gray-black powder, soluble in acids, insoluble in water, d 4.84, mp reduced to nickel oxide at 600C.
Derivation: By gentle heating of the nitrate or chlorate.
Hazard: See nickel.
Use: Storage batteries.

nickel iodide. (nickelous iodide).
  CAS: 13462-90-3. NiI_2 or $NiI_2 \cdot 6HOH$.
Properties: Black, crystalline powder or blue-green crystals; hygroscopic. Soluble in alcohol and water, d 5.834, sublimes at 797C without melting.
Derivation: Direct combination of nickel and iodine.
Hazard: Toxic by ingestion.
See nickel.

nickel-iron alloy. See iron-nickel alloy.

"Nickel-Lume."[288] TM for a bright nickel electroplating process; materials used are nickel sulfate, nickel chloride, boric acid, and organic addition agents.

nickel matte. See matte.

nickel nitrate. (nickelous nitrate).
  CAS: 13138-45-9. $Ni(NO_3)_2 \cdot 6HOH$.
Properties: Green, deliquescent crystals. Soluble in water, ammonium hydroxide, and alcohol. D 2.065, mp 55C, bp 136.7C.
Derivation: By the action of nitric acid on nickel, or on nickel oxide.
Grade: Technical, reagent.
Hazard: Dangerous fire risk, strong oxidizing agent. TLV (as nickel): 1 mg/m³ of air.
Use: Nickel plating, preparation of nickel catalysts, manufacture of brown ceramic colors.

nickel nitrate, ammoniated. (nickel nitrate tetrammine). $Ni(NO_3)_2 \cdot 4NH_3 \cdot 2HOH$.
Properties: Green crystals, soluble in water, insoluble in alcohol, decomposes in air.
Derivation: By adding ammonium hydroxide to a nitric acid solution of nickel oxide with subsequent crystallization.
Hazard: See nickel nitrate.
Use: Nickel plating.

nickelocene. (dicyclopentadienylnickel).
  $(C_5H_5)_2Ni$.
Properties: Dark green crystals, mp 171–173C, soluble in most organic solvents, insoluble in water. Decomposes in acetone, alcohol, and ether; highly reactive compound which decomposes rapidly in air.

Hazard: Toxic by inhalation and skin contact, a carcinogen (OSHA).
Use: Catalyst, complexing agent.
See also metallocene.

nickelous arsenate.   See nickel arsenate.

nickelous bromide.   See nickel bromide.

nickelous chloride.   See nickel chloride.

nickelous hydroxide.   (nickel hydroxide).
CAS: 12125-56-3. $Ni(OH)_2$.
Properties: Fine green powder, d 4.15, mp (decomposes) 230C, very slightly soluble in water, soluble in acids and ammonium hydroxide.
Derivation: By adding caustic soda to a solution of nickelous salt.
Hazard: See nickel.
Use: Nickel salts.

nickelous iodide.   See nickel iodide.

nickelous nitrate.   See nickel nitrate.

nickelous phosphate.   See nickel phosphate.

nickel oxide.   (nickelous oxide; nickel(II) oxide.; nickel protoxide; green nickel oxide).
CAS: 1313-99-1.   NiO.
Properties: Green powder becoming yellow, soluble in acids and ammonium hydroxide, insoluble in water and caustic solutions, d 6.6–6.8, absorbs oxygen at 400C forming nickelic oxide, which is reduced to nickel oxide at 600C.
Derivation: By heating nickel above 400C in presence of oxygen.
Hazard: See nickel.
Use: Nickel salts, porcelain painting, fuel cell electrodes.

nickel oxide, black.   See nickelic oxide.

nickel oxide, green.   See nickel oxide.

nickel peroxide.   See nickelic oxide.

nickel phosphate.   (nickelous phosphate; trinickelous-orthophosphate).   CAS: 10381-36-9.
$Ni_3(PO_4)_2 \cdot 7HOH$.
Properties: Light-green powder, soluble in acids and ammonium hydroxide, insoluble in water.
Use: Electroplating.

nickel potassium sulfate.   (potassium nickel sulfate). $NiSO_4 \cdot K_2SO_4 \cdot 6HOH$.
Properties: Blue-green crystals, soluble in water, d 2.124.

nickel protoxide.   See nickel oxide.

nickel-rhodium.   Alloys containing nickel and 25–80% of rhodium, but sometimes also some platinum, iridium, palladium, molybdenum, tungsten, copper, iron, or cobalt.
Use: Electrodes, chemical apparatus, reflectors.

nickel salt, double.   See nickel ammonium sulfate.

nickel salt, single.   See nickel sulfate.

nickel sesquioxide.   See nickelic oxide.

nickel silver.   Nonferrous alloy of nickel, copper, and zinc having a silver appearance.
Use: Etching, enameling, silver-plating, and chromium-plating.

nickel stannate. $NiSnO_3 \cdot 2HOH$.
Properties: Light colored, crystalline powder; approximate temperature of dehydration 120C.
Hazard: Toxic by ingestion and inhalation. TLV (as nickel): 2 mg/m^3 of air.
Use: Additive in ceramic capacitors.

nickel sulfate.   (nickel salts, single; blue salt).
CAS: 7786-81-4.   (a) $NiSO_4$,
(b) $NiSO_4 \cdot 6HOH$, (c) $NiSO_4 \cdot 7HOH$.
Properties: (a) Yellow-green crystals, (b) blue or emerald green crystals, (c) green crystals. All the sulfates are soluble in water, (b) and (c) are soluble in alcohol, (a) is insoluble in alcohol and ether; d (a) 3.68, (b) 2.031, (c) 1.98; mp (a) 840C (loses SO_3), (b) and (c) loses 6HOH at 103C.
Derivation: Action of sulfuric acid on nickel.
Grade: Technical, CP, single crystals.
Hazard: Toxic material. TLV (as nickel): 0.1 mg/m^3 of air.
Use: Manufacture of nickel ammonium sulfate, nickel catalysts, nickel plating, mordant in dyeing and printing textiles, coatings, ceramics.

nickel tetracarbonyl.   See nickel carbonyl.

nickel titanate.   See "Solfast."

nicol.   An optical material (Iceland spar) that functions as a prism, separating light rays that pass through it into two portions, one of which is reflected away and the other transmitted. The transmitted portion is called plane-polarized light.
See also calcite, optical isomerism.

"Nicomo-12."[241]   TM for a hydrodesulfurization catalyst of nickel, cobalt, and molybdenum on alumina.

"Nicon."[169]   Brand name for diethyldithiocarbamate used in the colorimetric determination of nickel.

nicotinamide. See niacinamide.

nicotinamide adenine dinucleotide phosphate.
Name recommended by the International Union of Biochemistry and IUPAC.
(NADP; diphosphopyridine nucleotide; DPN; coenzyme I; Co I; codehydrogenase I).
$C_{22}H_{27}O_{14}N_7P_2$.
A coenzyme necessary for the alcoholic fermentation of glucose and the oxidative dehydrogenation of other substrates. It occurs widely in living tissue, especially in liver.
Use: Biochemical research, chromatography.

nicotine. (β-pyridyl-α-n-methylpyrrolidine).
CAS: 54-11-5. $C_5H_4NC_4H_7NCH_3$.

Properties: Alkaloid from tobacco; thick, water-white, levorotatory oil turning brown on exposure to air; hygroscopic. Soluble in alcohol, chloroform, ether, kerosene, water, and oils. Bp 247C (decomposes), d 1.00924, also in form of dust or powder, autoign temperature 471F (243C). Combustible.
Derivation: By distilling tobacco with milk of lime and extracting with ether.
Hazard: Toxic by ingestion, inhalation, and skin absorption. TLV: 0.5 mg/m³ of air.
Use: Insecticide, fumigant, use as insecticide may be restricted. Available as the dihydrochloride, salicylate, sulfate, and bitartrate.

nicotine salts. (a) Hydrochloride, $C_{10}H_{14}N_2 \cdot HCl$; (b) salicylate, $C_{10}H_{14}N_2 \cdot C_7H_6O_3$; (c) sulfate, $(C_{10}H_{14}N_2)_2 \cdot H_2SO_4$
(40% solution = Black Leaf Forty); (d) tartrate, $C_{10}H_{14}N_2 \cdot 2C_4O_6H_6 \cdot HOH$.
Properties: (a) Colorless oil; (b) white crystals, mp 117.4C; (c) white crystals; (d) white plates, mp 89C. All the salts are soluble in water, alcohol, and ether.
Derivation: By the action of the respective acid on the alkaloid.

nicotinic acid. See niacin.

nicotinic acid amide. See niacinamide.

Niementowski quinazoline synthesis.
Formation of 4-oxo-3,4-dihydroquinazolines by cyclization of the reaction products of anthranilic acid and amides.

Niementowski quinoline synthesis. Formation of γ-hydroxyquinoline derivatives from anthranilic acids and carbonyl compounds.

Nierenstein reaction. Formation of omega-chloroacetophenones by reaction of diazomethane in dry ether with aroyl chlorides. Coumaranones are obtained if an ortho-hydroxy group is present.

Nieuland, Father J. A. (1878-1936). A Jesuit whose research on polymers of acetylene formed the basis for the development of polychloroprene (neoprene) in 1931.
See also Carothers.

nigrosine. A class of dark blue or black dyes, some soluble in water, some in alcohol, and some in oil.
Use: Manufacture of ink and shoe-polish and in dyeing leather, wood, textiles, etc. It is also used as a shark-repellent.

nigre. The dark-colored layer containing some soap as well as salts and impurities, formed in soap manufacture between the layers of soap proper and caustic solution.

NIH. Abbreviation for National Institutes of Health (Bethesda, MD).

ninhydrin. (triketohydrindene hydrate).
CAS: 485-47-2. $C_9H_4O_3 \cdot HOH$.
Properties: White crystals or powder, mp 240–245C, becomes red when heated above 100C, soluble in water and alcohol, slightly soluble in ether and chloroform.
Hazard: Irritant.
Use: Chemical intermediate, reagent for determination of amines, amino acids, ascorbic acid.

niobe oil. See methyl benzoate.

niobic acid. Any hydrated form of Nb_2O_5. It forms as a white, insoluble precipitate when a potassium hydrogen sulfate fusion of a niobium compound is leached with hot water or when niobium fluoride solutions are treated with ammonium hydroxide. Soluble in concentrated sulfuric acid, concentrated hydrochloric acid, hydrogen fluoride, and in bases. Important in analytical determination of niobium.

niobite. See columbite.

niobium. (columbium).   CAS: 7440-03-1.
Nb.   The name niobium is officially approved by chemical authorities, but columbium is still used chiefly by metallurgists. Metallic element, atomic number 41, Group VB of the Periodic

Table, aw 92.9064, valences = 2, 3, 4, 5, no stable isotopes.

Properties: Gray or silvery, ductile metal; does not tarnish or oxidize at room temperature; d 8.57; mp 2468C. Reacts with oxygen and halogens only when heated, less corrosion-resistant than tantalum at high temperature. Not attacked by nitric acid up to 100C, but vigorously attacked by a mixture of nitric and hydrofluoric acids. Hot concentrated hydrochloric acid, sulfuric acid, and phosphoric acid attack it, but hot concentrated nitric does not. Unaffected at room temperature by most acids and by aqua regia. It is attacked by alkaline solutions to some extent at all temperatures.

Occurrence: Found in two major ores, columbite and pyrochlore (a carbonate-silicate rock). Chief sources are Brazil, Nigeria, Canada.

Derivation: Niobium is so closely associated with tantalum that they must be separated by fractional crystallization or by solvent extraction with subsequent purification.

Grade: Plates, rods, powder, single crystals.

Use: Superconducting and magnetic alloys (with tin and titanium), cermets, missiles and rockets, cryogenic equipment, ferroniobium for alloy steels.

niobium carbide. NbC.
Properties: Lavender-gray powder, insoluble in water and in all acids except a mixture of nitric acid and hydrogen fluoride, mp approximately 3500C, d 7.82.
Derivation: By direct combination of niobium with carbon or by the reduction of niobium oxide with lampblack.
Use: Cemented carbide tipped tools, special steels, preparation of niobium metal, coating graphite for nuclear reactors.

niobium chloride. (niobium pentachloride).
CAS: 10026-12-7. $NbCl_5$.
Properties: Yellow, crystalline solid; soluble in carbon tetrachloride, hydrochloric acid, concentrated sulfuric acid. Mp 205C, bp 254C, d 2.75, deliquescent, decomposes in moist air with evolution of hydrogen chloride fumes, available in commercial quantities.
Derivation: Direct combination of niobium and chlorine, chlorination of niobium oxide in the presence of carbon.
Hazard: May evolve fumes of hydrogen chloride. Keep dry.
Use: Preparation of pure niobium, intermediate.

niobium diselenide. $NbSe_2$.
Properties: Gray-black solid, mp greater than 1316C, vacuum-stable from −430 to 2400F (−170 to +1315C). Has higher electrical conductivity than graphite.

Use: Lubricant and conductor at high temperatures and high vacuum.

niobium oxalate. $NbO(HC_2O_4)_3 \cdot 4HOH$.
Properties: White crystals, 99.99% pure, very soluble in water.
Use: Intermediate and in special catalysts.

niobium oxide. (niobium pentoxide).
CAS: 1313-96-8. Nb_2O_5.
Properties: White powder; insoluble in acids except hydrofluoric and hot sulfuric acids; insoluble in water; soluble in fused potassium hydrogen sulfate, or carbonates or hydroxides of the alkali metals; d 4.5–5.0; mp 1520C.
Derivation: Strong ignition of niobic acid.
Use: Intermediate, electronics.

niobium pentachloride. See niobium chloride.

niobium pentoxide. See niobium oxide.

niobium potassium oxyfluoride. (potassium niobium oxypentafluoride; potassium oxyfluoniobate). CAS: 17523-77-2.
$K_2NbOF_5 \cdot HOH$.
Properties: White, lustrous leaflets; greasy to touch; soluble in hot water.
Hazard: Toxic by ingestion, strong irritant. TLV (as F): 2.5 mg/m³ of air.
Use: Separation of niobium from tantalum, electrolytic preparation of niobium metal.

niobium silicide. $NbSi_2$.
Properties: Crystalline solid, mp 1950C.
Use: Refractory material.

niobium-tin. Nb_3Sn. Alloy used for special wire for super-conducting magnets to obtain high magnetic fields for use in communication and containment of thermonuclear fusion plasmas.

niobium-titanium. Alloy used for magnetic devices with fields up to 100,000 gauss.

niobium-uranium. Niobium alloyed with 20% uranium yields a nuclear fuel which maintains tensile strength and hardness at 871C.

NIOSH. Abbreviation for National Institute for Occupational Safety and Health.

NIRMS. Noble gas-ion reflection mass spectroscopy.

"Nirez."[36] TM for polyterpene, terpene, phenol, and resins used in adhesives, coatings, chewing gums, and printing inks.

nisin. Antibiotic containing 34 amino acid residues, produced by *Streptomyces lactis*.
Use: Food preservative, especially in canned products.

niter. (nitre; saltpeter).   CAS: 7757-79-1.
KNO_3.   A natural potassium nitrate.

niter cake. A common name for sodium bisulfate ($NaHSO_4$) because it was a product of the reaction by which nitric acid was first made: $NaNO_3 + H_2SO_4 \rightarrow NaHSO_4 + HNO_3$.
See sodium bisulfate.

niter, Chile. See sodium nitrate.

nitralloy. See nitriding.

nitralin. A herbicide used largely to control weeds in cotton and soybeans. Research indicates possible use in plant breeding; treatment of corn roots induces abnormal chromosome formation and cell wall deterioration.

nitramine. See tetryl.

nitranilic acid. (2,5-dihydroxy-3,6-dinitroquinone).   CAS: 479-22-1.
$C_6O_2(NO_2)_2(OH)_2$.
Properties: Flat, yellow crystals. Loses water at 100C, decomposes explosively at 170C, soluble in water and alcohol, insoluble in ether.
Hazard: Fire risk when heated. Evolves toxic fumes on decomposition.

nitraniline. See nitroaniline.

nitrating acid. Legal label name for mixed acid.

nitration. A reaction in which a nitro group (—NO_2) replaces a hydrogen on a carbon atom by the use of nitric acid or mixed acid. An example is the nitration of cellulose to nitrocellulose. It is widely used in aromatic reactions to form such compounds as nitrobenzene, trinitrotoluene, nitroglycerin, and other explosives. Aromatic nitrations are usually effected with mixed acid, a mixture of nitric and sulfuric acids, at 0–120C. Aliphatic nitration is less common than aromatic, but propane can be nitrated under pressure to yield nitroparaffins.

"Nitrelmang."[250] TM for a high purity nitrided manganese for use in the production of free-machining steels, high nitrogen tinplate, and special high temperature alloy steels.

"Nitrix."[248] TM for synthetic rubber latices of the butadiene-acrylonitrile type.
Use: Paper saturation, leather finishing and as a plasticizer for resin latices.

nitric acid. (aqua fortis; engraver's acid; azotic acid).   CAS: 7697-37-2. HNO_3.   11th highest-volume chemical produced in US (1985).
Properties: Transparent, colorless, or yellowish fuming, suffocating, hygroscopic, corrosive liquid. Will attack almost all metals. The yellow color is due to release of nitrogen dioxide on exposure to light; strong oxidizing agent, miscible with water, decomposes in alcohol, bp (decomposes) 78C, fp −42C, d 1.504 (25/4C), vap press 62 mm (25C), refr index 1.3970 (24C), viscosity 0.761 cp (25C).
Derivation: (1) Oxidation of ammonia by air or oxygen with platinum catalyst. (*Note*: A pelleted catalyst not containing platinum or other noble metals is available.) Air oxidation yields 60% acid; concentration is achieved by (a) distillation with sulfuric acid, (b) extractive distillation with magnesium nitrate, or (c) by neutralizing the weak acid with soda ash, evaporating to dryness, and treating with sulfuric acid. Method (c) yields synthetic niter cake ($NaHSO_4$) as a byproduct. (2) High-pressure oxidation of nitrogen tetroxide (yields 98% acid). (3) Reaction of nitrogen and oxygen in nuclear reactors; two tons of nitric acid are said to be produced from one gram of enriched uranium. Not in commercial use. Strength of solutions: 36, 38, 40, 42 degrees Bé; 58–63.5%; 95%.
Hazard: Dangerous fire risk in contact with organic materials. Highly toxic by inhalation, corrosive to skin and mucous membranes, strong oxidizing agent. TLV: 2 ppm in air.
Use: Manufacture of ammonium nitrate for fertilizer and explosives, organic synthesis (dyes, drugs, explosives, cellulose nitrate, nitrate salts), metallurgy, photoengraving, etching steel, ore flotation, urethanes, rubber chemicals, reprocessing spent nuclear fuel.

nitric acid, fuming. (1) White fuming nitric acid (WFNA) contains more than 97.5% nitric acid, less than 2% water, and less than 0.5% NO_x. It is a colorless or pale yellow liquid which fumes strongly. It is decomposed by light or elevated temperatures, becoming red in color from nitrogen dioxide.
(2) Red fuming nitric acid (RFNA) contains more than 85% nitric acid, approximately 6–15% NO_x (as nitrogen dioxide), and less than 5% water.
Derivation: From dilute nitric acid, nitrogen dioxide, and oxygen.
Hazard: Toxic by inhalation, corrosive to skin and mucous membranes. Strong oxidizing agent, may explode in contact with strong reducing agents. Dangerous fire risk.
Use: Preparation of nitro-compds, rocket fuels, laboratory reagent.

nitric oxide. CAS: 10102-43-9.   NO.
Properties: Colorless gas (readily reacts with oxygen at room temperature to form nitrogen dioxide, NO_2, a reddish brown gas), bp −152C, fp −164C, d at bp 1.27, slightly soluble in water. Noncombustible.

Derivation: Oxidation of ammonia above 500C, decomposition of nitrous acid (aqueous solution). Also from atmospheric oxygen and nitrogen in the electric-arc process for fixation of nitrogen.
Grade: Pure (99%).
Hazard: Supports combustion. Toxic by inhalation, strong irritant to skin and mucous membranes. TLV: 25 ppm in air.
Use: Intermediate in production of nitric acid from ammonia, preparation of nitrosyl carbonyls, bleaching rayon.
See also nitrogen dioxide.

nitriding. A process of case hardening in which a ferrous alloy, usually of special composition, is heated in an atmosphere of ammonia or in contact with nitrogenous material to produce surface hardening by absorption of the nitrogen without quenching. The alloys used for nitriding are known as nitroalloys. Several types are available with ranges of composition as follows: aluminum 0.85–11.2%, carbon 0.20–0.45%, chromium none to 1.8%, Mo 0.15–1.00%, Mn 0.4–0.7%, silicon 0.2–0.4%.

nitrile. An organic compound containing the $-C\equiv N$ grouping, for example, acrylonitrile $H_2C=CHC\equiv N$; it may also contain a triple bond as in acetonitrile ($CH_3C\equiv N$).
Hazard: Some organic cyanide compounds are flammable (acetonitrile, acrylonitrile). Most organic cyanide compounds are toxic (acetonitrile, acrylonitrile).

nitrile rubber. (acrylonitrile rubber; acrylonitrile-butadiene rubber; nitrile-butadiene rubber; NBR). A synthetic rubber made by random polymerization of acrylonitrile with butadiene by free radical catalysis. Alternating copolymers using Natta-Ziegler catalyst have been developed. approximately 20% of the total is used as latex, also available in powder form. Its repeating structure may be represented as $-CH_2CH=CHCH_2CH_2CH(CN)-$.
Typical properties: (Medium acrylonitrile): d (polymer) 0.98, tensile strength (psi) 1000–3000, elongation (%) 100–700, maximum service temperature 121–148C. Combustible.
Use: (High acrylonitrile): Oil well parts, fuel tank liners, fuel hose, gaskets, packing oil seals, hydraulic equipment.
    (Medium acrylonitrile): General-purpose oil-resistant applications, shoe soles, kitchen mats, sink topping, and printing rolls.
    (Low acrylonitrile): Gaskets, grommets and O-rings (flexible at a very low temperature), adhesives. Fuel binder in solid rocket propellants.

nitrile-silicone rubber. (NSR). Combines the characteristic properties of silicones with the oil resistance of nitrile rubber. Resistant to jet fuels, solvents, and hot oils.

nitrilotriacetic acid. (NTA; triglycine; TGA; triglycolamic acid). CAS: 139-13-9. $N(CH_2COOH)_3$.
Properties: White, crystalline powder; mp 240C (with decomposition). Insoluble in water and most organic solvents, forms mono-, di-, and tribasic salts which are water-soluble. Combustible. 70% biodegradable.
Use: Synthesis, chelating agent, eluting agent in purification of rare-earth elements, detergent builder. Also available as the di- and trisodium salts.

nitrilotriacetonitrile. (NTAN). $N(CH_2CH)_3$.
Properties: White, crystalline solid; mp 130–134C; insoluble in water; soluble in acetone.
Hazard: See nitriles.
Use: Intermediate and chelating agent.

p-nitroacetanilide. $NO_2C_6H_4NHCOCH_3$.
Properties: White crystals, soluble in alcohol and ether, very slightly soluble in cold water, soluble in hot water and in potassium hydroxide solution, mp 214–216C. Combustible.
Derivation: By acetylating aniline, then nitrating.
Use: Manufacture of nitraniline.

p-nitro-o-aminophenol. CAS: 99-57-0. $C_6H_3OHNH_2NO_2$.
Properties: Yellow-brown leaflets containing water of crystallization melting at 80–90C, anhydrous melts at 154C, soluble in acid.
Derivation: From dinitrophenol.
Use: Dyes.

m-nitroaniline. (m-nitraniline). CAS: 99-09-2. $NO_2C_6H_4NH_2$.
Properties: Yellow needles, d 1.43, mp 111.8C, bp 306C, soluble in alcohol and ether, slightly soluble in water.
Derivation: From aniline by nitration after acetylation, with subsequent removal of the acetyl group by hydrolysis, from m-nitrobenzoic acid.
Hazard: Moderate fire risk. Toxic when absorbed by skin.
Use: Dry intermediate.

o-nitroaniline. (o-nitraniline). CAS: 88-74-4. $NO_2C_6H_4NH_2$.
Properties: Orange-red needles, d 1.443, mp 69.7C, soluble in alcohol and ether, slightly soluble in water, not light-fast, flash p 335F (168C), autoign temperature 970F (521C).
Derivation: From aniline by nitration after acetylation with subsequent removal of the acetyl group by hydrolysis, from o-dinitrosobenzene.
Hazard: Explosion risk. Toxic when absorbed by skin.
Use: Dye intermediate, synthesis of photographic antifogging agent, o-phenylenediamine, coccidiostats.

p-nitroaniline. (p-nitraniline).
CAS: 100-01-6. $NO_2C_6H_4NH_2$.

$$O_2N\!-\!\langle\bigcirc\rangle\!-\!\ddot{N}H_2$$

Properties: Yellow needles, d 1.437, mp 148C, flash p 390F (198C), soluble in alcohol and ether, insoluble in water. Combustible.
Derivation: (a) From p-chloronitrobenzene, (b) from aniline by nitration after acetylation, (c) from acetanilide.
Hazard: Explosion risk. Toxic when absorbed by skin. TLV: 3 mg/m³ in air.
Use: Dye intermediate, especially p-nitraniline red, intermediate for antioxidants, gasoline gum inhibitors, corrosion inhibitor.

o-nitroanisole. CAS: 91-23-6.
$C_6H_4OCH_3NO_2$.
Properties: Light reddish or amber liquid, soluble in alcohol and ether, insoluble in water, d 1.255 (20/20C), fp 9.6C, boiling range 268–271C, refr index 1.5602 (20C). Combustible.
Derivation: From o-nitrophenol by methylation or from o-nitrochlorobenzene by action of methanol and caustic soda.
Grade: Technical.
Use: Organic synthesis, intermediate for dyes and pharmaceuticals.

p-nitroanisole. (1-methoxy-4-nitrobenzene).
CAS: 100-17-4. $NO_2C_6H_4OCH_3$.
Properties: Colorless crystals, mp 54C, bp 260C, insoluble in water, soluble in alcohol and ether. Combustible.
Grade: Technical.
Use: Intermediate for dyes.

5-nitrobarbituric acid. (dilituric acid).

CAS: 480-68-2. $O_2NHCCONHCONHCO$.
Properties: Prisms and leaflets from water, mp 176C (decomposes) slightly soluble in water, soluble in alcohol and sodium hydroxide solution, insoluble in ether.
Use: Microreagent for potassium.

3-nitrobenzaldehyde. (m-nitrobenzaldehyde).
$NO_2C_6H_4CHO$.
Properties: Yellowish, crystalline powder. Mp 58C, bp 164C (23 mm). Almost insoluble in water; soluble in alcohol, chloroform, ether.
Use: Synthesis of dyes, pharmaceuticals, surface active agents, vapor phase corrosion inhibitor, antioxidant for chlorophyll, mosquito repellent.

nitrobenzene. (oil of mirbane). CAS: 98-95-3.
$C_6H_5NO_2$.

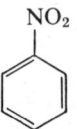

Properties: Greenish-yellow crystals or yellow, oily liquid; soluble in alcohol, benzene, and ether; slightly soluble in water. D 1.19867, mp 5.70C, bp 210.85C, flash p 190F (87.7C), autoign temperature 900F (482C). Combustible.
Derivation: From benzene by nitrating with nitric acid-sulfuric acid mix.
Method of purification: By washing and distilling with steam then redistilling.
Grade: Technical, redistilled, 97%.
Hazard: Toxic by ingestion, inhalation, and skin absorption. TLV: 1 ppm in air.
Use: Manufacture of aniline, solvent for cellulose ethers, modifying esterification of cellulose acetate, ingredient of metal polishes and shoe polishes, manufacture of benzidine, quinoline, azobenzene, etc.

p-nitrobenzeneazoresorcinol.
$NO_2C_6H_4N_2C_6H_3(OH)_2$.
Properties: Red crystals, slightly soluble in water, soluble in nitrobenzene, mp 198C. Combustible.
Derivation: Diazotized p-nitroaniline is coupled with resorcinol.
Grade: Analytical.
Use: Determination of magnesium.

m-nitrobenzenesulfonic acid. (3-nitrobenzenesulfonic acid). CAS: 5337-19-9.
$C_6H_4NO_2SO_3H$.
Properties: Crystals, mp 70C, soluble in water and alcohol.
Use: Organic synthesis. The sodium salt is a protective antireduction agent.

6-nitrobenzimidazole. CAS: 94-52-0.

$O_2NC_6H_4NHCH\!:\!N$
Properties: Solid; mp 203C; very soluble in alcohol; slightly soluble in water, ether, and benzene.
Use: Antifogging agent in photographic developers.

nitrobenzoic acid. (a) m-, (b) o-, (c) p-(nitrodracylic acid). CAS: (m-) 121-92-6; (p-) 62-23-7.
$C_6H_4(NO_2)COOH$.
Properties: Yellowish-white crystals; (a) soluble in alcohol and ether, slightly soluble in water; (b) soluble in water, alcohol, and ether; (c) solu-

ble in alcohol, sparingly soluble in water; (a) d 1.494, mp 140–141C; (b) d 1.575, mp 147.7C; (c) d 1.5497, mp 238C. Combustible.
Derivation: (a) Nitration of benzoic acid, (b) oxidation of o-nitrotoluene with MnO_2 and sulfuric acid, (c) oxidation of p-nitrotoluene by hot chromic acid mixture.
Use: (a) Dye intermediate, reagent for alkaloids, (b) and (c) organic synthesis.

m-nitrobenzotrifluoride. (3-nitrobenzotrifluoride; m-nitrotrifluoromethylbenzene).
CAS: 98-46-4. $NO_2C_6H_4CF_2$.
Properties: Pale straw, thin, oily liquid; aromatic odor. Distillation range 200.5–208.5C, fp −5.0C, d 1.437 (15.5C), bp 203C, flash p (OC) 214F (101C), 11.98 lb/gal (15.5C), viscosity 2.35 cp (37.7C), soluble in organic solvents, insoluble in water. Combustible.
Hazard: Toxic material.

m-nitrobenzoyl chloride.   CAS: 121-90-4.
$NO_2C_6H_4COCl$.
Properties: Yellow to brown liquid, partially crystallized at room temperature, mp approximately 34C, bp 278C, soluble in ether, decomposes in water and alcohol. Combustible.
Hazard: Toxic by ingestion.
Use: Manufacture of dyes for fabrics and color photography, intermediate in preparation of pharmaceuticals.

p-nitrobenzoyl chloride.   CAS: 122-04-3.
$NO_2C_6H_4COCl$.
Properties: Yellow, crystalline solid. Mp 72C, bp 154C (15 mm), decomposes in water and alcohol, soluble in ether. Combustible.
Use: Intermediate for procaine hydrochloride, dyestuffs.

p-nitrobenzyl cyanide. (p-nitro-α-tolunitrile).
CAS: 555-21-5. $NO_2C_6H_4CH_2CN$.
Properties: Crystals, mp 116–118C, insoluble in water, soluble in alcohol and ether.
Derivation: Action of concentrated nitric acid on benzyl cyanide.
Hazard: Toxic material. TLV (as CN): 5 mg/m³ of air.
Use: Intermediate for dyestuffs and pharmaceuticals, preparation of p-nitrophenylacetic acid.

o-nitrobiphenyl. (ONB; o-nitrodiphenyl).
CAS: 92-93-3. $C_6H_5C_6H_4NO_2$.
Properties: Light yellow to reddish solid or liquid, d 1.203 (25/25C), 10 lb/gal, crystallizing p 34.5C (min), refr index approximately 1.613 (25C), bp approximately 330C, flash p 290F (143C), autoign temperature 356F (180C). Soluble in carbon tetrachloride, mineral spirits, pine oil, turpentine, benzene, acetone, glacial acetic acid, and

perchloroethylene; insoluble in water. Combustible.
Derivation: By controlled nitration of biphenyl.
Hazard: Toxic by ingestion.
Use: Dye intermediate, fungicide, plasticizer for cellulosics, wood preservative.

nitrobromoform.   See bromopicrin.

2-nitro-1-butanol. $CH_3CH_2CHNO_2CH_2OH$.
Properties: Colorless liquid, soluble in water (20 g/100 cc at 20C), d 1.133 (20/20C), bp 105C (10 mm), fp −48 to −47C, bulk d 9.44 lb/gal (20C), refr index 1.4390 (20C), pH of 0.1M solution 4.51. Combustible.
Use: Organic synthesis.

nitro carbon nitrate.   A blasting agent consisting of ammonium nitrate sensitized with diesel oil. Will burn with explosive violence.
Hazard: Dangerous explosion risk, strong oxidizing agent.

nitrocellulose. (cellulose nitrate; nitrocotton; guncotton; pyroxylin).   CAS: 9004-70-0.

Properties: Pulpy, cotton-like, amorphous solid (dry); colorless liquid to semisolid (solution). Contains from 10–14% nitrogen. High-nitrogen form (explosives) is soluble in acetone, insoluble in ether-alcohol mixture. Low-N form (pyroxylin) is soluble in ether-alcohol mix and acetone (collodion and lacquers), d 1.66, flash p 55F (12.7C), autoign p 338F (170C).
Derivation: Treatment of cellulose (as cotton linters, wood pulp) with mixture of nitric and sulfuric acids. By varying strength of acids, temperature and time of reaction, and acid/cellulose ratio, widely different products are obtained.
Forms: Colloided, block, colloided, flake or granular, flakes, powder, solutions of several viscosities (from 1/4 to 1000 sec). May be dry or wet with alcohol or water.
Hazard: Flammable, dangerous fire and explosion risk. Somewhat less flammable when wet.
Use: Fast-drying automobile lacquers, high explosives, collodion, rocket propellant, printing ink base, flashless propellant powder, coating bookbinding cloth, leather finishing, manufacture of "Celluloid."
See also Hyatt.

nitrocellulose lacquer. See lacquer.

nitrochlorobenzene. Legal label name (Rail) for chloronitrobenzene.

nitrochloroform. See chloropicrin.

p-nitro-o-chlorophenyl dimethyl thionophosphate. See dicapthon.

nitrocobalamin. $C_{62}H_{90}N_{14}O_{16}PCo$.
One of the active forms of vitamin B_{12} in which a nitro group is attached to the central cobalt atom.

nitrocotton. See nitrocellulose.

2-nitro-p-cresol. (4-methyl-2-nitrophenol). $NO_2(CH_3)C_6H_3OH$.
Properties: Yellow crystals, d 1.24 (38/4C), mp approximately 35C, bp 234C, slightly soluble in water, soluble in alcohol and ether. Combustible.
Hazard: Toxic by ingestion, inhalation, and skin absorption.
Use: Intermediate.

nitrodichloro derivative. See the corresponding dichloronitro derivative.

o-nitrodiphenyl. See o-nitrobiphenyl.

o-nitrodiphenylamine. $C_6H_5NHC_6H_4NO_2$.
Properties: Red-brown, crystalline powder; mp 75–76C. Combustible.
Use: Stabilizer for nitroglycerin, chemical intermediate.

nitrodracylic acid. See p-nitrobenzoic acid.

nitro dye. A dye whose molecules contain the NO_2 chromophore group.

nitroethane. CAS: 79-24-3. $CH_3CH_2NO_2$.
A nitroparaffin.
Properties: Colorless liquid, solubility in water 4.5 cc/100 cc (20C), solubility of water in nitroethane 0.9 cc/100 (20C), d 1.052 (20/20C), fp −50C, bp 114C, vap press 15.6 mm (20C), flash p 106F (41C), autoign temperature 779F (415C), bulk d 8.75 lb/gal (20C), refr index 1.3917 (20C).
Derivation: By reaction of propane with nitric acid under pressure.
Hazard: Moderate fire risk. TLV: 100 ppm in air.
Use: Solvent for nitrocellulose, cellulose acetate, cellulose acetopropionate, cellulose acetobutyrate, vinyl, alkyd and many other resins, waxes, fats and dyestuffs, Friedel-Crafts synthesis, propellant research, fuel additive.

2-nitro-2-ethyl-1,3-propanediol.
$CH_2OHC(C_2H_5)NO_2CH_2OH$.
Properties: White, crystalline solid. Mp 56–75C, bp decomposes (10 mm), pH 0.1M aqueous solution 5.48, soluble in organic solvents, very soluble in water.
Use: Organic synthesis.

nitrofuran. Any of several synthetic antibacterial drugs used to treat mammary gland infections in cows and to inhibit disease in swine, chickens, etc. Among them are nitrofurazone, furazolidone, nihydrazone, and furaltadone. All of the latter have been found to cause cancer in laboratory animals, and their use has been discontinued.
See furazolidone.

nitrofurantoin. CAS: 67-20-9. $C_8H_6N_4O_5$.
(N-(5-nitro-2-furfurylidene)-1-aminohydantoin).
Properties: Yellow, bitter powder with slight odor. Mp (decomposes) 270–272C, very slightly soluble in alcohol and practically insoluble in ether and water.
Grade: USP.
Use: Medicine (antibacterial).

nitrogen. CAS: 7727-37-0.   N.   Gaseous element of atomic number 7 of group VA of the Periodic system, aw 14.0067, valences = 1, 2, 3, 4, 5. There are 2 stable and 4 radioactive isotopes, the molecular formula is N_2. Second highest-volume chemical produced in US (1985).
Properties: Colorless, odorless, tasteless, diatomic gas constituting approximately four-fifths of the air; colorless liquid, chemically unreactive, d 1.251 g/L (0C, 1 atm), d (gas) 0.96737 (air = 1.00), (liquid) 0.804, (solid) 1.0265, fp −210C, bp −195.5C, slightly soluble in water, slightly soluble in alcohol, an asphyxiant gas. Combustible.
Derivation: From liquid air by fractional distillation, by reducing ammonia.
Grade: USP, prepurified 99.966% min, extra dry 99.7% min, water pumped 99.6% min.
Use: Production of ammonia, acrylonitrile, nitrates, cyanamide, cyanides, nitrides; manufacture of explosives; inert gas for purging, blanketing, and exerting pressure; electric and electronic industries; in-transit food refrigeration and freeze drying; pressurizing liquid propellants; quick-freezing foods; chilling in aluminum foundries; bright annealing of steel; cryogenic preservation; food antioxidant; source of pressure in oil wells; inflating tires; component of fertilizer mixtures.

nitrogen-15. A stable isotope with an atomic mass of 15.00011 which is present in naturally occurring nitrogen to the extent of 0.37%. Many N-15 compounds are commercially available.

nitrogenase. Enzyme which fixes nitrogen and can be isolated from soil bacteria. It is possible to synthesize ammonia from nitrogen and hydrogen without high temperatures and pressures by means of nitrogenase. Pyruvic acid is an adjunct of the reaction.

nitrogen chloride.   See nitrogen trichloride.

nitrogen dioxide.   CAS: 10102-44-0. NO_2.
  Properties: Red to brown gas above 21.1C, brown liquid below 21.1C, colorless solid approximately $-11C$. The pressurized liquid is nitrogen tetroxide (dinitrogen tetroxide) because of admixture of N_2O_4 with NO_2, fp $-9.3C$, bp 21.1C. Noncombustible but supports combustion.
  Derivation: By oxidation of nitric acid, an intermediate stage in the oxidation of ammonia to nitric acid.
  Grade: Pure, 99.5% min.
  Hazard: Can react strongly with reducing materials. Inhalation instead of air may be fatal. TLV: 3 ppm in air.
  Use: Production of nitric acid, nitrating agent, oxidizing agent, catalyst, oxidizer for rocket fuels, polymerization inhibitor for acrylates.

nitrogen fixation.   Utilization of atmospheric nitrogen to form chemical compounds. In nature, this function is performed by bacteria located on the root hairs of plants, as a result, plants are able to synthesize proteins. Industrial nitrogen fixation is exemplified by the synthesis of ammonia from a gaseous mixture containing carbon monoxide, hydrogen, and nitrogen. Nitrogen can also be fixed by the electric arc process and the cyanamide process. The latter involves reaction of water and calcium cyanamide: $CaCN_2 + 3HOH \rightarrow CaCO_3 + 2NH_3$.
  See also synthesis gas.

nitrogen monoxide.   See nitrous oxide.

nitrogen mustard.   A class of compounds with fishy odor and lachrymatory properties. They are named from their similarity in structure to mustard gas (dichlorodiethyl sulfide). The sulfur of the mustard gas is replaced by an amino nitrogen. Typical nitrogen mustards are halogenated alkylamines, such as methyl bis(2-chloroethyl) amine: $(ClCH_2CH_2)_2NCH_3$. Other examples are triethylene melamine, triethylene thiophosphoramide, and triethylene phosphoramide.
  Hazard: Strong irritant to tissues, lachrymatory.
  Use: Medicine, military poison gas.
  See mechloroethamine hydrochloride.

nitrogen oxides. NO_x The nitrogen oxides provide an example of the Law of Multiple Proportions and the $+1$ to $+5$ oxidation states of nitrogen. They are as follows: N_2O (nitrous oxide), NO (nitric oxide), N_2O_3 (nitrogen trioxide or nitrogen sesquioxide), N_2O_4 (dinitrogen tetroxide or nitrogen peroxide), NO_2 (nitrogen dioxide), N_2O_5 (dinitrogen pentoxide), N_3O_4 (trinitrogen tetroxide), and NO_3, which is unstable.
  Hazard: Toxic by inhalation, especially NO_2.

nitrogen solutions.   Aqueous solutions of ammonium nitrate, ammonia, and/or urea. They are graded according to total nitrogen content and composition.
  Use: Direct application to soil as fertilizer, neutralizing superphosphate fertilizers.

nitrogen tetroxide.   See nitrogen dioxide.

nitrogen trichloride.   (nitrogen chloride).
  CAS: 10025-85-1. NCl_3.
  Properties: Yellow oil or rhombic crystals, d 1.653, bp below 71C, fp below $-40C$. Insoluble in cold water; decomposes in hot water; soluble in chloroform, phosphorus trichloride, and carbon disulfide.
  Hazard: Explodes when heated to approximately 200F (93C) or when exposed to direct sunlight. Toxic by ingestion and inhalation, strong irritant.
  Use: Flour bleach (no longer permitted in US).

nitrogen trifluoride.   (nitrogen fluoride).
  CAS: 7783-54-2. NF_3.
  Properties: Colorless, unstable gas having a moldy odor. D 1.537 ($-129C$), gas d 0.1864 lb/ft³ (70F), fp $-206.6C$, bp $-128.8C$, very slightly soluble in water. Nonflammable.
  Derivation: Electrolysis of ammonium acid fluoride.
  Hazard: Severe explosion hazard. Corrosive to tissue. TLV: 10 ppm in air.
  Use: Oxidizer for high-energy fuels, chemical synthesis.

nitrogen triiodide.   (nitrogen iodide).
  CAS: 13444-85-4. NI_3.
  Properties: Black, unstable crystals.
  Hazard: Explodes at slightest touch when dry; when handled it should be kept wet with ether. Too sensitive to be used as explosive, as it cannot be stored, handled, or transported safely.

nitrogen trioxide.   (dinitrogen trioxide; nitrogen sesquioxide).   CAS: 10544-73-7. N_2O_3.
  Properties: Blue liquid, d 1.447 (2C), bp 3.5C (1 atm), fp $-102C$ (1 atm). Partially dissociated into nitric oxide (NO) and nitrogen dioxide (NO_2).
  Derivation: Prepared by passing nitric oxide into nitrogen tetroxide at approximately 0C until the stoichiometric amount has been absorbed.
  Hazard: Toxic by inhalation, strong irritant.

Use: Oxidant in special fuel systems, identification of terpenes, preparation of pure alkali nitrites.

nitroglycerin. (glyceryl trinitrate; trinitroglycerin). CAS: 55-63-0. $CH_2NO_3CHNO_3CH_2NO_3$.
Properties: Pale yellow, viscous liquid. Soluble in alcohol and ether, slightly soluble in water, d 1.6009, fp 13.1C, explosion p 424F (218C).
Derivation: By dropping glycerol through cooled, mixed acid and stirring, followed by repeated washing with water.
Grade: Technical.
Hazard: Severe explosion risk, highly sensitive to shock and heat. Toxic by ingestion, inhalation, and skin absorption. TLV: 0.05 ppm in air.
Use: High explosive, production of dynamite and other explosives, medicine (vasodilator) combating fires in oil wells, rocket propellants.

nitroguanidine. CAS: 556-88-7. $H_2NC(NH)NHNO_2$. Exists in two forms, alpha and beta.
Properties: (a) Long, thin, flat, flexible lustrous needles; (b) small, thin, elongated plates. Both melting ranges are from 220–250C. The beta-form appears to be more soluble in water.
Derivation: (a) Results when guanidine nitrate is dissolved in concentrated sulfuric acid and the solution is poured into water. (b) Nitration of a mixture of guanidine sulfate and ammonium sulfate which results from the hydrolysis of dicyandiamide by sulfuric acid.
Hazard: May explode when shocked or heated.
Use: High explosives, especially flashless propellant powder (with nitrocellulose), chemical intermediate.

nitrohydrochloric acid. Legal label name for aqua regia.

3-nitro-2-hydroxybenzoic acid. See m-nitrosalicylic acid.

4-nitro-3-hydroxymercuri-o-cresol anhydride. See nitromersol.

3-nitro-4-hydroxyphenylarsonic acid. See 4-hydroxy-3-nitrobenzenearsonic acid.

nitromannite. See mannitol hexanitrate.

nitromersol. (4-nitro-3-hydroxymercuri-o-cresol anhydride). $C_6H_2(CH_3)(NO_2)(OHg)$.
Properties: Brownish-yellow or yellow granules or powder; odorless and tasteless. Insoluble in water; almost insoluble in alcohol, acetone, ether; soluble in solutions of alkalies, ammonia, by opening the anhydride ring and salt formation.

Grade: NF.
Hazard: Toxic by ingestion and inhalation.
Use: Disinfectant for skin in extremely dilute soluble (1 part in 1000 is maximum strength).

nitromethane. CAS: 75-52-5. CH_3NO_2. A nitroparaffin.
Properties: Colorless liquid, soluble in water and alcohol, solubility of water in nitromethane 2.2 cc/100 cc (20C), d 1.139 (20/20C), bp 101C, vap press 27.8 mm (20C), flash p 95F (35C), bulk d 9.5 lb/gal, refr index 1.3817 (20C), fp −29C, autoign temperature 785F (418C).
Derivation: By reaction of methane or propane with nitric acid under pressure.
Hazard: Dangerous fire and explosion risk, lower explosion limit 7.3% in air. Toxic by ingestion and inhalation. TLV: 100 ppm in air.
Use: Solvent for cellulosic compounds, polymers, waxes, fats, etc.; chemical synthesis; rocket fuel; gasoline additive.

2-nitro-4-methoxyaniline.
$NO_2C_6H_3(OCH_3)NH_2$. Orange-red powder, mp 118–120C, sparingly soluble in cold water and alcohol, soluble in hot water and dioxane.
Use: Chemical intermediate.

2-nitro-2-methyl-1,3-propanediol.
$CH_2OCH(CH_3)NO_2CH_2OH$.
Properties: White, crystalline solid. Solubility in water 80 g/100 cc (20C), mp 147–149C, bp decomposes (10 mm), pH 0.1M solution 5.42.
Hazard: Toxic by ingestion.
Use: Intermediate.

nitron. (1,4-diphenyl-3,5-endo-anilino-4,5-dihydro-1,2,4-triazole). CAS: 487-88-7. $C_{20}H_{16}N_4$.
Properties: Lemon-yellow, fine, crystalline needles; soluble in chloroform, acetone, and acetic acid ester; slightly soluble in ether and alcohol; insoluble in water.
Use: Reagent for detection of nitrate, perchlorate, boron, rhodium.

α-nitronaphthalene. (1-nitronaphthalene). CAS: 86-57-7. $C_{10}H_7NO_2$.
Properties: Yellow crystals, soluble in alcohol and ether, insoluble in water, d 1.331, mp 55–56C, bp 304C, flash p 327F (164C). Combustible.
Derivation: Action of a mixture of nitric and sulfuric acids on finely ground naphthalene.
Hazard: Irritant. The beta isomer is highly toxic by ingestion.
Use: Dyes, naphthylamine, added to mineral oils to mask fluorescence.

1-nitronaphthalene-5-sulfonic acid. See Laurent's alpha acid.

nitronium perchlorate. NO_2ClO_4.
Properties: White, crystalline solid; mp 120–140C; hygroscopic; noncorrosive. Soluble in water to form nitric and perchloric acids, highly reactive.
Derivation: From ozone, nitrogen dioxide, chlorine dioxide.
Hazard: Strong oxidizing agent, may explode in contact with organic materials. Irritant to skin and mucous membranes.
Use: Suggested as propellant oxidizer.

nitroparaffin. (nitroalkane).   Any of a homologous series of compounds whose generic formula is $C_n H_{2n+1}NO_2$, the nitro groups being attached to a carbon atom via the nitrogen.
Properties: Colorless, mobile liquids; pleasant odor. Bp range 101–131C, fp range −18 to −104C, d range: 0.983–1.131, dielectric constant range 23–35, slightly soluble in water, flash p range 75–100F (24–38C).
Hazard: Toxic by inhalation.
Derivation: Nitration of propane and other paraffin hydrocarbons under pressure.
Use: Solvents for dyes, waxes, resins, gums, and various polymeric substances; gravure and flexographic inks; propellants and fuel additives; chemical intermediates.
See also nitroethane, nitromethane, nitropropane.

p-nitrophenetole. CAS: 100-29-8.
$NO_2C_6H_4OC_2H_5$.
Properties: Crystals in prisms, soluble in alcohol and ether, mp 58C, bp 283C.
Derivation: By ethylation of p-nitrophenol with ethyl chloride.
Use: Dyes and other intermediates.

nitrophenide. [bis(3-nitrophenyl)disulfide].
CAS: 537-91-7. $(NO_2C_6H_4)_2S$.
Properties: Yellow, rhomboid crystals; mp 83C. Insoluble in water, soluble in ether, slightly soluble in alcohol.
Derivation: Reduction of m-nitrobenzenesulfonyl chloride with hydriodic acid.
Use: Pharmaceutical intermediate.

m-nitrophenol. CAS: 554-84-7.
$NO_2C_6H_4OH$.

Properties: Pale yellow crystals, d 1.485 (20C), mp 96–97C, bp 194C (70 mm), slightly soluble in water, soluble in alcohol. Combustible.
Derivation: Diazotized m-nitroaniline is boiled with water and sulfuric acid.

Hazard: Toxic by ingestion.
Use: Indicator.

o-nitrophenol. CAS: 88-75-5. $NO_2C_6H_4OH$.

Properties: Yellow crystals, d 1.295 (45C), 1.657 (20C), mp 44–45C, soluble in hot water, alcohol, ether.
Derivation: Action of dilute nitric acid on phenol at low temperature, p-nitrophenol formed at same time. They are separated by steam distillation.
Hazard: Toxic by ingestion.
Use: Intermediate in organic synthesis, indicator, reagent for glucose.

p-nitrophenol. CAS: 100-02-7.
$NO_2C_6H_4OH$.

Properties: Yellowish, monoclinic, prismatic crystals. D 1.479–1.495 (20C), mp 111.4–114C (sublimes), bp 279C (decomposes). Soluble in hot water, alcohol, ether.
Derivation: (a) From p-chloronitrobenzene, (b) as in o-nitrophenol.
Hazard: Toxic by ingestion.
Use: Intermediate in organic synthesis, production of parathion, fungicide for leather, indicator.
See o-nitrophenol.

p-nitrophenol, sodium salt. (sodium-p-nitrophenolate; p-nitro sodium phenolate).
$NO_2C_6H_4ONa$.   Yellow, crystalline solid; soluble in water. Intermediate.
Hazard: Toxic by ingestion.

p-nitrophenylacetic acid. (p-nitro-α-toluic acid).
CAS: 104-03-0. $NO_2C_6H_4CH_2COOH$.
Properties: Yellow needles, mp 152.3C, slightly soluble in cold water, soluble in alcohol and chloroform. Combustible.
Derivation: Hydrolysis of p-nitrobenzyl cyanide with 50% sulfuric acid.
Use: Intermediate for dyestuffs, pharmaceuticals, penicillin precursors, local anesthetics.

p-nitrophenylhydrazine. CAS: 100-16-3.
$C_6H_7N_3O_2$.
Properties: Reddish to orange crystals; mp 155C (decomposes); soluble in hot water, alcohol, and chloroform.
Use: Analytical reagent (aldehydes, ketones).

o-**nitrophenylpropiolic acid.**   CAS: 530-85-8.
$C_9H_5NO_4$.
  Properties: Yellow to brown crystals, mp 157C
  (decomposes), partially soluble in hot water and
  alcohol, insoluble in carbon disulfide.
  Hazard: Decomposition may occur with explosive
  violence.
  Use: Analytical reagent (alkaloids).

p-**nitrophenylazosalicylate sodium.**   See alizarin
  yellow R.

"Nitrophoska."[440]   TM for a group of complete
  (N-P-K) fertilizers for agricultural and horticul-
  tural crops.

nitrophosphate.   A nitrogen-phosphorus fertilizer
  produced by the action of nitric acid or a mixture
  of nitric and sulfuric or phosphoric acids on
  phosphate rock. Potassium salts usually are
  added to produce complete fertilizers. Typical
  analysis: Available nitrogen 15%, available P_2O_5
  15%, available K_2O 15%.
  Use: Fertilizer.
  See also superphosphate, triple superphosphate.

1-nitropropane.   CAS: 108-03-2.
  $CH_3CH_2CH_2NO_2$.   A nitroparaffin.
  Properties: Colorless liquid, soluble in water 1.4
  mL/100 mL (20C), solubility of water in 1-nitro-
  propane 0.5 cc/100 cc (20C), d 1.003 (20/20C),
  miscible with organic solvents, bp 132C, vap
  press 7.5 mm (20C), flash p (TOC) 93F (34C),
  bulk d 8.4 lb/gal (20C), refr index 1.4015 (20C),
  fp −108C, autoign temperature 789F (420C).
  Derivation: By reaction of propane with nitric acid
  under pressure.
  Hazard: Flammable, moderate fire risk, moderate
  explosion hazard when shocked or heated. TLV:
  25 ppm in air.
  Use: Solvent, chemical synthesis, rocket propellant, gasoline additive.

2-nitropropane.   CAS: 79-46-9.
  $CH_3CHNO_2CH_3$.   A nitroparaffin.
  Properties: Colorless liquid, soluble in water 1.7
  mL/100 mL (20C), solubility of water in 2-nitro-
  propane 0.6 cc/100 cc (20C), miscible with organic solvents, d 0.992 (20/20C), bp 120C, vap
  press 12.9 mm (20C), flash p (TOC) 75F (24C),
  bulk d 8.3 lb/gal (20C), refr index 1.3941 (20C),
  fp −93C, autoign temperature 802F (427C).
  Derivation: By reaction of propane with nitric acid
  under pressure.
  Hazard: Flammable, dangerous fire risk, moderate
  explosion hazard when shocked or heated. Carcinogen. TLV: CL 25 ppm in air.
  Use: Solvent especially for vinyl and epoxy coat-

ings, chemical synthesis, rocket propellant, gasoline additive.

m-**nitrosalicyclic acid.**   (3-nitro-2-hydroxybenzoic
  acid). $C_6H_3COOH(OH)NO_2$.
  Properties: Yellowish crystals, soluble in benzene
  and alcohol, mp 144C.
  Derivation: Nitration of salicylic acid.
  Use: Intermediate, azo dyes.

nitrosamine.   Any of a series of organic compounds in which $=N-N=O$ is attached to
  alkyl or aryl group, e.g., diphenylnitrosamine $(C_6H_5)_2NNO$ or dimethylnitrosamine
  $(CH_3)_2NNO$. Such compounds are formed by
  reaction between an amine and NO_x or nitrites.
  They occur in many food products, whisky, herbicides and cosmetics, as well as in industrial
  environments such as tanneries, rubber factories
  and iron foundries. They are also formed within
  the body by reaction of amine-containing drugs
  with the nitrites resulting from bacterial conversion of nitrates. Nitrosamines have been found
  to be strong carcinogens in experimental animals.

N-nitrosodimethylamine.   (dimethylnitrosamine;
  DMN).   CAS: 67-75-9. $(CH_3)_2N_2O$.
  Properties: Yellow liquid; soluble in water, alco-
  hol, and ether; bp 152C; d 1.006. Combustible.
  Hazard: A carcinogen, often present in fish meal.
  Use: Rocket fuels, solvent, rubber accelerator.

p-**nitrosodiphenylamine.**   (diphenylnitrosamine;
  nitrous diphenylamide).   CAS: 156-10-5.
  $(C_6H_5)_2NNO$.
  Properties: Greenish crystals; mp 145C; soluble
  in alcohol, benzene, ether, sulfuric acid.
  Hazard: Carcinogen.
  Use: Rubber accelerator.

p-**nitro sodium phenolate.**   See p-nitrophenol, sodium salt.

nitroso dyes.   (quinone oxime dyes).
  Dyes whose molecules contain the $-N=O$ or
  $=N-OH$ chromophore group.

nitroso ester terpolymer.   See nitroso polymer.

nitrosoguanidine.   CAS: 674-81-7.
  $ONNHC(NH)NH_2$.   Pale yellow, crystalline
  powder.
  Derivation: Cooling the reaction products of nitro-
  guanidine, ammonium chloride and zinc dust.
  Hazard: Severe explosion risk.
  Use: An initiating explosive.

1-nitroso-2-naphthol.   (naphthoquinone oxime;
  [one of several isomers] α-nitroso-β-naphthol).
  CAS: 131-91-9. $C_{10}H_7NO_2$.
  Properties: Yellow needles, soluble in alcohol and
  ether, insoluble in water, mp 110C. Combustible.

Derivation: Action of nitrous acid on β-naphthol.
Use: Organic synthesis, prevention of gum formation in gasoline, dye intermediate, analytical reagent.

p-nitrosophenol.   CAS: 104-91-6.
C_6H_4OHNO.
Properties: Crystals of light brown leaflets which decompose at 140C; soluble in alcohol, ether, and acetone; moderately soluble in water; mp 140C.
Derivation: From phenol by action of nitrous acid in the cold.
Hazard: Dangerous fire and explosion risk, reacts violently with acids and alkalies, may explode spontaneously, intraplant transport must be in tightly covered steel barrels.
Use: Dyes.

nitroso polymer. (nitroso rubber).   A flame- and heat-resistant copolymer derived from trifluoronitrosomethane (CF_3NO) and tetrafluoroethylene ($F_2C=CF_2$). A third monomer (methyl-4-nitrosoperfluorobutyrate) provides more easily crosslinked functional groups than heretofore available. This terpolymer called nitroso ester terpolymer (NET) is a notable improvement on the earlier copolymer and is especially effective for fireproof interiors of aerospace vehicles and airplanes. Crosslinking is by a peroxide mechanism. Properties of the terpolymer are a glass transition temperature of approximately −50C, decomposition temperature of 275C, tensile strength 750 psi, elongation 500%, Shore hardness 78. Nonflammable.

nitrostarch. (starch nitrate).   CAS: 9056-38-6.
$C_{12}H_{12}(NO_2)_8O_{10}$.
Properties: Orange-colored powder, contains 16.5% nitrogen, soluble in ether-alcohol.
Hazard: Flammable, dangerous fire risk, severe explosion risk when dry.
Use: Explosives.

β-nitrostyrene. $C_6H_5CH:CHNO_2$.
Properties: Liquid, mp 58C, available as a 30% solution in styrene.
Hazard: Moderate fire and explosion risk.
Use: Chain stopper in styrene type polymerization.

p-nitrosulfathiazole.   CAS: 473-42-7.
$NO_2C_6H_4SO_2NHC_3H_2NS$.
Properties: Pale yellow powder, odorless, slightly bitter taste, mp 258–266C. Slightly soluble in alcohol; very slightly soluble in chloroform, ether and water; practically insoluble in benzene.
Use: Medicine (antibacterial).

nitrosyl chloride.   CAS: 2696-92-6.   NOCl.
One of the oxidizing agents in aqua regia.
Properties: Yellow-red liquid or yellow gas, bp −5.5C, fp −61.5C, d (gas) 2.3 (0C, air = 1), d (liquid): 1.273 (20C), decomposed by water, dissociates into nitric oxide and chlorine on heating, soluble in fuming sulfuric acid. Nonflammable.
Derivation: By action of chlorine on sodium nitrate, also from nitrosylsulfuric acid and hydrochloric acid.
Grade: Pure, 93% min.
Hazard: Strong irritant, especially to lungs and mucous membranes.
Use: Synthetic detergents, catalyst, intermediate.

nitrosyl fluoride.   CAS: 7789-25-5.   NOF.
Properties: Colorless gas, fp −132C, bp −60C, d (liquid): 1.32, (solid) 1.72, attacks glass severely and corrodes quartz.
Hazard: Toxic by inhalation, irritant to skin and mucous membranes.
Use: Fluorinating agent, rocket fuel oxidizer.

nitrosylsulfuric acid.   CAS: 7782-78-7.
HNO_5S.
Properties: Straw-colored, oily liquid; furnished as a 40% solution in 87% sulfuric acid; stable at room temperature.
Hazard: Strong irritant to skin and mucous membranes.
Use: Diazotizing agent for dyes, chemical intermediate, drugs and pharmaceuticals.

m-nitrotoluene. (m-methylnitrobenzene).
CAS: 99-08-1. $NO_2C_6H_4CH_3$.
Properties: Yellow liquid, d 1.1571 (20/4C), fp 15C, bp 232.6C, refr index 1.5466 (20C), insoluble in water, soluble in alcohol and ether, flash p 223F (106). Combustible.
Hazard: Toxic by inhalation, ingestion, skin absorption. TLV: 2 ppm in air.
Use: Organic synthesis.

o-nitrotoluene. (o-methylnitrobenzene).
CAS: 88-72-2. $NO_2C_6H_4CH_3$.
Properties: Yellow liquid, d 1.1629 (20C), fp −9.3C, bp 220.4C, refr index 1.544 (25C), insoluble in water, miscible with alcohol and benzene, flash p 223F (106C). Combustible.
Derivation: From toluene by nitration and separation by fractional distillation.
Hazard: Toxic by inhalation, ingestion, skin absorption. TLV: 2 ppm in air.
Use: For production of toluidine, tolidine, fuchsin and various synthetic dyes.

p-nitrotoluene. (p-methylnitrobenzene).
CAS: 99-99-0. $NO_2C_6H_4CH_3$.
Properties: Yellow crystals, d 1.299 (0/4C), mp 51.7C, bp 238.3C, refr index 1.5382 (15C). In-

soluble in water; soluble in alcohol, ether, and benzene; flash p 223F (106C). Combustible.
Derivation: From toluene by nitration and separation by fractional distillation.
Hazard: Toxic by inhalation, ingestion, skin absorption. TLV: 2 ppm in air.
Use: For production of toluidine, fuchsin, and various synthetic dyes.

p-nitro-α-toluic acid. See p-nitrophenylacetic acid.

m-nitro-p-toluidine. (3-nitro-4-toluidine). $NO_2C_6H_3(CH_3)NH_2$.
Properties: Orange-red crystals, soluble in alcohol and concentrated sulfuric acid, mp 117C, flash p 315F (157C). Combustible.
Derivation: From acetyl-p-toluidine by nitration.
Hazard: Toxic by ingestion.
Use: Intermediate for dyes and pigments.

p-nitro-o-toluidine. (4-nitro-2-toluidine). $NO_2C_6H_3(CH_3)NH_2$.
Properties: Yellow, crystalline solid; soluble in alcohol and ether; mp 104C; flash p 315F (157C). Combustible.
Derivation: From o-toluidine by nitration.
Use: Intermediate for dyes and pigments.

p-nitro-α-tolunitrile. See p-nitrobenzyl cyanide.

nitrotrichloromethane. See chloropicrin.

m-nitrotrifluoromethylbenzene. See m-nitrobenzotrifluoride.

2-nitro-4-trifluoromethylbenzonitrile. $NO_2(CF_3)C_6H_3CN$.
Properties: Liquid, bp 156–158C (18–19 mm), mp 47–48C.
Derivation: From a cyclic halogen compound by heating with copper cyanide in the presence of amines.
Use: Dyes.

2-nitro-4-trifluoromethyl chlorobenzene. $NO_2C_6H_3(CF_3)Cl$.
Properties: Pale yellow oil; fp −6F; soluble in acetone, alcohol, and other solvents; insoluble in water.
Use: Intermediate for dyes and organic chemicals.

nitrourea. CAS: 556-89-8. $NH_2CONHNO_2$.
Properties: White, crystalline powder; slightly soluble in water; soluble in alcohol, acetone, and acetic acid; mp 158–159C.
Derivation: Reaction of urea nitrate and concentrated sulfuric acid.
Hazard: Severe explosion risk.
Use: Explosives.

nitrous acid. CAS: 7782-77-6. HNO_2.
Properties: A weak acid occurring only in the form of a light blue solution.
Derivation: Reaction of strong inorganic acids with nitrites, e.g., hydrochloric acid + $NaNO_2$.
Use: Formation of diazotizing compounds by reaction with primary aromatic amines, source of nitric oxide.

nitrous diphenylamide. See N-nitrosodiphenylamine.

nitrous oxide. (nitrogen monoxide). CAS: 10024-97-2. N_2O.
Properties: Colorless, sweet-tasting gas; d 1.52 referred to air, liquid 1.22 (−89C); fp −90.8C; bp −88.5C; soluble in alcohol, ether, and concentrated sulfuric acid; slightly soluble in water; an asphyxiant, anesthetic gas. Noncombustible.
Derivation: Thermal decomposition of ammonium nitrate, controlled reduction of nitriles or nitrates.
Grade: Pure, 98.0% min, USP (97% min).
Hazard: Supports combustion, can form explosive mixture with air. Narcotic in high concentration.
Use: Anesthetic in dentistry and surgery, propellant gas in food aerosols, leak detection.

nitroxanthic acid. See picric acid.

nitroxylene. (dimethylnitrobenzene). CAS: 25168-04-1. $C_6H_3(CH_3)_2NO_2$.
There are three isomers (2,4-; 3,4-; and 2,5-).
Properties: Yellow liquid (2,4- and 2,5-), or crystalline needles (3,4-), d 1.135, fp 2C, bp 246C, soluble in alcohol and ether, insoluble in water.
Derivation: By nitrating xylene and separating from the resulting mixture by rectification.
Hazard: Toxic by ingestion, inhalation, and skin absorption.
Use: Organic synthesis, gelatinizing accelerators for pyroxylin.

nitryl chloride. CAS: 13444-90-1. $ClNO_2$.
Properties: Colorless gas with odor of chlorine, bp −14C, fp −145C, d (liquid) 1.33, liquid and solutions have yellow tinge.
Derivation: Reaction of chlorosulfonic acid with nitric acid.
Hazard: May explode on contact with organic materials. Corrosive to tissue.
Use: Organic synthesis (nitrating and chlorinating agent).

nitryl fluoride. CAS: 10022-50-1. FNO_2.
Properties: Colorless gas, fp −165C, bp −72C, d (liquid): 1.80, strong oxidizing agent, hydrolyzes to form nitric and hydrogen fluoride acids.

Derivation: Reaction of fluorine with nitrogen dioxide.

Hazard: Ignites on contact with selenium, iodine, phosphorus, arsenic, antimony, boron, silicon, molybdenum. Corrosive to tissue.

Use: Rocket propellants, fluorinating agent.

"Nivco."[308]   TM for a precipitation-hardening cobalt-base alloy with good strength-to-weight ratio for use at temperatures up to 1200F. It contains somewhat greater than 20% nickel and small amount of titanium, zirconium, and iron. Damping material for steam turbine blade applications.

nm.   Abbreviation for nanometer.

NMR.   Abbreviation for nuclear magnetic resonance.

No.   Symbol for the element nobelium.

Nobel, Alfred B.   (1833–1896) A native of Sweden, Nobel devoted most of his life to a study of explosives and was the inventor of a mixture of nitroglycerin and diatomaceous earth which he called dynamite. He also invented blasting gelatin and smokeless powder. With the fortune he accumulated from his work, Nobel established the foundation that bears his name, which annually recognizes outstanding work in physics, chemistry, medicine, literature, and human relations. The Nobel prize is still the world's most valued scientific award.

nobelium.   CAS: 10028-14-5.   No.
Synthetic radioactive element no. 102, aw 254, one of the actinide series of elements; its discovery has been claimed by research groups in Sweden, USSR, and California. It can be produced in a cyclotron by bombarding copper with nuclei of carbon-13 accelerated to high energies. The name nobelium has been accepted by the IUPAC Commission on Nomenclature of Inorganic Chemistry. It has nine isotopes (251–259) that are so short-lived that its chemical properties have not been determined. It has no known uses or compounds.

noble.   In chemical terminology, this term describes an element which either is completely unreactive or reacts only to a limited extent with other elements. Six noble gases constitute a group in the Periodic Table that is variously called the zero group (as the first three of its members have a valence of 0), the inert gas group, and the noble gas group. The last is preferable, as three of the gases, though unreactive, are not inert. The gases of this group are helium, neon, argon, krypton, xenon, and radon. The noble metals are generally considered to be gold, silver, platinum, palladium, iridium, rhenium, mercury, ruthenium, and osmium. The term has no reference to their commercial value.
See also inert.

NODA.   Abbreviation for n-octyl-n-decyl adipate.

nodular iron.   A ductile form of cast iron.

nodules.   Aggregates of ferromanganese occurring in huge quantities on the ocean floors. It has been estimated that they cover approximately 10% of the bottom of the Pacific. They are also present in other oceans and even in large lakes. The size averages approximately 4 cm, varying from small pellets to masses several meters in diameter. Their composition is approximately 55% manganese and 35% iron, the balance being cobalt, copper, and nickel. Their origin has not been definitely established. Proposals for commercial exploitation have been made but no economically satisfactory method has been developed. Considerations of international law and "ownership" of the sea floor are also involved.

"Nolane."[342]   TM for 2,6-allyl-dichlorosilaneresorcinol.
Use: Priming glass fibers, i.e., prior to laminating.

nomenclature, chemical.   See chemical nomenclature.

"Nomex."[28]   TM for an aramid fiber or fabric.
Use: Bag filters for industrial dust control and insulation of spacecraft.
See also aramid.

nomograph.   A chart in which a straight line, either drawn or indicated by a ruler, intersects three scales at points that represent values that satisfy an equation or a given set of physical conditions. It thus enables one to determine the value of a dependent variable when that of the independent variable is known.
Use: Chemical engineers to determine relationships between the properties of materials.

nonadecane. $C_{19}H_{40}$ or $CH_3(CH_2)_{17}CH_3$.
Properties: Leaflets, soluble in alcohol and ether, insoluble in water, d 0.777, mp 32C, bp 330C. Combustible.
Use: Organic synthesis.

n-nonadecanoic acid. $CH_3(CH_2)_{17}COOH$.
A saturated fatty acid normally not found in natural vegetable fats or waxes. Combustible.
Properties: Colorless crystals, mp 68.7C, bp 297C (100 mm), soluble in alcohol and ether, insoluble

in water. Synthetic product available 99% pure for organic synthesis.

γ-nonalactone. See γ-nonyl lactone.

nonanal. (pelargonic aldehyde; n-nonyl aldehyde; aldehyde C-9).   CAS: 124-19-6. $C_8H_{17}CHO$.
Properties: Colorless liquid with an orange-rose odor, d 0.822–0.830, refr index 1.424–1.429, soluble in three volumes of 70% alcohol, in mineral oil; insoluble in glycerol. Combustible.
Grade: Technical, FCC.
Use: Perfumery, flavoring agent.

nonane. (nonyl hydride).   CAS: 111-84-2. C_9H_{20} or $CH_3(CH_2)_7CH_3$.
Properties: Colorless liquid, soluble in alcohol, insoluble in water, d 0.722, bp 150.7C, fp −54C, refr index 1.40561 (20C), flash p (CC) 86F (30C), autoign temperature 403F (206C).
Grade: Technical (95%), 99%, research.
Hazard: Flammable, moderate fire risk. Irritant, narcotic in high concentration. TLV: 200 ppm in air.
Use: Organic synthesis, biodegradable detergents, distillation chaser.

nonanedioic acid. See azelaic acid.

nonane-1,3-diol monoacetate. $C_9H_{18}OOCCH_3$.
Properties: A mixture of isomers, colorless to slightly yellow liquid, stable, soluble in alcohol. Combustible.
Use: Perfumery, flavoring.

n-nonanoic acid. See pelargonic acid.

nonanol. See nonyl alcohol.

nonallyl chloride. See pelargonyl chloride.

noncombustible material. A solid, liquid, or gas that will not ignite or burn, regardless of how high a temperature it is exposed to, e.g., silicon dioxide, water, carbon dioxide.
See also nonflammable material.

nondestructive testing. A test that does not involve destruction of the sample or material tested. Such tests are usually carried out by radiographic methods on large, finished metal products (castings and machine components) to determine the presence of internal defects likely to cause operational failure. An infrared camera scanning device for detection of internal weaknesses in tires is a recent development in nondestructive testing. X-radiation is used to determine the authenticity of paintings and other objects of art.
See also testing.

nondrying oil. See drying oil.

nonelectrolyte. A compound that resists passage of electricity (nonconductor) both in liquid form and in solution. Included in this classification are most organic compounds, with the exception of acids and amides, and such inorganic compounds as nonmetal halides. Nonelectrolytes are covalently bonded; some inorganic compounds having covalent bonds form electrolytic solutions in aqueous solution.
See also electrolyte.

nonene. See nonylene.

nonflammable material. A gas or liquid which will not burn under normal conditions. Do not confuse with combustible materials.
See combustible material and note under flammable material.

"Nonic."[204]   TM for a group of non-ionic surface-active agents based on polyethylene glycol tert-dodecylthioether or related compounds.
Use: Detergents, paper manufacture, grease removal, wetting agents, emulsifier, clouding agent.

nonmetal. (1) Any of a number of elements whose electronic structure, bonding characteristics, and consequent physical and chemical properties differ markedly from those of metals, particularly in respect to electronegativity and thermal and electrical conductivity. In general, nonmetals have very low to moderate conductivity and high electronegativity. The 25 elements classified as nonmetals may be considered in two groups: (a) those having moderate electrical conductivity (semiconductors), all of which are solids, and (b) those having very low conductivity, many of which are gases. The semiconductors of group (a) were formerly called metalloids since they more nearly resemble metals than those of group (b), but this term is no longer used by chemists. The nonmetals are given below based on this subgrouping, though any such list is open to challenge:

	(a)		(b)
arsenic	polonium	halogens
antimony	phosphorus	hydrogen
boron	selenium	nitrogen
carbon	silicon	oxygen
germanium	tellurium	noble gases
		sulfur

(2) Loosely, any material that is not a metal, e.g., petroleum, plastics, waxes, etc. The term

is widely used in this sense by engineers and specification writers.

n-nonoic acid. See pelargonic acid.

nonoic acid. An acid of the formula $C_8H_{17}COOH$ of which there are many possible isomers. Pelargonic acid is the normal or straight-chain acid. A mixture of various branched chain nonoic acids is recovered from products of the Oxo process.

nonviscous neutral oil. A neutral oil of viscosity less than 135 SUS at 37.7C.

nonwoven fabric. A fabric made from staple lengths of cotton, rayon, glass, or thermoplastic synthetic fiber mechanically positioned in a random manner, and usually bonded with a synthetic adhesive or rubber latex. The sheets thus formed can be pressed together to form porous mats of high absorptivity and good elastic recovery on deformation. Permanent bonds are formed where the fibers touch each other as a result of heat treatment when the fibers are thermoplastic, or by use of a high-polymer binder. Disposable filters of polyethylene and polyester can be made without a binder.

A specialty nonwoven (so-called "melded") fabric trademarked "Cambrelle" is a composite of two different polymers; heating the exterior layer to its melting point causes the fibers to fuse into a fabric.
Use: Applications are as a backing for plastic film, padding for surgical dressings, diapers, sanitary napkins, drapes and other decorative textile products, filtration, shoe liners, industrial wiping and polishing fabrics, disposable clothing, carpet backing.

n-nonyl acetate. (acetate C-9).
CAS: 143-13-5. $CH_3COO(CH_2)_8CH_3$.
Properties: Colorless liquid, strong pungent odor, d 0.864–0.868, refr index 1.422–1.426, soluble in four volumes of 70% alcohol, flash p 155F (68.3C), several isomers exist. Combustible.
Use: Perfumery, flavoring.

n-nonyl alcohol. (nonanol; alcohol C-9; octyl carbinol; pelargonic alcohol). CAS: 143-08-8. $CH_3(CH_2)_7CH_2OH$.
Properties: Colorless liquid with floral odor, d 0.826–0.829, refr index 1.431–1.435, fp −5C, bp 215C, flash p 165F (73.9C), soluble in seven volumes of 50% alcohol, insoluble in water, several isomers exist. Combustible.
Use: Perfumery, flavoring (lemon oil).

n-nonylamine. $C_9H_{19}NH$.
Properties: Colorless liquid, bp 75–85C (20 mm), d 0.798 (25C), refr index 1.4366 (20C), may be prepared from nonyl halides by conventional techniques. Combustible.
Hazard: Moderate fire risk. Skin irritant.

tert-nonylamine. Principally tert-$C_9H_{19}NH_2$ and tert-$C_{10}H_{21}NH_2$.
Properties: Colorless liquid, boiling range 160–174C, d 0.789 (25C), refr index 1.428 (25C), flash p 120F (48.9C), insoluble in water, soluble in common organic solvents, especially in petroleum hydrocarbons. Combustible.
Hazard: Moderate fire risk. Skin irritant.
Use: Intermediate for rubber accelerators, insecticides, fungicides, dyestuffs, pharmaceuticals.

nonylbenzene. (1-phenylnonane). $C_9H_{19}C_6H_5$.
Properties: Light-straw liquid, aromatic odor, boiling range 245–252C, d 0.864 (20/20C), refr index 1.488 (20C), viscosity 41.9 cp (20C), flash p (CC) 210F (98.9C). Combustible.
Hazard: Irritant; narcotic in high concentration.
Use: Manufacture of surface-active agents.

nonyl bromide. $C_9H_{19}Br$.
Properties: Liquid, bp 81–85C (10 mm), d 1.101 (25C), refr index 1.4583 (20C). Nonflammable.
Derivation: High yields of nonyl bromide are obtained by passing hydrogen bromide into the alcohol while heating or by refluxing with aqueous hydrogen bromide in the presence of an acid catalyst.

nonyl chloride. $C_9H_{19}Cl$.
Properties: Liquid, bp 58–63C (8 mm), d 0.878 (25C), refr index 1.4379 (20C). Nonflammable.
Derivation: Nonyl alcohol reacts with hydrochloric acid at elevated temperatures and pressures to give nonyl chloride. It can also be made by refluxing a mixture of concentrated hydrochloric acid with the alcohol in presence of zinc chloride.

1-nonylene. (1-nonene). C_9H_{18} or $CH_3(CH_2)_6CH:CH_2$.
Properties: Colorless liquid, soluble in alcohol, insoluble in water, d 0.7433, bp 149.9C. Combustible.
Derivation: From propylene.
Use: Organic synthesis, wetting agent, lube oil additive, polymer gasoline.

nonyl hydride. See nonane.

n-nonylic acid. See pelargonic acid.

γ-nonyl lactone. (aldehyde C-18; 4-hydroxynonanoic acid, gamma-lactone; gamma-nonalactone). $CH_3(CH_2)_4CHCH_2CH_2C(O)O$.
Properties: Yellowish to almost colorless liquid, coconut-like odor, d 0.956–0.963, refr index

1.447, soluble in five volumes of 50% alcohol, soluble in most fixed oils and mineral oil, practically insoluble in glycerol. Combustible.
Grade: Technical, FCC.
Use: Perfumery, flavors.

nonyl nonanoate. (nonyl pelargonate).
$C_9H_{19}OOCC_8H_{17}$.
Properties: Liquid, d 0.863 (25C), bp 315C, refr index 1.4419 (20C), floral odor. Combustible.
Use: Flavors, perfumes, organic synthesis.

nonylphenol. CAS: 25154-52-3.
$C_9H_{19}C_6H_4OH$.   A mixture of isomeric monoalkyl phenols, predominantly p-substituted.
Properties: Pale yellow, viscous liquid with a slight phenolic odor. Insoluble in water, soluble in most organic solvents, d 0.950 (20/20C), bp 293C, fp −10C (sets to glass below this temperature), viscosity 563 cp (20C), flash p 285F (140.5C). Combustible.
Grade: Technical.
Use: Non-ionic surfactant (nonbiodegradable), lube oil additives, stabilizers, petroleum demulsifiers, fungicides, antioxidants for plastics and rubber.

nonylphenoxyacetic acid. CAS: 3115-49-9.
$C_9H_{19}C_6H_4OCH_2COOH$.
Properties: Light amber liquid, miscible with organic solvents, d 1.02 (20C), insoluble in water, soluble in alkali, flow p 5C, viscosity 6500 cp (25C). Combustible.
Use: Corrosion inhibitor for turbine oils, lubricants, fuels, greases, antifoaming agent in gasoline, hydraulic fluids, cutting oils.

nonyl thiocyanate. $C_9H_{19}SCN$.
Properties: Bp 84–86.5C (1 mm), d 0.919 (25C), refr index 1.4696 (20C). Nonyl thiocyanate can be made from nonyl chloride by refluxing with alcoholic sodium thiocyanate solution using conventional techniques.

nonyltrichlorosilane. (trichlorononylsilane).
$C_9H_{19}SiCl_3$.
Properties: Water-white liquid with a pungent irritant odor. Fumes readily in moist air.
Hazard: Strong irritant to skin and mucous membranes.

"Nopalcol."[309]   TM for a series of diethylene glycol and polyethylene glycol fatty esters.
Use: General-purpose emulsifying agents for oils, solvents, essential oils, opacifying agents for cosmetics.

"Nopcodraw" Compounds.[309]   TM for a group of die box lubricants. Powdered dry compounds for drawing and heading ferrous, alloy, stainless steel, and refractory metals. Full range of formulations based on soda, potash, aluminum, zinc, calcium, and barium soaps combined with inorganic and organic lubricants plus anti-wear agents.

"Nopcogen."[309]   TM for a series of fatty acid derivatives, including alkylolamides, polyamine condensates, alkylolamine condensates and sulfonated amine condensates, used as softeners, detergents, wetting, or emulsifying agents.

"Nopcosulf."[309]   TM for a series of sulfated oils, fatty acids and fatty esters.
Use: Mineral oil emulsifiers, dyestuff dispersants, latex paint additives, cosmetic emulsifiers, textile wetting and re-wetting agents, paper wetting agents and softeners.

nopinene. See β-pinene.

NOPON. See p-bis[2-(5-α-naphthyloxazolyl)-benzene.

"Norad."[121]   TM for a water-soluble adhesive used to prevent slippage of palletized multiwall bags.

"Norane."[42]   TM for a group of water repellents used in the textile industry for a wide variety of fabrics. Include formulations of anionic wax emulsions, silicone resins, thermosetting resins, fluorophilic compounds, and fatty acid amide condensates.

norbormide.   Generic name for 5-(α-hydroxy-α-2-pyridylbenzyl)-1-(α-2-pyridylbenzylidene)-5-norbornene-2,3-dicarboximide.
CAS: 991-42-4.
Properties: White solid, mp 180–190C, insoluble in water, soluble in dilute acids.
Hazard: Toxic by ingestion.
Use: Rodenticide.

5-norbornene-2-methanol. (2-hydroxymethyl-5-norbornene; "Cyclol"). $C_7H_9CH_2OH$.
Properties: Stable, colorless liquid. D 1.022–1.024 (25/25C), refr index 1.4985–1.4995 (25C), a high-boiling solvent, bp 192–198C, miscible with most common organic solvents, slightly soluble in hot water, flash p (TOC) 183F (83.9C). Combustible.
Use: Monomer for the modification of condensation and addition polymers for coatings.

5-norbornene-2-methyl acrylate. CAS: 95-39-6.
$(C_3H_4O_2)_4$.
Properties: Colorless liquid, bp 103–105C, d 1.027–1.031 (25/25C), soluble in most organic solvents, inhibitor 0.02% MEHQ.

Use: A difunctional monomer particularly suited for in situ crosslinking of vinyl acetate emulsions polymer systems.

"Nordel."[28] TM for an elastomer based on an ethylene-propylene-hexadiene terpolymer, sulfur-curable.
Use: Automotive and appliance components, wire insulation: electrical accessories, belts, hose, mechanical products.

nordhausen acid. CAS: 7664-93-9. Fuming sulfuric acid of d 1.86–1.90.
Hazard: Corrosive to skin and tissue.
See sulfuric acid, fuming.

nordihydroguaiaretic acid. (NDGA; 4,4′-(2,3-dimethyltetramethylene)dipyrocatechol).
CAS: 500-38-9. $[C_6H_3(OH)_2CH_2CH(CH_3)]_2$.
Properties: Crystals from acetic acid; mp 184–185C; soluble in methanol, ethanol, ether; slightly soluble in hot water, chloroform; nearly insoluble in benzene, petroleum ether.
Grade: Technical, FCC.
Derivation: Extraction from guaiac, also synthetically.
Hazard: Use in foods prohibited by FDA.
Use: Antioxidant to retard rancidity of fats and oils.

norea. (3-(hexahydro-4,7-methanoindan-5-yl)-1,1-dimethylurea). CAS: 18530-56-8.
$C_{10}H_{14}NHCON(CH_3)_2$.
Properties: White solid; mp 168–169C. Very slightly soluble in water; soluble in polar solvents such as cyclohexanone, acetone, alcohol; less soluble in nonpolar solvents such as benzene, xylene, hexane, and kerosene.
Use: Selective herbicide.

norgestrel. CAS: 797-63-7. A synthetic steroid hormone used as ingredient of oral contraceptives. Approved by FDA.
See "Ovral."

"Norit."[107] TM for activated adsorption carbons of vegetable origin.
Use: Purification and decolorization of chemical solutions; bleaching agent for glycerol, vegetable oils, and fats; dechlorination, taste, and odor control in water treatment; catalyst and carrier for catalysts; recovery of volatile solvents; air purification; gas treatment and separation.

norleucine. (α-aminocaproic acid).
CAS: 327-57-1.
$CH_3(CH_2)_3CH(NH_2)COOH$.
A nonessential amino acid found naturally in the l(+) form.
Properties: Leaflets, crystals from water, dl-norleucine, soluble in water, slightly soluble in alcohol, soluble in acids, decomposes 327C, available commercially.
   l(+)-norleucine: slightly sweet, sublimes 275–280C, mp 301C (decomposes).
   d(−)-norleucine: bitter, partially sublimes 275–280C, mp 301C (decomposes).
Derivation: Found in traces in proteins, also synthetically.
Use: Biochemical research.

"Norlig."[121] TM for a line of unmodified or partially modified lignosulfonates derived from spent sulfite liquor. Available in both powder and liquid forms. Liquids are darker brown in color and more sticky as solids concentration increases.
Use: Road binder, linoleum paste, foundry products, leather tanning, gypsum board, soil stabilization, feed pellets.

normal. (1) Of hydrocarbon molecules: containing a single unbranched chain of carbon atoms, usually indicated by the prefix n-, as n-butane, n-propane, etc. (2) Of solutions: containing one equivalent weight of a dissolved substance per liter of solution; a standard measure of concentration, indicated by N, e.g., 2N, 0.5N, etc.
See also molar.
(3) Perpendicular or at right angles to, as for example, incident light rays normal to a surface.

normalize. To temper steels and other alloys by heating to a predetermined temperature and cooling at a controlled rate to relieve internal stresses and improve strength (analogous to annealing of glass).

Normant reagents. Vinylmagnesium halides that were previously unavailable are now prepared by reacting the parent vinyl halides with magnesium in tetrahydrofuran. These compounds are called Normant reagents and behave like typical Grignard reagents.

nornicotine. (3-(2-pyrrolidinyl)pyridine).
CAS: 479-97-3. $C_9H_{12}N_2$.
Properties: Moderately thick, hygroscopic liquid. Bp 270C, d 1.07, optical rotation −89 degrees, refr index 1.53. Soluble in water, alcohol, and kerosene.
Hazard: Toxicity about one-third that of nicotine.
Use: Insecticide.

"Norpar."[592] TM for normal paraffin solvents comprised of C_{11}—C_{14} hydrocarbons. They are characterized by chemical inertness, high selectivity, low volatility, and resistance to hydrolysis and redox reactions. They have a number of industrial uses, for example, in resin dispersions, cold-rolling aluminum foil, and such specialized formulations as waterless hand cleaners.

norphytane. See pristane.

Norrish, Ronald G. W. (1897-1978) An English physical chemist who was awarded a Nobel prize in 1967 with Eigen and Porter. His analysis of reactions of one ten- billionth of a second were made possible by disturbing the chemical equilibrium with short energy pulses. After receiving a doctorate, he went to the Sorbonne before returning to Cambridge to teach. His career was long and distinguished by many awards.

Norrish type cleavage. Homolytic cleavage of aldehydes and ketones originating from their excited n pi * state. Synthetically useful for the ring cleavage of cyclic ketones.

Northrup, John H. (1891-) An American chemist who won a Nobel prize in chemistry in 1946 along with Sumner and Stanley. His work was primarily concerned with isolation and crystallization of enzymes. Many firsts included the production of the enzyme trypsin in the laboratory and isolation of the first bacterial virus. He was also responsible for producing diptheria antitoxin in crystalline form. His education was at eastern schools including Harvard, Yale, and Princeton.

noruron. CAS: 2163-79-3. $C_{13}H_{22}N_2O$.
Properties: Colorless crystals, mp 175C, insoluble in water, soluble in acetone, slightly soluble in benzene.
Hazard: Highly toxic.
Use: Herbicide.

Norway saltpeter. See ammonium nitrate.

Norwegian saltpeter. See calcium nitrate.

NOS. Abbreviation for "not otherwise specified."
Use: In shipping regulations for classes of substances to which a restriction applies, individual members of which are not listed in the regulations.

noscapine. (l-α-narcotine; narcosine).
CAS: 128-62-1. $C_{22}H_{23}NO_7$. An isoquinoline alkaloid of opium.
Properties: Fine, white, crystalline powder. Mp 176C, sublimes at 150–160C, insoluble in water, practically insoluble in vegetable oils, slightly soluble in hot solutions of potassium hydroxide and sodium hydroxide, soluble in most organic solvents. Salts formed with acids are dextrorotatory and unstable in water.
Derivation: An opium alkaloid from the seed capsules of *Papaver somniferum*.

Grade: USP.
Hazard: Narcotic, use legally restricted.
Use: Medicine (cough control).

novobiocin. CAS: 303-81-1. $C_{31}H_{36}H_2O_{11}$.
Properties: A light yellow to white antibiotic produced by Streptomyces niveus. Available as calcium and sodium salts.
Hazard: May have damaging side effects.

"Novocain."[162] TM for a brand of procaine hydrochloride.
Use: Local anesthetic in dentistry and minor surgery.

novolak. (novolac). A thermoplastic phenol-formaldehyde type resin obtained primarily by the use of acid catalysts and excess phenol. Generally alcohol soluble and require reaction with hexamethylenetetramine, p-formaldehyde, etc. for conversion to cured, crosslinked structures by heating at 200–400F.
Use: Molding materials, bonding materials, bonding agent in brake linings, abrasive grinding wheels, electrical insulation, clutch facings, air-drying varnishes, reinforcing agent and modifier for nitrile rubber.

novoldiamine. (1-diethylamino-4-aminopentane).
CAS: 140-80-7. $C_9H_{22}N_2$.
Properties: Colorless liquid, d 0.82, bp 200C, refr index 1.44, soluble in water and alcohol.
Derivation: From 2-diethylaminoethanol and ethyl acetoacetate.
Use: Manufacturing antimalarials (quinacrine and related compounds).

noxious gas. Any natural or byproduct gas or vapor that has specific toxic effects on humans or animals (military poison gases are not included in this group). Examples of noxious gases are ammonia, carbon monoxide, nitrogen oxides, hydrogen sulfide, sulfur dioxide, ozone, fluorine, and vapors evolved by benzene, carbon tetrachloride and a number of chlorinated hydrocarbons. Gases which act as simple asphyxiants are not classified as noxious.
See also air pollution.

Np. Symbol for neptunium.

NPK mixtures. Fertilizers containing nitrogen, phosphorus, and potassium. These are usually characterized by numbers such as 5–10–10, meaning 5% nitrogen, 10% phosphorus as P_2O_5 and 10% potassium as K_2O. The percentages refer to the amount of nitrogen, phosphorus, or potassium present in available form rather than to the amount of the compounds.

NPN. Abbreviation for n-propyl nitrate.

NPO. See α-naphthylphenyloxazole.

NQR spectroscopy. See nuclear quadrupole resonance.

NR. Abbreviation for natural rubber.

NRC. Abbreviation of Nuclear Regulatory Commission.

NRDC. Abbreviation for National Resources Defense Council.

"n-Sol."[201] TM for a series of patented processes used in water treatment. Dilute alklai silicate is partially neutralized with an acid reactant to develop an activated silica for use as a coagulant aid in clarifying raw and waste waters.

NSR. Abbreviation for nitrile silicone rubber.

NTA. Abbreviation for nitrilotriacetic acid.

NTAN. Abbreviation for nitrilotriacetonitrile.

"Nuactant."[300] TM for a fiber reactant to improve crush resistance and dimensional control in cellulosic fabrics.

"Nucerite."[522] TM for a material of construction for high-temperature corrosive environments. A family of composites consisting of a *crystalline* glass ceramic bonded to steel, e.g., "Nucerite" 7000 ceramic.
Use: Nozzles, flange faces, reaction equipment, etc.
See also "Glasteel," glass ceramic.

nuclear chemistry. The division of chemistry dealing with changes in or transformations of the atomic nucleus. It includes spontaneous and induced radioactivity, the fission or splitting of nuclei, and their fusion, or union; also the properties and behavior of the reaction products and their separation and analysis. The reactions involving nuclei are usually accompanied by large energy changes, far greater than those of chemical reactions; they are carried out in nuclear reactors for electric power production and manufacture of radioactive isotopes for medical use, also (in research work) in cyclotrons.
See also fission, fusion, radiochemistry. See nucleus (1).

nuclear energy. The energy liberated by (1) the splitting or fission of an atomic nucleus, (2) the union or fusion of two atomic nuclei, and (3)

the radioactive decay of a nucleus (transmutation). For details refer to the following entries, nuclear reaction, fission, fusion, tokamak, radioactivity, ionizing radiation, nuclear chemistry, transmutation, breeder, uranium-235, plutonium, nuclear reactor.
See also acceptable risk.

nuclear fuel. A fissionable material which is the source of energy in a nuclear reactor, specifically Uranium-235, thorium, and Plutonium-239. For thermonuclear (fusion) reactions the "fuels" are the hydrogen isotopes deuterium and tritium.
See also breeder, fission, fusion.

nuclear fuel element. A rod, tube, plate, or other mechanical shape or form into which a nuclear fuel is fabricated for use in a reactor. The word "element" is used here in its electrical engineering rather than its chemical sense. Cladding materials are usually ceramics or zirconium alloys.

nuclear magnetic resonance. (NMR).
A type of radio-frequency spectroscopy based on the magnetic field generated by the spinning of electrically charged atomic nuclei. This nuclear magnetic field is caused to interact with a very large (10,000–50,000 gauss) magnetic field of the instrument magnet. The magnetic properties of atomic nuclei are the spin number and the magnetic moment. Hydrogen nuclei, fluorine, phosphorus, boron, nitrogen, carbon-13, and oxygen-17 have distinctive magnetic properties. the molecular or chemical environment of the nucleus produces characteristic shifts and fine structure in the NMR spectra. Because of its dependence on molecular structure, NMR has become a fundamental research tool for structure determinations in organic chemistry. Studies of hydrogen locations in crystals have been useful, as these cannot be determined directly by means of x-rays. NMR techniques have been applied to studies of electron densities and chemical bonding, the composition of mixtures, percentage purity determinations, and elemental hydrogen analyses.

nuclear quadrupole resonance. (NQR).
A spectroscopic technique related to nuclear magnetic resonance which utilizes the electric fields naturally present in crystals instead of a magnet. It is useful for studies of electrical field gradients around nuclei, chemical bonding, space groupings in crystals, and molecular structure.

nuclear reaction. A reaction that involves a change in the nucleus of an atom, as opposed to a chemical reaction in which only the electrons take part. Nuclear reactions usually result in release of tremendous amounts of energy; while the energy obtained from chemical reactions is slight; e.g., rupture of a chemical bond evolves approximately 5 eV compared with 200 MeV resulting from the fission of an atomic nucleus. There are three types of nuclear reactions, namely, (1) transmutation (radioactive disintegration), (2) fission, and (3) fusion (thermonuclear).

See these entries for details.

nuclear reactor. An assembly of fissionable material (uranium-235 or plutonium-239) designed to produce a sustained and controllable chain reaction for the generation of electric power. The first reactor (then called a "pile") was constructed at University of Chicago in 1942 under the leadership of Enrico Fermi. The essential components of a modern nuclear reactor are: (1) The core, composed of metal- or ceramic-clad rods containing enough fissionable material to maintain a chain reaction at the necessary power level; as much as 50 tons of uranium may be required (d: 19). (2) A source of neutrons to initiate the reaction, such as a mixture of polonium and beryllium. (3) A moderator to reduce the energy of fast neutrons for more efficient fission (called slow or thermal neutrons), materials such as graphite, beryllium, heavy water and light water are used. (4) A coolant to remove the fission-generated heat, water is generally used, which is converted to steam in heat exchangers and used to drive turbines. Sodium, helium, and nitrogen may also be used. The heating of streams and estuaries by reactor effluent is a serious environmental problem, which can be surmounted by construction of special cooling towers. (5) A control system to absorb neutrons rapidly when their concentration becomes too high, rods of boron or cadmium which have high capture cross-sections, are necessary for this purpose. (6) Adequate shielding, remote control equipment, and appropriate instrumentation are essential for personnel safety and efficient operation.

The primary use of nuclear reactors is for electric power generation. This has been a reasonably successful endeavor, for nuclear energy has made a notable contribution to the overall energy supply situation in the US. But the efficiency of reactors is often reduced far below their potential by failures resulting from the highly corrosive operating conditions to which the materials of construction of the reactor and the associated hardware are exposed. Frequent and extended shutdowns are commonplace. This, together with the environmental radiation hazard and escalating construction costs, has restricted the development of nuclear energy as a power source. Plans for construction of many plants have been de-

layed or cancelled. A further negative fact is that the estimated safety factor has changed radically since 1975, at which time the risk of serious accident that might result in core meltdown was reported to be 1 in 20,000 years of operation. A study conducted by ORNL based on accident data in the decade 1969–1979 concluded that this risk has risen to 1 in 1000 operating years --a 20-fold decrease in the safety factor.

See also fission, breeder, acceptable risk.

Nuclear Regulatory Commission. A Federal agency established in 1975 to regulate all commercial uses of atomic energy, including construction and operation of nuclear power plants, nuclear fuel reprocessing plants, and research applications of radioactive materials. It is also responsible for safety and environmental protection.

nuclear waste. See radioactive waste.

nucleation. The process by which crystals are formed from liquids, supersaturated solutions (gels), or saturated vapors (clouds). Crystals originate on a minute trace of a foreign substance acting as a nucleus. These are provided by impurities or by container walls in laboratory apparatus. Crystals form initially in tiny regions of the parent phase and then propagate into it by accretion. Rain and snow are formed in this way in moisture-laden air, the moisture condenses on minute particles of dust or ice. Cloud seeding with silver iodide or carbon dioxide is based on this principle. The former has a crystalline structure similar to that of ice and the latter causes rapid formation of ice nuclei by intense local lowering of temperature.

See also crystals, whiskers.

There are various methods for growing crystals including (a) evaporation of a solution, (b) cooling of a saturated solution or melt, (c) condensation of a vapor (plasmas are sometimes used to generate the vapor), (d) electrodeposition, (e) growth in gel media (in order to slow growth). Methods (b) through (e) are particularly suited for the growth of large crystals, which are widely used in modern technology.

nucleic acid. Any of several complex compounds occurring in living cells, usually chemically bound to proteins to form nucleoproteins. Nucleic acids are of high molecular weight and are easily changed by many mild chemical reagents. They contain carbon, hydrogen, oxygen, nitrogen (15–16%), and phosphorus (9–10%).

The fundamental units of nucleic acid are nucleotides, nucleic acids and polynucleotides in which the nucleotides are linked by phosphate bridges. Upon extensive heating in the presence

of water (hydrolysis), nucleic acids yield a mixture of purines and pyrimidines, d-ribose or d-deoxyribose, and phosphoric acid. Nucleic acids are subdivided into two types: ribonucleic acid (RNA), containing the sugar d-ribose; and deoxyribonucleic acid (DNA), containing the sugar d-deoxyribose.
See deoxyribonucleic acid, ribonucleic acid, nucleoside, nucleotide, nucleoprotein.
Good sources of nucleic acids are salmon, thymus, yeast, and wheat kernel embryo.
Use: Biochemical and medical research in genetics, virus diseases, and cancer.

nucleogenesis*. The original synthesis of the chemical elements resulting from a huge explosion of undifferentiated energy which astrophysicists believe initiated the universe approximately 20 billion years ago, at least 7 billion years before the formation of the Milky Way galaxy (the so-called "big bang" theory). Information obtained with radiotelescopes indicates that temperatures were so inconceivably high in this continuum of radiation that the elementary particles that were undoubtedly present were unable to combine. As the continuum expanded rapidly after the original eruption, it cooled within a few minutes to the point where protons and electrons could unite to form hydrogen. After further cooling to 10^7C, thermonuclear fusion of hydrogen to helium occurred in a matter of seconds. But no more complex combinations took place in the young universe for several million years, until the shrinking cores of giant stars, many times the mass of the sun (supernovae), raised their internal temperatures to 10^8C, high enough to fuse helium nuclei to form carbon. The heavier elements were eventually synthesized during explosions of supernovae, which produced temperatures in the range of 600×10^6C. Thus the sequence of creation of the elements is believed to be hydrogen and helium, then after a few million years came carbon, oxygen, magnesium, silicon, sulfur, and iron; after still another interval the heavy metals were formed in another round of stellar explosions. All this occurred over a period of 5–7 billion years.
See also fusion, astrochemistry; life, origin.
*Based on E. T. Chaisson, *Harvard Magazine*, Vol. 80, No. 2 (1977).

nucleon. General name applied to neutrons and protons, the essential constituents of atomic nuclei and also used as a class name for fundamental particles of that mass. The study of subatomic particles is often called nucleonics.

nucleophile. An ion or molecule that donates a pair of electrons to an atomic nucleus to form a covalent bond, the nucleus that accepts the electrons is called an electrophile. This occurs, for example, in the formation of acids and bases according to the Lewis concept, as well as in covalent carbon bonding in organic compounds.
See also Lewis base, donor.

nucleoprotein. A type of protein universally present in the nuclei and the surrounding cytoplasm of living cells. A nucleoprotein is composed of a protein, which is rich in basic amino acids, and a nucleic acid. The nucleic acid portions can be isolated and used in medical and biochemical research.
See also deoxyribonucleic acid, chromatin, virus.

nucleoside. A compound of importance in physiological and medical research, obtained during partial decomposition (hydrolysis) of nucleic acids, and containing a purine or pyrimidine base linked to either d-ribose, forming ribosides, or d-deoxyribose, forming deoxyribosides. They are nucleotides minus the phosphorus group. For specific nucleosides
See adenosine, cytidine, guanosine, and uridine.

nucleotide. A fundamental unit of nucleic acids, some are important coenzymes. The four nucleotides found in nucleic acids are phosphate monoesters of nucleosides, adenylic acid, guanylic acid, uridylic acid, and cytidylic acid. Great progress has been made in determining the nucleotide sequence in fundamental materials, such as yeast genes.
The term is also applied to compounds not found in nucleic acids and which contain substances other than the usual purines and pyrimidines. Such compounds are modified vitamins, and function as coenzymes; examples are riboflavin phosphate (flavin mononucleotide), flavin adenine nucleotide, nicotinamide adenine dinucleotides, nicotine adenine dinucleotide phosphate, and coenzyme A. The nucleotides inosine-5'-monophosphate and guanosine-5'-monophosphate are used as flavor potentiators.

nucleus. (1) The positively charged central mass of an atom, it contains essentially the total mass in the form of protons and neutrons. The nucleus of the hydrogen atom consists of one proton, while that of uranium is comprised of 93 protons and 146 neutrons.
(2) The central portion of a living cell, consisting primarily of nucleoplasm in which chromatin is dispersed. It is enclosed by a membrane that separates it from the surrounding cytoplasm. All the most important functions of the cell, including the mechanics of division (mitosis) and the programming of the genetic code, take place in the nucleus.
See also gene, chromosome.

(3) The characteristic structure of a group of chemical compounds, e.g., the benzene nucleus.

(4) Any small particle which can serve as the basis for crystal growth (See nucleation). *Note*: The multiple meanings of "nucleus" and "resonance" can be a source of confusion, especially when these terms are closely associated, as in nuclear magnetic resonance and resonance of a molecular nucleus. In the first of these expressions, nucleus is used in sense (1) under nucleus and resonance in sense (2) under resonance. In the second expression, nucleus is used in sense (3) under nucleus and resonance in sense (1) under resonance.

nuclide. A particular species of atom, characterized by the mass, the charge (number of protons), and the energy content of its nucleus. A radionuclide is a radioactive nuclide. Example: carbon-14 is a radionuclide of carbon.
See also isotope.

"Nuclon."[177] TM for a polycarbonate thermoplastic resin.
Properties: High impact resistance not appreciably reduced by temperature fluctuations, good dimensional stability, good electrical resistance. Naturally transparent of light straw color.
Use: Engineering plastic for "hard service" parts and components.

"Nucon."[466] TM for a magnesia-chrome refractory made by a high-fired direct bonding process.
Use: Copper and lead furnace roofs; linings of rotary kilns used to burn lime, dolomite, and magnesia; steel furnace sidewalls and roofs.

"Nuflo."[285] TM for a high quality, extra fine kaolin treated with a surfactant by an intimate mixing process. A number of grades are obtained by applying various percentages of additive.
Use: Conditioning prilled ammonium nitrate and other high analysis granular fertilizers.

nuisance particulate. Fine particles (dusts) that are not very toxic in low concentrations. Among them are clay, calcium carbonate, emery, glass fiber, silicon carbide, gypsum, starch, Portland cement, marble, and titanium dioxide.
See also dust, industrial.

Nujol. Mineral oil used to prepare mulls for infrared analysis.

numerals. For their use and meaning in chemical names see Geneva System; benzene; chemical nomenclature.

"Numet."[439] TM for depleted uranium.
Properties: Uranium metal wherein uranium-235 isotopic content has been reduced below 0.7% as found in normal uranium. Has high structural strength coupled with high density of 18.9 g/cc.
Derivation: Processed from uranium hexafluoride as tailing from gaseous diffusion plants.
Use: Balance weights for aircraft, high speed rotors in gyro-compasses, gamma radiation shielding, radio-isotope transportation casks and fuel element transfer casks, in general as structural material applicable to radiation shielding.

"Nuroz."[36] TM for a polymerized wood rosin.
Use: Adhesives, gloss oils, paper label coatings, oleoresinous varnishes, solder flux, spirit varnishes, waxed paper and hot melt compounds, synthetic resins.

Nusselt number. A value used in heat transfer studies and calculations to compare heat losses by conduction from various shaped objects under various conditions. It is combined into a single number the actual heat loss (Q) the temperature difference (ΔT) between the body and its surroundings, the size (d) and shape of the body and the thermal conductivity (k) of the fluid surrounding the object, in the equation $Nu = Qd/\Delta Tk$.

nutrient. Any element or compound that is essential to the life and growth of plants or animals either as such or as transformed by chemical or enzymatic reactions. In plants, nutrients include numerous mineral elements as well as nitrogen, carbon dioxide, and water. In animals and man, the primary nutrients are the proteins, carbohydrates, and fats obtained from plants, either directly or indirectly, supplemented by vitamins and minerals. Water and oxygen are included in this definition. All told, there are 43 basic nutrients.
See also food.

nutrient solution. A water solution of minerals and their salts necessary for plant growth which is used instead of soil, the plants being supported by mechanical means. Such solutions contain combined nitrogen, potassium, phosphorus, calcium, sulfur, and magnesium, together with traces of iron, boron, zinc, and copper. They are extensively used for commercial growing of flowers and vegetables particularly on islands, and also to some extent for house plants.

nutrification. Addition of nutrients to a food either to replace those lost in processing (restoration) to provide nutrients that are not normally present in the food (fortification), or to bring the food into conformity with a specific standard for that food.

nutrition. The effects of nutrients on living organisms and the biochemical mechanisms involved in bringing them about; also, that subdivision of biochemistry which deals specifically with these effects. In plant nutrition the essential requirements are carbon dioxide and water, from which the plant forms carbohydrates by photosynthesis, nitrogen, which is essential for the synthesis of proteins by the plant, with the aid of nitrogen-fixing bacteria, as well as phosphorus, calcium, potassium, and a number of trace elements (micronutrients). Besides proteins and carbohydrates, plants also synthesize vitamins and various fats and oils. Thus they provide a basis for human nutrition, both directly (grain and other vegetables) and indirectly (meats and dairy products), though the conversion to protein values for human nutrition is only approximately 10% for meats.

Human diet requires proteins (milk, eggs, fish, and some vegetables), carbohydrates (plants), fats (oils) from both plants and animals, minerals from milk and meats, salt (chloride), vitamins from green vegetables and citrus fruits, and water. Micronutrients are furnished by sea food, cereals, vegetables, and fruit.

Human digestive processes involve primarily the hydrolysis of complex carbohydrates to simple sugars, of proteins to a mixture of amino acids, and of fats to glycerol and higher fatty acids. Hydrolysis is catalyzed by various enzymes in the saliva and digestive tract. The end products of digestion are absorbed across a semipermeable membrane in the intestine and thus enter the blood stream, unusable products being eliminated. The efficiency of digestion plus absorption is approximately 92% for protein, 95% for fat, and 98% for carbohydrates.

See also metabolism, digestion (1), plant (1), nutrient; RDA.

nut shells. In a fine-ground state, the shells of coconuts and other nuts are a source of decolorizing carbon; the pits of peaches and similar fruits have been used for gas-adsorbent carbon.

NW acid. Abbreviation for Neville-Winter acid.

Nylander's reagent. Basic solution of bismuth subnitrate and Rochelle salt.
Use: To detect glucose in urine.

nylidrin hydrochloride. (p-hydroxy-α-[1-(1-methyl-3-phenyl-propylamino)-ethyl]benzyl alcohol hydrochloride.
$C_{19}H_{25}O_2N \cdot HCl$.

Properties: White, odorless, tasteless crystals or powder; slightly soluble in water and alcohol; very slightly soluble in chloroform, ether; pH of 1% solution is between 4.5 and 6.5.
Grade: NF.
Use: Medicine (treatment of heart disease).

nylon. CAS: 63428-83-1 $(C_6H_{11}NO)_n$.
Generic name for a family of polyamide polymers characterized by the presence of the amide group —CONH. By far the most important are nylon 66 (75% of US consumption) and nylon 6 (25% of US consumption). Except for slight difference in melting points the properties of the two forms are almost identical, though their chemical derivations are quite different. Other types are nylons 4, 9, 11, and 12 (See Grade).
Properties: Crystals, thermoplastic polymers. May be extruded as monofilaments over a wide dimensional range. Filaments are oriented by cold-drawing. Tensile strength (high-tenacity) up to 8 g/denier (approximately 100,000 psi); d 1.14; mp (66) 264C, (6) 223C; low water absorption. Good electrical resistance, but accumulates static charges. Highly elastic, with rather high percentage of delayed recovery at low strain values; low permanent elongation; moisture absorption 4% at 65% R.H. Wet strength approximately 90% of dry strength. Can be dyed with ionic and nonionic dyestuffs. Attacked by mineral acids, but resistant to alkalies and cold abrasion; soluble in hot phenols, cresols, and formic acid; insoluble in most organic solvents; difficult to ignite, self-extinguishing, melts forming beads; resistant to attack by moths, carpet beetles, etc.; compatible with wool and cotton, increases wear and crease resistance in 30% blends with natural fibers; rods and blanks are machinable.
Forms: Monofilaments, yarns, bristles, molding powders, rods, bars, sheets. Microcrystalline nylon is now available.
Grade: Nylon 66 is a condensation product of adipic acid and hexamethylenediamine developed by Carothers in 1935. Adipic acid is obtained by catalytic oxidation of cyclohexane. Nylon 6 is a polymer of caprolactam, originated by I. G. Farbenindustrie in 1940. Nylon 4 is based on butyrolactam (2-pyrrolidone); its tenacity, abrasion resistance and melting point are said to be about the same as for the 6 and 66 grades. It has excellent dyeability. Nylon 610 (TM "Tynex.") is obtained by condensation of sebacic acid and hexamethylenediamine and nylon 11 (TM "Rilsan") from castor bean oil (developed in France). Nylon 12 (also called "Rilsan" 12) is made from butadiene, also by a French process involving photonitrosation of cyclododecane by actinic light from mercury lamps. Its properties are similar to those of nylon 11. Nylon 9 can

be made from 9-aminononanoic acid, present in soybean oil. It has properties specifically desired in metal coatings and electrical parts; higher electrical resistance than 6 and 66; absorbs less moisture; and has better distortion resistance.

Use: Tire cord, hosiery, wearing apparel component, bristles for toothbrushes, hairbrushes, paint brushes (nylon 610); cordage and towlines for gliders; fish nets and lines; tennis rackets; rugs and carpets; molded products; turf for athletic fields; parachutes; composites; sails; automotive upholstery; film; gears and bearings; wire insulation; surgical sutures; artificial blood vessels; metal coating; pen tips; osmotic membranes; fuel tanks for automobiles.

See also polyamide, aramid, "Tajmir."

Note: Not all nylons are polyamide resins, nor are all polyamide resins nylons, e.g., "Versamide." One class of polyamide resins distinct from nylons is derived from ethylenediamine; they may be liquids or low-melting solids and have lower molecular weight than nylons. Another class, called aramids, is aromatic in nature.

"NyoGel."[483] TM for a series of low shear thixotropic greases and semifluid instrument lubricants for use where nonspreading properties are critical.

"NyoSil."[483] TM for a wide temperature silicon instrument oil halogenated for improved wear properties; viscosity 55 centistokes at 37.7C.

nystatin. (fungicidin). CAS: 1400-61-9. $C_{47}H_{75}NO_{17}$.

Properties: Yellow to light tan powder, odor suggestive of cereals, hygroscopic, affected by light, heat, air and moisture; sparingly soluble in methanol, ethanol; very slightly soluble in water, insoluble in chloroform, ether and benzene; in solution is rapidly inactivated by acids and bases.

Derivation: Produced by fermentation with Streptomyces noursei and aureus.

Grade: USP.

Use: Medicine (antifungal antibiotic), feed additive.

nytril. Generic name for a manufactured fiber containing at least 85% of a long-chain polymer of vinylidene dinitrile, $—CH_2C(CH)_2—$, where the vinylidene dinitrile content is no less than every other unit in the polymer chain (Federal Trade Commission).

Properties: Soft, resilient fabric is obtained, is easy to clean, does not pill, resists wrinkling and retains shape after pressing.

Use: Fur-like pile fabrics, sweaters, yarns, blended fabrics for coats and suits.

O

O. Symbol for oxygen; the molecular formula is O_2.

o-. Abbreviation for ortho-.

oakum. Hemp fiber impregnated with tar or pitch.
Use: Caulking.

Obermayer's reagent. Ferric chloride in concentrated hydrochloric acid (4 g $FeCl_3$ in 1 liter concentrated HCl).
Use: For determining indoxyl in urine.

OC. Abbreviation of oxygen consumed.

Occupational Safety and Health Administration. (OSHA). A Federal agency responsible for establishing and enforcing standards for exposure of workers to harmful materials in industrial atmospheres, and other matters affecting the health and well-being of industrial personnel. A number of OSHA regulations and proposals have been controversial.

Ocean Thermal Energy Conversion. (OTEC). Utilization of ocean temperature differentials between solar-heated surface water and cold deep water as a source of electric power. In tropical areas such differences amount to 35–40F. A pilot installation now operating near Hawaii utilizes a closed ammonia cycle as a working fluid, highly efficient titanium heat exchangers and a polyethylene pipe 2000 feet long and 22 inches inside diameter to handle the huge volume of cold water required. Alternative uses for such a system, such as electrolysis of water, ammonia production, and desalination are envisaged. There has been active interest in the possibilities of this energy source in France from the time of d'Arsonval (1885) which is currently continuing, and more recently in Japan (1970). Ongoing research indicates that OTEC should be able to produce substantial amounts of power by 1990.

ocean water. (sea water). A uniform solution of essentially constant composition containing 96.5% water and 3.5% ionized salts, associated compounds, elements and ionic complexes. Na^+ and Cl^- are completely ionized, but $MgSO_4$ and AuCl remain in bound form. Dissolved gases, e.g., oxygen and nitrogen are also present. Composition: Over 60 elements, by far the most important of which are (in tons/cubic mile): chlorine 89,500,000; sodium 49,500,000; magnesium 6,400,000; sulfur 4,200,000; calcium 1,900,000; potassium 1,800,000; bromine 306,000; carbon 132,000. Other elements present include (tons/cubic mile): copper 14, indium 47, silver 1, lead 0.1, gold 0.02. Total solids content/cubic mile is 166 million tons.
Properties: Colorless liquid, d 1.02, pH 7.8–8.2, fp −2.78C, bp 101.1C, average temperature about 5C, bitter taste, faint odor (depending on organic impurities).
Hazard: Ingestion of substantial amounts will create bodily chloride imbalance with harmful effects.
Use: Source of magnesium, bromine, sodium chloride, and fresh water; source of hydrogen.

"Ocenol."[28] TM for technical grades of oleyl alcohol rich in cetyl and unsaturated alcohols, used as antifoam agents.

ocher. Any of various colored earthy powders consisting essentially of hydrated ferric oxides mixed with clay, sand, etc. Some grades are calcined (burnt ocher). The colors are yellow, brown or red. Noncombustible.
Use: Paint pigments, cosmetics, theatrical makeup.
See also umber, sienna.

ocimene. (2,6-dimethyl-1,5,7-octatriene).
CAS: 25167-71-9.
$CH_2:C(CH_3)(CH_2)_2CH:C(CH_3)CH:CH_2$.
Mixture of isomers. A terpene obtained from sweet basil oil, bp 81C (30 mm), insoluble in water, soluble in alcohol, d 0.8031 (15C). Combustible.
Use: Flavors and perfumes.

ocotea oil. CAS: 68917-09-9. A volatile oil derived from Ocotea cymbarum used for its safrole content for the manufacture of heliotropin.

octabenzone. CAS: 1843-05-6. $C_{21}H_{26}O_3$.
Properties: Colorless crystals, mp 45C.
Use: UV absorber especially for polyethylene.

octadecadienoic acid. See linoleic acid.

n-octadecane. $C_{18}H_{48}$ or $CH_3(CH_2)_{16}CH_3$.
Properties: Colorless liquid, d 0.7767 (28/4C), bp 318C, mp 28.0C, refr index 1.4369 (28C), flash

p 200F (93C). Soluble in alcohol, acetone, ether, petroleum, and coal-tar hydrocarbons; insoluble in water. Combustible.
Use: Solvents, organic synthesis, calibration.

1,12-octadecanediol.   See 1,12-hydroxystearyl alcohol.

n-octadecanoic acid.   See stearic acid.

1-octadecanol.   See stearyl alcohol.

n-octadecanoyl chloride.   See stearoyl chloride.

octadecatrienol.   See linolenyl alcohol.

9-octadecen-1,12-diol.   See ricinoleyl alcohol.

1-octadecene. $C_{18}H_{36}$ or $CH_3(CH_2)_{15}CH:CH_2$.
Properties: Colorless liquid, d 0.7884 (20/4C), refr index 1.4456 (20C), bp 180C (15 mm), flash p 200F (93C). Soluble in alcohol, acetone, ether, petroleum, and coal tar solvents, insoluble in water. Combustible.
Grade: 95% min purity.
Use: Organic synthesis, surfactants.

octadecene-octadecadieneamine.   See oleyl-linoleylamine.

cis-9-octadecenoic acid.   See oleic acid.

trans-9-octadecenoic acid.   See elaidic acid.

octadecenol.   See oleyl alcohol.

cis-9-octadecenoyl chloride.   See oleoyl chloride.

octadecenyl aldehyde.   (oleyl aldehyde). $C_{17}H_{35}CHO$.
Properties: Liquid, bp 167C (20 mm), refr index 1.4620 (25C), d 0.847 (25C).
Hazard: Flammable, moderate fire risk. Irritant to skin and mucous membranes.
Use: Intermediate for vulcanization accelerators, rubber antioxidants, synthetic drying oils and pesticides.

octadecyl alcohol.   See stearyl alcohol.

octadecyldimethylbenzylammonium chloride. $C_{18}H_{37}(CH_3)_2(C_6H_5CH_2)NCl$.   A quaternary ammonium salt.
Properties: White, crystalline powder; soluble in water, chloroform, benzene, acetone, xylene.

octadecyl isocyanate.   A straight-chain, saturated monoisocyanate consisting principally of a mixture of C_{18} and C_{16} alkyls, $CH_3(CH_2)_{16}CH_2NCO$, and $CH_3(CH_2)_{14}CH_2NCO$.
Properties: Colorless, slightly cloudy liquid at room temperature. Fp 10–20C (2 mm), flash p (OC) 355F (179C). Combustible.

Hazard: By inhalation, irritant to skin.
Use: Intermediate in synthesis, water-repellent textiles, paper, and other surfaces.

octadecyl mercaptan.   See stearyl mercaptan, thiol.

octadecyltrichlorosilane.   (trichlorooctadecylsilane). $C_{18}H_{37}SiCl_3$.
Properties: Water-white liquid, pungent odor, d 0.984 (25C), refr index 1.4580 (25C), bp 380C, flash p 193F (89.4C). Soluble in benzene, ethyl ether, heptane, and perchloroethylene. Combustible.
Hazard: Strong irritant to skin and mucous membranes.
Use: Intermediate for silicones.

octafluoro-2-butene.   (perfluoro-2-butene).   CAS: 360-89-4. $CF_3CF:CFCF_3$.
Properties: Colorless, nonflammable gas or liquid. Density of liquid (at bp) 1.5297 g/cc, bp 1.2C, fp −136 to −134C, sp volume 2.7 cu ft/lb (70F, 1 atm).
Grade: 95%.
Use: Organic intermediate, fluorocarbon polymers.

octafluorocyclobutane.   CAS: 115-25-3. C_4F_8.
Properties: Colorless gas, fp −41C, bp −6C. Nonflammable.
Use: Refrigerant, heat transfer agent.

octafluoropropane.   (perfluoropropane).   CAS: 76-19-7. C_3F_8.
Properties: Colorless, nonflammable gas. D 1.29, bp −36.7C, fp −160C, specific volume 2.02 cu ft/lb (21C), 1 atm).
Grade: 98% min purity.
Derivation: Electrofluorination of various organic compounds.
Use: Refrigerant (when combined with chlorofluorohydrocarbons), gaseous insulator, especially for radar wave guides.

octakis(2-hydroxypropyl)sucrose.
Properties: Viscous, straw-colored liquid. D 1.170 (70/20C), refr index 1.485 (25C), pour point 38C, flash p 485F (251C). Soluble in water, methanol, and ether. Combustible.
Use: Crosslinking agent for urethane foams; plasticizer for cellulosics, glue, starch, and many resins.

octamethylcyclotetrasiloxane.   CAS: 556-67-2. $C_8H_{24}O_4Si_4$.
Properties: Smooth, viscous liquid. Bp 175C, mp 17C, d 0.95, refr index 1.4.

Derivation: Hydrolysis of dimethyldichlorosilane.
Use: Silicone oils and related products.

octamethyl pyrophosphoramide. See schradan.

octamethyltrisiloxane. CAS: 105-51-7.
$C_8H_{24}O_2Si_3$.
Properties: Colorless liquid, bp 151C, d 0.820, fp
−80C, soluble in light hydrocarbons.
Use: Silicone oils, antifoam agent in lube oils.

"Octamine."[248] TM for a reaction product of diphenylamine and diisobutylene.
Properties: Light brown, granular, waxy solid. D
0.99, mp 75–85C, good storage stability. Soluble
in gasoline, benzene, ethylene dichloride, and acetone; insoluble in water.
Use: Rubber antioxidant.

1-octanal. (n-octyl aldehyde; aldehyde C-8; ca-
prylic aldehyde). CAS: 124-13-0.
$CH_3CH_2)_6CHO$.
Properties: Colorless liquid with strong fruity
odor, d 0.820–0.830, refr index 1.418–1.425, bp
163S, soluble in 70% alcohol and mineral oil,
insoluble in glycerol, flash p (CC) 125F (51.6C).
Combustible.
Grade: Technical, FCC.
Hazard: Moderate fire risk.
Use: Perfumery, flavors.

n-octane. CAS: 111-65-9. $CH_3(CH_2)_6CH_3$.
Properties: Colorless liquid, d 0.7026 (20/4C), refr
index 1.39745 (20C), bp 125.6C, fp −56.798C,
flash p 56F (13.3C), autoign temperature 428F
(220C), soluble in alcohol and acetone; insoluble
in water.
Grade: 95%, 99% research.
Hazard: Flammable, dangerous fire risk. TLV:
300 ppm in air.
Use: Solvent, organic synthesis, calibrations, azeotropic distillations.

1,8-octanedicarboxylic acid. See sebacic acid.

octanedioic acid. See suberic acid.

octane number. A number indicating the antiknock properties of an automotive fuel mixture
under standard test conditions. Pure normal heptane (a very high-knocking fuel) is arbitrarily
assigned an octane number of zero, while 2,2,4-
trimethylpentane or isooctane (a branched-chain
paraffin) is assigned 100. Thus, a rating of 80
for a given fuel indicates that its degree of knocking in a standard test engine is that of a mixture
of 80 parts isooctane and 20 parts n-heptane.
    Octane ratings as high as 115 have been ob-
tained by addition of tetraethyllead to isooctane.

Premium leaded gasolines have a Research octane rating of about 100, but this value drops
to 85–90 for unleaded gasolines, though it may
be improved by addition of methyl-tert-butyl
ether. The octane rating scale ends at 125 and
any higher figure is meaningless.
    Research octane numbers are those obtained
under test or "laboratory" conditions; they generally run about ten points higher than the so-
called motor octane numbers, which represent
actual road operating conditions.

octanoic acid. (octylic acid; octoic acid; caprylic
acid). CAS: 124-07-2.
$CH_3(CH_2)_6COOH$.
Properties: Colorless, oily liquid; slight odor; rancid taste. D 0.9105 (20/4C), mp 16C, bp 237.9C,
147.9 (30 mm), refr index 1.4278 (20C), flash
p 270F (132C), slightly soluble in water, soluble
in alcohol and ether. Combustible.
Derivation: By saponification and subsequent distillation of coconut oil.
Method of purification: Crystallization or rectification.
Grade: Technical, 99%, FCC.
Use: Synthesis of various dyes, drugs, perfumes,
antiseptics and fungicides, ore separations, synthetic flavors.

octanol. See octyl alcohol.

2-octanone. See methyl hexyl ketone.

octanoyl chloride. (capryloyl chloride; sometimes
called caprylyl chloride). $CH_3(CH_2)_6COCl$.
Properties: Water-white to straw-colored liquid
with pungent odor, miscible with most common
solvents; reacts with alcohol and water, fp −70C,
distillation range 183–212C, d 0.9576 (15.5/
15.5C), refr index 1.4357 (20C), flash p 180F
(82.2C). Combustible.
Hazard: Irritant to eyes and skin.
Use: Organic synthesis.

1-octene. (1-octylene; 1-caprylene). C_8H_{16} or
$CH_3(CH_2)_5CH:CH_2$.
Properties: Colorless liquid, d 0.7160 (20/4C), bp
121.27C, fp −102.4C, refr index 1.4088 (20C),
flash p (TOC) 70F (21.1C). Soluble in alcohol,
acetone, ether, petroleum, and coal-tar solvents;
insoluble in water.
Grade: 95%, 99% research.
Hazard: Flammable, dangerous fire risk.
Use: Organic synthesis, plasticizer, surfactants.

2-octene. C_8H_{16} or $CH_3(CH_2)_4CH:CHCH_3$.
Cis and trans forms exist.
Properties: Colorless liquid; d (cis) 0.7243, (trans)
0.7199; commercial 0.7185–0.7200 (20.4C); bp

(cis) 125.6C, (trans) 125.0C, commercial 124–127C; fp −94.04C; refr index (cis) 1.4150, (trans) 1.4132, commercial 1.4120–1.4145 (20C); flash p (mixed isomers) (TOC) 70F (21.2C); soluble in alcohol, acetone, ether, petroleum, and coal-tar solvents; insoluble in water.
Grade: 95, 99 mole %.
Hazard: Flammable, dangerous fire risk.
Use: Organic synthesis, lubricants.

octoic acid. An eight-carbon acid, usually designates caprylic acid.

octyl. The general name describing all eight-carbon radicals having the formula C_8H_{17}, often used interchangeably for the 2-ethylhexyl isomer.

n-octyl acetate. (acetate C-8; caprylyl acetate). CAS: 103-09-3. $CH_3COO(CH_2)_7CH_3$.
Properties: Colorless liquid, floral-fruity odor, slightly soluble in water, soluble in alcohol and most other organic liquids, d 0.865–0.869, flash p (OC) 180F (82.2C), refr index 1.419–1.422, bp 199C. Combustible.
Use: Perfumery, flavors, solvent.

n-octyl alcohol, primary. (1-n-octanol; alcohol C-8; heptyl carbinol). CAS: 111-87-5. $CH_3(CH_2)_6CH_2OH$. In industrial practice, the term octyl alcohol has been used for both 1-octanol and 2-ethylhexanol. the latter is also sometimes called isooctanol. The term capryl alcohol has been used for both 1-octanol and 2-octanol. It therefore seems preferable to designate the normal primary alcohol as 1-n-octanol.
Properties: Colorless liquid with penetrating aromatic odor, d 0.826 (20C), bp 194–195C, 108.7C (30 mm), refr index 1.430 (20C), fp −16C, flash p 178F (81.1C). Miscible with alcohol, chloroform, mineral oil; immiscible with water and glycerol. Combustible.
Derivation: By reduction of caprylic acid.
Grade: Technical, CP, pure, perfume, FCC.
Use: Perfumery, cosmetics, organic synthesis, solvent manufacture of high-boiling esters, antifoaming agent, flavoring agent.

sec-n-octyl alcohol. (2-n-octanol; methyl hexyl carbinol). $CH_3(CH_2)_5CHOHCH_3$.
Frequently called capryl alcohol.
Properties: Colorless, oily liquid; refractive; aromatic odor; miscible with alcohol, ether; slightly soluble in water. D 0.825 (15C), bp 178–179C, fp −38C, refr index 1.437 (20C), flash p 140F (60C). Combustible.
Derivation: By distilling sodium ricinoleate with an excess of sodium hydroxide.
Grade: Technical 92–99%, pure.
Use: Solvent, manufacture of plasticizers, wetting

agents, foam control agents, hydraulic oils, petroleum additives, perfume intermediates, masking of industrial odors.

octyl aldehyde. See octanal and 2-ethylhexaldehyde.

octylamine. $CH_3(CH_2)_7NH_2$.
Properties: Water-white liquid, amine odor, boiling range 170–179C, d 0.779 (20/20C), refr index 1.431 (20C), flash p 140F (60C). Combustible.

tert-octylamine. $(CH_3)_3CCH_2C(CH_3)_2NH_2$.
Properties: Colorless liquid, amine odor, bp 137–143C, d 0.771 (25C), refr index 1.423 (25C), flash p (OC) 92F (33.3C), insoluble in water, soluble in common organic solvents, especially petroleum hydrocarbons.
Hazard: Flammable, moderate fire risk. Skin irritant.
Use: Intermediate for rubber accelerators, insecticides, fungicides, dyestuffs, pharmaceuticals.

n-octylbicycloheptene dicarboximide. (n-(2-ethyl hexyl)-bicyclo(2,2,1)-5-heptene-2,3-dicarboximide). $C_8H_{17}NC_9H_8O_2$.
Properties: Liquid, bp 158C (2 mm), d 1.05 (18C), refr index 1.505 (20C), miscible with most organic solvents and oils. Combustible.
Derivation: From maleic anhydride, cyclopentadiene and 2-ethylhexylamine.
Hazard: By ingestion and skin absorption.
Use: Insecticide and pesticide synergist.

n-octyl bromide. (capryl bromide; caprylic bromide). CAS: 111-83-1. $CH_3(CH_2)_6CH_2Br$.
Properties: Colorless liquid, miscible with alcohol and ether; immiscible with water. D 1.118 (15C), bp 202C, fp −55C, refr index 1.4503 (25C).
Grade: Technical.
Use: Synthesis of quaternary ammonium compounds, organometallics, vinyl stabilizers.

octyl carbinol. See nonyl alcohol.

n-octyl chloride. $CH_3(CH_2)_6CH_2Cl$.
Properties: Colorless liquid, soluble in most organic solvents, d 0.8697 (25/15.5C), refr index 1.4288 (25C), fp −62C, bp 181.6C, flash p 158F (70C). Combustible.
Use: Chemical intermediate, manufacture of organometallics.

n-octyl-n-decyl adipate. (NODA).
Properties: Liquid, mild odor, d 0.92–0.98 (20/20C), fp −50C, boiling range 220–254C (4 mm), refr index 1.447. Combustible.
Use: Low-temperature plasticizer.

n-octyl-decyl alcohol. A blend of alcohols, available in tank cars and trucks. Combustible.
Use: Intermediate for plasticizers.

n-octyl-n-decyl phthalate. CAS: 119-07-3. $C_{26}H_{42}O_4$.
Properties: Clear liquid, mild characteristic odor, d 0.972–0.976 (20/20C), fp −40C, boiling range 232–267C (4 mm), refr index 1.482 (25C), flash p 455F (235C). Combustible.
Use: Plasticizer for vinyl resins.

octylene. See octene.

octylene glycol titanate.
Properties: Light yellow solid.
Derivation: Reaction of butyl titanate with octylene glycol.
Use: Crosslinking agent, surface-active agent. See also titanium chelate.

octylene oxide. Mixed $CH_3(CH_2)_5\overline{CHCH_2}O$ and $CH_3(CH_2)_4\overline{CHCH(CH_3)}O$. Density of liquid 0.830 (25C). Combustible.
Use: Organic intermediate, epoxy resins.

octyl formate. CAS: 112-32-3. $C_8H_{17}OOCH$.
Properties: Colorless liquid, fruity odor, soluble in mineral oil, practically insoluble in glycerol, 1 mL dissolves in 5 mL of 70% alcohol, d 0.869–0.872 (25C), refr index 1.4180–1.4200 (20C). Combustible.
Grade: FCC.
Use: Flavoring agent.

octylic acid. See caprylic acid.

2-octyl iodide. (caprylic iodide; secondary capryl iodide). $CH_3(CH_2)_5CHICH_3$.
Properties: Oily liquid, d 1.318 (18C), bp 210C (decomposes).
Hazard: By ingestion and inhalation. Keep away from light and air.
Use: Organic synthesis.

octylmagnesium chloride. $C_8H_{17}MgCl$.
A Grignard reagent available in tetrahydrofuran solution.

n-octyl mercaptan. $C_8H_{17}SH$.
Properties: Water-white liquid with mild odor, bp 199C, d 0.8395 (25/4C), refr index 1.4497 (25C), flash p 115F (46.1C) (OC). Combustible.
Hazard: Moderate fire risk.
Use: Polymerization conditioner, synthesis.

tert-octyl mercaptan. $C_8H_{17}SH$.
Properties: Colorless liquid, boiling range 154–166C, d 0.848 (15.5C), refr index 1.454 (20C), flash p 105F (40.5C). Combustible.

Grade: 95%.
Hazard: Moderate fire risk.
Use: Polymer modification, lubricant additive.

n-octyl methacrylate.
$H_2C:C(CH_3)COOC_8H_{17}$.
Properties: Water-insoluble, colorless liquid; polymerizes to a resin if unstabilized.

octyl peroxide. (caprylyl peroxide).
Properties: Straw-colored liquid with sharp odor, immiscible with water.
Hazard: Dangerous fire risk, strong oxidizing agent.

octyl phenol. (diisobutyl phenol).
$C_8H_{17}C_6H_4OH$.   Probably a mixture of isomers.
Properties: White flakes, mp 72–74C, d 0.89 (90C), bp 280–302C, hydroxyl coefficient 259–275. Insoluble in hot and cold water; limited solubility in alkalies; soluble in 1:1 mixture of methanol and 50% aqueous potassium hydroxide, also in alcohol, acetone, fixed oils. Combustible.
Derivation: (p-tert-isomer): Catalytic alkylation of phenol with olefins.
Use: Nonionic surfactants, plasticizers, antioxidants, fuel oil stabilizer, intermediate for resins, fungicides, bactericides, dyestuffs, adhesives, rubber chemicals.

p-tert-octylphenoxy polyethoxyethanol.
CAS: 9002-93-1.
$(CH_3)_3CCH_2C(CH_3)_2C_6H_4O(CH_2CH_2O)xH$.
Anhydrous liquid mixture of mono-p-(1,1,3,3-tetramethylbutyl)phenyl esters of polyethylene glycols in which x varies from 5–15.
Properties: Yellow, viscous liquid. Faint odor, bitter taste, d 1.060, refr index 1.489 (25C). Soluble in water, alcohol, acetone, benzene, toluene; insoluble in hexane; pH 7–9. Combustible.
Grade: NF.
Use: Food packaging, probably as a plasticizer for films.

p-octylphenyl salicylate.
$C_6H_4OHCOOC_6H_4C_8H_{17}$.
Properties: White solid, mp 72–74C, bulk d 5.6 lb/gal (20C). Soluble in hexane, benzene, acetone, and ethanol; insoluble in water.
Use: Prevents photooxidation in polyethylene and polypropylene.

octyl phosphate. See trioctyl phosphate.

n-octyl sulfoxide isosafrole. See sulfoxide.

octyl trichlorosilane.   (trichlorooctylsilane).
CAS: 5283-66-9. $C_8H_{17}SiCl_3$.
Properties: Water-white liquid with pungent irritating odor, fumes readily in moist air to evolve corrosive vapors.
Hazard: Moderate fire risk in contact with oxidizing materials. Toxic by ingestion and inhalation, strong irritant to skin and mucous membranes.
Use: Intermediate for silicones.

ODPN.   Abbreviation for β,β'-oxydipropionitrile.

odor.   An important property of many substances, manifested by a physiological sensation due to contact of their molecules with the olfactory nervous system. Odor and flavor are closely related, and both are profoundly affected by submicrogram amounts of volatile compounds. Attempts to correlate odor with chemical structure have produced no definitive results. Objective measurement techniques involving chromatography are under development. Even potent odors must be present in a concentration of 1.7×10^7 molecules/cc to be detected. It has been authentically stated that the nose is 100 times as sensitive in detection of threshold odor values as the best analytical apparatus.

Many compounds have a characteristic odor that is an effective means of identification. Toxic and noxious gases have distinctive odors often utilized for warning purposes. An important exception is carbon monoxide, which is almost odorless. The penetrating banana-like odor of amyl acetate has been used in mine rescue work. Among the most powerful unpleasant odors are those of organic sulfur compounds, especially ethyl mercaptan (skunk). Organic substances having a pleasant odor are broadly designated as aromatic, regardless of chemical nature. The cyclic aromatic (benzene) series of hydrocarbons was so named for this reason. Most essential oils have a pleasant odor and are the basis of perfumes and fragrances. Odor research, including evaluation by test panels, is conducted at the Olfactronics and Odor Sciences Center at Illinois Institute of Technology, Chicago.

odorant.   A substance having a distinctive, sometimes unpleasant odor which is deliberately added to essentially odorless materials to provide warning of their presence. For example, mercaptan derivatives may be added to natural gas for this purpose. In a broad sense, perfumes are odorants which are added to cosmetics, toilet goods, etc., largely for consumer appeal.
See also odor, deodorant, perfume, fragrance.

Oenanthic acid.   See n-heptanoic acid.

oenology.   See enology.

oil.   The word "oil" is applied to a wide range of substances that are quite different in chemical nature. Oils derived from animals or from plant seeds or nuts are chemically identical with fats, the only difference being one of consistency at room temperature. They are composed largely of glycerides of the fatty acids, chiefly oleic, palmitic, stearic, and linolenic. As a rule the more hydrogen the molecule contains, the thicker the oil becomes. Petroleum (rock oil) is a hydrocarbon mixture comprising hundreds of chemical compounds. It is thought to be derived from the remains of tiny sea animals laid down in past geologic ages. Following is a classification of oils by type and function.
I. Mineral
  1. Petroleum
    (a) Aliphatic or wax-base (Pennsylvania)
    (b) Aromatic or asphalt-base (California)
    (c) Mixed-base (Midcontinent)
  2. Petroleum-derived
    (a) Lubricants: engine oil, machine oil, cutting oil
    (b) Medicinal: refined paraffin oil
II. Vegetable (chiefly from seeds or nuts)
  1. Drying (linseed, tung, oiticica)
  2. Semidrying (soybean, cottonseed)
  3. Nondrying (castor, coconut)
  4. Inedible soap stocks (palm, coconut)
III. Animal
  These usually occur as fats (tallow lard, stearic acid). The liquid types include fish oils, fish-liver oils, oleic acid, sperm oil, etc. They usually have a high fatty acid content.
IV. Essential
  Complex volatile liquids derived from flowers, stems and leaves, and often the entire plant. They contain terpenes (pinene, dipentene, etc.) and are used chiefly for perfumery and flavorings. Usually resinous products are admixed with them. Turpentine is a highly resinous essential oil.
V. Edible
  Derived from fruits or seeds of plants and used chiefly in foodstuffs (margarine, etc.). Most common are corn, coconut, soybean, olive, cottonseed, and safflower. They have varying degrees of saturation.

oil black.   A carbon black made from oil, usually an aromatic-type petroleum oil.
See furnace black.

oil cake. The residue obtained after the expression of vegetable oils from oil-bearing seeds, used as cattle feed and fertilizer. When ground they are known as meal.
See also cottonseed cake and meal, peanut oil meal, etc.

oil, chloronaphthalene.
Properties: Almost colorless, thin, mobile liquid. D 1.20–1.25 (68F), liquid down to −25F (−31.6C), congealing p −30F, flash p 350F (176C), volatile at 212F (100C), bp 480–550F (248–287C), soluble in practically all organic solvent liquids and oils. Combustible.
Derivation: By chlorinating naphthalene.
Hazard: By inhalation, strong skin irritant.
Use: Plasticizer, carbon softener and remover, heat-transfer medium, solvent for rubber, aniline and other dyes, mineral and vegetable oils, varnish gums and resins, waxes.

"Oildag."[46] TM for a dispersion of colloidal graphite in petroleum oil for general industrial and automotive lubricants.

"Oilfos."[58] TM for glassy sodium phosphate, a dry powder used exclusively for controlling viscosity of oil-well drilling muds.

oil gas. A gas made by the reaction of steam at high temperature on gas oil or similar fractions of petroleum, or by high-temperature cracking of gas oil. One typical analysis is: heating value 554 Btu/ft^3, illuminants 4.2%, carbon monoxide 10.4%, hydrogen 47.6%, methane 27.0%, carbon dioxide 4.6%, oxygen 0.4%, nitrogen 5.8%, autoign temperature 637F (336C).
Hazard: Flammable, dangerous fire and explosion risk. Toxic by inhalation.

oiliness. That property of a lubricant that causes a difference in coefficient of friction when all the known factors except the lubricant itself are the same. This concept is also expressed by the term lubricity.

oil of bitter almond. See almond oil.

oil of mirbane. See nitrobenzene.

oil of vitriol. See sulfuric acid.

oil of wintergreen. See methyl salicylate.

oil sands. (tar sands). Porous sandstone structures occurring on the surface and to depths of 100 meters or more in certain localities; they contain a high proportion of bitumen composed chiefly of asphaltenes and maltha, together with substantial percentage of sulfur and heavy metals. Its viscosity is about midway between that of crude oil and soft asphalt. The largest deposit in North America is in the Athabasca region of Alberta; there are smaller ones in the western US. Venezuela and Trinidad have large deposits. The Athabasca sands have been successfully mined and have made a substantial contribution to Canadian energy resources over the past decade.

oil shale. Extensive sedimentary rock deposits in the mountains of Colorado, Utah and Wyoming containing a high percentage of kerogen, which can be separated from the shale either by heating in retorts (surface mining) or by direct combustion in situ in interior excavations. The deposits range in thickness from 10–800 ft and yield from 25–30 gal of oil/ton of shale. Only 33% of the oil content is recoverable by present techniques.
See also shale oil, kerogen.

oil varnish. See varnish.

oil, vulcanized. See factice.

oil white. One of several mixtures of lithopone and white lead or zinc white. It may also contain gypsum, magnesia, whiting, or silica.
Use: White-lead substitute.

ointment. (salve). A semisolid pharmaceutical preparation based on a fatty material such as lanolin, and often containing petrolatum or zinc oxide together with specific medication for relief of rashes and other forms of dermatitis.

oiticica oil.
Derivation: By expression from the seeds of the Brazilian oiticica tree, Licania rigida. Chief constituents: Glycerides of α-licanic acid (4-keto-9,11,13-octadecatrienoic acid).
Use: Drying oil in paints, varnishes, etc.

-ol. A suffix indicating that one or more hydroxyl groups (OH) are present in an organic compound, e.g., alcohol, phenol, menthol. Thiol is an exception, the oxygen of the OH group being replaced by sulfur. There are a few other exceptions among the essential oils, e.g., eucalyptol.

oleamide.
cis-$CH_3(CH_2)_7CH{:}CH(CH_2)_7CONH_2$.

Properties: Ivory-colored powder, mp 72C, d 0.94. Combustible.

Grade: Refined.

Use: Slip-agent for extrusion of polyethylene, wax additive, ink additive.

olefin. (alkene). A class of unsaturated aliphatic hydrocarbons having one or more double bonds, obtained by cracking naphtha or other petroleum fractions at high temperatures (1500–1700F). Those containing one double bond are called alkenes, and those with two alkadienes, or diolefins. They are named after the corresponding paraffins by adding "-ene" or "-ylene" to the stem. α-olefins are particularly reactive because the double bond is on the first carbon. Examples are 1-octene and 1-octadecene, which are used as the starting point for medium-biodegradable surfactants. Other olefins (ethylene, propylene, etc.) are starting points for certain manufactured fibers.

See also diolefin.

oleic acid. (cis-9-octadecenoic acid; red oil).

CAS: 112-80-1.

$CH_3(CH_2)_7CH:CH(CH_2)_7COOH$. A monounsaturated fatty acid, a component of almost all natural fats as well as tall oil. Most oleic acid is derived from animal tallow or vegetable oils.

Properties: Commercial grades: Yellow to red oily liquid, lard-like odor, darkens on exposure to air. Insoluble in water; soluble in alcohol, ether, and most organic solvents, fixed and volatile oils. Solvent for other oils, fatty acids and oil-soluble materials. Purified grades: Water-white liquid, d 0.895 (20/4C), fp 4C, bp 286C (100 mm), 225C (10 mm), refr index 1.4599 (20C), acid value 196–204, iodine value 83–103, saponification value 196–206, flash p 372 (189C), Combustible.

Derivation: The free fatty acid is obtained from the glyceride by hydrolysis, steam distillation and separation by crystallization or solvent extraction. Filtration from the press cake results in the oleic acid of commerce (red oil) which is purified and bleached for specific uses.

Grade: Variety of technical grades, grade free from chick edema factor, USP, FCC, 99+%. A purified technical oleic acid containing 90% or more oleic, 4% maximum linoleic and 6% maximum saturated acids is available.

Use: Soap base, manufacture of oleates, ointments, cosmetics, polishing compounds, lubricants, ore flotation, intermediate, surface coatings, food-grade additives.

olein. (triolein; glyceryl trioleate).

$C_{17}H_{33}COO)_3C_3H_5$. The triglyceride of oleic acid, occurring in most fats and oils. It constitutes 70–80% of olive oil.

Properties: Yellow, oily liquid; d 0.915; mp -4 to $-5C$; soluble in chloroform, ether, carbon tetrachloride; slightly soluble in alcohol. Combustible. Impurities: Stearin, linolein.

Derivation: Refined natural oils.

Use: Textile lubricants.

oleoresin. Any of a number of mixtures of essential oils and resins characteristic of the tree or plant from which they are derived. Most types are semisolid and tacky at room temperature, becoming soft and sticky at high temperatures. They have various distinctive odors.

See also balsam, rosin.

oleoyl chloride. (cis-9-octadecenoyl chloride).

$CH_3(CH_2)_7CH:CH(CH_2)_7COCl$.

Properties: Liquid, bp 175–180C (3 mm), soluble in hydrocarbons and ethers, reacts slowly with water. Combustible.

Use: Chemical intermediate.

n-oleoylsarcosine.

$C_{17}H_{33}C(O)N(CH_3)CH_2COOH$.

Properties: Amber liquid, d 0.955 (20/20C), refr index 1.4703 (20C), 95% pure. Combustible.

Use: Surfactants.

oleum. The Latin word for oil, applied to fuming sulfuric acid. (Sulfuric acid was originally called oil of vitriol).

oleyl alcohol. (octadecenol). CAS: 143-28-2.

$CH_3(CH_2)_7CH:CH(CH_2)_7CH_2OH$. The unsaturated alcohol derived from oleic acid. Clear, viscous liquid at room temperature. Iodine value 88, cloud p $-6.6C$, bp 333C, fp $-75C$, d 0.84. Combustible. Impurities: Linoleyl, myristyl, and cetyl alcohols.

Derivation: Reduction of oleic acid, occurs in fish and marine mammal oils.

Grade: Technical, commercial (80–90% pure).

Use: Surfactants, metal cutting oils, printing inks, textile finishing, antifoam agent, plasticizer.

oleyl aldehyde. See octadecenyl aldehyde.

oleylhydroxamic acid. $C_{17}H_{33}CONHOH$.

Properties: Waxy solid, off-white color, d 0.897 (70/25C), insoluble in water, soluble in aqueous potassium hydroxide and organic solvents.

oleyl-linoleylamine. (octadecene-octadecadieneamine).

Properties: Highly unsaturated primary amine,

soluble in many organic solvents, insoluble in water, d 0.83, mp 19C, bp 198–209C, amine no. 200–210, iodine value 90 min.
Use: Organic intermediate.

oleyl methyl tauride. See sodium-N-methyl-N-oleoyl taurate.

oligo-. A prefix meaning "a few" or "very little." See following entries.

oligodynamic. Literally, active in small amounts. In technical literature, the term describes the sterilizing or purifying action of a substance, e.g., silver.

oligomer. A polymer molecule consisting of only a few monomer units (dimer, trimer, tetramer).

oligopeptide. A peptide made up of not more than 10 amino acids.

oligosaccharide. A carbohydrate containing from two to ten simple sugars linked together (e.g., sucrose, composed of dextrose and fructose). Beyond ten they are called polysaccharides.

olive oil. CAS: 8001-25-0.
Properties: Pale yellow or greenish-yellow liquid, a nondrying oil, slight odor and taste. Soluble in ether, chloroform, and carbon disulfide; sparingly soluble in alcohol. D 0.910–0.918, saponification value 188–196, iodine value 77–88, flash p 437F (225C), cloud p −6.6 to −1.1C. Combustible.
Use: Salad dressings and other foods; ointments, liniments, etc.; Castile soap, special textile soaps; lubricant; sulfonated oils; cosmetics.

olivetol. (5-pentylresorcinol).   CAS: 500-66-3.
$CH_3(CH_2)_4C_6H_3$-1,3-$(OH)_2$.   Store under nitrogen.
Properties: Off-white solid with mw 180.25, mp 42–44, flash p above 110C.
Hazard: Skin irritant and combustible.
Use: Pharmaceutical intermediate.

olivine. (chrysolite).   CAS: 1317-71-1.
$(Mg,Fe)_2SiO_4$.   Natural magnesium-iron silicate, found in igneous and metamorphic rocks, meteorites, and blast furnace slags. A complete series exists from Fe_2SiO_4 to Mg_2SiO.
Grade: Crude, 20 mesh, 100 mesh.
Use: Refractories, cements.

"Omadine."[84]   TM for a series of derivatives of pyridinethione [such as 1-hydroxy-2-pyridinethione, $C_5H_4NOH(S)$] having bactericide-fungicide properties.
Use: Cosmetics, textiles, cutting oils and coolant systems, vinyl films and rubber products.

"Omazene."[84]   TM for copper dihydrazinium sulfate. Available as 50% wettable powder.
Use: Foliage fungicide.

OMC. Abbreviation of oxidized microcrystalline waxes.

OMPA. Abbreviation for octamethyl pyrophosphoramide.
See schradan.

"Onamine."[328]   TM for a series of liquid cationic detergents.
Use: Intermediate, acid-stable emulsifier, corrosion inhibitor.

ONB. Abbreviation for o-nitrobiphenyl.

oncogen. Any substance that will cause tumors in test animals, either benign or malignant. EPA pesticide regulations use this term instead of "tumorogenic" and "carcinogenic."

one-step resin. See A-stage resin.

Onsager, Lars. (1903-1976) A Norwegian chemist who won the Nobel prize for chemistry in 1968. He studied and wrote on the theory of electrolytic conduction and theory of dielectrics. He also worked with superfluids and crystal statistics and reciprocal relations in irreversible processes. After receiving his doctorate in Norway, he came to the US and became a citizen.

"Onyx-ol."[328]   TM for a series of fatty acid ethanolamine and isopropanolamine condensates.
Use: Foam stabilizers, thickeners, nonionic detergents.

opacity. The optical density of material, usually a pigment; the opposite of transparency. A colorant or paint of high opacity is said to have good hiding power or covering power, by which is meant its ability to conceal another tint or shade over which it is applied. Apparatus for measuring opacity is available.

"Opax."[337]   TM for a brand of zirconium oxide containing 88% min of zirconium oxide and 7% max silicon dioxide, mp 2480C.
Use: Ceramic enamels and glazes.
See also zirconium oxide.

OPDN. Abbreviation for β,β-oxydipropionitrile.

operation. See unit operation.

OPG. Abbreviation for oxypolygelatin.

opium. A mixture of alkaloids.
Derivation: The air-dried, milky exudate obtained from the unripe capsules of Papaver somniferum.
Forms available: Deodorized, granulated, powdered.
Hazard: A habituating narcotic; importation and sale restricted by law in US.
Use: Source of morphine.

"Oppanol."[440] TM for a series of polyisobutylenes, varying from oily liquids through highly viscous materials to rubberlike solids according to degree of polymerization.

Oppenauer oxidation. The aluminum alkoxide-catalyzed oxidation of a secondary alcohol to the corresponding ketone (the reverse of the Meerwein-Ponndorf-Verley reduction).

optical activity. See optical rotation.

optical brightener. (optical bleach; colorless dye; fluorescent brightener). A colorless, fluorescent, organic compound which absorbs UV light and emits it as visible blue light. The blue light masks the undesirable yellow of textiles, paper, detergents, and plastics. Some examples are: derivatives of 4,4′-diaminostilbene-2,2′-disulfonic acid, coumarin derivatives such as 4-methyl-7-diethylaminocoumarin.

optical crystal. A comparatively large crystal, either natural or synthetic, used for infrared and ultraviolet optics, piezoelectric effects, and short wave radiation detection. Examples are sodium chloride, potassium iodide, silver chloride, calcium fluoride, and (for scintillation counters) such organic materials as anthracene, naphthalene, stilbene, and terphenyl.

optical fiber. See fiber, optical.

optical glass. See glass, optical.

optical isomer. Either of two kinds of optically active 3-dimensional isomers (stereoisomers). One kind is represented by mirror-image structures called enantiomers which result from the presence of one or more asymmetric carbon atoms in the compound (glyceraldehyde, lactic acid, sugars, tartaric acid, amino acids). The other kind is exemplified by diastereoisomers, which are not mirror images. These occur in compounds having two or more asymmetric carbon atoms; thus, such compounds have 2^n optical isomers, where n is the number of asymmetric carbon atoms.
See also enantiomer, diastereoisomer, optical rotation, asymmetry.

optical microscope. (light microscope). A magnifying lens system that utilizes light in the visible wavelength range of the electromagnetic spectrum (5000Å). A convex glass lens bends or focuses light waves because of the difference in density between glass and air. Invented in 1590 by the Janssen brothers and later improved by van Leeuwenhoek, the compound microscope has three lenses: a condenser lens, which concentrates the incident light; an objective lens, which gives an enlarged reverse image of the specimen, and a projector lens, which further enlarges the image and returns it to normal position. Its maximum resolving power is 0.5 micron, compared with 100 microns for the human eye. The compound microscope is particularly useful in studying bacteria and other microorganisms in their natural state without interfering with their behavior. It has been of untold benefit to biologists and bacteriologists, and also has innumerable uses in chemical and metallurgical research, as well as in forensic chemistry.
See also: resolving power, electron microscope, ultramicroscope, field-ion microscope, van Leeuwenhoek.

optical rotation. The change of direction of the plane of polarized light to either the right or the left as it passes through a molecule containing one or more asymmetric carbon atoms, e.g., sugars. The direction of rotation, if to the right, is indicated by either a plus sign (+) or an italic d; if to the left, by a minus sign (-) or an italic l. Molecules having a right-handed configuration (d) usually are dextrorotatory, $d(+)$, though they may be levorotatory, $d(-)$; those having a left-handed configuration (l) are usually levorotatory, $l(-)$, but may be dextrorotatory, $l(+)$. Compounds having this property are said to be optically active, and are isomeric. The amount of rotation varies with the compound, but is the same for any two isomers, though in opposite directions.
See also optical isomer, nicol.

optical spectroscopy. See spectroscopy, absorption spectroscopy, emission spectroscopy.

optrode. A component of fiber-optical analytical systems that is analogous to an electrode. Its function is to couple the laser beam to the sample solution being analyzed. There are three types: cuvette, sapphire ball, and membrane.

oral contraceptive. See antifertility agent.

orange III. See methyl orange.

orange IV. See tropaeolin 00.

orange cadmium. See cadmium sulfide.

orange mineral. A red lead oxide pigment made in a furnace by roasting lead carbonate or sublimed litharge; it is a very bright orange, but has low tinting strength.
Properties: Fine powder, −325 mesh, d 9.0, contains 95.5% red lead (Pb_3O_4).
Use: Pigment in printing inks and primers.

orange peel. A term used in the paint industry to refer to a roughened film surface due to too rapid drying.

orange peel oil. See citrus peel oils.

orange seed oil. See citrus seed oils.

"Orasol."[219] TM for solvent dyes for transparent coatings, inks and plastics.

orbital theory. The quantum theory of matter applied to the nature and behavior of the electron either in a single atom (atomic orbital) or combined atoms (molecular orbital). A combination of Schrodinger's wave mechanics and Heisenberg's uncertainty principle, the orbital theory was formulated in 1926. It has yielded a better understanding of the electron and its critical part in chemical bonding than is possible with Newtonian mechanics. In simple language, the orbital theory considers the electron not as a particle, but as a 3-dimensional wave that can exist at several energy levels; its exact location and position in the "shell" (which in most elements is a group of orbitals) cannot be precisely determined, but only predicted by the laws of mathematical probability. The orbital levels and the movement of electrons within them are expressed by wave functions and quantum numbers. The probability that an electron will be found in a given volume (i.e., the square of the one-electron wave function) is called the orbital of that electron, and the shape of the orbital is defined by surfaces of constant probability, (i.e., spheres and doughnut-shaped ellipses). The electron orbital, described in terms of probability, is like a cloud, with indefinite boundaries. The energy state of each electron is given by four quantum numbers which describe its principal level, its angular momentum, its magnetic moment, and its spin. This concept has exerted a profound effect on modern ideas about chemical bonding, transition metal complexes, semiconductors, and solid state physics.

orcin. (dihydroxytoluene; methylresorcinol; orcinol).   CAS: 504-15-4.
$CH_3C_6H_3(OH)_2$.

Properties: White, crystalline prisms becoming red in air; intensely sweet, unpleasant taste; soluble in water, alcohol, and ether; d 1.2895; mp (anhydrous) 107C, (hydrated) 56C; bp 287–290C.
Derivation: By fermentation of various species of lichens (rocella), and extraction.
Use: Reagent for certain carbohydrates (beet sugar, lignin, pentoses, etc.).

order of magnitude. A range of values applied to numbers, distances, dimensions, etc., which begins at any value and extends to 10 times that value; e.g., 2 is of the same order of magnitude as any number between itself and 20; 5 miles is of the same order of magnitude as any distance between 5 and 50 miles. The expression usually applies to extremely large or extremely small units, i.e., the size ranges of atoms, molecules, colloidal particles, etc., or astronomical distances.

ore. An aggregate of valuable minerals and gangue from which one or more metals can be extracted at a profit.

ore flotation. See flotation.

organelle. A portion of a cell having specific functions, distinctive chemical constituents and characteristic morphology; it is a unit subsystem of a cell. Examples are mitochondria and chromosomes. Organelles are often closely associated with enzymes. The lysosome (an enzyme-bearing organelle) has been synthesized.

organic chemistry. A major branch of chemistry which embraces all compounds of carbon except such binary compounds as the carbon oxides, the carbides, carbon disulfide, etc.; such ternary compounds as the metallic cyanides, metallic carbonyls, phosgene ($COCl_2$), carbonyl sulfide (COS), etc.; and the metallic carbonates, such as calcium carbonate and sodium carbonate. The total number of organic compounds is indeterminate, but some 6,000,000 have been identified and named. These fall into several structural groups as follows:

I. Aliphatic (straight-chain)
1. Hydrocarbons (petroleum and coal-derived)
   (a) paraffins or alkanes (saturated) ($C_n H_{2n+2}$) methane and homologs; halogen-substituted derivatives, e.g., fluorocarbons.
   (b) Olefins (unsaturated)
       (1) Alkenes (one double bond) ($C_n H_{2n}$) ethylene and homologs
       (2) alkadienes (2 double bonds) ($C_n H_{2n-1}$) butadiene, allene
   (c) acetylenes or alkynes (triple bond)
2. Alcohols (ROH): methanol and homologs
3. Ethers (ROR): methyl ether and homologs
4. Aldehydes (RCHO): formaldehyde, acetaldehyde
5. Ketones: (RCOR): acetone, methylethylketone
6. Carboxylic acids (RCOOH)
7. Carbohydrates ($C_n H_{2n} O_n$)
   (a) Sugars: glucose, fructose, sucrose, gums
   (b) Starches: wheat, corn, potato
   (c) Cellulose: cotton, plant fibers
II. Cyclic (closed ring)
1. Alicyclic hydrocarbons (properties similar to aliphatics):
   (a) cycloparaffins (naphthenes) (saturated): cyclohexane, cyclopentane, etc.
   (b) cycloolefins (unsaturated): cyclopentadiene, cyclooctatetraene
   (c) cycloacetylenes (triple bond)
2. Aromatic hydrocarbons (arenes): unsaturated compounds; hexagonal ring structure; single and multiple fused rings
   (a) benzene group (1 ring)
   (b) naphthalene group (2 rings)
   (c) anthracene group (3 rings)
   (d) polycyclic group (steroids, sterols)
3. Heterocyclic: unsaturated; usually pentagonal rings, containing at least one other element besides carbon
   (a) pyrroles
   (b) furans
   (c) thiazoles
   (d) porphyrins
III. Combinations of aliphatic and cyclic structures
1. terpene hydrocarbons
2. amino acids (some are aliphatic and others combinations)
3. proteins and nucleic acids (coiled or helical formations)

IV. Organometallic compounds
V. Synthetic high polymers, including silicones

Important areas of organic chemistry include polymerization, hydrogenation, isomerization, fermentation, photochemistry, and stereochemistry. There is no sharp dividing line between organic and inorganic chemistry, for the two often tend to overlap.
See also entry on inorganic chemistry.

organoborane. A compound comprised of an unsaturated organic group and a borane obtained by the hydroboration reaction. Such compounds are useful catalytic reagents in organic syntheses of some complexity, e.g., cis- or trans-olefins, optically pure alcohols, alkanes, and ketones. Prostaglandins and insect pheromones have been synthesized by this means. A particularly versatile example is triphenylboron, $B(C_6H_5)_3$.
See also hydroboration, carborane, borane.

organoclay. (organopolysilicate). A clay such as kaolin or montmorillonite, to which organic structures have been chemically bonded; since the surfaces of the clay particles, which have a lattice-like arrangement, are negatively charged, they are capable of binding organic radicals. When this type of structure is in turn reacted with a monomer such as styrene, a complex results that is known as a polyorganosilicate graft polymer.

organoleptic. A term widely used to describe consumer testing procedures for food products, perfumes, wines, and the like in which samples of various products, flavors, etc., are submitted to groups or panels. Such tests are a valuable aid in determining the acceptance of the products, and thus may be viewed as a marketing technique. They also serve psychological purposes and are an important means of evaluating the subjective aspects of taste, odor, color, and related factors. The physical and chemical characteristics of foods are stimuli for the eye, ear, skin, nose, and mouth, whose receptors initiate impulses that travel to the brain, where perception occurs.

organopolysilicate. See organoclay.

organometallic compound. An organic compound comprised of a metal or nonmetal attached directly to carbon (RM); such compounds have been prepared of practically all the metals as well as with such nonmetals as silicon and phosphorus. Metallic salts (soaps) of organic acids are excluded. Examples are diethylzinc (the first known organometallic), Grignard compounds such as methyl magnesium iodide (CH_3MgI), metallic alkyls such as butyllithium (C_4H_9Li),

tetraethyllead, triethyl aluminum, tetrabutyl titanate, sodium methylate, copper phthalocyanine, and metallocenes. Some are highly toxic or flammable, others are coordination compounds.
See specific compound for details.
Reactive and moderately reactive organometallic compounds will react with all functional groups, two major types of reaction in which they are involved are oxidation and cleavage by acids. Probably the most important organometallic reactions are those involving addition to an unsaturated linkage. Many of them are powerful catalysts and form useful coordination complexes.
See also catalysis, metallocene, coordination compound.

organophosphorus compound. Any organic compound containing phosphorus as a constituent. These fall into several groups, chief of which are the following: (1) phospholipids, or phosphatides, which are widely distributed in nature in the form of lecithin, certain proteins, and nucleic acids; (2) esters of phosphinic and phosphonic acids, used as plasticizers, insecticides, resin modifiers, and flame retardants; (3) pyrophosphates, for example, tetraethyl pyrophosphate, which are the basis for a broad group of cholinesterase inhibitors used as insecticides; (4) phosphoric esters of glycerol, glycol, sorbitol, etc., which are components of fertilizers. While many of these compounds play an important part in animal metabolism, those in group (3) are toxic and should be handled with extreme care.

organosilane. See organosilicon.

organosilicon. An organic compound in which silicon is bonded to carbon (organosilane). Such compounds were first made by Friedel and Crafts in 1863. Silicon was found to have a remarkable chemical similarity to carbon, which it can replace in organic compounds. the silicon-carbon bond is about as strong as the carbon-carbon bond, and compounds containing them are similar in properties to all-carbon compounds. Organosilicon oxides (organosiloxanes or silicones) were discovered by F. S. Kipping in England in 1900; he found that Grignard reagents would react with silicon tetrachloride to form silicon-carbon bonded polymers of both ring and chain types. These were named silicones because of the similarity of their empirical formula (R_2SiO) to that of ketones (R_2CO).
An organosilicon compound (tetramesityldisilene) containing a silicon- to silicon double bond has been synthesized. It is a crystalline solid, mp 176C and has reactive properties similar to olefins. Compounds of this type are called silylenes.
See also silicone.

organosol. Colloidal dispersion of any insoluble material in an organic liquid; specifically the finely divided or colloidal dispersion of a synthetic resin in plasticizer in which dispersion the volatile content exceeds 5% of the total.
See Plastisol.

organotin compound. A family of alkyl tin compounds widely used as stabilizers for plastics, especially rigid vinyl polymers used as piping, construction aids, and cellular structures. Some have catalytic properties. They include butyl tin trichloride, dibutyltin oxide, etc., and various methyltin compounds. They are both liquids and solids.
Hazard: All are highly toxic, with a TLV of 0.1 mg/m^3 of air.
See dibutyltin entries for specific data.

"Oriex."[482] TM for a bi-axially oriented vinyl. Available in roll and sheet form for shrink packaging.

origanum oil. An essential oil used in pharmacy and as a flavoring.

"Orlon."[28] TM for a copolymer containing at least 85% acrylonitrile. Available in various types of staple and tow.
Properties: Tensile strength (psi) 32,000–39,000, d 1.14–1.17, break elongation 20–28%, moisture regain 1.5% (21.2C, 65% RH), softens at 235C, soluble in butyrolactone (hot), dimethyl formamide (hot), ethylene carbonate (hot), resistant to mineral acids, fair to good resistance to weak alkalies. Insoluble in alcohol, acetone, benzene, carbon tetrachloride and petroleum ether; soluble in dimethyl sulfoxide, maleic anhydride, ethylene carbonate, nitriles, and nitrophenols.
Hazard: Combustible, burns freely and rapidly.
Use: In apparel, usually blended with wool or other fibers.

Orn. Abbreviation for ornithine.

ornithine. (2,5-diaminovaleric acid).
CAS: 70-26-8. $NH_2(CH_2)_3CH(NH_2)COOH$.
A nonessential amino acid produced by the body and important in protein metabolism.
Properties: l(+)-ornithine: Crystals from alcohol-ether; mp 140C, soluble in water and alcohol. dl-ornithine: Crystals from water, slightly soluble in alcohol.
Derivation: Isolated from proteins after hydrolysis with alkali.
Use: Biochemical research; medicine.

ORNL. Abbreviation for Oak Ridge National Laboratory.

"Orotan" TV[23]. TM for a synthetic tanning agent with attributes of vegetable tannins. Dark-red, viscous solution; 31% tannin. Imparts high degree of tannage, strength, fullness, and solidity to leather. Solubilizing, penetrating, and bleaching agent.

orotic acid. (uracil-6-carboxylic acid; 6-carboxyuracil).　　CAS: 65-86-1. $C_4N_2H_3(O)_2COOH$.　　Occurs in cow's milk and has also been isolated from certain strains of molds (Neurospora). A growth factor for certain microorganisms.
Properties: Crystals with mp 345–346C.
Use: Biochemical research, especially the biosynthesis of nucleic acids.

orpiment. Obsolete name for arsenic trisulfide.

orthamine. See o-phenylenediamine.

ortho-. (o-).　　A prefix meaning "straight ahead." Compare with meta- (m-) meaning "beyond, " para- (p-) meaning "opposite." These prefixes are used in organic chemistry in naming disubstitution products derived from benzene in which the substituent atoms or radicals are located in certain definite positions on the benzene ring. This is illustrated in the diagram, where A and B represent the substituent atoms or groups. When attached to adjoining carbon atoms, B is in the o- position in respect to A (also called the 1,2-position). If B is located on the third carbon atom in respect to A, it is in the m- position (also called 1,3-); when B is attached to the opposite carbon atom, it is the para position (1,4).

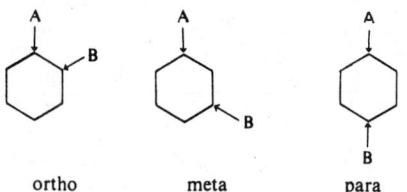

ortho　　　　　meta　　　　　para

In organic compounds, these prefixes can appear in italics (often abbreviated o-, m-, and p-), and are ignored in alphabetizing.
　　In inorganic chemistry, the prefix "ortho" designates the most highly hydrated acid, or its salt, to contrast with the "meta" or less hydrated acid or salt. For example, H_3PO_4 or $(P_2O_5 \cdot 3HOH)$ is orthophosphoric acid.

orthoarsenic acid. See arsenic acid.

orthoboric acid. See boric acid.

"Orthochrom."[23] TM for pigmented plasticized nitrocellulose lacquers and thinners. Produce

durable, washable, flexible, colored lacquer finishes of good light fastness.
Use: Finishing of belt, garment, upholstery, and other leathers.

"Orthoclear."[23] TM for permanently plasticized nitrocellulose binders and lacquers in various solvents. Produce clear, durable, flexible finishes.
Use: Topcoat finishes for glazing or high gloss leather coatings.

orthoform. See methyl-m-amino-p-hydroxybenzoate.

"Ortholite."[23] TM for clear and pigmented vinyl lacquers, binders, and solvents. Produce finishes of outstanding abrasion resistance and low-temperature flexibility.
Use: Finishes on upholstery, automotive, luggage, and case leathers.

"Orthophen" 278.[204] TM for a special blend of amyl phenol isomers.
Properties: Straw colored liquid, d 0.94–0.95, distillation range 75–270C.
Use: Antiskin agent for paint industry.

"Orthosil."[204] TM for a practically anhydrous water-soluble sodium orthosilicate in granular form. Quick-acting detergent used in heavy-duty metal cleaning. Special types made for still tank cleaning have suitable wetting and water softening additions.

Os. Symbol for osmium.

-ose. A suffix indicating a carbohydrate compound or polymer, usually a simple or complex sugar, for example, sucrose, fructose, glucose, maltose, etc., and also cellulose, cellobiose, amylose (starch).

OSHA. Abbreviation for Occupational Safety and Health Administration.

osmic acid. (osmium tetroxide; perosmic acid anhydride; perosmic oxide).　　CAS: 20816-12-0. OsO_4.
Properties: A dimorphic compound with both crystalline and amorphous forms. Colorless, pungent, disagreeable odor; soluble in water, alcohol, and ether; d 4.90, mp 40C, bp 130C.
Derivation: By heating powdered osmium in air, or by treating it with nitric acid, aqua regia, or chlorine.
Hazard: Toxic by inhalation; strong irritant to eyes and mucous membranes. TLV (as Os): 0.0002 mg/m^3 of air.
Use: Microscopic staining, photography, oxidation catalyst in organic synthesis.

osmiridium. See iridosmine.

osmium. CAS: 7440-04-2. Os. Metallic element having atomic number 75, in group VIII of the periodic system, aw 190.2, valences = 2, 3, 4, 6, 8; has seven stable isotopes.
Properties: Hard white metal of the platinum group. On heating in air gives off poisonous fumes of osmium tetroxide; insoluble in acids and aqua regia, attacked by fused alkalies, d 22.5, mp 3000C, bp 5500C, it has the highest specific gravity and melting point of the platinum metals, metallurgically unworkable.
Occurrence: Tasmania, South Africa, USSR, Canada.
Derivation: Occurs with platinum from which it is recovered during the purification process. Also occurs with iridium as a natural alloy, iridosmine.
Hazard: Highly toxic; irritant to skin.
Use: Hardener for iridium and platinum, pen points, instrument pivots, catalyst.

osmium ammonium chloride. See ammonium hexachloroosmate.

osmium chloride. (osmium dichloride; osmous chloride). $OsCl_2$.
Properties: Dark green needles, hygroscopic, keep away from air. Soluble in alcohol, ether; insoluble in water.

osmium dichloride. See osmium chloride.

osmium sodium chloride. See sodium hexachloroosmate.

osmium tetroxide. See osmic acid.

osmocene. (dicyclopentadienylosmium). $(C_5H_5)_2Os$.
Properties: Stable, white solid; mp 229–230C.
Use: Intermediate, high-temperature applications, derivatives used as an UV radiation absorber.
See metallocene.

osmometry. The measurement of osmotic pressure.

osmosis. Passage of a pure liquid (usually water) into a solution, (e.g., of sugar and water) through a membrane that is permeable to the pure water but not to the sugar in the solution. This passage can also occur when the two phases consist of solutions of different concentration. The membrane is called semipermeable when the molecules of the solvent, but not those of the solute, can penetrate it. This pushing of water through a membrane into a solution results from the greater tendency of water molecules to escape from water than from a solution. The term osmosis is usually restricted to movement through a solid or liquid barrier that prevents the phases from mixing rapidly. In test apparatus parchment or collodion membranes are used; in plants and animals the cell wall acts as a diffusion barrier. The pressure exerted by osmosis is substantial and accounts for the elevation of sap from root systems to the tops of trees. Osmosis is considered an essential characteristic of growth.

Reverse osmosis is utilized as a method of desalting sea water, recovering waste water from paper mill operations, pollution control, industrial water treatment, chemical separations, and food processing. This method involves application of pressure to the surface of a saline solution, thus forcing pure water to pass from the solution through a membrane that is too dense to permit passage of sodium and chlorine ions. Hollow fibers of cellulose acetate or nylon are used as membranes, since their large surface area offers more efficient separation.
See also dialysis, membrane, diffusion, desalination.

osmous chloride. See osmium chloride.

Ostromyslenskii (Ostromisslenskii) reaction. Dehydrogenation of ethanol over copper-containing catalysts and conversion of the acetaldehyde-ethanol mixture to butadiene by passage at high temperature over silica gel containing a small amount of tantalum oxide.

Ostwald, Wilhelm. (1853-1932) A German chemist who won the Nobel prize for chemistry in 1909. He was considered to be a founder of modern physical chemistry. His work involved research in catalysis, the rates of chemical reactions, equilibrium, and conductivity of organic acids. He was an admirer of Mach and did not readily accept the atomic theory. He was educated at the University of Dorpat.

OTEC. See Ocean Thermal Energy Conversion.

otto. See attar.

ouricury wax. A vegetable wax exuded by the leaves of Cocos coronapa (South America).
Properties: Brown, acid value 10, saponification value 80, d 0.970 (15C), mp 83C, foreign matter (dirt, etc) sometimes 18%. Combustible.
Grade: Crude, refined.
Use: Substitute for carnauba wax.

-ous. A suffix used in naming inorganic compounds which indicates that the central element is present in its lower oxidation state. For example, in ferrous chloride ($FeCl_2$), the iron atom is in its lower oxidation state of $+2$, equivalent to its valence: in an ionized state it would have two positive charges (Fe^{++}). (A recommended change in this system of nomenclature is to use the common name of the element (iron) together with a Roman numeral showing the oxidation number; thus, ferrous chloride would be iron(II) chloride).

outgassing. The removal of gas from a metal by heating at a temperature somewhat below melting, while maintaining a vacuum in the space around the metal. Usually done before melting but sometimes afterward.
Use: Manufacture of tubes and other vacuum devices.

ovalbumin. See albumin, egg.

oven. A heated chamber of varying sizes used for removing moisture from industrial products before or during processing, for example, finely divided solids, food products, tobacco, textiles, wood, etc. Laboratory sizes are used for testing the effect of heat exposure on such materials as rubber, plastics, fibers, paints, etc., as well as for sterilization, drying electronic components and curing encapsulation compounds.

ovex. (Generic name for p-chlorophenyl-p-chlorobenzenesulfonate).   CAS: 80-33-1. $ClC_6H_4OSO_2C_6H_4Cl$.
Properties: White, crystalline solid; mp 86.5C. Insoluble in water, soluble in acetone and aromatic solvents.
Hazard: Toxic by ingestion, strong irritant to skin.
Use: Insecticide and acaricide.

ovicide. A type of pesticide that kills the eggs of insects.

"Ovral."[24] TM for an oral contraceptive containing norgestrel, a synthetic progestogen. Contains 0.05 milligram norgestrel and 0.05 milligram ethinyl estradiol.

oxa-. Prefix indicating the presence of oxygen in a heterocyclic ring.

"OXAF."[248] TM for the zinc salt of 2-mercaptobenzothiazole ($Zn(SCNSC_6H_4)_2$).
Properties: White to pale yellow powder, d 1.63, melting range decomposes without melting when heated to 200C or over, excellent storage stability. Slightly soluble in ethylene dichloride and actone; insoluble in water, benzol, and gasoline. Available in pelletized form.
Use: Rubber accelerator, especially latex foam

sponge, wire insulation, air-cured footwear, druggist sundries, and specialties.

oxalic acid. HOOCCOOH·2HOH.
CAS: 144-62-7.
Properties: Transparent, colorless crystals; mp 187C anhydrous form, 101.5C for dihydrate.
Derivation: Occurs naturally in many plants (wood sorrel, rhubarb, spinach) and can be made by alkali extraction of sawdust. Now manufactured by reaction of carbon monoxide and sodium hydroxide or of sodium formate with sodium hydroxide, followed by distillation of the resulting dihydrate crystals.
Grade: Technical (crystals and powder), CP.
Hazard: Toxic by inhalation and ingestion, strong irritant. TLV: 1 mg/m³ of air.
Use: Automobile radiator cleanser, general metal and equipment cleaning, purifying agent and intermediate for many compounds, leather tanning, catalyst, laboratory reagent, stripping agent for permanent press resins, bleaching of textiles, rare-earth processing, printing and dyeing auxiliary.

oxalomolybdic acid. CAS: 53450-33-2. $[MoO_3(C_3O_4)]H_2$.   A water-soluble, crystalline compound used in invisible inks.

oxalonitrile. See cyanogen.

oxalyl chloride. (ethanedioyl chloride). CAS: 79-37-8. $(COCl)_2$.
Properties: Colorless liquid; if cooled to $-12C$, solidifies to a white, crystalline mass; gives off carbon monoxide on heating; decomposed by water and alkaline solutions; soluble in ether, benzene, chloroform; bp 64C; fp $-12C$; d 1.43.
Derivation: Interaction of oxalic acid and phosphorus pentachloride.
Hazard: Toxic by inhalation and ingestion.
Use: Military poison gas, chlorinating agent in organic synthesis.

oxamic acid. (aminooxoacetic acid). CAS: 471-47-6. H_2NCOCO_2H.
Properties: Crystalline powder with mw 89.05, mp 210C. Sparingly soluble in water.
Use: In medical diagnostic manufacture.

oxamide. CAS: 471-46-5. $NH_2COCONH_2$.
Properties: White, odorless powder; mp 419C (decomposes). Probably the highest melting organic compound, slightly soluble in water, very slightly soluble in alcohol and ether, not hygroscopic, decomposes to ammonia and carbonic acid.
Use: Stabilizer for nitrocellulose preparations, possible substitute for urea as fertilizer.

oxammonium. See hydroxylamine.

"Oxanal."[443] TM for dyes for coloring anodized aluminum.

oxazole. $\overline{OCH:NCH:CH}$. A five-membered heterocyclic compound valuable for its derivatives. The dihydro forms, 2- and 4-oxazoline, are the parents of increasingly useful commercial compounds, e.g., surface-active agents, detergents, etc.; 2-oxazoline is $\overline{OCH_2NHCH:CH}$.

oxazoline wax. A series of synthetic waxes having the oxazoline structure.
See oxazole. They can be made to fairly exact specifications; are miscible with most natural and synthetic waxes, and can be applied to the same uses.

oxetane. (trimethylene oxide).
CAS: 503-30-0. $\overline{CH_2OCH_2CH_2}$. An oxetane group ($=COCH_2C=$) is one kind of epoxy group.
See "Penton."

oxidase. An enzyme whose activity results in the transfer of electrons on the substrate; an oxidizing enzyme.

oxidation. The term "oxidation" originally meant a reaction in which oxygen combines chemically with another substance, but its usage has long been broadened to include any reaction in which electrons are transferred. Oxidation and reduction always occur simultaneously (redox reactions), and the substance which gains electrons is termed the oxidizing agent. For example, cupric ion is the oxidizing agent in the reaction: Fe (metal) $+ Cu^{++} \rightarrow Fe^{++} + Cu$ (metal); here, two electrons (negative charges) are transferred from the iron atom to the copper atom; thus, the iron becomes positively charged (is oxidized) by loss of two electrons while the copper receives the two electrons and becomes neutral (is reduced). Electrons may also be displaced within the molecule without being completely transferred away from it. Such partial loss of electrons likewise constitutes oxidation in its broader sense and leads to the application of the term to a large number of processes which at first sight might not be considered to be oxidations. Reaction of a hydrocarbon with a halogen, for example, $CH_4 + 2 Cl \rightarrow CH_3Cl + HCl$, involves partial oxidation of the methane; halogen addition to a double bond is regarded as an oxidation.
Dehydrogenation is also a form of oxidation, when two hydrogen atoms, each having one electron, are removed from a hydrogen-containing

organic compound by a catalytic reaction with air or oxygen, as in oxidation of alcohols to aldehydes.
See dehydrogenation.

oxidation number. The number of electrons that must be added to or subtracted from an atom in a combined state to convert it to the elemental form; i.e., in barium chloride ($BaCl_2$) the oxidation number of barium is $+2$ and of chlorine is -1. Many elements can exist in more than one oxidation state.
See also valence.

oxidation-reduction indicator. A substance that has a color in the oxidized form different from that of the reduced form, and can be reversibly oxidized and reduced. Thus, if diphenylamine is present in a ferrous sulfate solution to which potassium dichromate is being added, a violet color appears with the first drop of excess dichromate.
See also indicator.

oxidative coupling. A polymerization technique for certain types of linear high polymers. Oxidation of 2,6-dimethylphenol with an amine complex of a copper salt as catalyst forms a polyether, with splitting off of water. The product is soluble in aromatic and chlorinated hydrocarbons; insoluble in alcohols, ketones, and aliphatics. It is thermoplastic and unaffected by acids, bases, and detergents. It has a very broad useful temperature range (from -170 to $+190C$). It is also dimensionally stable and has good electrical resistance. Oxidative coupling of diacetylenes and dithiols also yields promising polymers.
See also "PPO."

oxide. A mineral in which metallic atoms are bonded to oxygen atoms.

"Oxidex."[188] TM for an antioxidant for soap fats and oils. It can be added to the oils before saponification or it can be added by milling into the finished soap in the same manner that perfume is incorporated. The correct proportion for solid soaps is 0.1%.

oxidizing material. Any compound that spontaneously evolves oxygen either at room temperature or under slight heating. The term includes such chemicals as peroxides, chlorates, perchlorates, nitrates, and permanganates. These can react vigorously at ambient temperatures when stored near or in contact with reducing materials such as cellulosic and other organic compounds. Storage areas should be well ventilated and kept as cool as possible.

oxine. See 8-hydroxyquinoline.

oxirane. CAS: 75-21-8.   H₂COCH.

A synonym for ethylene oxide. An oxirane group is one having the structure

$$=\!\!\overline{\text{COC}}\!\!=$$

and is one kind of epoxy group.
See ethylene oxide.

Oxirane process. A method of making ethylene glycol by catalytic oxidation of ethylene to the diacetate, which is then hydrolyzed to ethylene glycol.

oxirene. (oxacyclopropene).   An organic intermediate containing four pi electrons, reported to result from oxidation of acetylene.

2-oxohexamethylenimine. See caprolactam.

"Oxone."[28]   TM for an acidic, white, granular, free-flowing solid containing the active ingredient potassium peroxymonosulfate; readily soluble in water; 1% solution has pH of 2–3; minimum active oxygen content 4.5%; strong oxidizing agent.
Hazard: Fire risk in contact with organic materials.
Use: Manufacture of dry laundry bleaches, detergent-bleach washing compound, scouring powders, plastic dishware cleaners, and metal cleaners, hair wave neutralizers and pharmaceuticals, general oxidizing reactions.

oxonium ion. See hydronium ion.

2-oxopentanedioic acid. See α-ketoglutaric acid.

4-oxopentanoic acid. See levulinic acid.

Oxo process. Production of alcohols, aldehydes, and other oxygenated organic compounds by passage of olefin hydrocarbon vapors over cobalt catalysts in the presence of carbon monoxide and hydrogen. Aldehydes are formed as products, but in most cases these are hydrogenated at once to the corresponding alcohol. Propylene produces normal and isobutyraldehyde; higher olefins produce a mixture of aldehydes containing one more carbon atom than the olefins; n-butyl, isobutyl, amyl, isooctyl, decyl, and tridecyl alcohols are produced in large quantities.

oxosilane. See siloxane.

oxybenzoic acid. See hydroxybenzoic acid.

oxybenzone. (4-methoxy-2-hydroxybenzophenone).   CAS: 131-57-7.   C₁₄H₁₂O₃.
Properties: Colorless crystals, mp 65C, soluble in common organic solvents.
Use: Sunscreen lotions.

p,p′-oxybis(benzenesulfonylhydrazide).
[(4,4′-oxybis(benzenesulfonylhydrazide)].
H₂NNHSO₂C₆H₄OC₆H₄SO₂NHNH₂.
Properties: Fine, white, crystalline powder; odorless. D 1.52, mp decomposes at 150–160C, soluble in acetone, moderately soluble in ethanol and polyethylene glycols, insoluble in gasoline and water. Combustible.
Use: Blowing agent for sponge rubber and expanded plastics.

oxyconiine. See conhydrine.

oxydemetonmethyl. Generic for S-[2-(ethylsulfinyl)ethyl]-O,O-dimethylphosphorothioate.
See O,O-dimethyl-S-2-(ethylsulfinyl)ethyl phosphorothioate.

n-oxydiethylene-2-benzothiazolesulfenamide.
See 2-(morpholinothio)benzothiazole.

β,β′-oxydipropionitrile. (ODPN).
CAS: 1656-48-0.   (O(CH₂CH₂CH)₂.
Properties: Colorless liquid, fp −26.3C, bp 120C (1 mm), bp 155C (5 mm), d 1.0405 (30C), viscosity 8.00 cp (30C), refr index 1.4392 (25C), flash p (TOC) 180F (82.2C), soluble in water. It is thermally unstable, yielding acrylonitrile and water at above 175C. Hydrolyzed by strong acids and bases, quite immiscible with paraffin hydrocarbons, but dissolves aromatics. Combustible.
Derivation: From acrylonitrile.
Hazard: Toxic by inhalation and ingestion.
Use: Solvent in fractional extraction.

"Oxyfume."[214]   TM for a sterilant mixture of ethylene oxide and either carbon dioxide or dichlorodifluoromethane. Colorless and tasteless gas, excellent penetrating ability and ether odor. Available as: "Oxyfume" Sterilant-20, 20% ethylene oxide and 80% carbon dioxide by weight and gas volume; "Oxyfume" Sterilant-12, a mixture of 12% ethylene oxide and 88% dichlorodifluoromethane by weight.

oxygen. CAS: 7782-44-7.   O.   Nonmetallic gaseous element of atomic number 8; Group VIA of the Periodic Table, aw 15.9994, valence = 2, isotopes 16, 17, 18; molecular oxygen is O₂ and ozone O₃. Atmospheric oxygen is the result of photosynthesis. Oxygen was discovered by Priestley in England in 1774. Fifth highest-volume chemical produced in US (1985).
Properties: Colorless, odorless, tasteless, diatomic gas. Liquefiable at −183C to slightly bluish liq-

uid, solidifiable at $-218C$. It constitutes 20% by volume of air at sea level; d (gas) 1.429 g/L at 0C and 1 atm (air $= 1.29$); d (gas): 1.10535 (air $= 1.00$), (liquid) 1.14 ($-183C$); soluble in water and alcohol. Oxygen is noncombustible, but actively supports combustion.

Derivation: Prior to 1971 the only commercial method for large-scale oxygen production was fractionation of liquefied air; this has largely been replaced by a process which utilizes ambient temperature separation by means of a pressure cycle in which molecular sieves of synthetic zeolite preferentially adsorb nitrogen from air, giving 95% oxygen, and 5% argon. Electrolysis of water is used for small amounts and for laboratory demonstration.

Grade: Low purity, high purity, USP.

Hazard: Gaseous: Moderate fire risk as oxidizing agent; therapeutic overdoses can cause convulsions. Liquid: May explode on contact with heat or oxidizable materials. Irritant to skin and tissue.

Use: Blast furnaces, copper smelting, steel production (basic oxygen converter process); manufacture of synthesis gas for production of ammonia, methyl alcohol, acetylene, etc.; oxidizer for liquid rocket propellants; resuscitation, heart stimulant; decompression chambers; spacecraft; chemical intermediate; to replace air in oxidation of municipal and industrial organic wastes; to counteract effect of eutrophication in lakes and reservoirs; coal gasification.

See also ozone.

oxygen 18. (heavy oxygen).   Oxygen isotope of aw 18. Occurs in proportion of 8 parts to 10,000 of ordinary oxygen in water, air, rocks, etc. The proportion may be increased by passing carbon dioxide gas repeatedly through a packed column down which water is passed. The carbon dioxide leaving the top of tower is enriched in heavy oxygen and the water leaving the bottom is depleted.

Use: Tracer experimentation.

See also heavy water.

"Oxygenated Hydrocarbons."[569] TM for a series of petroleum-derived oxidates composed primarily of organic acids and esters. Designated by TC or TX followed by a four digit number, e.g., TC-5416, TC-6664.

Use: Corrosion inhibitors, surface-active components in wax emulsions, emulsifiable lubricants, plasticizer, intermediate, leather and cordage oils, lubricity agents and solubilizers.

oxygen consumed. (COD; OC; DOC).
A measure of the quantity of oxidizable components present in water. Since the carbon and hydrogen, but not the nitrogen, in organic matter are oxidized by chemical oxidants, the oxygen consumed is a measure only of the chemically oxidizable components and is dependent upon the oxidant, structure of the organic compound, and manipulative procedure. Since this value does not differentiate stable from unstable organic matter, it does not necessarily correlate with the biochemical oxygen demand value. It is also known as chemical oxygen demand (COD) and dichromate oxygen consumed (DOC).

See also biochemical oxygen demand; dissolved oxygen.

oxygen fluoride. (oxygen difluoride; fluorine monoxide).   CAS: 7783-41-7. OF_2.
An unstable, colorless gas; fp $-224C$; bp $-145C$; slightly soluble in water and alcohol. Suggested as oxidizer for rocket propellants.

Hazard: Explodes on contact with water, air, and reducing agents. Corrosive to tissue. TLV: CL of 0.05 ppm in air.

"Oxygen Scavenger K-91."[323] TM for a technical grade of sodium sulfite in a dry, tan, water-soluble powder form.

Use: To chemically remove dissolved oxygen left in the feed-water after mechanical deaeration. Specially catalyzed for quick reaction. Usable in cold water.

oxygen sink.   A reservoir consisting of a chemical element or compound that combines readily with oxygen and thus removes it from the atmosphere. During the early part of Precambrian time, sulfur, iron, and other elements and compounds served as important oxygen sinks, preventing oxygen from accumulating in the atmosphere.

oxyhemoglobin.   See hemoglobin.

oxyluminescence.   See chemiluminescence.

oxymethurea.   (1,3-bis(hydroxymethyl)urea).   CAS: 140-95-4. $(HOCH_2NH)_2CO$.
Properties: Crystalline solid, mp 137C, soluble in water, ethanol and methanol.

Use: Textile auxiliary (crease- and shrinkproofing agent for cotton), photographic developers, antiseptic.

oxymethylene.   See formaldehyde.

β-oxynaphthoic acid.   See 3-hydroxy-2-naphthoic acid.

oxyneurine.   See betaine.

oxyphosphorane.   One of a class of compounds derived from trialkyl phosphites and o-quinones.

Their molecules have a five atom ring

OCCOP(OR)$_3$ in which the two carbon atoms are part of an aromatic ring. They react by liberating a phosphate ester.

oxypolygelatin. (OPG). A purified gelatin treated with glyoxal, followed by oxidation with hydrogen peroxide. A possible plasma substitute.

oxyquinoline. See 8-hydroxyquinoline.

oxytetracycline. CAS: 79-57-2.
$C_{22}H_{24}N_2O_9 \cdot 2HOH$. An antibiotic obtained by fermentation from *Streptomyces rimosus*, an actinomycete. Its chemical structure is that of a modified naphthacene molecule having six assymetric centers; also recently synthesized.
Properties: Dull yellow, odorless, slightly bitter, crystalline powder; mp 179–182C (decomposes); soluble in acids and alkalies; very slightly soluble in acetone, alcohol, chloroform, and water; practically insoluble in ether; stable in air; affected by sunlight; deteriorates in solutions with pH less than 2; destroyed rapidly by alkali hydroxide solutions; pH (saturated solution) 6.115.
Grade: NF.
Use: Medicine (antibiotic). Inhibitor of lethal yellowing in coconut palm trees; feed additive; the hydrochloride is TM "Terramycin."

oxythioquinox. (6-methyl-2,3-quinoxalinedithiol cyclic carbonate). CAS: 2439-01-2.
Properties: Mp 172C, insoluble in water.
Use: Acaricide for tree fruits.

oxytocin. (α-hypophamine). CAS: 50-56-6.
$C_{43}H_{66}N_{12}O_{12}S_2$. A hormone secreted by the posterior lobe of the pituitary gland. Its chief action is stimulation of the contraction of the smooth muscle of the uterus. It contains eight different amino acids. In 1955 du Vigneaud elucidated its amino acid sequence, the first such determination ever made; it may be represented:

Cys—Tyr—Ileu
| |
Cys—Asp—Glu
|
Pro—Leu—Gly

It is available as a solution for injection (oxytocin injection, USP).

"Oxytrol."[492] TM for a chemically modified, cold-water-swelling starch that produces noncongealing pastes. Free-flowing, light cream, small granules that form straw-colored solutions that are mildly alkaline in water.

Use: Sizing cotton yarns and other fabrics. A low BOD in sizing wastes is claimed.

oyster shells. Shells of *Ostrea virginica*, taken from the Gulf of Mexico coast in Texas and Louisiana and from Chesapeake Bay. Average analysis: $CaCO_3$ 93–97%, $MgCO_3$ 1%, silica 0.5–2.0%, SO_4 (as $CaSO_4$) 0.3–0.4%, also miscellaneous substances.
Use: Source of lime, drilling muds, road beds, poultry and cattle feeds.

ozalid. Copying process which gives positive prints (dark on white).

"Ozene."[292] TM for an emulsifiable o-dichlorobenzene.

ozocerite. (mineral wax; fossil wax; ozokerite).
Properties: Wax-like, hydrocarbon mixture; yellow-brown to black or green, translucent when pure and having a greasy feel; soluble in light petroleum hydrocarbons, benzene, turpentine, kerosene, ether, carbon disulfide; slightly soluble in alcohol; insoluble in water; d 0.85–0.95; mp 55–110C (usually 70C). Combustible.
Occurrence: Utah, Australia, near the Caspian Sea.
Method of purification: Filtration.
Grade: Technical.
Use: Electric insulation, rubber products, paints, leather polish, lithographic and printing inks, electrotypers' wax, carbon paper, source of ceresin, floor polishes, impregnating furniture and parquet floor lumber, lubricating compositions, grease crayons, sizing and glossing paper, waxed paper, cosmetics, ointments, matrices for galvanoplastic work, textile sizings, waxed cloth, substitute for carnauba and beeswax.

ozone. CAS: 10028-15-6. O_3. An allotropic form of oxygen.
Properties: Unstable blue gas with pungent odor, liquefiable at −12C, more active oxidizing agent than oxygen. Contributes to formation of photochemical smog; deterioration of rubber is accelerated by traces of ozone; bp −112C, fp −192C, d (liquid) 1.6, more soluble in water than oxygen.
Occurrence: Formed locally in air from lightning, in stratosphere by UV radiation, inhibits penetration of UV. Also occurs in automobile engines and by electrolysis of alkaline perchlorate solutions. Commercial mixtures containing up to 2% ozone are produced by electronic irradiation of air. It is usually manufactured on the spot, as it is too expensive to ship. Tonnage quantities are used.
Hazard: Dangerous fire and explosion risk in con-

tact with organic materials. Toxic by inhalation, strong irritant. TLV: 0.1 ppm in air, EPA standard for ambient air is 0.12 ppm.

Use: Purification of drinking water, industrial waste treatment, deodorization of air and sewage gases, bleaching waxes, oils, wet paper and textiles, production of peroxides, bactericide: Oxidizing agent in several chemical processes (acids, aldehydes, ketones from unsaturated fatty acids), steroid hormones, removal of chlorine from nitric acid, oxidation of phenols and cyanides.

Note: Depletion of the ozone layer in the stratosphere which acts as a shield against penetration of UV light in the sun's rays is believed to be caused by light-induced chlorofluorocarbon decomposition resulting from increased use of halocarbon aerosol propellants. Their manufacture and use were prohibited in 1979, except for a few specialized items.

ozonide. A product of ozonolysis.

ozonolysis. (1) Oxidation of an organic material by means of ozone, i.e., tall oil, oleic acid, safflower oil, cyclic olefins, carbon treatment, peracetic acid production. (2) The use of ozone as a tool in analytical chemistry to locate double bonds in organic compounds, and a similar use in synthetic organic chemistry for preparing new compounds. Under proper conditions, ozone attaches itself at the double bond of an unsaturated compound to form an ozonide. Since many ozonides are explosive, it is customary to decompose them in solution and deal with the final product. See also ozone.

P

P. Symbol for phosphorus.

p-. Abbreviation for para-.

Pa. Symbol for protactinium.

PA. Abbreviation for phthalic anhydride and for polyamide.

Paal-Knorr pyrrole synthesis. Formation of pyrroles by heating 1,4-dicarbonyl compounds with ammonia or primary amines in a sealed tube.

PABA. Abbreviation for p-aminobenzoic acid.

PABA sodium. See sodium-p-aminobenzoate.

packaging. The operation of placing materials in suitable containers or protective covering for purposes of storage, distribution, and sale. Some packages act merely as containers but others protect perishable materials (especially foodstuffs) from environmental damage, contamination, and biological deterioration; in this respect the critical factor is exclusion of moisture vapor, bacteria, and oxygen. Some packages perform both functions simultaneously. Common packaging materials are:

(for nonperishable products)	(for perishable products)
wooden boxes, kegs, barrels	glass bottles
fiber drums	"tin" cans
glass bottles (perfumes, pharmaceuticals)	plastic film
polyvinyl chloride bottles (detergents)	cellophane (tobacco)
aluminum tubes (toothpaste)	polypropylene
paperboard cartons	polyvinylidene chloride
polyethylene film (textiles)	polyethylene
ceramic jars (cosmetic creams)	paraffin-coated paper and board
steel cylinders (gases)	aluminum cans (beer, soft drinks)

packing. (1) A collar or gasket used to seal mechanical devices to prevent leakage of oil or water; often made of specially compounded rubber or of a flexible plastic. (2) The operation of placing solid materials or objects in shipping containers in such a way as to secure maximum space economy and freedom from damage by vibration or impact. Barriers of paperboard, foamed plastic, or glass fiber are widely used. (3) An inert material used in distillation columns to baffle the downward flow of countercurrent liquid; it may be glass fiber or beads, metal tubes called Raschig rings, metal chains, or specially shaped devices of various kinds (saddles, helices, rings, etc.).
See tower, distillation.

padan. [S,S'(2-dimethylaminotrimethylenebis-thiocarbamate)]. CAS: 15263-52-2. $C_7H_{15}O_2N_2S_2Cl$.
Hazard: Toxic by ingestion.
Use: Insecticide.

PAHA. See p-aminohippuric acid.

paint. A uniformly dispersed mixture having a viscosity ranging from a thin liquid to a semisolid paste and consisting of (1) a drying oil, synthetic resin, or other film-forming component, called the binder; (2) a solvent or thinner; and (3) an organic or inorganic pigment. The binder and the solvent are collectively called the vehicle. Paints are used (1) to protect a surface from corrosion, oxidation, or other type of deterioration, and (2) to provide decorative effects.
Hazard: Flammable, dangerous fire risk (except water-based). Toxic if vapors are inhaled over a long period. The lead content of household paints is limited to 0.5%. For further information refer to National Paint, Varnish and Lacquer Association, 1500 R.I. Ave., Washington, DC.
See also paint, emulsion, vehicle, protective ocating, antifouling paint.

paint, emulsion. (latex paint). A paint composed of two dispersions: (1) dry powders (colorants, fillers, extenders) and (2) a resin dispersion. The former is obtained by milling the dry ingredients into water. The resin dispersion is either a latex formed by emulsion polymerization or a resin in emulsion form. The two dispersions are blended to produce an emulsion paint. Surfactants and protective colloids are necessary to stabilize the product. Emulsion paints are characterized by the fact that the binder is in a water-dispersed form, whereas in a solvent paint it is in soluble form. The principal latex paints are styrene-butadiene, polyvinyl acetate, and acrylic resins. Percentage composition may be 25–30% dry ingredients, 40% latex and 20–30% water, plus stabilizers. The unique properties of emulsion paints are ease of application, absence of disagreeable odor, and nonflammability. They can be used on both interior and exterior surfaces.

paint, inorganic. A potassium silicate-based corrosion-resistant coating designed for use on bridges and other metal work subject to marine environments.

paint, metallic. A paint in which the primary pigment is a finely divided metal dispersed in the vehicle. Most common is aluminum paint, but other metals are also used.

paint remover. (varnish remover). A mixture in liquid or paste form containing volatile solvents and nonvolatile components which retard evaporation of the solvent, thereby prolonging its action. Typical solvents are: methanol, denatured ethanol, methylene chloride, toluene, benzene, and ethyl acetate. Paraffin is often used as the retarder. Caustic removers contain sodium phosphate, sodium silicate, caustic soda, or the like.

paint, water-based. See paint, emulsion.

"Palacet."[203] TM for a series of organic pigments used for dyeing and printing on acetate, nylon, and polyester fibers.

"Palatin."[203] TM for metallized acid dye stuffs approaching the fastness of chrome colors.

palladium. CAS: 7440-05-3. Pd.
Metallic element of atomic number 46; aw 106.4; valences = 2, 3, 4; Group VIII of the Periodic Table; there are 6 stable isotopes.
Properties: Silver-white, ductile metal which does not tarnish in air. It is the least noble (most reactive) of the platinum group. Absorbs up to 800 times its own volume of hydrogen. Attacked by hot, concentrated nitric acid and boiling sulfuric acid, soluble in aqua regia and fused alkalies, insoluble in organic acids, good electrical conductor, d 12.0, mp 1554C, bp 2800C, Mohs hardness 4.8, Brinell 61, Vickers (annealed) 41. Noncombustible, except as dust.
Occurrence: Siberia, Ural Mountains (USSR), Ontario, South Africa.
Derivation: In ores with platinum, gold, copper, etc. Concentrated ores are dissolved in aqua regia; after gold and platinum are removed by chemical treatment, palladium is precipitated by ammonia, followed by hydrochloric acid. After

further purification treatment, ignition yields palladium metal.
Forms: Wire, leaf, powder, single crystals.
Grade: CP (99.99%), technical (99.0%).
Use: Alloys for electrical relays and switching systems in telecommunication equipment, catalyst for reforming cracked petroleum fractions and hydrogenation, metallizing ceramics, "white gold" in jewelry, resistance wires, hydrogen valves (in hydrogen separation equipment), aircraft spark plugs, protective coatings.

palladium chloride. (palladous chloride; palladium dichloride). CAS: 7647-10-1.
(a) $PdCl_2$ (b) $PdCl_2 \cdot 2HOH$.
Properties: Dark brown, deliquescent powder or crystals; soluble in water, hydrochloric acid, alcohol, and acetone; (a) d 4.0 (18C), mp 675C (decomposes).
Derivation: By solution of palladium in aqua regia and evaporation.
Grade: Technical, reagent.
Use: Analytical chemistry, "electroless" coatings for metals, photography, leak detection in gas lines, indelible inks, catalyst.

palladium diacetate. CAS: 3375-31-3.
$(CH_3COO)_2Pd$.
Properties: Reddish-brown, crystalline solid; decomposes at 200C; insoluble in water and alcohols; soluble in acetone, chloroform, acetonitrile.
Derivation: Reaction of palladium nitrate or palladium sponge with glacial acetic acid.
Use: Catalyst for organic reactions.

palladium iodide. (palladous iodide). PdI_2.
Properties: Black powder, d 6.003 (18C). Soluble in a solution of potassium iodide; insoluble in alcohol, water, and ether; decomposes at 350C.

palladium monoxide. See palladium oxide.

palladium nitrate. (palladous nitrate).
CAS: 10102-05-3. $Pd(NO_3)_2$.
Properties: Brown salt, deliquescent, decomposed by heat, soluble in water with turbidity, soluble in dilute nitric acid.
Hazard: Oxidizing agent, may react with organic materials.
Use: Analytical reagent, catalyst.

palladium oxide. (palladium monoxide).
CAS: 1314-08-5. PdO.
Properties: Black-green or amber solid, d 8.70 (20C), mp 750C (decomposes), soluble in dilute acids.
Derivation: Careful ignition of the nitrate or prolonged heating of the finely divided metal at 800C.
Use: Reduction catalyst in organic synthesis.

palladium potassium chloride. (palladous potassium chloride; potassium palladium chloride). PdCl$_2$·2KCl.
Properties: Reddish-brown crystals, soluble in water, slightly soluble in hot alcohol, d 2.67, mp 524C.
Use: Reagent for carbon monoxide determination.

palladium sodium chloride. (palladous sodium chloride; sodium palladium chloride). NaPdCl$_2$·3HOH.
Properties: Brown salt, hygroscopic, soluble in alcohol and water.
Use: Analysis (testing for carbon monoxide, ethylene, illuminating gas, iodine).

"Palldon."[169]   TM for p-nitrosodimethylaniline used in the colorimetric determination of palladium and platinum.

palladous chloride.   See palladium chloride.

palladous iodide.   See palladium iodide.

palladous nitrate.   See palladium nitrate.

palladous potassium chloride.   See palladium potassium chloride.

palladous sodium chloride.   See palladium sodium chloride.

pallet.   A low platform of wood or metal used for transportation or temporary storage of materials or semi-finished products, it stands on supports that are high enough to permit handling by forklift trucks.

palmarosa oil.   (geranium oil, Turkish).
CAS: 8014-19-5.   A light yellow essential oil consisting chiefly of geraniol, optically active.
Use: Source of geraniol, and in perfumes and flavors.

palm butter.   See palm oil.

"Palmetto."[84]   TM for agricultural dusting sulfur.
Use: Insecticide and fungicide.

"Palmex."[152]   TM for a series of metal processing oils available in several grades for hot dip thinning, pickling, and rolling of sheet, strip, and thin tinplate.

palmitic acid.   (hexadecanoic acid; cetylic acid).
CAS: 57-10-3.   CH$_3$(CH$_2$)$_{14}$COOH.
A saturated fatty acid. It occurs in natural fats and oils and in tall oil and in most commercial-grade stearic acid.

Properties: White crystals, soluble in hot alcohol and ether, insoluble in water, d 0.8414 (80/4C), mp 62.9, bp 351.5C, 271.5C (100 mm), 139.0C (1 mm), refr index 1.4309 (70C). Combustible.
Derivation: From spermaceti by saponification and from palm oil, hydrolysis of natural fats.
Method of purification: Crystallization.
Grade: Technical, 99.8%, FCC.
Use: Manufacture of metallic palmitates, soaps, lube oils, waterproofing, food-grade additives.

palmitic acid cetyl ester.   See cetin.

palmitin.   See tripalmitin.

palmitoleic acid.   (cis-9-hexadecenoic acid).
CH$_3$(CH$_2$)$_5$CH:CH(CH$_2$)$_7$COOH.   An unsaturated fatty acid found in nearly every fat, especially in marine oils (15–20%).
Properties: Colorless liquid, mp 1.0C, bp 140–141C (5 mm), insoluble in water, soluble in alcohol and ether. Combustible.
Grade: Purified product 99%.
Use: Organic synthesis, chromatographic standard.

palmitoyl chloride.   (hexadecanoyl chloride; palmityl chloride, so-called).   CH$_3$(CH$_2$)$_{14}$COCl.
Properties: Colorless liquid, soluble in ether, decomposes in water or alcohol, mp 11–12C, bp 194.5C (17 mm).

palmityl alcohol.   See cetyl alcohol.

palm nut cake.   (palm cake).   The cakes formed in the press when the palm nut kernels are expressed to obtain the oil. Contains various useful constituents, such as unexpressed oil, carbohydrates, proteins, and salts. Typical analysis: proteins 30.4%, fats 8.4%, fiber 41.0%, water 9.5%, ash 10.6%.
Use: Cattle-food, fertilizer ingredient.

palm oil.
Properties: Yellow-brown, buttery, edible solid at room temperature. D 0.952, mp 30C, iodine number 13.5, saponification number 247.6, soluble in alcohol, ether, chloroform, carbon disulfide. Combustible.
Occurrence: Oil palms are native to several countries in central Africa and are extensively cultivated in Malaysia, which is its chief commercial source. It is also produced in Indonesia.
Use: Soap manufacture, pharmacy, food shortening, cutting-tool lubricant, hot-dipped tin coating, terne plating, cosmetics, softener in rubber processing, cotton goods finishing, substitute for tallow as mold-release agent.

2-PAM. Abbreviation for 2-pyridine aldoxime methiodide.

"Pamak."[266] TM for various tall oil products including a series of tall oil fatty acids and distilled tall oils containing varying percentages of rosin acids. "Pamak" TP and WTP are residues from fractionation of crude tall oil in the manufacture of tall oil fatty acids.

pamaquine naphthoate. $C_{42}H_{45}N_3O_7$.
Properties: Yellow to orange-yellow, odorless, almost tasteless powder. Insoluble in water, soluble in alcohol and acetone.
Use: Medicine (antimalarial).

"Panasol."[216] TM for a petroleum aromatic solvent available in a variety of boiling ranges.
Use: Paint and varnish applications and in the formulation of insecticides.

pancreatin. A mixture of enzymes, principally pancreatic amylase, trypsin, and pancreatic lipase. Obtained from the pancreas of hog or ox.
Properties: Cream-colored amorphous powder, characteristic odor, acts upon starch and proteins, soluble in water, insoluble in alcohol, it changes protein into proteoses and derived substances, and starch into dextrins and sugars. Its greatest activity is in neutral or slightly alkaline media.
Derivation: Pancreas gland is extracted by macerating with chloroform, water, dilute boric acid, glycerol, or alcohol, filtered and evaporated.
Grade: NF.
Use: Preparation of so-called predigested protein nutrients, in bating compounds of leather, to remove starch and protein sizings from textiles.

Paneth technique. Method demonstrating the existence of free radicals (e.g., methyl) or atoms, which is based on the removal of a metallic "mirror" by a stream of gas containing the radicals. The reaction products can be collected and assayed.

"Pan Glaze."[149] TM for a resinous silicone coating developed to replace pan grease in the baking industry. It provides easy release for 100 or more bakes.

"Pano-drench."[401] TM for a liquid soil treatment concentrate containing 0.6% cyano(methylmercuri)guanidine.
Hazard: Toxic by ingestion.

pantethine. CAS: 16816-67-4. $C_{22}H_{42}N_4O_8S_2$.
The disulfide form of N-pantothenylthioethanolamine. *Lactobacilus bulgaricus* growth factor (LBF). A fragment of coenzyme A, a pantothenic acid derivative.
Use: Biochemical research.

panthenol. USAN name for pantothenol.

pantocaine. (4-butylaminobenzoic-beta-dimethylaminoethyl ester hydrochloride).
$C_4H_9NHCOOCH_2CH_2N(CH3)_2 \cdot HCl$.
Use: Local anesthetic.

pantolactone. (2,4-dihydroxy-3,3-dimethylbutyric acid; γ-lactone). CAS: 599-04-2.

$$\overline{HC(OH)C(CH_3)_2CH_2OCO}.$$

Properties: Crystals, d 1.180 (20/20C), mp 79.2C, soluble in water.
Grade: 80% aqueous solution.
Use: Preparation of pantothenic acid.

pantothenic acid. [N-(2-4-dihydroxy-3,3-dimethylbutyryl)-β-alanine]. CAS: 79-63-4.
$HOCH_2C(CH_3)_2CHOHCONH(CH_2)_2COOH$.

$$HOOC-CH_2-CH_2-NH-\overset{\displaystyle O}{\overset{\|}{C}}-\underset{\displaystyle OH}{\overset{}{\underset{|}{CH}}}-\underset{\displaystyle CH_3}{\overset{\displaystyle CH_3}{\underset{|}{\overset{|}{C}}}}-CH_2-OH$$

A member of the vitamin B complex; it is a component of coenzyme A and may be considered a β-alanine derivative with a peptide linkage. It is involved in the release of energy from carbohydrate utilization and is necessary for synthesis and degradation of fatty acids, sterols, and steroid hormones; it also functions in the formation of porphyrins. It occurs in all living cells and tissues. The natural product is dextrorotatory [d(+)] and is the only form having vitamin activity.
Properties: Viscous, hygroscopic liquid; soluble in water, ethyl acetate, glacial acetic acid; insoluble in benzene.
Sources: Food sources: liver, kidney, yeast, crude molasses, milk, whole grain cereals, rice. Commercial sources: produced synthetically from 2,4-dihydroxy-3,3-dimethylbutyric acid and β-alanine.
See calcium pantothenate.

pantothenol. (d(+)-pantothenyl alcohol; panthenol). CAS: 81-13-0.
$HOCH_2C(CH_3)_2CHOHCONH(CH_2)_2CH_2OH$.
The alcohol corresponding to pantothenic acid, with vitamin activity.
Properties: Viscous liquid; soluble in water, ethanol, methanol; specific rotation +28.36 to 30.7 degrees in water (c = 5); refr index 1.497 (20C).
Use: Biochemical research, food additive and dietary supplement.

papain. (papayotin). CAS: 9001-73-4.
Properties: White or gray, slightly hygroscopic powder; partially soluble in water and glycerol; insoluble in common organic solvents. The most thermostable enzyme known. Digests proteins.
Derivation: Obtained as dried and purified latex of Carica papaya.
Grade: Technical, purified. Technical grade is susceptible to decomposition in storage.
Use: Meat tenderizer, other food industries (mainly to prevent protein haze on chilling beer), tobacco, pharmaceutical, cosmetic, leather, textiles.

papaverine. (6,7-dimethoxy-1-veratrylisoquino-line). CAS: 58-74-2.
$(CH_3O)_2C_6H_3CH_2C_9H_4N(OCH_3)_2$. An alkaloid.
Properties: White, crystalline powder; soluble in chloroform, hot benzene, aniline, glacial acetic acid, and acetone; slightly soluble in alcohol and ether; insoluble in water; mp 147C.
Derivation: From opium or by synthesis.
Hazard: Toxic narcotic.
Use: A vasoldilator used for treatment of hypertension (also as the hydrochloride which is soluble in water).

paper. A semisynthetic product made by chemically processing cellulosic fibers. A wide variety of sources have been used for specialty papers (flax, bagasse, esparto, straw, papyrus, bamboo, jute, and others), but by far the largest quantity is made from softwoods (coniferous trees), such as spruce, hemlock, pine, etc.; some is also made from such hardwoods as poplar, oak, etc., as well as from synthetic fibers. Papermaking technology involves the following basic steps: (1) chipping or other subdivision of the logs (see groundwood); (2) manufacture of chemical or semichemical pulp by digestion in acidic or alkaline solutions, which separates the cellulose from the lignin (see pulp, paper); (3) beating the pulp to break down the fibers and permit proper bonding when the sheet is formed; (4) addition of starches, resins, clays, and pigments to the liquid stock (or "furnish"); (5) formation of the sheet continuously on a fourdrinier machine, where the water is screened out and the sheet dried by passing over a series of heated drums; (6) high-speed calendering for brightness and finish; (7) coating either by machine application or (for heavy finishes) by brushes. *Note*: Wet paper stock and waste are flammable and are considered as dangerous fire risks. Further information can be obtained from the Technical Association of the Pulp and Paper Industry, 155 E. 44th St., New York, NY, or from the Institute of Paper Chemistry, Appleton, Wisconsin.
See Appendix II for history of the industry.

paper chromatography. (PC). A micro type of chromatography. A drop of the liquid to be investigated is placed near one end of a strip of paper. This end is immersed in solvent, which travels down the paper and distributes the materials present in the original drop selectively. Comparison with known substances makes identification possible.

paper, coated. A paper that is covered on one or both sides with a suspension of clays, starches, casein, rosin, wax, or combinations of these to serve special purposes. Machine-coated paper is required for standard book printing; the rather light coating is applied by any of several devices (air knife, trailing blade, or roll coater). Heavier coatings are applied by means of brushes or spreading devices. These are required for high-grade printing of magazines, art books, etc., where excellent photographic reproduction is essential. Special-purpose coatings as for packaging are applied in a separate operation.

paper, synthetic. Paper or paper-like material made from a polyolefin, polypropylene is usually selected. A paper made from styrene copolymer fibers has been developed to production stage in Japan. Plastic-coated cellulosic papers are available for children's books, posters, and similar applications.

"Papi."[520] TM for a series of methylene diphenyl diisocyanate urethane polymers. Average viscosity 250 cp at 25C.
Use: One-shot rigid urethane foams.
"Papi"50. 50% solution in monochlorobenzene.
Use: in adhesives (rubber to metals and synthetic fabrics), coating intermediate.
"Papi"94. Light colored polymer.
Use: Foam seating and packaging.

"Papricol."[342] TM for an oleoresin of paprika for food coloring and flavoring.

para-. (p-). A prefix.
See ortho- for definition of para- compounds. For p-compounds, see specific compound; p-cresol is listed under cresol and p-dichlorobenzene under dichlorobenzene.

"Parabens."[583] TM for the methyl, propyl, butyl, and ethyl esters of p-hydroxybenzoic acid. Antimicrobial agents for foods and pharmaceuticals. Approved by FDA.

paracasein. See casein.

paracetaldehyde. See paraldehyde.

"Paracol."[266] TM for a series of wax and wax-rosin emulsions produced from paraffin waxes,

microcrystalline waxes, or combinations of these waxes with rosin.
Use: Impart water resistance to paper and allied materials.

"Paracril."[248]   TM for a group of synthetic rubbers of the Buna-N or nitrile type, produced by the copolymerization of butadiene and acrylonitrile. Resist deterioration by aliphatic hydrocarbon, mineral and vegetable oils, and animal fats and oils, and are particularly resistant to petroleum products. "Paracril" is also used as a plasticizer for vinyls and other thermoplastic and thermosetting resins.

"Paradene."[21]   TM for low-priced, dark, thermoplastic, coal-tar resins (coumarone-indene) available in low to high softening point ranges.
Use: Rubber compounding.

paraffin.   (1) Also called alkane. A class of aliphatic hydrocarbons characterized by a straight or branched carbon chain; generic formula $C_n H_{2n+2}$. Their physical form varies with increasing molecular weight from gases (methane) to waxy solids. They occur principally in Pennsylvania and Midcontinent petroleum. (2) Paraffin wax.

paraffin distillate.   A distilled petroleum fraction which when cooled consists of a mixture of crystalline wax and oil.

paraffin, chlorinated.   A paraffin oil or wax in which some of the hydrogen atoms have been replaced by chlorine atoms. Nonflammable.
Use: High-pressure lubricants, as flame retardants in plastics and textiles, as plasticizer for polyvinyl chloride in polyethylene sealants, and in detergents.

paraffin oil.   An oil either pressed or dry-distilled from paraffin distillate. Liquid petrolatum is also known as paraffin oil. Combustible.
Grade: By viscosity and color.
Use: Floor treatment, lubricant.

paraffin wax.   (paraffin scale; paraffin).
CAS: 8002-74-2.
Properties: White, translucent, tasteless, odorless solid consisting of a mixture of solid hydrocarbons of high molecular weight, e.g., $C_{36}H_{74}$; soluble in benzene, ligroin, warm alcohol, chloroform, turpentine, carbon disulfide, and olive oil; insoluble in water and acids; d 0.880–0.915; mp 47–65C; flash p 390F (198C); autoign temperature 473F (245C). Combustible.
Grade: Yellow crude scale, white scale, refined wax, ASTM, NF. Also graded by melting point in F and color. The higher-melting grades are the more expensive.

Hazard: Many waxes contain carcinogens. TLV: 2 mg/m³ of air.
Use: Candles; paper coating; protective sealant for food products, beverages, etc.; glass-cleaning preparations; hot-melt carpet backing; biodegradable mulch (hot melt-coated paper), impregnating matches, lubricants; crayons; surgery; stoppers for acid bottles; electrical insulation; floor polishes; cosmetics; photography; antifrothing agent in sugar refining; packing tobacco products; protecting rubber products from suncracking; chewing gum base (to ASTM specifications).

"Paraflint."   TM for a polymethylene wax. It is white, odorless, and has congealing point of 96C. Available in flaked form. Approved by FDA.

"Para-Flux."[94]   TM for a series of petroleum-based plasticizers for natural and synthetic rubbers.

paraformaldehyde.   (paraform).
CAS: 30525-89-4. $HO(CH_2O)_n H$.
A polymer of formaldehyde in which n equals 8 to 100. Not to be confused with the trimer, symtrioxane.
Properties: White solid with slight odor of formaldehyde, insoluble in alcohol and ether, soluble in strong alkali solution. The higher polymers are insoluble in water. Melting range 120–170C, flash p (CC) 160F (71C), autoign temperature 572F (300C). Combustible.
Derivation: By evaporating an aqueous solution of formaldehyde.
Forms: Flake, powder.
Grade: Bags, carlots.
Hazard: Toxic by ingestion.
Use: Fungicides, bactericides, and disinfectants; adhesives; hardener and waterproofing agent for gelatin; contraceptive creams.

paraldehyde.   (2,4,6-trimethyl-1,3,5-trioxane).
CAS: 123-63-7. $C_6H_{12}O_3$.   A cyclic polymer (trimer) of acetaldehyde. A depressant drug; may be addictive.
Properties: Colorless liquid, disagreeable taste, agreeable odor, decomposes on standing, stable toward alkalies but slowly decomposed to acetaldehyde when treated with a trace of mineral acid, miscible with most organic solvents and volatile oils, soluble in water, d 0.9960 at 20/20C, bp 124.5C, mp 12.6C, vap press 25.3 mm (20C), flash p (TOC) 96F (35.5C), specific heat 0.434, refr index 1.40–1.42 (20C), bulk d 8.27 lb/gal (20C), autoign temperature 460F (237C).
Derivation: Action of hydrochloric acid and sulfuric acid upon acetaldehyde.
Grade: Technical, USP.
Hazard: Flammable, moderate fire risk. Toxic by ingestion.

Use: Substitute for acetaldehyde, rubber accelerators, rubber antioxidants, synthetic organic chemicals, dyestuff intermediates, solvent for fats, oils, waxes, gums, resins; leather, solvent mixture for cellulose derivatives, sedative (hypnotic).

paralytic shellfish poisoning.   See red tide.

"Paramine."[300]   TM for a cationic finishing agent for textiles, an aminofatty condensation product solubilized with a volatile acid.

paranitraniline.   See p-nitroaniline.

paranitraniline red.   See para red.

para-oxon.   (diethyl para-nitrophenyl phosphate). CAS: 311-45-5. $(C_2H_5O)_2P(O)OC_6H_4NO_2$. Generic name for the oxygen analog of parathion.
Properties: Odorless, reddish-yellow oil; bp 148–151C (1 mm); d 1.269 (25/25C); refr index 1.5060 (25C). Slightly soluble in water, soluble in most organic solvents, decomposes rapidly in alkaline solutions.
Hazard: Poison by ingestion, inhalation, and skin absorption; cholinesterase inhibitor; use may be restricted.
Use: Insecticide.

"Paraplex."[94]   TM for polymeric plasticizers for polymers and resinous coatings. Primarily polyesters, but some are epoxidized oils which impart heat and light stability as well as plasticization. Supplied as viscous liquids in a range of molecular weights all at 100% solids. Compatible with polyvinyl chloride, polyvinyl butyral, cellulosics, and other high polymers and elastomers.
Use: Calendered sheet and film; extruded and molded items; electrical wire insulation; coatings for wood, metal, fabrics, and paper.

paraquat.   (generic name for 1,1'-dimethyl-4,4'-bipyridinium salt).   CAS: 4685-14-7. $[CH_3(C_5H_4N)_2CH_3] \cdot 2CH_3SO_4$.
Properties: Yellow solid, soluble in water.
Hazard: Highly toxic by ingestion, inhalation, and skin absorption; use is restricted. TLV: 0.1 mg/m³ of air.

para red.   (paranitraniline red). $C_{10}H_6(OH)NNC_6H_4NO_2$.   A pigment formed by coupling diazotized p-nitroaniline with β-naphthol. The term is also used to refer to a group of lakes based on this dye.
See para toner.

"Para Resin" 2457.[94]   TM for a petroleum base resin.
Use: Plasticizer and softener for rubbers.

pararosaniline.   (CI No 42500). CAS: 569-61-9. $HOC(C_6H_4NH_2)_3$.
A triphenylmethane dye. Component of fuchsin.
Properties: Colorless to red crystals, mp 205C, soluble in alcohol, very slightly soluble in water and ether. Combustible.
Use: Dye (usually as the hydrochloride).

pararosolic acid.   See aurin.

parathion.   (generic name for O,O-diethyl-O,p-nitrophenyl phosphorothioate; ethyl parathion; O,O-diethyl-p-nitrophenyl thiophosphate; AATP).   CAS: 56-38-2. $(C_2H_5O)_2P(S)OC_6H_4NO_2$.
Properties: Deep brown to yellow liquid, usually has faint odor, refr index 1.5367 (25C), d 1.26 (25/4C), bp 375C, fp 6C, vap press 0.003 mm (24C). very slightly soluble in water (20 ppm); completely soluble in esters, alcohols, ketones, ethers, aromatic hydrocarbons, animal and vegetable oils; insoluble in petroleum ether, kerosene, spray oils; stable in distilled water and in acid solution; hydrolyzed in the presence of alkaline materials; slow decomposition in air. Purity: Technical grade is 95% pure. Also supplied diluted with inert carriers of various types, and in various proportions.
Derivation: From sodium ethylate, thiophosphoryl chloride, and sodium p-nitrophenate.
Hazard: Highly toxic by skin contact, inhalation, or ingestion; cholinesterase inhibitor. Repeated exposure may, without symptoms, be increasingly hazardous. Fatalities have resulted from its accidental use; use may be restricted. TLV: 0.1 mg/m³ of air.
Use: Insecticide and acaricide.
See also methyl parathion.

para toner.   An insoluble red pigment derived from β-naphthol and p-nitroaniline. The former is sometimes partly replaced by mono-acid F, 2-naphthol-7-sulfonic acid. By varying the conditions of temperature and acid concentration, different shades may be obtained.
Use: Paint and printing ink pigments, making para lakes.

"Parco."[62]   TM for phosphoric acid and phosphate compounds for dissolving rust from the surface of metal.

"Parcolene."[62]   TM for chemicals for treating metal surfaces to remove extraneous matter and/or condition the surface prior to other finishing operations.

paregoric.   Contains a derivative of opium that is habituating on continued use. Its use is restricted by FDA.
Use: Medicine, especially for digestive disorders.

"Parez."[57] TM for a series of melamine-formaldehyde and urea-formaldehyde resins, designed for use in papermaking.

"Paricin."[302] TM for various alkyl hydroxystearates (soft, low melting point waxes useful as firming agents in cosmetics and specialty inks and as coupling agents for incompatible mixture of polar and non-polar materials in hot melt applications) and acetoxystearates (plasticizers for nitrocellulose, cellulose acetate butyrate, ethylcellulose, and vinyl resins).

"Pariflux."[250] TM for fluorspar.
Use: Welding electrode coatings for arc stabilization, fluxing weld metal, and forming protective slag.

Paris green. See copper acetoarsenite.

parison. (1) An unformed mass of molten glass from which finished products are manufactured. (2) An extruded tube of plastic from which toys and similar items are made by blow molding.

Paris white. See whiting.

Parkes process. A standard process for the separation of silver from lead. From 1 to 2% molten zinc is added to the lead-silver mixture, heated to above the melting point of zinc. A scum containing most of the silver and zinc forms on the surface. This is separated and the silver recovered. The separation of silver is not complete and the process is repeated several times.

"Parlon."[266] TM for chlorinated rubber. White, odorless, nonflammable, granular powder. Available in five viscosity grades.
Use: Film-former in paints, inks, adhesives and concrete-treating compounds, traffic and marine paints, corrosion-resistant coatings.

"Parmo."[51] TM for petrolatums meeting USP or NF requirements and having melting points, consistencies, and colors suitable for pharmaceuticals and cosmetics.

"Parolite."[159] TM for normal zinc sulfoxylate formaldehyde.
Use: Stripping of all forms of wool and synthetics (nylon, acetates, Dacron, etc.).

paromomycin sulfate. $C_{23}H_{47}N_5O_{18}S$.
Antibiotic from a strain of Streptomyces.
Properties: Creamy white, odorless, hygroscopic powder; soluble in water; insoluble in chloroform and ether.
Grade: ND.
Use: Medicine (antimicrobial).

partial pressure. The pressure due to one of the several components of a gaseous or vapor mixture. In general this pressure cannot be measured directly but is obtained by analysis of the gas or vapor and calculated by use of Dalton's law. See also Raoult's law.

particle. Any discrete unit of material structure; the particulate basis of matter is a fundamental concept of science. The size ranges of particles may be summarized as follows: (1) Subatomic: protons, neutrons, electrons, deuterons, etc. These are collectively called fundamental particles. (2) Molecular: includes atoms and molecules with size ranging from a few angstroms to half a micron. (3) Colloidal: includes macromolecules, micelles, and ultrafine particles such as carbon black, resolved via electron microscope, size ranges from 1 millimicron up to lower limit of the optical microscope (1 micron). (4) Microscopic: units that can be resolved by an optical microscope (includes bacteria). (5) Macroscopic: all particles that can be resolved by the naked eye.
See also fundamental particle, particle size.

particle accelerator. A device in which the speed of charged subatomic particles (protons, electrons) and heavier particles (deuterons, alpha particles) can be greatly increased by application of electric fields of varying intensity, often in conjunction with magnetic fields. It is possible to accelerate electrons and protons to speeds approaching the speed of light if sufficiently high voltage is used. Straight-line (linear) accelerators are used for protons, and doughnut-shaped betatrons for electrons; other types are the Van de Graaf electrostatic generator, the synchrotron, and the cyclotron. Before the development of nuclear reactors, the cyclotron was used to accelerate deuterons for use in bombarding stable nuclei to produce neutrons for inducing artificial radioactivity, fission, and formation of synthetic (transuranic) elements.
See also betatron, cyclotron.

particle size. This term refers chiefly to the solid particles of which industrial materials are composed (carbon black, zinc oxide, clays, pigments, and the like). The smaller the particle, the greater will be the total exposed surface area of a given mass. Activity is a direct function of surface area; that is, the finer a substance is, the more efficiently it will react, both chemically and physically. A colloidal pigment is a more effective colorant than a coarse one because of the greater surface area of its particles. A pound of channel carbon black has a surface area of 18 acres, which largely accounts for its powerful reinforcing effect in rubber. Thus, ultrafine grinding of powders is of utmost importance in such products as paints, cement, plastics, rubber, dyes, pharmaceuticals, printing inks, and numerous others.

See also particle, surface chemistry, colloid chemistry, sedimentation.

parting agent. See abherent.

partition chromatography. See liquid chromatography.

parylene. Generic name for thermoplastic film polymers based on p-xylylene and made by vapor-phase polymerization.

Derivation: p-Xylene, $CH_3C_6H_4CH_3$, is heated with steam at 950C to produce the cyclic dimer di-p-xylylene, a solid which can be separated in pure form. The dimer is then pyrolyzed at 550C to produce monomer vapor of p-xylylene, $CH_2:C_6H_4:CH_2$ which is then cooled below 50C and condenses on the desired object as a polymer having the repeating structure —($CH_2C_6H_4CH_2$—)$_n$, with n about 5000 and molecular weights of about 500,000. The polymer is used as a protective coating. Films as thin as 500Å to 5 mils are obtained.

Use: Thin coatings of high purity and uniformity on almost any substrate which will resist a high vacuum, as paper, fabric, polyethylene and polystyrene film, ceramics, metals, many solid chemicals; electronic miniaturization systems, capacitors, thin film circuits.

"Parzate."[28] TM for a series of fungicides. "Parzate" liquid is a solution containing 22% nabam to be combined with zinc sulfate in the spray tank. "Parzate" carbon is a wettable powder containing 75% zineb. "Parzate" D is a finely divided powder containing 85% zineb.

Hazard: Irritant to skin and mucous membranes.

PAS. Abbreviation for p-aminosalicylic acid.
See 4-aminosalicylic acid.

PAS sodium. See sodium-p-aminosalicylate.

Passerini reaction. Formation of 2-acyloxy amides on treatment of an isonitrile with a carboxylic acid and an aldehyde or ketone.

passivity. A property shown by iron, chromium, and related metals involving loss of their normal chemical activity in an electrochemical system or in a corrosive environment after treatment with strong oxidizing agents, like nitric acid, and

when oxygen is evolved upon them during electrolysis, forming an oxide coating.

paste. (1) An adhesive composition of semisolid consistency, usually water-dispersible. The common pastes are based upon starch, dextrin, or latex often with the addition of gums, glue, and antioxidants. They are widely employed for the adhesion of paper and paperboard. (2) More generally, a soft, viscous mass of solids dispersed in a liquid. For example, paste resins are finely divided resins mixed with plasticizers to form fluid or semifluid mixtures, without the use of low boiling solvents or water emulsions.

paste solder. A paste (2) containing flux, cleaner, tinning agent, and powdered metallic solder.

Pasteur, Louis. (1822–1895) A French chemist and bacteriologist who made three notable contributions to science: (1) As a result of extensive study of fermentation, which led him to conclude that it is caused by infective bacteria, he extended the work of Jenner on smallpox serum made from cowpox (1775) to development of the concept of immunizing serums and the antibody-antigen relationship (1880). Pasteur was the first to inoculate for rabies and anthrax, and suggested the term vaccination (from Latin vaccus = cow) in recognition of Jenner's achievement. (2) Initiation of the practice of heat-treating wine, and later milk and other food products, to kill or inactivate toxic microorganisms, especially the tuberculosis bacillus. (3) Discovery of the optical properties of tartaric acid, present in wine residues, which laid the basis for modern knowledge of optical isomers (right- and left-handed molecular structure), a phenomenon now often called chirality.

See also pasteurization, optical isomer.

pasteurization. Heat treatment of milk, fruit juices, canned meats, egg products, etc., for the purpose of killing or inactivating disease-causing organisms. For milk, the minimum exposure is 62C for 30 minutes or 72C for 15 seconds, the latter being called flash pasteurization. Although this treatment kills all pathogenic bacteria and also inactivates enzymes which cause deterioration of the milk, the shelf-life is limited; to prolong storage life, temperatures of 80–88C for 20–40 seconds must be used. Complete sterilization requires ultra-high pasteurization at from 94C for 3 seconds to 150C for 1 second, in-can heating at 116C for 12 minute and 130C for 3 minutes is also employed for maximum stability and long storage life. Some meat products are pasteurized by alpha-radiation.

PAT. Abbreviation for polyaminotriazole.

patchouli oil. A yellow-to-brownish essential oil used in perfumery and flavoring. It is strongly levorotatory.

patentability. The qualifications for obtaining a patent on an invention or chemical process. These are: (1) the invention must not have been published in any country or in public use in the US, in either case for more than one year prior to date of filing the application; (2) it must not have been known in the US before date of invention by the applicant; (3) it must not be obvious to an expert in the art; (4) it must be useful for a purpose not immoral and not injurious to the public welfare; (5) it must fall within the five statutory classes on which only patents may be granted, namely, (a) composition of matter, (b) process of manufacture or treatment, (c) machine, (d) design (ornamental appearance), or (e) a plant produced asexually. Special regulations relate to atomic energy developments and subjects directly affecting national security. (Robert Calvert) Note: In 1980, the Supreme Court in a landmark decision upheld the patentability of synthetic bacteria created by recombinant DNA techniques.

patent alum. See aluminum sulfate.

patent leather. Fashion leather used chiefly for formal shoes, bags, etc., characterized by a high, glossy finish applied as the final step in processing. The finish greatly reduces the poromeric nature of the leather.

Paterno-Buchi reaction. Formation of oxetanes by photochemical cycloaddition of carbonyl compounds to olefins.

pathfinder element. An element present in small proportions less than 1%, generally metallic in nature, associated with ore deposits at the time of formation. Mapping of the concentration variation of the selected element serves to locate the main ore deposit. Examples are zinc as the "pathfinder" for lead, copper, and silver ores, and molybdenum associated with copper deposits.

pathway. A sequence of reactions, usually of a biochemical nature, in which more complex substances are converted to simple end products, as in the degradation of the components of foods to carbon dioxide and water. Its course is largely determined by preferential factors involving coenzymes and other catalysts. An example is the TCA cycle, which is the common pathway in the degradation of foodstuffs and cell constituents to carbon dioxide and water.

patina. Variously used to refer to an ornamental and/or corrosion-resisting film on the surface of copper, copper alloys, including bronzes, and also sometimes of iron and other metals. Such a film is formed by exposure to the air, or by a suitable chemical treatment.

patronite. A mixture of vanadium-bearing substances with the formula VS_4 found in Peru.

Pattinson process. Process for the removal of silver from lead. The silver-lead mixture is melted in one of a series of pots and allowed to cool slowly. The lead which is free from silver or poorer in silver separates out as crystals which are removed, leaving the silver-rich lead in the molten state. From a number of such operations in series a lead rich in silver is obtained, collected, and the silver recovered.
See also Parkes process.

Pauli exclusion principle. A fundamental generalization concerning the energy relationships of electrons within the atom, namely that no two electrons in the same atom have the same value for all four quantum numbers; corollary to this is the fact that only two electrons can occupy the same orbital, in which case they have opposite spins, i.e., $+1/2$ and $-1/2$. This principle has an important bearing on the sequence of elements in the Periodic Table and on the limiting numbers of electrons in the shells (2 in the first, 8 in the second, 18 in the third, 32 in the fourth, etc.).
See also quantum number, shell, orbital theory.

Pauling, Linus. (1901-) An American chemist and physicist who won the Nobel prize for chemistry in 1954. By using x-ray diffraction analysis he determined the crystal structure of molecules. He made significant progress in the study of chemical bonds and discovered the atomic structure of many proteins including hemoglobin. He is known to the masses for his advocacy of vitamin C. It is interesting to note that Pauling was also a recipient of the Nobel Peace Prize in 1962.

Pb. Symbol for lead.

PBAA. Abbreviation for polybutadieneacrylic acid copolymer.

PBD. See 1,3,4-phenylbiphenylyloxadiazole.

PBI. See polybenzimidazole.

PBPB. Abbreviation for pyridinium bromide perbromide.

PCB. Abbreviation for polychlorinated biphenyl.

PCE. Abbreviation for pyrometric cone equivalent, a scale of melting or fusion points of refractory materials, based on comparison with the temperature at which pyrometric cones melt.

p-cellulose. See phosphorylation.

PCNB. Abbreviation for pentachloronitrobenzene.

PCP. (1) Abbreviation for pentachlorophenol. (2) Abbreviation for phenylcyclidene hydrochloride.

PCTFE. Abbreviation for polychlorotrifluoroethylene.
See chlorotrifluoroethylene resin.

pcu. Abbreviation for pound centigrade unit, the amount of heat needed to raise one pound of water from 15 to 16C.
See also chu.

Pd. Symbol for palladium.

PDB. Abbreviation for p-dichlorobenzene.

PDMS. See polydimethylsiloxane.

PE. Abbreviation for pentaerythritol and for polyethylene.

"PE-100."[210] TM for a cationic polyelectrolyte containing a quaternary ammonium group.
Use: Conductive coating, coagulant aid, and metal complexing agent.

peacock blue. (α,α-bis[N-ethyl-N-(4-sulfobenzyl)aminophenyl]-α-hydroxy-o-toluenesulfonic acid sodium salt). CAS: 2650-18-2.
$HSOC_6H_4COH[C_6H_4N(C_2H_5)CH_2C_6H_4SO_3Na]_2$.
A blue organic pigment used especially in inks for multicolor printing. It is a lake of acid glaucine blue dye on alumina hydrate, and is prepared from aniline, ethanol, benzyl chloride, o-chlorobenzaldehyde, sulfuric acid and sodium bisulfite. *Note*: The term peacock blue is sometimes applied to other pigments of similar color, such as Prussian blue which has been treated with phosphotungstic acid.

peanut cake. The press cake resulting from the extraction of oil from the peanut.
See peanut oil meal.

peanut oil. (arachis oil; groundnut oil).
CAS: 8002-03-7. A fixed, nondrying oil.
Properties: Yellow to greenish-yellow; soluble in ether, petroleum ether, carbon disulfide, and chloroform; insoluble in alkalies, but saponified by alkali hydroxides with formation of soaps; insoluble in water; slightly soluble in alcohol.

D 0.912–0.920 (25C), solidifying point −5 to +3C, saponification value 186–194, iodine number 88-98, refr index 1.4625–1.4645 (40C), flash p 540F (282C), autoign temperature 883F (472C). Combustible.
Use: Substitute for olive oil and other edible oils, both hydrogenated and unhydrogenated; soaps; vehicle for medicines; salad oil, mayonnaise, margarine.

peanut oil meal. The crushed form of peanut cake resulting from the extraction of oil from the seed. Prepared with or without the shells, the oil meal of commerce contains between 39–45% crude protein and is sold on that basis. Typical analysis of 39% protein meal: 39.1% crude protein, 5.3% crude fiber, 34.3% nitrogen-free extract, 6.2% ether-soluble (fats), 5.3% ash, total digestible nutrient 80%.
Use: Animal feeds, fertilizer ingredient.

pearl alum. See aluminum sulfate.

pearl ash. See potassium carbonate.

pearl essence. See nacreous pigment.

pearl pigment. See nacreous pigment.

pearl white. See bismuth oxychloride, bismuth subnitrate.

pear oil. See amyl acetate.

peat. Semicarbonized residue of plants formed in water-saturated environments (bogs and marshes). It occurs in surface layers 3–10 ft thick and has a water content of 85%. Before peat can be used for chemical or fuel purposes it must be field-dried to a water content of 30–40%. Since the dried product is susceptible to autoignition, storage conditions must be such as to minimize this risk. Peat is easily converted to hydrocarbons and is an excellent source of natural gas; when dry it can be used directly as a fuel. The US has peat sources second only to those of the USSR, located in Alaska, the north central states, and in Maine, where processing on a large scale is planned. Their total energy content is said to be equivalent to 240 billion barrels of petroleum. The peat can be gasified for production of methanol after mechanical dewatering. Experimental conversion studies have been under way for some time. Substantial quantities of oil, ammonia, and sulfur can be obtained as byproducts.

pebble. A piece of gravel between 4 and 8 millimeters in size.

pebble mill. A jacketed steel cylinder rotating on a horizontal axis and containing flint or porcelain pebbles as the grinding medium. Its operation is similar to that of a ball mill. It is used for grinding and mixing of dry chemicals, pigments, food products, and the like. Pebble mills are usually lined with alumina, buhrstone, or similar material to protect the walls from wear.

pebulate. (propyl ethyl-n-butylthiocarbamate). CAS: 1114-71-2. $C_{10}H_{21}NOS$.
Properties: Colorless liquid, bp 142C (20 mm), d 0.945, refr index 1.47, soluble in benzene, acetone, methanol, and xylene.
Hazard: Toxic by ingestion.
Use: Herbicide.

Pechmann pyrazole synthesis. Formation of pyrazoles from acetylenes and diazomethane. The analogous addition of diazoacetic esters to the triple bond yields pyrazolecarboxylic acid derivatives.

pectic acid. An acid derived from pectin by treating it with sodium hydroxide solution, washing with isopropyl alcohol, adding alcoholic hydrochloric acid, and finally washing again with isopropyl alcohol and drying.
Use: Acidulant in pharmaceuticals.

pectin. A high molecular weight hydrocolloidal substance (polyuronide) related to carbohydrates and found in varying proportions in fruits and plants. Pectin consists chiefly of partially methoxylated galacturonic acids joined in long chains.
Properties: White powder or syrupy concentration. Commonest characteristic of pectins is their property of jelling at room temperature, after addition of sugar to fruit juices in the preparation of jams or jellies. Soluble in water, insoluble in organic solvents.
Derivation: By dilute-acid extraction of the inner portion of the rind of citrus fruits, or of fruit pomaces, usually apple.
Method of purification: Following decolorization, the extracts are concentrated by evaporation or the pectins precipitated with alcohol or acetone.
Grade: Pure (NF) containing not less than 6.7% methoxy groups and not less than 74% galacturonic acid; 150-, 200-, 250-jelly grades, containing various diluents.
Use: Jellies, foods, cosmetics, drugs, protective colloids, emulsifying agents, dehydrating agents.
See also gel.

pectinase. An enzyme present in most plants. It catalyzes the hydrolysis of pectin to sugar and galacturonic acid.
Use: Biochemical research, juice and jelly industry.

"Pectinol."[23] TM for formulated enzyme concentrate of fungal origin with varying degrees of pectinase activity which hydrolyze pectic substances.
Use: Clarification of wines and fruit juices and processing of jellies.

pectin sugar. See l-arabinose.

PEG. Abbreviation for polyethylene glycol.

"Pegosperse."[73] TM for a series of polyglycol esters of fatty acids.
Use: Plasticizers, softeners, wetting agents, detergents, lubricants, emulsifying agents.

pelargonic acid. (n-nonoic acid; n-nonanoic acid; n-nonylic acid). CAS: 112-05-0.
$CH_3(CH_2)_7COOH$.
Properties: Colorless or yellowish oil with slight odor, d 0.9052 (20/4C), mp 12.5C, bp 255.6C, refr index 1.4322 (20C). Soluble in alcohol, ether, and organic solvents; almost insoluble in water. Combustible.
Derivation: By the oxidation of nonyl alcohol or nonyl aldehyde, the oxidation of oleic acid, especially by ozone.
Grade: Technical, 99%.
Hazard: Strong skin irritant.
Use: Organic synthesis, lacquers, plastics, production of hydrotropic salts, pharmaceuticals, synthetic flavors and odors, flotation agent, esters for turbojet lubricants, vinyl plasticizer, gasoline additive.
See also nonic acid.

pelargonic alcohol. See nonyl alcohol.

pelargonic aldehyde. See nonanal.

pelargonyl chloride. (n-nonanoyl chloride). $CH_3(CH_2)_7COCl$.
Properties: Bp 80–85C (5 mm), min assay 97%, soluble in hydrocarbons and ethers, decomposes in water.
Hazard: Skin irritant.
Use: Intermediate in organic synthesis.

pelargonyl peroxide. $(C_8H_{17}COO-)_2$.
Properties: Water-white liquid with a faint odor, d 0.926 min (25/25C), mp 10C, refr index 1.443 min (25C), insoluble in water and glycerol, soluble in alcohol and hydrocarbons.
Hazard: Dangerous fire risk in contact with organic materials. Strong skin irritant and oxidizing agent.
Use: Initiator of polymerization reactions.

"Pelaspan."[233] TM for a series of expandable polystyrenes in bead or pellet form. Each bead contains its own expanding agent, which is activated by heat.

Peligot's salt.   See potassium chlorochromate.

pellagra.   A disease caused by deficiency of niacin in the diet.

pellet.   A small unit of a light, bulky material compressed into any of several shapes and sizes, usually either spherical or rectangular. The operation is performed on a pellet mill, which consists essentially of a pair of steel rollers around which rotates a circular perforated metal die. Material is fed into the chambers above and below the inner face of the die. As the die turns in contact with the rollers, the latter also turns thus compressing the material and forcing it through the holes in the die at the point of tangency, where the extruded segment is sheared off by knives. Pelletizing is advantageous for fluffy particulates which are difficult to handle in loose form, e.g., carbon black, clays, plastic molding powders, etc. Binding materials called excipients are often used.

pelletierine.   (β-2(-piperidyl) propionaldehyde).
CAS: 4396-01-4. $C_5H_{10}N(CH_2)_2CHO$.
Properties: Liquid alkaloid from the root of the pomegranate; soluble in water, alcohol, ether, chloroform, benzene; d 0.988 (20/4C); bp 195C.
Use: Medicine (in form of its salts, sulfate, tannate, valerate).

Pellizzari reaction.   Formation of substituted 1,2,4-triazoles by the condensation of amides and acyl hydrazines. When the acyl groups of the amide and of the acylhydrazine are different, interchange of acyl groups may occur with formation of a mixture of triazoles.

Pelouze synthesis.   Formation of nitriles from alkali cyanides by alkylation with alkyl sulfates or alkyl phosphates.

"Peltex."[476]   TM for a ferrochrome complex of sodium lignosulfonate.
Use: Specialized dispersant in oil well drilling fluids.

"Pemco."[296]   TM for ceramic frits and coloring oxides.

"Penacolite."[11]   TM for resorcinolformaldehyde and resorcinol phenol formaldehyde adhesives and thermosetting resins in a variety of formulations.
Use: Bonding of wood and cellulosic products, some thermosetting and thermoplastic resins and plastics, natural and some synthetic rubbers.

"Penchlor."[204]   TM for cold-setting silicate-type acid-resistant cements. Completely stable in dry storage.
Grade: Acid-Proof, S-25, Fireproof, and FCC.
Use: Quick setting cement mortar used with acid-proof brick to resist acids and strong oxidizing agents, except hydrogen fluoride, up to 1093C.

"Pendane."[342]   TM for lindane insecticide concentration.
Hazard: Highly toxic. Use may be restricted. Tolerance, 0.5 mg/m^3 of air.

penetrant.   Any agent used to increase the speed and ease with which a bath or liquid permeates a material being processed by effectively reducing the interfacial tension between the solid and liquid. Penetrants are widely used in the textile, tanning, and paper industries for improving dyeing, finishing, etc., operations. Sulfonated oils, soluble pine oils, and soaps are popular among the older penetrants, and the salts of sulfated higher alcohols are typical of the synthetic organics developed for this purpose.
See also wetting agent.

"Penglo 65."[36]   TM for a pale maleic modified pentaerythritol ester of a special tall oil in mineral spirits.
Use: Paint and varnish.

penicillamine.   (USAN; d,3-mercaptovaline).
CAS: 52-67-5.
$(CH_3)_2C(SH)CH(NH_2)COOH$.
Properties: Crystals, decomposes at 178C.
Use: Medicine as a chelate for copper. Degradation product of antibiotics of the penicillin type and in treatment of rheumatoid arthritis.

penicillin.   CAS: 1406-05-9.
$(CH_3)_2C_5H_3NSO(COOH)NHCOR$ (bicyclic).
A group of isomeric and closely related antibiotic compounds with outstanding antibacterial activity, obtained from the liquid filtrate of the molds Penicillium notatum and Penicillium chrysogenum, or by a synthetic process which includes fermentation. Total synthesis of the penicillin molecule by J. C. Sheehan in 1957 was an outstanding achievement.
Derivation: The mold is grown in a nutrient solution such as corn steep liquor, lactose, or dextrose. After several days of cultivation the mold excretes penicillin into its liquid culture medium. This liquid is then filtered off and the penicillin extracted and purified by countercurrent extraction with amyl acetate, adsorption on carbon, or other methods. Different varieties of penicillin are produced biosynthetically by adding the proper precursors to the nutrient solution.

Grade: Aluminum penicillin G, benzathine penicillin G, benzylpenicillin G, chloroprocaine penicillin O, hydrabramine pencillin V, phenoxymethylpenicillin, potassium penicillin G, potassium α-phenoxyethylpenicillin, potassium phenoxymethylpenicillin, procaine penicillin G, sodium methicillin, sodium penicillin G.

Hazard: Strong allergen, reaction may be severe in susceptible people.
Use: Medicine (antibiotic).

penicillin V. See phenoxymethylpenicillin.

penicillinase. An enzyme which antagonizes the antibacterial action of penicillin. Such enzymes are found in many bacteria.
Use: Pharmaceutical, biological research.
See also antagonist, structural.

pennyroyal oil. (pulegium oil; hedeoma oil).
Yellow to reddish essential oil, strongly dextrorotatory.
Use: Manufacture of pulegone, flavoring alcoholic beverages, emmenagogue.

"Penros."[36] TM for polymerized wood rosin.
Use: Adhesives, gloss oils, paper label coatings, oleoresinous varnishes, solder flux, spirit varnishes, waxed paper and hot melt compounds, synthetic resins.

pentaborane. CAS: 19624-22-7. B_5H_9.
Properties: Colorless liquid with pungent odor, fp −46.6C, bp 58C, d 0.61, vap press (0C) 6 mm, decomposes at 150C, ignites spontaneously in air if impure, flash p 86F (30C), hydrolyzes slowly in water.
Derivation: Hydrogenation of diborane.
Grade: Technical 95%, high purity 99%.
Hazard: Highly flammable, dangerous fire and explosion risk. Toxic by ingestion and inhalation, strong irritant. TLV: 0.005 ppm in air.
Use: Fuel for air-breathing engines, propellant.

"Pentac."[62] TM for miticide whose active ingredient is (bis(pentachloro-2,4-cyclopentadien-1-yl)).

pentacene. (2,3,4,7-dibenzoanthracene).
CAS: 135-48-8. $C_{22}H_{14}$. Highly reactive aromatic compound consisting of five fused benzene rings.
Properties: Deep blue-violet solid, sublimes at 290–300C, decomposes in air above 300C, insoluble in water, slightly soluble in organic solvents.
Use: Suggested as organic photoconductor (instead of selenium) in copying systems.

"Pentacetate."[204] TM for synthetic amyl acetate.
Use: Extractant for antibiotics, lacquer solvent.

pentachloroethane. (pentalin). CAS: 76-01-7. $CHCl_2CCl_3$.
Properties: Dense, high-boiling, colorless liquid. D 1.685 (15/4C), bp 159.1C, fp −22C, refr index 1.503 (24C), insoluble in water.
Derivation: By chlorination of trichloroethylene, obtained by a two-step process involving chlorination of acetylene to obtain tetrachloroethane, and removal of hydrogen chloride by action of alkali.
Hazard: Moderate fire and explosion risk. Toxic by inhalation and ingestion.
Use: As solvent for oil and grease in metal cleaning. Also used for separation of coal from impurities by density difference.
See tetrachloroethane for other uses.

pentachloronaphthalene. CAS: 1321-64-8. $C_{10}H_3Cl_5$.
Properties: White powder.
Hazard: Action similar to chlorinated naphthalenes and chlorinated diphenyls. TLV: 0.5 mg/m³ of air.

pentachloronitrobenzene. (PCNB).
CAS: 82-68-8. $C_6Cl_5NO_2$.
Properties: Cream crystals, musty odor, d 1.718 (25/4C), mp 142–145C, bp 328C (some decomposition). Practically insoluble in water; slightly soluble in alcohols; somewhat soluble in carbon disulfide, benzene, chloroform.
Derivation: By reacting pentachlorobenzene with fuming nitric acid.
Grade: Dust, emulsion concentrate, wettable powder.
Hazard: Skin irritant.
Use: Intermediate, soil fungicide, slime prevention in industrial waters, herbicide.

pentachlorophenol. (PCP). CAS: 87-86-5. C_6Cl_5OH.
Properties: White powder or crystals, mp 190C, bp 310C with decomposition, d 1.978 (22/4C). Slightly soluble in water; soluble in dilute alkali, alcohol, ether, benzene.
Derivation: Chlorination of phenol.
Hazard: Toxic by ingestion, inhalation, and skin absorption; abuse may be fatal. TLV: 0.5 mg/m³ of air.
Use: Fungicide, bactericide, algicide, herbicide; sodium pentachlorophenate, wood preservative (telephone poles, pilings, etc.).

pentachlorothiophenol. See "RPA."

"Pentacite."[36] TM for pale colored rosin ester gums and modified rosin esters used in coating vehicles and in chewing gums and some rubber adhesives.

pentadecane. $CH_3(CH_2)_{13}CH_3$.
Properties: Colorless liquid, soluble in alcohol, insoluble in water, d 0.776, bp 270.5C, mp 10C. Combustible.
Grade: Technical.
Use: Organic synthesis.

n-pentadecanoic acid. (pentadecylic acid).
$CH_3(CH_2)_{13}COOH$. A saturated fatty acid normally not found in vegetable fats, but made synthetically.
Properties: Colorless crystals, d 0.8423 (80/4C), mp 51.8–52.8C, bp 339.1C, 212C (16 mm), refr index 1.4529 (60C). Insoluble in water, soluble in alcohols and ethers.
Grade: 99% pure.
Use: Organic synthesis, reference standard in gas chromatography.

pentadecanolide. (15-hydroxypentadecanoic acid lactone; pentadecalactone).

$CH_2(CH_2)_{13}C(O)O$.

Properties: Colorless liquid with a strong musky odor, congeals to white crystals at room temperature. Minimum congealing p 36C, soluble in equal volume of 90% ethanol. Combustible.
Derivation: Angelica root oil.
Grade: 98% min.
Use: Perfumery.

pentadecenyl phenol. See "Cardanol."

pentadecylic acid. See n-pentadecanoic acid.

1,3-pentadiene. See piperylene.

pentaerythrite tetranitrate. Legal label name for pentaerythritol tetranitrate.

pentaerythritol. (PE; tetramethylolmethane; monopentaerythritol). CAS: 115-77-5. $C(CH_2OH)_4$.
Properties: White, crystalline powder; readily esterified by common organic acids. Unaffected when boiled with dilute caustic alkali; soluble in water; slightly soluble in alcohol; insoluble in benzene, carbon tetrachloride, ether, and petroleum ether. Bp 276C (30 mm), mp 262C, refr index 1.54–1.56 (20C), d 1.399 (25/4C). Combustible. Technical grade is 88% monopentaerythritol and 12% dipentaerythritol.

Derivation: Reaction of acetaldehyde with an excess of formaldehyde in an alkaline medium.
Grade: Technical, nitration, CP.
Use: Alkyd resins, rosin and tall oil esters, special varnishes, pharmaceuticals, plasticizers, insecticides, synthetic lubricants, explosives, paint swelling agents.

pentaerythritol tetraacetate. CAS: 597-71-7.
$C(CH_2OOCCH_3)_4$.
Properties: White, crystalline powder; extremely stable in sunlight; soluble in water, alcohol, and ether; mp 84C; bp 225C (30 mm). Combustible.
Derivation: By the esterification of pentaerythritol with acetic acid.
Grade: Technical.

pentaerythritol tetrakis(diphenyl phosphite).
(tetra(diphenylphosphito)pentaerythritol).
$C[CH_2OP(OC_6H_5)_2]_4$.
Properties: A low-melting, white, waxy solid with a slight phenolic odor, d 1.24 (25/15.5C), mp 30–60C, ref index 1.5823 (25C). Combustible.
Use: Ingredient in stabilizer systems for resins.

pentaerythritol tetra(3-mercaptopropionate).
$C(CH_2OOCCH_2CH_2SH)_4$.
Properties: Liquid; d 1.28 (25C); refr index 1.5300 (25C); insoluble in water, alcohol, and hexane; soluble in acetone and benzene. Combustible.
Use: Curing or crosslinking agents for polymers, especially epoxy resins, intermediates for stabilizers and antioxidants.

pentaerythritol tetranitrate. (PETN).
CAS: 78-11-5. $C(CH_2ONO_2)_4$.
Properties: White, crystalline material; d 1.75; mp 138–140C; decomposes above 150C. Very soluble in acetone, slightly soluble in alcohol and ether, insoluble in water.
Derivation: Esterification of pentaerythritol with nitric acid.
Hazard: Shock-sensitive explosive. Detonates at 210C.
Use: Demolition explosive, blasting caps, detonating compositions ("Primacord.")

pentaerythritol tetrastearate.
$C(CH_2OOCC_{17}H_{35})_4$. Hard, high-melting wax, ivory-colored and essentially neutral, acid number 1, softening p 67C. Combustible.
Use: Polishes, coatings, textile finishes.

pentaerythritol tetrathioglycolate.
$C(CH_2OOCCH_2SH)_4$.
Properties: Liquid; d 1.385 (25C); refr index 1.5499 (25C); insoluble in water, alcohol, and hexane; partially soluble in benzene; soluble in acetone. Combustible.

Use: Curing or crosslinking agents for polymers, especially epoxy resins, intermediate for stabilizers and antioxidants.

pentaglycerine. See trimethylolethane.

pentahydroxycyclohexane. See quercitol.

2′,3,4′,5,7-pentahydroxyflavone. See morin.

pentalin. See pentachloroethane.

"Pentalyn."[266] TM for a series of nonreactive and heat-reactive pentaerythritol esters of rosin. Available in solid and/or flake form in various grades.
Use: Varnishes, floor polishes, inks, and adhesives.

1,1,3,3,5-pentamethyl-4,6-dinitroindane.
CAS: 116-66-5. $C_{14}H_{18}N_2O_4$.
Properties: Pale yellow crystals with a musk-type odor, mp min 132C, slightly soluble in alcohol, soluble in diethyl phthalate.
Use: Perfumery.

pentamethylene. See cyclopentane.

pentamethyleneamine. See piperidine.

pentamethylene-1,1-bis(1-methylpyrrolidinium bitartrate). See pentolinium tartrate.

pentamethylenediamine. See cadaverine.

pentamethylene dibromide. (1,5-dibromopentane). $BrCH_2(CH_2)_3CH_2Br$.
Properties: Colorless, aromatic liquid; fp −35C; bp 224C; insoluble in water.

pentamethylene glycol. See 1,5-pentanediol.

pentamethylpararosaniline chloride. See methyl violet.

pentanal. See n-valeraldehyde.

n-pentane. (amyl hydride). CAS: 109-66-0.
$CH_3(CH_2)_3CH_3$.
Properties: Colorless liquid, pleasant odor, fp −129.7C, bp 36.074C, refr index 1.35748 (20C), d 0.62624, soluble in alcohol and most organic solvents, insoluble in water, flash p −40F (−40C), autoign temperature 588F (308C).
Derivation: Fractional distillation from petroleum, purified by rectification.
Grade: Pure, technical, commercial.
Hazard: Highly flammable, dangerous fire explosion risk. Explosive limits 1.4–8% in air. Narcotic in high concentration. TLV: 600 ppm in air.

Use: Artificial ice manufacture, low-temperature thermometers, solvent extraction processes, blowing agent in plastics, e.g., expandable polystyrene, pesticide.

pentanedinitrile. See glutaronitrile.

pentanedioic acid. See glutaric acid.

pentanedioic acid anhydride. See glutaric anhydride.

1,5-pentanediol. (pentamethylene glycol).
CAS: 111-29-5. $HOCH_2(CH_2)_3CH_2OH$.
Properties: Viscous liquid, bp 240C, fp −15.6C, d 0.9921 (20/20C), bulk d 8.2 lb/gal (20C), flash p (OC) 265F (129C), miscible with water and alcohol, autoign temperature 633F (334C). Combustible.
Grade: Technical.
Use: Hydraulic fluid, lube oil additive, antifreeze, plasticizer and polyester resin intermediate.

2,3-pentanedione. See acetyl propionyl.

2,4-pentanedione. See acetylacetone.

pentanethiol. (amyl mercaptan).
CAS: 110-66-7. $C_5H_{11}SH$. A mixture of isomers.
Properties: Water-white to light yellow liquid, d 0.83–0.84 (20C), mercaptan content greater than 90.0%, initial bp above 104.0C, final bp below 130C, bulk d 6.99 lb/gal, insoluble in water, soluble in alcohol, flash p (OC) 65F (18.3C). These properties vary with proportions of isomers. Strong offensive odor.
Derivation: Mixing amyl bromide and potassium hydrosulfide in alcohol.
Hazard: Flammable, dangerous fire risk. Toxic by inhalation.
Use: Synthesis of organic sulfur compounds, chief constituent of odorant used in gas lines to locate leaks.

pentanoic acid. See n-valeric acid.

1-pentanol. See n-amyl alcohol, primary.

2-pentanol. (sec-n-amyl alcohol; sec-amyl alcohol, active; methyl propyl carbinol; 1-methylbutyl alcohol). CAS: 6032-29-7.
$CH_3CH_2CH_2CHOHCH_3$.
Properties: (racemic form) Colorless liquid, fp −75C, bp 119.3C, d 0.811 (20/20C), bulk d 6.75/gal (20C), refr index 1.4041 (40.5C), flash p (OC) 105F (40.5C), soluble in water, miscible with alcohol and ether, autoign temperature 657F (347C). Combustible.
Derivation: Fractional distillation of the mixed

alcohols resulting from the chlorination and hydrolysis of pentanes.
Hazard: Moderate fire risk. Irritant to eyes, nose, and throat.
Use: Solvent for paints and lacquers, pharmaceutical intermediate.

3-pentanol. (sec-n-amyl alcohol; 1-ethyl-1-propanol; diethyl carbinol). CAS: 584-02-1. $CH_3CH_2CHOHCH_2CH_3$.
Properties: Colorless liquid, d 0.82 (20C), fp below −75C, bp 115.6C, bulk d 6.81/gal, refr index 1.41 (20C), flash p (CC) 94F (34.4C), autoign temperature 650F (343C), soluble in alcohol and ether, slightly soluble in water.
Hazard: Moderate fire risk. Irritant to eyes, nose, and throat.
Use: Solvent, flotation agent, pharmaceuticals.

2-pentanone. See methyl propyl ketone.

3-pentanone. See diethyl ketone.

penta resin. Ester gum made from rosin and pentaerythritol.

pentasodium diethylenetriaminepentaacetate.
CAS: 140-01-2. $C_{14}H_{23}N_3O_{10} \cdot 5Na$.
A sodium salt of diethylenetriaminepentaacetic acid.
Use: A chelating agent.

pentasodium triphosphate. See sodium tripolyphosphate.

n-pentatriacontane. $C_{35}H_{72}$ or $CH_3(CH_2)_{33}CH_3$.
Properties: Crystals, d 0.782 at 75C, bp 331C at 15 mm, mp 75C. Combustible.
Use: Organic synthesis.

pentazocine. (2-dimethylallyl-5,9-dimethyl-2'-hydroxy benzomorphan). CAS: 359-83-1.
$C_{19}H_{27}NO$. A synthetic drug claimed to be as effective as morphine but without its addictive properties. Has a noncumulative effect. FDA approved.

"Pentecat L."[230] TM for a 50% aqueous solution of lithium naphthenate.
Use: Alcoholysis catalyst in alkyd varnish cooking.

1-pentene. (α-n-amylene; propylethylene).
CAS: 109-67-1. $CH_3CH_2CH_2CH:CH_2$.
Properties: Colorless liquid, soluble in alcohol, insoluble in water, fp −165C, bp 30C, d 0.6410 (20C), flash p (OC) 0F (−17.7C); autoign temperature 523F (272C).
Derivation: Natural gasoline.

Hazard: Flammable, dangerous fire risk. Toxic by ingestion, inhalation, and skin absorption.
Use: Organic synthesis, blending agent for high octane motor fuel, pesticide formulations.

2-pentene. (β-n-amylene; sym-methylethylethylene). CAS: 109-68-2.
$CH_3CH_2HC:CHCH_3$. Mixed cis- and trans-isomers are available commercially.
cis-isomer

$$CH_3\diagdown\diagup CH_2CH_3$$
$$C=C$$
$$H\diagup\diagdown H$$

Properties: Bp 37C, fp −180C, d 0.656 (20/4C), flash p 0F (−17.7C), soluble in alcohol, insoluble in water.
Grade: Technical, 95.0 mole %.
trans-isomer

$$CH_3\diagdown\diagup H$$
$$C=C$$
$$H\diagup\diagdown CH_2CH_3$$

Properties: Bp 36.4C, fp −139C, d 0.6482 (20C), soluble in alcohol, insoluble in water, flash p 0F (−17.7C).
Derivation: Natural gasoline.
Hazard: Flammable, dangerous fire risk.
Use: Polymerization inhibitor, organic synthesis.

"Pentex."[248] TM for tetrabutylthiuram monosulfide.
Use: Rubber accelerator. When mixed with 87.5% clay it is used for sponge rubbers and called "Pentex Flour."

"Pentite."[62] TM for tetra(diphenylphosphito) pentaerythritol.
See pentaerythritol tetrakis(diphenyl phosphite).

pentlandite. (Fe, Ni)S.
Properties: Light bronze-yellow mineral, metallic luster, contains 35.57% of nickel, soluble in nitric acid, d 4.6–5, hardness 3.5–4.
Occurrence: Canada (Ontario), Norway.
Use: Nickel ore.

pentobarbital. (5-ethyl-5-(1-methylbutyl)barbituric acid). CAS: 76-74-4. $C_{11}H_{18}O_3N_2$.
See barbiturate.

pentolinium tartrate. (pentamethylene-1,1-bis(1-methylpyrrolidinium bitartrate).
CAS: 52-62-0. $C_{23}H_{42}N_2O_{12}$.
Properties: White to light cream-colored, crystalline powder; slightly soluble in alcohol; insoluble

in ether, chloroform; very soluble in water; pH of 1% solution in water is 3.0–4.0, decomposes 203C.
Use: Medicine (antihypertensive).

pentolite. A high explosive consisting of equal parts of pentaerythritol tetranitrate and trinitrotoluene.
Hazard: Dangerous, explodes on shock or heating.

"Penton."[266] TM for a thermoplastic resin derived from 3,3-bis(chloromethyl)oxetane

$(CH_2Cl)_2\overline{CCH_2OCH_2})$.
A chlorinated polyether.
Properties: A linear polymer extremely resistant to chemicals and to thermal degradation at molding and extrusion temperature, d 1.4, self-extinguishing, dimensionally stable, very low water absorption, outstanding chemical resistance. Natural, black, or olive green molding powder. Finely divided powder for coatings.
Use: Solid and lined valves, pumps, pipe, and fittings; monofilament for filter supports and column packing.

pentosan. A complex carbohydrate (hemicellulose) present with the cellulose in many woody plant tissues, particularly cereal straws and brans, characterized by hydrolysis to give five-carbon-atom sugars (pentoses). Thus the pentosan xylan yields the sugar xylose $(HOH_2C \cdot CHOH \cdot CHOH \cdot CHOH \cdot CHO)$ which is dehydrated with sulfuric acid to yield furfural $(C_5H_4O_2)$.

pentose. General term for sugars with five carbon atoms per molecule.

"Pentothal."[3] TM for sodium thiopental, a barbiturate.
See thiopental sodium.

pentyl. Synonym for the amyl group, C_5H_{11}—.

pentyl acetate. See amyl acetate.

pentylamine. See n-amylamine.

α-pentylcinnamaldehyde. See α-amylcinnamic aldehyde.

p-tert-pentylphenol. (p-tert-amylphenol).
CAS: 80-46-6. $C_{11}H_{15}OH$.
Properties: Crystalline solid, mp 95C, bp 261C, d 0.962 (20/4C), insoluble in water, soluble in organic solvents.
Derivation: Condensation of tert-pentanol with phenol with aluminum chloride catalyst.

Use: Pesticide intermediate, oil-soluble resin manufacture, may be useful as germicide and fumigant.

Penzold's reagent. Solution of diazobenzosulfonic acid and potassium hydroxide.
Use: Testing for sugar in urine.

peppermint oil. Essential oil with strong aromatic odor and taste, levorotatory, chief component is menthol.
Use: To flavor mouth washes, chewing gum, liqueurs, toothpastes; source of menthol.

pepsin. (pepsinum).
Properties: A digestive enzyme of gastric juice. White or yellowish-white powder or lustrous transparent or translucent scales; should have no odor; converts proteins into albumoses and peptones; soluble in water; insoluble in alcohol, chloroform, and ether.
Derivation: From the glandular layer of fresh hog stomachs.
Grade: Technical, NF.
Use: Medicine (digestive ferment); substitute for rennet in cheese making.

pepsinogen. An inactive precursor of pepsin.

peptidase. See protease.

peptide. See polypeptide.

peptization. Stabilization of hydrophobic colloidal solutions by addition of electrolytes which provide the necessary electric double layer of ionic charges around each particle. Such electrolytes are known as peptizing agents. The ions of the electrolyte are strongly adsorbed on the particle surfaces. Stable solutions of non-ionizing substances acquire a charge in contact with water by preferential adsorption of the hydroxyl ions, which may be considered peptizing agents. The term is also loosely applied to the softening or liquefaction of one substance by trace quantities of another, analogous to the digestion of a protein by an enzyme (pepsin).

peptone.
Properties: (a) From albumin: white or pale yellow amorphous powder. (b) From meat: light brown amorphous powder. (c) From milk: light brown powder. Soluble in water, insoluble in alcohol or ether.
Derivation: (a) By digestion of egg albumin by pepsin and a small quantity of dilute hydrochloric acid at 38–40C (body temperature). (b) By digestion of red meat with pancreatin at body temperature. (c) By digestion of casein.

Grade: Technical, reagent.
Use: Preparation of nutrient media in bacteriology; nutrient.

per-. A prefix signifying complete or extreme and specifically denoting: (1) a compound containing an element in its highest state of oxidation, as perchloric acid; (2) presence of the peroxy group, —O—O—, as peracetic and perchromic acids; (3) exhaustive substitution or addition, as perchloroethylene.

peracetic acid. (peroxyacetic acid).
CAS: 79-21-0. CH_3COOOH.
Properties: Colorless liquid, strong odor, bp 105C, fp —30C, d 1.15 (20C), flash p (OC) 105F (40.5C). Soluble in water, alcohol, sulfuric acid, strong oxidizing agent.
Derivation: (a) Oxidation of acetaldehyde, (b) reaction of acetic acid and hydrogen peroxide with sulfuric acid catalyst.
Grade: Technical, 40% soluble in acetic acid.
Hazard: Oxidizing material, dangerous in contact with organic materials, explodes at 110C. Strong irritant.
Use: Bleaching textiles, paper, oils, waxes, starch; polymerization catalyst; bactericide and fungicide, especially in food processing; epoxidation of fatty acid esters and epoxy resin precursors; reagent in making caprolactam, synthetic glycerol.

per-acids. Derivatives of hydrogen peroxide, the molecules of which contain one or more directly linked pairs of oxygen atoms, —O—O—. Examples are persulfuric, perchromic, peracetic acids. Permanganic, perchloric, and periodic acids are not per-acids in this sense.
See per-(2). See per-(1).

"Perapret."[440] TM for finishing agents and binders (polymer dispersions), additives for wash-and-wear finishing.

"Perazil."[301] TM for chlorcyclizine hydrochloride, an antihistamine.

perbenzoic acid. See benzoyl hydroperoxide.

percarbamide. See urea peroxide.

perchloric acid. CAS: 7601-90-3. $HClO_4$.
Properties: Colorless, fuming, hygroscopic liquid, unstable in concentrated form, d 1.764, bp 19C (11 mm), fp —112C, evolves heat on combination with water. Commercial aqueous solutions contain 65–70% $HClO_4$.
Derivation: By distilling potassium perchlorate

with strong sulfuric acid (96%) under reduced pressure in an oil bath at 140–190C.
Method of purification: Rectification.
Grade: Technical, CP.
Hazard: Strong oxidizing agent, will ignite vigorously in contact with organic materials, or detonate by shock or heat. Toxic by ingestion and inhalation, strong irritant.
Use: Analytical chemistry, catalyst, manufacture of various esters, ingredient of electrolytic bath in deposition of lead, electro-polishing, explosives.

perchlorobenzene. See hexachlorobenzene.

perchlorocyclopentadiene. See hexachlorocyclopentadiene.

perchloroethane. See hexachloroethane.

perchloroethylene. (tetrachloroethylene).
CAS: 127-18-4. $Cl_2C:CCl_2$.
Properties: Colorless liquid, ether-like odor, extremely stable, resists hydrolysis, d 1.625 (20/20C), bp 121C, fp —22.4C, bulk d 13.46 lb/gal (26C), refr index 1.5029 (25C), flash p none. Miscible with alcohol, ether, and oils; insoluble in water. Nonflammable.
Derivation: (a) By chlorination of hydrocarbons and pyrolysis of the carbon tetrachloride also formed, (b) from acetylene and chlorine via trichloroethylene.
Method of purification: Distillation.
Grade: Purified, technical, USP, as tetrachloroethylene, spectrophotometric.
Hazard: Irritant to eyes and skin. TLV: 50 ppm in air.
Use: Dry-cleaning solvent, vapor-degreasing solvent, drying agent for metals and certain other solids, vermifuge, heat-transfer medium, manufacture of fluorocarbons.

perchloromethane. See carbon tetrachloride.

perchloromethyl mercaptan. Legal label name for trichloromethylsulfenyl chloride.

perchloropentacyclodecane. CAS: 2385-85-5.
$C_{10}Cl_{12}$.
Properties: White, crystalline solid; mp (sealed tube) 485C. Nonflammable.
Use: Fire-retardant additive in elastomeric resin systems.

perchloropropylene. See hexachloropropylene.

perchloryl fluoride. CAS: 7616-94-6. $ClFO_3$.
Properties: Colorless, noncorrosive gas or liquid with sweet odor. Fp —146C, bp —46.8C, d (liq-

uid) 1.434 (20C). Nonflammable but supports combustion.

Hazard: Dangerous in contact with organic materials, strong oxidizing agent. Toxic by inhalation. TLV: 3 ppm in air.

Use: Oxidant in rocket fuels, oxidizing and fluorinating agent in chemical reactions.

perchromic acid. Probably $(HO)_4Cr(OOH)_3$.

Properties: Unstable acid formed when a solution of chromic acid is added to hydrogen peroxide; forms deep blue crystals below $-15C$; the blue color can be extracted from solutions by ether; decomposes in acid solution to form chromic salts and in alkaline solution to form chromates; the blue color can be used as a test for chloride or chromate.

Hazard: Irritant.

"Perclene."[309] TM for dry cleaning composition consisting of perchloroethylene and surfactant additives.

"Peregal."[307] TM for a series of textile chemicals. O. Polyoxyethylated fatty alcohol, nonionic. OK. Methyl polyethanol quaternary amine, cationic. ST. Polymeric dye complexing agent. TW. Alkyl polyoxyethylene glycol amine.

perfect gas. See gas.

perfluidone. CAS: 37924-13-3. $C_{14}H_{12}F_3NO_4S_2$. A pre-emergence herbicide.

perfluoro-2-butene. Legal label name for octafluoro-2-butene.

perfluorobutyric acid. See heptafluorobutyric acid.

perfluorocarbon compound. A fluorocarbon in which the hydrogen directly attached to the carbon atoms is completely replaced by fluorine.

perfluorocyclobutane. See octafluorocyclobutane.

perfluorodimethylcyclobutane. An inert fluorocarbon liquid.

Properties: Bp 45C, stable to 260C.

Use: Evaporative coolant for electronic equipment.

perfluoroethylene. See tetrafluoroethylene.

perfluoropropane. Legal label name (Air) for octafluoropropane.

perfluoropropene. See hexafluoropropylene.

perfluorosulfonic acid. See "Nafion."

performic acid. (peroxyformic acid). CAS: 107-32-4.   HCOOOH.

Properties: Colorless liquid; miscible with water, alcohol, ether; soluble in benzene, chloroform; solutions are unstable.

Derivation: Mixture of formic acid, peroxide, and sulfuric acid is allowed to interact for two hours and then distilled.

Grade: 90% solution.

Hazard: Explodes when shocked or heated or in contact with reducing materials, metals, and metallic oxides. Strong irritant and oxidizing agent.

Use: Oxidation, epoxidation, and hydroxylation reactions.

perfume. A blend of pleasantly odorous substances (usually liquids) obtained from the essential oils of flowers, leaves, fruit, roots, or wood of a wide variety of plants, either by steam distillation or solvent extraction. Flower oils (rose, jasmine) are extracted with a non-polar solvent to give a waxy mixture called a concrete; the wax is then removed by a second solvent (an alcohol), which is then in turn removed to form an absolute. It is necessary that all the solvent be eliminated to obtain the finest perfumes. The center of this industry has long been in Grasse, France. Perfume materials are also derived from animal sources (musk, ambergris) and from resinous extracts (terpenes and balsams); they are also made synthetically. Cologne and toilet water are weak alcoholic solutions (5% or less) of perfumes. Fine perfumes may contain as many as 30 ingredients and their blending is an art rather than a science. The largest volume use of perfumes is in soaps, lotions, shaving creams, and cosmetics.

See also fragrance, odorant.

perhydrosqualene. See squalane.

peri acid. (1-naphthylamine-8-sulfonic acid).

Properties: White needles, slightly soluble in water.

Derivation: Nitration of naphthalene-α-sulfonic acid followed by reduction with iron and crystallization of the sodium salt.

Use: Azo dye intermediate, production of Chicago acid.

periclase. MgO.   A natural or calcined magnesium oxide used as a lining and maintenance material for basic oxygen steel-making furnaces and other refractories.

perilla oil.
Properties: Light yellow drying oil. Soluble in alcohol, ether, chloroform, and carbon disulfide; d 0.932–0.945; saponification value 191–193; iodine value 187–202; refr index 1.4841. Combustible. Subject to spontaneous heating.
Derivation: From the seeds of Perilla ocimoides, grown commonly in Japan and Korea. Chief constituents: Linoleic and linolenic acids.
Use: Substitute for linseed oil, edible oil in Asia, manufacture of varnishes.

periodic acid. CAS: 10450-60-9.
$HIO_4 \cdot 2HOH$.
Properties: White crystals, mp 122C, decomposes at 130C, loses 2HOH at 100C. Soluble in water, alcohol; slightly soluble in ether.
Derivation: By the interaction of iodine and concentrated perchloric acid, by low-temperature electrolytic oxidation of concentrated iodic acid.
Method of purification: Crystallization.
Hazard: Dangerous in contact with organic materials. Irritant and oxidizing material.
Use: Oxidizing agent, increasing wet strength of paper, photographic paper.

periodic law. Originally stated in recognition of an empirical periodic variation of physical and chemical properties of the elements with atomic *weight*, this law is now understood to be based fundamentally on atomic *number* and atomic *structure*. A modern statement is: *The electronic configurations of the atoms of the elements vary periodically with their atomic number. Consequently all properties of the elements that depend on their atomic structure (electronic configuration) tend also to change with increasing atomic number in a periodic manner* (R. T. Sanderson).

periodic table. An arrangement of the chemical elements by symbol in a geometric pattern designed to represent the periodic law by aligning the elements in periods so that the corresponding parts of the several periods are adjacent. When the elements are aligned in order of increasing atomic number, they constitute a succession of period, each beginning with an alkali metal (one electron in the outermost principal quantum level) and ending with an element of the helium family, helium, neon, argon, krypton, xenon, and radon (each having eight electrons in the outermost principal quantum level, except for helium which is limited to two). Each helium family element is followed directly in atomic number by an alkali metal which begins a new period.

The advantage of placing each successive period so that it is adjacent to the preceding period in its corresponding parts is that similar elements are thus brought together in groups. For exam-

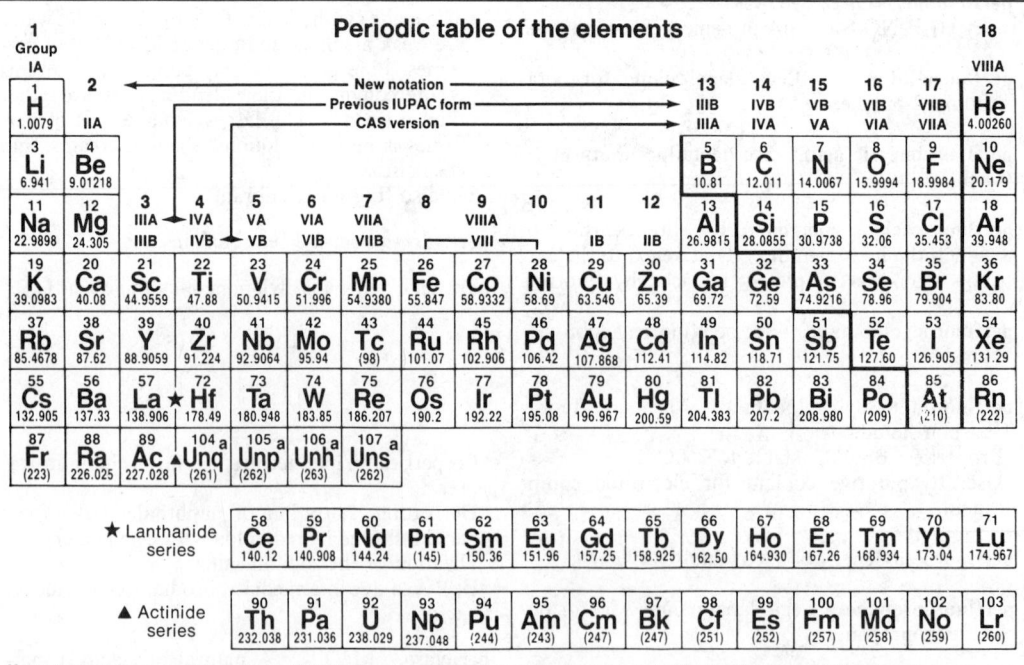

Note: Atomic masses shown here are the 1983 IUPAC values (maximum of six significant figures). **a** Symbols based on IUPAC systematic names.

ple, the alkali metals that begin each period have similar properties that correspond to their atomic structure, each having one outermost shell electron. In a periodic table, the elements exhibit a steady trend in properties from the beginning of each period, where they are metals, to the end, where they are nonmetals. Within each group the elements are quite similar. A given element can be predicted, on the basis of its position in a periodic table, to resemble the other elements of its group and be intermediate in properties between its adjacent neighbors within its period. The chief function of a periodic table is to serve as a fundamental framework for the systematic organization of chemistry. R. T. Sanderson.

See also Mendeleef.

Note: Dr. Glenn T. Seaborg predicted the existence of synthetic superheavy elements 107-168 on the basis of theoretical calculations. Numbers 122-153 he called 'superactinides'.

Perkin alicyclic synthesis. Synthesis of alicyclic compounds from alpha,omega-dihaloalkanes and compounds containing active methylene groups in the presence of sodium ethoxide.

Perkin reaction. Formation of α,β-unsaturated carboxylic acids by aldol condensation of aromatic aldehydes and acid anhydrides in the presence of an alkali salt of the acid.

Perkin rearrangement. Formation of benzofuran-2-carboxylic acids and benzofurans by heating 3-halocoumarins with alkali.

Perkin, Sir William Henry. (1838–1907). An English chemist who was the first to make a synthetic dyestuff (1856). He studied under Hofmann at the Royal College of London. Perkin's first dye was called mauveine, but he proceeded to synthesize alizarin and coumarin, the first man-made perfume. In 1907 he was awarded the first Perkin Medal, which has ever since been awarded by the American Division of the Society of Chemical Industry for distinguished work in chemistry. Notwithstanding the fact that Perkin patented and manufactured mauve dye in England, the center of the synthetic dye industry shifted to Germany where it remained till 1914. See also Hofmann.

Perkow reaction. Formation of enol phosphates on treatment of α-halocarbonyl compounds with trialkyl phosphites.

perlite. Eutectic between ferrite and cementite (steelmaking).

"Perluxe."[56] TM for a dry cleaning solvent consisting of perchloroethylene or tetrachloroethylene.

"Permachlor Red."[141] Azo pigments made from p-chloro-o-nitraniline and β-naphthol.
Use: Bright, light red and pink house paints.

"Permachrom Red."[141] Permanent red azo pigments derived from β-hydroxynaphthoic acid.
Grade: Medium red, dark red, maroon.
Use: Printing inks, paints, enamels, lacquers, plastics.

permafil. A mixture in which the liquid undergoes complete polymerization and hardens without the necessity of any evaporation. Anaerobic permafils harden out of contact with air.

"Permafresh."[42] TM for a series of resins used in the textile industry. They include cellulose reactants, melamine derivatives designed as a replacement for conventional trimethylol melamine and related products, partially condensed modified carbamide resins, cyclic nitrogenous reactant, bis-keto-triazine, methylol imidazolidone, modified carbamide resin, urea-formaldehyde liquid, durable thermosetting urea-formaldehyde reactive resin.
Use: Wash-and wear fabrics, crease resistance, shrinkage control, durable finishes, bodying agent.

"Permagreen Gold."[141] TM for a yellow nickel azo pigment.
Use: Paints, printing inks, plastics.

"Permalloy." TM for an alloy containing 78.5% nickel and 21.5% iron. It has high magnetic permeability and electrical resistivity.
Use: Transocean submarine cables.

"Permalume G."[288] TM for a semi-bright nickel electroplating process. The plating bath contains nickel sulfate, nickel chloride, boric acid, and organic addition agents.

"Permalux."[28] TM for a di-o-tolylguanidine salt of dicatechol borate.
$(HOC_6H_4O)_2B \cdot HNC(NHC_6H_4CH_3)_2$.
Properties: Light grayish-brown powder, d 1.27, fp greater than 165C.
Use: Rubber antioxidant, vulcanization of neoprene.

permanent magnet. A magnet whose magnetic properties are not affected by removing it from an external magnetic field.

Permanent Orange. See dinitraniline orange.

permanent-press resin. (durable-press resin).
A thermosetting resin used as a textile impregnant or fiber coating to impart crease resistance and permanent hot-creasing to suitings, dress

fabrics, etc. Chemicals such as formaldehyde and maleic anhydride are the basis of these products. The resin is "cured" after the fabric has been tailored into a garment. A permanent-press fabric that requires no resin has been developed (a blend of polyester with cotton or rayon).

Permanent Red 2B Amine. See 4-amino-2-chlorotoluene-5-sulfonic acid.

permanent set. (permanent elongation).
The extent to which vulcanized rubber and other elastomeric materials are permanently deformed after extension to break. In general it varies in the range of 5–10% of the original length, depending on the formulation and state of cure. See also elasticity.

permanent white. $BaSO_4$.   Precipitated barium sulfate.

"Permanickel" 300.[283]   TM for an alloy which contains 98.6% nickel and small quantities of other metals.

"Permanite."[250]   TM for purified natural magnetic iron ore for use in radio cores, shields, etc.

"Permansa."[266]   TM for a series of light and heat stable yellow, orange, and green azo pigments used primarily in paints.

"Permasep."[28]   TM for a method of desalinating water by reverse osmosis using membranes composed of hollow nylon fibers which provide maximum membrane surface.
See also osmosis.

permenorm.   Nickel-iron alloy produced by magnetic annealing and drastic cold reduction, and used for mechanical rectifiers and low-frequency amplifiers. This alloy has a rectangular hysteresis loop that eliminates arcing at the contacts of mechanical rectifiers, as well as other desirable properties.

permethrin.   [(3-phenoxyphenyl)methyl-3-(2,2-dichlorovinyl)-2,2-dimethylcyclopropanecarboxylate].   CAS: 52645-53-1.
Use: Insecticide, nematocide, acaricide.

permissible explosive. See explosive, permissible.

"Perone."[28]   TM for HOOH, 30, 35 and 50% solutions by weight.
See hydrogen peroxide.

perovskite.   A natural or synthetic crystalline mineral composed of calcium dioxide and tita-nium dioxide. The natural material was discovered in Russia in about 1850; the synthetic type can be so made as to incorporate particles of catalysts (platinum or palladium) which are protected by the crystalline structure from contamination by lead and other poisons, thus permitting use of leaded gasoline in cars equipped with emission-control devices.

peroxidase.   An enzyme found in most plant cells and some animal cells which promote the oxidation of various substrates, such as phenols, aromatic amines, etc., by means of hydrogen peroxide.

peroxide.   (1) any compound containing a bivalent O—O group, i.e., the oxygen atoms are univalent. Such compounds release atomic (nascent) oxygen readily. Thus they are strong oxidizing agents, and fire hazards when in contact with combustible materials, especially under high-temperature conditions. The chief industrial uses of peroxides are as oxidizing agents, bleaching agents, and initiators of polymerization. (2) Hydrogen peroxide.

peroxyacetic acid. See peracetic acid.

peroxybenzoyl nitrate.   A component of photochemical smog.
Hazard: Over 200 times as irritating to the eyes as formaldehyde. A concentration of 0.02 ppm in air causes moderate to severe conjunctival irritation.
See also smog.

peroxyformic acid. See performic acid.

peroxysulfuric acid. See Caro's acid.

perphenazine.   CAS: 58-39-9.
$C_{21}H_{26}ClN_3OS$.
Properties: Crystals, sensitive to light, mp 97–100C, insoluble in water, soluble in ethanol and acetone.
Grade: ND.
Use: Medicine.

perrhenic acid. $HReO_4$.   Exists only in solution, commercially available as aqueous syrup. Strong, very stable, monobasic acid; extremely soluble in water and organic solvents.

"Persistol."[440]   TM for agents for the washproof water-repellent finishing of textiles of natural and synthetic fibrous materials and fiber mixes.

perstoff See trichloromethylchloroformate.

persulfuric acid.   See Caro's acid.

"Perthane."[23] TM for an agricultural insecticide based on 1,1-dichloro-2,2-bis(p-ethylphenyl)ethane. Supplied as a wettable powder or emulsifiable concentration.

Peru balsam. See balsam.

Perutz, Max F. (1914-) An Austrian molecular biologist who was a recipient of the Nobel prize for chemistry in 1962 along with Kendrew. His work was concerned with crystalline protein structure, particularly the molecular structure of hemoglobin and myoglobin. His education was in England and Austria.

PES. Abbreviation for photoelectron spectroscopy.

pesticide. Any substance, organic or inorganic, used to destroy or inhibit the action of plant or animal pests; the term thus includes insecticides, herbicides, rodenticides, miticides, etc.
See specific entry. See note under methyl parathion.
Virtually all pesticides are toxic to man to some degree. They vary in biodegradability. The use of more toxic types have been restricted, especially DDT. Microencapsulated controlled-release forms are available.
See insecticide (toxic). See also biocide. For additional information consult International Pesticide Institute, Syracuse, NY.

"Pestmaster" Soil Fumigant-1[426]. TM for a mixture of methyl bromide 98%, chloropicrin 2%.
Properties: Colorless liquefied gas with a sharp pungent odor.
Hazard: As for methyl bromide.
Use: To control insects, weeds, nematodes, and certain fungi in the soil.

petalite. (lithium aluminum silicate).
CAS: 1302-66-5. $LiAl(Si_2O_5)_2$
Properties: Colorless, white, gray, or occasionally pink mineral, white streak, vitreous luster. Resembles spodumene in appearance. Contains up to 4.9% lithia, sometimes with partial replacement by sodium or less often by potassium. Insoluble in acids, d 2.39–2.46, Mohs hardness 6–6.5.
Occurrence: US (Massachusetts, Maine); Sweden.
Use: Source of lithium salts, in ceramics and glass.

Peterson reaction (olefination). Reaction of α-silylated carbanions with carbonyl compounds yielding β-hydroxyalkyl silanes which undergo instantaneous elimination to afford olefins.

petitgrain oil. See neroli oil, to which it is similar.

PETN. Abbreviation for pentaerythritol tetranitrate.

Petrenko-Kritschenko piperidone synthesis.
Formation of piperidones by cyclization from two moles of aldehyde and one mole each of acetonedicarboxylic ester and of ammonia or a primary amine. Used widely in synthesis of tropane derivatives.

petri dish. A small concave glass plate having an easily removable cover, used for the culture of bacteria. Named after its inventor, a German biologist.

"Petrobase."[25] TM for a series of emulsifying and rust-preventive bases. On dilution with suitable petroleum oils or light distillates the proper concentration provides soluble cutting and grinding oils; solvent emulsion cleaner, preservative, slushing and household oils, and water-displacing fluids.

"Petro Bond."[236] TM for a bonding agent in the preparation of waterless foundry sands, used with oil and catalyst. Fine sands giving excellent reproduction of detail can be used since high permeability of the foundry sand is not required with this binder.

petrochemical. An organic compound for which petroleum or natural gas is the ultimate raw material. Thus, cracking of petroleum produces ethylene which is converted to ethylene glycol, the latter being a typical petrochemical. The term is also applied to substances such as ammonia, because the hydrogen used to form the ammonia is derived from natural gas. Thus, synthetic fertilizers are considered to be petrochemicals. At least 175 substances are designated as petrochemicals including many paraffin, olefin, naphthene, and aromatic hydrocarbons (methane, ethane, propane, ethylene, propylene, butenes, cyclohexane, benzene, toluene, naphthalene, etc.) and their derivatives, even though some of their commercial production is from sources other than petroleum. The percentage of total hydrocarbon consumption represented by petrochemicals is steadily increasing; some authorities maintain that petroleum is too valuable to be used for fuel and that it should be conserved for future petrochemical development.

petrolatum. A semisolid or liquid mixture of hydrocarbons derived by distillation of paraffin-base petroleum fractions. The solid form (mineral jelly) may be either water-white or pale yellow. Its chief uses are in mild ointments, cosmetics, as softener in rubber mixtures and in food processing (release agent in bakery products, dehydrated fruits and vegetables), protective

coating (raw fruits and vegetables), and defoaming agent (beet sugar, yeast). The liquid form (white mineral oil) is used as a laxative, textile lubricant, and dispersing agent. There are three grades of both solid and liquid types with various specifications (USP, NF, and FCC).
See also ointment.

petrolatum wax. A microcrystalline wax containing hydrocarbons from $C_{33}H_{70}$ to $C_{43}H_{88}$. Solidifying range 71–83C.
See also wax, microcrystalline.

"Petrolene."[200] TM for a petroleum solvent prepared by straight-run distillation.
Properties: Water-white, initial bp 140–145F, 95% distills between 195–200F, d 0.701 (15.5C), flash p (TCC) −16F (−26.6C), mild, nonresidual odor.
Hazard: Highly flammable, dangerous fire risk.
Use: Rubber cements, sealers, fast-drying lacquers, lacquer dopes, roto inks used on high-speed presses.

petroleum. (crude oil). A highly complex mixture of paraffinic, cycloparaffinic (naphthenic), and aromatic hydrocarbons, containing a low percentage of sulfur and trace amounts of nitrogen and oxygen compounds. Said to have originated from both plant and animal sources 10–20 million years ago. The most important petroleum fractions, obtained by cracking or distillation, are various hydrocarbon gases (butane, ethane, propane), naphtha of several grades, gasoline, kerosene, fuel oils, gas oil, lubricating oils, paraffin wax, and asphalt. From the hydrocarbon gases, ethylene, butylene, and propylene are obtained; these are important industrial intermediates, being the source of alcohols, ethylene glycols and monomers for a wide range of plastics, elastomers, and pharmaceuticals. Benzene, phenol, toluene, and xylene can be made from petroleum, and hundreds of other products, including biosynthetically produced proteins, are petroleum-derived. About 5% of the petroleum consumed in the US is used as feedstocks by the chemical industries.
Occurrence: At present half of the world's proven resources are in the Middle East and North Africa, the other half being divided among the US (including Alaska), Canada, Venezuela, USSR, the North Sea area, Indonesia, Mexico, Rumania, and Australia.
Properties: Viscous, dark-brown liquid, unpleasant odor, d 0.78–0.97, flash p 20–90F.
Hazard: Flammable, moderate fire risk. Toxic by ingestion, local skin irritant.
For further information refer to American Petroleum Institute, 1271 Avenue of the Americas, NY, NY.

See also natural gas, petrochemical, See Appendix E.4 for history of the industry.

petroleum benzin. A special grade of ligroin.

petroleum coke. See coke.

petroleum ether. This term is used synonymously with petroleum naphtha. It is also sometimes used as a synonym for ligroin or petroleum spirits. It is technically a misnomer for it is not an ether in the chemical sense. For details about specified distillation ranges and other distinctive properties, consult ASTM and API specifications.
See naphtha (1).

petroleum gas, liquefied. See liquefied petroleum gas.

petroleum jelly. See petrolatum

petroleum naphtha. See naphtha (1).

petroleum spirits. In Great Britain the term "petroleum spirits" refers to a volatile hydrocarbon mixture having a flash p 32F (0C).
Hazard: Highly flammable, dangerous fire risk.
See naphtha (1a). See also spirits; petroleum ether.

petroleum, synthetic. See pyrolysis.

petroleum thinner. See naphtha (1a).

petroleum wax. A high molecular weight solid hydrocarbon derived from petroleum. There are three types: paraffin waxes, microcrystalline waxes, and petrolatum waxes. All are made mostly by solvent dewaxing, although pressing and sweating processes are still used.
See specific entry.

"Petronates."[45] TM for salts of petroleum sulfonic acids, varying in molecular weight and color.
Use: Emulsifying agents, dispersing agents, wetting agents, corrosion-preventive.

"Petrosul."[25] TM for a series of highly purified natural petroleum sulfonate products available in high, medium, and low molecular weight ranges. Useful in applications requiring the surface active functions of foaming, detergency, emulsibility, dispersion, solubilization, spreading, and rust protection.

"Petrotect."[25] TM for a series of rust-preventive and hydraulic fluids in general meeting military specifications. The rust preventives are classified as solvent cut backs, petrolatum barriers, general purpose preservatives, and engine perservative lubricants. Hydraulic fluids include both preservative and operational types.

"Petrothene."[192]   TM for polypropylene resins for blown, cast, and water-quenched films, substrate coating, wire and cable coating, injection molding, blow molding, thermoforming, pipe extrusion, calendering.
Available forms: Solid cubes and pellets in natural and black.

"Petrotone."[236]   TM for an organophilic clay that increases suspending properties of crude or refined oils and oil muds without appreciably increasing gels. Makes the use of weighted oil muds safer and more economical.

"Petrowet."R[28]   TM for a surface-active agent composed of saturated-hydrocarbon sodium sulfonate. A wetting and penetrating agent effective in high concentration of electrolytes and acids, suitable for use in acidizing of oil wells.

pewter.   Tin alloys with 5–15% tin, 0–3% copper, and 0–15% lead. White metal and britannia metal are also of this general composition.

peyote.   Root of cactus Lophophora williamsii (Mexico) known as a hallucinogen.

PF resins.   Abbreviation for phenol-formaldehyde resins.

Pfau-Plattner azulene synthesis.   Formation of azulenes by ring enlargement of indanes on addition of diazoacetic ester, hydrolysis, dehydrogenation, and decarboxylation of the resulting acid.

Pfitzinger reaction.   Formation of quinoline-4-carboxylic acids by condensation of isatic acids from isatin with α-methylene carbonyl compounds; subsequent decarboxylation yields quinolines.

Pfitzner-Moffatt oxidation.   Oxidation of alcohols to carbonyl derivatives with dimethyl sulfoxide and dicyclohexylcarbodiimide. The procedure is especially useful for the conversion of a primary alcohol to an aldehyde without further oxidation to the carboxylic acid.

PG.   Abbreviation for polypropylene glycol.

PGA.   Abbreviation for pteroylglutamic acid. See folic acid.

pH.   pH is a value taken to represent the acidity or alkalinity of an aqueous solution, it is defined as the logarithm of the reciprocal of the hydrogen-ion concentration of a solution:

$$pH = \log_{10} \frac{1}{[H+]}$$

Pure water is the standard used in arriving at this value. Under ordinary conditions water molecules dissociate into the ions H^+ and OH^-, with recombination at such a rate that with very pure water at 22C there is a concentration of oppositely charged ions of 1/10,000,000, or 10^{-7}, mole per liter. This is commonly expressed by saying that pure water has a pH of 7, which means that its concentration of hydrogen ions is expressed by the exponent 7, without its minus sign.

When acids or hydroxyl-containing bases are in water solution they ionize more or less completely, furnishing varying concentrations of H^+ and OH^- ions, respectively, to the solution. Strong acids and bases ionize much more completely than weak acids and bases, thus strong acids give solutions of pH 1–3, while solutions of weak acids have a pH of 6. Strong bases give solutions of pH 12 or 13, while weak bases give solutions of pH 8. As the pH scale is logarithmic, the intervals are exponential, and thus represent far greater differences in concentration than the values themselves seem to indicate. (See table)

	pH Value	Ratio of H^+ or OH^- Concentration to That of Pure Water at 22°C
Acid side (excess of H^+ ions)	1	1,000,000
	2	100,000
	3	10,000
	4	1,000
	5	100
	6	10
Neutrality	7	1
Alkaline side (excess of OH^- ions)	8	10
	9	100
	10	1,000
	11	10,000
	12	100,000
	13	1,000,000

Liquid	pH Value
Pure water	7
Sea water	7.8–8.2
Electroplating bath	6.5–5
0.01 N HCl	2
0.1 N HCl	1.08
0.1 N H_2SO_4	1.17
0.01 N NaOH	12
0.1 N acetic acid	3
0.1 N NH_4OH	11
Gastric juices	1.7
Urine	5–7
Blood	7.3–7.5
Milk	6.5–7
Soil (optimum for crops)	6–7

In acid-base titrations, changes in pH can be detected by indicators such as methyl orange, phenolphthalein, etc. Litmus paper can also be used as a rough indication of acidity or alkalinity. In carrying out titrations the point at which the solution is found to be neutral (pH 7) is not always the correct end point.

pH control is of critical importance in a large number of industrial operations such as water purification, chrome tanning process for leather, in preservation of food products, in electroplating baths, dyeing, agriculture and numerous other instances.
See also acid; base.

phaltan. See folpet.

"Phantolid."[105] TM for a standardized mixture of structural isomers of 1,1,2,3,3,5-hexamethylindan methyl ketone.
Properties: Waxy, white solid with musk-like odor that begins to sinter at 35C and becomes liquid at 40C; soluble in conventional solvents and essential oils.
Use: Perfumery.

pharaoh's serpent eggs. (mercuric thiocyanate). $(NCS)_2Hg$. Swells when heated.

pharmaceutical. A broad term that includes not only all types of drugs and medicinal and curative products but also ancillary products as tonics, dietary supplements, vitamins, deodorants, and the like.
See also drug.

phase. (1) One of the three states or conditions in which substances can exist, i.e., solid, liquid, or gas (vapor). The condition depends primarily on the concentration of atoms of molecules; solids are the most dense, gases the least, and liquids occupy the intermediate position. Solids are normally crystals, liquids are amorphous, and gases are without structure.
See also matter.
(2) A physically distinct and mechanically separable portion of a dispersion or solution. Phases may be either solid, liquid, or gaseous (vapor). In any mixture or solution the major component is called the continuous or external phase and the minor component the dispersed or internal phase. The latter may or may not be uniformly dispersed in the continuous phase.
See colloid chemistry, solution.

phase rule. Propounded by J. Willard Gibbs in 1877, the phase rule is a general system of equations of the form $F = C - P + 2$ stating the boundaries of thermodynamic equilibrium in a system

of chemical reactants. The number of degrees of freedom (F) allowed in a given heterogeneous system may be examined by analysis or observation and plotted on a graph by proper choice of the components (C), the phases (P), and the independently variable factors of temperature and pressure. The principles of the phase rule apply to all multicomponent systems, including solvent blends, glass, alloys, and plastics.

Phe. Abbreviation for phenylalanine.

α-phellandrene. (4-isopropyl-1-methyl-1,5-cyclohexadiene). CAS: 99-83-2.

$CH_3C:CHCH_2CH[CH(CH_3)_2]CH:CH$.
A monocyclic terpene occurring as (a) d- and (b) l-optical isomers.
Properties: Colorless oil; insoluble in water; soluble in ether; (a) d 0.8463 (25C), bp 66–68C (16 mm), refr index 1.4777; (b) d 0.8324 (20C), bp 58–59C (16 mm), refr index 1.4724. Combustible.
Derivation: (a) Found in ginger oil, Ceylon and Seychelles cinnamon oil. (b) Found in eucalyptus oil.
Hazard: Toxic by ingestion and skin absorption; strong irritant.
Use: Flavoring, perfumery.

β-phellandrene. (4-isopropyl-1-methylene-2-cyclohexene). CAS: 555-10-2.

$CH_2C:CH:CHCH[CH(CH_3)_2]CH_2CH_2$.
A monocyclic terpene occurring as (a) d- and (b) l-optical isomers.
Properties: (a) Mobile oil with pleasant odor and a burning taste, d 0.8520 (20C), bp 171–172C, refr index 1.4788; (b) mobile oil, d 0.8497 (15C), bp 178–179, flash p (TCC) 120F (48.9C), refr index 1.4800. Both are insoluble in water and alcohol, soluble in ether. Combustible.

phenacaine hydrochloride. (N,N'-bis(p-ethoxyphenyl)acetamidine hydrochloride). CAS: 620-99-5. $C_{18}H_{22}O_2N_2 \cdot HCl \cdot HOH$.
Properties: Small, white crystals; odorless; faintly bitter taste; incompatible with alkalies; mp 190C; soluble in alcohol, boiling water, and chloroform; less so in cold water; insoluble in ether.
Grade: NF, technical.
Hazard: Toxic by ingestion.
Use: Medicine (topical anesthetic).

phenacetin. USP name for acetophenetidin.

phenacyl chloride. See chloroacetophenone.

phenacyl fluoride. See fluoroacetophenone.

phenanthraquinone. See phenanthrenequinone.

phenanthrene. CAS: 85-01-8. $C_{14}H_{10}$.

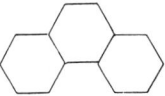

Properties: Colorless, shining crystals; soluble in alcohol, ether, benzene, carbon disulfide, and acetic acid; insoluble in water; d 1.063; mp 100.35C; bp 340C. Combustible.
Derivation: Fractional distillation of high-boiling coal-tar oils with subsequent recrystallization from alcohol.
Hazard: A carcinogen.
Use: Dyestuffs, explosives, synthesis of drugs, biochemical research, manufacturing phenanthrenequinone.

phenanthrenequinone. (not phenanthraquinone). CAS: 84-11-7. $C_{14}H_8O_2$.
Properties: Yellow-orange, needle-like crystals, soluble in sulfuric acid, benzene, glacial acetic acid and hot alcohol; insoluble in water, d 1.4045, mp 206–207C, bp sublimes above 360C.
Derivation: By oxidation of a boiling solution of phenanthrene in glacial acetic acid with chromic acid, soluble in sodium disulfite, precipitation by means of hydrochloric acid, and recrystallization.
Use: Organic synthesis, dyes.

1,10-phenanthroline. (4,5-phenanthroline; o-phenanthroline). CAS: 66-71-7. $C_{12}H_8N_2 \cdot HOH$. A heterotricyclic compound.
Properties: White, crystalline powder; mp 93–94C, anhydrous 117C; slightly soluble in water; soluble in alcohol, benzene, and acetone.
Derivation: Heating o-phenylenediamine with glycerol, nitrobenzene and concentrated sulfuric acid, also from 8-aminoquinoline.
Use: Forms a complex compound with ferrous ions used as an indicator, drier in coatings industry.

phenarsazine chloride. See diphenylaminechloroarsine.

phenazine. (azophenylene). CAS: 92-82-0. $C_6H_4N_2C_6H_4$. A tricyclic compound.
Properties: Yellow crystals, mp 170–171C, bp above 360C, very slightly soluble in water, partially soluble in alcohol and ether. Combustible.
Use: Organic synthesis, manufacturing of dyes, larvicide.

phenethicillin. See potassium-α-phenoxyethyl penicillin.

phenethyl acetate. See 2-phenylethyl acetate.

phenethyl alcohol. (phenylethyl alcohol; 2-phenylethanol; benzyl carbinol). CAS: 60-12-8. $C_6H_5CH_2CH_2OH$.
Properties: Colorless liquid, floral odor, sharp burning taste, d 1.017–1.020 (25C), refr index 1.5310–1.5340 (20C), fp −27C, bp 219C, flash p 216F (102C). Soluble in 50% alcohol; soluble 1 part in 50 parts of water; soluble in fixed oils, alcohol, and glycerol; slightly soluble in mineral oil. Combustible.
Derivation: (a) By reduction of phenylacetic ethyl ester by sodium in absolute alcohol. (b) By the action of ethylene oxide on phenylmagnesium bromide and subsequent hydrolysis. (c) Component of many essential oils.
Grade: Technical, NF, FCC.
Use: Organic synthesis, synthetic rose oil, soaps, flavors, antibacterial, preservative.

sec-**phenethyl alcohol.** See α-methylbenzyl alcohol.

phenethylamine. See 2-phenylethylamine.

phenethyl anthranilate. See 2-phenylethyl anthranilate.

phenethyl isobutyrate. See 2-phenylethyl isobutyrate.

phenethyl phenylacetate. See 2-phenylethyl phenylacetate.

phenethyl propionate. See 2-phenylethyl propionate.

phenethyl salicylate. See 2-phenylethyl salicylate.

o-phenetidine. (2-aminophenetole). CAS: 94-70-2. $NH_2C_6H_4OC_2H_5$.
Properties: Oily liquid, rapidly becomes brown on exposure to light or air, solidifies at about −20C, bp 228–230C, soluble in alcohol and ether, insoluble in water. Combustible.
Derivation: Reduction of o-nitrophenetole with iron filings and hydrochloric acid.
Hazard: Toxic by ingestion, inhalation, and skin absorption.
Use: Manufacture of dyes, laboratory reagent.

p-phenetidine. (4-aminophenetole). CAS: 156-43-4. $NH_2C_6H_4OC_2H_5$.
Properties: Colorless, oily liquid; becomes red to brown on exposure to air and light. D 1.0613 (15C), mp 2–4C, bp 253–255C, insoluble in water, soluble in alcohol. Combustible.

Derivation: Ethylating p-nitrophenol with ethyl sulfate or chloride in presence of sodium hydroxide followed by reduction with iron filings and hydrochloric acid.

Hazard: Toxic by ingestion, inhalation, and skin absorption.

Use: Dyestuffs intermediate, pharmaceuticals, lab reagent.

phenetole. (phenyl ethyl ether).
CAS: 103-73-1. $C_6H_5OC_2H_5$.

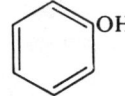

Properties: Colorless, oily liquid. Bp 172C, fp −30C, d 0.967 (20/4C), insoluble in water, soluble in alcohol and ether. Combustible.

phenetsal. See p-acetylaminophenyl salicylate.

"Phenex."[94] TM for α-ethyl-β-propylacrylaniline.

Use: Accelerator for natural and synthetic rubber and latexes.

phenic acid See phenol.

phenindione. (2-phenyl-1,3-indanedione).
CAS: 83-12-5. $C_{15}H_{10}O_2$.
Properties: Pale yellow crystals; practically odorless; insoluble in water; soluble in methanol, alcohol, ether, acetone, benzene; solutions in alkalies are red; in concentrated sulfuric acid blue.
Use: Medicine (blood anticoagulant).

pheniramine maleate. (prophenpyridamine maleate; 1-phenyl-1-(2-pyridyl)-3-di-methylaminopropane maleate). CAS: 132-20-7.
$C_{16}H_{20}N_2 \cdot C_4H_4O_4$.
Properties: White, crystalline powder with faint amine-like odor; mp 104–108C. Very soluble in alcohol and water, slightly soluble in benzene and ether, 1% solution has pH between 4.5 and 5.5.
Grade: NF.
Use: Medicine (antihistamine).

"Phenmad."[329] TM for a 10% phenylmercuric acetate aqueous solution.
Use: Turf fungicide.
Hazard: Highly toxic.

phenobarbital. (phenylbarbital; phenylethylmalonylurea; 5-ethyl-5-phenylbarbituric acid).
CAS: 50-06-6. $C_{12}H_{12}N_2O_3$.
Properties: White, shining, crystalline powder; odorless; stable; mp 174–178C; soluble in alcohol, ether, chloroform, alkali hydroxides, alkali carbonate solutions; sparingly soluble in water.

Derivation: Condensation of phenylethylmalonic acid derivatives and urea.
Grade: USP.
Hazard: May have damaging side effects. See barbituate.
Use: Medicine (sedative), laboratory reagent. Also available as the sodium salt which has good water-solubility.

phenocoll hydrochloride. (aminoacetophenetidide hydrochloride; glycocoll-p-phenetidine hydrochloride).
$C_2H_5OC_6H_4NHCOCH_2NH_2 \cdot HCl$.
Properties: Fine, white, crystalline powder; soluble in water and warm alcohol; slightly soluble in chloroform, ether, and benzene; mp 95C.
Derivation: By the action of aminoacetic acid upon phenetidine and acidifying.
Use: Medicine (analgesic).

phenol. (1) A class of aromatic organic compounds in which one or more hydroxy groups are attached directly to the benzene ring. Examples are phenol itself (benzophenol), the cresols, xylenols, resorcinol, naphthols. Though technically alcohols, their properties are quite different.

(2) phenol. (carbolic acid; phenylic acid; benzophenol; hydroxybenzene). CAS: 108-95-2. C_6H_5OH.

34th highest-volume chemical produced in US (1985).

Properties: White, crystalline mass which turns pink or red if not perfectly pure or if under influence of light; absorbs water from the air and liquefies; distinctive odor; sharp burning taste. When in very weak solution it has a sweetish taste. D 1.07, mp 42.5–43C, bp 182C, flash p (CC) 172.4F (78C). Soluble in alcohol, water, ether, chloroform, glycerol, carbon disulfide, petrolatum, fixed or volatile oils, and alkalies. Autoign temperature 1319F (715C). Combustible.

Derivation: Most of the phenol in the US is made by the oxidation of cumene, yielding acetone as a byproduct. The first step in the reaction yields cumene hydroperoxide, which decomposes with dilute sulfuric acid to the primary products, plus acetophenone and phenyl dimethyl carbinol. Several other benzene-based processes have been used in the past; derivation from benzoic acid is also possible.

Method of purification: Rectification.

Grade: Fused, crystals, or liquid, all as technical

(82%, 90%, 95%, other components mostly cresols), CP, and USP.

Hazard: Toxic by ingestion, inhalation, and skin absorption; strong irritant to tissue. TLV: 5 ppm in air.

Use: Phenolic resins, epoxy resins (bisphenol-A) nylon-6 (caprolactam), 2,4-D, selective solvent for refining lubricating oils, adipic acid, salicylic acid, phenolphthalein, pentachlorophenol, acetophenetidine, picric acid, germicidal paints, pharmaceuticals, laboratory reagent, dyes and indicators, slimicide, biocide, general disinfectant.

Note: High-boiling phenols are mixtures containing predominantly m-substituted alkyl phenols. Their boiling points range from 238 to 288C, they set to a glass at −30C.

Use: In phenolic resins, as fuel-oil sludge inhibitors, as solvents and as rubber chemicals.

phenolate process. A process for removing hydrogen sulfide from gas by the use of sodium phenolate, which reacts with the hydrogen sulfide to give sodium hydrosulfide and phenol. This can be reversed by steam heat to regenerate the sodium phenolate.

phenol coefficient. In determining the effectiveness of a disinfectant using phenol as a standard of comparison, the phenol coefficient is a value obtained by dividing the highest dilution of the test disinfectant by the highest dilution of phenol that sterilizes a given culture of bacteria under standard conditions of time and temperature. See also disinfectant.

phenoldisulfonic acid. CAS: 96-77-5. $C_6H_6O_7S_2$.

Properties: Deliquescent crystals, mp 90C, decomposes above 100C, soluble in water and alcohol.

Use: Manufacture of dye intermediates.

phenol-formaldehyde resin. The first synthetic thermosetting polymer, the reaction product of phenol with aqueous 37–50% formaldehyde at 50–100C, with basic catalyst, discovered by Baekeland in 1907 and trademarked "Bakelite" in 1911. Polymerization is of the condensation type, proceeding through three stages. With an acid catalyst "novolak" resins are produced, which are thermoplastic.

Properties: Gray to black, hard, infusible solid when cured, resistant to moisture, solvents, and heat up to 200C, dimensionally stable, good electrical resistance, noncombustible, sound- and noise-absorbent, decomposed by oxidizing acids, fair resistance to alkalies. Cannot be successfully colored.

Use: Molded and cast articles, bonding powders, ion exchange, laminating and impregnating, plywood and glass-fiber composites, ablative coatings for aerospace use, binder for oil-well sands, paint and baked enamel coatings, thermal and acoustic insulation, brake linings, clutch facings, shell molds, chemical equipment, machine and instrument housings, chemical-resistant mortars, machine parts, electrical devices.

See also A-, B-, and C-stage resin; novolak, Baekeland, phenolic resin.

phenol-furfural resin. A phenolic resin that has a somewhat sharper transition than phenol-formaldehyde from the soft, thermoplastic stage to the cured, infusible state and can be fabricated by injection molding since it has little tendency to harden before curing conditions are reached.

phenolic resin. Any of several types of synthetic thermosetting resin obtained by the condensation of phenol or substituted phenols with aldehydes such as formaldehyde, acetaldehyde, and furfural. Phenol-formaldehyde resins are typical and constitute the chief class of phenolics.

phenolphthalein. (3,3-bis(p-hydroxyphenyl)-phthalide).   CAS: 77-09-8. $(C_6H_4OH)_2C_2O_2C_6H_4$ (an approximation).

Properties: Pale yellow powder; forms an almost colorless solution in neutral or acid solution, pink to deep red in presence of alkali, but colorless in the presence of large amounts of alkali; soluble in alcohol, ether, and alkalies; insoluble in water; d 1.2765; mp 261C.

Derivation: Interaction of phenol and phthalic anhydride in sulfuric acid.

Grade: Technical, pure reagent, NF.

Use: Acid-base indicator, laboratory reagent, medicine (laxative).

phenol red. See phenolsulfonephthalein.

phenolsulfonephthalein. (phenol red). CAS: 143-74-8. $(C_6H_4OH)_2COSO_2C_6H_4$ (an approximation).

Properties: Bright to dark red, crystalline powder; stable in air; slightly soluble in water, alcohol, and acetone; almost insoluble in chloroform and ether; soluble in alkali hydroxides and carbonates.

Derivation: Reaction of phenol with o-sulfobenzoic acid anhydride. Differs from phenolphthalein in containing an SO_2 group in place of a CO group.

Grade: Technical, reagent, USP. The USP spelling is phenolsulfonphthalein.

Use: Acid-base indicator, diagnostic reagent in medicine, laboratory reagent.

phenolsulfonic acid. (sulfocarbolic acid). $HOC_6H_4SO_3H$.
Properties: Yellowish liquid, becoming brown on exposure to air. A mixture of o- and p-phenolsulfonic acids; soluble in water and alcohol.
Derivation: Action of sulfuric acid on phenol.
Grade: Technical, reagent.
Hazard: Irritant to skin and tissue.
Use: Water analysis, laboratory reagent, electroplated tin coating baths, manufacture of intermediates and dyes, pharmaceuticals.

phenol trinitrate. See picric acid.

phenosafranin. (3.7-diamino-5-phenylphenazinium chloride; CI 50200). CAS: 81-93-6. $C_{18}H_{15}ClN_4$.
Properties: Green, acicular crystals; soluble in water and alcohol; solutions have purple to red color.
Use: Staining tissues for microscopy.

phenothiazine. (thiodiphenylamine).
CAS: 92-84-2.　　$C_{12}H_9NS$.
Properties: Grayish-green to greenish-yellow powder, granules, or flakes; tasteless; slight odor; insoluble in water; soluble in benzene, ether, hot acetic acid; mp 175–185C; bp 371C; sublimes 130C (1 mm).
Derivation: By reaction of diphenylamine and sulfur in presence of an oxidizing catalyst.
Grade: Technical, NF.
Hazard: Toxic by ingestion and inhalation, absorbed by skin. TLV: 5 mg/m³ of air.
Use: Insecticide, manufacture of dyes, parent compound for chlorpromazine and related antipsychotic drugs, polymerization inhibitor, antioxidant.

phenothrin [(3-phenoxyphenyl)methyl-2,2-dimethyl-3-(2-methyl-1-propenyl)-cyclopropane carboxylate].　　CAS: 26002-80-2.
Use: Insecticide.

phenoxyacetic acid. CAS: 122-59-8. $C_6H_5OCH_2COOH$.
Properties: Light tan powder, bp 285C, mp 98C. Soluble in ether, water, alcohol, carbon disulfide, glacial acetic acid. Combustible.
Use: Intermediate for dyes, pharmaceuticals, pesticides, other organics, fungicides, flavoring, laboratory reagent, precursor in antibiotic fermentations, especially penicillin V.

phenoxydihydroxypropane. See phenoxypropanediol.

2-phenoxyethanol. See ethylene glycol monophenyl ether.

α-phenoxyethylpenicillin. See penicillin.

phenoxymethylpenicillin. (penicillin V). $C_{16}H_{18}N_2O_5S$.
Properties: White, odorless, crystalline powder; very slightly soluble in water; soluble in alcohol and acetone; insoluble in fixed oils, pH of saturated solution is 2.5–4.0, decomposes at 120C.
Grade: NF.
Use: Medicine (antibiotic for oral use). Available also as potassium salt.

phenoxypropanediol. (1-phenoxypropanediol-2,3).　　$C_6H_5OCH_2CHOHCH_2OH$.
Properties: White, crystalline solid; mp 53C; bp 150–155C (4 mm); soluble in water, alcohol, glycerol, carbon tetrachloride, warm benzene; insoluble in gasoline. Combustible.
Derivation: Phenol and glycerol.
Use: Plasticizer, resins, lacquers.

phenoxypropylene oxide. CAS: 122-60-1.

$C_6H_5OCH_2CHCH_2O$.
Properties: Practically colorless liquid with characteristic odor, very slightly soluble in water, d 1.1110 (20/20C), bp 244.2C, vap press less than 0.1 mm (20C), fp 2.8C, viscosity 6.93 cp (20C).

phenoxy resin. A high molecular weight thermoplastic copolymer of bisphenol A and epichlorohydrin having the basic molecular structure $—[OC_6H_4C(CH_3)_2C_6H_4OCH_2CH(OH)CH_2]_n—$ (n is about 100). It uses the same raw materials as epoxy resins, but contains no epoxy groups. It may be cured by reacting with polyisocyanates, anhydrides, or other crosslinking agent capable of reacting with hydroxyl groups.
　　The ductility of phenoxy resins resembles that of metals. They are transparent and are also characterized by low mold shrinkage, good dimensional stability, and moderate good resistance to temperature and corrosion. Phenoxy resins are soluble in methyl ethyl ketone and have been used for coatings and adhesives. A typical injection-molded specimen has a tensile strength of 9000 psi, heat distortion point 86.6C at 264 psi load, and d 1.18.
　　They may be extruded or blow-molded. Parts may be thermally formed and heat- or solvent-welded. Some applications are blow-molded containers, pipe, ventilating ducts, and molded parts.

phenyl. The univalent $C_6H_5^+$ group derived from benzene and characteristic of phenol and other derivatives.
See following entries.

phenylacetaldehyde. (α-toluic aldehyde). CAS: 122-78-1.　　$C_6H_5CH_2CHO$.

Properties: Colorless liquid, strong hyacinth-like odor, soluble in 2 parts of 80% alcohol, soluble in ether and most fixed oils, slightly soluble in water, d 1.023–1.030 (25C), fp −10C, bp 193–194C, refr index 1.520–1.530, becomes more viscous on aging. Combustible.
Derivation: From phenyl-α-chloroacetic acid, by action of alkalies, oxidation of phenylethyl alcohol.
Grade: Technical, 50% soluble in benzyl alcohol, FCC.
Use: Perfumes, flavoring, laboratory reagent.

phenylacetaldehyde dimethylacetal. (α-tolyl aldehyde dimethyl acetal).   CAS: 101-48-4. $C_6H_5CH_2CH(OCH_3)_2$.
Properties: Colorless liquid, strong odor, more stable than phenylacetaldehyde, not known to cause discoloration, d 1.000–1.004 (25/25C), refr index 1.493–1.496 (20C), soluble in 2 parts of 70% alcohol, flash p (TCC) 191F (88.3C). Combustible.
Grade: Technical, FCC.
Use: Perfumery, flavoring, laboratory reagent.

phenylacetamide. (α-toluamide).
CAS: 103-81-1. $C_6H_5CH_2CONH_2$.
Properties: White crystals, bp 280–290C (decomposes), mp 156–160C, soluble in hot water and alcohol, slightly soluble in cold water and ether. Combustible.
Derivation: From acetophenone or styrene by Willgerodt reaction, dehydration of ammonium phenyl acetate.
Use: Organic synthesis, pharmaceuticals, penicillin G precursor, laboratory reagent.

N-phenylacetamide.   See acetanilide.

phenyl acetate. (acetylphenol).
CAS: 122-79-2. $C_6H_5OOCCH_3$.

OCOCH$_3$

Properties: Water-white liquid, soluble in alcohol and ether, almost insoluble in water, d 1.073 (25/25C), bp 195–196C, flash p 176F (80C). Combustible.
Derivation: (a) From phenol and acetyl chloride. (b) By heating triphenyl phosphate with potassium acetate and alcohol.
Use: Solvent, organic synthesis, laboratory reagent.

phenylacetic acid. (α-toluic acid).
CAS: 103-82-2. $C_6H_5CH_2COOH$.

Properties: Shiny, white plate crystals. D 1.0809, floral odor, fp 76–78C, bp 262C. Soluble in alcohol, ether and hot water. Combustible.
Derivation: From benzyl cyanide refluxed with dilute hydrochloric acid.
Grade: Technical, FCC.
Use: Perfume, precursor in manufacture of penicillin G, fungicide, flavoring, laboratory reagent.

phenylacetonitrile.   See benzyl cyanide.

α-phenylacetophenone.   See deoxybenzoin.

phenylacetyl chloride.   CAS: 103-80-0.
$C_6H_5CH_2COCl$.
Properties: Colorless liquid, refr index 1.5320 (20C).
Hazard: Strong irritant.
Use: Acylating agent, including manufacture of esters for flavors, laboratory reagent.

β-phenylacrylic acid.   See cinnamic acid.

phenylacrylyl chloride.   See cinnamoyl chloride.

phenylalanine. (α-amino-β-phenylpropionic acid).   CAS: 63-91-2.
$C_6H_5CH_2CH(CH_2)COOH$.   An essential amino acid.
Properties: L(-)-phenylalanine: Plates and leaflets from concentrated aqueous solutions, hydrated needles from dilute aqueous solutions, decomposes 283C, soluble in water, slightly soluble in methanol and ethanol. D(+)-phenylalanine: Leaflets from water, decomposes 285C, soluble in water, slightly soluble in methanol. DL-phenylalanine: Leaflets or prisms from water or alcohol, sweet tasting, decomposes 318–320C, soluble in water.
Sources: (L-)Phenylalanine is isolated commercially from proteins (ovalbumin, lactalbumin, zein and fibrin). DL-Phenylalanine is synthesized from α-acetaminocinnamic acid.
Grade: Technical, FCC.
Use: (L isomer): medicine and nutrition, essential ingredients of Aspartame. Available commercially as DL-dihydroxyphenylalanine and as DL-phenylalanine.

phenylallylic alcohol.   See cinnamic alcohol.

phenylamine.   See aniline.

1-phenyl-2-aminopropane.   See amphetamine.

N-phenylaniline.   See diphenylamine.

o-phenylaniline.   See o-aminobiphenyl.

phenylarsonic acid.   CAS: 98-05-5.
$C_6H_5AsO(OH)_2$.
Properties: Crystalline powder, mp 160C with de-

composition, soluble in water and alcohol, insoluble in chloroform. Preparation: Reaction of the diazonium salt and sodium arsenite.
Hazard: Toxic by ingestion.
Use: Analytical reagent for tin.

phenylazoaniline. See aminoazobenzene.

1-(phenylazo)-2-naphthylamine. See yellow AB.

phenylbarbital. See phenobarbital.

phenylbenzamide. See benzanilide.

3-phenyl-3-benzoborepin. A synthetic carbon-boron heterocyclic compound with aromatic properties, stable in air. Key intermediate in its synthesis is the nonaromatic carbon-tin heterocycle 3,3-dimethyl-3-benzostannepin.

1,3,4-phenylbiphenylyloxadiazole. (PBD; phenylbiphenyloxadiazole). $C_{20}H_{14}N_2O$.
Properties: Crystals, mp 166–168C.
Grade: Purified.
Use: As primary fluors or as wavelength shifters in soluble scintillators.

phenylbis[1-(2-methyl)aziridinyl]phosphine oxide. $C_6H_5(C_3H_6N)_2PO$.
Properties: (technical) Straw-colored liquid, limited solubility in water, soluble in most organic solvents. The pure material is a low-melting solid. Combustible.
Use: Polymerization initiator.

phenyl bromide. See bromobenzene.

1-phenylbutane. See n-butylbenzene.

2-phenylbutane. See sec-butylbenzene.

phenylbutazone. (4-n-butyl-1,2-diphenyl-3,5-pyrazolidinedione). CAS: 50-33-9.
$C_{19}H_{20}N_2O_2$. A synthetic pyrazolone derivative, also occurs naturally in certain herbs.
Properties: White or very light yellow powder; slightly bitter taste and very slight aromatic odor; mp 103–106C; freely soluble in acetone, ether, and ethyl acetate; very slightly soluble in water; stable if stored at room temperature in closed containers in absence of moisture. Also available as sodium salt.
Grade: NF.
Hazard: May have harmful side-effects.
Use: Medicine, anti-inflammatory drug licensed for human use, veterinary medicine.

1-phenylbutene-2. $C_6H_5CH_2CH:CHCH_3$.
Properties: Colorless liquid, boiling range 174–176C, d 0.888 (15.5C), refr index 1.511 (20C), flash p (TOC) 160F (71.1C). Combustible.

Grade: Technical, 95 mole %.
Use: Organic synthesis, laboratory reagent.

phenylbutynol. (3-phenyl-1-butyn-3-ol).
$HC:CC(C_6H_5)(OH)CH_3$.
Properties: Crystals, camphor odor, mp 51–52C, bp 217–218C, d 1.0924 (20/20C). Slightly soluble in water; soluble in acetone, benzene, most organic solvents. Combustible.
Use: Acid inhibitor, organic synthesis.

2-phenylbutyric acid. See phenylethylacetic acid.

1-phenyl-3-carbethoxypyrazolone-5.
$C_{12}H_{12}N_2O_3$.
Properties: White to light buff powder, stable in aqueous solution, mp 182–188C. Combustible.
Use: Dyestuff intermediate, laboratory reagent.

phenyl carbimide. See phenyl isocyanate.

phenylcarbinol. See benzyl alcohol.

phenyl carbonate. CAS: 102-09-0.
$C_6H_5OCOOC_6H_5$.
Properties: Shiny crystals, mp 80C, bp 300C, insoluble in water, soluble in hot alcohol, benzene.
Use: Solvent for nitrocellulose.

phenylcarbylamine chloride. CAS: 622-44-6.
$C_6H_5NCCl_2$.
Properties: Pale yellow, oily liquid; onion-like odor; mildly volatile; soluble in alcohol, benzene, ether; insoluble in water; d 1.30 at 15C; bp 208–210C.
Derivation: Chlorination of phenylisothiocyanate.
Grade: Technical.
Hazard: Toxic by inhalation and skin absorption, strong irritant to skin and mucous membranes.
Use: Organic synthesis, military poison.

phenyl "Cellosolve."[214] TM for ethylene glycol monophenyl ether.

phenyl chloride. See chlorobenzene.

phenylchloroform. See benzotrichloride.

phenyl chloromethyl ketone. See chloroacetophenone.

2-phenyl-6-chlorophenol. CAS: 85-97-2.
$C_{12}H_9ClO$.
Properties: Yellow, oily liquid; slight odor. Mp 6C, bp 318C (decomposes), d 1.24, insoluble in water, soluble in organic solvents.
Use: Fungicide, disinfectant.

1-phenyl-3-chloropropane. See phenylpropyl chloride.

phenyl cyanide. See benzonitrile.

phenylcyclidene hydrochloride. (PCP).
$C_{17}H_{25}N \cdot HCl$.
Properties: White crystals, mp 243C.
Hazard: Toxic by ingestion and inhalation. A psychotropic drug which may cause serious mental disorders.
Use: Veterinary anesthetic (manufacture legally restricted to 5 kg annually).

phenylcyclohexane. (cyclohexylbenzene).
CAS: 827-52-1. $C_6H_5C_6H_{11}$.
Properties: Colorless, oily liquid; pleasant odor. D 0.938 (25/25C), mp 5C, bp 237.5C, refr index 1.523 (25C), flash p 210F (98C). Insoluble in water, glycerol; very soluble in alcohol, acetone, benzene, carbon tetrachloride, castor oil, hexane, xylene. Combustible.
Use: High-boiling solvent, penetrating agent, intermediate, laboratory reagent.

2-phenycyclohexanol.

$HOCH(CH_2)_4CH(C_6H_5)$.
Properties: Colorless to pale, straw-colored liquid, pour point −18C, bp 276–281C, d 1.033 (25/25C), refr index 1.536 (25C), flash p 280F (137C), very slightly soluble in water, soluble in methanol, ether. Combustible.
Use: Solvent, intermediate.

N-phenyl-N'-cyclohexyl-p-phenylenediamine.
$C_6H_5HNC_6H_4NHC_6H_{11}$.
Properties: Gray-violet powder, d 1.16, mp 103–107C. Soluble in acetone, benzene, MEK, ethyl acetate, ethylene dichloride; insoluble in water.
Use: Antioxidant-antiozonant for elastomers.

phenyldichloroarsine. CAS: 696-28-6.
$C_6H_5AsCl_2$.
Properties: Liquid, microcrystalline mass at the fp, decomposed by water. Soluble in alcohol, benzene, and ether; insoluble in water. D 1.654 (20C), bp 255–257C, fp −20C, vapor tension 0.014 mm (15C), volatility 404 mg/m³ (20C), coefficient of thermal expansion 0.00073.
Derivation: Arsenic trichloride and phenylmercuric chloride are heated together at 100C.
Grade: Technical.
Hazard: Toxic by ingestion and inhalation, strong irritant to eyes, skin, and tissue.
Use: Lachrymator poison gas, solvent for diphenylcyanoarsine.

phenyldidecyl phosphite. $C_6H_5OP(OC_{10}H_{21})_2$.
Properties: Nearly water-white liquid with an odor of alcohol, d 0.940 (25/15.5C), mp 0C, refr index 1.4785 (25C), flash p (COC) 425F (218C). Combustible.
Use: Chemical intermediate, antioxidant, ingredient in stabilizer systems for resins.

phenyl diethanolamine. CAS: 120-07-0.
$C_6H_5N(C_2H_4OH)_2$.
Properties: Colorless liquid, mp 58C, bp 190C (1 mm), vap press less than 0.01 mm (20C), bulk d 10.0 lb/gal (20C), d 1.1203 at 60/20C, viscosity 1.19 poise (20C), slightly soluble in water, soluble in ethanol and acetone, flash p (OC) 375F (190C). Combustible.
Grade: Technical.
Use: Organic synthesis, dyestuff, laboratory reagent.

phenyl diglycol carbonate. [diethyl glycol bis-(phenyl carbonate)].
$(C_6H_5OOCOCH_2CH_2)_2O$.
Properties: Colorless solid, d 1.23 (20/4C), mp 40C, bp 225–229C, refr index 1.525 (20C), evaporation rate 0.026 mg/sq cm/hr at 100C, insoluble in water (very stable to hydrolysis), widely soluble in organic solvents, compatible with many resins and plastics. Combustible.
Use: Plasticizer.

N-phenyl-N'-(1,3-dimethyl butyl)-p-phenylenediamine. $C_6H_5HNC_6H_4NHC_6H_{13}$.
Properties: Dark violet, staining, low-melting solid, mp 50C, d 1.07.
Use: Antiozonant, antioxidant, and polymer stabilizer.

3-phenyl-1,1-dimethylurea. See fenuron.

m-phenylenediamine. (m-diaminobenzene).
CAS: 108-45-2. $C_6H_4(NH_2)_2$.
Properties: Colorless needles becoming red in air, usually in the form of the stable hydrochloride, d 1.1389, mp 63C, bp 282–287C. Soluble in alcohol, ether, and water.
Derivation: Reduction of m-dinitrobenzene or nitroaniline with iron and hydrochloric acid; purified by crystallization.
Grade: Technical, 99% min purity.
Hazard: Toxic by ingestion and inhalation, strong irritant to skin.
Use: Dyestuff manufacture, detection of nitrite, textile developing agent, laboratory reagent, vulcanizing agent, ion-exchange resins, block polymers, corrosion inhibitors, photography.

o-phenylenediamine. (orthamine; o-diaminobenzene).   CAS: 95-54-5. $C_6H_4(NH_2)_2$.
Properties: Colorless, monoclinic crystals; darkens in air. Mp range 102–104C, bp 252–258C, soluble in alcohol, ether, and chloroform.
Derivation: Reduction of o-dinitrobenzene or nitoraniline with iron and hydrochloric acid. Purified by crystallization.
Grade: Technical, 99.0% min purity.
Use: Manufacture of dyes, photographic developing agent, organic synthesis, laboratory reagent.

p-**phenylenediamine.** (p-diaminobenzene).
CAS: 106-50-3. $C_6H_4(NH_2)_2$.
Properties: White to light purple crystals (oxidizes on standing in air to purple and black), mp 147C, bp 267C. Soluble in alcohol, ether and 100 parts water; affected by light; flash p 312F (155C). Combustible.
Derivation: Reduction of p-dinitrobenzene or nitroaniline with iron and hydrochloric acid. Purified by crystallization.
Grade: Technical, 99% min purity.
Hazard: Toxic by ingestion and inhalation, strong irritant to skin. TLV: (vapor) 0.1 mg/m³ of air.
Use: Azo dye intermediate, photographic developing agent, photochemical measurements, intermediate in manufacture of antioxidants and accelerators for rubber, synthetic fibers, laboratory reagent, dyeing hair and fur.

phenylethane. See ethylbenzene.

2-**phenylethanol.** See phenethyl alcohol.

phenylethanolamine. CAS: 7568-93-6.
$C_6H_5NHCH_2CH_2OH$.
Properties: Yellow crystals, mp 56C, bp 157C (17 mm), soluble in water.
Use: Shortstopping agent for SBR, intermediate, wax hardener.

phenyl ether. See diphenyl oxide.

2-**phenylethyl acetate.** (phenethyl acetate).
CAS: 103-45-7. $C_6H_5CH_2CH_2OOCCH_3$.
(Not the same as sec-phenylethyl acetate).
Properties: Colorless liquid, peach-like odor. Soluble in alcohol, ether, and most fixed oils. D 1.030–1.034, refr index 1.497–1.501 (20C), bp 226C, flash p 230F (110C). Combustible.
Derivation: (a) Interaction of ethyl acetate and aluminumphenyl ethylate. (b) Interaction of acetic anhydride and phenylethyl alcohol in the presence of sodium acetate.
Grade: Technical, FCC.
Use: Perfumery, laboratory reagent.

sec-**phenylethyl acetate.** See α-methylbenzyl acetate.

phenylethylacetic acid. (2-phenylbutyric acid).
$C_2H_5CHC_6H_5COOH$.
Properties: White crystals with aromatic odor, mp 41.0C (min). Insoluble in water; soluble in alcohol, ketones, and esters. Combustible.
Use: Organic synthesis, lab reagent.

2-**phenylethyl alcohol.** See phenethyl alcohol.

2-**phenylethylamine.** (phenethylamine; 1-amino-2-phenylethane). CAS: 64-04-0.
$C_6H_5C_2NH_2$.

Properties: Liquid with a fishy odor, absorbs carbon dioxide from the air, strong base, d 0.9640, bp 194.5C. Soluble in water, alcohol, and ether. Combustible.
Derivation: From phenylethyl alcohol and ammonia under pressure.
Grade: Technical, scintillation.
Hazard: Toxic by ingestion, skin irritant.
Use: Organic synthesis, lab reagent, scintillation counter (CO_2 absorber).

2-**phenylethyl anthranilate.** (phenethyl anthranilate). CAS: 133-18-6.
$H_2NC_6H_4COOC_2H_4C_6H_5$.
Properties: Colorless liquid which yellows with age, odor of grape and orange, d 1.14 (25/25C). Combustible.
Use: Perfumes, flavoring.

phenylethyl carbinol. See phenylpropyl alcohol.

phenylethylene. See styrene.

phenylethylene glycol. See styrene glycol.

N-phenylethylethanolamine.
$C_6H_5N(C_2H_5)C_2H_4OH$.
Properties: Solid, mp 37.2C, bp 268C (740 mm), d 1.04 (20/20C), very slightly soluble in water, flash p (COC) 270F (132C), soluble in alcohol, acetone, benzene. Combustible.
Use: Solvents, chemical intermediates, preparation of dyes for acetate rayons, laboratory reagent.

phenyl ethyl ether. See phenetole.

2-**phenylethyl isobutyrate.** (phenethyl isobutyrate). CAS: 103-48-0.
$(CH_3)_2CHCOOC_2H_4C_6H_5$.
Properties: Colorless liquid, pleasant odor, d 0.988 (25/25C), refr index 1.488 (20C), soluble in alcohol and ether. Combustible.
Use: Perfumes, flavoring.

phenylethylmalonylurea. See phenobarbital.

2-**phenylethyl mercaptan.** $C_6H_5CH_2CH_2SH$.
Properties: Liquid, boiling range 193–225C, unpleasant odor, d 1.0264 (15.5C), refr index 1.5582 (20C), flash p 160F (71.1C). Combustible.
Use: Organic synthesis, laboratory reagent.

2-**phenylethyl phenylacetate.** (phenethyl phenylacetate). CAS: 102-20-5.
$C_6H_5(CH_2)_2OOCCH_2C_6H_5$.
Properties: White crystals, hyacinth odor, d 1.080–1.082, congealing p 27C. Combustible.
Use: Perfumery, flavors.

2-**phenylethyl propionate.** (phenethyl propionate). CAS: 120-45-6.
$C_2H_5COOC_2H_4C_6H_5$.

Properties: Synthetic, colorless liquid; flower-fruit odor; miscible with alcohols and ether; d 1.012 (25/25C). Combustible.
Use: Perfumes, flavors.

2-phenylethyl salicylate. (phenethyl salicylate). $C_6H_5C_2H_4OOCC_6H_4OH$.
Properties: White crystals, faint aromatic odor, soluble in 14 parts of 95% alcohol, congealing point 41.5C. Combustible.
Use: Flavors.

phenyl ferrocenyl ketone. See benzoylferrocene.

phenyl fluoride. See fluorobenzene.

phenyl fluoromethyl ketone. See fluoroacetophenone.

phenylformamide. See formanilide.

phenylformic acid. See benzoic acid.

phenyl α-acid. (phenyl-2-amino-8-naphthol-6-sulfonic acid). $HOC_{10}H_5(NHC_6H_5)(SO_3H)$.
Properties: Gray crystals, soluble in alkali. Combustible.
Derivation: From α-acid and aniline (condensation with heat).
Use: Azo dye manufacture.

phenyl glycidyl ether. (1,2-epoxy-3-phenoxypropane; PGE). CAS: 122-60-1. $H_2COCHCH_2OC_6H_5$.
Properties: Colorless liquid, d 1.11, bp 245C, mp 3.5C.
Hazard: Toxic by skin absorption, moderate irritant to eyes and skin. TLV: 1 ppm in air.

phenylglycol. See styrene glycol.

phenylglycolic acid. See mandelic acid.

phenylhydrazine. CAS: 100-63-0. $C_6H_5NHNH_2$.
Properties: Colorless liquid, becomes red-brown on exposure to air. Soluble in alcohol, ether, chloroform, benzene, and dilute acids; slightly soluble in water. D 1.0978, mp 19.35C, bp 243.5C with decomposition, flash p (CC) 192F (88.9C), autoign temperature 345F (173C). Also available as the hydrochloride. Combustible.
Derivation: Reduction of diazotized aniline, followed by reaction with sodium hydroxide.
Grade: Commercial, CP, reagent.
Hazard: Toxic by inhalation, ingestion, and skin absorption; attacks red blood cells. A carcinogen. TLV: 5 ppm in air.
Use: Analytical chemistry (reagent for detecting aldehydes, sugars, etc.), organic synthesis (inter-

mediates, dyestuffs, pharmaceuticals). The hydrochloride is a strong reducing agent.

α-phenylhydroxyacetic acid. See mandelic acid.

phenylic acid. See phenol.

phenylindan dicarboxylic acid. See 1,1,3-trimethyl-5-carboxy-3-(p-carboxyphenyl)indan.

phenyl isocyanate. (phenyl carbimide; carbanil). CAS: 103-71-9. $C_6H_5N:C:O$.
Properties: Liquid, bp 165C, fp −30C, d 1.095 (20/4C), refr index 1.53684 (19.6C), decomposes in water and alcohol, very soluble in ether, flash p (TOC) 132F (55.5C). Combustible.
Hazard: Moderate fire risk. Strong irritant, a lachrymator.
Use: Reagent for identifying alcohols and amines, intermediate.

phenyl isothiocyanate. See phenyl mustard oil.

phenyl J acid. (phenyl-2-amino-5-naphthol-7-sulfonic acid). $HOC_{10}H_5(NHC_6H_5)(SO_3H)$.
Properties: Slate-colored crystals, soluble in alkali. Combustible.
Derivation: From hydrogen acid and aniline (condensation with heat).
Hazard: Toxic by ingestion.
Use: Azo dye manufacturing.

phenyllithium. C_6H_5Li. Available in a 20% by weight solution of a 70:30 volume percentage benzene-ether mixture.
Hazard: Flammable, dangerous fire risk.
Use: Organic synthesis.

phenylmagnesium bromide. C_6H_5MgBr.
A grignard reagent available as a solution in ether, d 1.14.
Derivation: From magnesium and bromobenzene.
Hazard: Dangerous fire risk, solution highly flammable.
Use: Arylating agent in organic synthesis.

phenylmagnesium chloride. CAS: 100-59-4. C_6H_5MgCl. Available dissolved in tetrahydrofuran.
Hazard: Dangerous fire risk, solution highly flammable.
Use: Grignard reagent, organic synthesis.

phenyl mercaptan. CAS: 108-98-5. C_6H_5SH.
Properties: Water-white liquid, offensive odor, oxidizes on exposure to air, supplied under nitrogen atmosphere, d 1.080 (15.5/15.5C), refr index 1.5815 (25C), fp −15C, bp 169C, insoluble in water, very soluble in aromatic and aliphatic hydrocarbons.
Hazard: Store out of contact with air and acids.

Toxic by inhalation. TLV: 0.5 ppm in air.
Use: Chemical intermediate, mosquito larvicide.

phenylmercuric acetate. CAS: 62-38-4.
$C_6H_5HgOCOCH_3$.
Properties: White to cream prisms, mp 148–150C. Slightly soluble in water; soluble in alcohol, benzene, and glacial acetic acid; slightly volatile at ordinary temperatures.
Derivation: Action of heat on benzene and mercuric acetate.
Grade: CP, technical, commercial.
Hazard: Toxic by ingestion, inhalation, and skin absorption; strong irritant.
Use: Fungicide, herbicide; mildewcide for paints, slimicide in paper mills.

phenylmercuric benzoate. $C_6H_5HgOOCC_6H_5$.
Properties: Crystals, mp 94C (min).
Hazard: Toxic by ingestion, inhalation and skin absorption.
Use: Bactericide, fungicide, alcohol denaturant.

phenylmercuric borate. CAS: 102-98-7.
$(C_5H_5Hg)_2HBO_3$.
Properties: White, crystalline powder; mp 120–130C; slightly soluble in water; soluble in alcohol.
Derivation: Reaction of phenylmercuric acetate with boric acid.
Hazard: Toxic by ingestion, inhalation, and skin absorption.
Use: Fungicide, bactericide.

phenylmercuric chloride. CAS: 100-56-1.
C_6H_5HgCl.
Properties: White, satiny crystals; mp 251C; insoluble in water; slightly soluble in hot alcohol; soluble in benzene, ether, pyridine.
Derivation: Reaction of phenylmercuric acetate and sodium chloride.
Hazard: Toxic by ingestion, inhalation, and skin absorption.
Use: Antiseptic, fungicide, germicide.

phenylmercuric hydroxide. CAS: 100-57-2.
C_6H_5HgOH.
Properties: Fine white to cream crystals, mp 197–205C, slightly soluble in water, soluble in acetic acid and alcohol.
Grade: Technical, pure.
Hazard: Toxic by ingestion, inhalation, and skin absorption.
Use: Manufacture of phenylmercuric salts, fungicide and germicide. Principal compound in manufacturing organic mercury derivatives, denaturant for alcohol.

phenylmercuric lactate.
$C_6H_5HgOOCCHOHCH_3$.

Hazard: Toxic by ingestion, inhalation, and skin absorption.
Use: Bactericide, fungicide.

phenylmercuric naphthenate. Prepared by interaction of phenylmercuric acetate and naphthenic acid, producing a colored solution.
Hazard: Toxic by ingestion, inhalation, and skin absorption.
Use: Wood preservative, mildew proofing agent.

phenylmercuric nitrate (basic). CAS: 8003-05-2.
$C_6H_5HgNO_3 \cdot C_6H_5HgOH$.
Properties: Fine white crystals or grayish powder, mercury content 63–65% (theoretical 63.2%), melting range 175–185C with decomposition, ash 0.1% max, very slightly soluble in water and alcohol, insoluble in ether, moderately soluble in glycerol.
Grade: NF.
Hazard: Toxic by ingestion, inhalation, and skin absorption.
Use: Germicide, fungicide.

phenylmercuric oleate.
$C_6H_5HgOOC(CH_2)_7CH:CHC_8H_{17}$.
Properties: White, crystalline powder; mp 45C. Insoluble in water, soluble in organic solvents and some oils.
Derivation: Reaction of phenylmercuric acetate with oleic acid.
Hazard: Toxic by ingestion, inhalation, and skin absorption.
Use: Mildewproofing agent for paints, fungicide and germicide.

phenylmercuric propionate. CAS: 103-27-5.
$C_6H_5HgOCOCH_2CH_3$.
Properties: Technical grade, white to off-white wax-like free flowing powder, mp 65–70C, stable to 200C for short periods, 57% min mercury content.
Hazard: Toxic by ingestion, inhalation, and skin absorption.
Use: Fungicide and bactericide for paints and industrial finishes.

phenylmercuric salicylate.
$C_6H_4(OH)(COOHgC_6H_5)$.
Hazard: Toxic by ingestion, inhalation, and skin absorption.
Use: Seed disinfectant.

phenylmercuriethanolammonium acetate.
$[(HOC_2H_4)NH_2(C_6H_5Hg)]OOCCH_3$.
Properties: White, crystalline solid; soluble in water.
Derivation: Reaction of phenylmercuric acetate with monoethanolamine.

Hazard: Toxic by ingestion, inhalation, and skin absorption.

Use: Insecticide and fungicide, may not be used on food crops.

phenylmercuritriethanolammonium lactate.
[tris(2-hydroxyethyl)(phenylmercuri)ammonium lactate].
$[(HOC_2H_4)_3NHgC_6H_5]OOCCHOHCH_3$.
Properties: White, crystalline solid; soluble in water.
Derivation: Reaction of phenylmercuric acetate with triethanolamine and lactic acid.
Hazard: Toxic by ingestion, inhalation, and skin absorption.
Use: Turf fungicide and eradicant fungicide for fruit trees.

phenylmercury formamide. $HCONHHgC_6H_5$.
Hazard: Toxic by ingestion, inhalation, and skin absorption.
Use: Seed disinfectant.

phenylmercury urea.   CAS: 2279-64-3.
$C_6H_5HgNHCONH_2$.
Hazard: Toxic by ingestion, inhalation, and skin absorption.
Use: Disinfectant and fungicide for seed treatment.

phenylmethane.   See toluene.

phenylmethanol.   See benzyl alcohol.

phenylmethyl acetate.   See benzyl acetate.

phenylmethylcarbinol.   See α-methylbenzyl alcohol.

phenylmethylcarbinyl acetate.   See α-methylbenzyl acetate.

N-phenylmethylethanolamine.
$C_6H_5N(CH_3)C_2H_4OH$.
Properties: Liquid which sets to a glass at $-30C$, bp 192C (100 mm), d 1.0661 (20/20C), slightly soluble in water, flash p (COC) 280F (137C). Combustible.
Use: Chemical intermediate, solvent for dyes for acetate fibers.

phenyl methyl ketone.   See acetophenone.

1-phenyl-3-methyl-5-pyrazolone.   See 3-methyl-1-phenyl-2-pyrazolin-5-one.

phenyl α-methylstyryl ketone.   See dypnone.

N-phenylmorpholine.

$C_6H_5NCH_2CH_2OCH_2CH_2$.
Properties: White solid, soluble in water, bp 268C, mp 57C, vap press less than 0.1 mm (20C), flash

p 220F (104C), d 1.06 (57/20C). Combustible.
Grade: Technical.
Hazard: Toxic by ingestion and skin absorption.
Use: Chemical intermediate for dyestuffs, rubber accelerators, corrosion inhibitors, and photographic developers; insecticide.

phenyl mustard oil.   (thiocarbanil; phenyl isothiocyanate; phenylthiocarbonimide).
C_6H_5NCS.
Properties: Pale yellow or colorless liquid; penetrating, irritating odor. Readily volatilized with steam, soluble in alcohol and ether, insoluble in water, d 1.1382, fp $-21C$, bp 221C. Combustible.
Derivation: (a) By action of concentrated hydrochloric acid and sulfocarbanilide, (b) by reaction of thiophosgene with aniline.
Hazard: Toxic by ingestion and inhalation, irritant to tissue.
Use: Medicine, organic synthesis.

N-phenyl-α-naphthylamine.   CAS: 90-30-3.
$C_{10}H_7NHC_6H_5$.
Properties: Crystallizes in prisms, white to slightly yellowish. Soluble in alcohol, ether, and benzene; mp 62C; bp 335C (260 mm). Combustible.
Derivation: From α-naphthylamine and aniline. Purified by distillation.
Use: Dyes and other organic chemicals, rubber antioxidant.

N-phenyl-β-naphthylamine.   CAS: 135-88-6.
$C_{10}H_7NHC_6H_5$.
Properties: Light gray powder, mp 107C, bp 395C, d 1.24. Insoluble in water; soluble in alcohol, acetone, benzene. Combustible.
Hazard: A carcinogen.
Use: Rubber antioxidant, lubricant, inhibitor (butadiene).

N-phenyl-1-naphthylamine-8-sulfonic acid.
(N-phenyl peri acid). $C_{16}H_{13}NO_3S$.
Properties: Greenish-gray needles, insoluble in water, soluble in alcohol.
Derivation: Arylation of 1-naphthylamine-8-sulfonic acid with aniline.
Grade: Technical, mostly as sodium salt.
Use: Azo dyes.

phenylneopentyl phosphite.
$(CH_3)_2C(CH_2O)_2POC_6H_5$.
Properties: Water-white liquid, mp 19C, bp 138–140C (10 mm), refr index 1.517 (25C), d 1.135 (25/15C).
Use: Chemical intermediate.

1-phenylnonane.   See nonylbenzene.

1-phenylpentane.   See n-amylbenzene.

2-phenylpentane.   See sec-amylbenzene.

o-**phenylphenol.** (o-hydroxydiphenyl; o-xenol).
CAS: 90-43-7. $C_6H_5C_6H_4OH$.
Properties: Nearly white or light buff crystals, mp
56–58C, bp 280–284C, d 1.217 (25/25C), flash
p 255F (124C). Soluble in alcohol, sodium hydroxide solution; insoluble in water. Combustible.
Derivation: From reaction of chlorobenzene and
caustic soda solution at elevated temperatures
and pressures.
Hazard: Toxic by ingestion.
Use: Intermediate for dyes, germicides, fungicides,
rubber chemicals, laboratory reagents, food
packaging.

p-**phenylphenol.** (p-hydroxydiphenyl; p-xenol).
CAS: 92-69-3. $C_6H_5C_6H_4OH$.
Properties: Nearly white crystals, mp 164–165,
bp 308C, flash p 330F (165.5C), soluble in alcohol, also in alkalies and most organic solvents,
insoluble in water. Combustible.
Derivation: From reaction of chlorobenzene and
caustic soda solution at elevated temperatures
and pressures.
Hazard: Toxic by ingestion.
Use: Intermediate for dyes and resins, rubber
chemicals, laboratory reagent, fungicide.

N-phenyl-p-phenylenediamine. See p-aminodiphenylamine.

phenylphosphine. (phosphaniline). $C_6H_5PH_2$.
Properties: Colorless liquid, bp 160C, d 1.001
(15C).
Hazard: Toxic by inhalation and ingestion. TLV:
CL of 0.05 ppm in air.

phenylphosphinic acid. See benzenephosphinic
acid.

phenylphosphonic acid. See benzenephosphonic
acid.

phenyl phthalate. $C_6H_4(COOC_6H_5)_2$.
Properties: White crystals, mp 70C, bp 255 (14
mm), d 1.57, soluble in ketones, insoluble in water.
Use: Plasticizer in nitrocellulose lacquers.

N-phenylpiperazine. CAS: 92-54-6.

$C_6H_5NCH_2CH_2NHCH_2CH_2$.
Properties: Pale yellow oil, insoluble in water,
soluble in alcohol and ether, d 1.0621 (20/4C),
bp 286.5C, 156–7C (10 mm), mp 18.8C, flash
p 285F (140.5C). Combustible.
Use: Intermediate for pharmaceuticals, anthelmintics, surface-active agents, synthetic fibers.

phenylpropane. See n-propylbenzene.

3-phenyl-1-propanol. See phenylpropyl alcohol.

1-phenylpropanone-1. See propiophenone.

3-phenylpropenal. See cinnamic aldehyde.

3-phenylpropenoic acid. See cinnamic acid.

3-phenylpropenol. See cinnamic alcohol.

1-(α-phenyl)-propenylveratrole. See benzyl isoeugenol.

3-phenylpropionaldehyde. See phenylpropyl aldehyde.

3-phenylpropionic acid. See hydrocinnamic acid.

phenylpropyl acetate. (hydrocinnamyl acetate).
CAS: 122-72-5.
$C_6H_5CH_2CH_2CH_2OOCCH_3$.
Properties: White crystals, soluble in 70% alcohol,
d 1.012–1.016, refr index 1.497. Combustible.
Use: Perfumery, flavoring, laboratory reagent.

phenylpropyl alcohol. (hydrocinnamic alcohol; 3-
phenyl-1-propanol; phenylethyl carbinol).
CAS: 122-97-4. $C_6H_5CH_2CH_2CH_2OH$.
Properties: Colorless liquid, floral odor, bp 219C,
soluble in 70% alcohol, insoluble in water, d
0.998–1.000, refr index 1.524–1.528 (20C). Combustible.
Grade: Technical, FCC.
Use: Perfumery, flavoring, laboratory reagent.

phenylpropyl aldehyde. (3-phenylpropionalde-
hyde; hydrocinnamic aldehyde).
CAS: 104-53-0. $C_6H_5CH_2CH_2CHO$.
Properties: Colorless liquid, floral odor, soluble
in 50% alcohol, d 1.010–1.020, refr index 1.520–
1.532. Combustible.
Grade: Chlorine-free.
Use: Perfumery, flavors.

phenylpropyl chloride. (hydrocinnamyl chloride;
1-phenyl-3-chloropropane).
$C_6H_5CH_2CH_2CH_2Cl$.
Properties: Colorless to pale yellow liquid, bp 219–
220C, d 1.056 (25/4C), refr index 1.5220 (20C).
Combustible.
Use: Organic synthesis, laboratory reagent.

1-phenyl-3-pyrazolidone.

$C_6H_5NNHC(O)CH_2CH_2$.
Properties: Crystals, mp 121C, soluble in water.
Combustible.
Use: Photographic developer, laboratory reagent.

4-phenylpropylpyridine. $C_6H_5(CH_2)_3C_5H_4N$.
Properties: Colorless liquid, bp 322C, soluble in
organic solvents.
Use: Heat-transfer agent, chemical intermediate.

phenyl salicylate.　(salol).　　CAS: 118-55-8.
$C_6H_4OHCOOC_6H_5$.
Properties: White, crystalline powder; faint aromatic odor and taste. Soluble in alcohol, ether, chloroform, benzene, and fixed or volatile oils; sparingly soluble in water. D 1.2614, mp 41.9C, bp 172–173C, absorbs light, especially at 290–330μ. Combustible.
Derivation: Heating salicylic acid and phenol with phosphorus pentachloride or other dehydrating agent.
Grade: NF, granular powder.
Hazard: Toxic by ingestion.
Use: Medicine, preservative, UV absorber in plastics, waxes, polishes, laboratory reagent.

phenylstearic acid.　An organic fatty acid having a high degree of fluidity and no definite melting point. Pour point is −26C.
Use: Lubricant stabilizer; its potential uses include corrosion inhibitor, plasticizer, and textile auxiliary. Use: Phenylstearic acid and its quaternary and ethoxylated derivatives are used in synthetic latices, as mineral oil emulsifiers, and in invert systems.

phenylsulfanilic acid.　CAS: 101-57-5.
$C_{12}H_{11}NO_3S$.
Properties: White plates turning blue in light, soluble in water and alcohol, decomposes above 200C.
Use: Indicator, colorimetry, determination of nitrates.

phenylsulfohydrazide.　$C_6H_5SO_2NHNH_2$.
Properties: Colorless crystals, decomposes at 100C, evolving nitrogen.
Use: Blowing agent for cellular rubber and plastics.

phenylsulfonic acid.　See benzenesulfonic acid.

4-phenyl-1,4-thiazane.

$\overline{SCH_2CH_2}N(C_6H_5)CH_2CH_2$.
Properties: White powder, soluble in hot toluene, mp 108–111C.
Derivation: Interaction of dichlorodiethyl sulfide and an aliphatic amine in the presence of alcohol and sodium carbonate.

phenylthiocarbonimide.　See phenyl mustard oil.

phenyl tolyl ketone.　(methylbenzophenone).
$C_{14}H_{12}O$.
Properties: (o-) Viscous liquid, mp below -18C, bp 309C; (p-) colorless crystals, mp 59C, bp 311C, soluble in alcohol, benzene, and common organic solvents.

Derivation: (o-) Benzene and o-toluic acid chloride (aluminum chloride catalyst); (p-) benzoyl peroxide + toluene (aluminum chloride catalyst).
Use: Perfume additive (fixative).

phenyltrichlorosilane.　CAS: 98-13-5.
$C_6H_5SiCl_3$.
Properties: Colorless liquid, bp 201C, d 1.321 (25/25C), refr index 1.5240 (25C), flash p (COC) 185F (85C). Soluble in benzene, ether, perchloroethylene; readily hydrolyzed by moisture with liberation of hydrogen chloride. Combustible.
Derivation: By Grignard reaction of silicon tetrachloride and phenylmagnesium chloride, reaction of benzene with trichlorosilane.
Hazard: Strong irritant to tissue.
Use: Intermediate for silicones, laboratory reagent.

1-phenyltridecane.　See tridecylbenzene.

phenyltrimethoxysilane.　CAS: 780-69-8.
$C_6H_5Si(OCH_3)_3$.
Properties: Liquid, d 1.063 (25C), refr index 1.4710 (25C), bp 211C. Soluble in acetone, benzene, perchloroethylene, methanol. Combustible.
Use: In polymers to be applied to powders, glass, paper, and fabrics.

phenylurethane.　See ethyl phenylcarbamate.

phenyl valerate.　$C_4H_9COOC_6H_5$.
Properties: Colorless liquid, slightly soluble in water, soluble in alcohol and ether.
Use: Flavors and odorants.

pheromone.　A group of organic compounds produced by insects which function as communication means and as sex attractants. Synthetic pheromones have been used experimentally to control insect pests by disrupting their mating behavior, e.g., 4-methyl-3-heptanone. Pheronomes have also been synthesized by use of organoborane reactions.
See phoromone.

"Philprene."[303]　TM for a series of styrene-butadiene rubbers, hot and cold, oil-extended, pigmented and non-pigmented, including special types for specific needs.
Use: Tire carcasses and treads, molded and extruded goods, sporting goods, footwear, coated fabrics, wire and cable jackets, hospital goods, floor tile, insulation.

phloroglucinol.　(phloroglucine; 1,3,5-trihydroxybenzene).　　CAS: 108-73-6.
$C_6H_3(OH)_3 \cdot 2HOH$.
Properties: White to yellowish crystals, odorless. Mp 212–217C if rapidly heated, 200–209C if

slowly heated; bp sublimes with decomposition. Soluble in alcohol, ether, and pyridine; soluble in 100 parts water.

Derivation: By fusion of resorcinol with caustic soda, by reduction of trinitrobenzene.

Use: Analytical chemistry (reagent for pentoses and with vanillin for determining the presence of free hydrogen chloride), decalcifying agent for bones, preparation of pharmaceuticals and dyes, resins, preservative for cut flowers, textile dyeing and printing.

phorate. (Generic name for O,O-diethyl-S-[(ethylthio)methyl]phosphorodithioate). CAS: 298-02-2. $(C_2H_5O)_2P(S)SCH_2SC_2H_5$.

Properties: Liquid, bp 118–120C (0.8 mm). Insoluble in water; miscible with carbon tetrachloride, dioxane, xylene.

Hazard: Toxic by skin contact, inhalation, or ingestion. Rapidly absorbed through skin. Repeated inhalation or skin contact may, without symptoms, progressively increase susceptibility. Use may be restricted. A cholinesterase inhibitor. TLV: 0.5 mg/m³ of air;

Use: Insecticide.

phorbol. (4,9,12-β-13,20-pentahydroxy-1,6-tigliadien-3-on). CAS: 17673-25-5.

The parent alcohol of tumor-producing compounds in croton oil.

Hazard: A carcinogen.

Use: Biochemical and medical research.

phoromone. (7-ethyl-5-methyl-6,8-dioxabicyclo-[3.2.1]-octane). Product excreted by bark beetles which acts as sex attractant. It has been isolated and synthesized for possible use in protection of forest timber.

phorone. (diisopropylidene acetone). CAS: 504-20-1. $(CH_3)_2CCHCOCHC(CH_3)_2$.

Properties: Yellow liquid or yellowish green prisms, d 0.8791 at 20/20C, bp 197.9C, fp 28.0C, vap press 0.38 mm (20C), flash p 185F (85C), bulk d 7.3 lb/gal (20C), slightly soluble in water, soluble in alcohol. Combustible.

Use: Solvent for nitrocellulose, coating compositions, stains, intermediate (organic synthesis).

phosalone. CAS: 2310-17-0. $C_{12}H_{15}ClNO_4PS_2$.

Properties: Colorless crystals, mp 48C, soluble in alcohols and aromatic solvents, insoluble in water and aliphatic solvents.

Hazard: Toxic by ingestion and inhalation.

Use: Insecticide, molluscicide.

"Phos-Chek."[58] TM for fire-fighting solutions of phosphorus pentoxide: 201 and 202 contain 46.5% and 47% min P_2O_5, designed for application from aircraft, 258 contains 47.7% min P_2O_5 applied ground equipment and helicopters. Said to be solutions of ammonium phosphate thickened with sodium methyl cellulose.

"Phosdrin."[125] TM for a mixture which contains more than 60% of the alpha isomer of 2-$(CH_3O)_2P(O)OC(CH_3):CHCOOCH_3$ (generic name mevinphos) and less than 40% of insecticidally active related compounds. It is 100% active.

See mevinphos.

"Phosflake."[177] TM for a uniform blend of caustic soda and trisodium phosphate prepared in flake form, especially for bottle-washing use.

"Phosgard" C-22-R[58]. TM for a flame retardant organophosphorus compound with high efficiency through high phosphorus and chlorine content.

Use: In urethane foams, phenolics, polymethacrylates, polyester resins and epoxies.

phosgene. (carbonyl chloride; carbon oxychloride; chloroformyl chloride). CAS: 75-44-5. $COCl_2$.

Properties: Liquid or easily liquefied gas, colorless to light yellow, odor varies from strong and stifling when concentrated to hay-like in dilute form, d 1.392 (19/4C), fp −128C, bp 8.2C, slightly soluble in water and slowly hydrolyzed by it, soluble in benzene and toluene, specific volume 3.9 cu ft/lb (21.1C). Noncombustible.

Derivation: By passing a mixture of carbon monoxide and chlorine over activated carbon.

Hazard: Very toxic via inhalation, strong irritant to eyes. TLV: 0.1 ppm in air.

Use: Organic synthesis, especially of isocyanates, polyurethane and polycarbonate resins, carbamates, organic carbonates and chloroformates, pesticides, herbicides, dye manufacture.

phosmet. CAS: 732-11-6. $C_{11}H_{12}NO_4PS_2$.

A dimethyl ester of phosphorodithioic acid.

Properties: Colorless crystals, mp 72C, partially soluble in water, decomposes on heating.

Hazard: Toxic by ingestion, may inhibit cholinesterase.

Use: Acaricide, insecticide.

phosphamidon. (2-chloro-2-diethylcarbamoyl-1-methylvinyl dimethyl phosphate). CAS: 13171-21-6. $(CH_3O)_2P(O)OC(CH_3):C(Cl)C(O)N(C_2H_5)_2$.

Properties: Colorless liquid, bp 162C (1.5 mm), soluble in water and organic solvents.

Hazard: Toxic by ingestion, inhalation, skin ab-

sorption; cholinesterase inhibitor; use may be restricted.
Use: Insecticide.

phosphatase, alkaline. An enzyme excreted into the bile by the liver and found in the blood. It is concerned with bone formation, probably being produced by osteoblasts. It hydrolyzes phosphoric acid esters at pH 7–8, liberating phosphate ions.
Use: Biochemical research.

phosphate, condensed. A phosphorus compound with two or more phosphorus atoms in the molecule. Examples are polyphosphates, pyrophosphates.
See polyphosphoric acid.

phosphate glass. A type of glass containing phosphorus pentoxide. Aluminum-metaphosphate is frequently the basic material. Such glasses have properties not attainable in silicate glasses, e.g., resistance to hydrogen fluoride.

phosphate rock. (phosphorite). A natural rock consisting largely of calcium phosphate and used as a raw material for manufacture of phosphate fertilizers, phosphoric acid, phosphorus, and animal feeds. Recovery of uranium from the manufacture of phosphoric acid and other phosphate chemicals is expected to become an important source of this metal. Phosphate rock is the primary source of superphosphate, prepared by treatment of the pulverized rock with sulfuric acid (superphosphate having 16–18% P_2O_5) or by acidifying with phosphoric acid (triple superphosphate having 40–48% P_2O_5). Nitric acid is sometimes used, i.e., nitrophosphate. Defluorinated phosphate rock is the source of phosphate used in animal feeds and feed concentrations. Important deposits are in US (Florida, North Carolina, Tennessee, California, Wyoming, Montana, Utah, Idaho), North Africa (Morocco, Libya, Algeria), USSR, and various islands in the Pacific.

phosphate slag. Glassy calcium silicate, byproduct of electric furnace phosphorus manufacture.
Properties: Lumps, loose bulk d 85 lb/ft³.

phosphatide. See phospholipid.

phosphatidyl choline. See lecithin.

phosphatidyl ethanolamine. See cephalin.

phosphatidyl serine. See cephalin.

phosphazene. (phosphonitrile). A ring or chain polymer that contains alternating phosphorus and nitrogen atoms with two substituents on each phosphorus atom. Characteristic structures are cyclic trimers, cyclic tetramers, and high polymers. The substituent can be any of a wide variety of organic groups, halogen, amino, etc. Most cyclic trimers are crystalline, solids, organo-soluble and stable to weather conditions; the high polymers (polyphosphazenes) are elastomeric or thermoplastic. A copolymer of phosphazene and styrene has been investigated for use as a flame-retardant.

phosphinate. A derivative of the hypothetical phosphinic acid, $H_2P(O)OH$.

phosphine. (hydrogen phosphide). PH_3.
Properties: Colorless gas, disagreeable, garlic-like odor. Soluble in alcohol, ether, and cuprous chloride solution; slightly soluble in cold water; insoluble in hot water. D 1.185, fp −133.5C, bp −85C, autoign temperature 100F (37.7C).
Derivation: By action of freshly formed hydrogen or of caustic potash on phosphorus.
Hazard: Spontaneously flammable. Toxic by inhalation, strong irritant. TLV: 0.3 ppm in air.
Use: Organic preparations, phosphonium halides, doping agent for n-type semiconductors, polymerization initiator, condensation catalyst.
Note: A synthetic dye, chrysaniline yellow, is sometimes called phosphine.

phosphodiesterase. An enzyme which cleaves hydrolytically the carboxy- and phosphoesters of phosphatides (phospholipids).

2-phosphoglyceric acid.
$HOCH_2CH[OPO(OH)_2]COOH$. An intermediate in the metabolism of carbohydrates in biological systems.
See enolase.

phosphoglyceride. See phospholipid.

phospholipid. (phosphatide). A group of lipid compounds that yield on hydrolysis phosphoric acid, an alcohol, fatty acid, and a nitrogenous base. They are widely distributed in nature and include such substances as lecithin, cephalin, and sphingomyelin.

phosphomolybdate. See heteromolybdate.

phosphomolybdic acid. (phospho-12-molybdic acid; PMA; 12-molybdophosphoric acid).
CAS: 11104-88-4. $H_3PMo_{12}O_{40} \cdot xHOH$.
Properties: Yellowish crystals. Soluble in water, alcohol, and ether. D 3.15 g/cc, mp 78–90C, strong oxidizing agent in aqueous solution, strong acid in free acid form.

Grade: Technical, CP, reagent.

Hazard (solution): May ignite combustible materials.

Use: Reagent for alkaloids; pigments; catalyst; fixing agent in photography; additive in plating processes; imparts water resistance to plastics, adhesives, and cement.

phosphomolybdic pigment. (molybdenum lake). A pigment made by precipitating a basic organic dye with phosphomolybdic acid or a mixture of phosphomolybdic and phosphotungstic acids. See also phosphotungstic pigment.

phosphonitrile. See phosphazene.

phosphonium iodide. (iodophosphonium). PH₄I.

PH_4I.

Properties: Colorless or slightly yellowish crystals, d 2.86, mp 18.5C, sublimes 61.8C, bp 80C, decomposed by water or alcohol evolving highly toxic phosphine.

Derivation: Hydrolysis of phosphorus tetraiodide and white phosphorus.

Hazard: Rapid heating can cause detonation. Toxic by ingestion.

phosphor. (fluor). A substance, either organic or inorganic, liquid or crystals, that is capable of luminescence, that is, of absorbing energy from sources such as x-rays, cathode rays, UV radiations, alpha particles and emitting a portion of the energy in the UV, visible, or infrared. When the emission of the substance ceases immediately or in the order of 10^{-8} second after excitation, the material is said to be fluorescent. Material that continues to emit light for a period after the removal of the exciting energy is said to be phosphorescent. The half-life of the afterglow varies with the substance and may range from 10^{-6} second to days.

Use: Fluorescent light tubes, television, radar, and cathode ray tubes, instrument dials, scintillation counters.

See also fluorescence, phosphorescence.

phosphor bronze. A tin bronze which has been deoxidized by the addition of up to 0.5% phosphorus. Relatively hard, strong, and corrosion-resistant. Has good cold work properties and high strength.

Grade: Grade A (5% tin), grade C (8% tin), grade D (10% tin), grade E (1.25% tin).

Use: Springs, electrical switches, contact fingers, chains, fourdrinier wire.

See brass and bronze.

phosphorescence. A type of luminescence in which the emission of radiation resulting from excitation of a crystalline or liquid material occurs *after* excitation has ceased and may last from a fraction of a second to an hour or more. This phenomenon is characteristic of some organic compounds, as in the firefly, and also of a number of inorganic solid materials, both natural and synthetic.

Use: Industrially as phosphors.

See also fluorescence.

phosphoric acid. (orthophosphoric acid).

CAS: 7664-38-2. H_3PO_4.   7th highest-volume chemical produced in US (1985).

Properties: Colorless, odorless, sparkling liquid or transparent, crystalline solid, depending on concentration and temperature. At 20C the 50 and 75% strengths are mobile liquids, the 85% is of a syrupy consistency, while the 100% acid is in the form of crystals, d 1.834 (18C), mp 42.35C, loses 1/2 water at 213C (to form pyrophosphoric acid), soluble in water and alcohol, corrosive to ferrous metals and alloys.

Derivation: (a) Action of sulfuric acid on pulverized phosphate rock; (b) action of hydrochloric acid on phosphate rock, with extraction by tributylphosphate; (c) by heating phosphate rock, coke, and silica in an electric furnace, burning the elemental phosphorus produced, and then hydrating the phosphoric oxide (furnace acid).

Grade: Agricultural, technical (50, 75, 85, 90, 100%), food (50, 75, 85%), NF 85–88%), FCC (75–85%). (Polyphosphoric acid is sometimes called 115% phosphoric acid).

Hazard: Toxic by ingestion and inhalation, irritant to skin and eyes. TLV: 1 mg/m³ of air.

Use: Fertilizers, soaps and detergents, inorganic phosphates, pickling and rust-proofing metals, pharmaceuticals, sugar refining, gelatin manufacture, water treatment, animal feeds, electropolishing, gasoline additive, conversion coatings for metals, catalyst for ethanol manufacture, lakes in cotton dyeing, yeasts, soil stabilizer, waxes and polishes, binder for ceramics, activated carbon, in foods and carbonated beverages as acidulant and sequestrant, laboratory reagent.

phosphoric acid, anhydrous. See phosphoric anhydride.

phosphoric acid, meta-. (metaphosphoric acid; phosphoric acid, glacial).   CAS: 37267-86-0. $(HPO_3)_x$.

Properties: Transparent, highly deliquescent, glassy mass; d 2.2–2.488; soluble in water slowly forming the ortho-acid. Also soluble in alcohol. Noncombustible. ·

Derivation: By heating orthophosphoric acid to redness, by treating phosphorus pentoxide with a calculated quantity of water, by heating diammonium phosphate.

Grade: Technical, CP.

Use: Phosphorylating agent, dehydrating agent, dental cements, laboratory reagent.

See also polyphosphoric acid.

phosphoric anhydride. (phosphorus pentoxide; phosphoric oxide; phosphoric acid, anhydrous). P_2O_5.

Properties: Soft, white powder; absorbs moisture from the air with avidity forming meta-, pyro-, or ortho-phosphoric acid, depending upon amount of water absorbed and upon conditions of absorption; d 2.30; mp 340C; bp sublimes at 360C.

Derivation: By burning phosphorus in dry air.

Grade: Technical, nitrogen-complex coated (for slow solution).

Hazard: Reacts violently with water to evolve heat; dangerous fire risk, keep tightly sealed or stoppered. Corrosive to skin and tissue.

Use: Preparation of phosphorus oxychloride and metaphosphoric acid, acrylate esters, surfactants, dehydrating agent, condensing agent in organic synthesis, sugar refining, laboratory reagent, fire extinguishing, special glasses.

phosphoric bromide. See phosphorus pentabromide.

phosphoric chloride. See phosphorus pentachloride.

phosphoric oxide. See phosphoric anhydride.

phosphoric perbromide. See phosphorus pentabromide.

phosphoric perchloride. See phosphorus pentachloride.

phosphoric sulfide. See phosphorus pentasulfide.

phosphorous acid, ortho-. (orthophosphorous acid; phosphonic acid). CAS: 13598-36-2. H_3PO_3.

Properties: White or yellowish, crystalline mass; very hygroscopic. Keep tightly sealed or stoppered, absorbs oxygen very readily with formation of orthophosphoric acid, soluble in alcohol and water, d 1.651, bp 200C (decomposes) mp 70C.

Grade: Reagent, technical, 70%.

Use: Analysis (testing for mercury), chemical reducing agent, phosphite salts.

phosphorus. CAS: 7723-14-0. P.

Nonmetallic element of atomic number 15; group VA of Periodic Table; aw 30.97376; valences = 1, 3, 4, 5; allotropes, white (or yellow), red, and black phosphorus. No stable isotopes, several artificial radioactive isotopes with mass numbers 29–34.

White phosphorus:

Properties: Crystals, wax-like, transparent solid, metastable with respect to red phosphorus, an impurity present in white allotrope, bp 280C, vapor density corresponds to formula P_4, mp 44.1C, d (solid 20C), 1.82 (liquid 44.5C) 1.745, Mohs hardness 0.5, high electrical resistivity, insoluble in water and alcohol, soluble in carbon disulfide, exhibits phosphorescence at room temperature, an essential dietary nutrient.

Derivation: (1) Produced in an electric furnace from phosphate rock, sand, and coke. The phosphorus vapor is driven off and condensed under water. (2) By reaction of phosphate rock with sulfuric acid, the resulting $CaSO_4$ being removed by filtration and the phosphoric acid concentrated by evaporation (wet process).

Grade: Technical 99.9%, electronic grade 99.9999%.

Occurrence: Occurs in nature in phosphate rock [impure $Ca_3(PO_4)_2$], in apatite [$Ca_5(PO_4)_3F$], in bones, teeth and in organic compounds of living tissue. Also as phosphorite nodules on ocean floor.

Hazard: Ignites spontaneously in air at 86F (30C), store under water and away from heat, dangerous fire risk. Toxic by ingestion and inhalation, skin contact causes burns. TLV: 0.1 mg/m^3 of air.

Use: Rodenticides, smoke screens, analytical chemistry.

Red phosphorus:

Properties: Violet-red amorphous powder, obtained from white phosphorus by heating at 240C with catalyst, sublimes at 416C, d 2.34, autoign temperature 500F (260C), high electrical resistivity, much less reactive than white phosphorus, insoluble in most solvents.

Hazard: Large quantities ignite spontaneously and on exposure to oxidizing materials. Reacts with oxygen and water vapor to evolve phosphine. Extinguish with foam or dry chemical (not water).

Use: Manufacture of phosphoric acid and other phosphorus compounds, phosphor bronzes, metallic phosphides, additive to semiconductors, electroluminescent coatings, safety matches, fertilizers.

Black phosphorus: Black solid resembling graphite, d 2.25–2.69, obtained by heating white phosphorus under pressure, insoluble in most solvents, electrically conducting,

phosphorus-32. Radioactive phosphorus of mass number 32.

Properties: Half-life, 14.3 days, radiation, beta.

Derivation: Pile irradiation of potassium dihydro-

gen phosphate or sulfur and sulfur compounds.
Forms available: Phosphate ion in weak hydrochloric acid solution, solid potassium dihydrogen phosphate, P-32 sterile solution, in tagged compounds such as hexaethyltetraphosphate, ribonucleic acid, triphenylphosphine, etc.

Grade: USP as sodium phosphate P-32 solution.

Use: Biochemical radioactive tracer studies, medical treatment of leukemia, skin lesions, etc., industrial measurements, e.g., tire tread wear, thickness of ink and paint films, lead detection.

phosphorus chloride. See phosphorus trichloride.

phosphorus heptasulfide. (tetraphosphorus heptasulfide). CAS: 12037-82-0. P_4S_7.
Properties: Light yellow crystals, d 2.19, mp 310C, bp 523C, slightly soluble in carbon disulfide.
Hazard: Flammable, dangerous fire risk.

phosphorus nitride. P_3N_5.
Properties: Amorphous white solid, insoluble in cold water, decomposes in hot water, soluble in common organic solvents. Nonhygroscopic and stable in air, decomposes at 800C.
Use: Doping semiconductors.

phosphorus oxybromide. CAS: 7789-59-5.
$POBr_3$.
Properties: Colorless crystals, d 2.82, mp 56C, bp 189C, decomposed by water, soluble in ether and benzene, reacts strongly with organic matter.
Hazard: Strong irritant to skin and tissue, store in sealed glass containers.
Use: Chemical intermediate.

phosphorus oxychloride. (phosphoryl chloride).
CAS: 10025-87-3. $POCl_3$.
Properties: Colorless, fuming liquid; pungent odor. D 1.675 (20/20C), mp 1.2C, bp 107.2C, refr index 1.460 (25C), decomposed by water and alcohol with evolution of heat.
Derivation: From phosphorus trichloride, phosphorus pentoxide, and chlorine.
Grade: Technical, 99.999+%.
Hazard: Toxic by inhalation and ingestion, strong irritant to skin and tissue.
Use: Manufacture of cyclic and acyclic esters for plasticizers, gasoline additives, hydraulic fluids and organophosphorus compounds, chlorinating agent and catalyst, dopant for semiconductor grade silicon, tricresyl phosphate, and fire-retarding agents.

phosphorus pentabromide. (phosphoric bromide; phosphoric perbromide). CAS: 7789-69-7.
PBr_5.
Properties: Yellow, crystalline mass; keep hermetically sealed! Soluble in water (decomposes), carbon disulfide, carbon tetrachloride, benzene; bp 106C (decomposes).
Grade: Technical.
Hazard: Corrosive to skin and tissue.
Use: Brominating agent.

phosphorus pentachloride. (phosphoric chloride; phosphoric perchloride). CAS: 10026-13-8.
PCl_5.
Properties: Slightly yellow, crystalline mass; irritating odor; fuming in moist air. D 3.60, mp (under pressure) 148C, ordinarily sublimes at 160–165C, soluble in carbon disulfide and carbon tetrachloride.
Derivation: By action of chlorine on phosphorus or phosphorus trichloride.
Grade: Technical, reagent.
Hazard: Flammable, reacts strongly with water, use carbon dioxide or dry chemical to extinguish, store in tightly closed containers. Corrosive to eyes and skin. TLV: 0.1 mg/m³ of air.
Use: Chlorinating and dehydrating agent, catalyst.

phosphorus pentafluoride. CAS: 7647-19-0.
PF_5.
Properties: Colorless, nonflammable gas;, fumes strongly in air. Fp −94C, bp −84.8C, decomposed by water, available in small cylinders.
Grade: 99%.
Hazard: Corrosive to eyes and skin. TLV (as F): 2.5 mg/m³ of air.
Use: Polymerization catalyst.

phosphorus pentasulfide. (phosphoric sulfide; phosphorus persulfide; thiophosphoric anhydride). CAS: 1314-80-3. P_2S_5.
Properties: Light-yellow or greenish-yellow crystalline mass, odor similar to hydrogen sulfide, keep in sealed containers, very hygroscopic, burns in air forming P_2O_5 and SO_2, decomposed by moist air, soluble in solutions of alkali hydroxides, soluble in carbon disulfide, mp 286–290C, bp 515C, d 2.03, vap press 1 mm (300C), autoign temperature 287F (141.6C).
Derivation: By reaction of phosphorus and sulfur.
Grade: Technical, distilled.
Hazard: Dangerous fire risk, ignites by friction, contact with water or acids liberates poisonous and flammable hydrogen sulfide. Toxic by inhalation, strong irritant. TLV: 1 mg/m³ of air.
Use: Intermediate for lube oil additives, insecticides (chiefly parathion and malathion), flotation agents, safety matches, ignition compounds, sulfonation.

phosphorus pentoxide. See phosphoric anhydride.

phosphorus persulfide. See phosphorus pentasulfide.

phosphorus salt. See sodium ammonium phosphate.

phosphorus sesquisulfide. (tetraphosphorus trisulfide). CAS: 1314-85-8. P_4S_3.
Properties: Yellow, crystalline mass. Soluble in carbon disulfide, insoluble in cold water, decomposed by hot water, d 2.00, mp 172C, bp 407.8C, autoign temperature 212F (100C) (solution).
Derivation: By gently heating phosphorus and sulfur.
Hazard: Dangerous fire risk, ignites by friction. An irritating poison.
Use: Organic synthesis, manufacture of matches.

phosphorus sulfide. See phosphorus trisulfide.

phosphorus tribromide. CAS: 7789-60-8. PBr_3.
Properties: Fuming, colorless liquid; penetrating odor; soluble in acetone, alcohol, carbon disulfide, hydrogen sulfide, water (decomposes); d 2.852 (15C); bp 175C; fp −40C.
Hazard: Corrosive to skin and tissue, store in tightly closed containers.
Use: Analysis (testing for sugar and oxygen), catalyst, synthesis.

phosphorus trichloride. (phosphorus chloride). CAS: 7719-12-2. PCl_3.
Properties: Clear, colorless, fuming liquid; decomposes rapidly in moist air; soluble in ether, benzene, carbon disulfide, and carbon tetrachloride; d 1.574; fp −111.8C; bp 76C.
Derivation: By passing a current of dry chlorine over gently heated phosphorus, which ignites. The trichloride admixed with some pentachloride distills over. A small amount of phosphorus is added and the whole distilled.
Grade: Technical, 99.9%.
Hazard: Corrosive to skin and tissue, reacts with water to form hydrochloric acid, store in tightly closed containers. TLV: 0.2 ppm in air.
Use: Making phosphorus oxychloride, intermediate for organophosphorus pesticides, surfactants, phosphites (reaction with alcohols and phenols), gasoline additives, plasticizers, dyestuffs, chlorinating agent, catalyst, preparing rubber surfaces for electrodeposition of metal, ingredient of textile finishing agents.

phosphorus triiodide. PI_3.
Properties: Red crystals, hygroscopic. Soluble in alcohol, carbon disulfide, water (decomposes). Mp 61C (decomposes); d 4.18.
Grade: Technical, reagent.
Hazard: Flammable, reacts with water. Irritating to skin and eyes.
Use: Organic synthesis.

phosphorus trisulfide. (phorphorus sulfide; tetraphosphorus hexasulfide; thiophosphorous anhydride). CAS: 12165-69-4. P_2S_3 or P_4S_6.
Properties: Grayish-yellow mass, tasteless, odorless, keep well stoppered, decomposes in moist air. Soluble in alcohol, carbon disulfide, ether; bp 490C; mp 290C.
Hazard: Flammable, highly dangerous fire risk, reacts with water.
Use: Organic chemistry (reagent).

phosphorylase. An enzyme occurring in muscle and liver which catalyzes the conversion of glycogen into glucose-1-phosphate.

phosphorylation. A reaction in which phosphorus combines with an organic compound, usually in the form of the trivalent phosphoryl group. It occurs naturally in cellular metabolism and is of particular importance in vitamin activity and enzyme formation. It is also used to produce a modified cellulose (P-cellulose) for cation exchange in chromatographic separations.

phosphoryl chloride. See phosphorus oxychloride.

phosphotungstic acid. (phospho-12-tungstic acid; phosphowolframic acid; 12-tungstophosphoric acid). CAS: 12067-99-1. $H_3PW_{12}O_{40} \cdot xHOH$.
Properties: Yellowish-white solid, mp (for $24H_2O$ of hydration) 89C. Soluble in water, acetone, and diethyl ether. Relatively insoluble in nonpolar organic solvents. Strong oxidizing agent in aqueous solution; strong acid in the free acid form.
Derivation: Addition of phosphates to sodium tungstate in the presence of hydrochloric acid.
Grade: Reagent, technical.
Hazard: Strong irritating to skin and eyes.
Use: Reagent in analytical chemistry and biology; manufacture of organic pigments; additive in plating industry; imparts water resistance to plastics, adhesives, and cement; catalyst for organic reactions; photographic fixing agent; textile antistatic agent.

phosphotungstic acid, sodium salt. See sodium-12-tungstophosphate.

phosphotungstic pigment. (tungsten lake).
A green or blue pigment manufactured by precipitating basic dyestuffs such as malachite green or Victoria blue with solutions of phosphotungstic acid, or phosphomolybdic acid, or mixture of both.
See also phosphomolybdic pigment.
Use: Printing inks, paper, paints, and enamels.

photochemistry. The branch of chemistry concerned with the effect of absorption of radiant energy (light) in inducing or modifying chemical changes. Photosynthesis is the most important example of a photochemical reaction; others are the photosensitization of solids, applied in photography, photocells, photovoltaic cells, and the formation of visual pigments; photochemical decomposition (photolysis), photo-induced polymerization, oxidation, and ionization; fluorescence and phosphorescence; and the reaction of chlorine with organic compounds. Free-radical chain mechanisms are usually involved.
See also free radical.

photochromism. The ability of a transparent material to darken reversibly when exposed to light.
Plastics can be made light-sensitive by certain aromatic organic nitro compounds such as 2-(2,4-dinitrobenzyl)pyridine. Such chemicals are compatible with most transparent plastics and are either blended with the base resin or applied as coatings.
See also glass, photochromic.

photodecomposition. See photolysis, photochemistry.

photo-glycin. See p-hydroxyphenylglycine.

photographic chemistry. In photographic films and papers the sensitive surface usually consists of microscopic grains of a silver halide, suspended in gelatin. Exposure to light renders the halide particles susceptible to reduction to metallic silver by developing agents containing a reducing agent, as well as an accelerator, preservative, and restrainer. The accelerator increases the activity of the reducing agent (due principally to ionization of the phenolic agents to their active form) and is usually an alkaline compound. The preservative, usually sodium sulfite, minimizes air oxidation. The restrainer helps to prevent "fog" (reduction of silver halide grains which have not been exposed to light) and is almost always potassium bromide.
Color sensitizers are dyes added to silver halide emulsions to broaden their response to various wavelengths. Unsensitized emulsions are most responsive in the blue region of the spectrum and thus do not correctly represent the light spectrum striking them. Widely used sensitizers include the cyanine dyes, the merocyanines, the benzooxazoles, and the benzothiazoles. Cryptocyanine sensitizes the extreme red and infrared.
In color photography diethyl-p-phenylenediamine is an important developer since its oxidation product readily couples with a large number of phenol and reactive methylene compounds to form indophenol and indoaniline dyes which are the basis of most of the current color processes.
See also holography.

photolysis. Decomposition of a compound into simpler units as a result of absorbing one or more quanta of radiation; examples are splitting of hydrogen iodide by the reaction $2HI + hv \rightarrow H_2 + I_2$; and of ketene ($H_2C{=}CO$) into CO and carbene (methylene) (${=}CH_2$). Photodecomposition may also occur with aldehydes, ketones, azo compounds, and organometallic compounds. Continuous generation of hydrogen by photolysis of water has been achieved using platinum catalyst in conjunction with ruthenium and rhodium. Similarly, hydrogen can be split from H_2S by photolysis with cadmium sulfide as catalyst, aided by ruthenium dioxide.
See also flash photolysis, photochemistry.

photometric analysis. Chemical analysis by means of absorption or emission of radiation, primarily in the near UV, visible, and infrared portions of the electromagnetic spectrum. It includes such techniques as spectrophotometry, spectrochemical analysis, Raman spectroscopy, colorimetry, and fluorescence measurements.
See also spectroscopy.

photon. The unit (quantum) of electromagnetic radiation. Light waves, gamma rays, x-rays, etc., consist of photons. Photons are discrete concentration of energy that seem to have no rest mass and move at speed of light. Their nature can be described only in mathematical terms. Photons are emitted when electrons move from one energy state to another, as in an excited atom.
See also radiation.

photophor. See calcium phosphide.

photopolymer. A polymer or plastic so made that it undergoes a change on exposure to light. Such materials can be used for printing and lithography plates, for photographic prints and microfilm copying. The effect of the light may be to cause further polymerization or crosslinking, or it may cause degradation. One application involves the use of esters of polyvinyl alcohol which crosslink and so become insoluble, whereas unexposed portions of the material remain soluble.

photosynthesis. The utilization of sunlight by plants as well as by bacteria to convert two inorganic substances (carbon dioxide and water) into carbohydrates. Chlorophyll acts as the energy-converter in this reaction, which is perhaps the most important on earth. The generalized reaction is: $6CO_2 + 6HOH + 672$ kcal $\rightarrow C_6H_{12}O_6$

+ $6O_2$. The significance of this process lies in the conversion of energy from radiant to chemical form. The chemical energy a green plant stores by photosynthesis provides the total energy requirement of the plant. Directly or indirectly plants supply the primary organic nutrient for most other living organisms. Most fossil fuels are storehouses of the radiant energy transformed by photosynthesis in earlier geologic eras.

Photosynthesis is the principal source of atmospheric oxygen. At least two-thirds of the total photosynthetic activity of the earth takes place in the oceans. Its exact chemical mechanism is extremely complex. Essential features are the reduction of carbon dioxide and utilization of the hydrogen of water to form carbohydrates, the oxygen being liberated; the nucleotides nicotinamide and adenosine triphosphate are involved in this conversion. Sugar (sucrose) is formed in the cytoplasm surrounding the chloroplasts. Photosynthesis has been shown to be substantially inhibited by air pollution to the extent of 20% in rural locations and 33% in urban areas.
See algae.

photovoltaic cell. See solar cell.

"PH 990 Resin."[469] TM for a phosphonitrilic modified phenolic resin; stable, off-white, free-flowing powder, soluble in most organic solvents, flame retardant, retains electrical and structural properties up to 260–426C.

"Phthalamaquin."[342] TM for an aureoquin preparation (4-(2-dimethylaminoethylamino)-6-methoxyquinoline diethylaminotetrahydrophthalate).
Use: Medicine.

phthalamide. CAS: 88-96-0. $C_6H_4(CONH_2)_2$. The double acid amide of phthalic acid.
Properties: Colorless crystals, mp 220C (decomposes into phthalimide and ammonia). Very slightly soluble in water and alcohol, insoluble in ether.
Derivation: By stirring phthalimide with cold concentrated ammonia solution; by the reaction of phthalyl chloride and ammonia; or from the addition of ammonia to phthalic anhydride under pressure.
Use: Intermediate in organic synthesis, laboratory reagent.

phthalic acid. (o-phthalic acid; o-benzene dicarboxylic acid).   CAS: 88-99-3.
$C_6H_4(COOH)_2$.
Properties: Colorless crystals, soluble in alcohol, sparingly soluble in water and ether, d 1.585, mp decomposes at 191C.

Derivation: Catalytic oxidation of o-toluic acid and oxidation of xylene.
Grade: Technical, reagent.
Use: Dyes, phenolphthalein, phthalmide, anthranilic acid, synthetic perfumes, laboratory reagent.

p-phthalic acid. See terephthalic acid.

phthalic anhydride.   CAS: 85-44-9.
$C_6H_4(CO)_2O$.

Properties: White, crystalline needles; sublimes below boiling point; mild odor; d 1.527 (4C); mp 131.16C; bp 285C; flash p (CC) 305F (151.6C); autoign temperature 1083F (583C); soluble in alcohol, carbon disulfide, and hot water. Combustible.
Derivation: Catalytic oxidation of naphthalene.
Method of purification: Sublimation.
Grade: Pure, technical.
Hazard: Skin irritant. TLV: 1 ppm in air.
Use: Alkyd resins, plasticizers, hardener for resins, polyesters, synthesis of phenolphthalein and other phthaleins, many other dyes, chlorinated products, pharmaceutical intermediates, insecticides, diethyl phthalate, dimethyl phthalate, laboratory reagent.

phthalimide.   CAS: 85-41-6. $C_6H_4(CO)_2NH$

Properties: White, crystalline leaflets; slightly soluble in water; insoluble in benzene; soluble in boiling alcohol or acetic acid and in aqueous alkalies; mp 233–238C; bp sublimes. Combustible.
Derivation: By dissolving phthalic anhydride in ammonium hydroxide, evaporating to dryness, and using the residue.
Use: Synthetic indigo via anthranilic acid, fungicide, organic synthesis, laboratory reagent.

phthalocyanine. Any of a group of benzoporphyrins which have strong pigmenting power, forming a family of dyes. The basic structure of the molecule comprises four isoindole groups $(C_6H_4)C_2N$ joined by four nitrogen atoms. Four commercially important modifications are: (1)

metal-free phthalocyanine $(C_6H_4C_2N)_4N_4$ having a blue-green color (structure shown below): (2) copper phthalocyanine, in which a copper atom is held by secondary valences of the isoindole nitrogen atoms; see Pigment Blue 15; (3) chlorinated copper phthalocyanine, green, in which 15-16 hydrogen atoms are replaced by chlorine; see Pigment Green 7; and (4) sulfonated copper phthalocyanine, water-soluble, green, in which two hydrogen atoms are replaced by sulfuric acid (HSO_3) groups.

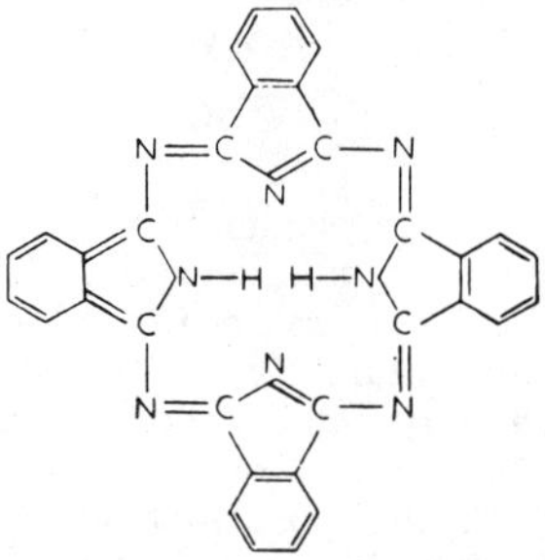

Use: Decorative enamels, automotive finishes, and similar applications where light fastness and chemical stability are required.

phthalonitrile. (o-dicyanobenzene).
CAS: 91-15-6. $C_6H_4(CN)_2$.
Properties: Buff-colored crystals, mp 138C, insoluble in water, soluble in acetone and benzene. Combustible.
Derivation: Vapor phase reaction of ammonia and phthalic anhydride over alumina catalyst at high temperature.
Hazard: Toxic by ingestion.
Use: Intermediate in organic synthesis, especially pigments and dyes, base material for high-temperature lubricants and coatings, insecticide.

phthaloyl chloride. (phthaloyl dichloride; phthalyl chloride). CAS: 88-95-9.
$C_6H_4(COCl)_2$.
Properties: Colorless, oily liquid, Mp 16C, bp 277C, refr index 1.568 (20C), decomposed by water or alcohol, soluble in ether. Combustible.
Derivation: By the action of phosphorus pentachloride on phthalic anhydride.
Hazard: Irritating by inhalation and skin contact.
Use: Chemical intermediate, especially for plasticizers and resins, laboratory reagent.
See also isophthaloyl chloride.

phycocolloid. One of several carbohydrate polymers (polysaccharides) occurring in algae (seaweed). They are hydrophilic colloids having a tendency to absorb water with swelling and to form gels of varying strength and consistency.

The chief types of phycocolloid are carrageenan from Irish moss, algin from brown algae, and agar from red algae. They contain complex galactose and mannose sugars and are sometimes considered as seaweed mucilages.
See specific entries for details.

"Phyllicin."[9] TM for theophylline-calcium salicylate.

physical chemistry. Application of the concepts and laws of physics to chemical phenomena in order to describe in quantitative (mathematical) terms a vast amount of qualitative (observational) information. A selection of only the most important concepts of physical chemistry would include: the electron wave equation and the quantum mechanical interpretation of atomic and molecular structure, the study of the subatomic fundamental particles of matter, application of thermodynamics to heats of formation of compounds and the heats of chemical reaction, the theory of rate processes and chemical equilibria, orbital theory and chemical bonding, surface chemistry, including catalysis and finely divided particles, the principles of electrochemistry and ionization. Although physical chemistry is closely related to both inorganic and organic chemistry, it is considered a separate discipline.

physiological salt solution. A solution of sodium chloride and water (0.9%) which is identical with the concentration found in the body. Also called isotonic salt solution.
Use: Medicine to replace acute loss of water as from burns, etc.

physostigmine. (eserine; calabarine).
CAS: 57-47-6. $C_{15}H_{21}O_2N_3$. An alkaloid.
Properties: Colorless or pinkish crystals, slightly soluble in water, soluble in alcohol and dilute acids, mp 86–87C and 105–106C (unstable and stable forms), specific rotation −119 to −121 degrees.
Derivation: By solvent extraction from the seeds of Physostigma venenosum.
Grade: USP.
Hazard: Toxic by ingestion.
Use: Medicine (anticholinesterase). Available as the salicylate and sulfate.

phytane. (2,6,10,14-tetramethylhexadecane).
$C_{20}H_{42}$. A hydrocarbon found in rock specimens 2.5–3 billion years old. Is known to be synthesized only by living organisms (is a derivative of chlorophyll) and to withstand heat and pressure, so helps to date the existence of life on earth.
See also pristane.

phytic acid. (inositolhexaphosphoric acid).
CAS: 83-86-3. $C_6H_6[OPO(OH)_2]_6$.
Occurs in nature in the seeds of many cereal grains, generally as the insoluble calcium-magnesium salt. It inhibits absorption of calcium in the intestine.
Properties: White to pale yellow liquid, odorless with acid taste, pH less than 1.0 (in 1% solution), soluble in water and alcohol, d 1.58, bulk d 13.1 lb/gal.
Derivation: From corn steep liquor.
Grade: Technical (as a 70% solution).
Use: Chelation of heavy metals in processing of animal fats and vegetable oils, rust inhibitor, preparation of phytate salts, metal cleaning, treatment of hard water, nutrient.

phytochemistry. That branch of chemistry dealing with (1) plant growth and metabolism and (2) plant products. The former includes the absorption of inorganic nutrients (nitrogen, phosphorus, potassium, carbon dioxide, water, etc.) to form sugars, starches, proteins, fats, vitamins, etc., and is closely associated with photosynthesis. Plant products comprise a vast group of natural materials and chemicals; besides those used directly as foods, these include alkaloids, cellulose, lignin, dyes, glucosides, essential oils, resins, gums, tannins, rubbers, terpene hydrocarbons, and glycerides (fats and oils). Some of these are basic raw materials for industry (paper, pharmaceuticals, food, paint, perfume, flavoring, leather, rubber); there are also many miscible plant products such as drugs, poisons, and pigments. Phytochemistry also embraces the study of plant hormones or growth regulators (auxin, gibberellin, synthetic types).

phytol. CAS: 150-86-7.
$C_{20}H_{40}O$.   An alcohol obtained by the decomposition of chlorophyll.
Properties: Odorless liquid; bp 202–204C (10 mm); d 0.8478 (25/4C); soluble in the common organic solvents; insoluble in water. Combustible.
Use: Synthesis of vitamins E and K.

phytonadione. (2-methyl-3-phytyl-1,4-naphthoquinone; vitamin K_1). $CH_3C_{10}H_4O_2C_{20}H_{39}$.
Properties: Clear, yellow, viscous, odorless liquid. D 0.967 (25/25C), refr index 1.5230–1.5252 (25C), stable in air. Protect from sunlight! Insoluble in water; soluble in benzene, chloroform, and vegetable oils; slightly soluble in alcohol.
Derivation: Synthetically from 2-methyl-1,4-naphthoquinone and phytol.
Grade: USP.
Use: Food supplement.

phytosterol. Sterol alcohols from plants. See sterol.

pi bond. A covalent bond formed between atoms by electrons moving in orbitals which extend above and below the plane of an organic molecule containing double bonds. A double bond consists of one pi and one sigma bond and a triple bond consists of one sigma and two pi bonds. See also metallocene, orbital theory.

pickle alum. See aluminum sulfate.

pickling. (1) Removal of scale, oxides, and other impurities from metal surfaces by immersion in an inorganic acid, usually sulfuric, hydrochloric, or phosphoric. Rate of scale removal varies inversely with concentration and temperature; the usual concentration is 15% at or above 100C. The rate is also increased by electrolysis. (2) A method of food preservation involving use of salt, sugar, spices, and organic acids (acetic). (3) Preserving or preparing hides for tanning by immersion in a 6–12% salt solution, together with enough acid to maintain pH at 2.5 or less.

picloram. (4-amino-3,5,6-trichloropicolinic acid).
CAS: 1918-02-1. $C_6H_3Cl_3N_2O_2$.
Properties: Crystalline solid, mp 218C.
Hazard: Toxic by ingestion and inhalation. Use has been restricted. TLV: 10 mg/m³ of air.
Use: Herbicide and defoliant.

pico-. Prefix meaning 10^{-12} unit (symbol = p) 1 pg = 1 picogram = 10^{-12} gram.

α-picoline. (2-methylpyridine; 2-picoline).
CAS: 109-06-8. $C_5H_4N(CH_3)$ or

$$\overline{NC(CH_3)CHCHCH}$$

Properties: Colorless liquid, strong unpleasant odor, d 0.952, bp 129C, fp −69.9C, refr index 1.4957 (20C), miscible with water and alcohol, flash p (OC) 102F (39C), autoign temperature 1000F (537C). Combustible.
Derivation: From cyclohexylamine plus ammonia and zinc chloride, also from coal tar and bone oil.
Hazard: Moderate fire risk. Irritant.
Use: Organic intermediate for pharmaceuticals, dyes, rubber chemicals, solvent, source for vinyl pyridine, laboratory reagent.

β-picoline. (3-methylpyridine; 3-picoline).
CAS: 108-99-6.

Properties: Colorless liquid, unpleasant odor, bp 143.5C, fp −18.3C, d 0.9613 (15/4C), refr index 1.5060 (20C). Soluble in water, alcohol, and ether. Combustible.

Derivation: From cyclohexylamine plus ammonia and zinc chloride, also from coal tar and bone oil.

Hazard: Moderate fire risk. Irritant.

Use: Solvent in synthesis of pharmaceuticals, resins, dyestuffs, rubber accelerators, insecticides, preparation of nicotinic acid and nicotinic acid amide, waterproofing agents, laboratory reagent.

γ-picoline. (4-methylpyridine; 4-picoline).

CAS: 108-89-4. $\overline{NCHCHC(CH_3)CHCH.}$

Properties: Colorless, moderately volatile liquid. D 0.957 (15/4C), bp 144.9C, refr index 1.5050 (20C), mp 3.8C. Soluble in water, alcohol, and ether. Flash p (OC) 134F (56.6C). Combustible.

Derivation: From cyclohexylamine plus ammonia and zinc chloride, also from coal tar and bone oil.

Hazard: Moderate fire risk. Irritant.

Use: Solvent in synthesis of pharmaceuticals, resins, dyestuffs, rubber accelerators, pesticides and waterproofing agents, laboratory reagent, making isoniazid, catalyst, curing agent.

picoline-N-oxide. (2-picoline-N-oxide).

$\overline{N(O)C(CH_3)CHCHCHCH.}$

Properties: Crystals, very soluble in water. Mp (2-isomer) 49.5C, (3-isomer) 40.5C, (4-isomer) 186.3C. Combustible.

Use: Organic synthesis.

4-picolylamine. $(CH_2NH_2)CHCHNCHCH.$

A heterocyclic compound, a pyridine derivative, highly reactive, strong base. Combustible.

Use: Manufacture of polyamides, epoxy curing agents, carbinols, and amine polyols.

picramic acid. (picraminic acid; 2-amino-4,6-dinitrophenol; dinitroaminophenol).

CAS: 96-91-3. $C_6H_2(NO_2)_2(NH_2)OH.$

Properties: Red crystals; soluble in alcohol, benzene, glacial acetic acid, aniline, and ether; sparingly soluble in water; mp 168C.

Derivation: By partial reduction of picric acid.

Hazard: May explode when shocked or heated, dangerous fire risk.

Use: Azo dyes, reagent for albumin.

picramide. Legal label name (Air) for trinitroaniline.

picraminic acid. See picramic acid.

picric acid. (picronitric acid; trinitrophenol; nitroxanthic acid; carbazotic acid; phenoltrinitrate).   CAS: 88-89-1. $C_6H_2(NO_2)_3OH.$

Properties: Yellow crystals; soluble in water, alcohol, chloroform, benzene, and ether. Very bitter taste, d 1.767, mp 122C, explodes at 300C.

Derivation: Nitration of phenolsulfonic acid, obtained by heating phenol with concentrated sulfuric acid.

Grade: Technical paste, pure paste.

Hazard: Severe explosion risk when shocked or heated, especially reactive with metals or metallic salts. Toxic by skin absorption. TLV: 0.1 mg/m³ of air.

Use: Explosives, matches, electric batteries, etching copper, mordant in textile dyeing, reagent, picrates.

picrolonic acid.

$NO_2C_6H_4NNC(CH_3)C(NO_2)COH.$

3-methyl-4-nitro-1-(p-nitrophenyl)-5-pyrazolone.

Properties: Yellow leaflets, mp 116–117C, decomposes 125C, slightly soluble in water, soluble in alcohol.

Use: Reagent for alkaloid identifications, for tryptophan and phenylalanine, for the detection and estimation of calcium.

pictronitric acid. See picric acid.

picrotoxin. (cocculin).   CAS: 124-87-8. $C_{30}H_{34}O_{13}.$ A glucoside.

Properties: Flexible, shining, prismatic crystals or microcrystalline powder; odorless; very bitter taste; stable in air; affected by light; mp 200C; soluble in boiling water, boiling alcohol, dilute acids, and alkalies; sparingly soluble in ether and chloroform.

Derivation: Derived from the fruit of Anamirta paniculata or Cocculus indicus, fishberries.

Hazard: Toxic in overdose.

Use: Medicine, as central nervous system stimulant and antidote for barbiturate poisoning.

picryl chloride. (2-chloro-1,3,5-trinitrobenzene). $C_6H_2(NO_2)_3Cl$.
Hazard: Severe explosion and fire risk. A high explosive.

Pictet-Gams isoquinoline synthesis. Formation of isoquinolines by cyclization of acylated aminomethyl phenyl carbinols or their ethers with phosphorus pentoxide in toluene or xylene.

Pictet-Hubert reaction. Phenanthridine cyclization by dehydrative ring closure of acyl-o-aminobiphenyls on heating with zinc chloride at 250-300 degrees, or with phosphorus oxychloride in boiling nitrobenzene.

Pictet-Spengler isoquinoline synthesis.
Formation of tetrahydroisoquinoline derivatives by condensation of β-arylethylamines with carbonyl compounds and cyclization of the Schiff bases formed.

"Pictol."[329]   TM for monomethyl-p-aminophenol sulfate, photo-developer.

PIDA. Abbreviation for phenylindane dicarboxylic acid.
See 1,1,3-trimethyl-5-carboxy-3-(p-carboxyphenyl)indane.

Pidgeon process. (ferrosilicon process; silicothermic process).   Process for the production of high-purity magnesium metal from dolomite or magnesium oxide by reduction with ferrosilicon at 1150C under high vacuum.

piezochemistry. Study of reactions occurring at very high pressures, e.g., in the interior of the earth's crust.

piezoelectricity. Electric energy created by application of pressure to ceramics or plastics. Devices utilizing this phenomenon are gas flame igniters, ultrasonic welding, and sonar navigation aids.

pig iron. Product of blast-furnace reduction of iron oxide in presence of limestone. About half the ore is converted to iron. Average analysis is: 1% silicon, 0.03% sulfur, 0.27% phosphorus, 2.4% manganese, 4.6% carbon, balance iron. Pig iron is the basic raw material for steel and cast iron. In metal terminology a "pig" is a bar or ingot of cooled metal.
See also iron.

pigment. Any substance, usually in the form of a dry powder, that imparts color to another substance or mixture. Most pigments are insoluble in organic solvents and water; exceptions are the natural organic pigments, such as chlorophyll, which are generally organosoluble. To qualify as a pigment, a material must have positive colorant value. This definition excludes whiting, barytes, clays, and talc.
See fillers, extenders.
Some pigments (zinc oxide, carbon black) are also reinforcing agents, but the two terms are not synonymous; in the parlance of the paint and rubber industries these distinctions are not always observed. Pigments may be classified as follows:
  I. Inorganic
      (a) metallic oxides (iron, titanium, zinc, cobalt, chromium).
      (b) metal powder suspensions (gold, aluminum).
      (c) earth colors (siennas, ochers, umbers).
      (d) lead chromates.
      (e) carbon black.
  II. Organic
      (a) animal (rhodopsin, melanin).
      (b) vegetable (chlorophyll, xanthophyll, indigo, flavone, carotene). See also pigment, plant.
      (c) synthetic (phthalocyanine, lithos, toluidine, para red, toners, lakes, etc). See also dye, natural and synthetic.

"Pigmentar."[296]   TM for tar products derived from the distillation and decomposition of oleoresinous southern pine. Produced to various viscosity grades.
Use: Rubber compounding and reclaiming, marine paints and roof coatings.

Pigment Blue 15. (CI No 74160).
$C_{32}H_{16}N_8Cu$.   A bright blue copper phthalocyanine pigment. Preparation: By heating phthalonitrile with cuprous chloride.
Use: In paints, alkyd resin enamels, printing inks, lacquers, rubber, resins, papers, tinplate printing, colored chalks and pencils.
See also phthalocyanine.

Pigment Blue 19. (CI No. 42750A).
$C_{32}H_{28}N_3O_4SNa$.   A bright blue to bright reddish-navy triphenylmethane pigment.
Use: Coloring for candles.

Pigment Blue 24. $C_{37}H_{34}N_2O_9S_3Na_2$.
A bright greenish-blue triarylmethane pigment. (CI No. 42090).
Use: In printing inks, especially for tinplate printing; in rubber; plastics; artist colors; lacquers.

Pigment E. See barium potassium chromate.

Pigment Green 7. (CI No. 74260).
$C_{32}O_{0-1}N_8Cl_{15-16}Cu$. A bright green chlorinated copper phthalocyanine pigment.
Derivation: Heating copper phthalocyanine in sulfur dichloride under pressure.
Use: Paints, printing inks, lacquers, leather and book cloth, paper surfacing, chalks, colored pencils.

pigment, plant. Any of a large number of organic natural colorants produced by living plants, with the exception of fungi and lichens. They may be classified into three groups, further information is given in specific entries.
(1) The chlorophylls (types a, b, and c): Green color. They are magnesium- containing porphyrins and are technically considered to be microcrystalline waxes.
(2) The carotenoids: yellow and orange colors.
  (a) carotene (straight-chain hydrocarbon)
  (b) xanthophyll (straight-chain hydrocarbon containing two oxygen molecules)
(3) The flavanoids: red, yellow, blue, orange, ivory colors. They are oxygen-containing heterocyclic compounds.
  (a) catechins
  (b) flavones, flavanols, anthocyanins
  (c) flavanones and leucoanthocyanidins
  (d) flavonols
Some of these pigments can be made synthetically. They have limited use as textile colorants and pharmaceutical products.

pigment, precipitated. See lake.

pigment volume concentration. See PVC (2).

Pigment Yellow 12. (CI No. 21090).
CAS: 6358-85-6. $C_{32}H_{26}Cl_2N_6O_4$.
A yellow diazo pigment. Preparation: Condensation of 3,3'-dichlorobenzidine di-diazotate with acetoacetanilide.
Use: Printing inks, lacquers resistant to heat and solvents, in rubber and resins, in paper coloring, textile printing.
See diazotization.

"Pilate" Fast Dyes.[440] TM for 1:1 metal complex dyes for dyeing and printing textiles of animal fibers and union materials of wool and nylon fibers.

pilchard oil. An oil expressed from the pilchard fish, a member of the herring family.
Properties: Pale yellow liquid, deposits stearin on long standing, d 0.931–0.933, saponification value 186–189.6, refr index 1.4751 (40C). Combustible.
Use: Potash soft soap, paints.

pilot plant. A trial assembly of small-scale reaction and processing equipment which is the intermediate stage between laboratory experiment and full-scale operation in the production of a new product. The functions of this stage are (1) to furnish chemical engineers with design data needed to construct a large-scale plant, (2) to resolve the many problems inherent in conversion from batch to continuous production, (3) to eliminate the differences that accompany change from constant laboratory conditions to a less closely controlled environment, and (4) to provide management with a basis for cost evaluation and estimation of the capital requirements of the new product. As the size of the pilot plant varies with the nature of the product, it must be determined on an individual basis.

Piloty-Robinson synthesis. Formation of pyrroles by heating azines of enolizable ketones with acid catalysts, usually zinc chloride or hydrochloric acid.

l-pimaric acid. (levopimaric acid).
CAS: 127-27-5. $C_{20}H_{30}O_2$.
Properties: Solid, mp 150C, optical rotation [a] 20/D −280 degrees (c = 0.7 in alcohol), soluble in most organic solvents, insoluble in water. Combustible.
Derivation: From pine gum.
Use: Resins.
See maleo-pimaric acid.

pimelic acid. (1,7-heptanedioic acid).
CAS: 111-16-0. $OOC[CH_2]_5COOH$.
Properties: Crystals, mp 105–106C, slightly soluble in water, soluble in alcohol and ether, nearly insoluble in cold benzene. Combustible.
Use: Biochemical research, polymers, plasticizers.

pimelic ketone. See cyclohexanone.

Pinacol rearrangement. Acid-catalyzed rearrangement of vicinal glycols to aldehydes and ketones.

pinacolone. (pinacoline; methyl-t-butyl ketone; 3,3-diemthyl-2-butanone). CAS: 75-97-8.
$CH_3COC(CH3)_3$.
Properties: Bp 106C, refr index 1.3964 (20C), d 0.801, flash p 75F (23C).

pindone. Coined name for 2-pivaloyl-1,3-indandione.

α-pinene. $C_{10}H_{16}$. A terpene hydrocarbon derived from sulfate wood turpentine.
Properties: Colorless, transparent liquid; terpene odor; insoluble in water; soluble in alcohol, chloroform, and ether. D 0.8620–0.8645 (15.5/

15.5C), refr index 1.4655–1.4670 (20C), boiling range 95% between 156–160C, fp −40C, flash p 90F (32.3C) (TCC), occurs in *d-*, *l-*, and racemic forms.
Hazard: Flammable, moderate fire risk. Skin irritant.
Use: Solvent for protective coatings, polishes, and waxes; synthesis of camphene, camphor, geraniol, terpin hydrate, terpineol, synthetic pine oil, terpene esters and ethers, lube oil additives, flavoring, odorant.

β-pinene. (nopinene). $C_{10}H_{16}$. A terpene hydrocarbon derived from sulfate wood turpentine.
Properties: Colorless, transparent liquid; terpene odor; insoluble in water; soluble in alcohol, chloroform, and ether. D 0.8740–0.8770 (15.5/15.5C), refr index 1.4775–1.4790 (20C), boiling range 95% between 164–169C, flash p 117F (47.2C), levorotatory. Combustible.
Hazard: Fire risk.
Use: Polyterpene resins, substitute for α-pinene, intermediate for perfumes and flavorings.

pinene hydrochloride. See bornyl chloride.

pine oil.
Properties: Colorless to light amber liquid having a strong, piny odor, miscible with alcohol, d 0.927–0.940, refr index 1.4780–1.4820 (20C), distilling range 200–225C, flash p (CC) 172F (77.7C). Combustible. Chief constituents: Tertiary and secondary terpene alcohols.
Derivation: From the wood of Pinus palustris by extraction and fractionation, or by steam distillation, also from turpentine.
Use: Odorant, disinfectant, penetrant, wetting agent, preservative (textile and paper industries), laboratory reagent, fragrances.

pine tar.
Properties: Sticky, viscous, dark brown to black liquid or semisolid with strong odor and sharp taste. Translucent in thin layers, hardens with aging, d 1.03–1.07, boiling range 240–400C, flash p (CC) 130F (54.4C). Soluble in alcohol, acetone, fixed and volatile oils, and in sodium hydroxide solution; slightly soluble in water. Combustible. Chief constituents: Complex phenols, turpentine, rosin, toluene, xylene, and other hydrocarbons.
Derivation: By destructive distillation of pine wood, especially Pinus palustris.
Grade: Kiln burnt, retort, NF.
Hazard: Fire risk, subject to spontaneous heating.
Use: Ore flotation, roofing compositions, paints and varnishes, softener in plastics and rubber processing, tar soaps, deKhotinsky cement, asphaltic compositions, marine preservative, medicine (cough syrups), laboratory reagent.

pine-tar oil. See tar oil, wood.

pine-tar pitch. The residue after distillation of practically all the volatile oils from pine tar. Similar to coal-tar pitch.

pinhole. (1) A small hole in electrical insulating tape caused by failure of the rubber coating to penetrate the fabric. The acceptable number of pinholes per unit area is subject to specification. (2) In paints, the presence of pimples or tiny holes in a coating.

Pinner reaction. Formation of imino esters (alkyl imidates) by addition of dry hydrogen chloride to a mixture of a nitrile and an alcohol. Treatment of alkyl imidates with ammonia or primary or secondary amines affords amidines, while treatment with alcohols yields ortho-esters.

Pinner triazine synthesis. Preparation of 2-hydroxy-4,6-diaryl-s-triazines by reaction of aryl amidines and phosgene.

pintsch gas. See oil gas.

pipecoline. See 2-methylpiperidine.

piperalin. [3-(2-methylpiperidino)propyl-3,4-dichlorobenzoate]. CAS: 3478-94-2. $CH_3C_5H_9N(CH_2)_3OC(O)C_6H_3Cl_2$.
Properties: Amber liquid, bp 156–157C (0.2 mm). Slightly soluble in water; miscible in paraffin hydrocarbon, aromatic hydrocarbon, and chlorinated hydrocarbon solvents.
Use: Fungicide.

piperazine. (diethylenediamine; pyrazine hexahydride; piperazidine). CAS: 110-85-0.

NHCH$_2$CH$_2$NHCH$_2$CH$_2$.

Properties: Colorless, deliquescent, transparent, needle-like crystals which absorb carbon dioxide from the air; soluble in water, alcohol, glycerol, and glycols; mp 104–107C; bp 145C; flash p 190F (87.7C). Combustible.
Derivation: Treatment of ethylene bromide or chloride with alcoholic ammonia at 100C.
Use: Corrosion inhibitor, anthelmintic, insecticide, accelerator for curing polychloroprene.

piperazine dihydrochloride. CAS: 142-64-3. $C_4H_{10}N_2 \cdot 2HCl$.
Properties: White needles, soluble in water.
Use: Fibers, insecticides, pharmaceuticals. The monochloride, $C_4H_{10}N_2 \cdot HCl$, is also commercially available.

piperazine hexahydrate. $C_4H_{10}N_2 \cdot 6HOH$.
Properties: White crystals, mp 44C, bp 125C, soluble in water and alcohol.

Use: Fibers, insecticides, pharmaceuticals, laboratory reagent, anthelmintic.

"Pipersin."[188] TM for a substitute for oleoresin of black pepper. Officially recognized by USDA Meat Inspection Division for use under its supervision.

piperidine. (hexahydropyridine; pentamethyleneamine). CAS: 110-89-4.

CH$_2$CH$_2$CH$_2$CH$_2$CH$_2$NH. Completely saturated ring compound.
Properties: Colorless liquid with odor of pepper, d 0.862, bp 106C, fp −7 to −9C, soluble in water, alcohol, and benzene; strong base. Combustible.
Derivation: By electrolytic reduction of pyridine.
Grade: 95 and 98% pure.
Hazard: Toxic by ingestion, strong irritant.
Use: Solvent and intermediate, curing agent for rubber and epoxy resins, catalyst for condensation reactions, ingredient in oils and fuels, complexing agent.

piperidine pentamethylene dithiocarbamate. ("Pippip"). C$_{11}$H$_{22}$N$_2$S$_2$.
Properties: White powder.
Hazard: Toxic by ingestion, strong irritant to eyes and skin.
Use: Ultra-accelerator for rubber.

2-piperidinoethanol. (n-2-hydroxyethylpiperidine). C$_5$H$_{10}$NCH$_2$OH.
Properties: Colorless liquid, d 0.972–0.974 (20/4C), bp 115–117C (45 mm), refr index 1.478–1.480 (20C), miscible with water and most organic solvents.
Use: Intermediate.

piperocaine hydrochloride. (3-(2-methyl-piperidyl)propyl benzoate hydrochloride). CAS: 533-28-8. C$_{16}$H$_{23}$NO$_2$·HCl.
Properties: White, crystalline powder; odorless and stable in air; mp 172–175C; bitter taste; solution (1 in 10) acid to litmus; soluble in water, alcohol, and chloroform; almost insoluble in ether and fixed oils.
Use: Medicine (local anesthetic).

piperonal. (heliotropin; piperonyl aldehyde; 3,4-methylenedioxybenzaldehyde). CAS: 120-57-0. C$_6$H$_3$(CH$_2$)OO)CHO (bicyclic).
Properties: White, shining crystals; turns red-brown on exposure to light. Floral odor, mp 35.5–37C, bp 263C, soluble in alcohol and ether, slightly soluble in water and glycerol. Combustible.
Derivation: By oxidation of isosafrole.
Grade: Technical, FCC.

Use: Perfumery, suntan preparations, mosquito repellent, laboratory reagent, flavoring.

piperonyl butoxide. (Generic name for α-[2-(2-butoxyethoxy)-ethoxy]-4,5-(methylene-dioxy)-2-propyltoluene). CAS: 51-03-6. C$_3$H$_7$C$_6$H$_2$(OCH$_2$O)CH$_2$OC$_2$H$_4$OC$_2$H$_4$OC$_4$H$_9$.
Properties: Light-brown liquid; mild odor; insoluble in water; soluble in alcohol, benzene, petroleum hydrocarbons. D 1.06 (25C), refr index 1.50 (20C), bp 180C (1 mm), flash p 340F (171C). Combustible.
Use: Synergist in insecticides in combination with pyrethrins in oil solutions, emulsions, powders or aerosols.

piperonyl cyclonene. (Generic name for a mixture of 3-alkyl-6-carbethoxy-5-(3,4-methylenedioxyphenyl)-2-cyclohexen-1-one and 3-alkyl-5-(3,4-methylenedioxyphenyl)-2-cyclohexen-1-one). CAS: 119-89-1.

(CH$_2$O$_2$)C$_6$H$_3$CHCH$_2$C(O)CH:CRCH$_2$.
R is usually C$_6$H$_{13}$.
Properties: Red liquid; insoluble in water, oils, and refrigerant 12; flash p 290F (143C). Combustible.
Use: Synergist in insecticides in combination with rotenone, pyrethrins, or rotenone-pyrethrin mixture in oil solutions, emulsions, or powders.

piperylene. (1,3-pentadiene). CH$_2$:CHCH:CHCH$_3$. Cis- and trans-forms.
Properties: Colorless liquid; d 0.693 (60/60F); fp (cis) −141C, (trans) −87C; bp (cis) 44C, (trans) 42C; refr index (cis) 1.43634 (20C), (trans) 1.43008 (20C); insoluble in water; soluble in alcohol and ether; flash p (mix) −20F (−28.8C).
Hazard: Highly flammable, dangerous fire risk.
Use: Polymers, maleic anhydride adducts, intermediate.

pipette. (pipet). A slender glass tube open at both ends and having an expanded area at or near the center designed to contain a specific volume of liquid, e.g., 5 ml. Liquid is drawn into the tube by oral or, for the sake of safety, some other form of suction.
Use: Transferring measured volumes of liquid from one container to another.

Piria reaction. Formation of arylsulfamic acids or sulfonation products or both by refluxing aromatic nitro compounds with a metal sulfite and boiling the mixture with dilute acid to yield the amines and sulfamic acids.

pitch. (1) A carbonaceous, tacky residue resulting from distillation of coal tar, petroleum, pine tar, and fatty acids. Some types, such as glance pitch, occur naturally. They are used chiefly as sealants, roofing compounds, and wood preservatives. Synthetic carbon fibers are made from petroleum pitch. (2) In papermakers' terminology, a mixture of calcium carbonate, calcium soaps from wood components, and miscellaneous residues from materials used in paper manufacture. Pitch of this type is a production nuisance which requires close control. (3) The degree of slope of an inclined plane as in a screw or auger, as measured by the distance between the flights or threads.

pitchblende. CAS: 1317-99-3. A massive variety of uraninite or uranium oxide found in metallic veins. Contains 55–75% UO_2, up to 30% UO_3, usually a little water, and varying amounts of other elements. Thorium and the rare earths are generally absent.
Properties: Color black, streak brownish-black, luster pitchy to dull, Mohs hardness 5.5, d 6.5–8.5.
Occurrence: Canada, Colorado, Europe, Zaire.
Hazard: Radioactive material.
Use: Most important ore of uranium, original source of radium.

pituitrin. (coluitrin; hypophysin; pitilobin).
Use: Antidiuretic hormone.

pivalic acid. See trimethylacetic acid.

2-pivaloyl-1,3-indandione. (pivalyl-1,3-indandione; pindone). CAS: 83-26-1.
$C_9H_5O_2C(O)C(CH_3)_3$.
Properties: Bright yellow powder or crystals, mp 109C, insoluble in water, soluble in most organic solvents.
Hazard: Toxic by inhalation and ingestion; inhibits blood clotting. TLV: 0.1 mg/m³ of air.
Use: Rodenticide, insecticide, pharmaceutical intermediate.

pK. A measurement of the completeness of an incomplete chemical reaction. It is defined as the negative logarithm (to the base 10) of the equilibrium constant, K for the reaction in question. The pK is most frequently used to express the extent of dissociation or the strength of weak acids, particularly fatty acids, and amino acids and also complex ions, or similar substances. The weaker an electrolyte the larger its pK. Thus at 25C for sulfuric acid (strong acid), pK is about −3.0, acetic acid (weak acid), pK = 4.76; boric acid (very weak acid) pK = 9.24. In a solution of a weak acid, if the concentration of undissociated acid is equal to the concentration of the anion of the acid, the pK will be equal to the pH.

planetology, chemical. See chemical planetology.

plankton. Microscopic plant and animal life that floats in the oceans or in lake waters.

plant. (1) A broad group of vegetable organisms comprised of all types of vegetation that contain chlorophyll (algae, mosses, grasses, vegetables, trees, etc., but excluding fungi). Their metabolic processes are vital to the maintenance of life on earth and result in the following products: (1) oxygen (from respiration), (2) carbohydrates (from photosynthesis), (3) amino acids and proteins (from nitrates and nitrogen-fixing bacteria), (4) fats and oils, (5) vitamins, (6) natural fibers, (7) coal, (8) various other substances of value such as alkaloid drugs, rubber, etc.
See also photosynthesis, phytochemistry.
(2) Any large-scale manufacturing unit including pipelines, reaction equipment, machinery, etc.

plant growth regulator. An organic compound either natural or synthetic that modifies or controls one or more specific physiological processes within the plant. If the compound is produced by the plant it is called a plant hormone e.g., auxin which regulates the growth of longitudinal cells involved in bending of the stem one way or another. Substances applied externally also bring about modifications such as improved rooting of cuttings, increased rate of ripening (ethylene), and easier scission (separation of fruit from stem). A large number of chemicals tend to increase the yield of certain plants such as sugar cane, corn, etc. All these, as well as plant-produced hormones, are included in the term plant growth regulator.
See dinitrobutylphenol, kinin, gibberellin, abscisic acid.

plant location. Selection of a site for a new chemical or process industry plant. The problem has been compounded in recent years by the increasing number of environmental regulations and by the energy shortage. Among the more important considerations are: (1) accessibility of essential materials, including water; (2) transportation of finished product (rail, air truck, barge); (3) reliability of fuel and power supply; (4) liquid and solid waste disposal restrictions; (5) commuting distance for employees in view of gasoline consumption; (6) availability of housing for employees; (7) state and local regulations (zoning, hazardous chemicals, building codes); (8) availability of qualified labor; (9) taxation; (10) weather factors (temperature range, severe storms, floods, etc.); (11) expansion possibilities.

"Plasdone."[307] TM for the pharmaceutical grade of polyvinylpyrrolidone.
Use: Tablet binding and coating agent, detoxicant and demulcent lubricant in ophthalmic preparations, film forming agent in medical aerosols.

"Plaskon."[175] TM for plastics and resins including alkyd, urea, melamine and nylon molding compounds; polyester, coating, foundry, bonding, impregnating, chlorotrifluoroethylene resins, adhesives, hardeners, phenolic laminating varnishes. 210–577A. A polyolefin designed for fabrication of tape for insulating computer cable, available in resin form or as specification product. FR 1050. A flame-retardant polypropylene resin, continuous-use at 100C.
Use: As TV tube sockets, structural parts for appliances, etc.

"Plaslube."[539] TM for unreinforced thermoplastics with lubricating additives, includes nylons, polycarbonates, and acetals with various concentrations of molybdenum disulfide and/or "Teflon."

plasma. (1) The portion of the blood remaining after removal of the white and red cells and the platelets; it differs from serum in that it contains fibrinogen which induces clotting by conversion into fibrin by activity of the enzyme thrombin. Plasma is made up of more than 40 proteins and also contains acids, lipids, and metal ions. It is an amber, opalescent solution in which the proteins are in colloidal suspension and the solutes (electrolytes and nonelectrolytes) are either emulsified or in true solution. The proteins can be separated from each other and from the other solutes by ultrafiltration, ultracentrifugation, electrophoresis, and immunochemical techniques.
(2) Two kinds of plasma are recognized by physicists, namely, a particle plasma and a reactor plasma. A particle plasma is a neutral mixture of positively and negatively charged particles interacting with an electromagnetic field, which dominates their motion. Temperatures of 10,000 to 15,000C can be reached. Such plasma formed by sudden energy releases, can be utilized as an energy source, as in magnetohydrodynamics. Reactor plasmas, on the other hand, are composed of positively charged ions of hydrogen isotopes (deuterium, tritium), the electric charge is the controlling factor. These are used in nuclear fusion devices, where temperatures of 74,000,000C have been attained and still higher temperatures are expected. These plasmas also respond to electromagnetic forces which are used to confine them.
See also magnetohydrodynamics, fusion, tokamak.

plasma volume expander. A substance used to partially or wholly replace blood plasma in treatment of the injured. Most important are gelatin, polyvinylpyrrolidone, and dextran.

plasmid. A strand or fragment of genetic material existing outside the chromosomes in certain types of bacteria. R-type plasmids, which are present in *E. coli*, impart resistance to antibiotics in organisms that are exposed to them. The plasmids can be transferred from animals to man, as well as to other, harmful bacteria which also become resistant to antibiotics. Feeding of traces of antibiotics to animals is believed to promote the growth of *E. coli* and, thus, to produce strains of pathogenic bacteria that are not amenable to antibiotic treatment. For this reason FDA has recommended elimination of certain antibiotics from animal feeds, e.g., penicillin, oxytetracycline, and chlortetracycline. Synthetic plasmids have been used successfully in recombinant DNA research.

plasmin. See fibrinolysin.

plasmoquin. (pamaquine; plasmochin; 8-dimethylamino-isoamyl-6-methoxyquinoline). $C_{19}H_{28}N_2O$.
Properties: Yellow powder, mw 300.2. Insoluble in water.
Use: Antimalarial.

"Plastacele."[28] TM for cellulose acetate flake, a fine white powder used for molding powders, films, sheets, rods, and tubes.

plaster of Paris. See calcium sulfate.

"Plasthall."[94] TM for a broad range of monomeric and polymeric plasticizers used in polymers and elastomers. Types include adipates, glutarates, trimellitates, azelates, sebacates, and tallates.

plastic. (1) Capable of being shaped or molded with or without the application of heat. Soft waxes and moist clay are good examples of this property.
See also plasticity.
(2) A high polymer, usually synthetic, combined with other ingredients, such as curatives, fillers, reinforcing agents, colorants, plasticizers, etc.; the mixture can be formed or molded under heat and pressure in its raw state and machined to high dimensional accuracy, trimmed, and finished in its hardened state. The thermoplastic type can be resoftened to its original condition by heat; the thermosetting type cannot.
Plastics in general (including all forms) are sensitive to high temperatures, among the more

resistant being fluorocarbon resins, nylon, phenolics, polyimides, and silicones, though even these soften or melt above 260C. Other types are combustible when exposed to flame for a short time (cellulosics, polyethylene, acrylic polymers, polystyrene), and still others burn with evolution of toxic fumes (polyurethane).

Engineering plastics are those to which standard metal engineering equations can be applied; they are capable of sustaining high loads and stresses, and are machinable and dimensionally stable. They are used in construction, as machine parts, automobile components, etc. Among the more important are nylon, acetals, polycarbonates, ABS resins, PPO/styrene and polybutylene terephthalate.

Fibers, films, and bristles are examples of extruded forms. Plastics may be shaped by either compression molding (direct pressure on solid material in a hydraulic press) or injection molding (ejection of a measured amount of material into a mold in liquid form). The latter is most generally used and articles of considerable size can be produced. Because of their dielectric properties, plastics are essential components of electrical and electronic equipment (especially for use within the human body).

Plastics can be made into flexible and rigid foams by use of a blowing agent; these foams are light and strong, and the rigid type is machinable. They are collectively called cellular plastics. Plastics can also be reinforced, usually with glass or metallic fibers for added strength. They are laminated to paper, cloth, wood, etc., for many uses in the packaging, electrical and furniture industries; they also can be metal-plated. Plastic pipe is widely used for underground transportation of gases and liquids over long distances as well as intraplant.

Several natural materials (waxes, clays, and asphalts) have rheological properties similar to synthetic products, but as they are not polymeric, they are not considered true plastics. Certain proteins (casein, zein) are natural high polymers from which plastics are made (buttons and other small items), but they are of decreasing importance.

Plastics have permeated industrial technology. Not only have they replaced and improved upon many materials formerly used, but also have made possible industrial and medical applications that would have been impracticable with older technologies. Use of plastics in the US has been authoritatively estimated at close to 60 billion pounds a year in 1980, which is twice the 1970 consumption. Their major application areas are: (1) automobile bodies and components, boat hulls (2) building and construction (siding, piping insulation, flooring) (3) packaging (vapor-proof barriers, display cartons, bottles, drum linings) (4) textiles (carpets, cordage, suiting, hosiery, drip-dry fabrics, etc.) (5) organic coatings (paint and varnish vehicles) (6) adhesives (plywood, reinforced plastics, laminated structures) (7) pipelines (8) electrical and electronic components (9) surgical implants (10) miscellaneous (luggage, toys, tableware, brushes, furniture, etc.). For additional information refer to Society of the Plastics Industry, 250 Park Ave, New York. See also polymer, high; cellular plastic; reinforced plastic; foam, plastic; plastic pipe. See Appendix III for history of the industry.

plastic flow. A type of rheological behavior in which a given material shows no deformation until the applied stress reaches a critical value called the yield value. Most of the so-called plastics do not exhibit plastic flow. Common putty is an example of a material having plastic flow.

plastic foam. See foam, plastic; cellular plastic.

plastic, reinforced. See reinforced plastic.

plasticity. A rheological property of solid or semi-solid materials expressed as the degree to which they will flow or deform under applied stress and retain the shape so induced, either permanently or for a definite time interval. It may be considered the reverse of elasticity. Application of heat and/or special additives is usually required for optimum results. See also thermoplastic; plasticizer.

plasticizer. An organic compound added to a high polymer both to facilitate processing and to increase the flexibility and toughness of the final product by internal modification (solvation) of the polymer molecule. The latter is held together by secondary valence bonds; the plasticizer replaces some of these with plasticizer-to-polymer bonds, thus aiding movement of the polymer chain segments. Plasticizers are classed as primary (high compatibility) and secondary (limited compatibility). Polyvinyl chloride and cellulose esters are the largest consumers of plasticizers; they are also used in rubber processing. Among the more important plasticizers are nonvolatile organic liquids and low-melting solids, e.g., phthalate, adipate, and sebacate esters, polyols such as ethylene glycol and its derivatives, tricresyl phosphate, castor oil, etc. Camphor was used in the original modification of nitrocellulose to "Celluloid" See also plastisol, softener.

"Plasticone Red."[141] TM for light, medium, and deep shades of pyrazolone red pigments. Use: Paints, enamels, lacquers, plastics, rubber, printing inks, textiles and floor coverings.

plastic pipe. Tubes, cylinders, conduits, and continuous length piping made (1) from thermoplastic polymers unreinforced (polyethylene, polyvinyl chloride, ABS polymers, polypropylene) or (2) from thermosetting polymers (polyesters, phenolics, epoxies) blended with 60–80% of such reinforcing materials as chopped asbestos or glass fibers to increase strength. The latter type is a reinforced plastic. In general the properties of plastic tubing or pipe are those of the polymers that comprise it. Most have good resistance to chemicals, corrosion, weathering, etc., combined with flexibility, light weight, and high strength. They are combustible, but generally slow-burning. The reinforced type is widely used as underground conduit for transportation of gases and fluids, including city water services, sewage disposal systems, etc. Its use in buildings is subject to local building codes.
See Plastic Pipe Institute, 250 Park Avenue, New York, NY, for further details.

"PlastifixPC."[470] TM for a polychloroprene-based resin for corrosion-resistant, elastic metal coatings.

"Plat-Iron."[296] TM for high purity electrolytic iron powder and reduced iron oxide powder, annealed and unannealed.
Use: Powder enrichment, catalyst, pole pieces, magnets, electronic cores, welding rod coatings, sintered structural parts, and oil-less bearings.

plastisol. A dispersion of finely divided resin in a plasticizer. A typical composition is 100 parts resin and 50 parts plasticizer, forming a paste that gels when heated to 150C as a result of solvation of the resin particles by the plasticizer. If a volatile solvent is included, the plastisol is called an organosol. Plastisols are used for molding thermoplastic resins, chiefly polyvinyl chloride.
See also plasticizer.

"Plast-Manganese."[296] TM for electrolytic manganese powder.
Use: Welding rod coatings, pyrotechnics, and fuses.

"Plast-Nickel."[296] TM for nickel powder.
Use: Welding rod coatings, sintered permanent magnets, filters and parts.

"Plastogen."[69] TM for a plasticizing agent.
Properties: Liquid, amber to mahogany, d 0.81–0.84, acid number 1.0–1.1.
Use: Plasticizer and softener in all elastomers, effective in sponge rubber.

"Plastolein."[242] TM for vinyl plasticizers, including di-n-hexyl azelate, diethylene glycol dipelar-gonate, diisooctyl azelate, di-2-ethylhexyl azelate, tetrahydrofurfuryl oleate, triethylene glycol dipelargonate.

"Plast-Silicon."[296] TM for silicon powder.
Use: Fuses and pyrotechnics.

platelet. (thrombocyte). A proteinaceous cellular structure occurring in blood in the amount of $150–500 \times 10^3$ units/mm^3. Platelets range from 2 to 4 microns in diameter and contain no nuclei. They are rich in amine compounds which constrict the blood vessels at the site of an injury, to which the platelets adhere; on dissolution they release thromboplastin, which initiates the coagulation mechanism.
See also blood, fibrinogen, thrombin.

platen. A vertically movable plate (deck) of a compression molding press.
See also hydraulic press.

platinic ammonium chloride. See ammonium hexachloroplatinate.

platinic chloride. See chloroplatinic acid, platinum chloride.

platinic oxide. See platinum dioxide.

platinic sal ammoniac. See ammonium hexachloroplatinate.

platinic sodium chloride. See sodium chloroplatinate.

platinic sulfate. See platinum sulfate.

platinous ammonium chloride. See ammonium chloroplatinate.

platinous chloride. See platinum dichloride.

platinous iodide. See platinum iodide.

platinum. CAS: 7440-06-4. Pt.
Metallic element of atomic number 78, group VIII of the periodic system, aw 195.09, valences = 2, 4. There are five stable isotopes.
Properties: Silvery, white, ductile metal. Insoluble in mineral and organic acids, soluble in aqua regia, attacked by fused alkalies, d 21.45, mp 1769C, bp 3827C. Brinell hardness 97, annealed (Vickers) 42. Does not corrode or tarnish; heated platinum absorbs large volumes of hydrogen. It is also a strong complexing agent. As a catalyst it is abnormally sensitive to poisons.
Occurrence: Canada (Ontario), South Africa, USSR, Alaska. Usually mixed with ores of copper, nickel, etc.

Derivation: By dissolving the ore concentrate in aqua regia, precipitating the platinum by ammonium chloride as ammonium hexachloroplatinate, igniting the precipitate to form platinum sponge. This is then melted in an oxyhydrogen flame or in an electric furnace.

Grade: Physically pure (99.99%), chemically pure (99.9%), crucible platinum (99.5%), commercial (99.0%).

Forms available: Powder (platinum black), single crystals, wire (2 by 0.05–0.005 inches diameter); special composition for electronics, metallizing and decorating ceramics and metals.

Hazard: Flammable in powdered form. Soluble salts are highly toxic by inhalation. TLV: 1 mg/m^3 of air.

Use: Catalyst (nitric acid, sulfuric acid, high-octane gasoline, automobile exhaust gas converters), laboratory ware, spinnerets for rayon and glass fiber manufacture, jewelry, dentistry, electrical contacts, thermocouples, surgical wire, bushings, electroplating, electric furnace windings, chemical reaction vessels, permanent magnets.

platinum ammonium chloride. See ammonium hexachloroplatinate and ammonium chloroplatinate.

platinum barium cyanide. See barium cyanoplatinite.

platinum black. Finely divided metallic platinum.
Properties: Black powder, exhibits a metallic luster when rubbed. Soluble in aqua regia, d 15.8–17.6 (apparent).
Derivation: Reduction of solution of a platinum salt with zinc or magnesium.
Hazard: Flammable when dispersed in air.
Use: Catalyst; to absorb gases (hydrogen, oxygen, etc.) which it again liberates at red-heat; gas ignition apparatus.

platinum chloride. (platinum tetrachloride; platinic chloride).   CAS: 10025-65-7.   (a) PtCl$_4$; (b) PtCl$_4 \cdot$5HOH.   The platinum chloride of commerce is usually chloroplatinic acid.
Properties: (a) Brown solid, (b) red crystals; soluble in water and alcohol; (a) d 4.30 (25C), decomposes at 370C; (b) d 2.43, mp loses 4H$_2$O at 100C.
Derivation: By solution of platinum in aqua regia and evaporation.
Hazard: TLV (as Pt): 0.002 mg/m^3 of air.
Use: See chloroplatinic acid.

platinum-cobalt alloy. A 76.7 platinum/23.3 cobalt alloy forms a more powerful permanent magnet than any other known.

platinum dichloride. (platinous chloride). PtCl$_2$.
Properties: Greenish-gray powder which forms double salts with the chlorides of the alkali metals, soluble in hydrochloric acid and ammonium hydroxide, insoluble in water, d 5.87, mp is decomposed at red heat, yielding platinum.
Derivation: (a) By heating platinum sponge in presence of dry chlorine, (b) by heating chloroplatinic acid to 200C.
Use: Platinum salts.

platinum dioxide. (platinic oxide).   PtO$_2$.
Properties: Black powder, soluble in concentrated acids and dilute solutions of potassium hydroxide.
Derivation: Reaction of platinic chloride with excess sodium hydroxide.
Use: Hydrogenation catalyst (forms platinum black when reduced by hydrogen).

platinum iodide. (platinous iodide; platinum diiodide).   PtI$_2$.
Properties: Heavy black powder, slightly soluble in hydriodic acid, insoluble in alkalies and water; d 6.4, mp 300–350C (decomposes)

platinum-iridium alloy. The most important platinum alloy. Commercial alloys contain 1–30% iridium. As the iridium is increased the hardness of the alloy increases, as does the resistance to chemical attack. The melting point of platinum is raised by the addition of iridium.
Use: Jewelry ("medium" platinum is 95% platinum, 5% iridium and "hard" platinum is 90% platinum, 10% iridium), electrical contacts (10–25% iridium), fuse wire (10–20% iridium), hypodermic needles (20–30% iridium), and in general where high corrosion resistance is needed.
See also iridium.

platinum-lithium.   LiPt$_2$.   Brittle solid, non reactive with water, made by direct combination at 540C. If the lithium and platinum are combined at 200C the product can be decomposed by water, hydrolyzing and dissolving the lithium and leaving unusually active platinum catalyst.

platinum metal. Any of a group of six metals, all members of group VIII of the periodic system: ruthenium, rhodium, palladium, osmium, iridium, and platinum. All of these are also transition metals.

platinum potassium chloride. See potassium chloroplatinate.

platinum-rhodium alloys. Alloys containing up to 40% rhodium. Such alloys are harder than plati-

num but not as hard as the corresponding platinum-iridium alloys. The addition of rhodium to
platinum increases the resistance to attack by
aqua regia. The melting points of the alloys are
higher than those of platinum.
Use: Catalyst in nitric acid production, high temperature vessels, furnace resistors, thermocouples and resistance thermometers, spinneret
nozzles, components of gas turbine aircraft engines.

platinum sodium chloride.   See sodium chloroplatinate.

platinum sponge.
Properties: Grayish-black, porous mass of finely
divided platinum; soluble in aqua regia.
Derivation: By ignition of ammonium hexachloro-
platinate or other salts.
Use: Catalyst.
See also platinum black.

platinum sulfate.   (platinic sulfate). $Pt(SO_4)_2$.
Properties: Greenish-black mass, hygroscopic.
Soluble in acids (dilute), alcohol, ether, water.
Hazard: Toxic by inhalation. TLV: 0.002 mg/m³
of air.
Use: Analysis (microtesting for bromine, chlorine,
iodine).

platinum tetrachloride.   See platinum chloride.

"Plexiglas."[23]   TM for thermoplastic poly(methyl
methacrylate)-type polymers. Available in bead
or granule form and sheets.
Use: Manufacture of lenses, ornaments, letters for
signs, aircraft canopies and windows, light diffus-
ers, industrial and architectural glazing, chalk-
boards, boat windshields, and similar products.

"Plexol."[23]   TM for synthetic lubricants and additives for petroleum oils. Most grades are diesters
of dibasic acids, some are polyesters or polyether
alcohols. The ester lubricants have very low
freezing points, high flash points, little change
of viscosity with temperature.
Use: Aircraft engine lubricants, hydraulic systems,
instrument oils, petroleum-base lubricant formulation.

pliofilm.   TM for rubber hydrochloride as transparent base.

"Plioflex."[265]   TM for a series of staining and
nonstaining synthetic dry elastomers produced
by emulsion polymerization. These rubbers are
either styrene/butadiene or polybutadiene
polymers. All are protected with antioxidants.

"Pliolite."[265]   TM for a series of styrene/buta-
diene polymers in both latex and solid form.
Use: Coatings, printing inks, adhesives, and rubber
reinforcement.

"Pliopave."[265]   TM for a series of rubber latices
used specifically as modifiers of bituminous paving materials. The series is composed of polymers
of styrene/butadiene, acrylonitrile/butadiene,
and others.

"Pliovic" Vinyl Resins.[265]   TM for a group of
thermoplastic resins composed of polymers and
copolymers consisting of 50% or more vinyl
chloride. Supplied in the form of fine white powders, the resins are easily compounded and
formed into finished goods by extruding, calendering, compression molding, injection molding,
and blow molding.

plow.   A scraping device of various contours used
for dislodging sediment or "mud" accumulated
at the bottom of clarifying and thickening tanks.
It is activated by a rotating arm which moves
it circumferentially around the tank riding close
to the bottom; it thus transfers the sedimented
material from the periphery to the center of the
tank where it is discharged into a hopper. Plows
are also used in mixing equipment of the Muller
type, where they serve to continuously rake the
material being mixed into the path of the Muller
wheels.

plumbic acid, anhydrous.   See lead dioxide.

plumboplumbic oxide.   See lead oxide, rėd.

"Plumb-O-Sil" B and C.[304]   TM for coprecipitates
of lead-orthosilicate and silica gel.
Properties: Soft, white powders; d (B) 3.9, (C)
3.1, refr index (B) 1.58–1.60, (C) 1.58.
Use: Translucent and colored vinyl film, sheeting,
and upholstery stocks, as vinyl stabilizers.

plumbous oxide.   See litharge.

plumbous sulfide.   See lead sulfide.

plumbum.   The Latin name for lead, hence the
symbol Pb and the names plumbic and plumbous.

"Pluracol."[203]   TM for a series of organic compounds used in hydraulic brake and other functional fluids, chemical intermediates, urethane
foams, elastomers, and coatings.

"Plurafac."[203]   TM for a series of 100% active,
nonionic biodegradable surfactants of straight
chain, primary aliphatic oxyethylated alcohols.
Available in liquid, paste, flake, and solid form.
Use: Range from light-duty hand dishwashing formulations to heavy-duty industrial detergents,
rinse aids, metal cleaners, etc.

"Pluronic."[203]   TM for a nonionic series of 28 related difunctional block-polymers terminating in primary hydroxyl groups with molecular weights ranging from 1,000 to above 15,000. They are polyoxyalkylene derivatives of propylene glycol. Available in liquid, paste, flake powder, and cast-solid forms.
Use: Defoaming agents, emulsifying and demulsifying agents, binders, stabilizers, dispersing agents, wetting agents, rinse aids, and chemical intermediates.

plutonium.   Pu.   CAS: 7440-07-5.
Synthetic radioactive metallic element with atomic number 94, first prepared in 1941, aw 239.11, valences = 3, 4, 5, 6, there are 15 isotopes (from 232–246), 6 allotropic forms. Plutonium-239 (half-life 24,360 years) is produced in a nuclear reactor by neutron bombardment of the nonfissionable isotope ^{238}U. Plutonium is readily fissionable with both slow and fast neutrons and can be used for either nuclear weapons or electric power production. The critical mass of pure ^{239}Pu is 10 lb. One lb of reactor-grade material contains a heat energy equivalent of 10^6 kilowatt hours. Weapons-grade plutonium contains up to 7% of ^{240}Pu; though reactor-grade plutonium is somewhat less pure, it can be used for weapons. The Safeguards Manual of the International Atomic Energy Agency states: "Plutonium of any grade, in either metal, oxide, or nitrate form, can be put in a form suitable for the manufacture of explosive devices in a matter of a few days or weeks."
According to Glenn T. Seaborg, "in breeder reactors it is possible to create more new plutonium from U-238 than that consumed in sustaining the fission chain reaction. Because of this, plutonium is the key to unlocking the enormous energy reserves in the nonfissionable U-238." Reactor fuels containing plutonium can be either liquid or solid; since plutonium forms low-melting alloys with a number of metals (gallium, bismuth, tin, iron, cobalt, and nickel) these are often used as liquid reactor fuels. Cerium also may be a component.
Hazard: The most radiotoxic of the elements and one of the most toxic substances known; dangerous ionizing radiation persists indefinitely; a powerful carcinogen. Must be handled by remote control and with adequate shielding.
See also breeder.

"Plyamule."[36]   TM for a series of vinyl acetate homopolomers.

"Plyamine."[36]   TM for a group of liquid water-soluble urea-formaldehyde adhesive resins
Use: Binders in the manufacture of plywood, furniture, wood particle products, etc.

"Plyophen."[36]   TM for a water-soluble impregnating resin. Penetrates deeply and quickly into wood, canvas, asbestos, paper and other laminating and molding stocks. Can be diluted as much as 8–10 parts water to 1 part resin for spraying glass fiber or rock wool.

plywood.   A composite composed of thin wood veneers (with grains placed at right angles to each other) bonded with a synthetic resin, usually phenol-formaldehyde or resorcinol-formaldehyde. It is superior to metals in strength-to-weight ratio, and has low thermal expansion, high heat capacity, and low water absorption.
See also laminate, composite.

Pm.   Symbol for promethium.

PMA.   Abbreviation for phosphomolybdic acid and for pyromellitic acid.

"PMA."[74]   TM for a series of fungicides containing phenyl mercury acetate.
Use: Emulsion paint and other aqueous systems.
Hazard: As for mercury compounds.

"PMD-10."[74]   TM for a mineral spirits solution of phenyl mercury oleate.
Use: Mildew-resistant oil, oleoresinous and alkyd paints.
Hazard: As for mercury compounds.

PMDA.   Abbreviation for pyromellitic dianhydride.

PMHP.   Abbreviation for p-menthane hydroperoxide. A polymerization catalyst.

PMP.   Abbreviation for 1-phenyl-3-methyl-5-pyrazolone.

PMTA.   Abbreviation for a mixture of phosphomolybdic and phosphotungstic acids.
Use: Making pigments.
See phosphotungstic pigment.

pn.   Abbreviation for propylenediamine, as used in formulas for coordination compounds.
See also dien, en, py.

"PNF."[278]   TM for a special-purpose synthetic rubber, a phosphonitrilic fluoroelastomer, said to be flexible at −56C and serviceable up to 176C. Resistant to oils over a wide temperature range. Can be processed on standard equipment.

Po.   Symbol for polonium.

"POCO."[486]   TM for a series of unique fine-grained, high strength, isotropic, formed graphite materials. Three basic density grades range between 1.50 and 1.88 g/cm³, the associated flexural strengths range from 4000 to 17,250 psi. Use: Jigs and fixtures for electronic components, electrodes for spectrographic use, electrical discharge machining, aerospace and nuclear fields.

POEMS. Abbreviation for polyoxyethylene monostearate,
See polyoxyl 40 stearate.

POEOP. See polyoxyethyleneoxypropylene.

POF. See *dl*-α-lipoic acid.

poise. See centipoise.

poison. (1) Any substance that is harmful to living tissues when applied in relatively small doses. The most important factors involved in effective dosage are (a) quantity or concentration, (b) duration of exposure, (c) particle size or physical state of the substance, (d) its affinity for living tissue, (e) its solubility in tissue fluids, and (f) the sensitivity of the tissues or organs. Sharp distinction between poisons and non-poisons is not always possible, as many variables must be taken into consideration in each case. Poisons are divided into four classes by the shipping regulatory agencies, as follows: Poison A: A gas or liquid so toxic that an extremely small amount of the gas or the vapor formed by the liquid is dangerous to life. Poison B: Less toxic liquids and solids that are hazardous either by contact with the body (skin absorption) or by ingestion. Poison C: Liquids or solids that evolve toxic or strongly irritating fumes when heated or when exposed to air (excluding Class A poisons). Poison D: Radioactive materials.
See also toxicity, toxic substances.
Note: A computerized poison information center is operated by the FDA in Washington, DC. The national clearinghouse for poison information centers is located at 5401 Westbard Ave., Bethesda, MD, 20016.
(2) In nuclear technology, any material with a high capture probability for neutrons that may divert an undesirable number of neutrons from the fission chain reaction.
(3) A substance that reduces or destroys the activity of a catalyst. carbon monoxide and phosphorus, arsenic, or sulfur compounds have this effect on the formation of ammonia from hydrogen and nitrogen gases and the gases must be highly purified to avoid this. Another example is the poisoning of the platinum catalysts used in emission-control devices by organic lead compounds.

poison gas. A toxic or irritant gas or volatile liquid designed for use in chemical warfare or riot control. They vary in toxicity from nerve gases, which are lethal, to tear gases (lachrymators), which cause only temporary disability.
See also noxious gas, chemical warfare.

polar. Descriptive of a molecule in which the positive and negative electrical charges are permanently separated, as opposed to nonpolar molecules in which the charges coincide. Polar molecules ionize in solution and impart electrical conductivity. Water, alcohol, and sulfuric acid are polar in nature; most hydrocarbon liquids are not. Carboxyl and hydroxyl groups often exhibit an electric charge. The formation of emulsions and the action of detergents are dependent on this behavior.
See also dipole moment.

polarimetry. Measurement of the degrees and direction of the plane of polarized light as it passes through an optically active compound. It is used in the investigation of optical isomers, especially in analysis of sugars.

"Polaris Red."[141] TM for precipitated azo pigments of a very bright, blue shade of red derived from β-hydroxynaphthoic acid.
Use: Printing inks, rubber, plastics.

polarized light. See nicol, polarimetry.

"Polectron."[307] TM for modified vinylpyrrolidone resins.
Properties: 40% active aqueous emulsion, stable to intense mechanical shear, freeze-thaw cycling. Compatible with other commercial latexes.
Use: Binding agents for wood, cotton, paper, glass fibers; stabilizer, opacifier and dyestuff; medium precoat for photosensitive papers. Adhesive for metal, glass, cotton, paper, and wood.

"Polidase."[91] TM for cultured vegetable enzyme product.
Use: Laboratory reagent, experimental nutrition, recovery of silver from photographic film, component of leather bates, conversion of starch in mashes used for alcohol production, low cost industrial enzyme.

"Polidene."[263] TM for a series of polyvinylidene chloride copolymer emulsions.
Use: Pigmented coating for textiles, paper, leather, etc.; fire retardant finishes.

polish. (1) A solid powder or a liquid or semiliquid mixture that imparts smoothness, surface protection, or a decorative finish. The most widely used solid polishing agent is fine-ground red iron oxide (rouge), applied to the surface of plate glass, backs of mirrors, and optical glass.

A wide variety of liquid and paste-like polishes are based on vegetable waxes (carnauba and candelilla), combined with softeners, fillers, and pigments or emulsified in alcohol or other solvent. Furniture polishes often contain red oil, lemon oil, and petroleum solvent; most types of metal and wood polish contain organic solvents, and hence are flammable liquids. Nail polishes are nitrocellulose lacquers, usually with amyl acetate solvent.

See also electro-polishing.

Hazard: May be flammable.

(2) The hard outer coating of cereal grains, especially rice, which is usually removed in processing. These coatings are rich in vitamin B_1. Their removal robs the cereal of much of its nutritive value.

pollucite. $Cs_4Al_4Si_9O_{26} \cdot HOH$.   A natural cesium aluminum silicate found in pegmatites.

Properties: Colorless, Mohs hardness 6.5, d 2.9.

Use: Source of cesium, catalyst, fluxes, welding materials, ion propulsion, thermocouple units.

pollution.   Introduction into any environment of substances that are not normally present therein and that are potentially toxic or otherwise objectionable. The most serious atmospheric contaminants have been (a) sulfur dioxide evolved from the fuels used in electric power production and industrial processing, and (b) automobile exhaust gases rich in carbon monoxide and tetraethyllead residues. The former is being alleviated by mandatory use of low-sulfur fuels and the latter by elimination of tetraethyllead from most gasoline and by use of catalytic converters.

Water pollution due to discharge of toxic chemical wastes is closely regulated by both EPA and FDA. Such substances are defined in the 1972 amendment of the Federal Water Pollution Control Act as those "which will cause death, disease, cancer, or genetic malfunctions in any organisms with which they come into contact." Substances added to water for purification purposes (chlorine, aluminum sulfate, etc.) are excluded from the category of pollutants.

See also Environmental Protection Agency, air pollution, water pollution.

polonium. Po.   Radioactive element of atomic number 84, member of Group VIA of the Periodic Table, aw 210, valences = 2, 4, 6. There are no stable isotopes. Polonium is a member of the uranium natural radioactive decay series, occurring naturally in uranium-bearing ores; it is produced artificially by bombarding bismuth with neutrons. It has been identified in cigarette smoke.

Properties: Similar to those of tellurium, mp 254C, bp 962C, d 9.4, dissolved by concentrated sulfuric and nitric acids and aqua regia, and by dilute hydrochloric acid.

Hazard: Dangerous radioactive poison.

Use: Source of alpha radiation and neutrons, instrument calibration, oil-well logging, moisture determination, power source.

See smoke, 4.

Polonovski reaction.   Demethylation of tertiary (or heterocyclic) amine N-oxides on treatment with acetyl chloride or acetic anhydride to give N-acylated secondary amines and formaldehyde, along with O-acylated aminophenols as a result of a side reaction.

poly-.   A prefix signifying many. For example, a polymer is an aggregate formed by combination of a number of single molecules.

See polymer, high.

"Polyac."[28]   TM for a butyl rubber conditioner containing 25% poly-p-dinitrosobenzene $[C_6H_4(NO)_2]_x$ with an inert wax. Dark brown, waxy pellets; d 0.96.

Use: Processing aid and accelerator of vulcanization for butyl rubber.

polyacetal.   See acetal resin.

polyacetylene.   A linear polymer of acetylene having alternate single and double bonds, developed in 1978. It is electrically conductive but this property can be varied in either direction by appropriate doping either with electron acceptors (arsenic pentafluoride or a halogen) or with electron donors (lithium, sodium). Thus, it can be made to have a wide range of conductivity from insulators to n- or p-type semiconductors, to strongly conductive forms. Polyacetylene can be made in both cis and trans modifications in the form of fibers and thin films, the conductivity of the fibers increasing with their degree of orientation. Films can be applied on glass or metal substrates. Though still in an experimental stage, these polymers have significant possibilities for industrial applications, e.g., in batteries.

See also cyclooctatetraene.

polyacrylamide. $(CH_2CHCONH_2)_x$.

White solid, water-soluble high polymer.

Derivation: Polymerization of acrylamide with N,N'-methylene bisacrylamide.

Use: Thickening agent, suspending agent, additive to adhesives. Permissible food additive.

See also acrylic resin.

polyacrylate.   See acrylic resin.

polyacrylic acid.   See acrylic acid, methacrylic acid.

polyacrylonitrile. A polymer of acrylonitrile which is the basic material used in the manufacture of a number of synthetic fibers, e.g., "Orlon" and "Dynel." When combined with other materials it produces a hard resin having high solvent resistance

$$—CH_2—CH_2$$
$$|$$
$$CN$$

and high-temperature stability; from these are made such items as moldings, shoe soling, wall panels, and the like.
See also acrylonitrile, acrylic resin, acrylic fiber.

polyalcohol. See polyol.

"Polyall."[175] TM for a series of alkyd resin-based thermosetting compounds.
Properties: Good electrical resistance, high strength, flame resistance, good dimensional stability, very fast cure, resistance to solvents and weak acids, good color fastness in sunlight and heat, fungus-resistant, d 2.1, heat distortion temperature 204C.

polyallomer. A copolymer which has a uniform crystalline structure but a mixed chemical composition. It is prepared by anionic coordination catalysis using a Ziegler catalyst. The best-known polyallomer is a copolymer of propylene and ethylene; it has the stereoregular crystalline structure of the homopolymers of these resins, but a variable chemical composition. Temperature range −40 to 99C. The physical properties of polyallomers are generally intermediate between those of the homopolymers of the component resins, but give a better balance of properties than blends of the homopolymers.
Use: Vacuum-formed, injection-molded, blow-molded, and extruded products; film; sheeting; wire cables.

polyamide. A high molecular weight polymer in which amide linkages (CONH) occur along the molecular chain. They may be either natural or synthetic. Important natural polyamides are casein, soybean and peanut proteins, and zein, the protein of corn, from all of which plastics, textile fibers, and adhesive compositions can be made. Synthetic polyamides are typified by the numerous varieties of nylon, though some, e.g., "Versamide", are quite different from nylon.
See also nylon, polypetide, protein, aramid.

polyamine-methylene resin. A polyethylene polyamine methylene substituted resin of diphe-nylol dimethylmethane and formaldehdyde in basic form.
Properties: Light amber, granular, freely flowing powder with appreciable odor. Insoluble in dilute acids and alkalies, alcohol, ether, and water.
Use: Medicine, ion-exchange resin, antacid.

polyaminotriazole. (PAT). A synthetic polymer made from sebacic acid and hydrazine with small amounts of acetamide. Poly-octamethylene-amino-triazole is a specific example.
See monobasic.

"Poly B-D"liquid Resins.[222] TM for low molecular weight liquid polymers based on butadiene containing controlled hydroxyl functionality. They consist of both homo- and terpolymers. Designed for general rubber products, coatings, adhesives, etc.

polybenzimidazole. (PBI). $(C_7H_6N_2)_n$.
A synthetic polymer designed for high-temperature space technology applications. Reputed to withstand temperature up to 260C for 1000 hours.
Derivation: Condensation of diphenyl isophthalate and 3,3'-diaminobenzidine.
Use: Fibers, composites, adhesives (high adhesion to steel, titanium, beryllium, and aluminum alloys), coatings, ablative materials.

polyblend. A combination in any proportion of either (1) two homopolymers (natural or synthetic), (2) a homopolymer and a copolymer, or (3) two copolymers. An example of (1) is rubber-polystyrene, of (2) is rubber and butadiene-styrene, and of (3) is a mixture of butadiene-acrylonitrile and isobutylene-isoprene. A polyblend is a mixture that is made after its components have been polymerized, and thus is different from a copolymer, which is made by chemical combination of two monomers.
See also homopolymer, copolymer, blend.

polybutadiene. A synthetic thermoplastic polymer made by polymerizing 1,3-butadiene with a stereospecific organometallic catalyst (butyl lithium), though other catalysts such as titanium tetrachloride and aluminum iodide may be used. The cis-isomer, which is similar to natural rubber, is used in tire treads due to its abrasion and crack resistance and low heat build-up. Large quantities are also used as blends in SBR rubber. The trans-isomer resembles gutta percha and has limited utility. Liquid polybutadiene, which is sodium-catalyzed, has specialty uses as a coating resin. It is cured with organic peroxides. Combustible.

Hazard (liquid): By ingestion and inhalation; skin irritant.
See also polymer, stereospecific.

polybutene.   See polybutylene.

polybutylene.   (polybutene; polyisobutylene; polyisobutene).   Any of several thermoplastic isotactic (stereo-regular) polymers of isobutene of varying molecular weight, also polymers of butene-1 and butene-2. Butyl rubber is a type of polyisobutene to which has been added 2% of isoprene, which provides sulfur linkage sites for vulcanization. Isobutane can be homopolymerized to various degrees in chains containing from 10 to 1000 units, the viscosity increasing with molecular weight. Combustible.
See also "Vistanex."
Use: Lubricating-oil additive, hot-melt adhesives, sealing tapes, special sealants, cable insulation, polymer modifier, viscosity index improvers, films and coatings.

polybutylene terephthalate.   An engineering plastic derived from 1,4-butanediol, it is a thermoplastic polyester with a broad spectrum of uses.

"Polycarbafil."[539]   TM for a glass fiber-reinforced polycarbonate.

polycarbonate. $(COOC_6H_5C(CH_3)_2C_6H_5O)_n$.
A synthetic thermoplastic resin derived from bisphenol A and phosgene, a linear polyester of carbonic acid: Can be formed from any dihydroxy compound and any carbonate diester, or by ester interchange. Polymerization may be in aqueous emulsion or in nonaqueous solution.
Properties: Transparent (90% light transmission), noncorrosive, weather and ozone-resistant, nontoxic, stain-resistant. Combustible but self-extinguishing, low water absorption, high impact strength, heat-resistant, high dielectric strength, dimensionally stable, soluble in chlorinated hydrocarbons and attacked by strong alkalies and aromatic hydrocarbons, stable to mineral acids, insoluble in aliphatic alcohols. Excellent for all molding methods, extrusion, thermoforming etc.; easily fabricated by all methods including thermoforming and fluidized bed coating.
Use: Molded products, solution-cast or extruded film, structural parts, tubes and piping, prosthetic devices, meter face plates, nonbreakable windows, street-light globes, household appliances.

polycarboxylic acid.   An organic acid containing two or more carboxyl (COOH) groups.

polychlor.   General name for synthetic chlorinated hydrocarbons.
Use: Pesticides.

polychlorinated biphenyl.   (PCB).
CAS: 1336-36-3.   One of several aromatic compounds containing two benzene nuclei with two or more substituent chlorine atoms. They are colorless liquids with d 1.4–1.5. Because of their persistance, toxicity, and ecological damage via water pollution their manufacture was discontinued in the US in 1976.
Hazard: Highly toxic.

polychloroprene.   See neoprene.

polychlorotrifluoroethylene.   (PCTFE).
See chlorotrifluoroethylene polymer.

"Polycin."[202]   TM for (1) an elastic, tacky, gel-like solid resulting from the polymerization of castor oil, used in rubber compounding, floor tile manufacture, and as a polymeric plasticizer; (2) a series of polyols used in the preparation and curing of urethane polymers for protective coatings, foamed insulation, and elastomers.

"Polyco."[65]   TM for a series of thermoplastic polymers in the form of water emulsions or solvent solutions, applied to vinyl acetate polymers and copolymers, butadiene-styrene copolymer latics, polystyrenes, vinyl and vinylidene chloride copolymers, acrylic copolymers and water-soluble polyacrylates.
Use: Adhesives and coatings, in paint, leather, textiles, paper, cosmetics, and construction fields.

polycondensation.   See condensation (1), polymerization.

polycoumarone resin.   See coumarone-indene resin.

polycyclic.   An organic compound having three or more ring structures, which may be the same or different, e.g., anthracene, naphthacene.
See polynuclear.

poly(1-4-cyclohexylenedimethylene)terephthalate.
TM "Kodel."   A linear polyester film or fiber obtained by condensation of terephthalic acid with 1,4-cyclohexanedimethanol. It has good electrical resistivity and hydrolytic stability.
Use: Carpet fibers and chemically resistant films.
See also terephthalic acid.

"Polycyclol 1222."[214]   TM for an intermediate for the preparation of alkyd-type resins used for coatings. These are known by the coined name "cyclyd."

poly-1,1-dihydroperfluorobutyl acrylate.
Properties: White, rubber-like polymer. D 1.5, be-

gins to degrade at 148C, retains strength and elastomeric properties in contact with synthetic lubricants, solvents, hydraulic fluids, oils, etc. at temperatures in the range 148–204C, has limited flexibility at temperatures below -17C. Nonflammable.
Use: O-rings, seals, gaskets, diaphragms, hose, sheets and coatings for fabrics and other surfaces.

polydimethylsiloxane. (PDMS). A silicone polymer developed for use as a dielectric coolant and in solar energy installations. It also may have a number of other uses. It is stated to be highly resistant to oxidation and to biodegradation by microorganisms. It is degradable when exposed to a soil environment by chemical reaction with clays and water, by which it is decomposed to silicic acid, carbon dioxide, and water.

poly-p-dinitrosobenzene. See "Polyac."

"Polydril."[233] TM for a synthetic water-soluble polymer.
Use: Flocculating agent in the oil industry.

polyelectrolyte. A high polymer substance, either natural (protein, gum arabic) or synthetic (polyethyleneimine, polyacrylic acid salts) containing ionic constituents; may be either cationic or anionic. The former type is widely used for industrial applications. Water solutions of both types are electrically conducting, some are effective in concentrations as low as 1 ppm. In a given polyelectrolyte, ions of one sign are attached to the polymer chain, while those of opposite sign are free to diffuse into the solution. Major uses are flocculation of solids (especially dissolved phosphates) in potable water, dispersion of clays in oil well drilling muds, soil conditioning, antistatic agents, and treatment of paper-mill waste water. Ion-exchange resins are cross-linked (stabilized) polyelectrolytes.
See also flocculant; "Purifloc"; "Cat-Floc."

polyene. Any unsaturated aliphatic or alicyclic compound containing more than four carbon atoms in the chain and having at least two double bonds. Examples are pentadiene, cyclooctatriene.

polyester fiber. Generic name for a manufactured fiber (either as staple or continuous filament) in which the fiber-forming substance is any long chain synthetic polymer composed of at least 85% by weight of an ester of a dihydric alcohol and terephthalic acid (Federal Trade Commission).
See "Dacron"; polyethylene terephthalate.
Properties: Strength (staple) 2.2–4.0 g per denier; (continuous filament) up to 9.5 g denier, mp 264C, water absorption 0.5%. Nonflammable.

Use: Tire fabric, seat belts, reinforcement of rubber hose for sea water cooling systems, as blend in clothing fabrics, fire-hose jackets.

polyester film. Continuously extruded polyester sheet of various thicknesses, especially useful in electrical equipment because of its high resistivity. Its tensile strength of 25,000 psi is much greater than that of other plastic films. Sensitized polyester film is used in magnetic tapes, in the photocopying technique known as reprography.

polyester resin. Any of a group of synthetic resins, which are polycondensation products of dicarboxylic acids with dihydroxy alcohols. They are thus a special type of alkyd resin but, unlike other types, are not usually modified with fatty acids or drying oils. The outstanding characteristics of these resins is their ability, when catalyzed, to cure or harden at room temperature under little or no pressure. Most polyesters now produced contain ethylenic unsaturation, generally introduced by unsaturated acids. The unsaturated polyesters are usually crosslinked through their double bonds with a compatible monomer, also containing ethylenic unsaturation, and thus become thermosetting. Flame resistance is imparted by using either acid or glycol ingredients having a high content of halogens, e.g., HET acid.
   The principal unsaturated acids used are maleic and fumaric. Saturated acids, usually phthalic and adipic, may also be included. The function of these acids is to reduce the amount of unsaturation in the final resin, making it tougher and more flexible. The acid anhydrides are often used if available and applicable. The dihydroxy alcohols most generally used are ethylene, propylene, diethylene, and dipropylene glycols. Styrene and diallyl phthalate are the most common crosslinking agents. Polyesters are resistant to corrosive, chemicals, solvents, etc.
Forms: Sheets, powder, chips.
Use: Reinforced plastics, automotive parts, boat hulls, foams, encapsulation of electrical equipment, protective coatings, ducts, flues and other structural applications, low-pressure laminates, magnetic tapes, piping, bottles, nonwoven disposable filters, low-temperature mortars.
See also alkyd resin, polyester fiber.

polyethenoid. Characterizing an aliphatic compound having more than one ethene group —CH=CH—. Linoleic acid is a polyethenoid fatty acid.

polyether. A polymer in which the repeating unit contains a C=O bond derived from aldehydes or epoxides or similar materials.
See also following entries.

polyether, chlorinated. A highly crystalline material that is 46% chlorine. Outstanding corrosion-resistance. Good electrical resistance. Readily processed and fabricated.
Use: Fluid-bed coating, tank linings, piping, valves, laboratory equipment, chemical processing equipment.

polyether, cyclic. See crown ether.

polyether foam. A polyurethane foam, either rigid or flexible, made by use of a polyether as distinct from a polyester or other resin component.
Hazard: As for polyurethane.

polyether glycol. A compound with a structural skeleton such as HO—C—C—O—C—C—O—C—C—O—C—C—OH. The length of the chain can vary widely and the number of consecutive carbon atoms may be greater than two. Examples are polyethylene glycol and polypropylene glycol.

polyethylene. CAS: 9002-88-4. $(H_2C{=}CH_2)_x$
chlorosulfonated. See "Hypalon."
crosslinked (XLPE).
Properties: Thermosetting white solid, high-temperature-resistant, excellent resistance to chemicals and to creep, high impact and tensile strength, high electrical resistivity, insoluble in organic solvents, does not stress-crack. Combustible.
Derivation: (a) By irradiating linear polyethylene with electron beam or gamma radiation, crosslinking taking place through a primary valence bond, as shown.

(b) By chemical crosslinking agent such as an organic peroxide (e.g., benzoyl peroxide). All grades of polyethylene and most copolymers can be chemically crosslinked.
Use: Wire and cable coatings and insulation (low-density grades), pipe and molded fittings (high-density grades). Special types having low electrical resistivity can be made; these can be regarded as semiconductors.
Note: In molding crosslinked polyethylene, the desired part must be formed before crosslinking is initiated, as material will not change its shape after crosslinking. The variations in composition and wide range of properties approach the ideal of a universal material more closely than most polymers.
Density
The density of polyethylene and other thermoplastic polymers is affected by the shape and spacing of the molecular chains; low-density materials have highly branched and widely spaced chains, whereas high-density materials have comparatively straight and closely aligned chains. Polymers of the latter type are called linear. The physical properties are markedly affected by increasing density.
low-density (branched chain)
Properties: Crystallinity 50–60%, d 0.915, mp 240F, tensile strength 1500 psi, impact strength above 10 ft-lb/inch/notch, thermal expansion 17 $\times$ 10^{-5}inch/inch/C, soluble in organic solvents above 200F, insoluble at room temperature.
Derivation: (1) Ethylene is polymerized in a free-radical-initiated liquid phase reaction at 1500 atmospheres (22,000 psi) and 375F with oxygen as catalyst (usually from peroxides). (2) A much more effective and cheaper process uses pressures of only 100–300 psi at less than 212F; the catalyst is undisclosed and reaction is vapor-phase.
Use: Packaging film, especially for food products, paper coating, liners for drums and other shipping containers, wire and cable coating, toys, cordage, refuse and waste bags, chewing-gum base, squeeze bottles, electrical insulation.
high-density (linear)
Properties: Crystallinity 90%, d 0.95, mp 275F, tensile strength 4000 psi, impact strength 8 ft-lb/in notch, high electrical resistivity, film is gas-permeable, hydrophobic, does not resist nitric acid.
Derivation: Ethylene polymerized by Ziegler catalysts at 1–100 atmospheres (15–1500 psi) at from room temperature to 200F. Catalyst is a metal alkyl, e.g., triethylaluminum plus a metallic salt $(TiCl_4)$ dissolved in a hydrocarbon solvent. A vapor-phase modification of this process was developed in 1965. Another method uses such metallic catalysts as Cr_2O_3 at 100–500 psi with solvents such as cyclohexane or xylene.
Use: Blow-molded products, injection-molded items, film and sheet, piping, fibers, gasoline and oil containers.
Note: Ethylene may be copolymerized with varying percentages of other materials, e.g., 2-butene or acrylic acid; a crystalline product results from copolymerization of ethylene and propylene. When butadiene is added to the copolymer blend, a vulcanizable elastomer is obtained.
low molecular weight
Properties: Molecular weight 2000–5000. Translucent white solids, excellent electrical resistance,

abrasion-resistant, resistant to water and most chemicals, d 0.92. Slightly soluble in turpentine, petroleum naphtha, xylene, and toluene at room temperature; soluble in xylene, toluene, trichloroethylene, turpentine, and mineral oils at 82.2C; practically insoluble in water; slightly soluble in methyl acetate, acetone, and ethanol up to the boiling points of these solvents. Available as emulsified and nonemulsified forms. Combustible.

Use: Mold-release agent for rubber and plastics, paper and container coatings, liquid polishes and textile finishing agents.

polyethylene glycol. (PEG; polyoxyethylene; polyglycol; polyether glycol).
CAS: 25322-68-3. Any of several condensation polymers of ethylene glycol with the general formula $HOCH_2(CH_2OCH_2)_n CH_2OH$ or $H(OCH_2CH_2)_n OH$. Average molecular weights range from 200 to 6000. Properties vary with molecular weight.
Properties: Clear, colorless, odorless, viscous liquids to waxy solids. Soluble or miscible with water and for the most part with alcohol and other organic solvents, heat-stable, inert to many chemical agents, do not hydrolyze or deteriorate, have low vapor pressure. Combustible.
Derivation: By condensation of ethylene glycol or of ethylene oxide and water.
Use: Chemical intermediates (lower molecular weight varieties), plasticizers, softeners and humectants, ointments, polishes, paper coating, mold lubricants, bases for cosmetics and pharmaceuticals, solvents, binders, metal and rubber processing, permissible additives to foods and animal feed, laboratory reagent.
See also "Carbowax."

polyethylene glycol chloride.
$H(OCH_2CH_2)_n Cl$. Any of a group of polymers, usually colorless liquids with very low vapor pressure at room temperature. Molecular weights from 100 to 600. Miscible with water, d for a low molecular weight polymer is 1.18 (20C), for a high molecular weight polymer 1.14 (10C). The former sets to a glass at −90C, the latter sets to a wax-like solid at 20C. Combustible.
Use: Solvents for cleaning, extracting, and dewaxing.

polyethylene glycol ester. A mono- or di- ester resulting from the interaction of an organic acid with one or both of the glycol ends of the polyethylene glycol polymer. These are also called polyoxyethylene esters, polyglycol esters, or by a coined generic name.

polyethylene imine. CAS: 26913-06-4.
$(CH_2CH_2NH)_x$. A synthetic polymer which is a highly viscous, hygroscopic liquid when anhydrous; completely miscible with water and lower alcohols; insoluble in benzene. Reactive toward cellulose. Combustible.
Use: Adhesive and anchoring agent for paper and cellophane, dewatering agent and wet strength improver in paper manufacture, fixative, levelling agent in textile fibers, antiblocking agent on plastic films, flocculating agent, ion exchange resins, complexing agents, disinfectant for textiles, skins, photographic chemistry, absorbent for carbon dioxide, water purification, polyelectrolyte.

polyethylene oxide. A plastic reported to be dimensionally stable at high and low temperatures and designed as a substitute for phenolics.

polyethylene oxide sorbitan fatty acid esters.
See polysorbate.

polyethylene terephthalate. CAS: 25038-59-9.
$(C_{10}H_8O_4)_x$. A thermoplastic polyester formed from ethylene glycol by direct esterification or by catalyzed ester exchange between ethylene glycol and dimethyl terephthalate. Offered as oriented film or fiber. It melts at 265C, tenacity is 2.2–4 g/denier (staple) and up to 9.0 g/denier as continuous filament; d 1.38. It has good electrical resistance and low moisture absorption. Resists combustion and is self-extinguishing.
Use: Blended with cotton, for wash-and-wear fabrics; blended with wool, for worsteds and suitings; packaging films, recording tapes, soft-drink bottles.

polyethylene thiuram sulfide.
Derivation: Oxidation of diammonium ethylene bisdithiocarbamate with calcium hypochlorite.
Grade: 50% vegetable powder, 95% technical powder.
Use: Fungicide.

polyformaldehyde. See p-formaldehyde.

polyfurfuryl alcohol. See furfuryl alcohol.

polyformaldehyde resin. See acetal resin.

"Poly-G."[94] TM for a series of polyethylene glycols, polypropylene glycols, and polyoxypropylene adducts of glycerol. G200, 300, 400, and 600 are liquid polyethylene glycols; G1000, 1500, GB-1530, and BG-2000 are waxy polyethylene glycols. The number indicates the molecular weight. G420P, 1020P, 2020P are propylene oxide condensation polymers of propylene glycol. G1030PG, 3030PG, 4030PG are propylene oxide condensation polymers of glycerol.

"Polygard."[248] TM for a mixture of alkylated aryl phosphites.
Properties: Liquid; clear amber; d 0.99; soluble in acetone, alcohol, benzene, carbon tetrachlo-

ride, solvent naphtha, and ligroin; insoluble in water, but can hydrolyze.
Use: Nondiscoloring stabilizer for rubber and plastics.

polyglycerol. One of several mixtures of ethers of glycerol with itself, ranging from diglycerol to triacontaglycerol. Some examples are: (a) diglycerol, possibly $(CH_2OHCHOHCH_2)_2$, mw 166, a liquid with 4 OH groups, viscosity of 287 cs at 65.5C; (b) hexaglycerol mw 462, a liquid with 8 OH groups, viscosity 1671 cs at 65.5C; (c) decaglycerol mw 758, a liquid with 12 OH groups, viscosity 3199 cs.
Properties: Viscous liquids to solids, soluble in water, alcohol, and other polar solvents. Act as humectants much like glycerol but have progressively higher molecular weight and boiling point. Combustible.
Derivation: Glycerol is heated with an alkaline catalyst (200–275C) at normal or reduced pressure. A stream of inert gas may be used to blanket the reaction and help remove the water of reaction.
Use: Surface-active agents, emulsifiers, plasticizers, adhesives, lubricants, and other compounds used for both edible and industrial applications.

polyglycerol ester. One of several partial or complete esters of saturated and unsaturated fatty acids with a variety of derivatives of polyglycerols ranging from diglycerol to triacontaglycerol. Prepared by (a) direct esterification and (b) transesterification reactions. Combustible.
Some examples of (a): decaglycerol monostearate, semisolid, d 1.04, mp 51.9C, decaglycerol monooleate, viscous liquid, d 1.13, mp around 0C; decaglycerol hexaoleate, liquid, d 0.97, mp -17.7C.
Examples of (b): triglycerol monolinoleate, viscosity 322 cp (75.5C); triglycerol trilinoleate, viscosity 30.1 cp (75.5C).
Use: Lubricants, plasticizers, paint and varnish vehicles, gelling agents, urethane intermediates, adhesives, crosslinking agents, humectants, textile fiber finishes, functional fluids, surface-active agents, dispersants and emulsifiers in foods, pharmaceuticals, cosmetic preparations.

polyglycol. See polyethylene glycol.

polyglycol amine H-163.
$HO[C_2H_4O]_2C_3H_6NH_2$.
Properties: Colorless liquid, d 1.0556, bp decomposes, fp 14.5C, soluble in water, bulk d 8.8 lb/gal, flash p 295F (146C). Combustible.

polyglycol distearate. (polyethylene glycol distearate).   CAS: 9005-08-7.

$C_{17}H_{35}COO(CH_2CH_2O)_x OCC_{17}H_{35}$.   Distearate ester of polyglycol.
Properties: A soft, off-white solid. D 1.04 (50C), mp 43C, pH of 10% dispersion 7.26, saponification number variable. Soluble in chlorinated solvents, light esters, and acetone; slightly soluble in alcohols, insoluble in glycols, hydrocarbons, and vegetable oils. Combustible.
Use: Plasticizer for various resins, component of grinding and polishing pastes to promote easy removal in water.

"Polygriptex." TM for an adhesive especially designed for bonding polyethylene sheeting to porous surfaces, used in making polyethylene-lined bags and multiwall kraft bags. Has good adhesion to waxed surfaces. Available in viscosities from 300 to 20,000 centipoises.

polyhalite. $2CaSO_4 \cdot MgSO_4 \cdot K_2SO_4 \cdot 2HOH$.
A naturally occurring potash salt found in Germany, Texas, and New Mexico.
Use: Source of potash for fertilizer.

polyhexamethyleneadipamide. Same as nylon 66.

polyhexamethylene sebacamide. Same as nylon 610.

polyhydric alcohol. See polyol.

polyimide. Any of a group of high polymers which have an imide group (—CONHCO—) in the polymer chain.
Properties: Tensile strength 13,500 psi, d 1.42, water absorption (24 hours at 77K) 0.3%, heat distortion point above 260C, dielectric constant at 2000 mc 3.2, and coefficient of linear expansion 28.4 × 10^{-6} inch/inch/foot. High-temperature stability (up to 370C), excellent frictional characteristics, good wear resistance at high temperatures, resists radiation, exhibits low outgassing in high vacuum, resistant to organic materials at quite high temperatures, not resistant to strong alkalies and to long exposure to steam. Flame retardant.
Use: High-temperature coatings, laminates and composites for aerospace vehicles, ablative materials, oil sealants and retainers, adhesives, semiconductors, valve seats, bearings, insulation for cables, printed circuits, magnetic tapes (high- and low-temperature-resistant), flame-resistant fibers, binders in abrasive wheels.

polyindene resin. See coumarone-indene resin.

"Poly-Ionic."[300] TM for a series of polyethylene emulsions.
Use: Softeners and lubricating agents for textiles.

polyisobutene. See polybutylene.

polyisobutylene. See polybutylene.

polyisocyanurate. See isocyanurate.

polyisoprene. $(C_5H_8)_n$. The major component of natural rubber, also made synthetically. Forms are stereo-specific cis-1,4- and trans-1,4-polyisoprene. Both can be produced synthetically by the effect of heat and press on isoprene in the presence of stereospecific catalysts. Natural rubber is cis-1,4-; synthetic cis-1,4- is sometimes called synthetic natural rubber. Trans-1,4-polyisoprene resembles gutta-percha. Polyisoprene is thermoplastic until mixed with sulfur and vulcanized. Supports combustion.
See rubber, natural and synthetic. See catalyst, stereospecific.

"Polylan."[493] TM for a polyunsaturated ester of linoleic acid and lanolin alcohols. An amber, viscous, oily liquid; soluble in mineral oil, castor oil, anhydrous ethanol, isopropanol, ethyl acetate; insoluble in water.
Use: Hydrophobic conditioner in cosmetics and pharmaceuticals.

"Poly-Lease."[175] TM for an aerosol mold release and parting agent for plastics and rubber materials based on a low molecular weight polyethylene lubricant. The usual precautions for shipping and handling aerosol containers apply.

"Polylite."[36] TM for a group of 100% reactive alkyd resins, dissolved in styrene and other monomers. Highly diversified applications both alone and in combination with such materials as fibrous glass. This group also includes resins for use with diisocyanate to form rigid or flexible polyurethane foams.

"Polymeg."[224] TM for polytetramethylene ether glycols. Available in three molecular weight ranges: 1000, 2000, and 3000.
Properties: Waxy solids which melt to clear, viscous liquids at 37C. On supercooling (or nucleation) the liquid resolidifies, d 0.985 (1000 mw) to 0.982 (3000 mw) at 35C, soluble in aromatic and chlorinated hydrocarbons, slightly soluble in water, solubility decreasing with increasing molecular weight.
Use: Polyurethane technology.

polymer. A macromolecule formed by the chemical union of five or more identical combining units called monomers. In most cases the number of monomers is quite large (3500 for pure cellulose), and often is not precisely known. In synthetic polymers this number can be controlled to a predetermined extent, e.g., by shortstopping agents. (Combinations of two, three, or four monomers are called, respectively, dimers, trimers, and tetramers and are known collectively as oligomers). A partial list of polymers by type is as follows:
  I. Inorganic siloxane, sulfur chains, black phosphorus, boron-nitrogen, silicones
  II. Organic
    1. Natural
      (a) Polysaccharides starch, cellulose, pectin, seaweed gums (agar, etc.), vegetable gums (arabic, etc.).
      (b) Polypeptides (proteins) casein, albumin, globulin, keratin, insulin, DNA
      (c) Hydrocarbons rubber and gutta percha (polyisoprene)
    2. Synthetic
      (a) Thermoplastic elastomers (unvulcanized), nylon, polyvinyl chloride, polyethylene (linear), polystyrene, polypropylene, fluorocarbon resins, polyurethane, acrylate resins
      (b) Thermosetting elastomers (vulcanized), polyethylene (crosslinked), phenolics, alkyds, polyesters
    3. Semisynthetic cellulosics (rayon, methylcellulose, cellulose acetate), modified starches (starch acetate, etc.)
See also following entries.

polymer, addition. See addition polymer.

polymer, atactic. See atactic.

polymer, block. See block polymer.

polymer, condensation. A polymer formed by a condensation reaction.

polymer, electroconductive. A polymer or elastomer made electrically conductive by incorporation of a substantial percentage of a suitable metal powder, (e.g., aluminum) or acetylene carbon black, the proportion used must be high enough to permit the particles to be in contact with one another in the mixture. Polyelectrolytes such as ion-exchange resins, salts of polyacrylic acid, and sulfonated polystyrene are electroconductive in the presence of water. Pyrolysis of polyacrylonitrile makes it electrically conductive without impairment of its structure. Polyacetylene and a few related polymers are made conductive by various doping agents such as arsenic pentafluoride and iodine.
See also polyacetylene.

polymer, graft. See graft polymer.

polymer, high. An organic macromolecule composed of a large number of monomers. The molecular weight may range from 5000 into the millions (for some polypeptides). Natural high polymers are exemplified by cellulose $(C_6H_{10}O)_n$ and rubber $(C_5H_8)_n$. Proteins are natural high polymer combinations of amino acid monomers. The dividing line between low and high polymers is considered to be in the neighborhood of 5000 to 6000 mw.

Synthetic high polymers (or "synthetic resins") include a wide variety of materials having properties ranging from hard and brittle to soft and elastic. Addition of such modifying agents as fillers, colorants, etc., yields an almost infinite number of products collectively called plastics. High polymers are the primary constituents of synthetic fibers, coating materials (paints and varnishes), adhesives, sealants, etc. Polymers having special elastic properties are called rubbers, or elastomers.

Synthetic polymers in general can be classified: (1) by thermal behavior, i.e., thermoplastic and thermosetting; (2) by chemical nature, i.e., amino, alkyd, acrylic, vinyl, phenolic, cellulosic, epoxy, urethane, siloxane, etc.; and (3) by molecular structure, i.e., atactic, stereospecific, linear, crosslinked, block, graft, ladder, etc. Copolymers are products made by combining two or more polymers in one reaction (styrene-butadiene).

See also crosslinking; and specific entries.

polymer, inorganic. A polymer in which the main chain contains no carbon atoms and in which behavior similar to that of an organic polymer can be developed, i.e., covalent bonding and crosslinking, as in silicone polymers. Here the element silicon replaces carbon in the straight chain; substituent groups are often present, forming highly useful polymers. Other inorganic high polymers are black phosphorus, boron, and sulfur all of which can form polymeric structures under special conditions. At present these have little or no commercial significance. *Note:* Some authorities consider silicone resins to be semi-organic, since their substituent groups are comprised of methyl groups.

polymer, isotactic. A type of polymer structure in which groups of atoms which are not part of the backbone structure are located either all above or all below the atoms in the backbone chain, when the latter are all in one plane. See polymer, stereospecific.

polymerization. A chemical reaction, usually carried out with a catalyst, heat, or light, and often under high pressure, in which a large number of relatively simple molecules combine to form a chain-like macromolecule. The combining units are called monomers, e.g., styrene is the monomer for polystyrene. The linear chains can be combined (crosslinked) by addition of appropriate chemicals.

The polymerization reaction occurs spontaneously in nature; industrially it is performed by subjecting unsaturated or otherwise reactive substances to conditions that will bring about combination. This may occur by *addition*, in which free radicals are the initiating agents that react with the double bond of the monomer by adding to it on one side, at the same time producing a new free electron on the other.

$$R + CH_2=CHX \rightarrow R-CH_2-CHX.$$

By this mechanism the chain becomes self-propagating. Polymerization may also occur by condensation, involving the splitting out of water molecules by two reacting monomers, and by so-called oxidative coupling. The degree of polymerization (DP) is the number of monomer units in an average polymer unit of a given sample.

Polymerization techniques may be: (1) in the gas phase at high pressures and temperatures (200C), (2) in solution at normal pressure and temperatures from -70 to $+70C$, (3) bulk or batch polymerization at normal pressure at 150C, (4) in suspension at normal pressure at 60–80C, (5) in emulsion form at normal pressure at -20 to $+60C$ (used for copolymers). Catalysts of the peroxide type are necessary with some of these methods. *Note:* Polymerization, like its handmaiden catalysis, has long been one of the most complex and productive areas of chemical research; from year to year new materials and reaction mechanisms are constantly being explored, sometimes with only marginal success. But it need only be recalled that such now commonplace materials as polyethylene, polycarbonate, nylon, neoprene, epoxies, acrylics, to mention only a few, as well as block, graft, and stereospecific polymers, have resulted from continuous and intensive research by many brilliant chemists over the last 60 years, and this research continues undiminished.

See also free radical, crosslinking.

polymer, ladder. See ladder polymer.

polymer, low. A polymer comprised of comparatively few monomer units and having a molecular weight from 300–5000.

polymer, natural. See polymer.

polymer, stereoblock. See stereoblock polymer.

$$-\overset{\displaystyle \overset{H}{|}}{\underset{\displaystyle \underset{H}{|}}{C}}-\overset{\displaystyle \overset{H}{|}}{\underset{\displaystyle \underset{O}{|}}{C}}-\overset{\displaystyle \overset{H}{|}}{\underset{\displaystyle \underset{H}{|}}{C}}-\overset{\displaystyle \overset{H}{|}}{\underset{\displaystyle \underset{O}{|}}{C}}-\overset{\displaystyle \overset{H}{|}}{\underset{\displaystyle \underset{H}{|}}{C}}-\overset{\displaystyle \overset{H}{|}}{\underset{\displaystyle \underset{O}{|}}{C}}-$$

polymer, stereospecific. (stereoregular).
A polymer whose molecular structure has a definite spatial arrangement, i.e., a fixed position in geometrical space for the constituent atoms and atomic groups comprising the molecular chain, rather than the random and varying arrangement that characterizes an amorphous polymer. Achievement of this specific steric (three-dimensional) structure (also called tacticity) requires use of special catalysts such as those developed by Ziegler and Natta about 1950. Such polymers are wholly or partially crystalline. Synthetic natural rubber, cis-polyisoprene, is an example of a stereospecific polymer made possible by this means. There are five types of stereospecific (or stereoregular) structures: cis, trans, isotactic, syndiotactic, and tritactic.
See also catalyst, stereospecific.

polymer, syndiotactic. See syndiotactic polymer.

polymer, synthetic. See polymer.

polymer, water-soluble. Any substance of high molecular weight that swells or dissolves in water at normal temperature. These fall into several groups, including natural, semisynthetic, and synthetic products. Their common property of water solubility makes them valuable for a wide variety of applications as thickeners, adhesives, coatings, food additives, textile sizing, etc.
See specific entries.
(1) *Natural*. This type is principally comprised of gums, which are complex carbohydrates of the sugar group. They occur as exudations of hardened sap on the bark of various tropical species of trees. All are strongly hydrophilic. Examples are arabic, tragacanth, karaya.
(2) *Semisynthetic*. This group (sometimes called water-soluble resins) includes such chemically treated natural polymers as carboxymethylcellulose, methylcellulose, and other cellulose ethers, as well as various kinds of modified starches (ethers and acetates).
(3) *Synthetic*. The principal members of this class are polyvinyl alcohol, ethylene oxide polymers, polyvinyl pyrrolidone, polyethyleneimine.

polymethacrylate resin. See acrylic resin, methyl methacrylate.

polymethylbenzene. See durene and pseudocumene, the two members of this group with some commercial production and use.

polymethylene polyphenylisocyanate.
A polymer of diphenylmethane-4,4′ diisocyanate.

polymethylene wax. See wax, polymethylene.

poly-4-methylpentene-1.
Properties: High resistance to all chemicals except carbon tetrachloride and cyclohexane, excellent heat resistance, high clarity and light transmittance. Temperature limit 170C, d 0.83.
Use: Laboratory ware (beakers, graduates, etc.), electronic and hospital equipment; food packaging, especially types subject to high temperature such as trays for TV dinners, etc.; light reflectors.

poly(methyl vinyl ether). See polyvinyl methyl ether.

polymorphism. See allotropy.

polymyxin. CAS: 1406-11-7. Generic term for a series of antibiotic substances produced by strains of *Bacillus polymyxa*. Various polymyxins are differentiated by the letters A, B, C, D, and E. All are active against certain gram-negative bacteria. Polymyxin B is most used.
Properties: All are basic polypeptides, soluble in water; the hydrochlorides are soluble in water and methanol, insoluble in ether, acetone, chlorinated solvents, and hydrocarbons. Permissible food additives.
Use: Medicine (antibiotic), beer production.

polynuclear. Descriptive of an aromatic compound containing three or more closed rings, usually of the benzenoid type, e.g., sterols.
See also polycyclic, nucleus (3).

polyol. A polyhydric alcohol, i.e., one containing three or more hydroxyl groups. Those having three hydroxyl groups (trihydric) are glycerols, those with more than three are called sugar alcohols, with general formula $CH_2OH(CHOH)_n CH_2OH$, where n may be from 2 to 5. These react with aldehydes and ketones to form acetals and ketals.
See also alcohol, glycerol.

polyolefin. A class or group name for thermoplastic polymers derived from simple olefins, among the more important are polyethylene, polypropylene, polybutenes, polyisoprene and their copolymers. Many are produced in the form of fibers. This group comprises the largest tonnage of all thermoplastics produced.

polyorganosilicate graft polymer. An organoclay to which a monomer or an active polymer has been chemically bonded, often by the use of ionizing radiation. An example is the bonding of styrene to a polysilicate containing vinyl radicals, resulting in the growth of polystyrene chains from the surface of the silicate. Such complexes are stable to organic solvents. They have consid-

erable use potential in the ion-exchange field, as ablative agents, reinforcing agents, and hydraulic fluids.
See also organoclay, graft polymer.

"Polyox."[214]   TM for a series of water-soluble ethylene oxide polymers with molecular weights in the 100,000 to several million range.
Use: Textile warp size, paper coatings, detergents, hair spray, toothpastes, water-soluble packaging film, adhesives.

polyoxadiazole.   A polymer of oxadiazole, cyclic C_2N_2O, prepared by cyclodehydration (ring formation from a chain with subsequent loss of water) of polyisophthalic hydrazide. Because of its high temperature tolerance (above 398C) fibers made from it may be useful for space vehicles.

polyoxamide.   A nylon-type material made from oxalic acid and diamines.

polyoxetane.   See oxetane, "Penton."

polyoxyethylene.   See polyethylene glycol.

polyoxyethylene fatty acid ester.   See polysorbate.

polyoxyethylene (40) monostearate.   (polyethylene glycol stearate).   CAS: 9004-99-3.
A mixture of the mono- and distearate esters of mixed polyoxyethylene diols and corresponding free glycols. The monostearate can be represented as: $H(OCH_2CH_2)_n OCOC_{17}H_{35}$ (n is about 40).
Properties: Waxy, light tan, nearly odorless solid; congealing range 39–45C; soluble in water, alcohol, ether and acetone; insoluble in mineral oil and vegetable oils.
Grade: USP.
Use: Ointments, emulsifier, surfactant, food additive.
See also polysorbate.

polyoxyethyleneoxypropylene.   (POEOP).
A polymer of ethylene and propylene glycols (ethylene oxide, propylene oxide).
Use: Solvent.

polyoxyethylene (8) stearate.   CAS: 9004-99-3.
A mixture of the mono- and diesters of stearic acid and mixed polyoxyethylene diols having an average polymer length of 7.5 oxyethylene units.
Properties: Cream-colored, soft, waxy, or pasty solid at 25C, faint, fatty odor and a slightly bitter, fatty taste. Soluble in toluene, acetone, ether, and ethanol.
Use: Emulsifier in bakery products.
See also polysorbate.

polyoxymethylene.   Any of several polymers of formaldehyde and trioxane.
See acetal resin.

polyoxypropylene diamine.   (POPDA).
Any of six high molecular weight amines of low viscosity and vapor pressure, high primary amine content and light color.
Use: As crosslinking agents in epoxy coatings, imparting high flexibility and adhesion at low temperatures. Other possible uses are in polyamide and polyurethane coatings, adhesives, elastomers and foams, as intermediates for textile and paper treatment, and viscosity index improvers in lube oils.

polyoxypropylene ester.   See polypropylene glycol ester.

polyoxypropylene-glycerol adduct.   One of several condensation polymers of propylene oxide and glycerol with molecular weights in the range 1000–4000. Clear, stable, almost colorless, noncorrosive liquids.
Use: Similar to those of polypropylene glycol.

"Poly-pale."[286]   TM for pale, hard, thermoplastic resins; 40% dimeric resin acids; acid number 145; USDA color WG; softening point 102C. Available in solid and flake forms.
Use: Adhesives, lacquers, varnishes, printing inks.

polypeptide.   (peptide).   The class of compounds composed of acid units chemically bound together with amide linkages (CONH) with elimination of water. A polypeptide is thus a polymer of amino acids, forming chains that may consist of several thousand amino acid residues. A segment of such a chain is as follows:

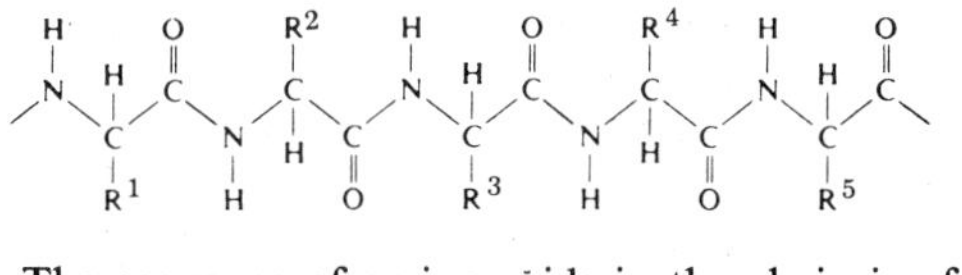

The sequence of amino acids in the chain is of critical importance in the biological functioning of the protein, and its determination is one of the most difficult problems in molecular biology. The chains may be relatively straight, or they may be coiled or helical. In the case of certain types of polypeptides, such as the keratins, they are crosslinked by the disulfide bonds of cystine. Linear polypeptides can be regarded as proteins. Synthesis of a 20-amino acid polypeptide that induces formation of antibodies for foot and mouth disease was announced in 1982. It is the first vaccine that has ever been synthesized.
See also protein, polyamide, keratin.

polyphenylene oxide. See "PPO."

polyphenylene triazole.
[—C_6H_4—$C_2N_3(C_6H_5)$—]$_n$. A polymer
stated to be serviceable up to 260C for films,
coatings, adhesives, and lamination.

"Polyphos."[84] TM for a water-soluble glassy sodium phosphate of standardized composition,
($Na_{12}P_{10}O_{31}$) analyzing 63.5% P_2O_5 (ratio of
$Na_2O:P_2O_5$ is 1.2:1). It is closely similar to a
sodium hexametaphosphate and sodium tetraphosphate; frequently the three names are used
interchangeably.
Grade: Ground, walnut-size to pea-size lumps.
Use: Boiler water compounds, detergents, textiles,
leather tanning, photographic film developing,
deflocculation of clays, flotation and desliming
of minerals, dispersion of pigments, paper processing, industrial and municipal water treatment.

polyphosphazene. See phosphazene.

polyphosphoric acid. $H_{n+2}P_nO_{3n+1}$, for n greater
than 1. Any of a series of strong acids, from
pyrophosphoric acid, $H_4P_2O_7$ ($n = 2$), through
metaphosphoric acid (large values of n).
Properties: Viscous, water-white liquid; water-
soluble; does not crystallize on standing; hygroscopic. The commercial acid is a mixture of
orthophosphoric acid with pyrophosphoric,
triphosphoric, and higher acids and is sold on
the basis of its calculated content of H_3PO_4, as
for example 115%. Superphosphoric acid is a
similar mixture sold at 105% H_3PO_4. These acids
revert slowly to orthophosphoric acid on dilution
with water.
Hazard: Toxic by ingestion; strong irritant.
Use: Dehydrating, catalytic, and sequestering
agents, for metal treating; many applications
where a concentrated monooxidizing acid is
needed; lab reagent.
See also phosphoric acid.

polypropylene. CAS: 9003-07-0. (C_3H_5)$_n$
A synthetic, crystalline, thermoplastic polymer
with molecular weight of 40,000 or more. *Note*:
Low molecular weight polymers are also known
which are amorphous in structure, and used as
gasoline additives, detergent intermediates,
greases, sealants, and lube oil additives. Also
available as a high melting wax.
Derivation: Polymerization of propylene with a
stereo-specific catalyst such as aluminum alkyl.
Properties: Translucent, white solid with d 0.90;
mp 168–171C; tensile strength 5000 psi; flexural
strength 7000 psi; usable up to 121C. Insoluble

in cold organic solvents; softened by hot solvents.
Maintains strength after repeated flexing. Degraded by heat and light unless protected by antioxidants. Readily colored; good electrical resistance; low water absorption and moisture
permeability; poor impact strength below −9.4C;
not attacked by fungi or bacteria; resists strong
acids and alkalies up to 60C, but is attacked
by chlorine, fuming nitric acid, and other strong
oxidizing agents. Combustible, but slow-burning.
Fair abrasion and good heat resistance if properly
modified. Can be chrome-plated, injection- and
blow-molded, and extruded.
Available forms: Molding powder: extruded sheet,
cast film (1–10 mils), textile staple and continuous filament yarn, fibers with diameters from
0.05 to 1 micron and fiber webs down to 2 microns thick, low-density foam.
Use: Packaging film; molded parts for automobiles, appliances, housewares, etc.; wire and cable
coating; food container closures; coated and
laminated products; bottles; artificial grass and
turfs; plastic pipe; wearing apparel (acid-dyed);
fish nets; surgical casts; strapping; synthetic paper; reinforced plastics; nonwoven disposable filters.

polypropylenebenzene. See dodecylbenzene.

polypropylene, chlorinated. White, odorless, nonflammable powder. A film-forming polymer used
in coatings, inks, adhesives, and paper coatings.

polypropylene glycol ester. Exactly analogous to
polyethylene glycol ester.

polypropylene glycol monobutyl ether.
See butoxy polypropylene glycol.

polypropylene glycol. (PG). HO(C_3H_6O)$_n$ H.
One of a group of compounds comparable to
polyethylene glycols but more oil-soluble and
substantially less water-soluble. Classified by molecular weight as 425, 1025, 2025. Non-volatile,
noncorrosive liquids; lower molecular weight
members are soluble in water. Solvents for vegetable oils, waxes, resins. Combustible.
Use: Hydraulic fluids, rubber lubricants, antifoam
agents, intermediates in urethane foams, adhesives, coatings, elastomers, plasticizers, paint formulations, lab reagent.

polypropyleneimine. Polymeric form of propyleneimine. Available in 50% aqueous solution.
Use: Textile, paper, and rubber industries.

polypropylene oxide. (C_3H_6O)$_n$. A derivative
of propylene.
Use: Intermediate for urethane foams.

polypyrrolidone. Synonym for nylon-4.

"Polyrad."[266] TM for reaction products of "Amine D" and ethylene oxide.
Grade: Various grades which differ in chain length of polyoxyethylene units and free amine content. Vary in viscosity at 25C from 0.5–24.8 poises.
Use: Corrosion inhibitors and detergents in petroleum processing equipment, wetting and emulsifying agents, inhibiting hydrogen chloride.

"Polyram."[55] TM for a wettable powder.
Hazard: Toxic by ingestion and inhalation.
Use: Fungicide and approved for many vegetables.

polysaccharide. A combination of nine or more monosaccharides, linked together by glycosidic bonds. Examples: starch, cellulose, glycogen.
See also carbohydrate, phycocolloid.

polysiloxane. See siloxane.

"Poly-Solv."[84] TM for a series of glycol ether solvents for paints, varnishes, dry cleaning soaps, cutting oils, insecticides.

D2M. Diethylene glycol dimethyl ether.
Use: Anhydrous reaction medium for organometallic syntheses.

polysorbate. (USAN name for a polyoxyethylene fatty acid ester). One of a group of nonionic surfactants obtained by esterification of sorbitol with one or three molecules of a fatty acid (stearic, lauric, oleic, palmitic) under conditions which cause splitting out of water from the sorbitol, leaving sorbitan. About 20 moles of ethylene oxide per mole of sorbitol are used in the condensation to effect water solution.
Properties: Lemon to amber, oily liquids; d 1.1; faint odor and bitter taste; most types are soluble in water, alcohol, and ethyl acetate. Combustible.
Grade: Polysorbate 20 (polyoxyethylene (20) sorbitan monolaurate). Polysorbate 60 (polyoxyethylene (20) sorbitan monostearate). Polysorbate 80 (polyoxyethylene (20) sorbitan monooleate). Polysorbate 65 (polyoxyethylene (20) sorbitan tristearate).
Use: Surfactant, emulsifying agent, dispersing agents, shortenings and baked goods, pharmaceuticals, flavoring agents, foaming and defoaming agents.
See also sorbitan fatty acid ester.

polystyrene. CAS: 9003-53-6.
$(C_6H_5CHCH_2)_n$. Polymerized styrene, a thermoplastic synthetic resin of variable molecular weight depending on degree of polymerization.
Properties: Transparent, hard solid; high strength and impact resistance; excellent electrical and thermal insulator. Attacked by hydrocarbon solvents but resists organic acids, alkalies, and alcohols. Not recommended for outdoor use, unmodified polymer yellows when exposed to light, but light-stable modified grades are available. Easily colored, molded, and fabricated. Copolymerization with butadiene and acrylonitrile and blending with rubber or glass fiber increase impact strength and heat resistance; autoign temperature 800F. Combustible.
Derivation: Polymerization of styrene by free radicals with peroxide initiator.
Forms: Sheet, plates, rods, rigid foam, expandable beads or spheres.
Hazard: As for foam, plastic.
Use: Packaging, refrigerator doors, air conditioner cases, containers and molded household wares, machine housings, electrical equipment, toys, clock and radio cabinets. (As foam): thermal insulations, light construction as in boats, etc., ice buckets, water coolers, fillers in shipping containers, furniture construction. (As spheres): radiator leak stopper.
See also "Styron", "Styrofoam."

polysulfide elastomer. A synthetic polymer in either solid or liquid form obtained by the reaction of sodium polysulfide with organic dichlorides such as dichlorodiethyl formal, alone or mixed with ethylene dichloride. Outstanding for resistance to oils and solvents and for impermeability to gases. Poor tensile strength and abrasion resistance but are resilient and have excellent low-temperature flexibility. Some grades have fairly strong odor, which is not objectionable in most applications. Sealant grades are furnished in two parts which cure at room temperature when blended.
Use: Gasoline and oil-loading hose, sealants and adhesive compositions, binder in solid rocket propellants, gaskets, paint spray hose.
See also "Thiokol."

polysulfone. A synthetic thermoplastic polymer.
Properties: Hard, rigid, transparent solid. Tensile strength 10,000 psi, d 1.24, flexural strength 15,000 psi, good electrical resistance, minimum creep, low expansion coefficient. Soluble in aromatic hydrocarbons, ketones, and chlorinated hydrocarbons; resistant to corrosive acids and alkalies, to heat and oxidation and to detergents, oils, and alcohols. Dimensionally stable over temperature range −100 to +148C; tends to absorb moisture, readily processed and fabricated. Combustible, but self-extinguishing.
Derivation: Condensation of bis-phenol A and dichlorophenyl sulfone.
Use: Power-tool housings, electrical equipment, extruded pipe and sheet, auto components, elec-

tronic parts, appliances, computer components, base matrix for stereotype printing plates.

polyterpene resin. A class of thermoplastic resins or viscous liquids of amber color, obtained by polymerization of turpentine in the presence of catalysts such as aluminum chloride or mineral acids. The resins consist essentially of polymers of α- or β-pinene and are soluble in most organic solvents.
Use: Paints, rubber plasticizers, curing concrete, impregnating paper, adhesives, hot-melt coatings, pressure-sensitive tapes.
See also pinene.

polytetrafluoroethylene. (PTFE; TFE; Teflon). CAS: 9002-84-0. $(C_2F_4)_n$. A polymer of tetrafluoroethylene; it is essentially a straight chain of the repeating unit $[-CF_2-CF_2-]_n$. Soft and waxy with a milk-white color; it can be molded by powder metallurgy techniques involving mixing with a diluent that is subsequently removed and sintering at 371C.
Properties: Highly resistant to oxidation and action of chemicals including strong acids, alkalies, oxidizing agents; resistant to nuclear radiation and UV rays, ozone, and weather; halogenated solvents at high temperatures and pressures have some adverse effect. Retains useful properties up to 287C and is strong and tough. Low coefficient of friction (0.05) and antistick properties; excellent resistance to electricity; coefficient of thermal expansion greater than other plastics and metals. Nonflammable.
Forms: Extrusion and molding powders, aqueous dispersion, film, multifilament fiber.
Hazard: Evolves toxic fumes on heating.
Use: Gaskets, liners, seals, flexible hose; ablative coatings for rockets and space vehicles, chemical process equipment, coatings in aerospace, coaxial spacers, insulators, wire coating and tape in electrical and electronic fields, bearings, seals, piston rings, antistick coatings for cooking vessels, felts, packings and bearings.
See also "Teflon", "Halon."

polytetramethylene ether glycol. See "Polymeg."

polythene. Generic name for polyethylene. No longer current in the US, but is still used in England.

polythiadiazole. $[-C_6H_4-C_2N_2S-]_n$. A polymer made from polyoxathiahydrazide. Can be converted to fibers; stated to retain properties to 398C and to have resistance to thermal degradation such that it retains 60 or 70% of original tenacity after 32 hours at 398C.

polythiazyl. $(SN)_n$. An experimental polymer of sulfur nitride with covalent linkages said to have the optical and electrical properties characteristic of metals. Thin films are reported to exhibit epitaxial growth.

polytrifluorochloroethylene resin. See chlorotrifluoroethylene resin.

polyunsaturated fat. A fat or oil based at least partly on fatty acids having two or more double bonds per molecule, such as linoleic and linolenic acids. Examples are corn oil and safflower oil.
Use: Margarine and dietary foods, salad dressings, etc.

polyurethane. CAS: 9009-54-5. A thermoplastic polymer (which can be made thermosetting) produced by the condensation reaction of a polyisocyanate and a hydroxyl-containing material, e.g., a polyol derived from propylene oxide or trichlorobutylene oxide. The basic polymer unit is formed as follows: $R_1NCO + R_2OH \rightarrow R_1NHCOOR_2$.
Fiber:
Properties: High elastic modulus, good electrical resistance, high moisture-proofness, crystalline structure. Combustible.
Derivation: Reaction of hexamethylene diisocyanate and 1,4-butanediol.
Use: Chiefly in so-called spandex fibers for girdles and other textile structures requiring exceptional elasticity, bristles for brushes, etc.
Coatings:
Properties: Excellent hardness, gloss, flexibility, abrasion resistance, and adhesion; resistant to impact, weathering, acids and alkalies; attacked by aromatic and chlorinated solvents. Applied by brush, spray, or dipping. Combustible.
Derivation: Formed from "prepolymers" containing isocyanate groups (toluene and 4,4'-diphenylmethane diisocyanates) and hydroxyl-containing materials such as polyols and drying oils.
Use: Baked coatings, two-component formulations, wire coatings, tank linings, maintenance paints, masonry coating.
Elastomers:
Properties: Good resistance to abrasion, weathering, and organic solvents, tend to harden and become brittle at low temperatures. Combustible.
Derivation: Reaction of polyisocyanates with linear polyesters or polyethers containing hydroxyl groups.
Use: Sealants and caulking agents, adhesives, films and linings, shoe uppers and heels, encapsulation of electronic parts, binders for rocket propellants, abrasive wheels and other mechanical items, auto bumpers, fenders, and other components.
Foams:
Properties: Both flexible and rigid foams are available. Density varies from 2–50 lbs/ft³, thermal conductivity as low as 0.11, excellent insulators. Combustible unless protected by effective thermal barrier.

Hazard: Evolves toxic fumes on ignition.

See also foam, plastic.

Derivation: A polyether such as polypropylene glycol is treated with a diisocyanate in the presence of some water and a catalyst (amines, tin soaps, organic tin compounds). As the polymer forms, the water reacts with the isocyanate groups to cause crosslinking and also produces carbon dioxide which causes foaming. In other cases, trifluoromethane or similar volatile material may be used as blowing agent.

Flexible foams are based on polyoxypropylenediols of 2000 molecular weight and triols up to 4000 molecular weight. Rigid foams are based on polyethers made from sorbitol, methyl glucoside, or sucrose.

Use: (Flexible) Furniture, mattresses, laminates and linings, flooring leveling, seat cushions and other automotive accessories, carpet underlays, upholstery, absorbent of crude oil spills on sea water. (Rigid) Furniture; packaging and packing; building insulation; marine flotation; auto components; cigaret filters; light structures (as boat hulls) soundproofing; salvaging operations; shipbuilding (for buoyancy); transportation insulation for box cars, refrigerated cars, tank and hopper cars, trucks and trailers (claimed to be twice as effective as glass fiber); insulation for storage tanks, ships' holds, and pipelines; auto bumpers.

polyvinyl acetal. One of the family of vinyl resins resulting from the condensation of polyvinyl alcohol with an aldehyde; acetaldehyde, formaldehyde and butyraldehyde are commonly used. The three main groups are polyvinyl acetal itself, polyvinyl formal, and polyvinyl butyral. These are all thermoplastic and can be processed by extruding, molding, coating, and casting. They are used chiefly in adhesives, paints and lacquers, and films. Polyvinyl butyral is used in sheet form as an interlayer in safety glass and shatter-resistant protection in aircraft.

$$-CH_2CH \underset{O\qquad O}{\overset{CH_2}{\vert\qquad\vert}} CH-$$
$$\underset{C_3H_7}{CH}$$

polyvinyl acetate. (PVAc).   CAS: 9003-20-7.
[—CH$_2$CH(OOCCH$_3$)—]$_x$ A thermoplastic high polymer.

Properties: Colorless, odorless, transparent solid; d 1.19 at 15C; insoluble in water, gasoline, oils, and fats; soluble in low molecular weight alcohols, esters, benzene, and chlorinated hydrocarbons. Resistant to weathering. Combustible.

Derivation: Polymerization of vinyl acetate with peroxide catalysts.

Use: Latex water paints; adhesives for paper, wood, glass, metals, and porcelain; intermediate for conversion to polyvinyl alcohol and acetals; sealant; shatterproof photographic bulbs; paper coating and paperboard; bookbinding; textile finishing; nonwoven fabric binder; component of lacquers, inks, and plastic wood; strengthening agent for cements.

polyvinyl alcohol. (PVA; PVOH).
CAS: 9002-89-5.   (—CH$_2$CHOH—)$_x$.
A water-soluble synthetic polymer made by alcoholysis of polyvinyl acetate.

Properties: White to cream-colored powder, d 1.27–1.31, refr index 1.49–1.53. Properties depend on degree of polymerization and the percentage of alcoholysis, both of which are controllable in processing. Water solubility increases as molecular weight decreases; strength, elongation, tear resistance, and flexibility improve with increasing molecular weight. Tensile strength up to 22,000 psi; decomposes at 200C. PVA has high impermeability to gases, is unaffected by oils, greases, and petroleum hydrocarbons. Attacked by acids and alkalies. It forms films by evaporation from water solution. Combustible.

Grade: Super high viscosity (molecular weight 250,000–300,000), high viscosity (molecular weight 170,000–220,000), medium-viscosity (molecular weight 120,000–150,000), low viscosity (molecular weight 25,000–35,000).

Use: Textile warp and yarn size, laminating adhesives, molding powders, binder for cosmetic preparations, ceramics, leather, cloth, nonwoven fabrics and paper, paper coatings, greaseproofing paper, emulsifying agent, thickener and stabilizer, photosensitive films, cements and mortars, intermediate for other polyvinyls, imitation sponges, printing inks (glass).

polyvinylbenzyltrimethyl ammonium chloride.
An electrically conductive polymer.

Use: Increase the conductivity of papers.

polyvinyl butyral. See polyvinyl acetal.

polyvinyl carbazole. A brown thermoplastic resin obtained by the reaction of acetylene with carbazole. It softens at 150C, has excellent dielectric properties, good heat and chemical stability, but poor mechanical strength.

Use: Substitute for mica in electrical equipment and as an impregnant for paper capacitors.

polyvinyl chloride. (PVC).   CAS: 9002-86-2.
(-H$_2$CCHCl-)$_x$.   A synthetic thermoplastic polymer.

Properties: White powder or colorless granules. Resistant to weathering and moisture; dimensionally stable; good dielectric properties; resistant to most acids, fats, petroleum hydrocarbons, and fungus. Readily compounded into flexible and rigid forms by use of plasticizers, stabilizers, fillers, and other modifiers. Easily colored and processed by blow molding, extrusion, calendering, fluid bed coating, etc. Available as film, sheet, fiber, and foam.

Derivation: Polymerization of vinyl chloride by free radicals with peroxide initiator. May be copolymerized with up to 15% of other vinyls.

Hazard: Decomposes at 148C evolving toxic fumes of hydrogen chloride.

See also vinyl chloride and note.

Use: Piping and conduits of all kinds; siding; gutters; window and door frames; officially approved for use in interior piping, plumbing, and other construction uses. Raincoats, toys, gaskets, garden hose, electrical insulation, shoes, magnetic tape, film and sheeting, containers for toiletries, cosmetics, household chemicals, fibers for athletic supports, sealant liners for ponds and reservoirs, adhesive and bonding agent, plastisols and organosols, tennis court playing surfaces, flooring, coating for paper and textiles, wire and cable protection, base for synthetic turf, phonograph records, fuel in pyrotechnic devices.

Note: Use of PVC in rigid and semirigid food containers such as bottles, boxes, etc., is under restriction by FDA as well as in coatings for fresh citrus fruits. Its use in thinner items such as films and package coatings is permissible. Possibility of migration of vinyl chloride monomer into food products is the critical factor; this tends to increase with the thickness of the material.

polyvinyl chloride-acetate. CAS: 34149-92-3. $(C_2H_4O \cdot C_2H_3Cl)_x$. A vinyl chloride and vinyl acetate copolymer that is more flexible than polyvinyl chloride. The copolymer usually contains 85–97% of the chloride. It generally has similar properties and uses as polyvinyl chloride.

polyvinyl dichloride. (PVDC). A chlorinated polyvinyl chloride. Has high strength and superior chemical resistance over a broad temperature range. Combustible but self-extinguishing.

Use: Pipe and fittings for hot corrosive materials up to 100C. Is immune to solvation or direct attack by inorganic reagents, aliphatic hydrocarbons, and alcohols.

polyvinyl ether. See polyvinyl ethyl ether, polyvinyl isobutyl ether, polyvinyl methyl ether, polyvinyl methyl ether-maleic anhydride.

polyvinyl ethyl ether. (PVE; polyvinyl ether). $[-CH(OC_2H_5)CH_2-]_n$.

Properties: Viscous gum to rubbery solid, depending on molecular weight; colorless when pure,

d 0.97 (20C), refr index 1.45 (25C), insoluble in water, soluble in nearly all organic solvents, stable toward dilute and concentrated alkalies and dilute acids, compatible with a limited number of commercial resins, including rosin derivatives and some phenolics.

Derivation: Polymerization of vinyl ethyl ether.

Use: Pressure-sensitive tape, to improve adhesion to porous surfaces, cellophane, cellulose acetate and vinyl sheet.

polyvinyl fluoride. $(-H_2CCHF-)_n$.

Polymer of vinyl fluoride. In film form it is characterized by superior resistance to weather, high strength, high dielectric constant, low permeability to air and water, as well as oil, chemical solvent, and stain resistance.

Hazard: Not recommended for food packaging; evolves toxic fumes on heating.

Use: Protective material for outdoor use, packaging, electrical equipment.

polyvinyl formal. See polyvinyl acetal.

polyvinylidene chloride. (saran). A stereoregular, thermoplastic polymer.

Properties: Tasteless, odorless, abrasion resistant, low vapor transmission, impermeable to flavor, highly inert to chemical attack, softened by chlorinated hydrocarbons and soluble in cyclohexanone and dioxane. Combustible but self-extinguishing.

Derivation: Polymerization of vinylidene chloride or copolymerization of vinylidene chloride with lesser amounts of other unsaturated compounds.

Forms: Extruded and molded products, oriented fibers, films.

Use: Packaging of food products, especially meats and poultry, insecticide-impregnated multiwall paper bags, pipes for chemical processing equipment, seat covers, upholstery, fibers, bristles, latex coatings.

See also saran fiber; "Cryovac."

polyvinylidene fluoride. $H_2C=CF_2$.

A thermoplastic fluorocarbon polymer suitable for compression and injection molding and extrusion.

Properties: Crystals, mp 171C, thermally stable from -62 to $+148C$; combustible, self-extinguishing and nondripping. Tensile strength 7000 psi at 25C, yield stress 5500 psi, elongation 300%, compression strength 10,000 psi, thermal conductivity 1.05 Btu/hr/sq ft/F/in., water absorption 0.04% in 24 hours, d 1.76, refr index 1.42. Resistant to oxidative degradation, electricity, acids, alkalies, oxidizers, halogens; somewhat soluble in dimethylacetamide, attacked by hot concentrated sulfuric acid or n-butylamine.

Forms: Powder, pellets, solution, and dispersion.

Use: Insulation for high-temperature wire, tank linings, chemical tanks and tubing, protective paints and coatings with exceptional resistance (30 years) to weathering and UV, valve and impeller parts, shrinkage tubing to encapsulate resistors, diodes, and soldered joints, sealant.

See also fluorocarbon polymer.

polyvinyl isobutyl ether. (PVI; polyvinyl ether). [—CH(OCH$_2$CH(CH$_3$)$_2$CH$_2$—]$_n$.

Properties: White, opaque elastomer or viscous liquid depending on molecular weight; almost odorless; d 0.91–0.93 (20C); refr index 1.45–1.46 (25C); insoluble in water, ethanol, and acetone; soluble in most other organic solvents; stable toward dilute and concentrated alkalies and dilute acids. Compatible with a limited number of commercial resins, including rosin derivatives and some phenolics. Combustible.

Derivation: Polymerization of vinyl isobutyl ether by peroxides or acid catalysts.

Grade: As 100% material in three molecular weight ranges.

Use: Adhesives, waxes, tackifiers, plasticizers, surface coatings, laminating agents, cable filling compositions, lubricating oils.

polyvinyl methyl ether. (PVM). (—CH$_2$CHOCH$_3$—)$_n$. A nonionic polymer of high molecular weight.

Properties: Colorless (when pure), tacky liquid; soluble in water at 32C above which it precipitates from water but redissolves on cooling. Soluble in most organic solvents except aliphatic hydrocarbons. Compatible with rubber latexes, rosin derivatives, and some phenolics; d 1.05; refr index 1.47; solidifies near 0C. Combustible

Derivation: Polymerization of vinyl methyl ether with peroxides or acid catalysts.

Use: Pressure-sensitive and hot-melt adhesives, bonding of paper to polyethylene, laminating adhesive, tackifier and plasticizer for coatings, heat sensitizer for rubber latex, pigment binder in textile finishing, printing inks, protective colloid in emulsions.

polyvinyl methyl ether-maleic anhydride. (PVM/MA). [—CH$_2$CHOCH$_3$CHCOOCOCH—]$_n$.

Water-soluble copolymer of vinyl methyl ether and maleic anhydride

Use: Protective colloid, dispersing agent, thickener, binder, adhesive and size in coatings, detergents, emulsions, paper, textiles, leather, latex, rust preventive, foam stabilizer.

poly-2-vinylpyridine. (—CH(C$_5$H$_4$N)—CH$_2$—)$_n$. A vinyl-type

polymer suggested for use as a photographic dye mordant, tablet coating material, antistat for textiles and paper, and textile dye receptor.

polyvinylpyrrolidone. (PVP). CAS: 9003-39-8. (C$_6$H$_9$NO)$_n$.

Properties: White, free-flowing, amorphous powder or aqueous solution, soluble in water and organic solvents, compatible with wide range of hydrophilic and hydrophobic resins, d 1.23–1.29, bulk d 15lb/ft^3, hygroscopic.

Grade: Various molecular weight: 10,000, 40,000, 160,000, 360,000.

Use: Pharmaceuticals; blood plasma expander; cast films adherent to glass, metals, and plastics; complexing agent; detoxification of chemicals such as dyes, iodine, phenol, and poisonous drugs. Tableting, photographic emulsions, cosmetics (hair sprays, shampoos, hand creams, skin lotions), dentifrices, dye-stripping, textile finishes, protective colloid, detergents, adhesives, beer and wine clarification.

polyvinyl resin. (vinyl plastic). Any of a series of polymers (resins) derived by polymerization or copolymerization of vinyl monomers (CH$_2$=CH—) including vinyl chloride and acetate, vinylidene chloride, methyl acrylate and methacrylate, acrylonitrile, styrene, the vinyl ethers, and numerous others. Specifically, polyvinyl chloride, acetate, alcohol, etc., and copolymers or closely related materials.

See under both vinyl and polyvinyl.

"Polywood."[448] TM for polyester coatings for wood furniture and other wood products.

"Polyzime" P.[212] TM for a product containing diastatic and proteolytic enzymes.

Properties: Dry, fine, white powder; fully water soluble; nonflammable; optimum pH for diastatic reaction, 7.0–7.2, for proteolytic reaction 7.5–8.0, optimum temperature 45C.

Use: Desizing of textile fabrics preparatory to dyeing, bleaching, mercerizing, printing, and finishing.

"Poly-Zole" AZDN.[511] TM for azodiisobutyronitrile, a blowing agent and catalyst for plastics.

Pomeranz-Fritsch reaction. Formation of isoquinolines by the acid-catalyzed cyclization of benzalaminoacetals prepared from aromatic aldehydes and aminoacetal.

Ponceau 3R. (3-Hydroxy-4-[2,4,5-trimethylphenyl)azo]-2,7-naphthalenedisulfonic acid, disodium salt; C.I. No. 16155).

CAS: 3564-09-8.
$(CH_3)_3C_6H_2NNC_{10}H_4(OH)(SO_3) \cdot 2Na$.
A water-soluble red dye.
Properties: Dark-red powder. Soluble in water and acids to form a cherry-red solution; slightly soluble in alcohol; insoluble in alkalies.
Hazard: Prohibited by FDA for use in foods, drugs, cosmetics.
Use: Dyeing wool; biological stain.

"Ponolith."[28] TM for a type of highly dispersed pigments used primarily for the coloring and tinting of paper.

pontianak. A type of Manila resin which is semifossil, hard, and soluble in oils and hydrocarbons.
Use: Adhesives; paints, varnishes, and lacquers.
See also copal resin.

pony. A small unit of equipment used for laboratory or plant experimental work.

Ponzio reaction. Formation of dinitrophenylmethanes from benzaldoximes by oxidation with nitrogen dioxide in ether.

POPDA. See polyoxypropylene diamine.

POPOP. Abbreviation for phenyl-oxazolyl-phenyloxazolyl-phenyl.
See 1,4-bis[2-(5-phenyloxazolyl)]benzene.

porcelain. (potassium aluminum silicate).
$(4K_2O \cdot Al_2O_3 \cdot 3SiO_2)$. A mixture of clays, quartz, and feldspar usually containing at least 25% alumina. Ball and china clays are ordinarily used. A slip or slurry is formed with water to form a plastic, moldable mass, which is then glazed and fired to a hard, smooth solid.
Properties: High impact strength, impermeable to liquids and gases, resistant to chemicals except hydrogen fluoride and hot, strong caustic solutions, usable up to 1093C but subject to heat shock, d 2.41, Mohs hardness 6–7, compression strength 100,000 psi.
Grade: Chemical and electrical.
Use: Reaction vessels, sparkplugs, electrical resistors; electron tubes, corrosion-resistant equipment, ball mills and grinders, food-processing equipment, piping, valves, pumps, tower packing, lab ware.
See also ceramics; porcelain enamel; porcelain, zircon.

porcelain enamel. A substantially vitreous inorganic coating bonded to metal by fusion above 426C (ASTM). Composed of various blends of low-sodium frit, clay, feldspar, and other silicates, ground in a ball mill and sprayed onto a metal surface (steel, iron, or aluminum) to which it bonds firmly after firing, giving a glass-like fire-polished surface.
Properties: High hardness, abrasion and chemical resistance except to hydrogen fluoride and hot strong caustic; low expansion coefficient. Corrosion-resistant, stable to heat-shock, easy to clean and decontaminate.
Use: Chemical reaction equipment, light reflectors, lab bench tops, storage tanks and containers, high-temperature process equipment, seawater treatment and marine engine parts, glazing frit.
See also porcelain, glaze.

porcelain, zircon. $(ZrO_2 \cdot SiO_2)$. A special high-temperature porcelain used for spark plugs and furnace trays because of its high mechanical strength and heat-shock resistance. Usable up to 1700C. High dielectric strength but rather lower power factor at high frequencies.

pore. (1) A minute cavity in epidermal tissue as in skin, leaves or leather, having a capillary channel to the surface which permits transport of water vapor from within outward but not the reverse. (2) A void or interstice between particles of a solid such as sand minerals or powdered metals, which permit passage of liquids or gases through the material in either direction. In some structures, such as gaseous diffusion barriers and molecular sieves, the pores are of molecular dimensions, i.e., 4–10 Å units. Such microporous structures are useful for filtration and molecular separation purposes in various industrial operations. (3) A cell in a spongy structure made by gas formation (foamed plastic) which absorbs water on immersion but releases it when stressed.
See also membrane, semipermeable; molecular sieve.

"Porocel."[99] TM for an activated bauxite. Supplied in various standard meshes, moisture contents, and in regular low iron and low silica grade.
Use: Absorbent catalysts.

poromeric. A term coined to describe the microporosity, air-permeability, and water- and abrasion resistance of natural and synthetic leather. The pores decrease in diameter from the inner surface to the outer and thus permit air and water vapor to leave the material while excluding water from the outside. Polyester-reinforced urethane resins have poromeric properties and vinyl resins have been used as leather substitutes with some success, primarily for shoe uppers.

porphyrin. Any of several physiologically active nitrogenous compounds occurring widely in nature. The parent structure is comprised of four pyrrole rings, shown as I, II, III, IV in the diagram below, together with four nitrogen atoms and two replaceable hydrogens, for which various metal atoms can be readily substituted. A metal-free porphyrin molecule has the structure:

Prophyrins of this type have been made synthetically by passing an electric current through a mixture of ammonia, methane, and water vapor. Biochemists believe that this phenomenon may account for the original formation of chlorophyll and other porphyrins which have been essential factors in the origin of life.

The most important porphyrin derivatives are characterized by a central metal atom; hemin is the iron-containing porphyrin essential to mammalian blood, and chlorophyll is the magnesium-containing porphyrin that catalyzes photosynthesis. Other derivatives of importance are cytochromes, which function in cellular metabolism, and the phthalocyanine group of dyes.

"Poroplastic."[578] TM for a microporous plastic material reported to be effective for the controlled release of fragrances, flavors, medicinals, pesticides, etc. It has the mechanical properties of a polymer with pore size of molecular dimensions, which enable it to carry a "payload" of up to 98% of a liquid or solid immobilized within its structure. Compounds can be selectively admitted into or excluded from the material on a precisely controlled basis.
Available forms: Microbeads, films, and granules.

porpoise oil. (dolphin oil).
Properties: Pale yellow liquid, d 0.926–0.929, saponification value 290, iodine value 10–30. Soluble in ether, chloroform, benzene, and carbon disulfide; the refined jaw and head oil has uniquely low pour point and high lubricity and is highly resistant to gumming, oxidation, and evaporation.
Use: Lubricant for watches and precision instruments.

Porter, George. (1920-) An English chemist who won the Nobel prize for chemistry in 1967 with Eigen and Norrish. His research concerned fast chemical reactions and the chemistry of photosynthesis. He was educated at Cambridge University and taught there before going on to other posts.

Portland cement. See cement, Portland.

positron. An antiparticle whose mass and spin are the same as those of an electron, but whose electric charge is positive. Positrons can be made in a linear particle accelerator for use in physical research.
See also antiparticle.

potasan. (4-methyl-7-hydroxycoumarin diethoxythiophosphate; hymecromone- O,O-diethylphosphorothioate).   CAS: 299-45-6.
$C_{14}H_{17}O_5PS$.
Properties: Acicular crystals, mp 38C, bp 210C (1 mm) (decomposes), d 1.26, refr index 1.567, soluble in common organic solvents, almost insoluble in water.
Hazard: Toxic by ingestion, cholinesterase inhibitor.
Use: Insecticide.

potash. See potassium carbonate. Also refer to American Potash Institute, 16th St., NW, Washington, DC.

potash alum. See aluminum potassium sulfate.

potash, caustic. See potassium hydroxide.

potash chrome alum. See chromium potassium sulfate.

potash feldspar. See feldspar.

potash magnesia double salt. A material containing potassium carbonate, magnesium sulfate, and a low proportion of chloride, containing 20–30% K_2O.
Use: Fertilizer.

potash, sulfurated. (potassium, sulfurated).
A mixture composed chiefly of potassium polysulfides and potassium thiosulfate, containing not less than 12.8% sulfur in combination as sulfide.
Properties: Liver brown when freshly made changing to a greenish-yellow; odor of hydrogen sulfide; bitter, acrid, and alkaline taste; decomposes upon exposure to air, soluble in water leaving a residue.
Grade: USP.
Use: Medicine; decorative color effects on brass, bronze, and nickel.

potassium. CAS: 7440-09-7.   K.
Metallic element of atomic number 19, group IA of the periodic system, an alkali metal, aw 39.098, valence = 1, potassium-40 is a naturally occurring radioactive isotope. There are also two stable isotopes. The synthetic isotope, potassium-42, is used in tracer studies, primarily in medicine. An essential element in plant growth and in animal and human nutrition; occurs in all soils.
Properties: Soft, silvery metal; rapidly oxidized in moist air; d 0.862; mp 63C; bp 770C; soluble in liquid ammonia, aniline, mercury, and sodium.
Derivation: Ores and deposits in Stassfurt (Germany), Carlsbad (New Mexico), Saskatchewan (Canada), Searles Lake (California), Great Salt Lake (Utah), Yorkshire (England), Ural Mountains (USSR), Israel, and eastern Mediterranean area. Most important ores are carnallite, sylvite, and polyhalite. Thermochemical distillation of potassium chloride with sodium is the chief US method of production.
Grade: Technical, 99.95% pure.
Hazard: Dangerous fire risk, reacts with moisture to form potassium hydroxide and hydrogen. The reaction evolves much heat, causing the potassium to melt and spatter. It also ignites the hydrogen. Burning potassium is difficult to extinguish; dry powdered soda ash or graphite or special mixture of dry chemicals are recommended. It can ignite spontaneously in moist air. Moderate explosion risk by chemical reaction. Potassium metal will form the peroxide and the superoxide at room temperature even when stored under mineral oil; may explode violently when handled or cut. Oxide-coated potassium should be destroyed by burning. Store in inert atmospheres, such as argon or nitrogen or under liquids which are oxygen-free, such as toluene or kerosene, or in glass capsules which have been filled under vacuum or inert atmosphere.
Use: Preparation of potassium peroxide, heat exchange alloys (see NaK); laboratory reagent, seeding of combustion gases in magnetohydrodynamic generators, component of fertilizers (as potassium chloride).

potassium abietate. $KO_2CC_{19}H_{29}$.   Water soluble soap resulting from action of rosin on potassium hydroxide.

potassium acetate. CAS: 127-08-2.
$KC_2H_3O_2$.
Properties: White, crystalline, deliquescent powder; saline taste. Soluble in water and alcohol, insoluble in ether, solutions alkaline to litmus but not to phenolphthalein, mp 292, d 1.57 (25C). Combustible.
Derivation: By the action of acetic acid on potassium carbonate.

Grade: Pure, pure fused, CP, NF reagent.
Use: Dehydrating agent, textile conditioner, reagent in analytical chemistry, medicine, cacodylic derivatives, crystal glass, synthetic flavors.

potassium acid carbonate.   See potassium bicarbonate.

potassium acid fluoride.   See potassium bifluoride.

potassium acid oxalate.   See potassium binoxalate.

potassium acid phosphate.   See potassium phosphate, monobasic.

potassium acid saccharate.
$HOOC(CHOH)_4COOK$.
Properties: Light off-white powder, pH of solution 3.5. Slightly soluble in cold water; soluble in hot water, acid, or alkaline solutions. Combustible.
Use: Chelating agent, rubber formulations, metal plating, soaps and detergents.

potassium acid sulfate.   See potassium bisulfate.

potassium acid sulfate, anhydrous.   See potassium pyrosulfate.

potassium acid sulfite.   See potassium bisulfite.

potassium acid tartrate.   See potassium bitartrate.

potassium alginate. (potassium polymannuronate). $(C_6H_7O_6K)_n$.   Hydrophilic colloid having a molecular weight of 32,000–250,000.
Properties: Occurs in filamentous, grainy, granular, and powdered forms. It is colorless or slightly yellow and may have a slight characteristic smell and taste. Slowly soluble in water forming a viscous solution; insoluble in alcohol.
Grade: Technical, FCC.
Use: Thickening agent and stabilizer in dairy products, canned fruits, and sausage casings; emulsifier.
See also alginic acid.

potassium alum.   See aluminum potassium sulfate.

potassium aluminate. CAS: 12003-63-3.
$K_2Al_2O_4 \cdot 3HOH$.
Properties: Hard crystals, lustrous, soluble in water with hydrolysis to form strongly alkaline solution, insoluble in alcohol.
Derivation: By fusing potassium hydroxide with aluminum oxide.
Grade: Technical.
Use: Dyeing, printing (mordant); lakes, paper sizing.

potassium aluminosilicate.   See feldspar.

potassium aluminum fluoride. K_3AlF_6.
Properties: White powder, slightly soluble in water.
Derivation: Aluminum fluoride, ammonium fluoride, and potassium chloride.
Hazard: Toxic by ingestion and inhalation, strong irritant. TLV (as fluorine): 2.5 mg/m³ of air.
Use: Insecticide.

potassium aluminum sulfate. See aluminum potassium sulfate.

potassium-p-aminobenzoate. CAS: 138-84-1. $C_7H_6KNO_2$.
Properties: Colorless crystals, soluble in water, partially soluble in alcohol, insoluble in ether.
Use: Condensation catalyst, mainly for polyglycol ether polymers.

potassium antimonyl tartrate. See antimony potassium tartrate.

potassium argentocyanide. See silver potassium cyanide.

potassium arsenate. (Macquer's salt).
CAS: 7784-41-0. KH_2AsO_4.
Properties: Colorless crystals, d 2.867, mp 288C, soluble in water, insoluble in alcohol.
Hazard: Toxic by ingestion and inhalation, strong irritant.
Use: Manufacture of fly paper, insecticidal preparations, preserving hides, printing textiles.

potassium arsenite. (potassium metaarsenite).
CAS: 10124-50-2. $KH(AsO_2)_2 \cdot HOH$.
Properties: White powder, hygroscopic, decomposes slowly in air, variable composition, keep well stoppered, soluble in water, slightly soluble in alcohol.
Grade: Technical, reagent.
Hazard: Toxic by ingestion and inhalation, strong irritant.
Use: Reducing agent in silvering mirrors.

potassium aurate. $KAuO_2 \cdot 3HOH$.
Properties: Yellow crystals, soluble in water and alcohol.
Derivation: Gold oxide dissolved in potassium hydroxide solution.
Use: To prepare other gold compounds.

potassium beryllium fluoride. See beryllium potassium fluoride.

potassium bicarbonate. (potassium acid carbonate; baking soda). CAS: 298-24-6. $KHCO_3$.
Properties: Colorless, odorless, transparent crystals or white powder; slightly alkaline, salty taste.

Soluble in water and potassium carbonate solution, insoluble in alcohol, d 2.17, mp decomposes between 100 and 120C, refr index 1.482.
Derivation: By passing carbon dioxide into a solution of potassium carbonate in water.
Grade: Commercial, highest purity, USP, reagent, FCC.
Use: Baking powders, soft drinks, medicine (antacid), manufacture of pure potassium carbonate, fire-extinguishing agent, low pH liquid detergents, laboratory reagent, food additive.

potassium bichromate. See potassium dichromate.

potassium bifluoride. (potassium acid fluoride; potassium hydrogen fluoride).
CAS: 7789-29-9. KHF_2.
Properties: Colorless crystals, decomposed by heat, soluble in alcohol (dilute) and water, insoluble in alcohol (absolute), d 2.37, mp 238C.
Grade: Technical.
Hazard: Corrosive to tissue. TLV (as F): 2.5 mg/m³ of air.
Use: Etching glass, flux in silver solders, alkylation catalyst, electrolyte in fluorine production.

potassium binoxalate. (potassium acid oxalate; acid potassium oxalate; sorrel salt).
CAS: 127-95-7. $KHC_2O_4 \cdot 1/2HOH$.
Properties: White crystals; bitter, sharp taste; somewhat hygroscopic. Soluble in water, insoluble in alcohol, density of the anhydrous salt 2.088, decomposes when heated.
Derivation: Neutral potassium oxalate and oxalic acid are dissolved in water and crystallized.
Hazard: Toxic by ingestion.
Use: Removing ink stains, scouring metals, cleaning wood, photography, laboratory reagent, mordant.

potassium biphthalate. See potassium hydrogen phthalate.

potassium bisulfate. (acid potassium sulfate; potassium hydrogen sulfate; potassium acid sulfate). CAS: 7646-93-7. $KHSO_4$.
Properties: Colorless crystals, the fused salt is deliquescent, soluble in water yielding a solution with acid reaction, decomposes in alcohol, d 2.245, mp 195 (decomposes).
Derivation: Heating potassium sulfate with sulfuric acid.
Use: Conversion of wine lees and tartrates into potassium bitartrate, flux, manufacture of mixed fertilizers, methyl acetate, ethyl acetate, lab reagent.

potassium bisulfide. See potassium hydrosulfide.

potassium bisulfite. (potassium acid sulfite; potassium hydrogen sulfite). CAS: 1310-61-8. $KHSO_3$.
Properties: White, crystalline powder; sulfur dioxide odor. Soluble in water, insoluble in alcohol, mp decomposes at 190C.
Derivation: Sulfur dioxide is passed through a solution of potassium carbonate until no more carbon dioxide is given off; the solution is concentrated and allowed to crystallize.
Grade: Commercial, reagent, medicinal.
Use: Reduction of various organic compounds; purification of aldehydes and ketones, iodine, sodium hydrosulfite; antiseptic; source of sulfurous acid, particularly in brewing; analytical chemistry; tanning; bleaching straw and textile fibers; chemical preservative in foods (except meats and other sources of vitamin B_1).

potassium bitartrate. (cream of tartar; potassium acid tartrate). CAS: 868-14-4. $KHC_4H_4O_6$.
Properties: White crystals or powder; soluble in boiling water; pleasant, slightly acid taste; insoluble in alcohol; d 1.984 (18C).
Derivation: From wine lees by extraction with water and crystallization.
Grade: Technical, NF, FCC.
Use: Baking powder, preparation of other tartrates, medicine, galvanic tinning of metals, food additive.

potassium borate. See potassium tetraborate.

potassium borofluoride. See potassium fluoborate.

potassium borohydride. CAS: 13762-51-1. KBH_4.
Properties: White, crystalline powder. Partially soluble in water and ammonia, insoluble in ethers and hydrocarbons, d 1.18, stable in moist and dry air, stable in vacuum to 500C, decomposed by acids with evolution of hydrogen.
Derivation: By reaction of sodium borohydride and potassium hydroxide.
Grade: Technical, powder, pellets.
Hazard: Flammable, dangerous fire risk. Toxic by ingestion.
Use: Source of hydrogen; reducing agent for aldehydes, ketones, and acid chlorides; foaming agent for plastics.

potassium bromate. CAS: 7758-01-2. $KBrO_3$.
Properties: White crystals or crystalline powder, soluble in boiling water, d 3.27, mp (decomposes) 370C, strong oxidizing agent.
Derivation: By passing bromine into a solution of potassium hydroxide.

Grade: Pure, CP, FCC.
Hazard: Dangerous fire risk in contact with organic materials, strong irritant.
Use: Laboratory reagent, oxidizing agent, permanent wave compounds, maturing agent in flour, dough conditioner, food additive. *Note*: Food additive uses restricted as to proportions used.

potassium bromide. CAS: 7758-02-3. KBr.
Properties: White, crystalline granules or powder; pungent, strong, bitter, saline taste; somewhat hygroscopic. Soluble in water and glycerol, slightly soluble in alcohol and ether, d 2.749, mp 730C, bp 1435C.
Derivation: Solutions of iron bromide and potassium carbonate are mixed and heated, the solution filtered and concentrated, and the bromide crystallized out.
Grade: Technical, CP, NF, reagent, single crystals.
Hazard: Toxic by ingestion and inhalation.
Use: Photography (gelatin bromide papers and plates), process engraving and lithography, special soaps, spectroscopy, infrared transmission, lab reagent.

potassium-tert-butoxide. See potassium-tert-butylate.

potassium-tert-butylate. (potassium-tert-butoxide). $(CH_3)_3COK$. White, hygroscopic powder; a strong organic base.
Hazard: Flammable. May cause severe caustic tissue burns.

potassium carbonate. (potash; pearl ash). CAS: 584-08-7. (a) K_2CO_3, (b) $K_2CO_3 \cdot 1.5HOH$. 32nd highest-volume chemical produced in the US (1985).
Properties: White, deliquescent, granular, translucent powder; alkaline reaction. Soluble in water, insoluble in alcohol; (a) d 2.428 (19C), mp 891C, bp decomposes. Noncombustible.
Derivation: (1) Engel-Precht process uses magnesium oxide, potassium chloride, and carbon dioxide, separating the Engels salt ($MgCO_3 \cdot KHCO_3 \cdot 4HOH$). Decomposition leaves potassium bicarbonate in solution, which can be processed to potassium carbonate. (2) Alkyl amines or ion-exchange resins can be used with potassium chloride and carbon dioxide to yield potassium bicarbonate, which is calcined to the carbonate. This is analogous to the Solvay process for sodium carbonate.
Grade: Crystals, technical, reagent, calcined 80–85%, 85–90%, 90–95%, 96–98%, hydrated 80–85%, FCC.
Hazard: Solutions irritating to tissue.

Use: Special glasses (optical and color TV tubes), potassium silicate, dehydrating agent, pigments, printing inks, lab reagent, soft soaps, raw wool washing, general-purpose food additive.

potassium chlorate. CAS: 3811-04-9. $KClO_3$.
Properties: Transparent, colorless crystals or white powder; cooling, saline taste. Soluble in boiling water, d 2.337, mp 368C, bp decomposes at 400C giving off oxygen gas.
Derivation: (a) By electrolyzing a hot concentrated alkaline solution of potassium chloride. Preferably (b) by interaction of solutions of potassium chloride and sodium chlorate or calcium chlorate.
Hazard: Forms explosive mixture with combustible materials (sulfur, sugar, etc.); strong oxidizing agent.
Use: Oxidizing agent, explosives, matches, source of oxygen, textile printing, pyrotechnics, percussion caps, disinfectant, bleaching.

potassium chloride. CAS: 7447-40-7. KCl.
Properties: Colorless or white crystals, strong saline taste, occurs naturally as sylvite, soluble in water, slightly soluble in alcohol, insoluble in ether and acetone, d 1.987, mp 772C, sublimes at 1500C. Noncombustible.
Derivation: (a) Mined from sylvite deposits in New Mexico and Saskatchewan, purified by fractional crystallization or flotation; (b) extracted from salt lake brines and purified by recrystallization.
Grade: Chemical (99.5 and 99.9%), agricultural grades sold as 60–62%, 48–52% and 22% K_2O, single crystals, USP, FCC.
Use: Fertilizer, source of potassium salts, pharmaceutical preparations, photography, spectroscopy, plant nutrient, salt substitute, lab reagent, buffer solutions, food additive.

potassium chloraurate. See potassium gold chloride.

potassium chlorochromate. (Peligot's salt). $KClCrO_3$.
Properties: Red crystals, liberates chlorine on heating, soluble in water, d 2.497.
Hazard: Toxic by ingestion and inhalation.
Use: Oxidizing agent.

potassium chloroiridate. See iridium potassium chloride.

potassium chloroplatinate. (platinum potassium chloride; potassium platinichloride).
CAS: 16921-30-5. K_2PtCl_6.
Properties: Small, orange-yellow crystals or powder. Insoluble in alcohol, very slightly soluble in water, decomposes when heated (250C), d 3.499 (24C).
Derivation: Adding platinic chloride to a solution of potassium salt and crystallizing.
Hazard: Toxic by ingestion and inhalation.
Use: Photography, lab reagent.

potassium chromate. (potassium chromate, yellow; neutral potassium chromate).
CAS: 7789-00-6. K_2CrO_4.
Properties: Yellow crystals, soluble in water, insoluble in alcohol, d 2.7319, mp 971C.
Derivation: Roasting powdered chromite with potash and limestone, treating the cinder with hot potassium sulfate solution and leaching.
Grade: Technical, reagent.
Hazard: Toxic by ingestion and inhalation.
Use: Reagent in analytical chemistry, aniline black, textile mordant, enamels, chromate pigments, inks.

potassium chromate, red. See potassium dichromate.

potassium chromium sulfate. See chromium potassium sulfate.

potassium citrate. CAS: 866-83-1. $K_3C_6H_5O_7 \cdot HOH$.
Properties: Colorless or white crystals or powder, deliquescent. Cooling, saline taste; odorless. Soluble in water and glycerol, almost insoluble in alcohol, d 1.98, loses water at 180C, decomposes at 230C.
Derivation: Action of citric acid on potassium carbonate.
Grade: Technical, CP, NF, FCC.
Use: Medicine (antacid), sequestrant, stabilizer, buffer in foods.

potassium cobaltinitrite. See cobalt potassium nitrite.

potassium columbate. Obsolete name for potassium niobate.

potassium copper cyanide. (copper potassium cyanide; potassium cuprocyanide).
CAS: 13682-73-0. $C_2CuN_2 \cdot K$.
Properties: White, crystalline double salt of copper cyanide and potassium cyanide; copper content min 25.8%; free potassium cyanide 1.25–3.0%.
Hazard: Highly toxic. TLV (as CN): 5 mg/m³ in air.
Use: For preparing and maintaining cyanide copper plating baths based on potassium cyanide.

potassium copper ferrocyanide. See copper potassium ferrocyanide.

potassium cupric ferrocyanide. See copper potassium ferrocyanide.

potassium cuprocyanide. Legal label name (Air) for potassium copper cyanide.

potassium cyanate. CAS: 590-28-3. KOCN.
Properties: Colorless crystals, d 2.05, decomposes at 700–900C, soluble in water, insoluble in alcohol.
Derivation: Heating potassium cyanide with lead oxide.
Use: Herbicide, manufacture of organic chemicals and drugs, treatment of sickle cell anemia.

potassium cyanide. CAS: 151-50-8. KCN.
Properties: White, amorphous, deliquescent lumps or crystalline mass; faint odor of bitter almonds. Soluble in water, alcohol, and glycerol; d 1.52 (16C); mp 634C.
Derivation: absorption of hydrogen cyanide in potassium hydroxide.
Grade: Commercial, pure, solution, reagent.
Hazard: A poison as absorbed by skin. TLV (as CN): 5 mg/m³ of air.
Use: Extraction of gold and silver from ores, reagent in analytical chemistry, insecticide, fumigant, electroplating.

potassium cyanoargentate. See silver potassium cyanide.

potassium cyanoaurite. See potassium gold cyanide.

potassium cyclamate. See sodium cyclamate.

potassium di-n-butyl dithiocarbamate.
(C₄H₉)₂NCSSK. A 50% aqueous solution in a straw-colored liquid, d 1.08–1.12.
Use: Ultraaccelerator for natural and synthetic latexes.

potassium dichloroisocyanurate. (potassium dichloro-s-triazinetrione). A cyclic compound. See also dichloroisocyanuric acid.
Properties: White, slightly hygroscopic, crystalline powder or granules; loose bulk density (approximate) powder 37 lb/cu ft, granular 64 lb/cu ft. Active ingredient 59% available chlorine; decomposes at 240C.
Hazard: Dangerous fire risk in contact with organic materials. Strong oxidizing agent. Toxic by ingestion.
Use: Household dry bleaches, dishwashing compounds, scouring powders, detergent- sanitizers; replacement for calcium hypochlorite.
See also dichloroisocyanuric acid.

potassium dichromate. (potassium bichromate; red potassium chromate). CAS: 7778-50-9. K₂Cr₂O₇.
Properties: Bright, yellowish-red, transparent crystals; bitter, metallic taste. Soluble in water, insoluble in alcohol, mp 396C, d 2.676 (25C), bp decomposes at 500C.
Derivation: Reaction of potassium chloride and sodium dichromate.
Grade: Commercial; highest purity; highest purity fused; reagent.
Hazard: Toxic by ingestion and inhalation. Dangerous fire risk in contact with organic materials. Strong oxidizing agent.
Use: Oxidizing agent (chemicals, dyes, intermediates); analytical reagent; brass pickling compositions; electroplating; pyrotechnics; explosives; safety matches; textiles; dyeing and printing; chrome glues and adhesives; chrome tanning leather; wood stains; poison fly paper; process engraving and lithography; synthetic perfumes; chrome alum manufacture; pigments; alloys, ceramic products; depolarizer in dry cell batteries; bleaching fats and waxes.

potassium dihydrogen phosphate. See potassium phosphate, monobasic.

potassium dimethyldithiocarbamate. See sodium dimethyldithiocarbamate.

potassium diphosphate. See potassium phosphate, monobasic.

potassium dithionate. (potassium hyposulfate). K₂S₂O₆.
Properties: Colorless crystals, soluble in water, insoluble in alcohol, d 2.27.
Use: Analytical reagent.

potassium endothall. (dipotassium salt of endothall). The water solution is an amber liquid, used as a contact herbicide.

potassium ethyldithiocarbonate. See potassium xanthate.

potassium ethylxanthate. See potassium xanthate.

potassium ethylxanthogenate. See potassium xanthate.

potassium feldspar. (feldspar). (K₂O·3SiO₂) + (Al₂O₃·3SiO₂). "Potassium" feldspar is usually the pink variety of feldspar.

potassium ferric oxalate.
K₃Fe(C₂O₄)₃·3HOH.
Properties: Monoclinic green crystals, loses three molecules water at 100C, decomposes 230C,

soluble in water and acetic acid, incompatible with alkali and ammonia since these react to precipitate ferric hydroxide.
Use: Photography and blue-printing.

potassium ferricyanide. (red prussiate of potash; red potassium prussiate).   CAS: 13746-66-2. $K_3Fe(CN)_6$.
Properties: Bright red, lustrous crystals or powder. Soluble in water, slightly soluble in alcohol, d 1.85 (25C).
Derivation: Chlorine is passed into a solution of potassium ferrocyanide, the ferricyanide separating out.
Grade: Pure crystals, pure powder, reagent, technical.
Hazard: Decomposes on strong heating to evolve highly toxic fumes, but the compound itself has low toxicity.
Use: Tempering steel, etching liquid, production of pigments, electroplating, sensitive coatings on blueprint paper, fertilizer compositions, laboratory reagent.

potassium ferrocyanide. (yellow prussiate of potash; yellow potassium prussiate).
CAS: 13943-58-3. $K_4Fe(CN)_6 \cdot 3HOH$.
Properties: Lemon yellow crystals or powder, mild saline taste, effloresces on exposure to air, soluble in water, insoluble in alcohol, d 1.853 (17C), loses its water of crystallization when heated to 70C, bp decomposes.
Derivation: From nitrogenous waste products, iron filings, and potassium carbonate.
Grade: Technical, CP.
Hazard: It evolves highly toxic fumes on heating to red heat, but the compound itself has low toxicity.
Use: Potassium cyanide and ferricyanide, dry colors, tempering steel, dyeing, explosives, process engraving and lithography, lab reagent.

potassium fluoborate. (potassium borofluoride). KBF_4.
Properties: Colorless crystals, d 2.498 (20C), decomposes at 350C. Very slightly soluble in water; insoluble in alcohol, ether, alkalies.
Derivation: By the reaction of boric acid, hydrogen fluoride, and potassium hydroxide.
Hazard: Toxic by ingestion.
Use: Sand-casting of aluminum and magnesium; grinding aid in resinoid grinding wheels, flux for soldering and brazing, electrochemical processes and chemical research.

potassium fluoride. CAS: 7789-23-3.   (a) KF; (b) $KF \cdot 2HOH$.
Properties: White, crystalline, deliquescent powder; sharp saline taste. Soluble in water and hy-

drogen fluoride, insoluble in alcohol. D (a) 2.48, (b) 2.454; mp (a) 846C; bp 1505C.
Derivation: Neutralizing hydrogen fluoride with potassium carbonate.
Grade: Technical, free of arsenic, CP.
Hazard: Toxic by ingestion and inhalation, strong irritant to tissue. TLV (as F): 2.5 mg/m^3 of air.
Use: Etching glass, preservative, insecticide, solder flux, fluorination reactions.

potassium fluosilicate. (potassium silicofluoride). CAS: 16871-90-2. K_2SiF_6.
Properties: White, odorless, crystalline powder; d 3.0; slightly soluble in water; soluble in hydrochloric acid.
Hazard: Toxic by ingestion and inhalation, strong irritant to tissue. TLV (as F): 2.5 mg/m^3 of air.
Use: Vitreous enamel frits, synthetic mica, metallurgy of aluminum and magnesium, ceramics, insecticide, opalescent glass.

potassium fluotantalate. See tantalum potassium fluoride.

potassium fluozirconate. See zirconium potassium fluoride.

potassium germanium fluoride. See germanium potassium fluoride.

potassium gibberellate. (salt of gibberellic acid).
Use: Promote and control development of malt in grain.

potassium gluconate. CAS: 299-27-4. $KC_6H_{11}O_7$.
Properties: Odorless, salty tasting, fine, white, crystalline powder. Soluble in water, insoluble in alcohol and benzene, mp 180C (decomposes).
Derivation: Reaction of potassium hydroxide or carbonate with gluconic acid.
Grade: Pharmaceutical.
Use: Medicine (vitamin tablets).

potassium glutamate. (monopotassium-*l*-glutamate).   CAS: 19473-49-5. $KOOC(CH_2)_2CH(NH_2)COOH \cdot HOH$.
Properties: White, practically odorless, free-flowing hygroscopic powder. Freely soluble in water, slightly soluble in alcohol, pH of a 2% solution 6.9–7.1.
Grade: FCC.
Use: Flavor enhancer, salt substitute.

potassium glycerophosphate. (potassium glycerinophosphate).   CAS: 1319-69-3. $K_2C_3H_5O_2 \cdot H_2PO_4 \cdot 3HOH$.
Properties: Pale yellow viscous mass or liquid (75% aqueous solution), acid taste, soluble in alcohol, miscible with water.

Derivation: Glycerol and phosphorus pentoxide or metaphosphoric acid are mixed, warmed, and exactly neutralized with potassium carbonate, then warmed and concentrated.

Grade: Technical, 50 or 75% solution, FCC (syrup or solution).

Use: Food additive and dietary supplement.

potassium gold chloride. (potassium aurichloride; potassium chloroaurate; gold potassium chloride). $KAuCl_4 \cdot 2HOH$.

Properties: Yellow crystals; soluble in water, alcohol, and ether.

Derivation: Neutralizing chloroauric acid with potassium carbonate.

Use: Photography, painting porcelain and glass, medicine.

potassium gold cyanide. (potassium cyanoaurite; gold potassium cyanide). $KAu(CN)_2$.

Properties: White, crystalline powder. Soluble in water, slightly soluble in alcohol, insoluble in ether.

Derivation: Action of hydrogen cyanide on potassium aurate.

Hazard: Highly toxic. TLV (as CN): 5 mg/m^3 of air.

Use: Electrogilding.

potassium guaiacol sulfonate. CAS: 1321-14-8. $C_6H_3OCH_3OHSO_3K \cdot 1/2HOH$.

Properties: White powder or crystals, gradually turns pink on exposure to air and light, bitter taste, afterward becoming sweetish; odorless, contains 60% guaiacol, soluble in water, sparingly soluble in alcohol.

Grade: NF.

Use: Medicine (expectorant), lab reagent.

potassium-2,4-hexadienoate. See potassium sorbate.

potassium hexafluorophosphate. KPF_6.

Properties: Solid, mp 575C, bp decomposes, soluble in water.

Grade: 98–100%.

Hazard: Toxic by ingestion.

Use: Maintenance of fluoride atmospheres, preparation of bactericides and fungicides, lab reagent.

potassium hexanitrocobaltate III. See cobalt potassium nitrite.

potassium hexyl xanthane. $C_6H_{13}OCSSK$.

Use: Flotation agent.

potassium hydrate. See potassium hydroxide.

potassium hydride. KH. Marketed as a semi-dispersion of gray powder in oil.

Properties: The solid decomposes on heating or in contact with moisture.

Hazard: Dangerous fire and explosion risk, evolves toxic and flammable gases on heating and on exposure to moisture.

Use: Organic condensations and alkylations.

See also hydride.

potassium hydrogen fluoride. See potassium bifluoride.

potassium hydrogen phosphate. See potassium phosphate dibasic.

potassium hydrogen phthalate. (potassium biphthalate). $HOOCC_6H_4COOK$.

Properties: Colorless crystals, soluble in water, d 1.636.

Derivation: Potassium hydroxide and phthalic anhydride.

Grade: CP, analytical.

Use: Alkalimetric standard, buffering agent.

potassium hydrosulfide. (potassium sulfhydrate; potassium bisulfide). KHS.

Properties: White to yellow crystals, hydrogen sulfide odor. Forms the polysulfide when exposed to air. Hygroscopic, soluble in alcohol and water, d 1.69, mp 455C.

Grade: Technical.

Use: Separation of heavy metals.

potassium hydroxide. (caustic potash; potassium hydrate; lye). CAS: 1310-58-3. KOH.

Properties: White, deliquescent pieces, lumps, sticks, pellets, or flakes having a crystalline fracture. Keep well stoppered, absorbs water and carbon dioxide from the air; soluble in water, alcohol, glycerol; slightly soluble in ether; d 2.044; mp 405C (varies with water content).

Derivation: Electrolysis of concentrated potassium chloride solution.

Method of purification: Sulfur compounds are removed by the addition of potassium nitrate to the fused caustic. The purest form is obtained by solution in alcohol, filtration, and evaporation.

Grade: Commercial, ground, flake, fused (88–92%), purified by alcohol (sticks, lumps, and drops), reagent, highest purity, USP, liquid (45%), FCC.

Hazard: Toxic by ingestion and inhalation, strong caustic, handle with gloves or tongs, corrosive to tissue. TLV: CL of 2 mg/m^3 of air.

Use: Soap manufacture, bleaching, manufacture of potassium carbonate and tetrapotassium pyrophosphate, electrolyte in alkaline storage batteries and some fuel cells, absorbent for carbon dioxide and hydrogen sulfide, dyestuffs, liquid fertilizers, food additive, herbicides, electroplating, mercerizing, paint removers, reagent.

potassium hypophosphite. (potassium hypophosphite, monobasic). CAS: 77-82-87-8. KH_2PO_2.
Properties: White, opaque crystals or powder with pungent saline taste; very deliquescent. Soluble in water and alcohol, decomposed by heat.
Derivation: Interaction of calcium hypophosphite and potassium carbonate.
Hazard: Moderate fire risk, may explode if ground with chlorates, nitrates, or other strong oxidizing agents.

potassium hyposulfate. See potassium dithionate.

potassium hyposulfite. See potassium thiosulfate.

potassium iodate. CAS: 7758-05-6. KIO_3.
Properties: White, crystalline powder; odorless. Soluble in water, sulfuric acid (dilute); insoluble in alcohol. D 3.9, mp 560C (partial decomposition).
Grade: Technical, CP, FCC.
Use: Analysis (testing for zinc and arsenic), iodometry, reagent, feed additive, in foods as maturing agent and dough conditioner, medicine (topical antiseptic).

potassium iodide. CAS: 7681-11-0. KI.
Properties: White crystals, granules, or powder; strong, bitter, saline taste; soluble in water, alcohol, acetone, and glycerol; d 3.123; mp 686C; bp 1330C.
Grade: Reagent, USP, single crystals, FCC.
Use: Reagent in analytical chemistry, photographic emulsions (precipitating Ag), feed additive, spectroscopy, infrared transmission, scintillation, dietary supplement (up to 0.01% in table salt).

potassium iridium chloride. See iridium potassium chloride.

potassium laurate. $KOOCC_{11}H_{23}$.
Properties: Light tan paste, soluble in water and alcohol.
Use: Emulsifying agent, base for liquid soaps and shampoos.

potassium linoleate. $KOOCC_{17}H_{31}$.
Properties: Light tan paste, soluble in water.
Use: Emulsifying agent.

potassium magnesium sulfate.
$K_2SO_4 \cdot 2MgSO_4$.
Properties: White, tetragonal crystals; d 2.829; mp 927.
Use: Fertilizer.

potassium manganate. CAS: 10294-64-1. K_2MnO_4.
Properties: Dark green powder or crystals, soluble in potassium hydroxide solution and water, decomposes in acid solution, mp 190C (decomposes).
Derivation: Fusion of pyrolusite with potassium hydroxide.
Grade: Technical.
Hazard: Dangerous fire risk in contact with organic materials, strong oxidizing agent.
Use: Bleaching skins, fibers, oils; disinfectants; mordant (wool); batteries; photography; printing; source of oxygen (dyeing); water purification; oxidizing agent.

potassium mercuric iodide. See mercuric potassium iodide.

potassium metaarsenite. See potassium arsenite.

potassium metabisulfite. (potassium pyrosulfite). CAS: 16731-55-8. $K_2S_2O_5$.
Properties: White granules or powder, pungent sharp odor, d 2.3, decomposes at 150–190C, oxidizes in air and moisture to sulfate, soluble in water, insoluble in alcohol.
Derivation: By heating potassium bisulfite until it loses water.
Grade: Technical, reagent, FCC.
Use: Antiseptic, reagent, source of sulfurous acid, brewing (cleaning casks and vats), wine-making (said to kill only undesirable yeasts and bacteria), food preservative, developing agent (photography), process engraving and lithography, dyeing, antioxidant, bleaching agent.

potassium, metallic, liquid alloy. A flammable mixture or alloy of potassium and another metal which is liquid at normal temperature.

potassium metaphosphate. (monopotassium metaphosphate). KPO_3 or $(KPO_3)_6$.
Properties: White powder, soluble in dilute acids, slightly soluble in water, d 2.45.

potassium molybdate. CAS: 13446-49-6. K_2MoO_4.
Properties: White, deliquescent, microcrystalline powder. Soluble in water, insoluble in alcohol, d 2.91 (18C), mp 919C.
Use: Reagent.

potassium monophosphate. See potassium phosphate, dibasic.

potassium naphthenate.
Properties: Gray paste, soluble in water. Combustible.
Derivation: From naphthenic acids.
Use: Driers, emulsifying agents.

potassium nickel sulfate. See nickel potassium sulfate.

potassium nitrate. (niter; nitre; saltpeter).
CAS: 7757-79-1. KNO_3.
Properties: Transparent, colorless or white, crystalline powder or crystals; slightly hygroscopic; pungent, saline taste. D 2.1062, mp 337C, bp decomposes at 400C, soluble in water and glycerol, slightly soluble in alcohol.
Grade: Commercial, CP, FCC.
Hazard: Dangerous fire and explosion risk when shocked or heated, or in contact with organic materials, strong oxidizing agent.
Use: Pyrotechnics, explosives, matches, specialty fertilizer, reagent, to modify burning properties of tobacco, glass manufacture, tempering steel, curing foods, oxidizer in solid rocket propellants.

potassium nitrite. CAS: 7758-09-0. KNO_2.
Properties: White or slightly yellowish prisms or sticks, deliquescent, d 1.915, bp explodes at 1000F (537C), mp 440C (decomposition starts at 350C), soluble in water, insoluble in alcohol.
Grade: CP, technical, reagent, FCC.
Hazard: Fire and explosion risk when shocked or heated, or in contact with organic materials, strong oxidizing agent.
Use: Analysis (testing for amino acids, cobalt, iodine, urea) food additive (curing agent).

potassium oleate. CAS: 143-18-0.
$C_{17}H_{33}COOK$.
Properties: Gray-tan paste, soluble in water and alcohol. Combustible.
Use: Textile soaps, emulsifying agent.

potassium orthophosphate. See potassium phosphate, monobasic, dibasic or tribasic.

potassium orthotungstate. See potassium tungstate.

potassium osmate. (potassium perosmate).
CAS: 19718-36-6. $K_2OsO_4 \cdot 2HOH$.
Properties: Violet crystals, hygroscopic, soluble in water, insoluble in alcohol and ether.
Hazard: Toxic by ingestion and inhalation.
Use: Determination of nitrogenous matter in water.

potassium oxalate. CAS: 583-52-8.
$K_2C_2O_4 \cdot HOH$.
Properties: Colorless, transparent crystals; odorless. Soluble in water, efflorescent in warm dry air, d 2.08, decomposes when heated.
Derivation: Potassium formate or carbonate mixed with a small quantity of oxalate and a slight excess of alkali is heated, the oxalate extracted with water and crystallized.
Grade: Technical, CP.
Hazard: Toxic by inhalation and ingestion.

Use: Reagent in analytical chemistry, source of oxalic acid, bleaching and cleaning, removing stains from textiles, photography.

potassium oxide. K_2O.
Properties: Gray, crystalline mass. Soluble in water, forms potassium hydroxide, soluble in alcohol and ether, d 2.32 (0C), decomposes 350C.
Derivation: By heating potassium nitrate and metallic potassium
Hazard: Corrosive to tissue.
Use: Reagent and intermediate.

potassium oxyfluoniobate. See niobium potassium oxyfluoride.

potassium palladium chloride. See palladium potassium chloride.

potassium penicillin G. (benzylpenicillin potassium). CAS: 113-98-4. $C_{16}H_{17}KN_2O_4S$.
Properties: Colorless or white crystals or powder, odorless, moderately hygroscopic, solutions dextrorotatory, relatively stable to air and light. Very soluble in water, in saline, and in dextrose solutions; moderately soluble in alcohol; pH of solution (30 mg/mL) is 5.0–7.5; mp 214–217C (decomposes)
Grade: USP.
Use: Medicine (antibiotic).

potassium penicillin V. See potassium phenoxymethylpenicillin.

potassium percarbonate. CAS: 589-97-9.
$K_2C_2O_6 \cdot HOH$.
Properties: Granular white mass, keep away from light and moisture, soluble in water (liberates oxygen), mp 200–300C.
Grade: Technical.
Hazard: Strong irritant to tissue. Fire risk in contact with organic materials, strong oxidizing agent.
Use: Analysis (testing for cerium, chromium, vanadium, titanium), photography, textile printing.

potassium perchlorate. CAS: 7778-74-7.
$KClO_4$.
Properties: Colorless crystals or white, crystalline powder; decomposed by concussion, organic matter, and agents subject to oxidation; more stable than potassium chlorate. Soluble in water, insoluble in alcohol, d 2.524, mp 400C (decomposes)
Grade: Technical, reagent, ordnance.
Hazard: Fire risk in contact with organic materials, strong oxidizing agent. Strong irritant.
Use: Explosives, oxidizing agent, photography, py-

rotechnics and flares, reagent, oxidizer in solid rocket propellants.

potassium periodate. CAS: 7790-21-8. KIO$_4$.
Properties: Small, colorless crystals or white, granular powder. Slightly soluble in water, d 3.168, mp 582C, explodes at 1076F (580C).
Grade: Technical, CP, reagent.
Hazard: Fire risk in contact with organic materials, strong oxidizing agent. Strong irritant to tissue.
Use: Analysis (oxidizing agent).

potassium permanganate. CAS: 7722-64-7. KMnO$_4$.
Properties: Dark purple crystals with blue metallic sheen; sweetish, astringent taste; odorless. Soluble in water, acetone, and methanol; decomposed by alcohol. D 2.7032, mp decomposes at 240C, oxidizing material.
Derivation: (a) By oxidation of the manganate in an alkaline electrolytic cell. (b) A hot solution of the manganate is treated with carbon dioxide; on cooling, the solution deposits crystals of the permanganate.
Grade: Technical, CP, USP.
Hazard: Dangerous fire and explosion risk in contact with organic materials; powerful oxidizing agent.
Use: Oxidizer, disinfectant, deodorizer, bleach, dye, tanning, radioactive decontamination of skin, reagent in analytical chemistry, medicine (antiseptic), manufacture of organic chemicals, air and water purification.

potassium perosmate. See potassium osmate.

potassium peroxide. CAS: 17014-71-0. K$_2$O$_2$.
Properties: Yellow, amorphous mass; decomposes in water evolving oxygen; mp 490C.
Derivation: Oxidation of potassium in oxygen.
Hazard: Dangerous fire and explosion risk in contact with organic materials, strong oxidizing agent. Irritant to skin and tissue.
Use: Oxidizing agent, bleaching agent, oxygen-generating gas masks.

potassium peroxydisulfate. See potassium persulfate.

potassium peroxymonosulfate. See "Oxone."

potassium persulfate. (potassium peroxydisulfate). CAS: 7727-21-1. K$_2$S$_2$O$_8$.
Properties: White crystals, soluble in water, insoluble in alcohol, d 2.477, mp decomposes below 100C.

Derivation: By electrolysis of a saturated solution of potassium sulfate.
Hazard: Fire risk in contact with organic materials. Strong irritant and oxidizing agent.
Use: Bleaching, oxidizing agent, reducing agent in photography, antiseptic, soap manufacture, analytical reagent, polymerization promoter, pharmaceuticals, modification of starch, flour-maturing agent, desizing of textiles.

potassium α-phenoxyethylpenicillin. (potassium penicillin 152; potassium phenethicillin; phenethicillin). KC$_{17}$H$_{19}$N$_2$O$_5$S. Synthetically prepared, a mixture of two stereoisomers.
Properties: White, crystalline solid; moderately hygroscopic. Decomposes above 220C, very soluble in water, resistant to acid decomposition.
Preparation: By N-acylation of α-phenoxypropionic acid and G-aminopenicillanic acid (produced by fermentation using Penicillium chrysogenum).
Grade: NF.
Use: Antibiotic.

potassium phenoxymethylpenicillin. (potassium penicillin V). CAS: 132-98-9. KC$_{16}$H$_{17}$N$_2$O$_5$S.
Properties: White, odorless, crystalline powder. Very soluble in water, slightly soluble in alcohol, insoluble in acetone.
Grade: USP.
Use: Antibiotic.

potassium phosphate, dibasic. (DKP; potassium hydrogen phosphate; potassium monophosphate; dipotassium orthophosphate). CAS: 7758-11-4. K$_2$HPO$_4$.
Properties: Hygroscopic, white crystals or powder; very soluble in water; converted to pyrophosphate by ignition.
Derivation: Action of phosphoric acid on potassium carbonate.
Grade: Commercial, pure, highest purity, NF, FCC.
Use: Buffer in antifreezes, ingredient of "instant" fertilizers, nutrient for penicillin culturing, humectant, pharmaceuticals. In foods as a buffer, sequestrant, and yeast food. Laboratory reagent.

potassium phosphate, monobasic. (MKP; potassium acid phosphate; potassium diphosphate; potassium orthophosphate; potassium dihydrogen phosphate). CAS: 7778-77-0. KH$_2$PO$_4$.
Properties: Colorless crystals, acid in reaction, soluble in water, insoluble in alcohol, d 2.338, mp 253C.

Derivation: Action of phosphoric acid on potassium carbonate.

Grade: Technical, CP, FCC.

Use: Baking powder, nutrient solutions, yeast foods, buffer and sequestrant, lab reagent.

potassium phosphate, tribasic. (potassium phosphate, neutral; potassium phosphate normal; tripotassium orthophosphate; potassium phosphate, tertiary; tripotassium phosphate).
CAS: 7778-53-2. $K_3PO_4 \cdot HOH$ or K_3PO_4.
Properties: Granular, white powder; deliquescent. Soluble in water giving strongly basic solution, insoluble in alcohol, mp (anhydrous) 1340C, d (anhydrous) 2.564 (17C).
Grade: Reagent, technical, FCC.
Use: Purification of gasoline, water-softening, liquid soaps, fertilizer, in foods as an emulsifier, laboratory reagent.

potassium phosphite, monobasic. KH_2PO_3.
Properties: White powder, hygroscopic, soluble in water, insoluble in alcohol, slowly oxidized by air to phosphate.

potassium platinichloride. See potassium chloroplatinate.

potassium polymetaphosphate. $(KPO_3)_n$.
The molecular weight may be as high as 500,000.
Properties: White, odorless powder. Insoluble in water, soluble in sodium salt solutions which may have high viscosity.
Derivation: Dehydration of monobasic potassium phosphate.
Grade: Technical, FCC.
Use: Fat emulsifier and moisture-retaining agent in foods.
See sodium metaphosphate.

potassium polysulfide. K_2S_x.
Properties: Crystals, soluble in water and alcohol. Hazard: Moderate fire risk. Toxic by ingestion, irritant to skin and eyes.
Use: Fungicide.

potassium prussiate, red. See potassium ferricyanide.

potassium prussiate, yellow. See potassium ferrocyanide.

potassium pyroantimonate. $K_2H_2SbO_7 \cdot 4HOH$.
Properties: White, crystalline powder or granules. Slightly soluble in cold water, readily soluble in hot water, insoluble in alcohol.
Grade: Reagent, technical.
Use: Starch sizes and flame-retarding compounds.

potassium pyroborate. See potassium tetraborate.

potassium pyrophosphate. (TKPP; tetrapotassium pyrophosphate; potassium pyrophosphate, normal).   CAS: 7320-34-5.
$K_4P_2O_7 \cdot 3HOH$.
Properties: Colorless crystals or white powder, somewhat hygroscopic in air (deliquescent at a relative humidity of above 40–45%). Similar to tetrasodium pyrophosphate except for greater solubility, d 2.33, dehydrates at about 300C, mp 1090C, soluble in water, insoluble in alcohol.
Grade: Technical, 99.4%, 60% solution, FCC.
Use: Soap and detergent builder, sequestering agent, peptizing and dispersing agent.

potassium pyrosulfate. (potassium acid sulfate, anhydrous).   CAS: 7790-62-7. $K_2S_2O_7$.
Properties: Colorless needles or white crystalline powder or fused pieces, soluble in water, converted to potassium bisulfate, d 2.25 (25/4C), mp 325C.
Use: Acid flux in analysis, laboratory reagent.

potassium pyrosulfite. See potassium metabisulfite.

potassium rhodanide. See potassium thiocyanate.

potassium ricinoleate. $C_{17}H_{32}OHCOOK$.
Properties: White paste, soluble in water. Combustible.
Use: Emulsifying agent.

potassium silicate. CAS: 1312-76-2.
Properties: (Solid) Weight ratio $SiO_2:K_2O$ varies with grade as 2.1:1; 2.5:1; colorless anhydrous lump, shattered or granular, soluble in water at high temperature and pressure, insoluble in alcohol. (Solution) Colorless liquid, Bé range 29–48 degrees.
Derivation: Supercooled melt of potassium carbonate and pure silica sand.
Use: (Solid) Manufacture of glass and refractory material, welding rods, high-temperature mortars, binder in carbon arc-light electrodes, detergents, catalyst, adhesives.

potassium silicofluoride. See potassium fluosilicate.

potassium sodium carbonate. See sodium potassium carbonate.

potassium sodium ferricyanide.
$K_2NaFe(CN)_6$.
Properties: Red crystals, over 99% pure, mp decomposes, nonhygroscopic and stable, easily soluble in water.
Derivation: From ferrocyanides.
Use: Blueprint paper and photography.

potassium-sodium phosphate. See sodium-potassium phosphate.

potassium-sodium tartrate. (Rochelle salt; sodium potassium tartrate). CAS: 304-59-6. $KNaC_4H_4O_6 \cdot 4HOH$. It is salt of l(+)-tartaric acid.
Properties: Colorless, transparent, efflorescent crystals or white powder having cool, saline taste. Soluble in water, insoluble in alcohol, loses water of crystallization at 140C, unstable above 225C, d 1.77, mp 70–80C.
Derivation: Potassium acid tartrate is dissolved in water, the solution saturated with sodium carbonate, concentrated after purification, and crystallized.
Grade: Highest purity, reagent, commercial crystals or powder, NF, FCC.
Use: Baking powders, medicine (cathartic), component of Fehling's solution, silvering mirrors.

potassium sorbate. (potassium-2,4-hexadienoate). CAS: 24634-51-5.
$CH_3CH:CHCH:CHCOOK$.
Properties: White powder, mp 270C (decomposes), 58.5% soluble in water (25C), d 1.36 (25/20C).
Grade: Technical, FCC.
Use: Bacteriostat and preservative in meats, sausage casings, wines, etc.

potassium stannate. CAS: 12142-33-5.
$K_2SnO_3 \cdot 3HOH$.
Properties: White to light tan crystals, soluble in water, insoluble in alcohol, d 3.197.
Grade: Technical.
Hazard: Highly toxic. TLV (as Sn): 2 mg/m³ of air.
Use: Textiles (dyeing and printing), alkaline tin-plating bath.

potassium stearate. CAS: 59329-3.
$C_{17}H_{35}COOK$.
Properties: White, crystalline powder; soluble in hot water and alcohol; slight odor of fat.
Grade: Commercial, contains considerable palmitate, FCC.
Use: Base for textile softeners.

potassium strontium chlorate. See strontium potassium chlorate.

potassium styphnate. $KC_6H_2N_3O_8 \cdot HOH$.
Properties: Yellow prisms, mp loses water at 120C.
Hazard: Explodes when shocked or heated.
Use: High explosive.

potassium sulfate. CAS: 7778-80-5. K_2SO_4.
Properties: Colorless or white, hard crystals or powder; bitter saline taste, soluble in water, insoluble in alcohol, d 2.66, mp 1072C.

Derivation: (a) By treatment of potassium chloride either with sulfuric acid or with sulfur dioxide, air, and water (Hargreaves process); (b) by fractional crystallization of a natural sulfate ore; (c) from salt lake brines.
Grade: Highest purity medicinal, commercial, crude, CP, agricultural, reagent, technical.
Use: Reagent in analytical chemistry, medicine (cathartic), gypsum cements, fertilizer for chloride-sensitive crops such as tobacco and citrus, alum manufacture, glass manufacture, food additive.

potassium sulfhydrate. See potassium hydrosulfide.

potassium sulfide. CAS: 1312-73-8. K_2S.
Properties: Red or yellow-red crystalline mass or fused solid; deliquescent in air; soluble in water, alcohol, and glycerol; insoluble in ether; d 1.75 (20/4C); mp 910C.
Grade: Technical.
Hazard: Flammable, dangerous fire risk, may ignite spontaneously, explosive in form of dust or powder.
Use: Reagent in analytical chemistry, depilatory, medicine.

potassium sulfite. CAS: 10117-38-1.
$K_2SO_3 \cdot 2HOH$.
Properties: White crystals or powder, soluble in water, sparingly soluble in alcohol, decomposes on heating and slowly oxidizes in air.
Grade: Technical, CP, FCC.
Use: Photographic developer, medicine (cathartic), food and wine preservative.

potassium sulfocarbonate. (potassium trithiocarbonate). K_2CS_3.
Properties: Yellowish-red crystals, very hygroscopic, soluble in alcohol and water.
Grade: Technical.
Hazard: Toxic by ingestion, strong irritant.
Use: Analysis (testing for cobalt, nickel), medicine, soil fumigant.

potassium sulfocyanate. See potassium thiocyanate.

potassium sulfocyanide. See potassium thiocyanate.

potassium tantalum fluoride. See tantalum potassium fluoride.

potassium tartrate. CAS: 921-53-9.
$K_2C_4H_4O_6 \cdot 1/2HOH$.
Properties: Colorless, crystalline solid; soluble in water; insoluble in alcohol; decomposed by heat (200–220C); d 1.98.

Grade: CP, technical.
Use: Manufacture of potassium salts, medicine (cathartic), lab reagent.

potassium tellurite. CAS: 15571-91-2.
K_2TeO_3.
Properties: Granular, white powder; hygroscopic; soluble in water; decomposes at 460–470C.
Use: Analysis (testing for bacteria).

potassium tetrathiocyanodiammonochromate.
See Reinecke salt.

potassium tetroxalate. CAS: 127-96-8.
$KHC_2O_4 \cdot H_2C_2O_4$.
Properties: White crystals, soluble in water, slightly soluble in alcohol.
Use: Metal polish, spot removal, analytical chemistry.

potassium thiocyanate. (potassium rhodanide; potassium sulfocyanate; potassium sulfocyanide). CAS: 333-20-0. KCNS.
Properties: Colorless, transparent, deliquescent, odorless crystals; soluble in water, alcohol, and acetone; saline cooling taste; d 1.88; mp 173C; turns brown, green, blue when fused, white again on cooling; bp decomposes at 500C.
Derivation: By heating potassium cyanide with sulfur.
Grade: Commercial, pure, reagent, ACS.
Hazard: Toxic by ingestion.
Use: Reagent; manufacture of sulfocyanides, thioureas; printing and dyeing textiles; photographic restrainer and intensifier; synthetic dyestuffs; medicine (hypotensive).

potassium thiosulfate. (potassium hyposulfite). CAS: 10294-66-3. $K_2S_2O_3$ (with varying proportions of water of crystallization).
Properties: Colorless crystals, hygroscopic, soluble in water.
Grade: Technical, CP.
Use: Analytical reagent.

potassium titanate. K_2TiO_3.
Properties: White salt, hydrolyzes in water to give a strongly alkaline solution.
Derivation: From titanic acid and potassium hydroxide.
Use: See potassium titanate fiber.

potassium titanate fiber. Approximate composition $K_2O \cdot (TiO_2)_n$ where n is 4–7.
Properties: High refr index, mp 1371C, can diffuse and reflect infrared radiation.
Use: Rockets, missiles, nuclear-powered applications as an insulator, especially for the range 1300–2100F.

potassium titanium fluoride. See titanium potassium fluoride.

potassium titanium oxalate. See titanium potassium oxalate.

potassium trichlorophenate. $Cl_3C_6H_2OH$.
Offered as a solution containing 47% potassium trichlorophenate, and 3% other potassium chlorophenates, d 1.3, fp −9C.
Use: Slime control agent for pulp and paper mill systems.

potassium tripolyphosphate. (KTPP).
$K_5P_3O_{10}$.
Properties: White, crystalline solid; mp 620–640C; d 2.54; loose bulk d 67 lb/cu/ft; solubility in water (26C) more than 140 g/100 mL water.
Use: Water-treating compounds, cleaners, specialty fertilizers, sequestrant.

potassium trithiocarbonate. See potassium sulfocarbonate.

potassium tungstate. (potassium orthotungstate; potassium wolframate). CAS: 7790-60-5.
$K_2WO_4 \cdot 2HOH$.
Properties: Heavy crystalline powder, d 3.1, mp 921C, deliquescent, soluble in water, insoluble in alcohol.

potassium undecylenate.
$CH_2{:}CH(CH_2)_8COOK$.
Properties: Finely divided white powder, decomposes above 250C, limited solubility in most organic solvents, soluble in water.
Hazard: Toxic in high concentration.
Use: Bacteriostat and fungistat in cosmetics and pharmaceuticals.

potassium wolframate. See potassium tungstate.

potassium xanthate. (potassium ethyldithiocarbonate; potassium xanthogenate; potassium ethyl xanthate; potassium ethylxanthogenate).
CAS: 140-89-6. $KS_2COC_2H_5$.
Properties: Colorless or light yellow crystals, soluble in water and alcohol, insoluble in ether, d 1.558 (21.5C).
Derivation: Reaction of potassium ethylate and carbon disulfide.
Hazard: Toxic by ingestion.
Use: Fungicide for soil treatment, reagent in analytical chemistry.

potassium zinc iodide. (zinc potassium iodide).
$ZnI_2 \cdot KI$.
Properties: Colorless crystals, very hygroscopic.
Use: Analysis (testing for alkaloids).

potassium zinc sulfate.   See zinc potassium sulfate.

potassium zirconifluoride.   See zirconium potassium fluoride.

potassium zirconium chloride.   See zirconium potassium chloride.

potassium zirconium sulfate.   See zirconium potassium sulfate.

potentiator.   A term used in the flavor and food industries to characterize a substance that intensifies the taste of a food product to a far greater extent than does an enhancer. The most important of these are the 5′-nucleotides. They are approved by FDA. Their effective concentration is measured in parts per billion, whereas that of an enhancer such as MSG is in parts per thousand. The effect is thought to be due to synergism. Potentiators do not add any taste of their own, but intensify the taste response to substances already present in the food.
See also enhancer, seasoning, flavor.

"Potter's" Compounds.[309]   TM for drawing lubricants and compounds for nonferrous and space age materials. Basic formulations vary from water-soluble soaps through fortified synthetic oil types.

potting compound.   See encapsulation.

pour point.   (1) The lowest temperature at which a liquid will flow when a test container is inverted. (2) The temperature at which an alloy is cast.

pour point depressant.   An additive for lubricating and automotive oils which lowers the pour point (or increases the flow point) by 11.0C. The agents now generally used are polymerized higher esters of acrylic acid derivatives. They are most effective with low-viscosity oils.

powder.   Any solid, dry material of extremely small particle size ranging down to colloidal dimensions, prepared either by comminuting larger units (mechanical grinding), by combustion (carbon black, lamp-black), or by precipitation via a chemical reaction (calcium carbonate, etc.). Powders that are so fine that the particles cannot be detected by rubbing between thumb and forefinger are called impalpable. Typical materials used in powder form are cosmetics, inorganic pigments, metals, plastics (molding powders), dehydrated dairy products, pharmaceuticals and explosives. Metal powders are used to make specialized equipment by sintering and pressing (powder metallurgy) as well as for sprayed coatings and paint pigments (aluminum, bronze). Thermoplastic polymers in powder form are used in a technology known as powder molding, and thermosetting polymers are used in the sprayed coatings field for autos, machinery, and other industrial applications, in which they have many advantages over sprayed solvent coatings.
See also metal, powdered; carbon black; black powder.

powder of Algaroth.   A mixture of SbOCl and Sb_2O_3.
Use: To prepare tartar emetic.

powder metallurgy.   See metal, powdered; sintering.

"Powdura blue."[266]   TM for a series of pigments and coloring agents for use with inks.

ppb.   Abbreviation for parts per billion.

ppm.   Abbreviation for parts per million.

"PPO."[245]   TM for polyphenylene oxide.
Properties: Engineering thermoplastic, light beige, opaque, d 1.06, Rockwell hardness (R scale) 118–120, tensile modulus at 22.7C 3.6–3.8 × 10⁵ psi, self-extinguishing, useful temperature range of greater than 332C, excellent mechanical properties and dielectric characteristics. Soluble in most aromatic and chlorinated hydrocarbons; insoluble in alcohols, ketones, aliphatic hydrocarbons, and water; highly resistant to hydrolysis, acids, bases, and detergents.
Derivation: Oxidative polymerization of 2,6-dimethylphenol in the presence of a copper-amine-complex catalyst.
Use: Dielectric components, hospital and lab equipment, pump housings, impellers, pipe, valves and fittings required by chemical and food industries, substitute for die-cast metals, nose cones for space vehicles.

"PQ Soluble Silicates."[201]   TM for a group of chemicals produced by varying the proportions of sodium or potassium oxide, silica and water. Includes sodium and potassium silicates, metasilicates, sesquisilicates, and orthosilicates.
Use: Adhesives, soap builders, detergents, cleaning compounds, cements, binders, sizes, protective coatings and films, coagulant aids for raw and waste water, rust inhibitors, catalyst base, and deflocculants.

Pr.   (1) Symbol for praseodymium. (2) Informal abbreviation for propyl.

pralidoxime methiodide.   See 2-pyridine aldoxime methiodide.

Prandtl number. For any substance, the ratio of the viscosity to the thermal conductivity. The lower the number, the higher is the convection capacity of the substance. This ratio is important in heat and chemical engineering calculations.

praseodymia. See praseodymium oxide; see also rare earths.

praseodymium. Pr.   Metallic element of atomic number 59, Group IIIB of the periodic table, one of the rare earth elements of the lanthanide group, aw 140.9077, valences = 3, 4. No stable isotopes.
Properties: Yellowish metal, tarnishes easily (color of salts green), d 6.78–6.81, mp 930C, bp 3200C, ignites to oxide (200–400C), liberates hydrogen from water, soluble in dilute acids, paramagnetic.
Form and grade: Ingots, rods, sheets, 98.8–99.9+% pure.
Source: Monazite, cerite, and allonite; also a fission product.
Derivation: Reduction of the trifluoride with an alkaline metal or by electrolysis of the fused halides.
Use: Praseodymium salts, ingredient of mischmetal, core material for carbon arcs, colorant in glazes and glasses, catalyst, phosphors, lasers. See also didymium.

paraseodymium oxalate. $Pr_2(C_2O_4)_3 \cdot 10HOH$.
Green powder, insoluble in water, slightly soluble in acids.
Use: Ceramics.

praseodymium oxide. (praseodymia). Pr_2O_3.
Yellow-green powder, d 7.07, insoluble in water, soluble in acids, hygroscopic, absorbs carbon dioxide from air, purities to 99.8% oxide. Combustible.
Use: Glass and ceramic pigment, laboratory reagent.

precipitate. ($\downarrow$).   Small particles that have settled out of a liquid or gaseous suspension by gravity, or that result from a chemical reaction. Precipitated compounds, such as blanc fixe (barium sulfate), are prepared in this way, for example, by the reaction $BaCl_2 + Na_2SO_4 \rightarrow NaCl + BaSO_4$. In formulas, a downward vertical arrow $\downarrow$ is sometimes used to indicate a precipitate. A class of organic pigments called lakes are made by precipitating an organic dye onto an inorganic substrate. Colloidal particles dispersed in a gas, as flue dust in industrial stacks, can be precipitated by introducing an electric charge opposite to that which sustains the particles.
See Cottrell. See also sedimentation.

precipitator, electrostatic. See Cottrell, precipitate.

precision investment casting. See investment casting.

precursor. In biochemistry, an intermediate compound or molecular complex present in a living organism; when activated physiochemically it is converted to a specific functional substance. The prefix "pro-" is usually used to indicate that the compound in question is a precursor. Examples are ergosterol (pro-vitamin D_2), which is activated by UV radiation to vitamin D; carotene (pro-vitamin A), a precursor of Vitamin A; prothrombin, which forms thrombin upon activation in the bloodclotting mechanism; and phenylacetic acid, a precursor in the biosynthesis of penicillin G.

prednisolone. ($\Delta^{1,4}$-pregnadiene-11β,17α,21-triol-3,20-dione).   CAS: 50-24-8.
$C_{21}H_{28}O_5$.
Generic name for an analog of hydrocortisone.
Properties: White to practically white, odorless, crystalline powder; very slightly soluble in water; soluble in alcohol, chloroform, acetone, methanol, dioxane; mp 235C with some decomposition.
Grade: USP.
Hazard: Causes sodium retention; may have side effects similar to cortisone.
Use: Medicine, also available as acetate.

prednisone. ($\Delta^{1,4}$-pregnadiene-17α,21-diol-3,-11,20-trione).   CAS: 53-03-2. $C_{21}H_{26}O_5$.
Generic name for an analog of cortisone.

preferential. Descriptive of the selectivity of action, either chemical or physiochemical, exhibited by a substance when in contact with two other substances; it may be due either to chemical affinity or to surface phenomena. An example of a preferential chemical combination is that of hemoglobin with carbon monoxide, with which it unites 200 times as readily as with oxygen when exposed to a mixture of the two. Such phenomena as adsorption, corrosion, and the wetting of dry powders by liquids are other examples.

Pregl, Fritz. (1869-1930) An Austrian chemist who won the Nobel prize in 1923. He was also a medical doctor who worked in micromechanical analysis and developed determinations for hydrogen, carbon, nitrogen, and organic groups using micromethods. He was educated at Tubingen, Leipzig, and Berlin.

pregnanediol. (5β-pregnane-3α,20α-diol).
$C_{21}H_{36}O_2$.   A steroid, the metabolic product of progesterone.
Properties: Crystallizes in plates from acetone, mp 238C, dextrorotatory in solutions, sparingly solu-

ble in organic solvents, not precipitated by digitonin.

Derivation: Isolation from urine of pregnant women, cows, mares, and chimpanzees; by reduction of pregnanedione.

Use: Synthesis of progesterone, medically as a pregnancy test.

pregnenedione. See progesterone.

pregneninolone. See ethisterone.

4-pregnen-21-ol-3,20-dione. See deoxycorticosterone.

pregnenolone. (Δ^5-pregnene-3β-ol-20-one). CAS: 145-13-1. $C_{21}H_{32}O_2$. A steroid which is a biologically active hormone similar to progesterone and the adrenal steroid hormones.

Properties: Crystals in needles from dilute alcohol, mp 193C. Slightly soluble in acetone, petroleum ether, benzene, and carbon tetrachloride.

Derivation: From stigmasterol or other steroids.

Use: Medicine, biochemical research. Also available as acetate salt.

Prelog, Vladimir. (1906-) A Swiss organic chemist who won the Nobel prize for chemistry in 1975 along with Cornforth for work on chemical synthesis of organic compounds. Although educated in Yugoslavia, his most recent years have been spent in Zurich.

premix molding. A mixture of plastic ingredients prepared in advance of the molding or extruding operation and stored in bags or bins until required. They are made by mixing the components (resin, filler, fibrous materials such as glass and necessary curatives) in a dough blender. Storage life may be from a few days to a year or more, depending on formulation. Such mixtures are then calendered or extruded after warming to suitable temperature.

prenitene. (1,2,3,4-tetramethylbenzene; prenitol). $(CH_3)_4C_6H_2$.

Properties: Colorless liquid, soluble in alcohol, insoluble in water, d 0.901, bp 204C, fp −7.7C.

prepolymer. An adduct or reaction intermediate of a polyol and a monomeric isocyanate, in which either component is in considerable excess of the other. A polymer of medium molecular weight having reactive hydroxyl and —NCO groups.

Use: Preparation of polyurethane coatings and foams.

prepreg. A term used in the reinforced plastics field to mean the reinforcing material containing

or combined with the full complement of resin before molding.

"Pre-San."[329] [N-(2-mercaptoethyl)benzenesulfonamide-S-(O,O-diisopropylphosphorodithioate)]. TM for a selective herbicide.

preservative. Any agent that prolongs the useful life of a material. Food products are preserved by (1) low temperature, (2) ionizing radiation (x- and gamma rays), and (3) antioxidants and similar additives. Antioxidants are also used in lube oils, rubber, and plastics; fungicides on textiles; aldehydes on biological specimens; paints on wood and metals.

See also protective coating; antioxidant; radiation, industrial.

press, hydraulic. See hydraulic press.

"Prestabit Oil" V. TM for an anionic textile chemical consisting of purified sulfated castor-oil fatty acids.

Use: Dyeing assistant for cotton and wool fiber, in viscose manufacture, clarifying agent to prevent milkiness of the yarn, antistatic agent for acetate and polyacrylonitrile fibers.

Prevost reaction. Hydroxylation of olefins with iodine and silver benzoate in an anhydrous solvent to give trans-glycols.

Priestly, Joseph. (1733–1804) Born near Leeds, England, Priestley originally planned to enter the ministry. As a youth he became interested in both physics and chemistry and his research soon established his position as a scientist. He was elected to the Royal Society in 1766. He discovered nitrous oxide in 1772 but his greatest contribution to science was his discovery of oxygen in 1774. He emigrated from England to Northumberland, PA, where he lived from 1784 to his death. His research in America resulted in the discovery of carbon monoxide (1799).

Prigogine, Ilya. (1917-) A Belgian chemist who won the Nobel prize for chemistry in 1977 for his contributions to nonequilibrium thermodynamics. His education was at the University of Brussels. The Center for Statistical Mechanics and Thermodynamics at the University of Texas bears his name.

Prilezhaev (Prileschajew) reaction. Formation of epoxides by the reaction of alkenes with peracids.

prills. Small round or acicular aggregates of a material, usually a fertilizer, that are artificially

prepared. In the explosives field, prills-and-oil consists of 94% coarse, porous ammonium nitrate prills and 6% fuel oil.
See also explosives, high.

"Primacord." TM for a detonating composition. See pentaerythritol tetranitrate.

"Primafloc."[23] TM for a series of organic polyelectrolyte products used for flocculating suspended solids in water, waste, and process streams.
See also polyelectrolyte.

"Primal."[23] TM for aqueous dispersions of acrylic resins, supplied in various grades that differ in hardness and flexibility, produced finishes which are water-insoluble, require no plasticizer for flexibility, are unimpaired by aging, and adhere tenaciously to leather and lacquer coats.

primary. (1) In reference to monohydric alcohols, amines, and a few related compounds, this term, together with secondary and tertiary, describes the molecular structure of isomeric or chemically similar individuals. Monohydric alcohols are based on the methanol group

$$-C-OH$$

in which three of the bonds of the methanol carbon may be attached either to hydrogen atoms or to alkyl groups. A primary alcohol has one alkyl group and two hydrogens,

$$CH_3-C-OH$$

except methanol, in which all three bonds are to hydrogen atoms. A secondary alcohol has two alkyl groups and one hydrogen, e.g.,

$$CH_3-C-OH$$

A tertiary alcohol has three alkyl groups, e.g.,

$$CH_3-C-OH$$

The three types can be readily identified by the number of hydrogen atoms attached to the central (methanol) carbon atom: if it is two or more, the alcohol is primary; if one, it is secondary; and if zero, it is tertiary. For example, CH_3CH_2OH is primary; $(CH_3)_2CHOH$ is secondary; $(CH_3)_3COH$ is tertiary.

Primary, secondary, and tertiary amines are formed from ammonia (NH_3) when one, two, or three hydrogen atoms, respectively, are replaced by alkyl groups.

These terms are also used to name salts of orthophosphoric acid (H_3PO_4) in which one, two, or three of the hydrogen atoms have been replaced by metal or radicals: NaH_2PO_4 is primary sodium phosphate, Na_2HPO_4 is secondary sodium phosphate. The same system of names is used for salts of other acids containing three replaceable hydrogen atoms.

(2) A type of battery that is irreversible in respect to power output; a voltaic cell. A secondary battery (storage battery) is reversible and can be recharged.

(3) In the terminology of minerals, primary (in the case of metals) refers to direct production from the ore; in the case of petroleum it refers to production from wells by direct means. This meaning contrasts with the term "secondary," used to denote recovery of metal from scrap and, for petroleum, recovery by means of special techniques such as flooding and hydraulic pressure.

primary azo dyes. Azo dyes derived from primary amines.

primary calcium phosphate. See calcium phosphate, monobasic.

"Primene JM-T."[23] TM for a solvent comprised of branched chain primary amines containing from 18–22 carbon atoms.

primuline dye. See thiazole dye.

Prins reaction. Acid catalyzed addition of olefins to formaldehyde to give 1,3-diols, allylic alcohols, or meta-dioxanes.

printing ink. A viscous to semisolid suspension of finely divided pigment in a drying oil such as heat-bodied linseed oil. Alkyd, phenol-formaldehyde, or other synthetic resins are frequently used as binders; and cobalt, manganese, and lead soaps are added to catalyze the oxidative drying reaction. Some types of inks dry by evaporation of a volatile solvent rather than by oxidation and polymerization of a drying oil or resin. Use distribution is: offset 40%, gravure 23%, flexographic 18%, letterpress 9%, screen 4%, other 6%. For further information refer to National Printing Ink Institute, Lehigh University, Bethlehem, PA.

pristane. (2,6,10,14-tetramethylpentadecane; norphytane). CAS: 1921-70-6. $C_{19}H_{40}$.
Found in rock specimens 2.5–3 billion years old. It is known to be synthesized only by living organisms and to withstand heat and pressure; thus it serves to date the existence of life on earth.
Properties: Colorless, faintly odored, transparent, stable liquid. Bp 290C, fp −60C, d 0.775–0.795 (20C), refr index 1.435–1.440 (20C), soluble in most organic solvents. Combustible.
Grade: 90% purity.
Use: Precision lubricant, chromatographic oil, anticorrosive agent.
See also phytane.

procaine hydrochloride. (procaine).
CAS: 51-05-8.
$C_6H_4NH_2COOCH_2CH_2N(C_2H_5)_2 \cdot HCl$.
Properties: Small, colorless crystals or white, crystalline powder; odorless; stable in air. Soluble in water and in alcohol at 25C, slightly soluble in chloroform, almost insoluble in ether, solutions acid to litmus, mp 153–156C.
Derivation: (a) By heating chloroethyl-p-nitrobenzoic ester with diethylamine for 24 hours under pressure at 120C. The product is then reduced with tin and hydrochloric acid. (b) By condensation of ethylene chlorohydrin with diethylamine. The chloroethyldiethylamine formed is heated with sodium-p-aminobenzoate.
Grade: USP.
Use: Medicine (local anesthetic).

procaine penicillin G. CAS: 6130-64-9.
$C_{16}H_{18}N_2O_4S \cdot C_{13}H_{20}N_2O_2 \cdot HOH$.
Properties: White, fine crystals or powder; odorless; relatively stable to air and light; solutions dextrorotatory. Sparingly soluble in water, slightly soluble in alcohol, fairly soluble in chloroform.
Grade: USP.
Use: Antibiotic, animal feed additive.

process. See unit process.

process industry. See chemical process industry.

"Prodag."[46] TM for a dispersion of graphite in water.
Use: Mold wash for aluminum permanent molds, ingot molds, and molds for mechanical rubber goods; stop-off coating.

prodrug. A term applied in pharmaceutical chemistry to a chemical compound that is converted into an active curative agent by metabolic processes within the body.
See also precursor.

producer gas. A gas obtained by burning coal or coke with a restricted supply of air, or by passing air and steam through a bed of incandescent fuel under such conditions that the carbon dioxide formed is converted into carbon monoxide. The water vapor reacts to form carbon monoxide and hydrogen. Producer gas is cheap but has low Btu and is used where transportation is not required.
Hazard: Highly flammable and toxic. Explosive range 20–73% in air.
See also water gas, synthesis gas.

"Profil."[539] TM for glass fiber-reinforced polypropylene.

profile. See soil.

progesterone. (Δ^4-pregnene-3,20-dione).
CAS: 57-83-0. $C_{21}H_{30}O_2$. The female sex hormone secreted in the body by the corpus luteum, by the adrenal cortex, or by the placenta during pregnancy. It is important in the preparation of the uterus for pregnancy and for the maintenance of pregnancy. It exists in two crystalline forms (α- and β-) of equal physiological activity. Progesterone is believed to be the precursor of the adrenal steroid hormones.
Properties: White, crystalline powder; odorless and stable in air but sensitive to light; mp (α form) 128–133C, (β form) 121C; practically insoluble in water; soluble in alcohol, acetone, and dioxane; sparingly soluble in vegetable oils.
Units: The international unit (IU) of progestational activity is expressed as 1 milligram of progesterone.
Derivation: Isolation from corpus luteum of pregnant sows, synthesis from other steroids such as stigmasterol.
Grade: NF.
Hazard: A carcinogen (OSHA).
Use: Oral contraceptive, lab reagent.

CH₃ — shown as structure with C=O and O= (steroid structure)

proguanyl. (chlorguanide; 1-p-chlorophenyl-5-isopropyl biguanide). An antimalarial drug said to be less toxic than others.

prolactin. See luteotropin.

prolamin. Any of a group of simple vegetable proteins, e.g., gliadin in wheat, zein in corn. When split by acids they give only amino acids.

prolan. (2-nitro-1,1-bis(p-chlorophenyl)propane). CAS: 117-27-1. $C_{15}H_{13}Cl_2NO_2$.
Properties: Thick, oily liquid; crystalline form melts at 80C.
Hazard: Toxic by ingestion.
Use: Insecticide; when mixed with bulan, the insecticide known as dilan is formed.
See also bulan, dilan.

proline. (2-pyrrolidinecarboxylic acid). CAS: 147-85-3. C_4H_8NCOOH. A nonessential amino acid found naturally in the l(-) form.

$$CH_2-CH_2$$
$$CH_2 \quad CH-COOH$$
$$NH$$

Properties: Colorless crystals, soluble in water and alcohol, insoluble in ether, optically active. dl-proline: mp 205C with decomposition. d(+) proline: mp 215–220C with decomposition. l(-)-proline: mp 220–222C with decomposition.
Derivation: Hydrolysis of protein, also synthetically and by recombinant DNA techniques.
Use: Biochemical and nutritional research, microbiological tests, culture media, dietary supplement, lab reagent. Available commercially as the l(-)-proline.

prolipin. A compound sterile solution of protein obtained from nonpathogenic bacteria, various animal fats and lipoids derived from bile.

promazine hydrochloride. (10-(3-dimethylaminopropyl)phenothiazine hydrochloride). $C_{17}H_{20}N_2S \cdot HCl$. Isomeric with promethazine hydrochloride.
Properties: White to slightly yellow, practically odorless, hygroscopic powder; oxidizes upon prolonged exposure to air and acquires a blue or pink color. Melts within a range of three degrees between 172–182C, pH of 5% solution between 4.2–5.2, soluble in water and alcohol, insoluble in benzene.
Grade: NF.
Use: Medicine (tranquilizer).

promecarb. (3-methyl-5-isopropyl-N-methyl-carbamate). CAS: 2631-37-0. $C_{12}H_{17}NO_3$.
Properties: Colorless, crystalline solid; mp 87C. Insoluble in water, soluble in alcohol.
Hazard: Toxic by ingestion.
Use: Insecticide.

promethium. Pm. Radioactive, rare-earth element, a member of the lanthanide series, atomic number 61, aw 147. The 145 isotope has a half-life of 18 years, 147 half-life 2.64 years; the latter is the only form available.
Properties: Silvery-white metal, mp 1160C, density 7.2.
Derivation: The 147 isotope is recovered from spent uranium fission products, also by reduction of the chloride or fluoride with an alkali metal.
Hazard: Strong radioactive poison, use requires shielding and glove-boxes.
Use: (147 isotope) Nuclear auxiliary power generators, special semiconductor battery, luminescent paint for watch dials, x-ray source, source of beta rays for thickness gauges, with tungsten cermet for space power system (withstands 2000C for 1000 hours).
Note: Some 30 compounds are known, but are not commercially available.

promoter. (1) A substance which, when added in relatively small quantities to a catalyst, increases its activity, e.g., aluminum and potassium oxide are added as promoters to the iron catalyst used in facilitating a combination of hydrogen and nitrogen to form ammonia. (2) In ore flotation, a substance which provides the minerals to be floated with a water-repellent surface that will adhere to air bubbles. Such reagents are generally more or less selective towards minerals of certain classes.

proof. The ethanol content of a liquid at 15.5C, stated as two times the percentage of ethanol by volume. One gallon of 95% alcohol is therefore equivalent to 1.9 gallons of proof alcohol. In the US, the alcohol tax is based on the number of proof gallons.

propachlor. CAS: 1918-16-7. $C_{11}H_{14}NOCl$.
Properties: Tan powder, mp 68C, soluble in alcohol, benzene.
Use: Selective weed killer.

propadiene. See allene.

"Propaloid-T."[236] TM for a creamy white, organic-modified, beneficiated magnesium montmorillonite with a rapid hydration characteristic.
Use: As a viscosity control, stabilizer for emulsions and suspensions in aqueous systems. Primarily for use in paints, cosmetics, and paper coatings.

propanal. See propionaldehyde.

propane. (dimethylmethane). CAS: 74-98-6. C_3H_8.
Properties: Colorless gas, natural gas odor, noncorrosive, bp −42.5C, fp −189.9C, density of liquid at 0C 0.531, density of vapor at 0C 1.56, flash p −156F (−105C), autoign temperature 874F (467C). Soluble in ether, alcohol; slightly soluble in water. An asphyxiant gas.
Grade: Technical, research (9.9%).
Derivation: From petroleum and natural gas.
Hazard: Flammable, dangerous fire risk, explosive limits in air 2.4–9.5%, for storage, see butane (note).
Use: Organic synthesis, household and industrial fuel, manufacture of ethylene, extractant, solvent, refrigerant, gas enricher, aerosol propellant, mixture for bubble chambers.
See butane (note).

1,3-propanediamine. See 1,3-diaminopropane.

1,2-propanediol. See 1,2-propylene glycol.

1,3-propanediol. See trimethylene glycol.

propane hydrate. See gas hydrate.

propanenitrile. See ethyl cyanide.

1-propanethiol. (n-propyl mercaptan).
CAS: 107-03-9. C_3H_7SH.
Properties: Offensive-smelling liquid, boiling range 67–73C, d 0.8408 (20/4C), refr index 1.4380 (20C), flash p −5F (−20.5C).
Grade: 95%.
Hazard: Highly flammable, dangerous fire risk.
Use: Chemical intermediate, herbicide.

propanil. (Generic name for 3,4-dichloropropionanilide) CAS: 709-98-8. $Cl_2C_6H_3NHCOCH_2CH_3$.
Properties: (Pure) Light brown solid, mp 85–89C. (Technical) Liquid bp 91–91.5C.
Hazard: Toxic by ingestion and inhalation.
Use: Post-emergence herbicide, especially for rice culture; nematocide.

propanoic acid. See propionic acid.

1-propanol. See propyl alcohol.

2-propanol. See isopropyl alcohol.

propanolamine. See 2-amino-1-propanol; 3-amino-1-propanol.

2-propanol nitrate. See isopropyl nitrate.

2-propanolpyridine. $C_5NH_4C_3H_6OH$.
Properties: Colorless liquid, bp 260.2C, d 1.060 (25C), refr index 1.5298 (20C), miscible with water at 20C.

4-propanolpyridine. $C_5NH_4C_3H_6OH$.
Properties: Colorless liquid, bp 289.0C, fp 36.7C, d 1.053 (40C), soluble in water.

2-propanone. See acetone.

2-propanone oxime. See acetoxime.

propanoyl chloride. See propionyl chloride.

propargite. (2-(p-tert-butylphenoxy)-cyclohexyl-2-propynylsulfite; Dark oil).
CAS: 2312-35-8.
Properties: D 1.1 g/cc at 25C, insoluble in water.
Use: Acaricide for fruits, vegetables, row crops.

propargyl alcohol. (2-propyn-1-ol).
CAS: 107-19-7. $HC\equiv CCH_2OH$.
Properties: Colorless liquid, geranium-like odor, d 0.971, fp −48C, bp 114C. Soluble in water, alcohol, and ether; immiscible with aliphatic hydrocarbons; flash p (OC) 97F (36C).
Derivation: From acetylene by high-pressure synthesis.
Grade: Technical, 75% solution.
Hazard: Flammable, moderate fire risk. Toxic by ingestion, inhalation, and skin absorption. TLV: 1 ppm in air.
Use: Chemical intermediate, corrosion inhibitor, lab reagent, solvent stabilizer, prevents hydrogen embrittlement of steel, soil fumigant.

propargyl bromide. (3-bromo-1-propyne).
CAS: 106-96-7. $HC\equiv CCH_2Br$.
Properties: Liquid, d 1.520, bp 88–90C, flash p (COC) 65F (18.3), sharp odor.
Derivation: From acetylene by high-pressure synthesis.
Hazard: Flammable, dangerous fire and explosion risk. Irritant.
Use: Chemical intermediate, soil fumigant.

propargyl chloride. (3-chloro-1-propyne).
$HC\equiv CCH_2Cl$.
Properties: Liquid, fp −76.9C, bp 57.1, refr index 1.4310 (25C), flash p 90F (32C). Soluble in ben-

zene, alcohol, carbon tetrachloride; insoluble in water.

Derivation: From acetylene by high-pressure synthesis.

Hazard: Flammable, moderate fire risk.

Use: Chemical intermediate, soil fumigant.

propazine 80W. CAS: 139-40-2. $C_9H_{16}ClN_5$. Generic name for a pre-emergence herbicide containing 80% 2-chloro-4,6-bis(isopropylamino)-s-triazine.

Use: Weed control in sorghum culture.

propellant. (1) A rocket fuel. (2) A compressed gas used to expel the contents of containers in the form of aerosols. Chlorofluorocarbons were once widely used because of their nonflammability. The strong possibility that they contribute to depletion of the ozone layer of the upper atmosphere has resulted in prohibition of their use for this purpose. Other propellants used are hydrocarbon gases, such as butane and propane, carbon dioxide, and nitrous oxide. The materials dispersed include insecticides, shaving cream, whipping cream, and cosmetic preparations. See also ozone (note).

2-propenal. See acrolein.

propene. See propylene.

propenenitrile. See acrylonitrile.

2-propene-1-thiol. See allyl mercaptan.

propene-1,2,3-tricarboxylic acid. See aconitic acid.

propenoic acid. See acrylic acid.

2-propen-1-ol. See allyl alcohol.

propenyl alcohol. See allyl alcohol.

2-propenylamine. See allylamine.

p-propenylanisole. See anethole.

α-propenyldichlorohydrin. See α-dichlorohydrin.

propenyl guaethol. (1-ethoxy-2-hydroxy-4-propenylbenzene). $C_2H_5OC_6H_3(OH)(C_3H_5)$.
Properties: Free-flowing, white powder; odor and taste similar to vanilla, but much more powerful; mp 85–86C; very soluble in fats, edible solvents, and essential oils; very slightly soluble in water.
Use: Artificial vanilla flavoring, flavor enhancer.

propenyl guaiacol. See methyl isoeugenol.

2-propenyl hexanoate. See allyl caproate.

2-propenyl isothiocyanate. See allyl isothiocyanate.

propineb. (zinc-1,2-propylene-bis-dithiocarbamate). CAS: 12071-83-9. $C_5H_8N_2S_4Zn$.
Properties: Faintly yellow crystals or powder becoming dark when heated, decomposes in presence of strong acids or bases, insoluble in most common solvents.
Hazard: Toxic by ingestion.
Use: Fungicide.

propiodal. (1,3-bis(trimethylamino)-2-propanol diiodide; iodisan; hexamethyldiaminoisopropanol diiodide). $[CH_2N(CH_3)_3I]_2CHOH$.
Properties: White, crystalline solid; mp 275C (decomposes); turns brown at 240C. Freely soluble in water, slightly soluble in alcohol, insoluble in ether and acetone.
Use: Medicine (iodine therapy).

β-propiolactone. (USAN) (BPL).
CAS: 57-57-8. OCH_2CH_2CO.
Properties: Colorless liquid, pungent odor, bp 155C with rapid decomposition, fp −33.4C, refr index 1.4131 (20C), d 1.1450 (20/4C). Soluble in water; miscible with ethanol, acetone, ether, and chloroform at 25C; reacts with alcohol. Flash p (OC) 167F (75C), stable when stored in glass at refrigeration temperature (+5 to +10C). Combustible.
Derivation: Direct combination of ketene and formaldehyde.
Hazard: Strong skin irritant, suspected human carcinogen, worker exposure should be minimized. TLV: 0.5 ppm in air.
Use: Organic synthesis, vapor sterilant, disinfectant.

propionaldehyde. (propanal; propyl aldehyde; propionic aldehyde). CAS: 123-38-6. C_2H_5CHO.
Properties: Water-white liquid with suffocating odor, soluble in water, flash p (OC) 15F (−9.4C), bp 48.8C, fp −81C, d 0.807 (20.4C), refr index 1.364 (20C), soluble in water and alcohol, autoign temperature 405F (207C).
Derivation: (a) Oxidation of propyl alcohol with dichromate, (b) by passing propyl alcohol over copper at elevated temperatures.
Hazard: Flammable, dangerous fire risk, explosive limits in air 3.0–16%. Irritant.
Use: Manufacture of propionic acid, polyvinyl and other plastics, synthesis of rubber chemicals, disinfectant, preservative.

propionamide nitrile. See cyanoacetamide.

propione. See diethyl ketone.

propionic acid. (methylacetic acid; propanoic acid). CAS: 79-09-4. $CH_3CH_2CO_2H$.
Properties: Colorless, oily liquid; rancid odor; d 0.9942 (20/4C); fp −20.8C; refr index 1.3862 (20C); bp 140.7C. Soluble in water, alcohol, chloroform, and ether. Flash p 130F (54.4C), autoign temperature 955F (512C). Combustible.
Derivation: By reaction of ethanol with carbon monoxide, using a boron trifluoride catalyst; also by the reaction of carbon monoxide with hydrogen and olefins or alcohols.
Method of purification: Rectification.
Grade: Technical, 99.0%, FCC.
Hazard: Moderate fire risk. Strong irritant. TLV: 10 ppm in air.
Use: Propionates, some of which are used as mold inhibitors in bread and fungicides in general, herbicides, preservative for grains and wood chips, emulsifying agents, solutions for electroplating nickel, perfume esters, artificial fruit flavors, pharmaceuticals, cellulose propionate plastics.

propionic aldehyde. See propionaldehyde.

propionic anhydride. CAS: 123-62-6. $(CH_3CH_2CO)_2O$.
Properties: Colorless liquid, pungent odor, fp −45C, bp 167–169C, d 1.0119 at 20C, vap press 1 mm at 20C, flash p 165F (73.9C) bulk d 8.4 lb/gal at 20C. Soluble in alcohol, ether, chloroform, and alkalies; decomposed by water. Combustible.
Hazard: Strong irritant to tissue.
Use: Esterifying agent for fats, oils, cellulose, dehydrating medium for nitrations and sulfonations, alkyd resins, dyestuffs, and pharmaceuticals.

propionitrile. See ethyl cyanide.

propionylbenzene. See propiophenone.

propionyl chloride. (propanoyl chloride). CAS: 79-03-8. CH_3CH_2COCl.
Properties: Colorless liquid with pungent odor, fp −94C, bp 80C, d 1.065 (20/4C), decomposes in water and alcohol.
Hazard: Strong irritant to skin.
Use: Chemical intermediate.

propionyl peroxide. CAS: 3248-28-0. $C_2H_5C(O)OOC(O)C_2H_5$. Available as a 25% solution in a high-boiling hydrocarbon, flash p 125F (51.6C).
Use: Initiator in polymerization reactions, such as the high pressure polymerization of ethylene.

Hazard: Strong oxidizing agent, may explode if shocked or heated.

propiophenone. (ethyl phenyl ketone; propionylbenzene; 1-phenylpropanone-1). CAS: 93-55-0. $C_6H_5COC_2H_5$.
Properties: Colorless to light amber liquid or crystals, strong persistent odor, d 1.012 (20/20C), refr index 1.527 (20C), congealing temperature 17.5–21C, bp 218C, flash p (TOC) 210F (99C). Insoluble in water, ethylene glycol, glycerol; miscible with alcohol, ether, benzene, and toluene. Combustible.
Use: Fixative in perfumes, starting material for synthesis of ephedrine and other important pharmaceuticals and synthetic organic chemicals, lab reagent.

"Propi-Rhap."[248] TM for the 2-ethylhexyl ester of 2-(2,4-dichlorophenoxy)propionic acid.
Use: Herbicide.

propoxur. See o-isopropoxyphenyl-N-methyl carbamate.

propoxyphene. (4-dimethylamino-3-methyl-1,2-diphenyl-2-diphenyl-2-butanol propionate). CAS: 469-62-5. $C_{22}H_{29}NO_2$. The α-diastereoisomers are optically active and are preferred for their greater pain-relieving ability. The drug has about the same analgesic effect as codeine. Abuse can cause addiction and overdosage can be fatal. Its use has been restricted by FDA.

n-propoxypropanol. $C_6H_{14}O_2$.
Properties: Liquid, d 0.8865 (20/20C), bp 149.8C, fp −80C (sets to glass below this), soluble in water, flash p 128F (53.3C).
Hazard: Moderate fire risk.
Use: Solvent for water-based enamel.

proppant. A term coined by petroleum engineers to refer to agents such as sand, sintered bauxite, etc., which are used in hydraulic fracturing of oil wells; they are so called because they prop open the minute cracks in rock formations created by hydraulic press.
See hydraulic fracturing.

propranolol. [1-(isopropylamino)-3-(1-naphthyloxy)-2-propanol]. CAS: 525-66-6. $C_{16}H_{21}NO_2$.
Properties: Colorless crystals, mp 96C, soluble in water and alcohol, insoluble in benzene and ether.
Hazard: Highly toxic.
Use: An adrenergic blocker used in treatment of hypertension and various forms of heart disease. Approved by FDA.

OH
|
OCH$_2$CHCH$_2$NHCH(CH$_3$)$_2$

n-propyl acetate. CAS: 109-60-4.
C$_3$H$_7$OOCCH$_3$.
Properties: Colorless liquid, pleasant odor. Slightly soluble in water; miscible with alcohols, ketones, esters, hydrocarbons. D 0.887, flash p 58F (14.4C), boiling range 96.0–102.0C, bulk d 7.36 lb/gal, autoign temperature 842F (450C), fp −92C.
Derivation: Interaction of acetic acid and n-propyl alcohol in the presence of sulfuric acid.
Grade: Technical.
Hazard: Flammable, dangerous fire risk, explosive limits in air 2–8%. TLV: 200 ppm in air.
Use: Flavoring agents, perfumery, solvent for nitrocellulose and other cellulose derivatives, natural and synthetic resins, lacquers, plastics, organic synthesis, lab reagent.

propyl acetone. See methyl butyl ketone.

propyl alcohol. (1-propanol; n-propyl alcohol).
CAS: 71-23-8. CH$_3$CH$_2$CH$_2$OH.
Properties: Colorless liquid, odor similar to ethanol, bp 97.2C, fp −127C, d 0.804 (20/4C), flash p (OC) 77F (25C), autoign temperature 700F (371C), refr index 1.385 (20C), viscosity 2.256 cp (20C). Soluble in water, alcohol, and ether.
Derivation: From oxidation of natural gas hydrocarbons, also from fusel oil.
Hazard: Flammable, dangerous fire risk. Explosive limits in air 2–13%. Toxic by skin absorption. TLV: 200 ppm in air.
Use: Organic synthesis and chemical intermediate, solvent for waxes, vegetable oils, natural and synthetic resins, cellulose esters and ethers, polishing compositions, brake fluids, solvent degreasing, antiseptic.

sec-propyl alcohol. See isopropyl alcohol.

propyl aldehyde. See propionaldehyde.

n-propylamine. CAS: 107-10-8. C$_3$H$_7$NH$_2$.
Properties: Colorless liquid, d 0.7182 (20C), bp 47.8C, vap press 248 mm (20C), fp −83C, odor amine. Soluble in water, alcohol, and ether. Flash p −35F (−37.2C), autoign temperature 604F (317C).
Hazard: Highly flammable, dangerous fire risk, explosive limits in air 2–10%, use alcohol foam to extinguish. Strong irritant to skin and tissue.
Use: Intermediate, lab reagent.

n-propylbenzene. (1-phenylpropane).
CAS: 103-65-1. C$_6$H$_5$CH$_2$CH$_2$CH$_3$.
Properties: Colorless liquid, bp 160C, fp −100C, d 0.862, flash p 86F (30C), refr index 1.49, soluble in alcohol and ether, sparingly soluble in water.
Derivation: Reaction of benzylmagnesium chloride and diethyl sulfate.
Hazard: Flammable, moderate fire risk.
Use: Solvent for cellulose acetate, dyeing textiles.

propyl butyrate. CAS: 105-66-8.
C$_3$H$_7$OOCC$_3$H$_7$.
Properties: Colorless liquid, d 0.8789 (15C), bp 142.7C, fp −95.2C, slightly soluble in water, soluble in alcohol and ether. Combustible.
Hazard: Irritant to mucous membranes, narcotic in high concentration.
Use: Solvent mixture for cellulose ethers.

propyl chloride. (1-chloropropane).
CAS: 540-54-5. CH$_3$CH$_2$CH$_2$Cl.
Properties: Liquid, fp −122.8C, bp 46.6C, refr index 1.3886 (20C), soluble in alcohol and ether, slightly soluble in water, flash p 0F (−17.7C).
Hazard: Highly flammable, dangerous fire risk, explosive limits in air 2.5–11%. Irritant and narcotic.
See also isopropyl chloride.

propyl chlorosulfonate. CAS: 109-61-5.
CH$_3$CH$_2$CH$_2$OSO$_2$Cl.
Properties: Liquid, bp 70–72C (20 mm).
Derivation: Interaction of n-propyl alcohol and sulfuryl chloride.
Hazard: Toxic by inhalation and ingestion, strong irritant to eyes.
Use: Organic synthesis, military poison gas (lachrymator).

n-propyl cyanide. See n-butyronitrile.

propyl-3,5-diiodo-4-oxo-1(4H)pyridineacetate.
See propyliodone.

propylene. (propene). CAS: 115-07-1.
CH$_3$CH:CH$_2$. 12th highest-volume chemical produced in US (1985).
Properties: Colorless gas, soluble in alcohol and ether, slightly soluble in water, bp −47.7C, fp −185.2C, d (liquid) 0.5139 (20/4C), vap d at 0C (air = 1) 1.46, flash p −162F (−108C), autoign temperature 927F (497C).
Derivation: Catalytic and thermal cracking of ethylene with zeolite catalyst, from naphtha.
Grade: 95%, 99%, and research.
Hazard: Highly flammable, dangerous fire risk, explosive limits in air 2–11%. An asphyxiant gas.

Use: Manufacture of isopropyl alcohol, polypropylene, synthetic glycerol, acrylonitrile, propylene oxide, heptene, cumene, polymer gasoline, acrylic acid, vinyl resins, oxo chemicals.

propylene carbonate. $C_4H_6O_3$ or

‾OCOCH$_2$CH(CH$_3$)O‾.
Properties: Odorless, colorless liquid; fp −49.2C (easily super-cooled); bp 241.7C; d 1.2057 (20/4C); bulk d 10 lb/gal (20C); refr index 1.4209 (20C); flash p 270F (132C). Miscible with acetone, benzene, chloroform, ether, ethyl acetate; moderately soluble in water and carbon tetrachloride. Combustible.
Use: Solvent extraction, plasticizer, organic synthesis, natural gas purification, synthetic fiber spinning solvent.

propylene chloride. See propylene dichloride.

propylene chlorohydrin. (chloro-isopropyl alcohol; 1-chloro-2-propanol). CAS: 78-89-7. CH$_2$ClCHOHCH$_3$.
Properties: Colorless liquid, mild odor, nonresidual, bp 127.5C, vap press 4.9 mm (20C), flash p (CC) 125F (51.6C), bulk d 9.3 lb/gal (20C), d 1.1128 at 20/20C, soluble in water and alcohol.
Grade: Technical.
Hazard: Moderate fire risk. Toxic by ingestion and skin absorption.
Use: Organic synthesis (introducing hydroxypropyl group), manufacture of propylene oxide.

propylenediamine. Legal label name for 1,2-diaminopropane.

propylene dichloride. (1,2-dichloropropane; propylene chloride). CAS: 78-87-5. CH$_3$CHClCH$_2$Cl.
Properties: Colorless, stable liquid; chloroform-like odor. Bp 96.3C, d 1.1583 at 20/20C, bulk d 9.6 lb/gal (20C), refr index 1.4068 (20C), flash p 61F (16.1C), soluble in water 0.26% by wt (20C), fp −80C, miscible with most common solvents, insoluble in water, autoign temperature 1035F (557C).
Derivation: Action of chlorine on propylene.
Grade: Refined.
Hazard: Flammable, dangerous fire risk, explosive limits in air 3.4–14.5%. Toxic by ingestion and inhalation. TLV: 75 ppm in air.
Use: Intermediate for perchloroethylene and carbon tetrachloride; lead scavenger for antiknock fluids; solvents for fats, oils, waxes, gums, and resins; solvent mixture for cellulose esters and ethers; scouring compounds; spotting agents;

metal degreasing agents; soil fumigant for nematodes.

propylene disulfate. $C_3H_8(SO_4)_2$.
Hazard: A carcinogen.

1,2-propylene glycol. (1,2-dihydroxypropane; 1,2-propanediol; methylene glycol; methyl glycol). CAS: 57-55-6. CH$_3$CHOHCH$_2$OH.
Properties: Colorless, viscous, stable, hygroscopic liquid; practically odorless and tasteless; miscible with water, alcohols, and many organic solvents. Bp 187.3C, fp −60C, d 1.0381 at 20/20C, bulk d 8.64 lb/gal (20C), refr index 1.4293 (27C), surface tension 40.1 dynes/cm (25C), viscosity 0.581 poise (20C), vap press 0.07 mm (20C), specific heat 0.590 cal/g (20C), latent heat of evaporation 168.6 cal/g at bp, flash p (OC) 210F (99C), autoign temperature 780F (415C), heat of combustion 431.0 kg cal/mole. Combustible.
Derivation: By hydration of propylene oxide.
Method of purification: By distillation.
Grade: Refined, technical, USP, FCC, feed.
Use: Organic synthesis, especially polypropylene glycol and polyester resins, cellophane, antifreeze solution. Solvent for fats, oils, waxes, resins, flavoring extracts, perfumes, colors, soft-drink syrups, antioxidants. Hygroscopic agent, coolant in refrigeration systems, plasticizers, hydraulic fluids, bactericide, textile conditioners. In foods as solvent, wetting agent, humectant. Emulsifier, feed additive, anticaking agent, preservative (retards molds and fungi), cleansing creams, sun tan lotions, pharmaceuticals, brake fluids, deicing fluids for airport runways, tobacco. See also polypropylene glycol.

1,3-propylene glycol. See trimethylene glycol.

propylene glycol alginate. (hydroxypropyl alginate). ($C_9H_{14}O_7$).
Properties: Vary with degree of esterification. White powder, practically tasteless and odorless, soluble in water and dilute organic acids.
Grade: FCC.
Use: Stabilizer, thickener, emulsifier, food additive.

propylene glycol distearate. See propylene glycol monostearate.

propylene glycol monomethyl ether. (polypropylene glycol methyl ether). CAS: 107-98-2. CH$_3$OCH$_2$CHOHCH$_3$.
Properties: Colorless liquid, fp −95C (sets to glass), bp 120.1C, d 0.9234 (20/20C) bulk d 7.65 lb/gal (25C), refr index 1.402 (25C), flash p 97F (36.1C). soluble in water, methanol, ether.

Hazard: Flammable, moderate fire risk. TLV: 100 ppm

Use: Solvent for celluloses, acrylics, dyes, inks, stains; solvent-sealing of cellophane.

propylene glycol monoricinoleate.
$C_{17}H_{32}(OH)COOCH_2CHOHCH_3$.
Properties: Pale yellow, moderately viscous, oily liquid; mild odor. D 0.960 (25/25C), saponification value 160, hydroxyl value 285, solidifies at −26C, soluble in most organic solvents, insoluble in water. Combustible.
Derivation: Castor oil and propylene glycol.
Grade: Technical.
Use: Plasticizer, dye solvents, lubricant, cosmetics, urethane polymers and hydraulic fluids.

propylene glycol monostearate. The FCC grade is a mixture of propylene glycol mono- and diesters of stearic and palmitic acids. White beads or flakes, bland odor and taste, insoluble in water, soluble in alcohol, ethyl acetate, chloroform and other chlorinated hydrocarbons. Combustible.
Use: Emulsifier, stabilizer.

propylene glycol phenyl ether.
$C_6H_5OCH_2CHOHCH_3$.
Properties: Colorless liquid, d 1.060–1.070 (25/25C), boiling range 5.95%, 237–242C, flash p 275F (135C). Combustible.
Use: High-boiling solvent, bactericidal agent, fixative for soaps and perfumes, intermediate for plasticizers.

propylene imine. (2-methylaziridine; propyleni-

mine). CAS: 75-55-8. $\overline{CH_2HCNHCH_2}$.
Properties: Water-white liquid, bp 66–67C, d 0.8039–0.8070 (25/25C), 1.4094–1.4109 (25C), soluble in water and most organic solvents.
Hazard: Flammable, dangerous fire risk. Toxic by ingestion, inhalation, and skin absorption. A suspected carcinogen. TLV: 2 ppm in air.
Use: Organic intermediate whose derivatives are used in the paper, textile, rubber and pharmaceutical industries.

propylene oxide. CAS: 75-56-9.

$$H_2C\!\!-\!\!CH\!\!-\!\!CH_3$$
$$\diagdown\!\!\diagup$$
$$O$$

41st highest-volume chemical produced in US (1985).
Properties: Colorless liquid, ethereal odor, d 0.8304 at 20/20C, bp 33.9C, vap press 445 mm (20C), flash p −35F (−37.2C), bulk d 6.9 lb/

gal (20C), fp −104.4C, partially soluble in water, soluble in alcohol and ether.
Derivation: (1) Chlorohydration of propylene followed by saponification with lime, (2) peroxidation of propylene, (3) epoxidation of propylene by a hydroperoxide complex with molybdenum catalyst.
Hazard: Highly flammable, dangerous fire risk, explosive limits in air 2–22%. TLV: 20 ppm in air. Irritant.
Use: Polyols for urethane foams, propylene glycols, surfactants and detergents, isopropanol amines, fumigant, synthetic lubricants, synthetic elastomer (homopolymer), solvent.

propyl formate. CAS: 110-74-7.
$CH_3CH_2CH_2COOH$.
Properties: Liquid, d 0.9006 (20/4C), fp −92.9C, bp 81.3C, refr index 1.3769 (20C), slightly soluble in water, miscible with alcohol and ether. Flash p (COC) 27F (−2.8C), autoign temperature 851F (455C).
Hazard: Flammable, dangerous fire risk.
Use: Flavoring.

propylformic acid. See butyric acid.

n-propyl furoate. $C_4H_3OCO_2C_3H_7$.
Properties: Colorless, fragrant liquid; becomes yellow in light. Practically insoluble in water, soluble in alcohol and ether, d 1.0745 (25.9/4C), bp 210.9C (corrected), refr index 1.4737 (25.9C). Combustible.
Use: Flavoring.

propyl gallate. CAS: 121-79-9.
$C_3H_7OOCC_6H_2(OH)_3$.
Properties: Colorless crystals, mp 150C, almost insoluble in water, soluble in alcohol (50/50), somewhat soluble in oils.
Hazard: Use in foods restricted to 0.02% of fat content.
Use: Food preservative and antioxidant for animal fats and oils, flavoring, transformer oils.

propyl-p-hydroxybenzoate. See propylparaben.

propyliodone. (propyl-3,5-diiodo-4-oxo-1(4H)-pyridineacetate). CAS: 587-61-1.
$I_2(O)C_5H_2NCH_2COOC_3H_7$.
Properties: White, crystalline powder; odorless or nearly so. Practically insoluble in water; soluble in acetone, alcohol, and ether; mp 187–190C.
Grade: USP.
Use: Medicine (radiopaque medium).

propyl isomer. Generic name for
CAS: 83-59-0. dipropyl-5,6,7,8-tetrahydro-

7-methylnaphtho[2,3-d]-1,3-dioxole-5,6-dicarboxylate. $C_{20}H_{26}O_6$.
Properties: Orange liquid, insoluble in water, slightly soluble in oils, soluble in most organic solvents.
Use: Insecticide synergist.

propylmagnesium bromide. C_3H_7MgBr.
Available as a solution in ether; a Grignard reagent.
Use: Alkylating agent in organic synthesis.

propylmalonic acid, diethyl ester. (propyl diethyl malonate). $C_3H_7CH(COOC_2H_5)_2$.
Properties: Colorless liquid; fragrant odor; soluble in water; soluble in alcohols, ethers, esters, and ketones; d 0.9860 (25C); bp 222C.
Use: Intermediate, tobacco flavoring.

n-**propyl mercaptan.** Legal label name for 1-propanethiol.

2-n-propyl-4-methylpyrimidyl-6-N,N-dimethylcarbamate. $C_3H_7C_4HN_2(CH_3)OOCN(CH_3)_2$.
Properties: A liquid, miscible with water and most organic solvents.
Use: Insecticide.

n-**propyl nitrate.** (NPN). CAS: 627-13-4. $C_3H_7NO_3$.
Properties: White to straw-colored liquid, ethereal odor, d 1.07 (20C), bp 110C, flash p 68F (20C), autoign temperature 350F (176C), fp −100C, refr index 1.3975 (20C), insoluble in water, soluble in alcohol and ether.
Grade: 96–98% pure.
Hazard: Flammable, severe fire and explosion risk, strong oxidizing material, explosive limits in air 2–100%. TLV: 25 ppm in air.
Use: Rocket fuel (monopropellant).

propylparaben. (propyl-p-hydroxybenzoate). CAS: 94-13-3. $C_{10}H_{12}O_3$.
Properties: Colorless crystals or white powder; slightly soluble in boiling water; soluble in alcohol, ether, and acetone; mp 95–98C.
Grade: USP, FCC.
Use: Food preservative, fungicide, mold control in sausage casings.
See also "Parabens."

propyl pelargonate. $C_3H_7OOCC_8H_{17}$.
Properties: Liquid, d 0.870 (15/15C), bp 237C, insoluble in water, soluble in alcohol and most organic solvents. Combustible.
Use: Flavors and perfumes, bactericides and fungicides.

propylpiperidine. See coniine.

6-propylpiperonyl butyl diethylene glycol ether. See piperonyl butoxide.

n-**propyl propionate.** CAS: 106-36-5. $CH_3CH_2COOCH_2CH_2CH_3$.
Properties: Colorless liquid, soluble in most organic solvents, slightly soluble in water, boiling range 122–124C, fp −76C, flash p (OC) 174F (78.9C), bulk d 7.31 lb/gal. Combustible.
Grade: Technical.
Use: Solvent for nitrocellulose, paints, varnishes, lacquers, coating agents.

4-n-propylpyridine. $C_8H_{11}N$ or $C_3H_7C_5H_4N$.
Not to be confused with coniine (propylpiperidine).
Properties: Bp 188C.
Use: Intermediate.

propylthiouracil. (6-propyl-2-thiouracil). CAS: 51-52-5. $C_7H_{10}N_2OS$.
Properties: White, powdery, crystalline substance; starch-like in appearance and to touch. Bitter taste, mp 218–221C, sensitive to light, very slightly soluble in water, sparingly soluble in alcohol, soluble in ammonia and alkali hydroxides.
Derivation: Condensation of β-oxocaproate with thiourea.
Grade: USP.
Use: Medicine (thyroid inhibitor).

n-**propyltrichlorosilane.** CAS: 141-57-1. $C_3H_7SiCl_3$.
Properties: Colorless liquid, bp 12.5, d 1.195 (25/25C), refr index 1.4292 (25C), flash p (COC) 100F (37.7C), readily hydrolyzed with liberation of hydrogen chloride.
Derivation: By Grignard reaction of silicon tetrachloride and propylmagnesium chloride.
Grade: Technical.
Hazard: Flammable, moderate fire risk. Strong irritant.
Use: Intermediate for silicones.

propyl xanthate. See xanthic acid.

propyne. See methylacetylene.

2-propyn-1-ol. See propargyl alcohol.

prostacycline. See prostaglandin.

prostaglandin. One of a group of physiologically active compounds derived from arachidonic acid, a 20-carbon fatty acid which occurs in glandular organs and the liver. It was named from the prostate gland, where it was originally found. Research on prostaglandins has been intense in recent years in view of their importance in various reproduction mechanisms and their effects on blood pressure. They are believed to have significant relationships to a number of hormones; they also affect the nervous system, inhibit production of gastric juice, stimulate smooth muscles, and induce labor. They occur naturally in body tissues and biological fluids (especially semen). Both the chemical structure and metabolic functions of these compounds have been established with considerable accuracy, and several types have been synthesized; one pathway uses norbornadiene as a starting point followed by a ten-step sequence of conversions to a diol which serves as a precursor; another starts with cyclopentadiene, followed by hydroboration, yielding two intermediates from which prostaglandin can be derived. The most prolific source of prostanglandin intermediates (called syntons) is a marine organism called a gorgonian sea whip, which occurs in great numbers in coral reefs, especially in the Caribbean area. Harvesting of these has made the production of prostaglandins much less expensive. The occurrence of prostaglandin A_1 in yellow onions has been confirmed. Important derivatives of prostaglandins are prostacyclins and thromboxanes; a closely related group of compounds derived directly from arachidonic acid are the leukotrienes, which occur in white blood cells.

prosthetic group. A chemical grouping in which a metal ion is associated with a large molecule or molecular complex, for example, coenzymes and metal-porphyrin complexes such as chlorophyll and hemin. Such groups activate metabolic mechanisms such as phosphorylation, decarboxylation, etc., by coordination reactions with amino acids, proteins, enzymes, and nucleic acids. The behavior of certain vitamins and other metabolites is due in part to prosthetic groups; catalysis is also involved.

protactinium. CAS: 7440-13-3. Pa.
A radioactive element of atomic number 91, a member of the actinide series, aw 231.0359, valences = 4, 5; 13 unstable isotopes, two of which occur naturally. Protactinium is a constituent of all uranium ores, 340 milligrams being extracted from one ton. Protactinium may also be produced by irradiation of thorium-230. Purification is carried out by ion exchange and solvent extraction techniques. The longest lived isotope, Pa-231, decays by alpha emission and has a half-life of 33,000 years. Protactinium may be precipitated as the double potassium fluoride K_2PaF_7 or the oxide Pa_2O_5. The metal may be prepared by reducing PaF_4 with barium or by heating PaI_4 in a vacuum. It is hard and white, melting near 1600C. It is too rare for commercial use. Forms several compounds with halogens.
Hazard: Highly toxic radioactive material.

protamines. Simplest proteins, without sulfur, molecular weights about 3000.
Properties: Water soluble, producing basic solutions.

protease. A proteolytic enzyme which weakens or breaks the peptide linkages in proteins. They include some of the more widely known enzymes such as pepsin, trypsin, ficin, bromelin, papain, and rennin. Being water soluble they solubilize proteins and are commercially used for meat tenderizers, bread baking, and digestive aids.

protective coating. A film or thin layer of metal glass or paint applied to a substrate primarily to inhibit corrosion and secondarily for decorative purposes. Metals such as nickel, chromium, copper, and tin are electrodeposited on the base metal; paints may be sprayed or brushed on. Vitreous enamel coatings are also used, which require baking. Zinc coatings are applied by a continuous bath process where a strip of ferrous metal is passed through molten zinc.
See also galvanizing, terne plate, electroplating, paint, corrosion, cladding.

protective colloid. See colloid, protective.

protein. A complex high polymer containing carbon, hydrogen, oxygen, nitrogen, and usually sulfur, and comprised of chains of amino acids connected by peptide linkages ($—CO \cdot NH—$). Proteins occur in the cells of all living organisms and in biological fluids (blood plasma, protoplasm). They are synthesized by plants largely because of the nitrogen fixing ability of certain soil bacteria. Their molecular weight may be as high as 40 million (tobacco mosaic virus). They have many important functional forms; enzymes, hemoglobin, hormones, viruses, genes, antibodies, and nucleic acids. They also comprise the basic component of connective tissue (collagen), hair (keratin), nails, feathers, skin, etc. Some have been synthesized in the lab.
The sequence of amino acids in the polypeptide chain is of critical importance in genetics. Proteins can be hydrolyzed to their constituent amino acids and can be broken down into simpler forms by proteolytic enzymes. They form colloidal solutions, and behave chemically as both acids and bases simultaneously (amphoteric). They are denatured by changes in pH, by heat, UV radiation, and many organic solvents.
Simple proteins contain only amino acids, conjugated proteins contain amino acids plus nucleic acids, carbohydrates, lipids, etc. On the basis of solubility, they can be classified as albumins (water-soluble), globulins (insoluble in water but soluble in aqueous salt solutions), and prolamins (soluble in alcohol-water mixture but not in alco-

hol or water alone). A number of proteins have been synthesized, notably the hormone insulin. Proteins are an essential component of the diet, occurring chiefly in meat, eggs, milk, and fish. Edible proteins suitable for human food as well as cattle feed can be produced from microorganisms grown in carbonaceous or nitrogenous media to form yeast-like materials. Paraffinic hydrocarbons (methane) and petroleum-derived ethanol can be used as growth media for protein biosynthesis.

Industrial applications of proteins include plastics, adhesives, and fibers derived from casein and soybean protein, but these have been declining in recent years. Special forms in which proteins are commercially available include textured proteins for food products, and protein hydrolyzate and liquid predigested protein, both for medical use.

See ribonuclease for structure; also see deoxyribonucleic acid. See also nutrition; amino acid; protein, textured; protein, single-cell; polypetide.

protein hydrolyzate. Solution of protein hydrolyzed into its constituent amino acids.
Use: Medicine and surgery. Usually administered by a stomach tube or intravenous injection.
Grade: USP.

protein, single-cell. (SCP). A protein nutrient derived from bacteria or yeast by hydrocarbon fermentation or from fungi by fermentation on foodplant waste. A process developed in West Germany during the 1970s utilizes the bacterium *Methylomanas clara* cultured in a mixture of methanol, ammonia, water, and air. The continuous fermentation process is followed by dewatering and spray drying. The product contains 70% protein, 10% nucleic acids, 8% fats, and 7% minerals; in this form it is suitable for animal feeds. A purified type (90% protein) is made by dissolving the product in an ammonia-methanol mixture, followed by filtration to remove the fats and then by water-washing to extract the nucleic acids. The product may prove to be satisfactory for use in human foods. The presence of nucleic acids is undesirable, as they may lead to metabolic disorders such as gout. Commercial production awaits further testing.

protein, spun. See protein, textured.

protein, textured. A meat extender or substitute made from defatted soybean flour or similar protein, usually by an extrusion process. Some types are used to fortify cereals and other food products. The filaments produced by extrusion are designed to stimulate the fibrous structure (texture) of meats. The term "spinning" is also used, and the products are often called spun proteins.

"Protek-Sorb."[241] TM for a group of silica gels.

proteolysis. The structural breakdown of proteins, usually by hydrolysis, as a result of the action of an enzyme, e.g., trypsin, pepsin, papain, etc.

prothrombin. The precursor of the enzyme thrombin, a proteinaceous component of blood plasma which is converted into thrombin by the blood-clotting mechanism when activated at the site of an open wound.
See also coagulation.

"Protina."[266] TM for line of products for cleaning, inhibiting corrosion, and coating copper, brass, or bronze artwork and statuary.

protocatechuic aldehyde, methyl ether.
See vanillin.

"Protolin."[23] TM for reducing agents based on zinc sulfoxylate and zinc formaldehyde sulfoxylate. Supplied as water-soluble white powder.
Use: Stripping colors from fabrics, chemical synthesis.

proton. A fundamental unit of matter having a positive charge and a mass number of 1, equivalent to 1.67×10^{-24} gram. Its mass is 1837 times that of the negatively charged electron, but is almost identical with that of the uncharged neutron. Protons are constituents of all atomic nuclei, their number in each nucleus being the atomic number of the element. An atom of normal hydrogen contains one proton and one electron. A proton is identical with a hydrogen ion (H^+).

"Protopet."[45] TM for petrolatum of medium consistency and ranging in color from pure white to amber, but meeting USP or NF purity requirements for petrolatum.

protoplasm. The total contents of the living cell, including both nucleus and cytoplasm. Predominantly a mixture of proteins, protoplasm is the physical basis of life. Most of its components are in the colloidal size range. This term is falling into disuse among modern biochemists.
See also cell (1).

"Protorez."[53] TM for finishing resins and chemicals for the textile industry.

protoveratrine. A substance isolated from the Veratrum album plant. It is a mixture of two alkaloids, designated protoveratrine A and protoveratrine B.
Use: Medicine to lower blood pressure.

"Protowet."[53] TM for scouring and wetting agents for general industrial use and finishing chemicals for the textile industry.

"Protox."[268] TM for a series of lead-free zinc oxides manufactured by the American Process (produced from zinc ore) and the French Process (produced from metal).
Available forms: Pellets and free flowing, nondusting, low bulk for easy handling.
Use: Reinforcing agent and accelerator activator for rubber.

"Provinite."[304] TM for a two part barium-cadmium organic vinyl stabilizer.
Properties: (a) Soft white powder, d 1.4, refr index 1.56; (b) clear, straw-colored liquid, d 0.91, refr index 1.45.
Use: Heat and light stabilizer for vinyl film, sheeting, and dispersion resin systems.

provitamin. The precursor of a vitamin. Examples are carotene and ergosterol, which upon activation become Vitamin A and Vitamin D, respectively.
See also specific compounds.

Prussian blue. The most common and best known name for blue iron ferrocyanide (iron blue) pigments made by a variety of procedures.
See iron blue.

prussic acid. See hydrocyanic acid.

"Prym."[42] TM for a group of colorless syrups of thermosetting carbamide fiber reactants.
Properties: Readily soluble in water at 25C.
Use: Textile finishes.

Pschorr reaction. Synthesis of phenanthrene derivatives from diazotized α-aryl-omicron-aminocinnamic acids by intramolecular arylation.

pseudobutylene glycol. See 2,3-butylene glycol.

pseudocumene. (1,2,4-trimethylbenzene; uns-trimethylbenzene). CAS: 95-63-6. $C_6H_3(CH_3)_3$.

Properties: Liquid, fp −43.91C, bp 168.89C, d 0.8758 (20/4C), refr index 1.5045 (20C), flash p 130F (54.4C). Insoluble in water; soluble in alcohol, benzene, and ether.
Derivation: From C_9 fraction of refinery reformate streams by fractional distillation.

Grade: 95%, 99% and research.
Hazard: Moderate fire risk. Central nervous system depressant, irritant to mucous membranes.
Use: Manufacture of trimellitic anhydride, dyes, pharmaceuticals, and pseudocumidine.

pseudocumidine. (2,4,5-trimethylaniline; 1,2,4-trimethyl-5-aminobenzene). CAS: 137-17-7. $C_6H_2(CH_3)_3NH_2$.
Properties: White crystals, d 0.957, mp 62C, bp 236C, soluble in alcohol and ether, insoluble in water. Combustible.
Derivation: From pseudocumene.
Use: Manufacture of dyes, organic synthesis.

pseudohexyl alcohol. See 2-ethylbutyl alcohol.

pseudoionone. (6,10-dimethyl-3,5,9-undeca-triene-2-one). CAS: 141-10-6. $(CH_3)_2C:CH(CH_2)_2C(CH_3):CHCH:CHCOCH_3$.
Properties: Pale yellow liquid, d 0.8984 (20C), bp 143–145C (12 mm), soluble in alcohol and ether. Combustible.
Use: Perfumery, cosmetics.

psi. Abbreviation for pounds per square inch.

psia. Abbreviation for pounds per square inch absolute.

psicain. Acid tartrate of d-psi-cocaine.
Use: Same as cocaine hydrochloride.

psig. Abbreviation for pounds per square inch gauge.

psilocin. CAS: 520-53-6.
$C_8H_5N(OH)C_2H_4N(CH_3)_2$. An indole derivative. An alkaloid from certain mushrooms; a hallucinogenic drug.

psilocybin. CAS: 520-52-5.
$C_8H_5N(OPO_3H_2)C_2H_4N(CH_3)_2$. An indole derivative. An alkaloid from certain mushrooms; hallucinogenic drug.

psilomelane. $BaMn_9O_{16}(OH)_4$. A natural oxide of variable composition. Calcium, nickel, cobalt, and copper frequently are present. The name sometimes refers to mixture of manganese minerals.
Properties: Color black, streak brownish black, luster submetallic, hardness 5–6, d 3.7–4.7.
Occurrence: USSR, India, South Africa, Cuba, US (Arkansas, Virginia, Georgia).
Use: Important ore of manganese.

psychotropic drug. Any of a number of therapeutic agents which affect the behavior, emotional state, or mental functioning of psychologically disturbed persons. They are widely known as "tranquilizers," but this term is no longer accepted as clinically accurate, as the minor tranquilizers (benzodiazepine and glycerol derivatives) act quite differently from the major tranquilizers. For this reason the latter are now classified as antipsychotics and antidepressants, and the term "antianxiety agent" is applied to the minor tranquilizers. Antipsychotic agents include phenothiazines (chlorpromazine), thioxanthenes, and butyrophenones; antidepressant agents have two major types, i.e., monoamine oxidase (MAO) inhibitors and several tricyclic compounds; antianxiety agents are glycerol derivatives (meprobamate) and benzodiazepine derivatives, e.g., oxazepam.

Pt. Symbol for platinum.

PTA. Abbreviation for phosphotungstic acid; also for purified terephthalic acid.

pteroylglutamic acid. See folic acid.

PTFE. Abbreviation for polytetrafluoroethylene.

PTMA. Abbreviation for a mixture of phosphotungstic and phosphomolybdic acids.
Use: Making pigments.
See phosphotungstic pigment.

ptomaine. A group of highly toxic substances (derivatives of ethers of polyhydric alcohols) resulting from the putrefaction or metabolic decomposition of animal proteins. Examples which have been isolated and prepared synthetically are cadaverine (1,5-diaminopentane), muscarine (hydroxyethyltrimethylammonium hydroxide), putrescine (tetraethylenediamine), and neurine (trimethylvinylammonium hydroxide). *Note*: The term "ptomaine poisoning" is usually a misnomer for other types of food poisoning.

PTSA. Abbreviation for p-toluenesulfonamide.

ptyalin. A salivery amylase which acts upon α-1,4-glycosidic linkages converting starch to various dextrins and maltose. It can act over a pH range of 4.0–9.0; optimum pH 5.6–6.5. It requires the presence of certain negative ions for activation; chlorides and bromides are the most effective.
Use: Biochemical research.

Pu. Symbol for plutonium.

pug mill. A comminuting or granulating machine whose essential components are a shaft equipped

with blades or arms with alloy-hardened tips rotating in a trough-like compartment.
Use: Grinding and amalgamating fertilizer ingredients, clay mixtures, cement components, and similar products.

pulegium oil. See pennyroyal oil.

pulegone. (1-isopropylidene-4-methyl-2-cyclohexanone). CAS: 89-82-7. $C_{10}H_{16}O$.
A ketone found in pennyroyal and hedeoma oil. Combustible.
Properties: Oily liquid with pleasant odor, d 0.9323 (20C), bp 221C, dextrorotatory, refr index 1.4894 (20C), insoluble in water, soluble in alcohol and ether.
Use: Chemical intermediate, flavoring.

"Pullulan." A biodegradable polysaccharide made by yeast fermentation originally developed in Japan. Its adhesive and oxygen impermeable properties enable it to be used to coat pharmaceutical products. It is water-soluble, odorless, and edible; these properties have led to its use admixed with foodstuffs for special-purpose applications. The mixture can be processed into a semi-rigid plastic sheet or film which can serve as an emergency food source. FDA approval is pending.

pulp, paper. Processed cellulosic fibers derived from hardwoods, softwoods, and other plants. There are two major types of pulp: (1) ground wood or mechanical pulp, which is merely finely divided wood without purification and is made into newsprint, cheap manila papers, and nonpermanent tissues; (2) chemical pulp, of which there are three kinds: (a) soda process pulp; obtained from the digestion of wood chips (mostly poplar) by caustic soda, (b) sulfite process pulp (mostly spruce and other coniferous woods) obtained by digestion with a solution of magnesium, ammonium, or calcium disulfite containing free sulfur dioxide; and (c) sulfate process (kraft) pulp, in which sodium sulfate is added to the caustic liquors but is reduced by the carbon present to the sulfide, which becomes a digesting agent. Sulfite and sulfate pulps (chiefly from softwoods) comprise the bulk of paper pulps. Sulfate pulps are known as kraft pulps because of their strength and are used for wrapping, packaging, container board, etc. A relatively new process called holopulping replaces sodium sulfate with oxidants. A synthetic pulp based on polyolefins (styrene copolymer fibers) has been developed to the production stage in Japan.
See also holopulping, paper, digestion.

pultrusion. A technique for making certain products from glass-reinforced plastics, such as rods, electrical insulators, etc. It involves passage of continuous bundles of glass fiber that have been impregnated with liquid resin through an oven at the rate of 18 inches per minute at 140C (285F).

pumice. A highly porous igneous rock, usually containing 67–75% SiO_2 and 10–20% Al_2O_3, glassy texture. Potassium, sodium, and calcium are generally present. Insoluble in water, not attacked by acids.

Occurrences: USA (Arizona, Oregon, California, Hawaii, New Mexico), Italy, New Zealand, Greece.

Grade: Lump, powdered coarse, medium, fine; NF, technical.

Use: Concrete aggregate, heat and sound insulation, filtration, finishing glass and plastics, road construction, scouring preparations, paint fillers, absorbents, support for catalysts, dental abrasive, abherent for uncured rubber products, possible substitute for asbestos.

Pummerer rearrangement. Rearrangement of sulfoxides to α-acyloxythioethers in the presence acyclic anhydrides.

punty. A solid or hollow iron rod 4–6 ft long, usually with an insulation covering on one end.

Use: By glass workers to remove molten material from the melt preparatory to shaping finished articles.

"Purafil."[525] TM for an odoroxidant consisting of activated alumina impregnated with potassium permanganate in pellet form. Destroys odors by oxidation.

Purdie (Irvine-Purdie) methylation. Exhaustive methylation of a methyl glycoside by repeated treatment with methyl iodide and silver oxide, followed by hydrolysis of the pentamethyl ether with dilute acid to yield the anomeric hydroxyl group.

Purex process. See reprocessing.

purification. Removal of extraneous materials (impurities) from a substance or mixture by one or more separation techniques. A pure substance is one in which no impurity can be detected by any experimental procedure. Though absolute purity is impossible to attain, a number of standard procedures exist for approaching it to the extent of 1 ppm of impurity or less. The following fractionation techniques are widely used: crystallization, precipitation, distillation, adsorption (various types of chromatography), extraction, electrophoresis and thermal diffusion.

See also purity, chemical.

"Purifloc."[233] TM for a polyelectrolyte.

Use: To flocculate solids in water and industrial waste treatment.

purine. (1) [imidazo(4,5-d)pyrimidine].

CAS: 120-73-0.

Properties: Colorless crystals; mp 217C; soluble in water, alcohol, toluene.

Derivation: Prepared from uric acid and regarded as the parent substance for compounds of the uric groups, many of which occur naturally in animal waste products.

Use: Organic synthesis, metabolism, and biochemical research.

(2) One of a number of basic compounds found in living matter and having a purine-type molecular structure.

See adenine, guanine, hypoxanthine, xanthine, uric acid, caffeine, and theobromine.

"Purite."[84] TM for a specially prepared fused soda ash furnished in the form of 2-pound cast pigs and stated to contain 98% sodium carbonate.

Use: Cupola flux, for refining and desulfuring iron, steel, and other metals.

purity, chemical. A substance is said to be pure when its phsyical and chemical properties coincide with those previously established and recorded in the literature, and when no change in these properties occurs after application of the most selective fractionation techniques. In other words, purity exists when no impurity can be detected by any experimental procedure. There are a number of recognized standards of purity.

See also grade.

puromycin. (USAN). CAS: 53-79-2. $C_{22}H_{29}N_7O_5$. An antibiotic which inhibits protein synthesis, prevents transfer of amino acid from its carrier to the growing protein. Produced by Streptomyces alboniger, effective against bacteria, protozoa, parasitic worms, and cancerous tumors.

Properties: Crystals, mp 176C.

Hazards: Toxic to living cells of all kinds.

purple of Cassius. See gold-tin purple.

purpurin. (1,2,4-trihydroxyanthraquinone; CI 58205). CAS: 81-54-9. $C_{14}H_5O_2(OH)_3$.
Properties: Reddish needles, mp 256C, slightly soluble in hot water, soluble in alcohol and ether.
Derivation: Occurs as a glucoside in madder root. Made synthetically by oxidation of alizarin.
Use: Dye for cotton, stain for microscopy, reagent for boron, manufacture of acid and chrome dyes.

purpurin red. See anthrapurpurin.

"PuTrol."[188] TM for a powerful aromatic compound.
Use: For masking the odors of putrefaction associated with the decomposition of proteins, as in fat-rendering operations, sewage disposal, and industrial wastes.

putty. A mixture of whiting (chalk) with 12–18% of linseed oil, with or without white lead or other pigment. Containers must be air-tight.
Use: Sealant, glass setting, caulking agent.

putty powder. A soft abrasive composed of tin oxide.

PVA. Abbreviation for polyvinyl alcohol.

PVAc. Abbreviation for polyvinyl acetate.

PVB. Abbreviation for polyvinyl butyral.

PVC. (1) Abbreviation for polyvinyl chloride. (2) Abbreviation for pigment volume concentration, a term used in paint technology to mean pigment volume divided by the sum of the pigment volume and the vehicle solids volume, multiplied by 100.

PVC, chlorinated. See polyvinyl dichloride.

PVDC. See polyvinyl dichloride.

PVE. Abbreviation for polyvinyl ethyl ether.

PVI. Abbreviation for polyvinylisobutyl ether.

PVM. Abbreviation for polyvinyl methyl ether.

PVM/MA. See polyvinyl methyl ether/maleic anhydride.

PVOH. Abbreviation for polyvinyl alcohol.

PVP. Abbreviation for polyvinylpyrrolidone.

PVT. Abbreviation for pressure-volume-temperature.
Use: Chemical engineering.

py. An Abbreviation for pyridine.
Use: As in formulas for coordination compounds. See also dien, en, pn.

"Pycal."[89] TM for a series of plasticizers.
Use: For cellulose acetate, cellulose acetate butyrate, nitrocellulose, vinyls, and synthetic rubber.

"Pyranol."[245] TM for a dielectric material, principally of the askarel type.

"Pyratex."[248] TM for a vinyl pyridine latex.
Properties: Total solids 40–42%, pH 10.4–11.5, d 0.96.
Use: To promote adhesion between rayon or nylon fibers and rubber, as in tire cord, belting, hose, etc.

"Pyrax."[69] TM for a ground pyrophyllite-aluminum silicate.
Use: Filler in paints, plastics, and for dusting unvulcanized rubber.

pyrazine hexahydride. See piperazine.

pyrazole. CAS: 288-13-1. $\overline{HNNCHCHCH}$.
Properties: Off-white, crystalline solid with pyridine odor; mp 68–70C; bp 186–188C; soluble in water and alcohol.
Hazard: Toxic by ingestion and inhalation, irritant to skin and eyes.
Use: Chemical intermediate, stabilizer for halogenated solvents and lubricating oils.

pyrazoline. $\overline{HNNCHCH_2CH_2}$.
Properties: Bp 144C.
Use: Organic synthesis.

pyrazolone dye. A dye whose molecules contain both the —N≡N— and the ≡C≡C≡ chromophore groups in their structure. These are acid dyes most used for silk and wool and to some extent for lakes. Tartrazine, CI 19140, is an important member of this group.
See dye.

"Pyre-ML."[28] TM for polyimide coated fabrics and laminates.
Use: Class H electrical insulating materials.

pyrene. CAS: 129-00-0. $C_{16}H_{10}$.
A condensed ring hydrocarbon.
Properties: Colorless solid (tetracene impurities give a yellow color), solutions have a slight blue fluorescence, mp 156C, d 1.271 (23C), bp 404C, insoluble in water, partially soluble in organic solvents.
Derivation: From coal tar.
Hazard: A carcinogen, absorbed by skin.
Use: Biochemical research.

pyrethroid. Any of a group of insect growth regulators that act as neurotoxins, analogous to juvenile hormones, restricting the development of insect larvae. Thus they are especially effective against insects that are destructive in the adult stage. They are considered nontoxic to animals and man.
See also juvenile hormone.

pyrethrin I. CAS: 8003-34-7. $C_{21}H_{28}O_3$.
Pyrethrolone ester of chrysanthemummonocarboxylic acid. Most potent insecticidal ingredient of pyrethrum flowers.
Properties: Viscous liquid, oxidizes readily in air, insoluble in water, soluble in other common solvents, incompatible with alkalies.
Hazard: Toxic by ingestion and inhalation.
Use: Household insecticide (flies, mosquitoes, garden insects, etc.), treatment of paper bags for shipping cereals, etc.
See also cinerin I and II and pyrethrin II.

pyrethrin II. $C_{22}H_{28}O_5$. Pyrethrolone ester of chrysanthemumdicarboxylic acid. One of the four primary active insecticidal ingredients of pyrethrum flowers.
Properties: Similar to those of pyrethrin I.
See also pyrethrin I, cinerin I and II.

pyrethrum. A natural insecticide obtained by extraction of chrysanthemum flowers native to Kenya, Ecuador, and Japan. The solvent used is a hydrocarbon of the kerosene type. Pyrethrum is also made synthetically. Not compatible with alkaline material. The chief constituents are pyrethrins I and II and cinerins I and II.
See also allethrin.
These compounds are nonvolatile and very slightly soluble in water.
Hazard: Toxic by ingestion and inhalation. TLV: 5 mg/m³ of air.
Use: Household insecticide (flies, mosquitoes, garden insects, etc.).

"Pyrex Glass Brand No. 7740."[20] TM for a borosilicate glass.
Properties: Linear coefficient of expansion, 32 × 10^{-7}/C, elasticity coefficient 6.230 kg/sq mm, hardness (scleroscope) 120, d 2.25, specific heat 0.20, refr index 1.474 dispersion 0.00738, light and heat transmission higher than the best plate glass, dielectric constant 4.5 (25C), upper working temperature (mechanical considerations only): annealed, extreme limit 490C, normal service 230C, thermal stress resistance 53C, impact abrasion resistance 3.1, softening temperature 820C.
Use: Laboratory and pharmaceutical glassware and apparatus, electrochemical equipment, fiber manufacture, domestic ovenware apparatus, and equipment for many processes.

pyridine. CAS: 110-86-1. $N(CH)_5$.

Properties: Slightly yellow or colorless liquid, nauseating odor, burning taste, slightly alkaline in reaction. Soluble in water, alcohol, ether, benzene, ligroin, and fatty oils. D 0.987, fp −42.0C, bp 115.5C, flash p (CC) 68F (20C), autoign temperature 900F (482C).
Derivation: (a) By coal carbonization and recovery both from coke-oven gases and the coal tar middle oil. (b) Also synthetically from acetaldehyde and ammonia.
Grade: Technical, as 20C, 2C, etc., (meaning distillation range), medicinal, CP, spectrophotometric.
Hazard: Flammable, dangerous fire risk, explosive limits in air 1.8–12.4% Toxic by ingestion and inhalation. TLV: 5 ppm in air.
Use: Synthesis of vitamins and drugs, solvent waterproofing, rubber chemicals, denaturant for alcohol and antifreeze mixtures, dyeing assistant in textiles, fungicides.

2-pyridine aldoxime methiodide. (2-PAM). CAS: 94-63-3. $C_5NH_4CHNOH \cdot ICH_3$.
Use: Antidote for cholinesterase-inhibiting pesticides of the parathion type because of its property of reactivating the cholinesterase by removal of phosphoryl groups. Also antidote for nerve gases.

pyridine-3-carboxylic acid. See niacin.

3-pyridine diazonium fluoborate. Intermediate in manufacture of 3-fluoropyridine.
Properties: Explodes when dry.

2,5-pyridinedicarboxylic acid. See isocinchomeronic acid.

pyridine-N-oxide. (pyridine-1-oxide). CAS: 694-59-7. C_5H_5NO.
Properties: Fp 67.0C, soluble in water.
Hazards: Probably flammable and toxic.
Use: Intermediate.

pyridine polymer. A polymer or copolymer of methylvinylpyridine and vinylpyridine.
Use: Corrosion inhibitors and as intermediates for coatings and printing inks.

2-pyridinethiol-1-oxide. See 1-hydroxy-2-pyridine thione.

pyridinium bromide perbromide. (PBPB).
CAS: 39416-48-3. $C_5H_6NBr \cdot Br_2$.
Properties: Red prismatic crystals, mp 135–137C (decomposes) with preliminary softening. The salt is stable in the dry state and can be used in glacial acetic acid, ethanol and related solvents. This compound has 45–50% available bromine.
Use: Brominating phenols and addition to double bonds; mono- and polybromination of ketones, including aliphatic, alicyclic, steroid, and amino carbonyls. Micro or semimicro quantitative analysis.

pyridoxal hydrochloride. CAS: 58-56-0.
$C_8H_9NO_3 \cdot HCl$.   An aldehyde derivative of pyridoxine, with vitamin B_6 activity.
Properties: Rhombic crystals, mp 165C (decomposes), soluble in water and 95% ethanol.
Use: Nutrition.

pyridoxal phosphate. (2-methyl-3-hydroxy-4-formyl-5-pyridylmethylphosphoric acid).
CAS: 54-47-7.
$CH_3C_5HN(OH)(CHO)CH_2PO_4H_2)$.   The coenzyme of amino acid metabolism which also is the active group of various decarboxylases and other types of enzymes. It is closely related to pyridoxine.
Derivation: Synthesized (a) through the action of adenosine triphosphate, or phosphorus oxychloride, on pyridoxal, and (b) by phosphorylation of pyridoxamine followed by oxidation with 100% H_3PO_4.
Use: Nutrition, biochemical research.

pyridoxine. (vitamin B_6; 3-hydroxy-4,5-dimethylol-2-methylpyridine).
$CH_3C_5HN(OH)(CH_2OH)_2$.

A group name designating the naturally occurring pyridine derivatives having vitamin B_6 activity. Essential for the dehydration and desulfhydration of amino acids and for the normal metabolism of tryptophan; appears to be related to fat metabolism. Pyridoxine is required in the nutrition of all species of animals.
Sources: (Food): Vegetable fats, whole grain cereals, legumes, yeast, muscle meats, liver and fish. (Commercial): Synthetic pyridoxine, pyridoxal

and pyridoxamine are produced by a complex series of reactions from isoquinoline.
Units: Amounts are expressed in micrograms.
Use: Medicine, nutrition (available as pyridoxine hydrochloride).

α-pyridylamine. See 2-aminopyridine.

β-pyridylamine. See 3-aminopyridine.

3-pyridylcarbinol. (3-pyridine methanol).
$C_5H_4NCH_2OH$.
Properties: Colorless liquid, bp 266C, fp −6.5C, d 1.131 g/mL (20C), refr index .1.5455 (20C), miscible in water at 20C, hygroscopic. Combustible.
Use: Intermediate.

pyrimidine. One of a group of basic compounds found in living matter. They may be isolated following complete hydrolysis of nucleic acids. They include uracil, thymine, cytosine, and methylcytosine. Thiamine is also a pyrimidine derivative. Other pyrimidines such as alloxan and thiouracil are important in medicine and biochemical research.
See also purine.

pyrimithate. CAS: 5221-49-8.
$C_{11}H_{20}N_3O_3PS$.   A diethyl ester of phosphoric acid.
Properties: Colorless liquid; d 1.16; soluble in acetone, alcohol, and benzene; almost insoluble in water.
Hazard: Cholinesterase inhibitor.
Use: Acaricide, insecticide.

pyrite. (iron pyrite; fool's gold).
CAS: 1309-36-0. FeS_2.   Often mixed with small amounts of copper, arsenic, nickel, cobalt, gold, selenium.
Properties: Brass-yellow or brown tarnished mineral, greenish or brownish-black streak, metallic luster, d 4.9–5.2, hardness 6–6.5.
Use: Manufacture of sulfur, sulfuric acid and sulfur dioxide, ferrous sulfate, cheap jewelry, recovery of metals.
See also ferrous sulfide.

pyrite, magnetic. See pyrrhotite.

pyrithiamine. (neopyrithiamine).
CAS: 534-64-5. $C_{14}H_{20}Br_2N_4O$.   A thiamine antagonist.
Properties: Crystallizes from acetone, mp 219C (decomposes), soluble in water.
Derivation: Synthetically from the condensation of 2-methyl-3-(β-hydroxyethyl)pyridine with the pyrimidine moiety of thiamine.
Use: Biochemical research.

pyro-. A prefix indicating formation by heat, specifically, an inorganic acid derived by loss of one molecule of water from two molecules of an orthoacid as pyrophosphoric acid.

pyroboric acid. (tetraboric acid). $H_2B_4P_7$. Vitreous or white powder, soluble in water and in alcohol.

"Pyrobrite."[288] TM for a bright leveling, pyrophosphate copper electroplating process. The materials used are copper pyrophosphate trihydrate, potassium pyrophosphate, ammonium hydroxide, and addition agents.

pyrocatechol. (1,2-benzenediol; o-dihydroxybenzene; catechol). CAS: 120-80-9. $C_6H_4(OH)_2$.
Properties: Colorless crystals, discolors to brown on exposure to air and light, especially when moist. D 1.371, mp 104C, bp 245C, sublimes. Soluble in water, alcohol, ether, benzene, and chloroform, also in pyridine and aqueous alkaline solutions; flash p 261 (127C) (CC). Combustible.
Derivation: (a) By fusion of o-phenolsulfonic acid with caustic potash at 350C. (b) By heating guaiacol with hydriodic acid.
Grade: Technical, CP, resublimed.
Hazard: Strong irritant. TLV: 5 ppm in air.
Use: Antiseptic, photography, dyestuffs, electroplating, specialty inks, antioxidants and light stabilizers, organic synthesis.

pyrocatechol dimethyl ether. See veratrole.

pyrocatechol methyl ester. See guaiacol.

pyrocellulose See guncotton.

"Pyroceram" Brand Cement.[20] TM for powdered glasses which are thermosetting and utilized for sealing inorganic materials. The resultant seals are crystalline and have service temperatures in excess of the sealing temperatures.

"Pyroceram" Brand 9608.[20] TM for a crystalline ceramic material made from glass by controlled nucleation.
Properties: Hard, brittle solid; d 2.5; softening temperature 1250C; specific heat 0.19 (25C); thermal conductivity 0.0047 cgs; dielectric constant 6.54 (25C); flexural strength 14,000–23,000 psi; Knoop hardness 703 (100g).
Use: Telescope mirrors, special-purpose ceramic products.

pyrochlore. $NaCaNb_2O_6F$. A complex oxide of sodium, calcium, and niobium. Tantalum, rare earth metals, and other elements may be present.

Color brown to black, streak light brown, hardness 5–5.5, d 4.2–6.4.
Occurrence: Canada (Quebec), Brazil, Africa.
Use: Ore of niobium.

pyrochroite. See manganous hydroxide.

pyrogallol. (pyrogallic acid; 1,2,3-trihydroxybenzene). CAS: 87-66-1. $C_6H_3(OH)_3$.
Properties: White, lustrous crystals; turn gray on exposure to light; d 1.463; mp 132.5; bp 309C. Soluble in water, alcohol, and ether. A solution of pyrogallol acquires a brown color on exposure to air. This absorption of oxygen and change of color take place rapidly when the solution is made alkaline.
Derivation: By heating gallic acid with three times its weight of water in an autoclave.
Hazard: Toxic by ingestion and skin absorption.
Use: Protective colloid in preparation of metallic colloidal solutions, photography, dyes, intermediates, synthetic drugs, medicine (antibacterial), process engraving, laboratory reagent, gas analysis (an oxygen absorber), reducing agent, antioxidant in lubricating oils.

pyrolan. (1-phenyl-3-methyl-5-pyrazolyl dimethylcarbamate). CAS: 87-47-8. $C_{13}H_{15}N_3O_2$.
Properties: Crystalline solid, mp 50C, soluble in water and fats.
Hazard: Toxic by ingestion. Cholinesterase inhibitor.
Use: Insecticide.

pyroligneous acid. (wood vinegar; pyroligneous liquor). CAS: 8030-97-5.
Properties: Crude yellow to red liquid, a mixture of materials from wood distillation. Crude product contains methanol, acetic acid, acetone, furfural, and various tars and related products; d 1.018–1.030; miscible with water and alcohol.
Use: Smoking meats.

pyrolusite. (manganese dioxide, black). MnO_2.
Properties: Iron-black to dark steel-gray or bluish mineral; streak, black or bluish-black; luster, metallic or dull. Soluble in hydrochloric acid, d 4.73–4.86, Mohs hardness 2–2.5.
Occurrence: US (Virginia, Georgia, Arkansas, Lake Superior region, Massachusetts, Vermont, New Mexico), Germany, Australia, India, Canada.
Use: Manganese ore. For many uses the ore and synthetic material are interchangeable, but pyrolusite is not usable for batteries.
See also manganese dioxide.

pyrolysis. Transformation of a compound into one or more other substances by heat alone, i.e., without oxidation. It is thus similar to destructive distillation. Though the term implies decomposition into smaller fragments, pyrolytic change may also involve isomerization and formation of higher molecular weight compounds. Hydrocarbons are subject to pyrolysis, e.g., formation of carbon black and hydrogen from methane at 1300C and decomposition of gaseous alkanes at 500–600C. The latter is the basis of thermal cracking (pyrolysis gasoline).

One application of pyrolysis is conversion of acetone into ketenes by decomposition at about 700C; the reaction is $CH_3COCH_3 \rightarrow H_2C{=}C{=}O + CH_4$. Pyrolysis of natural gas or methane at about 2000C and 100 mm Hg pressure produces a unique form of graphite. Synthetic crude oil can be made by pyrolysis of coal, followed by hydrogenation of the resulting tar. Large-scale pyrolysis of cellulosic wastes is being conducted for production of synthetic fuel oils and other products; the method is said to require only 30 minutes at about 537C (flash pyrolysis).
See also destructive distillation.

pyrolysis gasoline. See under gasoline.

pyromellitic acid. (PMA; 1,2,4,5-benzenetetracarboxylic acid). CAS: 89-05-4. $C_6H_2(COOH)_4$.
Properties: White powder, d 1.79, mp 257–265C, bp converts to dianhydride, bulk d 32 lb/cu ft. Absorbs moisture slowly if exposed to atmosphere; soluble in alcohol, slightly soluble in water.
Grade: 99% purity.
Hazard: Skin irritant, may be carcinogenic.
Use: Intermediate for polyesters and polyamides used in electrical and nonfogging plasticizers, lubricants, and waxes.

pyromellitic dianhydride. (PMDA). $C_6H_2(C_2O_3)_2$.

Properties: White powder, d 1.68, mp 286C, bp 397–400C, 305–310C (30 mm) bulk d 21 lb/cu ft, soluble in some organic solvents, hydrolyzes to the acid when exposed to moisture.
Derivation: Oxidation of durene, either in wet process by nitric acid or a dichromate or as direct air oxidation with catalyst.
Grade: 98+% purity.
Hazard: Skin irritant.
Use: Curing agent for epoxy resins used in high temperature laminates, molds, and coatings; crosslinking agent for epoxy plasticizers in vinyls, alkyd resins; intermediate for pyromellitic acid.

pyrometer. An instrument for measuring temperatures as high as 1800C or higher, for example, molten steels, hot springs, volcanoes, etc., there are three kinds: (1) thermocouples of the graphite to silicon carbide type; (2) optical, in which the indications depend on the brightness at some one wavelength, of the hot body whose temperature is being measured; and (3) radiation, in which the indications depend on the radiance of a source of radiant energy.
See also thermocouple.

pyrometric cone. (Seger cone). A small pyramid composed of mixture of oxides which melt at known temperatures.
Use: To measure temperatures in autoclaves, curing ovens, etc., in the 100–3700F range.

pyromucamide. See furoamide.

pyromucic acid. See furoic acid.

pyromucic aldehyde. See furfural.

pyrophoric material. Any liquid or solid that will ignite spontaneously in air at about 130F (54.4C). Titanium dichloride and phosphorus are examples of pyrophoric solids, tributylaluminum and related compounds are pyrophoric liquids. Sodium, butyllithium, and lithium hydride are spontaneously flammable in moist air, as they react exothermically with water. Such materials must be stored in an atmosphere of inert gas or under kerosene. Some alloys (barium, misch metal) are called pyrophoric because they spark when slight friction is applied.
Hazard: Dangerous fire risk near combustible materials.
Use: As tips on pocket lighters and similar devices.

pyrophosphoric acid. (diphosphoric acid). CAS: 2466-09-3. $H_4P_2O_7$.
Properties: A viscous, syrupy liquid which tends to solidify on long standing at room temperature. When diluted with water it is rapidly converted into orthophosphoric acid. Soluble in water, mp 54C.
Derivation: By heating phosphoric acid at 250–260C. Further heating produces metaphosphoric acid.

Grade: Technical.
Use: Catalyst, manufacture of organic phosphate esters, metal treatment, stabilizer for organic peroxides.
See also polyphosphoric acid.

pyrophyllite. (agalmatolite). $Al_2Si_4O_{10}(OH)$.
A natural hydrous aluminum silicate, found in metamorphic rocks.
Properties: Color white, green, gray, brown; luster pearly to greasy, good micaceous cleavage, d 2.8–2.9, Mohs hardness 1–2, similar to talc.
Occurrence: North Carolina, California, Newfoundland, Japan.
Use: Ceramics, insecticides, slate pencils, substitute for talc, extender in paints, wallboard, buffer in high-pressure equipment, permissible animal feed additive.

pyroracemic acid. See pyruvic acid.

pyrosulfuric acid. See sulfuric acid, fuming.

pyrosulfuryl chloride. (disulfuryl chloride).
CAS: 7791-27-7. $S_2O_5Cl_2$.
Properties: Colorless, mobile, very refractive, fuming liquid, d 1.819, bp 146C, fp −38C, refr index 1.449 (19C).
Hazard: Decomposes violently with water to sulfuric acid and hydrochloric acid. Corrosive to tissue.
Use: Organic synthesis.

pyrotartaric acid. (methylsuccinic acid).
$HOOCCH(CH_3)CH_2COOH$.
Properties: White or yellowish crystals; soluble in water, alcohol, and ether; d 1.4105; mp 111–112C.
Derivation: By distilling tartaric acid.
Use: Organic synthesis.

pyrotartaric acid, normal. See glutaric acid.

pyrotechnics. The formulation and manufacture of fireworks, signal flares and military warning devices. The industry is professionally represented by the Pyrotechnic Guild International, Inc. The primary ingredients of pyrotechnic products are as follows: (1) Oxidizers: potassium nitrate, potassium chlorate, or potassium perchlorate; ammonium perchlorate; barium chlorate and nitrate; and strontium nitrate. (2) Fuels: aluminum, magnesium, antimony sulfate, dextrin, sulfur, and titanium. (3) Binders: dextrin and various polymers. Colored flames are produced by strontium compounds (red); barium compounds (green); copper carbonate, sulfate, and oxide (blue); sodium oxalate and cryolite (yellow); and magnesium, titanium, or aluminum (white). Black powder is used as propellant.
See also chemiluminescence.

pyrovanadic acid. See vanadic acid.

"Pyrovatex" CP.[443] TM for a fiber-reactive phosphone alkyl amide.
Properties: Odorless, nontoxic, nonirritating, stable to laundering and dry-cleaning. Reduces tear and tensile strengths, no effect on "hand." Compatible with other finishing and proofing agents and precured durable press processes. From 25–35% by weight is required; imparts self-extinguishing properties to cotton.
Use: Flame retardance of 100% cotton fabrics, such as tenting, military uniforms, draperies, etc.

pyroxylin. See nitrocellulose.

pyrrhotite. (magnetic pyrites; pyrrhotine).
FeS. A natural iron sulfide. Frequently has a deficiency in iron. May contain small amounts of nickel, cobalt, manganese and copper.
Properties: Color brownish bronze, streak black, luster metallic, slightly magnetic, hardness 4, d 4.6.
Occurrence: Tennessee, Pennsylvania, Europe, Canada.
Use: Ore of iron, manufacture of sulfuric acid.

pyrrobutamine phosphate. (1-[4(p-chlorophenyl)-3-phenyl-2-butenyl]pyrrolidine diphosphate).
$ClC_6H_4CH_2C(C_6H_5):CHCH_2NC_4H_8 \cdot 2H_3PO_4$.
Properties: Cream to off-white powder with a slight odor and a bitter taste, soluble in water, slightly soluble in alcohol, almost insoluble in chloroform and ether, melting range 127–131C.
Grade: NF.
Use: Medicine (antihistamine).

pyrrole. CAS: 109-97-7. Pyrrole is regarded as a resonance hybrid and no one structure adequately represents it. An approximation is:

$$
\begin{array}{ccc}
HC & \!\!-\!\!\!-\!\! & CH \\
\| & & | \\
HC & & CH \\
 & \underset{H}{\overset{+}{N}} &
\end{array}
$$

Properties: Yellowish or brown oil; burning, pungent taste; odor similar to chloroform; readily polymerizes by the action of light and turns brown. Soluble in alcohol, ether and dilute acids; insoluble in water and dilute alkalies. Bp 130–131C, fp −24C, d 0.968 (20/4C), refr index 1.5091 (20C), flash p 102F (38.9C) (TCC), soluble in most organic chemicals. Combustible.
Derivation: Fractional distillation of bone-oil with sulfuric acid.
Method of purification: Conversion into the potassium compound (C_4H_4NK) washing with ether,

and treatment with water, followed by drying and distillation.

Grade: Technical.

Hazard: Moderate fire risk. Toxic by ingestion and inhalation.

Use: Manufacture of pharmaceuticals.

pyrrolidine. CAS: 123-75-1. C_4H_9N.

Properties: Colorless to pale yellow liquid, penetrating amine-like odor, d 0.8660 (20/20C), fp −60C, bp 87C, refr index 1.4425 (20C), flash p 37F (2.7C) (TCC), soluble in water and alcohol.

Grade: 95% min purity.

Use: Intermediate for pharmaceuticals, fungicides, insecticides, rubber accelerators, citrus decay control, curing agent for epoxy resins, inhibitor.

Hazard: Flammable, dangerous fire risk. Toxic by ingestion and inhalation.

2-pyrrolidinecarboxylic acid. See proline.

2-pyrrolidone. (2-pyrrolidinone; butyrolactam).

CAS: 616-45-5. $\overline{CH_2CH_2CH_2C(O)}NH$.

Properties: Light yellow liquid, d 1.1, bp 245C. Soluble in water, ethanol, ethyl ether, chloroform, benzene, ethyl acetate, carbon disulfide. High-boiling, noncorrosive, flash p 265F (129C). Combustible.

See polyvinylpyrrolidone.

Derivation: From acetylene and formaldehyde by high-pressure synthesis.

Use: Plasticizer and coalescing agent for acrylic latexes in floor polishes, solvent for polymers, insecticides, polyhydroxylic alcohols, sugar, iodine, specialty inks, monomer for nylon-4.

pyrrone. An aromatic heterocyclic polymer derived from a cyclic dianhydride and an aromatic o-tetramine or a derivative. Stable to 482C.

Use: Films, coatings, adhesives, laminates, and moldings.

pyruvaldehyde. See pyruvic aldehyde.

pyruvic acid. (α-ketopropionic acid; acetylformic acid; pyroracemic acid).   CAS: 127-17-3. $CH_3COCOOH$.   A fundamental intermediate in protein and carbohydrate metabolism in the cell.

Properties: Liquid with odor resembling acetic acid; d 1.2272 (20/4C); mp 13.6C; bp 165C; miscible with water, ether, and alcohol.

Derivation: Dehydration of tartaric acid by distilling with potassium acid sulfate.

Use: Biochemical research.

pyruvic alcohol. See hydroxy-2-propanone.

pyruvic aldehyde. (pyruvaldehyde; methyl glyoxal). CH_3COHCO.

Properties: Supplied commercially as a 30% aqueous solution, d 1.20 (20/20C), bulk d 10 lb/gal (20C).

Use: Organic synthesis, as of complex chemical compounds such as pyrethrins, tanning leather, flavoring.

Q

"Qiana"[28]. TM for a nylon-type synthetic fiber with properties similar to silk.

Q-lure. See cue-lure.

"QO"[224]. TM for a series of furan chemicals and derivatives.

"Q-Tac"[30]. TM for a brand of starch used as a fiber-bonding agent and retention aid for pigments and filler material in papermaking.

quad. An energy unit that has come into use in recent years in predicting future energy requirements on a national basis. One quad equals 10^{15} Btu, which is the energy equivalent of 10^{12} cubic feet of natural gas, or 182 million barrels of oil, or 42 million tons of coal, or 293 billion kilowatt-hours of electricity.

quadrupole resonance. See nuclear quadrupole resonance.

qualitative analysis. See analytical chemistry.

quantitative analysis. See analytical chemistry.

quantum number. The quantum is the basic unit of electromagnetic energy, it characterizes the wave properties of electrons, as distinct from their particulate properties. The quantum theory developed by Max Planck states that the energy associated with any quantum is proportional to the frequency of the radiation, that is e (energy) $= hv$, where v is the frequency and h is a universal constant.

An electron has four quantum numbers which define its properties. These are as follows: (1) The principal quantum number is a constant that can be any positive integer ($n = 1, 2, 3 . .$). It determines the principal energy level, or shell, of the electron, sometimes designated by letters such as K, L, or M, depending on the value of the principal quantum number. (2) The angular momentum constant l, also an integer, is related to n as: $l = 0, 1, . ., n-1$. *Here again, letter designations are often used: in s* electrons $l = 0$, *in p* electrons $l = 1$, *in d* electrons $l = 2$, and *in f* electrons $l = 3$. (3) The magnetic quantum number m is an integer related to l as: $m = -1, . ., -1, 0, +1, . ., +l$. (4) The spin quantum number is independent of the other three and has a value of either $+1/2$ or $-1/2$, depending on the direction of rotation of the electron on its axis in the atomic frame of reference.

See also orbital theory, electron, photon, radiation, Pauli exclusion principle.

quartz. CAS: 14808-60-7. SiO_2.
Crystallized silicon dioxide (silica).
Properties: Color white to reddish, luster vitreous, Mohs hardness 7, d 2.65, insoluble in acids except hydrogen fluoride, only slightly attacked by solutions of caustic alkali, piezoelectric and pyroelectric, mp 1713C. Noncombustible.
Derivation: Synthetic crystals of good size and purity are grown by mass production methods under very carefully regulated conditions of temperature and concentration.
Hazard: Avoid inhalation of fine particles. TLV: (for respirable dust) 10 mg/m³/% respirable quartz + 2.
Use: Electronic components, piezoelectric control in filters, oscillators, frequency standards, wave filters, radio and TV components; barrel-finishing abrasive.
See also silica.

quartz, fused. Pure silica that has been melted to yield a glass-like material on cooling.
Use: For apparatus and equipment (such as vacuum tubes) where its high melting point, ability to withstand large and rapid temperature changes, chemical inertness and transparency (including UV light), and electrical resistance are valuable. Produced as fibers and fabrics for heat resistance, low expansion coefficient, and insulating value.
See also glass.

quassia. (bitter ash; bitterwood).
Derivation: The wood or bark of Picrasma excelsa or Quassia amara, very bitter taste, white to bright yellow chips or shavings.
Use: Decoction or tincture as a fly poison, surrogate for hops, medicine (anthelminthic), hair lotion, flavoring, alcohol denaturant.

quaternary ammonium salt. A type of organic nitrogen compound in which the molecular structure includes a central nitrogen atom joined to four organic groups (the cation) and a negatively charged acid radical (the anion). The structure is indicated as:

$$\left[\begin{array}{c} R \\ | \\ R-N-R \\ | \\ R \end{array} \right]^{+} Z^{-}$$

Octadecyldimethylbenzyl ammonium chloride and hexamethonium chloride are examples. Pentavalent nitrogen ring compounds, such as lauryl pyridinium chloride, are also considered quaternary ammonium compounds. They are all cationic surface-active coordination compounds and tend to be adsorbed on surfaces.

Use: Disinfectant, cleanser and sterilizer, cosmetics (deodorants, dandruff removers, emulsion stabilizers), fungicides, mildew control, to increase affinity of dyes for film in photography, coating of pigment particles to improve dispersibility, to increase adhesion of road dressings and paints, antistatic additive, biocide.

See also detergent, synthetic, coordination compound.

p-quaterphenyl. $C_6H_5C_6H_4C_6H_4C_6H_5$.
Properties: Crystals, mp 316–318C, bp 428C (18 mm).
Grade: Purified.
Use: As primary fluor or as wavelength shifter in soluble scintillators.

quebrachine.   See yohimbine.

quebracho.
Properties: A wood-derived tannin, the most important tanning agent used in the American leather industry. Combustible.
Derivation: From *Aspidosperma quebracho* and *Quebracho lorentzi*, imported as logs from Argentina.
Grade: Liquid: 35–37% tannin. Solid: 65% tannin.
Use: Vegetable tanning, retanning of chrome-tanned upper leathers, dyeing, ore flotation, oil well drilling fluids, flavoring.

Quelet reaction.   Passage of dry hydrogen chloride through a solution in ligroin of a phenolic ether and an aliphatic aldehyde in the presence or absence of a dehydration catalyst to yield α-chloroalkyl derivatives by substitution in the para position to the ether group or in the ortho position in para-substituted phenolic ethers.

quench.   In the terminology of metallurgy, quick cooling of metals or alloys by immersion in cold water or oil. This is an essential part of the tempering process, especially for steels. If the metal or alloy is in the liquid (molten) state and the quench time is extremely short (less than a second), the product will have an amorphous or glass-like structure, since no crystallization occurs.
See also glass, metallic.

quercetin.   CAS: 117-39-5. $C_{15}H_{10}O_7$.
Properties: Yellow needles (dihydrate), anhydrous form decomposes at 315C, soluble in alcohol and glacial acetic acid, insoluble in water.
Derivation: Bark of fir trees, also synthetically.
Use: Medicine, reported formation of epoxy resins on mixing with epichlorohydrin.

"Questex"[1].   TM for ethylenediaminetetraacetic acid (EDTA) and derivatives, a group of polyamino-acid-based organic sequestering agents which complex or chelate multivalent, metallic cations (such as calcium, magnesium, copper and iron) into stable, coordinated anionic complexes.

quick-.   Prefix meaning alive or active, as in quicksilver (mercury), quicklime (unslaked lime), quicksand, quick (fingernail).

quicklime.   See calcium oxide.

quicksilver.   See mercury.

"Quilon"[28].   TM for a Werner-type chromium complex in isopropanol, 30% solution of stearatochromic chloride.
Use: Water repellent and sizing treatment of cellulosic materials, treatment of negatively charged surfaces, antiblocking agent, for insolubilizing various water-soluble or swellable coatings, improving grease-resistant coatings, treatment of feathers.

quinacridone.   A light-fast pigment used in paints, printing inks, plastics, etc.

quinacrine dihydrochloride.   (3-chloro-7-methoxy-9-(1-methyl-4-diethylamino-butylamino) acridine dihydrochloride).   CAS: 69-05-6. $C_{23}H_{30}ClN_3O \cdot 2HCl \cdot 2HOH$.
Properties: Bright yellow, crystalline powder; odorless and with a bitter taste; decomposes at 248–250C; soluble in hot water; pH of 1% water solution 4.5.
Derivation: Organic synthesis.
Grade: USP.
Use: Medicine (antimalarial and antihelminthic).

quinaldine.   (chinaldine; α-methylquinoline).   CAS: 91-63-4. $C_9H_6NCH_3$.
Properties: Colorless, oily liquid; odor of quinoline; darkens to reddish brown in air. Soluble

in alcohol, ether, and chloroform; insoluble in water. Bp 246–247C, mp −2C, d 1.51.

Derivation: (a) By the treatment of aniline and paraldehyde with hydrochloric acid and heat. (b) From coal tar.

Hazard: Strong irritant to mucous membranes.

Use: Manufacture of dyes, pharmaceuticals, fine organic chemicals, acid-base indicators.

quinaphthol. (quinine-β-naphthol α-sulfonate). $C_{20}H_{24}N_2O_2 \cdot (OHC_{10}H_6 \cdot SO_3H)_2$.

Properties: Yellow crystalline powder, contains 42% quinine, bitter taste, moderately soluble in hot water or alcohol, insoluble in cold water, mp 185–186C.

Derivation: Interaction of quinine and β-naphtholsulfonic acid.

Use: Medicine.

"Quindex"[74]. TM for solubilized form of copper-8-quinolinolate. Contains 1.8% copper.

Use: When nonmercurial fungicide is required.

quinhydrone. CAS: 106-34-3. $C_6H_4O_2C_6H_4(OH)_2$.

Properties: Dark green crystals; slightly soluble in water; soluble in alcohol, ether, hot water, ammonia; mp 171C; d 1.40.

Derivation: Oxidation of hydroquinone with sodium dichromate.

Grade: Reagent, technical.

Hazard: Toxic by ingestion.

Use: Electrode for pH determination.

quinine. CAS: 130-95-0. $C_{20}H_{24}N_2O_2 \cdot 3HOH$. An alkaloid.

Properties: Bulky, white, amorphous powder or crystalline alkaloid; very bitter taste; odorless and levorotatory; soluble in alcohol, ether, chloroform, carbon disulfide, glycerol, alkalies, and acids (with formation of salts); very slightly soluble in water.

Derivation: Finely ground cinchona bark mixed with lime is extracted with hot, high-boiling paraffin oil. The solution is filtered, shaken with dilute sulfuric acid, and latter neutralized hot with sodium carbonate; on cooling, quinine sulfate crystallizes out. The sulfate is then treated with ammonia, the alkaloid being obtained.

Source: Indonesia, Bolivia.

Hazard: Skin irritant, ingestion of pure substance adversely affects eyes.

Use: Medicine (antimalarial) as the alkaloid or as numerous salts and derivatives; flavoring in carbonated beverages.

β-quinine. See quinidine.

quinine acid sulfate. See quinine.

quinine bisulfate. See quinine.

quinizarin. (1,4-dihydroxyanthraquinone; CI 58050). $C_{14}H_6O_2(OH)_2$.

Properties: Red or yellow-red crystals; mp 194–195C; soluble in hot water, alcohol, ether, and benzene, and in potassium chloride and sulfuric acid.

Use: Antioxidant in synthetic lubricants, dyes.

quinol. See hydroquinone.

quinoline. (chinoline). CAS: 91-22-5. C_9H_7N.

Properties: A basic heterocyclic nitrogenous compound occurring in coal tar and obtained from it, but more frequently by synthesis; highly refractive, colorless liquid, darkens with age, hygroscopic, peculiar odor. Soluble in water, alcohol, ether, and carbon disulfide; d 1.0899; fp −15C; bp 238C; autoign temperature 896F (480C). Combustible.

Derivation: By treatment of aniline and nitrobenzene with glycerol and sulfuric acid and heat.

Use: Medicine (antimalarial), preserving anatomical specimens, manufacture of quinolinol sulfate, niacin and copper-8-quinolinolate, flavoring.

quinoline dye. See cyanine dye.

8-quinolinol. See 8-hydroxyquinoline.

quinomethionate. $C_{10}H_6N_2OS_2$.

Properties: Tan to yellow crystals; mp 170C; soluble in benzene, toluene, and dioxane; insoluble in water.

Use: Fungicide, acaricide.

quinone. (1,4-benzoquinone; chinone). CAS: 106-51-4. $C_6H_4O_2$.

Properties: Yellow crystals; irritant odor; soluble in alcohol, ether, and alkalies; slightly soluble in hot water; d 1.307; mp 115.7; bp sublimes; volatile with steam, being in part decomposed. Combustible.

Derivation: By oxidation of aniline with chromic acid, extraction with ether and distillation.

Hazard: Toxic by inhalation, strong irritant to skin and mucous membranes. TLV: 0.1 ppm in air.

Use: Manufacture of dyes and hydroquinone, fungicides, analytical reagent, photography, oxidizing agent.

p-quinonedioxime. See "GMF."

quinone oxime dye. See nitroso dye.

quinsol. $C_9H_6NOSO_3K \cdot HOH$.   Yellow crystalline powder. Combustible.
Use: Fungicide.

quinoxaline. (1,4-benzodiazine; benzo-p-diazine). $C_8H_6N_2$ (bicyclic). An organic base.

Properties: Colorless, crystalline powder; mp 30C; bp 229C; soluble in water and organic solvents.
Use: Organic synthesis.

quinoxalinedithiol cyclic trithiocarbonate. See thioquinox.

quintozene. See pentachloronitrobenzene.

"Quotane"[71]. TM for an ointment and lotion containing dimethisoquin hydrochloride.

"QUSO"[201]. TM for a series of microfine precipitated amorphous silicas having a fully hydroxylated surface.
Use: Thickening, flatting, anti-caking and reinforcing agent, adsorptive carrier.

R

R. (1) Symbol used to represent an organic group in a chemical formula, for example CH_3, C_2H_5, C_6H_5, etc. (2) A free radical (with superior dot, $R^{\bullet}$). (3) The gas constant, equal to $p_0v_0/273C$. (4) Abbreviation of Rankine temperature scale.

R&D. Abbreviation for Research and Development, usually referring to the department or division of a company whose major responsibility is applied research and creative development of new products and processes.

Ra. Symbol for radium.

R-acid. (2-naphthol-3,6-disulfonic acid; β-naphtholdisulfonic acid).

Properties: Deliquescent, colorless needles; soluble in water, alcohol, and ether.
Derivation: Sulfonation of β-naphthol. For details see Schaeffer acid.
Use: Azo dye intermediate. The disodium salt is used as a reagent in detection of nitrogen dioxide in the air.

2R acid. See RR acid.

racemization. Conversion, by heat or by chemical reaction (e.g., enolization) of an optically active compound into an optically inactive form, in which half of the optically active substance becomes its mirror-image (enantiomer). This change results in a mixture of equal quantities of dextro- and levorotatory isomers, as a result of which the compound does not rotate plane-polarized light to either right or left since the two opposite rotations cancel each other. This is sometimes referred to as external compensation, as opposed to the internal compensation exhibited by meso-compounds.
See also meso-(1), tartaric acid.

racephedrine. (racemic ephedrine; dl-ephedrine). $C_{10}H_{15}NO$.
Properties: Crystals; mp 79C; soluble in water, alcohol, chloroform, and oils.
Derivation: Synthetic.
Use: Medicine (also as hydrochloride and sulfate). See ephedrine for optically active form.

racking. Experimental cold-stretching of unvulcanized rubber, whose behavior under stress is unique among natural materials. A thin, narrow strip stretched, e.g., 500–600% at 0C will retain that extension indefinitely after release of stress as long as the low temperature persists. In this state it loses its elasticity and has virtually 100% permanent set. It also displays a crystalline x-ray pattern similar to that of a fiber, in contrast to the amorphous structure of the unstretched state. On exposure to room temperature it slowly retracts to its original length; higher temperature increases its rate of recovery. Tests made on racked rubber have shown that crude rubber can be exposed to any degree of low temperature for any length of time without impairment of its properties.

rad. That quantity of ionizing radiation that results in the absorption of 100 ergs of energy per gram of irradiated material, regardless of the source of the radiation. The Federal radiation safety standard is 0.17 rad per person per year, and even this is considered too high by some authorities. See also rem.

"Radel."[214] TM for a water-soluble polyethylene oxide film.
Use: Packaging powdered detergents, dyes, insecticides, fungicides and other household, industrial, and agricultural products that are dissolved in water prior to use.

radiation. Energy in the form of electromagnetic waves (also called radiant energy, or light). It is emitted from matter in the form of photons (quanta), each having an associated electromagnetic wave having frequency (ν) and wavelength (λ). The various forms of radiant energy are characterized by their wavelength, and together they comprise the electromagnetic spectrum, the components of which are as follows: (1) cosmic rays (highest energy, shortest wavelength); (2) gamma rays from radioactive disintegration of atomic nuclei; (3) x-rays; (4) UV rays; (5) visible light rays; (6) infrared; (7) microwave; and (8) radio (Hertzian) and electric rays. All these are identical in every way except wavelength, those having the shortest wavelength being the most penetrating. They are not electrically charged and have no mass, their velocity of propagation is the same, all display the properties characteristic of light and have a dual nature (wave-like and corpuscular). In a looser sense the term "radiation" also includes energy emitted in the form of particles which possess mass and may or may not be electrically charged, i.e., alpha (positive) and beta (negative), also neutrons. Beams of such particles may be considered as "rays." The charged particles may all be accelerated and high energy imparted to "beams" in particle accelerators such as cyclotrons, betatrons, synchrotrons, and linear accelerators.

Radiation is used in medicine in the form of x-rays and radioactive isotopes; it is used in industry in many ways, e.g., as vitamin activator, sterilizing agent, and polymerization initiator; it is also the basis of all types of spectroscopic analysis.
See also following entries.

radiation biochemistry. The study of substances having the ability to protect cells and body tissue against the deleterious effects of ionizing radiation. As one of these effects is to deprive proteins of sulfhydryl (-SH) groups necessary for cell division, the injection of compounds rich in this radical (notably cysteine) has been successfully tried with laboratory animals. Thiourea has been found to protect DNA from depolymerization by x-rays; enzymes containing -SH groups inactivated by radiation are reactivated by addition of glutathione. Some of the other radiochemically induced reactions that adversely affect biochemical activity are (1) formation of hydrogen peroxide (a biological poison) by free radical mechanism, (2) denaturation of proteins, (3) change in substituent groups of amino acids, (4) oxidation of hemoglobin, (5) depolymerization of DNA.
See also radiation, ionizing.

radiation curing. See radiation, industrial (6).

radiation, industrial. Chemical or physiochemical changes induced by exposure to various types and intensities of radiation. (1) Synthesis of ethyl bromide from hydrogen bromide and ethylene, using alpha radiation from cobalt-60. (2) Crosslinking of such polymers as polystyrene and polyethylene with either beta or gamma radiation. (3) Vulcanization of rubber with ionizing radiation. (4) Polymerization of methyl methacrylate monomer with cobalt-60 as source of gamma rays. Free radical formation is involved in both crosslinking and polymerization reactions. This technique is also being applied in the textile finishing field for grafting and crosslinking fibers with chemical agents for durable-press fabrics. (5) Processing of various foods (cooking, drying, pasteurizing, etc.) by electromagnetic energy in the microwave region of the spectrum; preservation and sterilization of food products

by ionizing radiation (gamma and x-rays). The dosage of radiation is strictly controlled, and FDA approval is required. Irradiation is also effective in inhibiting sprouting and preventing insect infestation of stored grains. (6) Curing or hardening of organic protective coatings (paints, inks) by exposure to infrared, UV or electron-beam radiation. Required are a monomer or oligomer and a photoninitiator, which induces polymerization by free radical formation.
See also "Electrocure."

radiation, ionizing. Extremely short-wavelength, highly energetic, penetrating rays of the following types: (a) gamma rays emitted by radioactive elements and isotopes (decay of atomic nuclei), (b) x-rays generated by sudden stoppage of fast-moving electrons, (c) subatomic charged particles (electrons, protons, deuterons) when accelerated in a cyclotron or betatron. The term is restricted to electromagnetic radiation at least as energetic as x-rays, and to charged particles of similar energies. Neutrons also may induce ionization.

Such radiation is strong enough to remove electrons from any atoms in its path, leading to the formation of free radicals. These short-lived but highly reactive particles initiate decomposition of many organic compounds. Thus ionizing radiation can cause mutations in DNA and in cell nuclei; adversely affect protein and amino acid mechanisms; impair or destroy body tissue; and attack bone marrow, the source of red blood cells. Exposure to ionizing radiation for even a short period is highly dangerous, and for an extended period may be lethal. The study of the chemical effects of such radiation is called radiation chemistry or (in the case of body reactions) radiation biochemistry.
See radiation, industrial for applications.

radical. (1) An ionic group having one or more charges, either positive or negative, e.g., OH^-, NH_4^+, SO_4^{--}.
(2) See free radical. See also group (2).

radioactive isotope. See radioisotope.

radioactive waste. Disposal of waste containing radioisotopes and of spent nuclear reactor fuel presents a serious problem for which there is as yet no completely satisfactory solution. Such wastes may remain radioactive for thousands of years and can constitute a long-term hazard which is restraining the development of nuclear-generated electric power. Safe disposal techniques are being intensively studied. Ocean dumping, practiced some years ago, is no longer permissible. Small amounts of low-level wastes containing radioisotopes can be diluted

with an inert material sufficiently to reduce its activity to an acceptable point. High-level reactor wastes, for example, at Hanford, are stored in concrete tanks lined with steel and buried under a foot of concrete and 5 or 6 ft of soil. Containers of compressed alumina (corundum) have been recommended, as this material remains impervious to water indefinitely. Storage in the form of calcine (granular solid) and in borosilicate glass is a promising possibility under active investigation. The DOE has recommended disposal in deep geologic formations, although the heat generated by the radioactivity could cause fracturing of the surrounding rock structures; this would admit water which would eventually rise to the surface after being contaminated. A test program involving storage in basalt is being conducted by DOE. Storage in salt formations is under serious consideration because they are self-sealing and free from water.
See also waste control.

radioactivity. Natural or artificial nuclear transformation; discovered by Becquerel in 1895. The energy of the process is emitted in the form of alpha, beta, or gamma rays. Thus radium-226 undergoes radioactive decay by the emission of an alpha particle, and the new product is radon-222. The decay series terminates in lead-206. Radioactivity is not affected by the physical state or chemical combination of the element. The radioactivity of a nuclide is characterized by the nature of the radiations, their energy, and the half-life of the process,, i.e., the time required for the activity to decrease to one-half of the original. Half-lives vary from microseconds to millions of years. Some radioactive elements occur in nature (radium, uranium). Radioactivity can be caused artifically in many stable elements by irradiation with neutrons in a nuclear reactor, or by charged particles from an accelerator.

Amounts of radioactive material are usually expressed in units of activity, the rate of radioactive decay. The accepted unit is the curie (Ci) and its metric multiples and fractions, the mega, kilo, milli-, and microcurie. A curie is 3.73×10^{10} disintegrations per sec. A common unit is millicuries per millimole. Packaging and shipment of radioactive materials, which are highly toxic, must be in accord with official requirements. Consult IATA and DOT shipping regulations for labeling and other instructions.
See also rad, rem.

radiocarbon. See carbon-14, chemical dating.

radiochemistry. That subdivision of chemistry which deals with the properties and uses of radio-

active materials in industry, biology, and medicine, including tracer research and radioactive waste disposal.
See also nuclear chemistry.

radioisotope. (radionuclide). An isotopic form of an element (either natural or artificial) that exhibits radioactivity. Radioisotopes are used as diagnostic and therapeutic agents in medicine, in biological tracer studies, and for many industrial purposes, from measurement of thickness to initiating polymerization. Artifical radioisotopes are made by neutron bombardment of stable isotopes in a nuclear reactor.
See also tracer; isotope.

radium. CAS: 7440-14-4. Ra. Radioactive element of Group IIA of the Periodic Table, atomic number 88, aw 226.0254, valence = 2. There are 14 radioactive isotopes but only radium-226 with half-life of 1620 years is usable. Discovered by the Curies in 1898.
Properties: Brilliant white solid, mp 700C, bp 1140C, d 5, luminescent, turns black on exposure to air, soluble in water with evolution of hydrogen, forms water-soluble compounds. Decays by emission of alpha, beta, and gamma radiation. Bone-seeking when taken into the body.
Occurrence: Colorado, Canada, Zaire, France, USSR.
Derivation: Uranium ores (pitchblende and carnotite). The method used for isolating radium is similar to that of Mme. Curie and involves coprecipitation with barium and lead, chemical separation with hydrochloric acid, and further purification by repeated fractional crystallization. The metal is separated from its salts by electrolysis and subsequent distillation in hydrogen. Dry salts are stored in sealed glass tubes, opened regularly by experienced workers to relieve pressure. The tubes are kept in lead containers.
Hazard: Highly toxic, emits ionizing radiation. Lead shielding should be used in storage and handling, adequate protective clothing and remote control devices are essential. Destructive to living tissue.
Use: Medical treatment for malignant growths, industrial radiography, source of neutrons and radon.

radium bromide. CAS: 10031-23-9. $RaBr_2$.
Properties: White crystals, becoming yellow or pink, radioactive, d 5.79, mp 728C, sublimes at 900C, soluble in water and alcohol.
Derivation: Freed from the ores as a bromide mixed with barium bromide.
Method of purification: Fractional crystallization.
Impurities: Barium salts.
Grade: Technical, pure. The purity is determined

by the strength of the ionizing power of the salt, i.e., the extent to which it causes air to conduct electricity.

Hazard: As for radium.

Use: Medicine (cancer treatment), physical research.

radium carbonate. $RaCO_3$.

Properties: Amorphous radioactive powder; white when pure, but sometimes yellow, orange, or pink due to impurities; insoluble in water; marketed as a mixture with barium carbonate.

Hazard: As for radium.

radium chloride.   CAS: 10025-66-8. $RaCl_2$.

Properties: Yellowish-white crystals becoming yellow or pink on standing, radiactive, soluble in water and alcohol, mp 1000C, d 4.91.

Derivation: Freed from the ores as a chloride mixed with barium chloride.

Method of purification: Fractional crystallization.

Grade: Technical, pure. The purity of radium salts is determined by the strength of their ionizing power, i.e., the extent to which they cause air to conduct electricity.

Hazard: As for radium.

Use: Medicine (cancer treatment), physical research.

radium sulfate. $RaSO_4$.

Properties: White crystals when pure, but sometimes yellow, orange, or pink due to impurities. Radioactive, insoluble in acids and water.

Hazard: As for radium.

radon.   CAS: 10043-92-2.   Rn.   Gaseous radioactive element. Atomic number 86; noble gas group of Periodic Table; aw 222; valences = 2, 4, (6); 18 radioactive isotopes, all short-lived. The radon-222 isotope has a half-life of 3.8 days, emits alpha radiation.

Properties: Colorless gas, density 9.72 g/L (0C), soluble in water, can be condensed to a colorless, transparent liquid (bp −61.8C) and to an opaque, glowing solid. The heaviest gas known.

Derivation: Radioactive decay of radium. Radon is obtained by bubbling air through a radon salt solution and collecting the gas plus air.

Hazard: As for radium.

Use: Medicine (cancer treatment), tracer in leak detection, flow rate measurement, radiography, chemical research.

raffinate.   The portion of an oil that is not dissolved in solvent refining of lubricating oil.

raffinose.   CAS: 512-69-6. $C_{18}H_{32}O_{16} \cdot 5HOH$. A trisaccharide composed of one molecule each of d(+)galactose, d(+)glucose, and d(-)-fructose.

Properties: White, crystalline powder; sweet taste. Soluble in water, very slightly soluble in alcohol, d 1.465, mp (anhydrous) 118–119C, bp decomposes at about 130C, optical rotation +104.5 degrees. Split by invertase to melibiose and saccharose. Combustible.

Derivation: Hydrolysis of cottonseed meal, from sugar beet concentrates.

Use: Bacteriology, preparation of other saccharides.

rainfall, induced.   See nucleation.

Raman spectroscopy.   An analytical technique discovered in 1928 by C. V. Raman, an Indian physicist.

See spectroscopy, photometric analysis.

Ramberg-Backlund reaction.   Reaction of α-halo sulfones with strong bases to yield alkenes.

ramie.   A natural fiber obtained from the stems of *Boehmeria nivea,* of the hemp family. High wet strength, absorbent but dries quickly, can be spun or woven. Wears well and has great rot and mildew resistance, tensile strength four times that of flax, elasticity 50% greater than flax.

Sources: Taiwan, Egypt, Brazil, Florida.

Hazard: Combustible, not self-extinguishing.

Use: High-grade paper (Europe), fabrics (wearing apparel and car seat covers), stern-tube packing in ships, patching watermains (Great Britain).

"Ramrod."[58]   TM for selective preemergence herbicide available as a wettable powder (contains 65% 2-chlor-N-isopropylacetanilide) and granular form (contains 20% 2-chloro-N-isopropyl-acetanilide).

Ramsay, Sir William.   (1852-1916) A British chemist born in Scotland who received the Nobel prize for chemistry in 1904. He participated in the discovery of helium, argon, neon, xenon, and krypton. Much of his work concerned investigations of inert gases. He was also known for studies in organic, physical, and inorganic chemistry. Ramsay was educated at the Universities of Glasgow, Heidelberg, and Tubingen, and was a Professor at the Universities of Bristol and London.

"Randox."[58]   TM for a liquid herbicide containing 47.1% α-l-chloro-N,N-diallylacetamide.

Raney nickel.

Properties: Dark gray powder or crystals, pyrophoric.

Derivation: By leaching the aluminum from an

alloy of 50% aluminum-50% nickel with 25% caustic soda solution.
Hazard: Ignites spontaneous in air, store under alcohol or water, dangerous fire risk.
Use: Catalyst for hydrogenation.

Rankine. A scale of absolute temperature based on Fahrenheit degrees. Temperatures on the Rankine scale are 9/5 (or 1.8) times those on the Kelvin scale.
See also absolute temperature.

Raoult's law. The vapor pressure of a substance in equilibrium with a solution containing the substance is equal to the product of the mole fraction of the substance in the solution and the vapor pressure of the pure substance at the temperature of the solution. The law is not applicable to most solutions, but is often approximately applicable to a mixture of closely similar substances, particularly the substance present in high concentration.

rapeseed meal. The ground press-cake left from expression of rapeseed oil. Its major use is as an animal feed ingredient, but the presence of harmful glucosides, which may be goiter-inducing, limits its application for this purpose. Research efforts are being directed to removal of this constituent from meal produced in Canada. The meal also has some use as a fertilizer. If fed to animals, it should be blended with other feeds.

rapeseed oil. A vegetable oil derived from rape by expression or solvent extraction; it is now produced chiefly in Canada. It is a viscous, brownish liquid, though when refined it is yellow. D 0.913–0.916, solidifies at 0C, flash p 325F (162C), autoign temperature 836F (446C), subject to spontaneous heating. It is high in unsaturated acids, especially oleic, linoleic, and erucic.
Use: Edible oil for salad dressings, margarine, etc.; lubricant additive; substitute for soybean oil; soft soaps; blown oils.

"Rapidase."[173] TM for a bacterial amylase for removing starch sizings from textiles.

rare earth. One of a group of 15 chemically related elements in Group IIIB of the Periodic Table (lanthanide series). Their names and atomic numbers are:

lanthanum	La	57
cerium	Ce	58
praseodymium	Pr	59
neodymium	Nd	60
promethium	Pm	61
samarium	Sm	62
europium	Eu	63
gadolinium	Gd	64
terbium	Tb	65
dysprosium	Dy	66
holmium	Ho	67
erbium	Er	68
thulium	Tm	69
ytterbium	Yb	70
lutetium	Lu	71

The elements 57–62 are known as the cerium subgroup, and 63–71 as the yttrium subgroup. Yttrium, atomic number 39, although not a rare earth element, is found associated with the rare earths and is only separated with difficulty. Source: Monazite, bastnasite, and related fluocarbonate minerals as well as minerals of the yttrium group. These ores contain varying percentages of rare earth oxides, which are often loosely called rare earths. Rare earth elements also occur as fission products of uranium and plutonium, and is the only source of promethium. Occurrence: (monazite) India, Brazil, Florida, Carolinas, Australia, South Africa; (bastnasite) California.
See also didymium. See specific entry for properties and uses.

rare gas. Any of the six gases comprising the extreme right-hand group of the Periodic Table, namely helium, neon, argon, krypton, xenon, and radon. They are preferably called noble gases or (less accurately) inert gases. The first three have a valence of 0 and are truly inert, but the others can form compounds to a limited extent.

rare metal. A loose term for the less common metallic elements. They include the alkaline-earth metals (barium, calcium, and strontium), beryllium, bismuth, cadmium, cobalt, gallium, germanium, hafnium, indium, lithium, boron, silicon, manganese, molybdenum, rhenium, selenium, tantalum, niobium, tellurium, thallium, thorium, titanium, tungsten, uranium, vanadium, zirconium, and the rare earths.

"Rareox."[241] TM for optical quality cerium oxide for high speed polishing.

Raschig phenol process. Commercial process for the production of phenol by the hydrolysis of chlorobenzene, produced by the chlorination of benzene with hydrochloric acid and air.

Raschig rings. Short sections of metal tubes.
Use: Packing in distillation towers.
See tower, distillation.

rate of recovery. Tests made on racked rubber have shown that crude rubber can be exposed to any degree of low temperature for any length of time without impairment of its properties.

"Raticate."[507] TM for 5-(α-hydroxy-α-2-pyridyl-benzyl)-7-(α)-2-pyridylbenzylidene-5 -norbornene-2,3-dicarboximide. A rodenticide specific for rats, supposedly nontoxic to other animals or to man (US Dept. of Agriculture).

"Raudixin."[412] TM for Rauwolfia serpentina whole root.

"Rauserfia."[342] TM for an extract of weakly basic alkaloids from Rauwolfia.

rauwolfia. The powdered whole root of *Rauwolfia serpentina* found in India and Indonesia. The plant is of value as a source of alkaloids, especially reserpine.
Properties: Light tan to light brown, bitter, fine, amorphous powder; slight aromatic odor, sparingly soluble in alcohol and very slightly soluble in water.
Grade: NF.
Use: Medicine (antihypertensive agent).

rayon. Generic name for a semisynthetic fiber composed of regenerated cellulose as well as manufactured fibers composed of regenerated cellulose in which substituents have replaced not more than 15% of the hydrogen of the hydroxyl groups. Rayon was first made by denitration of cellulose nitrate fibers (Chardonnet process), but most rayon is made from wood pulp by the viscose process. "Regular" viscose tenacity = 2 g/denier; "high-tenacity" = 3–6 g/denier (tire cord). Elongation 15–30% (dry) and 20–40% (wet). Swells and weakens when wet. Moisture regain 11–13%, d 1.50.
Modified rayon is made principally of regenerated cellulose and contains nonregenerated cellulose fiber-forming material; for example, a fiber spun from viscose containing casein or other protein (ASTM). This greatly increases both dry and wet strength and also permits mercerization. Rayon is readily dyed by standard methods.
Hazard: Flammable, not self-extinguishing, moderate fire risk.
Use: Nonwoven fabrics, surgical dressings, mechanical rubber goods, coated fabrics, felts and blankets, blends with cotton for home furnishings, etc.
See also cellophane, acetate fiber, viscose process.

rayon coning oil. An oil used to lubricate and reduce the static of yarns wound by a coning machine. Usually composed of mineral oils of low viscosity so compounded as to emulsify in water.

raw material. (1) The basic material from which one or more useful products are derived, e.g., bauxite is the raw material for aluminum, wood for paper, rayon, etc., petroleum for fuels and chemicals. (2) As commonly used, the term refers to any ingredient or component of a mixture or product before mixing and processing take place, e.g., fillers, colorants, antioxidants, etc. See also storage (1).

"Rayox." TM for titanium dioxide.

Rb. Symbol for rubidium.

"RC."[52] TM for a series of plasticizers.
Use: Electrical tapes, metal laminates, surgical tubing, gaskets, non-marring film and sheeting, organosols, plastisols, plasticizer for nitrocellulose and synthetic rubbers, mild lubricant.

"R-2" Crystals[58] TM for a reaction product of carbon disulfide and 1,1'-methylenedipiperidine.
Use: Vulcanization accelerator for natural and synthetic rubbers, mostly in cements and latex.

RDA. Abbreviation for Recommended Dietary Allowances of food requirements, including proteins, vitamins and minerals for infants, children and adults, established by the Food and Nutrition Board of the National Academy of Sciences-National Research Council. They are revised periodically, particularly in reference to certain vitamins (C, B_{12}, and E) and proteins.
See also USRDA.

RDGE. See resorcinol diglycidyl ether.

RDX. See cyclonite.

Re. Symbol for rhenium.

reaction, chemical. See chemical reaction.

reaction injection molding. See injection molding.

reaction, nuclear. See nuclear reaction.

reagent. Any substance used in a reaction for the purpose of detecting, measuring, examining, or analyzing other substances. High purity and high sensitivity are essential requirements of lab reagents. Over 8000 reagent chemicals are commercially available.
See also grade.

rearrangement. A type of chemical reaction in which the atoms of a single compound recom-

bine, usually under the influence of a catalyst, to form a new compound having the same molecular weight but different properties. Thus ammonium cyanate in solution will rearrange to form urea, in which the four hydrogen atoms are equally distributed between the two nitrogen atoms: $NH_4OCN \rightarrow (NH_2)_2C\!\!=\!\!O$. Many such rearrangements have been named for their discoverers, e.g., Beckmann rearrangement.
See also Wohler.

Reaumur. A temperature scale in which 0 is the freezing point of water and 80 is its boiling point.

recipe. A product formula.
Use: Food industry.

reclaiming. Recovery and reuse of scrap materials, either in low percentage in new product manufacture or in larger proportions in products in which the highest quality is not essential. Among the materials widely reclaimed in industry are aluminum, steel, paper, rubber, glass, crankcase oil, greases, etc. Solid materials are comminuted, the contaminants being removed with organic solvents or strong alkali solutions (paint from metals, ink from paper, fabric and metals from tires). In the case of crosslinked elastomers, more intensive solvent and heat treatment is necessary. The resulting product is used as an adulterant in low-quality items. Research on high-temperature conversion of scrap rubber to oil, with recovery of carbon black and metal inserts, indicates that substantial values may be obtained by this method. Devulcanization by means of microwave radiation is also under development. The term "reprocessing" refers specifically to the recovery of nuclear fuels from reactor waste.
See also recycling, reprocessing.

recombinant DNA. See genetic engineering, biotechnology.

reconstitution. In food technology, restoration of a dehydrated food product to its original edible condition by adding water to it at the time of use. It is also called rehydration.
See also dehydration.

rectification. The enrichment or purification of the vapor during the distillation process by contact and interaction with a countercurrent stream of liquid condensed from the vapor.
See also reflux.

recycling. The practice of returning a portion of the reaction products to the start of the system, either for the purpose of more efficient conversion of unreacted components or to reuse auxiliary materials that remain unchanged during processing. In the petroleum refining industry, some of the product stream may be recycled and blended with the fresh input materials to obtain a product of maximum value. In some industries, processing wastes are recycled.
See also reclaiming, reprocessing.

red acetate. See aluminum diacetate.

red arsenic. See arsenic disulfide.

"Redax."[69] TM for n-nitrosodiphenylamine.
Use: Pre-vulcanization retarder for natural and synthetic rubber compounds.

red brass. Copper-zinc (brass) alloys characterized by their red color and high copper content. The term is used for several different types of brass. One ASTM classification permits 2–8% zinc, tin less than zinc, and lead less than 0.5%. Other alloys referred to as red brass include those with 75–85% copper, up to 20% zinc and usually very small amounts of lead and tin. In one alloy possessing good machining qualities, the lead content may be as high as 10%, the tin as high as 5%.
Red brasses are widely used for decorative purposes and in plumbing and piping because of their resistance to atmospheric corrosive and dezincification.

Red Dye No. 2. See amaranth, FD&C colors.

red glass. A soda-zinc glass containing a small amount of cadmium and 1% selenium. Red may also be obtained by using cuprous oxide or gold chloride, the latter usually in form of purple of Cassius.

"Redicote."[15] TM for a series of compounded amines.
Use: Cationic emulsifiers for asphalt.

red iron oxide. See iron oxide, red.

red iron trioxide. See ferric oxide.

red lake C. One of a family of organic acid azo pigments prepared by coupling the diazonium salt of o-chloro-m-toluidine-p-sulfonic acid with β-naphthol. Both sodium and barium salts of the parent dye are used.
Use: Plastics, rubber products, printing inks.

red lead. See lead oxide, red.

red mud. A byproduct sludge from aluminum ore processing; it contains 30–60% iron oxide. It may be used for steel making.

red ocher.   See ocher.

red oil.   A commercial grade of oleic acid comprised of 70% oleic and 15% each of linoleic and stearic acids.

redox.   Short form of term oxidation-reduction, as in redox reactions, redox conditions, etc. See also oxidation.

red oxide.   See iron oxide red.

red phosphorus.   See phosphorus.

red tide.   Yellow-to-reddish discoloration of seawater due to rapid multiplication of various species of plant-like microorganisms, called dinoflagellates, which occurs seasonally in areas of warm water, especially off the coast of Florida and occasionally as far north as New England. Some, though by no means all, of the species are poisonous. Concentration of these organisms (phytoplankton) may be as high as 10^8 units/L. They are lethal to free-swimming fish. Shellfish are unharmed by them, but are able to store and concentrate the toxin which causes paralytic poisoning when they are eaten by man. So potent is this poison that death may result from ingestion of milligram amounts.

reduced states.   See corresponding states.

"Reducolite."[159]   TM for a sodium zinc sulfoxylate formaldehyde compound used in stripping dyed colors from wool and synthetics.

reduction.   (1) The opposite of oxidation. Reduction may occur in the following ways: (1) acceptance of one or more electrons by an atom or ion; (2) removal of oxygen from a compound; (3) addition of hydrogen to a compound. See oxidation. (2) See comminution.

Reed reaction.   Photochemical sulfonation of paraffins and cycloparaffins by sulfur dioxide and chlorine under irradiation with UV light.

refinery gas.   A mixture of hydrocarbon gases (and often some sulfur compounds) produced in large-scale cracking and refining of petroleum. The usual components are hydrogen, methane, ethane, propane, butanes, pentanes, ethylene, propylene, butenes, pentenes, and small amounts of other components such as butadiene.
Use: Source of raw material for petrochemicals, high-octane gasoline, and organic synthesis of alcohols.

"Refinex."[436]   TM for a specially processed grade of attapulgite clay.
Use: Bleaching, neutralizing, and deodorizing agent in the re-refining of engine and lubricating oils by the contact method.

refining.   Essentially a separation process whereby undesirable components are removed from various types of mixture to give a concentration and purified product. Such separation may be effected (a) mechanically, by pressing, centrifuging, filtering, etc.; (b) by electrolysis; (c) by distillation, solvent extraction, or evaporation; and (d) by chemical reaction. One or more of these operations is applied to (1) food products, (2) petroleum, (3) lubricating oils, and (4) metals. As regards petroleum, refining is generally understood to include not only fractional distillation of crude oils to naphthas, low-octane gasoline, kerosene, fuel oil, and asphaltic residues, but also the processes involved in thermal and catalytic cracking (hydroforming, reforming, etc.) for production of high-octane gasoline. See also specific entries.

reflux.   In distillation processes in which a fractionating column is used the term reflux refers to the liquid that has condensed from the rising vapor and allowed to flow back down the column toward the still. As it does so, it comes into intimate contact with the rising vapor resulting in improved separation of the components. The separation resulting from contact of the countercurrent streams of vapor and liquid is called rectification or fractionation.

Reformatskii reaction.   Condensation of carbonyl compounds with organozinc derivatives of α-halo esters to yield β-hydroxy esters.

reforming.   Decomposition (cracking) of hydrocarbon gases or low-octane petroleum fractions by heat and pressure, either without a catalyst (thermoforming) or with a specific catalyst (molybdena, platinum). The latter method is the more efficient and is almost exlcusively in the US. The chief cracking reactions are (1) dehydrogenation of cyclohexanes to aromatic hydrocarbons; (2) dehydrocyclization of certain paraffins to aromatics; (3) isomerization, i.e., conversion of straight-chain to branched-chain structures, as octane to isooctane. These result in substantial increase in octane number. Steam reforming of natural gas is an important method of producing hydrogen by the reaction $CH_4 + HOH \rightarrow 3H_2 + CO$; steam reforming of naphtha is used to produce synthetic natural gas. See also hydroforming; "Platforming."

"Refractaloy."[308]   TM for a nickel-cobalt-chromium-molybdenum-iron alloy. Type 26 is a precipitation-hardenable material using titianum as the hardening agent, and having high strength, high ductility, and corrosion resistance up to 1450F. Gas turbine discs, bolting, and blading are typical applications.

refraction. The change in direction (apparent bending) of a light ray passing from one medium to another of different density, as from air to water or glass. The ratio of the sine of the angle of incidence to the sine of the angle of refraction is the index of refraction of the second medium. Index of refraction of a substance may also be expressed as the ratio of the velocity of light in a vacuum to its velocity in the substance. It varies with the wavelength of the incident light, temperature, and pressure. The usual light source is the D line of sodium, the standard temperature being 20C, the expression of refractive index is 20/D.

refractive index. See refraction.

refractive. Descriptive of a substance having a high refractive index.

refractory. (1) An earthy, ceramic material of low thermal conductivity that is capable of withstanding extremely high temperature (1650–2200C) without essential change. There are three broad groups of these: (a) acidic (silica, fireclay), (b) basic (magnesite, dolomite), and (c) amphoteric (alumina, carbon and silicon carbide). Their primary use is for lining steel furnaces, coke ovens, glass lehrs, and other continuous high-temperature applications. They are normally cast in the form of brick and are sometimes bonded to assure stability. The outstanding property of these materials is their ability to act as insulators. The most important are fireclay (aluminum silicates), silica, high alumina (70–80% Al_2O_3), mullite (clay-sand), magnesite (chiefly MgO), dolomite (CaO-MgO), forsterite (MgO-sand), carbon, chrome ore-magnesite, zirconia, and silicon carbide.
See also specific entries.
(2) Characterizing the ability to withstand extremely high temperature, e.g., tungsten and tantalum are refractory metals, clay is a refractory earth, ceramics are refractory mixtures.

"Refrasil."[161] TM for a group of materials having outstanding high-temperature resistance. Composed of white, vitreous fibers having up to 99% silicon dioxide content. Available as bulk fiber, batt, cloth, tape, sleeving, yarn, cordage, and flakes.
Use: Aircraft and industrial applications, missiles.

"Refrax."[280] TM for silicon nitride-bonded silicon carbide refractories. Available in brick and precision-formed shapes and parts.
Use: Brazing and furnace fixtures; pumps and pump parts handling corrosive, abrasive slurries; rocket motor components; spray nozzles; burners; pyrometer protection tubes; sinker assemblies in wire aluminizing; bolts and nuts; valve parts; aluminum melting furnace linings and parts; conveyor parts.

refrigerant. Any substance which, by undergoing a change of phase (solid to liquid or liquid to vapor), lowers the temperature of its environment because of its latent heat. Melting ice, with latent heat of 80 calories per gram removes heat and exerts a considerable cooling effect. Most commercial refrigerants are liquids whose latent heat of vaporization results in cooling. Ammonia, sulfur dioxide, and ethyl or methyl chloride were once widely used. The flammablity and toxicity of these compounds led to a search for safer refrigerants that resulted in the discovery of halogenated hydrocarbons, especially fluorocarbons, which are nonflammable. Under various trademarks, these are now generally used for domestic refrigeration and air-conditioning. Ice and circulating brine are still used for preservation of fish at sea and ammonia systems are operated for seafood storage in warehouses.

regeneration. (1) Restoration of a material to its original condition after it has undergone chemical modification necessary for manufacturing purposes. The most common instance is that of cellulose for rayon production. The wood pulp used must first be converted to a solution by reaction with sodium hydroxide and carbon disulfide; in this form (cellulose xanthate) it can be extruded through spinnerets. After this operation it is regenerated to cellulose by passing it through acid (viscose process). Collagen can also be regenerated by acid treatment after it has been purified for use in food products by alkaline solution.
(2) Renewal or reactivation of a catalyst that has accumulated reaction residues such as coke, usually accomplished by passage of steam or reducing gases over the catalyst bed.
(3) Replenishing the sodium ions of a zeolite or similar ion-exchange agent by treatment with sodium chloride solution. Molecular sieves are regenerated by heat removal of the water (200C), followed by treatment with an inert gas.

rehydration. See reconstitution.

Reichert-Meissl number. A measure of volatile, soluble fatty acids derived under arbitrary conditions.

Reich process. A method of purifying carbon dioxide produced in fermentation. The small amounts of organic impurities are oxidized and absorbed and the gas is then dehydrated with chemicals.

Reimer-Tiemann reaction. Reaction for the formation of phenolic aldehydes by heating a phenol with chloroform in the presence of alkali.

Reinecke salt. (ammonium tetrathiocyanodiammonochromate; ammonium reineckate). CAS: 13573-16-5.
$NH_4[Cr(NH_3)_2(SCN)_4] \cdot HOH$.
Properties: Dark-red crystals or crystalline powder, moderately soluble in cold water, soluble in hot water and alcohol, decomposes in aqueous solution.
Derivation: From fusion of ammonium thiocyanate with ammonium dichromate.
Use: Precipitating agent for organic bases such as choline, amines, for certain amino acids; reagent for mercury.

reinforced plastic. A composite structure comprised of a thermosetting or thermoplastic resin and fibers, filaments, or whiskers of glass, metal, boron, or aluminum silicate. Unless otherwise indicated, this term refers to fiberglass-reinforced plastic (FRP).
Properties: Exceptionally high strength, good electrical resistivity, weather- and corrosion-resistance, low thermal conductivity, and low flammability.
Derivation: Basic acid glass in the form of fiber (0.0005 inch), strands of 50–200 fibers, filaments, or woven fabric is coated by passing through a bath of molten resin, which acts as a binder. The assembly can be either compression- or injection-molded. Large, high-strength parts are made by filament winding. Resins used are polyester, epoxy, phenolic, polypropylene, polystyrene, nylon, polycarbonate, and polyphenylene oxides.
Use: Automotive body components, ablative coatings on rockets and space vehicles, appliances (air-conditioning and refrigerator cases), electrical equipment, oil-well piping and tubing, large-diameter pipe, industrial piping systems, chemical storage and mixture tanks, unitized cargo containers, marine equipment, pressure vessels, prefabricated building panels and other structural components, blades for wind machines.
See also glass fiber, composite, fiber, whiskers.
For further information consult the Reinforced Plastics/Composites Institute of the SPI, 250 Park Ave, NY, NY.

reinforcing agent. (1) One of numerous fine powders used in rather high percentages to increase the strength, hardness, and abrasion resistance of rubber, plastics, and flooring compositions. The reinforcing effect is in general a function

of the particle size of the powder. The finest of all is channel carbon black, whose surface area may be as great as 18 acres per pound. Other widely used reinforcing agents are thermal and furnace blacks, magnesium carbonate, zinc oxide, hard clay (kaolin), and hydrated silicas. Though some reinforcing agents have positive coloring properties, the term should not be used as a general synonym for pigment.
(2) Fibers, fabric, or metal insertions in plastics, rubber, flooring, etc., for the purpose of imparting impact strength and tear resistance
See pigment, filler. See also whiskers.

Reissert indole synthesis. Condensation of an o-nitrotoluene with oxalic ester, reduction to the amine, and cyclization to the indole.

Reissert reaction. Formation of 1-acyl-2-cyano-1,2-dihydroquinoline derivatives (Reissert compounds) by reaction of acid chlorides with quinoline and potassium cyanide; hydrolysis of these compounds yields aldehydes and quinaldic acid.

rem. The unit of radiation dose equivalent, the dosage in rads multiplied by a factor representing the different biological effects of various types of radiation.
See also rad, dosimetry, radiation.

Remsen, Ira. (1846–1927) An American chemist born in New York. He began his career in medical practice, but abandoned it for chemistry, and went to Germany to study. He received his doctorate in chemistry from the University of Gottingen in 1870. Returning to the US he taught physics and chemistry at Williams and was later invited to join the staff of Johns Hopkins University where he became head of the department of chemistry. There he established the first graduate curriculum in chemistry in the US based on the system then in use in Germany. In 1879, saccharin was discovered in his research laboratory. He wrote several widely-used textbooks, and founded the American Journal of Chemistry which later merged with JACS. He was President of Johns Hopkins from 1901–1912, and is recognized as one of the great teachers of chemistry.

renewable resources. See biomass, waste control, gasohol.

release. (1) Separation of a cured or baked product from the metal mold or pan in which it is formed. Common release agents for rubber and plastics are waxy or fatty materials such as paraffin and tallow; vegetable oils are used in the baking industry. Such materials are collectively called abherents.
(2) Gradual diffusion of an active ingredient

through a permeable or soluble coating. Pelleted products such as fertilizers and medicinals are often covered with a layer of a substance which permits slow and uniform escape of the active principle. Sulfur is used to coat controlled-release fertilizers; gelatin and similar materials serve the same purpose in pharmaceuticals.

"Renex."[89] TM for a group of nonionic synthetic organic detergents, chiefly ethylene oxide esters and ethers.

"Renite."[577] TM for a series of high-temperature lubricants, swabbing compounds, release agents, and automatic spray equipment for hot-forming glass and hot-working metals.
Use: Improve the quality and production of glassware products, forgings, extrusions, and die castings.

rennet. See rennin.

rennin. (rennase; chymosin).
CAS: 9001-98-3. A digestive enzyme secreted by the glands of the stomach which causes curdling of milk. It has the power of coagulating 25,000 times its own weight of milk.
Properties: Yellowish-white powder or as yellow grains or scales. Characteristic and slightly salty taste and peculiar odor, slightly hygroscopic, partially soluble in water and dilute alcohol.
Derivation: From the glandular layer (inner lining) of the true stomach of the calf. It has been made experimentally by gene-splicing techniques.
Grade: Rennet is the dried commercial extract containing the rennin.
Use: Pharmacy, cheese-making, coagulation of casein for plastics, food additive.

repellent. (1) A substance that causes an insect or animal to turn away from it or reject it as food. Repellents may be in the form of gases (olfactory), liquids, or solids (gustatory). Standard repellents for mosquitos, ticks, etc., are citronella oil, dimethyl phthalate, n-butylmesityl oxide oxalate, and 2-ethyl hexanediol-1,3. Actidione is the most effective rodent repellent, but is too toxic and too costly to use. Thiuram disulfide, amino complexes with trinitrobenzene, and hexachlorophene are successfully used. Copper naphthenate and lime/sulfur mixtures protect vegetation against rabbits and deer. Shark repellents are copper acetate or formic acid mixed with ground asbestos. Bird repellents are chiefly based on taste, but this sense varies widely with the type of bird, so that generalization is impossible. α-naphthol, naphthalene, sandalwood oil, quinine, and ammonium compounds have been used, with no uniformity of result.

See also fumigant.
(2) A substance which, because of its physicochemical nature, will not mix or blend with another substance. All hydrophobic materials have water-repellent properties due largely to differences in surface tension or electric charges, e.g., oils, fats, and waxes and certain types of plastics. Silicone resin coatings can keep water from penetrating masonry by lining the pores, not by filling them; they will not exclude water under pressure.

replacement. See substitution.

replication. Making a reverse image of a surface by means of an impression on or in a receptive material; usually applied to microscopic techniques for obtaining plastic replicas of observed objects. In biochemistry, the term refers to reproduction of the DNA molecule, which is composed of two interlocking chains of nucleotides, the double helix structure elucidated by Watson and Crick in 1953. It reproduces itself by forming two identical daughter molecules, each of which receives one of the two chains of the original molecule, the other in each case being synthesized from nucleic acids by enzymes (DNA polymerases). In the oversimplified drawing below the solid lines are the strands of the original molecule and the broken lines are the synthesized strands:

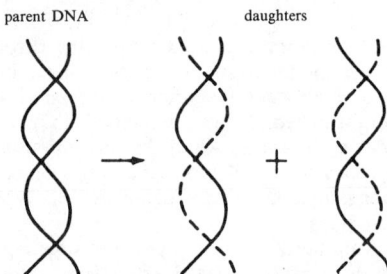

parent DNA   daughters

The chemical mechanisms in replication are more complex than previously thought; 12–15 proteins are involved as well as several enzymes.
See also deoxyribonucleic acid, genetic code.

Reppe process. Any of several processes involving reaction of acetylene: (a) with formaldehyde to produce 2-butyne-1,4-diol which can be converted to butadiene; (b) with formaldehyde under different conditions to produce propargyl alcohol and from this, allyl alcohol; (c) with hydrogen cyanide to yield acrylonitrile; (d) with alcohols to give vinyl ethers; (e) with amines or phenols to give vinyl derivatives; (f) with carbon monoxide and alcohols to give esters of acrylic acid; (g) by polymerization to produce cyclooctatetraene; and (h) with phenols to make resins. The use of catalysts, of pressures up to 30 atm, and of special techniques to avoid or contain explosions, are important factors in these processes. See also acetylene.

reprocessing. Treatment of spent nuclear reactor fuel to recover the unconsumed uranium-235 and plutonium by separating them from each other and from the fission products formed in the reactor. The Purex process is the accepted procedure used for this purpose. The spent fuel is dissolved in nitric acid; separation is effected by solvent extraction with tributyl phosphate, ion-exchange reactions, and precipitation. The reclaimed uranium-235 and plutonium are sent to fuel fabrication plants for reuse. The fission product waste is evaporated and stored. Another method, called the Civex process, has been proposed to prevent theft of plutonium; here the mixture of waste products, uranium isotopes, and plutonium is not separated. Since its plutonium content is only 20% it could not be used for weapons; the mixture is suitable for fast breeder reactors. Serious radiation hazards are involved in reprocessing which require use of appropriate shielding and remote-control handling procedures. Storage of the radioactive waste also presents a long-range problem which has not yet been satisfactorily solved. There are no commercial reprocessing plants operational in the US, though there are several in Europe.
See radioactive waste, breeder.

reprography. A coined name for the technique of reproducing drawings, blueprints, and typographic matter by the use of photosensitized papers, or more recently polyester sheeting, which may be coated with diazo dyes. The process has broad potential in the photocopying field and in communications technology. It involves colloid and surface chemistry, ink-paper interactions, and unusual imaging techniques.

resinamine. $C_{35}H_{42}N_2O_9$. Alkaloid from certain species of Rauwolfia.
Properties: White or pale buff to cream colored, odorless, crystalline powder; darkens slowly on exposure to light, more rapidly when in solution; partially soluble in organic solvents; insoluble in water; mp 238C (in vacuo).
Use: Medicine (antihypertensive).

Research octane number (RON). See octane number.

resene. The unsaponifiable component of rosin and other natural resins.

reserpine. CAS: 50-55-5. $C_{33}H_{40}N_2O_9$. An alkaloid.

Properties: White or pale buff to slightly yellowish, odorless powder; darkens slowly on exposure to light and darkens more rapidly in solution; insoluble in water, slightly soluble in alcohol, soluble in chloroform and benzene; mp 264–265C (decomposes).
Derivation: From Rauwolfia serpentina.
Grade: USP.
Hazard: Carcinogen in animals, potential cancer risk in humans.
Use: Antihypertensive agent, tranquilizer.

residual oil. A liquid or semiliquid product resulting from the distillation of petroleum and containing largely asphaltic hydrocarbons. Also known as asphaltum oil, liquid asphalt, black oil, petroleum tailings and residuum. Combustible.
Use: Roofing compounds, hot-melt adhesives, friction tape, sealants, heating oil for large buildings, factories, etc.
See also fuel oil.
Note: Gasoline of 94 octane can be produced from residual oil in a high-temperature catalytic process, thus increasing the yield of gasoline from a barrel of crude by 33% when full-scale production is achieved.

"Resilon."[326] TM for a bituminous base lining for chemical process equipment.
Use: To resist acids and alkalies up to 150F.

"Resimene."[58] TM for melamine and ureaformaldehyde resins. Supplied in organic liquid solutions. The melamine is also available in water-alcohol and soluble spray-dry powders.
Use: Paint, varnish, lacquer for automobiles, machinery, appliances, construction; electronics, missiles; chemicals, pulp and paper.

resinate. A salt of the resin acids found in rosin. They are mixtures rather than pure compounds.
Use: See soap (2).

"Resin C."[175] TM for a neutral synthetic coal-tar resin of high styrene content.
Properties: Light color, mp 115–123C, d 1.05, mineral oil cloud point 130–150C.
Use: To impart alkali-and grease-resistance to floor tile.

resin, ion-exchange. See ion-exchange resin.

resin, liquid. An organic polymeric liquid which, when converted to its final state for use, becomes solid (ASTM). Example linseed oil, raw or heat-bodied (partially polymerized).
See also drying oil, resinoid.

resin, natural. (a) Vegetable-derived, amorphous mixture of carboxylic acids, essential oils, and terpenes occurring as exudations on the bark of many varieties of trees and shrubs. They are combustible, electrically nonconductive, hard and glassy with conchoidal fracture when cold, and soft and sticky below the glass transition point. Most are soluble in alcohols, ethers and carbon disulfide, and insoluble in water. The best known of these are rosin and balsam, obtained from coniferous trees; these have a high acid content. Of more remote origin are such resins as kauri, congo, dammar, mastic, sandrac, and copal. Their use in varnishes, adhesives and printing inks is still considerable, though diminishing in favor of synthetic products. (b) Miscellaneous types. Shellac, obtained from the secretion of an Indian insect, is still in general use as a transparent coating. Amber is a hard, polymerized resin that occurs as a fossil. Ester gum is a modified rosin. Amorphous sulfur is considered an inorganic natural resin. Liquid resins, sometimes called resinoids, are represented by linseed and similar drying oils.
See also gum, natural (note); resin, synthetic (note).

resinoid. Any thermosetting synthetic resin, either in its initial temporarily fusible state or its final infusible state (ASTM). Heat-bodied linseed oil, partially condensed phenol-formaldehyde and the like, are also considered resinoids.

resinol. A coal-tar distillation fraction containing phenols. It is the fraction soluble in benzene but insoluble in light petroleum, obtained by solvent extraction of low temperature tars or similar materials. Resinols are very sensitive to heat and oxidation.

"Resinox."[58] TM for a series of phenolic resins, supplied in various forms suitable for applications as bonding agents for shell molding and as core binders for metal casting; impregnants or bonding materials for grinding wheels, brake linings, insulation, and similar industrial uses; as pipe linings, air conditioning equipment coatings, special primers; as laminating, bonding, impregnating resins for paper, fibers; for use in cans, drums, and tank car linings requiring a high degree of chemical and solvent resistance; and for heavy duty products, such as equipment housings. Special formulations for high temperature use of space technology.

resin, synthetic. A man-made high polymer resulting from a chemical reaction between two (or more) substances, usually with heat or a catalyst. This definition includes synthetic rubbers and silicones (elastomers), but excludes modified, water-soluble polymers (often called resins). Distinction should be made between a synthetic resin and a plastic, the former is the polymer itself, whereas the latter is the polymer plus such additives as filters, colorant, plasticizers, etc.

The first truly synthetic resin was developed by Baekeland in 1911 (phenol-formaldehyde). This was soon followed by a petroleum-derived product called coumarone-indene, which did indeed have the properties of a resin. The first synthetic elastomer was polychloroprene (1931) originated by Nieuwland, and later called neoprene. Since then many new types of synthetic polymers have been synthesized, perhaps the most sophisticated of which are nylon and its congeners (polyamides, by Carothers) and the inorganic silicone group (Kipping). Other important types are alkyds, acrylics, aminoplasts, polyvinyl halides, polyester, epoxies, and polyolefins.

In addition to their many applications in plastics, textiles and paints, special types of synthetic resins are useful as ion-exchange media.
See "Cumar." See also plastic, paint, fiber, film, elastomer.
Note: Because the term "resin" is so broadly used as to be almost meaningless, it would be desirable to restrict its application to natural organo-soluble, hydrocarbon-based products derived from trees and shrubs. But in view of the tendency of inappropriate terminology to "gel" irreversibly, it seems like a losing battle to attempt to replace "synthetic resin" with the more precise "synthetic polymer."
See also note under gum, natural.

resist. A material which will prevent the fixation of dye on a fiber, thus making color designs and pattern prints possible. The resist may act mechanically, as a wax, resin, or gel which prevents absorption of the dye, or its accompanying mordant. Citric acid, oxalic acid, and various alkalies are among the more common resists of the chemical type.

resistor composition. A specially treated semiconducting metal powder compounded with glass binders and temporary organic carriers. Can be applied to glass or ceramic surfaces by stenciling, spraying, brushing or dipping; firing range 704–760C. Compositions can be blended with mem-

bers of the same series to produce intermediate resistance values. Fired resistors have good reproducibility, low temperature and voltage coefficients, and stability to abrasion, moisture, and relatively high (125C) ambient temperature.
Use: To produce fired-on resistor components for electronic circuits.

"Resistox."[296]　TM for stabilized grades of copper powder assaying at greater than 99% copper with a density 8.9 and apparent density range of 2.0–3.5 g/cm³. Marketed in several grades of various particle sizes.
Use: Fabrication of porous bearings, sintered ferrous machine parts, catalysts, magnesium chloride cements, metal friction surfaces, electric brushes, electrical contacts, metallic paints.

resite.　See C-stage resin.

resitol.　See B-stage resin.

"Resloom."[58]　TM for a series of synthetic resins. E-50 Resins. Supplied at 50% solid cyclic ethylene urea-formaldehyde resin in liquid form. Melamine. A series of essentially monomeric melamine resins.
Use: Textile industry.

"Resmetal."[65]　TM for a resin-metal composition which when catalyzed converts to metal-like solid. Recommended for mold making, patching, forming, and general repair of metal surfaces and objects.

resol.　See A-stage resin.

"Resolube."[483]　TM for a wide temperature, high film strength lubricant (to −57C) which does not have the plastic-attacking properties of most synthetic low-temperature oils.

resolution.　See resolving power.

resolving power.　The extent to which a lens can distinguish small particles and minute distances, i.e., fine structure. The human eye can resolve objects of 1/250th inch (100 microns) in any dimension. The compound microscope has a resolving power of 0.5 micron; an electron microscope can resolve fine structure as small as 5 Å units, that is, in the molecular range. Two factors determine resolving power: the wavelength of the radiation utilized and the focal depth of the lens. The resolving power of a microscope is much more important than its ability to magnify, for no magnification, however, large, can add detail to an image that was not first discerned by the lens system.
See also optical microscope, electron microscope.

resonance.　(1) In chemistry, resonance (or mesomerism) is a mathematical concept based on quantum mechanical considerations (i.e., the wave functions of electrons); it is used to describe or express the true chemical structure of certain compounds which cannot be accurately represented by any one valence-bond structure. It was originally applied to aromatic compounds such as benzene, for which there are many possible approximate structures, none of which is completely satisfactory.
See benzene.
The resonance concept indicates that the actual molecular structure lies somewhere between these various approximations, but is not capable of objective representation. This idea can be applied to any molecule, organic or inorganic, in which an electron pair bond is present. The term "resonance hybrid" denotes a molecule which has the property. Such molecules do not vibrate back and forth between two or more structures, nor are they isotopes or mixtures; the resonance phenomenon is rather an idealized expression of an actual molecule which cannot be accurately pictured by any graphic device.
(2) In the terminology of spectroscopy, resonance is the condition in which the energy state of the incident radiation is identical with that of the absorbing atoms, molecules or other chemical entities. Resonance is applied in various types of instrumental analysis such as nuclear resonance absorption and nuclear magnetic resonance.
See also absorption spectroscopy.
Note: The multiple meanings of "nucleus" and "resonance" can be a source of confusion, especially when these terms are closely associated, as in nuclear magnetic resonance and resonance of a molecular nucleus. In the first of these expressions, nucleus is used in sense (1) under nucleus, and resonance in sense (2) under resonance. In the second expression, nucleus is used in sense (3) under nucleus and resonance in sense (1) under resonance.

resorcinol.　(resorcin; m-dihydroxybenzene; 3-hydroxyphenol).　CAS: 108-46-3. $C_6H_4(OH)_2$.

Properties: White crystals, becoming pink on exposure to light when not perfectly pure; sweet taste; d 1.2717; mp 110.7; bp 281C; soluble in water, alcohol, ether, glycerol, benzene, and amyl alcohol; slightly soluble in chloroform; flash p

261F (127C); autoign temperature 1126F (607C). Combustible.

Derivation: By fusing benzene-m-disulfonic acid with sodium hydroxide.

Grade: USP, powder, resublimed, pure, reagent, technical, crude.

Hazard: Irritant to skin and eyes. TLV: 10 ppm in air.

Use: Resorcinol-formaldehyde resins, dyes, pharmaceuticals, crosslinking agent for neoprene, rubber tackifier, adhesives for wood veneers and rubber-to-textile composites, manufacture of styphnic acid, cosmetics.

resorcinol acetate. (resorcinol monoacetate). CAS: 102-29-4. $HOC_6H_4OCOCH_3$.

Properties: Viscous, pale yellow or amber liquid with a faint odor and a burning taste, bp 283C (decomposes), boiling range (10 mm) 150–153C, saturated solution is acid to litmus, soluble in alcohol and most organic solvents, sparingly soluble in water, d 1.203–1.207. Combustible.

Derivation: Action of acetic anhydride on resorcinol.

Grade: CP, NF.

Use: Medicine, cosmetics.

resorcinol blue. See lacmoid.

resorcinol diglycidyl ether. (RDGE; 1,3-diglycidyloxybenzene; m-bis(2,3-epoxypropoxybenzene). CAS: 101-90-6.

$C_6H_4(OCH_2CHOCH_2)_2$.

Properties: Straw-yellow liquid, d 1.21 (25C), bp 172C (0.8 mm), refr index 1.541 (25C), viscosity 500 cp (25C), flash p (COC) 350F (176C), miscible with most organic resins. Combustible.

Use: Epoxy resins.

resorcinol dimethyl ether. (dimethyl resorcinol; 1,3-dimethoxy benzene). CAS: 151-10-0. $C_6H_4(OCH_3)_2$.

Properties: Pale straw-tint liquid, bp 204–212C, d 1.063–1.066 (25/25C), refr index 1.523–1.527 (20C). Combustible.

Use: Organic intermediate, flavoring.

resorcinol-formaldehyde resin. A type of phenol-formaldehyde resin. Permanently fusible; soluble in water, ketones, and alcohols. By dissolving and adjusting the pH to 7, an adhesive base is formed. These adhesives can be used whenever phenolics are used and where fast or room-temperature cure is desired. An important use of these adhesives is in wood gluing, particularly marine plywood.

resorcinol monobenzoate. CAS: 136-36-7. $C_6H_5COOC_6H_4OH$.

Properties: White, crystalline solid; mp 132–135C; bp 140C (0.15 mm); insoluble in water, benzene, di-(2-ethylhexyl)phthalate; soluble in acetone and ethanol. Combustible.

Use: Noncoloring UV inhibitor for various plastics, color stabilizer in cosmetic compositions.

resorcinolphthalein. See fluorescein.

resorcinolphthalein sodium. See uranine.

α-resorcylic acid. (3,5-dihydroxybenzoic acid). CAS: 99-10-5. $(OH)_2C_6H_3COOH$.

Properties: White crystals; mp 237C; soluble in water, ethanol, and ether. Combustible.

Grade: CP.

Use: Intermediate for dyes, pharmaceuticals, light stabilizers, resins.

β-resorcylic acid. (BRA; 2,4-dihydroxybenzene carboxylic acid; 2,4-dihydroxybenzoic acid; 4-hydroxysalicylic acid; 4-carboxyresorcinol). CAS: 89-86-1. $(OH)_2C_6H_3COOH$.

Properties: White needles; mp (decomposes) 219–220C; bp decomposes; almost insoluble in water and benzene; soluble in alcohol, ethyl ether. The sodium, potassium, ammonium, calcium, and barium salts are soluble in water; the silver, lead and copper salts are only slightly soluble. Combustible.

Use: Dyestuff and pharmaceutical intermediate, chemical intermediate in synthesis of fine organic chemicals, reagent for iron.

respiration. (1) In man and animals, inhalation of oxygen and exhalation of carbon dioxide; the oxygen supports the oxidation (combustion) of organic nutrients in the body, yielding energy, carbon dioxide, and water. (2) In growing plants, respiration occurs in both the presence and the absence of light. Some of the energy produced in respiration is used to form adenosine triphosphate, pyruvic acid and other metabolic intermediates. Fruits and vegetables continue to respire after harvest, a fact that must be taken into account in transportation and storage.

"Reswax."[65] TM for a series of wax-resin blends used as coatings and hot melt adhesives in paper conversion. The polymers used include butyl rubber, polyisobutylene, chlorinated rubber, polyethylene, and styrene co-polymers.

"Resyn."[53] TM for polyvinyl acetate latex for adhesive and binder uses.

ret. To reduce or digest fibers, especially linen, by enzymatic action.

retarder. See inhibitor.

"Reten" 205.[266] TM for a strongly cationic, high molecular weight, synthetic, water-soluble polymer. A finely divided white powder, dissolves in either hot or cold water to produce clear, smooth, viscous, nonthixotropic solutions; available in a variety of viscosity grades and cationic functionality. "Reten" 763 is an aqueous solution of a modified polyamide-epichlorohydrin resin.
Use: Flocculant, binder, and viscosifier.

retene. (7-isopropyl-1-methylphenanthrene). CAS: 483-65-8. $C_{18}H_{18}$.
Properties: Mw 234.36.

retinal. Preferred name for retinene.

retinene. (vitamin A aldehyde; retinal). $C_{20}H_{28}O$. A necessary component of rhodopsin, the light-sensitive pigment of the eye. Retinene is the aldehyde form of vitamin A, which is an alcohol.

retinol. CAS: 68-26-8. $C_{20}H_{29}OH$.
(1) A component of vitamin A.
See also carotene.
(2) A resin distillate similar to rosin oil.

retorting. A process much used in the early years of chemistry for destructive distillation of heavy organic liquids and for laboratory separations. It involves the use of a cylindrical vessel made of glass (for laboratory work), fireclay, or metal, with a neck bent at a downward angle to facilitate distillation. For gas manufacture the equipment is built on a heavier scale to handle destructive distillation of coal. Use of this term has been revived in current developments for processing shale oil.

retro Diels-Alder reaction. Thermal dissociation of Diels-Alder adducts, occurring most readily when one or both fragments are particularly stable.

retropinacol rearrangement. Conversion of an alcohol to the rearranged olefin on treatment with acid.

retrosynthesis. A computer-assisted analysis of an organic molecule that is to be synthesized, i.e., the target molecule, in which the computer works back through the precursors of the target substance to a group of possible starting materials that are readily available from natural sources or as commercial products. Retrosynthetic analysis is thus the opposite of the usual direct approach to laboratory synthesis.
See also computational chemistry.

reverberatory furnace. An ore-roasting kiln having a curved or sloping roof from which the heat is deflected onto the material being treated. The fuel and the charge occupy separate areas in the kiln so that there is no direct contact between them, thus avoiding contamination of the ore with fuel particulates. The heat rising from the ignited fuel impinges on the curved roof and is reflected downward onto the ore. After passing over the ore, the heat escapes through suitably located vents.

Reverdin reaction. Migration of iodine during nitration of iodophenolic ethers.

reverse osmosis. See osmosis, desalination.

reversible. (1) A chemical reaction which proceeds first to the right and then to the left when the ambient conditions are changed; the product of the first reaction decomposes to the original components as a result of different conditions of temperature or pressure. Examples are: $HOH + CO_2 \leftrightarrows H_2CO_3$ in which the carbonic acid reverts to water and carbon dioxide on heating; $NH_4Cl \leftrightarrows NH_3 + HCl$ in which the ammonium chloride decomposes on heating to ammonia and hydrochloric acid, which recombine on cooling.
(2) A colloidal system such as a gel or suspension that can be changed back to its original liquid form by heating, addition of water, or other method. For example, evaporated egg white can be restored (reconstituted) by addition of water.

"Reversil."[425] TM for an inert hydrophobic chromatography powder, designed as a reversed-phase column support or thin-layer powder.

reversion. The softening and weakening of a natural rubber vulcanizate when the curing operation has been too long continued.

"Rexforming."[416] A proprietary process combining certain of the elements of the "Platforming" process and "Udex" extraction to convert a naphtha fraction into a highly aromatic motor fuel blending component of high octane rating.

Reynold's number. The function DUP/v used in fluid flow calculations to estimate whether flow through a pipe or conduit is streamline or turbulent in nature. D is the inside pipe diameter, U is the average velocity of flow, P is density, and v is the viscosity of the fluid. Different systems of units give identical values of the Reynold's number, and values much below 2100 correspond to streamline flow, while values above 3000 correspond to turbulent flow.

"Rezipas."[412] TM for aminosalicylic acid resin.

"Rezklad."[41] TM for epoxy-based flooring compound, grout, and cements. Gives corrosion-proof, strong, abrasion-resistant surfaces.

"Rezyl."[11] TM for short, medium, and long oil semioxidizing or oxidizing types of alkyd resins, also short nonoxidizing types, nondrying plasticizing types, specialty formulations.
Use: Baking enamels, interior and exterior brushing enamels, air dry finishes, nitrocellulose lacquers.

RFNA. Abbreviation for red fuming nitric acid. See nitric acid, fuming.

Rh. Symbol for rhodium.

rhamnose. CAS: 3615-41-6. $C_6H_{12}O_5$.
A deoxyhexose monosaccharide found combined in the form of glycosides in many plants. Two forms exist, alpha- and beta-, of which the alpha- is the more stable.
Properties: White crystals, very soluble in water and in methanol, soluble in absolute alcohol, mp 82–92C (alpha- form converts partially to the beta- form on heating). For equilibrium mixture optical rotation +9.18 degrees.
Use: Synthetic sweetener research.

rhenium. CAS: 7440-15-5. Re. Metallic element, atomic number 75, group VIIB of the periodic table, aw 186.207. Valences = 1-7; 4, 6, and 7 are most common, the last being the most stable. There are two isotopes.
Properties: Silver-white solid or gray to black powder, d 21.02 (20C), mp 3180C, bp 5630C, tensile strength 80,000 psi, high modulus of elasticity, attacked by strong oxidizing agents (nitric and sulfuric acids), practically insoluble in hydrochloric acid, retains its crystalline structure all the way to its mp, has widest range of valences of any element. Rhenium-molybdenum alloys are superconductive at 10K. Not attacked by sea water.
Source: Principally molybdenite.
Derivation: Solutions from refinery residues (molybdenum ore flue dust, copper ore treatment) are (a) concentrated by a salting-out processes and reduced by hydrogen gas under press to give the metal, or (b) passed through an anionic resin from which pure rhenium can be extracted by a strong mineral acid.
Forms: Powder which can be consolidated into rods, wires, or strips by powder metallurgy. Annealed metal is very ductile and can be bent, coiled, or rolled. Single crystals 2 inches × 0.05-0.005 inch diameter.

Hazard: Flammable in powder form.
Use: Additive to tungsten-and molybdenum-based alloys, electronic filaments, electrical contact material, high-temperature thermocouples, igniters for flash bulbs, refractory metal components of missiles, catalyst, plating of metals by electrolysis and vapor-phase deposition.

rhenium heptasulfide. Re_2S_7.
Properties: Brown-black solid, d 4.87, decomposes to ReS_2 at 600C, insoluble in water, dissolves in solutions of alkali sulfides.
Hazard: Ignites on heating in air.
Use: Catalyst.

rhenium heptoxide. CAS: 1314-68-7. Re_2O_7.
Properties: Yellow crystals, d 6.103, mp 297C, dissolves in water to form perrhenic acid $HReO_4$, very soluble in alcohol.
Derivation: Oxidation of metallic rhenium at 400C.

rhenium pentachloride. $ReCl_5$.
Properties: Dark green to black solid, d 4.9, decomposes on heating, decomposes in water, soluble in hydrochloric acid and alkalies.
Derivation: By reacting rhenium heptoxide with carbon tetrachloride at 400C.

rhenium trichloride. $ReCl_3$.
Properties: Dark red solid, on heating emits green vapor from which the metal may be deposited, nonelectrolyte in solution, soluble in water and glacial acetic acid.
Derivation: Distillation of rhenium pentachloride.

rhenium trioxychloride. ReO_3Cl.
Properties: Colorless liquid, d 3.867 (20/4C), mp 4C, bp 131C, decomposes in water, soluble in carbon tetrachloride, reacts readily with organic substances.

rheology. Science of the deformation and flow of materials in terms of stress, strain, and time. Has important bearing on the behavior of viscous liquids in plastic molding.
See liquid, Newtonian; dilatancy; thixotropy; and viscosity.

"Rheolube."[483] TM for a series of low-shear thixotropic greases and semifluid instrument lubricants for use where nonspreading properites are critical.

rheometer. A device that continuously measures the viscosity and elasticity of resin solutions and polymer melts at high shear rates.

rhesus factor. (Rh factor). A substance present in the red blood cells of the rhesus monkey and of 85% of an average white American population. Those whose red cells contain this factor are termed Rh-positive; others, Rh-negative. A negative individual may develop anti-Rh antibodies, if Rh-positive red cells enter his blood; such antibodies can then agglutinate Rh-positive cells. Hemolytic reactions may thus occur following transfusion of Rh-positive blood cells into a recipient previously sensitized and having Rh antibodies in the serum. Likewise, an Rh-positive fetus may give rise to antibodies in the blood of an Rh-negative mother; the antibodies, returning into the fetus may then produce the disease erythroblastosis fetalis. There are many subtypes of the Rh factor; these can be distinguished by serologic tests, and the laws of their inheritance have been determined.

rhizobitoxin. A recently developed broad-spectrum herbicide which attacks young growth and new leaves but has little effect on older growth. Said to have low toxicity to humans (USDA).

rho acid. See anthraquinone-1,5-disulfonic acid.

rhodamine B. (CI No.45170). CAS: 81-88-9. $C_{28}H_{31}ClN_2O_3$. A basic red fluorescent dye, structurally related to xanthene.
Properties: Green crystals or reddish violet powder; very soluble in water and alcohol, forming bluish red, fluorescent solution; slightly soluble in acids or alkalies.
Derivation: By fusion of m-diethylaminophenol and phthalic anhydride followed by acidification with hydrochloric acid.
Use: Red dye for paper, also for wool and silk where brilliant fluorescent effects are desired and light-fastness is of secondary importance; analytical reagent for certain heavy metals, biological stain.

rhodamine toner. Red to maroon lakes of rhodamine dyes and phosphotungstic or phosphomolybdic acid. They have good lightfastness and are used principally in printing inks.
See also phosphomolybdic pigment, phosphotungstic pigment.

rhodanine. (2-thio-4-keto-thiazolidine).

CAS: 141-84-4. $\overline{SCH_2C(O)NHCS}$.
Properties: Finely crystalline, light yellow color, d 0.868, bulk d 0.617, mp decomposes often violently 166C, pure material 167–168.5C. Soluble in methanol, ethyl ether, and hot water.
Hazard: May explode when heated to 166C. Toxic by ingestion.
Use: Organic synthesis (phenylalanine), laboratory reagent.

rhodinol. CAS: 6812-78-8. A mixture of terpene alcohols consisting principally of l-citronellol.
Properties: Colorless liquid with a pronounced rose-like odor, d 0.860–0.880 (25C), refr index 1.4630–1.4730 (20C), optical rotation −4 to −9 degrees, soluble in alcohol and mineral oil, insoluble in water. Combustible.
Derivation: From Reunion geranium oil.
Grade: FCC.
Use: Perfumery, flavoring agent.

rhodinyl acetate. A mixture of terpene alcohol acetates consisting primarily of l-citronellyl acetate.
Properties: Colorless to slightly yellow liquid with a rose-like odor, d 0.895–0.908 (25C), refr index 1.4530–1.4580 (20C), optical rotation −2 to 6 degrees, soluble in alcohol and mineral oil, insoluble in glycerol. Combustible.
Derivation: Action of acetic anhydride on rhodinol in the presence of sodium acetate.
Grade: Technical, FCC.
Use: Perfumery, flavoring agent.

rhodium. CAS: 7440-16-6. Rh. Metallic element having atomic number 45, group VIII of the periodic system, aw 102.9055, no isotopes, valence = 3.
Properties: White solid of platinum group, d 12.41 (20C), mp 1966C, bp 4500C, insoluble in acids and aqua regia, soluble in fused potassium bisulfate. Harder and higher-melting than platinum or palladium, highest in electrical and thermal conductivity of the platinum group. High surface reflectivity. A strong complexing agent.
Occurrence: Ontario, South Africa, Siberia.
Derivation: Occurs with platinum, from which it is recovered during the purification process.
Forms: Produced as powder which can be fabricated by casting or powder metallurgy techniques. Single crystals are available.
Hazard: Flammable in powder form. TLV (insoluble rhodium compounds): 1 mg/m³ of air; (soluble rhodium salts): 0.01 mg/m³ of air.
Use: Alloy with platinum for high temperature thermocouples, furnace windings, laboratory crucibles, spinnerets in rayon industry. Electrical contacts, jewelry, catalyst, optical instrument mirrors, electro-deposited coatings for metals, vacuum-deposited glass coatings, headlight reflectors.

rhodium carbonyl chloride. CAS: 14523-22-9. $Rh(CO)_2Cl_2$.
Properties: Reddish crystals, mp 125C. Soluble in alcohol, benzene, acetone, with decomposition; solid material is stable in dry air.
Use: Catalyst for organic reactions.

rhodium chloride. (rhodium trichloride).
CAS: 10049-07-7. $RhCl_3$.
Properties: Reddish-brown powder, insoluble in water and acids, soluble in solutions of alkalies and cyanides, mp 450–500C, bp 800C (sublimes).
Hazard: Toxic by ingestion.
Use: Manufacture of rhodium trifluoride.

rhodochrosite. $MnCO_3$ with partial replacement by iron, calcium, magnesium, zinc.
Properties: Light pink, rose red, brownish red, or brown mineral; white streak; vitreous to pearly luster; photoluminescent; found in veins with ores of silver, lead, copper manganese; d 3.3–3.6; Mohs hardness 3–4.
Occurrence: US (Connecticut, New Jersey, Colorado, Montana, Nevada), Europe.
Use: Manganese ore.

rhodonite. An ore of manganese.
See manganous silicate.

rhodopsin. The red-light-sensitive pigment of the eye (visual purple) consisting of the proteins opsin and retinene (vitamin A aldehyde). It occurs in land and marine vertebrates.

rhodoxanthin. CAS: 116-30-3. $C_{40}H_{50}O_2$.
A natural carotenoid pigment used in the food, drug and cosmetic industries; soluble in benzene and chloroform, slightly soluble in alcohols.

"Rhoduline."[307] TM for basic dyestuffs, used for dyeing of silk, cotton, rayon, paper, leather, lacquers, and plastics. Also used as phosphotungstic toners or lakes for printing inks, wallpaper, and coated paper. Characterized by exceptional brilliancy.

"Rhonite."[23] TM for thermosetting modified and unmodified urea-formaldehyde condensates. Supplied as water-clear solutions and aqueous pastes. Reactive with cotton, various grades producing shrink-resistance, crease-proofing, or modification of hand.
Use: Finishing of natural and synthetic fabrics.

"Rhoplex."[23] TM for aqueous dispersions of acrylic copolymers. White opaque emulsions; various grades differing in hardness, flexibility, adhesion and tack of film; some thermosetting. Produce colorless, transparent films with outstanding permanence, durability, adhesion, and pigment-binding capacity.
Use: Emulsion paints, paper coatings and saturation, floor sealers and wax emulsions, textile-backing and finishing, bonding fibers and pigments, clear and pigmented coatings on wood and metals.

"Rhothane."[23] TM for an agricultural insecticide based on 1,1-bis(chlorophenyl)-2,2-dichloroethane and supplied as a wettable powder or emulsion concentrate.

"Rhozyme."[23] TM for enzyme concentrate with diastatic or proteolytic activity. Buff-colored powders or liquids of fungal or bacterial origin which hydrolyze and soluble proteins and starches.
Use: Desizing textile fabrics; drycleaning; liquefaction of starch paste; fermentation processes; manufacture of corn syrup, fish solubles, septic tank formulations; animal feed; meat tenderizer.

"Ria."[511] TM for a finely ground surface-treated urea; also available as paste.
Use: Cure accelerator and activator for resins.

Rib. Abbreviation for ribose.

ribbon mixer. A mixing or blending machine whose essential components are a steel bowl or trough, jacketed for temperature control, within which rotates an agitating device consisting of two or more metal strips (ribbons) pitched in opposite directions spirally arranged around a central shaft. The curved and reverse-pitched ribbons operate on the principle of the screw; the material is moved forward by one ribbon and backward by another, so that efficient mixing is effected. Continuous operation is possible in some types. Such equipment is used for mixing dry powders, viscous liquids, slurries, etc., as well as for drying, crystallizing, and deaerating. Large sizes have a bowl diameter of 5 ft and a length of 9.5 feet.

riboflavin. (vitamin B_2; 7,8-dimethyl-10-(1'-d-ribityl) isoalloxazine). CAS: 83-88-5. $C_{17}H_{20}N_4O_6$. A crystalline pigment, the principal growth-promoting factor of the vitamin B_2 complex. It functions as a flavoprotein in tissue respiration. A syndrome resembling pellagra is thought to be due to riboflavin deficiency.

Properties: Orange-yellow crystals, bitter taste, mp 282C (decomposes), slightly soluble in water and alcohols, insoluble in lipid solvents, stable to heat in dry form and in acid solution, stable to ordinary oxidation, unstable in alkaline solution and quite sensitive to light. In solution, riboflavin has an intense greenish yellow fluorescence.

Units: Amounts are expressed in milligram or μg of riboflavin.

Sources: (Food): Milk, green leafy vegetables, egg yolk, liver, enriched flour, yeast. (Commercial): Distiller's residues, fermentation solubles, synthetic production (indirectly from dextrose, lactose, yeast, and whey).

Grade: USP, FCC.

Use: Medicine, nutrition, animal feed supplement, enriched flours, dietary supplement.

riboflavin-5'-phosphate.　(FMN; flavin mononucleotide).　　The phosphate ester of riboflavin in which the phosphate is esterified to the ribityl portion of riboflavin. It functions as a coenzyme for many flavine enzymes. The riboflavin group has the ability to take up hydrogen atoms, thus oxidizing the substrate.

Properties: (sodium salt): Yellow crystals, much more soluble than riboflavin in water, quite sensitive to UV light.

Derivation: By treating riboflavin with chlorophosphoric acid.

Use: Dietary supplement, flavor potentiator.

9-β-D-ribofuranosyladenine.　See adenosine.

ribonuclease.　An enzyme which causes splitting of ribonucleic acid. Pancreatic ribonuclease for example, cleaves only phosphodiester bonds that are linked to pyrimidine-3'-phosphates. It is a critical regulator of life processes in the cell. The first enzyme to be synthesized (1969), it is composed of 124 amino acid residues. It is one of the proteins for which the sequence of amino acids has been elucidated (the order or sequence of amino acids is of critical importance in the functioning of enzymes, genes and nucleotides). See also genetic code.

ribonucleic acid.　(ribose nucleic acid; RNA). Generic term for a group of natural polymers consisting of long chains of alternating phosphate and d-ribose units, with the bases adenine, guanine, cytosine, and uracil bonded to the 1-position of the ribose. Ribonucleic acid is universally present in living cells and has a functional genetic specificity due to the sequence of bases along the polyribonucleotide chain.

Four types are recognized: (1) Messenger RNA, synthesized in the living cell by the action of an enzyme that carries out the polymerization of ribonucleotides on a DNA template region which carries the information for the primary sequence of amino acids in a structural protein. It is a ribonucleotide copy of the deoxynucleotide sequences in the primary genetic material. (2) Ribosomal RNA which exists as a part of a functional unit within living cells called the ribosome, a particle containing protein and ribosomal RNA in roughly 1:2 parts by weight, having a particle weight, of about three million.

See ribosome.

Messenger RNA combines with ribosomes to form polysomes containing several ribosome units, usually 5 (e.g., during hemoglobin synthesis), complexed to the messenger RNA molecule. This

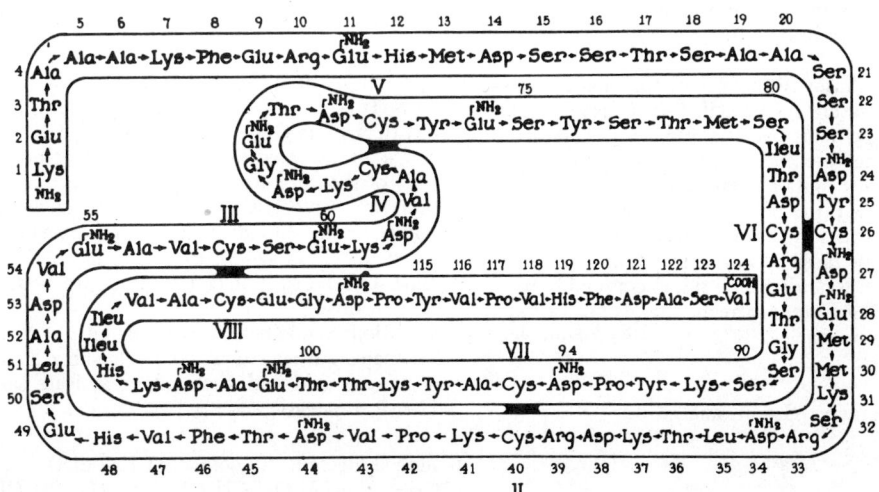

Ribonucleic acid

aggregate structure is the active template for protein biosynthesis. (3) Transfer RNA, the smallest and best characterized RNA class. Its molecules contain only 80 nucleotides per chain. Within the class of transfer RNA molecules there are probably at least 20 separate kinds, correspondingly related to each of the 20 amino acids naturally occurring in proteins. Transfer RNA must have at least two kinds of specificity: (a) It must recognize (or be recognized by) the proper amino acid activating enzyme so that the proper amino acid will be transferred to its free 2'- or 3'-OH group. (b) It must recognize the proper triplet on the messenger RNA-ribosome aggregate. Having these properties, the transfer RNA accepts or forms an intermediate transfer RNA-amino acid that finds its way to the polysome, complexes at a triplet coding for the activated amino acid, and allows transfer of the amino acid into peptide linkage. (4) Viral RNA isolated from a number of plant, animal and bacterial viruses, may be considered as a polycistronic messenger RNA. It has been shown to have molecular weights of one or two million. Generally speaking, there is one molecule of RNA per infective virus particle. The RNA of RNA virus can be separated from its protein component and is also infective, bringing about the formation of complete virus. From article by F. J. Bollum in "Encyclopedia of Biochemistry."
See also deoxyribonucleic acid.

D-ribose-5-phosphoric acid. CAS: 4300-28-1.
$C_5H_9O_4 \cdot H_2PO_4$. A constituent of nucleotides and nucleic acids.
Properties: The barium salt (5.5 H_2O) is sparingly soluble in cold water and crystallizes in hexagonal plates.
Derivation: From ionosinic acid.
Use: Biochemical research.

ribosome. A ribonucleoprotein, the smallest organized structure in the cell. Ribosomes occur in all cells including bacteria, fungi, algae, and protozoa. They are the central point of protein synthesis. They contain from 45–60% of ribonucleic acid (RNA), the balance being proteins. Ribosome crystals have been produced; in the electron microscope these appear as sheets of black dots, each sheet (one ribosome thick) containing hundreds of ribosomes in recurring groups of four.
See deoxyribonucleic acid. See also ribonucleic acid.

D-ribosyl uracil. See uridine.

rice paper. See straw.

Richards, Theodore W. (1868–1928) An American chemist born in Germantown, PA. He was the first American to receive the Nobel prize in chemistry. He studied chemistry at Haverford and Harvard with a doctorate in chemistry from Harvard where he later became Erving Professor of Chemistry. An outstanding experimental chemist, his major interests were atomic weights, thermochemistry, and thermodynamics. He was also a brilliant teacher. He was President of the ACS in 1914, and the recipient of many honorary awards, including the Davy, Faraday, and Gibbs medals.

"Ricilan."[493] TM for esters synthesized from castor oil and lanolin components. Amber, viscous liquids.
Use: Hydrophobic emollients; pigment dispersants.

ricin. White powder. The albumin of the castor oil bean is the toxic principle. Extracted from the pressed seeds with 10% solution of sodium chloride followed by precipitation with magnesium sulfate.
Hazard: Very highly toxic by ingestion; small particle in cut or abrasion, eye or nose may prove fatal.
Use: Reagent for pepsin and trypsin.

ricinine. (1,2-dihydro-4-methoxy-1-methyl-2-oxonicotinonitrile). CAS: 524-40-3.
$C_8H_8O_2N_2$.
Properties: White, crystalline alkaloid; slightly soluble in water, chloroform, and ether; mp 201.5C; sublimes at 170–180C at 20 mm pressure.
Derivation: From castor oil seeds and leaves.
Hazard: Toxic by ingestion.
Use: Biochemical research.

ricinoleic acid. (cis-12-hydroxyoctadec-9-enoic acid; 12-hydroxyoleic acid; castor oil acid). CAS: 141-22-0.
$CH_3(CH_2)_5CH(OH)CH_2CH:CH(CH_2)_7COOH$.
A C_{18} unsaturated fatty acid which comprises 80% of the fatty acid content of castor oil.
Properties: Colorless to yellow, viscous liquid; soluble in most organic solvents; insoluble in water; d 0.940 (27.4/4C); mp 5.5C; bp 226C (10 mm); refr index 1.4697 (20C); dextrorotatory. Combustible.
Derivation: Saponifiction of castor oil.
Use: Soaps, Turkey red oil, textile finishing, source of sebacic acid and heptanol, ricinoleate salts, 12-hydroxystearic acid.

ricinoleyl alcohol. (9-octadecen-1,12-diol).
$CH_3(CH_2)_5CH(OH)CH_2CH:CH(CH_2)_7CH_2OH$.
The fatty alcohol derived from ricinoleic acid. It has a long straight chain with one double bond

and one hydroxyl group in a secondary position besides the primary group on the end. Available as a 90% product.

Properties: Colorless non-drying liquid at room temperature, iodine value 91.8%, cloud point −12.2C, boiling range 170–328C, viscosity 51 (SSU/21C). Combustible.

Derivation: Reduction of acid made from castor oil.

Impurities: Oleyl and linoleyl alcohols.

Use: Protective coatings, polyesters, plasticizers, organic synthesis, pharmaceuticals, lubricants, surface active agents.

ricinus oil. See castor oil.

"Ridzlik."[15a] TM for an emulsifying agent developed for treatment of oil spills in both fresh and salt water. Claimed to have minimal toxicity to marine life and to be biodegradable.

Riehm quinoline synthesis. Formation of quinoline derivatives by prolonged heating of arylamine hydrochlorides with ketones with or without use of aluminum chloride or phosphorus pentachloride.

Riemschneider thiocarbamate synthesis. The action of concentrated sulfuric acid followed by treatment with ice-water serves to transform arylthiocyanates into the corresponding thiocarbamates.

"Rigortex."[11] TM for vinyl copolymer coating for severe corrosive conditions.

Riley oxidations. Oxidations of organic compounds with selenium dioxide, e.g., the oxidation of active methylene groups to carbonyl groups.

"Rilsan." TM for nylon 11.
See nylon.

RIM. Abbreviation for reaction injection molding.

ring compound. See cyclic compds, alicyclic, aromatic, heterocyclic.

"Rio Resin."[69,119] TM for a blend of resinous and protective materials.

Properties: Orange to dark red, d 1.13; softening point 54C minimum.

Use: Compounding heat-resistant copolymers and corona-resistant neoprene.

ripening. (cheese). See ageing, curing.

ristocetin. An antibiotic produced by the fermentation of *Nocardia lurida,* a species of Actinomycetes. The antibiotic has two components, risto-cetin A and ristocetin B. The commercial product is a lyophilized preparation representing a mixture of A and B, of which B comprises no more than 25%.

Properties: White or tan crystals or powder, practically odorless, freely soluble in water, practically insoluble in organic solvents.

Grade: USP.

Use: Medicine.

Ritter reaction. Formation of amides by addition of olefins or secondary and tertiary alcohols to nitriles in strongly acidic media.

Rittinger's law. The energy required for reduction in particle size of a solid is directly proportional to the increase in surface area.
See also Kick's law.

Rn. Symbol for radon.

RNA. Abbreviation for ribonucleic acid.

RNA polymerase. An enzyme essential in imparting the DNA genetic code to ribonucleic acid (RNA).

roasting. Heating in the presence of air or oxygen. Most commonly used in converting natural metal sulfide ores to oxides as a first step in recovery of metals such as zinc, lead, copper, etc. Roasting is an oxidation process.
See also smelting.

Robinson annellation reaction. Formation of six-membered ring α,β-unsaturated ketones by condensation of cyclohexanones with methyl vinyl ketone or its equivalents, followed by an intramolecular aldol condensation.

Robinson-Schopf reaction. Synthesis of tropinones from a dialdehyde, methylamine, and acetonedicarboxylic acid.

Robinson, Sir Robert. (1886-1975) An English chemist who won the Nobel prize for chemistry in 1947. His work began on plants which have biological significance, particularly alkaloids. He synthesized brazilin and haematoxylin. In organic chemistry he discovered an important qualitative electronic theory. He explicated the penicillin structure as well. His education, which began at the University of Manchester, took him throughout Europe.

Rochelle salt. See potassium-sodium tartrate.

rock crystal. See quartz.

rock flour. Mud-sized material ground from coarse grains by the movement of a glacier.

rocket fuel. (rocket propellant). A substance or mixture that has the capacity for extremely rapid but controlled combustion, which produces large volumes of gas at high pressure and temperature. They may be either liquid, solid, or combinations of both. Liquid monopropellants are hydrogen peroxide and hydrazine, catalyzed by finely divided metals to decompose them into gases. Liquid bipropellants consist of the fuel and an oxidizer; typical fuels of this type are hydrogen, hydrazine, ammonia, and boron hydride, the oxidizers being oxygen, nitric acid, ozone, hydrogen peroxide, and water.

Solid propellants include nitrocellulose, plasticized with nitroglycerin or various phthalates; inorganic salts suspended in a plastic or synthetic rubber (e.g., "Thiokol") and containing a finely divided metal. The inorganic oxidizers used are ammonium and potassium nitrates and perchlorates.

Rockwell hardness. See hardness.

"Rodar."[155] TM for an alloy composed of nickel 29%, cobalt 17%, manganese 0.30%, iron 53.7%.
Properties: Produces a permanent, vacuum-tight seal with simple oxidation procedure, resists mercury corrosion, readily machined and fabricated, can be welded, soldered, or brazed.
Forms: Wire, strip, bar.
Use: Sealing metal to hard glass.

rodenticide. A pesticide used to kill rats and other rodents.
See warfarin, squill.

"Rodine."[342] TM for red squill liquid extract rodenticides.

rod mill. A closed steel cylinder one-third filled with rods of about the same length as the cylinder and 1–2 inches in diameter. As the cylinder rotates the rods roll over one another, exerting a combination of impact and grinding action on the charge. It gives a product of 50–60 mesh with a minimum of fines. Rod mills are used for pulp grinding in the paper industry and for size reduction of ores, minerals, metal powders, etc.

Roentgen, W. K. (1845–1923) A German physicist who discovered x-rays in 1895 for which he was awarded the Nobel Prize in 1901. Application of these to a number of important problems in analytical chemistry was developed by the Braggs, Moseley, von Laue, and Debye and Sherrer.

roentgen. (r). The international unit of quantity or dose for both x-rays and gamma rays. It is defined as the quantity of x- or gamma rays which will produce as a result of ionization one electrostatic unit of electricity of either sign in 1 cc (0.001293 g) of dry air as measured at 0C and standard atmospheric pressure. The use of the roentgen unit has been extended to include particle radiation such as alpha and beta particles and protons and neutrons.
See also rad, curie.

Rohrbach solution.
Properties: Clear, yellow liquid, very refractive, d 3.5.
Derivation: An aqueous solution of mercuric barium iodide.
Hazard: Toxic by ingestion and inhalation.
Use: Separating minerals by their specific gravity, microchemical detection of alkaloids.

roll mill. Two chilled steel rolls 48–72 inches wide and 12–24 inches in diameter, turning in opposite directions at different speeds to exert shearing action; the separation (or nip) is adjustable by set screws. As the shearing friction generates considerable heat, the rolls are water-cooled. Such mills are standard equipment in the rubber, plastic and adhesives industries, several usually rotating on one shaft. They can be used for mixing, but their chief use is for prewarming calender and extruder feed. Mills with three rolls are used for mixing and grinding paints and printing inks. Laboratory sizes of all types are available.

"Romark."[448] TM for alkyd and chlorinated rubber type road-marking paints.

RON. Abbreviation for Research octane number.

ronnell. (O,O-dimethyl-O-(2,4,5- trichlorophenyl)phosphorothioate).   CAS: 299-84-3.
$(CH_3O)_2P(S)OC_6H_2Cl_3$.
Properties: Powder or granules, mp 41C, insoluble in water, soluble in most organic solvents.
Hazard: Toxic by ingestion and inhalation. Cholinesterase inhibitor, use may be restricted. TLV: 10 mg/m^3 of air.
Use: Insecticide.

room temperature. An ambient temperature from 20–25C (68–77F).

rosaniline. CAS: 632-99-5.
$HOC(C_6H_4NH_2)_2C_6H_3(CH_3)NH_2$.   A triphenylmethane dye.
Properties: Reddish-brown crystals, mp 186C (decomposes), soluble in acids and alcohol, slightly soluble in water.

Use: Dye (usually as the hydrochloride), fungicide.

roscoelite. $K_2V_4Al_2Si_6O_{20}(OH)_4$. A vanadium-bearing species of mica. Formula variable with V_2O_3 up to 28%. Occurs as minute scales with micaceous cleavage, dark green to brown in color, pearly luster, Mohs hardness 2.5, d 3.0.
Occurrence: Colorado, California, Australia.
Use: Source of vanadium.

Rose Bengal. CAS: 11121-48-5.
$C_{20}H_4Cl_4I_4O_5$.
Properties: Bluish-pink powder, soluble in water. CI 45440.
Use: Biological stain, colorant for inks, cellulosics, foods, cosmetics, medicine (diagnostic aid).

Rosenheim color test. On addition of nine parts of trichloroacetic acid in water to a solution of ergosterol in chloroform, an immediate red color develops, gradually changing to pure blue. The test is specific to sterols containing a diene system or capable of forming one on acid dehydration.

Rosenmund reaction. Formation of aromatic arsonic acids by heating equimolar amounts of sodium or potassium arsenite with aryl halides in aqueous solution. The reaction is an extension of the Meyer reaction.

Rosenmund reduction. Catalytic reduction of acid chlorides to aldehydes. To prevent further hydrogenation, a poison is added to the catalyst.

Rosenmund-von Braun synthesis. Formation of aromatic nitriles from aryl halides and cuprous cyanide.

rosin.
Properties: Angular, translucent, amber-colored fragments. D 1.08, mp 100–150C, acid number above 150, flash p 370F (187C). Insoluble in water; freely soluble in alcohol, benzene, ether, glacial acetic acid, oils, carbon disulfide, dilute solutions of fixed alkali hydroxides. Hard and friable at room temperature; soft and very sticky when warm. Combustible. Chief constituents: Resin acids of the abietic and pimaric types, having the general formula $C_{19}H_{29}COOH$ and having a phenanthrene nucleus.
See also turpentine.
Derivation: From pine trees, chiefly Pinus palustris and Pinus caribaea. (a) Gum rosin is the residue obtained after the distillation of turpentine oil from the oleoresin tapped from living trees. (b) Wood rosin is obtained by extracting pine stumps with naphtha and distilling off the volatile fraction. (c) Tall oil rosin is a byproduct of the fractionation of tall oil.

Grade: Virgin, yellow dip, hard, NF. Wood rosin is grades B, C, D, E, F, FF, G, H, I, J, K, L, M, N, W-G (window-glass), W-W (water-white). The grading is done by color, B being the darkest and W-W the lightest.
Hazard: Evolves irritating and suffocating fumes on heating.
Use: Hot-melt and pressure-sensitive adhesives, mastics and sealants, varnishes, ester gum, soldering compds, core oils, insulating compds, soaps, paper sizing, printing inks, polyesters (formed by reaction of the conjugated acids of rosin with acrylic acid, followed by reaction with a glycol).
See also abietic acid.

rosin oil.
Properties: Water-white to brown liquid, viscous, odorless, strong peculiar taste. Soluble in ether, chloroform, fatty oils, and carbon disulfide; slightly soluble in alcohol, insoluble in water. Essentially decarboxyalted rosin acids, d 0.980–1.110, iodine number 112–115.
Derivation: By fractional distillation of rosin, that portion distilling above 360C being rosin oil.
Hazard: Spontaneous heating, fire risk when heated.
Use: Lubricant, adulterant for boiled linseed oil, hot-melt adhesives, printing inks, impregnating paper for wrapping electric cables, core compounds, varnishes.

rosin soap. See sodium abietate and soap.

"Roskydal."[470] TM for a series of unsaturated polyesters curing with peroxides for coating applications.

rosolic acid. See aurin.

rotameter. An instrument for measuring the flow rates of liquids and gases.

"Rotax."[256] TM for a purified grade of 2-mercaptobenzothiazole.
Use: Primary accelerator for rubber.

rotenone. (tubatoxin). CAS: 83-79-4.
$C_{23}H_{22}O_6$. A pentacyclic compound.
Properties: White, odorless crystals; soluble in ether, alcohol, acetone, and other organic solvents; insoluble in water; not compatible with alkalies. D 1.27 at 20C, mp 163C, strongly levorotatory in solution, specific rotation for D line 230 degrees in benzene, 62 degrees in ethylene dichloride.
Grade: CP crystals, technical, also as extracts of derris and cube root.
Hazard: Toxic by ingestion, overexposure can be fatal, irritant to skin. TLV: 5 mg/m³ of air.

Use: Insecticide, flea powders, fly sprays, moth-proofing agents.

Rothemund reactions. Preparation of meso-tetra-substituted porphyrins by condensation of pyrrole with an aldehyde.

"Rotoclor."[56] TM for ferric chloride solution with other additives specially prepared for roto-gravure work. Contains 43% $FeCl_3$. Sold only in 155-lb carboys.

rottenstone. Soft, friable aluminum silicate.
Use: Polishing agent.

rouge. (1) A high-grade red pigment used as a polishing agent for glass, jewelry, etc.
See also iron oxide red.
  (2) A cosmetic prepared from dried flowers of the safflower.

"Rovana."[565] TM for a thermoplastic vinylidene chloride copolymer filament in the form of a folded flat tape, offered in 300, 400, and 550 deniers.

Rowe rearrangement. Intramolecular rearrangement of pseudo-phthalazones into phthalazones by heating at 180 degrees in the presence of dilute aqueous acids.

Roxel process. A process for increasing the flame resistance of cotton by chemically modifying the fiber, i.e., by applying a formulation consisting of tetrakis(hydroxymethyl)phosphonium chloride (THPC), triethylolamine, and urea in aqueous solution. These substances crosslink with the hydroxyl groups of the cellulose to form an efficient and durable protective medium. This process is used in the manufacture of blankets, curtains, bed linen, and industrial safety garments.

"Royalac."[248] TM for an accelerator for EPDM elastomers.

"Royalene."[248] TM for a low-cost high-performance ethylene-propylene terpolymer.
Properties: Resistance to ozone, heat, weather, sunlight aging, steam, and chemicals. Resilient at low temperatures, excellent electrical properties and can be easily colored.
Use: Tires, automotive parts, tank cars, hoses, and insulation.

royal jelly. A complex mixture secreted by "worker" honeybees which comprises the sole nutrient of the queen bee. It contains 31% protein, 15% carbohydrate, 15% lipid, plus vitamins. The free fatty acid portion of the lipid is a mixture of C_{10} acids.

"Royal Methyl Violet."[141] TM for violet pigment produced by precipitation of the basic methyl violet dyestuff with phosphomolybdic acid.
Use: Printing inks.

"Royal Spectra."[133] TM for an impingement carbon black.
Use: Specialty application requiring highest blackness and reinforcing power.

"Royal Victoria Blue."[141] TM for blue pigment produced by precipitation of basic Victoria Blue dye with phosphomolybdic acid.
Use: Printing inks and some paints.

RR acid. (2-amino-8-naphthol-3,6-disulfonic acid; 2R acid). $C_{10}H_4NH_2OH(SO_3H)_2$.
Derivation: Fusion of a naphthylamine trisulfonic acid with sodium hydroxide.
Use: Azo dye intermediate.

"RSR."[173] TM for a proteolytic enzyme preparation for removal of stains, available in powder form specifically designed for removing albuminous spots and stains from garments.

"RTV."[245] TM for a family of silicone rubber compounds. *Note:* "RTV" is reported to have been the adhesive used to bond refractory tiles to the body of "Columbia" space shuttle (1981).

RTV. (room-temperature-vulcanizing).
Rubbers that have good physical properties and electrical properties similar to silicone rubber.
Use: Sealing, caulking, encapsulating and flexible mold-making in electronic, aircraft, missile and building industries.
See also silicone.

Ru. Symbol for ruthenium.

"Rubanox Red."[143] TM for lithol rubine pigments of bright bluish-red shades. Composed of calcium salts of azo pigments formed when 4-aminotoluene-3-sulfonic acid is coupled with β-hydroxynaphthoic acid.
Use: Printing inks, paints, enamels, lacquers, rubber, plastics, wallpaper, floor coverings.

rubber. Any of a number of natural or synthetic high polymers having unique properties of deformation (elongation or yield under stress) and elastic recovery after vulcanization with sulfur or other crosslinking agent, which in effect changes the polymer from thermoplastic to thermosetting. The yield or stretch of the vulcanized material ranges from a few hundred to over 1000%. The deformation after break, called "permanent set," is usually taken as the index of recovery; it ranges from 5–10% for natural

rubber to 50% or more for some synthetic elastomers, and varies considerably with the state of vulcanization and the pigment loading.
See also elastomer and following entries.

rubber cement. See adhesive, rubber-based.

rubber, chlorinated. An elastomer (natural rubber or a polyolefin) to which 65% of chlorine has been added to give a solid, film-forming resin. White amorphous powder available in viscosity grades from 5–125 centipoises, the figures indicating viscosity of a 20% solution in toluene. Decomposes at 125C, soluble in aromatics, insoluble in aliphatics and alcohols. Compatible with almost all natural and synthetic resins. Chief use is in maintenance paints (marine, swimming pool, traffic, masonry, etc.).
See also "Parlon," "Hypalon," rubber hydrochloride.
Hazard: Do not dry-mill chlorinated rubber with zinc oxide; mixture reacts violently at 216C. Do not use in baked enamels.

rubber, cold. See cold rubber.

rubber fiber. Generic name for a manufactured fiber in which the fiber-forming substance is comprised of natural or synthetic rubber (Federal Trade Commission). Often the rubber is a core around which cotton or other fibers are wrapped to make an elastic yarn used for girdles, swimwear, elastic bands, and tapes.

rubber, hard. A rubber compounded with 30–50% by weight of sulfur and cured until an extremely hard, brittle product is formed. Lime or magnesia is used as activator. The theoretical maximum of sulfur that can combine chemically with rubber hydrocarbon is 32%. Combustible.
Hazard: Flammable in form of dust.
Use: Battery boxes, tank linings, acid- and alkali-resistant equipment, combs. As dust, filler for low-cost rubber products.

rubber hydrochloride. A hydrochloride derivative, as distinct from a chlorine derivative.
Properties: Thermoplastic white powder or clear film, odorless, tasteless, nonflammable, nontoxic, chlorine content 29–30.5%, soluble in aromatic hydrocarbons, softens at 110–120C. Films are highly resistant to moisture, oils, acids, and alkalies but tend to become brittle on exposure to sunlight. The life of such films is greatly extended by the incorporation of suitable stabilizers and plasticizers.
Derivation: A solution of natural rubber is treated with anhydrous hydrogen chloride under pressure and at low temperature. After neutralization of excess hydrochloric acid the product is precipitated by the addition of ethanol.
Use: Protective coverings for machinery, rainclothing, shower curtains, food packaging.

rubber latex. See latex

rubber, liquid. Any of several proprietary products consisting of high polymers in liquid form for use as coatings, adhesives, etc.

rubber, natural. (polyisoprene).
CAS: 9006-04-6. $(C_5H_8)_x$.

$$\left(\begin{array}{c} \quad H \quad CH_3 \\ \quad | \quad\ | \\ CH_2{=}C{-}C{=}CH_2 \end{array} \right)_x$$

(a) Crude (unvulcanized):
Properties: Chemically unsaturated, d 0.92; amorphous when unstretched, but has oriented crystalline structure on stretching; not stable to temperature changes (thermoplastic), readily oxidizable by mastication; soluble in acetone, carbon tetrachloride, and most organic solvents; refr index 1.52; dielectric constant 2.5. Processed by calenders and extruders; can be injection-molded with low sulfur and high accelerator. Cured by hot-molding or in open steam, at temperatures from 120–150C after addition of 3% sulfur, 1% organic accelerator, 3% zinc oxide, plus fillers or reinforcing agents. The only factors of significance in vulcanization are the time of exposure to heat and the temperature used.
Derivation: From latex obtained from Hevea trees, coagulated with acetic or formic acid. Also made synthetically.
See "Coral," "Natsyn."
Occurrence: Brazil, Malaysia, Indonesia.
Grade: Ribbed smoked sheets, pale (yellow) crepe, brown crepe.
Use: Cements, adhesives, electrical insulating tapes and cable wrapping. (b) Cured (vulcanized, i.e., sulfur crosslinkages):
Properties: High tensile strength; relatively low permanent set; insensitive to temperature changes; attacked by heat, atmospheric oxygen, ozone, hydrocarbons, and unsaturated fats and oils; insoluble in acetone; permeable to gases; supports combustion; abrasion resistance poor unless compounded with carbon black; dissipates vibration shock; high electrical resistivity.
Use: Vehicle tires, hose, conveyor belt covers, footwear, specialized mechanical products, drug sundries, foam rubber, electric insulation, etc.
Note: Gutta percha and balata have similar chemical composition (isomeric) but have very different properties and few commercial uses. Neither can be vulcanized.

See Appendix II for history of the industry. See also latex, guayule.

rubber sponge. (foam rubber; cellular rubber). A flexible foam produced by beating air into heat sensitized latex, with subsequent vulcanization, or by incorporating ammonium carbonate or sodium bicarbonate into a strongly masticated and highly accelerated rubber mixture. As the temperature rises to the curing range, ammonia or carbon dioxide is released, forming uniform pores throughout the mixture just before the onset of vulcanization.
Use: Vibration damping pads and inserts, rug and carpet underlays, mattresses and upholstery, seat cushions.

rubber, synthetic. Any of a group of man-made elastomers which approximate one or more of the properties of natural rubber. Some of these are: sodium polysulfide ("Thiokol"), polychloroprene (neoprene), butadiene-styrene copolymers (SBR), acrylonitrilebutadiene copolymers (nitrile rubber), ethylenepropylene-diene (EPDM) rubbers, synthetic polyisoprene ("Coral," "Natsyn"), butyl rubber (copolymer of isobutylene and isoprene), polyacrylonitrile ("Hycar"), silicone (polysiloxane), epichlorohydrin, polyurethane ("Vulkollan").
The properties of these elastomers are widely different. All require vulcanization. In general, sulfur is used only for unsaturated polymers; peroxides, quinones, metallic oxides, or diisocyanates effect vulcanization with saturated types. Many are special-purpose rubbers, some can be used in tires when loaded with carbon black, others have high resistance to attack by heat and hydrocarbon oils and thus are superior to natural rubber for steam hose, gasoline and oil-loading hose. Most are available in latex form.
See specific type for details.

rubber, thermoplastic. Any of several block copolymers of propylene/EPDM or styrene/ethylene-butylene. Crosslinking results from crystallization of polypropylene or polystyrene segments. Since this is reversible on heating, the product is thermoplastic. Its chief use is in oil-resistant wire and cable insulation.

ruberythric acid. $C_{26}H_{28}O_{14}$. An alizarin glucoside.

rubidium. Rb. CAS: 7440-17-7.
Metallic element of atomic number 37, group IA of the periodic table, aw 85.4678, valence = 1. One stable form, principal natural radioactive isotope is rubidium 87. It is the second most electropositive and the second most alkaline element, has lowest ionization potential. Highly reactive.

Properties: Soft, silvery-white solid, easily oxidized in air, soluble in acids and alcohol, decomposes water, d 1.532 (20C as solid), mp 39C, bp 688C, high heat capacity and heat transfer coefficient.
Derivation: (a) Thermochemical reduction of rubidium chloride with calcium, (b) electrolysis of fused cyanide or chloride.
Source: Lepidolite, carnallite, pollucite, mineral springs, and natural brines.
Grade: Technical, 99+%.
Hazard: Reacts vigorously with air and water, must be stored under kerosene or similar liquid, dangerous fire and explosion risk. Metal causes serious skin burns.
Use: Photocells, catalyst or catalyst promotor.

rubidium alum. See aluminum rubidium sulfate.

rubidium carbonate. Rb_2CO_3.
Properties: White powder, mp 837C, extremely hygroscopic, soluble in water, dissociates above 900C.
Hazard: Strong irritant to tissue.
Use: Special glass formulations.

rubidium chloride. CAS: 7791-11-9. RbCl.
Properties: White, crystalline powder; lustrous. When heated it decrepitates, melts, and volatizes. Soluble in water, almost insoluble in alcohols, d 2.76, mp 715C, bp 1390C.
Grade: Technical, CP, single crystals.
Use: Analysis (testing for perchloric acid), source of rubidium metal.

rubidium fluoride. RbF.
Properties: White crystals, d 3.557, mp 775C, bp 1410C, soluble in water, insoluble in alcohol, single large crystals available.
Hazard: Strong irritant to tissue. TLV (as F): 2.5 mg/m^3 of air.

rubidium hexafluorogermanate. Rb_2GeF_6.
Properties: White, crystalline solid; mp 696C; slightly soluble in cold water; very soluble in hot water.

rubidium hydroxide. (rubidium hydrate). CAS: 1310-82-3. RbOH.
Properties: Grayish-white mass; extremely hygroscopic; absorbs carbon dioxide from air; strong base; soluble in alcohol, water; d 3.2; mp 300C; attacks glass at room temperature.
Hazard: Strong irritant to tissue.
Use: Suggested for electrolyte in storage batteries for low-temperature use.

ruby. Synthetic rubies, usually in the form of rods, are made from aluminum oxide containing small amounts of other metals, such as 0.05% chromic oxide, by single-crystal-growing techniques. These are used extensively in masers and lasers.

See also corundum.

ruby glass. See glass, ruby.

Ruff-Fenton degradation. Shortening of the carbon chain of sugars by the oxidation of aldonic acids (as calcium salts) with hydrogen peroxide and ferric salts.

"Ruflux."[337]. TM for a series of titanium and zirconium products (potassium and sodium titanates, rutile, zirconium oxide, and zircon in specially prepared forms).

Use: Metallurgical and welding products.

runner. The secondary feed channel (usually circular) in a multiple-cavity injection mold that connects the sprue with the gate. In certain cases, molding can be performed satisfactorily without runners (runnerless molding).

rust. (1) The reddish corrosion product formed by electrochemical interaction between iron and atmospheric oxygen; ferric oxide (Fe_2O_3). The reaction occurs most rapidly in moist air, indicating the catalytic activity of water.

(2) A reddish-yellow fungus which attacks plants, especially cereal grains; it is also called smut. It can be controlled by treatment with formaldehyde.

"Rust-Ban."[29]. TM for a series of protective coatings which include alkyd-based, epoxy ester-based, converted epoxy-based, oil-based, phenolic-based, silicone-based, vinyl-based, and inorganic zinc coatings. The inorganic zinc coatings are primarily for heavy-duty protection of steel surfaces and are used on marine vessels and equipment, steel structures, offshore installations, and heavy industrial equipment.

ruthenium. CAS: 7440-18-8. Ru.
Metallic element of atomic number 44, group VIII of the periodic system, aw 101.07, valences = 3, 4, 5, 6, 8. Seven stable isotopes.
Properties: Silvery white solid of the platinum metal group, d 12.41 (20C), mp 2310C, bp 3900C, Brinell hardness 220 (cast), insoluble in acids and in aqua regia, attacked by alkaline oxidants (concentrated sodium hydroxide solution) and by fused alkalies.
Derivation: Occurs with platinum, from which it is recovered.
Grade: Technical, single crystals.
Use: Hardener for platinum and palladinum in jewelry, electrical contact alloys, catalyst, medical instruments, corrosion-resistant alloys, electrodeposited coatings, nitrogen-fixing agent (experimental), solar cells (experimental); the oxide

is used to coat titanium anodes in electrolytic production of chloride; the dioxide serves as an oxidizer in photolysis of hydrogen sulfide.

ruthenium chloride. (ruthenic chloride; ruthenium sesquichloride).   CAS: 10049-08-8. $RuCl_3$.
Properties: Black solid, deliquescent, d 3.11, decomposes above 500C, insoluble in cold water, decomposes in hot water, slightly soluble in alcohol.
Grade: Technical, CP.
Use: Analysis (testing for sulfur trioxide).

ruthenium red. (ammoniated ruthenium oxychloride).   CAS: 12790-48-6.
$Ru_2(OH)_2Cl_4 \cdot 7NH_3 \cdot 3HOH$.
Properties: Brownish-red powder, soluble in water.
Use: Microscopic stain and reagent for pectin, plant mucin, and gum.

ruthenium tetroxide. CAS: 20427-56-9.
RuO_4.
Properties: Yellow crystals, sublimes at room temperature, mp 25C, bp 100C, strong oxidizing agent, soluble in carbon tetrachloride.
Derivation: Acidic solutions of ruthenium compounds are heated with strong oxidizing agents.
Hazard: Fire risk in contact with organic materials.
Use: Oxidizing agent (carbon tetrachloride as solvent).

ruthenocene. $(C_5H_5)_2Ru$.   Dicyclopentadienylruthenium.
Properties: A stable, light yellow solid; mp 199C.
Use: Intermediate for high-temperature compounds and for UV radiation absorbers in paints.

Rutherford, Sir Ernest. (1871–1937) Born in New Zealand, Rutherford studied under J. J. Thomson at the Cavendish Laboratory in England. His work constituted a notable landmark in the history of atomic research as he developed Bacquerel's discovery of radioactivity into an exact and documented proof that the atoms of the heavier elements, which had been thought to be immutable, actually disintegrate (decay) into various forms of radiation. Rutherford was the first to establish the theory of the nuclear atom and to carry out a transmutation reaction (1919) (formation of hydrogen and an oxygen isotope by bombardment of nitrogen with alpha particles). Uranium emanations were shown to consist of three types of rays, alpha (helium nuclei) of low penetrating power, beta (electrons), and gamma, of exceedingly short wavelength and great energy. Rutherford also discovered the half-life of radioactive elements and applied this to studies of age determination of rocks by measuring the decay period of radium to lead-206. See Aston; Bohr.

rutile. TiO_2. Natural titanium dioxide. May contain up to 10% iron.
Properties: Color red, reddish-brown to black, luster adamantine to submetallic, streak pale brown, hardness 6–6.5, d 4.3, mp 1640C, refr index 2.7, insoluble in acids.
Occurrence: Virginia, West Africa, Australian beach sands.
Use: Source of titanium and titanium compounds; ceramics; steel deoxidizer; welding rod coatings; pigment for paints, enamels, and tile.

"Rutile, Ceramic."[317]. TM for 92% titanium oxide. Light brown powder, bulk d 98 lb/ft³, mp 1793C, particle size less than 44 microns.
Use: Stable stains for ceramic glazes and bodies.

Ruzicka large ring synthesis. Formation of large ring alicyclic ketones from dicarboxylic acids by thermal decomposition of salts with metals of the second and fourth groups of the periodic system (calcium, thorium, cerium).

Ruzicka, Leopold. (1887-1976) A chemist who won the Nobel prize in 1939 with Butenandt. His work involved research in organic synthesis including polymethylenes and higher terpenes.

He was the first chemist to synthesize musk, androsterone, and testosterone from cholesterol. His medical degree was awarded at the University of Basel, Switzerland, although he was born in Croatia and educated partially in Germany.

ryania. An insecticidal principle extracted from the wood of a family of tropical American trees (Ryania speciosa). In commercial practice the wood is fine-ground and used as a component of certain insecticide formulations.
Hazard: Moderately toxic.

"Rynite."[28]. TM for a glass-reinforced polyester engineering plastic; it is extremely stiff, with glass content of 45%. A modified polyethylene terephthalate featuring high temperature resistance, high tensile and impact strength, and good electrical resistance.

"Ryton."[303]. TM for a polyphenylene sulfide plastic.
Use: Coatings and molded parts.

"RZ-50-A."[58]. TM for a 50% solution of N,N-dimethylcyclohexylamine salt of dibutyldithiocarbamic acid. Also available as "RZ-50-B" with emulsifying agents for use in latex.
Use: Rubber accelerator.

S

S. Symbol for sulfur.

Sabatier-Senderens reduction. Catalytic hydrogenation of organic compounds in the vapor phase by passage over hot, finely divided nickel (the oldest of all hydrogenation methods).

Sabatier, Paul. (1854-1941) A French chemist who received the Nobel prize in chemistry in 1912 along with Victor Grignard. His work involved the behavior of oxides as oxidizing catalysts and as agents for dehydrating and dehydrogenating. He received his PhD in Nimes, France, and went on to become lecturer and faculty member in Toulouse, France.

d-saccharic acid. (2,3,4,5-tetrahydroxyhexanedioic acid; tetrahydroxyadipic acid).
COOH(CHOH)$_4$COOH. The 1,6-dicarboxylic acid formed by the oxidation of d-glucose.
Properties: White needles or syrup; very soluble in water, alcohol, or ether; deliquescent; mp 125–126C with decomposition. Combustible.
Derivation: Oxidation of cane sugar, glucose, starch by nitric acid.

saccharification. Conversion of wood and other cellulosics (biomass) to glucose by acid hydrolysis or enzymic hydrolysis. An experimental method using anhydrous hydrogen fluoride with vacuum distillation is under development.

saccharin. (o-benzosulfimide; gluside; benzoylsulfonic imide). CAS: 81-07-2. The anhydride of o-sulfimide benzoic acid. A non-nutritive sweetener.

Properties: White, crystalline powder; exceedingly sweet taste (500 times that of sucrose); mp 226–230C; soluble in amyl acetate, ethyl acetate, benzene, and alcohol; slightly soluble in water, chloroform, and ether.
Derivation: A mixture of toluenesulfonic acids is converted into the sodium salt then distilled with phosphorus trichloride and chlorine to obtain the o-toluene sulfonyl chloride which, by means of ammonia, is converted into o-toluenesulf-amide. This is oxidized with permanganate, treated with acid, and saccharin crystallized out. In food formulations, saccharin is used in the form of its sodium and calcium salts. Sodium bicarbonate is added to provide water solubility.
Grade: Commercial, CP, USP, FCC.
Hazard: The National Academy of Sciences has stated that saccharin is a weak carcinogen in animals and a potential human carcinogen. Products containing it must have warning label.

saccharolactic acid. See mucic acid.

saccharose. See sucrose, carbohydrate.

saccharose unit. See carbohydrate.

"SACI."[104] TM for a rust and corrosion-preventive product effective in coatings as thin as 0.5 mil.

Sachsse process. See BASF process.

S acid. (1-amino-8-naphthol-4-sulfonic acid).
Properties: Gray needles, white when pure, slightly soluble in water, insoluble in alcohol and ether.
Derivation: Fusion of 1-naphthylamine-4,8-disulfonic acid with sodium hydroxide.
Use: Azo dye intermediate.

2S acid. See Chicago acid.

"Sacon."[173] TM for a concentrated, water-soluble material used to restore sizing in dry-cleaned garments.

sacrificial protection. The preferential corrosion of a metal coating for the sake of protecting the substrate metal. For example, when zinc is in contact with a more noble (less reactive) metal in the electromotive series, such as steel, a galvanic cell is created in which electric current will flow in the presence of an electrolyte. Atmospherically contaminated moisture constitutes the electrolyte. Under these conditions, the zinc coating rather than the steel is affected. Thus, galvanic protection with zinc is sometimes called sacrificial protection.
See also galvanizing.

SAE. Abbreviation for Society of Automotive Engineers. The initials are applied to its specifications and test for motor fuels, oils, and steels.

SAE steel. A grade or type of steel indicated by a number system, they are principally plain carbon and of low to medium alloy content, used primarily in machine parts. The first two numbers designate either plain carbon or the alloy grouping and quantity, and the last two the mean carbon content in hundredths of 1%. Thus, 10 are carbon steels; 13 manganese steels; 40 and 44 molybdenum steels (the latter of higher alloy content); 50 and 51 chromium steels (the latter of higher alloy content); 41 chromium-molybdenum; 61 chromium-vanadium; 92 silicon and silicon-chromium; 46 and 48 nickel-molybdenum (the latter of higher nickel content); and 81, 94, 86, 87, 88, 47, 43, and 93 nickel-chromium-molybdenum in the order of increasing alloy content. The letter B between the first two and last two numerals indicates the presence of boron in amounts of 0.0005–0.003% as a depth-hardening addition.

SAF black. Abbreviation for super-abrasion furnace black.

safety engineering. Application of engineering principles to chemical plant safety by professionally trained personnel. Following is a check list of the more important items.

(1) Plant construction: separate buildings or outdoor location of hazardous reaction vessels, storage tanks, etc.; interior fire walls and doors, exterior blow-out walls, sprinkler systems, enclosed stairways, explosion vents and safety valves, scram alarm systems, color-coded pipelines.

(2) Fire and explosion prevention, dust control, proper storage of flammable liquids, grounding of electrical equipment, accessibility of extinguishers and hose lines, leak detection of reaction vessels, adequate ventilation of storage rooms, accumulation of solid wastes, static spark control (metal-free shoes, static bars on friction-generating machinery).

(3) Toxic hazards: workroom concentration of toxic agents must conform to OSHA and ACGIH tolerances.

(4) Protective equipment: goggles and gloves, acid-proof clothing, face masks and respirators, lifelines, eye-wash fountains, flooding showers; for hot labs and plants handling radioactive materials, decontamination equipment, glove boxes, and remote-control devices.

(5) Accident prevention: emergency shut-offs on machines with moving parts; housing on gears, lathes, rotors, etc.; operator-restraining devices; proper handling of chain hoists, carton stackers, pallets; training personnel in safety practice.

safety glass. (shatterproof glass). A composite or laminate consisting of two sheets of plate glass with an interlayer of polyvinyl butyral. The plastic is so bonded to the glass that shattering on impact is virtually eliminated. Required for automobile windshields, also used for bulletproof glass, train windows, etc.

"Saffil." TM for a group of synthetic inorganic fibers made from alumina and zirconia.

safflower oil. Drying oil from safflower (carthamus) seed, somewhat similar to linseed oil. It is non-yellowing. Contains 78% linoleic acid (unsaturated fatty acid).
Properties: Straw-colored liquid, d 0.923–0.927 (25/25C), refr index 1.4740–1.4745 (25C), acid value 0.6–1.5, iodine no. 140–152, saponification no. 186–193, unsaponifiable 0.3–1.0%. Combustible.
Derivation: Hydraulic or solvent extraction of seeds.
Use: Alkyd resins, paints, varnishes, medicine, dietetic foods, margarine, hydrogenated shortening.

"Saflex."[58] TM for polyvinyl butyral adhesive film supplied in clear, translucent, tinted, or graduated color, for the plastic interlayer in safety glass.

safranine. A family of dyes, some of which are useful as biological stains, based on phenazine, having CI Nos. 50200–50375

safrole. (4-allyl-1,2-methylenedioxybenzene).
CAS: 94-59-7. $C_3H_5C_6H_3O_2CH_2$.
Properties: Colorless or pale yellow oil, the odor giving constituent of sassafras, camphorwood, and other oils. D 1.100–1.107 (15C), mp 11C, bp 233C, optical rotation −0 degrees 30 minutes (15C), refr index 1.5363–1.5385 (20C), soluble in alcohol, slightly soluble in propylene glycol, insoluble in water and glycerol.
Derivation: From oil of sassafras or camphor oil.
Method of purification: Rectification or freezing.
Grade: Technical.
Hazard: Toxic by ingestion, may not be used in food products (FDA), a carcinogen.
Use: Perfumery and soaps, manufacture of heliotropin, medicine (antiseptic).

"SAG."[214] TM for silicone defoamers used in many industrial and chemical processes.

sage oil. A yellow-to-green essential oil used in perfumery and flavoring; dextrorotatory. There are two varieties (Clary and Dalmatian) which have different constituents.

SAIB. Abbreviation for sucrose acetate isobutyrate.

sal ammoniac.   See ammonium chloride.

salicin.   (salicyl alcohol glucoside).
CAS: 138-52-3. $HOCH_2C_6H_4OC_6H_{11}O_5$.
A glucoside obtained from several species of Salix and Populus.
Properties: Colorless crystals or white powder; soluble in water, alcohol, alkalies, glacial acetic acid; insoluble in ether; mp 199–202C; d 1.43.
Use: Medicine (analgesic), reagent for nitric acid.

salicylal.   See salicylaldehyde.

salicyl alcohol.   (o-hydroxybenzyl alcohol; α-2-dihydroxytoluene; saligenin).   CAS: 90-01-7.
$HOC_6H_4CH_2OH$.
Properties: White crystals; mp 86–87C; pungent taste; sublimes at 100C; very soluble in alcohol, chloroform, ether; soluble in propylene glycol, benzene, and fixed oils; sparingly soluble in cold water, soluble in hot water, d 1.16. Combustible.
Derivation: Hydrolysis of salicin, heating phenol and methylene chloride with caustic.
Use: Medicine (local anesthetic).

salicyl alcohol glucoside.   See salicin.

salicylaldehyde.   (salicylal; salicylic aldehyde; o-hydroxybenzaldehyde).   CAS: 90-02-8.
C_6H_4OHCHO.

Properties: Colorless, oily liquid or dark red oil; bitter, almond-like odor; burning taste. D 1.165–1.172, fp −7C, bp 196C. Soluble in alcohol, ether, and benzene; slightly soluble in water; flash p 172F (77.7C). Combustible.
Derivation: Interaction of phenol and chloroform in presence of aqueous alkali.
Use: Analytical chemistry, perfumery (violet), synthesis of coumarin, auxiliary fumigant, flavoring.

salicylamide.   (o-hydroxybenzamide).
CAS: 65-45-2. $C_6H_4(OH)CONH_2$.
Properties: White or slightly pink crystals; mp 139–142C; bp decomposes at 270C; soluble in hot water, alcohol, ether, chloroform; slightly soluble in cold water, naphtha, and carbon tetrachloride.
Derivation: Treatment of methyl salicylate with dry ammonia gas.
Grade: Technical, NF.
Use: Medicine (analgesic).

salicylanilide.   CAS: 87-17-2.
$HOC_6H_4CONHC_6H_5$.
Properties: Odorless, white, or slightly pink crystals. Mp 136–138C, bp decomposes, stable in air. Slightly soluble in water; freely soluble in alcohol, ether, chloroform and benzene.
Grade: NF.
Hazard: Toxic by ingestion, irritant to skin.
Use: Fungicide, slimicide, antimildew agent, intermediate.

salicylic acid.   (o-hydroxybenzoic acid).
CAS: 69-72-7. $C_6H_4(OH)(COOH)$.

Properties: White powder, acrid taste, stable in air but gradually discolored by light, d 1.443 (20/4C), mp 158–161C, bp 211C at 20 mm, sublimes at 76C. Soluble in acetone, oil of turpentine, alcohol, ether, benzene; slightly soluble in water. Combustible.
Derivation: Reacting a hot solution of sodium phenolate with carbon dioxide and acidifying the sodium salt thus formed.
Grade: Technical, USP, crude.
Hazard: Dust forms explosive mixture in air. Toxic by ingestion.
Use: Manufacture of aspirin and salicylates, resins, dyestuff intermediate, prevulcanization inhibitor, analytical reagent, fungicide.

salicylic acid dipropylene glycol monoester.
See dipropylene glycol monosalicylate.

salicylic aldehyde.   See salicylaldehyde.

saligenin.   (salicyl alcohol).
$C_6H_4(OH)CH_2OH$.
Use: Treatment for rheumatism.

saline water.   See brine, ocean water.

salinity.   The saltiness of natural water. The salinity of normal sea water is 35 parts salt per 1000 parts water.

salmine.   A protein specific to the salmon.
Use: Nutritional and biochemical research.

salol.   See phenyl salicylate.

saloquinine.   (salicyl quinine.)
$HOC_6H_4COOC_{20}H_{23}N_{20}$.
Use: Antipyretic, antiperiodic.

sal soda.   (washing soda; sodium carbonate decahydrate). $Na_2CO_3 \cdot 10HOH$.
Properties: White crystals, d 1.44, mp 32.5–34.5C, loses water at this temperature, easily soluble

in water, insoluble in alcohol, a pure form of sodium carbonate (soda ash).

Use: Washing textiles, bleaching linen and cotton, general cleanser.

salt. (1) The compound formed when the hydrogen of an acid is replaced by a metal or its equivalent, (e.g., an NH_4 radical). Example: $HCl + NaOH \rightarrow NaCl + HOH$. This is typical of the general rule that the reaction of an acid and a base yields a salt and water. Most inorganic salts ionize in water solution.

(2) Common salt, sodium chloride, occurs widely in nature, both as deposits left by ancient seas and in the ocean, where its average concentration is 2.6%.
See sodium chloride. See also soap.

salt bath. A molten mixture of sodium, potassium, barium, and calcium chlorides or nitrates, to which sodium carbonate and sodium cyanide are sometimes added. Used for hardening and tempering of metals and for annealing both ferrous and nonferrous metals. Temperatures used may be as high as 1315C for hardening high-speed steels. Commercial mixtures are available for a variety of specifications.
See also fused salt.

salt cake. Impure sodium sulfate (90–99%).
Properties: For properties and derivation see sodium sulfate.
Grade: Technical, glassmakers' (iron-free).
Use: Paper pulp, detergents and soaps, plate and window glass, sodium salts, ceramic glazes, dyes.
See sodium sulfate.

salting out. Reduction in the water-solubility of an organic solid or liquid by adding a salt (usually sodium chloride) to an aqueous solution of the substance. Ions of the dissolved salt attract and hold water molecules, thus making them less free to react with the solute. The result of this is to decrease the solubility of the solute molecules with consequent separation or precipitation. Colloidal suspensions of proteins, soaps, and similar substances are precipitated in this way.

salt, molten. See fused salt.

salt of tartar. See acid potassium tartrate.

saltpeter. See niter, potassium nitrate.

salt, rock. See sodium chloride.

salt, fused. See fused salt.

salvarsan. (dihydroxydiaminoarsenobenzene dihydrochloride).

$C_{12}H_{14}O_2N_2Cl_2As_2 \cdot 2HOH$.
Use: To treat syphilis.

salvia oil. The Dalmatian variety of sage oil.

samarium. CAS: 7440-19-9.   Sm.
A rare-earth metal of the lanthanide group (Group IIIB of the Periodic Table); atomic number 62; aw 150.4; valences = 2, 3; seven stable isotopes.
Properties: Hard, brittle metal which quickly develops an oxide film in air. An active reducing agent. Ignites at 150C, liberates hydrogen from water, d 7.53, mp 1072C, bp 1900C, hardness similar to iron, high neutron absorption capacity. Combustible.
Occurrence: Australia, Brazil, Southeastern US, South Africa; also from bastnasite ore in California.
Derivation: Reduction of the oxide with barium or lanthanum.
Use: Neutron absorber, dopant for laser crystals, metallurgical research, permanent magnets.

samarium chloride. $SmCl_3 \cdot 6HOH$.
Properties: Faintly yellow, hygroscopic crystals; d 2.383. Loses $5H_2O$ at 110C, soluble in water.
Derivation: By treating the carbonate or oxide with hydrochloric acid.

samarium oxide. Sm_2O_3.
Properties: Cream-colored powder, d 8.347, mp 2300C, insoluble in water, soluble in acids, absorbs moisture and carbon dioxide from the air.
Use: Catalyst in the dehydrogenation of ethanol, infrared-absorbing glass, neutron absorber, preparation of samarium salts.

sampling. The methods and the techniques used in obtaining representative test samples of quantity lots of raw materials, semiprocessed work, and finished product for production and quality control. Rules for sampling procedures for both solid and liquid materials have been established by the National Cottonseed Products Association, Memphis, TN, and by the National Institute of Oilseed Products, San Francisco, CA. The techniques of physical sampling are one application of statistical quality control.

SAN. Abbreviation for styrene-acrylonitrile polymer.
See polystyrene.

sand. Sediment particulates ranging in size from 1/16 to two millimeters.
See silica.

sandalwood oil. (santal oil).   A pale yellow, essential oil; strongly levorotatory.
Use: In fragrances, perfumes, and flavoring.

sandarac. A natural resin obtained from Morocco. Its commercial form is yellow, brittle, amorphous lumps or powder; soluble in alcohol; insoluble in benzene and water.
Use: Special types of varnishes and lacquers.

sand casting. See foundry sand.

Sandmeyer diphenylurea isatin synthesis.
Formation of a cyanoformamidine by treatment of a symmetrical diphenylthiourea with potassium cyanide in alcohol containing lead carbonate, reduction with ammonium sulfide, and ring-closure with concentrated sulfuric acid to isatin-2-anil; also formed smoothly by ring closure of the cyanoformamidine with aluminum chloride in benzene or carbon disulfide.

Sandmeyer isonitrosoacetanilide isatin synthesis.
Formation of isonitrosoacetodiphenylamidine by condensation of chloral hydrate, hydroxylamine, and aniline, cyclization with concentrated sulfuric acid, and quantitative hydrolysis to isatin on dilution.

Sandmeyer reaction. Replacement of diazonium groups in aromatic compounds by halo or cyano groups in the presence of cuprous salts, copper powder, or cupric salts.

"Sandrol."[462] TM for a group of detergents, foam boosters, foam stabilizers and thickeners. Made from lauric acid, coconut fatty acid, and coconut oil "Superamides," coconut and lauric acid alkanolamides, hydrogenated coconut alkanolamide, and lauric acid isopropanolamide.

sandwich molecule. See metallocene; ferrocene.

"Sangamo."[492] TM for snow-white powder with a clean, neutral, sweet taste; essentially, regular corn syrup with water removed.
Use: Sweetener in a variety of food products.

sandstone. A siliciclastic sedimentary rock consisting primarily of sand, usually sand that is predominantly quartz.

Sanger, Frederick. (1918-) An English biochemist who won the Nobel prize for chemistry in 1958. His research was on the protein structure. He identified the amino acid sequence of the protein insulin. His PhD was awarded from Cambridge University.

sanitizer. A special class of disinfectant designed for use on food-processing equipment, dairy utensils, dishes, and glassware in restaurants. Among them are the hypochlorites, chloramines, and other organic chlorine-liberating compounds, and quaternary ammonium compounds, many of which are proprietary.
See also antiseptic, disinfectant.

santal oil. See sandalwood oil.

santalol. CAS: (α) 115-71-9; (β) 77-42-9.
$C_{15}H_{24}O$. A sesquiterpene alcohol.
Properties: Colorless liquid, odor of oil of sandalwood, soluble in three parts of 70% alcohol, insoluble in water, d 0.971–0.973, refr index 1.504–1.508, bp 300C. Combustible.
Derivation: From sandalwood oil.
Use: Perfumery.

santalyl acetate. Acetic acid ester of a mixture of α- and β-santalols.
Properties: Colorless liquid, light odor of sandalwood, d 0.982–0.985, refr index 1.487–1.492.
Derivation: Treatment of sandalwood oil or santalol with acetic anhydride.
Use: Perfumery.

"Santicizer."[58] TM for a series of plasticizers, including various sulfonamides, phthalates and glycollates.

"Santocure."[58] TM for a series of accelerators for natural and synthetic rubbers.

"Santoflex."[58] TM for a series of rubber antioxidants and antiozonants.

"Santolite."[58] TM for a group of sulfonamide resins.

α-santonin. CAS: 481-06-1. $C_{15}H_{18}O_3$.
A tricyclic structure.
Properties: White powder turning yellow on exposure to light, odorless, tasteless at first, then bitter. Soluble in chloroform, alcohol, and most volatile and fatty oils; very slightly soluble in water; solutions are levorotatory. D 1.187, mp 170–173C, bp sublimes, specific rotation −170 to −175 degrees (2 g/100 mL alcohol).
Derivation: By extraction from Artemisia.
Hazard: Toxic by ingestion, affects color vision.
Grade: Technical.
Use: Medicine (anthelmintic).

"Santoquin."[58] TM for ethoxyquin.
Use: Antioxidant in animal feeds and dehydrated forage crops.

"Santovar A."[58] TM for 2,5-di-tert-amylhydroquinone.

"Santowax." R[58] TM for mixed terphenyls. Yellowish-white, noncrystalline, flaked solid.
Use: Extender for polystyrene.

saponification. The chemical reaction in which an ester is heated with aqueous alkali such as sodium hydroxide to form an alcohol (usually glycerol), and the sodium salt of the acid corresponding to the ester. The process is usually carried out on fats (glyceryl esters of fatty acids). The sodium salt formed is called a soap. A typical saponification reaction is: $(C_{17}H_{35}COO)_3C_3H_5 + 3NaOH \rightarrow 3C_{17}H_{35}COONa + C_3H_5(OH)_3$.
See also soap.

saponification number. The number of milligrams of potassium hydroxide required to hydrolyze 1 g of a sample of an ester (glyceride, fat) or mixture.

saponin. (1) A general term applied to two groups of plant glycosides that on shaking with water form colloidal solutions giving soapy lathers; they form oil/ester emulsions and are used as protective colloids. They also have the ability to hemolyze red blood corpuscles at very great dilutions. The two groups are triterpenoid and steroid saponins; the latter are used in research on sex hormones.
(2) Specific term: saponin derived from Saponaria or Quillaja.
Properties: White, amorphous glucoside; pungent, disagreeable taste and odor. It foams strongly when shaken with water; soluble in water.
Grade: Crude, purified, highest purity.
Hazard: Highly toxic by injection; destroys red blood cells. Moderately toxic by ingestion.
Use: Foam producer in fire extinguishers, detergent in textile industries, sizing, substitute for soap, emulsification agent for fats and oils.

"SAPP #4."[1] TM for a sodium acid pyrophosphate.
Grade: FCC.
Use: In leavening (has the slowest reaction rate of several phosphates which are otherwise chemically identical), in acid-type metal cleaners, oil well drilling muds, household and industrial cleaners.

sapphire. CAS: 1317-82-4. (Al_2O_3).
For natural material, see corundum. Synthetic sapphire is made by crystal-growing technique.
Properties: Hard, crystalline solid; d 3.98; Mohs hardness 9.0; mp 2040C. Dielectric strength 480 kV/cm, dielectric constant 9.0 (20C), coefficient of friction 0.05 micron, inert to strong acids and alkalies, excellent high-temperature stability, can be sealed to glass, high transmission in infrared and ultraviolet.
Forms: Rods, spheres, disks, whiskers, single crystals.
Use: Electron and microwave tubes, optical elements in radiation detectors, substrate for thin-film components and integrated circuits, abrasive, record needles, precision instrument bearings, aluminum composites, micromortars for hand-pulverizing chemicals.
See corundum.

saran. See polyvinylidene chloride, saran fiber.

saran fiber. Generic name for a manufactured fiber in which the fiber-forming substance is any long-chain synthetic polymer composed of at least 80% by weight of vinylidene chloride units $(-CH_2CCl_2-)$ (Federal Trade Commission).
Properties: Tenacity, 2–2.5 g/denier, elongation 15–30%, softens at 115–137C. Highly resistant to most chemicals and solvents, to weather, moths, and mildew. Combustible, but self-extinguishing.
Use: Screens, upholstery, curtain and drapery fabrics, rugs and carpets, awnings, filter cloth.
See also polyvinylidene chloride.

sarcolysin. See melphalan.

sarcosine. (methyl glycocoll; methylaminoacetic acid). CAS: 107-97-1.
CH_3NHCH_2COOH.
Properties: Deliquescent crystals with sweet taste, mp 210–215C (decomposes), very soluble in water, slightly soluble in alcohol. Combustible.
Derivation: Decomposition of creatine or caffeine.
Grade: Technical.
Use: Synthesis of foaming antienzyme compounds for toothpaste, cosmetics, and pharmaceuticals.

sardine oil. See fish oil.

Sarett oxidation. Oxidation of primary and secondary alcohols to aldehydes and ketones by means of CrO_3-pyridine complex.

sarin. (methylphosphonofluoride acid, isopropyl ester). CAS: 107-44-8.
$[(CH_3)_2CHO](CH_3)FPO$. A nerve gas.
Hazard: Toxic by inhalation and skin absorption, cholinesterase inhibitor.

"Sarkosyl."[219] TM for a series of surface-active N-acylated sarcosines.
Use: Tooth pastes, cosmetics, and pharmaceuticals (anti-enzyme); corrosion inhibitors; lubricants and greases; inks.
See also sarcosine.

SAS. (1) Abbreviation for sodium aluminum sulfate.
See aluminum sodium sulfate.
(2) Abbreviation for sodium alkane sulfonate.

SASOL process. See gasification.

satin spar. See calcite.

satin white. A high-bulking filler used in paper coating formulations; a mixture of hydrated lime, potash alum, and aluminum sulfate. Readily hydrated. Particle size range 0.2–2 microns.

saturation. (1) The state in which all available valence bonds of an atom (especially carbon) are attached to other atoms. The straight-chain paraffins are typical saturated compounds.
(2) The state of a solution when it holds the maximum equilibrium quantity of dissolved matter at a given temperature.
See unsaturation, solubility (true), supersaturation.

saveall. A device used in the papermaking industry to reclaim wastewater containing suspended solids (fiber, pigment, etc.) for further processing. There are three types: (1) sedimentation, (2) vacuum filter, and (3) flotation. Types (1) and (2) are large tanks made of tile or concrete, type (3) is a rotary vacuum filter drum which separates solids from wastewater.

savory oil. CAS: 8016-68-0. An essential oil used in flavoring, especially in stuffing for meats and poultry.

saxitoxin. CAS: 35523-89-8. $C_{10}H_{17}N_7O_4 \cdot 2HCl$. A toxic principle present in certain species of shellfish. It is a paralytic poison that attacks the central nervous system acting as a muscular nerve block.

Saybolt Universal viscosity. The efflux time in seconds (SUS) of 60 mL of sample flowing through a calibrated Universal orifice in a Saybolt viscometer under specified conditions.
See also viscosity.

Saytex.[313] TM for brominated flame retardants.

Saytzeff (Zaitsev) rule. The rule predicts that in elimination reactions the olefin predominantly formed will be the one with the largest possible number of substituents on the carbon-carbon double bond.

Sb. Symbol for antimony (from Latin stibium).

SBA. Abbreviation for sec-butyl alcohol.

SBG. Abbreviation for standard battery grade, highly purified chemicals manufactured for use in the battery industry.

SBR. Abbreviation for styrene-butadiene rubber.

Sc. Symbol for scandium.

scale. (1) A calcareous deposit in water tubes or steam boilers resulting from deposition of mineral compounds present in the water, e.g., calcium carbonate. (2) A type of paraffin or petroleum wax from which all but a few percent of oil has been removed by hydraulic pressing and subsequent processing. (3) A graduated standard of measurement in which the units (degrees) are defined in relation to some property of what is measured, e.g., temperature scale, Brix scale, Baumé scale. (4) The markings indicating such units as on a thermometer or graduate. (5) A weighing device which may be of the beam type in which weights (poises) and lever systems are used, or of the direct-reading spring type in which the gravitational pull of the object being weighed is counterbalanced by a known constant spring force.
See boiler scale. See also balance (2).

scale-up. A term used in chemical engineering to describe the calculations and planning involved in carrying a chemical processing operation from the pilot plant to large-scale production stage.

scandium. CAS: 7440-20-2. Sc. Metallic element of atomic number 21, Group IIIB of the Periodic Table, aw 44.9559, valence = 3, no stable isotopes.
Properties: Silvery-white solid, mp 1539C, bp 2727C, d 2.99. Does not tarnish in air, reacts rapidly with acids, strongly electropositive. Not attacked by 1:1 mixture of nitric and 48% hydrofluoric acids. It is chemically similar to the rare earths.
Source: Chief ores are wolframite and thortveitite (Norway, Madagascar).
Derivation: Reduction of scandium fluoride with calcium or with zinc or magnesium alloy. Purification by distillation gives purity of 99+%.
Use: No major industrial use. An artificial radioactive isotope has been used in tracer studies and leak detection, and there is some application in the semi-conductor field.

scandium fluoride. ScF_3.
Derivation: Reaction of scandium oxide with ammonium hydrogen fluoride.
Use: Preparation of scandium metal.

scandium oxide. Sc_2O_3.
Properties: White, amorphous powder; soluble in hot acids, less so in cold acids; d 3.864; specific heat 0.153; a weak base.
Source: Thortveitite.
Use: Preparation of scandium fluoride.

scavenger. (getter). (1) In chemistry, any substance added to a system or mixture to consume or inactivate traces of impurities. (2) In metallurgy, an active metal added to a molten metal or alloy which combines with oxygen or nitrogen in the melt and causes its removal into the slag. Alloys including such metals as thorium, zirconium, and cerium as well as carbon and misch metal are used in vacuum tubes to absorb traces of residual gases.

"Scav-Ox."[84] TM for a hydrazine oxygen scavenger.
Use: Prevent corrosion in boilers and oil wheel casings.

Schaeffer acid. (2-naphthol-6-sulfonic acid; β-naphtholsulfonic acid).

Properties: White leaflets, soluble in water and alcohol, mp 122C.
Derivation: Sulfonation of β-naphthol with 94% sulfuric acid at 95C, yielding a mixture comprised chiefly of Schaeffer acid (56%), R-acid (15%), and G-acid (10%). Separation is effected by dilution with water, boiling to hydrolyze the sulfonic acid group, and addition of metallic salts.
Use: Azo dye intermediate.

Schaeffer's salt. (sodium salt of 2-naphthol-6-sulfonic acid). $C_{10}H_6OHSO_3Na$.
Use: Intermediate for organic chemicals.

Scheele, C. W. (1742–1786) One of the outstanding early chemical thinkers and experimenters, Scheele was a Swedish scientist who discovered a number of previously unknown substances, among which were tartaric acid, chlorine, manganese salts, arsine, and copper arsenite (Scheele's green). He also noted the oxidation states of various metals; observed the nature of oxygen two years before Priestley's discovery, and discovered the chemical action of light on silver compounds, thus laying the foundation of photochemistry and photography. Scheelite, or natural calcium tungstate, is named after him.

Scheele's green. See copper arsenite.

scheelite. $CaWO_4$. A natural calcium tungstate found in igneous rocks, usually with granite. Some molybdenum may replace tungsten.
Use: Ore of tungsten; as a phosphor.

Schiemann reaction. Formation of diazonium fluoborates by diazotization of aromatic amines in the presence of fluoborates, followed by their thermal decomposition to aryl fluorides.

Schiff base. A class of compounds derived by chemical reaction (condensation) of aldehydes or ketones with primary amines. The general formula is RR'X = NR''
Properties: Usually colorless, crystalline solids, although some are dyes. Very weakly basic and hydrolyzed by water and strong acids to carbonyl compounds and amines.
Use: Rubber accelerators; dyes (phenylene blue and naphthol blue); chemical intermediate; liquid crystals in electronic display systems; perfume base.

Schmidlin ketene synthesis. Formation of ketene by thermal decomposition of acetone over electrically heated wire at 500–750 degrees by a reaction involving radical formation with generation of methane and carbon monoxide.

Schmidt reaction. Acid catalyzed addition of hydrazoic acid to carboxylic acids, aldehydes, and ketones to give amines, nitriles, and amides, respectively.

Scholl reaction. Coupling of aromatic molecules by treatment with Lewis acid catalysts.

Scholler saccharification process. Industrial saccharification of wood using 0.5% sulfuric acid at 170–180 degrees and 165–180 lb/sq inch pressures. Recovered sugars are fermented to produce about 40 gal alcohol per ton of dry wood.

Schorigin (Shorygin) reaction. Organometallic reactions of the Grignard-type, employing sodium in place of magnesium; the reaction of alkyl sodium compounds with carbon dioxide to give monobasic acids is sometimes known as the Wanklyn reaction.

Schotten-Baumann reaction. Acylation of alcohols with acyl halides in aqueous alkaline solution.

schradan. (Generic name for octamethyl pyrophosphoramide; OMPA). CAS: 152-16-9. $[(CH_3)_2N]_2P(O)OP(O)[N(CH_3)_2]_2$.
Properties: Viscous liquid, d 1.137, bp 120–125C (0.5 mm), refr index 1.462 (25C), miscible with water, soluble in most organic solvents, hydrolyzed in the presence of acids, but not by alkalies or water alone.
Hazard: Toxic by ingestion and inhalation, a cholinesterase inhibitor, use may be restricted.

Use: A systemic insecticide which is absorbed by the plant which then becomes toxic to sucking and chewing insects.

Schweitzer's reagent. A solution of copper hydroxide in strong ammonia used in analytical chemistry as a test for wool. It dissolves cotton, silk, and linen.

scintillation counter. A device used to detect a pulse of radiation by emitting a flicker of light. Alpha radiation is counted by inorganic detectors such as sodium iodide, while organic materials such as plastics may be used for beta and gamma particles.
See also phosphor.

scission. (1) The rupture of a chemical bond with production of 5-electron-volts of energy. (2) In agricultural technology, the separation of fruit or vegetable products from the tree or vine.

scleroprotein. Any of a large class of proteins that have a supporting or protective function in tendons, bones, cartilages, ligaments, and other hard or tough parts of the animal body. They include the collagens of skin, tendons, and bones, as well as the elastic proteins known as elastins and the keratins. Specific examples are the keratin of hair, hoofs, and horns and fibroin from silk.

scleroscope hardness. See hardness.

scopolamine. CAS: 51-34-3. $C_{17}H_{21}NO_4$.
A drug used to inhibit effects of acetylcholine; viscous liquid, soluble in water and alcohol.

scouring agent. A compound used to remove the natural oils and fats from raw wool, also used to remove lubricants applied to rayon yarns or fabrics during such operations as throwing, winding, weaving, knitting, etc.

SCP. Abbreviation for single-cell protein.
See protein, single-cell.

screen. A woven fabric-like structure made of intersecting strands of wire or plastic, usually mounted in a steel frame. They are available in a wide range of sizes, weaves, and meshes from as coarse as 25 to as fine as 400. The mesh is the number of apertures per square inch; it is the square of the number of strands of metal or plastic per linear inch. The strands can be made of any suitable metal (copper, nickel), alloy (steel, bronze), or synthetic (nylon, PVC). Some types of screen are mechanically vibrated or gyrated to facilitate solids separation; fine mesh screens require application of force, such as a stream of water, to effect separation. Screens are used for filtration, clarification of suspensions, separation and classification of solids, and removal of contaminants from semi-solid materials. Lab sizes are available. A special application is the wire of a fourdrinier papermaking machine; it may be 38–60 in. wide and 55–85 mesh; it moves continuously over return rolls, the sheet being formed upon it by filtration of wood pulp slurry.
See also filter media.

screw. (auger; worm). A simple machine employing the principle of the inclined plane, invented by the Greek scientist Archimedes. It consists of a central shaft around which winds a spiral of ribs (called "flights") that are integral with the shaft. The distance between the flights is the pitch, or angle of inclination of the screw. This distance may be uniform throughout the length of the screw or it may vary from one point to another, depending on use requirements. The shorter the distance between flights, the more pressure the screw will deliver.

Screws have a number of important applications in industrial operations. (1) Extrusion of rubber, plastics, and food products: the screw rotates in a chamber or barrel, the product being introduced through a port where the flights are farthest apart; it is forced through a die at the opposite end of the barrel, which molds it to the form desired. (2) Mixing of solids: continuous mixing is possible with screws having a wide variety of pitches and contours which impart a back-and-forth motion to the product being mixed without moving it to the discharge end until it is pushed along by added material. Two screws may operate in parallel, their pitches opposing each other in such a way as to effect maximum mixing. Some screws are cored for circulation of cooling water. (3) Conveying of solids: for this purpose screws with uniform and fairly wide pitch are used; as the screw turns the solids are passed along from one flight to the next. These are used for conveying wet and dry solids, wood chips, and similar particulates. (4) An engineering application of the screw is the mechanism known as a worm gear, in which the flights of the screw engage corresponding indentations or notches in a shaft or wheel, causing it to turn.

scrubbing. Process for removing one or more components from a mixture of gases and vapors by passing it upward and usually countercurrent to and in intimate contact with a stream of descending liquid, the latter being chosen so as to dissolve the desired components and not others. The gas or vapor may be broken into fine bubbles upon entering a tower filled with liquid,

but more frequently the tower is filled with coke, broken stone, or other packing, over which the liquid flows while exposing a relatively large surface to the rising gas or vapor.
See also absorption (1).

scruple. Unit of weight used in pharmacy equivalent to 20 grains or 1/3 dram.

"SD-75."[430] TM for alkenyl dimethyl ethyl ammonium bromide.
Use: Primarily as an algicide in water towers and swimming pools.

SDA. Abbreviation for specially denatured alcohol.

SDA No. 1. Specially denatured alcohol (government regulation formula). It consists of 5 gal methanol added/100 gal 95% ethanol.

SDDC. Abbreviation for sodium dimethyldithiocarbamate.

SDP. Abbreviation for 4,4'-sulfonyldiphenol.

Se. Symbol for selenium.

Seaboard process. Method of removing hydrogen sulfide from a gas by absorption in sodium carbonate solution. Sodium bicarbonate and sodium hydrosulfide are formed. By blowing air through this solution, hydrogen sulfide is released and carried off, and the sodium carbonate is regenerated.

Seaborg, Glen T. (1912-) An American chemist who won the Nobel prize for chemistry in 1951 along with McMillan. He did research in nuclear chemistry, physics, and artificial radioactivity. Discovered the elements plutonium, americium, berkelium, californium, einsteinium, fermium, and medelevium with his colleague. He co-discovered numerous isotopes and radioisotopes. His PhD came from the University of California at Berkeley.

sea coal. Finely ground bituminous coal used in sand mixtures for iron molds to prevent sticking.

sealant. Any organic substance that is soft enough to pour or extrude and is capable of subsequent hardening to form a permanent bond with the substrate. Most sealants are synthetic polymers (silicones, urethanes, acrylics, polychloroprene) which are semisolid before application and later become elastomeric. A few of the best known sealants are such natural products as linseed oil (putty), asphalt, and various waxes.
See also adhesive.

"Sealol."[51] TM for "white" mineral oil used as lubricant for sealed refrigeration compressor units.

"Sealstix." TM for a cement of the deKhotinsky type which adheres to almost any surface; insoluble in all common reagents except alcohol, strong caustics, and chromic acid cleaning solution.

"Sealvar."[209] TM for an iron-nickel-cobalt alloy used for making hermetic seals with the harder glasses and ceramic materials.

"Sealz."[248] TM for a line of thermoplastic rubber compounds which are used as additives for asphalt and tar products to raise softening point, improve low temperature flexibility, enhance adhesive qualities, and increase elasticity.

Searles Lake brine. See brine; Trona process.

seasoning. Any food additive used in low concentration to contribute to the taste of food. Best known is sodium chloride, but the term also includes various spices, onions, mustards, etc.
See also enhancer, potentiator.

sea water. See ocean water.

seaweed. See algae, phycocolloid, carrageenan, kelp.

sebacic acid. (1,8-octanedicarboxylic acid; sebacylic acid; decanedioic acid).
CAS: 111-20-6. $COOH(CH_2)_8COOH$.
Properties: White leaflets, mp 133C, d 1.110 (25C), bp 295.0C (100 mm), refr index 1.422 (133.3C), slightly soluble in water, soluble in alcohol and ether. Combustible.
Derivation: From butadiene via dichlorobutene and its nitrile derivatives, dry distillation of castor oil with alkali.
Grade: CP, purified.
Use: Stabilizer in alkyd resins, maleic and other polyesters, polyurethanes, fibers, paint products, candles and perfumes, low-temperature lubricants and hydraulic fluids, manufacture of nylon 610.

"Sebacol."[179] TM for a tanners' depilatory containing sodium hydrosulfite suitably stabilized for use in alkaline unhairing systems.

sebaconitrile. $NC(CH_2)_8CN$.
Properties: Straw-colored, oily liquid; bp 199C (15 mm).
Use: Chemical intermediate for drugs, dyes, and high polymers. ·

sebacoyl chloride. (n-octane-1,8-dicarboxylic acid dichloride). $ClOC(CH_2)_8COCl$.
Properties: Liquid, bp 137–140C (3 mm), 97%

pure, decomposes slowly in cold water, soluble in hydrocarbons and ethers.
Use: Organic intermediate.

sebacylic acid.   See sebacic acid.

sec-.   Abbreviation for secondary.

seconal sodium.   (sodium seconal).   Proprietary preparation of quinalbarbitone sodium.
Use: Sedative and hypnotic.

secondary.   (1) For chemical meaning refer to "primary." (2) Designates a reversible (rechargeable) electric battery, generally called a storage battery. (3) Designates recovery of metal values from scrap or waste products such as aluminum from cans or steel from car bodies; also recovery of petroleum by chemical flooding and hydraulic fracturing.

secondary calcium phosphate.   See phosphate, dibasic.

sedanolic acid. $C_{12}H_{20}O_3$.   A constituent of celery seed oil. Its lactone, also present, which is the source of the characteristic odor, is sedanolide.

sedative.   A natural or synthetic therapeutic agent having the property of inducing relaxation and varying degrees of depression of the central nervous system. The major types of sedatives are: (1) chlorine substitution products (chloral hydrate, chlorobutanol), (2) ethanol derivatives (ethyl carbamate, hedonal), (3) specific aldehydes and ketones (paraldehyde, amylene hydrate), (4) barbituric acid derivatives (barbital, barbiturates). Sedatives are almost always prescription drugs.
See also tranquilizer.

sedimentation.   The settling out by gravity of solid particles suspended in a liquid, the rate of settling being defined by Stoke's Law. This method is used industrially in water purification. It is also an analytical procedure for separation of solids of different particle size, as well as for molecular weight determination. The sedimentation of large molecules in a strong centrifugal field permits determination of both average molecular weights and the distribution of molecular weights in certain systems. When a solution containing polymer or other large molecules is centrifuged at forces up to 250,000 times gravity, the molecules begin to settle, leaving pure solvent above a boundary which progressively moves toward the bottom of the cell. An optical system is provided for viewing this boundary, and a study as a function of the time of centrifuging yields the rate of sedimentation for the single component, or for each of many components of a polydisperse system. These sedimentation rates may then be related to the corresponding molecular weights of the species present after the diffusion coefficients for each species are determined by independent experiments. The result of this work is the distribution of molecular weights in the sample which is attainable by few other methods.
See also precipitate.

seed.   (1) That part of a plant which includes the plant embryo itself, a quantity of stored food (fats, carbohydrates, and proteins in varying proportions), and the enclosing cellulosic coats. The food storage tissue is called the endosperm. Starch is an important food reserve in the endosperms of cereals and legumes; sugar in sweet corn; protein in soybean and wheat; and fats in such plants as coconut, cacao, castor bean, and most types of palms. Such growth substances as giberellin also occur in seeds; some also contain alkaloids (ricinine, hyoscine, and caffeine). The seeds or bulbs of some plants are highly toxic, e.g., hyacinth, larkspur, castor bean, cashew nut shells. Bitter almond contains as much as 10% of hydrogen cyanide which is removed in processing. Seeds (which in many cases are equivalent to nuts) are an important source of vegetable oils, which are used as food components, in paints and other industrial products, and in medicine. There has been considerable experimentation with sunflower seeds, coconuts, and soybeans as a source of diesel fuel.

(2) Trace proportions of a material introduced to initiate a desired reaction where the "seed" acts as a nucleating agent. An example is the seeding of a cloud with silver iodide to cause precipitation by providing means for the nucleation of droplets.
See nucleation.

seeding, cloud.   See nucleation.

Seger cone.   See pyrometric cone.

Seidlitz salt.   See epsom salts.

Seismotite.   Trade name for pumice.
Use: Abrasive in scouring agents.

selenic acid.   CAS: 7783-08-6. H_2SeO_4.
Properties: White, hygroscopic solid; d 3.004 (15/4C); mp 58C (easily undercools); decomposes at 260C; very soluble in water; decomposes in alcohol; usually available as a liquid; d 2.609 (15C).
Hazard: Strong irritant to skin and mucous membranes.

selenious acid.   See selenous acid.

selenium. CAS: 7782-49-2. Se. A nonmetallic element, atomic number 34, Group VIA of the Periodic Table, aw 78.96, valences = 2, 4, 6. These are 6 stable isotopes.

Properties: Amorphous, red powder becoming black on standing and crystalline on heating; vitreous and colloidal forms may be prepared. Crystalline form has d 4.5, mp 217C, bp 685C; amorphous form softens at 40C and melts at 217C. Crystalline selenium is a p-type semiconductor; electrically it acts as a rectifier and has marked photoconductive and photovoltaic action (converts radiant to electrical energy); the electrical conductivity increases with increasing light irradiation. Soluble in concentrated nitric acid and (in liquid form) in common alkalies; forms binary alloys with silver, copper, zinc, lead, etc.; a necessary nutritional factor for animals.

Occurrence: Canada, Japan, Yugoslavia, Mexico; also in certain soils.

Grade: Commercial (powder or lumps), high-purity up to 99.999%.

Use: Electronics, xerographic plates, TV cameras, photocells, magnetic computer cores, solar batteries (rectifiers, relays); ceramics (colorant for glass), steel and copper (degasifier and machinability improver), rubber accelerator, catalyst, trace element in animal feeds. For further information consult Selenium Tellurium Development Association, 11 Broadway, NY, NY.

selenium diethyldithiocarbamate.
CAS: 5456-28-0. $Se[SC(S)N(C_2H_5)_2]_4$.

Properties: Orange-yellow powder, d 1.32 (20/20C), melting range 63–71C, characteristic odor, soluble in carbon disulfide, benzene, chloroform, insoluble in water.

Hazard: Toxic by inhalation, ingestion, and skin absorption. TLV: (as Se): 0.2 mg/m³ of air.

Use: Vulcanization agent without added sulfur or as a primary or secondary accelerator with sulfur.

selenium dimethyldithiocarbamate. See "Methyl Selenac."

selenium dioxide. (selenous acid anhydride).
CAS: 7446-08-4. SeO_2.

Properties: White or yellowish-white to slightly reddish, lustrous, crystalline powder or needles. D 3.954 (15/15C), mp 340–350C (sublimes), soluble in alcohol, water.

Hazard: Toxic by inhalation, ingestion, and skin absorption. TLV: (as Se): 0.2 mg/m³ of air.

Use: Analysis (testing for alkaloids), oxidizing agent, antioxidant in lubricating oils, catalyst.

selenium sulfide. (selenium disulfide).
CAS: 7488-56-4. SeS_2.

Properties: Bright orange powder, mp less than 100C, practically insoluble in water and organic solvents.

Grade: USP.

Hazard: Toxic by ingestion, strong irritant to eyes and skin, an animal carcinogen. TLV: (as Se): 0.2 mg/m³ of air.

Use: Medicine (treatment of seborrhea, medicated shampoos).

selenium tetrafluoride. CAS: 10026-03-6. SeF_4.

Properties: Colorless, fuming liquid; fp −10C; bp 105C; d 2.75; reacts strongly with phosphorus and with water (hydrolysis). Soluble in alcohol, sulfuric acid, ether, and carbon tetrachloride.

Derivation: Reaction of selenium chloride and silver fluoride.

Hazard: Irritant.

Use: Fluorinating agent.

selenous acid. (selenious acid). H_2SeO_3.

Properties: Transparent, colorless, deliquescent crystals. Soluble in water and alcohol, insoluble in ammonia, d 3.0066, mp 70C (decomposes)

Derivation: Action of hot nitric acid on selenium.

Hazard: Toxic by inhalation, ingestion, and skin absorption. TLV: (as Se): 0.2 mg/m³ of air.

Use: Reagent for alkaloids.

selenous acid anhydride. See selenium dioxide.

Semenov, Nikolai N. (1896-) A Russian chemist and physicist who won the Nobel prize in 1956. He authored books on the chain reaction and problems of chemical kinetics and reactivity as well as many articles. His work concerning thermal combustion and explosion is utilized in rockets and jet engines. He received his doctorate at Leningrad State University.

"Semesan."[28] TM for a wettable powder containing 25.3% hydroxymercurichlorophenol.

Hazard: As for mercury compounds.

"Semesan Bel."[28] TM for a seed disinfectant containing 12.5% hydroxymercurinitrophenol and 3.8% hydroxymercurichlorophenol.

Hazard: As for mercury compounds.

semicarbazide hydrochloride. (carbamylhydrazine hydrochloride; aminourea hydrochloride). CAS: 563-41-7. $H_2NCONHNH_2HCl$.

Properties: White crystals, mp 172–175C (decomposes), soluble in water, insoluble in absolute alcohol and ether.

Derivation: From hydrazine sulfate, potassium or sodium cyanate, and sodium carbonate, or electrolytically by the reduction of nitrourea.

Grade: CP, technical.

Hazard: Toxic by ingestion.

Use: Reagent for aldehydes and ketones, isolation of hormones and isolation of certain fractions from essential oils.

semiconductor. An element or compound having an electrical conductivity intermediate between that of conductors and non-conductors (insulators). Most metals have quite high conductivity, while substances like diamond and mica have very low conductivity (high resistance). Between these extremes lie the semiconductors, of which germanium, silicon, silicon carbide, and selenium are examples, with resistivities in the range of 10^{-2} to 10^9 ohms/cm. Slight traces of impurities in the crystalline structure are essential for semiconduction; arsenic is a typical impurity in semiconductor crystals. These impurities function as electron donors or acceptors and the semiconductor is designated n-type or p-type, depending on the electrical nature of the "holes" or energy deficits in the crystalline lattice.

The functioning of semiconductors involves the science of solid state physics. Their discovery in the early 1940s made possible the development of transistors, with their manifold applications in electronic devices, in which they have largely replaced the vacuum tube.

There are a few organic semiconducting compounds which contain a significant amount of carbon-carbon bonding and are also capable of supporting electronic conduction. Anthracene and Ziegler-catalyzed acetylene polymers (conjugated polyolefins) are examples.
See also crystals, impurity, solid, solid state chemistry.

semimicrochemistry. Any chemical method (usually analytical) in which the weight of the sample used is from 10–100 mg.

semipermeable membrane. See membrane, semipermeable.

semisynthetic. A term often used to describe endproducts that are manufactured from natural materials but do not occur in the free state, for example, paper, glass, soap, cement, rayon, leather, etc.

Semmler-Wolff reaction. Rearrangement of α,β-unsaturated cyclohexenyl ketoximes into aromatic amines under acidic conditions.

"Sentry."[214] TM for sorbic acid and potassium sorbate.
Use: Fungistats for the control of certain molds and yeast in foods. Also TM for propylene glycol, USP. Solvent for flavors and colors; humectant for baked goods, and plasticizer for cork seals and crowns.

"Separan."[233] TM for a series of flocculating agents. AP30. Synthetic, high molecular weight, anionic polymer. C-90 and C-120. Synthetic,

high molecular weight, cationic polymers. MGL. Similar to NP10. Used in production of uranium. NP10. Synthetic, water-soluble, nonionic, high molecular weight polymer of acrylamide. "Separan" NP 10 potable water grade flocculant has been accepted, subject to maximum use concentration of 1 ppm, by the US Public Health Service. NP20 Nonionic polyacrylamide polymer. PG2 Similar to NP10. Used in paper manufacture.

separation. A collective term including a large number of unit operations which, in one way or another, isolate the various components of a mixture. Chief among these are evaporation, distillation, drying, gas absorption, sedimentation, solvent extraction, press extraction, adsorption, and filtration. Specialized methods include centrifugation, electromagnetic separation (mass spectrograph), gaseous diffusion, and various types of chromatography.
See specific entry for further information.

"Sephadex."[485] TM for a dry insoluble powder composed of microscopic beads which are synthetic, organic compounds derived from the polysaccharide dextran. The dextran chains are crosslinked to give a three dimensional network and the functional ionic groups are attached to the glucose units of the polysaccharide chains by ether linkages. Available in various forms for use in many different phases of chromatography.

septiphene. See o-benzyl-p-chlorophenol.

"Septo-Sour."[44] TM for a product consisting chiefly of zinc salts of fluorine compounds.
Properties: White, dustless crystals; readily soluble in water; neutralizing value 25.6 oz sodium bicarbonate/lb.
Use: Laundry sour, especially low temperatures.

sequestration. The formation of a coordination complex by certain phosphates with metallic ions in solution so that the usual precipitation reactions of the latter are prevented. Thus, calcium soap precipitates are not produced from hard water treated with certain polyphosphates and metaphosphates (these polyphosphates are often improperly referred to as hexametaphosphates). The term sequestration may be used for any instance in which an ion is prevented from exhibiting its usual properties due to close combination with an added material. Two groups of organic sequestering agents (chelates in these examples) of economic importance are the aminopolycarboxylic acids such as ethylenediaminetetraacetic acid and the hydroxycarboxylic acids such as gluconic, citric, and tartaric acids. In the food

industry sequestering agents aid in stabilizing color, flavor, and texture.
See also chelate, coordination compound, complex.

"Sequestrene."[219] TM for a series of complexing agents and metal complexes consisting of ethylenediaminetetraacetic acid and salts.

Ser. Abbreviation for serine.

serendipity. An unexpected scientific discovery which turns out to be more important than the project being researched. One example is the discovery of the sodium sulfide polymer, later known as "Thiokol." White, the researcher, was seeking to develop an improved automotive coolant. Another is the discovery of the reinforcing effect of carbon black on rubber while technologists were using it as a black pigment to counteract the whitening effect of zinc oxide.

serendipity berry. See sweetener, nonnutritive.

SERI. Abbreviation for Solar Energy Research Institute.
See solar energy.

serine. (β-hydroxyalanine; α-α-β-hydroxypropionic acid). $HOCH_2CH(NH_2)COOH$. A nonessential amino acid occurring naturally in the l(-) form.
Properties: Colorless crystals, soluble in water, insoluble in alcohol and ether, optically active; dl-serine, mp 246C with decomposition; d(+)-serine, mp 228C with decomposition. Available commercially in all three forms.
Derivation: Hydrolysis of protein (especially silk protein); synthesized from glycine.
Use: Biochemical research, dietary supplement, culture media, microbiological tests, feed additive.

Serini reaction. Zinc-promoted rearrangement of 17-hydroxy-20-acetoxysterol derivatives into C-20 ketones, the reaction is applicable to other cyclic, as well as open-chain alcohols.

"Serizyme."[173] TM for a proteolytic enzyme preparation for desizing textiles.

serotonin. (5-hydroxytryptamine; 5-hydroxy-3-(β-aminoethyl)indole). CAS: 50-67-9. $C_{10}H_{12}N_2O$. A powerful vasoconstrictor occurring in the brain and blood platelets, it is thought to play a part in regulation of blood pressure and also has muscle-contracting properties. It has been isolated from beef serum and may be synthesized from 5-benzyloxyindole. Serotonin is similar to the hallucinogenic drugs mescaline, LSD, psilocin and psilocybin. Its formation is inhibited by p-chlorophenylalanine.

serpentine. A type of asbestos.
See asbestos.

serum. (1) The continuous phase of a biocolloid after the solid or disperse phase has been removed by centrifugation, coagulation, or similar means. In the case of milk, for example, the serum, or whey, is a true solution of sugars, proteins, and mineral compounds in water. (2) Specifically blood from which corpuscles, platelets, etc., have been removed, especially when prepared with antigenic bacteria for inoculation to effect the cure of a disease.

serum albumin. Blood albumin comprising 60% of the plasma proteins.
Use: As a plasma volume expander.

SES. Abbreviation for sodium-2,4-dichlorophenoxyethyl sulfate.
See sesone.

sesame oil. (benne oil; teel oil).
CAS: 8008-74-0. A bland, yellowish vegetable oil.
Use: Shortenings, salad oil, margarine, and similar food products.

sesamex. (Generic name for 2-(2-ethoxyethoxy)ethyl-3,4-(methylenedioxy)phenyl acetal of acetaldehyde). CAS: 51-14-9.
$CH_2O_2C_6H_3OCH(CH_3)OC_2H_4OC_2H_4OC_2H_5$.
Properties: Liquid, bp 137–141C (0.08 mm), soluble in non-polar solvents.
Use: Synergist for insecticides.

sesamin. (Generic name for 2,6-bis[3,4-(methylenedioxy)phenyl]-3,7-dioxabicyclo-[3.3.0]octane). CAS: 607-80-7.
$C_{20}H_{18}O_6$.
Properties: Solid, mp 122.5C. Combustible.
Use: Synergist for insecticides and fungicides.

sesamolin. (Generic name for 6-[3,4-(methylenedioxy)phenoxy]-2-[3,4-(methylenedioxy)phenyl]3,7-dioxabicyclo[3.3.0]octane).
CAS: 526-07-8. $C_{22}H_{18}O_7$. Constituent of sesame oil which is a powerful synergist of pyrethrum. A 1:1 mixture of pyrethrum and sesamoline has 31 times the insecticidal activity of pyrethrum alone.

sesone. (Generic name for sodium-2,4-dichloro-phenoxy-ethyl sulfate; SES; "Crag").
CAS: 136-78-7. $C_6H_3Cl_2OCH_2CH_2SO_4Na$.
Properties: Crystalline solid, mp 245C, soluble in water.
Use: Pre-emergence herbicide.

sesqui. A prefix meaning one-and-a-half. Often used for salts in which the proportions of metal oxide to acid anhydride are 2:3, or vice-versa, as in sodium sesquicarbonate.

sesquiterpene. A terpene having the formula $C_{15}H_{24}$, one and a half times the standard terpene formula of $C_{10}H_{16}$. An example is cadinene, a constituent of various wood oils (cade oil, cedar-leaf oil, etc.).

"Setit."[337] TM for suspension aid used in the ceramic industry. Composed of a colloidal form of aluminum oxide.

"Setole."[43] TM for thermoplastic resin emulsions.
Use: As finishing agents for all types of textile fabrics.

"Sevin."[214] TM for a carbaryl insecticide (1-naphthyl-N-methyl carbamate).
Hazard: Toxic by inhalation. TLV: 5 mg/m³ of air.

sewage sludge. A mixture of organic materials resulting from purification of municipal waste. There are two types: (a) Imhoff sludge: a low grade sludge containing 2–3% ammonia and about 1% phosphoric acid.
See also Imhoff tank.
    (b) Activated sludge: a high-grade sludge containing 5.0–7.5% ammonia and 2.5–4.0% phosphoric acid. Derivation: (a) By running sewage through settling tanks without access of air. The sludge, or solid matter is decomposed by anaerobic bacteria. Lagooning may be used effectively following activated sludge treatment. (b) By running sewage through settling tanks and pumping air (or oxygen) through porous plates at the bottom of the tanks; 20% of the current "make" is also added. The waste acts as nutrient for aerobic bacteria, which consume the polluting organic matter. The resulting solids are filtered and dried.
Use: Fertilizer base.

"SF-100."[236] TM for a granular organic material used as the basic additive for preparing low filtrate casing packs and for spotting fluid to free differentially stuck pipe. It can be prepared with density of 7.6 lb/gal for use where formation is too weak to support a full column of mud or heavier packs.

"S-Fatty Acids."[487] TM for fatty acids derived from soybean oil. S-210 is a soy fatty acid developed primarily for alkyd resins. S-230 is a soy-type fatty acid having a lower unsaturated acids content and higher titer than S-210.
Properties: S-210 is a light yellow liquid and S-230 a light yellow semi-solid at ambient temperature.
Use: Chemical intermediate, paints and varnishes, alkyd resins, and soaps.

SFS. (1) Abbreviation for Saybolt Furol seconds. See Furol viscosity and Saybolt Universal viscosity.
    (2) Abbreviation for sodium formaldehyde sulfoxylate.

SGR. Abbreviation for steam gas recycle process for shale oil recovery.
See also shale oil.

shale oil. A mixed-base crude oil extracted from mountains of sedimentary shale in Colorado, Utah, and Wyoming by heating at 425–535C (approximately 800–1000F). Two methods can be used--surface mining and excavation. In the first, the shale is bulldozed from beds, crushed, and fed into retorts of either the vertical or horizontal type. In the second, shafts are driven into the mountain and the shale is heated *in situ* by direct combustion in an interior chamber excavated for the purpose. The *in situ* method is the more efficient. The oil-bearing component of the shale is called kerogen. Only 20–30 gallons (less than one barrel) of oil is obtained per ton of shale processed and less than 33% of the total oil content is recoverable. Major obstacles are the vast earth-moving operations necessary, the need for large volumes of cooling water, and disposal of the spent shale. Though rather extensive pilot plant operations have been carried out in recent years, no large-scale production is likely for the indefinite future for both economic and technical reasons.
See also kerogen, oil shale, synfuel.

shape-selective catalyst. See catalyst, shape-selective.

shark liver oil.
Properties: Yellow to red-brown liquid, strong odor, d 0.917–0.928. Soluble in ether, chloroform, benzene, and carbon disulfide.
Derivation: By expression from shark livers.
Method of purification: Chilling and filtration.
Grade: Crude, refined.
Use: Source of vitamin A and of squalene, biochemical research.

sharp. (1) In reference to cheeses, this term denotes length of the curing period, usually at least six months, twice as long as for mild cheeses. (2) Descriptive of a technique for quick-freezing of foods at −20F.

Sharpless reaction. Metal-catalyzed asymmetric epoxidation of allylic alcohols employing a system consisting of titanium tetraisopropoxide, (+)- or (-)-diethyl tartrate, and tert-butyl hydroperoxide. The epoxy alcohols are obtained with a high degree of optical purity (90%) and their absolute stereochemistry is predictable.

"Sharstop."[204] TM for a series of short-stopping agents used in polymerization of synthetic rubbers.

shear. The ratio between a stress (force per unit area) applied laterally to a material and the strain resulting from this force. Determination of this ratio is one method of measuring the viscosity of a liquid or semi-solid.
See also viscosity.

shell. (1) In physical chemistry, this term is applied to any of the several paths or orbits of the electrons in an atom as they revolve around its nucleus. They constitute a number of principal quantum paths representing successively higher energy levels. There may be from one to seven shells, depending on the atomic number of the element and corresponding to the seven periods of the Periodic Table. The shells are usually designated by number, though letter symbols have been used, i.e., K, L, M, N, O, P, Q. The laws of physics limit the number of electrons in the various shells as follows: two in the first (K), eight in the second (L), 18 in the third (M), and 32 in the fourth (N). With the exception of hydrogen and helium, each shell contains two or more orbitals, each of which is capable of holding a maximum of two electrons.
See also quantum number, orbital theory, Pauli exclusion principle.
(2) The hard integument of molluscs and crustaceans, consisting mostly of calcium carbonate, chitin, etc.
See nacre, oyster shells.
(3) The brittle covering of avian eggs, chiefly calcium carbonate, lime, etc. The formation of proper shell structures in certain species of birds is said to be adversely affected by DDT and similar insecticidal contaminants of their food.
(4) The shells of nuts are cellulosic in character, some contain industrially useful oils.

shellac. (lac; garnet lac; gum lac; stick lac).
Derivation: A natural resin secreted by the insect *Laccifer lacca* (*Coccus lacca*) and deposited on the twigs of trees in India. After collection, wash-

ing and purification by melting and filtering it is formed into thin sheets, which are later fragmented into flakes of orange shellac. This may be dewaxed and bleached to a transparent product. Soluble in alcohol, insoluble in water.
Grade: (1) Orange: TN (impure), fine, superfine, heart, superior; (2) Bleached and dewaxed (colorless).
Hazard (alcohol solution): Flammable, dangerous fire risk.
Use: Sealer coat under varnish; finish coat for floors, furniture, etc.; dielectric coatings, deKhotinsky cement.

"Sheliflex."[125] TM for a series of petroleum products largely composed of naphthenic and/or paraffinic hydrocarbons.
Properties: Colors ranging from near water-white to black, viscosity 3–55 cp at 99C, d 0.85–0.95, odor slight to none, very low volatilities.
Hazard: Combustible.
Use: Rubber processing and extending oils, miscellaneous process oil uses.

"Shell Sol."[125] TM for a series of aliphatic hydrocarbon solvents composed of mixture of paraffinic, naphthenic, and aromatic hydrocarbons.

"Shellwax."[125] TM for a series of paraffin waxes derived from distillate lube streams.

sherardizing. The process by which relatively small articles made of iron or steel are coated with zinc powder. The metal forms an alloy with the steel surface and produces a thin, tightly adherent, corrosion-resistant coating.

"Sherbelizer."[322] TM for an algin derivative-vegetable gum composition.
Use: Stabilizer for sherbets, water ices, frozen fruits, syrups, purees, chocolate ice cream.

"Sherbrite."[266] TM for brighteners for nickel plating. Active ingredient is 1,2-benzisothiazolin-3-one-1,1-dioxide sodium salt.

shielding. Protection of personnel from the harmful effects of ionizing radiation and/or neutrons by enclosure of the equipment (reactor, particle accelerator, x-ray generator, etc.) with an absorbing material. The most efficient of these are cadmium, lead, steel, and high-density concrete (3 ft of concrete equals about 1 ft of steel). Among the plastics, polyethylene affords reasonable protection in thick sheets. Water and paraffin wax are good neutron absorbers because of their high hydrogen content. Materials such as graphite and beryllium are able to deflect and retard neutrons and are used for this purpose as moderators in nuclear reactors. Heavy water is also used.

shoddy. Reclaimed scrap wool, rubber, leather, etc., often used in the manufacture of low-quality products.

Shoelkopf acid. (1-naphthol-4,8-disulfonic acid). $C_{10}H_5OH(SO_3H)_2$.
Properties: Colorless crystals.
Derivation: Decomposition of 1-naphthylamine-4,8-disulfonic acid by diazotization and acidifying with heat.
Use: Azo dye intermediate.

short. (Baked products): Crisp, flaky, e.g., shortcake; (steel): brittle, friable under certain temperature conditions, e.g., hot-short (above red heat) and cold-short (below red heat); (clay): dry, lacking plasticity.

shortening. See cooking (5).

shortstopping agent. A material used in a polymerization reaction to cut off the reaction at a predetermined point. Example: use of diethylhydroxylamine or sodium dimethyldithiocarbamate to control synthetic rubber polymerization; tetraethylsilanol is used in silicone polymers.

shot metal. Lead alloy with less than 3% arsenic.

SI. Abbreviation for International System of Units (metric system) now in use in most countries, but not yet officially adopted in US.

Si. Symbol for silicon.

side-chain. See chain.

siderite. (chalybite; spathic iron ore). $FeCO_3$ (usually with some calcium, magnesium, or manganese). The term siderite is also used for an iron alloy found in meteorites.
Properties: Gray, yellow, brown, green, white, or brownish-red mineral; vitreous inclining to pearly luster; white streak; d 3.83–3.88; Mohs hardness 3.5–4.
Occurrence: US (Vermont, Massachusetts, Connecticut, New York, North Carolina, Pennsylvania, Ohio), Europe.
Use: An ore of iron, when high in manganese used in the manufacture of spiegeleisen.

sienna. A yellowish hydrated iron oxide. Raw sienna is a brown-tinted yellow ocher occurring in Alabama, California, Pennsylvania, Cyprus, and Italy. Burnt sienna is an orange-brown pigment made by calcining raw sienna.
See also ocher, iron oxide reds.
Use: Colorant in oil paints, stains, pastels, etc.

sieve. See screen.

siglure. (Generic name for sec-butyl-6-methyl-3-cyclohexane-1-carboxylate). $CH_3C_6H_8COOCH(CH_3)C_2H_5$.
Properties: Liquid, bp 113–114C (15 mm), soluble in most organic solvents, insoluble in water. Combustible.
Use: Insect attractant.

sigma blade. A rotating agitator set horizontally in a kneading bowl or chamber used for mixing doughs and heavy pastes. The blade or arm is shaped somewhat like a Greek capital sigma (Σ) lying on its side; variations of this shape simulate horizontal letters S and Z. Some kneaders have two such blades which overlap as they turn to provide maximum mixing efficiency.
See also kneading.

sigma bond. A covalent bond directed along the line joining the centers of two atoms. They are the normal single bonds in organic molecules.
See also pi bond.

silane. (silicon tetrahydride).
CAS: 7803-62-5. SiH_4.
Properties: A gas with repulsive odor, solidifies at −200C, bp −112C, decomposes in water, insoluble in alcohol and benzene, d 0.68.
Hazard: Dangerous fire risk, ignites spontaneously in air. Strong irritant to tissue. TLV: 5 ppm in air.
Use: Doping agent for solid-state devices, production of amorphous silicon.

"Silaneal."[149] TM for organopolysiloxanes used for making materials water-repellent.

silane compounds. Gaseous or liquid compounds of silicon and hydrogen (Si_nH_{2n+2}), analogous to alkanes or saturated hydrocarbons. SiH_3 is called silyl (analogous to methyl) and Si_2H_5 is disilanyl (analogous to ethyl). A cyclic silicon and hydrogen compound having the formula (SiH_2) is called a cyclosilane. Organo-functional silanes are noted for their ability to bond organic polymer systems to inorganic substrates.
Hazard: Dangerous fire risk.
See also silicone, siloxane.

"Silastic."[149] TM for compositions in physical character comparable to milled and compounded rubber prior to vulcanization but containing organosilicon polymers. Parts fabricated of "Silastic" are serviceable from −73 to +260C, retain good physical and dielectric properties in such service, show excellent resistance to compression set, weathering, and corona. Thermal conductivity is high, water absorption low.
Use: Diaphragms, gaskets and seals, O-rings, hose, coated fabrics, wire and cable, and insulating components for electrical and electronic parts.

"Silbond."[1] TM for ethyl silicate, available as pure, condensed, prehydrolyzed and specialty formulations.

"Silflake."[31] TM for a commercially pure silver, containing 1% of organic lubricant and traces of iron. Particle size 2–10 microns.
Grade: 131, 135, and 850 (variable conductivities and covering power). "Wet" flake contains a flammable solvent.
Use: Conductive coatings and adhesives.

"Silfrax."[230] TM for bonded refractories containing 40–78% silicon carbide.
Properties: High refractoriness, great strength, high thermal conductivity, freedom from spalling, resistance to clinker adhesion, and resistance to mechanical and flame abrasion.
Use: Bricks for boiler and furnace installations, kiln furniture in ceramic kilns, shapes for boiler furnaces, air-cooled furnace linings, glass lehrs, pit furnaces, and enameling furnace-ware supports.

silica. (silicon dioxide). SiO_2. Occurs widely in nature as sand, quartz, flint, diatomite.
Properties: Colorless crystals or white powder, odorless and tasteless, d 2.2–2.6, insoluble in water and acids except hydrogen fluoride, soluble in molten alkali when finely divided and amorphous. Combines chemically with most metallic oxides; melts to a glass with lowest known coefficient of expansion (fused silica); thermal conductivity about half that of glass, mp 1710C, bp 2230C, high dielectric constant, high heat and shock resistance. Noncombustible.
Derivation: Can be made from a soluble silicate (water glass) by acidification, washing and ignition. Arc silica is made from sand, vaporized in a 3000C electric arc.
Grade: By purity and mesh size, silica aerogel, hydrated, precipitated.
Hazard: Toxic by inhalation, chronic exposure to dust may cause silicosis.
Use (powder): Manufacture of glass, water glass, ceramics, abrasives, water filtration, microspheres, component of concrete, source of ferrosilicon and elemental silicon, filler in cosmetics, pharmaceuticals, paper, insecticides, hydrated and precipitated grades as rubber reinforcing agent, including silicone rubber, anticaking agent in foods, flattening agents in paints, thermal insulator. (Fused): Ablative material in rocket engines, spacecraft, etc.; fibers in reinforced plastics; special camera lenses. (Amorphous): Silica gel.
See also quartz, silicic acid, silica gel.

silica, fumed. A colloidal form of silica made by combustion of silicon tetrachloride in hydrogen-oxygen furnaces. Fine white powder.
Use: Thickener, thixotropic, and reinforcing agent in inks, resins, rubber, paints, cosmetics, etc. Base material for high-temperature mortars.
See also "Aerosil," "Cab-O-Sil."

silica fused. See silica, quartz, fused.

silica gel. A regenerative adsorbent consisting of amorphous silica. Noncombustible.
Derivation: From sodium silicate and sulfuric acid.
Grade: Commercial grades capable of withstanding temperatures up to 260–315C are supplied in the following mesh sizes: 3–8, 6–16, 14–20, 14–42, 28–200, and through 325.
Use: Dehumidifying and dehydrating agent, air-conditioning, drying of compressed air and other gases, and liquids, such as refrigerants and oils containing water in suspension, recovery of natural gasoline from natural gas, bleaching of petroleum oils, catalyst and catalyst carrier, chromatography, anticaking agent in cosmetics and pharmaceuticals, in waxes to prevent slipping, in dietary supplements.
See also silicic acid.

"Silic AR."[329] TM for silica-gel-based formulations, suitable for various chromatographic applications. The numerical suffixes indicate the approximate pH of a 10% slurry. Letters F, G, or GF indicate that the product contains a fluorescent material, gypsum binder, or both. "TLC" indicates suitability for thin layer chromatography.

silicate. Any of the widely occurring compounds containing silicon, oxygen, and one or more metals with or without hydrogen. The silicon and oxygen may combine with organic groups to form silicate esters. Most rocks (except limestone and dolomite) and many mineral compounds are silicates. Typical natural silicates are gemstones (except diamond), beryl, asbestos, talc, clays, feldspar, mica, etc. Portland cement contains a high percentage of calcium silicates. Best known of the synthetic (soluble) silicates is sodium silicate (water glass).
Hazard (natural silicate dusts): Toxic by inhalation.
Use: Fillers in plastics and rubber, paper coatings, antacids, anticaking agents, cements.

silicic acid. (hydrated silica).
CAS: 7699-41-4. $SiO_2 \cdot nHOH$. The jelly-like precipitate obtained when sodium silicate solution is acidified. The proportion of water

varies with the conditions of preparation and decreases gradually during drying and ignition, until relatively pure silica remains. During drying the jelly is converted to a white amorphous powder or lumps.
Use: Laboratory reagent and reinforcing agent in rubber.
See silica gel.

silicochloroform. See trichlorosilane.

silicomanganese. Alloys consisting principally of manganese, silicon, and carbon.
Use: Low-carbon steel in which silicon is not objectionable. Silicon manganese steels are used for springs and high-strength structural steels.
See also manganese steels and ferromanganese.

silicomolybdic acid. See 12-molybdosilicic acid.

silicon. Si. Nonmetallic element Atomic number 14, Group IVA of the Periodic Table, aw 28.086, valence = 4, three stable isotopes. It is the second most abundant element (25% of the earth's crust) and is the most important semiconducting element; it can form more compounds than any other element except carbon.
Properties: Dark-colored crystals (the octahedral form in which the atoms have the diamond arrangement). The amorphous form is a dark brown powder (see silicon, amorphous). Soluble in a mixture of nitric and hydrofluoric acids and in alkalies; insoluble in water, nitric acid, and hydrochloric acid. D 2.33, mp 1410C, bp 2355C, Mohs hardness 7, dielectric constant 12, coordination number 6. Combines with oxygen to form tetrahedral molecules in which one silicon atom is surrounded by four oxygen atoms. In this respect it is similar to carbon. It is also capable of forming $\equiv Si\equiv Si\equiv$ double bonds in organosilicon compounds.
Occurrence: Does not occur free in nature, but is a major portion of silica and silicates (rocks, quartz, sand, clays, etc.).
Derivation: Crystalline silicon is made commercially (96–98% pure) in an electric furnace by heating SiO_2 with carbon, followed by zone refining. It can be purified to 99.7% by leaching. The ultrapure semiconductor grade (99.97%) is obtained by reduction of purified silicon tetrachloride or trichlorosilane with purified hydrogen; the silicon is deposited on hot filaments (800C) of tantalum or tungsten. In a one-step method, sodium fluorosilicate is reacted with sodium, the heat produced being sufficient to form silicon tetrafluoride; this, when reacted with sodium, yields high-purity silicon and sodium fluoride. The process requires no heat except that provided by the original reaction. Single crystals

of both n- and p-type are grown by highly specialized techniques.
Grade: Ferrosilicon, regular (97% silicon), semiconductor or hyperpure (99.97% silicon), amorphous.
Hazard: Flammable in powder form.
Use: Semiconductor in solid state devices (transistors, photovoltaic cells, computer circuitry, rectifiers, etc.); organosilicon compounds; silicon carbide; alloying agent in steels, aluminum, copper, bronze, and iron (ferrosilicon); cermets and special refractories; halogenated silanes; spring steels; deoxidizer in steel manufacture.
See also silica; silicate; silicone; ferrosilicon; silicon, amorphous.

silicon, amorphous. A non-crystalline allotrope of silicon which exists in the form of a dark-brown powder. It is made from silane (SiH_4) plus doping agents in a glow-discharge tube at low pressure. A film only a few microns in thickness is deposited on a glass or metal substrate. The amorphous product contains about 20% hydrogen. It has been found superior to crystalline silicon in the manufacture of solar cells.

silicon-bronze. An alloy of copper, tin, and silicon used for telephone and telegraph wires.

silicon carbide. CAS: 409-21-2. SiC.
Properties: Bluish-black, iridescent crystals. Mohs hardness 9, d 3.217, sublimes with decomposition at 2700C, insoluble in water and alcohol, soluble in fused alkalies and molten iron. Excellent thermal conductivity, electrically conductive, resists oxidation at high temperatures. Noncombustible, a nuisance particulate.
Derivation: Heating carbon and silica in an electric furnace at 2000C.
Forms available: Powder, filament, whiskers (3 million psi), single crystals.
Use: Abrasive for cutting and grinding metals, grinding wheels, refractory in nonferrous metallurgy, ceramic industry and boiler furnaces, composite tubes for steam reforming operations. Fibrous form used in filament-wound structures and heat-resistant, high-strength composites.
See also "Carborundum," "Carbofrax."

silicon-copper. (copper silicide).
Properties: A hard, tough, bronze-like alloy containing 10–30% silicon.
Derivation: From silica and copper electrolytically.
Use: Manufacture of silicon-bronze.

silicone. (organosiloxane). Any of a large group of siloxane polymers based on a structure consisting of alternate silicon and oxygen atoms with various organic radicals attached to the silicon:

$$-OSi\begin{bmatrix} CH_3 \\ | \\ | \\ CH_3 \end{bmatrix} -O-Si\begin{bmatrix} CH_3 \\ | \\ | \\ CH_3 \end{bmatrix}_n \begin{matrix} CH_3 \\ | \\ OSi- \\ | \\ CH_3 \end{matrix}$$

Discovered by Kipping in England in 1900.

Properties: Liquids, semisolids, or solids depending on molecular weight and degree of polymerization, viscosity ranges from less than 1 to more than 1 million centistokes. Polymers may be straight-chain, or crosslinked with benzoyl peroxide or other free radical initiator, with or without catalyst. Stable over temperature range from -50 to $+250C$. Very low surface tension; extreme water repellency; high lubricity; excellent dielectric properties; resistant to oxidation, weathering, and high temperatures; permeable to gases. Soluble in most organic solvents; unhalogenated types are combustible.

Derivation: (1) Silicon is heated in methyl chloride to yield methylchlorosilanes; these are separated and purified by distillation and the desired compound mixed with water. A polymeric silicone results. (2) Reaction of silicon tetrachloride and a Grignard reagent (RMgCl), with subsequent hydrolysis and polymerization.

Forms: fluids, powders, emulsions, solutions, resins, pastes, elastomers.

Use (Liquid): Adhesives, lubricants, protective coatings, coolants, mold-release agents, dielectric fluids, heat transfer, wetting agents and surfactants, foam stabilizer for polyurethanes, diffusion pumps, antifoaming agent for liquids, textile finishes, water repellent, weatherproofing concrete, brake fluids, cosmetic items, polishes, foam shields in solar energy collectors, rust preventives. (Resin): Coatings, molding compounds, laminates (with glass cloth), filament winding sealants, room-temperature curing cements, electrical insulation, impregnating electric coils, bonding agent, modifier for alkyd resins, vibration-damping devices. (Elastomer, or silicone rubber): Encapsulation of electronic parts; electrical insulation; gaskets; surgical membranes and implants; automobile engine components; flexible windows for face masks, air locks, etc.; miscellaneous mechanical products.

See organosilicon, "RIV."

silicone oil. See silicone properties and uses.

silicone rubber. See silicone properties and uses.

silicon-gold alloy. See gold-silicon alloy.

silicon monoxide. CAS: 10097-28-6.   SiO.

Properties: Amorphous, black solid. D 2.15–2.18, sublimes at high temperature, hard and abrasive. Noncombustible.

Grade: Lumps, powders, tablets; optical.

Use: To form thin surface films for protection of aluminum coatings, optical parts, mirrors, dielectrics or insulators.

silicon nitride. Si_3N_4.

Properties: Gray, amorphous powder (can be prepared as crystals). Sublimes at 1900C, d 3.44, bulk d 70–75 lb/cu ft depending on mesh, Mohs hardness 9+, thermal conductivity 10.83 Btu/in/sq ft/hr/F at 400–2400F. Resistant to oxidation, various corrosive media, molten aluminum, zinc, lead, and tin; soluble in hydrogen fluoride.

Derivation: Reaction of powdered silicon and nitrogen in an electric furnace at 1300C.

Use: Refractory coatings, bonding silicon carbide, mortars, abrasives, thermocouple tubes in molten aluminum, crucibles for zone-refining germanium, rocket nozzles, high-strength fibers and whiskers, insulator and passivating agent in transistors and other solid-state devices.

silicon tetrabromide. (tetrabromosilane). CAS: 7789-66-4. $SiBr_4$.

Properties: Fuming, colorless liquid which turns yellow in air; disagreeable odor. Decomposed by water with evolution of heat, d 2.82 (0C), bp 153C, mp 5C. Noncombustible.

Purity: 99.999%.

Hazard: Strong irritant to tissue.

silicon tetrachloride. (tetrachlorosilane; silicon chloride).   CAS: 10026-04-7. $SiCl_4$.

Properties: Colorless, exceedingly mobile, fuming liquid; suffocating odor. Corrosive to most metals when water is present; in the absence of water it has practically no action on iron, steel, or the common metals and alloys, and can be stored and handled in metal equipment without danger. D 1.483 (20C), bulk d 12.4 lb/gal, fp $-70C$, bp 57.6C, refr index 1.412 (20C). Miscible with carbon tetrachloride, tin tetrachloride, titanium tetrachloride, and sulfur mono- and dichlorides; decomposed by water and alcohol with evolution of hydrogen chloride. Noncombustible.

Derivation: Heating silicon dioxide and coke in a stream of chlorine.

Grade: Technical, 99.5%, CP (99.8%), semiconductor.

Hazard: Toxic by ingestion and inhalation, strong irritant to tissue.

Use: Smoke screens; manufacture of ethyl silicate and similar compounds; production of silicones; manufacture of high-purity silica and fused silica glass; source of silicon, silica, and hydrogen chloride; lab reagent.

silicon tetrafluoride. (tetrafluorosilane; silicon fluoride). CAS: 7783-61-1. SiF₄.
Properties: Colorless gas, suffocating odor similar to hydrogen chloride, fumes strongly in air, d 3.57 (gas, air = 1) (15C), fp −90C, bp −86C, absorbed readily in large quantities by water with decomposition, soluble in absolute alcohol. Noncombustible.
Derivation: (a) Action of hydrogen fluoride or concentrated sulfuric acid and a metallic fluoride on silica or silicates. (b) Direct synthesis.
Grade: Pure, 99.5% min.
Hazard: Toxic by inhalation, strong irritant to mucous membranes. TLV (as F): 2.5 mg/m³ of air.
Use: Manufacture of fluosilicic acid, intermediate in manufacture of pure silicon, to seal water out of oil wells during drilling.

silicon tetrahydride. See silane.

silicon tetraiodide. SiI₄.
Properties: White crystals, mp 120C, bp 290C. Noncombustible. Purity: Up to 99.999%.
Hazard: Toxic by ingestion and inhalation, irritant to tissue.

silicotungstic acid. (12-tungstosilicic acid; silicowolframic acid). CAS: 12027-38-2. H₄SiW₁₂O₄₀·5HOH.
Properties: White, crystalline powder. Very soluble in water and polar organic solvents, relatively insoluble in non-polar organic solvents, strong acid. Noncombustible.
Grade: Reagent, technical.
Use: Catalyst for organic synthesis, reagent for alkaloids, additive to plating processes, as precipitant and inorganic ion-exchanger, minerals separation, mordant.
See also sodium 12-tungstosilicate and tungstosilicates.

silicowolframic acid. See silicotungstic acid.

silk. A natural fiber secreted as a continuous filament by the silkworm, Bombyx mori; silk consists essentially of the protein fibroin and, in the raw state, is coated with a gum which is usually removed before spinning. D 1.25, elongation at rupture about 20%, tenacity 3–5 g/denier. Combustible but self-extinguishing.

sillimanite. An aluminum silicate, a high heat-resisting material containing a maximum amount of mullite, developed from the alteration of andalusite during firing. This necessitates firing at above 1550C for the development of a suitable crystalline structure.
Use: Spark plugs, chemical lab ware, pyrometer tubes, special porcelain shapes, furnace patch and refractories.

silo. See ensilage.

siloxane. (oxosilane). A straight-chain compound (analogous to paraffin hydrocarbons) consisting of silicon atoms single-bonded to oxygen and so arranged that each silicon atom is linked with four oxygen atoms.

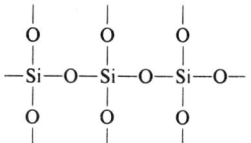

In some types, hydrogen may replace two or more of the oxygens. Disiloxane and trisiloxane are examples.
See also silicone.

"Silpowder."[31] TM for a commercially pure silver, metallic impurities 0.1% maximum. Particle size, 5–125 microns depending on grade.
Derivation: By precipitation from solutions of silver salts and by atomization.
Grade: 120, 130, 140, 150, and 160 (galvanic and chemical precipitates of variable size and density).
Use: Powder metallurgy parts such as electrical contacts, batteries, and brushes, pigments for conductive coatings and adhesives.

"Sil-Temp."[349] TM for a substantially pure fibrous silica for use in rocket and missile constructions and for high temperature insulation of motor components and similar applications. As a construction material it is used in laminate form impregnated and bonded with high-temperature resins.

"Sil-Trode."[407] TM for silicon bronze electrodes and filler rod for use in inert gas welding.

silumin. Aluminum-silicon alloys containing 12% silicon.
Properties: D 2.63–2.65.

"Silvacel."[129] TM for a series of wood fibers.
Use: Insulation, pulp molding, special papers and boards where high bulk and absorbent qualities are desirable, filter aid and filter media, treatment of oil well drilling muds.

"Silvacon."[129] TM for a series of products made from Douglas fir (Pseudotsuga taxifolia) bark.
Use: Phenolic adhesive extender, molding compound component, conditioner for fertilizer, insecticides, etc.; burnout material in foundry sands, rubber sponge manufacture, other vinyl

and rubber compounds, treatment of oil well drilling fluids, replacement for cork.

silver. CAS: 7440-22-4. Ag. Metallic element, atomic number 47, Group IB of the Periodic Table, aw 107.868, valence = 1, two stable isotopes.
Properties: Soft, ductile, lustrous, white solid; highest electrical and thermal conductivity of all metals. Excellent light reflector that resists oxidation, but tarnishes in air through reaction with atmospheric sulfur compounds. D 10.53, mp 961C, bp 2212C, thermal conductivity 1.01 cal/cm/sec/C, absorbs oxygen strongly at the melting point. Soluble in nitric acid, hot sulfuric acid, and alkali cyanide solutions; insoluble in water and alkalies. Noncombustible except as powder.
Derivation: Byproduct of operations on copper, zinc, lead, or gold ores, but some smelters still operate on native silver. The recovery ranges from 166 ounces to a few thousandths of an ounce per ton.
See Parkes process, The Pattinson process.
Source: Chief silver ores are native silver, argentite (silver sulfide), and cerargyrite (silver chloride).
Forms available: Pure ("fine"), sterling (7.5% copper), various alloys, plate; ingot, bullion, moss, sheet, wire, tubing, castings, powder, high purity (impurities less than 100 ppm), single crystals, whiskers.
Hazard: Toxic material. TLV (metal): 0.1 mg/m³ of air; (soluble compounds as silver): 0.01 mg/m³ of air.
Use: Manufacture of silver nitrate, silver bromide, photographic chemicals; lining vats and other equipment for chemical reaction vessels, water distillation, etc.; mirrors, electric conductors, such as bus bars; silver plating, electronic equipment; sterilant; water purification; surgical cements; hydration and oxidation catalyst; special batteries; solar cells; reflectors for solar towers; low-temperature brazing alloys; table cutlery; jewelry; dental, medical, and scientific equipment; electrical contacts; bearing metal; magnet windings; dental amalgams. Colloidal silver is used as a nucleating agent in photography and in medicine, often combined with protein.
See "Argyrol."
Note: A sandwich assembly consisting of a layer of silver between two layers of TiO_2 is used to coat the interior of light bulbs, it is said to reduce power consumption by more than 50% and triple the life of the bulb. Each layer of the coating, applied by the sputtering technique is 180 Å thick.

silver acetate. CAS: 563-63-3. CH_3COOAg.
Properties: White crystals or powder, d 3.26, moderately soluble in hot water, soluble in nitric acid.

Hazard: Toxic material. TLV: 0.01 mg/m³ of air.
Use: Lab reagent, oxidizing agent.

silver acetylide. CAS: 13092-75-6. Ag_2C_2.
Properties: White, unstable powder; a salt of acetylene.
Derivation: Reaction of acetylene with aqueous solution of argentous salts.
Hazard: Severe explosion risk when shocked or heated.
Use: Detonators.

silver bromate. $AgBrO_3$.
Properties: White powder, sensitive to light, keep in amber bottle, soluble in ammonium hydroxide, slightly soluble in hot water, decomposed by heat, d 5.2.

silver bromide. CAS: 7785-23-1. AgBr.
Properties: Pale yellow crystals or powder, darkens on exposure to light, finally turning black; d 6.473 (25C); mp 432C; bp decomposes at 700C; soluble in potassium bromide, potassium cyanide, and sodium thiosulfate solutions; very slightly soluble in ammonia water; insoluble in water; light-sensitive.
Derivation: Silver nitrate is dissolved in water and a solution of alkali bromide added slowly. The precipitated silver bromide is washed repeatedly with hot water, the operation must be carried on in a darkroom under a ruby red light.
Use: Photographic film and plates, photochromic glass, laboratory reagent.

silver carbonate. CAS: 534-16-7. Ag_2CO_3.
Properties: Yellow to yellowish-gray powder; contains 78% silver; light-sensitive; soluble in ammonium hydroxide, nitric acid; insoluble in alcohol and water; d 6.077; decomposes at 218C.
Use: Lab reagent.

silver chlorate. (argentous chlorate).
CAS: 7783-92-8. $AgClO_3$.
Properties: White, crystalline solid; mp 230C; decomposes at 270C. D 4.44, partially soluble in water, light-sensitive.
Derivation: Reaction of silver nitrate and sodium chlorate.
Hazard: Oxidizing agent, may react violently when shocked or heated, store away from combustible materials. Toxic by ingestion.
Use: Organic synthesis.

silver chloride. CAS: 7783-90-6. AgCl.
Properties: White, granular powder which darkens on exposure to light, finally turning black; exists in several modifications differing in behavior toward light and in their solubility in various solvents; soluble in ammonium hydroxide, concen-

trated sulfuric acid and sodium thiosulfate and potassium bromide solutions; very slightly soluble in water; can be melted, cast, and fabricated like a metal; d 5.56; mp 445C; bp 1550C.

Derivation: Silver nitrate solution is heated and hydrochloric acid or salt solution added. The whole is boiled, then filtered, all in the dark or under a ruby-red light.

Method of purification: Resolution in ammonium hydroxide and precipitation by hydrochloric acid.

Grade: Technical, CP, single pure crystals.

Hazard: As for silver.

Use: Photography, photometry and optics, batteries, photochromic glass, silver plating, production of pure silver, antiseptic. Single crystals are used for infrared absorption cells and lens elements, lab reagent.

silver chromate. CAS: 7784-01-2. Ag_2CrO_4.
Properties: Dark, brownish-red powder; soluble in acids, ammonium hydroxide, potassium cyanide, solutions of alkali chromates; insoluble in water; d 5.625.
Use: Lab reagent.

silver cyanide. CAS: 506-64-9. AgCN.
Properties: White, odorless, tasteless powder which darkens on exposure to light; soluble in ammonium hydroxide, dilute boiling nitric acid, and potassium cyanide and sodium thiosulfate solutions; insoluble in water; d 3.95; decomposes at 320C.
Derivation: By adding sodium cyanide or potassium cyanide to a solution of silver nitrate.
Hazard: Toxic by ingestion or inhalation. TLV: 0.01 mg/m³ of air.
Use: Silver plating.

silver dichromate. (silver bichromate).
$Ag_2Cr_2O_7$.
Properties: Dark red, almost black, crystalline powder; d 4.770, soluble in ammonium hydroxide and nitric acid, slightly soluble in water.
Hazard: Possible carcinogen. TLV: 0.01 mg/m³ of air.

silver fluoride. CAS: 7775-41-9. AgF·HOH.
Properties: Yellow or brownish crystalline mass, very hygroscopic, becomes dark on exposure to light, light-sensitive, soluble in water, d 5.852 (anhydrous), mp 435C, bp 1159C.
Hazard: Strong irritant. TLV: 0.01 mg/m³ of air.
Use: Medicine (antiseptic), substitution of fluorine for bromine and chlorine in organic compounds.

silver iodide. CAS: 7783-96-2. AgI.
Properties: Pale yellow, odorless, tasteless powder darkening on exposure to light; soluble in hydro-

gen iodide, potassium iodide, potassium cyanide, ammonium hydroxide, sodium chloride, and sodium thiosulfate solutions; insoluble in water; light-sensitive; d 5.675; mp 556C.
Derivation: Silver nitrate solution is heated, alkali iodide solution added, and the precipitate washed with boiling water in the dark or under ruby red illumination.
Use: Photography, cloud seeding for artificial rainmaking, lab reagent, antiseptic.

"Silver-Lume" A, B.[288] TM for a bright silver electroplating process for use by silversmiths and electronics manufacturers. The materials used are silver cyanide, potassium cyanide, potassium carbonate and addition agents.

silver mercury iodide. See mercuric silver iodide.

silver methylarsonate. (methanearsonic acid, disilver salt). $CH_3AsI_3Ag_2$.
Derivation: Reaction of disodium methylarsonate with silver salts.
Hazard: Very toxic. TLV: 0.01 mg/m³ of air.
Use: Algicides.

silver nitrate. CAS: 7761-88-8. $AgNO_3$.
Properties: Colorless, transparent, tabular, rhombic crystals becoming gray or grayish-black on exposure to light in the presence of organic matter; odorless; bitter, caustic metallic taste; strong oxidizing agent and caustic; soluble in cold water; more soluble in hot water, glycerol, and hot alcohol; slightly soluble in ether; d 4.328; mp 212C; bp decomposes.
Derivation: Silver is dissolved in dilute nitric acid and the solution evaporated. The residue is heated to a dull red heat to decompose any copper nitrate, dissolved in water, filtered, and recrystallized.
Grade: Technical, CP, USP.
Hazard: Strong irritant to skin and tissue. TLV: 0.01 mg/m³ of air.
Use: Photographic film, catalyst for ethylene oxide, indelible inks, silver plating, silver salts, silvering mirrors, germicide (as a wall spray), hair dyeing, antiseptic, fused form to cauterize wounds, lab reagent.

silver nitride. Ag_3N.
Properties: Colorless powder.
Derivation: Reaction of silver compounds with ammonia, with or without additives.
Hazard: Severe explosion hazard when shocked. It is unusually sensitive to mechanical action of any kind and can explode even in water suspension.

silver nitrite. CAS: 7783-99-5. $AgNO_2$.
Properties: Small yellow or grayish-yellow needles becoming gray on exposure to light; contains 70%

silver; decomposed by acids; soluble in hot water; insoluble in alcohol; decomposes at 140C; d 4.453 (26C).
Hazard: Highly toxic. TLV: 0.01 mg/m³ of air.
Use: Reagent for alcohols, preparation of aliphatic nitrogen compounds and standard solutions for water analysis.

silver orthophosphate. See silver phosphate.

silver oxide. (argentous oxide).
CAS: 20667-12-3. Ag₂O.
Properties: Dark brown or black, odorless powder; metallic taste. Soluble in ammonium hydroxide, potassium cyanide solution, nitric acid, and sodium thiosulfate solution, slightly soluble in water, insoluble in alcohol, d 7.14 (16C), mp decomposes when heated above 300C, strong oxidizing agent.
Derivation: Silver nitrate and alkali hydroxide solutions are mixed, the precipitate filtered and washed.
Grade: Technical, up to 99.6% pure, particle size 2–3 microns.
Hazard: Fire and explosion risk in contact with organic materials or ammonia.
Use: Polishing glass, coloring glass yellow, catalyst, purifying drinking water, lab reagent.

silver oxide battery. See zinc-silver oxide battery.

silver perchlorate. CAS: 7783-93-9. AgClO₄.
Properties: Colorless crystals, deliquescent, mp 485C (decomposes), d 2.80, soluble in water, reacts with explosive violence with many organic solvents, explodes on grinding.
Use: Manufacture of explosive compositions.

silver permanganate. CAS: 7783-98-4.
AgMnO₄.
Properties: Violet, crystalline powder; d 4.27 (25C); mp decomposes; contains 47.5% silver; decomposed by alcohol; light-sensitive, use dark-colored bottles.
Grade: Technical.
Hazard: Dangerous explosion risk, may detonate if shocked or heated.
Use: Gas masks, medicine (antiseptic).

silver peroxide. (argentic oxide). Ag₂O₂.
Properties: Grayish powder, d 7.44 (25C), decomposes above 100C. Insoluble in water; soluble in sulfuric acid, nitric acid, and ammonium hydroxide.
Hazard: Dangerous fire and explosion risk, strong oxidizing agent. Keep out of contact with organic materials.
Use: Manufacture of silver-zinc batteries.

silver phosphate. (silver orthophosphate).
CAS: 7784-09-0. Ag₃PO₄.
Properties: Yellow powder, turns brown when heated or on exposure to light, soluble in acids, potassium cyanide solutions, ammonium hydroxide, ammonium carbonate and acetic acid, very slightly soluble in water, d 6.370 (25C), mp 849C.
Derivation: Interaction of silver nitrate and sodium phosphate.
Use: Photographic emulsions, catalyst, pharmaceuticals.

silver picrate. CAS: 146-84-9.
C₆H₂O(N₂)₃Ag·HOH.
Properties: Yellow crystals containing 30% silver, soluble in water, slightly soluble in alcohol and acetone, insoluble in ether and chloroform.
Hazard: Severe explosion risk. Very toxic.
Use: Medicine (antimicrobial agent).

silver potassium cyanide. (potassium cyanoargentate; potassium argentocyanide).
CAS: 506-61-6. KAg(CN)₂.
Properties: White crystals, d 2.36 (25C), sensitive to light, soluble in water and alcohol, insoluble in acids.
Derivation: By adding silver chloride to a solution of potassium cyanide.
Hazard: Very toxic. TLV: 0.01 mg/m³ in air.
Use: Silver plating, bactericide, antiseptic.

silver sodium chloride. (sodium silver chloride).
AgCl·NaCl.
Properties: Hard, white crystals; decomposed by water; soluble in a concentrated solution of sodium chloride.

silver sodium thiosulfate. (sodium silver thiosulfate). Ag₂S₂O₃·2Na₂S₂O₃·2HOH.
Properties: White to gray, crystalline powder; sweet taste; soluble in water.

silver, sterling. See sterling silver.

silver sulfate. (silver sulfate, normal).
CAS: 10294-26-5. Ag₂SO₄.
Properties: Small, colorless, lustrous crystals or crystalline powder; contains about 69% silver, turns gray on exposure to light. Soluble in ammonium hydroxide, nitric acid, sulfuric acid, hot water; insoluble in alcohol. D 5.45 (29C), bp 1085C (decomposes), mp 652C.
Grade: Technical, CP.
Hazard: Highly toxic. TLV: 0.01 mg/m³ of air.
Use: Lab reagent.

silver sulfate, normal. See silver sulfate.

silver sulfide. CAS: 21548-73-2. Ag₂S.
Properties: Grayish black powder, soluble in concentrated sulfuric and nitric acids, insoluble in water, d 7.32 (25C), bp decomposes, mp 825C.

Derivation: By passing hydrogen sulfide gas into silver nitrate solution, washing and drying.
Use: Inlaying in niello metal-work, ceramics.

silvex. (fenoprop; generic name for 2-(2,4,5-trichlorophenoxy)propionic acid).
CAS: 93-72-1. $Cl_3C_6H_2OCH(CH_3)COOH$.
Properties: Solid, mp 180.4–181.6C, slightly soluble in water, freely soluble in acetone, methanol. Combustible.
Hazard: Use has been restricted.
Use: herbicide and plant growth regulator.

silvichemical. A chemical derived from wood, e.g., lignins, lignosulfonates (from spent sulfite liquor), vanillin, yeast (from fermentation of wood sugars), tall oil, sulfate turpentine, bark extracts, phenolic materials.

silvicide. A nonselective herbicide used to kill or defoliate bushes and small trees, e.g., ammonium sulfamate.

silylene. An organosilicon compound containing double-bonded silicon.

simazine. (2-chloro-4,6-bis(ethylamino)-s-triazine).   CAS: 122-34-9.
$ClC_3N_3(NHC_2H_5)_2$.
Properties: White solid, mp 225C, insoluble in water, slightly soluble in organic solvents. Combustible.
Use: Herbicide.

Simmons-Smith reaction. Stereospecific synthesis of cyclopropanes by treatment of olefins with methylene iodide and zinc-copper couple.

Simonini reaction. The preparation of aliphatic esters by the reaction of two moles of the silver salt of a carboxylic acid and one mole of iodine.

Simonis chromone cyclization. Formation of chromones from phenol and β-keto esters in the presence of phosphorus pentoxide, phosphorus oxychloride or sulfuric acid. Coumarins may also be formed.

Simons process. An electrochemical fluorination process which makes fluorocarbons by passing an electric current through a mixture of the organic starting compound and liquid anhydrous hydrogen fluoride. The products are hydrogen and the desired fluorocarbon.

simple distillation. Distillation in which no appreciable rectification of the vapor occurs, i.e., the vapor formed from the liquid in the still is completely condensed in the distillate receiver and does not undergo change in composition due to partial condensation or contact with previously condensed vapor.

SIMS. Abbreviation for secondary ion mass spectroscopy.

single-cell protein. See protein, single-cell.

sintering. The agglomeration of metal or earthy powders at temperatures below the melting point. Occurs in both powder metallurgy and ceramic firing. While heat and pressure are essential, decrease in surface area is the critical factor. Sintering increases strength, conductivity, and density. See Rittinger's law.

"Sipenol."[542] TM for ethoxylated fatty and short-chain amines.
Use: Cosmetics, textile industry, metal cleaning, agricultural emulsifiers, chemical and pharmaceutical intermediates.

"Sipex."[542] TM for industrial grade alcohol sulfates and alcohol ethoxylate sulfates.
Use: Detergents, emulsion, polymerization, and textile industry. Available are principally ammonium, magnesium, sodium, and triethanolamine lauryl sulfates, sodium tridecyl sulfate, sodium-2-ethylhexyl sulfate, and sodium lauryl ethoxylate sulfate.
See also "Sipon."

"Sipomer."[542] TM for a group of speciality monomers; they include dimethylaminoethyl methacrylate, hydroxyethyl methacrylate, dimethyl and diethyl maleates, and allyl glycolate.
Use: For polymerization.

"Sipon."[542] TM for a cosmetic grade of fatty alcohol sulfates and fatty alcohol ethoxylate sulfates. They include about the same materials as the "Sipex" grade and also diethanolamine lauryl sulfate, sodium cetyl sulfate, and ammonium lauryl ethoxylate sulfate.

"Siponate."[542] TM for purified alkylarylsulfonates, including sodium dodecyl benzene sulfonate (branched or linear) and sodium lauroyl monoglyceride sulfate.

SIPP. Abbreviation for sodium iron pyrophosphate.

sisal.
Properties: Hard, strong, light yellow to reddish fibers obtained from the leaves of Agave sisilana. Strength 4.5 g/denier, fineness ranges from 300–500 denier. Combustible, not self-extinguishing.

Source: Africa, Haiti, Bahama, Indonesia.

Hazard: Dust is flammable, may ignite spontaneously when wet. Toxic by inhalation.

Use: Twine, sacking, upholstery, life preservers, mattress liners, floor covering.

"Sitol."[28] TM for an oxidizing agent in flake form used in discharge printing.

β-sitosterol. (dihydrostigmasterol).
CAS: 83-46-5. $C_{29}H_{50}O$.

Properties: Waxy white solid, almost odorless and tasteless. Insoluble in water; soluble in benzene, chloroform, carbon disulfide, and ether. Can be crystallized from ether as anhydrous needles or from aqueous alcohol as leaflets with one molecule of water.

Derivation: Soybeans, tall oil.

Use: Biochemical research, anticholesteremic.

See also cholesterol.

size oil. See throwing oil.

sizing compound. (1) A material such as starch, gelatin, casein, gums, oils, waxes, asphalt emulsions, silicones, rosin, and water-soluble polymers applied to yarns, fabrics, paper, leather, and other products to improve or increase their stiffness, strength, smoothness or weight. (2) A material used to modify the cooked starch solutions applied to warp ends prior to weaving.

See also slashing compound.

"Skamex."[28] TM for a fluorocarbon plastic used as a metallurgical additive for the beneficiation of molten metal. Cakes and slugs of it are specifically designed for immersion in hot metal for removal of excessive quantities of dissolved hydrogen. Other forms are added to molds during casting to create a protective atmosphere against the effects of absorbed oxygen. Can be used for ferrous and nonferrous metals.

skatole. (3-methylindole). CAS: 83-34-1. C_9H_9N.

Properties: White, crystalline substance browning upon aging; fecal odor. Mp 93–95C, bp 265C, soluble in hot water, alcohol, benzene. Gives violet color in potassium ferrocyanide and sulfuric acid.

Derivation: Feces; African civet cat; *Celtis reticulosa,* a Javanese tree.

Use: Perfumery (fixative), artificial civet.

"Skellysolves."[409] TM for straight-run aliphatic naphthas having various boiling ranges, specific gravities, evaporation rates, and other properties, which make them suitable for a number of industrial uses.

Hazard: Flammable, dangerous fire risk.

Skraup synthesis. Synthesis of quinoline or its derivatives by heating aniline or an aniline derivative, glycerol and nitrobenzene in the presence of sulfuric acid.

"Skydrol."[58] TM for a series of fire-resistant aircraft hydraulic fluids. 500-A Used for hydraulic systems in turbo jet and turbo prop aircraft, which must operate at −54C. 7000 Used in aircraft cabin superchargers, expansion turbines for air-conditioning systems and the aircraft hydraulic system itself.

slack. (1) Descriptive of a soft paraffin wax resulting from incomplete pressing of the settlings from the petroleum distillate. Though it has some applications in this form, it is actually an intermediate product between the liquid distillate and the scale wax made by expressing more of the oil.
See also scale (2).

(2) Specifically, to react calcium oxide (lime) with water to form calcium hydroxide (slaked or hydrated lime), the reaction is CaO + HOH → $Ca(OH)_2$ + heat. The alternate spelling "slake" has the same meaning.

slaframine. (1-acetoxy-8-aminooctahydroindolizine). An alkaloid derived from a fungus that infests clover. It is under research development for use as an agent in retarding cystic fibrosis.

slag. (dross; cinder). (1) Fused agglomerate (usually high in silicates) which separates in metal smelting and floats on the surface of molten metal. Formed by combination of flux with gangue of ore, ash of fuel, and perhaps furnace lining. Slag is often the medium by means of which impurities may be separated from metal.

(2) The residue or ash from coal gasification processes, it may run as high as 40% depending on the rank of coal used.

Use: Railroad ballast, highway construction, cement and concrete aggregate, raw material for Portland cement, mineral wool, and cinder block.

slake. See slack.

slaked lime. See calcium hydroxide; lime, hydrated.

slashing compound. A textile sizing material applied to cotton or rayon warp ends by a special machine (slasher).

slash pine. A loblolly pine growing in swampy areas (slashes) in southeastern US.
Use: Primarily for manufacture of kraft paper pulp.

slate. A fine-grained metamorphic rock which cleaves into thin slabs or sheets. Color usually gray to black, sometimes green, yellow, brown, or red. Slates are composed of micas, chlorite, quartz, hematite, clays, and other minerals.
Occurrence: Pennsylvania, Vermont, Maine, Virginia, California, Colorado, Europe.
Use: Roofing, blackboards; (as powder) filler in paint, rubber, abrasive.

slate black. See mineral black.

slate flour. Finely divided slate used as a filler and dusting agent in rubber, plastics, etc.

slate, green. See slate flour.

slave. A remote-controlled mechanism or instrument that repeats the action of an identical mechanism that is controlled by an operator in another location; it may be activated by electromagnets or by electronic means. Such devices are used chiefly in handling or processing radioactive materials but also have communication uses, as in the Telautograph.

slimicide. A chemical which is toxic to the types of bacteria and fungi characteristic of aqueous slimes. Examples are chlorine and its compounds, organomercurial compounds, phenols, and related substances.
Use: Largely in paper mills and to some extent in textile and leather industries.
See also biocide.

slip clay. A type of clay containing such a high percentage of fluxing impurities and of such a texture that it melts at a relatively low temperature to a greenish or brown glass, thus forming a natural glaze. It must be fine-grained, free from lumps or concretions, show a low air-shrinkage, and mature in burning at as low as 704C.

"Slipicone." TM for fluid silicone compositions to prevent adhesion of materials to one another.
Use: Food-processing and packaging equipment.

sludge. Any thick, viscous mass, usually a sedimented or filtered waste product. Examples are (1) aluminum ore tailings (red mud), (2) asphaltic petroleum residues, and (3) municipal waste (Imhoff and activated sludge). All these may be used as components of useful products, e.g., (1) in steel manufacturing, (2) in roofing and road treatment compositions, and (3) as a base for fertilizers.
See also sewage sludge.

"Sludge Conditioner."[108] TM for a series of polyelectrolytes.
Use: Conditions sludge for dewatering and settling in municipal sewage treatment plants.

slurry. A thin, watery suspension; for example, the feed to a filter press or to a fourdrinier machine; also a stream of pulverized metal ore. A special use of this term refers to a type of explosives called "slurry blasting agents" based on gelatinized aqueous ammonium nitrate, sensitized with various other explosives.

slushing agent. A nondrying oil, grease, or similar material.
Use: Coat metals to afford temporary protection against corrosion.

slush molding. A method of molding certain toys such as doll parts in which a preheated mold is filled with liquid plastic composition and then heated until the required wall thickness has formed. The remaining liquid plastic is then poured out and the mold heated further at 200–220C until the product has completely set. The mold is then cooled and the product removed.

Sm. Symbol for samarium.

smalt.
Properties: Blue powder.
Derivation: A potash-cobalt glass made by fusing pure sand and potash with cobalt oxide, grinding, and powdering.
Use: Paint pigments, ceramic industries (pigment), coloring glass, bluing paper, starch and textiles, coloring rubber.

smectic. A molecular structure (layers or planes) occurring in some liquid crystals; it imparts a soft, soapy property. There are nine types of smectic orientation.

smelting. Heat treatment of an ore to separate the metallic portion with subsequent reduction.
See also roasting.

"Smentox."[236] TM for a chemical compound for reconditioning cement-contaminated drilling mud or for preventing mud from becoming cement-contaminated, since contact with cement flocculates untreated drilling mud, rendering it unfit for use.

Smiles rearrangement. Intramolecular nucleophilic aromatic substitution in alkaline solution resulting in the migration of an aromatic system from one heteroatom to another.

"Smite."[28] TM for emulsifiable insecticide containing 12.5% methoxychlor and 12.5% malathion.

"Smithol 26."[403] TM for a synthetic fatty alcohol ester of fatty acids to substitute for 45 degrees NW sperm oil. May replace sperm oil, lard oil, olive oil, and other natural fatty oils for industrial use. The lower cloud point and pour point and the fact that it does not have the objectionable odor of some natural oils makes it attractive for many applications where natural oils have limitations.

smog. A coined word denoting a persistent combination of smoke and fog occurring under appropriate meteorological conditions in large metropolitan or heavy industrial areas. The discomfort and danger of smog is increased by the action of sunlight on the combustion products in the air, especially sulfur dioxide, nitric oxide, and exhaust gases (photochemical smog). Strongly irritant and even toxic substances may be present, e.g., peroxybenzoyl nitrate. Fatalities have resulted from exposure to particularly severe photochemical smogs.
See also air pollution.

smoke. A colloidal or microscopic dispersion of a solid in gas, an aerosol. (1) Coal smoke: A suspension of carbon particles in hydrocarbon gases or in air, generated by combustion. The larger particles can be removed by electrostatic precipitation in the stack (Cottrell). Dark color, nauseating odor.
See also smog, air pollution, Cottrell.
(2) Wood smoke: Light-colored particles of cellulose ash, pleasant aromatic odor. Smoke from special kinds of wood, (e.g., hickory, maple) is used to cure ham, fish etc., also to preserve crude rubber.
(3) Chemical smoke: Generated by chemical means for military purposes (concealment, signaling, etc.).
(4) Metallic smoke (fume): An emanation from heated metals or metallic ores, the particles being of specific geometric shapes. Such smoke is particularly damaging to vegetation in the neighborhood of zinc and tin smelters.
(5) Cigarette smoke. There is conclusive evidence that the tars occurring in cigarette smoke

can lead to lung cancer; chief factors are age of individual at initiation of smoking, extent of inhalation, and amount smoked per day. Polonium, a radioactive element, is known to occur in cigarette smoke; more than 100 compounds have been identified, including nicotine, cresol, carbon monoxide, pyridene, and benzopyrene, the latter a carcinogen.
See also cigarette tar.

smokeless powder. Nitrocellulose containing about 13.1% nitrogen, produced by blending material of somewhat lower (12.6%) and slightly higher (13.2%) nitrogen content, converting to a dough with alcohol-ether mixture, extruding, cutting, and drying to a hard, horny product. Small amounts of stabilizers (amines) and plasticizers are usually present, as well as various modifying agents (nitrotoluene, nitroglycerin salts).
Hazard: A low explosive, dangerous fire and explosion risk when exposed to flame or impact.
Use: Sports ammunition, military purposes.

smudge oil. An oil burned in fruit orchards to prevent frost from injuring the trees. No. 3 fuel oil is typical of oils used.

smut. See fungus, rust (2).

Sn. Symbol for tin (from Latin stannum).

snake venom. There are two functional types (1) those that bring about blood coagulation either by direct action on fibrogen or by converting prothrombin to thrombin; (2) neurotoxins that act on the central nervous system, e.g., by inactivation of acetylcholine. Rattlesnake and moccasin venom are examples of (1) and cobra venom of (2). The enzymes of snake venoms are thought to be the actual toxic principles. Solutions of cobra venom have found use in treatment of arthritis and cancer. The chemistry and pharmacological properties of these poisons are not well understood. *Note*: A person bitten by a poisonous snake should be carried, *not walked*, to a hospital. *No alcohol* should be administered.

SNG. Abbreviation for synthetic or substitute natural gas.

sneezing gas. See diphenylchloroarsine.

snow, artificial. See artificial snow.

"Snowflake."[292] TM for sodium sesquicarbonate ($Na_2CO_3 \cdot NaHCO_3 \cdot 2HOH$), small, white, sparkling, needle-like crystals of uniform size which are free-flowing and non-caking under normal conditions.
Use: Base for household and industrial cleansers where mild alkaline action is desired, and as a

base for both foaming and non-foaming bath salts.

SNTA. Abbreviation for sodium nitrilotriacetate.

soap. (1) The water-soluble reaction product of a fatty acid ester and an alkali (usually sodium hydroxide), with glycerol as byproduct. For the reaction see saponification. A soap is actually a specific type of salt, the hydrogen of the fatty acid being replaced by a metal, which in common soaps is usually sodium. Soap lowers the surface tension of water and thus permits emulsification of fat-bearing soil particles. A typical commercial cleansing soap is made by reacting sodium hydroxide with a fatty acid. The lower the hydrogen content of the acid the thinner the soap. The byproduct of the reaction is glycerol. Many different carboxyl-containing substances are used, including rosin, tall oil, vegetable and animal oils and fats (stearic, palmitic, and oleic acids). Olive oil is used for Castile soap; transparent soaps are made from decolorized fats. The specific gravity of soaps is slightly more than 1.0; inclusion of air gives a floating product. Water solutions of sodium soaps in bar, chip, or powder form are universally used as mild emulsifying detergents for washing textiles, skin, paint, etc. Medically soap is used as an antidote for poisoning by ingestion of mineral acids or heavy metals. Liquid green soap is made with potassium hydroxide and a vegetable oil. For further information refer to Soap and Detergent Association, 485 Madison Ave., New York, NY.

(2) Heavy-metal soaps (loosely called metallic soaps) are those formed by metals heavier than sodium (aluminum, calcium, cobalt, lead, and zinc). These soaps are not water-soluble; specific types are used in lubricating greases, gel thickeners, and in paints as driers and flatting agents. Napalm is an aluminum soap.
See also saponification, detergent.

soap builder. Any material mixed with soap to improve the cleaning properties, modify the alkali content, or impart water-softening characteristics. The commonest are some of the sodium phosphates and rosin.
See also builder, detergent.

soap, heavy metal. See soap (2).

soap, soft. See green soap, soap.

soapstone. See talc.

Society of the Chemical Industry. This society was founded in London in 1881 to advance applied chemistry in all its branches. The American Section was established in 1894. In 1906 on the anniversary of the "birth of the coal-tar industry," it founded the famous Perkin Medal, which has since been awarded annually for outstanding achievement in chemistry in the US. It has also offered the Chemical Industry Medal since 1932, given for conspicuous service to applied chemistry. Perkin Medalists include L. Baekeland, F. G. Cottrell, Irving Langmuir, Herbert Dow, A. D. Little, R. R. Williams, Glenn T. Seaborg, and Roger Adams.

Society of Plastics Engineers. (SPE). An incorporated technical organization devoted primarily to the application of sound engineering principles to the manufacture and use of plastics, this rapidly growing organization (founded in 1942) has contributed much to the correct evaluation of these versatile materials in many fields. It publishes a monthly journal and sponsors a series of technical books on both theoretical and applied aspects of plastics technology. Its headquarters is at Greenwich, Connecticut.

Society of the Plastics Industry. (SPI). An incorporated technical organization serving the needs of the entire plastics industry in the US. It establishes standards for the properties and selection of materials and for product design and engineering. Its two major publications are the "Plastics Engineering Handbook" and the "Reinforced Plastics Handbook." With it are associated the Plastics Pipe Institute and the Reinforced Plastics/Composites Institute. Its offices are at 250 Park Ave., New York.

SOCMA. Abbreviation for Synthetic Organic Chemical Manufacturers Association.

soda. Any one of the forms of sodium carbonate, also used loosely as equivalent to the word sodium in compounds.

soda alum. See aluminum sodium sulfate.

soda ash. (soda, calcined; sodium carbonate, anhydrous). CAS: 497-19-8. Na_2CO_3.
The crude sodium carbonate of commerce, 10th highest-volume chemical produced in the US (1985).
Properties: Grayish-white powder or lumps containing up to 99% sodium carbonate, soluble in water, insoluble in alcohol. Noncombustible.
Derivation: For nearly a century most of the soda ash produced in the US has been made synthetically by the ammonia soda (Solvay) process. Largely because of high energy costs and strict pollution controls, most of the synthetic production is being abandoned in favor of the natural

product obtained from deposits in Utah, California, Wyoming, etc. One of the largest producing sites of the natural product is in Green River, Wyoming.

Impurities: Sodium chloride, sodium sulfate, calcium carbonate and magnesium carbonate, sodium bicarbonate.

Grade: Dense 58%, light 58%, extra light, natural, refined.

Use: Glass manufacture, chemicals, pulp and paper manufacture, sodium compounds, soaps and detergents, water treatment, aluminum production, textile processing, cleaning preparations, petroleum refining, sealing ponds from leakage (sodium ions bind to clay particles, which swell to seal leaks), catalyst in coal liquefaction.

soda, baking. See sodium bicarbonate.

soda, calcined. See soda ash.

soda, caustic. See sodium hydroxide.

soda crystals. See sodium carbonate monohydrate.

soda lime. CAS: 8006-28-8. A mixture of calcium oxide with sodium hydroxide or potassium hydroxide intended for the absorption of carbon dioxide gas and water vapor.

Properties: White or grayish-white granules unless colored by a specified indicator. Must be kept in air-tight containers.

Grade: Technical, reagent. Usually percentage moisture and mesh size are stated.

Hazard: Toxic by ingestion and inhalation, strong irritant to tissue.

Use: Drying agent and carbon dioxide absorbent, lab reagent.

soda lime glass. See glass.

sodalite. See cage, zeolite.

sodamide. See sodium amide.

soda, modified. (neutral soda). A combination of soda ash and bicarbonate of soda in definite proportions for purposes where an alkali is needed, ranging in causticity between bicarbonate of soda and soda ash. White, crystalline powders; water-soluble and possessing valuable cleansing and purifying properties. Prepared in various strengths.

Use: Washing powders, laundering, wool scouring powders, bottle cleansers, textile cleaners, mild detergents.

soda monohydrate. See sodium carbonate, monohydrate.

soda, natural. See trona.

soda niter. See sodium nitrate.

"Sodaphos."[55] TM for glassy sodium tetraphosphate.

soda pulp. See pulp, paper.

soda, washing. See sal soda.

α-sodio-sodium acetate. (sodium α-sodioacetate). $NaCH_2COONa$.

Properties: Free-flowing powder, stable in dry air, decomposes slowly in moist air, decomposes 280C without melting, insoluble in ethers and hydrocarbons, reacts mildly with water.

Grade: 80–85% pure. Impurities are sodium acetate, sodium amide, and sodium hydroxide.

Hazard: Toxic by inhalation, irritant to skin and mucous membranes.

Use: Organic intermediate, drying agent for organic solvents.

Soddy, Frederick (1877-1965) A British physicist who won Nobel prize in chemistry in 1921. Work was concerned with radioactive elements and atomic energy. His concept of isotopes and displacement law of radioactive change is basic to nuclear physics. His education was at Oxford and Glasgow. He later worked in Canada and Australia.

sodium. (natrium). CAS: 7440-23-5. Na. Metallic element, atomic number 11, group IA of Periodic Table, aw 22.98977, valence = 1, no stable isotopes but several radioactive forms, extremely reactive.

Properties: Soft, silver-white solid oxidizing rapidly in air; wax-like at room temperature, brittle at low temperatures. Store in air-tight containers or in naphtha or similar liquid that does not contain water or free oxygen. D 0.9674 (25C), mp 97.6C, bp 892C. Decomposes water on contact, with evolution of hydrogen to form sodium hydroxide; insoluble in benzene, kerosene, and naphtha. Has excellent electrical conductivity and high heat-absorbing capacity.

Derivation: Electrolysis of a fused mixture of sodium chloride and calcium chloride.

Method of purification: Distillation.

Grade: Commercial, technical, brick, amalgam, coated powders, dispersions (sodium dispersion), reactor (99.99% pure).

Hazard: Severe fire risk in contact with water in any form, ignites spontaneously in dry air when heated; to extinguish fires use *dry* soda ash, salt, or lime. Forms strong caustic irritant to tissue.

Use: Tetraethyl and tetramethyl lead, titanium re-

duction, sodium peroxide, sodium hydride, polymerization catalyst for synthetic rubber, lab reagent, coolant in nuclear reactors, electric power cable (encased in polyethylene), non-glare lighting for highways, radioactive forms in tracer studies and medicine, heat transfer agent in solar-powered electric generators.
See sodium dispersions.

sodium abietate.   (rosin soap; sodium resinate). $C_{19}H_{29}COOONa$.
Properties: White powder, soluble in water. Combustible.
Derivation: By leaching rosin with sodium hydroxide solution.
Use: Soap making, paper coating.

sodium acetate.   CAS: 127-09-3.
(a) $NaC_2H_3O_2$, (b) $NaC_2H_3O_2 \cdot 3HOH$.
Properties: Colorless, odorless crystals; efflorescent; soluble in water; slightly soluble in alcohol; soluble in ether. (a) D 1.528, mp 324C; (b) d 1.45, mp 58C; autoign temperature 1125F (607C). Combustible.
Grade: Highest purity, pure fused, CP, NF, technical, FCC.
Use: Dye and color intermediate, pharmaceuticals, cinnamic acid, soaps, photography, purification of glucose, meat preservation, medicine, electroplating, tanning, dehydrating agent, buffer, lab reagent, food additive.

sodium acetone bisulfate.   (acetone-sodium bisulfite). $(CH_3)_2CONaHSO_3$.
Properties: Crystalline material, soluble in water, decomposed by acids, slightly soluble in alcohol. Combustible.
Derivation: Interaction of sodium bisulfite and acetone.
Use: Chemical (pure acetone), photography, textile (dyeing and printing).

sodium acetylformate.   See sodium pyruvate.

sodium acid carbonate.   See sodium bicarbonate.

sodium acid methanearsonate.   See sodium methanearsonate.

sodium acid phosphate.   See sodium phosphate, monobasic.

sodium acid pyrophosphate.   See sodium pyrophosphate, acid.

sodium acid sulfate.   See sodium bisulfate.

sodium acid sulfite.   See sodium bisulfite.

sodium acid tartrate.   See sodium bitartrate.

sodium alginate.   (sodium polymannuronate). CAS: 9005-38-3. $(C_6H_7O_6Na)$.
Properties: Colorless or slightly yellow solid occurring in filamentous, granular, and powdered forms. Forms a viscous colloidal solution with water; insoluble in alcohol, ether, and chloroform. Combustible.
Derivation: Extracted from brown seaweeds (alginic acid).
Grade: NF, FCC, technical.
Use: Thickeners, stabilizers and emulsifiers in foods, especially ice cream, boiler compounds, medicine, experimental ocean-floor covering, textile printing, cement compositions, paper coating, food and pharmaceutical preparations, water-base paints.

sodium alkane sulfonate.   (SAS). RSO_3Na.
The sodium salt of an alkane sulfonic acid of linear paraffins having chain lengths of 14–18 carbon atoms. Preparation: By reaction of paraffins with sulfur dioxide and oxygen in the presence of gamma radiation from a cobalt-60 source.
Use: Biodegradable detergent intermediate.

sodium alum.   See aluminum sodium sulfate.

sodium aluminate.   CAS: 1302-42-7. $NaAlO_2$.
Properties: White powder, soluble in water, insoluble in alcohol, aqueous solution strongly alkaline, mp 1800C.
Derivation: By heating bauxite with sodium carbonate and extracting the sodium aluminate with water.
Grade: Technical, reagent, also 27 degrees Bé solution.
Hazard (solution): Strong irritant to tissue.
Use: Mordant, zeolites, water purification, sizing paper, manufacture of milk-glass, soap and cleaning compounds.

sodium aluminosilicate.   (sodium silicoaluminate).
A series of hydrated sodium aluminum silicates having a $Na_2O:Al_2O_3:SiO_2$ mole ratio of approximately 1:1:13.2.
Properties: Fine white amorphous powder or beads, odorless and tasteless. Insoluble in water and in alcohol and other organic solvents, but at 80–100C is partially soluble in strong acids and solutions of alkali hydroxides. pH of a 20% slurry is between 6.5 and 10.5. Noncombustible.
Grade: Technical, FCC.
Use: Anticaking agent in food preparations (up to 2%).

sodium aluminum hydride.   CAS: 13770-96-2. $NaAlH_4$.
Properties: White, crystalline material; d 1.24; stable in dry air at room temperature but very sensi-

tive to moisture; begins to melt at 183C with decomposition to evolve hydrogen. Soluble in tetrahydrofuran, dimethyl "Cellosolve."

Derivation: By reaction of aluminum chloride with sodium hydride.

Hazard: Severe fire and explosion risk in contact with oxidizing agents and water, forms caustic and irritating compounds.

Use: Reducing agent similar to lithium aluminum hydride.

sodium aluminum phosphate. (sodium aluminum phosphate, acidic). $NaAl_3H_{14}(PO_4)_8 \cdot 4HOH$ or $Na_3Al_2H_{15}(PO_4)_n$.

Properties: White, odorless powder; insoluble in water; soluble in hydrochloric acid.

Grade: Technical, FCC.

Use: Food additive (baked products).

sodium aluminum silicofluoride. (sodium aluminum fluosilicate). $Na_5Al(SiF_6)_4$.

Properties: White powder, somewhat soluble in cold water, corrosive to galvanized iron.

Hazard: Avoid prolonged skin contact. TLV (as F): 2.5 mg/m³ of air.

Use: Moth proofing and insecticides, also to obtain acid medium in dyebath.

sodium aluminum sulfate. See aluminum sodium sulfate.

sodium amalgam.   CAS: 11110-52-4. $Na_x Hg_y$.

Properties: A silver-white, porous, crystalline mass containing 2–20% of metallic sodium, decomposes water.

Derivation: Mercury is heated to about 200C and sodium in small pieces, added slowly. Also formed at one stage of process for making chlorine and sodium hydroxide by mercury cell process.

Grade: 2, 3, 4, 5, 6, 7, 8, 9, 10, and 20%.

Hazard: Flammable, dangerous fire risk.

Use: Preparation of hydrogen, reduction of metal halogen compounds and organic compounds, reagent in analytical chemistry.

sodium amide. (sodamide).   CAS: 7782-92-5. $NaNH_2$.

Properties: White, crystalline powder with ammonia odor; decomposes in water and hot alcohol; mp 210C; bp 400C.

Derivation: Dry ammonia gas is passed over metallic sodium at 350C.

Hazard: Flammable, dangerous fire risk.

Use: Manufacture of sodium cyanide, organic synthesis, lab reagent, dehydrating agent.

sodium aminophenylarsonate. See sodium arsanilate.

sodium ammonium hydrogen phosphate. See sodium ammonium phosphate.

sodium ammonium phosphate. (microcosmic salt; sodium ammonium hydrogen phosphate; phosphorus salt). $NaNH_4HPO_4 \cdot 4HOH$.

Properties: Transparent, colorless, odorless, efflorescent, monoclinic crystals; gives off water and ammonia on heating, leaving $NaPO_3$; soluble in water; insoluble in alcohol; d 1.57; mp about 79C with decomposition.

Derivation: Mixing solutions of sodium phosphate and ammonium chloride.

Grade: Granular, CP, technical.

Use: Analytical reagent.

sodium amytal. (sodium isoamyl ethyl barbiturate).

Use: Hypnotic.

sodium anilinearsonate.   See sodium arsanilate.

sodium antimonate. (antimony sodiate).

CAS: 11112-10-0. $NaSbO_3$.   Other forms are sodium metaantimonate $2NaSbO_3 \cdot 7HOH$ and sodium pyroantimonate $Na_2H_2Sb_2O_7 \cdot HOH$.

Properties: White, granular powder; slightly soluble in water and alcohol; insoluble in dilute alkalies, mineral acids but soluble in tartaric acid. Noncombustible.

Grade: Technical, glassmakers' grade.

Hazard: Toxic by ingestion and inhalation. TLV (As Sb): 0.5 mg/m³ of air.

Use: Opacifier in enamels for cast iron and glass, ingredient of acid resisting sheet steel enamels, high-temperature oxidizing agent.

sodium arsanilate. (sodium anilinearsonate; sodium aminophenylarsonate).   CAS: 127-85-5. $C_6H_4NH_2(AsO \cdot OH \cdot ONa)$ Often with one or more water.

Properties: White, crystalline, odorless powder; faint salty taste; soluble in water; slightly soluble in alcohol.

Derivation: By dissolving arsanilic acid in sodium carbonate solution and crystallizing.

Grade: Technical, medicinal.

Hazard: Highly toxic by ingestion and inhalation, may cause blindness.

Use: Medicine (anthelmintic), organic synthesis.

sodium arsenate.   CAS: 7778-43-0. $Na_3AsO_4 \cdot 12HOH$.

Properties: Clear, colorless crystals; mild alkaline taste. Soluble in water, slightly soluble in alcohol and glycerol, insoluble in ether, d 1.7593, mp 86C.

Derivation: Reaction of arsenic trioxide and sodium nitrate.

Grade: Pure crystals, CP, technical (60% arsenic pentoxide).
Hazard: Toxic by ingestion and inhalation.
Use: Mordant and assist in dyeing and printing, other arsenates, germicide.

sodium arsenite. (sodium metaarsenite).
CAS: 7784-46-5. $NaAsO_2$.
Properties: Grayish-white powder which absorbs carbon dioxide from the air, soluble in water, slightly soluble in alcohol, d 1.87.
Derivation: Arsenic trioxide is dissolved in a solution of sodium carbonate or hydroxide and boiled.
Grade: Crude, pure, 75% arsenious oxide (94% solution).
Hazard: Toxic by ingestion and inhalation.
Use: Arsenical soaps for taxidermists, antiseptic, dyeing, insecticides, hide preservation, herbicide.

sodium arsphenamine.   See arsphenamine.

sodium ascorbate.   CAS: 134-03-2.
$C_6H_7NaO_6$.   The sodium salt of ascorbic acid.
Use: Antioxidant in food products.

sodium azide. CAS: 26628-22-8. NaN_3.
Properties: Colorless, hexagonal crystals; decomposes at about 300C; d 1.846 (20C). Soluble in water and in liquid ammonia, slightly soluble in alcohol, hydrolyzes to form hydrazoic acid. Combustible.
Hazard: Highly toxic. TLV: CL of 0.1 ppm in air.
Use: Air bag inflation, preservative in diagnostic medicinals, intermediate in explosive manufacture.

sodium barbiturate. $C_4H_3N_2O_3Na$.
Properties: White to yellow tinted powder, pH of a 1% aqueous solution 10.5, soluble in water and dilute mineral acid.
Hazard: Toxic by ingestion.
See barbiturate.
Use: Synthetic intermediate, catalyst in ammonium nitrate propellants, wood impregnating solutions.

sodium benzenesulfonchloramine.   See chloramine-B.

sodium benzoate. CAS: 532-32-1.
C_6H_5COONa.
Properties: White crystals or granular, odorless powder; sweetish, astringent taste, soluble in water and alcohol. Combustible.
Derivation: Benzoic acid is neutralized with sodium bicarbonate solution, the solution filtered, concentrated, and allowed to crystallize.
Grade: USP, FCC, technical.

Hazard: Use in foods limited to 0.1%.
Use: Food preservative, antiseptic, medicine, tobacco, pharmaceutical preparations, intermediate for manufacture of dyes, rust and mildew inhibitor.

sodium benzosulfimide.   See saccharin.

sodium benzylpenicillin.   See penicillin.

sodium benzyl succinate.
$C_6H_5CH_2O_2C(CH_2)_2COONa$.
Properties: White amorphous or crystalline powder having a slight benzyl odor and cool, salty taste; soluble in water.

sodium beryllium fluoride.   See beryllium sodium fluoride.

sodium bicarbonate. (baking soda; sodium acid carbonate).   CAS: 144-55-8. $NaHCO_3$.
Properties: White powder or crystalline lumps; cooling, slightly alkaline taste; soluble in water; insoluble in alcohol; stable in dry air, but slowly decomposes in moist air; d 2.159; mp loses carbon dioxide at 270C. Noncombustible.
Derivation: Principally by treating a saturated solution of soda ash with carbon dioxide to precipitate the less soluble bicarbonate; also by purifying the crude product from the Solvay process.
Grade: Commercial, pure, highest purity, CP, USP, FCC.
Use: Manufacture of effervescent salts and beverages, artificial mineral water, baking powder; other sodium salts, pharmaceuticals, sponge rubber, gold and platinum plating, treating wool and silk, fire extinguishers, prevention of timber mold, cleaning preparations, lab reagent, antacid, mouthwash.

sodium bichromate.   See sodium dichromate.

sodium bifluoride. (sodium acid fluoride).
CAS: 1333-83-1. $NaHF_2$.
Properties: White, crystalline powder; soluble in water; decomposes on heating; d 2.08.
Hazard: Highly toxic, strong irritant to tissue.
TLV (as F): 2.5 mg/m³ of air.
Use: Tin plate production, neutralizer in laundry rinsing operations, preservative for zoological and anatomical specimens, etching glass, antiseptic.

sodium biphosphate.   See sodium phosphate, monobasic.

sodium bis(2-methoxyethoxy)-aluminohydride.
An organometallic metal hydride, the alkoxy adduct of sodium aluminum hydride, soluble in

benzene and other hydrocarbons; reacts less strongly with water than lithium aluminum hydride.

Use: For reduction of organic compounds.

sodium bismuthate.   CAS: 12232-99-4. $NaBiO_3$.
Properties: Yellow or brown, amorphous powder; slightly hygroscopic; insoluble in cold water; decomposes in hot water and acids.
Use: Analysis (testing for manganese in iron, steel, and ores), reagent, pharmaceuticals.

sodium bisulfate.   (sodium acid sulfate; niter cake; sodium hydrogen sulfate).   CAS: 7681-38-1.
(a) $NaHSO_4$ or (b) $NaHSO_4 \cdot HOH$.
Properties: Colorless crystals or white fused lumps, aqueous solution is strongly acid, soluble in water (a) d 2.435 (13C), mp above 315C, (b) d 2.103 (13C), mp 58.5C. Noncombustible.
Derivation: A byproduct in the manufacture of hydrochloric acid and nitric acid.
Method of purification: Recrystallization.
Grade: Pure crystals, pure fused, pure dry, reagent, crude, CP, technical, FCC.
Forms available: Cakes, powder, prills, pearls.
Hazard: Strong irritant to tissue.
Use: Flux for decomposing minerals; substitute for sulfuric acid in dyeing; disinfectant; manufacture of sodium hydrosulfide, sodium sulfate, and soda alum; liberating CO_2 in carbonic acid baths, in thermophores; carbonizing wool; manufacture of magnesia cements, paper, soap, perfumes, foods, industrial cleaners, metal pickling compounds; lab reagent.

sodium bisulfide.   See sodium hydrosulfide.

sodium bisulfite.   (sodium acid sulfite; sodium hydrogen sulfite).   CAS: 7631-90-5. $NaHSO_3$.
Properties: White crystals or crystalline powder, slight sulfurous odor and taste, soluble in water, insoluble in alcohol, unstable in air, d 1.48, mp decomposes. Noncombustible.
Derivation: Sodium carbonate solution is saturated with sulfur dioxide and the solution crystallized.
Grade: Crystals, pure dry, commercial dry, reagent, commercial solution 35 degrees Bé, powder (67% SO_2), CP, USP, FCC.
Hazard: Not permitted in meats and other sources of vitamin B_1, strong irritant to skin and tissue. TLV: 5 mg/m³ of air.
Use: Chemicals (sodium salts, cream of tartar), vat dye preparation, textiles (antichlor, mordant, discharge), food preservative. Bleaching groundwood, wool, etc. Reducing agent, fermentation, antiseptic, cask sterilization (brewing), copper

and brass plating, color preservative for pale crepe rubber, wood pulp digestion, analytical reagent, dietary supplement.

sodium bitartrate.   (acid sodium tartrate). CAS: 526-94-3. $NaHC_4H_5O_6 \cdot HOH$.
Properties: White, crystalline powder; soluble in water (aqueous solution in acid); slightly soluble in alcohol; mp loses water at 100C; bp 219C (decomposes). Combustible.
Grade: CP, technical, reagent.
Use: Analysis (testing for potassium), effervescing mixture, nutrient media.

sodium bithionolate.   (USAN; (thiobis[4,6-dichloro-o-phenylene)oxy]disodium; disodium 2,2′-thiobis-[4,6-dichlorophenoxide]).
$C_{12}H_4Cl_4Na_2O_2S$.
Properties: Solid, soluble in water.
Use: Self-sanitizing polishes, waxes, and cleaners; shoe polish and leather conditioners.

sodium borate.   (sodium tetraborate; sodium pyroborate; borax).   CAS: 1303-96-4.
$Na_2B_4O_7 \cdot 10HOH$.
See also borax, anhydrous; borax pentahydrate.
Properties: White crystals or powder, odorless, d 1.73 (20/4C). Loses water of crystallization when heated to melting, between 75C and 320C; fuses to a glassy mass at red heat (borax glass); effloresces slightly in warm, dry air. Soluble in water and glycerol, insoluble in alcohol. Noncombustible.
Derivation: Fractional crystallization from Searles Lake brine, solution of kernite ore followed by crystallization. Also from colemanite, natural borax, uxelite, and other borates.
Grade: Crystals, granulated, powdered (refined, USP), CP, technical.
Hazard: Toxic by inhalation. TLV: 5 mg/m³ of air.
Use: Heat-resistant glass, porcelain enamel, detergents, herbicides, fertilizers, rust inhibitors, pharmaceuticals, leather, photography, bleaches, paint, boron compounds, flux for smelting, flame-retardant fungicide for wood, soldering flux, cleaning preparations, lab reagent.

sodium borate, anhydrous.   See borax, anhydrous.

sodium boroformate.
$NaH_2BO_3 \cdot 2HCOOH \cdot 2HOH$.
Properties: White,.crystalline solid; soluble in water; mp 110C.
Use: Buffering agent toward both acid and alkali in the range of pH 8.5, textile treating and tanning baths.

sodium borohydride.   CAS: 16940-66-2.
  $NaBH_4$.
  Properties: White, crystalline powder; soluble in
  water, ammonia, amines, pyridine, and dimethyl
  ether of diethylene glycol; insoluble in other
  ethers, hydrocarbons, and alkyl chlorides; d 1.07;
  hygroscopic; stable in dry air to 300C; decomposes slowly in moist air or in vacuum at 400C.
  Derivation: By reaction of sodium hydride and
  trimethyl borate at 250C.
  Grade: Technical, powdered, pellets.
  Hazard: Reacts with water to evolve hydrogen
  and sodium hydroxide; flammable, dangerous
  fire risk, store out of contact with moisture.
  Note: An explosion can occur by spontaneous
  ignition of the gases released from a saturated
  solution of sodium borohydride in dimethylformamide at 17C.
  Use: Source of hydrogen and other borohydrides.
  Reduces aldehydes, ketones and acid chlorides.
  Bleaching wood pulp, blowing agent for plastics,
  precipitation of mercury from waste effluent (by
  reduction), decolorizer for plasticizers, recycling
  of gold and platinum group metals, inorganic
  and organometallic reductions, organic synthesis.

sodium bromate.   CAS: 7789-38-0. $NaBrO_3$.
  Properties: White, odorless crystals or powder;
  soluble in water; insoluble in alcohol; d 3.339; mp
  381C (decomposes).
  Derivation: By passing bromine into a solution
  of sodium carbonate; sodium bromide and sodium
  bromate being formed.
  Hazard: Oxidizing material, dangerous fire risk
  near organic materials. Toxic by ingestion.
  Use: Analytical reagent.

sodium bromide.   CAS: 7647-15-6.   (a) NaBr,
  (b) $NaBr \cdot 2HOH$.
  Properties: White, crystalline powder or granules;
  saline and somewhat bitter taste. Absorbs moisture from the air, becoming very hard. Soluble
  in water; moderately soluble in alcohol; d (a)
  3.208, (b) 2.176; mp (a) 757.7C; bp 1390C.
  Derivation: Occurs naturally in some salt deposits.
  Made synthetically by first causing iron to react
  with bromine and water. The resulting ferroso-
  ferric bromide is dissolved in water, sodium carbonate added, the solution filtered and evaporated.
  Grade: CP, crystals, powdered, commercial, pure,
  highest purity, NF.
  Hazard: Toxic by inhalation and ingestion.
  Use: Photography, medicine (sedative), preparation of bromides.

sodium bromite. $NaBrO_2$.
  Use: A desizing compound for textiles.

sodium cacodylate.   (sodium dimethylarsenate).
  CAS: 124-65-2. $(CH_3)_2AsOONa \cdot 3HOH$.
  Properties: White, amorphous crystals or powder;
  deliquescent; melts about 60C; loses water at
  120C; soluble in water and alcohol.
  Derivation: Oxidation and neutralization of caco-
  dyl oxide.
  Hazard: Toxic by inhalation and ingestion.
  Use: Herbicide.

sodium carbolate.   See sodium phenate.

sodium carbonate.   (soda).   See soda ash, sal
  soda, sodium bicarbonate, sodium carbonate
  monohydrate, sodium sesquicarbonate.

sodium carbonate monohydrate.   (crystal carbonate; soda monohydrate; soda crystals).
  CAS: 497-19-8. $Na_2CO_3 \cdot HOH$.
  Properties: White, odorless, small crystals or crystalline powder; alkaline taste. D 1.55, soluble
  in water and glycerol, insoluble in alcohol, mp
  109C (loses water) 851C. Noncombustible.
  Grade: USP, technical, FCC.
  Use: Photography, cleaning and boiler compounds, pH control of water, food additive, intermediate in thermochemical reactions, manufacture of glass, cleaning and bleaching textiles,
  analytical reagent.

sodium carbonate peroxide.
  $2Na_2CO_3 \cdot 3HOOH$.
  Properties: White, crystalline powder; soluble in
  water; stable at room temperature when dry; decomposes rapidly at 100C evolving oxygen; active oxygen content 14% min.
  Derivation: Crystallization from solution of soda
  ash and hydrogen peroxide.
  Grade: Technical.
  Hazard: Oxidizing agent, dangerous near organic
  materials.
  Use: Source of hydrogen peroxide, household detergents, dental cleansers, bleaching and dyeing,
  modification of starch.

sodium carbonate peroxohydrate.   See sodium
  percarbonate.

sodium carboxymethylcellulose.   See carboxymethylcellulose.

sodium caseinate.   CAS: 9005-46-3.
  Properties: White, coarse powder; odorless; tasteless. Contains 65% proteins, soluble in water
  (usually with turbidity).
  Derivation: By dissolving casein in sodium hydroxide and evaporating.
  Grade: Edible.
  Use: Food additive (binder and extender in sausage, soups, etc.), emulsifier and stabilizer.

sodium catechol disulfonate. See disodium-1,2-dihydroxybenzene-3,5-disulfonate.

sodium cellulosate.
Properties: A cellulosic fiber.
Derivation: Reaction of sodium methoxide with cotton or rayon swollen in methanol.
Use: Intermediate in preparing grafted fibers when combined with polyacrylonitrile, there is an average substitution of one polyacrylonitrile chain for every 100 to 300 anhydroglucose units of the cellulose.

sodium chlorate. CAS: 7775-09-9. $NaClO_3$.
Properties: Colorless, odorless crystals; cooling, saline taste; must not be triturated with any combustible substance; soluble in water and alcohol; d 2.490; mp 255C; bp decomposes.
Derivation: A concentrated acid solution of sodium chloride is heated and electrolyzed, the chlorate crystallizing out.
Grade: Technical, CP, crystals, powder.
Hazard: Dangerous fire risk, strong oxidant, contact with organic materials may cause fire.
Use: Oxidizing agent and bleach (especially to make chlorine dioxide) for paper pulps; ore processing; herbicide and defoliant; substitute for potassium chlorate, being more soluble in water; matches, explosives, flares, and pyrotechnics; recovery of bromine from natural brines; leather tanning and finishing; textile mordant; to make perchlorates.

sodium chloraurate. See sodium gold chloride.

sodium chloride. (table salt; sea salt; halite; rock salt). NaCl.
Properties: Colorless, transparent crystals or white, crystalline powder; somewhat hygroscopic. Soluble in water and glycerol, very slightly soluble in alcohol, d 2.165, mp 801C, essential in diet to maintain chloride balance in body. Noncombustible.
Occurrence: Ocean water (2.6% concentration), deposits in central New York, Southern Michigan, Gulf Coast, Great Salt Lake, Newfoundland.
Derivation: (a) Evaporation and crystallization of natural brines, (b) solar evaporation of sea water, (c) direct mining from underground or surface deposits.
Method of purification: Recrystallization. Impurities: Sulfates, heavy metals, alkaline earths, magnesium salts, ammonium salts.
Grade: Highest purity medicinal, crystals; highest purity, dried; highest purity, fine powder; highest purity, fused; reagent; reagent, fused; sea evaporated; ground; microsized; powdered; table salt; rock salt; CP; USP; FCC; single pure crystals.

Use: Chemical (sodium hydroxide, soda ash, hydrogen chloride, chlorine, metallic sodium), ceramic glazes, metallurgy, curing of hides, food preservative, mineral waters, soap manufacture (salting out), home water softeners, highway deicing, regeneration of ion-exchange resins, photography, food seasoning, herbicide, fire extinguishing, nuclear reactors, mouthwash, medicine (heat exhaustion), salting out dyestuffs, supercooled solutions. Single crystals are used for spectroscopy, UV and infrared transmissions. See also fused salt.

sodium chlorite. CAS: 7758-19-2. $NaClO_2$.
Properties: White crystals or crystalline powder, slightly hygroscopic, soluble in water, mp 180–200C (decomposes)
Grade: Technical, reagent.
Hazard: Flammable, strong oxidizing agent, dangerous fire and moderate explosion risk. (Solution): Strong irritant to skin and tissue.
Use: For improving taste and odor of potable water (as an oxidizing agent); bleaching agent for textiles, paper pulp, edible and inedible oils, shellacs, varnishes, waxes and straw products; oxidizing agent; reagent.

sodium chloroacetate. CAS: 3926-62-3. $ClCH_2COONa$.
Properties: White, nonhygroscopic powder; odorless; easier to handle than chloroacetic acid; soluble in water; slightly soluble in methanol; insoluble in acetone benzene, ether, and carbon tetrachloride. Noncombustible.
Use: Manufacture of weed killers, dyes, vitamins, pharmaceuticals; also a defoliant.

sodium chloroaluminate. See sodium tetrachloroaluminate.

sodium chloroaurate. See sodium gold chloride.

sodium-p-chloro-m-cresolate. A water-soluble preservative for cutting oils.

sodium-4-chlorophthalate. (monosodium-4-chlorophthalate). $ClC_6H_3(COOH)COONa$.
Properties: White to light gray powder.
Grade: Commercial.
Use: Modifying agent in the manufacture of phthalocyanine pigments.

sodium chloroplatinate. (platinic sodium chloride; platinum sodium chloride; sodium platinichloride; sodium hexachloroplatinate). CAS: 1307-82-0. $Na_2PtCl_6 \cdot 4HOH$.
Properties: Yellow powder, soluble in alcohol, water. Noncombustible.
Grade: Technical, CP.

Use: Etching on zinc, indelible ink, microscopy, mirrors, photography, plating, catalyst, determination of potassium.

sodium-o-chlorotoluene-p-sulfonate.
$NaSO_3C_6H_3(CH_3)Cl$.
Properties: Gray to light tan powder, soluble in water and organic solvents.
Derivation: Sulfonation of o-chlorotoluene, neutralized to form the sodium salt.
Use: Synthesis of dyes, intermediates and drugs.

sodium chromate.   CAS: 7775-11-3.
$Na_2CrO_4 \cdot 10HOH$.
Properties: Yellow, translucent, efflorescent crystals; soluble in water; slightly soluble in alcohol. D 1.483, mp 19.92C, anhydrous sodium chromate is also available commercially.
Derivation: Chrome iron ore is melted in a reverberatory furnace with lime and soda, in presence of air. The melt is dissolved in water, a small amount of sodium carbonate added, the solution decanted, acidified with acetic acid, concentrated, and crystallized.
Grade: Pure neutral, highest purity, technical, CP, reagent.
Hazard: Toxic material. TLV (as chromium): 0.5 mg/m³ of air.
Use: Inks, dyeing, paint pigment, leather tanning, other chromates, protection of iron against corrosion, wood preservative.

sodium chromate tetrahydrate.
$Na_2CrO_4 \cdot 4HOH$.
Properties: Yellow crystals, deliquescent, soluble in water.
Grade: Technical.
Hazard: As for sodium chromate.
Use: Pigment manufacture, corrosion inhibition, leather tanning, other chromium compounds.

sodium citrate.   (trisodium citrate).
CAS: 68-04-2. $C_6H_5O_7Na_3 \cdot 2HOH$.
Properties: White crystals or granular powder, odorless, stable in air, pleasant acid taste, soluble in water, insoluble in alcohol, mp loses 2HOH at 150C, bp decomposes at red heat. Combustible.
Derivation: Sodium sulfate solution is treated with calcium citrate, filtered, concentrated and crystallized.
Grade: Highest purity, medicinal; pure; commercial; CP; USP; FCC.
Use: Soft drinks, photography, frozen desserts, meat products, detergents, special cheeses, electroplating, sequestrant and buffer, nutrient for cultured buttermilk, removal of sulfur dioxide from smelter waste gases, medicine (diuretic, expectorant), anticoagulant for blood withdrawn from body.

sodium copper chloride.   See copper sodium chloride.

sodium copper cyanide.   (copper sodium cyanide; sodium cyanocuprate). $NaCu(CN)_2$.
Properties: White, crystalline, double salt of copper cyanide and sodium cyanide; d 1.013 (20C); decomposes at 100C; soluble in water.
Hazard: Toxic material. TLV (as CN): 5 mg/m³ of air.
Use: For preparing and maintaining cyanide copper plating baths based on sodium cyanide.

sodium cyanate.   CAS: 917-61-3.   NaOCN.
Properties: White, crystalline powder; soluble in water; insoluble in alcohol and ether; d 1.937 (20C); mp 700C.
Use: Organic synthesis, heat treating of steel, intermediate for manufacture of medicinals.

sodium cyanide.   CAS: 143-33-9.   NaCN.
Properties: White, deliquescent, crystalline powder; soluble in water; slightly soluble in alcohol. Mp 563C, bp 149C, the aqueous solution is strongly alkaline and decomposes rapidly on standing.
Derivation: By absorption of hydrogen cyanide in a solution of sodium hydroxide with subsequent vacuum evaporation.
Grade: 30% solution, 73–75%, 96–98%, reagent, technical, briquettes, granular.
Hazard: Toxic by ingestion and inhalation. TLV (as CN): 5 mg/m³ of air.
Use: Extraction of gold and silver from ores, electroplating, heat treatment of metals (case-hardening), making hydrogen cyanide, insecticide, cleaning metals, fumigation, manufacture of dyes and pigments, nylon intermediates, chelating compounds, ore flotation.

sodium cyanoaurite.   See sodium gold cyanide.

sodium cyanocuprate.   See sodium copper cyanide.

sodium-2-cyanoethanesulfonate.   (sodium-2-sulfopropionitrile). $NaO_3S(CH_2)_2CN$.
Properties: Colorless crystals; mp 240C; soluble in water and glacial acetic acid; insoluble in benzene, alcohol, and ether.
Use: Introduction of sulfonic acid group into surfactants and other organics.

sodium cyclamate.   (sodium cyclohexylsulfamate).   CAS: 139-05-9.
$C_6H_{11}NHSO_3Na$.
Properties: White crystals; practically odorless powder; sweet taste; freely soluble in water; practically insoluble in alcohol, benzene, chloroform,

and ether; pH (10% solution) 5.5–7.5; sweetening power about 30 times that of sucrose.
Grade: NF, FCC.
Hazard: Some evidence of causing cancer in laboratory animals. Prohibited by FDA for food use.
Use: Nonnutritive sweetener.

sodium decametaphosphate. See metaphosphate.

sodium dehydroacetate. CAS: 4418-26-2.
$C_8H_7NaO_4 \cdot HOH$.
Properties: Tasteless, white powder; soluble in water and propylene glycol; insoluble in most organic solvents.
See also dehydroacetic acid.
Grade: FCC.
Use: Fungicide, plasticizer, toothpaste, pharmaceutical, preservative in food, mold inhibitor for strawberries and similar fruits.

sodium deoxycholate. See deoxycholic acid.

sodium dextran sulfate. (dextran sulfate).
Properties: Solid, soluble in water.
Derivation: Derivatives of dextran having a molecular weight of 500,000 to 2,000,000.
Use: Fractionation and separation of biological preparations.

sodium-3,5-diacetamido-2,4,5-triiodobenzoate.
See sodium diatrizoate.

sodium diacetate. CAS: 126-96-5.
$CH_3COONa \cdot x(CH_3COOH)$, anhydrous or $CH_3COONa \cdot x(CH_3COOH) \cdot yHOH$, technical.
Properties: White crystals with an acetic acid odor, soluble in water, slightly soluble in alcohol, insoluble in ether, decomposes above 150C. Combustible.
Grade: FCC.
Use: Buffer, mold inhibitor, souring agent, intermediate for acid salts, mordants, varnish hardeners, antitarnishing agents, sequestrant and preservative in foods.

sodium diatrizoate. (sodium-3,5-diacetamido-2,4,6-triiodobenzoate). CAS: 737-31-5.
$C_6I_3(COONa)(NHCOCH_3)_2$.
Properties: White crystals, soluble in water, solutions are radiopaque.
Grade: USP (as solution for injection).
Use: Radiopaque medium, medicine.

sodium-1-diazo-2-naphthol-4-sulfonate.
See 1-diazo-2-naphthol-4-sulfonic acid.

sodium dibutyldithiocarbamate. See "Trepidone."

sodium-α,β-dichloroisobutyrate. A plant growth regulator.

sodium dichloroisocyanurate. (sodium salt of dichloro-s-triazine-2,4,6-trione).

CAS: 2244-21-5. $NaNC(O)NClC(O)NClCO$.
Properties: White, slightly hygroscopic, crystalline powder; loose bulk d about 37 lb/ft³; granulated 57 lb/ft³. Active ingredient about 60% available chlorine; decomposes at 230C.
Hazard: Strong oxidizing material, fire risk near organic materials. Toxic by ingestion.
Use: Active ingredient in dry bleaches, dishwashing compounds, scouring powders, detergent-sanitizers, swimming pool disinfectants, water and sewage treatment, replacement for calcium hypochlorite.

sodium-2,4-dichlorophenoxyacetate. (2,4-D, sodium salt). CAS: 2702-72-9.
$C_6H_5(OCH_2COONa)Cl_2$.
Properties: Crystalline solid, decomposes at 215C, slightly soluble in water.
Hazard: Irritant by inhalation.
Use: Herbicide.
See 2,4-D.

sodium-2,4-dichlorophenoxyethyl sulfate.
See sesone.

sodium-2,2-dichloropropionate. See dalapon.

sodium dichromate. (sodium bichromate).
CAS: 10588-01-9. $Na_2Cr_2O_7 \cdot 2HOH$.
Properties: Red or red-orange deliquescent crystals, d 2.52 (13C), mp 357C, decomposes at 400C, loses 2HOH on prolonged heating at 100C, soluble in water, insoluble in alcohol. Noncombustible.
Derivation: (a) From chromite ore by alkaline roasting and subsequent leaching, (b) action of sulfuric acid on sodium chromate.
Grade: Technical crystals, technical liquor containing 69–70% $Na_2Cr_2O_7 \cdot 2HOH$, anhydrous 95C, soluble in water and alcohol.
Hazard: A toxic material.
Use: Colorimetry (copper determination), complexing agent, oxidation inhibitor in ethyl ether.

sodium dihydrogen phosphate. See sodium phosphate, monobasic.

sodium dihydroxyethylglycine. (N,N-bis[2-hydroxyethyl]-glycine).
$NaOOCCH_2N(CH_2CH_2OH)_2$.
Properties: Clear, straw-colored liquid, d 1.204 (25C), fp about −10C.
Use: Complexing agent for the transition metals.

sodium dimethylarsenate. See sodium cacodylate.

sodium dimethyldithiocarbamate. (SDDC).
CAS: 148-18-5. $(CH_3)_2NCS_2Na$.

Properties: 40% solution is amber to light green, d 1.17–1.20 (25/25C).

Derivation: Reaction of dimethylamine, carbon disulfide, and sodium hydroxide.

Use: Fungicide, corrosive inhibitor, rubber accelerator, intermediate, polymerization shortstop.

sodium dinitro-o-cresylate. CAS: 25641-53-6. $CH_3C_6H_2(NO_2)_2ONa$.

Properties: Brilliant orange-yellow dye which may stain clothing and wood.

Derivation: By treating 4,6-dinitro-o-cresol with sodium hydroxide.

Hazard: Toxic by ingestion and inhalation.

Use: Herbicide (control of mustard and other susceptible weeds), fungicide.

See also 4,6-dinitro-o-cresol.

sodium dioctyl sulfosuccinate. See dioctyl sodium sulfosuccinate.

sodium dioxide. NaO_2. Exists as impurity (about 10%) in sodium peroxide, obtained by heating sodium peroxide in oxygen, reacts with water to yield hydrogen peroxide, oxygen, and sodium hydroxide.

sodium dispersion. A stable suspension of microscopic sodium particles in a hydrocarbon or other medium, boiling at temperatures above the melting point of sodium (97.5C), e.g., heptane, n-octane, toluene, xylene, naphtha, kerosene, mineral oil, n-butyl ether, etc. Particles range in size from submicron to 30 μ depending on the method of preparation. Dispersions contain up to 50% (by weight) of sodium metal.

Hazard: Flammable, dangerous fire risk. To extinguish use dry soda ash, sodium chloride, or graphite, followed by carbon dioxide. Do not use carbon tetrachloride.

Use: For chemical reactions where advantages of controlled reaction rate, lower reaction temperature, increased yields, or substitution for more expensive reagent can be achieved, as, removal of sulfur from hydrocarbons and petroleum, metal powders, sodium hydride, alcohol-free alcoholates, phenylsodium.

sodium dithionate. (sodium hyposulfate). CAS: 7631-94-9. $Na_2S_2O_6 \cdot 2HOH$.

Properties: Large, transparent crystals; bitter taste. Soluble in water, insoluble in alcohol and concentrated hydrochloric acid, d 2.189 (25C), loses 2HOH at 110C, decomposes at 267C. Combustible.

Use: Chemical reagent.

sodium dithionite. (sodium hydrosulfite). CAS: 7775-14-6. $Na_2S_2O_4$.

Properties: Light lemon-colored solid in powder or flake form, mp 55C (decomposes), also available in liquid form, partially soluble in water, insoluble in alcohol, strong reducing agent.

Derivation: (1) Zinc is dissolved in a solution of sodium bisulfite, the zinc-sodium sulfite precipitated by milk of lime, leaving the hydrosulfite in solution. (2) Reaction of sodium formate with sodium hydroxide and sulfur dioxide.

Grade: Technical, reagent.

Hazard: Fire risk in contact with moisture. Use dry sand, soda ash, or carbon dioxide to extinguish fires.

Use: Vat dyeing of fibers and textiles; stripping agent for dyes; reagent; bleaching sugar, soap, oils, groundwood; oxygen scavenger for synthetic rubbers.

sodium diuranate. (uranium yellow). $Na_2U_2O_7 \cdot 6HOH$.

Properties: Yellow-orange solid, insoluble in water, soluble in dilute acids.

Derivation: By treating a solution of uranyl salt with sodium hydroxide.

Use: Ceramics to produce colored glazes, manufacture of fluorescent uranium glass.

sodium dodecylbenzenesulfonate. CAS: 25155-30-0. $C_{12}H_{25}C_6H_4SO_3Na$.

Properties: White to light yellow flakes, granules, or powder; biodegradable. The dodecyl radical may have many isomers and the benzene may be attached to it in many positions. Combustible.

Derivation: Benzene is alkylated with dodecene, to which it attaches itself in any secondary position; the resulting dodecylbenzene is sulfonated with sulfuric acid and neutralized with caustic soda. For ABS (branched-chain alkyl) the dodecene is usually a propylene tetramer, made by catalytic polymerization of propylene. For LAS (straight-chain alkyl), the dodecene may be removed from kerosene or crudes by molecular sieve, may be formed by Ziegler polymerization of ethylene, or by cracking wax paraffins to α-olefins.

See also biodegradability; alkyl sulfonate, linear; detergent; alkylate (3).

Use: Anionic detergent.

sodium dodecyldiphenyl oxide disulfonate. $C_{24}H_{34}O(SO_3)_2Na$.

Properties: Dry form 90% min active solution (45%); mp 150C; d 1.1; very soluble in water, strong acids, bases, and electrolytes; stable to oxidation.

sodium edetate. See tetrasodium EDTA.

sodium erythorbate. (sodium isoascorbate). $NaC_6H_7O_6$. White free-flowing crystals, soluble in water.

Grade: Technical, FCC.
Use: Antioxidant and preservative.

sodium ethoxide. See sodium ethylate.

sodium ethylate. (sodium ethoxide; caustic alcohol). C_2H_5ONa.
Properties: White powder, sometimes having brownish tinge; readily hydrolyzes to alcohol and sodium hydroxide.
Derivation: By carefully adding small amounts of sodium to absolute alcohol kept at a temperature of 10C, heating carefully to 37.7C, again carefully adding sodium, cooling to 10C and adding the same quantity of absolute alcohol as was used originally.
Hazard: As for caustic soda, ethanol.
See sodium hydroxide, which is formed when exposed to moisture.
Use: Organic synthesis.

sodium ethylenebisdithiocarbamate. See nabam.

sodium-2-ethylhexylsulfoacetate.
$C_8H_{17}OOCCH_2SO_3Na$.
Properties: Light cream-colored flakes, water-soluble, good foaming properties and excellent resistance to hard water; solutions practically neutral and stable to mineral acids. Combustible.
Use: Solubilizing agent, particularly for soapless shampoo compositions, electroplating detergent.

sodium ethylmercurithiosalicylate. See thimerosal.

sodium ethyl oxalacetate. See ethyl sodium oxalacetate.

sodium ethylxanthate. (sodium xanthogenate; sodium xanthate). $C_2H_5OC(S)SNa$.
See also xanthic acids.
Properties: Yellowish powder, soluble in water, alcohol.
Use: Ore flotation agent.

sodium ferricyanide. (red prussiate of soda). CAS: 14217-21-1. $Na_3Fe(CN)_6 \cdot HOH$.
Properties: Ruby-red, deliquescent crystals; soluble in water; insoluble in alcohol.
Derivation: Chlorine is passed into sodium ferrocyanide solution.
Grade: Technical, CP.
Use: Production of pigments, dyeing, printing.

sodium ferrocyanide. (yellow prussiate of soda). CAS: 13601-19-9. $Na_4Fe(CN)_6 \cdot 10HOH$.
Properties: Yellow, semitransparent crystals; partially soluble in water; insoluble in organic solvents; d 1.458.
Grade: Technical, FCC.

Use: Manufacture of sodium ferricyanide, blue pigments, blueprint paper, anticaking agent for salt, ore flotation, pickling metals, polymerization catalyst, photographic fixing agent.

sodium fluoborate. See sodium fluoroborate.

sodium fluophosphate. See sodium fluorophosphate.

sodium fluorescein. USP name for uranine.

sodium fluoride. CAS: 7681-49-4. NaF.
Properties: Clear, lustrous crystals or white powder; insecticide grade frequently dyed blue. Soluble in water, very slightly soluble in alcohol, d 2.558 (41C), mp 988C, bp 1695C.
Derivation: By adding sodium carbonate to hydrogen fluoride.
Grade: Pure, CP, USP, insecticide, technical, single pure crystals.
Hazard: Toxic by ingestion and inhalation, strong irritant to tissue. TLV (as F): 2.5 mg/m³ of air.
Use: Fluoridation of municipal water (1 ppm), degassing steel, wood preservative, insecticide (not to be used on living plants), fungicide and rodenticide, chemical cleaning, electroplating, glass manufacture, vitreous enamels, preservative for adhesives, toothpastes, disinfectant (fermentation equipment), dental prophylaxis, cryolite manufacture, single crystals used as windows in UV and infrared radiation detecting systems.

sodium fluoroacetate. (also known as 1080). CAS: 62-74-8. FCH_2COONa.
Properties: Fine, white, odorless powder. Soluble in water, nonvolatile, decomposes at 200C, soluble in water, insoluble in most organic solvents.
Derivation: Ethyl chloroacetate and potassium fluoride form ethyl fluoroacetate, which is then treated with a methanol solution of sodium hydroxide.
Hazard: Toxic by ingestion, inhalation, and skin absorption. TLV: 0.05 mg/m³ of air. For use by trained operators only; use has been restricted.
Use: Predator elimination (coyotes), rodenticide.

sodium fluoroborate. $NaBF_4$.
Properties: White powder, bitter acid taste, slowly decomposed by heat, d 2.47 (20C), mp 384C, soluble in water, slightly soluble in alcohol.
Derivation: By heating sodium fluoride and hydrofluoboric acid and cooling slowly.
Use: Sand casting of aluminum and magnesium, electrochemical processes, oxidation inhibitor, fluxes for nonferrous metals, fluorinating agent.

sodium fluorophosphate. (sodium fluophosphate). Na_2PO_3F.
Properties: Colorless crystals, mp about 625C, soluble in water.

Grade: 97%.
Hazard: Toxic by ingestion, strong irritant.
Use: Preparation of bactericides and fungicides.

sodium fluorosilicate. (sodium silicofluoride; sodium hexafluorosilicate; sodium fluosilicate).
Na_2SiF_6.
Properties: White, odorless, tasteless, free-flowing powder. D 2.7, mp decomposes at red heat, partially soluble in cold water, more soluble in hot water, insoluble in alcohol.
Derivation: From fluosilicic acid and sodium carbonate, or sodium chloride.
Grade: Technical, CP.
Hazard: Toxic by ingestion and inhalation, strong irritant to tissue. TLV (as F): 2.5 mg/m³ of air.
Use: Fluoridation, laundry sours, opalescent glass, vitreous enamel frits, metallurgy (aluminum and beryllium), insecticides and rodenticides, chemical intermediate, glue, leather and wood preservative, moth repellent, manufacture of pure silicon.

sodium formaldehyde bisulfite.
$HOCH_2SO_3Na$.
Properties: White water-soluble solid.
Derivation: Action of sodium bisulfite, formaldehyde, and water.
Use: Textile stripping agent.
See also hydrosulfite-formaldehyde compounds.

sodium formaldehyde hydrosulfite.
Use: Synthetic rubber polymerization and textile dyeing and stripping agent.
See hydrosulfite-formaldehyde compounds.

sodium formaldehyde sulfoxylate. (sodium sulfoxylate; sodium sulfoxylate formaldehyde; SFS).
$HCHO \cdot HSO_2Na \cdot 2HOH$.
Properties: White solid, mp 64C, soluble in water and alcohol. Purity: Usually admixed with a sulfite.
Use: Stripping and discharge agent for textiles, bleaching agent for molasses, soap.
See also hydrosulfite-formaldehyde preparations.

sodium formate. CAS: 141-53-7.   HCOONa.
Properties: White, slightly hygroscopic, crystalline powder; slight odor of formic acid. Soluble in water and glycerol, slightly soluble in alcohol, insoluble in ether, d 1.919, mp 253C.
Derivation: Carbon monoxide and sodium hydroxide are heated under pressure; also from pentaerythritol manufacture.
Grade: Technical, CP.
Use: Reducing agent, manufacture of formic acid and oxalic acid, organic chemicals, mordant, manufacture of sodium dithionite, complexing agent, analytical reagent (noble metal precipitant), buffering agent.

sodium glucoheptonate.
$HOCH_2(CHOH)_5COONa$.
Properties: Light tan, crystalline powder; a sequestering agent for polyvalent metals.
Use: Metal cleaning, bottle washing, kier boiling, mercerizing, caustic boiloff, paint stripping, aluminum etching.

sodium gluconate. CAS: 527-07-1.
$NaC_6H_{11}O_7$.
Properties: White to yellowish crystalline powder, soluble in water, sparingly soluble in alcohol.
Derivation: From glucose by fermentation.
Grade: Purified, technical, FCC.
Use: Foods and pharmaceuticals, sequestering agent, metal cleaners, paint stripper, aluminum deoxidizer, bottle-washing preparations, rust removal, chrome tanning, metal plating, mordant in dyeing.

sodium glucosulfone. (p′-diaminodiphenylsulfone-N,N′-di-dextrose sodium sulfonate).
$C_{14}H_{34}N_2Na_2O_{18}S_3$ The USP grade is an aqueous solution (for injection), clear and pale yellow, pH 5.0–6.5.
Use: Medicine.

sodium glutamate. (monosodium glutamate; MSG).   CAS: 142-47-2.
$COOH(CH_2)_2CH(NH_2)COONa$.   Sodium salt of glutamic acid, one of the common naturally occurring amino acids.
Properties: White, crystalline powder; mp decomposes; soluble in water and alcohol; shows optical activity; most effective between pH 6 and 8.
Derivation: (a) Alkaline hydrolysis of the waste liquor from beet sugar refining, (b) a similar hydrolysis of wheat or corn gluten, (c) organic synthesis based on acrylonitrile.
Grade: Technical, 99%, ND, FCC.
Use: Flavor enhancer for foods in concentration of about 0.3%.
See also flavor, glutamic acid.

sodium glycolate. (sodium hydroxyacetate).
$NaOOCCH_2OH$.
Properties: White powder.
Use: Buffer in electrodeless plating and textile finishing.

sodium gold chloride. (sodium aurichloride; sodium chloraurate; sodium chloroaurate; gold sodium chloride; gold salts).
$NaAuCl_4 \cdot 2HOH$.
Properties: Yellow crystals, soluble in water and alcohol.
Derivation: By neutralizing chloroauric acid with sodium carbonate.
Use: Photography, staining fine glass, decorating porcelain, medicine.

sodium gold cyanide. (sodium cyanoaurite; sodium aurocyanide; gold sodium cyanide). NaAu(CN)$_2$.
Properties: Yellow powder, soluble in water, contains 46% gold (min).
Hazard: Toxic. TLV (as CN): 5 mg/m^3 of air.
Use: For gold-plating electronic components.

sodium guanylate. (GMP; disodium guanylate). CAS: 5550-12-9. Na$_2$C$_{10}$H$_{12}$N$_5$O$_8$P·2HOH. A 5'-nucleotide.
Properties: Crystals, soluble in cold water, very soluble in hot water.
Derivation: From a seaweed or from dried fish.
Use: Flavor potentiator in foods.
See also guanylic acid.

sodium heparin. See heparin.

sodium heptametaphosphate. See sodium metaphosphate.

sodium hexachloroosmate. (osmium-sodium chloride; sodium-osmium chloride). Na$_2$OsCl$_6$.
Properties: Orange, rhombic prisms, contains 40.3% osmium, unstable, soluble in alcohol, water.
Grade: Technical.
Use: Catalyst (oxidation).

sodium hexachloroplatinate. See sodium chloroplatinate.

sodium hexylene glycol monoborate. C$_6$H$_{12}$O$_3$BNa.
Properties: Amorphous, white solid; bulk d 0.25; mp 426C; soluble in nonpolar solvents. Purity: Min 98%.
Use: Corrosion inhibitor in organic systems, additive to lubricating oils, flame-retardant, siloxane resin additive.

sodium hydrate. See sodium hydroxide.

sodium hexachloroplatinate. See sodium chloroplatinate.

sodium hexafluorosilicate. See sodium fluorosilicate.

sodium hexametaphosphate. See "Calgon."

sodium hydride. CAS: 7646-69-7. NaH.
Properties: Practically odorless powder, d 0.92, mp 800C (decomposes), must be kept cool and dry, particle size range 5–50 μ, starts to decompose with evolution of hydrogen at about 255C.
Preparation: Reaction of sodium metal with hydrogen. A microcrystalline dispersion of gray powder in oil containing 50 or 25% by weight. Packaging ranges from one pint to 55-gal drums.
Hazard: Dangerous fire risk, reacts violently with water evolving hydrogen. Irritant.
Use: Condensing or alkylating agent, especially for amines, descaling metals.

sodium hydrogen sulfide. See sodium hydrosulfide.

sodium hydrogen sulfite. See sodium bisulfite.

sodium hydrosulfide. (sodium sulfhydrate; sodium bisulfide; sodium hydrogen sulfide). CAS: 7775-14-6. NaSH·2HOH.
Properties: Colorless needles to lemon-colored flakes; soluble in water, alcohol, and ether; 70–72% NaSH; mp 55C; water of crystallization 26–28%.
Derivation: From calcium sulfide by treating it in the cold with sodium bisulfate.
Grade: Technical, flake, 70–72%, soluble 40–44%.
Hazard: Contact with acids causes evolution of toxic gases.
Use: Paper pulping dyestuffs processing, rayon and cellophane desulfurizing, unhairing hides, bleaching reagent.

sodium hydrosulfite. See sodium dithionite.

sodium hydroxide. (caustic soda; sodium hydrate; lye; white caustic). The most important commercial caustic, eighth highest-volume chemical produced in US (1985).
Properties: White, deliquescent solid; chiefly in form of beads or pellets, also 50% and 73% aqueous solutions. Absorbs water and carbon dioxide from the air, d 2.13, mp 318C, bp 1390C, soluble in water, alcohol, and glycerol.
Derivation: Electrolysis of sodium chloride (brines) (electrolytic cell), reaction of calcium hydroxide and sodium carbonate.
Grade: Commercial, ground, flake, beads, FCC, granulated (60% and 75% Na$_2$O), rayon (low in iron, copper, and manganese), purified by alcohol (sticks, lumps, and drops), reagent, highest purity, CP, USP.
Hazard: Corrosive to tissue in presence of moisture, strong irritant to tissue (eyes, skin, mucous membranes), by ingestion. TLV: CL of 2 mg/m^3 of air.
Use: Chemical manufacture, rayon and cellophane, neutralizing agent in petroleum refining,; pulp and paper, aluminum, detergents, soap, textile processing, vegetable oil refining, reclaiming rubber, regenerating ion exchange resins, organic fusions, peeling of fruits and vegetables in food

industry, lab reagent, etching and electroplating, food additive.

sodium hypochlorite. CAS: 7681-52-9.
NaOCl·5HOH.
Properties: Unstable in air unless mixed with sodium hydroxide. Strong oxidizing agent, usually stored and used in solution, disagreeable, sweetish odor and pale greenish color, soluble in cold water, decomposed by hot water, mp 18C.
Derivation: Addition of chlorine to cold dilute solution of sodium hydroxide.
Grade: Technical.
Hazard: Fire risk in contact with organic materials. Toxic by ingestion, strong irritant to tissue.
Use: Bleaching paper pulp, textiles, etc., intermediate, organic chemicals, water purification, medicine, fungicides, swimming pool disinfectant laundering, reagent, germicide.

sodium hypophosphite. CAS: 7681-53-0.
NaH$_2$PO$_2$·HOH.
Properties: Colorless, pearly, crystalline plates or white granular powder; saline taste; deliquescent. Soluble in water, partially soluble in alcohol.
Derivation: By neutralizing hypophosphoric acid with sodium carbonate.
Grade: Technical, CP.
Hazard: Explosion risk when mixed with strong oxidizing agents, decomposes to phosphine on heating, store in cool, dry place, away from oxidizing materials.
Use: Pharmaceuticals, reducing agent in electrodeless nickel plating of plastics and metals, lab reagent, substitute for sodium nitrite in smoked meats.

sodium hyposulfate. See sodium dithionate.

sodium inosinate. CAS: 4691-65-0.
C$_{10}$H$_{11}$Na$_2$N$_4$O$_8$P. A 5'-nucleotide derived from seaweed or dried fish. Sodium guanylate is a byproduct.
Use: Flavor potentiator in foods.
See also inosinic acid.

sodium iodate. CAS: 7681-55-2. NaIO$_3$.
Properties: White crystals, d 4.28, soluble in water and acetone, insoluble in alcohol.
Derivation: Interaction of sodium chlorate and iodine in presence of nitric acid.
Grade: CP, reagent, technical.
Hazard: Oxidizing agent, fire risk near organic materials.
Use: Antiseptic, disinfectant, feed additive reagent.

sodium iodide. CAS: 7681-82-5. (a) NaI (b) NaI·2HOH.
Properties: White, cubical crystals or powder, or colorless, odorless crystals; slowly becomes

brown in air, deliquescent, saline, somewhat bitter taste; soluble in water, alcohol, and glycerol; d (a) 3.665, (b) 2.448 (21C), mp (a) 653C, bp (a) 1304C.
Grade: Technical, CP, USP, single crystals.
Use: Photography, solvent for iodine, organic chemicals, reagent, feed additive, cloud seeding, scintillation (thallium-activated form), expectorant.

sodium iodide (I-131). (sodium radio-iodide).
A radioactive form of sodium iodide containing iodine-131 which can be used as a tracer.
Grade: USP (as capsules or solution).
See iodine 131.

sodium iodipamide. C$_{20}$H$_{12}$I$_6$N$_2$NaO$_6$.
N,N'-adipolybis(3-amino-2,4,6-triiodobenzoic acid) disodium salt.
Properties: Radiopaque, water-soluble, available as a 20% solution for injection as a clear, colorless to pale yellow, slightly viscous liquid.
Derivation: By dissolving the free acid in dilute sodium hydroxide and buffering to pH 6.5–7.7.
Grade: USP.
Use: X-ray contrast medium.

sodium iodomethanesulfonate. See sodium methiodal.

sodium iothalamate. (sodium-5-acetamido-2,4,6-triiodo-N-methylisophthalamate).
CAS: 1225-20-3.
C$_6$I$_3$(CONHCH$_3$)$_2$COONa.
Grade: USP (for injection).
Use: Medicine (radiopaque medium).

sodium ipodate. (USAN; sodium-3-(dimethylaminomethyleneamino)-2,4,6-triiodohydrocinnamate). CAS: 1221-56-3.
NaOOCCH$_2$CH$_2$C$_6$H(I$_3$)N:CHN(CH$_3$)$_2$.
Use: Radiopaque agent.

sodium iron pyrophosphate. (SIPP).
Na$_8$Fe$_4$(P$_2$O$_7$)$_5$·xHOH.
Properties: Tan powder, insoluble in water but soluble in dilute acid, min 14.5% iron. Iron is in complex form and will not catalyze oxidation reactions.
Derivation: By reacting tetrasodium pyrophosphate with a soluble iron salt.
Use: For iron enrichment, particularly in flours and cereals.

sodium isoascorbate. See sodium erythorbate.

sodium isobutylxanthate.
CH(CH$_3$)$_2$CH$_2$OC(S)SNa.
Use: Ore flotation agent.
See xanthic acid.

sodium isopropylxanthate. CAS: 140-93-2. $(CH_3)_2CHOC(S)SNa$.
Properties: Light-yellow crystals soluble in water, deliquescent, decomposes 150C.
Hazard: Moderate fire risk. Irritant to skin and mucous membranes.
Use: Chemical weed killer, fortifying agent for certain oils, ore flotation.

sodium lactate. CAS: 72-17-3. $CH_3CHOHCOONa$.
Properties: Colorless or yellowish syrupy liquid, very hygroscopic, soluble in water, mp 17C, decomposes 140C. Combustible.
Grade: Technical, USP (solution with pH 6.0–7.3).
Use: Hygroscopic agent, glycerol substitute, plasticizer for casein, corrosion inhibitor in alcohol antifreeze.

sodium-n-lauroyl sarcosinate. $C_{15}H_{28}NO_4Na$.
Use: Dentrifices, hair shampoos, rug shampoos.

sodium lauryl sulfate. CAS: 151-21-3. $NaC_{12}H_{25}SO_4$.
Properties: Small white or light yellow crystals, slight characteristic odor, soluble in water, forming an opalescent solution.
Grade: USP, technical, FCC.
Use: Wetting agent in textile industry, detergent in toothpaste, food additive (emulsifier and thickener).

sodium-lead alloy. One of several alloys as follows: (1) usually containing 10% sodium and 90% lead, used in the manufacture of lead tetraethyl, (2) containing 2% sodium used as a deoxidizer and homogenizer in nonferrous metals where lead is a component, (3) used as a stabilizer and deoxidizer for lead in cable sheathing.
Hazard: Moderate fire and explosion risk, reacts with moisture, acids, and oxidizing agents.

sodium lead hyposulfite. See lead sodium thiosulfate.

sodium lead thiosulfate. See lead sodium thiosulfate.

sodium lignosulfonate.
Properties: Tan, free-flowing, spray-dried powder, containing 70–80% total lignin sulfonates, balance wood sugars. Combustible.
Use: Dispersant, emulsion stabilizer, chelating agent.
See also lignin sulfonate.

sodium liothyronine. (sodium-l-3[4-(4-hydroxy-3-iodophenoxy)-3,5-diiodophenyl]alanine). $NaOOCCH(NH_2)CH_2C_6H_2I_2OC_6H_3IOH$.
Properties: Light-tan, odorless, crystalline powder; very slightly soluble in water; slightly soluble in alcohol; insoluble in most other organic solvents.
Grade: USP.
Use: Medicine (a thyroid hormone).

sodium MBT. (NaMBT). $C_7H_4NS_2Na$.
A 50% aqueous solution of sodium mercaptobenzothiazole, light-amber liquid, bulk d 10.5 lb/gal.
Use: Corrosion inhibitor for nonferrous metals, antifreeze, paper mill systems.

sodium mercaptoacetate. See sodium thioglycolate.

sodium-2-mercaptobenzothiazole. See sodium MBT.

sodium metabisulfite. (sodium pyrosulfite). CAS: 7681-57-4. $Na_2S_2O_5$. Chief constituent of commercial dry sodium bisulfite with which most of its properties and uses are practically identical.
Grade: FCC.
Hazard: Toxic by inhalation. TLV: 5 mg/m³ of air.
Use: In foods, as preservative, lab reagent.

sodium metaborate. CAS: 7775-19-1. $NaBO_2$.
Properties: White lumps, d 2.464, soluble in water, mp 966C, bp 1434C. Noncombustible.
Derivation: By fusing sodium carbonate and borax.
Use: Herbicide. Also available commercially as octahydrate and tetrahydrate.

sodium metanilate. $NaSO_3C_6H_4NH_2$.
Derivation: The metasodium sulfonate of aniline sold as a solid or 20% aqueous solution prepared by neutralizing metanilic acid.
Grade: Technical, 99%, also 20% solution.
Use: Manufacture of synthetic dyestuffs and drugs.

sodium metaperiodate. See sodium periodate.

sodium metaphosphate. CAS: 10361-03-2. $(NaPO_3)_n$ The value of n ranges from 3–10 (cyclic molecules) or may be much larger number (polymers). Cyclic sodium metaphosphate, based on rings of alternating phosphorus and oxygen atoms, range from the trimetaphosphate $(NaPO_3)_3$ to at least the decametaphosphate. So-called sodium hexametaphosphate is probably a polymer where n is between 10 and 20 ("Calgon").

The vitreous sodium phosphates having a Na_2O/P_2O_5 mole ratios near unity are classified

as sodium metaphosphates (Graham's salts). The average number of phosphorus atoms per molecule in these glasses ranges from 25 to infinity. The term sodium metaphosphate has also been extended to short-chain vitreous compositions, the molecules of which exhibit the polyphosphate formula $Na_{n+2}P_nO_{3n+1}$ with n as low as 4–5. These materials are more correctly called sodium polyphosphates.
Use: Dental polishing agents, detergent builders, water softening, sequestrants, emulsifiers, food additives, textile processing laundering.

sodium metasilicate, anhydrous.   CAS: 6834-92-0. Na_2SiO_3.   A crystalline silicate.
Properties: Dustless, white granules; mp 1089C; total Na_2O content 51.5%; total Na_2O in active form 48.6%; bulk d 2.61 or 75 lb/cu ft. Soluble in water, precipitated by acids and by alkaline earths and heavy metal ions, pH of 1% solution 12.6, Noncombustible.
Derivation: Crystallized from a melt of Na_2O and SiO_2 at approximately 1089C.
Use: Laundry, dairy and metal cleaning; floor cleaning, base for detergent formulations, bleaching aid, deinking paper.
Also available as the pentahydrate, whose properties are mp 72.2C, total Na_2O content 29.3%, total Na_2O in active form 27.8%, d 1.75, or 55 lb/cu ft.

sodium metavanadate.   CAS: 13718-26-8. $NaVO_3$, often with 4HOH.
Properties: Colorless, monoclinic, prismatic crystals or pale green crystalline powder; soluble in water; mp 630C. Noncombustible.
Derivation: Sodium hydrate and vanadium pentoxide in water solution.
Grade: Technical, CP.
Hazard: Toxic by ingestion.
Use: Inks, fur dyeing, photography, inoculation of plant life, mordants and fixers, corrosive inhibitor in gas-scrubbing systems.

sodium methacrylate. $CH_2:C(CH_3)COONa$.
Properties: Water-soluble monomer.
Use: Resins, chemical intermediate.

sodium methanearsonate.   (disodium methylarsonate; monosodium methanearsonate; sodium acid methanearsonate; sodium methylarsonate). CAS: 2163-80-6. $(CH_3AsO(OH)ONa)$.
Properties: White solid, mp 130–140C, very soluble in water.
Hazard: Toxic by ingestion and inhalation.
Use: Post-emergence herbicide for grassy weeds.

sodium methiodal.   (sodium iodomethanesulfonate).   CAS: 126-31-8. ICH_2SO_3Na.

Properties: A white, crystalline powder; odorless, with slight salty taste followed by sweetish aftertaste. Decomposes on exposure to light, solutions are neutral to litmus, soluble in water, very soluble in methanol, slightly soluble in ethanol, practically insoluble in acetone, ether, and benzene.
Derivation: From sodium sulfite and methylene iodide.
Grade: NF.
Use: Radiopaque contrast medium.

sodium methylate.   (sodium methoxide). CAS: 124-41-4. CH_3ONa.
Properties: White, free-flowing powder; sensitive to oxygen; decomposed by water; soluble in methanol and ethanol; decomposes in air above 126C.
Hazard: (Solid) Flammable when exposed to heat or flame. (Solution) Flammable, moderate fire risk.
Use: Condensation reactions, catalyst for treatment of edible fats and oils, especially lard, intermediate for pharmaceuticals, preparation of sodium cellulosate, analytical reagent.

sodium methyl carbonate. $CH_3OCOONa$.
Properties: White powder, mp 330C (decomposes), d 1.66, purity 90% min.

sodium-N-methyldithiocarbamate dihydrate.
CAS: 137-42-8. $CH_3NHC(S)SNa \cdot 2HOH$.
Properties: White, crystalline solid; readily soluble in water; moderately soluble in alcohol; stable in concentrated aqueous solution but decomposes in dilute aqueous solution; unstable in moist soil.
Hazard: Irritant to tissue, toxic to plants and vegetation.
Use: Fungicide, nematocide, weed killer, and insecticide; soil fumigant.

sodium-N-methyl-N-oleoyl taurate.   (oleyl methyl tauride).   CAS: 137-20-2.
$C_{21}H_{40}O_4NSNa$
Properties: Fine, white powder; sweet odor.
Grade: Technical, 32% purity (remainder is mainly sodium sulfate).
Use: Detergent, pesticide aid.

sodium methyl siliconate.   Most effective water-repellent and cleaner for limestone, concrete and similar masonary. Reaction product of aqueous sodium hydroxide and a resinous silicone. Total solids about 30%.

sodium-N-methyltaurate.   See   N-methyltaurine.

sodium molybdate.   CAS: 7631-95-0.
Commercially, the normal molybdate Na_2MoO_4 or its dihydrate (called sodium molybdate crystals). Chemically, a wide variety of complex mo-

lybdenum and sodium compounds are known.
Properties: Small, lustrous, crystalline plates; soluble in water; mp 687C; d 3.28 (18C). Noncombustible.
Derivation: By the action of sodium hydroxide on molybdenum trioxide. Complex molybdates are prepared by dissolving large amounts of molybdenum trioxide in solutions of normal molybdates.
Grade: Anhydrous, crystals.
Hazard: Irritant.
Use: Reagent in analytical chemistry, paint pigment, production of molybdated toners and lakes, metal finishing, brightening agent for zinc plating, corrosion inhibitor, catalyst in dye and pigment production, additive for fertilizers and feeds, micronutrient.

sodium-12-molybdophosphate. (sodium phospho-12-molybdate).   CAS: 1313-30-0.
$Na_3PMo_{12}O_{40}$.
Properties: Yellow crystals, d 2.83, soluble in water, although less soluble than the free acid, strong oxidizing action in aqueous solution.
Grade: Technical.
Use: Analysis, neuromicroscopy, catalysts, additives in photographic processes, imparting water resistance to plastics, adhesives and cements, pigments.
See also heteromolybdate.

sodium-12-molybdosilicate. (sodium silico-12-molybdate). $Na_4SiMo_{12}O_{40} \cdot xHOH$.
Properties: Yellow crystals; d 3.44; soluble in water, acetone, alcohol, ethyl acetate; insoluble in ether, benzene, and cyclohexane.
Grade: Reagent.
Hazard: Water solution is strong oxidizer, store away from organic materials.
Use: Catalysts, reagents, fixing and oxidizing agents in photography, precipitants and ion exchangers in atomic energy, plating processes, imparting water resistance to plastics, adhesives, and cement.

sodium monoxide. (sodium oxide).
CAS: 12401-86-4. Na_2O.
Properties: White powder, d 2.27, sublimes at 1275C, soluble in molten caustic soda or potash, converted to sodium hydroxide by water.
Hazard: Caustic and strong irritant when wet with water.
Use: Condensing or polymerizing agent in organic reactions, dehydrating agent, strong base.

sodium myristyl sulfate.   CAS: 1191-50-0.
$NaC_{14}H_{29}SO$.   Anionic detergent.
Use: Foaming, wetting, and emulsifying in cosmetic, household, and industrial uses.

sodium naphthalenesulfonate. $C_{10}H_7SO_3Na$.
Properties: Yellowish, crystalline plates or white odorless scales; soluble in water; insoluble in alcohol. Combustible.
Grade: Technical.
Hazard: Toxic by ingestion and inhalation.
Use: Organic synthesis, liquefying agent in animal glue preparations, naphthols.

sodium naphthenate.   A white paste, the most important of the naphthenic acid salts, commercial samples have consistency of grease, but this will vary with source and manner of processing. Excellent emulsifying and foam-producing properties, low hydrolytic dissociation.
Use: Detergent, emulsifier, disinfectant, manufacture of paint driers.
See also naphthenic acid.

sodium naphthionate. (sodium α-naphthyl-aminesulfonate). $NH_2C_{10}H_6SO_3Na \cdot 4HOH$.
Properties: White crystals, turns violet on exposure to light, soluble in water, insoluble in ether. Combustible.
Grade: Technical (paste, crystals).
Use: Riegler's reagent for nitrous acid, manufacture of dyestuffs.

sodium naphthylaminesulfonate.   See sodium naphthionate.

sodium niobate. (sodium columbate).
$Na_2Nb_2O_6 \cdot 7HOH$.   Important in the purification of niobium materials. The crystalline compound forms when a niobium compound is treated with hot concentrated sodium hydroxide. It is sparingly soluble in water.

sodium nitrate. (soda niter).   CAS: 7631-99-4.
$NaNO_3$.   Chile saltpeter (caliche) is impure natural sodium nitrate.
Properties: Colorless, transparent, odorless crystals; saline, slightly bitter taste. D 2.267, mp 308C, explodes at 1000F (537C), decomposes at 380C, soluble in water and glycerol, slightly soluble in alcohol.
Derivation: From nitric acid and sodium carbonate and from Chile saltpeter.
Grade: Granular, sticks, powder; crude, 99.5%, double refined, recrystallized, CP, technical, reagent, diuretic, FCC.
Hazard: Fire risk near organic materials, ignites on friction and explodes when shocked or heated to 1000F (537C). Toxic by ingestion, content in cured meats, fish, and other food products restricted.
Use: Oxidizing agent, solid rocket propellants, fertilizer, flux, glass manufacture, pyrotechnics,

reagent, refrigerant, matches, dynamites, black powders, manufacture of sodium salts and nitrates, dyes, pharmaceuticals, anaphrodisiac. Color fixative and preservative in cured meats, fish, etc.; enamel for pottery; modifying burning properties of tobacco.

sodium nitrilotriacetate. See nitrilotriacetic acid.

sodium nitrite. CAS: 7632-00-0. $NaNO_2$.
Properties: Slightly yellowish or white crystals, pellets, sticks, or powder. Oxidizes on exposure to air, soluble in water, slightly soluble in alcohol and ether, d 2.157, mp 271, explodes at 1000F (537C), decomposes at 320C.
Grade: Reagent, technical, USP, FCC.
Hazard: Dangerous fire and explosion risk when heated to 537C (1000F) or in contact with reducing materials; a strong oxidizing agent. Carcinogen in test animals; its use in curing fish and meat products is restricted to 100 ppm.
Use: Diazotization (by reaction with hydrochloric acid to form nitrous acid), rubber accelerators, color fixative and preservative in cured meats, meat products, fish; pharmaceuticals, photographic and analytical reagent, dye manufacture, antidote for cyanide poisoning.

sodium nitroferricyanide. (sodium nitroprussiate; sodium nitroprusside).
$Na_2Fe(CN)_5NO \cdot 2HOH$.
Properties: Red, transparent crystals; d 1.72; soluble in water with slow decomposition; slightly soluble in alcohol.
Grade: Reagent, technical.
Use: Analytical reagent.

sodium-p-nitrophenolate. See p-nitrophenol, sodium salt.

sodium novobiocin. See novobiocin.

sodium octyl sulfate. CAS: 142-31-4.
$C_8H_{17}OSO_3Na$. Anionic detergent, available commercially as a 35% solution.
Use: Wetting, dispersing, and emulsifying agent.

sodium oleate. CAS: 143-19-1.
$C_{17}H_{33}COONa$.
Properties: White powder, slight tallow-like odor, mp 232–235C, soluble in water with partial decomposition, soluble in alcohol. Combustible.
Derivation: Action of alcoholic sodium hydroxide on oleic acid.
Use: Ore flotation, waterproofing textiles, emulsifier of oil/water systems.

sodium orthophosphate. See sodium phosphate, mono-, di-, and tribasic.

sodium orthosilicate. $Na_2SiO_3 \cdot 2NaOH$ or other proportions such as $2Na_2O \cdot SiO_2$ (anhydrous) or $2Na_2O \cdot SiO_2 \cdot 5.4HOH$.
Properties: (Composition $2Na_2O \cdot SiO_2$) Dustless, white, flaked product; bulk d 75/lb/cu ft. Total Na_2O content 60.8%, percent of total Na_2O in active form 59.0, soluble in water, pH of a 1% solution 13.0.
Hazard: Strong irritant to skin, eyes, and mucous membranes.
Use: Commercial laundries, metal cleaning, heavy duty cleaning.

sodium oxalate. CAS: 62-76-0. $Na_2C_2O_4$.
Properties: White, crystalline powder; d 2.34; mp 250–270C (decomposes); soluble in water; insoluble in alcohol.
Grade: Reagent, technical, 88%, 99%.
Hazard: Toxic by ingestion.
Use: Reagent, textile finishing, pyrotechnics, leather finishing, blue printing.

sodium oxide. See sodium monoxide.

sodium palconate. The sodium salt of an acid that may be extracted with alkali from redwood dust. The dark reddish-brown material consists mainly of a partially methylated phenolic acid containing aliphatic hydroxyls, phenolic hydroxyls, and carboxyl groups in the ratio of 2:4:3. The viscosity of aqueous solutions rises rapidly with concentration.
Use: To control viscosity and water loss in drilling muds and as a dispersing agent.

sodium palladium chloride. See palladium sodium chloride.

sodium palmitate. $CH_3(CH_2)_{14}COONa$.
The sodium salt of palmitic acid. Combustible.
Use: Polymerization catalyst for synthetic rubbers, laundry and toilet soaps, detergents, cosmetics, pharmaceuticals, printing inks, emulsifier.

sodium paraperiodate. (sodium triparaperiodate).
$Na_3H_2IO_6$.
Properties: White, crystalline solid; very slightly soluble in water; soluble in concentrated sodium hydroxide solutions.
Use: Selectively oxidizes specific carbohydrates and amino acids, wet-strengthens paper, aids combustion of tobacco.

sodium pentaborate decahydrate.
CAS: 11139-65-4. $Na_2B_{10}O_{16} \cdot 10HOH$.
Properties: White crystals; free-flowing; stable under ordinary conditions; solubility in water 15.4% (20C), increasing with temperature; d 1.72; pH of solution 7.5. Noncombustible.

Use: Weed killer, cotton defoliant, fireproofing compositions, glass manufacture, B supplement for tree fruit and truck crops.

sodium pentachlorophenate. CAS: 131-52-2. C_6Cl_5ONa.
Properties: White or tan powder; soluble in water, ethanol, and acetone; insoluble in benzene.
Grade: Technical, powder, pellets, or briquettes.
Hazard: Toxic by ingestion, inhalation; skin irritant.
Use: Fungicide, herbicide, slimicide; fermentation disinfectant, especially in finishes and papers.

sodium pentobarbital. See barbiturate.

sodium "Pentothal." See "Pentothal," thiopental sodium.

sodium perborate, anhydrous. CAS: 7632-04-4. $NaBO_3$.
Properties: White, amorphous powder of unknown constitution containing active oxygen. Evolves oxygen gas when dissolved in water; hygroscopic.
Derivation: By heating sodium perborate tetrahydrate.
Hazard: Strong oxidizing agent, fire risk in contact with organic materials. Toxic by ingestion.
Use: Denture cleaner, oxygen source.

sodium perborate monohydrate.
$NaBO_3 \cdot HOH$, better represented as $Na_2[B_2(O_2)_2(OH)_4]$.
Properties: White, amorphous powder; rapidly soluble in water giving a solution of HOOH and sodium borate.
Derivation: By partial dehydration of sodium perborate tetrahydrate.
Hazard: Strong oxidizing agent, fire risk in contact with organic materials. Toxic by ingestion.
Use: Denture cleaner, bleaching agent in special detergents.

sodium perborate tetrahydrate.
$NaBO_3 \cdot 4HOH$ better represented as $Na_2[B_2(O_2)_2(OH)_4] \cdot 6HOH$.
Properties: Colorless crystals, mp 63C, loses water at 130–150C, soluble in water 21.5 g/L at 18C, giving a solution of HOOH and sodium borate.
Derivation: By crystallization from solutions of borax and HOOH.
Hazard: Oxidizing agent. Toxic by ingestion.
Use: Bleaching agent for domestic detergents and industrial laundries, mild antiseptic, mouthwash (under medical supervision).

sodium percarbonate. CAS: 4452-58-8.
$Na_2CO_3 \cdot 1.5HOH$.   Commercial name for sodium carbonate peroxohydrate.
Properties: Stable, microcrystalline powder; soluble in water 120 g/kg at 20C; giving a solution of HOOH and sodium carbonate.
Derivation: By solution crystallization or by a fluid bed process, from concentrated solutions of sodium carbonate and hydrogen peroxide, with stabilizers.
Hazard: Oxidizing agent; when intimately mixed with certain organic substances, it may initiate combustion. Toxic by ingestion.
Use: Bleaching agent for domestic and industrial use, denture cleaner, mild antiseptic.

sodium perchlorate. CAS: 7601-89-0.
$NaClO_4$ sometimes with an HOH.
Properties: White, deliquescent crystals; soluble in water and alcohol; mp 482C; bp decomposes; d 2.02.
Derivation: (a) Sodium chlorate and sodium chloride are mixed and heated until fused. The unchanged chloride is leached out. (b) A cold solution of sodium chlorate is electrolyzed, the solution concentrated and crystallized.
Hazard: Dangerous fire and explosion risk in contact with organic materials and sulfuric acid.
Use: Explosives, jet fuel, analytical reagent.

sodium periodate. (sodium metaperiodate).
(a) $NaIO_4$, (b) $NaIO_4 \cdot 3HOH$.   See also sodium-para-periodate.
Properties: Colorless crystals, d (a) 3.865 (16C), (b) 3.219 (18C), mp (a) 300C (decomposes), (b) 175C (decomposes), very soluble in water.
Hazard: Fire risk in contact with organic materials. Toxic by ingestion.
Use: Source of periodic acid, analytical reagent, oxidizing agent.

sodium permanganate. CAS: 10101-50-5.
$NaMnO_4 \cdot 3HOH$.
Properties: Purple to reddish-black crystals or powder, soluble in water, d 2.47, mp 170C (decomposes).
Derivation: Sodium manganate is dissolved in water and chlorine or ozone passed in. The solution is concentrated and crystallized.
Grade: Technical, sold commercially in solution.
Hazard: Dangerous fire risk in contact with organic materials, strong oxidizing agent.
Use: Oxidizing agent, disinfectant, bactericide, manufacture of saccharin, antidote for poisoning by morphine, curare, and phosphorus.

sodium peroxide. CAS: 1313-60-6. Na_2O_2.
Properties: Yellowish-white powder, turning yellow when heated, absorbs water and carbon diox-

ide from air, active oxygen content about 20% by weight, d 2.805, mp 460C (decomposes), bp 657 (decomposes), soluble in cold water with evolution of heat.
Derivation: Metallic sodium is heated at 300C in aluminum trays in a retort in a current of dry air, from which the carbon dioxide has been removed.
Grade: Technical, reagent.
Hazard: Dangerous fire and explosion risk in contact with water, alcohols, acids, powdered metals, and organic materials; keep dry; strong oxidizing agent. Irritant.
Use: Oxidizing agent, bleaching of miscellaneous materials including paper and textiles, deodorant, antiseptic, organic chemicals, water purification, pharmaceuticals, oxygen generation for diving bells, submarines, etc.; textile dyeing and printing, ore processing, analytical reagent, calorimetry, germicidal soaps.

sodium persulfate.   (sodium peroxydisulfate).
CAS: 7775-27-1. $Na_2S_2O_8$.
Properties: White, crystalline powder; soluble in water; decomposed by alcohol; decomposes in moist air.
Hazard: By ingestion, strong irritant to tissue.
Use: Bleaching agent (fats, oils, fabrics, soap), battery depolarizers, emulsion polymerization.

sodium phenate.   (sodium phenolate; sodium carbolate).   CAS: 139-02-6. C_6H_5ONa.
Properties: White, deliquescent crystals; soluble in water and alcohol; decomposed by carbon dioxide in the air.
Derivation: Phenol is dissolved in caustic soda solution, concentrated and crystallized.
Hazard: Strong irritant to skin and tissue.
Use: Antiseptic, salicylic acid, organic synthesis.

sodium phenobarbital.   (phenobarbital, solution).
See barbiturate.

sodium phenolate.   Legal label name for sodium phenate.

sodium phenolsulfonate.   (sodium sulfocarbolate).
CAS: 1300-51-2. $HOC_6H_4SO_3Na \cdot 2HOH$.
Properties: Colorless crystals or granules; slightly efflorescent; chars at high temperature, evolving phenol; soluble in water, hot alcohol, and glycerol.
Use: Medicine (intestinal antiseptic).

sodium phenylacetate.   (sodium α-toluate).
$C_6H_5CH_2 \cdot COONa$.
Properties: Soluble in water; insoluble in alcohol, ether, and ketones; 50% aqueous solution has pH 7.0–8.5 and is pale yellow, solution tends to crystallize at 15C.

Grade: 50% solution, dry salt.
Use: Precursor in production of penicillin G, intermediate for producing heavy metal salts which act as fungicides.

sodium-N-phenylglycinamide-p-arsonate.
See tryparsamide.

sodium-o-phenylphenate.   (sodium-o-phenylphenolate). $C_6H_4(C_6H_5)ONa \cdot 4HOH$.
Properties: Practically white flakes, bulk d 38–43 lb/cu ft, pH of saturated solution in water 12.0–13.5, soluble in water, methanol, acetone.
Use: Industrial preservative (bactericide and fungicide), mold inhibitor for apples and other fruit (post-harvest).

sodium phenylphosphinate. $C_6H_5PH(O)(ONa)$.
Properties: Crystals, mp 355C (decomposes to give phenylphosphine), stable at room temperature, soluble in water.
Use: Antioxidant, heat and light stabilizer.

sodium phosphate.   See sodium metaphosphate, sodium phosphate, dibasic; sodium phosphate, monobasic; sodium phosphate P-32; sodium phosphate, tribasic; sodium polyphosphate; sodium pyrophosphate; sodium pyrophosphate, acid; sodium tripolyphosphate.

sodium phosphate, dibasic.   (DSP; disodium phosphate; sodium orthophosphate, secondary; disodium orthophosphate; disodium hydrogen phosphate).   CAS: 7558-79-4.
(a) Na_2HPO_4, (b) $Na_2HPO_4 \cdot 2HOH$,
(c) $Na_2HPO_4 \cdot 7HOH$, (d) $Na_2HPO_4 \cdot 12HOH$.
The dihydrate (b) is also marketed as the duohydrate.
Properties: Colorless, translucent crystals or white powder; saline taste. Soluble in water, very soluble in alcohol. (a) Hygroscopic, converted to sodium pyrophosphate at 240C; (b) mp loses water at 92.5C, d 2.066 (15C); (c) d 1.679, loses 5HOH at 48C, (d) mp 35C, d 1.5235, readily loses 5HOH on exposure to air at room temperature, loses 12HOH at 100C; pH of 1% solution 8.0–8.8. Nonflammable.
Derivation: (1) By treating phosphoric acid with a slight excess of soda ash, boiling the solution to drive off carbon dioxide and cooling to permit the dodecahydrate to crystallize; (2) by precipitating calcium carbonate from a solution of dicalcium phosphate with soda ash.
Grade: Commercial, NF (a) and (c), FCC (a) or (b).
Use: Chemicals, fertilizers, pharmaceuticals, textiles (weighting silk, dyeing and printing), fireproofing wood and paper; ceramic glazes, tanning, galvanoplastics, soldering enamels,

analytical reagent, cheese, detergents, boiler water treatment, dietary supplement, buffer, sequestrant in foods.

sodium phosphate, monobasic. (sodium acid phosphate; sodium biphosphate; sodium orthophosphate, primary; MSP, sodium dihydrogen phosphate).   CAS: 7558-80-7.
(a) NaH_2PO_4, (b) $NaH_2PO_4 \cdot HOH$.
Properties: (a) White, crystalline powder; slightly hygroscopic. Very soluble in water, has acid reaction, forms sodium acid pyrophosphate at 225–250C and sodium metaphosphate at 350–400C; (b) large transparent crystals, mp loses water at 100C, d 2.040, very soluble in water, insoluble in alcohol, pH of 1% solution 4.4–4.5. Nonflammable.
Derivation: By treating disodium phosphate with proper proportion of phosphoric acid.
Grade: Commercial, food, (b) NF, (a) FCC.
Use: Boiler water treatment, electroplating, dyeing, acid cleansers, baking powders, cattle feed supplement, buffer, emulsifier, nutrient supplement in food; lab reagent, acidulant.

sodium phosphate (P-32). (sodium radio-phosphate).   A radioactive form of sodium phosphate (which phosphate is not specified) containing phosphorus-32 which can be used as a tracer. See phosphorus 32.
Grade: USP, as solution.
Use: Biochemical research, medicine (diagnostic aid, antineoplastic).

sodium phosphate, tribasic. (TSP; trisodium orthophosphate; trisodium phosphate; tertiary sodium phosphate; sodium orthophosphate, tertiary).   CAS: 7601-54-9. $Na_3PO_4 \cdot 12HOH$.
Properties: Colorless crystals, soluble in water, d 1.62 (20C), mp 75C (decomposes), loses 12HOH at 100C, pH of 1% solution is 11.8–12.0. Nonflammable.
Derivation: By mixing soda ash and phosphoric acid in proper proportions to form disodium phosphate and then adding caustic soda.
Grade: Commercial, high purity, CP, FCC (anhydrous), anhydrous salt also available.
Hazard: Toxic by ingestion, irritant to tissue.
Use: Water softeners, boiler water compounds, detergent, metal cleaner, textiles, manufacture of paper, laundering, tanning, sugar purification, photographic developers, paint removers, industrial cleaners, dietary supplement, buffer, emulsifier, food additive.

sodium phosphide.   CAS: 12058-85-4. Na_3P.
Properties: Red solid, decomposes on heating and in water, forming phosphine.

Hazard: Dangerous fire risk, reacts with water and acids to form phosphine.

sodium phosphite.   CAS: 13708-85-5. $Na_2HPO_3 \cdot 5HOH$.
Properties: White, crystalline powder; hygroscopic; soluble in water; insoluble in alcohol; mp 53C; bp 200–250C (decomposes).
Use: Antidote in mercuric chloride poisoning.

sodium phosphoaluminate.   White powder composed primarily of sodium aluminate (hydrated), sodium phosphate (ortho), and small amounts of sodium carbonate and sodium silicate.
Use: Paper industry as a sizing adjunct, as an aid in retention of filler and fiber and in pH control; boiler feed water treatment, and as a food additive.

sodium phospho-12-molybdate.   See sodium-12-molybdophosphate.

sodium phospho-12-tungstate.   See sodium-12-tungstophosphate.

sodium phytate.   (USAN; insitol hexaphosphoric ester, sodium salt). $C_6H_9O_{24}P_6Na_9$.
Properties: Hygroscopic powder, water-soluble.
Use: Chelating agent for trace (heavy metals) color improvement, medicine.

sodium picramate.   CAS: 831-52-7. $NaOC_6H_2(NO_2)_2NH_2$.
Derivation: Yellow water-soluble salt resulting from neutralization of picramic acid with caustic soda.
Hazard: Dangerous fire and explosion hazard when dry. Toxic by ingestion and skin absorption.
Use: Manufacture of dye intermediates, organic synthesis.

sodium platinichloride.   See sodium chloroplatinate.

sodium plumbate.   CAS: 12201-47-7. $Na_2PbO_3 \cdot 3HOH$.
Properties: Fused, light yellow lumps; hygroscopic; decomposed by water and acids; soluble in alkalies.
Hazard: As for lead.

sodium plumbite. Na_2PbO_2.
Derivation: Solution of PbO (litharge) in sodium hydroxide.
Hazard: Highly toxic, corrosive.
See lead.
Use: Doctor solution for improving the odor of gasoline and other petroleum distillates.

sodium polyphosphate. $Na_{n+2}P_nO_{3n+1}$.
The two most important crystalline sodium poly-

phosphates are the pyrophosphate ($n = 2$) and the tripolyphosphate, also called the triphosphate ($n = 3$). The term sodium polyphosphate also includes the system of vitreous sodium phosphates for which the mole ratio of Na_2O/P_2O_5 is between 1 and 2.

Hazard: As for sodium phosphate.

Use: Sequestering and deflocculating agents, primarily in water treatment, food processing and cleaning compounds; heavy-set detergent builders.

See sodium meta-, pyro-, and tripolyphosphates.

sodium polysulfide. Na_2S_x.
Properties: Yellow-brown, granular, free-flowing polymer; bulk d 56 lb/cu ft. Combustible.
Use: Manufacture of sulfur dyes and colors, insecticides, oil-resistant synthetic rubber ("Thiokol"), petroleum additives, electroplating.

sodium-potassium alloy. Legal label name for NaK.

sodium-potassium carbonate. (potassium-sodium carbonate). $NaKCO_3 \cdot 6HOH$.
Properties: Colorless crystals; the double salt fuses more readily than the single salts; d 1.6344; mp 135C (decomposes); soluble in water.
Derivation: Mixture of potassium and sodium carbonates.
Use: Analysis (flux).

sodium-potassium phosphate. (potassium-sodium phosphate). $NaKHPO_4 \cdot 7HOH$.
Properties: White powder, stable in air, soluble in water.

sodium-potassium tartrate. See potassium-sodium tartrate.

sodium propionate. CAS: 137-40-6.
CH_3CH_2COONa or $C_2H_5COONa \cdot xHOH$.
Properties: Transparent crystals or granules, almost odorless, deliquescent in moist air, soluble in water and alcohol. Combustible.
Grade: NF, FCC.
Use: Fungicide, mold-preventive, food preservative (bread and other bakery products).

sodium prussiate, red. See sodium ferricyanide.

sodium prussiate, yellow. See sodium ferrocyanide.

sodium pyroantimonate. See sodium antimonate.

sodium pyroborate. See sodium borate.

sodium pyrophosphate. (tetrasodium pyrophosphate; sodium pyrophosphate, normal; TSPP).

CAS: 7722-88-5.   (a) $Na_4P_2O_7$,
(b) $Na_4P_2O_7 \cdot 10HOH$ (one of the sodium polyphosphates).
Properties: Colorless, transparent crystals or white powder; (a) mp 880C, d 2.45, soluble in water, decomposes in alcohol; (b) mp 94C (loses water), d 1.8, soluble in water, insoluble in alcohol and ammonia.
Derivation: By fusing sodium phosphate, dibasic.
Grade: Pure crystals, dried, fused, CP, FCC.
Hazard: Toxic by inhalation.
Use: Water softener, soap and synthetic detergent builder, dispersing and emulsifying agent, metal cleaner, boiler water treatment, viscosity control of drilling muds, de-inking news print, synthetic rubber manufacture, textile dyeing, scouring of wool, buffer, sequestrant, nutrient supplement, food additive.

sodium pyrophosphate, acid. (disodium pyrophosphate; sodium acid pyrophosphate; disodium diphosphate; disodium dihydrogen pyrophosphate; SAPP). $Na_2H_2P_2O_7 \cdot 6HOH$.
Properties: White, crystalline powder; mp (decomposes) 220C; d 1.862; soluble in water.
Derivation: Incomplete decomposition of monobasic sodium phosphate.
Grade: Technical, food, FCC.
Use: Electroplating, metal cleaning and phosphatizing, drilling muds, baking powders and leavening agent, buffer, sequestrant, peptizing agent in cheese and meat products, frozen desserts.

sodium pyrophosphate, normal. See sodium pyrophosphate.

sodium pyrophosphate peroxide.
$Na_4P_2O_7 \cdot 2HOOH$.
Properties: White powder, bulk d 73 lb/cu ft, active oxygen min 9.0% by weight, water-soluble, mildly alkaline.
Hazard: Fire risk in contact with organic materials, oxidizing agent.
Use: Denture cleansers, dentrifices, household and laundry detergents, antiseptic.

sodium pyroracemate. See sodium pyruvate.

sodium pyrosulfite. See sodium metabisulfite.

sodium pyrovanadate. $Na_4V_2O_7 \cdot 8HOH$.
Properties: Colorless, six-sided plates; soluble in water; insoluble in alcohol; mp (anhydrous) 654C.
Derivation: sodium hydroxide and vanadium pentoxide in water solution.

sodium pyruvate. (sodium pyroracemate; sodium acetylformate). $NaOOCCOCH_3$.   White powder, apparent mp 205C, very soluble in water.

Use: Biochemical research.

sodium resinate. See sodium abietate.

sodium rhodanate. See sodium thiocyanate.

sodium rhodanide. See sodium thiocyanate.

sodium ricinoleate. CAS: 5323-95-5.
$C_{17}H_{32}OHCOONa$.
Properties: White or slightly yellow, nearly odorless powder; soluble in water or alcohol. Combustible.
Derivation: Sodium salt of the fatty acids from castor oil.
Use: Emulsifying agent in special soaps.

sodium saccharin. (sodium benzosulfimide; gluside, soluble; soluble saccharin).
CAS: 128-44-9. $C_7H_4NNaO_3S \cdot 2HOH$.
The sodium salt of saccharin.
Properties: White crystals or crystalline powder, odorless or with a faint aromatic odor, in dilute solutions has an intensely sweet taste (500 times sweeter than sugar), very soluble in water, slightly soluble in alcohol.
Grade: NF, FCC.
Use: Foods (non-nutritive sweetener).
See also saccharin.
Hazard: The use of saccharin is being limited due to possible carcinogenicity.

sodium salicylate. CAS: 54-21-7.
HOC_6H_4COONa.
Properties: Lustrous, white, crystalline scales or amorphous powder; saline taste; soluble in water, alcohol, and glycerol. Combustible.
Grade: Technical, CP, USP.
Use: Production of salicylic acid, preservative for paste, mucilage, glue and hides, medicine (analgesic).

sodium sarcosinate. (sodium sarcosine).
CH_3NHCH_2COONa.
Grade: 33% aqueous solution.
Use: Intermediate, stabilizer for diazonium salts, chelating agent.

sodium secobarbital. See barbiturate.

sodium selenate. CAS: 13410-01-0.
$Na_2SeO_4 \cdot 10HOH$.
Properties: White crystals, d 1.603–1.620, soluble in water.
Hazard: Toxic by ingestion. TLV (as Se): 0.2 mg/m³.
Use: Reagent, insecticide for nonedible plants.

sodium selenite. CAS: 10102-18-8.
$Na_2SeO_3 \cdot 5HOH$.
Properties: White crystals, soluble in water, insoluble in alcohol.
Derivation: By neutralizing selenious acid with sodium carbonate and crystallizing.
Hazard: Toxic by ingestion. TLV (as Se): 0.2 mg/m³.
Use: Glass manufacture (color control), reagent in bacteriology, testing germination of seeds, decorating porcelain.

sodium sesquicarbonate. CAS: 533-96-0.
$Na_2CO_3 \cdot NaHCO_3 \cdot 2HOH$.
Properties: White, needle-shaped crystals; d 2.112; mp decomposes; soluble in water; less alkaline than sodium carbonate. Noncombustible.
Derivation: Crystallization of a solution containing equimolar quantities of sodium carbonate and sodium bicarbonate, also occurs native (as trona) in desert areas and in Searles Lake brine.
Grade: Technical, FCC.
Hazard: Irritant to tissue.
Use: Detergent and soap builder, mild alkaline agent for general cleaning and water softening, bath crystals, alkaline agent in leather tanning, food additive.

sodium sesquisilicate. $Na_6Si_2O_7$ (anhydrous).
Properties: White, granular powder; soluble in water; pH of 1% solution 12.7. Noncombustible.
Derivation: Crystals from solutions obtained by heating silica or sodium metasilicate with sodium hydroxide. Intermediate in composition between ortho- and metasilicates, less alkaline than sodium orthosilicate.
Hazard: Toxic by ingestion, protective clothing required.
Use: Heavy-duty cleaning (metals, laundries), textile processing.

sodium silicate. (water glass).
CAS: 1344-09-8.   Formulas: Vary from $Na_2O \cdot 3.75SiO_2$ to $2Na_2O \cdot SiO_2$ and with various proportions of water. 47th highest-volume chemical produced in US (1985). The simplest form of glass.
Properties: Lumps of greenish glass soluble in steam under pressure, white powders of varying degrees of solubility, or liquids cloudy or clear and varying from highly fluid to extreme viscosity, viscosity range from 0.4 to 600,000 poise, fp slightly below water, miscible with some polyhydric alcohols, partially miscible with primary alcohols and ketones, gels form with acids between pH 3 and pH 9, coagulated by brine, precipitated by alkaline earth and heavy metal ions. Noncombustible.
Derivation: By fusing sand and soda ash.

Grade (liquid): 40, 47, 52 degrees Bé.

Use: Catalysts and silica gels, soaps and detergents, adhesives (especially sealing and laminating paper board containers), water treatment, bleaching and sizing of textiles and paper pulp, ore treatment, soil solidification, glass foam, pigments, drilling fluids, binder for abrasive wheels, foundry cores and molds, waterproofing mortars and cements, flame retardant, chemical equipment lining, enhanced oil recovery.

See also the other soluble sodium silicates: sodium metasilicate, anhydrous; sodium metasilicate pentahydrate; sodium sesquisilicate; sodium orthosilicate.

sodium silicoaluminate.   See sodium aluminosilicate.

sodium silicofluoride.   See sodium fluorosilicate.

sodium silico-12-molybdate.   See sodium-12-molybdosilicate.

sodium-12-silicotungstate.   See sodium-12-tungstosilicate.

sodium silver chloride.   See silver sodium chloride.

sodium silver thiosulfate.   See silver sodium thiosulfate.

sodium-α-sodioacetate.   See α-sodiosodium acetate.

sodium sorbate.   CAS: 7757-81-5.
$CH_3CH:CHCH:CHCOONa$.   Combustible.
Use: Food preservative.

sodium stannate.   CAS: 12058-66-1.
$Na_2SnO_3 \cdot 3HOH$ or $Na_2Sn(OH)_6$.
Properties: White to light tan crystals, soluble in water, insoluble in alcohol, decomposes in air, aqueous solution slightly alkaline, loses 3HOH at 140C.
Derivation: (a) By fusion of metastannic acid and sodium hydroxide, (b) by boiling tin scrap and sodium plumbate solution.
Hazard: Toxic material. TLV: 2 mg/m³ of air.
Use: Mordant in dyeing, ceramics, glass, source of tin for electroplating and immersion plating, textile fireproofing, stabilizer for hydrogen peroxide, blueprint paper, laboratory reagent.

sodium stearate.   CAS: 822-16-2.
$NaOOCC_{17}H_{35}$.
Properties: White powder with fatty odor, soluble in hot water and hot alcohol, slowly soluble in cold water and cold alcohol, insoluble in many organic solvents. Impurities: Varying quantities of sodium palmitate.

Grade: Technical.
Use: Waterproofing and gelling agent, toothpaste and cosmetics, stabilizer in plastics.

sodium stearoyl-2-lactylate.
Properties: White powder, melting range 46–52C.
Derivation: Sodium salt of reaction product of lactic and stearic acids.
Use: Emulsifier, dough conditioner, whipping agent in baked products, desserts and mixes, complexing agent for starches and proteins.

sodium styrenesulfonate.
$CH_2:CH_2C_6H_4SO_3Na$.   White, free-flowing powder.
Use: Reactive monomer.

sodium subsulfite.   See sodium thiosulfate.

sodium sulfate, anhydrous.   CAS: 7757-82-6.
Na_2SO_4.   37th highest-volume chemical produced in US (1985).
Properties: White crystals or powder, odorless, bitter saline taste, d 2.671, mp 888C, soluble in water and glycerol, insoluble in alcohol. Noncombustible.
Derivation: (a) Purification of natural sodium sulfate from deposits or brines. (b) Byproduct of hydrochloric acid manufacture from salt and sulfuric acid, $2NaCl + H_2SO_4 \rightarrow 2HCl + Na_2SO_4$. (c) Byproduct of phenol manufacture (caustic fusion process). (d) Hargreaves process.
Grade: Technical, CP, detergent, rayon, glass makers.
Use: Manufacture of kraft paper, paperboard and glass, filler in synthetic detergents, sodium salts, ceramic glazes, processing textile fibers, dyes, tanning, pharmaceuticals, freezing mix, laboratory reagent, food additive.
See also salt cake.

sodium sulfate decahydrate.   (sodium sulfate, crystals; Glauber's salt).   CAS: 7727-73-3.
$Na_2SO_4 \cdot 10HOH$.
Properties: Large, transparent crystals, small needles, or granular powder. D 1.464 (crystals), mp 33C (liquefies), loses water of hydration at 100C, energy storage capacity is more than seven times that of water, soluble in water and glycerol, insoluble in alcohol, solutions neutral to litmus. Nonflammable.
Derivation: Crystallization of sodium sulfate from water solutions. (Glauber's salt): also occurs in nature as mirabilite.
Grade: Technical, NF.
Use: Solar heat storage, air conditioning.

sodium sulfhydrate.   See sodium hydrosulfide.

sodium sulfide. CAS: 16721-80-5; (anhydrous) 1313-82-2.   (a) Na_2S, (b) $Na_2S \cdot 9HOH$.
Properties: Yellow or brick red lumps or flakes or deliquescent crystals; (a) d 1.856 (14C), mp 1180C; (b) d 1.427 (16C), decomposes at 920C; soluble in water, slightly soluble in alcohol, insoluble in ether, largely hydrolyzed to sodium acid sulfide and sodium hydroxide.
Derivation: By heating sodium acid sulfate with salt and coal to above 950C, extraction with water and crystallization.
Grade: Flake, fused, chip sulfide (60% Na_2S), 60% fused and broken, 30% crystals, liquid.
Hazard: Flammable, dangerous fire and explosion risk. Strong irritant to skin and tissue, liberates toxic hydrogen sulfide on contact with acids.
Use: Organic chemicals, dyes (sulfur), intermediates, viscose rayon (sulfur removal), leather (depilatory), paper pulp, hydrometallurgy of gold ores, sulfiding oxidized lead and copper ores preparatory to flotation, sheep dips, photographic reagent, engraving and lithography, analytical reagent.

sodium sulfite. CAS: 7757-83-7.   (a) Na_2SO_3, (b) $Na_2SO_3 \cdot 7HOH$.
Properties: White crystals or powder; saline, sulfurous taste; soluble in water; sparingly soluble in alcohol; d (a) 2.633, (b) 1.539; mp (a) decomposes, (b) loses 7HOH at 150C.
Derivation: (a) Sulfur dioxide is reacted with soda ash and water and a solution of the resulting sodium bisulfite is treated with additional soda ash; (b) byproduct of the caustic fusion process for phenol.
Grade: Reagent, technical, FCC.
Hazard: Use prohibited in meats and other sources of Vitamin B_1.
Use: Paper industry (semichemical pulp), reducing agent (dyes), water treatment, photographic developer, food preservative and antioxidant, textile bleaching (antichlor).

sodium sulfobromophthalein.
$C_{20}H_8Br_4O_{10}S_2Na_2$.
Properties: White, crystalline powder; odorless with a bitter taste; hygroscopic; soluble in water; insoluble in alcohol and acetone.
Derivation: From phenol and tetrabromophthalic acid or anhydride.
Grade: USP, technical.
Use: Medicine (diagnostic aid).

sodium sulfocarbolate.   See sodium phenolsulfonate.

sodium sulfocyanate.   See sodium thiocyanate.

sodium sulfocyanide.   See sodium thiocyanate.

sodium sulfonate.   Class name for various sulfonates derived from petroleum e.g., sodium dodecylbenzenesulfonate, sodium xylenesulfonate, sodium toluenesulfonate.
See also sodium alkanesulfonate.
Use: Textile processing oils, oils for metal working (emulsifying and antirust agents), lubricating oils, emulsifiers for insecticides, herbicides, fungicides, preparation of dyes and intermediates, hydrotropic solvent, coatings in food packaging.

sodium-2-sulfopropionitrile.   See sodium-2-cyanoethanesulfonate.

sodium sulforicinoleate.
Derivation: Product of successive sulfonation (partial) and saponification of castor oil. Composition indefinite.
Use: Emulsifying and wetting agent.

sodium sulfoxylate.   See sodium formaldehyde sulfoxylate.

sodium tartrate.   (sal tartar; disodium tartrate). CAS: 868-18-8. $Na_2C_4H_4O_6 \cdot 2HOH$.
Properties: White crystals or granules, soluble in water, insoluble in alcohol, d 1.794, loses 2HOH at 150C.
Derivation: Neutralization of tartaric acid with sodium carbonate, concentration, and crystallization.
Grade: Technical, CP, reagent, FCC.
Use: Reagent, food additive, as sequestrant and stabilizer.

sodium tartrate, acid.   See sodium bitartrate.

sodium TCA.   See sodium trichloroacetate.

sodium tellurite.   CAS: 10102-20-2.
Na_2TeO_3.
Properties: White powder, soluble in water.
Hazard: Toxic by ingestion.
Use: Bacteriology, medicine.

sodium tetraborate.   See sodium borate.

sodium tetrachloroaluminate.   CAS: 7784-16-9. (sodium chloroaluminate). $AlCl_4Na$.
A water-soluble powder used as catalyst in organic reactions.

sodium-2,3,4,6-tetrachlorophenate.
$C_6HCl_4ONa \cdot HOH$.
Properties: Buff to light brown flakes, bulk d 26–29 lb/cu ft, pH of water-saturated solution 9.0–13.0. Soluble in water, methanol and acetone.
Hazard: Toxic by ingestion.
Use: Industrial preservative (bactericide and fungicide).

sodium tetradecyl sulfate. (sodium-7-ethyl-2-methyl-4-hendecanol sulfate). CAS: 139-88-8. $C_{14}H_{29}SO_4Na$.
Properties: White, waxy, odorless solid; soluble in alcohol, ether, and water; 5% solution is clear and colorless; pH (5% solution) 6.5–9.0.
Use: Wetting agent, detergent.

sodium tetraphosphate. See sodium polyphosphate.

sodium tetrasulfide. Na_2S_4.
Properties: Yellow, hygroscopic crystals or clear, dark red liquid; mp of crystals 275C.
Grade: Aqueous solution containing 40% by weight of compound.
Hazard: Fire risk when exposed to flame. Irritant to skin and tissue.
Use: Reducing organic nitro compounds, manufacture of sulfur dyes, insecticides and fungicides, ore flotation, soaking hides and skins, preparation of metal sulfide finishes.

sodium thiocyanate. (sodium sulfocyanate; sodium sulfocyanide; sodium rhodanate; sodium rhodanide). CAS: 540-72-7. NaSCN.
Properties: Colorless, deliquescent crystals or white powder; mp 287C; soluble in water and alcohol; hygroscopic and affected by light.
Derivation: By boiling sodium cyanide with sulfur.
Grade: Technical; pure, crystals or dried; CP; reagent; ACS.
Use: Analytical reagent, dyeing and printing textiles, black nickel plating, manufacture other thiocyanate salts and artificial mustard oil, solvent for polyacrylates, medicine (antihypertensive).

sodium thioglycolate. (sodium mercaptoacetate). CAS: 367-51-1. HSCH$_2$COONa. The sodium salt of thioglycolic acid.
Properties: Crystals, characteristic odor, hygroscopic, discolors on exposure to air or iron, soluble in water, slightly soluble in alcohol. Combustible.
Hazard: Yields toxic hydrogen sulfide on decomposition, may be toxic by skin absorption.
Use: Bacteriology, cold waving hair, depilatory, analytical reagent.

sodium thiosulfate. (sodium subsulfite; hypo). CAS: 7772-98-7. $Na_2S_2O_3 \cdot 5HOH$.
The anhydrous salt is also commercially available.
Properties: White, translucent crystals or powder; cooling taste and bitter aftertaste. Soluble in water and oil of turpentine, insoluble in alcohol, deliquescent in moist air, efflorescent above 33C

in dry air, d 1.729 (17C), mp 48C, bp decomposes.
Derivation: Heating a solution of sodium sulfite with powdered sulfur.
Grade: Technical, crystals, granulated, photographic, CP, pure, USP, FCC.
Hazard: Use in foods restricted to 0.1%.
Use: Photography (fixing agent to dissolve unchanged silver salts from exposed negatives), chrome tanning, removing chlorine in bleaching and papermaking, extraction of silver from its ores, dechlorination of water, mordant, reagent, bleaching, reducing agent in chrome dyeing, sequestrant in salt (up to 0.1%), antidote for cyanide poisoning.

sodium titanate. (sodium tritanate). $Na_2Ti_3O_7$.
Properties: White crystals, insoluble in water, d 3.35–3.50, mp 1128C, Combustible.
Use: Welding.

sodium-α-toluate. See sodium phenylacetate.

sodium toluenesulfonate. (p-toluenesulfonic acid, sodium salt). CAS: 657-84-1. $CH_3C_6H_4SO_3Na$.
Properties: Crystals, soluble in water.
Use: Dye chemistry, hydrotropic solvent.

sodium-p-toluenesulfonchloramine. See chloramine.

sodium trichloroacetate. (sodium TCA). CAS: 650-51-1. CCl_3COONa.
Hazard: Toxic by ingestion, irritant to skin and eyes.
Use: Herbicide, pesticide.

sodium-2,4,5-trichlorophenate. $C_6H_2Cl_3ONa \cdot 1.5HOH$.
Properties: Buff to light brown flakes, bulk d 28–33 lb/cu ft, pH of water-saturated solution 11.0–13.0. Soluble in water, methanol, acetone.
Use: Industrial preservative (bactericide and fungicide).

sodium tridecylbenzenesulfonate. Not a true compound but a mixture of C_{12} and C_{15} alkyl benzene sulfonates which approximates C_{13}.
See sodium dodecylbenzene sulfonate for derivation and use.

sodium tri-metaphosphate. See sodium metaphosphate.

sodium tri-para-periodate. See sodium-para-periodate.

sodium triphenyl-p-rosaniline sulfonate. See methyl blue.

sodium triphosphate. See sodium tripolyphosphate.

sodium tripolyphosphate. (STFF; sodium triphosphate, tripoly; pentasodium triphosphate). CAS: 7758-29-4. $Na_5P_3O_{10}$.   49th highest-volume chemical produced in US (1985).
Properties: White powder, two crystalline forms of anhydrous salt (transition pt 417C, mp 622C to give melt and sodium pyrophosphate), and a hexahydrate.
Derivation: Controlled calcination of sodium orthophosphate mixture from sodium carbonate and phosphoric acid. May contain up to 10% pyrophosphate and up to 5% tri-metaphosphate.
Grade: Powdered and granular, FCC, food.
Use: Water softening; sequestering, peptizing, or deflocculating agent; food additive and texturizer.

sodium trititanate. See sodium titanate.

sodium tungstate. (sodium wolframate). CAS: 13472-45-2. $Na_2WO_4 \cdot 2HOH$.
Properties: Colorless crystals, soluble in water, insoluble in alcohol and acids, d 3.245, mp loses 2HOH at 100C and then melts at 692C. Noncombustible.
Derivation: By dissolving tungsten trioxide or the ground ore in caustic soda solution, concentration, and crystallization.
Grade: Technical, CP, crystals.
Use: Intermediate in preparation of tungsten compounds, (e.g., phosphotungstate), reagent, fireproofing fabrics and cellulose, alkaloid precipitant.

sodium-12-tungstophosphate. (sodium phosphotungstate; phosphotungstic acid, sodium salt). CAS: 51312-42-6.
$Na_4O_2 \cdot O_5P_2 \cdot O_{36}W_{12} \cdot 18HOH$.
Properties: Yellowish-white powder, very soluble in water and alcohols.
Grade: Reagent, technical.
Hazard: An oxidizing agent.
Use: Reagent; manufacture of organic pigments; treatment of furs; antistatic agent for textiles; leather tanning; making plastic films, cements, and adhesives water-resistant.

sodium-12-tungstosilicate.
$Na_4SiW_{12}O_{40} \cdot 5HOH$.
Properties: White, crystalline powder; soluble in water and alcohols, although less so than the free acid. Noncombustible.
Grade: Reagent, technical.

Use: Catalyst for organic synthesis, precipitant and inorganic ion-exchanger, additive in plating processes.

sodium undecylenate.
$(CH_2{:}CH(CH_2)_8COONa$.
Properties: White powder, decomposes above 200C, limited solubility in most organic solvents, soluble in water. Combustible.
Use: Bacteriostat and fungistat in cosmetics and pharmaceuticals.

sodium uranate. See sodium diuranate.

sodium valproate. A drug that has proved effective in treating epilepsy. Approved (1978) by FDA, successfully used in Europe since 1970.

sodium vanadate. See sodium orthovanadate, sodium pyrovanadate, sodium metavanadate.

sodium-p-vinylbenzene sulfonate. See sodium styrene sulfonate.

sodium warfarin. (sodium-(3-α-acetonylbenzyl)-4-hydroxycoumarin). $C_{19}H_{15}NaO_4$.
See warfarin.

sodium xanthate. See sodium ethylxanthate.

sodium xanthogenate. See sodium ethylxanthate.

sodium xylenesulfonate. (dimethylbenzenesulfonic acid, sodium salt).   CAS: 1300-72-7.
$(CH_3)_2C_6H_3SO_3NaHOH$.
Use: Hydrotropic solvent, used in detergents.

sodium zinc hexa-metaphosphate. See sodium metaphosphate.

sodium zirconium glycolate.
$NaH_3ZrO(H_2COCOO)_3$.
Properties: Clear, light straw-colored solution; d 1.28–1.30; containing 35.7–38.6% solids, 12.5–13.5% ZrO_2.
Use: Deodorant, astringent, germicide, sequestrant, fire retardant.

sodium zirconium lactate. CAS: 10377-98-7.
$NaH_3ZrO(CH_3CHOCOO)_3$.
Properties: Clear, straw-colored solution; d 1.28–1.30; containing 12.5–13.5% ZrO_2, equivalent to 42.5–45.9% sodium zirconium lactate; pH 7.5–8.0.
Use: Deodorant and antiperspirant.

sodium zirconium sulfate. See zirconium sodium sulfate.

soft. A nontechnical word used by chemists in several senses, it describes the following: (1) an acid having little or no positive charge and whose valence electrons are easily excited (see acid); (2) water which is relatively free from calcium compounds (see water, hard); (3) wood from coniferous trees (see softwood).

softener. (1) A substance used when dry powders are added to a polymeric material (e.g., rubber or plastic) to reduce the friction of mechanical mixing and to facilitate subsequent processing. They exert both lubricating and dispersing action, often by means of emulsification. Examples are vegetable oils, asphaltic materials, and stearic acid, the latter being especially effective with carbon black. It is difficult to distinguish precisely between softeners and plasticizers; in general, softeners do not enter into chemical combination with the polymer and their softening effect tends to be temporary. (2) A fatliquoring agent used to soften leather. (3) A sulfonated oil, fatty alcohol, or quaternary ammonium compound used in textile finishing to impart superior "hand" to the fabric and facilitate mechanical processing. (4) A substance that reduces the hardness of water by removing or sequestering calcium and magnesium ions; among those used are various sodium phosphates and zeolites.
See water, hard.

softwood. In papermaking terminology, an arbitrary name for the wood from coniferous trees (pine, spruce, fir, hemlock) regardless of the hardness or softness of the wood itself. Softwoods are used for almost all commercial grades of paper.
See also pulp, paper.

soil. (1) A mixture of inorganic matter derived from weathered rocks and organic components resulting from decay of prior vegetation. Eight elements are present in the inorganic component in excess of 1% (oxygen, silicon, aluminum, iron, calcium, potassium, magnesium, and sodium), most in the ionized state. Water and air are also present, either in the voids between the particles or adsorbed on their surfaces. Many other elements occur in lower percentages, including trace elements in concentration of less than 1000 ppm. Some of these, e.g., boron (about 20 ppm), are essential for plant nutrition. Both nitrogen and phosphorus are associated with the organic content. The concentration of these is a fraction of 1% of each, but they play a vital part in plant and animal life. The pH of soils varies widely with location; some soils are as low as pH 4.5 (very acid) and others as high as pH 10 (strongly alkaline). For most crops the pH ranges around the neutral point (6.5–7.5). Texturally, soils are classified on the basis of their content of sand,

silt, and clay. Those having 45–50% sand and 20–28% clay are called loams, those with more than 50% sand are called sandy, and those with more than 28% clay are in the clay group. Technologists consider soil as being made up of layers, known as horizons, each having a characteristic composition and physical properties; the spectrum of these horizons is called the soil profile. Organic matter is usually excluded from the profile. (2) In textile literature, any foreign matter present in or on fiber or fabric, i.e., dirt, oil, grease, etc. These are usually removable by the action of soap, synthetic detergents, or organic solvents.

soil conditioner. (1) A synthetic, long-chain organic molecule having carboxyl groups along the chain whose charges react with the positive charges on the soil particles (aluminum and iron). The conditioners affect the anion exchange capacity of a soil. (2) Loosely, any material added to topsoil to reduce acidity (lime) and promote growth (bone meal).

"Soilfume."[55] TM for a soil fumigant whose active ingredient is ethylene dibromide.
Hazard: As for ethylene dibromide.

sol. An abbreviation for soluble.
See solution, colloidal.

solan. (Generic name for 3′-chloro-2-methyl-p-valerotoluidide or N-(3-chloro-4-methylphenyl)-2-methylpentanamide).
CAS: 2307-68-9.
$H_3CC_6H_3(Cl)NHCOCH(CH_3)CH_2CH_2CH_3$.
Properties: Solid, mp 86C, insoluble in water, soluble in pine oil, diisobutyl ketone, isophorone, and xylene. Combustible.

solanine. CAS: 20562-02-1. $C_{45}H_{73}NO_{15}$.
An alkaloid found in low percentages in potato and other plants, insoluble in water, soluble in alcohol, decomposes at 285C.

solar cell. (photovoltaic cell). A battery-like device in which the radiant energy of the sun is converted to electrical energy by means of a semiconductor. The essential component of a solar cell is a thin sheet or wafer of crystalline or amorphous silicon, plus doping agents. The crystalline form is used in 3-inch squares, but the hydrogenated amorphous form is effective in 1-foot squares which are produced as vapor-deposited coatings on glass only a few microns in thickness. A notable advance in the technology of solar cells is the experimental development of a photoelectrochemical cell, which utilizes silicon electrodes by which incident light is converted to energy. The conversion efficiency is sufficient to indicate that this type of cell may have an important future. Its success is due to microthin films of doped oxides of alumina or magnesia deposited on the electrode surfaces, which in turn are coated with platinum a few angstroms thick. The cell can also be operated as a battery for use in remote power sources.

solar collector. A device for utilizing solar energy either by absorption or reflection of the incident radiation. Several types have been developed for both domestic and industrial application. Some are in the experimental stage. (1) Collection units consisting of a blackened absorber surface enclosed by one or two layers of silvered glass or polished aluminum, which admit the incident radiation but prevent the escape of heat; the energy absorbed is transferred to air or water circulating between the glass plates and the absorber. (2) Energy-focusing systems (solar towers) in which the incident radiation is concentrated on a central receiver from an array of flat or parabolic mirrors. This method is utilized for evaporating low-level radioactive waste water at Los Alamos, as well as in a power generating plant.

Solar energy has important possibilities for chemical synthesis. A prototype collector for producing ammonia from the reduction of nitrogen and water at low temperature and pressure has been reported. This reaction is

$$N_2 + 3H_2O \xrightarrow[TiO_2]{hv} 2NH_3 + 1.5\ O_2$$

solar energy. See solar cell, solar collector, solar furnace. For the problem of heat and power storage when sunlight is not available; see storage (4). The Solar Energy Research Institute, a division of Midwest Research Institute, is located in Golden, Colorado (1617 Cole Blvd).

solar furnace. An experimental device for attaining extremely high temperatures for physical and chemical research. Solar radiation is directed by an array of 63 heliostats onto a parabolic reflector which is 2000 m² in area and composed of more than 9000 curved mirrors. By this means the radiation is focused on a target area of 0.3 m². Three furnaces, each of 1000 kw capacity, have been constructed; one is located in New Mexico, the others are in France and Japan. Calcium carbide is reported to have been made in the French solar furnace from lime and coke at 1980C (3600F), under the sponsorship of the Institute of Gas Technology.

solar pond. An experimental means of storing solar energy in either fresh or salt water. The "pond" is actually a large, suitably lined open container of varying dimensions. In the freshwater type, the heat absorbed is retained at night by covering the pond with an inflated plastic film, supplemented by a layer of foam applied directly to the water. In the salt-water type, heat loss is prevented by a salt gradient, which minimizes convection currents that would cool the water. The sodium chloride or magnesium chloride used is not uniformly dispersed, but is most highly concentrated in the darker bottom layers where most of the solar radiation is absorbed. Thus the water is warmest and densest at the bottom, where it remains, and no convection occurs. The salt concentration ranges from zero at the surface to 18% or more at the bottom. An experimental salt-gradient pond at Argonne National Laboratory 100 × 100 × 12 ft containing 2800 tons of water and 700 tons of sodium chloride has attained a bottom temperature approaching 180F.

solder. A low-melting alloy usually of the lead-tin type used for joining metals at temperatures below 425C. The solder acts as an adhesive and does not form an intermetallic solution with the metals being joined.
See also brazing, welding.

"Solfast."[266] TM for a series of lightfast pigments used in paints, printing inks, plastics, rubber, floor coatings, and textiles.

solid. Matter in its most highly concentrated form, i.e., the atoms or molecules are much more closely packed than in gases or liquids and thus more resistant to deformation. The normal condition of the solid state is crystalline structure—the orderly arrangement of the constituent atoms of a substance in a framework called a lattice (see crystal). Crystals are of many types and normally have defects and impurities that profoundly affect their applications, as in semiconductors. The geometric structure of solids is determined by x-rays which are reflected at characteristic angles from the crystalline lattices, which act as diffraction gratings.
See crystallography.

Some materials that are physically rigid, such as glass, are regarded as highly viscous liquids because they lack crystalline structure. All solids can be melted (i.e., the attractive forces acting between the crystals are disrupted) by heat and are thus converted to liquids. For ice, this occurs at 0C; for some metals the melting point may be as high as 3300C.

solid state chemistry. Study of the exact arrangement of atoms in solids, especially crystals, with particular emphasis on imperfections and irregularities in the electronic and atomic patterns in a crystal and the effects of these on electrical and chemical properties.
See also crystal, semiconductor, impurity.

"Solphenyl."[443] TM for fast-to-light direct dyes for cellulosic fibers.

"Solprene."[303] TM for a solution of polymers of butadiene/styrene.
Use: Footwear, wire and cable, sponge, floor tile and cove base and other molded and extruded goods.

"Solricin" 135.[202] TM for 35% aqueous solution of potassium ricinoleate; mild germicide, synergizes phenol coefficients of disinfectants.

"Solricin" 285.[202] TM for an 85% aqueous solution of ammonium ricinoleate.
Use: Rustproofing agent.

solubility. The ability or tendency of one substance to blend uniformly with another, e.g., solid in liquid, liquid in liquid, gas in liquid, gas in gas. Solids vary from 0–100% in their degree of solubility in liquids depending on the chemical nature of the substances; to the extent that they are soluble they lose their crystalline form and become molecularly or ionically dispersed in the solvent to form a true solution. Examples are sugar/water, salt/water. Liquids and gases are often said to be miscible in other liquids and gases rather than soluble. Thus nitrogen, oxygen, and carbon dioxide are freely miscible in each other and air is a solution (uniform mixture) of these gases.
    The physical chemistry of solutions is an extremely complex mathematical subject in which the principles of electrolytic dissociation, diffusion, and thermodynamics play controlling parts. Raoult's Law and Henry's Law are also involved. See also miscibility; solution, true.

soluble oil. An oil (also called emulsifying oil) which, when mixed with water, produces milky emulsions. In some soluble oils the emulsion is so fine that instead of milky solutions in water, amber colored, transparent solutions are formed. Typical examples are sodium and potassium petroleum sulfonates.
Use: Metal cutting lubricants, textile lubricants, metal boring lubricants, emulsifying agents.

soluble starch. See starch, modified.

"Solulan."[493] TM for ethoxylated derivatives of lanolin and lanolin components. Some are also acetylated. "Solulan" C-24 is a polyethoxylated cholesterol.
Use: Pharmaceuticals and cosmetics.

solute. One or more substances dissolved in another substance, called the solvent; the solute is uniformly dispersed in the solvent in the form of either molecules (sugar) or ions (salt), the resulting mixture comprising a solution.
See solution, true; solvent.

solution, true. A uniformly dispersed mixture at the molecular or ionic level, of one or more substances (the solute) in one or more other substances (the solvent). These two parts of a solution are called phases. Common types are:
    liquid/liquid: alcohol/water
    solid/liquid: salt/water
    solid/solid: carbon/iron
    Solutions that exhibit no change of internal energy on mixing and complete uniformity of cohesive forces are called *ideal*; their behavior is described by Raoult's Law. Solutions are involved in most chemical reactions, refining and purification, industrial processing, and biological phenomena.
    The proportion of substances in a solution depends on their limits of solution. The solubility of one substance in another is the maximum amount that can be dissolved at a given temperature and pressure. A solution containing such a maximum amount is *saturated*. A state of supersaturation can be created, but such solutions are unstable and may precipitate spontaneously.

solution, colloidal. A liquid colloidal dispersion is often called a solution. Since colloidal particles are larger than molecules it is strictly incorrect to call such dispersions solutions; however, this term is widely used in the literature. *Note*: Wolfgang Ostwald stated, "...There are no sharp differences between mechanical suspensions, colloidal solutions, and molecular [true] solutions. There is a gradual and continuous transition from the first through the second to the third."
See also colloid chemistry.

solutrope. A ternary mixture having two liquid phases between which one component is distributed in an apparent ratio varying with concentration from less than 1 to greater than 1. In other words, the solute may be selectively dissolved in one or the other of the phases or solvents depending on the concentration. This phenomenon has been compared to azeotropic behavior.

"Solvat."[243] TM for leuco esters of vat dyes used for wool, cellulose, and synthetic fibers.

solvation. In the parlance of colloid chemistry, the adsorption of a microlayer or film of water or other solvent on individual dispersed particles of a solution or dispersion. The term "solvated hulls" has been used to describe such particles. It is also applied to the action of plasticizers on resin dispersions in plastisols.
See also hydration (2).

Solvay process. (ammonia soda process).
Manufacture of sodium carbonate (soda ash, Na_2CO_3) from salt, ammonia, carbon dioxide, and limestone by an ingenious sequence of reactions involving recovery and reuse of practically all the ammonia and part of the carbon dioxide. Limestone is heated to produce lime and carbon dioxide. The latter is dissolved in water containing the ammonia and salt, with resultant precipitation of sodium bicarbonate. This is separated by filtration, dried, and heated to form normal sodium carbonate. The liquor from the bicarbonate filtration is heated and treated with lime to regenerate the ammonia. Calcium chloride is a major byproduct. *Note*: Because this process requires much energy and pollutes streams and rivers with chloride effluent many plants using it have closed, production being obtained from the natural deposits in the Western US.

"Solvenol."[266] TM for a group of monocyclic terpene hydrocarbons with minor amounts of terpene alcohols and ketones.
Use: General solvent, rubber reclaiming.

solvent. A substance capable of dissolving another substance (solute) to form a uniformly dispersed mixture (solution) at the molecular or ionic size level. Solvents are either polar (high dielectric constant) or non-polar (low dielectric constant). Water, the most common of all solvents, is strongly polar (dielectric constant 81), but hydrocarbon solvents are non-polar. Aromatic hydrocarbons have higher solvent power than aliphatics (alcohols). Other organic solvent groups are esters, ethers, ketones, amines, and nitrated and chlorinated hydrocarbons.
The chief uses of organic solvents are in the coatings field (paints, varnishes and lacquers), industrial cleaners, printing inks, extractive processes, and pharmaceuticals. Since many solvents are flammable and toxic to varying degrees, they contribute to air pollution and fire hazards. For this reason their use in coatings and cleaners has declined in recent years.
See individual compounds.

solvent, aprotic. A solvent that cannot act as a proton acceptor or donor i.e., as an acid or base.

solvent drying. Removal of water from metal surfaces by means of a solvent that displaces it preferentially, as on precision equipment, electronic components, etc. Examples of solvents used are acetone, 1,1,2-trichloro-1,2,2-trifluorethane, 1,1,1-trichloroethane.

solvent dye. See dye, solvent.

solvent extraction. A separation operation which may involve three types of mixture: (a) a mixture composed of two or more solids, such as a metallic ore; (b) a mixture composed of a solid and a liquid; (c) a mixture of two or more liquids. One or more components of such mixture are removed (extracted) by exposing the mixture to the action of a solvent in which the component to be removed is soluble. If the mixture consists of two or more solids, extraction is performed by percolation of an appropriate solvent through it. This procedure is also called leaching, especially if the solvent is water; coffee-making is an example. Synthetic fuels can be made from coal by extraction with a coal-derived solvent followed by hydrogenation.
In liquid-liquid extraction one or more components are removed from a liquid mixture by intimate contact with a second liquid which is itself nearly insoluble in the first liquid and dissolves the impurities and not the substance that is to be purified. In other cases the second liquid may dissolve, i.e., extract from the first liquid, the component that is to be purified, and leave associated impurities in the first liquid. Liquid-liquid extraction may be carried out by simply mixing the two liquids with agitation and then allowing them to separate by standing. It is often economical to use counter-current extraction, in which the two immiscible liquids are caused to flow past or through one another in opposite directions. Thus fine droplets of heavier liquid can be caused to pass downward through the lighter liquid in a vertical tube or tower.
The solvents used vary with the nature of the products involved. Widely used are water, hexane, acetone, isopropyl alcohol, furfural, xylene, liquid sulfur dioxide, and tributyl phosphate. Solvent extraction is an important method of both producing and purifying such products as lubricating and vegetable oils, pharmaceuticals and nonferrous metals.

solvent, latent. (co-solvent). An organic liquid that will dissolve nitrocellulose in combination with an active solvent. Latent solvents are usually alcohols and are used widely in nitrocellulose lacquers in a ratio of 1 part alcohol to 2 parts active solvent.

solvent naphtha. See naphtha (2b).

Solvent Red 73. See 4',5'-diiodofluorescein.

solvent refining.   See solvent extraction.

Solvent Yellow 3.   See o-aminoazotoluene.

solvolysis.   A reaction involving substances in solvent, in which the solvent reacts with the dissolved substance (solute) to form a new substance. Intermediate compounds are usually formed in this process.
See also hydrolysis.

soman.   (methylphosphonofluoridic acid-1,2,2-trimethylpropylester).   CAS: 96-64-0. $(CH_3)_3CCH(CH_3)OPF(O)CH_3$.   A nerve gas.
Properties: Colorless liquid, evolves odorless gas, bp 167C, fp −70C, d 1.026 (20C).
Hazard: Highly toxic by ingestion, inhalation, and skin absorption; may be fatal on short exposure; cholinesterase inhibitor; military nerve gas; fatal dose (man) 0.01 mg/kg.

somatotropic hormone.   (STH; somatotropin).
CAS: 9002-72-6.   Hormone secreted by the anterior lobe of the pituitary. It causes an increase in general body growth and also affects carbohydrate and lipid metabolism.

Sommelet reaction.   Preparation of aldehydes from aralkyl or alkyl halides by reaction with hexamethylenetetramine followed by mild hydrolysis of the formed quaternary salt.

Sommelet-Hauser rearrangement.   Rearrangement of benzyl quaternary ammonium salts to ortho-substituted benzyldialkylamines on treatment with alkali metal amides.

Sonn-Muller method.   Preparation of aromatic aldehydes from anilides by conversion of an acid anilide with phosphorus pentachloride to an imido chloride, reduction of the imido chloride with stannous chloride, and hydrolysis of the obtained anil.

sonolysis.   The breaking-up (molecular fragmentation) of molecules by ultrasonic radiation. Examples: sonolysis in pure water produces hydrogen atoms, hydroxyl radicals, molecular hydrogen, oxygen and hydrogen peroxide; acetonitrile, in an argon atmosphere, produces molecular hydrogen, nitrogen, and methane.

"Sonostat."[309]   TM for a series of antistatic agents based on nitrogenous compounds for use on natural and synthetic fibers.

"Sonowax."[45]   TM for a series of wax emulsions based on microcrystalline wax for use with resins or as top finish on various textile fabrics.

sorb.   See sorption.

sorbic acid.   (2,4-hexadienoic acid).
CAS: 110-44-1. $CH_3CH:CHCH:CHCOOH$.
Properties: White, crystalline solid; mp 134.5C; bp 228C (decomposes); 153C (50 mm); flash p (OC) 260F (126C); slightly soluble in water and many organic solvents. Combustible.
Derivation: Trimerization of acetaldehyde and catalytic air oxidation of the resulting hexadienal. Found in berries of mountain ash.
Grade: FCC, technical.
Use: Fungicide, food preservative (mold inhibitor), alkyd resin coatings, upgrading of drying oils, cold rubber additive, intermediate for plasticizers and lubricants.

sorbide.   (dianhydrosorbitol). $C_6H_8O_2(OH)_2$.
Generic name for anhydrides (dicyclic ether dihydric alcohols) derivable from sorbitol by removal of two molecules of water. The name is also applied to specific commercial varieties.

sorbitan.   (monoanhydrosorbitol; sorbitol anhydride). $C_6H_8O(OH)_4$.   Generic name for anhydrides (cyclic ether tetrahydric alcohols) derivable from sorbitol by removal of one molecule of water.

sorbitan fatty acid esters.   Mixture of partial esters of sorbitol and its anhydrides with fatty acids.
Properties: Sorbitan monolaurate and sorbitan monooleate are amber liquids; sorbitan monostearate, sorbitan monopalmitate, and sorbitan tristearate are cream-colored waxy solids with slight odor and bland taste; d of both liquid and solid esters is about 1.0 (25C), mp of solid esters is 54C; they are insoluble in water, somewhat soluble in organic solvents. Combustible.
Derivation: Esterification of sorbitol with fatty acids.
Grade: Technical, FCC (for sorbitan monostearate).
Use: Emulsifiers and stabilizers in foods, cosmetics, drugs, textiles; plastics, agricultural chemicals.
See also polysorbate.

sorbitol.   (d-sorbite; d-sorbitol; hexahydric alcohol).   CAS: 50-70-4. $C_6H_8(OH)_6$.
Properties: White, odorless, crystalline powder; hygroscopic; faint sweet taste; soluble in water, glycerol, and propylene glycol; slightly soluble in methanol, ethanol, acetic acid, phenol, and acetamide; almost insoluble in most other organic solvents; d 1.47 (−5C); mp (metastable form) 93C, (stable form) 97.5C; approved by FDA for food use.

Derivation: By pressure hydrogenation of dextrose with nickel catalyst. Occurs in small amounts in various fruits and berries.

Grade: Crystals, technical, 70% aqueous solution (USP), resin, powder, FCC (solid and solution).

Use: Ascorbic acid fermentation. In solution form, for moisture-conditioning of cosmetic creams and lotions, toothpaste, tobacco, gelatin; bodying agent for paper, textiles, and liquid pharmaceuticals; softener for candy; sugar crystallization inhibitor; surfactants; urethane resins and rigid foams; plasticizer, stabilizer for vinyl resins; food additive (sweetener, humectant, emulsifier, thickener, anticaking agent); dietary supplement.

sorbitol anhydride. See sorbitan.

"Sorbo-Cell."[247] TM for a chemical-coated diatomite filter aid used for selectively removing traces of oil from oil-in-water emulsions. Effective for removing other trace components free of emulsified systems.

Use: Conditioning of boiler feed water.

l(-)-**sorbose.** CAS: 87-79-6.

$HOCH_2CO(CHOH)_3CH_2OH$.

Properties: White, crystalline powder; sweet taste; mp 159–161C; soluble in water; almost insoluble in organic solvents. Combustible.

Derivation: From sorbitol by submerged culture aerobic fermentation.

Grade: Technical, reagent.

Use: Manufacture of ascorbic acid (vitamin C), preparation of special diets, and media for the study of metabolism in animals and microorganisms.

Sorel cement. See magnesium oxychloride cement.

sorghum. A cereal plant cultivated in the temperate zones; its stems are rich in sucrose and can be processed in much the same way as sugarcane and used as a source of sugar and syrups. It is grown in the US in the Midwest, especially Texas.

sorption. A surface phenomenon which may be either absorption or adsorption, or a combination of the two. The term is often used when the specific mechanism is not known.

sorrel salt. See potassium binoxalate.

sour. (1) Any substance used in textile or laundry operations to neutralize residual alkali or decompose residual hypochlorite bleach. The commonly used sours are sodium bifluoride and sodium fluosilicate. (2) Contaminated with sulfur

compounds, e.g., gasoline or natural gas. (3) Taste characteristic of an acidic or fermented substance.

See doctor treatment.

"Source."[523] TM for a biconstituent fiber composed of 70% nylon and 30% polyester by weight. The fiber is a dispersion of polyester fibrils within and oriented to the axis of the nylon fiber. Chief use is for luxury carpet fiber.

soybean flour. A fine-ground powder having a particle size of 100-mesh or less, made by steaming soybeans to inactive enzymes, followed by removal of hulls and mechanical grinding. It contains 40–50% protein, about 20% fat, and 5% moisture. Defatted flours are made by extraction with hexane to remove the oil. The flakes produced are used chiefly for animal feeds but the full-fat flour has become an increasingly important factor in high-protein food products, especially textured proteins and meat analogs.

soybean meal. The crushed residue from the extraction of soybeans. Extraction by the hydraulic or expeller process produces normally a meal with 6% residual oils, while the solvent process yields meal with 1% residual oil. Typical analysis; crude protein 43%, crude fiber 5.5%, nitrogen-free extract 30%, ash 6%, oil content 1–6%. Total digestible nutrients about 75%.

Use: Animal feeds, adhesives, medium for bacitracin production.

soybean oil. (soya bean oil; chinese bean oil; soy oil).

Properties: Pale yellow, fixed drying oil; soluble in alcohol, ether, chloroform, and carbon disulfide. D 0.924–0.929, mp 22–31C, refr index 1.4760–1.4775, solidifying p −15 to −8C, Hehner value 94–96, saponification value 190–193, iodine value 137–143, flash p 540F (282C), autoign temperature 833F (445C), moderate spontaneous heating. Combustible.

Use: Soap manufacture, high-protein foods, paints varnishes, cattle feeds, margarine and salad dressings, printing inks, source of nylon 9, plasticizer (epoxidized), alkyd resins. *Note*: Soybean oil is the most widely used vegetable oil for both edible and industrial use in the US.

Soxhlet extractor. A laboratory apparatus consisting of a glass flask and condensing unit used for continuous reflux extraction of alcohol- or ether-soluble components of food products. Named after its inventor, a German chemist.

space, chemistry in. Experiments carried out on the space shuttle in the early 1980s indicate that unique types of chemical reactions occur in outer space, and that actual products may result that are not achievable under terrestrial environment. Several factors are believed to account for this, primarily zero gravity, though absence of oxygen and enhanced magnetic effects may also play a part. Several encouraging results have already been obtained, though until further experiments and operating data have been investigated the conclusions must be considered tentative. Among projects that have been carried out or are contemplated are the following: (1) Uniform polymer microspheres that are over twice as large as possible on earth have been made due to zero gravity. (2) More effective electrophoresis reactions for making biological materials have been discovered, probably also because of zero gravity. (3) Possibilities exist for (a) making unique alloys in space that are not possible on earth, for example lead-copper, lead-zinc, and aluminum-indium; (b) purer crystals for microelectronics; (c) better glass for fiber optics; (d) new drugs and pharmaceuticals. Future experiments will involve human cells, enzymes, and hormones.

space velocity. The volume of gas or liquid, measured at specified temperature and pressure (usually standard conditions) passing through unit volume in unit time.
Use: Comparing flow processes involving different conditions, rates of flow and sizes or shapes of containers.

spalling. Chipping an ore for crushing, or the cracking, breaking, or splintering of materials due to heat.

"Span."[89] TM for each member of a series of general-purpose emulsifiers and surface-active agents. They are fatty acid partial esters of sorbitol anhydrides (or sorbitan). Generally insoluble in water and soluble in most organic solvents.

spandex. Generic name for a fiber in which the fiber-forming substance is a long-chain synthetic polymer comprised of at least 85% of a segmented polyurethane (Federal Trade Commission). Imparts elasticity to garments such as girdles, socks, special hosiery.

spanish white. (1) chalk, $CaCO_3$; (2) bismuth white, $BiO(NO_3)$, basic bismuth white.

spar. (1) A type of crystalline material such as Iceland spar or feldspar, usually containing calcium carbonate or an aluminum silicate; fluorspar is calcium fluoride. Iceland spar has unique optical properties. (2) A weather-resistant varnish originally used for coating wooden decks

of ships, which may be the reason for its name. See spar varnish.

sparger. A perforated pipe through which steam, air or water is sprayed into a liquid during a fermentation reaction.
Use: Brewing industry to remove traces of wort from the mash.

spar, Greenland. See cryolite.

spar, heavy. See barite.

spar, Iceland. See calcite.

sparking metal. See pyrophoric alloy.

spar, satin. See calcite; gypsum.

spar varnish. A durable, water-resistant varnish for severe service on exterior exposure. It consists of one or more drying oils (linseed, tung, or dehydrated castor oil), one or more resins (rosin, ester gum, phenolic resin, or modified phenolic resin), one or more volatile thinners (turpentine or petroleum spirits), and driers (linoleates, resinates, or naphthenates of lead, manganese, and cobalt). It is classed as a long-oil varnish and generally consists of 45–50 gals of oil for each 100 lb of resin.
See also varnish.

SPE. Abbreviation for Society of Plastics Engineers.

spearmint oil. A yellowish essential oil, strongly levorotatory.
Use: Source of carvone and as flavoring for medicines, chewing gum, etc.

specification. A schedule of minimum performance requirements for specialized products such as those established by the various committees of the American Society for Testing and Materials and the Underwriters Laboratories. Such products are subject to inspection and test before acceptance.
See also testing.

specific gravity. The ratio of the density of a substance to the density of a reference substance; it is an abstract number that is unrelated to any units. For solids and liquids, specific gravity is numerically equal to density, but for gases it is not, because of the difference between the densities of the reference substances, which are usually water (1 g/cc) for solids and liquids and air (0.00129 g/cc, or 1.29 g/L at 0C and 760 mm) for gases. The specific gravity of a gas is the ratio of its density to that of air; since the *specific gravity* of air = 1.0 (1.29/1.29), this is usually stated to indicate the comparison with the gas under consideration. For example, the density of hydrogen is 0.089 g/L but its specific gravity is 0.069, (i.e., 0.089/1.29). The specific gravity of solids and liquids is the ratio of their density to that of water at 4C, taken as 1.0, as 1 cc of water weighs 1 gram. Thus a solid or liquid with a density of 1.5 g/cc has a specific gravity of 1.5/1 or 1.5.

Since weights of liquids and gases vary with temperature, it is necessary to specify both temperatures involved, except for rough or approximate values. Thus the specific gravity of alcohol should be given as 0.7893 at 20/4C, the first temperature referring to the alcohol and the latter to the water. At 15.56C the specific gravity of alcohol is 0.816.
See also density, API gravity, Baumé.

specific heat. The ratio of the heat capacity of a substance to the heat capacity of water, or the quantity of heat required for a 1 degree temperature change in a unit weight of material. Commonly expressed in Btu/lb/degree F or in cal/g/degree F. For gas the specific heat at constant pressure is greater than that at constant volume by the amount of heat needed for expansion.

specific volume. The volume of unit weight of a substance, as cubic feet per pound, or gallons per pound, but more frequently milliliters per gram. The reciprocal of density

specific weight. The weight per unit volume of a substance.

"SpectrAR."[229] TM for a group of spectrophotometric-grade solvents developed for the extremely high purity required by modern analytical techniques.

spectrophotometry. See absorption spectroscopy.

spectroscopy. (instrumental analysis).
A branch of analytical chemistry devoted to identification of elements and elucidation of atomic and molecular structure by measurement of the radiant energy absorbed or emitted by a substance in any of the wavelengths of the electromagnetic spectrum in response to excitation by an external energy source. The types of absorption and emission spectroscopy are usually identified by the wavelength involved, namely, gamma ray, x-ray, UV, visible, infrared, microwave, and radiofrequency. The technique of spectroscopic analysis was originated by Fraunhofer

who in 1814 discovered certain dark (D) lines in the solar spectrum, which were later identified as characterizing the element sodium. In 1861 Kirchhoff and Bunsen produced emission spectra and showed their relationship to Fraunhofer lines. X-ray spectroscopy was utilized by Moseley (1912) to determine the precise location of elements in the Periodic system. Since then, a number of sophisticated and highly specialized techniques have been developed including Raman spectroscopy, nuclear magnetic resonance, nuclear quadropole resonance, dynamic reflectance spectroscopy, microwave and gamma ray spectroscopy, and electron paramagnetic resonance.

spectrum. The radiant energy emitted by a substance as a characteristic band of wavelengths by which it can be identified.
See radiation, spectroscopy.

speculum metal. (1) 66% copper, 34% tin with trace of arsenic; (2) 64% copper, 32% tin, 4% nickel.
Properties: D 8.6, mp 750C.
Use: For mirrors for reflecting telescopes.

spelter. Relatively pure zinc as encountered in industrial operations such as galvanizing. Lead and/or iron are common impurities.

spent mixed acid. Mixed acid which has given up part of its nitric acid.
Hazard: Dangerous fire risk. Strong irritant to tissue.

spent oxide. See iron sponge, spent.

Sperry process. An electrolytic process for the manufacture of lead carbonate, basic (white lead) from desilverized lead containing some bismuth. The impure lead forms the anode. A diaphragm separates anode and cathode compartments, and carbon dioxide is passed into the solution. Impurities, including bismuth, remain on the anode as a slime blanket.

sphalerite. (blende, zinc blende). ZnS. Natural zinc sulfide, usually containing some cadmium, iron, and manganese.
Properties: Color yellow, brown, black, or red; luster resinous; hardness 3.5–4; d 3.9–4.1; good cleavage; soluble in hydrochloric acid.
Occurrence: Missouri, Kansas, Oklahoma, Colorado, Montana, Wisconsin, Idaho, Australia, Canada, Mexico.
Use: Most important ore of zinc, also a source of cadmium; phosphor; source of sulfur dioxide for production of sulfuric acid.

sphingomyelin. Diaminophosphatides occurring primarily in nervous tissue and containing a fatty acid, phosphoric acid, choline, and sphingosine. They are soluble in hot absolute alcohol and insoluble in ether, acetone and water.

sphingosine. (1,3-dihydroxy-2-amino-4-octadecene). CAS: 123-78-4. $CH_3(CH_2)_{12}CH:CHCHOHCH(NH_2)CH_2OH$. Forms part of certain phosphatides, such as cerebrosides and sphingomyelins.
Properties: Waxy crystals, mp 67C, soluble in ether.

SPI. Abbreviation for Society of the Plastics Industry.

spider. (1) A component of the ejector mechanism of a compression molding press. (2) A supporting device used in the die assembly of an extrusion machine.

spindle oil. Low-viscosity lubricating oil for textile and other high-speed machinery.

spinel. CAS: 1302-67-6. $MgAl_2O_4$.
A natural oxide of magnesium and aluminum with replacement of magnesium by iron, zinc, and manganese and of aluminum by iron and chromium. There are also synthetic spinels, e.g., as magnesia-alumina or magnesia-chromia. Their structure is similar to ferrites.
Properties: Color, various shades of red, grading to green, brown, and black; luster vitreous; hardness 8; there are many varieties.
Occurrence: New York, New Jersey, Massachusetts, Virginia, North Carolina, Ceylon, Thailand, Mozambique.
Use: Crystallography; synthetic spinel is used as a refractory and in electronic applications.

spinneret. An extrusion device shaped somewhat like a thimble and containing a number of holes of exceedingly small diameter through which are forced solutions of viscose rayon, nylon, molten glass, and various other materials. It is made of precious metals such as gold or platinum. Spinnerets enable extrusion of filaments of one denier or less, one denier filament has a diameter of 40 microns. For commercial work 12–15 denier fiber is generally used.
See denier.

"Spiralloy."[266] TM for glass- and other filament-wound, resin-bonded internal and external pressure vessels.
Use: Rocket motor cases, underwater and space structures, pressure vessels, radomes, torpedo cases, booms, and tubes.

spirits. An obsolescent and ambiguous term usually taken to mean the distilled essence of a substance. Mineral or petroleum spirits is a volatile hydrocarbon distillate similar to naphtha; in pharmacy, the term refers to an alcoholic solution of volatile principles, e.g., spirits of ammonia, niter, camphor, etc.
See also tincture.

spirochete. A type of bacteria characterized by a spiral shape; the infective organism of syphilis.

spirocyclane. See spiropentane.

spironolactone. (17-hydroxy-7-α-mercapto-3-oxo-17-α-pregn-4-ene-21-carboxylic acid-γ-lactone-7-acetate). CAS: 52-01-7. $C_{24}H_{32}O_4S$.
Properties: Light cream-colored to light tan, crystalline powder; mild mercaptan-like odor; stable in air. Practically insoluble in water, soluble in ethyl acetate and ethanol, slightly soluble in methanol, mp 198–207C (decomposes).
Grade: USP.
Use: Medicine (diuretic).

spiropentane. (spirocyclane; cyclopropanespiro-cyclopropane). $H_2CH_2CCCH_2CH_2$.
Properties: Colorless liquid, refr index 1.41220 (20C), d 0.7551 (20/4C), fp −107.05C, bp 39.03C.
Derivation: Heating pentaerythrityl tetrabromide in ethanol with zinc dust.

spiro system. A structural formula consisting of two rings having one carbon atom in common. Most bicyclic compounds such as naphthalene have two carbons in common.
See for example spiropentane.

spodumene. $LiAl(SiO_3)_2$.
Properties: White, pale green, emerald green, pink, or purple mineral; white streak; vitreous luster. Contains up to 8% lithium oxide with some replacement by sodium. Insoluble in acids, d 3.13–3.20, Mohs hardness 6.5–7.
Occurrence: US (North Carolina, California, Massachusetts, South Dakota), Brazil, Mozambique.
Use: Source of lithium, in ceramics and glass as a source of lithia and alumina.

sponge. (1) Metal: A finely divided and porous form of metal such as iron, platinum, or titanium. May be used in sponge form as a catalyst or pressed into metal ingots. (2) Plastic:
See cellular plastic, foam, rubber sponge. (3) Natural: Siliceous cells of the Porifera group of sessile sea animals.

sponge iron. See iron sponge.

spontaneous combustion. See combustion.

"Spotleak."[204] TM for a group of natural gas odorants based on mercaptans.

sputtered coating. A protective metallic coating applied in a vacuum tube and consisting of metal ions emanating from a cathode and deposited as a film on objects within the tube. The process involves three phases: generation of metal vapor, diffusion of the vapor, and its condensation. Paper, plastics, and similar materials can be coated in this way.

sprue. (1) The main feed channel that runs from the outer face of an injection mold to the gate in a single-cavity mold or to the runners of a multiple-cavity mold. The liquid polymer is forced through this orifice from a nozzle till the mold is filled to capacity. Some polymer remains in the sprue after the mold is closed, leaving a projecting piece which must be removed after the product is ejected. The viscosity of the polymer must be low enough to permit it to pass through the sprue readily.
(2) A disease attacking the digestive organs as a result of inadequate nutrition, especially in tropical areas.

spun protein. See protein, textured.

squalane. (perhydrosqualene; 2,6,10,15,19,23-hexamethyltetracosane; spinacane; dodecahydrosqualene). CAS: 111-01-3. $C_{30}H_{62}$.
A saturated hydrocarbon.
Properties: Colorless, odorless, tasteless liquid; miscible with vegetable and mineral oils; organic solvents, and lipophilic substances. Nondrying, nonoxidizing, noncongealing. D 0.805–0.812 (20C), bp 350C, fp −38C, refr index 1.4520–1.4525 (20C). Combustible.
Derivation: Hydrogenation of squalene; may occur naturally in sebum.
Use: High-grade lubricating oil, vehicle for externally applied pharmaceuticals and cosmetics, perfume fixative, gas chromatographic analysis, transformer oil.

squalene. (spinacene).   CAS: 7683-64-9.
2,6,10,15,19,23-hexamethyl-2,6,10,14,18,22-tetracosahexaene) $C_{30}H_{50}$.   A natural raw material found in human sebum (5%) and in shark liver oil. An unsaturated aliphatic hydrocarbon (carotenoid) with six unconjugated double bonds.

Properties: Oil with faint odor, d 0.858–0.860 (20C), bp 285C (25 mm), fp −60C, refr index 1.49–1.50, iodine no 360–380, saponification value 0–5, insoluble in water, slightly soluble in alcohol, soluble in lipids and organic solvents. Combustible.
Grade: 90% min.
Use: Biochemical and pharmaceutical research, a precursor of cholesterol in biosynthesis, chemical intermediate.

squirrel-cage disintegrator. See cage mill.

Sr. Symbol for strontium.

"SR-173."[245] TM for a silicone intermediate resin for reaction with alkyd coating resin constituents for air-dry and baking co-polymer high-performance coatings. Also available with special solvent to comply with air pollution requirements.

SRF black. Abbreviation for semireinforcing furnace black.
See furnace black.

SS acid. See Chicago acid.

"Stabilide."[329] TM for potassium iodide stabilized with calcium stearate.
Use: Animal feeds.

"Stabilisal C."[230] TM for an organic compound used as an antigellation agent for nitrocellulose-based bronzing lacquers and for stabilizing dipping lacquers.

stabilizer. Any substance which tends to keep a compound, mixture, or solution from changing its form or chemical nature. Stabilizers may retard a reaction rate, preserve a chemical equilibrium, act as antioxidants, keep pigments and other components in emulsion form, or prevent the particles in a colloidal suspension from precipitating.
See also inhibitor.

"Staclipse."[492] TM for specially processed, acid-modified, thin-boiling starches. Highly converted products having relatively high percentage of cold water-soluble fractions.

Use: Finishes and binders in textile wet processing; adhesives;and paper industry for surface sizing, calender sizing, and coating.

"Stacolloid."[492]   TM for a group of noncongealing sizing and finishing agents derived from corn starch.

Use: Sizing of spun synthetic, combed cotton, worsted yarns, and paper.

Staedel-Rugheimer pyrazine synthesis.
Formation of pyrazines by high temperature autoclave reaction of α-halogenomethyl ketones with ammonia.

"Starcrylic."[492]   A pale yellow, cloudy solution of an acrylate terpolymer.

Use: Sizing polyester filament yarns of all types.

"Stadex."[492]   TM for products of the partial hydrolysis of corn starch produced by heating the starch in a dry atmosphere in the presence of acid.

Use: Adhesives, binders, thickeners, sizing ingredients, stabilizers, bodying agents, and in edible products as carriers for color and flavor additives.

stain. (1) An organic protective coating similar to a paint, but with much lower solids content (pigment loading).

Use: Exterior and interior coating of wood, furniture, flooring, etc. (2) Any compound used to color bacteria for microscopic examination, e.g., osmium tetroxide, phosphotungstic acid, uranyl acetate, and certain chromium compounds.

stainless steel.   See steel, stainless.

"Staleydex."[492]   TM for a crystalline dextrose produced by an enzyme conversion process. Several types of dextrose available varying from coarse particles to very fine dextrose monohydrate crystals, as well as liquid dextrose.

Use: Canning, brewing, confections, beverages, prepared mixes, pharmaceuticals, chemicals, and other industrial uses.

"Stamere."[406]   TM for a series of edible hydrocolloids extracted from seaweeds. Various products are available which are standardized and recommended for particular applications.

standard cell.   See Weston cell.

Stanley, Wendell M.   (1904-1971) An American biochemist who won the Nobel prize for chemis-

try in 1946 along with Northrop and Sumner. His work on virus research resulted in isolation of crystals proving the virus to be proteinaceous. In the 1930's, he was concerned with isolating nucleic acid from crystallized virus and the reproduction of influenza virus. His doctorate was from University of Illinois. His many accomplishments included membership in the National Advisory Cancer Council of the United States Public Health Service in the 1950's.

stannic acid.   See stannic oxide.

stannic anhydride.   See stannic oxide.

stannic bromide.   (tin bromide; tin tetrabromide). CAS: 7789-67-5. $SnBr_4$.

Properties: White, crystalline mass; fumes when exposed to air; soluble in water, alcohol, and carbon tetrachloride; d 3.3; bp 203C; mp 31C.

Hazard: Irritant to skin and eyes. TLV (as Sn): 2 mg/m³ of air.

Use: Mineral separations.

stannic chloride.   (tin chloride; tin tetrachloride; tin perchloride).   CAS: 7646-78-8. $SnCl_4$. Often sold in the form of the double salt with sodium chloride: $Na_2SnCl_6 \cdot HOH$.

Properties: Colorless, fuming, caustic liquid, which water converts into a crystalline solid, $SnCl_4 \cdot 5HOH$; keep well stoppered; d 2.2788; fp −33C; bp 114C; soluble in cold water, alcohol, carbon disulfide; decomposed by hot water.

Derivation: Treatment of tin or stannous chloride with chlorine.

Grade: Technical, CP.

Hazard: Evolves heat on contact with moisture. Corrosive liquid. TLV (as Sn): 2 mg/m³ of air.

Use: Electroconductive and electroluminescent coatings, mordant in dyeing textiles, perfume stabilization, manufacture of fuchsin, color lakes, ceramic coatings, bleaching agent for sugar, stabilizer for certain resins, manufacture of blueprint and other sensitized papers, other tin salts, bacteria and fungi control in soaps.

stannic chloride pentahydrate.   CAS: 10026-06-9. $SnCl_4 \cdot 5HOH$.

Properties: White solid, mp 56C, soluble in water or alcohol.

Hazard: Toxic material. TLV (as Sn): 2 mg/m³ of air.

Use: Substitute for anhydrous stannic chloride where the presence of water is not objectionable.

stannic chromate.   (tin chromate). CAS: 38455-77-5. $Sn(CrO_4)_2$.

Properties: Brownish-yellow, crystalline powder; partially soluble in water.

Derivation: Action of chromic acid on stannic hydroxide.

Hazard: Toxic material. TLV (as Sn): 2 mg/m³ of air.

Use: Coloring porcelain and china.

stannic oxide. (stannic anhydride; tin peroxide; tin dioxide; stannic acid). CAS: 18282-10-5. SnO_2 or $SnO_2 \cdot nHOH$.

Properties: White powder; anhydrous or containing variable amounts of water; d 6.6–6.9; mp 1127C; sublimes at 1800–1900C; soluble in concentrated sulfuric acid, hydrochloric acid; insoluble in water. Noncombustible.

Derivation: (a) Found in nature as in the mineral cassiterite. (b) Precipitated from stannic chloride solution by ammonium hydroxide.

Grade: White, pure; white; gray; CP.

Use: Tin salts; catalyst; ceramic glazes and colors; putty; perfume preparations and cosmetics; textiles (mordant, weighting); polishing powder for steel, glass, etc.; manufacture of special glasses.

stannic phosphide. (tin phosphide).
CAS: 25324-56-5. Sn_2P_2 or SnP.

Properties: Silver white, hard mass or lumps, soluble in acids, d 6.56, mp forms Sn_4P_3 at 415C.

Derivation: By heating together tin and phosphorus.

Hazard: Flammable, dangerous fire risk. TLV (as Sn): 2 mg/m³ of air.

stannic sulfide. (artificial gold; mosaic gold; tin bronze; tin disulfide). CAS: 1315-01-1. SnS_2.

Properties: Yellow to brown powder, soluble in concentrated hydrochloric acid and alkaline sulfides, insoluble in water, d 4.42–4.60, mp decomposes at 600C.

Derivation: (a) Action of sulfur on a solution of stannic chloride. (b) By heating tin amalgam with sulfur and ammonium chloride, distilling off the mercury sulfide and ammonium chloride.

Grade: Technical, reagent.

Hazard: Irritant to skin and eyes. TLV (as Sn): 2 mg/m³ of air.

Use: Imitation gilding, pigment.

stannous acetate. CAS: 638-39-1.
$Sn(C_2H_3O_2)_2$.

Properties: Gray to yellow crystals, mp 182C, d 2.30, soluble in dilute hydrochloric acid.

Derivation: Refluxing stannous oxide with 50% acetic acid in an atmosphere of nitrogen.

Hazard: Toxic by ingestion.

Use: Reducing agent.

stannous bromide. (tin bromide; tin dibromide).
CAS: 10031-24-0. $SnBr_2$.

Properties: Yellow powder; soluble in hydrochloric acid (dilute); soluble in water, alcohol,
ether, and acetone; oxidizes and turns brown in air; d 5.117 (17C); bp 619C; mp 215C.

Hazard: Skin irritant. TLV (as Sn): 2 mg/m³ of air.

stannous chloride. (tin crystals; tin salt; tin dichloride; tin protochloride).
CAS: 7772-99-8. (a) $SnCl_2$,
(b) $SnCl_2 \cdot 2HOH$.

Properties: White, crystalline mass which absorbs oxygen from the air and is converted into the insoluble oxychloride; soluble in water, alkalies, tartaric acid, and alcohol; (a) d 3.95 (25/4C), mp 246.8C, bp 652C; (b) d 2.71, mp 37.7C, bp decomposes.

Derivation: By dissolving tin in hydrochloric acid.

Grade: Technical, CP, reagent, anhydrous, hydrated.

Hazard: Irritant to skin, use in foods restricted to 0.0015%, as tin. TLV (as Sn): 2 mg/m³ of air.

Use: Reducing agent in manufacture of chemicals, intermediates, dyes, polymers, phosphors; manufacture of lakes; textiles (reducing agent in dyeing, discharge in printing); tin galvanizing; reagent in analytical chemistry; silvering mirrors; revivication of yeast sown in must (accelerator); antisludging agent for lubricating oils; food preservative; stabilizer for perfume in soaps; catalyst; soldering flux; sensitizing agent for glass, paper, plastics.

stannous chromate. (tin chromate). $SnCrO_4$.

Properties: Brown powder, almost insoluble in water.

Derivation: Interaction of stannous chloride and sodium chromate.

Hazard: Toxic material. TLV (as Sn): 2 mg/m³ of air.

Use: Decorating porcelain.

stannous-2-ethylhexoate. (stannous octoate; tin octotate). $Sn(C_8H_{15}O_2)_2$.

Properties: Light yellow liquid; insoluble in water, methanol; soluble in benzene, toluene, petroleum ether; hydrolyzed by acids and bases; d 1.25; Gardner color 3 (max).

Hazard: Toxic material. TLV (as Sn): 0.1 mg/m³ of air.

Use: Polymerization catalyst for urethane foams, lubricant, addition agent, stabilizer for transformer oils.

stannous fluoride. (tin fluoride; tin difluoride).
CAS: 7783-47-3. SnF_2.

Properties: White, lustrous, crystalline powder with bitter, salty taste; mp 212–214C; practically insoluble in alcohol, ether, and chloroform; slightly soluble in water.

Grade: NF.
Hazard: Toxic by ingestion, strong irritant to skin and tissue. TLV (as Sn): 2 mg/m³ of air.
Use: Fluoride source in toothpastes.   Note: Stannous hexafluorozirconate is said to be more effective than the fluoride in preventing dental caries.

stannous octoate.   See stannous-2-ethylhexoate.

stannous oleate.   (tin oleate).
    CAS: 1912-84-1. $Sn(C_{18}H_{33}O_2)_2$.
Properties: Light yellow liquid; insoluble in water and methanol; soluble in benzene, toluene, petroleum ether; hydrolyzed by acids and bases.
Hazard: Absorbed by skin. TLV (as Sn): 0.1 mg/m³ of air.
Use: Polymerization catalyst, inhibitor.

stannous oxalate.   (tin oxalate).
    CAS: 814-94-8. SnC_2O_4.
Properties: Heavy, white, crystalline powder; d 3.56; mp decomposes at 280C; soluble in acids; insoluble in water and acetone.
Derivation: By the action of oxalic acid on stannous oxide.
Grade: Technical, CP, reagent.
Hazard: Absorbed by skin. TLV (as Sn): 0.1 mg/m³ of air.
Use: Dyeing and printing textiles, catalyst for esterification reactions.

stannous oxide.   (tin oxide; tin protoxide).
    CAS: 21651-19-4.   SnO.
Properties: Brownish-black powder, unstable in air, reacts with acids and strong bases, insoluble in water, d 6.3, mp 1080C (600 mm) (decomposes) a nuisance particulate.
Derivation: By heating stannous hydroxide in a current of carbon dioxide.
Grade: Technical, CP.
Use: Reducing agent, intermediate in preparation of stannous salts as used in plating and glass industries, pharmaceuticals, soft abrasive (putty powder).

stannous pyrophosphate.   CAS: 15578-26-4.
    $Sn_2P_2O_7$.
Properties: White, free-flowing crystals; insoluble in water; d 4.009 (16C).
Use: Toothpaste additive.

stannous sulfate.   (tin sulfate).
    CAS: 7488-55-3. $SnSO_4$.
Properties: Heavy white or yellowish crystals, soluble in water and sulfuric acid, water solution decomposes rapidly, mp loses sulfur dioxide at 360C.
Derivation: Action of sulfuric acid on stannous oxide.

Hazard: Toxic material. TLV (as Sn): 2 mg/m³ of air.
Use: Dyeing, tin-plating, particularly for plating automobile pistons and steel wire.

stannous sulfide.   (tin monosulfide; tin protosulfide; tin sulfide).   CAS: 1314-95-0.   SnS.
Properties: Dark gray or black crystalline powder, d 5.080, bp 1230C, mp 880C, soluble in concentrated hydrochloric acid (decomposes), insoluble in dilute acids and water.
Hazard: Toxic material. TLV (as Sn): 2 mg/m³ of air.
Use: Making bearing material, catalyst in polymerization of hydrocarbons, analytical reagent.

stannous tartrate.   (tin tartrate).
    CAS: 815-85-0. $SnC_4H_4O_6$.
Properties: Heavy, white, crystalline powder; soluble in water; dilute hydrochloric acid.
Derivation: Action of tartaric acid on stannous oxide.
Hazard: Toxic material. TLV (as Sn): 0.1 mg/m³ of air.
Use: Dyeing and printing fabrics.

stannum.   The Latin name for tin, hence the symbol Sn in chemical nomenclature.

staple.   A cotton fiber usually in reference to length, i.e., short or long staple cotton.

starch.   CAS: 9005-84-9.   A carbohydrate polymer having the following repeating unit:

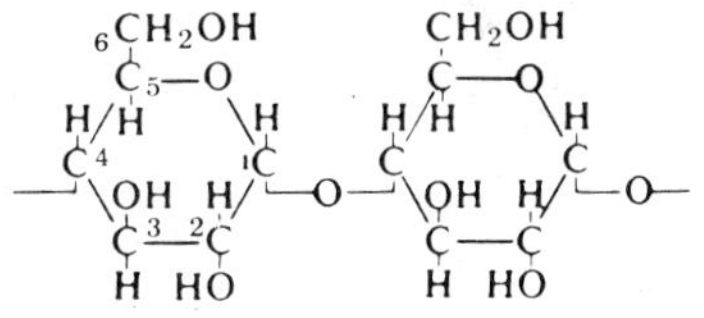

It is composed of about 25% amylose (anhydroglucopyranose units joined by glucosidic bonds) and 75% amylopectin, a branched-chain structure.
Properties: White, amorphous, tasteless powder or granules; various crystalline forms may be obtained, including microcrystalline. Irreversible gel formation occurs in hot water; swelling of granules can be induced at room temperature with such compounds as formamide, formic acid, and strong bases and metallic salts.
Occurrence: Starch is a reserve polysaccharide in plants (corn, potatoes, tapioca, rice, and wheat are commercial sources).
Grade: Commercial, powdered, pearl, laundry, technical, reagent, edible, USP.
Use: Adhesive (gummed paper and tapes, cartons, bags, etc.), machine-coated paper, textile filler

and sizing agent, beater additive in papermaking, gelling agent and thickener in food products (gravies, custards, confectionery), oil-well drilling fluids, filler in baking powders (cornstarch), fabric stiffener in laundering, urea-formaldehyde resin adhesives for particle board and fiberboard, explosives (nitrostarch), dextrin (starch gum), chelating and sequestering agent in foods, indicator in analytical chemistry, anticaking agent in sugar, face powders, abherent and mold-release agent, polymer base.
See starch-based polymer.

starch-based polymer. (1) A reactive polyol derived from a mixture of a starch with dibasic acids, hydrogen-donating compounds, and catalysts dissolved in water; the slurry is subjected to high temperatures and pressures, yielding a low-viscosity polymer in a 50% solids aqueous solution. A molecular rearrangement takes place, and the polymer formed is completely different from starch in structure and properties. It can be further reacted with acids, bases, and crosslinking agents. Suggested uses are in high wet-strength papers, as binders in paper coatings, as moisture barriers in packaging, and as water-resistant adhesives.

(2) Yeast fermentation of starch to form a biodegradable plastic called "pullulan" (a trigluco polysaccharide) has been reported to be commercially feasible.
See "Pullulan."

starch dialdehyde. Starch in which the original anhydroglucose units have been partially changed to dialdehyde form by oxidation, for example, the product of the oxidation of cornstarch by periodic acid. Available in cationic dispersions up to 15% solids for mixing with paper pulp.
Use: Thickening agent, tanning agent, binder for leaf tobacco, adhesives, wet-strength additive in paper.

starch-iodide paper. Indicator paper made by dipping paper in starch paste containing potassium iodide.
Use: To test for halogens and oxidizers such as hydrogen peroxide.

starch, modified. Any of several water-soluble polymers derived from a starch (corn, potato, tapioca) by acetylation, chlorination, acid hydrolysis, or enzymatic action. These reactions yield starch acetates, esters, and ethers in the form of stable and fluid solutions and films. Modified starches are used as textile sizing agents and paper coatings. Thin-boiling starches have high gel strength, oxidized starches made with sodium

hypochlorite have low gelling tendency. Introduction of carboxyl, sulfonate, or sulfate groups into starch gives sodium or ammonium salts of anionic starches, yielding clear, non-gelling dispersions of high viscosity. Cationic starches result from addition of amino groups.

The glucose units of starch can be crosslinked with such agents as formaldehyde, soluble metaphosphates, and epichlorohydrin; this increases viscosity and thickening power for adhesives, canned foods, etc.

starch phosphate. An ester made from the reaction of a mixture of orthophosphate salts (sodium dihydrogen phosphate and disodium hydrogen phosphate) with starch.
Properties: Soluble in cold water (unlike regular starch) and has high thickening power. Can be frozen and thawed repeatedly without change in physical properties.
Use: Thickener for frozen foods; taconite ore binder; in adhesives, drugs, cosmetics; substitute for arabic gum, locust bean gum, and carboxymethyl cellulose.

starch syrup. See glucose.

starch, thin-boiling. See starch, modified.

starch xanthate. A water-insoluble synthetic polysaccharide made by reacting starch with sodium hydroxide and carbon disulfide; biodegradable.
Use: To encapsulate pesticides; the coating, though insoluble, is permeable to water, thus slowly releasing the pesticide. Rubber reinforcing agent.

starch xanthide. Starch xanthate which has been crosslinked with oxygen.
Use: Strengthening agent in paper and manufacture of powdered rubbers.

"Starfol."[589] TM for various grades of glyceryl monostearate and butyl stearate. Most grades have been approved for food additives.

"Starwax" 100.[128] TM for a hard petroleum microcrystalline wax, minimum mp 82.2C.

"Sta-set."[50] TM for a trichlorophenoxypropionic acid product for use on apples to reduce preharvest drop.

"Sta-Sol."[492] TM for mixture of naturally occurring phosphatides or phospholipids, derived from soybean oil. They contain some residual oil as a solvent or carrier. Available in regular (plastic) or fluid (pourable liquid) form, bleached and unbleached. They act as emulsifiers, stabilizers, antioxidants.

Use: Foods, especially chocolate and compound coatings, candies, bakery products, margarine, industrial uses; paints and printing inks, soaps and cosmetics, textile compounds, leather tanning, petroleum lubricants; animal feeds and pet foods.

"Sta-Tac."[36]   TM for olefinic hydrocarbon resins used as coating vehicles.

"Stat-Eze."[430]   TM for a specially prepared quaternary ammonium compound.
Use: Externally applied antistatic agent on plastics, textiles, paper, and carpeting.

Staudinger, Hermann.   (1881–1965) A German chemist, winner of a Nobel Prize in 1953 for his pioneer work on the structure of macromolecules and polymerization. A large part of modern high-polymer chemistry is based on his original research.

Staudinger reaction.   Synthesis of phosphazo compounds by the reaction of tertiary phosphines with organic azides.

"Staybelite."[266]   TM for hydrogenated rosin, a pale, thermoplastic resin. Acid number 165, USDA color X, softening point 75C, saponification number 167.
Use: Adhesives and protective coatings.

"Stayco."[492]   TM for a family of oxidized starches. Available in five viscosities: S, A, G, C, and M.
Use: Paper industry for wet-end addition, in tub, press, and calender sizing, and as coating adhesives.

"Stayrite."[104]   TM for stabilizers used to provide increased heat and light stability to vinyl resins. Available in range of products, both liquid and solid.

"Staysize" 109.[492]   TM for a white, highly uniform, chemically modified corn starch in pearl form developed for surface sizing.

"Stayzyme."[492]   TM for a thick-boiling corn starch buffered to a pH of 6.3–7.0 to produce the proper pH for enzyme conversion in the paper mill.

steam.   An allotropic form of water formed at 212F (100C) and having a latent heat of condensation of 540 calories per gram. It has a number of industrial uses, one of the most important being the production of hydrogen by the steam-hydrocarbon gas process (reforming), by the steam-water gas process, the steam-iron process, and the steam-methanol process. It is also used in steam cracking of gas oil and naphtha; in food processing, as a cleaning agent; in rubber, vulcanization; as a source of heat and power; in distillation of plants for production of essential oils and perfumes; and in secondary oil recovery.

Steam from geothermal sources such as hot springs and fumaroles is utilized as an energy source. Steam for industrial processing is also being generated by solar energy techniques.
See geothermal energy, latent heat.

steam distillation.   See hydrodistillation.

steam reforming.   See reforming.

steapsin.   A lipase in the pancreatic juice. See enzyme.

stearamide.   (octadecanamide).
CAS: 124-26-5. $CH_3(CH_2)_{16}CONH_2$.
Properties: Colorless leaflets, mp 109C, bp 251C (12 mm), insoluble in water, slightly soluble in alcohol and ether.
Use: Corrosion inhibitor in oil wells.

stearato chromic chloride.   A polynuclear complex in the form of a six-membered ring. The two chromium atoms are bridged on one side by a hydroxyl group, and on the other side by the carboxyl oxygens of the stearic acid. The water-soluble complex results from the neutralization of stearic acid with basic chromic chloride. It acts as a water repellent and abherent.

stearic acid.   (n-octadecanoic acid).
CAS: 57-11-4. $CH_3(CH_2)_{16}COOH$.
The most common fatty acid occurring in natural animal and vegetable fats. Most commercial stearic acid is 45% palmitic acid, 50% stearic acid and 5% oleic acid, but purer grades are increasingly used.
Properties: Colorless, wax-like solid; odor and taste slight suggesting tallow. Soluble in alcohol, ether, chloroform, carbon disulfide, carbon tetrachloride; insoluble in water. D 0.8390 (80/4C), mp 69.6C, bp 361.1C, refr index 1.4299 (80C), flash p 385F (196C), autoign temperature 743F (395C). Combustible.
Derivation: (a) From high-grade tallows and yellow grease stearin by washing, hydrolysis with the Twitchell or similar reagent, boiling, distilling, cooling, and pressing; (b) from oleic acid by hydrogenation.
Grade: Saponified, distilled, single-pressed, double-pressed, triple-pressed, USP, FCC, 90%

stearic with low oleic, grade free from chick edema factor, 99.8% pure.

Use: Chemicals, especially stearates and stearate driers, lubricants, soaps, pharmaceuticals and cosmetics, accelerator activator, dispersing agent and softener in rubber compounds, shoe and metal polishes, coatings, food packaging, suppositories and ointments.

stearin.   (tristearin; glyceryl tristearate).
$C_3H_5(C_{18}H_{35}O_2)_3$.

Properties: Colorless crystals or powder, odorless, tasteless, mp 71.6C, d 0.943 (65C). Insoluble in water; soluble in alcohol, chloroform, carbon disulfide. Combustible.

Derivation: Constituent of most fats.

Grade: Technical, also graded as to source.

Use: Soap, candles, candies, adhesive pastes, metal polishes, water-proofing paper, textile sizes, leather stuffing, manufacture of stearic acid.

stearin pitch.   See fatty acid pitch.

stearone.   An aliphatic ketone, insoluble in water, stable to high temperatures, acids and alkalies; compatible with high-melting vegetable waxes, paraffins, and fatty acids; incompatible with resins, polymers and organic solvents at room temperature but compatible with them at high temperature. Combustible.

Use: Antiblocking agent.

N-stearoyl-p-aminophenol.
$HO(C_6H_4)NHCOCH_2(CH_2)_{15}CH_3$.

Properties: White to off-white powder, mp 131–134C. Insoluble in water; soluble in polar organic solvents (especially when hot) such as alcohol, dioxane, acetone, and dimethylformamide. Combustible.

Use: Antioxidant.

stearoyl chloride.   (n-octadecanoyl chloride).
$CH_3(CH_2)_{16}COCl$.

Properties: Soluble in hydrocarbons and ethers, mp 23C, bp 174–178C (2 mm). Combustible.

Use: Preparation of substituted amines and amides, acid anhydrides, esterification of alcohols, synthesis of other organic compounds.

stearyl alcohol.   (1-octadecanol; octadecyl alcohol).   CAS: 112-92-5.
$CH_3(CH_2)_{16}OH$.

Properties: Unctuous, white flakes or granules with faint odor and bland taste; d 0.8124 (59/4C); bp 210.5C (15 mm); mp 59C; soluble in alcohol, acetone, and ether; insoluble in water. Combustible.

Derivation: Reduction of stearic acid.

Grade: Commercial, technical, USP.

Use: Perfumery, cosmetics, intermediate, surface active agents, lubricants, resins, antifoam agent.

stearyl mercaptan.   (octadecyl mercaptan).
$CH_3(CH_2)_{17}SH$.

Properties: Solid, mp 25C, bp 205–209C (11 mm), d 0.8420 (25/4C), refr index 1.4591 (34C).

Grade: 95% (min) purity.

Hazard: Strong irritant.

Use: Organic intermediate, synthetic rubber processing.

stearyl methacrylate.   Group name for
$CH_2{:}C(CH_3)COO(CH_2)_n CH_3$ (n = 13 to 17).

Properties: Bulk d 0.864 (25/15C), boiling range 310–370C.

Use: Monomer for plastics, molding powders, solvent coatings, adhesives, oil additives; emulsions for textile, leather, paper finishing.

stearyl monoglyceridyl citrate.

Properties: A soft, practically tasteless, off-white to tan, waxy solid having a lard-like consistency; insoluble in water, soluble in chloroform and ethylene glycol.

Derivation: Reaction of citric acid, monoglycerides of fatty acids (obtained by the glycerolysis of edible fats and oils or derived from fatty acids) and stearyl alcohol.

Grade: FCC.

Use: Emulsion stabilizer in foods (not over 0.15%).

steatite.   A mixture of talc, clay, and alkaline-earth oxides.

Use: Chiefly as a ceramic insulator in electronic devices.

"Stedbac."[430]   TM for stearyl dimethyl benzyl ammonium chloride, a hair conditioner used primarily in after-shampoo hair rinses.

steel.   An alloy of iron and 0.02–1.5% carbon; it is made from molten pig iron by oxidizing out the excess carbon and other impurities. The open-hearth process, which uses air for this purpose, has been supplanted by the basic oxygen process in which pure oxygen is injected into molten iron. A small percentage of steel is also made in the electric furnace, with iron ore as oxidant. Low-carbon (mild) steels contain 0.02–0.3% carbon; medium-carbon grades from 0.3–0.7% carbon; and high-carbon grades 0.7–1.5%.

There are many special-purpose types of steel in which one or more alloying metals are used, with or without special heat treatment. The most common additives are chromium and nickel, as in the 18-8 stainless steels; these add greatly to corrosion resistance. High-speed and tool steels,

designed primarily for efficient cutting, contain such alloying metals as tungsten, molybdenum, manganese, and vanadium as well as chromium. Cobalt and zirconium are used for construction steels. For further information refer to American Iron and Steel Institute, New York, NY.
Available forms: Rods, bars, sheet, strips, wire, wool.
Use: Construction, ship hulls, auto bodies, machinery and machine parts, cables, abrasive, chemical equipment, belts for tire reinforcement.

steel, stainless. Alloy steels containing high percentages of chromium, from less than 10% to more than 25%. There are three groups: (1) austenitic, which contain both chromium (16% min) and nickel (7% min); a stress-corrosion resistant type contains 2% silicon; (2) ferritic, which contain chromium only and cannot be hardened by heat treatment; (3) martensitic, which contain chromium and can be hardened by heat treatment; these last are ferromagnetic. A subgroup of the martensitic steels comprises the precipitation-hardening types.
See also SAE steel.

Steffen process. A process used in beet sugar manufacture to separate residual sugar from molasses. Based on the formation of insoluble tricalcium saccharate and its subsequent decomposition to sugar in the presence of a weak acid such as carbonic.

Steinbuhl yellow. See barium chromate.

stellarator. A type of torus developed for fusion research. It differs from tokamaks in respect to the method of confining the plasma.
See also torus, fusion, tokamak.

"Stellite."[214] TM for a series of cobalt-chromium-tungsten alloys. There are nine alloys available in the wear-resistant group.
Use: Special castings, hard-facing rods, and metal-cutting tools; high-temperature grades are also produced by the investment casting process for jet engine turbine blades and vanes.

Stengel process. A method of making ammonium nitrate fertilizer from anhydrous ammonia and nitric acid. The fertilizer particles can be made in different sizes according to the use.

Stephen aldehyde synthesis. Preparation of aldehydes from nitriles by reduction with stannous chloride in ether saturated with hydrochloric acid. The intermediate aldimine salts have to be hydrolyzed. The best results are obtained in the aromatic series.

stereoblock polymer. A polymer whose molecule is made of comparatively long sections of identical stereospecific structure, these sections being separated from one another by segments of different structure, as for example, blocks of an isotactic polymer interspersed with blocks of the same polymer with a syndiotactic structure.
See polymer, stereospecific; block polymer.

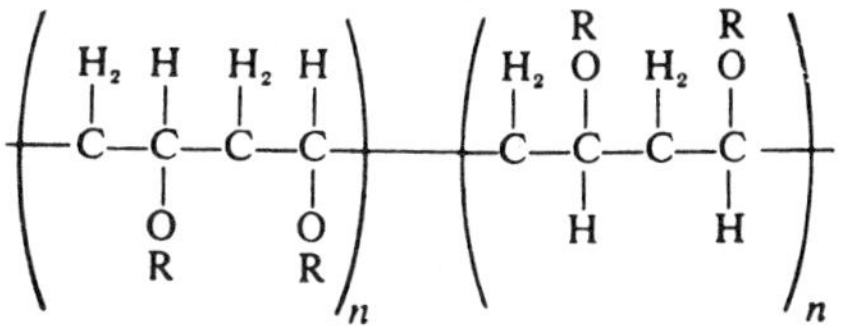

stereoblock polymer

stereochemistry. A subdiscipline of organic chemistry devoted to study of the three-dimensional spatial configurations of molecules. One aspect of the subject deals with stereoisomers-compounds which have identical chemical constitution, but differ as regards the arrangement of the atoms or groups in space. Stereoisomers fall into two broad classes: optical isomers and geometric (cis-trans) isomers. Another aspect of stereochemistry concerns control of the molecular configuration of high polymer substances by use of appropriate (stereospecific) catalysts.
See also optical isomer; enantiomer; asymmetry; polymer, stereospecific; Ziegler catalyst; geometric isomer.

stereoisomer. See stereochemistry.

stereoregular polymer. See polymer, stereospecific.

stereospecific catalyst. See catalyst, stereospecific.

stereospecific polymer. See polymer, stereospecific.

steric hindrance. A characteristic of molecular structure in which the molecules have a spatial arrangement of their atoms such that a given reaction with another molecule is prevented or retarded.

sterilization. (1) Complete destruction of all bacteria and other infectious organisms in an industrial, food or medical product; it must be followed by aseptic packaging to prevent recontamination, usually by hermetic sealing. The methods used involve either wet or dry heat, use of chemicals such as formaldehyde and ethylene oxide filtration (pharmaceutical products), and irradiation by UV or gamma radiation. Milk is sterilized by heating for 30 seconds at 135C, followed by in-can heating at 115C for several minutes and aspetic packaging.

(2) Rendering a lifeform incapable of reproduction by radiation or chemical treatment. See chemosterilant.

sterling silver. According to US law, this term is restricted to silver alloys containing at least 92.5% pure silver and 7.5% maximum other metal (usually copper). See also silver.

steroid. One of a group of polycyclic compounds closely related biochemically to terpenes. They include cholesterol, numerous hormones, precursors of certain vitamins, bile acids, alcohols (sterols), and certain natural drugs and poisons (such as the digitalis derivatives). Steroids have as a common nucleus a fused, reduced 17-carbon-atom ring system, cyclopentanoperhydrophenanthrene. Most steroids also have two methyl groups and an aliphatic side-chain attached to the nucleus. The length of the side-chain varies and generally contains 8, 9, or 10 carbon atoms in the sterols, 5 carbon atoms in the bile acids, 2 in the adrenal cortical steroids, and none in the estrogens and androgens. Steroids are classed as lipids because of their solubility in organic solvents and insolubility in water.

Most of the naturally occurring steroids have been synthesized and many new steroids unknown in nature have been synthesized for use in medicine, such as the fluorosteroids (dexamethasone).

sterol. A steroid alcohol. Such alcohols contain the common steroid nucleus, plus an 8 to 10-carbon-atom side-chain and a hydroxyl group. Sterols are widely distributed in plants and animals, both in the free form and esterified to fatty acids. Cholesterol is the most important animal sterol, ergosterol is an important plant sterol (phytosterol).

"Sterox."[58] TM for a series of nonionic surface-active agents including polyoxyethylene ethers and polyoxyethylene thioethers.

Stevens rearrangement. Migration of an alkyl group from a quaternary ammonium salt to an adjacent carbanionic center on treatment with strong base. The product is a rearranged tertiary amine, or sulfonium or a sulfide.

STH. See somatotropic hormone.

stibic anhydride. See antimony pentoxide.

stibine. See antimony hyride.

stibium. The Latin name for the element antimony; hence the symbol Sb.

stibnite. (gray antimony; antimony glance; antimonite). Sb_2S_3.
Properties: Lead gray mineral, subject to blackish tarnish, metallic luster, soluble in concentrated boiling hydrochloric acid with evolution of hydrogen sulfide, d 4.52–4.62, Mohs hardness 2.
Occurrence: Japan, China, Mexico, Bolivia, Peru, South Africa.
Use: The most important ore of antimony.

Stieglitz rearrangement. Rearrangement of trityl hydroxylamines to Schiff bases on treatment with phosphorous pentachloride.

stigmasterol. $C_{29}H_{48}O \cdot HOH$.   A plant sterol.

Properties: Insoluble in water, soluble in usual organic solvents. Combustible.
Derivation: From soy or calabar beans.
Use: Preparation of progesterone and other important steroids.

stilbene. (toluelene; trans form of α,β-diphenylethylene).   CAS: 103-30-0.
$C_6H_5CH{:}CHC_6H_5$.
Properties: Colorless or slightly yellow crystals, d 0.9707, mp 124–125, bp 306–307C, soluble in benzene and ether, slightly soluble in alcohol, insoluble in water. Combustible.
Derivation: By passing toluene over hot lead oxide.
Method of purification: Crystallization, zone melting used for very pure crystals.
Grade: Technical, pure.
Use: Manufacture of dyes and optical bleaches, crystals are used as phosphors and scintillators.
Note: The *cis* form of α,β-diphenylethylene (isostilbene) is a yellow oil, bp 145C (13 mm), mp 1C.

stilbene dye. A dye whose molecules contain both the —N=N— and the >C=C< chromophor groups in their structure and whose CI numbers range from 40000 to 40999. These are direct cotton dyes.

stilbestrol. See diethylstilbestrol.

stillage. The grain residue from alcohol production used in feeds and feed supplements.

stirrer. See impeller, agitator.

STM. Abbreviation for scanning tunneling microscope.

Stobbe condensation. Condensation of aldehydes or ketones with diethyl succinate in the presence of a strong base to form monoesters of α-alkylidene (or arylidene) succinic acids.

Stock system. See chemical nomenclature.

Stoddard solvent. A widely used dry-cleaning solvent. US Bureau of Standards and ASTM D-484-52 define it as a petroleum distillate clear and free from suspended matter and undissolved water, and free from rancid and objectionable odor. The minimum flash p is 100F (37.7C). Distillation range, more than 50% over at 350F (177C), 90% over at 375F (190C), and the end point below 410F (210C); autoign temperature 450F (232C). Combustible.
Hazard: Fire risk. Toxic by ingestion. TLV: 100 ppm in air.
Use: Dry cleaning, spot and stain removal.

stoichiometry. The branch of chemistry and chemical engineering that deals with the quantities of substances that enter into and are produced by chemical reactions. For example, when methane unites with oxygen in complete combustion, 16 g of methane require 64 g of oxygen. At the same time 44 g of carbon dioxide and 36 g of water are formed as reaction productions. Every chemical reaction has its characteristic proportions. The methods of obtaining these from chemical formulas, equations, atomic weights, and molecular weights and determination of what and how much is used and produced in chemical processes is the major concern of stoichiometry.

Stokes' law. (1) The rate at which a spherical particle will rise or fall when suspended in a liquid medium varies as the square of its radius; the density of the particle and the density and viscosity of the liquid are essential factors. Stokes' law is used in determining sedimentation of solids, creaming rate of fat particles in milk, etc.
(2) In atomic processes, the wavelength of fluorescent radiation is always longer than that of the exciting radiation.

Stolle synthesis. Formation of indole derivatives by the reaction of arylamines with α-haloacid chlorides or oxalyl chloride, followed by cyclization of the resulting amides with aluminum chloride.

"storage." Any method of keeping raw materials, chemicals, food products, and energy while awaiting use, transportation or consumption. The term "storage" is often applied to various types of wastes, but the more accurate word is "disposal."
See radioactive waste, chemical waste, waste control.
(1) *Raw materials.* Normally storage is in suitably protected and well-ventilated interior areas at ambient temperature. Outdoor storage is practicable in some cases, e.g., logs for pulpwood, certain bulk solids and liquids received in metal or fiber drums. Storage of flammable liquids in large underground tanks is standard practice (gasoline, fuel oil). Hygroscopic materials (paper, textiles) should be in a humidity-controlled environment. Combustible materials that tend to build up internal heat on long standing at high ambient temperature (cellulosics such as paper, hay, grain; bulk wool; and certain vegetable oils) should be stored in well-ventilated areas.
(2) *Chemicals.* Materials which may react to form hazardous products in case of spillage should be kept well separated, oxidizing agents (nitrates, peroxides, etc.) should not be stored near reducing or combustible materials. Heat-sensitive materials should be kept away from hot pipes or other heat sources, especially in the case of flammable liquids. Chemicals that will ignite spontaneously in air or react with water vapor require special storage conditions to keep them out of contact with air.
See pyrophoric.
Reactive organic monomers that tend to polymerize at room temperature, e.g., styrene, must contain an inhibitor when stored or shipped.
(3) *Food Products.* Long shelf-life at or near room temperature is highly desirable for processed foods. This is achieved partly by the use of antioxidants and other preservatives and partly by processing techniques. Much experimental work has been done in this field. Refrigerated storage at temperatures near 4.5C is used for meats, eggs, and other dairy products. Meats and quick-frozen foods can be stored indefinitely at or below −18C. Controlled-atmosphere storage to retard post-harvest ripening is used for unprocessed fruits and vegetables.
See also ageing (2). See atmosphere, controlled.
(4) *Energy.* The conventional method of storing energy is by means of primary and secondary batteries.
See dry cell, storage battery.
The growing need for energy conservation has stimulated research on new and more effective methods, especially in regard to solar and wind energy, the collection of which is intermittent or non-uniform. Two such methods are in limited

use. One (for electric power plants) involves compressing air with off-peak electricity and storing it in subterranean cavities from which it can be withdrawn when needed. The other (for domestic use) involves electrically heating refractory bricks at night at off-peak rates; the stored heat is given up during the day with 90% energy recovery. A number of other techniques are in the experimental stage: use of Glauber's salt, which has seven times the heat capacity of water, for storing solar energy; specialized batteries; so-called solar ponds; groundwater heated by solar or industrial process heat and returned to underground storage; and mechanical devices such as flywheel technology.

(5) For information on storage, see chemical data storage.

storage battery. A secondary battery, so called because the conversion of chemical to electrical energy is reversible and the battery is thus rechargable. An automobile battery usually consists of 12–17 cells with plates (electrodes) made of sponge lead (negative plate or anode) and lead dioxide (positive plate or cathode) which is in the form of a paste. The electrolyte is sulfuric acid. The chemical reaction that yields electric current is $Pb + PbO_2 + 2H_2SO_4 \leftrightarrows 2PbSO_4 + 2HOH + 2e$. More complicated and expensive types have nickel-iron, nickel-cadmium, silver-zinc, and silver-cadmium as electrode materials. A sodium-liquid sulfur battery for high-temperature operation as well as a chlorine-zinc type using titanium electrodes have also been developed.

As part of the US effort to replace gasoline with another form of energy, DOE is supporting short- and long-term research on batteries for electric vehicles at Argonne National Laboratories. Three types intended to deliver 20–30 kwh are in the short-term program: improved lead/acid, nickel/zinc and nickel/iron. The long-term program includes lithium/metal sulfide, sodium/sulfur (beta battery), zinc/chlorine, and metal/air. No conclusive results are expected before the late eighties. Independent research indicates that a zinc/nickel oxide system has encouraging possibilities. A lead-acid battery for storing energy from solar cells has been reported to have a life of 5–7 years.

storax. USP name for styrax.

Stork enamine reaction. Synthesis of α-alkyl or α-acyl carbonyl compounds from enamines and alkyl or acyl halides.

STP. Abbreviation for standard temperature and pressure, i.e., 0C and one atmosphere pressure.

STPP. Abbreviation for sodium tripolyphosphate.

straight-chain. See aliphatic.

straight-run. See gasoline.

strain. The resistance to deformation, that is, the tendency to resume its original shape, of a material subjected to a static or dynamic load. Strain increases as a function of stress; at rupture it represents the maximum strength of the material, usually measured in pounds per square inch, or kilograms per square centimeter. The relationship of stress to strain is indicated by a curve obtained by stretching a sample of standard dimensions in a device designed for that purpose. See also stress, modulus of elasticity.

Straits tin. 99.895% pure tin.

strangeness. See antiparticle.

"Stratabond."[526] TM for organic phosphate coatings used for the preparation of steel, galvanized steel, and aluminum for painting. Application by spray, dip, flow-coat, or roller coat.

strategy. A computer-aided approach to a problem in organic synthesis, e.g., in biomimetic chemistry or recombinant DNA. Strategies may be group-oriented, non-oriented, or long-range.

straw. A fibrous, cellulosic component of cereal plants (wheat, rice, etc.). Its fibers are 1–1.5 mm long, similar to those of hardwoods. Straw can be pulped by the alkaline process to yield specialty papers of high quality. Use of straw for papermaking is of limited importance in the US due to the abundance of pulpwood.

streaming potential. A potential resulting from the friction of a liquid passing through a tube or pipe; its strength depends on the viscosity of the liquid, diameter of the pipe, and rate of flow. This can be a factor in pipelines conveying oil and chemicals.

Strecker amino acid synthesis. Synthesis of α-amino acids by simultaneous reaction of aldehydes with ammonia and hydrogen cyanide followed by hydrolysis of the resulting amino nitriles.

Strecker degradation. Interaction of an alpha-amino acid with a carbonyl compound in aqueous solution or suspension to give carbon dioxide and an aldehyde or ketone containing one less carbon atom. Inorganic oxidizing agents can also be used to bring about the reaction.

Strecker sulfite alkylation. Formation of alkyl sulfonates by reaction of alkyl halides with alkali or ammonium sulfites in aqueous solution in the presence of iodide.

streptolin.
Properties: An antibiotic isolated as the hydrochloride. Gummy mass, soluble in water, most stable at pH 3.0–3.5.
Derivation: Produced by *Streptomyces* # 11.
Use: Antibiotic, possible rodenticide.

streptomycin. CAS: 57-92-1. $C_{21}H_{39}N_7O_{12}$.
A specific antibiotic, but the term is also used loosely to designate several chemically related antibiotics produced by actinomycetes. Streptomycin is produced by *Streptomyces griseus* and consists of streptidine attached in glycosidic linkage to the disaccharide, streptobiosamine. It is active against gram-negative bacteria and the tubercle bacillus. Usually available as trihydrochloride, phosphate or sesquisulfate.
Properties: A base which readily forms salts with anions. Quite stable but very hygroscopic.
Units: One unit equals one microgram of pure crystalline streptomycin base.
Derivation: From *Streptomyces griseus* by aerobic fermentation. The streptomycin is then concentrated by adsorption on activated carbon and purified.
Hazard: Damage to nerves and kidneys may result from ingestion. Use restricted by FDA.
Use: Medicine (antibacterial).
See also dihydrostreptomycin.

streptonigrin. (USAN) CAS: 3930-19-6.
$C_{25}H_{22}N_4O_8$. An antibiotic derived from *Streptomyces flocculus*. Dark brown, rectangular crystals.

stress. The deformation undergone by a material when subjected to a definite load (the force applied per unit area). The load may be static (constant) or dynamic (increasing at a uniform rate). In either case it induces a strain in the material which results in rupture if the deforming force exceeds its strength.
See also strain, modulus of elasticity.

stress cracking. (tension cracking).
Development of transverse cracks in a rubber or plastic product exposed to atmospheric oxygen at low (5–10%) elongation for long periods of time, for example, coiled hose, packaging materials, etc., both in service and during storage. Cracking will occur in the absence of light. It can be minimized in the case of a plastic such as polyethylene by lowering the density and the melt index, and in rubber by use of antioxidants.

"Stripolite."[159] TM for a modified sodium sulfoxylate formaldehyde.
Use: Powerful reduction agent to strip dyed colors from wool and synthetics.

stripping. (1) Removal of relatively volatile components from a gasoline or other liquid mixture by distillation, evaporation, or by passage of steam, air or other gas through the liquid mixture. (2) Rapid removal of color from an improperly dyed fabric or fiber by a chemical reaction. Compounds used for this purpose in vat dyeing or in discharge printing are termed discharging agents. Substances commonly used as strippers are sodium hydrosulfite, titanous sulfate, sodium and zinc formaldehyde sulfoxylates.

strontia. See strontium oxide.

strontianite. $SrCO_3$. Natural strontium carbonate.
Properties: Color white, gray, yellow, green; luster vitreous; Mohs hardness 3.5–4; d 3.7.
Occurrence: California, New York, Washington, Germany, Mexico.
Use: Source of strontium chemicals.

strontium. CAS: 7440-24-6. Sr.
Metallic element of atomic number 38, Group IIA of Periodic Table, aw 87.62, valence = 2, radioactive isotopes strontium-89 and strontium-90. There are four stable isotopes.
Properties: Pale yellow, soft metal; chemically similar to calcium. Soluble in alcohol and acids, decomposes water on contact, d 2.54, mp 752, bp 1390C.
Occurrence: Ores of strontianite and celestite (Mexico, Spain).
Derivation: (a) Electrolysis of molten strontium chloride in a graphite crucible with cooling of the upper, cathodic space; (b) thermal reduction of the oxide with metallic aluminum (strontium aluminum alloy formed), and distilling the strontium in a vacuum.
Grade: Technical.
Hazard: Spontaneously flammable in powder form; ignites when heated above its mp; reacts with water to evolve hydrogen; store under naphtha.
Use: Alloys, "getter" in electron tubes.

strontium-90. Radioactive strontium isotope.
Properties: Half-life = 38 years; radiation: beta.
Derivation: From the fission products of nuclear reactor fuels.
Forms available: A mixture containing strontium-90, yttrium-90, and strontium-89 chlorides in hydrochloric acid solution; also as the carbonate and titanate.

Hazard: Highly toxic radioactive poison; present in fallout from nuclear explosions. Absorbed by growing plants; when ingested attacks bone marrow with possibly fatal results. It may be partially removed from milk by treatment with vermiculite.

Use: Radiation source in industrial thickness gauges; elimination of static charge, treatment of eye diseases, in radio-autography to determine the uniformity of material distribution, in electronics for studying strontium oxide in vacuum tubes, activation of phosphors, source of ionizing radiation in luminous paint, cigarette density control, measuring silk density, atomic batteries, etc.

strontium acetate.   CAS: 543-94-2.
$Sr(C_2H_3O_2)_2 \cdot 1/2HOH$.
Properties: White, crystalline powder; soluble in water; loses $1/2HOH$ at 150C.
Derivation: Interaction of strontium hydroxide and acetic acid, followed by crystallization.
Use: Intermediate for strontium compounds, catalyst production.

strontium bromate.   CAS: 14519-18-7.
$Sr(BrO_3)_2 \cdot HOH$.
Properties: Colorless or yellowish crystals, lustrous powder, hygroscopic, soluble in water, d 3.773, loses water at 120C, decomposes at 240C.
Hazard: Strong oxidant, fire risk in contact with organic materials.

strontium bromide.   CAS: 10476-81-0.
$SrBr_2 \cdot 6HOH$.
Properties: White, hygroscopic crystals or powder; soluble in water, alcohol, and amyl alcohol; insoluble in ether; d 2.386 (25/4C); loses 4HOH at 89C, losing remaining water by 180C; mp anhydrous salt 643C.
Derivation: Strontium carbonate is treated with bromine or hydrobromic acid.
Grade: Anhydrous powder, crystals, technical, CP.
Hazard: Toxic by ingestion and inhalation.
Use: Medicine (sedative), lab reagent.

strontium carbonate.   CAS: 1633-05-2.
$SrCO_3$.
Properties: White, impalpable powder; soluble in acids, carbonated water, and solutions of ammonium salts; slightly soluble in water; d 3.62; loses carbon dioxide at 1340C.
Derivation: Celestite is boiled with a solution of ammonium carbonate or is fused with sodium carbonate.
Grade: Precipitated, technical, natural, reagent.
Use: Catalyst, in radiation-resistant glass for color television tubes, ceramic ferrites, pyrotechnics.

strontium chlorate.   CAS: 7791-10-8.
(a) $Sr(ClO_3)_2$; (b) $Sr(ClO_3)_2 \cdot 8HOH$.
Properties: White, crystalline powder; soluble in water; slightly soluble in alcohol. (a) D 3.152, mp decomposes at 120C.
Derivation: Strontium hydroxide solution is warmed and chlorine passed in, with subsequent crystallization.
Grade: Technical, reagent.
Hazard: Dangerous explosion risk in contact with organic materials; highly sensitive to shock, heat, and friction; strong oxidizing agent.
Use: Manufacture of red-fire and other pyrotechnics, tracer bullets.

strontium chloride.   CAS: 101476-85-4.
(a) $SrCl_2$, (b) $SrCl_2 \cdot 6HOH$.
Properties: White, crystalline needles; odorless; sharp, bitter taste; soluble in water and alcohol; (a) d 3.054, mp 872C, bp 1250C; (b) d 1.964, mp loses 6HOH at 150C.
Derivation: Strontium carbonate is fused with calcium chloride, the melt extracted with water, the solution concentrated and crystallized.
Grade: Reagent, technical, anhydrous.
Use: Strontium salts, pyrotechnics, electron tubes.

strontium chromate.   CAS: 7789-06-2.
$SrCrO_4$.
Properties: Light yellow pigment, d 3.84, rust-inhibiting and corrosion-resistant properties, has good heat and light resistance and low reactivity in highly acid vehicles.
Hazard: Toxic by ingestion, a carcinogen.
Use: Metal protective coatings to prevent corrosion, colorant in polyvinyl chloride resins; pyrotechnics, sulfate ion control in electroplating baths.

strontium dioxide.   See strontium peroxide.

strontium fluoride.   CAS: 7783-48-4. SrF_2.
Properties: White powder, soluble in hydrochloric acid and hydrogen fluoride, insoluble in water, d 4.2, mp 1463C, bp 2489C.
Grade: Crystals, 99.2% pure.
Hazard: Toxic material. TLV (as F): 2.5 mg/m³ of air.
Use: Substitute for other fluorides, electronic and optical applications, single crystals for lasers, high-temperature dry-film lubricants.

strontium hydrate.   See strontium hydroxide.

strontium hydroxide.   (strontium hydrate).
CAS: 18480-07-4.   (a) $Sr(OH)_2$, (b) $Sr(OH)_2 \cdot 8HOH$.
Properties: Colorless, deliquescent crystals; soluble in acids and hot water; slightly soluble in

cold water; absorbs carbon dioxide from air; (a) d 3.625, mp 375C, decomposes 710C; (b) d 1.90.
Derivation: Heating the carbonate or sulfide in steam.
Grade: Technical, CP, reagent.
Use: Extraction of sugar from beet sugar molasses, lubricant soaps and greases, stabilizer for plastics, glass, adhesives.

strontium hyposulfite. See strontium thiosulfate.

strontium iodide. CAS: 10476-86-5.
(a) $SrI_2 \cdot 6HOH$. (b) SrI_2.
Properties: (a) White, crystalline plates; decomposes in moist air; becomes yellow on exposure to air or light. (b) White crystals, soluble in water and alcohol. (b) D 4.549, mp 515C, bp decomposes. (a) D 2.67 (25C), mp 90C (decomposes).
Derivation: By treating strontium carbonate with hydriodic acid.
Hazard: Toxic by ingestion and inhalation.
Use: Medicine (source of iodine), intermediate.

strontium lactate. CAS: 29870-99-3.
$Sr(C_3H_5O_3)_2 \cdot 3HOH$.
Properties: White crystals or granular powder; soluble in water, alcohol, ether; odorless.

strontium molybdate. $SrMoO_4$.
Properties: Crystalline powder, d 4, mp 1600C, insoluble in water.
Grade: 99.39% pure, single crystals.
Use: Anticorrosion pigments, electronic and optical applications, single crystals, solid-state lasers.

strontium monosulfide. See strontium sulfide.

strontium nitrate. CAS: 10042-76-9.
$Sr(NO_3)_2$.
Properties: White powder, soluble in water, slightly soluble in absolute alcohol, d 2.98, mp 570C, bp 645C.
Derivation: A concentrated solution of strontium chloride is treated with a solution of sodium nitrate.
Grade: Technical, reagent.
Hazard: Strong oxidizing agent, fire risk in contact with organic materials, may explode when shocked or heated.
Use: Pyrotechnics, marine signals, railroad flares, matches.

strontium nitrite. $Sr(NO_2)_2$.
Properties: White or yellowish powder or hygroscopic needles, soluble in water, insoluble in alcohol, d 2.8, decomposes at 240C.

strontium oxalate. CAS: 814-95-9.
$SrC_2O_4 \cdot HOH$.
Properties: White, odorless powder; loses HOH at 150C; insoluble in cold water.

Hazard: Toxic by ingestion.
Use: Pyrotechnics, catalyst manufacture, tanning.

strontium oxide. (strontia). CAS: 1314-11-0.
SrO.
Properties: Grayish-white, porous lumps. D 4.7, mp 2430C, bp 3000C, soluble in fused potassium hydroxide, converted to hydroxide by water. Combustible.
Derivation: Decomposition of strontium carbonate or hydroxide.
Grade: Lump, powder, porous carbide-free, reagent.
Use: Manufacture of strontium salts, pyrotechnics, pigments, greases and soaps, a major desiccant.

strontium perchlorate. CAS: 13450-97-0.
$Sr(ClO_4)_2$.
Properties: Colorless crystals, very soluble in water and alcohol.
Hazard: Fire and explosion risk in contact with organic materials.
Use: Oxidizing agent, pyrotechnics.

strontium peroxide. (strontium dioxide).
CAS: 1314-18-7. (a) SrO_2 (b) $SrO_2 \cdot 8HOH$.
Properties: White powder; odorless and tasteless; (a) d 4.56, mp decomposes at 215C; (b) d 1.951, loses 8HOH at 100C and decomposes when heated to a higher temperature; soluble in ammonium chloride solution and alcohol; decomposes in hot water; slightly soluble in cold water.
Derivation: (a) By passing oxygen over heated strontium oxide, (b) Reaction of strontium hydroxide and hydrogen peroxide.
Hazard: Dangerous fire and explosion risk in contact with organic materials; sensitive to heat, shock, catalysts, reducing agents; oxidizing agent.
Use: Bleaching, fireworks, antiseptic.

strontium-potassium chlorate. (potassium-strontium chlorate). $Sr(ClO_3)_2 \cdot 2KClO_3$.
Properties: White, crystalline powder; soluble in water.
Grade: Technical.
Hazard: Fire and explosion risk in contact with organic materials; sensitive to heat, shock, catalysts, reducing agents; oxidizing agent.
Use: Pyrotechnics.

strontium salicylate. $Sr(C_7H_5O_3)_2 \cdot 2HOH$.
Properties: White crystals or powder, odorless with sweet saline taste, soluble in water and alcohol, decomposes when heated, protect from light.
Derivation: Interaction of strontium carbonate and salicylic acid.
Use: Pharmaceutical and fine chemical manufacture.

strontium stearate. $Sr(OOCC_{17}H_{35})_2$.
Properties: White powder, mp 130–140C, insoluble in alcohol, soluble (forms gel) in aliphatic and aromatic hydrocarbons.
Use: Grease and wax compounding.

strontium sulfate. CAS: 7759-02-6. $SrSO_4$.
Properties: White precipitate or crystals of the mineral celestite, odorless, slightly soluble in concentrated acids, very slightly soluble in water, insoluble in alcohol and dilute sulfuric acid, d 3.71–3.97, mp 1605C.
Derivation: (a) Celestite is ground, (b) precipitation of any soluble strontium salt with sodium sulfate.
Grade: Natural, precipitated, air-floated, 90%, 325 mesh, free from sodium salts, CP.
Use: Pyrotechnics, ceramics and glass, paper manufacture.

strontium sulfide. (strontium monosulfide).
CAS: 1314-96-1. SrS.
Properties: Gray powder, has hydrogen sulfide odor when in presence of moist air, soluble in acids (decomposes), slightly soluble in water, d 3.7, mp 2000C.
Derivation: Reduction of celestite with coke at high temperatures.
Grade: Technical, high purity.
Hazard: Moderate fire and explosion risk. Irritant to skin and tissues.
Use: Depilatory, luminous paints, manufacture of strontium chemicals.

strontium tartrate. $SrC_4H_4O_6 \cdot 4HOH$.
Properties: White crystals, d 1.966, slightly soluble in water. Combustible.
Use: Pyrotechnics.

strontium thiosulfate. (stontium hyposulfite).
$SrS_2O_3 \cdot 5HOH$.
Properties: Fine needles, soluble in water, insoluble in alcohol, d 2.17, mp loses 4HOH at 100C.

strontium titanate. $SrTiO_3$.
Properties: Powder, d 4.81, mp 2060C, insoluble in water and most solvents.
Grade: Technical, single crystals.
Use: Electronics, electrical insulation.

strontium zirconate. $SrZrO_3$.
Properties: Powder, mp 2600C.
Use: Electronics.

"Structo-Foam."[1] TM for expanded polystyrene foam products used for insulation in buildings, refrigerator cars and trucks and for bouyancy in floats, moorings, etc.

structural antagonist. See antagonist, structural.

structural formula. See formula, chemical.

structure. (1) In the terminology of the textile industry, any woven fabrics. (2) In chemistry, the arrangement of atoms and groups in a molecule.

strychnidine. $C_{21}H_{24}ON_2$.
Properties: Colorless crystals, mp 256C.
Use: Reagent for nitrate determination.

strychnine. CAS: 57-24-9. $C_{21}H_{22}N_2O_2$.
An alkaloid.

Properties: Hard white crystals or powder, bitter taste, mp 268–290C, bp 270C (5 mm), soluble in chloroform, slightly soluble in alcohol and benzene, slightly soluble in water and ether.
Derivation: Extraction of the seeds of *Nux vomica* with acetic acid, filtration, precipitation by alkali and filtration.
Grade: Crystals, powder, technical.
Hazard: Toxic by ingestion and inhalation. TLV: 0.15 mg/m^3 of air.
Use: Poison.

stuff. Term used by paper makers to refer to the aqueous pulp suspension fed onto the fourdrinier wire from the headbox.
See also furnish.

Stuffer disulfone hydrolysis rule. 1. All sulfones containing vicinal sulfone group substituted carbon atoms (γ-disulfones) are hydrolyzed by dilute alkali with formation of β-hydroxysulfone and sulfinate. 2. All sulfones in which adjacent carbon atoms carry a strongly negative group and one or two sulfone residues can be cleaved into a sulfinic acid and an unsaturated compound.

"Stymer" Vinyl Styrene.[58] TM for resins used as sizes for filament acetates. L.F. A vinyl resin, soluble with ammonium hydroxide. sulfur. Styrene copolymer resin, soluble in water.

S-type synthetic elastomer. See styrene-butadiene rubber.

styphnic acid. (2,4,6-trinitroresorcinol).
CAS: 82-71-3. $C_6H(OH)_2(NO_2)_3$.
Properties: Yellow crystals, astringent taste, an initiating explosive, mp 179–180C, forms addi-

tion compounds with many hydrocarbons, soluble in alcohol and ether, slightly soluble in water.
Derivation: Nitration of resorcinol.
Hazard: Severe explosion risk when heated.
Use: Priming agents in the explosives industry.

styracin. See cinnamyl cinnamate.

"Styrafil."[539] TM for polystyrene with glass fiber reinforcement.

styralyl acetate. See α-methylbenzyl acetate.

styralyl alcohol. See α-methylbenzyl alcohol.

styrax. A type of balsam found in Central America and the Near East.
See balsam.

styrenated oil. A drying oil whose drying and hardening characteristics have been modified by incorporation of styrene or a similar monomer.

styrene. See polystyrene, styrene monomer.

styrene monomer. (vinylbenzene; phenylethylene; cinnamene). CAS: 100-42-5.
$C_6H_5CH:CH_2$. 20th highest-volume chemical produced in US (1985).
Properties: Colorless, oily liquid; aromatic odor; fp −30.63C; bp 145.2C; d 0.9045 (25/25C); bulk d 7.55 lbs/gal (20C); flash p 88F (31.1C); autoign temperature 914F (490C). Insoluble in water, soluble in alcohol and ether, readily undergoes polymerization when heated or exposed to light or a peroxide catalyst. The polymerization releases heat and may become explosive.
Derivation: From ethylene and benzene in the presence of aluminum chloride to yield ethylbenzene, which is catalytically dehydrogenated at 630C to form styrene.
Grade: Technical 99.2%, polymer 99.6%.
Hazard: Flammable, moderate fire risk, explosive limits in air 1.1–6.1%, must be inhibited during storage. Toxic by ingestion and inhalation. TLV: 50 ppm in air.
Use: Polystyrene; SBR, ABS, and SAN resins; protective coatings (Styrene-butadiene latex, alkyds); styrenated polyesters; rubber-modified polystyrene; copolymer resins; intermediate.

styrene-acrylonitrile. See polystyrene.

styrene-butadiene rubber. (SBR). By far the most widely used type of synthetic rubber, its

consumption for all applications is four times that of polybutadiene, its nearest competitor, and 1.5 times that of all other elastomers combined. Its manufacture involves copolymerization of three parts butadiene with one part styrene. These materials are suspended in finely divided emulsion form in a large proportion of water, in the presence of a soap or detergent. Also present in small amounts are an initiator or catalyst which is usually a peroxide, and a chain-modifying agent such as dodecyl mercaptan.
Use: Tires, footwear, mechanical goods, coatings, adhesives, solvent-release sealants, carpet backing.
See also rubber, synthetic, polymerization, free radical.

styrene glycol. CAS: 93-56-1. $C_8H_{10}O_2$.
Properties: Acicular crystals, mp 67C, bp 272C, soluble in water and organic solvents.
Use: Plasticizers.

styrene nitrosite. A compound resulting from the reaction between styrene and nitrogen dioxide and used as a qualitative or quantitative specific test for monomeric styrene in mixtures with other hydrocarbons.

styrene oxide. CAS: 96-09-3.

$C_6H_5CHOCH_2$.
Properties: Colorless to pale straw-colored liquid, boiling range 194.2–195C (5–95%), fp −36.6C, flash p (COC) 180F (82.2C), refractive index 1.5328 (25C), d 1.0469 (25/4C). Miscible with benzene, acetone, ether, and methanol. Combustible.
Hazard: Toxic by ingestion and inhalation.
Use: Highly reactive organic intermediate.

"Styresol."[36] TM for a group of styrenated alkyd resins with air-drying and baking properties and high resistance to gasoline, alkalies, acids, and water.

"Styrofoam."[233] TM for expanded cellular polystyrene (available in colors).
Use: Insulating materials; light-weight materials for boats, toys, etc.; separators in packing containers; airport runways; highway construction; battery cases.

"Styron."[233] TM for polystyrene resins, general purpose, medium and high impact, heat and impact-heat resistant, and light-stabilized resins ("Styron Verelite"). Available in wide range of translucent and opaque colors, as well as natural and crystal.
Use: Packaging, toys, appliance parts, bottle closures and containers, hot and cold drinking cups,

television cabinet backs, automotive components and machine housings, lighting equipment.

styryl carbinol.   See cinnamic alcohol.

suberane.   See cycloheptane.

suberic acid.   (octanedioic acid).
   CAS: 505-48-6. $HOOC(CH_2)_6COOH$.
   Properties: Colorless crystals from water, mp 143C, bp 279C (100 mm) partially soluble in water and ether, soluble in alcohol. Combustible.
   Derivation: Oxidation of oleic acid with nitric acid.
   Use: Intermediate for the synthesis of drugs, dyes, and high polymers.

suberone.   See cycloheptanone.

sublimation.   The direct passage of a substance from solid to vapor without appearing in the intermediate (liquid) state. An example is solid carbon dioxide which vaporizes at room temperature; the conversion may also be from vapor to solid under appropriate conditions of temperature.

subnuclear particle.   A particle either found in the nucleus or observed coming from the nucleus as the result of nuclear reaction or rearrangement, i.e., neutrons, mesons, etc.

substance.   Any chemical element or compound. All substances are characterized by a unique and identical constitution and are thus homogeneous. "A material of which every part is like every other part is said to be homogeneous and is called a substance." (Black and Conant, "Practical Chemistry.")
   See also homogeneous.

substantive dye.   See direct dye.

substituent.   An atom or radical that replaces another in a molecule as the result of a reaction. See substitution.

substitute natural gas.   See synthetic natural gas.

substitution.   The replacement of one element or radical by another as a result of a chemical reaction. Chlorination of benzene to produce chlorobenzene is a typical example; in this case a chlorine atom replaces a hydrogen atom in the benzene molecule.

substrate.   (1) A substance upon which an enzyme or ferment acts. (2) Any solid surface on which a coating or layer of a different material is deposited.

subtilin.   An antibiotic produced by the metabolic processes of a strain of Bacillus subtilis. It is a cyclic polypeptide similar to bacitracin in chemical structure and antibiotic activity, but not as important clinically. Subtilin is active against many gram-positive bacteria, some gram-negative cocci, and some species of fungi. It is a surface tension depressant and its antibiotic action is increased by use of wetting agents.
   Properties: Soluble in water in pH range 2.0–6.0; soluble in methanol and ethanol (up to 80%), insoluble in dry ethanol or other common organic solvents. Relatively stable in acid solutions. Inactivated by pepsin and trypsin and destroyed by light.
   Use: Seed disinfectant, bacteriostat in foods.

succinaldehyde.   (butanedial).   CAS: 638-37-9.
   $OHCCH_2CH_2CHO$.
   Properties: Liquid, d 1.064 (20/4C), bp 169–170C, refr index 1.4254, soluble in water, alcohol, and ether. The name succinaldehyde is often incorrectly used in commerce as a synonym for succinic anhydride.

succinic acid.   (butanedioic acid).
   CAS: 110-15-6. $CO_2H(CH_2)_2CO_2H$.
   Properties: Colorless crystals, slightly soluble in water, soluble in alcohol and ether, odorless, acid taste, d 1.552, mp 185C, bp 235C. Combustible.
   Derivation: Fermentation of ammonium tartrate.
   Grade: Technical, CP, FCC.
   Use: Organic synthesis, manufacture of lacquers, dyes, esters for perfumes, photography, in foods as a sequestrant, buffer, neutralizing agent.

succinic acid-2,2-dimethylhydrazide.
   (diaminozide).
   $(CH_3)_2NNHCOCH_2CH_2COOH$.
   Properties: White crystals, mp 155C, pH 3.8 (500 ppm), soluble in water, insoluble in simple hydrocarbons.
   Use: Growth retardant used in greenhouses, retards premature fruit drop.

succinic acid peroxide.   CAS: 123-23-9.
   $(HOOCCH_2CH_2CO)_2O_2$.
   Properties: Fine, white, odorless powder; mp 125C (decomposes). Moderately soluble in water, insoluble in petroleum solvents and benzene.
   Hazard: Fire risk in contact with combustible materials, oxidizing agent. Toxic by ingestion and inhalation, irritant to skin.
   Use: Polymerization catalyst, deodorants, antiseptics.

succinic anhydride. (2,5-diketotetrahydrofurane; succinyl oxide; butanedoic anhydride).

CAS: 108-30-5. $H_2\overline{CC(O)OC(O)C}H_2$.
Properties: Colorless or light-colored needles or flakes, d 1.104 (20/4C), mp 120C, bp 261C, soluble in alcohol and chloroform, insoluble in water, sublimes at 115C at 5 mm pressure. Combustible.
Grade: Distilled.
Use: Manufacture of chemicals, pharmaceuticals, esters; hardener for resins, starch modifier in foods.

succinimide. (2,5-diketopyrrolidine).

CAS: 123-56-8. $H_2\overline{CC(O)NHC(O)C}H_2$ or $C_4H_5O_2N \cdot HOH$.
Properties: Colorless crystals or thin, light tan flakes, nearly odorless, sweet taste, mp 125–127C, bp 287–288C, d 1.41, soluble in hot and cold water and in sodium hydroxide solution, slightly soluble in alcohol, insoluble in ether and chloroform.
Derivation: Conversion of succinic acid to succinamide, followed by heating; ammonia splits off to give a diacyl-substituted derivative (succinimide).
Grade: Purified, technical.
Use: Growth stimulants for plants, organic synthesis.

succiniodimide. See N-iodosuccinimide.

succinonitrile. See ethylene cyanide.

succinyl oxide. See succinic anhydride.

sucrase. See invertase.

sucroblanc. A mixture used to defecate and bleach sugar solution in one operation. Contains high-test calcium hypochlorite, calcium superphosphate, lime, and Filtercel (a grade of "Celite").

sucrose. (table sugar; saccharose).
CAS: 57-50-1. $C_{12}H_{22}O_{11}$.

Properties: Hard, white, dry crystals, lumps, or powder; sweet taste; odorless. Soluble in water, slightly soluble in alcohol, solutions are neutral to litmus, d 1.5877, decomposes 160–186C, optical rotation +33.6 degrees. Combustible.
Derivation: By crushing and extraction of sugar cane (Saccharum officinarum) with water or extraction of the sugar-beet (beta vulgaris) with water, evaporating, and purifying with lime, carbon, and various liquids. Also obtainable from sorghum by conventional methods. Occurs in low percentages in honey and maple sap.
Grade: Reagent, USP, technical, refined.
Use: Sweetener in foods and soft drinks, manufacture of syrups, source of invert sugar, confectionery, preserves and jams, demulcent, pharmaceutical products, caramel, chemical intermediate for detergents, emulsifying agents, and other sucrose derivatives.
See also sugar.

sucrose acetate isobutyrate. (SAIB).
$(CH_3COO)_2C_{12}H_{14}O_3[OOCCH(CH_3)_2]_6$.
Properties: Clear sucrose derivative available either as a semi-solid (100%) or as a 90% solution in ethanol, d 1.146 (25/25C). Combustible.
Derivation: By controlled esterification of sucrose with acetic and isobutyric anhydrides.
Use: Modifier for lacquers, hot-melt coating formulations, extrudable plastics.

sucrose monostearate.
Properties: Odorless, tasteless, white powder.
Derivation: By reaction of sugar and methyl stearate in a suitable solvent and with potassium carbonate catalyst.
Use: Low foam nonionic detergent, surfactants.

sucrose octaacetate. CAS: 126-14-7.
$C_{12}H_{14}O_3(OOCCH_3)_8$.
Properties: White, hygroscopic, crystalline material; bitter taste. When once melted does not recrystallize on cooling but becomes a transparent film. Molten film is very adhesive but this property is lost on cooling. Rate of hydrolysis, practically nil. Gives no action with Fehling's solution. Soluble in acetic acid, acetone, benzene; slightly soluble in water. D (fused) 1.28 (20/20C), bp 260C (0.1 mm), mp 79–84C, refr index 1.4660 (20C), decomposes at 285C, viscosity 29.5 poises (100C), specific rotation +54.96 degrees (CCl4). Combustible.
Grade: Technical, reagent, denaturing.
Use: Plasticizer for cellulose esters and plastics, adhesive and coating compositions, insecticide, termite repellent, denaturant in rubbing alcohol formulas, lacquers, flavoring.

sucrose polyester. An experimental, synthetic, noncaloric fat substitute comprised of a mixture of 6-, 7-, and 8-fatty acid esters of sucrose. These esters are not decomposed by the digestive enzymes, as are the triesters in natural fats; thus they are not metabolized and can be recovered from the excreta. The physical properties of sucrose polyester are similar to those of fats occurring in milk, margarine, and vegetable oils. It has been found effective in notably reducing body weight in preliminary trials when substituted for natural dietary fats in salad dressings, spreads, and milk mixture. Sucrose polyester also has a beneficial effect on serum cholesterol levels. It has met all FDA test requirements satisfactorily.

"Suganilla."[342] TM for concentrated extractives of vanilla beans adsorbed on sugar for food-flavoring.

sugar. A carbohydrate product of photosynthesis comprised of one, two, or more saccharose groups. The monosaccharide sugars (often called simple sugars) are composed of chains of 2–7 carbon atoms. One of the carbons carries aldehydic or ketonic oxygen which may be combined in acetal or ketal forms. The remaining carbons usually have hydrogen atoms and hydroxyl groups. Chief among the monosaccharides are glucose (dextrose) and fructose (levulose). These are optical isomers of formula $C_6H_{10}O_5$ i.e., their crystals have the property of rotating to either left or right under polarized light. Hence the alternate names dextrose and levulose; see also D and optical rotation.

Among the disaccharides are sucrose (cane or beet sugar); lactose, found in milk; maltose, obtained by hydrolysis of starch; and cellobiose from partial hydrolysis of cellulose. High-polymer sugars occur as water-soluble gums such as arabic, tragacanth, etc. Hydrolysis of sucrose yields invert sugar composed of equal parts fructose and glucose. Sugar is an important source of metabolic energy in foods and its formation in plants is an essential factor in the life process.

sugar-cane wax. A hard wax varying from dark green to tan and brown, produced by solvent extraction of cane, mp 76–79C. Combustible.
Grade: Domestic refined, in slabs.
Use: Carbon paper, pigment disperser, castings, lubricant for plastics in food wrappers.

sugar, corn. See dextrose.

sugar, grape. See dextrose.

sugar, invert. See invert sugar, sugar.

sugar of lead. See lead acetate.

sugar of milk. See lactose.

sugar, reducing. Sugars that will reduce Fehling's solution or similar test liquids, with conversion of the blue soluble copper salt to a red, orange or yellow precipitate of cuprous oxide. Glucose and maltose are typical examples of reducing sugars, their molecules containing an aldehyde group that is the basis for this type of reaction.

sugar substitute. See sweetener, non-nutritive.

sulfa drug. Any of a group of 50 or more compounds characterized by the presence of both sulfur and nitrogen, with high specificity for certain bacteria. Because of their toxicity and often serious side effects their use in treating disease is limited, since the less toxic antibiotics are generally available. Many are used in veterinary medicine. Among the best known are sulfanilamide, sulfadiazine, sulfapyridine, sulfamerazine, sulfasoxazole, and gantricin.
See the Merck Index for a detailed description.

sulfamic acid. CAS: 5329-14-6.

$$H-O-\overset{\displaystyle O}{\underset{\displaystyle O}{\overset{\displaystyle \|}{\underset{\displaystyle \|}{S}}}}-NH_2$$

Properties: White, crystalline solid; non-volatile; non-hygroscopic. D 2.1, mp 205C (decomposes), soluble in water, slightly soluble in organic solvents, odorless, aqueous solutions are highly ionized giving pH values lower than solutions of formic, phosphoric, and oxalic acids; all the common salts (including calcium, barium, and lead) are extremely soluble in water. Combustible.
Grade: Reagent, crystalline, granular.
Hazard: Toxic by ingestion.
Use: Metal and ceramic cleaning, nitrite removal in azo dye operations, gas-liberating compositions, organic synthesis, analytical acidimetric standard, amine sulfamates used as plasticizers and fire retardants, stabilizing agent for chlorine and hypochlorite in swimming pools, bleaching paper pulp and textiles, catalyst for urea-formaldehyde resins, sulfonating agent, pH control, hard-water scale removal, electroplating.

"Sulfan."[50] TM for stabilized sulfuric anhydride, a colorless liquid, mp 17C, bp 45C.
Use: Sulfonating agent used in the production of water-soluble and oil-soluble detergents; also used in the preparation of organic intermediates for pharmaceuticals, dyestuffs, and other products.
Grade: 99.5% of SO_3 min, stabilized to retain a mp of 17C.

sulfanilamide. See sulfa drug.

sulfanilic acid. (p-aminobenzenesulfonic acid; p-anilinesulfonic acid). CAS: 121-57-3. $H_2NC_6H_4SO_3H \cdot HOH$.
Properties: Grayish-white, flat crystals. Soluble in fuming hydrochloric acid, slightly soluble in water, very slightly soluble in alcohol and ether, mp chars at 280–300C. Combustible.
Derivation: By heating aniline with weak fuming sulfuric acid and pouring the reaction product into water.
Grade: Technical, pure, reagent.
Use: Dyestuffs, organic synthesis, medicine, reagent.

m-sulfanilic acid. See metanilic acid.

"Sulfanole."[42] TM for a group of sulfonated detergents used in the textile industry.

sulfate mineral. A mineral in which the basic building block is a sulfur atom linked to four oxygen atoms. Most sulfate minerals are highly soluble in water and form by the evaporation of natural waters.

sulfate pulp. See pulp, paper.

sulfate-reducing bacteria. Fermenting bacteria that obtain energy by converting sulfate compounds into sulfide compounds. These bacteria, which cannot tolerate oxygen, are common in the muds of swamps, ponds, and lagoons. Bacteria of this type seem to have flourished in Archean time, when there was little atmospheric oxygen.

sulfenamide. A compound having the typical structure $RSNH_2$.

sulfide dye. (sulfur dye). One of a group of water-insoluble dyes produced by heating various organic compounds with sulfur. The characteristic chromophore groupings are $\equiv$C—S—C$\equiv$ and $\equiv$C—S—S—C$\equiv$; CI numbers range from 53000–54999. Like vat dyes, sulfide dyes are reduced to a water-soluble, colorless form before application and are then oxidized to their original colored state. Application is usually to cotton from a sodium sulfide bath. Sulfur black (CI No.53185) is an important example.

sulfite acid liquor. An aqueous solution of calcium bisulfite or calcium and magnesium bisulfites containing a large amount of free sulfur dioxide. It is prepared from sulfur dioxide and limestone, dolomite, or lime by countercurrent extraction.
Use: Manufacture of sulfite pulp for paper.

sulfite pulp. See pulp, paper.

sulfite waste liquor. A waste liquor produced in the sulfite paper process. Sold in variations from a dilute solution to a solid.
Use: Foam producer, emulsifier, adhesives, tanning agent, binder for briquets, cores, unpaved roads, source of torula yeast.
See also lignin sulfonate, waste disposal.

m-sulfobenzoic acid.
$HO_3SC_6H_4COOH \cdot 2HOH$.
Properties: Grayish-white solid; stable but hygroscopic in air; mp 98C; anhydrous form melts at 141C; soluble in water, alcohol; insoluble in benzene.
Derivation: Direct sulfonation of benzoic acid with sulfur trioxide.
Use: Derivative for surface-active agents.

o-sulfobenzoic acid. $HO_3SC_6H_4COOH$.
Properties: White needles, soluble in water and alcohol, insoluble in ether, mp 68–69C (with 3HOH of crystallization), mp 134C (dry).
Derivation: (a) From saccharin and concentrated hydrochloric acid, (b) by the oxidation of thiosalicylic acid with potassium permanganate in alkaline solution.
Use: Manufacture of sulfonaphthalein indicators, dyes.

o-sulfobenzoic anhydride. $C_6H_4COOSO_2$.
Properties: Solid, mp 129.5C, bp 184–186C (18 mm). Soluble in hot water, ether, and benzene.
Use: Polymerization inhibitor.

"Sulfobetaines."[590] TM for a group of ionic neutral surfactants, compatible with all other classes of surface-active agents. The "zwitterion" of the unusual series of compounds contains a quaternary ammonium ion and a sulfonic acid ion, arranged in a betaine-like inner salt structure.
Use: Cosmetic soap bars, liquid hand soap, soap-based shampoos, germicidal soap preparations and soap-based laundry preparations.

sulfocarbanilide. See thiocarbanilide.

sulfocarbolic acid. See phenolsulfonic acid.

sulfoderm. Silicic acid with 1% colloidal sulfur.

1-(4-sulfo-2,3-dichlorophenyl)-3-methylpyrazolone.

$(Cl_2C_6H_2SO_3H)NNC(CH_3)CH_2CO$.
Properties: White or yellowish powder or crystals, very soluble in water, soluble in alkalies.
Derivation: By condensation of dichlorophenyl-

hydrazine sulfonic acid with ethylacetoacetate.
Use: Intermediate for dyes.

"Sulfogene."[28] TM for a series of sulfur colors.
Use: Extensively on cotton work-clothing and similar fabrics and to a limited extent on rayon and other materials.

sulfolane. (dapsone; tetrahydrothiophene-1,1-dioxide; tetramethylene sulfone).

CAS: 126-33-0. $CH_2CH_2CH_2CH_2SO_2$.
Properties: White or creamy white, crystalline powder; odorless; slightly bitter taste. Very slightly soluble in water, freely soluble in alcohol, soluble in acetone and dilute mineral acids, melting range 175 to 181C.
Grade: USP (as dapsone, the USAN name), technical.
Hazard: Toxic by ingestion.
Use: Curing agent for epoxy resins, medicine (antibacterial).

4,4'-sulfonyldiphenol. (SDP).
$(C_6H_5OH)_2SO_2$.
Properties: White, crystalline powder; mp above 247C; 99.5% pure.
Derivation: Oxidation of 4,4'-thiodiphenol.
Use: Intermediate, product modification.

p-1-sulfophenyl-3-methyl-5-pyrazolone.

$(C_6H_4SO_3H)NNC(CH_3)CH_2CO$.
Properties: White or yellowish powder, very slightly soluble in water, soluble in alkalies.
Derivation: By condensation of phenylhydrazine sulfonic acid with ethylacetoacetate.

4-sulfophthalic anhydride. $HO_3SC_6H_3(CO)_2O$.
Properties: Reddish-brown syrup, crystallizes partially on long standing, hygroscopic, fluorescent in solution under UV radiation, d 1.62 (25C). Very soluble in water, alcohol; insoluble in ether, benzene. Combustible.
Use: Esters of 4-sulfophthalic acid used in wetting, cleansing, emulsifying, softening, and equalizing agents with textiles. Derivatives have application as surface-active agents.

5-sulfosalicylic acid.
$C_6H_3OH(SO_2OH)COOH \cdot 2HOH$.
Properties: Colorless crystals, colored pink by traces of iron, very soluble in water, mp 120C (anhydrous), decomposes at higher temperatures.
Derivation: Action of sulfuric acid and salicylic acid.
Use: Reagent for albumin, colorimetric reagent for ferric ion, intermediate for dyes, surfactants, catalysts, grease additives.

sulfotepp. (ethyl thiopyrophosphate; tetraethyl dithiopyrophosphate). CAS: 3689-24-5.
$(C_2H_5O)_2P(S)OP(S)(OC_2H_5)_2$.
Properties: Yellow liquid, bp 136–139C (2 mm), slightly soluble in water, soluble in most organic solvents.
Hazard: Toxic by ingestion, inhalation, and skin absorption; cholinesterase inhibitor; use may be restricted. TLV: 0.2 mg/m³ of air.
Use: Insecticide.

sulfoxide. (Generic name for 1,2-(methylenedioxy)-4-[2-(octylsulfinyl)propyl]benzene(n-octyl sulfoxide isosafrole)). CAS: 120-62-7.
$C_{18}H_{28}O_3S$.
Properties: Brown liquid, insoluble in water, slightly soluble in oils, miscible with most organic solvents.
Use: Insecticide synergist.

sulfur. CAS: 7704-34-9. S. Nonmetallic element, atomic number 16, group VIA of periodic system, aw 32.06, valences = 2, 4, 6; four stable isotopes.
Properties: Pure sulfur exists in two stable crystalline forms, α and β, and at least two amorphous (liquid) forms. (a) α-Sulfur: rhombic, octahedral, yellow crystals; stable at room temperature. D 2.06, transition to β form 94.5C, mp (rapid heating) 112.8C, refr index 1.957. (b) β-Sulfur: monoclinic, prismatic, pale yellow crystals slowly changing to α form below 94.5C. D 1.96, mp 119C, bp 444.6C, flash p 405F (207C), autoign temperature 450F (232C), refr index 2.038. Both forms are insoluble in water, slightly soluble in alcohol and ether, soluble in carbon disulfide, carbon tetrachloride, and benzene. Combustible.
Occurrence: Native in Texas, Louisiana, Sicily, Canada (Alberta), Poland, Saudi Arabia, Mexico, Iraq, offshore deposits in Gulf of Mexico, USSR, Japan.
Derivation: Direct mining by Frasch process, low-grade ores by Chemico process, smelter waste gas, sour natural gas, coal, iron pyrites, gypsum, solvent extraction of volcanic ash, petroleum, coke oven gas, photolysis of hydrogen sulfide.
Grade: Technical (lumps, roll, flour), rubber makers, NF (sublimed), crude, refined, precipitated (milk of sulfur), high purity (impurities less than 10 ppm), also available in prilled form.
Hazard: Fire and explosion risk in finely divided form.
Use: sulfuric acid manufacture, pulp and paper manufacture, carbon dilsulfide, rubber vulcanization, detergents, petroleum refining, dyes and chemicals, drugs and pharmaceuticals, explosives, insecticides, rodent repellents, soil conditioner, fungicide, coating for controlled-release

fertilizers, nucleating agent for photographic film, cement sealant, binder and asphalt extender in road paving (up to 40%), base material for low-temperature mortars.

For further information refer to Sulfur Institute, 1725 P St., N.W., Washington, DC.

sulfur-35. Radioactive sulfur of mass number 35.
Properties: Half-life: 87.1 days, radiation: beta.
Derivation: By pile irradiation of elemental sulfur or of various chlorides.
Available forms: Solid elemental sulfur, sulfate in weak hydrochloric acid, barium sulfide in barium hydroxide solution, elemental sulfur in benzene solution, in tagged compounds such as carbon disulfide, chlorosulfonic acid, thiourea, sulfanilamide, thiamine, heparin, insulin, "Sucaryl," etc.
Hazard: Radioactive poison.
Use: Research tool in studying mechanism of rubber vulcanization and polymerization of synthetic rubber, role of sulfur in the coking process and in steel, effect of sulfur on engine wear, sulfur removal in the viscose process, behavior of detergents during ashing, sulfur deposition in diesel engines, action of sulfur in silver plating solutions, protein metabolism, surface active agents and surface phenomena, drug actions, etc.

sulfurated lime. See lime, sulfurated.

sulfurated potash. See potash, sulfurated.

sulfur bromide. (sulfur monobromide). S_2Br_2.
Properties: Yellow liquid, becomes red when exposed to air, decomposed by water, soluble in carbon disulfide, d 2.6 (15C), fp −40C, bp 54C (0.2 mm).
Hazard: Avoid inhalation of fumes, strong irritant to tissue.

sulfur chloride. (sulfur subchloride; sulfur monochloride). CAS: 10025-67-9. S_2Cl_2.
Properties: Amber to yellowish-red, oily, fuming liquid; penetrating odor; soluble in alcohol, ether, benzene, carbon disulfide, and amyl acetate; decomposes on contact with water; d 1.690 (15.5C); fp −80C; bp 138C; flash p 266F (130C). Combustible.
Derivation: By passing chlorine over molten sulfur.
Method of purification: Distillation.
Hazard: Reacts violently with water in closed vessel. Toxic by inhalation and ingestion, strong irritant to tissue. TLV: CL of 1 ppm.
Use: Chemicals (sulfur solvent, acetic anhydride, thionyl chloride, carbon tetrachloride from carbon disulfide, various chlorohydrins from gly-

cerol, glycol, etc.), analytical reagent, rubber vulcanizing, vulcanized oils, purifying sugar juices, military poison, insecticide, hardening soft woods (by treatment with sulfur chloride dissolved in carbon disulfide), pharmaceuticals, sulfur dyes, extraction of gold from its ores.

sulfur dichloride. SCl_2.
Properties: Reddish-brown, fuming liquid with pungent chlorine odor. D 1.638 (15.5C), fp −78C, bp decomposes above 59C on rapid heating, boils near 60C, decomposes in water and alcohol, soluble in benzene, refr index 1.567 (20C).
Derivation: Chlorine is passed into sulfur monochloride to saturation at 6–10C followed by carbon dioxide to drive off the excess chlorine.
Grade: Technical.
Hazard: Toxic by inhalation and ingestion, strong irritant to tissue.
Use: Chlorine carrier or chlorinating agent, rubber vulcanizing, vulcanized oils, purifying sugar juices, sulfur solvent, chloridizing agent in metallurgy, manufacture of organic chemicals and insecticides.

sulfur dioxide. CAS: 7449-09-5. SO_2.
Properties: Colorless gas or liquid with sharp, pungent odor; soluble in water, alcohol, and ether; forms sulfurous acid (H_2SO_3). D 1.4337, liquid at 0C, fp −76.1C, bp −10C, vap press 3.2 atmosphere at 20C, refr index (liquid) 1.410 (24C), an outstanding oxidizing and reducing agent. Noncombustible.
Derivation: (a) By roasting pyrites in special furnaces. The gas is readily liquefied by cooling with ice and salt or at a pressure of three atmospheres. (b) By purifying and compressing sulfur dioxide gas from smelting operations. (c) By burning sulfur.
Grade: Commercial, USP, technical, refrigeration, anhydrous 99.98% min.
Hazard: Toxic by inhalation, strong irritant to eyes and mucous membranes, especially under pressure. Dangerous air contaminant and constituent of smog. Not permitted in meats and other sources of vitamin B_1. TLV: 2 ppm in air, US atmospheric standard 0.140 ppm.
Use: Chemicals (H_2SO_4, salt cake, sulfites, hydrosulfites of potassium and sodium, thiosulfates, alum from shale, recovery of volatile substances), sulfite paper pulp, ore and metal refining, soybean protein, intermediates, solvent extraction of lubricating oils, bleaching agent for oils and starch, sulfonation of oils, disinfecting and fumigating, food additive (inhibition of browning, of enzyme-catalyzed reactions, bacterial growth), reducing agent, antioxidant.

sulfur dye. See sulfide dye.

sulfuretted hydrogen. See hydrogen sulfide.

sulfur hexafluoride. CAS: 2551-62-4. SF_6.
Properties: Colorless gas, fp $-64C$ (sublimes), d (gas) 6.5 g/L, d (liquid) 1.67, specific volume 2.5 cu ft/lb (21.1C), slightly soluble in water, soluble in alcohol and ether, odorless. Noncombustible.
Hazard: TLV: 1000 ppm in air.
Use: Dielectric (gaseous insulator for electrical equipment and radar wave guides).

sulfuric acid. (hydrogen sulfate; battery acid; electrolyte acid).   CAS: 7664-93-9.
H_2SO_4.   By far the most widely used industrial chemical, its production was 79.23 billion pounds in 1985. Highest-volume chemical produced in US (1985).
Properties: Strongly corrosive, dense, oily liquid; colorless to dark brown depending on purity; miscible with water. Very reactive, dissolves most metals; concentrated acid oxidizes, dehydrates, or sulfonates most organic compounds, often causes charring; d of pure material 1.84; mp 10.4C; bp varies over range 315–338C due to loss of sulfur trioxide during heating to 300C or higher.
Note: Use great caution in mixing with water due to heat evolution that causes explosive spattering. Always add the acid to water, *never* the reverse.
Derivation: From sulfur, pyrite (FeS_2), hydrogen sulfide, or sulfur-containing smelter gases by the contact process (vanadium pentoxide catalyst). The first step is combustion of elemental sulfur or roasting of iron pyrites to yield sulfur dioxide. Then follows the critical reaction--catalytic oxidation of sulfur dioxide to sulfur trioxide

$$(SO_2 + \tfrac{1}{2}O_2 \xrightarrow[425°C]{V_2O_5} SO_3)$$

in a four-stage converter at 425–450C with evolution of heat. After cooling to 160C, the sulfur trioxide is absorbed in a circulating stream of 98–99% sulfuric acid, where it unites with the small excess of water in the acid to form more sulfuric acid.

$$(SO_3 + H_2O \xrightarrow{260°C} H_2SO_4)$$

Sulfuric acid can also be made by the "Cat-Ox" process and from gypsum (calcium sulfate).
Grade: Commercial 60 degrees Bé: (d 1.71, 77.7% sulfuric acid); 66 degrees Bé (d 1.84, 93.2% sulfuric acid); 98% (d 1.84); 99% (d 1.84). 100% (d: 1.84) depending on supplier; reagent ACS, CP.
Hazard: Strong irritant to tissue. TLV: 1 mg/m^3 of air.
See note above.

Use: Fertilizers, chemicals, dyes and pigments, etchant, alkylation catalyst, electroplating baths, iron and steel, rayon and film, industrial explosives, lab reagent, nonferrous metallurgy.
See also sulfuric acid, fuming.

sulfuric acid, fuming. (oleum).
CAS: 8014-95-7. $xH_2SO_4 \cdot ySO_3$.   A solution of sulfur trioxide in sulfuric acid.
Properties: Heavy, oily liquid; colorless to dark brown depending on purity; fumes strongly in moist air. Extremely hygroscopic.
Derivation: Sulfur trioxide produced by the contact process is absorbed in concentrated sulfuric acid.
Grade: Commercial (10–70% free SO_3 depending on supplier), CP.
Hazard: Reacts violently with water. Strong irritant to tissue.
Use: Sulfating and sulfonating agent, dehydrating agent in nitrations, dyes, explosives, lab reagent.

sulfuric anhydride. See sulfur trioxide.

sulfuric chloride. See sulfuryl chloride.

sulfuric oxychloride. See sulfuryl chloride.

sulfur monobromide. See sulfur bromide.

sulfur monochloride. See sulfur chloride.

sulfurous acid. CAS: 7782-99-2.   A solution of sulfur dioxide in water. The formula H_2SO_3 is used, but the acid is known only through its salts.
Properties: Colorless liquid, suffocating sulfur odor, d 1.03, unstable, soluble in water.
Derivation: Absorption of sulfur dioxide in water.
Grade: Technical, CP.
Hazard: Toxic by ingestion and inhalation, strong irritant to tissue.
Use: Organic synthesis, bleaching straw and textiles, etc.; paper manufacture, wine manufacture, brewing, metallurgy, ore flotation, medicine (antiseptic), reagent in analytical chemistry, sulfites, as preservative for fruits, nuts, foods, wines, disinfecting ships.

sulfur oxychloride. See thionyl chloride.

sulfur subchloride. See sulfur chloride.

sulfur tetrafluoride. CAS: 7783-60-0. SF_4.
Properties: A gas, fp $-124C$, bp $-40C$, decomposes in water. Noncombustible.
Grade: Available in cylinders at 90–94% purity.
Hazard: High by inhalation, strong irritant to eyes and mucous membranes. TLV: CL of 0.1 ppm in air.

Use: Fluorinating agent for making water and oil repellents and lubricity improvers.

sulfur trioxide. (sulfuric anhydride).
CAS: 7449-11-9. SO_3, $(SO_3)_n$.
Properties: Exists in three solid modifications; α mp 62C, β mp 32.5C, γ mp 16.8C. The α form appears to be the stable form but the solid transitions are commonly slow; a given sample may be a mixture of the various forms and its melting point not constant. The solid sublimes easily.
Derivation: Passing a mixture of SO_2 and oxygen over a heated catalyst such as platinum or vanadium pentoxide.
See sulfuric acid.
Hazard: Oxidizing agent, fire risk in contact with organic materials, an explosive increase in vapor pressure occurs when the α form melts. The anhydride combines with water, forming sulfuric acid and evolving heat. Highly toxic, strong irritant to tissue.
Use: Sulfonation of organic compounds, especially nonionic detergents, solar energy collectors. It is usually generated in the plant where it is to be used.

sulfuryl chloride. (chlorosulfuric acid; sulfonyl chloride; sulfuric chloride; sulfuric oxychloride).
CAS: 7791-25-5. SO_2Cl_2.
Properties: Colorless liquid, pungent odor, rapidly decomposed by alkalies and by hot water, soluble in glacial acetic acid, d 1.667 at 20C, bp 69.2C, fp −54C, vap d 4.6.
Derivation: (a) By heating chlorosulfonic acid in the presence of catalysts. (b) From sulfur dioxide and chlorine in the presence of either activated carbon or camphor.
Grade: Technical.
Hazard: Strong irritant to tissue.
Use: Organic synthesis (chlorinating agent, dehydrating agent, acylating agent) pharmaceuticals, dyestuffs, rubber-base plastics, rayon, solvent, catalyst.

sulfuryl fluoride. CAS: 2699-79-8. SO_2F_2.
Properties: Colorless gas, fp −136.7C, bp −55.4C, slightly soluble in cold water and most organic solvents. Noncombustible.
Hazard: Toxic by inhalation. TLV: 5 ppm in air.
Use: Insecticide, fumigant.

sulphenone. (p-chlorophenyl phenyl sulfone).
CAS: 80-30-2.
Properties: White solid with mp 98C; insoluble in water; soluble in hexane, xylene, carbon tetrachloride, and acetone.
Use: A miticide.

"Sulphon."[203] TM for acid dyestuffs used on wool and silk, fair to good fastness to light, good fast-

ness to washing, etc.; can be used on leather and paper.

sultam acid. (1,8-naphthosultam-2,4-disulfonic acid). $C_{10}H_4(SO_3H)_2NHSO_2$.
Properties: White solid, soluble in water, slightly soluble in alcohol.
Derivation: Sulfonation of 1-naphthylamine-8-sulfonic acid or 1-naphthylamine-4,8-disulfonic acid.
Use: Intermediate for Chicago acid.

sulvanite. (ethyl sulfuryl chloride).
$ClSO_3C_2H_5$.
Properties: Colorless liquid with bp 135C.
Use: Poison gas.

Sumner, James B. (1887-1955) An American biochemist who won the Nobel prize in chemistry in 1946 along with Northrop and Stanley. His work was mainly concerned with enzymes, and he was the first in the field to isolate and crystallize an enzyme and determined it was a protein. He inspired followers to continue research in virus and enzyme study. His education was at Harvard and he taught for many years at Cornell.

sump. A chamber or large container into which waste liquids from processing or mechanical operations are allowed to collect and from which they are removed by a pump.

"Sumstar" 190[212]. TM for a polymeric dialdehyde; a fine, nonvolatile, virtually odorless powder, produced by highly specific periodate oxidation of the anhydroglucose units of starch; bulk d 40–42 lb/cu ft.
Use: Film-forming agents, adhesives, leather, and resins.

"Sundex" Oils[494]. TM for a series of aromatic process and extender oils. Recommended for rubber processing where color is not a problem.
Use: Plasticizers, softeners.

sunflower cake. The presscake resulting from hydraulic press expression of sunflower seeds. Typical analysis: proteins 21.0%, fats 8.5%, fiber 48.9%, water 10.2%, ash 11.4%.
Use: Cattle food, fertilizer ingredient.

sunflower meal. The mealy form assumed by sunflower seeds after crushing and heating. If the oil cake is ground the product again is in this mealy form. It contains 55–60% protein.
Use: Animal feed, fertilizers.

sunflower oil.
Properties: Pale yellow, semidrying oil; mild taste; pleasant odor; soluble in alcohol, ether, chloro-

form, and carbon disulfide; d 0.924–0.926; iodine value 130–135; refr index 1.4611. Combustible. Chief constituents: Mixed triglycerides of linoleic, oleic, and other fatty acids.
Derivation: By expression from the seeds of *Helianthus annuus*.
Method of purification: Filtration.
Grade: Crude, refined.
Use: Modified alkyd resins, soap, edible oil, margarine, shortening, dietary supplement.

"Sunolox."[204] TM for a formulation designed to promote the stability of hydrogen peroxide bleach solutions.
Use: Bleaching of textiles near the neutral point, softness is improved by low silicate content.

"Sunpar."[494] TM for a series of nonstaining paraffinic oils. Recommended for use with ethylene-propylene copolymers.
Use: Plasticizers and softeners.

"Sunproof."[248] TM for a series of protective waxes to prevent static atmospheric cracking in all types of synthetic and natural rubbers.

"Sunsize" 536A.[42] TM for a complex of fiber-reactive, alkyl-substituted, synthetic nitrogen derivatives.
Use: Internal size in textile industry in place of rosin-alum or rosin-alum-wax.

superactinide. See periodic table.

"Super-Alkali."[292] TM for combinations of caustic soda and soda ash with varying percentages of the two basic ingredients.
Use: Heavy-duty cleaning such as in metal cleaning and meat-packing houses.

superalloy. An iron-base, cobalt-base, or nickel-base alloy which combines high-temperature mechanical properties, oxidation resistance and creep resistance to an unusual degree. Constitutions of these alloys are as follows: iron-base: 10–45% nickel, 13–19% chromium, 1.3–6% molybdenum, balance iron. Cobalt-base: 0–26% nickel, 0–26% chromium, 0–15% tungsten, balance cobalt. nickel-base: 55–75% nickel, 10–20% chromium, 0–6% aluminum, 0–5% titanium. All contain less than 0.5% carbon, plus other special ingredients. Superalloys can be used up to 2500F.
Use: Jet engine parts, turbo-superchargers, extreme high-temperature applications.

"Super-Backacite."[36] TM for pure phenolic resins used as coating vehicles.

"Super-Beckamine."[36] TM for melamine-formaldehyde resins used as coating vehicles.

"Super-Sta-Bac."[36] TM for hydrocarbon resins which are polymers of mixed olefins.

supercalender. A machine used for finishing paper. It is a vertical stack of from 3 to 12 or more rolls through which the sheet is threaded. The top and bottom rolls are of chilled cast iron and are larger than the intermediate rolls, several of which are of the fiber type. Overall widths may be as much as 15 ft. Operating speeds range from 600 to 2000 ft per minute. The supercalender imparts gloss and smoothness to high-grade papers, both coated and uncoated, as a result of the interaction of the metal and resilient fiber rolls.
See also fiber roll, calender, paper.

superconductivity. The phenomenon in which certain metals, alloys, and compounds at temperatures usually near absolute zero lose both electrical resistance and magnetic permeability, i.e., have infinite electrical conductivity. Depending upon the substance, the maximum temperature (transition temperature) for the phenomenon is now (1985) 0.5–28K. Superconductivity does not occur in alkali metals, noble metals, ferro- and antiferromagnetic metals. It is well-known in elements having 3, 5, or 7 valence electrons per atom, and is associated with high room-temperature resistivity. A system for transmitting electric current underground by means of superconducting cables has been developed.
See also cryogenics.

supercritical fluid. A dense gas that is maintained above its critical temperature (the temperature above which it cannot be liquefied by pressure). Such fluids are less viscous and diffuse more readily than liquids, and are thus more efficient than other solvents in liquid chromatography.

"Super Dylan."[11] TM for high-density polyethylene, made by the Ziegler or low pressure process.
Use: Molded items, extruded sheets, film, pipe, bottles, wire insulation, filaments, etc.

"Superfine."[1] TM for a flour sulfur air-classified for delivery in two particle sizes, one 93% and the other 98% through a 325 US sieve.
Use: Manufacture of glass, matches, chrome oxide pigments, sulfur cements, stock foods, pyrotechnics and explosives; refining of petroleum; casting magnesium and aluminum.

"Superglo."[108] TM for a completely neutral liquid synthetic detergent, combination of highly active wetting agents and foam stabilizers, soluble in hot and cold water.

"Superloid."[322] TM for ammonium alginate, a hydrophilic colloid.
Use: Suspending, thickening, emulsifying, and stabilizing agent in creaming and bodying of rubber latex products; protective colloid in resin emulsion paints, adhesives, fire-retarding compositions, ceramics, etc.

"Superlume."[248] TM for a super-leveling bright nickel electroplating process on steel stampings, brass, copper, zinc die castings, etc. The materials used are nickel sulfate, nickel chloride, boric acid, and addition agents.

supernatant. A liquid or fluid forming a layer on the surface of another liquid.

superoxide. A compound characterized by the presence in its structure of the O_2^- ion. The O_2^- ion has an odd number of electrons (13) and as a result all superoxide compounds are paramagnetic. At room temperature they have a yellowish color. At low temperature many of them undergo reversible phase transitions accompanied by a color change to white. The stable superoxides are:

sodium superoxide	NaO_2
potassium superoxide	KO_2
rubidium superoxide	RbO_2
cesium superoxide	CsO_2
calcium superoxide	$Ca(O_2)_2$
strontium superoxide	$Sr(O_2)_2$
barium superoxide	$Ba(O_2)_2$
tetramethylammonium superoxide	$(CH_3)_4NO_2$

In these compounds each oxygen atom has an oxidation number of $-1/2$ instead of -2, as a normal oxide.

superpalite. (diphosgene; green cross gas; trichloromethylchloroformate).
$ClCOOCCl_3$.
Properties: D 1.6525, bp 127.5–128C.

"Superpax."[327] TM for 92–94.5% zirconium silicate with bulk d 68 lbs/cu ft, average particle size 5 microns max.
Use: Ceramic glazes and as a filler for resins and rubbers.

superphosphate. (acid phosphate). The most important phosphorus fertilizer, made by the action of sulfuric acid on insoluble phosphate rock (essentially calcium phosphate, tribasic) to form a mixture of gypsum and calcium phosphate, monobasic. A typical composition is $CaH_4(PO_4)_2 \cdot HOH$ 30%, $CaHPO_4$ 10%, $CaSO_4$ 45%, iron oxide, alumina, silica 10%, water 5%.

Typical analysis: Moisture 10–15%, available phosphoric acid (as P_2O_5) 18–21%, insoluble phosphoric acid 0.3–2%, total phosphoric acid (as P_2O_5) 19–23%.
Grade: Based on available P_2O_5.
Use: Fertilizer.
See also triple superphosphate and nitrophosphate.

superphosphoric acid. See polyphosphoric acid.

supersaturation. The condition in which a solvent contains more dissolved matter (solute) than is present in a saturated solution of the same components at equivalent temperature. Such solutions may occur, or can be made, when a saturated solution cools gradually so that nucleating crystals do not form. They are extremely unstable and will precipitate upon addition of even one or two crystals of the solute or upon shaking or other slight agitation. Supersaturated solutions occur in the confectionery industry, e.g., in fudges, maple sugar, etc.

"Supersheen."[292] TM for caustic soda solution containing chelating agent and wetting agent.
Hazard: Strong irritant to skin and tissue.
Use: Bottle washing and food plant sanitation.

"Super-sol."[25] TM for an odorless petroleum naphtha, a rapid-drying, highly purified solvent.
Use: Carrier for insecticides, preparation of odorless paints, cleaning compositions.

"Supralan."[203] TM for metallized acid colors of good fastness and level dyeing properties.

"Supramine." XA.[203] TM for a leather chemical, solubilized sulfur phenol condensate, 75% active.

"Supranol."[203] TM for dyestuffs used on wool and silk; good fastness to light, washing, and sea water; can also be used on leather.

surface. In physical chemistry the area of contact between two different phases or states of matter, e.g., finely divided solid particles and air or other gas (solid-gas); liquids and air (liquid-gas); insoluble particles and liquid (solid-liquid). Surfaces are the sites of the physicochemical activity between the phases that is responsible for such phenomena as adsorption, reactivity, and catalysis. The depth of a surface is of molecular order of magnitude. The term interface is approximately synonymous with surface, but it also includes dispersions involving only one phase of matter, i.e., solid-solid or liquid-liquid.
See also interface, surface area, surface chemistry.

surface-active agent. (surfactant). Any compound that reduces surface tension when dis-

solved in water or water solutions, or which reduces interfacial tension between two liquids, or between a liquid and a solid. There are three categories of surface active agents: detergents, wetting agents, and emulsifiers; all use the same basic chemical mechanism and differ chiefly in the nature of the surfaces involved.
See under above three entries for further information. See also interface, surface chemistry.

surface area. The total area of exposed surface of a finely divided solid (powder, fiber, etc.) including irregularities of all types. Since activity is greatest at the surface, that is, the boundary between the particle and its environment, the larger the surface area of a given substance the more reactive it is. Thus reduction to small particles is a means of increasing the efficiency of both chemical and physical reactions; for example, the coloring effect of pigments is increased by maximum size reduction. Carbon black is notable among solids for its huge surface area (as much as 18 acres/lb for some types); the activity of its surface accounts for its outstanding ability to increase the strength and abrasion resistance of rubber. The capacity of activated carbon to adsorb molecules of gases is due to this factor. Surface area is measured most accurately by nitrogen adsorption techniques.

surface chemistry. The observation and measurement of forces acting at the surfaces of gases, liquids and solids or at the interfaces between them. This includes the surface tension of liquids (vapor pressure, solubility); emulsions (liquid/liquid interfaces); finely divided solid particles (adsorption, catalysis); permeable membranes and microporous materials; and biochemical phenomena such as osmosis, cell function, and metabolic mechanisms in plants and animals. Surface chemistry has many industrial applications, a few of which are air pollution, soaps and synthetic detergents, reinforcement of rubber and plastics, behavior of catalysts, color and optical properties of paints, aerosol sprays of all types, monolayers and thin films, both metallic and organic. Outstanding names in the development of this science are Graham, Freundlich, and W. Ostwald in the 19th Century, and Harkins, Langmuir, LaMer, and McBain in the 20th. See also colloid chemistry.

surface tension. In any liquid, the attractive force exerted by the molecules below the surface upon those at the surface/air interface, resulting from the high molecular concentration of a liquid compared to the low molecular concentration of a gas. An inward pull, or internal pressure, is thus created which tends to restrain the liquid from flowing. Its strength varies with the chemical nature of the liquid. Polar liquids have high surface tension (water = 73 dynes/cm at 20C); nonpolar liquids have much lower values (benzene = 29 dynes/cm, ethanol = 22.3 dynes/cm), thus they flow more readily than water. Mercury, with the highest surface tension of any liquid (480 dynes/cm) does not flow, but disintegrates into droplets.
See also interface, surface-active agent.

surfactant. See surface-active agent.

"Surfactol."[202] TM for a series of castor oil-derived, nonionic surfactants.
Use: Emulsifiers, defoamers, plasticizers, solubilizers for oils, dyes, lubricants, in emulsion paints, pigment dispersions, cosmetics, and polishes.

"Surflo."[236] TM for a series of bactericides and scale inhibitors. "Surflo-B11" is a film-forming amine that acts as a corrosion inhibitor and bactericide for low pH water-base drilling muds. "Surflo-B33" is a bactericide for treating water-base packer fluids or oil field waters.

"Surlyn."[28] TM for a group of ionomer resins.
Properties: ("Surlyn" A) Thermoplastic produced as a granular material; flexible, transparent, grease-resistant; very light-weight but tough. Izod impact strength 5.7–14.6 ft-lb/in (higher than any other polyolefin), tensile strength 3,500–5,500 psi, elongation 300–400%, softening point 71C, insoluble in any commercial solvent, subject to slow swelling by hydrocarbons, to slow attack by acids.
Use: Coatings, packaging films, products made by injection or blow molding, or by thermoforming.

SUS. Abbreviation for Saybolt Universal Seconds.
See Saybolt Universal viscosity.

suspension. A system in which very small particles (solid, semisolid, or liquid) are more or less uniformly dispersed in a liquid or gaseous medium. If the particles are small enough to pass through filter membranes, the system is a colloidal suspension (or solution). Examples of solid-in-liquid suspensions are comminuted wood pulp in water, which becomes paper on filtration; the fat particles in milk; and the red corpuscles in blood. A liquid-in-gas suspension is represented by fog or by an aerosol spray. If the particles are larger than colloidal dimensions they will tend to precipitate, if heavier than the suspending medium, or to agglomerate and rise to the sur-

face, if lighter. This can be prevented by incorporation of protective colloids. Polymerization is often carried out in suspension, the product being in the form of spheres or beads.
See also solution, colloidal; dispersion; emulsion; colloid chemistry.

Svedberg, Theodor. (1884-1971) A Swedish chemist who won the Nobel prize in 1926. Author of "Die Methoden zur Herstellung Kolloider Losungen anorganischer Stoffe". His work included research in colloidal chemistry, molecular size determination, and methods of electrophoresis as well as the development of the ultrcentrifuge for separation of colloidal particles in solution. His education was in Sweden with later work done at the University of Wisconsin before returning to Uppsalla.

Swarts reaction. Fluorination of organic polyhalides with antimony trifluoride (or zinc and mercury fluorides) in the presence of a trace of a pentavalent antimony salt.

sweeten. (1) To add sugar or a synthetic product to foods or beverages to provide a sweet taste (flavor). (2) To deodorize and purify petroleum products by removing sulfur compounds (doctor treatment). (3) In industrial slang, to increase the quality of a low-cost product by adding more expensive ingredients.

sweetener, nonnutritive. A food additive, either natural or synthetic, usually having much greater sweetness intensity than sugar (sucrose), but without its caloric value. In some cases they act as enhancers or potentiators of sweetness. Chief among them is saccharin, a benzoic acid derivative. The cyclamate group was removed from the market by FDA in 1970 because of animal carcinogenicity, though the evidence in respect to human toxicity is controversial.

Increasing research has developed several new noncaloric sweeteners: the dihydrochalcone group of disaccharides, glycyrrhizin (licorice extract), especially its ammoniated derivative, dulcin (4-ethoxyphenylurea) the glycoprotein of "miracle fruit," and a polypeptide from a tropical fruit called the "serendipity berry" said to be 3000 times sweeter than sucrose. "Aspartame" ($C_{14}H_{18}N_2O_5$) has been approved by FDA.

sweet oil. See olive oil.

sweet water. (1) The glycerin and water mixture obtained when fats are split (or hydrolyzed) with water to give fatty acid and glycerin. (2) The washings from char used in sugar refining. (3) In engineering terminology, plain water cooled to just below the freezing point and used to preserve milk and other food products.

swep. (Generic name for methyl-3,4-dichlorocarbanilate; methyl-n-3,4-dichlorophenylcarbamate). $CH_3OOCNHC_6H_3Cl_2$.
Properties: Crystals, mp 113C, insoluble in water and kerosene, soluble in acetone and dimethylformamide.
Hazard: Toxic by ingestion.
Use: Herbicide.

"Sylfat."[296] TM for high-purity, tall oil fatty acids used in protective coatings, soaps, detergents, disinfectants, intermediates, and flotation chemicals.

"Sylgard."[149] TM for a series of silicone resin encapsulants used in electronic assemblies.

"Sylkyd."[149] TM for organosiloxy compositions for use as intermediates in the production of resins and formulating varnishes, paints, and enamels.

"Sylmer."[149] TM for silicone compositions used in finishing textile fabrics and in the preparation and treatment of leather to impart permanent water repellence.

"Syloid."[241] TM for a series of micron-sized silicas.
Use: Flatting agents, lacquer, baking finishes, inks, paper reproduction, adhesives, pharmaceuticals, insulation.

"Sylvamix."[296] TM for an internal additive for concrete masonry. A noncorrosive, nonflammable resin derivative of tall oil, developed principally to reduce water absorption, improve surface texture, and inhibit efflorescence in all types of concrete masonry units and in mortar joints.

sylvan. (alpha-methylfuran). C_5H_6O.
Constituent of wood tar.

"Sylvaros."[296] TM for high purity tall oil rosins.
Use: Sizing of paper and manufacture of resins for protective coatings and food containers.

sylvic acid. See abietic acid.

sylvite. (sylvine). KCl. A natural potassium chloride, contains 43% potassium chloride, 57% sodium chloride, sometimes with up to 0.26% bromide.
Properties: Colorless or white, bluish or reddish in color, streak white, vitreous luster, resembles

rock salt in appearance, d 1.97–1.99, Mohs hardness 2.

Occurrence: West Texas, New Mexico, Europe.

Use: Major source of potassium compounds in the US, fertilizers.

symmetrical compound. A compound derived from benzene in which the hydrogen atom on three alternate carbon atoms has been replaced with another element or group, for example, at the 1, 3, and 5 positions.

symmetry. Arrangement of the constituents of a molecule in a definite and continuously repeated space pattern or coordinate system. It is described in terms of three parameters called elements of symmetry, namely: (1) The center of symmetry, around which the constituent atoms are located in an ordered arrangement; there is only one such center in a molecule, which may or may not be an atom. (2) Planes of symmetry, which represent division of a molecule into mirror-image segments. (3) Axes of symmetry, represented by lines passing through the center of symmetry; if the molecule is rotated it will have the same position in space more than once in a complete 360 degree turn, e.g., the benzene molecule with 6 axes of symmetry requires a 60 degree rotation to return to its identical position.

See also stereochemistry, asymmetry.

"Symmetrel."[28] (amantadine hydrochloride). TM for a synthetic antiviral drug specific for A_2 type virus infection (one type of Asian flu). It is taken orally and acts by preventing the virus from entering cells. It is inactive against B type viruses. Reported to be effective in treatment of Parkinson's disease.

synaptase. See emulsin.

"Synasol."[214] TM for solvent composed of 100 gallons of S.D.1 ethanol denatured with one gallon of methyl isobutyl ketone, one gallon of ethyl acetate (87%), and one gallon of aviation gasoline.

Use: Wherever an alcohol-type solvent is required, except for anti-freeze, where its use is prohibited by the Bureau of Internal Revenue.

"Syncal."[266] TM for synthetic sweetener, saccharin, including sodium and calcium salts and insoluble form, as well as several different particle size distributions.

syndet. Abbreviated term for synthetic detergent. See detergent.

syndiotactic. A type of polymer molecule in which groups of atoms that are not part of the backbone structure are located in some symmetrical and recurring fashion above and below the atoms in the backbone chain, when the latter are arranged so as to be in a single plane.

See polymer, stereospecific.

$$\begin{array}{cccccccccccc} & & & & & & R & & & \\ & H_2 & H & H_2 & O & H_2 & H \\ & | & | & | & | & | & | \\ -C & -C & -C & -C & -C & -C- \\ & | & | & | \\ & O & & H & & O \\ & R & & & & R \end{array}$$

Syndiotactic polymer

syneresis. The contraction of a gel on standing, with exudation of liquid. The separation of serum from a blood clot, or of the whey in milk souring or cheese making are examples.

synergism. A chemical phenomenon (the opposite of antagonism) in which the effect of two active components of a mixture is more than additive, i.e., it is greater than the equivalent volume or concentration of either component alone. A substance that induces this result when added to another substance is called a synergist.

synfuel. Any gaseous or liquid fuel made by the synthesis gas reaction; synthetic natural gas (SNG) and methanol are examples, the latter being the only synfuel commercially available in the US (Othmer, 1983).

See also semisynthetic.

syngas. Shortened term for synthesis gas.

Synge, Richard L. M. (1914-) An Irish mathematician and physicist who won the Nobel prize for chemistry in 1952 along with Archer. His research was on the application of methods of physical chemistry to isolate and analyze proteins with special attention to antibiotic peptides and higher plants. He received his Doctorate from Cambridge.

"Synoca."[108] TM for a group of cleaning compounds including wetting agents, detergents, or alkaline hexametaphosphate.

Use: Continuous conditioning and batch washing of paper machine felts. Paper mill system boilouts. Deinking of secondary fiber.

"Synotol."[555] TM for a series of alkanolamides of fatty acids.

Use: Foam builders, viscosity improvers, wetting agents, detergents, and opacifiers.

syntan. A synthetic organic tanning material made from phenolsulfonic acids and formaldehyde.

Use: Chrome and vegetable tanned leathers. See also tanning.

synthane process. See Fischer-Tropsch process, gasification.

"Synthe-Copal."[36] TM for pale colored rosin ester gums used as coating vehicles and in chewing gums and rubber adhesives.

"Synthemul."[36] TM for a series of acrylate-acrylonitrile; acrylic and styrene acrylic emulsions.

"Synthenol."[64] TM for a series of dehydrated castor oils with fast, controllable rate of bodying, excellent color retention and water resistance. Use: Varnishes, enamels, house paints.

synthesis. Creation of a substance which either duplicates a natural product or is a unique material not found in nature, by means of one or more chemical reactions, or (for elements) by a nuclear change. High temperature and pressures as well as a catalyst are usually required. Though syntheses are more readily achieved with organic compounds because of the great versatility of the carbon atom, extremely important syntheses of inorganic compounds were made in the early years of chemistry. A list of noteworthy syntheses with approximate dates would include the following:

Inorganic		Organic	
sulfuric acid		urea (Wohler)	1828
(chamber)	1746	mauveine (Perkin)	1857
(contact)	1890	Celluloid (Hyatt)	1869
soda ash	1800	ethylbenzene	
(LeBlanc)			
(Solvay)	1861	(Friedel Crafts)	1877
ammonia rayon			
(Haber-Bosch)	1912	(Chardonnet)	1884
plutonium	1940	phenolic resins	
		(Baekeland)	1910
		neoarsphenamine	
		(Ehrlich)	1910
		aldehydes, alcohols	
		(Oxo synthesis)	1920
		methanol	1927
		neoprene	
		(Nieuwland)	1930
		nylon (Carothers)	1935
		SBR rubber	1940
		polyisoprene	1950

The tremendous proliferation of synthetic materials in recent years, especially in the high-polymer field was made possible by the increasingly sophisticated use of catalysts, particularly the Ziegler and Natta stereospecific type.

Synthesis of elements has also occurred; since 1940 all the transuranic elements from 93 to 106,

as well as a vast array of radioisotopes of natural elements, have been created by nuclear bombardment of various types.

See also following entries; resin, synthetic; semisynthetic.

synthesis gas. Any of several gaseous mixtures used for synthesizing a wide range of compounds, both organic and inorganic, especially ammonia. Such mixtures result from reacting carbon-rich substances with steam (steam reforming) or steam and oxygen (partial oxidation); they contain chiefly carbon monoxide and hydrogen, plus low percentages of carbon dioxide and usually less than 2.0% nitrogen. The organic source materials may be biomass, natural gas, methane, naphtha, heavy petroleum oils or coke (coal). The reactions are nickel-catalyzed steam-cracking (reforming) of methane or natural gas ($CH_4 + HOH \rightarrow CO + 3H_2$); partial oxidation of methane, naphtha, or heavy oils; and the water-gas reaction with coke ($C + HOH \rightarrow CO + H_2$). With transition-metal catalysts synthesis gas yields alcohols, aldehydes, acrylic acid, etc., and has many useful reactions with acetylene; it is the basis of the Oxo and Fischer-Tropsch reactions. Synthesis gas from which the carbon monoxide has been removed and which has been adjusted to a ratio of 3 parts hydrogen and 1 part nitrogen is used for ammonia synthesis (nitrogen fixation). Processing with a catalyst at high temperatures and pressures yields ammonia. See also ammonia, anhydrous; Oxo process; Haber; water gas; gasification; Fischer-Tropsch process.

synthetic natural gas. (SNG; syngas). Any manufactured fuel gas of approximately the same composition and Btu value as that obtained naturally from oil fields (85% methane and 15% ethane). There are two major methods of synthesis involving catalysts, high temperature, and high pressure: (1) direct hydrogenation of coal and (2) methanation of synthesis gas, obtained by hydrogenolysis of coal or steam-reforming of naphtha or similar petroleum distillate. Other methods involve pyrolysis of solid wastes and extraction from oil shale and manures. See also gasification, hydrogenolysis.

synthetic resin. See resin, synthetic.

synthine process. See Fischer-Tropsch process.

synthol. Liquid fuel containing fatty acids, water-soluble alcohols, aldehydes, ketones, and esters. Obtained from reduction of carbon monoxide in water-gas at high temperature and pressure with iron as catalyst.

synton. Any of several isomers of prostaglandin from which prostaglandin analogs and intermediates may be derived.

syrosingopine. CAS: 84-36-6. $C_{35}H_{42}N_2O_{11}$. Methyl reserpate ester of syringic acid ethyl carbonate. An analog of reserpine.
Properties: White or slightly yellowish powder, practically insoluble in water, slightly soluble in methanol, soluble in chloroform and acetic acid.
Grade: NF.
Use: Medicine.

syrup. Commercial name for an aqueous solution of cane or beet sugar (sucrose) sold in tank car lots to manufacturers of candy, soft drinks, soda-fountain goods, etc. USP grade is an aqueous solution of cane sugar (85g/100 mL). A viscous liquid with d 1.313.

"Systox."[181] TM for demeton.

"Syton."[58] TM for a series of colloidal silicas dispersed in water.
Use: Anti-soil and anti-slip agents.

T

T. Symbol for tritium, also for tera-.

2,4,5-T. Abbreviation for 2,4,5-trichlorophenoxyacetic acid.

2,4,6,-T. Abbreviation for 2,4,6-trichlorophenol.

Ta. Symbol for tantalum.

tabun. (dimethylphosphoramidocyanidic acid, ethyl ester). CAS: 77-81-6.
$(CH_3)_2NP(O)(C_2H_5O)(CN)$. A nerve gas.
Properties: Liquid, fp −50C, bp 240C, flash p 172F (77.7C), d 1.4250 (20/4C), readily soluble in organic solvents, miscible with water but readily hydrolyzed; destroyed by bleaching powder, generating cyanogen chloride. Combustible.
Hazard: Very toxic by inhalation, cholinesterase inhibitor, a military nerve gas, fatal dose (man) 0.01 mg/kg.

"TAC."[329] TM for tested additive chemical items, satisfactory for food additives and medical uses.

tachysterol. CAS: 115-61-7. $C_{28}H_{44}O$.
Properties: Oil, levorotatory, insoluble in water, soluble in most organic solvents, protect from air.
Use: Medicine, as the dihydrotachysterol.

tackiness. (tack). Property of being sticky or adhesive.

taconite. A low-grade iron ore consisting essentially of a mixture of hematite and silica. It contains 25% iron. Found in the Lake Superior district and western states.

tacticity. The regularity or symmetry in the molecular arrangement or structure of a polymer molecule. Contrasts with random positioning of substituent groups along the polymer backbone, or random position with respect to one another of successive atoms in the backbone chain of a polymer molecule.
See polymer, stereospecific, isotactic.

Tafel rearrangement. Rearrangement of the carbon skeleton of substituted acetoacetic esters to hydrocarbons with the same number of carbon atoms by electrolytic reduction at a lead cathode in alcoholic sulfuric acid.

tagetes. A permissible food additive used to increase the yellow color of the skin and eggs of poultry. It is made from the petals of the Aztec marigold (*Tagetes erecta L.*), either ground to a meal or extracted with hexane, with addition of up to 0.3% ethoxyquin.

tagged atom. A radioactive isotope used in tracing the behavior of a substance in both biochemical and engineering research, e.g., C-14 or I-131. See tracer, label (2).

Tag Closed Cup. See TCC.

Tag Open Cup. See TOC.

tailings. (1) In flour-milling, the product left after grinding and bolting middlings. (2) Impurities remaining after the extraction of useful minerals from an ore. (3) In general, any residue from a mechanical refining or separation process.

"Takamine" Pectinase.[212] TM for an enzyme system which hydrolyzes and depolymerizes pectins over a wide range of conditions.
Use: To produce clear fruit juices, wines, and jellies.

"Takimerse."[212] TM for a product containing diastatic and proteolytic enzymes.
Properties: Dry, fine, white powder; water-soluble; nonflammable; optimum pH 7.0–8.0; optimum temperature 40–45C.
Use: Digestion of albumin and starch-containing stains by immersion, in commercial drycleaning plants.

"Talase."[212] TM for a special enzyme formulation having both amylase and protease activity.
Use: Textile industry.

talc. (talcum; soapstone; steatite).
CAS: 14807-96-6. $Mg_3Si_4O_{10}(OH)_2$ or $3MgO \cdot 4SiO_2 \cdot HOH$. A natural hydrous magnesium silicate. Compact, massive varieties may be called steatite in distinction from the foliated varieties, which are called talc. Soapstone is an impure variety of steatite.
Properties: White, apple green, gray powder; luster pearly or greasy, feel greasy, Mohs hardness 1–1.5 (may be harder when impure), high resistance to acids, alkalies and heat; d 2.7–2.8.
Grade: Crude, washed, air-floated, USP, fibrous (99.5%, 99.95%).
Hazard: Toxic by inhalation.

Use: Ceramics; cosmetics and pharmaceuticals; filler in rubber, paints, soap, putty, plaster, oilcloth; abherent; dusting agent; lubricant; paper; slate pencils and crayons; electrical insulation.
See also magnesium silicate.

tall oil. (tallol; liquid rosin).   A mixture of rosin acids, fatty acids, and other materials obtained by acid treatment of the alkaline liquors from the digesting (pulping) of pine wood; flash p 360F (182C). Combustible.
Derivation: The spent black liquor from the pulping process is concentrated until the sodium salts (soaps) of the various acids separate out and are skimmed off. These are acidified by sulfuric acid. Composition and properties vary widely, but average 35–40% rosin acids, 50–60% fatty acids.
Grade: Crude, refined.
Use: Paint vehicles, source of rosin, alkyd resins, soaps, cutting oils and emulsifiers, driers, flotation agents, oil-well drilling muds, core oils, lubricants and greases, asphalt derivatives, rubber reclaiming, synthesis of cortisone and sex hormones, chemical intermediates.

tallow.   An animal fat containing C_{16} to C_{18}.
Properties: The solidifying points of the different tallows are as follows: from 20–45C for horse fat, 27–38C for beef tallow, 54–56C for stearin and oleo, 32–41C for mutton tallow; d 0.86; refr index 46–49 (40C) (Zeiss); iodine value 193–202; flash p 509F (265C). Combustible.
Derivation: Extracted from the solid fat or "suet" of cattle, sheep or horses by dry or wet rendering.
Chief constituents: Stearin, palmitin, and olein.
Grade: Edible; inedible; beef tallow; mutton tallow; horse fats; acidless; edible, extra.
Use: Soap stock, leather dressing, candles, greases, manufacture of stearic and oleic acids, animal feeds, abherent in tire molds.

"Tamax."[408]   TM for a premium grade, bonded mullite refractory available as brick and special shapes. Manufactured from calcined Indian kyanite or other high purity aluminum silicate materials.
Use: Construction of glass-melting furnaces.

"Tamol."[23]   TM for anionic, polymer-type dispersing agents. Supplied as light-colored powders or aqueous solutions. Effective dispersant for aqueous suspensions of insoluble dyestuffs, polymers, clays, tanning agents, and pigments.
Use: Manufacture of dyestuff pastes, textile printing and dyeing, pigment dispersion in textile backings, latex paints and paper coatings, retanning and bleaching of leather, dye resist in leather dyeing, dispersion of pitch in paper manufacture, pre-floc prevention in the manufacture of synthetic rubber.

"Tamul."[408]   TM for a synthetic mullite produced by sintering selected grades of bauxite. Used in the manufacture of a wide variety of special refractory materials. Also the brand name of some of the products manufactured from "Tamul" synthetic mullite grain.

"Tanak" MRX.[57]   TM for melamine-formaldehyde resin tanning agent used to make pure white leather and for bleaching and filling chrome leather.

"Tanamer."[57]   TM for sodium polyacrylate adhesive for use during the drying of leather.
See acrylate.

tangerine oil.   See citrus peel oil.

tankage. (animal tankage; tankage rough ammoniate).   The product obtained in abattoir byproduct plants from meat scraps and bones, which are boiled under pressure and allowed to settle. The grease is removed from the top and the liquor drawn off. The scrap is then pressed, dried, and sold for fertilizer.
Grade: Based on percentage of ammonia and bone phosphate. A medium grade has 10% ammonia and 20% bone phosphate. Concentrated tankage has had the boiled down tank liquor and press water added to it before drying and runs 15–17% ammonia.
Hazard: Flammable, may ignite spontaneously.

tankage, garbage. (tankage fertilizer).   Garbage treated with steam under pressure, the water and some of the grease removed by pressing, and further grease removed by solvent extraction. Contains 3–4% ammonia, 2–5% phosphoric acid, and 0.50–1.00% potash.
Hazard: Flammable, may ignite spontaneously.
Use: Fertilizer.

"Tannex."[236]   TM for a full substitute for quebracho in the thinning of drilling muds.
Use: Mud thinning when exposed to the high bottom-hole temperatures of deep wells. May be added directly to the mud system, but gives better results when dissolved in caustic soda solution before adding to the mud system.

tannic acid. (gallotannic acid; described as a penta-(m-digalloyl)-glucose).   CAS: 1401-55-4.
$C_{76}H_{52}O_{46}$.   Natural substance widely found in nutgalls, tree barks, and other plant parts. Tannins are known to be gallic acid derivatives. A solution of tannic acid will precipitate albumin.
Tannins are classified according to their behavior on dry distillation into two groups, (1)

condensed tannins, which yield catechol, and (2) hydrolyzable tannins, which yield pyrogallol; (2) comprises two groups on the basis of its products of hydrolysis, glucose and (a) ellagic acid or (b) gallic acid.

Properties: Lustrous, faintly yellowish, amorphous powder, glistening scales, or spongy mass; darkens on exposure to air; odorless; strong, astringent taste; mp decomposes at 210C; soluble in water, alcohol, and acetone; almost insoluble in benzene, chloroform, and ether; flash p 390F (198C); autoign temperature 980F (526C). Combustible.

Derivation: Extraction of powdered nutgalls with water and alcohol.

Grade: Technical, CP, NF, fluffy, FCC.

Hazard: Toxic by ingestion and inhalation.

Use: Chemicals (tannates, gallic acid, pyrogallic acid, hydrosols of the noble metals); alcohol denaturant; tanning; textiles (mordant and fixative); electroplating; galvanoplastics (gelatin precipitant); clarification agent in wine manufacture, brewing and foods, writing inks; pharmaceuticals; deodorization of crude oil; photography; paper (sizing, mordant for colored papers); treatment of minor burns.

tannin. Any of a broad group of plant-derived phenolic compounds characterized by their ability to precipitate proteins. Some are more toxic than others, depending on their source. Those derived from nutgalls are believed to be carcinogens, while those found in tea and coffee may be virtually nontoxic.

See also tannic acid.

tanning. The preservation of hides or skins by use of a chemical, which (1) makes them immune to bacterial attack, (2) raises the shrinkage temperature, and (3) prevents the collagen fibers from sticking together on drying, so that the material remains porous, soft, and flexible. Vegetable tanning is used mostly for sole and heavy-duty leathers. The chief vegetable tannins are water extracts of special types of wood or bark, especially quebracho and wattle. The main active constituent is tannic acid. The tannins penetrate the skin or hides after long periods of soaking, during which the molecular aggregates of the tannin form crosslinks between the polypeptide chains of the skin proteins; hydrogen bonding is an important factor.

In mineral or chrome tanning, the sulfates of chromium, aluminum, and zirconium are used (the last two for white leather); here the reaction is of a coordination nature between the carboxyl groups of the skin collagen and the metal atom. Syntans are also used; these are sulfonated phenol or naphthols condensed with formaldehyde.

Condensation products other than phenol having strong hydrogen-bonding power have been developed. Tannage by any method is a time-consuming and exacting process, requiring careful control of pH, temperature, humidity, and concentration factors.

See also leather. For further information refer to Tanners' Council, University of Cincinnati, Cincinnati, Ohio.

"Tanolin."[582] TM for basic chromium sulfate.

Use: Chrome tanning of leather, colors, compounding.

"Tansul 7."[216] TM for beneficiated magnesium montmorillonite.

Use: Flocculating agent and selective absorbent, primarily in the beverage industry.

"Tansul."[236] TM for a mixture of plant-derived proteolytic enzymes in white powder form or clear liquid form with a hydrolysis potency of 60–75 Activity Units/gram.

Use: Protein hydrolyzing agent in beverages.

tantalic acid anhydride. See tantalum oxide.

tantalic chloride. See tantalum chloride.

tantalite. $(Fe, Hg)(TaNb)_2O_6$. The most important ore of tantalum, found in Canada, Africa, Brazil, Malaysia. When niobium content exceeds that of tantalum, the ore is called columbite.

tantalum. CAS: 7440-25-7. Ta. Element of atomic number 73 in group VB of the periodic system, aw 180.9479, valences = 2, 3, 5, no stable isotopes.

Properties: (a) Black powder. (b) Steel-blue-colored metal when unpolished, nearly a platinum-white color when polished, soluble in fused alkalies, insoluble in acids except hydrofluoric and fuming sulfuric acids. D (a) 14.491, (b) 16.6 (worked metal). Mp 2996C, bp 5425C, tensile strength of drawn wire may be as high as 130,000 psi, refr index 2.05, expansion coefficient 8×10^{-6} over range 20–1500C. Electrical resistance 13.6 microhm-cm (0C), 32.0 (500C).

Occurrence: Canada, Thailand, Malaysia, Brazil.

Derivation: From tantalum potassium fluoride by heating in an electric furnace, by sodium reduction, or by fused salt electrolysis. The powdered metal is converted to a massive metal by sintering in a vacuum. Foot-long crystals can be grown by arc fusion.

Corrosion resistance: 99.5% pure tantalum is resistant to all concentrations of hot and cold sulfuric acid (except concentrated boiling), hydrochloric acid, nitric and acetic acids, hot and cold dilute sodium hydroxide, all dilutions of hot and cold ammonium hydroxide, mine and sea waters, moist sulfurous atmospheres, aqueous solutions of chlorine.

Forms and grades available: Powder 99.6% pure, sheet, rods, wire, ultrapure, single crystals.

Hazard: Dust or powder may be flammable. Toxic by inhalation. TLV: 5 mg/m^3 of air.

Use: Capacitors, chemical equipment, dental and surgical instruments, rectifiers, vacuum tubes, furnace components, high-speed tools, catalyst, getter alloys, sutures and body implants, electronic circuitry, thin-film components.

tantalum alcoholate. (pentaethoxytantalum). ($C_2H_5O)_5Ta$.
Use: Catalyst, intermediate for pure tantalates, preparing thin dielectric films.

tantalum carbide. TaC
Properties: Hard, heavy, refractory, crystalline solid. Mp 3875C, bp 5500C, d 14.5, hardness 1800 kg/sq mm, resistivity 30 microohm cm (room temperature), extremely resistant to chemical action except at elevated temperature.
Derivation: Tantalum oxide and carbon heated at high temperatures.
Use: Cutting tools and dies, cemented carbide tools.

tantalum chloride. (tantalic chloride; tantalum pentachloride).   CAS: 7721-01-9. $TaCl_5$.
Properties: Pale yellow, crystalline powder; highly reactive; decomposed by moist air; keep well stoppered. D 3.7, bp 242C, mp 221C, soluble in alcohol and potassium hydroxide solution.
Grade: Technical.
Use: Chlorination of organic substances, intermediate, production of pure metal.

tantalum disulfide. TaS_2.
Properties: Black powder or crystals, mp above 3000C, insoluble in water, available in 40 micron size.
Use: Solid lubricant.

tantalum fluoride. (tantalum pentafluoride).
CAS: 7783-71-3. TaF_5.
Properties: Deliquescent crystals, mp 97C, d 4.74, soluble in water and nitric acid.
Use: Catalyst in organic reactions.

tantalum nitride. TaN.
Properties: Hexagonal, brown, bronze, or black crystals, d 16.3, mp 3310–3410C, insoluble in water, slightly soluble in aqua regia, nitric acid, hydrogen fluoride.
Grade: Technical, powder.

tantalum oxide. (tantalic acid anhydride; tantalum pentoxide).   CAS: 1314-61-0. Ta_2O_5.
Properties: Rhombic, crystalline prisms; d 7.6; mp 1800C; insoluble in water and acids except hydrogen fluoride.
Derivation: From tantalite, by removal of other metals.
Grade: Technical, optical (99.995% pure), single crystals.
Use: Production of tantalum; "rare-element" optical glass; intermediate in preparation of tantalum carbide; piezoelectric, maser, and laser applications; dielectric layers in electronic circuits.

tantalum pentachloride. See tantalum chloride.

tantalum pentoxide. See tantalum oxide.

tantalum pentafluoride. See tantalum fluoride.

tantalum potassium fluoride. (potassium tantalum fluoride; potassium fluotantalate).
CAS: 16924-00-8. K_2TaF_7.
Properties: White, silky needles; slightly soluble in cold water; quite soluble in hot water.
Hazard: Toxic by inhalation. TLV (as F): 2.5 mg/m^3 of air.
Use: Intermediate in preparation of pure tantalum.

tantiron. A ferrous alloy containing 84.8% iron, 13.5% silicon, 1% carbon, 0.4% manganese, 0.18% phosphorus, 0.05% sulfur. It is resistant to acids.
Use: Chemical equipment.

"Tanz."[48] TM for an ammonium lignin sulfonate used as an extender for vegetable tannin extracts.

tapioca. A product rich in starch, obtained from the tuberous roots of the cassava plant. It is used both directly as a food and as a thickening agent in emulsions for food-products and industrial applications (lithographic inks). In many uses it is a satisfactory substitute for gum arabic.

TAPPI. Abbreviation for Technical Association of the Pulp and Paper Industry.

tar. See coal tar; cigarette tar; pine tar; tar, wood.

tar acid. Any mixture of phenols present in tars or tar distillates and extractable by caustic soda solutions. Usually refers to tar acids from coal tar and includes phenol, cresols, and xylenols. When applied to the products from other tars it should be qualified by the appropriate prefix, e.g., wood-tar acid, lignite tar acid, etc.
Properties: Soluble in alcohol and coal-tar hydrocarbons. Combustible.

Grade: 15–18%, 25–28% and 50–53% phenol.
Hazard: Toxic by ingestion, inhalation, and skin absorption; strong irritant to tissue.
Use: Wood preservative, insecticide for cattle and sheep dipping (dip oils or sheep dip), manufacture of disinfectants.

tar base. A basic nitrogen compound derived from coal tar, such as pyridine, picoline, lutidine, and quinoline.

tar camphor. See naphthalene.

tar camphor, chlorinated. See chloronaphthalene.

tar, dehydrated. (tar, refined).
Properties: Dark brown, thick, viscid liquid. Combustible.
Derivation: Tar from which the water has been driven off.
Grade: Technical.
Hazard: Strong irritant, absorbed by skin.
Use: Water-proofing compounds, roads, medicine.

tare. (1) The weight of a container, wrapper or liner which is deducted in determining the net weight of a material. (2) A weight used in analytical work to offset the weight of a container.

"Tarene."[235] TM for mixed naval stores (pine tar type).
Properties: Dark brown, viscous liquid; d 0.99–1.02; ash 0.5%; benzene-soluble. Combustible.
Use: Plasticizer and softener.

Tarmac. Proprietary preparation of blast furnace slag, refined tar, and other materials.
Use: Road dressing.

tarnish. A reaction product that occurs readily at room temperature between metallic silver and sulfur in any form. The well-known black film that appears on silverware results from reaction between atmospheric sulfur dioxide and metallic silver, forming silver sulfide. It is easily removable with a cleaning compound and is not a true form of corrosion. Plating with a mixture of silver and indium will increase tarnish resistance. Gold will also tarnish in the presence of a high concentration of sulfur in the environment.

tar oil. See creosote, coal-tar.

tar-oil, wood. (pine-tar oil).
Properties: Almost colorless liquid when freshly distilled, turns dark reddish-brown; strong odor and taste; d 0.862–0.872; soluble in ether, chloroform, alcohol, and carbon disulfide; flash p (CC) 144F (62.2). Combustible.

Derivation: Obtained by the destructive distillation of the wood of *Pinus palustris*.
Method of purification: Rectification.
Grade: Technical, rectified, refined.
Use: Paints, waterproofing paper, rubber reclaiming, varnishes, stains, ore flotation, cattle dips, insecticides.

tarragon oil. See estragon oil.

tar, refined. See tar, dehydrated.

tar sands. See oil sands.

tar soap. Soap containing juniper tar oil.

tartar, cream of. See potassium bitartrate.

tartar emetic. See antimony potassium tartrate.

tartaric acid. (dihydroxysuccinic acid).
CAS: 133-37-9. $HOOC(CHOH)_2COOH$.
Properties: Colorless, transparent crystals or white, crystalline powder; odorless; acidic taste; stable in air; soluble in water, alcohol, and ether; d 1.76; mp 170C. It has unusual optical properties. The molecule has two asymmetric carbon atoms which result in four isomeric forms, three of which are the natural d-form and the corresponding l-form, which are optically active, and the inactive meso form:

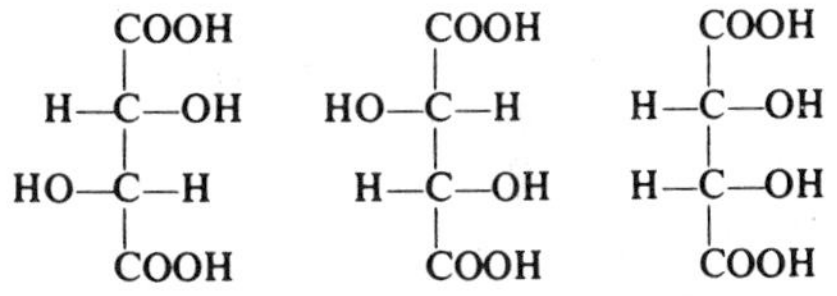

The occurrence of some dl-tartaric acid along with natural d-tartaric acid in the wine industry is explained on the basis of partial racemization. The l-form is the commercial product.
Derivation: Occurs naturally in wine lees; made synthetically from maleic anhydride and hydrogen peroxide and by an enzymatic reaction with a succinic acid derivative.
Grade: Technical, CP, crystals, powder, granular, NF, FCC.
Use: Chemicals (cream of tartar, tartar emetic, acetaldehyde), sequestrant, tanning, effervescent beverages, baking powder, fruit esters, ceramics, galvanoplastics, photography (printing and developing, light-sensitive iron salts), textile industry, silvering mirrors, coloring metals, acidulant in foods.

tartrazine. (FD&C Yellow No. 5; (3-carboxy-5-hydroxy-1-p-sulfophenyl-4-p-sulfophe-

nylazopyrazole trisodium salt; CI No. 19140).
CAS: 1934-21-0.　　$C_{16}H_9N_4O_9S_2$.
Properties: Bright orange-yellow powder, soluble in water.
Hazard: An allergen.
Use: Dye, especially for foods, drugs, and cosmetics.

tar, wood. (tar, hardwood).
Properties: Black, syrup-like, viscous fluid.
Derivation: A byproduct of the destructive distillation of wood.
Grade: Technical.
Hazard: Flammable, moderate fire risk.
Use: Hardwood pitch, wood creosote, heavy high-boiling wood oils, wood preserving oils, paint thinners.

"Tasil."[408] TM for a refractory brick, usually manufactured of calcined Indian kyanite.
Use: Construction of glass and metallurgical furnaces.

taste. See flavor.

Taube, Henry. (1915-) A Canadian born chemist who won the Nobel prize for chemistry in 1983 for his pioneering work in inorganic chemistry and the study of electron transfer reactions, particularly of metal complexes. Known as an outstanding teacher, he is admired and respected by students and colleagues for work at Stanford University.

taurine. (2-aminoethanesulfonic acid).
CAS: 107-35-7.　　$NH_2CH_2CH_2SO_3H$.
A crystallizable amino acid found in combination with bile acids; its combination with cholic acid is called taurocholic acid.
Properties: Solid, decomposes 300C, soluble in water, insoluble in alcohol.
Derivation: Isolated from ox bile, organic synthesis.
Hazard: Toxic by ingestion.
Use: Biochemical research, pharmaceuticals, wetting agents.

taurocholic acid. (cholaic acid; cholyltaurine).
CAS: 81-24-3.　　$C_{26}H_{45}NO_7S$.　　Occurs as sodium salt in bile. It is formed by the combination of the sulfur-containing amino acid taurine, and cholic acid as the sodium salt. It aids in digestion and absorption of fats.
Properties: Crystals, stable in air, mp 125C (decomposes) freely soluble in water, soluble in alcohol, almost insoluble in ether and ethyl acetate.
Derivation: Isolation from bile.
Use: Biochemical research, emulsifying agent in foods (not over 0.1%).

tautomerism. A type of isomerism in which migration of a hydrogen atom results in two or more structures, called tautomers. The two tautomers are in equilibrium. For example, aceto acetic ester has the properties of both an unsaturated alcohol and a ketone. The tautomers are called enol and keto.
See also enol, isomer (1).

"Taycor."[408] TM for a group of corundum-based refractory products. Manufactured of high purity alumina which has been sintered to form tabular corundum.

Tb. Symbol for terbium.

TBH. Abbreviation for technical benzene hexachloride.
See 1,2,3,4,5,6-hexachlorocyclohexane.
Use: An insecticide.

TBP. Abbreviation for tributyl phosphate.

TBT. Abbreviation for tetrabutyl titanate.

TBTO. Abbreviation for bis(tributyltinoxide).

Tc. Symbol for technetium.

TC. Abbreviation for trichloroacetic acid or its sodium salt.

TCA cycle. (tricarboxylic acid cycle; Krebs cycle; citric acid cycle).　　A series of enzymatic reactions occurring in living cells of aerobic organisms, the net result of which is the conversion of pyruvic acid, formed by anaerobic metabolism of carbohydrates, into carbon dioxide and water. The metabolic intermediates are degraded by a combination of decarboxylation and dehydrogenation. It is the major terminal pathway of oxidation in animal, bacterial and plant cells. Recent research indicates that the TCA cycle may have predated life on earth and may have provided the pathway for formation of amino acids.

TCB. Abbreviation for tetracarboxybutane.

TCBO. See trichlorobutylene oxide.

TCC. Abbreviation for Tagliabue Closed Cup, a standard method of determining flash points.

TCDD. See dioxin.

TCP. Abbreviation for tricresyl phosphate.

TDE. (Generic name for 1,1-dichloro-2,2-bis (p-chlorophenyl)ethane; tetrachlorodiphenyleth-

ane; DDD).   CAS: 72-54-8.
$(ClC_6H_4)_2CHCHCl_2$.
Properties: Colorless crystals, mp 109–110C, soluble in organic solvents, insoluble in water, not compatible with alkalies. Combustible.
Derivation: Chlorination of ethanol and condensation with chlorobenzene.
Grade: Technical.
Hazard: Toxic by ingestion, inhalation, and skin absorption; use restricted in some states.
Use: Dusts, emulsions, and wettable powders for contact control of leaf rollers and other insects.

TDI.   Abbreviation for toluene diisocyanate.

TDP.   Abbreviation for 4,4'-thiodiphenol.

TDQP.   Abbreviation for trimethyldihydroquinoline polymer.

Te.   Symbol for tellurium.

TEA.   (1) Abbreviation for triethanolamine. (2) Abbreviation for triethylaluminum.

TEAC.   Abbreviation for tetraethylammonium chloride.

technetium.   CAS: 7440-26-8.   Tc.
Element with atomic number 43 of group VIIB of the periodic system, aw 98.9062, valences = 4, 5, 6, 7, three radioactive isotopes of half-life more than 105 years, also several of relatively short half-life, some of which are beta emitters. Technetium was first obtained by the deuteron bombardment of molybdenum, but since has been found in the fission products of uranium and plutonium.
    The chemistry of technetium has been studied by tracer techniques and is similar to that of rhenium and manganese. The free metal is obtained from reactor fission products by solvent extraction followed by crystallization as ammonium pertechnetate, which is reduced with hydrogen. The metal is silver gray in appearance, mp 2200C (4000F), d 11.5, slightly magnetic. Compounds of the types TcO_2, Tc_2O_7, NH_4TcO_4, etc. have been prepared. The pertechnetate ion has strong anticorrosive properties. Technetium and its alloys are superconductors and can be used to create high-strength magnetic fields at low temperature. Tc-99 (metastable) is the most widely used isotope in nuclear medicine.
Use: Metallurgical tracer, cryochemistry, corrosion resistance, nuclear medicine.

Technical Association of the Pulp and Paper Industry.   (TAPPI).   A professional group of scientists devoted to the interests of pulp and paper chemistry and technology. Founded in 1915, it has seven sections, each concerned with a specific phase of the industry. It also has 11 local sections which hold monthly meetings. The Association publishes its own journal, as well as industry data sheets, bibliographies, technical monographs on subjects relating to the paper industry. It establishes standards of quality and testing procedures. Its headquarters is at 155 E 44th Street, New York, NY, 10017.

"Technoscents."[188]   TM for odor modifiers designed to cover a disagreeable odor associated with solvents, cutting oils, paints, lubricating oils, detergents, scrub soaps and powders, fuel oils.

"Tecmangam."[236]   TM for free-flowing granular solid containing 75–78% manganese sulfate.
Use: Fertilizers and feeds.

"Tedlar."[28]   TM for polyvinylfluoride film.

TEDP.   Abbreviation for tetraethyl dithiopyrophosphate.
See sulfotepp.

"Teflon."[28]   TM for tetrafluoroethylene (TFE) fluorocarbon polymers available as molding and extrusion powders, aqueous dispersion, film, finishes and multifilament yarn or fiber. The name also applies to fluorinated ethylene-propylene (FEP) resins available in the same forms. The no-stick cookware finishes may be of either type. Fibers are monofilaments made from copolymer of TFE and FEP.
Use: Packing, bearings, filters, electrical insulation, high-temperature industrial plastics, cooking utensils, plumbing sealants, coating glass fiber for architectural structure composites, bonding industrial diamonds to metal in manufacture of grinding wheels.
See also fluorocarbon polymer.

TEG.   Abbreviation for tetraethylene glycol and triethylene glycol.

"Tego."[23]   TM for thin tissue impregnated with heat-convertible phenol-formaldehyde resin, supplied in rolls. Produces waterproof bond with plywood veneers.
Use: Hot-press bonding of furniture veneers, premium wall panelling.

TEL.   Abbreviation for tetraethyl lead.

"Telloy."[69]   TM for finely ground tellurium.
Use: Secondary vulcanizing agent for rubber.

"Tellurac."[69]   TM for tellurium diethyldithiocarbamate $[(C_2H_5)_2NC(S)S]_4Te$.
Properties: Orange-yellow powder; d 1.44; melting range 108–118C; soluble in benzene, carbon di-

sulfide, chloroform; slightly soluble in alcohol, gasoline; insoluble in water.
Use: Primary or secondary (with thiazoles) rubber accelerator.

telluric acid. (hydrogen tellurate).
CAS: 7803-68-1.
$H_2TeO_4 \cdot 2HOH$ or H_6TeO_6.
Properties: White, heavy crystals. Soluble in hot water and alkalies, slightly soluble in cold water, d 3.07, mp 136C.
Derivation: Action of sulfuric acid on barium tellurate.
Hazard: As for tellurium.
Use: Chemical reagent.

telluric bromide. See tellurium tetrabromide.

tellurium. CAS: 13494-80-9. Te.
A nonmetallic element with many properties similar to selenium and sulfur. Atomic number 52, group VIA of the period system, aw 127.60, valences = 2, 4, 6, eight stable isotopes.
Properties: Silvery-white, lustrous solid with metal characteristics; d 6.24 g/cc (30C); Mohs hardness 2.3; mp 450C; bp 990C; soluble in sulfuric acid, nitric acid, potassium hydroxide, and potassium cyanide solutions; insoluble in water. Imparts garlic-like odor to breath, can be depilatory. It is a p-type semiconductor and its conductivity is sensitive to light exposure.
Source: From anode slime produced in electrolytic refining of copper and lead.
Derivation: Reduction of telluric oxide with sulfur dioxide; by dissolving the oxide in a caustic soda solution and plating out the metal.
Grade: Powder, sticks, slabs, and tablets, 99.5% pure, crystals up to 99.999% pure.
Hazard (metal and compounds as tellurium): Toxic by inhalation. TLV: 0.1 mg/m³ of air.
Use: Alloys (tellurium-lead, stainless steel, iron castings), secondary rubber vulcanizing agent, manufacture of iron and stainless steel castings, coloring agent in glass and ceramics, thermoelectric devices, catalysts, with lithium in storage batteries for space craft. For further information refer to Selenium-Tellurium Development Assoc., 11 Broadway, New York, NY.

tellurium bromide. See tellurium dibromide and tellurium tetrabromide.

tellurium chloride. See tellurium dichloride.

tellurium dibromide. (tellurium bromide; tellurous bromide). CAS: 7789-54-0. $TeBr_2$.
Properties: Blackish-green, crystalline mass or gray to black needles; very hygroscopic, decomposed by water, soluble in ether, violet vapor, mp 210C, bp 339C.
Hazard: As for tellurium.

tellurium dichloride. (tellurium chloride; tellurous chloride). CAS: 10025-71-5. $TeCl_2$.
Properties: Amorphous, black mass, greenishyellow when powdered, decomposed by water, d 6.9, bp 327C, mp 209C.
Hazard: As for tellurium.

tellurium diethyldithiocarbamate. See "Tellurac."

tellurium dioxide. (tellurous acid anhydride). CAS: 7446-07-3. TeO_2.
Properties: Heavy, white, crystalline powder; odorless; soluble in concentrated acids, alkalies; slightly soluble in dilute acids, water; d 5.89; mp 733C; bp 1245C.
Hazard: As for tellurium.

tellurium disulfide. (tellurium sulfide). TeS_2.
Properties: Red powder, turns in time to a dark brown amorphous powder, fuses to gray lustrous mass, soluble in alkali sulfides, insoluble in acids, water.
Hazard: As for tellurium.

tellurium lead. See lead, tellurium.

tellurium sulfide. See tellurium disulfide.

tellurium tetrabromide. (telluric bromide; tellurium bromide). CAS: 10031-27-3. $TeBr_4$.
Properties: Yellow crystals, soluble in a little water (decomposes in excess water), d 4.3, mp 363C, bp 420C (decomposes into bromine and dibromide).
Hazard: As for tellurium.

tellurous acid. CAS: 10049-23-7. H_2TeO_3.
Properties: White, crystalline powder; soluble in dilute acids, alkalies; slightly soluble in water, alcohol; d 3.053; mp 40C (decomposes).
Hazard: As for tellurium.

tellurous acid anhydride. See tellurium dioxide.

tellurous bromide. See tellurium dibromide.

tellurous chloride. See tellurium dichloride.

telodrin. (1,3,4,5,6,7,8,8-octachloro-3a,4,7,7a-tetrahydro-4,7-methanonaphthalan).
CAS: 297-78-9.
Properties: Vap press 3 mm at 20C; soluble in acetone, benzene, carbon tetrachloride, fuel oil, toluene, xylene.

telomerization reactions. In telomerization reactions, a polymerizable unsaturated compound (the taxogen) is reacted under polymerization conditions in the presence of radical-forming catalysts or promoters with a so-called telogen. During the reaction, the telogen is split into radicals which attach to the ends of the polymerizing taxogen and in some instances add on to the double bond of the taxogen and thereby form chains whose terminal groups are formed of the radicals formed from the telogen. Organic compounds containing an olefinic double bond, such as ethylene, propylene, hexene, octene, or styrene, are normally employed as taxogens. Many different types of compounds can be employed as telogens, for example, halogenated hydrocarbons, such as chloroform or carbon tetrachloride, halogen derivatives of cyanogen, such as cyanogen chloride, aldehydes, alcohols and the like. Radical-forming catalysts, such as organic peroxides, hydrogen peroxide, aliphatic azo compounds of the type of azoisobutyric acid nitrile, and redox systems are employed for telomerization reactions. Telomerization reactions are as a rule carried out at an elevated temperature up to 250 degrees. When volatile reactants are used, the reaction is carried out under elevated pressures, for example, between 20 and 1000 atmospheres.

"Telone."[233] TM for fumigants containing 1,3-dichloropropene and related C_3 hydrocarbons.

"Tel-Tale."[241] TM for silica gel which is impregnated with cobalt chloride and turns from blue to pink as the relative humidity increases.

"Telvar."[29] TM for wettable powder containing 80% monuron, for weed control. "Telvar ML" is a stable liquid suspension containing 28% monuron, used for pre-emergence weed control in cotton.

TEM. Abbreviation for triethylene melamine.

temephos. (O,O,O',O'-tetramethyl-O,O'-thio-di-p-phenylene phosphorothionate).
CAS: 3383-96-8.
Properties: Mp 30–30.5C.
Use: Larvicide for mosquito and blackfly.

"Temex" 15.[304] TM for a barium-cadmium organic vinyl stabilizer.
Properties: Fine, white powder; d 1.3; good dispersion characteristics.
Use: Vinyl phonograph records and other rigid stocks, based on vinyl copolymer resins.

temper. To increase the hardness and strength of a metal by quenching or heat treatment.

temperature. The thermal state of a body considered with reference to its ability to communicate

heat to other bodies (J. C. Maxwell). There is a distinction between temperature and heat, as is evidenced by Helmholtz' definition of heat as "energy that is transferred from one body to another by a thermal process," where by a thermal process is meant radiation, conduction, and/or convection.
Temperature is measured by such instruments as thermometers, pyrometers, thermocouples, etc., and by scales such as centigrade (Celsius), Fahrenheit, Rankine, Reaumur, and absolute (Kelvin).
See also absolute temperature, thermodynamics.

tenacity. Strength per unit weight of a fiber or filament, expressed as grams per denier. It is the rupture load divided by the linear density of the fiber.
See also tensile strength, denier.

"Ten-Cem."[480] TM for a group of neodecanoates of cobalt, copper, calcium, lead, manganese, and zinc.
Use: Driers for printing inks, activators for protective coatings and plastics.

tenderization. (1) Treatment of meats with UV radiation or with certain enzyme preparations to accelerate softening of the collagen fibers and thus reduce the time necessary to "hang" the meat. (2) The degradation and mechanical weakening undergone by textile fibers under excessive wet abrasion during laundering.

"Tenex."[79] TM for an extremely pale heat-treated wood rosin.
Use: Adhesive tape, artificial Burgundy pitch, core coil, cosmetics, disinfectants, electric insulating compounds, emulsions, ester gum, printing ink, rubber cement, shellac diluent, solder, spirit varnishes, synthetic resins, varnishes, Venice turpentine.

"Tenite."[236] TM for cellulose esters and polyolefin thermoplastics. "Tenite" acetate, "Tenite" butyrate, "Tenite" propionate are made from cellulose acetate, cellulose acetate butyrate, and cellulose acetate propionate, respectively. "Tenite" polyethylene and "Tenite" polypropylene are made by the polymerization of ethylene and propylene. "Tenite" polyallomer is a block copolymer of propylene and ethylene; the lightest solid plastic offered commercially. Available in formulations meeting FDA requirements for use in contact with food.

"Tenlo"70.[309] TM for a blend of fatty amides and polyethylene glycol ester.
Use: Pigment grinding oil.

"Tenox."[256] TM for food grade antioxidants containing one or more of the following ingredients:

butylated hydroxyanisole, butylated hydroxytoluene, and/or propyl gallate with or without citric acid. Some formulas are supplied in solvents such as propylene glycol.

"Texo 120."[84] TM for a nonionic surface active agent composed of polyethenoxy tallate.
Properties: Light straw-colored liquid, bland odor, d 1.065–1.070, refr index 1.4762, viscosity 190 cp, heat-stable at 644F (340C), very slightly hygroscopic, pH 7.2 (0.2% solution). Soluble in water, acetone, benzene, ethyl ether, carbon tetrachloride, ethanol, methanol, and xylene.
Use: Detergents, surface cleaners, textile cleaning and dyeing, emulsion formulations.

tensile strength. The rupture strength (stress-strain product at break) per unit area of a material subjected to a specified dynamic load; it is usually expressed in pounds per square inch (psi). This definition applies to elastomeric materials as well as to certain metals.
See tenacity.

tension cracking. See stress-cracking.

tenter. A machine used for holding a processed fabric taut as it is fed into a wind-up or to a cutter. It consists of a frame along the inner sides of which travel continuous chains to which gripping devices are attached at intervals of a few inches; these may be either hooks or clamps of various kinds. As the fabric moves into the machine, the edges are engaged by the grippers and are automatically released at the end of the frame.

tepa. Generic name for tris(1-aziridinyl)phosphine oxide.
See triethylenephosphoramide.

"Tepidone."[28] $(C_4H_9)_2NC(S)SNa$. TM for a water solution of sodium dibutyldithiocarbamate.
Properties: An amber liquid, d 1.08.
Use: To accelerate vulcanization of natural and synthetic rubber and latex compounds.

TEPP. (ethyl pyrophosphate; tetraethyl pyrophosphate). CAS: 107-49-3. $(C_2H_5)_4P_2O_7$.
Properties: Water-white to amber liquid depending on purity; hygroscopic; miscible with water and all organic solvents except aliphatic hydrocarbons; hydrolyzed in water with formation of mono-, di-, and triethyl orthophosphates; water solutions attack metals; commercial material contains 40% tepp. D 1.20, refr index 1.420, bp (pure compound) 135–138C (1 mm).
Derivation: From phosphorus oxychloride and ethanol or phosphorus oxychloride and triethyl phosphate.
Grade: 40%.
Hazard: Toxic by skin contact, inhalation, or ingestion; rapidly absorbed through skin; repeated exposure may, without symptoms, be increasingly hazardous; cholinesterase inhibitor, use may be restricted. TLV: 0.05 mg/m³ of air.
Use: Insecticide for aphids and mites, rodenticide.

tera-. Prefix meaning 10^{12} units (symbol T), 1 Tg = 1 teragram = 10^{12} grams.

teratogen. An agent that causes growth abnormalities in embryos, genetic modifications in cells, etc.; ionizing radiation can have this effect.

terbacil. CAS: 5902-51-2. $C_9H_{13}ClN_2O_2$.
Properties: Colorless crystals, mp 175C, soluble in dimethylacetamide and cyclohexanone, partially soluble in xylene and butyl acetate.
Use: Herbicide.

"Terbec."[233] TM for a wet strength improver for soils; based on 4-tert-butylcatechol.

terbia. See terbium oxide.

terbium. CAS: 7440-27-9. Tb. Atomic number 65, Group IIIB of the periodic table, a rare-earth element of the yttrium subgroup (lanthanide series), aw 158.9254, valences = 3, 4, no stable isotopes.
Properties: Metallic luster, reacts slowly with water and is soluble in dilute acids, d 8.332, mp 1356C, bp 2800C, salts colorless, highly reactive, handled in inert atmosphere or vacuum.
Source: See rare-earths.
Derivation: Reduction of fluoride with calcium.
Grade: Regular, 99.9+% purity (ingots, lumps), single crystals.
Use: Phosphor activator, dope for solid state devices.

terbium chloride, hexahydrate. CAS: 13798-24-8. $TbCl_3 \cdot 6HOH$.
Properties: Transparent, colorless, prismatic crystals; readily soluble in water or alcohol; d 4.35; mp (anhydrous) 588C; very hygroscopic.
Derivation: By treatment of carbonate or oxide with hydrochloric acid in an atmosphere of dry hydrogen chloride.

terbium fluoride. $TbF_3 \cdot 2HOH$.
Properties: Solid, mp 1172C, bp 2280C, insoluble in water.

Hazard: Strong irritant.
Use: Source of terbium.

terbium nitrate. $Tb(NO_3)_3 \cdot 6HOH$.
Properties: Colorless, monoclinic needles or white powder; soluble in water; mp 89.3C.
Derivation: By treatment of oxide, carbonate, or hydroxide with nitric acid.
Hazard: Strong oxidant, fire risk in contact with organic materials.

terbium oxide. (terbia). Tb_2O_3.
Properties: Dark brown powder, soluble in dilute acids, slightly hygroscopic, absorbs carbon dioxide from air.
Derivation: By ignition of hydroxides or salts of oxy-acids.
Grade: 98–99%.
See also rare earth.

terbium sulfate. $Tb_2(SO_4)_3 \cdot 8HOH$.
Properties: Colorless crystals which lose 8HOH at 360C, soluble in water.

terbufos. (O,O-diethyl-S-(tert-butyl)methyl phosphorodithioate). CAS: 13071-79-9. $C_9H_{21}O_2PS_3$. An ester of phosphoric acid.
Properties: Yellowish liquid, d 1.10, bp 70C (0.01 mm), fp −29C, flash p (TOC) 88C, soluble in alcohol, acetone.
Hazard: Moderate fire risk. Toxic by ingestion.
Use: Soil insecticide.

terebene. A mixture of terpenes, chiefly dipentene and terpinene.
Properties: Colorless liquid, d 0.862–0.866, optical rotation inactive; bp 160–172C, soluble in alcohol, insoluble in water.
Derivation: From oil of turpentine.
Hazard: Flammable, moderate fire risk. Toxic by ingestion and inhalation.
Use: To impart water and oil resistance to cellulosics.

terephthalic acid. (p-phthalic acid; TPA; benzene-p-dicarboxylic acid). CAS: 100-21-0. $C_6H_4(COOH)_2$. 21st highest-volume chemical produced in US (1985).
Properties: White crystals or powder; insoluble in water, chloroform, ether, acetic acid; slightly soluble in alcohol; soluble in alkalies; d 1.51; sublimes above 300C. Combustible.
Derivation: (1) Oxidation of p-xylene or of mixed xylenes and other alkyl aromatics (phthalic anhydride). (2) reacting benzene and potassium carbonate over a cadmium catalyst.
Grade: Commercial, fiber.
Use: Production of linear, crystalline polyester resins, fibers, and films by combination with glycols; reagent for alkali in wool; additive to poultry feeds.

2-terephthaloylbenzoic acid.
$HOOCC_6H_4COC_6H_4COOH$.
Properties: White to gray powder, mp 233–237C.
Use: Chemical intermediate.

terephthaloyl chloride. (1,4-benzenedicarbonyl chloride). $C_6H_4(COCl)_2$.

Properties: Colorless needles, mp 82–84C, bp 259C, decomposes in water and alcohol, soluble in ether, flash p 356F (180C). Combustible.
Hazard: Skin irritant.
Use: Dye manufacture; synthetic fibers, resins, films; UV absorption; pharmaceuticals; rubber chemicals; cross-linking agent for polyurethanes and polysulfides.

"Tergitol."[214] TM for a series of nonionic and anionic surfactants.
Use: Detergents, wetting agents, emulsifiers in water systems, leveling and spreading agents.

"Ternalloy."[453] TM for a series of four aluminum-zinc-magnesium alloys. Ternalloys 5 and 6 are self-aging, while Ternalloys 7 and 8 are amenable to heat treatment. Nominal compositions are 0.5% manganese, 0.3% chromium, and 3.0–4.8% zinc, 1.6–2.3% magnesium, dependent on particular alloy. Alloys have good corrosion resistance and excellent machinability in both sand and permanent mold type castings. Supplied in ingot form.

ternary. Descriptive of a solution or alloy having three components, or of a chemical compound having three constituent elements or groups.

tereplate. (roofing tin). A lead-tin alloy used for coating iron or steel; its composition is 75% lead and 25% tin. It has a dull finish.
Use: Roofing, also for deep stamping, gasoline tanks, etc.

1,4(8)-terpadiene. See terpinolene.

terpene. $C_{10}H_{16}$. An unsaturated hydrocarbon occurring in most essential oils and oleoresins of plants. The terpenes are based on the isoprene unit

C_5H_8, and may be either acyclic or cyclic with one or more benzenoid groups. They are classified as monocyclic (dipentene), dicyclic (pinene), or acyclic (myrcene), according to the molecular structure. Many terpenes exhibit optical activity. Terpene derivatives (camphor, menthol, terpineol, borneol, geraniol, etc.) are called terpenoids; many are alcohols.
See also polyterpene resin.

terpeneless oil. An essential oil from which the terpene components have been removed by extraction and fractionation, either alone or in combination. The optical activity of the oil is thus reduced. The terpeneless grades are much more highly concentrated than the original oil (15–30 times). Removal of terpenes is necessary to inhibit spoilage, particularly of oils derived from citrus sources; on atmospheric oxidation the specific terpenes form compounds that impair the value of the oil; for example, d-limonene oxidizes to carvone and γ-terpinene to p-cymene. Terpeneless grades of citrus oils are commercially available.

terpenoid. See terpene.

"Terpex."[296] TM for a polymerized turpentine modified with various additives.
Use: Component of coatings, especially aluminum paints.

"Terpex" Extra.[296] TM for a low molecular weight liquid terpene resin.
Use: Rubber tackifier and plasticizer, adhesives, resin and oil extender.

p-terphenyl. (1,4-diphenylbenzene).
CAS: 92-94-4. $(C_6H_5)_2C_6H_4$.
Properties: Liquid, d 1.234 (0C), mp 213C, bp 405C, flash p 405F (207C). Combustible.
Derivation: From p-dibromobenzene or bromobenzene and sodium.
Method of purification: Zone-melting.
Grade: Technical, scintillation.
Hazard: Toxic by ingestion and inhalation. TLV: CL of 0.5 ppm in air.
Use: Polymerized with styrene to make a plastic phosphor. Single crystals used as scintillation counters.

terpilenol. See terpineol.

terpinene. $C_{10}H_{16}$. A mixture of three isomeric cyclic terpenes, α-, β-, and γ-terpinene. α-terpinene has a bp of 180–182C and d 0.8484 (14C). It is found in cardamom, marjoram, and coriander oils. β-terpinene has a bp of 183C and d 0.853 (15C). It is found in coriander, lemon, cumin, and ajawan oils.
Use: Synthetic flavors.

terpineol. (α-terpineol; β-terpineol; γ-terpineol; terpilenol). CAS: (α) 98-55-5.
$C_{10}H_{17}OH$.
Properties: Colorless liquid or low melting transparent crystals, lilac odor, usually sold commercially as a mixture of the three isomers, d 0.930–0.936 (25C), solidification point 2C, optical rotation between −0 degrees 10 minutes and +0 degrees 10 minutes; boiling range between 214 and 224C, 90% within 5C, refr index 1.4825–1.4850 (20C), soluble in two volumes 70% alcohol, slightly soluble in water, glycerol. Combustible.
Derivation: By heating terpin hydrate with phosphoric acid and distilling or with dilute sulfuric acid, using an azeotropic separation; fractional distillation of pine oil. Occurs naturally in several essential oils.
Grade: Technical, perfumery, extra, prime, FCC.
Use: Solvent for hydrocarbon materials, mutual solvent for resins and cellulose esters and ethers, perfumes, soaps, disinfectant, antioxidant, flavoring agent.

"Terpineol 318."[266] TM for a highly refined mixture of tertiary terpene alcohols, predominantly α-terpineol; 97% total terpene alcohols; colorless liquid; fp below 10C; d 0.937 (15.6/15.6C); ASTM distillation range 5–95%, 216–220C.

terpin hydrate. (dipentene glycol).
$CH_3(OH)C_6H_9C(CH_3)_2OH \cdot HOH$.
Properties: Colorless, lustrous, rhombic, crystalline prisms or white powder. Slight characteristic odor, slightly bitter taste, efflorescent, mp 115–117C, anhydrous mp 105C, bp 258C, soluble in alcohol and ether, slightly soluble in water. Combustible.
Derivation: From turpentine oil or d-limonene.
Grade: Technical, NF.
Use: Pharmaceuticals, source of terpineol, medicine (expectorant).

terpinolene. (1,4(8)-terpadiene).
CAS: 586-62-9. $C_{10}H_{16}$.
Properties: Water-white to pale amber liquid; insoluble in water; soluble in alcohol, ether, glycol. D 0.864 (15.5/15.5C), bp 183–185C, flash p (CC) 99F (37.2C), bulk d 7.2 lbs/gal (15.5C).
Derivation: Fractionation of wood turpentine.
Hazard: Flammable, moderate fire risk.

Use: Solvent for resins, essential oils, manufacture of synthetic resins, synthetic flavors.

terpinyl acetate. CAS: 80-26-2. $C_{10}H_{17}OOCCH_3$.
Properties: Colorless liquid, odor suggestive of bergamot and lavender, d 0.958–0.968 (15C), refr index 1.4640–1.4660 (20C), optical rotation varies around 0 degrees, fp −50C, bp 220C, soluble in five or more volumes of 70% alcohol, slightly soluble in water and glycerol. Combustible.
Derivation: By heating terpineol with acetic acid or anhydride in the presence of sulfuric acid and subsequent distillation.
Grade: Technical, prime, extra, FCC.
Use: Perfumes, flavoring agent.

terpolymer. A polymer made from three monomers, e.g., ABS polymers.

"Terraclor."[84] TM for pentachloronitrobenzene. Available as technical grade and formulations of 75% wettable powder, 2 lb concentrate, 20% dust, and 10% granular.
Use: Soil sterilant and fungicide.

Terramycin. Proprietary preparation containing oxytetracycline and oxytetracycline hydrochloride.
Use: Antibiotic.

"Tersan" 75.[28] TM for a turf fungicide containing 75% thiram. "Tersan" OM contains 45% thiram and 10% hydroxymercurichlorophenol.
Hazard: Toxic by ingestion and inhalation.

tert-. Abbreviation for tertiary.

tertiary. (1) For chemical meaning see primary. (2) In Petroleum extraction technology the term refers to recovery of petroleum by pumping detergents, high polymers, silicates, etc., into the rock structure. These techniques are generally known as enhanced oil recovery.
See also chemical flooding, hydraulic fracturing.

testing, chemical. Identification of a substance by means of reagents, chromatography, spectroscopy, melting and boiling point determination, etc.
See also analytical chemistry.

testing, physical. Application of any procedure whose object is to determine the physical properties of a material. There are four major categories of tests: (1) Those that are direct measurements of a property e.g., tensile strength. (2) Those that subject the material to actual service conditions;

these often require a long period of time, e.g., shelf-life of foods and corrosion of metals. (3) Accelerated tests, which require specially designed equipment that simulates service conditions on an exaggerated scale; in these, only a few hours are necessary to duplicate years of service life, e.g., oxygen bomb ageing of elastomers. (4) Nondestructive testing by x-ray or radiography. Elaborate standard testing procedures are established by the American Society for Testing and Materials. The more common types of test are as follows:
abrasion (elastomers, textiles)
adhesion (glues, resins)
ageing (elastomers, plastics, leather, food products)
color stability (pigments, organic dyes) (exposure)
corrosion (metals, alloys) (exposure)
dielectric (electrical tapes, plastics, glass)
flammability (textiles, fibers, paper, plastics)
flash point of combustible liquids--Tag closed cup (TCC), Cleveland open cup (COC), open cup (OC).
hardness (metals, elastomers, plastics) (Brinell, Rockwell, Shore penetration)
high temperature (elastomers, adhesives)
impact strength (composites, glass cement)
suncracking (paints, varnishes, elastomers) (exposure)
tear (paper, rubber, textiles)
tensile strength (fibers, elastomers, paper, textiles, metals)
viscosity (lubricants) (Saybolt, Engler)
See also exposure testing, nondestructive testing, ageing.

testing, physiological. Determination of the toxicity of a substance or product by administering it to laboratory animals in controlled dosages, either by mouth, by skin application, or by injection. Materials commonly subjected to such evaluation are pharmaceuticals, pesticides, and foods. Extensive testing programs are required before such products are approved for human use.
See also LD_{50}.

testosterone. CAS: 58-22-0. $C_{19}H_{28}O_2$.

An androgenic steroid; the male sex hormone produced by the testis. It has six times the androgenic activity of its metabolic product, androsterone.

Properties: White or slightly cream-white crystals or crystalline powder; odorless and stable in air; mp 153–157C; dextrorotatory in dioxane solution; very soluble in chloroform; soluble in alcohol, dioxane, and vegetable oils; slightly soluble in ether; insoluble in water.

Derivation: Isolation from extract of testis, synthesis from cholesterol or from the plant steroid diosgenin.

Grade: NF.

Hazard: A carcinogen (OSHA).

Use: Medicine, biochemical research.

See also methyltestosterone.

TETD. Abbreviation for tetraethylthiuram disulfide.

tetraamminecopper sulfate. CAS: 14283-05-7. $N_4H_{12}CuSO_4$.

Derivation: Dissolving copper sulfate in ammonia water, with precipitation by alcohol.

Use: Mordant in textile printing, fungicide.

tetrabromo-o-cresol. CAS: 576-55-6. $C_7H_4Br_4O$.

Properties: Fine, white crystals; mp 205C (decomposes); insoluble in water; soluble in alcohol and ether.

Derivation: Bromination of o-cresol.

Hazard: Irritant to skin and mucous membranes.

Use: Fungicide.

sym-tetrabromoethane. See acetylene tetrabromide.

tetrabromoethylene. C_2Br_4.

Properties: Colorless crystals, mp 55–56C, bp 227C.

Derivation: Bromination of dibromoacetylene.

Use: Organic synthesis.

tetrabromofluorescein. See eosin.

tetrabromomethane. See carbon tetrabromide.

tetrabromophthalic anhydride. $C_6Br_4C_2O_3$.

Properties: Pale yellow, crystalline solid; mp 280C.

Use: Flame retardant for plastics, paper, and textiles.

tetrabromosilane. See silicon tetrabromide.

tetra-n-butylammonium chloride. $(C_4H_9)_4NCl$.

Properties: Deliquescent, light tan powder, mp 50C.

Use: Substitute in solution for glass-calomel electrode systems, coagulant for silver iodide solutions, stereospecific catalyst.

tetrabutylthiuram disulfide. $[(C_4H_9)_2NCH]_2S_2$.

Properties: Liquid; amber color; d 1.03–1.06 (20/20C); solidifies at −30C; slight sweet odor; soluble in carbon disulfide, benzene, chloroform, and gasoline; insoluble in water and 10% caustic. Combustible.

Use: Vulcanizing and accelerating agent.

tetrabutylthiuram monosulfide. $[(C_4H_9)_2NCS]_2S$.

Properties: Brown, free-flowing liquid; d 0.99; soluble in acetone, benzene, gasoline, and ethylene dichloride; insoluble in water. Combustible.

Use: Rubber accelerator.

tetrabutyltin. CAS: 1461-25-2. $(C_4H_9)_4Sn$.

Properties: Colorless or slightly yellow, oily liquid. D 1.0572 (20/4C), fp −97C, bp 145C (10 mm), decomposes at 265C, insoluble in water, soluble in most common organic solvents. Combustible.

Derivation: Reaction of tin tetrachloride with butyl magnesium chloride.

Hazard: Irritant. TLV (as Sn): 0.1 mg/m³ of air.

Use: Stabilizing and rust-inhibiting agent for silicones, lubricant and fuel additive, polymerization catalyst, hydrogen chloride scavenger.

tetrabutyl titanate. (TBT; butyl titanate; titanium butylate). CAS: 5593-70-4. $Ti(OC_4H_9)_4$.

Properties: Colorless to light yellow liquid, bp 310–314C, forms a glass −55C, d 0.996, refr index 1.486, flash p 170F (76.6C), decomposes in water, soluble in most organic solvents except ketones. Combustible.

Derivation: Reaction of titanium tetrachloride with butanol.

Use: Ester exchange reactions; heat-resistant paints (up to 500C); improving adhesion of paints, rubber, and plastics to metal surfaces; crosslinking agent; condensation catalyst.

tetrabutyl urea. $(C_4H_9)_2NCON(C_4H_9)_2$.

Properties: Liquid, d 0.880, refr index 1.4535, vap press below 0.01 mm, bp 305C, fp below −60C, flash p 200F (93.3C), insoluble in water. Combustible.

Use: Plasticizer.

tetrabutyl zirconate. $(C_4H_9O)_4Zr$.

Properties: White solid from reaction of zirconium tetrachloride with butanol.

Use: Condensation catalyst and crosslinking agent.

tetracaine. (2-dimethylaminoethyl-p-butylamino-benzoate).   CAS: 94-24-6.
$CH_3(CH_2)_3NHC_6H_4COOCH_2CH_2N(CH_3)_2$.
Properties: White or light yellow, waxy solid; very slightly soluble in water; soluble in alcohol, ether, benzene, chloroform; mp 41–46C.
Grade: USP.
Use: Medicine (anesthetic).

tetracalcium aluminoferrate. An ingredient of Portland cement.

1,2,3,4-tetracarboxybutane. (TCB; 1,2,3,4-butanetetracarboxylic acid).
$HOOCCH_2CH(COOH)CH(COOH)CH_2COOH$.
Use: Alkyd resins, epoxy curing agent, sequestrant.

tetracene. See naphthacene.

tetracine. See tetrazene.

1,2,3,4-tetrachlorobenzene. CAS: 634-66-2.
$C_6H_2Cl_4$.
Properties: White crystals, mp 46.6, bp 254C, insoluble in water. Combustible.
Use: Component of dielectric fluids, synthesis.

1,2,4,5-tetrachlorobenzene. CAS: 95-94-3.
$C_6H_2Cl_4$.
Properties: White flakes, mp 137.5–140C, flash p 311F (155C), distillation range 240–246C. Combustible.
Use: Intermediate for herbicides and defoliants, insecticide, impregnant for moisture resistance, electrical insulation.

tetrachloro-p-benzoquinone. See chloranil.

tetrachlorobisphenol. $(C_6H_2Cl_2OH)_2C(CH_3)_2$.
A monomer for flame-retardant epoxy, polyester, and polycarbonate resins.

2,3,7,8-tetrachlorodibenzo-p-dioxin. See dioxin.

sym-tetrachlorodifluoroethane. (1,2-difluoro-1,1,2,2-tetrachloroethane; freon 112).
CAS: 76-12-0. CCl_2FCCl_2F.
Properties: White solid or colorless liquid with slightly camphor-like odor when concentrated, bp 92.8C, mp 26C, critical temperature 278C, d 1.6447 (25C), refr index 1.413 (25C), bulk d 13.8 lb/gal, insoluble in water, soluble in alcohol. Nonflammable.
Grade: Purified, solvent.
Hazard: Toxic by inhalation. TLV: 500 ppm in air.
Use: Degreasing solvent.

tetrachlorodinitroethane. $O_2NCCl_2CCl_2NO_2$.
Properties: White crystals, decomposes at 130C to nitrogen peroxide.

Hazard: Toxic by ingestion and inhalation, strong irritant.

tetrachlorodiphenylethane. See TDE.

tetrachlorodiphenyl sulfone. See tetradifon.

sym-tetrachloroethane. (acetylene tetrachloride).
CAS: 79-34-5. $CHCl_2CHCl_2$.
Properties: Heavy, colorless, corrosive liquid; chloroform-like odor. Soluble in alcohol and ether, slightly soluble in water, d 1.593 (25/25C), bp 146.5C, fp −43C, bulk d 13.25 lb/gal (25C), refr index 1.4918 (25C), flash p none. Nonflammable.
Derivation: Reaction of acetylene and chlorine and subsequent distillation.
Grade: Technical.
Hazard: Toxic by ingestion, inhalation, skin absorption. TLV: 1 ppm in air.
Use: Solvent, cleansing and degreasing metals, paint removers, varnishes, lacquers, photographic film, resins and waxes, extraction of oils and fats, alcohol denaturant, organic synthesis, insecticides, weed killer, fumigant, intermediate in manufacture of other chlorinated hydrocarbons.

tetrachloroethylene. See perchloroethylene.

cis-n-[1,1,2,2-tetrachloroethyl)thio]-4-cyclohexene-1,2-dicarboximide. ("Difolatan").
$C_{10}H_9Cl_4NO_2S$.
Properties: White solid, mp 160C, slightly soluble in most organic solvents, insoluble in water.
Hazard: Absorbed by skin.
Use: Fungicide.

tetrachloroisophthalonitrile. (1,3-dicyano-2,-4,5,6-tetrachlorobenzene). $C_8Cl_4N_2$.
Properties: Colorless crystals, mp 245C, bp 350C, d 1.70, insoluble in water, almost insoluble in organic solvents.
Use: Bactericide, nematocide.

tetrachloromethane. See carbon tetrachloride.

tetrachloronaphthalene. See chlorinated naphthalene.

2,3,4,6-tetrachlorophenol. CAS: 58-90-2.
C_6HCl_4OH.
Properties: Brown flakes or sublimed mass with a strong odor, mp 69–70C, bp 164C (23 mm), d 1.839 (25/4C), soluble in acetone, benzene, ether, and alcohol. Nonflammable.
Hazard: Toxic by ingestion and inhalation, strong irritant.
Use: Fungicide.

2,4,5,6-tetrachlorophenol.　CAS: 58-90-2.
C_6HCl_4OH.
Properties: Brown solid, phenol odor, d 1.65 at 60/4C, mp above 50C, soluble in sodium hydroxide solutions and most organic solvents, insoluble in water.
Hazard: Toxic by ingestion and inhalation, strong irritant.
Use: Fungicide, wood preservative.

tetrachlorophthalic acid.　$C_6Cl_4(CO_2H)_2$.
Properties: Colorless, crystalline plates; soluble in hot water; sparingly soluble in cold water.
Derivation: By passing a stream of chlorine through a mixture of phthalic anhydride and antimony pentachloride.
Use: Dyes, intermediates.

tetrachlorophthalic anhydride.　CAS: 117-08-8.
$C6Cl_4(CO)_2O$.
Properties: White, odorless, free-flowing, nonhygroscopic powder; slightly soluble in water; mp 254–255C; bp 371C.
Use: Intermediate in dyes, pharmaceuticals, plasticizers, and other organic materials; flame-retardant in epoxy resins.

tetrachloroquinone.　See chloranil.

tetrachlorosalicylanilide.　$C_{13}H_7Cl_4NO_2$.
Properties: Crystalline solid, mp 160C, insoluble in water, soluble in common organic solvents and in alkaline solutions.
Use: Bacteriostat, preservative (textile finishes, cutting oils, plastics); use in foods, drugs, and cosmetics may be restricted.

tetrachlorosilane.　See silicon tetrachloride.

tetrachlorothiophene.　CAS: 6012-97-1.

CClCClCClCClS.
Properties: Liquid; mp 29–30; bp 104C (10 mm); soluble in benzene, hexane, alcohols, chlorocarbons.
Use: Agricultural chemicals, lubricants.

tetrachlorvinphos.　CAS: 961-11-5.
$C_{10}H_9Cl_4O_4P$.
Properties: Powder, mp 97C, partially soluble in chloroform, slightly soluble in water.
Hazard: Cholinesterase inhibitor.
Use: Insecticide.

tetracosamethylhendecasiloxane.　CAS: 107-53-9.
$C_{24}H_{72}O_{10}Si_{11}$.
Properties: Colorless liquid, bp 200C (5 mm), d 0.924, refr index 1.399, flash p 188C, soluble in benzene and low molecular weight hydrocarbons, viscosity unaffected by temperature. Combustible.
Use: Silicone oils, foam inhibitor in lube oils.

tetracosane.　$C_{24}H_{50}$ or $CH_3(CH_2)_{22}CH_3$.
Properties: Crystals, soluble in alcohol, insoluble in water, d 0.779 (51/4C), bp 324.1C, mp 51.5C. Combustible.
Use: Organic synthesis.

n-tetracosanoic acid.　See lignoceric acid.

tetracyanoethylene.　CAS: 670-54-2.
$(CN)_2C:C(CN)_2$.　The first member of a class of compounds called cyanocarbons.
Properties: Colorless crystals, sublimes above 120C, mp 198–200C, bp 223C, high thermal stability, burns in oxygen with a hotter flame than acetylene.
Hazard: Hydrolyzes in moist air to hydrogen cyanide.
Use: Organic synthesis, dyes, to make colored solutions with aromatics.

tetracycline.　CAS: 60-54-8.　$C_{22}H_{24}N_2O_8 \cdot 0$ to $6HOH$.　An antibiotic obtained from certain *Streptomyces* species. It can be prepared from chlortetracycline or oxytetracycline, which it resembles in its actions and uses. It has been synthesized. Its chemical structure is that of a modified naphthacene molecule.
See naphthacene.
Properties: Yellow, odorless, crystalline powder; stable in air; affected by strong sunlight; potency affected in solutions with pH below 2 and destroyed in alkali hydroxide solutions. Practically insoluble in chloroform and ether, very slightly soluble in water, slightly soluble in alcohol, very soluble in dilute hydrochloric acid and alkali hydroxide solutions, pH (saturated solution) 3.0–7.0, induces fluorescence in mitochondria of living cells in tissue culture or in fresh preparations from various organs.
Grade: USP.
Use: Medicine (antibacterial).

tetradecamethylhexasiloxane.　CAS: 107-52-8.
$C_{14}H_{42}O_5Si_6$.
Properties: Colorless liquid, bp 140C (20 mm), fp below −100C, d 0.89, refr index 1.4, flash p 118C, viscosity unaffected by temperature, soluble in benzene and low molecular weight hydrocarbons, slightly soluble in alcohol. Combustible.
Use: Silicone oil base, foam inhibitor in lube oils.

n-tetradecane.　CAS: 629-59-4.　$C_{14}H_{30}$ or $CH_3(CH_2)_{12}CH_3$.
Properties: Colorless liquid, d 0.7653 (20/4C), mp 5.5C, bp 253.5C, refr index 1.4302 (20C), flash

p 212F (100C), autoign temperature 396F (202C), soluble in alcohol, insoluble in water. Combustible.
Grade: 95%, 99%.
Hazard: A fire hazard. Lower flammable limit in air 0.5%.
Use: Organic synthesis, solvent standardized hydrocarbon, distillation chaser.

tetradecanoic acid.   See myristic acid.

tetradecanol.   See myristyl alcohol, see also 7-ethyl-2-methyl-4-undecanol.

1-tetradecene.   (α-tetradecylene).
$CH_2:CH(CH_2)_{11}CH_3$.
Properties: Colorless liquid, d 0.775 (20/4C), fp −12C, bp 256C, flash p 230F (110C), insoluble in water, very slightly soluble in alcohol and ether. Combustible.
Use: Solvent in perfumes, flavors, medicines, dyes, oils, resins.

cis-tetradec-9-enoic acid.   See myristoleic acid.

tetradecylamine. $C_{14}H_{29}NH_2$.
Properties: White solid with odor of ammonia, mp 37C, bp 291.2C, insoluble in water, soluble in alcohol and ether. Combustible.
Grade: 90% purity.
Use: Intermediate for manufacture of cationic surface-active agents, germicides.

tetradecylbenzyldimethylammonium chloride monohydrate.
$[C_{14}H_{29}N(CH_3)_2CH_2C_6H_5]Cl \cdot HOH$.   A quaternary ammonium salt.
Properties: 50% solution in aqueous isopropanol: Viscous liquid which may gelatinize on standing; d 0.978 (16C); miscible with water, alcohol, glycerol and acetone; pH of a 10% solution in distilled water 7–8.
Grade: 50% solution in aqueous isopropanol.
Use: Production of bacteriostatic and fungistatic paper.

tetradecyl chloride.   (myristyl chloride).
$CH_3(CH_2)_{13}Cl$.
Properties: Water-white distilled liquid, mild odor, d 0.8590, fp −0.2C, bp 154–155C (15 mm), 15.2% chloride, subject to mild hydrolysis on standing.
Grade: 97% min.

α-tetradecylene.   See 1-tetradecene.

tetradecyl thiol.   (myristyl mercaptan).
$CH_3(CH_2)_{13}SH$.
Properties: Liquid, mp 6.5C, bp 176–180C (22 mm), d 0.8398 (25/4C), refr index 1.4612 (20C), strong odor. Combustible.

Grade: 95% (min) purity.
Hazard: Toxic by inhalation, strong irritant.
Use: Organic intermediate, synthetic rubber processing.

tetradifon.   (chlorophenyl-2,4,5-trichlorophenyl sulfone; 2,4,4′,5-tetrachlorodiphenylsulfone).
CAS: 116-29-0. $Cl_3C_6H_2SO_2C_6H_4Cl$.
Properties: White, crystalline powder; mp 147C; soluble in chloroform and aromatic hydrocarbons; insoluble in water.
Use: Insecticide, acaricide, ovicide.

tetra(diphenylphosphito)pentaerythritol.
See pentaerythritol tetrakis(diphenyl phosphite).

tetraethanolammonium hydroxide.   CAS: 77-98-5.
$(HOCH_2CH_2)_4NOH$.
Properties: White, crystalline solid; mp 123C; vap press below 0.01 mm (20C); completely soluble in water. A strong base, approaching sodium hydroxide in alkalinity. Aqueous solutions are stable at room temperature but decomposes on heating to weakly basic polyethanolamines.
Grade: Commercial grade is a 40% water solution.
Hazard: Strong irritant to skin and tissue.
Use: Alkaline catalyst, solvent for certain types of dyes, metal-plating solutions.

1,1,3,3-tetraethoxypropane.   CAS: 122-31-6.
$[(C_2H_5O)_2CH]_2CH_2$.
Properties: Liquid, bp 105C, refr index 1.410 (20C), d 0.920 (20/20C), slightly soluble in water, soluble in ether and alcohol, flash p 190F (87.7C). Combustible.
Use: Organic synthesis.

tetraethylammonium chloride.   (TEAC; TEA chloride). $(C_2H_5)_4NCl$.
Properties: Anhydrous: Colorless, odorless, hygroscopic crystals; d 1.080; freely soluble in water, alcohol, chloroform, acetone; slightly soluble in benzene and ether. Tetrahydrate: $(C_2H_5)_4NCl \cdot 4HOH$, crystals; mp 37.5C, d 1.084.
Use: Medicine (nerve-blocking agent).

tetraethylammonium hexafluorophosphate.
$(C_2H_5)_4NPF_6$.
Properties: Solid, mp 255C, nonhygroscopic but soluble in water, stable to heat, can be stored in solution without decomposition of the PF_6 ion.
Hazard: Irritant to skin.
Use: Maintenance of fluoride atmospheres; preparation of bactericides and fungicides.

tetra-(2-ethylbutyl) silicate.
$[(C_2H_5)C_4H_8O]_4Si$.
Properties: Colorless liquid, d 0.8920–0.9018 (20/20C), fp −100C, bp 238C (50 mm), insoluble

in water, slightly soluble in methanol, miscible with most organic solvents.
Use: Heat-transfer medium, hydraulic fluid, wide temperature-range lubricant.

tetraethyl dithiopyrophosphate. (TEDP).
See sulfotepp.

tetraethylene glycol. (TEG).
$HO(C_2H_4O)_3C_2H_4OH$.
Properties: Colorless liquid; hygroscopic; soluble in water; insoluble in benzene, toluene, or gasoline. D 1.1248 at 20/20C, fp −4C, bp 327.3C, vap press above 0.001 mm (20C), refr index 1.4577 (20C), flash p 345F (174C), bulk d 9.4 lb/gal (20C). Combustible.
Use: Solvent for nitrocellulose, plasticizer, lacquers, coating compositions.

tetraethylene glycol dibutyl ether. See dibutoxytetraglycol.

tetraethylene glycol dimethacrylate.
Properties: Water-white to pale straw liquid, bp 200C (1 mm), d 1.075 (20/20C), refr index 1.4620 (20C), viscosity 12 cp. Insoluble in water; soluble in styrene, many esters and aromatics; limited solubility in aliphatic hydrocarbons. Combustible.
Hazard: Irritant to skin and eyes.
Use: Plasticizer.

tetraethylene glycol dimethyl ether. See dimethoxytetraglycol.

tetraethylene glycol distearate.
$(C_{17}H_{35}COOCH_2CH_2OCH_2CH_2)_2O$.
Properties: Liquid, mp 32–33C, insoluble in water. Combustible.
Use: Plasticizer.

tetraethylene glycol monostearate.
$C_{17}H_{35}COO(CH_2CH_2O)_4H$.
Properties: Liquid, d 0.971, mp 30–31C, insoluble in water. Combustible.
Use: Plasticizer.

tetraethylenepentamine. CAS: 112-57-2.
$NH_2(CH_2CH_2NH)_3CH_2CH_2NH_2$.
Properties: Viscous, hygroscopic liquid. D 0.9980 at 20/20C, fp −30C, bp 333C, vap press above 0.01 mm (20C), bulk d 8.3 lb/gal (20C), flash p 325F (162.7C), soluble in most organic solvents and water. Combustible.
Hazard: Strong irritant to eyes and skin.
Use: Solvent for sulfur, acid gases, various resins and dyes; saponifying agent for acidic materials; manufacture of synthetic rubber; dispersant in motor oils; intermediate for oil additives.

tetra-(2-ethylhexyl) silicate.
$[C_4H_9CH(C_2H_5)CH_2O]_4Si$.
Properties: Colorless liquid, d 0.8838, bp 350–370C, fp −90C, solubility in water below 0.01, 7.4 lbs/gallon, flash p 390F (198C). Combustible.
Use: Synthetic lubricants and functional fluids.

tetra-(2-ethylhexyl) titanate.
(tetrakis(2-ethylhexyl)titanate).
$[C_4H_9CH(C_2H_5)CH_2O]_4Ti$.
Properties: Light yellow, viscous liquid from the transesterification of isopropyl titanate with 2-ethylhexanol. Combustible.
Use: Crosslinking agent, condensation catalyst, adhesion promoter, water repellents.

tetraethyl lead. (TEL).   CAS: 78-00-2.
$Pb(C_2H_5)_4$.
Properties: Colorless, oily liquid; pleasant odor. Soluble in all organic solvents, insoluble in water and dilute acids or alkalies, d 1.65, bp 198–202C, 75–85C (13–14 mm), fp −136C, decomposes slowly at room temperature, rapidly at 125–150C. Combustible.
Derivation: (a) Alkylation of lead-sodium alloy with ethyl chloride; (b) electrolysis of an ethyl Grignard reagent with an anode of lead pellets.
Grade: One grade only, about 98% pure.
Hazard: Toxic by ingestion, inhalation, and skin absorption. TLV (as Pb): 0.1 mg/m³ of air.
Use: Antiknock agent. Leaded gasoline contains 1.10 g of lead per gal. TEL has been largely replaced by MBTE.

tetraethyl orthosilicate. See ethyl silicate.

tetraethyl pyrophosphate. Legal label name for tepp.

tetraethylthiuram disulfide. (disulfiram; TTD; TETD; bis[diethylthiocarbamyl] disulfide).
CAS: 97-77-8. $[(C_2H_5)_2NCS]_2S_2$.
Properties: Light gray powder; d 1.27; melting range 65–70C; slight odor; soluble in carbon disulfide, benzene, and chloroform; insoluble in water.
Hazard: Toxic symptoms when ingested with alcohol; animal teratogen. TLV: 2 mg/m³ of air.
Use: Fungicide, ultra-accelerator for rubber.

tetraethylthiuram sulfide.
[bis-(diethylthiocarbamyl)sulfide]
$[(C_2H_5)_2NCS]_2S$.
Properties: Dark brown powder, slight odor, d 1.12 (20/20C), boiling range 225–240C (3 mm).
Hazard: Toxic by ingestion and inhalation.
Use: Pharmaceutical ointments, fungicide, insecticide.

tetraethyltin.   CAS: 597-64-8. $Sn(C_2H_5)_4$.
Properties: Colorless liquid, d 1.187 (23C), bp 181C, fp −112C, insoluble in water, soluble in alcohol and ether.
Hazard: Toxic material. TLV (as Sn): 0.1 mg/m³ of air.

"Tetra-Flex."[511] TM for internally plasticized resins based on polymethylene polyphenol. Permanent flexibility makes them suitable for indus- trial and electrical laminates.

tetrafluorodichloroethane.   See dichlorotetrafluo- roethane.

tetrafluoroethylene.   (perfluoroethylene; TFE). CAS: 116-14-3. $F_2C:CF_2$.
Properties: Colorless gas, fp −142C, bp −78.4C, insoluble in water, much heavier than air.
Derivation: By passing chlorodifluoromethane through a hot tube.
Hazard: Flammable, dangerous fire risk.
Use: Monomer for polytetrafluoroethylene poly- mers.

tetrafluoroethylene epoxide.   (TFEO).

$$F_2C \overset{O}{\underset{\textstyle \diagdown}{-\!\!-\!\!-}} CF_2$$

Derivation: Oxidation of tetrafluoroethylene at 120C with UV light; reaction proceeds by free- radical mechanism.
Use: Monomer for products ranging from dimers to polymers of dp 35;
See also "Freon E" and "Krytox."

tetrafluorohydrazine.   CAS: 13847-65-9. F_2NNF_2.
Properties: Colorless, mobile liquid or colorless gas; bp (calc) −73C; heat of vaporization 3170 cal/mole; critical temperature 36C.
Hazard: Explodes on contact with reducing agents and at high pressures. Irritant.
Use: Organic synthesis, oxidizer in fuels for rock- ets, missiles, etc.

tetrafluoromethane.   (fluorocarbon 14; carbon tetrafluoride).   CAS: 75-73-0. CF_4.
Properties: Colorless gas, fp −184C, bp −128C, slightly soluble in water, density of liquid 1.96 at −184C, sp vol 4.4 cu ft/lb (70F). Nonflamma- ble.
Grade: 95% min purity.
Hazard: Toxic by inhalation.
Use: Refrigerant, gaseous insulator.

tetrafluorosilane.   See silicon tetrafluoride.

tetraglycol dichloride.
$(ClCH_2CH_2OCH_2CH_2)_2O$.
Properties: Colorless liquid, slightly soluble in wa- ter, d 1.186, bp 114 (2 mm). Combustible.
Use: High-boiling solvent and extractant for oils, fats, waxes, and greases; chemical intermediate.

tetraglyme.   See dimethoxytetraglycol.

1,2,3,6-tetrahydrobenzaldehyde.   Legal label name for 3-cyclohexene-1-carboxaldehyde.

1,2,3,4-tetrahydrobenzene.   See cyclohexene.

tetrahydrocannibol. $C_{21}H_{30}O_2$.   The active principle of marijuana, a hallucinatory drug. It has been synthesized and is available in lab quan- tities, subject to legal restrictions. Animal tests have indicated that it can retard cancer growth and also may promote the acceptance of organ transplants in the human body.

tetrahydrofuran.   (THF).   CAS: 109-99-9.

$\overline{CH_2CH_2CH_2CH_2O}$.
Properties: Water-white liquid with ethereal odor, d 0.888 (20C), refr index 1.4070 (20C), fp −65C, bp 66C, flash p (OC) 5F (−15C), autoign tem- perature 610F (321C), soluble in water and or- ganic solvents.
Derivation: (1) Catalytic hydrogenation of furan with nickel catalyst. (2) Acid-catalyzed dehydra- tion of 1,4-butanediol.
Grade: Technical, spectrophotometric.
Hazard: Flammable, dangerous fire risk. Flamma- ble limits in air 2–11.8%. Toxic by ingestion and inhalation. TLV: 200 ppm in air.
Use: Solvent for natural and synthetic resins, par- ticularly vinyls, in topcoating solutions, polymer coating, cellophane, protective coatings, adhe- sives, magnetic tapes, printing inks, etc. Grignard reactions, lithium aluminum hydride reductions, and polymerizations; chemical intermediate and monomer.

2,5-tetrahydrofurandimethanol.
(2,5-bis[hydroxymethyl]tetrahydrofuran).
CAS: 104-80-3. $C_6H_{12}O_3$.
Properties: Colorless liquid, d 1.154, mp below −50C, bp 265C, refr index 1.47. Soluble in water, alcohol, acetone, benzene; hygroscopic.
Hazard: Strong irritant to tissue.
Use: Organic synthesis, humectant, solvent.

tetrahydrofurfuryl acetate.
$C_4H_7OCH_2OOCCH_3$.
Properties: Colorless liquid; soluble in water, alco- hol, ether, and chloroform; d 1.061 (20/0C); bp 194–195C (753 mm). Combustible.

Derivation: By treatment of tetrahydrofurfuryl alcohol with acetic anhydride.

Use: Flavoring.

tetrahydrofurfuryl alcohol. (tetrahydrofuryl carbinol). CAS: 97-99-4. $C_4H_7OCH_2OH$.

$$CH_2-CH_2$$
$$|\qquad|$$
$$CH_2\quad CHCH_2OH$$
$$\diagdown O \diagup$$

Properties: Colorless liquid, mild odor, hygroscopic, miscible with water, d 1.0543 (20/20C), bp 178C, refr index 1.4520 (20C), flash p (OC) 167F (75C), viscosity 5.49 cp (25C), autoign temperature 540F (282C). Combustible.

Derivation: Catalytic hydrogenation of furfural.

Grade: Commercial, industrial (80%).

Use: Solvent for vinyl resins, dyes for leather; chlorinated rubber; cellulose esters; solvent-softener for nylon; vegetable oils; coupling agent; organic synthesis.

tetrahydrofurfurylamine. CAS: 4795-29-3. $C_4H_7OCH_2NH_2$.

Properties: Colorless to light yellow liquid, refr index 1.4520–1.4535 (25C), distilling range 150 to 156C. d 0.977 (20C/20C), refr index (20C) 1.4551. Combustible.

Use: Chemical intermediate, fine grain photographic development, vulcanization accelerator.

tetrahydrofurfuryl benzoate. $C_4H_7OCH_2OOCC_6H_5$.

Properties: Colorless liquid; insoluble in water; soluble in alcohol, ether, and chloroform; d 1.137 (20/0C); bp 300–302C; 138–140C (2 mm). Combustible.

Derivation: Tetrahydrofurfuryl alcohol and benzoic acid by esterification.

Use: Chemical intermediate.

tetrahydrofurfuryl laurate. $C_4H_7OCH_2OOCC_{11}H_{23}$.

Properties: Colorless liquid, d 0.930 (25C), insoluble in water. Combustible.

Use: Plasticizer.

tetrahydrofurfuryl levulinate. $CH_3CO(CH_2)_2COOCH_2C_4H_7O$.

Properties: Colorless liquid, mp 59–62C, soluble in water. Combustible.

Use: Plasticizer.

tetrahydrofurfuryl oleate. $C_{17}H_{33}COOCH_2C_4H_7O$.

Properties: Colorless liquid, d 0.923 (25C), bp 240C (5 mm), fp −30C, insoluble in water, flash p 329F (165C). Combustible.

Use: Plasticizer.

tetrahydrofurfuryl phthalate. $C_6H_4(COOCH_2C_4H_7O)_2$.

Properties: Colorless liquid, d 1.194 (25C), mp below 15C, insoluble in water. Combustible.

Use: Plasticizer.

tetrahydrofuryl carbinol. See tetrahydrofurfuryl alcohol.

tetrahydrogeraniol. See 3,7-dimethyl-1-octanol.

tetrahydrolinalool. (3,7-dimethyl-3-octanol). CAS: 78-69-3. $C_{10}H_{21}OH$.

Properties: Colorless liquid, floral odor, d 0.832–0.837, optically inactive. Combustible.

Use: Perfumery, flavoring.

1,2,3,4-tetrahydro-6-methylquinoline. $C_{10}H_{13}N$.

Properties: Yellowish crystals, strong civet-like odor, mp 33C, soluble in 2 parts 80% alcohol. Combustible.

Use: Perfumery.

tetrahydronaphthalene. CAS: 119-64-2. $C_{10}H_{12}$.

$$
\begin{array}{c}
CH_2 \\
\diagup \quad \diagdown CH_2 \\
| \\
CH_2 \diagdown \quad | \\
\quad CH_2
\end{array}
$$

Properties: Colorless liquid, pungent odor, miscible with most solvents and compatible with natural and synthetic vehicles, insoluble in water, d 0.981 at 13C, bp 206C, refr index 1.540–1.547, flash p 160F (71.1C), fp −25C, moisture content none, residue on evaporation none, acidity neutral, bulk d 8 lbs/gal, autoign temperature 723F (384C). Combustible.

Derivation: Hydrogenation of naphthalene in the presence of a catalyst at 150C.

Grade: Technical.

Hazard: Irritant to eyes and skin; narcotic in high concentration.

Use: Chemical intermediate; solvent for greases, fats, oils, waxes; substitute for turpentine.

tetrahydrophthalic anhydride. CAS: 85-43-8. $C_6H_8(CO)_2O$.

Properties: White, crystalline powder. Solidification point 99–101C, d 1.20 (105C), slightly soluble in petroleum ether and ethyl ether, soluble in benzene, flash p (OC) 315F (157C). Combustible.

Derivation: Diels-Alder reaction of butadiene and maleic anhydride.

Use: Chemical intermediate for light-colored alkyds, polyesters, plasticizers, and adhesives; intermediate for pesticides; hardener for resins.

tetrahydropyran-2-methanol. CAS: 100-72-1.

$\overline{OCH_2CH_2CH_2CH_2CHCH_2OH}$.

Properties: Liquid, d 1.0272 (20C), bp 187.2C, fp sets to glass below −70C, miscible with water, flash p (OC) 200F (93.3C). Combustible.

Use: Chemical intermediate.

1,2,5,6-tetrahydropyridine. CAS: 694-05-3. C_5H_9N.

Properties: Colorless liquid, d 0.912–0.914 (20/4C), fp −44C, bp 115.5–120.0C. Combustible. Purity: 96% min.

Use: Organic intermediate.

tetrahydrothiophene. (thiophane).

CAS: 110-01-0. $\overline{CH_2CH_2CH_2CH_2S}$

Properties: Water-white liquid, d 1.00 (15.6/15.6C), boiling range 115–124.4C. Combustible.

Use: Solvent, intermediate, fuel gas odorant.

tetrahydrothiophene-1,1-dioxide. See sulfolane.

tetrahydroxybutane. See erythritol.

tetrahydroxydiphenyl. See diresorcinol.

tetrahydroxyethylethylenediamine. [N,N,N′,N′-tetrakis-(2-hydroxyethyl)ethylenediamine]. $(HOCH_2CH_2)_2NCH_2CH_2N(CH_2CH_2OH)_2$

Properties: Clear, viscous liquid; good heat stability. Combustible.

Use: Organic intermediate, crosslinking of rigid polyurethane foams, chelating agent, humectant, gas absorbent, resin formation, detergent processing.

2,3,4,5-tetrahydroxyhexanedioic acid. See saccharic acid.

tetraiodoethylene. (iodoethylene). $I_2C{:}CI_2$.

Properties: Light-yellow, odorless crystals; on exposure to light turns brown. Mp 187C, d 2.98, insoluble in water, soluble in most organic solvents.

Derivation: Iodine on diiodoacetylene obtained from calcium carbide and iodine.

Use: Surgical dusting powder, antiseptic ointment, fungicide.

tetraiodofluorescein. See iodeosin.

tetraisopropylthiuram disulfide. $[(CH_3CH_3CH)_2NCS]S_2$.

Properties: Tan powder; d 1.12 (20/20C); melting range 95–99C; amine odor; soluble in benzene, chloroform, gasoline; insoluble in water, 10% caustic, carbon disulfide.

Use: Rubber accelerator.

tetraisopropyl titanate. (TPT; titanium isopropylate; isopropyl titanate). $Ti[OCH(CH_3)_2]_4$.

Properties: Light-yellow liquid which fumes in moist air, bp 102–104C (10 mm), mp 14.8C, d 0.954, refr index 1.46, apparent viscosity 2.11 cp (25C), decomposes rapidly in water, soluble in most organic solvents.

Derivation: Reaction of titanium tetrachloride with isopropanol

Use: Ester exchange reactions; adhesion of paints, rubber, and plastics to metals; condensation catalyst.

tetraisopropyl zirconate. $Zr[OCH(CH_3)_2]_4$.

Properties: White solid, decomposes before melting.

Derivation: By reaction of zirconium tetrachloride with isopropanol.

Use: Condensation catalyst, crosslinking agent.

tetrakis(hydroxymethyl)phosphonium chloride. (THPC). $(HOCH_2)_4PCl$. A crystalline compound made by the reaction of phosphine, formaldehyde, and hydrochloric acid.

Use: Flame-retarding agent for cotton fabrics. May be used in combination with triethylolamine and urea (Roxel process) or with triethanolamine and tris(1-aziridinyl) phosphine oxide.

N,N,N′N′-tetrakis(2-hydroxypropyl)ethylenediamine. (ethylenedinitrilotetra-2-propanol). $[-CH_2N(CH_2CHOHCH_3)_2]_2$.

Properties: Viscous, water-white liquid; miscible in water; soluble in ethanol, toluene, ethylene glycol, and perchloroethylene. Combustible.

Use: Crosslinking agent and catalyst in urethane foams, epoxy resin curing, metal complexes, intermediate.

"Tetralin."[28] TM for tetrahydronaphthalene.

tetralite. See tetryl.

1-tetralone. $\overline{CHCHCHCHCCCH_2CH_2CH_2CO}$

A ketone of tetrahydronaphthalene.

Properties: Liquid, d 1.090–1.095 (20/20C), bp 120–125C (10 mm), vap press 0.02 mm (20C),

mp 5.3–6.0C, insoluble in water, flash p 265F (129.5C). Combustible.
Grade: Solvent and intermediate.

tetram. (O,O-diethyl-S-(β-diethylamino)-ethyl phosphorothioate hydrogen oxalate).
CAS: 78-53-5.
$(C_2H_5O)_2POSCH_3CH_2N(C_2H_5)_2$.
Hazard: Cholinesterase inhibitor, use may be restricted.
Use: Insecticide.

tetramer. An oligomer whose molecule is composed of four molecules of the same chemical composition.
See also polymer.

1,1,3,3-tetramethoxypropane.
$[(CH_3O)_2CH]_2CH_2$.
Properties: Liquid, bp 183C, refr index 1.408 (20C), d 0.995 (20/20C), soluble in water, hexane, ether, and alcohol. Combustible.
Use: Organic synthesis.

tetramethylammonium chloride. $(CH_3)_4NCl$.
A quaternary ammonium compound.
Properties: White, crystalline solid; d 1.1690 (20/4C); mp decomposes. soluble in water and alcohol, insoluble in ether, also available as a 50% solution.
Use: Chemical intermediate, catalyst, inhibitor.

tetramethylammonium chlorodibromide.
$(CH_3)_4NClBr_2$.
Properties: Powder, mp 118–126C, soluble in water and other polar solvents.
Hazard: Evolves bromine on contact with water.
Use: Dry brominating agent, ingredient in formulation of sanitizers.

tetramethylammonium hydroxide. CAS: 75-59-2.
$(CH_3)_4NOH$.
Properties: A strong base available in 10% solution.
Hazard: Strong irritant to skin and tissue.

1,2,3,5-tetramethylbenzene. See isodurene.

sym-**tetramethylbenzene.** See durene.

N,N,N′,N′-tetramethyl-1,3-butanediamine.
$CH_3CHN(CH_3)_2CH_2CH_2N(CH_3)_2$.
Properties: Colorless stable liquid, fp −100C, bp 165.0C, d 0.8020 (20/20C), vap press 1.64 mm (20C), miscible with water, viscosity 1.0 cp (20C), flash p (TOC) 114F (45.5C). Combustible.
Use: Catalyst for polyurethane foams and epoxy resins, high-energy fuels.

2,2,4,4-tetramethyl-1,3-cyclobutanediol.
$(CH_3)_4C_4H_2(OH)_2$.
Properties: White solid, mp 124–135C, bp 220–225C, isomer composition 50% cis, 50% trans, flash p 125F (51.6C).
Hazard: Moderate fire risk. Irritant.
Use: Chemical intermediate, lubricants.

tetramethyldiamidophosphoric fluoride. See dimefox.

tetramethyldiaminobenzhydrol. (tetramethyldiaminodiphenylcarbinol; Michler's hydrol; hydrol).
$(CH_3)_2NC_6H_4CH(OH)C_6H_4N(CH_3)_2$.
Properties: Colorless prisms; forms a colorless solution in ether or benzene and a blue solution in alcohol or acetic acid; soluble in alcohol, ether, benzene, and acetic acid; mp 96C. Combustible.
Derivation: Reaction of tetramethyldiaminodiphenylmethane, hydrochloric acid, and glacial acetic acid; oxidized with lead peroxide.
Grade: Technical.
Use: Dye intermediate, organic synthesis.

tetramethyldiaminobenzophenone. (Michler's ketone; 4,4′-bis[dimethylamino]benzophenone).
CAS: 90-94-8. $CO[C_6H_4N(CH_3)_2]_2$.
Properties: Crystalline leaflets; mp 172C; bp decomposes at 360C; soluble in alcohol, ether, and water. Combustible.
Derivation: From dimethylaniline by reaction with phosgene.
Use: Synthesis of dyestuffs, especially auramine derivatives.

4,4′-tetramethyldiaminodiphenylmethane.
(tetra base). CAS: 101-61-1.
$H_2C[C_6H_4N(CH_3)_2]_2$.
Properties: Yellowish leaflets or glistening plates; mp 90–91C; sublimes with decomposition; bp 390C; insoluble in water; soluble in benzene, ether, carbon disulfide, and acids.
Derivation: By heating dimethylaniline with hydrochloric acid and formaldehyde.
Use: Dye intermediate.

tetramethyldiaminodiphenylsulfone.
(4,4′-bis(dimethylamino)diphenylsulfone).
$[(CH_3)_2NC_6H_4]_2SO_2$.
Properties: Solid, mp 259–260C. Combustible.
Grade: Technical, reagent.
Use: Intermediate in making dyestuffs and medicinal chemicals; analytical reagent for lead.

tetramethylene. See cyclobutane.

tetramethylenediamine. CAS: 110-60-1.
$H_2N(CH_2)_4NH_2$.
Properties: Colorless crystals with strong odor,

mp 27C, bp 158–159C, soluble in water with strongly basic reaction. Combustible.

Use: Chemical intermediate, complexing agent, catalyst in resin technology, synthesis of quaternary ammonium compounds.

tetramethylene dichloride. See 1,4-dichlorobutane.

tetramethylene glycol. See 1,4-butylene glycol.

tetramethylene sulfone. See sulfolane.

1,1,4,4-tetramethyl-6-ethyl-7-acetyl-1,2,3,4-tetrahydronaphthalene. $C_{18}H_{26}O$. A polycyclic musk.

Properties: Colorless crystals, mp 45C, bp 130C (2 mm), insoluble in water, soluble in alcohol.

Use: Perfumes, cosmetics, soaps.

tetramethylethylenediamine. (TMEDA; N,N,N′,N′-tetramethylethylenediamine). CAS: 110-18-9. $(CH_3)_2NCH_2CH_2N(CH_3)_2$.

Properties: Colorless liquid with slight ammoniacal odor, soluble in water and most organic solvents, bp 121–122C, d 0.7765 (20/4C), refr index 1.4170 (25C), fp −55.1C. Combustible.

Grade: Anhydrous (100%), aqueous (65%).

Use: Preparation of epoxy curing agents, polyurethane formation, corrosion inhibitor, textile finishing agents, intermediate for quaternary ammonium compounds.

tetramethylguanidine.
$(CH_3)_2NC(NH)N(CH_3)_2$.

Properties: Liquid with slight ammoniacal odor, bp 159–160C, soluble in both water and organic solvents. A strong base. Combustible.

2,6,10,14-tetramethylhexadecane. See phytane.

tetramethyl lead. (TML). CAS: 75-74-1.
$(CH_3)_4Pb$.

Properties: Colorless liquid; d 1.995; fp −27.5C; bp 110C (10 mm); insoluble in water; slightly soluble in benzene, petroleum ether, alcohol; flash p 100F (37.7C).

Derivation and grade: As for tetraethyl lead. Methyl reagents used instead of ethyl.

Hazard: Flammable, moderate fire risk. Toxic by ingestion, inhalation, and skin absorption. Lower explosion level 1.8%. TLV (as Pb): 0.15 mg/m³ of air.

tetramethylmethane. See neopentane.

3,3′-tetramethylnonyl thiodipropionate. See ditridecyl thiodipropionate.

tetramethylsilane. CAS: 75-76-3. $(CH_3)_4Si$.

Properties: Colorless, volatile liquid; bp 26.5C; d 0.646 (20/4C). insoluble in water and cold concentrated sulfuric acid, soluble in most organic solvents, flash p 0F (−17.7C).

Derivation: By Grignard reaction of silicon tetrachloride and methylmagnesium chloride.

Grade: Technical, purified.

Hazard: Flammable, high fire risk.

Use: Aviation fuel, internal standard for NMR analytical instruments.

tetramethylthiuram disulfide. See thiram.

tetramethylthiuram monosulfide. (bisdimethylthiocarbamyl sulfide). CAS: 97-74-5. $[(CH_3)_2NCH]_2S$.

Properties: Yellow powder; d 1.40; mp 104–107C; soluble in acetone, benzene, and ethylene dichloride; insoluble in water and gasoline. Combustible.

Use: Ultra-accelerator for rubber, fungicide, insecticide.

tetramethylurea. CAS: 632-22-4. $C_5H_{12}N_2O$.

Properties: Liquid, bp 176C, flash p 75C (167F), d 1.45, soluble in water and organic solvents.

Use: Solvent, analytical reagent.

tetranitroaniline. (TNA). CAS: 53014-37-2. $C_6H(NO_2)_4NH_2$. A nitration product of aniline.

Properties: Mp at 170C and explodes at 237C.

Hazard: Dangerous fire and explosion risk.

Use: Manufacture of detonators and primers.

tetranitromethane. CAS: 509-14-8. $C(NO_2)_4$.

$$O_2N - \underset{\underset{NO_2}{|}}{\overset{\overset{NO_2}{|}}{C}} - NO_2$$

Properties: Colorless liquid, pungent odor, bp 125.7C, mp 12.5C, d 1.650 (13C), miscible with alcohol and ether, insoluble in water, decomposed by alcoholic solution of potassium hydroxide, powerful oxidizing agent.

Derivation: By action of fuming nitric acid on benzene, acetic anhydride, or acetylene.

Hazard: Dangerous fire and explosion risk. Toxic by ingestion, inhalation, skin absorption. TLV: 1 ppm in air.

Use: Rocket fuel, as an oxidant or monopropellant; qualitative test for unsaturated compounds; diesel fuel booster; organic reagent.

"Tetranol."[300] TM for a highly sulfated fatty ester of the oleic type with wetting and dye-leveling properties.

tetra(octylene glycol) titanate. See octylene glycol titanate.

tetradotoxin. A highly toxic venom found in puffer-like fish.

1,1,4,4-tetraphenylbutadiene. (TPB). $(C_6H_5)_2C:CHCH:C(C_6H_5)_2$.
Properties: White crystals, two forms, mp 194–196C and mp 202–204C, insoluble in water, soluble in most organic solvents. Combustible.
Grade: Purified.
Use: Primary fluor or wavelength shifter in soluble scintillators.

tetraphenylsilane. $(C_6H_5)_4Si$.
Properties: White solid, mp 237C, bp 428C, very stable and inert. Combustible.
Derivation: By Grignard reaction of silicon tetrachloride and phenylmagnesium chloride.
Grade: Technical.
Use: Heat-transfer medium, polymers.

tetraphenyltin. $(C_6H_5)_4Sn$.
Properties: White powder, d 1.490, mp 225–228C, bp above 420C. Insoluble in water; soluble in hot benzene, toluene, xylene.
Derivation: Reaction of tin tetrachloride with phenylmagnesium bromide.
Hazard: Skin irritant.
Use: Stabilizer in chlorinated transformer oils, mothproofing agent, scavenger in dielectric fluids, intermediate.

tetraphosphoric acid. See polyphosphoric acids.

tetraphosphorus heptasulfide. See phosphorus heptasulfide.

tetraphosphorus hexasulfide. See phosphorus trisulfide.

tetraphosphorus trisulfide. See phosphorus sesquisulfide.

tetrapotassium ethylenediaminetetraacetate. See ethylenediaminetetraacetic acid (note).

tetrapotassium pyrophosphate. (TKPP). See potassium pyrophosphate.

tetrapropenylsuccinic anhydride. See dodecenylsuccinic anhydride. Many isomers are possible.

tetra-n-propyl dithionopyrophosphate.
CAS: 3244-90-4.
$(C_3H_7O)_2P(S)OP(S)(OC_3H_7)_2$.
Properties: Amber liquid, bp 148C (2 mm), miscible with most organic solvents, insoluble in water.
Hazard: Highly toxic.
Use: Insecticide, acaricide.

tetrapropylene. (dodecene; propylene tetramer). CAS: 6842-15-5. $C_{12}H_{24}$. A mixture of C_{12} monoolefins.
Properties: Liquid, d 0.770 (20/20C), boiling range 183–218C, bulk d 6.44 lb/gal (15.5). Combustible.
Derivation: Olefin fraction obtained from catalytic polymerization of propylene.
Use: Detergents (dodecylbenzene), lubricant additives, plasticizers.

tetrapropylthiuram disulfide.
$[(C_3H_7)_2NCS]_2S_2$.
Properties: Light cream color; d 1.13 (20/20C); melting range 49–51.5C; musty odor; soluble in carbon disulfide, benzene, chloroform, and gasoline; insoluble in water and 10% caustic. Combustible.

tetrasilane. CAS: 7783-29-1. Si_4H_{10}.
Properties: Colorless liquid, fp −93.5C, bp 109C, d 0.825 (0C).
Hazard: Severe fire and explosion risk, can ignite or explode in air.

tetrasodium diphosphate. See sodium pyrophosphate and sodium polyphosphate.

tetrasodium EDTA. (ethylenediaminetetraacetic acid, tetrasodium salt; EDTA Na₄; sodium edetate). CAS: 64-02-8. $C_{10}H_{12}N_2Na_4O_8$ anhydrous or 2HOH.
Properties: White powder, freely soluble in water.
Use: General-purpose chelating agent.

See also ethylenediaminetetraacetic acid (note).

tetrasodium monopotassium tripolyphosphate.
$Na_4KP_3O_{10}$.
Properties: White, crystalline solid; mp 580–600C; d 2.55; solubility in water 30 g/100 mL (26C).
Use: Sequestrant.

tetrasodium pyrophosphate. See sodium pyrophosphate and sodium polyphosphate.

tetrastearyl titanate. Organic intermediate, adhesion promoter, pigment dispersant.

tetrazene. (4-amidino-1-[nitrosamino-amidino]-1-tetrazene). CAS: 27120-23-6.
$H_2NC(:NH)NHNHN:NC(:NH)NHNHNO.$
Properties: Colorless or pale yellow fluffy solid, apparent d 0.45 but yields a pellet of d 1.05 under pressure of 3000 psi. Practically insoluble in water, alcohol, ether, benzene, and carbon tetrachloride; slightly hygroscopic.
Derivation: Interaction of an aminoguanidine salt with sodium nitrite in absence of free mineral acid.
Use: Initiating explosive.

tetrazolium chloride. (tetrazolium salt; TTC; 2,4,5-triphenyltetrazolium chloride).
$CN_4Cl(C_6H_5)_3.$
Properties: White to pale-yellow crystalline powder which darkens on exposure to light. Readily soluble in water, mp (with decomposition) 245C.
Use: In germination and viability tests. Viable parts of seed are stained red by deposition of red insoluble triphenyl formazan.

tetrol. See furan.

"Tetron."[88] TM for tetraethyl pyrophosphate (tepp), technical and in liquid formulations for insecticidal use.
Hazard: Highly toxic.

"Tetrone" A.[28] TM for dipentamethylenethiuram tetrasulfide rubber accelerator.

"Tetronic."[203] TM for a nonionic tetrafunctional series of polyether block-polymers ranging in physical form from liquids through pastes to flakable solids. They are polyoxyalkylene derivatives of ethylenediamine. Physical state varies with molecular weight and oxyethylene content, 100% active.
Use: Low-foaming detergent formulations; defoaming agents, flexible and rigid polyurethane foams, emulsifying and demulsifying agents, textile processing.

tetryl. (tetralite and nitramine are common commercial names for trinitrophenylmethylnitramine).
CAS: 479-45-8. $(NO_2)_3C_6H_2N(NO_2)CH_3.$
Properties: Yellow crystals; mp 130–132C; d 1.57 (19C); explodes at 187C; insoluble in water; soluble in alcohol, ether, benzene, glacial acetic acid.
Hazard: Dangerous fire and explosion risk. Skin irritant, absorbed by skin. TLV: 1.5 mg/m³ of air.
Use: Detonating agent for less sensitive high explosives, indicator (colorless at pH 10.8, dull red at pH 13.0).

"Texicote."[263] TM for a series of vinyl, acrylic, and styrene polymer and copolymer emulsions, plasticized and unplasticized; available in many grades.
Use: Adhesives, paint, carpet backsizing, pigment binder in clay coatings, self-polishing finishes.

"Texicryl."[263] TM for a series of acrylic and methacrylic polymer and copolymer emulsions.
Use: Paper coating, paint, leather finishes, textiles, adhesives, and wall paper.

"Texigel."[263] TM for a series of polyacrylic thickening agents, water-soluble anionic colloids.
Use: Stabilizers, protective colloid and thickeners for natural and synthetic latices and other aqueous dispersions, binders, and flocculants.

"Texilac."[263] TM for a series of vinyl and acrylic polymer and copolymer solutions in various solvents, plasticized and unplasticized.
Use: Wet-bond and heat-seal adhesives, impregnants, specialty lacquers, paper coating, nitrocellulose finishes, and printing inks for plastic films.

textile oil. Any of various specially compounded oils used to condition raw fibers, yarns, or fabric for manufacturing, bleaching, dyeing, and finishing operations.

"Textolite."[245] TM for laminated plastic sheets, tubes, and rods used as insulating materials. Includes laminations of various combinations of phenolic, melamine, silicone, and epoxy resins with such base materials as paper, asbestos, cotton, linen, and nylon. Also some of the above materials with copper foil cladding for printed circuit applications.

texture. The physical structure of a solid or semisolid material that results from the shape, arrangement, and proportions of its components. The term is used in the textile industry to characterize fabrics of various types and in the food industry to describe quality characteristics of bakery products, margarines, meats, spun proteins, etc. It is also regarded by geologists as a property of rocks and soils.
See also protein, textured.

textured protein. See protein, textured.

textryl. Generic name for nonwoven structures which may be manufactured by wet-processing from staple fibers and fibrid binder.

"Texzyme."[114] TM for a proteolytic enzyme which digests blood, albumin, food, beverage, urine, perspiration, etc.
Use: Silk and filament rayon fabric to accelerate the rate of desizing and insure good results in dyeing, bleaching, and finishing.

TFE. Abbreviation for tetrafluroethylene. See also polytetrafluorethylene.

TFEO. Abbreviation for tetrafluoroethylene epoxide.

TGA. Abbreviation for triglycollamic acid. See nitrilotriacetic acid.

Th. Symbol for thorium.

thallic oxide. See thallium peroxide.

thallium. CAS: 7440-28-0. Tl. Metallic element of atomic number 81, Group IIIA of the periodic table, aw 204.37, valences = 1, 3, two stable isotopes.
Properties: Bluish-white, lead-like metal, d 11.85, mp 302C, bp 1457C, oxidizes in air at room temperature, soluble in nitric and sulfuric acids, insoluble in water but readily forms soluble compounds when exposed to air or water.
Derivation: Flue dusts from lead and zinc smelting. The thallium compounds recovered are treated to obtain the metal by electrolysis, precipitation, or reduction.
Grade: Technical, high-purity.
Hazard: Forms toxic compounds on contact with moisture; keep from skin contact. TLV (as Tl): For soluble salts, 0.1 mg/m³ of air.
Use: Thallium salts, mercury alloys, low-melting glasses, rodenticides, photoelectric applications, electrodes in dissolved oxygen analyzers.

thallium acetate. (thallous acetate).
CAS: 563-68-8. TlOCOCH$_3$.
Properties: White, deliquescent crystals;, soluble in water and alcohol. Constants: mp 131, d 3.68.
Derivation: Interaction of acetic acid and thallium carbonate.
Hazard: As for thallium.
Use: High specific gravity solutions used to separate ore constituents by flotation.

thallium bromide. (thallous bromide).
CAS: 7789-40-4. TlBr.
Properties: Yellowish-white, crystalline powder. Soluble in alcohol, slightly soluble in water, insoluble in acetone, d 7.557, bp 815C, mp 460C.
Hazard: As for thallium.
Use: Mixed crystals with thallium iodide for infrared radiation transmitters used in military detection devices.

thallium carbonate. (thallous carbonate).
CAS: 6533-73-9. Tl$_2$CO$_3$.
Properties: Heavy, shiny, colorless or white crystals. Highly refractive, melts to dark gray mass, slightly alkaline taste, soluble in water, insoluble in alcohol, d 7, mp 272C.

Hazard: As for thallium.
Use: Analysis (testing for carbon disulfide), artificial diamonds.

thallium chloride. (thallous chloride).
CAS: 7791-12-0. TlCl.
Properties: White, crystalline powder; becomes violet on exposure to light; slightly soluble in water; insoluble in alcohol, ammonium hydroxide; d 7.004 (30/4C); mp 430C; bp 720C.
Hazard: As for thallium.
Use: Catalyst (chlorination), sun-tan lamp monitors.

thallium hydroxide. (thallous hydroxide).
CAS: 12026-06-1. TlOH·HOH.
Properties: Yellow needles; soluble in alcohol, water; bp (dehydrated) 139C, decomposes.
Hazard: As for thallium.
Use: Analysis (testing for ozone), indicator.

thallium iodide. (thallous iodide).
CAS: 7790-30-9. TlI.
Properties: Yellow powder, insoluble in alcohol, slightly soluble in water, soluble in aqua regia, d 7.09, bp 824C, mp 440C, becomes red at 170C.
Hazard: As for thallium.
Use: Mixed crystals with thallium bromide for infrared radiation transmitters.

thallium-amalgam.
Properties: Reported to have fp −60C (30C below that of mercury).
Hazard: As for thallium.
Use: Substitute for mercury in electrical switches, thermometers, etc., for extreme low-temperature service.

thallium monoxide. (thallium oxide; thallous oxide). CAS: 1314-32-5. Tl$_2$O.
Properties: Black powder, oxidizes when exposed to air, keep well stoppered, soluble in alcohol, water (decomposes). Mp 300C, bp 1080C, d 9.52 (16C).
Hazard: As for thallium.
Use: Analysis (testing for ozone), artificial gem, optical glass of high refractive index.

thallium nitrate. (thallous nitrate).
CAS: 10102-45-1. TlNO$_3$.
Properties: Colorless crystals, soluble in hot water, insoluble in alcohol, d 5.5, mp 206C (solidifies to a glass-like solid), decomposes at 450C.
Grade: Technical.
Hazard: A poison. Strong oxidizing agent, fire and explosion risk. TLV: 0.1 mg/m³ of air.
Use: Analysis, pyrotechnics (green fire).

thallium oxide. (thallous oxide).
CAS: 1314-12-1. Tl_2O.
Properties: Finely divided black solid, mp about 300C, soluble in water and alcohol, oxidizes when exposed to air.
Hazard: Toxic by ingestion.
Use: Optical glass, synthetic gemstones.

thallium sesquichloride. (thallo-thallic chloride).
$TlCl_3 \cdot 3TlCl$ or Tl_2Cl_3.
Properties: Yellow, crystalline powder; slightly soluble in water; d 5.9; mp 400–500C.
Hazard: As for thallium.

thallium sulfate. (thallous sulfate).
CAS: 7446-18-6. Tl_2SO_4.
Properties: Colorless crystals, soluble in water, d 6.77, mp 632C.
Grade: Technical, 99%.
Hazard: As for thallium.
Use: Analysis (testing for iodine in the presence of chlorine), ozonometry, rodenticides, pesticide.

thallium sulfide. (thallous sulfide).
CAS: 1314-97-2. Tl_2S.
Properties: Blue-black, lustrous, microscopic crystals or amorphous powder; soluble in mineral acids; insoluble in water, alcohol, or ether; d 8.46; mp 448C.
Hazard: As for thallium.
Use: Infrared-sensitive photocells.

thallous compound. See corresponding thallium compound.

THAM. See tris(hydroxymethyl)aminomethane.

"Thancat."[569] TM for a series of catalysts for use in urethane resins.

thebaine. (p-morphine).   CAS: 115-37-7.
$C_{19}H_{21}NO_3$.
Properties: White, crystalline alkaloid; slightly soluble in water; soluble in alcohol and ether; mp 193C; d 1.30. It is extracted from poppies of a different species from morphine-producing opium poppies.
Hazard: Toxic by ingestion, may induce addiction.

theine. See caffeine.

Thenard's blue. See cobalt blue.

thenyl alcohol. (2-thienylmethanol; 2-hydroxymethylthiophene; 2-thiophenecarbinol).
$C_4H_4SCH_2OH$.
Properties: Colorless liquid, bp 207C, insoluble in water, soluble in alcohol and ether.
Derivation: A heterocyclic alcohol made by reac-

tion of 2-thienylmagnesium iodide and formaldehyde.
Use: No commercially developed applications.

thenyldiamine. (Coined name for 2-[2-dimethylaminoethyl]-3-thenylaminopyridine).
CAS: 91-79-2.
$(C_4H_3SCH_2)N(C_5H_4N)CH_2CH_2N(CH_3)_2$.
Properties: Liquid, bp 169–172C (1.0 mm).
Derivation: Condensing N,N-dimethylaminoethyl-α-aminopyridine with 3-thenyl bromide.
Use: Medicine (as base for various salts, especially the hydrochloride).

theobroma oil. (cacao butter; cocoa butter).
Properties: Yellowish-white solid with chocolate-like taste and odor, d 0.858–0.864 (100/25C), mp 30–35C, refr index 1.4537–1.4585 (40C), saponification number 188–195, iodine number 35–43, insoluble in water, slightly soluble in alcohol, soluble in boiling dehydrated alcohol, freely soluble in ether and chloroform. Combustible.
Derivation: From the cacao bean, by expression, decoction, or extraction by solvent. Chief constituents: Glycerides of stearic, palmitic, and lauric acids.
Grade: Crude, refined, USP.
Use: Confectionery, suppositories and pharmaceuticals, soaps, cosmetics.

theobromine. (3,7-dimethylxanthine).
CAS: 83-67-0. $C_7H_8N_4O_2$.   The alkaloid found in cocoa and chocolate products. A purine base closely related to caffeine. Also occurs in tea and cola nuts.

Hazard: Toxic by ingestion.

theophylline. (1,3-dimethylxanthine).
CAS: 58-55-9. $C_7H_8N_4O_2$.

Properties: White, crystalline alkaloid; odorless; bitter taste. mp 270–274C, slightly soluble in water and alcohol, more soluble in hot water, soluble in alkaline solutions.

Derivation: (a) By extraction from tea leaves, (b) synthetically from ethyl cyanoacetate.
Grade: Technical, NF.
Use: Medicine (diuretic, muscle relaxant).

theoretical plate. Any contacting device in a fractionating column, such as packing, grids, or screens, that effects the same degree of separation of vapor from liquid as one simple distillation. A column that gives the same separation as 10 successive simple distillations is considered to have 10 theoretical plates. The effectiveness of a fractionating column is measured in terms of theoretical plates. As many as 100 theoretical plates are used in laboratory and industrial operations. The total column height divided by the number of theoretical plates is known as HETP (height equivalent to a theoretical plate). This concept is also used in chromatographic techniques.

therm. A unit of heat equal to 100,000 Btu. It has also been used to mean one Btu or one small calorie, but these uses have been abandoned in the US.

thermal black. Carbon black made from natural gas by the thermatomic process or by pyrolysis of bituminous coal. It is no longer widely used.

thermal cracking. See cracking.

thermal decomposition. See decomposition (6), pyrolysis.

thermal expansion coefficient. The change in volume per unit volume per degree change in temperature (cubical coefficient). For isotropic solids the expansion is equal in all directions, and the cubical coefficient is about three times the linear coefficient of expansion. These coefficients vary with temperature, but for gases at constant pressure the coefficient of volume expansion is nearly constant and equals 0.00367 for each degree Celcius at any temperature.

thermal neutron. A slow neutron.
See neutron.

thermal pollution. Heat introduced into rivers or estuaries by power plants or other industrial cooling waters or chemical wastes, which has adverse effects on estuarine ecology.
See also water pollution.

thermatomic process. Methane or natural gas is cracked over hot bricks at 870C to form amorphous carbon (carbon black) and hydrogen.
See thermal black, carbon black.

"Thermax."[69] TM for a medium thermal carbon black used for compounding with rubber.

"Thermid."[53] TM for polyamide resins.

"Therminol"FR.[58] TM for heat-transfer liquids which are thermally and oxidatively stable to 315C and extremely fire-resistant.

thermite. A mixture of ferric oxide and powdered aluminum, usually enclosed in a metal cylinder and used as an incendiary bomb, invented by the German chemist Hans Goldschmidt around 1900. On ignition by a ribbon of magnesium, the reaction produces a temperature of 2200C, which is sufficient to soften steel. This is typical of some oxide/metal reactions which provide their own oxygen supply and thus are very difficult to stop.
Hazard: Dangerous fire risk.

thermochemistry. That branch of chemistry comprising the measurement and interpretation of heat changes that accompany changes of state and chemical reactions. It is closely related to chemical thermodynamics. The heat of formation of a compound is the heat absorbed when it is formed from its elements in their standard states. An exothermic reaction evolves heat, an endothermic reaction requires heat for initiation.

Application of thermochemical principles to generation of hydrogen by water splitting are being extensively researched. The advantage of this method over electrolysis is its greater net efficiency. Many techniques have been explored, but only a few have practical potential. One of the more promising is the S-I cycle under study at Lawrence Livermore Laboratory. The heat for this would be obtained from either solar receivers or nuclear reactors. The reactions involved are

$$2HOH + SO_2 + I_2 \rightarrow H_2SO_4 + 2HI$$
$$H_2SO_4 \rightarrow HOH + SO_2 + 1/2O_2$$
$$2HI \rightarrow H_2 + I_2$$

See also hydrogen (Note 2).

thermocouple. (thermoelectric thermometer). An instrument comprised of two wires made of dissimilar metals or semiconducting materials that are joined at one end (the measuring junction), the other end being the reference junction, which is maintained at a known temperature (usually 0C). The difference in temperature between the measuring junction and the reference junction generates an electromotive force that is proportional to the temperature difference. Thermocouples are applicable over a range from −200C to 1800C. The most suitable conducting materials are iron-constantan, platinum-platinum-rhodium, copper-constantan, and Chromel-Alumel; graphite-silicon carbide is used in the metallurgical field. Thermocouples are essential for determinations of extreme temperatures that are beyond the range of liquid-in-glass thermometers. Their industrial applications include molten metals, fuel beds, ceramic kilns, furnaces, etc.; in laboratories they are used for both high-temperature and cryogenic research. They are also applicable to intermediate temperatures in cases where conventional thermometers are impracticable.

See also thermoelectricity.

thermodynamics. A rigorously mathematical analysis of energy relationships (heat, work, temperature, and equilibrium), the principles of which were first elaborated by J. Willard Gibbs in the mid-19th Century. It describes systems whose states are determined by thermal parameters, such as temperature, in addition to mechanical and electromagnetic parameters. A *system* is a geometric section of the universe whose boundaries may be fixed or varied, and which may contain matter or energy or both. The *state* of a system is a reproducible condition, defined by assigning fixed numerical values to the measurable attributes of the system. These attributes may be wholly reproduced as soon as a fraction of them have been reproduced. In this case the fractional number of attributes determines the state, and is referred to as the *number of variables of state* or *the number of degrees of freedom* of the system.

The concept of *temperature* can be evolved as soon as a means is available for determining when a body is "hotter" or "colder." Such means might involve the measurement of a physical parameter such as the volume of a given mass of the body. When a "hotter" body, A, is placed in contact with a "colder" body, B, it is observed that A becomes "colder" and B "hotter." When no further changes occur, and the joint system involving the two bodies has come to equilibrium, the two bodies are said to have the same temperature. Thus temperature can only be measured at equilibrium. Therefore thermodynamics is a science of equilibrium and a thermodynamic state is necessarily an equilibrium state. Thermodynamics is a macroscopic discipline, dealing only with the properties of matter and energy in bulk, and does not recognize atomic and molecular structure. Although severely limited in this respect, it has the advantage of being completely insensitive to any change in our ideas concerning molecular phenomena, so that its laws have broad and permanent generality. Its chief service is to provide mathematical relations

between the measurable parameters of a system in equilibrium so that, for example, a variable like pressure may be computed when the temperature is known, and *vice versa*.

The three laws of thermodynamics involve concepts too involved to be described here. For a more detailed explanation, see the article by Howard Reiss in "Encyclopedia of Chemistry," 3rd ed. (Hampel and Hawley), from which this entry was adapted.

thermodynamics, chemical. See chemical thermodynamics.

thermoelectricity. Electricity produced directly by applying a temperature difference to various parts of electrically conducting or semiconducting materials. Usually two dissimilar materials are used, and the points of contact are kept at different temperatures (Peltier effect). Many temperature-measuring devices (thermocouples, thermopiles) work on this principle, since the voltage is proportional to the temperature difference. Metallic conductors are usually used for these "thermometers, " which produce a rather small current. A newer use for the effect is as a source of electrical energy, i.e., a means of direct conversion of heat into electricity (or vice versa) without the use of a generator (or motor). The materials used for these thermoelectric couples are semiconductors, (e.g., tellurium; zinc antimonide; lead, bismuth, and germanium tellurides; samarium sulfide) or thermoelectric alloys, all of which produce relatively large currents. Several of these "cells" are then hooked in series much like the cells of a battery.

"Thermoflex" A[28]. TM for a rubber antioxidant containing 25% di-p-methoxydiphenylamine $(CH_3OC_6H_4)_2NH$, 25% diphenyl-p-phenylenediamine $C_6H_4(NHC_6H_5)_2$ and 50% phenyl-β-naphthylamine $C_{10}H_7NHC_6H_5$.

Properties: Dark gray pellets, d 1.21, fp above 67C.

Use: Tire carcasses, transmission belts, etc., to promote resistance to flexing at operating temperatures.

See also antioxidant.

thermofor. A heat-transfer medium.
See coolant.

thermoforming. (1) See reforming. (2) Forming or shaping a thermoplastic sheet by heating the sheet above its melting point, fitting it along the contours of a mold with pressure supplied by vacuum or other force, and removing it from the mold after cooling below its softening point.

The method is applied to polystyrenes, acrylics, vinyls, polyolefins, cellulosics, etc.

Thermofor process. A moving-bed catalytic cracking process in which petroleum vapor is passed up through a reactor countercurrent to a flow of small beads or catalyst. The deactivated catalyst then passes through a regenerator and is recirculated.

"Thermoguard."[288] TM for antimony-based materials for incorporation in PVC and other chlorine-containing plastics for flameproofing properties.

thermometer. An instrument for measuring temperature. The liquid-in-glass thermometer consists of a graduated glass tube and a bulb containing a suitable liquid whose expansion and contraction indicates the temperature. Its range is from −130 to 600C. For scientific purposes the most widely used liquid is mercury down to its freezing point at −40C; below this, alcohol gives readings to −100C and pentane to −130C. Colored alcohol is generally used in household thermometers. Mercury thermometers ranging up to 600C are available; the mercury is prevented from vaporizing by a pressurized inert gas inserted above the mercury column. Metal protection tubes for stem and bulb are necessary. The softening point of the glass is of primary importance; borosilicate glasses are satisfactory up to 500C, but Jena glass is required for higher temperatures. Minimum and maximum thermometers are so made as to retain their lowest and highest readings indefinitely; the latter are used for oil-well and other geothermal measurements.

There are several other types of thermometers: (1) Gas, in which either the pressure at constant volume or the volume at constant pressure measured the temperature; these are used for extremely accurate thermodynamic determinations. The gases used are helium, nitrogen, and hydrogen. (2) Bimetallic, in which the sensing element consists of two strips of metals having different expansion coefficients; its range is from −185 to 425C. (3) Thermoelectric (thermocouple), in which measurement is made by the electromotive force generated by two dissimilar metals; its range is from −200 to 1800C. (4) Resistance, in which temperature is measured by change in the electrical resistance of a metal, usually platinum; its range is from −163 to 660C. (5) An optical fiber thermometer recently developed by NBS Center for Chemical Engineering has a range of up to 2000C. It is made from a single crystalline sapphire and is much more accurate than the existing standard. Based on fun-

damental radiation principles, it measures thermodynamic temperatures directly.
See also thermocouple; bimetal.

thermonuclear reaction. See fusion.

thermoplastic. A high polymer that softens when exposed to heat and returns to its original condition when cooled to room temperature. Natural substances that exhibit this behavior are crude rubber and a number of waxes; however, the term is usually applied to synthetics such as polyvinyl chloride, nylons, fluorocarbons, linear polyethylene, polyurethane prepolymer, polystyrene, polypropylene, and cellulosic and acrylic resins. See also thermoset.

thermoset. A high polymer that solidifies or "sets" irreversibly when heated. This property is usually associated with a crosslinking reaction of the molecular constituents induced by heat or radiation, as with proteins, and in the baking of doughs. In many cases, it is necessary to add "curing" agents such as organic peroxides or (in the case of rubber) sulfur. For example, linear polyethylene can be crosslinked to a thermosetting material either by radiation or by chemical reaction. Phenolics, alkyds, amino resins, polyesters, epoxides, and silicones are usually considered to be thermosetting, but the term also applies to materials where additive-induced crosslinking is possible, e.g., natural rubber.

THF. Abbreviation for tetrahydrofuran.

thia-. Prefix indicating the presence of sulfur in a heterocyclic ring.

thiabendazole. (4-[2-benzimidazolyl]thiazole). CAS: 148-79-8.　$C_{10}H_7N_3S$.
Properties: White to tan crystals; mp 304C; slightly soluble in water, alcohols, and chlorinated hydrocarbons; soluble in diemthylformamide.
Use: Fungicide effective on citrus fruits, anthelmintic.

thiamine. (vitamin B_1). $C_{12}H_{17}ClN_4OS$. (3-(4-amino-2-methylpyrimidyl-5-methyl)-4-methyl-5, β-hydroxy-ethylthiazolium chloride). The antineuritic vitamin, essential for growth and the prevention of beriberi. It functions in intermediate carbohydrate metabolism in coenzyme form in the decarboxylation of α-keto acids. Deficiency symptoms: emotional hypersensitivity, loss of appetite, susceptibility to fatigue, muscular weakness, and polyneuritis.

Sources: Enriched and whole grain cereals, milk, legumes, meats, yeast. Most of the thiamine commercially available is synthetic.

Use: Medicine, nutrition, enriched flours. Isolated usually as the chloride (see formula above). Available as thiamine hydrochloride and thiamine mononitrate.

"Thiate."[69] TM for a series of accelerators for neoprene. "Thiate A" (thiohydropyrimidine), "Thiate B" (trialkyl thiourea), "Thiate E" (trimethyl thiourea).

1,4-thiazane. $CH_2SCH_2CH_1NHCH_2$.
Properties: Colorless liquid; pyridine-like odor; fumes in air; absorbs carbon dioxide from the air; soluble in alcohol, benzene, ether, water; bp 169C (758 mm). Combustible.
Derivation: Interaction of alcoholic ammonia and dichlorodiethyl sulfide.
Grade: Technical.
Use: Organic synthesis.

thiazole. CAS: 288-47-1. $SCH:NCH:CH$.
Properties: Colorless or pale yellow liquid, d 1.18, bp 116.8C, soluble in alcohol and ether, slightly soluble in water, odor resembles that of pyridine.
Use: Organic synthesis of fungicides, dyes, and rubber accelerators.

thiazole dye. A dye whose molecular structure contains the thiazole ring. The chromophore groups are $=C=N-$, $-S-C=$, but the conjugated double bonds are also of importance. The members of the class are mainly used as direct or developed dyes for cotton, though some find use as union dyes. One example is primuline, CI No. 49000.
See thiazole.

2-thiazylamine. See 2-aminothiazole.

thickener. A circular or cylindrical tank equipped with revolving rakes or plows so designed as to increase the concentration of solids in a suspension. The slurry enters through a feed trough or well, which distributes it uniformly around the circumference of the tank; the rakes move the thickened sediment toward the center where it is discharged through a collecting cone. Thickeners are applicable to the metallurgical and food industries, ore flotation, and municipal water treatment.
See also clarification.

thickening agent. Any of a variety of hydrophilic substances used to increase the viscosity of liquid mixture and solutions and to aid in maintaining stability by their emulsifying properties. Four classifications are recognized: (1) starches, gums casein, gelatin, and phycocolloids; (2) semisynthetic cellulose derivatives (carboxymethyl-cellulose, etc.); (3) polyvinyl alcohol and carboxyvinylates (synthetic); and (4) bentonite, silicates, and colloidal silica. The first group is widely used in the food industries, especially in ice creams, confectionery, gravies, etc.; other major consumers are the paper, adhesives, textiles, and detergent fields.
See also emulsion; colloid, protective; gel.

Thiele reaction. Formation of triacetoxy aromatic compounds by the reaction of quinones with acetic anhydride catalyzed by sulfuric acid or boron trifluoride.

thimerosal. (sodium ethylmercurithiosalicylate; ethylmercurithiosalicylic acid, sodium salt).
CAS: 54-64-8. $NaOOCC_6H_4SHgC_2H_5$.
Properties: Light cream-colored, crystalline powder with slight characteristic odor, pH (1% solution) 6.7, affected by light, soluble in water and alcohol, almost insoluble in ether and benzene.
Derivation: Reaction between ethylmercuric chloride and thiosalicylic acid in alcoholic sodium hydroxide.
Grade: NF.
Hazard: Toxic by ingestion, inhalation.
Use: Medicine, bacteriostat, fungistat.

thin. A nontechnical word used by scientists with a variety of meanings. (1) In electronic metallurgy, a thin film is a vapor-deposited coating having a thickness of only a single atom; such *monatomic* films, e.g., thorium on tungsten, are used in electronic devices such as cathodes. (2) A coating or film of a fatty acid on water which is one molecule thick (about 200 Å) is called a *monomolecular* film. (3) In thin-layer chromatography the term applies to a specially prepared mixture of adsorbents spread on a glass slide to a thickness of 1/100 in. (4) The word is also used in the sense of a liquid of low viscosity, as in paint thinner and thin-boiling starch.

thin-boiling starch. See starch, modified; thin.

thin film. In electronic engineering a film having a "thickness" of a single atom and consisting

of a metal deposited on a metallic substrate either externally by vapor deposition or internally by diffusion. The base metal is usually tungsten (for a cathode), the film being any of a number of other metals (thorium, cesium, zirconium, barium, or cerium). "The greatest benefit is obtained when the films are of a monatomic nature; in the case of thorium on tungsten, the optimum coverage is 0.67 monatomic layer, i.e., the thorium atoms do not completely cover the tungsten surface. Such films are very tightly bound to the base metal by atomic forces." (W. H. Kohl.)
See also thin; film.

thin-layer chromatography. (TLC).
A micro type of chromatography. The thin layer (0.01 in.) is the adsorbent, usually a special silica gel spread on glass, or incorporated in a plastic film. Single drops of the solutions to be investigated are placed along one edge of the glass plate, and this edge then dipped into a solvent. The solvent carries the constituents of the original test drops up the thin layer in a selective separation, so that a comparison with known standards and various identifying tests may be made on the spots formed.
See also thin.

thinner. A hydrocarbon (naphtha) or oleoresinous solvent (turpentine) used to reduce the viscosity of paints to appropriate working consistency usually just prior to application. In this sense a thinner is a liquid diluent, except that it has active solvent power on the dissolved resin.

thio-. A prefix used in chemical nomenclature to indicate the presence of sulfur in a compound, usually as a substitute for oxygen.
See also thiol.

thioacetamide. CAS: 62-55-5. CH_3CSNH_2.
Properties: Colorless leaflets; stable in solution; mp 115C; soluble in water, alcohol, ether, benzene. Combustible.
Hazard: Toxic by ingestion and inhalation, a carcinogen (OSHA).
Use: To replace gaseous hydrogen sulfide in qualitative analysis.

thioacetic acid. (thiacetic acid; ethanethiolic acid).
CAS: 507-09-5. CH_3COSH.
Properties: Clear, yellow liquid; strong unpleasant odor; d 1.05 (25C); fp $-17C$; bp 81.8C (630 mm); soluble in water, alcohol, and ether. Combustible.
Derivation: By heating glacial acetic acid and phosphorus pentasulfide with subsequent distillation.

Hazard: Toxic by ingestion and inhalation.
Use: Chemical reagent, lachrymator.

thioallyl ether. See allyl sulfide.

thioanisole. $C_6H_5CH_3$.
Properties: Colorless liquid with strong unpleasant odor, d 1.053 (25C), fp $-15.5C$, bp 188C, refr index 1.5842 (25C), insoluble in water, soluble in most organic solvents. Combustible.
Use: Intermediate, solvent for polymeric systems.

thiobenzoic acid. C_6H_5COSH.
Properties: Yellow oil or crystals, d 1.1825–1.1835 (20/4C), mp 24C, bp 77.5C (5 mm), 122C (30 mm), refr index 1.602–1.604 (20C), insoluble in water, miscible with organic solvents. Combustible.
Grade: 95% min.
Use: Organic intermediate.

4,4′-thiobis(6-tert-butyl-m-cresol). CAS: 96-69-5.
Properties: Light gray to tan powder, mp 150C min, d 1.10 (25C).
Hazard: Toxic by inhalation. TLV: 10 mg/m³ of air.
Use: Protection of light-colored rubber from oxidation and of non-staining neoprene compounds against deterioration.

2,2′-thiobis(chlorophenol). CAS: 97-18-7.
$[ClC_6H_3(OH)]_2S$.
Properties: White, crystalline solid; mp 175.8–186.8C; odorless; insoluble in water.
Hazard: Toxic by ingestion.
Use: Bacteriostat for cosmetics, fungicide.

2,2′-thiobis(4,6-di-sec-amylphenol).
(2,2′-thiobis[4,6-bis-(1-methylbutyl)phenol]).
$[(CH_3[CH_2]_2CH[CH_3])_2OHC_6:H_2]_2S$.
Properties: A dark, viscous liquid; softening point 0C; d 0.99 (50C).
Use: Rubber antioxidant.

thiocarbamide. See thiourea.

thiocarbanil. See phenyl mustard oil.

thiocarbanilide. (N,N′-diphenylthiourea; sulfocarbanilide).   CAS: 102-08-9.
$CS(NHC_6H_5)_2$.
Properties: Gray powder, mp 148C, d 1.32, soluble in alcohol and ether, insoluble in water. Combustible.
Derivation: Interaction of aniline and carbon disulfide and alcohol in the presence of sulfur.
Use: Intermediates, dyes (sulfur colors, indigo, methyl indigo), vulcanization accelerator, synthetic organic pharmaceuticals, flotation agent, acid inhibitor.

thiocarbonyl chloride. See thiophosgene.

thioctic acid. See dl-α-lipoic acid.

thiodiethylene glycol. See thiodigylcol.

thiodiglycol. (thiodiethylene glycol; β-bis-hydroxyethyl sulfide; dihydroxyethyl sulfide). $(CH_2CH_2OH)_2S$.
Properties: Syrupy, colorless liquid; characteristic odor. D 1.1852 at 20C, bp 283C, fp −10C, viscosity 0.652 poise (20C), flash p 320F (160C), bulk d 9.85 lb/gal, refr index 1.5217 (20C). Soluble in acetone, alcohol, chloroform, water; slightly soluble in benzene, carbon tetrachloride, and ether. Combustible.
Derivation: Hydrolysis of dichloroethyl sulfide, interaction of ethylene chlorohydrin and sodium sulfide.
Hazard: Do not use with hydrochloric acid.
Use: Intermediate for elastomers and antioxidants; solvent for dyes in textile printing.

thiodiglycolic acid. CAS: 123-93-3.
$HOOCCH_2SCH_2COOH$.
A dicarboxylic acid.
Properties: Colorless crystals, mp 128C, soluble in water and alcohol. Combustible.
Use: Analytical reagent.

4,4′-thiodiphenol. (TDP). $(C_6H_5OH)_2S$.
Properties: White, crystalline powder; mp above 151C; 99.5% pure.
Use: Intermediate, flame retardant, antioxidant, engineering plastics.

thiodiphenylamine. See phenothiazine.

thiodipropionic acid. CAS: 111-17-1.
$HOOCCH_2CH_2 \cdot S \cdot CH_2CH_2COOH$. A dicarboxylic acid.
Properties: Leaflets, mp 135, soluble in water and alcohol.
Hazard: Use in foods restricted to 0.02% of fat and oil content, including essential oils.
Use: Antioxidant in food packaging, soaps, plasticizers, lubricants, fats and oils.

β,β-thiodipropionitrile. CAS: 111-97-7.
$S(CH_2CH_2CN)_2$.
Properties: White crystals or light yellow liquid, d 1.1095 (30C), mp 28.65C. Slightly soluble in water and alcohol; soluble in acetone, chloroform, and benzene.
Use: Preservative, selective solvent, chromatography.

thio-1,3-dithio[4,5-b]quinoxaline. See thioquinox.

thioethanolamine. See 2-aminoethanethiol.

"Thiofast."[438] TM for a thio-indigo deep maroon pigment with a red-violet undertone.
Use: Paints, printing inks, and plastics.

thioflavine T. $CH_3C_6H_3N(HCl)SCC_6H_4N(CH_3)_2$.
Properties: A yellow basic dye of the thiazole class, CI No. 49005, fluoresces yellow to yellowish-green when excited by UV.
Derivation: By heating p-toluidine with sulfur in the presence of lead oxide.
Use: Textile dyeing, fluorescent sign paints, in combination with green or blue pigments to produce brilliant greens, phosphotungstic pigments.

thiofuran. See thiophene.

"Thiogel."[91] TM for gelatin into which thio or -SH groups have been introduced by an improved thiolation process.
Use: Vehicle for gradual release of active compounds, enzyme activity measurement, tissue culture medium, photography and graphic arts, gel filtration, surface coatings.

1-thioglycerol. CAS: 96-27-5.
$CH_2(OH)CH(OH)CH_2SH$.
Properties: Water-white liquid, bp 118C at 5 mm, d 1.295 (14.4C), soluble in water, alcohol, and ether. Combustible.
Use: Reducing agent for cystine molecule in human hair and wool, for stabilization of acrylonitrile polymers, medicine.

thioglycolic acid. (mercaptoacetic acid). CAS: 68-11-1. $HSCH_2COOH$.
Properties: Colorless liquid, strong unpleasant odor, d 1.325, fp −16.5C, bp 123C (29 mm). Miscible with water, alcohol, or ether. Combustible.
Derivation: Heating chloracetic acid with potassium hydrogen sulfide.
Hazard: Toxic by ingestion and inhalation, strong irritant to tissue. TLV: 1 ppm in air.
Use: Reagent for iron, manufacture of thioglycolates, permanent wave solutions and depilatories, vinyl stabilizer, manufacture of pharmaceuticals.

2-thiohydantoin. (glycolylthiourea).
$NHC(S)NHC(O)CH_2$.
Properties: Crystals or tan powder, mp 230C, slightly soluble in water, insoluble in alcohols and ethers. Purity: 99% min.
Use: Intermediate for pharmaceuticals, rubber accelerators, copper plating brighteners and dyestuffs.

2-thio-4-keto-thiazolidine. See rhodanine.

thiol. (mercaptan). Any of a group of organic compounds resembling alcohols, but having the oxygen of the hydroxyl group replaced by sulfur, as in ethanethiol (C_2H_5SH). Many thiols are characterized by strong and repulsive odors.
Hazard: Aliphatic thiols are flammable. Toxic by inhalation.
Use: Warning agents in fuel gas lines, chemical intermediates.
See also specific compound.
Note: Adoption of the name "thiol" to replace "mercaptan" has been officially approved as more consistent with the molecular constitution of these compounds. The older term, which literally means "mercury seizing, " is inappropriate.

thiolactic acid. (2-mercaptopropionic acid).
CAS: 79-42-5. $CH_3CH(SH)COOH$.
Properties: Oily liquid with unpleasant odor, becomes crystalline at 10C, d 1.22, bp 116C (16 mm), refr index 1.482. Soluble in water, alcohol, and acetone. Readily forms salts with numerous metals which have quite different properties.
Derivation: Reaction of sodium sulfide, sulfur, and bromopropionic acid.
Use: Depilatory, hair-waving preparations.

thiomalic acid. (mercaptosuccinic acid).
CAS: 70-49-5. $HOOCCH(SH)CH_2COOH$.
Properties: White crystals or powder; sulfuric odor; mp 149–150C; soluble in water, alcohol, acetone, and ether; slightly soluble in benzene. Combustible.
Use: Biochemical research, intermediate, rust inhibitor, antidarkening agent for crepe rubber, tackifier for synthetic rubber.

thionazin. (Generic name for O,O-diethyl-O-2-pyrazinyl phosphorothioate).
CAS: 297-97-2.

$\overset{\lceil\qquad\rceil}{NCHCHNCHCOPS(C_2H_5O)_2}$.

Properties: Amber liquid, mp −1.7C, bp 80C (.001 mm), slightly soluble in water, miscible with most organic solvents.
Hazard: Toxic by ingestion, inhalation, and skin absorption; cholinesterase inhibitor.
Use: Insecticide, fungicide, nematocide.

"Thionex."[28] TM for tetramethylthiuram monosulfide.

thionyl chloride. (sulfurous oxychloride; sulfur oxychloride). CAS: 7719-09-7. $SOCl_2$.
Properties: Pale yellow to red liquid with suffocating odor, d 1.638, fp −105C, bp 79C, decomposes at 140C, decomposes (fumes) in water, soluble in benzene, carbon tetrachloride.
Grade: 93%, 97.5%.

Hazard: Strong irritant to skin and tissue.
Use: Pesticides, engineering plastics, chlorinating agent, catalyst.

thiopental sodium. ("Pentothal;" "Sodium Pentothal"; sodium 5-ethyl-5(1-methylbutyl)-2-thiobarbiturate). ($C_{11}H_{17}N_2O_2SNa$).
A rapidly acting barbiturate administered intravenously for general anesthesia and hypnosis. Commonly known as "truth serum."
Hazard: May cause respiratory failure, use only with physician in attendance.

thiophane. See tetrahydrothiophene.

thiophene. (thiofuran). CAS: 110-02-1.

$$\begin{array}{ccc} HC&\!\!\!\!-\!\!\!\!&CH \\ \| & & \| \\ HC & & CH \\ & S & \end{array}$$

Properties: Colorless liquid, refr index 1.5285 (20C), d 1.0644 (20/4C), fp −38.5C, bp 84C, flash p 30F (−1.1C), soluble in alcohol and ether, insoluble in water.
Derivation: From coal tar (benzene fraction) and petroleum, synthetically from heating sodium succinate with phosphorus trisulfide.
Hazard: Flammable, dangerous fire risk.
Use: Organic synthesis (condenses with phenol and formaldehyde, copolymerizes with maleic anhydride), solvent, dye, and pharmaceutical manufacture.

α-thiophenealdehyde. C_4H_3SCHO. (2-thiophenecarboxaldehyde).
Properties: Oily liquid with almond-like odor; bp 198C; 90C (20 mm); d 1.210–1.220; very soluble in alcohol, benzene, ether; slightly soluble in water. Combustible.
Grade: 95%.
Use: Thiophene derivatives, introducing thenyl group into organic compounds.

thiophenol. (phenyl mercaptan).
CAS: 108-98-5. C_6H_5SH.
Properties: Water-white liquid, repulsive odor, bp 168.3C, bp 71C (15 mm), refr index 1.5891, d 1.075 (25/25C), insoluble in water, soluble in alcohol and ether. Combustible.
Derivation: Reduction of benzenesulfonyl chloride with zinc dust in sulfuric acid.
Grade: 99%.
Hazard: Skin irritant.
Use: Pharmaceutical synthesis.

thiophosgene. (thiocarbonyl chloride).
CAS: 463-71-8. CSCl$_2$.
Properties: Reddish liquid, d 1.5085 (15C), bp 73.5C, decomposes in water and alcohol, soluble in ether.
Hazard: Toxic by ingestion and inhalation.
Use: Organic synthesis.

thiophosphoryl chloride. CAS: 3982-91-0.
PSCl$_3$.
Properties: Colorless liquid, penetrating odor, d 1.635, bp 126C, fp −35C, decomposed by water. Soluble in carbon disulfide, carbon tetrachloride; flash p none. Nonflammable.
Hazard: Strong irritant to skin and tissue.

thioquinox. (Generic name for 2,3-quinoxalinedithiol cyclic trithiocarbonate). CAS: 93-75-4.
C6H$_4$N$_2$C$_2$S$_2$CS (tricyclic).
Properties: Yellow solid; mp 180C; insoluble in water; slightly soluble in acetone, kerosene, and alcohol.
Hazard: Toxic by ingestion.
Use: Fungicide, acaricide.

thioridazine. (USAN name for 2-methylthio-10-[2-(N-methyl-2-piperidyl)-ethyl]phenothiazine).
CAS: 50-52-2. C$_{21}$H$_{26}$N$_2$S$_2$.
Properties: Colorless crystals, mp 158C, soluble in water and alcohol.
Grade: ND (as the hydrochloride).
Use: Medicine (tranquilizer).

thiosalicylic acid. (2-mercaptobenzoic acid).
CAS: 147-93-2. HOOCC6H$_4$SH.
Properties: Yellow solid; mp 164–165C; sublimes; slightly soluble in hot water; soluble in alcohol, ether, and acetic acid. Combustible.
Grade: Reagent, technical 80%, pharmaceutical.
Use: Dyes, reagent for iron determination, intermediate.

thiosemicarbazide. (aminothiourea).
CAS: 79-19-6. NH$_2$CSNHNH$_2$.
Properties: White, crystalline powder; no odor; soluble in water and alcohol; mp 180–184C.
Derivation: From potassium thiocyanate and hydrazine salts.
Grade: Technical and pure.
Use: Reagent for ketones and certain metals, photography, rodenticide.

thiosinamine. See allyl thiourea.

thiosorbitol. CAS: 24531-57-5.
CH$_2$OH(CHOH)$_4$CH$_2$SH.
Properties: Colorless, nonhygroscopic crystals; mp 92C; strong reducing agent; soluble in water, ethylene glycol, and formamide; insoluble in ben-

zene, carbon tetrachloride and carbon disulfide.
Use: Corrosion protection in pickling and plating baths.

"Thiostat-B."[248] TM for 40% aqueous solution of sodium dimethyldithiocarbamate.
Use: Bactericide in oil-well water flooding, paper slimicide, specialty bactericide.

"Thiostop."[248] TM for aqueous solutions of the sodium and potassium salts of dimethyl dithiocarbamate.
Properties: Clear yellow to amber liquid, d 1.18–1.23, good storage stability, potassium salt will crystallize out at −6.6C and the sodium salt at 0C. Avoid long storage in partially filled containers.
Use: Shortstop in SBR polymerization.

thiotepa. USP name for triethylenethiophosphoramide.

thiourea. (thiocarbamide). CAS: 62-56-6.
(NH$_2$)$_2$CS.

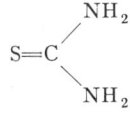

Properties: White, lustrous crystals; bitter taste. D 1.406, mp 180–182C, bp sublimes in vacuo at 150–160C. Soluble in cold water, ammonium thiocyanate solution, and alcohol, nearly insoluble in ether.
Derivation: (1) By heating dry ammonium thiocyanate, extraction with a concentrated solution of ammonium thiocyanate with subsequent crystallization; (2) action of hydrogen sulfide on cyanamide.
Grade: Technical, reagent.
Hazard: A carcinogen (OSHA), may not be used in food products (FDA), skin irritant (allergenic).
Use: Photography and photocopying papers, organic synthesis (intermediate, dyes, drugs, hair preparations), rubber accelerator, analytical reagent, amino resins, mold inhibitor.

thioxylenol. (mixed isomers).
C6H$_3$(CH$_3$)$_2$SH.
Properties: Colorless liquid, d 1.03 (15.6/15.6C), bp 122–133C (50 mm), oxidizes on exposure to air, supplied under nitrogen atmosphere. Combustible.
Use: Chemical intermediate, peptizer for natural and SBR rubbers.

thiram. (tetramethylthiuram disulfide; bis-(di-methylthiocarbamyl)disulfide; thiuram;

TMTD). CAS: 137-26-8.
$[(CH_3)_2NCH]_2S_2$.
Properties: White, crystalline powder with a characteristic odor; soluble in alcohol, benzene, chloroform, carbon disulfide; insoluble in water, dilute alkali, gasoline; d 1.29 (20C); melting range 155–156C.
Grade: 75% wettable powder, 95% technical powder, ND.
Hazard: Toxic by ingestion and inhalation, irritant to skin and eyes. TLV: 5 mg/m³ of air.
Use: Vulcanizing agent for rubber, especially for steam hose and other heat-resistant uses; fungicide; insecticide; seed disinfectant; lube oil additive; bacteriostat; animal repellent.

"Thiramad."[329] TM for a 75% thiram formulation. A turf fungicide.
Hazard: See thiram.

thiuram. A compound containing the group R_2NCS. Most are disulfides, The most common monosulfide is tretramethylthiuram monosulfide.
See thiram.

"Thixon."[465] TM for a family of adhesives or cements composed of mixtures of rubber and other bonding agents in solvents.
Use: Bonding natural and synthetic rubber to metals or plastics and for splicing various elastomers to themselves or each other.

thixotropy. The ability of certain colloidal gels to liquefy when agitated (as by shaking or ultrasonic vibration) and to return to the gel form when at rest. This is observed in some clays, paints, and printing inks which flow freely on application of slight pressure, as by brushing or rolling. Suspensions of bentonite clay in water display this property, which is desirable in oil-well drilling fluids.
See also rheology, dilatancy.

Thomson, Sr., J. J. (1856–1940) A native of England, Thomson entered Cambridge University in 1876 and remained there permanently as a professor of physics, especially in the field of electrical phenomena. His observations and calculations of cathode ray experiments led to proof of the existence of the electron as the lightest particle of matter (1896). This proof was announced at the Royal Institution in the following year. This was the keystone of the theory of atomic structure and one of the most notable discoveries in the history of science.
See Rutherford.

thomsonite. $(NaCa_2Al_5(SiO_4)_5 \cdot 6HOH)$.
A mineral, one of the zeolites.

thonzylamine hydrochloride. (2-[(2-di-methylaminoethyl)(p-methoxybenzyl)amino]-pyrimidinehydrochloride). CAS: 63-56-9.
$C_{16}H_{22}N_4O \cdot HCl$.
Properties: White, crystalline powder with faint odor. Mp 173–176, very soluble in water, freely soluble in alcohol and chloroform, practically insoluble in ether and benzene, pH 5.0–6.0 (2% solution).
Grade: NF.
Use: Medicine (antihistamine).

thoria. See thorium dioxide.

thorin. $HOC_{10}H_4(SO_3H)_2NNC_6H_4AsO_3H_2$.
A reagent for the colorimetric determination of microgram quantities of thorium.
Hazard: Toxic by ingestion.

thorite. $ThSiO_4$. A natural thorium silicate, usually impure, found in pegmatites.
Properties: Color black to orange, luster vitreous to resinous, hardness 4.5–5, d 4.4–5.2, radioactive.
Occurrence: Norway, Ceylon.
Use: Source of thorium.

thorium. CAS: 7440-29-1. Th. Metallic element of atomic number 90, a member of the actinide series (Group IIIB of Periodic Table), aw 232.0381, valence = 4, radioactive, no stable isotopes.
Properties: Soft metal with bright silvery luster when freshly cut, similar to lead in hardness when pure. Can be cold-rolled, extruded, drawn, and welded. D about 11.7, mp 1700C, bp 4500C, soluble in acids, insoluble in alkalies and water, some alloys may ignite spontaneously, the metal in massive form is not flammable.
Source: Monazite, thorite. It is about as abundant as lead.
Derivation: (a) Reduction of thorium dioxide with calcium, (b) fused salt electrolysis of the double fluoride $ThF_4 \cdot KF$. The product of both processes is thorium powder, fabricated into the metal by powder metallurgy techniques. Hot surface decomposition of the iodide produces crystal bar thorium.
Forms available: Powder, unsintered bars, sintered bars, sheets.
Hazard: Flammable and explosive in powder form. Dusts of thorium have very low ignition points and may ignite at room temperature. Radioactive decay isotopes are dangerous when ingested.
Use: Nuclear fuel (thorium-232 is converted to uranium-233 on neutron bombardment after several decay steps), sun lamps, photoelectric cells, target in x-ray tubes, alloys.

thorium acetylacetonate.
Th[OC(CH$_3$):CHCOCH$_3$]$_4$.
Properties: Crystalline powder, slightly soluble in water, resistant to hydrolysis, a chelating, non-ionizing compound.

thorium anhydride. See thorium dioxide.

thorium carbide. ThC$_2$.
Properties: Yellow solid, d 8.96 (18C), mp 2630–2680C, bp 5000C, decomposes in water.
Use: Nuclear fuel.

thorium chloride. (thorium tetrachloride).
CAS: 10026-08-1.   ThCl$_4$.
Properties: Colorless or white, lustrous needles (light yellow color caused by iron trace); hygroscopic; partially volatile; crystallizes with variable water of crystallization; soluble in alcohol, water; d 4.59; bp 928C (decomposes); mp 820C.
Grade: Technical; as 50% ThO$_2$.
Use: Incandescent lighting.

thorium decay series. The series of radioactive elements produced as successive intermediate products when thorium undergoes spontaneous natural radioactive disintegration into lead. Many of these are severe radioactive poisons when ingested or inhaled as thorium dust particles.

thorium dioxide. (thorium anhydride; thorium oxide; thoria).   ThO$_2$.
Properties: Heavy white powder, d 9.7, mp 3300C (highest of all oxides), bp 4400C, hardness 6.5 (Mohs), very refractory, soluble in sulfuric acid, insoluble in water.
Derivation: Reduction of thorium nitrate.
Grade: Technical and purities to 99.8% ThO$_2$; granular particles, crystals.
Use: Ceramics (high temperature), gas mantles, nuclear fuel, flame spraying, crucibles, medicine, nonsilica optical glass, catalyst, thoriated tungsten filaments.

thorium disulfide. ThS$_2$.
Properties: Dark brown crystals, d 7.30 (25/4C), mp 1875–1975C (in vacuo), insoluble in water.
Use: Solid lubricant.

thorium fluoride. ThF$_4$.
Properties: White powder, dehydrated between 200–300C, d (anhydrous) 6.32 (24C), mp 1111C, above 500C reacts with atmospheric moisture to form thorium oxyfluoride, ThOF$_2$, and finally the oxide, ThO$_2$. Forms a series of compounds with other metallic fluorides such as NaF and KF.
Grade: 79–80% ThO$_2$.

Hazard: Toxic material. TLV (as F): 2.5 mg/m^3 of air.
Use: Production of thorium metal and magnesium-thorium alloys, high temperature ceramics. ThOF$_2$ is used as a protective coating on reflective surfaces.

thorium nitrate. CAS: 13823-29-5.
Th(NO$_3$)$_4$·4HOH.
Properties: White, crystalline mass; soluble in water and alcohol; mp 500C (decomposes).
Grade: Technical, CP.
Hazard: Dangerous fire and explosion risk in contact with organic materials, strong oxidizing agent. Radioactive.
Use: Reagent for determination of fluorine, thoriated tungsten filaments.

thorium oxalate. Th(C$_2$O$_4$)$_2$·2HOH.
Properties: White powder, insoluble in water and most acids, d (anhydrous) 4.637 (16C), soluble in solutions of alkali and of ammonium oxalates, above 300–400C decomposes to thorium oxide, ThO$_2$.
Grade: Purities to 99.9%; as 59% ThO$_2$.
Use: Ceramics.

thorium oxide. See thorium dioxide.

thorium sulfate. (thorium sulfate, normal).
CAS: 103810-37-0.   Th(SO$_4$)$_2$·8HOH.
Properties: White, crystalline powder; d 2.8; mp loses 4HOH at 42C; remainder at 400C; slightly soluble in water; soluble in ice water.
Grade: As 43% ThO$_2$.

thorium tetrachloride. See thorium chloride.

"Thornel."[214] TM for a graphite yarn.
Properties: Tensile strength 200,000 psi, elastic modulus 25 million psi, d 1.5.
Use: Structural components for high-performance aircraft, solid-propellant-rocket motor casings, re-entry vehicles, and deep submergence vessels; high-strength composites.
See also graphite.

Thorpe reaction. Base-catalyzed condensation of two molecules of nitrile to yield imines which tautomerize to enamines.

thortveitite. An ore containing 37–42% scandium oxide. A basic material in the production of scandium.

Thr. Abbreviation of threonine.

threonine. (α-amino-β-hydroxybutyric acid).
CAS: 72-19-5.

$CH_3CH(OH)CH(NH_2)COOH$). An essential amino acid.
Properties: Colorless crystals, soluble in water, optically active. dl-threonine, mp 228–229C with decomposition; l(-)-threonine (naturally occurring), mp 255–257C with decomposition; dl-allothreonine, mp 250–252C.
Derivation: Hydrolysis of protein (casein), organic synthesis.
Use: Nutrition and biochemical research, dietary supplement.

threshold limit value. (TLV). A set of standards established by the American Conference of Governmental Industrial Hygienists for concentrations of airborne substances in workroom air. They are time-weighted averages based on conditions which it is believed that workers may be repeatedly exposed to day after day without adverse effects. The TLV values are revised annually and provide the basis for the safety regulations of OSHA. They are intended to serve as guides in control of health hazards, rather than definitive marks between safe and dangerous concentration. In this book, these are indicated by "TLV."
See also air pollution.

thrombin. A proteolytic enzyme which catalyzes the conversion of fibrinogen to fibrin and thus is essential in the clotting mechanism of blood. It is present in the blood in the form of prothrombin under normal conditions; when bleeding begins, the prothrombin is converted to thrombin, which in turn activates the formation of fibrin.

thrombocyte. See platelet.

thromboxane. See prostaglandin.

throwing oil. An oil applied to prepare raw silk and filament rayon for "throwing," the operation by which the filaments are twisted into threads. Applied by a bath, the oils condition the filaments and yarns for subsequent weaving or knitting. Usually compounded to be self-emulsifying and may contain a sizing agent such as dextrin, gelatin, etc.
See also slashing compound.

throwing power. A term denoting the effectiveness of an electrolytic cell for depositing metal uniformly over a surface being electroplated, particularly in irregular and recessed areas. The throwing power is the weight of deposition per unit distance between the electrodes.

THPC. See tetrakis(hydroxymethyl)phosphonium chloride.

thuja oil. (arbor vitae oil). CAS: 8007-20-3.
Properties: Pale yellow essential oil, camphor-like odor, d 0.910–0.920, refr index 1.459 (20C), optical rotation −10 to −13 degrees in 100 mm tube. Soluble in alcohol, ether, chloroform, carbon disulfide, fixed oils, and mineral oil. Combustible.
Chief known constituents: Dextro-pinene, levofenchone, thujone, should contain more than 60% ketones calculated as thujone. Derivation: Distilled from the leaves of the white cedar, *Thuja occidentalis*.
Grade: Technical, FCC (as cedarleaf oil).
Use: Perfumery, flavoring.

thujone. CAS: 546-80-5. $C_{10}H_{16}O$.
A terpene-type ketone contained in thuja oil and the oils of sage, tansy, and wormwood.
Properties: Colorless liquid, d 0.915–0.919 (20/20C), bp 203C, insoluble in water, soluble in alcohol. Combustible.
Hazard: Toxic by ingestion.
Use: Solvent.

thulia. See thulium oxide.

thulium. CAS: 7440-30-4. Tm. Atomic number 69, Group IIIB of the periodic table, a rare earth element of the lanthanide groups, aw 168.9342, valence = 3, no stable isotopes.
Properties: Metallic luster, reacts slowly with water, soluble in dilute acids, salts colored green, d 9.318, mp 1550C, bp 1727C.
Derivation: Isolated by reduction of the fluoride with calcium.
See rare earth minerals.
Grade: Regular high purity (ingots, lumps).
Hazard: Fire risk in form of dust.
Use: Ferrites, x-ray source.

thulium 170. Radioactive thulium of mass number 170.
Use: X-ray source in portable units.

thulium chloride. $TmCl_3 \cdot 7HOH$.
Properties: Green, deliquescent crystals; mp 824C; bp 1440C; very soluble in water and alcohol.

thulium oxalate. $Tm_2(C_2O_4)_3 \cdot 6HOH$.
Properties: Greenish-white precipitate, loses H_2O at 50C, soluble in aqueous alkali oxalates.
Derivation: Precipitation of a solution containing a thulium salt and a mineral acid by addition of oxalic acid.
Use: Analytical separation of thulium (and other rare-earth metals) from the common metals.

thulium oxide. (thulia). Tm_2O_3.
Properties: Dense white powder with greenish tinge, slightly hygroscopic, absorbs water and

carbon dioxide from the air, d 8.6. Exhibits a reddish incandescence on heating, changing to yellow and then white on prolonged heating. Slowly soluble in strong acids.
Derivation: By ignition of thulium oxalate, salt of other oxyacids, or hydroxide.
Grade: 99–99.9%.
Use: Source of thulium metal.

"Thylate."[28] TM for a wettable off-white powder containing 65% thiram.

thymic acid. See thymol.

thymidine. (thymine-2-deoxyriboside).
CAS: 50-89-5. $C_{10}H_{14}N_2O_5$. The nucleoside (deoxyriboside) of thymine. Occurs in DNA.
Properties: Crystalline needles; mp 185C; dextrorotatory in solution; soluble in water, methanol, hot ethanol, hot acetone, and hot ethyl acetate; sparingly soluble in hot chloroform; soluble in pyridine and glacial acetic acid.
Use: Biochemical research. Also available as trityl thymidine and as tritiated thymidine in a radioactive form.

thymidylic acid. The nucleotide of thymine, i.e., the phosphate ester of thymidine.

thymine. (5-methyluracil; 5-methyl-2,4-dioxypyrimidine). CAS: 165-71-4.

$CH_3CC(O)NHC(O)NHCH$. One of the pyrimidine bases of living matter.
Properties: White, crystalline powder; decomposes at 335–337C; slightly soluble in hot water; insoluble in cold water, alcohol; sparingly soluble in ether; readily soluble in alkalies.
Derivation: Hydrolysis of deoxyribonucleic acid, from methylcyanoacetylurea by catalytic reduction.
Use: Biochemical research.

thymine-2-deoxyriboside. See thymidine.

thymol. (isopropyl-m-cresol; thyme camphor; thymic acid). CAS: 89-83-8.
$(CH_3)_2CHC_6H_3(CH_3)OH$.

Properties: White crystals with aromatic odor and taste. Soluble in alcohol, carbon disulfide, chloro-

form, glacial acetic acid, ether, and fixed or volatile oils; slightly soluble in water and glycerol. D 0.979, mp 48–51, bp 233C. Combustible.
Derivation: From thyme oil or other oils, synthetically from m-cresol and isopropyl chloride by the Friedel-Crafts method at −10C.
Grade: Technical, NF, reagent.
Use: Perfumery, mold and mildew preventive, microscopy, preservative, antioxidant, flavoring, lab reagent, synthetic menthol.

thymol blue. (thymolsulfonphthalein).
CAS: 76-61-9.

$C_6H_4SO_2OC[C_6H_2(CH_3)(OH)CH(CH_3)_2]_2$.
Properties: Brown green powder or crystals, insoluble in water, soluble in alcohol or dilute alkali, mp 223C (decomposes)
Use: Acid base indicator in pH range 1.5 (pink) to 2.8-8 (yellow) to 9.6 (blue).
See indicator.

thymol iodide. CAS: 552-22-7.
$[C_6H_2(CH_3)(OI)(C_3H_7)]_2$. Principally dithymol diiodide.
Properties: Red-brown powder or crystals; slight aromatic odor; affected by light; soluble in ether, chloroform, and fixed or volatile oils; slightly soluble in alcohol; insoluble in water. Combustible.
Derivation: Interaction of thymol and potassium in alkaline solution.
Grade: Technical.
Use: Feed additive, antifungal agent.

thymolphthalein. CAS: 125-20-3.

$C_6H_4COOC[C_6H_2(CH_3)(OH)CH(CH_3)_2]_2$.
Properties: White powder, mp 245C, insoluble in water, soluble in alcohol and acetone and in dilute alkali and acids.
Use: Acid-base indicator in pH range 9.3 (colorless) to 10.5 (blue).
See indicator.

thymolsulfonphthalein. See thymol blue.

p-thymoquinone. (2-isopropyl-5-methylbenzoquinone). CAS: 490-91-5.
$C_6H_2O_2(CH_3)CH(CH_3)_2$.
Properties: Bright yellow crystals, penetrating odor, mp 45.5C, bp 232C, slightly soluble in water, soluble in alcohol and ether. Combustible.
Derivation: From diazonium salt of aminothymol and nitrous acid.
Use: Fungicide.

"Thyrite."[245] TM for composite dielectric or resistance material in molded form.

thyrocalcitonin. A thyroid hormone having significant effect on the calcium content of bone and blood. It can retard bone damage and may be involved in the incidence of dental caries and the form of dental calculus known as tartar.

thyroid. See thyroxine.

thyronine. (desiodothyroxine).
$HOC_6H_4OC_6H_4CH_2CH(NH_2)COOH$.
The p-hydroxyphenyl ether of tyrosine. Thyronine and its iodinated derivatives are used in biochemical research on the thyroid gland and its activity.

thyrotropic hormone. (TSH; thyrotropin).
CAS: 9002-71-5. A hormone secreted by the anterior lobe of the pituitary gland. It increases the rate of removal of iodine from the blood by the thyroid gland, of synthesis of the thyroid hormone, and of its release into the bloodstream. The thyrotropic hormone is a protein which has a low molecular weight (about 10,000) and which contains some carbohydrate.

thyroxine. (3,5,3',5'-tetraiodothyronine).
CAS: 51-48-9.
$HOC_6H_2I_2OC_6H_2I_2CH_2CH(NH_2)COOH$.

The hormone produced by the thyroid gland. It is an amino acid and a derivative of tyrosine. It increases the metabolic rate and oxygen consumption of animal tissues.
Properties: Optically active, the l-isomer is the natural and physiologically active form. dl-thyroxine: Needles; decomposes at 231–233C; insoluble in water, alcohol, and the common organic solvents; soluble in alcohol in the presence of mineral acids or alkalies. l-thyroxine: Crystals, decomposes 235–236C. d-thyroxine: Crystals, decomposes 237C.
Derivation: Obtained from the thyroid glands of animals, also made synthetically.
Use: Medicine (hormone), biochemical research. See also triiodothyronine.

Ti. Symbol of titanium.

TIBAL. Abbreviation for triisobutylaluminum.

"Ticon."[337] TM for a series of metal salts used as additives in ferroelectric and piezoelectric devices. They include barium, bismuth, calcium, cerium, lead, magnesium, strontium, and zinc titanates; barium, bismuth, calcium, lead, magne-

sium, and strontium zirconates; barium, calcium, and strontium stannates.

Tiemann rearrangement. Rearrangement of amide oximes (available from nitriles and hydroxylamine) to monosubstituted ureas by treatment with benzenesulfonyl chloride and water.

Tiffeneau-Demjanov ring expansion.
Rearrangement of beta-amino alcohols on diazotization with nitrous acid to give ring expanded carbonyl compounds.

tiglic acid. (methylcrotonic acid; crotonolic acid; trans-2-methyl-2-butenoic acid).
CAS: 80-59-1. $CH_3CH:C(CH_3)COOH$.
The trans isomer of angelic acid.
Properties: Thick, syrupy liquid or colorless crystals, spicy odor, soluble in alcohol and ether, soluble in hot water, d 0.9641, mp 65C, bp 198.5C. Combustible.
Derivation: From croton oil. Also occurs in English chamomile oil.
Use: Perfumes, flavors, emulsion breakers.

tiglium oil. See croton oil.

"Ti-Kote."[537] TM for bonded coating of a silicon carbide material completely impermeable to all gases. Has unusual resistance to abrasion, thermal shock, oxidation, mineral acid corrosion.
Use: Mechanical seals of pumps, ejector nozzles, thermocouple protection tubes, molds and mandrels, rocket nozzle inserts, nuclear fuel elements, missile leading edges, spacecraft re-entry surfaces and deflection elements.

timolol. CAS: 26839-75-8. $C_{13}H_{24}N_4O_3S$.

Properties: Colorless crystals, mp 72C.
Use: Adrenergic blocker used in treatment of hypertension, glaucoma, and for reducing risk of second heart attack. FDA approved.

tin. (stannum). CAS: 7440-31-5. Sn.
Metallic element of atomic number 50, group IVA of the periodic system, aw 118.69, valences = 2, 4. Ten isotopes.
Properties: Silver-white, ductile solid (beta form), d 7.29 (20C), mp 232C, bp 2260C. Changes to brittle gray (alpha) tin at temperature of 18C but the transition is normally very slow. Soluble in acids and hot potassium hydroxide solution,

insoluble in water. Elemental tin has low toxicity but most of its compounds are toxic.

Derivation: By roasting the ore (cassiterite) to oxidize sulfates and to remove arsine, then reducing with coal in a reverberatory furnace, or by smelting in an electric furnace; secondary recovery from tin plate.

Occurrence: Malaysia (called Straits tin), Indonesia, Thailand, Bolivia.

Grade: By percentage purity; spectrographic grade is 99.9999%; block tin is a common designation for pure tin.

Forms: Sheet, wire, tape, pipe, bar, ingot, powder, single crystals.

Hazard: All organic tin compounds are toxic. TLV (inorganic compounds except SnH_4 and SnO_2 as Sn): 2 mg/m^3 of air; (organic compounds as Sn): 0.1 mg/m^3 of air.

Use: Tin plate, terneplate, Babbitt metal, pewter, bronze, corrosion-resistant coatings, collapsible tubes, anodes for electrotin plating, hot-dipped coatings, cladding, solders, low-melting alloys for fire control, organ pipes, dental amalgams, die-casting. White, type, and casting metal. Manufacture of chemicals, tinned wire (all copper wire which is to be rubber covered). Block tin is used to coat copper cooking utensils and lead sheet, or to line lead pipe for distilled water, beer, carbonated beverages, and some chemicals.

Note: In speaking of fabricated articles "tin" is often used when tinplate (thin sheets of steel coated with tin) is meant, e.g., "a tin can." To distinguish, articles (such as condenser coils) made of solid tin are said to be made of "block tin." For further information refer to Tin Research Institute, Inc., 483 West Sixth Ave., Columbus, OH.

tin anhydride.　See stannic oxide.

tin ash.　See stannic oxide.

tin bisulfide.　See stannic sulfide.

tin bronze.　See stannic sulfide.

tincal.　See borax.

tincture.　An alcoholic or aqueous alcoholic solution of an animal or vegetable drug or a chemical substance. The tincture of potent drugs is essentially a 10% solution. Tinctures are more dilute than fluid extracts and less volatile than spirits.

"Tinopal."[443]　TM for a group of fluorescent whitening agents, which absorb UV light in the near visible range and re-emit the energy as visible light.

Use: In heavy duty detergents to whiten fabrics, paper, plastics, fibers, coatings, waxes, etc.

tin oxide.　See stannous oxide.

tin perchloride.　See stannic chloride.

tin plating.　The process of covering steel, iron, or other metal with a layer of tin by dipping it in the molten metal, by electroplating, or by immersion in solutions which deposit tin by chemical action of their components. Ingredients for the chemical process are cream of tartar and stannous chloride. The metal being plated also takes part in the process. The objective is to utilize the superior corrosion resistance of tin and in some cases to improve appearance. See also terneplate.

tin protochloride.　See stannous chloride.

tin protosulfide.　See stannous sulfide.

tin protoxide.　See stannous oxide.

tin resinate.　See soap (2).

tin salt.　See stannous chloride.

tin spirits.　Solutions of tin salts used in dyeing.

tin tetrabromide.　See stannic bromide.

tin tetrachloride.　Legal label name (Rail) for stannic chloride.

tin tetraiodide.　See stannic iodide.

"Tinuvin."[219]　TM for a group of UV absorbers, substituted hydroxyphenyl benzotriazoles. For example: 2-(2'-hydroxy-5'-methylphenyl)benzotriazole.

"Tipersul."[28]　TM for fibrous potassium titanate, crystalline fibers one micron in diameter melting at 1371C, useful to 1204C.

Use: High temperature thermal, acoustical, and electrical insulation; filter media. Available as lumps, blocks, and loose fibers.

"Tiron."[169]　TM for disodium-1,2-dihydroxybenzene-3,5-disulfonate.

Use: Colorimetric determination of ferric iron, titanium, or molybedenum.

Tiselius, Arne W. K.　(1902-1971) A Swedish biochemist who won the Nobel prize for chemistry in 1948. Renowned for research in separation methods of biochemical matter, in particular electrophoresis and chromatography. Work also involved virus isolation and synthesis of blood plasma. Degrees from University of Upsala and

Princeton University as well as a multitude of honorary degrees.

Tishchenko reaction. Formation of esters from aldehydes by an oxidation-reduction process in the presence of aluminum or sodium alkoxides.

titanellow. See titanium trioxide (not to be confused with titan yellow, an organic dye containing no titanium).

titania. See titanium dioxide.

titanic acid. (titanic hydroxide; metatitanic acid). CAS: 20338-08-3. H_2TiO_3 or $Ti(OH)_4$. Water content variable.
  Properties: White powder, insoluble in mineral acids and alkalies except when freshly precipitated, insoluble in water.
  Derivation: From hydrochloric acid solution of titanates by treating with ammonia and then drying over concentrated sulfuric acid or by boiling titanium sulfate solution.
  Grade: Technical.
  Use: Mordant.

titanium. CAS: 7440-32-6. Ti. Metallic element of atomic number 22, Group IVB of Periodic Table, aw 47.90, valences = 2, 3, 4; five isotopes.
  Properties: Silvery solid or dark gray amorphous powder, d 4.6 (20C), mp 1675C, bp 3260C, specific heat 0.13 Btu/lb/F, thermal conductivity 105 Btu/ft²/F/hour, as strong as steel but 45% lighter, Vickers hardness 80–100, excellent resistance to atmospheric and seawater corrosion and to corrosion by chlorine, chlorinated solvents, and sulfur compounds; reactive when hot or molten. Insoluble in water, inert to nitric acid but attacked by concentrated sulfuric acid and hydrochloric acid. Unaffected by strong alkalies.
  Sources: Ilmenite, rutile, titanite, titanium slag from certain iron ores.
  Derivation: (a) Reduction of titanium tetrachloride with magnesium (Kroll process) or sodium (Hunter process) in an inert atmosphere of helium or argon. The titanium sponge is consolidated by melting. (b) Electrolysis of titanium tetrachloride in a bath of fused salts (alkali or alkaline-earth chlorides).
  Grade: Technical (powder), commercially pure (sheets, bars, tubes, rods, wire, and sponge), single crystals.
  Hazard: Flammable, dangerous fire and explosion risk. (Metal) Ignites in air at 1200C, will burn in atmosphere of nitrogen. Do not use water or carbon dioxide to extinguish.
  Use: Alloys (especially ferrotitanium); as structural material in aircraft, jet engines, missiles,

marine equipment, textile machinery, chemical equipment (especially as anode in chloride production), desalination equipment, surgical instruments, orthopedic appliances, food-handling equipment; x-ray tube targets; abrasives; cermets; metal-ceramic brazing, especially in nickel-cadmium batteries for space vehicles; electrodeposited and dipped coatings on metals and ceramics; electrodes in chlorine battery.

titanium acetylacetonate. See titanylacetylacetonate.

titanium ammonium oxalate. See ammonium titanium oxalate.

titanium boride. (titanium diboride). TiB_2.
  Properties: Extremely hard solid with oxidation resistance up to 1400C, mp 2980C, d 4.50, hardness 9+ (Mohs), low electrical resistivity.
  Use: Metallurgical additive, high-temperature electrical conductor, refractory, cermet component, coatings resistant to attack by molten metals, aluminum manufacture, super alloys.

titanium butylate. See tetrabutyl titanate.

titanium carbide. TiC.
  Properties: Extremely hard, crystalline solid with gray metallic color, mp 3140C, bp 4300C, d 4.93, resistivity 60 micro-ohm-cm (room temperature), insoluble in water, soluble in nitric acid and aqua regia.
  Use: Additive (with tungsten carbide) in making cutting tools and other parts subjected to thermal shock, arc-melting electrodes, cermets, coating dies for metal extrusion (.2 mil film by vapor deposition).

titanium chelate. $(HOYO)_2Ti(OR)_2$ or $(H_2NYO)_2Ti(OR)_2$ A series of titanium compounds where Y and R are hydrocarbon groups, e.g., octylene glycol titanate, triethanolamine titanate.
  Use: Surface-active agents, corrosive inhibitors, crosslinking agents.

titanium diboride. See titanium boride.

titanium dichloride. CAS: 10049-06-6. $TiCl_2$.
  Properties: Black powder; d 3.13; decomposed by water; decomposes at 475C in vacuo; hygroscopic; soluble in alcohol; insoluble in chloroform, ether, carbon disulfide.
  Derivation: Reduction of titanium tetrachloride with sodium amalgam, dissolving titanium metal in hydrochloric acid.
  Hazard: Flammable, dangerous fire risk, ignites in air, store under water or inert gas.

titanium dioxide. (titanic anhydride; titanic acid anhydride; titanic oxide; titanium white; titania). CAS: 13463-67-7. TiO_2.
45th highest-volume chemical produced in US (1985).
Properties: White powder in two crystalline forms, anatase and rutile. It has the greatest hiding power of all white pigments. Noncombustible.
Derivation: From ilmenite or rutile. (a) Ilmenite is treated with sulfuric acid and the titanium sulfate further processed. The product is primarily the anatase form. (b) Rutile is chlorinated and the titanium tetrachloride converted to the rutile form by vapor-phase oxidation. Papermakers are using this form to an increasing extent in preference to the anatase form.
Grade: Technical, of many variations, pure, USP, single crystals, whiskers.
Use: White pigment in paints, paper, rubber, plastics, etc.; opacifying agent, cosmetics, radioactive decontamination of skin, floor coverings, glassware and ceramics, enamel frits, delustering synthetic fibers, printing inks, welding rods. Single crystals are high-temperature transducers.

titanium disilicide. $TiSi_2$.
Use: In special alloy applications, as a flame or blast impingement-resistant coating material.

titanium disulfide. TiS_2.
Properties: Yellow solid, d 3.22 (20C), decomposed by steam.
Use: Solid lubricant.

titanium ester. A series of titanium compounds whose general formula is $Ti(OR)_4$ where R is a hydrocarbon group, e.g., tetraisopropyl titanate.
Use: Adhesion promoters, ester exchange catalysts, crosslinking agents, heat-resistant paints.
See also tetrabutyl titanate, tetra(2-ethylhexyl) titanate.

titanium ferrocene. See titanocene dichloride.

titanium hydride. CAS: 7704-98-5. TiH_2.
Properties: Black, metallic powder which dissociates above 288C. The evolution of hydrogen is gradual and practically complete at 650C, d 3.8, attacked by strong oxidizing agents.
Derivation: Direct combination of titanium with hydrogen, reduction of titanium oxide with calcium hydride in the presence of hydrogen above 600C.
Hazard: Flammable, dangerous fire risk, dust may explode in presence of oxidizing agents.
Use: Powder metallurgy, production of pure hydrogen (can contain 1800 cc (STP) H/cc of hydride), production of foamed metals, solder for metal-glass composites, electronic getter, reduc-

ing atmosphere for furnaces, hydrogenation agent, refractories.

titanium isopropylate. See tetraisopropyl titanate.

titanium monoxide. TiO. A weakly basic oxide of titanium, prepared by reduction of the dioxide at 1500C; it has no important industrial uses.

titanium nitride. TiN.
Properties: Golden-brown, hard, brittle plates, mp 2927C, d 5.24, specific heat 8.86 cal/mole at 25C, electrical resistivity 21.7 micro-ohm-cm.
Use: High-temperature bodies, cermets, alloys, rectifiers, semiconductor devices.

titanium ore. See rutile and ilmenite.

titanium oxalate. (titanous oxalate). $Ti_2(C_2O_4)_3 \cdot 10HOH$.
Properties: Yellow prisms, soluble in water, insoluble in alcohol and ether.
Derivation: Action of oxalic acid on titanous chloride.

titanium peroxide. See titanium trioxide.

titanium potassium fluoride. (potassium titanium fluoride). TiK_2F_6.
Properties: White leaflets, soluble in hot water.
Hazard: Toxic material. TLV (as F): 2.5 mg/m³ of air.
Use: Titanic acid, titanium metal.

titanium potassium oxalate. (potassium titanium oxalate). $TiO(C_2O_4K)_2 \cdot 2HOH$.
Properties: Colorless, lustrous crystals, soluble in water.
Derivation: By treating titanium hydroxide with potassium oxalate and oxalic acid.
Grade: Technical, pure, 22% TiO_2 (min).
Use: Mordant in cotton and leather dyeing, sensitization of aluminum for photography.

titanium sesquisulfate. See titanous sulfate.

titanium sponge. The metal in crude form as reduced from the tetrachloride, ingots are made from it by consumable-electrode refining (arc-melting).
See titanium.

titanium sulfate. (titanic sulfate; basic titanium sulfate; titanyl sulfate). CAS: 13825-74-6. $Ti(SO_4)_2 \cdot 9HOH$, also $TiOSO_4 \cdot H_2SO_4 \cdot 8HOH$. A commercial material, possibly a mixture.
Properties: White, cake-like solid, highly acidic, similar to 50% sulfuric acid, typical composition

20% TiO_2, 50% sulfuric acid, 30% water, hygroscopic, d 1.47, soluble in water, solutions hydrolyze readily unless protected from heat and dilution.
Derivation: Action of sulfuric acid on ilmenite ore.
Hazard: Irritant to skin and tissue.
Use: Treatment of chrome yellow and other colors, production of titanous sulfate used as reducing agent or stripper for dyes, laundry chemical.

titanium tetrachloride.   (titanic chloride).
  CAS: 7550-45-0. $TiCl_4$.
Properties: Colorless liquid, fumes strongly when exposed to moist air, forming a dense and persistent white cloud. Pure: d 1.7609 at 0C, bp 136.4C, fp −30C. Soluble in dilute hydrochloric acid, soluble in water with evolution of heat, concentrated aqueous solutions are stable and corrosive, dilute solutions precipitate insoluble basic chlorides.
Derivation: By heating titanium dioxide or the ores and carbon to redness in a current of chlorine.
Grade: Technical, CP.
Hazard: Toxic by inhalation, strong irritant to skin and tissue.
Use: Pure titanium and titanium salts, iridescent effects in glass, smoke screens, titanium pigments, polymerization catalyst.

titanium trichloride.   (titanous chloride).
  CAS: 7705-07-9. $TiCl_3$.
Properties: Dark violet, anhydrous, deliquescent crystals; d 2.6; decomposes above 440C; decomposes in air and water with heat evolution; soluble in alcohol, acetonitrile, certain amines; slightly soluble in chloroform; insoluble in ether and hydrocarbons.
Hazard: Fire risk in presence of oxidizing materials. Irritant to skin and tissue; open container only in oxygen free or inert atmosphere.
Use: Reducing agent, organic synthesis, cocatalyst for polyolefin polymerization, organometallic synthesis involving titanium, laundry stripping agent.

titanium trioxide.   (titanium peroxide; titanello; pertitanic acid). TiO_3.
Properties: Yellow powder, soluble in acids.
Use: Dental porcelain and cements, yellow tile.

titanocene dichloride.   (dicyclopentadienyltitanium dichloride; titanium ferrocene).
  $(C_5H_5)_2TiCl_2$.
Properties: Red crystals; mp 287–289C; moderately soluble in toluene, chloroform; sparingly soluble in ether and water; stable in dry air, slowly hydrolyzed in moist air.

Hazard: Toxic by inhalation, irritant to skin and mucous membranes.
Use: Ziegler polymerization catalyst (with aluminum alkyls).
See also metallocene.

titanous chloride.   See titanium trichloride.

titanous oxalate.   See titanium oxalate.

titanous sulfate.   (titanium sesquisulfate).
  $Ti_2(SO_4)_3$.
Properties: Green, crystalline powder; insoluble in water, alcohol, concentrated sulfuric acid; soluble in dilute hydrochloric acid or sulfuric acid, giving violet solutions.
Grade: Commercial grade made and supplied as a dark purple solution containing 15% $Ti_2(SO_4)_3$.
Use: Textile industry as reducing agent for stripping or discharging colors.

"Titanox."[316]   TM of a series of white pigments comprising titanium dioxide in both anatase and rutile crystalline forms and also extended with calcium sulfate (titanium dioxide-calcium pigment). Noncombustible.
Use: Paints, paper, rubber, plastics, leather and leather finishes, inks, coated textiles, textile delustering, ceramics, roofing granules, welding rod coatings, and floor coverings.

titanylacetylacetonate.   (titaniumacetylacetonate). $TiO[OC(CH_3):CHCOCH_3]_2$.
Properties: Crystalline powder, slightly soluble in water, resistant to hydrolysis, a chelating, nonionizing compound.
Derivation: Reaction of titanium oxychloride with acetylacetone and sodium carbonate.
Use: Crosslinking agent for cellulosic lacquers.

titanyl sulfate.   See titanium sulfate.

titer.   In solutions (1) the concentration of a dissolved substance as determined by titration; (2) the minimum amount or volume needed to bring about a given result in titration; or (3) the solidification point of fatty acids which have been liberated from the fat by hydrolysis.

titration.   Any of a number of methods for determining volumetrically the concentration of a desired substance in solution by adding a standard solution of known volume and strength until the reaction is completed, usually as indicated by a change in color due to an indicator. Organic solvents are used in titrating water-insoluble substances (non-aqueous titration). Several types of electrical measurement techniques are used, among which are amperometric, conductometric, and coulometric titration methods and spectrophotometry is a further possible procedure. Titration has been successfully automated, the extent depending on the type of measurement used.

TKP. Abbreviation for tripotassium phosphate. See potassium phosphate, tribasic.

TKPP. Abbreviation for tetrapotassium pyrophosphate.
See potassium pyrophosphate.

Tl. Symbol for thallium.

TLV. See threshold limit value.

Tm. Symbol for thulium.

TMA. Abbreviation for trimethylamine and trimellitic anhydride.

TMEDA. Abbreviation for tetramethylethylenediamine.

TML. Abbreviation for tetramethyllead.

TMTD. Abbreviation for tetramethylthiuram disulfide.
See thiram.

TNA. Abbreviation for tetranitroaniline.

TNB. Abbreviation for trinitrobenzene.

TNT. Abbreviation for trinitrotoluene.

tobacco. Cured leaves of plants of the family *Solanaceae*, genus *Nicotiniana;* the species *N. tabacum* is the most important domestic source. Curing consists of drying and long ageing. The leaves are dehydrated by hanging in warm air to terminal moisture content of 20–30%, during which time starches reduce to sugars, the chlorophyll is discharged, and the color darkens. The product is then aged from one to five years to remove unpleasant odor; cigar tobacco is "fermented" with water with resulting hydrolysis, deamination, and decarboxylation. Aging and fermentation and oxidizers improve taste, aroma, and smoking qualities. Humectants and flavoring agents are added to some cigarette blends.

Scores of chemical compounds have been identified in unburned tobacco. Basically it is cellulose. The cured product contains acids (citric, oxalic, formic); alkaloids (nicotine, anabasine, myosmine); carbohydrates (lignin, pentosans, starch, sucrose), as well as tannin, ammonia, glutamine and micro amounts of zinc, iodine, copper, manganese, and polonium-210.

See cigarette tar, smoke (4).
A nicotine-free tobacco substitute TM "Cytrel" has been developed.
See "Cytrel."
Hazard: Carcinogenesis, many other deleterious effects from chronic inhalation of cigarette smoke.

tobacco mosaic virus. The first virus to be obtained in crystalline form (from diseased tobacco plants). The protein portion of this virus (95% of each particle) contains about 2300 peptide chains, each having 150 amino acids and ending with threonine. Molecular weight of each chain is about 17,000, total molecular weight 40 million. The complete sequence of the amino acids has been determined.

tobermorite gel. $3CaO \cdot 2SiO_2 \cdot 3HOH$.
A calcium silicate hydrate, a main cementing ingredient of concrete because of its great surface area.

Tobias acid. (2-naphthylamine-1-sulfonic acid; 2-amino-1-naphthalene-sulfonic acid).

Properties: White needles, soluble in hot water.
Derivation: Sulfonation of β-naphthol with chlorosulfonic acid in nitrobenzene at 0C followed by heating the resulting 1-sulfonic acid with ammonium hydrogen sulfite and ammonia at 145–150C. Purification is by precipitation from dilute solution of the sodium salt.
Use: Azo dye intermediate, optical brighteners.

TOC. Abbreviation for Tagliabue Open Cup, a standard method for determining flash points.

dl-α-tocopherol. (dl-2,5,7,8-tetramethyl-2-(4′,8′,12′-trimethyltridecyl)-6-chromanol).
CAS: 59-02-9. $C_{29}H_{49}OOH$. The most important of the vitamin E group.

Properties: Clear yellow, viscous oil; d 0.947–0.958 (25/25C); refr index 1.5030–1.5070 (20C).
See also tocopherol.
Derivation: Synthetic.

Grade: NF, FCC.
Use: Biological antioxidant, meat curing (nitrosamine blocker), nutrient, medicine. Also available as the acetate.

tocopherol. (vitamin E). Any of a group of related substances (alpha-, beta-, gamma-, and delta-tocopherol) which constitute vitamin E. The alpha-form (which occurs naturally as the d-isomer) is the most potent. Occurs naturally in plants, especially wheat germ. All are derivatives of dihydrobenzo-gamma-pyran and differ from each other only in the number and position of methyl groups. Vitamin E is required by certain rodents for normal reproduction. Muscular and central nervous system depletion along with generalized edema are deficiency symptoms in all animals.
Properties: Viscous oils; soluble in fats; insoluble in water; stable to heat in the absence of oxygen, to strong acids, and to visible light; unstable to UV light, alkalies, and oxidation.
Use: Medicine, nutrition, antioxidants for fats, animal feed additive.

tocophersolan. (USAN name for d-α-tocopheryl polyethylene glycol 1000 succinate).
$C_{29}H_{49}O \cdot OOC(CH_2)_2COO(CH_2CH_2O)_n H.$
($n = 22$.) A water-miscible vitamin E.
Use: Dietary food supplement, medicine.

Todd, Sir Alexander R. (1907-) A British chemist who won the Nobel prize for chemistry in 1957. His diverse research and accomplishments involving phosphorylation and mechanisms of biological reactions concerning phosphates. Many of his studies concerned the structure of nucleic acids, nucleotides, nucleotidic coenzymes as well as vitamins B_1, B_{12}, and E. Work in biological organic chemistry indicated that hemp plant could be used for production of narcotics. Todd had degrees awarded from Oxford, Frankfurt, and Glasgow among others.

Tofranil. Proprietary preparation of imipramine hydrochloride.
Use: Antidepressant.

toiletries. Liquid cosmetic preparations usually consisting of an essential oil-alcohol mixture used chiefly for application to hair and scalp, as aftershave lotions, etc.

"Tok."[23] TM for 2,4-dichlorophenyl-4-nitrophenyl ether product.
Use: Herbicide.

tokamak. A transliterated Russian word for an assembly for producing nuclear fusion. It consists essentially of a doughnut-shaped evacuated chamber called a torus, of 18-inch cross-section, through which the plasma moves. Surrounding the torus is an electromagnetic field powerful enough to confine the energized plasma sufficiently to achieve the required density of 10^{14} particles/cc/sec and a temperature well above 44 million centigrade. Up to 74 million centigrade has been obtained experimentally, but 100 million centigrade will be necessary for power production. The vacuum chamber and the magnetic field simulate conditions on the sun, i.e., absence of air and immense gravitational forces. Several experimental tokamaks are in operation in the US, the largest being at Princeton, NJ. It is utilizing hydrogen and deuterium as a test plasma, and thus can attain temperatures of only 100,000C. After further experimentation and modification, it should be capable of reaching the 100 million degrees necessary for power production when tritium is used.
See also fusion.

tolan. (diphenylacetylene). CAS: 501-65-5.
$C_6H_5C:CC_6H_5.$
Properties: Monoclinic crystals, mp 59–61C, bp 300C, 170C (19 mm), d 0.966 (100/4C), insoluble in water, soluble in ether or hot alcohol.
Grade: Technical, purified.
Use: Organic synthesis, purified grade, primary fluor or wavelength shifter in soluble scintillators.

o-tolidine. (3,3'-dimethylbenzidine; diaminoditolyl). CAS: 119-93-7.
$[C_6H_5(CH_3)NH_2]_2.$
Properties: Glistening plates, white to reddish, mp 129–131C, soluble in alcohol and ether, sparingly soluble in water, affected by light. Combustible.
Derivation: Reduction of o-nitrotoluene with zinc dust and caustic soda and conversion of the hydrazotoluene by boiling with hydrochloric acid.
Grade: Technical, dry, or paste.
Use: Dyes, sensitive reagent for gold (1:10 million detectable) and for free chlorine in water, curing agent for urethane resins (also available as the dihydrochloride).

Tollens' reagent. Solution of ammonaical silver nitrate containing free sodium bicarbonate.
Use: To test for aldehydes.

tolnaftate. (USAN name for o-2-naphthyl-meta-N-dimethylthiocarbanilate).
CAS: 2398-96-1.
$C_{10}H_7OC(S)N(CH_3)C_6H_4CH_3.$
Use: Antifungal treatment (medicine).

3-o-toloxy-1,2-propanediol. See mephenesin.

"Tolpit."[79] TM for a tall oil pitch used in roofing cements, caulking compounds, shoemaker's wax, asphalt emulsions and anti-stripping agents, coal dust briquetting.

Topfer's reagent. Dimethylaminoazobenzene 0.5 g in 100 cc 95% ethanol.
Use: To test acidity of stomach contents.

tolualdehdye. See tolyl aldehydes.

toluazotoluidine. See o-aminoazotoluene.

tolu balsam. See balsam.

toluene. (methylbenzene; phenylmethane).
CAS: 108-88-3. $C_6H_5CH_3$ 27th highest-volume chemical produced in US (1985).

Properties: Colorless liquid, benzene-like odor, d 0.866 (20/4C), fp −94.5C, bp 110.7C, refr index 1.497 (20C), aniline equivalent 15, flash p (CC) 40F (4.4C), autoign temperature 997F (536C). Soluble in alcohol, benzene and ether; insoluble in water.
Derivation: (a) By catalytic reforming of petroleum. (b) By fractional distillation of coal-tar light oil.
Method of purification: Rectification.
Grade (usually defined in terms of boiling ranges): Pure, commercial, straw-colored, nitration, scintillation, industrial.
Hazard: Flammable, dangerous fire risk. Explosive limits in air 1.27–7%. Toxic by ingestion, inhalation, and skin absorption. TLV: 100 ppm in air.
Use: Aviation gasoline and high-octane blending stock; benzene, phenol, and caprolactam; solvent for paints and coatings, gums, resins, most oils, rubber, vinyl organosols; diluent and thinner in nitrocellulose lacquers; adhesive solvent in plastic toys and model airplanes; chemicals (benzoic acid, benzyl and benzoyl derivatives, saccharin, medicines, dyes, perfumes); source of toluenediisocyanates (polyurethane resins); explosives (TNT); toluene sulfonates (detergents); scintillation counters.

toluene-2,4-diamine. (m-tolylenediamine; MTD; m-toluylenediamine; diaminotoluene).
CAS: 95-80-7. $CH_3C_6H_3(NH_2)_2$.
Properties: Colorless crystals; soluble in water, alcohol, and ether; mp 99C; bp 280C.
Derivation: Reduction of m-dinitrotoluene with iron and hydrochloric acid.

Hazard: A carcinogen.
Use: Chain extender and crosslinker, intermediate in organic synthesis of dyes, polymers, especially polyurethanes.

toluene-2,4-diisocyanate. (2,4-tolylene diisocyanate; m-tolylene diisocyanate; TDI).
CAS: 584-84-9. $CH_3C_6H_3(NCO)_2$.

Properties: Water-white to pale yellow liquid, sharp pungent odor, bp 251C, 120C (10 mm), flash p 270F (132C), mp 19.4–21.5C (pure isomer), d 1.22 (25/15.5C), vap press 0.01 mm at 20C. Reacts with water producing carbon dioxide; reacts with compounds containing active hydrogen (may be violent); soluble in ether, acetone, and other organic solvents. Combustible.
Derivation: Reaction of 2,4-diaminotoluene with phosgene.
Method of purification: Distillation to remove hydrogen chloride.
Grade: 100% 2,4-isomer; 80% and 65% 2,4-isomer both mixed with 2,6-isomer.
Hazard: Toxic by ingestion and inhalation, strong irritant to skin and tissue, especially to eyes. TLV: 0.005 ppm in air.
Use: Polyurethane foams, elastomers and coatings, crosslinking agent for nylon 6.

toluene-2,6-diisocyanate. See toluene-2,4-diisocyanate.

p-toluenesulfamine. See p-toluenesulfonamide.

p-toluenesulfanilide. $CH_3C_6H_4SO_2C_6H_4NH_2$.
Properties: White to pink crystalline solid, mp 103C, soluble in most lacquer solvents. Combustible.
Derivation: p-Toluene sulfonchloride treated with aniline in presence of lime or carefully regulated amounts of alkalies.
Grade: Technical.
Use: Softener for acetylcellulose in proportions up to 50%, dyestuff intermediate.

o-toluenesulfonamide. CAS: 88-19-7. $CH_3C_6H_4SO_2NH_2$.
Properties: Colorless crystals, mp 156.3C, soluble in alcohol, slightly soluble in water and ether. Combustible.
Use: Plasticizer.

p-toluenesulfonamide. (p-toluenesulfamine; PTSA). CAS: 70-55-3. $CH_3C_6H_4SO_2NH_2$.
Properties: White leaflets, soluble in alcohol, very slightly soluble in water, mp 137C. Combustible.
Derivation: By amination of p-toluene sulfonchloride.
Use: Organic synthesis, plasticizers and resins, fungicide and mildewicide in paints and coatings.

p-toluenesulfondichloroamide.
See dichloramine-T.

o-toluenesulfonic acid. (o-toluenesulfonate). CAS: 104-15-4. $C_6H_4(SO_3H)(CH_3)$.
Properties: Colorless crystals, mp 67.5C, bp 129C, soluble in alcohol, water, and ether. Combustible.
Derivation: By sulfonating toluene with concentrated sulfuric acid at 100C.
Grade: Anhydrous, monohydrate, 40% aqueous solution.
Hazard: Toxic by ingestion and inhalation, strong irritant to tissue.
Use: Dyes, organic synthesis, acid catalyst.

p-toluenesulfonic acid. (p-toluenesulfonate). CAS: 104-15-4. $C_6H_4(SO_3H)(CH_3)$.
Properties: Colorless leaflets, mp 107C, bp 140C (20 mm), soluble in alcohol, ether, and water. Combustible.
Derivation: By action of chlorosulfonic acid on toluene at a low temperature.
Grade: Anhydrous, monohydrate; 40% aqueous solution.
Hazard: Skin irritant.
Use: Dyes, organic synthesis, organic catalyst.

o-toluenesulfonyl chloride. (o-toluenesulfochloride; o-toluenesulfonchloride). CAS: 98-59-9. $H_3CC_6H_4SO_2Cl$.
Properties: Oily liquid, d 1.3383 (20/4C), mp 10.2C, bp 154C (36 mm), insoluble in water, soluble in hot alcohol and in ether and benzene. Combustible.
Derivation: Action of chlorosulfonic acid on toluene.
Use: Intermediate in the synthesis of saccharin and dyestuffs.

p-toluenesulfonyl chloride. (tosyl chloride; p-toluenesulfochloride; p-toluenesulfonchloride). $H_3CC_6H_4SO_2Cl$.
Properties: Solid, mp 71C, bp 145–146C (15 mm), insoluble in water, soluble in alcohol, ether, and benzene. Combustible.
Use: Organic synthesis.

p-toluenesulfonyl semicarbazide.
$C_8H_{11}N_3O_3S$.
Properties: Fine, white powder; d 1.428; decomposes at 440F (226C dry); 415–430F (212–221C compounded).
Use: Blowing agent for polyolefins, impact polystyrene, polypropylene, ABS, etc.

toluenethiol. (thiocresol; tolyl mercaptan). CAS: 100-53-8. $CH_3C_6H_4SH$.
Properties: Cream to white, moist crystals; musty odor; bp 195C. Insoluble in water, soluble in alcohol or ether. There are three isomers with different boiling points.
Hazard: Skin irritant.
Use: Intermediate, bacteriostat.

α-toluenethiol. See benzyl thiol.

toluene trichloride. See benzotrichloride.

toluene trifluoride. See benzotrifluoride.

toluhydroquinone. $CH_3C_6H_3(OH)_2$.
Properties: Pink to white crystals, mp 125–127C, slightly soluble in water, soluble in alcohol and acetone.
Use: Antioxidant, polymerization inhibitor.

α-toluic acid. See phenylacetic acid.

m-toluic acid. (m-toluylic acid; 3-methylbenzoic acid). $C_6H_4CH_3COOH$.
Properties: White to yellowish crystals, slightly soluble in water, soluble in alcohol and ether, d 1.0543, mp 109C, bp 263C, ionization constant 5.3×10^{-5}. Combustible.
Derivation: Oxidation of m-xylene with nitric acid.
Use: Organic synthesis, to form N,N-diethyl-m-toluamide, a broad-spectrum insect repellent.

o-toluic acid. (o-toluylic acid; 2-methylbenzoic acid). $C_6H_4CH_3COOH$.
Properties: White crystals, slightly soluble in water, soluble in alcohol and chloroform, d 1.0621, mp 103.5–104C, bp 256C, refr index 1.512 (114.6C), ionization constant 1.2×10^{-5}. Combustible.
Derivation: Oxidation of o-xylene with dilute nitric acid.
Use: Bacteriostat.

p-toluic acid. (p-toluylic acid; 4-methylbenzoic acid). $C_6H_4CH_3COOH$.

Properties: Transparent crystals, slightly soluble in water, soluble in alcohol and ether, mp 180C, bp 275C, ionization constant 4.3×10^{-5}. Combustible.

Derivation: By treating cymene or turpentine with nitric acid.

Use: Agricultural chemicals, animal feed supplement.

α-toluic aldehyde.   See phenylacetaldehyde.

m-toluidine.   (m-aminotoluene).
CAS: 108-44-1. $CH_3C_6H_4NH_2$.
Properties: Colorless liquid, d 0.980, fp −31.5C, bp 203.3C, slightly soluble in water, soluble in alcohol or ether, flash p 187F (86.1C). Combustible.

Derivation: Reduction of m-nitrotoluene.

Hazard: Toxic by inhalation and ingestion, absorbed by skin.

Use: Dyes, manufacture of organic chemicals.

o-toluidine.   (o-aminotoluene).   CAS: 95-53-4. $CH_3C_6H_4NH_2$.

Properties: Light yellow liquid, becomes reddish-brown on exposure to air and light, volatile with steam, d 1.008 (20/20C), fp −16C, bp 200C, flash p 185F (85C), soluble in alcohol and ether, very slightly soluble in water. Combustible.

Derivation: By the reduction of o-nitrotoluene or obtained mixed with p-toluidine by the reduction of crude nitrotoluene.

Grade: Technical.

Hazard: Toxic by inhalation and ingestion, absorbed by skin, a carcinogen. TLV: 2 ppm in air.

Use: Textile printing dyes, vulcanization accelerator, organic synthesis.

p-toluidine.   (p-aminotoluene).
CAS: 106-49-0. $CH_3C_6H_4NH_2$.
Properties: White lustrous plates or leaflets, d 1.046 (20/4C), mp 45C, bp 200.3C, soluble in alcohol and ether, very slightly soluble in water, flash p 189F (87.2C). Combustible.

Derivation: By the reduction of p-nitrotoluene with iron and hydrochloric acid.

Grade: Technical, flake, or cast.

Hazard: Toxic by inhalation and ingestion, absorbed by skin.

Use: Dyes, organic synthesis, reagent for lignin, nitrite, phloroglucinol.

toluidine maroon.
$CH_3C_6H_3NO_2N_2C_{10}H_5OHCONHC_6H_4NO_2$.
An organic azo pigment obtained by the azo coupling of m-nitro-p-toluidine with the m-nitroanilide of β-hydroxynaphthoic acid.

Properties: Good lightfastness and weather resistance, excellent acid and alkali resistance, poor resistance to bleeding.

Use: Air-dried and baked enamels, truck body finishes.

toluidine red.   Any of a class of pigments based on couplings of β-naphthol and m-nitro-p-toluidine.
See toluidine maroon.

tolu oil.   (tolu balsam oil).
Properties: Yellow liquid; hyacinth-like odor; soluble in alcohol, ether, chloroform, and carbon disulfide; d 0.945–1.09.
Chief known constituents: A terpene, $C_{10}H_{16}$, and esters of cinnamic and benzoic acid. Derivation: From tolu balsam by distillation.
Grade: Technical.
Use: Medicine (expectorant), cough syrups. See also balsam.

toluol.   Obsolete name for toluene.

toluqinone.   (2-methylquinone; p-toluqinone). $CH_3C_6H_3O_2$.
Properties: Yellow leaflets or needles; mp 65–67C; soluble in hot water; very soluble in alcohol, ether, acetone, ethyl acetate, and benzene.

"Tolu-Sol."[125]   TM for a series of solvents, predominantly C_7 hydrocarbons, low in naphthenic hydrocarbons and containing 3–50% aromatics, the balance being essentially paraffinic.
Use: Lacquer diluents and gravure ink solvents.
Hazard: May be flammable.

toluyl aldehyde.   See tolylaldehyde.

toluylene.   See stilbene.

m-toluylenediamine.   See toluene-2,4-diamine.

toluylene red.   See neutral red.

m-, o-, and p-toluylic acid.   See corresponding toluic acid.

tolyl acetate.   See cresyl acetate.

p-tolylaldehyde.   (p-toluylaldehyde; p-tolualdehyde; p-methylbenzaldehyde).
CAS: 1334-78-7. $CH_3C_6H_4CHO$.
Properties: Colorless liquid, refr index 1.54693 (16.6C), d 1.020, bp 204C, slightly soluble in water, soluble in alcohol and ether, there are also o- and m- isomers. Combustible.
Grade: Technical, pure.
Use: Perfumes, pharmaceutical and dyestuff intermediate, flavoring agent.

α-tolylaldehyde dimethylacetal.   See phenylacetaldehydedimethylacetal.

4-o-tolylazo-o-diacetotoluide. See diacetylaminoazotoluene.

o-tolyl biguanide. $NH_2(CNHNH)_2C_6H_4CH_3$.
Properties: White to off-white powder, mp 138C (min). Combustible.
Use: Antioxidant for soaps produced from animal or vegetable oil.

p-tolyldiethanolamine.
$(HOC_2H_4)_2NC_6H_4CH_3$.
Properties: Crystals, mp 62C, bp 297.1C, vap press less than 0.1 (20C), d 1.0723 (20/20C), solubility in water 1.67% by weight (20C), viscosity 155 cp (20C). Very soluble in acetone, ethanol, ethyl acetate, benzene; flash p 385F (196C). Combustible.
Use: Emulsifier, dyestuff intermediate.

m-tolylenediamine. See toluene-2,4-diamine.

m-tolylenediaminesulfonic acid. [4,6-diamino-m-toluenesulfonic acid. $(SO_3H = 1)$].
$CH_3C_6H_2(NH_2)_2SO_3H$.
Properties: White, crystalline solid; soluble in alkalies.
Derivation: By addition of m-toluylenediamine sulfate to oleum and heating.
Use: Dyes.

m-tolylenediisocyanate. See toluene-2,4-diisocyanate.

p-tolyl isobutyrate. See p-cresyl isobutyrate.

p-tolyl-1-naphthylamine-8-sulfonic acid.
(tolylperi acid). $C_{17}H_{15}NO_3S$.
Properties: Greenish-gray needles, soluble in alcohol, rather insoluble in water. Combustible.
Derivation: Arylation of 1-naphthylamine-8-sufonic acid with p-toluidine.
Grade: Technical, mostly as sodium salt.
Use: Azo colors.

p-tolyl phenylacetate. See p-cresyl phenylacetate.

tolyl-p-toluenesulfonate. See cresyl-p-toluenesulfonate.

tomatidine. CAS: 77-59-8. $C_{27}H_{45}NO_2$.
A steroid secondary amine, the nitrogenous aglycone of tomatine. Isolated from the roots of the Rutgers tomato plant as the hydrochloride, $C_{27}H_{45}NO_2 \cdot HCl$.
Properties: Crystals, decomposes at 275–280C.

tomatine. CAS: 17406-45-0. $C_{50}H_{83}NO_{21}$.
A glycosidal alkaloid prepared from the dried leaves and stems of the tomato plant. White crystals, used as plant fungicide and as a specific precipitating agent for cholesterol and other sterols. The crude extract is referred to as tomatin.

"Tona."[173] TM for a proteolytic enzyme meat tenderizer. Supplied as powder and used as liquid.

"Tonalid."[105] TM for 1,1,2,2,3,3,5-heptamethylindan-6-methyl ketone. Readily soluble in all common perfumery raw materials, solvents, and aerosol propellants. Stable and non-discoloring in soaps and cosmetics.
Use: Musk odorant.

toner. An organic pigment which does not contain inorganic pigment or inorganic carrying base. (ASTM) See also lake. Important toners are Pigment Green 7, Pigment Blue 15, Pigment Yellow 12, and Pigment Blue 19.

tonka. (tonka bean; coumarouna bean; dipteryx).
Properties: Black-brownish seeds with wrinkled surface and brittle shining or fatty skins, aromatic bitterish taste, balsamic vanilla-like odor, efflorences of coumarin often appear on the surface. Combustible.
Use: Production of natural coumarin, flavoring extracts, toilet powders.

tonka bean camphor. See coumarin.

"Tonox."[248] TM for a blend of aromatic primary amines, the main component of which is p,p-diaminodiphenylmethane.
Use: Epoxy and urethane curative.

topical. A medical term meaning "applied to the surface of the skin."

topochemical reaction. Any chemical reaction that is not expressible in stoichiometric relationships. Such reactions are characteristic of cellulose; they can take place only at certain sites on the molecule where reactive groups are available, i.e., in the amorphous areas or on the surfaces of the crystalline areas.

"Tophel."[155] TM for a nickel-base thermocouple alloy of 10% chromium. Contains a number of minor elements to give proper electromotive force against pure platinum and produces a black oxide whose structure matches the alloy.
Properties: Mp 2600F, d 8.63, sp heat at 20C 11 cal/g, tensile strength at 20C 88,000 psi.

TOPO. Abbreviation for trioctylphosphinic oxide.

"Toranil."[168] TM for a desugared calcium lignosulfonate produced from a desugared extract of coniferous woods.
Use: Adhesives, binding agents, dispersants, refractory materials, boiler water treating com-

pounds, oil well drilling muds, dye mordants, emulsifiers and foaming agents, concrete, leather, agricultural sprays.

"Tordon."[233]   TM for picloram.

torr.   A pressure unit used chiefly in vacuum technology; it is the pressure required to support 1 mm of mercury at 0C.

torula yeast.   A yeast that utilizes fermentable sugar in industrial wastes, such as fruit cannery refuse and sulfite liquor from pulp mills. The dried yeast is high in protein and vitamin content, enabling it to be used for enriching animal feeds. The enzymes present are destroyed during drying. It is now being made by a new process utilizing petroleum-derived ethanol.

torus.   A doughnut-shaped vacuum chamber which is an essential part of a nuclear fusion reactor.
See tokamak, JET.

tosyl. (Ts).   CAS: 302-17-0.
$CH_3C_6H_4SO_2^-$.   The p-toluenesulfonyl radical. Esters of p-toluenesulfonic acid are known as tosylates.

Toth process.   A process under development for production of aluminum metal which utilizes kaolin and other high-alumina clays. The clay is chlorinated after calcination and the aluminum chloride resulting is reacted with metallic manganese to yield aluminum and manganese chloride. The reaction occurs at the comparatively low temperature of 260C. The manganese chloride is recovered as manganese metal and chlorine by oxidation and subsequent reduction, the manganese being recycled. This is a much cheaper and more efficient method than the Hall process, as it requires less energy input and does not utilize imported bauxite.

tourmaline.
$(Na,Ca)(Al,Fe)B_3Al_3(AlSi_2O_9)(O,OH,F)_4$.
A complex borosilicate of aluminum.
Use: Pressure gauges, optical equipment, oscillator plates, source of boric acid.

tower, distillation.   A metal cylinder from 6 inches to 20 ft in diameter located between the boiler and the condenser in distillation units. The vapors rise through the tower, some of the liquid condensate flowing back down through the tower (reflux). Horizontal plates at intervals of 2 ft are used to achieve contact between countercurrent liquid and the vapor stream. Vapor passes through the liquid on the plate through several

apertures, each covered by a cap, an arrangement called a bubble cap plate. Various types of packings are used in smaller columns, e.g., metal chains, tubes, glass beads, Raschig rings, etc. See HETP (2).

Possibly supplanting distillation towers in the future is a circular metal drum which rotates at 1800 rpm and contains conventional packing materials. The centrifugal force separates the liquid and vapor components as they pass through the drum. This novel distillation technique has been developed by ICI in England and is called Higee, i.e., high gravity.

toxaphene.   (Generic name for technical chlorinated camphene).   CAS: 8001-35-2.
$C_{10}H_{10}Cl_8$.   Formula is approximate, contains 67–69% chlorine.
Properties: Amber, waxy solid with a mild odor of chlorine and camphor, melting range 65–90C, d 1.66 (27C), soluble in common organic solvents.
Hazard: Toxic by ingestion, inhalation, skin absorption; most uses prohibited. TLV: 0.5 mg/m^3 of air.

toxicity.   The ability of a substance to cause damage to living tissue, impairment of the central nervous system, severe illness or, in extreme cases, death when ingested, inhaled, or absorbed by the skin. The amounts required to produce these results vary widely with the nature of the substance and the time of exposure to it. "Acute" toxicity refers to exposure of short duration, i.e., a single brief exposure, "chronic" toxicity refers to exposure of long duration, i.e., repeated or prolonged exposures.

The toxicity hazard of a material may depend on its physical state and on its solubility in water and acids. Some metals that are harmless in solid or bulk form are quite toxic as fume, powder, or dust. Many substances that are intensely poisonous are actually beneficial when administered in micro amounts, as in prescription drugs, e.g., strychnine.

Toxicity is objectively evaluated on the basis of test dosages made on experimental animals under controlled conditions. Most important of these are the LD_{50} (lethal dose, 50%) and the LC_{50} (lethal concentration, 50%) tests, which include exposure of the animal to oral ingestion and inhalation of the material under test. A substance having an LD_{50} of less than 400 mg/kg of body weight is considered very toxic.

toxic substances.   The following list includes a number of chemical individuals and groups that are generally regarded as having toxic properties by either ingestion, inhalation, or absorption via

the skin. There is considerable variation in the degree of toxicity among these and the listing is by no means exhaustive.
See also toxicity.

Individuals	Groups
aniline	aldehydes
asbestos (carcinogen)	alkaloids
benzidine (carcinogen)	allyl compounds
benzpyrene (carcinogen)	arsenic and compounds
	barium and soluble compounds
carbon monoxide	barbiturates
chlorine	beryllium and soluble compounds
coal-tar pitch (carcinogen)	chlorinated hydrocarbons
cresol	chromium (hexavalent carcinogenic compounds)
hydrogen peroxide	
hydrogen sulfide	
methanol	corrosive materials
nickel carbonyl	cyanides
osmium tetroxide	cadmium and compounds
oxalic acid	
ozone	fluorine and compounds
phenol	
pyrene	lead compounds
pyridine	mercury and compounds
phosphine	
stibine	organic phosphate esters
sulfur dioxide	
vinyl chloride nonomer (carcinogen)	radioactive substances
	selenium and compounds
	thallium and compounds
	tin (organic compounds)

The Toxic Substances Control Act (TSCA) passed by Congress in 1976 provides the legal basis for regulations concerning all aspects of the manufacture of such products. Establishment and enforcement of such regulations is carried out by the Environmental Protection Agency (EPA). The American Conference of Governmental Industrial Hygienists issues a periodically revised list of Threshold Limit Values for substances in workroom air, upon which the industrial safety standards of the Occupational Health and Safety Administration (OSHA) are based. The Food and Drug Administration (FDA) is responsible for the enforcement of the Food, Drug, and Cosmetic Act. Decisions made by these agencies are arrived at only after extensive testing by both manufacturers and independent groups. Effective control of toxic materials has assumed increasing importance in recent years and may be expected to become still more rigorous.

See also poison (1), NIOSH.
A Toxicology Data Bank (TDB) has been established by the National Library of Medicine of NIH. The data are set up in a computer-based on-line file available for public use. Extensive physical data are included. The file contains over 5000 toxic substances.
See also Dangerous Properties of Industrial Materials, 6th Edition, N. Irving Sax, published by Van Nostrand Reinhold Company, 115 5th Avenue, New York, NY, 10003. This book contains 20,000 entries, each of which gives physical, chemical, and toxicological data about potentially hazardous materials.

toxicology. The branch of medical science devoted to the study of poisons, including their mode of action, effects, detection, and countermeasures. The subject may be subdivided into (1) clinical, (2) environmental, (3) forensic, and (4) occupational (industrial). The Institute of Chemical Toxicology was formed in 1975, its members comprising a number of the larger chemical companies.
See also toxicity; poison (1).

Toxic Substances Control Act. (TSCA).
See toxic substances; Environmental Protection Agency.

TPA. Abbreviation for terephthalic acid.

TPB. Abbreviation for tetraphenylbutadiene.

TPG. Abbreviation for triphenylguanidine.

TPN. Abbreviation for triphosphopyridine nucleotide.
See nicontinamide adenine dinucleotide phosphate.

TPO rubber. Abbreviation for thermoplastic polyolefin rubber.
See elastomer.

Tpp. (1) Abbreviation for triphenyl phosphate. (2) Abbreviation for thiamine pyrophosphate.
See cocarboxylase.

TPT. (1) Abbreviation for triphenyltetrazolium chloride.
See tetrazolium chloride.
(2) Abbreviation for tetraisopropyl titanate.

trace element. (micronutrient). An element essential to plant and animal nutrition in trace concentration, i.e., minute fractions of 1% (1000 ppm or less). Plants require iron, copper, boron, zinc, manganese, potassium, molybedenum, sodium, and chlorine. Animals require iron, copper, manganese, cobalt, selenium, and potassium. Such elements are also called micronutrients. Do not confuse with "tracer."

tracer. A chemical entity (almost invariably radioactive and usually an isotope) added to the reacting elements or compounds in a chemical process, which can be traced through the process by appropriate detection methods, e.g., Geiger counter. Compounds containing tracers are often said to be "tagged" or "labelled." Carbon-14 is a commonly used tracer and radioactive forms of iodine and sodium are also used. Many complex biochemical reactions have been examined in this way, (e.g., photosynthesis). Nonradioactive deuterium (hydrogen isotope) is sometimes used, the detection being by molecular weight determination. Radioactive enzymes are also available for tracer studies, e.g., ribonuclease, pepsin, trypsin, and others.
See also labeling (2).

trade mark. (TM). A word, symbol, or insignia designating one or more proprietary products or the manufacturer of such products, which has been officially registered with the government trade mark agency. The accepted designation is a superior capital R enclosed in a circle; however, quotation marks may also be used, as in this Dictionary. The term "trade name," though widely used, is not applicable to such products; according to the US Trade Mark Association, a trade name is the name under which a company does business, e.g., the Blank Chemical Company. Use of a trade mark without proper indication of its proprietary nature places the name in jeopardy; a number of trade marks have been invalidated as a result of this practice.

trade name. See trade mark.

trade sales. In the paint industry this term is applied to paints intended for sales to the general public, as in hardware stores and similar outlets.

tragacanth gum. CAS: 9000-65-1.
Properties: Dull white, translucent plates or yellowish powder; soluble in alkaline solutions, aqueous hydrogen peroxide solution; strongly hydrophilic; insoluble in alcohol. Combustible.
Constituents: Polysaccharides of galactose, fructose, xylose, and arabinose with glucuronic acid.
Occurrence: Southwestern Europe, Greece, Turkey, Iran.
Grade: USP, FCC, No. 1, 2, 3.
Use: Pharmacy (emulsions), adhesives, leather dressing, textile printing and sizing, thickener and emulsifier, dyes, food products (ice cream, desserts), toothpastes, coating soap chips and

powders, hairwave preparations, confectionery, printing inks, tablet binder.

tranquilizer. See psychotropic drug

trans-. See cis-.

transalkylation. A type of disproportionation reaction by which toluene is hydrogenated to benzene and mixed xylene isomers free from ethylbenzene, avoiding the formation of methane resulting from the conventional hydrodealkylation process. Transalkylation of toluene to benzene involves the use of a catalyst; the yield is claimed to be 97%, based on toluene feed.

transferase. An enzyme whose activity causes a transfer of a radical from one molecule to another. Examples are transaminases, transacetylases, and transmethylases, which effect the transfer of amino, acetyl, and methyl groups respectively.
See enzyme.

transfer mold. A chamber in which a thermosetting plastic is softened by heat and pressure, and from which it is forced by high pressure through a suitable orifice into a closed mold for final curing.

transfer RNA. See ribonucleic acid.

transference number. That portion of the total current carried by any species of ion in an electrolyte in the fluid state. The symbol t^+ is usually used for a positive ion and t^- for a negative ion.

transformer oil. A liquid having the property of insulating the coils of transformers, both electrically and thermally. There are two broad classes, (1) natural and (2) synthetic. The natural type includes refined mineral oils (petroleum fractions), which have low viscosities and high chemical and oxidative stability. The synthetic types are (a) chlorinated aromatics (chlorinated biphenyls and trichlorobenzene) known collectively as askarels; (b) silicone oils; and (c) ester liquids such as dibutyl sebacate. All these types are nonflammable, but are combustible. Flash points are 250–300F (121–148C).
See also dielectric.
Note: Use of PCB (chlorinated biphenyls) has been discontinued because of their ecologically damaging effects.

"Transist AR."[329] A grade of reagent quality chemicals, specially controlled for use in the manufacture of semiconductors and other electronic devices and precision instruments.

transition element. (transition metal). Any of a number of elements in which the filling of the outermost shell to eight electrons within a period is interrupted to bring the penultimate shell from 8 to 18 or 32 electrons. Only these elements can use penultimate shell orbitals as well as outermost shell orbitals in bonding. All other elements, called "major group" elements, can use only outermost shell orbitals in bonding. Transition elements include elements 21 through 29 (scandium through copper), 39 through 47 (yttrium through silver), 57 through 79 (lanthanum through gold), and all known elements from 89 (actinium) on. All are metals. Many are noted for exhibiting a variety of oxidation states and forming numerous complex ions as well as possessing extremely valuable properties in the metallic state.
See also orbital theory. (R. T. Sanderson).

transmutation. The natural or artificial transformation of atoms of one element into atoms of a different element as the result of a nuclear reaction. The reaction may be one in which two nuclei interact, as in the formation of oxygen from nitrogen and helium nuclei (beta particles), or one in which a nucleus reacts with an elementary particle such as a neutron or a proton. Thus a sodium atom and a proton form a magnesium atom. Radioactive decay, e.g., of uranium can be regarded as a type of transmutation. The first transmutation was performed by the English physicist Rutherford in 1919.

transportation label. See label (1).

transuranic element. An element of higher atomic number than uranium, not found naturally, and produced by nuclear bombardment. The most recently discovered are 105 and 106, although there is as yet no confirmation. They are all radioactive.
See actinide, Periodic Table.

Traube purine synthesis. Preparation of an appropriate 4,5-diaminopyrimidine by introduction of the amino group into the 5-position of 4-amino-6-hydroxy- or 4,6-diaminopyrimidines by nitrosation and ammonium sulfide reduction, followed by ring closure with formic acid or chlorocarbonic ester.

tremolite. CAS: 1332-21-4. $Ca_2Mg_5Si_8(OH)_2$. A variety of asbestos. Some tremolite is sold as "fibrous talc."
Properties: Color white to light green, luster vitreous to silky, hardness 5–6, d 3.0–3.3, resistant to acids. Noncombustible.
Occurrence: New York, California, Maryland, South Africa.

Hazard: Inhalation of dust or fine particles is dangerous. Carcinogenic.
Use: As asbestos, particularly in acid-resisting applications, ceramics, paint.

"Treopax."[337] TM for 90% minimum purity form of zirconium oxide. Contains 5% max silicon dioxide, mp 2480C.
Use: Source for zirconium oxide for welding rod coatings and to stabilize color in titanium enamels.

tretamine. Generic name for 2,4,6-tris(1-aziridinyl)-s-triazine.
See under triethylenemelamine.

"Trevira 271."[589] TM for a durable flame retardant fiber in which the retarding agent is a constituent of the polymer structure and thus is impossible to remove by laundering or dry-cleaning. The polymer used is polyethylene terephthalate; the nature of the retardant has not been disclosed. The fiber is stated to be self-extinguishing and will melt when subjected to a direct flame. It is intended for use primarily in children's sleepwear.

triacetate fiber. See acetate fiber.

triacetin. (glyceryl triacetate).
CAS: 102-76-1. $C_3H_5(OCOCH_3)_3$.
Properties: Colorless liquid with slight fatty odor and a bitter taste, d 1.160 (20C), bp 258–260C, sets to a glass −37C, refr index 1.4312 (20C), flash p 300F (149C), bulk d 9.7 lbs/gal. Slightly soluble in water; very soluble in alcohol, ether, and other organic solvents. Combustible.
Derivation: Action of acetic acid on glycerol.
Method of purification: Vacuum distillation.
Grade: Technical, CP, ND, FCC.
Use: Plasticizer, fixative in perfumery, manufacture of cosmetics, specialty solvent, to remove carbon dioxide from natural gas, medicine (topical antifungal).

triacontanoic acid. See melissic acid.

1-triacontanol. CAS: 593-50-0.
$CH_3(CH_2)_{28}CH_2OH$.   A 30-carbon, straight-chain fatty alcohol.
Properties: Colorless needles from ether, mp 85−88C, soluble in most organic solvents, insoluble in water, d at mp 0.777. Combustible.
Occurrence: Beeswax, carnauba wax, leaf wax.
Use: Biochemical research, growth promoter, fertilizer supplement.

"Triafil."[470] TM for electric insulating film based on cellulose triacetate.

trialkyl boranes. R_3B. (where R = alkyl radical) See tributylborane, triethylborane.

trialkylsilanol. An alcohol derivative of silane.
Hazard: May be flammable.
Use: Short stopping agent for silicone polymers.

triallylamine. CAS: 102-70-5.
$(H_2C:CHCH_2)_3N$.
Properties: Liquid, d 0.800 (20/4C), fp -70C, bp 150–151C, refr index 1.4501 (20C), flash p (TOC) 103F (39.4C).
Hazard: Fire risk. Irritant.
Use: Intermediate.

triallyl cyanurate. CAS: 101-37-1.
$(CH_2:CHCH_2OC)_3N_3$.
Properties: Colorless liquid or solid, mp 27.32C, flash p (TOC) above 176F (80C), d 1.1133 (30C), refr index 1.5049 (25C). Miscible with acetone, benzene, chloroform, dioxane, ethyl acetate, ethanol, and xylene. Combustible.
Hazard: Toxic by ingestion and inhalation.
Use: Polymers as monomer and modifier, organic intermediate.

triallyl phosphate. $(CH_2:CHCH_2O)_3PO$.
Properties: Water-white liquid, fp −50C, bp 80C (0.5 mm), refr index 1.448 (25C), d 1.064 (25/15C). Combustible.
Use: Intermediate.

triamcinolene. (α-fluoro-16(α)-hydroxyprednisolone). CAS: 124-94-7. $C_{21}H_{27}FO_6$.
Properties: White, crystalline powder; mp 264–268C. Insoluble in water, slightly soluble in usual organic solvents, soluble in dimethylformamide.
Grade: ND.
Use: Medicine. Also available as the acetonide.

1,3,5-triaminobenzene. CAS: 108-72-5.
$C_6H_3(NH_2)_3$.
Properties: Solid; mp (anhydrous) 129C, hydrate 84–86C (1.5 moles water); soluble in water, acetone, and alcohol; insoluble in ether, cold benzene, carbon tetrachloride, and petroleum ether. Combustible.
Use: Ion exchange resin intermediate, wetting and frothing agents, photographic developers, organic reactions.

2,4,6-triaminotoluene trihydrochloride.
$C_6H_2(NH_2HCl)_3CH_3 \cdot HOH$.
Properties: Light tan to cream crystals, very soluble in water, soluble in alcohol and acetone, insoluble in benzene, mp 119C (free base)
Use: Nongelatin photographic emulsion with ethylenediamine for fixation, ion exchange resins, wetting and frothing agents, photographic developers, intermediate varnishes and rubber chemicals.

2,4,6-triamino-sym-triazine. See melamine.

triamylamine. $(C_5H_{11})_3N$.
Properties: Colorless to yellow liquid, d 0.79–0.80 (20C), triamylamine content at least 98.0%, 95% boils between 225–260C, flash p 215F (101.6C), refr index 1.4374 (18C), insoluble in water, soluble in gasoline. Combustible.
Derivation: Reaction of amyl chloride and ammonia.
Hazard: Irritant.
Use: Corrosion inhibitor, insecticidal preparations.

triamylbenzene. $(C_5H_{11})_3C_6H_3$.
Properties: Colorless liquid, d 0.87 (20C), boiling range 300–320C, odor faintly aromatic, flash p 270F (132.2C). Combustible.
Use: Chemical intermediate.

triamyl borate. $(C_5H_{11})_3BO_3$.
Properties: Colorless liquid, d 0.845 (20C), boiling range 220–280C, flash p 180F (82.2C), odor faintly alcoholic, soluble in alcohol and ether. Combustible.
Derivation: direct heating of boric acid and amyl alcohol.
Use: Varnishes.

tri-p-tert-amylphenyl phosphate.
$(C_5H_{11}C_6H_4)_3PO$.
Properties: Liquid, boiling range 305–345C at 5 mm, mp 62–63C, white solid, odorless, insoluble in water. Combustible.
Use: Plasticizer.

tri-p-anisylchloroethylene. See chlorotrianisene.

triarylmethane dye. Any of a group of dyes whose molecular structure involves a central carbon atom joined to three aromatic nuclei. CI numbers range from 42000–44999. The color is due in part to the aromatic rings and to the chromophore groups $=C=NH$ and $=C=N—$. The members of this class function as basic dyes for cotton, using tannin as a mordant or if they contain sulfonic acid groups, as acid dyes for wool and silk. Examples are malachite green and methyl violet.
See also triphenylmethane dyes.

s-triazine derivatives. See ammelide, ammeline, melamine.

s-triazine-3,5(2H,4H)dione riboside. See 6-azauridine.

s-triazine-2,4,6-triol. See cyanuric acid.

triazole. C$_2$H$_3$N$_3$. A five-membered ring compound containing three nitrogens in the ring.
Use: Suggested as photoconductors in copying systems.

triazone resin. One of a class of amino resins produced from urea, formaldehyde, and a primary amine.
Use: Textile and fabric treatment.
See dimethylolethyltriazone.

"Tribase."[304] TM for lead sulfate, tribasic. "Tribase-E." Basic lead silicate sulfate.
Use: Stabilizer for vinyl resins.

tribasic. See monobasic.

tribromoacetaldehyde. (bromal). CBr$_3$CHO.
Properties: Oily, yellowish liquid; d 2.66; bp 174C; soluble in water, alcohol, or ether. Combustible.
Derivation: (a) By adding bromine to a solution of paraldehyde in ethylacetate. (b) By adding bromine to absolute alcohol, fractionating, treating the fraction boiling at 165C with water, and distilling.
Hazard: As for bromine.
Use: Organic synthesis.

tribromoacetic acid. CAS: 75-96-7.
CBr$_3$COOH.
Properties: Colorless crystals; soluble in water, alcohol, or ether; mp 135C; bp 245–250C. Combustible.
Derivation: By oxidizing tribromoacetaldehyde with nitric acid.
Hazard: As for bromine.
Use: Organic synthesis.

tribromo-tert-butyl alcohol. (acetone-bromo-form). CAS: 76-08-4. CBr$_3$C(CH$_3$)$_2$OH.
Properties: Fine, white, prismatic crystals, camphor odor and taste, mp 176C, slightly soluble in water, soluble in alcohol and ether. Combustible.
Derivation: Reaction of acetone and bromoform with solid potassium hydroxide.
Hazard: As for bromine.
Use: Vinyl chloride polymerization.

tribromoethanol. (1,1,1-tribromoethyl alcohol).
CAS: 75-80-9. CBr$_3$CH$_2$OH.
Properties: White crystals or powder with slight aromatic odor and taste, mp 79–82C, bp 94C (11 mm), unstable in air and light. Slightly soluble in water, soluble in alcohol, ether, benzene, and amylene hydrate; aqueous and alcoholic so-lutions decompose on exposure to light. Combustible.
Grade: NF.
Derivation: By reduction of tribromoacetaldehyde with aluminum isopropylate.
Use: Medicine (basal anesthetic).

tribromomethane. See bromoform.

1,1,1-tribromo-2-methyl-2-propanol.
CBr$_3$C(CH$_3$)$_2$OH.
Properties: Fine, white crystals; mp 176–177C; soluble in water, methanol, ether. Combustible.
Use: Organic synthesis.

tribromonitromethane. See bromopicrin.

tribromophenol. See bromol.

1,2,3-tribromopropane. (allyl tribromide).
CAS: 96-11-7. BrCH$_2$CHBrCH$_2$Br.
Properties: Colorless liquid, d 2.43, mp 16C, bp 220C, refr index 1.584, soluble in alcohol and ether, insoluble in water.
Derivation: Gamma-ray initiated reaction of bro-motrichloromethane with allyl bromide.
Use: Nematocide.

3,4',5-tribromosalicylanilide. (tribromasalan).
CAS: 87-10-5.
Br$_2$C$_6$H$_2$(OH)C(O)NHC$_6$H$_4$Br. An active antiseptic.
Use: Soaps.
Hazard: A suspected carcinogen, use in cosmetics prohibited (FDA).

tributoxyethyl phosphate.
[CH$_3$(CH$_2$)$_3$O(CH$_2$)$_2$O]$_3$PO.
Properties: Slightly yellow, oily liquid; insoluble or limited solubility in glycerol, glycols, and certain amines; soluble in most organic liquids. D 1.020 (20C), fp −70C (viscous liquid), boiling range 215–228C (4 mm), flash p 435F (223C), refr index 1.434 (25C). Combustible.
Use: Primary plasticizer for most resins and elastomers, floor finishes and waxes, flame-retarding agent.

tri-n-butyl aconitate. C$_3$H$_3$(COOC$_4$H$_9$)$_3$.
Properties: Colorless, odorless liquid. D 1.018 (20C), refr index 1.4500–1.4530 (25C), bp 190C (3 mm), insoluble in water, soluble in organic solvents. Combustible.
Use: Plasticizer-stabilizer for vinylidene chloride polymers, synthetic rubbers and cellulosic lacquers, insecticides.

tri-n-butylaluminum. CAS: 1116-70-7.
(CH$_3$CH$_2$CH$_2$CH$_2$)$_3$Al.
Properties: Colorless, pyrophoric liquid. Bulk d 0.823 g/mL (20C), fp −26.7C.

Derivation: Exchange reaction of butene-1 and isobutyl aluminum.

Hazard: Highly flammable, dangerous fire risk, ignites spontaneously.

Use: Production of organo-tin compounds.

tri-n-butylamine. CAS: 102-82-9. $(C_4H_9)_3N$.
Properties: Pale yellow liquid with amine odor, bp 214C, fp −70C, d 0.8 (20/20C), bulk d 6.5 lbs/gal, refr index 1.4297 (20C), flash p (OC) 185F (85C), slightly soluble in water, soluble in most organic solvents. Combustible.
Derivation: By reaction of butanol or butyl chloride with ammonia.
Grade: Technical.
Hazard: Skin irritant, central nervous system stimulant.
Use: Solvent, inhibitor in hydraulic fluids, intermediate.

tri-n-butylborane. (tri-n-butylborine). $(CH_3CH_2CH_2CH_2)_3B$.
Properties: Colorless pyrophoric fluid, fp −34C, bp 170C (222 mm), d 0.747 (25C), vap press 0.1 mm (20C), refr index 1.4285 (20C), insoluble in water, soluble in most organic solvents, flash p −32F (−35.5C).
Hazard: Flammable, dangerous fire risk, store and use in inert atmosphere, ignites spontaneously in air.
Use: Petrochemical industry, organic reactions, catalyst.

tributyl borate. (butyl borate). $(C_4H_9)_3BO_3$.
Properties: Water-white liquid, d 0.8550–0.8570, bp 232.4C, distillation range, 85% distills between 135C and 140C (40 mm), refr index 1.4071 (25C), fp below −70C, viscosity 1.601 cp (25C), flash p (OC) 200F (93.3C), hydrolyzes rapidly, miscible with common organic liquids. Combustible.
Derivation: From butanol and boric acid.
Use: Welding fluxes, intermediate in preparation of borohydrides, flame retardant for textiles (with boric acid).

tri-n-butylborine. See tri-n-butylborane.

tri-n-butylchlorostannate. See tributyltin chloride.

tributyl citrate. (butyl citrate). $C_3H_5O(COOC_4H_9)_3$.
Properties: Colorless or pale yellow, stable, odorless, nonvolatile liquid. Practically insoluble in water, fp −20C, bp 233.5C at 22.5 mm, flash p (COC) 315F (157C), refr index 1.4453 at 20C, d 1.042 (25/25C), bulk d 8.7 lbs/gal at 20C, pour point viscosity 31.9 cp (25C). Combustible.

Grade: Technical.
Use: Plasticizer, antifoam agent, solvent for nitrocellulose.

tributyl(2,4-dichlorobenzyl)phosphonium chloride. CAS: 115-78-6. $Cl_2C_6H_3CH_2P(C_4H_9)_3Cl$.
Properties: White, crystalline solid with a mild aromatic odor; soluble in water, acetone, ethanol, isopropanol and hot benzene; insoluble in hexane and ether; technical grade melts 114–120C.
Use: Growth retardant for ornamental plants.

2,4,6-tri-tert-butylphenol. $[(CH_3)_3C]_3C_6H_2OH$.
Properties: Solid, mp 131C, insoluble in water, soluble in most organic solvents. Combustible.
Use: Permissible antioxidant for aviation gasolines (when mixed with other butylphenols) (ASTM).

tri-p-tert-butylphenyl phosphate. $[(CH_3)_3CC_6H_4O]_3PO$.
Properties: Solid, bp 320C (5 mm), mp 102–105C, flash p 545F (285C), insoluble in water. Combustible.
Use: Plasticizer.

tributyl phosphate. (TBP). CAS: 126-73-8. $(C_4H_9)_3PO_4$.
Properties: Stable, colorless, odorless liquid. Miscible with most solvents and diluents, soluble in water, refr index 1.4226 at 25C, bp 292C, latent heat of vaporization 55.1 cals/g at 289C, fp below −80C, flash p (COC) 295F (146C), Saybolt viscosity 38.6 sec at 29.4C, bulk d 8.19 lb/gal, d 0.978 (20/20C). Combustible.
Grade: Technical.
Hazard: Toxic by ingestion and inhalation, irritant to skin. TLV: 2.5 mg/m³ of air.
Use: Heat-exchange medium, solvent extraction of metal ions from solution of reactor products, solvent for nitrocellulose, cellulose acetate, plasticizer, pigment grinding assistant, antifoam agent, dielectric.

tri-n-butyl phosphine. $(CH_3CH_2CH_2CH_2)_3P$.
Properties: Colorless liquid with garlic odor, d 0.8100 (min at 25/4C), fp −60 to -65C, bp 240C, flash p 104F (40C), autoign temperature 392F (200C), refr index 1.463 (20C). Almost insoluble in water; miscible with ether, methanol, ethanol, benzene. An organic base and strong reducing agent. Combustible.
Use: Polymerization crosslinking catalyst, organic intermediate, fuels.

tributyl phosphite. CAS: 102-85-2. $(C_4H_9O)_3P$.
Properties: Water-white liquid, bp 120C (8 mm), d 0.911 (25C), refr index 1.4301 (25C), flash p 250F (121C), decomposes in water, soluble in common organic solvents. Combustible.

Use: Additive for greases and extreme-pressure lubricants, stabilizer for fuel oils and polyamides, gasoline additive.

O,O,O-tributyl phosphorothioate. (tributyl thiophosphate). $(C_4H_9O)_3PS$.
Properties: Colorless liquid with characteristic odor, bp 142–145C (4.5 mm), d 0.987, flash p (COC) 295F (146C), insoluble in water, soluble in most organic solvents. Combustible.
Hazard: Highly toxic, cholinesterase inhibitor.
Use: Plasticizer, lubricant additive, antifoam agent, hydraulic fluid, intermediate.

S,S,S-tributyl phosphorotrithioate. ("DEF"). CAS: 78-48-8. $(C_4H_9S)_3PO$.
Properties: Liquid, bp 150C (0.3 mm), insoluble in water, soluble in aliphatic, aromatic and chlorinated hydrocarbons.
Hazard: Cholinesterase inhibitor.
Use: Cotton defoliant.

tributylphosphorotrithioite. CAS: 150-50-5. $(C_4H_9S)_3P$.
Properties: Nearly colorless liquid, insoluble in water, soluble in a variety of organic solvents, bp 115–134C (0.08 mm), d 1.02 (20C), refr index 1.542 (25C).
Derivation: Reaction of butyl mercaptan with phosphorus trichloride.
Hazard: Cholinesterase inhibitor.
Use: Cotton defoliant.

tributyl thiophosphate. See tributyl phosphorothioate.

tributyltin acetate. CAS: 56-36-0. $(C_4H_9)_3SnOOCCH_3$.
Properties: White, crystalline solid.
Derivation: Reaction of sodium acetate with tributyltin chloride.
Hazard: Toxic material. TLV (as Sn): 0.1 mg/m³ of air.
Use: Fungicide and bactericide.

tributyltin chloride. (tri-n-butylchlorostannate). $(C_4H_9)_3SnCl$.
Properties: Colorless liquid; soluble in the common organic solvents, including alcohol, heptane, benzene, and toluene; insoluble in cold water, but hydrolyzes in hot water; d 1.20 (20/4C); refr index 1.4903 (25C); bp 145–147C (5 mm).
Derivation: Reaction of tetrabutyltin with dibutyltin chloride.
Hazard: Toxic material. TLV (as Sn): 0.1 mg/m³ of air.
Use: Rodenticide, intermediate, rodent-repellent cable coatings.

tributyltin oxide. CAS: 56-35-9. $(C_4H_9)_3SnO$.
Properties: Colorless to pale yellow liquid, soluble in many organic solvents, practically insoluble in water, bp 180C (2 mm).
Hazard: Toxic material. TLV (as Sn): 0.1 mg/m³ of air.
Use: Bactericide, fungicide, intermediate.

tri-n-butyl tricarballylate. $(C_4H_9OCOCH_2)_2CHCOOC_4H_9$.
Properties: Liquid, d 1.004 (24C), refr index 1.4388 (26.5C), insoluble in water. Combustible.
Use: Plasticizer.

tributyrin. See glyceryl tributyrate.

tricalcium aluminate. See calcium aluminate.

tricalcium citrate. See calcium citrate.

tricalcium orthoarsenate. See calcium arsenate.

tricalcium orthophosphate. See calcium phosphate, tribasic.

tricalcium phosphate. See calcium phosphate, tribasic.

tricalcium silicate. See cement, Portland. Also used as an anticaking agent in foods (up to 2%).

tricamba. (3,5,6-trichloro-o-anisic acid). CAS: 2307-49-5. $C_6HCl_3(COOH)(OCH_3)$.
Properties: Crystals, mp 137–139C, very slightly soluble in water, moderately soluble in xylene, freely soluble in alcohol.
Hazard: Toxic by ingestion.
Use: Herbicide.

tricarbimide. See cyanuric acid.

tricarboxylic acid cycle. See TCA cycle.

trichlorfon. (O,O-dimethyl[2,2,2-tri-chloro-1-hydroxyethyl]phosphonate). CAS: 52-68-6. $(CH_3O)_2P(O)CH(OH)CCl_3$.
Properties: White, crystalline solid; mp 83–84C; bp 100C (1 mm); d 1.73 (20/4C); soluble in water, benzene, chloroform, ether; insoluble in oils.
Hazard: Cholinesterase inhibitor, absorbed by skin, use may be restricted.
Use: Systemic insecticide, medicine (anthelmintic).

trichloroacetaldehyde. See chloral.

trichloroacetaldehyde, hydrated. See chloral hydrate.

trichloroacetic acid. (TCA).   CAS: 76-03-9.
CCl_3COOH.
Properties: Deliquescent, colorless crystals; sharp, pungent odor; d 1.6298; mp 57.5C; bp 197.5C; soluble in water, alcohol, and ether; flash p none. Nonflammable.
Derivation: (a) Treating chloral hydrate with fuming nitric acid, (b) from glacial acetic acid by the action of chlorine in presence of sunlight, UV radiation, or catalysts.
Grade: Technical, CP, USP.
Hazard: Toxic by ingestion and inhalation, strong irritant to skin and tissue. TLV: 7 mg/m³ of air.
Use: Organic synthesis, reagent for detection of albumin, medicine, pharmacy, herbicides.

trichloroacetonitrile.   CAS: 545-06-2.
CCl_3CN.
Properties: Colorless liquid, bp 85C, d 1.44, refr index 1.440.
Derivation: Reaction of methylnitrile, hydrochloric acid, and chlorine.
Hazard: Strong irritant to tissue.
Use: Insecticide.

trichloroacetyl chloride. CCl_3COCl.
Properties: Liquid, d 1.654 (0/4C), bp 118C, decomposes in water, soluble in alcohol.
Hazard: Toxic by ingestion and inhalation, strong irritant to skin and tissue.

S-2,3,3-trichloroallyl-N,N-diisopropylthiolcarbamate.   CAS: 2303-17-5.
$[(CH_3)_2CH]_2NC(O)SCH_2CCl:CCl_2$.
Properties: Oily liquid; bp 148–149C (9 mm); mp 29–30C; practically insoluble in water; soluble in alcohol, acetone, ether, and heptane. Combustible.
Use: Herbicide.

trichloro-o-anisic acid.   See tricamba.

2,4,6-trichloroanisole.   CAS: 87-40-1.
$C_7H_5Cl_3O$.
Properties: Acicular crystals; mp 60C; bp 240C (238 mm); insoluble in water; soluble in benzene, methanol, and dioxane; gradually sublimes at room temperature.
Use: Dyeing auxiliary for polyester fabrics.

1,2,3-trichlorobenzene.   CAS: 87-61-6.
$C_6H_3Cl_3$.
Properties: White crystals, d (solid) 1.69, refr index 1.5776 (19C), bp 221C, mp 52.6C, insoluble in water, slightly soluble in alcohol, soluble in ether, flash p 235F (112.7C) (CC). Combustible.
Hazard: Toxic by ingestion and inhalation.
Use: Organic intermediate.

1,2,4-trichlorobenzene.   CAS: 120-82-1.
$C_6H_3Cl_3$.
Properties: Colorless, stable liquid, odor similar to that of o-dichlorobenzene, miscible with most organic solvents and oils, insoluble in water, d 1.4634 (25C), bp 213C, mp 17C, flash p 210F (98.9C). Combustible.
Derivation: Chlorination of monochlorobenzene.
Grade: Technical, 99%, mixture of 1,2,4- and 1,2,3-isomers distilling at 213–219C.
Hazard: Toxic by ingestion and inhalation. TLV: CL of 5 ppm in air.
Use: Solvent in chemical manufacturing, dyes and intermediates, dielectric fluid, synthetic transformer oils, lubricants, heat-transfer medium, insecticides.

2,3,6-trichlorobenzyloxypropanol.
$(Cl)_3C_6H_2CH_2OCH_2CH(CH_3)OH$.
Properties: Liquid, bp 121–124C (0.1 mm), almost insoluble in water, soluble in most organic solvents. Combustible.
Hazard: Toxic by ingestion and inhalation.
Use: Herbicide.

1,1,1-trichloro-2,2-bis(chlorophenyl)ethane.
See DDT.

1,1,1-trichloro-2,2-bis(p-methoxyphenyl)ethane.
See methoxychlor.

B-trichloroborazole.   BClNHBClNHBClNH.
Properties: White, crystalline solid; mp 84.5–85.5C; bp 96.5–98C/37 mm; soluble in many organic solvents; highly reactive.
Use: Intermediate, gelling agent, catalyst complexing agent.

trichlorobromomethane.   See bromotrichloromethane.

trichlorobutylene oxide.   (TCBO).

$$Cl-\underset{\underset{Cl}{|}}{\overset{\overset{Cl}{|}}{C}}-CH_2-CH-CH\diagdown\!\!\!\diagup O$$

A reactive liquid epoxide used as an organic solvent and surfactant intermediate; its polymers can be used for polyester, polyurethane, and polyacrylic resins, polyether polyols, flame retardants, etc.

trichlorobutyraldehyde hydrate.   See butyl chloral hydrate.

3,4,4'-trichlorocarbanilide.
$C_6H_3Cl_2NHCONHC_6H_4Cl$.
Properties: Heat-resistant white powder, mp 250C.

Use: Bacteriostat in soaps and detergents, plastics.

1,1,1-trichloroethane. (methyl chloroform).
CAS: 71-55-6. CH_3CCl_3.
Properties: Colorless liquid, d 1.325, bp 75C, fp −38C, insoluble in water, soluble in alcohol and ether, flash p none. Nonflammable.
Hazard: Irritant to eyes and tissue. TLV: 350 ppm in air.
Use: Solvent for cleaning precision instruments, metal degreasing, pesticide, textile processing.

1,1,2-trichloroethane. (vinyl tichloride; β-trichloroethane).   CAS: 79-00-5.
$CHCl_2CH_2Cl$.
Properties: Clear, colorless liquid, characteristic sweet odor, bp 113.7C, d 1.4432 (20C/4C), refr index 1.4458, vap press 16.7 mm (20C), bulk d 12.0 lbs/gal (20C), fp −36.4C, flash p none. Miscible with alcohols, ethers, esters and ketones; insoluble in water. Nonflammable.
Grade: Technical.
Hazard: Irritant, absorbed by skin. TLV: 10 ppm in air.
Use: Solvent for fats, oils, waxes, resins, other products; organic synthesis.

trichloroethanol.   CAS: 115-20-8.
CCl_3CH_2OH.
Properties: Viscous liquid, ether-like odor, hygroscopic. Slightly soluble in water, miscible with alcohol, ether, and carbon tetrachloride. Bp 150C, fp 13C, d 1.541 (25/4C). Combustible.
Use: Intermediate, anesthetic.

trichloroethylene. (tri).   CAS: 79-01-6.
$CHCl:CCl_2$.
Properties: Stable, low-boiling, colorless, photoreactive liquid; chloroform-like odor; will not attack the common metals even in the presence of moisture. Bp 86.7C, fp −73C, d 1.456–1.462 (25/25C), refr index 1.4735 (27C), miscible with common organic solvents, slightly soluble in water. Nonflammable.
Derivation: From tetrachloroethane by treatment with lime or alkali in the presence of water, or by thermal decomposition, followed by steam distillation.
Grade: USP, technical, high purity, electronic, metal degreasing, extraction.
Hazard: Toxic by inhalation. Use as solvent not permitted in some states. FDA has prohibited its used in foods, drugs, and cosmetics. TLV: 50 ppm in air.
Use: Metal degreasing; extraction solvent for oils, fats, waxes; solvent dyeing; dry cleaning; refrigerant and heat exchange liquid; fumigant; cleaning and drying electronic parts; diluent in paints and adhesives; textile processing; chemical intermedi-

ate; aerospace operations (flushing liquid oxygen).

trichlorofluoromethane. (fluorotrichloromethane; fluorocarbon-11).   CAS: 75-69-4.
CCl_3F.
Properties: Colorless, nearly odorless, volatile liquid. Bp 23.7C, fp −111C, d 1.494 (17.2C), critical pressure 43.2 atmospheres. Noncombustible.
Derivation: From carbon tetrachloride and hafnium, in the presence of fluorinating agents such as antimony tri- and pentafluoride.
Grade: Technical, 99.9% min.
Hazard: TLV: CL of 1000 ppm in air.
Use: Solvent, fire extinguishers, chemical intermediate, blowing agent.

trichloroisocyanuric acid. (1,3,5-trichloro-s-triazine-2,4,6-trione).   CAS: 87-90-1.

$\overline{OCNClCONClCONCl}$.
Properties: White, slightly hygroscopic ,crystalline powder or granules; loose bulk d 31 lbs/cu ft; granular 60 lbs/cu ft; available chlorine 85%; decomposes 225C.
Hazard: Fire risk in contact with organic materials, strong oxidizing agent. Toxic by ingestion.
Use: Active ingredient in household dry bleaches, dishwashing compounds, scouring powders, detergent-sanitizers, commercial laundry bleaches, swimming pool disinfectant, bactericide, algicide, bleach, and deodorant.

trichloroisopropyl alcohol.   See isopral.

trichloromelamine. (N,N',N'-trichloro-2,4,6-triamine-1,3,5-triazine).

$\overline{NC(NHCl)NC(NHCl)NC(NHCl)}$.
Properties: Fine, white powder, slightly soluble in water and glacial acetic acid, insoluble in carbon tetrachloride and benzene, pH of saturated aqueous solution 4, autoign temperature 320F (160C).
Derivation: By chlorination of melamine.
Grade: 89% available chlorine.
Hazard: Dangerous fire risk, can ignite spontaneously in contact with reactive organic materials.
Use: Chlorine bleach and bactericide.

trichloromethane.   See chloroform.

α-(trichloromethyl)benzyl acetate.   See trichloromethylphenylcarbinol acetate.

trichloromethyl chloroformate. (diphosgene).
$ClCOOCCl_3$.
Properties: Colorless liquid; odor similar to phosgene (new mown hay); decomposed by heat, po-

rous substances, activated carbons (with evolution of phosgene), also by alkalies, hot water; soluble in alcohol, benzene, and ether. D 1.65 (15C), bp 127–128C, fp −57C, vap density 6.9 (air = 1), refr index 1.45664 (22C). Noncombustible.
Derivation: (a) By chlorinating methyl formate, (b) by chlorinating methyl chloroformate. In both methods the mixture of chloro-derivatives is then separated by fractionation.
Grade: Technical.
Hazard: Toxic by inhalation and ingestion, strong irritant to tissue.
Use: Organic synthesis, military poison gas.

trichloromethyl ether. CHCl$_2$OCH$_2$Cl.
Properties: Liquid; pungent odor; d 1.5066 (10C); bp 130–132C; soluble in alcohol, benzene, and ether; insoluble in water.
Hazard: Strong irritant to eyes and skin, evolves lachrymatory fumes.

N-(trichloromethylmercapto)-tetrahydrophthalimide. See captan.

trichloromethylphenylcarbinyl acetate.
(α-[trichloromethyl]benzyl acetate).
CAS: 90-17-5.   C$_6$H$_5$CH(CCl$_3$)OOCCH$_3$.
Properties: White, crystalline solid; intense rose odor; mp 86–88C; soluble in 18 parts of 95% alcohol.
Use: Perfumes, fixative for essential oils and perfumes.

trichloromethylphosphonic acid.
CCl$_3$PO(OH)$_2$.   Strong dibasic acid.
Properties: Soluble in water and alcohol, insoluble in benzene and hexane.
Use: Catalyst and condensation agent.

1,1,1-trichloro-2-methyl-2-propanol. See chlorobutanol.

trichloromethylsulfenyl chloride. (perchloromethyl mercaptan).   CAS: 594-42-3.
ClSCCl$_3$.
Properties: Yellow, oily liquid; disagreeable odor. Mildly decomposed by moist air, subject to the action of oxidizing agents, reducing agents, chlorine, etc. D 1.722 (0C), bp 148–149C (decomposes), vap d 6.414, volatility 18,000 mg/cu m (20C), insoluble in water. Nonflammable but supports combustion.
Derivation: Chlorination of carbon disulfide, thiophosgene, or methyl thiocyanate.
Grade: Technical.
Hazard: Toxic by ingestion and inhalation, strong irritant to eyes and skin.
Use: Organic synthesis, dye intermediate, fumigant.

trichloronaphthalene.   See chlorinated naphthalene.

trichloronitromethane.   See chloropicrin.

trichloronitrosomethane.   CCl$_3$NO.
Properties: Dark blue liquid; unpleasant odor; slowly decomposes, but is more stable in solution; soluble in alcohol, benzene, ether; insoluble in water; d 1.5 (20C); bp 5C (70 mm).
Derivation: Interaction of sulfuric acid, sodium trichloromethylsulfinate, and sodium nitrate.
Grade: Technical.
Hazard: Strong irritant to eyes and tissue.
Use: Organic synthesis, military poison gas (lachrymator).

trichlorononylsilane.   See nonyl trichlorosilane.

trichlorooctadecylsilane.   See octadecyltrichlorosilane.

trichlorooctylsilane.   See octyl trichlorosilane.

2,4,5-trichlorophenol.   CAS: 95-95-4.
C$_6$H$_2$Cl$_3$OH.
Properties: Gray flakes in sublimed mass with a strong phenolic odor, d 1.678 (25/4C), bp 252C, mp 68–70C, no flash p, soluble in alcohol, ether, and acetone. Nonflammable.
Hazard: May cause skin irritation.
Use: Fungicide, bactericide.

2,4,6-trichlorophenol.   (2,4,6-T).
CAS: 88-06-2.   C$_6$H$_2$Cl$_3$OH.
Properties: Yellow flakes with strong phenolic odor, d 1.675 (25/4C), fp 61C, bp 248–249C, soluble in acetone, alcohol, and ether. Nonflammable.
Hazard: May cause skin irritation.
Use: Fungicide, herbicide, defoliant.

2,4,5-trichlorophenoxyacetic acid.   (2,4,5-T).
CAS: 93-76-5.   C$_6$H$_2$Cl$_3$OCH$_2$CO$_2$H.
Properties: Light tan solid, mp 151–153C, soluble in alcohol, insoluble in water, available as sodium and amine salts.
Hazard: Use has been restricted.
Use: Plant hormone, herbicide, defoliant. See also dioxin.

2-(2,4,5-trichlorophenoxy)ethyl-2,2-dichloropropionate.   See erbon.

2-(2,4,5-trichlorophenoxy)propionic acid.
See silvex.

2,4,5-trichlorophenyl acetate.   CAS: 5393-75-9.
C$_6$H$_2$Cl$_3$OOCCH$_3$.
Use: As fungicide, especially on cotton seed.

1,2,3-trichloropropane.   CAS: 96-18-4.
  $CH_2ClCHClCH_2Cl$.
  Properties: Colorless liquid, d 1.3888 (20/4C), fp
  —15C, bp 156.17C, refr index 1.4822 (20C), flash
  p (COC) 180F (82.2C). Slightly soluble in water;
  dissolves oils, fats, waxes, chlorinated rubber,
  and numerous resins. Autoign temperature 580F
  (304C). Combustible.
  Derivation: Chlorination of propylene.
  Hazard: Toxic by inhalation and skin absorption;
  strong irritant. TLV: 10 ppm in air.
  Use: Paint and varnish remover, solvent, degreasing agent.

trichlorosilane.   (1) (silicochloroform).
  CAS: 10025-78-2. $SiHCl_3$
  Properties: Colorless volatile liquid, d 1.336, fp
  —127C, bp 32C, refr index 1.3990. Soluble in
  benzene, ether, heptane, perchloroethylene; decomposed by water. Flash p 7F (—13.9C), purity
  of 99.9999% is commercially attainable.
  Hazard: Flammable, dangerous fire risk.
  Use: Intermediate, purification of silicon. (2) Generic name for compounds of the formula $RSiCl_3$
  of which methyl trichlorosilane, CH_3SiCl_3 is
  most important.

N,N',N''-trichloro-2,4,6-triamine-1,3,5-triazine.
  See trichloromelamine.

2,4,6-trichloro-1,3,5-triazine.   See cyanuric chloride.

1,3,5-trichloro-s-triazine-2,4,6-trione.
  See trichloroisocyanuric acid.

trichlorotrifluoroacetone.   (1,1,3-trichloro-1,3,3-
  trifluoroacetone). $CCl_2FCOCClF_2$.
  Properties: Colorless liquid, bp 84.5C, fp —78C,
  miscible with water and most organic solvents,
  stable to acid but not alkalies. Nonflammable.
  Hazard: Strong irritant to eyes.
  Use: Solvent in acid media, complexing agent.

1,1,2-trichloro-1,2,2-trifluoroethane.
  (trifluorotrichloroethane).   CAS: 76-13-1.
  CCl_2FCClF_2.
  Properties: Colorless nearly odorless, volatile liquid, bp 47.6C, fp —35C, critical pressure 33.7
  atmosphere, d 1.42 (25C). Noncombustible.
  Derivation: From perchloroethylene and hafnium.
  Grade: Technical, spectrophotometric.
  Hazard: TLV: 1000 ppm in air.
  Use: Dry-cleaning solvent, fire extinguishers, to
  make chlorotrifluoroethylene, blowing agent,
  polymer intermediate, solvent drying, drying
  electronic parts and precision equipment.

tricholine citrate.   (tris[2-hydroxyethyl]trimethyl-
  ammonium citrate).
  $[(CH_3)_3NCH_2CH_2OH]_3 \cdot C_6H_5O_7$.
  Use: Medicine, nutrition.

trichotecene.   See mycotoxin.

tricobalt tetraoxide.   See cobalto-cobaltic oxide.

tricosane. $CH_3(CH_2)_{21}CH_3$.
  Properties: Glittering leaflets, soluble in alcohol,
  insoluble in water, d 0.779 (48C), bp 234C (15
  mm), mp 48C. Combustible.
  Grade: Technical.
  Use: Organic synthesis.

n-tricosanoic acid. $CH_3(CH_2)_{21}COOH$.
  A saturated fatty acid not normally found in
  natural fats or oils.
  Properties: Synthetic compound is a white crystalline solid, mp 79.1C.
  Use: Purified product is used in medical research
  and as reference standard for gas chromatography.

tri-m,p-cresyl borate. $(CH_3C_6H_4)_3BO_3$.
  Properties: Light amber liquid, d 1.065 (25C), bp
  385–395C, refr index 1.5480 (24C), flash p
  (COC) 240F (115.5C). Miscible in acetone, benzene, chloroform; hydrolyzes on contact with
  water. Combustible.
  Use: Plasticizer, organic synthesis.

tricresyl phosphate.   (tritolyl phosphate; TCP).
  CAS: 78-30-8. $(CH_3C_6H_4O)_3PO$.
  A mixture of isomers.
  Properties: Practically colorless, odorless liquid;
  stable, nonvolatile. Bp 420C, refr index 1.556
  (25C), d 1.162 (25/25C), bulk d 9.7 lb/gal, crystallizing point —35C, miscible with all the
  common solvents and thinners, also with
  vegetable oils, insoluble in water, flash p 437F
  (225C), autoign temperature 770F (410C).
  Combustible.
  Derivation: From cresol and phosphorus oxychloride.
  Hazard: Toxic by ingestion and skin absorption.
  The o- isomer is highly toxic.
  Use: Plasticizer for polyvinyl chloride, polystyrene, nitrocellulose; fire retardant for plastics;
  air filter medium; solvent mixtures; waterproofing; additive to extreme pressure lubricants; hydraulic fluid; heat exchange medium.

tricresyl phosphite. $(CH_3C_6H_4O)_3P$.
  Properties: Colorless liquid, slight phenolic odor,
  bp 191C (0.11 mm), d 1.115 (20/4C), flash p
  (OC) 440F (226.6C). Insoluble in water; miscible
  with acetone, alcohol, benzene, ether, and kerosene. Combustible.
  Grade: Technical.
  Use: Stabilizer and plasticizer for plastics and
  resins.

tricyanic acid.   See cyanuric acid.

tricyclic. An organic compound comprised of only three ring structures, which may be identical or different, e.g., anthracene.

sym-tricyclodecane. See adamantane.

tricyclohexyl borate. See boric acid ester.

n-tridecane. $CH_3(CH_2)_{11}CH_3$.
Properties: Colorless liquid, soluble in alcohol, insoluble in water, d 0.755 (20/4C), bp 225.5C, fp −5.45C, refr index 1.4250 (20C), flash p 175F (79.4C). Combustible.
Grade: 95%, 99%, research.
Use: Organic synthesis, distillation chaser.

n-tridecanoic acid. (tridecylic acid; tridecoic acid). $CH_3(CH_2)_{11}COOH$. A saturated fatty acid usually prepared synthetically.
Properties: Colorless crystals, mp 44.5C, d 0.8458 (80/4C), bp 312.4C, 192.2C (16 mm), refr index 1.4328 (50C), slightly soluble in water, soluble in alcohol and ether. Combustible.
Grade: 99% pure.
Use: Organic synthesis, medical research.

tridecanol. See tridecyl alcohol.

tridecoic acid. See n-tridecanoic acid.

tridecyl alcohol. (tridecanol).
$C_{12}H_{25}CH_2OH$. A commercial mixture of isomers.
Properties: Low-melting white solid with pleasant odor, bp 274C, mp 31C, d 0.845 (20/20C), bulk d 7.0 lb/gal, flash p (TOC) 180F (82.2C). Combustible.
Derivation: Oxo process from C_{15} hydrocarbons.
Grade: Technical.
Use: Esters for synthetic lubricants, detergents, antifoam agent, other tridecyl compounds, perfumery.

tridecylbenzene. (1-phenyltridecane).
$C_6H_5(CH_2)_{12}CH_3$.
Properties: Colorless liquid, d 0.85–0.86 (60/60F), refr index 1.4815–1.4830. Combustible.
Use: Detergent intermediate.

tridecylic acid. See n-tridecanoic acid.

tri(decyl)orthoformate. $CH(OC_{10}H_{21})_3$.
Properties: Liquid, bp 194C, fp −15 to −20C, refr index 1.448. Insoluble in water; soluble in benzene, naphtha, ether, and alcohol.
Use: To remove small quantities of water from ethers or other solvents where acid catalysts can be employed.

tri(decyl) phosphite. $(C_{10}H_{21}O)_3P$.
Properties: Water-white liquid, decyl alcohol odor, d 0.892 (25/15.5C), mp below 0C, refr index 1.4565 (25C), flash p 455F (235C). Combustible.
Use: Chemical intermediate, stabilizer for polyvinyl and polyolefin resins.

2,4,6-tri(dimethylaminomethyl)phenol.
$[(CH_3)_2NCH_2]_3C_6H_2OH$.
Properties: Liquid, refr index 1.5181. Combustible.
Use: Antioxidants, acid neutralizers, stabilizers, and catalysts for epoxy and polyurethane resins.

tri(dimethylphenyl)phosphite. (trixylenyl phosphate). $[(CH_3)_2C_6H_3O]_3PO$.
Properties: Liquid, d 1.155, refr index 1.5535, bp 243–265C (10 mm), flash p 450F (232C), solubility in water 0.002% (85C) by weight. Combustible.
Use: Plasticizer.

tridodecyl amine. See trilauryl amine.

tridodecyl borate. See boric acid ester.

tridymite. SiO_2. A vitreous, colorless, or white native form of pure silica, found variously but not so commonly as quartz. Quartz will change into tridymite with a 16.2% increase in volume at 870C. Unlike quartz, it is soluble in boiling sodium carbonate solution; d 2.28–2.3; Mohs hardness 7.

trietazine. (Generic name for 2-chloro-4-diethylamino-6-ethylamino-s-triazine).
CAS: 1912-26-1.
$ClC_3N_3[N(C_2H_5)_2]NHC_2H_5$.
Properties: Solid, practically insoluble in water, partially soluble in benzene and chloroform.
Use: Herbicide, plant growth regulator.

triethanolamine. (TEA; tri[2-hydroxyethyl] amine). CAS: 102-71-6.
$(HOCH_2CH_2)_3N$.
Properties: Colorless, viscous, hygroscopic liquid with slight ammoniacal odor. Mp 21.2C, bp 335C (decomposes), vap press below 0.01 mm (20C), d 1.126, flash p (OC) 375F (190.5C), bulk d 9.4 lb/gal. Miscible with water, alcohol; soluble in chloroform; slightly soluble in benzene and ether; slightly less alkaline than ammonia, commercial product contains up to 25% diethanolamine and up to 5% monoethanolamine. Combustible.
Derivation: Reaction of ethylene oxide and ammonia.
Grade: Technical, regular, 98%, USP.
Use: Fatty acid soaps used in drycleaning, cosmetics, household detergents, and emulsions. Wool scouring, textile antifume agent and water-repellent, dispersion agent, corrosion inhibitor, soften-

ing agent, emulsifier, humectant and plasticizer, chelating agent, rubber accelerator, pharmaceutical alkalizing agent.

triethanolamine lauryl sulfate. CAS: 139-96-8. $(HOC_2H_4)_3NOS(O)_2OC_{12}H_{25}$. A liquid or paste.
Use: Detergent; wetting, foaming and dispersing agent for industrial, cosmetic and pharmaceutical applications, especially shampoos.

triethanolamine methanarsonate.
CAS: 5902-97-6.
$CH_3As(O)[ONH(C_2H_4OH)_3]_2$.
Hazard: Toxic by ingestion.
Use: Herbicide.

triethanolamine oleate. See trihydroxyethylamine oleate.

triethanolamine stearate. See trihydroxyethylamine stearate.

triethanolamine titanate. See titanium chelate.

1,1,3-triethoxyhexane.
$CH(OC_2H_5)_2CH_2CH(OC_2H_5)C_3H_7$.
Properties: Colorless liquid, d 0.8746 (20/20C), bp 133C (50 mm), fp −100C, bulk d 7.3 lb/gal, flash p 210F (98.9C), insoluble in water. Combustible.
Use: Synthesis of aldehydes, acids, esters, chlorides, amines, etc.

triethoxymethane. See triethyl-o-formate.

1,1,3-triethoxy-3-methoxypropane. (triethylmethyl malonaldehyde diacetal).
$(CH_3O)(C_2H_5O)CHCH_2CH(OC_2H_5)_2$.
Properties: Colorless liquid, d 0.9300 (25/4C), bp 86C (6 mm). Combustible.
Grade: 99%.
Use: Intermediate, crosslinking and insolubilizing agent.

triethylaconitate.
$C_2H_5OOCCHC(COOC_2H_5)CH_2COOC_2H_5$.
Properties: Liquid, d 1.096 (25C), refr index 1.4517 (26C), bp 154–156 (5 mm). Combustible.
Use: Plasticizer.

triethylaluminum. (ATE; TEA; aluminum triethyl). CAS: 97-93-8. $(C_2H_5)_3Al$.
Properties: Colorless liquid, d 0.837, fp −52.5C, bp 194C, specific heat 0.527 (33C), miscible with saturated hydrocarbons, flash p −63F (−53C).
Derivation: By introduction of ethylene and hydrogen into an autoclave containing aluminum. The reaction proceeds at moderate temperature and varying pressures.

Grade: 88–94%.
Hazard: Flammable, dangerous fire risk, ignites spontaneously in air, reacts violently with water, acids, alcohols, halogens, and amines. Destructive to tissue.
Use: Catalyst intermediate for polymerization of olefins, especially ethylene; pyrophoric fuels; production of alpha-olefins and long chain alcohols; gas plating of aluminum.

triethylamine. CAS: 121-44-8. $(C_2H_5)_3N$.
Properties: Colorless liquid, strong ammoniacal odor, bp 89.7C, fp −115.3C, 0.7293 (20/20C), bulk d 6.1 lb/gal, flash p (OC) 10F (−6.67C), soluble in water and alcohol.
Derivation: From ethyl chloride and ammonia with heat and pressure.
Hazard: Flammable, dangerous fire risk, explosive limits in air 1.2–8.0%. Toxic by ingestion and inhalation, strong irritant to tissue. TLV: 10 ppm in air.
Use: Catalytic solvent in chemical synthesis; accelerator activators for rubber; wetting, penetrating, and waterproofing agents of quaternary ammonium types; curing and hardening of polymers (e.g., corebinding resins); corrosion inhibitor; propellant.

triethylborane. (triethylborine; boron triethyl). $(C_2H_5)_3B$.
Properties: Colorless liquid, d (25C), flash p −32F (35.5C), fp −93C, bp 95C, refr index 1.3971, heat of combustion 20,000 Btu/lb, miscible with most organic solvents, immiscible with water.
Derivation: Reaction of triethylaluminum and boron halide or diborane and ethylene.
Hazard: Flammable, dangerous fire risk, ignites spontaneously in air, reacts violently with water and oxidizing materials. Toxic by inhalation, strong irritant.
Use: Igniter or fuel for jet and rocket engines, fuel additive, olefin polymerization catalyst, intermediate.

triethyl borate. (ethyl borate). $(C_2H_5)_3BO_3$.
Properties: Colorless liquid, mild odor, hydrolyzes rapidly depositing boric acid in finely divided crystalline form, bp 120C, d 0.863–0.864 (20/20C), flash p (CC) 51.8F (11C), bulk d 7.20 lb/gal (20C), refr index 1.37311 (20C).
Hazard: Flammable, dangerous fire risk.
Use: Antiseptics, disinfectants, antiknock agent.

triethylborine. See triethylborane.

triethyl citrate. (ethyl citrate). CAS: 77-93-0.
$C_3H_5(COOC_2H_5)_3$.
Properties: Colorless, mobile liquid, bitter taste, bp 294C, bp 126–127C (1 mm), d 1.136 (25C),

pour point −46C, solubility in water 6.5 g/100cc, solubility in oil 0.8 g/100c, flash p (COC) 303F (150.5C), Combustible.
Derivation: Esterification of citric acid.
Grade: Technical, refined, FCC.
Use: Solvent and plasticizer for nitrocellulose and natural resins, softener, paint removers, agglutinant, perfume base, food additive (not over 0.25%).

triethylene diamine.   (1,4-diazobicyclo[2,2,2] octane).   CAS: 280-57-9. $N(CH_2CH_2)_3N$.
Properties: Colorless, hygroscopic crystals; mp 158C; bp 174C.
Hazard: Skin irritant.
Use: Catalyst for polyurethane foams, oxidation and polymerization catalyst, chemical intermediate (metal complexes, quaternary ammonium compounds, etc.), bromine and iodine addition compounds.

triethylene glycol.   (TEG).   CAS: 112-27-6.
$HO(C_2H_4O)_3H$.
Properties: Colorless, hygroscopic, practically odorless liquid. D 1.1254 (20/20C), bp 287.4C, vap press below 0.01 mm (20C), flash p (CC) 350F (176.6C), bulk d 9.4 lb/gal (20C), fp −7.2C, viscosity 0.478 poise (20C), autoign temperature 700F (371C), soluble in water, immiscible with benzene, toluene, and gasoline. Combustible.
Derivation: From ethylene and oxygen as a byproduct of ethylene glycol manufacture.
Grade: Technical, CP.
Use: Solvent and plasticizer in vinyl, polyester, and polyurethane resins; dehydration of natural gas; humectant in printing inks; extraction solvent.

triethylene glycol diacetate. CAS: 111-21-7.
$CH_3COOCH_2CH_2OCH_2CH_2OCH_2CH_2OOCCH_3$
Properties: Colorless liquid, d 1.112 (25C), refr index 1.437 (25C), bp 300C, fp below −60C. Combustible.
Use: Plasticizer.

triethylene glycol dibenzoate.
$C_6H_5CO(OCH_2CH_2)_3OOCC_6H_5$.
Properties: Crystals, bp 210–223C, mp 46C, flash p 457F (236C) (TOC), d 1.168. Combustible.
Use: Plasticizer for vinyl resins, adhesives.

triethylene glycol dicaprylate.   (triethylene glycol dioctoate).
$C_7H_{15}COO(CH_2CH_2O)_3OCC_7H_{15}$.
Properties: Clear liquid, d 0.973 (20C), acidity 0.3% max (caprylic); moisture 0.05% max, fp −3C, bp 243C (5 mm), soluble in most organic solvents. Combustible.
Use: Low-temperature plasticizer for elastomers.

triethylene glycol dichloride.   See triglycol dichloride.

triethylene glycol didecanoate.
$C_9H_{19}COO(C_2H_4O)_3OCC_9H_{19}$.
Properties: Colorless liquid, bp 237 (2.0 mm), d 0.9584 (20/20C), viscosity 28.6 cp (20C). Combustible.
Use: Plasticizer.

triethylene glycol di(2-ethylbutyrate).
$C_5H_{11}OCOCH_2(CH_2OCH_2)_2CH_2OCOC_5H_{11}$.
Properties: Light-colored liquid, d 0.9946 (20/20C), bulk d 8.3 lb/gal (20C), bp 196C (5 mm), vap press 5.8 mm (200C), solubility in water 0.02% by weight (20C), viscosity 10.3 cp (20C), flash p 385F (196C). Combustible.
Use: Plasticizer.

triethylene glycol di(2-ethylhexoate).
$C_7H_{15}OCOCH_2(CH_2OCH_2)_2CH_2OCOC_7H_{15}$.
Properties: Light-colored liquid, d 0.9679 (20/20C), bulk d 8.1 lb/gal (20C), bp 219C (5 mm), vap press 1.8 mm (200C), insoluble in water, viscosity 15.8 cp (20C), flash p 405F (207C). Combustible.
Use: Plasticizer.

triethylene glycol dihydroabietate.
$C_{19}H_{31}COO(C_2H_4O)_3OCC_{19}H_{31}$.
Properties: Liquid, d 1.080–1.090 (25C), refr index 1.5180 (20C), vap press 2.5 (225C), flash p 438F (226C), insoluble in water. Combustible.
Use: Plasticizer.

triethylene glycol dimethyl ether.   (triglyme).
$CH_3(OCH_2CH_2)_3OCH_3$.
Properties: Water-white liquid, mild ether odor, d 0.9862 (20/20C), refr index 1.4233 (20C), flash p 232F (111C), bp 216.0C (760 mm), 153.6C (100 mm), fp −46C, completely soluble in water and hydrocarbons at 20C, may contain peroxides. Combustible.
Use: Solvent for gases, coupling immiscible liquids.

triethylene glycol dioctoate.   See triethylene glycol dicaprylate.

triethylene glycol dipelargonate.
$C_8H_{17}COO(C_2H_4O)_3OCC_8H_{17}$.
Properties: Clear liquid, d 0.964 (20/20C), bp 251C (5 mm), fp +1 to −4C, refr index 1.4470 (23C), flash p 410F (210C), almost insoluble in water, soluble in most organic solvents. Combustible.
Use: Plasticizer.

triethylene glycol dipropionate.
$C_2H_5CO(OCH_2CH_2)_2OOCC_2H_5$.
Properties: Colorless liquid, d 1.066 (25C), refr index 1.436 (25C), bp 138–142C (2 mm), fp be-

low −60C, solubility in water 6.70% by weight. Combustible.
Use: Plasticizer.

triethylene glycol monobutyl ether. See butoxytriglycol.

triethylenemelamine. (tretamine; TEM; 2,4,6-tris(1-aziridinyl)-s-triazine). CAS: 51-18-2.

$NC[N(CH_2)_2]NC[N(CH_2)_2]NC[N(CH_2)_2]$.

Properties: White, crystalline, odorless powder; mp 160C (polymerizes); polymerizes readily with heat or moisture; soluble in alcohol, water, methanol, chloroform, and acetone.
Grade: NF.
Hazard: Highly toxic.
Use: Medicine (antineoplastic), insecticide, chemosterilant.

triethylenephosphoramide. (tepa; tris-(1-aziridinyl)-phosphine oxide; APO).

CAS: 545-55-1C010. $(NCH_2CH_2)_3PO$

Properties: Colorless crystals, mp 41C, soluble in water, alcohol, and ether. Combustible.
Derivation: From ethyleneimine.
Hazard: Highly toxic, strong irritant to skin and tissue.
Use: Medicine (antineoplastic), insect sterilant. Also used with tetrakis(hydroxymethyl)phosphonium chloride (THPC) to form a condensation polymer suitable for flameproofing cotton. See also tris[1-(2-methyl)aziridinyl]phosphine oxide.

triethylenetetramine. CAS: 112-24-3.
$NH_2(C_2H_4NH)_2C_2H_4NH_2$.
Properties: Moderately viscous, yellowish liquid, less volatile than diethylenetriamine but resembles it in many other properties, soluble in water, bp 277.5C, d 0.9818 (20/20C), mp 12C, flash p (CC) 275F (135C), bulk d 8.2 lb/gal (20C), autoign temperature 640F (337.7C). Combustible.
Grade: Technical, anhydrous.
Hazard: Strong irritant to tissue, causes skin burns and eye damage.
Use: Detergents and softening agents; synthesis of dyestuffs, pharmaceuticals, and rubber accelerators.

tri(2-ethylhexyl)phosphate. CAS: 78-42-2.
$[C_4H_9CH(C_2H_5)CH_2]_3PO_4$.
Properties: Light-colored liquid, d 0.9260 (20/20C), bulk d 7.70 lb/gal (20C), bp 220C (5 mm), vap press 1.9 mm (200C), insoluble in water, viscosity 14.1 cp (20C), pour point −74C, flash p 405F (207C). Combustible.
Use: Plasticizer.

tri(2-ethylhexyl) phosphite. $(C_8H_{17}O)_3P$.
Properties: Straw-colored liquid, d 0.897 (25/15C), mp sets to a glass at low temperature, refr index 1.451 (25C), flash p (COC) 340F (171C). Combustible.
Use: Plasticizer, intermediate.

tri(2-ethylhexyl) trimellitate.
$C_6H_3(COOC_8H_{17})_3$.
Properties: Clear liquid, mild odor, d 0.992 (20/20C), distillation range at 3 mm 278–284C (5–95%), fp a gel at −35C, refr index 1.4846 (23C), bulk d 8.26 lb/gal (20C). Combustible.
Use: Plasticizer.

triethylmethane. See 3-ethylpentane.

triethylmethyl malonaldehyde diacetal.
See 1,1,3-triethoxy-3-methoxypropane.

triethylorthoformate. (triethoxymethane).
CAS: 122-51-0. $CH(OC_2H_5)_3$.
Properties: Colorless liquid, pungent odor, bp 145.9C, refr index 1.39218 (18.8C), d 0.895 (20/20C), soluble in alcohol, ether; decomposes in water. Flash p (CC) 86F (30C).
Derivation: Reaction of sodium ethylate with chloroform or reaction of hydrochloric acid with hydrocyanic acid in ethanol solution.
Hazard: Flammable, moderate fire risk. Toxic.
Use: Organic synthesis, pharmaceuticals.

triethyl phosphate. (TEP). CAS: 78-40-0.
$(C_2H_5)_3PO_4$.
Properties: Colorless, high-boiling liquid; mild odor. Very stable at ordinary temperatures, compatible with many gums and resins, soluble in most organic solvents, miscible in water. When mixed with water is quite stable at room temperature, but at elevated temperatures it hydrolyzes slowly. Fp −56.4C, bp 216C, flash p 240F (115.5C), refr index 1.4055 (20C), bulk d 8.90 lb/gal (20C). Combustible.
Grade: Technical, 97%.
Hazard: May cause nerve damage but to less extent than other cholinesterase-inhibiting compounds.
Use: Solvent; plasticizer for resins, plastics, gums; manufacture of pesticides; catalyst; lacquer remover.

triethyl phosphite. CAS: 122-52-1.
$(C_2H_5)_3PO_3$.
Properties: Colorless liquid, d 0.9687 (20C), bp 156.6C, refr index 1.413 (25C), flash p 130F (54.4C), insoluble in water, soluble in alcohol and ether. Combustible.
Hazard: Moderate fire risk.
Use: Synthesis, plasticizers, stabilizers, lube and grease additives.

O,O,O-triethyl phosphorothioate. CAS: 126-68-1. (triethyl thiophosphate). $(C_2H_5O)_3PS$.
Properties: Colorless liquid with characteristic odor, bp 93.5–94C (10 mm), d 1.074, flash p (COC) 225F (107.2C). Combustible.
Hazard: Toxic by ingestion, cholinesterase inhibitor.
Use: Plasticizer, lubricant additive, antifoam agent, hydraulic fluid, intermediate.

triethyl tricarballylate.
$(C_2H_5OCOCH_2)_2CHCOOC_2H_5$.
Properties: Colorless liquid, d 1.087 (20C), refr index 1.4234 (26C), bp 158–160C (5 mm), solubility in water 0.62% (20C) by weight. Combustible.
Use: Plasticizer.

"Triexcel."[342] TM for rotenone and isomesynergized pyrethrin extracts in concentrated form for insecticidal formulations.

trifluoroacetic acid. CAS: 76-05-1.
CF_3COOH.
Properties: Colorless, fuming liquid; hygroscopic; pungent odor. Bp 72.4C, d 1.535, fp −15.25C, index of refr 1.2850 (20C), very soluble in water. Nonflammable.
Hazard: Irritant to skin.
Use: Strong non-oxidizing acid, lab reagent, solvent, catalyst.

trifluoroamine oxide. CAS: 13847-65-9.
(F_3NO). A perfluorated amine obtained by fluorination of nitrosyl fluoride with UV light or high temperature or pressure. An alternative process is by burning nitric oxide and fluorine, rapidly quenching the gaseous mixture as it leaves the flame zone.

trifluorochloroethylene. Legal label name (Rail) for chlorotrifluoroethylene.

trifluorochloromethane. See chlorotrifluoromethane.

1,1,1-trifluoro-2,6-dinitro-N,N-dipropyl-p-toluidine. See trifluralin.

trifluoromethane. See fluoroform.

trifluoromethylbenzene. See benzotrifluoride.

3-trifluoromethyl-4-nitrophenol. (1,1,1-trifluoro-4-nitro-m-cresol). $CF_3C_6H_3(NO_2)OH$.
Properties: Crystals, mp 74–76C.
Use: To exterminate lampreys, especially in the Great Lakes. It is placed in tributary streams where it kills the lamprey larvae.

trifluoronitrosomethane. CF_3NO.
Properties: Midnight blue, fairly stable gas, disagreeable odor, bp −84C, fp −150C. Nonflammable.
Derivation: (a) Interaction of fluorine and silver cyanide in the presence of silver nitrate; (b) from nitric oxide and iodotrifluoromethane or bromotrifluoromethane in the presence of UV light.
Hazard: Strong irritant to mucous membranes and tissue.
Use: Monomer for nitroso rubber.

trifluorostyrene. $C_8H_5F_3$. A monomer designed for the production of polytrifluorostyrene and for copolymerization with vinyl monomers.
Properties: Liquid, bp 68C, fp −23C, refr index 1.474, d 1.22, dipole moment 1.98 (D). The polymer is soluble in toluene, chloroform, and methyl ethyl ketone and has dielectric constant of 2.56. Nonflammable.
Use: Membranes for fuel tanks and water purification.

trifluorotrichloroethane. See trichlorotrifluoroethane.

trifluorovinylchloride. See chlorotrifluoroethylene.

trifluralin. (Generic name for 1,1,1,-trifluoro-2,6-dinitro-N,N-dipropyl-p-toluidine).
CAS: 1582-09-8. $F_3C(NO_2)_2C_6H_2N(C_3H_7)_2$.
Properties: Yellowish-orange solid, mp 48.5–49C, bp 139–140C (4.2 mm), insoluble in water, soluble in xylene, acetone, and ethanol.
Hazard: Toxic by ingestion.
Use: Herbicide, especially for cotton plant.

triformol. See sym-trioxane.

triglyceride. Any naturally occurring ester of a normal acid (fatty acid) and glycerol. The chief constituents of fats and oils, they have the general formula:
$CH_2(OOCR_1)CH(OOCR_2)CH_2(OOCR_3)$,
where R_1, R_2, and R_3 are usually of different chain length. Refining processes often yield commercial products in which the R chain lengths are the same.
Derivation: Extraction from animal, vegetable, and marine matter.
Use: Fatty acids and derivatives, manufacture of edible oils and fats, manufacture of monoglycerides.

triglycerol monolinolenate. See polyglycerol ester.

triglycerol trilinoleate. See polyglycerol ester.

triglycine. See nitrilotriacetic acid.

triglycol dichloride. (triethylene glycol dichloride). $Cl(C_2H_4O)_2C_2H_4Cl$.
Properties: Colorless liquid, d 1.1974 (20/20C), bp 241.3C, flash p 250F (121C), bulk d 10.0 lb/gal (20C), fp −31.5C, insoluble in water. Combustible.
Grade: Technical.
Use: Solvent for hydrocarbons, oils, etc.; extractant; intermediate for resins and insecticides; organic synthesis.

triglycollamic acid. See nitrilotriacetic acid.

triglyme. See triethylene glycol dimethyl ether.

trigonelline. (coffearine; caffearine; gynesine; N-methylnicotinic acid betaine).
CAS: 535-83-1. $C_5H_4NCOOCH_3 \cdot HOH$.
A base formed in the seeds of many plants.
Properties: Colorless prisms; mp 218C (decomposes); very soluble in water; soluble in alcohol; nearly insoluble in ether, benzene, and chloroform.
Derivation: Plant seeds, coffee beans, synthetically by heating nicotinic acid with methyl iodide and treatment with silver oxide.
Use: Biochemical research.

tri-n-hexylaluminum. $(C_6H_{13})_3Al$.
Properties: Colorless, pyrophoric liquid; bp 105C (0.001 mm).
Derivation: Exchange reaction between hexene and isobutyl aluminum.
Hazard: Ignites in air at room temperature.
Use: Polyolefin catalyst.

trihexylene glycol biborate. $(C_6H_{12}O_2)_3B_2$.
A cyclic borate.
Properties: Colorless liquid, d 0.982 (21C), boiling range 314–326C, refr index 1.4375 (25C), flash p 345F (173.9C), soluble in most organic solvents, hydrolyzes slowly in water. Combustible.
Derivation: Reaction of hexylene glycol with boric oxide.
Use: Gasoline additive, chemical intermediate.

trihexyl phosphite. $(C_6H_{13}O)_3P$.
Properties: Mobile, colorless liquid with characteristic odor. D 0.897 (20/4C), bp 135–141C (0.2 mm), flash p (COC) 320F (160C), miscible with most common organic solvents, insoluble in water, hydrolyzes very slowly, high degree of thermal stability, exposure to air should be minimum. Combustible.
Use: Intermediate for insecticides, component of vinyl stabilizers, lubricant additive, specialty solvent.

trihydric. Any alcohol in which three hydroxyl groups are present.
See also polyol, glycerol.

1,2,3-trihydroxyanthraquinone. See anthragallol.

1,2,4-trihydroxyanthraquinone. See purpurin.

1,2,7-trihydroxyanthraquinone. See anthrapurpurin.

1,2,3-trihydroxybenzene. See pyrogallol.

1,3,5-trihydroxybenzene. See phloroglucinol.

3,4,5-trihydroxybenzoic acid. See gallic acid.

2,4,5-trihydroxybutyrophenone.
$C_6H_2(OH)_3COC_3H_7$.
Properties: Yellow-tan crystals, mp 149–153C, bulk d 6.0 lb/gal (20C), very slightly soluble in water, soluble in alcohol and propylene glycol.
Use: Antioxidant for polyolefins and paraffin waxes, food additive.

tri(2-hydroxyethyl)amine. See triethanolamine.

trihydroxyethylamine oleate. (triethanolamine oleate). $(HOCH_2CH_2)_3 \cdot HOOCC_{17}H_{33}$.
Surfactant made by reaction of triethanolamine with oleic acid. Combustible.
Use: Emulsifying agent.

trihydroxyethylamine stearate. (triethanolamine stearate). $(HOCH_2CH_2)_3N \cdot HOOCC_{17}H_{35}$.
Properties: Cream-colored, wax-like solid; faint fatty odor; soluble in methanol, ethanol, mineral oil, vegetable oil; dispersible in hot water. D 0.968, pH 8.8–9.2 (25C) (5% aqueous dispersion), mp 42–44C. Combustible.
Use: Emulsifying agent for cosmetic and pharmaceutical industries.

tri(hydroxymethyl)aminomethane. See tris(hydroxymethyl)aminomethane.

1,3,8-trihydroxy-6-methylanthraquinone.
See emodin.

2,4,6-trihydroxytoluene. See methylphloroglucinol.

2,3,5-triiodobenzoic acid. $C_6H_2I_3COOH$.
A plant growth regulator used as a growth retardant.

triiodomethane. See iodoform.

triiodothyronine. (liothyronine; 3,5,3'-triiodothyronine).

$HOC_6H_3IOC_6H_2I_2CH(NH_2)COOH$.
Either a derivative or precursor of thyroxine. Triiodothyronine increases the metabolic rate and oxygen consumption of animal tissues.
Use: Biochemical research, medicine (metabolic insufficiency).

triisobutylaluminum. (TIBAL).
$[(CH_3)_2CHCH_2]_3Al$.
Properties: Colorless liquid, d 0.7876 (20C), fp −5.6C, bp 114C (30 mm), flash p 32F (0C), autoign temperature 39F (4C).
Derivation: Reaction of isobutylene and hydrogen with aluminum under moderate temperature and varying pressures.
Hazard: Highly toxic, destroys tissue. Flammable, dangerous fire risk, ignites spontaneously in air, reacts violently with water, acids, alcohols, amines, and halogens.
Use: Polyolefin catalyst, manufacture of primary alcohols and olefins, pyrophoric fuel.

triisobutylene. $(C_4H_8)_3$. A mixture of isomers readily prepared by polymerizing isobutylene. A typical mixture is 2,2,4,6,6-pentamethylheptane-3 and 2-neopentyl-4,4-dimethylpentene-1. May be depolymerized to simpler isobutylene derivatives.
Properties: Liquid, d 0.764 (60F), boiling range 175.5–178.9C. Combustible.
Use: Synthesis of resins, rubbers and intermediate organic compounds, lubricating oil additive, raw material for alkylation in producing high octane motor fuels.

triisooctyl phosphite. $(C_8H_{17}O)_3P$.
Properties: Colorless liquid with characteristic odor, miscible with most common organic solvents, insoluble in water, hydrolyzes very slowly, exposure to air should be minimum, high thermal stability, d 0.891 (20/4C), bp 161–164C (0.3 mm), flash p (COC) 385F (196C). Combustible.
Use: Intermediate for insecticides, component of vinyl stabilizers, lubricant additive, specialty solvent.

O,O,O-triisooctyl phosphorothioate. (triisooctyl thiophosphate). $(C_8H_{17}O)_3PS$.
Properties: Colorless liquid with characteristic odor, bp 160–170C (0.2 mm), d 0.933, flash p (COC) 410F (210C), insoluble in water, soluble in most organic solvents. Combustible.
Hazard: Highly toxic, cholinesterase inhibitor.
Use: Plasticizer, lubricant additive, hydraulic fluid, intermediate.

triisooctyl trimellitate. $C_6H_3(COOC_8H_{17})_3$.
Properties: Clear liquid, mild odor, d 0.992 (20/20C), distillation range 272–286C (5–95%), fp

gel at −45C, refr index 1.4852 (23C), practically insoluble in water. Combustible.
Use: Plasticizer.

triisopropanolamine. $N(C_3H_6OH)_3$.
Properties: Crystalline, white solid. Mild base (a mixture of isopropanolamines which has density of 1.004–1.010 and is liquid at room temperature is also marketed), d 0.9996 (50/20C), mp 45C, bp 305C, vap press below 0.01 mm (20C), viscosity 1.38 poise (60C), soluble in water, flash p (OC) 320F (160C). Combustible.
Grade: Technical.
Hazard: Irritant to skin and eyes.
Use: Emulsifying agents.

triisopropyl borate. $[(CH_3)_2CH]_3BO_3$.
Properties: Colorless liquid, bp 138–140C, fp −59C, d 0.8138, flash p (TCC) 82F (27.7C).
Derivation: Reaction of isopropyl alcohol with boric oxide.
Hazard: Flammable, moderate fire risk.

triisopropyl phosphite. $[(CH_3)_2CH]_3PO_3$.
Properties: Colorless liquid, characteristic odor, d 0.914 (20/4C), bp 94–96C (50 mm), flash p (COC) 165F (73.9C), miscible with most common organic solvents, insoluble in water, hydrolyzes slowly in water, exposure to air should be minimum, high thermal stability. Combustible.
Use: Intermediate for insecticides, component of vinyl stabilizers, lubricant additive, specialty solvent.

triketohydrindene hydrate. See ninhydrin.

trilaurin. The glyceride of lauric acid (glyceryl trilaurate).

trilaurylamine. (tridodecyl amine).
$(C_{12}H_{25})_3N$.
Properties: Colorless liquid, d 0.82, mp 14C, soluble in organic solvents, insoluble in water. Combustible.
Use: Chemical intermediate, metal complexes.

trilauryl phosphite. $(C_{12}H_{25}O)_3P$.
Properties: Water-white liquid, d 0.866 (25/15C), refr index 1.456 (25C), mp 10C. Combustible.
Use: Stabilizer in polymers, chemical intermediate.

trilauryl trithiophosphite. $(C_{12}H_{25}S)_3P$.
Properties: Pale yellow liquid, d 0.915 (25/15C), mp 20C, refr index 1.502 (25C), flash p (COC) 430F (221C). Combustible.
Use: Stabilizer, lubricant, chemical intermediate.

trimagnesium phosphate. See magnesium phosphate, tribasic.

"Tri-Mal."[304] TM for tribasic lead maleate vinyl stabilizer.

trimedlure. (Generic name for tert-butyl 4(or 5)-chloro-2-methylcyclohexanecarboxylate). $H_3C(Cl)C_6H_9COOC(CH_3)_3$.
Properties: Liquid, bp 90–92C (0.6 mm), soluble in most organic solvents, insoluble in water.
Use: Insect attractant.

trimellitic acid. (1,2,4-benzenetricarboxylic acid). CAS: 528-44-9. $C_9H_6O_6$.
Properties: Colorless crystals, mp 220C, partially soluble in DMF and alcohol, insoluble in benzene and carbon disulfide, slightly soluble in water.
Derivation: Oxidation of pseudocumene.
Use: Organic synthesis (plasticizers, polymers, and similar products).

trimellitic anhydride. (TMA; 1,2,3-benzenetricarboxylic acid-1,2-anhydride).
CAS: 552-30-7. $HOCOC_6H_3COOCO$.
Properties: Solid, mp 164–166C. Combustible.
Derivation: From pseudocumene.
Hazard: Toxic by inhalation. TLV: 0.04 mg/m³ of air.
Use: Plasticizer for polyvinylchloride, alkyd coating resins, high-temperature plastics, wire insulation, gaskets, automotive upholstery.

"Trimene Base."[248] TM for a reaction product of ethyl chloride, formaldehyde, and ammonia.
Properties: Dark brown, viscous liquid; d 1.10; soluble in water and acetone; insoluble in gasoline and benzene.
Use: Rubber accelerator.

trimer. An oligomer whose molecule is comprised of three molecules of the same chemical composition. Examples are trioxane and tripropylene. See also polymer, dimer.

trimercuric orthophosphate. See mercuric phosphate.

trimercurous orthophosphate. See mercurous phosphate.

trimesoyl trichloride. (benzene-1,3,5-tricarboxylic acid chloride). $C_6H_3(COCl)_3$.
Use: Specialty organic.

trimethadione. (3,5,5-trimethyl-2,4-oxazolidinedione). CAS: 127-48-0.
C_6H_9NO or $OC(O)N(CH_3)C(O)C(CH_3)_2$.
Properties: White, granular, crystalline substance; camphor-like odor; mp 45–47C; soluble in water; freely soluble in alcohol, chloroform, and ether; pH 6.0 (5% solution).
Grade: USP.
Hazard: May have adverse side effects; toxic in overdose.
Use: Medicine (anticonvulsant).

trimethoxyborine. See trimethyl borate.

trimethoxyboroxine. (methyl metaborate). $(CH_3O)_3B_3O_3$. A cyclic compound.
Properties: Colorless liquid, mp 10–11C, bp dissociates, d 1.216 (25C), refr index 1.3986. Nonflammable.
Derivation: Reaction of methyl borate with boric acid.
Grade: 99%.
Use: Metal-fire extinguishing fluid.

trimethoxymethane. See methyl orthoformate.

3,4,5-trimethoxyphenethylamine. See mescaline.

2,4,5-trimethoxy-1-propenylbenzene. (asarone). $(CH_3O)_3C_6H_2CH:CHCH_3$.
Properties: Crystals, mp 67C, insoluble in water, soluble in alcohol.
Derivation: Either extracted from calamus oil or synthesized (Wittig reaction).
Use: Grain fumigant, insect chemisterilant.

trimethylacethydrazide ammonium chloride. See Girard's T Reagent.

trimethylacetic acid. (pivalic acid; neopentanoic acid). CAS: 72-98-9. $(CH_3)_3CCOOH$.
Properties: Colorless crystals, d 0.905 (50C), refr index 1.3931 (36.5C), mp 35.5C, bp 163.8C. Soluble in water, alcohol, and ether. Combustible.
Use: Intermediate, as a replacement for some natural materials.

trimethylaluminum. (aluminum trimethyl; ATM). CAS: 75-24-1. $(CH_3)_3Al$.
Properties: Colorless, pyrophoric liquid; bp 126C; mp 15.4C; d 0.752.
Derivation: By sodium reduction of dimethylaluminum chloride.
Hazard: Highly flammable, dangerous fire risk, flames instantly on contact with air, reacts violently with water, acids, halogens, alcohols, and amines.
Use: Catalyst for olefin polymerization, pyrophoric fuel, manufacture of straight-chain primary alcohols and olefins, to produce luminous trails in upper atmosphere to track rockets.

trimethyladipic acid. $C_9H_{16}O_4$.
Properties: Powder. Combustible.

Use: Esterification agent for plasticizers, lubricants, alkyd resins, polyurethane, polyester, special polyamides, intermediate for the production of glycols.

trimethylamine. (TMA).   CAS: 75-50-3. $(CH_3)_3N$.
Properties: Colorless gas at room temperature, readily liquefied, anhydrous form shipped as liquefied compressed gas, fishy ammoniacal odor, d 0.6621 (−5C), bp −4C, fp −117.1C, flash p of 25% solution (TOC) 38F (3.3C). Soluble in water, alcohol, and ether; autoign temperature 374F (190C); flash p (CC) 10F (−12.2C).
Derivation: Interaction of methanol and ammonia over a catalyst at high temperature. The mono-, di-, and trimethylamines are produced and yields are regulated by conditions.
Method of separation: Azeotropic or extractive distillation.
Grade: Anhydrous 99% min, aqueous solution 25, 30, 40%.
Hazard: Flammable, dangerous fire risk, explosive limits in air 2–11.6%. Toxic by inhalation, vapor highly irritant. TLV: 10 ppm.
Use: Organic synthesis, especially of choline salts, warning agent for natural gas, manufacture of disinfectants, flotation agent, insect attractant, quaternary ammonium compounds, plastics.

trimethylamine sulfur trioxide. $(CH_3)_3N \cdot SO_3$.
Properties: White powder; mp 232–238C (decomposes); soluble in hot water, ethanol; soluble with difficulty in cold water and acetone; not dissociated in benzene and chloroform solutions; distinctly different from the isomeric adduct of trimethylamine oxide and sulfur dioxide.
Use: Separation of isomers, soil sterilant, catalyst for thermosetting resins.

2,4,5-trimethylaniline. See pseudocumidine.

1,2,3-trimethylbenzene. See hemimellitene.

1,2,4-trimethylbenzene. See pseudocumene.

1,3,5-trimethylbenzene. See mesitylene.

(trimethylbenzyl)dodecyldimethylammonium chloride.
$[(CH_3)_3C_6H_2CH_2N(CH_3)_2C_{12}H_{25}]Cl$.   A quaternary ammonium salt.
Properties: White to slightly yellow, crystalline powder; mild odor and taste. Mp 162–163C, bulk density 4.62 lbs/gal. Soluble in water, alcohol, glycerol, acetone; pH of 10% solution in distilled water 4.4.
Use: Germicide.

trimethyl borate. (methyl borate; trimethoxyborine).   CAS: 121-43-7. $(CH_3O)_3B$.
Properties: Water-white liquid; bp 67–68C; d 0.915; fp −29C; miscible with ether, methanol, hexane, tetrahydrofuran; decomposes in presence of water; flash p 80F (26.6C).
Derivation: Reaction of boric acid and methanol.
Hazard: Flammable, fire risk, reacts with water and oxidizing agents.
Use: Solvent, dehydrating agent, fungicide for citrus fruit, neutron scintillation counters, brazing flux, boron compounds, catalyst.

2,2,3-trimethylbutane. (isopropyltrimethylmethane; triptane). $CH_3C(CH_3)_2C(CH_3)CH_3$.
Properties: Colorless liquid, soluble in alcohol, insoluble in water, d 0.691, bp 81.0C, fp −24.96C, refr index 1.3895 (20C).
Hazard: Flammable, moderate fire risk.
Use: Organic synthesis, aviation fuel.

trimethyl carbinol. See tert-butyl alcohol.

1,1,3-trimethyl-5-carboxy-3-(p-carboxyphenyl)indane. An aromatic di-acid used as intermediate in manufacture of polyester fibers, polyamides and alkyd resins; hot-melt adhesives, engineering thermoplastics.

trimethylchlorosilane. CAS: 75-77-4. $(CH_3)_3SiCl$.
Properties: Colorless liquid, bp 57C, d 0.854 (25/25C), refr index 1.3893 (25C), flash p −18F (−27.7C), readily hydrolyzed with liberation of hydrogen chloride, soluble in benzene, ether and perchloroethylene.
Derivation: By Grignard reaction of silicon tetrachloride and methylmagnesium chloride.
Hazard: Flammable, dangerous fire risk, reacts violently with water. Strong irritant to tissue.
Use: Intermediate for silicone fluids, as a chain terminating agent, imparting water repellency.

trimethylcyclododecatriene. (TMCDT).
A cyclic hydrocarbon.
Use: Intermediate in making derivatives useful in the perfume and pharmaceutical industries as well as catalysts.

3,3,5-trimethylcyclohexanol-1. CAS: 116-02-9. $C_6H_8(CH_3)_3OH$.
Properties: Colorless liquid, d 0.878 (40/20C), mp 35.7C, bp 198C, soluble in most organic solvents, hydrocarbons, oils; insoluble in water, flash p (OC) 165F (73.9C). Combustible.
Hazard: Toxic by inhalation, strong irritant.
Use: Menthol and camphor substitute, antifoaming agent, hydraulic fluids and textile soaps, odor masking, esterification agent, pharmaceuticals, wax additive, printing inks.

3,5,5-trimethyl-2-cyclohexen-1-one. See isophorone.

3,3,5-trimethylcyclohexyl salicylate. See homomenthyl salicylate.

trimethyl dihydroquinoline polymer. (TDQP). $(C_{12}H_{15}N)_n$ (probably three or more quinoline groups).
Properties: Amber pellets; d 1.08; softening point 75C; insoluble in water; miscible with ethanol, acetone, benzene, monochlorobenzene, isopropyl acetate, and gasoline.
Use: Antioxidant, stabilizer or polymerization inhibitor.

3,7,11-trimethyl-1,6,10-dodecatrien-3-ol. See nerolidol.

3,7,11-trimethyl-2-6-10-dodecatrien-1-ol. See farnesol.

trimethylene. See cyclopropane.

trimethylene bromide. (1,3-dibromopropane).
CAS: 109-64-8. $CH_2BrCH_2CH_2Br$.
Properties: Colorless liquid, sweet odor, d 1.979 (20/4C), bp 166C, insoluble in water, soluble in organic solvents, fp −34.4C. Combustible.
Grade: Technical, CP.
Use: Intermediate for dyestuff and pharmaceutical industries, cyclopropane manufacture.

trimethylene chlorobromide. See 1-bromo-3-chloropropane.

trimethylene chlorohydrin. (3-chloro-1-propanol). $ClCH_2CH_2CH_2OH$.
Properties: Colorless to pale yellow liquid, characteristic odor, d 1.130–1.150 (25/25C), refr index 1.445–1.447 (25C). Soluble in water, alcohols, and ethers; insoluble in hydrocarbons. Combustible.
Use: Intermediate.

trimethylenedicyanide. See glutaronitrile.

trimethylene glycol. (1,3-propylene glycol; 1,3-propanediol). CAS: 504-63-2. $CH_2OHCH_2CH_2OH$.
Properties: Colorless, odorless liquid; d 1.0537 (25C); bp 210–211C; soluble in water, alcohol, and ether; autoign temperature 752F (400C). Combustible.
Derivation: From acrolein.
Grade: Technical 95%, pure 99%.
Use: Intermediate, especially for polyesters.

trimethylene oxide. See oxetane.

sym-trimethylene trinitramine. See cyclonite.

trimethylethylene. See 3-methyl-2-butene.

trimethylglycine. See betaine.

trimethylheptanoic acid. See isodecanoic acid.

2,2,5-trimethylhexane.
$(CH_3)_3CCH_2CH_2CH(CH_3)_2$.
Properties: Colorless liquid, fp −105.84C, bp 124.06C, d 0.711 (15.5/15.5C), refr index 1.399 (20C), flash p 55F (12.7C).
Grade: 95%, 99%; research.
Hazard: Flammable, dangerous fire risk; the 2,3,3-2,3,4-, and 3,3,4-isomers are less flammable.
Use: Synthesis, motor fuel additive.

3,5,5-trimethylhexan-1-ol. $C_9H_{20}O$.
Properties: Colorless liquid of mild odor, bp 194C, d 0.8236 (25/4C), bulk d 6.86 lbs/gal (25C), refr index 1.4300 (25C), flash p (OC) 200F (93.3C), insoluble in water. Combustible.
Derivation: High-pressure synthesis.
Use: Synthetic lubricants, additives to lubricating oils, wetting agent, softener in manufacture of various plastics, disinfectants and germicides.

trimethylmethane. See isobutane.

trimethyl nitrilotripropionate.
$N(CH_2CH_2COOCH_3)_3$.
Properties: Weakly basic tertiary amine and organic ester.
Use: Plasticizer for PVC, intermediate for pharmaceutical and agricultural chemicals, high-boiling solvents, low-temperature lubricants.

trimethylnonanol. See 2,6,8-trimethylnonyl-4 alcohol.

2,6,8-trimethyl-4-nonanone. (isobutyl heptyl ketone). CAS: 123-17-1.
$(CH_3)_2CHCH_2COCH_2CH(CH_3)CH_2CH(CH_3)_2$.
Properties: Water-white liquid with pleasant odor; high solvent power for vinyl resins, the cellulose esters and ethers and many difficultly soluble substances; insoluble in water. D 0.8165 (20/20C), bulk d 6.8lbs/gal, fp −75C, bp 211–219C, viscosity 1.91 cp (20C), flash p (COC) 195F (90.5C). Combustible.
Use: Solvent, dispersant, intermediate, lube oil dewaxing.

2,6,8-trimethylnonyl-4-alcohol.
(trimethylnonanol). $C_{12}H_{26}O$.
Properties: Colorless liquid with characteristic odor, d 0.8913 (20/20C), bulk d 6.9 lbs/gal

(20C), bp 225.2C, fp −60C (sets to a glass), viscosity 21.4 cp, flash p (COC) 200F (93.3C), insoluble in water. Combustible.
Derivation: Oxo process.
Use: Surface-active and flotation agents, lube additives, rubber chemicals.

trimethylolethane. (pentaglycerine; methyltrimethylolmethane). $CH_3C(CH_2OH)_3$.
Properties: Colorless, hygroscopic crystals; soluble in water and alcohol. Combustible.
Use: Conditioning agent, manufacture of varnishes, alkyd and polyester resins, synthetic drying oils.

trimethylolmelamine. CAS: 1017-56-7.
$C_3N_3(NHCH_2OH)_3$.   The first stage in making melamine resins.

trimethylolpropane. (hexaglycerol).
$C_2H_5C(CH_2OH)_3$.
Properties: Colorless, hygroscopic crystals; soluble in water and alcohol. Combustible.
Use: Conditioning agent, manufacture of varnishes, alkyd resins, synthetic drying oils, urethane foams and coatings, silicone lube oils, lactone plasticizers, textile finishes, surfactants, epoxidation products.

trimethylolpropane monooleate.
$C_2H_5C(CH_2OH)_2CH_2OOCC_{17}H_{33}$
(theoretically).   The commercial product is a mixture of mono-, di-, and tri-esters, free polyol, and free oleic acid. Combustible.
Properties: Oily liquid, d 0.954 (25C), fp less than −20C, insoluble in water, soluble in most organic solvents.
Use: Water-in-oil emulsifier, corrosion inhibitor, low-temperature plasticizer, deicing agent for gasoline.

trimethylolpropane tris(mercaptopropionate).
$C_2H_5C(CH_2OOCCH_2CH_2SH)_3$.
Properties: Liquid; d 1.21 (25C); refr index 1.5151 (25C); insoluble in water and hexane; soluble in acetone, benzene, and alcohol. Combustible.
Use: Curing or crosslinking agents for polymers, especially epoxy resins, intermediate for stabilizers and antioxidants.

trimethylolpropane trithioglycolate.
$CH_3CH_2C(CH_2OOCCH_2SH)_3$.
Properties: Liquid, d 1.28, refr index 1.5292 (25C). Combustible.
Use: Curing or crosslinking agent for polymeric systems, especially epoxy resins; intermediate for stabilizers and antioxidants.

3,5,5-trimethyl-2,4-oxazolidinedione.
See trimethadione.

2,2,4-trimethylpentane.   CAS: 540-84-1.
$(CH_3)_2CHCH(CH_3CH(CH_3CH_3$.
Properties: Liquid, fp −109.43C, bp 113C, d 0.723 (60/60F), refr index 1.4042 (20C), flash p 41F (5C).
Grade: 95%, 99%; research.
Hazard: Flammable, dangerous fire risk.
Use: Intermediate, azeotropic distillation entrainer.

2,2,4-trimethyl-1,3-pentanediol.
$(CH_3)_2CHCH(OH)C(CH_3)_2CH_2OH$.
Properties: (96% pure): White solid, d 0.928 (55/15C), mp 46–55C, bp 215–235C, flash p 235F (112.7C). Slightly soluble in water; soluble in alcohol, acetone, ether, and benzene. Combustible.
Use: Polyester resins, plasticizers, lubricants, surface coatings and printing inks, insect repellent.

2,2,4-trimethyl-1,3-pentanediol monoisobutyrate.
$(CH_3)_2CHCH(OH)C(CH_3)_2CH_2OOCCH(CH_3)_2$.
Properties: Liquid, d 0.945–0.955 (20/20C), bp 180–182C (125 mm), pour point −57C, refr index 1.4423 (20C), flash p (COC) 245F (118.3C). Insoluble in water; soluble in benzene, alcohol, acetone, and carbon tetrachloride. Combustible.
Use: Intermediate in the manufacture of plasticizers, surfactants, pesticides, and resins.

2,4,4-trimethylpentene-1 (α-diisobutylene).
$H_2C:C(CH_3)CH_2C(CH_3)_3$.
Properties: Colorless liquid, bp 101.44C, fp −93.5C, refr index 1.4086 (20C), d 0.7150 (20C), flash p 35F (1.67C).
Derivation: Polymerization of isobutene.
Grade: 95%, 99%; research.
Hazard: Flammable, dangerous fire risk.
Use: Organic synthesis, motor fuel synthesis, particularly isooctane; peroxide reactions.

2,4,4-trimethylpentene-2. (β-diisobutylene).
$H_3CC(CH_3):CHC(CH_3)_3$.
Properties: Colorless liquid, bp 104.55C, d 0.724 (60/60F), fp −106.4C, refr index 1.416 (20C), flash p (TOC) 35F (1.67C).
Grade: 95%.
Hazard: Flammable, dangerous fire risk. Irritant and narcotic in high concentration.
Use: Organic synthesis.

tri-2-methylpentylaluminum.
$[(CH_3)_2CH(CH_2)_3]_3Al$.
Properties: Colorless liquid.
Derivation: Reaction of 2-methylpentene and isobutylaluminum.
Hazard: Flammable.
Use: Polyolefin catalyst.

trimethyl phosphate.   CAS: 512-56-1.
  $(CH_3O)_3PO$.
Properties: Colorless liquid, 22.1% phosphorus,
density 1.210 at 68F, flash p above 300F (148C),
bp 193C, refr index 1.397 (20C), pour point
−46C, soluble in both gasoline and water. Combustible.
Hazard: Toxic by ingestion and inhalation, strong
irritant to skin and eyes.
Use: For controlling spark plug fouling, surface
ignition and rumble in gasoline engines.

trimethyl phosphite.   CAS: 121-45-9.
  $(CH_3O)_3P$.
Properties: Colorless liquid, bp 108–108.5C, pour
point below −60C, d 1.046 (20/4C), flash p
(COC) 100F (37.7C). Insoluble in water; soluble
in hexane, benzene, acetone, alcohol, ether, carbon tetrachloride, and kerosene.
Hazard: Flammable, moderate fire risk. TLV: 2
ppm in air.
Use: Chemical intermediate, especially for insecticides.

trimethyl phosphorothionate. $(CH_3O)_3PS$.
Properties: Water-white liquid, bp 78C (12 mm).
Hazard: Toxic by ingestion and inhalation.
Use: Extraction of mineral salts from alkyl acid
phosphate solvent solutions, plasticizer.

1,2,4-trimethylpiperazine. $(CH_3)_3C_4H_7N_2$.
Properties: Colorless liquid, d 0.851 (25/25C), fp
−44C, bp 151C, refr index 1.4480 (20C), flash
p (OC) 257F (125C), miscible in water. Combustible.
Use: Polymerization catalyst.

trimethylpropylmethane.   See 2,2-dimethylpen-
tane.

2,4,6-trimethylpyridine.   See 2,4,6-collidine.

2,4,6-trimethyl-1,3,5-trioxane.   See paraldehyde.

**1,3,5-trimethyl-2,4,6-tris(3,5-di-tert-butyl-4-hy-
droxybenzyl)benzene.**
Properties: Free-flowing, white, crystalline powder; mp 244C; no odor. Partially soluble in benzene and methylene chloride, insoluble in water,
permissible in contact with food products. Combustible.
Use: Antioxidant for polypropylene, high-density
polyethylene, spandex fibers, polyamides and
specialty rubbers.

2,6,10-trimethyl-9-undecen-1-al.
  $(CH_3)_2C:CHC_2H_4CH(CH_3)C_3H_6CH(CH_3)CHO$.
Properties: Clear, yellow liquid; strong, pungent,
ozone-like odor. D 0.850–0.860 (25/25C), refr
index 1.4530–1.4630 (20C). Combustible.
Use: Perfume.

trimethylvinylammonium hydroxide.   See neurine.

trimethylxanthine.   See caffeine.

"Trimulso."[236]   TM for a liquid synthetic surfactant used in the preparation of oil-in-water emulsion drilling muds, effective in both fresh water
and brine muds.

trimyristin.   The glyceride of myristic acid, glyceryl trimyristate.

trinickelous orthophosphate.   See nickel phosphate.

trinitite.   Green glazed glass material produced by
the 100 million degree Fahrenheit heat from an
atomic explosion. It was named for Trinity, New
Mexico, a sand covered site 80 miles from the
Almagordo atom bomb test site.

trinitroaniline.   (picramide).
  CAS: 26952-42-1. $C_6H_2NH_2(NO_2)_3$.
Properties: Orange-red crystals, mp 188C, bp ex-
plodes, d 1.762.
Derivation: Nitrating aniline in glacial acetic acid
solution or by the use of mixed nitric-sulfuric
acid in limited amounts.
Hazard: Dangerous, explodes by heat or shock.
Use: Explosive compositions.

trinitroanisole.   (methyl picrate; 2,4,6-trinitrophe-
nyl methyl ether). $CH_3OC_6H_2(NO_2)_3$.
Properties: Crystals, mp 68.4C, d 1.408 (20/4C).
Hazard: High. Dangerous, explodes by heat or
shock.
Derivation: Interaction of methyl iodide and silver
picrate, nitration of anisic acid.
Use: Explosive compositions.

1,3,5-trinitrobenzene.   (TNB).   CAS: 99-35-4.
  $C_6H_3(NO_2)_3$.
Properties: Yellow crystals, d 1.688 (20/4C), mp
122C, soluble in alcohol and ether, insoluble in
water.
Derivation: From trinitroluene by removal of
methyl group.
Hazard: Dangerous, explodes by heat or shock.
Use: Explosive compositions.

2,4,6-trinitrobenzoic acid.   (trinitrobenzoic acid).
  CAS: 129-66-8. $C_6H_2(NO_2)_3COOH$.
Properties: Orthorhombic crystals, mp 228.7C,
sublimes with decomposition forming carbon dioxide and trinitrobenzene, slightly soluble in water and benzene, soluble in alcohol, ether, and
acetone.
Derivation: Oxidation of 2,4,6-trinitrotoluene with
chromic acid.

Hazard: Dangerous, explodes by heat or shock.
Use: Explosive compositions.

2,4,6-trinitro-m-cresol. (cresolite; cresylite). $(NO_2)_3C_6H(CH_3)OH$.
Properties: Yellow crystals; mp 106C; readily soluble in alcohol, ether, and acetone.
Derivation: Prepared from m-cresol by a process similar to which picric acid is prepared from phenol.
Hazard: Explodes at 300F (148.8C), severe explosion risk when shocked or heated.
Use: Bursting charges and other high explosive uses.

trinitroglycerin. See nitroglycerin.

trinitromethane. CAS: 517-25-9. $CH(NO_2)_3$.
Properties: White crystals, mp 15C, d 1.469 (25C), decomposes above 25C, heat of combustion 746 cal/g, soluble in water.
Derivation: Reaction of acetylene with nitric acid.
Hazard: Explodes on heating, concentrations above 50% in air may explode.
Use: Manufacture of propellants and explosives.

1,3,5-trinitronaphthalene. (naphite). $C_{10}H_5(NO_2)_3$. Commercial preparation is a mixture of isomers which melts at 110C.
Hazard: Explosion risk when shocked or heated.
Use: Explosive, stabilizer for nitrocellulose.

trinitrophenol. See picric acid.

2,4,6-trinitrophenyl methyl ether. See trinitroanisole.

trinitrophenylmethylnitramine. See tetryl.

2,4,6-trinitroresorcinol. Legal label name for styphnic acid.

2,4,6-trinitrotoluene. (TNT; methyltrinitrobenzene). CAS: 118-96-7. $CH_3C_6H_2(NO_2)_3$.

Properties: Yellow, monoclinic needles; d 1.654; mp 80.9C; soluble in alcohol and ether; insoluble in water.
Derivation: Nitration of toluene with mixed acid.

Small amounts of the 2,3,4- and 2,4,5-isomers are produced which may be removed by washing with aqueous sodium sulfite solution.
Grade: Technical.
Hazard: Flammable, dangerous fire risk, moderate explosion risk, will detonate only if vigorously shocked or heated to 450F (232C). Toxic by ingestion, inhalation, and skin absorption. TLV: 0.5 mg/m^3 of air.
Use: Explosive, intermediate in dyestuffs and photographic chemicals.

trinitrotrimethylenetriamine. See cyclonite.

trioctadecyl phosphite. $(C_{18}H_{37}O)_3P$.
Properties: White, waxy solid; mp 45–47C; d 0.940 (25/25C).
Use: Stabilizer in polymers and an intermediate.

tri-n-octylaluminum. $(C_8H_{17})_3Al$.
Properties: Colorless, pyrophoric liquid.
Derivation: Reaction between octene and isobutylaluminum.
Hazard: Flammable, ignites in air.
Use: Polyolefin catalyst.

trioctyl phosphate. (octyl phosphate). $(C_8H_{17})_3PO_4$.
Properties: Liquid, d 0.924 (26C), bp 220–30 (8 mm). Soluble in alcohol, acetone, and ether. Combustible.
Hazard: Toxic by ingestion and inhalation.
Use: Solvent, antifoaming agent, plasticizer.

trioctylphosphinic oxide. (TOPO). $(C_8H_{17})_3PO$.
Properties: Solid, mp 55C, min purity 95%.
Use: Reagent for extraction of metals from aqueous and nonaqueous solutions, including fissionable actinide elements.

trioctyl phosphite. See tris-2-ethyl-hexyl phosphite.

triolein. See olein.

tri-o-cresylphosphate. See tricresyl phosphate.

sym-trioxane. (triformol; trioxin; metaformaldehyde). $(CH_2O)_3$ or $\overline{CH_2OCH_2OCH_2O}$. A trimer of formaldehyde, not to be confused with paraformaldehyde, which consists of eight or more formaldehyde units.
Properties: White crystals, formaldehyde odor, mp 62C, sublimes at 115C, flash p 113F (45C) (OC). Soluble in water, alcohol and ether; autoign temperature 777F (413C).

Derivation: By distillation of formaldehyde with an acid catalyst and extraction with solvent.

Hazard: Moderate fire risk, explosive limits in air 3.6–29%.

Use: Organic synthesis; disinfectant; nonluminous, odorless fuel.

See also formaldehyde.

2,6,8-trioxypurine. See uric acid.

tripalmitin. (palmitin; glyceryl tripalmitate).
CAS: 555-44-2. $C_3H_5(OOCC_{15}H_{31})_3$.
Properties: White, crystalline powder; soluble in ether and chloroform; insoluble in water. Mp 65.5C, d 0.866 (80/4C). Combustible.
Derivation: From glycerol and palmitic acid.
Grade: Technical.
Use: Soap, leather dressing.

tripelennamine citrate. CAS: 91-81-6.
$C_{16}H_{21}N_3 \cdot C_6H_8O_7$.   2-[benzyl-(2-dimethyl-aminoethyl)amino]pyridine dihydrogen citrate.
Properties: White, bitter, crystalline powder; solutions are acid to litmus. Mp 107C, soluble in water and alcohol, very slightly soluble in ether, practically insoluble in chloroform and benzene, 1% solution in water has a pH of 4.3.
Grade: USP.
Hazard: Toxic by ingestion.
Use: Medicine (antihistamine, sunburn treatment).

tripentaerythritol. $C_{15}H_{35}O_8$
Properties: White to ivory powder; has eight primary hydroxyl groups, all esterifiable; melting range 225–240C. Combustible.
Use: Hard resins, varnishes and fast-drying tall oil vehicles.

triphenol phosphorus. See 1,1,3-tris(hydroxyphenyl)-propane.

triphenylantimony. (triphenylstibine).
CAS: 603-36-1. $Sb(C_6H_5)_3$.
Properties: White, crystalline solid. D 1.434 (25C), mp 46–53C, bp below 360C, insoluble in water, slightly soluble in alcohol, soluble in most organic solvents. Combustible.
Derivation: Reaction of antimony trichloride with phenyl magnesium bromide or phenyl sodium.
Use: Stibonium salts, co-catalyst in converting trienes to aromatics and hydroaromatics, reacts with nitric-sulfuric acid to give trinitro derivatives, polymerization inhibitor, lubricating oil additive.

triphenylboron. $B(C_6H_5)_3$.   A type of Lewis acid used as catalyst and intermediate.

triphenylcarbinol, bis(4-dimethylamino).
$C_6H_5C(OH)[C_6H_4N(CH_3)_2]_2$.
Properties: Solid, mp 121–123C, very soluble in ether and hot benzene, soluble in acids.
Use: Dyestuffs
See malachite green.

triphenyl formazan. $CN_4H(C_6H_5)_3$.   Red, insoluble derivative of tetrazolium chloride, formed when the latter comes into contact with viable portions of a seed.
Use: Germination and viability tests.

triphenylguanidine. (TPG).
$C_6H_5NC(C_6H_5NH)_2$.
Properties: White, crystalline powder; soluble in alcohol; d 1.10; mp 144C. Combustible.
Derivation: Desulfurization of thiocarbanilide in presence of aniline.
Use: Accelerator for vulcanization of rubber.

triphenylmethane dyes. Any of a group of dyes whose molecular structure is basically derived from $(C_6H_5)_3CH$, usually by substitution of NH_2, OH, HSO_3, or other groups or atoms for some of the hydrogen of the C_6H_5 groups. Many coal tar and synthetic dyes are of this class, including rosaniline, fuchsin, malachite green, and crystalline violet.
See also triarylmethane dye.

triphenylmethane triisocyanate. Available as a brown 20% solution in methylene chloride.
Use: Bonding uncured rubber to metal or other surfaces.

triphenylmethylhexafluorophosphate.
Properties: Orange-colored, free-flowing, crystalline powder.
Use: Catalyst in manufacture of polyoxymethylenes (polymers of formaldehyde and trioxane).
See also acetal resin.

triphenyl phosphate. (TPP).   CAS: 115-86-6.
$PO(OC_6H_5)_3$.
Properties: Colorless, odorless, crystalline powder; soluble in most lacquers, solvents, thinners, oils; insoluble in water. Mp 50C, bp 245C (11 mm mercury), d 1.268 (60C), bulk d 10.5 lb/gal, refr index 1.550 (60C), flash p (CC) 428F (220C). Combustible.
Derivation: Interaction of phenol and phosphorus oxychloride.
Grade: Technical.
Hazard: Toxic by inhalation. TLV: 3 mg/m³ of air.
Use: Fire-retarding agent, plasticizer for cellulose acetate and nitrocellulose.

triphenylphosphine. See triphenylphosphorus.

triphenyl phosphite.   CAS: 101-02-0.
$(C_6H_5O)_3P$.
Properties: Water-white to pale yellow solid or oily liquid, pleasant odor, d 1.184 (25/25C), mp 22–25C, bp 155–160C (0.1 mm), refr index 1.589 (25C), flash p (COC) 425F (218.3C). Combustible.
Use: Chemical intermediate, stabilizer systems for resins, metal scavenger, diluent for epoxy resins.

triphenylphosphorus.   (triphenylphosphine).
$(C_6H_5)_3P$.
Properties: White, crystalline solid; mp 79–82C; bp above –360C; d 1.132 (25C); insoluble in water; slightly soluble in alcohol; soluble in benzene, acetone, carbon tetrachloride; flash p (OC) 356F (180C). Combustible.
Derivation: By a modified Grignard synthesis.
Use: Synthesis of organic compounds, phosphonium salts, other phosphorus compounds, polymerization initiator.

triphenylstibine.   See triphenylantimony.

triphenyltetrazolium chloride.   See tetrazolium chloride.

triphenyltin acetate.   CAS: 9000-95-8.
$(C_6H_5)_3SnOOCCH_3$.   An agricultural biocide, white crystalline solid, made by reaction of sodium acetate with triphenyltin chloride.
Hazard: Irritant to skin. TLV (as Sn): 0.1 mg/m^3 of air.

triphenyltin chloride.   CAS: 639-58-7.
$(C_6H_5)_3SnCl$.
Properties: White, crystalline solid; mp 106C; bp 240C (13.5 mm); insoluble in water; soluble in organic solvents.
Derivation: Reaction of tin tetrachloride with phenylmagnesium bromide.
Hazard: An irritant to skin. TLV (as Sn): 0.1 mg/m^3 of air.
Use: Biocidal intermediate.

triphenyltin hydroxide.   CAS: 76-87-9.
$(C_6H_5)_3SnOH$.
Properties: White solid; mp 118–120C; insoluble in water; soluble in ether, benzene, and alcohol.
Hazard: Irritant to skin. TLV (as Sn): 0.1 mg/m^3 of air.
Use: Insect chemisterilant, fungicide.

"Triphosaden."[91]   TM for adenosine triphosphate used for biochemical and clinical research and medicine.

triphosgene.   See hexachloromethylcarbonate.

triphosphoric acid.   See polyphosphoric acid.

triple bond.   A highly unsaturated linkage between the two carbon atoms of acetylenic compounds (alkynes), typified by acetylene ($HC{\equiv}CH$).
See also double bond, chemical bonding.

triple point.   The temperature and pressure at which the solid, liquid, and vapor of a substance are in equilibrium with one another. Also applied to similar equilibrium between any three phases, i.e., two solids and a liquid, etc. The triple point of water is +0.072C at 4.6 mm Hg; it is of special importance because it is the fixed point for the absolute scale of temperature.

triple superphosphate.   A dry, granular, free-flowing product, gray in color, produced by addition of phosphoric acid to phosphate rock, thus avoiding formation of insoluble gypsum, as in superphosphate, and achieving three times the amount of available phosphate (as P_2O_5). Typical analysis: Moisture 2%, available P_2O_5 50%, water solution P_2O_5 45%, free phosphoric acid 1%, also minor ingredients.
Use: Fertilizer.
See also superphosphate and nitrophosphate.

tripoli.   (rottenstone).   Amorphous silica.
Properties: Soft, porous granules resulting from natural decomposition of siliceous rock.
Grade: Various grades according to fineness for polishing; rose, cream, white.
Use: Abrasive, polishing powder, filtering material, absorbent for insecticidal chemicals, paints (inert filler, wood filler), rubber filler, base for scouring soaps and powders, oil-well drilling muds.

tripolyphosphate.   See sodium tripolyphosphate.

tripotassium orthophosphate.   See potassium phosphate, tribasic.

tripotassium phosphate.   See potassium phosphate, tribasic.

tripropionin.   See glyceryl tripropionate.

tri-n-propylaluminum. $(C_3H_7)_3Al$.
Properties: Colorless, pyrophoric liquid; d 0.820; fp –84C.
Derivation: Reaction of propylene and isobutylaluminum.
Hazard: Flammable, dangerous fire risk, ignites spontaneously in air.
Use: Polyolefin catalyst.

tripropylamine.   CAS: 102-69-2.
$(CH_3CH_2CH_2)_3N$.
Properties: Water white liquid, amine odor, fp –94C, boiling range 150–156C, d 0.754 (20/

20C), refr index 1.417 (20C), flash p (OC) 105F (40.5). Combustible.
Hazard: Moderate fire risk. Toxic by inhalation and ingestion.

tripropylene. (propylene trimer). $(C_3H_6)_3$.
Properties: Colorless liquid, d 0.738 (20/20C), boiling range 133.3–141.7C, bulk d 6.17 lb/gal (60F), flash p (TOC) 75F (23.9C).
Derivation: Catalytic polymerization of propylene.
Hazard: Flammable, dangerous fire risk.
Use: Oxo feed stock, lubricant additive, plasticizers, nonyl phenol.

tripropylene glycol. CAS: 13987-01-4.
$HO(C_3H_6O)_2C_3H_6OH$.
Properties: Colorless liquid, supercools instead of freezing, bp 268C, d 1.019 (25/25C), bulk d 8.51 lb/gal, refr index 1.442 (25C), flash p 285F (140.5C). Soluble in water, methanol, ether. Combustible.
Use: Intermediate in resins, plasticizers, pharmaceuticals, insecticides, dyestuffs, mold lubricants.

tripropylene glycol monomethyl ether.
CAS: 10213-77-1. $HO(C_3H_6O)_2C_3H_6OCH_3$.
Properties: Colorless liquid, d 0.961 (25C), bp 242C, 116C (10 mm), viscosity 5.5 cp (25C), refr index 1.427 (25C), flash p 250F (121C). Miscible with water, VMP naphtha, acetone, ethanol, benzene, carbon tetrachloride, ether, methanol, and monochlorobenzene. Combustible.
Use: Ingredient in hydraulic fluids.

triptane. See 2,2,3-trimethylbutane.

"Triptide."[91] TM for lyophilized monosodium glutathione used for biochemical and clinical research and medicine.

tris-. A prefix indicating that a certain chemical grouping occurs three times in a molecule, e.g., tris(hydroxymethyl)aminomethane $(CH_2OH)_3CNH_2$.
See also bis-.

tris amine buffer. See tris(hydroxymethyl)aminomethane.

tris(1-aziridinyl)phosphine oxide. Legal label name for triethylenephosphoramide.

2,4,6-tris(1-aziridinyl)-s-triazine. See triethylenemelamine.

trisazo dye. One of the four kinds of azo dyes, characterized by the presence of three azo couplings ($-N=N-$) in each molecule. (CI 30000 to 34999).

tris(2-chloroethyl) phosphate. CAS: 115-96-8.
$(ClC_2H_4O)_3PO$.
Properties: Clear, transparent liquid; d 1.425 (20/20C); bp 214C (25 mm); refr index 1.4721 (20C); flash p (COC) 421F (216C). Combustible.
Use: Flame-retardant plasticizer.

tris(2-chloroethyl) phosphite. CAS: 140-08-9.
$(ClC_2H_4O)_3P$.
Properties: Colorless liquid with characteristic odor, d 1.353 (20/4C), bp 119C (0.15 mm), flash p (OC) 280F (137.7C), miscible with most common organic solvents, undergoes intramolecular isomerization at higher temperatures, exposure to air should be minimum, insoluble in water and hydrolyzes in water. Combustible.
Use: Intermediate, component of vinyl stabilizers, grease additives, flameproofing compositions.

tris(2-chloroisopropyl) thionophosphate.
$[CH_3(CH_2Cl)CHO]_3PS$.
Properties: Liquid, phosphorus content 9.0%, d 1.282 at (20C), flash p (OC) above 347F (175C), pour point below −50C, readily soluble in gasoline, insoluble in water. Combustible.
Use: To extend spark plug life, to control deposit-induced knocking in gasoline engines.

tris(2,3-dibromopropyl) phosphate.
$(CH_2BrCHBrCH_2O)_3PO$.
Properties: Viscous, pale yellow liquid; bulk d 18.5 lb/gal; refr index 1.5772 (20C). Combustible.
Hazard: A carcinogen; use restricted.
Use: Flame retardant for plastics and synthetic fibers.

tris(2,3-dichloropropyl) phosphate.
$(CH_2ClCHClCH_2O)_3PO$. Combustible.
Use: Flame retardant in plastics and as a secondary plasticizer.
Hazard: A carcinogen.

tris(diethylene glycol monoethyl ether) citrate.
$C_{19}H_{42}O_{13}$
Properties: Solid, d 1.28 (25C), mp 16–19C, soluble in water. Combustible.
Use: Plasticizer.

tris(2-ethylhexyl) phosphate.
$[C_4H_9CH(C_2H_5)CH_2O]_3PO$.
Properties: Colorless liquid, d 0.9260 (20/20C), fp −90C (sets to glass), pour point −74C, mid-boiling point 216C (4 mm), refr index 1.441 (25C), insoluble in water, soluble in mineral oil and gasoline, flash p 420F (215.5C). Combustible.
Use: Low-temperature plasticizer for PVC resins, imparting flame and fungus resistance.

tris-2-ethylhexyl phosphite. (trioctyl phosphite). [C$_4$H$_9$CH(C$_2$H$_5$)CH$_2$O]$_3$P.
Properties: Colorless liquid, d 0.902, bp 163–164C (0.3 mm), insoluble in water, soluble in alcohol and ether, flash p 340F (171C). Combustible.
Use: Synthesis, plasticizers, stabilizers, lube and grease additives, flameproofing compositions.

tris(2-hydroxyethyl)isocyanurate. (THEIC; tris(2-hydroxyethyl)-s-triazine-2,4,6-trione). C$_3$N$_3$O$_3$(CH$_2$CH$_2$OH)$_3$.
Properties: White solid, mp 135C, bp dissociates at 180C (3 mm), very soluble in water, somewhat soluble in alcohol and acetone, insoluble in chloroform and benzene. Combustible.
Use: Additive to plastics, especially to impart thermal stability.

tris(hydroxymethyl)acetic acid.
HOCH$_2$C(CH$_2$OH)$_2$COOH. A photographic chemical made by bacterial oxidation of pentaerythritol.

tris(hydroxymethyl)aminomethane.
(tri[hydroxymethyl]-aminomethane; THAM; 2-amino-2-hydroxymethyl-1,3-propanediol; tris amine buffer). (CH$_2$OH)$_3$CNH$_2$.
Properties: White, crystalline solid; solubility in water 80 g/100cc at 20C. Mp 171–172C, bp 219–220C (10 mm), pH 0.1M aqueous solution 10.36, corrosive to copper, brass, aluminum. Combustible.
Hazard: Irritant to eyes and skin.
Use: Emulsifying agent (in soap form) for oils, fats, and waxes; absorbent for acidic gases; chemical synthesis; buffer; medicine.

tris(hydroxymethyl)nitromethane.
(2-hydroxymethyl-2-nitro-1,3-propanediol). CAS: 126-11-4. (CH$_2$OH)$_3$CNO$_2$.
Properties: White crystals or amorphous solid, mp 175C (decomposes), soluble in water and alcohol.
Hazard: Moderate fire risk. Irritant to skin and eyes.
Use: Bactericide and slimicide for aqueous systems, cutting oil emulsions, industrial water systems, drilling muds.

1,1,3-tris(hydroxyphenyl)propane. (triphenol P). (C$_6$H$_4$OH)$_2$CHCH$_2$CH$_2$C$_6$H$_4$OH.
Properties: White solid, mp 84C, d 1.226 (20/20C), fp 90–110C, sets to glass below this temperature. Combustible.
Use: Antioxidant, intermediate for polyester and alkyd resins.

trisilane. CAS: 7783-26-8. Si$_3$H$_8$.
Properties: Colorless liquid, bp 53C, mp −117C, d 0.74.

Hazard: Explodes on contact with air, reacts violently with carbon tetrachloride and chloroform.

trisiloxane. See siloxane.

tris[1-(2-methyl)aziridinyl]phosphine oxide. (C$_3$H$_6$N)$_3$PO. See metepa.

trisodium citrate. See sodium citrate.

trisodium dipotassium tripolyphosphate.
Na$_3$K$_2$P$_3$O$_{10}$.
Properties: White, crystalline solid; mp 620–640C; d 2.48; solubility in water 80 g/100mL (26C).
Use: Sequestrant.

trisodium EDTA. (ethylenediaminetetraacetic acid trisodium salt). CAS: 150-38-9.
C$_{10}$H$_{13}$N$_2$Na$_3$O$_8$·HOH.
Properties: White powder, freely soluble in water.
Use: Chelating agent.

trisodium hydroxyethylethylenediaminetriacetate.
(NaOOCCH$_2$)$_2$NC$_2$H$_4$N(CH$_2$COONa)C$_2$H$_4$OH.
Properties: Liquid, d 1.285, fp below −5C.
Use: Chelating agent.

trisodium nitrilotriacetate. CAS: 5064-31-3.
N(CH$_2$COONa)$_3$. A sodium salt of nitrilotriacetic acid.
Use: Sequestrant.

trisodium orthophosphate. See sodium phosphate, tribasic.

trisodium phosphate. See sodium phosphate, tribasic.

trisodium phosphate, chlorinated.
4(Na$_3$PO$_4$·12HOH)·NaOCl. Active ingredients 3.25% min sodium hypochlorite and 91.75% min trisodium phosphate dodecahydrate. Inert ingredients below 5% sodium chloride.
Properties: White, crystalline, water-soluble solid; stable under normal storage conditions; in solution has the properties of both trisodium phosphate and sodium hypochlorite.
Derivation: By reacting sodium phosphate, caustic soda, and sodium hypochlorite.
Hazard: Irritant to skin and eyes.
Use: Cleaner and bactericide in dairies, food plants, dishwashing compounds, and scouring powders.

trisodium phosphate monohydrate. See sodium phosphate, tribasic, monohydrate.

tristearin. See stearin.

tris(tetrahydrofurfuryl) phosphate.
$(C_4H_7OCH_2O)_3PO$.
Properties: Yellow liquid, fp −75C, refr index 1.4759 (20C), soluble in water and aromatic solvents.
Use: Plasticizer, solvent.

"Tritac."[62]   TM for a group of herbicides whose active ingredient is 2,3,6-trichlorobenzyloxypropanol.

"Trithion."[1]   TM for an organic phosphate insecticide-acaricide available as aqueous emulsion, dust, and wettable powder containing various percentages of carbophenothion.
Hazard: As for carbophenothion.

tritium.   CAS: 10028-17-8.   T.   Radioactive isotope of hydrogen, mass number 3, isotopic weight 3.017 (two neutrons and one proton in the nucleus).
Properties: Half-life 12.5 years, radiation beta, radiotoxic material.
Derivation: Bombardment of lithium with low-energy neutrons.
Forms available: Gas packaged in ampules and in tagged compounds such as water, streptomycin, cortisone, epinephrine, octadecane, stearic acid, etc.
Use: Bombarding particle in cyclotrons, activator in self-luminous phosphors, in cold cathode tubes, tracer in biochemical research and various special problems in chemical analysis, luminous instrument dials, thermonuclear power research
See fusion.

tritolyl phosphate.   See tricresyl phosphate.

triton.   The nucleus of the T atom (1 proton, two neutrons).

tritopine.   See laudanidine.

triturate.   To reduce to a powder by rubbing or grinding, to pulverize.

triuranium octoxide.   (uranous-uranic oxide; uranyl uranate). U_3O_8.
Properties: Olive-green to black solid, crystals, or granules. Insoluble in water, soluble in nitric acid and sulfuric acid, d 8.39, decomposes when heated to 1300C to uranium dioxide.
Source: The naturally occurring uranium oxide found in pitchblende.
Derivation: (a) As one of the forms of uranium produced from the ores, often by a solvent extraction process. The solvent used is dodecylphosphoric acid. (b) A common form of triuranium octoxide is yellow cake, the powder obtained by evaporating an ammonia solution of the oxide.
Hazard: Radioactive poison. Use appropriate protection in handling.
Use: Nuclear technology, preparation of other uranium compounds.

trivial name.   The name applied by early chemists to a number of simple organic compounds, usually based on their sources or properties, e.g., acetone and acetic acid, from Latin *acetum* (vinegar), urea from urine, glucose and glycerol from Greek *glyc*- (sweet). Such names remained in common use regardless of the systematic nomenclature later developed.

trixylenyl phosphate.   See tri(dimethylphenyl) phosphate.

trona.   (urao).   CAS: 497-19-8.
$Na_2CO_3 \cdot NaHCO_3 \cdot 2HOH$.   A natural sodium sesquicarbonate and the most important of the natural sodas.
Properties: White, gray, or yellow with vitreous, glistening luster. Noncombustible.
See Trona process.
Occurrence: Hungary, Egypt, Africa, Venezuela, and US (Wyoming, California, especially Searles Lakes, Owens Lake).
Use: Source of sodium compounds, especially the sodium carbonates.

"Tronabor."[88]   TM for crude borax pentahydrate, from Searles Lake brines.

Trona process.   The method used for separation and purification of soda ash, anhydrous sodium sulfate, boric acid, borax, potassium sulfate, bromine, and potassium chloride from Searles Lake (California) brine.

tropacocaine hydrochloride. $C_{15}H_{19}NO_2 \cdot HCl$.
An alkaloidal salt.
Properties: White crystals, soluble in water, alcohol and ether; mp 271C.
Derivation: From a variety of Erythroxylon coca.
Hazard: Toxic by ingestion and inhalation.
Use: Medicine (spinal anesthetic).

3-tropanol.   See tropine.

tropaeolin D.   See methyl orange.

tropaeolin OO.   (Orange IV).
$NaSO_3C_6H_4NNC_6H_4NHC_6H_5$.   (p-diphenylaminoazobenzene-sodium sulfonate). A biological stain and acid-base indicator, red at pH 1.4, yellow at pH 2.6. CI No. 13080.

Trouton's rule.   The molal heat of vaporization of normal liquids, at the boiling point and under atmospheric pressure, divided by the absolute boiling temperature is a constant, about 22.

Trp.   Abbreviation for tryptophan.

true solution.   See solution (true).

truth serum.   See scopolamine.

truttine.   A protein obtained from fish of the trout family.

"Trycite."[233]   TM for an oriented polystyrene film.
Use: Packaging and in envelope windows.

tryparsamide.   (sodium-N-phenylglycineamide-p-arsonate).   CAS: 554-72-3.
$NaOAs(OOH)C_6H_4NHCH_2CONH_2 \cdot 1/2HOH$.
Properties: White, crystalline powder; odorless. Contains 24.6% arsenic, may affect eyes. Soluble in water; almost insoluble in alcohol; insoluble in ether, chloroform, benzene.
Grade: Medicinal, USP.
Hazard: Highly toxic by ingestion and inhalation.
Use: Medicine (treatment of syphilis).

trypsin.   CAS: 9002-07-7.   The proteolytic enzyme of pancreatic juice, yellow to grayish powder, soluble in water, insoluble in alcohol or glycerol. It acts on albuminoid material producing amino acids. The maximum result is obtained in a neutral or slightly alkaline medium. Trypsins or similar materials are found not only in the pancreas but also in the spleen, leucocytes, and urine, as well as in beer yeast, molds, and bacteria.
Grade: NF (crystallized).
Hazard: Irritant to skin.
Use: Dehairing of hides.

trypsinogen.   An inactive precursor of trypsin.

tryptophan.   (indole-α-aminopropionic acid; 1-α-amino-3-indolepropionic acid).
CAS: 54-12-6.

One of the essential amino acids occurring naturally in the l-(-)-form.
Properties: dl-: White crystals, slightly soluble in water, stable in alkaline solution, decomposed by strong acids. $d(+)$-: Characteristic sweet taste, mp 275–290C (decomposes), soluble in water, hot alcohol, alkali hydroxides, insoluble in chlo-

roform. l(-)-: Flat taste (other properties identical with $d(+)$-tryptophan).
Derivation: (a) Synthetic tryptophan can be made by the conversion of indole to gramine followed by methylation, interaction with acetylaminomalonic ester and hydrolysis, (b) hydrolysis of proteins.
Grade: Reagent, technical, FCC.
Use: Nutrition and research, medicine, dietary supplement, cereal enrichment.
Available commercially in all three forms, as well as acetyl-dl-tryptophan.

"Trysben" 200.[28].   TM for a weed killer based on aqueous solution of the dimethylamine salt of trichlorobenzoic acid, containing two pounds of acid equivalent per gallon.
Use: Control of broadleaf weeds.

Ts.   Abbreviation for tosyl.

TSA.   Abbreviation for toluenesulfonic acid.

Tscherniac-Einhorn reaction.   Introduction of the amidomethyl group into aromatic rings or activated methylene groups in the presence of sulfuric acid.

TSH.   See thyrotropic hormone.

TSP.   Abbreviation for trisodium phosphate. See sodium phosphate, tribasic.

TSPA.   Abbreviation for triethylenethiophosphoramide.

TSPP.   Abbreviation for tetrasodium pyrophosphate.
See sodium pyrophosphate.

TTC.   Abbreviation for tetrazolium chloride.

TTD.   Abbreviation for tetraethylthiuram disulfide.

"Tuads."[69]   TM for thiuram; tetramethylthiuramdisulfide.
See also "Methyl Tuads."

tubatoxin.   See rotenone.

tube mill.   A fine-grinding machine comprised of a rotating steel cylinder which may be from 15–50 ft long and 6 to 8 ft in diameter, within which are steel balls from 1 to 5 inches in diameter. Depending on its construction it may be either batch or continuous. Some types have several compartments, each containing balls of different sizes. Its function is finish-grinding of particu-

lates; it is usually fed with 20-mesh material which it reduces to 325-mesh.
See also ball mill.

tuberose oil.
Properties: Colorless to very light-colored oil, d 1.007–1.035 at 15C, taken from Polianthes tuberosa, by enfleurage.
Use: Perfume, flavoring.

***d*- tubocurarine chloride.**
$C_{38}H_{44}Cl_2N_2O_6 \cdot 5HOH$.
Properties: White to light tan, odorless, crystalline alkaloid; mp 270C with decomposition; soluble in water and alcohol; insoluble in acetone, chloroform, and ether; aqueous solution is strongly dextrorotatory (specific rotation for 1% solution of anhydrous −208 to +218 degrees).
Grade: USP.
Hazard: Highly toxic.
Use: Medicine (muscle relaxant).

"Tuex."[248]   TM for tetramethylthiuram disulfide. Available in pellets as "Tuex Naugets."
See thiram.

"Tumerol."[342]   TM for oleoresin of tumeric for food coloring and flavoring.

tung oil. (China-wood oil).
Properties: Yellow drying oil; soluble in chloroform, ether, carbon disulfide, and oils. D 0.9360–0.9432, saponification value 193, iodine value 150–165, refr index 1.5030, flash p 552F (288.9C), autoign temperature 855F (457C). Combustible.
Derivation: From the seeds of *Aleurites cordata*, a tree indigenous to China. It is now produced in China, Argentina, and Paraguay; US production has been virtually discontinued. Chief constituent: eleostearic acid (80%).
Use: Exterior paint, varnishes.

tungstated pigment.   See phosphotungstic pigment.

tungstate white.   See barium tungstate.

tungsten. (wolfram).   CAS: 7440-33-7.
W.   Metallic element, atomic number 74, group VIB of periodic system, aw 183.85, valences = 2, 4, 5, 6; five isotopes.
Properties: Hard, brittle, gray solid; not found native; the ores are scheelite and wolframite. D 19.3 (20C), mp 3410C (highest of all the metals), bp 5927C, high electrical conductivity, soluble in a mixture of nitric acid and hydrogen fluoride, corroded by seawater, oxidizes in air at 400C, the rate increasing rapidly with temperature.
Derivation: (a) Aluminothermic reduction of tungstic oxide, (b) hydrogen reduction of tungstic acid or its anhydride. Fabrication: The metal can be plated onto objects by vapor deposition from tungsten hexafluoride or hexacarbonyl and can bond metal parts together. Tungsten powder (produced by carbon reduction) is converted into solid metal by powder metallurgical techniques. Large single crystals are grown by an arc-fusion process. Granules obtained by reduction of the hexafluoride.
Occurrence: Canada, Bolivia, Peru, Thailand, China, USSR, US (Arizona, California, Colorado, Nebraska, Nevada, New Mexico, and Texas).
Grade: Technical, powder, single crystals, ultrapure granules of 50–600 microns.
Hazard: Finely divided form is highly flammable and may ignite spontaneously. TLV (as W) (insoluble compounds): 5 mg/m³ of air; (soluble compounds) 1 mg/m³ of air.
Use: High-speed tool steel, ferrous and nonferrous alloys, ferrotungsten, filaments for electric light bulbs, contact points, x-ray and electron tubes, welding electrodes, heating elements in furnaces and vacuum-metallizing equipment, rocket nozzles and other aerospace applications, shell steel, chemical apparatus, high speed rotors as in gyroscopes, solar energy devices (as vapor-deposited film which retains heat at 500C).

tungsten boride. WB_2.
Properties: Silvery solid, d 10.77, mp 2900C, insoluble in water, soluble in aqua regia and concentrated acids, decomposed by chlorine at 100C.
Derivation: Heating tungsten and boron in electric furnace.
Use: Refractory.

tungsten carbide.   CAS: 12070-12-1.   WC.
A ditungsten carbide, W_2C, with similar physical properties is also known. WC is said to be the strongest of all structural materials.
Properties: Gray powder, d 15.6, mp 2780C, bp 6000C, hardness of 9+ (Mohs) in solid form, insoluble in water but readily attacked by a nitric acid-hydrofluoric acid mixture, stable to 400C with chlorine, burns in fluorine at room temperature, oxidizes on heating in air.
Derivation: Heating tungsten and lamp black at 1500–1600C.
Hazard: Toxic by inhalation. TLV (as W): 5 mg/m³ of air.
Use: Cemented carbide, dies and cutting tools, wear-resistant parts, cermets, electrical resistors. An abrasive in liquids.

tungsten carbide, cemented.   A mixture consisting of tungsten carbide 85–95% and cobalt 5–15%.
Properties: Hardness about that of corundum, not affected by high temperatures, d 12–16.

Derivation: Ball milling of powdered tungsten carbide with metallic cobalt, followed by sintering.

Hazard: As for tungsten carbide.

Use: Machine tools and abrasives for machining and grinding metals, rocks, molded products, porcelain, and glass; in gages, blast nozzles, knives, drill bits.

tungsten carbonyl. See tungsten hexacarbonyl.

tungsten diselenide. WSe_2. Lamellar-structured, dry, solid lubricant with exceptional high temperature and high vacuum stability, retains its lubricity to temperatures as low as $-265C$.

Grade: 1–2 micron and 40 micron.

Hazard: TLV (as W): 5 mg/m³ of air.

tungsten disulfide. WS_2.

Properties: Grayish-black solid, mp above 1480C, can lubricate at temperatures above 2400F (1316C). Attacked by fluorine and hot sulfuric and hydrogen fluoride.

Grade: 0.40, 0.70 and 1–2 micron grades.

Use: Solid lubricant for many applications, including use as an aerosol.

tungsten hexacarbonyl. (tungsten carbonyl). $W(CO)_6$.

Properties: White, volatile, highly refractive, crystalline solid; decomposes without melting at 150C; one of the more stable carbonyls. D 2.65, vapor pressure 0.1 mm (20C), insoluble in water, soluble in organic solvents.

Derivation: Reaction of tungsten with carbon monoxide at high pressures, reduction of tungsten hexachloride with iron alloy powders in carbon monoxide atmosphere.

Hazard: TLV (as W): 5 mg/m³ of air.

Use: Tungsten coatings on base metals by deposition and decomposition of the carbonyl.

tungsten hexachloride. WCl_6.

Properties: Dark blue or violet hexagonal crystals, volatile, mp 275C, bp 347C, d 3.52, vap press 43 mm (215C), electrical conductivity (fused state) poor, soluble in organic solvents including ligroin and ethanol, decomposed by moist air and water, reduced by hydrogen to the metal.

Derivation: Heating tungsten with dry chlorine at red heat.

Hazard: TLV (as W): 5 mg/m³ of air.

Use: Tungsten coatings on base metals, vapor deposition for bonding metals, single crystal tungsten wire, additive to tin oxide to produce electrically conducting coating for glass, catalyst for olefin polymers.

tungsten hexafluoride. CAS: 7783-82-6. WF_6.

Properties: Colorless gas or light yellow liquid, d (liquid) 3.44, mp 2.5C, bp 19.5C, decomposes in water.

Derivation: Direct fluorination of powdered tungsten, purified by distillation under pressure.

Hazard: Toxic by ingestion and inhalation, strong irritant to tissue. TLV (as W): 1 mg/m³ of air.

Use: Vapor phase deposition of tungsten, fluorinating agent.

tungsten lake. See phosphotungstic pigment.

tungsten oxychloride. $WOCl_4$.

Properties: Dark red, acicular crystals. Decomposed by water and moist air, keep in sealed glass container, bp 227.5C, mp 211C, d 11.92, soluble in carbon disulfide.

Derivation: By the action of chlorine on tungsten or tungstic oxide at elevated temperatures.

Purification: Vacuum distillation.

Grade: Technical.

Hazard: Irritant. TLV: 1 mg/m³ of air.

Use: Incandescent lamps.

tungsten silicide. A ceramic, probably WSi_2.

Properties: Blue-gray, very hard solid, d 9.4, mp above 900C, insoluble in water, attacked by fused alkalies and mixture of nitric and hydrofluoric acids.

Grade: Cylindrical shapes, lumps, standard sieve sizes.

Hazard: Dust flammable. TLV (as W): 5 mg/m³ of air.

Use: Oxidation-resistant coatings, electrical resistance and refractory applications.

tungsten steel. In many of its alloying effects, tungsten is similar to molybdenum. Tungsten increases the density of alloys to which it is added.

Use: To obtain steels with great wear resistance and special resistance to tempering, as in high-speed steels, hot-work steels, finishing steels that maintain keen cutting edge and great wear resistance, creep-resisting steels, and oxidation-resistant, high-temperature, high-strength alloys.

tungsten trioxide. See tungstic oxide.

tungstic acid. (wolframic acid; orthotungstic acid). CAS: 7783-03-1. H_2WO_4.

Properties: Yellow powder, d 5.5, insoluble in water, soluble in hydrogen fluoride and alkalies. A white form of tungstic acid exists, having the formula $H_2WO_4 \cdot HOH$. This is formed by acidifying tungsten solutions in the cold.

Derivation: Decomposition of sodium tungstate with hot sulfuric acid.

Grade: Technical, CP, reagent.

Hazard: Toxic material. TLV (as W): 5 mg/m³ of air.

Use: Textiles (mordant, color resist), plastics, tungsten metal, wire, etc.

tungstic acid anhydride. See tungstic oxide.

tungstic anhydride. See tungstic oxide.

tungstic oxide. (tungstic acid anhydride; tungstic anhydride; tungsten trioxide; wolframic acid, anhydrous). WO_3.
Properties: Canary yellow, heavy powder, dark orange when heated and regains original color on cooling, mp 1473C, d 7.16, insoluble in water, soluble in caustic alkalies, soluble with difficulty in acids. Noncombustible.
Derivation: Scheelite ore is treated with hydrochloric acid and the resulting product dissolved out with ammonia. The complex ammonium tungstate can then be ignited to tungstic oxide.
Hazard: Toxic material. TLV (as W): 5 mg/m³ of air.
Use: To form metal by reduction, alloys, preparation of tungstates for x-ray screens, fireproofing fabrics, yellow pigment in ceramics.

"Tungstide."[289] TM for a series of coatings consisting of metallic tungsten of near-colloidal particle-size suspended in a liquid plastic and incorporated with a form of "Liqui-Moly," (a molybdenum disulfide product) to produce a hard, abrasion-resistant, self-lubricating coating.

tungstophosphate. Structure and properties are similar to molybdophosphate.

12-tungstophosphoric acid. See phosphotungstic acid.

tungstosilicate. One of a group of complex inorganic compounds of high molecular weight, containing a central silicon atom surrounded by tungsten oxide octahedra. They have high molecular weight, high degree of hydration, strong oxidizing action in aqueous solutions; they decompose in strongly basic aqueous solutions to give simple tungstate solutions; they form highly colored anions or reaction products.
See silicotungstic acid and sodium-12-tungstosilicate.

12-tungstosilicic acid. See silicotungstic acid.

"Turgum S."[285] TM for a resin acid; d 1.06–1.07; an odorless tackifier, retarder, and processing aid; slightly retarding in cure.

turkey brown. (turkey umber). Natural earth used as permanent pigment. Contains iron and manganese oxide with some clay.

Turkey red. See iron oxide red.

Turkey red oil. (castor oil, sulfonated; castor oil, soluble). CAS: 72-48-0. It is also known as alizarin assistant and alizarin oil because of its use in dyeing with alizarin.
Properties: Viscous liquid, d 0.95, iodine no. 82.1, acid no 174.3, saponification no. 189.3, soluble in water, autoign temperature 833F (445C). Combustible.
Derivation: By sulfonating castor oil with sulfuric acid and washing.
Grade: Sulfonated castor oil graded as to moisture and color.
Use: Textiles, leather, manufacture of soaps, alizarin dye assistant, paper coatings.

Turnbull's Blue. An inorganic blue pigment made by the reaction of a ferrous salt and potassium ferricyanide $[Fe_3Fe(CN)_6]_2$. One of its important uses is in making blueprints, in which sensitized paper containing ferric ammonium citrate and potassium ferricyanide is exposed to light, the ferric ions being thus reduced to ferrous ions.
See also Iron Blue.

turpentine (gum). The oleoresin or pitch obtained from living pine trees. Sticky, viscous, balsamic liquid comprising a mixture of rosin and turpentine oil; strong piny odor; soluble in alcohol, ether, chloroform, and glacial acetic acid. Combustible.
Use: Source of turpentine oil and gum rosin.
See also rosin.

turpentine (oil). CAS: 8006-64-2. $C_{10}H_{16}$
A volatile essential oil whose chief constituents are pinene and diterpene.
Derivation: Steam-distillation of the turpentine gum exuded from living pine trees (gum turpentine), or naphtha-extraction of pine stumps (wood turpentine), destructive distillation of pine wood.
See also rosin, rosin oil.
Properties: Colorless liquid, penetrating odor, immiscible with water, lighter than water. Considerable variation appears in constants reported; the following are based on tests made by the Forest Products Laboratory: d 0.860–0.875 (15C), refr index 1.463–1.483 (20C), flash p (CC) 90–115F (32–46C) acidity none, bulk d 17.18 lb/gal, autoign temperature 488F (253C).
Hazard: Moderate fire risk. Toxic by ingestion. TLV: 100 ppm in air.
Use: Solvent; thinner for paints, varnishes, and lacquers; rubber solvent and reclaiming agent; insecticide; synthesis of camphor and menthol; wax-based polishes; medicine (liniments); perfumery.

"Tutane."[530] TM for 2-aminobutane.
Use: Fungicide.

tutocaine hydrochloride. (gamma-dimethylmino-alpha,beta-dimethylpropyl-p-aminoben-

zoate hydrochloride.) CAS: 532-62-7. ($C_{14}H_{22}O_2N_2 \cdot HCl$.
Use: Local anesthetic.

tuyere. A duct or pipe through which a stream of hot air is introduced into a blast-furnace or cupola to support combustion.

Tw. Abbreviation for Twaddell, used in reporting specific gravities for densities greater than water, as degrees Tw. A twaddell reading, multiplied by five and added to 1000, gives specific gravity with reference to water as 1000.

Twitchell process. Commercial process for splitting fats to glycerol and fatty acids by heating the sulfuric-acid-washed fat 20-48 hours in an open tank with steam in a mixture of 25-50% water, 0.5% sulfuric acid, and 0.75-1.25% Twitchell reagent. The original Twitchell catalyst was prepared by sulfonation of a mixture of fatty acid and benzene, but nowadays sulfonated petroleum products are used.

Twitchell reagent. Catalyst for the Twitchell process (acid hydrolysis of fats). It is a sulfonated addition product of naphthalene and oleic acid, a naphthalenestearosulfonic acid.

"Tygobond."[326] TM for a series of vinyl- and rubber-based adhesives for cementing porous and semiporous materials to each other.

"Tygofil."[326] TM for a modified epoxy base metal filler.

"Tygon."[326] TM for a series of vinyl compounds used as linings, coatings, adhesives, tubing, and extruded shapes applied to chemical process equipment as corrosion protection.

"Tygorust."[326] TM for a vinyl-based primer for application to damp or dry rusted steel.

"Tygoweld."[326] TM for a modified epoxy-based structural adhesive for bonding both similar and dissimilar materials.

"Tygozinc."[326] TM for a series of inorganic zinc-rich protective coatings of both self-curing and post-cured types.

"Tylac."[36] TM for a series of synthetic latexes and elastomers. NBL High-strength, film-forming, nitrile rubber latexes and rubbers and carboxylated polymers characterized by exceptional oil, solvent, and abrasion resistance. SBL Butadiene/styrene and carboxylated butadiene/styrene and high styrene latexes of various monomer ratios. Modified versions available.
Use: Paper, textiles, adhesives, Portland cement, molded products, polyvinyl chloride and phenolic resin blends.

"Tylenol."[595] TM for p-acetylaminophenol (acetaminophen).

"Tylose."[450] TM for a wide range of water-soluble cellulose ethers.
Use: Thickeners, binders, dispersing agent, emulsifiers, protective colloids, lubricants, and film forming materials; drilling muds; detergents; textile, paper, print, and varnish industries; ceramics; cosmetics and pharmaceuticals.
See methylcellulose, carboxymethylcellulose; numerous grades and viscosities are available.

tyloxin. Antibiotic substance isolated from a strain of *Streptomyces fradiae*.
Properties: Crystals, mp 128–130C, solutions are stable between pH 4–9.
Use: Veterinary medicine.

tyloxapol. (USAN). CAS: 25301-02-4.
An oxyethylated-tert-octylphenolpolymethylene polymer and low toxicity, nonionic surfactant.

Tyndall effect. A colloidal phenomenon in which particles too small to be resolved in an optical microscope suspended in a gas or liquid reveal their presence by scattering a beam of light as it passes through the suspension, the extent of reflection being dependent on the position of the irregularly shaped particles relative to the incident light. The effect causes the appearance of a visible cone of light through the suspension. This principle is utilized in the ultra-microscope.

"Tynex."[28] TM for nylon filament. Available tapered with an essentially uniform taper from butt to tip, and also level, i.e., in a wide range of constant diameters. The tapered form is used primarily in paint brushes, the level form in other brushes.

type metal. Alloy of 75–95% lead, 2.5–18% antimony, with a little tin and sometimes copper, which expands slightly upon solidification and produces sharp castings.

"Ty-Ply."[525] TM for a family of nontacky vulcanizing adhesives for bonding natural and synthetic rubber compounds.

Tyr. Abbreviation for tyrosine.

Tyrer sulfonation process. Sulfonation of benzene in the vapor phase with sulfuric acid at 170-180 degrees with passage of benzene vapor and azeotropic removal of the water of reaction.

"Tyrez."[36] TM for a series of impact modifiers for plastics.

"Tyril."[233] TM for a group of styrene-acrylonitrile copolymers.

"Tyrilfoam" 80.[233] TM for expanded styreneacrylonitrile for flotation uses under conditions of gasoline spillage, petroleum scum from outboard motors, and stagnant water.

tyrocidine. CAS: 8011-61-8. Antibiotic produced by the metabolic processes of the bacteria, *Bacillus brevis*. It is a cyclic polypeptide which is active against most gram positive pathogenic bacteria. It is one of the two antibiotic components of tyrothricin but has been isolated and used alone.
Properties: (Probably the hydrochloride): Fine, crystalline needles which decomposes at 240C; soluble in 95% alcohol, acetic acid, and pyridine; slightly soluble in water, acetone, and absolute alcohol; insoluble in ether, chloroform, and hydrocarbons; depresses surface tension; forms fairly stable colloidal emulsion in distilled water.
Use: Medicine (usually as component of tyrothricin), possible fungistat and bacteriostat.

tyrosinase. CAS: 9002-10-2. An enzyme containing copper which occurs in plant and animal tissue and is responsible for turning peeled potatoes black when exposed to air.
Use: Medicine (antihypertensive).

tyrosine. (β-p-hydroxyphenylalanine; α-amino-β-p-hydroxyphenylpropionic acid).

CAS: 60-18-4.
$C_6H_4OHCH_2CHNH_2COOH$. A nonessential amino acid.
Properties: White crystals, readily oxidized by the animal organism, soluble in water, slightly soluble in alcohol, insoluble in ether, optically active. dl-tyrosine mp 316C. d(+)-tyrosine mp 310–314C; l(-)-tyrosine mp 295C with decomposition, d 1.456 (20/4C).
Derivation: Hydrolysis of protein (casein), organic synthesis.
Grade: FCC.
Use: Growth factor in nutrition, biochemical research, dietary supplement.
Available commercially as dl-tyrosine.

tyrothricin. An antibiotic produced by growth of *Bacillus brevis*. It consists of a mixture of antibiotics, principally gramicidin and tyrocidine. Gramicidin is the more active component. Use is generally limited to local external applications. It is active against some gram-positive bacteria, including species of pneumococci, streptococci and staphylococci.

"Tyzor."[28] TM for a group of simple and chelated esters of ortho-titanic acid, such as tetrabutyltitanate, of varying reactivity.
Use: Chemical intermediates, primers for adhesion promotion in extrusion coating, dispersants, scratch-resistant finishes on glass, masonry water repellents, crosslinking agents, esterification and olefin polymerization catalyst.

U

U. Symbol for uranium.

"Ucane."[214] TM for soft detergent alkylates which are anionic in character after sulfonation; derived from the chlorination of n-paraffins and the subsequent alkylation of benzene with these n-paraffin chlorides. Show almost complete biodegradability.

"Ucar."[214] TM for various synthetic organic chemicals including butylene oxide, butylphenol, nonylphenol, triphenol phosphorus, bisphenol A, p-tert-amylphenol, ethylene and propylene glycols; also applied to synthetic latexes and water-soluble polymers.

"Ucon."[214] (1) TM for a series of nonflammable fluorocarbon solvents and solvent blends with high chemical stability and extremely low residue levels.
Use: Cleaning electronic and mechanical instruments and controls, degreasing motors, cleaning liquid oxygen equipment, and motion picture and television film and magnetic tape.
(2) TM for polyalkylene glycols and diesters. Available as water-soluble or insoluble products.
Grade: LB-Series, 50-HB Series, 75-H Series, DLB Series and Hydrolube Series.
Use: High-temperature lubricants, low temperature fluids, compressor lubricants, hydraulic brake fluids, quenchant fluid, heat transfer fluids, textile lubricants, rock drill lubricants, leather and paper-treating compounds, rubber lubricants, plasticizers and solvents, chemical intermediate; stationary phase in gas chromatography.
See fluorocarbon.

"Udel."[214] TM for biaxially oriented polypropylene film used for overwraps, laminations, box windows, and shrink wraps; also for polysulfone resins.

UDP. Abbreviation for uridine diphosphate.
See uridine phosphate.

UDPG. Abbreviation for uridine diphosphate glucose.

"Uformite."[23] TM for synthetic resins based on urea-formaldehyde, melamine-formaldehyde, and triazine condensates. Supplied as colorless or light-colored aqueous solutions or solutions in volatile solvents. Solvent type produces hard, alkali-resistant, colorless coatings on curing with adhesion to a variety of surfaces.
Use: With alkyd resins in coatings, industrial finishes on appliances, automobiles, etc; adhesives for paper-board boxes, paper coatings, wet-strength paper, textile pigment binding.

ulexite. (cotton balls).   CAS: 1319-33-1.
$NaCaB_5O_9 \cdot 8HOH$.   A natural hydrated borate of sodium and calcium.
Properties: Color white, luster silky, Mohs hardness 1–2.5, d 1.96, usually found as rounded, loose-textured masses of fine crystals.
Occurrence: Chile, Argentina, California, Nevada.
Use: Source of borax.

Ullman reaction. Synthesis of biaryls by copper-induced coupling of aryl halides. Similar coupling of aryl halides with aroxides yields diaryl ethers. (A modification of the Fittig synthesis in which copper powder is used instead of sodium.)

ulmin brown. See Van Dyke brown.

ulmin. One of a class of amorphous substances resulting from the decomposition of the cellulose and lignite tissues of plants. Ulmins represent one of the initial changes by which vegetable matter is converted into coal.

"Ultem."[245] A series of engineering plastics based on polyetherimide resins. Said to be intermediate between thermoplastic polyesters, nylons, and acetals, and such specialty polymers as fluorocarbons. Performance in a broad range of properties is excellent, including heat resistance, nonflammability, and ease of processing.

ultra-accelerator. An unusually powerful accelerator of rubber vulcanization, typified by thiuram sulfides and dithiocarbamates.

ultracentrifuge. A high-speed rotational separating device, usually of lab size, capable of developing a force of 250,000 times gravity. Its major uses are in research on molecular weight distribution, macromolecular structure and properties (proteins, nucleic acids, viruses) and separation of solutes from solutions.
See also centrifugation.

"Ultra-Clor."[329] TM for a turf fungicide whose active ingredients are mercuric dimethyldithio-

carbamate, potassium chromate, and cadmium succinate.
Hazard: Highly toxic by ingestion.

ultramarine blue.
Properties: Inorganic pigment, blue powder, good alkali and heat resistance, low hiding power, poor acid resistance, poor outdoor durability. Noncombustible. CI 77007.
Derivation: Heating a mixture of sulfur, clay, alkali, and a reducing agent to high temperatures.
Use: Colorant for machinery and toy enamels, white baking enamels, printing inks, rubber products, soaps and laundry blues, cosmetics, textile printing. *Note*: Used in very low percentages to intensify whiteness of white enamels, rubber compounds, laundered clothing, etc. by offsetting yellowish undertones, gives a "blue" rather than a "yellow" white.

ultramicroscope. A development of the compound optical microscope invented in 1903 by Zsigmondy and Siedentopf. Its essential feature is a strong light beam from an arc lamp, focused by passing through two lenses, which illuminates the specimen at right angles to the axis of observation. The presence of suspended colloidal particles as small as 5 microns is detectable because of the light-scattering effect of the particles as they move about in the suspension (Tyndall effect). Since the light reflected or scattered by the particles is the only light that enters the microscope, the particles appear as points of light against a dark background (dark-field illumination). There is no resolution of individual particle shape or size; the instrument shows only that particles are present. The ultramicroscope has been of great value in the study of colloidal suspensions, such as rubber latex and of various biological phenomena; its usefulness has diminished since the advent of the electron microscope.
See also Tyndall effect.

ultrasonics. The science of effects of sound vibrations beyond the limit of audible frequencies. Used for dust, smoke, and mist precipitation; preparation of colloidal dispersions; cleaning of metal parts, precision machinery, fabrics, etc. Friction welding, formation of catalysts, degassing and solidification of molten metals, extracting flavor oils in brewing, electroplating, drilling hard materials, fluxless soldering, nondestructive testing. Also used for investigation of physical properties, determination of molecular weights of liquid polymers, degree of association of water, and for inducing chemical reaction. A developing application is the use of ultrasonic vibration in diagnostic medicine.

"Ultrathene."[192] TM for a series of ethylenevinyl acetate copolymer resins for adhesives, conversion coatings and thermoplastic modifiers. Wide range of melt indexes. Improves specific adhesion of hot-melt, solvent-based and pressure-sensitive adhesives.

ultraviolet. (UV). Radiation in the region of the electromagnetic spectrum including wavelengths from 100 to 3900 Å.
See radiation.
Hazard: Dangerous to eyes, overexposure causes severe skin burns (sunburn).
Use: Air sterilization in hospitals, microscopy.

ultraviolet absorber. A substance which absorbs radiant energy in the wavelength of UV. The radiant energy absorbed is converted to heat (thermal energy). UV absorbers are added to unsaturated substances (plastics, rubbers, etc.) to decrease light sensitivity and consequent discoloring and degradation. Among compounds used are benzophenones, benzotriazoles, substituted acrylonitriles, and phenol-nickel complexes.
See also absorption (2).

"Ultrawet."[104] TM for a series of biodegradable linear alkylate sulfonate (LAS) anionic detergents or surface-active agents (linear dodecylbenzene type). Available forms: Sodium or triethanolamine salts in liquid, slurry, flake, or bead form.

"Ultratex."[443] TM for a group of silicone elastomeric finishes for natural and synthetic textile fabrics.

umbellic acid. See anisic acid.

umbelliferone. CAS: 93-35-6. $C_9H_6O_3$.

Properties: Colorless needles; mp 225C; soluble in alcohol, acetic acid, weak alkalies; also in boiling water.
Derivation: Distillation of plant resins (umbelliferae).
Use: Cosmetics, sun-screen preparations.

umber. A naturally occurring brown earth containing ferric oxide together with silica, alumina, manganese oxides, and lime. Raw umber is umber which is ground and then levitated. Burnt umber is umber calcined at low heat.
Use: Paint pigment, lithographic inks, wall paper (pigment), artists' color.

UMP. Abbreviation for uridine monophosphate. See uridine phosphates and also uridylic acid.

uncertainty principle. The conclusion of Heisenberg based on quantum mechanical theory that the precise position of a specific electron in an atomic orbit cannot be determined and that consequently the ultimate nature of matter is not susceptible to objective measurement. The result of this concept was development of the orbital theory in which electron behavior is dealt with on a statistical basis. Its validity was confirmed by 1930 by the work of other mathematical physicists such a DeBroglie, Fermi, and Schrodinger. See also orbital theory; Heisenberg.

γ-undecalactone. (peach aldehyde; γ-undecyl lactone; 4-hydroxy-undecanoic acid, gamma-lactone). CAS: 710-04-3.

$$CH_3(CH_2)_6\overline{CHCH_2CH_2COO}.$$

Properties: Colorless to light-yellow liquid, peach-like odor, d 0.941–0.944, refr index 1.450–1.454. Soluble in 4–5 volumes of 60% alcohol; soluble in benzyl alcohol, benzyl benzoate, and most fixed oils. Combustible.
Derivation: By heating undecylenic acid in the presence of sulfuric acid.
Grade: Chlorine-free, FCC.
Use: Perfumery, flavoring agent.

undecanal. (n-undecylic aldehyde; hendecanal). CAS: 112-44-7. $CH_3(CH_2)_9CHO$.
Properties: Colorless liquid, sweet odor, d 0.825–0.832 (25C), refr index 1.4310–1.4350 (20C), soluble in oils and alcohol, insoluble in glycerol and water, flash p 235F (112.7C). Combustible.
Derivation: By oxidation of 1-undecanol or reduction of undecanoic acid.
Grade: FCC.
Hazard: Toxic by ingestion and inhalation, irritant to tissue.
Use: Perfumery, flavors.

n-undecane. (hendecane). CAS: 1120-21-4. $CH_3(CH_2)_9CH_3$.
Properties: Colorless liquid, d 0.7402 (20/4C), fp −25.75C, bp 195.6C, refr index 1.41725 (20C), flash p 149F (65C). Combustible.
Grade: 95%, 99%, research.
Use: Petroleum research, organic synthesis, distillation chaser.

undecanoic acid. (n-undecylic acid; hendecanoic acid). $CH_3(CH_2)_9COOH$. Small amounts occur in castor oil. It is best derived from undecylenic acid by hydrogenation.
Properties: Colorless crystals, d 0.8505 (80/4C), mp 28.5C, bp 284.0C, 222.2C (128 mm), refr in-

dex 1.4319 (40C), insoluble in water, soluble in alcohol and ether.
Grade: Technical, 99%.
Use: Organic synthesis.

1-undecanol. (n-undecyl alcohol; decyl carbinol; 1-hendecanol). CAS: 112-42-5. $CH_3(CH_2)_9CH_2OH$.
Properties: Colorless liquid with a citrus odor, d 0.829–0.834, refr index 1.435–1.443, mp 19C, soluble in 60% alcohol, flash p 200F (93.3C). Combustible.
Use: Perfumery, flavoring.

2-undecanol. (2-hendecanol). $CH_3(CH_2)_8CHOHCH_3$.
Properties: Colorless liquid, d 0.8363 (20C), mp 12C, bp 228–229C, insoluble in water, soluble in alcohol and ether, flash p 235F (112.7C). Combustible.
Use: Antifoaming agent, intermediate, perfume fixatives, plasticizer.

2-undecanone. See methyl nonyl ketone.

undecenal. (undecylenic aldehyde; hendecen-1-al). Listed by different authorities as 10-undecenal, $CH_2{:}CH(CH_2)CHO$, and 9-undecenal, $CH_3CH{:}CH(CH_2)_7CHO$.
Properties: Colorless liquid, strong odor suggesting rose, d 0.840–0.850 (25/25C), refr index 1.4410–1.4470 (20C), soluble in 80% alcohol. Combustible.
Use: Perfumery, flavoring.

10-undecenoic acid. See undecylenic acid.

10-undecen-1-ol. See undecylenic alcohol.

n-undecyl alcohol. See 1-undecanol.

undecylenic acid. (10-undecenoic acid). CAS: 112-38-9. $CH_2{:}CH(CH_2)_8COOH$.
Properties: Light-colored liquid; fruity-rosy odor; almost insoluble in water, miscible with alcohol, chloroform, ether, benzene and with fixed and volatile oils. Congealing point 21C, d 0.910–0.913 (25/25C), refr index 1.4475–1.4485 (25C), flash p 295F (146C). Combustible.
Derivation: Destructive distillation of castor oil.
Grade: Technical, NF.
Use: Perfumery, flavoring, plastics, modifying agent (plasticizer, lubricant additive, etc.), medicine (antifungal agent).

undecylenic alcohol. (n-undecylenic alcohol; 10-undecen-1-ol; alcohol C-11). CAS: 112-43-6. $CH_2{:}CH(CH_2)_8CH_2OH$.
Properties: Colorless liquid, citrus odor, d 0.842–

0.847, refr index 1.449–1.454; fp −3.0C, soluble in 70% alcohol. Combustible.
Use: Perfumes.

undecylenic aldehyde. See undecenal.

undecylenyl acetate. (10-hendecenyl acetate). CAS: 112-19-6. $CH_3COO(CH_2)_9CH:CH_2$.
Properties: Colorless liquid, floral-fruity odor, d 0.876–0.883, refr index 1.438–1.442, soluble in 80% alcohol. Combustible.
Use: Perfumery, flavoring.

n-undecylic acid. See undecanoic acid.

γ-undecyl lactone. See γ-undecalactone.

UNH. Abbreviation for uranyl nitrate hydrated. See uranyl nitrate.

unhairing. (dehairing). Removal of hair from hides and skins as practiced on a commercial scale in the leather industry. Several methods are used, involving application of hydrated lime, dimethylamine, trypsin, and other enzymes.

"Unicel."[28] TM for blowing agents for natural and synthetic rubber sponge.

"Unitol."[420] TM for a series of tall oil fatty acids, distilled tall oil, tall oil rosin, acid-refined tall oils, synthetic crude tall oils, and tall oil pitch.
Use: Alkyd resins, oleoresinous vehicles, esters, maleic-modified resins, fumaric-modified resins, gloss oils, epoxy resins, polyurethanes, soaps, synthetic detergents, ore flotation, stabilizer-plasticizers, metallic soaps, paper size, rubber processing aids, and others.

unit operation. A particular kind of physical change used in the industrial production of various chemicals and related materials. Filtration, evaporation, distillation, fluid flow and heat transfer are examples.
See also chemical engineering.

unit process. A process characterized by a particular kind of chemical reaction; oxidation, hydrolysis, esterification, and nitration are examples.
See also kinetics, chemical; chemical engineering.

"Univis."[51] TM for a series of power transmission or hydraulic oils. They have viscosity indexes of 150 or higher and pour points of −46C or lower, permitting wide temperature ranges in operation.

"Univolt."[51] TM for oils used as electrical insulating mediums in transformers, switches, and some electrical cables. Suitable for transformers,

whether used indoors or outdoors or under low-temperature conditions.
See also transformer oil.

uns-. (unsym). Abbreviation for unsymmetrical. A prefix denoting the structure of organic compounds in which substituents are disposed unsymmetrically with respect to the carbon skeleton or to a functional group, such as a double bond. For example, unsdichloroethane is CH_3CHCl_2.

unsaturation. Of a chemical compound, the state in which not all the available valence bonds along the alkyl chain are satisfied; in such compounds the extra bonds usually form double or triple bonds (chiefly with carbon). Thus unsaturated compounds are more reactive than saturated compounds, as other elements readily add to the unsaturated linkage. An unsaturated compound (ethylene, C_2H_4; butadiene, C_4H_6; benzene, C_6H_6) has fewer hydrogen atoms or equivalent groups than the corresponding saturated compound (ethane, C_2H_6; butane, C_4H_{10}; cyclohexane, C_6H_{12}).
In structural formulas unsaturation may be represented by parallel lines joining the carbon atoms (ethylene, $H_2C=CH_2$; butadiene $H_2C=CHCH=CH_2$) or by colons or triple dots, $H_2C:CH_2 H_2C:CH_2$ (ethylene) and $HC \vdots CH$ (acetylene).
See also carboxyl group.

UPS. Ultraviolet photoelectron spectroscopy.

"Urab."[50] TM for a complex of fenuron and TCA, available in liquid concentrations, granular and pelleted formulations. A brush and weed killer for control of woody plants and deep-rooted weeds in noncrop land.

"Urac."[57] TM for products based on ureaformaldehyde condensates used mainly as adhesives for the production of moisture-proof bonds in plywood manufacture, plywood assembly, and furniture manufacture.

uracil. (2,4-dioxypyrimidine). CAS: 66-22-8. $HNC(O)NHC(O)CHCH$. A pyrimidine that is a constituent of ribonucleic acids and the coenzyme, uridine diphosphate glucose.
Properties: Crystalline needles, mp 335C (decomposes), soluble in hot water, ammonium hydroxide, and other alkalies; insoluble in alcohol and ether.
Derivation: Hydrolysis of nucleic acids, precipitation from urea and ethyl formylacetate. Radioactive forms available.
Use: Biochemical research.

uracil-6-carboxylic acid. See orotic acid.

"Uracil Mustard."[327] TM for 5-[bis(2-chloroethyl)amino]uracil. An antineoplastic.

uracil, D-ribosyl. See uridine.

"Uramite."[28] TM for a fertilizer derived from urea-formaldehyde containing 38% nitrogen.

"Uramon."[28] TM for solutions of urea in aqueous ammonia.
Use: Manufacture of mixed fertilizers.

urania. See uranium dioxide.

urania-thoria. Crystals of the mixed oxides of uranium and thorium are available. The crystals are denser and cheaper than the pellet form.
Use: Nuclear fuel.

uranic chloride. See uranium tetrachloride.

uranic oxide. See uranium dioxide.

uranine. (uranine yellow; sodium fluorescein; resorcinolphthalein sodium; Color Index No. 45350). CAS: 518-47-8. $Na_2C_{20}H_{10}O_5$.
Properties: Orange red, odorless powder, hygroscopic, soluble in water and sparingly soluble in alcohol.
Derivation: By treatment of fluorescein with sodium carbonate solution and crystallizing.
Method of purification: Recrystallization.
Grade: Technical, USP (as sodium fluorescein).
Use: Dyeing silk and wool yellow; tracing subterranean waters, marking water for air-sea rescues; clinical test solution.

uraninite. CAS: 1317-99-3. UP_2. A natural phosphide of uranium usually partly oxidized to UO_3 with variable amounts of lead, radium, thorium, rare earth metals, helium, argon, nitrogen. Pitchblende is an important variety.
Occurrence: Colorado, Utah, South Africa, Canada, Europe, USSR, Australia, Zaire.
Use: Source of uranium and radium.

uranium. CAS: 7440-61-1. U. Metallic element number 92; a member of the actinide series; aw 238.029; valences = 3, 4, 6; three natural radioactive isotopes: uranium-234 (0.006%), uranium-235 (0.7%), and uranium-238 (99%).
See separate entries for details.
Properties: Dense, silvery, solid, strongly electropositive, ductile and malleable, poor conductor of electricity, d 19.0, mp 1132C, bp 3818C, heat of fusion 4.7 kcal/mole, heat capacity 6.6 cal/mole/C. Forms solid solutions (for nuclear reactors) with molybdenum, niobium, titanium, and zirconium. The metal reacts with nearly all nonmetals. It is attacked by water, acids, and peroxides, but is inert toward alkalies. Green tetravalent uranium and yellow uranyl ion (UO_2^{++}) are the only species which are stable in solution.
Occurrence: Pitchblende (essentially UO_2), a variety of uraninite, coffinite $(USiO_4)$ and carnotite (Colorado, New Mexico, France, Zaire, Canada, South Africa, Australia, USSR). US resources of uranium oxide in 1970s were estimated at 150,000–175,000 tons; there is about the same amount in Canada.
Derivation: Finely ground ore is leached under oxidizing conditions to give uranyl nitrate solution. The uranyl nitrate, purified by solvent extraction (ether, alkyl phosphate esters), is then reduced with hydrogen to uranium dioxide. This is treated with hydrogen fluoride to obtain uranium tetrafluoride, followed by either electrolysis in fused salts or by reduction with calcium or magnesium. Uranium can also be recovered from phosphate sand.
Forms available: Solid pure metal, alloys, powder (99.7%).
Hazard: (Powder) Dangerous fire risk, ignites spontaneously in air. Highly toxic, radioactive material, source of ionizing radiation. TLV: (including metal and all compounds, as uranium) 0.2 mg/m^3 of air.
Use: Source of fissionable isotope uranium-235, source of plutonium by neutron capture, electric power generation.
See also enrichment (2), uranium compounds.

uranium-233. (U-233). A fissionable isotope of uranium produced artificially by bombarding thorium-232 with neutrons. Used as an atomic fuel in molten salt reactor and is a possible fuel in breeder reactors. Half-life 1.62×10^5 years.

uranium-234. (U-234). A natural isotope of uranium with half-life of 2.48×10^5 years; it is separated by extraction with trioctylphosphine oxide.
Use: Nuclear research, with potential use in fission detectors for counting fast neutrons.

uranium-235. (U-235). The readily fissionable isotope of uranium used to enrich natural uranium in nuclear fuels. It is present in uranium only to the extent of 0.7% and can be separated from it by any of several methods: the gaseous diffusion process using uranium hexafluoride, the gas centrifuge process, and the electromagnetic separation method. Its half-life is 7.13×10^8 years. It was the energy source used in the original atom bomb. Its critical mass is about 33 lbs.
See also fission, uranium.

uranium-238. (U-238).   The abundant isotope of uranium of which it comprises 99%. It is not fissionable, but will form plutonium-239 as a result of bombardment by neutrons in a reactor. Its half-life is 4.51×10^9 years. It will be used in breeder reactors, together with plutonium where its energy potential will be exploited by transmuting it to fissionable plutonium.
See also breeder.

uranium carbide.   See uranium dicarbide.

uranium compounds.   Before the advent of nuclear energy, uranium had very limited uses. It had been suggested for filaments of lamps. A small tube of uranium dioxide, UO_2, connected in series with the tungsten filaments of large incandescent lamps used for photography and motion pictures, tends to eliminate the sudden surge of current through the bulbs when the light is turned on, thereby extending their life. Compounds of uranium have been used in photography for toning and in the leather and wood industries uranium compounds have been used for stains and dyes. Uranium salts are mordants of silk or wool. In making special steels, a little ferrouranium has been utilized, but its value is questionable in this connection. Such alloys have not proved commercially attractive. In the production of ceramics, sodium and ammonium diuranates have been used to produce colored glazes.

Uranium carbide has been suggested as a good catalyst for the production of synthetic ammonia. Uranium salts in small quantities are claimed to stimulate plant growth, but large quantities are clearly poisonous to plants.

By far the most important use of uranium lies in its application for nuclear (or atomic) energy. This use, in fact, has so increased the value of uranium as to eliminate its use for many of the purposes mentioned above (Glenn T. Seaborg, "Encyclopedia of Chemical Elements, " ed. by C. A. Hampel).

uranium decay series. (uranium-radium series). The series of elements produced as successive intermediate products when the element uranium undergoes spontaneous natural radioactive disintegration into lead. Radium and radon are members of this series.

uranium, depleted.   Uranium from which most of the uranium-235 isotope has been removed.
See uranium-238.

uranium dicarbide. (uranium carbide). UC_2.
Properties: Gray crystals, d 11.28 (18C), mp 2350C, bp 4370C, decomposes in water, slightly soluble in alcohol.

Hazard: Highly toxic, radiation risk.
Use: As crystals, pellets, or microspheres for nuclear reactor fuel.

uranium dioxide. (uranium oxide; uranic oxide; urania; yellowcake).   CAS: 1344-57-6. UO_2.
Properties: Black crystals, insoluble in water, soluble in nitric acid and concentrated sulfuric acid, d 10.9, mp 3000C.
Derivation of pure oxide: Powdered uranium ore is digested with hot nitric-sulfuric acid mixture and filtered to remove the insoluble portion. Sulfate is precipitated from the solution with barium carbonate, and uranyl nitrate is extracted with ether. After re-extraction into water, it is heated to drive off nitric acid, leaving uranium trioxide. The latter is reduced with hydrogen to the dioxide. Can be prepared from uranium hexafluoride by treating with ammonia and subsequent heating of the ammonium diuranate. It is also recovered from phosphoric acid.
Hazard: High radiation risk. Ignites spontaneously in finely divided form.
Use: A crystalline (or pellet) form is used to pack nuclear fuel rods.

uranium, enriched.   Natural uranium to which a few percent of the fissionable 235-isotope has been added.
See also enrichment.

uranium hexafluoride.   CAS: 7783-81-5. UF_6.
Properties: Colorless, volatile crystals; sublimes; triple point 64.0C (1134mm); mp 64.5C (2 atmospheres); d 5.06 (25C); soluble in liquid bromine, chlorine, carbon tetrachloride, sym-tetrachloroethane, and fluorocarbons. Reacts vigorously with water, alcohol, ether, and most metals. Vapor behaves as nearly perfect gas.
Derivation: (1) Triuranium octoxide (U_3O_8) and nitric acid react to form a solution of uranyl nitrate; this is decomposed to UO_3 and reduced to the dioxide with hydrogen. The dioxide as a fluidized bed is reacted with hydrogen fluoride. The resulting tetrafluoride is fluorinated to the hexafluoride. (2) Triuranium octoxide is converted directly to the hexafluoride with hydrogen fluoride and fluorine, then purified by fractional distillation.
Hazard: Highly corrosive, radiation risk.
Use: Gaseous diffusion process for separating isotopes of uranium.

uranium hydride. UH_3.
Properties: Brown gray to black powder, d 10.92, conductor of electricity.
Derivation: Action of hydrogen on hot uranium.
Hazard: Highly toxic. Ignites spontaneously in air.

Use: Preparation of finely divided uranium metal by decomposition, separation of hydrogen isotopes, reducing agent, lab source of pure hydrogen.

uranium monocarbide. UC.
Properties: Lumps or powder that can be formed into desired shapes by powder metallurgy or arc-melt casting, mp 2375C, density 13.63, thermal conductivity 0.08 cal/sec/cm$_2$/C/cm, must be stored in inert atmosphere.
Hazard: Radioactive poison.

uranium oxide. See uranium dioxide.

uranium tetrafluoride. (green salt).
CAS: 10049-14-6.   UF$_4$.
Properties: Green, nonvolatile, crystalline powder; d 6.70; mp 1036C; insoluble in water.
Derivation: Treatment of uranium dioxide with hydrogen fluoride. See uranium hexafluoride.
Hazard: Highly corrosive, radioactive poison.
Use: Intermediate in preparation of uranium metal.

uranocene. (bis-cyclooctatetraenyluranium).
A synthetic metallocene "sandwich" molecule whose unique feature is that the pi-molecular orbitals of the large organic rings share electrons with f-atomic orbitals of the uranium.
See also metallocene.

uranyl acetate. CAS: 541-09-3.
UO$_2$(C$_2$H$_3$O$_2$)$_2$.
Properties: (Dihydrate): Yellow crystals, d 2.89, soluble in water plus acetic acid, slightly soluble in alcohol.
Use: Bacterial oxidation activator, copying inks, reagent.

uranyl nitrate. (uranium oxynitrate; UNH; yellow salt).   CAS: 10102-06-4.
UO$_2$(NO$_3$)$_2$·6HOH.
Properties: Yellow, rhombic crystals; d 2.807; mp 60.2C; bp 118C; soluble in water, alcohol, and ether.
Derivation: Action of nitric acid on uranium octoxide (U$_3$O$_8$).
Hazard: Strong oxidizer. Corrosive and irritating.
Use: Source of uranium dioxide, extraction of uranium into non-aqueous solvents.

urban waste. Solid waste materials including garbage, cellulosics, glass, metals, etc., but not sewage (municipal waste). Garbage is being used both directly as fuel and fermented to yield proteins; the cellulosic portion can be hydrolyzed to glucose, which is in turn converted to methane by anaerobic fermentation. A continuous process

for this operation utilizing newspapers or sawdust has been reported.
See also waste control; biomass.

urea. (carbamide).   CAS: 57-13-6.
CO(NH$_2$)$_2$.

$$O{=}C{\overset{\textstyle NH_2}{\underset{\textstyle NH_2}{\big<}}}$$

Occurs in urine and other body fluids. The first organic compound to be synthesized (Wohler 1824). 15th highest-volume chemical produced in US (1985) (as primary solution).
Properties: White crystals or powder, almost odorless, saline taste, d 1.335, mp 132.7C, decomposes before boiling. Soluble in water, alcohol, and benzene; slightly soluble in ether; almost insoluble in chloroform. Noncombustible.
Derivation: Liquid ammonia and liquid carbon dioxide at 1750–3000 psi and 160–200C react to form ammonium carbamate, NH$_4$CO$_2$NH$_2$, which decomposes at lower pressure (about 80 psi) to urea and water. Several variations of the process include once-through, partial recycle, and total recycle.
Method of purification: Crystallization.
Grade: Technical, CP, USP, fertilizer (45–46% nitrogen), feed grade (about 42% nitrogen).
Use: Fertilizer, animal feed, plastics, chemical intermediate, stabilizer in explosives, medicine (diuretic), adhesives, separation of hydrocarbons (as urea adducts), pharmaceuticals, cosmetics, dentrifices, sulfamic acid production, flameproofing agents, viscosity modifier for starch or casein-based paper coatings, preparation of biuret.
See also urea-formaldehyde resin.

urea adduct. See inclusion complex.

urea ammonia liquor. A solution of crude urea in aqueous ammonia containing ammonium carbamate.
Use: Reaction with superphosphate in preparation of fertilizers, furnishing combined nitrogen.

urea-ammonium orthophosphate. A fertilizer developed especially for food-deficient regions, especially rice-dependent areas. Several grades contain all three primary plant nutrients (nitrogen, phosphorus, and potassium). Contains up to 60% nitrogen, phosphoric anhydride, and potassium oxide.

urea-ammonium polyphosphate. A fertilizer similar to urea-ammonium orthophosphate except that about half the phosphorus is in polyphosphate form, which gives improved sequestering action and solubility. It is excellent for use as a liquid fertilizer.

ureaform. A urea-formaldehyde reaction product that contains more than one molecule of urea per molecule of formaldehyde. It can be used as a fertilizer because of its high nitrogen content, its insolubility in water, and its gradual decomposition in the soil during the growing season to yield soluble nitrogen.

urea-formaldehyde resin. An important class of amino resin. Urea and formaldehyde are united in a two-stage process in the presence of pyridine, ammonia, or certain alcohols with heat and control of pH to form intermediates (methylolurea, dimethylolurea) that are mixed with fillers to produce molding powders. These are converted to thermosetting resins by further controlled heating and press in the presence of catalysts. These were the first plastics that could be made in white, pastel, and colored products.
Use: Scale housings, dinnerware, interior plywood, foundry core binder, flexible foams, insulation. See amino resin, melamine resin.

urea hydrogen peroxide. See urea peroxide.

urea nitrate. CAS: 124-47-0.
$CO(NH_2)_2 \cdot HNO_3$.
Properties: Colorless crystals, decomposes 152C, slightly soluble in water, soluble in alcohol.
Derivation: By adding an excess of nitric acid to a strong aqueous solution of urea.
Hazard: Dangerous fire and explosion risk.
Use: Explosives, manufacture of urethane.

urea peroxide. (urea hydrogen peroxide; percarbamide; carbamide peroxide).
CAS: 124-43-6. $CO(NH_2)_2 \cdot HOOH$.
Properties: White crystals or crystalline powder, mp (decomposes) 75–78C, decomposed by moisture at temperatures around 40C. Soluble in water, alcohol, and ethylene glycol; solvents such as ether and acetone extract hydrogen peroxide and may form explosive solutions. Active oxygen (min) 16%.
Grade: Technical, sometimes compounded with waxes in pellet form.
Hazard: Dangerous fire risk in contact with organic materials; strong oxidizing agent. Irritant.
Use: Source of water-free hydrogen peroxide, bleaching disinfectant, cosmetics, pharmaceuticals, blue print developer, modification of starches.

urea phosphoric acid. See carbamide phosphoric acid.

urease. Enzyme present in low percentages in jack bean and soybean, water-soluble, its action is inhibited by heavy-metal ions. Its principal use is in the determination of urea in urine, blood, and other body fluids; it splits urea into ammonia and carbon dioxide or ammonium carbonate.

Urech cyanohydrin method. Cyanohydrin formation by addition of alkali cyanide to the carbonyl group in the presence of acetic acid (Urech); or by reaction of the carbonyl compound with anhydrous hydrogen cyanide in the presence of a basic catalyst (Ultee).

Urech hydantoin synthesis. Formation of hydantoins from alpha-amino acids by treatment with potassium cyanate in aqueous solution and heating of the salt of the intermediate hydantoic acid with 25% hydrochloric acid.

"Urepan."[470] TM for urethane rubbers which have to be crosslinked with isocyanates or peroxides.

urethane. (ethyl carbamate; ethyl urethane).
CAS: 51-79-6. $CO(NH_2)OC_2H_5$.   Its structure is typical of the repeating unit in polyurethane resins.
Properties: Colorless crystals or white powder, odorless, saltpeter-like taste, solutions neutral to litmus. Soluble in water, alcohol, ether, glycerol, and chloroform; slightly soluble in olive oil. D 0.9862, mp 49C, bp 180C. Combustible.
Derivation: (a) By heating ethanol and urea nitrate to 120–130C; (b) by action of ammonia on ethyl carbonate or ethyl chloroformate.
Grade: Technical, NF.
Hazard: Toxic by ingestion.
Use: Intermediate for pharmaceuticals, pesticides, and fungicides; biochemical research; medicine (antineoplastic).
See also polyurethane.

urethane alkyd. (uralkyd). A urethane resin modified with a drying oil (linseed or safflower) for use in high-quality paints.

Urey, Harold C. (1894–1981) An American chemist who received the Nobel prize in chemistry in 1934 for his discovery of the heavy isotopes of hydrogen and oxygen. His discovery which became an important factor in the development of nuclear fission and fusion and made possible the production of the first transuranic element Pu. He was one of the leaders of the Manhattan Project, which constructed the first nuclear reactor at University of Chicago and eventually produced the first atomic bomb. Obtaining his doctorate at University of California in 1923, he taught at several leading universities, including Columbia, where he discovered deuterium oxide (heavy water), used as a moderator in early types of nuclear reactors. Later he devoted much study to the origin of the universe and the origin of life on earth. He was author of many scientific treatises and made notable contributions to the cosmological theories of recent years.

uric acid.   (lithic acid; uric oxide; 2,6,8-trioxypurine).   CAS: 69-93-2.

OCNHC(O)NHCCNHC(O)NH   (keto   form).

May also be written in enolic form. The end-product of purine metabolism in man and other primates, birds, and some dogs and reptiles.
Properties: Odorless, tasteless, white crystals; soluble in hot concentrated sulfuric acid; very slightly soluble in water; insoluble in alcohol and ether; soluble in glycerol, solutions of alkali hydroxides, sodium acetate, and sodium phosphate; d 1.855–1.893; mp decomposes.
Derivation: From guano.
Grade: Technical, reagent.
Hazard: Evolves highly toxic hydrogen cyanide when heated.
Use: Organic synthesis.
See also pyrine.

uridine.   (d-ribosyl uracil).   CAS: 58-96-8. $C_9H_{12}N_2O_6$.   The nucleoside of uracil. It is a constituent of ribonucleic acid and some coenzymes (such as uridine diphosphate glucose).
Properties: White, odorless powder of slightly acrid and faintly sweet taste. Mp 165C, soluble in water, slightly soluble in dilute alcohol, insoluble in strong alcohol.
Derivation: From nucleic acid hydrolyzates from yeast. Radioactive forms available.
Use: Biochemical research.

uridine diphosphate glucose.   (UDPG).
CAS: 133-89-1.   A coenzyme which acts in the transfer of glucose from the coenzyme to another chemical compound during the reaction for which the coenzyme is a catalyst.
Use: Biochemical research.

uridine monophosphate.   See uridylic acid.

uridine phosphate.   A nucleotide used by the body in growth processes, important in biochemical and physiological research. Those isolated and commercially available (as sodium salts) are the monophosphate (UMP), the diphosphate (UDP), and the triphosphate (UTP).
See also uridine diphosphate glucose (UDPG).

uridylic acid.   (uridine phosphoric acid; UMP; uridine monophosphate).   CAS: (5') 58-97-9. $(C_9H_{13}N_2O_9P)$.   The monophosphoric ester

of uracil, i.e., the nucleotide containing uracil-d-ribose and phosphoric acid. The phosphate may be esterified to either the 2, 3, or 5 carbon of ribose, yielding uridine-2'-phosphate, uridine-3'-phosphate, and uridine-5'-phosphate, respectively.
Properties: (Uridine-3'-phosphate): Crystallizes in prisms from methanol, mp 202C (decomposes), freely soluble in water and alcohol, dextrorotatory in solution.
Derivation of commercial product: From yeast ribonucleic acid. Also made synthetically; radioactive forms available.
Use: Biochemical research.

uroformine.   (urotropin, formin, crystogen, cystamine).   See hexamethyltetramine.

uronic acid.   Any of a class of compounds similar to sugars but differing from them in that the terminal carbon has been oxidized from an alcohol to a carboxyl group. The most common are galacturonic acid and glucuronic acid.

"Urox."[50]   TM for a soil sterilant herbicide developed for general weed control in non-crop areas, contains monuron trichloroacetate. Properties: Miscible with kerosene, fuel oil, diesel oil, aromatic weed oil, liquid asphalts and tars; available in granular and liquid oil concentrated form.

ursin.   See arbutin.

urushiol.   Mixture of catechol derivatives.
Properties: Pale yellow liquid; d 0.968; bp 200C; soluble in alcohol, ether, and benzene.
Derivation: Poison ivy (*Rhus toxicodendron*).
Hazard: The toxic principle of poison ivy. Causes severe allergic dermatitis.

urylon.   A polyurea synthetic fiber made by condensation of nonamethylenediamine and urea, d 1.07, softening point 205C, mp 235C, weakens on heating to 150C or long exposure to light.
Use: Fiber blends, fishing gear, and the like.

USAN.   Abbreviation for United States Adopted Name, a nonproprietary name approved by the American Pharmaceutical Association, American Medical Association, and the US Pharmacopeia. Such names applied to pharmaceutical products do not imply endorsement; their use in advertising and labeling is required by law.

USDA.   Abbreviation for United States Department of Agriculture, the Federal regulatory authority for meats and meat products. This department maintains Regional Laboratories for agricultural and food research and development located in Philadelphia, Peoria, New Orleans, and Albany (CA). Its central Office is in Washington, DC.

usnic acid. (usninic acid). CAS: 125-46-2. $C_{18}H_{16}O_7$. A tricyclic compound, a constituent of many lichens, known in d-, l-, and dl-forms.

Properties: Crystalline, yellow solid, melting range 192–203C, insoluble in water, slightly soluble in alcohol and ether.

Derivation: From *Usnea barbata*, a lichen growing on trees.

Use: Medicine (antibiotic).

USP. Abbreviation for United States Pharmacopeia, the official publication for drug product standards.

USRDA. Abbreviation for United States Recommended Dietary Allowances for food and nutrition established by the Food and Drug Administration (FDA) in 1975 to serve as a basis for regulations on nutritional labeling of food products. They are based on the Recommended Dietary Allowances previously established by the Food and Nutrition Board, National Research Council of the National Academy of Sciences.

See also RDA.

UTP. See uridine phosphate.

"Uvitex."[219] TM for a group of fluorescent whitening agents used on natural and synthetic textile fabrics and yarn.

V

V. Symbol for vanadium.

vacancy. (1) A defect in a crystal due to the absence of an atom in the lattice. See also hole. (2) The absence of one or more electrons in the outer shell of an atom.

vacuum deposition. The process of coating a base material by evaporating a metal under high vacuum and condensing it on the surface of the material to be coated, which is usually another metal or a plastic. Aluminum is most commonly used for this purpose. The coatings obtained range in thickness from 0.01 to as much as 3 mils. A vacuum of about one-millionth atmosphere is necessary. The process is used for jewelry, electronic components, decorative plastics, etc. Thermally evaporated metals and dielectric coatings can be effectively applied to glass by this method. It is also called vacuum coating and vacuum metallizing.

vacuum distillation. Distillation at a pressure below atmospheric but not so low that it would be classed as molecular distillation. Since lowering the pressure also lowers the boiling point, vacuum distillation is useful for distilling high-boiling and heat-sensitive materials such as heavy distillates in petroleum, fatty acids, vitamins, etc.

vacuum forming. See thermoforming (2).

val. Abbreviation for value.

valacidin. CAS: 53762-92-8. $C_{26}H_{24}N_4O_8$.
Properties: Brown-to-red solid, soluble in polar solvents and alkaline solutions, insoluble in most nonpolar solvents and acid solutions.
Use: An antibiotic used as a preservative for biological specimens, and in embalming fluids.

"Valclene."[28] TM for clear fluorocarbon formulations with slight ethereal odor. Nonflammable.
Use: Drycleaning fluids.

"Valdet."[496] TM for a series of detergents. 561. Ethylene oxide condensate of nonyl phenol. AC-40, CC Amine condensate.
Use: Wetting agent; emulsifying, leveling, dispersing, antistatic agent; multipurpose detergent for dyeing, bleaching, scouring, washing after printing; anionic synthetic detergents; kier boiling aids.

valence. A whole number which represents or denotes the combining power of one element with another. By balancing these integral valence numbers in a given compound, the relative proportions of the elements present can be accounted for. If hydrogen and chlorine both have a valence of 1, oxygen a valence of 2 and nitrogen 3, the valence-balancing principle gives the formulas HCl, H_2, NH_3, Cl_2O, NCl_3 and N_2O_3, which indicate the relative numbers of atoms of these elements in compounds which they form with each other. In inorganic compounds it is necessary to assign either a positive or negative value to each valence number, so that valence-balancing will give a zero sum by algebraic addition. Negative numbers are called polar valence numbers (-1 and -2). The valence of chlorine may be -1, $+1$, $+3$, $+5$, and $+7$, depending on the type of compound in which it occurs. In organic chemistry only nonpolar valence numbers are used.
See also chemical bonding, oxidation, oxidation number, coordination number.

valentinite. (antimony trioxide [ortho-rhombic]; white antimony). CAS: 1209-64-4. Sb_2O_3. White or gray mineral, sometimes pale red, white streak and adamantine or silky luster, d 5.57–5.76, Mohs hardness 2–3.
Occurrence: Algeria, Yugoslavia, Italy, Germany.
Use: Ore of antimony.

valeral. See valeraldehyde.

n-valeraldehyde. (valeric aldehyde; valeral; amyl aldehyde; pentanal). CAS: 110-62-3. $CH_3(CH_2)_3CHO$.
Properties: Colorless liquid, d 0.8095 (20/4C), fp $-91C$, bp 102–103C, refr index 1.3944 (20C), flash p (CC) 54F (12.2C), slightly soluble in water, soluble in alcohol and ether.
See also isovaleraldehyde.
Derivation: Oxidation of amyl alcohol, also by the Oxo process.
Hazard: Flammable, dangerous fire risk. TLV: 50 ppm in air.
Use: Flavoring, rubber accelerators.

valerianic acid. See valeric acid.

valerian oil.
Properties: Yellowish or brownish liquid; penetrating odor; soluble in alcohol, ether, chloro-

form, acetone, benzene, and carbon disulfide; d 0.903–0.960; refr index 1.486 (20C). Combustible.

Chief constituents: Pinene, camphene, borneol, and esters of borneol and valeric acid. Derivation: Distilled from roots and rhizome of Valeriana officinalis.

Use: Tobacco perfume, industrial odorant, flavors.

valeric acid. (valerianic acid; n-pentanoic acid). CAS: 109-52-4. $CH_3(CH_2)_3COOH$.

Properties: Colorless liquid, penetrating odor and taste, d 0.9394 (20/4C), bp 185.4C, refr index 1.4081 (20C), vap press 0.08 mm (20C), fp −34C, flash p (OC) 205F (96C), soluble in water, soluble in alcohol and ether, undergoes reactions typical of normal monobasic organic acids. Combustible.

Derivation: With other C_5 acids by distillation from valerian, by oxidation of n-amyl alcohol, numerous essential oils.

Grade: Technical, reagent.

Hazard: Toxic by ingestion, strong irritant to skin and tissue.

Use: Intermediate for flavors and perfumes, ester-type lubricants, plasticizers, pharmaceuticals, vinyl stabilizers.

See also isovaleric acid.

valeric aldehyde. See n-valeraldehyde.

γ-valerolactone. CAS: 108-29-2.

$$H_2C-CH_2-CH_2-CH-CH_3$$
$$O \overline{\qquad\qquad} C=O$$

Properties: Colorless liquid, d 1.0518 (25/25C), bp 205–206.5C, crystallizing point −37C, flash p (COC) 205F (96C), refr index 1.4301 (25C), surface tension 30 dynes/cm (25C), viscosity 2.18 cp (25C), pH (anhydrous) 7.0 (pH of 10% solution in distilled water 4.2). Miscible with water and most organic solvents, resins, waxes, etc.; slightly miscible with zein, beeswax, petrolatum; not miscible with anhydrous glycerin, glue, casein, arabic gum, and soybean protein. Combustible.

Use: In dye baths (coupling agent), brake fluids, cutting oils, and as solvent for adhesives, insecticides, and lacquers.

valine. (α-aminoisovaleric acid). CAS: (l-) 72-18-4. $(CH_3)_2CHCH(NH_2)COOH$.

An essential amino acid.

Properties: White, crystalline solid; soluble in water; very slightly soluble in alcohol; insoluble in ether; shows the following optical isomers: dl-valine: mp 298C with decomposition. d-valine

(natural isomer): mp 315C with decomposition. l-valine: mp 293C with decomposition.

Derivation: Hydrolysis of proteins, synthesized by the reaction of ammonia with α-chloroisovaleric acid. Available commercially as d-, l-, or dl-valine.

Use: Dietary supplement, culture media, biochemical and nutritional investigations.

"Valium."[180] TM for diazepam.

valone. (2-isovaleryl-1,3-indanedione). $C_{14}H_{14}O_3$.

Properties: Yellow, crystalline solid; mp 68C; insoluble in water; soluble in common organic solvents; blood anticoagulant.

Hazard: Toxic by ingestion.

Use: Pesticide, rodenticide.

"Valrez."[496] TM for a series of wash and wear resins, many of which are urea derivatives or modified urea-formaldehyde resins.

"Valstat"E.[496] TM for a modified fatty condensate.

Use: Antistatic agent for synthetics, cotton, and as a napping assistant.

"Valzopon ON."[496] TM for a polyoxyethylated fatty alcohol.

Use: Dispersing agent to improve crockfastness of naphthol dyeings, stabilizer for color salts.

vanadic acid. (a) meta-HVO_3, (b) ortho-H_3VO_4, (c) pyro-$H_4V_2O_7$. These acids apparently do not exist in the pure state, but are represented in the various alkali and other metal vanadates. Ordinarily, vanadic acid implies vanadium pentoxide (vanadic acid anhydride).

vanadic acid anhydride. See vanadium pentoxide.

vanadic sulfate. See vanadyl sulfate.

vanadic sulfide. See vanadium sulfide.

vanadinite. $Pb_5Cl(VO_3)$. A natural chlorovanadate of lead.

Properties: Ruby red, orange red, brown, yellow; luster resinous to adamantine; Mohs hardness 3; d 6.7–7.1; soluble in strong nitric acid.

Occurrence: New Mexico, Arizona, Africa, Scotland, USSR.

Use: Ore of vanadium and lead.

vanadium. CAS: 7440-62-2. V. Metallic element having atomic number 23, Group VB of the periodic system, aw 50.9414, valences = 2, 3, 4, 5,; two natural isotopes.

Properties: Silvery-white ductile solid; insoluble in water; resistant to corrosion, but soluble in nitric, hydrofluoric, and concentrated sulfuric acids; attacked by alkali, forming water-soluble vanadates. D 6.11, mp 1900C, bp 3000C, acts as either a metal or a nonmetal and forms a variety of complex compounds.

Source: Not found native: principal ores are patronite, roscoelite, carnotite and vanadinite. Also from phosphate rock (Idaho, Montana, Arkansas).

Occurrence: Colorado, Utah, New Mexico, Arizona, Mexico, and Peru.

Derivation: (a) Calcium reduction of vanadium pentoxide yields 99.8+% pure ductile vanadium; (b) aluminum, cerium, etc., reduction produces a less pure product; (c) solvent extraction of petroleum ash or ferrophosphorus slag from phosphorus production; (d) electrolytic refining using a molten salt electrolyte containing vanadium chloride.

Grade: 99.99% pure (electrolytic process), single crystals.

Use: Target material for x-rays, manufacture of alloy steels, vanadium compounds, especially catalyst for sulfuric acid and synthetic rubber. See ferrovanadium.

vanadium acetylacetonate.
Properties: Blue to blue-green crystals, decomposes before melting.
Derivation: Reaction of vanadyl sulfate with acetylacetone and sodium carbonate.
Use: Catalyst.

vanadium carbide. VC.
Properties: Crystals with hardness 2800 kg/sq mm, d 5.77, mp 2800C, bp 3900C, resistivity 150 micro-ohm cm (room temperature).
Use: Alloys for cutting tools, steel additive.

vanadium dichloride. (vanadous chloride). CAS: 10213-09-9. VCl_2.
Properties: Apple-green hexagonal plates, soluble in alcohol and ether, decomposes in hot water, d 3.23 (18C).
Derivation: From vanadium trichloride by heating in atmosphere of nitrogen.
Method of purification: Sublimation in nitrogen.
Grade: CP.
Hazard: Strong irritant to tissue.
Use: Strong reducing agent, purification of hydrogen chloride from arsenic.

vanadium disulfide. V_2S_2. Solid, d 4.20, mp decomposes, soluble in hot sulfuric or nitric acids, insoluble in alkalies.
Use: Solid lubricant, electrode in lithium-based batteries (experimental).

vanadium ethylate. $(C_2H_5O)_4V$.
Properties: Dark reddish-brown solid.

Derivation: Reaction of vanadium chloride with sodium ethylate.
Use: Polymerization catalyst.

vanadium hexacarbonyl. $V(CO)_6$.
Properties: Blue-green powder, sublimes easily at 50C (15 mm), paramagnetic, decomposes without melting at 60–70C.
Hazard: Strong irritant to tissue; store under inert gas.
Use: Chemical intermediate, production of plating compounds and fuel additives.

vanadium hexacarbonyl, sodium salt. (bisdiglymesodium hexacarbonyl vanadate). $Na(C_6H_{14}O_3)_2V(CO)_6$. A convenient though incomplete name for the complex compound.
Properties: Yellow solid; mp 173–176C (with decomposition); soluble in water, alcohol, and ether; slightly soluble in hydrocarbons; relatively inert to air but is stored and shipped under nitrogen.
Hazard: Strong irritant to tissue.
Use: Source of vanadium hexacarbonyl by treatment with phosphoric acid under special conditions.

vanadium nitride. VN.
Properties: Black solid, d 6.13, mp 2320C, insoluble in water, slightly soluble in aqua regia.
Use: Refractory.

vanadium oxydichloride. See vanadyl chloride.

vanadium oxytrichloride. CAS: 7727-18-6. $VOCl_3$.
Properties: Lemon-yellow liquid, d 1.811 (32C), fp −78.9C, bp 125–127C, nonionizing solvent, dissolves most nonmetals, dissolves and/or reacts with many organic compounds, hydrolyzes in moisture.
Hazard: Strong irritant to tissue.
Use: Catalyst in olefin polymerization (ethylenepropylene rubber) organovanadium synthesis.

vanadium pentoxide. (vanadic acid anhydride). CAS: 1314-62-1. V_2O_5.
Properties: Yellow to red crystalline powder, d 3.357 (18C), mp 690C, bp decomposes at 1750C, soluble in acids and alkalies, slightly soluble in water.
Derivation: (a) Alkali or acid extraction from vanadium minerals. (b) By igniting ammonium metavanadate. (c) From concentrated ferrophosphorus slag by roasting with sodium chloride, leaching with water and purification by solvent extraction followed by precipitation and heating.
Method of purification: Alkali solution, precipitation as ammonium metavanadate and ignition to V_2O_5.

Grade: Commercial air-dried, commercial fused, CP air-dried, CP fused.

Hazard: Toxic by inhalation. TLV (as vanadium dust and fume): 0.05 mg/m^3 of air.

Use: Catalyst for oxidation of sulfur dioxide in sulfuric acid manufacture ferrovanadium, catalyst for many organic reactions, ceramic coloring material, vanadium salts, inhibiting UV transmission in glass, photographic developer, dyeing textiles.

See also contact process.

vanadium sesquioxide.   See vanadium trioxide.

vanadium sulfate.   See vanadyl sulfate.

vanadium sulfide.   (vanadium pentasulfide; vanadic sulfide). V_2S_5.

Properties: Black-green powder, soluble in acids, alkali-metal sulfides and alkalies, insoluble in water, d 3.0, decomposes on heating.

Derivation: Action of hydrogen sulfide on vanadium chloride solution.

Hazard: Toxic by inhalation (especially to animals).

Use: Vanadium compounds.

vanadium tetrachloride.   CAS: 7632-51-1. VCl_4.

Properties: Red liquid, soluble in absolute alcohol and ether, decomposes slowly to vanadium trichloride and chlorine below 63C, d 1.816 (20C), fp −28C, bp 154C. Nonflammable.

Derivation: Chlorination of ferrovanadium.

Method of purification: Distillation and fractionation.

Hazard: Toxic by ingestion, inhalation, and skin absorption. Open containers only in dry, oxygen-free atmosphere or inert gas; wear goggles and protective clothing; corrosive to tissue.

Use: Preparation of vanadium trichloride, vanadium dichloride, and organovanadium compounds.

vanadium tetraoxide. V_2O_4.

Properties: Blue-black powder, d 4.339, mp 1967C, insoluble in water, soluble in alkalies and acids.

Derivation: (1) From vanadium pentoxide by oxalic acid reduction. (2) From vanadium pentoxide by carbon reduction.

Hazard: Toxic and irritating.

Use: Catalyst at high temperature.

vanadium trichloride.   CAS: 7718-98-1. VCl_3.

Properties: Pink, deliquescent crystals; soluble in absolute alcohol and ether; decomposes in water; d 3.0 (18C); decomposes on heating.

Derivation: From vanadium tetrachloride boiling under reflux condenser.

Hazard: Irritant.

Use: Preparation of vanadium dichloride and organo-vanadium compounds.

vanadium trioxide.   (vanadium sesquioxide).   CAS: 1314-34-7. V_2O_3.

Properties: Black crystals, soluble in alkalies and hydrogen fluoride, slightly soluble in water, d 4.87 (18C), mp 1970C.

Derivation: From vanadium pentoxide by either hydrogen or carbon reduction.

Hazard: Irritant.

Use: Catalyst for conversion of ethylene to ethanol.

vanadous chloride.   See vanadium dichloride.

vanadyl chloride.   (vanadium oxydichloride; vanadyl dichloride; divanadyl tetrachloride).   CAS: 10213-09-9. $V_2O_2Cl_4 \cdot 5HOH$.

Properties: Green, deliquescent crystals slowly decomposed by water; usual technical product is a dark green syrupy mass 76–82% pure, or a solution; soluble in water, alcohol, and acetic acid; may react violently with potassium.

Grade: Technical.

Hazard: Irritant.

Use: Mordanting textiles.

vanadyl sulfate.   (vanadic sulfate; vanadium sulfate).   CAS: 27774-13-6. $VOSO_4 \cdot 2HOH$.

Properties: Blue crystals, soluble in water.

Derivation: Reduction of cold solution of concentrated sulfuric acid and vanadium pentoxide by sulfur dioxide gas.

Hazard: Irritant.

Use: Mordant, catalyst, aniline black preparation, reducing agent, colorant in glasses and ceramics.

"Vancide 89."[69]   TM for n-(trichloromethylmercapto)-4-cyclohexene-1,2-dicarboximide.

Use: Fungicide for vinyl compositions.

See captan.

"Vancide 26EC."[69]   TM for lauryl pyridinium-5-chloro-2-benzothiazyl sulfide.

Use: Preservative for cotton fabrics used in rubber structures. Applied from an aqueous solution.

"Vancide 51Z."[69]   TM for zinc dimethyldithiocarbamate $[(CH_3)_2NC(S)S]_2Zn$ with a small proportion of zinc-2-mercaptobenzothiazole.

Use: Fungicide for neoprene compositions, ultra-accelerator for rubber.

vancomycin hydrochloride.

Properties: Tan to brown powder; odorless; with bitter taste; soluble in water; moderately soluble

in dilute methanol; insoluble in higher alcohols, acetone, ether.
Derivation: Produced by *Streptomyces orientalis* from Indonesian and Indian soil.
Grade: USP.
Use: Medicine (antibiotic).

van der Waals' forces. Weak attractive forces acting between molecules. They are somewhat weaker than hydrogen bonds and far weaker than interatomic valences. They are involved in the van der Waals equation of state for gases which compensates for the actual volume of the molecules and the forces acting between them. "Information regarding the numerical values of van der Waals forces is mostly semiempirical, derived with the aid of theory from an analysis of chemical or physical data. Attempts to calculate the forces from first principles have had a measure of success only for the simplest systems, such as H—H, He—He and a few others. When judging the difficulties of such calculations, one must bear in mind that the energies sought are of the same order of magnitude as those in the best atomic energy calculations." (Henry Margenau).
See also hydrogen bond, chemical bonding.

"Vandex."[69] TM for a finely ground selenium.
Properties: Dark gray metallic powder, d 4.80, fineness through 200 mesh 99%, melting range above 217C, slightly soluble in benzene, soluble in carbon disulfide, insoluble in water.
Use: Rubber vulcanization.

Van Dyke brown. (Cassel brown; Cologne brown; Cologne earth; ulmin brown). A naturally occurring pigment.
Derivation: Indefinite mixture of iron oxide and organic matter. Obtained from bog-earth, peat deposits, or from ochers containing bituminous matter.
Use: Pigment for artists' colors and stains.

Van Dyke red. A brownish red pigment consisting of copper ferrocyanide; sometimes used to refer to red varieties of ferric oxide.
Use: Pigment.
See iron oxide red.

vanillin. (3-methoxy-4-hydroxybenzaldehyde; vanillic aldehyde). CAS: 121-33-5.
$(CH_3O)(OH)C_6H_3CHO$. The methyl ether of protocatechuic aldehyde.
Properties: White, crystalline needles; sweetish smell; d 1.056; mp 81–83C; bp 285C. Soluble in 125 parts water, in 20 parts glycerol, and in two parts 95% alcohol; soluble in chloroform and ether. Combustible.
Derivation: (a) By extraction from the vanilla

bean, (b) from lignin contained in sulfite waste pulp liquor.
Method of purification: Crystallization.
Grade: Technical, USP, FCC.
Use: Perfumes, flavoring, pharmaceuticals, laboratory reagent, source of L-dopa.

van't Hoff, Jacobus H. (1852-1911) A Dutch chemist who received the first Nobel prize for chemistry in 1901. A father of physical chemistry, he did research on decomposition and formation of double salts. He related optically active carbon compounds to three dimensional and asymmetrical molecular structure. This led to the development of stereochemistry. Educated at the Universities of Paris and Utrecht, where he received a Doctorate in 1874. He was a Professor at Amsterdam, Leipzig, and Berlin.

van Leeuwenhoek, Anton. (1632–1723) A native of Holland, van Leeuwenhoek was a professional lens grinder. He developed the compound optical microscope, which had been invented in 1590, to a point where he was able to obtain magnifications up to 275 times. He was the first to observe bacteria, spermatozoa, and other unicellular animals which he described to the Royal Society of London.
See also optical microscope.

"Vanplast 125."[69] TM for cresyl diphenyl phosphate.
Use: Flame-retardant plasticizer for vinyl formulations, including foam.

Van Slyke determination. Treatment of primary aliphatic amines and α-amino acids with nitrous acid and volumetric determination of evolved nitrogen.

"Vanstay."[69] TM for a series of heat and light stabilizers for polyvinyl chloride resins.

"Vapam."[1] TM for a soil fumigant of sodium-N-methyldithiocarbamate dihydrate.

"Vapona."[125] TM for an insecticide which contains more than 93% 2,2-dichlorovinyl dimethyl phosphate and less than 7% active, related compounds.
See DDVP.
Hazard: As for dichlorovos.

vapor. An air dispersion of molecules of a substance that is liquid or solid in its normal state, i.e., at standard temperature and pressure. Examples are water vapor and benzene vapor. Vapors of organic liquids are also loosely called fumes.
See also evaporation, gas.

vapor-phase chromatography. See gas chromatography.

vapor pressure. (vap press; vap p; v.p.).
The pressure (often expressed in millimeters of mercury, mm Hg) characteristic at any given temperature of a vapor in equilibrium with its liquid or solid form.

vapor tension. See vapor pressure.

"Varkon."[300] TM for sequestering agents of the ethylenediaminetetraacetate type.
Use: Kier boiling, peroxide bleaching, dyeing, and stripping operations. Available as both powder and liquid.

varnish. (1) An organic protective coating, similar to a paint except that it does not contain a colorant. It may be comprised of a vegetable oil (linseed, tung, etc.) and solvent or of a synthetic or natural resin and solvent. In the first case the formation of the film is due to polymerization of the oil and in the second to evaporation of the solvent. "Long-oil" varnishes such as spar varnish have a high proportion of drying oil; "short-oil" types have a lower proportion, i.e., furniture varnishes. Spirit varnishes contain such solvents as methanol, toluene, ketones, etc., and often also thinners such as naphtha or other light hydrocarbon. Flammable. (2) A hard, tightly adherent deposit on the metal surfaces of automobile engines resulting from resinous oxidation products of gasoline and lubricating oils.

varnish remover. See paint remover.

"Varnon."[446] TM for hard-fired super-duty fireclay brick which resists the destructive action of carbon monoxide and other reducing gases.
Use: Glass tank regenerator checkers; also to line various metallurgical furnaces, rotary kilns, shaft kilns, carbon baking furnaces, and incinerators.

"Varox."[69] TM for a 50% active blend of 2,5-bis(tert-butylperoxy)-2,5-dimethylhexane with an inert mineral carrier.
Properties: White powder, d 1.43.
Use: Crosslinking agent for polymers such as polyethylene.

Varrentrapp reaction. Cleavage of oleic acid into palmitic and acetic acids by heating with molten potassium hydroxide. The procedure has been extended to olefinic acids in general.

"Varsol."[51] TM for straight petroleum aliphatic solvents used as paint and varnish thinners, for dry cleaning, and for general plant machinery cleaning. Conform to CS3-40, the US Dept. of Commerce commercial standard for Stoddard Solvent and have minimum TCC flash points of 100F (37.7C). Combustible.

"Varsoy."[221] TM for a prebodied chemically modified soybean oil for varnish cooking or bodied drying oil.

vasopressin. (β-hypophamine; antidiuretic hormone). One of the hormones secreted by the posterior lobe of the pituitary gland. It causes an increase in blood pressure and an increase in water retention by the kidney. Vasopressin is an octapeptide consisting of eight amino acids.
Derivation: Synthetic, or from the posterior lobe of the pituitary of food animals.
Grade: USP as an aqueous solution for injection.
Use: Medicine (antidiuretic).

vat dye. A class of water-insoluble dyes that can be easily reduced, i.e., vatted to a water-soluble and usually colorless leuco form in which they can readily impregnate fibers. Subsequent oxidation then produces the insoluble colored form that is remarkably fast to washing, light and chemicals. Examples are indigo (CI# 73000) and Indanthrene Blue BFP (CI# 69825). The reducing agents are usually an alkaline solution of sodium hydrosulfite ($Na_2S_2O_4$). Oxidation is by air, perborate, dichromate, etc.
Use: For cotton, wool and cellulose acetate.

vat printing assistant. A mixture of gums and reducing and wetting agents used to carry the dye in printing fabrics with vat dyes. They assist in securing penetration of the fabric and in converting the dyes from a semi-leuco to a leuco state.

"Vatrolite."[159] TM for a group of concentrated sodium hydrosulfite preparations.
Use: Bleaching textiles and paper, oxygen scavenger in synthetic rubber production.

"Vatsol."[57] TM for a series of wetting agents made in several different grades and types: OS, sodium isopropyl naphthalene sulfonic acid; OT, sodium dioctyl sulfosuccinate.

vauquelin's salt. $(Pd(NH_3)_4)Cl_2 \cdot PdCl_2$.
A compound obtained by treating palladium chloride with ammonia.

VC. Abbreviation for vinyl chloride or vinylidene chloride.

vector. In biochemistry, an animal (insect, rodent, etc.) that carries or transports infectious microorganisms. Typical vectors are rats and mosquitoes.

vegetable black. In general, any form of more or less pure carbon produced by incomplete combustion or destructive distillation of vegetable matter, wood, vines, wine lees.

vegetable dye. A colorant derived from a vegetable source, i.e., logwood, indigo, madder, etc.

vegetable gum. See dextrin.

vegetable oil. An oil extracted from the seeds, fruit, or nuts of plants and generally considered to be a mixture of mixed glycerides, (e.g., cottonseed, linseed, corn, coconut, babassu, olive, tung, peanut, perilla, oiticica, etc.). Many types are edible. Being plant-derived products, vegetable oils are a form of biomass. Some are reported to be convertible to liquid fuels by passing them over zeolite catalysts.
Use: Paints (as drying oils), shortenings, salad dressings, margarine, soaps, rubber softeners, dietary supplements, pesticide carriers.
See also specific entry.

vegetable tanning. The tanning of leather by plant extracts.
See tannic acid, tanning, wattle bark, quebracho.

vehicle. A term used in paint technology to indicate the liquid portion of a paint, comprised of drying oil or resin, solvent, and thinner in which the solid components are dissolved or dispersed.
See also paint.

Venetian red. A high-grade ferric oxide pigment of a pure red hue. It is obtained either native as a variety of hematite red or more often artificially, by calcining copperas (ferrous sulfate) in the presence of lime. The composition ranges from 15–40% ferric oxide and from 60–80% calcium sulfate. The 40% ferric oxide is the "pure" grade and has a d 3.45.
Grade: 20–40% ferric oxide.
See also iron oxide red.

venturi. A type of flow meter used for liquids or fine particulates. It is a tube-like device having broad, flaring ends and a narrow central portion, or throat; this so constricts the passage of the fluid that its rate of flow increases while the pressure decreases. The difference in pressure thus created is a measure of the flow. Venturis are used in scrubbers, liquid and solid conveying systems, pipelines, and aircraft instrument control, as well as in numerous chemical process techniques such as hydrogenation, chlorination, and oxidation. The hydrogenation technology in-

volves hydrogen entrainment by rapid flow of a liquid catalyst through the venturi nozzle.

veratraldehyde. (vertraldehyde; 3,4-dimethoxybenzaldehyde).   CAS: 120-14-9.
$(CH_3O)_2C_6H_3CHO$.
Properties: Mp 42–45C, bp 281C, flash p 235F(112C), mw 166.18.

veratric acid. (3,4-dimethoxybenzoic acid).
CAS: 93-07-2. $(CH_3O)_2C_6H_3COOH$.
Properties: Mw 182.18, mp 179–82C.

veratrole. (1,2-dimethoxybenzene; pyrocatechol dimethyl ether).   CAS: 91-16-7.
$C_6H_4(OCH_3)_2$.
Properties: Colorless crystals or liquid, mp 21–22C, bp 206–207C, d 1.084 (25/25C), soluble in alcohol and ether, slightly soluble in water.
Derivation: Treatment of catechol in methanol with dimethyl sulfate and caustic.
Use: Medicine (antiseptic).

veratrum alkaloid. One of a group of alkaloids used in medicine to relieve hypertension. They include *Veratrum viride* (American hellebore).
Hazard: May have severe side effects.
See also alkaloid.

verdigris. See copper acetate, basic.

vermiculite. Hydrated magnesium-iron-aluminum silicate capable of expanding 6–20 times when heated to 1093C. The platelets exhibit an active curling movement when heated, hence the name.
Occurrence: Montana, North Carolina, South Carolina, Wyoming, Colorado, South Africa.
Properties: Platelet-type crystalline structure, high porosity, high void volume to surface area ratio, low density, large range of particle size, insoluble in water and organic solvents, soluble in hot concentrated sulfuric acid. Water vapor adsorption capacity of expanded vermiculite less than 1%, liquid adsorption dependent on conditions and particle size, ranges 200–500%. Noncombustible.
Grade: Unexpanded (ore concentrate), expanded (also called exfoliated), flake, activated.
Use: Lightweight concrete aggregate, insulation, sound conditioning, fireproofing, plaster, soil conditioner, additive for fertilizers, seed bed for plants, refractory, lubricant, oil well drilling mud. Filler in rubber, paint, plastics. Wall paper printing, removal of strontium-90 from milk, absorption of oil spills on seawater, animal feed additive, packing, carrier for insecticides, catalyst and catalyst support, litter for hatcheries, adsorbent.
See also verxite.

vermifuge. An agent used in veterinary medicine to eliminate intestinal worms; an anthelmintic.

vermilion, natural. See cinnabar.

vernolepin. A sesquiterpene dilactone extracted from leaves of an African plant *Veronia hymenolepis*.
Use: Biochemical research (inhibits tumor growth in rats and reversibly retards plant growth).

"Versadyme."[590] TM for dimerized vegetable fatty acids.
Use: Production of polymers, petroleum additives, corrosion inhibitors, paint additives, and surfactants.

"Versalon."[590] TM for thermoplastic polyamide resins of high strength, good adhesion, and flexibility.
Use: Hot-melt adhesives, heat-seal coatings, electrical encapsulation.

"Versamid."[590] TM for thermoplastic and reactive polyamide resins. The reactive resins copolymerize with epoxy resins.
Use: Heat-seal coatings, hot-melt adhesives, varnishes, paints, and inks; coatings, structural adhesives.

"Versa-TL."[53] TM for sulfonated polystyrene polymers.

"Versatyl."[141] TM for a series of printing ink vehicles and pigment colors flushed in the vehicles.
Flushed Colors: Wide range of yellows, reds, blues, and greens.
Use: Letterpress, lithographic, offset, tin lithography, quick-set, gloss, multicolor, and snap-dry inks.

"Versene."[233] TM for a series of chelating agents based on ethylenediaminetetraacetic acid.

"Versene Fe-3 Specific."[233] TM for an iron specialty chelating agent; active ingredient is sodium dihydroxyethylglycine; available as straw-colored liquid or white powder.

"Versenex" 80.[233] TM for pentasodium salt of diethylenetriaminepentaacetic acid.

"Versenol" 120.[233] TM for the trisodium salt of N-hydroxyethylethylenediaminetriacetic acid ($C_{10}H_{15}O_7N_2Na_3$).
Use: Organic chelating agent.

"Versilate."[244] TM for a composition of fortified sodium silicate.
Properties: Viscous liquid; opalescent.

"Versilube."[245] TM for a series of silicone hydraulic fluids, lubricating fluids, and greases.

verxite. (exfoliated hydrobiotite).
Properties: Thermally expanded (exfoliated) magnesium-iron-aluminum silicate having a minimum of 98% purity and a bulk d of 5–7 lb/cu ft. Expansion occurs by heating at 793C. Sponge-like structure which absorbs liquids and permits reexpansion after compression to 70–80% of the original heat-expanded volume.
Use: Poultry feed in quantities less than 5% as a non-nutritive bulking agent, pelleting or anti-caking agent and nutrient carrier in dog and ruminant feeds. For dog feeds the maximum permitted is 1.5%. Labeling must state content when in excess of 1%.
See also vermiculite.

"Vespel."[28] TM for fabricated parts, based on polyimide resin and diamond abrasive wheels formulated with a high-temperature polyimide binder.

vetiver oil. An essential oil with violet-like odor, strongly dextrorotatory.
Use: Perfumery and fragrances.

"Vexar."[28] TM for both low and high density polyethylene and polypropylene plastic netting. Available in a wide variety of forms and colors for use in packaging.

"V-G-B."[248] TM for reaction product of acetaldehyde and aniline.
Properties: Brown resinous powder; d 1.152; mp 60–80C; soluble in acetone, benzene, and ethylene dichloride; insoluble in water and gasoline.
Use: Rubber antioxidant.

"Vibrathane."[248] TM for a group of polyurethane raw materials for manufacture of foam and elastomers. Includes isocyanates, polyesters, polyether glycols, polyester and polyether prepolymers, liquid casting resins, and gums and catalysts.

"Vibrin."[248] TM for resin compositions of polyesters and crosslinking monomers which, when catalyzed, will polymerize to infusible solid resins without evolving water or other byproducts.
Use: Molding, laminating, impregnating, casting, automotive and aircraft structural parts, wall panels, table tops, coating for paper, boat hulls, chemically inert tanks, large-diameter pipe.

vic-. Prefix meaning vicinal.

"Vicalloy."[155] TM for a magnetic alloy composed of iron 38%, cobalt 52%, and vanadium 10%.
Properties: Density 8.09, tensile strength 100,000 psi.

Use: Magnetic recording tape, machinable permanent magnets.

vicinal. (abbreviated as vic-). Neighboring or adjoining positions on a carbon ring or chain; the term is used in naming derivatives with substituting groups in such locations in a structural formula or molecule. For example, vicinal locations in the molecule shown are occupied by the hydrogen atoms and the hydroxyl groups:

$$\begin{array}{ccc} H & H \\ | & | \\ -C & -C- \\ | & | \\ HO & OH \end{array}$$

"Victacid 105."[1] TM for a high-strength product of approximately equal quantities of orthophosphoric and polyphosphoric acids.
Use: Dehydrating agent for organic reactions, sequestrant for heavy metals, and control of P_2O_5 content of bath process for electropolishing and bright-dipping metals.

"Victamide."[1] TM for exceedingly fine particles of ammonium salt of an amido polyphosphate, practically all below 5 microns. Slowly soluble in cold water, more rapidly in hot water.
Use: Sequestering agent for metallic ions, flameproofing agent, deflocculating agent for oil-drilling muds, paint pigments, and clay slips.

"Victawet."[1] TM for surface-active phosphorus compounds. Anionic and nonionic wetting agents, used as penetrants, dye carriers and dispersing agents.

"Victor Cream."[1] TM for sodium acid pyrophosphate. Purity meets all requirements of federal and state pure food laws; FCC grade.
Use: Baking acid in doughnut and prepared flours, manufacture of commercial baking powders and instant puddings, for conditioning oil-well drilling muds, formulation of acid-type metal cleaners.

Victoria blue. (CI 44045). $C_{33}H_{31}N_3 \cdot HCl$.
Properties: Crystalline powder; bronze colored; soluble in hot water, alcohol, or ether.
Derivation: Michler's ketone is condensed with phenyl-α-naphthylamine.
Use: Dyeing silk, wool, and cotton; biological stain, dye intermediate for complex acid pigment toners.

Victoria green. See malachite green.

vicuna. A soft wool-like fiber obtained from a South American animal similar to the llama.

Use: Specialty high-grade coats, sweaters, etc. Combustible.

vidarabine. Generic name for ara-A.

"Vigofac 6."[299] TM for an unidentified growth factor for addition to animal feeds. Derived from dried streptomyces fermentation solubles.

"Vikane."[233] TM for sulfuryl fluoride.

Vilsmeier-Haack reaction. Formylation of activated aromatic or heterocyclic compounds with disubstituted formamides and phosphorus oxychloride.

"Vinac."[472] TM for polyvinyl acetate emulsions, powders, beads and solutions.
Emulsions:
Properties: (range): 48–57% solids, 80–4200 cp, 4.0–7.0 pH, 0.15 to 2.5 microns average particle size, nonionic.
Use: Binder in pigmented paper coatings, adhesive bases, concrete adhesive, textile finishes.
Powders:
Properties: White, free-flowing powder; bulk d 33 lbs/cu ft; 98% through 100 mesh; viscosity 5000–7000 cp in 50% dispersion; particle size 2 to 6 microns in dispersion.
Use: Adhesive for concrete and joint cements.
Beads:
Properties: (Range): Glass-like spheres; low, medium, and high molecular weight grades; soluble in organic solvents except aliphatic hydrocarbons; ASB-grade is soluble in alkalies in addition to organic solvents.
Use: Hot melt and solvent adhesives, solvent paints, specialty paper coatings, binders in inks.
Solutions:
Properties: Methanol solvent, 50–51% polymer, 900–1100 cp.
Hazard: Flammable.
Use: Heat-seal adhesives.

vinal fiber. Generic name for a manufactured fiber in which the fiber-forming substance is any long-chain synthetic polymer composed of at least 50% by weight of vinyl alcohol units, —CH_2CHOH— and in which the total of the vinyl alcohol units and any one or more of the various acetal units is at least 85% by weight of the fiber (Federal Trade Commission). It has good chemical resistance, low affinity for water, good resistance to mildew and fungi. Combustible.
Use: Fishing nets, stockings, gloves, hats, rainwear, swimsuits.

"Vinamyl."[53] TM for dry starch for use as a carrier in adhesive used in the manufacture of corrugated board.

"Vin-Clad."[326] TM for a series of dry, free-flowing, polyvinyl chloride powders formulated for application by the fluidized bed process to provide protective and decorative coatings.

vinegar.
Properties: Brownish or colorless liquid, dilute aqueous solution containing 4–8% acetic acid, depending on source. Legal minimum is 4%. Also contains low percentages of alcohols and mineral salts. Nonflammable.
Derivation: (a) Bacterial fermentation of apple cider, wine, or other fruit juice. (b) Fermentation of malt or barley. The fermenting agent is usually a mold, e.g., Mycoderma aceti, generally known as "mother." Either type can be distilled to remove color and other impurities and is then called white vinegar.
Use: Mayonnaise, salad dressings, pickled foods, preservative, medicine (antifungal agent), latex coagulant.

"Vinoflex."[440] TM for vinyl chloride homo- and copolymers.

"Vinsol."[266] TM for a series of low-cost, dark, brittle, thermoplastic resins, ruby red by transmitted light, dark brown by reflected light. Available in solid form, flakes, fine powder and aqueous dispersion.
Use: Adhesives, emulsions, electrical insulation, inks, plastics.

vinyl. See vinyl compound.

vinyl acetate. CAS: 108-05-4.
$CH_3COOCH:CH_2$. 39th highest-volume chemical produced in US (1985). Raw material for polyvinyl resins.
Properties: Colorless liquid stabilized with either hydroquinone or diphenylamine inhibitors. The HQ-stabilized material can be polymerized without redistillation. The DPA-stabilized material must be distilled prior to polymerization; d 0.9345 (20/20C), fp −100.2C, bp 73C, refr index 1.3941, bulk d 7.79 lb/gal, flash p (TOC) 30F (−1.1C), soluble in most organic solvents including chlorinated solvents, insoluble in water, autoign temperature 800F (426.6C).
Derivation: (a) Vapor-phase reaction of ethylene, acetic acid, and oxygen, with a palladium catalyst. (b) Vapor-phase reaction of acetylene, acetic acid, and oxygen, with zinc acetate catalyst. (c) From synthesis gas.
Grade: Technical.
Hazard: Flammable, dangerous fire risk. Flammable limits in air 2.6–13.4%. Toxic by inhalation and ingestion. TLV: 10 ppm in air.
Use: Polyvinyl acetate, polyvinyl alcohol, poly-

vinyl butyral, and polyvinyl chloride-acetate resins, used particularly in latex paints, paper coating, adhesives, textile finishing, safety glass interlayers. A vinyl acetate-ethylene copolymer is available for specialty products.
See also polyvinyl acetal resins, polyvinyl acetate, polyvinyl alcohol, polyvinyl chloride-acetate resins.

vinylacetonitrile. See allyl cyanide.

vinylacetylene. C_4H_4 or $H_2C:CHC:CH$.
The dimer of acetylene, formed by passing it into a solution of cuprous and ammonium chlorides in hydrochloric acid.
Properties: Colorless gas or liquid, d 0.6867 (20C), bp 5C. Combustible.
Use: Intermediate in manufacture of neoprene and for various organic syntheses.
See also divinyl acetylene.

vinyl alcohol. (ethenol). $CH_2:CHOH$.
Unstable liquid; isolated only in form of its esters, or the polymer, polyvinyl alcohol.

vinylation. The formation of a vinyl derivative by reaction with acetylene. Thus vinylation of alcohols yields vinyl ethers such as vinyl ethyl ether, $C_2H_5OC_2H_3$.

vinyl bromide. CAS: 593-60-2. CH_2CHBr.
Properties: Gas, fp −138C, bp 15.6C, d 1.51.
Hazard: A carcinogen. TLV: 5 ppm in air.
Use: Flame-retarding agent for acrylic fibers.

vinyl-n-butyl ether. (n-butyl vinyl ether; BVE).
CAS: 111-34-2. $CH_2:CHOC_4H_9$.
Properties: Liquid, d 0.7803 (20C), bp 94.1C, fp −113C, refr index 1.3997, flash p (OC) 15F (−9.4C), bulk d 7.45 lb/gal (20C), slightly soluble in water, soluble in alcohol and ether.
Derivation: Reaction of acetylene with n-butyl alcohol.
Grade: Technical (98%).
Hazard: Flammable, dangerous fire risk.
Use: Synthesis, copolymerization.

vinyl butyrate. CAS: 123-20-6.
$CH_2:CHOOCC_3H_7$.
Properties: Liquid, d 0.9022 (20/20C), bp 116.7C, fp −86.8C, very slightly soluble in water, flash p (OC) 68F (20C).
Hazard: Flammable, dangerous fire risk.
Use: Polymers, emulsion paints.

n-vinylcarbazole. $C_2H_8NHC:CH_2$.
Properties: Liquid. Combustible.
Derivation: From acetylene and carbazole.
Use: Polymerizes to form heat-resistant and insu-

lating resins somewhat similar to mica in dielectric properties.
See polyvinyl carbazole.

vinyl cetyl ether. See cetyl vinyl ether.

vinyl chloride. (VC; chloroethene; chloroethylene). CAS: 75-01-4.
CH_2:CHCl. The most important vinyl monomer; 19th highest-volume chemical produced in US (1985).
Properties: Compressed gas, easily liquefied, ethereal odor, usually handled as liquid, phenol is added as a polymerization inhibitor, d 0.9121 (liquid at 20/20C), bp −13.9C, fp −159.7C, vap press 2300 mm (20C), flash p −108F (−77C), autoign temperature 882F (472C), slightly soluble in water, soluble in alcohol and ether.
Hazard: Explosive limits in air 4–22% by volume. An extremely toxic and hazardous material by all avenues of exposure. A carcinogen. TLV: TWA = 5 ppm in air. Workroom exposure established by OSHA is 1 ppm (average over an 8-hr period), exposure times up to 15 minutes are permissible up to 5 ppm. Respirators must be used in cases where this concentration is exceeded. Use in aerosol sprays prohibited.
Derivation: (a) Dehydrochlorination of ethylene dichloride; (b) reaction of acetylene and hydrogen chloride, either as liquids or gases.
Grade: Technical, pure 99.9%.
Use: Polyvinyl chloride and copolymers, organic synthesis, adhesives for plastics.

vinyl-2-chloroethyl ether. CAS: 110-75-8.
CH_2:CHOCH$_2$CH$_2$Cl.
Properties: Liquid, d 1.0498 (20C), bp 109.1C, fp −69.7C, very slightly soluble in water, flash p (OC) 80F (26.6C).
Hazard: Flammable, dangerous fire risk.

vinyl compound. A compound having the vinyl grouping (CH_2=CH−), specifically vinyl chloride, vinyl acetate, and similar esters but also referring more generally to compounds such as styrene C_6H_5CH=CH_2, methyl methacrylate CH_2=C(CH$_3$)COOCH$_3$, and acrylonitrile CH_2=CHCN. The vinyl compounds are highly reactive, polymerize easily and are the basis of a number of important plastics.

vinyl cyanide. See acrylonitrile.

vinylcyclohexene. (1-vinylcyclohexene-3; 4-vinylcyclohexene-1; cyclohexenylethylene).
CAS: 100-40-3.
CH_2:CHCHCH$_2$CH:CHCH$_2$CH$_2$. A butadiene dimer.
Properties: Liquid, d 0.8303 (20/4C), fp −108.9C, bp 128C, refr index 1.464 (20C), flash p (TOC)

70F (21.2C), autoign temperature 517F (269C), temperatures above 80F (26.6C) and prolonged exposure to oxygen-containing gases should be avoided as these conditions lead to discoloration and gum formation.
Grade: Technical 95%, pure 99%, research.
Hazard: Flammable, dangerous fire risk. Narcotic in high concentration.
Use: Polymers, organic synthesis.

vinylcyclohexene dioxide. (vinylcyclohexane dioxide). CAS: 106-87-6.
$\overline{CH_2CHOC_6H_9O}$.
Properties: Colorless liquid, d 1.098 (20/20C), bp 227C, refr index 1.4782 (20C), viscosity 7.77 cp (20C), flash p 230F (110C). Combustible.
Hazard: Toxic by ingestion and skin absorption, strong irritant to skin and tissue. TLV: 10 ppm in air.
Use: Polymers, organic synthesis.

vinylcyclohexene monoxide. (vinylcyclohexane monoxide). CH_2:CHC$_6$H$_9$O.
Properties: Liquid, d 0.9598 (20/20C), bp 169C, fp −100C, flash p 136F (57.8C), viscosity 1.69 cp (20C), very slightly soluble in water. Combustible.
Hazard: Moderate fire risk. Irritant.
Use: Polymers, organic synthesis.

vinyl ether. (divinyl ether; divinyl oxide).
CAS: 109-93-3. CH_2:CHOCH:CH_2.
Properties: Colorless liquid with characteristic odor, d 0.769, bp 39C, refr index 1.3989 (20C). Slightly soluble in water; miscible with alcohol, acetone, chloroform, and ether; must be protected from light. Flash p −22F (−30C), autoign temperature 680F (360C).
Derivation: Treatment of dichloroethyl ether with alkali.
Grade: NF (contains 96–97% vinyl ether, remainder dehydrated alcohol).
Hazard: Flammable, severe fire and explosion risk; explosive limits in air 1.7–27%. Toxic by inhalation, overexposure may be fatal.
Use: Copolymer with 3–5% polyvinyl chloride for plastic products such as clear plastic bottles, medicine (anesthetic, for brief operations only).

vinyl-β-ethoxyethyl sulfide.
CH_2:CHSCH$_2$CH$_2$OC$_2$H$_5$.
Properties: Colorless liquid, pungent camphor-like odor, d 0.9532 (15C), bp 65C (8 mm).
Use: Organic synthesis.

vinylethylene. See butadiene.

vinyl ethyl ether. (ethyl vinyl ether; EVE).
CAS: 109-92-2. $CH_2:CHOC_2H_5$.
Properties: Colorless liquid, extremely reactive, can be polymerized in liquid or vapor phase, slightly soluble in water (0.9% by weight), d 0.754 (20/20C), bulk d 6.28 lb/gal (20C), fp −115.0C, viscosity 0.22 degrees cp (20C), refr index 1.3739, flash p −50F (−46C), autoign temperature 395F (201.6C). Commercial material contains inhibitor to prevent premature polymerization. Often stored underground to minimize vapor losses.
Derivation: Reaction of acetylene with ethanol.
Grade: Technical.
Hazard: Highly flammable, severe fire and explosion risk, explosive limits in air 1.7–28%.
Use: Copolymerization, intermediate.

vinyl-2-ethylhexoate. CAS: 94-04-2.
$CH_2{=}CHOOCCH(C_2H_5)C_4H_9$.
Properties: Liquid, d 0.8751 (20/20C), bp 185.2C, fp −90C, flash p (OC) 165F (73.9C), insoluble in water. Combustible.
Use: Polymers, emulsion paints.

vinyl-2-ethylhexyl ether. CAS: 103-44-6.
$CH_2{=}CHOCH_2CH(C_2H_5)C_4H_9$.
Properties: Liquid, d 0.8102 (20/20C), bp 177.7C, fp −100C, flash p (OC) 135F (57.2C), autoign temperature 395F (201.6C), insoluble in water. Combustible.
Hazard: Moderate fire and explosion risk.
Use: Intermediate for pharmaceuticals, insecticides, adhesives, viscosity index improver.

2-vinyl-5-ethylpyridine.
$(CH_2{=}CH)C_5H_3N(C_2H_5)$.
Properties: Liquid, d 0.9449 (20/20C), bp 138C (100 mm), vap press 0.1 mm (20C), fp −50.9C, insoluble in water, flash p (COC) 200F (93.3C). Combustible.
Use: Copolymer, synthesis.

vinyl fluoride. (fluoroethylene).
CAS: 75-02-5. $CH_2{=}CHF$.
Properties: Colorless gas, bp −72C, insoluble in water, soluble in alcohol and ether.
Hazard: Flammable, dangerous fire and explosion risk. Toxic by inhalation. TLV (as F): 2.6 mg/ m^3 of air.
Use: Monomer.
See polyvinyl fluoride.

vinylidene chloride. (VC). CAS: 75-35-4.
$CH_2{=}CCl_2$.
Properties: Colorless liquid, fp −122.53C, bp 37C, flash p (OC) 14F (−10C), insoluble in water, autoign point 856F (457C), readily polymerizes. Commercial product contains small proportion of inhibitor.

Hazard: Flammable, dangerous fire risk, explosive limits in air 5.6–11.4%. Toxic by inhalation. TLV: 5 ppm in air.
Use: Copolymerized with vinyl chloride or acrylonitrile to form various kinds of saran. Other copolymers are also made. Adhesives; component of synthetic fibers.
See also saran.

vinylidene fluoride. (1,1-difluoroethylene).
CAS: 75-38-7. $H_2C{=}CF_2$. A monomer.
Properties: Colorless gas with faint ethereal odor, bp −83C (1 atmosphere), fp −144C (1 atmosphere), d (liquid) 0.617 (24C), slightly soluble in water, soluble in alcohol and ether.
Derivation: Interaction of hydrogen with dichlorodifluoroethane.
Grade: 99% min purity.
Hazard: Flammable, dangerous fire risk, explosive limits in air 5.5–21%. Toxic by inhalation. TLV (as F): 2.5 mg/m^3 of air.
Use: Polymers and copolymers, chemical intermediate.
See polyvinylidene fluoride.

vinylidene resin. (polyvinylidene resin).
A polymer in which the unit structure is ($-H_2CCX_2-$), in which X is usually chlorine, fluorine, or cyanide radical. Examples are saran and "Vitron" A.

vinyl isobutyl ether. (isobutyl vinyl ether; IVE).
CAS: 109-53-5. $CH_2{=}CHOCH_2CH(CH_3)_2$.
Properties: Colorless liquid, d 0.7706 (20/20C), bp 83.3C, vap press 68 mm (20C), fp −132C, refr index 1.3938, flash p (OC) 15F (−9.4C), very slightly soluble in water, soluble in alcohol and ether, easily polymerized.
Derivation: Catalytic union of acetylene and isobutyl alcohol.
Method of purification: Washing with water, drying in presence of alkali and distillation from metallic sodium.
Grade: Technical.
Hazard: Flammable, dangerous fire risk.
Use: Polymer and copolymers used in surgical adhesives, coatings, and lacquers; modifier for alkyd and polystyrene resins; plasticizer for nitrocellulose and other plastics; chemical intermediate.

vinylmagnesium chloride. $CH_2{=}CHMgCl$.
Usually supplied dissolved in tetrahydrofuran.
Use: Grignard reagent.

vinyl methyl ether. (methyl vinyl ether; methoxyethylene; MVE). CAS: 107-25-5.
$CH_2{=}CHOCH_3$.
Properties: Colorless compressed gas, or colorless liquid, d 0.7500 (20/20C), bp 6.0C, vap press

1052 mm (20C), flash p −60F (−51C), fp −121.6C, slightly soluble in water, soluble in alcohol and ether, easily polymerized, commercial material contains a small proportion of polymerization inhibitor.
Derivation: Catalytic reaction of acetylene and methanol.
Grade: Technical (95% min), pure.
Hazard: Highly flammable, severe fire and explosion risk, explosive limits in air 2.6–39%.
Use: Copolymers used in coatings and lacquers; modifier for alkyl, polystyrene, and ionomer resins; plasticizer for nitrocellulose and adhesives. See polyvinyl methyl ether.

vinyl methyl ketone. (3-buten-2-one; methyl vinyl ketone). CAS: 78-94-4.
$CH_3COCH=CH_2$.
Properties: Colorless liquid, d 0.8636 (20/4C), bp 80C, soluble in water and alcohols, flash p 20F (−6.6C) (CC).
Hazard: Flammable, dangerous fire risk. Skin and eye irritant.
Use: Monomer for vinyl resins, component of ionomer resins, intermediate in steroid and vitamin A synthesis, alkylating agent.

vinyl plastics. See polyvinyl resins.

vinyl propionate. $CH_2=CHOOCC_2H_5$.
Properties: Liquid, d 0.9173 (20/20C), bp 95.0C, fp −81.1C, flash p (OC) 34F (1.1C), almost insoluble in water.
Hazard: Flammable, dangerous fire risk.
Use: Polymers, emulsion paints.

vinylpyridine. $C_5H_4NCH=CH_2$.
Properties: Colorless liquid, boils with resinification at 159C, d 0.9746 (20C), refr index 1.5509 (20C). Dissolves in water to extent of 2.5%; water dissolves in it to 15%; soluble in dilute acids, hydrocarbons, alcohols, ketones, esters. Commercial material contains inhibitor. Combustible.
Hazard: Irritant to skin and eyes.
Use: Elastomers and pharmaceuticals. Latex used as tire cord binder (41% solids). The latex is copolymerized with butadiene-styrene.

n-vinyl-2-pyrrolidone.

$CH_2=CHNCH_2CH_2CH_2CO$. CAS: 88-12-0.
Properties: Colorless liquid, bp 148C (100 mm), mp 13.5C, flash p (COC) 209F (98.3C), d 1.04. Combustible.
Derivation: From acetylene and formaldehyde by high-pressure synthesis.
Hazard: Irritant and narcotic.
Use: Polyvinylpyrrolidone, organic synthesis.

vinyl stabilizer. A substance added to vinyl chloride resins during compounding, to retard the rate of deterioration due to formation of hydrogen chloride. Many combine readily with hydrogen chloride but do not otherwise interfere with the properties and uses of the final product. Amines, basic oxides, and metallic soaps are commonly used.

vinyl stearate. $CH_3(CH_2)_{16}COOCH=CH_2$.
Properties: White waxy solid, mp 28–30C, bp 175C (3 mm), d 0.9037 (20/20C), refr index 1.4355–1.4362 (55C), iodine no. 80–82, insoluble in water and alcohol, moderately soluble in ketones and vegetable oils, soluble in most hydrocarbon and chlorinated solvents. Combustible.
Use: Plasticizer (copolymerizer), lubricant.

vinylstyrene. See divinyl benzene.

vinyltoluene. (methyl styrene).
CAS: 25013-15-4. $CH_2=CHC_6H_4CH_3$.
Properties: Colorless liquid, fp −76.8C, bp 170–171C, d 0.890 (25/25C), bulk d 7.41 lb/gal, refr index 1.534 (34C), flash p 130F (54.4C) (CC), autoign temperature 921F (494C), very slightly soluble in water, soluble in methanol, ether. Combustible.
Hazard: Moderate fire risk. TLV: 50 ppm in air.
Use: Solvent, intermediate.

vinyl trichloride. See 1,1,2-trichloroethane.

vinyltrichlorosilane. CAS: 75-94-5.
$CH_2CHSiCl_3$.
Properties: Colorless or pale yellow liquid, bp 90.6C, d 1.265 (25/25C), refr index 1.432 (20C), flash p 16F (−8.89C), readily hydrolyzed with liberation of hydrogen chloride, polymerizes easily, soluble in most organic solvents, reacts with alcohol.
Derivation: Reaction of acetylene and trichlorosilane (peroxide catalyst), reaction of trichlorosilane with vinyl chloride.
Grade: Technical.
Hazard: High by ingestion and inhalation, strong irritant to tissue. Flammable, dangerous fire risk.
Use: Intermediate for silicones, coupling agent in adhesives and bonds.

vinyon. Generic name for a manufactured fiber in which the fiber-forming substance is any long chain synthetic polymer composed of at least 85% by weight of vinyl chloride units, —CH_2CHCl— (Federal Trade Commission). It has good resistance to chemicals, bacteria, moths; is unaffected by water and sunlight; low softening point. Tenacity 3.1 g/denier, difficult to ignite, self-extinguishing.
See polyvinyl chloride.

"Vinyzene."[8] TM for series of fungicides and bactericides formulated as additives to vinyls. Typical formulation is a condensation product of epoxidized soybean oil and 10,10'-oxybisphenoxarsine, functioning as both primary plasticizer-stabilizer and biocide.

Hazard: Toxic by ingestion and inhalation.

vioform. (iodochloroxyquinoline). Odorless, non-irritant, sterilizable substitute for iodoform.

violanthrone. See dibenzanthrone.

Violet #1. An FD&C color used for meat grading, cosmetics, beverages, etc. It is a triphenylmethane dye banned by the FDA in 1973 due to carcinogenic risk.

violet gentian. See methyl violet.

viomycin. An antibiotic produced by *Streptomyces puniceus*. Unique among antibiotics in that it is more active against acid-fast organisms than against other groups. Mycobacteria are most sensitive to viomycin and the antibiotic is active against strains of Mycobacterium tuberculosis which are resistant to other antibiotics. Available commercially as sulfate.

viosterol. Irradiated ergosterol.

Virtanen, Arrturi I. (1895-1973) A Finnish biochemist who won the Nobel prize in 1945. His work was primarily concerned with research in nutrition and agriculture. He made important discoveries regarding prevention of fodder spoilage and bacterial fermentation as well as nitrogen metabolism in plants. His PhD was awarded at the University of Helsinki followed by an illustrious career that included awards throughout Scandanavia.

virus. An infectious agent composed almost entirely of protein and nucleic acids (nucleoprotein). Viruses can reproduce only within living cells and are so small that they can be resolved only with an electron microscope. Since they pass through filters that retain bacteria, they are often called filterable viruses. Tobacco mosaic was the first virus to be crystallized and isolated (Dr. W. M. Stanley, 1935); it contains some 2000 protein molecules in a sequence of 158 amino acids (molecular weight 40,000,000). Bushy stunt virus found in tomato plants has a molecular weight of 7,600,000. First synthesis of a virus was reported in 1967.

"Viruses differ from organisms in that they are only half alive; they lack metabolism, are unable to utilize oxygen, to synthesize macromolecules, to grow, or to die. They are parasites, relying on a living host cell." They account for many diseases, including mumps, measles, scarlet fever, smallpox, influenza, and possibly the common cold. Their shapes are similar to those of bacteria (rods, spheres, filaments). They have the ability to mutate; they are also antigenic and thus initiate formation of antibodies. Some act as bacteriophages. A direct relation between virus and cancer has been shown, the DNA of the virus becoming irreversibly bound to the DNA of the affected cells.

See also bacteria, deoxyribonucleic acid.

"Viscarin."[124] TM for carrageenan extractives from Irish moss. Dry, free-flowing powder packed in fiber drums.

Use: Gelling, suspending, binding, and viscosity-building in food products.

"Viscasil."[245] TM for a series of polish grade dimethyl silicone fluids. Available in viscosities from 50 to 1,000 centistokes at 25C. They are not compatible with waxes and act as lubricants, producing a polish which is more easily rubbed out and gives a higher gloss.

viscometer. (viscosimeter). A device for measuring the viscosity of a liquid. The types most widely used are the Engler, Saybolt, and Redwood, which indicate viscosity by the rate of flow of the test liquid through an orifice of standard diameter or the flow rate of a metal ball through a column of the liquid; other types utilize the speed of a rotating spindle or vane immersed in the test liquid. The liquids commonly measured are lubricating oils and the like; heavier (non Newtonian) liquids such as paints and paper coatings require more complex devices, e.g., Brookfield and Krebs-Stormer.

See also viscosity.

viscose process. The best-known process for making regenerated cellulose (rayon) by converting cellulose to the soluble xanthate, which can be spun into fibers and then reconverted to cellulose by treatment with acid. Wood pulp is steeped with 17–20% caustic soda; the resulting alkali cellulose is pressed to remove excess liquor and the soluble beta- and gamma-cellulose, and then shredded and aged. It is then treated with carbon disulfide and sodium hydroxide to form an orange, viscous solution of cellulose xanthate. After filtration and deaeration, this solution (viscose) is forced through minute spinneret openings (or long slit dies in the case of cellophane) into a bath containing sulfuric acid and various salts such as sodium and zinc sulfate. The salts cause the viscose to gel immediately, forming a fiber or film of sufficient strength to permit it to be drawn through the bath under tension. At the same time the sulfuric acid decomposes the xanthate, converting the fibers to cellulose, in which form they are washed and dried.

See also rayon, cellophane.

viscosimeter. See viscometer.

viscosity. The internal resistance to flow exhibited by a fluid, the ratio of shearing stress to rate of shear. A liquid has a viscosity of one poise if a force of 1 dyne/square centimeter causes two parallel liquid surfaces one square centimeter in area and one centimeter apart to move past one another at a velocity of 1 cm/second. One poise equals 100 centipoises. Viscosity in centipoises divided by the liquid density at the same temperature gives kinematic viscosity in centistokes (cs). One hundred centistokes equal one stoke. To determine kinematic viscosity, the time is measured for an exact quantity of liquid to flow by gravity through a standard capillary.

Water is the primary viscosity standard with an accepted viscosity at 20C of 0.01002 poise. Hydrocarbon liquids such as hexane are less viscous. Molasses may have a viscosity of several hundred centistokes, while for a very heavy lubrication oil the viscosity may be 100 centistokes. There are many empirical methods for measuring viscosity.

See Saybolt University Viscosity; viscometer.

viscosity index improver. A lubricating oil additive that has the effect of increasing the viscosity of the oil in such a way that it is greater at high temperature than at low temperature. Agents used for this purpose are polymers of alkyl esters of methacrylic acid, polyisobutylenes, etc.

viscosity, kinematic. See viscosity.

"VisKing."[313] TM for plastic film and sheets for wrapping, packaging, and containers; rods, bars, and tubing.

"Vistac."[230] TM for a series of synthetic hydrocarbon polymers used in rubber-base adhesives and cements, waxes and asphalts, paints and caulking compounds and in latex and asphalt emulsions.

"Vistalon."[592] TM for an ethylene-propylene rubber, claimed to have excellent resistance to ozone and weathering, heat resistance, low-temperature flexibility, and resistance to acids and detergents.

Use: Automotive appliances, industrial hose, cable insulation, electrical equipment.

visual purple. See rhodopsin.

vitamin. Any of a number of complex organic compounds, present in natural products or made synthetically, which are essential in small propor-

tions in the diet of animals and man. Some are fat-soluble (A, D, K); others are water-soluble (B complex, C). Their precursors are called provitamins. A normal diet usually contains sufficient vitamins for health, although older people, the ill, young, or infirm may require different standards. Their usual use in medicine is restricted to correction of specific metabolic deficiencies. Some authorities believe that habitual intake of standard vitamin preparations readily available on the market is of little if any nutritional benefit. The following list of cross-references will serve to locate technical information about the various vitamin entries in this book:

Vitamin A
See carotene, cryptoxanthin, retinol, 3-dehydroretinol, provitamin.
Vitamin B
See Vitamin B complex, also thiamine, riboflavin, niacin, panthothenic acid, biotin, cyanocobalamin, pyridoxine, folic acid, inositol.
Vitamin C
See ascorbic acid.
Vitamin D
See ergosterol, ergocalciferol, cholecalciferol.
Vitamin E
See tocopherol.
Vitamin K
See phytonadione, menadione, phthiocol.

vitamin B complex. A group of closely interrelated vitamins found in rice bran, yeast, wheat germ, etc., originally thought to be one substance. Studies carried out by R. J. Williams and associates later revealed the astonishing complexity of this group. He states as follows: "The physiological activity originally observed was due to the *additive* effect of a considerable number of substances, *each one* of which is of itself essential. If and when the designations B_1, B_2, B_3, etc., are used, they have an entirely different meaning from the parallel use of D_1, D_2, D_3 or K_1 and K_2 because in the case of the D and K vitamins one form can replace another. In the case of the B vitamins each form is a distinctly different substance with different functions, and each member of the family is separately indispensable. No one B vitamin can replace any other."

"Vitel."[265] TM for a series of polyester resins, for spinning into polyester staple and high tenacity fibers for wearing apparel, industrial fabrics, tire cord; for extrusion of high strength monofilaments and film; for extrusion coating on paper, textiles, foils, plastics; also used as hot melt resins.

"Viton."[28] TM for a series of fluoroelastomers based on the copolymer of vinylidene fluoride

and hexafluoropropylene with the repeating structure possibly
—CF_2—CH_2—CF_2—$CF(CF_3)$—.
Properties: White transparent solid; d 1.72–1.86; resistant to corrosive liquids and chemicals up to 315C; useful continuous service at 204–232C; resistant to ozone, weather, flame, oils, fuels, lubricants, many solvents; radiation resistance good. Nonflammable.
Use: Gaskets, seals, diaphragms, tubing, aerospace and automotive components, high vacuum equipment, low-temperature and radiation equipment.

"Vitrafos."[1] TM for a clear, glassy, granular sodium phosphate.
Properties: Phosphorus pentoxide, 63% min; sieve size 10- to 80-mesh screen, pH (1% solution) 7.8; bulk d 79 lb/cu ft.
Use: Detergent builder, deflocculating agent in oil-well drilling muds, sequestering agent in the textile industry.

"Vitresil."[220] TM for a high-quality natural silica, fused into a pure and uniform woven fabric containing more than 99.8% SiO_2. Available in translucent and transparent forms.
Use: Filter medium.

vitreous. Descriptive of a material having the appearance and properties of a glass, i.e., a hard, amorphous, brittle structure, as in porcelain enamel.
See vitrification, glass.

vitreous enamel. See porcelain enamel.

"Vitrex."[41] TM for an acid-proof silicate cement which sets by chemical action. Inert to acids, except hydrogen fluoride, up to 1148C.

"Vitric 10."[326] TM for a silicate base, acid proof, chemical setting cement suitable for use with strong oxidizing acids such as nitric and chromic.

vitrification. The process of converting a siliceous material into an amorphous, glassy form by melting and cooling. As applied to radioactive waste disposal, it refers to incorporation of the waste in glassy materials for permanent storage.

vitriol. An obsolete term once used to refer to a number of sulfates (lead, copper, zinc) because of their glassy appearance. Sulfuric acid was called oil of vitriol. Derived from vitrum (glass).

"Vitrobond."[41] TM for a plasticized sulfur cement, resistant to nonoxidizing acids, oxidizing acids, acid and neutral salts, up to 93C.

"Vitroplast."[41] TM for a polyester resin cement, resistant to nonoxidizing acids, acid and neutral

salts, some organic solvents and weak alkalies up to 135C.

"Vituf."[265] TM for polyester extrusion resins designed for jacket applications over electrical wire insulations as a replacement for nylon over vinyl and polyethylene.
Use: Aircraft, auto and electronic hook-up wire, building wire, missile ground cable.

"Vivana."[565] TM for fibers, filaments and yarn of polymers and interpolymers of vinylidene chloride.

"Viz-Thin."[48] TM for a ferro-chrome lignin sulfonate.
Use: Thinner for gypsum base oil well drilling muds.

VM&P naphtha. See naphtha (la).

voids. Empty spaces of molecular dimensions occurring between closely packed solid particles, as in powder metallurgy. Their presence permits barriers made by powder metallurgy techniques to act as diffusion membranes for separation of uranium isotopes in the gaseous diffusion process.
See diffusion, gaseous.

Voight amination. Amination of benzoins with amines in the presence of phosphorus pentoxide or hydrochloric acid.

"Volan."[28] TM for a Werner type chromium complex (methacrylatochromic chloride) in isopropanol.
Use: Bonding agent for glass fibers, paper, wood, and polymeric coatings.

volatility. The tendency of a solid or liquid material to pass into the vapor state at a given temperature. Specifically the vapor pressure of a component divided by its mole fraction in the liquid or solid.

Volhard-Erdmann cyclization. Synthesis of alkyl and aryl thiophenes by cyclization of disodium succinate or other 1,4-difunctional compounds (gamma-oxo acids, 1,4-diketones, chloroacetyl-substituted esters) with phosphorus heptasulfide.

Volhard's solution. A solution of potassium thiocyanate.
Use: Analytical chemistry.

voltaic cell. Two conductive metals of different potentials in contact with an electrolyte which generate an electric current. The original voltaic

cell was comprised of silver and zinc, with brine-moistened paper as electrolyte. Semisolid pastes are now used; electrodes may be lead, nickel, zinc, or cadmium.
See also solar cell.

volumetric analysis. See titration.

Von Baeyer, Adolf. (1835-1917) A German chemist who received the Nobel prize for chemistry in 1905. This work concerned organic dyes and hydroaromatic compounds. He was educated in Berlin under the direction of Bunsen and Kekule. He was professor in Strasbourg and Munich. Many discoveries included barbituric acid and molecular structure of indigo.

von Richter reaction. Carboxylation of substituted aromatic nitro compounds with ethanolic potassium cyanide at 120-270 degrees, the carboxyl group entering a position ortho to that previously occupied by the eliminated nitro group.

von Richter synthesis. Formation of cinnoline derivatives by diazotization of o-aminoarylpropiolic acids or o-aminoarylacetylenes followed by hydration and cyclization. The method is applicable to preparation of cinnolines substituted in the benzenoid ring.

"Voranate."[233] TM for a series of urethane intermediates which are the reaction products of polyols and isocyanates. They are adducts or quasiprepolymers to be used in combination with "Voranol" products to obtain rigid urethane foams.

"Vorane."[233] TM for a group of polyurethane chemicals, raw materials for polyurethane elastomers, coatings, and foams.

"Voranol."[233] TM for a series of polyols used as intermediates in urethane elastomers, coatings, flexible and rigid foams.

VPC. Abbreviation for vapor-phase chromatography.
See gas chromatography.

"VPM."[28] TM for a solution containing 32.7% sodium methyldithiocarbamate used as a soil fumigant.

"Vulca."[53] TM for a series of ether derivatives of ungelatinized starch; graded by resistance to swelling in boiling water. "Vulca" 100 is completely nonswelling, resistant to agents that gelatinize starch, retains stability during autoclaving and pressure cooking, and may be steam sterilized without any appreciable change in its powdery appearance.
Use: Thickeners, cosmetics, dusting powders.

"Vulcalock."[119] TM for rubber-based adhesive especially designed for bonding rubber to metals.

vulcanization. A physicochemical change resulting from cross-linking of the unsaturated hydrocarbon chain of polyisoprene (rubber) with sulfur, usually with application of heat. The precise mechanism which produces the network structure of the cross-linked molecules is still incompletely known. Sulfur is also used with unsaturated types of synthetic rubbers; some types require use of peroxides, metallic oxides, chlorinated quinones, or nitrobenzenes. Natural rubber can be vulcanized with selenium, organic peroxides, and quinone derivatives, but these have limited industrial use; high-energy radiation curing is an important innovation.

Vulcanization can be effected with sulfur alone in high percentage, but the time required is too long to be economic and the properties obtained are poor. Inorganic accelerators and metallic oxides (usually zinc) are essential for satisfactory cure. Organic accelerators, introduced in the early 1920s, notably shortened vulcanization time.

Three factors affect the properties of a vulcanizate: (1) the percentage of sulfur and accelerator used, (2) the temperature, and (3) time of cure. Sulfur is usually from 1 to 3%; with strong acceleration the time can be as short as three minutes at high temperature (150C). Vulcanization can also occur at room temperature with specific formulations (self-curing cements).

Vulcanization was discovered in 1846 by Charles Goodyear in the US and simultaneously by Thomas Hancock in England. Its overall effect is to convert rubber hydrocarbon from a soft, tacky, thermoplastic to a strong, temperature-stable thermoset having unique elastic modulus and yield properties.
See also rubber; rubber, synthetic.

"Vulcosal."[233] TM for the industrial grade of salicylic acid.
Use: Stabilizer and retarder of vulcanization.

"Vulklor."[248] TM for tetrachloro-p-benzoquinone.
Use: Vulcanizing agent.

"Vulkollan."[470] TM for urethane elastomers.
See polyurethane.

"Vultac."[204] TM for a series of vulcanizing agents composed chiefly of alkyl phenol disulfides.

"Vultex."[572]   TM for a vulcanized rubber or synthetic rubber latex.
Use: Surgeons' gloves, drug sundries, dipped products, adhesives.

"Vycor" Brand Glass No. 7900.[20]   TM for a glass made by a process in which an article fabricated by conventional methods is chemically leached to remove substantially all of the ingredients except silica. When fired at high temperatures a transparent glass of high softening point and extremely low expansion coefficient is produced.
Properties: Softening point 1500C, temperature limit in service 900C, linear coefficient of expansion per degree C = 0.0000008, d 2.18, refr index 1.458 (a similar glass, No. 7910, will transmit over 60% radiation at 254 millimicrons in 2 mm section).
Use: Laboratory and industrial glassware, including beakers, crucibles, flasks, dishes, tubes, cylinders, containers, flat glass rods.

"Vynolour."[438]   TM for pigments dispersed on vinyl copolymer for use in vinyl chloride and vinyl chloride-vinylidene chloride types for inks, calendering solution resin coatings, extrusion, and molding.

W. Symbol for tungsten.
See tungsten, wolfram.

Wacker reaction. The oxidation of ethylene to acetaldehyde in the presence of palladium chloride and cupric chloride.

Wagner-Jauregg reaction. Addition of maleic anhydride to diarylethylenes with formation of bis adducts which can be converted to aromatic ring systems.

Wagner-Meerwin rearrangement. Carbon-to-carbon migration of alkyl, aryl, or hydride ions. The original example is the acid catalyzed rearrangement of camphene hydrochloride to isobornyl chloride.

Waksman, Selman A. (1888–1973) American microbiologist (Nobel Prize 1952) and Professor at Rutgers University. He was the first to use the term "antibiotic" to designate the mold-produced antibacterial substances discovered by Fleming in 1928. He became the outstanding authority in this field.

Wallach, Otto. (1847-1931) A German chemist who received the Nobel prize for chemistry in 1910 for his work in alicylic compounds. His mentors were Hofmann and Wahler and worked at the University of Bonn under Kekule. He studied pharmacy and did work on terpenes, camphors, and essential oils. This was followed by research in aromatic oils, perfumes, and spices. He studied thuja oil and fenchone. His research on terpenes revealed their significance in sex hormones and vitamins. Ethereal oils, industrial uses were made possible by his work.

Walden inversion. Inversion of configuration of a chiral center in bimolecular nucleophilic substitution reactions.

Wallach rearrangement. Acid-catalyzed rearrangement of azoxybenzenes to p-hydroxyazobenzenes.

"Wallpol."[36] TM for a series of vinyl acetate/acrylate and vinyl acetate emulsions.

"Warfarin."[316] (3-(α-acetonylbenzyl)-4-hydroxycoumarin).
$C_6H_4C_3O(OH)(O)CH(CH_2COCH_3)C_6H_5$.

Hazard: TLV 0.1 mg/m^3.
Use: Rodenticide.

warning odor. A distinctive odor imparted to fuel gases for safety purposes, as they have little or no odor of their own. Ethanethiol and other malodorous mercaptan derivatives are added to natural gas, SNG, LPG, etc., for this purpose.

warp. The lengthwise threads or strands of a textile fabric.

wash-and-wear fabric. See aminoplast resin.

washing soda. (soda crystals).
$Na_2CO_3 \cdot 10HOH$.   See soda ash, sodium carbonate monohydrate.

waste, chemical. See chemical waste.

waste control. Waste materials can be classified by type, as gaseous, liquid, solid, and radioactive; and by source, as chemical, municipal, agricultural, urban, and nuclear. Many methods of treating wastes, either by converting them to useful byproducts or by disposing of them, are in operation or under experimentation. Dumping in streams and rivers has long been illegal and ocean dumping has been prohibited since 1973. Methods of disposal or treatment include incineration (garbage, paper, plastics), precipitation (smoke, solid-in-liquid suspensions), adsorption (gases and vapors), chemical treatment (neutralization, ion-exchange, chlorination), reclamation of sulfite liquors (paper industry), reclaiming (paper, metals, rubber), compaction (urban wastes), bacterial digestion (sewage), comminution, and melting (glass, metals). Some cases may involve a combination of these.
  The following procedures should also be mentioned: (1) Flash pyrolysis of certain solid wastes yields synthetic fuel oil and other useful products. (2) Urban, animal, and agricultural wastes can be fermented by anaerobic bacteria to yield proteins, fuel oil, etc.; the possibilities of converting cannery and other food-processing wastes into protein-rich foods and feeds are being investigated. (3) Catalytic oxidation of exhaust emission gases; devices have been installed in cars since 1973 for air-pollution control. (4) Recovery of methane from manures. (5) Incorporation of special additives to certain plastics (polystyrene bottles) to render them biodegradable. (6) High-pressure hydrogenation of garbage to yield a low-

sulfur combustible oil. (7) Deactivating radioactive wastes by adsorption or ion exchange, as well as by solidification and hydraulic fracturing; high-activity wastes are buried in steel-lined concrete tanks. (8) Catalytic oxidation of waste chlorinated hydrocarbons, with partial recovery of chlorine. (9) Incineration of semisolid and liquid wastes at sea in ships designed for that purpose. See also radioactive waste, chemical waste, sewage sludge, urban waste.

waste, hazardous.   See chemical waste, radioactive waste.

waste, radioactive.   See radioactive waste.

waste wool, wet.   See wool waste.

"Watchung."[28]   TM for precipitated diazo red pigments.

water.   (ice, steam).   HOH, H_2O.
Properties: Colorless, odorless, tasteless liquid. Allotropic forms are ice (solid) and steam (vapor). Water is a polar liquid with high dielectric constant (81 at 17C), which largely accounts for its solvent power. It is a weak electrolyte, ionizing as H_3O^+ and OH^-. At atmospheric pressure it has d 1.00 (4C), fp 0C (32F) and expands about 10% on freezing. Viscosity 0.01002 poise (20C), sp heat 1 calorie/g, vap press 760 mm (100C), triple point 273.16K at 4.6 mm, surface tension 73 dynes/cm at 20C, latent heat of fusion (ice) 80 cal/g, latent heat of condensation (steam) 540 cal/g. Bulk d 8.337 lbs/gal, 62.3/lb/cu ft. Refr index 1.333. Water may be superheated by enclosing in an autoclave and increasing pressure; it may be supercooled by adding sodium chloride or other ionizing compound. It has definite catalytic activity, especially of metal oxidation. Physiologically water is classed as a nutrient substance.
Derivation: (1) Oxidation of hydrogen, (2) end product of combustion, (3) end product of acid-base reaction, (4) end product of condensation reaction.   Purification: (1) Distillation, (2) ion exchange reaction (zeolite), (3) chlorination, (4) filtration.
Use: Suspending agent (papermaking, coal slurries), solvent (extraction, scrubbing), diluent, beer and carbonated beverages, hydration of lime, paper coatings, textile processing, moderator in nuclear reactors, debarking logs, industrial coolant, filtration, washing and scouring, sulfur mining, hydrolysis, Portland cement, hydraulic systems, power source, steam generation, food industry, source of hydrogen by electrolysis and thermochemical decomposition.
See hydrogen. See also ice; steam; heavy water; ocean water; water, hard.

water, bound.   See bound water.

water gas.   (blue gas).   A mixture of gases made from coke, air, and steam. The steam is decomposed by passing it over a bed of incandescent coke, or by high-temperature reaction with natural gas or similar hydrocarbons. Approximate composition: carbon monoxide 40%, hydrogen 50%, carbon dioxide 3% and nitrogen 3%.
Hazard: Flammable, dangerous fire and explosion risk. Explosive limits 7–72% in air. Toxic by inhalation.
Use: Organic synthesis, fuel gas, ammonia synthesis.
See also synthesis gas.

water glass.   See sodium silicate.

water, hard.   Water containing low percentages of calcium and magnesium carbonates, bicarbonates, sulfates, or chlorides, as a result of long contact with rocky substrates and soils. Degree of hardness is expressed either as grains per gallon or parts per million (ppm) of calcium carbonate (1 grain of $CaCO_3$ per gal is equivalent to 17.1 ppm). Up to 5 grains is considered soft, over 30 grains is very hard. Hardness may be temporary (carbonates and bicarbonates) or permanent (sulfates, chlorides). Treatment with zeolites is necessary to soften permanently hard water. Temporary hardness can be reduced by boiling. These impurities are responsible for boiler scale and corrosion of metals on long contact. Hard waters require use of synthetic detergents for satisfactory "sudsing."
See also zeolite.

watermark.   See dandy roll.

water of crystallization.   Water chemically combined in many crystallized substances; it can be removed at or near 100C, usually with loss of crystalline properties.

water pollution.   Contamination of fresh or salt water with materials that are toxic, noxious, or otherwise harmful to fish and other animals and to man, including thermal pollution. Disposal of untreated chemical and municipal wastes in streams and rivers has been illegal since the early 1900s; in 1973 the EPA prohibited dumping of all types of wastes into the ocean. Unintentional pollution results from run-off containing toxic insecticidal residues. Oil spills at sea are a continual problem and probably will remain so.
See also waste treatment, oil spill treatment, Environmental Protection Agency, environmental chemistry.

waterproofing agent.   (1) Any film-forming substance which coats a substrate with water-repellent layer, such as paint, a rubber or plastic film, a wax, or an asphaltic compound. These are used on a wide variety of surfaces, including cement, masonry, metals, textiles, etc.

(2) Any metal salt or other chemical which impregnates textile fibers in such a way as to give an air-permeable, water-resistant product. There are three types of these: renewable, semi-durable, and durable. Renewable repellents are water dispersions containing aluminum acetate or formate, emulsifying agents and protective colloids in the continuous phase, and a blend of waxes in the disperse phase. Semidurable repellents involve precipitation of insoluble metal salts on the fibers; water-soluble soaps and waxes are usually added to the mixture, which is especially effective on synthetic fibers. Durable repellents coat each fiber with a protective film without bonding them together or sealing the apertures.

water purification. (water conditioning).
Any process whereby water is treated in such a way as to remove or reduce undesirable impurities. The following methods are used: (1) sedimentation, in which coarse suspended matter is allowed to settle by gravity in special tanks or reservoirs; (2) coagulation of aggregates (called "floc") by means of aluminum sulfate, ferric sulfate, or sodium aluminate (the aggregates are formed from colloidally dispersed impurities activated by the coagulant); (3) filtration through a bed of fine sand, either by gravity flow or by pressure, to remove suspended particles; (4) chlorination, which is effective in sterilizing potable water, swimming pools, etc.; (5) adsorption on activated carbon for removal of organic contaminants causing unpleasant taste and odor; (6) hardness removal by ion-exchange or zeolite process. The USDA has reported that mercury can be removed from water by treatment with low concentrations of black liquor from kraft paper-making.

water-soluble gum. See gum, natural.

water-soluble oil. Ammonia, potash, or sodium soaps of oleic, rosin, or naphthenic acids dissolved in mineral oils.
Use: Boring, lathe-cutting, milling, polishing lubricants, dressing textile fibers, dust laying.
See also soluble oil.

water-soluble resin. See polymer, water-soluble.

wattle bark. (Australian bark; mimosa bark).
Derivation: From the Australian wattles and South African wattles and South African acacias.
Bark contains 25–35% tannin.
Grade: Based on tannin content.

Use: Source of wattle bark extract, used in vegetable tanning of leather, especially retannage of upper leathers and production of heavy leathers.

wax. A low-melting organic mixture or compound of high molecular weight, solid at room temperature and generally similar in composition to fats and oils except that it contains no glycerides. Some are hydrocarbons, others are esters of fatty acids and alcohols. They are classed among the lipids. Waxes are thermoplastic, but since they are not high polymers, they are not considered in the family of plastics. Common properties are water repellency, smooth texture, low toxicity, freedom from objectionable odor and color. They are combustible and have good dielectric properties; soluble in most organic solvents, insoluble in water. The major types are as follows:
I. Natural
  1. Animal (beeswax, lanolin, shellac wax, Chinese insect wax)
  2. Vegetable (carnauba, candelilla, bayberry, sugar cane)
  3. Mineral
    (a) Fossil or earth waxes (ozocerite, ceresin, montan)
    (b) Petroleum waxes (paraffin, microcrystalline) (slack or scale wax)
II. Synthetic
  1. Ethylenic polymers and polyol ether-esters ("Carboxwax," sorbitol)
  2. Chlorinated naphthalenes ("Halowax")
  3. Hydrocarbon type, i.e., Fischer-Tropsch synthesis
Use: Polishes, candles, crayons, sealants, sun-cracking protection of rubber and plastic products, abherent, cosmetics, paper coating, packaging food products, electrical insulation, water-proofing and cleaning compounds, carbon paper, precision investment casting.
See also following entries.

wax, chloronaphthalene.
Properties: Translucent, black, light and varied colors, d 1.40–1.7 (300F), mp 87.7–129.4C, bp 287–371C, soluble in many organic solvent liquids and oils (when heated together).
Derivation: By chlorinating naphthalene.
Hazard: Toxic by ingestion and skin contact.
Use: Condenser impregnation, moisture-, flame-, acid, insect-proofing of wood, fabric, and other fibrous bodies; moisture- and flame-proofing covered wire and cable; solvent (for rubber, aniline and other dyes, mineral and vegetable oils, varnish gums and resins, and other waxes when mixed in the molten state).

wax, microcrystalline. A wax, usually branched-chain paraffins, characterized by a crystalline structure much smaller than that of normal wax and also by much higher viscosity. Obtained by dewaxing tank bottoms, refinery residues, and other petroleum waste products; they have an average molecular weight of 500–800 (twice that of paraffin). Viscosity 45–120 seconds (SUS at 98.9C), penetration value 3–33. Petroleum-derived products are used for adhesives, paper coating, cosmetic creams, floor wax, electrical insulation, heat-sealing, glass fabric impregnation, leather treatment, emulsions, etc. Some natural products, notably chlorophyll are classed as microcrystalline waxes.

wax, polymethylene. White, odorless solid with congealing point of 96.1C. Offered in flaked form. Approved by FDA.

wax tailings. Brown, sticky, semiasphalt product obtained in the destructive distillation of petroleum tar just prior to formation of coke.
Use: Wood preservative, roofing paper.

"Weatherometer." See ageing (c).

web. A roll of paper as it comes from the Fourdrinier machine and used to feed a rotary printing press.

weed-killer. See herbicide.

Weerman degradation. Formation of an aldose with one less carbon atom from an aldonic acid by a Hoffmann-type rearrangement of the corresponding amide. This is a general reaction of alpha-hydroxy carboxylic acids.

weight. See mass.

weighting agent. (1) In soft drink technology, an oil or oil-soluble compound of high specific gravity, such as a brominated olive oil, which is added to citrus flavoring oils to raise the specific gravity of the mixture to about 1.00, so that stable emulsions with water can be made for flavoring. (2) In the textile industry a compound used both to deluster and lower the cost of a fabric, at the same time improving its "hand" or feeling. Zinc acetylacetonate, clays, chalk, etc., are used.

welding. Joining or bonding of metals or thermoplastics by application of temperatures high enough to melt the materials so that they fuse to a permanent union on cooling. In general, the temperatures used for thermoplastics are considerably lower than required for metals. The following methods are used for metals: (1) An oxyacetylene flame is applied with a torch to the butted ends or edges of the pieces to be

joined. (2) A method called brazing is similar to (1), except that a nonferrous filler alloy is inserted between the pieces. A number of alloys are used, e.g., Ag/Cu/Zn; the filler cannot be remelted. It forms an intermetallic compound at the interfaces. (3) In resistance welding, the heat is provided by the resistance to an electric current as it passes through the material. No filler metal is used. (4) In ultrasonic welding, the heat source is the friction resulting from ultrasonic vibrations. It is a type of friction welding. (5) Electron-beam welding is a comparatively recent technique in which energy is supplied by a stream of electrons focused by a magnetic field under high vacuum. It is used for complicated weldments of tool steels.
The following methods are used for welding such thermoplastics as polyvinyl chloride, HDPE, polypropylene, and polycarbonates: (1) Hot gas technique, in which an electrically or gas-heated "gun" melts a rod of the same material as the parts to be joined. (2) Friction welding, in which heat is generated by rapid rubbing together of the two surfaces, one of which is held stationary while the other is rubbed against it at a speed great enough to cause softening. (3) Ultrasonic welding, which is also used for thermoplastics. See (4) above.
See also solder.

Werner, A. (1866–1919) A native of Switzerland, Werner was awarded the Nobel Prize for his development of the concept of the coordination theory of valence, which he advanced in 1893. His ideas revolutionized the approach to the structure of inorganic compounds and in recent years have permeated this entire area of chemistry. The term "Werner complex" has largely been replaced by "coordination compound."

Wessely-Moser rearrangement. Rearrangement of flavones and flavanones possessing a 5-hydroxyl group, through fission of the heterocyclic ring and reclosure of the intermediate diaroylmethanes in the alternate direction.

Weston cell. An electrical cell used as a standard which consists of an amalgamated cadmium anode covered with crystals of cadmium sulfate dipping into a saturated solution of the salt, and a mercury cathode covered with solid mercury sulfate.

Westphalen-Lettre rearrangement. Dehydration of 5-hydroxycholesterol derivatives accompanied by C-10 to C-5 methyl migration in compounds with a beta-substituent in C-6.

wet deposition. See acid precipitation.

wetting agent. A surface-active agent which, when added to water causes it to penetrate more easily into, or to spread over the surface of, another material by reducing the surface tension of the water. Soaps, alcohols, and fatty acids are examples.
See also detergent.

WFNA. Abbreviation for white fuming nitric acid.
See nitric acid, fuming.

Wharton reaction. Reduction of α,β-epoxy ketones by hydrazine to allylic alcohols.

wheat germ oil. Light-yellow, fat-soluble oil extracted from wheat germ. Dietary supplement.
See also tocopherol.

whey. The serum remaining after removal of the solids (fat and casein) from milk. Dried whey contains about 13% protein, 71% lactose, 2.3% lactic acid, 4.5% water, and 8% ash, including a low concentration of phosphoric anhydride. Besides its value as an inexpensive source of protein for animal feeds, whey is used as a source of lactose and lactic acid, as well as for the synthesis of riboflavin, acetone, butanol, and fuel-grade ethanol by fermentation processes. Some types of cheese are made from whey and it is also a possible culture medium. Dried whey may be used to replace up to 75% of the polyol component of rigid polyurethane foams.
See also lactose.

whiskers. Single, axially-oriented, crystalline filaments of metals (iron, cobalt, aluminum, tungsten, rhenium, nickel, etc.), refractory materials (sapphire, aluminum oxide, silicon carbide), carbon, boron, etc. They have tensile strengths of 3–6 million psi and very high elastic moduli. Their upper temperature limit in oxidizing atmospheres may be as high as 1700C and in inert atmospheres up to 2000C. Length may be up to two inches, with diameter up to 10 microns. Their chief use is in the manufacture of composite structures with plastics, glass, or graphite which have many applications in the aircraft and space vehicle field, where their high heat capacity and tremendous strength are invaluable, especially as ablative agents.

whiskey.
Properties: Light yellow to amber liquid, d 0.923–0.935 (15.56C), 47–53% alcohol by volume, flash p (CC) 26.6C.
Derivation: Distillation of fermented malted grains (corn, rye, or barley). After distillation whiskey is aged in wooden containers for several

years. The following changes occur during ageing: extraction of wood components (acids and esters), oxidation of the components of the liquid, and reaction between organic compounds in the liquid, forming new flavors.
Hazard: Flammable, moderate fire risk. A noncumulative poison, usually harmless in moderate amounts, but may be toxic when habitually taken in large amounts.
Use: Beverage, medicine (stimulant, antiseptic, vasodilator).
See also ethanol, proof.

white acid. A mixture of ammonium bifluoride and hydrogen fluoride used for etching glass.
Hazard: Strong irritant to skin and tissue.

white arsenic. See arsenic trioxide.

white copperas. See zinc sulfate.

white dye. An optical bleach or, in general, any substance, such as bluing, which may be added to a white article to increase its apparent whiteness.

white gasoline. See gasoline.

white gold. Alloy of 90% gold, 10% palladium, or 59% nickel and 41% gold.

white lead. Name primarily applied to lead carbonate, basic, but also used for lead sulfate, basic (white lead sulfate) and lead silicate, basic (white lead silicate).
Hazard: Toxic by ingestion and skin absorption. Content in paints limited to 0.05% (FDA).
Use: Paint pigment.

white liquor. See liquor (c).

white metal. (1) Any of a group of alloys having relatively low melting points. They usually contain tin, lead, or antimony as the chief component. Type metal, Babbitt, pewter, and Britannia metal are of this group. (2) Copper matte containing about 75% copper, as obtained in copper smelting operations.
See copper.

whitener. (1) Any of several oil-in-water emulsions in powder form that are dried and concentrated; when added to an aqueous medium they form stable emulsions, giving a white color and a cream-like body to coffee. A typical formulation contains vegetable fat, protein, sugar, corn syrup solids, plus emulsifier and stabilizers. (2) A white pigment or colorant used in the paper and textile industries.

white oil. Any of several derivatives of paraffinic hydrocarbons having moderate viscosity, low

volatility, and high flash point. Available in USP grade. Combustible.
Use: Textile lubricant and finishing agent, plastics modifier.

white phosphorus. See phosphorus.

white precipitate. See mercury, ammoniated.

white vitriol. See zinc sulfate.

whitewash. A suspension of hydrated lime or calcium carbonate in water used for temporary antiseptic coatings of interiors of chicken houses and the like.
See also calcimine.

whiting. Finely ground, naturally occurring calcium carbonate, $CaCO_3$, 98% pure, contaminated with silica, iron, aluminum, or magnesium.
Properties: White or off-white powder, d 2.7, insoluble in water, soluble in acids. It has no tinctorial power and hence is not a pigment.
Derivation: Traditionally from chalk, obtained from England, France, Belgium. A pure limestone or calcite is the principal commercial source. Crude chalk or limestone is ground dry or wet, air- or water-floated, and sieved. Grades are based on particle size, softness, and light reflectance. Dry ground, air-floated limestone whiting can be as fine as 99% through 300 mesh.
Grade: Various. Paris white is the finest; coarser grades are extra gilders whiting, gilders whiting, and commercial, the last being quite coarse and of poor color. A putty grade is also sold.
Use: Filler in rubber, plastics, and paper coatings; putty (with linseed oil); whitewashes; sealants.
See also calcium carbonate; chalk.

Whiting reaction. Alkynediols are reduced by lithium aluminum hydride in ether or tertiary amines to dienes.

Wichterle reaction. Modification of the Robinson annellation reaction in which 1,3-dichloro-cis-2-butene is used instead of methyl vinyl ketone.

Widman-Stoermer synthesis. Synthesis of cinnolines by cyclization of diazotized o-aminoaryl-ethylenes at room temperature.

Wieland, Heinrich O. (1877-1957) A German chemist who won the Nobel prize in chemistry in 1927. His research included work on bile acids, organic radicals, nitrogen compounds, toxic substances, and chemical oxidation as well as the discovery of the structure of cholesterol. He received his PhD at the University of Munich.

Wij's solution. ICl_3 and I_2.
Source: Iodine chloride (9.4 g) and iodine (7.2 g) are dissolved separately in glacial acetic acid and the solutions added together.

Use: Determination of iodine values of fats and oils.

Wilkinson, Geoffrey. (1921-) A British organic chemist who won the Nobel prize for chemistry in 1973 with Fischer. Their work involved research on organometallic compounds. He was a professor at the Universities of California and Harvard before returning as Professor of Inorganic Chemistry at the University of London.

willemite. Zn_2SiO_4. Natural zinc orthosilicate. Troostite is a manganese-bearing variety.
Properties: Color yellow, green, red, brown, white; luster vitreous to resinous; sometimes fluoresces in UV light; Mohs hardness 5.5; d 3.3.
Occurrence: New Jersey, New Mexico, Africa, Greenland.
Use: Ore of zinc, a phosphor.

Willgerodt reaction. Discovered in 1887, this reaction involves heating a ketone, e.g., $ArCOCH_3$ with an aqueous solution of yellow ammonium sulfide (sulfur dissolved in ammonium sulfide). It results in formation of an amide derivative of an arylacetic acid and in some reduction of the ketone: $ArCOCH_3 + (NH)_4S_x \rightarrow ArCH_2CONH_2 + ArCH_2CH_3$. The dark reaction mixture usually is refluxed with alkali to hydrolyze the amide, the arylacetic acid being recovered from the alkaline solution. (L. & M. Fieser).

Willgerodt-Kindler reaction. Conversion of aryl alkyl ketones to amides and/or the ammonium salts of the corresponding acids by aqueous ammonium polysulfide or by sulfur and a primary or secondary amine.

Williamson synthesis. An organic method for preparing ethers by the interaction of an alkyl halide with a sodium alcoholate (or phenolate).

Willstater, Richard. (1872-1942) A German chemist who won the Nobel prize for chemistry in 1915 for his work with plant pigmentation. His education was at the University of Munich where he studied and taught before going to Zurich, Switzerland. He researched chlorophylls and pigments of plants and the relationship of cornflower blue to rose red. Work included the study of alkaloids including cocaine, tropine, and atropine. His work perfected the process of chromatographic partition.

Wilzbach procedure. Exposure of organic compounds to tritium gas yields tritiated products of high activity without extensive radiation damage. Concentrations of tritium ranging from one to ninety millicuries per gram have been obtained with very varied compounds.

Windaus, Adolf. (1876-1959) A German chemist who won the Nobel prize in chemistry in 1928. His work involved the study of steroids and the effect of ultraviolet light activity, ergosterol, and vitamin D_2. He also researched digitalis and histamine. Although he studied medicine he received his doctorate in chemistry at the University of Freiburg.

wine. The fermented juice of grapes or other fruits or plants. Contains 7–20% alcohol (by volume). The higher percentages are obtained by addition of pure alcohol (fortifying). Coloring matter, sugars, and small amounts of acetic acid, salts, higher fatty acids, etc., give wines their distinctive appearance and flavor.
Hazard: Flammable, moderate fire risk.

wine ether. See ethyl pelargonate.

wine gallon. Same as US gallon.

wine lees. A deposit or sediment formed in the bottom of wine casks during fermentation. Wine lees vary greatly in quality, but usually contain 20–35% potassium acid tartrate and up to 20% calcium tartrate. They also contain yeast cells, proteins, and other solid matter which was suspended in the grape juice.
Use: Source of tartaric acid and tartarates.
See also Pasteur, tartaric acid.

"Wing-Stay."[265] TM for a series of p-phenylenediamine derivatives used as stabilizers, antioxidants, and antiozonants for natural and synthetic rubbers and latices.

"Wing-Stop."[265] TM for sodium dimethyldithiocarbamate used in terminating the polymerization reactions of SBR elastomers, in "hot" and "cold" reactions.

wintergreen oil. See methyl salicylate.

winterize. A process of refrigerating edible and lubricating oils to crystallize the saturated glycerides, which are then removed by filter pressing.

wire cloth. See screen, filter media.

witherite. $BaCO_3$. A natural barium carbonate usually found in veins with lead ores.
Properties: White, yellowish, or grayish; vitreous, inclining to resinous luster; d 4.27–4.35; Mohs hardness 3–3.75.
Occurrence: US (Kentucky, Lake Superior region), England (most important source).

Hazard: Toxic by ingestion, skin irritant.
Use: Chemicals (barium dioxide, barium hydroxide, blanc fixe), plate glass and porcelain, brick making, rodenticide.

Wittig, George. (1897-) A University of Heidelberg professor who won the Nobel prize for chemistry in 1979 along with Brown of Purdue. Wittig's research showed that phosphorous ylids react with ketones and aldehydes to form alkenes. This reaction is used a great deal in the synthesis of pharmaceuticals and other complex organic substances.

Wittig reaction. Preparation of olefins from alkylidene phosphoranes (ylids) and carbonyl compounds.

Wittig rearrangement. Rearrangement of ethers with alkyl lithiums to yield alcohols via a 1,2-shift.

Wohl degradation. Method for the conversion of an aldose into an aldose with one less carbon atom by the reversal of the cyanohydrin synthesis. In the Wohl method, the nitrile group is eliminated by treatment with ammoniacal silver oxide.

Wohl-Ziegler reaction. Allylic bromination of olefins with N-bromosuccinimide. Peroxides or ultraviolet light are used as initiators.

Wohler, Friedrich. (1800–1882) A native of Germany, Wohler, working with Berzelius, placed the qualitative analysis of minerals on a firm foundation. In the early 19th Century chemists still thought it impossible to synthesize organic compounds. In 1828, Wohler publicized his laboratory synthesis of urea, which he had obtained four years previously by heating the inorganic substance ammonium cyanate, the synthesis is a result of intramolecular rearrangement:

$$(NH_4)OCN \longrightarrow NH_2-\overset{\displaystyle O}{\overset{\|}{C}}-NH_2$$

This was the beginning of the science of synthetic organic chemistry.

Wohler synthesis. Classical synthesis of urea by heating an aqueous solution of ammonium cyanate extended to preparation of urea derivatives.

Wohlwill process. The official process of the US mints for refining gold. It consists in subjecting gold anodes to electrolysis in a hot solution of hydrochloric acid containing gold chloride, the solution being continuously agitated with compressed air.

Wolff rearrangement. Rearrangement of diazoketone to ketenes by action of heat, light or some metallic catalysts. The rearrangement is the key step in the Arndt-Eistert synthesis.

Wolff-Kishner reaction. This reduction reaction was discovered independently in Germany (Wolff, 1912) and in Russia (Kishner, 1911). A ketone (or aldehyde) is converted into the hydrazone, and this derivative is heated in a sealed tube or an autoclave with sodium ethoxide in absolute ethanol.

$$\diagdown \atop {C=O} \diagup \quad \xrightarrow{H_2NNH_2} \quad \diagdown \atop {C=NNH_2} \diagup \quad \xrightarrow{NaOC_2H_5,\ 200°}$$

$$\diagdown \atop {CH_2} \diagup \ +\ N_2$$

After preliminary technical improvements, Huang Minlon (1946) introduced a modified procedure by which the reduction is conducted on a large scale at atmospheric pressure with efficiency and economy. (L. & M. Fieser).

Wolffenstein-Boters reaction. Simultaneous oxidation and nitration of aromatic compounds to nitrophenols with nitric acid or the higher oxides of nitrogen in the presence of a mercury salt as catalyst. Hydroxynitration of benzene yields picric acid.

wolfram. See tungsten. Wolfram is the official international alternate name for tungsten. Tungsten is preferred in the US.

wolframic acid, anhydrous. See tungstic oxide.

wolframite. $(Fe,Mn)WO_4$. A natural tungstate of iron and manganese. Ferberite is the iron-rich member of the series, and huebnerite is the manganese-rich member.
Properties: Color black to brown, luster sub-metallic to resinous, streak black to brown, Mohs hardness 5–5.5, d 7.0–7.5.
Occurrence: Colorado, South Dakota, Nevada, Australia, Bolivia, Europe.
Use: Chief ore of tungsten.

wolfram white. See barium tungstate.

wollastonite. CAS: 13983-17-0. $CaSiO_3$.
A natural calcium silicate found in metamorphic rocks.
Properties: Color white to brown, red, gray, yellow; luster vitreous to pearly; Mohs hardness 4.5–5; d 2.8–2.9.

Occurrence: New York, California.
Grade: Fine, medium paint grades.
Use: Ceramics, paint extender, welding rod coatings, rubber filler, silica gels, paper coating, filler in plastics, cements and wallboard, mineral wool, soil conditioner.

"Wolman."[11] TM for a wood preservative comprising a mixture of chlorinated arsenate, fluoride and phenolic salts used in aqueous solutions to impregnate lumber to protect it against decay and termites.

wood. A mixture composed of 67–80% holocellulose and 17–30% lignin, together with low percentages of resins, sugars, a variable amount of water and potassium compounds. Its fuel value varies widely around 3000–6000 Btu/lb according to variety, moisture, etc. Combustible.
Use: Pulp and paper, construction, packaging and cooperage, furniture, destructive distillation products (charcoal), extraction products (turpentine, rosin, tall oil, pine oil, etc.), methanol, plywood, fuel, rayon and cellophane, flock.
Note: Catalytic conversion of wood chips to fuel oil and methanol is under development.
See also biomass.

wood alcohol. See methanol.

wood ash. The inorganic residue of wood combustion.
Use: Fertilizer for its potash content, which averages about 4% K_2O.

wood flour. (wood flock). Pulverized dried wood from either soft or hardwood wastes. Graded according to color and fineness.
Grade: Domestic standard, domestic fine, imported 40–60 mesh, 70–80 mesh, etc.
Hazard: Flammable, dangerous fire risk, especially suspended in air.
Use: Extender and filler in dynamite, plastics, rubber, paperboard; fur cleaning; polishing agents; Sorel cement.

wood, indurated. A wood hardened by impregnation with a phenol-formaldehyde product.
Use: Storage batteries.

wood, petrified. Wood in which the original chemical components have been replaced by silica. The change occurs in such a way that the original form and structure of the wood are preserved. The most famous instance is the Petrified Forest of Arizona.

wood pulp. See pulp, paper.

wood rosin. See rosin.

Wood's metal. A four-component fusible alloy used largely in sprinkler systems, it melts at 70C, the composition being bismuth 50%, cadmium 10%, tin 13.3%, lead 26.7%.
See also alloy, fusible; eutectic.

wood sugar. See $d(+)$-xylose.

wood tar. See pine tar; creosote, wood-tar.

wood turpentine. See turpentine (oil).

Woodward cis-hydroxylation. The hydroxylation of an olefin with iodine and silver acetate in acetic acid to give cis-glycols. The method involves the trans addition of iodine and silver acetate to give a trans-iodo-acetoxy derivative. The latter is hydrolyzed to a cis-mono-glycol acetate with acetic acid and water. Alkaline hydrolysis results in the final cis-glycol.

Woodward-Hoffmann rules. Rules that predict the stereochemical course of concerted reactions in terms of the symmetry of the interacting molecular orbitals.

Woodward, Robert B. (1917–1979) An American chemist born in Quincy, MA, and widely regarded as one of the world's leading synthetic organic chemists. After receiving his doctorate from MIT (the youngest student in the history of the Institute to do so), he joined the Harvard faculty in 1937 as instructor and attained full professorship in 1950. He was recipient of the Nobel Prize in 1965 for his brilliant work in synthesizing complex organic compounds, among them quinine, cholesterol, chlorophyll, reserpine, and cobalamin (vitamin B_{12}). When he died in 1979, his synthesis of the antibiotic erythromycin was virtually complete; it was finished by his associates two years later. He was director of the Woodward Research Institute in Basel, Switzerland and a member of the governing board of the Weizmann Institute in Israel.

wool. Staple fibers, usually 2–8 in. long, obtained from the fleece of sheep (and also alpaca, vicuna, and certain goats). Physically, wool differs from hair in fineness and by the presence of prominent cortical scales and a natural crimp. The latter properties are responsible for the felting properties of wool and the ability of the fibers to cling together when spun into yarns. Chemically, wool consists essentially of protein chains (keratin) bound together by disulfide cross linkages.
Properties: Tenacity ranges from 1–2 g/denier; elongation 25–50%, d 1.32, moisture regain 16% (21.2C, 65% relative humidity), decomposes at 126C, scorches at 204C, resistant to most acids except hot sulfuric, destroyed by alkalies and chlorine bleach, resistant to mildew but attacked by insects, amphoteric to dyes. Combustible.
Sources: Australia, Argentina, US, New Zealand, Uruguay, USSR, England.
Use: Outerwear, blankets, carpets, upholstery, felt, clothing source of lanolin.
For additional information refer to Wool Bureau, 360 Lexington Avenue, New York, NY.

wool fat. See lanolin.

wool waste. Wool from scrap materials cut up for remaking into cloth. Used as a fertilizer. Wool waste usually contains from 4–7% ammonia. Pure wool shoddy may contain as much as 15% nitrogen and is particularly valued by hop growers.
Hazard: Wet wool waste is flammable and a dangerous fire risk.

wormseed oil. See chenopodium oil.

wormwood. See absinthium.

wort. A clear infusion of grain extract (usually malt) used as the basis of fermentation in the brewing industry.
See also brewing.

"Worthite."[269] TM for an austenitic stainless steel, containing 3% silicon, 20% chromium, 24% nickel, 2.75% molybdenum, 1.75 copper. Available in castings and wrought bars.
Use: Pumps for corrosive environments (paper industry, citrus fruit canners, sea water, sulfuric acid plants, etc.).

wove. (paper). See dandy roll.

writing ink. A solution of colorant in water, usually containing also low concentrations of tannic or gallic acid. Washable inks contain glycerol. Various water-soluble dyes are used for colored inks; dispersions of carbon black stabilized with a protective colloid are used for drawing inks. Fountain pen inks retain the fluidity of water; for ball-point pens the mixture is of a paste-like consistency.

wrought iron. A ferrous material, aggregated from a solidifying mass of pasty particles of highly refined metallic iron into which 1–4% of slag is uniformly dispersed without subsequent melting. Wrought iron is distinguished by its low carbon and manganese contents. Carbon seldom exceeds 0.035% and manganese content is held at 0.06% maximum. Phosphorus usually ranges from 0.10% to 0.15%; it adds about 1000 psi for each 0.01% above 0.10%. Sulfur content is normally low, ranging from 0.006% to less than 0.015%. Silicon content ranges from 0.075% to 0.15%; silicon content of base metals is 0.015% or less. Residuals such as chromium, nickel, cobalt, copper, and molybdenum are generally low, totaling less than 0.05%. Wrought iron is readily fabricated by standard methods and is quite corrosion-resistant.

wulfenite. $PbMoO_4$, sometimes with calcium, chromium, vanadium.

Properties: Yellow, orange, or bright orange-red mineral of resinous luster; d 6.7–7.0; Mohs hardness 2.75–3.

Occurrence: Found in veins with ores of lead in US (Massachusetts, New York, Pennsylvania, Nevada, Utah, New Mexico, Arizona), Europe, Australia.

Use: Ore of molybdenum.

Wulff process. Production of acetylene by treating a hydrocarbon gas with superheated steam in a regenerative type of refractory furnace at 1148–1371C. Contact times are very short. Acetylene and ethylene are produced in 1:1 ratio. See also acetylene.

Wurtz reaction. A method of synthesizing hydrocarbons discovered by Wurtz in 1855. It consists in treatment of an alkyl halide with metallic sodium, which has a strong affinity for bound halogen and acts on methyl iodide in such a way as to strip iodine from the molecule and produce sodium iodide. The reaction involves two molecules of methyl iodide and two atoms of sodium.

The reaction probably proceeds through the formation of methylsodium, which interacts with methyl iodide:

$$CH_3I \xrightarrow{Na} CH_3Na \xrightarrow{CH_3I} CH_3CH_3$$

The Wurtz reaction can be applied generally to synthesis of hydrocarbons by the joining together of hydrocarbon residues of two molecules of an alkyl halide (usually the bromide or iodide). With halides of high molecular weight the yields are often good and the reaction has been serviceable in the synthesis of higher hydrocarbons starting with alcohols found in nature.

X

xanthan. A synthetic, water-soluble bipolymer made by fermentation of carbohydrates; it is a thickening and suspending agent that is heat-stable, with good tolerance for strongly acid and basic solutions. Viscosity remains stable over wide temperature range.
Use: In drilling fluids, ore flotation, and the food and pharmaceutical fields, especially in prepared mixed and high-protein fast foods. An experimental use is to flood oil-fields as an aid to enhanced recovery.

xanthate. A salt (usually potassium or sodium) of a xanthic acid. Available as yellow pelletized solids having a pungent odor, soluble in water. Examples are potassium amyl xanthate, potassium ethyl xanthate, sodium isobutyl xanthate, sodium isopropyl xanthate.
See also cellulose xanthate, viscose process.
Use: Collector agents in the flotation of sulfide minerals, metallic elements such as copper, silver, gold, and some oxidized minerals of lead and copper.

xanthene. (dibenzopyran, tricyclic).
$CH_2(C_6H_4)_2O$. The central structure of the fluoroscein, eosin, and rhodamine dyes.
Properties: Yellowish, crystalline leaflets, mp 100.5C, bp 315C, soluble in ether, slightly soluble in alcohol, very slightly soluble in water.
Derivation: By the condensation of phenol and o-cresol by means of aluminum chloride.
Use: Organic synthesis, fungicide.

xanthene dye. A group of dyes whose molecular structure is related to that of xanthene. The aromatic (C_6H_4) groups constitute the chromophore. Color Index number ranges from 45000–45999. The dyes are closely related structurally to diaryl methane dyes. Eosin (CI# 45380) is an example.

xanthenol. See xanthydrol.

xanthic acid. (xanthogenic acid). A substituted dithiocarbonic acid of the type ROC(S)SH, in which R is ordinarily an alkyl radical. Xanthic acid salts are called xanthates. Unless otherwise designated, xanthic acid is understood to be the ethyl derivative $(C_2H_5OC(S)SH)$, also called ethylxanthic acid.
Properties: Liquid melting at −53C and decomposing at room temperature.

xanthine. (dioxopurine). CAS: 69-89-6.
$C_5H_4N_4O_2$. A purine base occurring in the blood and urine and in some plants. Theophylline and theobromine, the alkaloids of tea and cocoa, respectively, are both dimethylxanthines; caffein in coffee is a trimethylxanthine.
Properties: Yellowish-white powder, sublimes with partial decomposition, soluble in potassium hydroxide, insoluble in water and acid.
Derivation: By the action of nitrous acid on guanine.
Grade: Technical, CP, monohydrate, sodium salt, radioactive forms available.
Hazard: Toxic by ingestion.
Use: Organic synthesis, medicine.

xanthine oxidase. An enzyme found in animal tissues which acts upon hypoxanthine, xanthine, aldehydes, reduced coenzyme I, etc., producing, respectively, xanthine, uric acid, acids, oxidized coenzyme I, etc.
Use: Biochemical research.

xanthogenic acid. See xanthic acid.

xanthone. (benzophenone oxide; dibenzopyrone; xanthene ketone). CAS: 90-47-1.
$CO(C_6H_4)_2O$. Tricyclic; occurs in some plant pigments.
Properties: White needles or crystalline powder, mp 173–4C, bp 350C, sublimes. Insoluble in water; soluble in alcohol, chloroform, and benzene, especially when hot.
Use: Larvicide; intermediate for dyes, perfumes, and pharmaceuticals.

xanthophyll. CAS: 127-40-2. $C_{40}H_{56}O_2$.
Properties: Yellow pigment, an oxygenated carotenoid occurring in green vegetation and in some animal products, notably egg yolk; mp 190–3C; insoluble in water; slightly soluble in alcohol and ether.
See also lutein, carotenoid.

xanthopterin. (2-amino-4,6-dihydroxypteridine). CAS: 119-44-8. $C_6H_5N_5O_2 \cdot HOH$.
Pigment found in the wings of butterflies, can be converted by yeast into folic acid.
Properties: Orange-yellow crystals, sinters 360C, decomposes above 410C, practically insoluble in water, freely soluble in dilute ammonium or sodium hydroxide giving yellow solutions, and in 2N hydrochloric acid giving colorless solutions.
Use: Biochemical research.

xanthydrol. (xanthenol). CAS: 90-46-0. $HCOH(C_6H_4)_2O$. A derivative of xanthene.
Properties: White powder, mp 123C, insoluble in water, soluble in alcohol.
Derivation: Reduction of xanthone with alcohol and sodium.
Grade: CP (analytical).
Use: Determination of urea and DDT.

"X-Cor."[236] TM for an ethoxylated diamine liquid, added to drilling and packer fluids to protect drill pipe, casing, and tubing from hydrogen sulfide attack, oxidation, or the corrosive effects of electrolytes.

Xe. Symbol for xenon.

xenon. Xe. Element of atomic number 54, noble gas group of the periodic table, aw 131.30, valences = 2, 4, 6, 8, nine stable isotopes.
Properties: Colorless, odorless gas or liquid; gas (at STP) has density of 5.8971 g/L (air = 1.29 g/L) and dielectric constant (25C) at 1 atmosphere of 1.0012; liquid has bp of −108.12C at one mm, density (at bp) 1.987 g/cc; liquefaction temperature −106.9C. Chemically unreactive but not completely inert. Noncombustible.
Derivation: Fractional distillation of liquid air.
Use: Luminescent tubes, flash lamps in photography, fluorimetry, lasers, tracer studies, anesthesia.
See also noble gas, xenon compounds.

xenon compounds. Xenon tetrafluoride XeF_4, is easily prepared by mixing fluorine and xenon in gaseous form, heating in a nickel vessel to 400C, and cooling. The product forms large colorless crystals. The difluoride and hexafluoride, XeF_2 and XeF_6, also colorless crystals, can be obtained somewhat similarly. The hexafluoride melts to a yellow liquid at 50C and boils at 75C. Many xenon fluorine complexes with other compounds are also known. Xenon oxytetrafluoride, $XeOF_4$, a volatile liquid at room temperature, is obtained from the reaction of xenon hexafluoride and silica. Gram amounts have been isolated and studied.
All these fluorides must be protected from moisture to avoid formation of xenon trioxide, XeO_3, a colorless, nonvolatile solid which is dangerously explosive when dry. Its solution, the so-called xenic acid, is a stable weak acid but a strong oxidizing agent, which will even liberate chlorine from hydrochloric acid.
In an alkaline solution, xenon trioxide reacts to give free xenon and perxenate, such as $Na_4XeO_6 \cdot 8HOH$. Perxenates are probably the most powerful oxidizing agents known, just as the xenon fluorides are extremely effective fluorinating agents.

Complex compounds containing nitrogen bonded to xenon have also been prepared.
Hazard: Toxic by inhalation; oxidizing agents, strong irritant.

xenyl. The biphenyl group $C_6H_5C_6H_4-$.

p-xenylamine. See p-aminodiphenyl.

xerography. A "dry" method of photography or photocopying. A metal plate is covered with a layer of photoconductive powder, such as selenium; the surface of this plate is given an electric charge by passing it under a series of charged wires. An image of the material to be photographed is projected onto the charged plate through a camera lens. The electric charges disappear in the areas exposed to light, but elsewhere the surface retains its charge. A powder consisting of a coarse carrier and a fine developing resin is then spread over the plate. Adhesion between powder and plate occurs only at the charged areas. Elsewhere developing resin and carrier are not retained on the plate, which thus has become a negative of the original image. A positive is obtained by placing a piece of paper against the plate, and applying an electric charge as in the first stage of the process. This causes adhesion of developing resin and its carrier to the paper. This positive print is fixed by heating in a press for a few seconds to melt the developing resin and fuse it to the paper. Colored prints are possible by use of suitable developing resins. Various materials other than paper can thus be printed.

xibornol. (6-isobornyl-3,4-xylenol).
CAS: 34632-99-0. $C_{18}H_{26}O$.
Properties: Crystalline solid or viscous, yellowish liquid; d 1.02; refr index 1.538.
Use: Antibacterial agent, rubber antioxidant.

XLPE. Trade abbreviation for crosslinked polyethylene.

XPS. Abbreviation for X-ray photoelectron spectroscopy.

x-radiation. (Roentgen rays; x-rays). Electromagnetic radiation of extremely short wavelength (0.06–120 Å), emitted as the result of electron transitions in the inner orbits of heavy atoms bombarded by cathode rays in a vacuum tube. Those of the shortest wavelength have the highest intensity, and are called "hard" x-rays. X-radiation was discovered by Roentgen in 1898. Its properties are: (1) Penetration of solids of moderate density, such as human tissue; they are retarded by bone, barium sulfate, lead, and other dense materials. (2) Action on photographic plates and fluorescent screens. (3) Ionization of the gases through which they pass. (4) Ability to damage or destroy diseased tissue; there is also a cumulative deleterious effect on healthy tissue.

Hazard: Overexposure can permanently damage cells and tissue structures; effect is cumulative.

Use: Spectrometry; structure determination of molecules, cancer therapy, diagnostic medicine, nondestructive testing of metals, identification of original paintings, preservation of foods.

See also radiation, ionizing, diffraction, Roentgen.

xylene. (dimethylbenzene).   CAS: 1330-20-7. $C_6H_4(CH_3)_2$.   A commercial mixture of the three isomers, o-, m-, and p-xylene. The last two predominate. 26th highest-volume chemical produced in US (1975) (all grades).

Properties: Clear liquid, soluble in alcohol and ether, insoluble in water, d about 0.86. See under Grade for boiling range; flash p (TOC) from 81–115F (27.2–46.1C).

Derivation: (a) Fractional distillation from petroleum (90%), coal tar or coal gas, (b) by catalytic reforming from petroleum, followed by separation of p-xylene by continuous crystallization, (c) from toluene by transalkylation.

Grade: Nitration (bp range 137.2–140.5C), 4 degrees (bp range 138–134C), 5 degrees (bp range 137–142C, high in m- isomer), 10 degrees (bp range 135–145C), industrial (bp 90% 40C, complete 160C). Also other grades depending upon use. In some cases one or another of the industrial isomers are partially removed for use in chemical production.

Hazard: Flammable, moderate fire risk. Toxic by ingestion and inhalation. TLV: 100 ppm in air.

Use: Aviation gasoline; protective coatings; solvent for alkyd resins, lacquers, enamels, rubber cements; synthesis of organic chemicals.

See also following entries.

m-xylene. (1,3-dimethylbenzene).
CAS: 108-38-3. $1,3\text{-}C_6H_4(CH_3)_2$.

Properties: Clear, colorless liquid; soluble in alcohol and ether; insoluble in water. D 0.8684 (15C), fp −47.4C, bp 138.8C, refr index 1.4973 (20C), flash p 85F (29.4C), autoign temperature 982F (527.7C).

Derivation: Selective crystallization or solvent extraction of m-p- mixture.

Grade: 95% (technical), 99%, 99.9% (research).

Hazard: Flammable, moderate fire risk. TLV: 100 ppm in air.

Use: Solvent; intermediate for dyes and organic synthesis, especially isophthalic acid; insecticides; aviation fuel.

o-xylene. (1,2-dimethylbenzene).
CAS: 95-47-6. $1,2\text{-}C_6H_4(CH_3)_2$.

Properties: Clear, colorless liquid; soluble in alcohol and ether; insoluble in water. D 0.880 (20/4C), fp −25C, bp 144C, refr index 1.505 (20C), flash p (TOC) 115F (46.1C), autoign temperature 867F (463.8C). Combustible.

Grade: 99%, free of hydrogen sulfide and sulfur dioxide, technical 95%, research 99.9%.

Hazard: Moderate fire risk. TLV: 100 ppm in air.

Use: Manufacture of phthalic anhydride, vitamin and pharmaceutical syntheses, dyes, insecticides, motor fuels.

p-xylene. (1,4-dimethylbenzene).
CAS: 106-42-3. $1,4\text{-}C_6H_4(CH_3)_2$.   28th highest-volume chemical produced in US (1985).

Properties: Colorless liquid, crystallizes at low temperature, soluble in alcohol and ether, insoluble in water, d 0.8611 (20C), mp 13.2C, bp 138.5C, refr index 1.5004 (21C), flash p (TOC) 81F (27.2C).

Derivation: Selective crystallization or solvent extraction of m-, p- mixture; separation from mixed-xylene feedstocks by adsorption (parex process[416]).

Hazard: Flammable, dangerous fire risk. TLV: 100 ppm in air.

Use: Synthesis of terephthalic acid for polyester resins and fibers ("Dacron", "Mylar," "Terylene"), vitamin and pharmaceutical syntheses, insecticides.

p-xylene-α,α'-diol. $C_6H_4(CH_2OH)_2$.

Properties: White, crystalline solid; mp 118C; bp 138–144C (0.8–1.0 mm). Slightly soluble in water (25C). Purity: 98%.

Use: Polyester resins.

xylenol. (dimethylphenol; hydroxydimethylbenzene; dimethylhydroxybenzene).
CAS: 1300-71-6. $(CH_3)_2C_6H_3OH$.
There are five isomers (2,4-; 2,5-; 2,6-; 3,4-; 3,5-). This entry describes commercially offered mixture.

Properties: White, crystalline solid; d 1.02–1.03 (15C); mp 20–76C; bp 203–225C. Only slightly soluble in water, soluble in most organic solvents and in caustic soda solution. Combustible.
Derivation: Cresylic acid or tar acid fraction of coal tar.
Hazard: Toxic by ingestion and skin absorption.
Use: Disinfectants, solvents, pharmaceuticals, insecticides and fungicides; plasticizers, rubber chemicals, additives to lubricants and gasolines, manufacture of polyphenylene oxide (2,6-isomer only), wetting agents, dyestuffs.

xylidine. (aminodimethylbenzene; aminoxylene).
CAS: 1300-73-8. $(CH_3)_2C_6H_3NH_2$.
A varying mixture of isomers (2,3-; 2,4-; 2,5-; 2,6-).

Properties: Liquid, d 0.97–0.99, bp 213–226C, slightly soluble in water, soluble in alcohol and ether, flash p (CC) 206F (96.6C). Combustible.
Derivation: Nitration of xylene and subsequent reduction.
Hazard: Toxic by ingestion, inhalation, and skin absorption. TLV: 2 ppm in air.

Use: Dye intermediate, organic syntheses, pharmaceuticals.

X-linked. Shortened form for crosslinked, e.g., X-linked polyethylene.

D(+)-xylose. (wood sugar). CAS: 58-86-6.
$C_5H_{10}O_5$. (not to be confused with phenylosazone of the same name).
Properties: White, crystalline, dextrorotatory powder; sweet taste. D 1.525 (20C), mp 144C (also given as 153C), soluble in water and alcohol. Combustible.
Derivation: Hydrolysis with hot dilute acids of wood, straw, corn cobs, etc.; wood pulp wastes.
Grade: Reagent, technical.
Use: Dyeing, tanning, diabetic food, source of ethanol.

xylocaine. Preparation of lignocaine hydrochloride.
Use: Local anesthetic.

xylol. Commercial grade xylene, a mixture of isomers and other benzene derivatives.

xylyl bromide. (α-bromoxylene).
CAS: 28258-59-5. $CH_3C_6H_4CH_2Br$.
Mixed o-, m-, and p-isomers.
Properties: Colorless liquid, pleasant aromatic odor, decomposed slowly by water, d 1.4, bp 210–220C. Combustible.
Derivation: Bromination of xylene.
Grade: Technical.
Hazard: Toxic by inhalation and ingestion; strong irritant to eyes, skin, and tissue.
Use: Organic synthesis, tear gas.

m-xylyl chloride. $CH_3C_6H_4CH_2Cl$.
Properties: Colorless liquid, bp 196, d 1.064. Combustible.
Hazard: Toxic by ingestion and inhalation, strong irritant to eyes and skin.
Use: Intermediate.

p-xylylene. $CH_2{:}C_6H_4{:}CH_2$. A monomer.
See parylene.

Y

Y. Symbol for yttrium.

Yankee machine. A papermaking machine similar to a fourdrinier, but with much shorter wire. It is designed for light-weight papers such as toilet and facial tissues and operates at comparatively high speeds. It has a specialized cylindrical drying roll that effects drying of the web much more efficiently than felts.
See also fourdrinier.

"Yarmor."[266] TM for series of pine oils for widely varied uses. Total terpene alcohol contents from 55–91%.
Use: Disinfectants, textile specialties, wetting and flotation agents, special solvents, household and industrial cleaners; odorants.

Yb. Symbol for ytterbium.

yeast. (barm). Unicellular organisms known as *saccharomycetaceae*. The following description applies to the cultured commercial product and not to various wild varieties.
Properties: Yellowish-white, viscid liquid or soft mass, flakes, or granules, consisting of cells and spores of *Saccharomyces cerevisiae*.
Derivation: A ferment obtained in brewing. Yeasts induce fermentation by enzymes (zymases) which convert glucose and other carbohydrates into carbon dioxide and water in the presence of oxygen, or into alcohol and carbon dioxide (or lactic acid) in the absence of oxygen.
Grade: Technical, brewers', cooking, compressed (contains about 74% moisture), dried, NF (contains no starch or filler, not more than 7% moisture nor more than 8% ash). Also graded according to vitamin B_1 content.
Use: Fermentation of sugars, molasses, and cereals for alcohol; brewing; baking; food supplement; protein biosynthesis from many carbonaceous and nitrogenous materials, including petroleum; source of vitamins, enzymes, nucleic acids, etc.; biochemical research.
See also bacteria; fermentation.

yeast adenylic acid. See adenylic acid.

yellow AB. (1-(phenylazo)-2-naphthylamine; Color Index No. 11380). CAS: 85-84-7. $C_6H_5N_2C_{10}H_6NH_2$.
Properties: Orange or red platelets, mp 102–104C, insoluble in water, soluble in alcohol and oils.

Hazard: Consult regulations before using in food products.
Use: Biological stain.

yellow brass. A brass containing 34–37% zinc; it has excellent fabrication properties and corrosion resistance.
Use: Structural and decorative purposes.
See also brass.

yellowcake. See uranium dioxide.

yellow enzyme. See flavin enzyme.

yellow glass. A soda-lime glass.

yellow lake. Any of several pigments made by precipitating soluble yellow dyes on an aluminum hydrate base. They are transparent in oil and lacquer vehicles and are used for metal decorating.

yellow OB. (1-o-tolueneazonaphthylamine-2; Color Index No. 11390). CAS: 131-79-3. $CH_3C_6H_4N_2C_{10}H_6NH_2$.
Properties: Orange or yellow powder, mp 122–125C, insoluble in water, soluble in alcohol and oils.
Hazard: Consult regulations before using in food products.
Use: Biological stain.

yellow phosphorus. See phosphorus.

yellow precipitate. See mercuric oxide, yellow.

yellow prussiate of potash. See potassium ferrocyanide.

yellow prussiate of soda. See sodium ferrocyanide.

yellow rain. See mycotoxin.

yellow salt. See uranyl nitrate.

ylang ylang oil. An essential, yellowish, volatile oil distilled from the flowers of *Cananga odorata*. Strongly levorotatory.
Use: Perfumery.
See also cananga oil.

ylid. A substance in which a carbanion is attached to a heteratom with a high degree of positive charge, i.e., $>C^- — X^+$. It is similar to a zwitter-ion and related to the Wittig reaction.

yohimbine. (aprodine; corynine; quebrachine). CAS: 146-48-5. $C_{21}H_{26}O_3N_2$. Properties: Glistening, needle-like alkaloid, mp 234C, soluble in alcohol and ether, very slightly soluble in water.
Derivation: By extraction from the bark of *Corynanthe yohimbe*, found in the Cameroons.
Hazard: Said to be an aphrodisiac. Toxic by ingestion.

Young's modulus. See modulus of elasticity.

ytterbium. CAS: 7440-64-4. Yb.
A metallic element. A rare-earth metal of yttrium subgroup, atomic number 70, aw 173.04, valence = 2, 3, exists in alpha and beta forms, the latter being semiconductive at pressures above 16,000 atmospheres. There are seven natural isotopes.
Properties: Metallic luster, malleable, mp 824C, bp 1427C, d 7.01, reacts slowly with water, soluble in dilute acids and liquid ammonia.
Source: See rare earth metals.
Derivation: Reduction of the oxide with lanthanum or misch metal.
Grade: Regular high purity (ingots and lumps).
Use: Lasers, dopant for garnets, portable x-ray source, chemical research.

ytterbium chloride. $YbCl_3 \cdot 6HOH$.
Properties: Green crystals, d 2.575, loses 6HOH at 180C, mp 865C, very soluble in water, hygroscopic.

ytterbium fluoride. YbF_3.
Properties: Solid, mp 1157C, bp 2200C, insoluble in water, hygroscopic.
Hazard: TLV (as F): 2.5 mg/m³ of air.

ytterbium oxide. (ytterbia). Yb_2O_3.
Properties: Colorless mass when free of thulia, tinted brown or yellow when containing thulia. The weakest base of the yttrium group save scandia and lutetia. Slightly hygroscopic, absorbs water and carbon dioxide from the air, d 9.2, mp 2346C, soluble in hot, dilute acids, less so in cold acids.
Use: Special alloys, dielectric ceramics, carbon rods for industrial lighting, catalyst, special glasses.

ytterbium sulfate. $Yb_2(SO_4)_3 \cdot 8HOH$.
Properties: Crystalline solid, d 3.286, soluble in water.

yttria. See yttrium oxide.

yttrium. CAS: 7440-65-5. Y. Metallic element of atomic number 39, group IIIB of the periodic table, aw 88.9059, valence = 3, no stable isotopes.
Properties: Dark gray metal, d 4.47, mp 1500C, bp 2927C, soluble in dilute acids and potassium hydroxide solution, decomposes water, known only in the tripositive state, low neutron capture cross section.
Source: See rare-earth metals.
Derivation: Reduction of the fluoride with calcium.
Grade: Regular high purity (ingots, lumps, turnings), metallurgical, low-oxygen, crystal sponge, powder.
Hazard: Flammable in finely divided form. TLV: 1 mg/m³ of air.
Use: Nuclear technology, iron and other alloys, deoxidizer for vanadium and other nonferrous metals, microwave ferrites, coating on high-temperature alloys, special semiconductors.

yttrium acetate. $Y(C_2H_3O_2)_3 \cdot 9HOH$.
Properties: Colorless crystals, soluble in water.
Derivation: Action of acetic acid on yttrium oxide.
Use: Analytical chemistry.

yttrium antimonide. YSb. A high-purity binary semiconductor.

yttrium arsenide. YAs. A high-purity binary semiconductor.
Hazard: Highly toxic.

yttrium bromide. $YBr_3 \cdot 9HOH$.
Properties: Colorless crystals, hygroscopic, mp (anhydrous) 904C, soluble in water, slightly soluble in alcohol, insoluble in ether.

yttrium chloride. CAS: 10361-92-9. $YCl_3 \cdot 6HOH$.
Properties: Reddish-white, transparent, deliquescent prisms; soluble in water and alcohol; insoluble in ether; d 2.18; decomposes at 100C.
Derivation: By the action of hydrochloric acid on yttrium oxide.
Grade: Purities to 99%.
Use: Analytical chemistry.

yttrium oxide. (yttria). CAS: 1314-36-9. Y_2O_3.
Properties: Yellowish-white powder, soluble in dilute acids, insoluble in water, d 4.84, mp 2410C.
Derivation: By the ignition of yttrium nitrate.
Grade: Purities to 99.8%, electronic grade 99.999%.
Use: Phosphors for color TV tubes (alloy with europium), yttrium-iron garnets for microwave filters, stabilizer for high-temperature service materials (zirconia and silicon nitride refractories).

yttrium phosphide.   YP.
Used as a high-purity binary semiconductor.

yttrium sulfate. $Y_2(SO_4)_3 \cdot 8HOH$.
Properties: Small, reddish-white, monosymmetric crystals. Soluble in concentrated sulfuric acid, sparingly soluble in water, insoluble in alkalies, d 2.558, loses 8HOH at 120C, decomposes 700C.

Derivation: Action of sulfuric acid on monazite sand.
Method of purification: Fractional crystallization.
Use: Reagent.

yttrium vanadate. YVO_4.
Properties: White, crystalline solid.
Use: With europium vanadate as red phosphor in color television tubes.

Z

"Zaclon."[28] TM for a series of fluxes based on active zinc ammonium chloride combined with additives; available in various forms.

"Zalba."[28] TM for a rubber antioxidant containing a hindered phenol.
Properties: Yellow cream-colored powder, d 1.30.
Use: Nondiscoloring antioxidant for natural and synthetic rubbers and latex; stabilizer in SBR manufacture.

Zanzibar gum. A hard, usually fossil type of copal.
Source: Found on the island of Zanzibar and the adjacent African mainland.
Properties: D 1.062–1.068, mp 240–250C, insoluble in most solvents. Combustible.
Use: Varnishes.

ZDP. Abbreviation for zinc dithiophosphate.

"Zectran."[233] TM for insecticides containing 4-dimethylamino-3,5-xylyl-n-methylcarbamate.

"Zefran."[565] TM for an acrylic fiber in white staple form, based on polyacrylonitrile and supplemented with a dye-receptive component. Has higher strength than other acrylics and different dyeing properties.

zein. The protein of corn, a prolamin.
Properties: White to slightly yellow powder; odorless, nontoxic protein of the prolamine class, derived from corn; contains 17 amino acids; tasteless; free of cystine, lysine, and tryptophane. A resinous material dispersible in water with neutral sulfonated castor oil; soluble in dilute alcohol; insoluble in water, dilute acids, anhydrous alcohols, turpentine, esters, oils, fats; d 1.226. Combustible.
Derivation: Byproduct of corn processing, by extraction of gluten meal with 85% isopropanol, extraction of the zein from the extract with hexane, precipitation by water, and spray drying.
Use: Paper coating, grease-resistant coating, label varnishes, laminated board, solid color prints, printing inks, food coatings, microencapsulation, fibers.

Zeisel determination. Cleavage of methoxy and ethoxy groups with boiling hydriodic acid, distillation of the alkyl iodide into alcoholic silver nitrate and gravimetric determination as silver iodide. Alternatively, the alkyl iodide is oxidized to form iodate and the iodine liberated by addition of potassium iodide is titrated.

"Zelan."[28] TM for a line of durable, water-repellent textile finishes.

"Zelcon" SL.[28] TM for a quaternized, long-chain, complex amine condensation product used as a fabric softener and conditioner for compounding into home laundry softeners.

"Zelec."[28] TM for a series of antistatic agents, both durable and nondurable.

"Zendel."[214] TM for polyethylene films; available both cast and blown.
Use: Packaging products and netting for agricultural and construction industries.

"Zenite."[28] TM for a group of rubber accelerators based on zinc salt of 2-mercaptobenzothiazole, with or without various modifying agents.

"Zeo."[285] TM for a series of hydrous silicas produced by precipitation.
Use: Reinforcing agent for rubber compounds, adsorbent carrier for insecticides, flatting agent for lacquers and varnishes, anticaking agent in foods, feeds and chemicals.

"Zeodur."[184] TM for a processed greensand used as a low capacity inorganic cation exchanger suitable for salt cycle, primarily water softening.

"Zeogel."[236] TM for a special clay used in drilling fluids to give an equally high yield and stable viscosity and gel characteristics in either fresh or salt water, regardless of the concentration of salt in the latter.

"Zeokarb."[184] TM for a commercial grade cation exchanger of sulfonated coal; available in two ionic forms, sodium and hydrogen; intermediate acidity, used for metal recovery, deionization, removal of cations, purification and esterification catalyst.

Zeise's salt. $(Pt(C_2H_4)Cl_3)K$. Formed by adding potassium chloride to a solution of platinous chloride saturated with ethylene.

zeolite. A natural hydrated silicate of aluminum and either sodium or calcium or both, of the

1248

type $Na_2O \cdot Al_2O_3 \cdot xSiO_2 \cdot xHOH$. Both natural and artificial zeolites are used extensively for water softening, as detergent builders, and cracking catalysts. For the former purpose the sodium or potassium compounds are required, since their usefulness depends on the cationic exchange of the sodium of the zeolite for the calcium or magnesium of hard water. When the zeolite has become saturated with calcium or magnesium ion it is flooded with strong salt solution, a reverse exchange of cations takes place and the material is regenerated. The natural zeolites are analcite, chabazite, heulandite, natrolite, stilbite, and thomosonite.

Synthetic zeolites are made either by a gel process (sodium silicate and alumina) or a clay process (kaolin), which form a matrix to which the zeolite is added. These processes are quite complex, involving substitution of various rare-earth oxides. The effectiveness of zeolites depends upon their pore size, which may be as small as 4–5 Å. Other applications of zeolites are as adsorbents, desiccants, and in solar collectors, where they function as both heating and cooling agent. See also molecular sieve; aluminosilicate; ion-exchange resin; cage, zeolite.

"Zepar."BP[28] TM for a reducing agent based on sodium hydrosulfite.
Use: Bleaching agent for ground wood pulp, unbleached pulp, and old papers.

"Zepel."[28] TM for a fluorocarbon textile finish used as a durable oil- and water-repellent.

Zerewitinoff determination. The reaction of methylmagnesium iodide with an active hydrogen-containing compound provides methane, which can be measured quantitatively by the increase in volume of the system at constant pressure.

Zerewitinoff reagent. Solution of methylmagnesium iodide in purified n-butyl ether. A clear, light-colored liquid which reacts rapidly with moisture and oxygen.
Use: Analytical reagent for active hydrogen atoms in organic compounds, also to determine water, alcohols, and amines in inert solvents.

"Zerlate."[28] TM for a wettable fungicide powder containing 75% ziram.
Hazard: Irritant to eyes and mucous membranes.

"Zerlon."[233] TM for a methyl methacrylate-styrene copolymer.
Use: Plastic molding material.

zero gravity. See space, chemistry in.

zero group. See group (1).

"Zerok"101.[41] TM for a synthetic resin coating of the vinyl type which is resistant to oxidizing acids and useful for protection against fumes and splashing, up to 65.5C.

zeroth law (of thermodynamics). Two bodies which have been shown to be individually in equilibrium with a third body will be in equilibrium when placed in contact with each other, that is, they will have the same temperature (H. Reiss).

"Zeset."[28] TM for a fiber-reactant resin for crease resistance and dimensional stabilization of textiles.

zeta potential. (electrokinetic potential). The potential across the interface of all solids and liquids. Specifically, the potential across the diffuse layer of ions surrounding a charged colloidal particle, which is largely responsible for colloidal stability. Discharge of the zeta potential, accompanied by precipitation of the colloid, occurs by addition of polyvalent ions of sign opposite to that of the colloidal particles. Zeta potentials can be calculated from electrophoretic mobilities, i.e., the rates at which colloidal particles travel between charged electrodes placed in the solution.
See also electric double layer.

"Zetax."[265] TM for zinc-2-mercaptobenzothiazole.
Properties: Pale yellow powder, d 1.70; mp above 300C, zinc content 15–18%.
Use: Rubber accelerator.

Ziegler catalyst. A type of stereospecific catalyst, usually a chemical complex derived from a transition metal halide and a metal hydride or a metal alkyl. The transition metal may be any of those in groups IV to VIII of the periodic table; the hydride or alkyl metals are those of groups I, II, and III. Typically, titanium chloride is added to aluminum alkyl in a hydrocarbon solvent to form a dispersion or precipitate of the catalyst complex. These catalysts usually operate at atmospheric pressure and are used to convert ethylene to linear polyethylene and also in stereospecific polymerization of propylene to crystalline polypropylene (Ziegler process).
See also Natta catalyst.

Ziegler, Karl. (1898-1973) A German chemist who won the Nobel prize for chemistry in 1963 with Natta. A great deal of work was concerned with the chemistry of carbon compounds and development of plastics. A recipient of the Swinburne medal from the Plastics Institute of London in 1964. After studying at Marburg he was a professor at Heidelberg.

Ziegler method. Cyclization of dinitriles at high dilution in dialkyl ether in the presence of ether-soluble metal alkylanilide and hydrolysis of the resultant imino-nitrile with formation of macrocyclic ketones in good yields.

Ziegler-Natta polymerization. Polymerization of vinyl monomers under mild conditions using aluminum alkyls and $TiCl_4$ (or other transition element halide) catalyst to give a stereoregulated, or tactic, polymer. These polymers, in which the stereochemistry of the chain is not random, have very useful physical properties.

"Zimate." TM for zinc dimethyldithiocarbamate. See ziram.

Zimmermann reaction. The reaction that occurs between methylene ketones and aromatic polynitro compounds in the presence of alkali. When applied to 17-oxosteroids, the colored compounds formed can be used for the quantitative determination of 17-oxosteroids.

zinc. CAS: 7440-66-6.   Zn.   Metallic element of atomic number 30, Group IIB of periodic table, aw 65.38, valence = 2, five stable isotopes.
Properties: Shining white metal with bluish-gray luster (called spelter). Not found native, soluble in acids and alkalies, insoluble in water, d 7.14, mp 419C, bp 907C, malleable at 100–150C, strongly electropositive, zinc foil will ignite in presence of moisture. Ores and minerals:
See calamine, franklinite, hydrozincite, smithsonite, sphalerite, willemite, wurtzite.
Source: British Columbia, Mexico, US (Colorado), Australia, Belgium.
Derivation: Extracted from ores by two distinct methods, both starting with zinc oxide formed by roasting the ores: (1) the pyrometallurgical or distillation process wherein the zinc oxide is reduced with carbon in retorts from which the resultant zinc is distilled and condensed; and (2) the hydrometallurgical or electrolytic process wherein the zinc oxide is leached from the roasted or calcined material with sulfuric acid to form zinc sulfate solution which is electrolyzed in cells to deposit zinc on cathodes.
Grade: Special high-grade (99.990%), high-grade (99.95%), intermediate (99.5%), brass special (99%), prime western (98%).
Forms available: Slab, rolled (strip, sheet, rod, tubing), wire, mossy zinc, zinc dust powder (99% pure), single crystals, zinc anodes.

Hazard (dust): Flammable, dangerous fire and explosion risk.
See also zinc dust.
Use: Alloys (brass, bronze, and die-casting alloys), galvanizing iron and other metals, electroplating, metal spraying, auto parts, electrical fuses, storage and dry cell batteries, fungicides, nutrition (essential growth element). Roofing, gutters, engravers' plates, cable wrappings, organ pipes. For further information refer to American Zinc Institute, 292 Madison Ave., New York, NY.

zinc-65. Radioactive isotope of mass number 65.
Properties: Half-life 250 days, radiation beta and gamma.
Derivation: Pile irradiation of zinc metal and, in the cyclotron, by bombarding copper-65 with deuterons.
Forms available: Zinc metal and zinc chloride in hydrochloric acid solution.
Hazard: A radioactive poison.
Use: Tracer nuclide in study of wear in alloys, the nature of phosphor activators, galvanizing, function of traces of zinc in body metabolism, the functions of oil additives in lubricating oils, etc.

zinc abietate. See zinc resinate.

zinc acetate. CAS: 557-34-6.
$Zn(C_2H_3O_2)_2 \cdot 2HOH$.
Properties: White, monoclinic, crystalline plates; pearly luster; faint acetous odor; astringent taste. Soluble in water and alcohol, d 1.735, loses 2HOH at 100C, mp 200C (decomposes).
Derivation: Action of acetic acid on zinc oxide.
Use: Medicine (astringent), preserving wood, textile dyeing (mordant and resist), zinc chromate, laboratory reagent, crosslinking agent for polymers, ingredient of dietary supplements (up to one milligram daily), feed additive, ceramic glazes.

zinc acetylacetonate.
$Zn[OC(CH_3):CHCO(CH_3)]_2$.
Properties: Crystalline solid; mp 138C; bp sublimes; very soluble in benzene, acetone; decomposes in water.
Use: Catalyst in synthesis of long-chain alcohols and aldehydes, textile weighting agent.

"Zincalume."[288] TM for a bright zinc electroplating process for consumer's goods and military materials. The bath contains zinc cyanide, sodium cyanide, sodium hydroxide, and addition agents.

zinc ammonium chloride. $ZnCl_2 \cdot 2NH_4Cl$.
A complex salt; double salts with 3–6 molecules of ammonium chloride have also been prepared.
Properties: White powder or crystals, soluble in water, d 1.8.
Grade: Technical (foaming and nonfoaming).
Use: Welding, soldering flux, dry batteries, galvanizing.

zinc ammonium nitrite. CAS: 63885-01-8. $ZnNH_4(NO_2)_3$.
Properties: White powder, strong oxidizing agent.
Hazard: Dangerous fire risk in contact with organic materials.

zinc antimonide. $ZnSb_2$. Silvery-white crystals, d 6.33, mp 570C, decomposes in water.
Use: Thermoelectric devices.
Hazard: May be irritant to skin.

zinc arsenate. Various forms are known. The description following is for zinc orthoarsenate, $Zn_3(AsO_4)_2 \cdot 8HOH$.
Properties: White, odorless powder; d 3.31 (15C); loses 1HOH at 100C; insoluble in water; soluble in acids and alkalies.
Derivation: (a) Occurs in nature as mineral koettigite, (b) by reaction of a solution of sodium arsenate and a soluble zinc salt.
Hazard: Toxic by ingestion and inhalation.
Use: Insecticide, wood preservative.

zinc arsenite. (zinc metaarsenite; ZMA). CAS: 10326-24-6. $Zn(AsO_2)_2$.
Properties: Colorless powder, soluble in acids, insoluble in water, federal specification TT-W-581 describes the composition of the solution used for wood preservation.
Hazard: Toxic by ingestion and inhalation.
Use: Timber preservative, insecticide.

zinc bacitracin.
Properties: Creamy-white powder, almost insoluble in water, good thermal stability, usually has 50–60 units/mg of bacitracin activity.
Derivation: Action of zinc salts on bacitracin broth.
Grade: USP.
Use: Antibacterial agent in ointments, suppositories, etc.

zinc borate. Typical composition: zinc oxide 45%, B_2O_3 34%, may have 20% water of hydration.
Properties: White, amorphous powder; soluble in dilute acids; slightly soluble in water; zinc borate of composition $3ZnO \cdot 2B_2O_3$ has a d 3.64; mp 980C. Nonflammable.
Derivation: Interaction of the oxides at 500–1000C or of zinc oxide slurries with solutions of boric acid or borax.
Use: Medicine, fireproofing textiles, fungistat and mildew inhibitor, flux in ceramics.

zinc bromate. $Zn(BrO_3)_2 \cdot 6HOH$.
Properties: White solid, d 2.566, mp 100C, deliquescent, loses 6HOH at 200C, very soluble in water.
Hazard: Dangerous fire risk in contact with organic materials, strong oxidizing agent.

zinc bromide. CAS: 7699-45-8. $ZnBr_2$.
Properties: White, hygroscopic, crystalline powder; soluble in water, alcohol, and ether; d 4.219; mp 394C; bp 650C.
Derivation: Interaction of solutions of barium bromide and zinc sulfate with subsequent crystallization.
Use: Photographic emulsions, manufacture of rayon. A solution of 80% zinc bromide is used as a radiation viewing shield.

zinc butylxanthate. $Zn(C_4H_9OCS_2)_2$.
Properties: White powder, d 1.45, decomposes when heated, moderately soluble in benzene and ethylene dichloride, slightly soluble in acetone, insoluble in water and gasoline.
Use: Ultra-accelerator used in self-curing rubber cements.
See also xanthate.

zinc cadmium sulfide. A fluorescent pigment, a phosphor.
Hazard: As for cadmium.

zinc caprylate. (zinc octanoate). CAS: 557-09-5. $Zn(C_8H_{15}O_2)_2$.
Properties: Lustrous scales, slightly soluble in boiling water, fairly soluble in boiling alcohol, mp 136C, decomposes in moist atmosphere giving off caprylic acid.
Derivation: By precipitation from a solution of ammonium caprylate with zinc sulfate.
Use: Fungicide.

zinc carbolate. See zinc phenate.

zinc carbonate. CAS: 3486-35-9. $ZnCO_3$.
Properties: White, crystalline powder; soluble in acids, alkalies, and ammonium salt solutions; insoluble in water; d 4.42–4.45; evolves carbon dioxide at 300C.
Derivation: (a) Grinding the mineral smithsonite, (b) action of sodium bicarbonate on a solution of a zinc salt.
Use: Ceramics, fire-proofing filler for rubber and plastic compositions exposed to flame temperature, cosmetics and lotions, pharmaceuticals (ointments, dusting powders), zinc salts, medicine (topical antiseptic).

zinc chlorate. CAS: 10361-95-2.
$Zn(ClO_3)_2 \cdot 4HOH$.
Properties: Colorless to yellowish crystals, deliquescent, d 2.15, decomposes at 60C, soluble in water and alcohol, soluble in glycerol and ether.
Hazard: Dangerous fire risk in contact with organic materials, strong oxidizing agent.

zinc chloride. CAS: 7646-85-7. $ZnCl_2$.
Properties: White, granular, deliquescent crystals or crystalline powder; soluble in water, alcohol, glycerol, and ether; d 2.91; mp 290C; bp 732C; a 10% solution is acid to litmus.
Derivation: Action of hydrochloric acid on zinc or zinc oxide.
Method of purification: Recrystallization.
Grade: CP, technical; fused, crystals, granulated; 62–1/2% solution, 50% solution, USP.
Hazard: (Solid) skin irritant; (solution) severe irritant to skin and tissue. TLV (as fume): 1 mg/m^3 of air.
Use: Catalyst, dehydrating and condensing agent in organic synthesis, fireproofing and preserving food, soldering fluxes, burnishing and polishing compounds for steel, electroplating, antiseptic and deodorant preparations (up to 2% solution), textiles (mordant; carbonizing agent; mercerizing, sizing, and weighting compositions; resist for sulfur colors, albumin colors, and para red), adhesives, dental cements, glass etching, petroleum refining, parchment, dentrifices, embalming and taxidermists' fluids, medicine (astringent), antistatic, denaturant for alcohol.

zinc chloride, chromated. A mixture of zinc chloride and sodium dichromate used as a wood preservative. Federal Specification TT-W-551 requires that it contain no less than 77.5% zinc chloride and 17.5% sodium dichromate dihydrate.

zinc chloroiodide. Mixture of zinc chloride and iodide.
Properties: White powder, soluble in water.
Use: Disinfectant, pharmaceutical preparations.

zinc chromate. $ZnCrO_4 \cdot 7HOH$
Properties: Solid yellow pigment; mw 307.6

zinc cyanide. CAS: 557-21-1. $Zn(CN)_2$.
Properties: White powder, d 1.852, mp 800C (decomposes), soluble in dilute mineral acids with production of hydrogen cyanide, soluble in alkalies, insoluble in water and alcohol.
Derivation: By precipitation of a solution of zinc sulfate or chloride with potassium cyanide.
Grade: Technical.
Hazard: Toxic by ingestion and inhalation. TLV (as CN): 5 mg/m^3 of air.

Use: Metal plating, chemical reagent, insecticide.

zinc dialkyldithiophosphate. A lube oil additive for corrosion resistance, wear resistance, antioxidant.

zinc dibenzyldithiocarbamate.
$Zn[SCSN(C_7H_7)_2]_2$.
Properties: White powder; d 1.41; melting range 165–175C; moderately soluble in benzene and ethylene dichloride; insoluble in acetone, gasoline, and water.
Use: Accelerator for latex dispersions and cements.

zinc dibutyldithiocarbamate. CAS: 136-23-2.
$Zn[SC(S)N(C_4H_9)_2]_2$.
Properties: White powder; d 1.24 (20/20C); melting range 104–108C; pleasant odor; soluble in carbon disulfide, benzene, and chloroform; insoluble in water.
Use: Accelerator for latex dispersions and cements, etc; ultra-accelerator for lubricating oil additive.

zinc dichromate. CAS: 7789-12-0.
$ZnCr_2O_7 \cdot 3HOH$.
Properties: Orange-yellow powder, soluble in acids and hot water, insoluble in alcohol and ether.
Derivation: Action of chromic acid on zinc hydroxide.
Hazard: Toxic by ingestion and inhalation.
Use: Pigment.

zinc diethyl. See diethylzinc.

zinc diethyldithiocarbamate. CAS: 14324-55-1.
$Zn[SC(S)N(C_2H_5)_2]_2$.
Properties: White powder; d 1.47 (20/20C); melting range 172–176C; soluble in carbon disulfide, benzene, and chloroform; insoluble in water.
Hazard: Strong irritant to eyes and mucous membranes.
Use: Rubber vulcanization accelerator, especially latex foam, heat stabilizer for polyethylene.

zinc dimethyldithiocarbamate. See ziram.

zinc dimethyldithiocarbamatecyclohexylamine complex. (zinc dithioamine complex).
Properties: White powder or slurry of low solubility.
Hazard: Toxic by ingestion.
Use: Fungicide, rodent poison, deer and rabbit repellent.

zinc dioxide. See zinc peroxide.

zinc dithionite. See zinc hydrosulfite.

zinc dust. A gray powder.
Grade: Commercial, pigment.
Hazard: Dangerous fire risk; may form explosive mixture with air; in bulk when damp, may heat and ignite spontaneously on exposure to air.
Use: Zinc salts, other zinc compounds, reducing agent, precipitating agent, purifier, catalyst; rust-resistant paints, bleaches, pyrotechnics, soot-removal, pipe-thread compounds, sherardizing, decorative effect in resins, autobody coatings.

zinc ethyl. See diethylzinc.

zinc ethylenebisdithiocarbamate. See zineb.

zinc-2-ethylhexoate. See zinc octoate.

zinc ethylsulfate. $Zn(C_2H_5SO_4)_2 \cdot 2HOH$.
Properties: Colorless, hygroscopic, crystalline leaflets; soluble in water and alcohol.
Derivation: Interaction of zinc hydroxide and diethyl sulfate.
Use: Organic synthesis.

zinc fluoride. CAS: 7783-49-5. ZnF_2.
Properties: White powder, soluble in hot acids, slightly soluble in water, insoluble in alcohol, d 4.84 (15C), mp 872C, bp about 1500C.
Derivation: (a) Action of hydrogen fluoride on zinc hydroxide, (b) addition of sodium fluoride to a solution of zinc acetate.
Grade: Technical, 95% pure.
Hazard: Toxic material. TLV (as F): 2.5 mg/m^3 of air.
Use: Phosphors, ceramic glazes, wood preserving, electroplating, organic fluorination.

zinc fluoroborate. $Zn(BF_4)_2$.
Properties: Colorless liquid, handled as 40 or 48% solution.
Use: Plating and bonderizing, resin curing.

zinc fluorosilicate. (zinc silicofluoride).
CAS: 16871-71-9. $ZnSiF_6 \cdot 6HOH$.
Properties: White crystals, d 2.104, decomposes on heating, soluble in water.
Derivation: Reaction of zinc oxide and fluosilicic acid.
Use: Concrete hardener, laundry sour, preservative, mothproofing agents.

zinc formaldehyde sulfoxylate.
$Zn(HSO_2 \cdot CH_2O)_2$ (normal);
$Zn(OH)(HSO_2 \cdot CH_2O)$ (basic).
Properties: Rhombic prisms; very soluble in water (normal) (basic is insoluble in water); insoluble in alcohol; decomposes in acid.
Derivation: Reaction of formaldehyde and zinc sulfoxylate.

Grade: Basic, normal.
Hazard: Toxic by ingestion.
Use: Stripping and discharging agent for textiles. See also hydrosulfite-formaldehyde preparations.

zinc formate. CAS: 557-41-5.
$Zn(CH_2O)_2 \cdot 2HOH$.
Properties: White crystals, d 2.207 (20C), loses 2HOH at 140C, soluble in water, insoluble in alcohol.
Derivation: Action of formic acid on zinc hydroxide.
Hazard: Toxic by ingestion.
Use: Catalyst for production of methanol, waterproofing agent, textiles, antiseptic.

zinc gluconate. Dietary supplement and food additive, vitamin tablets.

zinc green. One of a group of brilliant green pigments consisting of mixture of Prussian blue and zinc yellow. They are permanent to light but not to alkali or water.
Use: Flat wall paints and interior work.

zinc hydrosulfite. (zinc dithionite).
CAS: 7779-86-4. (ZnS_2O_4).
Properties: White, amorphous solid; soluble in water.
Grade: Technical.
Use: Brightening groundwood, kraft, and other paper pulps; to treat beet and cane sugar juices; depressant in mining flotations; bleaching textiles, vegetable oils, straw, hemp, vegetable tannins, animal glue, etc.

zinc hydroxide. $Zn(OH)_2$.
Properties: Colorless crystals, d 3.053, decomposes 125C, almost insoluble in water, forms both zinc salts and zincates.
Derivation: Addition of a strong alkali to a solution of a zinc salt.
Use: Intermediate, absorbent in surgical dressings, rubber compounding.

zinc hypophosphite. $Zn(H_2PO_2)_2 \cdot HOH$.
Properties: White, hygroscopic crystals; soluble in water and alkalies.
Derivation: Action of hypophosphorous acid on zinc hydroxide.

zinc iodide. CAS: 10139-47-6. ZnI_2.
Properties: Hygroscopic, white, crystalline powder; sharp saline taste; turns brown on exposure to light or air; soluble in water, alcohol, and alkalies; d 4.67; mp 446C; bp 625C (decomposes).
Derivation: Interaction of barium iodide and zinc sulfate with subsequent crystallization.

Use: Medicine (topical antiseptic), analytical reagent.

Zincke disulfide cleavage. Formation of sulfenyl halides by three essentially similar methods involving the action of chlorine or bromine on aryl disulfides, thiophenols, or arylbenzyl sulfides.

Zincke nitration. Replacement of ortho- or para-bromine or iodine atoms (but not fluorine or chlorine atoms) in phenols by a nitro group on treatment with nitrous acid or a nitrite in acetic acid.

Zincke-Suhl reaction. Phenol-dienone rearrangement of p-cresols by addition of carbon tetrachloride in the presence of aluminum chloride with formation of 4-methyl-4-trichloromethylcyclohexa-2,5-dienone.

zinc lactate. CAS: 16039-53-5.
$Zn(C_3H_5O_3)_2 \cdot 3HOH$.
Properties: White crystals, soluble in water. Combustible.
Derivation: Action of lactic acid on zinc hydroxide.

zinc laurate. $Zn(C_{12}H_{23}O_2)_2$.
Properties: White powder, mp 128C, almost insoluble in water and alcohol. Combustible.
Derivation: Precipitation of a soluble coconut oil soap with a solution of a zinc salt.
Use: Paints, varnishes, rubber compounding (softener and activator).

zinc linoleate. $Zn(C_{17}H_{31}COO)_2$. Brown solid containing 8.5–9.5% zinc. Combustible.
Derivation: Precipitation from solutions of sodium linoleate and soluble zinc salt, or by fusion of the fatty acid and zinc oxide.
Use: Paint drier, especially with cobalt and manganese soaps.

zinc malate.
$Zn(OOCCH_2CHOHCOO) \cdot 3HOH$.
Properties: White, crystalline powder; soluble in water. Combustible.
Derivation: Action of malic acid on zinc hydroxide.

zinc-2-mercaptobenzothiazole. $Zn(C_7H_4NS_2)_2$.
Use: Rubber accelerator, fungicide.

zinc molybdate. $ZnMoO_4 \cdot 2HOH$.
Properties: Solid, d 3.3, mp 1650C, insoluble in water.
Use: Anticorrosion agent, starting material for growing single crystals.

zinc naphthenate. $Zn(C_6H_5COO)_2$.
Properties: Amber, viscous, basic liquid or basic solid; the liquid contains 8–10% zinc, the solid contains 16% zinc; very soluble in acetone. Combustible.
Derivation: Fusion of zinc oxide or hydroxide and naphthenic acid or precipitation from mixture of soluble zinc salts and sodium naphthenate.
Use: Drier and wetting agent in paints, varnishes, resins; insecticide, fungicide, and mildew preventive; wood preservative, waterproofing textiles, insulating materials.

zinc nitrate. CAS: 7779-88-6.
$Zn(NO_3)_2 \cdot 6HOH$.
Properties: Colorless lumps or crystals, soluble in water and alcohol, d 2.065 (13C), mp 36.4C, loses water of crystallization between 105 and 131C.
Derivation: Action of nitric acid on zinc or zinc oxide.
Hazard: Dangerous fire and explosion risk, strong oxidizing agent.
Use: Acidic catalyst, latex coagulant, reagent, intermediate, mordant.

zinc octanoate. See zinc caprylate.

zinc octoate. (zinc-2-ethylhexoate).
$Zn(OOCCH(C_2H_5)C_4H_9)_2$.
Properties: Light straw-colored, viscous liquid; d 1.16. Insoluble in water, soluble in hydrocarbon solvents. Combustible.
Use: Catalyst.

zinc oleate. CAS: 557-07-3.
$Zn(C_{17}H_{33}COO)_2$.
Properties: Dry, white to tan, greasy, granular powder containing 8.5–10.5% zinc; mp 70C; soluble in alcohol, ether, carbon disulfide, ligroin; insoluble in water. Combustible.
Derivation: Interaction of solutions of zinc acetate and sodium oleate, or by fusion of zinc oxide and oleic acid.
Use: Paints, resins and varnishes (drier).

"Zincon."[169] TM for 2-carboxy-2'-hydroxy-5'-sulfoformazylbenzene.
Use: In colorimetric determination of zinc and copper.

zinc orthoarsenate. See zinc arsenate.

zinc orthophosphate. See zinc phosphate.

zinc orthosilicate. See zinc silicate.

zinc oxalate. CAS: 547-68-2.
$ZnC_2O_4 \cdot 2HOH$.
Properties: White powder, soluble in acids and alkalies, slightly soluble in water, d 2.562 (24C), mp 100C (decomposes), combustible.

Derivation: Interaction of zinc sulfate and sodium oxalate.

Use: Zinc oxide, organic synthesis.

zinc oxide. (Chinese white; zinc white).
CAS: 1314-13-2.   ZnO.
Properties: Coarse white or grayish powder, odorless, bitter taste, absorbs carbon dioxide from the air, has greatest UV absorption of all commercial pigments, d 5.47, mp 1975C, soluble in acids and alkalies, insoluble in water and alcohol. Noncombustible.
Derivation: (a) Oxidation of vaporized pure zinc (French process), (b) roasting of zinc oxide ore (franklinite) with coal and subsequent oxidation with air, (c) similar treatment starting with other ores, (d) oxidation of vapor-fractionated die castings.
Grade: American process, lead-free; French process, lead-free, green seal, red seal, white seal (according to fineness); leaded (white lead sulfate); USP; single crystals.
Hazard: zinc oxide fume is harmful by inhalation. Zinc oxide powder reacts violently with chlorinated rubber at 215C. TLV (fume): 5 mg/m³ in air.
Use: Accelerator activator, pigment and reinforcing agent in rubber, ointments, pigment and mold-growth inhibitor in paints, UV absorber in plastics, ceramics, floor tile, glass, zinc salts, feed additive, dietary supplement, seed treatment, cosmetics, photoconductor in office copying machines and in color photography, piezoelectric devices, artists' colorant.

zinc oxychloride. A saturated solution of zinc chloride and zinc oxide.
Use: Dentistry.

zinc palmitate. $Zn(C_{16}H_{31}O_2)_2$.
Properties: White, amorphous powder; d 1.121; mp 100C; insoluble in water and alcohol; slightly soluble in benzene and toluene. Combustible.
Use: Flatting agent in lacquer, pigment suspending agent for paints, rubber compounding, lubricant in plastics.

zinc perborate. $Zn(BO_3)_2$ with water of hydration.
Properties: Amorphous white powder, insoluble in water but slowly decomposed by it, liberating hydrogen peroxide.
Derivation: Interaction of sodium peroxide, boric acid, and zinc salt, or of boric acid and zinc peroxide.
Hazard: Fire risk when wet, in contact with organic materials.
Use: Medicine, oxidizing agent.

zinc permanganate. CAS: 23414-72-4.
$Zn(MnO_4)_2 \cdot 6HOH$.
Properties: Violet-brown or black, hygroscopic crystals, d 2.47, loses 5HOH at 100C, decomposes on exposure to light and air, soluble in water and acids, decomposes in alcohol.
Grade: Technical (95% pure).
Hazard: Dangerous fire risk in contact with organic materials, strong oxidizing agent.
Use: Oxidizing agent, medicine (antiseptic).

zinc peroxide. (zinc dioxide).
CAS: 1314-22-3. ZnO_2.
Properties: White powder containing 45–60% ZnO_2, balance ZnO; d 1.571; decomposes rapidly above 150C; decomposes in acids, alcohol, acetone; insoluble in water but decomposed by it.
Derivation: Action of barium peroxide on zinc sulfate solution, followed by filtration.
Grade: USP (mixture of peroxide, carbonate, and hydroxide), technical 50–60%.
Hazard: Severe explosion risk when heated; explosive range 190–212C. Fire risk in contact with organic materials; strong oxidizing agent.
Use: Curative for rubber and elastomers, pharmaceuticals, high-temperature oxidation.

zinc phenate. (zinc carbolate; zinc phenolate).
$Zn(C_6H_5O)_2$.   (May be only a mixture of zinc oxide and phenol).
Properties: White powder, soluble in alcohol, slightly soluble in water. Combustible.
Derivation: By heating zinc hydroxide with phenol and extracting with alcohol.
Hazard: Toxic by ingestion.
Use: Insecticide.

zinc-1,4-phenolsulfonate. (zinc sulfophenate; zinc sulfocarbolate).   CAS: 127-82-2.
$Zn(SO_3C_6H_4OH)_2 \cdot 8HOH$.
Properties: Colorless, transparent crystals or white granular powder; odorless; astringent metallic taste; effloresces in air; turns pink on exposure to air and light; loses water of crystallization at 120C; soluble in water and alcohol.
Derivation: By heating zinc hydroxide with p-phenolsulfonic acid.
Grade: Technical.
Hazard: Toxic by ingestion.
Use: Insecticide, medicine (antiseptic).

zinc phosphate. (zinc orthophosphate; zinc phosphate, tribasic).   CAS: 7779-90-0.
$Zn_3(PO_4)_2$.
Properties: White powder, soluble in acids and ammonium hydroxide, insoluble in water, d 3.998 (15C), mp 900C.
Derivation: Interaction of zinc sulfate and trisodium phosphate.

Grade: Technical, 98% pure.
Use: Dental cements, phosphors, conversion coating of steel, aluminum and other metal surfaces.

zinc phosphide.   CAS: 1314-84-7. Zn_3P_2.
Properties: Dark gray, gritty powder. D 4.55 (15C), mp above 420C, stable if dry, insoluble in alcohol, soluble in acids, decomposes in water.
Derivation: By passing phosphine into a solution of zinc sulfate.
Grade: Technical, 80–85% pure.
Hazard: Reacts violently with oxidizing agents, produces toxic and flammable phosphine by reaction with acids. Toxic by ingestion. A deadly poison.
Use: Rodenticide.

zinc potassium chromate.   See zinc yellow.

zinc potassium iodide.   See potassium zinc iodide.

zinc propionate.   CAS: 557-28-8.
$Zn(OOCC_2H_5)_2$.
Properties: Platelets, tablets, or needlelike crystals; fairly soluble in water, slightly soluble in alcohol, decomposes in moist atmosphere liberating propionic acid. Combustible.
Derivation: By dissolving zinc oxide in dilute propionic acid and concentrating the solution.
Use: Fungicide on adhesive tape.

zinc-1,2-propylene bisdithiocarbamate.
See propineb.

zinc pyrophosphate.   CAS: 7446-26-6.
$Zn_2P_2O_7$.
Properties: White powder, d 3.756 (23C), soluble in acids and alkalies, insoluble in water.
Use: Pigment.

zinc resinate.
Properties: Powder, clear amber lumps, or yellowish liquid; may be acid, basic, or neutral; soluble in some organic solvents (ether, amyl alcohol). Combustible.
Chief constituent: zinc abietate. Derivation: By fusion of zinc oxide and rosin, or by precipitation from solutions of zinc salts and sodium resinate.
Use: Wetting, dispersing, and hardening agent; drier in paints, varnishes, and resins.

zinc rhodanide.   See zinc thiocyanate.

zinc ricinoleate.
$Zn[CH_3CH_2)_5CHOHCH_2CH:CH(CH_2)_7CO_2]_2$.
Properties: Fine white powder with faint fatty acid odor, mp 92–95C, d 1.10 (25/25C). Combustible.
Use: Fungicide, emulsifier, greases, lubricants, waterproofing, lubricating-oil additive, stabilizer in vinyl compounds.

"Zincrometal."[579]   TM for a precoated, highly corrosion-resistant steel designed chiefly for use in automobile bodies. The coating is stated to be comprised of two parts: an undercoat of zinc-chromium solution and a topcoat of zinc-rich, epoxy-based resin. The combination of coatings is applied to the steel before shaping and other mechanical processing. The effectiveness of this product is said to be greater than that of other anticorrosion systems.

zinc salicylate.   CAS: 16283-36-6.
$Zn[C_6H_4(OH)COO]_2 \cdot 3HOH$.
Properties: White crystalline needles or powder; soluble in water and alcohol. Combustible.
Derivation: By heating zinc hydroxide and salicylic acid.
Use: Medicine (antiseptic).

zinc selenide.   CAS: 1315-09-9.   ZnSe.
Properties: Yellowish to reddish crystals, d 5.42 (15/4C), mp above 1100C, insoluble in water.
Hazard: Fire risk in contact with water or acids.
Use: Windows in infrared optical equipment, phosphor.

zinc silicate.   (zinc orthosilicate).
CAS: 13597-65-4. Zn_2SiO_4.
Properties: White crystals, d 4.103, mp 1509C, insoluble in water.
Use: Phosphors, spray ingredients, to remove traces of copper from gasoline.
See also willemite.

zinc silicofluoride.   See zinc fluorosilicate.

zinc-silver oxide battery.
Primary or secondary battery used where space and weight are critical, i.e., in missiles. The battery has large energy output for its weight, but the components are expensive and the cycle life is short. To avoid deterioration, potassium hydroxide electrolyte is added just before use.
See also battery.

zinc stearate.   CAS: 557-05-1.
$Zn(C_{18}H_{35}O_2)_2$.   Percentage of zinc may vary according to intended use, some products being more basic than others.
Properties: (pure substance) White, hydrophobic powder free from grittiness; faint odor; d 1.095; mp 130C; soluble in acids; soluble in common solvents when hot; insoluble in water, alcohol, and ether. Combustible.
Derivation: Action of sodium stearate on solution of zinc sulfate.
Grade: USP, technical, available free from chick edema factor.
Use: Cosmetics, lacquers, ointments; dusting pow-

der, lubricant, mold-release agent, filler, anti-foamer, heat and light stabilizer, medicine (dermatitis), tablet manufacture, dietary supplement.

zinc sulfate. (white vitriol; white copperas; zinc vitriol).   CAS: 7733-02-0. $ZnSO_4 \cdot 7HOH$.
Properties: Colorless crystals, small needles, or granular, crystalline powder without odor; astringent, metallic taste; efflorescent in air; solutions acid to litmus; d 1.957 (25/4C); mp 100C; loses 7HOH at 280C; soluble in water and glycerol; insoluble in alcohol.
Derivation: (a) Roasting zinc blende and lixiviating with subsequent purification, (b) action of sulfuric acid on zinc or zinc oxide.
Grade: Technical, USP, reagent.
Use: Rayon manufacture, dietary supplement, animal feeds, mordant, wood preservative, analytical reagent.

zinc sulfate monohydrate. $ZnSO_4 \cdot HOH$.
Properties: White, free-flowing powder; soluble in water; insoluble in alcohol.
Use: Rayon manufacture, agricultural sprays, chemical intermediate, dyestuffs, electroplating.

zinc sulfide. CAS: 1314-98-3.   ZnS.   Exists in two crystalline forms, alpha (wurtzite) and beta (sphalerite).
Properties: Yellowish-white powder, stable if kept dry. Alpha: d 3.98. Beta: d 4.102, changes to alpha form at 1020C. Sublimes at 1180C, soluble in acids, insoluble in water.
Derivation: By passing hydrogen sulfide gas into a solution of a zinc salt.
Grade: Technical, CP, fluorescent or luminous, single crystals.
Use: Pigment, white and opaque glass, base for color lakes, rubber, plastics, dyeing (hydrosulfite process), ingredient of lithopone, phosphor in x-ray and television screens, luminous paints, fungicide.

zinc sulfite. $ZnSO_3 \cdot 2HOH$.
Properties: White, crystalline powder; absorbs oxygen from the air to form sulfate; loses 2HOH at 100C; decomposes at 200C. Soluble in sulfurous acid, insoluble in cold water and alcohol, decomposes in hot water.
Derivation: Action of sulfurous acid on zinc hydroxide.
Use: Preservative for anatomical specimens.

zinc sulfocarbolate.   See zinc phenolsulfonate.

zinc sulfocyanate.   See zinc thiocyanate.

zinc sulfophenate.   See zinc phenolsulfonate.

zinc sulfoxylate. $ZnSO_2$.
Properties: White, crystalline powder; decomposed by heat; salt of unstable sulfoxylic acid; H_2SO_2; a strong reducing agent.
Derivation: Action of zinc and sulfuryl choride in ethereal solution, or of sulfur dioxide on granulated zinc in absolute alcohol.
Use: Stripping agent in dyeing.

zinc telluride. CAS: 1315-11-3.   ZnTe.
Properties: Reddish crystals, d 6.34 (15C), mp 1238C, decomposes in water, single crystals available for phosphors.
Derivation: Reaction of zinc oxide and tellurium powder in alkaline solution.
Use: Semiconductor research, photoconductor.

zinc thiocyanate. (zinc rhodanide; zinc sulfocyanate).   CAS: 557-42-6. $Zn(CNS)_2$.
Properties: White, hygroscopic powder or crystals; soluble in water, alcohol, and ammonium hydroxide.
Derivation: Interaction of zinc hydroxide and ammonium thiocyanate.
Grade: Technical, solution, reagent, ACS.
Use: Analytical chemistry, swelling agent for cellulose esters, dyeing assistant.

zinc thiophenate. A coined name for a class of peptizing agents for natural and synthetic rubbers; a typical example is the tert-butylphenyl sulfide $[(CH_3)_3CC_6H_4S]_2Zn$.

zinc undecylenate. $[CH_2{:}CH(CH_2)_8COO]_2Zn$.
Properties: White, amorphous powder; nearly insoluble in water and alcohol; mp 115–116C. Combustible.
Grade: NF.
Use: Medicine (fungistat), cosmetics, chemical intermediate.

zinc white.   See zinc oxide.

zinc yellow. (citron yellow; buttercup yellow; zinc potassium chromate; zinc chrome).
CAS: 11103-86-9.
$4ZnO \cdot 4CrO_3 \cdot K_2O \cdot 3HOH$.
Properties: Greenish-yellow pigment of comparatively low tinting strength, partially water-soluble.
Derivation: Reaction of a solution of potassium dichromate with zinc oxide and sulfuric acid.
Hazard: Toxic by ingestion.
Use: Rust-inhibiting paints, artists' color.

zinc zirconium silicate. $ZnO \cdot ZrO_2 \cdot SiO_2$.
Properties: White powder, d 4.8, bulk d 115 lbs/cu ft, mp 2100C, soluble in hydrogen fluoride, insoluble in water and alkalies, slightly soluble

in mineral acids and hot concentrated sulfuric acid. Noncombustible.
Use: Opacifier for ceramic glazes.

zineb. (zinc ethylenebis[dithiocarbamate]).
CAS: 12122-67-7. $Zn(CS_2NHCH_2)_2$.
Properties: Light-tan solid, insoluble in water, soluble in pyridine, decomposes on heating.
Derivation: Reaction of sodium ethylenebisdithiocarbamate with zinc sulfate or other zinc salts. In practical application as a fungicide these reactants are mixed in the presence of lime; the zineb is not formed until after reaction of the carbon dioxide of the air with the film of the other chemicals on the leaf or fruit.
Grade: Commercial dusts and wettable powders usually contain 65% active material.
Hazard: Toxic by inhalation and ingestion; irritant to eyes and mucous membraness.
Use: Insecticide and fungicide.

zingerone. (4-[4-hydroxy-3-methoxyphenyl]-2-butanone). CAS: 122-48-5.
$HOC_6H_3(OCH_3)CH_2CH_2COCH_3$.
Properties: Crystals, mp 40–41C, soluble in ether, sparingly soluble in water and petroleum ether.
Use: Flavoring.

"Zinol."[79] TM for a special zinc resinate in solution in mineral spirits.
Hazard: Flammable, moderate fire risk.
Use: Printing ink, paint and varnish.

"Zin-O-Lyte."[28] TM for a series of electroplating products for use in zinc cyanide plating baths.

"Zinophos."[57] TM for a soil insecticide and nematocide whose active ingredient is thionazin.

"Zinros."[79] TM for a pale, high-melting zinc resinate.
Use: Adhesives, printing ink, rubber compounding.

"Zinstabe."[268] TM for a series of zinc-based products used as heat stabilizers for PVC foams and as activators for azodicarbonamide and similar chemical blowing agents for PVC and rubber foams.

ziram. (zinc dimethyldithiocarbamate).
CAS: 137-30-4. $Zn(SCSNCH_3CH_3)_2$.
Properties: White and odorless when pure; d 1.71; mp 246C; almost insoluble in water; soluble in acetone, carbon disulfide, chloroform, in dilute alkalies and concentrated hydrochloric acid.
Derivation: Reaction of sodium dimethyldithiocarbamate with a soluble zinc salt in aqueous solution.

Grade: 76% wettable powder, 90% technical powder.
Hazard: Strong irritant to eyes and mucous membranes.
Use: Fungicide, rubber accelerator.

"Zircaloy." TM for alloys of zirconium with low percentages of antimony, iron, chromium, and nickel.
Use: Cladding for nuclear fuel elements and other reactor applications.

"Zirco."[230] TM for an oil-soluble polymeric zirconyl complex in odorless mineral spirits. Not a paint drier in itself, has synergistic action on metallic driers.

"Zircofrax."[280] TM for super-refractory products from zirconium oxide and zirconium silicate.
Properties: High heat resistance, great strength, high thermal conductivity, high resistance to attack by acids and acid slags, porosity about 25%, permeability low.
Use: Bricks and special shapes for ceramic kiln furniture and in chemical and metallurgical furnaces.

zircon. CAS: 14940-68-2. $ZrSiO_4$ or $ZrO_2 \cdot SiO_2$. A natural zirconium silicate, represents 60–70% of all zirconium used.
Properties: Color brown, gray, red, colorless; luster adamantine; hardness 7.5; d 4.68; insoluble in acids.
Occurrence: Georgia, Florida, Australia, Brazil.
Use: Source of zirconium oxide, metallic zirconium, and hafnium; abrasive; refractories; enamels; refractory porcelain; catalyst; silicone rubbers; foundry cores.

"Zircon H-W."[446] TM for a compound made from the purified mineral by impact pressing, or by air ramming or slip casting.
Properties: High density (230 lbs/cu ft), helps resist the wetting and penetration of molten glass. Resistant also to thermal spalling, fluxing conditions; has constancy of volume under soaking heat in excess of 1593C.
Use: To pave glass tank bottoms and floors, line sodium metaphosphate and sodium silicate furnaces; tap hole blocks for aluminum and nonferrous melting, nozzles for continuous steel casting.

zirconia. See zirconium oxide.

zirconic anhydride. See zirconium oxide.

"Zirconite."[337] TM for foundry materials containing 97–98% zircon minimum, mp 2220–2260C, available as sand and flour.
Use: Foundries for cores, molds, and facings; re-

duces cleaning costs and produces smoother castings.

zirconium. CAS: 7440-67-7. Zr.
Metallic element of atomic number 40, group IVB of the periodic table, aw 91.22, valences = 2, 3 (halogens only), 4, five stable isotopes.
Properties: Hard, lustrous, grayish, crystalline scales or gray amorphous powder; d 6.4; mp 1850C; bp 4377C; soluble in hot, very concentrated acids; insoluble in water and cold acids. Corrosion-resistant, low neutron absorption.
Sources: Zircon, baddeleyite (zirconia).
Derivation: The ore is converted to a cyanonitride and is chlorinated to obtain zirconium tetrachloride. This is reduced with magnesium (Kroll process) in inert atmosphere. The metal can be prepared in a highly pure and ductile form by vapor phase decomposition of the tetraiodide. Hafnium must be removed for uses in nuclear reactors See hafnium.
Grade: Plate, strip, bars, wire, sponge and briquettes, powder, foil, technical, pure (hafnium-free), single crystals.
Hazard: Flammable and explosive as dust or powder, and in form of borings, shavings, etc. A suspected carcinogen. Not permitted in cosmetics (FDA). TLV (as Zr): 5 mg/m^3 of air (applies to all Zr compounds). Powder should be kept wet in storage and protective clothing should be worn.
Use: Coating nuclear fuel rods, corrosion-resistant alloys, photoflash bulbs (foil), pyrotechnics, metal-to-glass seals, special welding fluxes, getter in vacuum tubes, explosive primers, acid manufacturing plants, deoxidizer and scavenger in steel manufacturer, lab crucibles, spinnerets.

zirconium-95. Radioactive zirconium of mass number 95.
Properties: Half-life 63 days, radiation beta and gamma.
Derivation: Obtained in a mixture with niobium from the fission products of nuclear reactor fuels.
Forms available: Zirconium oxalate complex in oxalic acid solution.
Hazard: Radioactive poison.
Use: To trace the flow of petroleum products in pipelines, to measure rate of catalyst circulation in petroleum cracking plants, to study the cracking and polymerization of hydrocarbons with various catalysts, etc.

zirconium acetate. H$_2$ZrO$_2$(C$_2$H$_3$O$_2$)$_2$.
Properties: (a) Available as aqueous solution, 22% ZrO$_2$. Clear to pale amber liquid, d 1.46, pH 3.8–4.2 (20C), fp −7C, stable at room temperature. (b) Available as 13% ZrO$_2$ (aqueous solution), d 1.20 (approx.), pH 3.3–4.0 (20C), stable

at room temperature but temperature of hydrolysis decreases with pH, undergoes exchange with anion exchange resins but not with cation exchangers.
Hazard: TLV (as Zr): 5 mg/m^3 of air.
Use: Preparation of water repellents, other chemicals.

zirconium acetylacetonate. See zirconium tetraacetylacetonate.

zirconium ammonium fluoride. (ammonium zirconifluoride). Zr(NH$_4$)$_2$F$_6$.
Properties: White crystals, soluble in water.
Hazard: Irritant. TLV (as Zr): 5 mg/m^3 of air.

zirconium anhydride. See zirconium oxide.

zirconium boride. (zirconium diboride).
ZrB$_2$.
Properties: Gray metallic crystals or powders, d 6.085, mp 3000C, Mohs hardness 8, electrical resistivity 9.2 micro-ohm cm at 20C, excellent thermal shock resistance, poor oxidation resistance above 1100C.
Hazard: TLV (as Zr): 5 mg/m^3 of air. Toxic.
Use: Refractory for aircraft and rocket applications, thermocouple protection tubes, high temperature electrical conductor, cutting tool component, coating tantalum, cathode in high-temperature electrochemical systems.

zirconium carbide. ZrC.
Properties: Gray, crystalline solid; d 6.78; hardness 8+; mp 3400C; bp 5100C; insoluble in water and hydrochloric acid; soluble in oxidizing acids and attacked by oxidizers.
Derivation: By heating zirconium oxide and coke in an electric furnace.
Grade: Technical.
Hazard: Powder or dust will ignite spontaneously. TLV (as Zr): 5 mg/m^3 of air.
Use: Incandescent filaments, abrasive, cermet component, high temperature electrical conductor, refractory, metal cladding, cutting tool component.

zirconium carbonate, basic. (zirconyl carbonate; zirconium carbonate). ZrOCO$_3$ or ZrOCO$_3$·xHOH.
Properties: White, amorphous powder; soluble in acids; insoluble in water.
Derivation: By adding sodium carbonate to a solution of zirconium salt.
Hazard: TLV (as Zr): 5 mg/m^3 of air.
Use: Preparation of zirconium oxide.

zirconium chloride. See zirconium tetrachloride.

zirconium chloride, basic. See zirconium oxychloride.

zirconium diboride. See zirconium boride.

zirconium dioxide. See zirconium oxide.

zirconium disilicide. (zirconium silicide). $ZrSi_2$.
Properties: Gray solid, d 4.88 (22C), soluble in hydrogen fluoride, insoluble in water and aqua regia.
Hazard: TLV (as Zr): 5 mg/m³ of air.
Use: Coatings resistant to flame or blast impingement, special alloys.

zirconium disulfide. ZrS_2.
Properties: Gray, crystalline solid; d 3.87; mp 1550C; insoluble in water.
Use: A solid lubricant.
Hazard: TLV (as Zr): 5 mg/m³ of air.

zirconium fluoride. See zirconium tetrafluoride.

zirconium glycolate. $H_2ZrO(C_2H_2O_3)_3$.
Properties: Solid, decomposes without melting on heating to 220C, insoluble in water and organic solvents, soluble in alkali and sulfuric acid solutions. One or more of the acidic hydrogens may be replaced by alkali metals or ammonium to give water-soluble salts.
Hazard: TLV (as Zr): 5 mg/m³ of air.
Use: Cosmetic (deodorant), medicine, sequestrant.

zirconium hydride. CAS: 7704-99-6. ZrH_2.
Contains 1.7–2.1% combined hydrogen which can be driven off in a vacuum above 600C.
Properties: Gray-black metallic powder, stable toward air and water, d 5.6, autoign temperature 518F (270C).
Derivation: Reduction of zirconia with calcium hydride or magnesium in the presence of hydrogen, direct combination of hydrogen and zirconium.
Grade: Commercial (contains hafnium), reactor (hafnium-free), electronic.
Hazard: Flammable, dangerous fire risk, especially in presence of oxidizers. TLV (as Zr): 5 mg/m³ of air.
Use: Vacuum-tube getter, powder metallurgy, source of hydrogen, metal-foaming agent, nuclear moderator, reducing agent, hydrogenation catalyst.

zirconium hydroxide. CAS: 14475-63-9. $Zr(OH)_4$.
Properties: White, bulky, amorphous powder; soluble in dilute mineral acids; insoluble in water and alkalies; d 3.25; decomposes to ZrO_2 at 550C.
Derivation: Action of sodium hydroxide (solution) on a zirconium salt solution.
Hazard: TLV (as Zr): 5 mg/m³ of air.

Use: Source of zirconium oxide and zirconium sulfate, glass colorants.

zirconium hydroxychloride solution. See zirconyl hydroxychloride solution.

zirconium lactate. CAS: 63919-14-2. $H_4ZrO(CH_3CHOCO_2)_3$.
Properties: White, slightly moist pulp; decomposes without melting; very slightly soluble in water and the common organic solvents; soluble in aqueous alkalies with formation of salts, decomposes to hydrous zirconia above pH 10.5, efficient odor absorber. Combustible.
Grade: Zirconia 25% (min).
Hazard: TLV (as Zr): 5 mg/m³ of air.
Use: Body deodorants, source of zirconia.

zirconium naphthenate.
Properties: Amber-colored, heavy, transparent liquid, viscosity equivalent to that of heavy lubricating oil, very stable, unlike other metallic naphthenates, possesses no drying properties, soluble in all common solvents, d 1.05. Combustible.
Derivation: By heating a mixture of naphthenic acid and zirconium sulfate.
Hazard: TLV (as Zr): 5 mg/m³ of air.
Use: Ceramics (enamels, glazes); lubricants; paints and varnishes (anti-chalking agent, minimizer of moisture and solar radiation effects).

zirconium nitrate. CAS: 13746-89-9. $Zr(NO_3)_2 \cdot 5HOH$.
Properties: White, hygroscopic crystals; water and alcohol soluble; decomposes at 100C.
Derivation: Action of nitric acid on zirconium oxide.
Hazard: Dangerous fire and explosion risk in contact with organics, strong oxidizing agent. Hazard: TLV: (as Zr) 5 mg/m³ of air.
Use: Preservative.

zirconium nitride. ZrN.   A brassy-colored powder produced by heating the metal in nitrogen.
Properties: Hardness Mohs 8+, d 7.09, mp 2930C, slightly soluble in dilute hydrochloric acid and sulfuric acid, soluble in concentrated acids.
Hazard: TLV (as Zr): 5 mg/m³ of air.
Use: Special crucibles, cermets, refractories.

zirconium orthophosphate. See zirconium phosphate.

zirconium oxide. (zirconia; zirconium dioxide; zirconic anhydride; zirconium anhydride).
CAS: 1314-23-4. ZrO_2.   Occurs in nature as baddeleyite.
Properties: Heavy, white, amorphous powder; d 5.73; mp 2700C; Mohs hardness 6.5; refractive

index 2.2. Insoluble in water and most acids or alkalies at room temperature, soluble in nitric acid and hot concentrated hydrochloric, hydrofluoric, and sulfuric acids. Most heat-resistant of commercial refractories; dielectric.
Derivation: By heating zirconium hydroxide or zirconium carbonate.
Grade: Reagent, technical, crystals, fused, whiskers, CP (99% zirconia), hydrous. The fused grade is reported to be harder than diamond (11 on Mohs scale).
Hazard: TLV (as Zr): 5 mg/m³ of air.
Use: (Unstabilized) production of piezoelectric crystals, high-frequency induction coils, colored ceramic glazes, special glasses, source of zirconium metal, heat-resistant fibers, (hydrous) odor absorbent, to cure dermatitis caused by poison ivy. (Stabilized with CaO) refractory furnace linings, crucibles, etc., solid electrolyte for batteries operating at high temperature.

zirconium oxychloride. (zirconium chloride, basic; zirconyl chloride). $ZrOCl_2 \cdot 8HOH$.
Properties: White, silky crystals; loses 6HOH at 150C; 8HOH at 210C; density 44 lb/cu ft; soluble in water, methanol, and ethanol; insoluble in other organic solvents; aqueous solutions are acidic.
Derivation: Action of hydrochloric acid on zirconium oxide.
Grade: Technical, 36% ZrO_2, HP.
Use: Textile, cosmetic and grease additive, antiperspirant, water repellents, chemical reagent, zirconium salts, in lakes and toners of acid and basic dyes, oil-field acidizing aid.

zirconium phosphate. (zirconium phosphate, basic; zirconium orthophosphate).
$ZrO(H_2PO_4)_2 \cdot 3HOH$.
Properties: White, dense, amorphous powder; decomposes on heating. Soluble in acids, insoluble in water and organic solvents, extensively hydrolyzed in basic solution.
Derivation: Action of phosphoric acid on zirconium hydroxide.
Hazard: TLV (as Zr): 5 mg/m³ of air.
Use: Chemical reagent, cation scavenger, coagulant, carrier for radioactive phosphorus.

zirconium potassium chloride. (potassium zirconium chloride). $ZrCl_4 \cdot KCl$. A source of zirconium for magnesium alloys, to remove iron in an insoluble form.

zirconium potassium fluoride. (potassium fluozirconate; potassium zirconifluoride).
CAS: 16923-95-8. ZrK_2F_6.
Properties: White crystals, soluble in water (hot).
Hazard: Irritant. TLV (as F): 2.5 mg/m³ of air.

Use: Grain refiner in magnesium and aluminum, welding fluxes, catalyst, optical glass.

zirconium potassium sulfate. (potassium zirconium sulfate). $2K_2SO_4 \cdot Zr(SO_4)_2 \cdot 3HOH$.
Properties: White, crystalline powder; slightly soluble in water.
Hazard: TLV: (as Zr) 5 mg/m³ in air.

zirconium pyrophosphate. ZrP_2O_7.
Properties: White solid, stable to 1550C, insoluble in water and dilute acids other than hydrogen fluoride, coefficient of thermal expansion 5 × 10⁻⁶ at 1000C.
Hazard: TLV (as Zr): 5 mg/m³ in air.
Use: Refractory, olefin polymerization catalyst, phosphor.

zirconium silicate. See zircon.

zirconium silicide. See zirconium disilicide.

"Zirconium Spinel."[337] Trade designation for a synthetic complex containing 39–41% zirconium oxide, 20–22% silicon dioxide, 18.5–20.5% aluminum oxide and 17–21% zinc oxide; mp 1704C.
Use: Glaze opacifier in the ceramic industry.

zirconium sulfate. CAS: 14644-61-2. $Zr(SO_4)_2 \cdot 4HOH$.
Properties: White, crystalline powder; bulk density 70 lb/cu ft; decomposes to monohydrate at 100C. Soluble in water, slightly soluble in alcohol, insoluble in hydrocarbons. Aqueous solutions are strongly acidic, will precipitate potassium ions and amino acids from solution, are decomposed by bases and heat.
Derivation: Action of sulfuric acid on zirconium hydroxide.
Hazard: TLV (as Zr): 5 mg/m³ in air.
Use: Chemical reagent, lubricants, catalyst support, protein precipitation, tanning of white leather.

zirconium sulfate, basic. (zirconyl sulfate).
$Zr_5O_8(SO_4)_2 \cdot xHOH$. Similar in properties to the oxychloride and is prepared in a similar fashion, the end result being in cake form.
Use: Textile treatment and white leather tanning and retannage.

zirconium tetraacetylacetonate. (zirconium acetylacetonate). $Zn[OC(CH_3):CHCO(CH_3)]_4$.
Properties: A colorless, crystalline tetrachelate; d 1.415; mp 194–5C (decomposition begins at 125C); soluble in pyridine, acetone, benzene, and other organic solvents having some polarity; slightly soluble in water.
Derivation: Reaction between zirconyl chloride, acetylacetone and sodium carbonate.

Hazard: TLV (as Zr): 5 mg/m³ in air.
Use: Crosslinking agent for polyol, polyester, and polyalkyloxy resins; lubricant and grease additive; reagent; catalyst.

zirconium tetrachloride. (zirconium chloride). CAS: 10026-11-6. $ZrCl_4$.
Properties: White, lustrous crystals; soluble in alcohol; decomposes in water; d 2.8; sublimes above 300C.
Derivation: Action of hydrochloric acid on zirconium hydroxide.
Grade: Technical.
Hazard: Irritant. TLV (as Zr): 5 mg/m³ in air.
Use: Source of the pure metal (formed as intermediate in process), analytical chemistry, water repellents for textiles, tanning agent, zirconium compounds, special catalysts of the Friedel-Crafts and Ziegler types.

zirconium tetrafluoride. (zirconium fluoride). CAS: 7783-64-4. ZrF_4.
Properties: White powder, d 4.43, mp 600C (sublimes), slightly soluble in water and hydrogen fluoride.
Hazard: Strong irritant. TLV (as F): 2.5 mg/m³ of air.
Use: Component of molten salts for nuclear reactors.

zirconocene dichloride. (dicyclopentadienylzirconium dichloride). $(C_5H_5)_2ZrCl_2$.
Properties: White crystals, mp 244C, soluble in polar organic solvents, stable in dry air, very slowly hydrolyzes in moist air.
Hazard: Toxic by inhalation and skin contact, irritant to eyes and mucous membranes.
Use: Rubber accelerator, component of a catalyst system for polymerization of vinyl monomers, curing agent for water-repellent silicone materials, agent for plating with zirconium.
See also metallocene.

zircon, porcelain. See porcelain, zircon.

zircon sand. A sand containing considerable zirconium, titanium, and related metals.
Use: Source of these elements and also in foundries for casting of alloys.

zirconyl acetate. CAS: 14311-93-4. $Zr(OH)_2(C_2H_3O_2)_2$.
Properties: (22% ZrO_2 solution) Tacky, resinous, amorphous mass; d 1.46; fp −7C; becomes solid on heating.
Derivation: Addition of acetic acid to water suspension of carbonated zirconia.
Use: Waterproofing textiles, precipitation of proteins, starch, etc., for textile and paper coatings.

zirconyl carbonate. See zirconium carbonate, basic.

zirconyl chloride. See zirconium oxychloride.

zirconyl hydroxychloride. CAS: 10119-31-0. $ZrOOHCl \cdot nHOH$.
Properties: Colorless or slightly amber liquid (aqueous solution), d 1.26, forms a soluble glass on evaporation, pH of solution 0.8, reacts with alkalies to form hydrous zirconia; contains 20% zirconia.
Use: Pharmaceuticals, deodorants, precipitation of acid dyes, water repellents for textiles.

zirconyl nitrate. (basic). (zirconyl hydroxynitrate). $ZrO(OH)NO_3$.
Properties: Aqueous solution, d 1.35 (25C).
Hazard: Fire risk in contact with organic materials.
Use: Gelatins and improving lamination bonds of polyvinyl alcohol.

zirconyl sulfate. See zirconium sulfate, basic.

"Zircopax."[337] TM for zirconium silicate, 94–96.5% min purity.
Use: Opacifier for ceramic glazes.

"Zircotan."[23] TM for zirconium tanning agents which produce through-white leather.
Use: Tannage of white kid suede, glove leathers, retannage of chrome leather.

"Zirmul."[408] TM for a bonded alumina-zirconia-silica refractory containing a minimum of glassy phase. It has broad applications in construction of glass melting furnaces, both above and below the glass line.

"Zirox" B.[337] TM for a polishing compound of zirconium oxide.
Use: Ophthalmic lenses, precision glass, marble and granite.

ZMA. Abbreviation for zinc metaarsenite. See zinc arsenite.

Zn. Symbol for zinc.

zoalene. (dinitolmide; 3,5-dinitro-o-toluamide). CAS: 148-01-6. $(O_2N)_2C_6H_2(CH_3)CONH_2$.
Properties: Yellowish solid; mp 177C; very slightly soluble in water; soluble in acetone, acetonitrile, dioxane, and dimethylformamide.
Use: Coccidiostat, food additive.

"Zoamix."[233] TM for poultry coccidiostats containing 3,5-dinitro-o-toluamide.

"Zobar."[28]   TM for a weed killer based on an aqueous solution of the dimethylamine salts of polychlorobenzoic acids, containing four pounds of acid equivalent per gallon.

"Zonarez."[252]   TM for a series of polyterpene resins. Thermoplastic polymers produced by the polymerization of terpene hydrocarbons consisting primarily of beta-pinene and dipentene.
Use: Manufacture of pressure-sensitive adhesives for tapes and labels, rubber cements, solvent-based adhesives, emulsion adhesives, hot-melt adhesives and coatings, can sealants, caulking and general sealant compounds, inks, investment casting waxes, paints, concrete waterproofing agents, varnishes, chewing and bubble gum bases, moisture resistant soft gelatin capsules and powders of ascorbic acid and its salts, and other applications which require low molecular weight, linear polymers that promote tack and adhesion and have excellent aging properties.

zone refining.   A purification process that involves repeated melting and crystallization. The sample to be purified is placed in a relatively long narrow tube and then passed slowly through a furnace having short, alternate hot and cold zones. Melting occurs opposite the hot zones and crystallization opposite the cold zones. As the rod moves through the furnace the zones move along the rod. Impurities remain in the molten zones and so are carried to one end of the rod. The process has been most used for relatively high cost materials used in small quantities at very high purities, as for solid state electronic purposes.

"Zonolite."[241]   TM for verxite (expanded hydrobiotite).

"Zonyl."[28]   TM for a fluorosurfactant wetting agent that is superior to hydrocarbon surfactants due to greater surface tension reduction.

Zsigmondy, Richard.   (1865–1929) A native of Austria, Zsigmondy received the Nobel Prize in chemistry in 1925 for his work in the field of colloid chemistry, which was initiated by his interest in ruby glass (a colloidal gold suspension). His most important contribution to chemistry was his invention of the ultramicroscope (with Siedentopf) in 1903.
See also ultramicroscope; Tyndall effect.

Zr.   Symbol for zirconium.

zwitterion.   See isoelectric point.

zymase.   The enzyme present in yeast which converts sugars to alcohol and carbon dioxide.

zymohexase.   See aldolase.

"Zymol."[221]   TM for a chemically modified linseed oil for part or complete replacement for tung oil in varnish manufacture.

zymose.   Enzyme found in yeast, hydrolyzes glucose to alcohol (fermentation).

"Zytel."[28]   TM for a nylon resin available as molding powder, extrusion powder, and soluble resin.

APPENDIX I

Origin of Some Chemical Terms

The etymology of chemical terms is not only of historical interest but is often of value in illuminating their meaning. Those listed here were selected chiefly on the basis of these two criteria. The names of many elements (tin, sulfur, gold, zinc, lead) are merely transliterations from older languages; others are in honor of famous scientists (curium, fermium, einsteinium) or of countries (germanium, francium, polonium); still others are from their place of discovery (berkelium, yttrium, strontium). The predominance of Latin (L.) and Greek (Gk.) origins and the many references to mythology reflect the classical background of the nineteenth century.

The dates given in parentheses indicate the approximate year in which the term was introduced. Many common chemical words were in literary use long before the development of modern chemistry in the late 1700s. For dates prior to 1750 the Editor is indebted to the **Oxford Universal Dictionary**; however, some of these (e.g., petroleum, 1526) appear open to question.

abrasive (1656) L. *ab*-(away) + *ras*-(scrape). From the same root as **erase**.

abscission L. *ab*-(from + *sciss*-(cut). To detach, as fruit from a tree. From the same root as **abscissa**.

absorption See **sorb**.

accelerator (1900) L. *ad*-(*ac*-) (to) + *celer*-(speed), i.e., an agent that adds speed to a reaction.

acetic (1808) L. *acetum* (vinegar), i.e., a dilute solution of acetic acid. **Acetone** is from the same root.

acid (1626) L. *acidum*, from *acer* (sour, sharp-tasting), one of the characteristic properties of acids. Acrid is from the same root.

adhesive (1670) L. *ad*-(to) + *haes*-(stick, cling), referring to two dissimilar materials. See also **coherent**.

adsorption See **sorb**.

agglomeration L. *ad*-(*ag*-) + *glomer*-(ball, sphere), i.e., a material made up of small, spherical aggregates.

aggregation (1547) L. *ad*-(to, together) + *greg*-(group, herd), i.e., of particulates. From the same root are **gregarious** (group-seeking) and **egregious** (out of the herd), i.e., noteworthy, outstanding in a deprecatory sense, as an egregious error.

albumin L. *alba* (white). In the 1600s this term (spelled albumen) referred to egg-white as well as to a substance found in plants and seeds. Modern use as a class of proteins began in 1869.

alchemy (1514) Arabic *al*(the) + Gk. *kemia*-(transmutation of lead to gold, which was the primary concern of the alchemists). See also **chemistry**.

alcohol (1543) Arabic *al* (the) + *koh'l*, a cosmetic paste (antimony sulfide) used to paint the eyelids, later the term referred to any fine metal powder. By some inexplicable semantics probably originating with the alchemists, it came to mean distilled spirits of wine (1753). The present sense originated about 1850.

aldehyde (1850) Coined from *al*(cohol) and *dehyd* (rogenation), describing the chemical derivation.

aliphatic Gk. *aliphatos* (oil, fat) from *aleiphein* (to rub or smear with oil). Originated with early chemists who were familiar with animal and vegetable fats whose properties were determined by high molecualar weight hydrocarbon groups.

alkali (1813) Arabic *al* (the) + *qalay* (pan-roast, calcine). As early as 1612 it referred to any soda-like substance obtained by calcining plants and reducing them to ashes.

alkaloid (1831) See **alkali**. So-called because of their superficial resemblance to alkalis, i.e., bitter taste.

aluminum L. *alumen* (alum). So named by Sir Humphry Davy in 1812.

aromatic (1869) L. *aroma* (spice, fragrance). Applied to benzenoid compounds derived from balsams because of their characteristic odor.

alkyd (1847) Coined from the words *al*(cohol) and *acid* (cid = kyd), describing their chemical derivation. First prepared by Berzelius.

amalgam (1471) L. *amalgma*, from Gk. malessein (a soft mixture of poultice); hence a mix softened with mercury.

ammonia (1799) From the name of the Egyptian deity Ammon to whom a temple was built in the North African desert. It refers to the gas evolved by camel excreta in the vicinity of the temple. The ammonium group NH_4 was named by Berzelius in 1808.

amorphous (1801) Gk. *a*-(not, without) +

morph-(shape), referring to a material that lacks crystal structure, usually a liquid.

amphoteric (1849) Gk. *ampho*-(both), i.e., both acidic and alkaline.

analysis (1581) Gk. *ana*-(up) + *lyein* (loosen, set free), i.e., to separate. It was applied to chemistry in 1655. **Dialysis** is from the same root.

anesthesia (1846) Gk. *an*-(not) + *aisthesis* (feeling, sense) This term was coined by Dr. Oliver Wendell Holmes soon after the first successful use of ether.

anhydride (1863) Gk. *an*-(not, without) + *hydr*-(water), in the sense of a compound that forms an acid on combining with water, e.g., SO_3 is the anhydride of H_2SO_4.

anhydrous (1819) Same roots as for anhydride, but used in the sense of absence of water of crystallization in a salt or similar compound.

aniline (1850) Arabic *anil* (indigo), from which aniline was originally made.

anthracene (1863) *anthrac*-(coal), from which it is obtained by distillation.

anthracite See **anthracene**.

antibiotic Term coined by Selman A. Waksman for mold-derived antibacterial agents.

argon (1894) Gk. *argos* (inactive), referring to the chemical nature of this element.

asbestos Gk. *a*-(not) + *sbestos* (quench, extinguish). The meaning as derived is the exact opposite of the unique property of this material.

asphalt (1650?) Gk. *asphaltos*, used in ancient times as an illuminant.

atom Gk. *a*) + *tom* (divide, cut) The original use goes back to the Greek philosopher Democritus (465 BC) who first conceived of matter as composed of small indivisible particles. Boyle about 1650 came to a similar conclusion, but it was not till 1807 that Dalton advanced his atomic theory and the word entered the scientific vocabulary. However, the etymology was proved to be incorrect by Rutherford's discovery of atomic disintegration in 1910.

auxiliary L. *auxilia* (help, aid).

azo Gk. *a* + *zo*-(alive), referring to the presence of nitrogen, which Lavoisier had named azote in 1791 because of its nonreactive nature. Related words are zoology, zoo.

barbiturate L. *barba* (beard), presumably in reference to the original plant source of these substances, namely, a lichen characterized by a beard-like structure (*Usnea barbata*).

barium, barytes (1808, 1789) Gk. *barus* (heavy), referring to the characteristic property of these substances. The same occurs in **barometer**.

base (1810) Gk. *basis* from root *ba*-(go). The reason for the selection of this term in its present chemical sense is not clear.

bauxite Named from a town in southern France, Les Baux, where it was discovered.

benzene See **benzoin**.

bentonite Named from Fort Benton, Wyoming, where it was discovered.

benzoin (1560) Probably from French *benjoin*, said to be derived from Arabic *ban jawi*, an aromatic gum of Java (frankincense?). The term was later corrupted to *benjamin*. *Benz*-is the root responsible for benzol (Liebig, 1838), later modified to benzene, and for benzoic, an acid present in gum benzoin.

bromine (1827) Gk. *bromos* (stench), after one of its characteristic properties.

butyric (1826) L. *butyrum* (butter). The term butyl from the same source, was introduced in 1868.

cadmium (1917) L. *cadmia* (calamine = zinc ore); this in turn was derived from the legendary hero Cadmus, founder of the Greek city of Thebes, though the connection is obscure.

calcium (1808) L. *calx* (lime). Related is the term *calcine*, used by the alchemists to mean reduction of a metal to a powder (calx) by heat; also to desiccate by fire (1640).

calender (1688) L. *cylindrus* (cyliner), descriptive of the major feature of this machine.

carbon (1789) L. *carbo* (coal); so named by Lavoisier, as the chief constituent of coal.

casein (1841) L. *caseus* (cheese), of which it is a major component.

catalysis Gk. *kata*-(completely) + *lyein* (loosen, set free). Adopted by Berzelius (1836) to describe this type of reaction. The term dates back to 1660 in the nontechnical sense of dissolution or destruction.

cement (1490) L. *caed*-(cut), in reference to the comminuted or crushed materials that are the basis of this product. **Cementation** (finely divided metal powders) has the same origin.

ceramic (1850) Gk. *keramos* (pottery, potters' clay) Contrary to general belief, this term is *not* derived from Latin *cera* (wax).

ceresin See **ozocerite**.

cerium (1804) Named by Berzelius from the then recently discovered asteroid Ceres, which in turn was named after the Greek goddess of harvest. Cereal has the same derivation.

cesium (1860) L. *caesius* (sky-blue). So named by Kirchhof and Bunsen because of the blue lines in its spectrum.

chelate (1930) Gk. *chela* (claw of a crab or lobster). So names by G. T. Morgan and H. D. K. Drew because of the tenacity with which the coordinating group holds the metal ion.

chemistry Gk. *kemia* (transmutation). *Oxford Universal Dictionary* states that this relates to "the Egyptian art" associated with the "black earth" (*khem*) of the Nile delta, but the exact etymology is obscure. Early writers referred to chemistry as an art rather than a science. Use in its present sense probably began with Robert Boyle about 1650. See also **alchemy.**

chiral (1950?) Gk. *chiro* (hand), referring to the right-and left-handed structure of optical isomers. See also **enantiomorph.**

chlorine (1810) Gk. *chloros* (green). So named from its greenish-yellow color by Sir Humphry Davy.

chlorophyll (1819) Gk. *chloros* (green) + *phyl-* (leaf).

cholesterol (1827) Gk. *chol-*(bile) + *stereo-*(hard, solid).

chromatography (1906) Gk. *chroma* (color) + *graph-*(write, draw). So named by Tswett, its inventor, who used it to analyze plant pigments, which produced bands of characteristic color.

chromium (1797) Gk. *chroma* (color), in reference to the distinctive colors of its compounds. Many other chemical terms have the same derivation, e.g., **chromosome, chromatin, chromatography**; also **chromatic.**

chromophore Gk. *chroma* (color) + *phor-*(carry, bear), i.e., a chemical group that imparts color to a compound.

chromosome (1890) Gk. *chroma* (color) + *soma* (body), the reference to color being due to the extreme ease with which this nucleo-protein accepts biological stains. The sense of the word "body" here is probably the same as that in "antibody."

clathrate Gk. *klathra* (a grating of metal bars), suggestive of the cage-like structure of these compounds (also called inclusion compounds). The term was known in the 19th century in a botanical sense, but its chemical use did not appear till about 1940.

coagulate (1610) L. *co-*(together) + *ag-*(drive), i.e., a natural forcing together of dissolved substances to form a clot.

cobalt (1735) German *kobold* (goblin). The extreme difficulty of separating Co from Ni ore during extraction operations led miners in Saxony to endow it with an evil spirit that interfered with obtaining the nickel, which they considered the valuable material. See also **nickel.**

coherent L. *co-*(together) + *haer-*(stick, cling), referring to sticking together of identical or closely similar materials, e.g., coherent light. See also **adhesive.**

colloid (1861) Gk. *kolla* (glue). Introduced by Thomas Graham to describe suspensions whose disperse phase did not pass readily through a parchment membrane. **Collagen** and **collodion** have the same derivation.

component L. *con-*(together) + *pon-*(place), i.e., a material that is blended with other materials in a mixture. Compound has the same derivation but a different chemical meaning.

compound See **component.**

composite L. *con-*(together) + *pos-*(place), i.e., a combination or placing together of two or more materials, usually in a laminar structure. See also **component.**

conjugated L. *con-*(together) + *jugum* (yoke for oxen). An organic compound "yoked together" by alternate double bonds.

copper L. *cuprum* or *Cyprium*, because originally found in Cyprus. Also German *Kupfer.*

corrode (1610) L. *con-*(completely) + *rod-* (gnaw). From the same root as erode and rodent.

crucible (1460) L. *cruc-*(cross), referring to the practice of the alchemists of placing a cross on their experimental equipment to ward off evil spirits. Crucial has the same derivation, e.g., a crossroad.

desiccate L. *de-*(out, away) + *siccus* (dry). From the same root as siccative, a drying agent.

detergent (1620) L. *de-*(away) + *terg-*(wipe), no doubt because of the cleaning action of these compounds.

dextrose See **levulose.**

dialysis See **analysis.**

diffusion L. *dis-(dif-)* (apart) + *fusus* (melt), i.e., gradual dispersal of one substance in another.

disintegrate L. *dis-*(apart) + *in-*(not) + *teg-* (touch), i.e., division of an untouched whole (integer) into many parts. An integer is an "untouched" whole number.

dissociation L. *dis-*(apart) + *socius* (friend, ally), i.e., a separation of substances previously united. Association is from the same root, but with opposite meaning.

distillation L. *de-*(down) + *still-*(to fall as drops). The exact date is unknown but the term was familiar to the alchemists in the Middle Ages. It was also used by poets in the 16th century, e.g., "leperous distillment" (Shakespeare).

dysprosium (1886) Gk. *dysprositos* (hard to get at), because of the difficulty of its isolation.

electrode Gk. *elektron* (amber) + *-ode* (road, pathway), i.e., a highway for electricity.

electron (1890) Gk. *elektron* (amber), referring to the ability of amber to accumulate a static charge by being rubbed. Similarly, electric and related terms.

enantiomorph (about 1900) Gk. *enantio-*(opposite, contrary) + *morph-*(form, shape), descriptive of the mirror-image structure of optical isomers. Also called enantiomer.

enzyme (1881) Gk. *zyme* (ferment, leaven), in reference to the catalytic activity of these compounds.

equilibrium (1608) L. *aequus* (the same) + *libra-* (weight, balance); hence any stabilized condition, whether or not weight is involved.

ether (1587) Gk. *aether* (the sky or upper atmosphere), from a root meaning to glow or burn. The term was long used to mean simply "the air;" it was also applied to a supposed elastic medium that filled all space and by which light was transmitted. This was disproved by Michelson and Morley in 1887. The chemical sense originated about 1838, doubtless because of the highly volatile (air-like) nature of ether.

ethyl   See ether.

evolve, evolution   L. *e*-(out) + *volv*-(roll), e.g., a gas given off by a chemical reaction. In general, the "rolling out" of any development over a long time span.

eutectic   (1884) Gk. *eu*-(well) + *tek*-(melt), i.e., well or ideally melting.

extract   L. *ex*-(out) + *tract*-(draw, drag). From the same root as contract (draw together) and retract (draw back).

extrude   L. *ex*(out) + *trud*-(thrust), descriptive of the action of an extrusion machine.

ferment   (1605) L. *fermentum* (Yeast), from the root *ferv*-(boil), in reference to the apparent similarity to boiling. The chemical meaning did not appear till Pasteur's studies about 1850. From the same root as effervescent.

filter   (1791) L. *filtrum* (felt), which was originally used for liquid-solid separations.

flavone   L. *flavus* (yellow), the predominant color produced in flowers by this class of plant pigments.

flocculation   (1877) L. *floccus* (a tuft of wool), referring to the appearance of such aggregates.

fluorine   (1813) L. *flu*-(flow). So named by Ampere, after Scheele's earlier naming of hydrofluoric acid.

flux   (1800) L. *flu*-(flow), originally applied to excretory processes of the body; later it referred to flow-inducing materials in ceramics and metallurgy, as well as to neutron emanations.

fossile   L. *foss*-(dig), e.g., fossil fuels.

fractionate   L. *fract*-(break), i.e., separation of a liquid mixture into its components by distillation. From the same root as refraction ("breaking" of a light ray).

fumigant, fume   L. *fumus* (smoke).

fuse, fusion   L. *fusus* (melt). The original meaning (1681) was to melt by application of heat, e.g., fused salts, electric fuse. This was later (1817) extended to include blending or joining, as in welding. Since 1940 the term has been applied to the union of hydrogen nuclei with release of energy. In this sense the concept of melting has disappeared.

gas   (1600) Adapted by van Helmont from Gk. *chaos*, the letter *g* replacing the *ch*. The Gk. term meant "a yawning gulf or abyss; the formless void of primordial matter" (*Oxford Universal Dictionary*). Modern usage dates from the late 1700s.

gasoline   (1871) Standard dictionaries give the etymology as gas + ol + ine, but no clue as to how the term originated. It was probably coined from "gas oil," a heavy petroleum fraction from which the first gasoline was made.

glycerol   (1838) Gk. *glyc*-(sweet). From the same root are glycol, glycogen (sweet-maker), glycine, etc.

graphite   (1789) Gk. *graph*-(write, draw); so named by Werner in reference to its marking ability, as in pencils. Graph is from the same root.

halogen   Gk. *halo*-(salt) + L. *gen*-(create), from the tendency of these elements to form salts with metals.

helium   (1868) Gk. *helios* (the sun), where it was originally discovered by spectrographic analysis.

heterogeneous   (1630) Gk. *hetero*-(different) + L. *gen*-(kind). Descriptive of mixtures and solutions, which are made up of two or more different components. Its opposite is homogeneous (*homo*- = the same).

homologous   (1850) Gk. *homo*-(the same) + *log*-(proportion), i.e., compounds having the same proportion of certain constituents in an ascending series.

hormone   (1906) Gk. *ormon* (to excite, stimulate), in reference to the physiological activity of these compounds.

hydrogen   (1783) Gk. *hydro*-(water) + L. *gen*-(create), i.e., water-producer. Named by Lavoisier.

hydrolysis   (1880) Gk. *hydro*-(water) + *lysis* (loosen, dissolve), in reference to the cleavage of the water molecule in this reaction.

hydrophilic   Gk. *hydro*-(water) + *phil*-(love). The term describes the chief property of vegetable gums, seaweeds, gelatin, etc., which absorb many times their weight of water. Hydrophobic (*phob* = hate) has the opposite meaning.

hygroscopic   (1775) Gk. *hygro*-(moisture) + *skopos* (observe, measure). Hygroscopic materials "observe" the presence of atmospheric moisture by absorbing it.

-iferous   L. *fer*-(carry, bear). A common suffix as in carboniferous, coniferous, etc.

inhibitor   L. *in*-(in) + *hab*-(*hib*-)(have, hold), i.e., an agent that holds in or restrains a reaction.

insulin   (1925) L. *insula* (island). So named by Banting, its discoverer, because it is formed in the isles of Langerhans in the pancreas. From the same word as insulation.

iodine   (1814) Gk. *iodos* (violet-colored), so named from the color of its vapor by Gay-Lussac.

ion   (1834) L. *i*-(go, move), in reference to the extreme mobility of these particles.

iridium   (1804) Named by Tennant from Iris, goddess of the rainbow because of "the striking color it gives while dissolving in marine acid."

isomer (1838) Gk. *iso*-(the same + *mer*-(part, share), i.e., compounds that (usually) share their constituent atoms equally, though their arrangement is different.

isotope (1913) Gk. *iso*-(the same + *topos*(place), i.e., in the Periodic Table.

kerosene (1854) L. *cer*-(wax), referring to the Pennsylvania paraffinic crude from which it was derived.

ketone (1851) A german modification of **acetone.**

krypton (1898) Gk. *kryptos* (hidden). Named by Sir William Ramsay in reference to the difficulty of its separation.

laser Coined from the first letters of "light amplification by stimulated emission."

latent L. *lat*-(lie hidden), e.g., a latent solvent.

levulose L. *laev*-(the left) + *-ose* (sugar suffix), descriptive of optical isomers that rotate light to the left. The opposite meaning is dextrose from L. *dextro*-(the right).

ligand L. *lig*-(bind), referring to the capacity of certain molecules to attach themselves to metal atoms in coordination compounds. Discovered by Werner in 1893.

lignin, lignite (1820) L. *lignus* (wood). Lignin is a component of wood, and lignite is formed from it.

linoleic L. *linum* (flax) + *oleum* (oil). The principal fatty acid in linseed (flaxseed) oil.

litharge (1800) Gk. *lithos* (stone) + *argyros* (silver); so named because it was once thought to be a product of the separation of silver from lead.

lithium (1818) Gk. *lithos* (stone), probably because of the rock-like nature of petalite, the mineral complex in which it was discovered.

-lysis Gk. *lysis* (decompose, loosen, break down). A suffix in many chemical terms, e.g., analysis, hydrolysis, catalysis, etc.

magnesia (1610) Gk. for Magnesian Stone, which was thought to have magnetic properties and which the alchemists believed to be a component of the Philosopher's Stone. The element magnesium was so named by Sir Humphry Davy in 1808. See also **manganese.**

manganese (1775) L. *magnes* (magnet) = German Mangan, because of the magnetic nature of its ore, pyrolusite. A corrupted form of magnesium.

mercaptan (1835) L. *mer*(curium)(mercury) + *captans* (seizing), in reference to their ability to form relatively insoluble salts of mercury and other heavy metals. Has been replaced by thiol.

mercury L. *Mercurium*, in Roman mythology the messenger of the gods, noted for his quickness and agility.

microscope (1656) Gk. *micro*-(small) + *skopos* (observe).

molecule (1794) L. *moles* (mass) + diminutive suffix *-cule*, i.e., a little mass.

molybdenum (1816) Gk. *molybdos* (a soft, lead-like mineral). Scheele had used the term "molybic acid" in 1778.

mordant L. *morsus* (bite), i.e., an agent that bites (binds) a dye to a fiber.

naphtha (1572) Transliteration of Gk. *naphtha*; a component of asphalt. Both naphtha and asphalt were used as illuminants in ancient times.

nascent (1790?) L. *nasc*-(arise, be born), referring to the active state of an element as it is liberated from a compound, e.g., nascent oxygen. Now obsolescent.

neon (1898) Named by Sir William Ramsay from Gk. *neos* (new).

neutron (1932) L. *neuter* (neither) So named by its discoverer Chadwick because it has neither a positive nor a negative charge. Neutral has the same derivation (neither acidic nor basic).

nickel (1751) Swedish *koppernickel*, from German *Kupfer* (copper) + *nickel* (demon, devil). So called because the ore niccolite (NiAs), being reddish in color, resembled copper, and was thus supposed the trick of an evil spirit. See also **cobalt.**

niobium (1644) Named by H. Rose after the lachrymose goddess Niobe, though his reason for doing so is not clear.

nitrogen (1790) L. *nitrum*(nitre) + *gen*-(create). The term was used by Chaptal to indicate that the element is a constituent of nitre. It had previously been recognized as an element by Lavoisier.

nucleus (1762) L. *nuc*-(nut, kernel).

olefin (1860) L. *ole*-(oil). Ethylene was originally called "olefin gas" because on combination with chlorine it formed an oily liquid, ethylene dichloride, also called Dutch oil. See also **aliphatic.**

oleum L. *ole*-(oil) Fuming H_2SO_4, probably so called because of its oily appearance.

oligomer Gk. *oligo*-(a few) + *mer*-(part). Originally used in botany; adopted by chemists to describe polymers composed of less than five monomer units.

organic (1807) First used by Berzelius to designate substances derived from living organisms. The Greek word *organon* meant implement, instrument, machine, part of the body. The semantics of its transfer from inanimate matter to living entities is not clear.

oxygen (1790) Gk. *oxy*-(sharp, acidic) + L. *gen*-(create). Thus the literal but erroneous meaning is "acid-maker" because of the belief of Lavoisier, who named it, that oxygen is the characteristic element of acids. In 1774 it had been discovered by Priestly who called it dephlogisticated air.

ozocerite (1834) Gk. *ozo*-(bad odor) + *keros*

(wax), i.e., a malodorous earth wax. From the latter root is ceresin.

ozone (1840) Gk. *ozo*-(bad odor) a characteristic property of this gas.

palladium (1803) Named by its discoverer, Wollaston, after the then recently discovered asteroid Pallas, which in turn was so called because of its pale-white color associated with the goddess Pallas.

paper (1670) Gk. and L. *papyrus*, a plant used by the ancients for writing. Thin strips were cut from the stem, soaked in water and pressed together after drying.

paraffin (1830) L. *parum* (little + L. *affinis* (relationship). The literal meaning is thus "having little affinity," in reference to the low reactivity of saturated hydrocarbons. Introduced by the German chemist Reichenbach.

periodic (1870) Gk. *peri*-(around) + *odos* (way, road), i.e., a circuit; any system having repeated beginnings and endings, e.g., the Periodic Table.

petroleum (1526) L. *petr*-(rock) + *ole*-(oil), from its early discovery in rock formations.

pH (1909) Abbreviation of *pouvoir hydrogene* (hydrogen power), the term used by Sorenson, who invented this scale. The word "power" is used in its mathematical sense.

phen- The root that characterizes derivatives of benzene. Gk. *phaino* (shine, show, appear). First used by Laurent (1841) in "hydrate de phenyle" and "acid phenique," which were names for a coal-tar product (later called phenol) from which illuminating gas could be made. Hence this root was used to name all compounds derived from benzene, e.g., phenanthrene (*phen* + *anthrac*-(coal); phenol; phenyl; phenacetin, etc. Phenomenon (i.e. what appears) is from the same root.

phosphorus (1669) Gk. *Phos*-(light) + *phor*-(bear, carry). So named by H. Brand because of its property of glowing in the dark.

phthalic (1857) Coined from na*phthal*ene, from which phthalic anhydride is made.

pigment L. *pigmentum* (paint) from the root *ping*-(paint a picture), implying use of a colorant.

platinum (1803) Originally called *platina* from the Spanish *plata* (silver), because of its color.

plutonium (1941) Named after the most remote planet Pluto, since plutonium was the last element in the Periodic System at the time of its discovery. See also **uranium.**

polymer (1866) Gk. *poly*-(many) + *mer*-(part); similarly for monomer, dimer, etc.

precipitation L. *prae*-(first) + *caput* (head). Literally, headfirst, in the sense of falling directly downward.

polyunsaturated See **saturated.**

protein (1868) Gk. *proteios* (first, primary), in reference to its fundamental part in nutrition. Proton is from the same root.

racemic (1890) L. *racemus* (grapes) referring to Pasteur's discovery of optical isomers in the tartaric acid of wine lees.

radical (1560) L. *radic*-(root) First used in the chemical sense in 1816, but the connection is obscure.

reaction, reagent (1797) L. *re*-(back) + *act*-, *ag*-(do, drive). The same root is responsible for such common words as action, activity, agent, etc. See also coagulate.

retort (1605) L. *re*-(back) + *tort*-(twist), referring to the shape of a type of simple distillation unit characterized by a glass neck twisted downward.

rhodium (1803) Gk. *rhodon* (rose), so named by Wollaston from the rosy color of its chloro salt. Rhodopsin, the red pigment of the eye, is from the same root.

rubber (1788) So named by Joseph Priestley, when he found that the crude product obtained by coagulation of latex was able to erase pencil marks by rubbing.

salt L. *sal*; present chemical meaning (acid-base reaction), 1790. Its original use cannot be accurately dated; it goes at least back to the alchemists and Chaucer (salte teeres) about 1400. Historically its economic value was unique. Roman soldiers accepted salt as payment for their services. Hence the word *salary*, and the expression "earn your salt."

saponification (1821) L. *sapo*-(soap). The term soap has the same derivation, with vowels reversed.

saturated L. *satis* (enough), i.e., an element with which enough constituents are united to fill (satisfy) its valence bonds. Thus polyunsaturated literally means "many times not enough" constituents (i.e. two or more double bonds). Similarly, any material that has absorbed all the liquid it can retain.

scan L. *scand*-(climb over), i.e., as a light beam moves over an irregular surface.

scintillation (1809) L. *scintilla* (spark).

scleroscope Gk. *sclero*-(hard) + *skopos* (measure), (observe), i.e., a type of metal-hardness tester.

selenium (1818) Named by Brezelius from Gk. *selenon* (the moon), because its chemical similarity to tellurium (from L. *tellus*, the earth) suggested an analogical relationship between the two elements similar to that of the moon to the earth.

silicon (1823) L. *silic*-(flint).

silvicide L. silva (trees, forest) + *caed*-(kill), i.e., a special type of herbicide.

solder L. *solid*-(to make firm), descriptive of the metal-joining function of these alloys.

solvent L. *solv*-(free, loosen), i.e., a liquid that "frees" solids from their confined state. Related words from the same root are solve, solution, dissolve, resolve.

sorb, sorption L. *sorb*-, *sorp*-(suck in, swallow).

Absorption describes the swallowing action of a porous material; adsorption means attraction to the surface only (*at, ad* = to, toward). "Sorb" is used when the distinction is not clear.

sparger   (1839) L. *sparg*-(sprinkle). Sparse (thinly scattered) is from the same root.

ster-, stereo-   Gk. *stereo* (solid), i.e., three-dimensional. This root occurs in many chemical terms, e.g., sterol (solid alcohol), stearic, steroid, steric, stereospecific, cholesterol (hard bile alcohol).

synthesis   (1611) Gk. *syn*-(together) + *tithenai*-(place). The chemical meaning did not appear until 1733.

tacticity   Gk. *tax*-(arrange), referring to the arrangement of atoms in the molecular structure of high polymers. This and related terms (syndiotactic, atactic, etc.) were introduced in the 1930s when stereospecific catalysts were developed by Ziegler and Natta.

tall   Derived from the Swedish word for pine, to distinguish tall oil from the U.S. meaning of pine oil.

tall   oil See **tall**.

tantalus   (1802) From the mythological punishment of Tantalus for revealing the secrets of the gods. He was condemned to stand in water, which always receded as he tried to drink. Isolation of the element was apparently attended by the same frustration.

tellurium   See **selenium**.

tension, tensile   L. *ten*-(hold, as against strain). From the same root as retension (holding back) and tenter (a machine for holding fabric taut).

tetrahedral   Gk. *tetra* (four) + *hedron* (side), i.e., a pyramidal structure characteristic of certain molecules.

thallium   (1861) L. *thallus* (budding leaf). Named by Sir William Crookes because of a pronounced green line in its spectrum.

thiol   (1899) Gk. *thio*-(sulfur) + *-ol*, a sulfur alcohol in which SH has replaced the OH group. From the same root as thiamine, dithiocarbamate, etc.

tincture   (1648) L. *tinct*-(color, dye) The term is obsolete in this sense; the connection with alcoholic solutions is not clear, though possibly some were colored. From the same root are tinctorial and tint.

titanium   (1800) So named by Klaproth after the Titans, legendary giants supposed to have originally inhabited the earth. His reason for naming this substance thus is not clear.

transfer, transference   L. *trans*-(across) + *fer*-(carry, bear). There is no connection between this root and the L. *ferrum* (iron) from which ferric is derived.

transition   L. *trans*-(across + *i*-(go, move), e.g., transition elements which "move across" the Periodic Table, making a bridge from electropositive to electronegative.

transmutation   L. *trans*-(across) + *mut*-(change), i.e., changing one element into another. In the Middle Ages alchemists tried to change lead into gold with little success.

tungsten   (1781) Swedish *tung*(heavy) + *sten*-(stone); named in honor of Swedish chemist Scheele.

turpentine   Gk. *terebenthos*(turpentine tree). Terpene is undoubtedly related.

uranium   (1789) Named by Klaproth after the then recently discovered planet Uranus, believed to be the most distant member of the solar system. This planet in turn was named from the Greek word for sky (heavens). See also **plutonium.**

valence   (1869) L. *val*-(be strong), referring to the binding strength of interatomic attraction. From the same root as valid.

vanadium   (1830) From the Scandanavian goddess Vanadis, because of its multicolored compounds. So named by its discoverer, Sefstrom.

vermiculite   (1824) L. *vermis*(worm), so called because of the worm-like writhing motion of the material as it expands on exposure to high temperature.

vitamin   (1921) L. *vita* (life) + amin(e). Originally named vitamine by Funk (1911); the terminal -e was later dropped when the non-nitrogenous carotene (vitamin A) was discovered.

vitriol   L. *vitrum* (glass) An old name for concentrated H_2SO_4, because of its glassy appearance. Vitrification (to convert to a glassy state) is from the same root.

vulcanization   (1846) Named by Charles Goodyear after Vulcan, the Roman god of fire and forging, to describe his process for heat-curing of rubber with sulfur.

xenon   (1898) Gk. *xenos*(strange); so named by Sir William Ramsay.

x-ray   (1896) Named by Roentgen, the discoverer of this form of radiation, because its nature was at that time unknown.

xylene   Gk. *hyle*, a suffix meaning stuff or material, as in ethyl = ether-material (Liebig).

zein   L. *zea*(grain) The protein occurring in corn.

zymase   See **enzyme.**

APPENDIX II

Highlights in the History of Chemistry

A. Chronology of Notable Achievements

Democritus (465 BC) First to conceive matter in the form of particles, which he called atoms.

alchemists (about 1000–1650) Attempted to (1) change lead and other base metals to gold; (2) discover a universal solvent; and (3) discover a life-prolonging elixir. Used plant products and arsenic compounds to treat diseases.

Boyle, Sir Robert (1637–1691) Formulated fundamental gas laws. First to conceive the possibility of small particles combining to form molecules; distinguished between compounds and mixtures; studied air and water pressures, desalination, crystals and electrical phenomena.

Priestley, Joseph (1733–1804) Discovered oxygen, carbon monoxide and nitrous oxide.

Scheele, C. W. (1742–1786) Discovered chlorine, tartaric acid, sensitivity of silver compounds to light (photochemistry); and oxidation of metals.

Le Blanc, Nicholas (1742–1806) Invented a process for making soda ash from sodium sulfate, limestone and coal.

Lavoisier, A. L. (1743–1794) Discovered nitrogen; studied acids and described composition of many organic compounds. Generally regarded as the father of chemistry.

Volta, A. (1745–1827) Invented the electric battery, a series of "piles" or stacks of alternating layers of silver and zinc, or copper and zinc, separated by paper soaked in brine (electrolyte). See activity (1).

Berthollet, C. L. (1748–1822) Corrected Lavoiser's theory of acids; discovered bleaching power of chlorine; studied combining weights of atoms (stoichiometry).

Jenner, Edward (1749–1823) Discoverer of vaccination for prevention of smallpox (1776).

Dalton, John (1766–1844) The first great chemical theorist; proposed atomic theory (1807); stated law of partial pressure of gases. His ideas led to laws of multiple proportions, constant composition and conservation of mass.

Avogadro, A. (1776–1856) Proposed principle that equal volumes of gases contain the same number of molecules. The number (6.02×10^{23} for 22.41 liters of any gas) is a fundamental constant that applies to all chemical units.

Davy, Sir Humphry (1778–1829) Laid foundation of electrochemistry; studied electrolysis of salts in water and other electrochemical phenomena; isolated Na and K.

Gay-Lussac, J. L. (1778–1850) Discovered boron and iodine; studied acids and bases and discovered indicators (litmus); improved production method for H_2SO_4, did basic research on behavior of gases versus temp and on the ratios of gas volumes in chemical reactions.

Berzelius, J. J. (1779–1848) Classified minerals chemically; discovered and isolated many elements (Se, Th, Si, Ti, Zr); coined the terms *isomer* and *catalyst*; noted existence of radicals; anticipated discovery of colloids.

Faraday, Michael (1791–1867) Extended Davy's work in electrochemistry. He developed theories of electrical and mechanical energy, electrolysis, corrosion, batteries, and electrometallurgy.

Wohler, F. (1800–1882) First to synthesize an organic compound (urea, 1828) (a rearrangement reaction). This discovery was the beginning of synthetic organic chemistry.

Goodyear, Charles (1800–1860) Discovered vulcanization of rubber (1844) by sulfur, inorganic accelerator, and heat. Hancock in England made a parallel discovery.

Liebig, J. von (1803–1873) Fundamental investigation of plant life (photosynthesis) and soil chemistry; first to propose use of fertilizers. Discovered chloroform and cyanogen compounds.

Graham, Thomas (1805–1869) Studied diffusion of solutions through membranes; established principles of colloid chemistry.

Pasteur, Louis (1822–1895) (1) First to recognize infective bacteria as disease-causing agents; (2) developed concept of immunochemistry; (3) initiated heat-sterilization of wine and milk (pasteurization); (4) observed optical isomers (enantiomers) in tartaric acid.

Lister, Joseph (1827–1912) Initiated use of antiseptics in surgery, e.g., phenols, carbolic acid, cresols.

Kekulé, A. (1829–1896) Laid foundations of aromatic chemistry; conceived of four-valent carbon and structure of benzene ring; predicted isomeric substitutions (ortho-, meta-, para-).

Nobel, Alfred (1833–1896) Invented dynamite, smokeless powder, blasting gelatin. Established international awards for achievements in chemistry, physics and medicine.

Mendeleev, D. I. (1834–1907) Discovered periodicity of the elements and compiled the first Periodic Table.

Hyatt, J. W. (1837–1920) Initiated plastics industry (1869) by invention of Celluloid (nitrocellulose modified with camphor).

Perkin, Sir W. H. (1838–1907) Synthesized first organic dye (mauveine, 1856) and first synthetic perfume (coumarin). His work on dyes was continued and expanded by Hofmann in Germany.

Beilstein, F. K. (1838–1906) Compiled *Handbuch der Organischen Chemie*, a multi-volume compendium of properties and reactions of organic chemicals.

Gibbs, Josiah W. (1839–1903) Stated three principal laws of thermodynamics; expounded nature of entropy and phase rule and the relation between chemical, electric and thermal energy.

Chardonnet, H. (1839–1924) First to produce a synthetic fiber (nitrocellulose) with properties similar to rayon.

Boltzmann, L. (1844–1906) Developed kinetic theory of gases; their viscosity and diffusion properties are summarized in Boltzmann's Law.

Roentgen, W. K. (1845–1923) Discovered x-radiation (1895). Awarded Nobel Prize in 1901.

Le Chatelier, H. L. (1850–1936) Fundamental research on equilibrium reactions (Le Chatelier's Law), combustion of gases, and metallurgy of iron and steel.

Becquerel, H. (1851–1908) Discovered radioactivity, deflection of electrons by magnetic fields and gamma radiation. Nobel Prize 1903 (with the Curies).

Moisson, H. (1852–1907) Developed electric furnace for making carbides and preparing pure metals; isolated fluorine (1886). Nobel Prize 1906.

Fischer, Emil (1852–1919) Basic research on sugars, purines, uric acid, enzymes, nitric acid, ammonia. Pioneer work in stereochemistry. Nobel Prize 1902.

Thomson, Sir J. J. (1856–1940) Research on cathode rays resulted in proof of existence of electrons (1896). Nobel Prize 1906.

Arrhenius, Svante (1859–1927) Fundamental research on rates of reaction versus temperature, expressed by the Arrhenius equation; and on electrolytic dissociation. Nobel Prize 1903.

Hall, Charles Martin (1863–1914) Invented method of aluminum manufacture by electrochemical reduction of alumina. Parallel discovery by Heroult in France.

Baekeland, Leo H. (1863–1944) Invented phenol-formaldehyde plastic (1907), the first completely synthetic resin (Bakelite).

Nernst, Walther Hermann (1864–1941) Awarded Nobel Prize in 1920 for his work in thermochemistry; did basic research in electrochemistry and thermodynamics.

Werner, A. (1866–1919) Introduced concept of coordination theory of valence (complex chemistry). Nobel Prize 1913.

Curie, Marie (1867–1934) Discovered and isolated radium; research on radioactivity of uranium. Nobel Prize 1903 (with Becquerel) in physics; in chemistry 1911.

Haber, F. (1868–1934) Synthesized ammonia from nitrogen and hydrogen, the first industrial fixation of atmospheric nitrogen (the process was further developed by Bosch). Nobel Prize 1918.

Rutherford, Sir Ernest (1871–1937) First to prove radioactive decay of heavy elements and to carry out a transmutation reaction (1919). Discovered half-life of radioactive elements. Nobel Prize 1908.

Lewis, Gilbert N. (1875–1946) Proposed electron-pair theory of acids and bases; authority on thermodynamics.

Aston, F. W. (1877–1945) Pioneer work on isotopes and their separation by mass spectrograph. Nobel Prize 1922.

Fischer, Hans (1881–1945) Basic research on porphyrins, chlorophyll, carotene; synthesized hemin. Nobel Prize 1930.

Langmuir, Irving (1881–1957) Fundamental research on surface chemistry, monomolecular films, emulsion chemistry. Also electric discharges in gases, cloud seeding, etc. Nobel Prize 1932.

Staudinger, Hermann (1881–1965) Fundamental research on high-polymer structure, catalytic synthesis, polymerization mechanisms, resulting eventually in development of stereospecific catalysts by Ziegler and Natta (stereoregular polymers). Nobel Prize 1963.

Fleming, Sir Alexander (1881–1955) Discovered penicillin (1928); initiated antibiotics. Nobel Prize 1945. The science was developed in the U.S. by Selman A. Waksman.

Moseley, Henry G. J. (1887–1915) discovered the relation between frequency of x-rays emitted by an element and its atomic number, thus indicating the element's true position in the Periodic Table.

Adams, Roger (1889–1971) Noted educator and contributor to industrial research in catalysis and structural analysis. Priestley Medal.

Midgley, Thomas (1889–1944) Discovered tetraethyllead and antiknock treatment for gasoline (1921) and fluorocarbon refrigerants early research on synthetic rubber.

Ipatieff, Vladimir N. (1890?–1952) Basic research and development of catalytic alkylation and isomerization of hydrocarbons (with Herman Pines).

Banting, Sir Frederick   (1891–1941) Isolated the insulin molecule. Nobel Prize 1923.

Chadwick, Sir James   (1891–1974) Discovered the neutron (1932). Nobel Prize 1935.

Urey, Harold C.   (1894–1981) Discovered heavy isotope of hydrogen (deuterium). Nobel Prize 1934. A leader of the Manhattan Project. Made original contributions to theories of the origin of the universe and of life processes.

Carothers, Wallace   (1896–1937) Polymerization research resulting in synthesis of neoprene (polychloroprene) and of nylon (polyamide)

Kistiakowsky, George B.   (1900–1982) Developed the detonating device used in first atomic bomb.

Heisenberg, W. K.   (1901–1976) Research in quantum mechanics resulting in development of the orbital theory of chemical bonding. Stated Uncertainty Principle. Nobel Prize 1932.

Fermi, Enrico   (1901–1954) First to achieve a controlled nuclear fission reaction (1939); basic research on subatomic particles. Nobel Prize 1938.

Lawrence, Ernest O.   (1901–1958) Invented the cyclotron in which first synthetic elements were created. Nobel Prize 1939.

Libby, Willard F.   (1908–1980) Developed radiocarbon dating technique based on carbon-14. Nobel Prize 1960.

Crick, F. H. C.   (1916—) (with Watson, J. D.) Elucidated structure of DNA molecule (1953) resulting in development of gene-splicing (recombinant DNA) techniques.

Woodward, Robert W.   (1917–79) Nobel prize 1965 for his brilliant syntheses of such compounds as cholesterol, quinine, chlorophyll and cobalamin.

B. American Chemical Society

The following is based in part on an article prepared by Alden H. Emery, then Secretary of the ACS, for the first edition of *Encyclopedia of Chemistry*.

At a meeting held at Priestley's old home in Northumberland, Pennsylvania in 1874 to celebrate the centennial of the discovery of oxygen, plans were laid to organize an American Chemical Society. The first official meeting was held two years later in New York City. It was attended by 35 chemists out of a membership list of 133. Contrasted with these figures are the current attendance of about 15,000 at national meetings from a membership of over 125,000.

The Society was originally incorporated in New York State; a national charter was granted by Congress as of January 1, 1938. The objectives of the Society are "to encourage in the broadest and most liberal manner the advancement of chemistry in all its branches; the promotion of research in chemical science and industry; the improvement of the qualifications and usefulness of chemists . . .; the increase and diffusion of chemical knowledge; and to promote scientific interests and inquiry; thereby fostering public welfare and education, aiding the development of our country's industries, and adding to the material prosperity and happiness of our people."

These objectives are pursued by several means: (1) Meetings at the national, regional, and local levels, at which papers devoted to advanced chemical theory and practice are presented; several hundred of these are delivered at a single national meeting. (2) Publication of 24 journals in major areas of chemistry, including *Chemical Abstracts, Journal of the American Chemical Society, Industrial and Engineering Chemistry, Chemical Reviews,* and *Chemical and Engineering News.* The Society also publishes reports of chemical symposia, *Advances in Chemistry;* and numerous chemical treatises, including the Monograph Series. (3) Strong and positive interest in chemical education, evidenced by a series of intensive short courses, radio broadcasts on special topics, and tape recordings of symposia. (4) Maintenance of the general well-being of its membership by conservative management and the highest ethical and professional standards, as well as by many adjunctive functions such as employment service: "it has protected its members against legal encroachments on their rights and has acted positively to gain recognition for their proper status."

The Priestley Medal, considered the highest award in chemistry, was established by the ACS in 1923. The ACS also administers or sponsors many other awards.

C. Chemical Abstracts. (CA)

The following is adapted from a *Chemical and Engineering News* report of the 75th anniversary of the first issue of *CA*, January 1, 1907. *CA* came into being primarily because U.S. chemists felt that European abstracting journals were neglecting U.S, chemical research.

From the beginning, the American Chemical Society charged *CA* with abstracting the complete world's literature of chemistry. The exponential growth of chemical research and publication in the ensuing years led to a parallel growth in *CA* and the organization that produces it.

The first years issues of *CA* contained just under 12,000 abstracts. It now abstracts, indexes, or cites more than half a million scientific papers and patents annually, and each weekly issue contains almost 9,000 abstracts. The nine millionth abstract appeared in the first issue in 1982.

The early issues were edited by William A.

Noyes, Sr., Chairman of the ACS committee on papers and publications. Noyes was assisted by two part-time editors, a secretary, and 129 unpaid volunteer abstractors (there are now over 3,300). The original staff of four has grown to nearly 1,200, located in two large buildings near Ohio State University in Columbus, Ohio.

Staff members monitor some 12,000 scientific journals and other periodicals from more than 150 nations, patents issued by 26 nations and two international bodies, and conference proceedings, reports, dissertations and books from around the world in search of new information of chemical interest.

Document analysts abstract papers and patents published in more than 50 languages. They also thoroughly index each paper or patent using both key words and phrases from the titles and abstracts of the document and highly controlled and structured index entries to subjects covered and chemical substances mentioned.

Abstracts fill almost 35,000 pages in *CA* annually. Weekly indexes add another 8,000 pages each year, and the more detailed and comprehensive volume indexes published every six months total almost 30,000 pages per year.

CA has become thoroughly international both in terms of the information it covers and the audience it serves. Nearly two-thirds of its circulation is abroad.

Noyes was succeeded as editor of *CA* in 1908 by Austin M. Patterson. Patterson retired in 1914 but made a major contribution after his retirement. When it was decided to publish a 10-year collective index to *CA* in 1916, it became evident that some systematic means of naming and indexing chemical substances was necessary. Patterson and Carleton E. Curran devised a naming scheme that had a profound effect on chemical nomenclature in general.

In 1915, E. J. Crane took over as editor and remained at the helm for 43 years, becoming the first director of Chemical Abstracts Service when the *CA* editorial organization was renamed and made a division as ACS in 1956. Crane made *CA* the model and pacesetter for all scientific abstracting and indexing services. He was succeeded as director of CAS in 1958 by Dale B. Baker, who led it through a difficult transition to financial self-sufficiency and a major growth in staff and facilities and moved it into the computer age. An extensive, in-depth systems approach to the storage and handling of chemical information has enabled CAS to assemble a vast computerized body of knowledge, typified by its Chemical Compound Registry, a collection of some six million structural formulas, any of which can be instantaneously retrieved and displayed. This unique system is known as CAS On-Line.

D. Center for History of Chemistry

The Center for History of Chemistry was established in January 1982, under the joint sponsorship of the American Chemical Society and the University of Pennsylvania. It is located at the Van Pelt Library on the campus of the University in Philadelphia. The long-range purpose of the Center is to increase general appreciation of the significance of chemistry as a creative enterprise resulting from a more adequate history of American chemistry. The short-range goal of the Center is to locate, catalog, preserve, publicize, and make available the historical records that illuminate the development of chemistry, chemical engineering, and the chemical industry.

E. History of Five Major Industries

1. Drug and Pharmaceutical

The use of drugs and medicines as curative agents began with the Hindus, who successfully practiced vaccination techniques as early as 550 AD, and continued with the alchemists in the Middle Ages. They experimented with a wide range of plant principles as well as with arsenic compounds and various metals. The first compilation of drugs, the *Nuremburg Pharmacopeia*, appeared in 1542, just a year after the death of Paracelsus, a German physician and chemist. He modernized the approach of the alchemists to diseases and ridiculed the ancient ideas that were still current in the 16th century. Plant products such as cinchona bark, ipecac, and chaulmoogra oil were used for many years. In 1805, morphine was obtained from the opium poppy, paving the way for the introduction of an extensive series of alkaloid drugs, many of which are still in use.

The first notable breakthrough in disease treatment was Jenner's successful inoculation for smallpox (1775) which initiated the science of immunochemistry. Since this disease had ravaged Europe for centuries, Jenner's brilliant work is a landmark in medical history. Pasteur adopted this technique; he was the first to inoculate for anthrax and rabies (1880) and to establish the antigen-antibody relationship. Still more important was his identification of many types of infective microorganisms (germs) as disease-causing agents. Almost simultaneously, Lister introduced surgical antiseptics, such as mercury compounds, carbolic acid and phenol, and Robert Koch discovered the germs of anthrax, cholera, and tuberculosis.

Chemotherapy may be said to have begun when Ehrlich developed his arsphenamine treatment of syphilis (1910), the so called "magic bullets" that killed microorganisms with minimum damage to the host. This practice has been broadly applied to cancer therapy in recent years, but without positive results. Another significant event was Funk's discovery (1911) of the ability of certain plant products to cure a disease called beri-beri; these soon were named vitamins. Their classification, metabolic functions, and curative properties were established by many researchers, including McCollum, Szent-Gyorgyi, Sherman, and R. J. Williams (B complex).

The science of endocrinology was developing rapidly at this time. Outstanding was the isolation of insulin for treatment of diabetes by Banting in the early 1920s. Continuing research in this field led to development of many hormones for treatment of a wide vareity of disorders, especially the adrenal cortical steroid hormone (ACTH), cortisone, and thyroxine.

In 1928, Fleming in England noted the antimicrobial action of certain plant molds that produced penicillin. Some years later, Waksman in the U.S. initiated the development of these mold products, which he called antibiotics. These were so effective against many infectious diseases that they soon displaced the sulfa drugs that had been introduced a few years earlier. Antibiotics have proliferated in recent years and have been the most significant addition to curative drugs in the history of medicine. Coincidental with this development was discovery of the tobacco mosaic virus by Stanley in 1935, which resulted in new concepts of the nature and behavior of infective microorganisms. The early 1940s saw the introduction of antimalarials as substitutes for quinine, and of antihistamines for treatment of allergic diseases.

Significant breakthroughs are by no means confined to the past. Current research on gene-splicing techniques has already resulted in "programming" bacteria to produce insulin and interferon, as well as the creation of wholly new forms of bacteria that have great potential not only in medicine but in chemistry and agriculture.

2. Paper

Paper owes its name to papyrus, a plant from which the ancient Egyptians made rolls of tissue on which they inscribed historical records. The actual manufacture of paper from plant fibers originated in China about 100 AD; the primitive technology traversed Asia and the Middle East over a period of several centuries. By 1400, it had reached Europe. The first use of paper for bookmaking was in the Gutenberg Bible (1455). Papermaking began in England about 1500. The first mill in the U.S. was installed in 1690 in Philadelphia. Scheele's discovery of the bleaching power of chlorine (1774) was an important advance; another was the invention of vat sizing with alum and rosin (1800).

On the mechanical side, the invention of the fourdrinier machine was largely responsible for the future development of the industry. Patented in 1807, it was built in England and installed in the U.S. in 1827. Soda pulp was first made in 1854. The principles of sulfite pulping were disclosed by Tilghman in America (1867), and the process was developed in Germany by Mitscherlich in 1874. Groundwood pulp entered commercial use for newsprint about 1870. The sulfate or kraft process, invented by Dahl in Germany in 1884, was first used in the U.S. in 1909. The technique of machine coating was developed in 1875. Continual improvement of the fourdrinier and similar machines by the 1920s resulted in paper speeds of 1000 feet a minute and continuous operation over a period of several days.

Developments since 1920 include multistage bleaching of kraft; bleaching with peroxides, chlorine dioxide and sodium hypochlorite; use of synthetic resins for wet-strength paper for bags, maps, etc.; and the use of soluble bases for sulfite pulping, which allow the use of more species of wood.

The official organization of papermakers in the U.S. is the Technical Association of the Pulp and Paper Industry (TAPPI), founded in 1915. Basic research is carried out at the Institute of Paper Chemistry in Appleton, Wisconsin.

3. Plastics

The first plastic was a mixture of cellulose nitrate and camphor invented in the 1860's by John Wesley Hyatt, it was given the TM "Celluloid." In 1899, Spitteler developed a method of hardening casein with formaldehyde and thus founded the casein plastics industry (small items such as buttons). The earliest high-volume plastic, a condensation product of phenol and formaldehyde, was introduced by Leo Baekeland in 1907. Trademarked "Bakelite," it was the first truly synthetic high polymer. Its chief uses were as engineering material since its dark color limited its application to items in which color was not a factor.

In 1927, cellulose returned to the scene with the development of cellulose esters in the form of the acetate and butyrate produced as sheet, fibers, and molded items. At this time occurred an important mechanical development: compression molding began to give way to injection molding, a greatly improved technique in which the pelleted plastic is introduced to the mold in semiliquid form under pressure. This was a landmark in the growth of the industry.

During the period 1920–1930, the German chemist Hermann Staudinger was deeply involved in fundamental research on polymerization mechanisms which was to profoundly affect the future of plastics. The importance of his work can hardly be overestimated.

Several notable events took place during the

1930s, when the field was crowded with significant developments that occurred almost simultaneously. One of these was the introduction of amino resins (catalyzed reaction products of urea or melamine with formaldehyde). These had a great advantage over phenol-formaldehyde because they made possible white or light-colored products suitable for scale housings and domestic items such as dinnerware. Two other achievements were due to the genius of Wallace Carothers, from whose research came neoprene (polychloroprene), the first approach to synthetic rubber (1930), and soon after the polyamide nylon, the first wholly synthetic fiber. Two other notable polymers were polystyrene (about 1932) and polyethylene, the latter developed in England (1939).

Other important developments were the technique of emulsion polymerization, in which two or more monomers can be co-polymerized; reinforced plastics (glass fibers impregnated with phenolic and other resins); foamed or cellular plastics, both flexible and rigid; silicone resins originally researched by Kipping and developed in the 1940s; the invention of block and graft polymers; and the extremely important stereospecific catalysts devised by Ziegler and Natta and based on the earlier research of Staudinger. These permit selective control of polymer structure and resulted among other products in the development of polypropylene and polyisoprene (the first actual synthetic rubber).

The industry is represented by the Society of the Plastics Industry (250 Park Avenue, New York); the Society of Plastics Engineers (1942) located in Greenwich, Connecticut, is chiefly concerned with the engineering aspects of plastics technology.

4. Petroleum

Petroleum was found in many locations throughout the world before its discovery in the U.S. The first well was drilled by Drake in Pennsylvania in 1859; its depth was a mere 70 feet, in contrast to the 20,000 feet that can be attained with modern equipment. For many years, production was confined to the eastern states, the oil being used chiefly for lubrication and illumination (kerosene). Around the turn of the century, the great strikes in the southwest and California coincided with the expanding demand for motor fuels. Production figures (in millions of gallons) were 1880, 26; 1900, 64; 1920, 443; 1940, 1,300; 1960, 2,470; and 1970, 3,327. Since 1970, U.S. production has declined substantially in spite of new methods of secondary extraction, such as chemical flooding and hydraulic fracturing.

Notable advances in refining techniques occurred at an increasing rate after 1910, when thermal cracking was introduced as a result of research by Burton. This did not yield gasoline of high quality.

Its performance was greatly improved by Midgley's discovery in 1921 of the octane-boosting power of tetraethyllead from about 60 to 100 octane or more. An outstanding achievement was the development of catalytic methods over the next 20 years. Early catalysts were activated clays or silica-alumina. In recent years, these have given way to zeolites (molecular sieves) and to metals of the platinum group. The processing techniques include the fixed-bed method, later replaced by moving-bed and fluidized-bed continuous crackers. The exhaustive study and evaluation of catalysts largely accounted for the present sophisticated techniques that characterize the industry. Many catalytic reforming processes (alkylation, isomerization, hydrocracking, etc.) were researched and developed between 1940 and 1960, important contributions to this development were being made by Ipatieff and Egloff.

All questions relating to the industry, including production statistics, reserves and exploitation of alternative sources, can be referred to the American Petroleum Institute, 1271 Avenue of the Americas, New York.

5. Rubber

While exploring the Amazon in 1735, LaCondamine, a French scientist, noted a resilient material that the natives obtained by tapping certain trees (Hevea) and coagulating the sap (latex) over fires. This they used for making waterproof shoes and large balls with which they played games. He took samples of this substance to Europe, where it aroused great interest, especially among chemists such as Priestly. Its ability to erase pencil marks by rubbing was the origin of the word *rubber*. The Hevea forests of Brazil remained the primary source of rubber for 150 years. In 1875, an English botanist, Henry Wickham, was requested by his government to obtain a supply of Hevea seeds. He helped himself to several thousand of these; they were planted in Kew Gardens in London, and the seedlings were transplanted in Malaya. Thus began the commercial production of plantation rubber which soon displaced Brazilian sources and gave Great Britain control of the world rubber market until World War II.

The rubber industry began in 1844 with the simultaneous discovery of vulcanization with S and inorganic accelerators (metallic oxides) by Goodyear in America and Hancock in England. But it was not until the early years of the 20th century that it became a major industry as a result of three important developments: (1) the discovery in 1906 of the much more efficient organic accelerators (nitrogenous compounds), which were used for some years by one company before being widely accepted (1920); (2) fortuitous discovery of the reinforcing and abrasion-resistant properties of carbon black; and (3) the introduction of age-resisters (antioxidants), which greatly prolonged the life of rubber by inhibiting oxidation. As a result of these advances, the mileage of tires suddenly increased by a factor of 10 with comparable reduction in risk of failure, and thus made possible the vast growth of the automobile industry.

An efficient method of reclaiming vulcanized scrap was developed about 1900, involving digestion of the scrap with NaOH plus a hydrocarbon softener. This separated the cotton insertions from the rubber and partially reversed the effect of vulcanization. Use of reclaim has long been an important adjunct to the industry, especially for low-cost, high-volume products.

Methods of concentrating latex sufficiently by centrifugation to make long distance transportation economically possible were introduced in the early 1930s, thus simplifying the manufacture of thin and irregularly shaped items, either by dipping or electrodeposition, as well as of cellular rubber (foamed or sponge products).

Significant developments in synthetic rubber began at this time. Outstanding were the introduction of polychloroprene (neoprene) by Carothers and of the oil-resistant polysulfide rubber Thiokol by Patrick. These were soon followed by styrene-butadiene copolymers, nitrile rubber, butyl rubber, and various other types, some of which were rushed into production for the war effort in the early 1940s. The stereospecific catalysts researched by Ziegler and Natta aided this development, including synthesis of true rubber hydrocarbon (polyisoprene). Since 1935, synthetic rubbers have been referred to as elastomers.

APPENDIX III

Manufacturers of Trademarked Products
Numerical List

(for addresses, see Alphabetical List, Appendix IV)

1. Stauffer Chemical Company
2. Calgon Carbon Corporation
3. Abbott Laboratories
4. Crown Zellerbach, Chemical Products Division
5. Firestone Tire and Rubber Company
6. Laticrete International, Inc.
7. Parker Chemical Company
8. Scientific Chemicals, Inc.
9. Knoll Pharmaceutical Company
10. Van Dyk & Company, Inc.
11. Koppers Company, Inc.
12. Zoecon Corporation
13. Formica Corporation
14. Occidental Petroleum Corporation
15a. Reheis Chemical Company
16. Thomas, Arthur. H. Company
17. Heresite - Saekaphen, Inc.
18. Morflex Chemical Company
19. Inmont Corporation
19a. Interpace Corporation
20. Corning Glass Works
21. Neville Chemical Company
22. Indian Gum Industries Ltd.
23. Rohm and Haas Company
24. Wyeth Laboratories
25. Pennsylvania Refining Company
26. PMC Specialties Group, Inc.
27. Thiokol Corp., Specialty Chemicals Div.
28. Du Pont de Nemours, E. I. & Company
30. CPC International, Inc.
31. Handy & Harman
35. Firestone Plastics Company
36. Reichhold Chemicals, Inc.
41. Atlas Minerals & Chemicals, Inc.
42. Sun Chemical Corporation
45. See 104
46. Acheson Colloids Company
47. Duriron Company, Inc.
48. Cleary, W. A. Corporation
50. Allied Chemical Corporation (all divisions)
51. Humble Oil & Refining Company
53. National Starch and Chemical Corporation
54. Carrier Corporation
55. FMC Corporation
56. C-I-L, Inc.
57. American Cyanamid Company
58. Monsanto Company

62. Occidental Chemical Corporation
63. Richardson Company, The
64. Spencer Kellogg Products, NL Chemicals
65. Borden Chemical
67. Climax Molybdenum Company
69. Vanderbilt Company, R. T., Inc.
70. Searle, G. D., & Company
71. Smith Kline Corporation
73. Glyco, Inc.
74. Nuodex, Inc.
79. Nuodex, Inc.
81. Morton Thiokol, Inc., Ventron Division
82. Graphite Metallizing Corporation
84. Olin Chemicals
85. Shulton, Inc.
88. American Potash & Chemical Corporation
89. ICI United States, Inc.
91. Schwarz/Mann
92. Masonite Corporation
93. Tennessee Chemical Co.
94. Hall, C. P., Company
97. Chesebrough-Pond's, Inc.
99. Engelhard Corporation
100. Lilly, Eli, and Company
103. Hoyt, Arthur S., Company, Inc.
104. Witco Chemical Company, Inc.
106. American Mineral Spirits Company
107. American Norit Company, Inc.
108. See 323.
114. Premier Malt Products, Inc.
115. Eastman Kodak Company
116. Siemens-Allis, Inc.
117. Alox Corporation
118. California Industrial Minerals Company
119. Goodrich, B. F., Chemical Group
121. American Can Company
122. G & A Laboratories, Inc.
123. Merck & Company, Inc.
124. Marine Colloids, Inc.
125. Shell Chemical Company
128. Petrolite Corporation, Bareco Division
129. Weyerhaeuser Company
134. Harshaw/Filtrol Partnership
136. Atlantic Refining Company
138. Tenneco Chemicals, Inc.
140. Pennsylvania Industrial Chemical Corporation
141. Sherwin-Williams Chemicals

142. Enthone, Inc.
145. Lithcote Corporation
148. National Lead Company, DeLore Division
149. Dow Corning Corporation
152. Swift Edible Oil Company
154. Pennwalt Corporation, Lucidol Division
155. Driver, Wilbur B., Company
158. 3M
159. Royce Chemical Company
161. Hitco
162. Winthrop Laboratories, Sterling Drug Company
165. Apex Chemical Company, Inc.
166. Hoskins Mfg. Company
167. Kendall Refining Company
168. Lake States Chemical Division, St. Regis Paper Company
169. LaMotte Chemical Products Company
173. Wallerstein Company
175. See 50.
177. PPG Industries, Inc.
179. Diversitech General
181. Chemagro Agricultural Division, Mobay Chemical Corporation
182. Nalco Chemical Company
184. Permutit Company
188. Fritzsche Dodge & Olcutt, Inc.
189. Wallace & Tiernan, Inc. Harchem Division
190. Hoffmann - La Roche, Inc.
191. Owens-Corning Fiberglas Corporation
192. U. S. Industrial Chemicals Company
196. American Agricultural Chemical Company
197. See 50.
200. Total Petroleum
201. Philadelphia Quartz Company
202. Baker Castor Oil Company
203. BASF-Wyandotte Corporation
204. Pennwalt Corporation
205. Grace & Company, W. R. Construction Products Division
206. See 89.
210. Ionac Chemical Company
212. Miles Laboratories, Inc.
216. Amoco Chemicals Corporation
217. Filtrol Corporation
219. Ciba-Geigy Corporation
229. Thermal American Fused Quartz Company
221. Archer Daniels Midland Company
222. Arco Chemical Company
223. Osborn, C. J., Chemicals, Inc.
224. Quaker Oats Company
226. Aluminum Company of America
227. Givaudan Corporation
230. Carlisle Chemical Works, Inc., Advance Division
232. Holliday Dyes and Chemicals, Ltd.
233. Dow Chemical Company
235. National Rosin Oil Products, Inc.

236. National Lead Company, Baroid Division
238. Chemical Development Corporation
241. Grace, W. R., & Company, Davison Chemical Division
242. Emery Industries, Inc.
243. See 50.
245. General Electric Company, Polymers Product Dept.
247. Manville Products Corporation
248. Uniroyal Chemical
249. Norton Company, Metals Division
250. Foote Mineral Company
251. Republic Steel Corporation
252. Arizona Chemical Company
253. See 151.
256. Tennessee Eastman Company
259. General Mills Chemicals, Inc.
261. FMC Corporation, Food & Pharmaceutical Products Division
263. Scott Bader & Company, Ltd.
265. Goodyear Tire & Rubber Company
266. Hercules, Inc.
267. Kenrich Petrochemicals, Inc.
268. The New Jersey Zinc Co., Inc.
269. Worthington Corporation
270. Mearl Corporation
271. American Lignite Products Company
274. Beckman Instruments, Inc.
275. Cabot Corporation
277. Xylos Rubber Company
278. Firestone Synthetic Rubber and Latex Company
280. See 442.
282. See 154, 189.
283. Huntington Alloys, Inco Alloys International, Inc.
285. Huber, J. M., Corporation
287. Beaunit Corporation
288. M & T Chemicals, Inc.
289. Lockrey Company
292. See 50.
293. BASF Inmont Corporation
296. Glidden-Durkee Division, SCM Corporation
299. Pfizer, Inc., Chemicals Division
300. Arkansas Company, Inc.
303. Phillips Petroleum Company
304. National Lead Company
305. Ciba Pharmaceutical Company
307. GAF Corporation
308. Westinghouse Electric Corporation
309. Diamond Shamrock Corporation, Nopco Division
310. Norac Company, Inc.
311. Grace, W. R., & Company, Organic Chemicals Division
312. Evans Chemetics, Inc.
312a. Fabric Research Laboratories
313. Ethyl Corporation

315. American Cyanamid Company, Fine Chemicals Dept.
316. Velsicol Chemical Corporation
321. Schering Corporation
322. Kelco Company
324. Anaconda American Brass Company
326. U. S. Stoneware Company
327. Upjohn Company
328. Onyx Chemical Company
329. Mallinckrodt, Inc.
331. Mobil Chemical Company
336. See 304.
337. See 304.
342. Penick, S. B., & Company
343. Parker Chemical Company
345. Hammond, W. A., Drierite Company
346. Sterling Drug, Inc. Consumer Products Division
347. Kennametal, Inc.
348. Beckton Dickinson & Company
349. Haveg Industries, Inc.
350. Driver-Harris Company
351. Celotex Corporation
352. Celanese Chemical Company
354. See 346.
400. Aceto Chemical Company, Inc.
401. Morton Chemical Company
403. Smith, Werner G., Inc.
405. Mona Industries, Inc.
406. Meer Corporation
407. Ampco Metal, Inc.
408. Taylor Sons, Charles, Company
409. Skelly Oil Company
410. Gross, A., & Company
412. Squibb, E. R., & Sons, Inc.
415. Robeco Chemicals, Inc.
418. Roussel Corporation
419. Cadet Chemical Corporation
420. Union Camp Corporation
422. Verona Division, Baychem Corporation
424. Johnson, S. C. & Son, Inc.
425. Applied Science Laboratories, Inc.
430. Fine Organics, Inc.
431. Biddle Instruments
434. National Lead Company, Evans Lead Division
436. Pennsylvania Glass Sand Corp.
438. See 50.
439. See 304.
440. See 203.
441. U. S. Borax & Chemical Corporation
442. Harbison-Carborundum Corporation
443. See 219.
445. Devcon Corporation
446. Harbison-Walker Refractories Company
448. See 296.
449. See 104.
450. American Hoeschst Corporation
451. Stepan Chemical Company

453. Apex Smelting Company
455. Babcock & Wilcox Company
457. Brown, R. J., Company
459. Carus Chemical Company
461. Chemtree Corporation
463. Interplastic Corporation
464. Consolidated Astronautics Division (United Guardian, Inc.)
465. Dayton Chemicals Division, Whittaker Corporation
466. King, Robert J., Company
467. Eastman Kodak Company
468. Eagle-Picher Industries, Inc.
470. Farbenfabriken Bayer AG
472. Air Products & Chemicals, Inc.
473. Humko Chemical
474. Jones Dabney Division, Devoe & Raynolds Company, Inc.
475. Kaiser Chemicals
476. Reed, Inc., Chemical Division
477. Loctite Corporation
480. Mooney Chemicals, Inc.
481. National Gypsum Company
482. Nixon-Baldwin Chemicals, Inc.
483. Nye, William F., Inc.
484. Ozark-Mahoning Company
485. Pharmacia Biotechnology Group
486. Poco Graphite, Inc.
487. Proctor & Gamble Company
488. Puero Rico Chemical Company, Inc.
489. Reynolds Metal Company
491. Stainless Foundry & Engineering, Inc.
492. Staley, A. E., Mfg. Company
493. Amerchol
494. Sun Refining and Marketing Company
496. Valchem, Chemical Division
497. Wilson Pharmaceutical & Chemical Corporation
498. See 104.
499. See 104.
500. See 104.
504. Reactor Experiments, Inc.
505. See 416.
506. General Adhesives & Chemical Company
507. McDanel Refractory Company
508. Breon Laboratories
510. Kay Fries, Inc., Chemical Division
511. National Polychemicals, Inc.
512. Hodag Chemical Corporation
513. Tylac, Standard Brands Chemical Industries, Inc.
516. Polyvinyl Chemical Industries
517. Organic Chemicals Division, W. R. Grace & Company
518. Upjohn Company
519. Upjohn Company
520. Upjohn Company
522. Pfaudler Company
523. See 50.

APPENDIX IV

Manufacturers of Trademarked Products
Alphabetical List

A

Abbott Laboratories
  Chemical Division
  1400 Sheridan Road
  North Chicago, IL 60064
Aceto Chemical Co., Inc.
  126-02 Northern Blvd.
  Flushing, NY 11368
Acheson Colloids Co.
  Port Huron, MI
Adria Laboratories, Inc.
  PO Box 16529
  Columbus, OH 43216
Air Products & Chemicals, Inc.
  PO Box 538
  Allentown, PA 18105
Alcolac Inc.
  3440 Fairfield Road
  Baltimore, MD 21226
Aldrich Chemical Co., Inc.
  PO Box 355
  Milwaukee, WI 53201
Allied Chemical Corp.
  Morristown, NJ 07960
Alox Corp.
  PO Box 517
  Niagara Falls, NY 14302
Aluminum Company of America
  Pittsburgh, PA 15219
Amerace Corp., Penetone Div.
  74 Hudson Avenue
  Tenafly, NJ 07670
Amerchol
  136 Talmadge Road
  CN 4051
  Edison, NJ 08818
American Can Co.
  247 American Lane
  Greenwich, CT 06830
American Cyanamid Co.
  One Cyanamid Plaza
  Wayne, NJ 07470
American Cyanamid Co.
  Industrial Chemicals Division
  Wayne, NJ 07470
American Hoechst Corp.
  Dyes & Pigments Div.
  Box 2500
  Somerville, NJ 08876
American Lignite Products Co.
  PO Box 1066
  Ione, CA 95640
American Mineral Spirits Co.
  Division of Union Oil Co. of CA
  Palatine, IL 60067
American Norit Co., Inc.
  420 Agmac Avenue
  Jacksonville, FL 32205

American Potash & Chemical Corp.
  3000 West 6th Street
  Los Angeles, CA 90054
AMETEK Inc.
  Haveg Division
  900 Greenbank Road
  Wilmington, DE 19808
Amoco Chemical Corp.
  200 E. Randolph Drive
  PO Box 8640-A
  Chicago, IL 60680
Ampco Metal, Inc.
  Box 2004
  Milwaukee, WI 53201
Anaconda American Brass Co.
  Waterbury, CT 06720
Ansul Fire Protection
  1 Stanton Street
  Marinette, WI 54143
Apex Chemical Co., Inc.
  200 So. First Street
  Elizabethport, NJ 07206
Apex Smelting Co.
  Division of American Metal
  Climax, Inc.
  6700 Grant Avenue
  Cleveland, OH 44105
Applied Science Laboratories, Inc.
  Box 440
  Stat College, PA 16801
Archer Danials Midland Co.
  Box 1470
  Decatur, IL 62525
Arco Chemical Co.
  1500 Market Street
  Philadelphia, PA 19101
Arizona Chemical Co.
  200 S. Sudduth Place
  Panama City, FL 32404
Arkansas Co., Inc.
  Box 210
  Newark, NJ 07101
Atlantic Refining Co.
  Box 7258
  Philadelphia, PA 19101
Atlas Minerals & Chemicals Co.
  Farmington Road
  Mertztown, PA 19539
Alvisun Corp.
  21 South 12th Street
  Philadelphia, PA 19107

B

Babcock & Wilcox Co.
  Refractories Div.
  Augusta, GA 30903
Badische Corp.

PO Drawer "D"
  Williamsburg, VA 23185
Baker Castor Oil Co.
  Bayonne, NJ 07002
Barium and Chemicals, Inc.
  Box 218
  Steubenville, OH 43952
Barker, R. W. & Co.
  13 Charterhouse Sq.
  London, England
BASF Inmont Corp.
  1255 Broad Street
  PO Box 6001
  Clifton, NJ 07015-6001
BASF Wyandotte
  Intermediate Chemicals Dept.
  Box 181
  Parsippany, NJ 07054
Beaunit Corp.
  261 Madison Avenue
  New York, NY 10016
Beckman Instruments, Inc.
  2500 harbor Blvd.
  Box 3100
  Fullerton, CA 92634
Becton Dickinson & Co.
  Mack Centre Drive
  Paramus, NJ 07652
Biddle Instruments
  510 Township Line Road
  Blue Bell, PA 19422
Biddle, James G. Co.
  Plymouth Meeting, PA 19462
Borden Chemical
  Division Borden, Inc.
  180 E. Broad Street
  Columbus, OH 43215
Borg-Warner Chemicals
  International Center
  Parkersburg, WV 26101
Breon Laboratories
  90 Park Avenue
  New York, NY 10016

C

Cabot Corp.
  1020 West Park Avenue
  PO Box 9013
  Kokoma, IN 46902-9013
Cadet Chemical Corp.
  Burt, NY 14028
Calgon Carbon Corp.
  PO Box 717
  Pittsburgh, PA 15230
California Industrial Minerals Co.
  Friant, CA 93626
C-I-L, Inc.
  PO Box 200, Station A

North York
Ontario, Canada
Carlisle Chemical Works, Inc.
Advance Division
New Brunswick, NJ 08903
Carrier Corp.
Syracuse, NY 13201
Carus Chemical Co.
1500 Eighth Street
LaSalle, IL 61301
Celanese Chemical Co.
1211 Avenue of Americas
New York, NY 10036
Celotex Corp.
1500 North Dale Mabry Hwy.
Tampa. FL 33607
Chemagro Agricultural Div.
Mobay Chemical Corp.
Box 4913
Kansas City, MO 64120
Chemical Development Corp.
Danvers, MA 09123
Chemtree Corp.
Central Valley, NY 10917
Chesebrough-Ponds, Inc.
485 Lexington Avenue
New York, NY 10017
CIBA-GEIGY
Agricultural Division
PO Box 18300
Greensboro, NC 27419
Ciba-Geigy Corp.
444 Saw Mill River Road
Ardsley, NY 10502
Ciba Pharmaceutical Co.
556 Morris Avenue
Summit, NJ 07901
Cleary, W. A., Corp.
Box 10
Somerset, NJ 08873
Climax Molybdenum Co.
A Division of AMAX
1600 Huron Parkway
PO Box 1658
Ann Arbor, MI 48106
Consolidated Astronautics Div.
(United Guardian Inc.)
230-T Marcus Blvd.
Hauppage, NY 11787
Continental Oil Co.
Petrochemicals Dept.
Saddle Brook, NJ 07662
Corning Glass Works
Corning, NY 14831
CPC International, Inc.
Englewood Cliffs, NJ 07632
Crown Zellerbach Corp.
Chemical Products Div.
PO Box 4524
Vancouver, WA 98662

D

Dawe's Laboratories
7100-TN Tripp
Chicago, IL 60646

Dayton Chemicals Division
Whittaker Corp.
PO Box 27
West Alexandria, OH 45381
Degussa, Inc.
Chemicals Division
2 Penn Plaza
New York, NY 10001
Devcon Corp.
Endicott Street
Danvers, MA 01923
Diamond Shamrock Corp.
Chemetals Division
711 Pittman Road
Baltimore, MD 21226
Diamond Shamrock Corp.
Nopco Division
Morristown, NJ 07960
Diversitech General
Polymers Division
One General Street
Akron, OH 44329
Dow Chemical Co.
PO Box 2166
Midland, MI 48640
Dow-Corning Corp.
Midland, MI 48640
Drew Chemical Corp.
Boonton, NJ 07065
Driver-Harris Co.
PO Box 31
Harrison, NJ 07029
Driver, Wilbur B., Co.
1875 McCarter Hwy.
Newark, NJ 07104
Du Pont de Nemours, E. I., Co.
Wilmington, DE 19898
Duriron Co., Inc.
PO Box 1145
Dayton, OH 45401

E

Eagle-Picher Industries, Inc.
580 Walnut Street
Cincinnati, OH 45202
Eastman Chemical Products, Inc.
Box 431
Kingsport, TN 37662
Eastman Kodak Co.
Organic Chemicals Dept.
343 State Street
Rochester, NY 14650
Elanco Products Co.
Box 1750
Indianapolis, IN 46206
Emery Industries, Inc.
4300 Carew Tower
Cincinnati, OH 45202
Engelhard Corp.
70 Wood Avenue South CN 770
Iselin, NJ 08830
Enthone Inc.
Box 1900
New Haven, CT 06508
Essex Chemical Corp.

1401 Broad Street
Clifton, NJ 07015
Ethyl Corp.
451 Florida Blvd.
Baton Rouge, LA 70801
Evans Chemetics, Inc.
90 Tokeneke Road
Darien, CT 06820

F

FMC Corp.
Food and Pharmaceutical
Products Div.
2000 Market Street
Philadelphia, PA 19103
FMC Corp.
Agricultural Chemicals Div.
Middleport, NY 14105
Fabric Research Laboratories
Dedham, MA
Fairmount Chemical Co.
117 Blanchard Street
Newark, NJ 07105
Farbenfabriken Bayer AG
425 Park Avenue
New York, NY 10022
Filtrol Corp.
5959 W. Century Blvd.
Los Angeles, CA 90045
Fine Organics, Inc.
205 Main Street
Lodi, NJ 07644
Firestone Plastics Co.
Bo 699
Pottstown, PA 19464
Firestone Synthetic Rubber
& Latex Co.
381 West Wilbeth Road
Akron, OH 44301
Firestone Tire & Rubber Co.
Akron, OH 44301
Fitzpatrick Co.
832 Industrial Drive
Elmhurst, IL 60126
Foote Mineral Co.
Exton, PA 19341
Ford Motor Co.
Dearborn, MI
Formica Corp.
4614 Spring Grove Avenue
Cincinnati, OH 45232
Fritzsche Dodge & Olcutt, Inc.
76 Ninth Avenue
New York, NY 10011

G

G & A Laboratories, Inc.
Box 1217
Savannah, GA 31402
GAF Corp. Chemical Products
140 W. 51st Street
New York, NY 10020
General Adhesives & Chemical Co.
6100 Centennial Blvd.
Nashville, TN 36202

General Electric Co.
  Polymers Product Dept.
  Pittsfield, MA 01201
General Latex & Chemical Corp.
  666 Main Street
  Cambridge, MA 02139
General Mills Chemicals, Inc.
  4620 West 77th Street
  Minneapolis, MN 55435
General Refractories Co.
  225 City Avenue
  Bala Cynwyd, PA 19004
Givaudan Corp.
  100 Delawanna Avenue
  Clifton, NJ 07014
Glidden-Durkee Division
  SCM Corp.
  Organic Chemicals Dept.
  Box 389
  Jacksonville, FL 32201
Glyco, Inc.
  488 Main Avenue (Route 7)
  PO Box 5100
  Norwalk, CT 06856
B. F. Goodrich Chemical Group
  6100 Oak Tree Blvd.
  Cleveland, OH 44131
Goodyear Tire & Rubber Co.
  Akron, OH 44316-0001
Grace, W. R., & Co.
  Davison Chemical Div.
  Charles and Baltimore Sts.
  Baltimore, MD 21203
Grace, W. R. & Co.
  Construction Products Div.
  62 Whittemore Avenue
  Cambridge, MA 02140
Grace, W. R., & Co.
  Organic Chemicals Div.
  55 Hayden Avenue
  Lexington, MA 02173
Graphite Metallizing Corp.
  1050 Nepperhan Avenue
  Yonkers, NY 10702
Gross, A. & Co.
  Box 818
  Newark, NJ 07101
Gulf Oil Chemicals Co.
  Industrial & Specialty Chemicals
  Div.
  Box 3766
  Houston, TX 77001

H

Hall, C. P. Co.
  7300 S. Central Avenue
  Chicago, IL 60638
Hamblet & Hayes Co.
  Colonial Road
  Salem, MA 01970
Hammond, W. A. Drierite Co.
  138 Dayton Avenue
  Xenia, OH 45385
Handy & Harman
  850 Third Avenue
  New York, NY 10022

Harbison-Carborundum Corp.
  Box 337
  Niagara Galls, NY 14302
Harbison-Walker Refractories Co.
  Division of Dresser Industries
  One Gateway Center
  Pittsburgh, PA 15222
Harshaw-Filtrol Pertnership
  30100 Chagrin Blvd.
  Cleveland, OH 44124
Hercules Inc.
  Hercules Plaza
  Wilmington, DE 19894
Heresite-Saekaphen, Inc.
  822 S. 14th Street
  Manitowoc, WI 54420
Hitco
  Box 1097
  Gardena, CA 90249
Hodag Chemical Corp.
  7247 N. Central Park Avenue
  Skokie, IL 60076
Hoechst Fibers Industries
  Spartanburg, SC
Hoffman-La Roche, Inc.
  340 Kingsland Street
  Nutley, NJ 07110
Holliday Dyes and Chemicals, Ltd.
  PO Box B22
  Leedo Road
  Huddersfield HD2 1UH England
Hoskins Mfg. Co.
  4445 Lawton Avenue
  Detroit, MI 48208
Hoyt, Arthur S., Co., Inc.
  Box 24
  Hicksville, NY 11802
Huber, J. M., Corp.
  Box 831
  Borger, TX 79007
Hughson Chemical Co.
  Erie, PA 16512
Humble Oil & Refining Co.
  Box 2180
  Houston, TX 77001
Humko Chemical
  Div. Witco Corp.
  PO Box 125
  Memphis, TN 38101
Humphrey Chemical Co.
  7261 Devine Street
  North Haven, CT 06473
Hunter Engineering Co.
  Div. American Metal Climax,
  Inc.
  Box 5437
  Riverside, CA 92507
Huntington Alloys
  Inco Alloys International, Inc.
  Huntington, WV 25720

I

ICI United States, Inc.
  Wilmington, DE 19897
International Rustproof Co.

Div. of Lubrizol Corp.
  Box 17100
  Cleveland, OH 44117
Interpace Corp.
  Box 785
  Ione, CA 95640
Interplastic Corp.
  2015 NE Broadway
  Minneapolis, MN 55413
Ionac Chemical Co.
  Div. Pfaudler Permutit, Inc.
  Birmingham, NJ 08011

J

Johnson, S. C., and Son, Inc.
  1525 Howe Street
  Racine, WI 53403
Jones-Dabney Division
  Devoe & Raynolds Co., Inc.
  Box 1863
  Louisville, KY 40201

K

Kaiser Chemicals
  30100 Chagrin Blvd.
  Cleveland, OH 44124
Kay Fries, Inc.
  Chemical Division
  Dynamit Noble of America
  10 Link Drive
  Rockleigh, NJ 07647
Kelco Co.
  Box 23076
  San Diego, CA 92123
Kendall Refining Co.
  Bradford, PA 16701
Kennametal Inc.
  Latrobe, PA 15650
Kenrich Petrochemicals, Inc.
  140 East 22nd Street
  PO Box 32
  Bayonne, NJ 07002-0032
King, Robert J., Co.
  Box 588
  Norwalk, CT 06856
Knoll Fine Chemicals, Inc.
  120 E. 56 Street
  New York, NY 10022
Kohnstamm, H., & Co., Inc.
  161 Avenue of Americas
  New York, NY 10013
Kolene Corp.
  12890 Westwood Avenue
  Detroit, MI 488223
Koppers Co., Inc.
  Organic Materials Div.
  1900 Koppers Bldg.
  Pittsburgh, PA 15219

L

Lake States Chemical Div.
  St. Regis Paper Co.
  Rhinelander, WI 54501

La Motte Chemical Products Co.
  Chestertown, MD 21620
Laticrete Internatinoal, Inc.
  1 Laticrete Park North
  Bethany, CT 06525
Lilly, Eli, & Co.
  Box 618
  Indianapolis, IN 46206
Lithcote Corp.
  5000 West Lake Street
  Milrose Park, IL 60161
Lockrey Co., Inc.
  2517-T Finlaw Avenue
  Pennsauken, NJ 08109
Loctite Corp.
  705 N. Mountain Road
  Newington, CT 06111

M

3M
  3M Center
  St. Paul, MN 55144-1000
M & T Chemicals, Inc.
  Rahway, NJ 07065-0970
Magnesol
  Pilot Engineering Div. of
  Reagent Chemical & Research
    Inc.
  Middlesex, NJ 08846
Mallinckrodt Inc.
  Box 5439
  St. Louis, MO 63160
Mallinckrodt Inc.
  Washine Div.
  165 Main Street
  Lodi, NJ 07644
Manville Products Corp.
  Ken-Caryl Ranch
  Denver, CO 80217
Marine Colloids, Inc.
  Rockland, ME 04102
Masonite Corp.
  1 S. Wacker Drive
  Chicago, IL 60606
McDanel Refractory Co.
  Box 560
  Beaver Falls, PA 15010
Mearl Corp.
  41 East 42nd Street
  New York, NY 10017
Meer Corp.
  North Bergen, NJ 07047
Merck & Co., Inc.
  PO Box 2000
  Rahway, NJ 07065-0900
Miles Laboratories, Inc.·
  PO Box 40
  Elkhart, IN 46515
Minerals & Chemicals Div.
  Englehard Minerals & Chemicals
  Corp.
  Menlo Park, NJ 00817
Mobay Chemical Corp.
  Mobay Road
  Pittsburgh, PA 15205-9741

Mobile Chemical Co.
  Petrochemicals Div.
  120 E. 42nd Street
  New York, NY 10017
Mona Industries, Inc.
  76 E. 24th Street
  PO Box 425
  Paterson, NJ 07544
Monsanto Co.
  800 N. Lindbergh Blvd.
  St. Louis, MO 63167
Mooney Chemicals, Inc.
  2301 Scranton Road
  Cleveland, OH 44113
Morflex Chemical Co.
  2110 High Point Road
  Greensboro, NC 27043
Morton Chemical Co.
  110 N. Wacker Drive
  Chicago, ILL 60606
Morton Thiokol, Inc.
  Ventron Division
  150 Andover Street
  Danvers, MA 01923

N

Nalco Chemical Co.
  1601 Diehl Road
  Naperville, IL 60566
Nalge Co.
  Division of Sybron Corp.
  65 Panarama Creek Drive
  Box 365
  Rochester, NY 14602
Napp Chemicals
  199 Main Street
  Lodi, NJ 07644
National Gypsum Co.
  Buffalo, NY 14202
National Lead Co.
  Baroid Division
  Box 1675
  Houston, TX 77001
National Lead Co.
  111 Broadway
  New York, NY 10006
National Lead Co.
  Evans Lead Division
  Box 1467
  Charleston, WV 25325
National Polychemicals, Inc.
  Wilmington, MA 01887
National Rosin Oil Products, Inc.
  Box 1205
  Savannah, GA 31498
National Starch & Chemical Corp.
  10 Finderne Avenue
  Bridgewater, NJ 08807
Neville Chemical Co.
  Neville Island
  Pittsburgh, PA 15225
The New Jersey Zinc Co., Inc.
  Palmerton, Pa. 18071
Nixon-Baldwin Chemicals, Inc.
  Nixon, NJ 08818

Norac Co., Inc.
  Azusa, CA 91702
Norton Co.
  Metals Division
  45 Industrial Place
  Newton, MA
Nuodex, Inc.
  Turner Place
  PO Box 365
  Piscataway, NJ 08854
Nye, William F., Inc.
  Box 927
  New Bedford, MA 02742

O

Occidental Chemical Corp.
  PO Box 189
  Niagara Falls, NY 14302
Occidental Petroleum Corp.
  10889 Wilshire Blvd.
  Los Angeles, CA 90024
Olin Chemicals
  120 Long Ridge Road
  Stamford, CT 06904
Onyx Chemical Co.
  Division of Millmaster Onyx
  Group, Inc.
  190 Warren Street
  Jersey City, NJ 07302
Osborn, C. J. Chemicals, Inc.
  Pennsauken, NJ 08109
Osmonics, Inc.
  5951 Clearwater Drive
  Minnetonka, MN 55343
Owens-Corning Fiberglas Corp.
  Fiberglas Tower
  Toledo, OH 43659
Ozark-Mahoning Co.
  1870 S. Boulder
  Tulsa, OK 74119

P

Parker Chemical Co.
  32100 Stephenson Highway
  Madison Heights, MI 48071
Penick, S. B., & Co.
  1050 Wall Street
  W. Lyndhurst, NJ 07071
Pennwalt Corp.
  Lucidol Div.
  1740 Military Road
  Buffalo, NY 14240
Pennwalt Corp.
  3 Parkway
  Philadelphia, PA 19102
Pennsylvania Glass Sand Corp.
  PO Box 187
  Berkeley Springs, WV
Pennsylvania Industrial Chemical
  Corp.
  131 State Street
  Clairton, PA 15025
Pennsylvania Refining Co.
  Butler, PA 16001

Permutit Co.
  Div. of Pfaudler-Permutit, Inc.
  Paramus, NJ 07632
Petrolite Corp. Bareco Div.
  6910 E. 14 Street
  Tulsa, OK 74115
Pfaudler Co.
  A Sohio Company
  Rochester, NY 14603
Pfizer Inc., Chemicals Div.
  235 E. 42nd Street
  New York, NY 10017
Pharmacia Biotechnology Group
  800 Centennial Avenue
  Piscataway, NJ 08854
Philadelphia Quartz Co.
  Box 840
  Valley Forge, PA 19482
Phillips Petroleum Co.
  Rubber Chemicals Div.
  Bartlesville, OK 74004
Pitt-Consol Chemical Co.
  Teterboro, NJ 07608
PMC Specialties Group, Inc.
  101 Prospect Avenue, NW
  Cleveland, OH 44115
Poco Grahite, Inc.
  Box 1524
  Garland, TX 75041
Polyvinyl Chemical Industries
  Wilmington, MA 01887
PPG Industries, Chemical Div.
  1 Gateway Center
  Pittsburgh, PA 15222
Premier Malt Products, Inc.
  1137 N. Eighth Street
  Milwaukee, Wis. 52301
Proctor & Gamble Co.
  Industrial Chemicals Div.
  Box 599
  Cincinnati, OH 45201
Puerto Rico Chemical Co., Inc.
  277 Park Avenue
  New York, NY 10017
Purafil, Inc.
  Box 80434
  Atlanta, GA 30366

Q

Quaker Oat Co.
  Merchandise Mart Plaza
  Chicago, IL 60654

R

Reactor Experiments, Inc.
  963 Terminal Way
  San Carlos, CA 94070-3278
Reed, Inc.
  Chemical Division
  10-16 boul des Capucins
  PO Box 2025
  Quebec, Quebec Canada G1K
  7N1
Reheis Chemical Co.

Div. of Armour Pharmaceutical
  Co.
  Phoenix, Ariz 85077
Reichhold Chemicals, Inc.
  Resins and Binders Division
  407 South Pace Blvd.
  PO Box 1433
  Pensacola, FL 32596
Renite Co.
  2500 East Fifth Avenue
  Columbus, OH 43219
Republic Steel Corp.
  1441 Republic Bldg.
  Cleveland, OH 44101
Reynolds Metals Co.
  6601 West Broad Street
  Richmond, VA 23261
Richardson Co.
  Organic Chemicals Div.
  2400 E. Devon Street
  Des Plaines, IL 60018
Robeco Chemicals, Inc.
  51 Madison Avenue
  New York, NY 10010
Rohm and Haas Co.
  Independence Mall West
  Philadelphia, PA 19105
Rorer-Amchem Products, Inc.
  Ambler, PA 19002
Roussel Corp.
  155 E. 44th Street
  New York, NY 10017
Royce Chemical Co.
  Box 237
  E. Rutherford, NJ 07073
Ruetgers Nease Chemical Co., Inc.
  Box 221
  State College, PA 16801

S

Schering Corp.
  60 Orange St.
  Bloomfield, NJ 07003
Schwarz/Mann
  Orangeburg, NJ 10962
Scientific Chemicals, Inc.
  1637 S. Kilbourn Avenue
  Chicago, IL 60623
Scott Bader & Co., Ltd.
  Wollaston, Wellingborough
  Northamptonshire, England
Searle, G. D., & Co.
  Box 1045
  Skokie, IL 60076
Shell Chemical Co.
  Box 2463
  Houston, TX 77001
Sherwin Williams Chemicals
  Box 6520
  Cleveland, OH 44101
Shulton, Inc.
  Clifton, NJ 07015
Skelly Oil Co.
  1437 Boulder Street
  Tulsa, OK 74102

SmithKline Corp.
  1500 Spring Garden St.
  Philadelphia, PA 19101
Smith, Werner G., Inc.
  1730 Train Avenue
  Cleveland, OH 44113
Spencer Kellogg Products
  NL chemicals
  120 Delaware Avenue
  Box 807
  Buffalo, NY 14240
Squibb & Sons, E. R., Inc.
  PO Box 4000
  Princeton, NJ 08540
Stainless Foundry & Engineering,
  Inc.
  5132 No. 35th St.
  Milwaukee, WI 53209
Staley, A. E. Mfg. Co.
  Box 151
  Decatur, IL 62525
Stauffer Chemical Co.
  Industrial Chemical Div.
  Westport, CT 06880
Stepan Chemical Co.
  Industrial Chemicals Div.
  Northfield, IL 60093
Sterling Drug, Inc.
  Consumer Products Division
  Montvale, NJ 07645
Sun Chemical Corp.
  Box 70
  Chester, SC 29706
Sun Co., Inc.
  1608 Walnut St.
  Philadelphia, PA 19103
Sun Refining and Marketing Co.
  Ten Penn Center
  1801 Market Street
  Philadelphia, PA 19103-1699
Swift Edible Oil Co.
  115 W. Jackson Blvd.
  Chicago, IL 60604

T

Taylor Sons Co., Charles
  Box 14058
  Cincinnati, OH 45214
Tennessee Chemical Co.
  347 Lenox Road
  Suite 670
  Atlanta, GA 30326
Tennessee Eastman Co.
  Kingsport, TN 37662
Texaco, Inc.
  Petrochemical Dept.
  Box 52332
  Houston, TX 77052
Textilana Corp.
  12607 Cerise Avenue
  Hawthorne, CA 90250
Thermal American Fused Quartz
  Co.
  Montville, NJ 07045

Thiokol Corp.
  Specialty Chemicals Division
  930 Lower Ferry Road
  Box 8296
  Trenton, NJ 08650
Thomas, Arthur H., Co.
  99 High Hill Road
  PO Box 99
  Swedesboro, NJ 08085-0099
Tylac, Standard Brands Chemical
  Industries, Inc.
  Dover, DE 19901

U

Union Camp Corp.
  1600 Valley Road
  Wayne, NJ 07470
Uniroyal Chemical
  Div. Uniroyal Inc.
  Naugatuck, CT 06770
UOP Chemical Co.
  Division Universal Oil Products
  Co.
  Des Plaines, IL 60016
Upjohn Co.
  Kalamazoo, MI 49001
U. S. Borax & Chemical Corp.
  3075 Wilshire Blvd.
  Los Angeles, CA 90010
U. S. Industrial Chemicals Co.
  Div. National Distillers &
  Chemical Corp.
  11500 Northlake Drive
  Cincinnati, OH 45249
U. S. Stoneware Co.

5300 E. Tallmadge Avenue
Akron, OH 44310

V

Valchem, Chemical Div.
  Box 38
  Langley, SC 29834
Vanderbilt, R. T., & Co.
  Norwalk, CT 06855
Van Dyk & Co., Inc.
  Main and William Sts.
  Belleville, NJ 07109
Velsicol Chemical Corp.
  341 E. Ohio St.
  Chicago, IL 60611
Ventron Corp.
  Beverly, MA 01915
Verona Division
  Baychem Corp.
  Box 385
  Union, NJ 07083

W

Wallace & Tiernan, Inc.
  Harchem Division
  25 Main Street
  Belleville, NJ 07101
Wallerstein Co.
  Div. Travelol Laboratories
  Morton Grove, IL 60053
Warner-Lambert Co.
  201 Tabor Road
  Morris Plains, NJ 07950
Westinghouse Electric Corp.
  E. Pittsburgh, PA 15235

Weyerhaeuser Co.
  Tacoma, WA 98401
Wilson-Fiberfil International
  PO Box 3333
  Evansville, IN 47732
Wilson Pharmaceutical & Chemical
  Corp.
  Jackson and Swanson Sts.
  Philadelphia, PA 19148
Winthrop Laboratories
  Div. Sterling Drug Co.
  90 Park Avenue
  New York, NY 10016
Witco Chemical Corp.
  Organics Division
  Box 45296
  Houston, TX 77245
Worthington Corp.
  Harrison, NJ 07029
Wyeth Laboratories
  Div. American Home Products
  Corp.
  Box 8299
  Philadelphia, PA 19101

X

Xylos Rubber Co.
  Div. Firestone Synthetic Rubber
  & Latex Co.
  1300 Emerling Avenue
  Akron, OH 44301

Z

Zoecon Corp.
  975 California Avenue
  Palo Alto, CA 94304